GREEN CHEMISTRY

GREEN CHEMISTRY

AN INCLUSIVE APPROACH

Edited by

Béla Török
Timothy Dransfield
University of Massachusetts Boston, Boston, MA, United States

ELSEVIER

Elsevier
Radarweg 29, PO Box 211, 1000 AE Amsterdam, Netherlands
The Boulevard, Langford Lane, Kidlington, Oxford OX5 1GB, United Kingdom
50 Hampshire Street, 5th Floor, Cambridge, MA 02139, United States

Library of Congress Cataloging-in-Publication Data
A catalog record for this book is available from the Library of Congress

British Library Cataloguing-in-Publication Data
A catalogue record for this book is available from the British Library

ISBN: 978-0-12-809270-5

Publisher: Candice Janco
Acquisition Editor: Laura Kelleher; Kathryn Morrissey; Anneka Hess
Editorial Project Manager: Emily Thomson
Production Project Manager: Anitha Sivaraj
Designer: Christian J. Bilbow

Back cover image: Photo courtesy: Harry Brett

Typeset by TNQ Books and Journals

Contents

3.7 Contaminants of Emerging Concern, With an Emphasis on Nanomaterials and Pharmaceuticals

HELEN C. POYNTON, WILLIAM E. ROBINSON

3.8 Pollution in the Coastal Zone: A Case Study of Wastewater on Cape Cod, MA

AMY COSTA, LAUREL SCHAIDER, PATRICIA HUGHES

3.9 The Composition of Soils and Sediments

STEFAN D. KALEV, GURPAL S. TOOR

3.10 Heavy Metal Pollution and Remediation

ABHISHEK ROYCHOWDHURY, RUPALI DATTA, DIBYENDU SARKAR

3.11 Application of Green Chemistry in Homogeneous Catalysis

ANNE KOKEL, CHRISTIAN SCHÄFER

3.12 Heterogeneous Catalysis: A Fundamental Pillar of Sustainable Synthesis

ISTVÁN PÁLINKÓ

3.13 Phase Transfer Catalysis: A Tool for Environmentally Benign Synthesis

MANISHA MISHRA, BÉLA TÖRÖK

3.14 Biocatalysis: Nature's Chemical Toolbox

JULIE A. HIMMELBERGER, KATHRYN E. COLE, DANIEL P. DOWLING

3.21 Principles of Photochemical Activation Toward Artificial Photosynthesis and Organic Transformations

KEN T. NGO, JONATHAN ROCHFORD

3.22 Biopolymers: Biodegradable Alternatives to Traditional Plastics

CHRISTOPHER BRIGHAM

3.23 Modern Applications of Green Chemistry: Renewable Energy

JOHN COLLINS, GERALD GOURDIN, DEYANG QU

3.24 From Ethanol to Biodiesel: A Survey of Green Fuels

CHRISTOPHER BRIGHAM, JOHANNES BADER

3.25 Solar Energy Conversion

NICHOLAS A. LEE, GERALD E. GILLIGAN, JONATHAN ROCHFORD

3.26 Toward a Sustainable Carbon Cycle: The Methanol Economy

ALAIN GOEPPERT, GEORGE A. OLAH, G.K. SURYA PRAKASH

3.27 Natural and Nature-Inspired Synthetic Small Molecule Antioxidants in the Context of Green Chemistry

WILLIAM HORTON, MARIANNA TÖRÖK

3.28 The Value-Adding Connections Between the Management of Ecoinnovation and the Principles of Green Chemistry and Green Engineering

PHILIP COISH, ENDA MCGOVERN, JULIE B. ZIMMERMAN, PAUL T. ANASTAS

3.29 The International Chemicals Regime: Protecting Health and the Environment

NATALIA ESCOBAR-PEMBERTHY, MARIA IVANOVA, GABRIELA BUENO

List of Contributors

Steven Ackerman University of Massachusetts Boston, Boston, MA, United States

Karelle Aiken Georgia Southern University, Statesboro, GA, United States

Nicholas D. Anastas United State Environmental Protection Agency, Cincinnati, OH, United States

Paul T. Anastas Yale University, New Haven, CT, United States

Gopalakrishnan Aridoss LG Life Sciences Ltd, Daejeon, South Korea

Johannes Bader Beuth University of Applied Sciences, Berlin, Germany

Nadine Borduas ETH Zurich, Zurich, Switzerland

Christopher Brigham University of Massachusetts Dartmouth, North Dartmouth, MA, United States

Gabriela Bueno University of Massachusetts Boston, Boston, MA, United States

Timothy P. Canty University of Maryland, College Park, MD, United States

Philip Coish Yale University, New Haven, CT, United States

Kathryn E. Cole Christopher Newport University, Newport News, VA, United States

John Collins IBM Thomas J Watson Research Center, Yorktown Heights, NY, United States

Amy Costa Center for Coastal Studies, Provincetown, MA, United States

Levente Cseri The University of Manchester, Manchester, United Kingdom

Rupali Datta Michigan Technological University, Houghton, MI, United States

Neil M. Donahue Carnegie Mellon University, Pittsburgh, PA, United States

Daniel P. Dowling University of Massachusetts Boston, Boston, MA, United States

Timothy Dransfield University of Massachusetts Boston, Boston, MA, United States

Clifford J. Ellstrom University of Massachusetts Boston, Boston, MA, United States

Natalia Escobar-Pemberthy University of Massachusetts Boston, Boston, MA, United States

Daniel M. Genest University of Massachusetts Boston, Boston, MA, United States

Debanjana Ghosh Georgia Southern University, Statesboro, GA, United States

Gerald E. Gilligan University of Massachusetts Boston, Boston, MA, United States

Alain Goeppert University of Southern California, Los Angeles, CA, United States

Gerald Gourdin Georgia Institute of Technology, Atlanta, GA, United States

Robyn E. Hannigan University of Massachusetts Boston, Boston, MA, United States

Julie A. Himmelberger DeSales University, Center Valley, PA, United States

William Horton University of Massachusetts Boston, Boston, MA, United States

Patricia Hughes Center for Coastal Studies, Provincetown, MA, United States

Maria Ivanova University of Massachusetts Boston, Boston, MA, United States

Stefan D. Kalev Gulf Coast Research and Education Center, University of Florida, Wimauma, FL, United States

Daniel Kirk-Davidoff University of Maryland, College Park, MD, United States

Anne Kokel University of Massachusetts Boston, Boston, MA, United States

Kenneth K. Laali University of North Florida, Jacksonville, FL, United States

Shainaz Landge Georgia Southern University, Statesboro, GA, United States

Nicholas A. Lee University of Massachusetts Boston, Boston, MA, United States

Alexandra Maertens Johns Hopkins Bloomberg School of Public Health, Baltimore, MD, United States

Enda McGovern Sacred Heart University, Fairfield, CT, United States

Meaghan McKinnon University of Massachusetts Boston, Boston, MA, United States

Manisha Mishra University of Massachusetts Boston, Boston, MA, United States

Ken T. Ngo University of Massachusetts Boston, Boston, MA, United States

James Noblet California State University San Bernardino, San Bernardino, CA, United States

George A. Olah University of Southern California, Los Angeles, CA, United States

István Pálinkó University of Szeged, Szeged, Hungary

Peter Pogany Gedeon Richter Plc., Budapest, Hungary

Helen C. Poynton University of Massachusetts Boston, Boston, MA, United States

Deyang Qu University of Wisconsin Milwaukee, Milwaukee, WI, United States

Mayamin Razali The University of Manchester, Manchester, United Kingdom

William E. Robinson University of Massachusetts Boston, Boston, MA, United States

Jonathan Rochford University of Massachusetts Boston, Boston, MA, United States

Abhishek RoyChowdhury Stevens Institute of Technology, Hoboken, NJ, United States

Heather A. Rypkema Heritage Strategies, INTL, Washington, DC, United States

Ross J. Salawitch University of Maryland, College Park, MD, United States

Dibyendu Sarkar Stevens Institute of Technology, Hoboken, NJ, United States

Christian Schäfer University of Massachusetts Boston, Boston, MA, United States

Laurel Schaider Silent Spring Institute, Newton, MA, United States

Linda Schweitzer Oakland University, Rochester, MI, United States

Abid Shaikh Georgia Southern University, Statesboro, GA, United States

G.K. Surya Prakash University of Southern California, Los Angeles, CA, United States

Gyorgy Szekely The University of Manchester, Manchester, United Kingdom

Gurpal S. Toor University of Maryland, College Park, MD, United States

Béla Török University of Massachusetts Boston, Boston, MA, United States

Marianna Török University of Massachusetts Boston, Boston, MA, United States

David M. Wilmouth Harvard University, Cambridge, MA, United States

Julie B. Zimmerman Yale University, New Haven, CT, United States

Preface

The concept of this book was born in 2005 when we (along with Deyang Qu, who wrote Chapter 3.23) taught *Introduction to Green Chemistry* at UMass Boston. We team-taught the course because none of us individually had expertise in everything we wanted to cover. We were lucky to have faculty whose specialties included air pollution, novel battery technologies, and green synthesis, but even with our combined backgrounds, we felt unprepared to show the true breadth of the field of Green Chemistry. We searched for a textbook to help us fill in the gaps, but we could not find one. We successfully ran the course using an environmental chemistry textbook and a lot of primary literature, but from the outset, we realized that the course deserved a textbook designed around it.

This, then, is the textbook we wish had been available to us. It is intended for a broad audience, including industry and academia. It is aimed to be a contemporary and inclusive Green Chemistry text that can be used in undergraduate and graduate education and as a resource for researchers. The main goal of the work was to be as broad as possible, including many aspects of Green Chemistry. The book includes three main parts. The first two parts are intended for those who teach Green Chemistry: it covers the basic definitions, environmental chemistry, renewable energy, sustainable synthesis, fundamental chemical toxicology, and the effect of environmental factors on our genetic information. These chapters follow a textbook style, providing examples, recommended reading, and problem sets. The third part of the work is designed for researchers, as it contains in-depth reviews on selected topics. Our intention is that educators, after covering the fundamentals laid down in the first part, may select some of the specialized research chapters as case studies to further illustrate the state-of-the-art practice of Green Chemistry. Although the book includes more topics than could be covered in a typical undergraduate or even graduate class, the variety of topics will provide opportunity for the faculty instructors to select topics they are comfortable covering and can fit into their schedules.

The centerpiece of the third part of the book is Green Chemistry in practice. The topics included in this part focus on special areas of the field. Every chapter in this part provides an up-to-date reference section, together comprising thousands of original papers and review articles. We believe that by including experts in many of the fields discussed, the book provides the readers with "insider information": the aspects or challenges of a given field that the specialists consider the most important and urgent. This way, we hope that the book will serve as a primary resource for those who are new to Green Chemistry or those who intend to branch out and discover other topics that are related to their own research fields.

We would like to thank our distinguished colleagues and authors, experts in their fields, for contributing to this unique endeavor. We also thank Kathryn Morrissey and Laura Kelleher, who helped us through the proposal phase of the book, and Anitha Sivaraj, who handled the galley proofs. We are indebted to Emily Thomson, our Editorial Project Manager, for her enormous help and continuous encouragements throughout the process.

Béla Török and Timothy Dransfield
Boston, May 1, 2017

PART 1

INTRODUCTION

CHAPTER

1.1

Green Chemistry: Historical Perspectives and Basic Concepts

Béla Török, Timothy Dransfield

University of Massachusetts Boston, Boston, MA, United States

Chemists and chemistry, in general, have made an enormous contribution to the history of humankind. Beginning with the early alchemists, these contributions include several developments that changed the course of history for the better or, in some cases, for the worse. Many of them may seem to be simple by today's standards, but at the time, they were groundbreaking discoveries and inventions. The fabrication of simple soaps made the formation of large cities possible by improving the personal hygiene. The production of dyes and paints contributed significantly to fashion and art over the centuries. As the usefulness of chemistry became clear, more people decided to pursue such endeavors, which brought exponential growth in this field. As Dalton, Avogadro, and Lavoisier made their famous discoveries, chemistry became viewed more and more as science than as black magic. With significant developments in chemical theory came the increased pace of new applications that inspired yet further progress. However, not every step along the way was problem free. Several inventions that were made with the best intentions backfired and caused health or environmental issues. The use of freons as inflammable carrier gases in all sorts of sprays in the 1960s–70s seemed to be perfect, until Rowland and Molina published their findings on the terrible effect of these chemicals on the ozone layer that protects the earth from harmful ultraviolet radiation. Antibiotics were hot commodities after World War II, until it was found that bacteria can develop resistance toward these compounds making them more difficult to fight against. Plastics seemed like a blessing until it was found that their degradation takes over a thousand years. Dichlorodiphenyltrichloroethane (DDT) appeared to be an effective agent to fight malaria-spreading mosquitos, until it was found in fish around Antarctica, proving that it lingers for a long time. Contemporary pesticides can leach into natural waters and cause gender change in frogs. Many drugs have unintended harmful side effects. The list is long. What is common in all these cases is that a product was developed for a certain purpose without a careful analysis of its broader impact on the ecosystem. All these disasters initiated a different chemical thinking that now we call green chemistry.

Green Chemistry
http://dx.doi.org/10.1016/B978-0-12-809270-5.00001-7

1.1.1 EMERGENCE OF GREEN CHEMISTRY

Early in the evolution of the chemical industry, scientists were already, although unconsciously, applying some of the much later formulated principles of green chemistry. For instance, the development of heterogeneous catalytic petrochemical processes dates back to the 1930s. The more conscious development of such thinking began after several major environmental disasters and industrial accidents occurred. The aforementioned problems are just a few examples to demonstrate the potentially harmful nature of chemicals if applied and introduced to the biosphere without sufficiently careful forward thinking. Rachel Carson's book, *Silent Spring* (1962), which described the destruction of local ecosystems by toxic chemicals, likely was a wake-up call for the society to address the issues or face grave consequences. As the first important step to address these issues the US Congress passed the *National Environmental Policy Act* in 1969. The US Environmental Protection Agency (US EPA) was established by President Nixon in 1970. Since the 1970s several environmental legislations have been implemented, such as the *Clean Air Act* of 1970 and the *Safe Drinking Water Act* of 1974, that signaled the government's intention to solve the problems via regulations. The US *Toxic Substances Control Act* was passed in 1976, and now it has over 80,000 chemicals in its listings. Later the more comprehensive *Clean Air Act* and the *Pollution Prevention Act* were enacted, both in 1990. The term *Green Chemistry* was coined by the EPA Office of Pollution Prevention and Toxins in the early 1990s. In 1995, the US EPA established an annual awards program called the *Presidential Green Chemistry Awards* to recognize the leaders of innovation from both industry and academia. In 1997, the first *PhD in Green Chemistry* program was established at the University of Massachusetts Boston. In the same year the Green Chemistry Institute was founded, which later became the American Chemical Society Green Chemistry Institute. Starting with the 1990s, several scientific journals devoted to green chemistry research began publishing original research and review articles in the field. Today, all major publishers have at least one journal devoted to green or sustainable chemistry research with several books and textbooks published to aid research and education.

1.1.2 SUSTAINABLE PRODUCTION OF COMMODITIES: PRINCIPLES AND BASIC CONCEPTS

Several tools and methods that are now considered as part of sustainable synthesis (e.g., catalysis) were developed much earlier than the formal green chemistry movement began. It took concerted and conscious efforts to envision and design a framework that included the earlier developments and initiated further progress in this field. The basic principle, *benign by design*, emphasized that both the product and the process used to produce it should conform to the basic rules of sustainability. In their seminal book in 1998, Anastas and Warner established the major principles of green chemistry. Although since then several "new principles" have been added to the list, the original list is still applicable.

1.1.2.1 Principles of Green Chemistry

1. *Prevention*: It is better to prevent waste than to treat or clean up waste after it has been created.
2. *Atom economy*: Synthetic methods should be designed to maximize the incorporation of all materials used in the process into the final product.

3. *Less hazardous chemical syntheses*: Wherever practicable, synthetic methods should be designed to use and generate substances that possess little or no toxicity to human health and the environment.
4. *Designing safer chemicals*: Chemical products should be designed to affect their desired function while minimizing their toxicity.
5. *Safer solvents and auxiliaries*: The use of auxiliary substances (e.g., solvents, separation agents, etc.) should be made unnecessary wherever possible and innocuous when used.
6. *Design for energy efficiency*: Energy requirements of chemical processes should be recognized for their environmental and economic impacts and should be minimized. If possible, synthetic methods should be conducted at ambient temperature and pressure.
7. *Use of renewable feedstocks*: A raw material or feedstock should be renewable rather than depleting whenever technically and economically practicable.
8. *Reduce derivatives*: Unnecessary derivatization (use of blocking groups, protection/deprotection, temporary modification of physical/chemical processes) should be minimized or avoided if possible, because such steps require additional reagents and can generate waste.
9. *Catalysis*: Catalytic reagents (as selective as possible) are superior to stoichiometric reagents
10. *Design for degradation*: Chemical products should be designed so that at the end of their function they break down into innocuous degradation products and do not persist in the environment.
11. *Real-time analysis for pollution prevention*: Analytical methodologies need to be further developed to allow for real-time, in-process monitoring and control prior to the formation of hazardous substances.
12. *Inherently safer chemistry for accident prevention*: Substances and the form of a substance used in a chemical process should be chosen to minimize the potential for chemical accidents, including releases, explosions, and fires.

Since products that involve any chemistry during their preparation are all manufactured by industry, the aforementioned list had to be amended to include specific issues that chemical engineers face while transitioning a laboratory process to the industrial setting. Hence, Anastas and Zimmerman developed a similar set of principles for engineering.

1.1.2.2 Principles of Green Engineering

1. *Inherent rather than circumstantial*: Designers need to strive to ensure that all materials and energy inputs and outputs are as inherently nonhazardous as possible.
2. *Prevention instead of treatment*: It is better to prevent waste than to treat or clean up waste after it is formed.
3. *Design for separation*: Separation and purification operations should be designed to minimize energy consumption and materials use.
4. *Maximize efficiency*: Products, processes, and systems should be designed to maximize mass, energy, space, and time efficiency.
5. *Output pulled versus input pushed*: Products, processes, and systems should be "output pulled" rather than "input pushed" through the use of energy and materials.

6. *Conserve complexity*: Embedded entropy and complexity must be viewed as an investment when making design choices on recycle, reuse, or beneficial disposition.
7. *Durability rather than immortality*: Targeted durability, not immortality, should be a design goal.
8. *Meet need, minimize excess*: Design for unnecessary capacity or capability (e.g., "one size fits all") solutions should be considered a design flaw.
9. *Minimize material diversity*: Material diversity in multicomponent products should be minimized to promote disassembly and value retention.
10. *Integrate material and energy flows*: Design of products, processes, and systems must include integration and interconnectivity with available energy and materials flows.
11. *Design for commercial "afterlife"*: Products, processes, and systems should be designed for performance in a commercial "afterlife."
12. *Renewable rather than depleting*: Material and energy inputs should be renewable rather than depleting.

Several other sets of principles have been developed by different groups, for example, Poliakoff's mnemonic *PRODUCTIVELY*, that similarly summarizes the basic concepts (**P***revent waste*, **R***enewable materials*, **O***mit derivatization*, **D***egradable products*, **U***se of safe methods*, **C***atalysis*, **T***emperature, pressure ambient*, **I***n-process monitoring*, **V***ery few auxiliaries*, **E***-factor*, **L***ow toxicity*, **Y***es, it is safe*). Certainly, the aforementioned principles were not developed overnight. Several research groups have contributed to the development of the major concepts that guided the growth of green chemistry. Here, we list these basic concepts and definitions that will be used in the later chapters of this book.

E-factor: The *E (environmental)-factor*, developed by Roger Sheldon, is one of the most practical descriptors of the efficiency; it is the mass ratio of waste to the target product. For instance, $E = 20$ means that 20 kg waste is produced to every kilogram of product. Obviously, the smaller the number, the better; in the best possible circumstances (0 kg waste is generated with the product), $E = 0$. The *E-factor* is a commonly accepted and applied measure to describe the efficiency of processes in the chemical and pharmaceutical industry for the assessment of the overall environmental impact.

Atom economy (AE): *Atom economy* (atom efficiency is also used), first described in 1991 by Trost, is defined by the ratio of the molecular weight of the product and the sum of the molecular weights of all substances consumed in the stoichiometric equation of a reaction. Commonly it is expressed as a percentage. It is important to highlight that *AE* is based on the theoretical reaction (i.e., no unexpected by-products are factored in) and 100% theoretical yield. Therefore the AE is the best possible scenario and can be used to assess a reaction at the theoretical level. For example, if a reaction scheme does not involve the formation of any expected by-product, the AE is 100% (Scheme 1.1.1).

However, it is worth noting that it cannot be used exclusively to describe the environmental impact of a reaction: it may be that a reaction with 100% AE yields unexpected by-products (e.g., stereo- or regioisomers) that would decrease the actual AE. Thus a highly selective reaction with 80% theoretical AE and no unexpected by-products may have less

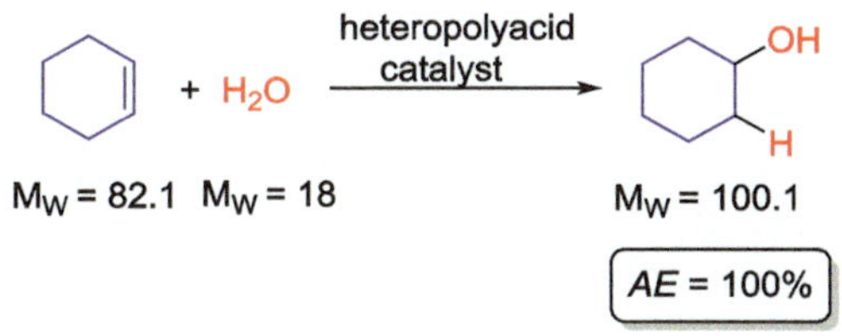

SCHEME 1.1.1 Hydration of cyclohexene to cyclohexanol; a 100% atom economic process.

environmental impact than a 100% AE reaction that is accompanied by extensive unexpected by-product formation.

Although these are the two most important measures to describe the efficiency of a chemical process, several alternative metrics have been proposed, such as the *reaction mass efficiency* (*RME*) defined as the mass ratio of the obtained product to the total mass of the reactants, thus incorporating the actual percent yield into traditional *AE* calculations. *Carbon efficiency* is similar to *RME*, but only considers carbon as a part of the product or starting materials and reagents. *Mass efficiency* (total mass of the materials used divided by the mass of product obtained given as a percentage) and *effective mass yield* (the ratio of the mass of the desired product and the total mass of nonbenign reactants) are other available ways to describe the environmental impact of processes. There is a general agreement in the literature that the *AE* and *E-factor* are the most applicable and widely used measures in many industries.

Although the aforementioned metrics are able to estimate the impact of a process, there are other considerations to discuss, such as the chemical characteristics of the waste. Obviously, the ultimate process occurs with 100% *AE* and 0 *E-factor*; however, practical processes are different and mostly produce either expected or unexpected by-products that are considered waste. The nature of that waste is highly important. It is easy to realize that if the waste is water (e.g., dehydration reactions) or sodium chloride (nucleophilic substitutions) that are considered harmless, the process is quite different from, to choose one example from many, the Jones oxidation of alcohols to ketones that generates a significant amount of chromium salt waste (Scheme 1.1.2).

To consider this highly important aspect, Sheldon introduced the *environmental quotient* (*EQ*), which is calculated by multiplying the *E-factor* by an arbitrarily assigned *environmental unfriendliness quotient*, *Q*. As an example, one can assign 1 to benign chemicals (such as water or NaCl) and a large number (e.g., 100 or 1000) to chromium sulfate. Although as of yet a clear definition or individual assignment of *Q* values to chemicals is not available, theoretically it is possible to quantify the environmental impact of chemical waste based on its amount and toxicity. It is, however, a difficult task as compounds exhibit all sorts of harmful biological activities, and although it is possible to rank them in terms of one effect (e.g., carcinogenicity or mutagenicity), the ranking (and the *Q* value) could be significantly different if another type of toxicity is considered. In addition, when a process is generating a compound that is not yet known, its biological effect only can be estimated, e.g., by quantitative structure-activity relationship models.

An even more extended approach that embraces the holistic evaluation of a process or product is the *life cycle assessment (LCA)*. The *LCA* considers a broad range of issues that

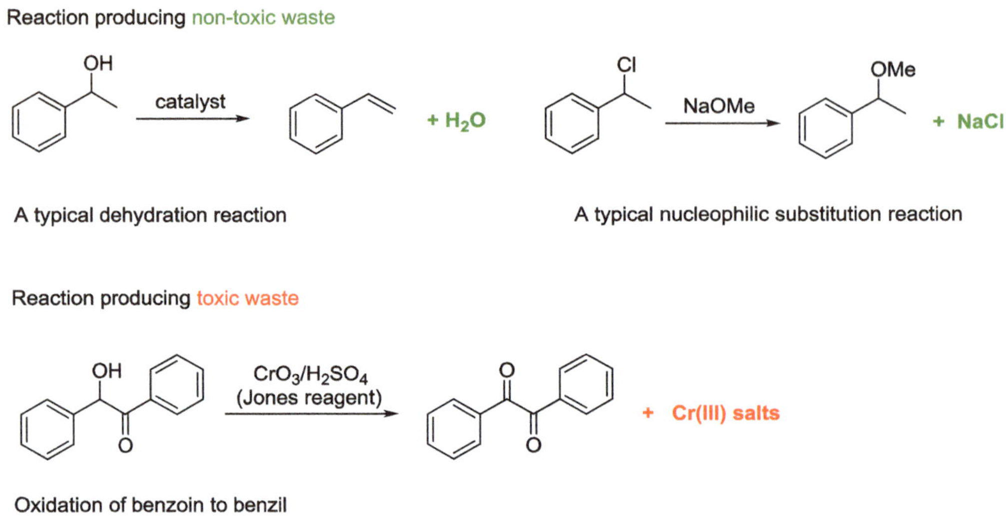

SCHEME 1.1.2 Examples of chemical reactions producing toxic (red) and non-toxic (green) waste.

can be quantified by environmental friendliness metrics. These indicators include energy consumption, carbon footprint and emission of other greenhouse gases, potential contributions to ozone depletion or smog formation, amount of waste generation, and toxicity of the waste generated. The application of *EQ* or *LCA*, however, requires extended analysis before a process can be implemented, and this often contributes to the cost of a product. Nonetheless, as highlighted by the original 12 principles, it is always better (and likely less expensive) to prevent problems by a thorough analysis than to clean up after an environmental disaster.

1.1.3 GREEN CHEMISTRY AND THE ENVIRONMENT

It is not within the scope of this text to provide a comprehensive examination of environmental chemistry. Indeed, many fine textbooks exist on that subject, and our goal is not to reduplicate such works. Rather, the goal of the various environmental chemistry chapters in this text is to put into context the impact of human society on the natural world. While the precise placement of the dividing line between green chemistry and environmental chemistry can be debated, it is clear that the entire purpose of green chemistry is to minimize that impact. This is most evident in principles 1 and 10: waste prevention and the design of chemicals such that they degrade harmlessly in the environment. Obviously, then, the practice of green chemistry requires an understanding of those degradation pathways and an understanding of what happens to the waste and by-products when they are emitted. Although various textbooks exist that provide greener pathways for industrial synthesis, for example, there is an alarming shortage of texts that *train* green chemists to think about the chemistry of their products in the wild. With this work, we hope to begin to bridge that gap.

It may come as a surprise to some readers that society has long been aware of the impacts of science and industry on the environment. The Mishnah laws of first- and second-century Israel specified that threshing floors, leather tanneries, and lime kilns be removed 50 cubits from the city to reduce exposure to airborne pollutants, and that flax stems be soaked at least 4–5 m from any vegetable fields to prevent water pollution from affecting the neighbor's food crops. The impetus for the construction of the Roman aqueducts was to transport clean drinking water, as the Tiber had become so fouled with human waste. English laws attempted to curb pollution from the burning of coal as early as 1273, but the wave of the Industrial Revolution overwhelmed those early efforts.

Even if we as a species were unaware of our effect on the environment, there has been pollution as long as there has been civilization. There is evidence in ice core data that the expansion of agriculture by the Romans and the cultivation of rice by the Han dynasty in China, both in 1st century BCE, led to measurable increases in global methane concentrations. Looking even further into the past, there is clear evidence of heavy metal pollution of soil and water arising from metallurgy as long ago as 1500 BCE and continuing for nearly 2000 years. Indeed, significant levels of these pollutants measured in modern lakes may in fact be sourced to ancient industry rather than to more modern endeavors.

Ancient impacts aside, it is clear that since the Industrial Revolution the scale and the character of the pollution has fundamentally changed. Perhaps we can date this to 17th century England, when John Evelyn wrote of the damage caused by London's coal smoke in his pamphlet, *Fumifugium*, although even this document refers to the history of England's problems with coal dating back to the middle ages. Perhaps it dates to the cholera outbreaks around the world in the 19th century, caused by contaminated drinking water in the growing cities; or perhaps the burning of the Cuyahoga River, most notably in 1952, a result of the accumulation of oil slicks on its surface; or the smog incidents of the mid-20th century in London, and Pennsylvania and Belgium, in which thousands of people died; or the widespread use of DDT after World War II, now found in animal tissue samples from the most remote locations on Earth; or the tragedy of Love Canal, where people living in houses built on a landfill were exposed to toxic waste as the containers leeched into the soil; or the 1984 Union Carbide disaster in Bhopal, India, where 4000 people died from exposure to methyl isocyanate; or the photochemical smog in Los Angeles during the 1980s, or that of Mexico City, Delhi, and Beijing today; or Three Mile Island, Chernobyl, and Fukushima, reminding us that the wonders of the nuclear age bring with them new dangers. It is true that human science and technology have combined to produce a society that is awe inspiring. However, it is also true that this society has wreaked havoc on our natural environment.

According to most geology textbooks, the most recent geological epoch began nearly 12,000 years ago with the dawn of the Holocene. However, the conversation in recent years has recognized the shift in mankind's ability to harm our planet on a global scale. In 2000, Nobel laureate Paul Crutzen argued for the use of the term Anthropocene, referencing that our current period is defined by our species more than by any other characteristic. Crutzen did not coin the phrase—it had been used by Soviet scientists as early as the 1960s, and perhaps even predates them. However, the dawn of the 21st century was also the dawn of widespread acceptance of the harsh realities of global warming and its associated climate changes. Already the pH of the oceans is dropping, leading to the distinct possibility of mass extinction of life in the sea. The ongoing "Holocene extinction" is removing up to 140,000 species from our planet each year, a rate that rivals that of the extinction that took the dinosaurs. Left unchecked, the modern tail of the Industrial Revolution threatens all life on the planet. Often doomsayers will claim that we are on the verge of destroying the planet itself. This is presumably hyperbole—the planet has seen mass extinction before, and will again. However, the same cannot be said of the human race.

To be clear, the authors do not believe that we are witnessing the end of human civilization. Because one way or another, our destruction of the environment *will* be checked. At some point, the financial arguments against change will fall by the wayside as the need for change becomes more urgent. Whether driven by government intervention or an industrial recognition of the bottom line, at some point the solutions to these problems will become financially competitive with the cost of doing nothing. It falls to practitioners of green chemistry to provide the solutions at a cost that obtains that result as soon as possible.

1.1.4 REGULATORY AGENCIES

Environmental regulatory agencies are part of life in most countries. They provide guidance for new developments and oversight for existing industrial technologies as well as common aspects of life, from the disposal of restaurant waste to application of cosmetics. Since the detailed discussion of environmental law and regulatory agencies is far beyond the scope of this introductory chapter, here we describe several of most visible regulatory agencies that are charged with managing the environmental issues in the largest economies.

1.1.4.1 International: The United Nations

The United Nations (UN) took a leadership role in facilitating discussions, organizing international conferences where the member nations could develop strategies to combat environmental issues. Per the suggestion of Sweden, the UN organized the *United Nations Conference on Human Environment* in Stockholm in 1972. The assembly agreed upon the *Stockholm Declaration,* which put forth 26 principles to guide environmental protection and development. The *World Commission for Environment and Development,* first chaired by Gro Harlem Brundtland (former Prime Minister of Norway), was established in 1983. It was tasked with generating a report on the environment and with making recommendations for a worldwide sustainable and environmentally benign economic development to 2000 and beyond. The assessment and recommendations of the commission were published in a book entitled *Our Common Future* (Oxford University Press) in 1987. Since then, the UN has spearheaded efforts on various environmental issues.

One enormously successful early agreement was the *Montreal Protocol* (1987) to combat the depletion of ozone over the poles in spring, commonly referred to as the "ozone hole." The global ban on chlorofluorocarbons and related compounds succeeded in stopping the deterioration, and the atmosphere continues to recover slowly but surely. The success of the Montreal Protocol was remarkable; former UN Secretary-General Kofi Annan hailed it as "perhaps the single most successful international agreement to date." This is in marked contrast to the international community's much slower and more contested response to global warming. The most important and well-known treaties to combat climate change are the *Kyoto Protocol* (1997) and the *Paris Agreement* (2015). The Kyoto Protocol described the commitment of the participating countries to reduce the emission of greenhouse gases. The Paris Agreement, the result of the *UN Framework Convention on Climate Change*, while aiming for a similar goal, made recommendations on how much to limit the annual temperature increase of the planet. As of today, 131 (out of 197) parties have ratified the Paris Agreement.

This leads us to mention the weaknesses of some of the UN-facilitated agreements. Given the nature of the UN, these treaties are negotiated by the governments of the participating countries. However, once the agreement is signed, the law-making bodies of the nations have to ratify it; thus the countries essentially commit themselves by their own laws to uphold the agreement. That has been so far the Achilles' heel of many such agreements: several countries ratified it, whereas many others did not. The reasons for not ratifying vary from country to country and depend on economic development, energy/fuel production and use, and many other socioeconomic factors. Unfortunately, many of these agreements are purely political and do not include the development of actual technologies as a response to global warming.

1.1.4.2 International: International Organization for Standardization

The International Organization for Standardization (ISO) is a nongovernmental international organization with 161 members that are commonly the similar standards bodies of the member countries. In 1946, 25 countries decided to establish the ISO to provide unified industrial standards. It is a forum to share knowledge and develop consensus-based international standards that help innovation and offer solutions to global problems. Although the ISO is not specifically an environmental organization, many of its approximately 21,000 international standards are related to environmental and safety issues. The most relevant of these are ISO1400-Environmental Management, ISO45001-Occupational Health and Safety, ISO50001-Energy Management, ISO22000-Food Safety Management, and ISO31000-Risk Management. However, it is important to note that similar to the above UN recommendations the ISO standards are applied voluntarily by the individual organizations (e.g., corporations) of the member countries.

1.1.4.3 United States

We have already introduced the US EPA earlier in this chapter, which takes the lead on many issues from toxic waste cleanup to mitigation of global warming. The EPA fulfills the roles of a regulatory agency and recommends changes and additions to current environmental laws that Congress considers. The EPA, however, is not just an agency that regulates and enforces regulations. It has a broad network of research facilities, institutes in which far-reaching scientific research is conducted in several areas of environment-related sciences, from sustainable synthesis to atmospheric chemistry. In addition, all US states have their own EPA-like agencies, sometimes just one umbrellalike agency; however, in many states, there are separate entities to deal with separate issues (water, agriculture, etc.).

The *National Oceanic and Atmospheric Administration* (*NOAA*) (part of the Department of Commerce) also concerns itself with environmental issues. It grew out of some of the oldest government branches dedicated to the environment, including the United States Coast and Geodetic Survey, which was first established in 1807 by Thomas Jefferson. In 1970, President Nixon created NOAA from several other agencies, including the Weather Bureau (1870) and the Bureau of Commercial Fisheries (1871). Its current substructure includes the National Weather Service, the National Ocean Service, and the Environmental Satellite, Data and Information Service, among others. Through these divisions it contributes by observing and communicating data related to global warming and atmospheric and water pollution. Similar to the EPA, the NOAA has facilities that conduct active research on the environment.

1.1.4.4 Canada

The Environment and Climate Change Canada is the major government agency that regulates and enforces environmental protection in Canada. It was established in 1971 by the Department of Environment Act to assess, monitor, and protect the environment, including providing basic weather and meteorological services to the citizens of Canada. The agency's responsibilities are those of a typical environmental agency: making environmental decisions/regulations based on available evidence, especially with regard to pollution prevention and the like. In addition to its support to policy making, the agency is a supporter of a broad variety of environment-related research through many funding initiatives. Just as in the United States, the Canadian Provinces have their own environmental agencies.

1.1.4.5 European Union

The European Union established its main environmental agency, the European Environment Agency (EEA), in 1990, which became operational in 1994. The agency has 33 member states (28 EU members and Norway, Iceland, Lichtenstein, Switzerland, and Turkey). Six additional countries from the Balkans work with the Agency as cooperating countries. Given the nature of the European Union, most member countries had established their own environmental agencies long before the EEA, such as the Federal Environmental Agency (and others) in Germany (1974), the Environment Agency in the United Kingdom (1995), the Ministry of Ecology, Sustainable Development and Energy in France (since 1974 under various names), just to name a few. Thus the EEA's role is mainly to provide independent information on the environment. They are the major EU information source for policy makers as well as the general public, integrating the principals of sustainability into political and economic decisions.

1.1.4.6 Russia

The *Ministry of Natural Resources and the Environment* (*MNRE*) is the main policy-making and enforcing body in Russia. Russia has had some sort of natural resources governmental unit since the 18th century. The *MNRE* was created in 2008 by merging the former Ministry of Environment and Ministry of Natural Resources (both founded in 1996, after the collapse of the Soviet Union). It has several agencies such as the Federal Service for the Supervision of Natural Resources, the Service for Hydrometeorology and Environmental Monitoring, and separate agencies for subsoil, water, and forestry management.

1.1.4.7 China

Immediately after the Stockholm Declaration, China created its Environmental Protection Leadership Group (1974) and environmental protection became a state policy in 1983 with the Environmental Protection Commission. After several upgrades and name changes, the current Ministry of Environmental Protection (MEP) was established in 2008. China, the most populous country in the world, has seen an unprecedented industrial growth and urbanization since the 1980s, which has brought with it significant environmental problems, including water and air pollution. The major mandates of the MEP include the design, organization, and implementation of national policies, programs, and plans for environmental protection, policy making, and regulations and leading the response to major environmental problems. In addition, it carries out environmental protection science and technological activities including the organization of projects on engineering, and facilitates the development of environmental technology management systems as well as conducts and organizes environmental education.

1.1.4.8 India

India, as the second most populous country in the world, with rapidly growing industrial activity has its fair share of environmental problems. In addition to industrial accidents, water shortages, soil problems (e.g., exhaustion, erosion), deforestation, and, especially in the major metropolitan areas, air and water pollution affect many areas in the country. To address these problems, the Central Pollution Control Board (founded in 1974) created the National Air Quality Monitoring Program. Due to the inspirational power of the Stockholm Declaration, in 1972, the National Council for Environmental Policy and Planning within the Department of Science and Technology was established to protect the environment. This council later became the Ministry of Environment and Forest, which is India's most important governmental agency for environmental protection. The Environment Protection Act of 1986 is one of the major early milestones in its actions. The current legislative framework includes climate change, deforestation management, coastal regulations zones, and pollution control, among many other tasks.

1.1.4.9 Japan

Many countries started seriously considering environmental protection only after the Stockholm Declaration (1972), whereas Japan was among the first countries to enact environmental regulations after four major pollution outbreaks occurred in the country in the 1950s–60s. In 1958, a water quality conservation law was passed, which was followed in

1962 by a smoke and soot emission control law. In 1971, the first Environmental Agency was established in Japan. Currently, the Ministry of the Environment is the branch of the Government of Japan that is responsible for activities related to the environment. This agency has broad responsibilities, as it is involved in policy making concerning waste management and recycling, pollution control, nature conservation, wildlife protection, air quality and transportation, as well as health and chemicals, and is also charged with the care of Japan's national parks.

1.1.4.10 Australia

The Department of Environment and Energy (DEE) is the major government agency in Australia in charge of environmental protection. It designs and implements government policies and programs to protect and conserve the environment, water, and heritage, and promotes climate action. DEE deals with a broad array of activities, such as far-reaching environmental protection of Australian air, land, and water; managing the national parks; conducting research on environmental problems; as well as acting as a funding agency for environmental research. Similar to the United States and Canada, the Australian states also have their own individual environmental agencies.

1.1.5 CLOSING THOUGHTS

The nature and scope of the field of chemistry has changed dramatically since the days of the alchemists. Chemistry impacts nearly every aspect of modern life. The progression has not always been smooth, and there have been significant missteps along the way. However, the harnessing of chemistry in the interest of society has brought us to a modern age unimaginable by our ancestors. Toward the end of the 20th century, the field of green chemistry was born from a recognition of those missteps and a desire to minimize the impact of human society, and especially of human industry, on our natural environment. The core philosophy can be expressed in various ways, with lists of principles and metrics. The goal is always the same: benign by design.

Societal impact on the environment presents us with multiple pressing problems, which can be addressed by various means. All things considered, it would be preferable if chemical solutions could be found before political solutions are required. This could entail, for example, a reimagining of an existing technology (e.g., the electric car, biodegradable plastics), a means to reduce the environmental impact (e.g., solvent-free synthesis, exhaust plume scrubbers), or a technology to remove existing pollutants (e.g., CO_2 sequestration, bioreactor landfills). By finding a way for companies to profit from the mitigation of pollution, green chemists can encourage society to adopt these changes much more quickly and completely than would a reluctant culture enforced by governmental policies. This is not to minimize the importance of effective policies to control pollution; as we have seen, these efforts are critical to solving national- and global-scale problems. However, one of the goals of green chemistry must be to better facilitate widespread adoption of these new technologies and new methodologies by making them acceptable for everyone.

PROBLEMS

1. Calculate the atom economy and E-factor of the following processes:

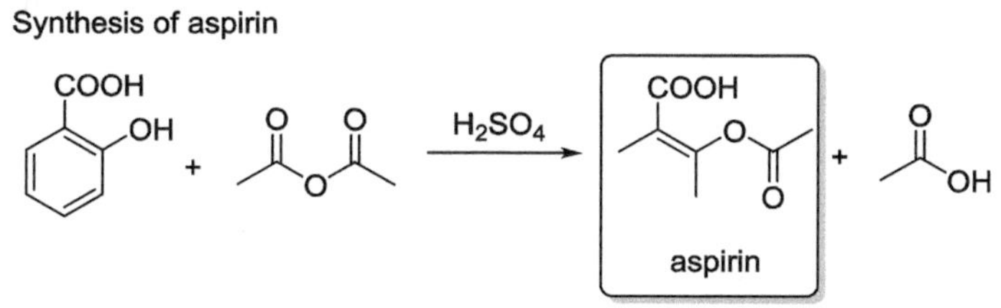

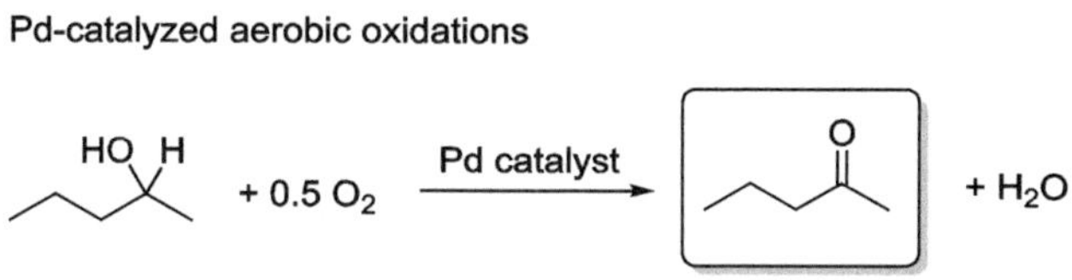

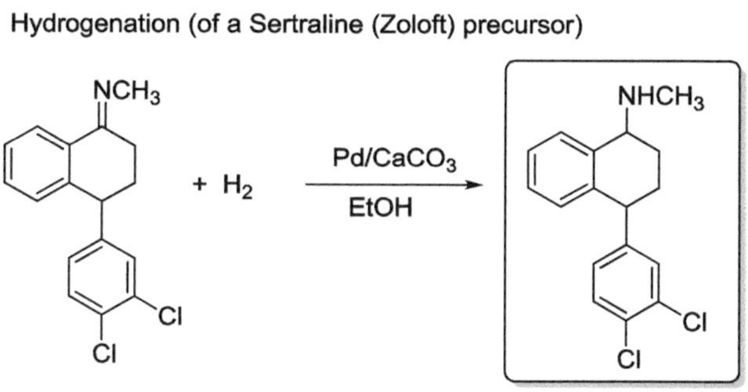

2. Reactions A and B are two different methods producing the same product P. Reaction A has a theoretical atom economy of 100% with an actual yield of 65% for the product. Reaction B only possesses 85% atom economy, with 95% actual yield for the product. Which reaction is greener, that is, generates less waste considering 100% conversion of A and B?
3. Research the similarities and differences between environmental quotient and life cycle assessment.
4. Select any chemical product and design a green pathway for it considering the complete process (all details from finding/manufacturing starting materials, other components, minimizing environmental impact, etc.). Show your work in a flowchart from the beginning of the manufacturing phase to the likely fate of the product/postproduct in the environment.

5. Solar panels are a popular product marketed widely to offer consumers the opportunity to harness the energy of the sun. Research the manufacturing process of solar panels and prepare an in-depth analysis on the cost and environmental impact of their production. Compare the environmental impact of generating energy using solar panels through the complete lifetime of these panels with that of the same amount energy generated by a coal-based power plant.
6. Several times in this chapter, the smoke from coal burning was mentioned as being especially damaging to the environment and to human health. Research what compounds in coal smoke are primarily responsible for these effects.
7. John Evelyn's *Fumifugium* detailed the health impacts of coal burning in 17th century London and was written to request intervention by the Parliament and the King. Investigate the alarming health statistics that led to his appeal.
8. How does an increase in pastoral agriculture result in an increase in methane emissions? What about rice cultivation?
9. The history of environmental pollution is often the history of unanticipated consequences. Two excellent examples of this are the stories of DDT and CFCs.
   a. In 1948, the Nobel Prize in medicine was awarded for the invention of DDT to combat disease in the aftermath of World War II. Why was DDT so effective in this role? What is it about DDT that made its widespread use so problematic?
   b. CFCs were one of the "miracle" inventions of the 20th century, replacing toxic and/or flammable gases in a variety of applications. What were some of these applications and what were the gases that they replaced?
10. A staggering variety of organic pollutants were found in the soil, water, and air samples taken from Love Canal in 1977. One of the highest concentrations was of benzene. Research the toxicity of benzene, including its mode of action and its mandated concentration limits in the environment.
11. Analyze the 1997 Kyoto Protocol and 2015 Paris Agreement and describe the major points in both. By comparing the two, summarize the development that occurred during the nearly 20 years that passed between the two agreements.

Recommended Reading

1. Anastas PT, Warner JC. *Green chemistry: theory and practice*. New York: Oxford University Press; 1998.
2. Lancaster M. *Green chemistry – an introductory text*. 3rd ed. Cambridge: RSC; 2016.
3. Matlack AS. *Introduction to green chemistry*. 2nd ed. New York: CRC Press, Taylor & Francis; 2010.
4. Kovacs L, Csupor D, Lente G, Gunda T. *100 chemical myths: misconceptions, misunderstandings, explanations*. Heidelberg, New York, Dordrecht, London: Springer Cham; 2014.
5. Li C-J, Anastas PT. Green chemistry: present and future. *Chem Soc Rev* 2012;**41**(4):1413–4.
6. vanLoon GW, Duffy SJ. *Environmental chemistry: a global perspective*. 3rd ed. Oxford: Oxford University Press; 2010.
7. Spiro TG, Purvis-Roberts KL, Stigliani WM. *Chemistry of the environment*. 3rd ed. California: University Science Books; 2012.
8. Baird C, Cann M. *Environmental chemistry*. 5th ed. New York: WH Freeman; 2012.

PART 2

CHEMICAL ISSUES OF MODERN SOCIETY

CHAPTER

2.1

Environmental Chemistry, Renewable Energy, and Global Policy

Heather A. Rypkema

Heritage Strategies, INTL, Washington, DC, United States

2.1.1 INTRODUCTION

The future welfare of our planet depends on a thorough understanding of our current technology and its effects on the environment. Earlier manifestations of mankind, in which the height of environmental impact consisted of the occasional slaughtering of a large mammal, had little or no impact on the world that surrounded it. Today, however, almost every aspect of our lifestyle has a profound effect on the macrocosm of Earth's ecosystems. Most daily activities contribute to this impact in varied, and often unexpected, ways.

Most people realize that the mundane act of driving a car contributes to one's carbon footprint, thereby pouring carbon dioxide into the atmosphere, which contributes to global warming, more currently enveloped into the phenomenon of climate change. But what impact does this really have? What is the implicit environmental effect of trading in a traditional vehicle for a hybrid—does the increased fuel economy and reduced carbon footprint counterbalance the addition of an otherwise usable vehicle to the exploding mass of metal-contaminated landfill? And, more generally, what is the difference between global warming and climate change?

There are national movements to reduce the consumption of bottled water, but why? Most obviously, there is the economic element of paying $1 or more for a bottle of water, when the equivalent amount of tap water costs 0.1 cents. Additionally, there is the environmental waste impact of the plastic container. What are the long-term environmental impacts of the plastic bottle? In 2015, 11.7 billion gallons of bottled water were consumed in the United States, a volume that equates to more than 75 billion individual 20-oz bottles, more than 60% of which ultimately end up in landfills. Furthermore, how much energy is expended just to produce this packaging? The annual energy cost of producing plastic bottles to contain commercial

Green Chemistry
http://dx.doi.org/10.1016/B978-0-12-809270-5.00002-9

water is enough to fuel 1.3 million cars for a year. It is easy to muster outrage at a cartoon image of a glowing, gelatinous goop being dumped into a pristine river full of happy, jumping fish, but too few people realize that even the most innocent and cleanest of man-made products can carry a hidden environmental threat.

The ambiguous phrase, "pollution," has been used for decades to excoriate emissions from industry and other sources, but it is rarely explained in detail through mainstream media coverage because to truly understand the threat of pollution, one must first understand the chemistry of the pollutants being expelled. Yes, pollutants are bad, but what precisely are the climatological, ecological, and economic impacts of their presence? How and why are they so detrimental to ecosystems and human life? Without understanding these concepts, one is poorly armed to argue against them. And without well-researched dissent, there can be no change to the status quo of environmental pollution.

Today's technology offers myriad pathways for improving the life of any human being with access to it, but these advancements come with a cost. Almost any piece of technology you use on a daily basis has a hidden price of environmental degradation. Cars, as previously mentioned, contribute to your carbon footprint and add dangerous radicals to the atmosphere. Smartphones are now ubiquitous in the developed world and beyond, but the disposal of faulty or unwanted phones provides a new challenge toward chemical remediation in landfills. Nearly every industry associated with natural resources, mining, agriculture, drilling, etc., has a deleterious environmental impact that must ultimately be eliminated if we are to cooperatively maintain the health of our planet.

The fundamental purposes of green chemistry are to:

1. Identify environmental threats.
2. Understand the chemical processes leading to environmental threats.
3. Analyze how and why these threats are occurring.
4. Devise a way to alter current technology to avoid these problems.
5. Determine how to remediate damage already inflicted.

Each of these points is equally important, and the solution to any one of them will require a wide variety of chemical and other scientific expertise. Nevertheless, a thorough understanding of green chemistry can allow a scientist to initiate the process of resolution for any given step.

2.1.2 ENVIRONMENTAL CHALLENGES

2.1.2.1 Challenges by Air

Air pollution is one of the most immediate and visible forms of environmental harm enacted by human technological advances and the source of many of the early environmental disasters in human history. While far from being the earliest incident, the London killer fog is one of the most notorious. In the winter of 1952, an unexpected cold snap hit London, causing a rise in the level of coal burning, which infused the city's air with sulfurous gas. An inversion layer, in which higher altitude air is warmer than that nearer the ground, trapped the pollutants at ground level, resulting in a toxic miasma throughout the city. Exposure to sulfurous compounds such as sulfuric acid (H_2SO_4) and sulfur trioxide (SO_3) resulted in damage to the eyes, lungs, and gastrointestinal tracts of thousands of people. This convergence of human pollution and meteorological happenstance is estimated to have caused the death of more than 12,000 people, with young children and the elderly being particularly susceptible targets. This disaster led to the Clean Air Act of 1956, which was passed by the parliament to impose restrictions on urban coal burning, and encouraged homeowners to upgrade to more modern technologies for climate control (Fig. 2.1.1).

Although the killer fog is an exceptional example, it highlights the dangers of anthropogenic chemical emissions, which have increased exponentially in the half century since the event itself. Furthermore, direct toxicity represents only a small portion of the environmental threats caused by human production. Environmental scientists must also consider more subtle effects, such as the ultimate fate of a chemical in the atmosphere—how it will react and what by-products it will produce—as well as its interaction with the bombardment of radiation—infrared radiation (IR) from the earth and visible and ultraviolet (UV) from the sun—that form a ceaseless flux of energy through our atmosphere.

Another significant example is the Bhopal, India, disaster of 1984, in which a leak from a pesticide plant resulted in the death of at least 4000 people and injury to more than 500,000. In this incident, at least 30 tons of methyl isocyanate (CH_3NCO) was released from the plant to the surrounding area. Additional toxins are believed to have included phosgene ($COCl_2$), chloroform ($CHCl_3$), hydrogen chloride (HCl), and further organic pollutants, which were either included in the tanks or were formed subsequently in the atmosphere. Children were blinded and thousands of animals were killed, and Union Carbide, the company that operated the plant, resumed operations without cleaning up the site (Fig. 2.1.2).

The challenges of environmental scientists in the atmospheric arena are multifold. They must identify potentially toxic emissions, predict the fate of these emissions after exposure to the chemical context of the atmosphere, develop alternative technologies to supplant

FIGURE 2.1.1 Westminster, London, 1952. A toxic fog has enveloped London due to increased coal burning in an unexpected cold snap. Thousands of people died as a result of this deadly miasma. *Image credit: George Tsiagalakis.*

FIGURE 2.1.2 Remains of the Bhopal chemical plant, following the 1984 disaster in which more than 4000 people died as a result of substandard safety precautions. *Image credit: Julian Nitzsche.*

harmful substances, and advocate for policy changes, both local and international, that might help prevent further environmental disasters from taking place.

2.1.2.2 Challenges by Sea

The oceans and natural waterways of our planet are particularly vulnerable to the damage posed by environmental contaminants. Exposure to pollutants can occur through runoff from common industries such as mining, direct dumping from industrial plants, pesticides transported through rainwater, or multitudinous other mechanisms (Fig. 2.1.3).

This waste poses a threat to both humans and wildlife, poisoning the species it encounters as well as the humans who harvest them or drink the polluted water. The environmental impact of water pollutants can affect humans and animals directly, or indirectly through the transformations that occur via the chemical or biological environment. Microbes can transform neutral species into toxic ones, and the acidity or basicity of waterways can alter the reactivity of the otherwise harmless chemical compounds.

Although nearly every human industry contributes to water pollution, the most infamous ecological disasters have come from the petrochemical industry. In 1989, the Exxon Valdez oil tanker struck a reef off the coast of Alaska, spilling almost 11 million gallons of oil into open waters (Fig. 2.1.4).

FIGURE 2.1.3 Waste being dumped into a natural waterway. Chemical waste deposited directly into natural waters is an efficient way to spread ecological harm. *Image Credit: Frank J. Aleksandrowicz.*

FIGURE 2.1.4 The Exxon Valdez oil tanker ran aground in 1989. At the time, this was one of the worst ecological disasters on record, resulting in the death of between 100,000 and 250,000 animals, including sea birds, seals, orcas, and otters. *Photo courtesy of the Office of Response and Restoration, National Ocean Service, National Oceanic and Atmospheric Administration.*

This disaster was in part remediated by the dumping of solvents and surfactants into the affected area, although these chemical dispersants had not been thoroughly tested for toxicity to humans or wildlife. In fact, one of the components, 2-butoxyethanol, was later associated with organ, nervous system, and blood disorders among the cleanup crew. As recently as 2014, scientists reported ongoing disruption to the wildlife. Later remediation attempts involved the dispersal of hot water, which devastated the population of marine microorganisms at the base of the local food chain. A better strategy for the isolation and removal of this chemical threat might have saved the lives of thousands of animals (Fig. 2.1.5).

A more recent ecological disaster associated with oil is the BP oil spill in the Gulf of Mexico, also known as the Deepwater Horizon oil spill. In this 2010 disaster, an offshore oil rig caught fire, resulting in the largest ocean contamination event in human history. Over 130 million gallons of oil were leaked into the gulf, with devastating ecological effects. Although the amount of wildlife fatalities is difficult to pin down, it is estimated that hundreds of thousands of sea birds were killed, along with innumerable invertebrates, and the dolphin fatality rate more than doubled in the years following the spill. Remediation was accomplished by physical barriers and chemical dispersants, whose purpose is to break the hydrophobic oil down into smaller droplets that could be more readily incorporated into water, so as to be more easily dispersed by tides and natural diffusion. Still, our knowledge of the long-term ecological effects of these dispersants remains limited (Fig. 2.1.6).

Sea-based pollution is an ongoing threat to our ecological welfare as human technology expands and its population increases. Oil spills threaten a wide range of ocean-based biota, and

FIGURE 2.1.5 Birds killed by oil contamination from the Exxon Mobil oil spill in 1989. This disaster resulted in the death of between 100,000 and 250,000 sea birds and other animals.

FIGURE 2.1.6 The Deepwater Horizon oil rig being extinguished by a fleet of boats. This spill, which contaminated the Gulf of Mexico and a large proportion of the US southern coast, was the largest oil spill in American history. *Photo courtesy of the U.S. Coast Guard.*

the sanctity of our drinking water is constantly at risk. Humans, animals, and ecosystems are under threat, and we must take every precaution to protect them from chemical pollutants.

2.1.2.3 Challenges by Land

Human refuse, especially in first-world countries, is now at an all-time high. In the past century, the human population has exploded, increasing by more than 5 billion people. This population boom, combined with the development and fast integration of plastics into consumer technology, has produced a massive increase in the amount of landfill. In the United States alone, more than 250 million tons of waste are generated annually, which is the equivalent of almost 30 million *Tyrannosaurus rex* or more than 80 million elephants. This figure has tripled in the past 50 years, with only a small fraction (10% in the United States) of the refuse being recycled (Fig. 2.1.7).

However, landfill alone is far from the only environmental threat posed to the land by human activity. In 1986, a nuclear power plant in Chernobyl, Ukraine, suffered a catastrophic chain reaction, causing an explosion in which plumes of radioactive material were sent into the atmosphere, to precipitate onto much of the landmass of the then USSR and Europe. This event resulted in the evacuation of 43,000 residents from the nearby city of Pripyat, caused the death of dozens of plant and first-response workers, and remains the largest nuclear disaster in history (Fig. 2.1.8).

FIGURE 2.1.7 A landfill site. The average waste generated per US citizen is almost 5 pounds a day.

FIGURE 2.1.8 School hallway in Pripyat, near the location of the Chernobyl nuclear disaster. A catastrophic explosion left the region riddled with nuclear debris, making it a dangerous habitat for humans, plants, and animals.

The greatest threat to human health from this disaster has heretofore been increased cancer rates. Researchers estimates that as of 2006, around 5000 cases of cancer in Europe can be attributed to the Chernobyl event. However, the impact of this catastrophe on the soil has been longer in manifesting. Radionucleotides, such as Cs-137, Sr-90, and Pu-239, exhibit slow diffusion rates in soil, so the radiation levels remain high even 30 years after the incident. A 30-km-radius exclusion zone has been in place since the disaster to prevent habitation or agricultural efforts, although a few pre-Chernobyl villages do remain occupied. A pine forest in the region surrounding the site was killed off by radioactive dust immediately following the explosion and was subsequently bulldozed to prevent further dissemination of radioactive materials, but the associated soil still shows high levels of radiation. A reduction of regional fauna has also been observed, particularly among birds. Natural grass and forest fires are a continuing threat, as they can release radioactive dust into the atmosphere and further contaminate the surrounding area (Fig. 2.1.9).

One of the most pervasive threats to the land and its biota is from pesticides, a chemical treatment vital to the sustenance of our agricultural economy. The problem is that although pests that destroy crops must be eliminated, most pesticides are nontargeting, meaning that

FIGURE 2.1.9 A radioactivity warning sign in the "Red Forest" near Chernobyl. Pine trees were killed and turned red by radioactive fallout. *Image courtesy of Timm Suess.*

they do not only kill the creatures they are designed to eliminate. Pesticides sprayed on plants can make direct contact with insects, insinuate into the soil and remain pervasive pollutants, be transported into the ground water system, or, of course, be taken up by local biota. Each of these chemical fates poses an environmental threat, and each one must be evaluated.

A recent concern over pesticides is their damaging effect on bee populations. Bees account for a crucial role in sustaining local flora populations through the distribution of pollen, which allows plants to reproduce. In the United States alone, bee pollination efforts have an estimated value of 14 billion dollars a year, due to their continual maintenance of crops including fruits, vegetables, and nuts. Neonicotinoids, a reasonably new class of pesticides, cause acute toxicity in bees, resulting in either death or behavioral changes that impair a bee's ability to pollinate efficiently. The phenomenon of colony collapse disorder involves the sudden death of adult honey bees within a hive and is still under scientific inquiry. Whether this is caused by pesticides, pathogens, or parasites is not yet understood, but it has been found that even small traces of neonicotinoid in an adult bee can induce sublethal effects, in which the bee is disoriented and deprived of essential learning functions (Fig. 2.1.10).

Chemical pollution on land is a triple threat. It can directly affect plant growth in the deposition area and amplify up the food chain via bioaccumulation; it can enter the atmosphere through volatility, or as the result of natural fires; and it can enter the water system through groundwater distribution, where, again, species low on the food chain can absorb the pollutant, which is subsequently accumulated in higher order predators. Each one of these mechanisms presents an ecological threat that must be met by all the tools that science can provide.

FIGURE 2.1.10 Dead bees, overtaken by colony collapse disorder, in which whole colonies of bees are killed to the detriment of local flora, both natural and agricultural. *Image courtesy of Skinkie.*

2.1.3 TOPICS IN ENVIRONMENTAL CHEMISTRY

2.1.3.1 Toxicology

Toxicology is the study of the effect of chemical and biological substances on human health. It is the means by which scientists can evaluate the risks posed by the introduction of man-made or man-imposed species into the environment. These species can emerge through the disposal of industrial waste, chemical runoff from agricultural or other treatment processes, human excretion or disposal of pharmaceuticals, or the collateral deposition of substances such as heavy metals through mining or smelting. Obviously, toxicology is a pivotal field in the advancement of environmental chemistry because understanding what chemicals might lead to negative health consequences, and at what level of exposure, is at the very heart of strategizing the replacement and remediation of such substances. Toxicological studies can involve direct testing on animals, epidemiological studies on humans, or in vitro studies, in which chemicals and potential target species represented by tissue cultures are mixed directly in the laboratory.

However, a further challenge in this field is understanding *why* certain species harm biota the way they do. With the myriad of molecules deposited into the environment by human activity every day, it would be impossible to perform case studies on each one, as each case study is labor and expertise intensive as well as expensive to perform. Instead, toxicologists must turn to a toolbox of methods to help them identify potential threats. One method is the analysis of quantitative structure-activity relationships, wherein specific component substructures in moderate to large molecules can be associated with a particular interaction with a biological substructure resulting in an adverse reaction. A compound's reactive tendencies, including acidity, nucleo- or electrophilicity, and oxidizing capacity all give clues about how it might behave within a biological matrix. This type of technique makes toxicology a predictive as well as a preventative science. An understanding of the toxic behavior in particular molecular sub-structures, or moieties, allows for the identification of potentially problematic moieties in new molecules, targeting a source for research on new drugs and pinpointing the areas for concern that are irrelevant to a drug's functional attributes (Fig. 2.1.11).

2.1.3.2 Soil Chemistry

Soil plays a crucial role in the continuing survival of all plants and animals on the earth. It is responsible for nourishing plants, which form the foundation of most food chains; is the foundation of the agricultural industry; and is the core of the nitrogen cycle, in which nutrients are taken up by the plants and then returned to the soil through the decay of organic material. Soil is also an immensely complex venue for chemical reactivity, composed as it is of colloidal suspensions, humic acids, microbes and other biota, and both organic and inorganic components. Given the wide range of water saturation, acidity, and chemical and biological composition that soils can attain, they present a rich and varied medium for chemical interaction. Chemical pollutants that are not released directly into the air are generally deposited into the soil, where they can be retained, enter the atmosphere through volatility, or be transported into water systems. Thus pollutants in the soil have the potential for exposure to all aspects of our biosphere.

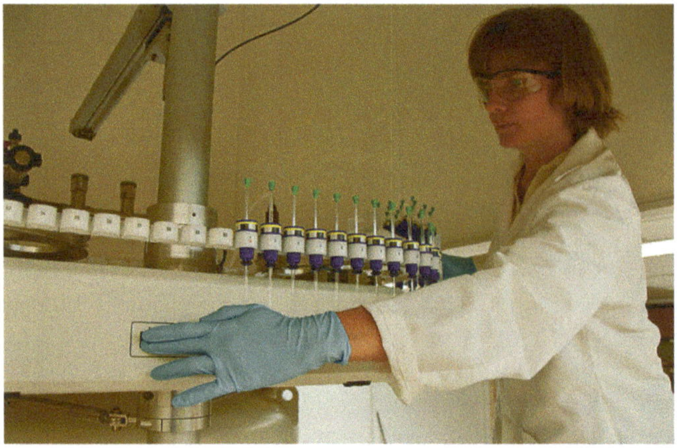

FIGURE 2.1.11 A scientist at the Food and Drug Administration prepares a sample for nuclear magnetic resonance analysis as part of a toxicological study.

Soil contamination can emerge from multiple sources. Mining and pesticides are obvious examples, but contamination can also occur through petrochemical spills or the breakdown of refuse in landfills. In addition to organic materials from chemical runoff such as naphthalene ($C_{10}H_8$), and species associated with the petrochemical industry, heavy metals, such as aluminum (Al), copper (Cu), and lead (Pb), can also pose a threat to natural soils. These contaminants can affect plant growth in the immediate area, as well as induce further environmental harm to animals and humans through further distribution via air and groundwater channels (Fig. 2.1.12).

2.1.3.3 Atmospheric Chemistry

The study of the atmosphere is a key branch of environmental chemistry in that it seeks to explain the multitudinous and complex chemical processes that occur in the mantle of air surrounding our planet. While the most abundant species in the atmosphere are nitrogen (N_2), oxygen (O_2), and argon (Ar), there are innumerable trace species, both natural and anthropogenic, that can initiate and/or participate in reactions deleterious to our natural environment. A constant flux of solar radiation provides a ceaseless source of energy by which reactions can be initiated to produce highly reactive species that could not be formed in a closed thermodynamic system.

One of the most famous problems in atmospheric chemistry is that of stratospheric ozone (O_3) depletion. The ozone layer is a band of unusually concentrated ozone that serves to absorb much of the UV solar radiation impinging the planet. Intense UV light is hazardous to humans and animals, as it can cause DNA damage, which leads to mutation and cancer. The ozone layer provides a protective shield to reduce the amount of UV reaching the earth's surface, thereby protecting the organisms that live on it. However, a number of anthropogenic chemicals, especially oxygenated nitrogen and chlorine compounds, can act as catalysts to destroy ozone and thereby the protection it affords. A mechanism for this catalysis was first proposed by Rowland and Molina and led to the regulation of chlorofluorocarbons

FIGURE 2.1.12 Barrels of chemical waste stored in swampland. These barrels leak not only their potentially toxic contents but also the heavy metals that they are composed of, including aluminum and lead. *Image courtesy of John Messina.*

(CFCs), which were commonly used as refrigerants at that time, in the Montreal Protocol of 1987. The authors received the Nobel Prize for chemistry in 1995. Reduction in CFC emissions has resulted in a reduction of ozone loss, and atmospheric models project a slow but gradual recovery of stratospheric ozone (Fig. 2.1.13).

The high-profile nature of the ozone hole leads many to believe that ozone is a "good" chemical, but in fact it is a perfect example of why context is so important when we evaluate the chemical behavior with respect to the environment. In the stratosphere, which is the layer

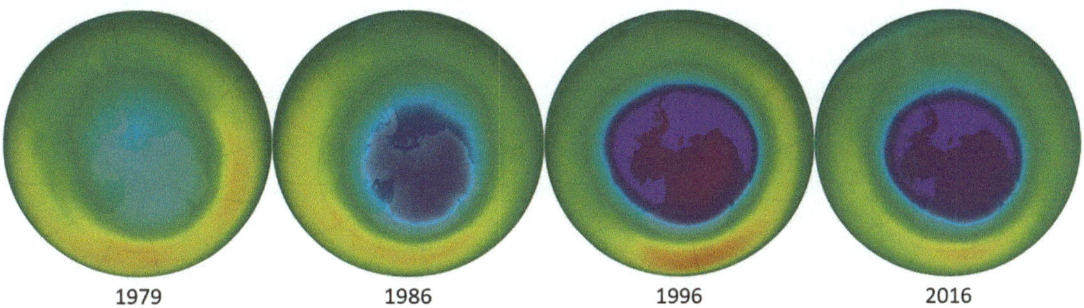

FIGURE 2.1.13 Ozone hole progress over the years. National Aeronautics and Space Administration images of the hole in the ozone layer over Antarctica in September of 1979–2016. Blue (dark gray in print versions) color represents very low ozone concentrations and green (gray in print versions) and yellow (light gray in print versions) show normal "healthy" concentrations. The progression shows the gradual enhancement of the ozone hole over the decades and indicates slowed growth over the past decade. *Satellite images courtesy of NASA.gov.*

of the atmosphere located roughly between 10 and 50 km above sea level, ozone is indeed a beneficent species. However, nearer the earth's surface, it is far from benign. In the troposphere (0–10 km above sea level) it serves as a pollutant, a major constituent of photochemical smog. Ozone can negatively affect the respiratory system of humans and animals and can also be damaging to the growth of crops.

Another famous challenge in atmospheric chemistry is the greenhouse effect. Many people do not understand the difference between the greenhouse effect, global warming, and climate change. Although they are all interrelated, it is important to understand the distinctions to fully grasp the immense threat that they pose to our planet. The greenhouse effect is the mechanism by which certain gases in the atmosphere absorb IR. When sunlight strikes the earth's surface, the energy is transformed and reradiated in the form of IR, which we would generally interpret as heat. Gases in the atmosphere, most notably carbon dioxide (CO_2), water (H_2O), and methane (CH_4), absorb this energy, trapping it within the atmosphere and causing it to increase in temperature. Global warming refers to the increase in mean global temperature caused by an increased greenhouse effect. This increase in temperature is the underlying cause of climate change, which describes the change in weather patterns and environmental conditions that have been increasingly prominent in recent years. The melting of the polar ice caps is an obvious consequence—as the ice melts, the sea levels rise, and more and more land is reclaimed by the sea. This obviously affects both humans and animals through the loss of low-altitude settled land and habitats. However, a particularly cold winter often prompts people to deride the idea of global warming. If the earth is getting warmer, why are the winters colder? This misapprehension is understandable, but dangerous. The answer lies in the excess of energy now roiling through the atmosphere. This energy translates not only as warmth but also as an engine for extreme weather—stronger temperature gradients mean stronger winds and thereby stronger storms: hurricanes, tornadoes, and tsunamis. This excess energy drives more extreme weather, which has devastating economic impact through the loss of property and lives. It is therefore critical that governments continue pushing forward on the restriction of greenhouse gases.

The debate about whether global warming is a man-made phenomenon is debated much more frequently in the media than it is in the scientific literature. It is clear from more than a century of global temperature data that the CO_2 emission surge sparked by the industrial revolution of the 19th century corresponds with a startling rise in Earth's temperature. Although natural cycles and solar phenomena can be used to explain a certain degree of fluctuation, they cannot fully account for the changes that Earth has experienced since man first started expanding his carbon footprint.

Acid rain is another famous phenomenon that has been explained through research in atmospheric chemistry. Many industrial processes result in the expulsion of sulfurous and/or nitrous compounds. When exposed to water, these compounds react to produce sulfuric or nitric acid, both strong acids that can acidify the environment with no energy barrier. They can therefore easily be integrated into the pure water in clouds, imbuing them with an unnaturally high acidity (which corresponds to low pH). The inclusion of a strong acid into clouds results in lowering of the pH of any resultant precipitation, and thereby the production of acid rain. This phenomenon has had a profound impact on ecosystems as well as man-made structures, particularly those made of limestone, also known as calcium carbonate ($CaCO_3$) (Fig. 2.1.14).

FIGURE 2.1.14 A gargoyle damaged by acid rain—the pockmarks represent the dissolution of the stone from acidic precipitation. *Image courtesy of Nino Barbier.*

Because of the delicate chemical and biological balance of many ecosystems, a sudden drop in pH can have devastating effects on the biota that it harbors. As for man-made structures, calcium carbonate ionizes into a weak base (which strongly attracts protons) and acid rain is able to rip apart this type of stone and dissolve it into the accompanying rainwater. The pock marks in the statue in Fig. 2.1.13 provide a visual example of the impact on both the beauty and structural integrity of limestone structures.

Fortunately, regulations have been enacted that have significantly reduced the cultural and ecological impacts of acid rain. Coal plants, which formerly supplied an inordinate amount of sulfurous pollution, are now regularly equipped with scrubbers that remove sulfur compounds before they are emitted into the atmosphere, and catalytic converters on vehicles have substantively reduced the emission of NO_2 and other nitric acid precursors.

2.1.3.4 Water Pollution

Water pollution comes in several forms: organic (carbon based), inorganic (specifically toxic metals), and bacterial. Pollutants can reach natural waterways through a number of sources, including runoff from agricultural pesticides, leaking from waste disposal facilities, wind- and water-borne distribution from landfills, and detritus from ocean-going ships. These sources pose a multitude of threats to the ecosystems that they penetrate, presenting health hazards to both humans and animals.

Some chemicals are soluble in water, which means that they can be quickly taken up into the organisms that encounter them, and, if toxic, they present an immediate impact on the health of those organisms. Others are lipid-soluble, which means that they can be stored in the fat of an organism, remaining nascent until the fat is burned for energy. The most severe effect of this type of toxin is that it can engage in the process of bioaccumulation, whereby many small organisms have a small store of the toxin in their fat. Then a larger organism eats many of these smaller organisms, but cannot process the chemical and ends up storing it in a magnified amount. This pattern continues up the food chain until the apex predators in an ecosystem end up with a level of toxic exposure that either proves fatal or has devastating consequences, such as the impact of dichlorodiphenyltrichloroethane (DDT) on the bald eagle population. DDT is a powerful insecticide, but one that has profound effects on the local ecosystem due to its persistence in readily being absorbed into sediments and soils. While in use, this toxin worked its way up through the food chain through bioaccumulation, resulting in the thinning of eggshells in birds of prey, including the bald eagle, peregrine falcon, and osprey. Eggshell thinning resulted in a significant drop in the reproductive capacity of these and other birds and contributed to the decline in their populations.

Attention to the deleterious effects of DDT was brought to the public by Rachel Carson's *Silent Spring*, a book that addressed the ecological pitfalls associated with mankind's technology. Developed countries began enacting bans on the substance as early as 1968, whereas national bans rolled out slowly, their enactment persisting into the 1980s. Finally, in 2004, an international accord was reached via the Stockholm Convention on Persistent Organic Pollutants, in which severe restrictions were imposed, excepting only those countries with severe malaria outbreaks (Fig. 2.1.15).

One of the most prominent sources of water contamination is from biocides, chemicals designed to kill insects and other pests that range from economically devastating to merely inconvenient. The agricultural industry relies heavily on such compounds to protect crops from invasive insects, thereby ensuring the food supply on which we live. However, insecticides are not specific to the pervasive species and often lead to collateral damage of benign or even beneficent species such as bees, which provide a critical role in the sustainability of local

FIGURE 2.1.15 A bald eagle pair in their nest. This species, the symbol of the United States, was brought to the brink of extinction because of human activity, most significantly the use of dichlorodiphenyltrichloroethane (DDT) as an agricultural pesticide. Since the banning of DDT, the species has rebounded. *Image courtesy of National Park Service.*

flora. Because these insecticides are sprayed onto the crops in an open environment, the chemicals easily make their way into the soil and waterways, and are thereby distributed far beyond the region in which they were applied.

Another major environmental threat to the marine environment is plastics, which are notable for being nonbiodegradable. This means that they cannot be broken down in the natural environment and so persist, accumulating more and more (Fig. 2.1.16).

Another well-known crisis in US water pollution is the lead contamination disaster in Flint, MI (discussed in detail in Chapter 3.6). With their drinking water supply insufficiently treated, city residents suffered the effects of lead poisoning, which include neurological damage that can be particularly devastating among children, among whom the city counts thousands of citizens. Of course, while the Flint crisis made headlines in the United States, it is only due to its rarity in the country. Globally, the problem is much greater. At least 98% of adults and 99% of children adversely affected by lead exposure live in developing countries. People all over the world are exposed to drinking water contaminated with lead and other heavy metals (or metalloids), including arsenic (As), cadmium (Cd), mercury (Hg), copper (Cu), and chromium (Cr). Although these compounds occur naturally in the earth, they are overwhelmingly most often released into the environment by human activities, particularly mining and industrial waste (Fig. 2.1.17).

In addition to heavy metals, biological contaminants such as feces serve as a threat to human water supplies, especially in low- and middle-income countries. The World Health Organization estimates that as of 2016, at least 1.8 billion people, more than 25% of the world's population, must rely solely on drinking water contaminated with human and animal excrement, and the Centers of Disease Control and Prevention reports that more

FIGURE 2.1.16 A seal entangled in ocean trash. The contents of landfills, particularly plastics, which are light, easily dispersed, and nonbiodegradable, have been polluting the oceans in increasing amounts over the past 20 years. *Image courtesy of Pixabay.*

FIGURE 2.1.17 A stream in Australia turned red by contamination from the runoff of mining activities.

than 10% of worldwide fatalities for children younger than 5 years can be attributed to intestinal infections caused by contaminated drinking water.

Nontoxic chemicals can also play a part in damaging the environment. Nutrient-containing compounds such as nitrates that find their way into natural waters can help some species while hindering others, disrupting the delicate balance of the ecosystem. One such example is that of algal blooms. If an excess of nutrients is deposited into a body of water, it can result in a population explosion of microflora such as algae, which reside on or near the surface of the water. The unusually high level of algae blocks sunlight from reaching deeper underwater flora and fauna, to their detriment. As deeper water plant life dies, so do the organisms that depend on them for sustenance, a deprivation that can continue up the food chain to affect both natural predators through the reduction of prey and the welfare of humans, whose fishing targets experience a severe drop in population (Fig. 2.1.18).

2.1.3.5 Emerging Contaminants

New chemicals are developed every day, and it is the onus of environmental scientists to pursue ongoing study into their effects on the environment. One category of emerging chemical species is that of nanoparticles, particles of the scale of nanometers or micrometers that are an aggregation of small molecules. These are typically inorganic and can have shapes ranging from simple spheres to elaborate trees. Nanoparticles have a wide range of valuable technological uses. In the medical field alone, they are used in drug delivery systems; as fluorescent biomarkers to detect damaged tissue due to cancer and other diseases; as an anchor and/or target in photodynamic therapy, in which tumors are burned away without damage to healthy tissue; in gene therapy delivery systems; as biodetectors; and in many more novel

FIGURE 2.1.18 Algal blooms can occur naturally, or be caused by a surfeit of nutrients from industrial waste. This phenomenon can produce a severe disruption in the delicate balance of an ecosystem.

applications. They are also used extensively in renewable energy development, both as semiconductor foundations and as a means to tie organic sensitizers to the semiconducting lattice. However, much of their deposition into the environment comes from less vital applications, such as cosmetics, textiles, and paint pigments. Because these substances are relatively new in the timescale of environmental exposure and ecological impact, little is known about their bioavailability, toxicity, or overall benignity (or otherwise) to the natural world (Fig. 2.1.19).

Another flavor of emerging contaminant is pharmaceuticals. The medicines we ingest are not fully metabolized, and can be emitted into the environment through urine, waste, refuse, and even common body fluids such as sweat. Although the medications that humans take are carefully monitored, dosed, and examined for conflicting effects, there is no oversight into drug interaction once the accumulated pharmaceuticals of an entire community are released into the natural environment. Most drugs are targeted to produce a specific effect in humans, but it is not feasible to test every drug on every animal, let alone chronicle the near-infinite permutations of drug interactions and their effects on the many species inhabiting the local

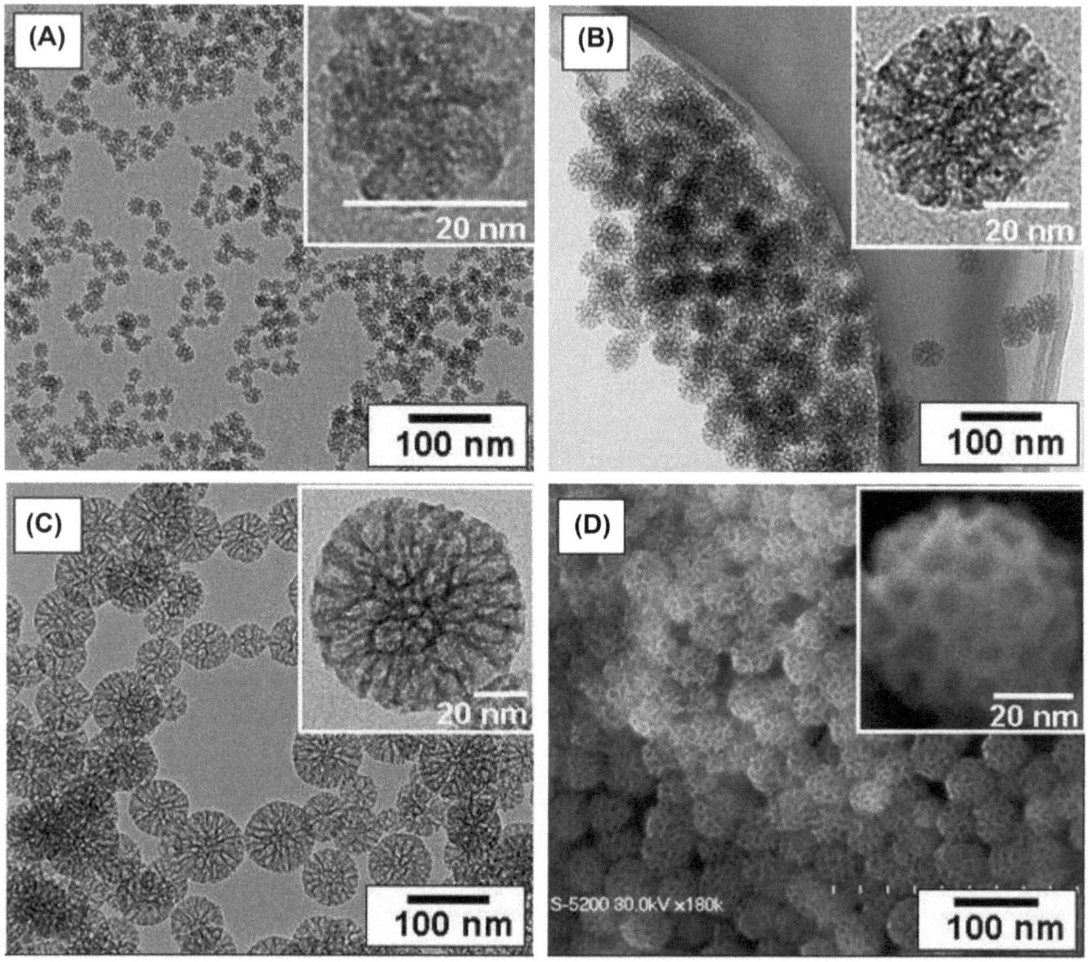

FIGURE 2.1.19 Mesoporous silica nanoparticles of varying diameter as measured by tunneling electron microscopy (A–C) and (D) scanning electron microscopy. Nanoparticles can be simple spheres or far more complex structures. They are used in a variety of applications including catalysis, as antimicrobial agents, and as carriers for chemotherapy drugs, among varied other applications.

environment. Here, the aspect of green chemistry by which harmful substances are identified must come into play, and it may take a long time—a generation or two for wild animals—to even observe any adverse effect. By their very nature, pharmaceuticals have a biological impact, and only time will tell how the introduction of these drugs into an ecosystem will affect the organisms within it.

2.1.3.6 Energy

Renewable energy is one of the largest challenges facing scientists today. Our reliance on fossil fuels means that billions of kilograms of carbon dioxide are released into the atmosphere each year. Natural reservoirs of fossil fuels pose an environmental hazard, but they are finite, meaning that in a few generations they will not be an option as an energy source. Furthermore, it is often overlooked that renewable energy does not merely mean clean energy, but a source of power that can sustain the existence of our grandchildren's grandchildren. It is vitally important that we develop a source of energy that can be sustained and improved over the decades, centuries, and, indeed, millennia to come. In this section, we will discuss some of the alternatives to fossil fuels that have the potential to heat and cool our homes, power our computers, and generally maintain our technology-based lifestyles in the generations to come.

Several sources of renewable energy exist, but the most prominent among these is solar energy in its various forms. The influx of solar radiation to our planet provides a source of fuel that can meet our energy needs millions of times over. Moreover, with the sun as the source of this power, the fuel of this resource will not be expended until a time at which, billions of years from now, the sun itself is expended, making life on Earth unviable in any case.

It might be surprising to learn that solar energy is encapsulated not only by the technology of solar cells but also in wind and hydroelectric power, both of which rely on natural processes whose driving mechanism can be traced back to the sun. The principal hurdle in each of these technologies is their cost-prohibitive nature relative to fossil fuels. Although the extraction of coal, oil, and gas from the earth is expensive and often onerous, the expense is far eclipsed by that required for alternate forms of energy production. So the problem faced by industry is to balance short-term versus Long-term advantages, and with the expenditure of remaining fossil fuels hundreds of years away, there is little motivation for companies to shift their energy sources at the expense of profits, when a monetary loss would engender disappointment and mistrust among their shareholders. One of the principal targets of alternative energy researchers is to bring down the cost and/or increase the efficiency of renewable energy sources to the point that they are competitive or at least nearly competitive with the cost of fossil fuels.

2.1.3.6.1 Solar

Solar energy is the cleanest and possibly most promising of all alternative energy sources. The annual flux of solar energy is almost 100 times larger than the annual global energy consumption, which means that harvesting only 1% of incident solar radiation could serve to satisfy the energy needs of every person on the earth. However, there are challenges. Energy scientists face the twin problems of economics and efficiency. High-functioning solar cells are expensive to produce and replace, so efforts are underway to develop cells that are both more efficient and cheaper to produce. Novel nanotechnologies are at the heart of this effort, in which new nanostructures, easily synthesized, can divert electrons more efficiently to the electrodes attached to the cells.

Solar energy has virtually no adverse environmental effects and is infinitely renewable. There are multiple mechanisms by which this energy can be harvested, including photovoltaics, in which light can be turned directly into electricity; solar thermal technologies, in which the incident radiation can be used to heat water and subsequently turned into other forms of energy; and concentrated solar power, in which solar energy is concentrated by lenses to produce intense energy beams, which can be used to heat water to provide conventional energy solutions (Fig. 2.1.20).

FIGURE 2.1.20 A solar panel array. Recent advancements in solar harvesting technology are bringing the immense potential of incident radiation from the sun into economic viability for the future. *Image courtesy of Mark Buckawick.*

Chemistry plays a surprisingly prominent role in the emerging solar power industry. One of the most promising technologies is photovoltaic cells, in which a chemical dye is used to absorb light, the energy of which is transferred to a semiconductor lattice. These chemical sensitizers can be organic in nature, or composed of nanoparticles engineered to maximize light absorption from the solar spectrum. The role of chemists is to develop new sensitizers, synthesize more efficient nanoparticles, and innovate semiconducting materials, which can serve as a platform for transferring photonic energy into electrical energy.

2.1.3.6.2 Organic Fuels

The area of organic fuels is one alternative that could provide a feasible gap between fossil fuels and truly clean energy. These fuels include methanol and ethanol and could be argued to be a form of solar energy as the plant matter from which they originate uses energy from the sun to grow. Although these sources have the advantage of being cleaner than fossil fuels in the sense that they combust almost completely into carbon dioxide and water without radical by-products that are damaging to the atmosphere and humans, the CO_2 will still contribute to the increasing anthropogenic carbon footprint.

Furthermore, the production of these fuels is an agricultural challenge, involving fertilizers, pesticides, shipping requirements, and other components that could contribute to environmental pollution. Ethanol as a prominent fuel source would magnify the need for large-scale farming, increasing land use, and reducing the viable land available for food production. Indeed, with the shortage of food in developing countries, it is single minded

FIGURE 2.1.21 Corn, a precursor to organic fuels. Although corn can produce ethanol, it is questionable whether ethanol as a fuel is an adequate substitute for oil and coal. Although it is a renewable resource, the relative carbon footprint provides little advantage over traditional fossil fuels. *Image courtesy of Maryam Afeefa.*

to utilize land for a deceptively clean but ultimately regressive-to-neutral energy source when it could be used to provide food for millions of starving people. In short, ethanol as a fuel can represent only a short-term solution to our energy needs if we are to mitigate carbon emissions and ultimately hampers our ability to provide the less fortunate with the basic necessities of life (Fig. 2.1.21).

2.1.3.6.3 Wind

In wind power, the mechanical forces induced in the atmosphere by solar radiation are harnessed to turn turbines, which generate electricity. This is an infinitely renewable resource, as the energy flux from the sun is constant and will continue to be so for billions of years, at which point the planet will be uninhabitable. This type of power is highly economical, often competitive with the generation of electricity through fossil fuels, and currently represents roughly 4% of the global power generation. One downside of this form of energy is that it inconsistent—although annual averages are typically reliable, day-to-day outputs are unpredictable, necessitating the supplementation of the power grid with alternative sources. Development of new energy storage technologies could vastly increase the viability of this form of energy.

Both land-based and offshore wind farms are feasible with our current level of technology. Land-based turbines are much less expensive and easier to connect to existing power grids, whereas offshore ones (in which the wind farms are situated over the ocean) are more productive energetically, but are much more expensive to construct. Energy transmission is a major challenge to the wind power industry, as geographical properties such as altitude and vicinity to oceans has a major impact on the energetic potential. Therefore, an infrastructure of electrical transmission, predominantly from coastal regions to inland locations, is a necessary undertaking to make wind power a viable alternative to fossil fuels.

Another major infrastructure improvement that could enhance the contribution of wind power to global energy needs would be the introduction of "smart grids," which would allow a higher time resolution of energy usage on a house-by-house basis to reduce the logistical challenge of diverting energy to the location where it is most needed (see Chapter 3.23.3). This technology would include a two-way flow of both information and energy, which would further enable the coupling of two or more alternative energy sources. For example, a house fitted with solar cells could "request" energy on a cloudy day, when battery stores are low, or contribute energy into the grid when its batteries were at capacity and no longer able to utilize accumulated energy from the sun (Fig. 2.1.22).

FIGURE 2.1.22 A wind farm at sunset. While "a visual blight on the environment" is an oft-used argument against arrays of turbines, this is certainly not always the case. *Image courtesy of Dirk Goldhahn.*

2.1.3.6.4 Geothermal

Geothermal energy harvesting is the technique of capturing energy produced in the heart of the earth. Due to its temperature, our planet produces a constant flux of IR, and the hot magma of the mantle powers many potential sources of renewable energy, including hot springs, hot dry rocks, and even volcanoes. The majority of geothermal technology in current use comes from hot springs, as they are generally shallow and more easily accessible. Deeper sources require expensive drilling, which can also release greenhouse gases such as CH_4 and CO_2 as well as other pollutants, although the net carbon footprint resulting from such activities still falls well below that of fossil fuel consumption.

The generation of geothermal energy works much like a steam engine, in which water is heated into vapor, which expands to induce mechanical motion in turbines. This mechanical energy can then be converted into electrical energy. The principal difference between a traditional steam engine and a geothermal energy plant is that the driving energy source comes from heat harvested from within the earth rather than from the burning of coal.

Geothermal energy is considered a renewable resource because the energy extracted from the earth, even if utilized to fulfill 100% of our energy needs, would represent only a miniscule fraction of the planet's total energy store. The global supply of geothermal energy comes from a combination of radioactive emissions from compounds embedded in the soil and latent energy leftover from the planet's formation. As with solar energy, the supply would be likely to last until after a time when the earth could be considered habitable (Fig. 2.1.23).

FIGURE 2.1.23 A geothermal energy plant in Iceland. Geothermal energy utilizes the planet's embodied energy to generate electricity as a cleaner alternative to fossil fuel burning. *Image courtesy of Debivort Work.*

2.1.3.6.5 Nuclear

Nuclear power must be mentioned as a promising alternative energy source to fossil fuels, although whether it can be considered a form of renewable energy, or whether it is truly cleaner than traditional energy, is a matter of debate. Energy from nuclear power is generated through the fission of radioactive isotopes, particularly U-238. In this process, nuclei are broken apart with a bombardment of neutrons, initiating a chain reaction in which more neutrons are released with each fission event. Ultimately, this means that a minimal bombardment can initiate the fission of an entire mass of uranium, resulting in a massive energy output. As with previous technologies, the generated energy is used to vaporize water and power turbines to generate electricity.

With proper engineering and oversight, nuclear power has immense potential to serve as an alternative to fossil fuels, but after the disasters of Three Mile Island (1979), Chernobyl (1986), and Fukushima (2011), public confidence in the safety of nuclear facilities is understandably low. These catastrophes have led to a warranted increase in regulations for the construction of nuclear facilities, which have decreased their economic viability significantly. Another consideration in the pursuance of nuclear power is the environmental threat posed by an unforeseen, calamitous event. Some of the worst environmental disasters in history, such as those mentioned earlier, have emerged from the improper oversight of nuclear energy facilities, and these have wreaked untold damage on the environment and human life.

Then there is the question of whether nuclear energy can be considered renewable. It is estimated that the accessible supply of U-238 would only power nuclear reactors for the next 100 years or so. Although there have been proposals for the development of "breeder reactors," which could generate fissile material to produce an ongoing source of fuel, this technology has yet to eventuate. Ultimately, nuclear energy has a significantly lower carbon footprint than fossil fuels, but has many associated drawbacks, including economic restrictions, environmental hazards, and a potentially limited span of viability. Under the current state of knowledge, nuclear power can provide an excellent bridge between our fossil fuel dependence and a more permanent, renewable, and clean solution to global energy needs (Fig. 2.1.24).

In summary, we currently have at our disposal several viable alternatives to the burning of fossil fuels for our principal source of energy. As technology continues to develop, these alternatives will become more and more economically viable, even outside of the consideration of their considerably lower environmental cost. Even so, it will take ongoing research and public pressure upon policy makers to demand a changeover to greener energy. The impetus of stagnation in customary energy technologies is high, and the best weapon against further anthropogenic damage to our climate is education of the public as to the full severity of the problem. This will ensure that the ongoing damage wrought by climate change is taken seriously by those equipped to make change, helping to provide funding, both monetary and intellectual, for the vital pursuit of alternative energy technologies.

2.1.3.7 Environmental Policy

Regulation through government policy is critical to the protection of the future welfare of our planet and the health of its citizens. Central to this is not only the enactment of local and

FIGURE 2.1.24 A nuclear power facility. Nuclear energy is a viable alternative to fossil fuels, but inadvisable as a permanent environmentally friendly replacement. *Image courtesy of Stefan Kühn.*

national regulations but also international cooperation to ensure consistent standards throughout the planet. Global threats including climate change and chemical waste can only be mitigated through the mutual efforts of developed and developing countries all over the world. Although great progress has been made toward this end in recent years, it is only through the coupling of science and policy that further advancements can be made. In particular, focus on green chemistry can provide the opportunity to develop alternate chemicals to be used in industrial processes that produce a large amount of hazardous waste. This research can not only mitigate negative environmental effects but also alleviate the economic burden that regulations impose upon the industry in both developed and developing countries.

Because of the dilemma presented by short-term economic gain vs. long-term environmental damage, which are almost always at odds, environmental policy can often be difficult to enact, particularly in the hands of policy makers whose very livelihood depends on the economic welfare of their jurisdiction. It is vital, therefore, that ongoing research should be conducted to improve the cost-benefit ratio of industrial processes in favor of more environment-friendly outcomes. Although it is often the burden of developed nations to take the lead in the innovation of greener chemical methods at the expense of economic gain, it is also the responsibility of developing nations to incorporate emerging technologies into their industrial economies. Policy regulations can help ensure that these responsibilities are met by all parties, thereby protecting the future welfare of our environment.

2.1.4 CONCLUSIONS

As you have seen from this chapter, environmental chemistry is not a mere subfield of chemistry but rather a vast and intricate mélange of fields, ranging from geology to medicine, that requires the contribution and collaboration of a wide variety of scientists to succeed. Although individual research studies may be conducted by a single specialist, the big picture requires environmental scientists of every stripe to remain informed about innovations in those fields adjacent to their own. Think of the variety of experts involved in the production of a new drug: synthetic chemists to design and build the molecule, biologists or medical scientists to perform animal trials, toxicologists to assess potential harm, doctors to conduct clinical trials, and statisticians to analyze them. However, assessing the full environmental impact requires even more expertise: geologists and soil scientists to understand its fate in the soil, microbiologists to assess how it might be taken up or transformed by microorganisms, biologists to analyze the effect of the drug and its by-products on other living creatures, and chemists and atmospheric scientists to predict further transformations in the water and air and project their implications down the line. And pharmaceuticals are just a small fraction of the chemical deposition for which humans are now responsible. It is only through collaboration and study that we as a society can work to limit and mitigate the damage our recent leaps in technological development have wrought upon our natural environment, damage that has already rebounded upon us in myriad ways. These efforts serve not only to protect us and the other living residents of planet Earth but also propagate forward to offer a scientific foundation for the self-protection of generations to come.

PROBLEMS

These questions are designed to encourage independent research. Please use the resources of scholarly articles as well as the Internet to provide detailed answers.

1. Select one of the environmental disasters presented in this chapter, and discuss its effects on the ecology and human habitation of the region.
2. Explain how soil pollution is a triple threat, and give an example of how a specific chemical, when introduced into the soil, can affect all elements of the biosphere.
3. Choose any renewable energy source and discuss its advantages and disadvantages (in comparison to fossil fuel combustion) from an environmental and economic standpoint.
4. The two greatest disasters with nuclear energy thus far have been Chernobyl and Fukushima, and the two greatest associated with fossil fuels have been the Exxon Valdez and Deepwater Horizon. Compare the human and animal death tolls, and discuss the environmental implications of each.
5. What was the environmental impact of microbeads in personal hygiene products, and why were they discontinued in the United States?
6. Present a particular chemical banned by the Montreal Protocol and explain why it was expunged from industrial use—what were the particular environmental consequences of this chemical?

7. Identify a drug that had vastly varying effects on different species. Explain the implications of this example on the field of toxicology.
8. What molecular properties contribute to water solubility versus fat solubility? How does this difference affect the environmental impact of a chemical when introduced into a natural water supply?
9. Research international environmental agreements that have taken place in the past 50 years. Select one, and briefly explain its purpose and the outcome.
10. Imagine the invention of a "miracle chemical" that solved all human energy needs, but killed all nonhuman animals that were exposed to it. What would be some of the challenges faced by legislators seeking to regulate this compound?

Recommended Reading

1. vanLoon GW, Duffy SJ. *Environmental chemistry: a global perspective*. 3rd ed. Oxford: Oxford University Press; 2010.
2. Spiro TG, Purvis-Roberts KL, Stigliani WM. *Chemistry of the environment*. 3rd ed. California: University Science Books; 2012.
3. Baird C, Cann M. *Environmental chemistry*. 5th ed. New York: WH Freeman; 2012.

CHAPTER

2.2

Sustainable Synthesis

Béla Török

University of Massachusetts Boston, Boston, MA, United States

Following the turn of the 20th century, a plethora of major breakthroughs were made in chemistry, including the discovery of radioactivity, new elements, and the three-dimensional structure of carbon. All these discoveries paved the way for the golden age of chemical synthesis. Several developments of military nature, such as chemical weapons, certainly did not serve the betterment of humanity; however, chemists had endorsed the idea that through chemistry everything was possible. While the preparation and application of dynamite and Haber's development of ammonia synthesis during World War I were primarily for military purposes, these inventions later were repurposed in mining and road constructions and generated artificial fertilizers. Simultaneously, the discovery of natural gas and petroleum resources in the late 1800s brought about what was presumed to be an unlimited flow of raw materials for chemical, particularly organic, synthesis that revolutionized society. Paul Ehrlich's systematic development of Salvarsan, the first systematic drug, established the field of medicinal chemistry, and scientists believed that by the application of chemistry, remedies for many diseases could be developed, thus reinvigorating the science of medicine. Rubber vulcanization reformed the then humble tire, and with that, the entire car industry. Thus developments in chemistry had far-reaching effects. All these scientific developments and the increasing demand for the manufacturing of products initiated major shifts resulting in the birth of the modern chemical industry and the large-scale production of chemicals. This had an immediate impact on the society and improved the standards of living, life expectancy, and many other societal factors. The growing chemical industry certainly played a major role in preparations for World War II, and also accelerated developments in many fields. For example, the large number of wounded soldiers suffering from infections on the frontlines inspired a strong push in medicinal chemistry. Ultimately, we can thank the US war efforts for the development of modern antibiotics. Following the war, the chemical and pharmaceutical industries were booming, and by the 1950s and the 1960s reached previously unimaginable heights. However, there was a dark side to these extensive developments. One by one, a series of unexpected negative effects began to surface. The list is long, ranging from industrial accidents (e.g., Bhopal, India), to harmful drugs (e.g., thalidomide), or other chemicals [e.g., freons, chlorobiphenyls, and bisphenol A (BPA)], all with detrimental

Green Chemistry
http://dx.doi.org/10.1016/B978-0-12-809270-5.00003-0

consequences. This phenomenon was mirrored in popular culture in the form of books and films focused on environmental topics, the most famous of which is perhaps Rachel Carson's 1962 book *Silent Spring* chronicling the damaging impact of dichlorodiphenyltrichloroethane pesticides on birds, while a more recent film *Erin Brokovich* features a legal aide working on finding the relationship between heavy metal contamination in drinking water and cancer. These examples raised public awareness of the potentially harmful nature of chemicals, whether they are damaging the environment (e.g., the role of freons in depleting the ozone layer) or toxic to living organisms (e.g., BPA plasticizers in plastic bottles). As a result of newfound knowledge on these issues, government regulations ensued all over the world and the chemical industry had to adapt. These harmful products have been replaced by others considered to be environmentally friendly, and the technologies employing dangerous conditions or producing toxic waste have been improved. The demand for environmentally friendly processes is expected to extend well into the foreseeable future as very few people are willing to give up the convenience of chemistry-based products, namely, plastics, drugs, cosmetics, and detergents. This chapter describes the tool set that is applied in the development of green synthetic methods for the preparation of chemical products. Here, we will focus on the discussion of the basic features of these methods, while the state-of-the-art current applications will be listed in Chapters 3.11–3.14. The experimental tools discussed in this chapter incorporate the major points of green chemistry and green engineering as previously discussed. The major contributors are as such: the various forms of catalysis, the selection of environmentally compatible solvents, and the application of nonconventional activation methods.

2.2.1 CATALYSIS

The major reason for catalysis being an invaluable tool for sustainable synthesis is the ability to substitute nonrenewable, low-atom economy reagents with an efficient catalyst that is able to connect the reactants to form the desired product. In general, the application of catalytic methods incorporates points 1, 2, 4–6, 8, 9, and 12 from the 12 points of green chemistry discussed in Chapter 1. A catalyst enables a reaction to occur without being consumed in the process; each catalytic molecule or particle is able to participate in multiple cycles. For a compound to undergo a reaction, chemical bonds need to be broken and new chemical bonds need to be formed. The initial energy investment to break the first chemical bond is called the activation energy (E_A). The major role of a catalyst in a reaction is to lower the required activation energy for the reaction. Often, the noncatalytic reaction requires significantly higher activation energy than the catalyzed process. Fig. 2.2.1 illustrates this phenomenon by the comparison of the energy profiles of a noncatalyzed reaction and a catalyzed one accompanied by the schematic representation of the elementary reaction steps that the reaction goes through with the aid of the catalyst. Often a catalyst breaks up one-step noncatalyzed reactions into a series of elementary steps that require only a fraction of the noncatalytic E_A.

However, Fig. 2.2.1 represents another very important aspect of catalysis. The Gibbs free energy of the reactants or the products remain the same regardless of the presence or absence of a catalyst. This highlights the fact that catalysis is a kinetic phenomenon; a catalyst does

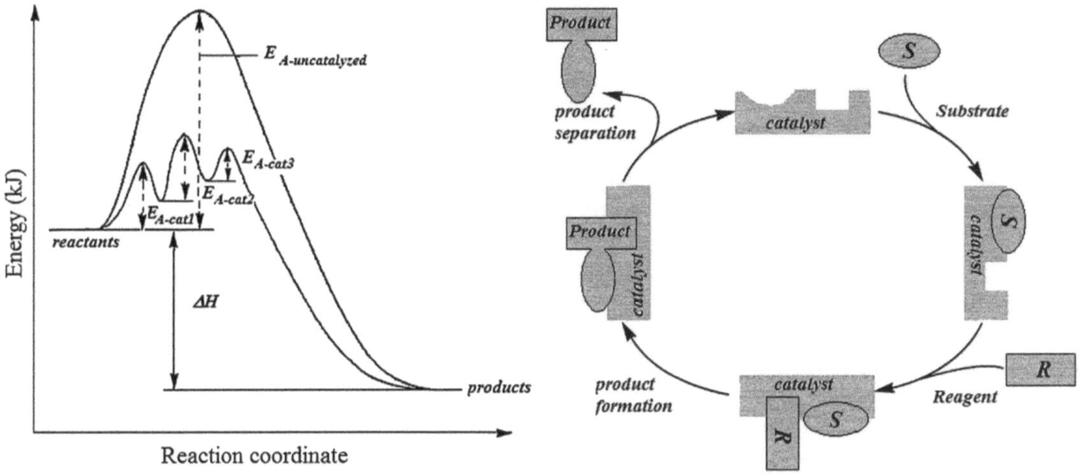

FIGURE 2.2.1 Energy profiles of noncatalytic and catalytic reactions and the scheme of a catalytic cycle.

not change the basic thermodynamic character of a reaction. Thus by the application of a catalyst, we can reach equilibrium faster, but the ratio of products to reactants will remain the same when the system reaches its thermodynamic equilibrium. Common chemical catalysts increase the reaction rate by a factor of 1000, and some enzymes are capable of increasing rates by a factor 10^{20}.

Catalysts are able to decrease the activation energy of the reactants by forming a bond with one or more participating molecules. They can have multiple roles, including breaking certain chemical bonds, physically anchoring and bringing the reactants together, and placing them in the right orientation to facilitate the collisions that more likely lead to product formation. As shown in Fig. 2.2.1, a catalyst initiates new elementary steps in a reaction when compared with the uncatalyzed pathway. Thus by changing the pathway, it changes the mechanism of the reaction.

There are several important terms that describe the efficiency of catalysts. While theoretically the catalytic material remains unchanged during the catalytic cycle that results in the formation of one product molecule, in real life, it undergoes deactivation. Thus, it is important to know how many such cycles or turnovers one catalytic unit (a molecule or an active center) can carry out in its lifetime, and this is described by the *turnover number* (TON), a description of the lifetime of the catalyst. The occurrence of these turnovers in a given time unit, essentially describing the reaction rate, is the *turnover frequency* (TOF = TON/time).

These characteristics can be significantly modified by certain compounds that would bind very strongly to the catalyst, partially or completely inhibiting its activity. Such compounds are called *catalyst poisons*. Although catalyst poisoning is usually considered to be an undesirable phenomenon, it cannot be avoided completely. Thus it is useful to take into account if the catalyst itself can be regenerated. This means that once its activity is diminished, the catalyst can be periodically treated by physical or chemical means to regain its original activity. Selective catalyst poisoning can be considered a positive phenomenon whereby selectively inhibiting a potentially undesirable pathway increases the likelihood of the desired reaction.

The accuracy with which a catalyst can carry out the reaction it is applied for is also of vital importance and is described by the *selectivity*. Selectivity comes in many forms. When the starting material possesses multiple different functional groups that can undergo reaction, *chemoselectivity* describes the excess of one product to the other. When the reaction can occur with the same functional groups in a compound but at different positions, the *regioselectivity* defines how well a catalyst can carry out the reaction at a specific position. When the product is chiral, the ratio of one enantiomer to the other is expressed as *enantioselectivity*. When the starting material already has a chiral center and the reaction generates another, the products are diastereomers and their abundance is described by the *diastereoselectivity*. While there are several examples of catalysts that are concurrently highly reactive and selective, we should consider the reaction conditions when designing a process. Harsher conditions or more powerful catalysts usually result in higher reaction rates but diminished selectivity, thus the reactivity and selectivity usually are in an inverse relationship.

Traditionally, catalysts are categorized based on whether the catalytic material is soluble in the reaction medium or not. The major classes are *homogeneous*, *heterogeneous*, and *phase transfer catalysts*. This traditional classification will be followed in our discussion. However, given the green chemistry focus of this book, we will further divide the methods, when applicable, to metal-containing catalysts and nonmetallic catalysts with a strong focus on heavy metals. The large majority of these metals are toxic, and as traces of the catalyst could remain as a product contaminant, the ultimate goal, theoretically, is to avoid using such catalysts. However, in a multitude of practical examples, these catalysts are far superior to anything else currently available, thus the benefits far outweigh the risk factors. Nonetheless, it is worth highlighting the difference between these catalysts and nonmetallic alternatives. We will discuss biocatalysis separately; these methods apply biomolecules such as enzymes or nucleic acids, often whole organisms (e.g., cells) when the isolation of the active catalyst (e.g., an enzyme) is impossible or at least highly impractical.

2.2.1.1 Homogeneous Catalysis

In homogeneous catalysis the catalyst is in the same phase as the reactants. In practical examples this means the liquid phase, in which both the catalyst and the reactants are soluble. There are several advantages to using homogeneous catalysis. First, since the catalyst and the reactants are in the same phase, the possibility of contact between them is the greatest of all types of catalysis. We do not have to deal with the serious mass transfer limitations that often decrease the efficiency of solid catalysts. Due to this inherent feature, these systems can achieve the same rate as other systems under milder conditions, which often result in greater selectivity. At the research and development phase of a catalytic process, it is quicker and simpler to work with well-defined homogeneous catalysts rather than their heterogeneous counterparts. It must be noted, however, that the scale-up of a process to industrial quantities may present significant technological problems. Thus with these advantages also come several disadvantages. Given that the catalyst is dissolved in the reaction mixture, the separation of the catalyst and the product may prove to be difficult. Tedious, solvent-heavy purification steps certainly undermine the sustainable character of a process. The separation problems often lead to another issue: catalyst recovery and recycling may present significant difficulties. When a catalyst (e.g., chiral metal-complexes or synthetic organocatalysts) is quite expensive, these difficulties could unnecessarily add to the cost of the product and decrease the green character of the method. As the majority of homogeneous catalysts are metal complexes, their stability and handling are also problematic. Many of these complexes are air or moisture sensitive and require special storage conditions, lest they decompose. This decreases the shelf-life of the catalyst, and ensuring the air/moisture-free environment on an industrial scale can be a serious detractor. Finally, complexes of transition metals are often toxic. Other homogeneous catalysts, such as strong mineral acids/bases require neutralization. Even organocatalysts can have toxicity issues.

2.2.1.1.1 Homogeneous Catalysis by Metal Complexes

While certainly the soluble acid/base-catalyzed reactions are the oldest class of homogeneous catalysis applications, the most common one is the application of soluble organometallic complexes of transition metals in a broad variety of reactions, where the liquid phase is an organic, commonly noncoordinating solvent. The first practical application of such compounds was the use of the Wilkinson's catalyst in hydrogenation reactions in 1965 by Wilkinson (Nobel Prize in Chemistry, 1973) (Fig. 2.2.2).

The central component of these reactions is the organometallic complex, which can be prepared from salts or covalent compounds, often halides, of transition metals and organic compounds (ligands). The ligand itself usually contains a strong electron pair donor, such as phosphorous, oxygen, or nitrogen (donor-acceptor complexes). However, there are a large number of organometallic compounds in which the ligand connects to the central metal

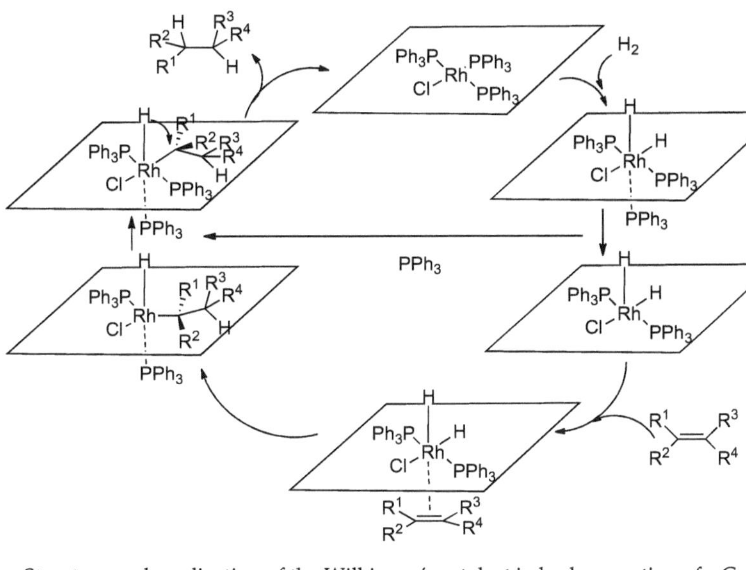

FIGURE 2.2.2 Structure and application of the Wilkinson's catalyst in hydrogenation of a C=C double bond. The catalytic cycle describes the proposed elementary steps of the reaction.

atom via a covalent carbon-metal bond. The basic mechanism of the reaction is relatively similar to that of the aforementioned example. The catalyst coordinates both reactants, first the hydrogen molecule and then the alkene. During the coordination of the reactants, chemical bonds are cleaved, thus increasing the reactivity of the individual components. The coordination of the alkene occurs via an exchange with a ligand molecule by its coordinating electron-rich π-bond. With the hydrogen atoms and the double bond now adjacent to each other, the migration of the hydrogen occurs and the double bond becomes saturated. In this transformation, the alkene (now alkane) loses its strong coordination ability and will be replaced by the free ligand, thus producing the hydrogenated product and re-forming the original catalyst. Although the mechanism varies for individual reactions, the major steps are similar to this process. The ultimate advantage of using organometallic catalysts is in the high reaction rates and selectivity, despite the mild reaction conditions. Their moisture and air sensitivity, however, often limit their applications. Since Wilkinson's original discovery, the development of metal complex catalysts has generated widespread attention. A broad array of methods have been developed and applied even under industrial conditions. The first such application was asymmetric homogeneous catalysis. The thalidomide tragedy of the 1950 and the 1960s focused attention to the important differences between the enantiomers of chiral drugs. Thalidomide, a sedative compound, was administered in the form of a racemate to expectant mothers to prevent morning sickness. One enantiomer was a powerful sedative, whereas the other enantiomer was a teratogen and caused serious birth defects resulting in underdeveloped limbs in the affected newborns. Thus it became evident that chiral drugs had to be synthesized separately and the effects of the enantiomers determined respectively. This can be achieved in two ways as illustrated in Fig. 2.2.3.

The first route is via chiral resolution. This process is based on the different character of enantiomers and diastereomers. Except for the optical rotation of polarized light, enantiomers possess the same physical properties. In contrast, diastereomers have different properties and can be separated based on those differences. To produce diastereomers, the enantiomers should be derivatized with optically pure chiral derivatizing agents. The obtained diastereomers are then separated, and the chiral derivatizing agent removed. From a green chemistry point of view, this is not a recommended process: it uses derivatization/removal of the agent (two extra reaction steps!), requires extensive amount of solvents for separation, displays low atom economy, and potentially results in a significant amount of waste. In the best case scenario, we obtain 50% yield and 100% optical purity. The application of chiral reagents via

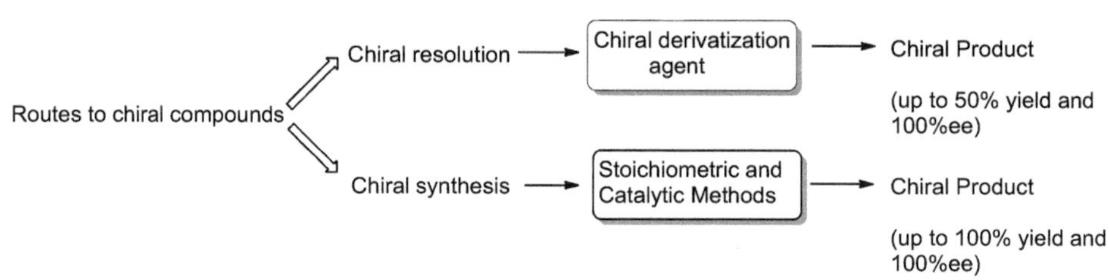

$$\text{enantiomeric excess (ee\%)} = \frac{|[R] - [S]|}{[R] + [S]} \times 100$$

FIGURE 2.2.3 Generation of optically pure products.

synthesis also carries similar problems including low atom economy and the nonrecyclable nature of the reagent.

Asymmetric catalysis is the most elegant and effective way of producing chiral compounds. In this case, we use a chiral catalyst that is recyclable and theoretically can produce 100% yield as well as 100% optical purity. Based on the principles that Wilkinson established, Knowles (Nobel Prize in Chemistry, 2001) was the first to design metal complex catalysts with chiral ligands and produced chiral L-DOPA (Monsanto). Some years later, Noyori (Nobel Prize in Chemistry, 2001) has developed the chiral binaphthyl-based Ru-BINAP catalyst that was found to be extremely broadly applicable in many industrial processes (Takasago). There are several other application possibilities, such as oxidations developed by Sharpless (Nobel Prize in Chemistry, 2001) and others. A few examples, shown in Fig. 2.2.4, only illustrate the widespread applicability of chiral homogeneous catalysis.

Many other applications were developed parallel to the ones shown previously. For establishing the mechanism of alkene metathesis and developing practical metal complex catalysts for such metathesis reactions, Chauvin, Grubbs, and Schrock shared the Nobel Prize in Chemistry in 2005. Now, both the Schrock and Grubbs catalysts, molybdenum and ruthenium complexes, respectively, are commercially available and allow for the synthesis of important products (Fig. 2.2.5).

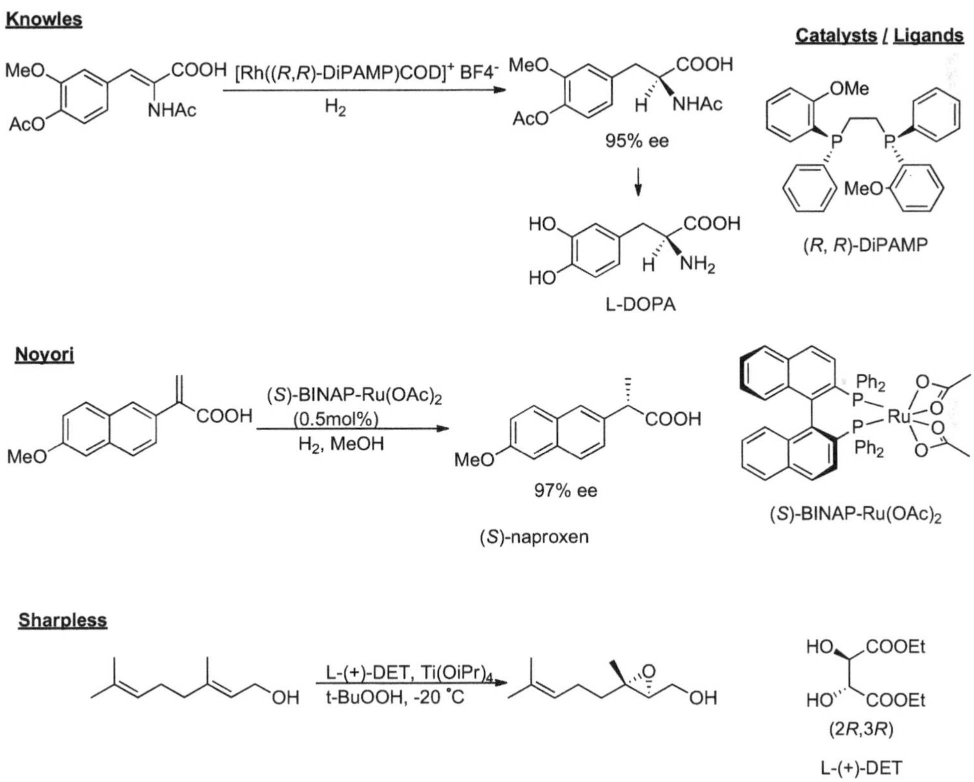

FIGURE 2.2.4 Application of chiral metal complexes in asymmetric synthesis.

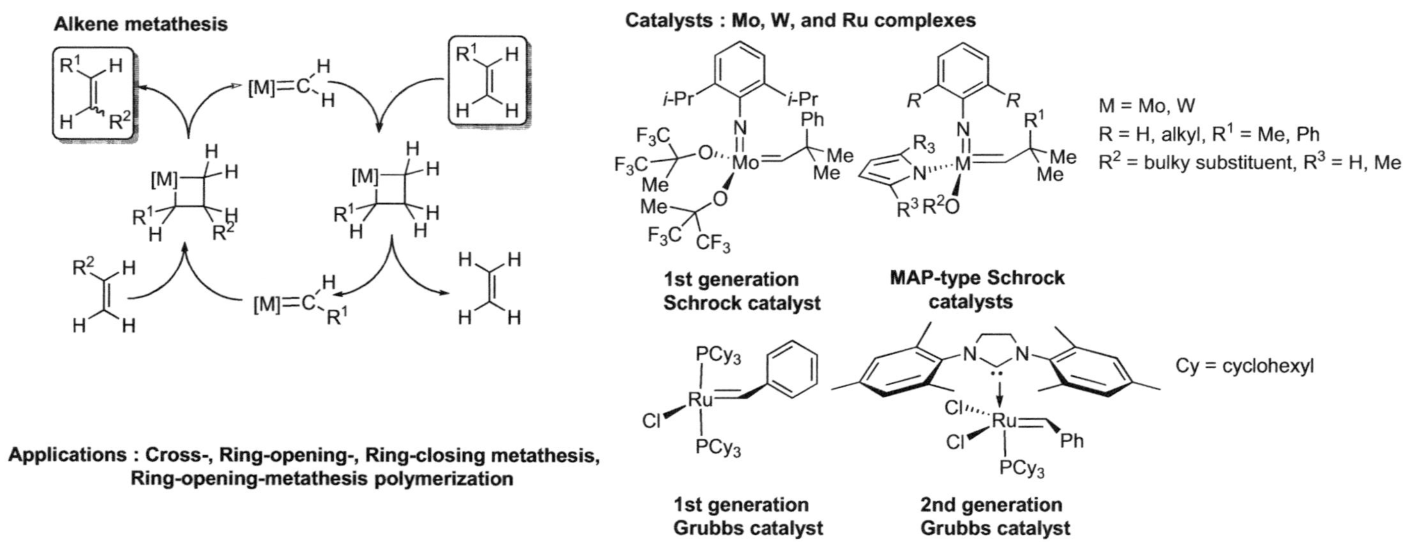

FIGURE 2.2.5 The mechanism of alkene metathesis and selected catalysts.

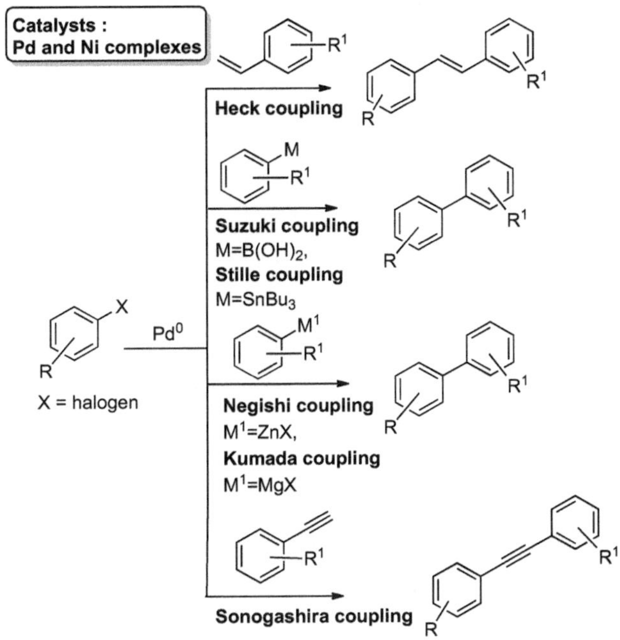

FIGURE 2.2.6 Selected examples of palladium-catalyzed C–C coupling reactions.

Other parallel developments in the 1970s established a new class of reactions known as metal-catalyzed coupling reactions. Mostly Pd and Ni complexes were used for these processes established by Heck, Suzuki, and Negishi (cowinners of the Nobel Prize in Chemistry 2010). Many other variations, such as the Sonogashira, Kumada, Stille, Hiyama, or Tsuji-Trost and Buchwald couplings, were developed on the basis of the same principles. These reactions made possible the direct coupling of aromatic rings that was previously only possible through tedious and long processes (Fig. 2.2.6).

Although the above examples only represent the most well-known cases, they illustrate the widespread applicability of these metal complexes. Further examples are provided in Chapter 3.11.

2.2.1.1.2 *Catalysis by Soluble Acids and Bases*

Classic acid-base catalysis has a long history and extensive application possibilities, including esterification, ester hydrolysis, aromatic electrophilic substitution, and electrophilic addition, just to name a few. Traditionally, these reactions were catalyzed by conventional mineral acids, such as HCl or H_2SO_4. Many of these acids are highly corrosive, and their reactions release gases that pose safety and health hazards. Therefore, their application is undesirable under the green chemistry principles. To address this problem, several new alternatives have been developed, mostly in the realm of heterogeneous catalysis (see Chapter 3.12.). There have been several attempts in applying water-soluble acids that do not hydrolyze under aqueous conditions and therefore are recyclable. The most typical examples are heteropoly acids (HPAs) such as the commercially available $H_3[PMo_{12}O_{40}]$,

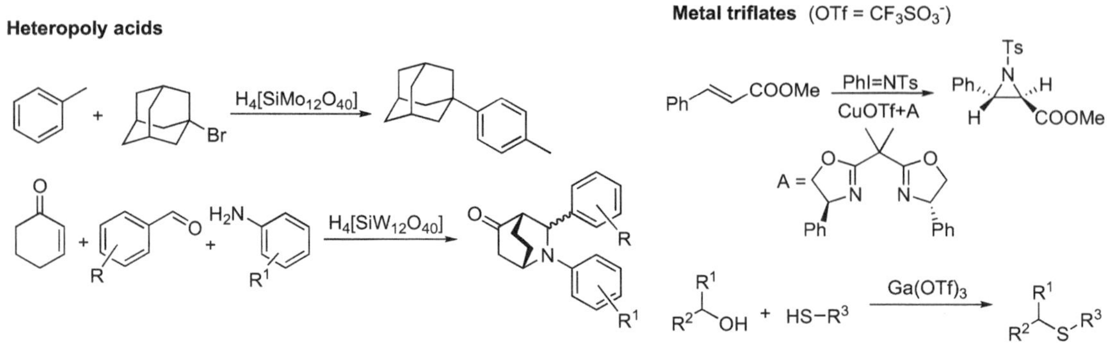

FIGURE 2.2.7 Selected examples of heteropolyacid and metal triflate-catalyzed reactions.

$H_4[SiMo_{12}O_{40}]$, $H_4[SiW_{12}O_{40}]$, or $H_3[PW_{12}O_{40}]$. These Keggin-type HPAs are soluble in water as well as in several organic solvents and readily initiate typical Brønsted acid-catalyzed reactions (Fig. 2.2.7). However, they can be regenerated from the reaction mixture, unlike the common mineral acids that release gases upon removal of the solvent. Another group is that of the metal triflates, common salts of the metals and trifluoromethanesulfonic acid (triflic acid) that function as Lewis acids (Fig. 2.2.7). In contrast to the traditional Lewis acids (such as $AlCl_3$), these acids resist aqueous hydrolysis and can be recovered from aqueous solutions. Due to their considerable cost, however, their widespread applications, particularly at the industrial scale, are not yet established.

2.2.1.1.3 Organocatalysis

Several organic compounds, which possess at least one of the most electronegative and coordinating heteroatoms (N, O, P, S), are also capable of catalyzing a broad range of reactions. Harnessing their power carries several benefits. First and foremost, products synthesized by organocatalytic approaches do not have metal impurities in them, which is a major advantage compared to the previously described transition metal complexes. Organocatalysts are, in general, less toxic than organometallic catalysts. While there are several new and expensive synthetic organocatalysts, the common examples are mostly natural products that are available in large quantities for a modest price. As Nature is highly selective, these catalysts, when chiral, are available in virtually 100% optical purity, which is a major benefit in the case of chiral catalysts. Due to the heteroatoms, it is relatively easy to remove them by simple salt formation, thus allowing for easy separation and recycling. The most common organocatalysts are amino acids and alkaloids and their derivatives. Representative groups and applications are depicted in Fig. 2.2.8.

In addition to natural product-based organocatalysts, synthetic alternatives have also garnered extended attention. In this group, the most popular catalysts are the chiral Brønsted acids that are based on Noyori's chiral binaphthyl unit. The generic catalyst structure is simple: these catalysts are partial esters of the diol BINOL with phosphoric acid. A broad variety of these compounds are already commercially available and are used in many different reactions (Fig. 2.2.9).

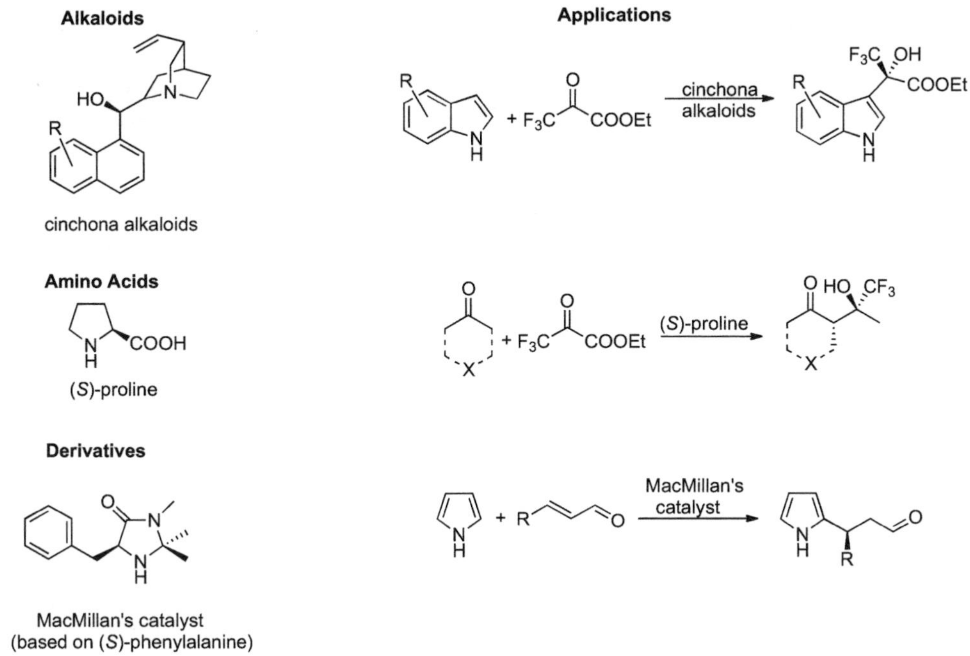

FIGURE 2.2.8 Selected examples of natural organocatalysts and a derivative and reactions catalyzed by them.

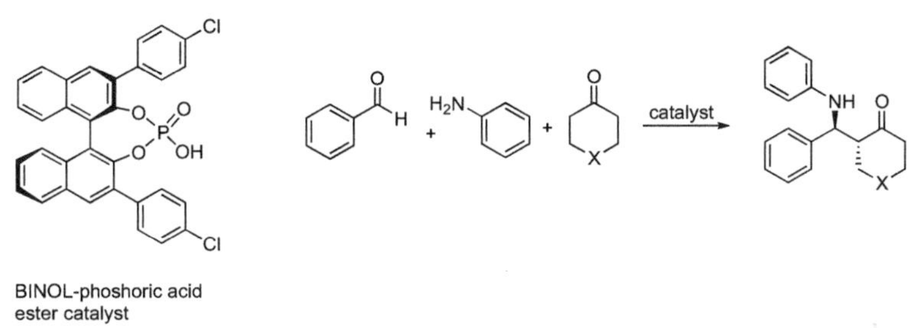

FIGURE 2.2.9 Structure of a representative example of chiral phosphoric acid catalysts and the typical Mannich reaction catalyzed by them.

Despite the major advantages, organocatalysts are not without problems. The most prevalent drawback is that in most cases the reactions require relatively high catalyst loading (20 mol% relative to the reactants). Another issue is that other than the natural compounds (e.g., proline or cinchona alkaloids) these catalysts are quite expensive due to their multistep preparation processes. A more detailed discussion about organocatalysts can be found in Chapter 3.11.

2.2.1.2 Heterogeneous Catalysis

In heterogeneous catalytic reactions, the catalyst and the reaction mixture are present in different phases. The catalyst (S) is usually a solid material and the reaction mixture can be in a liquid (L) (S/L reactions) or in the gas (G) phase (S/G reactions). In certain cases, all three phases are included in the system (S/L/G systems): in heterogeneous catalytic hydrogenation reactions the catalyst is a solid, the reactants are dissolved in a solvent, and the hydrogen is in the gas phase. Many industrial plants are based on an S/G system, including several processes in the petrochemical industry (e.g., isomerization, cracking) or the production of sulfuric acid (sulfur oxidation to SO_3).

As a classification, we distinguish between (1) metal-based heterogeneous catalysts and (2) transition metal-free catalysts, including solid acids and bases and polymer-bound organocatalysts. The bifunctional combinations of metal and solid acid catalysts were also developed for multistep reactions. In addition to traditional heterogeneous catalysts, recent developments in nanotechnology have presented us with a completely new type of catalytic materials, nanoparticle-based catalysts. These catalysts represent a transition between the traditional definition of homogeneous and heterogeneous catalysis. While the nanoparticles are very small (usually in the low nanometer size range), producing mixtures that resemble solutions, the particles are still solid particles within a liquid medium. They can be characterized by microscopic techniques, thus we will discuss their potential contribution to green chemistry under heterogeneous catalysis.

Just as its homogeneous counterpart, the application of heterogeneous catalysis has multiple benefits. Primarily, these catalysts are easily separated from the liquid (or the gas) phase by a simple filtration. This is a great advantage over soluble catalysts, as the removal of the catalyst from the product mixture, its recovery, and its potential reuse is made much simpler and more sustainable due to the elimination of several steps. Heterogeneous catalysts are also more stable than organometallic complexes, thus their easy handling and high stability are definite advantages compared to their homogeneous counterparts. They are readily applicable catalytic materials for continuous flow systems. The chemical industry prefers flow systems to static reactors due to the uninterrupted production it provides. Having a well-defined catalytic-bed makes catalyst recycling and recovery process much simpler, saving time and energy and reducing waste significantly. Last but not the least, heterogeneous catalysts often possess specific structural features (e.g., channels and cages in zeolites) that offer the opportunity to achieve special selectivity in a number of reactions.

However, heterogeneous catalysts are far from perfect. Given that the bulk of the material is hidden under the surface and the reactions are usually limited to surface phenomena, the lower effective concentration of the active centers decreases the activity (per unit weight) of these catalysts compared to the molecular level catalysts in homogeneous counterparts. Also, as the reactants are in a different phase, solid catalysts are usually hindered by mass transport limitations, which could also decrease reaction rates. This is particularly true for three-phase systems, such as hydrogenations. In addition, the preparation of effective catalysts, especially those used in industry require special preparation techniques. Some say that catalyst preparation is more black magic than science. While this is certainly not completely true, there is an art to it with special skills; expertise and intuition play significant roles in designing a catalyst preparation process. Since the preparation conditions have strong influence on the structure and thus the performance of the catalyst, minor changes in preparation conditions might result in notable changes in performance.

Regardless of the type of catalyst, adsorption-desorption phenomena play a crucial role in heterogeneous catalysis. Since the reaction occurs on the surface of the catalyst, the first key step is adsorption. A compound must possess reasonable adsorption capability to enter a surface-mediated reaction. Adsorption can be driven by physical or chemical forces, effectively defining physisorption or chemisorption, respectively. Whether adsorption occurs by physical or chemical bonding, it is important that it is of significant strength while remaining a reversible process. Irreversible adsorption of compounds on a catalyst's surface gradually eliminates the active centers and ultimately leads to the deactivation of the catalyst. Compounds that facilitate such phenomenon are called catalyst poisons.

Often, the catalyst poison is only present as a small impurity in the major reactant. For this reason, it is important to consider the potential impurities in a reactant before deciding on the type of catalyst to use. For example, sulfur-containing compounds are poisons for all noble metal catalysts. Thus, if we want to carry out a hydrogenation reaction, we have to remove all sulfur-containing impurities from the starting material. This is a common problem, as most reactants of petrochemical origin contain sulfur. Catalyst poisons, however, can play a positive role as well. By selectively poisoning a catalyst, it is possible to ensure that certain undesirable reactions will not occur, thus increasing the selectivity of the process for the target product. The Lindlar catalyst [$Pd/CaCO_3$, partially poisoned with $Pb(OAc)_2$ and quinoline] is a selective hydrogenation catalyst that is able to hydrogenate alkynes to alkenes, a practically impossible process, as the alkene otherwise would undergo a second hydrogenation to alkane (Fig. 2.2.10).

In addition to catalyst poisons, there are other additives that, when applied in small amounts, can significantly activate the originally inactive or low activity catalysts. These are called *promoters*. Chemical promoters improve the performance of catalysts usually by altering the electron density/distribution on the surface of the catalyst (see later at ammonia synthesis). Structural promoters improve the mechanical properties of catalysts and make sure that the catalyst is stable in its active form for an extended period of time. For example, they prevent sintering, which would decrease the active surface of the catalyst by compressing the catalyst particles to larger units.

2.2.1.2.1 Metal Catalytic Processes

Heterogeneous metal catalysts are employed in a broad range of processes, from catalytic hydrogenation and oxidations to the catalytic converters in cars. The major component of these catalysts is a transition metal, most commonly a noble metal (Pt, Pd, Rh, Ru, Ir, Au), although other transition metals such as Ni, Cu, Os, or Fe also possess remarkable catalytic activities. Most of these metals are expensive, and thus their use in the form of solid metal powders is not economic. Consider that even when the metal catalyst particles are small, the majority of the metal is inside the particle and does not participate as the reactions occur

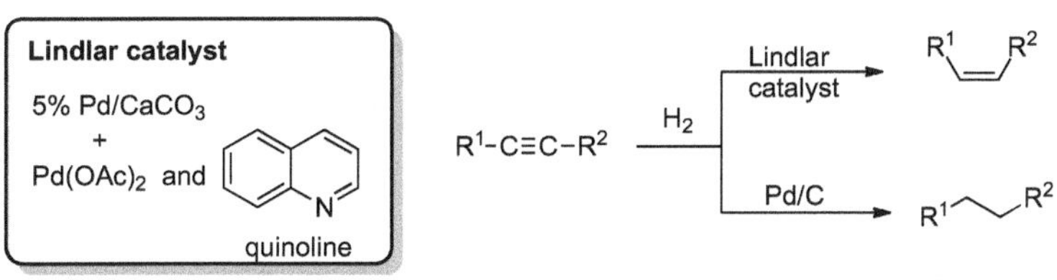

FIGURE 2.2.10 The Lindlar catalyst and its application in the hydrogenation of alkynes to alkenes.

on the surface. In addition, small metal particles are often dangerous (e.g., Raney Ni is pyrophoric), and they also undergo sintering to decrease their own surface tension.

To make such catalysts more economic and stable, *catalyst supports* were introduced. A catalyst support most commonly is an inert material that does not participate in the reaction, with few exemptions of bifunctional catalysts, when the support is often acidic. The major role of the support is to keep the catalytic metal in high dispersion (or small particle size) and stable so that the small particles do not sinter. There are several important characteristics that supports need to possess. High surface area ensures that the active centers are more or less evenly distributed on the catalyst's surface; usually, larger surface area results in higher reaction rates. For porous supports the pore volume and diameter are also important factors. Unless the catalyst is intended for bifunctional purpose the support should be inert. The support also has to possess good mechanical properties. It should be robust, so that the catalyst particles remain the same size over an extended period of time, even under harsh reaction conditions. Depending on the intended use, catalysts come in different shapes and sizes. They are formulated in powder, granules, beads, nets, and many other forms. Typically, mostly inert catalyst supports are activated carbon [Brunauer-Emmett-Teller (BET) surface $\sim 300-1000\ m^2/g$], silica ($\sim 300-600\ m^2/g$), or alumina ($\sim 100-300\ m^2/g$), while solid acids such as zeolites ($\sim 300-600\ m^2/g$) or clays ($\sim 50-300\ m^2/g$) are applied for the preparation of bifunctional catalysts.

The first, and probably the most dominant, application of heterogeneous metal catalysis is in catalytic hydrogenations. In 1897, Paul Sabatier (Nobel Prize in Chemistry, 1912) hydrogenated alkenes over nickel catalysts, which was the first report on catalytic hydrogenation reactions. Since then, the field expanded significantly and several books have been published in this area. The elementary steps of a typical hydrogenation process are illustrated in Fig. 2.2.11 and are described below.

While Fig. 2.2.11 shows a hydrogenation process, the major elementary steps can be seen as generic ones for common heterogeneous catalytic reactions. The first step is the mass transport, during which the reactant(s) reach the active surface. Once near it, the compounds make contact with the surface and adsorb on to it. Often this process starts with a weak electrostatic attraction. Metal atoms can be considered electron deficient and usually attract electron-rich substrates such as C=C or C=O double bonds. Once the adsorption is complete, some bonds undergo cleavage and the species are ready to react by forming new chemical bonds. The new compounds remain adsorbed on the surface; however, their adsorption capability is diminished by the saturation of the C=C bond. Since adsorption-desorption is a dynamic process, the remaining alkene reactants with stronger adsorption capability will compete for the active centers on the surface and replace the saturated products. The goal is to have the reactive (bond-breaking and/or bond-forming) processes (the actual reaction) to be the rate-determining step, thus the success of the process is highly dependent on the reactor design. Ideally, the mass transfer to or away from the catalyst should not play a significant role in the reaction kinetics.

There are many applications of hydrogenations in the chemical industry, such as alkene or alkyne hydrogenation and reduction of carbonyl compounds, in the pharmaceutical and fine chemical industries. Another common application can be found in the food industry in the preparation of hydrogenated oils. Here the goal is to saturate some of the C=C double bonds in the structure of vegetable oils to elevate their melting and boiling points (fried food

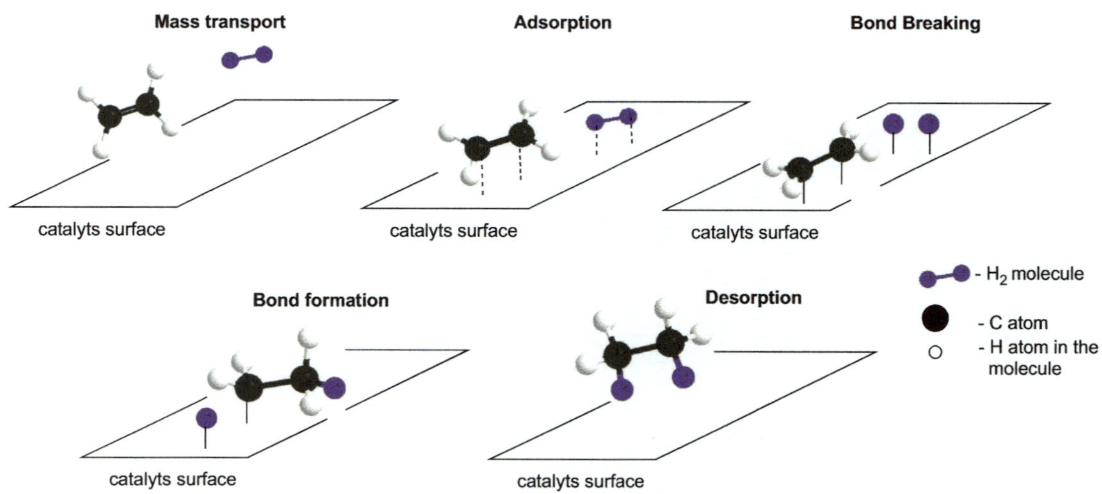

FIGURE 2.2.11 Elementary steps of heterogeneous catalytic hydrogenation of ethene. For clarity the hydrogen atoms introduced are shown in blue.

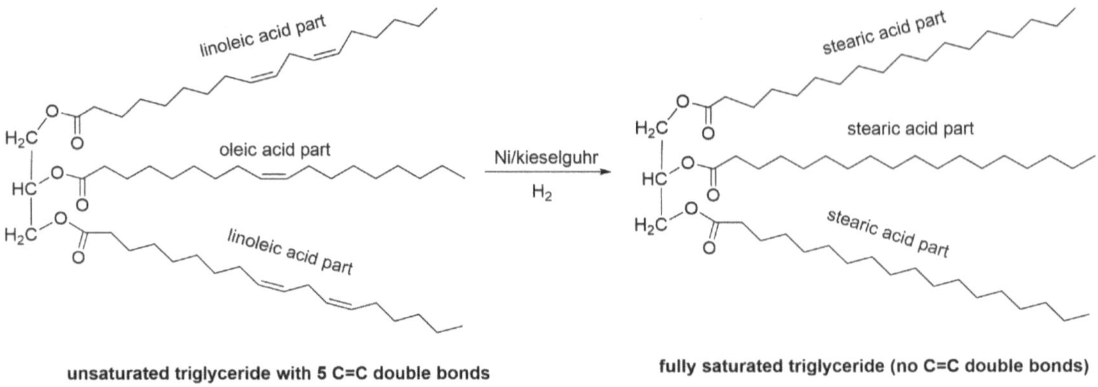

FIGURE 2.2.12 Schematic representation of the heterogeneous catalytic hydrogenation of vegetable oils.

becomes crispier if made at higher temperature) or to produce a solid vegetable spread (Fig. 2.2.12). These products were particularly popular from the late 1980s to the early 1990s, when consumers were frightened by the cholesterol and saturated fat content of animal fats. Although that trend is fading, the utility of the process remains.

Another green application of heterogeneous catalysis is in the automotive industry. All cars sold in the United States, Europe and most of Asia in the past 25 years have a catalytic converter built into their exhaust system. Catalytic metals are placed inside the converter on a ceramic or metallic support protecting it from vibration and shock. The converter itself fulfills the role of a small catalyst-filled reactor. Platinum, palladium, and rhodium are used as metals for the catalyst (Fig. 2.2.13).

The engine, while working, burns the hydrocarbon fuel in the presence of air. There are several gases that form in the engine: CO, CO_2, H_2O, partially oxidized hydrocarbons, nitrous oxides (NO_x), and unused hydrocarbons. The NO_x formation is particularly high in older engines that run at higher temperatures. All other gases except CO_2 and H_2O are toxic or harmful for the environment. Even CO_2 is problematic; it is a greenhouse gas directly promoting the global warming process. The complete unit is made of two different converters. The principle role of the catalytic converter is twofold: first to decompose and reduce the nitrous oxides to N_2 (front converter) and, second, to oxidize hydrocarbons and CO that remain in the

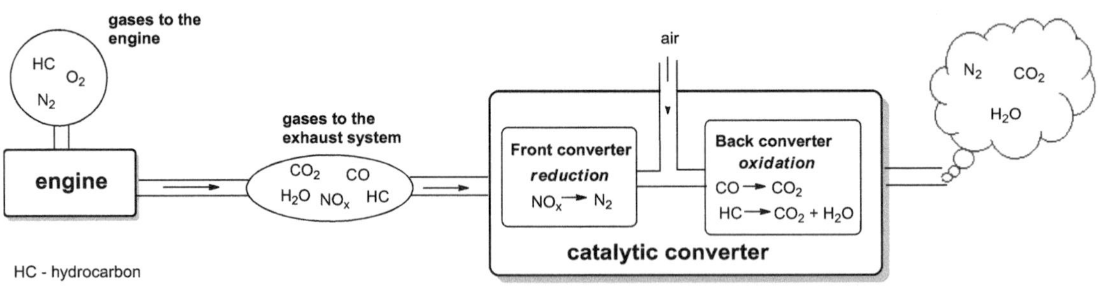

FIGURE 2.2.13 Model of the exhaust system with the catalytic converter and the oxidation/reduction processes.

exhaust to water and the less dangerous CO_2 (back converter). CO_2 is not a completely innocuous product, but it binds to hemoglobin in a reversible way and is thus much less toxic than carbon monoxide. The use of the converters can reduce the CO and hydrocarbon emissions by an order of magnitude, and the amount of NO_x, by two orders of magnitude.

For more applications of heterogeneous metal catalysts, please, see Chapter 3.12.

2.2.1.2.2 *Catalysis by Solid Nonmetal Catalysts: Metal Oxides, Solid Acids, Bases, and Surface-Bound Organocatalysts*

The family of solid nonmetal-based catalysts is quite large and diverse both in terms of the materials and the processes that they catalyze. Traditional acid/base catalysis was carried out using corrosive and harmful mineral acids (e.g., sulfuric acid or hydrochloric acid) or bases (NaOH, KOH, etc.). In addition to this obvious problem, these species are not applicable in flow reactors, which are the reactor of choice in industry. Thus almost a century before the green chemistry movement began, solid acid/base catalysts were introduced to industrial processes. Since then, a broad variety of catalysts have been developed. These catalysts can be classified by many characteristics. The least controversial method is to characterize them by their materials: (1) metal oxides (e.g., NiO, ZnO, MgO, SiO_2, Al_2O_3), (2) natural or synthetic aluminosilicates (i.e., zeolites, clays), and (3) acidic or basic ion exchange resins (e.g., Nafion, Amberlyst, Dovex). The application possibilities are broad and include oxidation, dehydration, polymerization, isomerization, cracking, addition, alkylation, and acylation.

The first industrial application of solid nonmetal catalysts was developed in the early 1900s in pre-World War I Germany by Haber (Nobel Prize in Chemistry, 1918) and was later scaled up to industrial production by Bosch (Nobel Prize in Chemistry, 1931). Due to the geographic isolation of the prewar Germany, its access to Chilean saltpeter, a common component of explosives at the time, was limited. Haber was entrusted with developing a chemical synthesis method for ammonia production from nitrogen. While Haber's original laboratory size apparatus used an osmium catalyst, the process was later scaled-up and put into industrial use by Bosch without an elemental noble metal. Instead, the Bosch system was based on magnetite (Fe_3O_4) using 1% potassium and calcium oxides (K_2O, CaO) as chemical promoters and 3% silica and alumina (SiO_2, Al_2O_3) as structural promoters (Fig. 2.2.14). This ammonia synthesis is now employed for the production of fertilizers rather than explosives, but it still remains one of the largest scale industrial operations in the world.

Another typical application of nonmetallic heterogeneous catalysts can be found in the petrochemical industry, which also dates back to the mid-20th century. Even then, there was a significant mismatch between the available petroleum resources and the needed hydrocarbon supplies. A large part of crude oil is composed of hydrocarbons that are straight chain and relatively large, while the growing chemical industry required smaller hydrocarbons as raw materials, e.g., for the synthesis of polymers such as poly(ethylene) and poly(propylene). One procedure that was developed to address this need is alkene cracking, a reaction catalyzed by zeolites. The best-performing catalysts among these was the ZSM-5 zeolite developed by Mobil in 1975. The process is depicted in Fig. 2.2.15.

The reaction proceeds via protonated intermediates called carbocations. The zeolites are strong solid acids at high reaction temperatures (around 500 °C) and protonate the alkenes. The carbocations undergo a series of scissions, isomerizations, and hydride shifts to produce the product C_2-C_4 alkenes. Carbocationic intermediates play a vital role in a large number of

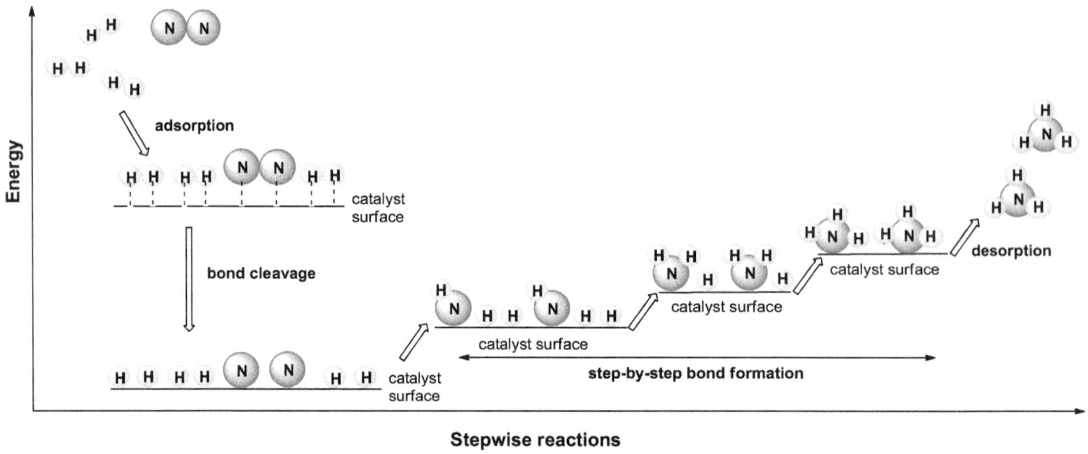

FIGURE 2.2.14 The Haber-Bosch ammonia synthesis.

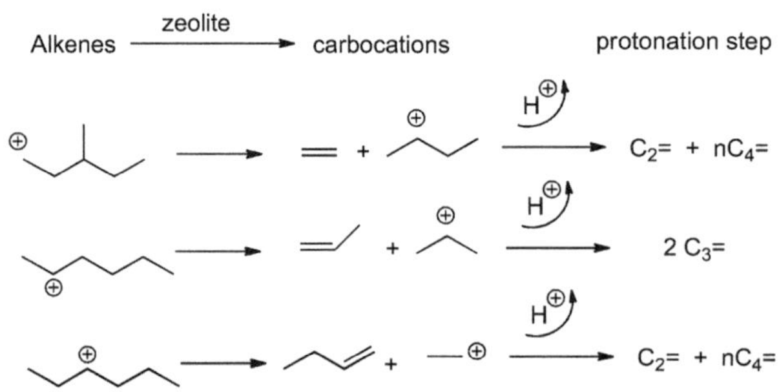

FIGURE 2.2.15 Schematic description of catalytic cracking using hexenes as an example.

reactions. Olah's (Nobel Prize in Chemistry, 1994) pioneering work on these reactive species allowed for a better understanding of these reaction mechanisms and the design of new processes.

Solid base catalysts are made of materials similar to those that compose solid acids, but they have different metal oxides and other synthetic or natural products with altered structures when compared with solid acids. They include alkaline metal oxides (MgO, CaO, etc.), rare earth oxides (e.g., La_2O_3), ZnO, Al_2O_3, layered double hydroxides (e.g., hydrotalcites), some zeolites (e.g., faujasites), and anionic exchange resins (e.g., Amberlyst-OH). These catalysts readily catalyze the conventional base-catalyzed reactions, such as alkene isomerizations, nucleophilic additions, condensations, and alkylations of phenols and the like (Fig. 2.2.16).

The previous paragraphs gave an overview of solid acid and base catalysis. While these catalysts usually catalyze the same type of reactions that their homogeneous counterparts do and are considered to be replacements of the corrosive and harmful liquid mineral acids, it is worth mentioning that in many processes, due to their specific structures, they are much more than simple replacements. Zeolites have a structure containing channels and cages that

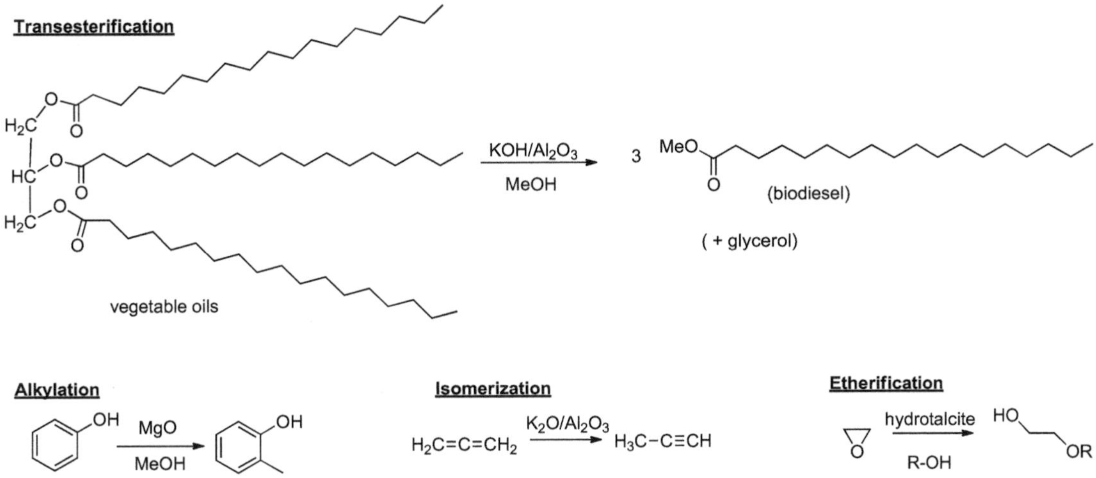

FIGURE 2.2.16 Application of solid base catalysts.

allows only certain reactants to enter their cavities and thus have been applied in shape-selective catalysis. Clays and layered double hydroxides possess a structure similar to that of graphite, that is, aluminosilicate layers separated by water or other molecules. By inserting inorganic or organic additives into these structures, the interlayer distance can be adjusted to the needs of the target reactions. The representation of these structures and their participation in shape-selective catalysis is illustrated in Fig. 2.2.17.

2.2.1.2.3 *Nanoparticle Catalysis*

The new developments in catalysis have certainly not been immune to the recent evolution of nanoscience. The application of nanoparticles in catalytic reactions falls on the borderline of homogeneous and heterogeneous catalysis. The research in *nanoparticle catalysis* currently focuses on transition metal nanoparticles, as the new catalysts would be able to carry the high activity of homogeneous transition metal complexes, without the risk of metal contamination or phosphine-based ligands. These nanoparticles vary from few tens to several thousands of metal atoms with a size of 1 nm to a few hundreds of nanometers. Certainly, the most active samples are the small nanoparticles usually below 10 nm. Such small particles are, however, highly unstable on their own and combine together to form much larger units, decreasing their catalytic activity. To prevent this, some form of stabilizer is applied during the preparation of nanoparticle catalysts. In one approach, the nanoparticles are deposited onto the surface of inorganic or polymer surfaces, essentially mimicking common supported metal catalysts. Other forms of stabilization are the application of surface-active agents, such as tetrabutylammonium chloride; soluble organic polymers, such as poly(vinylpyrrolidone); or carbohydrates, such as cyclodextrins (Fig. 2.2.18.) that are not only able to stabilize the particles but also can solubilize them under aqueous conditions.

The applications of nanoparticle catalysis combine the opportunities offered by both homogeneous as well as heterogeneous catalysis as previously discussed. The extremely small particle size ensures high reaction rates, yet the presence of solid particles preserve

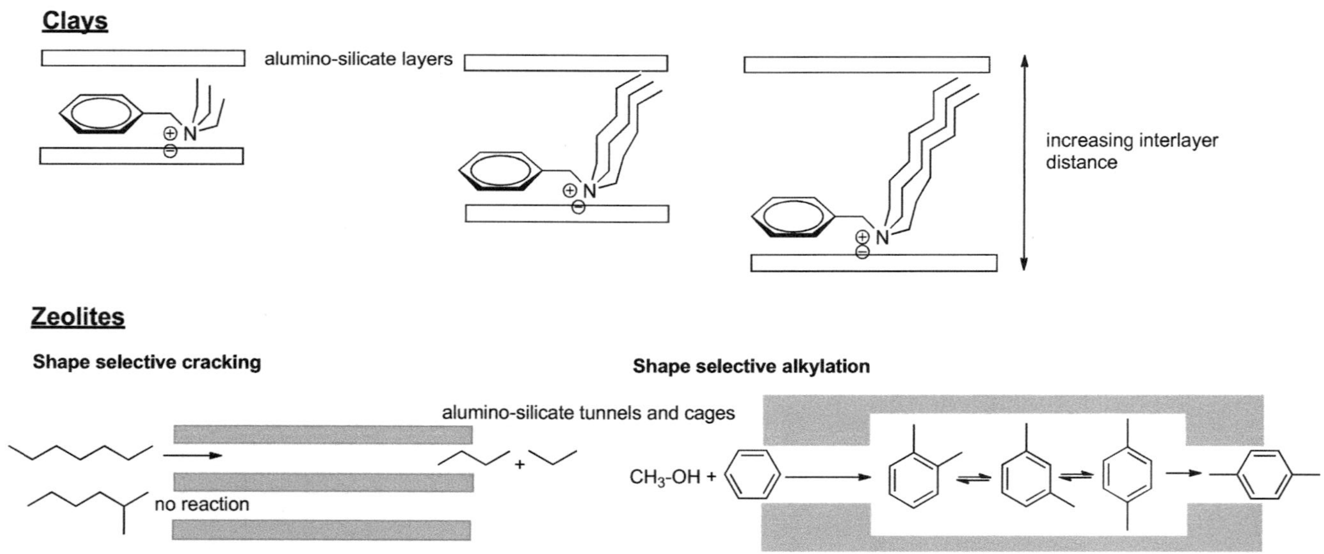

FIGURE 2.2.17 Increasing interlayer distances in clays upon the introduction of quaternary ammonium salts of varying chain length, and examples of shape-selective catalysis in zeolites.

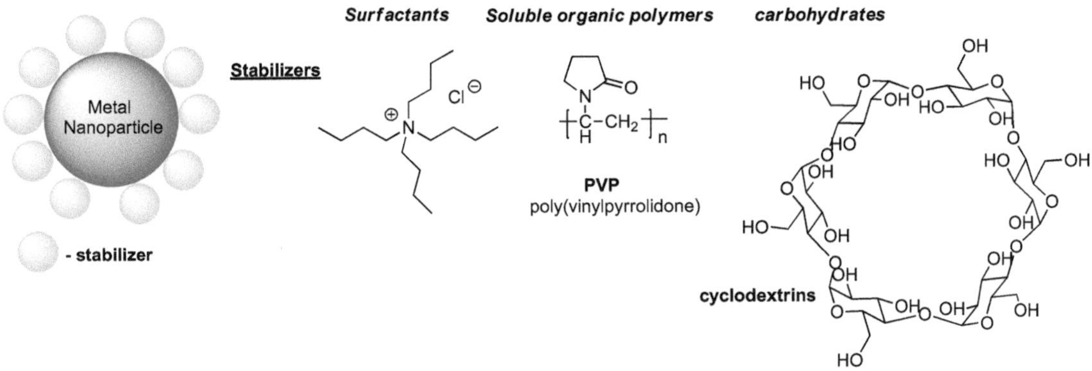

FIGURE 2.2.18 Stabilization of metal nanoparticles in the liquid phase.

the selectivity benefits of traditional heterogeneous catalysis. They are used in a broad variety of hydrogenation reactions, metal-catalyzed couplings (Heck, Suzuki, Sonogashira, Negishi, Kumada, Stille, etc.), hydroxycarbonylations, cycloadditions, oxidations, hydrocarbon conversions, and methanol reforming, just to name a few application possibilities.

Although at the laboratory or exploratory level of the reactions these catalysts appear to show outstanding performance, the application of nanoparticles in general, not just nanoparticle catalysts, brings up questions regarding their environmental impact. Due to their extremely small sizes, some of them are comparable to biomolecules, as these particles could pass through cell membranes directly, reaching targets where they can prove to be harmful (e.g., ZnO nanoparticles in sunscreens). Thus besides their positive contributions in catalysis or drug delivery, the potential negative health effects are not yet clear and sufficient precautionary measures should be applied during their large-scale applications.

2.2.1.3 Phase Transfer Catalysis

Although *phase transfer catalysis* (PTC) occurs in heterogeneous systems; the catalyst is dissolved in one, and thus it belongs to neither of the major classes discussed earlier. PTC reactions are carried out in two-phase systems including one organic and one aqueous phase. The role of the phase transfer agents is to ensure proper mass transport over the phase boundary, enabling the reactants to undergo reactions. The most common phase transfer catalysts are quaternary ammonium salts, cyclodextrin derivatives, or crown ethers. Their structure contains hydrophilic and hydrophobic units. Some common phase transfer catalysts and the depiction of the process are summarized in Fig. 2.2.19.

The application of these catalysts increases the rate of the mass transport through the phase boundary, thus increasing the reaction rates. The increased rates allow for the use of lower reaction temperatures, which is a key to ensuring high selectivities. Given the presence of the aqueous phase, we can avoid the application of anhydrous organic solvents, which are not only expensive but also often require harmful chemicals or energy-demanding setup for their preparation. The use of water as part of the reaction mixture also makes these processes greener compared with the organic-only alternatives. The obvious disadvantage is that

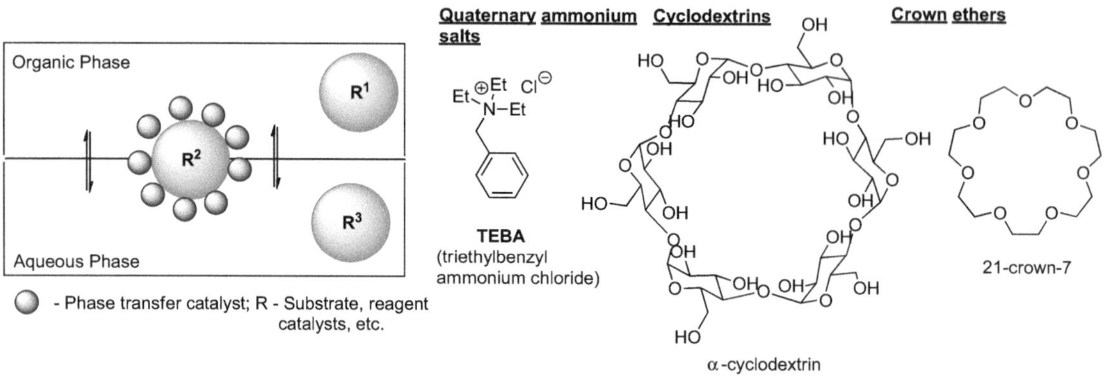

FIGURE 2.2.19 Phase transfer catalysts and the mechanism of the process.

moisture-sensitive compounds cannot be applied under PTC conditions. The application of PTC is widespread, and here we only list a few examples for the use of different catalysts (Fig. 2.2.20).

2.2.1.4 Biocatalysis

The term biocatalysis describes catalytic processes in which the applied catalysts are based on biological sources, usually enzymes or nucleic acids. The first application of biocatalysis goes back millennia: the fermentation of natural sugar-containing sources (grape juice, malt, potato, milk, etc.) to alcohol-containing drinks has been known for at least 3000 to 5000 years. The modern-day applications of biocatalysis are quite broad, ranging from the food industry to detergents, fine chemicals/pharmaceuticals synthesis, or bioremediation. The application of biocatalysts has several significant advantages over the chemical processes mentioned earlier. Biocatalysts are inherently green. They are biodegradable molecules and nontoxic, thus even if small portions of the catalyst appear in the product, it usually does not cause health problems. Biocatalysts are highly active and selective, thus the removal of the remaining starting material and undesirable by-products is not a significant issue. Therefore

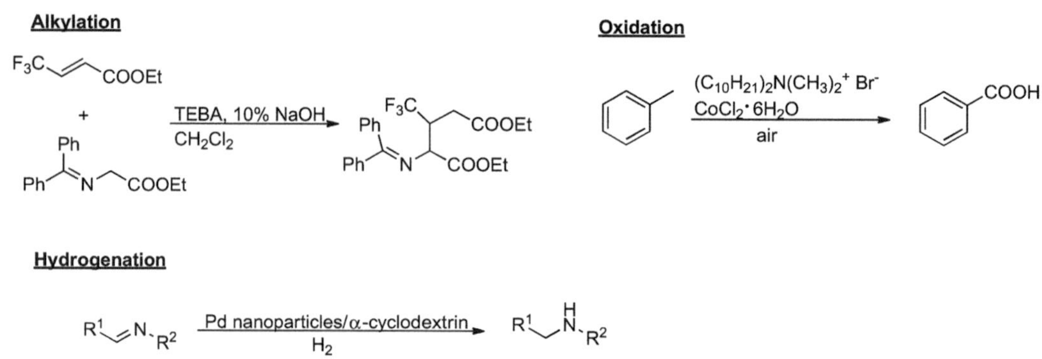

FIGURE 2.2.20 Examples for the application of different phase transfer catalysts.

the processes do not generate waste. It has to be mentioned that to reach yield/selectivity values similar to those the biocatalyst would produce in its natural environment is not always easy in chemical reactors where the ideal conditions are difficult to replicate. Other disadvantages are related to the preparation, handling, and stability of the biocatalysts. Enzymes in particular, are highly sensitive species and often denature at higher temperatures or irregular chemical conditions such as varying pH. Their optimized working conditions are very limited; a slight deviation from the optimum conditions could result in a significant rate/selectivity drop or loss of the catalyst itself. As mentioned earlier, enzymes are highly specific regarding their substrate, which often translates to narrow applicability. Furthermore, their performance is highly solvent dependent, as they usually do not tolerate organic solvents well.

At the moment, the large majority of biocatalytic processes are carried out by enzymes. The discussion of the basic description of enzymes, their structures, and their classification is beyond the scope of this book; for details on these topics the reader should consult biochemistry textbooks.

One of the most well-known applications of biocatalysis is the production of high fructose corn syrup (HFCS). The widespread use of HFCS as a large-scale industrial sweetener is rooted in its very low price, much lower than that of sucrose (table sugar) and is based on the fact that it is sweeter than regular sucrose. HFCS is produced from corn starch, a polymeric energy storage material made of glucose monomers. If a simple chemical hydrolysis were to be carried out, then the product would simply be glucose. Using the enzymatic process by applying an enzyme isomerase, however, a large part (over 40%) of the glucose is isomerized to fructose. Since fructose is about three times sweeter than glucose the product is much sweeter than pure glucose would be. The process, an astonishing example of productivity and catalyst stability, was developed in Japan in 1966. It applies a soluble enzyme, isomerase, that is immobilized on poly(acrylamide) gels. Using just 1 kg of enzyme, about 12,000 kg of HFCS can be produced. The reactor filled with the immobilized enzyme is active for 687 days, a little less than 2 years. The use of HFCS as a sweetener has been a controversial subject lately, and has been blamed to be responsible for the obesity epidemic. Without takings sides, here is an unbiased overview. Regular table sugar (sucrose) contains about 50% fructose, whereas general HFCS contains about 45% (there are examples with fructose content as high as 90%). Also, while our capacity to digest fructose is lower than that for glucose, common fruits also contain fructose, and hence fructose itself is not the problem. We believe that the problem is that HFCS is too cheap, thus the food industry employs it both liberally and ubiquitously even in food products that do not warrant the use of a sweetener. Therefore people overconsume it (e.g., free refills of soft drinks are common in many restaurants), leading to its unfavorable effects on public health.

Other examples of biocatalysis include the use of whole cell catalysis in which the enzymes are used in their natural environment through utilizing living organisms. The large-scale industrial production of antibiotics, particularly penicillin derivatives, is carried out by fermentation. In this case, an antibiotic-producing microorganism is cultured in large steel tanks, and once the process is complete the product is isolated. For example, penicillin is produced by *Penicillium chrysogenum* strain.

Propenamide is an important raw material in a broad range of processes, making its synthesis a priority. The traditional synthesis involves a copper-catalyzed oxidation of propionitrile. The major problems with this process are the regeneration and recycling of the catalyst, the formation of acrylic acid as by-product resulting in low yield, and the removal of the catalyst from the product. The biocatalytic method applies the enzyme nitrile hydratase in the form of whole cells from *Rhodococcus rhodochrous* immobilized on poly(propenamide) gel. This process has many advantages, namely, the immobilized cells can be used multiple times and the chemical yield of the process is over 99.99%, making the purification of the product or the recovery of the unreacted starting material unnecessary.

2.2.2 SOLVENTS

Many chemical reactions are carried out in solvents. The typical solvents for organic synthesis, which dominate the industrial production in a variety of reactions, are the common organic solvents, while many inorganic processes are carried out in water or solvent-free molten phases. Some of these solvents fulfill the majority of the requirements for a green solvent; however, many do not conform to environmental and safety standards. There are several alternative solvents to consider when selecting a medium for a reaction. We will consider the available options in the following sections.

2.2.2.1 Organic Solvents

As mentioned earlier, whether an organic solvent is an acceptable choice for a reaction is determined by several factors. One of the major factors is that the reactants should be soluble in the solvent, at least partially, so the reaction can occur. Based on the like-dissolves-like principle, most organic solvents fulfill this criterion, although polarity issues could rule out certain solvents. For example, whereas aromatic compounds are good solvents, saturated counterparts such as hexanes are usually poor solvents for organic reactions. The classification of solvents is based on their functional groups.

The first major group is the *halogenated hydrocarbons*. There are many solvents in this category, although the most common ones are the chlorinated methane derivatives, such as dichloromethane (CH_2Cl_2), chloroform ($CHCl_3$), and carbon tetrachloride (CCl_4). As a rule of thumb, these solvents are typical examples of good solvents that are not green due to several shortcomings. First, their material safety data sheets (MSDSs) reveal several health problems that they cause; among others they are carcinogenic and mutagenic, cause kidney and liver problems, and CH_2Cl_2 is metabolized in the body to CO, which is a poison for red blood cells. The list is long. Aside from their negative health effects, they are also relatively volatile (especially dichloromethane), and so it is almost impossible to recycle them without loss. With that, they enter the atmosphere and eventually could become catalyst for ozone decomposition, thus potentially contributing to the deterioration of the planet's ozone layer. Despite their inflammable nature, their use should be avoided as much as possible. A special group of halogenated solvents is the so-called fluorous (or perfluorinated) solvents that have specific effects and can be useful in multiple applications. The positions on their application in green synthesis are controversial, having both proponents and opponents. In any case, due to their perfluorinated structure, they are quite expensive, and thus their industrial-scale applications are strongly limited by cost issues as well.

The second group is the *ethers*. Ethers are also good solvents for many organic reactions. They are invaluable coordinating solvents for complex hydride reductions. The most common examples are diethyl ether and tetrahydrofuran. Again, their MSDSs reveal several problems; however, the most serious concern is that these solvents are highly flammable and potentially explosive. Ethers are known to slowly undergo oxidation under air. The products of this oxidation are peroxides that make these solvents explosive. Thus they have to be kept either in dark bottles or in metal cans to avoid the peroxidation and a spark by electric discharge. Due to this problem, they have a relatively short "expiration date" and the disposal of expired samples is inconvenient. Additionally, they have a relatively low boiling point and high vapor pressure at room temperature increasing the risk of fire. Although they are environmentally more beneficial than halocarbon solvents, the safety concerns largely temper any enthusiasm regarding their use in industrial-scale applications.

Alcohols, the parent compounds of ethers, are quite polar but still good solvents for a large group of organic compounds. The most commonly applied alcohols are methanol, ethanol, and isopropanol. As the carbon chain grows, the hydrophilicity decreases. There are several advantages of using alcohol-based solvents. Although they do not lack potential toxic effects (methanol may cause blindness and in higher amounts death), their relative toxicity is mild when compared with other solvents. Ethanol is even suitable for human consumption. Due to the extensive network of hydrogen bonds, their vapor pressure is relatively low if we consider their carbon number. Alcohols are flammable solvents, but not explosive on mechanical contact. During their disposal, they simply burn to CO_2 and water.

Carbonyl group-containing solvents are common in several processes. The commonly applied ones are acetone (CH_3COCH_3) or methylethylketone ($CH_3COCH_2CH_3$). Both are relatively harmless solvents; acetone is included in nail polish removers. However, they are also volatile and flammable. Given the strong reactivity of the carbonyl group in a plethora of transformations, they are considered reactive solvents and thus their application possibilities in reactions are limited.

Carboxylic acids and esters: Acetic acid (AcOH) is one of many organic compounds produced at large scale industrial level. Besides being used in the food industry, it is also a good solvent. It dissolves ionic compounds as well as many organic reactants. One of its major advantages is that it is nontoxic even at relatively high concentrations (up to 20% solution is sold in supermarkets). Due to its high resistance to oxidation, it is a frequently used solvent in oxidation reactions. Several oxidizing agents are routinely prepared in AcOH [Br_2, Cl_2, ICl, $Pb(OAc)_4$, ClO_2, etc.]. One of its most well-known applications is in the oxidation of *p*-xylene to terephthalic acid catalyzed by $CoBr_3$. However, it is an organic acid and thus is not inert in many reactions. Its ethyl ester, ethyl acetate, is also a nontoxic solvent at lower concentrations. It is also used in common nail polish removers. It is often the solvent of choice for chromatography or recrystallization. It is a good solvent for a broad variety of organic compounds and is mostly inert in reactions.

Hydrocarbon-based solvents can be separated into aromatic and aliphatic solvents. Historically, benzene was considered one of the best solvents, as it dissolves a broad variety of compounds, has a manageable boiling point, and thus is easy to remove from the products. In the 1990s, however, extensive studies reported its confirmed carcinogenic and mutagenic effects, and it was phased out from the line of solvents. Given its highly stable aromatic structure, the metabolic enzymes can do little to remove it from the body.

Currently, benzene can only be used as a reactant, as it is a frequent substrate in aromatic electrophilic substitution reactions (alkylation, acylation). Its "replacement," toluene is also a good solvent, although due to its high boiling point (110 °C), it is not a preferred one. The methyl group in toluene can be oxidized and converted to benzoic acid, which can be excreted by the body. Some halogenated benzene derivatives were also used as solvents; however, their use faces similar criticism as the previously mentioned halogenated solvents. The aliphatic hydrocarbon-based solvents are hexanes or octanes. They are not considered good solvents; however, they are often applied in chromatography. They are flammable, and especially, hexanes have a low boiling point for which they are not frequently used solvents in reactions.

Other miscellaneous solvents that are relatively rarely used include dimethylformamide (DMF), dimethylsulfoxide (DMSO), acetonitrile, and special solvents such as liquid ammonia or hydrogen fluoride. DMF and DMSO are popular solvents for nucleophilic substitution reactions, although their high boiling points ($bp_{DMF} = 153$ °C, $bp_{DMSO} = 189$ °C) makes them difficult to remove from products, which limits their applications. Both of these solvents are also mutagenic and have carcinogenic effects. Liquid NH_3 and HF are used for special purposes, including dissolving alkaline metal reactions (e.g., NH_3 for Birch reduction) or alkylate gasoline production, during which liquid HF is used both as a solvent and as a catalyst. These solvents are highly toxic and should be avoided (Table 2.2.1).

TABLE 2.2.1 Common Organic Solvents and the Benefits and Disadvantages of Their Application

Solvent	Examples	Application	Benefits	Disadvantages
Halogenated solvents	CH_2Cl_2, $CHCl_3$, CCl_4, etc.	Organic synthesis, chromatography	Good solvents for broad applicability, not flammable	Health hazards, environmental problems (ozone hole)
Ethers	Et_2O, tetrahydrofuran, MetBu ether	Organic synthesis	Only solvents for specific reactions, Et_2O is of low toxicity	Moderate health hazards especially MetBu ether, flammable, explosive
Alcohols	MeOH, EtOH, iPrOH, etc.	Organic synthesis, chromatography etc.	Good solvents, miscible with water, low toxicity (except MeOH), EtOH is a sustainable solvent	Flammable, not inert in several reactions
Ketones	Acetone, methylethyl ketone	Synthesis	Broad applicability, miscible with water, low toxicity	Flammable, not inert in many reactions
Esters, acids	Ethyl acetate, acetic acid, etc.	Chromatography, synthesis	Low toxicity, miscible with water, good solvent	Flammable, not inert in many reactions
Hydrocarbons	Hexanes, benzene, toluene, etc.	Chromatography, synthesis	Good solvents (aromatics)	Health hazard (carcinogenic aromatics), flammable, nonsustainable from petroleum sources
Miscellaneous	DMSO, DMF, etc.	Specific reactions	Miscible with water, good solvents	High boiling point, health hazards

DMF, dimethylformamide; *DMSO*, dimethylsulfoxide; *MetBu*, methyl-*t*-butyl.

2.2.2.2 Reactions in Aqueous Medium

The use of water as an abundant and nontoxic solvent in chemical reactions has generated much interest over the years. In addition to these advantages, water as a solvent is not flammable, which is a significant safety benefit compared with traditional organic solvents. It is abundant in nature and as such is considered to be a sustainable chemical that is inexpensive. The fact that its heat capacity is unusually high is both a benefit and a disadvantage. The temperature of aqueous systems is easier to control; however, in case of an accident it is difficult to cool it down. In addition, it takes a longer time and much more energy to heat the reaction mixtures to the desired temperature. Its purification and recycling is also difficult due to its high boiling point and can pose significant problems when the waste water is contaminated with organic impurities. Furthermore, water is a poor solvent for most organic compounds, thus mass transport limitations between the organic and aqueous phase may occur. Lastly, water is a well-known strong nucleophile; in fact, in many reactions (e.g., ester hydrolysis or halogen-OH exchange) it acts as a reactant, thus it cannot be considered as inert for many organic reactions. Despite these drawbacks, extended efforts have been made to replace harmful and/or simply just flammable organic solvents with water. Several examples of reactions occurring in aqueous medium are illustrated in Fig. 2.2.21.

2.2.2.3 Supercritical Fluids

A compound reaches its supercritical state when both its actual pressure and temperature exceeds its critical pressure and critical temperature (Fig. 2.2.22).

Although based on the $p_{critical}$ and $T_{critical}$ values several materials can be transformed into the critical state, the most common example is supercritical carbon dioxide ($scCO_2$) and water (scH_2O). The application of $scCO_2$ is particularly advantageous. The critical state of CO_2 is relatively easily reachable, with $p_{critical,CO_2} = 72.8$ bar (1071 psi) and $T_{critical,CO_2} = 31.1\,°C$. For comparison, the pressure of a common CO_2 tank (liquid/gas equilibrium) is about 830 psi at 25 °C. Thus transferring the liquid CO_2 into an appropriate vessel and heating it above its critical temperature generates the $scCO_2$. This also means that the properties of $scCO_2$ can be easily manipulated by the appropriate selection of the pressure and temperature. sCO_2 is not corrosive unlike many other such fluids (e.g., scH_2O). It is considered to be a nonpolar solvent, which makes it an ideal replacement for many halogenated solvents in several applications. In contrast to the halogenated solvents it is nontoxic, although in high concentrations it could cause suffocation by simply depleting oxygen. As it has the highest oxidation state of carbon, it is nonflammable. A gas under normal conditions, it can be easily removed from the reaction mixture and can also be recycled. It is also readily available in large quantities and is inexpensive. Since the formation and sequestration of carbon dioxide is an everyday process, it can be considered a sustainable material. The disadvantages that come with the application of $scCO_2$ include the cost of the high-pressure equipment that is necessary for working with $scCO_2$. Additionally, it is a relatively poor solvent and is not inert under multiple conditions, it reacts well with nucleophiles (water, alkaline metals, etc.). Nonetheless, $scCO_2$ has become part of well-known applications in the industry, as well as in our everyday life.

One of the earliest applications of supercritical fluids is the decaffeination of coffee. The early process used dichloromethane, which is not only a toxic and environmentally harmful

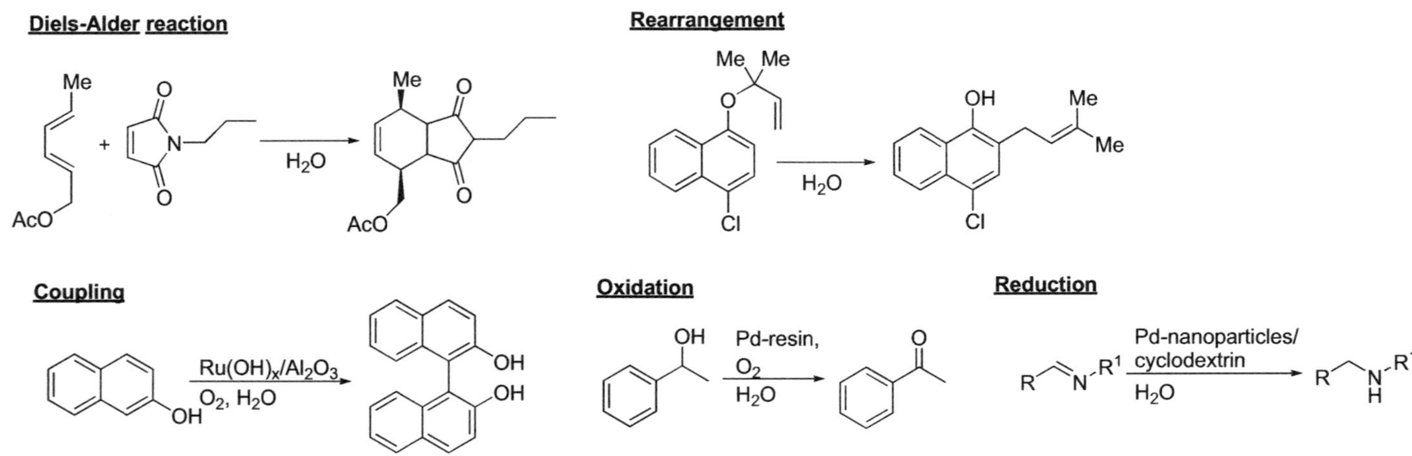

FIGURE 2.2.21 Reactions in aqueous medium.

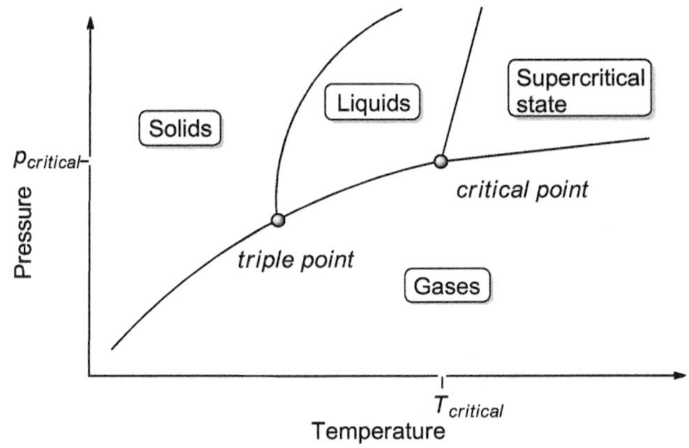

FIGURE 2.2.22 A generic phase diagram depicting the states of matter; $p_{critical}$ and $T_{critical}$ are the critical pressure and temperature of the material, respectively.

solvent but also left a residue in the coffee beans that resulted in a less-than-desirable taste. Currently, the industry is applying sCO_2, which appears to be an acceptable solvent to extract caffeine effectively, without the harmful effects of dichloromethane. The process is carried out in a flow system, where the coffee beans are loaded in a fixed bed through which $scCO_2$ flows continuously extracting the caffeine from the beans. Since it is a poor solvent, it removes relatively little of the other aroma and flavor materials, leaving an enjoyable product behind.

Other applications include dry cleaning, in which $scCO_2$ replaces tetrachloroethylene and several reactions, such as hydrogenation and polymerization, among others. Fig. 2.2.23. summarizes a few applications of $scCO_2$ in chemical reactions.

There are other materials that can be applied in supercritical state; however, their success do not match that of $scCO_2$. Either the harsh reaction condition (e.g., scH_2O) or thermal stability problems were not conducive to the widespread use of these solvents.

2.2.2.4 Ionic Liquids

While ionic compounds are usually known for their high melting point, there is a rapidly growing group of compounds with large organic cations and complex anions that are liquids

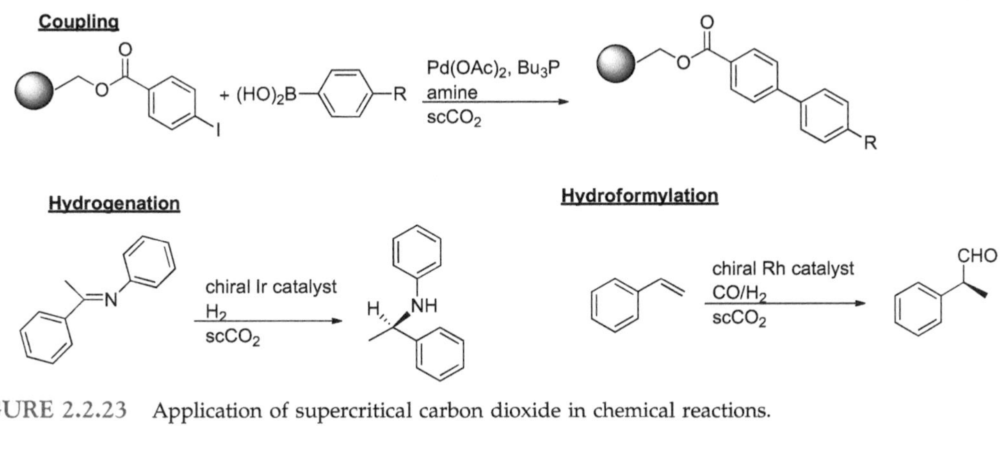

FIGURE 2.2.23 Application of supercritical carbon dioxide in chemical reactions.

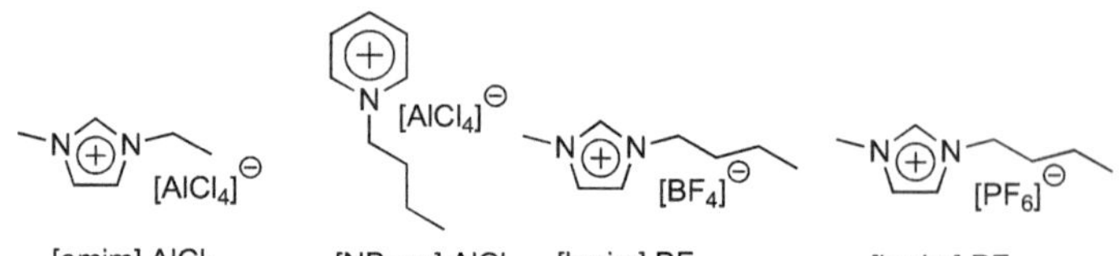

FIGURE 2.2.24 Selected commercially available ionic liquids.

at low (often as low as ambient) temperatures. These compounds are called nonaqueous ionic liquids (ILs). Since their melting point is due to their unique structure (large, nonsymmetric cations that do not form stable crystals well), the appropriate selection of the cation and counterion allows the design of solvents with broad application possibilities (Fig. 2.2.24).

ILs possess many advantages, including very low volatility and significant catalytic activity, mostly as Lewis acid catalysts, in a variety of reactions. As a disadvantage, they are often moisture sensitive, corrosive, and their toxicity has not been well studied. Their MSDSs usually include the "substance not yet fully tested" disclaimer. It is also of concern in green applications that although ILs are generally considered greener solvents than the common organic alternatives, a process or product always should be evaluated as a whole. From this point of view, the preparation of ILs involves harmful solvents (halogenated) and chemicals ($AlCl_3$, PF_5, BF_3, etc.). Another disadvantage is that while many ILs are commercially available, they are expensive (e.g., [bmim]PF_6; 50 g for $1280) and thus their use in large-scale processes is very costly. Nonetheless, ILs have been studied and found to be efficient in a variety of reactions (Fig. 2.2.25) that prove encouraging for further exploring their potential in green chemistry applications.

2.2.2.5 Solvent-Free Reactions

Solvent-free reactions represent a significant improvement over any reaction with added solvent. Despite the advantage of completely eliminating the use of solvent, not every reaction can be carried out this way. However, there are already many reactions in the industry

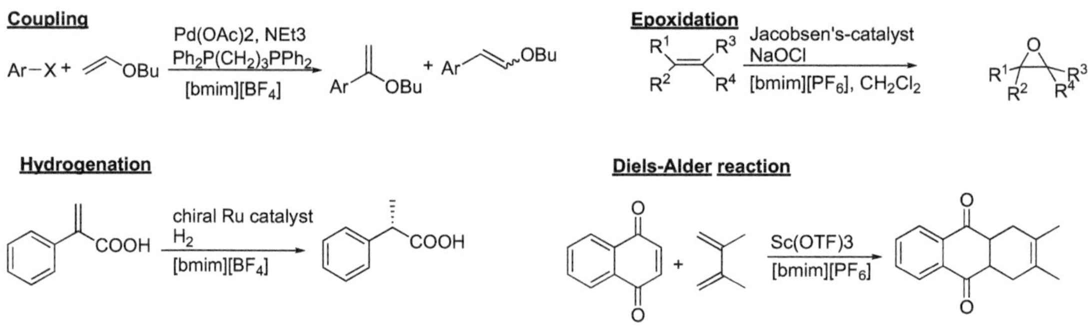

FIGURE 2.2.25 Selected applications of ionic liquids in synthesis.

that are executed without the use of solvent. Solvent-free reactions can occur in different setups. A large number of processes are carried out with solid/gas systems such as using a heterogeneous catalyst with gas-phase reactants. Most of the petrochemical processes are driven in this manner, including cracking or the isomerization of hydrocarbons. In solid/gas reactions, the reactants must be either gases or materials that can be evaporated into the gas phase without compromising the material, e.g., they should not decompose at the evaporation temperature. However, as previously mentioned, every process should be separately assessed to determine its sustainability. Although the solvent-free nature of these reactions is commendable, we should also consider how much energy is used to vaporize the reactants or liquefy the products after the reactions.

Liquid-phase reactions refer to when the reactants are miscible liquids, or at least one of the reactants is a liquid and dissolves the other reactant, reagent, or catalyst, thus a reactant also acts as a solvent. In these cases, a reactant is often used in excess and recycled after the reaction. It also allows for the complete consumption of the other reactant by forcing the equilibrium toward product formation (Le Chatelier-Brown principle). There are several industrial reactions that use this basic idea. We must note that although these reactions do not use extra or added solvent, we still have a solvent, that is, the solvent and reactant in one. This certainly may increase the recycling cost of the solvent-reactant; however, the technological problems are much less significant as compared to a case when we need to recycle the added solvent and the unreacted starting materials separately.

The third possibility is using a solid catalyst/solid (S/S) or liquid (S/L) reactant system. These reactions are quite frequently applied in organic synthesis when some special activation method (e.g., microwave irradiation or mechanochemical activation) is used. These reaction setups sound very similar to solid/gas reactions; however, while these S/S or S/L systems do not use solvent during the reactions, one is used to isolate the product. The compounds can be added to the solid catalyst as solids or liquids without added solvent; however, the usually solid product cannot be separated from the catalyst without using a solvent to dissolve it. The dissolved product is isolated by removing the solvent by evaporation that can be recycled. Therefore, the reactions are solvent-free; however, the process as a whole is not. Despite this drawback, we still believe that using a solvent in a limited role contributes to making a process more sustainable and significantly safer. Avoiding the use of a flammable and potentially explosive solvent during the reaction when extended period of heating is used still provides a major benefit. A short summary of solvent-free reactions with a few examples is shown in Fig. 2.2.26.

2.2.3 ACTIVATION AND ENERGY EFFICIENCY OF CHEMICAL PROCESSES

As discussed earlier, chemical reactions require the investment of their activation energy to occur. For some reactions, the ambient temperature provides this energy, whereas most reactions require an additional investment of energy. There are several reactions that only proceed at quite high temperatures of around 300–500 °C. Other reactions need subambient temperatures to ensure safety or high selectivity. Either way, investing energy for heating or cooling is a necessary part of designing chemical processes. Thus the cost of a product and the sustainability of a process depends not only on the chemicals used but also on the

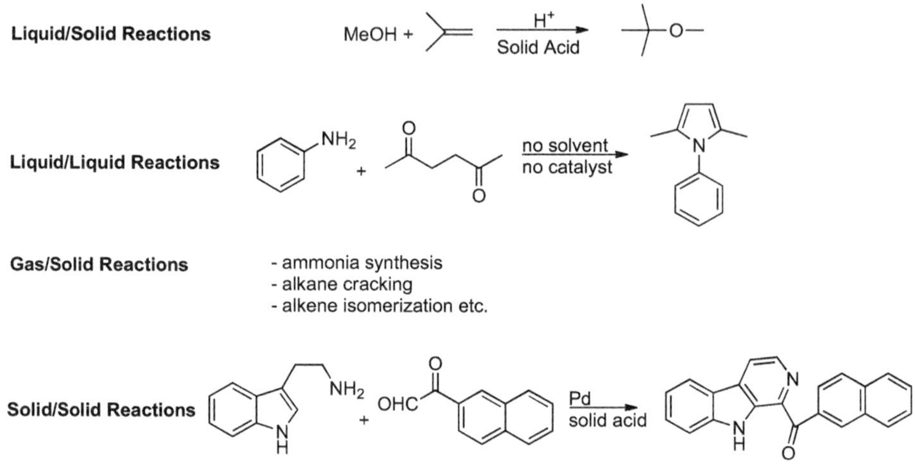

FIGURE 2.2.26 Examples of solvent-free reactions.

energy consumed. As the energy we use is currently produced mainly by burning fossil fuels, using more energy for a process indirectly promotes global warming by emitting a large amount of greenhouse gases. The design of better reactors and industrial plants to avoid unnecessary heat loss or recovering the lost heat is one way of addressing the problems. Moreover, to directly alleviate the amount of energy used, several nontraditional activation methods have been developed, which are predominantly based on the direct energy transfer from the source to the reaction mixture, in contrast to the traditional heating method that is based on external heating units and convective heat transfer.

2.2.3.1 Microwave-Assisted Organic Synthesis

The application of microwave irradiation for heating dates back to the 1940s. Randall and Both designed the so-called magnetron that is able to generate microwave radiation, as a part of radar during World War II. Even at this early stage it was noted that microwaves heated water rapidly. Since the 1950s, microwave equipments have been used in the food industry in the United States. The real breakthrough occurred in the 1970s and the 1980s when several companies began producing microwave ovens in mass quantity at affordable prices. Its first application in synthesis was published in 1969, for the aqueous emulsion polymerization of acrylate derivatives, but went largely unnoticed. The real breakthrough in synthesis occurred in 1986, when Gedye and Giguere independently published microwave-activated synthetic reactions, using household appliances. Since then, thousands of articles have been published and a new industry was developed focusing on the production of microwave instruments that were designed to fulfill the requirements of synthetic chemistry, including temperature control, reliability, reproducibility, and scalability. The efficiency of microwave irradiation is based on its unique ability to directly change the state of the matter and thus result in the heating of a reaction mixture internally. Microwaves occupy a section of the electromagnetic spectrum (Fig. 2.2.27) located between the infrared and radiowaves. Their frequency range is between 1 and 300 GHz.

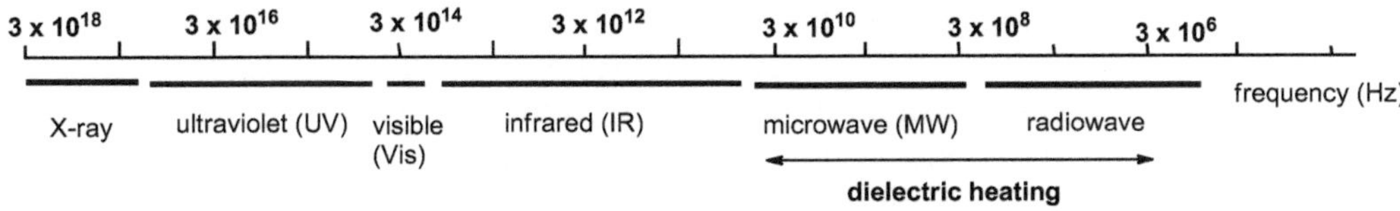

FIGURE 2.2.27 The position of microwaves in the electromagnetic spectrum.

Based on their energy, microwaves are able to affect the rotation of molecules. In low-pressure systems (gases), the material absorbs the energy and shows sharp peaks in a fingerprint-like spectrum (microwave spectroscopy). In liquids and solids, where the molecules are not capable of free rotation, we would observe broad peaks; however, due to the restricted movement, dielectric polarization will occur, which is the major source of microwave heating. Without further delving into the physical fundamentals in detail, it is worth noting that the ratio of the dielectric constant (ε') and dielectric loss (ε''), the so-called loss tangent ($\mathrm{tg}\,\delta = \varepsilon''/\varepsilon'$), describes the ability of materials to transform the microwave energy to heat. In general, materials with significant dipole moments (water, salts, metal oxides, etc.) are active in microwave heating; nonpolar materials (hydrocarbons, most plastics) are transparent to microwaves. The data in Table 2.2.2 illustrate how materials respond to microwave irradiation.

As the data show, solvents of polar character absorbed microwave irradiation better and generated more heat than those with low polarity. The same is true for the applied solids; however, in this case the differences are much more striking: whereas some solids only reached moderate temperatures, some are very strong microwave absorbers reaching over 1000 °C. Based on this, it is safe to say that the reason for the success that microwave reactors enjoy in the synthetic community is of thermal nature. Yet since the development of the first

TABLE 2.2.2 Temperature Change of Liquids and Solids After Microwave Irradiation (2.45 GHz) (Original Temperature: 25 °C, 50 mL Liquid, 25 g (1000 W) and 5 g (500 W) Solids)

Material	P (W)	t (min)	T (°C)	Material	P (W)	t (min)	T (°C)
H_2O	560	1	81	Al	1000	6	577
CH_3–OH	560	1	56	C	1000	1	1283
C_2H_5–OH	560	1	78	CO_2O_3	1000	3	1290
CH_3–COOH	560	1	110	$FeCl_3$	1000	4	41
$CHCl_3$	560	1	49	NiO	1000	6.2	1305
CCl_4	560	1	28	CaO	500	30	83
CH_3COCH_3	560	1	56	CuO	500	0.5	701
$(CH_3)_2NCH{=}O$	560	1	131	WO_3	500	0.5	532
Hexane	560	1	25	V_2O_5	500	9	701

Data from McGill SL, Walkiewicz JW. J Microw Power Electromagn Energy Symp Summ *1987:175 and Gedye RN, Smith FE, Westaway KC.* Can J Chem *1988;* ***66****:17.*

microwave reactors speculation has been ongoing in the literature about the so-called nonthermal nature of microwaves in reactions. Since a compelling evidence for this effect has not been presented yet, the majority of experts support that the microwave effect is of thermal nature. Microwave energy is not able to directly break chemical bonds. The special effects can be explained by the different E_A need of certain competing processes. Often, the traditional heating process is unable to provide the proper E_A for one of the processes and product A forms. In the microwave process, however, product B forms since much higher activation energy is provided and the product is much more stable thermodynamically, thus it will be the major one. This rings true especially in the light of recent investigations that point out that the real, internal temperature in microwave reactions is much higher than the value measured at the wall of the vessel by built-in infrared detectors.

The major advantage of microwave activation is the significantly increased reaction rates, or the short reaction times. Some of the traditionally heated reactions require long periods (sometimes days) of heating, whereas the microwave equivalent of the same reaction might be complete in an hour. Several reactions take only a few minutes. With the increased rates often comes higher selectivity. Overall, the obtained products are isolated with much higher purity. This could mean that the normally necessary purification step can be eliminated, thus a large amount of waste generation can be avoided. Microwave systems allow for the application of flow reactors, which enables continuous preparation of a product. Currently, microwave reactors are becoming common in laboratories, but less so in large-scale applications. The major reason for this is the limited penetration depth. Generally, depending on the power of the instrument, the microwaves are not able to penetrate to the reaction mixture more than about 1 inch. As long as one uses laboratory-scale instruments, this is not a problem; most instruments do not have larger than 1-inch-diameter sample tube. However, even a 10- to 20-kg-size batch reactor has much greater diameter (about 15–20 inch). Thus it is essential that the reactor is equipped with a powerful mixing system. Due to the short reaction times, it is commonly claimed that all microwave reactions are energy efficient, hence, green. Recent studies both in homogeneous and heterogeneous reactions pointed out that although this is mostly correct, it is not an automatic fact as exceptions were found. Thus the green label for microwave-assisted reactions is not automatically justified; every system needs individual evaluation.

The popularity of microwave activation is due to the short reaction times and its versatility. It can be applied for virtually any reaction that requires heat and has at least one component that is able to absorb microwave energy. Even when the components are not active, special graphite reaction vessels can be used that absorb the microwaves well and heat up rapidly. Even metal catalytic reactions can be carried out provided that the particle size of the metal is not too large. An overwhelming majority of commercial metal catalysts fulfill this requirement. Fig. 2.2.28 illustrates several types of reactions to which microwave activation was applied. These are only examples selected from the thousands of applications.

2.2.3.2 Ultrasonic Activation

The piezoelectric phenomenon, the first step toward the generation of ultrasounds, was discovered by the Curie brothers in the 1880s. In 1893, Galton developed the ultrasonic whistle; however, the first practical application of ultrasounds did not occur until 1917.

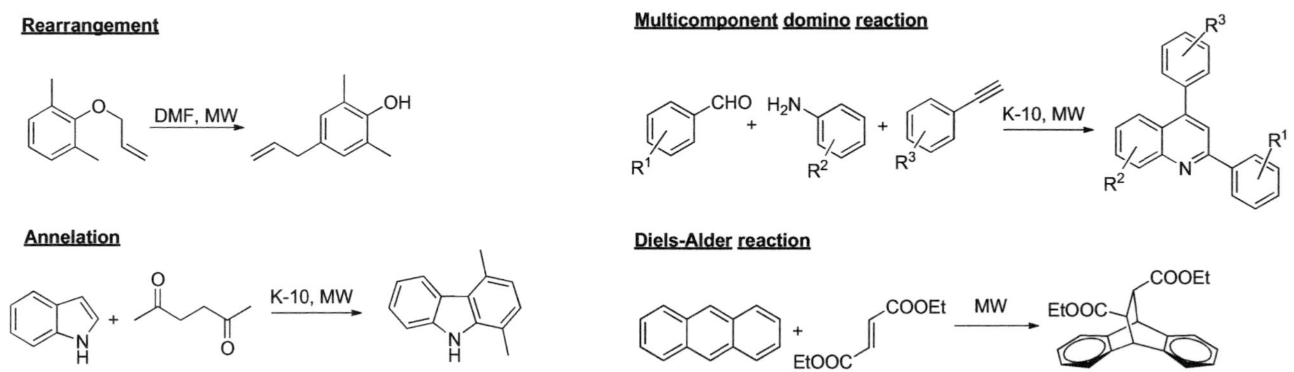

FIGURE 2.2.28 Selected applications of microwave (MW) activation in synthesis. *DMF*, dimethylformamide.

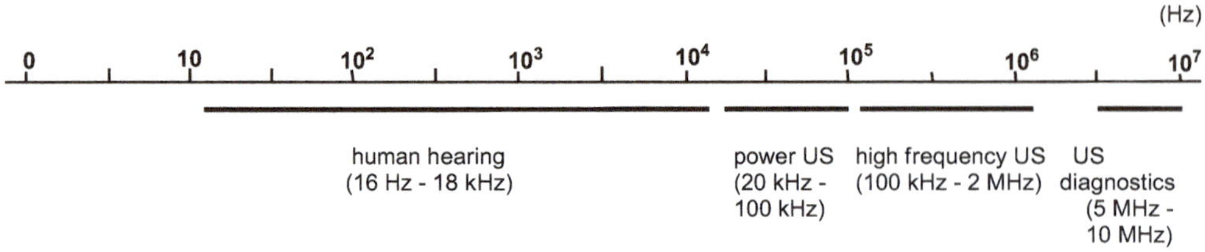

FIGURE 2.2.29 The sonic spectrum. *US*, ultrasound.

The tragedy of the Titanic in 1912 prompted a strong push for the detection of the size of icebergs, which was achieved by the application of the ultrasonic echo technique.

It has been known since 1927, that ultrasonic irradiation affects chemical reactions; however, a mechanistic explanation was not available until 1945. Just as in the case of microwaves, the developments in electronics resulted in the manufacturing of reliable ultrasound generators, which induced further developments on the application side. Probably the most well-known application is the medical ultrasound that made noninvasive detection of diseases such as cancer or the observation of fetus development possible. The beginning of wide-scale applications in chemical synthesis dates back to the 1980s. Since then, sonochemistry became a mainstream activation method.

Using the ultraviolet (UV) light, which we cannot see, as an analogy, ultrasounds refer to sound waves that are outside of the range of human hearing. Commonly, the 20–10,000 kHz frequency range is referred to as ultrasounds (Fig. 2.2.29)

Since the application of ultrasounds occurs exclusively in liquids, we will only focus on this medium. When ultrasounds travel through a liquid, they will excite the longitudinal motion of molecules, which will make them oscillate. Since the sound wave is periodic, it results in a series of compressions and rarefactions in the liquid. When the motion of the molecules is at such an extent that the secondary bonds cease to exist between them small bubbles or cavities will form in the liquid. The pressure inside the bubbles is significantly below the normal vapor pressure. This phenomenon is called acoustic cavitation. Depending on the size of the bubbles, we can distinguish between stable cavitation (large bubbles) and transient cavitation (small bubbles). The small bubbles and the transient cavitation is responsible for the ultrasonic effect (Fig. 2.2.30)

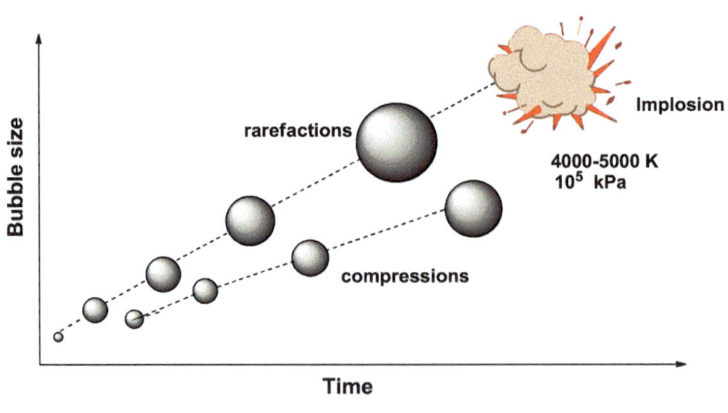

FIGURE 2.2.30 Transient cavitation.

As Fig. 2.2.30 illustrates the size of the bubbles grows up to a point of collapse. During the collapse, extreme temperature (1000–3000K) and pressure (10^5 kPa) were suggested to develop. In the early 2000s, Suslick experimentally determined these parameters and observed similar values, even higher temperatures (~4–5000K). This explanation is the hot spot model. Although there are other theories, this model can best explain reactions that occur both in polar and nonpolar solvents.

The experimental parameters significantly influence the sonochemical activation. The increasing frequency, as expected, increases the reaction rates. However, as described in several studies, the reactivity often shows an optimum as a function of frequency. The power of the ultrasound source also increases the reaction rates; however, at some point this effect turns to saturation and further increase in the power will not result in faster reactions. The temperature of the solvent is expected to have the opposite effect. The increase in vapor pressure inside the bubble would decrease the efficiency of the collapse. However, this is likely not a crucial issue as most studies report an optimum in temperature. The role of the external, or static pressure is not clear. In general, higher static pressure increases the rate of bubble formation on the surface of small particles in the solvent. If these particles are removed by ultrafiltration the reactivity significantly drops. The role of the protecting gas is twofold; monoatomic gases usually improve the rates, while certain gases (O_2, H_2, N_2, etc.) are not inert in sonochemical reactions. The solvents must be inert; however, this is difficult to maintain due to the radical formation that occurs with most solvents.

Taking into account these facts, we can conclude that just like in the case of microwaves ultrasonic activation is a process that would heat the reaction mixture internally resulting in faster reactions, shorter reaction times, and often higher yields and selectivity. Further advantages that the use of ultrasounds bring about are the extremely effective mixing of the system and the powerful surface cleaning effect of cavitation that will further enhance heterogeneous, particularly metal-catalyzed reactions. Selected applications of ultrasounds in synthesis are summarized in Fig. 2.2.31.

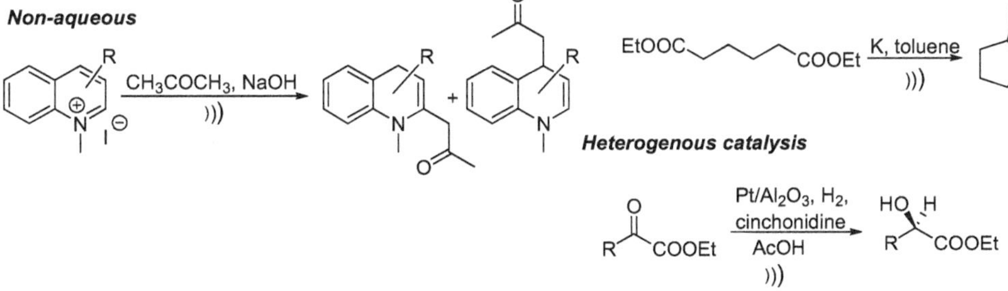

FIGURE 2.2.31 Selected applications of ultrasonic activation in chemical synthesis.

2.2.3.3 Photochemical Activation

Photochemical activation has been known since the early 1900s, when Ciamician used sunlight to initiate chemical reactions. As different light sources became available much earlier than microwave or ultrasound generators, many principles of photochemistry have been developed by the mid-20th century. Just as in the case of the other two previously described methods, a material must absorb the light to undergo a photochemical transformation. This is the first law of photochemistry, otherwise known as the Grotthuss-Draper law. Based on the dual wave/particle nature of light, Einstein described the law of photoequivalence, which states that each photon that is absorbed in a system can activate one molecule only (Stark-Einstein law). The efficiency of photochemical processes is described by quantum yield, which gives the ratio of the absorbed photons that generate a reaction to all absorbed photons. Theoretically, the best-case scenario is when every photon induces a reaction. In reality, many reactions have quantum yields much higher than one, even close to 1 million, which is indicative of a chain reaction with the photons likely only participating in the first step, the activation (e.g., chlorination of methane). In practice, photochemical reactions involve the application of UV and visible (Vis) light (see Fig. 2.2.27), thus narrowing the scope of potential substrates to compounds that can absorb UV-Vis light.

From a green chemistry perspective, these processes have several advantages. First, the photons are clean reagents that do not result in any waste. Depending on the energy of the photons, we are able to activate specific chemical bonds, thus, the reactions usually have high chemoselectivity. Additionally, the lack of conventional heating means low temperatures, so all other potential thermal processes are diminished resulting in often exclusive stereoselectivity.

Although laboratory applications are common, scaled up processes are rare at the industrial level. This is due to several disadvantages, such as the easy and significant loss of energy, due to several different reasons; ineffective photon absorption (e.g., luminescence), the wall of the reactor if transparent, and the broad spectrum of the light sources can result in often over 50% loss. Similarly to microwave applications, the penetration depth of the light is limited, thus if a large apparatus is used multiple light sources are required. This presents another problem, as the sources are delicate and expensive. Adding to the safety problem, all light sources will generate heat in addition to light and the extra heat requires constant cooling of the reactor when immersion lamps are used; otherwise conditions become hazardous, especially when using low boiling point solvents.

Despite its shortcomings, photochemical activation is applicable to many radical and pericyclic reactions using radical initiators or compounds with double bonds (alkenes, dienes, carbonyl compounds) as common substrates. Several applications are depicted in Fig. 2.2.32. As shown, many transformations such as pericyclic reactions occur readily. These processes are often difficult to perform by other methods and/or result in low stereoselectivity.

More recently an exciting new application of photochemistry has been developed and has generated significant attention, in the field of photochemical catalysis when the light activates the catalyst. While large-scale applications are yet to be developed, certainly this method carries much promise for future green industrial technologies. Some reactions are summarized in Fig. 2.2.33.

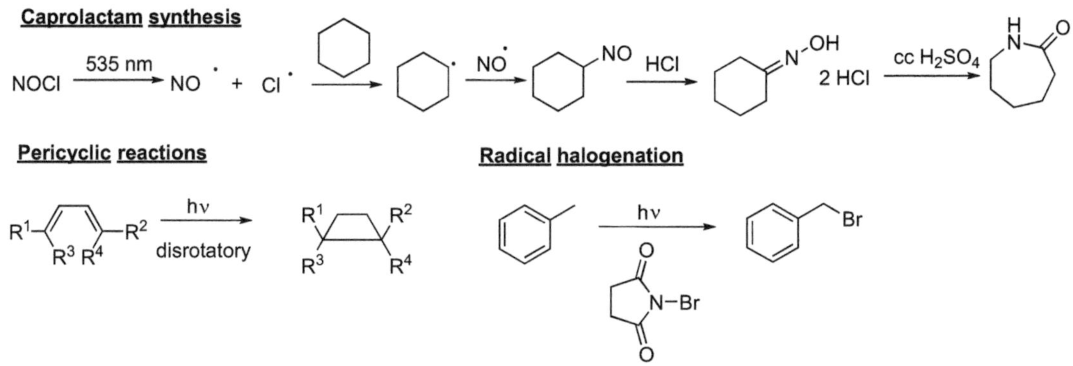

FIGURE 2.2.32 Selected applications of photochemical activation in chemical synthesis.

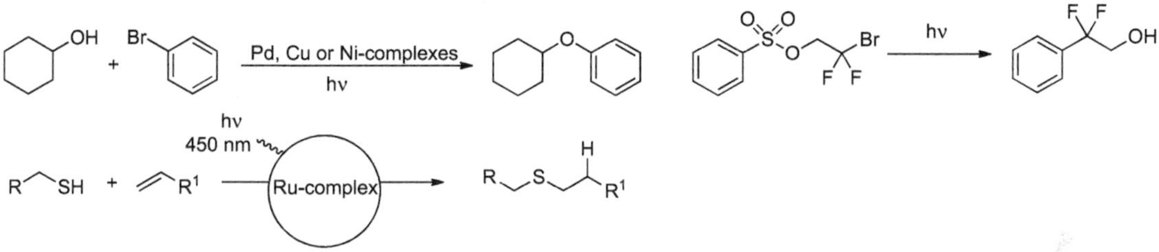

FIGURE 2.2.33 Selected applications of photocatalysis.

2.2.3.4 Electrochemical Activation

Among all activation methods electrochemistry has the most extended background. The electrochemical activation involves a direct transmission/removal of electrons to/from a substrate through metal or graphite electrodes. Electrochemical synthesis is involved in major industrial processes, such as chlorine generation or aluminum manufacturing. In addition, it is also involved in energy production (batteries, fuel cells), toxic material destruction, soil remediation, etc.

Electrochemical synthesis possesses major benefits. Just like photons mentioned earlier, electrons are also clean reagents; they do not generate any waste formation. Since reactions occur by a direct electron transfer at the electrode, the systems usually do not require any other activation, thus the conditions are mild resulting in high selectivities. The atom economy is high compared with processes that use moderate- to high-molecular-weight reagents. Although the electrochemical cells are not simple, due to the extensive history of reactor design in this field, they are surprisingly inexpensive. As a disadvantage, several systems include mercury-containing amalgam-based electrodes; however, these electrodes are phased out in most countries. Another significant drawback is that if the redox potential of the substrate is close to that of the solvent then the process will waste significant amount of energy and will partially transform the solvent, which may generate waste. While some of the major industrial technologies involve inorganic products, electrochemical synthesis of organic compounds is also common. Selected examples are shown in Fig. 2.2.34.

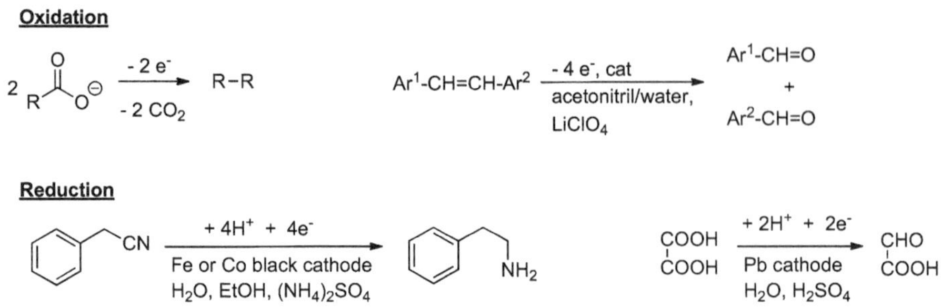

FIGURE 2.2.34 Selected applications of electrochemical activation in organic synthesis.

2.2.4 CONCLUSIONS

We have provided an overview of the synthetic toolbox that is available for the design of sustainable chemistry applications. As discussed, each of the contributing techniques has several benefits as well as disadvantages. It is important to understand that none of the methods can be considered the ideal green synthetic method: we should select the appropriate tool for the target synthesis we want to develop. One has to be careful to put the green label on any process automatically, just because one or the other listed methods are used. Each method and process should be holistically evaluated. The use of reagents, catalysts, and solvent or activation method should be evaluated as a whole considering not only the species but also the way they were prepared: designing a green process is a complex assignment. While in this chapter we have described the basics of each of these methods, the reader is encouraged to explore more about state-of-the-art case studies in the second part of the book as well as in the recommended further reading.

PROBLEMS

1. The use of catalyst poisons are quite common in medicinal chemistry. Name a certain type of drugs that acts as a catalyst poison.
2. Explain how catalytic hydrogenation is responsible for the formation of *trans*-fats in many oils and other hardened fats (margarines).
3. Compare chiral metal complex-catalyzed hydrogenations with available chiral stoichiometric methods regarding C=O reduction. Briefly discuss the advantages disadvantages.
4. Discuss how a metal catalyst decreases the activation energy of a hydrogenation reaction. Detail elementary steps and provide a hypothetical energy diagram for it. Compare homogeneous and heterogeneous alternatives.
5. Discuss the driving force of the carbocationic rearrangements that take place in hydrocarbon isomerization/cracking (stability of these cations vs. the structure of final products).

6. Design a hypothetical reaction system in the most ecofriendly manner (by combining knowledge we covered in the chapter) for the synthesis of the following materials:
Phenol
Ethylbenzene
(*R*)-ethyl lactate
Think carefully about the possibilities and explain your suggestions in detail (what and why).

7. Reactions (**A**) and (**B**) occur under similar conditions. Both reactions were carried out using microwave irradiation and traditional oil bath heating.

(A)

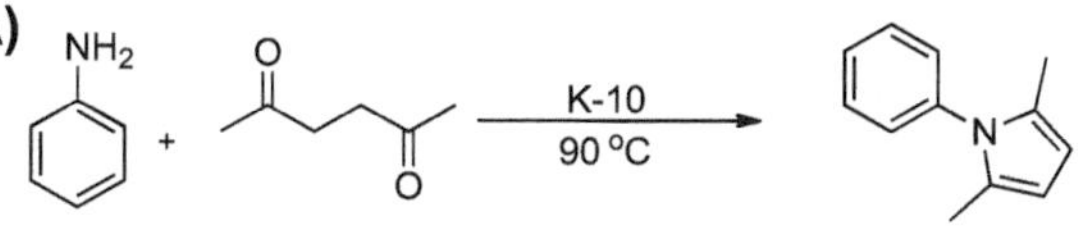

(B)

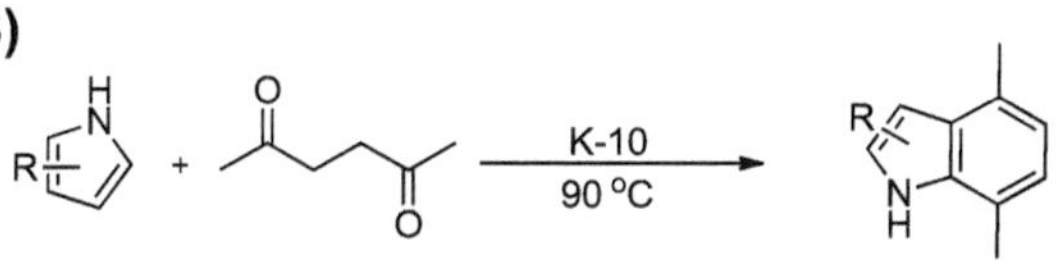

Reaction (**A**) consumed 0.01 kWh energy in a 1 min microwave assisted reaction yielding 99% product, while the conventionally heated reaction took 50 min, yielding 97% product and consuming 0.06 kWh energy.
Carrying out reaction (**B**) with microwave irradiation took only 50 min yielding 81% product while using 0.45 kWh energy compared to the 5 h of the traditional heating that produced 84% yield and consumed 0.28 kWh energy.
Which heating source is more energy efficient in these two reactions? Explain your answer with a brief calculation.

Recommended Reading

1. Anastas PT, Warner JC. *Green chemistry: theory and practice*. New York: Oxford University Press; 1998.
2. Lancaster M. *Green chemistry – an introductory text*. 3rd ed. Cambridge: RSC; 2016.
3. Matlack AS. *Introduction to green chemistry*. 2nd ed. New York: CRC Press, Taylor & Francis; 2010.
4. Rothenberg G. *Catalysis. Concepts and green applications*. Wiley-VCH Weinheim; 2008.
5. Horvath IT. *Encyclopedia of catalysis*. Wiley; 2003.
6. Somorjai G, Li Y. *Introduction to surface chemistry and catalysis*. Hoboken (NJ): Wiley; 2010.
7. Prins R, Wang A, Li X. *Introduction to heterogeneous catalysis*. London: Imperial College Press; 2016.
8. Astruc D, editor. *Nanoparticles and catalysis*. Weinheim: Wiley-VCH; 2008.
9. Leitner W, Jessop PG, Li CJ, Wasserscheid P, Stark A, editors. *Handbook of green chemistry – green solvents*. Weinhheim: Wiley-VCH; 2010.
10. Reichardt C, Welton T. *Solvents and solvent effects in organic chemistry*. 4th ed. Weinheim: Wiley-VCH; 2011.
11. Carrea G, Riva S, editors. *Solvent-free organic synthesis*. Weinheim: Wiley-VCH; 2009.
12. Kappe CO, Dallinger D, Murphree SS. *Practical microwave synthesis for organic chemists*. Weinheim: Wiley-VCH; 2009.
13. Grieser F, Choi PK, Enomoto N, Harada H, Okitsu K, Yasui K, editors. *Sonochemistry and the acoustic bubble*. Amsterdam: Elsevier; 2015.
14. Evans RC, Douglas P, Burrows HD, editors. *Applied photochemistry*. Dodrecht: Springer; 2013.
15. Fuchigami T, Atobe M, Inagi S. *Fundamentals and applications of organic electrochemistry. Synthesis, materials, devices*. Hoboken (NJ): Wiley; 2014.
16. http://www.nobelprize.org/nobel_prizes/chemistry/laureates/.

CHAPTER

2.3

Integrating the Principles of Toxicology Into a Chemistry Curriculum

Nicholas D. Anastas[1], *Alexandra Maertens*[2]

[1]United State Environmental Protection Agency, Cincinnati, OH, United States; [2]Johns Hopkins Bloomberg School of Public Health, Baltimore, MD, United States

2.3.1 AN INTRODUCTION TO THE PRINCIPLES OF TOXICOLOGY

Toxicology embodies the primordial nature of science. It is the science of survival, hard-wired not only within the DNA of human beings but also in every living organism on Earth. Plants and microorganisms invented and perfected chemical warfare, earning the well-deserved reputation for constructing some of the most lethal compounds ever tested with the tools of modern toxicology. Prolific and efficient, creating compounds with molecular structures so complex, the brightest and most skilled chemists are humbled by the elegance of the expertise possessed by these organisms to create defensive systems without the aid of a central nervous system or access to the Internet. Microorganisms synthesize antibiotics that are extremely complex structurally and have been essential to treating and preventing infectious diseases in humans and animals for decades. Strychnine is an acutely toxic poison produced by the plant *Strychnos nux-vomica*. Well-known synthetic chemists labored for years developing multistep total synthesis for strychnine and were universally lauded as pioneers and individuals of great insight. Hopefully, examining potential synthesis with modern philosophy, we now appreciate the relative irrationality of intentionally synthesizing a known toxic compound. Green chemistry and green toxicology strive to emphasize this very point.

Modern toxicology is the study of adverse effects of biological, chemical, and radiological agents on living organisms. This vast and very complex subject area requires a multidisciplinary, holistic, systems approach to fully grasp the breadth and potential of this field. Biological, chemical, and radiological agents have the potential to perturb normal biological and physiological processes, resulting in adverse consequences that may be reversible or

Green Chemistry
http://dx.doi.org/10.1016/B978-0-12-809270-5.00004-2

irreversible. The extent of perturbation of homeostasis depends on many factors associated with the agent including the intrinsic molecular structure and properties, the concentration or dose, exposure route, duration, timing of exposure, and the health status of the organism.

The structure-toxicity nexus provides common ground upon which to build the lexicon and skill set that allows chemists and toxicologists to discover, describe, and predict the potential for a toxicant to manifest harm. The arrangement of atoms in a molecule ultimately determines the pathway to the site of action, in other words, the processes including absorption, distribution, metabolism and excretion (ADME), which are collectively referred to as toxicokinetics. In common language, toxicokinetics describes what the body does to a toxicant or xenobiotic.

The magnitude and duration of activity of a toxicant at its site of action is determined by several fundamental chemical reactions that may influence the binding to macromolecules, which in turn may activate or inhibit the normal function, change molecular geometry of a receptor or membrane, or change the activity of the cellular architecture thereby permitting passage of molecules into a cell that would otherwise be excluded. This is referred to as the toxicodynamics, or what the toxicant does to the body. The mechanisms of toxicity are often organ specific, (e.g., liver, lung, kidney, nervous system), and one compound is most often associated with a spectrum of effects. Fortunately, organisms have developed ingenious defense mechanisms designed to combat the effects of xenobiotics using the same fundamental biological and chemical processes. Whether harm to the organism manifests is determined by a delicate, complex balance of kinetic, dynamic, and temporal processes.

Some of the molecular attributes that determine adverse outcomes include (1) molecular size, shape, and symmetry; (2) electronic properties and geometric configuration; (3) lipophilicity; and (4) frontier molecular orbitals, among others. Elucidating the mechanism of toxic action of xenobiotic requires linking fundamental molecular features and properties with molecular initiating events and all resulting adverse outcome pathways. This central structure-toxicity relationship forms a fertile nexus between chemists and toxicologists to work together to advance the understanding of molecular toxicology.

2.3.2 CURRENT STATUS OF TOXICOLOGY IN GREEN CHEMISTRY

Green chemistry is the design of products and processes that reduce or eliminate the use or generation of hazardous products. Twelve principles guide this challenging endeavor employing a systematic, holistic approach that includes pollution prevention, hazard reduction, energy efficiency, and atom economy. Inherent in these principles and practice of green chemistry are the principles and practice of toxicology, which represents a specific category of hazard.

Historically, advancements in green chemistry have been realized in a few focused areas, for example, catalysis, atom economy, and substituting known toxic solvents and auxiliaries in synthetic reactions with "safer" substitutes. The benefits of these improvements are laudable and have clearly demonstrated the benefits of employing the principles of green chemistry as a pollution prevention and environmental health strategy. However, demonstrated success in designing safer chemicals, often referred to as Principle 4, has lagged behind the other principles especially in the case of industrial chemicals and consumer products. There may be several reasons for this deficit including the complexity of the task of designing toxicologically benign chemicals, or a lack of experience with the principles of toxicology among chemistry educators and practitioners to address the issue of safer molecular design in an effective and efficient manner. Expanding opportunities for cross-disciplinary research, communication, and especially education are critical to reducing this knowledge gap.

2.3.3 CURRENT STATUS OF TOXICOLOGY IN THE CHEMISTRY CURRICULUM

The vast majority of colleges and universities neither incorporate toxicology as a specific requirement in the chemistry curriculum nor do they embed toxicology principles into the existing course material. This deficiency has resulted in a knowledge gap in the education of practicing and academic chemists that is truly unconscionable and must be addressed with a level of urgency concomitant with the need to vanquish the adverse consequences of chemical hazards to human health and the environment that have manifested as a result of this educational omission. Perhaps anecdotally students will hear about the dangers of acutely toxic chemicals used in the laboratory. For example, many students enrolled in a chemistry laboratory course have recounted the advice of laboratory instructors advising them to move to fresh air immediately if the smell of burnt almonds is discernable because it means cyanide is released. How many undergraduates have ever smelled burnt almonds, let alone cyanide? Likely equal numbers! Explosivity and flammability are chemical events that are often self-evident, exuberant, and memorable. Perhaps the time is right to convince chemists that understanding and applying the principles of toxicology to protect themselves, coworkers, families, fellow citizens, and future generations is equally or more important as the more familiar chemical hazards.

Recently, efforts to bridge this historical omission have been gaining momentum at the university secondary and primary levels of education. Science need not be taught in a dry, boring manner, unhinged from practical application. The material can be taught in an engaging manner by showing common application of toxicology and chemistry in everyday situations. String the baseball glove—make the parts work as one.

Incorporating toxicology into a mature chemistry curriculum presents a number of challenges at the scientific, pedagogical, and political levels of communication. The existing curriculum is already filled with fundamental principles and concepts that students must grasp to advance to a career in the chemical enterprise, so there is the issue of syllabus space. This will always be the case especially in this age of "big data." Chemistry departments and individual faculty members must determine innovative approaches to demonstrate the complementary nature of chemistry and toxicology as part of graduation requirement for chemists who are fully prepared to address the needs of a chemical industry that emphasizes safety and sustainability of their products and processes.

Even though the concepts of toxicology and chemistry overlap in a number of areas, there is a challenge of language, concept, and ontological differences between the two fields that must be reconciled. Often, chemistry instructors have not been exposed to these nuances in their training, which creates an issue of deciding how to bring toxicology into the discussion and who is best suited to teach the material most effectively. A stand-alone course in toxicology, special topics courses, seminars, and continuing education for chemists are all potential approaches to ensure that all chemists appreciate the implications of chemistry on the manifestation of toxicity.

Another major challenge is inertia. Chemistry has been taught in a common way for over a century, so why change now? This question has been expressed often in the past but less so recently. Science and technology have changed dramatically over the past decade pulling the two fields of chemistry and toxicology closer and expanding the amount of common material and objectives. Green chemistry has served as a uniting philosophy around which inclusive and diverse course have been designed to join chemistry and toxicology in academia.

2.3.4 TOXICOLOGY AS A CORE COMPONENT OF A COMPLETE CHEMIST'S EDUCATION

Teaching the principles and practices of toxicology to chemistry students requires that an instructor introduce a new lexicon of terms, present new concepts, and describe relationships among seemingly unrelated subject matter material. Effectively, embedding toxicology principles into the education of current and future chemistry students, as well as practicing chemists, must strive to soften the potentially harsh transition to a new paradigm and endeavor to make gentle the language of toxicology.

Begin a discussion of toxicology with the fact that everyone is an "apprentice toxicologist," even if they are unaware of that fact. As human beings we know that eating too much candy will often result in a belly ache. Insufficient consumption of too little food or vitamins will leave us hungry and vitamin deficient. The spectrum of effects is demonstrated with a recitation of the ecstasy and agony that manifests with the increasing ingestion of ethanol on a short- and long-term basis.

We seek common ground among chemists, toxicologists, environmental scientists, and other allied partners. Finding commonalities among fields is challenging but not unprecedented. Today's transdisciplinary science is merely acting as our forbearers did only with 21st century tools and knowledge. Many reasons, circumstances, and theories can be put forth regarding why the trend toward specialization in the sciences occurred, but that discussion will be left for another time. Society and science have forfeited a once treasured appreciation of the intrinsic interconnected components of Nature. Let us seize this opportunity to realign the future by embracing the holistic approaches embraced by our science mentors.

2.3.5 TOXICOLOGY, HAZARD, AND RISK ASSESSMENT

Hazard characterization requires that we understand what can potentially go wrong; risk is defined as the probability that someone will actually experience a hazard. Within the context of toxicology, to understand a chemical's risk we have to understand hazard. We need to know what health problems a chemical can potentially cause.

How do we assess a new chemical for potential hazard? There are several types of data we can use—animal (or in vivo) studies; in vitro studies, which typically use tissue culture;

human data (often from epidemiology studies); and quantitative structure-activity relationships (QSAR).

Animal studies are currently the main source of data for hazard assessment. Animal tests are used to understand acute toxicity (what happens when an animal is briefly exposed to a chemical), subchronic toxicity, and chronic toxicity (what happens after longer spans or close to lifetime exposure). Some tests are specific for a target organ (such as skin or eye exposure) or a specific process (reproductive and developmental toxicity). New chemicals with unknown hazard will typically be examined with a battery of tests to clarify potential concerns; for example, the US Environmental Protection Agency requires pesticides to have six different tests. Acute toxicity is studied through three different exposure routes—oral, dermal, and inhalation. Typically, this involves giving a test animal (usually a rodent) a high, one-time dose to quantify how much of a chemical it takes to cause lethality, which is reported as the LD_{50}, or the dose required to kill 50% of the animals. Eye and dermal irritation are both assessed by scoring the severity of irritation in rabbit eye and skin, respectively, and additionally, dermal sensitization is assayed via exposure of a guinea pig and testing immune response to a subject. Eye and dermal irritation are local effects, that is, they are limited to the area of exposure. Dermal sensitization, on the other hand, is a systemic effect. Finally, there is a test to establish a dose-response relationship, which is key because it helps to quantify the relationship between a given exposure and the incidence of a given adverse health effect, which is essential for risk characterization. Although animal models are often a "good enough" proxy for human results, there are often species differences that can complicate the results. Because of this, some tests have to be repeated on two species; for some types of hazard identification, for example, identifying the potential of a new drug to cause birth defects, often a primate species will be used.

Because of the difficulty of extrapolating from rodent species to humans, and because animal tests are both expensive and time consuming, it is often desirable to substitute an in vitro test. One of the more common in vitro tests is the Ames test, which used the ability of bacteria to provide a rapid and reliable read-out of mutations and coupled that with a mammalian metabolic system to provide a rapid way of assessing a chemical's potential to cause a DNA mutation. The Ames test not only drastically reduced the number of necessary cancer in vivo studies, which take 2 years and require many animals, but also demonstrated that many normal components of the environment (combustion products, cooked meat) were mutagens. In vitro tests are often most useful to understand molecular mechanisms.

Human data, which often come from epidemiological studies, can either look directly at the effects that an exposure has on humans (by using biomarkers of exposure and disease) or they can look indirectly, for example, by looking at the relationship between fluctuations in exposure and disease rates. Sometimes, human data come from case reports after a spill or misuse of a substance. Because of ethical considerations, there are rarely any controlled studies done on humans for toxicity testing. QSARs are often used for instances in which the data available for similar chemicals can establish a clear structure-activity relationship.

Ideally, whether a study is an animal study or an epidemiological study, there should be a relationship between the dose and the response, i.e., with increased dose, the effect increases, and most importantly, there is a threshold—often called a no observed adverse effect level (NOAEL)—below which adverse effects are not seen. In the absence of such dose-response information, it is possible that the findings were simply due to chance and we cannot adequately quantify the hazard.

Once we understand what specific effect a chemical can cause, and the dose necessary to cause that effect, we can do a risk assessment. Risk is a function of hazard and exposure; a highly hazardous chemical may have no risk if there is no exposure, therefore we cannot assess risk by understanding the potential for a population to be exposed. Risk assessment typically consists of quantifying whether an exposure is high enough to cause concern, after accounting for uncertainty in the data, usually by adding in several "uncertainty factors." For example, if a chemical caused an adverse effect at 150 mg/kg in rats in a subchronic test, what dose is problematic for occupational exposed humans? Typically, a regulatory agency might divide by 10 to account for the fact that we are extrapolating from rats to humans, and another 10 to account for the fact that we are using a subchronic study instead of a chronic study, depending on the quality of the data and the concern about the adverse outcome. The better we understand the health hazards a chemical can pose, the more accurate will the risk assessment be.

For the purpose of risk assessment, carcinogens and noncarcinogens are treated differently. For noncarcinogenic substances, it is usually assumed that there is an NOAEL, a dose below which an exposure causes no adverse outcome, and that an exposed individual can cope with or recover from the exposure. For carcinogens, on the other hand, it is assumed that there is no threshold—if a chemical can cause a mutation in DNA, then any exposure, however small, has some probability of causing cancer. To see how this works in practice, consider arsenic, which has both noncancer and cancer effects.

Arsenic can cause gastrointestinal effects, anemia, peripheral neuropathy, skin lesions, hyperpigmentation, gangrene of the extremities, vascular lesions, and liver or kidney damage in humans when there is chronic oral exposure. Since arsenic is present naturally in drinking water at different concentrations throughout the world, epidemiological studies have established a dose-response relationship for arsenic and what is called a "reference dose" (RfD). The RfD is an estimate, often an estimate that may be off by an order of magnitude, of a daily oral exposure to the human population that is likely to be without significant risk. In the case of arsenic, an RfD of 0.0003 mg As/kg/day for inorganic was derived from an NOAEL of 0.0008 mg As/kg/day for dermal effects and possible vascular complications in a Taiwanese farming population exposed to arsenic in well water. An uncertainty factor of 3 (to account for uncertainty in the data) was applied; note that this is a much smaller uncertainty factor than would typically have been used if extrapolating from animal data to humans.

Let us say some well water was recently sampled and the concentration of arsenic was determined to be 6.5 μg/L. Is this safe? To answer this question, we need to calculate exposure. For many activities, including water consumption, we can make some default assumptions that regulatory agencies typically use, for example, the average person drinks 2 L of water per day, weighs 70 kg, and will drink the well water for 350 days a year for 30 years. From this information, we can calculate the dose: it works out to be 1.8×10^{-4} mg/kg/day. When we compare this to the oral RfD, we see that it is above that level, and therefore not a safe amount.

Arsenic is also believed to cause cancer when individuals are exposed via inhalation, a scenario that can occur in occupational settings. Since carcinogens have no RfD, cancer risk assessment depends on the slope factor (SF). The SF is an upper-bound estimate of the probability that an exposed individual would develop cancer over a 70-year exposure; in other words, the SF represents an estimate of the potency of a given carcinogen.

Typically, the SF is derived from the potency observed at higher doses and extrapolated to lower doses, as such effects are difficult to detect with animal or epidemiological studies. Cancer risk is typically evaluated as individual excess life time cancer risk, which is the exposure multiplied by the SF; any risk above the threshold of one in a million is considered too high.

Arsenic has an oral SF of 1.5 mg/kg/day. Going to back to our previous estimate of oral ingestion from the well of 1.8×10^{-4} mg/kg/day, this gives us an excess risk of 2.7×10^{-4}. We multiply the ingested amount by the SF for an estimate of the excess risk of 2.7×10^{-3}. This is well above the target and therefore an unacceptable risk. Note that the acceptable and unacceptable risks are defined by regulations, and, ultimately therefore, political and other nonscience factors are part of the final decision making. We could decide that a risk of one in 10 million is acceptable for arsenic exposure, although doing so might involve undesirable trade-offs.

2.3.6 EXAMPLES OF CONNECTING CHEMISTRY AND TOXICOLOGY PRINCIPLES

2.3.6.1 Chemistry Concept: Nucleophilic Substitution

Nucleophilic substitution mechanisms are taught in fundamental organic chemistry courses, and this nexus of chemical mechanism and a biological mechanism can be illustrated with an example. These specific mechanisms are the central learning objectives in all organic, biochemistry, and many general chemistry courses. In a traditional curriculum, students are presented with a theoretical scheme for S_N1 and S_N2 reactions. An example for a general idealized S_N2 reaction may be presented in a typical chemistry lecture is as follows.

$$R{-}X + Y \rightarrow R{-}Y + X^-$$

After presenting this reaction mechanism illustrating how general S_N2-type reactions proceed, the instructor often provides actual examples using well-described, commonly used reactions represented by various R-groups.

2.3.6.1.1 Toxicology Concept Bridge

Instructors can introduce to students how this same reaction mechanism influences biological activity, including toxicity. Using the nucleophilic attack of a toxicant at the N^7 of guanine (Scheme 2.3.1), the instructor not only reinforces the central organic chemistry principle but also shows its fundamental link with biological and toxicological outcomes.

Alkylation of biological molecules, including DNA, RNA, and proteins, is a class of reactions involved in normal physiological processes that are necessary for maintaining homeostasis. For example, binding of topoisomerases to DNA is part of the natural replication process of genetic material. However, this same class of reactions provide molecular initiating events for adverse outcomes such as organ damage, reproductive and developmental deficits, neurotoxicity, developmental and reproductive effects, and cancer. Alkylation of any hetero atom within the DNA structure is possible; however, the N^7 of

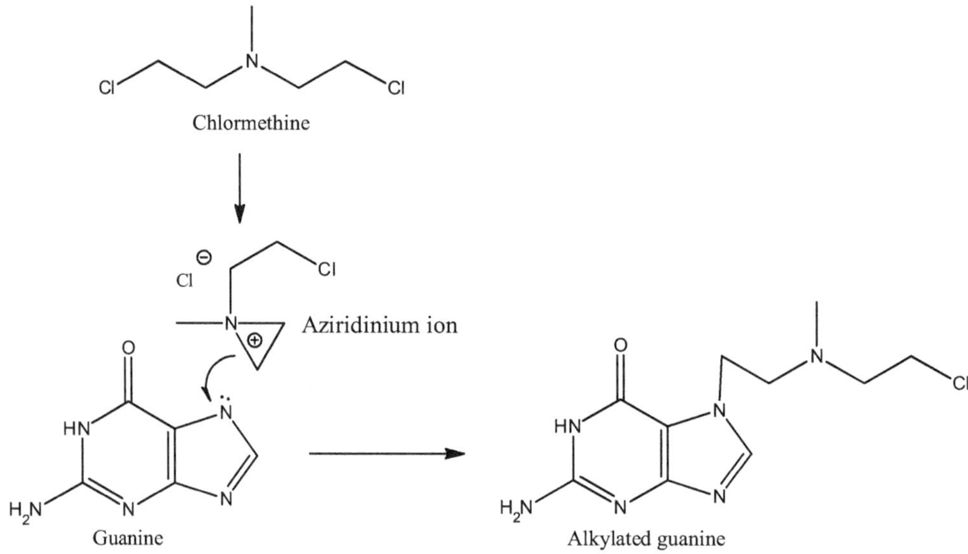

SCHEME 2.3.1 Mechanism of DNA alkylation at the N^7 of guanine for chlomethine.

guanine is the most electrophilic and reactive followed by the O^8 of guanine. Chlormethine is a compound that functions as a mustard gas, and as a medicine to treat certain cancers and some skin diseases. The instructor can use this opportunity to enlighten the students that the same compound can be used as a cure or to kill, depending on the dose, the timing of exposure, and other pharmacokinetic and pharmacodynamic properties. In the body, chlormethine is initially metabolized to an aziridinium ion, which increases its electrophilicity, making it a better electrophile and more likely to react with the N^7 of guanine-forming adducts. Using adduct formation exclusively to predict toxicity has not proven reliable. These types of molecular measurements must be supported with other data from high-throughput screening from animal testing in an integrated approach to describing and forecasting adverse outcomes.

An important opportunity for the instructor to realize is that while this concept introduces students to central chemistry concepts within a context of biochemistry, toxicology, and public health, it is also an opportunity to introduce the student to the concept of using mechanistic knowledge to design safer, more sustainable products and materials. At this point in the lecture, the instructor often compares the strength and properties of various nucleophiles including the nature of leaving groups. By showing how these factors affect the velocity, stereochemistry, and reactivity of a reaction, the instructor can present a set of toxicological data to introduce the concept of potency for various chemicals acting through this mechanism of action. Students will hopefully realize that understanding related mechanisms of action in both a chemical and toxicological sense empowers them to apply this knowledge to design chemicals that are less likely to initiate the process of toxic action.

2.3.6.2 Chemistry Concept: pH/pKa/Ionization

In a traditional chemistry course, the concept of pH is taught as a core learning objective. The hydrogen ion concentration is measured by determining the negative logarithm of the molar concentration of hydrogen ions and is used as a relative measure of acidity or basicity on a scale from 0 to 14.

The availability of the hydrogen or hydronium ions influences a myriad of chemical phenomena including rates of reactions, availability and accessibility of active sites, sorption phenomena, and corrosivity. As part of the traditional presentation of this material an instructor often provides examples of well-known chemicals and products along with their pH values. The ionization state of a chemical profoundly influences many biological reactions that can be beneficial biochemistry pathways or toxicology pathways of toxicity.

2.3.6.2.1 Toxicology Concept Bridge

A well-told story often engages students more effectively than a didactic presentation of material following traditional pedagogy. In short, stimulating rather than mundane. A real-world example of the link between the core chemistry concept of pH and toxicity is essential to show students that the availability of hydrogen ions can profoundly affect the extent of toxicity.

Curare is an extract of a mixture of natural products derived from a number of plant sources depending on the background, need, and mood of the local medicine man (shaman).

The active ingredient of these complex potions is tubocurarine (Fig. 2.3.1) and is so named because the original material obtained from a crude botanical extract of the plants used to make an arrow poison was transported in a tube.

In the Amazon rainforest, and other venues, hunters regularly use preparations of curare to coat blowgun darts used for killing prey that often winds up as a tasty meal. The poison-tipped dart enters the muscle of the animal, often a monkey lounging in a tree, and the monkey succumbs to the adverse (to the monkey) effects of the toxin, falling to the ground dead. The shaman approaches the animal, cuts a nice fillet from the flesh, and proceeds to chew, enjoy, swallow, and digest the curare-laden meat without any apparent adverse effect to the shaman/medicine person. Why? At this point hopefully the class is engaged,

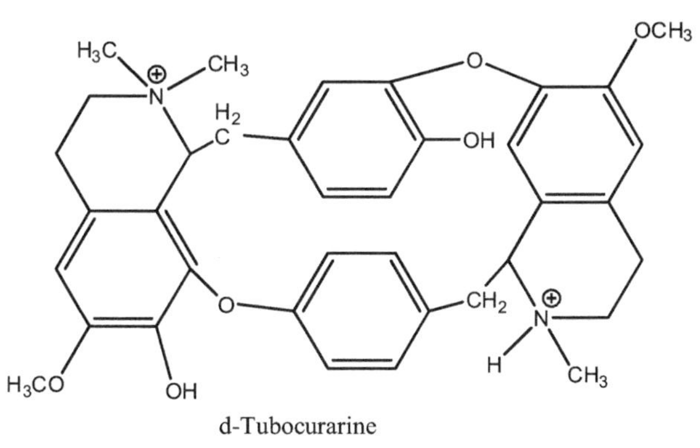

FIGURE 2.3.1 Structure of D-tubocurarine (curare).

scholastically and cerebrally salivating for the reasons why this seemingly incongruous event is possible. Prototoxicologists want to know!

It is the pH obviously.

A molecule must react at its site of action to initiate a biological response, in this context toxicity. Reaching the site of action may be simple as a single step as in the case of caustic acids or bases where the toxicity is immediate and at the site of exposure or it may be a multi-stage event that can occur immediately (acute effects) or on a delayed basis (chronic toxicity). The factors governing these steps, or the "molecular itinerary," are ADME.

The process of absorption can be introduced to illustrate the influence of physicochemical processes on toxicity. The molecular scaffolding of curare imparts a permanent positive charge on the nitrogen atom at a pH of approximately 7–7.4. However, we understand that the availability of the hydrogen ion is affected by the surrounding medium, in this case monkey muscle or shaman gut. What are the differences?

Here is an opportunity for the instructor to introduce, or reintroduce, the Henderson-Hasselbach equation for acidic (Eq. 2.3.1) and basic compounds (Eq. 2.3.2) and then link it with the toxicology concept of bioavailability. The toxicant, in this case the natural product toxin, cannot reach the site of action through the gut because curare is ionized and less likely to be absorbed.

$$\mathrm{pH} = \mathrm{pKa} + \log(\text{ionized}/\text{un-ionized}) \tag{2.3.1}$$

$$\mathrm{pH} = \mathrm{pKa} + \log(\text{un-ionized}/\text{ionized}) \tag{2.3.2}$$

The instructor expands on the topic by providing examples of the utility of this equation in a normally functioning situation (e.g., physiological homeostasis) versus a perturbed state (i.e., toxicity). Know the character of your molecule and the influence of the neighborhood it visits on its journey to the site of action and you will understand its toxic potential more completely.

2.3.6.3 Chemistry Concept: Electrophiles

Electrophiles contain electron-deficient atoms with a full or partial positive charge. Electrophiles react with nucleophiles in limitless reactions and are used as a core tool of synthetic chemistry. Electron-poor centers seek out atoms, molecules, or pieces of molecules that have a relative abundance of electrons or nucleophiles. Many nucleophiles of concern to toxicologist are those associated with biological systems including proteins, nucleic acids (e.g., DNA), and endogenous metabolic substances such as glutathione.

Introducing an electronegative atom or group into a molecule rearranges the distribution of electronic charge. These electron-withdrawing groups are found in a myriad of biological substances, natural products, and toxicants. Some examples are displayed in Table 2.3.1.

Aldehydes are a specific class of electrophilic compounds that are ubiquitous in our natural and built environments. Essential oils, fragrances, flavorings, adhesives, and other industrial chemicals have members in good standing of the aldehyde and ketone families. Aldehydes can be subdivided into groups based on the substitution of functional moieties to the carbonyl carbon. The nature of these entities will affect the physicochemical properties and reactivity including toxicity.

TABLE 2.3.1 Examples of General and Specific Electrophilic Entities

Electrophilic Entity	Specific Example
Aldehydes and ketones	Formaldehyde, acrolein, citral, vanillin
Nitrogen containing	Dimethylnitrosamine, *N*-methyl-*N*-nitrosourea
Sulfur containing	Hydrogen sulfide, carbon disulfide, sulfur mustards
Oxygen containing	Methanol, ethanol, epoxides

2.3.6.3.1 Toxicology Concept Bridge

Electrophiles can initiate a toxic response without any transformation (parent is the ultimate toxicant), or can be formed following single or multiple transformations (biotransformation). Compounds that are intrinsically toxic without any further transformation are often very reactive. Examples of substances where the parent compound is the ultimate toxicant are as follows:

- Strong acids and bases
- Nicotine
- Carbon monoxide
- Cyanide
- Methyl isocyanates

Biotransformation can result in a product that is more toxic (toxication) or in a species that is less toxic (detoxication) than the parent molecule. Insertion of an oxygen atom by the enzyme cytochrome P450 into a molecule is termed phase I metabolism, designed to render a molecule more water soluble and more easily excreted by the urine. These changes can occur by adding or exposing a hydrophilic functional group and involve hydrolysis, reduction, and oxidation. Some examples are provided in Table 2.3.2.

TABLE 2.3.2 Examples of Bioactivation of a Parent Compound to an Ultimate Toxicant

Parent Compound	Ultimate Toxicant	Toxicating Process
Acetaminophen	*N*-acetyl-*p*-quinoneimine	CYP
BP	BP 7,8-diol-9,10 epoxide	CYP
Ethylene glycol	Oxalic acid	ADH, oxidized
Hexane	2,5-Hexanedione	CYP
Hydrogen peroxide	Hydroxyl radical	Hydrolysis
Phosgene	Chloroform	CYP

ADH, alcohol dehydrogenase; *BP*, benzo[a]pyrene; *CYP*, cytochrome P450.

Phase II metabolism involves conjugation of hydrophilic groups to phase I metabolites or directly to parent molecules that are inherently hydrophilic, thereby increasing the water solubility in both cases. Phase II reactions include glucuronidation, conjugation with glutathione, sulfation, and methylation.

2.3.6.4 Chemistry Concept: The Process of Oxidation and Reduction (Redox)

Reduction and oxidation reactions must occur together and are therefore referred to as redox reactions. The oxidation of one species must be accompanied by the reciprocal reduction of another species, which can be an element, ion, molecule, or compound. This process essentially keeps track of where the electrons are going and what they are doing.

Oxidation can be thought of in a number of equivalent ways including the gain of oxygen and positivity or the loss of electrons. Reduction is the gain of electrons or hydrogen and the loss of electrons. One way to remember this process is to use the pneumonic **OIL RIG** that stands for oxidation *is* loss and reduction is gain of electrons. Oxidizing agents donate oxygen to another species, whereas reducing agents remove oxygen or add hydrogen to another species.

Redox reactions are central to biological and toxicological reactions including glycolysis, lipid peroxidation, photosynthesis, and a myriad of other mechanisms of toxicity that can provide chemistry students with connecting examples for making linkages between chemistry and toxicity.

2.3.6.4.1 Toxicology Concept Bridge

Redox property influences toxicity in a number of ways through a variety of mechanisms. Two specific examples will be presented here that can be used by instructors to illustrate the influence of redox chemistry in toxicology. Chromium is a naturally occurring metal most often complexed to other atoms to form ores and minerals that can assume several oxidation states. The two most important in toxicology are the trivalent (+3) and the more toxic hexavalent (+6) species. The differences in toxicity are associated with differences in the oxidation state stability and solubility. Trivalent chromium is not very soluble and is not absorbed efficiently by any route of exposure, whereas hexavalent chromium is well absorbed through the lungs, gastrointestinal tract, and skin. Trivalent chromium is an essential trace nutrient necessary for glucose metabolism, for growth, and for supporting immune system health.

Once absorbed, hexavalent chromium can be reduced by a number of endogenous molecules including hydrogen peroxide, ascorbic acid, and the detoxicating tripeptide glutathione. During this reduction process, reactive intermediates are produced that interact with proteins and lipids in target tissues causing damage. These reactive metabolites can also interact with DNA resulting in cancer in multiple organs including the lung, liver, kidney, and bladder. Hexavalent chromium is classified as a human carcinogen. This example demonstrates the potential adverse consequence that altering the redox cycle can have on the ability of living systems to maintain homeostasis and engage a proper detoxication response.

Nitrate and nitrite are part of the environmental redox cycle from atmospheric molecular nitrogen (N_2) to maximally reduced ammonium ion (NH^{+4}) to the fully oxidized nitrite (NO_2^-) ion. Reduction from nitrate to nitrite occurs physiologically from the breakdown of nitrogenous proteins and nucleic acids. The ratio of nitrate to nitrite is maintained at relatively constant levels; however, imbalances in metabolism or exposure to high levels from external sources can tip this balance in a direction that may lead to toxicity.

Excessive levels of nitrite in the blood can lead to a condition called methemoglobinemia or "blue baby syndrome," where the distribution of oxygen is compromised by a reduction in the binding capacity of oxygen caused by nitrite oxidation of iron on the hemoglobin molecule. Intake of drinking water containing elevated concentrations of nitrate can cause methemoglobinemia by being reduced to nitrite in the body. This is a second example where redox conditions dictate an adverse outcome. There are hundreds of examples that can be used to make this chemistry-toxicity link available for the curious chemist.

2.3.6.5 Chemistry Concept: Molecular Size and Charge Influence Reactivity

As a chemistry educator begins to explain how the intrinsic properties of shapes (molecular geometry) and sizes of atoms and molecules influence their reactivity, toxicology as a type of reactivity can be included. Size and geometry affect toxicity. Gaining access to the site of action represents the most important aspect of a toxicity pathway. The influence of size, shape, and charge of a toxicant is essential not only to understand how adverse biological outcomes occur but also to provide knowledge critical to designing molecules that have limited access to biologically critical sites in pursuit of reducing hazard.

2.3.6.5.1 Toxicology Concept Bridge

Toxic compounds that are foreign to an organism (i.e., xenobiotics) can mimic natural ions and molecules leading to disruption of homeostatic pathways. Molecular charge and size are often the key physicochemical attributes that are exploited in certain pathways. This concept can be demonstrated using metals (described below), small organic molecules, or inorganic compounds as examples.

Small molecules can pass through membrane pores gaining access to the internal matrix of a cell.

Hydrophilic molecules up to a molecular weight of approximately 600 permeate through pores located in the membranes. Some xenobiotic molecules can use the same transport systems as those that exist for endogenous molecules.

Lipid-soluble, nonpolar compounds are generally taken up through the process of simple diffusion. Toxicity for these types of molecules have been modeled using the partition coefficient (log P) as a surrogate in QSAR particularly to model narcosis in aquatic organisms. Narcosis represents nonspecific, baseline toxicity that depends on a compound's lipophilicity. Generally, this narcosis-log P relationship is parabolic peaking at log P of approximately 6, reflecting the reduced absorption of superlipophilic compounds (log $P > 6$) due to the tendency of these compounds to remain in the hydrophobic cellular membrane without crossing into the interior of the cell.

2.3.6.6 Chemistry Concept: Solvents

Solvents comprise a class of liquid compounds that are of variable lipophilicity and volatility used ubiquitously in undergraduate and graduate chemistry laboratories. Organic solvents are predominantly low molecular weight, lipophilic, and volatile, making them a concern for all routes of exposure including inhalation, ingestion, and dermal contact. These properties depend on the number of carbons, the type and number of halogen atoms attached to the hydrocarbon, and several other molecular structural features. Solvents are used as degreasers, paints, inks, adhesives, diluents or reactants in molecular reactions, and fuels.

2.3.6.6.1 Toxicity Concept Bridge

Hydrocarbon solvents are thought to be of low acute toxicity to humans because of the low reactivity associated with C–H bonds. The toxicity of solvents depends on (1) the number of carbon atoms, (2) the degree of saturation, (3) the number and type of functional groups, and (4) the configuration (branched or straight chain).

Chronic exposure to hexane is a concern for individuals exposed in the workplace or through other long-term use of products containing hexane. It is a widely used solvent, having molecular formula C_6H_{14}, found in adhesives, paints, and other household items. Sniffing glue or "huffing" leads to inhalation of very high levels of hexane on a short-term basis and can mimic the effects of chronic exposure. A spectrum of effects have been seen after long-term exposure to hexane including dizziness, blurred vision, fatigue, weight loss, malaise, and other nonspecific symptoms. More serious effects resulting from exposure to higher concentrations include numbness, paresthesia, and muscle weakness.

Hexane is an example of a molecule that must be bioactivated (metabolized) in the liver by cytochrome P450 enzymes to a γ-diketone 2,5-hexanedione, a reactive nucleophile, to be neurotoxic. This metabolite is the "ultimate toxicant," meaning that it is the molecule that can interact with that target site to cause toxicity, in this case with proteins in the nervous system. Molecules that contain the γ-diketone structure react with the amino acid lysine to produce a 2,5-dimethylpyrole adduct in neurofilaments inducing neuropathies. Steric preference is demonstrated by observing that 2,5-heptanedione is equitoxic with 2,5-hexanedione and 2,6-hexanedione is not neurotoxic.

2.3.6.7 Chemistry Concept: Metals

Metals comprise the largest group of elements on the periodic table, yet they are not unequivocally defined among scientists often because of their complex behavior and ability to assume numerous valence states and to complex with a variety of inorganic and inorganic partners. Among the entities of toxicological interest are elemental metals, metalloids, and organometallics, each exhibiting toxicity of unexpected magnitude and effect. Metals make up 75% of the periodic table including eight metalloids including arsenic.

Metals share common characteristics such as being highly reflective and thermally stable and possess high electroconductivity. The fact that they are elements means that they are not metabolized any further as organic compounds can be. Some metals like sodium, calcium, and selenium are essential for normal physiological function but can also become toxic at elevated concentrations.

2.3.6.7.1 Toxicology Concept Bridge

Metals and metalloids have been used for millennia as poisons (e.g., arsenic and lead) and as cures. For example, arsenic was used until relatively recently, to treat cases of syphilis, and lithium is still used as a therapy to manage some forms of bipolar disorders. Their biological and toxicological activity is often based on their proclivity to lose electrons and become positively charged electrophiles that can interact with biological nucleophiles such as proteins and nucleic acids possibly resulting in the disruption of biochemical pathways in cells, dysfunction of integral proteins, or alteration of DNA structure and function. Certain metals can mimic the behavior and action of endogenous metals by hijacking the same uptake processes or distribution pathways or act as substitutes in enzymatic reactions. Most metals are detoxicated by circulating nucleophiles such as glutathione.

Heavy metals are a subset of particular concern to toxicologists and risk assessors because of their well-documented potential hazards to organisms. There is no uniform definition of heavy metals, but the rule of thumb begins with metals with a molecular weight of 40 or greater. Among the more well studied are arsenic, antimony, cadmium, chromium, lead, silver, and zinc.

Lead has been used for thousands of years for many applications including pottery, ammunition, use in alloys and of course in pipes used to deliver water, or plumbing, from the Latin word for lead, *plumbus*. Biologically, lead follows the pathway of calcium and sequesters in bone, nervous tissue, kidneys, and other vital organs leading to toxicity at multiple sites similar to sites where calcium concentrates. Lead has been linked with peripheral nerve damage, deficits in the central nervous system leading to developmental and cognitive deficiencies, as well as adverse effects on the blood, kidney, and heart. Sequestration and storage in bone leads to a long biological half-life resulting in a body burden that can act as a source of lead in blood for many years.

Metalloids occupy a nebulous position between metals and nonmetals leaving room for judgment and interpretation on their identities. Arsenic, antimony, boron, and silicon are examples of toxicologically interesting metalloids. Organometallics are compounds containing at least one bond between a carbon atom of an organic compound and a metal. Their chemistry is highly complex because of the coordination possibilities and an essentially infinite possible number of combinations of organic molecule and metal. Several natural organometallic compounds should be familiar to us including hemoglobin, myoglobin, and chlorophyll, which is the core of photosynthesis. Vitamin B12 is actually called methyl cobalamin and contains an atom of cobalt conjugated to a complex organic molecule. When visiting a physician for a yearly B12 injection, the patient generally receives the dose in the form of cyanocobalamin that includes a cyano group attached to the cobalt atom to impart stability to the molecule and prolong potency. Even a cyanide moiety can be present in a molecule and not cause adverse effects at the appropriate dose and under appropriate circumstances. In this case, the cyano group is bound in a molecular cage and therefore unavailable to any target sites at sufficient concentrations.

Methylmercury is a very toxic organometallic compound and can be used to demonstrate that a methyl group moiety can increase toxicity of a metal tremendously by influencing access to the biological target site. Mercury exists naturally in several forms in the environment including elemental mercury (Hg^0), Hg II and IV, and several complexes including cinnabar (HgS). Exposures to humans occur primarily through industrial exposures and from the emissions from burning fossil fuels. Ingestion of fish containing methylmercury is a major source of exposure for certain populations.

Mercury is methylated both biotically (by organisms) and abiotically. Methylmercury is absorbed more efficiently then metallic mercury and tends to bioaccumulate, biomagnify, and persist in the environment and in organisms. More efficient absorption results in a higher concentration of a toxicant at the site of action resulting in greater toxicity. Elevated levels of mercury can cause neurotoxicity, especially to the developing fetus; kidney effects; and cancer. These effects can occur from acute exposures if the concentration is high enough or on a longer term (chronic) basis from lower exposures.

A tragic example of the extreme toxicity of organometallics is illustrated by examining the case of a chemistry professor using dimethylmercury as part of her research. A few drops of pure dimethylmercury [$Hg(CH_3)_2$] leaked from the end of a pipette, falling on the researcher's protective glove. Since the experiment was performed under a hood using protective gloves, there was no immediate consideration of the tragedy that was about to unfold.

Dimethylmercury is more lipophilic than methylmercury and passed through the "protective glove" as if unobstructed passage were guaranteed and entered the individual through the skin; made its journey to the circulatory system, crossing the blood-brain barrier with little resistance; and began its insidious assault on the scientist's nervous system. Plasma mercury levels peaked at day 39 at 4000 μg/L; the typical level in an average adult is between 1 and 8 μg/L. Abdominal pain was attributed to a virus, her vision blurred, and her speech was slurred. She was in irrevocable trouble. Chelation therapy was ineffective. This world lost a renowned researcher only 298 days after exposure to a drop or two of an organometallic compound.

This case illustrates the obligation to use and design chemicals and products that are intrinsically toxicologically benign. Exposure controls will ultimately fail whether those technologies involve state-of-the-art hoods, protective gloves, eye protection, or full body shields. The obligation chemists and allied scientists have to the students of chemistry is to ensure that the principles of toxicology are well known to them long before they enter a laboratory. There is also an obligation to all people living now and to those not yet here, to ensure that we design a sustainable, safer future.

2.3.7 TOXICOLOGY AND MOLECULAR DESIGN

Toxicology is required to design safer chemicals, for planning safer chemical synthesis, and to ensure that industrial processes employ safer, healthier, and sustainable chemistries. The principles of green chemistry provide a comprehensive framework to guide this vision by emphasizing a holistic, systems-based approach that acknowledges the inherent hazard associated with all molecules and products. Toxicology provides the theoretical and practical tools to profile the likely behavior of a molecule in living systems based on its physicochemical attributes using computational in silico approaches, in vitro assays at the molecular and cellular levels, and information gleaned from in vivo testing in appropriate test organisms.

Molecular design for hazard reduction requires incorporating 21st century toxicology insight to inform synthetic chemists, process engineers, and all scientists, to approach a utopian dream of zero risk from chemicals, realizing that zero hazard is only asymptotically approachable. The development and advancement of computational toxicology, characterizing mechanisms of toxicity using genomics and high-throughput assays, coupled with advancements in chemoinformatics, has augmented the ability to identify and document adverse outcome pathways for known and putative atoms, molecules, compounds, and materials. A more detailed discussion of these emerging technologies is beyond the scope of this chapter, but they are discussed in detail in the suggested references.

2.3.8 CONCLUSIONS

The goal of designing safer molecules is to make them intrinsically less capable of causing toxicity. This can be achieved in various ways using the tools of a chemist. The literature is rich with examples of the structure-toxicity relationship for many apical endpoints (e.g., liver and kidney toxicity, endocrine activity, mutagenicity and carcinogenicity, developmental and reproductive toxicity). QSAR models have been used to identify molecular features that are associated with toxicity. For example, structural alerts, toxicophores, and pharmacophores are available for many classes of compounds and can be searched through interactive, web-based resources.

The next frontier embraces these same cutting-edge toxicological techniques and tools at the advent of molecular design to ensure that every avenue has been examined to the best of our present knowledge and abilities, to design safer, sustainable chemistries. Toxicologists and chemists must work in the same plane, sharing knowledge, working collaboratively to identify safer products and processes. Training our next generation of chemists and toxicologists to work collaboratively at the interface of these two inextricably linked disciplines is our best opportunity to realize the sustainable future we envision. This education must begin at the grade K-12 level, continue through undergraduate and graduate education, and be part of the continuing education requirement of all professional and practicing chemists.

PROBLEMS

1. Choose a chemistry topic that interests you, and develop an example that aligns a parallel toxicology concept.
2. Is it true that if a chemical is more reactive it is always more toxic? Think about reasons why reactivity could be beneficial and or hazardous. What factors could contribute to the ability of a single compound to harm or to heal?
3. In the context of risk, list the opportunities to influence both toxicity and exposure through molecular design.

Disclaimer

The views and opinions represented in this chapter represent those solely of the authors and do not necessarily those of the United States Environmental Protection Agency (USEPA) or the Johns Hopkins Bloomberg School of Public Health.

Recommended Reading

1. Klassen C, editor. *Casarett and Doull's toxicology: the basic science of poisons*. 8th ed. McGraw-Hill Medical; 2013. 1454 pp.
2. Boelsterli U, editor. *Mechanistic toxicology: the molecular basis of how molecules disrupt biological targets*. 2nd ed. CRC Press; 2007. 416 pp.
3. Williams DA, editor. *Foye's principles of medicinal chemistry*. 7th ed. LWW Publishers; 2012. 1520 pp.
4. Wallace Hayes A, Kruger CL, editors. *Hayes' principles and methods of toxicology*. 6th ed. CRC Press; 2014. 2184 pp.

CHAPTER

2.4

Effects of Environmental Factors on DNA: Damage and Mutations

Steven Ackerman, William Horton

University of Massachusetts Boston, Boston, MA, United States

2.4.1 DNA MUTATIONS

2.4.1.1 Base Changes

The four bases comprising DNA (A, C, G, and T) (Fig. 2.4.1) may undergo mutation by chemicals or radiation [ultraviolet (UV), X-ray, etc.] or tautomerization. The modifiction may result in the change of a purine into a different purine derivative (or change of a pyrimidine into a different pyrimidine derivative) (transition) or the change of a purine into a pyrimidine (or change of a pyrimidine into a purine) (transversion). These are considered

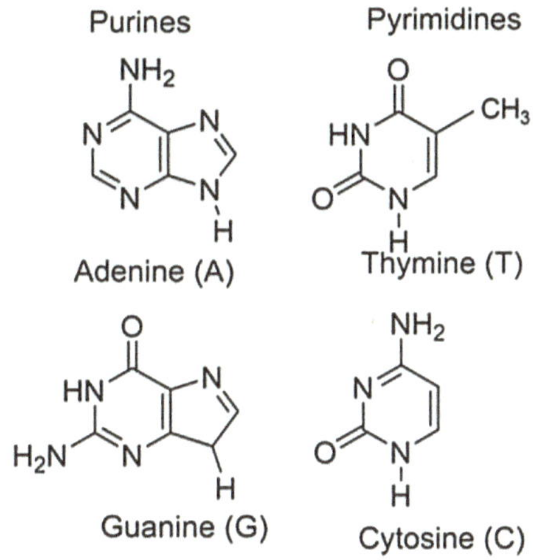

FIGURE 2.4.1 The four nitrogenous bases of DNA. A (amino form) base pairs with T (keto form) and G (keto form) base pairs with C (amino form).

Green Chemistry
http://dx.doi.org/10.1016/B978-0-12-809270-5.00005-4

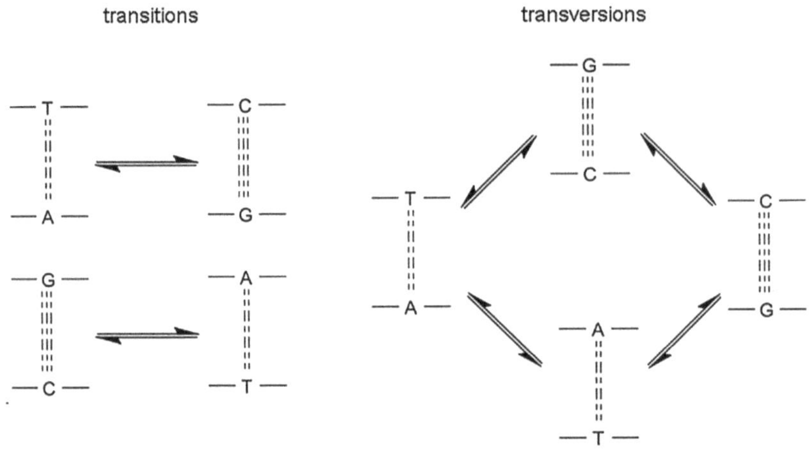

FIGURE 2.4.2 Transition and transversion mutations. Transitions are purine-to-purine (A to G, G to A) or pyrimidine-to-pyrimidine (C to T or T to C) changes. Transversions are changes of purine to pyrimidine or pyrimidine to purine (G to C, C to T, A to C, A to T, C to G, C to A, T to A, T to G).

substitutions (Fig. 2.4.2) When these mutations occur, they can be classified based on whether they change the amino acid that they code for in the protein.

2.4.1.2 Genetic Code

The 20 amino acids are coded by the DNA sequence; three bases specify an amino acid. The DNA is transcribed into RNA and the RNA is translated by the ribosome into a protein using the three-base code, which lacks punctuation (no commas, colons, or semicolons but there are periods) and is nonoverlapping (meaning that three bases specify only one amino acid) (Fig. 2.4.3). Hence, there are 64 possible codon combinations for 20 amino acids. Three

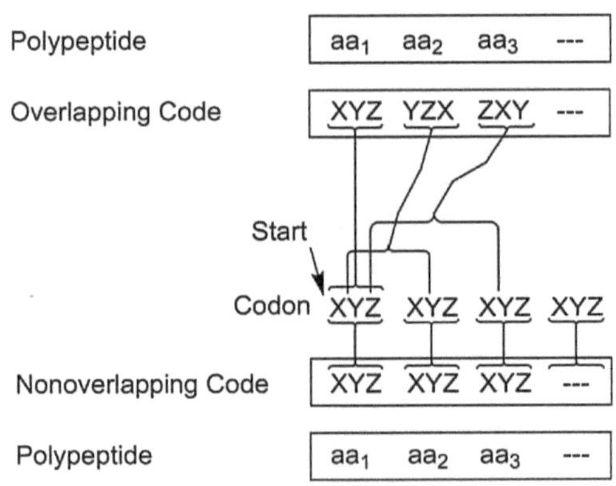

FIGURE 2.4.3 The genetic code is nonoverlapping. For simplification the code is represented as XYZ, repeating. If read nonoverlapping then the same amino acid is repeatedly incorporated by the translation machinery to form a homopolymer, whereas if the code is overlapping then it is read as XYZ, YZX, ZXY and three different amino acids are incorporated. Experimentally it was demonstrated that the code is nonoverlapping.

At the Protein Level

Neutral Mutation — Functionally equivalent but different amino acids:

AGA ⟶ AAA
Arginine to Lysine
(many times this will not change the protein's main function)

Silent Mutation — The same amino acid:

UCU ⟶ UCA
both code for Serine

Nonsense Mutation — Chain termination:

AAG ⟶ UAG
Lysine to an amber termination codon

Missense Mutation — Different and non-functional amino acid.

FIGURE 2.4.4 The different types of changes that occur from a single base change. The silent mutation is so named because the degeneracy allows some amino acids to have more than two codons. Even though the same amino acid is incorporated, the use of an alternative codon may affect the translation process due to less efficiency recognizing the alternate codon. This can affect how the polypeptide is folded, and changes in folding may affect protein function. A neutral mutation allows an amino acid of the same type [polar charged (acidic, basic), polar uncharged, nonpolar, special] to be inserted due to the mutation. This can cause a change of protein function by structural changes in the protein, or it changes how an amino acid is posttranslationally modified leading to gene expression changes. A missense mutation results in a different category of amino acid being inserted into the protein, and it may be nonfunctional because of a change in the structure, catalysis, or posttranslational modification. The nonsense mutation places a stop codon within the reading frame, thereby producing a truncated protein, which is either nonfunctional or partially functional (depending on where the stop codon is located).

codons, UAG, UAA, UGA, are stop codons (or periods) that end the translation process. Thus there are 61 remaining codon possibilities, and this means that the genetic code (really a genetic cipher) is degenerate because many amino acids have multiple codons. In every organism, however, there are codon preferences meaning that, for example, the six codons for leucine are not all used equivalently. A few may be used in preference to the others.

The amino acids are classified as acidic, basic, or neutral (polar and nonpolar). If a base change results in no change of the amino acid (due to degeneracy), this is referred to as a silent mutation (also called a synonymous mutation) (although the effect may be deleterious as discussed later), whereas if the base change results in an amino acid change but to the same type of amino acid (an acidic amino acid for an acidic amino acid, etc.), this is referred to as a neutral mutation (also called a nonsynonymous mutation) (again, the effect may be deleterious despite the name). If the base change results in a changed amino acid with different properties (acidic to basic, basic to acidic, basic to polar, etc.), it is referred to as a missense mutation, especially if the new amino acid causes a change in the protein's function (nonfunctional amino acid) (Fig. 2.4.4).

2.4.1.3 Diseases Associated With Base Changes

One example of how a silent mutation can wreak havoc for the individual is the MDR1 gene in which a rare codon for the same amino acid changes the rate of translation elongation. This change of rate results in a change of the folding of the polypeptide and an altered protein structure is formed. This causes the protein to not function correctly, resulting in multidrug resistance of cancer cells. The rate of translation, therefore, depends on the codon, and rare codons slow this rate because the cell lacks sufficient specific transfer RNAs associated with their cognate amino acid for this codon. A synonymous mutation can change the sequence of the messenger RNA (mRNA), and this can cause a change in the splicing pattern of the pre-mRNA to the mRNA, thereby either producing an altered protein sequence or even no protein at all.

A second example is the catechol-*O*-methyltransferase gene: when it has a silent mutation the base change in the mRNA changes the folding (secondary structure) of the mRNA, which changes the translation initiation rate and thereby produces less protein. This can cause a form of early-onset antisocial behavior, which may have associated attention-deficit/hyperactivity disorder. A neutral mutation that changes the amino acid (lysine changed to arginine) can lead to a change in the protein's posttranslational modification pattern, affecting function.

2.4.1.4 Mutations

Each cell has 10,000 insults (base changes, translesions) per day that need to be repaired. Some occur "naturally." DNA replication itself results in a few base changes [estimated to be between 1.20×10^{-8} and 1.0×10^{-9} every cycle (mutations per base pair per generation)]. To protect itself from accumulating mutations, cells typically divide (mitosis) between 40 and 70 times (the so-called Hayflick number) before they undergo programmed cell death (apoptosis). Hence, it is the accumulation of mutations in the DNA due to chemicals in the environment that seems to be the causative agent of maladies, including cancer. Some examples of base changes are listed in the following discussion.

2.4.1.5 Base Deletions

The deletion of a base, caused by certain chemicals such as ethidium or acridine orange, will change the reading frame of the code by the ribosome during translation of the RNA to a protein (Fig. 2.4.5).

2.4.1.6 Base Insertions

Similarly, the insertion of a base, caused by certain chemicals such as ethidium or acridine orange, will also change the reading frame of the code (Fig. 2.4.5).

2.4.1.7 Deamination

This is the removal of an amino ($-NH_2$) group as shown in Fig. 2.4.6. Methyl C (meC) (a major epigenetic mark for gene regulation) predisposes the meC residue for mutagenesis to T; these meC residues are considered hot spots for mutation.

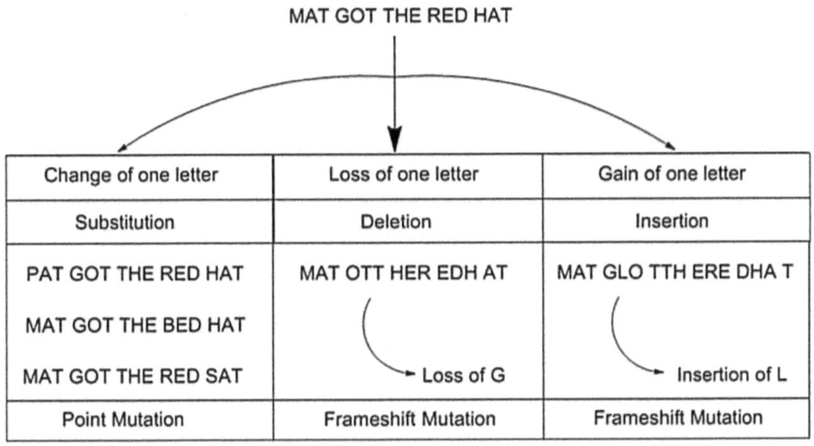

FIGURE 2.4.5 The effect of substitutions, insertions, or deletions in the code. The substitutions have been described (see Fig. 2.4.4). Deletions or insertions change the way the code is read (a reading frame change), so the incorporated amino acids after the insertion or deletion are wholly changed and result in a nonfunctional protein.

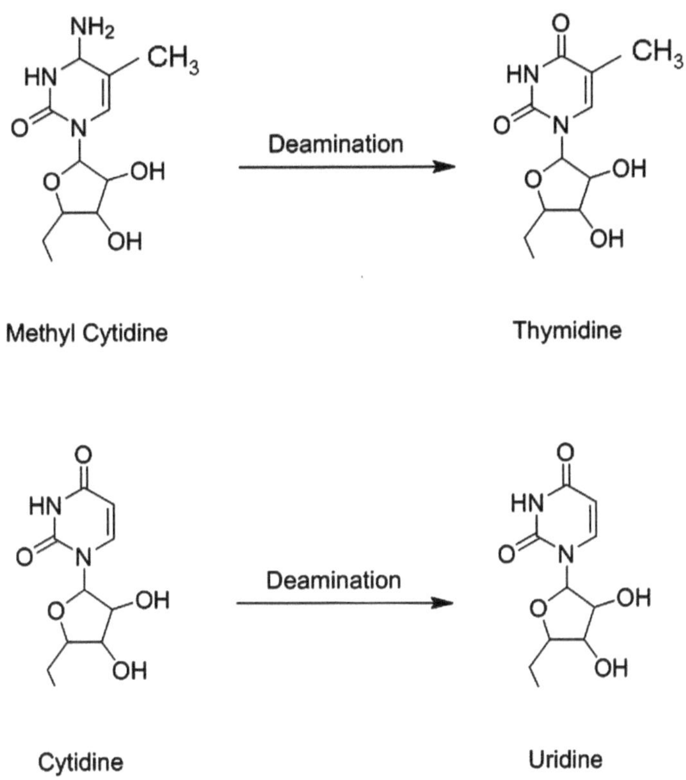

FIGURE 2.4.6 Base changes from chemical modification. One example is depicted, deamination, where C is changed to T or C is changed to U (found in RNA).

2.4.1.8 Tautomerization

Tautomers are isomers of a molecule that exist in solution or in a cell. They are interchangeable forms because chemical bonds are rearranged many times spontaneously. This is different from chirality, where molecules are mirror images (or enantiomers) of each other (D- and L-glucose, etc.). The DNA bases can undergo tautomeric shifts. This can lead to mutagenesis if it occurs during DNA replication such that the new daughter strand incorporates the incorrect base pair to complement the tautomer. James Watson's book *The Double Helix* discusses how tautomers had a role in elucidating the structure of DNA: both Watson and his colleague Francis Crick were building models using the wrong tautomers of the bases (since those were the structures in textbooks) until the organic chemist Jerry Donohue (a laboratory partner) told them to use the keto and amino forms. Watson adapted those forms in his modeling and soon thereafter realized that A=T and G≡C pairs "fit" together via hydrogen bonding. This allows a double helix to form with the strands antiparallel, and it also satisfies Erwin Chargaff's observation that the molar amounts of A and T are always equal, as are G and C. A summary of some of the base targets of mutagenesis are shown in Fig. 2.4.7.

2.4.1.9 Chemical Mutagens

One can easily write a huge tome about the toxic mutagenic chemicals that we have created for industrial and consumer processes that previously and presently contaminate our environment and bodies. A few examples will be considered here (Fig. 2.4.8).

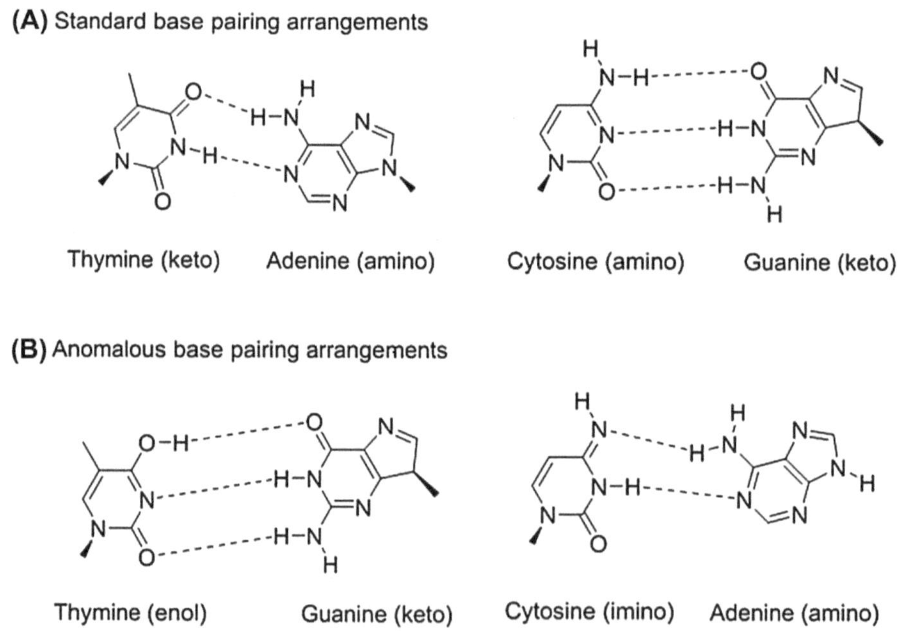

FIGURE 2.4.7 Tautomerization changes the base pairing abilities of the base as depicted in the figure.

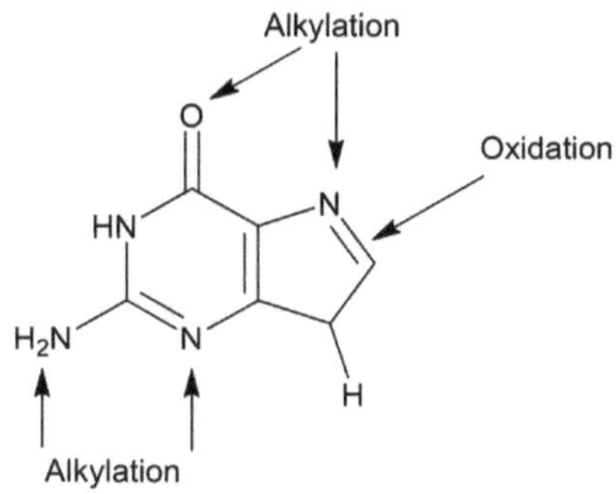

FIGURE 2.4.8 Some of the places on a base that can be affected by chemical processes in the cell.

1. Alkylating agents are of interest since some have been used in warfare. Mustard gas, also known as sulfur mustard, was used in World War I. It causes blisters on exposed skin and in the lungs. Alkylating agents can form cyclic "onium" ions (sulfonium, ammoniums, etc.) that modify DNA bases. This leads to base mutagenesis and point mutations, which can lead to diseases such as cancer later in life. The mustards are a class of compounds, and each has its own specificity. They were banned from warfare use in 1933, although similar chemical weapons are still illicitly used today. Conversely, in 1919, it was recognized that mustard agents suppressed hematopoiesis and autopsies of soldiers who had been exposed to, and died of, mustard agent during World War I had decreased counts of white blood cells. Later it was established that nitrogen mustard could be a therapy for Hodgkin's lymphoma, other types of lymphoma, and leukemia. One of these mustard agents was named "HN2," and it was one of the first chemotherapy drugs to be utilized; it is called mustine.
2. Ethyl methanesulfonate (EMS) is another alkylating agent that is mutagenic by producing random mutations in DNA via, generally, guanine alkylation, which produces point mutations. The ethyl group of EMS reacts with the O6 of guanine, forming the abnormal base O-6-ethylguanine. Because this changes the hydrogen bonding of guanine, DNA polymerase replicates the DNA by placing thymine, rather than cytosine, opposite O-6-ethylguanine. Ethyl ethanesulfonate too is an alkylating agent that is mutagenic, and also alkylates guanine creating O-6-ethylguanine. These alkylating agents can cause DNA interstrand links via G bases on opposite strands (Fig. 2.4.9).
3. Agent Orange, also referred to as Herbicide Orange, was a herbicide/defoliant used in the Vietnam conflict to defoliate forest/underbrush/swamp areas. The name is derived from the orange-striped barrels that contained the chemicals. It is a mixture of equal parts of two herbicides, 2,4,5-trichlorophenoxyacetic acid (2,4,5-T) and 2,4-dichlorophenoxyacetic acid (2,4-D). It also contains dioxin (tetrachlorodibenzodioxin), an impurity found in Agent Orange. There is evidence that Agent Orange causes chromosome damage including chromosome breaks. Dioxin is reported to attack the mitochondria and also cause cancer, and may have links to, at least, breast cancer. People exposed to Agent Orange (including those who handled the chemical), and their children, have medical issues related to their exposure to this chemical. These issues include leukemia, Hodgkin lymphoma, non-Hodgkin lymphoma, throat cancer, prostate cancer, lung cancer, colon cancer, ischemic heart disease, soft tissue sarcoma, and liver cancer. Further, an impurity in Agent

Orange is dioxin, a known DNA mutagen. The case for Agent Orange causing DNA damage is not causal, yet we know that some offspring of exposed individuals have congenital medical issues, implying DNA damage and mutagenesis. Laboratory experiments using herbicides such as 2,4,5-T and 2,4-D found that the highly purified forms of these herbicides (minus dioxins) are not very toxic compounds by themselves, but very high doses can cause effects in laboratory animals. The molecular mechanism has yet to be investigated. These pure compounds have not been linked with cancer in animal studies. The combination of these two with dioxin is harmful, however.

Additional chemical mutagens will be discussed in the following sections in relation to water contamination (triclosan, plastics, phthalates, lead, etc.).

2.4.1.10 Intercalating Agents

Intercalating agents are chemicals that "slide" between the DNA base pairs because they are planar, and they mutagenize because they can cause insertions, deletions, and (rarely)

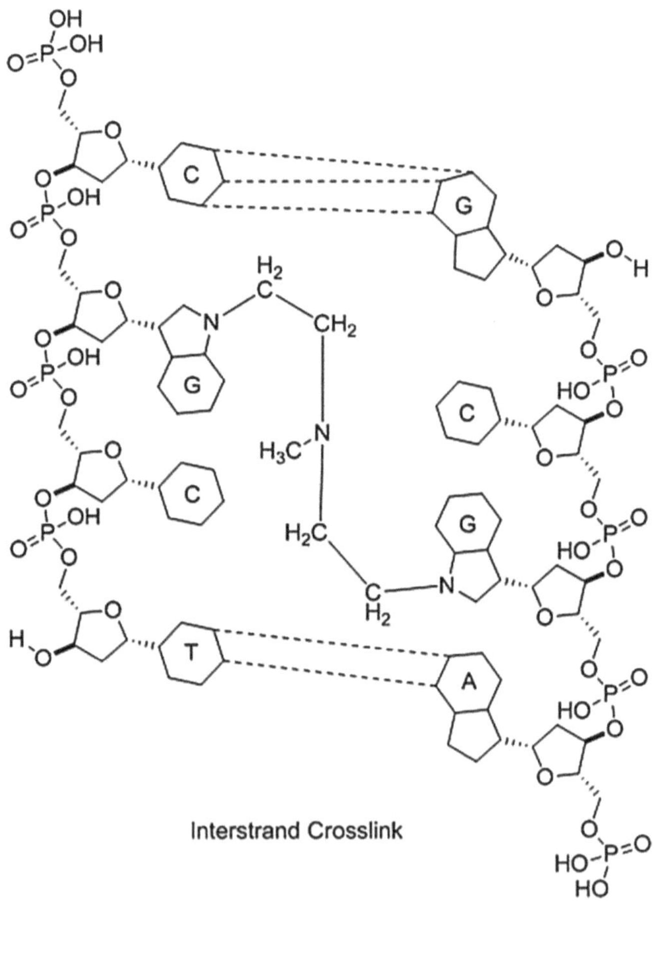

FIGURE 2.4.9 Chemical mutagens can cause cross-linking between DNA bases that are not part of a base pair. These type of cross-links block the transcription and replication processes until they are removed. Multiple cross-links can lead to extensive DNA damage, which may not be completely repaired or repaired incorrectly, leading to either cell death or uncontrolled cell growth (oncogenesis).

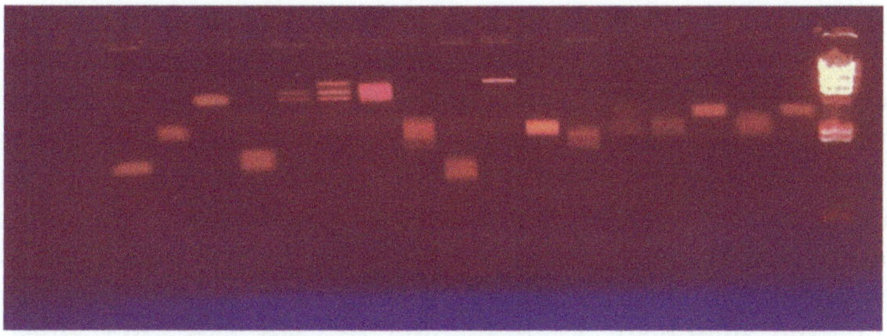

FIGURE 2.4.10 A typical agarose gel where DNA of different lengths has been electrophoresed to separate the strands by size. Ethidium bromide was added to the gel solution (0.1/μg/mL) prior to pouring into the mold. The ethidium inserts between the DNA base pairs and concentrates so that ultraviolet illumination under the gel allows the DNA fragments to fluoresce orange. The ethidium migrates opposite the DNA (note the bottom of the gel has a cleaner background than the upper half).

base pair changes. These intercalating agents were used extensively in research (ethidium, acridine orange) or as a disinfectant (proflavin). Ethidium bromide is used often to visualize DNA: it intercalates into DNA between the bases and fluoresces orange when UV light is shone on the sample. One example for its use is electrophoresis of DNA through an agarose matrix that separates the DNA molecules by size. To visualize the DNA in the agarose gel the gel may have ethidium bromide added to the sample, or to the liquid agarose before it gels, or the gel can be stained after electrophoresis. Upon exposure to UV light via a transilluminator the DNA bands fluoresce orange. For example, the ethidium is added to the liquid agarose to a final concentration of 0.1 μg/mL. This low concentration and is relatively nonfluorescing, but accumulation and concentration of the ethidium in the DNA allows us to visualize the DNA as orange bands (Fig. 2.4.10). Note that the places where there is no DNA do not fluoresce. Further, the ethidium electrophoreses in the opposite direction to the DNA, so when added to a gel the background remains dark.

These intercalating agents have also found use in, for example, deciphering the genetic code where the use of proflavin (an acridine) by Francis Crick and Leslie Barnett and Sydney Brenner and Richard Watts-Tobin in 1961 established that the genetic code is a triplet. An oversimplification is that using proflavin they could induce single deletions or insertions, rendering a protein inactive. Similar results were observed for double or quadruple deletions. Three deletions or insertions, however, restored the protein to a functional state, inferring that the code for an amino acid is three bases. Since the protein function was restored, this also proved that the code is nonoverlapping (one triplet is read as a triplet and the next code is a separate triplet). Further, a deletion subsequently followed by an addition elsewhere also restored the "reading frame"; hence the code lacks punctuation.

2.4.2 MUTAGENIC AGENTS THAT MAY AFFECT DNA SEQUENCE OR EPIGENETICS

2.4.2.1 Epigenetics

Epigenetics refers to above genetics meaning regulatory change that occurs without the DNA sequence changing. Epi means "on top" of the genetic code. The term was invented by Waddington from the Greek word epigenesis. This was originally a hypothesis for development that proposed that the early embryo was undifferentiated. Waddington changed epigenesis to epigenetics. He also coined the term epigenotype meaning "total developmental system consisting of interrelated developmental pathways through which the adult form of the organism is realized." The first level of epigenetics is methylation of DNA, generally at the 5 position of C, and when the C is followed by G (5′…CpG…3′; p = the phosphate between nucleosides). Embryonic cells, some insects, and plants have additional sequence variations that can be methylated. The modification of C by DNA methylase is a signal that recruits other enzymes that modify proteins bound to the DNA (specifically, the histones). Each histone protein can, and does, have multiple different modifications (acetylation, methylation, phosphorylation, ubiquitination, ADP-ribosylation, sumoylation, biotinylation, formylation, succinylation, citrullination, glycosylation, *N*-acetylglucosamine, propionylation, butyrylation, hydroxylation, crotonylation, malonylation, etc., as well as proline *cis-trans* isomerization, etc.) most often on arginines and lysines (but some modifications are on other amino acids), which can also be removed by other enzymes. These modifications occur in waves by "writers" and/or "erasers," and the "histone code" is deciphered by "readers" to determine the expression or repression of gene expression. DNA methylation is NOT a lock that prevents reprogramming and does not lead to irreversible gene silencing. DNA methylation is a signal that frequently is used to recruit other proteins that modify the histones with modifications that signal gene expression repression. Conversely, removal of these modifications (and sometimes consequent demethylation of the DNA) leads to gene expression activation.

These epigenetic modifications are often altered by mutagenic agents that do not necessarily change the DNA sequence. Hence, like the genetic examples aforementioned, epigenetic effects too are harmful. Some examples are fleetingly discussed later.

2.4.3 TRANSGENERATIONAL INHERITANCE

One consequence of these epigenetic changes is that the expression patterns are changed in the individual, but we can find the same changes in the offspring, for multiple generations. Although this sounds Lamarckian in principle, transgenerational inheritance has become a well-characterized phenomenon involving at least DNA methylation. It is transgenerational because the altered expression (often as a medical condition) occurs at least in the children, grandchildren, and great grandchildren of a parent exposed to an environmental factor while the offspring were in utero. This can, however, occur through maternal and paternal lineages, indicating that the female need not be pregnant at the time and that transmission is through the germline cells. Transgenerational inheritance is found in unicellular yeast and in plants, indicating that it is not a mammal-specific phenomenon.

2.4.3.1 Water Contamination

Contamination of the water supply today is both a concern and a controversial topic. This applies not only to the municipal sources of water but also to the various forms of bottled water. The latter is targeted as containing chemicals that are removed from municipal water as well as the plastic container contaminants [bisphenol A (BPA)], phthalates, etc., discussed later.

Contaminants in the water include agricultural components (fertilizers, pesticides), industrial components (chemicals to synthesize compounds, reaction by-products, impurities, etc.), and urban area components (discarded items that leach pollutants into the water, sewage contaminants, etc.). The agricultural contaminants have received attention as there is a realization that not only pesticides but also fertilizers have entered the water table. Indeed, twice as many are unregulated as are regulated (similarly urban pollutants are also mainly unregulated). An example is the organochlorines [dichlorodiphenyltrichloroethane (DDT), polychlorinated biphenyls (PCBs), etc.], which were used for their pesticide ability, in insulation, in vinyl chloride, etc. (PCBs are oils that do not catch fire under conditions of extreme pressure or temperature and were widely used in transistors, capacitors, and other electronics equipment during the 1950s to the 1970s in the United States.) Organochlorine pesticides are environmental contaminants that were banned in developed countries 20–30 years ago. Although largely banned in the United States, these chemicals persist in the environment, especially in the rivers and streams where they are found in sediments and fish. DDT and PCBs are extremely persistent in the environment, lasting for years or even decades in soil and lake sediment, where DDT often breaks down into dichlorodiphenyldichloroethylene (DDE), a toxic compound that is also extremely persistent. DDT, DDE, and PCBs are also are very long lasting in the human body, accumulating in fat (including breast milk), and most Americans have detectable levels of DDE and PCBs in their bodies. The compounds become concentrated as they move up through the food chain, and have been detected in polar bears, marine mammals, and humans in regions where neither compound has been used, due to their powerful resistance to degradation.

The landmark 1962 book, *Silent Spring* by Rachel Carson, focused on DDT and other pollutants as a source of environmental and public health concern because they remain in the environment for long periods. This outcry from the public caused the United States to ban the use of DDT and helped to initiate the environmental movement. Individuals subjected to chronic low-dose exposure to organochlorine pesticides show an increased risk to obtain a future diagnosis of cognitive impairment. Since they accumulate through the food chain and in water remain for a very long time in the human body, especially adipose tissue, high levels can still be found in a majority of the population.

The persistence of these contaminants in the water and the soil, combined with their possible health effects and mutagenic ability, raise serious concerns for the population. The concern about health effects are complex because unequivocal cause and effect either lack sufficient long-term data (or a large data set), may be controversial due to competing interests (industry vs. consumer protection), do not have a large affected group (resulting in insufficient data or deficiency of resources), etc. Fortunately, many (but not all) of these compounds are removed by water treatment plants that provide tap water to households.

Water contamination is often "stopped" by the use of disinfectants. This leaves residual disinfectant in your water, but this may lead to carcinogens (by-products) and also pipe corrosion and a bad taste in the water. Interestingly, these disinfectants are unnecessary; there are other safeguards such as filters, chlorine, etc. Note that chlorine reacts with organic materials in the water to produce chloroform. Disinfectants also lead to pipe corrosion and leaching of lead from pipes. This leaves chlorine and its derivatives as residuals to protect against contamination. Europe does not use disinfectants. Does the use of disinfectants lead to fewer disease outbreaks? There is no evidence that Europe has more disease outbreaks than the United States.

2.4.3.2 Triclosan

One of the more insidious contaminants is triclosan (Fig. 2.4.11). Triclosan has been used since the 1960s and has been on the consumer market since 1972, usually in disinfectant soaps because it is antibacterial, somewhat antiviral, and/or antifungal. Its disinfectant properties arise because it is a fatty acid biosynthesis inhibitor that blocks bacterial lipid biosynthesis. By 1988 it was known that it is harmful. In animal models, numerous reports indicate that 5-chloro-2-(2,4 dichlorophenoxy) phenol (TCS) adversely affects endocrine function, thyroid hormone function, and antibiotic resistance. Triclosan is very hydrophobic and is lipid (fat) soluble. It is a phenolic: phenoxyphenol (TCS) and 3,4,4′-trichlorocarbanilide. It also amends to soils. Of all of the TCS in consumer products, it is estimated that 96% is rinsed down the drain, leading to the concentration of TCS in untreated wastewater ranging from 1 to 10 mg/L. Triclosan is only partially degraded in sanitary facilities despite reports to the contrary. It is about 95% degraded under certain sanitary facility conditions (aerated bacterial treatment), and direct sunlight degrades triclosan. However, most treatment plants do not use direct sunlight or aerated bacterial treatment: treatment of the water causes the triclosan to precipitate and concentrate in the solid sludge residue. This residue is then used as fertilizer on farms, and triclosan can persist in the environment due to the chlorine atoms in the structure. Indeed, analyses of the triclosan levels after use of the triclosan-laden sludge for agriculture have been embarrassing. Analyses sampled the levels in the surface soil immediately after application of this fertilizer, and the subsequent measurements of the surface soil indicated a decline in triclosan contamination. However, neither were deeper soil samples examined subsequently (irrigation would cause the triclosan to migrate downward) nor were the crops examined for triclosan contamination. By 2008, it was found that 75% of the US population had triclosan in their urine! Prenatal exposure has also been established. During wastewater treatment processes, TCS may convert to other derivatives because it can be biologically methylated into methyltriclosan and/or be transformed using wastewater disinfectants with free chlorine into chlorinated TCS derivatives. Chlorinated TCS derivatives are more toxic than TCS itself. These possess a longer environmental persistence than TCS because they are lipophilic and they have resistance to biodegradation.

Triclosan has also been linked as an endocrine disruptor: it amplifies the effects of sex hormones and (although the evidence is weak) some have even suggested that it causes learning disabilities, affects thyroid hormone and testosterone levels, affects homeostasis, causes altered behavior, and may induce an increased allergy risk in children. Animal studies have made it evident that TCS acts as a thyroid-disrupting chemical. While the cause and effect for these claims are weak, the paucity of investigations and additional data beckon

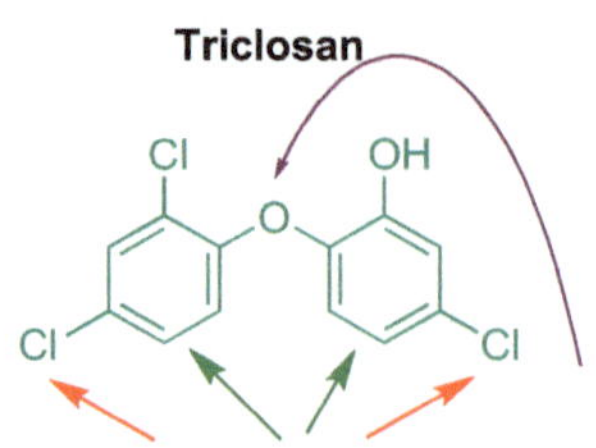

FIGURE 2.4.11 The structure of triclosan. A chlorinated biphenyl ether, 5-chloro-2-(2,4 dichlorophenoxy) phenol (TCS) is similar to polychlorinated biphenyls, bisphenol A, dioxins, and thyroid hormones. TCS is resistant to degradation. *Red arrows* point to the chlorine atoms, *green arrows* indicate the phenyl groups, and the *purple arrow* points to the ether bridge.

further investigation. Further, there is now concern that triclosan can cause changes in the body's microbiome (although these results, too, have been controversial).

Another use of triclosan has been in toothpastes. It is supposedly an antigingivitic—in 1997 the US Food and Drug Administration approved TCS in toothpaste at 0.3%. It is also used as an antibacterial in liquid soaps/bar soaps. Based on the mounting evidence of TCS detection in human body fluids, humans are unequivocally exposed to significant and potentially unsafe levels of TCS. TCS causes liver pathogenesis including tumor formation in mice.

In a clinical study, when men applied cream containing 2% TCS on their skin, the absorption of TCS, calculated from urinary excretion, was estimated to be ~10% in all individuals. TCS retention rate from mouthwash containing 0.03% TCS was ~7%. Following the application of 1% TCS in a soap formulation to the skin of rats and guinea pigs, TCS glucuronide was detected as the major urinary metabolite. These results attest to the body contamination by TCS. Following absorption, TCS is metabolized primarily through conjugation reactions to glucuronide and sulfate conjugates. These are eliminated in feces and urine.

2.4.3.3 Plasticizers

Bottled water has also received much "bad press" due to the plastic containers used, but there is a great deal of research to support the leaching of plasticizers into the water. Two main culprits are bisphenyl A and phthalates. Plastics are a range of materials, made by addition of chemical additives, prepared by the polymerization of monomers derived from oil or gas. During their manufacture the polymers are capable of flow (liquid) so that they can be extruded, molded, cast, spun, or applied as a coating. There are about 20 groups of plastics. Polystyrene was made in 1839 but not commercialized until 1930, whereas polyvinyl chloride (PVC) was synthesized in 1872. First, chlorine gas is used to produce ethylene dichloride, which is then converted to the vinyl chloride monomer, which in turn is converted to PVC. Burning PVC produces the carcinogenic group of dioxins. The first synthetic polymer, Bakelite, was developed by the Belgian chemist Leo Baekeland in 1907, but it was not until the 1940s to the 1950s that it was mass produced. Polyethylene was first made in 1933, and polypropylene in 1954. Of concern with plastics are the chemical additives used in their manufacture. Two examples are BPA and phthalates. Safety concerns about plastics began in 1987 when Dr. Theo Colborn investigated health problems with wildlife at the Great Lakes. She found many publications about wildlife problems that indicated that the youngest were most severely affected and that most of the issues involved the endocrine system. Dr. Colborn suspected that there were synthetic hormones in plastics, pesticides, etc. She obtained tissue samples and determined that, indeed, there are estrogenlike chemicals in the plastics that contaminate the environment. Her 1996 book *Our Stolen Future* was a monumental opus detailing her research, and it alerted the public to the danger of plastics to their health and environmental contamination.

2.4.4 BISPHENOL A (4,4′-ISOPROPYLIDENEDIPHENOL)

Frederick von Saal studied the effect of synthetic estrogens on fetal mouse development. Starting with BPA, he found that unlike estrogen it does not bind to blood proteins and instead locates directly to the cell interior. These studies found that male mice from mothers exposed to very low doses of BPA during pregnancy have enlarged prostates and low sperm counts. Many other subsequent studies have demonstrated the effects of BPA on mice and other animals, and there is growing evidence for its effect on humans. BPA exposure in utero also has transgenerational effects on the brain and social behavior in mice. However, using rodents to assess human susceptibility to BPA may be inaccurate as human testes are 100 times more sensitive to BPA than those of rodents.

BPA was first synthesized 1891, and by the 1930s the search for a cheap treatment for post-menopausal women led to the use of BPA. In 1936, commercial synthesis of BPA allowed it to be used as a synthetic estrogen. By 1938, diethylstilbestrol was found to be more effective, but it is very mutagenic. BPA was "rediscovered" in the 1940s for plastics: BPA facilitates the formation of the plastic and is polymerized into a resin or plastic where, for example, it hardens polycarbonate. BPA + phosgene (a World War I toxic gas) produces a clear plastic, polycarbonate, which is shatter resistant. BPA was used then for the synthesis of epoxyresins, polyester, styrene, etc. For example, BPA is used in making the epoxy resin in can liners: canned foods, sodas, etc. The problem arises because during manufacture not all of the BPA is locked into chemical bonds; residual BPA leaches into solution at elevated temperature, in a temperature dependent manner. Polycarbonate bottles leach 1 μg/mL at room temperature over 7 days. It is also used in some dental sealants and fillings, adhesives, flame retardants, water storage tanks, etc. BPA is metabolized via the liver (hepatic system) and enters the overall metabolism very efficiently and rapidly: it is detected in urine in less than 1 day. Indeed, it has been demonstrated that BPA is found in the urine of >93% people in the United States. There is evidence that BPA is dangerous in utero; it crosses the placenta resulting in prenatal exposure, and there is a growing body of evidence that this may cause medical issues for the newborn.

Often plastics are labeled BPA free, but they can contain a derivative of BPA called BPS (bisphenol S), which may be as dangerous as BPA. Investigations about the effect of BPS on animals have demonstrated impact congenic to those of BPA. Similarly, a BPA/BPS substitute known as Tritan is a plastic marketed as BPA free. Tritan is a clear, sturdy, heat-resistant plastic made from triphenyl phosphate, which is more estrogenic than BPA. There is leaching here too. One example of the use of Tritan is coffee take-out cups. BPA breakdown products also bind to the estrogen receptor, with a greater affinity (k_a) than BPA itself. Not surprisingly, there have been many disorders linked to BPA and its derivatives, from autism to zebrafish hyperactivity, as many researchers try to link a cause and effect to research grants. Until more conclusive evidence is produced, only the aforementioned confirmed effects should be accepted. The effect of BPA may last for generations (transgenerational inheritance).

2.4.4.1 Phthalates

These compounds, too, are endocrine disruptors that are similar to phytoestrogens. Phthalates are made from phthalic acid (1,2-benzene dicarboxylic acid); diesters of phthalic acid produce phthalates. Phthalates make PVC pliable, and they are a chemical additive in soft pliable plastics (pacifiers, bottle nipples). Phthalates have been linked to suppression of male hormones, effects on sperm count and quality, and mimicking of female hormones, and the male fetus in the first trimester may be most vulnerable. Phthalates leach out of products and are rapidly metabolized. They are found as well in personal care products, lacquers, varnishes, and even in some sunscreens. PVC floors contain phthalates. Some phthalates are used as solvents (dissolving agents) for other materials.

Phthalates are divided into two distinct groups, with very different applications, toxicological properties, and classification, based on the number of carbon atoms in their alcohol chain. Low-molecular-weight phthalates are being gradually replaced in many products. High-molecular-weight phthalates have increased permanency and durability. Phthalates are used in many personal care products and are found in 97% of people. Similar to BPA they may be the causative agent for myriad maladies.

2.4.4.2 Volatiles

Volatiles have long been recognized as having negative effects on people, from dizziness and a nauseated feeling to more manifest disorders.

Plastic shower curtains contain 108 different volatile organic chemicals that undergo aerosolization; this is the same event as new car odor. For the shower curtains, even after 1 week of "airing out" up to 40 different volatiles are still detected and after 2, 3, and 4 weeks, 16 different volatiles are still detected in each week. These volatiles can cause headaches, and some of the chemicals are endocrine disruptors. The aerosolization is often referred to as outgassing (sometimes called offgassing, particularly when in reference to indoor air quality).

2.4.5 REPAIR OF DNA DAMAGE

This discussion is limited to the repair of the damaged DNA bases and not to the removal of epigenetic modifications to ameliorate causative effects of epigenetics. There is a very nice, short YouTube video of this process (https://www.youtube.com/watch?v=HYS6EKnQcv0). There are several repair pathways in procaryotes and eucaryotes, and some will be discussed in a succinct manner, while others are omitted due to the extensive discussion required. The repair pathways include photoreactivation repair, base excision repair, nucleotide excision repair, mismatch repair, single- and double strand DNA break repair, recombination repair, and several other pathways (including transcription-coupled repair). Photoreactivation is used to remove thymine-thymine dimers by the photolyase enzyme in the presence of blue light. Base excision repair fixes base changes from tautomerization, chemical damage, etc. Nucleotide base excision replaces a larger segment of the DNA to repair a section of the DNA helix. Mismatch repair is used to replace incorrect base incorporation. Single- and double-strand DNA break repair generally uses recombination repair pathways, and will not be discussed further. Each repair system uses a pathway that includes proteins that detect/sense the DNA damage, proteins that receive the signal, and proteins that initiate the repair process. In bacteria there are multiple DNA polymerases (enzymes that synthesize DNA using a DNA template). In *Escherichia coli* DNA polymerase III is used for replication, whereas DNA polymerases I and II are general repair enzymes. DNA polymerases IV and V are inducible repair enzymes that are activated when there is extensive DNA damage or when the other repair DNA polymerases cannot complete a repair. The hallmark of DNA polymerases IV and V is their high error rate; they are actually mutagenic. Evolutionarily the reason for this is that when the DNA cannot be repaired by the other repair enzymes it is feasible to insert (almost) any base to allow the DNA to be replicated. Since a bacterial colony is unicellular in nature, even if 99.99% of the cells die due to the random nature of the "fix" those that do survive continue the culture. Eucaryotes have a panoply of DNA polymerases, some for initiation of replication, some for replication elongation, some are mitochondrial (or chloroplastic), and the remainder are for DNA repair. Although the total number of DNA polymerases varies among species a reasonable expectation is that there are ~15 DNA polymerases in a eukaryotic cell. Also, some of these are error-prone DNA polymerases as found in bacteria.

2.4.5.1 Photoreactivation

On exposure to UV light, the base pairing of adjacent thymines to a complementary adenine is disrupted and the thymines form covalent bonds between themselves (Fig. 2.4.12). Removal of thymine-thymine dimers by a repair mechanism is necessary for replication and transcription of the DNA. This was the first repair system identified and is used to repair thymine dimers, formed by UV exposure. This repair is done by photoreversal (Fig. 2.4.12). The photolyase enzyme binds to the DNA and in blue light repairs the DNA (Fig. 2.4.13).

FIGURE 2.4.12 Thymine dimers caused by ultraviolet (UV) exposure can occur in either of two forms.

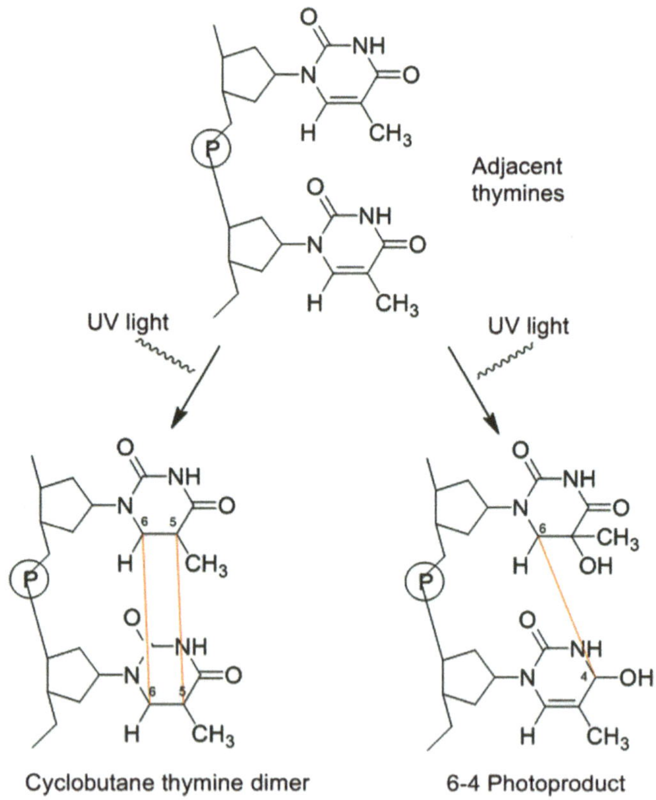

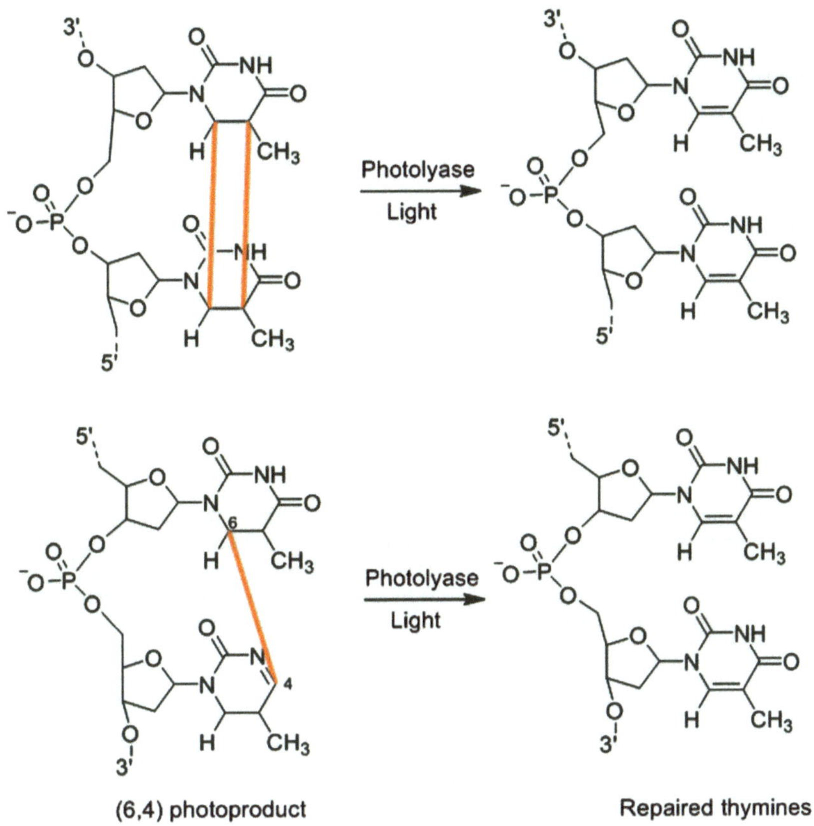

FIGURE 2.4.13 Repair of thymine dimers (both types) occurs using the enzyme photolyase and blue light.

2.4.5.2 Base Excision Repair

The general mechanism for base excision repair is shown in Fig. 2.4.14. A glycolase removes the damaged base, a DNA endonuclease "cuts" the DNA at the 3′ end of the preceding nucleoside, and then a DNA exonuclease removes the abasic sugar-phosphate group. The repair can be "short patch" or "long patch."

2.4.5.3 Nucleotide Excision Repair

The overall mechanism for nucleotide excision repair includes numerous enzymes used for the process and is similar to base excision repair. This process removes a much longer stretch of the DNA helix to repair large sections that have sustained damage.

2.4.5.4 Mismatch Repair

This process is used for incorrect base incorporation occurring during DNA replication, as well as incorrect insertions and deletions. This mechanism must be DNA strand specific and occurs on the newly synthesized DNA strand when replication has introduced errors

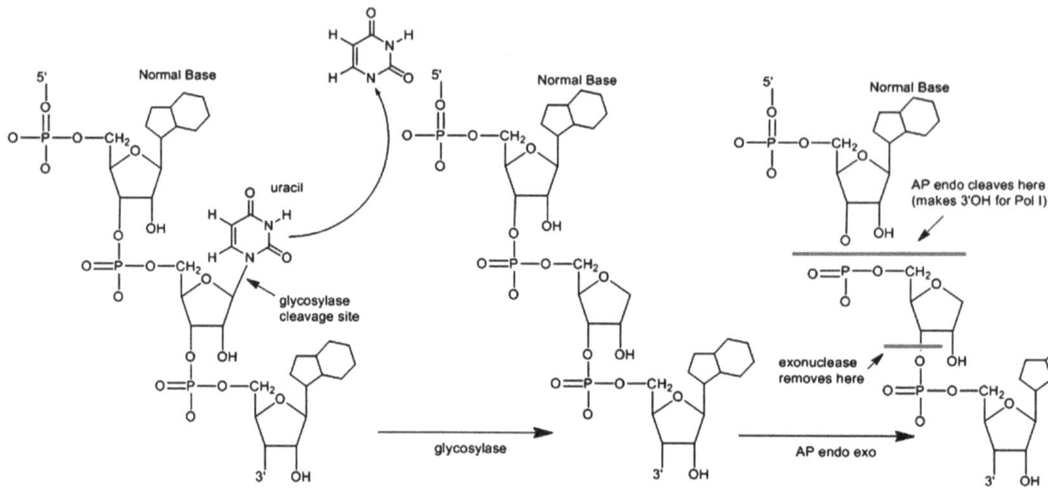

FIGURE 2.4.14 The mechanism for base excision repair. In this example, a uracil is in the DNA sequence, and once recognized, it is removed by an enzyme that breaks the glycosidic bond between the base and sugar, a glycolase. This results in an abasic site in the DNA (also called an AP site), and then an endonuclease (AP endo) cleaves the 3′-5′ phosphodiester bond, allowing an exonuclease (AP exo) to cleave the other side of the abasic site at the glycosidic bond. A repair DNA polymerase then fills in (adds) the complementary base to effect the repair. Ligase then forms a new 3′-5′ bond to seal the DNA strand break.

as detected by mismatched base pairing between the newly synthesized strand and the parental strand. The strands are specified easily in bacteria since the parental strand contains methylated C, whereas the newly synthesized strand lacks the methylation. It has also been suggested that since on one of the newly synthesized DNA strands (the lagging strand) the DNA is synthesized as short (Okazaki) pieces (about 1000 bases long) that are eventually ligated together, these short pieces discriminate the newly synthesized strand from the parental strand.

2.4.5.5 Transcription-Coupled Repair

This system works with the RNA polymerase, especially RNA polymerase II. Eukaryotic RNA polymerases are multisubunit enzymes that additionally have myriad associated proteins. One of these associated proteins TF IIH (itself a multisubunit protein) contains subunits that are also part of the repair system. Hence, when the RNA polymerase II encounters DNA damage the polymerase halts and a signal is sent to initiate the repair process, after which the RNA polymerase can proceed through the repaired region. In effect, this pathway is an offshoot of the nucleotide excision pathway. A limitation is that only transcribed genes are recognized by this pathway.

2.4.5.6 Single- and Double-Strand DNA Break Repair, Recombination Repair

These pathways will repair scissions in the backbone of the DNA, either one or both strands of the DNA, frequently using a recombination mechanism following chromatid separation.

PROBLEMS

1. It is well known that parental exposure to certain chemicals causes gene expression changes in the individual and these can be passed on to subsequent generations (transgenerational inheritance). Although the DNA sequence remains unchanged the epigenetic alterations persist over several generations. Do you think that if the exposure to these chemicals persisted for the parent and the offspring (P, F1, F2) generations, this would lead to a change in the DNA sequence?
2. What advantage is it to the cell to have some of the DNA damage sensors associated with the transcription machinery?
3. Thalassemias are often caused by a premature stop codon in the mRNA that leads to an incorrect mRNA that is detected by the cell and destroyed so that there is an absence of hemoglobin. There are drugs that can suppress the destruction of the mRNAs that have these premature stop codon mutations so that a partial protein is produced, and this can result in the partial or complete reversal of the disease. Is it possible, however, for these patients to have a higher incidence of cancer? Why?
4. (a) If BPA is harmful to the fetus, discuss the evidence you would seek to find out how the BPA caused genomic harm. (b) If we assume that the DNA base sequence has not changed due to exposure to BPA, what other changes would you investigate?
5. If mutations occur in a DNA repair protein
   a. why can that lead to oncogenesis?
   b. why is this likely the reason for the Hayflick number of cell divisions?
6. A neutral mutation changes a codon from lysine to arginine, both of which are basic amino acids. This is considered relatively harmless—a neutral mutation.
   a. Why, however, would such a change be particularly problematic with regard to posttranslational (epigenetic) modifications of proteins?
   b. Why could such a change affect a protein's structure?
7. Why is it not surprising that a particular cancer, say colon cancer, has different gene mutations for each patient?
8. If a single base change causes a phenotypic change, does that define a separate species or race? Why?
9. DNA base changes leading to mutations are often viewed as harmful and possibly cancer causing. Discuss why base changes that cause amino acid changes can be evolutionarily advantageous.
10. If you were designing a sunscreen product what qualities would you assay to demonstrate that the sunscreen is effective against DNA damage?
11. Many people have the expectation that we will eventually colonize planets in other solar systems. We have evolved in the presence of our cosmic radiation and our planet's light (UV, X-rays, gamma rays, etc.). For colonizers of a foreign Solar System what concerns would there be about mutations due to a different light spectrum?
12. Early in Earth's history the first genetic molecules may have been RNA that formed by chemical reactions in shallow pools. The intensity of our sun was much greater than today. How would this have affected mutation rates of these RNAs when they were replicated and how would this have provided genetic variation?

Recommended Reading

1. Chamary JV, Hurst LD. The price of silent mutations. *Sci Am* 2009;**300**(6):46–53.
2. Kimchi-Sarfaty C, Oh JM, Kim I-W, Sauna ZE, Calcagno AM, Ambudkar SV, et al. A "silent" polymorphism in the MDR1 gene changes substrate specificity. *Science* 2007;**315**:525–8.
3. Schmidt C. The fog of agent orange. *Sci Am* 2016;**314**(6):70–5.
4. Andrews FH, Strahl BD, Kutateladze TG. Insights into newly discovered marks and readers of epigenetic information. *Nat Chem Biol* 2016;**12**:662–8.
5. Rogers JA, Metz L, Yong VW. Review: endocrine disrupting chemicals and immune responses: a focus on bisphenol-A and its potential mechanisms. *Mol Immunol* 2013;**53**:421–30.
6. Giulivo M, Alda ML, Capri E, Barcelo D. Human exposure to endocrine disrupting compounds: their role in reproductive systems, metabolic syndrome and breast cancer. A review. *Environ Res* 2016;**151**:251–64.
7. Blake GET, Watson ED. Unravelling the complex mechanisms of transgenerational epigenetic inheritance. *Curr Opin Chem Biol* 2016;**33**:101–7.
8. Grossniklaus U, Kelly B, Ferguson-Smith AC, Pembrey M, Lindquist S. Transgenerational epigenetic inheritance: how important is it? *Nat Rev Genet* 2013;**14**:228–35.
9. Nestler EJ. Transgenerational epigenetic contributions to stress responses: fact or fiction? *PLoS Biol* 2016;**14**:e1002426.
10. Friedberg EC. DNA damage and repair. *Nature* 2003;**421**:436–40.
11. Peterson CL, Côté J. Cellular machineries for chromosomal DNA repair. *Genes Dev* 2004;**18**:602–16.
12. Laine J-P, Egly J-M. When transcription and repair meet: a complex system. *Trends Genet* 2006;**22**:430–5.
13. Ataian Y, Krebs JE. Five repair pathways in one context: chromatin modification during DNA repair. *Biochem Cell Biol* 2006;**84**:490–504.
14. Wood JL, Chen J. DNA-damage checkpoints: location, location, location. *Trends Cell Biol* 2008;**18**:451–5.

PART 3

GREEN CHEMISTRY IN PRACTICE

CHAPTER

3.1

The Natural Atmosphere

Nadine Borduas[1], *Neil M. Donahue*[2]

[1]ETH Zurich, Zurich, Switzerland; [2]Carnegie Mellon University, Pittsburgh, PA, United States

3.1.1 INTRODUCTION TO THE ATMOSPHERE

Imagine that the atmosphere is a large round-bottom flask. Wind and turbulence are the stir bars and mix the complex gas mixture inside. The starting materials are emissions from anthropogenic and biogenic sources and are typically small molecules with high vapor pressures that can readily evaporate and/or partition to the gas phase. Sunlight is the source of energy input into the mixture and acts as the hot plate. Since gas-phase molecules are few and far apart compared with the liquid or solid phases, only the most reactive species take part in chemical reactions. Therefore, gas-phase chemistry is governed by radicals, often initiated by photochemistry, because high-energy reactants are required to overcome the low collision frequency of molecules in the gas phase. The round-bottom flask also contains water, a crucial molecule to atmospheric chemistry. Water transports heat through the atmosphere, serves as a solvent for atmospheric aqueous reactions, and cleans the atmosphere through wet deposition. Water is also a precursor to the major oxidant in the atmosphere: the hydroxyl (OH) radical, the so-called atmospheric detergent. Note that it is traditional for atmospheric scientists not to write the radical symbol for small molecules where the odd electron is self-evident. The reaction mixture inside the round-bottom flask may also contain catalysts (e.g., NO_x for O_3 production), heterogeneous surfaces (e.g., polar stratospheric clouds for Cl production), and metals (e.g., Hg^0). When daily, seasonal, and annual variations in weather and climate are incorporated, not to mention pollution, the atmospheric system becomes quite complicated. The real atmosphere is intricate and thus chemists, physicists, engineers, geographers, biologists, meteorologists, climatologists, computer scientists, and researchers from many more disciplines study the atmosphere from different perspectives. This chapter reviews the chemical and physical processes that govern and dictate chemical reactions in the natural, unperturbed atmosphere.

3.1.2 LAYERS OF THE ATMOSPHERE

The earth's atmosphere is a thin layer of gases surrounding the planet. Yet this layer is vital for life on Earth as it regulates its surface temperature and holds oxygen and water,

Green Chemistry
http://dx.doi.org/10.1016/B978-0-12-809270-5.00006-6

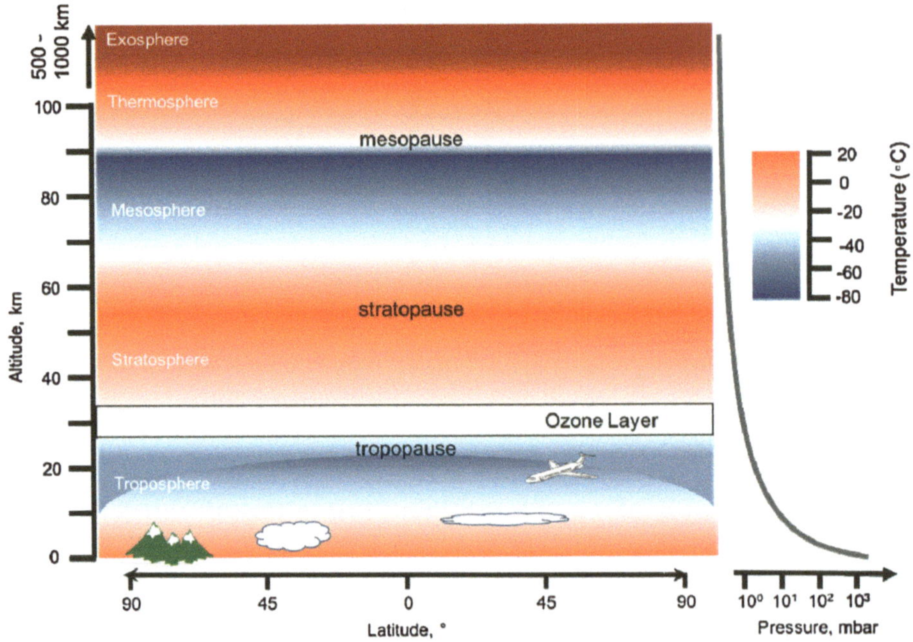

FIGURE 3.1.1 The five layers of the atmosphere are presented as a function of altitude: the exosphere, the thermosphere, the mesosphere, the stratosphere, and the troposphere. The altitude of each layer is labeled on the *y*-axis. The inset on the right shows the atmospheric pressure (mbar) as a function of altitude, which approaches zero in the exosphere. Note that the exosphere is at times omitted from the description of the layers of the atmosphere for the simple reason that the atmospheric pressure at 500–1000 km in altitude is extremely low and approaches vacuum. The temperature color scale ranges from −80 to 20 °C, where blue is coldest and red is hottest. In fact, changes in temperature gradients distinguish the atmospheric layers from one another, and are called the mesopause, the stratopause, and the tropopause. The *x*-axis is shown to illustrate the difference in thickness of the troposphere at different latitudes, where it is largest at the equator and smallest at the poles. The ozone layer is also outlined between 25 and 35 km. The airplane, the mountains, and the clouds are shown to give reference points to the altitude scale.

necessary for life. Pressure in the atmosphere ranges from 10^3 mbar at sea level to 10^{-4} mbar at the top of the atmosphere, and to eventually 0 in space (Fig. 3.1.1 right inset). Pressure decreases exponentially with altitude, dropping by e^{-1} of the surface pressure every 7.6 km, the approximate scale height of the atmosphere. 7.6 km is also the thickness the atmosphere would have if air were an incompressible fluid like water. However, if air did indeed have the density of water, the atmosphere would be about 1000 times thinner, or just 7.6 m thick. The atmosphere is truly the thin outer skin of the planet!

Temperature varies significantly throughout the atmosphere ranging from −80 to +30 °C and even higher in the uppermost parts of the atmosphere. In fact, changes in temperature gradients distinguish the five atmospheric layers from one another. From top to bottom, the layers are: the exosphere, the thermosphere, the mesosphere, the stratosphere, and the troposphere (Fig. 3.1.1). Note that the exosphere is at times omitted from the description of the layers of the atmosphere for the simple reason that atmospheric pressure at 500–1000 km altitude is extremely low.

The incoming solar radiation reaches first the exosphere, which begins at an altitude of approximately 500 km and extends into space and vacuum. The exosphere consists of the lightest elements, like H and He. Collisions between atoms are rare due to the low pressure in the exosphere, that is, due to the very low concentrations of these gases. Yet chemistry still happens! Indeed, ion chemistry dominates this region and is initiated by the extreme ultraviolet (UV) energy. Ions like H^+ and He^+ form when H and He absorb high-energy photons matching their high ionization energies at 13.60 eV (91 nm) and at 24.59 eV (50 nm), respectively. The exosphere also hosts the Hubble telescope and most orbiting satellites.

The thermosphere spans from 90 to approximately 500 km over the earth's surface and is often termed the ionosphere, because of the presence of ions produced by photoionization. N_2 and O_2 molecules absorb short-wavelength radiation with ionization energies of 15.58 eV (80 nm) and 12.07 eV (103 nm), respectively. The thermosphere is thus characterized by high temperatures, which increases with altitude. Following the logic that cold air sinks and hot air rises, because of their respective densities, the thermosphere is a stable layer of the atmosphere. Atoms and molecules separate by diffusion, and thus the ratio of lighter species increases with altitude. The higher pressure, relative to the exosphere, also implies that collisions between molecules are more frequent, resulting in chemical reactions mostly through charge transfer mechanisms. Interestingly, the presence of ions, but especially of electrons, in the earth's atmosphere is crucial to our telecommunication system. The electron density peaks at 10^6 electrons cm^{-3} at around 300 km and is the reason why the highest radio frequency to reflect from the ionosphere back to receivers on the ground is approximately 10−20 MHz. In addition, the International Space Station also resides in the thermosphere, orbiting at an altitude between 330 and 435 km. The bottom of the thermosphere is delimited by the mesopause, the coldest part of the earth's atmosphere.

The mesosphere lies below the mesopause and extends from about 50 to 90 km above sea level. The majority of the incoming solar irradiation is absorbed above the mesosphere, resulting in much lower ion densities in this atmospheric region. In the mesosphere, the temperature decreases with increasing altitudes, which leads to a somewhat less stratified layer (Fig. 3.1.1). Yet still low pressures exist in the mesosphere and few molecular transformations occur.

The stratopause divides the mesosphere from the stratosphere as well as the upper atmosphere from the lower atmosphere. The temperature in the stratosphere increases with altitude, which leads to a temperature inversion and to a stratified layer with low vertical mixing, because hot air rises and cool air sinks. The ozone layer is located in the stratosphere between 25 and 35 km at temperatures around −10 °C. The higher temperature in the upper stratosphere is due primarily to the photolysis of molecular oxygen, which is exothermic. Indeed, molecular oxygen absorbs UV radiation and consequently breaks apart homolytically and exothermically into two atoms of oxygen, which may then combine with molecular oxygen to form ozone (further discussed in Section 3.1.4.4 and depicted in Fig. 3.1.8). In the lower stratosphere, ozone molecules absorb UV radiation and break apart into atomic oxygen and molecular oxygen, also exothermically. The ozone layer is crucial to life on Earth as it prevents harmful UV radiation from reaching the surface of the planet. The chemistry of the ozone layer is further discussed in Chapter 3.3.

Finally, the sun's photons with wavelengths greater than roughly 290 nm penetrate the tropopause, which can be as high as 18 km in altitude at the tropics and as low as 8 km in altitude at the poles, before arriving at the earth's surface (Fig. 3.1.1). The troposphere's temperature decreases with altitude, creating a turbulent atmosphere. In fact, airplanes generally fly close to the tropopause to avoid the majority of the tropospheric turbulence. The troposphere contains about 80% of the total mass of the atmosphere and the majority of water vapor. It is divided into the free troposphere and the planetary boundary layer (PBL), approximately identified as the first kilometer above the earth's surface. The PBL is delimited by a change in temperature gradient and a change in aerosol concentrations, detected by Light Imaging, Detection, and Ranging instruments. The PBL is an important section of the atmosphere, since it contains the air we breathe. The discussion of air pollution is typically centered in the PBL region and is addressed in Chapter 3.2.

3.1.3 ENERGY IN THE ATMOSPHERE

The ability of the earth's atmosphere to absorb and reflect energy is related to its blackbody characteristics. A blackbody is a hypothetical physical body described as a perfect absorber with no capacity to reflect sunlight. Planets and stars are approximated to be blackbodies. Thus the earth approaches a blackbody and absorbs solar irradiation, energy transmitted as electromagnetic waves, in the atmosphere and at the surface. However, to maintain equilibrium between incoming and outgoing energy, the earth also emits longwave radiation. This radiative balance allows the planet to stay at a constant surface temperature. The energy balance is intrinsic yet intricate and has large implications in understanding and predicting the global climate change. Refer to Chapter 3.4 for the explicit equations governing climate.

3.1.3.1 Solar Irradiation

The sun's energy travels 149.6 million km before reaching the earth's atmosphere. Solar irradiance, which is measured as the power (Watts) per surface area (m^{-2}) per wavelength (nm^{-1}), enters the earth's atmosphere and may be absorbed and/or reflected either in the atmosphere or at the surface. Note that the word "irradiance" specifically refers to incoming radiation to the earth. The solar irradiance spectrum is composed of photons in the wavelength range from 200 to 3000 nm. It includes UV radiation (200–400 nm), visible light (400–700 nm), and near-infrared (IR) radiation (700–3000 nm), accounting for 5%, 43%, and 52%, respectively, of the total solar irradiance entering the top of the earth's atmosphere (Fig. 3.1.2). Indeed, this spectrum matches closely that of a blackbody at 5800K, which is inferred as the temperature at the surface of the sun (*gray outline* in Fig. 3.1.2). The solar irradiation that reaches the surface of the planet, however, is significantly less because of processes in the atmosphere that reflect and/or absorb energy. The amount of reflected and absorbed solar irradiance by the atmosphere is illustrated in Fig. 3.1.2 as the gray area. O_2 and O_3 in the stratosphere are responsible for absorbing irradiation below 290 nm, whereas CO_2 and water in the troposphere absorb in the near-IR region (between 1000 and 3000 nm) (see Fig. 3.1.4). Another distinct feature of the solar irradiance spectrum at the surface of the earth is an atmospheric "window" between 300 and 800 nm where the atmosphere is almost transparent and where visible light comes through (Fig. 3.1.2).

There are four possible fates for incoming solar energy (Fig. 3.1.3). It can be absorbed by particles and gases or reflected by particles and clouds. When the solar irradiation reaches the

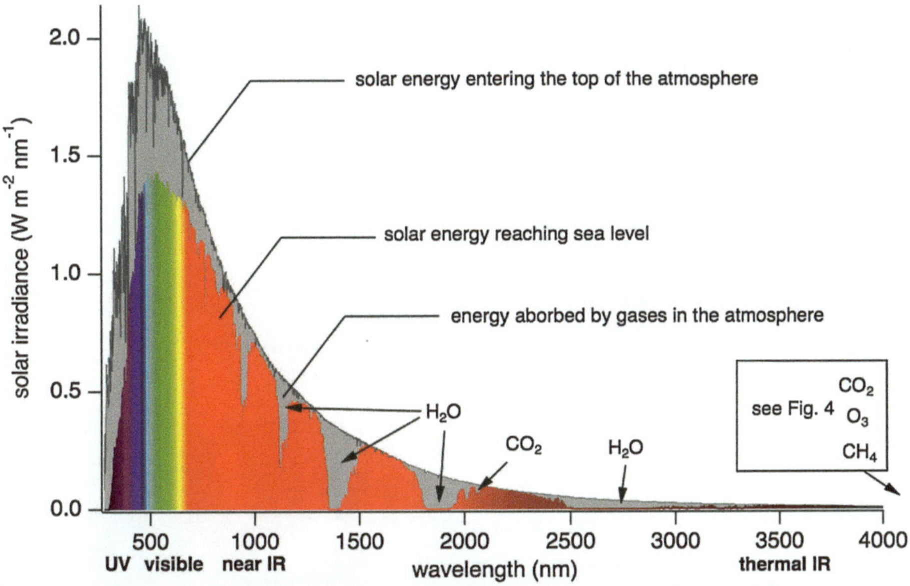

FIGURE 3.1.2 The solar irradiance is a function of wavelength spanning the ultraviolet (UV), visible, and infrared (IR) regions. The *gray line* is the sun's blackbody emission spectrum, which matches 5800K. The *gray shaded regions* represent the solar irradiance absorbed by the atmosphere, and the colored regions identify the solar irradiance that reaches the surface of the earth. The difference in the solar irradiance at the top of the atmosphere and at sea level is a result of atmospheric absorption by particles and gases like H_2O, CO_2, O_3, CH_4 and of reflection by particles and clouds. Terrestrial radiation is emitted in the thermal IR region and is further illustrated in Fig. 3.1.4. *Data from ASTM G173-03 Reference Spectra Derived from SMARTS v. 2.9.2.*

surface of the planet, it can once again be absorbed or reflected. The albedo effect is described as the ability of an environmental surface to reflect sunlight from a range of 0–1, where 0 is for a perfectly absorbing surface (black) and 1 is for a perfectly reflecting surface (white). For example, a dense forest may have an albedo of 0.10, while a white glacier may have an albedo of 0.90. On average, the planetary albedo of the earth is 0.30. The average intensity of sunlight over the earth is 340 W/m^2, where 102 W/m^2 are reflected back to space, while the remaining 238 W/m^2 are absorbed. The majority of incoming solar radiation is absorbed by the surface of the planet, consistent with the approximation of Earth as a blackbody (Fig. 3.1.3).

3.1.3.2 Terrestrial Radiation

Viewed from space, the earth emits 238 W/m^2 of thermal energy to balance the absorption of solar energy of the same magnitude. Because the earth can be approximated as a black body, this corresponds to an average temperature of 255K (−18 °C). Terrestrial radiation is longwave low-energy radiation and is emitted in the range of 6000–20,000 nm (Fig. 3.1.4 in micrometers). The terrestrial emission spectrum matches that of a combination of blackbody spectra of temperatures between 220 and 320K (Fig. 3.1.4). The gases that absorb terrestrial radiation (e.g., H_2O, CO_2, O_3, CH_4) are called greenhouse gases. Effectively, these gases prevent outgoing energy emitted from the surface from being released back into space, thereby trapping this energy

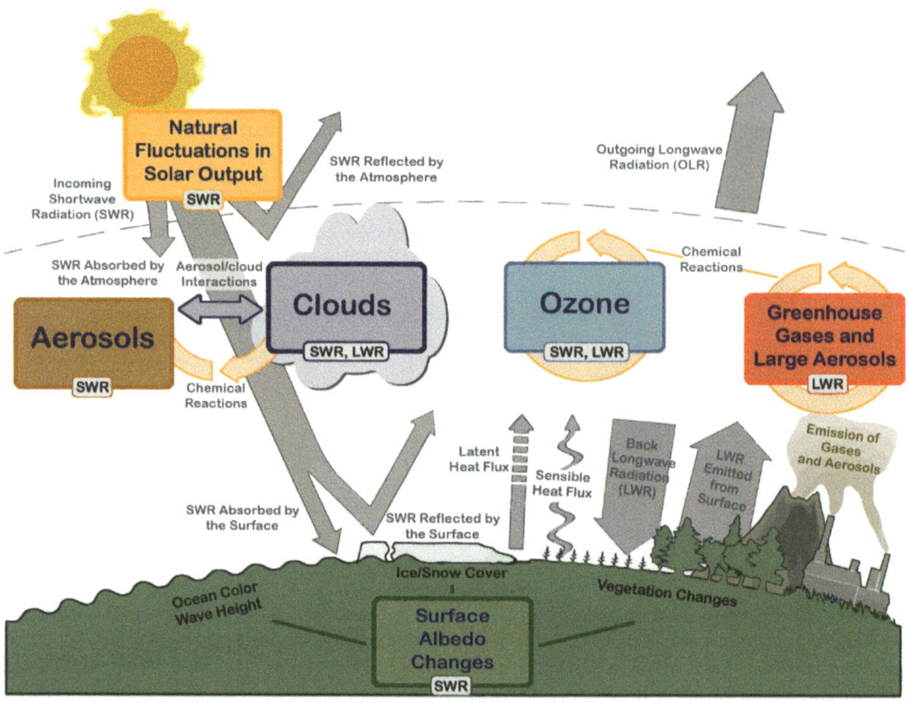

FIGURE 3.1.3 The main drivers of climate are represented while highlighting the energy balance of the earth's atmosphere. *Reproduced from the Intergovernmental Panel on Climate Change report 5th assessment Chapter 1, where the original caption reads as follows: Main drivers of climate change. The radiative balance between incoming solar shortwave radiation (SWR) and outgoing longwave radiation (OLR) is influenced by global climate 'drivers'. Natural fluctuations in solar output (solar cycles) can cause changes in the energy balance (through fluctuations in the amount of incoming SWR). Human activity changes the emissions of gases and aerosols, which are involved in atmospheric chemical reactions, resulting in modified O_3 and aerosol amounts. O_3 and aerosol particles absorb, scatter and reflect SWR, changing the energy balance. Some aerosols act as cloud condensation nuclei modifying the properties of cloud droplets and possibly affecting precipitation. Because cloud interactions with SWR and LWR are large, small changes in the properties of clouds have important implications for the radiative budget. Anthropogenic changes in GHGs (e.g., CO_2, CH_4, N_2O, O_3, CFCs) and large aerosols (>2.5 μm in size) modify the amount of outgoing LWR by absorbing outgoing LWR and re-emitting less energy at a lower temperature. Surface albedo is changed by changes in vegetation or land surface properties, snow or ice cover and ocean color. These changes are driven by natural seasonal and diurnal changes (e.g., snow cover), as well as human influence (e.g., changes in vegetation types).*

and releasing it in the form of heat. This heat release forces upward the average altitude, which relates to a lower temperature, from which outgoing energy escapes to space, leading to a temperature of 255K observed at the top of the atmosphere. This greenhouse effect occurs naturally in our atmosphere, yet human activity has significantly increased the amount of certain greenhouse gases (e.g., CO_2, N_2O, and CH_4) in the atmosphere since the industrial revolution.

3.1.3.3 Photolysis

The sun's photons include low-frequency, high-energy radiation capable of breaking chemical bonds through photolysis and of generating highly reactive chemical species. The

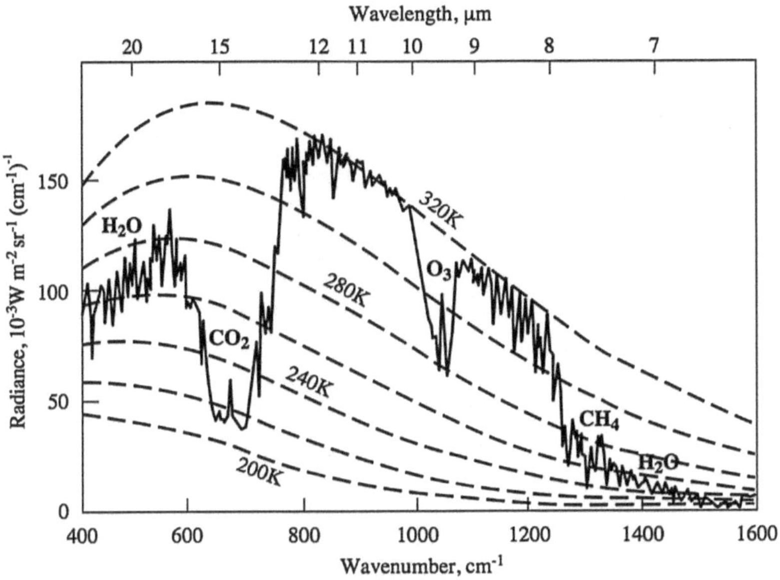

FIGURE 3.1.4 The actinic flux, or radiance, as a function of wave number, bottom x-axis, or wavelength, top x-axis, is depicted. The terrestrial emission spectrum is a combination of blackbody spectra of temperatures between 220 and 320K (*dotted lines*). The *black outlined spectrum* indicates the terrestrial radiation measured from a satellite at noon over the Niger valley in Northern Africa. The major gases responsible for atmospheric absorption are H_2O, CO_2, CH_4, and O_3. *From Hanel RA, Conrath BJ, Kunde VG, Prabhakara C, Revah I, Salomonson VV, Wolford G.* The Nimbus four infrared spectroscopy experiment: 1. Calibrated thermal emission spectra, volume 77, Oceans and atmospheres. *1972. p. 2629–41.*

energy of a photon is dictated by its frequency, according to Planck's law: $E = h\nu$. Ranges of frequencies (and wavelengths, related by $\lambda = c/\nu$, where c is the speed of light) relevant to atmospheric chemistry are depicted in Fig. 3.1.2. In particular, the wavelengths of the visible spectrum range from 700 to 420 nm, equivalent to photon energies from 170 to 280 kJ/mol, and UV spectrum wavelengths range from 420 to 290 nm, equivalent to photon energies from 280 to 410 kJ/mol. The UV spectrum is often subdivided into UVA (315–400 nm), UVB (280–315 nm), and UVC (100–280 nm). Since UVC are absorbed in the ozone layer (Fig. 3.1.5), UVB are the most harmful photons that reach the surface of the planet. Indeed, extended UVB exposure leads to sunburns and repeated prolonged UVB exposure may lead to DNA damage and skin cancer in humans.

Photolysis occurs when a molecule absorbs a photon that has the same energy as the gap between two of its electronic states. The absorbed photon then excites electrons to a higher energy orbital, as long as the dictated selection rules allow the transition. The molecule can subsequently release this energy through five possible relaxation pathways. The excited molecule, often denoted with a superscript asterisk (*), can undergo dissociation, chemical reaction, fluorescence, collisional deactivation, and/or ionization. The quantum yield (ϕ) describes the efficiency of an excited molecule's relaxation process through each of the five possible pathways. It is defined as the ratio of the number of molecules undergoing one relaxation process to the total number of photons absorbed and has a value between 0 and 1.

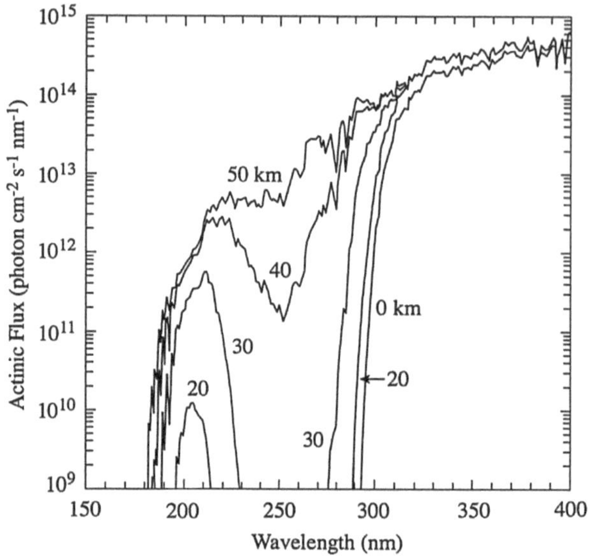

FIGURE 3.1.5 The actinic flux measures incoming solar radiation in units of photons per surface area, per second, per wavelength. The actinic flux is plotted here as a function of wavelength to show how much energy reaches altitudes of 0, 20, 30, and 50 km. The altitude contour lines are remarkably different because of the local concentrations of different key absorbers. This graph illustrates the role of the ozone layer at absorbing much of the harmful UVC radiation coming into the atmosphere from the sun between 20 and 40 km. *From DeMore.* Chemical kinetics and photochemical data for use in stratospheric modeling. *Pasadena (CA): JPL Publication; 1997.*

Actinic flux is an important term for atmospheric photochemical reactions; it defines the amount of sunlight available at a local point in the atmosphere for a specific wavelength. Fig. 3.1.5 depicts the actinic flux in units of photons per surface area per wavelength as a function of wavelength, where each contour represents the actinic flux at different altitudes. The high-energy wavelengths are typically absorbed at higher altitudes, in part by the ozone layer at around 30 km. Indeed, notice the stark difference in the actinic flux between 40 and 20 km altitudes. The ozone layer is key in the atmosphere to prevent UVC exposure at the surface of the planet.

Photochemical reaction rates are calculated by knowing the number of photons absorbed by a fixed concentration of a molecule in a specific volume of air. Photolysis rate constants (J) in the atmosphere depend on the absorption cross-section (σ), defined as the ability of a molecule to absorb a photon of a certain wavelength, the quantum yield (ϕ), and the actinic flux (I). All three parameters depend on the wavelength of the photon. To precisely calculate the photolysis rate of atmospheric gases, one must integrate the following equation as a function of wavelength at a known temperature, where λ_1 for the troposphere is 290 nm for example. The integral is often approximated to be the sum of the product of the cross-section, the quantum yield, and the actinic flux over small intervals of wavelength values.

$$J = \int_{\lambda_1}^{\lambda_2} \sigma(\lambda)\phi(\lambda)I(\lambda)\, d\lambda$$

The photolysis of O_2 and NO_2 are major precursors of ozone in the stratosphere and in the troposphere, respectively (see Fig. 3.1.8 in Section 3.1.4.4). Ozone itself can absorb photons with a wavelength of less than 305 nm to form O* and O_2. This reaction is particularly important as O* reacts with water in the troposphere to form the free radical hydroxyl (OH), the main atmospheric oxidant (Fig. 3.1.8). Photolysis is an important reaction-initiating process and the focus of much atmospheric research.

3.1.4 GASES IN THE ATMOSPHERE

The field of atmospheric chemistry is centered on the study of the lower atmosphere, where air density is largest and where molecular collisions are most frequent. Tropospheric chemistry is complex and heterogeneous. It contains anthropogenic and biogenic emissions of a huge range of molecules, mixed and transported by wind and turbulence across the surface of the planet. The troposphere also contains water in all three phases, serving as a medium for chemical reactions and as a controller for heat absorption and release. Fig. 3.1.6 gives an overview of the sources and sinks of tropospheric constituents as well as the impact of atmospheric processes such as photooxidation and aqueous-phase processing. Despite the lesser air density, stratospheric chemistry is important for life on Earth and for climate and is dominated by ozone reactivity, since 90% of all ozone molecules are found in the ozone layer centered at 30 km in altitude (see Chapter 3.3).

3.1.4.1 Measuring Atmospheric Composition

The amount of gases in the atmosphere can be expressed in absolute or relative concentrations. The absolute concentration is the number density (n), defined as the number of molecules per unit volume of air. Number densities are often expressed in units of molecules per cubic centimeter and are important for calculating gas-phase reaction rates. The relative concentration is given by a mixing ratio, and represents the mole (or molecule) fraction of a gas over the total number of moles (or molecules) of gases present in a volume. It is advantageous to work with mixing ratios when studying air mass parcels, as the mixing ratio does not change with pressure or temperature, whereas the absolute concentrations do change with volume. In other words, the ratio of a gas to the total amount of gases remains the same when an air mass parcel is carried upward in the atmosphere during expansion. A mixing ratio is expressed as a unitless value, and since the ratios are small, atmospheric scientists express them as parts per million/billion/trillion, written as ppmv/ppbv/pptv. The "v" emphasizes that the concentration is based upon gas-phase molecules, which occupy a volume rather than a mass. For example, the ambient concentration of CO_2 has recently surpassed 400 ppmv, or 400 molecules in a mixture of 1 million molecules of air.

The number density (absolute concentration) and the mixing ratio (relative concentration) are related by the number density of air, that is, the total number of air molecules at a given pressure and temperature. For example, carbon dioxide's mixing ratio (C_{CO_2}) is equal to its number density (n_{CO_2}) divided by the number density of air (n_{air}):

$$C_{CO_2} = n_{CO_2}/n_{air}$$

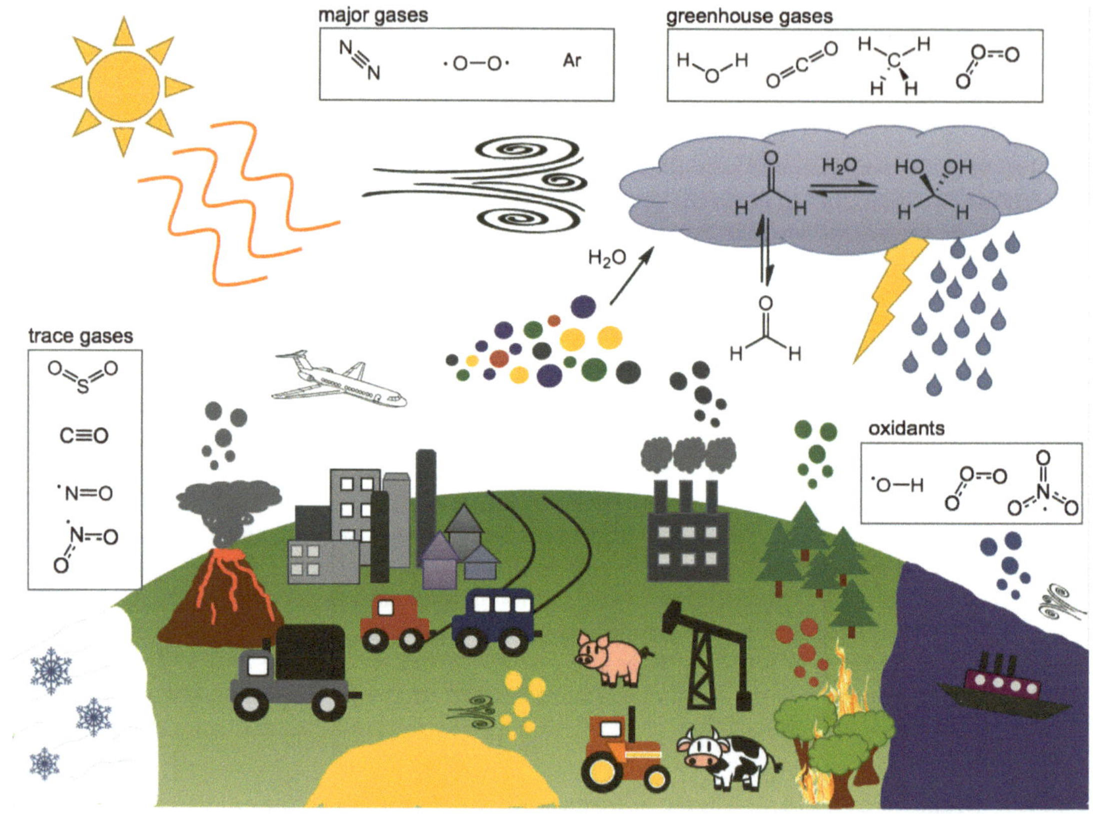

FIGURE 3.1.6 The composition of the atmosphere depends on emissions from different ecosystems including ocean, desert, cities, fields, ice-covered land, etc. Biogenic emission sources of aerosols (color coded) and of gases include volcanoes, deserts, forests, forest fires, and sea spay. Anthropogenic sources of particles (color coded) and of gases include urban areas, modes of transportation (cars, trucks, ships, and airplanes), agricultural activity, fossil fuel burning, and industrial plants. Key gas-phase species present in the troposphere can be categorized as major gases, like N_2, O_2, and Ar; as greenhouse gases, like H_2O, CO_2, CH_4, and O_3; as oxidants, like OH radicals, O_3, and NO_3 radicals; as trace gases, like SO_2, CO, NO radicals, and NO_2 radicals; as well as volatile organic compounds like formaldehyde. Chemical and physical atmospheric processing such as sunlight exposure, aerosol partitioning, and cloud processing also impact the composition of the atmosphere in important ways.

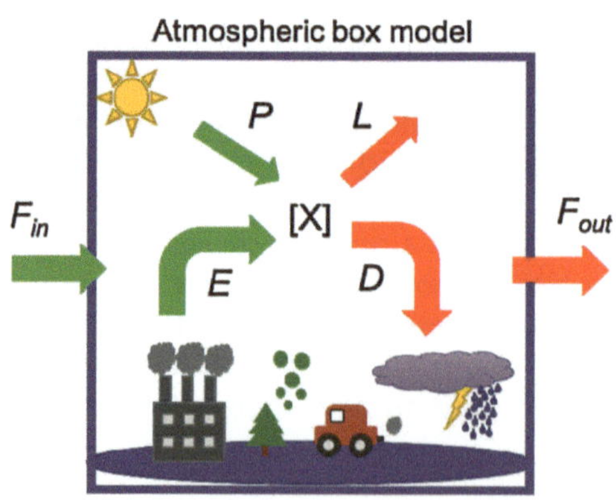

FIGURE 3.1.7 A hypothetical and simple box model for the atmospheric species X. There are three sources of X to the atmosphere (colored in green): F_{in}, inflow air; P, chemical production; E, emission. The fate of X may depend on three sinks (colored in red): F_{out}, outflow; L, chemical loss; D, deposition. The concentration of X as a function of time therefore depends on the rates of its sources and sinks over time. *Modified from Jacobs.* Introduction to atmospheric chemistry; *[chapter 3, Fig. 3.1].*

The number density of air (n_{air}) can be calculated using the ideal gas law, $PV = NRT$, where P is the pressure, V is the volume in m^3, N is the number of moles, R is the gas constant, and T is temperature in K. To convert into units of molecule cm^{-3}, the number density of air can be calculated with the following equation, which uses Avogadro's number A_v:

$$n_{air} = \frac{A_v P}{RT} \times 10^{-6} = \text{units of molecule cm}^{-3}$$

3.1.4.2 Fate of Chemical Species in the Atmosphere

To calculate the rate of appearance or disappearance of a species, information about its sources to the atmosphere and its sinks out of the atmosphere is needed. The following equation, known as the continuity equation, states that the change in concentration of species X with time must be equal to the sum of its sources minus the sum of its sinks. A simple box model is obtained by solving this equation. The different types of sources include inflow (F_{in}) of X from transport often due to meteorology, emission (E) of X, and chemical production (P) of X (Fig. 3.1.7). The different types of sinks include outflow (F_{out}) of X, deposition (D) of X, and chemical loss (L) of X (Fig. 3.1.7).

$$\frac{d[X]}{dt} = \sum \text{sources}_x - \sum \text{sinks}_x = (F_{in} + E + P) - (F_{out} + D + L)$$

The loss process by chemical reactions (L) is a major sink for gases in the atmosphere and is determined using chemical kinetics. The major loss processes of X, if X were a volatile organic compound (VOC), for example, are through reactions with oxidants and through photolysis. The following equations calculate the loss of X as a function of time due to three chemical loss processes, assuming a one-time injection of X (in other words, disregarding its sources for simplicity). The derivative is integrated as follows and leads to the time-dependent concentration equation of X, assuming a fixed amount of starting material. This example of chemical kinetics is common in laboratory experiments.

$$-\frac{d[X]}{dt} = k[X][OH] + k[X][O_3] + J[X]$$

$$-\frac{d[X]}{[X]} = (k[OH] + k[O_3] + J)dt$$

$$-\int_{[X]0}^{[X]t} \frac{d[X]}{[X]} = \int_0^t (k[OH] + k[O_3] + J)dt$$

$$\ln[X]_t - \ln[X]_0 = k[OH]t + k[O_3]t + Jt$$

$$[X]_t = [X]_0 + e^{k[OH]t + k[O_3] + Jt}$$

Atmospheric lifetime and/or residence time of a species is necessary information to evaluate the fate of a chemical or particle. Both terms are often used interchangeably, but lifetime is typically used for a chemical loss, like a chemical decomposition, whereas residence time is typically used for a physical loss, like deposition. The lifetime (τ) of a species is evaluated as the e-folding time and represents the time at which 1/e or ~36.8% of the starting concentration is left.[a] For the chemical loss of X in a one-time injection scenario, the lifetime is calculated by the inverse of its loss processes according to the following equation.

$$\tau = \frac{[X]}{k[OH][X] + k[O_3][X] + J[X]} = \frac{1}{k[OH] + k[O_3] + J}$$

3.1.4.3 Major Gases in the Atmosphere (N_2, O_2, Ar, O_x, H_2O)

Nitrogen is the most common element found in the atmosphere. Dinitrogen, N_2, makes up 78% of our atmosphere and is an inert molecule due primarily to its strong triple bond. Besides N_2, the largest components of the atmosphere are molecular oxygen, O_2 (21%), and argon, Ar (0.9%). Argon is a noble gas and is thus unreactive in the atmosphere. Note that these percentages are for dry air since mixing ratios of water vapor are extremely variable and range between 1 and 10,000 ppmv depending on the time of day, location, and altitude. In addition, there are countless trace gases in the atmosphere.

Atmospheric oxygen exists in four primary forms: diatomic molecular oxygen (O_2), ozone (O_3), ground-level triplet state atomic oxygen [O, or $O(^3P)$], and electronically-excited singlet state atomic oxygen [O* or $O(^1D)$]. Molecular oxygen is the most stable form of oxygen and is a by-product of life, cycled through living organisms on a thousand-year timescale. O_2 is also a reactive molecule, as it exists as a biradical with two unpaired electrons. Ozone has an intermediate reactivity, with a chemical lifetime ranging from a few days to a month and an abundance varying from 0.01 to 10 ppmv, depending on the location and mostly altitude. Both O and O* are present in very low quantities and are short lived. O* is the O atom in its excited singlet state and rapidly stabilizes to the triplet state atomic oxygen O through collisions with an M body (see Fig. 3.1.8). M is a third body molecule, and is typically an N_2 molecule, required to dissipate excess energy from the recombination reaction of two molecules. Ozone and O are often discussed as a chemical family known as odd oxygen (O_x), in other words, oxygen molecules with an odd number of oxygen atoms, which evidently excludes O_2.

Water is a key resource for life, but it also influences atmospheric circulation, controlling the way heat is released and transported in the atmosphere and establishing rates and depths of atmospheric mixing. Water serves as a precursor to the atmosphere's main oxidant, the OH radical, and also serves as a solvent for aqueous-phase reactions in aerosols and cloud droplets. In addition, rain water cleanses the atmosphere of many soluble and oxidized compounds and thus represents an important sink, through wet deposition, for assessing the atmospheric fate of compounds.

[a]Note that the notion of lifetime in atmospheric chemistry is different than the notion of half-life in radiochemistry, when half of the starting material is left.

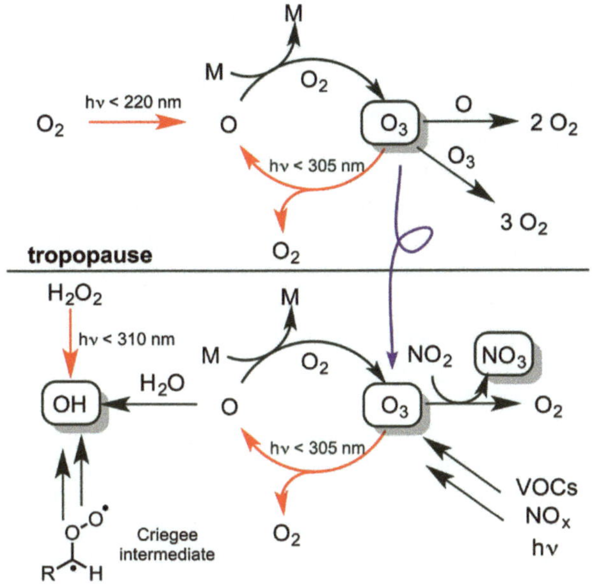

FIGURE 3.1.8 Chemical pathways of the sources and sinks of oxidants in the stratosphere and in the troposphere. The set of reactions in the stratosphere, above the tropopause, are referred to as the Chapman cycle. The *red arrows* indicate photolysis reactions, and the curly *blue arrow* indicates transport from the stratosphere to the troposphere. M is a third body molecule, and would typically be an N_2 molecule, required here to dissipate excess energy from the recombination reaction. O_2 photolysis only occurs in the stratosphere because of the presence of high-energy photons with wavelengths below 220 nm. Reactions of O_x with NO_x and with water occur solely in the troposphere because concentrations of NO_x or water are small in the stratosphere. The photochemical production of O_3 via reactions of volatile organic compounds and NO_x with sunlight are unique to polluted regions and are major components of smog. Finally, the production of OH radicals occurs predominantly in the troposphere through three different mechanisms: $O + H_2O$, photolysis of H_2O_2, and Criegee intermediate decomposition. The Criegee intermediate is formed from the ozonolysis of gas phase alkenes.

3.1.4.4 Oxidants (OH, O_3, NO_3)

The troposphere is an oxidative environment, and its major oxidants are hydroxyl radicals (OH), ozone molecules (O_3), and nitrate radicals (NO_3) (see Fig. 3.1.6). These three oxidants have unique reactivity, and oxidize gas-phase molecules, aqueous-phase molecules (e.g., cloud processing), and particle surfaces through different mechanisms.

Ozone chemistry consists of a series of reactions describing the production, cycling, and loss of odd oxygen, O_x, calculated as the sum of O atom and O_3 concentrations. The basic Chapman cycle reactions that describe the production of ozone in the stratosphere are depicted in the top panel of Fig. 3.1.8. The major source of stratospheric ozone is the photolysis of molecular oxygen with UV radiation with wavelengths below 220 nm and the subsequent reaction of the triplet state O atom with molecular oxygen. The cycling of odd oxygen between O and O_3 occurs with no net loss of ozone. Yet, odd oxygen loss occurs via the reaction of ozone with atomic O to form two O_2 and via the reaction of ozone with itself to form three O_2. O_x is being continually produced via the photolysis of O_2 and continually lost via the bimolecular reactions of ozone with O and with O_3 (Fig. 3.1.8). Ozone concentrations remain relatively constant in the atmosphere, whereas the flux through the system is large. The Chapman cycle qualitatively describes the observed ozone distribution in the stratosphere. However, it predicts more ozone than is actually observed because it omits catalytic ozone destruction. Small concentrations of ozone catalysts such as Cl atoms can greatly influence ozone levels and are further discussed in Chapter 3.3.

Ozone in the troposphere may come from downward transport from the stratosphere, but it may also be produced photochemically in the presence of NO_x, VOCs, and sunlight (Fig. 3.1.8). Tropospheric ozone production in the context of air pollution is discussed in Chapter 3.2. Ozone chemistry is dominated by cycloaddition reactions. The electron-poor ozone molecules are attracted to electron-rich double bonds, and will, for example, readily react with isoprene, terpenes, sesquiterpenes, and other unsaturated biogenic hydrocarbons, through an ozonide intermediate.

The most important oxidant in the troposphere, in terms of reactivity, is the OH radical, despite its very low concentrations. OH radicals are consumed as quickly as they are produced and thus have very short lifetimes, from a few milliseconds in polluted regions to 1 s in the free troposphere. Because of its short lifetime, the OH radical is almost always in a steady state, with concentrations ranging from 10^5 to 10^7 molecules cm^{-3} (or 0.004 to 0.4 pptv at sea level). The OH radical is produced via three dominant pathways: bimolecular reaction of water with O* originating from ozone photolysis, photolysis of hydrogen peroxide (H_2O_2), and decomposition of carbonyl oxides (i.e., Criegee intermediates) produced via reactions of ozone with alkenes (Fig. 3.1.8). The OH radical oxidizes molecules typically via H-abstraction and double bond addition mechanisms. As the OH radical is an excellent electrophile, it reacts preferentially with electron-rich C–H bonds and/or C–C bonds.

Ozone and OH radicals require sunlight for their production and typically have maximum concentrations during peak sunlight hours. On the other hand, NO_3 radicals are nighttime oxidants. They accumulate in the atmosphere solely at night, since during the day, NO_3 radicals are quickly photolyzed into NO_2 and O or into NO and O_2. During the night, NO_3 radicals may also react with NO_2 to form N_2O_5. N_2O_5 can readily decompose back to NO_3 and NO_2, but in the presence of liquid water, N_2O_5 can hydrolyze to form two HNO_3 molecules. This irreversible formation of HNO_3 is one of the dominant removal pathways of NO_x in the atmosphere. At nighttime, NO_3 radicals may also oxidize organic molecules by H-abstraction mechanisms or by addition mechanisms and are most reactive with molecules containing heteroatoms.

3.1.4.5 Volatile Organic Compounds

There are an endless number of VOCs in the atmosphere. Some are emitted directly into the troposphere from natural and industrial sources like oceans, forests, farms, coal power plants, and transport vehicles. Others are formed in the troposphere through direct and indirect photochemistry and through cloud processing. VOCs are generally defined as carbonaceous molecules, other than methane, that have large enough vapor pressures to partition to the gas phase. Thermodynamics, in particular partitioning coefficients, are used to predict the concentration of VOCs found in the gas, particle, or aqueous phases.

Biogenic VOCs include small hydrocarbons synthesized through biochemical pathways and subsequently emitted by plants. Isoprene, a small five-carbon unsaturated hydrocarbon, is the largest nonmethane VOC emitted to the atmosphere, with a global average emission of 500 million metric tons per year. It is emitted from a number of different trees such as oaks, poplars, and eucalyptus, and thus forests like the Amazon act as large sources of isoprene. Since isoprene accounts for about one-third of the VOC emissions globally, its atmospheric chemistry is well studied. It is known to be a key VOC in regulating the oxidative capacity of the troposphere and the lifetime of OH radicals by acting as one of its major sinks. In addition to isoprene, other common biogenic VOCs include terpenes and sesquiterpenes which are 10-carbon and 15-carbon unsaturated molecules, respectively, often containing cyclic structures. Terpenes and sesquiterpenes are emitted from forested areas covering large portions of the continents, particularly in the Northern Hemisphere. They impact the oxidative capacity of the troposphere by acting as a sink for oxidants and play an important role in secondary aerosol formation, growth, and processing. The plant-based VOCs are often hydrocarbons, but many biogenic VOCs also contain heteroatoms. Volatile fragments of proteinaceous material, for example, may contain nitrogenated and oxygenated residues. Moreover, the ocean is a source of many small molecules volatilizing as a result of the degradation of a multitude of organic matter substances.

Major anthropogenic VOCs include benzene, toluene, ethylbenzene, and xylene, often referred to as BTEX. They are present in natural gas and petroleum deposits and are emitted as combustion by-products. The concentrations of BTEX measured in ambient air are often used in source apportionment studies as they tend to correlate with anthropogenic activity, especially power stations and vehicle exhaust. VOCs in urban areas can be dominated by industrial sources and are ingredients required for the formation of tropospheric ozone via photochemistry, discussed in Chapter 3.2. Polycyclic aromatic hydrocarbons (PAHs) are less volatile than their monocyclic counterparts, yet they are commonly found in the particle phase of incomplete combustion emissions. PAHs are toxic, and consequently their atmospheric transport and chemistry is the subject of ongoing research.

Furthermore, a wide range of molecular functionalities are present in the gas and particle phases. Carboxylic acids, aldehydes, ketones, and alcohols are common functionalities found in the atmosphere, often produced by the oxidation of hydrocarbon precursors. Amines, amides, nitriles, and cyanates are also present and may be used to identify origins of air masses in source apportionment studies, as they often correlate with animal husbandry and/or biomass burning, for example. Sulfur-containing organic molecules are also common in the atmosphere but are more typically found in the aerosol phase, where they may be formed from substitution reactions by sulfates. Some functional groups are quite unique to the atmosphere, including peroxyacetyl nitrate molecules, a secondary pollutant formed in polluted regions.

3.1.4.6 Greenhouse Gases

Greenhouse gases are capable of trapping the earth's emitted radiation, which otherwise escapes back to space. Major greenhouse gases at all layers of the atmosphere are H_2O, CO_2, CH_4, O_3, and N_2O. These molecules have vibrational modes that match the frequency of the radiation emitted from the earth's surface. Greenhouse gases specifically trap photons of wavelengths in the IR region (see Fig. 3.1.4) and are therefore important temperature regulators of our atmosphere. However, since the industrial revolution, we have substantially increased the concentrations of most greenhouse gases in the atmosphere, thereby increasing the amount of trapped heat. Greenhouse gases and their impact on climate are further discussed in Chapter 3.4.

3.1.5 PARTICULATE MATTER

Aerosols are a collection of solid and liquid suspended particles of all shapes, sizes, and compositions in the air. Hazy conditions and reduced visibility are a result of high mass concentrations of particulate matter (PM). There are two types of sources of PM: primary PM, directly emitted particles from sources such as dust, sea spray, and volcanic and mine ash (also examples of coarse mode PM in Fig. 3.1.9), and secondary PM, formed from the condensation of gases in the atmosphere, producing particles such as sulfate and organic aerosol. Both primary and secondary PMs have biogenic and anthropogenic sources; examples are depicted in Fig. 3.1.6.

Secondary PM are created in the atmosphere when acids (e.g., H_2SO_4), bases (e.g., NH_3), as well as VOCs bind together to form a cluster held by weak intermolecular forces. The formation of these clusters is reversible, as they can dissociate. When the cluster reaches a critical size, typically around a few nanometers in diameter, a particle is nucleated and forms a

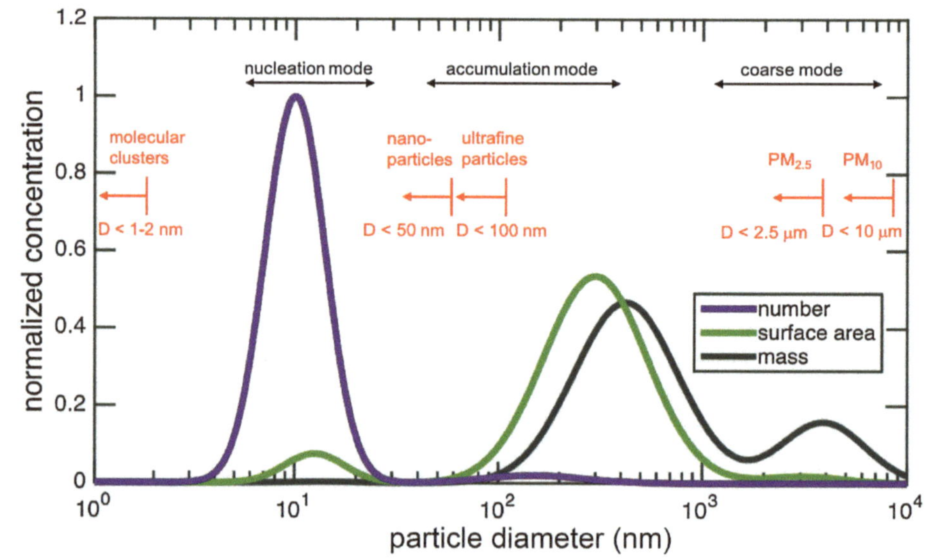

FIGURE 3.1.9 Number, surface area, and mass distributions of particulate matter in ambient air as a function of their diameter (in nm). The traces are log normal distributions for the number, surface area, and mass of particles from a range of 1–10,000 nm. Atmospheric particles are categorized into three modes based on particle diameter: the nucleation mode, the accumulation mode, and the coarse mode. The largest number of particles is found in the nucleation mode. However, when considering mass and surface volume of particles, the accumulation mode is the most important. Types of particles are also indicated in red, where particles smaller than 10 μm are PM_{10}, smaller than 2.5 μm are $PM_{2.5}$, smaller than 100 nm are ultrafine particles, smaller than 50 nm are nanoparticles, and finally smaller than 1–2 nm are dissociable molecular clusters.

nanoparticle (Fig. 3.1.9). The nanoparticle then may grow by agglomeration with other particles and/or by condensation of other vapors onto the existing nanoparticle. For example, a forest may emit large quantities of terpenes, which, upon reaction with OH radicals and ozone, can form oxygenated molecules. These molecules have lower vapor pressures and in the presence of atmospheric acids and/or bases, subsequently condense to nucleate new secondary particles.

Particles are classified based on their size and measured based on their number, surface area, and volume of particles. Fig. 3.1.9 depicts the different size modes of particles typically encountered in the atmosphere. The nucleation mode is composed of the smallest particles, 4–20 nm, and account for the majority of the particles in terms of number, yet a tiny proportion in terms of mass. The nucleation mode particles may grow in the atmosphere through deposition and partitioning of compounds and reach the accumulation mode. The accumulation mode ranges from tens to hundreds of nanometers and includes nanoparticles and ultrafine particles, which are particularly toxic as they can penetrate into the lungs. Finally, the largest particles, typically primary particles, make up the coarse mode. Particles with diameters less than 10 and 2.5 μm are designated as PM_{10} and $PM_{2.5}$, respectively. These size categories are important for evaluating health impacts. Indeed, $PM_{2.5}$, which constitutes the majority of PM by mass, has known serious health effects by causing oxidative stress in the lungs, leading to a variety of lung-related illnesses and cardiovascular complications. The World Health Organization and the United States Environmental Protection Agency, for example, have regulations requiring control of $PM_{2.5}$ particles, further described in Chapter 3.2.

Secondary organic aerosols (SOAs) are important types of particles where interesting chemistry can occur. Ambient SOA particles typically have sizes in the accumulation mode and thus account for an important fraction of aerosol surface area and mass. The presence of organic molecules in SOAs lends itself to a number of chemical processes, often initiated by sunlight. Photochemical processes, including photolysis, oxidation, and fragmentation, lead to changes in the chemical composition of the aerosols as they are transported in the atmosphere during their 1-week lifetime. The chemical composition of organic aerosols therefore changes as a function of time, complicating the evaluation of the impacts of SOAs on health, climate, and air pollution.

3.1.6 CLOUDS

Clouds play an important role in the radiative balance of the atmosphere. Depending on their altitude and their physical and chemical properties, clouds may scatter, reflect, and/or absorb solar irradiance and terrestrial radiation. Clouds do not form from the homogeneous condensation of water, but rather from the condensation of water onto existing particles. These particles can be of any size, origin, or composition. Particles efficient at growing into water droplets, and eventually into clouds, are called cloud condensation nuclei (CCN), and particles efficient at nucleating ice clouds are called ice nucleation particles (INPs). Understanding how clouds are activated (the term used for clouds containing liquid water) and nucleated (the term used for clouds containing solid water, or ice) is necessary for climate models to better predict the impacts of climate change on the earth's radiative balance.

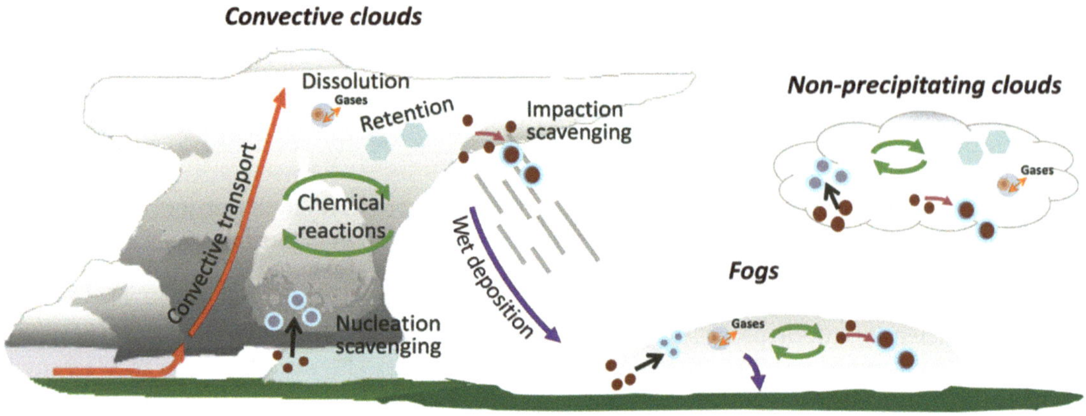

FIGURE 3.1.10 Schematic of transport, mass formation, and chemical processing and removal in clouds and fogs. In convective clouds, the upward movement of air masses generate supersaturation conditions with respect to water where a cloud condensation nuclei (CCN) may grow to droplet sizes. Particles may also get scavenged into the updraft and act as CCN. Water acts as a medium for aqueous-phase reactions as well as for dilution of gases. Convective clouds may also lead to precipitation, or wet deposition, by which many particles and water-soluble molecules are scavenged. Radiative cooling in fog also creates supersaturation conditions where CCN grow to droplet sizes. Chemical reactions, scavenging, and wet deposition also occur in fog. Nonprecipitation clouds have the same processes in addition to having cold enough conditions for ice nucleation. *Reprinted with permission from Ervens B.* Chem Rev *2015;**115**:4157–98. Copyright 2015 American Chemical Society.*

Unfortunately, aerosol-cloud interactions remain poorly constrained and currently represent the largest uncertainty in climate models.

Fig. 3.1.10 highlights the dominant processes in cloud formation: convective transport, CCN or INP scavenging, chemical reactions, and wet deposition. Convective clouds are typically storm clouds, which cover high altitudes and in which water exists in the gas, liquid, and solid phases. In convective clouds, the upward movement of air masses generate supersaturation conditions with respect to water, in other words, relative humidity greater than 100%. Particles can then grow into water droplets and/or get scavenged into the updraft and act as CCN. Water acts as a medium for aqueous-phase reactions as well as for dilution of gases. Convective clouds often lead to precipitation and wet deposition, through which many particles and water-soluble molecules are scavenged. Nonprecipitating clouds and fog are important media for aqueous-phase transformations of particles and lead to aging of particles. The process of particle aging refers to a change in chemical composition of the particle because of chemical reactions such as oxidation and fragmentation reactions. Radiative cooling in fog also creates supersaturation conditions where CCN grow to droplet sizes. Chemical reactions, scavenging, and wet deposition also occur in fog. Nonprecipitation clouds have the same processes in addition to having cold enough conditions for ice nucleation.

3.1.6.1 Warm Clouds

In the atmosphere, water condenses onto particles suspended in the atmosphere, which go on to grow as cloud droplets, reaching sizes in the micrometer size range. The process of activating clouds is initiated by specific types of CCN particles, but they may have biogenic or anthropogenic sources, and may be primary or secondary aerosols. Clouds form when particles experience supersaturation. This physical condition may occur during the updraft of an air parcel, which experiences expansion and thus cooling, or in radiative cooling like in fog, when the terrestrial surface cools the bottom of an air parcel until water vapor begins to condense. The CCN activity of particles is dictated by their size and chemical composition. Indeed, a large particle will activate as CCN at lower supersaturations than a small particle of the same composition, because the Gibbs free energy barrier of water cluster formation is lower for larger particles. Furthermore, the more hydrophilic is the particle's surface, the better it will act as a CCN via intermolecular interactions with water molecules. Chemical and physical atmospheric processes impact the ability of particles to act at CCN, and the major processes are depicted in Fig. 3.1.10. Chemical modifications like aging and deposition at the surface of a CCN increase the particle's size and hygroscopicity and thus lower the critical supersaturation necessary for droplet growth.

3.1.6.2 Ice Clouds

Ice clouds, also called cirrus clouds, are made up of ice crystals and start to form at altitudes of 5.5 km in temperate regions and of 6.5 km in tropical regions, making them the highest clouds in the troposphere. A small seed particle, or INP, is needed for heterogeneous ice nucleation. Indeed, the spontaneous homogeneous nucleation of an ice crystal from pure water droplets of 5 μm requires temperatures below −38 °C. Mineral dust, metallic particles, and biological material are the best-known ice nuclei to date and contribute to the formation of ice clouds. The amount of mineral dust and metallic dust in the atmosphere has increased due to human activity via deforestation, land use change, fossil fuel burning, and other industrial activities. Nonetheless, a large portion of mineral dust particles suspended in the air has naturally occurring sources such as the Sahara and Gobi deserts. Cirrus clouds play an important role in the global climate context by impacting the earth's radiative balance. In fact, thin cirrus clouds may contribute to net warming by absorbing the earth's thermal radiation, but thick cirrus clouds may contribute to net cooling by reflecting solar radiation. In other words, modeling cirrus clouds correctly, through accurate understanding of the physical processes of ice nucleation, is crucial to predicting the impacts of a changing climate and validates the ongoing research on ice nucleation mechanisms.

3.1.7 RESEARCH IN ATMOSPHERIC CHEMISTRY

The field of atmospheric chemistry research is composed of three pillars: laboratory, modeling, and field studies. These disciplines are all essential to understanding and characterizing the atmosphere. Atmospheric chemists study the chemical processes that occur in the gas and particle phases to better predict their impact on air pollution and global climate change. Gases and particles are emitted to the atmosphere, often as pollutants, and they may undergo a wide range of chemical processes. Field studies directly measure the composition of the atmosphere, whether it be in outdoor air or in indoor air. Laboratory studies allow chemists to replicate controlled environments, typically based on field observations, to study, for example, chemical kinetics and mechanisms. Experimental values determined in the laboratory are then used as parameters and inputs into chemical transport models. These models range from simple box models to complex global climate models and attempt to predict the impact of the changing atmospheric composition on air quality and on global climate change. Finally, there is substantial work involved in reconciling field observations with experimental results and model outputs.

Recommended Reading

1. Abbatt JPD, Lee AKY, Thornton JA. Quantifying trace gas uptake to tropospheric aerosol: recent advances and remaining challenges. *Chem Soc Rev* 2012;**41**(19):6555–81. http://dx.doi.org/10.1039/C2CS35052A.
2. Baird C, Cann M. *Environmental chemistry*. 5th ed. New York: WH Freeman; 2012.
3. Ervens B. Modeling the processing of aerosol and trace gases in clouds and fogs. *Chem Rev* 2015;**115**(10):4157–98. http://dx.doi.org/10.1021/cr5005887.
4. Farmer DK, Cappa CD, Kreidenweis SM. Atmospheric processes and their controlling influence on cloud condensation nuclei activity. *Chem Rev* 2015;**115**(10):4199–217. http://dx.doi.org/10.1021/cr5006292.
5. Finlayson-Pitts BJ, Pitts JNJ. *Chemistry of the upper and lower atmosphere*. San Diego: Academic Press; 2000.
6. Jacobs DJ. *Introduction to atmospheric chemistry*. Princeton (NJ): Princeton University Press; 1999.
7. Lohmann U, Lüönd F, Mahrt F. *An introduction to clouds from the microscale to climate*. 1st ed. Cambridge, UK: Cambridge University Press; 2016.
8. Seinfeld JH, Pandis SN. *Atmospheric chemistry and physics: from air pollution to climate change*. New Jersey: John Wiley & Sons, Inc.; 2006.
9. Shuman NS, Hunton DE, Viggiano AA. Ambient and modified atmospheric ion chemistry: from top to bottom. *Chem Rev* 2015;**115**(10):4542–70. http://dx.doi.org/10.1021/cr5003479.
10. Zhang R, Wang G, Guo S, Zamora ML, Ying Q, Lin Y, Wang W, Hu M, Wang Y. Formation of urban fine particulate matter. *Chem Rev* 2015;**115**(10):3803–55. http://dx.doi.org/10.1021/acs.chemrev.5b00067.

CHAPTER

3.2

Air Pollution and Air Quality

Neil M. Donahue

Carnegie Mellon University, Pittsburgh, PA, United States

3.2.1 INTRODUCTION

Bad air quality refers to any condition that reduces the healthfulness or any other desired property of the air such as visibility and smell. Air pollution, on the other hand, is the presence in air due to human activity of substances that are harmful to humans or other living organisms. Here we shall focus on how human activity influences the air quality and measures that can be taken to reduce air pollution.

Air pollution represents an enormous health hazard. Three of the top 10 sources of global mortality [measured in disability-adjusted life years (DALYs)] are attributable to air pollution, namely, tobacco smoking (including second-hand exposure), household air pollution from solid fuels, and exposure to ambient particulate matter pollution.[1] Of these, ambient pollution is the most ubiquitous, as ambient particulate matter has an atmospheric lifetime of approximately one week and is often transported hundreds of kilometers or more.[2] Smoking is a direct and severe hazard to smokers and those around them, while smoke from solid fuel combustion is a common and serious hazard in the developing world, especially for women and children, where solid fuel combustion dominates food preparation.

Air pollutants can be placed into two groups: primary pollutants and secondary pollutants. Primary pollutants are constituents that are emitted as such, whereas secondary pollutants form via chemical reactions in the atmosphere. In general, primary pollutants are somewhat easier to control than secondary pollutants because controls amount to reducing emissions, either by substituting a different process or activity for the one producing the emissions (via green design) or by adding "end-of-pipe" treatment to the existing process to reduce the emissions. Most pollutants are individual molecules [e.g., carbon monoxide (CO)] or closely linked groups of molecules (e.g., mercury-containing compounds), with the notable exception of atmospheric particulate matter, which consists of an enormous and varied mix of constituents residing in a (nonaqueous) condensed phase. A major classification for particulate matter is its size (equivalent spherical diameter), with fine particulate matter, generally known as $PM_{2.5}$, or particles with diameters less than or equal to 2.5 μm, being of special concern because it is transported over long distances and penetrates deep into the human lung when respired.

Green Chemistry
http://dx.doi.org/10.1016/B978-0-12-809270-5.00007-8

3.2.2 LONG-RANGE TRANSPORT

Air pollution legislation and policy has gone through several cycles of recognizing a simple fact of meteorology—air moves. Secondary pollutants take time to form, often hours to days, and many pollutants persist in the atmosphere until they are removed via wet or dry deposition (although many others are lost to oxidation chemistry). Many secondary pollutants are highly water soluble, as oxidation chemistry tends to produce polar compounds, and so wet deposition via rainfall is a common loss process. Over much of the earth, especially in the midlatitudes, the time between rainfall events is of the order of 1 week, so in very general terms the lifetime of water-soluble pollutants is also roughly 1 week. However, this is approximate and variable, and in any event an e-folding timescale.

Weather systems can travel thousands of kilometers in a week, and in a month they typically circumnavigate the globe. This means that regional transport of pollutants is ubiquitous and intercontinental transport cannot be ignored. Unfortunately, air quality legislation and policy has often been formulated with local emissions and local solutions in mind, and the lesson of long-range transport has been repeatedly learned, forgotten, and relearned.

The most notorious early example of long-range transport was acid rain, or acid deposition. This is in retrospect a form of fine particulate matter pollution, in conditions in which emissions of acid precursors (SO_2 and NO_2, leading to sulfuric acid, H_2SO_4, and nitric acid, $HONO_2$, respectively) overwhelm base molecules capable of neutralizing the acid products (principally ammonia from agricultural emissions). Before the onset of clean air regulations and widespread scrubbing of sulfur from coal-fired power plants (and/or utilization of low-sulfur fuels), and before nitrogen-reduction catalysts ("three-way catalysts") were deployed on spark engine (gasoline) vehicles, SO_x and NO_x emissions greatly exceeded ammonia emissions from agriculture.

Measurements at sites distant from local sources such as Vermont, Norway, and the Black Forest in Germany began to show a sharp drop in pH (increasing acidity) in surface waters and corresponding ecological damage.[3] This was especially true where the surface geology was characterized by silicate bedrock (granites and basalts) as opposed to carbonates, which tend to buffer surface water. It was quickly established that precipitation with low pH (acid rain) was responsible, and the sources of this acidification were sought out.

The sources are self-evident, although at the time they were controversial. The SO_2 and NO_2 emitted principally by combustion sources are readily oxidized in the atmosphere. SO_2 and NO_2 react in the gas phase with hydroxyl radicals to form H_2SO_4 (sulfuric acid) and $HONO_2$ (nitric acid). The gas-phase lifetime of NO_2 is roughly one day,[4] whereas that of SO_2 is several days.[5] However, SO_2 also reacts in cloud droplets with hydrogen peroxide[6] and ozone,[7] and globally, about 80% of the SO_2 oxidation occurs via cloud processing.[8] The bottom line is that both acid precursor gases are converted to their acid oxidation products after a day or two of atmospheric transport. Whether they are produced in the gas phase or directly in the aqueous phase, these acids are also highly water soluble and so remain in rain drops when they precipitate, driving acid rain.

In the United States, early efforts under the Clean Air Act emphasized regional acid deposition, and that was a major reason for implementation of sulfur emissions controls and a secondary reason (along with ozone abatement) for NO_x emissions controls. It is notable that at the time of issuance in 1990 the power industry had no proven technology for sulfur scrubbing from power plant plumes and consequently objected strenuously to the proposed (and promulgated) standards. However, chemical engineers are very good at what they do, and sulfur scrubbing was soon routine.[9] Coal combustion has remained a major source of electricity in the United States until the present day, although price pressure from cheap unconventional natural gas has put far more pressure on the domestic coal industry than sulfur scrubbing ever did.

3.2.3 OZONE

Ground-level ozone, O_3, is a powerful oxidant that acts as a respiratory irritant and can also cause extensive crop damage. Ozone is a secondary pollutant, and unique among air pollutants, it is completely a consequence of atmospheric oxidation chemistry. None of the three oxygen atoms are generally emitted with the primary species that lead to ozone formation. Instead, it is the oxidation of organic compounds catalyzed by nitrogen oxides that is the principal culprit, with the oxygen atoms coming from molecular oxygen (O_2) in air. In addition to being a pollutant, ozone is also a major atmospheric oxidant and a major source of other oxidants including the hydroxyl radical, OH. Consequently, when human activity perturbs ozone levels, it perturbs essentially all of atmospheric chemistry.

3.2.3.1 Ozone Formation

Ozone is produced when organic compounds are oxidized (principally by OH) in the presence of nitrogen oxides. The simplest sequence involves oxidation of carbon monoxide to carbon dioxide as follows:

$$CO + OH \rightarrow HOCO \tag{3.2.1}$$

$$HOCO + O_2 \rightarrow HO_2 + CO_2 \tag{3.2.2}$$

$$HO_2 + NO \rightarrow OH + NO_2 \tag{3.2.3}$$

$$NO_2 + h\nu \rightarrow NO + O \tag{3.2.4}$$

$$O + O_2 \rightarrow O_3 \tag{3.2.5}$$

$$\text{Net}: \; CO + 2O_2 \rightarrow CO_2 + O_3 \tag{3.2.6}$$

In this sequence, both the odd hydrogen radicals ($HO_x = OH + HO_2$) and the odd nitrogen radicals ($NO_x = NO + NO_2$) act as catalysts, but because they are highly reactive, there are many competing reactions that can spoil the chemistry and thus prevent ozone formation.

The symbol hν refers to a photon driving photolysis—in this case of NO_2. Most notably, when NO_2 is excessively high, formation of nitric acid can deplete the HO_x:

$$OH + NO_2 \rightarrow HONO_2 \tag{3.2.7}$$

and when NO is scarce, peroxy radicals (HO_2 in this case) can react with each other to terminate the chemistry:

$$HO_2 + HO_2 \rightarrow HOOH + O_2 \tag{3.2.8}$$

Oxidation of actual hydrocarbons—with C–H bonds—is best understood in terms of methane oxidation, although methane itself is important to hemispheric background ozone production and not local pollution due to its long (6.5 *y*) lifetime.[10] The hydrocarbons add an extra loop of peroxy radical chemistry to the CO example:

$$CH_4 + OH \rightarrow CH_3 + H_2O \tag{3.2.9}$$

$$CH_3 + O_2 \rightarrow CH_3O_2 \tag{3.2.10}$$

$$CH_3O_2 + NO \rightarrow CH_3O + NO_2 \tag{3.2.11}$$

$$CH_3O + O_2 \rightarrow CH_2O + HO_2 \tag{3.2.12}$$

The subsequent HO_2 and NO_2 chemistry follows the CO sequence. Thus oxidation of hydrocarbons most simply yields an aldehyde (formaldehyde in this case) and generates two molecules of ozone, although larger hydrocarbons have interesting nuances that have been extensively reviewed in the literature.[11–13] The aldehydes in turn can be important radical sources and also serve as precursors to key organonitrates—the peroxyacyl nitrates (PANs)—as discussed later. The hydrocarbon sequence can also terminate via hydroperoxide formation

$$RO_2 + HO_2 \rightarrow ROOH + O_2 \tag{3.2.13}$$

Because of the competition in termination processes, with too much OH leading to nitric acid formation and too much HO_2 leading to peroxide formation, there is an optimal (undesirable) balance of organic compounds and NO_x at which ozone production will be most efficient. This is usually described in terms of the mixing ratio of carbon atoms (ppmC, e.g., 1 ppm of propane = 3 ppmC) and the mixing ratio of NO_x, with a ratio of roughly 8:1 leading to the most rapid and efficient ozone formation.

This chemistry is often represented in an "ozone isopleth" diagram, once known as an "EKMA" diagram (Empirical Kinetic Modeling Approach), as depicted in Fig. 3.2.1. Before efficient chemical transport models could simulate ozone production with reasonable accuracy, EKMA diagrams were used by policy makers to assess ozone mitigation strategies.[14] The diagram shows contours of constant ozone, generally in parts per million of ozone as a maximum hourly average given the initial (near dawn) levels of volatile organic

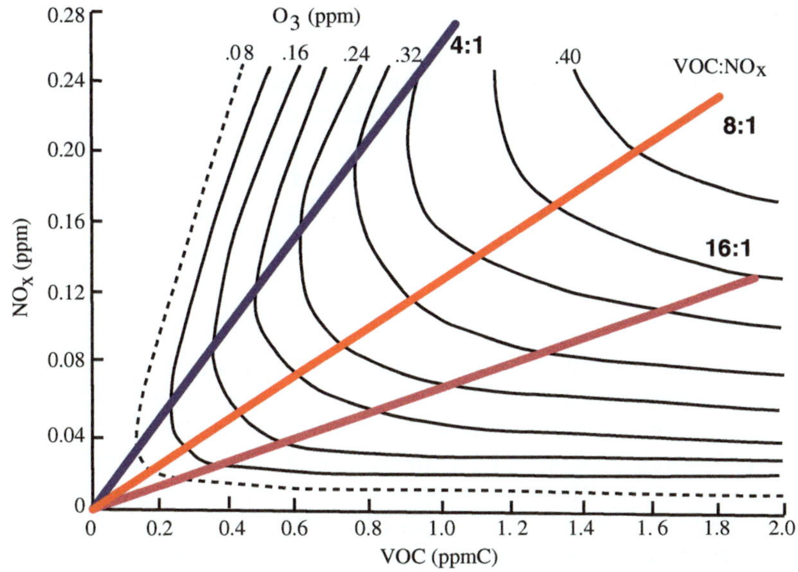

FIGURE 3.2.1 Ozone isopleth diagram. The contours (isopleths) show the expected afternoon ozone mixing ratio in parts per million (ppm) under typical conditions with different levels of volatile organic compounds (VOCs, shown in ppmC or the mixing ratio of the total number of carbon atoms) on the *x*-axis and nitrogen oxides (NO_x, in ppm) on the *y*-axis. A common air quality target of 80 ppb is shown with a dashed contour, so any situation to the upper right of the contour would be out of compliance. When VOC:NO_x is approximately 8:1 [*red line*], conditions are optimally poor. In the "NO_x-limited" regime to the lower right, NO_x reductions are highly effective in reducing ozone, whereas in the "VOC-limited" regime to the upper left, NO_x reductions can actually increase ozone and VOC reductions will reduce ozone.

compounds (VOCs) and NO_x. The highly curved contours of maximum ozone reveal three broad regions. First is a region of high VOC levels and (relatively) low NO_x (VOC:NO_x > 16:1) where maximum ozone is nearly independent of the VOC levels but increases sharply with rising NO_x. This is known as the "NO_x-limited" region where most VOC oxidation results in the formation of peroxides and only a small fraction (proportional to the NO concentration) results in ozone formation. Second is a central region (4:1 < VOC:NO_x < 16:1) known as the "spine" where conditions are optimal for ozone formation and peak ozone reaches a maximum for a given VOC and NO_x level. This is a bad place to be. Third is a region of high NO_x and (relatively) low VOCs (VOC:NO_x < 4:1) where the maximum ozone rises sharply with rising VOC levels but actually decreases with increasing NO_x. This is known as the "VOC-limited" region but could also be described as the "radical-limited" because reaction 2 effectively starves the system of radicals. Because the VOC-limited region is radical starved, there will be little oxidation of VOCs in air masses under these conditions. This in turn means that the air will tend to be exported from NO_x-rich source regimes (typically urban centers or the middle of power plant plumes) and diluted with air that is relatively free of NO_x but often still laden with substantial VOC levels. For this reason, ozone formation in these air masses is still often a problem as they are transported away from source regions.[15]

The general behavior captured in the EKMA diagrams has substantial policy implications. When regions have both high VOCs and high NO_x, they can often be in the spine portion of the EKMA phase space. Under these conditions, simultaneous reductions in VOCs and NO_x will tend to have modest effects because the backbone of the spine has the smallest slope. In extreme cases, such as aggressive NO_x reductions in strongly VOC-limited areas, the control measures can even be (locally) counterproductive. Much early policy focused on VOC reductions, but as it became evident that many regions had background VOCs from biogenic sources, mostly from trees, mitigation strategies were refocused on NO_x reductions.[16]

3.2.3.2 Ozone Control

Broadly speaking, the history of ozone control is a success story with an ambiguous end game. Ozone control measures largely began in the Los Angeles basin in the United States. Ozone levels were extremely high because the basin combined high population, extensive vehicle use, and unfavorable topography with encircling mountains to the leeward of the prevailing southwesterly winds. There is strong evidence that ozone is unhealthy down to the current National Ambient Air Quality Standard (NAAQS) of 70 ppbv (for an 8-h average).[17] During the late 1950s, peak ozone levels routinely exceeded 500 ppbv, with a maximum value of 680 ppbv in 1955.[16] Due to a succession of control measures, most notably tight emissions standards for motor vehicles, ozone levels in the Los Angeles basin now rarely exceed 100 ppbv, although the region remains out of attainment of the US Environmental Protection Agency (EPA) NAAQS.

Ozone is a global issue, with formation often depending on conditions specific to local sources and practices. For example, ozone formation in Mexico City has been associated with high usage of (and leakage from) propane distributed in small reusable tanks,[18] whereas the chemistry of smog in Sao Palo, Brazil, is significantly affected by the large and variable use of ethanol in vehicle fuel.[19] In Houston, extremely rapid ozone formation was associated with large, unintended emissions of ethene and propene from petrochemical facilities.[20]

3.2.3.3 Regional Ozone

Despite success in controlling ozone in large metropolitan areas, several factors have simultaneously contributed to a steady increase in background ozone levels throughout the Northern Hemisphere. At the extreme, air flowing into Europe, downwind of North America, can enter the European basin with ozone levels close to the EU ambient air quality standard. For intercontinental transport, methane also becomes a VOC capable of ozone production, in spite of its very long atmospheric lifetime.[21] These hemispheric increases are associated with increasing NO_x and also the temperature effect, which is largely due to the temperature dependence of PAN decomposition.[22]

Once again, air moves. Starting with long-range transport and acid deposition, this issue emerged again with ozone formation and transport. In the United States, early legislation and policies were developed under the assumption that local emissions largely controlled local ozone levels, and thus controls were based on state implementation plans designed to bring areas out of compliance with air quality standards back into compliance. However, the oxidation chemistry leading to ozone formation can take several days to occur, and air pollution events, especially in the summer, are often associated with stable high-pressure systems characterized by anticyclonic circulation and clear skies (and thus high photolysis and minimal precipitation removal) that can extend over 1000 km and more. Ozone episodes are thus often regional in extent, and local controls may have a minimal effect. For this reason, ozone control policy has begun to focus more on long-range transport of ozone precursors.

The most dramatic example of long-range transport of ozone and ozone precursors is the hemispheric background and associated intercontinental transport. Evidence from Paris near the turn of the last century suggests that background ozone in the preindustrial era may have been as low as 10 ppbv during much of the year.[16] Today, air entering Europe from North America frequently has levels above 30 ppbv and can often exceed the EU ozone standards, making it impossible for even continental-scale control measures to bring air quality into attainment. Long-range transport of air from eastern Asia across the Pacific has similar although less severe effects on air quality in western North America. Altogether, ozone in the Northern Hemisphere is permanently elevated over natural concentrations by a large factor; because ozone is central to atmospheric photochemistry, it is clear that human activity has significantly perturbed the oxidation behavior of Earth's atmosphere, at least in the Northern Hemisphere.

3.2.3.4 Climate Change

Ozone production increases sharply with temperature.[16,22] This is in large part because of a single process: the reversible formation of PANs, such as peroxyacetyl nitrate, $CH_3C(O)O_2NO_2$. Additional factors include strongly temperature-dependent emissions of biogenic VOCs such as isoprene and monoterpenes.[23] In a warming world, this temperature dependence to ozone production means that exceedances are likely to be much more frequent, all else being equal.

PANs form when aldehydes (RCHO) such as acetaldehyde (CH_3CHO) are oxidized via abstraction of the aldehydic hydrogen atom, typically by OH radicals. In the presence of molecular oxygen this immediately leads to the formation of a peroxyacyl radical. That radical can, in turn, react with NO_2 to generate a PAN:

$$RCHO + OH \xrightarrow{O_2} RC(O)O_2 \tag{3.2.14}$$

$$RC(O)O_2 + NO_2 \rightleftarrows RC(O)O_2NO_2 \tag{3.2.15}$$

The PANs differ from organonitrates ($RONO_2$) formed via isomerization from the $RO_2 + NO$ reaction[11,24] as well as peroxynitrates formed from the $RO_2 + NO_2$ reaction involving peroxy radicals other than peroxyacyl radicals. The $RONO_2$ species are thermally stable under atmospheric conditions, whereas RO_2NO_2 molecules are very unstable and decompose in seconds. PANs, on the other hand, have strongly temperature-dependent lifetimes, with a value near 1 h at 20 °C[25] that drops to minutes when it is warmer and increases dramatically when it is colder. Because of this, PAN formation is an effective sink of NO_2 under cool conditions but not an effective sink under warm conditions; furthermore, PAN formation and transport serves as an efficient NO_x transport mechanism.[26]

3.2.4 FINE PARTICULATE MATTER

Fine particulate matter has an outsized role in air pollution and air quality. It appears to be uniquely potent in causing death at very low concentrations;[27,28] it also plays a major role in climate, especially in the uncertainty of the total climate forcing since the Industrial Revolution.[29] Specifically, three of the top 10 causes of premature death or DALYs globally are forms of fine particles: ambient fine particles, fine particles in smoke from cooking fires, and fine particles in tobacco smoke.[1] Of these, ambient fine particles are ubiquitous and disproportionately affect people in developing countries, especially in rapidly growing urban areas.[30] The haze covering large parts of the developing world is also hard to see through and hard for the sun to penetrate. Consequently, fine particles mask up to half of the total greenhouse forcing associated with infrared-absorbing greenhouse gases by scattering sunlight back into space.[31] The extent of this masking is, however, highly uncertain because it depends on the change in fine particle levels between the present day and the preindustrial time and we have a poor understanding of pre-industrial fine-particle levels. Because of the severe health consequences associated with the fine particles and because a large portion of them are associated with fossil fuel combustion, it is likely that decarbonization associated with climate change mitigation will also lead to air quality improvements that might unmask some or all of this hidden climate forcing. Therefore, improvements to air quality serve in part as a much more localized incentive to take action to mitigate climate change, yet on the other hand the air quality improvements may partially exacerbate climate forcing by unmasking existing, latent forcing.

3.2.4.1 Fine Particulate Matter Definitions

Unlike most pollutants, which are defined by a chemical composition (most often they are a single chemical compound and even sometimes a specific enantiomer), fine particulate matter is defined by a state (any condensed phase) and a size (particles smaller than 2.5 μm diameter, $PM_{2.5}$). The motivation for using $PM_{2.5}$ is that fine particles are uniquely capable of penetrating deep into the human lung, where they initiate a cascade of responses resulting in the observed health effects, and also that fine particles are relatively mobile in the environment, with wet deposition (precipitation) being the major loss mechanism and residence times thus extending to more than 1 week[2] as opposed to coarse particles, which, for the most part, settle due to gravitation relatively quickly. $PM_{2.5}$ is also practically measurable, with straightforward filter weighing after sampling in a flow-controlled filter housing with a 2.5-μm cyclone removing any particle larger than 2.5 μm. Because of this, a relatively large data set of $PM_{2.5}$ mass concentrations existed for epidemiological studies during the 1990s, and still the most consistent associations for adverse health effects are with the $PM_{2.5}$ mass concentration.

A case could be made for 1.0 μm (PM_1) being a more logical dividing line from the perspective of particle microphysics and even physiology. This is closer to the dividing line where wet deposition dominates for smaller particles and dry deposition (gravitational settling) dominates for larger ones,[32] and the fine particle modes are for the most part submicron and share a common set of sources, whereas the sources for supermicron particles have more in common with the coarse modes between 2.5 and 10 μm. Furthermore, most regulatory standards apply at a fixed relative humidity (RH) because absorbed liquid water can constitute a large fraction of the total $PM_{2.5}$ mass concentration. The US EPA standard operating procedure specifies that filters must be preweighed and used within 30 days following detailed handling procedures, along with a series of handling and field blanks, and then weighed again, with both pre- and postsampling weighing carried out at 20–23 °C and at 30%–40% RH.[33]

$PM_{2.5}$ consists of a suspension of particles ranging from a few particles of a few nanometers diameter to more than 10 billion molecules 2.5 μm in diameter. Under most circumstances most of this mass is nonrefractory, meaning that it evaporates from the particles after heating. Consequently, instruments such as the aerosol mass spectrometer, which measures nonrefractory PM_1 when particles strike a 600 °C heated tungsten element and are subsequently ionized by 70 eV electrons,[34] generally measure 80%–90% of the total PM_1 signal. The major refractory component in PM_1 is elemental carbon, which in some circumstances near flaming, fuel-rich combustion can comprise a large fraction of the PM mass but under typical ambient conditions is 5%–10% of the total mass.[32]

The environmental effects of fine particles arise from multiple properties, including the composition, total mass, and total number. Health effects are in general associated with total mass, although it is clear that different constituents (i.e., composition) have different dose responses. Some materials, including trace metals such as nickel, comprise only a small fraction of the fine particle mass yet may have a disproportionate health effect[35]; however, although it is clear that health effects do depend on the particle composition, the overall role of composition on health effects remains a topic of current research. The climate effects of fine particles are caused both by direct scattering, which scales with the particle surface area, and the total number of cloud condensation nuclei (CCN), which are particles larger than 50–100 nm that have a sufficient number of soluble molecules to enable cloud droplet formation. Because the health effects depend on the total mass and the climate effects depend on the mass and size, the environmental effects of fine particles are largely driven by particles larger than 50–100 nm in diameter, although the role of nanoparticles is of great interest for a number of reasons.

Because particles are aggregates there are many different classification schemes beyond mere size. Furthermore, we care about many properties of particles, beginning with particle number and particle size. These are defined by distributions such as the one shown in Fig. 3.1.9, but in addition, the number distribution is characterized by the total number of particles *above* a certain size cutoff, whereas the total mass of particles is defined by the total mass on particles *below* a certain cutoff. For example, N_{50} refers to the total number of particles larger than 50 nm in diameter, whereas $PM_{2.5}$ refers to the total mass of particles smaller than 2.5 μm in diameter. Because we care (at least) about number and mass, care must be taken to distinguish between terminology referring to each. For example, "secondary particles" are particles formed via new particle formation (nucleation) in the atmosphere, directly from gas-phase precursors, whereas "primary particles" are particles that are emitted directly into the atmosphere from sources such as sea spray, wind-blown dust, pollen, and combustion (in the case of combustion, particles that nucleate within the cooling plume of a fire, stack, or tailpipe are generally considered to be primary particles, as the condensible material was not formed via gas-phase atmospheric chemistry). On the other hand, "secondary particle mass" refers to any condensed-phase mass that was formed as a consequence of atmospheric chemistry driving gas-phase precursors to the condensed phase. This can include gas-phase formation of sulfuric acid, nitric acid, and oxidized organics, or condensed-phase formation of these and other compounds.

Most of the mass comprising fine particles is secondary, including secondary organics,[36,37] secondary sulfate,[32] and secondary nitrate.[32] However, roughly half of the CCN are secondary particles, meaning that they formed via atmospheric nucleation,[38–40] whereas the other half are primary particles arising from combustion as well as sea spray and other primary sources.[41] Even for the primary particles, the large majority of CCN reach CCN size via condensational growth.[41] Over much of the earth, most accumulation-mode particles arise from growth of nanoparticles; because nanoparticles are also vulnerable to loss via collisional coalescence with larger particles, nanoparticles must "grow or die," and their survival probability is a key property to gauge their environmental effects.[42] Nucleated particles are the most vulnerable, because they have the farthest to grow and also have the highest Brownian diffusivity, but most primary nanoparticles also are lost via coagulation.[41]

Other definitions of note are "refractory" mass and particles, which refers to material (or the totality of particles) that survive above some high temperature (often several hundred degrees Celsius), whereas "nonrefractory" material is material that evaporates at or below this temperature. Inorganic mass refers to inorganic salts such as chlorides, nitrates, and sulfates as well as elemental carbon and nitrates, although there is some ambiguity regarding such compounds as covalently bound organosulfates and salts involving amines. Organics are sometimes constrained by their total mass concentration, in which case they are referred to as "organic aerosol" (OA) or "organic mass (OM), or sometimes by their carbon mass alone, in which case they are referred to as "organic carbon" (OC). The conversion factor is the organic-mass to organic-carbon ratio (OM:OC), which is thought to range between 1.7 and 2.2 in the atmosphere, with higher values associated with secondary OA.

3.2.4.2 Organic Aerosol

OA can comprise the large fraction of fine particle mass,[43] and it is generally split into "primary" organic aerosol (POA) and "secondary" organic aerosol (SOA). OA constitutes an extremely rich and complex mixture of thousands of different specific compounds, most of which have never been synthesized or isolated.[44,45] Because of this richness, representations of OA behavior almost universally involve some degree of simplification, either by representing the OA mixture with a suite of model compounds,[46] or pseudo compounds with specified properties (volatility and concentration at least),[47] or by describing a grid or phase space of properties across which OA constituents are distributed.[45,48,49] At least one model has been developed that attempts to completely enumerate the organic compounds involved in atmospheric chemistry, including those associated with OA.[50]

The extent to which OA concentrations have changed over the course of human civilization and especially since the industrial revolution is unclear.[36] On the one hand, most (80%–90%) of the emissions of organic material into the atmosphere globally arise from the biosphere,[44,51] although the majority of trees emitting organics are controlled through forestry. The "blue haze" of forested mountains such as the Great Smoky Mountains almost certainly predates human activity.[52] On the other hand, combustion, first of wood to clear land and to extract energy, has been a defining characteristic of humans since prehistory,[53] and the highest concentrations of OA today, by far, are found in cities[43,54] or downwind of fires associated with agriculture[55] or forest fires.[56] Consequently, OA is for the most part a pollutant caused by human activity.

OA concentrations in the atmosphere range from 1 μg/m^3 or even less in the remote troposphere[57] to well over 100 μg/m^3 in polluted cities.[43] However, 1 m^3 of air at STP (the concentrations are generally corrected to standard conditions) weighs roughly 1 kg, so 1 μg/m^3 is a mass fraction of 1 ppb. Furthermore, the molecules comprising OA tend to have molar weights roughly 10 times higher that of air, so the OA mole fraction is as low as 100 parts per trillion (pptv), or 10^{-10}. When there are multiple phases present in a system (e.g. OA suspended in air, meaning at least one condensed phase and the gas phase) there will always be some of a given substance in each phase. At equilibrium, the activity of any substance will be the same in each phase. In practice, this means that to be in OA, an organic compound must be in a form with a saturation vapor pressure $\lesssim 10^{-10}$ atm (hPa) or somewhat higher under polluted conditions. The compound must either have such a low saturation vapor pressure in its pure form, or else be in a form (such as a salt for volatile acids like oxalic acid or volatile bases like dimethyl amine) with a low saturation vapor pressure. The required vapor pressure is surprisingly low, as illustrated in the two-dimensional volatility basis set in Fig. 3.2.2, one of the spaces currently used to represent OA behavior. Compounds such as

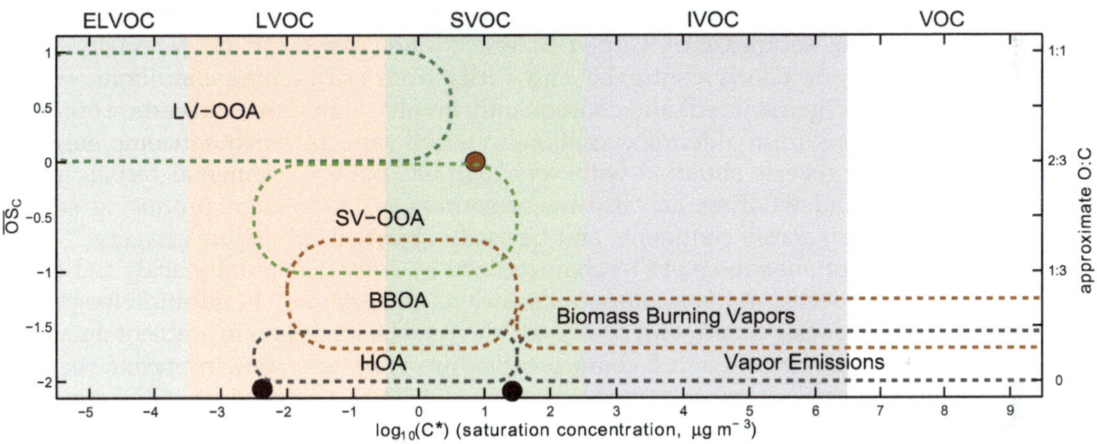

FIGURE 3.2.2 The two-dimensional volatility basis set used to describe organic aerosol behavior and evolution. Volatility (vapor pressure on a mass concentration basis) is expressed in logarithmic units on the *x*-axis, in μg/m^3. The degree of oxidation or oxidation state of carbon ($\overline{OS}_C$ = 2O:C−H:C) forms a linear *y*-axis. In addition, organics are grouped in broad volatility classes, including "extremely low-volatility organic compounds" (ELVOCs), "low-volatility organic compounds" (LVOCs), "semivolatile organic compounds" (SVOCs), "intermediate-volatility organic compounds (IVOCs), and finally "volatile organic compounds" (VOCs). Organics tend to be emitted as relatively reduced compounds, with $\overline{OS}_C \simeq -2$, shown with the regions marked "vapors" as well as "hydrocarbon-like organic aerosol" (HOA) and "biomass-burning organic aerosol" (BBOA). Oxidation tends to functionalize material and drive it upward and to the left, forming classes of "oxidized organic aerosol" (OOA) including "semivolatile" OOA (SV-OOA) and "low-volatility" OOA (LV-OOA). Extensive oxidation causes fragmentation and drives material toward the upper right. Typical SVOCs include tricosane ($C_{23}H_{48}$), shown by a *black circle* with logC* ≃ 1.5, and levoglucosan ($C_6H_{10}O_5$), a pyrolysis product of cellulose, shown by a *brown circle* with logC* ≃ 1. A typical primary LVOC is squalane ($C_{30}H_{62}$), shown by a *black circle* with logC* ≃ −2.5.

docosane ($C_{22}H_{46}$), oleic acid ($C_{18}H_{34}O_2$), or levoglucosan ($C_6H_{10}O_5$) are functionally "semivolatile" with saturation vapor pressures of the order 10^{-10} hPa, corresponding to saturation concentrations of 10 μg/m^{-3}, with significant fractions in the gas phase under ambient conditions[58–60] and thus important gas-phase oxidation chemistry.[61]

3.2.4.2.1 Primary Organic Aerosol

POA arises mostly from combustion, where high temperatures vaporize organic constituents in fuels or lubricants and incomplete combustion results in emission of combustion products other than CO_2 and H_2O. There are other sources of OA, including most notably pollen and other biogenic OAs, generally in the supermicron mode,[62] as well as "leaf detritus" associated with mechanical wearing (or possibly evaporation) of waxy leaf surfaces.

Smoke from wood burning is the most effusive source of POA in terms of emission per kilogram of fuel burned.[63] Broadly, wood burning tends to produce smoke containing pyrolysis products from cellulose and lignin, resulting in derivatives of sugars and aromatic compounds, respectively. Levoglucosan is an especially common pyrolysis product as it arises from the cleavage of a glucose unit from a cellulose polymer. Levoglucosan and associated sugars are commonly used as molecular markers of biomass burning.[64,65] Altogether, wood burning produces mass spectra with a characteristic appearance in the aerosol mass spectrometer, leading to a factor known as "biomass-burning organic aerosol" shown in Fig. 3.2.2.[63,66] Open burning includes burning for agriculture, which is widespread across the globe and contributes significantly to poor air quality in regions as diverse as Africa, the Amazon, eastern China, India, and the southeastern United States.[55,67] Emissions can be especially large because open burning is poorly controlled and varies from hot flaming conditions, which tend to be fuel rich (oxygen starved) and consequently involve high elemental carbon but relatively low OC emissions, to smoldering conditions in which vaporization but incomplete combustion results in the reverse situation with very high OC but low elemental carbon in the smoke. Forest fires and wildfires are also major sources of POA, often producing smoke plumes that can stretch across continents and be easily observed in satellite imagery.[68]

Cooking is an important source of OA characterized by emissions of fatty acids and other vegetable and meat-specific marker compounds such as cholesterol.[69] In addition to specific molecular markers, OA associated with cooking (COA) can be identified in ambient data by a modest average oxygen content and a characteristic diurnal pattern showing peaks near the lunch and dinner hours.[70] Because fatty acids tend to exist as triglycerides, COA tends to have a very low volatility; for example, olive oil consists mostly of glyceryl trioleate.

Emissions from controlled internal combustion, notably from vehicles, can also be a substantial source of POA. Vehicle emissions arise from three main sources: volatilized lubricating oil, incompletely burned fuel, and vaporization of fuel or other organics. These emissions tend to be rich in hydrocarbons.[71,72] Tailpipe emissions (e.g., the first two sources) typically dominate. Without emissions controls, engines can be substantial sources of organics, with organic emissions as high as 10,000 mg/kg-fuel (1%) and POA emissions as high as 100 mg/kg-fuel.[73] However, catalytic converters and particle filters have sharply reduced emission rates, and modern vehicles (both gasoline spark engines and diesel engines) have emissions rates as much as 100 times lower than for uncontrolled engines.[73,74] Electric vehicles are considered "zero emission vehicles," although the pollution associated with producing the electricity for the vehicle must be considered in a life cycle assessment of the pollution.

Gasoline consists of hydrocarbons with carbon numbers predominantly ranging from 5 to 12, while diesel fuel consists of hydrocarbons predominantly ranging from 12 to 20. The saturation concentration of even eicosane ($C_{20}H_{42}$) is greater than 100 μg/m^3, and so in the atmosphere, vehicle fuels themselves will be almost entirely in the gas phase. However, the larger carbon numbers in diesel fuel compared with gasoline correspond to a lower overall volatility, and especially the highest molecular weight fraction of diesel fuel is prone to solidification at low temperature in the fuel itself, which can cause difficulty in diesel engines when it is extremely cold. Most of the hydrocarbons in lubricating oil have carbon numbers between 22 and 32, and so when oil is emitted and diluted to ambient conditions, a fraction will stay in the condensed phase and a fraction will evaporate[75,76]; even with lubricating oil, up to 80% of the mass will evaporate from droplets under ambient conditions.[77] However, a fraction of any vaporized motor oil, and possibly pyrolysis products that did not aggregate to form refractory elemental carbon, constitute POA emissions from vehicles.

3.2.4.2.2 *Secondary Organic Aerosol*

SOA forms when chemistry transforms a gas-phase organic precursor into a condensible (or condensed) organic product. SOA formation can occur in the gas phase, for example, via oxidation of monoterpenes,[51,78] or it can occur in the condensed phase, often via aqueous oxidation involving hydrogen peroxide.[79,80] Condensed-phase association reactions (colloquially known as "oligomerization") also play a still uncertain role in the overall production of SOA mass.[51,81,82] In general, organic oxidation reactions in the atmosphere are complex, producing many products that vary with atmospheric conditions such as the VOC to NO_x ratio shown in Fig. 3.2.1.[11] Because SOA is produced by oxidation of a precursor, it is conventional to consider a mass yield of the SOA given after oxidation of a certain mass of that precursor: $y_{SOA} = \Delta C_{SOA}/\Delta C_{prec}$.

Global emissions of VOCs (in the far-right white-shaded region of Fig. 3.2.2) far exceed emissions of less volatile compounds[44,51,66] and so for many years most of the research into the formation of SOA focused on oxidation products of VOCs.[47,54,83,84] Although early studies considered the SOA mass yield as a more or less constant (typically small) value,[85] it became clear that a substantial portion of the SOA products were semivolatile organics (SVOCs) and that the SOA should be described in terms of semivolatile partitioning thermodynamics.[47,86]

While early studies focused on gas-phase oxidation chemistry as a major source of SOA, product molecules with sufficiently low volatility to condense to particles need at a minimum five or six carbon atoms unless they form ionic bonds (salts) as is the case with oxalic acid (oxalate) and dimethyl amine (dimethyl amminium). However, two pathways for new SOA formation chemistry involving association reactions emerged from recent studies.

First, even relatively large carbon number precursors such as α-pinene and trimethyl benzene can lead to "oligomers"[81,82] via reactions either in the condensed phase[82] or the gas phase.[87,88] The exact mechanisms and extent of these oligomerization reactions are topics of active research.[89]

Second, aqueous reactions involving the oxidation of abundant small organics, most notably glyoxal and oxidation products of isoprene, can also lead to larger carbon number association products.[79,90] The aqueous chemistry also can include important contributions from inorganics such as ammonia (e.g., in "browning" reactions[91]) or formation of organosulfates.[92–94] Again, the overall extent of these processes remains a topic of active research.[89,90]

Because SOA formation involves the oxidation of organic compounds and depends on the molecular identity (and volatility) of the products, SOA formation depends strongly on NO_x levels.[95–97] As with ozone formation in Fig. 3.2.1, RO_2 chemistry is a critical branch point for SOA production.[98] Recently, strong evidence has emerged that highly oxidized RO_2 radicals can also undergo unimolecular "autooxidation"[99] to generate highly functionalized, extremely low volatility products with many hydroperoxide functional groups.[100,101]

In addition to SOA formation from relatively volatile precursors, intermediate volatility organic compound (IVOC) and SVOC species emitted into the atmosphere can oxidize rapidly to form SOA.[61,102] In general, emissions are a form of volatilization, and so natural emissions from biogenic sources occur largely near ambient temperatures and are dominated by relatively volatile species, whereas emissions associated with combustion can include a higher fraction of less volatile species.[66,103] These IVOCs and SVOCs are often associated with POA sources such as wood burning, cooking, and vehicles.[77] Because these precursors have a much higher carbon number than VOCs, they can be very efficient sources of SOA[102] and are likely to be major contributors to both the ubiquity of SOA over POA[104] as well as the strong association between SOA and human activity.[105] IVOC emissions may play a major role in high OA and SOA concentrations observed during extreme haze events in eastern China.[106]

Because a substantial fraction of both POA and SOA consists of semivolatile species, rapid oxidation in the gas phase, known as "aging," is an important part of the overall OA life cycle in the atmosphere.[107] In some cases aging can dramatically increase OA concentrations, when air contains a large fraction of SOA, whereas in others aging can increase the oxidation state without significantly changing the total OA mass.[108] Ultimately, oxidation will in theory completely remove OA because the thermodynamic end product of oxidation is CO_2,[45,66] but there is little evidence that oxidative loss is a significant term on the atmospheric residence timescale for fine particles of roughly 1 week.[2]

3.2.4.3 New Particle Formation

In addition to secondary aerosol mass formation, production of secondary particles—nucleation—may play a major role in humanity's effect on the earth. Nucleation dominates the total number of fine particles, and even though nucleated particles must grow to 50–100 nm in diameter to significantly affect climate,[42] nucleation still contributes roughly half of the total number of CCN.[38,109] The total number of CCN in turn controls the number of droplets in clouds and thus their reflectivity and lifetimes.[110] Uncertainty in the *change* in CCN production between the preindustrial and the present day is one of the largest sources of uncertainty in overall climate forcing.[29]

Human activity influences new particle formation in myriad ways. In the present atmosphere it appears that sulfuric acid is almost always involved in new particle formation,[111,112] and the large majority of gas-phase sulfuric acid (at least over the continents) arises from SO_2 emissions associated with coal combustion.[32] Furthermore, the precursors and oxidation mechanisms of organics involved in new particle formation are strongly affected by human activity, just as with SOA formation. This in turn influences the growth rate, but anthropogenic aerosol mass also dominates the gas and particle condensation sink in most cases.[41] The competition between growth and coagulational loss in turn controls the survival probability of nucleated particles.[42,113]

Because the aerosol effect on clouds is highly nonlinear, with a very strong effect at low concentrations that saturates with increasing CCN concentrations, climate simulations are especially sensitive to the preindustrial CCN levels.[29] Recent experiments have shown that oxidation of biogenic precursors such as monoterpenes can lead to new particle formation in the absence of sulfuric acid[88] and that the subsequent growth of those particles can also be rapid under conditions typical of the background atmosphere.[114] This in turn suggests that the change in CCN (and thus the cooling associated with new particle formation) may be somewhat less than otherwise thought,[88] although the climate simulations associated with the Intergovernmental Panel on Climate Change (IPCC) climate estimates have traditionally assumed some baseline source of CCN in the preindustrial atmosphere.[31] What is clear is that human activity has profoundly influenced the overall atmospheric new particle formation rate and thus the overall budget of CCN.

3.2.4.4 Light-Absorbing Carbon

One important class of carbonaceous aerosols is light-absorbing carbon.[115] Most aerosol materials do not absorb visible light and so interact with light only because the index of refraction of the particles is different from the surrounding air. Especially if the particle diameter is similar to that of visible light (400–600 nm), the particles will scatter light via Mie scattering. When particles are in this size range, the scattering cross-section (which determines the probability that a given particle will scatter a given photon of light) increases dramatically with particle diameter, approximately with the sixth power. This is why it is important to know the extent to which particles absorb water as RH increases—their hygroscopic growth; as particles swell with rising RH, their cross-sections can rise dramatically, and thus the overall aerosol optical depth (which determines the probability that a given photon of light will be scattered while traveling through a given distance of air) will rise as well. Consequently, dry air masses with high (but small) particle number counts often retain relatively high visibility, but if humidity rises they can become extremely hazy.

Some particles do absorb light in addition to scattering it. The ratio of the absorption cross-section to the overall interaction cross-section (extinction cross-section) of a particle is known as the "single-scattering albedo." When this falls below 1.0, instead of visible light scattering off of haze and partly back into space, causing an overall cooling, light will be absorbed in the atmosphere, causing an overall warming. This dramatically changes the climate effect of the particles.

Molecules interact with light when the energy of the photon (which is directly proportional to the frequency, ν, and inversely proportional to the wavelength, λ, $E = h\nu = hc/\lambda$) matches the differences in energy between two quantum states of the molecule or ensemble of molecules. For visible light this corresponds to an electronic transition, where the electron distribution in the molecule is altered by the added energy. Transitions with a high degree of "delocalization" (a long wavelength) have a relatively low electronic energy difference and absorb in the visible range of the spectrum.

3.2.4.4.1 *Black Carbon*

For organic systems (with carbon-carbon bonds) this delocalization is accomplished by conjugation, a succession of carbon-carbon and carbon-oxygen double bonds (C=C–C=C... or O=C–C=C...) including cyclic (aromatic) structures. Other chemical moieties also absorb light; broadly they are known as "chromophores." When organic compounds with a high carbon number are heated in the absence of oxygen, these conjugated structures tend to form. Because the ensemble of molecules have different degrees of conjugation (different chromophores), together they absorb at all wavelengths, making the ensemble black. The molecules also tend to be large and resemble graphite, so these particles are known as "refractory" or "graphitic" carbon as well as "black" carbon.[116] In practice different measurement methods are sensitive to each of these properties and so represent different but substantially overlapping subsets of the same material.

Because black carbon causes greenhouse warming, is associated with aerosol health effects (mortality), and also has a very short atmospheric lifetime, it has received considerable attention as a short-lived climate forcer with potential large "cobenefit" in that reductions would simultaneously reduce climate forcing and pollution mortality.[115] However, particles containing black carbon also almost always contain water-soluble material, and thus many will serve as CCN. Sources of black carbon are thus sources of particle number and so cause a *cooling* effect on the climate due to the aerosol indirect effect; clouds formed from air masses with relatively more particles will have finer cloud droplets and thus be whiter, reflecting more light to space. When all these effects are considered together, it is not clear that reductions in sources will reduce warming, although they certainly will reduce fine particle pollution.[39]

3.2.4.4.2 *Brown Carbon*

Organic compounds that are not so refractory or heavily conjugated to be essentially nonvolatile and black can still absorb light, and they tend to absorb more light at shorter wavelengths (in the blue) and less light at longer wavelengths (in the red). The light that passes through will thus be reddish orange, or brown, and this is known as brown carbon. Unlike black carbon, brown carbon can be semivolatile and it can either consist of primary material, often from the same combustion sources producing black carbon,[117–119] or it can be produced via atmospheric chemistry as secondary brown carbon.[91,120,121] Some of these secondary reactions involve bases such as ammonia and constitute the classic "browning reactions" from organic chemistry that also cause fruits to brown when exposed to air.

Like any chemically active atmospheric constituent, brown carbon has a more complex life cycle than its relatively nonreactive black carbon cousin, and it is less well understood. The global budget for black carbon is uncertain and controversial, however, and it is possible that regional differences in brown carbon sources may partly explain differences between bottom-up and top-down emissions estimates for total light-absorbing carbon.[116,119]

3.2.4.5 Exposure

Exposure to ambient fine particulate matter is a worldwide public health risk. It is especially severe in urban areas of the developing world but remains a serious issue through much of the developed world as well. Estimates based on the 2010 Global Burden of Disease report constrained by satellite observations and self-consistent emissions inventories revealed extremely high exposure in much of the developing world, with especially high exposure in China and India, as shown in Fig. 3.2.3A.[30] This shows that almost 2.5 billion people are exposed to more than 25 μg/m^3 on an average each year, compared with the US NAAQS target of 12 μg/m^3 or the World Health Organization target of 10 μg/m^3. Overall, exposure to fine particles is associated with three of the top 10 sources of mortality globally (the air pollution shown here, exposure to smoke from open cooking fires, and exposure to cigarette smoke).[1]

Exposure in the developed world is by and large lower than in the developing world, although a large fraction of people are exposed to levels between 10 and 20 μg/m^3. Furthermore, exposure in the developed world has for the most part declined over the past decades, whereas exposure in the developing world is increasing.[30] The consequences of this exposure are severe. Fig. 3.2.3B shows the estimated mortality from the exposure estimates shown in Fig. 3.2.3A, revealing 3.24 million attributable deaths per year from $PM_{2.5}$ exposure. This results in significant reductions in life expectancy, with 72.4 million years of life lost, and a life expectancy reduction of 1.4 years on an average. Air pollution is thus neither a first- or a third-world problem but a global problem.

Although the specifics of $PM_{2.5}$ dose responses and especially composition response functions remain uncertain, broadly the risk of mortality is thought to follow a logarithmic (concave) form, rising sharply from low values near 5 μg/m^3 to roughly $10^{-3}\,y^{-1}$ at 50 μg/m^3 and further to $1.5 \times 10^{-3}\,y^{-1}$ at 150 μg/m^{3}[122] The predominant causes of death are heart attacks and strokes, both associated with systemic inflammation driven by $PM_{2.5}$ deposition in the lungs.[123] With a 70-year life expectancy, exposure to 50 μg/m^3 equates to a 7% increase in the risk of death. There is no evidence for or against a low concentration threshold because there are very few if any exposure or epidemiological studies at such low levels.[122]

The exposure and mortality estimates in Fig. 3.2.3 are almost certainly lower limits. A more recent analysis in support of the 2013 Global Burden of Disease found considerably higher exposure in India, with the mode for northern India coincident with the high values in China.[124] This is supported by in situ measurements showing that air quality in Beijing and New Delhi are comparably poor. This largely reflects improved measurements and not dramatically worsening air quality, although air quality is getting worse throughout much of the developing world.[124]

3.2.4.6 Control Efficacy

Control of fine particulate matter pollution constitutes some of the most effective public health measures in history, and air quality improvements throughout the world remain a high-priority objective. Historical data in regions where air quality policy has led to $PM_{2.5}$ reductions reveal substantial public health benefits consistent with the harm illustrated in Fig. 3.2.3. Fig. 3.2.4, based on data presented by Pope et al.,[125] shows the relationship between $PM_{2.5}$ concentrations in 51 municipal areas across the United States near 1980 and again near 2000. The data reveal both strong associations between $PM_{2.5}$ and life expectancy and a strong relationship between improved life expectancy and $PM_{2.5}$ reductions; approximately 15% of the total increase in life expectancy shown in Fig. 3.2.4 is attributable to the $PM_{2.5}$ reductions.[125]

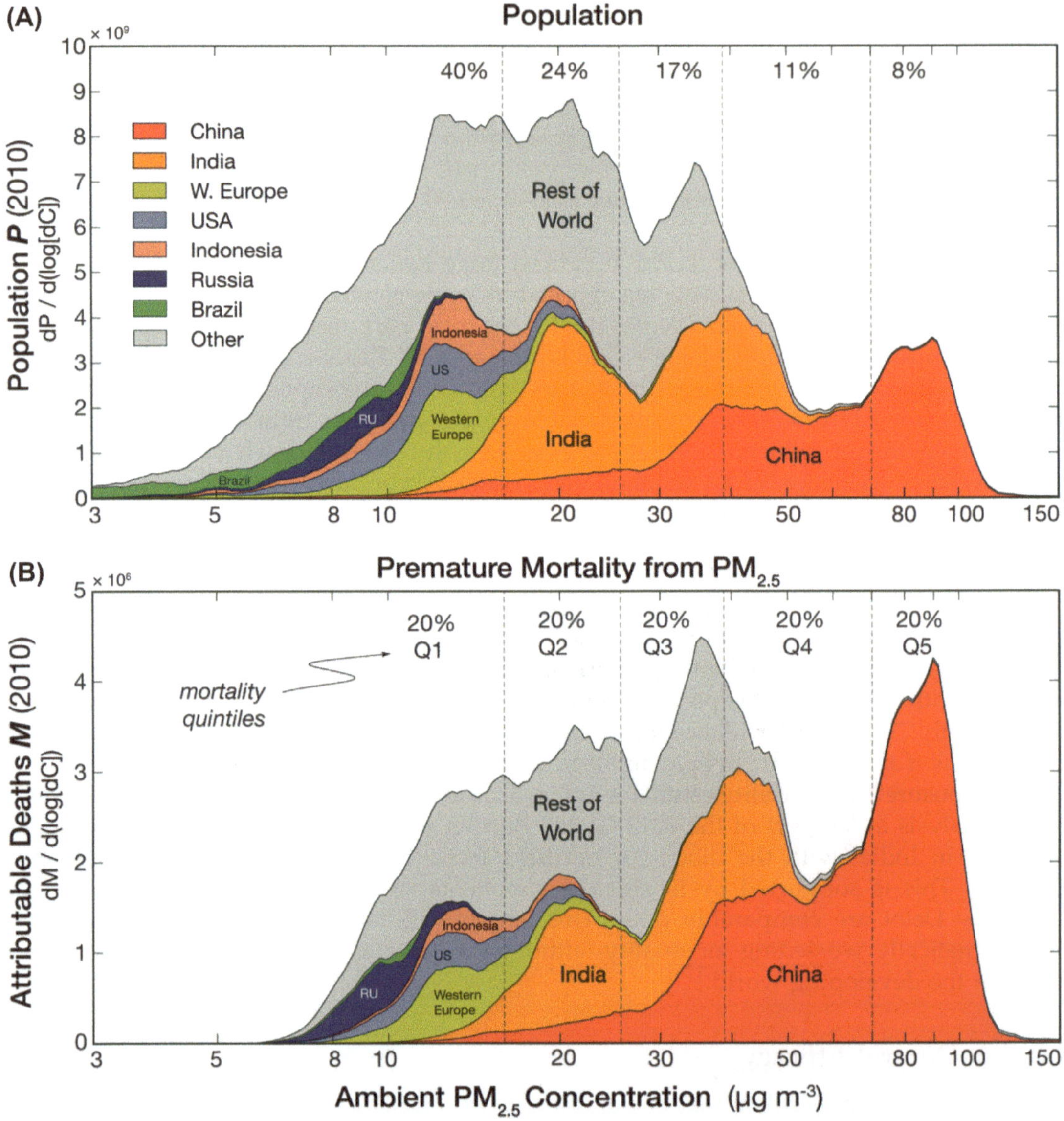

FIGURE 3.2.3 (A) Estimated worldwide exposure to 2.5-μm fine particle air pollution in 2010. The *x*-axis shows the $PM_{2.5}$ mass concentration on a log scale, whereas the *y*-axis shows the distribution of population exposed to those $PM_{2.5}$ levels. The integral gives total population—the 2010 population was approximately 7 billion, and the *y*-value over 1 $\log_{10}$ unit (10–100 $\mu g/m^3$) is $\sim 7 \times 10^9$. Colored patches form a stack showing population exposure for the seven most populous countries (or regions in the case of western Europe), with the rest of the world shown in *gray*. (B) Excess mortality (attributable deaths) derived from the population exposure data, with a total of 3.24 million annual deaths. The *dashed vertical lines* on both panels are quintiles (20%) of the mortality distribution, and the percentages in (A) are the corresponding percentage of total world population in each mortality quintile. Premature mortality adds to 72.4 million years of life lost, with life expectancy reduced by 1.4 years on average. *(A) From Apte JS, Marshall JD, Cohen AJ, Brauer M. Addressing global mortality from ambient $PM_{2.5}$.* Environ Sci Technol *2015;**49**(13):8057–66. PMID: 26077815. Available from: http://dx.doi.org/10.1021/acs.est.5b01236.*

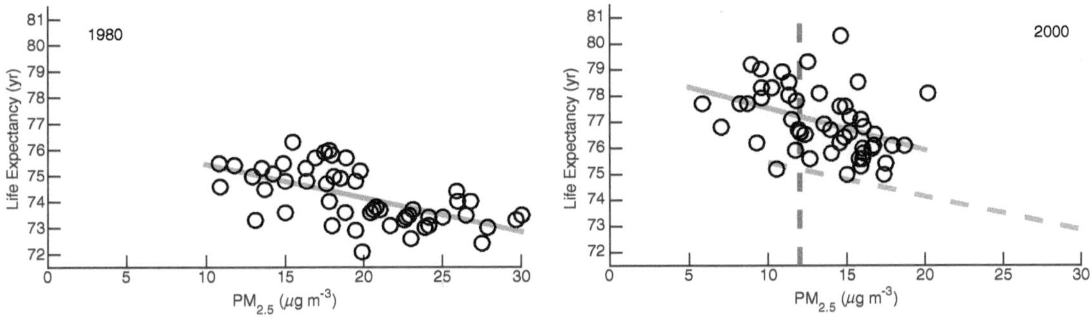

FIGURE 3.2.4 Life expectancy vs. $PM_{2.5}$ concentrations in 51 municipal areas across the United States near 1980 and 2000, based on Pope et al.[125] Concentrations near 1980 ranged between 10 and 30 $\mu g/m^3$ with a median life expectancy of roughly 74 years and a clear association between life expectancy and $PM_{2.5}$ (shown as a *straight-line regression*). By 2000, improved air quality revealed concentrations ranging between 5 and 20 $\mu g/m^3$ and a median life expectancy of roughly 77 years. The relationship from 1980 is the *dashed line* under the 2000 data, and that from the US annual NAAQS of 12 $\mu g/m^2$ is a *dashed vertical line.*

In simple, stark terms, the US EPA estimates the statistical value of a human life as \$9.1 million, based on an assessment of individuals' willingness to pay for interventions that would reduce their risk of death. The data above suggest that air quality improvements between 1980 and 2000 prevented roughly 26,000 deaths per year, with a public health value of more than \$200 billion annually. Although assigning a monetary value to avoided death can seem troubling, the figures illustrate the cost of interventions that are easily justified for reducing air pollution. For example, a significant (though uncertain) portion of the $PM_{2.5}$ improvements in Fig. 3.2.4 were caused by installation of catalytic converters and associated emissions standards for automobiles, although substantial improvements in SOA production from vehicle emissions are largely an unintended (fortunate) cobenefit of controls designed to reduce emissions of ozone precursors.[73] With roughly 17 million car sales in the United States annually, even assuming replacement costs for catalytic converters of roughly \$1000 (the increased price of a new car is probably more like \$100) and completely ignoring the benefits to employment and manufacturing from catalytic converter development and production, the annual costs borne by consumers for internalizing the social cost of air pollution via the catalytic converter is \$17 billion, less than 10% of the (total) benefit from reduced $PM_{2.5}$ alone.

There is strong evidence that straightforward interventions can substantially reduce the worldwide exposure and mortality shown in Fig. 3.2.3.[30] Even if regions simply meet the next most stringent WHO intermediate air quality target (i.e., regions with $PM_{2.5} > 35\ \mu g/m^3$ improve to that level, etc.), the world could avoid almost 1 million deaths annually. Bringing all regions to the WHO air quality goal of 10 $\mu g/m^3$ would avoid more than 2 million deaths annually.[30] Because the concentration-response function is concave, significant benefits for air quality improvements persist to relatively low concentrations. As the data in Fig. 3.2.4 show, the United States is already well along the journey, although cities such as Pittsburgh were once famously in a situation akin to modern-day Beijing and New Delhi.[126]

3.2.5 CONCLUSION

Most, however, by no means all of the sources of air pollution described earlier, including ozone and $PM_{2.5}$, derive from combustion of fossil fuels. Most NO_x production is from high-temperature combustion, most SO_2 emissions are from coal and high-sulfur petroleum, and a substantial share of OA is fuel derived from vehicle emissions.[32] For example, a large portion of $PM_{2.5}$ in eastern China during high-haze events may be formed from IVOCs associated with vehicle emissions.[106]

History shows that both $PM_{2.5}$ and ozone can be effectively controlled, and that the public health benefits are enormous. The benefits for reductions (or continued reductions) apply in both the developing and the developed world, and many remain associated with fossil fuel combustion in both places. Moreover, the compelling arguments for air quality improvement are relatively local in scope, as the harm from air pollution is much more localized in space and time than the harm from climate change.

Policy interventions can usefully focus on connections between air quality and climate benefits, but it may be far more resonant to convey the climate benefits of air quality improvements as a cobenefit, rather than the reverse, where air quality improvements are presented as a cobenefit. As an example, decarbonization of the US economy would plausibly reduce $PM_{2.5}$ concentrations by a factor of 3, saving at least 40,000 lives per year with an associated public health benefit of nearly \$400 billion annually. Given the strong relationship between energy use and prosperity, it is thus crucial to couple pollution reduction or prevention—both conventional air pollution and climate pollution—with energy development pathways that avoid both air pollution and fossil carbon emissions.

References

1. Lim SS, Vos T, Flaxman AD, Danaei G, Shibuya K, Adair-Rohani H, et al. A comparative risk assessment of burden of disease and injury attributable to 67 risk factors and risk factor clusters in 21 regions, 1990–2010: a systematic analysis for the Global Burden of Disease Study 2010. *Lancet* 2012;**380**(9859):2224–60. http://dx.doi.org/10.1016/S0140-6736(12)61766-8.
2. Wagstrom KM, Pandis SN. Determination of the age distribution of primary and secondary aerosol species using a chemical transport model. *J Geophys Res Atmos* July 23, 2009;**114**:D14303.
3. Likens GE, Bormann FH. Acid rain: a serious regional environmental problem. *Science* 1974;**184**(4142):1176–9. Available from: http://science.sciencemag.org/content/184/4142/1176.
4. Dransfield TJ, Perkins KK, Donahue NM, Anderson JG, Sprengenther MM, Demerjian KL. Temperature and pressure dependent kinetics of the gas-phase reaction of the hydroxyl radical with nitrogen dioxide. *Geophys Res Lett* 1999;**26**:687–90. Available from: http://www.agu.org/pubs/crossref/1999/1999GL900028.shtml.
5. Blitz MA, Hughes KJ, Pilling MJ. Determination of the high-pressure limiting rate coefficient and the enthalpy of reaction for $OH + SO_2$. *J Phys Chem A* 2003;**107**(12):1971–8. http://dx.doi.org/10.1021/jp026524y.
6. Martin LR, Damschen DE. Aqueous oxidation of sulfur dioxide by hydrogen peroxide at low pH. *Atmos Environ1967* 1981;**15**(9):1615–21. Available from: http://www.sciencedirect.com/science/article/pii/0004698181901463.
7. Hoyle CR, Fuchs C, Järvinen E, Saathoff H, Dias A, El Haddad I, et al. Aqueous phase oxidation of sulphur dioxide by ozone in cloud droplets. *Atmos Chem Phys* 2016;**16**:1693–712. Available from: http://www.atmos-chem-phys.net/16/1693/2016/.
8. Faloona I. Sulfur processing in the marine atmospheric boundary layer: a review and critical assessment of modeling uncertainties. *Atmos Environ* 2009;**43**:2841–54.

9. Taylor MR, Rubin ES, Hounshell DA. Effect of government actions on technological innovation for SO_2 control. *Environ Sci Technol* 2003;**37**(20):4527–34. http://dx.doi.org/10.1021/es034223b.
10. Prinn R, Huang J, Weiss R, Cunnold D, Fraser P, Simmonds P, et al. Evidence for variability of atmospheric hydroxyl radicals over the past quarter century. *Geophys Res Lett* April 8, 2005;**32**:7.
11. Atkinson R. Gas phase tropospheric chemistry of organic compounds. *J Phys Chem Ref Data* 1997;**26**:215–90.
12. Calvert JG, Atkinson R, Kerr JA, Madronich S, Moortgat GK, Wallington TJ, et al. *The mechanisms of atmospheric oxidation of the alkenes*. Oxford University Press; 2000.
13. Jordan CE, Ziemann PJ, Griffin RJ, Lim YB, Atkinson R, Arey J. Modeling SOA formation from OH reactions with C–8–C–17 n-alkanes. *Atmos Environ* 2008;**42**:8015–26.
14. Kinoslan JR. Ozone-precursor relationships from EKMA diagrams. *Environ Sci Technol* 1982;**16**:880–3.
15. Ehlers C, Klemp D, Rohrer F, Mihelcic D, Wegener R, Kiendler-Scharr A, et al. Twenty years of ambient observations of nitrogen oxides and specified hydrocarbons in air masses dominated by traffic emissions in Germany. *Faraday Discuss* 2016;**189**:407–37. http://dx.doi.org/10.1039/C5FD00180C.
16. Seinfeld JH, Atkinson R, Berglund RL, Chameides WL, Cotton WR, Demerjian KL, et al. *Rethinking the ozone problem in urban and regional air pollution*. National Academy of Sciences; 1991. http://dx.doi.org/10.17226/1889.
17. Bell ML, Peng RD, Dominici F. The exposure–response curve for ozone and risk of mortality and the adequacy of current ozone regulations. *Environ Health Perspect* 2006;**114**:532–6. Available from: https://www.ncbi.nlm.nih.gov/pmc/articles/PMC1440776/.
18. Molina LT, Madronich S, Gaffney JS, Apel E, de Foy B, Fast J, et al. An overview of the MILAGRO 2006 campaign: Mexico City emissions and their transport and transformation. *Atmos Chem Phys* 2010;**10**(18):8697–760. Available from: http://www.atmos-chem-phys.net/10/8697/2010/.
19. Salvo A, Geiger FM. Reduction in local ozone levels in urban Sao Paulo due to a shift from ethanol to gasoline use. *Nat Geosci* 2014;**7**(6):450–8. http://dx.doi.org/10.1038/ngeo2144.
20. Parrish DD, Allen DT, Bates TS, Estes M, Fehsenfeld FC, Feingold G, et al. Overview of the second Texas air quality study (TexAQS II) and the Gulf of Mexico atmospheric composition and climate study (GoMACCS). *J Geophys Res Atmos* 2009;**114**(D7):D00F13. Available from: http://dx.doi.org/10.1029/2009JD011842.
21. West JJ, Fiore AM, Naik V, Horowitz LW, Schwarzkopf MD, Mauzerall DL. Ozone air quality and radiative forcing consequences of changes in ozone precursor emissions. *Geophys Res Lett* 2007;**34**(6):L06806. http://dx.doi.org/10.1029/2006GL029173.
22. Racherla PN, Adams PJ. The response of surface ozone to climate change over the Eastern United States. *Atmos Chem Phys* 2008;**8**(4):871–85. Available from: http://www.atmos-chem-phys.net/8/871/2008/.
23. Guenther AB, Jiang X, Heald CL, Sakulyanontvittaya T, Duhl T, Emmons LK, et al. The model of emissions of gases and aerosols from nature version 2.1 (MEGAN 2.1): an extended and updated framework for modeling biogenic emissions. *Geosci Model Dev* 2012;**5**:1471–92. Available from: www.geosci-model-dev.net/5/1471/2012/.
24. Zhang J, Dransfield TJ, Donahue NM. On the mechanism for nitrate formation via the peroxy radical + No reaction. *J Phys Chem A* 2004;**108**:9082–95. Available from: http://pubs.acs.org/doi/abs/10.1021/jp048096x.
25. Talukdar RK, Burkholder JB, Schmoltner AM, Roberts JM, Wilson RR, Ravishankara AR. Investigation of the loss processes for peroxyacetyl nitrate in the atmosphere: UV photolysis and reaction with OH. *J Geophys Res Atmos* 1995;**100**(D7):14163–73. Available from: http://dx.doi.org/10.1029/95JD00545.
26. Singh HL, Salas LJ, Ridley BA, Shetter JD, Donahue NM, Fehsenfeld FC, et al. Relationship between peroxyacetyl nitrate (PAN) and nitrogen oxides in the clean troposphere. *Nature* 1985;**318**:347–9. Available from: http://www.nature.com/nature/journal/v318/n6044/abs/318347a0.html.
27. Pope CA. Respiratory disease associated with community air pollution and a steel mill, Utah Valley. *Am J Public Health* 1989;**79**:623–8.
28. Dockery DW, Pope CA, Xu X, Spengler JD, Ware JH, Fay ME, et al. An association between air pollution and mortality in six U.S. Cities. *N Engl J Med* 1993;**329**(24):1753–9. http://dx.doi.org/10.1056/NEJM199312093292401.
29. Carslaw KS, Lee LA, Reddington CL, Pringle KJ, Rap A, Forster PM, et al. Large contribution of natural aerosols to uncertainty in indirect forcing. *Nature* 2013;**503**:67–71. http://dx.doi.org/10.1038/nature12674.

30. Apte JS, Marshall JD, Cohen AJ, Brauer M. Addressing global mortality from ambient PM 2.5. *Environ Sci Technol* 2015;**49**(13):8057–66. http://dx.doi.org/10.1021/acs.est.5b01236.
31. IPCC. *Climate change 2013: the physical science basis*. 32 Avenue of the Americas, New York, NY 10013-2473, USA: Cambridge University Press; 2013.
32. Seinfeld JH, Nohl LE, Pandis SN. *Atmospheric chemistry and physics: from air pollution to climate change*. 3rd ed. Wiley; 2016.
33. Environmental, Division IS. *Standard operating procedure for particulate matter (PM) gravimetric analysis*. RTI International; 2008.
34. Jayne JT, Leard DC, Zhang X, Davidovits P, Smith KA, Kolb CE, et al. Development of an aerosol mass spectrometer for size and composition analysis of submicron particles. *Aerosol Sci Technol* 2000;**33**:49–70.
35. Bell ML. *Assessment of the health impacts of particulate matter characteristics*. Boston, MA: Health Effects Institute; 2012.
36. Donahue NM, Robinson AL, Pandis SN. Atmospheric organic particulate matter: from smoke to secondary organic aerosol. *Atmos Environ* 2009;**43**:94–106. http://dx.doi.org/10.1016/j.atmosenv.2008.09.055.
37. Donahue NM, Posner LN, Westervelt D, Li Z, Presto AA, Sullivan RC, et al. Where did this particle come from? Sources of particle number and mass for human exposure estimates. In: *Airborne particulate matter: sources, atmospheric processes and health*. The Royal Society of Chemistry; 2016. p. 35–71. http://dx.doi.org/10.1039/9781782626589-00035. ISBN : 978-1-78262-491-2.
38. Spracklen DV, Carslaw KS, Kulmala M, Kerminen VM, Sihto SL, Riipinen I, et al. Contribution of particle formation to global cloud condensation nuclei concentrations. *Geophys Res Lett* March 29, 2008;**35**(6):L06808.
39. Pierce JR, Adams PJ. Can cosmic rays affect cloud condensation nuclei by altering new particle formation rates? *Geophys Res Lett* May 13, 2009;**36**.
40. Dunne EM, Gordon H, Kürten A, Almeida J, Duplissy J, Williamson C, et al. Global atmospheric particle formation from CERN CLOUD measurements. *Science* 2016;**354**:1119–24. Available from: http://science.sciencemag.org/content/early/2016/10/26/science.aaf2649.
41. Westervelt DM, Pierce JR, Riipinen I, Trivitayanurak W, Hamed A, Kulmala M, et al. Formation and growth of nucleated particles into cloud condensation nuclei: model-measurement comparison. *Atmos Chem Phys Discuss* 2013;**13**(3):8333–86. Available from: http://www.atmos-chem-phys-discuss.net/13/8333/2013/.
42. Pierce JR, Adams PJ. Efficiency of cloud condensation nuclei formation from ultrafine particles. *Atmos Chem Phys* February 27, 2007;**7**:1367–79.
43. Jimenez JL, Canagaratna MR, Donahue NM, Prévôt ASH, Zhang Q, Kroll JH, et al. Evolution of organic aerosols in the atmosphere. *Science* 2009;**326**:1525–9. Available from: http://www.sciencemag.org/cgi/content/abstract/326/5959/1525.
44. Goldstein AH, Galbally IE. Known and unexplored organic constituents in the Earth's atmosphere. *Environ Sci Technol* 2007;**41**:1515–21.
45. Kroll JH, Donahue NM, Jimenez JL, Kessler S, Canagaratna MR, Wilson K, et al. Carbon oxidation state as a metric for describing the chemistry of atmospheric organic aerosol. *Nat Chem* 2011;**3**:133–9. Available from: http://www.nature.com/nchem/journal/vaop/ncurrent/full/nchem.948.html.
46. Pankow JF, Chang EI. Variation in the sensitivity of predicted levels of atmospheric organic particulate matter (OPM). *Environ Sci Technol* October 2008;**42**(19):7321–9.
47. Odum J, Hoffmann T, Bowman F, Collins D, Flagan R, Seinfeld J. Gas/particle partitioning and secondary organic aerosol yields. *Environ Sci Technol* August 1996;**30**(8):2580–5.
48. Donahue NM, Epstein SA, Pandis SN, Robinson AL. A two-dimensional volatility basis set: 1. Organic mixing thermodynamics. *Atmos Chem Phys* 2011;**11**:3303–18. Available from: http://www.atmos-chem-phys.net/11/3303/2011/.
49. Pankow JF, Barsanti KC. The carbon number-polarity grid: a means to manage the complexity of the mix of organic compounds when modeling atmospheric organic particulate matter. *Atmos Environ* June 2009;**43**(17):2829–35.
50. Aumont B, Szopa S, Madronich S. Modelling the evolution of organic carbon during its gas-phase tropospheric oxidation: development of an explicit model based on a self generating approach. *Atmos Chem Phys* September 22, 2005;**5**:2497–517.

51. Hallquist M, Wenger JC, Baltensperger U, Rudich Y, Simpson D, Claeys M, et al. The formation, properties and impact of secondary organic aerosol: current and emerging issues. *Atmos Chem Phys* 2009;**10**:5155–236. Available from: http://www.atmos-chem-phys.net/9/5155/2009/.
52. Went FW. Blue hazes in the atmosphere. *Nature* 1960;**187**:641–3. Available from: http://www.nature.com/nature/journal/v187/n4738/pdf/187641a0.pdf.
53. Hughes JD. *Environmental problems of the Greeks and Roman*. 2nd ed. Johns Hopkins University Press; 2014.
54. Haagen-Smit AJ. Chemistry and physiology of Los Angeles smog. *Indus Eng Chem* 1952;**44**(6):1342–6. http://dx.doi.org/10.1021/ie50510a045.
55. Hobbs PV, Sinha P, Yokelson RJ, Christian TJ, Blake DR, Gao S, et al. Evolution of gases and particles from a savanna fire in South Africa. *J Geophys Res Atmos* 2003;**108**:8485.
56. Yokelson RJ, Andreae MO, Akagi SK. Pitfalls with the use of enhancement ratios or normalized excess mixing ratios measured in plumes to characterize pollution sources and aging. *Atmos Measure Tech* 2013;**6**(8):2155–8. Available from: http://www.atmos-meas-tech.net/6/2155/2013/.
57. Heald CL, Kroll JH, Jimenez JL, Docherty KS, DeCarlo PF, Aiken AC, et al. A simplified description of the evolution of organic aerosol composition in the atmosphere. *Geophys Res Lett* April 2010;**37**.
58. Williams BJ, Goldstein AH, Kreisberg NM, Hering SV. In situ measurements of gas/particle-phase transitions for atmospheric semivolatile organic compounds. *Proc Nat Acad Sci* 2010;**107**:6676–81.
59. May AA, Saleh R, Hennign CJ, Donahue NM, Robinson AL. Volatility of organic molecular markers used for source apportionment analysis: measurements and atmospheric implications. *Environ Sci Technol* 2012;**46**:12435–44. Available from: http://pubs.acs.org/doi/abs/10.1021/es302276t.
60. Ye P, Ding X, Hakala J, Hofbauer V, Robinson ES, Donahue NM. Vapor wall loss of semi-volatile organic compounds in a Teflon chamber. *Aerosol Sci Technol* 2016;**50**:822–34. http://dx.doi.org/10.1080/02786826.2016.1195905.
61. Donahue NM, Chuang WK, Epstein SA, Kroll JH, Worsnop DR, Robinson AL, et al. Why do organic aerosols exist? Understanding aerosol lifetimes using the two-dimensional volatility basis set. *Environ Chem* 2013;**10**:151–7. http://dx.doi.org/10.1071/EN13022.
62. Després VR, Huffman JA, Burrows SM, Hoose C, Safatov AS, Buryak G, et al. Primary biological aerosol particles in the atmosphere: a review. *Tellus B Chem Phys Meteorol* 2012;**64**(1):15598. http://dx.doi.org/10.3402/tellusb.v64i0.15598.
63. Grieshop AP, Logue JM, Donahue NM, Robinson AL. Laboratory investigation of photochemical oxidation of organic aerosol from wood fires 1: measurement and simulation of organic aerosol evolution. *Atmos Chem Phys* 2009;**8**:1263–77. Available from: http://www.atmos-chem-phys.net/9/1263/2009/acp-9-1263-2009.pdf.
64. Simoneit BRT, Schauer JJ, Nolte CG, Oros DR, Elias VO, Fraser MP, et al. Levoglucosan, a tracer for cellulose in biomass burning and atmospheric particles. *Atmos Environ* 1999;**33**:173–82.
65. Hennigan CJ, Sullivan AP, Jeffrey L, Collett J, Robinson AL. Levoglucosan stability in biomass burning particles exposed to hydroxyl radicals. *Geophys Res Lett* 2010:L09806.
66. Donahue NM, Kroll JH, Robinson AL, Pandis SN. A two-dimensional volatility basis set - part 2: diagnostics of organic-aerosol evolution. *Atmos Chem Phys* 2012;**12**:615–34. Available from: http://www.atmos-chem-phys.net/12/615/2012/.
67. Lack DA, Langridge JM, Bahreini R, Cappa CD, Middlebrook AM, Schwarz JP. Brown carbon and internal mixing in biomass burning particles. *Proc Nat Acad Sci* 2012;**109**(37):14802–7. Available from: http://www.pnas.org/content/109/37/14802.abstract.
68. Ahern AT, Subramanian R, Saliba G, Lipsky EM, Donahue NM, Sullivan RC. Effect of secondary organic aerosol coating thickness on the real-time detection and characterization of biomass burning soot. *Aerosol Measure Tech* 2016;**9**:6117–37. Available from: http://www.atmos-meas-tech.net/9/6117/2016/.
69. Rogge WF, Hildemann LM, Mazurek MA, Cass GR, Simoneit BRT. Sources of fine organic aerosol. 1. Charbroilers and meat cooking operations. *Environ Sci Technol* 1991;**25**(6):1112–25.
70. Sun Y, Du W, Fu P, Wang Q, Li J, Ge X, et al. Primary and secondary aerosols in Beijing in winter: sources, variations and processes. *Atmos Chem Phys* 2016;**16**(13):8309–29. Available from: http://www.atmos-chem-phys.net/16/8309/2016/.
71. Schauer JJ, Rogge WF, Hildemann LM, Mazurek MA, Cass GR. Source apportionment of airborne particulate matter using organic compounds as tracers. *Atmos Environ* 1996;**30**(22):3837–55.

72. Schauer JJ, Kleeman MJ, Cass GR, Simoneit BRT. Measurement of emissions from air pollution sources. 2. C1 through C30 organic compounds from medium duty diesel trucks. *Environ Sci Technol* 1999;**33**(10):1578–87.
73. Jathar SH, Gordon TD, Hennigan CJ, Pye HOT, Pouliot G, Adams PJ, et al. Unspeciated organic emissions from combustion sources and their influence on the secondary organic aerosol budget in the United States. *Proc Nat Acad Sci* 2014;**111**:10473–8. Available from: http://www.pnas.org/content/111/29/10473.
74. Gordon TD, Presto AA, May AA, Nguyen NT, Lipsky EM, Donahue NM, et al. Secondary organic aerosol formation exceeds primary particulate matter emissions for light-duty gasoline vehicles. *Atmos Chem Phys* 2014;**14**(9):4661–78. Available from: http://www.atmos-chem-phys.net/14/4661/2014/.
75. Lambe AT, Miracolo MA, Hennigan CJ, Robinson AL, Donahue NM. Effective rate constants and uptake coefficients for the reactions of organic molecular markers (n-alkanes, hopanes and steranes) in motor oil and diesel primary organic aerosols with hydroxyl radicals. *Environ Sci Technol* 2009;**43**:8794–800. Available from: http://pubs.acs.org/doi/abs/10.1021/es901745h.
76. Grieshop AP, Miracolo MA, Donahue NM, Robinson AL. Constraining the volatility distribution and gas-particle partitioning of combustion aerosols using isothermal dilution and thermodenuder measurements. *Environ Sci Technol* 2009;**43**:4750–6. Available from: http://pubs.acs.org/doi/abs/10.1021/es8032378.
77. Robinson AL, Grieshop AP, Donahue NM, Hunt SW. Updating the conceptual model for fine particle mass emissions from combustion systems. *J Air Waste Manage Assoc* 2010;**60**:1204–22. Available from: http://secure.awma.org/journal/Abstract.aspx?id=2307.
78. Hoyle CR, Boy M, Donahue NM, Fry JL, Glasius M, Guenther A, et al. A review of the anthropogenic influence on biogenic secondary organic aerosol. *Atmos Chem Phys* 2011;**11**:321–43. Available from: http://www.atmos-chem-phys.net/11/321/2011/.
79. Carlton AG, Turpin BJ, Lim HJ, Altieri KE, Seitzinger S. Link between isoprene and secondary organic aerosol (SOA): Pyruvic acid oxidation yields low volatility organic acids in clouds. *Geophys Res Lett* 2006;**33**:L06822.
80. Ervens B, Turpin BJ, Weber RJ. Secondary organic aerosol formation in cloud droplets and aqueous particles (aqSOA): a review of laboratory, field and model studies. *Atmos Chem Phys* 2011;**11**(21):11069–102. Available from: http://www.atmos-chem-phys.net/11/11069/2011/.
81. Kalberer M, Paulsen D, Sax M, Steinbacher M, Dommen J, Prévôt ASH, et al. Identification of polymers as major components of atmospheric organic aerosols. *Science* 2004;**303**:1659.
82. Tolocka MP, Jang M, Ginter JM, Cox FJ, Kamens RM, Johnston MV. Formation of oligomers in secondary organic aerosol. *Environ Sci Technol* 2004;**38**:1428.
83. Kamens RM, Jeffries HE, Gery MW, Wiener RW, Sexton KG, Howe GB. The impact of α-pinene on urban smog formation: an outdoor smog chamber study. *Atmos Environ* 1981;**15**(6):969–81. Available from: http://www.sciencedirect.com/science/article/pii/0004698181900974.
84. Paulson SE, Pandis SN, Baltensperger U, Seinfeld JH, Flagan RC, Palen EJ, et al. Characterization of photochemical aerosols from biogenic hydrocarbons. *J Aerosol Sci.* 1990;**21**:S245–8.
85. Pandis SN, Paulson SE, Seinfeld JH, Flagan RC. Aerosol formation in the photooxidation of isoprene and beta-pinene. *Atmos Environ* 1991;**25**:997–1008.
86. Pankow JF. An absorption model of gas/particle partitioning of organic compounds in the atmosphere. *Atmos Environ* 1994;**28**:185.
87. Hall IV WA, Johnston MV. Oligomer content of α-pinene secondary organic aerosol. *Aerosol Sci Technol* 2011;**45**:37–45.
88. Kirkby J, Duplissy J, Sengupta K, Frege C, Gordon H, Williamson C, et al. Ion-induced nucleation of pure biogenic particles. *Nature* 2016;**530**:521–6. http://dx.doi.org/10.1038/nature17953.
89. Carlton AG, Bhave PV, Napelenok SL, Edney EO, Sarwar G, Pinder RW, et al. Model representation of secondary organic aerosol in CMAQv4. 7. *Environ Sci Technol* 2010;**44**(22):8553–60.
90. Volkamer R, Jimenez JL, San Martini F, Dzepina K, Zhang Q, Salcedo D, et al. Secondary organic aerosol formation from anthropogenic air pollution: rapid and higher than expected. *Geophys Res Lett* September 12, 2006;**33**(17).
91. Updyke KM, Nguyen TB, Nizkorodov SA. Formation of brown carbon via reactions of ammonia with secondary organic aerosols from biogenic and anthropogenic precursors. *Atmos Environ* 2012;**63**(C):22–31.
92. Iinuma Y, Boge O, Miao Y, Sierau B, Gnauk T, Herrmann H. Laboratory studies on secondary organic aerosol formation from terpenes. *Faraday Discuss* 2005;**130**:279–94. http://dx.doi.org/10.1039/B502160J.

93. Surratt JD, Gomez-Gonzalez Y, Chan AWH, Vermeylen R, Shahgholi M, Kleindienst TE, et al. Organosulfate formation in biogenic secondary organic aerosol. *J Phys Chem A* September 2008;**112**(36):8345–78.
94. Farmer DK, Matsunaga A, Docherty KS, Surratt JD, Seinfeld JH, Ziemann PJ, et al. Response of an aerosol mass spectrometer to organonitrates and organosulfates and implications for atmospheric chemistry. *Proc Nat Acad Sci* 2010;**107**(15):6670–5. Available from: http://www.pnas.org/content/107/15/6670.abstract.
95. Presto AA, Huff Hartz KE, Donahue NM. Secondary organic aerosol production from terpene ozonolysis: 1. Effect of UV radiation. *Environ Sci Technol* 2005;**39**:7036–45. Available from: http://pubs.acs.org/doi/abs/10.1021/es050174m.
96. Zhang J, Donahue NM. Constraining the mechanism and kinetics of OH + NO_2 and HO_2 + NO using the multiple-well master equation. *J Phys Chem A* 2006;**110**:6898–911. http://dx.doi.org/10.1021/jp0556512.
97. Ng NL, Chhabra PS, Chan AWH, Surratt JD, Kroll JH, Kwan AJ, et al. Effect of NOx level on secondary organic aerosol (SOA) formation from the photooxidation of terpenes. *Atmos Chem Phys* 2007;**7**(19):5159–74.
98. Presto AA, Donahue NM. Investigation of α-pinene + ozone secondary organic aerosol formation at low total aerosol mass. *Environ Sci Technol* 2006;**40**:3536–43. http://dx.doi.org/10.1021/es052203z.
99. Crounse JD, Nielsen LB, Jørgensen S, Kjaergaard HG, Wennberg PO. Autoxidation of organic compounds in the atmosphere. *J Phys Chem Lett* 2013;**4**(20):3513–20. Available from: http://pubs.acs.org/doi/abs/10.1021/jz4019207.
100. Ehn M, Thornton JA, Kleist E, Sipila M, Junninen H, Pullinen I, et al. A large source of low-volatility secondary organic aerosol. *Nature* 2014;**506**:476–9.
101. Rissanen MP, Kurtén T, Sipilä M, Thornton JA, Kangasluoma J, Sarnela N, et al. The formation of highly oxidized multifunctional products in the ozonolysis of cyclohexene. *J Am Chem Soc* 2014;**136**(44):15596–606. http://dx.doi.org/10.1021/ja507146s. 25283472.
102. Robinson AL, Donahue NM, Shrivastava MK, Sage AM, Weitkamp EA, Grieshop AP, et al. Rethinking organic aerosols: semivolatile emissions and photochemical aging. *Science* 2007;**315**:1259–63. Available from: http://www.sciencemag.org/cgi/content/full/315/5816/1259.
103. Donahue NM, Robinson AL, Stanier CO, Pandis SN. Coupled partitioning, dilution, and chemical aging of semivolatile organics. *Environ Sci Technol* 2006;**40**:2635–43. http://dx.doi.org/10.1021/es052297c.
104. Zhang Q, Jimenez JL, Canagaratna MR, Allan JD, Coe H, Ulbrich I, et al. Ubiquity and dominance of oxygenated species in organic aerosols in anthropogenically-influenced Northern Hemisphere midlatitudes. *Geophys Res Lett* July 7, 2007;**34**(13).
105. Spracklen DV, Jimenez JL, Carslaw KS, Worsnop DR, Evans MJ, Mann GW, et al. Aerosol mass spectrometer constraint on the global secondary organic aerosol budget. *Atmos Chem Phys* 2011;**11**(23):12109–36. Available from: http://www.atmos-chem-phys.net/11/12109/2011/.
106. Zhao B, Wang S, Donahue NM, Jathar SH, Huang X, Wu W, et al. Quantifying the effect of organic aerosol aging and intermediate-volatility emissions on regional-scale aerosol pollution in China. *Sci Rep* 2016;**6**:28815. http://dx.doi.org/10.1038/srep28815.
107. Donahue NM, Henry KM, Mentel TF, Scharr AK, Spindler C, Bohn B, et al. Aging of biogenic secondary organic aerosol via gas-phase OH radical reactions. *Proc Nat Acad Sci* 2012;**109**:13503–8. Available from: http://www.pnas.org/content/109/34/13503.abstract.
108. Kroll JH, Smith JD, Che DL, Kessler SH, Worsnop DR, Wilson KR. Measurement of fragmentation and functionalization pathways in the heterogeneous oxidation of oxidized organic aerosol. *Phys Chem Chem Phys* 2009;**11**(36):8005–14.
109. Pierce JR, Chen K, Adams PJ. Contribution of primary carbonaceous aerosol to cloud condensation nuclei: processes and uncertainties evaluated with a global aerosol microphysics model. *Atmos Chem Phys* 2007;**7**(20):5447–66.
110. Twomey S. The influence of pollution on the shortwave albedo of clouds. *J Atmos Sci* 1977;**34**(7):1149–52.
111. Riipinen I, Sihto SL, Kulmala M, Arnold F, Dal Maso M, Birmili W, et al. Connections between atmospheric sulphuric acid and new particle formation during QUEST III-IV campaigns in Heidelberg and Hyytiala. *Atmos Chem Phys* 2007;**7**(8):1899–914.
112. Kulmala M, Kontkanen J, Junninen H, Lehtipalo K, Manninen HE, Nieminen T, et al. Direct observations of atmospheric aerosol nucleation. *Science* 2013;**339**(6122):943–6. Available from: http://www.sciencemag.org/content/339/6122/943.abstract.
113. Riipinen I, Yli-Juuti T, Pierce JR, Petäjä T, Worsnop DR, Kulmala M, et al. Contribution of organics to atmospheric nanoparticle growth. *Nature Geoscience* 2012;**5**:453–8. Available from: http://www.nature.com/ngeo/journal/v5/n7/full/ngeo1499.html.

114. Tröstl J, Chuang WK, Heinritzi M, Yan C, Molteni U, Ahlm L, et al. Low-volatility organic compounds are key to initial particle growth in the atmosphere. *Nature* 2016;**530**:527–31. http://dx.doi.org/10.1038/nature18271.
115. Jacobson MZ. Strong radiative heating due to the mixing state of black carbon in atmospheric aerosols. *Nature* 2001;**409**(6821):695–7. http://dx.doi.org/10.1038/35055518.
116. Bond TC, Doherty SJ, Fahey DW, Forster PM, Berntsen T, DeAngelo BJ, et al. Bounding the role of black carbon in the climate system: A scientific assessment. *J Geophys Res Atmos* 2013;**118**(11):5380–552. http://dx.doi.org/10.1002/jgrd.50171.
117. Andreae MO, Gelencsér A. Black carbon or brown carbon? The nature of light-absorbing carbonaceous aerosols. *Atmos Chem Phys* 2006;**6**(10):3131–48. Available from: http://www.atmos-chem-phys.net/6/3131/2006/.
118. Saleh R, Robinson ES, Tkacik DS, Ahern AT, Liu S, Aiken A, et al. Brownness of organics in aerosols from biomass burning linked to their black carbon content. *Nat Geosci* 2014;**7**:647–50. Available from: http://www.nature.com/ngeo/journal/vaop/ncurrent/full/ngeo2220.html.
119. Saleh R, Marks M, Heo J, Adams PJ, Donahue NM, Robinson AL. Contribution of brown carbon and lensing to the direct radiative effect of carbonaceous aerosols in biomass and biofuel burning emissions. *J Geophys Res Atmos* 2015;**120**:10285–96. Available from: http://onlinelibrary.wiley.com/doi/10.1002/2015JD023697/abstract.
120. Sareen N, Moussa SG, McNeill VF. Photochemical aging of light-absorbing secondary organic aerosol material. *J Phys Chem A* 2013;**117**(14):2987–96. http://dx.doi.org/10.1021/jp309413j.
121. Laskin A, Laskin J, Nizkorodov SA. Chemistry of atmospheric brown carbon. *Chem Rev* 2015;**115**(10):4335–82. http://dx.doi.org/10.1021/cr5006167.
122. Burnett RT, Pope III CA, Ezzati M, Olives C, Lim SS, Mehta S, et al. An integrated risk function for estimating the global burden of disease attributable to ambient fine particulate matter exposure. *Environ Health Perspect* 2014;**122**:397–403. Available from: https://ehp.niehs.nih.gov/1307049.
123. Grunig G, Marsh LM, Esmaeil N, Jackson K, Gordon T, Reibman J, et al. Perspective: ambient air pollution: inflammatory response and effects on the lung's vasculature. *Pulmonary Circ* 2014;**4**(1):25–35. http://dx.doi.org/10.1086/674902.
124. Brauer M, Freedman G, Frostad J, van Donkelaar A, Martin RV, Dentener F, et al. Ambient air pollution exposure estimation for the global burden of disease 2013. *Environ Sci Technol* 2016;**50**(1):79–88. http://dx.doi.org/10.1021/acs.est.5b03709.
125. Pope CA, Ezzati M, Dockery DW. Fine-particulate air pollution and life expectancy in the United States. *N Engl J Med* 2009;**360**:376–86.
126. Davidson CI, Phalen RF, Solomon PA. Airborne particulate matter and human health: a review. *Aerosol Sci Technol* 2005;**39**:737–49.

CHAPTER

3.3

Stratospheric Ozone Depletion and Recovery

David M. Wilmouth[1], *Ross J. Salawitch*[2], *Timothy P. Canty*[2]

[1]Harvard University, Cambridge, MA, United States; [2]University of Maryland, College Park, MD, United States

Although present in only relatively small concentrations in Earth's atmosphere, ozone provides critical protection to life on this planet. Indeed, ozone is the only molecule in our atmosphere capable of significantly absorbing harmful radiation from the sun across the broad range of ultraviolet wavelengths. Ozone is also centrally important in defining the temperature structure of the atmosphere, which in turn impacts the chemical, radiative, and dynamical processes that occur. Because of its significance in protecting life on Earth as we know it, even small reductions in the ozone layer are noteworthy, but the dramatic loss of ozone first observed in the mid-1980s following decades of chlorofluorocarbon (CFC) and halon emissions had the potential to be catastrophic. Swift global action in the form of international agreements designed to protect Earth's ozone layer prevented much more significant damage from occurring, but full ozone recovery is still decades away. From a green chemistry perspective, perhaps there is no better example of unintended consequences from chemical use than the stratospheric ozone depletion that resulted from emissions of manmade halogen source gases, which were originally thought to be environmentally benign.

3.3.1 STRATOSPHERIC OZONE

The vertical structure of Earth's atmosphere is marked by changing temperature trends with increasing altitude. Temperatures decrease from the surface up to ~10–16 km in the troposphere (exact height varies with latitude and season) and then increase with altitude up to ~60 km in the stratosphere. Approximately 90% of atmospheric ozone is present in the stratosphere; hence, the stratosphere is referred to as the location of the ozone layer. Fig. 3.3.1 shows the vertical distribution of ozone in the atmosphere. For reference, commercial aircraft typically fly at 9–12 km (30,000–40,000 ft). Very little ozone is present in regions

Green Chemistry
http://dx.doi.org/10.1016/B978-0-12-809270-5.00008-X

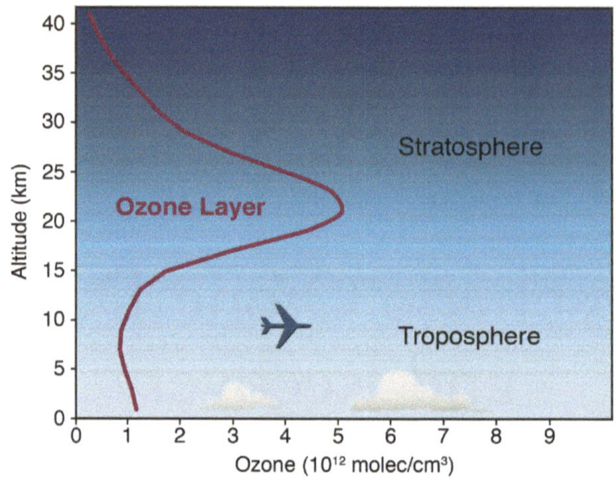

FIGURE 3.3.1 Vertical profile of ozone in the atmosphere. Approximately 90% of Earth's protective ozone layer is located in the stratosphere, the region of the atmosphere that extends from ~10 to 16 km above the surface up to ~60 km. The example profile shown here was acquired over the United States in summer.

of the upper atmosphere above the stratosphere. Even at its peak abundance in the stratosphere, ozone is a trace gas with a volume mixing ratio measuring only a few parts per million. Elevated ozone in the troposphere near the surface results from anthropogenic pollution and can be harmful to human health.

Ozone in the stratosphere absorbs much of the ultraviolet radiation from the sun, which is more energetic and at shorter wavelengths than visible light (400–740 nm). Within the ultraviolet, the wavelengths of solar light are further categorized as UV-A (315–400 nm), UV-B (280–315 nm), and UV-C (100–280 nm), as shown in Fig. 3.3.2. UV-C is almost completely absorbed by molecular oxygen and ozone, making penetration to the surface essentially negligible. While UV-B is mostly absorbed by stratospheric ozone, a small fraction does reach the surface. UV-A is only weakly affected by ozone, with most light in this wavelength region reaching the ground. UV-A accounts for more than 95% of the UV radiation from the sun that reaches Earth's surface.

The large difference in ozone absorption of UV-A relative to the shorter UV wavelengths is due to the dramatic variation of the ozone absorption cross sections over this spectral region. For example, from the peak ozone absorption at around 255 nm to the near midpoint of UV-A at 355 nm, the ozone absorption cross sections fall by approximately five orders of magnitude.[1,2] This means that ozone attenuates a factor of ~100,000 more sunlight at 255 nm than at 355 nm.

Ozone absorption of ultraviolet radiation in the stratosphere, particularly at the shorter wavelengths, is critical for human health because UV light damages DNA in living cells. Replication of these damaged DNA molecules can lead to melanoma skin cancers, nonmelanoma basal cell carcinoma, and squamous cell carcinoma. In addition, solar UV radiation can induce damage to the human eye, cause premature aging of the skin, suppress the immune

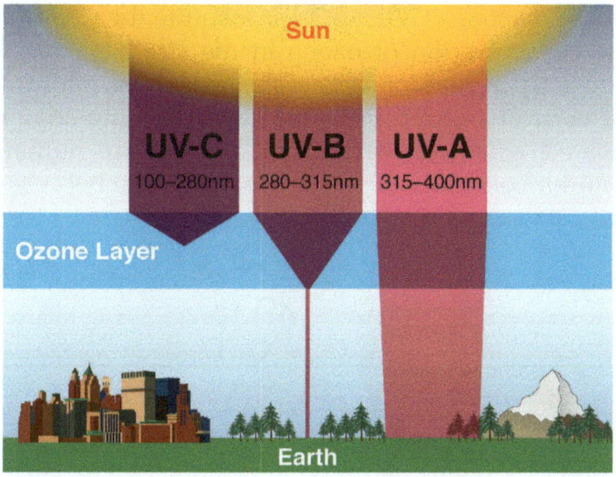

FIGURE 3.3.2 Absorption of ultraviolet radiation from the sun by the stratospheric ozone layer. The width of the bars schematically represents the amount of energy at that altitude. There is essentially no surface penetration of UV-C radiation, the most harmful to life on Earth. UV-B radiation from the sun is mostly absorbed by stratospheric ozone, but a small fraction does reach the surface. The vast majority of UV solar radiation that reaches Earth's surface is UV-A. Biological sensitivity to the shorter UV-B wavelengths is orders of magnitude more than the sensitivity to the less energetic UV-A wavelengths.

system, inhibit plant growth, and harm animals and marine life.[3–7] Biological organisms have been shown to be orders of magnitude more sensitive to erythema induction (sunburn), DNA damage, and plant damage at the shorter UV-B wavelengths than at the longer UV-A wavelengths.[8] This biological sensitivity is especially noteworthy because UV-B is also the wavelength region where the transmission of solar radiation is most impacted by the stratospheric ozone concentration. Clearly, it is important to understand the chemistry controlling production and loss of stratospheric ozone.

Production of ozone in the stratosphere occurs primarily by photolysis of molecular oxygen at UV-C wavelengths ($\lambda < 242$ nm). The O atoms thus produced each react with molecular oxygen to form ozone.[9]

$$O_2 + h\nu \rightarrow 2O \qquad (3.3.1)$$

$$2 \times (O + O_2 + M \rightarrow O_3 + M) \qquad (3.3.2)$$

$$\textbf{Net: } \mathbf{3O_2 \rightarrow 2O_3}$$

Reaction (3.3.2) is termolecular, where air (N_2 or O_2) is a nonreactive collision partner represented as "M" in the equation.

Destruction of ozone in the stratosphere was first proposed by Sydney Chapman to occur by photolysis and reaction with O atoms.[9]

$$O_3 + h\nu \rightarrow O + O_2 \qquad (3.3.3)$$

$$O + O_3 \rightarrow 2O_2 \qquad (3.3.4)$$

$$\textbf{Net: } \mathbf{2O_3 \rightarrow 3O_2}$$

These production and loss reactions (3.3.1–3.3.4) constitute the Chapman cycle and naturally produce a steady-state concentration of ozone in the stratosphere. These reactions alone, however, do not sufficiently represent ozone chemistry, as column ozone (i.e., the total amount of ozone in a vertical column from the surface to the edge of the atmosphere) is significantly overestimated with this mechanism.[10] There are numerous other ozone loss reactions that involve minor constituents, which are significant because of the catalytic nature of these reaction cycles. For example,

$$O_3 + h\nu \rightarrow O + O_2 \quad (3.3.3)$$

$$X + O_3 \rightarrow XO + O_2 \quad (3.3.5)$$

$$XO + O \rightarrow X + O_2 \quad (3.3.6)$$

$$\textbf{Net: } 2O_3 \rightarrow 3O_2$$

In this reaction cycle, the catalyst, X, represents either the radical NO, OH, Cl, or Br. Note that X is consumed in reaction (3.3.5) and reformed in reaction (3.3.6), such that its concentration does not change. Even though the radical concentrations are typically several orders of magnitude less than the ozone concentration, the catalytic nature of the reaction sequence enables significant ozone destruction.

Reactions involving O atoms are most important in the upper and middle stratosphere because of increasing O atom concentrations with altitude.[11] Reactions that destroy ozone in the lower stratosphere often involve more complex cycles that engage more than one radical family, such as coupled ClO_x–HO_x, BrO_x–HO_x, ClO_x–NO_x, and ClO_x–BrO_x cycles.[12–15] (The radical families are defined as: $ClO_x = Cl + ClO$, $BrO_x = Br + BrO$, $HO_x = OH + HO_2$, and $NO_x = NO + NO_2$.) Some removal of stratospheric ozone is also due to transport of air down to the troposphere.

In short, stratospheric ozone is continually produced and then destroyed by photolysis from sunlight along with a number of chemical reactions primarily involving hydrogen and nitrogen oxides, as well as reactive chlorine and bromine. Despite this complexity, ozone production and loss reactions tend to be in a balance (which varies with location, time of year, and other factors) that serves to maintain global ozone at particular levels when coupled with atmospheric air motions that transport and mix air of varying ozone concentrations. However, if the atmosphere were perturbed in such a way as to significantly enhance ozone loss rates relative to production, this natural balance would be distorted, and health risks to life on Earth would follow. Indeed, this is exactly what happened in past decades because of the emission of manmade halogen source gases (particularly CFCs and halons) that dramatically increased chlorine and bromine concentrations in the stratosphere. The effect of anthropogenic halogens on ozone is the primary focus of the remainder of this chapter.

3.3.2 OZONE-DEPLETING SUBSTANCES

During the mid-twentieth to the late twentieth century, a number of halogen source gases were in widespread use by industrialized nations, resulting in emissions of these chemicals into the atmosphere. Chlorine-containing CFC gases were used in refrigeration and air conditioning systems, in aerosol spray cans, and as blowing agents for foams. Unlike the toxic chemicals they replaced (e.g., ammonia, methyl chloride, and sulfur dioxide), CFCs in many ways were considered ideal chemicals; they are very stable and unreactive, nonflammable, and generally nontoxic. Bromine-containing halons similarly found widespread use as fire suppressants. Emissions of these halogen source gases increased dramatically in the decades following the 1950s.

Once emitted into the atmosphere, CFCs accumulate in the troposphere and eventually rise into the stratosphere at the tropics because of natural atmospheric circulation. There is essentially no chemical or photolytic removal of CFCs in the troposphere because of their low reactivity and significant absorption of sunlight only at UV-C wavelengths. Molina and Rowland,[16] who were awarded the Nobel Prize in Chemistry, were the first to recognize that the lack of tropospheric sinks for CFCs meant that these molecules would reach the stratosphere and eventually photolyze (i.e., decompose upon absorption of UV photons), yielding chlorine atoms capable of destroying stratospheric ozone. Indeed, one of the properties most attractive for industrial use—low reactivity—would prove to be the same property that made CFCs quite problematic for the environment.

Despite the concern following the publication by Molina and Rowland[16] in 1974, definitive observational evidence of ozone destruction due to chlorine from CFCs was lacking in the subsequent decade because of a high degree of natural variability in the abundance of column ozone. Industrial production and consumer use of CFCs increased in the late 1970s and early 1980s, although Canada, Norway, Sweden, and the United States did ban the use of CFCs as propellants in consumer products in the late 1970s. In 1985 a remarkable discovery was reported by Farman and coworkers: large losses in total column ozone had been occurring over Antarctica in the springtime.[17] Increased concentrations of CFCs were well correlated with the declining abundance of ozone.[17] The scientific details of seasonal polar ozone loss would be uncovered in the following years and decades (see Section 3.3.4), but the discovery of the Antarctic ozone hole was pivotal in motivating international agreements to regulate industrial production and emissions of CFCs and halons.

An example of the early atmospheric measurements showing large ozone losses over Antarctica is shown in Fig. 3.3.3. These observations of ClO and O_3 were acquired in situ in the stratosphere by scientific instruments flying onboard the NASA ER-2 aircraft in September 1987. As the aircraft approached the chemically perturbed air over Antarctica, ClO concentrations rose dramatically, O_3 concentrations fell, and a stark anticorrelation emerged in the abundance of these two species. Using their aircraft data from this field campaign, Anderson et al. published definitive observational evidence linking chlorine from CFC emissions to Antarctic ozone destruction.[18]

With the objective of protecting Earth's stratospheric ozone layer, the Montreal Protocol was signed in 1987. This international treaty placed limits on the production and consumption of halogen source gases such as CFCs and halons, which are known together as ozone-depleting substances (ODSs).[19] The provisions of the Montreal Protocol were vital for slowing the pace of industrial emissions of ODSs, but it was soon realized that further restrictions were necessary. Aided by the development of substitute chemicals to replace ODSs, subsequent amendments to the Montreal Protocol were implemented, which strengthened the provisions and accelerated the timing of ODS reductions and phaseouts. The impact of the Montreal Protocol and the amendments is displayed in Fig. 3.3.4. The EESC unit stands for equivalent effective stratospheric chlorine and is a means to represent both chlorine and bromine, which have different ozone-depletion capabilities, in terms of a single amount of chlorine available in the stratosphere for ozone destruction. As is evident from the

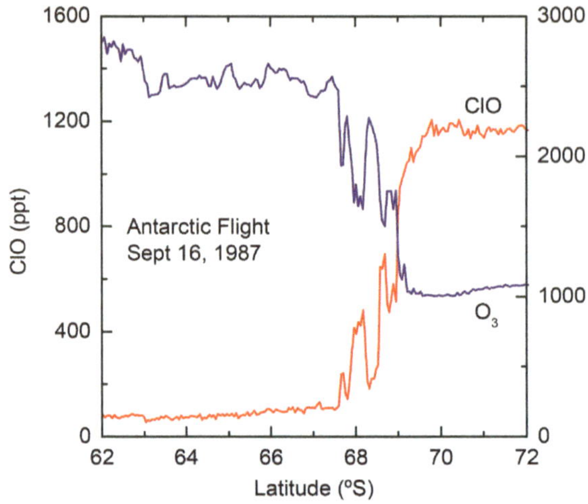

FIGURE 3.3.3 Aircraft measurements linking chlorine from the industrial release of CFCs to stratospheric ozone destruction over Antarctica. Measurements of ClO and O_3 were acquired by scientific instruments flying onboard a NASA ER-2 research aircraft at ~18 km altitude on September 16, 1987. The gradient inside the chemically perturbed air over Antarctica (>69.5°S) relative to outside (<67.5°S) is remarkable: ClO is a factor of ~20 higher and ozone is a factor of ~2.5 lower. In the transition region (67.5–69.5°S), there is a stark anticorrelation in the abundances of ClO and O_3.

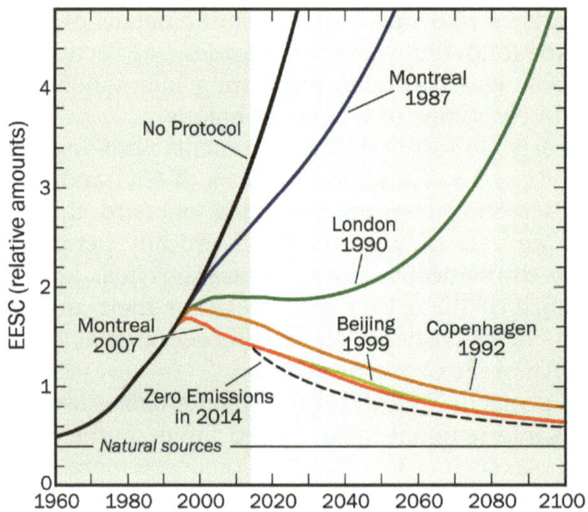

FIGURE 3.3.4 Impact of the Montreal Protocol and subsequent amendments. Projections of future halogen abundances expressed as equivalent effective stratospheric chlorine (EESC) are shown for (1) no Montreal Protocol provisions; (2) provisions of the 1987 Montreal Protocol and subsequent amendments, labeled by the city and year in which they were agreed upon; and (3) zero emissions of ozone-depleting substances beginning in 2014. EESC is a relative measure of ozone depletion potential that combines the influence of chlorine and bromine into a single unit. In the absence of the Montreal Protocol and subsequent amendments that strengthened its provisions, EESC would have increased dramatically in the twenty-first century. *Figure from Hegglin MI, Fahey DW, McFarland M, Montzka SA, Nash ER.* Twenty questions and answers about the ozone layer: 2014 update, scientific assessment of ozone depletion: 2014. *Geneva, Switzerland: World Meteorological Organization; 2015. 84 pp.*

projections in Fig. 3.3.4, in the absence of the Montreal Protocol and subsequent amendments, EESC values would have increased substantially in the twenty-first century rather than being in their current state of decline. In 2009, the Montreal Protocol became the first international treaty to be ratified by all 196 members of the United Nations.

Despite the success of international agreements to regulate ODS emissions, the long atmospheric lifetimes of CFCs and halons mean that stratospheric ozone levels will be impacted for many decades. The atmospheric lifetime of a gas is the time taken for the emitted concentration to decay to $1/e$ or approximately 37% of its starting value. The lifetimes of the most important CFCs range from around 50–100 years and for the halons from around 15–70 years.[19] The CFCs and halons are not very soluble in water and are not removed by precipitation. Some CFCs cycle through the atmosphere from troposphere to stratosphere and back (via the Hadley circulation) multiple times, over a period of many years, before ultimately being lost by photolysis in the stratosphere. This is in stark contrast to halogen source gases that are emitted naturally, such as methyl chloride, methyl bromide, and very-short-lived halogens, which typically have lifetimes less than 1 year in the atmosphere.[19]

The most significant ODSs containing chlorine are CFC-11 ($CFCl_3$), CFC-12 (CF_2Cl_2), CFC-113 ($CFCl_2CF_2Cl$), carbon tetrachloride (CCl_4), and methyl chloroform (CH_3CCl_3). The most significant ODSs with bromine are Halon-1211 (CF_2ClBr), Halon-1301 (CF_3Br), and methyl bromide (CH_3Br, which also has a significant natural source). Fig. 3.3.5 shows the mixing ratios of chlorine and bromine source gases entering the stratosphere, in units of parts per trillion by volume (pptv). The left column in each graph shows the abundance for the year in which the total amount peaked, 1993 for chlorine and 1998 for bromine, and the right

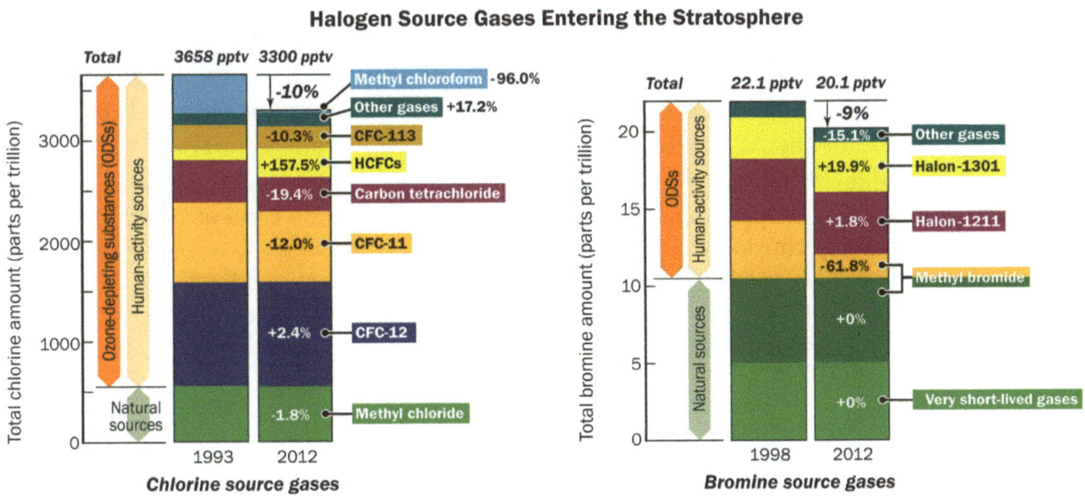

FIGURE 3.3.5 Chlorine and bromine source gases entering the stratosphere. The primary source gases are indicated for the year the abundances of chlorine and bromine entering the stratosphere peaked, 1993 and 1998 respectively, and are compared with the source gas mixing ratios in 2012. Total reductions in each halogen of approximately 10% were realized. Note that the abundance of total chlorine entering the stratosphere is approximately 150 times more than bromine. Human activities are responsible for the vast majority of chlorine and approximately half the bromine entering the stratosphere, as indicated. *Figure from Hegglin MI, Fahey DW, McFarland M, Montzka SA, Nash ER.* Twenty questions and answers about the ozone layer: 2014 update, scientific assessment of ozone depletion: 2014. *Geneva, Switzerland: World Meteorological Organization; 2015. 84 pp.*

column shows the abundance in 2012 for comparison. At the peak, approximately 85% of stratospheric chlorine and 60% of stratospheric bromine were from anthropogenic sources.[19,20] In 2012, the total chlorine and total bromine abundances entering the stratosphere were both reduced ~10% from their respective peaks, as shown in Fig. 3.3.5. That there was a reduction at all points to the effectiveness of the Montreal Protocol; that the reduction was relatively small points to the long atmospheric lifetimes of these chlorine and bromine ODSs.

3.3.3 HALOGEN CHEMISTRY IN THE STRATOSPHERE

Halogen source gases emitted at the surface both naturally and from human activities supply chlorine and bromine to the stratosphere. The vast majority of halogen-containing molecules emitted at the surface are organic, that is, contain carbon. As these organic molecules rise to higher altitudes in the atmosphere, they are eventually photolyzed by sunlight or react chemically to convert their halogen content into inorganic forms. CFCs and halons themselves do not destroy stratospheric ozone; it is their conversion to reactive forms of inorganic chlorine and bromine in the stratosphere that enables CFC and halon emissions to subsequently lead to ozone destruction.

Inorganic chlorine and bromine in the stratosphere may be grouped into either reservoir or active forms. Reservoir forms do not directly destroy ozone, whereas active forms participate in reaction cycles that cause ozone destruction. For inorganic chlorine, the vast majority (~97%) in the lower stratosphere is present in reservoir forms, namely, HCl and $ClONO_2$. The remainder is primarily Cl, ClO, HOCl, OClO, Cl_2, BrCl, and ClOOCl. Outside of polar winter-spring, total inorganic chlorine may be approximated as $Cl_y \approx HCl + ClONO_2 + ClO + HOCl$. Similarly, CCl_y is the total organic chlorine present in the form of CFCs, CCl_4, CH_3CCl_3, CH_3Cl, and other chlorocarbons. Fig. 3.3.6 shows the typical chlorine partitioning as a function of altitude in the northern midlatitude stratosphere (30–60°N), as measured primarily by the ACE-FTS (Atmospheric Chemistry Experiment-Fourier Transform Spectrometer) satellite.[21] Organic chlorine (CCl_y) decreases dramatically with altitude, whereas inorganic chlorine (Cl_y) increases in concert, such that total chlorine content (sum of organic and inorganic) is nearly constant.

As evident in Fig. 3.3.6, nearly all inorganic chlorine is in the form of either HCl or $ClONO_2$ in the midlatitude lower stratosphere. The primary production reactions in the stratosphere for HCl and $ClONO_2$, respectively, are

$$Cl + CH_4 \rightarrow HCl + CH_3 \quad (3.3.7)$$

$$ClO + NO_2 + M \rightarrow ClONO_2 + M \quad (3.3.8)$$

The mixing ratio of ClO peaks at midlatitudes in the upper stratosphere (~40 km altitude), which is where the Molina and Rowland paper[16] focused on the erosion of O_3 due to CFCs.

Stratospheric inorganic bromine is far less partitioned into reservoir forms than inorganic chlorine because bromine reservoirs are much less stable than chlorine reservoirs. The molecule HBr does not form via the reaction of Br with CH_4, as occurs for HCl via the analogous reaction (3.3.7), because $Cl + CH_4$ is exothermic, whereas $Br + CH_4$ is endothermic. The

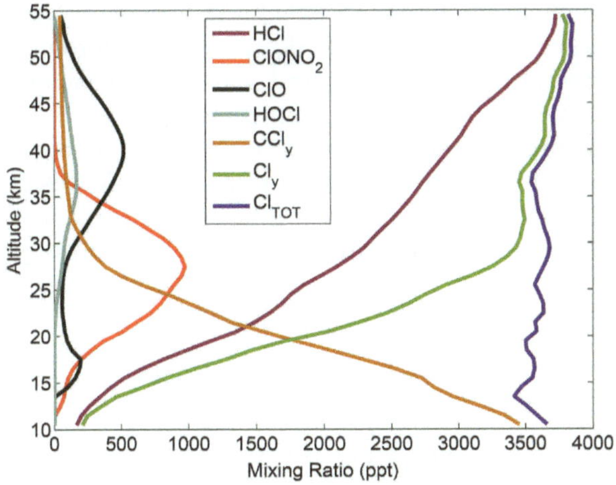

FIGURE 3.3.6 Vertical profile of chlorine species in the atmosphere. Shown are average stratospheric chlorine mixing ratios for northern midlatitudes (30–60°N) from Atmospheric Chemistry Experiment-Fourier Transform Spectrometer (ACE-FTS) satellite measurements and Michelson Interferometer for Passive Atmospheric Sounding (MIPAS) measurements (HOCl). CCl_y represents the total organic chlorine, Cl_y represents the total inorganic chlorine, and Cl_{TOT} represents the sum of organic and inorganic chlorine. As altitude increases, organic chlorine decreases and inorganic chlorine increases dramatically. Most inorganic chlorine is in the form of HCl or $ClONO_2$ at stratospheric altitudes where the ozone layer is most significant. *Data are from 2004, as published in Nassar R, Bernath PF, Boone CD, Clerbaux C, Coheur PF, Dufour G, et al. A global inventory of stratospheric chlorine in 2004.* J Geophys Res *2006;**111**:D22312, auxiliary material.*

reaction of Br with HO_2 does produce HBr, but it is slower than the corresponding chlorine atom reaction. Both $BrONO_2$ and HOBr are formed, but they photolyze relatively fast. The lack of a stable reservoir means that bromine spends more time in its active radical forms (i.e., Br and BrO), so that on a per atom basis, bromine is more efficient than chlorine at destroying ozone by approximately a factor of 60.[22,23] Therefore, despite the stratospheric abundance of inorganic bromine being less than that of inorganic chlorine by more than two orders of magnitude, the relative contributions of chlorine and bromine to stratospheric ozone destruction are somewhat similar.

Heterogeneous reactions, which involve interactions between gas-phase molecules and stratospheric aerosols, also play an important role in stratospheric halogen chemistry. This is especially true over the poles in winter and early spring, as detailed in Section 3.3.4. Here it is simply noted that heterogeneous reactions have the potential to alter the partitioning of inorganic halogens in the stratosphere by enabling reactions to occur that otherwise would not proceed in the gas phase. For example, the two main inorganic chlorine reservoirs, HCl and $ClONO_2$, do not react with each other in the gas phase but can participate in the following heterogeneous reactions to shift chlorine to its active form:

$$ClONO_2 + HCl + \text{aerosol} \rightarrow HNO_3 + Cl_2 \quad (3.3.9)$$

$$ClONO_2 + H_2O + \text{aerosol} \rightarrow HNO_3 + HOCl \quad (3.3.10)$$

$$HOCl + HCl + \text{aerosol} \rightarrow H_2O + Cl_2 \quad (3.3.11)$$

The Cl_2 and HOCl products in reactions (3.3.9)–(3.3.11) are readily photolyzed in the stratosphere to produce Cl atoms capable of ozone destruction. Heterogeneous reaction rates are typically extremely temperature dependent and sensitive to factors that limit the stratospheric conditions under which the reactions can occur.[2]

Of the important gas-phase chemical reaction cycles involving halogens presented in Section 3.3.1, none are capable of producing the large ozone losses observed annually over the Antarctic and Arctic poles. There are two other catalytic reaction cycles, uniquely important at high latitudes in late winter and spring, that control the rate of polar ozone loss. First is the ClOOCl mechanism[24]:

$$\mathrm{ClO + ClO + M \rightleftharpoons ClOOCl + M} \tag{3.3.12}$$

$$\mathrm{ClOOCl} + h\nu \rightarrow \mathrm{Cl + ClOO} \tag{3.3.13}$$

$$\mathrm{ClOO + M \rightarrow Cl + O_2 + M} \tag{3.3.14}$$

$$\mathrm{2(Cl + O_3 \rightarrow ClO + O_2)} \tag{3.3.15}$$

$$\textbf{Net: } \mathbf{2O_3 \rightarrow 3O_2}$$

Second is a coupled bromine and chlorine mechanism[13]:

$$\mathrm{BrO + ClO \rightarrow Br + Cl + O_2} \tag{3.3.16}$$

$$\mathrm{Br + O_3 \rightarrow BrO + O_2} \tag{3.3.17}$$

$$\mathrm{Cl + O_3 \rightarrow ClO + O_2} \tag{3.3.18}$$

$$\textbf{Net: } \mathbf{2O_3 \rightarrow 3O_2}$$

The majority of all polar ozone loss is due to these two reaction cycles.

The ClOOCl mechanism involves the self-reaction (dimerization) of ClO radicals at cold temperatures[24–29] and subsequent photolysis to produce a Cl atom and ClOO. The ClOO product is weakly bound and quickly dissociates to produce a second Cl atom. The net impact of ClOOCl photolysis is therefore production of two Cl atoms that react to destroy two O_3 molecules and reform ClO. The ClO products can then dimerize again, continuing the catalytic cycle that enables each chlorine atom to have the capability to destroy thousands of ozone molecules. Of critical importance in the ClOOCl cycle is that photolysis (3.3.13) breaks the Cl–O bond. If the O–O bond were broken instead, the photolysis reaction would produce two ClO molecules with no destruction of ozone. This indeed is believed to happen ~10%–20% of the time, but quantum yield studies of ClOOCl photolysis indicate much higher production of Cl radicals than of ClO,[2,30,31] consistent with the ClOOCl mechanism being the most important reaction cycle in polar ozone loss.[32] Thermal dissociation of ClOOCl (reverse reaction of 3.3.12) is much slower than photolysis at typical wintertime polar temperatures of the Antarctic stratosphere.[27,33]

The rate-limiting step for the loss of ozone by the ClOOCl mechanism is the photolysis reaction (3.3.13), which makes it the most important reaction for constraining ozone loss rates in the polar stratosphere.[34] The photolysis rate is determined from the absorption cross sections of ClOOCl and the quantum yield of the Cl-production channel over the wavelength range of incident sunlight. The absorption cross sections of ClOOCl, along with those of ClO and Cl_2, are shown in Fig. 3.3.7. The ClOOCl spectrum is continuous with a peak around 245 nm followed by steadily decreasing cross sections with increasing wavelength.

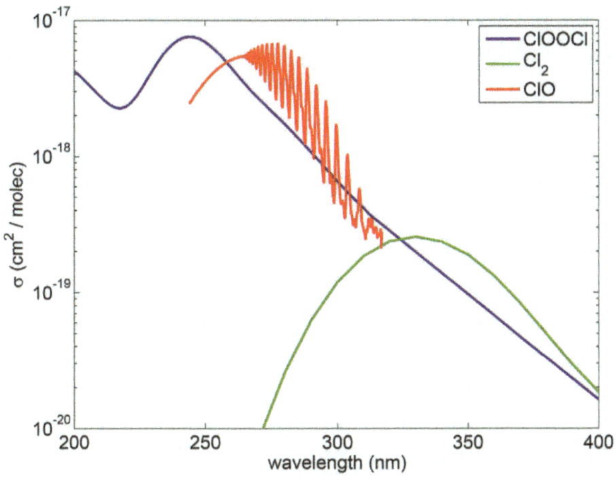

FIGURE 3.3.7 UV absorption cross sections of ClOOCl. Overlapping absorbers ClO and Cl_2 are shown for reference. Significant uncertainty in many laboratory determinations of ClOOCl cross sections has resulted from Cl_2, which has larger cross sections than ClOOCl at atmospherically important wavelengths >324 nm. Cross section data are from JPL-15.[2]

Although critical for calculating polar ozone loss rates, the laboratory-determined ClOOCl absorption spectrum historically suffered from large uncertainties due to experimental challenges in correcting for chemical interferences. The most important spectral interference is Cl_2, which has an unstructured absorption that peaks at approximately 330 nm (Fig. 3.3.7). A 2007 study that targeted reduction of this uncertainty reported cross sections for ClOOCl that were lower than previous studies by more than an order of magnitude in atmospherically relevant wavelength regions.[35] This study generated significant attention in both the scientific community and the mainstream media, implying that catalytic ozone destruction in the polar winter stratosphere via the ClOOCl mechanism was much less significant than previously believed.[36] Three new studies followed shortly thereafter, employing new experimental techniques to determine ClOOCl laboratory cross sections, while significantly reducing the uncertainty in correcting for spectral interferences.[37–39] All three studies found that ClOOCl cross sections at atmospherically relevant wavelengths were at least as large as had been believed before the 2007 study, solidifying the laboratory foundation underpinning the importance of ClOOCl photolysis in the catalytic destruction of ozone. In 2016, the authors of this chapter published a comprehensive update on the current understanding of ClOOCl spectroscopy and kinetics, which reinforces the view that the ClOOCl reaction cycle is the dominant polar ozone loss mechanism.[33]

The BrO + ClO reaction cycle[13] is also responsible for a substantial fraction of the observed polar stratospheric ozone loss.[32] Reaction (3.3.16) as written earlier is actually a simplification, as there are three product channels of the BrO + ClO reaction

$$BrO + ClO \rightarrow ClOO + Br \qquad (3.3.16a)$$

$$\rightarrow BrCl + O_2 \qquad (3.3.16b)$$

$$\rightarrow OClO + Br \qquad (3.3.16c)$$

Because ClOO thermally dissociates almost instantly and BrCl is photolyzed in sunlight, channels (3.3.16a and 3.3.16b) readily provide both Br and Cl atoms to react with O_3, as shown in reaction (3.3.16). Canty et al.[40] showed that ozone loss calculations are highly sensitive to the relative rate coefficients for the BrO + ClO channels. For example, changing the BrCl yield from 7% to 11% for channel (3.3.16b) results in a ~10% increase in calculated ozone loss due to the BrO + ClO reaction and a ~5% increase in total polar ozone loss.

Finally, given the importance of halogen chemistry involving chlorine and bromine in stratospheric ozone destruction, fluorine and iodine must also be considered. Like chlorine and bromine, fluorine has been released in large quantities in the stratosphere because of the photolysis of ODSs. However, unlike chlorine and bromine, fluorine radicals in the stratosphere react rapidly to form stable compounds such as hydrogen fluoride that do not lead to significant depletion of stratospheric ozone. On a per atom basis, fluorine is 1000–10,000 times less effective at ozone destruction than chlorine.[41–43] Iodine, on the other hand, does participate in reactions that lead to stratospheric ozone destruction, but it is present in only extremely small concentrations in the stratosphere. While some iodine source gases are released at the surface from human activities as well as naturally from the oceans, these gases have very short lifetimes. Therefore, iodine is believed to not play a significant role in the depletion of stratospheric ozone.[19,44]

3.3.4 POLAR OZONE LOSS

The most severe losses of stratospheric ozone have occurred over Earth's polar regions. As the high latitudes are sparsely populated, it may seem counterintuitive that ozone losses due to human activity would be largest there. Polar ozone loss is also not continuous year round; rather, there is a strong seasonal cycle, with peak losses occurring during late winter and early spring in each hemisphere. In this section, polar ozone depletion is addressed, with particular focus on explaining why ozone loss is more severe over the poles despite the presence of CFCs and halons at all latitudes. Ozone depletions in the Antarctic and Arctic are compared, with an emphasis on factors that control both the seasonal behavior and year-to-year variability.

As introduced in Section 3.3.2, Farman et al.[17] reported the steady, long-term decline of total column ozone over Antarctica, which they related to the rising burden of CFCs. Fig. 3.3.8A shows October average values of total column ozone over the Halley Bay Research Station at 75.6°S on Antarctica as published by Farman et al. (black points) and a continuation of this time series based on subsequent satellite observations (blue points). Ozone column is reported here in Dobson units, where 100 DU is equivalent to a layer of pure ozone of thickness 1 mm at standard temperature and pressure. Fig. 3.3.8B shows a time series of EESC for the polar stratosphere. The relation between the ozone column and EESC time series is remarkable, with the only break occurring during 2002, a year marked by extremely unusual meteorological conditions.[46]

Satellite observations of total column ozone over the entire southern hemisphere, published soon after the appearance of the paper by Farman et al.,[17] showed an extensive region of low-ozone values centered over the continent of Antarctica.[47] This low-ozone feature is evident in the images shown in the first two rows of Fig. 3.3.9. The region of low total column ozone had grown larger each austral winter, reaching the size of the continental United States

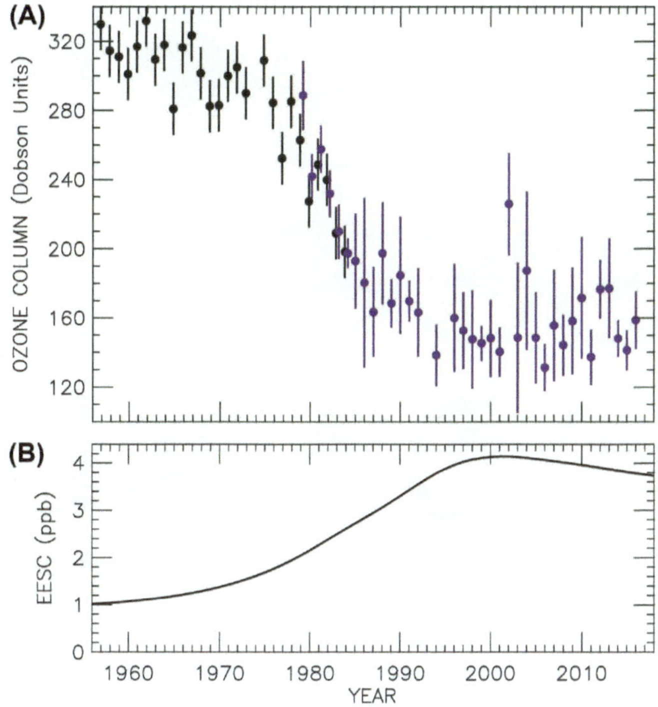

FIGURE 3.3.8 Total ozone and equivalent effective stratospheric chlorine (EESC) over Halley Bay, Antarctica. (A) Total ozone column in Dobson Units over Halley Bay, Antarctica (76°S, 26°W) during October (mean and standard deviation) as reported by Farman et al.[17] from ground-based observations, shown in black, and as subsequently measured by the Total Ozone Mapping Spectrometer and Ozone Monitoring Instrument satellite instruments, shown in blue (dark gray in print versions). (B) Time series of EESC over Antarctica, computed assuming complete conversion of organic chlorine and bromine to inorganic forms, age of air of 5.5 years with an age spectrum width of 3 years[45] and a scaling factor for bromine of 65. The bromine scaling factor of 65, which is appropriate for the polar stratosphere, differs slightly from the scaling factor of 60 for the midlatitude stratosphere.[22]

by the middle 1980s. Images such as those in Fig. 3.3.9 led to the coining of the term "Antarctic ozone hole" that soon became part of the public lexicon.

At the time of the discovery of the Antarctic ozone hole, it was known that the region of low column ozone was confined to a circulation system in the stratosphere termed the polar vortex. As the South Pole is not illuminated by sunlight during winter, stratospheric air above this region becomes exceedingly cold. A latitudinal temperature gradient ensues and as air parcels flow in response to this gradient, the transit of air parcels through the atmosphere is deflected by the Coriolis force. This results in a clockwise circulation system (when viewed looking down on the South Pole) that is known as the Antarctic polar vortex. Early satellite observations revealed that the region of low ozone was confined within the polar vortex boundaries.

Other satellite observations published a few years earlier showed that conditions within the polar vortex are cold enough to promote the formation of clouds.[48] The stratosphere is extremely arid; typically clouds form only in the much moister troposphere. Clouds at the

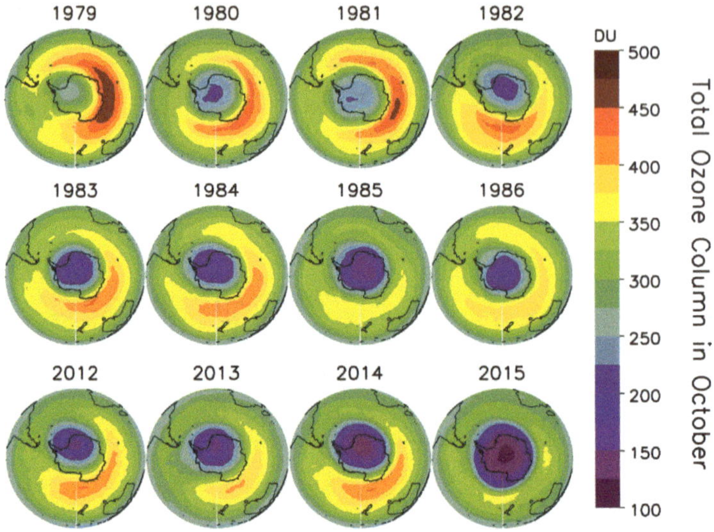

FIGURE 3.3.9 Total ozone over Antarctica during various Octobers. Measurements of the monthly mean abundance of total column ozone are shown over the high-latitude regions of the Southern Hemisphere, as measured by the Total Ozone Mapping Spectrometer and Ozone Monitoring Instrument satellite instruments. (Top row) Evolution of total ozone before the discovery of the Antarctic ozone hole. (Middle row) Total ozone during the time of discovery of the ozone hole. (Bottom row) Total ozone during four recent Octobers.

high altitudes of the stratosphere were first described nearly a century ago as "mother of pearl" clouds by Størmer[49] because of their striking appearance (Fig. 3.3.10). The advent of the modern satellite era provided for a more quantitative description of the characteristics of these clouds and the emergence of the phrase polar stratospheric clouds (PSCs) to describe this phenomenon.[48]

Three theories soon emerged to explain the occurrence of the Antarctic ozone hole. One involved anthropogenic halogens,[13,14] another involved naturally occurring nitrogen oxides,[50] and the third involved upward transport of ozone-poor air from the underlying troposphere.[51] All three theories seemed implausible to many other scientists at the time of these initial papers. The anthropogenic halogens theory, which is of course now known to be correct, required a mechanism for transformation of chlorine from the inactive reservoir species HCl and $ClONO_2$ to active forms, such as ClO and ClOOCl, that catalyze the loss of O_3. Solomon et al.[14] and McElroy et al.[13] proposed that reactions occurring on the surfaces of PSCs, that is, reactions such as (3.3.9)–(3.3.11), could cause this transformation.

Fortunately, each of these three theories posed distinctly different aspects of atmospheric chemistry and meteorology that could be tested by field observations. The ground-based National Ozone Expedition based at McMurdo Station in 1986 provided the first set of comprehensive observations, such as the presence of enhanced OClO,[52] suppressed column HCl and $ClONO_2$,[53] and elevated ClO,[54,55] that supported only the anthropogenic halogens theory. Balloon-borne observations obtained in 1986 revealed a region of extensive ozone loss between altitudes of about 12 and 20 km.[56] The Airborne Antarctic Ozone Experiment, conducted in 1987 using a NASA ER-2 aircraft, provided definitive evidence that industrially

FIGURE 3.3.10 Photograph of a polar stratospheric cloud as viewed from outside the Arena Arctica airplane research hangar in Kiruna, Sweden (68°N, 20°E), during January 2000. The photograph was taken by Ross Salawitch during the NASA-sponsored SOLVE (SAGE III Ozone Loss and Validation Experiment) mission, which used the NASA ER-2 and DC-8 research aircraft to study Arctic ozone depletion.

produced halogens were the cause of the Antarctic ozone hole. Not only did a strong anticorrelation exist between in situ observations of ClO and O_3[18] as shown in Fig. 3.3.3, but also Anderson et al.[57] showed that measured levels of ClO were quantitatively consistent with the rapid removal of O_3 by the ClO + ClO and BrO + ClO reaction cycles (3.3.12–3.3.18).

Fig. 3.3.11 provides a conceptual view of polar ozone depletion. This understanding was primarily developed during a flurry of atmospheric, laboratory, and modeling activities in the late 1980s, soon after the discovery of the ozone hole.[58] Initially, on formation of the wintertime polar vortex, chlorine is present primarily as HCl and $ClONO_2$.[59,60] Once temperatures drop below a certain threshold, PSCs form. Heterogeneous reactions such as (3.3.9)–(3.3.11) take place on the surfaces of PSCs[13,14] and on cold binary liquid aerosols,[61] initially transforming reservoir forms of inorganic chlorine into Cl_2 and HOCl. Upon the return of sunlight in late winter, the photolabile Cl_2 and HOCl molecules are lost by photolysis to form Cl atoms, which react with ozone to form ClO. For the subsequent period of weeks to months in the late winter and spring, ClO and ClOOCl are the dominant inorganic chlorine species. Satellite[62] as well as in situ[60] observations of ClO, such as the aircraft data plotted in Fig. 3.3.12, show that the activation of inorganic chlorine occurs rapidly once temperatures drop below a threshold of $\sim$195K in the polar vortex.

Under normal circumstances in the stratosphere, there is a strong tendency for ClO to react with NO_2, reforming $ClONO_2$ via reaction (3.3.8). However, PSC particles are composed of

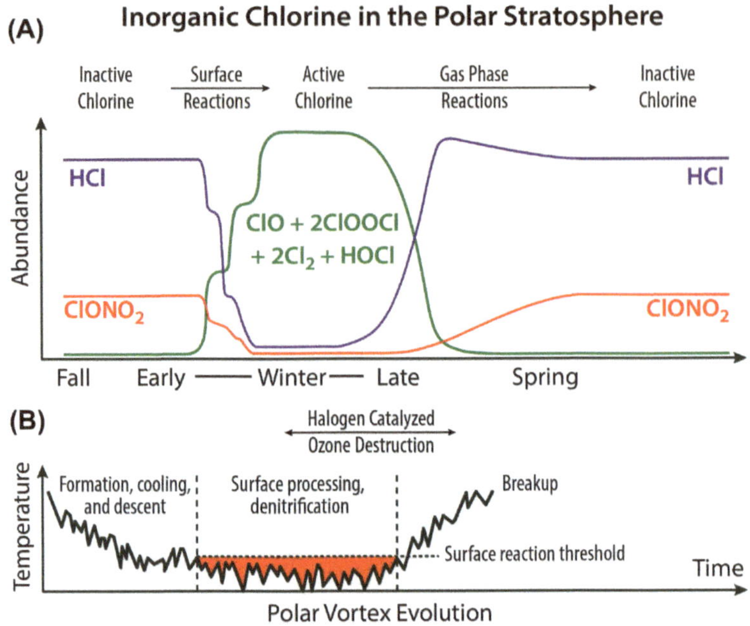

FIGURE 3.3.11 Seasonal evolution of inorganic chlorine and temperature in the Antarctic lower stratosphere. Most of the inorganic chlorine is resident as HCl and $ClONO_2$ as the Antarctic vortex starts to form during fall (A). Once temperature falls below a critical threshold (B), these species are converted to active chlorine and denitrification occurs. Upon return of sunlight, rapid loss of ozone occurs, driven by the ClO + ClO and the BrO + ClO cycles. Once ozone is dramatically reduced, active chlorine is converted to HCl. After the vortex circulation is disrupted and levels of ozone are replenished via mixing with extravortex air, the partitioning of inorganic chlorine back to HCl and $ClONO_2$ is eventually reestablished. *Figure adapted from WMO (World Meteorological Organization).* Scientific assessment of ozone depletion: 1994. *Global Ozone Research and Monitoring Project—Report No. 37. Geneva, Switzerland; 1995.*

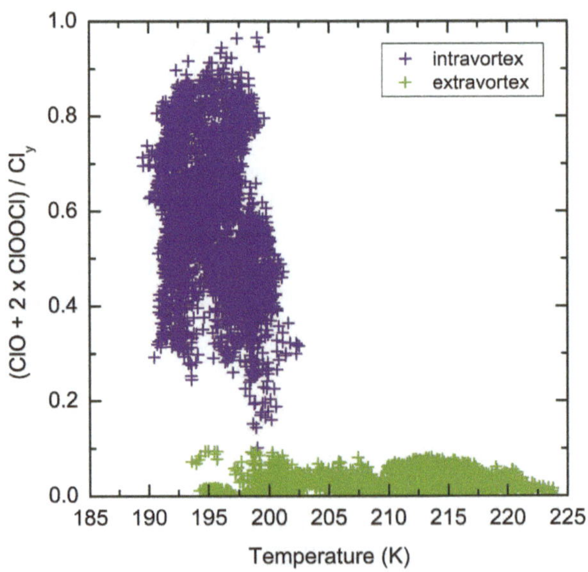

FIGURE 3.3.12 Atmospheric measurements of inorganic chlorine activation as a function of temperature. Data were acquired in situ in the lower stratosphere onboard the NASA ER-2 aircraft as part of the SOLVE mission January-March 2000. Measurements acquired inside the Arctic polar vortex are in blue (dark gray in print versions) and outside the polar vortex are in green (gray in print versions). Activated chlorine is dramatically higher at the colder temperatures inside the polar vortex. *Figure adapted from Wilmouth DM, Stimpfle RM, Anderson JG, Elkins JW, Hurst DF, Salawitch RJ, et al. Evolution of inorganic chlorine partitioning in the Arctic polar vortex.* J Geophys Res *2006;**111**:D16308.*

HNO_3 hydrates. The sedimentation of these PSC particles, termed denitrification, results in extremely low levels of NO_2 throughout the Antarctic polar vortex.[63] The suppression of NO_2 by the sedimentation of PSCs allows ClO concentrations to increase dramatically, greatly enhancing chemical removal of O_3 via the aforementioned ClO + ClO and BrO + ClO reaction cycles (3.3.12–3.3.18). Eventually, ClO is converted back to HCl and $ClONO_2$ during mid-spring,[60,64] but not until a large fraction of the ambient O_3 has been chemically removed.

The size and depth of the Antarctic ozone hole continued to expand until the late 1990s, which was coincident with the peak of EESC in the polar stratosphere (Fig. 3.3.8).[65] Since the late 1990s, variations in the size and depth of the ozone hole have been controlled more by the interannual fluctuations in both meteorological conditions and the influence of minor volcanic eruptions[65–67] than by the slight decline in EESC. As is apparent from the bottom row of Fig. 3.3.9, the Antarctic ozone hole during this decade remains much larger and deeper than it was during the years before its discovery.

The interplay between ozone and large-scale dynamics in the Arctic stratosphere is distinctly different than that of the Antarctic stratosphere, ultimately because of differences in the underlying surface. As Antarctica is somewhat uniformly surrounded by ocean, there is little to impede the zonal circulation of winds in the polar jet stream, resulting in the annual appearance of a much stronger, more isolated wintertime stratospheric vortex. This prevents warmer, ozone-rich midlatitude air from mixing with the considerably colder polar air that has experienced halogen-induced ozone loss. Conversely, the Arctic is encompassed by land and water. Mountain ranges, such as the Rockies, and convection over open ocean water, which by definition must be at a temperature above freezing, deflect the path of the northern polar jet. These undulations, called Rossby waves, weaken the polar circulation, transport warmer midlatitude air into the Arctic region, and decrease the intensity and duration of the northern hemisphere polar vortex.

Fig. 3.3.13 shows the annual cycle of minimum temperature at a pressure of 50 hPa (~20 km altitude) in both hemispheres. These temperatures are compared to thresholds for inorganic chlorine activation and for formation of pure ice clouds. Temperatures in the Antarctic lower stratosphere fall below the threshold for chlorine activation during many months each winter. In contrast, the Arctic lower stratosphere teeters around this threshold: minimum temperature drops below Cl activation conditions for weeks or even months during some winters and mostly remains above the threshold for the duration of other winters.

As a result of the contrasting meteorological conditions in the two hemispheres, total column ozone in the Arctic winter-spring is always greater than that in the Antarctic (Fig. 3.3.14). Even when unusually large chemical losses of Arctic ozone have occurred during winter-spring seasons that were abnormally cold (e.g., 1997 and 2011), total column ozone in the Arctic was still larger than that in the Antarctic. Although the term "ozone hole" has been used in some papers that describe chemical loss of Arctic ozone during winters with particularly cold conditions, most scientists point to comparisons such as those in Fig. 3.3.14 as support for their sentiment that an ozone hole has not yet occurred in the north. This is a debatable point, as there is not a clear quantitative definition of what is meant by the term ozone hole.

To conclude this section, two examples are shown of surprising simplifications to the highly complicated set of chemical and meteorological controls on polar ozone. Fig. 3.3.15 shows

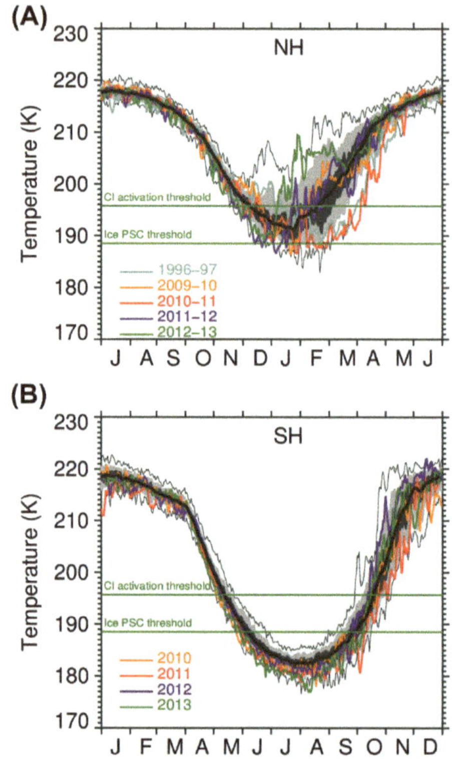

FIGURE 3.3.13 Seasonal evolution of minimum temperature in the lower stratosphere for the Northern Hemisphere (NH) and Southern Hemisphere (SH). Annual cycle of minimum temperature at a pressure of 50 hPa (~20 km altitude) in the lower stratosphere for 50–90°N (A) and 50–90°S (B). The black line shows the climatological mean annual cycle; light- and dark-gray shading indicate the 30%–70% and 10%–90% probabilities, respectively; and the thin dark-gray lines indicate the record maximum and minimum values. The thresholds for chlorine (Cl) activation and for condensation of pure ice polar stratospheric clouds (PSCs) are denoted. Finally, minimum temperatures for various winters are shown by the colored lines, as indicated. *Figure from WMO (World Meteorological Organization).* Scientific assessment of ozone depletion: 2014. *Global Ozone Research and Monitoring Project–Report No. 55. Geneva, Switzerland; 2014. 416 pp.*

chemical loss of column O_3 in the Arctic polar vortex plotted against the volume of the vortex exposed to temperatures cold enough to allow for the formation of PSCs.[68] The surfaces of PSCs have various phases and chemical compositions depending on the ambient conditions in which they form. The existence of a relatively small patch of PSCs at the edge of the vortex, perhaps resulting from a persistently cold region driven by upward motion over an underlying mountain, can in theory result in the activation of large amounts of chlorine as air flows through this region. On the other hand, PSCs in the center of the vortex likely activate chlorine for a much smaller volume because of stagnation of air near the vortex center. Despite these and other complexities, the decade-long series of atmospheric observations in Fig. 3.3.15 reveal that the chemical loss of Arctic ozone varies in a nearly linear manner with the volume of the vortex exposed to PSC temperatures, throughout any given winter. As stratospheric halogen levels decline due to the Montreal Protocol, the relation between

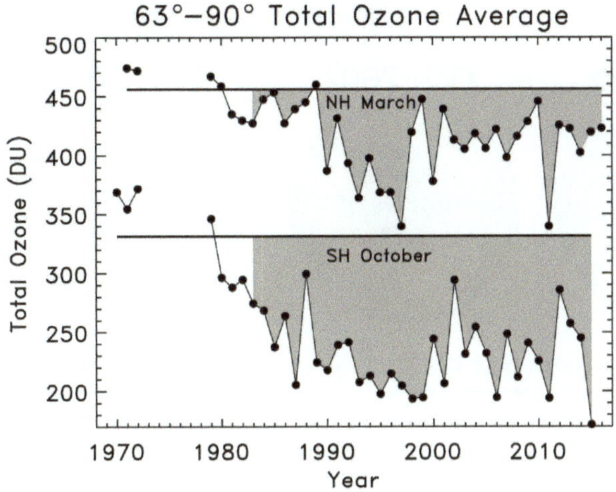

FIGURE 3.3.14 Long-term evolution of total ozone over polar regions, both hemispheres. (Top panel) Total ozone column in Dobson units (DU) measured during March averaged over 63–90°N. (Bottom panel) Total ozone observed during October averaged over 63–90°S. The horizontal black lines represent the average total ozone for the years before 1983 in the respective regions, and the gray shading is used to represent departure from these pre-1983 average values. *Figure adapted from WMO (World Meteorological Organization).* Scientific assessment of ozone depletion: 2014. *Global Ozone Research and Monitoring Project–Report No. 55. Geneva, Switzerland; 2014. 416 pp., updated to include more recent data.*

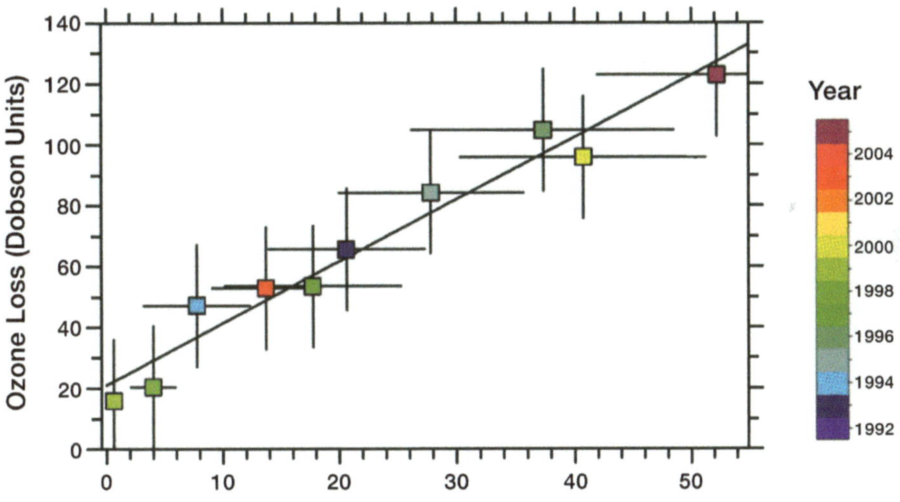

FIGURE 3.3.15 Ozone loss versus polar stratospheric cloud (PSC) conditions in the Arctic. The chemical loss of column ozone in the Arctic is plotted versus the volume of air exposed to PSC formation temperatures throughout the ozone loss season for specific years. *Figure adapted from Rex M, Salawitch RJ, Deckelmann H, von der Gathen P, Harris NRP, Chipperfield MP, et al. Arctic winter 2005: implications for stratospheric ozone loss and climate change.* Geophys Res Lett *2006;**33**:L23808.*

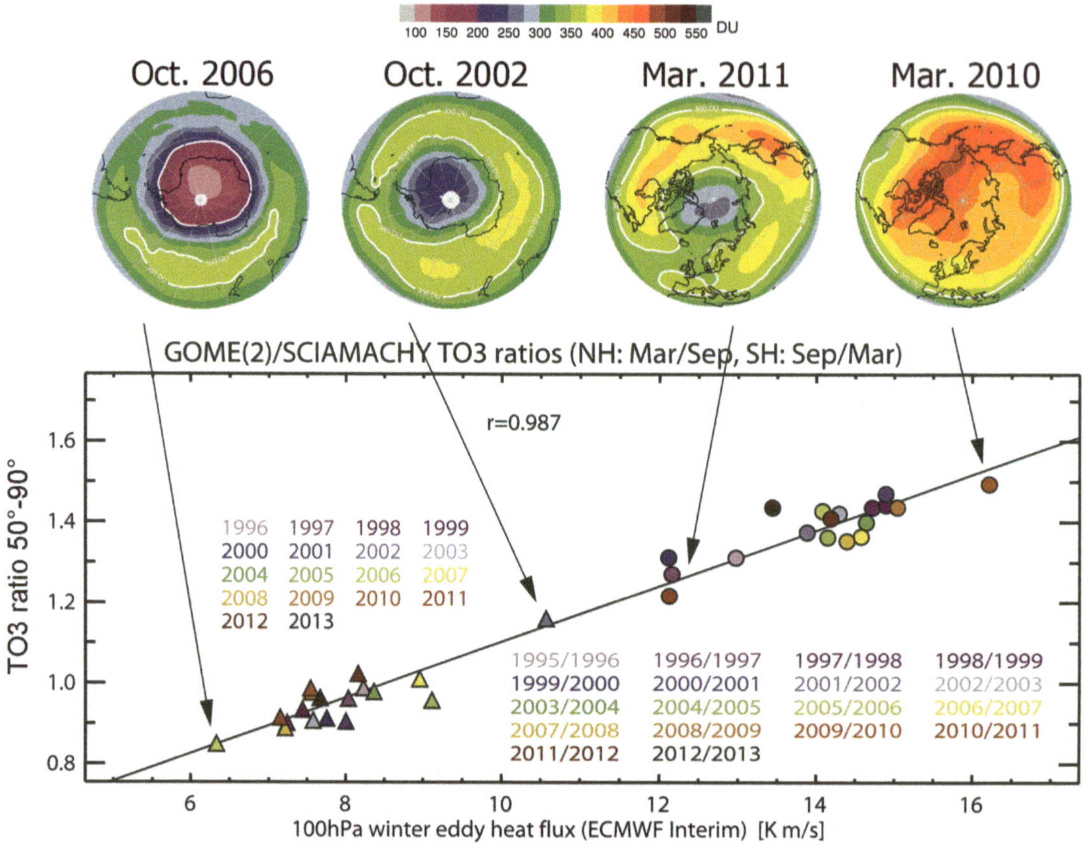

FIGURE 3.3.16 Spring-to-fall ratio of total ozone versus eddy heat flux. Total column ozone observed poleward of 50 degrees latitude in spring divided by total column ozone poleward of 50 degrees in fall is plotted versus the eddy heat flux at 100 hPa (~15 km altitude) during winter. Eddy heat flux is a measure of the strength of the dynamical conditions that impact the polar vortex; the higher the heat flux, the weaker the vortex circulation. The Northern Hemisphere (NH) analysis (circles) uses March as spring and September as fall; for the Southern Hemisphere (SH, triangles), September is spring and March is fall. Spring-to-fall ozone ratios greater than 1 indicate that ozone transport outweighs polar ozone loss. The distribution of polar ozone is shown at the top for selected months. *Figure from WMO (World Meteorological Organization).* Scientific assessment of ozone depletion: 2014. *Global Ozone Research and Monitoring Project—Report No. 55. Geneva, Switzerland; 2014. 416 pp.*

chemical loss and PSC temperatures shown in Fig. 3.3.15 is expected to eventually change; that is, as chlorine levels decline toward the natural background, chemical loss of Arctic ozone will eventually become decoupled from temperature. Currently, relations such as those in Fig. 3.3.15 provide an important test for the computer models that are used to simulate the effect of human activity on polar ozone.[69]

Fig. 3.3.16, an update to Weber et al.,[66] unifies polar ozone in both hemispheres. The figure shows the spring-to-fall ratio of total column ozone in the polar region versus eddy heat flux at 100 hPa (~15 km altitude). Eddy heat flux is a measure of the strength of the dynamical conditions (e.g., aforementioned convection) that affect the polar vortex; the higher the heat flux, the weaker the vortex circulation. The right grouping of points is for the northern hemisphere, which as noted earlier is characterized by more tropospheric stirring, and hence weaker vortices. The left grouping is for the southern hemisphere, and the point in the middle shows Antarctic conditions for the highly unusual winter-spring of 2002, during which the polar vortex was disrupted by an unusual burst of convective activity.[46] Two key themes of this section are reinforced in Fig. 3.3.16: (1) the stratospheric polar vortex in the southern hemisphere is more stable than that in the northern hemisphere, resulting in a larger amount of chemical loss of ozone during late winter and early spring and (2) within the northern hemisphere, those winters with a stronger, more isolated polar vortex (i.e., relatively low heat flux) tend to experience more chemical loss of polar ozone, ultimately because of colder conditions that promote higher levels of chlorine activation compared to Arctic winters with weaker vortices.

The relation between seasonal change of polar ozone and eddy heat flux shown in Fig. 3.3.16, which is connected to chemistry via the temperature dependence of chlorine activation, serves as another important test of the accuracy of computer model simulations. Rarely in nature does such a complex system simplify to provide the type of orderly behavior as that shown in Figs. 3.3.15 and 3.3.16.

3.3.5 MIDLATITUDE OZONE LOSS

Although stratospheric ozone losses are most severe over the poles, these regions contain only a small fraction of the global population. Ozone losses at midlatitudes are much smaller than at the poles, but they are concerning because of the potential impact on the far greater population in these regions. The midlatitudes are defined as the latitude bands between 35 and 60 degrees in both the northern and southern hemispheres. The tropics and subtropics are represented by the large region from 35°S to 35°N. Although most global ozone production occurs in the tropics (23.5°S–23.5°N) where the solar ultraviolet radiation is the highest, column ozone levels in the tropics are actually the lowest globally. This is due to natural meteorological motions in the atmosphere, in which air rises in the tropics and moves out toward the poles in the stratosphere in both hemispheres (the Brewer–Dobson circulation). Variations in this atmospheric circulation, along with other natural influences, can lead to significant variability in stratospheric ozone concentrations at midlatitudes. Quantitative analyses of midlatitude ozone records must first remove the naturally occurring seasonal and solar effects to facilitate more accurate trend detection.

Fig. 3.3.17 shows the changes in global total ozone using an average of measurements from the years 1964 to 1980 (before detectable ozone loss had occurred) as the point of reference. As shown in Fig. 3.3.17A, decreases in ozone were observed every year in the 1980s. The eruption of Mt. Pinatubo in 1991 led to additional ozone losses that year and in the subsequent few years.[70] The most recent years plotted, 2008–12, are on average about 2.5% below the 1964–80 average. Fig. 3.3.17B shows the percentage change in ozone for each latitude between a 1964–80 data average and a 2008–12 average. As is clear in Fig. 3.3.17B, the observed global ozone losses were very unevenly distributed latitudinally on the earth, with many areas experiencing losses far greater than the global average. As expected, polar losses due to winter-spring depletion in the Antarctic and Arctic resulted in the largest ozone

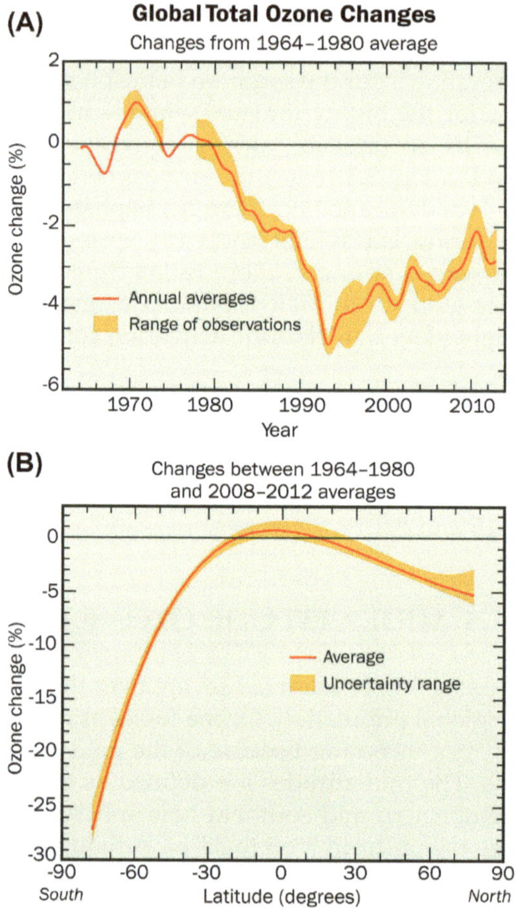

FIGURE 3.3.17 Percentage changes in global total ozone from satellite observations. (A) Annual average global ozone each year is compared to an average of ozone measurements from 1964 to 1980, a period before detectable ozone loss had occurred. A steady decline in global ozone is apparent throughout the 1980s, followed by a more significant drop due to the eruption of Mt. Pinatubo in 1991. (B) The percentage change in global ozone at each latitude between a 1964–80 average and a 2008–12 average. Ozone losses for each hemisphere are greatest in the polar regions, but significant losses are also evident at midlatitudes (35–60 degrees). *Figure from Hegglin MI, Fahey DW, McFarland M, Montzka SA, Nash ER.* Twenty questions and answers about the ozone layer: 2014 update, scientific assessment of ozone depletion: 2014. *Geneva, Switzerland: World Meteorological Organization; 2015. 84 pp.*

reductions in each hemisphere. What is also evident, however, are significant midlatitude ozone losses. Average northern midlatitude reductions of about 3.5% and southern midlatitude reductions of about 6% were observed.[19,20]

The extensive data record of midlatitude measurements also shows that stratospheric ozone losses have been distributed unevenly with respect to altitude.[71,72] Fig. 3.3.18A shows an ozone trend profile for northern midlatitudes (35–60°N) using an average of available observations for the period 1979–97. Here, 1997 is chosen because this year coincided with the peak of EESC. There are two distinct regions of ozone loss: the upper stratosphere at ~40 km

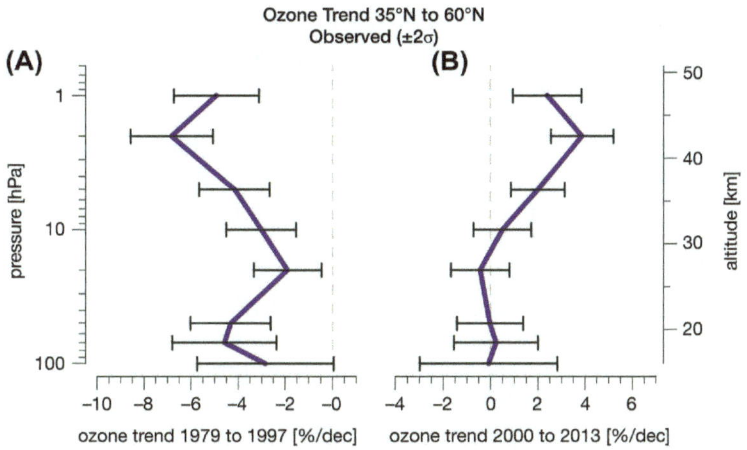

FIGURE 3.3.18 Observed ozone trend profiles for northern midlatitudes (35–60°N). (A) For the period 1979–97, ozone losses were unevenly distributed in the stratosphere, peaking in the upper stratosphere at ~40 km and in the lower stratosphere at ~20 km. (B) For the period 2000–13, midlatitude ozone increased in the upper stratosphere because of reductions in the atmospheric abundance of ozone-depleting substances as well as colder temperatures, with no statistically significant trend evident in the lower stratosphere. *Figure adapted from WMO (World Meteorological Organization).* Scientific assessment of ozone depletion: 2014. *Global Ozone Research and Monitoring Project–Report No. 55. Geneva, Switzerland; 2014. 416 pp.*

peaking around 6% loss per decade and the lower stratosphere at ~20 km peaking around 4% loss per decade. Because the ozone concentration is much greater in the lower stratosphere, reductions there result in greater declines in total column ozone.

In one regard, explaining this observed midlatitude ozone depletion is straightforward. The majority of observed losses have been unequivocally shown to be the result of increases in stratospheric chlorine and bromine resulting from anthropogenic halogen source gas emissions.[19] In the details, however, the causes of midlatitude ozone loss are complex and involve multiple factors, which vary with altitude, hemisphere, and time.

Focusing first on the upper stratosphere, where the photochemical lifetime of ozone is short (days to weeks), ozone depletion is dominated by photochemical processes. Specifically, gas-phase reactions such as (3.3.5) and (3.3.6), where X is Cl, lead to reductions in upper stratospheric ozone. As evident in Fig. 3.3.6, while the chlorine reservoir species HCl still dominates the partitioning of inorganic chlorine in the upper stratosphere, the mixing ratio of ClO is significantly larger there than in the lower stratosphere, facilitating a local maximum in ozone loss rates. The attribution of cause (rising CFCs) and effect (reduced upper stratospheric O_3) has been clear for the past several decades.[72] This connection played an important role in drafting the aforementioned amendments to the Montreal Protocol, which resulted in a near-complete global cessation of the industrial production of CFCs and other ODSs.

In the lower stratosphere the photochemical lifetime of ozone is much longer (months to years) than in the upper stratosphere, such that both chemical and dynamical processes are important. While less significant than at the poles in winter and spring, heterogeneous reactions on sulfate aerosols, in particular N_2O_5 hydrolysis, can impact the chemistry of the lower stratosphere at midlatitudes. Reduced concentrations of NO_x and increased concentrations of ClO_x, BrO_x, and HO_x from heterogeneous reactions can lead to enhanced ozone destruction in the lower stratosphere primarily through coupled radical reaction cycles such as those involving ClO_x–HO_x and ClO_x–BrO_x.[73]

Because the heterogeneous chlorine reactions that lead to halogen activation are sensitive to stratospheric aerosol loading, an increase in sulfates in the stratosphere due to a major volcanic eruption can lead to dramatic increases in the heterogeneous reaction rates.[74] Indeed this was observed following the 1991 eruption of Mt. Pinatubo, where reductions in northern midlatitude ozone were observed over several years.[73] The impact of increased stratospheric aerosol loading from Mt. Pinatubo is evident in Fig. 3.3.17A as the precipitous drop in global ozone levels in 1991. Likewise, significant quantities of very-short-lived bromocarbons (4–8 pptv) entering the stratosphere will impact lower stratospheric ozone losses, particularly at a time of high aerosol loading.[73,75]

In addition to chemical reactions that occur locally having an impact on midlatitude chemistry in the lower stratosphere, the perturbed chemistry in the polar vortices in winter and spring can impact midlatitudes. Export of ozone-depleted air from the polar vortex that mixes with midlatitude air has a dilution effect, which serves to reduce midlatitude ozone levels. Likewise, export of chemically activated air masses from the polar vortex that mix with midlatitude air may result in the catalytic destruction of midlatitude ozone. Because the Antarctic vortex is larger and ozone losses are greater than in the Arctic, the dilution effect at midlatitudes is more significant in the southern hemisphere. This is evident in Fig. 3.3.17B, which exhibits larger depletion of midlatitude ozone in the southern hemisphere.

While the chemistry associated with elevated halogen source gases dominates midlatitude ozone loss, as much as one-third of the trend in northern midlatitude ozone could be due to dynamical changes.[76] For example, changes in tropopause height[77,78] and in the strength of the Brewer-Dobson circulation[79,80] are important in influencing trends in midlatitude ozone. The latitudinal expansion of the Hadley cell in the troposphere, which causes the Brewer-Dobson circulation to expand, also plays a role in driving the observed changes in midlatitude ozone.[81]

Despite the significantly elevated halogen concentrations still present in the stratosphere, there are signs of midlatitude ozone recovery. Fig. 3.3.18B shows an ozone trend profile for northern midlatitudes (35–60°N) using an average of available observations for the period 2000–13. What was once a 6% ozone loss per decade in the upper stratosphere for 1979–97 (Fig. 3.3.18A) now appears as a 4% per decade increase. This change is statistically significant. The 4% loss per decade for 1979–97 in the lower stratosphere now appears much smaller, but there is not yet statistically significant ozone recovery at these altitudes.[19] The reason for the trend change in the upper stratosphere is two-fold: the decline in the atmospheric abundance of ODSs and a cooling of the upper stratosphere due to rising levels of greenhouse gases (GHGs). A colder upper stratosphere enhances ozone production because of the temperature dependence of reaction (3.3.2) and slows the chemical reactions that lead to ozone destruction[82]; this cooling has accelerated the recovery of upper stratospheric ozone due to decreasing ODSs.

Finally, given the detailed considerations of polar and midlatitude ozone, it should be noted why the tropics have been significantly less impacted by the dramatic increases in ODSs. As shown in Fig. 3.3.17B, ozone levels in the tropics have shown little change over the past half century. This relative stability is primarily because there is insufficient time for ODSs that were recently transported from the troposphere to the tropical stratosphere to decompose and produce reactive halogen gases before being transported to higher latitudes. In the lower stratosphere, air in the tropics has been transported from the troposphere within the previous 18 months, whereas at the poles, air has been in the stratosphere for an average of 4–7 years, allowing much more time for the conversion of ODSs to reactive halogen forms. This phenomenon, combined with the greater ozone production rates at the tropics and the physical separation of the tropics from the large ozone losses at the poles, has resulted in only small long-term changes in tropical ozone levels.

3.3.6 FUTURE OF STRATOSPHERIC OZONE

The rapidity with which irrefutable scientific evidence emerged that the Antarctic ozone hole, Arctic ozone depletion, and the slow erosion of the midlatitude ozone layer were all caused by CFCs and halons is, in retrospect, quite remarkable. The high level of scientific understanding that these chemicals cause ozone depletion, coupled with the fact that all ecosystems and peoples of the world are adversely affected by depletion of Earth's protective ozone layer (Section 3.3.1), played important roles in the international cooperation that led to the rapid phaseout of CFCs, halons, and other ODSs.

Following the Montreal Protocol, a class of chemicals called hydrochlorofluorocarbons (HCFCs) initially replaced CFCs as a transitional substitute. Chemically HCFCs are similar to CFCs except that one or more halogens is replaced by a hydrogen atom. The presence of hydrogen leads to the decomposition of HCFCs in Earth's troposphere primarily via the reaction of HCFCs with the hydroxyl radical (OH). The chlorine released from the tropospheric decomposition of HCFCs typically never reaches the stratosphere because it reacts to form compounds that dissolve in water and ice and are removed from the atmosphere by precipitation. Therefore, even though HCFCs contain chlorine, they pose less of a threat to the ozone layer than CFCs because of the much greater probability (88%–98%) of tropospheric decomposition.[20] Stratospheric halogen levels (i.e., EESC in Figs. 3.3.4 and 3.3.8) are in decline as a result of the banning of the industrial production of CFCs established by the Montreal Protocol and the initial replacement of these gases with HCFCs. Hydrofluorocarbons (HFCs) are now being used as replacement chemicals for CFCs and HCFCs. Because HFCs do not contain chlorine, bromine, or iodine, they pose no threat to the ozone layer.

Fig. 3.3.19 shows a time series of EESC at midlatitudes from 1960 to the end of the century, along with computer model estimates of future changes in total column ozone. The EESC time series for years before present is based on analyses of a host of atmospheric observations; the EESC projection is based on the assumption of compliance with the Montreal Protocol and subsequent amendments, combined with atmospheric decay of a suite of long-lived CFC and HCFC compounds in the current atmosphere. The computer model estimates of total column ozone shown in gray are based on ensemble forecasts of many models that consider a single future scenario for EESC as well as atmospheric CO_2, CH_4, and N_2O, which was termed the Special Report on Emissions Scenarios (SRES) A1B scenario.

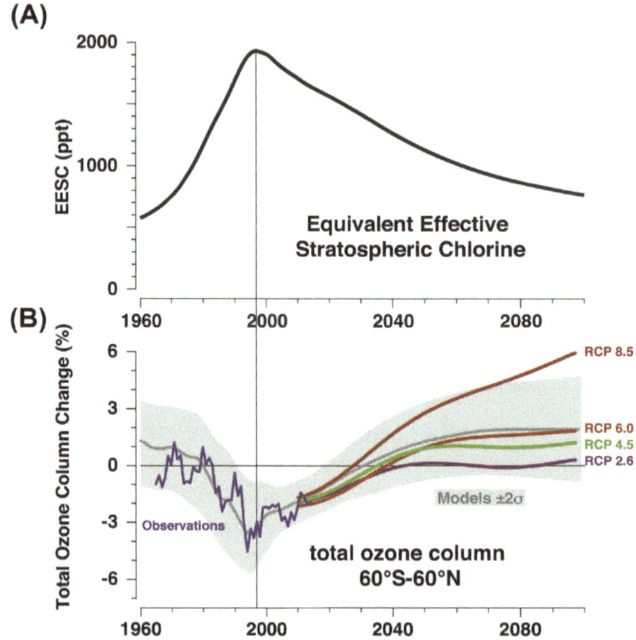

FIGURE 3.3.19 Equivalent effective stratospheric chlorine (EESC) and total column ozone change, 1960–2100. (A) Temporal variation of EESC for the midlatitude lower stratosphere. (B) Percentage change in total column ozone relative to 1980, between 60°S and 60°N, found by numerous chemistry-climate model (CCM) simulations that used the Special Report on Emissions Scenarios A1B scenario for greenhouse gases and the adjusted A1 scenario for ozone-depleting substances (multimodel mean shown by the dark-gray line; shading represents modeled range). The observed total ozone anomaly is shown in blue (dark gray in print versions). The other colored lines show simulations of total ozone by CCMs constrained using future levels of ODSs, CO_2, CH_4, and N_2O from the four Representative Concentration Pathway (RCP) scenarios considered by the most recent Intergovernmental Panel on Climate Change report. All the simulations used essentially the same constraint for ODSs. *Figure from WMO (World Meteorological Organization).* Assessment for decision-makers: scientific assessment of ozone depletion: 2014. *Global Ozone Research and Monitoring Project—Report No. 56. Geneva, Switzerland; 2014. 92 pp.*

As shown in Fig. 3.3.19, over the rest of this century, EESC is expected to remain higher than the level that was present in the stratosphere in 1960.[83] At the end of the century, EESC does approach the background level, which is larger than zero because of the production of CH_3Cl (methyl chloride) by ocean organisms. Methyl chloride partially decomposes in the troposphere, but a portion of the naturally produced burden reaches the stratosphere. The simulations of total column ozone in Fig. 3.3.19B show that a slow, steady, long-term recovery of the ozone layer is expected to occur as EESC declines.

It must be noted, however, that there are large uncertainties in this modeled picture of ozone recovery. Much of this uncertainty is not related to the decline of EESC, but to other factors. The computer model estimates of total column ozone shown by the colored lines in Fig. 3.3.19B rely on a new set of emission scenarios for EESC, CO_2, CH_4, and N_2O developed for the 2013 Intergovernmental Panel on Climate Change (IPCC) report, termed Representative Concentration Pathways (RCPs).[84] The number after each RCP represents the increase of radiative forcing of climate, in units of watt per square meter, which will occur in the year 2100. The RCP 8.5 and RCP 6.0 scenarios are driven by larger, future-projected atmospheric abundances of GHGs than the RCP 4.5 and RCP 2.6 scenarios.

Fig. 3.3.19B shows that should the future growth of GHGs be exceptionally high (i.e., RCP 8.5), the ozone layer will recover to significantly higher column values than existed in 1960. Scientists use the term "super recovery" of the ozone layer to describe this modeled scenario. Super recovery occurs for two reasons. One is that as rising GHGs cause the lower atmosphere to warm, radiative effects of the GHGs (i.e., the trapping of heat near the surface) cause the stratosphere to cool. The cooling of the stratosphere, particularly the middle and upper stratosphere as discussed in Section 3.3.5, causes ozone levels to increase. Also, RCP 8.5 is characterized by exceptionally high levels of CH_4 at the end of the century. High CH_4 levels are the other cause of the super recovery of the ozone layer because, as detailed by Revell et al.,[85] rising CH_4 causes a decline in ClO because of the reaction of Cl with CH_4 (3.3.7).

The dual role of N_2O as a GHG and as the source of stratospheric NO_y further complicates our ability to accurately project the future recovery of the ozone layer. Ravishankara et al.[86] showed that if emissions of N_2O were not controlled, this compound could emerge as the largest human threat of this century to the ozone layer. They showed that the supply of stratospheric NO_y by N_2O would cause the global ozone layer to erode, due to loss of ozone by reactions (3.3.5) and (3.3.6), where X represents NO.

Fig. 3.3.20 shows an updated evaluation of the role of future levels of GHGs on the ozone layer.[87] Fig. 3.3.20A shows a calculation of global mean column ozone, in the year 2100, as a function of future levels of CO_2 and CH_4. The end of the century values of the two GHGs in the four RCP scenarios are marked; the white band denotes total column ozone for the year 2000. Fig. 3.3.20B shows how the effect of N_2O on the ozone layer in the year 2100, expressed using a metric called ozone depletion potential (ODP), varies as a function of CO_2 and CH_4. This figure illustrates the synergistic nature of the effect of the three most important anthropogenic GHGs on Earth's ozone layer: N_2O will cause the most destruction for a future with exceptionally high levels of atmospheric CH_4 and modest to little growth of atmospheric CO_2. On the other hand, should future growth of CO_2 be large relative to that of CH_4, the impact of N_2O on the ozone layer is expected to be considerably smaller.

Figs. 3.3.19 and 3.3.20 illustrate that as EESC declines due to the Montreal Protocol, the future evolution of the ozone layer will be determined by the anthropogenic emission of GHGs. However, there are other potential risks to stratospheric ozone not included in the modeled future scenarios, such as volcanic eruption,[88] increased stratospheric water vapor,[89] and climate engineering.[90] Climate engineering (or geoengineering) would represent an intentional human effort to mitigate global warming by injecting chemicals into the stratosphere to reflect ultraviolet radiation back to space. The intentional injection of sulfates into the stratosphere would cool the climate in a manner analogous to the response to a major volcanic eruption and appears to be an economically feasible means to slow global warming, albeit not necessarily with the efficiency sometimes assumed.[91] Crutzen,[90] in a heavily cited paper about geoengineering, concludes with important perspective: "the very best would be if emissions of the greenhouse gases could be reduced so much that the stratospheric sulfur release experiment would not need to take place. Currently, this looks like a pious wish." Should the stratospheric sulfur release experiment ever occur, it would pose a new threat to the ozone layer at midlatitudes[92] as well as polar regions,[93] because of the enhancements in ClO that result from the presence of excess stratospheric sulfate aerosol.

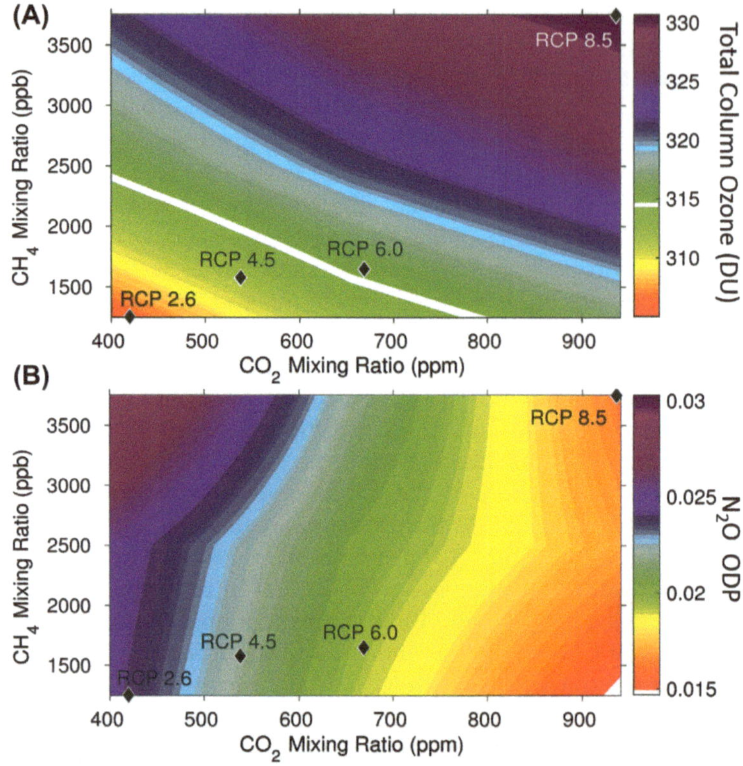

FIGURE 3.3.20 Projections of total ozone and ozone depletion potential (ODP) of N_2O, year 2100. (A) Calculation of global mean total column ozone in 2100 as a function of CO_2 and CH_4 found using constraints from Representative Concentration Pathway (RCP) 6.0 for all other quantities (i.e., ozone-depleting substances, N_2O). (B) Estimate of the ODP of N_2O in 2100 as a function of CO_2 and CH_4, again found using constraints from RCP 6.0 for other quantities. The white contours show the values of total column ozone (314 DU) and ODP of N_2O (0.015) for year 2000, which are provided to give an indication of how these quantities could evolve over the next century. The ODP of N_2O was found based on the decline in total ozone for a perturbation to N_2O at the lower boundary of the model, relative to the change in total ozone induced by a perturbation to CFC-11. *Figure adapted from Revell LE, Tummon F, Salawitch RJ, Stenke A, Peter T. The changing ozone depletion potential of N2O in a future climate.* Geophys Res Lett *2015;42:10047–10055.*

3.3.7 SUCCESS OF THE MONTREAL PROTOCOL

The efforts to save Earth's protective ozone layer from the threat posed by CFCs and other ODSs is a remarkable success story on many levels: human health, economics, science, cooperation of industry, and international participation. The Farman et al. paper[17] announcing the discovery of what would soon be termed the Antarctic ozone hole was published on May 16, 1985. The Montreal Protocol on Substances that Deplete the Ozone Layer was agreed to on September 16, 1987, and went into effect a few years later. The atmospheric abundance of EESC peaked in polar regions during the late 1990s, and a few years earlier at midlatitudes, owing to the longer time for surface air to be transported to the poles. Had the Montreal

Protocol and subsequent amendments not occurred or not been followed, massive depletion of Earth's ozone layer would have ensued.[94] Undoubtedly the early and prescient understanding that CFCs and other ODSs were the cause of polar ozone depletion, established by a concerted series of field observations (satellite, airborne, and ground-based), laboratory studies, and modeling efforts, played a role in the success of the Montreal Protocol.

The compounds that constitute CFCs and other ODSs also trap infrared radiation emitted by Earth's surface, acting as GHGs. The sum of the radiative forcing of climate due to the historical buildup of all anthropogenic halogens is nearly the same as the radiative forcing of climate due to CH_4.[95] Had the Montreal Protocol not been enacted—in other words, had CFCs not posed a threat to the ozone layer—the economic forecast of the atmospheric growth of CFCs (and others ODSs) indicates that these compounds would today have a radiative forcing of climate nearly equal to that of CO_2. The Montreal Protocol not only saved the ozone layer, it is also the single most effective piece of legislation ever enacted for mitigating the rapid rise of global temperatures.[95]

As noted earlier, the Montreal Protocol replaced CFCs with HCFCs, and subsequently with HFCs that pose no risk to stratospheric ozone. However, the global warming potential (GWP) of HFCs generally far exceeds the GWP of HCFCs.[96,97] The Montreal Protocol had, for a long time, lacked regulatory authority for HFCs, as these compounds pose no harm to ozone. Many scientists expressed their concern over this circumstance because some HFCs have a much larger GWP than other HFCs, and the GWP of HFCs was not factored into the industrial production of these compounds in a formal, legislative manner. On October 15, 2016, at the 28th Meeting of the Parties of the Montreal Protocol held in Kigali, Rwanda, negotiators from 197 countries reached an agreement to regulate the future production of HFCs under the Montreal Protocol. This marks the first time that the Montreal Protocol has had direct authority over a class of chemical compounds that pose no threat to the ozone layer.

Looking ahead, the green chemistry story of CFCs and their substitutes is still being written. While stratospheric ozone appears to be on a long trajectory toward recovery, important questions regarding CFC substitutes and climate change are still emerging. The passage of the Kigali Amendment to the Montreal Protocol represents an important extension of this legislation not only to protect stratospheric ozone but also to now regulate future industrial production of some potent GHGs. The Kigali Amendment will guide the future development of replacement compounds for HFCs that pose no threat to either the ozone layer or Earth's climate.

References

1. Burrows JP, Richter A, Dehn A, Deters B, Himmelmann S, Voigt S, et al. Atmospheric remote-sensing reference data from GOME: Part 2. Temperature-dependent absorption cross-sections of O_3 in the 231-794 nm range. *J Quant Spectrosc Radiat Transf* 1999;**61**:509–17.
2. Burkholder JB, Sander SP, Abbatt J, Barker JR, Huie RE, Kolb CE, et al. *Chemical kinetics and photochemical data for use in atmospheric studies*. Evaluation No. 18. JPL Publication 15-10. Pasadena: Jet Propulsion Laboratory; 2015. http://jpldataeval.jpl.nasa.gov.
3. Tevini M, Teramura AH. UV-B effects on terrestrial plants. *Photochem Photobiol* 1989;**50**:479–87.
4. Weinstock MA. Overview of ultraviolet radiation and cancer: what is the link? How are we doing? *Environ Health Perspect* 1995;**103**:251–4.

5. Duthie MS, Kimber I, Norval M. The effects of ultraviolet radiation on the human immune system. *Br J Dermatol* 1999;**140**:995–1009.
6. Matsumura Y, Ananthaswamy HN. Toxic effects of ultraviolet radiation on the skin. *Toxicol Appl Pharmacol* 2004;**195**:298–308.
7. Parkin DM, Mesher D, Sasieni P. Cancers attributable to solar (ultraviolet) radiation exposure in the UK in 2010. *Brit J Cancer* 2011;**105**:S66–9.
8. Madronich S. The atmosphere and UV-B radiation at ground level. In: Young A, et al., editors. *Environmental UV photobiology*. New York: Plenum Press; 1993. p. 1–39.
9. Chapman S. A theory of upper atmospheric ozone. *R Meteorol Soc Mem* 1930;**3**:103–25.
10. Crutzen PJ. Ozone production rates in an oxygen-hydrogen-nitrogen oxide atmosphere. *J Geophys Res* 1971;**76**:7311–27.
11. Osterman GB, Salawitch RJ, Sen B, Toon GC, Stachnik RA, Pickett HM, et al. Balloon-borne measurements of stratospheric radicals and their precursors: implications for the production and loss of ozone. *Geophys Res Lett* 1997;**24**:1107–10.
12. Crutzen PJ, Arnold F. Nitric acid cloud formation in the cold Antarctic stratosphere: a major cause for the springtime 'ozone hole'. *Nature* 1986;**324**:651–5.
13. McElroy MB, Salawitch RJ, Wofsy SC, Logan JA. Reductions of Antarctic ozone due to synergistic interactions of chlorine and bromine. *Nature* 1986;**321**:759–62.
14. Solomon S, Garcia RR, Rowland FS, Wuebbles DJ. On the depletion of Antarctic ozone. *Nature* 1986;**321**:755–8.
15. Wennberg PO, Cohen RC, Stimpfle RM, Koplow JP, Anderson JG, Salawitch RJ, et al. Removal of stratospheric O_3 by radicals: in situ measurements of OH, HO_2, NO, NO_2, ClO, and BrO. *Science* 1994;**266**:398–404.
16. Molina MJ, Rowland FS. Stratospheric sink for chlorofluoromethanes: chlorine atom-catalysed destruction of ozone. *Nature* 1974;**249**:810–2.
17. Farman JC, Gardiner BG, Shanklin JD. Large losses of total ozone in Antarctica reveal seasonal ClO_x/NO_x interaction. *Nature* 1985;**315**:207–10.
18. Anderson JG, Brune WH, Proffitt MH. Ozone destruction by chlorine radicals within the Antarctic vortex: the spatial and temporal evolution of ClO-O_3 anticorrelation based on *in situ* ER-2 data. *J Geophys Res* 1989a;**94**:11465–79.
19. WMO (World Meteorological Organization). *Scientific assessment of ozone depletion: 2014*. Global Ozone Research and Monitoring Project–Report No. 55. 2014. p. 416. Geneva, Switzerland.
20. Hegglin MI, Fahey DW, McFarland M, Montzka SA, Nash ER. *Twenty questions and answers about the ozone layer: 2014 update, scientific assessment of ozone depletion: 2014*. Geneva, Switzerland: World Meteorological Organization; 2015. p. 84.
21. Nassar R, Bernath PF, Boone CD, Clerbaux C, Coheur PF, Dufour G, et al. A global inventory of stratospheric chlorine in 2004. *J Geophys Res* 2006;**111**:D22312.
22. WMO (World Meteorological Organization). *Scientific assessment of ozone depletion: 2006*. Global Ozone Research and Monitoring Project–Report No. 50. 2007. p. 572. Geneva, Switzerland.
23. Sinnhuber B-M, Sheode N, Sinnhuber M, Chipperfield MP, Feng W. The contribution of anthropogenic bromine emissions to past stratospheric ozone trends: a modeling study. *Atmos Chem Phys* 2009;**9**:2863–71.
24. Molina LT, Molina MJ. Production of Cl_2O_2 by the self-reaction of the ClO radical. *J Phys Chem* 1987;**91**:433–6.
25. Sander SP, Friedl RR. Kinetics and product studies of the reaction ClO + BrO using flash photolysis-ultraviolet absorption. *J Phys Chem* 1989;**93**:4764–71.
26. Trolier M, Mauldin III RL, Ravishankara AR. Rate coefficient for the termolecular channel of the self-reaction of ClO. *J Phys Chem* 1990;**94**:4896–907.
27. Nickolaisen SL, Friedl RR, Sander SP. Kinetics and mechanism of the ClO + ClO reaction: pressure and temperature dependence of the bimolecular and termolecular channels and thermal decomposition of chlorine peroxide. *J Phys Chem* 1994;**98**:155–69.
28. Bloss WJ, Nickolaisen SL, Salawitch RJ, Friedl RR, Sander SP. Kinetics of the ClO self-reaction and 210 nm absorption cross section of the ClO dimer. *J Phys Chem* 2001;**105**:11226–39.
29. Stimpfle RM, Wilmouth DM, Salawitch RJ, Anderson JG. First measurements of ClOOCl in the stratosphere: the coupling of ClOOCl and ClO in the Arctic polar vortex. *J Geophys Res* 2004;**109**:D03301.
30. Moore TA, Okumura M, Seale JW, Minton TK. UV photolysis of ClOOCl. *J Phys Chem* 1999;**103**:1691–5.

31. Huang W-T, Chen AF, Chen I-C, Tsai C-H, Lin JJ-M. Photodissociation dynamics of ClOOCl at 248.4 and 308.4 nm. *Phys Chem Phys* 2011;**13**:8195–203.
32. Frieler K, Rex M, Salawitch RJ, Canty T, Streibel M, Stimpfle RM, et al. Toward a better quantitative understanding of polar stratospheric ozone loss. *Geophys Res Lett* 2006;**33**:L10812.
33. Canty T, Salawitch RJ, Wilmouth DM. The kinetics of the ClOOCl catalytic cycle. *J Geophys Res Atmos* 2016;**121**:13768–83.
34. Kawa SR, Stolarski RS, Newman PA, Douglass AR, Rex M, Hofmann DJ, et al. Sensitivity of polar stratospheric ozone loss to uncertainties in chemical reaction kinetics. *Atmos Chem Phys* 2009;**9**:8651–60.
35. Pope FD, Hansen JC, Bayes KD, Friedl RR, Sander SP. Ultraviolet absorption spectrum of chlorine peroxide, ClOOCl. *J Phys Chem A* 2007;**111**:4322–32.
36. von Hobe M. Revisiting ozone depletion. *Science* 2007;**318**:1878–9.
37. Chen H, Lien C, Lin W, Lee YT, Lin JJ. UV absorption cross sections of ClOOCl are consistent with ozone degradation models. *Science* 2009;**324**:781–4.
38. Papanastasiou DK, Papadimitriou VC, Fahey DW, Burkholder JB. UV absorption spectrum of the ClO dimer (Cl_2O_2) between 200 and 420 nm. *J Phys Chem A* 2009;**113**:13711–26.
39. Wilmouth DM, Hanisco TF, Stimpfle RM, Anderson JG. Chlorine-catalyzed ozone destruction: Cl atom production from ClOOCl photolysis. *J Phys Chem A* 2009;**113**:14099–108.
40. Canty T, Rivière ED, Salawitch RJ, Berthet G, Renard J-B, Pfeilsticker K, et al. Nighttime OClO in the winter Arctic vortex. *J Geophys Res* 2005;**110**:D01301.
41. Sehested J, Sehested K, Neilsen OJ, Wallington TJ. Atmospheric chemistry of FO_2 radicals: reaction with CH_4, O_3, NO, NO_2, and CO at 295K. *J Phys Chem* 1994;**98**:6731–9.
42. Ravishankara AR, Turnipseed AA, Jensen NR, Barone S, Mills M, Howard CJ, et al. Do hydrofluorocarbons destroy stratospheric ozone? *Science* 1994;**263**:71–5.
43. Li Z, Friedl RR, Sander SP. Kinetics of FO_2 with NO, NO_2, O_3, CH_4, and C_2H_6. *J Phys Chem* 1995;**99**:13445–51.
44. Bösch H, Camy-Peyret C, Chipperfield MP, Fitzenberger R, Harder H, Platt U, et al. Upper limits of stratospheric IO and OIO inferred from center-to-limb-darkening-corrected balloon-borne solar occultation visible spectra: implications for total gaseous iodine and stratospheric ozone. *J Geophys Res* 2003;**108**:4455.
45. Waugh DW, Hall TM. Age of stratospheric air: theory, observations, and models. *Rev Geophys* 2002;**40**:1010.
46. Allen DR, Bevilacqua RM, Nedoluha GE, Randall CE, Manney GL. Unusual stratospheric transport and mixing during the 2002 Antarctic winter. *Geophys Res Lett* 2003;**30**:1599.
47. Stolarski RS, Krueger AJ, Schoeberl MR, McPeters RD, Newman PA, Alpert JC. Nimbus 7 satellite measurements of the springtime Antarctic ozone decrease. *Nature* 1986;**322**:808–11.
48. McCormick MP, Steele HM, Hamill P, Chu WP, Swissler TJ. Polar stratospheric cloud sightings by SAM II. *J Atmos Sci* 1982;**39**:1387–97.
49. Størmer C. Remarkable clouds at high altitudes. *Nature* 1929;**123**:260–1.
50. Callis LB, Natarajan M. Ozone and nitrogen dioxide changes in the stratosphere during 1979–84. *Nature* 1986;**323**:772–7.
51. Tung KK, Ko MKW, Rodriguez JM, Sze ND. Are Antarctic ozone variations a manifestation of dynamics or chemistry? *Nature* 1986;**333**:811–4.
52. Solomon S, Mount GH, Sanders RW, Schmeltekopf AL. Visible spectroscopy at McMurdo station, Antarctica. 2. Observations of OClO. *J Geophys Res* 1987;**92**:8329–38.
53. Farmer CB, Toon GC, Schaper PW, Blavier J-F, Lowes LL. Stratospheric trace gases in the spring 1986 Antarctic atmosphere. *Nature* 1987;**329**:126–30.
54. De Zafra RL, Jaramillo M, Parrish A, Solomon P, Connor B, Barrett J. High concentrations of chlorine monoxide at low altitudes in the Antarctic spring stratosphere: diurnal variation. *Nature* 1987;**328**:408–11.
55. Solomon PM, Connor B, De Zafra RL, Parrish A, Barrett J, Jaramillo M. High concentrations of chlorine monoxide at low altitudes in the Antarctic spring stratosphere: secular variation. *Nature* 1987;**328**:411–3.
56. Hofmann DJ, Harder JW, Rolf SR, Rosen JM. Balloon-borne observations of the temporal development and vertical structure of the Antarctic ozone hole in 1986. *Nature* 1987;**326**:59–62.
57. Anderson JG, Brune WH, Lloyd SA, Toohey DW, Sander SP, Starr WL, et al. Kinetics of ozone destruction by chlorine oxide and bromine oxide within the Antarctic vortex: an analysis based on *in situ* ER-2 data. *J Geophys Res* 1989b;**94**:11480–520.

58. WMO (World Meteorological Organization). *Scientific assessment of ozone depletion: 1994*. Global Ozone Research and Monitoring Project—Report No. 37. 1995. Geneva, Switzerland.
59. Bonne GP, Stimpfle RM, Cohen RC, Voss PB, Perkins KK, Anderson JG, et al. An examination of the inorganic chlorine budget in the lower stratosphere. *J Geophys Res* 2000;**105**:1957–71.
60. Wilmouth DM, Stimpfle RM, Anderson JG, Elkins JW, Hurst DF, Salawitch RJ, et al. Evolution of inorganic chlorine partitioning in the Arctic polar vortex. *J Geophys Res* 2006;**111**:D16308.
61. Drdla K, Müller R. Temperature thresholds for chorine activation and ozone loss in the polar stratosphere. *Ann Geophys* 2012;**30**:1055–73.
62. Santee ML, Read WG, Waters JW, Froidevaux L, Manney GL, Flower DA, et al. Interhemispheric differences in polar stratospheric HNO_3, H_2O, ClO, and O_3. *Science* 1995;**267**:849–52.
63. Toon OB, Hamill P, Turco RP, Pinto J. Condensation of HNO_3 and HCl in the winter polar stratosphere. *Geophys Res Lett* 1986;**13**:1284–7.
64. Santee ML, Froidevaux L, Manney GL, Read WG, Waters JW, Chipperfield MP, et al. Chlorine deactivation in the lower stratospheric polar regions during late winter: results from UARS. *J Geophys Res* 1996;**101**:18835–59.
65. Yang E-S, Cunnold DM, Newchurch MJ, Salawitch RJ, McCormick MP, Russell III JM, et al. First stage of Antarctic ozone recovery. *J Geophys Res* 2008;**113**:D20308.
66. Weber M, Dikty S, Burrows JP, Garny H, Dameris M, Kubin A, et al. The Brewer-Dobson circulation and total ozone from seasonal to decadal time scales. *Atmos Chem Phys* 2011;**11**:11221–35.
67. Solomon S, Ivy DJ, Kinnison DE, Mills MJ, Neely RR, Schmidt A. Emergence of healing in the Antarctic ozone layer. *Science* 2016;**353**:269–74.
68. Rex M, Salawitch RJ, Deckelmann H, von der Gathen P, Harris NRP, Chipperfield MP, et al. Arctic winter 2005: implications for stratospheric ozone loss and climate change. *Geophys Res Lett* 2006;**33**:L23808.
69. SPARC report on the evaluation of chemistry-climate models (SPARC CCMVal). In: Eyring V, Shepherd T, Waugh D, editors. *SPARC Report No. 5, WCRP-30/2010, WMO/TD – No. 40*; 2010. www.sparc-climate.org/publications/sparc-reports.
70. Kinnison DE, Grant KE, Connell PS, Rotman DA, Wuebbles DJ. The chemical and radiative effects of the Mount Pinatubo eruption. *J Geophys Res* 1994;**99**:25705–31.
71. Jackman CH, Fleming EL, Chandra S, Considine DB, Rosenfield JE. Past, present, and future modeled ozone trends with comparisons to observed trends. *J Geophys Res* 1996;**101**:28753–67.
72. WMO (World Meteorological Organization). *Scientific assessment of ozone depletion: 1998*. Global Ozone Research and Monitoring Project—Report No. 44. 1999. Geneva, Switzerland.
73. Salawitch RJ, Weisenstein DK, Kovalenko LJ, Sioris CE, Wennberg PO, Chance K, et al. Sensitivity of ozone to bromine in the lower stratosphere. *Geophys Res Lett* 2005;**32**:L05811.
74. Fahey DW, Kawa SR, Woodbridge EL, Tin P, Wilson JC, Jonsson HH, et al. In situ measurements constraining the role of sulphate aerosols in mid-latitude ozone depletion. *Nature* 1993;**363**:509–14.
75. Salawitch RJ, Canty T, Kurosu T, Chance K, Liang Q, da Silva A, et al. A new interpretation of total column BrO during Arctic spring. *Geophys Res Lett* 2010;**37**:L21805.
76. Chipperfield MP. Mid-latitude ozone depletion. In: Müller R, editor. *Stratospheric ozone depletion and climate change*. Cambridge, UK: RSC Publishing; 2012. p. 169–89.
77. Steinbrecht W, Claude H, Köhler U, Winkler P. Interannual changes of total ozone and northern hemisphere circulation patterns. *Geophys Res Lett* 2001;**28**:1191–4.
78. Forster PMDF, Tourpali K. Effect of tropopause height changes on the calculation of ozone trends and their radiative forcing. *J Geophys Res* 2001;**106**:12241–51.
79. Fusco AC, Salby ML. Interannual variations of total ozone and their relationship to variations of planetary wave activity. *J Clim* 1999;**12**:1619–29.
80. Harris NRP, Kyrö E, Staehelin J, Brunner D, Andersen S-B, Godin-Beekmann S, et al. Ozone trends at northern mid- and high latitudes – a European perspective. *Ann Geophys* 2008;**26**:1207–20.
81. Hudson RD, Andrade MF, Follette MB, Frolov AD. The total ozone field separated into meteorological regimes – Part II: Northern Hemisphere mid-latitude total ozone trends. *Atmos Chem Phys* 2006;**6**:5183–91.
82. Rosenfield JE, Douglass AR, Considine DB. The impact of increasing carbon dioxide on ozone recovery. *J Geophys Res* 2002;**107**:4049.
83. WMO (World Meteorological Organization). *Assessment for decision-makers: scientific assessment of ozone depletion: 2014*. Global Ozone Research and Monitoring Project—Report No. 56. 2014. p. 92. Geneva, Switzerland.

84. Eyring V, Arblaster JM, Cionni I, Sedláček J, Perlwitz J, Young PJ, et al. Long-term ozone changes and associated climate impacts in CMIP5 simulations. *J Geophys Res* 2013;**118**:5029–60.
85. Revell LE, Bodeker GE, Huck PE, Williamson BE, Rozanov E. The sensitivity of stratospheric ozone changes through the 21st century to N_2O and CH_4. *Atmos Chem Phys* 2012;**12**:11309–17.
86. Ravishankara AR, Daniel JS, Portmann RW. Nitrous oxide (N_2O): the dominant ozone-depleting substance emitted in the 21st century. *Science* 2009;**326**:123–5.
87. Revell LE, Tummon F, Salawitch RJ, Stenke A, Peter T. The changing ozone depletion potential of N_2O in a future climate. *Geophys Res Lett* 2015;**42**:10047–55.
88. Klobas JE, Wilmouth DM, Weisenstein DK, Anderson JG, Salawitch RJ. Ozone depletion following future volcanic eruptions. *Geophys Res Lett* 2017;**44**:7490–9.
89. Anderson JG, Wilmouth DM, Smith JB, Sayres DS. UV dosage levels in summer: increased risk of ozone loss from convectively injected water vapor. *Science* 2012;**337**:835–9.
90. Crutzen PJ. Albedo enhancement by stratospheric sulfur injections: a contribution to resolve a policy dilemma? *Clim Change* 2006;**77**:211–20.
91. Canty T, Mascioli NR, Smarte MD, Salawitch RJ. An empirical model of global climate – Part 1: a critical evaluation of volcanic cooling. *Atmos Chem Phys* 2013;**13**:3997–4031.
92. Tilmes S, Kinnison DE, Garcia RR, Salawitch RJ, Canty T, Lee-Taylor J, et al. Impact of very short-lived halogens on stratospheric ozone abundance and UV radiation in a geo-engineered atmosphere. *Atmos Chem Phys* 2012;**12**:10945–55.
93. Tilmes S, Müller R, Salawitch RJ. The sensitivity of polar ozone depletion to proposed geoengineering schemes. *Science* 2008;**320**:1201–4.
94. Newman PA, Oman LD, Douglass AR, Fleming EL, Frith SM, Hurwitz MM, et al. What would have happened to the ozone layer if chlorofluorocarbons (CFCs) had not been regulated? *Atmos Chem Phys* 2009;**9**:2113–28.
95. Velders GJ, Andersen SO, Daniel JS, Fahey DW, McFarland M. The importance of the Montreal Protocol in protecting climate. *Proc Natl Acad Sci USA* 2007;**104**:4814–9.
96. Velders GJ, Fahey DW, Daniel JS, McFarland M, Andersen SO. The large contribution of projected HFC emissions to future climate forcing. *Proc Natl Acad Sci USA* 2009;**106**:10949–54.
97. Velders GJ, Ravishankara AR, Miller MK, Molina MJ, Alcamo J, Daniel JS, et al. Preserving Montreal Protocol climate benefits by limiting HFCs. *Science* 2012;**335**:922–3.

CHAPTER

3.4

The Greenhouse Effect, Aerosols, and Climate Change

Daniel Kirk-Davidoff

University of Maryland, College Park, MD, United States

3.4.1 FUNDAMENTALS

Over the past 50 years, the earth's climate has changed at a rapid pace relative to the observed climate variability of the past 5000 years. The global mean surface air temperature has increased by about 0.8 °C, sea level has risen by about 12 cm, and precipitation patterns have shifted. This global warming and associated climate change has been driven by increasing concentrations of gases that absorb and emit infrared radiation in the atmosphere, principally carbon dioxide and methane. The warming has been reduced to some degree by increasing aerosol pollution, which has its own associated impacts on precipitation.[1] Continued emissions of greenhouse gases at present rates can be expected to produce continued warming, which would also continue after emissions have ceased, until thermodynamic equilibrium is achieved. In this chapter, we will discuss the chemistry of these gases and particles, focusing on the processes that determine their lifetime in the atmosphere and their fate when they are removed from the atmosphere, and then review the physics of climate, focusing on the role of greenhouse gases and aerosols. We will also discuss some proposed technological interventions to remove greenhouse gases from the atmosphere or counteract their impact on climate.

Fig. 3.4.1 shows the annual mean global mean surface temperature change from 1880 through 2016, as a difference from the average over the years 1950–80, on the left and the rise in greenhouse gas concentration since the Industrial Revolution on the right. There is nothing particularly surprising about the agreement between the two figures: the idea that increases in carbon dioxide would lead to warming of approximately the observed magnitude has been around for over a century. However, the complexity of the climate system, and the presence of year-to-year and decade-to-decade ups and downs in the observed temperature record makes prediction of the exact magnitude of future change for a given amount of greenhouse gas emissions difficult to predict in advance, and the accompanying changes to precipitation and sea level have also required much scientific work to monitor, explain, and predict.

Green Chemistry
http://dx.doi.org/10.1016/B978-0-12-809270-5.00009-1

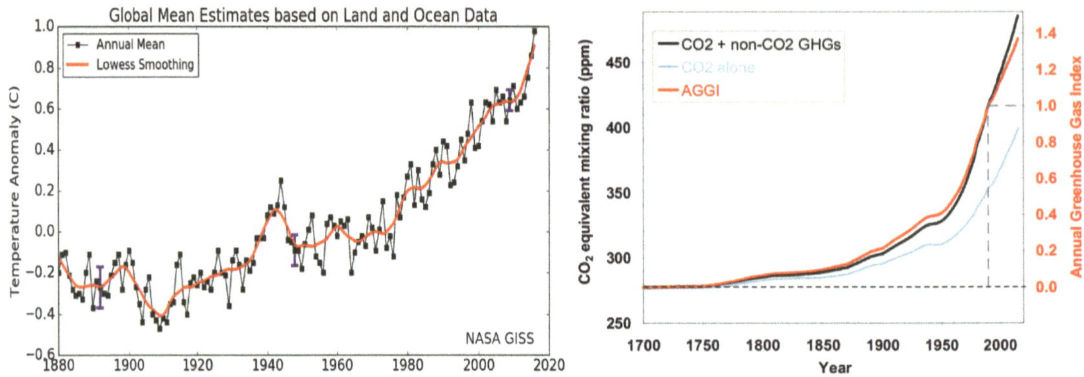

FIGURE 3.4.1 Left: Global average near-surface air temperature, 1880 to present, expressed as the difference from the temperature averaged from 1950 to 1980. The solid *black line* is the global annual mean and the solid *red line* (gray in print versions) is the 5-year LOWESS smoothed curve. The blue (dark gray in print versions) uncertainty bars (95% confidence limit) account only for incomplete spatial sampling.[38,39] Right: In red (gray in print versions), the annual greenhouse gas index, the ratio of the total direct radiative forcing due to long-lived greenhouse gases for any year for which adequate global measurements exist to that which was present in 1990.[40]

3.4.2 SOURCES AND SINKS OF GREENHOUSE GASES

Atmospheric composition changes that influence climate can be divided into two main categories: changes in molecular gases that interact with infrared radiation and changes in aerosols (small liquid or solid particles) that absorb or reflect visible radiation. In addition, ozone (O_3), which forms naturally in the atmosphere from molecular oxygen (O_2), interacts with both ultraviolet and infrared radiation, and its concentration is controlled by chemistry that has been changed substantially by human emissions of various pollutants.

Concentrations of CO_2, CH_4, and N_2O have all been rising rapidly since before 1900 and are now more abundant than their preindustrial levels by factors of 1.4, 2.5, and 1.5, respectively.[2] The impact of human emissions of these gases on climate depends on two main factors: the strength of interaction of gases with infrared radiation and the lifetime of gases in the atmosphere. For example, a ton of CH_4 added to the atmosphere increases the atmosphere's interaction with infrared radiation much more strongly than a ton of CO_2 added to the atmosphere [mostly because CH_4 is only present at a concentration of 1.8 parts per million by volume (ppmv), whereas CO_2 is present at about 400 ppmv], but CO_2 is much more persistent, so the overall impact of the CH_4 on climate will be about 110 times larger than that of the CO_2 in the first year after emission, about 30 times larger 40 years after emission, and only about 4 times larger 100 years after emission, when most of the CH_4 will have been lost to chemical processes.[3]

Anthropogenic greenhouse gas emissions arise from a range of human activities: industrial, agricultural, and domestic. Carbon dioxide emissions derive primarily from combustion of fossil fuels, in nearly equal parts each from burning coal, oil, and natural gas,[4] and secondarily from deforestation. Deforestation results in emission of carbon dioxide (and methane) because much of the woody biomass cut down when forests are cleared for conversion to agricultural land is either burnt or decomposes, converting carbon stored as cellulose in

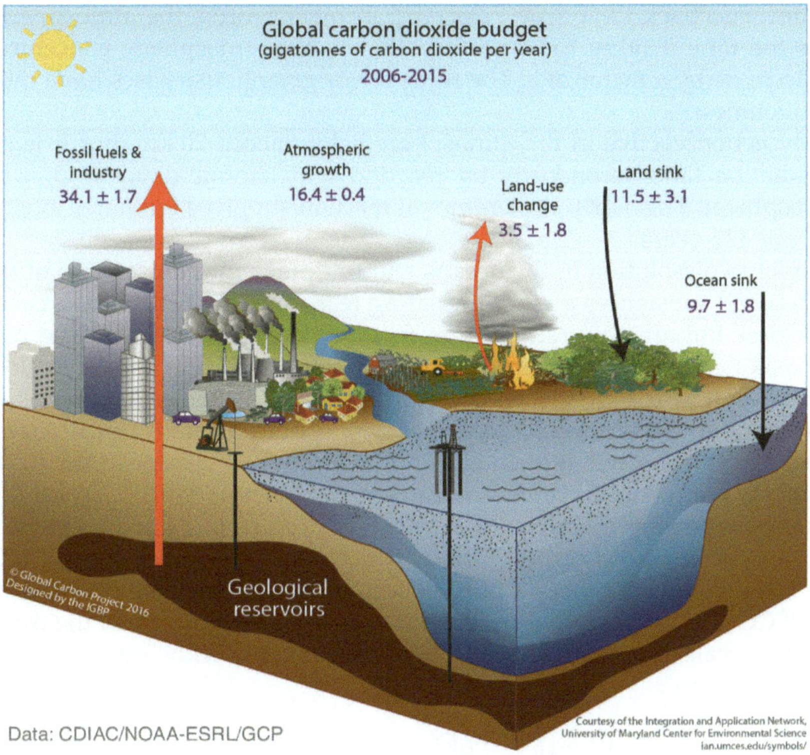

FIGURE 3.4.2 Fluxes of CO_2 averaged over the past decade.[41] Burning of oil, coal, and natural gas is the predominant cause of increasing atmospheric CO_2, with some additional CO_2 added by deforestation, especially in tropical rainforest areas. Of the CO_2 emitted from this combustion and deforestation, a bit less than half remains in the atmosphere, whereas the rest is either absorbed by the oceans or is taken up in forests and soils.

wood into carbon dioxide. In addition, chemical processes such as cement manufacturing, in which $CaCO_4$ is heated to form CaO_2, release CO_2. Fig. 3.4.2 shows the relative contribution of these sources to global CO_2 emissions. Methane emissions arise from fermentation either in the rumens of livestock or in anoxic agricultural environments such as rice paddies and manure piles or human waste treatment facilities, from leakage of natural gas (which is mostly methane) from wells and pipelines, and from burning of forests and fields. Methane emissions and concentrations rose exponentially through the 20th century along with the population, but leveled off in the late 20th century with improvements in rice farming practices that involved briefer periods of flooded fields, and reduction in leakage from natural gas drilling, before resuming their increase in the 2010s as these technological improvements were completed while population and wealth continued to increase.[5] Nitrous oxide emissions arise principally from soil bacteria, and are enhanced by the heavy use of nitrogen fertilizers.[6]

Greenhouse gases, once emitted to the atmosphere, do not remain there forever. We define the "atmospheric lifetime" of a chemical in the atmosphere as follows: if we add an increment of that chemical to the atmosphere so that its concentration increases from x kg/m^2 to $(x + y)$ kg/m^2, then the atmospheric lifetime is the time it would take for the concentration of that chemical to decrease back to $(x + ye^{-1})$ kg/m^2. In other words, the atmospheric lifetime of a chemical is the time it takes for a perturbation to the atmospheric concentration of that chemical to decrease by a factor of *e*. The three major greenhouse gases have rather different atmospheric lifetimes.

N_2O largely is nonreactive in the atmosphere below about 20 km, but it is destroyed by ultraviolet radiation (and secondarily by reaction with atomic oxygen, O, when it mixes into the stratosphere). This fairly slow removal mechanism gives it a fairly long atmospheric lifetime of about 116 years.[7]

CO_2 is subject to dissolution in the oceans and to uptake into those parts of the biosphere where carbon-based substances are accumulating (e.g., land areas that were cleared for agriculture in the past, but are now returning to forest). The fraction of CO_2 that remains in the atmosphere, making up the increment in concentration each year, had been close to 46% of what is emitted each year for the past several decades, with the remaining 54% being split between absorption in the ocean and in the land biosphere. There is much ongoing research about whether this fraction will remain constant over time—a change in the fraction remaining in the atmosphere would have major implications for how much CO_2 can be emitted in the future while preventing the global mean temperature from exceeding some threshold.[8]

The initial oceanic uptake of carbon dioxide takes places very quickly. One can think of the ocean mixed layer as simply a second box, adjacent to the atmosphere into which CO_2 flows. The amount of CO_2 that flows into this "box" is limited by the fraction of the ocean that turns over on the time scale of a year or so. CO_2 in the ocean quickly reacts with water in the following two reactions[9]:

$$CO_{2(aq)} + H_2O \rightleftharpoons HCO_3^- + H_3O^+$$

$$HCO_3^- \rightleftharpoons CO_3^{-2} + H_3O^+$$

The concentration of CO_2, HCO_3^-, and CO_3^{-2}, referred to collectively as dissolved inorganic carbon (DIC), is close to equilibrium with atmospheric CO_2, so as atmospheric CO_1 rises, the concentration of DIC and the concentration of hydronium ions (H_3O^+) increases. This implies an increase in the acidity of the ocean, as has been observed (Fig. 3.4.3). Thus the fraction of CO_2 emitted by people that fills the ocean mixed layer "box" does not influence the earth's climate, so the ocean absorption reduces the climate impact of a given amount of CO_2 emitted. However, the acidification of the ocean accompanying increasing DIC has its own set of ecological consequences, since it makes it more difficult for the microscopic animals and plants that form the base of the ocean food chain to make their shells.[10] This latter process leads to a slow sink of carbon from the ocean mixed layer to the ocean sediment: some fraction of those bits of plankton and dead plants and animals that are not decomposed and recycled within the mixed layer may eventually settle to the ocean floor and be buried, removing the carbon they contain completely from the ocean-atmosphere system, a process referred to as the "biological pump." If human emissions of carbon dioxide stop completely, over thousands of years this process will return atmospheric CO_2 concentrations to preindustrial levels.[1]

On land, CO_2 can be taken up in woody biomass (trees and bushes) or in soils (leaf litter that accumulates without decomposing completely), whereas it can be released when forests

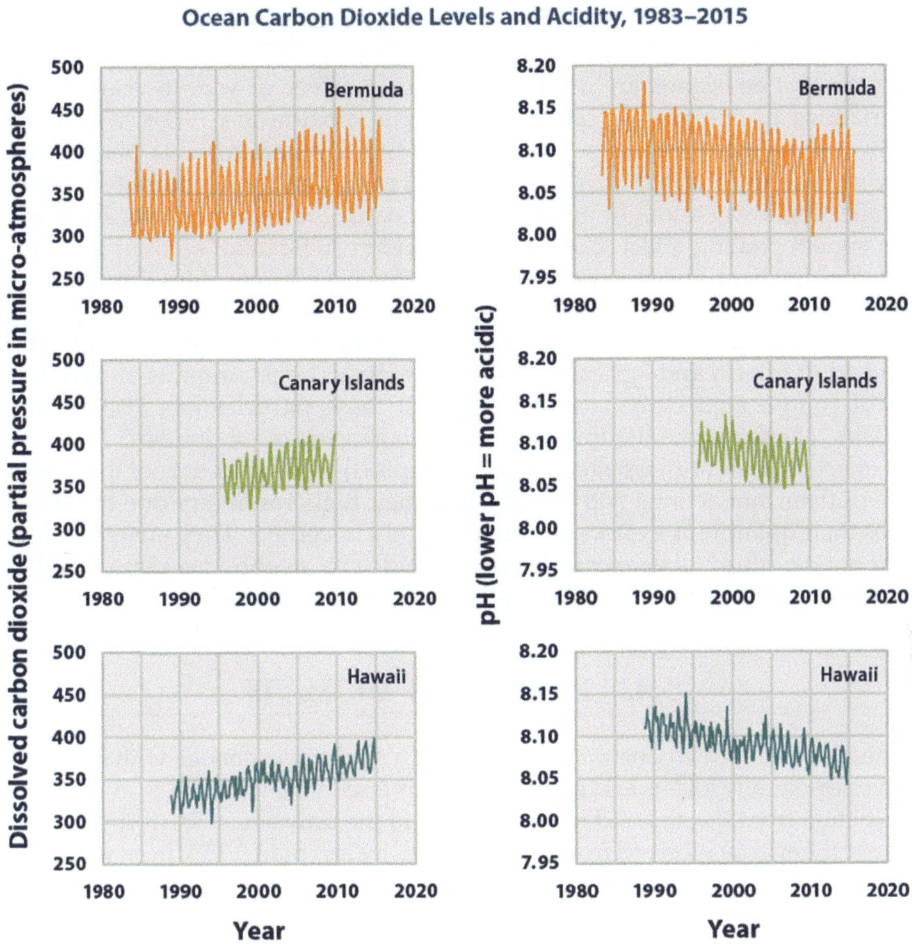

FIGURE 3.4.3 Direct observations of dissolved carbon dioxide concentration and pH of seawater sampled and the locations indicated in each graph. Trends toward greater acidity (lower pH) and higher carbon dioxide concentration are visible in each location.[42] A strong seasonal cycle is also present, associated with the seasonal cycle in atmospheric concentration of CO_2.

are burned or cut and allowed to rot, when organic soils are tilled and exposed to air, or when frozen peatlands thaw. While the tropical landmasses are significant net sources of CO_2 due to forest clearing (South America: 1.8 Pg CO_2/year, Africa: 0.4 Pg CO_2/year, Asia: 2.0 Pg CO_2/year), North America is a significant net sink (1.7 Pg CO_2/year) due to regrowth of forests cut down in the 19th and 20th centuries and due to growth of existing forests partly because of fire suppression.[11]

CH_4, unlike CO_2, is subject to chemical loss in the atmosphere. Most importantly, reaction with the OH (hydroxyl) radical yields CO_2 and water. This reaction yields an atmospheric lifetime for CH_4 of only about 10 years. Thus sustained high concentrations of CH_4 over more than a few decades requires sustained high emissions of CH_4.[5]

3.4.3 AEROSOLS AND CLIMATE

Volcanic aerosol emissions are a powerful natural driver of year-to-year and decade-to-decade variations in climate. Volcanoes emit sulfur dioxide gas (SO_2), which reacts with water in the atmosphere to form sulfuric acid (H_2SO_4). When volcanic plumes are emitted powerfully enough to reach the stratosphere,[a] the H_2SO_4 can form a persistent haze of liquid droplets, reflecting away sunlight and cooling the earth for a year or two. Human activities can have a similar cooling effect. Coal tends to contain substantial amount of sulfur, so that burning of coal for heat and power releases SO_2. Such pollution does not reach the stratosphere, so the added SO_2 has a fairly short atmospheric lifetime; however, the mass of emissions is large enough that a substantial cooling effect by H_2SO_4 droplets results. Burning of fossil fuels and of forests and agricultural fields results in large amounts of particle emission consisting of complicated carbon compounds. All these particles can also act to nucleate cloud particles, changing the lifetime and radiative properties of clouds.

The net effect of these anthropogenic (human-caused) aerosols is to cool the planet, but the complexity of their interactions with clouds, and their high variability due to their short lifetime, makes their quantitative effect on climate highly uncertain. They are estimated to cause a cooling about one-third as strong as the warming due to greenhouse gases, but the range of uncertainty is anywhere from zero to two-thirds of the greenhouse gas warming.[1]

3.4.4 PHYSICS OF CLIMATE

The climate of the earth is constrained by the earth's energy exchange with the universe. The earth receives essentially all[b] its energy from the universe in the form of ultraviolet, visible, and infrared radiation from the sun and returns energy to the universe in the form of infrared radiation. Because the energy received from the sun varies by less than 0.15% from year to year (most variability being associated with the approximately 11-year-long sunspot cycle),[12] the earth's climate is typically close to equilibrium: the annually averaged incoming and outgoing energy are equal to within about 1%. The bulk of yearly averaged net outward or inward energy transfer is caused by changes in atmospheric composition. These changes are large enough to be directly observed in satellite observations of the earth's outgoing radiative flux following volcanic eruptions, when increased aerosol concentrations in the stratosphere result in increased reflection of solar radiation to space.[13] For the somewhat smaller, but sustained and therefore more consequential, radiative response to increasing greenhouse gas concentrations, the energy imbalance can be calculated from the increase in heat stored in the oceans[14] and from radiative transfer models of the atmosphere.[15]

[a]The stratosphere is the layer of the atmosphere above 12–18 km in which the temperature rises with height, due to the absorption of solar ultraviolet radiation by ozone. This temperature structure suppresses strong vertical motions, so that air that reaches the stratosphere tends to remain there for a few years. Air enters the stratosphere in the tropics and returns to the troposphere (the layer between the stratosphere and the ground) in the polar regions.

[b]The next largest source of energy is the moon, from which the earth never receives more than about $1/(5 \times 10^4)$ as much energy as it receives from the sun (and that only when the moon is full).

3.4.4.1 Radiative Balance

Consider a small metal sphere orbiting the sun at about the earth's distance from the sun. We are using a metal sphere because it conducts temperature well, so we can assume that it has a single temperature. Its energy budget could be expressed as the change in heat energy resulting from the difference between the absorption of energy from the sun and the loss of energy by radiation emitted by the sphere:

$$\frac{4}{3}\pi r^3 \rho C \frac{dT}{dt} = (1-\alpha)S\pi r^2 - 4\pi r^2 \varepsilon\sigma T^4 \tag{3.4.1}$$

where ρ is the density of the metal, C is its heat capacity per unit mass, r is the radius of the sphere, T is its temperature, t is time, S is the flux of radiation from the sun that would fall on a plane facing the sun, α is the albedo of the sphere (the fraction of solar radiation reflected away from it, equal to 1 for a white sphere and 0 for a black sphere), σ is the Stefan-Boltzmann constant, and ε is the emissivity of the sphere (equal to 1 for a perfect blackbody).

Simplifying, we get:

$$\frac{r}{3}\rho C \frac{dT}{dt} = (1-\alpha)\frac{S}{4} - \varepsilon\sigma T^4 \tag{3.4.2}$$

If we assume the sphere's temperature starts at zero, the second term on the right-hand side of Eq. (3.4.2) will also be zero, so the right-hand side will be positive and the temperature will start to rise due to the absorption of solar radiation. As its temperature rises, the sphere will radiate ever more energy according to the second term on the right (the Stefan-Boltzmann law). Eventually the sphere will approach equilibrium, when the temperature will change only infinitesimally over time, so that the left-hand side is near zero. Then we can write

$$(1-\alpha)\frac{S}{4} = \varepsilon\sigma T^4 \text{ or}$$

$$T = \left[(1-\alpha)\frac{S}{4\varepsilon\sigma}\right]^{1/4} \tag{3.4.3}$$

If we assume that the sphere can radiate as well as a blackbody (so that $\varepsilon = 1$), and has the same albedo as the earth (so that $\alpha = 0.3$), and noting that the flux of radiation falling on a plane facing the sun at the earth's distance from the sun is $S = 1360\ \mathrm{W/m^2}$, we get a temperature $T = 255\mathrm{K} = -18\ ^\circ\mathrm{C}$, which is in the ballpark of the earth's temperature: it is the temperature of the air near 5 km elevation in the atmosphere.

However, the average temperature at the earth's surface is about 15 °C or 288K. What explains the difference? The peak wavelength of radiation emitted by a blackbody can be derived from Planck's law to be approximately $\lambda = 2900\ \mu\mathrm{m\ K}/T$, which gives about 11 μm for radiation emitted by our imaginary sphere. Substances like rock and water have emissivities of about 0.9, whereas snow, ice, and vegetation have emissivities of 0.97 or higher at wavelengths near 11 μm,[16] but even plugging a value of $\varepsilon = 0.9$ into Eq. (3.4.3) gives a temperature of just 260K. Assuming the earth's actual temperature and solving for the emissivity gives $\varepsilon = 0.61$. What is it that reduces the ability of the earth's surface to radiate infrared radiation to this level, resulting in its higher temperature?

The answer turns out to be the atmosphere. Because there are substances in the atmosphere, including water vapor, clouds, and a number of trace gases, that absorb and emit infrared radiation, an instrument placed outside the earth's atmosphere and looking toward the earth's surface will measure radiation that, depending on its wavelength, arises in part from the surface of the earth and in part from gases and clouds at various levels in the atmosphere. The average temperature at these levels in the atmosphere, weighted by the radiation they emit, corresponds approximately to the temperature we derived above, 255K.

One way to build intuition about this concept is to consider a two-level model of the earth's climate.[c] We can approximate the earth's surface as a perfect emitter (a blackbody) with $\varepsilon = 1$ and the earth's atmosphere as a partially absorbing layer with a finite emissivity, $\varepsilon = \varepsilon_a$. If we assume that the atmosphere is transparent to radiation from the sun, except for some that is reflected away by clouds and does not act to warm the climate, we can write two new energy balance equations:

$$\frac{4}{3}\pi r^2 D\rho_s C_s \frac{dT_s}{dt} = (1-\alpha)S\pi r^2 - 4\pi r^2\sigma T_s^4 + 4\pi r^2\varepsilon_a\sigma T_a^4 \quad (3.4.4)$$

$$4\pi r^2 H\rho_a C_a \frac{dT_a}{dt} = 4\pi r^2\varepsilon_a\sigma T_s^4 - 2\times 4\pi r^2\varepsilon_a\sigma T_a^4 \quad (3.4.5)$$

The terms on the right-hand sides of these equations are shown in the diagram on the left of Fig. 3.4.4. Eq. (3.4.4) represents the energy budget of the earth's surface: any positive change in the thermal energy of the surface (taken here to be a layer of water of depth D, density ρ_s, and heat capacity C_s spread on a sphere—the earth—of radius r), is equal to the solar heating of the surface minus the radiative energy emitted by the earth's surface plus the radiation emitted downward by the atmosphere. Eq. (3.4.5) represents the energy budget of the entire atmosphere, represented, contrary to the fact, as having a single temperature, a single density, and a finite thickness H. Any positive change in the thermal energy of the atmosphere is equal to energy radiated by the earth's surface (σT_s^4) and absorbed by the atmosphere with its characteristic emissivity ε_a (which we take to equal to its absorptivity, its fractional ability to absorb infrared radiation), minus the energy radiatively emitted by the atmosphere, both upward and downward (thus the factor of 2 in the second term on the left).

Simplifying, and assuming a steady solution with time derivatives equal to zero, we get from Eq. (3.4.5) that $T_a^4 = \frac{1}{2}T_s^4$. Substituting this back into Eq. (3.4.4) yields:

$$\sigma T_s^4 - \varepsilon_a\frac{1}{2}T_s^4 = \left(1-\frac{\varepsilon_a}{2}\right)\sigma T_s^4 = \frac{S}{4}(1-\alpha). \quad (3.4.6)$$

[c]A slightly more complicated version of this model that includes the possibility of some atmospheric absorption of solar radiation can be found in Marshall and Plumb.[36]

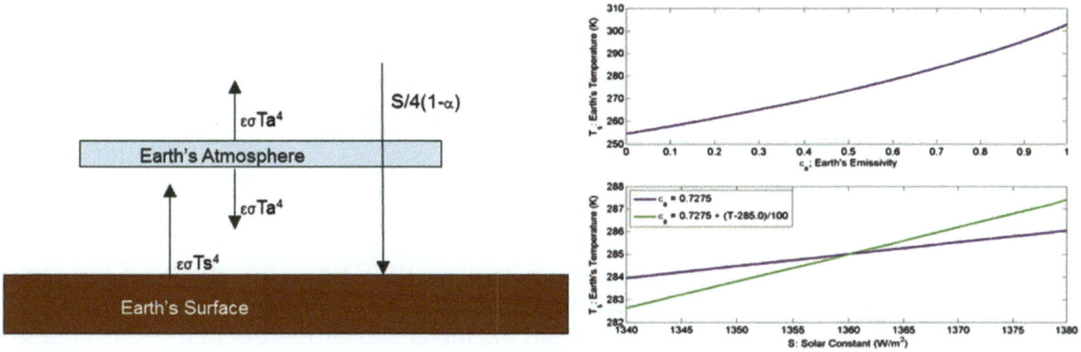

FIGURE 3.4.4 The diagram on the left shows the conceptual model of the earth's climate expressed in Eqs. (3.4.4) and (3.4.5). The top figure on the right shows the resulting steady-state temperature T_s as a function of the value of ε_a, the emissivity of the earth's atmosphere. The bottom figure on the right shows the T_s for a range of values of S, and again for the case where ε_a is a function of T_s. This shows the effect of a climate feedback: if the emissivity increases as a function of temperature, then the planet will respond more to an increase in solar forcing than it does for constant emissivity. This concept will be explored further in Section 3.4.4.4.

Solving for T_s gives

$$T_s = \sqrt[4]{\frac{S(1-\alpha)}{4\sigma\left(1-\frac{\varepsilon_a}{2}\right)}}. \tag{3.4.7}$$

Note that Eq. (3.4.6) implies that the emissivity of the earth-atmosphere system can be written as $\varepsilon = 1 - \frac{\varepsilon_a}{2}$: as the emissivity of the atmosphere increases, the emissivity of the earth as a whole decreases. Increasing atmospheric opacity to infrared radiation makes the earth's surface less able to shed radiation to space. So we should expect that as the atmosphere's emissivity increases, the temperature will need to increase, so that the system can shed the heat received from the sun and remain in equilibrium. The graph on the right of Fig. 3.4.4 shows the dependence of T_s on ε_a: as the atmosphere's ability to absorb and emit infrared radiation increases, the temperature of the surface does indeed increase. Thus gases such as carbon dioxide (CO_2), methane (CH_4), nitrous oxide (N_2O), and ozone (O_3) that absorb and emit infrared radiation are referred to as greenhouse gases, because, like the glass of a greenhouse, they reduce the net transfer of energy from the earth to space, warming the earth below.[d]

The upper limit of about 303K applies only to this model with its single-layer atmosphere with uniform temperature: a more realistic model with a continuous atmosphere can achieve much higher surface temperatures, as in fact occurs on Venus.[17] To understand why higher emissivity leads to higher temperature from a dynamical point of view, we can inspect the differential Eqs. (3.4.4) and (3.4.5) directly. If the emissivity increases, Eq. (3.4.4) tells us

[d]Actual greenhouse glass keeps greenhouses warm mostly by reducing *convection*, air motions that transfer heat away from the earth's surface into the atmosphere above,[37] whereas *greenhouse gases* warm the planet through radiative interactions, as we will discuss later.

that we should expect the resulting increase in downward radiation from the atmosphere to warm the surface. Eq. (3.4.5) tells us that we should expect the increase in radiative heat loss due to the atmosphere's emission of infrared radiation to be compensated by an increase in absorption of infrared radiation from the surface. The eventual increase in surface temperature leads to net warming of the atmosphere.

3.4.4.2 Radiative Transfer

The real atmosphere does not have a single emissivity. Rather, the ability of the atmosphere to absorb and emit radiation depends strongly on the wavelength of the radiation. This wavelength dependence in turn arises from the molecular properties of the atmosphere's various constituents, and from bulk properties of the atmosphere: temperature, density, and pressure. The phenomena associated with emission, absorption, and transmission of radiation in the atmosphere are collectively referred to as radiative transfer. Absorption and emission of energy by molecules is quantized. Radiative energy arrives at and departs from molecules in the form of photons with particular energies, and only those photons whose energy corresponds to a permitted energy transition of the molecule can be absorbed or emitted. For energies corresponding to infrared radiation, these transitions correspond to changes in the rotational and vibrational energy states of molecules in the atmosphere. Thus constituents like argon, which is present as individual atoms, and oxygen and nitrogen, which are present as homonuclear diatomic molecules, do not interact with infrared radiation: their symmetry means they do not present an electric dipole to the oscillating electromagnetic field associated with infrared radiation. Carbon dioxide obtains a dipole moment when it is warm enough to vibrate, as it is at terrestrial temperatures, so it has a rotational-vibrational set of absorption lines centered on the frequency of its bending mode at 667 cm^{-1}. Methane has similar rotational vibrational-rotational absorption bands. Water vapor has these, but, in addition, it has a strong pure rotational mode that extends over a broad region of the infrared spectrum below 600 cm^{-1}, because it possesses a dipole moment in its ground state, due to its bent geometry.

Fig. 3.4.5 shows the spectrum of radiation emitted by the earth to space as a function of location and season. Regions of the spectrum where atmospheric gases have high emissivity appear as low points in the spectrum (e.g., around 650 cm^{-1}), because radiation emitted by the warm surface is absorbed strongly by constituents of the atmosphere (in this case carbon dioxide, CO_2), whereas the radiation observed from space is emitted from the highest parts of the atmosphere, which are cold, and do not emit much radiation there. Regions of the spectrum where atmospheric gases have low emissivity (e.g., between 800 and 950 cm^{-1}) appear to touch up against a smooth curve, which would be close to the emission by a blackbody at a temperature close to that of the earth's surface, as predicted by Planck's law. The dips in radiation emitted to space in Fig. 3.4.5 are referred to as absorption lines and collectively as bands. For example, the region between 570 and 750 cm^{-1} is referred to as the CO_2 absorption band. Increasing the concentration of carbon dioxide increases the strength of its band, increasing the effective emissivity of the atmosphere as a whole.

Carbon dioxide is relatively abundant in the atmosphere at 400 ppmv, so the carbon dioxide band is strong, and increasing the concentration of the gas in the atmosphere mostly acts to increase the emissivity at the edges of the band, since the atmosphere is already nearly

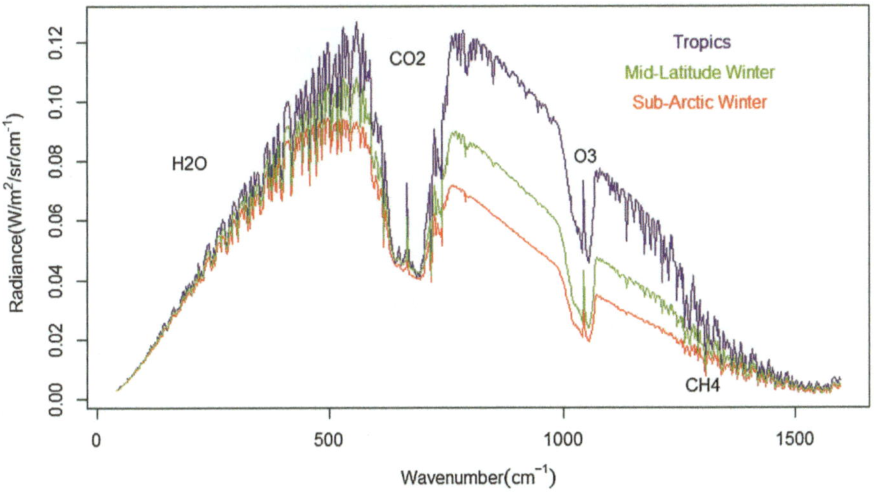

FIGURE 3.4.5 Overview of the earth's outgoing infrared radiation as a function of wave number (the inverse of wavelength) and latitude.[43] Radiances for this figure were calculated using Modtran and a web interface developed by David Archer available here: http://climatemodels.uchicago.edu/modtran/.

opaque to radiation at wavelengths in the center of the band. As a result, the radiative forcing of carbon dioxide (the extra heating at the surface due the presence of the gas, equivalent to $\varepsilon_a \sigma T^4$ in Eq. 3.4.5) goes as the natural logarithm of the concentration of the gas in the atmosphere. This is why one often hears about climate change referenced to a doubling of CO_2: for any doubling of the concentration of the gas, whether from 275 to 550 ppmv or from 300 to 600 ppmv, the natural log of the concentration increases by a fixed amount and the radiative forcing increases by about 3.7 W/m^2. On the other hand, a greenhouse gas like sulfur hexafluoride (SF_6), whose concentration is very low (8.3 parts per trillion), has very narrow, weak absorption lines. The impact of each additional molecule is very large (23,000 times larger than CO_2) because the lines can get deeper and broader with each additional molecule.[1] Its radiative forcing increases linearly with concentration. Methane, at 1.8 ppmv, is an intermediate case: its radiative forcing increases as the square root of its concentration.[18]

We humans are emitting carbon dioxide in large amounts (about 36×10^{12} kg, or 36 GT CO_2 per year from fossil fuel combustion and an additional 6 GT CO_2 that results from the burning or decomposition of trees when forecasts are cleared for agriculture). The *rate* of emissions has been increasing over the past 150 years, so that the concentration of CO_2 has increased exponentially, by about 0.5% per year since the mid-20th century. This exponential increase in concentration, combined with a logarithmic dependence of radiative forcing on concentration has resulted in a near-linear increase in radiative forcing due to CO_2. Expressed mathematically, if R is the warming due to CO_2 in W/m^2, a is a constant of proportionality, C is the present concentration of CO_2, and C_0 is the concentration at t_0, b is the exponential growth rate of the gas concentration, and t is the time in years, then:

$$R = a \ln (C/C_0) = a \ln \left(C_0 e^{b(t-t_0)}/C_0\right) = a(bt - bt_0) \tag{3.4.8}$$

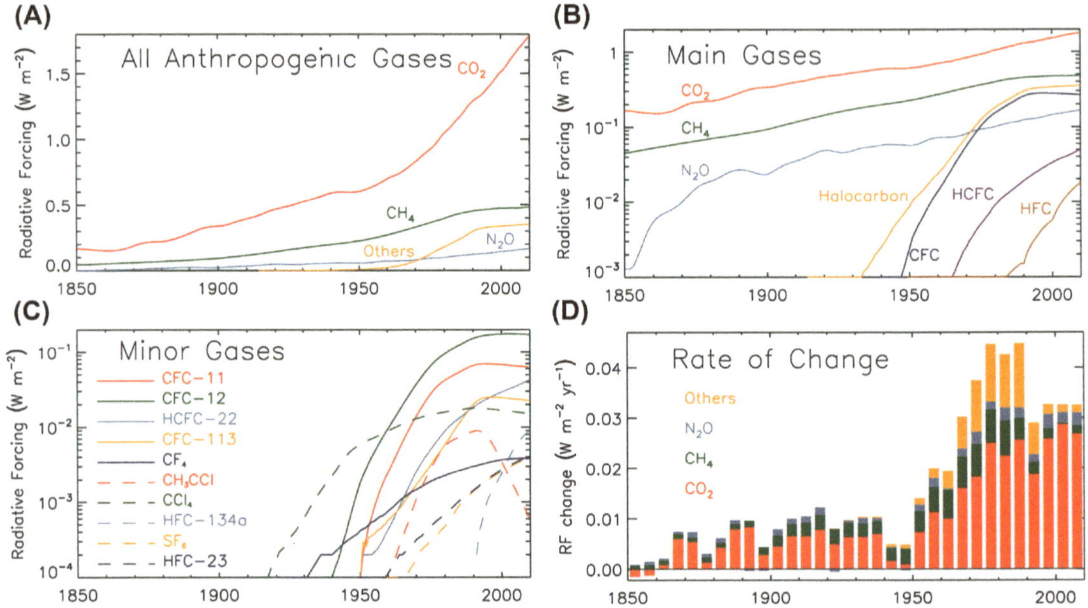

FIGURE 3.4.6 (A) Radiative forcing (RF) from the major well-mixed greenhouse gases (WMGHGs) and the sum of the minor gases such as chlorofluorocarbons and SF_6,1850 to 2011. (B) as (A) but with a logarithmic scale, (C) RF from the minor WMGHGs from 1850 to 2011 (logarithmic scale). (D) Rate of change in forcing from the major WMGHGs and groups of halocarbons from 1850 to 2011 (Fig. 8.6 in the IPCC Fifth Assessment Report[1]).

Fig. 3.4.6 shows the radiative forcing and its annual rate of change for the major greenhouse gases emitted by human activities: CO_2, CH_4, N_2O, and the sum of all other anthropogenic greenhouse gases. Since about 1960 the increase in radiative forcing due to CO_2 has been nearly linear, whereas the rate of increase for CH_4 has slowed and the rate for N_2O has accelerated, but from a low level. The large rates of change in the "other" category in the 1970s and the 1980s arose from the chlorofluorocarbons, gases that were found to damage the stratospheric ozone layer, and were banned by the 1989 Montreal Protocol: their rate of increase decreased sharply thereafter.

3.4.4.3 Anthropogenic Climate Change in Space and Time

Comparing the left panel of Fig. 3.4.6 with the left panel of Fig. 3.4.1, we note that as radiative forcing has increased in a nearly linear fashion over the past 50 years, temperature has also increased nearly linearly with time. We will now consider some of the questions this relationship suggests. Should we expect the temperature to continue to rise linearly as the forcing increases? What do we observe to be the pattern of temperature and other climate changes associated with this global mean temperature rise? And how should we expect these patterns to change over time?

In the following discussion, we will refer often to the "climate system," by which we mean the atmosphere, the oceans, and the parts of the land surface (including the living things that grow on or in it) that can interact with the atmosphere and ocean over time scales of days to thousands of years: soil, forests, fields, lakes, rivers, glaciers, and ice sheets. We often talk about the climate system responding to "external forcing," by which we mean either things not included in the earlier description (the sun, the orbital change caused by the earth's gravitational interaction with other planets, volcanoes, continental drift) or things done by human beings. We humans are not, of course, truly external to the climate system, but we will respect this useful fiction for the sake of clarity.

Understanding the response of the climate system to external forcing requires that we write down the equations that govern the system and then solve them. These equations involve derivatives of climate variables in space and time.[e] Since they do not have simple analytic solutions, we typically solve them in approximate form as difference equations using computers. Having expressed and assembled our understanding of the climate system in this way, we can pose questions to these climate or Earth system models: what do we expect to happen to temperatures in Siberia when a powerful volcanic eruption occurs in the Philippines? What do we expect to happen to rainfall in Iowa when carbon dioxide doubles? We can compare the answers that climate models give with what has been observed to occur in the historical record and gain a sense of how predictable the impacts of various forms of external forcing may be. For example, the volcanic eruption of Mount Pinatubo in 1991 resulted in a global averaged negative radiative forcing of about $-4\ W/m^2$, about the same magnitude as a doubling of carbon dioxide, but of opposite sign and lasting for only about 18 months.[19] Volcanoes act to cool the earth mostly by injecting sulfur dioxide into the stratosphere, where it is converted to a haze of sulfuric acid droplets. This haze was thick enough to be visible to the naked eye as whitening around the sun on clear days in the year following the eruption. This forcing resulted in global average cooling of about $-0.4\ ^\circ C$, peaking about 1 year after the eruption. This global average cooling was also produced by climate models, and the range of cooling observed in the models can be compared to the models' predicted warming under increasing carbon dioxide.[18] Other effects of such large climate forcing events, such as changes in wind and precipitation patterns, can also be compared in models and observations.[20]

3.4.4.4 Feedbacks and Climate Sensitivity

One of the primary questions that climate science has sought to address over the past few decades is: how much warmer will the earth get as a result of an increase in carbon dioxide? To answer this question, climate scientists first needed to state it carefully. This led to a few important definitions. The earth's *equilibrium climate sensitivity* (*ECS*) is the global average change in temperature of the air 2 m above the surface between two long periods (e.g., a few centuries), of which during one period the atmosphere's CO_2 concentration was close to its preindustrial value of 275 ppmv, and during the other it was twice that value, 550 ppmv. The *transient climate response* (*TCR*) is the change in temperature of the earth during a period when the CO_2 concentration is rising steadily from its preindustrial value of 275 ppmv, between the preindustrial average temperature and the temperature averaged over a few years centered on the time when the CO_2 concentration reaches two times its preindustrial values (550 ppmv). The TCR and the ECS are different because the earth's heat capacity, C_s in Eq. (3.4.4), is not zero, which means that the earth's temperature takes some time to respond to a radiative forcing (e.g., from a change in CO_2 concentration).

[e]For a thorough introduction to these equations, see Marshall and Plumb.[36]

The right-hand panels of Fig. 3.4.4 help to illustrate the concept of ECS. The upper panel shows the response of the surface temperature in the simple model of Eqs. (3.4.4) and (3.4.5) to changing emissivity (the simple model equivalent of changing the CO_2 in the real atmosphere). As the emissivity increases, the equilibrium temperature increases, by a nearly fixed amount for each increase in emissivity. The lower panel shows two different curves. In this case, the sensitivity being measured is to a change in the solar constant (S, the radiation received from the sun by a flat plane facing the sun at the position of the earth's orbit) rather than a change in emissivity. The blue and green curves show the sensitivity of two different model planets. The blue curve shows how the temperature changes as S increases for the model in Eqs. (3.4.4) and (3.4.5). The green curve shows how the temperature changes for a modified model in which the emissivity itself increases when the temperature increases:

$$\varepsilon_a = \varepsilon_0 + (T_s - 285\text{K})/(100\text{K}) \tag{3.4.9}$$

where ε_0 is the emissivity such that $T_s = 285\text{K}$ in the upper panel (~ 0.7275). In this model, a little extra heat from the sun (higher S), results in a larger increase in temperature than in the original model, because the warmer temperature also goes along with a higher emissivity, which means a little more radiation coming down from the atmosphere in Eq. (3.4.5), and thus an extra increase in surface temperature. This kind of mechanism, which increases the sensitivity of the model to the same change in energy from the sun, is referred to as *positive feedback*. This extra dependency of ε_a on T_s makes the model a bit harder to solve: one has to replace the derivatives on the left-hand sides of Eqs. (3.4.4) and (3.4.5) with time differences (setting $dT_s = T_{st+\Delta t} - T_{st}$, where Δt is a finite amount of time), solve for $T_{st+\Delta t}$ in terms of T_{st}, and then step through time until T_s stops changing.

Note that the existence of the aforementioned positive feedback does not *necessarily* imply an unstable climate, where any initial push causes a response that magnifies the initial push, so that the temperature rises or falls infinitely. Such a system is possible: simply replace the term (100K) in Eq. (3.4.8) with (50K) and the model climate will quickly shift to a state where either $\varepsilon_a = 0$ or $\varepsilon_a = 1$, assuming ε_a is required to lie in that range. And indeed, there have been times in the earth's history, most recently about 800 million years ago, when, because the sun emitted less radiation, a run-away to an "icehouse" climate occurred. As a result, the entire planet became covered with ice, and remained so until enough CO_2 was emitted by volcanoes so that the run-away operated in reverse, warming the planet swiftly and exposing enough ocean water to allow the CO_2 to be drawn down into the ocean and ocean floor deposits of calcium carbonate.[21] As the sun continues to brighten over the coming hundreds of millions of years, it is anticipated that we will again enter an unstable regime, when all the earth's oceans will evaporate into the atmosphere, leaving the earth with a Venus-like climate (unless of course intelligent life remains on the planet and is able to intervene effectively in this process with its own stabilizing feedback).[22]

However, in the current climate, we are in a regime where the existence of positive feedbacks merely weakens the strong negative feedback of the Stefan-Boltzmann law, so that warming temperatures result in increasing outgoing radiation, but the outgoing radiation increases with temperature *at a lower rate* than it would in the absence of positive feedbacks. Note that these feedbacks need not be active only for infrared radiation. The *ice-albedo feedback* describes the climate impact of the fact that warmer temperatures tend to result in smaller areas of ice and snow on Earth, so that as the temperature increases and the snow cover decreases, less solar radiation is reflected away from the earth and the earth warms more in response to a change in greenhouse gases or in solar radiation than it would otherwise. Cloudiness changes in response to warming temperatures can also cause changes in both long-wave emissions (since clouds act to increase infrared emissivity just as greenhouse gases do) and in short-wave emissions (since increasing cloud cover increases reflection of solar radiation). Changes in cloud properties with global temperature are among the most uncertain aspects of climate dynamics.[1]

The various climate feedback mechanisms can be compared quantitatively if we introduce a simplified form of the climate energy balance expression, in which we add a "forcing" term F representing some external change in the earth's radiative balance, for example, a change in the solar constant or a change in greenhouse gas concentration, that results in a measurable change in the balance of radiation at the top of the earth's atmosphere, expressed in W/m^2, and a "response" term, R, that represents the change of radiative balance at the top of the atmosphere associated with the warming or cooling of the earth in response to the forcing F, due to feedbacks other than blackbody. In equilibrium, the response will equal the forcing. In the "normal" climate, $\alpha = 0.31$, $\varepsilon_a = 0.61$, $S = 1360\ W/m^2$, and $T_0 = 288K$. We can write:

$$\frac{S}{4}(1-\alpha) + F = \varepsilon_a \sigma (T_0 + T)^4 + R,$$

where $\frac{S}{4}(1-\alpha) = \varepsilon_a \sigma T_0^4$, and since $\Delta T << T$, we can write:

$$F = 4\varepsilon_a \sigma T_0^3 \Delta T + R$$

Next we expand R into various responses to changing global temperatures:

$$F = 4\varepsilon_a \sigma T_0^3 \Delta T + R_{wv} + R_{\text{cloud}} + R_{\text{snow}} + R_{\text{lapserate}}$$

and express these responses as linear functions of the global mean temperature:

$$F = 4\varepsilon_a \sigma T_0^3 \Delta T - \lambda_{wv}\Delta T - \lambda_{\text{cloud}}\Delta T - \lambda_{\text{snow}}\Delta T - \lambda_{\text{lapserate}}\Delta T$$

The feedbacks are subtracted from the primary Planck feedback to preserve the convention that a *negative* feedback will result in more outgoing radiation (or less absorbed incoming radiation), and thus act to make the equilibrium temperature *lower*.

Rearranging and solving for ΔT, we get:

$$\Delta T = \frac{F}{4\varepsilon_a \sigma T_0^3 + \lambda_{wv} + \lambda_{\text{cloud}} + \lambda_{\text{snow}} + \lambda_{\text{lapserate}}} \tag{3.4.10}$$

For a doubling of CO_2, $F = 4\ W/m^2$, and since $4\varepsilon_a \sigma T_0^3 = 3.3\ W/m^2\ K$, Eq. (3.4.10) indicates that a doubling of CO_2 would lead, in the absence of any other feedbacks, to a warming of $\Delta T = 4.0\ W/m^2/3.3\ W/m^2\ K = 1.2\ K$. However, the other feedback mechanisms, when evaluated in climate models and in observational studies, are found on balance to be negative, meaning that warming temperature is associated with either reduced outgoing infrared radiation (e.g., because of increased atmospheric water vapor) or increased absorbed solar radiation (e.g., because of reduced snow cover and thus lower albedo). According to recent estimates from climate models, typical values of the climate feedback parameters are $\lambda_{wv} = 1.5\,W/m^2$, $\lambda_{cloud} = 0.5\,W/m^2$, $\lambda_{snow} = 0.3\,W/m^2$, and $\lambda_{lapserate} = -0.3\,W/m^2$. Subtracting these from the Planck feedback gives a denominator of $1.3\,W/m^2$ in Eq. (3.4.9) resulting in an ECS of about 3K, typical of many climate model estimates of the ECS.[23,24]

The ECS is one important aspect of the earth's climate system; it tells us how *much* the average temperature of the air just above the earth's surface will warm in response to a given radiative forcing, if we wait for the system to adjust itself completely. For practical purposes, however, it is important to understand how *fast* this adjustment will be. If all the extra energy from doubling CO_2 were mixed immediately throughout the oceans, the adjustment to equilibrium would take thousands of years and we might conclude that we could afford to do nothing to reduce emissions for hundreds of years, waiting for a perfect technological solution that would allow us to suck CO_2 out of the atmosphere and store it somewhere away from the atmosphere. However, in fact, the observed behavior of the climate system, both in response to greenhouse gas accumulation and in response to volcanic eruptions, shows that the leakage of heat into the deep oceans is quite slow. As a result, much of the approach to equilibrium is accomplished in the first tens of years, and involves only the land surface and the upper few hundred meters of the ocean. One way to quantify simultaneously the rate and magnitude of the climate response to greenhouse warming is the TCR. The TCR can be estimated from the rate of change of the global mean temperature observed so far, in combination with an estimate of the change in radiative forcing due to both greenhouse gases and aerosol pollution, compared to the preindustrial period. It can also be modeled using coupled climate models. Current (2016) global mean surface air temperature is about 1K warmer than in preindustrial times, and is warming at about 0.2K/decade, and CO_2 is currently increasing at 30 ppmv/decade from a value of 404 ppmv. A very simple approach to estimating the TCR would be to extrapolate these rates of increase to find the expected time of CO_2 doubling and the temperature at that time. At the present rate of increase, we expect that CO_2 would rise from its preindustrial value of 275 to −550 ppmv by about the year 2065, and if we extrapolate the present rate of global temperature increase, we would expect to reach a warming of 2K at that time. So the present trends are consistent with a TCR of 2K, assuming that aerosol cooling has approximately canceled warming from the greenhouse gases other than CO_2.[1] More detailed recent estimates of the TCR that account in detail for all known changes in radiative forcings do in fact result in a TCR of $2K \pm 0.8K$.[25]

3.4.4.5 Consequences of Global Warming

When the earth's average temperature changes, other climate system changes should be expected. For example, when the earth was about 4K colder at the coldest point of the last ice age, some 20,000 years ago, ice sheets expanded far equatorward, the sea level fell by about 125 m, precipitation was reduced over many midlatitude continent regions, and global dustiness increased dramatically.[26] What changes in climate variables other than temperature should we expect in a warmer world?

We will start from the most obvious consequences of a globally warmer world and move to consequences that are more complicated and more uncertain. In the first place, it has turned out to be the case that as our world has warmed globally by a degree or so, almost all locations have experienced warming. This need not have been the case: because of changes in ocean circulation that have accompanied global warming, a few locations such as the North Atlantic Ocean in the area around Greenland and the Southern Ocean off Antarctica have experienced some cooling. At times it has been speculated that these areas might have been larger, and might have included large section of Northern Europe, but the observational record has shown that these areas of cooling have been very limited.

However, the distribution of warming has not been, and is not expected to be, uniform. Warming is larger over land than over oceans for two reasons. As greenhouse gases increase and the infrared radiation received by the earth increases, the land and the ocean respond differently. First, the heat capacity of the ocean is larger than that of the land, both because water has a larger heat capacity than soil or rock, and more importantly, water can move and mix heat down from the surface, whereas heat can only diffuse slowly through rock and soil. So the extra heat received from the atmosphere is shared over a large depth of the ocean, but only a small volume of rock and soil on land. Second, evaporation from the ocean allows more heat to be carried away from the ocean (in the form of the latent heat of evaporation) than by turbulent mixing of sensible heat (proportional to temperature) alone. While this also occurs over moist soil, most land surfaces are less than fully saturated with water, so less evaporation can occur, and the land surface must warm to a higher temperature to balance the increase in downwelling infrared radiation.[27]

Warming is also larger over the high latitudes than in the tropics or over the South Pole. This is largely due to the ice-albedo and snow-albedo feedback, where warming results in less ice and snow, which in turn results in a darker surface of the earth that can absorb more radiation from the sun. The feedback does not function very strongly in Antarctica, which is so cold that no melting occurs (except on the Antarctic Peninsula), but has been very effective over the Arctic Ocean, where melting sea ice replaces a very reflective sea ice surface with very dark ocean water.[28] These effects are visible for both observed warming and warming projected by climate models in Fig. 3.4.7: warming is greater over the land than over the sea and greater at high latitudes than in the tropics.

Warming has occurred and is expected to occur in all seasons of the year and all times of the day, but in most locations, warming is stronger at night than in the day and slightly stronger in the winter than in the summer. This is because daytime conditions and summer conditions experience more *deep convection*, by which air rises vertically in thunderstorms. This process moves heat from the surface to the upper atmosphere more efficiently than radiation or the slanted motions that occur in wintertime storms.[1]

3.4.4.6 Sea Level Rise

Beyond warmer surface temperature, a globally warmer world implies that the upper layers of the ocean are warmer, and warmer water takes up more space than colder water. The density of seawater decreases by approximately 0.34 kg/m^3K. Since a cubic meter of sea water has a mass of about 1017 kg, for every 1K of warming of a 100-m-thick ocean layer, we expect about

$$\left(\frac{1\ \mathrm{m}^3}{1016.66\ \mathrm{kg}} - \frac{1\ \mathrm{m}^3}{1017.0\ \mathrm{kg}}\right) \times 1017\ \mathrm{kg} \times 100\ \mathrm{m} = 0.036\ \mathrm{m}$$

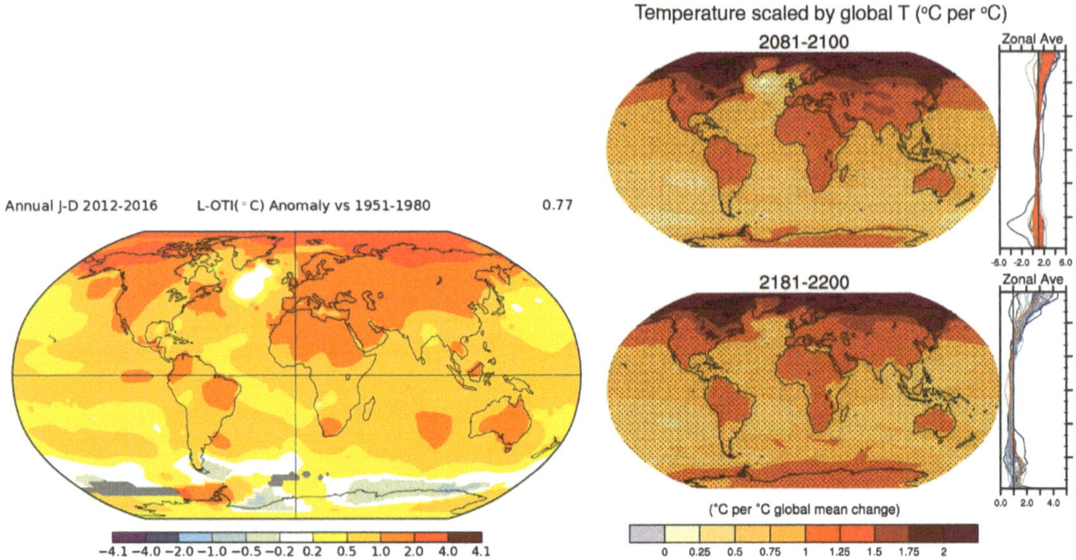

FIGURE 3.4.7 Warming observed to date (left)[38] and the distribution of warming expected for each degree of global average warming.[1] That is, the maps on the right show the amount of warming at each location in the world, divided by the global mean warming, averaged over the predictions of a large number of climate models, for two different times in the future: the end of the 21st century (top) and the end of the 22nd century (bottom). This emphasizes the strong agreement among the models that high latitude and land areas warm more than low latitude and ocean areas. The agreement with the observed pattern is also good, except for the observed cooling near the coast of Antarctica.

of sea level rise. The planet has warmed about 1K since the middle of the 20th century, and during this time, the sea level has risen by about 15 cm. The temperature of the ocean has been measured carefully and thoroughly down to about 1000 m since the turn of the 21st century; it turns out that the warming of the upper levels of the ocean accounts for about one-sixth of the observed sea level rise since 2003, as shown in Fig. 3.4.8. Other factors must be contributing to this rise. A warmer world is a world in which the snow line on mountains is higher up and glaciers are typically arranged with half their mass above the snow line and half their mass below. As the snow line rises, the total mass of the glacier will decrease (since the glacier cannot retreat above the top of the mountain!). Mountain glaciers are observed to be shrinking worldwide; this process has contributed a little more than 1 cm of sea level since 1960. Finally, warming air and sea temperatures are acting to increase melting of the continental ice sheets in Greenland and Antarctica, and to accelerate the flow of ice from these ice sheets into the ocean. This process has been especially strong in Greenland, whereas in Antarctica it has been at least partially counteracted by an increase in snowfall.[36] Although the contribution from this process has been a bit less than 1 cm of sea level rise since 1960, it is being closely monitored by satellite observations and direct measurement because the potential for sea level rise is so large. Melting the Greenland Ice Sheet would result in about 7 m of sea level rise, and we know that this has occurred in the past, as recently as the last warm period before the last ice age, 125,000 years ago. Such a continental-scale melting could not occur very rapidly, but

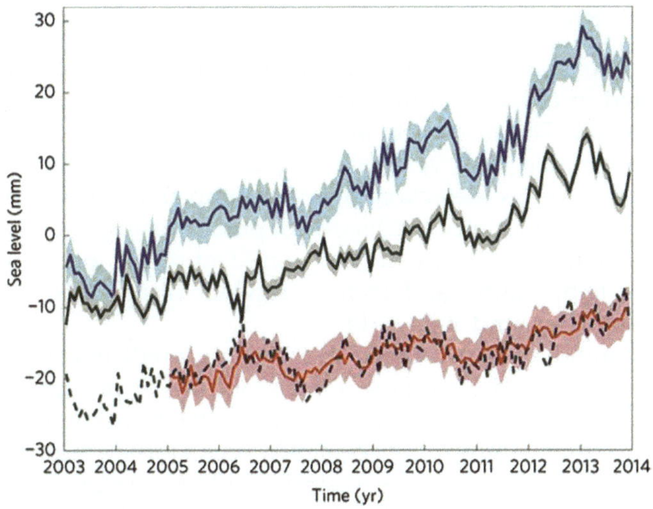

FIGURE 3.4.8 Contributions to sea level rise from reduced density of warming water [red (gray in print versions) curve, with pink (lightest gray in print versions) uncertainty bounds], from loss of ice on land (black curve with gray uncertainty bounds) and total sea level rise [blue (dark gray in print versions) curve, with light blue uncertainty bounds]. The *black (dark gray in print versions) dashed line* is a consistency check: it shows the difference of the *blue* (dark gray in print versions) and *black curves* (observed by different satellite methods) and agrees well with the *pink curve* (lightest gray in print versions) (calculated from observed temperature changes).[44]

up to 1 m of sea level rise due to melting of continental glaciers alone is thought to be possible by the year 2100, an amount sufficient to inundate large regions of coastal lands with high populations and large agricultural productivity.[1]

3.4.4.7 Ecological Consequences of Warming

Warming temperatures are causing changes in the range of various animal and plant species and stressing some populations. For example, warming of wintertime temperatures in western North America has led to a wider range of the Mountain Pine Beetle, which has led to blight and increased tree mortality over the Rocky Mountains of British Columbia, Canada.[29] Similarly, higher wintertime temperatures have led to the expansion of insect predators and forest damage to New Jersey pinelands and spruce forests in Alaska.[30] On the other hand, rising CO_2 concentrations themselves make it easier for plants to generate biomass via photosynthesis. Plants obtain CO_2 from the atmosphere by opening pores in their leaves called *stomates*, but when the stomates are open, they allow water to be lost from the leaves. With higher CO_2 concentration, plants can gain the needed carbon from the atmosphere with less loss of water vapor, allowing for better tolerance of high temperatures and water stress. In addition, warmer temperatures themselves extend plants' growing season in cool climates. As a result, the overall forest growth appears to be increasing over time in regions where forests are not being cut back.[31]

Warming is also associated with changes in atmospheric chemistry that can increase tropospheric ozone formation, a process discussed in more detail in Chapter 3.2.

3.4.4.8 Changes in Precipitation

Besides higher temperatures and their immediate consequences, a warming climate is observed and expected to differ from the preindustrial climate in its patterns of storminess and precipitation as well. These impacts are more uncertain, for two main reasons. First, precipitation is more variable from year to year than temperature in many locations, so it is harder to be confident of the detection of a long-term trend: the signal-to-noise ratio is lower. Second, precipitation is more difficult to predict using the weather models that form the core of climate models, so we are less confident of the mechanisms for changes in precipitation than for temperature and there is a wider disagreement among the various climate models developed around the world.

With this caveat in mind, there are some observed and predicted changes in precipitation of which we are reasonably confident. First, the distribution of precipitation is shifting from more frequent light events to less frequent and heavier events. Second, models and observations tend to show increases in precipitation in cold high latitude regions, and a decrease in precipitation in regions on the edge of the Tropics: the Mediterranean region in Europe and North Africa, Mexico, the southern United States, Southern Africa, and Australia. Third, changes are expected in the distribution and intensity of hurricanes and typhoons. These storms draw their energy from the contrast in temperature between the warm ocean surface and the cold upper atmosphere, and climate models generally predict that some regions of the planet, especially in the Pacific Ocean, will be able to support more intensive hurricanes in the future. The models generally predict a larger number of more intense storms, and a smaller number of less intense storms.[32] Fig. 3.4.9 shows an average over the models consulted by the Intergovernmental Panel on Climate Change (IPCC) of changes in precipitation, scaled by the change in temperature in each model. A general pattern of increasing precipitation near the equator, reduced precipitation in the subtropics (and extending notably into the Mediterranean region) and increased precipitation in the high northern and southern latitudes is evident.

3.4.5 TECHNOLOGY TO REDUCE GREENHOUSE GAS EMISSIONS

In light of the impacts discussed in the previous sections and reviewed in detail by the IPCC,[33] economists have estimated the damage done to the global human economy for each ton of CO_2 emitted today. A recent review calculated a range of values (due to uncertainty in economic damage of given effects, together with uncertainty in the amount of climate change for given level of greenhouse gas emissions) of \$7 to \$77 per ton of CO_2, with a most likely value of \$31 per ton.[34] The figure is often referred as the *social cost of carbon*. A consequence of this calculation is that an effort to reduce emission of CO_2 or an equivalent amount (in terms of climate impact averaged over some period of time) of another greenhouse gas, would be likely to have a net benefit to the human economy if it could be accomplished for less than \$32 per ton of reduced emissions.

Many such measures can be imagined. For example, it may be possible to add insulation to poorly insulated houses in such a way that the owner saves more money from reduced heating and cooling costs than the cost of the insulation and the labor to put it in place.

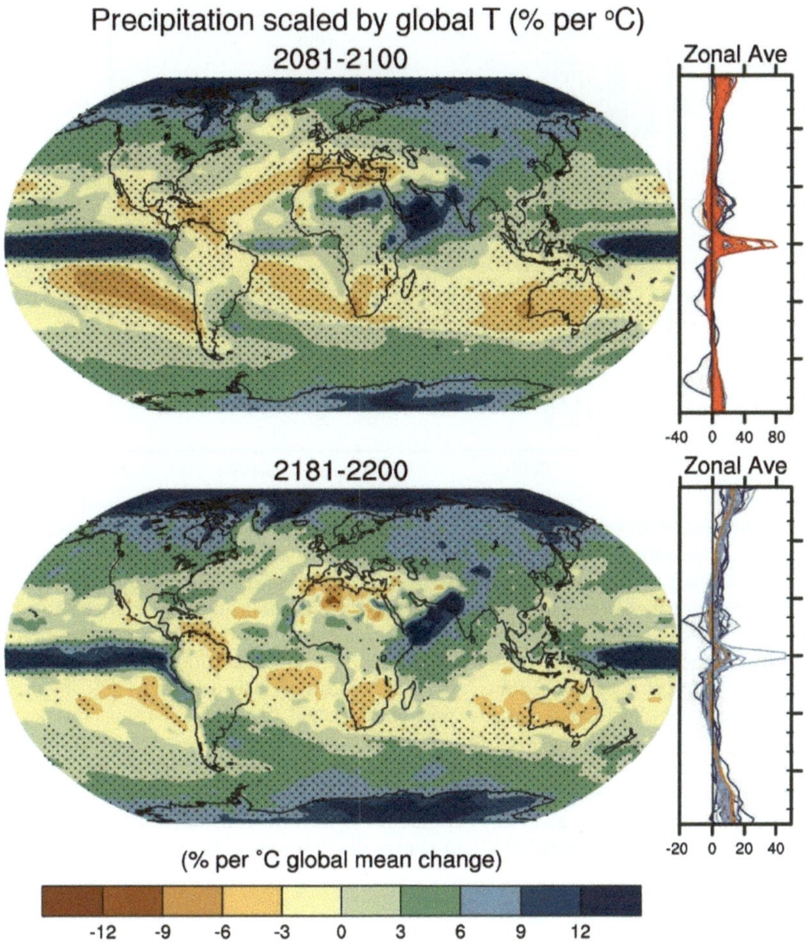

FIGURE 3.4.9 Patterns of observed precipitation change since 1970 and predicted patterns of precipitation changes per degree of global warming.[1]

In that case, CO_2 emissions are reduced for zero cost. Alternatively, one could design a device to remove CO_2 from the air and condense it into liquid form, using electricity generated by a hydroelectric dam. A demonstration facility to do just this was put in operation in Squamish, BC, in 2016.[35] The cost of such a process would depend on the success of a long process of learning by doing, but at present is likely much higher than $31 per ton CO_2. Other means of reducing CO_2 emissions are further along toward industrial-scale development. In locations that are windy or sunny, electricity generated from wind by wind turbines or from sunlight by photovoltaic panels is as cheap as any other source of electricity. A tax imposed on CO_2 emissions, set at something close to the estimated social cost of carbon, is often discussed as a simple measure that would tend to drive the economy toward lower carbon modes of production. Depending on the size of the tax, and the

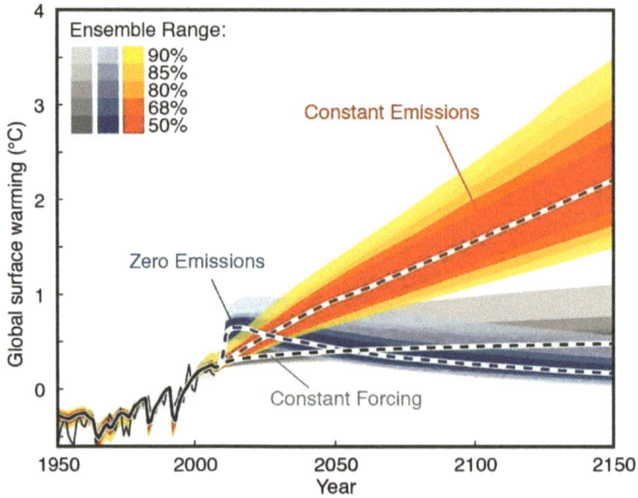

FIGURE 3.4.10 Probable results for global average temperature from various future greenhouse gas emission pathways.[1] The colors indicate the range of climate model uncertainty for each of three different emissions pathways. The red (gray in print versions) and yellow (light gray in print versions) colors show the range of climate outcomes for emissions of greenhouse gases frozen at 2011 levels, with the *dashed line* being the most probable outcome. The blue (dark gray in print versions) colors show the range of climate outcomes if emissions were suddenly halted. The gray colors show the outcome if emissions were halted in such a way as to produce constant climate forcing after 2011. The blue (dark gray in print versions) trajectories jump suddenly because emissions of greenhouse gases are accompanied by emissions of short-lived aerosols that tend to cool the planet. When emissions are halted, the aerosols decline sharply to preindustrial levels over a few years, resulting in a sudden jump in temperature, which decays as methane is converted to CO_2 in the atmosphere and CO_2 is drawn down into the ocean. Note that greenhouse gas emissions have been increasing steadily for decades, so all these pathways imply some social effort to reduce emissions, or at least to halt the growth of emissions. "Business as usual" with continued increasing emissions, would result in even warmer conditions than shown for the red (gray in print versions)-shaded emission scenario.

fraction of the world's economic activity on which it could be imposed, and on the response of human scientific, engineering, and economic activity, such a tax might be expected to slow the growth of emissions, to freeze emissions, or to actually reduce or even reverse emissions, with the accompanying range of results for climate shown in Fig. 3.4.10 below.

The human economy is far more complex and far more difficult to predict than the climate system. How rapidly we will be able and willing to reduce greenhouse gas emissions will determine how dramatic the consequences of their rising concentration on natural environments and human communities will become. Climate science has achieved a remarkable success in correctly predicting, well in advance, that fossil fuel combustion would lead to large changes in the earth's climate. Political and diplomatic action in response to this knowledge has led to the rapid growth in new non-fossil fuel energy industries. Our descendants will judge whether our response meets the demands of this knowledge and opportunity.

References

1. Myhre G, Shindell D, Bréon FM, Collins W, Fuglestvedt J, Huang J, et al. Anthropogenic and natural radiative forcing. In: Stocker TF, Qin D, Plattner GK, Tignor M, Allen SK, Boschung J, et al., editors. *Climate change 2013: The physical science basis. Contribution of working Group I to the Fifth Assessment Report of the Intergovernmental Panel on Climate Change*. Cambridge, UK: Cambridge University Press; 2013.
2. Blasing TJ. *Recent greenhouse gas concentrations*. Carbon Dioxide Information Analysis Center; 2016. http://dx.doi.org/10.3334/CDIAC/atg.032 [Internet] Available from:http://cdiac.ornl.gov/pns/current_ghg.html.
3. Allen MR, Fugelstvedt JS, Shine KP, Reisinger A, Pierrehumbert RT, Forster PM. New use of global warming potentials to compare cumulative and short-lived climate pollutants. *Nat Clim Change* 2016;**6**:773–6. http://dx.doi.org/10.1038/nclimate2998.
4. International Energy Agency. *CO_2 emissions from fuel combustion highlights*. Geneva: International Energy Agency; 2015 [Internet] Available from: https://www.iea.org/publications/freepublications/publication/CO2EmissionsFromFuelCombustionHighlights2015.pdf.
5. Kirschke S, Bousquet P, Ciais P, Saunois M, Canadell JG, Dlugokencky EJ, et al. Three decades of global methane sources and sinks. *Nat Geosci* 2015;**6**:813–23. http://dx.doi.org/10.1038/ngeo1955.
6. Tian H, Chen G, Lu C, Xu X, Ren W, Zhang B, et al. Global methane and nitrous oxide emissions from terrestrial ecosystems due to multiple environmental changes. *Ecosyst Health Sustainability* 2015. http://dx.doi.org/10.1890/EHS14-0015.1.
7. Prather MJ, Hsu J, DeLuca NM, Jackman CH, et al. Measuring and modeling the lifetime of nitrous oxide including its variability. *J Geophys Res Atmos* 2015;**120**:5693–705. http://dx.doi.org/10.1002/2015JD023267.
8. Ciais P, Sabine C, Bala G, Bopp L, Brovkin V, et al. Carbon and other biogeochemical cycles. In: Stocker TF, Qin D, Plattner GK, Tignor M, Allen SK, Boschung J, et al., editors. *Climate change 2013: The physical science basis. Contribution of working Group I to the Fifth Assessment Report of the Intergovernmental panel on climate change*. Cambridge, United Kingdom and New York (NY), USA: Cambridge University Press; 2013.
9. Feely RA, Doney SC, Cooley SR. Ocean acidification: present conditions and future changes in a high-CO_2 world. *Oceanography* 2009;**22**:36–47. http://dx.doi.org/10.5670/oceanog.2009.95.
10. Sabine CL, Feely RA, Gruber N, Key RM, Lee K, Bullister JL, et al. The oceanic sink for antrhopogenic CO_2. *Science* 2004;**305**:367–71. http://dx.doi.org/10.1126/science.1097403.
11. United States Department of Agriculture. *US agricultural and forestry greenhouse gas inventory: 1990–2013*. Technical Bulletin No. 1943. Office of the Chief Economist, Climate Change Program Office; 2016. http://dx.doi.org/10.15482/USDA.ADC/1264344.
12. Coddington O, Lean JL, Pilewski P, Snow M, Lindholm D. A solar irradiance climate data record. *Bull Am Met Soc* 2016. http://dx.doi.org/10.1175/BAMS-D-14-00265.1.
13. Minnis P, Harrison EF, Stowe LL, Gibson GG, Denn FM, et al. Radiative climate forcing by the Mount Pinatubo eruption. *Science* 1993;**259**:1411–5.
14. Trenberth KE, Fasullo JT, Balmaseda MA. Earth's energy imbalance. *J Climate* 2014. http://dx.doi.org/10.1175/JCLI-D-13-00294.1.
15. Liou KN. *An introduction to atmospheric radiation*. Academic Press; May 09, 2002.
16. Feldman DR, Collins WD, Pincus R, Huang X, Chen X. Far-infrared surface emissivity and climate. *Proc Nat Acad Sci* 2014;**111**:16297–302. http://dx.doi.org/10.1073/pnas.1413640111.
17. Pollack JB, Toon O, Boese R. Greenhouse models of Venus' high surface temperature, as constrained by Pioneer Venus measurements. *J Geophys Res Space Phys* 1980;**85**(A13):8223–31. http://dx.doi.org/10.1029/JA085iA13p08223.
18. Ramaswamy V, Boucher O, Haigh J, Hauglustaine D, Haywood J, Myhre G, et al. Radiative forcing of climate change. In: Houghton JT, Ding Y, Griggs DJ, Noguer M, van der Linden PJ, Dai X, et al., editors. *Climate change 2001: The scientific basis. Contribution of working Group I to the third Assessment Report of the Intergovernmental panel on climate change*. Cambridge, UK: Cambridge University Press; 2001.
19. Bender FAM, Ekman AML, Rodhe H. Response to the eruption of Mount Pinatubo in relation to climate sensitivity in the CMIP3 models. *Clim Dyn* 2010;**35**:875. http://dx.doi.org/10.1007/s00382-010-0777-3.
20. Barnes EA, Solomon S, Polvani L. Robust wind and precipitation responses to the Mount Pinatubo eruption, as simulated in the CMIP5 models. *J Clim* 2015. http://dx.doi.org/10.1175/JCLI-D-15-0658.1.

21. Benn DI, Le Hir G, Bao H, Donnadieu Y, Dumas C, et al. Orbitally forced ice sheet fluctuations during the Marinoan Snowball Earth glaciation. *Nat Geosci* 2015;**8**:704–7. http://dx.doi.org/10.1038/ngeo2502.
22. Schröder KP, Smith RC. Distant future of the Sun and Earth revisited. *Mon Not R Astron Soc* 2008;**386**:155–63. http://dx.doi.org/10.1111/j.1365-2966.2008.13022.x.
23. Gettelman A, Kay JE, Shell KM. The evolution of climate sensitivity and climate feedbacks in the community atmosphere model. *J Clim* 2012. http://dx.doi.org/10.1175/JCLI-D-11-00197.1.
24. Flato G, Marotzke J, Abiodun B, Braconnot P, Chou SC, et al. Evaluation of climate models. In: Stocker TF, Qin D, Plattner GK, Tignor M, Allen SK, et al., editors. *Climate change 2013: The physical science basis. Contribution of working Group I to the Fifth Assessment Report of the Intergovernmental Panel on Climate Change*. Cambridge, United Kingdom and New York (NY), USA: Cambridge University Press; 2013.
25. Storelvmo T, Leirvik T, Lohmann U, Phillips PCB, Wild M. Disentangling greenhouse warming and aerosol cooling to reveal Earth's climate sensitivity. *Nat Geosci* 2016;**2016**(9):286–9. http://dx.doi.org/10.1038/ngeo2670.
26. Albani S, Delmonte B, Maggi V, Baroni C, Petit JR, et al. Interpreting last glacial to Holocene dust changes at Talos Dome (East Antarctica): implications for atmospheric variations from regional to hemispheric scales. *Clim Past* 2012;**8**:741–50. http://dx.doi.org/10.5194/cp-8-741-2012.
27. Berg A, Findell K, Lintner B, Giannini A, Seneviratne SI, et al. Land–atmosphere feedbacks amplify aridity increase over land under global warming. *Nat Clim Change* 2016;**6**:869–74. http://dx.doi.org/10.1038/nclimate3029.
28. Pistone K, Eisenman I, Ramanathan V. Observational determination of albedo decrease caused by vanishing Arctic sea ice. *Proc Nat Acad Sci* 2014;**111**:3322–6. http://dx.doi.org/10.1073/pnas.1318201111.
29. Fauria MM, Johnson EA. Large-scale climatic patterns and area affected by mountain pine beetle in British Columbia, Canada. *J Geophys Res Biogeosci* 2009. http://dx.doi.org/10.1029/2008JG000760.
30. Weed AS, Ayres MP, Hicke JA. Consequences of climate change for biotic disturbances in North American forests. *Ecol Monogr* 2013;**83**:441–70. http://dx.doi.org/10.1890/13-0160.1.
31. McMahon SM, Parker GG, Miller DR. Evidence for a recent increase in forest growth. *Proc Nat Acad Sci* 2010;**107**:3611–5. http://dx.doi.org/10.1073/pnas.0912376107.
32. Trenberth KE. Changes in precipitation with climate change. *Clim Res* 2011;**47**:123–38. http://dx.doi.org/10.3354/cr00953.
33. Field CB, Barros VR, Dokken DJ, Mach KJ, Mastrandrea MD, et al., editors. *Climate change 2014: impacts, adaptation, and vulnerability. Contribution of working Group II to the Fifth Assessment Report of the Intergovernmental Panel on Climate Change.* Cambridge, United Kingdom and New York (NY), USA: Cambridge University Press; 2014. 1132 pp.
34. Nordhaus WD. Revisiting the social cost of carbon. *Proc Nat Acad Sci* 2017;**114**:1518–23. http://dx.doi.org/10.1073/pnas.1609244114.
35. Carbon Engineering. *CE demonstration plant-a year in review*. [Internet] Available from: http://carbonengineering.com/updates/.
36. Marshall J, Plumb RA. *Atmosphere, ocean, and climate dynamics: an introductory text*. Burlington (MA): Elsevier Academic Press; 2008.
37. Wood RW. Note on the theory of the greenhouse. *Philos Mag* 1909;**17**:319–20. http://dx.doi.org/10.1080/14786440208636602.
38. GISTEMP Team. *GISS surface temperature analysis (GISTEMP)*. NASA Goddard Institute for Space Studies. Available from: https://data.giss.nasa.gov/gistemp/.
39. Hansen J, Ruedy R, Sato M, Lo K. Global surface temperature change. *Rev Geophys* 2010:48. http://dx.doi.org/10.1029/2010RG000345. RG4004.
40. Butler JH, Montzka SA. *The NOAA annual greenhouse gas index (AGGI)*. 2016 [Internet] Available from: http://www.esrl.noaa.gov/gmd/aggi/aggi.html.
41. Le Quéré C, Andrew RM, Canadell JG, Sitch S, Korsbakken JI, Peters GP, et al. Global carbon budget 2016. *Earth Syst Sci Data* 2016;**8**:605–49. http://dx.doi.org/10.3334/CDIAC/GCP_2016 [Internet] Figure 2 is used courtesy of the Integration and Application Network, University of Maryland Center for Environmental Science. Available at: ian.umces.edu/symbols.
42. Environmental Protection Agency. *Climate change indicators: ocean acidity*. 2016 [Internet] Available from: https://www.epa.gov/climate-indicators/climate-change-indicators-ocean-acidity.
43. European Space Agency. Earth observation portal. *Far-infrared spectroscopy of the troposphere*. [Internet] Available from: https://directory.eoportal.org/web/eoportal/airborne-sensors/first.
44. Llovel W, Willis JK, Landerer FW, Fukumori I. Deep-ocean contribution to sea level and energy budget not detectable over the past decade. *Nat Clim Change* 2014;**4**:1031–5. http://dx.doi.org/10.1038/nclimate2387.

CHAPTER

3.5

Chemistry of Natural Waters

Robyn E. Hannigan, Daniel M. Genest, William E. Robinson

University of Massachusetts Boston, Boston, MA, United States

3.5.1 INTRODUCTION

As green chemists work on the synthesis of new chemicals and the redesign of existing chemical compounds, they learn a great deal about the chemical and physical properties of the chemical they are working on as well as the feedstocks and solvents used in its synthetic pathway. They also learn how to manipulate these properties by altering the basic structure of their chemical, adding and/or removing various functional groups, and altering the chemical environment in which these reactions take place. These processes are all part of the green chemists' "tool box," and are explored in more detail in the later chapters of this text. The present chapter, however, addresses what green chemists often may not fully consider (although they really should)—how their chemical reacts (physically, chemically, and biologically) when it is released into the environment, regardless of whether this release is through normal usage, industrial discharge during the manufacturing process, or by accident. The *fate* and *effect* of their chemical need to be considered prior to release, after the synthetic pathway has been worked out, but well before scale-up for manufacturing begins. The "effects" of synthetic compounds and molecules on biological systems (e.g., toxicological effects on individual organisms, ecological communities, wildlife, and humans) are discussed in other chapters (e.g., Chapters 2.4 and 3.7). This chapter deals with the "fate," specifically the natural processes that occur in aqueous media that will govern the fate of a compound. To understand the fate, we need to better understand the physical, chemical, and biological environment and how these vary across spatial and temporal scales.

Inorganic and organic compounds make their way to soils, sediments, groundwater, and surface waters through a variety of intermediate steps (Fig. 3.5.1). Use of these chemicals by the public may introduce them to the environment either directly (e.g., sunblocks, insect repellants, fragrances) or via landfills, household septic systems, or wastewater treatment facilities (e.g., pharmaceuticals, personal care products). Industrial wastes may be disposed in landfills, wastewater treatment facilities, or (in a few instances) via direct discharge of effluent.

Green Chemistry
http://dx.doi.org/10.1016/B978-0-12-809270-5.00010-8

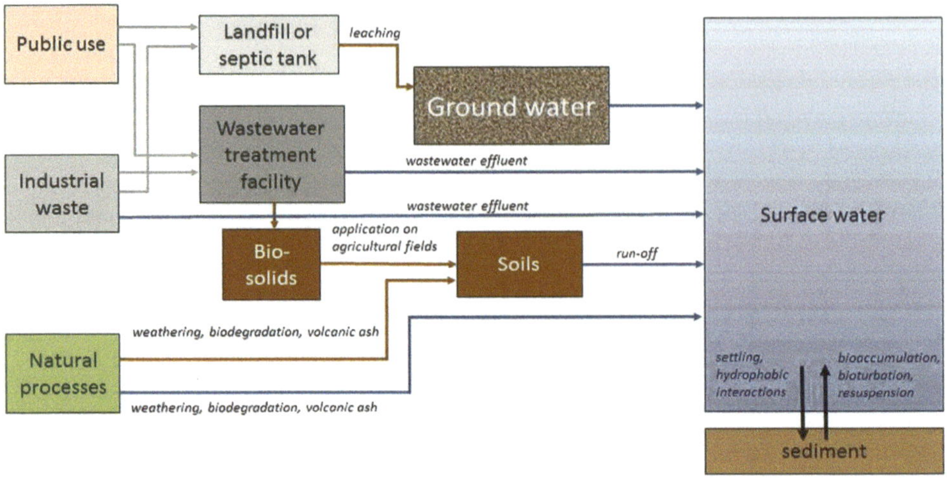

FIGURE 3.5.1 Sources of contaminants and transport to environmental reservoirs through water pathways. Industrial waste and the use of a product by the general public could disperse chemical compounds into the environment either directly or via landfills, septic tanks, or wastewater treatment facilities. Natural processes such as weathering, biodegradation, and volcanism also disperse chemical compounds into the environment. All avenues lead to surface waters (streams, rivers, lakes, oceans, and surface runoff after rainstorms) and ultimately to the sediments at the bottom of these water bodies. *Diagram created by Helen C. Poynton and used by permission.*

Chemicals designated as "hazardous waste" (e.g., pesticides, oils, grease, metals), whether generated by the public or by industry, are typically dealt with individually at the site of production and/or during wastewater treatment to prevent their transport into the environment. Volatiles, semivolatiles, and native metals can make their way directly onto soils and surface waters by both dry and wet deposition.[1]

Water, whether fresh or salt, plays a crucial role in the transport of chemical compounds through these various environmental compartments (Fig. 3.5.1). Leaching from septic tanks and landfills, direct discharge from wastewater treatment plants, runoff and infiltration into soils, and the mixing of groundwater with fresh and marine surface waters all involve water. The ultimate sink for all inorganic and organic compounds and molecules is oceanic sediment. The final form a compound takes when deposited depends on water column processes along the transport route as well as the mineralogy and pore water chemistry of the sediments themselves. Once deposited, biological (e.g., bioturbation, deposit feeding) and physical (e.g., resuspension) processes within the surface sediments may remobilize materials back into the water column, their chemistry possibly changing again before finally being deposited in a stable form, buried, and ultimately bound in the sediments during compaction and lithification into sedimentary rock over geologic time scales.

Water is an extremely important driver of Earth systems (i.e., weather, climate, carbon cycle). Almost 71% of the earth's surface is covered by water, most of which is in the oceans (96.5%) and only 3.5% of which is freshwater (i.e., ice, lakes, ponds, rivers, and permafrost) and thus available for plant and animal consumption.[2] Life, as we know it, depends on water. Approximately 50%–60% of the human body, for example, is water.[3] Most of this water is contained in our cells (~55%), although a significant amount is held in the blood and urine (~45%).[3] Water is the medium in which all our biochemical and molecular processes (anabolic and catabolic) occur. It allows for the gas exchange (CO_2 and O_2) needed for respiration, is the carrier of our excretory products, and maintains the permeability of our membranes. Given water's importance for life, its great abundance on the earth, and its crucial role in the transport of pollutants within the environment, we focused this chapter on discussing what green chemists need to anticipate will happen to their newly synthesized compounds in aquatic environments.

We have identified six categories of fundamental chemical processes to discuss (Table 3.5.1). All of them intimately involve water, and the role that water plays as the medium of chemical transport through the various compartments of the environment

TABLE 3.5.1 Chemical Processes That Occur in Aqueous Media That Are Important to Predict How a Newly Synthesized Chemical Will Move Through the Environment and/or to Be Taken up and Distributed in Biological Tissues

Chemical Process	Transport Through the Environment via Water	Transport Into Biological Tissues	Important Characteristics to Measure
Acid-base interactions	Interactions with POM/DOM	Bioavailability; binding/ repulsion by cell membranes; interactions with cellular components (protein, lipid, nucleic acid)	Acid-base equilibrium constants
Solubility and saturation	Movement among and within water, air, soil compartments; hydrophilicity versus lipophilicity	Bioavailability; passage through biological membranes	Solubility and saturability, polarity, K_{SP}, ΔG_r, BCF
Complexation (organic and inorganic)	Movement among and within water, air, soil compartments	Bioavailability	Stability constants (binding constants)
Ionization	Interactions with POM/DOM	Bioavailability; interactions with membranes and cellular constituents	pK_a
Redox	Interactions with POM/DOM	Bioavailability; interactions with membranes and cellular constituents	Pe, ΔG_r
Persistence	Movement among and within water, air and soil compartments	Bioavailability; amount of time a compound is available	Half-life ($t_{1/2}$), residence time

Each process should be considered in light of changing temperature, alkalinity, salinity, oxygenation, etc. Key characteristics that should be measured for every newly synthesized compound are listed.
BCF, bioconcentration factor; *DOM*, dissolved organic matter; *POM*, particulate organic matter.

(i.e., air, soil, surface, and groundwater) as well as the transport across biological membranes and into biological tissues. To truly understand environmental fate, green chemists should assess how these six chemical processes will affect their chemical of interest.

3.5.2 FUNDAMENTAL CHEMISTRY OF WATER

Although water is a neutral molecule, there is an uneven distribution of its electrons, making the molecule polar (dipole moment of 1.8546 d). This is a fortuitous characteristic since the dipole interactions (2 δ^-oxygen and each hydrogen being δ^+) allow the molecule to attract both negatively and positively charged ions. It also explains its ability to dissolve a great number of substances (to greater or lesser extent) than any other solvent and is the reason why water is the "universal solvent." The low vapor pressure of water is caused by the strength of its hydrogen bonding to other water molecules. Since these hydrogen bonds hold water molecules together tightly, the energy required for a molecule to "escape" is high and so evaporation is relatively slow. Because of these hydrogen bonds, as water begins to cool it initially behaves like a typical liquid with the density increasing as the molecules are more closely packed. However, when the temperature falls below 4 °C, the density of water begins to decrease. This means that chemists must be careful when working with water as a solvent as the density of water changes with temperature (the density of pure liquid water is 1.000 g/cm^3 at 4 °C). Above 4 °C, the density of pure water drops to 0.9584 g/cm^3 just prior to boiling (100 °C) and down to 0.9998 g/cm^3 at 0 °C just prior to freezing. Although the density of saline waters ($\sim$1.025 g/cm^3 at room temperature; estuaries, oceans, brackish lakes) is higher than that of freshwater due to dissolved salts (higher mass per volume results in a higher density), the changes in density with warming and cooling are the same as those for pure water. Thus ice floats on liquid water, effectively insulating the subsurface waters of our lakes and ponds to keep these waters well above 0 °C. The freezing temperature is lower for saline water than for freshwater (−2 vs. 0 °C).

When present in a basic environment, water may donate a proton to another compound and is considered a Brønsted-Lowry acid. Alternatively, if in an acidic environment, water may accept a proton and be considered a Brønsted-Lowry base. This is another reason why water is considered the universal solvent. This ability to act as either a donor or acceptor of protons (amphoteric) indicates that water itself has very little intrinsic buffering capacity. In the absence of buffering compounds, pH will quickly change if something is added to water. The pH of aqueous solutions can therefore be altered readily by both inorganic and organic compounds.

3.5.3 ACID-BASE INTERACTIONS

There are a wide range of H^+ concentrations in natural waters and biological tissues, and the difference in pH ($-\log [H_3O^+]$) is one of the most important factors governing the chemistry of organic compounds in its environment. The pH of freshwater ponds, lakes, and streams is typically in the neutral range (pH 6–8), depending on the geology, biota, and pH of the precipitation in the region. Seawater pH is slightly higher (currently averaging

~8.1, but is in the process of acidifying to <7.5 by the year 2300 due to the increase in CO_2 emissions from the burning of fossil fuels; see Chapter 3.4).[4] The pH of the pore waters of both marine and freshwater sediments is often identical to the overlying water column under aerobic conditions but acidifies under anaerobic conditions. Biological tissues generally have pH values in the neutral range, although the stomach of some animals may have lower pH. The human stomach, for example, may attain a pH of 1–2, while pH may change from 5.7 to 7.4 in various parts of the intestinal tract. The pH of natural waters can be readily altered by the presence of naturally occurring humic and fulvic acids, as well as by limestone dissolution, which introduces carbonate ion, producing carbonic acid, into the water. As discussed later, the pH of natural waters directly affects the chemical structure of the compounds transported in it.

In addition to the traditional Brønsted-Lowry description, acid-base reactions can be described in terms of the sharing of electron pairs (Fig. 3.5.2). Lewis acids are electron pair acceptors, whereas Lewis bases are electron pair donors. In the case of organic molecules, acidity/basicity is defined by the functional groups. Organic molecules with amine groups, for example, may act as Lewis bases in aqueous solutions. Those with carboxyl groups are, by definition, Brønsted-Lowry acids because rather than donating electrons, the functional group, COOH, donates a proton.

The acid-base behavior of a synthetic molecule and its influence on its fate in the environment are important and should be considered during synthesis as well as when determining the fate of a compound in natural waters. Take dioxins as an example. This class of molecules contains intermediaries in a number of industrial and synthetic processes, ranging from fossil fuel burning to the synthesis of chlorinated pesticides. Dioxins are polyaromatic chlorinated molecules including polychlorinated dibenzodioxins and polychlorinated dibenzofurans. Toxicity is determined by the number and position of chlorine atoms (Fig. 3.5.3). These molecules are relatively insoluble in water but have a high affinity to organic matter in natural waters and for lipids in aquatic organisms. Since industrial dioxins are sparingly dissolved in water, they are transported in the surface water and groundwater in the particulate load and will eventually be either sequestered in the sediment or accumulated in the food chain.

While industrial dioxins are well monitored, an emerging class of dioxins, those produced by the photodegradation of other man-made chemicals, has renewed concern about these highly toxic compounds. Triclosan, an antimicrobial banned in 2016 from soaps in the United

$$H^+(aq) + NH_3(aq) \rightarrow NH_4^+$$

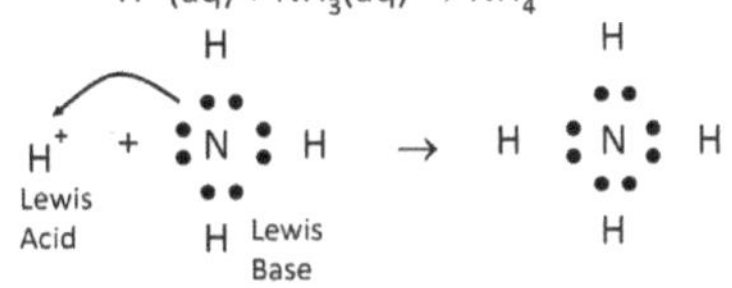

FIGURE 3.5.2 An example of a Lewis acid and base. Lewis acids are electron pair acceptors, whereas Lewis bases are electron pair donors. The hydrogen ion (Lewis acid) in aqueous solution accepts an electron pair from ammonia (Lewis base) to form the ammonium ion.

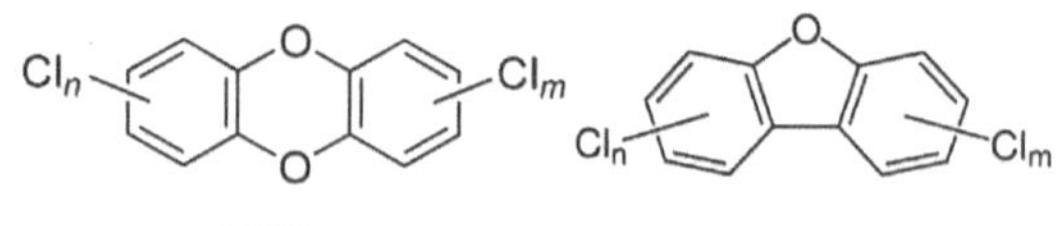

FIGURE 3.5.3 General structure of polychlorinated dibenzodioxins (PCDDs) and polychlorinated dibenzofurans (PCDFs) where *n* and *m* can range from 0 to 4 chlorine ions. The number and position of the substituted chlorines determines the nomenclature for each individual congener (75 individual congeners for PCDD and 135 congeners for PCDF).

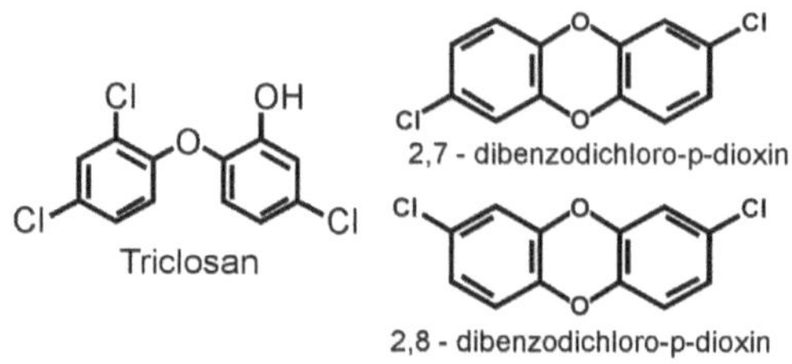

FIGURE 3.5.4 Structure of triclosan and its photodegradation dioxin products. *After Mezcua M, Gómez MJ, Ferrer I, Aguera A, Hernando MD, Fernández-Alba AR. Evidence of 2,7/2,8-dibenzodichloro-p-dioxin as a photodegradation product of triclosan in water and wastewater samples.* Anal Chim Acta *2004;***524***:241–47.*

States but still present in toothpaste and other consumer products, photodegrades to 2,7- and 2,8-dibenzodichloro-*p*-dioxin (Fig. 3.5.4) at neutral to basic pH.[5] Triclosan acts as a Lewis acid because of the phenolic OH groups, but can also act as a Lewis base due to the oxygen atom. Photodegradation under higher pHs, such as those encountered in the ocean (~pH 8.1), produces dioxin by leveraging both these characteristics to produce a fat-soluble Lewis base.

The strength of an acid is expressed by the acid dissociation constant:

$$K_a = \frac{[A^-][H^+]}{[HA]} \tag{3.5.1}$$

where HA is the acid such that strong acids have higher K_a values than weaker acids. In some cases, the K_a can be quite large and is, instead, expressed as pK_a (−log K_a). A final important factor to consider in acid-base reactions is that stronger acids (i.e., ones with lower pK_a values) are more readily ionized in aqueous solution, forming higher concentrations of reaction products compared with weaker acids. As the molar concentration of an acid increases, less is ionized (dissociated) (Fig. 3.5.5). This reflects Le Châtelier's principle, which states that high concentrations of products, the anion and hydrogen ion, push the equilibrium toward favoring the nonionized salt (acid).

3.5.4 SOLUBILITY AND SATURATION

Although chemical equilibrium is rarely attained in natural waters, the equilibrium approach allows us to approximate the fate of chemicals in natural waters and to estimate

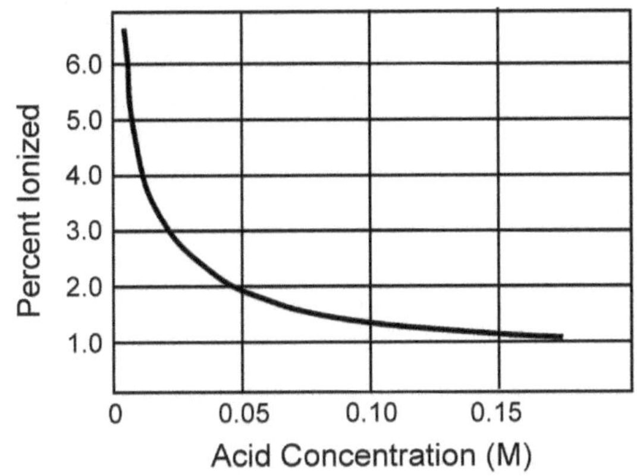

FIGURE 3.5.5 Acetic acid concentration and relation to percent ionization showing Le Châtelier's principle that as the concentration of the acid increases, the dissociation (ionization) decreases as equilibrium favors the nonionized salt (acid).

the rates of natural processes. Of course, equilibrium tracks back to the basics of Gibbs free energy, enthalpy, and entropy, which govern whether a system moving toward equilibrium will release heat, increase in entropy, or in some combination of these cause a net decrease in the energy of a system:

$$\Delta G = \Delta H - T\Delta S \tag{3.5.2}$$

where ΔG is the change in the Gibbs free energy, H is the enthalpy, T is the temperature, and S is the entropy. In natural waters, we are dealing with multiple phases of material, some dissolved (e.g., ions or organic molecules) and others solid (e.g., minerals and organic colloids). When thinking of how synthetic compounds will interact with natural waters we must remember that all components coexist at equilibrium, that the system is always striving to reach equilibrium, and that means that the chemical potential (μ_i) of each phase must be identical (Eq. 3.5.3). To achieve this, there will be exchange of component i (e.g., the synthetic molecule) between the phases until the chemical potential of phase A is the same as that of phase B. If, for example, $(\mu_i)_A$ is greater than $(\mu_i)_B$, some portion of i will be transferred from A to B resulting in a release of energy ($-\Delta G$) and so that the system cannot be in equilibrium. For green chemists, it is important to remember that for stoichiometric solids (ones that have a unique and invariable chemical composition) defining μ for components of a system is practically difficult. However, the principle still holds that, at equilibrium, the chemical potential of each component is the same for all phases:

$$\mu_i = \left(\frac{\delta G}{\delta n_i}\right)_{T,P} \tag{3.5.3}$$

where n is the number of moles and P is the pressure.

Connected to the exchange of compounds between phases in natural waters is the solubility of the compound/molecule in water. Aqueous solubility is the maximum quantity of a substance that can completely dissociate in water of given volume and temperature. Stated another way, it is the concentration of solute that can saturate a given volume of water at a given temperature. ΔG can be calculated as the difference in chemical potential, μ_i, between the reactants and products. Knowing that the chemical potential of a species is directly related to the activity of a species, a_i, we can calculate the equilibrium constant, K_{eq}, which is an estimate of the solubility of the products (Eq. 3.5.4):

$$\ln K_{eq} = \frac{-\Delta G^0}{RT} \tag{3.5.4}$$

Since temperature impacts solubility significantly, green chemists need to understand how their target compound's solubility in water will change across the temperature gradients it will encounter as it moves though the water cycle. We cannot assume that standard state conditions will apply. If temperature is close to 25 °C, we can calculate K_{eq} as follows:

$$\ln K_{eq} = \frac{-\Delta H_R^0}{RT} + \frac{\Delta S_R^0}{R} \tag{3.5.5}$$

This method assumes that ΔH_R^0 and ΔS_R^0 are independent of temperature, which is reasonable for temperatures within 20 °C or so of standard state. However, this is not true for more extreme temperatures such as those encountered in natural environments. To evaluate changes in K_{eq} across temperatures commonly encountered in natural waters we differentiate Eq. (3.5.5):

$$\frac{d \ln K_{eq}}{dT} = \frac{\Delta H_R^0}{RT^2} \tag{3.5.6}$$

Integrating from T_1 (reference temperature) to T_2 (temperature of interest) we get

$$\ln K_{T_2} - \ln K_{T_1} = \frac{\Delta H_R^0}{R}\left(\frac{1}{T_1} - \frac{1}{T_2}\right) \tag{3.5.7}$$

Of course, the trick to all the estimates of solubility is having good thermodynamic data for the target compound. In the case of novel synthetics, this is far easier said than done.

To be water soluble, a molecule or compound must be ionizable or must either be polar or have polar functional groups attached to its backbone. Polarity will allow the dipolar water molecule to attract opposite charges and "pull" individual molecules away from its solid mass. To solubilize a compound, the water-solute interaction must be stronger than the interactions between water molecules. Importantly, a green chemist's organic molecule is not the only organic molecule in water (Fig. 3.5.6). There are naturally occurring organics that contain aliphatic, aromatic, olefinic, and heterocyclic skeletons with different functional groups. The most important of these is the carboxylic acid group, $-COOH$, which behaves as an acid. Most carboxylic acid groups have pK_a values around 4 to 5 meaning that they are soluble in water. Carboxylic acids and their corresponding anions complex metals, especially trace metals and aluminum. This property means that a synthetic molecule with carboxylic

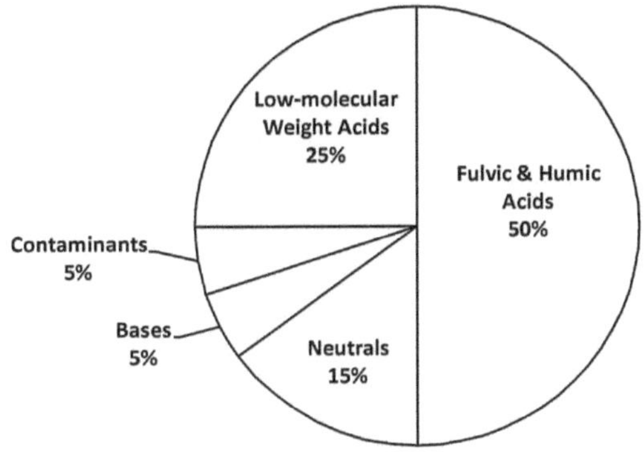

FIGURE 3.5.6 Distribution of different classes of dissolved organic carbon (DOC) compounds in the surface water of rivers in the United States. Half of the DOC is attributed to fulvic and humic acids, with an additional 25% attributed to low-molecular-weight acids. Neutral compounds (15%) and bases (5%) make up most of the remainder, with the remaining 5% being attributed to uncharacterized contaminants. *After Malcolm RL. Geochemistry of stream fluvic and humic substances. In: Aiken GR, McKnight DM, Wershaw RL, McCarthy P, editor.* Humic substances in soil, sediment, and water. *New York: Wiley; 1985. p. 181–209 used by permission of Wiley & Sons, New York.*

functional groups will dissociate in most natural waters and the carboxylate will bind metals, potentially essential metals, reducing their bioavailability and possibly negatively impacting the ecology of a receiving body of water.

A good example of this phenomenon is the class of emerging contaminants known as perfluorinated carboxylic acids (PFCAs).[6] Perfluorooctanoic acid (also known as C8, $C_8HF_{15}O_2$) now occurs in streams and groundwaters of several regions in the United States. C8, first manufactured in the 1940s, is a synthetic PFCA and surfactant used in emulsion polymerization of fluoropolymers. It is also used in the manufacture of Teflon. It is a toxicant (possible carcinogen and endocrine disrupter) and persists in the environment for long periods of time as it is highly soluble and non-particle reactive. Because of the metal binding efficiency of the carboxylic acid and carboxylate, however, novel methods for the removal of these compounds from water have been developed, such as electrocoagulation using an Fe anode with intermittent H_2O_2 addition, which facilitates the oxidative transformation of ferrous to ferric ion in the anode, which is then bound by the carboxylate in solution.[7]

This discussion centers on the assumption of equilibrium, and for most chemical questions this works fine. However, the reality is that most systems are not at equilibrium in the short term and that, in many cases, the chemicals synthesized by a green chemist may be stable in a beaker but will be subject to disequilibrium when in natural water. Therefore, nonequilibrium thermodynamics will govern solubility. We then must consider the activity product (AP) (the right side of Eq. 3.5.8) of a reaction (if the compound is ionic then the product is called the ion activity product, IAP). Consider the simple reaction of $aA + bB = cC + dC$. At equilibrium:

$$K_{eq} = \frac{a_C^c \cdot a_D^d}{a_A^a \cdot a_B^b} \tag{3.5.8}$$

The reaction will proceed to the left if $IAP/K_{eq} > 1$ and will proceed to the right if $IAP/K_{eq} < 1$. These expressions enable us to describe the extent to which a solution is supersaturated or undersaturated with respect to a compound using these equations, which are based on Le Châtelier's principle. The equilibrium constant for a dissolution reaction is called the solubility product, K_{sp}, and is calculated and employed just the same as any equilibrium constant:

$$\log K_{sp} = \frac{-\Delta G_R^0}{2.303\,RT} \tag{3.5.9}$$

where 2.303 is a conversion factor from natural log to $\log_{10}$.

For newly synthesized compounds, aqueous solubility must be determined experimentally and the data should be used to calculate the solubility constant (K_{sp}). Thermodynamically, K_{sp} could also be calculated if Gibbs free energy reference values (ΔG_r) have been determined for products and reactants. Of course, K_{sp}, being an equilibrium constant, ties directly to the AP. Take, for example, a water sample in which the calcium activity is $10^{-3.5}$ and the sulfate activity is $10^{-1.5}$. The reaction:

$$Ca^{2+}(aq) + SO_4{}^{2-}(aq) \rightleftharpoons CaSO_4(s) \tag{3.5.10}$$

produces solid gypsum. We can calculate the saturation (K_{sp}, which is the solubility product, the equilibrium constant for a reaction, at 25 °C, wherein the product is a solid) of the water by gypsum as well as how much the solution would have to be concentrated or diluted for it to be in equilibrium with gypsum. Since this is an ionic reaction we will calculate the IAP, which is $10^{-3.5} \times 10^{-1.5} = 10^{-5.0}$. K_{sp} is calculated as:

$$\log K_{eq} = -\frac{\Delta G^0}{5.708} \tag{3.5.11}$$

with ΔG in kJ/mol. The standard free energy of the reaction of Ca^{2+} ($\Delta_f G^0 = -553.58$ kJ/mol) and SO_4^{2-} ($\Delta_f G^0 = -744.53$ kJ/mol) in water to produce gypsum ($\Delta_f G^0 = -1797.28$ kJ/mol) is $[-553.58 - 744.53 - (2 \times 237.13)] - (-1797.28) = 24.91$ kJ/mol $= \log\ K_{sp} = -4.36$, thus $K_{sp} = 10^{-4.36}$. From this we can calculate the saturation index (SI) which determines whether the solution is undersaturated, saturated, or supersaturated with respect to the solid:

$$SI = \log\left(\frac{IAP}{K_{sp}}\right) = \log\left(\frac{10^{-5.0}}{10^{-4.36}}\right) = -0.64 \tag{3.5.12}$$

If $IAP = K_{sp}$ then $SI = 0$ and the water is saturated with respect to the solid. If $IAP < K_{sp}$ then $SI < 0$ and the water is undersaturated with respect to the solid. If $IAP > K_{sp}$ then $SI > 0$ and the water is supersaturated with respect to the solid. In this example, SI is less than 0, so the solid is undersaturated and will dissolve. The amount of energy released per mole of solid dissolved is:

$$RT \ln\frac{IAP}{K_{sp}} = -3.65 \tag{3.5.13}$$

Dissolution of the solid would increase the activities of Ca^{2+} and SO_4^{2-}, ultimately reducing the energy released per mole of gypsum dissolved until it became zero, at which time the solution would be saturated with respect to gypsum. What does all this mean to a green chemist? These reactions enable the prediction of the fate, with respect to dissolution, of a chemical in natural waters. Whether you are dealing with an ionic compound or a polar molecule, the thermodynamic equations of solubility and saturation still govern the fate of the chemical.

Recall that K_{sp} is a special type of equilibrium constant associated with dissolution reactions. When dealing with the complexation reactions, the formation of a solid complex, the extent to which a complexation reaction proceeds to the right, is expressed by an equilibrium constant called the stability constant, K_{stab}, and is calculated as

$$K_{\text{stab}} = \frac{aC}{aAaB} \tag{3.5.14}$$

where A and B are reactants, C is the product, and a is the activity. The magnitude of K_{stab} is proportional to the strength (stability) of the complex. K_{stab}, like K_{eq} and K_{sp}, can be manipulated as mentioned earlier to explore the effect of changes in state on complex stability.

So far, we have been assuming that the activity of a species is the same as its concentration, but we know that it is not. In natural waters, as in any chemical system, the ratio of the activity of a species to its concentration is called the activity coefficient (γ_i). For dilute waters, i.e., those with low ionic concentrations like streams and nonsaline groundwater, the activity of a species is approximately the same as its concentration, but this is not true for waters with high concentrations of charged solutes like marine waters, saline groundwater, and industrial and natural brackish waters such as lagoons.

There is no way, from a purely thermodynamic point of view, to measure the activity coefficient of an ion without making unverifiable assumptions. We cannot accurately quantify the amount of free energy change attributable to a single ion in aqueous solution, and so when dealing with activity coefficients, it is best to deal with the uncharged components rather than the ions themselves. We quantify the concentration of ions in a solution, the ionic strength, I, as:

$$I = \tfrac{1}{2}\sum M_i z_i^2 \tag{3.5.15}$$

where M is the molar concentration of an ion and z is the charge on the ion. Importantly, for dilute solutions ($I < 0.1$ mol/L) one can use single-ion activity coefficients even if there is little thermodynamic reality about them. In brackish and saline waters, one should focus on the uncharged species given the complexity of the water's chemistry. The activity coefficient, γ_i, is the ratio of activity divided by molal concentration, a_i/m_i. As the activity coefficient γ_i approaches 1, the concentration of all dissolved species, Σm_i, approaches zero. We know that in complex solutions such as natural waters, a multitude of cations and anions are interacting through dissolution-complexation reactions, acid-base reactions, and other reactions. Therefore the more complex the solution (e.g., the higher the ionic strength), the less an ion's activity will be directly proportional to its concentration.

Activity coefficients of uncharged species are near unity in dilute solutions and rise above unity (activity is higher than concentration) in concentrated solutions in large part because much of the water in concentrated solutions is involved in the hydration shells of ions and less water is available to solvate uncharged species. The activity coefficient of uncharged species can be represented, approximately, by $\gamma = 10^{0.1I}$. Single ion activity coefficients in dilute natural waters ($I < 0.1$ M) can be calculated using the Debye-Hückel model, which uses the effect of ionic interactions on free energy (Table 3.5.2). In an ionic solution, the positive ions will tend to be surrounded by a cloud of negative ions, and vice versa. If we assume that the ions are point charges, the interactions between the ions are electrostatic and the cloud of ions around any ion will follow a Boltzmann distribution:

$$F(\text{state})\alpha\, e^{-\left(\frac{E}{kT}\right)} \tag{3.5.16}$$

where E is the state energy and k is a constant of the distribution. From the Boltzmann distribution, we can derive the Debye-Hückel activity coefficient equation:

$$\log \gamma_i = -Az_i^2\sqrt{I} \tag{3.5.17}$$

where A is a constant depending only on pressure and temperature, z_i is the charge on the particular ion, and I is the ionic strength. At ionic strengths above 0.1 M, the model fails as it requires an impossibly high concentration of anions around each cation, and vice versa. This equation can be modified to consider the finite size of ions:

$$\log \gamma_i = \frac{-Az_i^2\sqrt{I}}{1 + Ba_0\sqrt{I}} \tag{3.5.18}$$

TABLE 3.5.2 Parameters for the Debye-Hückel Equation at 1 atm Pressure

T (°C)	A	B($\times10^8$)	Ion	a_0($\times10^{-8}$)	B
0	0.4883	0.3241	Ca^{2+}	5.0	0.165
5	0.4921	0.3249	Mg^{2+}	5.5	0.20
10	0.4960	0.3258	Na^+	4.0	0.075
15	0.5000	0.3262	K^+, Cl^-	3.5	0.015
20	0.5042	0.3273	SO_4^{2-}	5.0	−0.04
25	0.5085	0.3281	HCO_3^-, CO_3^{2-}	5.4	0.0
30	0.5130	0.3290	NH_4^+	2.5	
40	0.5221	0.3305	Sr^{2+}, Ba^{2+}	5.0	
50	0.5319	0.3321	Fe^{2+}, Mn^{2+}, Li^+	6.0	
60	0.5425	0.3338	H^+, Al^{3+}, Fe^{3+}	9.0	

Data from Manov GG, Bates RG, Hamer WJ, Acree SF. Values of the constants in the Debye—Hückel equation for activity coefficients. J Am Chem Soc *1943;***65***:1765–67; Klotz IM.* Chemical thermodynamics. *Englewood Cliffs (NJ): Prentice-Hall; 1950; Truesdell AH, Jones BF. WATEQ, a computer program for calculating chemical equilibria of natural waters.* J Res US Geol Surv *1974;***2***:233–48.*

where B is a constant depending on the pressure and temperature and a_0 is, theoretically, the hydrated radius of the particular ion. Since the value of B is close to 1 for most common ions, the Ba_0 term can be dropped from the equation.

If the ionic strength of solution is between 0.1 and 0.5 M and the temperature is close to 25 °C, the Davies equation is often used, as it expresses all dependence on the solution composition through ionic strength:

$$\log \gamma_i = -A_{\gamma,10} z_i^2 \left(\frac{\sqrt{I}}{1+\sqrt{I}} + 0.2I \right) \tag{3.5.19}$$

where the Debye-Hückel A_γ parameter bears the additional label "10" to ensure consistency with the fact that the activity coefficient is given in terms of the base 10 logarithm and not the natural logarithm.

For ionic strengths above 0.5 M, such as is observed in seawater, the Pitzer equation is used to describe the activity coefficients of ions.[8] The Pitzer equation is based on a semitheoretical interpretation of ionic interactions and is written in terms of interaction coefficients (and parameters from which such coefficients are calculated). There are two main categories of such coefficients: (1) "primitive" ones which appear in the original theoretical equation, but most of which are only observable in certain combinations, and (2) others which are "observable" by virtue of corresponding to observable combinations of the primitive coefficients or by virtue of certain arbitrary conventions. Only the observable coefficients are reported in the literature. These equations are based on a virial expansion for the excess of Gibbs free energy[8] and derivation to arrive at the following equations for the activity coefficients:

$$\ln \gamma_i = \left(\frac{z_i^2}{2}\right) f'(I) + 2\sum_i \lambda_{ij}(I) m_j + \sum_{jk} \left(\left(\frac{z_i^2}{2}\right) \lambda'_{jk}(I) + 3\mu_{ijk} \right) m_j m_k \tag{3.5.20}$$

where subscripts i, j, and k denote aqueous solution species; $f'(I)$ is the derivative of df/dI [$f(I)$ is a Debye-Hückel function describing the long-range electrical interactions to the first order]; $\lambda'_{ij}(I)$ is similarly the derivative of $d\lambda_{ij}/dI$ (λ_{ij} are second-order interaction coefficients); and μ_{ijk} are third-order interaction coefficients. In Pitzer calculations the activity of the water itself is considered and is assumed to be closely related to the osmotic coefficient, ϕ:

$$\ln a_w = -\left(\frac{\sum m}{\Omega}\right) \phi \tag{3.5.21}$$

where Ω is the molality of water and m is the molality of dissolved ions in solution.

When modeling activity coefficients other models exist and it is up to the chemist to decide which best reflects the condition of the system they are working in. Fortunately, several software packages allow for rapid calculation of g based on a multitude of parameters allowing for selection of the best model for the data. Recall Eqs. (3.5.15) and (3.5.17). The trick for green chemists—Deybe-Hückel activity coefficients are sufficient for calculating activities ($\gamma_i m_i$, where $m =$ concentration) in dilute waters (e.g., freshwater), but it falls apart for saline waters. Therefore the ionic strength is critical. We can use the Davies equation (Eq. 3.5.19) for higher ionic strength solutions ($I < 0.5$ mol/L).

For neutral species, we can use:

$$\log \gamma = 10^{0.1I} \tag{3.5.22}$$

For waters with an ionic strength greater than 0.5 mol/L, we must use the Pitzer equation, which uses the Debye-Hückel osmotic model such that:

$$f(I) = -\left(\frac{4A_\phi I}{b}\right)\ln\left(1 + b\sqrt{I}\right) \tag{3.5.23}$$

and

$$A_\phi = \frac{2.303A_\gamma}{3} \tag{3.5.24}$$

You may remember that the 2.303 in this expression is an approximation of log10. The complexity of this equation and its integration of a new term, A_Φ, accounts for both ionic size as well as the fact that numerous ions reacting within the solution shifts the chemical equilibrium.

Solubility and stability are important properties for green chemists to understand since highly soluble chemicals will likely be transported more easily from one environmental compartment to another, whereas insoluble compounds would have a higher probability of precipitating out of solution, either as a chemical precipitate or when bound to particulate organic matter (POM). As noted in Fig. 3.5.1, the movement of pollutants from one place to another is primarily driven by water. In addition, the ease with which compounds can cross biological membranes and are bioconcentrated is inversely related to their solubility (Fig. 3.5.7). A chemical's bioconcentration factor (BCF) is the ratio of the concentration of a chemical compound in the tissues of an aquatic organism versus the concentration of that chemical in the surrounding water. Compounds that are highly soluble (hydrophilic or lipophobic) do not easily cross biological membranes and typically do not bioaccumulate significantly in animal or plant tissues. Compounds with low aqueous solubility on the other hand (hydrophobic or lipophilic) can diffuse across biological membranes (if they are not too lipophilic and get caught up in the membrane itself). The degree of lipophilicity can be approximated by the solubility of the chemical compound in n-octanol [for a graph of BCF vs. the octanol-water partition coefficient (K_{ow}), see Chapter 3.7].

As ions and molecules move through aquatic systems, the dissolved species form complexes, which are defined as forming from two or more similar species each existing in aqueous solution. These can range from simple ionic complexes such as $Al(OH_2)^+$ to organic species such as C_2CuO_4. The equilibrium expression defines the stability of the product of the reaction (Eq. 3.5.14). For metal-organic ligand complexes, the stability equation is:

$$K_{\text{stab}} = \frac{aML_x}{aMaL^x} \tag{3.5.25}$$

where x is the stoichiometric coefficient, *aMLx* is the activity of the metal-ligand complex, *aM* is the activity of the metal, and *aL* is the activity of the ligand. Other processes, such as diffusion (the migration of a species because of a gradient in its concentration) and dispersion

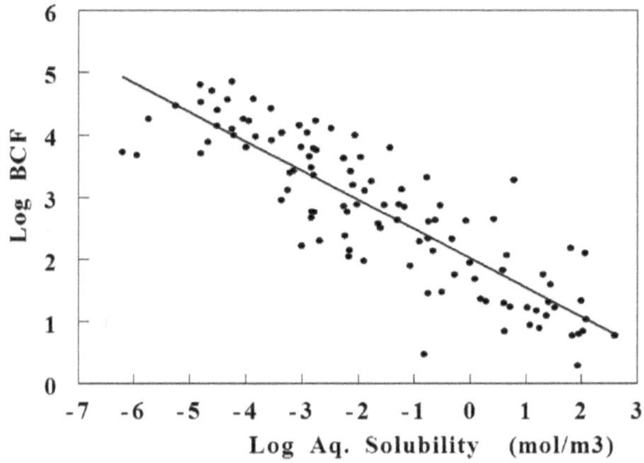

FIGURE 3.5.7 Log bioconcentration factor (BCF) versus aqueous solubility (mol/m^3) for 107 organic compounds. BCF is the ratio of the concentration of a compound in biological tissues to the concentration of the compound in aqueous solution. Thus BCF is a partitioning coefficient, which is inversely related to the aqueous solubility of compounds. Lipophilic compounds (with low aqueous solubility) tend to be more easily bioconcentrated by aquatic organisms. BCF = (−0.47 × Sol) + 2.01; $R^2 = 0.740$; $N = 107$. *Data taken from Table IV in Isnard P, Lambert S. Estimating bioconcentration factors from octanol-water partition coefficient and aqueous solubility.* Chemosphere *1988;***17***:21–34.*

(the mixing process that occurs when a fluid flows through a porous medium such as soils and sediments), influence the concentration of a chemical. In addition, reactions occur along the flow path, altering the chemical composition and thus changing a chemical's solubility and stability over time.

Critically, one must bear in mind that the pH and concentration influence the solubility and saturation (Fig. 3.5.8) and that the resulting changes in solution chemistry will further

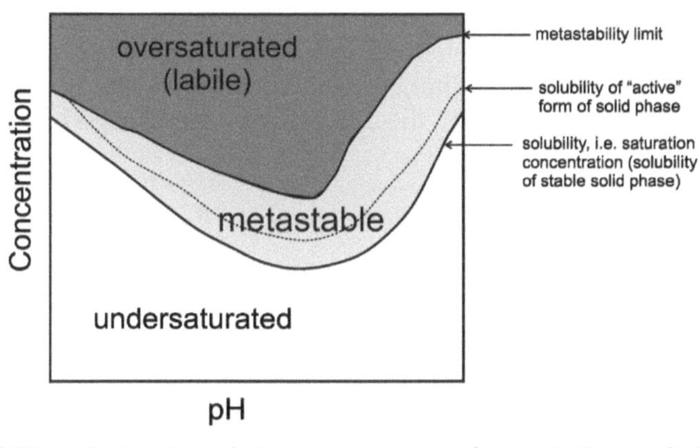

FIGURE 3.5.8 Solubility and saturation relations across a range of concentrations and pH are complex and vary for every chemical compound. Low concentrations of a compound typically form an undersaturated solution, whereas high concentrations may lead to oversaturation. However, the metastable region (where the saturated solution is stable) displays a curvilinear relationship that displays a metastability limit above which it is oversaturated and unstable. *After Stumm W, Morgan JJ.* Aquatic chemistry. Chemical equilibrium and rates in natural waters. *3rd ed. New York: John Wiley and Sons; 1996.*

drive reactions as the system moves toward equilibrium. Solubility and saturation relations vary across a range of concentration and pH (Fig. 3.5.8). A supersaturated solution in the "labile" concentration range may form no precipitates. Once the concentration of an ion and/or the pH of the solution moves into the metastable region, precipitation is initiated (the "active" form of the solid complex). This precipitate is usually an amorphous crystalline solid that is very fine grained and thermodynamically unstable. This active precipitate may persist until it undergoes a conversion to a more stable phase. An excellent example of this transition is the polymorphs of calcium carbonate. Calcium carbonate will first precipitate from a fluid as amorphous calcium carbonate, and, depending on the pH of the solution, will transform to aragonite or calcite, with the latter being more stable and less soluble than the former.

3.5.5 COMPLEXATION

The solubility calculations described earlier neglect one factor, namely, the formation of complexes or ion pairs. Complexes are dissolved species consisting of two or more simpler species. The formation of complexes is represented by an equilibrium constant for a reaction of this type. For example, recall the reaction between calcium and sulfate ion that produced calcium sulfate (gypsum) (Eq. 3.5.10). Earlier in this chapter we calculated the solubility product of gypsum. We can also calculate the stability of that complex, K_{stab} (Eq. 3.5.14).

The magnitude of a stability constant is proportional to the strength or stability of the complex. Complexes can be either weak or strong. Weak complexes are easily disrupted in the presence of higher strength complexes.

The complex $CaSO_4^0$ (where 0 denotes a neutral aqueous species), for example, has a stability constant of $10^{2.23}$. The existence of this aqueous complex affects the solubility of the mineral gypsum, $CaSO_4$(s). We will assume that the solution is in equilibrium with gypsum, which means that the activity of the neutral species is the product of the stability constant (Eq. 3.5.14) and the solubility product that we calculated above (Eq. 3.5.12). We can rearrange Eq. (3.5.14) and then substitute K_{sp} for the activities of the charged ions and solve for the activity of the complex:

$$a_{CaSO_4{}^0} = K_{stab} \cdot a_{Ca^{2+}} \cdot a_{SO_4{}^{2-}} = K_{stab} \cdot K_{sp} \tag{3.5.26}$$

$$a_{CaSO_4{}^0} = K_{stab} \cdot K_{sp} = 10^{2.23} \times 10^{-4.36} = 10^{-2.13} = 7.4 \times 10^{-3} \tag{3.5.27}$$

Since the complex is uncharged, its concentration is approximately equal to its activity, so the concentration of the complex is 7.4×10^{-3} mol/kg. The presence of this complex impacts the solubility of the solid. Recall that K_{sp} for this reaction (Eq. 3.5.10) is $10^{-4.36}$. If we assume that the concentration of Ca^{2+} and SO_4^{2-} are equal, then the concentration of both is 6.61×10^{-3}. The ionic strength (Eq. 3.5.15) is 2.64×10^{-2}. To calculate the activities we need to calculate the activity coefficients using Debye-Hückel (Eq. 3.5.18) where $A = 0.5085$, $B = 0.3281 \times 10^8$, and $a_0 = 5 \times 10^{-8}$. Substituting these values, we get the activity coefficient for both ions as

0.548. We can then use Eq. (3.5.1) to calculate the concentration of the ions, which must be equal when gypsum dissolves,

$$m_{Ca^{2+}} = m_{SO_4{}^{2-}} = \left(\frac{10^{-4.36}}{0.548 \times 0.548}\right)^{1/2} = 1.205 \times 10^{-2} \quad (3.5.28)$$

From this we calculate the ionic strength as 5.99×10^{-2} and iterate again new activity coefficients to get 0.441 for both ions. We would keep calculating new γ_i until no significant difference is found. Luckily, in this example, we only had to repeat the γ_i calculation once. Using these new activity coefficients, we calculate the solubility of gypsum in water, K_{sp}, which is now 15.0×10^{-3} mol/kg[1]. Compare this to the result for the complex, 7.4×10^{-3} mol/kg. The effect of the complex was to increase the solubility of gypsum by 50% over what it was without the complex. The impact of complexation on solution chemistry and on the fate of solids is, obviously, important to predicting the fate of a chemical in natural waters.

There are myriad possible inorganic and organic species that can form complexes in the waters of the environment (e.g., multiple inorganic species, humic and fulvic acids, dissolved organic matter, POM; Fig. 3.5.6). In general, complexation to these materials reduces the

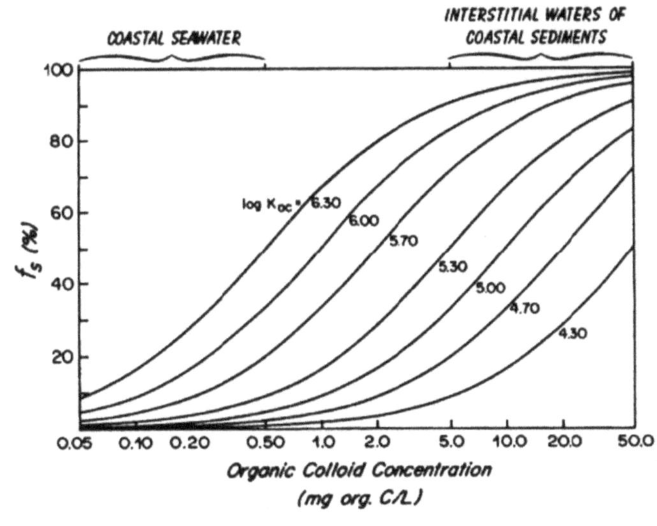

FIGURE 3.5.9 Nomograph depicting the relationship between the fraction of a hydrophobic organic compound that sorbs to organic colloids [f_s (%)] and the concentration of organic colloids, for various estimates of log K_{oc} of the compound (over the range from 4.30 to 6.30). Organic matter in surface waters can be characterized as dissolved, colloidal, or particulate. All three play important roles in the complexation and bioavailability of chemical compounds. Colloids are submicroscopic particles that do not settle out of solution. Depending on their makeup, colloids exhibit a range of organic carbon binding affinity (log K_{OC}) values. Typical colloidal concentrations in coastal waters range from about 0.05 to 0.5 mg of organic carbon per liter, whereas concentrations are much higher in the pore waters (interstitial waters of sediments). Organic chemical compounds bind to these organic colloids in direct proportion to the concentration of colloids present in the surface water and pore water. In pore waters, almost all of the chemical compound will bind to colloids, whereas in coastal seawater, up to 50% of the organic compounds may be bound. *From Farrington JW. Biogeochemical processes governing exposure and uptake of organic pollutant compounds in aquatic organisms.* Env Health Perspect *1991;***90***:75–84. Figure 2, page 78 of the EHP paper, with permission from Bruce J. Brownawell.*

concentration of the species that is bioavailable. Binding to POM or to the organic coating of inorganic particles can quickly reduce the total concentration of an organic compound in a water body, due to the settling out of these particles, depositing these compounds on the sediment surface, depending on the size of the particle. Even colloidal-size particles can be active in complexation. Brownawell[9] showed that colloidal particles reduced the bioavailability of polycyclic aromatic hydrocarbons (PAHs) and that the amount of bound PAH depended on the strength of the binding of the PAH to the colloidal materials (Fig. 3.5.9). In addition, Burgess and Lohmann[10] presented a multicompartment model regarding how organic contaminants can bind to various types of organic carbon films covering an inorganic particle. They included three types of organic carbon films: rubbery carbon (which behaved similarly to colloidal carbon), black carbon (from soot), and glassy carbon (from plant material). PAHs bound to rubbery colloidlike carbon, and bound strongly to both black and glassy carbon, whereas polychlorinated biphenols (PCBs) did not bind strongly to black or glassy carbon.[11]

Complex formation can also occur within organisms, either in the cytosol or in the blood plasma. For example, a variety of metals form complexes with serum albumen in human blood[12] and with histidine-rich glycoprotein (HRG) in the plasma of several bivalve mollusks.[13] Serum albumen and HRG are the major metal-binding proteins in the blood of these two groups, respectively, and are thought to act as metal transport proteins.

3.5.6 IONIZATION

As noted previously, ionic compounds tend to be readily soluble in water and so may be transported great distances in the environment. The ionized form may also be more chemically active than the nonionized form, making it easier to form new complexes with compounds that are also present in the water.

Ionization of an organic compound is very important biologically, since ionized forms do not tend to readily diffuse through biological membranes, but are either repelled by the slight negative charge of most membranes or bind to the outside of the membrane and do not pass through it.[14] From a biological activity point of view, the active form of the organic compound is usually the neutral molecule (i.e., HA in the case of acid compounds and A in the case of bases, as shown in the equations that follow), since this is the form that is most easily taken up by passive diffusion and is then bioaccumulated. For acids:

$$HA \rightleftharpoons H^+ + A^- \qquad (3.5.29)$$

whereas, for bases:

$$A + H^+ \rightleftharpoons HA^+ \qquad (3.5.30)$$

where H^+ is a dissociated proton. The degree of ionization depends primarily on the pH of the surrounding environment and can be easily calculated using the Henderson-Hasselbach equations:
For acids:

$$pH = pK_a + \log([A^-]/[HA]) \qquad (3.5.31)$$

and for bases:

$$pH = pK_a + \log([A]/[HA^+]) \quad (3.5.32)$$

where [....] denotes concentration and pK_a is the compound-specific pH at which 50% of the compound is ionized. The pK_a must be determined for new compounds experimentally. Within an acidic aqueous environment, acidic compounds remain nonionized and are thus likely to diffuse through biological membranes. Basic compounds, on the other hand, will primarily be ionized under acidic conditions. In basic aqueous solutions, acidic compounds will be ionized, whereas basic compounds will remain nonionized and can be absorbed by tissues by passive diffusion. It should be noted that once a nonionized compound diffuses through a biological membrane, its new environment (either intracellular cytosol or extracellular blood plasma) may have a pH that is very different from that used in the aforementioned calculations (often a neutral pH). With changes in pH come changes in complexation (e.g., Fig. 3.5.8). In natural waters, ionization of both inorganic and organic compounds also depends on other factors in the surrounding water, such as the presence of particulate and dissolved organic carbon, humic and fulvic acids, and the presence of black carbon (soot).

3.5.7 REDOX REACTIONS

Of the many kinds of reactions that occur in natural waters, oxidation-reduction reactions are critically important to the fate of compounds and to the formation of or dissociation of toxic reaction products. Oxidation-reduction (redox) reactions involve the transfer of electrons from one species to another, with the resultant change in valence or oxidation state. This oxidation state (also referred to as the oxidation number) is a hypothetical charge that an atom would have if the molecule in which the atom resides were to dissociate. Redox reactions always occur in pairs, where one species is oxidized and the other is reduced, since there are no free electrons present in aqueous solutions. Oxidation is the loss of an electron from an atom or a molecule and so results in an increase of positive charge of that oxidized species. Reduction is the gain of an electron and thus leads to the increase of negative charge by the reduced species. For example, a common redox reaction occurs with elemental iron (Fe^0) and the chlorine in seawater:

$$2Fe + 3Cl_2 \rightarrow 2Fe^{3+} + 6Cl^- \quad (3.5.33)$$

This can be analyzed as two half-reactions:

$$2Fe \rightarrow 2Fe^{3+} + 6e^- \quad (3.5.34)$$

$$3Cl_2 + 6e^- \rightarrow 6Cl^- \quad (3.5.35)$$

Although no free electrons exist in solution, it is useful to define a quantity called the electron activity of a solution:

$$pe = -\log[e] \quad (3.5.36)$$

that is analogous to the definition of pH as the negative log of the hydrogen ion concentration. Stumm and Morgan[15] refer to *pe* as "redox intensity," since *pe* indicates the tendency of a solution to donate or accept electrons. If *pe* is low, there is a strong tendency for the solution to donate electrons—the solution is reducing. If *pe* is high, there is a strong tendency for the solution to accept electrons—the solution is oxidizing.

Calculation of the *pe* of a half reaction provides a way to determine which way a reaction will proceed. However, the *pe* of a half reaction is a relative value, unless it is compared with a reference value. The usual choice of reference is the half reaction involving hydrogen gas:

$$\tfrac{1}{2}H_{2(g)} \rightleftharpoons H^+ + e^- \tag{3.5.37}$$

This is convenient because $H_{2(g)}$ can be easily measured by using a standard hydrogen electrode in the laboratory or in the field. The measured potential (measured in volts or millivolts) is an alternative to *pe* as an indicator of redox properties of a solution. Thus although *pe* is not directly measured, the potential of the solution relative to the production of H_2 gas can be measured, and is known as *Eh*, the "hydrogen scale potential." The *pe* and the measured *Eh* of a solution are essentially the same, since:

$$pe = \frac{F}{2.030RT} Eh \tag{3.5.38}$$

where $F = 96.42$ kJ/V eq (Faraday's constant), R is the gas constant, and T is the absolute temperature.

It is important for green chemists to understand what happens to their chemical of interest under both oxidizing and reducing conditions since redox reactions can change the species of ions and complexes as well as modify the structure of organic compounds (thereby changing their bioavailability to biota). Natural waters display a broad array of redox environments. Redox reactions can be very different at the air/water interface than in the sediments at the bottom of lakes, rivers, estuaries, and oceans.[15] Equilibria in redox reactions are often not achieved in natural systems because of the constantly varying conditions and the relative slowness of some processes. Waters can be classified with respect to redox as being oxic, suboxic, or anoxic.[16] Oxic waters contain measurable dissolved oxygen and are an oxidizing environment. Suboxic waters lack measurable oxygen or sulfide but have significant concentrations of iron ($\geq$0.1 mg/L), so are reducing waters. Anoxic waters contain both dissolved iron and sulfide and are strongly reducing.[16] A variety of factors control *pe* (and *Eh*) in natural waters. These factors are sometimes hard to predict. For example, in oxic groundwater, *Eh*, in some natural waters, does not correspond to the O_2/H_2O redox couple, but rather to the O_2/H_2O_2 couple.[17] *Eh* often corresponds to the concentration of POM, with high concentrations of POM in sediments producing reducing conditions.

The chemical properties of compounds can change dramatically depending on their redox state. For example, metals are far more mobile in one redox state than another, since their solubility changes. When a metal is more soluble, it is more mobile in the environment. Cr(VI) is more soluble than the less toxic Cr(III).[18] Organic contaminants may only slowly be broken down by the aerobic bacteria found in oxic waters, but are more readily broken down under anaerobic or suboxic conditions, because of the presence of anaerobic bacteria. Vinyl chloride, for example can be oxidized to CO_2 by anaerobic bacteria that use NO_3^- as an electron source.[14] These examples point out just how important redox chemistry is for understanding the fate of chemicals.

3.5.8 PERSISTENCE

There are a number of contaminants (e.g., DDT and several other cyclic pesticides, PAHs, PCBs, dibenzodioxins, and dibenzofurans) that are highly persistent in the environment. They are considered "PBT" chemicals because they are persistent, bioaccumulative, and/or toxic. The same chemicals are also considered as "POPs" (persistent organic pollutants). The chemicals on these two lists are of considerable concern to environmental toxicologists because their persistence increases their chance of being widely distributed in the environment and the fact that they bioaccumulate means that they are present in a form that is bioavailable to living organisms. Most of the chemicals on these lists are either known or suspected carcinogens and mutagens. Green chemists should work to avoid having to add new chemicals to these lists.

Persistence of a chemical can be established by calculating the half-life of the chemical in an environmental reservoir. For example, the declining concentration of a chemical in the outflow of a lake could be measured over time, following the elimination of a pollution source into the lake. A good example of this is mercury, which in lake sediments can become methylated by bacteria, changing a species of lower toxicity, Hg^{2+}, into a PBT, CH_3Hg^+. The primary source of mercury to lakes is atmospheric deposition of Hg^{2+} in rainwater, where the mercury itself comes from incinerators that emitted Hg^0 to the atmosphere where it is transformed to aqueous Hg^{2+}. Reduction of Hg^0 emissions through scrubber technologies and, in some regions, elimination of incinerators altogether drastically reduced mercury deposition to lakes and, as a result, reduced the formation of methylated mercury. Nevertheless, the remaining coal-powered power plants are still one of the major sources of Hg contamination. For animal populations (or various tissues within these animals), the concentration (body burden) of methylmercury can be measured at intervals of time after animals are moved to a clean environment. In both these cases, where the source of the contaminant has been reduced and/or eliminated, removal of the contaminant by natural processes (referred to as depuration—the action or process of freeing something of impurities) should, if of first order, follow an exponential decay mode.[19] From these data, a conditional depuration decay constant (k_d) can be calculated, and with this value the half-life ($t_{1/2}$) can be calculated[19]:

$$t_{1/2} = 0.693/k_d \tag{3.5.39}$$

Half-lives are independent of the initial concentration of the chemical in the reservoir.[19] A comparison of half-lives of various chemicals will indicate the persistence of that chemical, under a particular set of circumstances. Given that half-lives are calculated based on an exponential model, the concentration of chemical may never reach zero, regardless of how long the reservoir is monitored. In natural aquatic systems, whether saline or freshwater, the dissolved mass of chemical constituents is representative of the flux of material in and out of any given reservoir. The rate of change (or depuration rate) is simply the difference between the source(s) and the sink(s) such that if Q is the input rate (moles/year), S is the output rate (moles/year), and M is the total dissolved mass of chemical of interest in the reservoir, then the rate of change of the dissolved mass is given by:

$$dM/dt = Q - S \tag{3.5.40}$$

Over time, the system will approach steady state. At that point, $dM/dt = 0$ and $Q = S$. Once the system is at steady state, the residence time, τ, of a chemical is the ratio of the mass of the chemical to the flux of that chemical into or out of the reservoir:

$$\tau = M/Q = M/S \tag{3.5.41}$$

It important for green chemists to understand that the residence time and depuration rate of a chemical of concern depend on the residence time of water within a reservoir and on the biochemistry of an individual organism, respectively. For example, contamination of groundwater by triazine, a class of herbicides banned in the European Union in 2003, was found to be increasing at depth in the groundwater of the Vistrenque aquifer in southern France. This persistence of the chemical 13 years since being banned was due to the residence time of the water itself within this aquifer (12.32 years).[20] The fact that triazine concentrations increased with groundwater residence time highlights the fact that it is not only the absolute concentration of a chemical and the aquatic chemistry of the reservoir (e.g., pH, dissolved oxygen, etc.) but also the residence time of the water itself that are important considerations in determining the fate of a chemical in natural waters.

These examples can reveal how quickly a chemical moves out of an aquatic reservoir, but they say nothing about whether the parent compound is simply transported out of the system or whether the compound is degraded or destroyed. There are several processes, both chemical and biological, that may be involved in the destruction of organic or organometallic compounds:

Photolysis: A number of the PBT chemicals in aqueous solutions can be modified and even photolyzed by ultraviolet radiation. For example, benzo[a]pyrene and chrysene are photodegraded by a mechanism involving the creation of reactive oxygen species, whereas fluorene is photodegraded by a different, unknown mechanism.[21] Interestingly, many of the oxygenated PAHs develop a greater toxicity than the parent compounds when exposed to ultraviolet radiation, a process called photoactivation.[22] Photoactivation has also been reported in toxicity tests using *Daphnia magna* using several sulfonamide antibiotics,[23] raising questions as to how widespread this phenomenon is. Oxygenated PAHs have been measured in the soils of contaminated sites,[22] raising a warning that an increase in these persistent, more toxic metabolites (which are not routinely measured in monitoring programs) may be occurring as the parent compounds are being degraded.

Bacterial redox reactions: As noted earlier, chemical degradation of contaminants may occur via redox reactions. Anaerobic bacteria can oxidize vinyl chloride to CO_2, using NO_3^- as an electron acceptor.[24] Anaerobic bacteria can also dechlorinate perchloroethene.[25] Some aerobic bacteria use phase I metabolism (i.e., cytochrome P450 enzymes or reductases) to reduce compounds with nitro groups to primary amines.[14]

Invertebrate and vertebrate phase I enzymes: Phase I enzyme systems include cytochrome P450 and flavin adenine dinucleotide mixed function oxidase systems that insert a single oxygen molecule into organic compounds. These invertebrate and vertebrate systems therefore cannot break apart the benzene rings of PAHs. However, they can modify the structure of many of the PAHs, by adding oxygen-related functional groups. Several of these modified PAHs have higher toxicities than the parent compound. For example, benzo[a]pyrene is not particularly mutagenic, but is highly mutagenic after three rounds of phase I metabolism when benzo[a]pyrene-7,8-dihydrodiol-9,10 epoxide is produced,[14] which forms a covalent adduct with DNA. This process is referred to as metabolic activation. Although recognized for the PAHs, whether this is a more widespread phenomenon is unknown.

Bacterial dioxygenase systems: Some sedimentary bacteria have a dioxygenase phase I metabolism system that can insert two oxygen molecules into one of the benzene rings of a PAH, thereby splitting the ring apart. This system is not present in invertebrates and vertebrate tissues.[14]

Anaerobic bacterial systems: In aquatic systems, as much as 99% of organic matter (mostly protein, carbohydrate, and lipid) is expected to be recycled in the aerobic water column, leaving only 1% of the total material to be stored in the sediments.[26] Under low primary production conditions, bottom sediments remain aerobic. However, in highly productive and eutrophic areas, there is a rapid buildup of organic material in the sediments where aerobic bacteria use up the available oxygen faster than it can be replaced by diffusion from the overlying water, leaving the sediments anaerobic. In these sediments, various anaerobic bacteria outcompete aerobes, since they can use nitrate, sulfate (generally in non-freshwater situations), Mn(IV) oxides, and Fe(III) oxides in place of O_2, while using the organic matter as carbon sources. Often, these anaerobic bacteria can use some of the PBT compounds as carbon sources.

Although it is likely that many organic chemicals can be rapidly lysed or incrementally degraded by the processes discussed earlier, most of the in-depth research has been done on those chemicals that are highly persistent and only degrade slowly. Although this points out the characteristics that should not be incorporated into newly designed chemicals, research is needed to determine how better to design chemicals that can easily be broken down once their useful function has ended.

3.5.9 FINAL REMARKS

In this chapter, we have attempted to instill in green chemists an appreciation of how the chemistry of natural waters (both fresh and salt) impacts the fate and transport of both organic and inorganic compounds through various reservoirs of the environment. This transport occurs primarily via the association with water. Upon release into the environment, a newly synthesized compound can move from one aquatic reservoir to another and may be taken up by living organisms (e.g., plants, animals, bacteria) along the way. The compound will likely be changed during its transport, either slightly or considerably depending on aqueous solution chemistry and biological processes. To better predict the environmental fate of this new compound, we have discussed six important chemical processes: acid-base interactions, solubility and saturation, complexation, ionization, redox, and persistence. We hope that by carefully considering these six chemical processes together with knowledge specific to their compound of interest, green chemists will be better able to predict the extent to which the compounds they are working with may affect the environment and the organisms (including humans) that live in it.

References

1. Jurado E, Jaward F, Lohmann R, Jones KC, Simo R, Dachs J. Wet deposition of persistent organic pollutants to the global oceans. *Environ Sci Technol* 2005;**39**:2426–35.
2. USGS (United States Geological Survey), The USGS Water Science School. https://water.usgs.gov/edu/earthhowmuch.html; 2017.
3. West JB. *Best and Taylor's physiological basis of medical practice*. 11th ed. Baltimore (MD): Williams and Wilkins; 1985.
4. Tyrrell T. Calcium carbonate cycling in future oceans and its influence on future climates. *J Plankton Res* 2008;**30**:141–56.
5. Mezcua M, Gómez MJ, Ferrer I, Aguera A, Hernando MD, Fernández-Alba AR. Evidence of 2,7/2,8-dibenzodichloro-p-dioxin as a photodegradation product of triclosan in water and wastewater samples. *Anal Chim Acta* 2004;**524**:241–7.
6. Ahrens L, Bundschuh M. Fate and effects of poly-and perfluoroalkyl substances in the aquatic environment: a review. *Environ Toxicol Chem* 2014;**33**:1921–9.
7. Yang B, Han Y, Deng Y, Li Y, Zhuo Q, Wu J. Highly efficient remove of perfluorooctanoic acid from aqueous solution by H_2O_2-enhanced electrocoagulation-electroflotation technique. *Emerg Contam* 2016;**2**:49–55.
8. Pitzer KS, Mayorga G. Thermodynamics of electrolytes, II. Activity and osmotic coefficients with one or both ions univalent. *J Phys Chem* 1973;**77**:2300–8.
9. Farrington JW. Biogeochemical processes governing exposure and uptake of organic pollutant compounds in aquatic organisms. *Env Health Perspect* 1991;**90**:75–84.
10. Burgess RM, Lohmann R. Role of carbon in the partitioning and bioavailability of organic pollutants. *Environ Toxicol Chem* 2004;**23**:2531–3.
11. McGroddy SE, Farrington JW, Gschwend PM. Comparison of the *in situ* and desorption sediment – water partitioning of polycyclic aromatic hydrocarbons and polychlorinated biphenyls. *Environ Sci Technol* 1996;**30**:172–7.
12. Trisak ST, Doumgdee P, Rode BM. Binding of zinc and cadmium to human serum albumin. *Int J Biochem* 1990;**22**:977–81.
13. Abebe AT, Devoid SJ, Sugumaran M, Etter R, Robinson WE. Identification and quantification of histidine-rich glycoprotein (HRG) in the blood plasma of six marine bivalves. *Comp Biochem Physiol* 2007;**147B**:74–81.
14. Timbrell J. *Principles of biochemical toxicology*. 3rd ed. London: Taylor & Francis; 2000.
15. Stumm W, Morgan JJ. *Aquatic chemistry. Chemical equilibrium and rates in natural waters*. 3rd ed. New York: John Wiley and Sons; 1996.
16. Barcelona MJ, Holm TR. Oxidation-reduction capacities of aquifer solids. *Environ Sci Technol* 1991;**25**:1565–72.
17. Barcelona MJ, Holm TR, Schock MR, George GK. Spatial and temporal gradients in aquifer oxidation-reduction conditions. *Water Resour Res* 1989;**25**:991–1003.
18. Greenwood NN, Earnshaw A. *Chemistry of the elements*. Oxford (UK): Pergamon Press; 1984.
19. Barron MG, Stehly GR, Hayton WL. Pharmacokinetic modeling in aquatic animals. I. Models and concepts. *Aquat Toxicol* 1990;**17**:187–212.
20. Sassine L, La Salle CLG, Khaska M, Verdoux P, Meffre P, Benfodda Z, Roig B. Spatial distribution of triazine residues in a shallow alluvial aquifer linked to groundwater residence time. *Environ Sci Pollut Res* July 2016;**22**. http://dx.doi.org/10.1007/S11356-016-7224-x. Berlin: Springer-Verlag. 11 pp.
21. Miller JS, Olejnik D. Photolysis of polycyclic aromatic hydrocarbons in water. *Water Res* 2002;**35**:233–43.
22. Lundstedt S, White PA, Lemieux CL, Lynes KD, Lambert IB, Oberg L, et al. Sources, fate, and toxic hazards of oxygenated polycyclic aromatic hydrocarbons (PAHs) at PAH-contaminated sites. *Ambio* 2007;**36**:475–85. 23.
23. Jung J, Kim Y, Kim J, Jeong D-H, Choi K. Environmental levels of ultraviolet light potentiate the toxicity of sulfonamide antibiotics in *Daphnia magna*. *Ecotoxicology* 2008;**17**:37–45.

24. Coleman NV, Mattes TE, Gossett JM, Spain JC. Phylogenetic and kinetic diversity of aerobic vinyl chloride-assimilating bacteria from contaminated sites. *Appl Environ Microbiol* 2002;**68**:6162–71.
25. Fathepure BZ, Nengu JP, Boyd SA. Anaerobic bacteria that dechlorinate perchloroethene. *Appl Environ Microbiol* 1987;**53**:2671–4.
26. Drever JI. *The geochemistry of natural waters*, vol. 437. Englewood Cliffs (NJ): Prentice Hall; 1988.
27. Manov GG, Bates RG, Hamer WJ, Acree SF. Values of the constants in the Debye—Hückel equation for activity coefficients. *J Am Chem Soc* 1943;**65**:1765–7.
28. Klotz IM. *Chemical thermodynamics*. Englewood Cliffs (NJ): Prentice-Hall; 1950.
29. Truesdell AH, Jones BF. WATEQ, a computer program for calculating chemical equilibria of natural waters. *J Res US Geol Surv* 1974;**2**:233–48.
30. Malcolm RL. Geochemistry of stream fulvic and humic substances. In: Aiken GR, McKnight DM, Wershaw RL, McCarthy P, editors. *Humic substances in soil, sediment, and water*. New York: Wiley; 1985. p. 181–209.
31. Isnard P, Lambert S. Estimating bioconcentration factors from octanol-water partition coefficient and aqueous solubility. *Chemosphere* 1988;**17**:21–34.

CHAPTER

3.6

Water Contamination and Pollution

Linda Schweitzer[1], *James Noblet*[2]

[1]Oakland University, Rochester, MI, United States; [2]California State University San Bernardino, San Bernardino, CA, United States

3.6.1 INTRODUCTION

3.6.1.1 What is Water Pollution?

Water pollution is the presence of chemical, physical, or biological components or factors producing a condition of impairment of a given water body with respect to some beneficial use. The level of contamination necessary to render a water body impaired is highly dependent on the type of water body, its location, and the types of beneficial uses it supports. A water deemed unfit for drinking by humans may be suitable for other uses, such as habitat, irrigation, or recreation. Although certain natural events can cause water pollution, we will focus herein on the anthropogenic sources of pollution, that is, pollution arising from human activities.

3.6.1.2 Types of Water

Natural water is typically divided into surface water and groundwater. Surface water includes lakes, reservoirs, ponds, rivers, and streams, which each have their own dynamics and are exposed to both the underlying terrestrial surfaces and the atmosphere. Groundwater exists in porous rock units beneath the earth's surface and may be in contact with surface water or relatively isolated. The different pathways and mechanisms by which surface water and groundwater are exposed to contaminants and thus become polluted will be discussed throughout this section. Any of these waters may be used as a source of potable water for humans. Drinking water is subject to strict regulatory requirements and will be discussed separately.

3.6.1.3 Sources of Water Pollution

The sources of pollution in general are divided into **point sources** and **nonpoint sources**. Point sources are localized identifiable sources of contaminants, such as power plants,

Green Chemistry
http://dx.doi.org/10.1016/B978-0-12-809270-5.00011-X

refineries, mines, factories, wastewater treatment plants, etc. Nonpoint sources are those that are distributed over a wide geographic area, such as a watershed. Nonpoint sources can also include mobile sources such as cars, buses, and trains. Although each of these is a point source, they are moving and thus spread their cumulative impact over a large geographic area. A typical nonpoint source of pollution would be urban runoff, where the contaminant load may be the sum of thousands of small point sources within the watershed.

3.6.2 WATER QUALITY AND SUSTAINABILITY

The goal of Green Chemistry is to sustain a healthy environment in order to sustain the human population and healthy economies that are based on the extraction and utilization of natural resources. As stated by Hall and Klitgaard (2012), "All forms of economic production and exchange involve the transformation of materials, which in turn requires energy."[1] Energy and water are inextricably linked. For example, California uses 20% of its energy on moving drinking water and treating wastewater.[2] The procurement, treatment, and transmission of water require energy and petroleum-based chemicals. The sustainability of water is both a water quantity issue and a water quality issue and is dependent on the sustainability of our infrastructure (pipes and treatment plants), chemicals used in water treatment, and energy resources and associated technologies.

An irony is that the procurement of energy resources can lead to water pollution, such as the Gulf of Mexico oil spill from the Deepwater Horizon oil rig in 2010. Although British Petroleum (BP) was held responsible for the spill, we should all appreciate the value of oil and take responsibility for our contribution to the problem. The procurement of oil, natural gas, and coal is not without environmental risk and will inevitably result in occasional accidents such as an explosion in a coal mine, damage to aquifers, truck accidents, or pipeline bursts that spill oil, etc. As a society, we need to develop safer and more sustainable uses of energy and conserve energy wherever possible, while working to safeguard water resources.

Society also needs to develop safer chemicals and more biodegradable products. Many anthropogenic chemicals have been found in the environment, including in water or sediments at the bottom of some water bodies.[3–7] For example, think about what happens to a used computer monitor when it is discarded. Some of it may be recycled, and some of these materials may make their way to a landfill where they enter the environment. Parts of it may be combusted, for example, in an incinerator. Unless completely mineralized to carbon dioxide and water plus trace elements, these incompletely combusted materials become air pollutants. A portion of these will wash out of the air into water (i.e., wet deposition during rain or snowfall) or adsorb to particles in air that can fall by gravity (i.e., dry deposition). Atmospheric deposition is considered to be nonpoint source pollution because it is widely distributed. Chemicals deposited on land can also be washed into streams, rivers, lakes, and oceans, or make their way into groundwater.

An essential part of Green Chemistry teaching includes the fundamentals of fate-and-transport processes; how chemicals are transported both locally and globally into air, water, soil, sediments, and biota; and how chemicals are eventually broken down and eliminated. Fig. 3.6.1 shows the distribution (transport) of chemicals within an ecosystem. Some

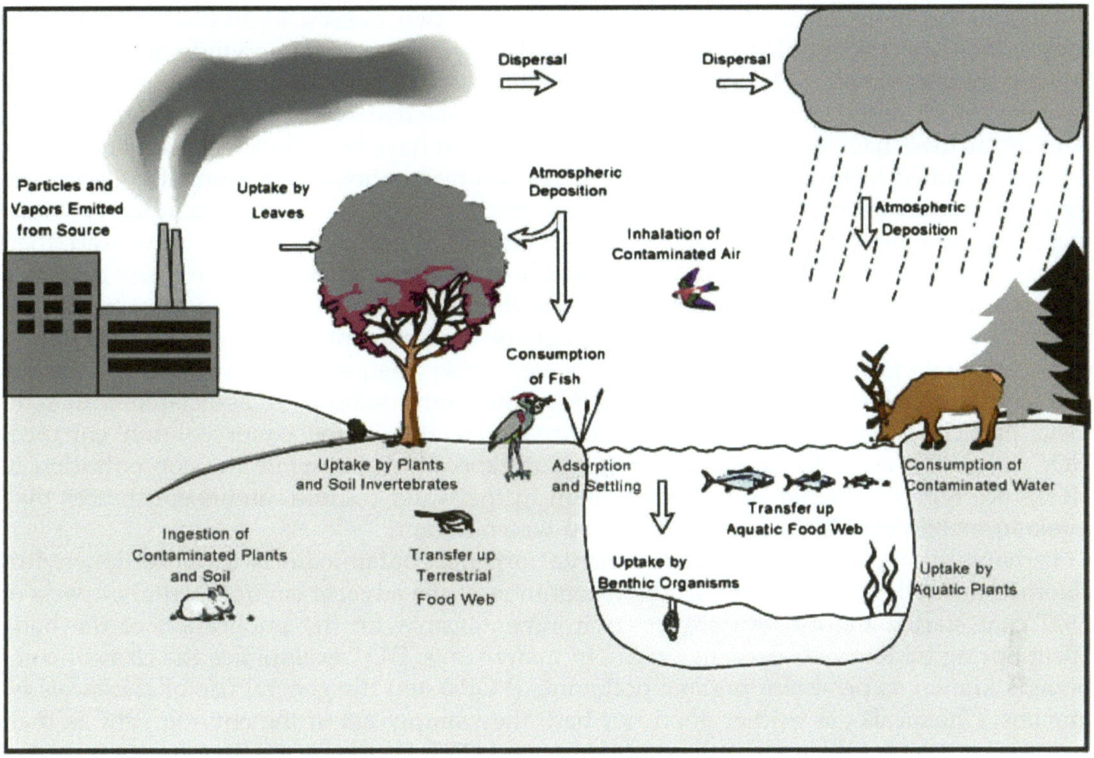

FIGURE 3.6.1 Transport of chemicals within an ecosystem. Chemicals are released from a source (in this case, a factory smoke stack) and are distributed throughout the environment. Chemicals move between air, water, soil, sediments, and biota. *From https://www.epa.gov/fera/multimedia-fate-and-transport-modeling-general.*

chemicals are highly persistent, whereas others more readily break down in the environment by processes such as photolysis, hydrolysis, and biodegradation (metabolism in organisms).

3.6.3 TYPES OF CONTAMINANTS

3.6.3.1 Anthropogenic Sources of Organic Chemical Pollutants

Organic compounds contain carbon and commonly other elements such as hydrogen, oxygen, nitrogen, sulfur, phosphorus, and halogens. Not all carbon-containing compounds are organic; exceptions include carbonates, cyanides, and oxalates. Many organic compounds occur naturally in water and may be considered contaminants if their concentrations are such that they adversely affect an aquatic system. However, the focus here will be on anthropogenic and largely synthetic organic compounds that occur in aquatic systems from use by humans. Such compounds include pesticides, herbicides, numerous industrial chemicals, and compounds derived from energy production and other combustion processes.

Organic compounds can be broadly divided into two classes: hydrophobic (nonpolar) compounds and hydrophilic (polar) compounds. Hydrophobic compounds are those that are not soluble or sparingly soluble in water, whereas hydrophilic compounds are soluble in water. Much of the focus of this chapter will be on the hydrophobic organic contaminants (HOCs), as these have been the class of compounds that have been the most problematic and the most studied historically. The past emphasis on hydrophobic contaminants is primarily for two reasons. First, owing to their hydrophobicity, they tend to bioaccumulate in biota, including humans, and thus reach elevated concentrations upon which they exert their adverse effects.[8] Second, until recently, hydrophobic compounds were more easily extracted and concentrated from environmental media, and thus more easily detected, quantified, and studied by available analytical methods. In aquatic environments, HOCs are often stored in sediments, with lower concentrations in the water phase. Their storage in sediments is mainly due to hydrophobic interactions of the HOCs with humic substances associated with sediment particles.[9] The presence of natural organic matter in the water column enhances HOC solubility, as HOCs adsorb to humic substances.[10] Thus, while the concentration of HOCs are typically higher in sediments than in the water column, humic substances play a role in mobilizing HOCs from sediment to water column.[11,12]

Perhaps the most notorious environmental organic contaminant is dichlorodiphenyltrichloroethane (DDT). Indeed, it was the recognition of the adverse environmental impacts of DDT that started the modern era of environmentalism with the publication of the book Silent Spring by Rachel Carson in 1962.[13] In many ways, DDT exemplifies the class of compounds known as persistent organic pollutants (POPs) and the general use of chemicals by humans. Chemicals are neither good nor bad, they simply act in the environment as their properties dictate. Like many other contaminants, DDT was very effective for its intended use. It was widely used in agriculture in the United States after World War II and was extremely effective as a pesticide. Its global use for mosquito control was also widespread. There is no doubt that DDT has saved millions of people worldwide from death due to malaria.

Although DDT has been blamed for causing many adverse impacts, there is one impact that has been truly documented and devastating. DDT interferes with calcium metabolism in birds and thus causes their eggshells to become thinner.[14,15] The eggs are thus more susceptible to breakage and greatly reducing fecundity. The book Silent Spring documents the reduction in songbird populations in the Midwestern United States. Also, DDT in the southern California Bight was responsible for the complete devastation of the bald eagle populations on the Channel Islands. Owing to its demonstrated impacts, DDT has been banned in most countries of the world. A few countries still use DDT for malaria control under the supervision of the World Health Organization.

As is often the case, the adverse effects of DDT became apparent when it was transported beyond its intended area of application. It is persistent in the aquatic environment because of its resistance to degradation by natural processes. It is transformed by many organisms into a slightly different compound through biotransformation. The product of the biotransformation of DDT is dichlorodiphenyldichloroethylene (DDE). The compounds produced through biotransformation are often called metabolites. Also DDT is subject to abiotic reductive transformation in reducing environments (e.g., aquatic sediments) to DDD (dichlorodiphenyldichloroethane). Fig. 3.6.2 shows the structures of DDT and its metabolites. The term DDT is often meant to refer to the sum of DDT and all of its isomers and transformation

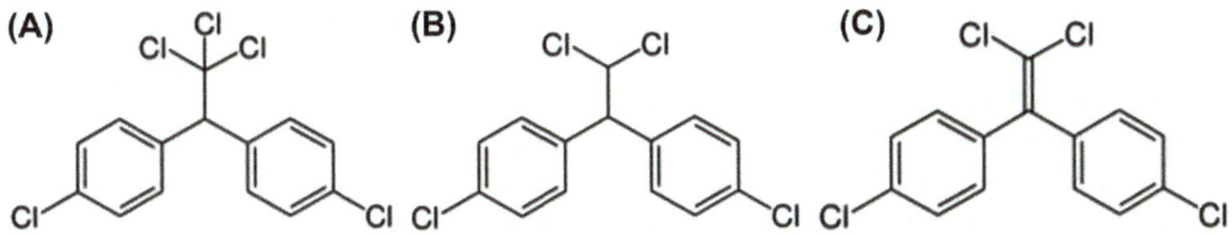

FIGURE 3.6.2 (A) Dichlorodiphenyltrichloroethane (DDT), and its metabolites, (B) dichlorodiphenyldichloroethane and (C) dichlorodiphenyldichloroethylene. DDT causes eggshell thinning in birds and is also an endocrine-disrupting chemical that has been associated with breast cancer.[14–16]

products. Indeed, most of the total DDT in the environment today is in the form of DDE. Because of its hydrophobicity, DDT can bind to suspended particles and molecules in water and be transported far away from its area of application. Moreover, its hydrophobicity allows it to bioaccumulate in organisms and even increases its concentration up the aquatic food chain through a process called biomagnification (Fig. 3.6.3).

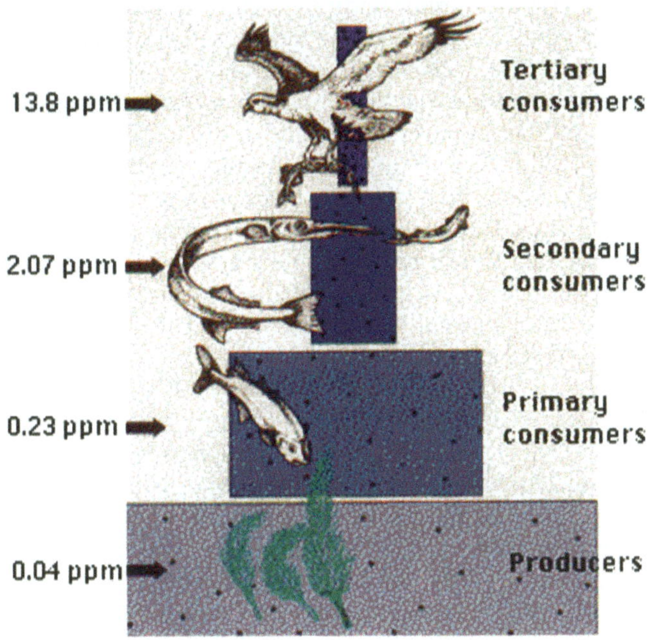

FIGURE 3.6.3 Biomagnification of dichlorodiphenyltrichloroethane (DDT), an organochlorine, is revealed in representative concentrations of DDT found in tissue samples (in parts per million, ppm). Hydrophobic chemicals build up in the food chain; the concentration of DDT increases in tissues of organisms at successively higher levels in the food chain. The mechanisms for this observation are still a matter of debate, but it is multifactorial including differences among organisms in metabolic rates, diet, and dietary assimilation efficiencies. DDT is metabolized and excreted more slowly than the nutrients that are passed up the food chain. Fat-soluble chemicals more readily diffuse into the digestive tracts of predatory organisms from their lipid-rich prey. Before DDT was banned in the United States, DDT levels in some wild bird eggs were significant enough to cause eggshell thinning and reduced reproductive success. *Image from http://bwbearthenviro2011.*

Most chemicals have some capacity to bioaccumulate, particularly hydrophobic chemicals that are lipid soluble. The process of bioaccumulation is as follows: upon repeated exposure to a chemical, the concentration of the chemical in the body increases over time up to a point that is a thermodynamic equilibrium between the body (usually fat compartments in which the chemical is soluble) and the ambient medium such as water, air, soil, sediments, or food to which the organism is exposed. The distribution of a chemical between these environmental compartments depends on the relative "solubility" of the chemical in each phase (water, air, fat, soil/sediment particle). An increase in tissue concentration of contaminants can also occur by a process called bioamplification when an organism loses weight and fat stores at a rate faster than it can eliminate the chemical, causing an increase in chemical concentration in its tissues.[17] Bioaccumulation should not be confused with toxicity. In theory, the chemical does not need to bioaccumulate to be toxic, as long as the tissue dose exceeds a threshold of toxicity.

Biomagnification occurs for several reasons, one main reason being that predators bioaccumulate chemicals more than their prey because chemicals more readily diffuse into the digestive track from lipid-rich prey compared to nonlipid food types such as fiber-rich foods.[18] Organisms at the top of the food chain are slower to depurate (metabolize and excrete) chemicals and often have higher body fat content compared to organisms on the bottom of the food chain, and they also tend to live longer, so they have more time to accumulate higher body burdens of hydrophobic organic chemicals. Biomagnification is observed in chemicals that have a log octanol-water (K_{ow}) partition coefficient greater than 4, but there are exceptions, as some chemicals bioaccumulate by a different mechanism than partitioning into fat stores, for example, by binding to specific tissues. Hydrophobic contaminants stay in the body longer, so they can become a problem even years after being exposed to them, and they build up to higher concentrations, being more likely to exceed a toxic dose than most hydrophilic contaminants that do not bioaccumulate as much. For example, HOCs accumulate in eggs and can impair fertility. The chemical may not do any harm while it is parked in some lipid compartment such as ova, but then poses a risk to a newly formed zygote, or when mobilized into the blood stream.

Another infamous class of water pollutants that are HOCs are polychlorinated biphenyls (PCBs). These compounds consist of a biphenyl backbone, with 1–10 chlorines replacing the hydrogens on the two phenyl groups. The various possibilities of chlorines positioned on the two phenyl groups lead to 209 possible discrete compounds called congeners. PCBs were first produced as complex commercial mixtures, primarily used as dielectric fluids in capacitors and transformers. They were also used widely as plasticizers in the plastics industry. They were produced in large quantities by the Monsanto Corporation and sold under the trade name Aroclor from 1930 to 1977. Aroclors were different mixtures of the congeners used for specific purposes.

Because of their chemical and physical properties, PCBs, like other HOCs, are prone to long-range atmospheric transport. They are now widely dispersed globally as evidenced by their presence in the fat of polar bears in the Arctic and penguin eggs in the Antarctic.[19] The toxicity of PCBs varies widely among congeners. Like DDT, they can cause eggshell thinning in birds and deformities such as crossed bills in cormorants of the Great Lakes region.[20,21] They can also inhibit reproduction in aquatic species. Another recently discovered impact of PCBs is their ability to bind to estrogen receptor sites and thus exert hormonal activity in organisms and interfere with normal biochemical processes.[22]

Another group of contaminants analogous to PCBs are polybrominated diphenyl ethers (PBDEs). Their structure differs from that of PCBs by the placement of an oxygen between the two phenyl groups (thus making it an ether). By analogy to PCBs, 1–10 of the hydrogens on the phenyl groups can be replaced by bromine atoms, and thus there are 209 possible congeners. Like PCBs, commercial formulations of PBDEs are mixtures of congeners rather than a single compound. Similarly, their toxicity differs considerably among different congeners. PBDEs were used extensively as flame retardants since the 1970s in plastics, in electronic devices such computers and TVs, and in polyurethane foam furniture padding and mattresses. Like other POPs, they are hydrophobic, tend to bioaccumulate in organisms, and have been distributed globally by long-range atmospheric transport. They are found in lipids of aquatic organisms such as salmon and in the milk of cows and humans.[23]

Chemicals of concern that are formed from the incomplete combustion of organic wastes or burning of hydrocarbons (oil, coal, and natural gas) include polychlorinated dibenzo-*p*-dioxins (PCDDs), polychlorinated dibenzofurans (PCDFs), and polyaromatic hydrocarbons (PAHs). Fig. 3.6.4 shows their structures. The major concern with these chemicals is their ability to cause cancer. Each chemical has a different carcinogenic potency. 2,3,7,8-Tetrachlorodibenzo[*p*]dioxin is considered to be the most potent carcinogen.[24]

PAHs consist of one to five fused benzene rings. They are produced from incomplete combustion of organic materials such as fossils fuels. PAHs also enter the aquatic environment via oil spills and refinery and drilling operations. Many of these compounds are potent carcinogens. The compounds were first discovered in soot because of an increased incidence of cancer in chimney sweeps. The cancer was eventually linked to a compound called benzo(*a*)pyrene, which has since been identified as one of the most potent carcinogens known. PAHs also bioaccumulate in biota because of their hydrophobicity.

Chemicals that are persistent in the environment on the order of years are called POPs. An attempt to limit the direct production and use of some of the worst POPs, termed the "Dirty Dozen," was ratified into an international treaty at the Stockholm Convention on Persistent Organic Pollutants in 2001. These include the PCDDs and PCDFs as well as organochlorine pesticides and industrially produced PCBs and hexachlorobenzene. See Table 3.6.1 and Fig. 3.6.5. In 2009, perfluorinated alkylated substances such as the perfluorooctane sulfonate (PFOS) mentioned earlier, as well as perfluorooctanoic acid and perfluorononanoic acid, were added as POPs under the Stockholm convention.

POPs will eventually break down in the environment by processes of photolysis (in air as well as in surface water and soil to a depth that sunlight can penetrate), chemical reactions in water such as hydrolysis, and biodegradation. But some of their transformation by-products may also be toxic and can be made more water soluble. Such is the case for nitro-polyaromatic hydrocarbons that arise primarily as products of incomplete combustion and from atmospheric transformation of PAHs. Atmospheric deposition of these chemicals leads to water contamination.[25] Other sources include industrial effluents and other air emissions from various industries.

Some organic chemicals are volatile, having low boiling points and high vapor pressures at room temperature; these constitute a category of chemicals called volatile organic chemicals (VOCs). Notable examples include BTEX (benzene, toluene, ethylbenzene, and xylenes) present in petroleum products, chloroform and bromoform as well as other low-molecular-weight chlorinated aliphatic and aromatic hydrocarbons, and lower-molecular-weight aliphatic and aromatic compounds including some PAHs.

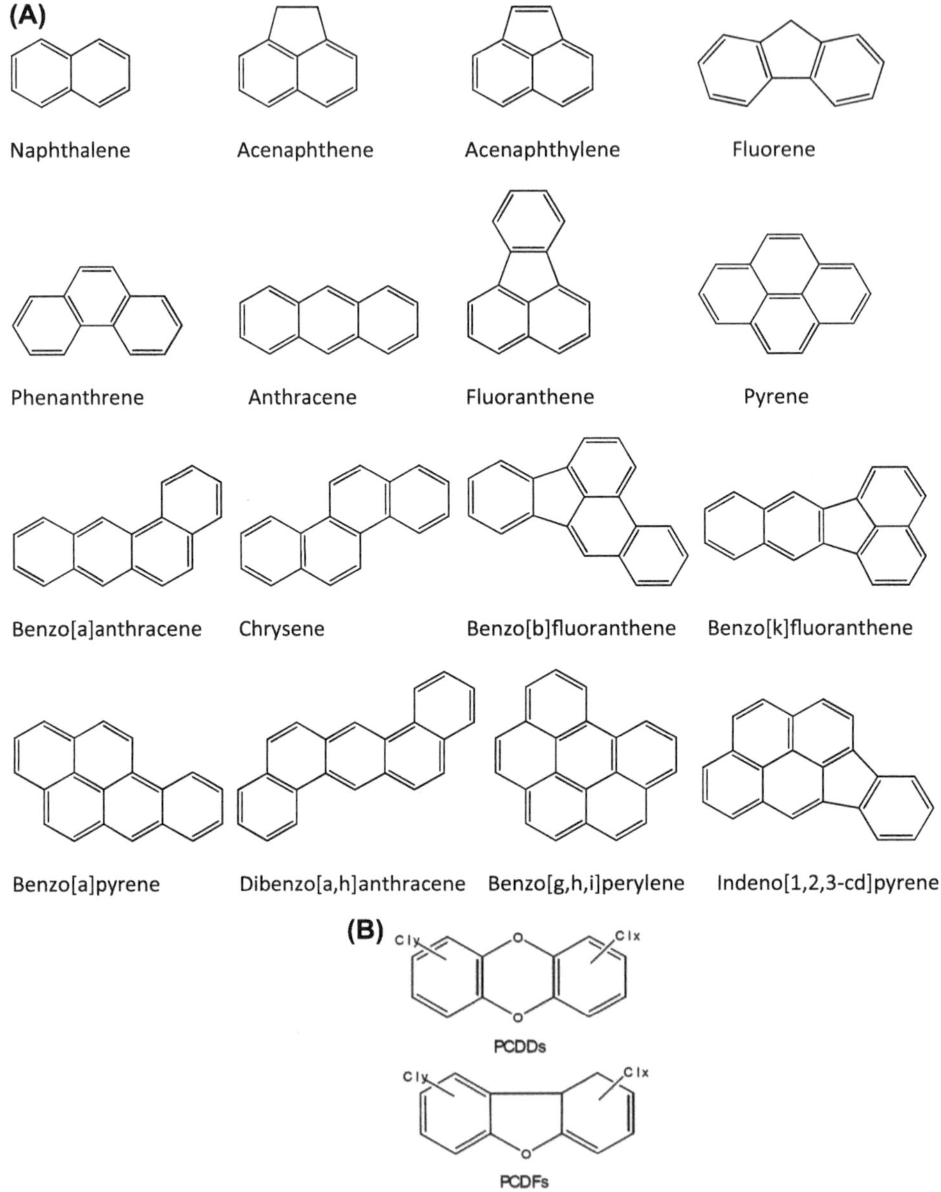

FIGURE 3.6.4 Environmental Protection Agency priority: (A) polyaromatic hydrocarbons (PAHs) and (B) polychlorinated dibenzofurans and dioxins (polychlorinated dibenzofurans (PCDFs) and polychlorinated dibenzo-*p*-dioxins (PCDDs)). All are endocrine-disrupting chemicals that have various carcinogenic potencies.[24]

TABLE 3.6.1 The "Dirty Dozen" Persistent Organic Pollutants Listed in the 2004 Stockholm Convention to Be Restricted or Eliminated in Their Production and Use

Chemical	Class of Chemical
Aldrin	Insecticide
Chlordane	Insecticide
DDT	Insecticide
Dieldrin	Insecticide
Dioxins	Industrial by-product
Endrin	Insecticide
Furans	Industrial by-product
Heptachlor	Insecticide
Hexachlorobenzene	Insecticide, industrial by-product
Mirex	Insecticide, flame retardant
PCBs	Industrial chemical
Toxaphene	Insecticide

PCB, polychlorinated biphenyl; *DDT*, dichlorodiphenyltrichloroethane.

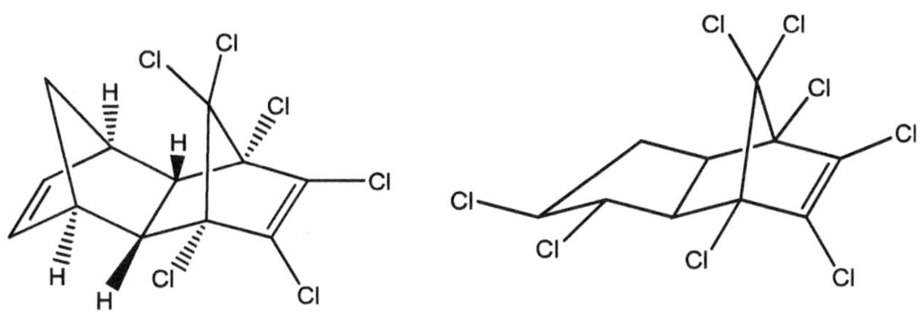

Aldrin Chlordane

FIGURE 3.6.5 The structures of the first two organochlorine persistent organic pollutants listed in Table 3.6.1. Heavily chlorinated organic chemicals tend to bioaccumulate and are persistent in the environment.

Most PAHs, however, fall into the category of "semivolatile" organics (SVOCs). (Sorry for the acronyms, but they are important to learn if you want to be employed in an environmental job because everyone uses them.) There are lists of various categories of chemicals; often, a chemical list will either refer to an analytical method, for example, Environmental Protection Agency (EPA) Method 524,[26] for determining VOCs in drinking water or a regulatory statute such as the Clean Water Act[27] that sets standards for chemicals in drinking water, called maximum contaminant levels (MCLs). An MCL is a level of contaminant that cannot be exceeded in a daily dose without potential unacceptable health risk and is a primary drinking water standard that requires monitoring as well as corrective action when exceeded.

There are only about 50 organic chemicals for which MCLs exist. This is a small fraction of the over 100,000 organic artificial chemicals. The MCL list is short because any MCL has to be defensible from a toxicological standpoint. Years of research and data evaluation go into the process of establishing an MCL. Also, MCLs were initially developed to target chemicals that are ubiquitous in the environment and/or are high-usage chemicals, such as pesticides and herbicides and some industrial chemicals. The handful of chemicals tested are surrogates for many of the others, such that if treated waters do not contain elevated levels of the chemicals on the MCL list, the water treatment process is probably sufficient enough to protect against other chemicals. In itself, this is not always adequate, but water treatment facilities have other monitoring safeguards, such as testing for total organic carbon or direct toxicity testing of water samples. Thus, a chemical contaminant that gets into the drinking water that is not on the MCL list could be detected by indirect means first, after which samples will be tested by mass spectrometry to identify the chemical(s) of interest.

Another way to categorize chemical pollutants other than based on their physiochemical properties (e.g., HOC vs. hydrophilic; VOC vs. SVOC or nonvolatile) is by their intended use, such as pesticides. Pesticide is a broad category that includes herbicides, fungicides, rodenticides, and others specific to what is intended to be killed. A better term encompassing all of the above is "biocide." As is clear from the etymology, biocides are intended to kill something. The problem with biocides is that they are often not perfectly target-specific, thus causing collateral damage to "nontarget" species such as the honey bee, *Apis mellifera*. Honey bees are often a nontarget species affected by pesticide spraying. Other than the obvious exposure from nectar and pollen collected from sprayed crops, a study by Samson-Robert et al.[28] showed that honey bees are getting exposed to neonicotinoid pesticides, which are highly toxic to bees, from puddle water near agricultural fields. Pollutants respect no boundaries. In case of the honey bee, water had been overlooked as an exposure route and transport pathway.

Owing to the prolific use of pesticides, surface water runoff can pollute surface water and groundwater, especially near agricultural areas. Pesticides are also used on golf courses and just about anywhere—in urban and suburban landscapes. Yet, levels are usually below MCLs. However, guidelines have not been established for the majority of the herbicides found in drinking water or for mixtures of pesticides.[29]

There are many different types of organic chemicals used as biocides. The neonicotinoids are structurally similar to nicotine, which is an effective pesticide that can be extracted from the nightshade family (Solanaceae) of plants. Neonicotinoids are highly water soluble and have low volatility, and being semipersistent, they can contaminate water supplies from agricultural runoff.[28] As they are used as systemic pesticides (the chemical is found within the plant or plant seed as opposed to being applied to a plant externally such as by spraying), environmental contamination can occur from seed dust during processing of seeds.[30] Pyrethrins are another class of pesticides produced or inspired by nature; the natural product is extracted from *Chrysanthemum* flowers. Pyrethrins target sodium channels of nerve cells, whereas neonicotinoids target the enzyme, acetylcholinesterase. Pyrethroids are artificial chemicals with the same type of structure and mode of action as pyrethrins (Fig. 3.6.6).

Organophosphates are esters of phosphoric acid; they include natural biological molecules, but insecticidal analogs are nerve agents acting on acetylcholinesterase with a wide range of

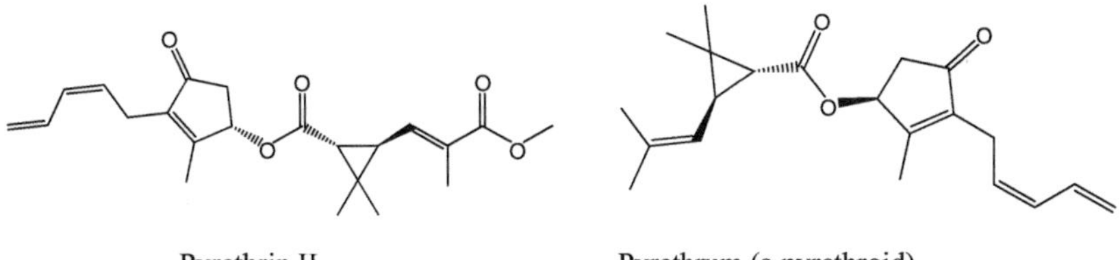

FIGURE 3.6.6 Pyrethrins and pyrethroids: a natural pyrethrin, pyrethrin II, and a synthetic pyrethroid, pyrethrum. Pyrethroids are synthetic analogs of pyrethrins (cyclopropane carboxylic acid esters). Their mode of action is through voltage-gated sodium channels in the axonal membranes (axonic excitotoxins).

acute toxicities depending on the structure. Although they are subject to hydrolysis and break down quickly when enhanced by photolysis, some analogs are dangerous from the standpoint of acute toxicity.[31] Carbamates, such as aldicarb, carbaryl, and sevin, have a similar mode of action as organophosphates but are more degradable. Water pollution of carbamates and organophosphates is mainly via agricultural runoff from spraying of fruits. Water quality criteria, both acute and chronic based, have been developed for some of these compounds to protect aquatic life, including for the organophosphates diazinon, chlorpyriphos, and parathion, as well as the carbamate carbaryl (Fig. 3.6.7).

One of the world's most widely used herbicides is glyphosate. Glyphosate is an organophosphate containing the amino acid glycine. It kills plants by interfering with the synthesis of aromatic amino acids. It adsorbs to soil and is broken down by soil microbes to aminomethylphosphonic acid (AMPA); its half-life is highly dependent on environmental conditions: 2–144 days.[32] The United States Geological Survey did an occurrence study of 154 water samples during 2002 in the Midwest; glyphosate was found in 36% of the samples and AMPA was found in 69% of the samples.[33] In 2015 the World Health Organization labeled it as a probable carcinogen, but probably not by a mutagenic mechanism. In mammals, glyphosate's mechanism of toxicity is believed to be by the uncoupling of oxidative phosphorylation.[34] Adjuvants including surfactants are added to glyphosate in commercial herbicide products, which enhance bioavailability and the potential for acute toxicity. A study exposing rats to environmentally relevant exposures (ultralow chronic doses) resulted in liver and kidney damage.[35]

Atrazine is an herbicide used widely on corn and sugarcane and less on golf courses and residential lawns. It was banned in Europe in 2004 because of widespread groundwater contamination. It blocks photosynthesis and was thought to be safe for nontarget organisms until it was found to be an endocrine-disrupting chemical. Simazine is a similar analog. Atrazine alone and combined with simazine were reviewed by US EPA and deemed to be safe at levels at which humans may be exposed, but further review is being conducted on their impact on amphibians, which has been a concern.[36]

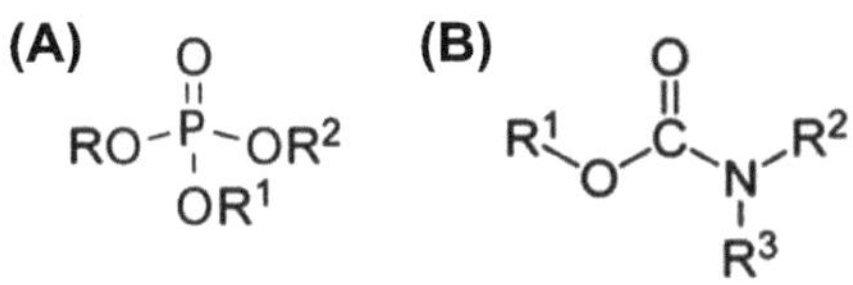

FIGURE 3.6.7 General structures of (A) organophosphates and (B) carbamates. These types of pesticides are neurotoxins that inhibit acetylcholinesterase.

Diquat and paraquat are herbicides identified chemically as dipyridils. They are highly toxic and if ingested by a person, all vital organs are affected. Because of the cations in the quaternary ammonium structure of the molecule, the chemical binds to tissues readily. Exposure to paraquat can lead to acute respiratory distress syndrome. A study by the National Institutes of Health linked farm workers' exposures to paraquat with Parkinson's disease via mitochondrial dysfunction and oxidative stress.[37] Both diquat and paraquat are potential drinking water contaminants (Fig. 3.6.8).

In recent years, the focus of contaminant research has shifted to hydrophilic contaminants because of the availability of new analytical techniques, primarily liquid chromatography-mass spectrometry, which allows for the analysis of compounds in water directly without prior extraction or concentration. This new generation of contaminants, mostly hydrophilic compounds, is collectively referred to as "emerging contaminants" and includes such chemicals as the ever-growing array of pharmaceuticals and personal care products. Emerging contaminants are the subject of Chapter 3.7.

An example of a hydrophilic water contaminant is methyl tertiary butyl ether (MTBE). MTBE was used extensively in gasoline from the 1980s until around 2000. MTBE was added to gasoline as an oxygenate to lower emissions and improve the burning of fuel as a requirement of the 1990 Clean Air Act Ammendments.[38] Oxygenating compounds are hydrocarbons that contain oxygen and thus boost the oxygen content of gasoline. Higher oxygen content ensures complete combustion and the subsequent lowering of tailpipe emissions. In addition, compounds like MTBE have the added benefit of burning very well in internal combustion engines, thereby increasing the octane rating of the gasoline. The EPA allowed up to 15% by volume of MTBE in gasoline in the late 1980s. Unfortunately, like so many other chemicals, there were unforeseen problems due to its properties, namely, the high water solubility of MTBE (48 g/L). Many gasoline stations had what came to be known as leaking underground storage tanks (or LUSTs). Most of the components of gasoline are nonpolar or hydrophobic and thus when they leaked from the tanks, they adsorbed to the surrounding soil and did not leach very far away from the LUSTs. However, for gasolines containing MTBE, the

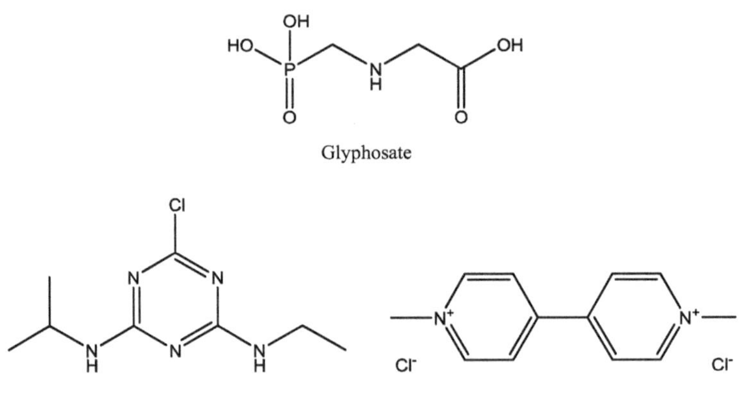

FIGURE 3.6.8 Common classes of herbicides. Glyphosate and atrazine are the most commonly used in the United States. Paraquat is an electrophile that readily binds to tissues, primarily targeting the lungs and kidney and has been associated with Parkinson's disease.

high water solubility led to significant leaching of the compound into underlying or nearby aquifers, thus contaminating the groundwater.

Significant contamination of drinking water resources was detected in Santa Monica, CA, in the mid-1990s.[39] Because of this and other contamination incidents throughout the state, California banned gasolines containing MTBE as of January 1, 2003. In 2008, Chevron, BP, and other major oil companies settled numerous lawsuits with California and 19 other states for damages due to drinking water supplies for $423 million.

The irony is that in all the examples discussed so far, the chemicals were perfect for their intended use. Part of the goal of Green Chemistry is to better understand, anticipate, and mitigate the adverse impacts of chemicals beyond their intended use and sphere of influence and thus avoid the need for costly and damaging lawsuits and remediation projects in the future.

Hydrophilic contaminants are less often discussed, monitored, and studied because they do not bioaccumulate readily, although there are exceptions. Some hydrophilic chemicals can bioaccumulate by mechanisms other than lipid partitioning, such as PFOS, which binds to proteins and other tissues such as serum albumin, liver tissue, and DNA.[5,40] PFOS was a main ingredient of the commercial product, Scotchgard, a fabric protector made by the company 3M (Minnesota Mining and Manufacturing Company). When PFOS was detected in wildlife, 3M invested research money into environmental monitoring of PFOS and the development of safer products that have much lower potential for tissue binding, bioaccumulation, and toxicity. The company began to phase out the chemical in 2002, partly due to its own concerns about the chemical having been detected in wildlife as well as concerns from the EPA regarding the chemical's high environmental persistence. Solutions do not always have to come from top-down regulations but also from the chemical industry itself if there is a desire to maintain a reputation of integrity and prevent fines and/or future litigation.

3.6.3.2 Marine Debris and Plastic in the Environment

One of the most alarming concerns for the environment is the growing amount of plastic and other trash in the world's oceans and lakes. Trash is sometimes illegally or accidentally dumped directly into the ocean, such as from ships and offshore platforms, but more often is carried by rivers and streams, washed overland by stormwater runoff or via raw sewage overspills, or blown in by wind. Many developing countries cannot afford wide-scale trash pickup and safe landfilling, and instead, they resort to open dumping. Improper handling of waste occurs in all countries, however, and is a global problem.

Millions of tons of plastics and other marine debris in the ocean get concentrated in giant gyres that swirl in circular patterns created by winds and the Coriolis effect from the rotation of the earth. One such gyre or giant vortex that accumulates trash is called the Great Pacific Garbage Patch. It stretches from California to Japan in the gyre of the North Pacific Subtropical Convergence Zone (Fig. 3.6.9). There are presently five major garbage patches that are located in the North Pacific, South Pacific, North Atlantic, South Atlantic, and Indian Oceans. The marine debris will move in and out of these gyres, but the sea floors accumulate more garbage below the middle of these gyres than elsewhere. Natural materials such as cotton can biodegrade rapidly, but plastics can take possibly hundreds of years to completely break down. Sunlight and physical stressors can break down plastic trash such as water bottles into smaller and smaller bits, such that the soup of marine debris in the gyres consists of tiny

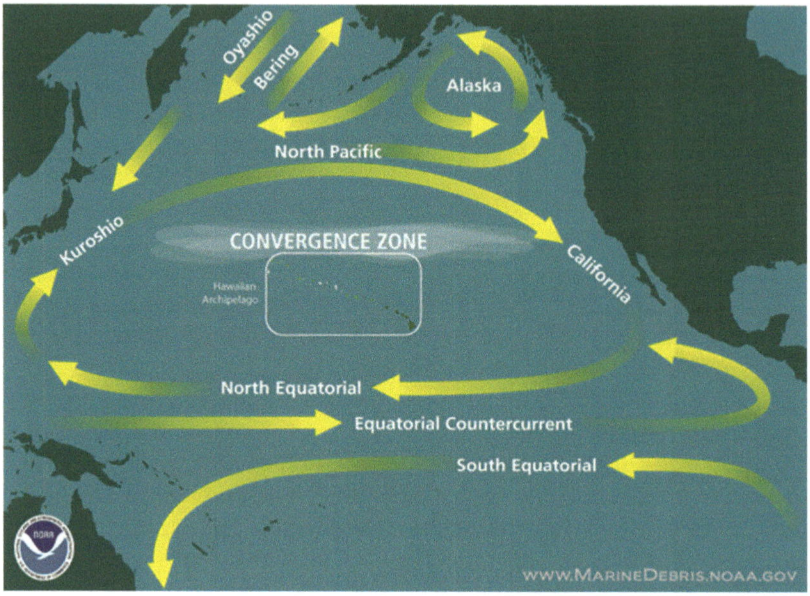

FIGURE 3.6.9 Pacific Ocean gyre (*circulation pattern*) that accumulates trash, otherwise known as a trash vortex or "Great Pacific Garbage Patch."

pieces of plastic. Birds and fish mistakenly ingest the plastic as food. In the most concentrated parts of these gyres, there is more plastic than biomass. Pieces of plastic debris also adsorb other pollutants in the ocean, such as PCBs. Thus, when an organism ingests the plastic, it is also ingesting contaminants that have collected on the plastic. Plastics may themselves contain toxins such as bisphenol A (an endocrine-disrupting chemical), phthalates, or vinyl chloride, depending on the type of plastic. The most persistent and bioaccumulating of chemicals will be transferred throughout the food chain. Humans should be very concerned because it is probable if not likely that all fish contain some measure of these chemicals.

When pieces of plastic break down into particles of less than 1 mm in diameter, they are called microplastics. In addition to the microplastics created from the decomposition of trash, microplastics enter the environment from sewage discharges carrying personal care products. Tiny microplastic beads make good facial scrubs and have been used in various cosmetics and toothpastes. But now the growing concern and awareness that marine biota including zooplanktons ingest microplastics[41] has led to a federal ban of microplastics in personal care products in the United States, starting in 2017. Canada and the European Union appear to be following suit. With the Netherlands leading this campaign, they are also trying to address the trash problem in the ocean gyres. The Ocean Cleanup[42] is a nonprofit organization out of the Netherlands founded by Boyan Slat, who was just 16 years at the time he envisioned designing a solution to the problem after seeing more trash than fish on a vacation dive. After a TED talk and crowd funding campaign, his organization started studying the marine debris problem and is designing technology that will aid in ocean cleanup. The organization believes it can remove half the plastic in the Great Pacific Garbage Patch in 10 years. Other organizations around the world are contributing to beach cleanup activities to collect marine debris when it comes ashore, and other volunteers are collecting trash from surface water offshore. Everyone can contribute to the solution by increasing recycling efforts and promoting biodegradable materials to reduce plastic dependency.

3.6.3.3 Metals and Metalloids

Many infamous organic pollutants are synthetic organic chemicals and as such any detected amount is an indication of anthropogenic contamination. Metals and metalloids differ from many organic contaminants in that they occur naturally. However, they can become enriched in a particular environment from anthropogenic activities related to their extraction or use by various industries and may even naturally exceed water quality or safe drinking water guidelines. An example of this would be arsenic: most groundwaters in the United States naturally exceed the MCL for arsenic in drinking water of 10.0 μg/L. An MCL is a chemical-specific concentration that is allowed in drinking water. A maximum contaminant level goal (MCLG) is the level of a contaminant in drinking water below which there is no known or expected risk to health. MCLGs allow for a margin of safety. The MCL is set as close to the MCLG as is feasible using the best available treatment technology. The MCLG sends a signal to water treatment industries to develop technology or water treatment approaches to eventually meet that goal. Table 3.6.2 shows MCLs and MCLGs for metals and metalloids, potential sources of enrichment from anthropogenic activity, and potential health effects. Industrial sources of these pollutants can be released directly to surface water, through sewage discharges, or to air which then deposit on surface water.

Trace elements are nutrients in our diets in small amounts (<100 mg/day), found in tissues at microgram per liter (μg/L) levels, and have a narrow therapeutic range. Too great or too little of trace elements such as copper, manganese, selenium, iron, zinc, fluoride, iodine, chromium, molybdenum, nickel, cobalt, and possibly boron and vanadium (the essentiality of these is still being evaluated) will affect human health; too low a dose induces a deficiency and too high a dose constitutes toxicity. Calcium and sodium are essential minerals found at higher levels in tissues. They, along with potassium and phosphorus, are needed at a daily intake of at least 1 g. Aluminum is one of the most common elements in soil, so it would not be expected to be a potent toxin. There is no MCL for aluminum. There are, however, ambient water quality criteria (AWQC) for aluminum. Striped bass and brook trout are two species shown to be sensitive to aluminum at μg/L concentrations.[43] AWQC were developed by EPA to protect aquatic organisms; they are benchmark concentrations that should not be exceeded. AWQC for aquatic life were developed for most metals and for some organic chemicals, some of which are on the MCL list. Metals that are affected by water quality parameters such as pH or hardness (divalent cations, mainly Ca^{2+} from calcium carbonate) have AWQC that are adjusted by the water quality parameter value. AWQC are developed for chronic and acute exposures.

The fate and toxicity of metals are highly dependent on their speciation, or the form in which they exist in a given aquatic system. It is the freely dissolved form of a metal that is typically the most toxic. The EPA has a list of priority water pollutants that includes organics, metals, and metalloids. The metals and metalloids include antimony, arsenic, beryllium, cadmium, chromium, copper, lead, mercury, nickel, selenium, silver, thallium, and zinc. Note that many of the metals listed are transition metals, such as cadmium and copper, or heavy metals, such as lead and silver, all of which are prone to forming complexes with organic or inorganic ligands. Ligands are organic or inorganic chemical species, elements,

TABLE 3.6.2 Maximum Contaminant Levels (MCLs) and MCL Goals, Potential Health Effects, and Potential Sources for Metals and Metalloids

Contaminant	MCLG (mg/L Unless Noted)	MCL (mg/L Unless Noted)	Potential Health Effects From Chronic Exposure Above the MCL (Unless Specified as Short Term)	Sources of Contamination in Drinking Water
Antimony	0.006	0.006	Increase in blood cholesterol levels, decrease in blood sugar levels	Discharge from petroleum refineries, fire retardants, ceramics, electronics, solder
Arsenic	0	0.01 as of January 23, 2006	Skin damage or problems with circulatory systems and may have increased risk of getting cancer	Erosion of natural deposits, runoff from orchards, runoff from glass and electronics production wastes
Asbestos (fiber >10 μm)	7 million fibers per liter	7 MFL	Increased risk of developing benign intestinal polyps	Erosion of natural deposits, runoff from orchards, runoff from glass and electronics production wastes
Barium	2	2	Increase in blood pressure	Discharge of drilling wastes, discharge from metal refineries, erosion of natural deposits
Beryllium	0.004	0.004	Intestinal lesions	Discharge from metal refineries and coal-burning factories; discharge from electrical, aerospace, and defense industries
Cadmium	0.005	0.005	Kidney damage	Corrosion of galvanized pipes, erosion of natural deposits, discharge from metal refineries, runoff from waste batteries and paints
Chromium (total)	0.1	0.1	Allergic dermatitis	Discharge from steel and pulp mills, erosion of natural deposits
Copper	1.3	Action level: 1.3[a]	Long-term exposure: Liver or kidney damage, especially among susceptible (Wilson's disease). Short-term exposure: gastrointestinal distress	Corrosion of household plumbing systems, erosion of natural deposits
Cyanide (as free cyanide)	0.2	0.2	Nerve damage or thyroid problems	Discharge from steel/metal factories, discharge from plastic and fertilizer factories
Fluoride	4	4	Bone disease (pain and tenderness of the bones), children may get mottled teeth	Water additive that promotes strong teeth, erosion of natural deposits, discharge from fertilizer and aluminum factories

TABLE 3.6.2 Maximum Contaminant Levels (MCLs) and MCL Goals, Potential Health Effects, and Potential Sources for Metals and Metalloids—cont'd

Contaminant	MCLG (mg/L Unless Noted)	MCL (mg/L Unless Noted)	Potential Health Effects From Chronic Exposure Above the MCL (Unless Specified as Short Term)	Sources of Contamination in Drinking Water
Lead	Zero	Action level: 0.015[a]	Infants and children: delays in physical or mental development, children could show slight deficits in attention span and learning abilities. Adults: kidney problems high blood pressure	Corrosion of household plumbing systems, erosion of natural deposits
Mercury (inorganic)	0.002	0.002	Kidney damage	Erosion of natural deposits, discharge from refineries and factories, runoff from landfills and croplands
Nitrate (measured as nitrogen) and nitrite	10 nitrate; 1 nitrite	10 nitrate; 1 nitrite	Infants below the age of 6 months who drink water containing nitrate in excess of the MCL could become seriously ill and, if untreated, may die. Symptoms include shortness of breath and blue baby syndrome	Runoff from fertilizer use, leaching from septic tanks, sewage, erosion of natural deposits
Selenium	0.5	0.5	Hair or fingernail loss, numbness in fingers or toes, circulatory problems	Discharge from petroleum refineries, erosion of natural deposits, discharge from mines
Thallium	0.0005	0.002	Hair loss, changes in blood, and kidney, intestine, or liver problems	Leaching from ore-processing sites, discharge from electronics, glass, and drug factories

[a]Lead and copper are regulated by a treatment technique that requires systems to control the corrosiveness of their water (EPA). If more than 10% of tap water samples exceed the action level, water systems must take additional steps. The action level for copper is 1.3 mg/L and for lead is 0.015 mg/L.
From EPA.

or compounds that act as Lewis bases or electron donors. Ligands readily donate a pair of electrons to positively charged metal ions in solution (i.e., cations) and thus create complex ions with one to six ligands bound to the central metal atom. In fact, metal cations in water are rarely present as just the freely dissolved ions. If nothing else, water molecules can act as ligands and bind to metal ions forming aquo-complexes.

This ability to form different complex species as the concentration of ligands in the water changes is a key characteristic of metals. As a metal makes the transition from freshwater to seawater in an estuary, the form of the dissolved metal can change dramatically. As the chloride ion concentration increases from zero to that typical of seawater (~0.6 mol/L), complexation increases. The dominant form of cadmium in pure water would be Cd^{2+} (actually the aquo-complexes are bound to humic substances in freshwater), and at high salinity, it exists as a mixture of chloro-complexes, with $CdCl_3^-$ predominating in seawater.

The speciation of metals also impacts their toxicity, so that just knowing the total metal concentration in an aquatic system does not allow for the accurate prediction of the associated toxicity. The binding of metals by various ligands inhibits bioavailability (the susceptibility to uptake by an organism), and thus toxicity. Metal toxicity is also affected by other factors such as pH and hardness (i.e., Ca^{2+} concentration). Thus, the same metal concentration in one aquatic system may be toxic, but not in another with different conditions.

Some metals and metalloids do not occur as the elemental cation, but rather as polyatomic anions. An example is arsenic, which occurs in water as either arsenite (AsO_4^{3-}) that contains As(III) or as arsenate (AsO_4^{3-}) that contains As(V). The oxidation-reduction conditions will also have an impact on metal speciation. Under reducing conditions, As(III) species dominate, whereas under oxidizing conditions, the As(V) species are more stable.

Arsenic in the form of lead arsenate was used for many years as a wood preservative, or generically as an insecticide. Wood pieces in soil will decompose, releasing the arsenic that can wash into surface water or make its way to groundwater. Smelting operations are another major source of arsenic and many other metals, especially lead, which can be either discharged directly into water or be atmospherically transported. Most arsenic poisonings today are the result of drinking well water in locations where groundwater is naturally enriched in inorganic arsenic. Other trace elements that are most often associated with toxicity include mercury, lead, and cadmium. As such, the MCLs for these are low.

Another infamous example of the importance of metal speciation and redox conditions is hexavalent chromium or chromium(VI). This is the carcinogenic form. This is another metal that does not occur as the free cation in water, appearing instead as the polyatomic anions chromate and dichromate (CrO_4^{2-} and $Cr_2O_7^{2-}$, respectively). This form of chromium was made famous in the movie *Erin Brockovich*, where it was found that in Hinkley, California, Pacific Gas and Electric (PG&E) had contaminated the groundwater of local residents, which led to an outbreak of cancer. The movie would lead one to believe that the contamination problem had been resolved by the famous lawsuit, but in fact many issues are still being debated. The remediation approach being used by PG&E to remove the Cr(VI) is to not remove it at all, but to create reducing conditions in the contaminant plume and reduce the Cr(VI) to Cr(III) that then forms insoluble Cr_2O_3 that precipitates out of solution. However, unintended consequences soon became apparent. Some local residents began observing black water coming from their wells that were near the plume. A sample of this black water was taken to California State University, San Bernardino, and the colloidal black material was isolated and analyzed and found to contain mostly manganese. It is suspected that the reducing conditions used to precipitate the Cr in the aquifer had also reduced Mn(IV) to Mn(II), the soluble form of manganese. The manganese became dissolved in the water and when it was drawn up into a well and reoxygenated, Mn(II) was oxidized back to Mn(IV) and formed a colloidal precipitate of manganese(IV)oxide.

Mercury as an inorganic metal is not nearly as toxic as its organic forms because of the differences in bioavailability. When mercury in aquatic environments is transformed by bacteria to methyl mercury, it readily bioaccumulates in fish and biomagnifies through the food chain. The EPA reference dose for mercury via fish consumption is a daily dose of 0.1 μg/kg, which is associated with threshold health effects. A major source of mercury to the environment is from coal burning plants. Smelting, alkali processing, and other industrial activities are other sources. Historically, mercury had many uses until the hazards became apparent.

"Mad hatter disease" was a neurological condition among hat makers who used mercury in the form of mercuric nitrate on felt. Mercury in the environment will cycle between air, water, soil, sediment, and biota, so it can be difficult to determine the original source of mercury. Mercury can also be naturally occurring and, for example, may be enriched in a layer of sediment after a volcanic eruption caused its release to the atmosphere. Moreover, mercury is a common component of fly ash from power plants and is subject to long-range transport in the atmosphere from its point of origin. In fact, it is suspected that a significant fraction of the mercury found in waters of the western United States came from power plants as far away as China.

3.6.3.4 Nutrients

"Nutrients" is the general term used to describe chemicals in aquatic systems necessary for the general health and growth of plants. Some nutrients are necessary for plants to thrive in a healthy aquatic system to provide food, habitat, oxygen, etc. One can think of nutrients in aquatic systems as being analogous to fertilizer in the terrestrial environment. Just the right amount is good, but too much can cause excessive plant growth and lead to serious water quality problems.

In aquatic systems, the term usually refers to various forms of nitrogen and phosphorus that are bioavailable to plants. These forms include nitrates and phosphates and related species. Sometimes cationic species such as ammonium ions and dissolved iron, both Fe^{2+} and Fe^{3+}, are important as well. The excess input of nutrients into an aquatic system leads to a process known as eutrophication (see Chapter 3.8 for a discussion of how this impacts coastal water).

Nitrates and nitrites may be naturally occurring, but excess levels are often the result of fertilizer use that washes into surface water and groundwater. Other sources may be animal manure or other chemical discharges. Infants younger than 6 months who drink water containing nitrate in excess of the MCL could become seriously ill and, if untreated, may go into a coma and possibly die. Symptoms include shortness of breath and blue baby syndrome, also known as methemoglobinemia, wherein a small percentage of the hemoglobin in the blood is oxidized and unable to carry oxygen. The mechanism of this effect is not completely understood in great detail, but a secondary or alternative mechanism involves a role of methemoglobin (MetHb) in reducing the binding of nitric oxide (NO). This allows NO to readily diffuse to neighboring cells and activate the NO signaling pathways of vasodilation leading to hypoxia.[44]

3.6.3.5 Radionuclides

A special subset of inorganic contaminants is radionuclides. Radionuclides are unstable isotopes that undergo radioactive decay to become more stable. A nuclide of an element, also called an isotope of an element, is an atom of that element that has a specific number of nucleons (protons and neutrons). All chemical elements have radionuclides; many radionuclides are naturally occurring, but they can become enriched in the environment from human activity such as from nuclear power plants, mining, medical wastes, or atmospheric deposition from nuclear bomb testing. Different nuclides of an element differ in the number of neutrons they contain. For example, hydrogen-1 (^{1}H) and hydrogen-2 are both nuclides of the element hydrogen, but the nucleus of hydrogen-1 contains only a proton, whereas the nucleus of hydrogen-2 contains a proton and a neutron. ^3H is called tritium; it rarely exists in nature and is found as a contaminant in water from nuclear power plant activity.

Uranium and thorium are primordial elements, formed in the beginning of the universe. These also may become enriched in the environment from human activity related to nuclear bombs and nuclear energy (including mining for the raw resources). The EPA drinking water MCL for uranium is 30 μg/L. The MCL for combined radium (Ra-226 and -228) is 5 picocuries per liter (5 pCi/L). A curie is a unit of ionizing radiation equal to 3.7×10^{10} disintegrations per second, roughly the amount of radiation emitted by 1 g of radium-226. All other radionuclides are regulated by their designation of either alpha or beta particle; MCLs are 15 pCi/L for gross alpha particles, and 4 mrem/year exposure to beta/photon emitters. A rem (acronym of roentgen equivalent man) is a measure of the radiation dose. An alpha particle consists of two protons and two neutrons, identical to that of a helium nucleus, and is produced in the process of alpha decay. Beta particles are either electrons or positrons. Beta decay occurs in a nucleus with too many protons or too many neutrons, resulting in the transformation of one of the protons or neutrons into the other. Groundwater supplies near nuclear power plants may have elevated (enriched) levels of several radionuclides, including tritium, radium-228/226, and strontium-90. Owing to the ubiquity of radionuclides, organisms have adapted the ability to repair tissue damage upon exposure to low levels of ionizing radiation. Only when they are enriched in water are they expected to pose a significant risk.

3.6.3.6 Bacterial Contamination and Other Water Pathogens

There are many different types of microorganisms that occur naturally in aquatic systems, including bacteria, protozoans, fungi, and viruses. Sometimes the presence or excess concentration of certain microbes, namely, pathogenic organisms capable of causing disease, will render a particular water body impaired with respect to certain beneficial uses, and thus polluted. Microbial contamination can render a water body unsuitable for use as a source of drinking water, irrigation water, or water for recreational activities.

Pathogenic organisms are difficult to isolate and analyze directly. Moreover, pathogenic organisms often stem from fecal contamination in water. Therefore, the use of fecal indicator bacteria (FIB) has been commonplace for investigating bacterial water quality since the 1920s. For example, coliform bacteria occur in the intestines of warm-blooded animals. Their presence in water is an indicator of fecal pollution and thus the potential presence of pathogenic enteric bacteria. Conversely, as FIB tend to be more persistent than pathogenic organisms, their absence is a good indicator that water is safe to drink with respect to pathogenic organisms. The most common metrics used in microbial water quality analyses are total coliforms, namely, *Escherichia coli*, and *Enterococcus* sp. These organisms can be easily and fairly rapidly measured using simple techniques in the laboratory.

Water treatment by filtration and chlorine disinfection is said to have reduced urban mortality by about one-half in the 19th century in the United States and saved more lives than any other single health development in human history.[45] The United States has one of the safest drinking water systems in the world, yet the Centers for Disease Control (CDC) estimates that there are between 4 and 32 million cases of acute gastrointestinal illnesses per year from public drinking water systems.[46] Water treatment processes are not 100% efficient; a small portion of pathogens can survive the process, so the cleaner the water is to begin with, the cleaner the water will be post treatment. It is therefore imperative to safeguard the source water. This is true for both surface water and groundwater. The MCLs for most microorganisms are based on treatment technique for disinfection. Inactivation or removal of pathogens is quantified in terms of a log-removal. Log reduction refers to the percentage of microorganisms physically removed or inactivated by a treatment process. For example, a log-4 removal means 99.99% reduction in the number of microorganisms in the water.

Two modern cases of waterborne disease outbreaks in North America include a cryptosporidiosis (from *Cryptosporidium* sp., a protozoan) outbreak in Milwaukee, WI, in 1993, which to date is the largest documented waterborne outbreak, and the 2000 Walkerton, Ontario *E. coli* outbreak. The Milwaukee "crypto" came from sewage discharges and was resolved with the implementation of ozone gas as an extra disinfectant. It is a classic case to demonstrate that the dirtier the water, the higher the contaminants are in the processed water, even when the treatment process is working as intended. The Walkerton case was partly a dirty source water problem as well as insufficient treatment. Groundwater was contaminated by agricultural waste. Then the utility manager and his brother, who was the foreman, delayed notification and falsified reports to hide the data. It was later found that they were negligent in their disinfection practices. Part of the problem was their lack of training. As a result, Ontario Clean Water Agency was put in charge to remediate the situation, and the Ontario provincial government established the Walkerton Clean Water Centre to train utility operators and other water professionals.

According to the World Health Organization, there are approximately 1.4 million preventable childhood deaths per year from diarrhea due to drinking unsafe water. Globally, waterborne disease has been and continues to be a major source of morbidity and mortality. Open defecation is practiced by about a billion people in the world. In 2010, only about 63% of the world's population used toilets or other sanitation facilities[47] and 2.5 billion lacked improved sanitation. While the number of sewage and drinking water treatment plants, septic systems, and other technologies are increasing, so is the human population, so the percentages belie the progress made in treatment and infrastructure.

Shigella dysenteriae is one of the species of bacteria that most often causes dysentery, along with bacteria of the genus *Salmonella*. In addition to *E. coli* and *Cryptosporidium* mentioned earlier, other pathogens that are common contributors to waterborne diseases include the protozoan *Giardia lamblia*, the hepatitis A virus, and Norwalk-type viruses. *Vibrio cholerae* is a bacterium that causes the disease cholera, which has been greatly reduced worldwide by chlorination. One of the worst modern cases occurred after the 7.0 earthquake in Haiti; genomic sequencing linked the epidemic to a Nepalese origin, thought to have been introduced into Haiti by UN relief workers.[48] The outbreak was once again made worse by Hurricane Matthew (2016).

Often overlooked as a source of illness is Legionnaires' disease, caused by a bacteria called *Legionella* sp. The disease was so named after an outbreak that occurred at the American Legions Convention in Philadelphia in 1976. *Legionella* exposure usually occurs from aspirating or breathing microdrops of contaminated water. Flint, MI, had one of the nation's worst *Legionella* outbreak that caused 12 deaths and 78 illnesses, now thought to be linked to the water debacle (see Section 3.6.4), and is now facing a rise in shigellosis. Shigellosis is usually caused by poor sanitation, including not washing one's hands, and Flint residents' fear of their own tap water has led to less handwashing; some residents were using baby wipes to clean their hands. It is not clear whether *Shigella* has increased in Flint's drinking water. However, it is known that damaged pipes can lead to pathogen intrusion and/or sloughing of biofilm off pipes into the water.[49]

3.6.3.7 Algal Toxins

Nature produces toxins that we can loosely call "pollutants" when they are regulated as such. The EPA is in the process at this time of developing standards for algal toxins. Algae is a broad term that encompasses cyanobacteria (i.e., blue-green algae). Cyanobacteria, sometimes called by the nickname "cyanos," are photosynthetic bacteria that share some properties with algae. Some species occur mainly in freshwater and some in seawater, or both. *Anabaena* is a genus of cyanos that occurs at the surface, and other types such as *Planktothrix* can be found in bottom sediments or can float to the surface water when disturbed. Excesses of nitrogen and phosphorus can cause these cyanos to grow prolifically, resulting in an algal "bloom." Cyanos can produce toxins called cyanotoxins. The most widespread type of toxins is peptides called microcystins, such as microcystin-LR, which is one of the most cyanotoxins. There may be toxins yet to be identified, but over 80 have been identified thus far.[50] Most are hepatotoxins (i.e., the toxin targets the liver), but the same toxins that injure the liver can also be associated with other health effects. Other cyanotoxins are neurotoxins; anatoxin can cause respiratory paralysis. Its mode of action is via the nicotinic acetylcholine receptor, where it is a competitive agonist for the receptor's natural ligand, acetylcholine.

3.6.4 CASE STUDY OF LEAD (Pb) IN DRINKING WATER—FLINT, MI

The drinking water crisis in Flint, MI, is a case of multiple failure modes and human negligence that resulted in lead contamination of drinking water from corrosion of water pipes and associated fixtures, starting in 2013. The damage to the infrastructure was then later implicated as the source of an outbreak of Legionnaires' disease because of which 9 people died and 87 were infected. But this is also a story of community resiliency and heroic actions by scientists who volunteered their expertise to uncover the problem and provide solutions.

Flint is a city of about 100,000 people, 40% of which live in poverty. About 60% of the population is black/African American, thus it is also an environmental justice issue that some critics argue would not have occurred in a more affluent town. In fact, a situation not unlike this did occur in 2001–04 in Washington, DC, one of the most affluent and powerful cities in the country, and in that case the federal government not only treated the water but the US EPA was also the primary agency. This is not to say that systemic racism did not play a role in the Flint crisis, considering that there was a motivation to save money and a scandal (cover-up) to hide the crisis when it occurred.

Before the crisis, Flint got its water from Detroit Water and Sewerage (DWS), which is a high-quality source from Lake Huron and Detroit River. The river has a high volume and discharge rate (diluting any potential contaminant sources) and DWS has a proven record of effective treatment. Flint, being distal to the Detroit River, was paying a premium for its water and future rate increases were projected. In an effort to avoid these higher water costs, state and Flint officials decided in 2013 they would switch to the Karegnondi Water Authority that was in the process of building a new pipeline to Lake Huron, one of the state's gold standards of water quality. The pipeline would, however, take a few years to be built, and the state neglected to renegotiate with DWS, instead opting for a cheaper alternative in the interim, by drawing water from the local Flint River. This decision was in part by the state's emergency manager, Darnell Earley, authorized under Public Act 72 of 1990, in cases of local governments that experience financial emergencies.

Flint's water treatment plant was upgraded in preparation for the switch, but the plant manager complained that they were not given adequate time to make the change to the new water source and the water started running by April 2014 (Fig. 3.6.10). For reasons that are not completely clear, federal corrosion control law was not followed, in that the practice of dosing a corrosion inhibitor, orthophosphate, that had been used in Flint for decades was not continued. Almost immediately, consumers started to complain about the taste and odor of the water. Within 4 months of running the new water supply through Flint's distribution system, the City of Flint had to issue two consecutive boil alerts in response to high bacterial counts, and consumer complaints increased regarding foul odors, yellow/brown or red colors, and particulates in their water. By October of that year, GM announced it could no longer use the public water supply in its manufacture of auto parts due to particulates in the water and signs it was corrosive. The University of Flint tested the water out of their water fountains and found elevated levels of lead, but they discounted the results as due to bad plumbing.

The Flint River had a legacy of industrial dumping that made environmentalists suspicious. But a second problem with Flint River water may have been the high concentrations

FIGURE 3.6.10 The Flint (MI, United States) Water Plant is front and center for a drinking water contamination crisis. A shortsighted economic decision was made by state politicians to switch from Detroit's water supply to the local Flint River for 2 years while a new pipe could be hooked up to Lake Huron. The Flint River is 19 times more corrosive than Detroit's water. Then the municipality failed to treat with orthophosphate resulting in lead leaching from distribution system materials into the drinking water. The crisis was made worse by the delayed and insufficient responses of the Michigan Department of Environmental Quality, local officials, and state government.

of chloride from road salt—nine times that of Lake Huron, as pointed out by Dr. Marc Edwards, a professor and water expert from Virginia Tech who was a key investigator of the Flint water crisis. Again, large rivers and lakes get the benefit of a dilution factor compared to small urban rivers. High levels of chloride corrode pipes if a corrosion inhibitor is not used. Ironically, Flint used iron chloride instead of aluminum sulfate as its coagulant to remove particles and organic matter from water; this turned out to be a poor choice of coagulant given that it further elevated chlorine levels. Although the US EPA Lead and Copper Rule and Clean Water Act requires monitoring for lead and copper throughout the distribution system and corrective action taken if lead or copper exceed a certain concentration (15 ppb for Pb and 1.3 ppm for Cu) in more than 10% of samples taken, the highest risk homes were not sampled as required, and some samples with high lead were invalidated. This led to claims that the water was meeting federal standards when it was not. The corrosion problem went from bad to worse when the water utility used extra chlorine to try and maintain required levels of disinfectant throughout the system. Chlorine is consumed by reacting with copper pipes, so adding more chlorine dioxide could further degrade the pipes. At that point the extra chlorine levels created higher concentrations of disinfection byproducts (DBPs) (e.g., chloroform, formed by the reaction of chlorine with natural organic matter in the water). DBPs are also required to be monitored and there are MCLs for them. By January 2015, Flint's water plant notified the public that the water was exceeding the DBP MCLs, prompting immediate action. The EPA was now aware of the lead problem, and discussions ensued with Michigan Department of Environmental Quality (MDEQ). EPA asked MDEQ if there could be a link between Legionnaires' cases that appeared in June 2014 out of McLaren hospital in Genesee County and the problems with its water and suggested they request help from the CDC because the state itself would be required to make that request, not EPA. MDEQ declined the suggestion. After the scandal was exposed, Governor Snyder appointed Professor Edwards to a task force to study the problem. Some Flint residents called for Edwards because he had similar experience on a case in Washington, DC, and had collaborated with them to expose the nature of the problem in Flint. The next month (February 2015), the governor approved 2 million dollars in emergency funding to address the widespread problems with the drinking water.

Leeanne Walters was one of many Flint residents who complained to the EPA and state officials about the water. Her child showed evidence of lead poisoning in March 2015, yet the state was still not using corrosion control at that point. Pediatrician Dr. Mona Hanna-Attisha was noticing elevated lead levels in several of her patients, but did not know the source of the exposures. In the meanwhile, Professor Edwards was testing Flint's water. By August, he was getting results that showed lead levels that exceeded drinking water standards in 20% of Flint households. EPA then required MDEQ to use corrosion control in the water treatment process. By mid-September, the residents and Edwards announced that there was a city-wide lead contamination problem. Dr. Hanna-Attisha mentioned her concerns about her patients to a friend who was a water engineer who had heard of Edwards' study of lead in the water supply. Dr. Mona, as she is affectionately called, went public with her data in September 2015. Soon thereafter, Governor Snyder announced that the water would be switched back to the Detroit system.

The Flint River water, improperly treated, ran through the distribution system for 18 months, all the while corroding the pipes, before actions were taken to address elevated lead levels and DBPs. Residents during that time were witnessing particulate matter, foul smells, and brown/yellow/ruddy colors in their water. They were told by local and state officials that the water was safe to drink in spite of health complaints of hair loss and skin rashes, followed by neurological effects such as nausea and nervous tremors, immune effects, bone pain, and growth problems in developing children. Lead is known to cause long-term problems such as IQ loss, behavioral problems, and bone damage. Epigenetic effects of lead can affect future generations, said Dr. Mona in the BBC television program, Our World, "The Poisoning of Flint," which aired on October 29, 2016.

Governor Snyder apologized for the whole fiasco. Protestors called for the governor's resignation, but the governor refused to resign. He argued that it was his responsibility to fix the problem and that he would have a lot of motivation because his reputation was on the line. He called for federal aid for funding to replace pipes, which was slow in coming. The community organized a program to distribute bottled water along with the state and local governments; little was done house-to-house, and there was hardship for people without cars to go pick up free bottles of water that had to cover cooking and bathing needs too. On November 3, 2016, a federal judge ordered the state to deliver bottles of water to individual residences. Finally, on March 11, 2017, EPA announced that it was awarding $100 million to replace damaged water pipes, with the state of Michigan matching with $20 million, with an expectation that all pipes to be fixed by January 2020.

Professor Edwards has been an advocate trying to express his concern about the potential for lead leaching into water supplies from distribution systems and trying to prevent another case like Washington, DC or Flint from happening. Common sense did not prevail soon enough in terms of recognizing the problem and mobilizing to fix it, and the MDEQ officials are under criminal investigation for denying the problem and delaying the use of corrosion control. There were many lessons to learn, litigation is pending, and legal investigations continue.

3.6.5 CASE STUDY—THE ST. CLAIR RIVER AND CHEMICAL VALLEY SARNIA

The St. Clair River is a designated "Area of Concern" because of the several Beneficial Use Impairments (BUIs) including restrictions on drinking water consumption due to historic and ongoing chemical pollutants that taint the water. The river divides the United States (Michigan) and Canada (Ontario) in the Great Lakes Region, flowing from Lake Huron to Lake St. Clair, which then empties into the Detroit River and on to Lake Erie. There are 14 drinking water plants along this waterway (ironic given the BUI) serving about 4 million people. The St. Clair River also serves the petrochemical industry known as Chemical Valley Sarnia, home to more than 60 oil refineries and chemical plants. The smell is evident, and locals give "toxic tours" in order to raise awareness, especially for the plight of Aamjiwnaang First Nation people living on reservation lands, a tiny fraction of their original territory. The Aamjiwnaang are completely surrounded by chemical plants and refineries, and many are now abandoning their land, sick of the chemical stench and health problems it brings. Chemical Valley Sarnia has the highest quantities of particulates and toxic air pollutants in Canada and the highest rates of pulmonary disease and cancers.[51] Gender-bending chemicals called endocrine disruptors are possibly contributing to an altered sex ratio of male to female, and a decline in the number of male babies born was documented from 1994 to 2003, with the greatest decline in the most recent years of the study.[52]

A 2006 report[53] identified over 700 chemical spills into the St. Clair-Detroit River corridor from 1990 to 2004, but the rate had been declining. One spill in 2002 occurred as a result of a blackout from a storm; in this event, a carcinogen, vinyl chloride, was accidentally released by a company that makes PVC for pipes, etc. They did not notify the public, including drinking water plants, for several days. It raised suspicions that the spills were still occurring with frequency, but just not being reported. Another major chemical spill occurred in 2003 on Super Bowl Sunday, this time involving methyl ethyl ketone (a reproductive toxicant) and methyl isobutyl ketone. These and other incidents drew the ire of local environmental advocates who demanded a better reporting system and the development of a drinking water quality monitoring system. Citizen advocate and St. Clair Channelkeeper Douglas Martz worked with his representative, Candice Miller (R-MI), to get bipartisan support for $1 million in federal funding through EPA (House Bill 2668; 2003) for an experimental drinking water monitoring system, patterned after a system on the Ohio River known as ORSANCO (Ohio River Valley Water Sanitation Commission). With matching funding from Homeland Security and local municipalities, the project funding grew to $3.5 million. The monitoring system was finally implemented from 2009 to 2011. During the 2-year monitoring program, only one major spill occurred; an acid was detected with a change in the pH in the river and the drinking water intakes were alerted. The project was deemed a success, but funding was not made available to continue with the monitoring.

Just as the monitoring system was being shelved, news hit that there had been an "outbreak" of five childhood Wilms kidney cancers within or near the small town (population < 5000) of Marine City along the St. Clair River in St. Clair County (Fig. 3.6.11). The global rate for Wilms is about 1 in 10,000.[54] A history of the disease revealed that there were three cases in St. Clair County in the years between 1990 and 1999 and eight cases in

FIGURE 3.6.11 Marine City, MI, along the St. Clair River; location of a Childhood Wilms Cancer Cluster due to an unknown source. *Courtesy of Richard Olawoyin.*

the next decade, from 2000 to 2009. The Health Department of St. Clair County conducted a study of the problem, reporting a standardized incidence ratio of Wilms for the whole county of 2.86, which was statistically significant, meaning that there was roughly three times the regional incidence (St. Clair County 2012).[55] However, the reality for Marine City residents was much worse, as the Health Department's statistics were in effect "diluted" in county-wide numbers. Meanwhile, across the river in Ontario, Canada, statistics were harder to obtain for Sarnia residents because, before 2010, they would normally have been hospitalized in London, over an hour's drive away; statistics were available by the hospital, but not by patients' residences. The St. Clair County Health Department tried to determine a potential source for the cancer, and the St. Clair Channelkeeper requested the Health Department investigate the river as a potential source. Indeed, the only correlation they found was how close residents had lived to a river—mainly the St. Clair River or nearby Black River at Port Huron, not any other lifestyle factors or occupational links to the cancer. This suggested an environmental link for the Marine City cases.

Some Wilms occurrences are known to have a heritable link, but others are thought to be linked to environmental exposures. A single dose of an alkylating agent, *N*-nitroso-methyl urea, administered to rodents can create a kidney tumor with the pathology and mutation pattern of human Wilms.[56] The potency of an alkylating agent is made much worse by folic acid deficiency.[57,58] In Ontario, a folic acid grain supplementation program was found to reduce the Wilms rate.[59] If there are chemical exposures that contribute to the disease, folic acid supplementation by pregnant mothers can aid in reducing the chance that their child develops the disease, but does not eliminate all risk. The St. Clair Channelkeeeper is now calling for the return of the monitoring system as a tool to prevent exposure to chemicals in the drinking water. He likens it to a cop on the beat, watching for violators. On the Canadian side, an effort is being made to delist the BUI on drinking water consumption. The Ontario Ministry of the Environment stepped up its inspections of chemical facilities and is working with the chemical companies to find technical and management solutions for preventing spills. Some of the mothers of the kids who got cancer started a nonprofit group called "Mothers Against Childhood Cancer" (MACC) as a support group and to raise money for a cancer investigation. The solution to the problem of chemical contamination is coming from citizen advocacy, politicians, local municipalities, and state and federal agencies working together.

References

1. Hall C, Klitgaard K. *Energy and the wealth of nations – understanding the biophysical economy*. Springer; 2012.
2. Bloomberg. http://www.bloomberg.com/news/articles/2013-01-14/water-and-energy-are-inextricably-linked-grace-report-says; 2013 [Internet].
3. Kolpin DW, Furlong ET, Meyer MT, Thurman EM, Zaugg SD, Barber LB, Buxton HT. Pharmaceuticals, hormones, and other organic wastewater contaminants in U.S. streams, 1999-2000: a national reconnaissance. *Environ Sci Technol* 2002;**36**(6):1202–11. http://dx.doi.org/10.1021/es011055j [Internet] Available from: ACS Publications Open Access.
4. Tanabe S. PCB problems in the future: foresight from current knowledge. *Env Pollut* 2008;**50**(1):5–28. http://dx.doi.org/10.1016/0269-7491(88)90183-2 [Internet] Available from: Science Direct.
5. Giesy JP, Kannan K. Global distribution of perfluorooctane sulfonate in wildlife. *Environ Sci Technol* 2001;**35**(7):1339–42. http://dx.doi.org/10.1021/es001834k [Internet] Available from: ACS Publications Open Access.

6. Toccalino PL, Norman JE, Scott JC. Chemical mixtures in untreated water from public-supply wells in the U.S. — occurrence, composition, and potential toxicity. *Sci Total Environ* 2012;**431**(1):262–70. http://dx.doi.org/10.1016/j.scitotenv.2012.05.044 [Internet] Available from: Science Direct.
7. Gatto NM, Henderson VW, Hodis HN, St. John JA, Lurmann F, Chen JC, Mack WJ. Components of air pollution and cognitive function in middle-aged and older adults in Los Angeles. *NeuroToxicology* 2014;**40**:1–7. http://dx.doi.org/10.1016/j.neuro.2013.09.004 [Internet] Available from: NCBI.
8. Mackay D, Fraser A. Bioaccumulation of persistent organic chemicals: mechanisms and models. *Environ Pollut* 2000;**110**(3):375–91. http://dx.doi.org/10.1016/S0269-7491(00)00162-7 [Internet] Available from: Science Direct.
9. Karickhoff SW, Brown DS, Scott TA. Sorption of hydrophobic organic pollutants on natural sediments. *Water Res* 1979;**13**:241–8. http://dx.doi.org/10.1016/0043-1354(79)90201-X [Internet] Available from: Science Direct.
10. Chiou CT, Malcolm RL, Brinton TI, Kile DE. Water solubility enhancement of some organic pollutants and pesticides by dissolved humic and fulvic acids. *Environ Sci Technol* 1986;**20**:502–8. http://dx.doi.org/10.1021/es00147a010 [Internet] Available from: SciTech Connect.
11. Caron G, Suffet IH, Belton T. Effect of dissolved organic carbon on the environmental distribution of nonpolar organic compounds. *Chemosphere* 1985;**14**:993–1000. http://dx.doi.org/10.1016/0045-6535(85)90020-7 [Internet] Available from: Science Direct.
12. Suffet IH, Jafvert CT, Kukkonen J, Servos MR, Spacie A, Williams LL, Noblet JA. Synopsis of discussion session: influences of particulate and dissolved material on the bioavailability of organic compounds. In: Hamelink JL, Landrum PF, Bergman HL, Benson WH, editors. *Bioavailability: physical, chemical, and biological interactions.* Ann Arbor (MI): Lewis Publishers; 1994 [chapter 3].
13. Carson R. *Silent spring*. Houghton Mifflin Company; 1962.
14. Hickey JJ, Anderson DW. Chlorinated hydrocarbons and eggshell changes in raptorial and fish-eating birds. *Science* 1968;**162**:271–3 [Internet] JSTOR.
15. Bitman J, Cecil HC, Harris SJ, Fries GF. DDT induces a decrease in eggshell calcium. *Nature* 1969;**224**:44–6. http://dx.doi.org/10.1038/224044a0 [Internet].
16. Cohn BA, La Merrill M, Kirgbaum NY, Yeh G, Park JS, Zimmermann L, Cirillo PM. DDT exposure in utero and breast cancer. *J Clin Endocrinol Metab* 2015;**100**(8):2865–70. http://dx.doi.org/10.1210/jc.2015-1841 [Internet] Available from: Oxford Academic.
17. Daley JM, Leadley TA, Drouillard KG. Evidence for bioamplification of nine polychlorinated biphenyl (PCB) congeners in yellow perch (*Perca flavescens*) eggs during incubation. *Chemosphere* 2009;**75**:1500–5. http://dx.doi.org/10.1016/j [Internet] Available from: Science Direct.
18. Gobas FAPC, Wilcockson JB, Russel RW, Haffner GD. Mechanism of biomagnification in fish under laboratory and field conditions. *Environ Sci Technol* 1999;**33**(1):133–41. http://dx.doi.org/10.1021/es980681m [Internet] Available from: ACS Publications Open Access.
19. Kumar KS, Kannan K, Corsolini S, Evans T, Giesy JP, Nakanishi J, Masunaga S. Polychlorinated dibenzo-p-dioxins, dibenzofurans and polychlorinated biphenyls in polar bear, penguin and south polar skua. *Environ Pollut* 2002;**119**(2):151–61. http://dx.doi.org/10.1016/S0269-7491(01)00332-3 [Internet] Available from: Science Direct.
20. Gilbertson M, Kubiak T, Ludwig J, Fox G. Great Lakes embryo mortality, edema and deformities syndrome (GLEMEDS) in clonial fish-eating birds: similarity to chick-edema disease. *J Toxicol Environ Health* 1991;**33**:455–520. http://dx.doi.org/10.1080/15287399109531538 [Internet] Available from: Taylor and Francis Online.
21. Fox GA, Collins B, Hayakawa E, Weseloh DV, Ludwig JP, Kubiak TJ, Erdman TC. Reproductive outcomes in colonial fish-eating birds: a biomarker for developmental toxicants in Great Lakes food chains: II. Spatial variation in the occurrence and prevalence of bill defects in young double-crested cormorants in the Great Lakes, 1979–1987. *J Gt Lakes Res* 1991;**17**:158–67. http://dx.doi.org/10.1016/S0380-1330(91)71353-1 [Internet] Available from: Science Direct.
22. Salama J, Chakraborty TR, Ng L, Gore AC. Effects of polychlorinated biphenyls on estrogen receptor-β expression in the anteroventral periventricular nucleus. *Environ Health Perspect* 2003;**111**(10):1278–82 [Internet] Available from: NCBI NIH.
23. De Wit CA. An overview of brominated flame retardants in the environment. *Chemosphere* 2002;**46**:583–624. http://dx.doi.org/10.1016/S0045-6535(01)00225-9 [Internet].

24. Van den Berg M, Birnbaum LS, Denison M, De Vito M, Farland W, Feeley F, Fiedler H, Hakansson H, Hanberg A, Haws L, Rose M, Safe S, Schrenk D, Tohyama C, Tritscher A, Tuomisto J, Tysklind M, Walker N, Peterson RE. The 2005 World Health Organization reevaluation of human and mammalian toxic equivalency factors for dioxins and dioxin-like compounds. *Toxicol Sci* 2006;**93**:223–41. http://dx.doi.org/10.1093/toxsci/kfl055 [Internet].
25. Perrini G, Tomasello M, Librando V, Minniti Z. Nitrated polycyclic aromatic hydrocarbons in the environment: formation, occurrences and analysis. *Ann Chim* 2005;**95**(7–8):567–77 [Internet] Available from: Research Gate.
26. *Methods for the determination of organic compounds in drinking water. Method 524.2.* US Environmental Protection Agency; 1988. EPA600/4–88/039 Rev. July 1991.
27. Federal Water Pollution Control Act (Clean Water Act) 33 U.S.C. §1251 et seq.; 1972.
28. Samson-Robert O, Labrie G, Chagnon M, Fournier V. Neonicotinoid-contaminated puddles of water represent a risk of intoxication for honey bees. *PLoS One* 2014;**9**:e108443. http://dx.doi.org/10.1371/journal.pone.0108443 [Internet] Open Access.
29. Donald DB, Cessna AJ, Sverko E, Glozier NE. Pesticides in surface drinking-water supplies of the northern Great Plains. *Environ Health Perspect* 2007;**115**(8):1183–91 [Internet].
30. Nuyttens D, Devarrewaere W, Verboven P, Foque D. Pesticide-laden dust emission and drift from treated seeds during seed drilling: a review. *Pest Manag Sci* 2013;**69**:564–75. http://dx.doi.org/10.1002/ps.3485. wileyonlinelibrary.com [Internet].
31. Ragnarsdottir KV. Environmental fate and toxicology of organophosphate pesticides. *J Geol Soc* 2000;**157**:859–76. http://dx.doi.org/10.1144/jgs.157.4.859 [Internet] Available from: Lyell Collection.
32. De Andréa MM, Peres TB, Luchini LC, Bazarin S, Papini S, Matallo MB, Savoy VL. Influence of repeated applications of glyphosate on its persistence and soil bioactivity. *Pesqui Agropecu Bras* 2003;**38**(11):1329–35. http://dx.doi.org/10.1590/S0100-204X2003001100012 [Internet] Available from: Scielo.
33. Scribner EA, Battaglin WA, Dietze JE, Thurman EM. *Reconnaissance data for glyphosate, other selected herbicides, their degradation products, and antibiotics in 51 streams in nine Midwestern States, 2002.* U.S. Geological Survey Open-File Report 03-217, 101. 2003.
34. Bates N, Campbell A. Glyphosate. In: Campbell A, Campbell A, editors. *Handbook of poisoning in dogs and cat.* 1st ed. England: Blackwell Science Ltd.; 2000. p. 135–8.
35. Mesnage R, Arno M, Costanzo M, Malatesta M, Séralini GE, Antoniou MN. Transcriptome profile analysis reflects rat liver and kidney damage following chronic ultra-low dose Roundup exposure. *Environ Health* 2015;**14**:70. http://dx.doi.org/10.1186/s12940-015-0056-1 [Internet] Available from: BioMed Central.
36. Draft ecological risk assessments: atrazine, simazine, and propazine registration review. US Environmental Protection Agency: EPA-HQ-OPP-2013-0266-0343, 2013.
37. Tanner CM, Kamel F, Ross GW, Hoppin JA, Goldman SM, Korell M, Marras C, Bhudhikanok GS, Kasten M, Chade AR, Comyns K, Richards MB, Meng C, Priestley B, Fernandez HH, Cambi F, Umbach DM, Blair A, Sandler DP, Langston JW. Rotenone, paraquat, and Parkinson's disease. *Environ Health Perspect* 2011;**119**(6):866–72 [Internet].
38. EPA Webpage. https://archive.epa.gov/mtbe/web/html/gas.html.
39. Andrews C. MTBE – a long-term threat to ground water quality. *Groundwater* 1998;**36**(5):705 [Internet] Academica OneFile.
40. Zhang Z, Stout JE, Yu VL, Vidac R. Effect of pipe corrosion scales in chlorine dioxide consumption in drinking water distribution systems. *Water Res* 2008;**42**:129–36. http://dx.doi.org/10.1016/j.watres.2007.07.054 [Internet] Available from: Science Direct.
41. Cole M, Lindeque P, Fileman E, Halsband C, Goodhead R, Moger J, Galloway TS. Microplastic ingestion by zooplankton. *Environ Sci Technol* 2013;**47**(12):6646–55. http://dx.doi.org/10.1021/es400663f [Internet] Available from: ACS Publications Online.
42. The Ocean Cleanup. https://www.theoceancleanup.com/.
43. Gostomski F. The toxicity of aluminum to aquatic species in the US. *Environ Geochem Health* 1990;**12**(1–2):51–4. http://dx.doi.org/10.1007/BF01734047 [Internet] Available from: NCBI.
44. Straub AC, Lohman AW, Billaud M. Endothelial cell expression of haemoglobin alpha regulates nitric oxide signalling. *Nature* 2012;**491**:473–7 [Internet].

45. McGuire MJ. *The chlorine revolution – water disinfection and the fight to save lives*. American Water Works Association; 2013.
46. CDC. *Magnitude & burden of waterborne disease in the U.S.* 2013. http://www.cdc.gov/healthywater/burden.
47. World Health Organization. *Safer water, better health: costs, benefits, and sustainability of interventions to protect and promote health*, 2008
48. Orata FD, Keim PS, Boucher Y. The 2010 cholera outbreak in Haiti: how science solved a controversy. *PLoS Pathog* 2014;**10**(4):1–5. http://dx.doi.org/10.1371/journal.ppat.1003967 [Internet].
49. Ashbolt NJ. Microbial contamination of drinking water and human health from community water systems. *Curr Environ Health Rep* 2015;**2**(1):95–106. http://dx.doi.org/10.1007/s40572-014-0037-5 [Internet] Available from: Springer Link.
50. Westrick J, Szlag D, Southwell B, Sinclair J. A review of cyanobacteria and cyanotoxins removal/inactivation in drinking water treatment. *Anal Bioanal Chem* 2010;**397**:1705–14.
51. Keith MM, Brophy JT. Participatory mapping of occupational hazards and disease among asbestos-exposed workers from a Foundry and Insulation Complex in Canada. *Int J Occup Environ Health* 2004;**10**:144–53 [Internet].
52. Mackenzie CA, Lockridge A, Keith M. Declining sex ratio in a first nation community. *Environ Health Perspect* 2005;**113**(10):1295–8.
53. IJC. *Report on spills in great lakes basin with a special focus on the St. Clair-Detroit River corridor*. International Joint Commision USA and Canada; July 2006, ISBN 1-894280-58-X.
54. Sharma PM, Bowman M, Yu BF, Sukumar S. A rodent model for Wilms tumors: embryonal kidney neoplasms induced by N-nitroso-N′-methylurea. *Proc Natl Acad Sci* 1994;**91**:9931–5.
55. St. Clair County Health Department. *Investigation summary report Wilms tumor cluster St. Clair County Health Department*. July 2012.
56. Terracini B, Testa MC. Carcinogenicity of a single administration of N-nitrosomethylurea: a comparison between newborn and 5-week-old mice and rats. *Br J Cancer* 1970;**24**(3):588–98 [Internet] Available from: NCBI.
57. Raykov ZZ, Ivanov VA, Raikova ET, Galabov AS. Folic acid role in mutagenesis, carcinogenesis, prevention and treatment of cancer. *Biotechnol Eq* March 18, 2004;**18**(3):125–35. http://dx.doi.org/10.1080/13102818.2004.10817133 [Internet] Available from: Taylor & Francis Online.
58. Branda RF, O'Neill JP, Brooks EM, Powden C, Naud SJ, Nicklas JA. The effect of dietary folic acid deficiency on the cytotoxic and mutagenic responses to methyl methanesulfonate in wild-type and in 3-methyladenine DNA glycosylase-deficient Aag null mice. *Mut Res* 2007;**615**(1–2):12–7. http://dx.doi.org/10.1016/j.mrfmmm.2006.09.007 [Internet] Available from: Science Direct.
59. Grupp SG, Greenberg ML, Ray JG, Busto U, Lanctot KL, Nulman I, Koren G. Pediatric cancer rates after universal folic acid flour fortification in Ontario. *J Clin Pharmacol* 2011;**51**(1):60–5. http://dx.doi.org/10.1177/0091270010365553 [Internet] Available from: Wiley Online Library.

CHAPTER

3.7

Contaminants of Emerging Concern, With an Emphasis on Nanomaterials and Pharmaceuticals

Helen C. Poynton, William E. Robinson

University of Massachusetts Boston, Boston, MA, United States

3.7.1 INTRODUCTION

Since the early 1960s, the majority of the environmental toxicological research has focused on what are now referred to as "legacy" contaminants. These included toxic chemicals that were produced in high volumes and have generally persisted after being released into the environment: metals (e.g., Cd, Cu, Pb, Sn, Ni, Cr), polycyclic aromatic hydrocarbons (such as naphthalene and benzo[*a*]pyrene), polychlorinated biphenyls, dioxins and dibenzofurans, and chlorinated pesticides [such as dichlorodiphenyltrichloroethane (DDT), dieldrin, and chlordane]. These legacy contaminants have been highly regulated. There has been an emphasis, especially with respect to the pesticides, to replace many of these persistent, bioaccumulating chemicals with substances that are less toxic to wildlife and humans and that have much shorter environmental half-lives. This has obvious relevance to green chemists.

Since around the turn of the current century, however, another group of chemicals, contaminants of emerging concern (CECs), has been receiving particular attention from environmental toxicologists, government regulators, and the general public.[1,2] The phrase "emerging contaminants" had been used by the US Environmental Protection Agency in the mid-1990s to focus on chemicals that were "recently discovered" in the environment, had no regulatory standards associated with them, and yet were at least potentially toxic to wildlife and even humans.[3] This group of chemicals is now widely recognized as CECs (Table 3.7.1). The US Geological Survey (USGS) (2016) defines CECs as "any synthetic or naturally occurring chemical or any microorganism that is not commonly monitored in the environment but has the potential to enter the environment and cause known or suspected adverse ecological and (or) human health effects."[4] Their "discovery" is primarily due to recent advances in analytical detection limits of environmental samples. However, this "discovery" also refers to the

Green Chemistry
http://dx.doi.org/10.1016/B978-0-12-809270-5.00012-1

TABLE 3.7.1 A List of the Current Groups of Major Contaminants of Emerging Concern

CHEMICAL CONTAMINANTS
Alkylphenol ethoxylate surfactants (e.g., nonylphenol ethoxylates)
Anticorrosive agents (e.g., benzotriazoles)
Brominated flame retardants (e.g., PBDEs)
Chemical warfare agents (e.g., sarin, cyclosarin)
Crude oil extraction chemicals (e.g., naphthenic acids)
Fluorinated surfactants (e.g., PFOA and PFOS)
Gasoline additives (e.g., MTBE, EDB)
Hormones & endocrine disruptors (e.g., estradiol)
Nanoparticles (e.g., fullerenes, Cd Te quantum dots)
New classes of pesticides (e.g., atrazine, acetanilide/acetamide herbicides)
Munition components/by-products (e.g., nitroaromatics, nitramines, and nitrate esters; perchlorates)
Personal care products (e.g., DEET, oxybenxone, siloxanes)
Pharmaceuticals (e.g., ethynyl estradiol, phenobarbital)
Stabilizers (e.g., dioxane)
MICROBIOLOGICAL COMPOUNDS
Infectious bacteria (e.g., botulin toxin)
Toxic diatoms (e.g., domoic acid)
Toxic cyanobacteria (e.g., nodularins)
Toxic dinoflagellates (e.g., saxitoxin, okadaic acid, ciguatoxin)
Viruses (e.g., viral RNA)

DEET, diethyltoluamide; *EDB*, ethylene dibromide; *MTBE*, methyl tertiary-butyl ether; *PBDE*, brominated diphenyl ether; *PFOA*, perfluorooctanoate; *PFOS*, perfluorooctane sulfonate.
Entries compiled from Richardson SD, Ternes TA. Water analysis: emerging contaminants and current issues. Anal Chem *2005;**77**(12):3807–38; Richardson SD. Water analysis: emerging contaminants and current issues.* Anal Chem *2009;**81**(12):4645–77; and USGS. Contaminants of emerging concern in the environment; 2016. Available from: http://toxics.usgs.gov/regional/emc/index.html.*

surprise expressed by toxicologists (who, in hindsight, should probably have known better) that some CECs could have deleterious effects on wildlife at very low environmental concentrations.

CECs comprise an exceptionally broad category of synthetic chemicals (Table 3.7.1)[4–6] that are artificially grouped together, even though they have a limited number of common properties and characteristics. A number of these chemicals (e.g., the pharmaceuticals and personal care products) have been used for years but have only recently been detectable in water, soil, and tissues, whereas others (e.g., nanomaterials) have only now been manufactured in sizable quantities. Note that natural chemical compounds produced by microbial entities are also included in the USGS definition of CECs.

Although CECs are detectable in environmental samples, even today they are not routinely monitored in the environment, due to the relatively high cost of analysis and the limited pool of well-trained analytical facilities. Their release into the environment is generally not as a by-product of manufacturing, but rather due to their use and disposal by industries and the general public.[4] They are often found in municipal, agricultural, and industrial wastewaters and even in drinking water. This type of non-point source release is typically unregulated. CECs can be measured in a broad range of environmental compartments (fresh and marine waters, soils and sediments, and tissues) and are now considered to be ubiquitously distributed.[7,8] Although scientists are gathering more and more data on the presence of CECs in various environmental compartments, determining whether these chemicals are having an adverse impact on organisms is much more complex. Their health risk, determined as the combination of hazard (which includes "inherent toxicity" or potency) and exposure (concentration and time), to flora, fauna, and even to humans is less known.

3.7.2 THE TOXICOLOGY OF CONTAMINANTS OF EMERGING CONCERNS

CECs, like the legacy chemicals, generally conform to the basic tenets of toxicology. These tenets are important for green chemists to consider when synthesizing new chemicals. The first tenet, originally articulated by the Swiss physician Paracelsus and published in 1564 in his Third Defense,[9] is that all chemicals are toxic; it is the dose or concentration of the chemical that determines whether it exhibits toxicity or not. This is not to say that all chemicals are toxic at the same concentration. Each chemical has its own unique potency. For example, a human would have to ingest several liters of ethanol to be lethal, whereas a micromolar concentration of strychnine would lead to the same outcome, although not through the same biochemical pathway.[10] A comparison of LD_{50}s of various chemicals (the lethal dose that kills 50% of the organisms tested), calculated from toxicity tests conducted on mice, rats, and other wildlife, can be used as an index of the toxic potencies of different chemicals (Fig. 3.7.1). Overall, this range spans over nine orders of magnitude.[11] Ames and Gold documented that the vast number of chemicals, both natural and synthetic, have been shown to be carcinogenic when tested at high concentrations.[12] This appears to be because high concentrations of most chemicals induce mitogenesis or cell division, and it is this increase in cell division rates that increases the chances of erroneous DNA replication, cancer, and thus an increased mortality rate. The finding that synthetic chemicals do not exhibit higher potencies than natural substances,[13] and that high concentrations of even those chemicals considered to be benign can induce mitogenesis, and thus carcinogenesis,[12,14] refutes the myth that natural chemicals are "good" and synthetic chemicals are "bad." This is good news for green chemists.

A second tenet of toxicology is that toxicity is not simply lethality. Lethality is only one endpoint of toxicity testing. There are a multitude of sublethal effects that may also affect organisms. The most important of these sublethal effects are those adversely affecting growth and reproduction. These effects on individual organisms may directly translate into adverse impacts on the population of organisms, and indirectly on the community in which this organism lives. Lethal and sublethal effects may differ depending on biological factors such as age, life stage (embryo, larva, juvenile, or adult), physiological condition of the organism, and reproductive state, as well as on environmental factors such as temperature, salinity,

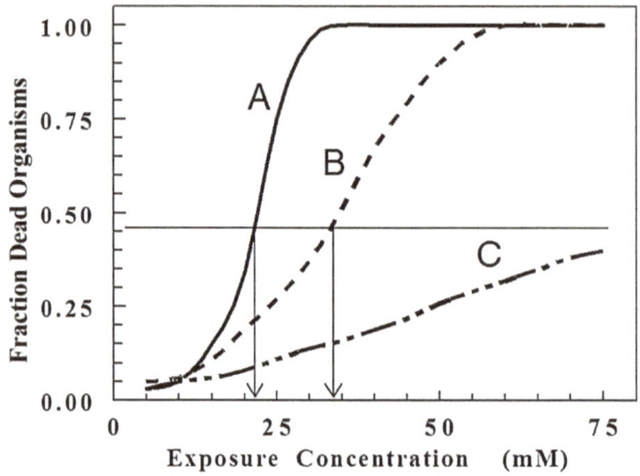

FIGURE 3.7.1 **Typical dose-response curve for three chemicals, A, B, and C.** The LD_{50} (lethal dose that kills 50% of the organisms tested) over a set exposure time (e.g., 24 h) for chemical A = 22 mM, for chemical B = 34 mM, and for chemical C = >75 mM. The potencies of the three chemicals are ranked A > B > C. While potencies in this example are within one order of magnitude, when all chemicals are considered, potencies span over nine orders of magnitude.

location (e.g., subtidal or tidal, shading), dissolved oxygen levels, and presence of other contaminants (chemical mixtures).

A third tenet of toxicology is that chemicals exhibit toxicity when the dose or concentration of the chemical reaches a critical concentration at a specific biochemical target site (Fig. 3.7.2). The target site for a particular chemical or class of chemicals refers to the protein, enzyme, or macromolecule (e.g., DNA, lipid, ribosome) that the chemical interacts with in such a way as

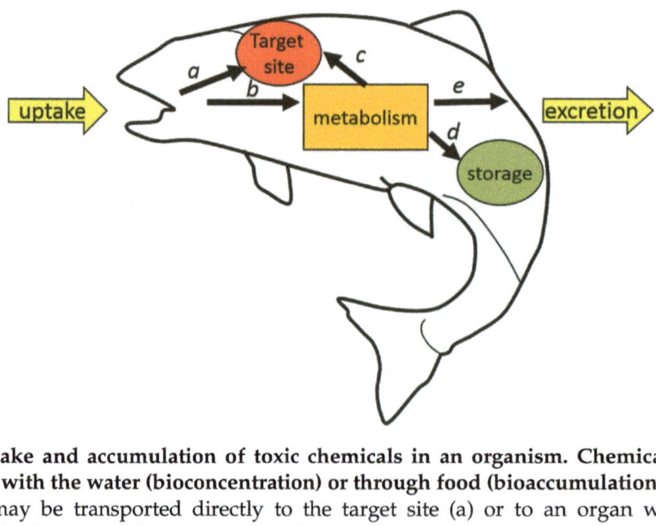

FIGURE 3.7.2 **Uptake and accumulation of toxic chemicals in an organism. Chemicals are taken up either through direct contact with the water (bioconcentration) or through food (bioaccumulation).** Once absorbed by the organism, chemicals may be transported directly to the target site (a) or to an organ where the compound is metabolized (b), possibly the liver or digestive organ. There are many examples where the metabolized compound is more toxic (e.g., polyaromatic hydrocarbons), and this metabolized chemical may be transported to the target site (c) causing toxicity. In other cases, metabolized compounds can be stored in adipose tissue, sequestered away from the target site in other organs (d), or are directly excreted (e).

to disrupt its usual cellular function. For example, a particular enzyme target site could be inhibited by a chemical, not allowing the enzyme to bind its substrate and catalyze a particular reaction. This target site will be different depending on the biochemical pathway that is adversely impacted. Different target sites are attacked by different chemicals, and the degree of chemical response can vary depending on the potency of the chemical.

In order for a chemical to become concentrated at the target site, it must first be taken up and accumulated within the target organism (Fig. 3.7.2). The extent that a chemical is taken up by an organism is related to its bioavailability. Many chemicals are "bioaccumulated" by organisms; chemicals are absorbed from the food, from the surrounding water (in aquatic organisms) or air (via the lungs of terrestrial organisms), or via direct contact (dermal exposure). Uptake of a chemical from the aqueous dissolved phase by an aquatic organism is referred to as "bioconcentration," whereas uptake from either the dissolved or particulate phase of a chemical in food is referred to as "bioaccumulation." Bioconcentration is generally proportional to a physicochemical property of the chemical, the octanol-water partitioning coefficient (usually designated as log K_{ow}, Fig. 3.7.3). This is particularly important for aquatic organisms where the bioconcentration of dissolved organic chemicals is generally directly related to log K_{ow} values up to approximately 5.0–5.5, above which body burdens may continue to increase through bioaccumulation via the organism's food (Fig. 3.7.3).[11,15] Regardless of how it occurs, or how much it occurs, the resulting bioburden of a chemical in an organism's body does not

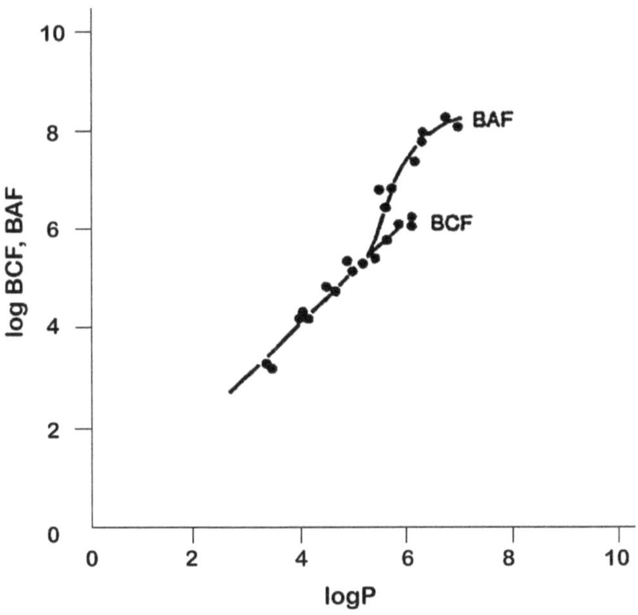

FIGURE 3.7.3 **The relationship between bioconcentration factor and the octanol-water partitioning coefficient (K_{OW}).** LogP = log K_{OW}; *BAF*, bioaccumulation factor, based on additional uptake of contaminant from food; *BCF*, bioconcentration factor, based on uptake of contaminant dissolved in water. Data obtained from experiments where guppies, goldfish, and rainbow trout were exposed to polychlorinated biphenyls and organochlorine compounds. *Reprinted by permission from Noegrohati S, Hammers W. Regression models for octanol-water partition coefficients, and for bioconcentration in fish.* Toxicol Environ Chem *1992;***34***(2–4):155–73. Copyright 1992 Taylor and Francis.*

necessarily mean that the chemical concentration at the target site is great enough to induce toxicity. In many cases, the bioaccumulated chemical is sequestered away in cellular compartments, preventing appreciable buildup at the target site until such time as the capacity of sequestration is exceeded (Fig. 3.7.2). Thus a chemical's whole body tissue concentration may not directly relate to the chemical's toxicity.

In many ways, CECs are harder to deal with today than legacy contaminants, which have been highly regulated. Legacy contaminants' distribution and fate in the environment have been well-studied, as have their bioaccumulation pathways, modes of toxicity, and mechanisms of detoxification and elimination.[11] Containment and cleanup procedures have been worked out, and best practices have been implemented to reduce unnecessary environmental release. For many CECs, however, we do not know if they are present or if they could be detected using current analytical techniques and instrumentation.[16] Even if they are present in measurable concentrations in the environment, we often do not know if they are actually inducing toxicity in any organism. Since many CECs were specifically designed to have biological effects at very low concentrations (e.g., pharmaceuticals), should we assume that the same biological activity will manifest itself unintentionally in wildlife? Are they persistent? Will they gradually build up in the environment and lead to chronic toxicity? Clearly, there are more "unknowns" than "knowns" concerning CECs.

The toxicology of these CECs is complex. Not all of these compounds necessarily bioconcentrate or bioaccumulate in the tissues of organisms, but they may still induce a toxic reaction at very low concentrations at the target site. When the chemical interacts with its target site, a biochemical or molecular pathway is disrupted causing the pathway to malfunction [e.g., Fig. 3.7.4A displays the interactions of an estrogenic compound such as 17-α-ethinyl estradiol (EE2), with a receptor site and its subsequent disruption of the normal endocrine system]. This adverse response at the molecular or biochemical level of biological organization (both are subcellular effects) can lead to impacts on the whole cell (cellular level of biological organization) such as an increase or decrease in mitosis, reduction in cell life, or even apoptosis (programmed cell death). These cellular effects may in turn lead to damage at the tissue level of biological organization (e.g., necrosis, carcinogenesis), leading to damage at the individual level of organization (e.g., reduced organism growth, survival and reproduction), which then leads to impacts at the population level (e.g., reductions in population numbers or the intrinsic rate of population increase). While these impacts of chemical substances at the various levels of biological organization have been studied by toxicologists for years (e.g., Fig. 3.7.4A), they have recently received renewed emphasis as an integrated series of tiered processes called adverse outcome pathways (AOPs; see Fig. 3.7.4B).

The concept of the AOP has provided a framework to organize information across levels of biological organization and also highlights the importance of key events (KEs) and key event relationships (KERs) critical for translating effects from one level to the next.[17] As shown in Fig. 3.7.4, a chemical will interact with a target biomolecule leading to inhibition or altered function in a step called the molecular initiating event (MIE). When concentrations of the chemical reach a critical level at the target site, the MIE will lead to impacts in the cell and tissue, possibly inhibiting the normal function of that tissue. These impacts will alter normal organ function, resulting in a disease state for that individual. In human toxicology, adverse effects on an individual person are enough to warrant regulatory action. However, in ecotoxicology, effects are not seen as a concern until impacts to populations or communities of

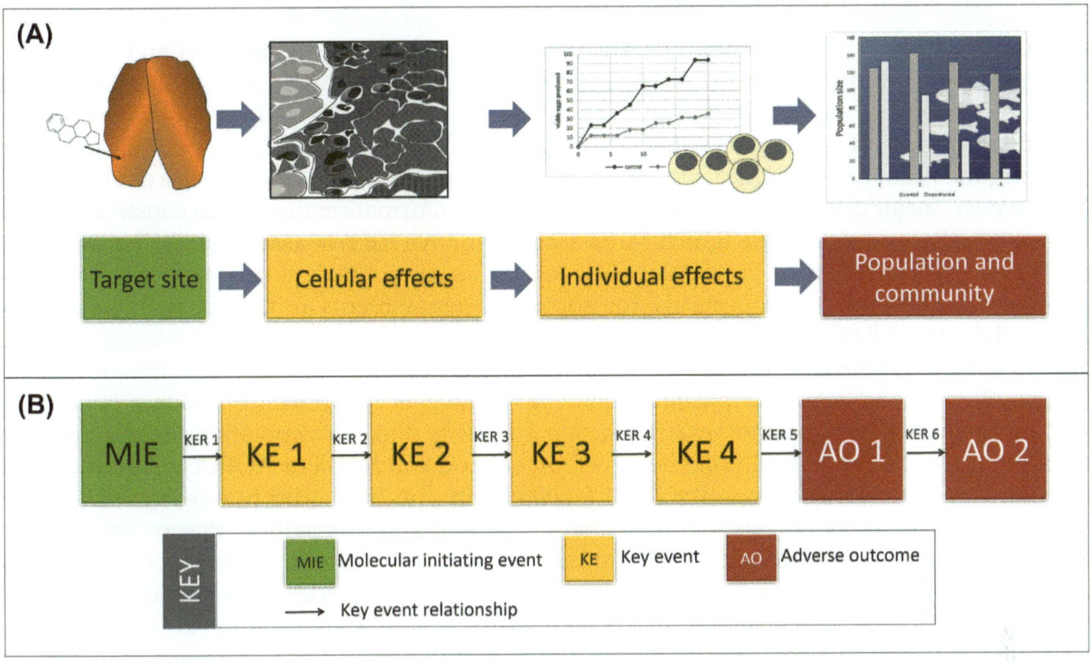

FIGURE 3.7.4 **The Adverse Outcome Pathways integrate impacts across levels of biological organization.** (A) The impact of estrogenic endocrine disrupting compounds on small fish species. The estrogenic compound binds to its target site, the estrogen receptor. This binding has been shown to induce a condition called intersex where male gonadal tissue contains both male and female reproductive tissue.[76] Matings of exposed fish produce a lower level of viable eggs,[68] which has also been shown to reduce the overall population size.[73] (B) The adverse outcome pathway describes these same impacts for estrogenic endocrine disruption in small fish, but in greater detail to establish relationships between each effect. The molecular initiating event (MIE) occurs when the estrogenic compound binds to the estrogen receptor. This binding causes the transactivation (or upregulation) of genes involved in the estrogen response (key event 1: KE 1). The upregulation of estrogen regulated genes causes some of the cells in the male gonad to differentiate into female reproductive tissue causing the intersex condition in male fish (KE 2). The change in the structure of male reproductive tissue interferes with the production of sperm (KE 3), resulting in reduced fertilization and production of viable eggs (KE 4). The reduction in viable eggs causes the first adverse outcome (AO 1) at the population level causing a decrease in population size. In the experimental lakes study by Kidd et al., the reduced population size of small fish led to a decrease in the size of its predators' populations (AO 2).[73] KE relationships between each KE and AO help to define the linkages between each level of biological organization and the threshold levels at each stage that will result in the impacts being translated to the next level.

organisms are realized and therefore population models are needed to relate individual-level impacts to population effects.[18] Two important considerations in the AOP are the threshold levels that (1) lead to initial inhibition at the target site (i.e., MIE) and (2) cause effects to reach the next level of biological organization (i.e., KE). Recent effort has focused on creating quantitative AOPs that emphasize the concentrations or accumulation of effects that initiate each of the KEs.[19]

The AOP concept was initially developed in direct response to the National Research Council report, *Toxicity Testing for the 21st Century*.[20] This report calls for an emphasis on high-throughput, predictive assays that can screen large numbers of chemicals for a particular adverse effect. By employing an AOP approach, assays can be developed to target MIEs and KEs and thus predict the potential for a toxicological pathway to be initiated and lead to an adverse effect. High-throughput screening assays that target MIEs and KEs and predict AOPs have obvious value for attacking the problem of CECs. Using these assays will aid in the identification of potential adverse effects early in chemical development, therefore reducing the number of chemicals that slip through and end up as CECs in the environment.

However, major challenges remain in acquiring the information needed to construct AOPs, establish KERs, and determine which MIEs are informative for risk assessments. For example, to connect mechanistic data to population-level risk assessments, it is necessary to first determine which endpoints are informative for population assessments and then provide data that define the casual linkages within these pathways.[18,21] In some cases, mechanistic data are almost completely lacking for toxicological pathways and this is especially true for CECs and many environmentally important organisms. However, genomic approaches have been helping to fill in these gaps and are a promising avenue to make extrapolations across different taxonomic groups.[22,23] In addition, the development and validation of AOPs can be conducted through a collaborative community effort facilitated by the AOP-wiki (www.aopwiki.org).[24]

Given that there are currently 85,000 chemicals used commercially,[25] few of which have been examined toxicologically, it is likely that the list of CECs will continue to grow even if no additional chemicals are synthesized and manufactured. This new "emergence" will likely be due to (1) further advances in the ability to detect these compounds at very low environmental concentrations, (2) the discovery of novel AOPs at low concentrations, (3) a recent release or a projected introduction into the environment, and/or (4) a ramp up of a chemical's production volume.

3.7.3 TWO CONTAMINANTS OF EMERGING CONCERN CASE STUDIES

3.7.3.1 Case Study #1: Nanomaterials

Nanomaterials are a unique class of emerging contaminants. They are a class of materials that were made possible by advancements in chemical synthesis that allowed for the bottom-up production of novel structures [e.g., carbon nanotubes (CNTs)[26] and quantum dots[27]], and the top-down production of ultrafine particles.[28] In the development of new nanotechnologies, it was recognized early on that nanomaterials have unique properties compared to their bulk counterparts. Therefore toxicological investigations were called for prior to large releases of these contaminants into the environment.[29,30] According to the Project on Emerging Nanotechnologies Consumer Products Inventory, there were over 1800 commercial products using nanotechnology in 2015 with market research suggesting a steady growth of the industry over the next few years reaching a value of $75 billion by 2020.[31,32] With this exponential growth of the industry, toxicological research has not been able to keep up with all the novel nanomaterials being developed.[33] However, the industry is far from mature and in most cases concentrations in the environment have not reached levels of concern (but see examples given later). This provides an almost unprecedented opportunity for green chemists to play a leading role in the development of sustainable nanotechnologies, and indeed, many investigators are exploring methods of creating nanomaterials in a manner that reduces the energy needs and/or materials with high toxicological potency.[34]

Nanomaterials are defined as molecular or macromolecular structures with one or more dimensions between 1 and 100 nm.[35] There are a diversity of different particles and applications (Table 3.7.2 and Fig. 3.7.5) that are grouped together based on this size definition. In addition to intentionally designed nanomaterials, there are natural nanomaterials, and

TABLE 3.7.2 Diversity of Common Engineered Nanomaterial Classes and Potential Toxicity Mechanisms

Nanoparticle Class	Examples	Applications	Toxic Mode of Action	References
Metal-Based				
Metal oxide	ZnO	Sunscreens, paints	Genotoxicity	50,51,53
	CuO	Electronics, sensors	Genotoxicity	50,51,53
	TiO_2	Cosmetics, sunscreens	Inflammation, oxidative damage, genotoxicity	50,53,90
	CeO_2	Diesel fuel additive, redox catalyst, medical applications	Low potency, inhibition of photosynthesis in algae, feeding inhibition in nematodes	91–93
Zero-valent	Ag	Antimicrobial agent, biosensors	DNA damage, oxidative stress	50,51,94–96
	Fe (nZVI)	Environmental remediation	Generation of reactive oxygen species	39
Carbon-Based				
Fullerenes	C_{60} C_{70}	Electronics, solar cells, optical applications	Genotoxicity	50
Carbon nanotubes	MWCNT SWCNT	Films and coatings, electronics, light-weight nanocomposites, energy storage	Asbestos-like toxicity, membrane leakage, apoptosis, DNA double-strand breakage	50,52,90
Graphene	Graphene oxide	Films and coatings, electronics	DNA damage	96
Quantum dots	CdSe/ZnS CdTe	Biomedical imaging, electronics	Cd-related toxicity, cell death, apoptosis	97
Dendrimers	PAMAM PPI	Drug delivery, antiviral agents, cosmetics, textiles	Membrane destabilization, apoptosis	54,98
Future classes	Nanosized robotic devices	Numerous, especially medical	Uncertain	99

MWCNT, multiwalled carbon nanotube; *nZVI*, nano zero-valent iron; *PAMAM*, polyamindoamine; *PPI*, polypropylenimine; *SWCNT*, single-walled carbon nanotube.

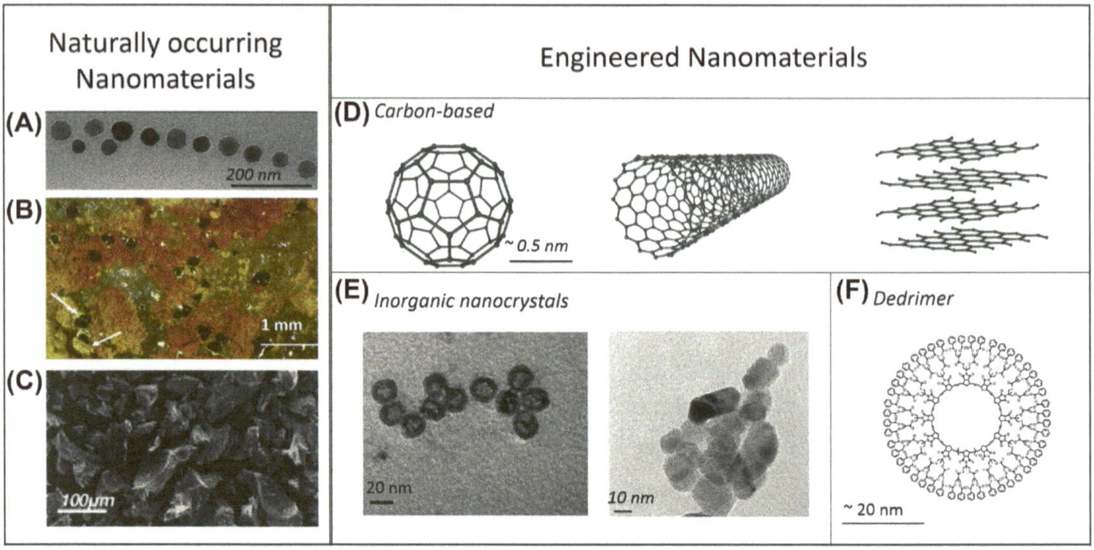

FIGURE 3.7.5 **Examples of naturally occurring (A–C) and engineered nanomaterials (D–F).** (A). Transmission electron microscopic image of nanosized magnetite crystals produced by the bacterium *Magnetospirillum gryphiswaldense.* (B) Optical microscope image of weathered quartz exposing gold "plates" (highlighted with white arrows) composed of gold nanoparticles. (C) Nanoparticulate ash from Mount Etna. (D) Structures of carbon-based nanomaterials including (from left to right) C_{60} fullerene, carbon nanotube, and graphene sheets. (E) Engineered core-shell semiconductor nanocrystals and ZnO nanoparticles. (F) Dendronized polymer synthesized using ring expansion metathesis polymerization. *Image of volcanic ash from Mount Edna (C) from Kadar E, Fisher A, Stolpe B, Calabrese S, Lead J, Valsami-Jones E, et al. Colloidal stability of nanoparticles derived from simulated cloud-processed mineral dusts.* Sci Total Environ *2014;***466***:864–70, Creative commons license. Image of gold nanoparticles (B) reprinted with permission from Hough R, Noble R, Reich M. Natural gold nanoparticles.* Ore Geol Rev *2011;***42***:55–61. Copyright 2011 Elsevier. (A) By Caulobacter subvibrioides, Wikimedia commons. (D) From Michael Ströck, Wikimedia commons. (E) From Qing Qing Dou et al., Wikimedia commons. (F) By Andrew Boydstone, Wikimedia commons.*

similar to other natural chemicals, some of these are harmful to organisms, whereas others are relatively inert. Some examples include ultrafine atmospheric dust and particles produced through geological and biological processes (e.g., natural weathering processes, combustion particulates from volcanoes and wildfires, and biologically derived colloids).[36] Although there are nanomaterials that exist in nature, most engineered nanomaterials (ENMs) are novel and do not have a natural analog. Despite the vast differences in particles and their chemical properties, their small size and particulate state create similar challenges across the spectrum of nanomaterials including (1) complex environmental interactions leading to changes in bioavailability and (2) high reactivity due to a high surface area to volume ratio.

3.7.3.1.1 Complex Environmental Interactions

ENMs will enter the environment through several routes based on their application. For example, TiO_2 nanoparticles (NPs) and ZnO NPs used as sunscreens have been shown to accumulate in recreational lakes.[37] Other ENMs used in textiles (e.g., Ag NPs), and those used in medical and health and beauty products will enter the environment through waste water streams, whereas ENMs found in nanocomposites (e.g., CNTs) or paints are subject to weathering and can enter aquatic systems through runoff.[38] Finally, ENMs used for environmental remediation (e.g., zero-valent iron NPs) may be directly injected into contaminated ground water.[39] ENMs that enter wastewater treatment facilities may undergo transformation reactions under the conditions of anaerobic digestion. For example, Ag NPs are quickly transformed into Ag_2S, a nonbioavailable form of Ag that accumulates in sewage sludge, effectively removing them from wastewater treatment plant (WWTP) effluent.[40,41]

Once they enter the environment, ENMs exhibit more complex behaviors compared with conventional chemicals due to not only chemical transformations that occur but also physical transformations. Particles in aqueous environments will generally follow colloidal chemistry as described by Derjaguin, Landau, Verwey, and Overbeek (DLVO theory) where the aggregation of particles is governed by competing forces, van der Waals forces, and electrostatic repulsion.[42] However, as illustrated in Fig. 3.7.6, environmental transformations (e.g., oxidation, reduction, reactions with natural organic matter, aggregation, and dissolution) can occur that alter the surface charge and particle size, which in turn affects fate and transport, bioavailability, and toxicity of the ENMs.[43,44] Metal-based nanomaterials may dissolve in aqueous environments. Therefore understanding the extent of dissolution, whether there is particle-specific toxicity, and/or whether the particles are more potent compared with the metal ions are major research emphases for metal-based nanomaterials.[45] However, it is also important to realize that environmental factors such as pH and salinity play a large role in the extent of dissolution of ENMs.[44] In addition, ENMs are being produced with

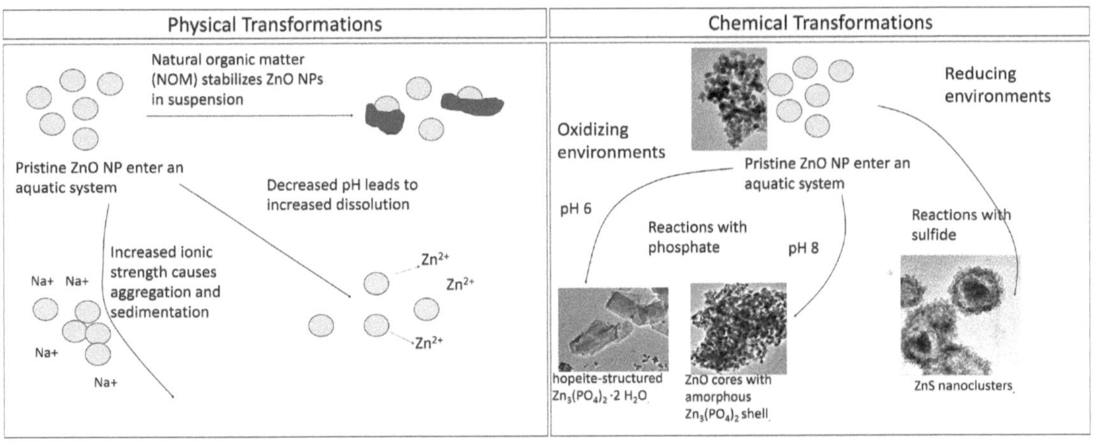

FIGURE 3.7.6 **Physical and chemical transformations of ZnO nanoparticles (NPs) influence their fate and transport in the environment.** ZnO NPs undergo a number of physical transformations including dissolution, which releases the free metal ion, aggregation, and sedimentation, or stabilization in the presence of humic acid.[44,100] The two major chemical transformations include sulfidation in reducing conditions[101] and reactions with phosphate to $Zn_3(PO_4)_2$ structures.[102] These chemical transformations also impact the stability of ZnO NPs in aqueous environments. *Nanoparticle images (pristine, hopeite-structured, and ZnO core with amphorous shell) reprinted with permission from Rathnayake S, Unrine JM, Judy J, Miller A-F, Rao W, Bertsch PM. Multitechnique investigation of the pH dependence of phosphate induced transformations of ZnO nanoparticles.* Environ Sci Technol *2014;**48**:4757–64. Copyright 2014 American Chemical Society. ZnS nanocluster image reprinted with permission from Ma R, Levard C, Michel FM, Brown Jr GE, Lowry GV. Sulfidation mechanism for zinc oxide nanoparticles and the effect of sulfidation on their solubility.* Environ Sci Technol *2013;**47**:2527–34. Copyright 2013 American Chemical Society.*

different functional groups, which have the potential to affect their fate in aqueous environments.[43] Overall, the complexity of understanding ENMs in the natural environment creates a number of questions regarding the adequacy of current test methods to truly predict toxicity of ENMs to aquatic organisms (see Handy et al. for a review on modifications needed for test methods[36]).

3.7.3.1.2 *Toxicological Considerations*

Through the transformation and transport mechanisms mentioned previously, ENMs will ultimately enter an ecosystem where there is potential to interact with the organisms present. At this time, their toxicity will be governed in part by their bioavailability, or the ability of the ENMs to be taken up by an organism. A lot of research has been dedicated to determining if ENMs have a distinct mechanism of uptake compared with conventional chemicals and their bulk counterparts. Many studies have found that ENM are less bioavailable, or are taken up by organisms and cells to a lesser degree than bulk counterparts. However, in some circumstances, the nanoform increases their uptake compared with metal ions (e.g., Qdots[46]; ZnO NPs[47]), and this may be through a "Trojan horse-type mechanism."[48] Increased uptake of ENMs has also been shown in marine bivalves. In natural marine waters, NPs form aggregates that are readily taken up by the gills of suspension feeding bivalves (i.e., blue mussels and oysters) and ingested into the digestive tract.[49] Other studies have shown that environmental factors, such as the presence of algae, can impact bioavailability and toxicity of ENMs.[46]

One of the major concerns with nanomaterials relates to the increased surface-area-to-volume ratio of these particles, such that a small mass or volume of substance has much enhanced reactivity, which may translate directly into enhanced toxicity.[33,50] Some nanomaterials are designed to be toxic, exploiting the antimicrobial properties of silver, copper, and zinc.[51] The enhanced surface area of these ENMs makes them not only more effective antimicrobial agents but also more likely to be toxic to other organisms. Although nanomaterials may have more potent toxicity due to the high surface-area-to-volume ratio, in general, they are not unique from conventional chemicals in their mechanisms of action (MOA), which activates the initial stages of an AOP (i.e., MIE and cellular- and tissue-level effects). They show a diverse suite of potential MOAs including genotoxicity, oxidative stress, and disruption of ion regulation, to name a few (see Table 3.7.2).[36] Many ENMs exhibit genotoxicity, but it is unclear whether this mechanism is particle specific or a generality of NMs.[50] One noteworthy MOA is characteristic of fibrous nanomaterials (e.g., rods and tubes), which exhibit asbestos-like effects. For example, cells lining the human intestine and lung cavity attempt to engulf CNTs through a process called phagocytosis. However, because of the length and shape of the fibrous particles, the cells are not able to engulf them, which leads to inflammation and ultimately cancer.[52,53] Finally, dendrimers that are being actively developed for drug delivery systems are often cationic and will interact with negatively charged biological molecules. Their primary target for cellular toxicity is the cell membrane where they form "nanoholes," leading to leakage of cellular contents or complete cell lysis.[54]

3.7.3.1.3 *Current Status and Future Outlook*

As mentioned earlier, only a few studies have detected ENMs in the environment, suggesting that the release of ENMs into the environment is minimal, as of yet. For example, Gondikas et al. detected TiO_2 NPs in Old Danube Lake, Austria, and found concentrations up to 1.38 μg/L.[37] Ferre et al. developed a method for the quantification of fullerenes and reported the first environmental concentrations of these nanomaterials in WWTP effluent. They found levels in the parts per billion range including *N*-methylfulleropyrrolidine C_{60}, an industrial intermediate, indicating that the levels represented ENMs and were not primarily derived from natural processes.[55] The levels of ENMs found in the aforementioned studies are below the levels shown to cause adverse effects in aquatic life; however, this does not mean that other ENMs could not be present in the environment at toxic levels. Cost-effective detection methods are still not available for monitoring the majority of ENMs in the environment, and therefore it has been difficult to attain accurate environmental concentrations. A number of modeling studies have suggested relatively low risk for most nanomaterials; however, there is a probable risk associated with the most potent NPs (e.g., Ag NPs) in some environmental matrices.[56] The limited information available on environmental concentrations of ENMs suggests that levels are quickly approaching concentrations that will pose a risk for environmental health.

It is highly likely that increasing amounts of nanomaterials will find their way into the environment in the future. New nanomaterials are being synthesized and applied to commercial products at an ever-increasing rate. Chronic toxicological studies and even environmental monitoring are lagging further behind the expanding nanomaterial universe, and there are currently no plans to bring these two divergent trends into alignment.

3.7.3.1.4 Green Chemistry's Approach to Nanomaterials

There is a critical need to apply green chemistry principles to design less toxic NMs and to limit their release and transport in the environment. In many ways, nanotechnology is a great example of the adoption of green chemistry principles, as many methods of creating nanomaterials in a manner that reduces energy needs and/or toxic materials are being explored.[34] High-throughput screening methods are being investigated to develop ENMs with reduced toxicity. For example, Nel et al. have created large libraries of ENMs of different sizes, shapes, and composition and that incorporate different coatings and functional groups. These libraries are screened through a set of high-throughput cellular assays to determine which parameters contribute most to the ENM's toxicity. Their results can then directly support green chemistry approaches to design NMs with the least toxic characteristics.[53,57] These strategies can also be employed to design highly effective and lower toxicity dendrimers for targeted drug delivery.[54]

In a similar way, it should be possible to design high-throughput assays to screen ENMs for environmental persistence and removal efficiency in WWTPs. As mentioned previously, some ENMs are designed to be antimicrobial agents and pesticides, and are therefore toxic by their design. In this case, creating NMs through Green Chemistry Principle #10 is key: "Chemical products should be designed so that at the end of their function they do not persist in the environment and break down into innocuous products."[58] Therefore, green chemists can use information about the environmental transformation, fate, and transport to design NMs that can easily be removed from the environment or transformed into inert forms. For example, ZnO NPs have been developed that are doped with iron, reducing their dissolution and toxicity to many aquatic organisms.[57] The challenge will be to develop high-throughput assays that not only report on the toxicity of the NMs but also predict environmental transformation and removal efficiency. With this information, green chemists can engineer safer NMs with limited transport and bioavailability.

3.7.3.2 Case Study #2: Pharmaceuticals

Pharmaceuticals comprise a large group of CECs, of rather dissimilar chemical structures and characteristics. Kinch et al. tabulated 1453 "new molecular entities" (i.e., drugs) that were registered as therapeutics for use in the United States by the US Food and Drug Administration (FDA) between 1827 and 2013. Only a small number of entries predate the formation of the FDA in 1927, and a steady, almost linear increase in the annual number of registrations began around 1945 and has continued through 2013.[59] From the standpoint of CECs, this number of drugs (i.e., 1453) is a minimum, since the number of drugs used globally could likely double this number. In addition, the number of FDA-registered drugs does not include metabolites of these drugs, which may exhibit either lesser or greater potency, or may act through a completely different MOA.

Due to current analytical limitations, not all of these 1453 drugs have been detected in environmental samples (water, sediments, tissues, etc.). Nevertheless, pharmaceuticals are now considered to have a ubiquitous distribution worldwide.[60–62] They were detected in 31 of the 71 countries sampled from all five of the United Nations global regions.[61] In general, contamination is higher in urban areas compared with rural ones.[63,64] Table 3.7.3 presents examples of drugs, arranged by therapeutic group, that have been classified as CECs simply because they have been detected in one or more environmental compartments, but are not necessarily toxic to humans or wildlife at these concentrations. This table presents examples, so should not be considered comprehensive.

3.7.3.2.1 Predictable Environmental Interactions

The distribution and fate of various pharmaceuticals is closely related to each compound's physicochemical properties. In most cases, environmental contamination by pharmaceuticals is caused by domestic, commercial, and agricultural waste disposal. Compounds that are hydrophilic are readily excreted by humans in urine, feces, and even perspiration, sometimes as parent compounds (i.e., without metabolism). These compounds make their way to either household septic systems[65] or to municipal wastewater treatment plants.[61] Since they are hydrophilic, they tend to be discharged as soluble compounds in effluent water. They are often not appreciably bioaccumulated by flora or fauna due to their hydrophilicity. Lipophilic compounds, however, tend to adsorb to particles and settle out during sewage treatment, often reaching relatively high concentrations in sewage sludge, which is often processed into fertilizer for croplands. They are also readily bioaccumulated in the lipid-rich tissues of organisms where they can remain stored for significant amounts of time before being metabolized to a more water-soluble metabolite and excreted. Changes in the salinity and pH, often found at the mouths of rivers and in estuaries, can have a profound effect on pharmaceutical compounds that have multiple ionizable functional groups. The "salting out" of these compounds as freshwater meets seawater reduces the dissolved concentration of these compounds and increases their concentrations in the sediments. Because of this salting out, in addition to the considerable dilution of seawater and effluent, concentrations of pharmaceuticals are expected to be very low in seawater and likely close to detection limits. The higher concentrations that are found in sediments and tissues make the analyses of these compartments more useful for monitoring programs.

TABLE 3.7.3 Therapeutic Groups of Pharmaceutical Contaminants of Emerging Concern, With Examples of Drugs That Have Been Measured in Water, Sediments, or Tissues

Pharmaceutical Class	Examples
Analgesics/antiinflammatories	Codeine, diclofenac, indomethacin, NSAIDs
Antibiotics	Sulfamethoxazole, triclosan, piperacillin
Anticoagulant	Warfarin, methylparaben
Anticonvulsives	Phenobarbital, primidone
Antidepressants	Fluoxetine, paroxetine
Antidiabetics	Metformin, glibenclamide
Antiepileptics	Carbamazepine, dilantin, gabapentin
Antifungals	Miconazole, thiabendazole
Antihelmintics	Levamisole, salbutamol
Antiplatelets	Clopidogrel
β-Blockers	Atenolol, nadolol, propranolol
Decongestants	Albuterol
Diuretics	Hydrochlorothiazide
Hemorrheologic agent	Pentoxifylline
Hormones (synthetic)	Ethinylestradiol, estrogen, estrone
Hypolipidemics/blood lipid regulators	Clofibric acid, fenofibrate, gemfibrozil
Laxatives	Anthraquinone
Sedatives	Phenobarbital, secobarbital, diphenhydramine
Steroids	Digoxigenin, progesterone
SSRIs	Fluoxetine
Stimulants	Dimethylxanthine, caffeine

NSAIDs, nonsteroidal antiinflammatory drugs (e.g., ibuprofen); *SSRIs*, selective serotonin reuptake inhibitors (popular antidepressants).
Data extracted from Kolpin DW, Furlong ET, Meyer MT, Thurman EM, Zaugg SD, Barber LB, et al. Pharmaceuticals, hormones, and other organic wastewater contaminants in US streams, 1999–2000: a national reconnaissance. Environ Sci Technol *2002;***36***(6):1202–11; Aus der Beek T, Weber FA, Bergmann A, Hickmann S, Ebert I, Hein A, et al. Pharmaceuticals in the environment–global occurrences and perspectives.* Environ Toxicol Chem *2015. http://dx.doi.org/10.1002/etc.3339; Fabbri E, Franzellitti S. Human pharmaceuticals in the marine environment: focus on exposure and biological effects in animal species.* Environ Toxicol Chem *2016;***35***(4):799–812; and Llorca M, Farré M, Eljarrat E, Díaz-Cruz S, Rodríguez-Mozaz S, Wunderlin D, et al. Review of emerging contaminants in aquatic biota from Latin America: 2002–2016.* Environ Toxicol Chem *2016. http://dx.doi.org/10.1002/etc.3626.*

3.7.3.2.2 Toxicological Considerations

It is important to note that pharmaceuticals have been designed to have specific biochemical effects. Their therapeutic effects in humans or livestock have been maximized, whereas their toxic impacts to humans (i.e., side effects) have simultaneously been minimized during the design process. Drug potencies are typically rather high, so that only small amounts are needed to be efficacious. Since they have been designed to target a specific biochemical pathway and receptor in humans at very low concentrations, it should not have surprised us to now discover that the low concentrations of these drugs in the environment are reacting with similar biochemical pathways and receptors in wildlife, often with adverse effects (i.e., AOPs). Pharmaceuticals do not always follow a typical dose-response curve, particularly if they are receptor specific. Instead, low concentrations can cause an effect once their concentration reaches a threshold value, causing their available receptors to become saturated. Higher concentrations would not therefore increase the degree of response.

As with other CECs, measuring a pharmaceutical in the environment is not necessarily an indication that the compound is actually having an adverse effect on any organism in the community or ecosystem. As analytical detection limits decrease, scientists are identifying more and more compounds that are then added to the CEC list. For many of these compounds, even basic toxicological data are lacking. However, for preliminary screening and prioritization, it may be possible to apply an acute-to-chronic ratio and also extrapolate from the relatively high doses used in toxicity tests to the low doses measured in nature, to assess the potential risk of newly detected environmental exposures.[66] While more widespread toxicity testing is clearly needed,[20] toxicity tests alone will not provide the biological response data necessary to assess low level, chronic toxicity. Toxicity tests, on both adults and early life stages, will have to be augmented with a suite of sublethal biomarkers (various physiological, behavioral, cellular, biochemical, and molecular measurements that can be interpreted to indicate if the organism is experiencing stress).[67] These biomarkers, however, must be chosen carefully if they are to be applied to low-dose exposures for extended periods of time (chronic exposures) and will be most informative if they are based on MIEs and KEs within the AOP for that chemical (Fig. 3.7.4).

Unlike well-controlled laboratory toxicity tests, animals in the field are almost never exposed to just a single potential toxicant at a time. Instead, exposure to multiple chemicals, at a wide range of concentrations, is more usual and especially common with pharmaceuticals. This makes extrapolating from laboratory studies to field situations extremely difficult if not currently impossible. The combination of chemical exposures may result in additive, synergistic, or antagonistic impacts, which are often difficult to predict just by knowing the concentrations of the chemicals present.[68] An additive response implies that a pair or suite of chemicals all behave independently of the others, yet may target the same biochemical pathway. Antagonistic responses imply that some chemicals in a mixture counteract the effect of other chemicals present. Toxicity is therefore at least reduced and at best negated completely. Synergism produces a greater-than-additive response for a group of chemicals, which may result if one chemical enhances the toxicity of another possibly through the inhibition of detoxification mechanisms. There are a few examples of chemical mixtures that are synergistic (e.g., hepatotoxicity of carbon tetrachloride and ethanol, lung carcinogenicity of cigarette smoke and asbestos).[10] Nevertheless, these three types of mixture responses need to be assessed for mixtures of pharmaceuticals.[69]

3.7.3.2.3 *Endocrine Disrupting Compounds*

A particularly interesting subset of the pharmaceuticals is the endocrine disrupting compounds (EDCs). However, although the synthetic hormones listed in Table 3.7.3 are clearly pharmaceuticals, not all EDCs are hormones. For example, the herbicide atrazine, phthalates, polychlorinated biphenols, chlorinated pesticides, dioxins and dibenzofurans, alkylphenol ethoxylates, bisphenol A, and organotins are all EDCs, but were not synthesized by the pharmaceutical industry. EDCs are compounds that interfere with the endocrine, or hormone, system of animals. Hormones are chemical messengers that relay signals from one tissue or organ to another. They are involved in a vast array of processes including growth, reproduction, appetite, and stress responses including the "fight-or-flight" response, to name a few. Hormones are synthesized in one tissue, but relay their messages through the circulatory system where they bind to protein receptors in a different tissue. Once they bind to a receptor, the chemical message is amplified in the target tissue, but the specific response of that tissue depends on the hormone. For example, when an estrogen, or female reproductive hormone, binds to its receptor, it may induce changes in the tissue to prepare the animal for pregnancy.

EDCs can interfere with hormones at any stage during their synthesis, during transport to a different tissue, or as they bind to their receptor. They may behave like a natural hormone (i.e., a hormone agonist, used as an example in Fig. 3.7.4A), inhibit a natural hormone (i.e., a hormone antagonist), interfere with normal hormonal receptor-mediated pathways, or interfere with the cycling, synthesis, or breakdown of natural hormones. EDCs have been shown to disrupt the endocrine system of humans and other animals, resulting in developmental and reproductive malfunctions, carcinogenesis,[7,70–72] and adverse impacts on animal populations and ecosystems.[73] Some of the earliest examples of EDCs involved masculinization of marine snails (imposex) due to the use of tributyltin-laced antifouling paint,[74] feminization of alligators due to DDT,[75] and the inappropriate synthesis of vitellogenin, an egg yolk protein produced by reproductively mature females, in marine and freshwater fish.[76] Additional work on fish demonstrated a wide range of impacts on sex determination, and reproductive pathways.[68,76] In addition, embryos and juveniles are often more sensitive to endocrine disruptors than are adults, and these effects may be delayed, not surfacing until the organism reaches reproductive maturity.

The synthetic estrogen EE2 is an excellent example of a hormone EDC. EE2 acts as a hormone agonist, mimicking the natural hormone estradiol (E2) by binding to the estrogen receptor (ER). It is a primary component of birth control pills and is also used in humans for estrogen replacement therapy, in livestock for enhancing muscle growth and in reproductive disorders, and in mariculture to produce single-sex fish populations.[77] EE2 and E2 are the most widely detected hormone CECs, having been quantified in fresh water, marine water, soils, sediments, and tissue samples from various parts of the globe.[62,78] Interestingly, EE2 has a higher affinity for the ER than E2, meaning that it will readily bind to the receptor and elicit a strong hormone response at even lower concentrations than the natural hormone.

EE2 is responsible for a variety of deleterious effects in biota, including feminization, altered population sex ratios, reduction in gonad development, reduced fecundity, and thyroid dysfunction. In both freshwater and marine fish, EE2 binding to the ER has been shown to feminize males and initiate the production of vitellogenin, as well as other reproduction-related effects.[68,79] As little as 0.1–1 ng/L EE2 was enough to initiate vitellogenin induction in fathead minnows, whereas 25–30 times more E2 was needed to produce the same result.[68,80] In a study that spiked EE2 into a lake in Ontario's Experimental Lakes Region, Kidd and colleagues found that the feminization of fathead minnows was so pervasive that the population of fathead minnows almost collapsed within 2–3 years.[73] This loss of forage fish led to reductions in the numbers of lake trout, their predators. The experiment was stopped to prevent further damage to the community.

Caldwell and coworkers summarized the effects of EE2 on aquatic organisms, as reported in over 100 studies between 1994 and 2007.[81] The most sensitive physiological process that was adversely impacted was reproduction in fish. The investigators focused on the "no observed effect concentrations" (NOECs), or the highest concentrations that animals could be exposed to without observation of any adverse effects, from 39 studies covering 26 aquatic vertebrate and invertebrate species. These values were arranged in a species sensitivity distribution (Fig. 3.7.7), and the hazardous concentration of EE2 that resulted in an adverse impact on 5% of all the species tested (HC_5) was estimated from the Weibull distribution fit to the data. This HC_5, 0.35 ng/L, was recommended as the predicted no effect concentration (PNEC), or regulatory threshold value, for EE2 exposure to aquatic organisms.[81] The recommended PNEC is lower than the method detection limits of many of the studies on EE2 in environmental samples. However, the authors argue that it is a realistic, albeit conservative, concentration to use for regulatory purposes, since it is below 95% of existing NOECs that are based on reproductive effects on aquatic organisms.

Because of the potency of EE2, detecting it at the very low levels responsible for adverse effects, such as the PNEC described in the previous paragraph, is not trivial, especially in coastal and marine waters. An alternative is to study the accumulation of EE2 in biota, such as the blue mussel *Mytilus edulis*, which bioaccumulates EE2 in its tissues from a 100 to just over a 1000-fold higher than the seawater.[82,83] Therefore, when water concentration reaches the PNEC of 0.35 ng/L, concentrations in mussel tissue would be between 35 and 350 ng/L, well within existing analytical detection limits.

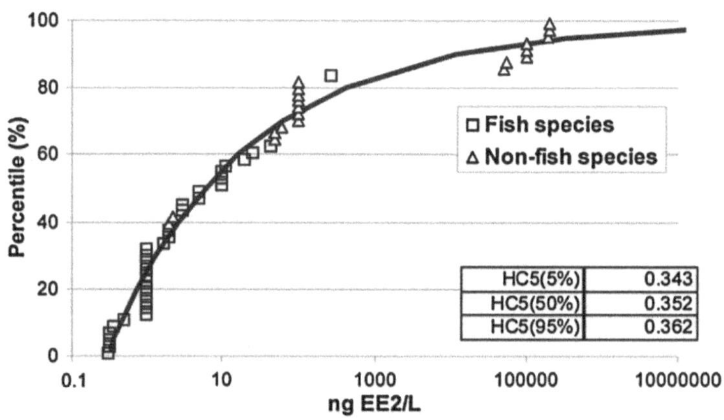

FIGURE 3.7.7 **Species sensitivity distribution for 26 aquatic vertebrate and invertebrate species (N = 52 data points; 39 studies) exposed to 17-α-ethinyl estradiol (EE2) in the laboratory for which no observed effect concentrations were calculated.** HCx (y) = hazardous concentration in which x % of all the species tested are adversely affected, and y = the 5%, 50%, and 95% confidence limits around the calculated HC5 value. *Reprinted from Open Access material from Caldwell DJ, Mastrocco F, Hutchinson TH, Länge R, Heijerick D, Janssen C, et al. Derivation of an aquatic predicted no-effect concentration for the synthetic hormone, 17α-ethinyl estradiol.* Environ Sci Technol *2008;**42**(19):7046–54. Copyright 2008 American Chemical Society. http://pubs.acs.org/doi/abs/10.1021/es800633q.*

3.7.3.2.4 *Current Status and Future Outlook*

As quantitative detection limits continue to be lowered and new pharmaceuticals make their way onto the global market, it is clear that pharmaceuticals will continue to be added to the growing list of CECs. Many pharmaceuticals appear to be relatively long-lived; therefore, it is likely that concentrations of these chemicals and their metabolites will increase in sediments over time. Since wastewater treatment plants are the primary point source whereby pharmaceuticals enter the environment, a number of researchers are investigating novel ways, such as using manganese oxide reactors, membrane bioreactors, and ozonation, to remove pharmaceuticals, especially EE2, from wastewater effluent[84,85] and drinking water.[86] Although this may not be practical for all pharmaceuticals, advanced water treatment practices may prove effective in reducing the amount of pharmaceuticals that enter the environment. This may, however, come at a significant financial cost.

3.7.3.2.5 *Green Chemistry's Approach to Pharmaceuticals*

In designing a new pharmaceutical, green chemists need to consider how the chemical and its metabolites will likely enter the environment when excreted or defecated by humans. One way to reduce a chemical's environmental influx would be to reduce its use by humans. If chemicals that were more potent than existing pharmaceuticals were designed, then less chemical would be needed for each therapeutic dose, and less chemical would enter the wastewater stream. Similarly, if a chemical was designed to be more rapidly and/or efficiently absorbed by the human digestive tract, then less overall chemical would be needed for a therapeutic dose. However, designing new pharmaceuticals that are either more potent or more easily absorbed is likely to be a false victory, since it may later prove to have adverse environmental effects at lower environmental concentrations and set up the need for environmental chemists to develop analytical techniques with even lower detection limits to monitor the chemical in the environment. A pharmaceutical "arms race" would ensue.

A more promising approach might be to design a new pharmaceutical that is effective, but that rapidly breaks down in the environment once it is excreted or defecated. Alternatively, it may be possible to design a pharmaceutical that will be rapidly broken down or removed in our existing secondary treatment plants. For example, the compound may be designed with a backbone that is easily targeted by aerobic bacteria present in the WWTP. Alternatively, the molecule may be "tagged" with a structure that can assist with its separation from the effluent stream (e.g., an iron NP that can be used to pull the molecule out of the WWTP wastestream using a magnet).[57]

3.7.4 CONCLUSIONS

CECs will continue to be a major environmental issue. Given the tremendous number of chemicals that are of commercial use, and the fact that some of these chemicals elicit toxic effects at very low environmental concentrations (in many cases, below the detection limits of currently used analytical techniques), it is likely that the list of CECs will continue to grow as analytical measurements improve. While a number of suggestions have been made on how medical and veterinary practices can reduce the quantities of the pharmaceuticals that are released into the environment,[87] these recommendations are generally not applicable to the vast majority of the other CECs. In the long term, the green chemistry principle of "benign by design" holds the best promise of curtailing CECs.[87]

In the case studies on nanomaterials and pharmaceuticals presented in this chapter, we discussed a number of roles that green chemists could play during the synthesis and design of new chemicals that would help prevent them from ending up on the list of CECs. To preclude new CECs from suddenly appearing in the future, green chemists need to design chemicals that are (1) inherently less toxic, (2) broken down immediately after use or by exploiting WWTP's typical microbial degradation pathways, and/or (3) modified structurally to be more easily separated from wastewater effluent.[88,89] One or more of these processes should be targeted during each new chemical's design.

Unfortunately, the benign synthesis of new chemicals does not solve the problem of current CECs. Green chemists can also play a role here, by redesigning existing CECs to substitute for some of the more egregious CECs currently in use. However, green chemists need access to more complete "life cycle" (i.e., "cradle to grave") data on each of the currently listed CECs. Some of these data are likely held by the chemical manufacturers, including data on the multiple pathways through which these chemicals make their way into the environment, where they will end up, and how fast they will cycle among environmental compartments. Basic toxicological information on uptake rates and pathways, bioaccumulation, metabolism, and excretion needs to be provided as well. This will allow green chemists to hone in on critical points in the life cycle of the chemical where they can intercede to make the greatest mitigation impact. Unfortunately today, these important life cycle data on the current list of CECs either do not exist or have not been made available.

Green chemists can play other important roles as well. They can base their designed synthetic pathways on less toxic feedstock chemicals, thus reducing the risk of toxic substances being released into the environment from manufacturing processes. They can also help develop assays that can be used to quantify existing CECs in various environmental compartments. Plus, for new or redesigned chemicals, they could consider the issue of detection and quantification during the chemical design process, so that the new chemical contains modifications that will not only make the chemical less toxic, but also easy to separate from environmental compartments and quantify. Finally, green chemists in industry will be in a position to advocate to their employers for full disclosure of any or all of the life cycle data that they have collected on current CECs and on all new chemicals. Although the Toxic Substances Control Act requires toxicity data on new chemicals that are produced commercially in amounts over 1 million pounds,[25] there are currently no regulations in the United States that require the collection or disclosure of life cycle data.

Philosophically, one might argue that a green chemist's job description can be summarized simply as "prevent any additional chemicals from being added to the current list of CECs, and remove chemicals currently listed." Given the increasing pace with which many new chemicals are being developed and marketed, green chemistry's task appears to be almost herculean. Nevertheless, adherence to the principles of green chemistry should lead to the eventual phase-out of CECs. The success of green chemistry will be measured by the rate at which the list of CECs shrinks.

References

1. Halling-Sørensen B, Nielsen SN, Lanzky P, Ingerslev F, Lützhøft HH, Jørgensen S. Occurrence, fate and effects of pharmaceutical substances in the environment-A review. *Chemosphere* 1998;**36**(2):357–93.
2. Daughton CG, Ternes TA. Pharmaceuticals and personal care products in the environment: agents of subtle change? *Environ Health Persp* 1999;**107**(Suppl. 6):907–38.
3. USEPA. *White paper: aquatic life criteria for contaminants of emerging concern. Part I. General challenges and recommendations*. U.S. Environmental Protection Agency, OW/ORD Emerging Contaminants Workgroup; 2008.
4. USGS. *Contaminants of emerging concern in the environment*. 2016. Available from: http://toxics.usgs.gov/regional/emc/index.html.
5. Richardson SD, Ternes TA. Water analysis: emerging contaminants and current issues. *Anal Chem* 2005;**77**(12):3807–38.
6. Richardson SD. Water analysis: emerging contaminants and current issues. *Anal Chem* 2009;**81**(12):4645–77.
7. Kolpin DW, Furlong ET, Meyer MT, Thurman EM, Zaugg SD, Barber LB, et al. Pharmaceuticals, hormones, and other organic wastewater contaminants in US streams, 1999–2000: a national reconnaissance. *Environ Sci Technol* 2002;**36**(6):1202–11.
8. Maruya KA, Dodder NG, Sengupta A, Smith DJ, Lyons JM, Heil AT, et al. Multi-media screening of contaminants of emerging concern (CECs) in coastal urban watersheds in Southern California (USA). *Environ Toxicol Chem* 2016;**35**:1986–94.
9. Deichmann W, Henschler D, Holmstedt B, Keil G. What is there that is not poison? A study of the Third Defense by Paracelsus. *Arch Toxicol* 1986;**58**(4):207–13.
10. Amdur MO, Doull J, Klaassen CD. *Casarett and Doull's toxicology: the basic science of poisons*. New York: McGraw Hill; 1991.
11. Timbrell J. *Principles of biochemical toxicology*. 3rd ed. CRC Press; 2000.
12. Ames BN, Gold LS. Chemical carcinogenesis: too many rodent carcinogens. *Proc Natl Acad Sci USA* 1990;**87**(19):7772–6.
13. Ames BN, Profet M, Gold LS. Nature's chemicals and synthetic chemicals: comparative toxicology. *Proc Natl Acad Sci USA* 1990;**87**(19):7782–6.
14. National Research Council. *Carcinogens and anticarcinogens in the human diet: a comparison of naturally occurring and synthetic substances*. US: National Academies Press; 1996.
15. Meylan WM, Howard PH, Boethling RS, Aronson D, Printup H, Gouchie S. Improved method for estimating bioconcentration/bioaccumulation factor from octanol/water partition coefficient. *Environ Toxicol Chem* 1999;**18**(4):664–72.
16. Muir DC, Howard PH. Are there other persistent organic pollutants? A challenge for environmental chemists. *Environ Sci Technol* 2006;**40**(23):7157–66.
17. Ankley GT, Bennett RS, Erickson RJ, Hoff DJ, Hornung MW, Johnson RD, et al. Adverse outcome pathways: a conceptual framework to support ecotoxicology research and risk assessment. *Environ Toxicol Chem* 2010;**29**(3):730–41.
18. Kramer VJ, Etterson MA, Hecker M, Murphy CA, Roesijadi G, Spade DJ, et al. Adverse outcome pathways and ecological risk assessment: bridging to population-level effects. *Environ Toxicol Chem* 2011;**30**(1):64–76.
19. Garcia-Reyero N. Are adverse outcome pathways here to stay? *Environ Sci Technol* 2015;**49**(1):3–9.
20. National Research Council. *Toxicity testing for the 21st century: a vision and strategy*. Washington, DC: National Academies Press; 2007.
21. Villeneuve DL, Garcia-Reyero N. Vision & strategy: predictive ecotoxicology in the 21st century. *Environ Toxicol Chem* 2011;**30**(1):1–8.
22. Perkins EJ, Chipman JK, Edwards S, Habib T, Falciani F, Taylor R, et al. Reverse engineering adverse outcome pathways. *Environ Toxicol Chem* 2011;**30**(1):22–38.
23. Poynton HC, Vulpe CD. Ecotoxicogenomics: emerging technologies for emerging contaminants. *J Am Water Resour Assoc* 2009;**45**(1):83–96.
24. Villeneuve DL, Crump D, Garcia-Reyero N, Hecker M, Hutchinson TH, LaLone CA, et al. Adverse outcome pathway (AOP) development I: strategies and principles. *Toxicol Sci* 2014;**142**(2):312–20.
25. USEPA. *Toxic substances control act (TSCA) chemical substance inventory*. 2016. Available from: www.epa.gov/tsca-inventory.
26. Iijima S. Helical microtubules of graphitic carbon. *Nature* 1991;**354**(6348):56–8.

27. Dabbousi BO, Rodriguez-Viejo J, Mikulec FV, Heine JR, Mattoussi H, Ober R, et al. (CdSe) ZnS core-shell quantum dots: synthesis and characterization of a size series of highly luminescent nanocrystallites. *J Phys Chem B* 1997;**101**(46):9463–75.
28. Klaine SJ, Alvarez PJJ, Batley GE, Fernandes TF, Handy RD, Lyon DY, et al. Nanomaterials is the environment: behavior, fate, bioavailability, and effects. *Environ Toxicol Chem* 2008;**27**(9):1825–51.
29. Roco MC. Environmentally responsible development of nanotechnology. *Environ Sci Technol* 2005;**39**(5):106A–12A.
30. Maynard AD, Aitken RJ, Butz T, Colvin V, Donaldson K, Oberdörster G, et al. Safe handling of nanotechnology. *Nature* 2006;**444**(7117):267–9.
31. Vance ME, Kuiken T, Vejerano EP, McGinnis SP, Hochella Jr MF, Rejeski D, et al. Nanotechnology in the real world: redeveloping the nanomaterial consumer products inventory. *Beilstein J Nanotechnol* 2015;**6**:1769–80.
32. Mulvaney P, Weiss PS. Have nanoscience and nanotechnology delivered? *ACS Nano* 2016;**10**(8):7225–6.
33. Klaine SJ, Koelmans AA, Horne N, Carley S, Handy RD, Kapustka L, et al. Paradigms to assess the environmental impact of manufactured nanomaterials. *Environ Toxicol Chem* 2012;**31**(1):3–14.
34. Albrecht MA, Evans CW, Raston CL. Green chemistry and the health implications of nanoparticles. *Green Chem* 2006;**8**(5):417.
35. USEPA. *Nanotechnology white paper*. Report No: EPA100/B-07/001. Washington, DC: U.S. EPA; 2007.
36. Handy RD, Cornelis G, Fernandes T, Tsyusko O, Decho A, Sabo-Attwood T, et al. Ecotoxicity test methods for engineered nanomaterials: practical experiences and recommendations from the bench. *Environ Toxicol Chem* 2012;**31**(1):15–31.
37. Gondikas AP, von der Kammer F, Reed RB, Wagner S, Ranville JF, Hofmann T. Release of TiO_2 nanoparticles from sunscreens into surface waters: a one-year survey at the old Danube recreational lake. *Environ Sci Technol* 2014;**48**(10):5415–22.
38. Ging J, Tejerina-Anton R, Ramakrishnan G, Nielsen M, Murphy K, Gorham JM, et al. Development of a conceptual framework for evaluation of nanomaterials release from nanocomposites: environmental and toxicological implications. *Sci Total Environ* 2014;**473–474**:9–19.
39. Crane RA, Scott TB. Nanoscale zero-valent iron: future prospects for an emerging water treatment technology. *J Hazard Mater* 2012;**211–212**:112–25.
40. Kaegi R, Voegelin A, Sinnet B, Zuleeg S, Hagendorfer H, Burkhardt M, et al. Behavior of metallic silver nanoparticles in a pilot wastewater treatment plant. *Environ Sci Technol* 2011;**45**(9):3902–8.
41. Levard C, Hotze EM, Colman BP, Dale AL, Truong L, Yang XY, et al. Sulfidation of silver nanoparticles: natural antidote to their toxicity. *Environ Sci Technol* 2013;**47**(23):13440–8.
42. Christian P, Von der Kammer F, Baalousha M, Hofmann T. Nanoparticles: structure, properties, preparation and behaviour in environmental media. *Ecotoxicology* 2008;**17**(5):326–43.
43. Lowry GV, Gregory KB, Apte SC, Lead JR. Transformations of nanomaterials in the environment. *Environ Sci Technol* 2012;**46**(13):6893–9.
44. Keller AA, Wang H, Zhou D, Lenihan HS, Cherr G, Cardinale BJ, et al. Stability and aggregation of metal oxide nanoparticles in natural aqueous matrices. *Environ Sci Technol* 2010;**44**(6):1962–7.
45. Notter DA, Mitrano DM, Nowack B. Are nanosized or dissolved metals more toxic in the environment? A meta-analysis. *Environ Toxicol Chem* 2014;**33**(12):2733–9.
46. Jackson BP, Bugge D, Ranville JF, Chen CY. Bioavailability, toxicity, and bioaccumulation of quantum dot nanoparticles to the amphipod *Leptocheirus plumulosus*. *Environ Sci Technol* 2012;**46**(10):5550–6.
47. Poynton HC, Lazorchak JM, Impellitteri CA, Blalock B, Smith ME, Struewing K, et al. Toxicity and transcriptomic analysis in *Hyalella azteca* suggests increased exposure and susceptibility of epibenthic organisms to zinc oxide nanoparticles. *Environ Sci Technol* 2013;**47**(16):9453–60.
48. Park EJ, Yi J, Kim Y, Choi K, Park K. Silver nanoparticles induce cytotoxicity by a Trojan-horse type mechanism. *Toxicol In Vitro* 2010;**24**(3):872–8.
49. Ward JE, Kach DJ. Marine aggregates facilitate ingestion of nanoparticles by suspension-feeding bivalves. *Mar Environ Res* 2009;**68**(3):137–42.
50. Magdolenova Z, Collins A, Kumar A, Dhawan A, Stone V, Dusinska M. Mechanisms of genotoxicity. A review of in vitro and in vivo studies with engineered nanoparticles. *Nanotoxicology* 2014;**8**(3):233–78.
51. Bondarenko O, Juganson K, Ivask A, Kasemets K, Mortimer M, Kahru A. Toxicity of Ag, CuO and ZnO nanoparticles to selected environmentally relevant test organisms and mammalian cells in vitro: a critical review. *Arch Toxicol* 2013;**87**(7):1181–200.

52. Poland CA, Duffin R, Kinloch I, Maynard A, Wallace WA, Seaton A, et al. Carbon nanotubes introduced into the abdominal cavity of mice show asbestos-like pathogenicity in a pilot study. *Nat Nanotechnol* 2008;**3**(7):423–8.
53. Zhao F, Zhao Y, Liu Y, Chang X, Chen C, Zhao Y. Cellular uptake, intracellular trafficking, and cytotoxicity of nanomaterials. *Small* 2011;**7**(10):1322–37.
54. Jain K, Kesharwani P, Gupta U, Jain NK. Dendrimer toxicity: let's meet the challenge. *Int J Pharm* 2010;**394**(1–2):122–42.
55. Farré M, Pérez S, Gajda-Schrantz K, Osorio V, Kantiani L, Ginebreda A, et al. First determination of C60 and C70 fullerenes and *N*-methylfulleropyrrolidine C60 on the suspended material of wastewater effluents by liquid chromatography hybrid quadrupole linear ion trap tandem mass spectrometry. *J Hydrol* 2010;**383**(1–2):44–51.
56. Gottschalk F, Sun T, Nowack B. Environmental concentrations of engineered nanomaterials: review of modeling and analytical studies. *Environ Pollut* 2013;**181**:287–300.
57. Nel A, Xia T, Meng H, Wang X, Lin S, Ji Z, et al. Nanomaterial toxicity testing in the 21st century: use of a predictive toxicological approach and high-throughput screening. *Acc Chem Res* 2013;**46**(3):607–21.
58. Anastas PT, Warner JC. *Green chemistry: theory and practice*. New York: Oxford University Press; 1998. 135 p.
59. Kinch MS, Haynesworth A, Kinch SL, Hoyer D. An overview of FDA-approved new molecular entities: 1827–2013. *Drug Discov Today* 2014;**19**(8):1033–9.
60. Daughton CG. Pharmaceuticals in the environment: sources and their management. In: *Analysis, removal, effects and risk of pharmaceuticals in the water cycle—occurrence and transformation in the environment*. New York: Elsevier; 2013. p. 37–69.
61. Aus der Beek T, Weber FA, Bergmann A, Hickmann S, Ebert I, Hein A, et al. Pharmaceuticals in the environment–global occurrences and perspectives. *Environ Toxicol Chem* 2015. http://dx.doi.org/10.1002/etc.3339.
62. Fabbri E, Franzellitti S. Human pharmaceuticals in the marine environment: focus on exposure and biological effects in animal species. *Environ Toxicol Chem* 2016;**35**(4):799–812.
63. Dodder NG, Maruya KA, Ferguson PL, Grace R, Klosterhaus S, La Guardia MJ, et al. Occurrence of contaminants of emerging concern in mussels (*Mytilus* spp.) along the California coast and the influence of land use, storm water discharge, and treated wastewater effluent. *Mar Pollut Bull* 2014;**81**(2):340–6.
64. Granek EF, Conn KE, Nilsen EB, Pillsbury L, Strecker AL, Rumrill SS, et al. Spatial and temporal variability of contaminants within estuarine sediments and native Olympia oysters: a contrast between a developed and an undeveloped estuary. *Sci Total Environ* 2016;**557**:869–79.
65. Phillips P, Schubert C, Argue D, Fisher I, Furlong ET, Foreman W, et al. Concentrations of hormones, pharmaceuticals and other micropollutants in groundwater affected by septic systems in New England and New York. *Sci Total Environ* 2015;**512**:43–54.
66. Guo J, Sinclair CJ, Selby K, Boxall A. Toxicological and ecotoxicological risk-based prioritization of pharmaceuticals in the natural environment. *Environ Toxicol Chem* 2016. http://dx.doi.org/10.1002/etc.3319.
67. Schoenfuss HL, Furlong ET, Phillips PJ, Scott TM, Kolpin DW, Cetkovic-Cvrlje M, et al. Complex mixtures, complex responses: assessing pharmaceutical mixtures using field and laboratory approaches. *Environ Toxicol Chem* 2015. http://dx.doi.org/10.1002/etc.3147.
68. Brian JV, Harris CA, Scholze M, Kortenkamp A, Booy P, Lamoree M, et al. Evidence of estrogenic mixture effects on the reproductive performance of fish. *Environ Sci Technol* 2007;**41**(1):337–44.
69. Rajapakse N, Silva E, Kortenkamp A. Combining xenoestrogens at levels below individual no-observed-effect concentrations dramatically enhances steroid hormone action. *Environ Health Persp* 2002;**110**(9):917.
70. Ketata I, Denier X, Hamza-Chaffai A, Minier C. Endocrine-related reproductive effects in molluscs. *Comp Biochem Physiol C Toxicol Pharmacol* 2008;**147**(3):261–70.
71. Soto AM, Sonnenschein C. Environmental causes of cancer: endocrine disruptors as carcinogens. *Nat Rev Endocrinol* 2010;**6**(7):363–70.
72. Newbold RR. Developmental exposure to endocrine-disrupting chemicals programs for reproductive tract alterations and obesity later in life. *Am J Clin Nutr* 2011;**94**(6 Suppl):1939S–42S.
73. Kidd KA, Blanchfield PJ, Mills KH, Palace VP, Evans RE, Lazorchak JM, et al. Collapse of a fish population after exposure to a synthetic estrogen. *Proc Natl Acad Sci USA* 2007;**104**(21):8897–901.
74. Matthiessen P. An assessment of endocrine disruption in mollusks and the potential for developing internationally standardized mollusk life cycle test guidelines. *Integr Environ Assess Manag* 2008;**4**(3):274–84.

75. Guillette Jr LJ, Gross TS, Masson GR, Matter JM, Percival HF, Woodward AR. Developmental abnormalities of the gonad and abnormal sex hormone concentrations in juvenile alligators from contaminated and control lakes in Florida. *Environ Health Persp* 1994;**102**(8):680.
76. Jobling S, Nolan M, Tyler CR, Brighty G, Sumpter JP. Widespread sexual disruption in wild fish. *Environ Sci Technol* 1998;**32**(17):2498–506.
77. Aris AZ, Shamsuddin AS, Praveena SM. Occurrence of 17α-ethynylestradiol (EE2) in the environment and effect on exposed biota: a review. *Environ Int* 2014;**69**:104–19.
78. Llorca M, Farré M, Eljarrat E, Díaz-Cruz S, Rodríguez-Mozaz S, Wunderlin D, et al. Review of emerging contaminants in aquatic biota from Latin America: 2002–2016. *Environ Toxicol Chem* 2016. http://dx.doi.org/10.1002/etc.3626.
79. Peters RE, Courtenay SC, Cagampan S, Hewitt ML, MacLatchy DL. Effects on reproductive potential and endocrine status in the mummichog (*Fundulus heteroclitus*) after exposure to 17α-ethynylestradiol in a short-term reproductive bioassay. *Aquat Toxicol* 2007;**85**(2):154–66.
80. Pawlowski S, Ternes T, Bonerz M, Rastall A, Erdinger L, Braunbeck T. Estrogenicity of solid phase-extracted water samples from two municipal sewage treatment plant effluents and river Rhine water using the yeast estrogen screen. *Toxicol Vitro* 2004;**18**(1):129–38.
81. Caldwell DJ, Mastrocco F, Hutchinson TH, Länge R, Heijerick D, Janssen C, et al. Derivation of an aquatic predicted no-effect concentration for the synthetic hormone, 17α-ethinyl estradiol. *Environ Sci Technol* 2008;**42**(19):7046–54.
82. Pojana G, Gomiero A, Jonkers N, Marcomini A. Natural and synthetic endocrine disrupting compounds (EDCs) in water, sediment and biota of a coastal lagoon. *Environ Int* 2007;**33**(7):929–36.
83. Ricciardi KL, Poynton HC, Duphily BJ, Blalock BJ, Robinson WE. Bioconcentration and depuration of (14)C-labeled 17alpha-ethinyl estradiol and 4-nonylphenol in individual organs of the marine bivalve *Mytilus edulis* L. *Environ Toxicol Chem* 2016;**35**(4):863–73.
84. Liu J, Lu G, Xie Z, Zhang Z, Li S, Yan Z. Occurrence, bioaccumulation and risk assessment of lipophilic pharmaceutically active compounds in the downstream rivers of sewage treatment plants. *Sci Total Environ* 2015;**511**:54–62.
85. De Rudder J, Van de Wiele T, Dhooge W, Comhaire F, Verstraete W. Advanced water treatment with manganese oxide for the removal of 17α-ethynylestradiol (EE2). *Water Res* 2004;**38**(1):184–92.
86. Clouzot L, Marrot B, Doumenq P, Roche N. 17α-Ethinylestradiol: an endocrine disrupter of great concern. Analytical methods and removal processes applied to water purification. A review. *Environ Prog* 2008;**27**(3):383–96.
87. Klatte S, Schaefer H-C, Hempel M. Pharmaceuticals in the environment–a short review on options to minimize the exposure of humans, animals and ecosystems. *Sustain Chem Pharm* 2016. http://dx.doi.org/10.1016/j.scp.2016.07.001.
88. Schug T, Abagyan R, Blumberg B, Collins T, Crews D, DeFur P, et al. Designing endocrine disruption out of the next generation of chemicals. *Green Chem* 2013;**15**(1):181–98.
89. Leder C, Rastogi T, Kümmerer K. Putting benign by design into practice-novel concepts for green and sustainable pharmacy: designing green drug derivatives by non-targeted synthesis and screening for biodegradability. *Sustain Chem Pharm* 2015;**2**:31–6.
90. Lan J, Gou N, Gao C, He M, Gu AZ. Comparative and mechanistic genotoxicity assessment of nanomaterials via a quantitative toxicogenomics approach across multiple species. *Environ Sci Technol* 2014;**48**(21):12937–45.
91. Taylor NS, Merrifield R, Williams TD, Chipman JK, Lead JR, Viant MR. Molecular toxicity of cerium oxide nanoparticles to the freshwater alga *Chlamydomonas reinhardtii* is associated with supra-environmental exposure concentrations. *Nanotoxicology* 2016;**10**(1):32–41.
92. Hoecke KV, Quik JTK, Mankiewicz-Boczek J, Schamphelaere KACD, Elsaesser A, Meeren PV, et al. Fate and effects of CeO2 nanoparticles in aquatic ecotoxicity tests. *Environ Sci Technol* 2009;**43**(12):4537–46.
93. Arnold M, Badireddy A, Wiesner M, Di Giulio R, Meyer J. Cerium oxide nanoparticles are more toxic than equimolar bulk cerium oxide in *Caenorhabditis elegans*. *Arch Environ Contam Toxicol* 2013;**65**(2):224–33.
94. Gagne F, Auclair J, Turcotte P, Gagnon C. Sublethal effects of silver nanoparticles and dissolved silver in freshwater mussels. *J Toxicol Environ Health A* 2013;**76**(8):479–90.
95. Poynton HC, Lazorchak JM, Impellitteri CA, Blalock BJ, Rogers K, Allen HJ, et al. Toxicogenomic responses of nanotoxicity in *Daphnia magna* exposed to silver nitrate and coated silver nanoparticles. *Environ Sci Technol* 2012;**46**(11):6288–96.

96. Ivask A, Voelcker NH, Seabrook SA, Hor M, Kirby JK, Fenech M, et al. DNA melting and genotoxicity induced by silver nanoparticles and graphene. *Chem Res Toxicol* 2015;**28**(5):1023–35.
97. Hardman RA. Toxicologic review of quantum dots: toxicity depends on physicochemical and environmental factors. *Environ Health Persp* 2006;**114**(2):165–72.
98. Bodewein L, Schmelter F, Di Fiore S, Hollert H, Fischer R, Fenske M. Differences in toxicity of anionic and cationic PAMAM and PPI dendrimers in zebrafish embryos and cancer cell lines. *Toxicol Appl Pharmacol* 2016. http://dx.doi.org/10.1016/j.taap.2016.06.008.
99. Patel GM, Patel GC, Patel RB, Patel JK, Patel M. Nanorobot: a versatile tool in nanomedicine. *J Drug Target* 2006;**14**(2):63–7.
100. Bian SW, Mudunkotuwa IA, Rupasinghe T, Grassian VH. Aggregation and dissolution of 4 nm ZnO nanoparticles in aqueous environments: influence of pH, ionic strength, size, and adsorption of humic acid. *Langmuir* 2011;**27**(10):6059–68.
101. Ma R, Levard C, Michel FM, Brown Jr GE, Lowry GV. Sulfidation mechanism for zinc oxide nanoparticles and the effect of sulfidation on their solubility. *Environ Sci Technol* 2013;**47**:2527–34.
102. Rathnayake S, Unrine JM, Judy J, Miller A-F, Rao W, Bertsch PM. Multitechnique investigation of the pH dependence of phosphate induced transformations of ZnO nanoparticles. *Environ Sci Technol* 2014;**48**:4757–64.

CHAPTER

3.8

Pollution in the Coastal Zone: A Case Study of Wastewater on Cape Cod, MA

Amy Costa[1], *Laurel Schaider*[2], *Patricia Hughes*[1]

[1]**Center for Coastal Studies, Provincetown, MA, United States;** [2]**Silent Spring Institute, Newton, MA, United States**

Nature is a part of our humanity, and without some awareness and experience of that divine mystery man ceases to be man. ***Henry Beston, writing about Cape Cod in The Outermost House***

3.8.1 CAPE COD: AN IMPERILED NATURAL TREASURE

3.8.1.1 A Land Surrounded by Water

Cape Cod is a sandy peninsula in the Commonwealth of Massachusetts that extends out into the Atlantic Ocean. Cape Cod Bay forms its northern coast, Nantucket Sound its southern coast, and its east coast faces the Atlantic. The man-made Cape Cod Canal separates it from the mainland. The Cape is approximately 339 square miles in area (216,960 acres), of which 11,000 acres are freshwater ponds. It was formed approximately 15,000 years ago during the last glacial period, as the ice sheets retreated and left sand and rocky material behind. Wave and wind action moved and redeposited this material, establishing Cape Cod as a landform.[1]

The United States Geological Survey (USGS) describes Cape Cod as having "absorbent geology," meaning that the glacial deposits of sand and rock are porous and that the groundwater held in these deposits is recharged by precipitation and directly affected by land-based discharges.[2] The Cape's groundwater is its sole source of potable water. This sole source aquifer (see glossary) consists of six separate lenses, each having a distinct flow behavior.[3] About 450 million gallons per day (Mgal/d) of water flow through the Cape Cod aquifer, of which approximately 69% discharges at the coast,[4] inextricably linking the Cape's freshwater to its surrounding coastal waters.

Cape Cod is composed of a total of 101 watersheds (see glossary), 53 of which discharge directly into coastal embayments (Fig. 3.8.1). Coastal embayments—semienclosed coastal

Green Chemistry
http://dx.doi.org/10.1016/B978-0-12-809270-5.00013-3

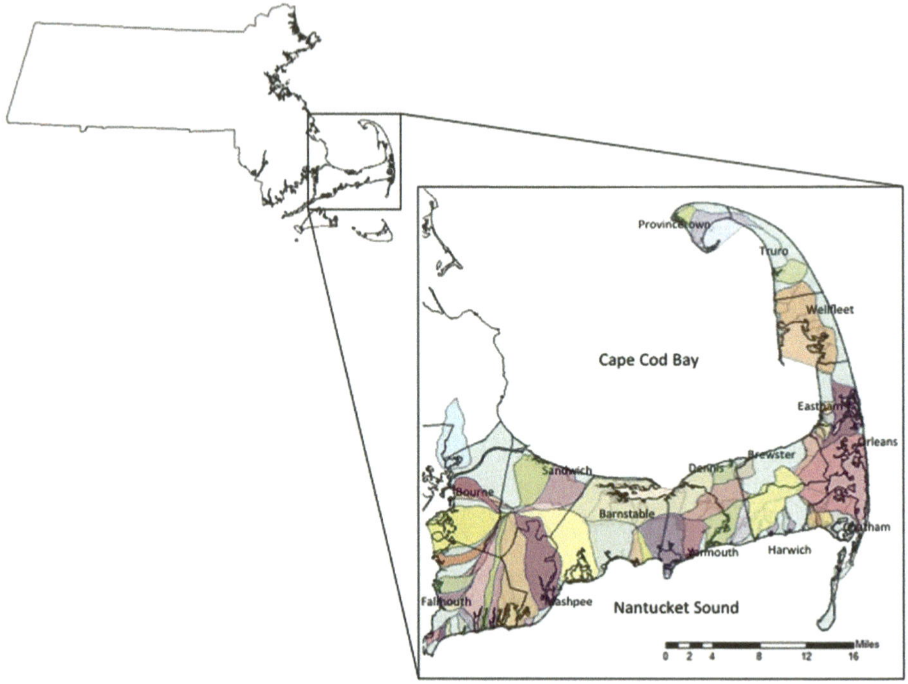

FIGURE 3.8.1 There are 101 delineated watersheds on Cape Cod. Of these, 53 discharge directly into coastal embayments. Watersheds often fall within more than one town, complicating management decisions. For example, the Pleasant Bay watershed falls within the towns of Orleans, Brewster, Chatham, and Harwich.

water bodies that ultimately discharge into open waters—provide valuable spawning and nursery habitats for fish and shellfish. They are also a significant recreational resource for residents and visitors. Each watershed, with its associated coastal embayment, extends from the top of the water table lens to the coastline, covering nearly 79% of the land area of the Cape.[5]

3.8.1.2 Changes in Population and Housing

Cape Cod has experienced enormous population growth and development pressures since 1970 (Fig. 3.8.2). The population more than doubled between 1970 (96,656 persons) and 2000 (222,230 persons).[6] Concomitant with this doubling of population was a more than twofold increase in housing units: 65,676 in 1970 compared with 160,281 in 2010.[7] Approximately one-third of the housing stock on Cape Cod is dedicated to seasonal use, which is much more prevalent in coastal areas than inland areas.[5]

The wastewater generated as a result of this explosive growth in population and development is treated almost exclusively by septic systems and cesspools, designed primarily to capture bacteria and viruses that are harmful to humans. Cape Cod's porous, sandy soils are highly absorbent, making the underground water supply vulnerable to contamination; toxic substances released into the ground can travel through the soil quickly and can move

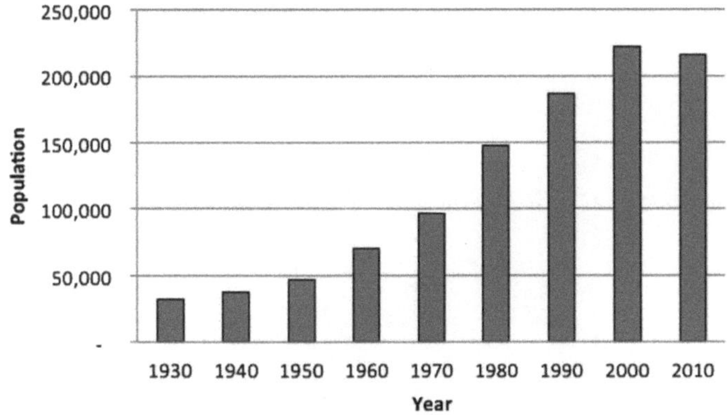

FIGURE 3.8.2 Cape Cod has experienced tremendous growth in population. Since the 1930s, the population has more than quadrupled. The highest rate of population growth occurred from the 1970s to the 1980s and continued through 2000 with only a slight dip in the population between 2000 and 2010. This explosion in the population has had ramifications on land use, development, and resource management on Cape Cod.

great distances. The travel time of groundwater to the Cape's coastal waters is less than 10 years for almost half of Cape Cod's land area.[5]

3.8.1.3 Groundwater and Coastal Pollution

From a distance, Cape Cod's coastal waters appear healthy. However, according to the Cape Cod Commission (CCC), "Cape Cod has a water problem. The saltwater border that has defined our peninsula is being poisoned by nitrogen."[5] High nutrient inputs impact coastal ecosystems by promoting excessive growth of algae and other aquatic plants (Fig. 3.8.3). This process, known as eutrophication, is occurring to some degree in Cape Cod Bay and Nantucket Sound and more noticeably in the coastal ponds and embayments that discharge into them. In coastal systems, inputs of nitrogen are primarily responsible for eutrophication.

About 80% of the nitrogen that enters Cape Cod's watersheds is from septic systems. Septic systems are onsite wastewater treatment systems that are designed mainly to remove pathogens and solids from wastewater and generally are not effective at removing nitrogen. Other sources of nitrogen pollution include fertilizers and other lawn/garden chemicals, wastes from pets, runoff from roadways, and atmospheric deposition. This nonpoint source (see glossary) pollution is carried into ponds, streams, and coastal waters by rain and snow melt.

Eutrophication has been linked to a number of harmful processes in coastal waters. Two symptoms of eutrophication that have been extensively documented are harmful algal blooms (HABs) and hypoxia. HABs are overgrowths of algae in the water. Although some are toxic (e.g., red tide, rust tide), nontoxic blooms are still classified as HABs if they hurt the environment or local economy. HABs are often a precursor to hypoxia, which occurs when large amounts of algae and other organic material decompose, depleting the water of oxygen and often leading to periods of low (hypoxic) or no (anoxic) oxygen. These conditions can cause physiological stress and even death of marine organisms.

FIGURE 3.8.3 Eutrophication is one of the most serious threats to coastal ecosystems. It occurs when excessive amounts of nitrogen enter our coastal waters, leading to prolific algal growth and deterioration of water quality. Eutrophication impacts the esthetics, ecosystem services, and habitat functions of our coastal waters.

3.8.2 REGULATORY FRAMEWORK FOR ADDRESSING COASTAL POLLUTION

3.8.2.1 Clean Water Act

Section 208 of the US Federal Clean Water Act of 1972 (33 U.S.C. §1251 et seq.) required states to identify areas that had substantial water quality problems, to delineate the areas' boundaries, and to designate a single representative organization "capable of developing effective area-wide waste treatment management plans for such an area." The Cape Cod Planning and Economic Development Commission (CCPEDC), the precursor to the current CCC, was so designated by the Commonwealth of Massachusetts and, in 1978, developed and gained approval for the first Cape Cod Area Wide Water Quality Management Plan, commonly referred to as the 208 Plan. In 1978, the 208 Plan identified increasing residential densities and a threefold summer population influx as the cause of isolated water quality and wastewater management problems. It anticipated that future growth threatened to cause more serious groundwater contamination and increased eutrophication in surface waters. At that time, the CCPEDC focused on better understanding and raising public awareness of Cape Cod's sole source aquifer and proper maintenance of on-site septic systems. In 1992, the CCC established a more restrictive limit on nitrogen concentrations in wastewater than state standards. This limit was based on research to characterize groundwater flow and the impacts of nitrogen in coastal waters conducted by the USGS, research scientists, and the United States Environmental Protection Agency (EPA)-funded National Estuaries Programs.

In the years and decades that followed the initial 208 Plan, water quality in Cape Cod's coastal waters steadily declined. Results of monitoring programs on the Cape have indicated an overall degradation of environmental conditions at many locations, primarily in the inshore areas. This gradual decline is thought to be due primarily to nitrogen pollution, and, tellingly, is occurring in the waters that are in the highest demand for human activities such as swimming, boating, fishing, and shellfishing. In estuarine systems, excessive inputs of nitrogen lead directly to thick mats of algae that replace eelgrass, diminish shellfisheries, and decrease dissolved oxygen concentrations—occasionally leading to fish and shellfish kills, odor, and frequent violations of water quality standards.[5]

Many of these degraded symptoms are apparent in the embayments and estuaries on Cape Cod. A 2011 areal study of eelgrass abundance in the coastal waters of Massachusetts indicated that the shallow water embayments are losing eelgrass habitat at a rate of approximately 3% per year. Some embayments have suffered a complete loss of eelgrass (Fig. 3.8.4). Of the embayments that were quantitatively assessed, there was a total loss of 20% of eelgrass acreage between 1994 and 2007.[8]

Algal blooms and low oxygen conditions have been documented in several estuaries over the years. In 2012, a coastal pond in Falmouth experienced a severe fish kill, likely due to septic system discharges.[9] In August 2016, North, West, and Cotuit Bays in Osterville and Cotuit experienced one of the worst algal blooms in twenty years and a minor fish kill, also thought to be the result of nutrient inputs from septic systems and runoff.[10]

Frustration at this lack of progress to limit nitrogen in wastewater and remediate coastal waters affected by excessive nitrogen prompted the Conservation Law Foundation (CLF) and the Buzzards Bay Coalition (BBC) to file suit against the US EPA in 2011 for not enforcing the implementation of the 208 Plan of 1978. As a consequence of this lawsuit, the CCC was directed and funded by the Massachusetts Department of Environmental Protection (MassDEP) to prepare an update to the 1978 Section 208 Water Quality Management Plan for Cape Cod to address the degradation of Cape Cod's coastal waters from excessive inputs of nitrogen (described in Section 3.8.4 Addressing Coastal Water Pollution on Cape Cod).

3.8.2.2 Massachusetts Estuaries Project

In December 2001 the Massachusetts Estuaries Project (MEP) was established as a collaboration between MassDEP and the University of Massachusetts Dartmouth with two primary goals: to study nitrogen loading in the coastal embayments and to develop regulatory limits on nitrogen [known as total maximum daily loads (TMDLs)] for 89 coastal embayments in Massachusetts.[11] The MEP developed a watershed embayment model that links watershed inputs with water circulation and nitrogen characteristics. MEP also established and conducted a standardized monitoring program in each of the embayments. By integrating the monitoring data with the model, the MEP developed watershed-specific recommendations for nitrogen loading. These recommendations provide the basis for the state-mandated TMDLs for nitrogen for many of the embayments that empty into the Nantucket Sound and Cape Cod Bay. The MEP monitoring program designated sentinel stations for each embayment. These MEP sentinel stations are important for monitoring the progress of actions taken to

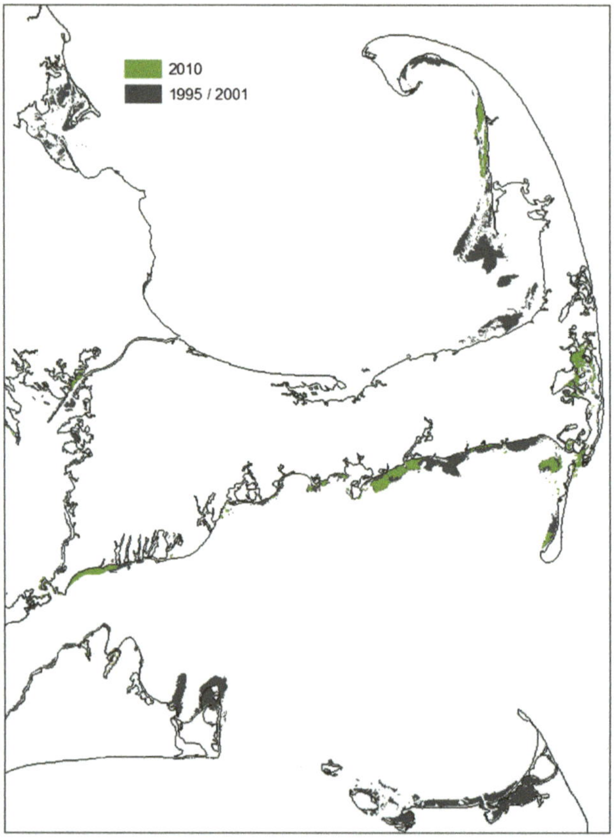

FIGURE 3.8.4 Eelgrass can be a key indicator of water quality. If water quality becomes impaired, eelgrass habitat will disappear. The Massachusetts Department of Environmental Protection (MassDEP) maps eelgrass habitat in the coastal waters of Massachusetts every 3–5 years. This figure shows the distribution of eelgrass beds in the coastal waters of Cape Cod.[38] Gray patches indicate where eelgrass once grew in 1995 and 2001. Green patches indicate the extent of eelgrass beds in 2010.

reduce nitrogen in an embayment's watershed and coastal waters. The MEP makes the assumption that by meeting the water quality standards at these stations, the water quality goals will be met throughout the entire system.

3.8.2.3 Coastal Water Quality Monitoring

To extend the work started by the MEP, the waters surrounding Cape Cod (Cape Cod Bay, Nantucket Sound, Vineyard Sound, and Buzzards Bay) and the many embayments, estuaries, and coastal ponds that empty into these waters are being monitored as a collaborative effort among the Commonwealth of Massachusetts, Barnstable County, towns on Cape Cod and the Islands, and environmental organizations. This monitoring effort is being led by the Center for Coastal Studies in Cape Cod Bay and Nantucket Sound and the BBC in Buzzards Bay and Vineyard Sound. Fig. 3.8.5 shows the locations of the stations currently being

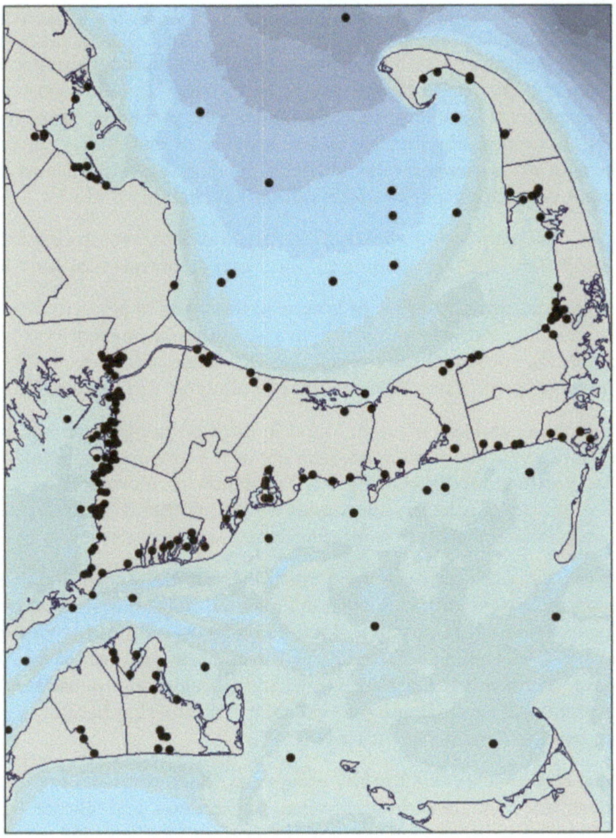

FIGURE 3.8.5 Locations of stations being monitored for water quality on Cape Cod. The monitoring of these stations is conducted through a collaboration among environmental organizations, towns on Cape Cod, Barnstable County, and the Commonwealth of Massachusetts. The majority of these stations have been monitored since the 1990s or early 2000s. These extensive data sets allow scientists to detect long-term trends in water quality and elucidate possible associations with resource management practices.

monitored for water quality. The information collected through this monitoring program is expanding our understanding of how human activities and management actions affect the surrounding water bodies. Many of the stations established as sentinel stations during the MEP are included in this monitoring effort. The key parameters monitored are surface and bottom temperatures, salinity, dissolved oxygen, chlorophyll, turbidity, and nutrients (mainly nitrogen and phosphorus). A description of these parameters is given in Table 3.8.1.

Nitrogen is often the limiting nutrient in coastal waters, so coastal systems are more sensitive to inputs of nitrogen than to inputs of other nutrients. Nitrogen can be found in the marine environment in a variety of inorganic compounds including nitrate (NO_3^-), nitrite (NO_2^-), nitrous oxide (N_2O), molecular nitrogen (N_2), and ammonia (NH_4^+). There are also many different types of organic compounds that contain nitrogen, such as amino acids and proteins. These types of compounds are released by living organisms and decaying organic matter.

TABLE 3.8.1 Key Water Quality Parameters Monitored by the Center for Coastal Studies

Water Temperature	An important property of the marine environment, influencing many physical properties (e.g., density), chemical properties (e.g., the capacity to hold dissolved oxygen, sensitivity to toxic wastes), and biological processes (metabolic processes, photosynthesis) as well as dictating the types, distribution, and abundance of marine flora and fauna. Measuring surface and bottom water temperatures indicates the degree of stratification in the water column, an important characteristic that affects nutrient availability and productivity.
Salinity	The total concentration of all dissolved salts in water. Monitoring fluctuations in salinity can give insight into the surrounding physical and environmental conditions.
Dissolved Oxygen (DO)	Concentrations are a measure of how well the water is aerated. This parameter is one of the most immediate indicators of a system's health. Because oxygen is needed to support animal and plant life, consequences of declining DO levels will set in quickly; under low oxygen conditions (hypoxia), nutrients and other pollutants can be released from sediments.
Chlorophyll a	A green photosynthetic pigment found in most phytoplankton and plant cells. Its concentrations can be used as an estimate of the amount of organic matter produced (i.e., primary production). By keeping track of trends in chlorophyll a, it is possible to assess the effects of nutrients entering the coastal waters and better understand the delicate balance between photosynthetic rates, nutrient inputs, and oxygen levels in coastal waters.
Turbidity	A measure of water clarity or how much the material suspended in the water column decreases light penetration. Suspended material may consist of both inorganic sediment and organic matter such as phytoplankton. High levels of turbidity can result from natural disturbances, as well as anthropogenic disturbances such as urban runoff, waste discharge, dredging, and boating. High turbidity can be ecologically detrimental by inhibiting photosynthesis by reducing light availability, smothering benthic organisms, and altering bottom material and sediment size.
Nutrients	Essential components of a healthy ecosystem. Macronutrients are those needed in larger quantities, and include carbon, nitrogen, phosphorus, and silicate. Micronutrients are those needed in smaller quantities, such as iron, copper, and zinc. Monitoring nutrient levels can indicate when excessive inputs of nutrients such as nitrogen and phosphorus can cause excessive growth of algae and other aquatic plants.

Most of the nitrogen that enters coastal waters on Cape Cod is from anthropogenic inputs and is in the form of dissolved inorganic nitrogen such as ammonia, nitrate, and nitrite. These inputs originate primarily from wastewater, runoff, and atmospheric deposition. Dissolved inorganic forms of nitrogen are biologically available for uptake by primary producers such as algae and phytoplankton. Excessive amounts of these forms of nitrogen entering coastal waters are the primary cause of most major problems affecting coastal ecosystems, such as eutrophication, algal blooms, and hypoxia.

3.8.3 NEW CHALLENGES: CONTAMINANTS OF EMERGING CONCERN

3.8.3.1 Sources and Concerns

In addition to nutrients, there are other potentially harmful contaminants associated with wastewater discharges broadly classified as contaminants of emerging concern (CECs).

CECs encompass a variety of natural and synthetic chemicals including pharmaceuticals, hormones, personal care products, household cleansers, industrial chemicals, and detergents.

Many Americans use over-the-counter and prescription drugs on a daily basis. According to the US Department of Health and Human Services, in 2009–12, almost half of all Americans took at least one prescription drug in the past 30 days, and one in five took three or more medications.[12] Excess pharmaceuticals that are not absorbed by the human body are excreted and end up in the wastewater stream. In addition, people may dispose of leftover pharmaceuticals by flushing them down the toilet or pouring them down the drain. Other CECs, such as personal care products, cleaners, detergents, and flame retardants, can end up in wastewater through washing and cleaning activities. Some of these chemicals are removed or degraded in septic systems and wastewater treatment plants, whereas many are discharged into the environment.

CECs are introduced into the environment through various pathways but are predominantly associated with wastewater. A 2002 study conducted by the USGS analyzed water from 139 streams in 30 states, including Massachusetts, specifically for the presence of pharmaceuticals, hormones, and other CECs. An astounding 80% of these 139 streams contained at least one of the contaminants included in the study.[13] Seven chemicals were found in at least half of the streams tested: *N,N*-diethyl-*meta*-toluamide (DEET, used in insect repellents), caffeine, triclosan (used in antibacterial soaps and toothpaste), tris(2-chloroethyl) phosphate (TCEP, a flame retardant used in foam), 4-nonylphenol (a breakdown product of detergents), and two chemicals found in human waste (cholesterol and coprostanol). Other commonly detected chemicals included over-the-counter medications, antibiotics, and other prescription drugs.

CECs have also been found in coastal waters. A 2010 study in an embayment in Puget Sound, Washington, detected pharmaceuticals, personal care products, and herbicides. These chemicals included DEET, caffeine, TCEP, ibuprofen (an over-the-counter pain medication), and several prescription medications.[14] In some locations, discharges from septic systems, which serve 70% of the surrounding community, were the likely source. Similarly, in Jamaica Bay, a wastewater-impacted estuary in New York, several prescription and over-the-counter medications, caffeine, and nicotine were detected.[15] As of 2016, a total of 113 pharmaceuticals and metabolites of pharmaceuticals have been detected in coastal waters.[16] The majority of those reported were antibiotics, followed by nonsteroidal antiinflammatory drugs and painkillers.

The presence of CECs in aquatic systems raises concerns about ecological health effects. In particular, endocrine disrupting compounds (EDCs)—chemicals that can mimic or alter the behavior of natural hormones—can affect reproduction and growth of fish and other aquatic organisms. A 2007 study in an experimental lake in Canada found that adding 5–6 parts per trillion of a synthetic hormone used in birth control, 17α-ethynylestradiol, led to a crash in the fathead minnow population.[17] Male fish had intersex (both male and female) characteristics and elevated levels of vitellogenin, a protein found in female fish associated with egg production. In the Potomac River, a 2007 study found over 80% of male smallmouth bass with intersex characteristics at some sampling locations.[18]

3.8.3.2 Contaminants of Emerging Concern and Cape Cod Water Quality

3.8.3.2.1 Drinking Water

The presence of CECs has been documented in public and private drinking water wells throughout Cape Cod.[19,20] The most frequently detected chemicals included sulfamethoxazole (an antibiotic), carbamazepine (an epilepsy medication also used as a mood stabilizer), perfluorooctane sulfonic acid (PFOS) and perfluorohexane sulfonic acid (highly fluorinated chemicals used in firefighting foams and nonstick and stain-resistant products), triethyl phosphate (a flame retardant), and acesulfame (an artificial sweetener) (Table 3.8.2). Wells in areas with more residential development—and thus more septic systems—in their zones of contribution had higher levels of CECs. Wells with higher levels of nitrate, even far below the federal drinking water standard for nitrate, also had higher levels of CECs, which is consistent with septic systems as their primary source.

TABLE 3.8.2 Examples of CECs Detected in Cape Cod Public and Private Drinking Water Wells, Ponds, and Estuaries.[19,20,24,25]

Chemical	Use
Acesulfame	Artificial sweetener
Acetaminophen	Pain medication
Caffeine	Stimulant
Carbamazepine	Epilepsy medication
Clofibric acid	Metabolite of lipid regulator
Cotinine	Metabolite of nicotine
DEET	Insect repellent
Dilantin	Epilepsy medication
Estrone	Endogenous hormone
Gemfibrozil	Lipid regulator
Meprobamate	Antianxiety medication
Nicotine	Stimulant
Paraxanthine	Metabolite of caffeine
PFOS	Perfluorosurfactant
Primidone	Epilepsy medication
Progesterone	Endogenous hormone
Sulfamethoxazole	Antibiotic
Triethyl phosphate	Flame retardant, plasticizer
Trimethoprim	Antibiotic

CECs, contaminants of emerging concern; *DEET*, *N,N*-diethyl-*meta*-toluamide; *PFOS*, perfluorooctane sulfonic acid.

The presence of CECs in drinking water, even at parts per trillion levels, raises potential health concerns, although we have little information about the health effects of long-term exposures to the pharmaceuticals and other CECs in drinking water. Most CECs are not currently regulated in drinking water, and their presence alone does not mean they are harmful. However, the US EPA is considering some CECs for future regulation and in 2016 released health advisory levels for two highly fluorinated chemicals, perfluorooctanoate and perfluorooctane sulfonate, in drinking water.[21] The typical amounts of pharmaceuticals ingested through drinking water are orders of magnitude lower than those found in a therapeutic dose prescribed for health conditions. Nevertheless, pharmaceuticals are designed to be biologically active, and certain individuals may be especially sensitive, as in the case of antibiotics. We do not yet understand how exposures to low levels of many different compounds may interact with one another in the human body, especially during sensitive stages of development before birth and in children. Certain EDCs can show effects at lower doses that are not apparent at higher doses.[22] Further research is needed to evaluate potential long-term health effects and prioritize chemicals for future drinking water standards.

3.8.3.2.2 *Ponds*

Cape Cod has 365 freshwater kettle ponds, depressions formed by ice blocks embedded in glacial deposits around 15,000 years ago that eventually melted and joined with groundwater. These ponds mostly lack surface inputs like streams, and instead are mainly fed by groundwater. Because of this close connection, they have been called a "window" on the aquifer.[23] Many ponds are exhibiting signs of eutrophication due to high levels of nutrient inputs—mainly phosphorus—from septic systems and other sources.

CECs from septic systems and other sources can move through groundwater and end up in ponds on Cape Cod (Fig. 3.8.6). A 2008 study found pharmaceuticals and hormones in ponds on Cape Cod.[24] These included sulfamethoxazole, carbamazepine, and several endogenous hormones (androstenedione, estrone, and progesterone) (Table 3.8.2). Ponds impacted by more densely developed areas had more frequent detections and higher levels of these CECs. The highest detected levels of hormones approached levels previously shown to cause endocrine disruption in fish according to laboratory studies. Further study is needed to assess whether endocrine disruption is occurring in freshwater fish on Cape Cod.

3.8.3.2.3 *Coastal Waters*

Based on prior work showing the presence of CECs in groundwater and ponds impacted by septic systems on Cape Cod, additional monitoring was implemented to address the question of whether these same CECs were ending up in coastal waters impacted by septic systems. Beginning in 2010, water samples from Cape Cod Bay, Nantucket Sound, and many of the harbors, estuaries, and coastal ponds that empty into the Bay and the Sound have been analyzed for CECs. The majority of the locations tested had detectable levels of at least one of the CECs that were tested for. This work was the first to document the presence of these types of contaminants in the coastal waters of the Cape and Islands. CEC levels were typically highest within embayments where tidal flushing was limited and where the surrounding area was more densely developed. Surprisingly, some CECs were detected several miles offshore in the open waters of Nantucket Sound and Cape Cod Bay. Of the 73 contaminants tested for in the coastal waters of Cape Cod, 21 were detected. Of the 21 detectable contaminants, 7 were

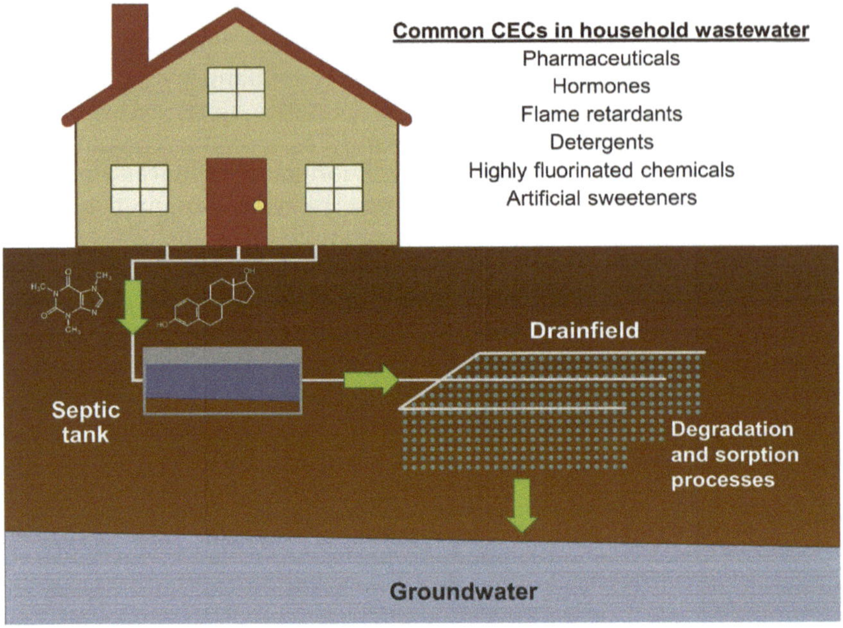

FIGURE 3.8.6 Schematic of a conventional septic system, consisting of a septic tank and a drainfield. Contaminants of emerging concern (CECs) from household products that are not fully removed in the septic tank or drainfield can percolate through the underlying soil and enter the groundwater. Almost 85% of Cape residences rely on septic systems and other onsite wastewater treatment systems.

hormones and 14 were pharmaceuticals or personal care products (Table 3.8.2). Several of the CECs detected in this study were the same as those found in the ponds and drinking water studies. Of the hormones, estrone was found in highest concentrations, but progesterone was detected at the most number of locations. Of the remaining CECs, DEET was found in highest concentrations, but cotinine (a metabolite of nicotine) was found at the most locations.[25] In 2014, this research was expanded to examine the ecological impacts by investigating the bioaccumulation of some CECs in the shellfish that live in these waters.[26] Oysters harvested from seven different embayments located on Cape Cod and the Islands were tested for 64 CECs. A total of six CECs were detected in the oyster tissue. Although the concentrations of the CECs found in the oyster tissue were only on the order of parts per trillion, the potential for some of these contaminants to increase in concentration up the food chain is of concern.

Spatial and temporal analysis of the concentrations of CECs in coastal waters indicates that the concentrations of these contaminants are linked to the degree of human use. For example, areas that were more developed generally had higher concentrations of CECs, and samples collected during the summer, when human activity on Cape Cod peaks, had higher concentrations than those collected during the spring or fall.

The strong correlation observed in these studies between concentrations of nitrate and CECs supports the conclusion that CECs are primarily associated with wastewater, and, correspondingly, that wastewater is a key source of nitrogen to the coastal waters of Cape Cod. Therefore, addressing the problems associated with wastewater, as is currently being done Capewide, could also lessen some of the potential impacts of CECs. Some of the nontraditional methods for wastewater treatment and their potential to remove CECs are discussed in Section 3.8.4.1.

3.8.4 ADDRESSING COASTAL WATER POLLUTION ON CAPE COD

As noted earlier, the CCC was required to update its 208 Plan, which was completed and approved in 2015. This new 208 Plan takes a watershed-based approach to restore the water quality of the embayments, estuaries, and coastal waters of Cape Cod. It recommends a mixture of wastewater management techniques including traditional (sewers connected to centralized treatment facilities) and alternative technologies [e.g., constructed wetlands, permeable reactive barriers (PRBs), shellfish growing and harvesting]. As individual towns on the Cape move forward with their wastewater management plans, many are incorporating these suggested approaches. Because the majority of these strategies are not yet approved by the MassDEP, the agency responsible for setting and enforcing TMDLs, rigorous testing and monitoring are required to evaluate the effectiveness of the various strategies. Thus, many towns have adopted an adaptive management approach in their wastewater management plans, allowing them to adjust the plan to take into consideration new information, new technologies, and progress over time.

3.8.4.1 Nontraditional Wastewater Management Strategies

The updated 208 Water Quality Management Plan for Cape Cod includes an analysis of a range of nontraditional methods of wastewater management designed either to remediate the impacts of nitrogen by treatment in the groundwater, restore an area impacted by wastewater by treatment within the affected water body, or reduce the amount of nitrogen entering groundwater at the source. The majority of strategies will impact only the watershed or water body within which they are located, so site selection is a crucial part of the planning and implementation process. A description of all the potential methods is beyond the scope of this chapter. Some of the primary methods that have already been implemented or being considered for use on Cape Cod are described.

PRBs are designed to remediate plumes of contaminants such as nitrogen. PRBs are installed directly in the ground to intercept nitrogen-enriched groundwater around the edge of an impacted water body by supplying a carbon source to the microbes that live in the groundwater. The carbon source provides the energy needed for the microbes to complete the denitrification cycle and convert bioavailable nitrogen to inert nitrogen gas, thereby effectively removing it from the groundwater.

Five towns on Cape Cod were selected during the spring of 2016 as part of EPA's Southeast New England Program (SNEP) for coastal watershed restoration to test the efficacy of PRBs. With funding from SNEP, EPA is collaborating with the USGS and the CCC to investigate the hydrogeological characteristics of these five sites to determine which is the most suitable site for installation of a PRB. The results of this work will be used to determine other locations on Cape Cod where this technology will be effective in reducing nitrogen impacts on coastal waters.

Inlet modification as a strategy to improve water quality of estuaries and semienclosed bays is based on the assumption that inlets limit tidal exchange between these systems and the open oceanic waters. Increased tidal flushing should therefore reduce nitrogen within these systems by flushing out nitrogen-rich water and replacing it with nitrogen-poor offshore water.

There are many examples of inlet modifications on Cape Cod. The Association to Preserve Cape Cod has been involved in a number of these projects through their Salt Marsh Monitoring Program, which monitors the impacts of culvert widening/modification on tidally restricted salt marshes. However, most of these studies were conducted with the goal of habitat restoration rather than nitrogen reduction, so data addressing the impact of inlet modification on nitrogen levels are lacking. The majority of these projects have shown a significant improvement in coastal habitat quality, so it is expected that nitrogen levels would likely be reduced. Several sites across the Cape are being considered for future pilot projects to further test this strategy of wastewater mitigation.

Shellfish aquaculture and habitat restoration as wastewater management strategies are gaining a great deal of interest on the Cape. These strategies are based on the premise that shellfish are filter feeders that remove phytoplankton and other organic matter from the water column. Because phytoplankton take up nitrogen, and organic matter contains nitrogen, the amount of nitrogen in the water body will be reduced when shellfish consume this material and incorporate the nitrogen into their tissue and shell as they grow. Harvesting these shellfish will then completely remove the incorporated nitrogen from the system. Additionally, studies have shown that denitrification rates in sediments around shellfish are typically higher.[27] Microbial processes that break down the waste products from shellfish convert dissolved forms of nitrogen to inert nitrogen gas, effectively removing it from the system.

Two different methods of using shellfish as a wastewater management strategy have been investigated on Cape Cod: aquaculture and habitat restoration. In 2013, the Town of Falmouth, in collaboration with the University of Massachusetts Dartmouth's School for Marine Science and Technology, started a 3-year aquaculture demonstration project to measure the impact of growing oyster seed in floating bags on water quality in Little Pond (Fig. 3.8.7). Several million oyster seed (<10 mm) were grown to a size of about 5 cm in Little Pond, a highly eutrophic coastal pond, and then transferred to a clean, grow-out site in another water body from which they can then be harvested for consumption once they reach marketable size. This demonstration project ended in 2015. The results of this project indicated that oysters did produce modest improvements in water quality. By removing phytoplankton via filtering, the oysters lowered the levels of chlorophyll and particulate carbon and nitrogen, which resulted in improved water clarity.[28] However, growing the number of oysters that would be required to meet the state-mandated TMDLs for nitrogen in this coastal pond was not feasible. Therefore, although the town continues to use this pond for shellfish propagation, they are also moving forward with plans to sewer the surrounding area.

The Town of Mashpee, working with the Mashpee Wampanoag Indian Tribal Council, has been relying on shellfish aquaculture (oyster and quahog) to improve the water quality of Mashpee waterways for over a decade. In 2015, their Comprehensive Wastewater Management Plan was the first plan to be approved by the Massachusetts Executive Office of Energy and Environmental Affairs that incorporated aquaculture as a

FIGURE 3.8.7 Shellfish aquaculture is one approach being implemented on Cape Cod as a nutrient reduction strategy in coastal waters. Using floating aquaculture bags, several million oyster seed are grown from a size of <1 cm to about 5 cm during the summer months in Little Pond. The filter feeding of the oysters helps remove organic material, improving water clarity and reducing nitrogen. In the fall, these oysters are transferred to a grow-out site where they can be harvested for consumption once they reach the legal size.

means to improve the health of its estuaries, and thus has set a precedent, which other Cape Cod towns are hoping to follow.

The second method, shellfish habitat restoration, is being implemented in the Town of Wellfleet. Beginning in 2012, the Town of Wellfleet, in collaboration with the University of Massachusetts Boston, and the Center for Coastal Studies, established a 2-acre propagation area for oysters. The goals of this project were to improve water quality and to expand the oyster industry. Approximately 5 million oysters were grown in the propagation area, and their impact on water quality and species diversity has been monitored since 2012. Initial results of this project indicated significant improvements in water quality within the oyster propagation area. Unfortunately, several winter storms during 2014 resulted in the severe accretion of sediment within the oyster bed, leading to high mortality and low spat recruitment. Ongoing monitoring will reveal if the oyster bed is able to rebound from this environmental setback.

The key difference between these two methods of wastewater management is that aquaculture involves directly removing nitrogen from the system by harvesting the shellfish, whereas habitat restoration involves leaving the shellfish in place to filter the water, provide habitat, increase biodiversity, and serve as a spawning sanctuary, among other benefits. Removal of nitrogen by habitat restoration, although potentially more significant than aquaculture, is much harder to quantify.

Innovative/Alternative (I/A) septic systems are an advanced type of onsite wastewater treatment system in which removal of solids, nitrogen, and other pollutants is better than in conventional septic systems. Many of these systems enhance nitrogen removal by shifting oxygen conditions to first promote microorganisms that carry out nitrification (conversion of nitrogen species to nitrate) and then denitrification (conversion of nitrate to N_2 gas, which is inert). Some I/A systems distribute effluent in shallow soil zones that are more biologically active and can promote nitrogen removal via uptake by plants.

There are approximately 50 approved I/A technologies in Massachusetts and more than 1500 I/A systems installed on Cape Cod to reduce the amount of nitrogen and other pollutants discharged into the groundwater. An example of one type of I/A system is a Residential RUCK system (Fig. 3.8.8). This system uses two septic tanks connected by a filter of alternating layers of sand and stone that promotes nitrification of effluent from the first tank. Effluent from the filter then enters a second, anaerobic septic tank that has a carbon source to promote denitrification of the effluent before it is discharged to the soil absorption system.[29]

The Massachusetts Alternative Septic System Test Center (MASSTC) on Cape Cod, operated by Barnstable County Department of Health and Environment, has conducted third-party testing and scientific research on different types of advanced onsite wastewater treatment systems since 1999. In addition to investigating nutrient reduction technologies, studies at MASSTC are investigating the removal of pharmaceuticals and other CECs in conventional and I/A septic systems.[30] The addition of the GPC Filter Process to a RUCK system has been shown to reduce some CECs.[31]

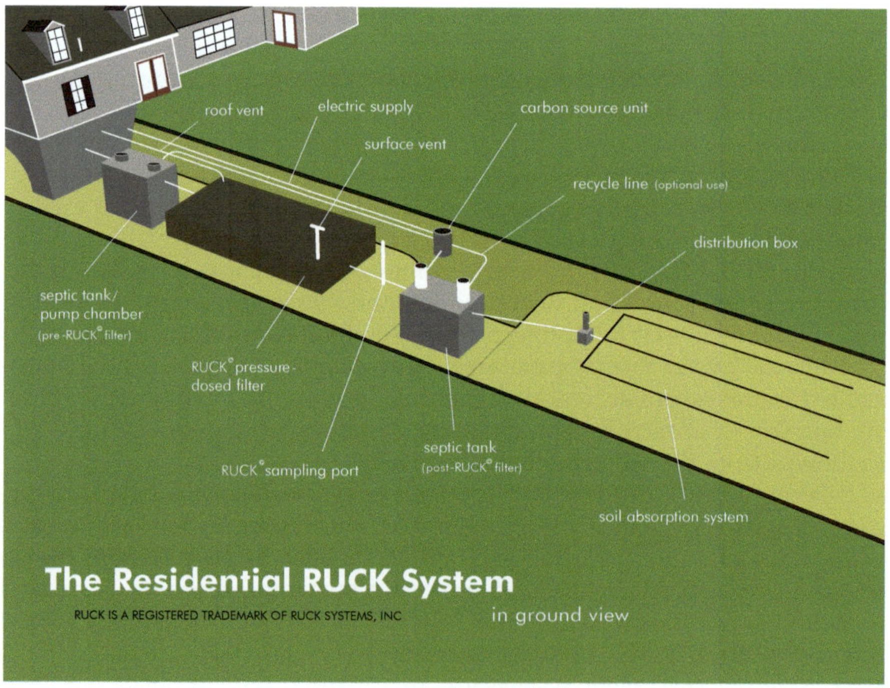

FIGURE 3.8.8 Schematic of a residential RUCK system. This system was granted approval by the Massachusetts Department of Environmental Protection for a General Use Certificate, which allows local boards of health to permit the construction and operation of this system in nitrogen-sensitive areas in Massachusetts. Unlike conventional septic systems, the RUCK system includes an aerobic filter that promotes microbial conversion of various nitrogen species to nitrate (a process called nitrification). Effluent from the filter then enters an anaerobic septic tank that promotes microbial conversion of nitrate to inert nitrogen gas (a process called denitrification). Since 1995, this system has been revised twice, and each version has shown improved removal of total nitrogen from wastewater.[29]

Ecotoilets include a number of different strategies to divert wastewater from the conventional wastewater stream to other methods of storage and handling. Although ecotoilets have been implemented in many institutional settings, their use as a nutrient reduction strategy for homes in impacted watersheds has not been investigated. Two of the more common systems are urine-diverting toilets and composting toilets. In a urine-diverting toilet, the toilet bowl has two chambers, one for urine and the other for solid waste. The urine is sent to a tank and pumped out periodically, and the solid waste goes to the existing septic system. In a composting toilet system, liquid and solid waste goes straight from the toilet into a composting tank, usually located in a home's basement. In addition to reducing or avoiding nutrient inputs into groundwater, ecotoilets save water and offer the potential for beneficial reuse of nitrogen and phosphorus contained in the waste. As an incentive to implement this type of alternative technology, the Town of Falmouth started a pilot program that offered a $5000 subsidy to homeowners who installed ecotoilets and participated in a study to monitor their performance. Additionally, in 2014, an ecotoilet betterment exemption program was approved by the Town of Falmouth that allowed homeowners in one highly impacted watershed to avoid sewer hookup fees by installing an ecotoilet system in their home.

3.8.4.2 Implications for Contaminants of Emerging Concern

The implications of these and other alternative wastewater management strategies on CECs in the waters of Cape Cod have mainly not been explored. The extent to which these strategies may affect CEC inputs into Cape Cod groundwater depends on the type of treatment being considered and the sources and chemical characteristics of each individual CEC. For instance, ecotoilets may prevent pharmaceuticals and hormones, which mainly enter wastewater via excretion, from entering groundwater, whereas CECs in detergents and household cleaners would not be diverted. By contrast, implementing PRBs will not prevent CECs from entering groundwater, but may alter the extent of CEC inputs into the water body by enhancing microbial activity. Strategies that focus on the water body itself will not alter groundwater concentrations, but may reduce CEC concentrations in the water column. Enhanced tidal exchange through inlet widening may increase transport of CECs into coastal waters. Shellfish aquaculture may lower the concentrations in water of certain CECs that are effectively removed by filter feeding and/or metabolized by the shellfish; however, some CECs may bioaccumulate within the food webs of these water bodies.

3.8.4.3 Strategies to Address Other Sources of Nitrogen Pollution

Beyond wastewater management, other strategies to reduce nutrient inputs, such as fertilizer management and stormwater remediation, are being implemented Capewide. Although wastewater is considered the main source of nitrogen in Cape Cod coastal waters, fertilizers and stormwater runoff are also substantial contributors. The contribution from these sources varies by watershed, for example, 23% of the total N loading to Oyster Pond, in Falmouth,[32] and 9% in Pleasant Bay, a watershed shared among Orleans, Brewster, Harwich, and Chatham.

Because fertilizer and stormwater remediation strategies are not directly linked to a targeted watershed or water body like those previously discussed, they can be implemented anywhere fertilizers are used (residential lawns and gardens, golf courses, municipal and commercial properties, ball fields) or for stormwater, anywhere there are impervious surfaces. Therefore, the impacts of these fertilizer management and stormwater remediation efforts are anticipated to be widespread.

Fertilizer management strategies are being implemented Capewide. In September 2013, the Barnstable County Commissioners adopted a Capewide Fertilizer Management District of Critical Planning Concern, noting that coastal and pond watersheds do not follow town boundaries and that nitrogen in these watersheds has increased over time as Cape Cod's population and development has increased. Since its passage, nine of the 15 Cape Cod towns adopted town-wide fertilizer bylaws.[33] In addition to these regulations, other best management practices can be followed such as replacing turf and other plants that require higher amounts of fertilizer with native vegetation with lower nutrient requirements or using slow-release fertilizers.

Stormwater remediation strategies can range from simple approaches such as street sweeping and proper maintenance of utilities to more involved scenarios of land use planning or construction to prevent stormwater runoff. Examples of stormwater remediation described in the *Massachusetts Stormwater Handbook* include the construction of deep sump catch basins, oil/grit separators, proprietary separators, sediment forebays, and vegetated filter strips.[34]

Another approach to stormwater management makes use of natural systems such as plants, soils, landscape design and engineered techniques to capture, cleanse, and reduce stormwater runoff. This "green infrastructure" uses practices that are designed to maintain natural hydrologic functions by absorbing and infiltrating precipitation where it falls.[35] One example of the use of green infrastructure to remediate stormwater is the use of permeable pavement. Permeable (or porous) pavement allows water to drain through it. Typically this is put over a reservoir bed or aggregate storage bed, which aids in filtration. This type of construction will reduce stormwater runoff volume, rate, and pollutant load. The Town of Provincetown has begun the process of repaving compromised streets identified in their stormwater assessment with permeable asphalt. This project is scheduled to be completed in May 2017. So far, early indications are that the installation of permeable asphalt is having beneficial impacts on the water quality.[36]

3.8.4.4 How Else Can Green Chemistry be Part of the Solution?

To address concerns about CEC pollution, wastewater management plans can take steps to reduce inputs of wastewater into drinking water and aquatic systems. For instance, minimizing discharges from septic systems and wastewater treatment plants in areas that recharge drinking water wells and in areas close to surface waters, as well as adapting advanced onsite and centralized wastewater treatment systems that enhance removal of certain CECs, can reduce effects of CEC inputs on water quality.

In the long run, preventing harmful, bioaccumulative, and toxic chemicals from contaminating surface waters and drinking water supplies will require a concerted effort to avoid using chemicals with these undesirable properties in the first place. Consumers can take steps to avoid purchasing household items that contain harmful chemicals and dumping hazardous products down the drain or into the ground. In Sweden, the Stockholm County Council publishes a guide that rates each pharmaceutical on the basis of its persistence, bioaccumulation potential, and toxicity to allow physicians to select more environment-friendly medications when multiple options are available.[37] Ultimately, these and many other green chemistry solutions offer the greatest promise to reduce the amount of harmful chemicals that are released into the environment.

Glossary of Key Terms

Watershed "Watersheds are drainage basins. They define flow paths for more than water; they are paths for nutrient cycling, vegetation changes, and geomorphic processes. Watersheds define the limits of aquatic habitats, and of the fish, reptiles, insects and other invertebrates endemic to those habitats. Watersheds collect and transmit rainfall with entrained atmospheric chemicals. And, when altered by human action, watersheds change their'-natural' flow paths accordingly." (excerpt from A Watershed Assessment Primer, US EPA)
Euphrat FD, Warkentin BP. A Watershed Assessment Primer. Region 10, U.S. Environmental Protection Agency (EPA) Seattle WA; 1994. EPA 910/B-94/005. 198 p. https://nepis.epa.gov/.

Aquifer "The name for a rock or soil which contains and transmits water and thus is a source for underground water is *aquifer.Aqua* means water in Latin and *fer*comes from the word meaning to yield. An aquifer is an underground zone or layer which is a relatively good source of water." (Excerpt from A Primer on Water, USGS)
Leopold LB, Langbein WB. 1960. A primer on water. U.S. Government Printing Office, Washington; 1960. 50 p. Available from: https://water.usgs.gov/edu/files/A-Primer-on-Water-USGS.pdf.

Point source pollution "The term 'point source' means any discernible, confined and discrete conveyance, including but not limited to any pipe, ditch, channel, tunnel, conduit, well, discrete fissure, container, rolling stock, concentrated animal feeding operation, or vessel or other floating craft, from which pollutants are or may be discharged. This term does not include agricultural storm water discharges and return flows from irrigated agriculture." (Section 502(14) of the Clean Water Act)
U.S. Environmental Protection Agency (EPA) Clean Water Act, Section 502 General Definitions. Available from: https://www.epa.gov/cwa-404/clean-water-act-section-502-general-definitions.

Nonpoint source pollution "Nonpoint source pollution generally results from land runoff, precipitation, atmospheric deposition, drainage, seepage or hydrologic modification. Nonpoint source (NPS) pollution, unlike pollution from industrial and sewage treatment plants, comes from many diffuse sources. NPS pollution is caused by rainfall or snowmelt moving over and through the ground. As the runoff moves, it picks up and carries away natural and human-made pollutants, finally depositing them into lakes, rivers, wetlands, coastal waters and ground waters." (Polluted Run-off: Nonpoint Source Pollution, EPA)
U.S. Environmental Protection agency (EPA) Polluted Runoff: Nonpoint Source Pollution. Available from: https://www.epa.gov/nps.

References

1. Oldale RN. *Cape Cod, Martha's Vineyard and Nantucket: The geologic story*. Yarmouth Port, MA: On Cape Publications; 2001. 224 p.
2. U.S. Geological Survey. *Cape Cod's unique, "absorbent" geology*. Science Information and Library Services; 2015. Available from: http://online.wr.usgs.gov/outreach/landpeople/students/cc_ccarea.html.
3. Masterson JP, Walter DA. *Hydrogeology and groundwater resources of the coastal aquifers of Southeastern Massachusetts*. Reston, VA: U.S. Geological Survey; 2009. Circular 1338. 16 p.
4. Walter DA, Masterson JP, Hess KM. *Ground-water recharge areas and traveltimes to pumped wells, ponds, streams, and coastal water bodies, Cape Cod, Massachusetts*. Reston, VA: Survey USG; 2004. Scientific Investigations Map I-2857.
5. Cape Cod Commission. *208 Plan-Cape Cod area wide water quality management plan update*. Barnstable, MA,. 2015. 254 p.
6. Massachusetts Ocean Management Task Force. *Waves of changes, appendix, trends in the demographics of human population and the Massachusetts marine economy*. Boston, MA,. 2004.

7. U.S. Census Bureau. *2010 Census of population and housing units*. CPH-2-23. Washington, DC: Massachusetts; 2012. 87 p.
8. Costello CT, Kenworthy WJ. Twelve-year mapping and change analysis of eelgrass (*Zostera marina*) areal abundance in Massachusetts (USA) identifies statewide declines. *Estuar Coast* 2011;**34**(2):232–42.
9. Teehan S. *Falmouth official: Fish kills a 'wake-up call'*. Cape Cod Times; July 25, 2012.
10. Spillane G. *Algae bloom outbreak in Cotuit*. Osterville: Cape Cod Times; August 16, 2016.
11. Buzzards Bay National Estuary Program. Massachusetts Estuaries Project information. Available from: http://buzzardsbay.org/mass-estuaries-project.htm.
12. National Center for Health Statistics. *Health, United States, 2015, with special feature on racial and ethnic health disparities*. Washington, DC,. 2015.
13. Kolpin DW, Furlong ET, Meyer MT, Thurman EM, Zaugg SD, Barber LB, et al. Pharmaceuticals, hormones, and other organic wastewater contaminants in U.S. streams, 1999–2000: A national reconnaissance. *Environ Sci Technol* 2002;**36**(6):1202–11.
14. Dougherty JA, Swarzenski PW, Dinicola RS, Reinhard M. Occurrence of herbicides and pharmaceutical and personal care products in surface water and groundwater around Liberty Bay, Puget Sound, Washington. *J Environ Qual* July–August 2010;**39**(4):1173–80.
15. Benotti MJ, Brownawell B. Distributions of pharmaceuticals in an urban estuary during both dry- and wet-weather conditions. *Environ Sci Technol* 2007;**41**:5795–802.
16. Gaw S, Thomas KV, Hutchinson TH. Sources, impacts and trends of pharmaceuticals in the marine and coastal environment. *Philos Trans Roy Soc B* November 19, 2014;**369**(1656).
17. Kidd KA, Blanchfield PJ, Mills KH, Palace VP, Evans RE, Lazorchak JM, et al. Collapse of a fish population after exposure to a synthetic estrogen. *Proc Natl Acad Sci USA* May 2007;**104**(21):8897–901.
18. Blazer VS, Iwanowicz LR, Iwanowicz DD, Smith DR, Young JA, Hedrick JD, et al. Intersex (testicular oocytes) in smallmouth bass from the Potomac River and selected nearby drainages. *J Aquat Anim Health* 2007;**19**:242–53.
19. Schaider LA, Rudel RA, Ackerman JM, Dunagan SC, Brody JG. Pharmaceuticals, perfluorosurfactants, and other organic wastewater compounds in public drinking water wells in a shallow sand and gravel aquifer. *Sci Tot Environ* September 17, 2014;**468**:384–93.
20. Schaider LA, Ackerman JM, Rudel RA. Septic systems as sources of organic wastewater compounds in domestic drinking water wells in a shallow sand and gravel aquifer. *Sci Tot Environ* 2016;**547**:470–81.
21. U.S. EPA (U.S. Environmental Protection Agency). *PFOA & PFOS drinking water health advisories*. Washington, DC,. 2016. EPA 800-F-16–003.
22. Vandenberg LN, Colborn T, Hayes TB, Heindel JJ, Jacobs Jr DR, Lee DH, et al. Hormones and endocrine-disrupting chemicals: Low-dose effects and nonmonotonic dose responses. *Endocr Rev* 2012;**33**(3):378–455.
23. Association to Preserve Cape Cod, Cape Cod Commission. *Cape Cod: a community connected by water*. 2000.
24. Standley LJ, Rudel RA, Swartz CH, Attfield KR, Christian J, Erickson M, et al. Wastewater-contaminated groundwater as a source of endogenous hormones and pharmaceuticals to surface water ecosystems. *Environ Toxicol Chem* December 2008;**27**(12):2457–68.
25. Costa A. *Testing Nantucket Sound and its embayments for contaminants of emerging concern, final report to Massachusetts Environmental Trust*. Provincetown, MA,. Center for Coastal Studies; 2014.
26. Costa A, Costa M, Benotti M. Evaluating the potential of Atlantic oysters (*Crassostrea virginica*) to bioaccumulative contaminants of emerging concern in coastal embayments. Charleston, SC. In: *16th International Conference on shellfish restoration*; 2014.
27. Humphries AT, Ayvazian SG, Carey JC, Hancock BT, Grabbert S, Cobb D, et al. Directly measured denitrification reveals oyster aquaculture and restored oyster reefs remove nitrogen at comparable high rates. *Front Mar Sci* 2016;**3**:74.
28. Howes B, Schlezinger D, Samimy R, Benson J, Sampieri S, Bartlett M, et al. *Shellfish aquaculture demonstration project, Little Pond year 2 monitoring, summer/fall 2014 oyster deployment*. New Bedford, MA: Coastal Systems Program, School of Marine Science and Technology, University of Massachusetts-Dartmouth; 2015. 169 pp.
29. McGrath MB, Santos TM, Kubrick JL. *A report on the operation of Residential Ruck^{TM} systems in Massachusetts*. 2009.
30. Barnstable County Department of Health and Environment. *Effectiveness of selected onsite wastewater-treatment systems in removing pharmaceutical and personal-care products*. Barnstable, MA,. 2010. Available from: http://www.barnstablecountyhealth.org/wp-content/uploads/downloads/2011/07/GrantFinalPharms.pdf.

31. McGrath MB, Hughes BT. *Ground penetrating carbon performance summary*. Falmouth, MA,. 2016. Available from: http://www.groundpenetratingcarbon.com/wp-content/uploads/GPC_Performance_Summary-March-11-2016_5-PPCP_PRINTOUT-mbm.pdf.
32. Massachusetts Department of Environmental Protection. *Oyster Pond embayment system: Total maximum daily loads for total nitrogen*. Boston, MA. 2008. 96-TMDL-7 Control #245.
33. Cape Cod Commission. *Cape Cod districts of critical planning concern*. 2013. Available from: http://www.capecodcommission.org/index.php?id=139.
34. Massachusetts Department of Environmental Protection. *Structural BMP specifications* [Chapter 2]. *Massachusetts stormwater Handbook*, vol. 2. Boston, MA: Executive Office of Energy and Environmental Affairs; 2008. 132 p.
35. U.S. EPA (U.S. Environmental Protection Agency). Different shades of green. Washington, DC: Office of Research & Development, Safe and Sustainable Water Resources Research Program; Available from: https://www.epa.gov/sites/production/files/2016-10/documents/green_infrastructure_brochure_final.pdf.
36. Roseen R, Waldo R, Janney J, Tripp S. Provincetown porous asphalt keeps beaches open. *Asph Pavement* 2014;**19**(5):19–24.
37. Stockholm County Council. *Environmentally classified pharmaceuticals*. Stockholm, Sweden. 2014. Available from: https://noharm-europe.org/documents/environmentally-classified-pharmaceuticals-2014-2015.
38. MassGIS Datalayers. MassDEP eelgrass mapping project. Available from: http://www.mass.gov/anf/research-and-tech/it-serv-and-support/application-serv/office-of-geographic-information-massgis/datalayers/layerlist.html.

CHAPTER

3.9

The Composition of Soils and Sediments

Stefan D. Kalev[1], *Gurpal S. Toor*[2]

[1]Gulf Coast Research and Education Center, University of Florida, Wimauma, FL, United States; [2]University of Maryland, College Park, MD, United States

3.9.1 INTRODUCTION

Soil is a precious natural resource that affects our ecosystems in numerous ways and therefore is an integral component of life on Earth. Soil is responsible for nutrient cycling, plant growth, gas exchange, carbon storage, and waste disposal. Soil consists of a mixture of weathered or eroded rocks, nutrients, living organisms, organic matter, water, and air. A soil in optimum condition contains approximately 45% solid material, 5% organic matter, and 50% pore spaces, which are occupied by an equal proportion of water and air (Fig. 3.9.1).

Soil formation is remarkably a slow process. It takes several hundred years for a centimeter of topsoil to form. This formation occurs through rock weathering and sediment deposition.

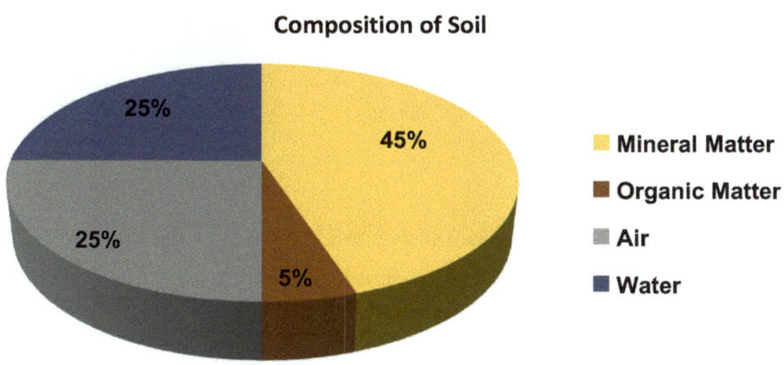

FIGURE 3.9.1 Components of soil in an ideal proportion. Mineral and organic matter make up the solid fraction of the soil; air and water comprise soil pore space. The proportion of these four components impact soil texture, porosity, and pore space. *Courtesy Stefan Kalev.*

http://dx.doi.org/10.1016/B978-0-12-809270-5.00014-5

Soil is derived from different sources of parent material and forms under diverse environmental conditions. This is the reason for the remarkable variety of different soils in the world. Even within the United States, there are thousands of different types of soils.

The size, shape, and clumping of the soil particles determine the volume of air and water in soil. For example, soils with larger pore spaces are more likely to contain more air and be more permeable to water, which may influence the type of crops that can be grown. In general, soil quality is based on different properties that determine the capacity to support plants, animals, and humans. These properties make soil a complex and durable entity that is essential for our quality of life. The ever-growing demands on the environment, be it climate change or waste recycling, require us to understand soils for the success and survival of our species on this planet.

3.9.2 ORIGIN OF SOILS AND SEDIMENTS

Soil formation begins with the physical and chemical breakdown of the earth's rocks, caused by atmospheric agents. These processes, known as *weathering*, chip away rock fragments and thus modify its inherent physical and chemical characteristics. Weathering can also synthesize new minerals that are essential for the soil formation process.

Two main pathways of weathering include *physical disintegration* and *chemical decomposition*. These physical and chemical processes act simultaneously on the parent material and are essential for soil formation. Both physical disintegration and chemical decomposition act differently on the parent material, creating recognizable features (Fig. 3.9.2). Physical disintegration breaks down the rock into smaller pieces and eventually into sand, silt, and clay particles. It predominates in dry and cold environments where the heating and cooling of the exposed rocks create physical stress and cracking (Fig. 3.9.2A). Other forms of physical weathering come from abrasion by water, ice, or wind and are just as significant to the origin of soils

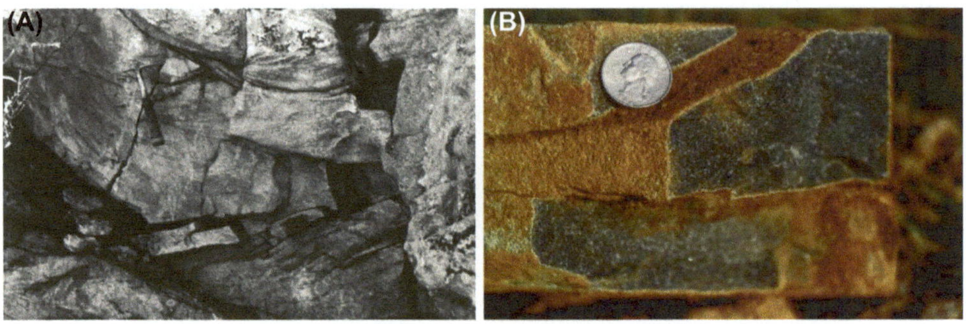

FIGURE 3.9.2 Physical and chemical weathering act simultaneously on the parent material and are essential for soil formation. (A) Physical weathering is a geological process that breaks rocks apart without changing their chemical composition. This image shows weathering caused by expansion and contraction due to temperature changes. Other forms of weathering include abrasion by water, ice, or wind. (B) Chemical weathering alters the structure of the parent material by chemical reactions instead of mechanical processes. It is enhanced by water, hot temperature, and biological agents. This image shows chemical weathering that has occurred from the outside inward, changing the color of this dark-colored igneous rock. *Reproduced from http://pubs.usgs.gov/of/2002/of02-437/gallery.htm.*

and sediments. For example, some of the most productive soils in the world are located in river valleys, where historically different civilizations developed, prospered, and vanished.

Simultaneously with the physical disintegration, chemical processes release soluble materials and minerals. Chemical forms of weathering alter the structure of the parent material by chemical reactions (Fig. 3.9.2B). Unlike physical degradation, *chemical decomposition* is more pronounced in hot and wet climatic regions. This form of weathering is impacted by geological and biological processes and therefore is also known as *biogeochemical weathering*. Water is an essential component for each of these forms of chemical weathering reactions. Some of these reactions include (1) *hydration*, in which water molecules bind to minerals; (2) *hydrolysis*, in which the water molecule splits and the hydrogen replaces a cation from the mineral structure; (3) *dissolution*, in which water dissolves some minerals; and (4) *oxidation-reduction*, which occurs when minerals, often containing iron, manganese, and sulfur, are affected by a loss or gain of an electron.

3.9.3 FACTORS AFFECTING COMPOSITION OF SOILS AND SEDIMENTS

Soil comprises a mixture of individual components that give each type of soil distinctive characteristics and structure. Historical observations and research show that similar soil profiles (*layers*) were present in areas located hundreds of kilometers apart. These studies led to the recognition of five major contributors of soil formation: parent material, climate, topography, biota or living organisms, and time. These factors are discussed in detail in the following sections.

3.9.3.1 Parent Material

Composition of parent materials determines the characteristics of the soil. For example, a sandy textured soil is usually derived from a parent material rich in quartz bedrock. There are different groups of parent materials that vary widely in properties, especially after being modified by physical and chemical processes. Parent material can be classified into several different groups, as described in the following:

- *Residual material* forms in stable landscapes from a weathered or underlying rock.
- *Colluvium* is composed of poorly sorted fragments that are detached from high areas and carried downslope by gravity.
- *Alluvial deposits* are materials carried by streams, rivers, and deltas.
- *Coastal sediments* have been deposited in marine waters and coastal areas.
- *Glacial ice and meltwater* materials have been deposited directly by ice, or vast loads of sediments, carried by torrential water from melting glaciers.
- *Wind* is another major transporter of material, as it can pick an enormous quantity from one area and deposit it to another.
- *Organic material* can collect at any areas with a high amount of accumulation and low rate of decomposition of organic material.

3.9.3.2 Climate

Climate is essential for soil formation, as it determines the intensity and type of weathering that occurs on the parent material. The two main components of climate influence are precipitation and temperature. Both affect soil formation on a physical, chemical, and biological level. *Precipitation* is necessary because water is required for all the major chemical reactions during the weathering process. For maximum impact on soil formation, precipitation must get into the loose earth material above solid rock, known as *regolith*. The water then transports soluble and suspended materials and stimulates soil-forming chemical reactions. *Temperature* is another major contributor, as soil formation reaction rates doubles for every 10 °C increase in temperature. It also influences organic matter content through plant growth and decomposition. If warm temperature is combined with abundant precipitation and leaching, weathering and plant growth will be maximized.

3.9.3.3 Topography

Topography influences formation, loss, and various properties of the soil. For example, steep slope and reduced vegetative cover result in higher rates of *erosion* (the wearing away of land surface by water, wind, or ice). This eroded material may become a source of sediments in different water bodies. On the other hand, low-lying areas exhibit soil accumulation and more pronounced soil development.

3.9.3.4 Biota

Plants have an important role in soil formation, as they influence organic matter accumulation, profile mixing, biogeochemical weathering, nutrient cycling, and aggregate stability. They affect both physical and chemical processes in the soil and interact with other soil organisms in multiple ways. Plants are an important factor with regard to soil porosity and the development of organic matter in soil. As roots decompose, the channels in the soil remain open for air and water and later mix within the soil. Roots also act as stabilizers, as intact root systems are effective in reducing erosion, and therefore improve the quality of the soil. Animals also have a significant impact on soil establishment. Large animals, such as gophers, moles, and prairie dogs, create tunnels and facilitate movement of water and air. Earthworms, termites, and ants also mix the soil, which stimulates soil formation. Some organisms, such as earthworms, enhance plant nutrient availability through ingestion, mineral enrichment, and subsequent excretion of the soil particles.

Human activities affect soil formation in numerous ways. Some of the *anthropogenic* soil alteration practices include the destruction of natural vegetation, tillage, mixing, and compaction. Although these effects in the soil are predominantly negative, humans can also accelerate natural soil restoration processes and minimize soil losses.

3.9.3.5 Time

Soil-forming processes require time, which does not necessarily refer to a specific period, such as months or years, but a degree of soil weathering. It takes hundreds or even thousands of years for soil to develop, making it extremely difficult to measure soil formation processes over time. Time is required for the formation of soil, as it interacts with the other soil-forming processes. Despite its importance, two soils located at different regions may form at a completely different rate. For example, a soil in a warm climate with high rate of precipitation and high parent material permeability will form much more rapidly than that located in a cool and dry climate with a resistant parent material.

3.9.4 PROPERTIES OF SOILS

Physical, chemical, and biological properties affect soil in numerous ways, making it an effective medium with different functions. Factors such as erosion degrade the soil and make it less capable of remaining as a plant medium. The economic implications of erosion can be huge, as in the United States alone, anthropogenic activities erode nearly 4 billion metric tons of soil each year.

There are many different soil types, with various properties that are linked to each other, making it challenging to view their functions separately. Instead, it is important to recognize multiple properties to inform reasonable soil management decisions.

3.9.4.1 Physical Properties

Soil is composed of four major constituents: mineral matter, organic matter, air, and water (Fig. 3.9.1). The proportion of these constituents is important in determining the soil's physical properties. These properties include texture, color, porosity, and density. These determine air and water movement, plant growth, microorganism activity, and nutrient availability.

3.9.4.1.1 Soil Texture

Knowledge of soil texture is critical in managing soils and is among the most important physical properties. The three main individual fractions that make up soil texture are sand, silt, and clay (Fig. 3.9.3). Individual soil particles vary greatly in size, but particles larger than 2 mm in diameter are not considered a part of the soil texture. *Sand* particles vary in size between 0.05 and 2 mm and are visible to the naked eye. Sand has relatively large pores in between individual particles. This is why sandy soils have poor water-holding properties, as water drains out relatively quickly. *Silt* particles vary in size between 0.002 and 0.05 mm, have smaller pores, and are more numerous compared to sand particles in the same volume of soil. The small size of silt particles makes them susceptible to wind and water erosion. *Clay* particles are smaller than 0.002 mm and have a large surface area. They have an incredible capacity to retain water due to the extremely small pore size, which causes water to move very slowly in clay soils. Some of the properties dictated by clay particles include shrinking and swelling, plasticity, and the water-holding capacity of the soil.

The proportion of sand, silt, and clay in a given soil results in 12 different types of soil textural classes (Fig. 3.9.4). These are defined according to the relative amount of particles in the soil and are named depending on the predominant particle, for example, "sand,"

FIGURE 3.9.3 Relative diameter, size, and classification of soil particles: (1) sand particles are classified as smaller than 2 mm and larger than 0.05 mm, (2) silt particles vary in size between 0.05 and 0.002 mm, and (3) clay particles are smaller than 0.002 mm. The proportions of sand, silt, and clay are not to scale. *This figure follows guidelines by the United States Department of Agriculture. Courtesy Stefan Kalev.*

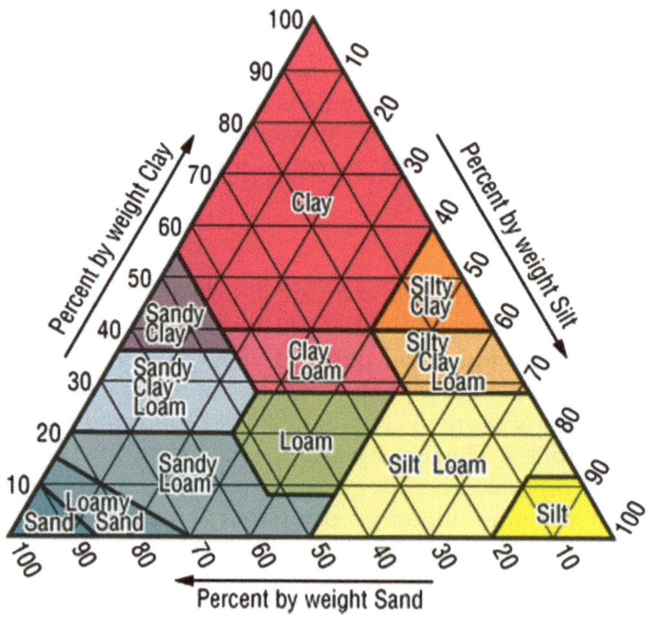

FIGURE 3.9.4 The major soil textural classes are defined by different proportions of sand, silt, and clay. Each class is confined by a corresponding color inside the textural triangle. The percentage of sand, silt, and clay must be known for the specific soil before being classified. If this is the case, a horizontal line corresponding to the percentage of clay is drawn across the triangle. Next a *line* is drawn from the right side toward the base. This *line* must correspond to the percentage by weight of silt and should be parallel to the left side of the textural triangle. The area at which the two *lines* intersect indicates the textural class of the soil. For example, a soil with 20% clay, 40% silt, and 40% loam is classified as a loam. *Courtesy of Prof. Douglas A. Miller.*

"clay," or "sandy clay." Another term, *loam*, describes similar concentration of sand, silt, and clay in the soil, for example, "clay loam" or "silt loam." The percentage of sand, silt, and clay must be known for the specific soil to be classified.

3.9.4.1.2 Soil Color

Soil color is a clear way to see and distinguish specific soil properties, especially for the trained soil scientist. Different soil colors are grouped in the *Munsell color chart* (Fig. 3.9.5), which recognizes different soil types by color. There are three main components of soil color: (1) *hue* is the soil's resemblance to red or yellow, (2) *value* describes the relative darkness and lightness of the soil, and (3) *chroma* is how weak or strong a color appears. A weak chroma can be a grayish orange soil, whereas a strong chroma can be a pronounced bright orange. In general, there are two factors that contribute toward soil color: *humus* (organic material) and the type of *iron compounds*. Humus adds a dark brown, almost black, appearance to the soil. Iron is an important contributor to soil color, especially in weathered forms (Table 3.9.1). Many other minerals also affect soil color and appearance.

Soil forms in a surface downward direction and consists of two or more layers with different characteristics and properties, known as *horizons* (Fig. 3.9.6). The accumulation of organic material on the uppermost part of the soil is called the *O horizon*. It comprises organic materials such as leaves, twigs, animal waste, and other materials in different stages of decomposition. A layer that underlines the O horizon, or lies beneath the surface, is called the *A horizon*. The accumulation of humus gives the *A* horizon a dark color. A whitish to gray *E horizon* may lay beneath the *A horizon*. Such a layer is commonly found in acidic soils of evergreen forests. Underneath the *A horizon* is the *B horizon*, a brownish to reddish horizon, enriched with clay, aluminum, iron, and some organic matter. The *C horizon* is the deepest

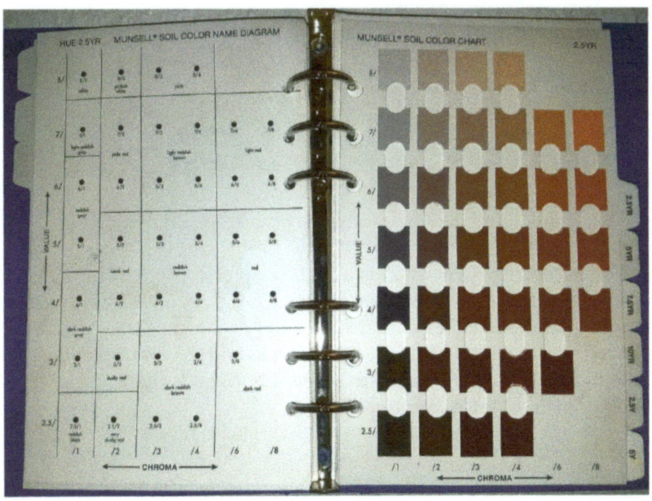

FIGURE 3.9.5 The Munsell Soil Color Chart allows for a standard system of comparing a given soil sample to the three components of soil color, namely, hue, value, and chroma. The book has arranged these components in different color variations. When a soil is held next to a potential color match, it can receive an appropriate color notation, which can assist in correctly identifying the sampled soil and its properties. *Courtesy Stefan Kalev.*

TABLE 3.9.1 Common Oxidation Forms of Iron (Fe) in the Soil and Their Corresponding Color

Form	Color
Iron(III) oxide (Fe_2O_3)	Red
Iron(II) oxide (FeO)	Gray
Iron(III) hydroxide ($Fe(OH)_3$)	Yellow

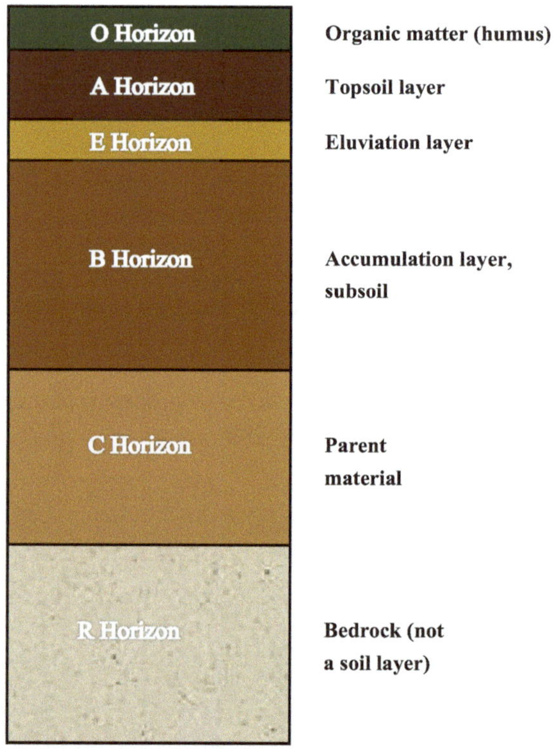

FIGURE 3.9.6 Basic soil horizons: (1) O horizon is a surface layer composed of organic matter, (2) A horizon is underlying the O horizon and is usually rich in organic matter, (3) E horizon is made up mostly of sand and silt and has lost its clays and minerals, (4) B horizon accumulates clay and carbonates that have leached downward from the above layers, (5) C horizon consists of slightly broken-up bedrock, and (6) R horizon comprises unweathered parent material (solid rock), which underlies the parent material. The R horizon is not a soil layer. Not all soils have each of the listed horizons and subhorizons. *Courtesy Stefan Kalev.*

true soil layer. It usually lies underneath the *B horizon* and consists of weathered rock material. The bedrock *R horizon* lies beneath the *C horizon* and is not considered a true part of the soil profile. Nevertheless, the *R horizon* consists of parent material that is essential for all soil formation on the planet.

3.9.4.1.3 Soil Density

Soil density provides important information about the different qualities of the soil. There are two major forms of density: *bulk density* and *particle density* (Fig. 3.9.7). Neither bulk density nor particle density considers the presence of any water. Particle density is the mass of a unit of soil solids divided by the volume of the soil solids only. It does not consider compaction and the pore spaces between the soil particles. A typical soil particle density is about 2.65 Mg/m^3, but it may narrowly vary from 2.60 to 2.75 Mg/m^3 depending on the composition of the soil. This is due to the similar density of the minerals present in different types of soils. These types of soils are usually low in organic matter (1%–5%) and may reach up to 3.0 Mg/m^3 or more particle density. On the other hand, particle density near 1.3 Mg/m^3 is an indicator of a soil with high organic matter content. The particle density of these soils may be as low as 0.9 Mg/m^3. Bulk density is another important measurement that is given by the mass of the dry soil divided by the volume of the soil. Bulk density takes into consideration both the mass of the soil particles and the pore space between these particles, which gives it a lower value.

A major reason for measuring bulk density is to find *porosity*, or the percentage of total pore space. Soil pores exhibit variability in size and shape that determine what role the pores will play in the soil. There are *macropores* (larger than 0.8 mm) and *micropores* (smaller than 0.8 mm). Macropores allow for the relatively easy movement of air and water. Micropores, on the other hand, are usually filled with water and are limited to permitting air movement. By measuring bulk density and porosity, it is possible to predict whether plants will successfully grow in a given soil.

3.9.4.2 Chemical Properties

Most chemical properties of soils are because of the *colloid fraction*, or the clay and humus. The vast majority of chemical reactions occur at the colloid surfaces because of their

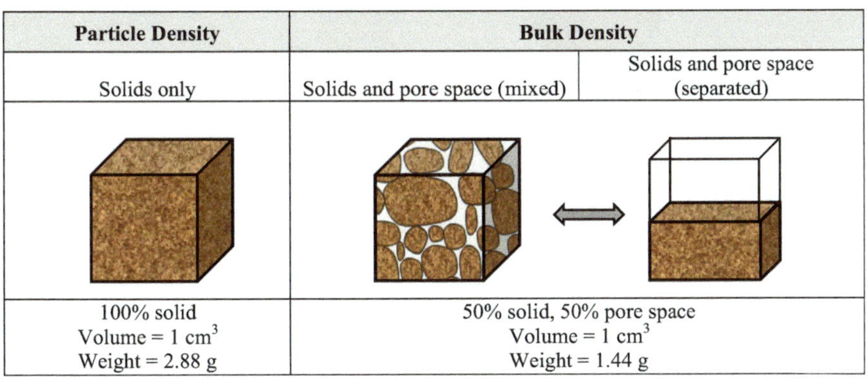

FIGURE 3.9.7 Bulk and particle densities of the soil. Bulk density considers the weight of the soil particles and pore spaces for a standard unit of soil. Particle density calculates soil particles per volume of these particles only and does not consider pore space. For example, if the particle density for a selected soil is twice as larger than its bulk density, then the bulk density of this soil has 50% pore space. *Courtesy Stefan Kalev.*

microscopic size, predominantly negative charge, and very large surface area. These chemical properties, in turn, control nutrient availability and transformations and can affect physical properties and thus plant growth. The concentration, charge, and size of various elements determine the amount of ions in the soil that can be adsorbed or exchanged, also known as *exchange capacity*. The following section briefly describes chemical properties of soils.

3.9.4.2.1 pH

pH is a term that measures the hydrogen ion (H^+) activity in the soil, and therefore the acidity or alkalinity of a soil. A high proportion of H^+ corresponds to an acidic soil and a lower pH value. A low H^+ means an alkaline soil with a high pH value. The pH scale typically ranges from 0 to 14, where a pH of 7 is neutral, pH higher than 7 is alkaline, and pH lower than 7 is acidic. Soils throughout the world exhibit different pH values (Fig. 3.9.8) because of the variations in climatic conditions, geology, land use, fluctuations in groundwater tables, etc. pH is an important factor for the chemical properties of the soil and can impact cation- and anion-exchange capacity by altering the charge on the soil colloids. Although some soils may exhibit electropositive charge, electronegative charge predominates in most soils.

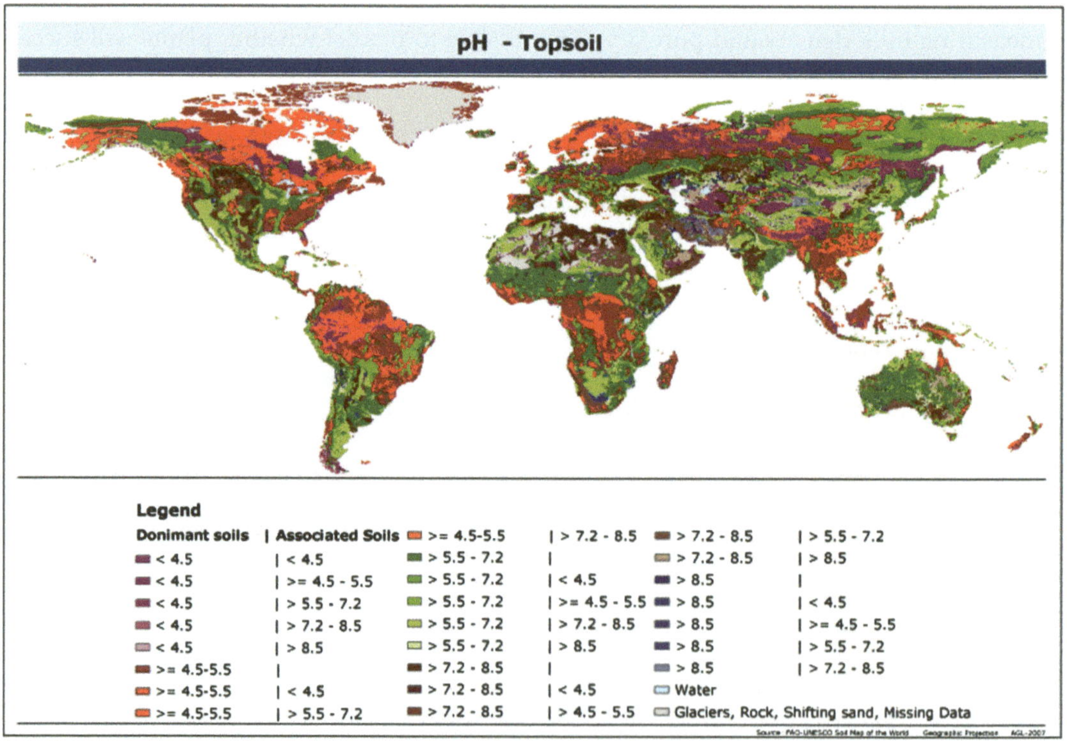

FIGURE 3.9.8 pH (topsoil) of the soils in the world. Soil pH varies greatly between 4.5 and 8.5 in different regions of the world. *Reproduced from FAO/UNESCO soil map of the world. http://www.fao.org/geonetwork/srv/en/main.home.*

The optimum pH ranges for different crops in different soils vary significantly. Many legume crops, for example, grow near neutral or alkaline soils. Most of the grain and vegetable crops grow at a soil range with a pH of 5.5–7.0.

Soil pH can be increased by using calcitic limestone, burned lime, wood ashes, and other materials if the soil is too acidic. The amount of lime that is used depends on the desired change of pH, depth of soil, soil buffer capacity, and the composition of the liming material. The majority of liming materials dissolve in water to produce bicarbonate, which in turn makes the soil less acidic. Lowering the pH of the soil can be accomplished by adding acidic material, but is rarely done in practice. The variation in pH controls the behavior of different elements in the soil. For example, trace metals are more soluble and easily available at low pH, whereas phosphorus—the second most important nutrient (see later discussion)—has higher availability in the neutral pH range.

3.9.4.2.2 *Plant Nutrients*

For plants to grow, they need to obtain nutrients from soil (Fig. 3.9.9). There are 15 nutrients that are needed in different quantities and are equally important to the plants. Plants extract essential carbon, oxygen, and hydrogen from air and water. The three most essential macronutrients obtained from the soil are nitrogen, phosphorus, and potassium. These nutrients are required in larger amounts than the rest. Other nutrients are needed in

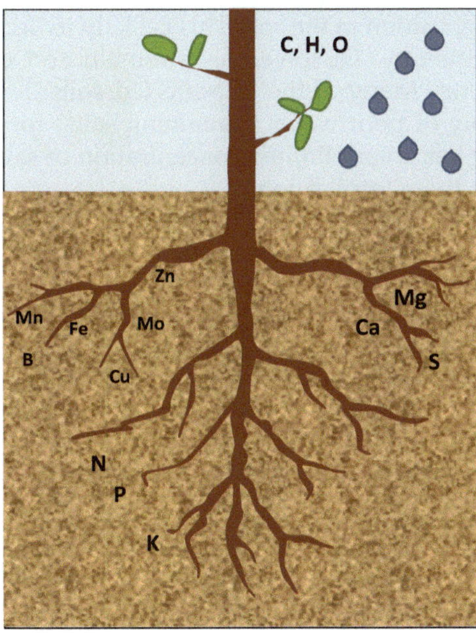

N - Nitrogen P - Phosphorus K - Potassium	**Macro nutrients**
Ca - Calcium Mg - Magnesium S - Sulphur	**Secondary Nutrients**
Fe - Iron B - Boron Zn - Zinc Cu - Copper Mn - Manganese Mo - Molybdenum	**Micro nutrients**

FIGURE 3.9.9 Plant essential nutrients: plants require *macronutrients* from the soil in the largest quantity; *secondary nutrients* are required in lesser amounts; *micronutrients* are also essential, but in relatively small quantities; and carbon, hydrogen, and oxygen are extracted by plants from air and water and make up the bulk of their weight. *Courtesy Stefan Kalev.*

lesser amounts but are just as important. Plant growth is limited by the element in the shortest supply, which may require amending soil with fertilizers and organic manures. Nutrient addition to the soil promotes root and shoot growth, helps establish newly planted plants, and enhances the flower and fruit of the plant.

3.9.4.2.3 Soil Carbon

Soil organic matter is a complex mixture of organic substances. All these substances contain the element carbon, which is the foundation for all living organisms and therefore is essential to life. Soil organisms obtain carbon either by breaking down organic materials or from the carbon dioxide in the air.

Carbon dioxide is considered a significant contributor toward global warming and soil plays an important part in storing it. On average, carbon comprises more than half of the mass in soil organic matter. Soil contains nearly twice as much carbon as the world's atmosphere and vegetation combined.

3.9.4.2.4 Soil Salinity and Sodicity

Soil *salinity* is the amount of dissolved salts in the soil solution (the aqueous phase in the soil). The process of accumulating soluble salts in the soil is known as *salinization*. Salts in the soil have an important effect on the functions and management. For example, high concentrations of soluble salts lower the pH and may limit plant growth, in some cases by more than 90%. Salt-affected soils are distributed throughout the world, especially in arid and semiarid regions. Salts are common in areas where the input of water containing salts exceeds the drainage, resulting in increased salt concentration in the soil. This is likely to occur in low-lying places, or areas that are irrigated frequently. Dissolved salts remain and eventually build up in these areas after water evaporates. Many of the salt-affected soils develop due to human activities that use excessive supply of poor water (containing salts) for irrigation purposes. Although irrigation water usually has a very limited concentration of salts, continuous use over a prolonged period can have huge implications.

Sodicity describes the amount of sodium that is present in the soil. A *sodic soil* contains enough sodium to interfere with the growth of plants. Sodic soils have relatively high levels of sodium and pH, usually over 8.5. High pH and sodicity cause dispersal of organic matter in the soil, which affect physical properties such as water movement and restrict plant growth. Furthermore, these types of soils have some of the poorest soil physical conditions and have slow permeability to air and water, resulting in less gas exchange needed for plants to grow.

3.9.4.3 Biological Properties

There is a fascinating biodiversity of organisms in the soil, and their vast variety and interactions enrich the growth properties of a high-quality soil. Soil organisms are classified based on the similarities in the genetic material, size, and ecological functions (Fig. 3.9.10; Table 3.9.2). *Flora* (general classification of true plants, and nonanimal organisms) and *fauna* (general classification of animals) interact with each other and affect the soil in numerous ways.

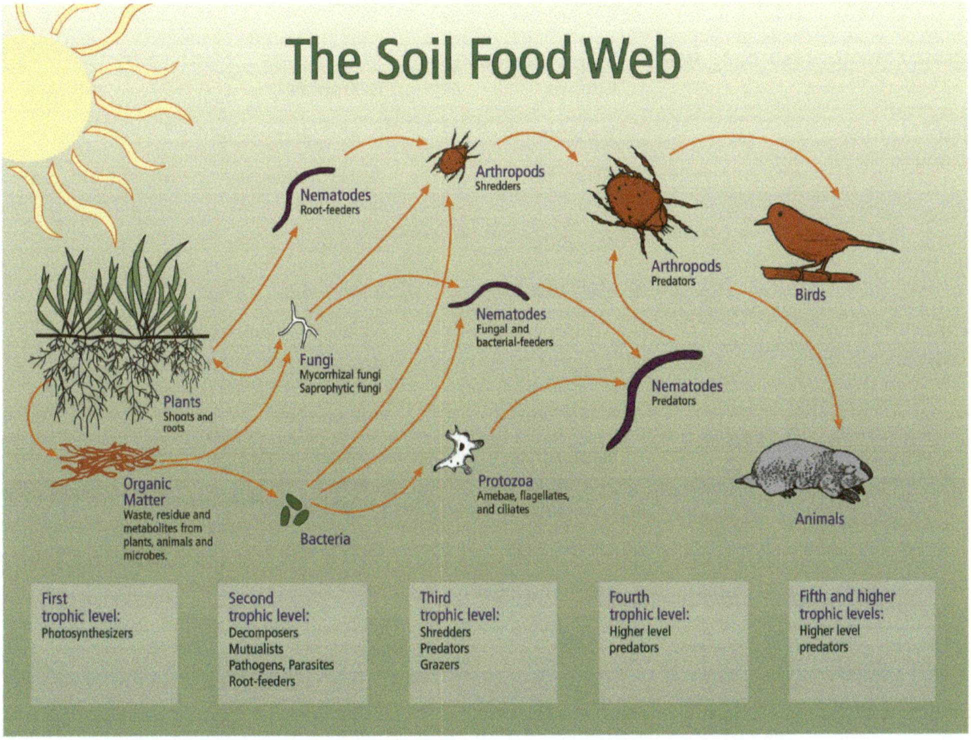

FIGURE 3.9.10 The soil food web. *From Soil and Water Conservation Society (SWCS). 2000. Soil Biology Primer. Rev. ed. Ankeny, IA: Soil and Water Conservation Society*

3.9.4.3.1 Classification of Organisms According to Size

Animals in the soil can be classified according to their size (Table 3.9.2): *macrofauna* are larger than 2 mm, *mesofauna* are between 0.1 and 2 mm, and *microfauna* are smaller than 0.1 mm. Plant species include vascular plants, mosses, microscopic algae, and diatoms. Other microorganisms include bacteria, fungi, and actinomycetes, which dominate the soil in terms of mass and quantity.

3.9.4.3.2 Classification According to Genetic Similarities

Biologists classify living organisms into three domains. The first domain is *eukarya*, a group of organisms whose cells contain organelles enclosed in membranes and a nucleus. This domain includes all plants, animals, and fungi. *Bacteria* is another domain that is among the first life forms on Earth. They are present in vast quantities in most habitats. *Archaea*, just like *bacteria*, is another single-celled prokaryotic microorganism group with similar appearance, but very different evolutionary origins.

3.9.4.3.3 Classification According to Ecological Function

Soil organisms are also classified according to the way in which they obtain food from their environment. These organisms can be either *autotrophs*, obtaining carbon from air

TABLE 3.9.2 General Classification of Organisms by Size

Generalized Grouping and Body Width (mm)	Major Taxonomic Groups	Examples
Macrofauna (2 mm) Heterotrophs	Vertebrates	Gophers, mice, moles
	Arthropods	Ants, beetles, centipedes, spiders, termites, woodlice
	Annelids	Earthworms
	Mollusks	Snails, slugs
Mesofauna (0.1–2 mm) Heterotrophs	Arthropods	Mites, protura, springtails
	Annelids	Pot worms
Microfauna Detritivores, fungivores, bacterivores, predators	Nematodes	Nematodes
	Rotifera	Rotifers
	Protozoa	Amebae, ciliates, flagellates
	Tardigrades	Water bears
Macroflora (>0.1 mm) Mainly autotrophs	Vascular plants	Feeder roots
	Bryophytes	Mosses
Microflora (<0.1 mm)		
Mainly autotrophs	Vascular plants	Root hairs
	Algae	Greens, yellow greens, diatoms
Mainly heterotrophs	Fungi	Yeasts, mildews, molds, rusts, mushroom
Autotrophs and heterotrophs	Bacteria	Aerobes, anaerobes
	Cyanobacteria	Blue-green algae
	Actinomycetes	Actinomycetes
	Archaea	Methanotrophs

and energy by photosynthesis or oxidation, or *heterotrophs*, obtaining carbon by breaking down organic materials. When one organism is eaten by another, nutrients and energy are transferred from one *trophic level* (position that organism occupies in the food web) to a higher one. Organisms in the soil food web are classified according to the following trophic levels:

- *First trophic level* organisms use the energy from the sun to make organic molecules and living tissues by combining carbon dioxide and water. These include vascular plants, mosses, lichens, algae, and some photosynthesizing bacteria.
- *Second trophic level* comprises primary consumers that include mites, woodlice, springtail insects, and earthworms. They consume plants or dead plant tissue.

- *Third trophic level* organisms use the primary consumers as their main food source. These include an array of soil predators and parasites, and their activities may influence the growth of fungal and bacterial populations in the soil.
- *Fourth trophic level* organisms include higher levels of predators, such as arthropods and predatory nematodes. These consume and influence the populations of the previous trophic levels.
- *Fifth and higher trophic levels* include ants, spiders, birds, and large animals that feed on the abovementioned lower groups.

The diversity of these organisms gives each soil a different set of characteristics, which, in turn, affect a soil's ability to provide various functions—be it nutrients for plant growth or recycling of waste.

3.9.5 PROPERTIES OF SEDIMENTS

The *benthic*, or bottom, layers of water bodies contain particle matter, known as *sediment*, which shares many similarities with the upland soils. Sediments are transported and deposited as particles of aggregates, which are derived from soil, rocks, or biological materials.

Sediments share many properties with soils, but also have some distinct differences. Unlike soils, sediments are always submerged under water and therefore have a high *redox potential*. Redox potential determines the sediment's tendency to acquire electrons and to become reduced in charge. This is what makes sediments likely to adsorb contaminants, which, in turn, cause different environmental concerns. One example of sediment contamination can be high nitrogen and phosphorus concentrations, which can provide a long-term source of nutrients for algae and eventually cause algal blooms. This type of pollution is known as *eutrophication* and can lead to oxygen depletion in the water bodies, which can kill fish and impact the entire aquatic ecosystem. Another area of concern is caused by metals that bind to sediments and cause toxicity to aquatic organisms. In the following section, we briefly discuss unique properties of sediments.

3.9.5.1 Physical Properties

Once soil is transported into a water body, such as stream, river, or lake, it becomes *sediment*. Humans have increased sediment concentration in the water bodies through different activities. Examples of such anthropogenic activities include runoff, logging, removal of *riparian* (from an area that borders a river) vegetation, construction, and other methods of water basin alterations. Sediments that enter streams because of watershed disturbance can remain there for decades and prolong water contamination.

A large amount of the sediments is carried away by the major rivers of the world. For example, the Mississippi River in the United States carries approximately 300 million Mg of sediment into the Gulf of Mexico annually. Fig. 3.9.11 illustrates the vast farmland (green) and city (red) land development in the Mississippi drainage basin, which contributes to significant amounts of sediment transport and nutrient runoff. More than 17 million tons of

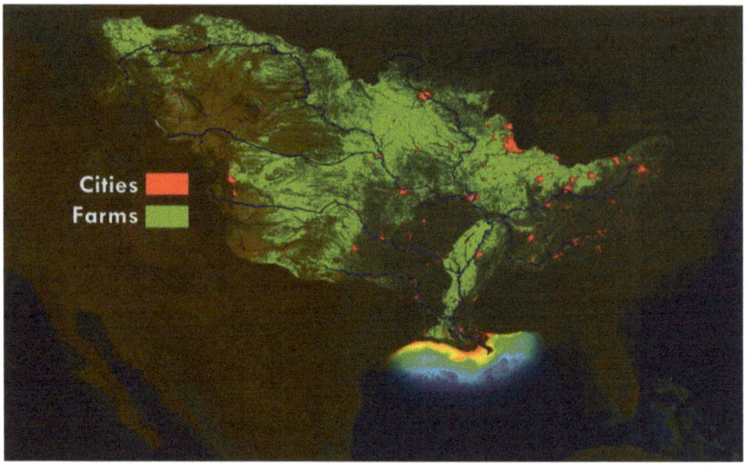

FIGURE 3.9.11 Anthropogenic activities in the Mississippi drainage basin of the United States results in vast amounts of sediment and nutrient runoff. This leads to extensive algal bloom areas (orange and yellow), resulting in *hypoxic zones* in the Gulf of Mexico. Hypoxic zones have extremely low oxygen concentration that suffocates aquatic life. *Reproduced from National Oceanic and Atmospheric Administration (NOAA). http://www.nnvl.noaa.gov/MediaDetail2.php?MediaID=1062&MediaTypeID=3&ResourceID=104616.*

combined nitrogen and phosphorus are carried into the Gulf of Mexico. Because of this huge influx of nutrients, massive phytoplankton blooms occur (yellow). This result in *hypoxic zones*, or areas with extremely low oxygen concentration, where only a few species can survive.

Globally, many river systems are impacted by dam structures. It is estimated that in the United States alone, there are nearly 75,000 dams, which impound about 17% of rivers. These structures cause sediments to settle and therefore leave rivers downstream to become significantly more erosive due to lack of sediments. This reduces the amount of natural sandbars that would form in such river systems. An example of downstream sediment limitation is the construction of the Aswan Dam at the Nile River, Egypt. This large-scale project has limited the sediment supply and has caused erosion and loss of important agricultural land near the Nile's delta.

Sediments may form sedimentary rocks, which are formed by erosion, and later accumulation and cementation of natural materials. The main forms of sedimentary rocks are limestone, sandstone, shale, and conglomerates. Different types of sediment deposition may occur by water, wind, and glaciers. Another example is quartz sand, which deposits near the shore and may cement into sandstone. Similarly, deposited clay particles may be compacted into shale.

3.9.5.2 Chemical Properties

As organic matter decomposes, it shows high oxygen demand, which releases nutrients into the environment and adds gases such as methane and hydrogen sulfide to the water column. Some of the chemical parameters that are important and frequently measured in sediment are pH, total organic carbon, total phosphorus, ammonia, nitrates, and metals. Some sediment contaminants and their pollution sources include (1) *ammonia*, associated with stormwater, combined sewer overflows, and aquaculture; (2) *toxic metals*, associated with mining waste and metal-processing facilities; (3) *pesticides* and *herbicides*, from agriculture or agricultural chemical companies; (4) *petroleum compounds*, from refineries, fuel storage facilities, marinas, and gas stations; (5) *polychlorinated dibenzo-p-dioxins* and *polychlorinated dibenzofurans*, from pulp and paper mills, waste facilities, cement factories, coal, wood and petroleum burning, and metals smelting; (6) *volatile organic compounds*, produced by solvents and in chemical manufacturing operations; and (7) *radioactive substances*, from nuclear power and processing plants, medical wastes, and military installations.

Areas prone to significant sediment pollution may occur in proximity to mining areas that have contaminated nearby sediments. After heavy rainfall, massive fish kills may occur in these areas if toxic sediments are deposited into the water.

3.9.5.3 Biological Properties

Sediments can have significant effects on life in the water, such as blocking sunlight for primary producers, which will inevitably impact the rest of the aquatic ecosystem. Sediments can carry several contaminants such as nutrients, metals, and organic contaminants that are adsorbed onto them into the water bodies. Sediment deposition may damage entire waterways by burying habitats and clogging fish gills. Sediment deposition may cause sensitive aquatic species to relocate and others to move in. Sediments likely determine the type of microorganisms that will thrive on the bottom of the water body. For example, a shifting sand bed does not offer rich opportunities for animals to thrive. On the other hand, rich organic sediment offers a diverse ecosystem for biota and various microorganisms. Benthic primary production is important for the lake ecosystem, as half of the sediment production in lakes can be attributed to *littoral* (near the shore) algae or macrophytes.

Bottom-dwelling organisms can have an important role in material exchange between the benthic zone and the water column. Therefore, these organisms are vital for energy flow and nutrient cycling, through processes such as digging, rowing, burrowing, stirring of the sediments, and processing these materials. Movement of these organisms can increase chemical transport, disturb sediment mass, and increase diffusion of materials. Migration of planktonic organisms and fish can also dramatically increase nutrient redistribution in water bodies.

3.9.6 IMPORTANCE OF SOILS AND SEDIMENTS

Individual soils and sediments are characterized by a specific set of properties, which are determined by the natural and environmental conditions at the specific site. Soil serves as a medium for growth by providing these six essential functions to both plants and organisms: physical support, water, air, temperature moderation, nutrients, and protection from toxins.

- Soil provides *support* that anchors the root system and prevents the plants from falling over or being blown away.

- Soil absorbs and holds *water* in pores and then makes it available to plants. This water-holding capacity is essential for their survival, as plants need a continuous water supply to transport nutrients, perform cooling, and undergo photosynthesis.
- Like our own respiration, plant roots also use oxygen and produce carbon dioxide. Soil provides *air* and *ventilation* for the roots and serves as a medium that allows for fresh oxygen to enter the root zone and for carbon dioxide to escape.
- Soil is an important *moderator of temperature* through its excellent insulation functions. These functions protect plant roots from extreme temperatures and allow them to function normally.
- Soils also supply plants with the necessary *mineral nutrients*, which they need to grow.
- Soils protect plants from *toxins* by adsorbing and suppressing organisms producing toxins.

Soil also provides vital function as a regulator of *water supply* and *purification*. It serves as a *natural recycling unit* by assimilating great amounts of organic waste and plays a vital role in major biogeochemical cycles. Certain soils also accumulate a large amount of carbon and thus play an important role in sequestering it in soil and assisting in climate change mitigation. Natural sediments can help in the attenuation of various chemical compounds by adsorbing and transforming them into harmless compounds. A moist and well-vegetated soil plays a key role in *air quality* by preventing dust pollution, which is a significant health hazard. It is also a vital habitat for soil organisms by providing a tremendous number of niches where millions of organisms live, work, and play.

The natural process of erosion turning soil into sediment serves to fill downstream valleys, plains, and deltas. Common natural geological erosion wears down the land relatively slowly and new soil is able to form from the underlying regolith, or loose earth material above solid rock. Accelerated erosion, on the other hand, is one of the most crucial problems facing humanity. It degrades the soil and causes great damage to reservoirs, rivers, and other water bodies. Water and wind erosion are the main causes for such erosion.

The anthropogenic processes of deforestation, improper soil management, and construction contribute to accelerated erosion. Rivers with contaminated sediments lead to pollution, as many pollutants such as metals and pesticides bind to the surface of sediment particles. Erosion control systems need to be developed and better conservation practices must be implemented to reduce soil loss.

3.9.7 CONCLUSIONS

The earth's soil comprises numerous individual soils, each with its own unique properties and horizons. A soil in good plant-growing condition consists of approximately 50% solid material (mainly mineral and some organic matter), water (~25%), and air (~25%). These components create a complex set of soil functions and interactions. The vast surface area of the soil colloids determines both chemical and physical properties. As colloids are negatively charged and colloid particles predominate, positive ions are usually attracted to their surface. This attraction is an essential mechanism for nutrient retention and other chemical properties of the soil. The origin, transport, and weathering of parent material are important factors in determining soil functions and properties. Soil formation depends on the climate, biota, parent material, time, and topography. The physical, chemical, and biological characteristics of soil are also influenced by hydrology, plant growth, and human soil management and engineering. Soil provides six main ecological functions: plant growth environment, recycling materials, regulating water supplies, modifying the atmosphere, providing habitat, and serving as engineering medium. Soil's importance will continue to increase in the future, as it will continue to supply us with food and vast amounts of goods and services. As a growing and urbanizing society, we should greatly improve our understanding and management of soils so they do not become sediments and pollute our water bodies. As new challenges continue to emerge, better understanding and research will allow societies to make decisions that are protective of the environment by reducing soil loss from the terrestrial landscapes.

Recommended Reading

1. Brady N, Weil R. *The nature and properties of soils*. 14th ed. New Jersey: Prentice Hall; 2007.
2. Eash N, Sauer T, O'Dell D, Odoi E, Bratz M. *Soil science simplified*. 6th ed. New Jersey: Wiley-Blackwell; 2015.
3. Paul E. *Soil microbiology, ecology and biochemistry*. 4th ed. Massachusetts: Academic Press; 2014.
4. Plaster E. *Soil science and management*. 6th ed. Massachusetts: Cengage Learning; 2013.
5. Nash D, McLaren S. *Geochemical sediments and landscapes*. New Jersey: Wiley Blackwell; 2007.

CHAPTER

3.10

Heavy Metal Pollution and Remediation

Abhishek RoyChowdhury[1], Rupali Datta[2], Dibyendu Sarkar[1]

[1]Stevens Institute of Technology, Hoboken, NJ, United States; [2]Michigan Technological University, Houghton, MI, United States

3.10.1 INTRODUCTION

Heavy metals are commonly defined as a set of metals/metalloids that possess densities at least 5 times higher than that of water.[1,2] Heavy metals are natural constituents of the earth's crust, and natural phenomena such as weathering and volcanic activity contribute significantly to heavy metal pollution. Anthropogenic activities such as mining, smelting, industrial, and agricultural use increase the abundance of heavy metals in the environment by either releasing or concentrating them into the environment.[2,3] Heavy metals are widely used in several industrial, agricultural, medical, and technological applications. They can be toxic to living organisms at very low levels of exposure. Excessive usage of heavy metals has raised concerns for human health over time, and their impacts on the overall environment are being studied extensively. Based on their immediate threat to human health and the environment, four of the top 10 chemicals of major public concern reviewed by the World Health Organization (WHO) are heavy metals (arsenic, lead, mercury, and cadmium). In addition, chromium is also considered as a very toxic heavy metal. This chapter will discuss the environmental occurrence, production and anthropogenic use, exposure, as well as toxicity of some of the major heavy metals, and their remediation approaches in detail. Table 3.10.1 summarizes the regulatory guidelines for some of the major heavy metals in soil in the United States.

3.10.1.1 Arsenic

Arsenic (As) is a metalloid found in almost all types of environmental matrices. Arsenic ranks No. 1 in the Agency for Toxic Substances and Disease Registry's (ATSDR) Priority List of Hazardous Substances,[6] and is a known human carcinogen. Soluble inorganic arsenic is acutely toxic. Arsenic can be found in four valence states: elemental arsenic (As 0), arsenate

Green Chemistry
http://dx.doi.org/10.1016/B978-0-12-809270-5.00015-7

TABLE 3.10.1 USEPA Regulatory Guidelines for Heavy Metals in Soil[4,5]

Metal	Regulatory Limits (mg/kg)
As	0.39
Pb	400
Hg	260
Cd	85
Cr	100

USEPA, United States Environmental Protection Agency.

(As V), arsenite (As III), and arsine gas (AsH_3-III). Arsenate is predominant in aerobic environments, whereas arsenite dominates under reducing conditions. Elemental arsenic and arsine can be found under extreme reducing conditions.

Inorganic arsenic intake over a long period can lead to arsenicosis. Arsenic poisoning can impact the nervous system, gastrointestinal system, and renal system, and can create skin lesions, diabetes, cardiovascular diseases, and cancer.[2,3] Organic arsenic compounds are commonly found in seafood, but they are less harmful, as the human body can rapidly eliminate them. Prolonged human exposure to As-contaminated groundwater has occurred in countries such as Bangladesh, India, Thailand, and Vietnam. A 2010 WHO report estimated 9100 deaths and 125,000 disability-adjusted life years in Bangladesh due to the consumption of As-contaminated drinking water in 2001.[2] In countries like the United States, inorganic arsenic was used as a wood preservative in the form of copper chromate arsenic (CCA), which became a predominant source of As contamination of soil. As of 2003, CCA is banned from any form of residential use in the United States. Inorganic arsenical pesticides were used in apple orchards, vegetable fields, cotton fields, and in cattle dipping vat sites to control ticks in livestock. Organic arsenic compounds were also used as pesticides, mainly for cotton crops. These historical uses of arsenic compounds resulted in widespread contamination of soil and water matrices, and posed a threat to human health. In 2001, the United States Environmental Protection Agency (USEPA) reduced the standard for As in drinking water in the United States from 50 to 10 μg/L.[7]

3.10.1.2 Lead

Lead (Pb) is the second most hazardous metal according to the ATSDR's Priority List of Hazardous Substances,[6] and is a "probable human carcinogen."[8] Both natural and anthropogenic sources play major roles in Pb exposure in the environment. Some of the major sources of Pb contamination in the environment include mining and smelting, welding of metals with Pb paint, batteries, leaded gasoline, paper and pulp, and explosives.[9–11] Lead-based paints are one of the major sources of Pb contamination in residential units. Although the USEPA banned the use of Pb-based paints in the United States in 1978, still more than 21 million homes currently exist throughout the country with Pb-based paints.[12] Due to the natural and anthropogenic activities like weathering, flaking, scrapping, chipping, and waste disposal, Pb from the exterior paints leaches out and tracks its way to the interior of the house as Pb-rich dust.[13,14] Children younger than 6 years are more susceptible to chronic exposure of Pb due to their hand-to-mouth activities.[15,16] Over 50% of ingested Pb is absorbed via the gastrointestinal tract and penetrates the immature blood-brain barrier of children, causing severe damage in their developing neurological systems.[9,13]

An incident of extensive Pb poisoning in residents of Flint, MI, indicates that Pb in water distribution pipes could also be a major health hazard. In the summer of 2014, the city of Flint began using water from the Flint River as a cost-saving measure, switching from Lake Huron, which had been their water source for the past 50 years. The Huron Lake water had low chloride and low chloride-to-sulfate mass ratio, and orthophosphate was added as corrosion inhibitor.[17] However, the Flint River water had high chloride and high chloride-to-sulfate mass ratio, and no orthophosphate was added. The water distribution system includes a large proportion of lead pipes, which corroded due to the change in water chemistry, leaching out Pb. Very high Pb levels were detected in the water, and the percentage of children with elevated blood lead levels increased two- to threefold.[17] Since water distribution systems in many cities have not been updated, this incident has raised widespread concern about Pb poisoning from public water systems in the United States (see Chapter 3.6).

3.10.1.3 Mercury

Mercury (Hg) ranked No. 3 in ATSDR's Priority List of Hazardous Substances.[6] Mercury causes severe risk to human health, particularly to the children in their early developing stages. Mercury can be found in the environment in different forms, such as elemental (or metallic), inorganic (such as $HgCl_2$), and organic (such as methyl- and ethyl-mercury). Mercury toxicity impacts the nervous system, digestive system, immune system, renal system, and ocular system, but the most prevalent impact is caused during early organ development periods in children. A major source of environmental release of Hg is caused by coal-fired power plants. Other sources include waste incineration, oil pipelines, manometers from pressure measuring stations, residential heating systems, and mining. After its release into the environment, elemental mercury is naturally transformed into methylmercury, which bioaccumulates in fish and shellfish. Inhalation of elemental mercury and consumption of mercury-contaminated fish and shellfish are the major sources of human exposure to mercury pollution.[2,3,18]

3.10.1.4 Cadmium

Cadmium (Cd) is a known human carcinogen, and it is ranked No. 7 in ATSDR's Priority List of Hazardous Substances.[6] Although Cd is present in the environment naturally at a lower concentration, anthropogenic activities have elevated its level significantly over time. Major sources of Cd exposure are Ni/Cd batteries, Zn and Pb refineries, pigments and stabilizers for polyvinyl chloride, alloys, electronic products, fertilizers, pesticides, and disposal of industrial wastes. Cadmium coatings are used in marine vessels, and other vehicles, for corrosion prevention. Cadmium toxicity impacts the kidneys leading to kidney dysfunction. It also impacts skeletal and respiratory systems and affects several enzymes.

Cadmium is very persistent in the environment and bioaccumulates in mollusks, crustaceans, and vegetables over time. Once it enters the human body it is very difficult to remove.[2,3,18]

3.10.1.5 Chromium

Chromium (Cr) is known to be a human carcinogen.[6] It is a naturally occurring element found in rocks, soils, and volcanic dust. Chromium is also found in plant and animal bodies as a trace metal, which means that small quantities of chromium are necessary for growth of living organisms but that excessive amounts can be extremely toxic to them. Anthropogenic activities that are responsible for elevated chromium levels in the environment are electroporating processes, disposal of industrial wastes, wood preservatives, textile dying, and leather tanning. Chromium can be found in three different forms, namely, trivalent (Cr III), hexavalent (Cr VI), and metallic (Cr 0). Metallic Cr is used for steel production. Chromium toxicity impacts the immune system, renal system, and respiratory system and can cause cancer.[3,18]

3.10.2 REMEDIATION OF HEAVY METALS

The main purpose of heavy metal remediation is to minimize the risk of these toxic compounds to human and ecological health. The selection of an appropriate remediation method depends on the characteristics of the site, extent of the heavy metal contamination, and regulatory limits for the heavy metal(s) of concern in that regulatory domain. The remediation methods can be broadly divided into three major categories: (1) physical methods, (2) chemical methods, and (3) biological methods.

3.10.2.1 Physical Methods

Physical remediation methods are very expensive and time consuming. Commonly used physical remediation methods are as follows.

3.10.2.1.1 Soil Replacement

This method involves the partial or complete removal of metal-contaminated soil and subsequent replacement with clean soil, often called the "dig-and-haul" method. The main challenges of this process are that it is only feasible for a smaller area that might be severely contaminated and is very expensive.[19,20] This method only moves the contaminated medium to a different location, transferring the contaminant from a higher risk area to one of lower risk.

3.10.2.1.2 Soil Washing

Soil washing helps to minimize the level of heavy metal contamination in severely contaminated soil. This process can be applied in-situ by forcing washing solution through the soil matrix, or ex situ by physically excavating the soil and washing it in reactors.[18] The main purpose of this method is to make the heavy metals soluble in the washing reagents and to separate them out from the soil. Soil washing is more effective in sandy and granular soils than in clay-rich soil as clay creates problems with solid-liquid separation after washing.[21]

Depending on the nature of the targeted heavy metals, a range of chemicals are used in the soil washing process, including surfactants, chelating agents, organic acids, and cosolvents.[18,22] The nature of the washing agents is determined based on the physicochemical characterization of the area and the type of heavy metal(s) present in it. Use of strong acids can destroy the crystalline structure of soil, so natural and low-molecular-weight acids such as oxalic, fumaric, acetic, formic, and lactic acids are used in this process. Use of chelating agents such as citric acid, tartaric acid, and ethylenediaminetetraacetic acid (EDTA) has also been evaluated in soil washing process. However, this method is expensive and proper handling and treatment of the washing solutions are required afterward. Removal of heavy metals like Pb, Ni, Zn, Cu, and Cd from contaminated soils in a significant amount using soil washing methods has been reported by several researchers.[18,23–25] Fig. 3.10.1 shows the schematic design of the soil washing technology.

3.10.2.1.3 Vitrification

This technology involves heating the soil at high temperature (1400–2000 °C) to decompose or volatilize organic matter and to form vitreous materials such as oxide solid. This technology can be applied both in ex situ or in situ processes. In ex situ vitrification processes, soils are excavated and subjected to pretreatment (such as screening, dewatering, and blending) and are fed to the reactors for melting. Collection of volatile metal species (such as Hg) is very important during this process. Cooling of melted soil produces vitreous materials, which are even stronger than concrete. For in situ vitrification processes, electrodes are inserted into the contaminated area vertically to heat the soil. The soil needs to be a little wet to provide the necessary conductance. The resistive heating melts the soil and moves both outward and vertically. Therefore, soil properties play significant roles in the vitrification

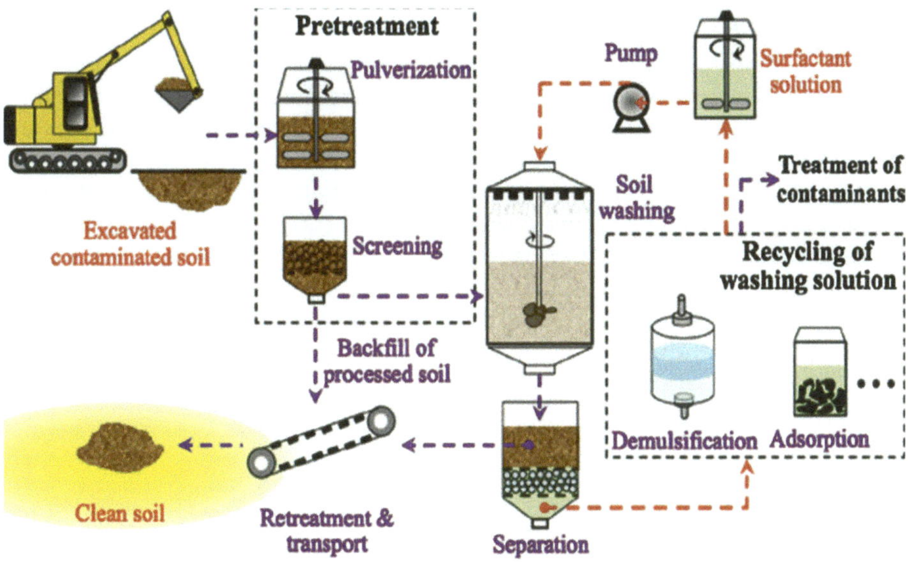

FIGURE 3.10.1 Schematic design of soil washing technology. *From Mao X, Jiang R, Xiao W, Yu J. Use of surfactants for the remediation of contaminated soils: a review. J Haz Mat 2015;**285**:419–35.*

process. Soil with high alkalinity is not suitable for this process as it fails to provide sufficient conductance to carry the current.[18,20] A large amount of energy investment is the main disadvantage of this process.

3.10.2.2 Chemical Methods

Chemical remediation methods involve the immobilization or extraction of heavy metals from the soil using different chemical agents.

3.10.2.2.1 Immobilization

Heavy metals partition in the soil in different phases such as soluble, exchangeable, Fe/Al bound, Ca/Mg bound, organic matter bound, and residual. Soluble forms of metals are readily available for uptake by biological organisms and are considered the most toxic to the environment. The immobilization process reduces the metal solubility and hence their mobility in the soil. Several organic and inorganic immobilizing amendments are used to reduce metal toxicity through precipitation, sorption, and complexation.[26] Commonly used immobilizing agents include clay, organic composts, cement, minerals, phosphates, and zeolites.[18,27] Several studies showed that other unconventional low-cost industrial residues such as red mud and termitaria can also be used to immobilize heavy metals in soil.[18,28,29] Immobilization processes are often coupled with soil capping methods to encapsulate the metals in a specific area. For this purpose, inorganic (such as fly ash, bentonite, kaolinite, charcoal, Fe/Mn oxides, and calcium carbonate) and organic (such as bitumen, compost, and manures) binders are used. The main objective of this process is to precipitate metals as metal hydroxides within the soil matrix. The presence of organic contaminants makes this method ineffective due to the possibilities of formation of toxic organic vapors.[30,31]

3.10.2.2.2 Extraction

This process involves the addition of chemical agents to bring the metals into the aqueous phase (by increasing the metal solubility) to extract/remove the metals from the system. Different types of inorganic eluents, chelating agents, and surfactants are commonly used in this method. Chelating agents possess the ability to form bonds with various metals, and hence they are widely used in physical, chemical, and biological remediation processes. Synthetic chelating agents such as EDTA, diethylenetriaminepentaacetic acid (DTPA), N-2-hydroxyethyl-ethylene-diaminetriacetic acid (HEDTA), nitrilotriacetate (NTA), and natural chelating agents such as ethylenediamine-N,N′-disuccinic acid (EDDS) have been successfully used to enhance soil Pb extraction by several researchers.[32–34] The role of potassium phosphate for arsenic extraction from soil has also been studied.[35,36] Chemically induced metal extraction processes are often coupled with phytoremediation methods and are found to be more effective. A detailed discussion of this part appears in the following phytoremediation section.

3.10.2.3 Biological Remediation

Biological remediation involves the use of living organisms for remediation of heavy metals in soil. In this section, we mainly focus on the use of plants as heavy metal remediation agents.

3.10.2.3.1 Phytoremediation

Phytoremediation is one of the emerging technologies for heavy metal remediation. The term phytoremediation comes from the Greek word *phyto* (meaning plant) and Latin word *remedium* (meaning to cure an evil). The success of phytoremediation depends on the presence of the metal of concern in a plant-available form in soil. Plants are only able to uptake soluble fractions of metals, hence, metal fractions that are strongly bound to soil particles are unavailable for plant uptake.[37] Some heavy metals such as Cd, Ni, Zn, As, and Cu are readily available to the plant, whereas others such as Pb and Cr are less available.[38] Although heavy metals like As, Cd, Hg, Pb, and Se are nonessential for plant growth, there are essential heavy metals such as Co, Cu, Ni, Zn, and Fe that plants need for their growth and development.

Excessive uptake of heavy metals can lead to a series of plant developmental problems and heavy metal poisoning. There are plants that can actively uptake heavy metals in large amounts from soil without showing any traits of heavy metal toxicity. This group of plants is known as metal hyperaccumulators, which account for only 0.2% of all known plant species.[39] Success of phytoremediation of heavy metals depends on the proper selection of the metal hyperaccumulator plants. Hyperaccumulator plants can accumulate heavy metals into their aboveground biomass (e.g., shoots and leaves) at concentrations 100- to 1000-fold higher than the plants belonging to the nonhyperaccumulator group. Generally, these plants can accumulate up to or more than 0.1% of metals like Cu, Pb, Cd, Cr, Ni, or Co or 1% of metals like Zn and Mn in their dry biomass.[40] Other factors such as the accumulation factor (AF) and the translocation factor (TF) also play determining roles for heavy metal hyperaccumulation inside plant tissues. AF is the ratio of metal concentration in plant tissue to the metal concentration in the soil, whereas TF is the ratio of metal concentration in the plant shoots to that in the plant roots. Heavy metal hyperaccumulator plants are able to rapidly translocate metals from root to shoot via the xylem. It was observed that hyperaccumulator plants such as *Thlaspi caerulescens* and *Sedum alfredi* can translocate metals from root to shoot within 48–72 h.[39] About 450 hyperaccumulator plant species belonging to 45 families have been identified so far, but there is a possibility that more unidentified plant species belonging to the hyperaccumulator category might exist in nature.[39,41,42] Fig. 3.10.2 shows the schematic diagram of the heavy metal phytoremediation process.

Apart from the aforementioned characteristics, it is desirable that plants used in phytoremediation should have other characteristics such as high growth rate, high aboveground biomass, a strong root system, resistance to toxic effects of target heavy metals, and the ability to translocate accumulated heavy metals from roots to shoots.[43] Some plants have developed certain mechanisms for solubilizing heavy metals in soil by making them more bioavailable. Plant roots secrete metal-mobilizing substances in the rhizosphere (the zone of soil surrounding a plant root) to increase metal dissolution. Root exudates reduce the pH of the rhizosphere region, which increases the solubility of heavy metals along with their availability for plant uptake.[43,44] Several studies also reported that the presence of rhizospheric microorganisms such as bacteria and mycorrhizal fungi can significantly increase the availability of heavy metals in soil. Root exudates provide a nutrient source for rhizospheric microorganisms, which stimulates microbial activities 10–100 times higher than the regular conditions.[1,43,45] Studies showed that symbiotic effect of the mycorrhizal

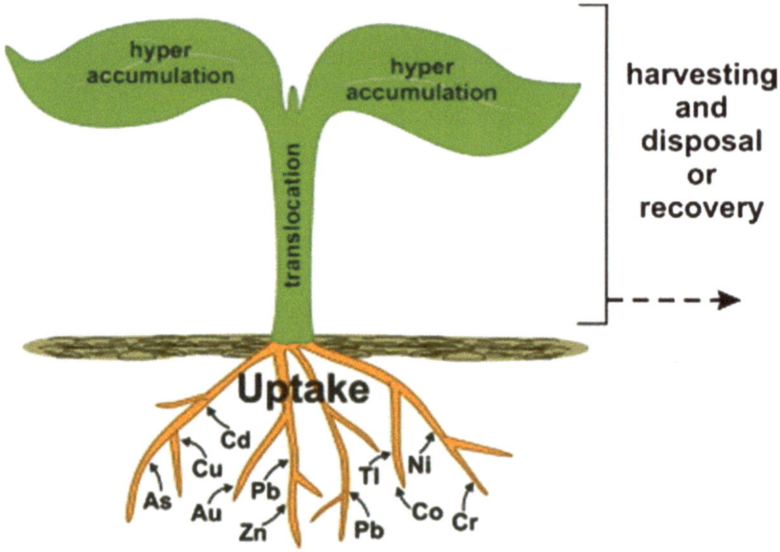

FIGURE 3.10.2 Phytoremediation of heavy metals using plants hyperaccumulator plants. Hyperaccumulators translocate metals to their aboveground parts, which can be periodically harvested.

fungi *Glomus mosseae* increases the phytoextraction of Pb by vetiver grass [*Chrysopogon zizanioides* (L.)].[46] Li et al. (2007) also reported that bacterial presence in the rhizosphere region enhances plant growth and metal uptake from soil by the plant species *S. alfredi*. Thus plant-microbe interactions play a significant role in phytoremediation.[47]

Fig. 3.10.3 shows the advantage of phytoremediation over other physical or chemical heavy metal remediation technologies such as soil excavation. For 10 acres of land,

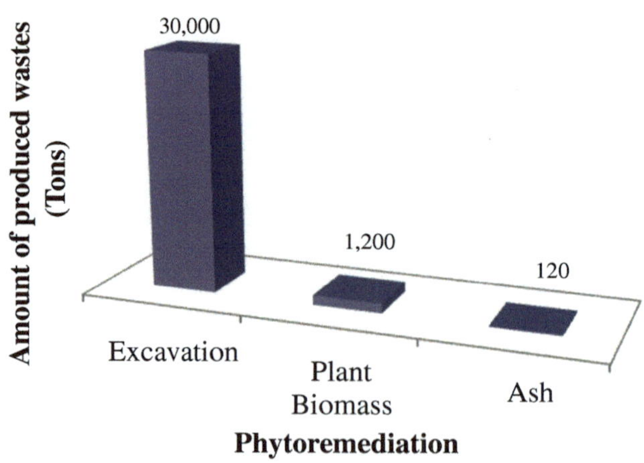

FIGURE 3.10.3 Comparative mass disposal (10 acres land) of hazardous wastes between different heavy metal remediation technologies. The amount of hazardous waste generated by phytoremediation process is <1% of the amount generated by conventional means.

TABLE 3.10.2 Plant Species Commonly Used for Phytoremediation of Heavy Metals

Plant Species	Metals of Concern	References
Atriplex halimus L.	Cd, Zn	48
Brassica spp.	Cd, Cr, Cu, Ni, Pb, Zn	49
Cichorium intybus	Pb	50
Cynodon dactylon	Cd, Cu, Pb, Zn	50,51
Cyperus alternifolius	Cd, Cu, Mn, Pb, Zn	52
Pteris vittata	As, Pb	53
Salix viminalis	Cd, Zn	54
Thlaspi caerulescens	Cd, Zn	55,56
Triticum aestivum L.	Cu, Cd, Cr, Zn, Fe, Ni, Mn, Pb	57
Chrysopogon zizanioides	Al, As, Cd, Cu, Fe, Hg, Mn, Pb, Zn	52,58–61

excavation processes require a safe disposal of 30,000 tons of heavy metal-contaminated soil. Application of phytoremediation technique on the same area can reduce the amount of heavy metal containing waste (in this case plant biomass) to 1200 tons. Incineration of the same plant biomass will generate only 120 tons of ash that needs to be disposed as hazardous waste. So, a significant reduction (more than 99%) of the total amount of hazardous waste is possible by applying phytoremediation in contrast to conventional heavy metal remediation technologies.

Table 3.10.2 lists some of the commonly used hyperaccumulator plants for phytoremediation of heavy metals. Photographs of some of these species are shown in Fig. 3.10.4. Plants such as *Brassica campestris* L., *Brassica carinata*, *Brassica juncea* L., and *Brassica nigra* L. can accumulate significant amounts of Pb in their dry biomass.[62] Studies showed that species such as *Cynodon dactylon* can accumulate metals like Cd (14 mg/kg), Pb (658 mg/kg), and Zn (828 mg/kg) in their biomass. It was also observed that species such as *Juncus usitatus* and *Lomandra longifolia* can also accumulate high amount of Cd in their biomass: 26 and 21 mg/kg, respectively.[5] Another potential plant species for remediation of Cd- and Zn-contaminated sites is *T. caerulescens*. Some studies reported that *T. caerulescens* is an excellent hyperaccumulator of Cd and Zn, and it can accumulate as much as 50–250 mg/kg Cd and 13,000–19,000 mg/kg Zn while growing in heavy metal-contaminated sites.[55,56] Studies also reported that *T. caerulescens* displays preferential root growth in areas that contain metals, thus increasing the efficiency of metal uptake processes.[1,63] However, due to its low biomass production, *T. caerulescens* is not considered as a preferred plant for phytoremediation technology. On the other hand, plants like *Cichorium intybus* L. and *C. dactylon* are potential phytoremediation candidates for Pb-contaminated sites. *C. intybus* and *C. dactylon* can accumulate as much as 800–1500 mg/kg and 400–1200 mg/kg Pb in their biomass, respectively.[50] In a similar study, Gonzalez and Gonzalez-Chavez[47] reported that *Atriplex*

FIGURE 3.10.4 Plants widely used for phytoremediation. (A) *Chrysopogon zizanioides* L. (vetiver grass); (B) *Pteris vittata* (Chinese Brake fern); (C) *Thlaspi sp.*; (D) *Brassica sp.*; (E) *Helianthus sp.*; and (F) *Polygonum sp.* *(A) From Abhishek RoyChowdhury; (C) From Brej T, Fabiszewski J. Plants accumulating heavy metals in the Sudety Mts. Acta Soc Bot Pol 2006;**75**(1):61–8; (D) From Felger RS, Rutman S, Salywon A, Malusa J. Ajo Peak to Tinajas Altas: a flora of southwestern Arizona. Part 11. Eudicots: Brassicaceae and Burseraceae. Phytoneuron 2015;**2015-6**:1–48; (E) From van der Kooi CJ. Plant biology: flower orientation, temperature regulation and pollinator attraction. Current Biology 2016;**26**(21):R1143–5; (F) From Lin L, Ni B, Lin H, Zhang M, Li X, Yin X, Qu C, Ni J. Traditional usages, botany, phytochemistry, pharmacology and toxicology of Polygonum multiflorum Thunb.: a review. Journal of Ethnopharmacology 2015;**159**:158–83.*

halimus L. can accumulate as much as 830 mg/kg of Cd and 440 mg/kg of Zn in its biomass while growing on contaminated soil under greenhouse conditions. The hyperaccumulator fern *Pteris vittata* was reported to be able to accumulate 27,000 mg/kg of As in its aboveground tissue.[64] *P. vittata* demonstrates the greatest root-to-shoot translocation of As, when compared with other nonhyperaccumulator ferns. Due to its high TF, *P. vittata* is one of the best performing As hyperaccumulator plants.[34,61,65] Another plant species that is highly suited for heavy metal remediation is *C. zizanioides*, commonly known as vetiver grass, due to its physiological characteristics and high tolerance to heavy metals such as As, Pb, Hg, and Cd.[33,60,65] Vetiver can tolerate Pb concentrations up to 3125 mg/kg, As concentrations up to 250 mg/kg, Cr concentrations up to 600 mg/kg, and Hg concentrations up to 6 mg/kg.[34,51,56,60,66,67] Vetiver can remove as much as 10.6% of total soil As while growing in a soil with 45 mg/kg As concentration.[67] Vetiver can remediate iron ore tailings contaminated with high concentrations of heavy metals such as Fe, Zn, Mn, and Cu and can accumulate 545–1197 mg/kg Fe, 302–531 mg/kg Zn, 415–648 mg/kg Mn, and 13–66 mg/kg Cu in its root and shoot tissues. High mean TFs for Mn (0.86), Fe (0.71), Zn (0.69), and Cu (0.55) were observed in vetiver tissue.[58]

As discussed earlier, the success of phytoremediation depends on the availability of the target metals. Hence, the overall efficiency of this technology can be improved by enhancing metal availability by various chemical or biological means, such as the addition of chelating agents, plant growth-promoting bacteria, mycorrhizal fungi, etc., to the soil. The use of chelating agents such as EDTA or citric acid increases the mobility of soil heavy metals and thus their plant availability. Chelating agents induce desorption of heavy metals that are strongly bound to the soil particles and form water-soluble complexes with the heavy metals in soil to facilitate the heavy metal uptake by plants. However, overuse of nonbiodegradable chelating agents such as EDTA can lead to metal leaching out of the soil and contaminating surface and groundwater. Excessive mobilization of metals can also result in plant toxicity. Therefore proper monitoring is required while applying these agents in the soil.[43,68–70] Use of natural and biodegradable chelating agents such as EDDS has been suggested to prevent excessive mobilization of metals. Several studies have reported successful use of EDDS to enhance soil Pb extraction by plants.[32–34] Moreover, introduction of mycorrhizal or plant growth-promoting bacterial associations with suitable hyperaccumulators would help improve mobilization and uptake of heavy metals.

3.10.2.3.2 Limitations of Phytoremediation

Although phytoremediation is a promising technology, it is not devoid of drawbacks. The major disadvantage of this technology is that it is a time-consuming process. Compared with other conventional physical and chemical remediation processes that might take weeks to months to clean up a heavy metal-contaminated site, phytoremediation may need several years to achieve the same result.[53,71] Another limitation of this technology is that the depth of the root system dictates the extent of remediation possible, which makes this technology effective only for sites with shallow contamination. In addition, the biomass of the plant plays a significant role in the extent of metals extracted. Plants with low biomass are not effective for this technology. Plant age also determines the success of the overall project. As young plants tend to uptake more metals than older ones, it is very important to maintain a healthy plant population in the target site throughout the process. [62,71] The presence of very high levels of heavy metals in soil can induce phytotoxicity, so the level of contaminants plays a significant role in the applicability of this technology. Other factors such as climatic conditions and soil properties also need to be considered for the successful application of this technology.[53,62,71]

3.10.3 CONCLUSIONS

Heavy metal contamination is a serious environmental problem, which is widespread due to various anthropogenic activities and which impacts both human and ecological health. Several remediation technologies have been developed over the years to deal with this problem. As physical and chemical remediation processes for heavy metal cleanup are highly expensive and sometimes generate a significant amount of secondary pollutants, the focus has now shifted toward developing "greener" and ecofriendly approaches such as phytoremediation. Several candidate plants suitable for phytoremediation have shown promising results, but this technology is highly site and target metal specific. Several factors such as soil properties, climate, and target metal concentrations need to be considered. Selection of proper plant species also plays a significant role in achieving a high performance of this remediation technology. Combination with chelating agents and microorganisms have been suggested to enhance metal mobilization and uptake. Further research to better understand rhizosphere interactions with microorganisms and the molecular basis of metal tolerance and hyperaccumulation needs to be carried out to maximize the applicability of phytoremediation.

References

1. Vamerali T, Bandiera M, Mosca G. Field crops for phytoremediation of metal-contaminated land. A review. *Environ Chem Lett* 2010;**8**:1–17.
2. World Health Organization (WHO). *Preventing diseases through healthy environment: action needed on chemicals of major public health concern.* 2010. Available from: http://www.who.int/ipcs/assessment/public_health/chemicals_phc/en/.
3. Agency for Toxic Substances, Disease Registry. *U.S. department of health and human services (ATSDR). Toxic substances portal.* 2011. Available from: https://www.atsdr.cdc.gov/substances/toxsubstance.asp.
4. USEPA. *Supplemental guidance for developing soil screening levels for superfund sites.* Washington, DC: Office of Solid Waste and Emergency Response; 2002. Available from: http://www.epa.gov/superfund/health/conmedia/soil/index.htm.
5. USEPA. *Regional screening levels table.* May 19, 2009.
6. ATSDR. *CERCLA priority list of hazardous substances.* 2015. Available from: https://www.atsdr.cdc.gov/spl/.
7. USEPA. *Implementation guidance for the arsenic rule; implementation guidance for the arsenic rule; EPA-816-K-02-018.* Washington, DC: Office of Ground Water and Drinking Water; 2002.
8. USEPA. *Integrated risk information system (IRIS).* Washington, DC: Environmental Criteria and Assessment Office; 1996.
9. Jarup L. Hazards of heavy metal contamination. *Br Med Bull* 2003;**68**:167–82.
10. Nriagu JO, Pacyna JM. Quantitative assessment of worldwide contamination of air, water, and soils by trace metals. *Nature* 1988;**333**:134–9.
11. USEPA. *Learn about lead (the National Lead Information Center).* August 2014. Available from: http://www.epa.gov.
12. CDC. *Impact of lead-contaminated soil on public health.* Publication Date: 5/1/1992. 1992. Available from:, http://wonder.cdc.gov.

13. CDC. *Preventing lead poisoning in young children: a statement by the U.S. department of health and human services.* Public Health Service, Centers for Disease Control; 1991.
14. Gulson BL, Davis JJ, Bawdensmith J. Plant as a source of recontamination of houses in urban environments and its role in maintaining elevated blood leads in children. *Sci Total Environ* 1995;**164**:221–35.
15. Pirkle JL, Kaufmann RB, Brody DJ, Hickman T, Gunter EW, Paschal DC. Exposure of the U.S. population to lead, 1991–1994. *Environ Health Perspect* 1998;**106**:745–50.
16. Andra SS, Sarkar D, Saminathan SKM, Datta R. Predicting potentially plant-available lead in contaminated residential sites. *Environ Monit Assess* 2011;**175**:661–76.
17. Hanna-Attisha M, LaChance J, Sadler RC, et al. Elevated blood lead levels in children associated with the Flint drinking water crisis: a spatial analysis of risk and public health response. *Am J Public Health* 2016;**106**:283–90.
18. Zhou DM, Hao XZ, Xue Y, et al. Advances in remediation technologies of contaminated soils. *Ecol Environ Sci* 2004;**13**(2):234–42.
19. Yao Z, Li J, Xie H, Yu C. Review on remediation technologies of soil contaminated by heavy metals. *Procedia Environ Sci* 2012;**16**:722–9.
20. Pearl M, Wood P. *Review of pilot and full scale soil washing plants. Warren Spring Laboratory Report LR 1018.* Department of the Environment, AEA Technology National Environmental Technology Centre; 1994. B551 Harwell, Oxfordshire, OX11 0RA.
21. Maturi K, Reddy KR. Extractants for the removal of mixed contaminants from soils. *Soil Sediment Contam* 2008;**17**(6):586–608.
22. Chen TC, Hong A. Chelating extraction of lead and copper from an authentic contaminated soil using *N*- (2-acetamido) iminodiacetic acid and S-carboxymethyl-L-cysteine. *J Hazard Mater* 1995;**41**(2–3):147–60.
23. Khodadoust AP, Reddy KR, Maturi K. Effect of different extraction agents on metal and organic contaminant removal from a field soil. *J Hazard Mater* 2005;**117**(1):15–24.
24. Wuana RA, Okieimen FE, Imborvungu JA. Removal of heavy metals from a contaminated soil using organic chelating acids. *Int J Environ Sci Technol* 2010;**7**(3):485–96.
25. Hashimoto Y, Matsufuru H, Takaoka M, Tanida H, Sato T. Impacts of chemical amendment and plant growth on lead speciation and enzyme activities in a shooting range soil: an X-ray absorption fine structure investigation. *J Environ Qual* 2009;**38**(4):1420–8.
26. Finžgar N, Kos B, Leštan D. Bioavailability and mobility of Pb after soil treatment with different remediation methods. *Plant Soil Environ* 2006;**52**(1):25–34.
27. Lombi E, Zhao FJ, Zhang G, et al. In situ fixation of metals in soils using bauxite residue: chemical assessment. *Environ Pollut* 2002;**118**(3):435–43.
28. Anoduadi CO, Okenwa LB, Okieimen FE, Tyowua AT, Uwumarongie-Ilori EG. Metal immobilization in CCA contaminated soil using laterite and termite mound soil. Evaluation by chemical fractionation. *Niger J Appl Sci* 2009;**27**:77–87.
29. Fawzy EM. Soil remediation using in situ immobilization techniques. *Chem Ecol* 2008;**24**(2):147–56.
30. Farrell M, Perkins WT, Hobbs PJ, Griffith GW, Jones DL. Migration of heavy metals in soil as influenced by compost amendments. *Environ Pollut* 2010;**158**(1):55–64.
31. Grčman H, Vodnik D, Velikonja-Bolta Š, Leštan D. Ethylenediaminedisuccinate as a new chelate for environmentally safe enhanced lead phytoextraction. *J Environ Qual* 2003;**32**:500–6.
32. Kos B, Leštan D. Induced phytoextraction/soil washing of lead using biodegradable chelate and permeable barriers. *Environ Sci Technol* 2003;**37**:624–9.
33. Sarkar D, Andra SS, Saminathan SKM, Datta R. Chelant-aided enhancement of lead mobilization in residential soils. *Environ Pollut* 2008;**156**:1139–48.
34. Alam GM, Tokunaga S, Maekawa T. Extraction of arsenic in a synthetic arsenic contaminated soil using phosphate. *Chemosphere* 2001;**43**(8):1035–41.
35. Lee M, Paik IS, Do W, et al. Soil washing of As-contaminated stream sediments in the vicinity of an abandoned mine in Korea. *Environ Geochem Health* 2007;**29**(4):319–29.
36. Datta R, Sarkar D. Effective integration of soil chemistry and plant molecular biology in phytoremediation of metals: an overview. *Environ Geosci* 2004;**11**(2):53–63.
37. Prasad MNV. Phytoremediation of metal-polluted ecosystems: hype for commercialization. *Russ J Plant Physiol* 2003;**50**:686–700.
38. Rascio N, Navari-Izzo F. Heavy metal hyperaccumulating plants: how and why do they do it? And what makes them so interesting? *Plant Sci* 2011;**180**:169–81.

39. Baker AJM, Brooks RR. Terrestrial higher plants which hyperaccumulate metallic elements. A review of their distribution, ecology and phytochemistry. *Biorecovery* 1989;**1**:12–81.
40. Baker AJM, Reeves RD, Hajara ASM. Heavy metal accumulation and tolerance in British populations of the metallophyte *Thlaspi caerulescens* J. & C. Presl (Brassicaceae). *New Phytol* 1994;**127**:61–8.
41. RoyChowdhury A, Sarkar D, Datta D. Remediation of acid mine drainage-impacted water. *Curr Pollut Rep* 2015;**1**:131–41.
42. Ali H, Khan E, Sajad MA. Phytoremediation of heavy metals – concepts and applications. *Chemosphere* 2013;**91**:869–81.
43. Thangavel P, Subbhuraam C. Phytoextraction: role of hyperaccumulators in metal contaminated soils. *Proc Indian Natl Sci Acad Part B* 2004;**70**:109–30.
44. Sheoran V, Sheoran A, Poonia P. Role of hyperaccumulators in phytoextraction of metals from contaminated mining sites: a review. *Crit Rev Env Sci Tec* 2011;**41**:168–214.
45. Punamiya P, Datta R, Sarkar D, Barber S, Patel M, Das P. Symbiotic role of *Glomus mosseae* in phytoextraction of lead in vetiver grass [*Chrysopogon zizanioides* (L.)]. *J Hazard Mater* 2010;**177**(1–3):465–74.
46. Li WC, Ye ZH, Wong MH. Effects of bacteria on enhanced metal uptake of the Cd/Zn hyperaccumulating plant, *Sedum alfredii*. *J Exp Bot* 2007:1–10.
47. Gonzalez RC, Gonzalez-Chavez MCA. Metal accumulation in wild plants surrounding mining wastes: soil and sediment remediation (SSR). *Environ Pollut* 2006;**144**:84–92.
48. Van Ginneken L, Meers E, Guisson R, et al. Phytoremediation for heavy metal-contaminated soils combined with bioenergy production. *J Environ Eng Landsc Manag* 2007;**15**(4):227–36.
49. Archer MJG, Caldwell RA. Response of six Australian plant species to heavy metal contamination at an abandoned mine site. *Water Air Soil Pollut* 2004;**157**:257–67.
50. Shu W. Exploring the potential utilization of Vetiver in treating acid mine drainage (AMD). In: *Proceedings of the 3rd international vetiver conference. Guangzhou (China)*; October 6–9, 2003. Available from: www.vetiver.org.
51. Knight B, Zhao FJ, McGrath SP, Shen ZG. Zinc and cadmium uptake by the hyperaccumulator *Thlaspi caerulescens* in contaminated soils and its effects on the concentration and chemical speciation of metals in soil solution. *Plant Soil* 1997;**197**:71–8.
52. Salido AL, Hasty KL, Lim JM, Butcher DJ. Phytoremediation of arsenic and lead in contaminated soil using Chinese Brake ferns (*Pteris vittata*) and Indian mustard (*Brassica juncea*). *Int J Phytoremediation* 2003;**5**(2):89–103.
53. Hammer D, Kayser A, Keller C. Phytoextraction of Cd and Zn with *Salix viminalis* in field trials. *Soil Use Manag* 2003;**19**(3):187–92.
54. Lutts S, Lefevre I, Delperee C, Kivits S, Dechamps C, Robledo A, et al. Heavy metal accumulation by the halophyte species mediterranean saltbush. *J Environ Qual* 2004;**33**:1271–9.
55. Du L, Truong PNV. Soil in Southern Vietnam. In: *Proceedings of the 3rd international vetiver conference, Guangzhou (China)*; October 6–9, 2003. Available from: www.vetiver.org.
56. Chandra R, Bharagava RN, Yadav S, Mohan D. Accumulation and distribution of toxic metals in wheat (*Triticum aestivum* L.) and Indian mustard (*Brassica campestris* L.) irrigated with distillery and tannery effluents. *J Hazard Mater* 2009;**162**(2–3):1514–21.
57. Yang ZY, Yuan JG, Xin GR, Chang HT, Wong MH. Germination, growth and nodulation of *Sesbania rostrata* grown in Pb/Zn mine tailings. *Environ Manage* 1997;**21**:617–22.
58. Truong P, Carlin G, Cook F, Thomas E. Vetiver grass hedges for water quality improvement in acid sulfate soils, Queensland, Australia. In: *Proceedings of the 3rd international vetiver conference. Guangzhou (China)*; October 6–9, 2003. p. 182–93. Available from: www.vetiver.org.
59. Roongtanakiat N, Tangruangkiat S, Meesat R. Utilization of vetiver grass (*Vetiveria zizanioides*) for removal of heavy metals from industrial wastewaters. *ScienceAsia* 2007;**33**:397–403.
60. Roongtanakiat N, Osotsapar Y, Yindiram C. Effects of soil amendment on growth and heavy metals content in vetiver grown on iron ore tailings. *Kasetsart J Nat Sci* 2008;**42**:397–406.
61. Tangahu BV, Sheikh Abdullah SR, Basri H, Idris M, Anuar N, Mukhlisin M. A review on heavy metals (As, Pb, and Hg) uptake by plants through phytoremediation. *Int J Chem Eng* 2011;**2011**. http://dx.doi.org/10.1155/2011/939161.
62. Whiting SN, Leake JR, McGrath SP, Baker AJM. Positive response to Zn and Cd by roots of the Zn and Cd hyperaccumulator *Thlaspi caerulescens*. *New Phytol* 2000;**145**:199–210.

63. Wang J, Zhao F-J, Meharg AA, Raab A, Feldmann J, McGrath SP. Mechanisms of arsenic hyperaccumulation in *Pteris vittata*. Uptake kinetics, interactions with phosphate, and arsenic speciation. *Plant Physiol* 2002;**30**(3):1552–61.
64. Poynton CY, Huang JWW, Blaylock MJ, Kochian LV, Ellass MP. Mechanisms of arsenic hyperaccumulation in *Pteris* species: root as influx and translocation. *Planta* 2004;**219**:1080–8.
65. Truong PN, Hart B. *Vetiver system for wastewater treatment. Technical bulletin no. 2001/2. Pacific Rim vetiver Network.* Bangkok (Thailand): Office of the Royal Development Projects Board; 2001.
66. Datta R, Quispe MA, Sarkar D. Greenhouse study on the phytoremediation potential of vetiver grass, *Chrysopogon zizanioides* L., in arsenic-contaminated soils. *Bull Environ Contam Toxicol* 2011;**86**:124–8.
67. Zhou QX, Cui S, Wei SH, Zhang W, Cao L, Ren LP. Effects of exogenous chelators on phytoavailability and toxicity of Pb in *Zinnia elegans* Jacq. *J Hazard Mater* 2007;**146**:341–6.
68. Freitas EV, Nascimento CW, Souza A, Silva FB. Citric acid-assisted phytoextraction of lead: a field experiment. *Chemosphere* 2013;**92**:213–7.
69. Paz-Ferreiro J, Lu H, Fu S, Mendez A, Gasco G. Use of phytoremediation and biochar to remediate heavy metal polluted soils: a review. *Solid Earth* 2014;**5**:65–75.
70. Mwegoha WJS. The use of phytoremediation technology for abatement soil and groundwater pollution in Tanzania: opportunities and challenges. *J Sustain Dev Afr* 2008;**10**(1):140–56.
71. Tu S, Ma LQ, Fayiga AO, Zillioux EJ. Phytoremediation of arsenic-contaminated groundwater by the arsenic hyperaccumulating fern *Pteris vittata* L. *Int J Phytoremediation* 2004;**6**(1):35–47.

Recommended Reading

1. Su YH, McGrath SP, Zhu YG, Zhao FJ. Highly efficient xylem transport of arsenite in the arsenic hyperaccumulator *Pteris vittata*. *New Phytol* 2008;**180**:434–41.
2. Truong PN, Baker D. *Technical bulletin No. 1998/1. Pacific Rim Vetiver Network.* Bangkok (Thailand): Office of the Royal Development Projects Board; 1998.

CHAPTER

3.11

Application of Green Chemistry in Homogeneous Catalysis

Anne Kokel, Christian Schäfer

University of Massachusetts Boston, Boston, MA, United States

3.11.1 INTRODUCTION

The concept of homogeneous catalysis is among the most used in organic synthesis making a broad variety of transformations possible by lowering the activation energies and thus opening new pathways for reactions to proceed. Although catalysis itself is one of the principles of green chemistry[1] and catalyzed reactions usually run under milder conditions than the corresponding noncatalyzed ones, from the point of view of green chemistry there are still several issues that need to be addressed. While several more general ideas as preventing waste (e.g., by being more atom efficient) and using less harmful chemicals and solvents are obvious for all types of reactions, special attention should be turned to the catalyst when considering green concepts in homogenous catalysis. Due to the large variety of catalysts applied, ranging from simple acids or bases to highly sophisticated metal complexes or organocatalysts,[2] there is no single approach valid for all these reactions. Instead, each type of catalyst should be considered separately. With regard to metal catalysts the identity of the metal is of importance. Transitioning from rare and often toxic heavy metals to more abundant and less toxic metals is desirable. An often overlooked aspect is the environmental impact of the frequently needed ligand for a metal catalyst. Although it would be ideal to perform ligand-free metal catalysis, ligands are often needed to stabilize the catalyst and achieve the desired activity. Still, the synthesis of the ligand should be factored into the overall green balance of a reaction, with highly engineered, complex ligands that require a multistep synthesis having a notable impact. Thus, even if the actual catalytic step may be considered green, the ligand synthesis can make the whole process less attractive from a sustainable point of view. The same argument is true when considering organocatalysis. Organocatalytic reactions are, by nature, metal free. Still, the design of the catalyst can have consequences on the overall environmental impact of the reaction, in particular with regard to the often high catalyst loadings needed for organocatalytic reactions. An

Green Chemistry
http://dx.doi.org/10.1016/B978-0-12-809270-5.00016-9

organocatalyst should therefore be ideally an easily obtained natural compound or a closely related molecule that is able to perform the desired transformation with low catalyst loadings. Unfortunately, many transformations require the use of a specifically designed organocatalyst to be efficiently performed.

Since their establishment in 1998, the implication of green chemistry principles has become more and more important, resulting in an increase in the number of reported homogeneous catalytic reactions that apply these principles. Here we will give an overview of the different approaches, illustrated by selected examples.

3.11.2 METAL-BASED CATALYSIS

3.11.2.1 Oxidation Reactions

The oxidation of alcohols to the corresponding carbonyl compounds is among the most important transformations in organic synthesis. The conventional catalytic reactions conducting such oxidation involve toxic metals (Pd, Ir, Ru, Mo, etc.) and hazardous solvents and/or oxidants. Some green pathways have been developed using less toxic catalysts and/or more environmentally benign solvents and oxidants.

Methods with lower potential environmental impact include aqueous reactions using water-soluble catalysts. The use of hazardous organic solvents is therefore eliminated and the ease for products isolation by simple extraction is increased. Several examples using palladium in water have been reported. Buffin et al. designed a water-soluble palladium-biquinoline complex for the aerobic oxidation of alcohols (Scheme 3.11.1).[3] Depending on the substrate, the temperature applied was in the range of 100–125 °C and an air pressure of 390–600 psi was applied. All conversions were performed at a pH of approximately 11 coupled with a catalyst loading of 1 mol%. Under these conditions, secondary alcohols afforded ketones with moderate to excellent yields. While aliphatic primary alcohols were fully oxidized to the corresponding carboxylic acid, substituted benzyl alcohols did not undergo overoxidation and yielded the corresponding aldehyde with only a small percentage

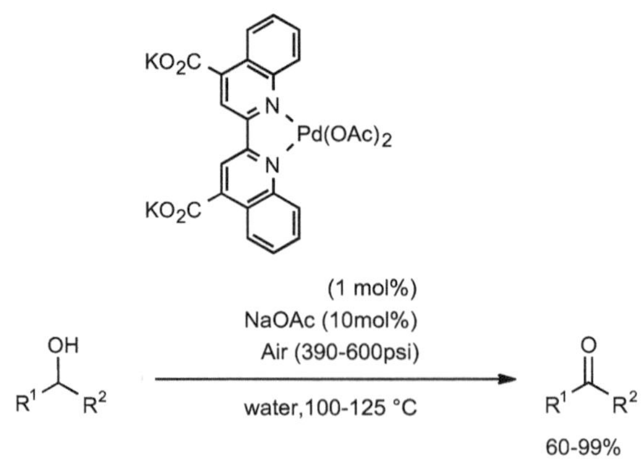

SCHEME 3.11.1

of the acid by-product. The catalytic system was tolerant toward a variety of functional groups, unlike other known palladium/water systems.[4]

Among others, Gorden et al. proved that nonprecious metals, such as copper, were able to catalyze the oxidation of alcohols in aqueous media. The copper(II) complex of sulfonated-2-quinoxalinol salen achieved the oxidation of different types of substrates from propargylic to benzylic and allylic alcohols with high yield and high selectivity using *t*-butyl hydroperoxide as oxidant.[5] The system required only 1% of catalyst loading and was performed under mild conditions (Scheme 3.11.2).

Oxygen and hydrogen peroxide are the greenest oxidants that can be used. They avoid the release of toxic by-products from the reduced oxidative agent. While the combination of molybdenum/hydrogen peroxide showed efficient oxidation activity (Scheme 3.11.3),[6] the copper-based reaction has more advantages because of the low toxicity and the affordable price of this metal (Scheme 3.11.4). The dimeric copper complex di-μ-chloro-bis[chloro(di-3,5-dimethylpyrazole)copper(II)] was shown to catalyze the oxidation of benzyl alcohols in the presence of hydrogen peroxide.[7] The conversion occurred at room temperature in acetonitrile with 1% catalyst to result in the formation of the aldehydes in moderate to excellent yields. The use of hydrogen peroxide regenerates the copper(II) species after the oxidation and releases water as the only by-product.

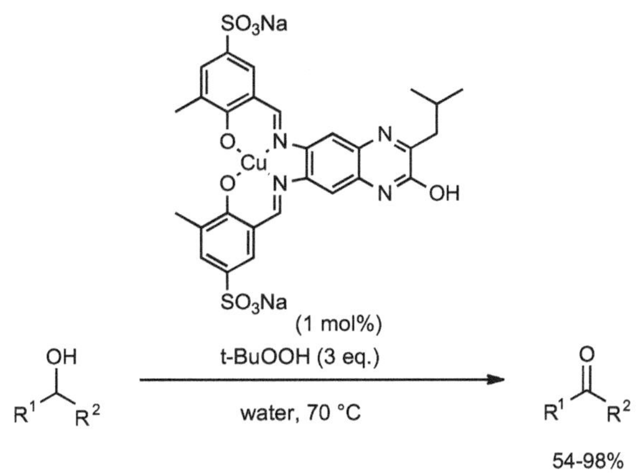

SCHEME 3.11.2

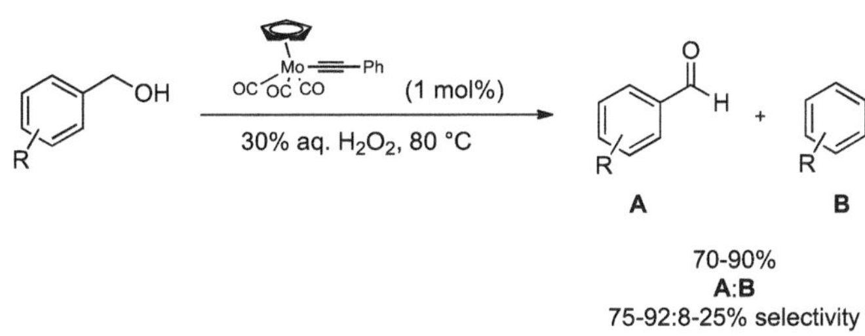

SCHEME 3.11.3

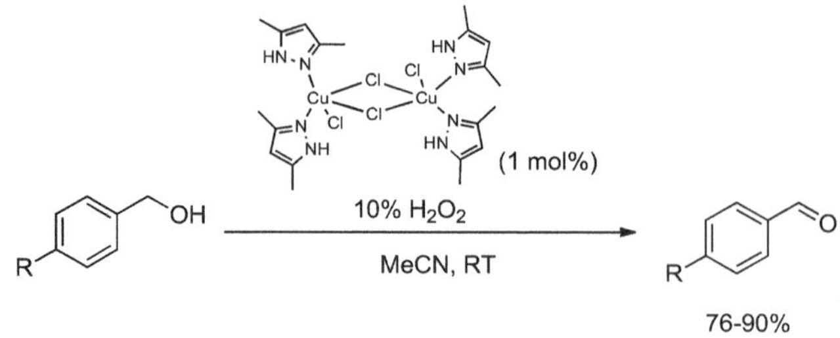

SCHEME 3.11.4

Another system using the easy-to-handle $CuBr_2$/2,2′-bipyridine ligand in combination with 2,2,6,6-tetramethylpiperidyl-1-oxyl (TEMPO) was conducted under air at atmospheric pressure (Scheme 3.11.5).[8] This Cu-TEMPO-catalyzed aerobic oxidation allowed the formation of aldehydes with excellent yield at room temperature using 5 mol% of catalyst and cocatalysts in a water/acetonitrile solvent mixture. The overoxidation to the acid was not observed, and no conversion occurred without the presence of TEMPO. It is believed that TEMPO acts as a hydrogen acceptor during the catalytic cycle.

Both water and hydrogen peroxide or molecular oxygen can be combined in the same system, further decreasing the environmental impact of the oxidation reaction. Repo et al. showed that $CuSO_4$-catalyzed oxidation of secondary benzylic alcohols with H_2O_2 in water at 100 °C occurred with high efficiency.[9] A second system with a lower temperature and the addition of a base and the ligand TEMPO improved the selectivity (Scheme 3.11.6).[9] Due to the high water solubility of the catalyst, when the oxidation products and the unreacted substrate were solid, the catalyst was recovered after precipitation of the solid compounds.

The readily available catalyst bismuth tribromide was also able to achieve oxidation of alcohols in aqueous 30% hydrogen peroxide. A large range of benzylic and aliphatic alcohols

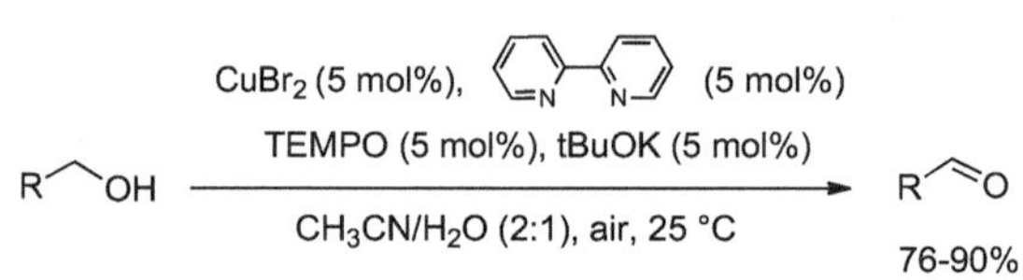

SCHEME 3.11.5

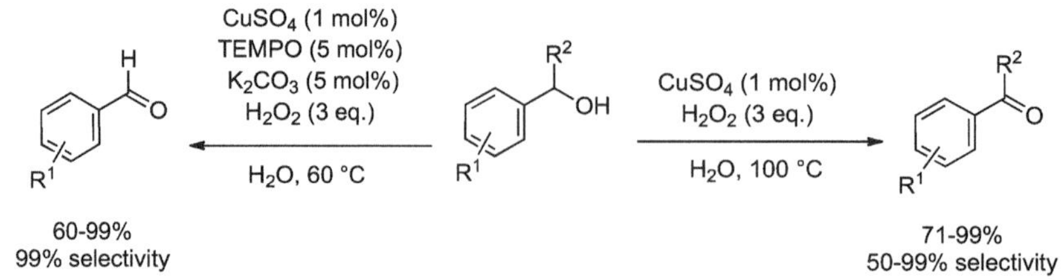

SCHEME 3.11.6

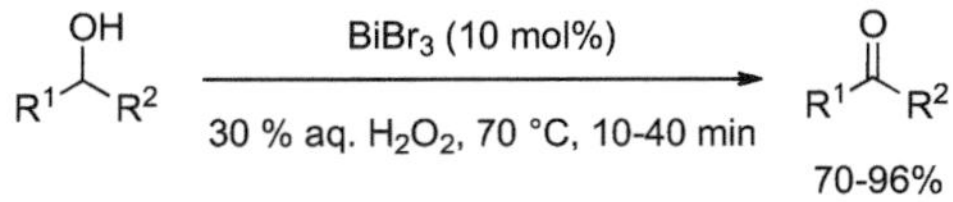

SCHEME 3.11.7

were converted to the corresponding carbonyl compounds in no more than 40 min at 70 °C (Scheme 3.11.7).[10] Neither the position of the substituent on the aromatic ring nor the nature of the substituent affected the yields.

In 2004, Ogawa et al. developed a novel and efficient system for the oxidation of benzylic alcohol derivatives catalyzed by vanadium-bipyridyl complexes in water using molecular oxygen as oxidant (Scheme 3.11.8).[11] Vanadium is an attractive catalyst because it is abundant, inexpensive and has low toxicity. Good yields were obtained for multiple substrates at 90 °C under 0.1 MPa oxygen pressure. It was shown for some examples that the reaction could be performed in an open flask under air, albeit the reaction time increased significantly. Furthermore, $MgSO_4$ was needed to improve the contact between the substrate and the catalyst; low conversions were observed otherwise.

A similar study combined a vanadium-phosphorus catalyst with the ligand TEMPO for the oxidation of a large range of substrates with molecular oxygen in aqueous medium as well (Scheme 3.11.9).[12] While benzyl alcohol derivatives led to moderate to excellent yields of the aldehyde product, aliphatic alcohols were hardly converted to the carbonyl compound. It is worth noting that the addition of phosphoric acid, which acts as an inorganic ligand for the vanadium species, was crucial to ensure high efficiency of the catalytic reaction; no conversion was observed without it.

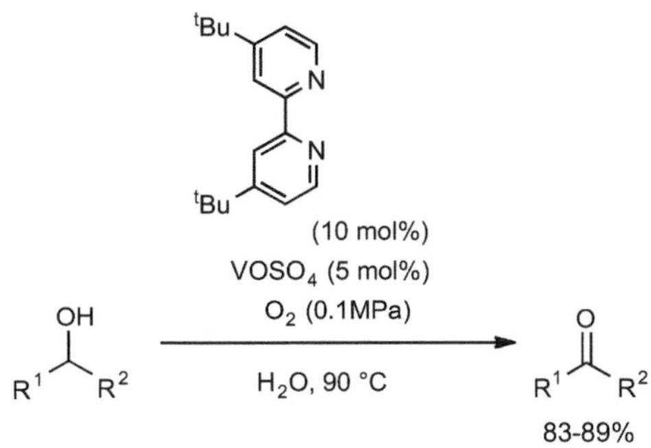

SCHEME 3.11.8

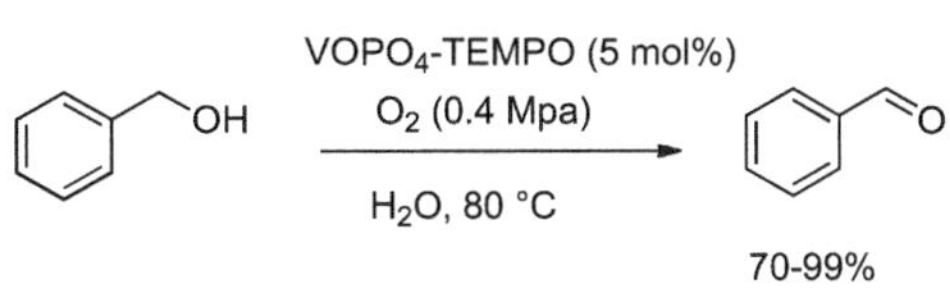

SCHEME 3.11.9

The imine formation is of key importance in organic and medicinal chemistry. Indeed, imines are used as versatile intermediates for the synthesis of biologically active compounds and fine chemicals. Traditionally, they are prepared by the condensation of carbonyl compounds with primary amines under harsh conditions using a Lewis acid catalyst, often a rare transition metal. Different approaches have been studied using abundant metals as catalyst under pure oxygen or air and under mild conditions. Using aerobic copper catalysis, high activity was observed for the oxidation of various primary benzylic and cyclic amines employing copper chloride.[13] The reaction was performed at 100 °C under neat conditions. The proposed method not only worked with the green oxidant air but also was performed without solvent, which is the best option from a sustainable point of view. The yields obtained were up to 90% (Scheme 3.11.10). It was found that the moisture in the air played a role in both avoiding the overoxidation of the amine and promoting the reaction, since poor selectivity was observed while performing the reaction under dry oxygen atmosphere. The authors proposed two pathways for the formation of the final imine. Both involve the formation of an initial imine by oxidation of the amine. It is then possible to proceed either via transamination with the unreacted amine or via hydrolysis, wherein the aldehyde formed reacts with the amine to give the final imine product.

Zhang et al. proposed another method using $Cu(ClO_4)_2$ to achieve the oxidative coupling of an imine from benzyl alcohol and amines.[14] Atmospheric pressure of molecular oxygen was employed as a green oxidant in this transformation (Scheme 3.11.11). The remarkably low catalyst loading (0.005%) is another advantage of the method. However, the catalytic system failed to produce imines from aliphatic alcohols.

As with alcohol oxidation, the use of hydrogen peroxide as an oxidant was explored for amine oxidation. H_2O_2 in water at room temperature using copper sulfate as catalyst constitutes a simple and energy-efficient method to afford imines (Scheme 3.11.12).[15] The reaction mechanism proposed is similar to that of the aforementioned study. The formation of amine/$CuSO_4$ complex occurs first, which reacts with hydrogen peroxide to produce an intermediate imine; second, the final product is obtained by transamination of the intermediate imine with the starting amine. A pathway involving the hydrolysis of the amine followed by condensation of the resulting aldehyde with the amine substrate is considered as well.

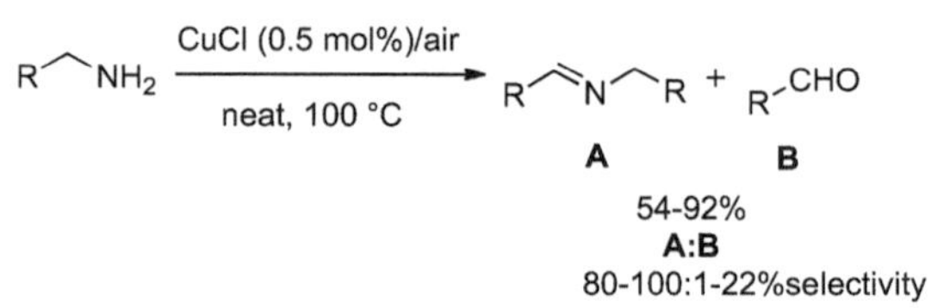

SCHEME 3.11.10

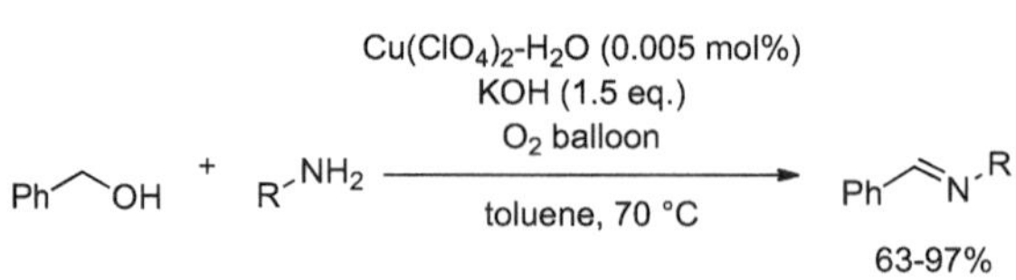

SCHEME 3.11.11

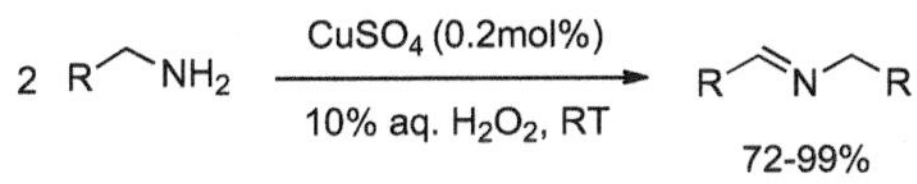

SCHEME 3.11.12

3.11.2.2 Reduction of C=C and C=X Double Bonds

Hydrogenation is an atom-efficient and essential step in organic synthesis. Many different catalysts are able to perform the hydrogenation reaction. Transition metals such as palladium, ruthenium, and rhodium complexes in combination with organic solvents are widely used in this field. However, these conventional catalysts and reaction media create health, safety, and environmental issues. Organic solvents should be avoided due to their flammability, toxicity, and volatility. Despite the noble metals' price and potential toxicity, some catalytic systems using these metals offer interesting reaction conditions toward greener synthesis. Grosselin et al. reported an aqueous/organic biphasic system allowing the recovery of the catalyst, which was either ruthenium or rhodium coordinated to tris(*m*-sulfophenyl)phosphine trisodium salt.[16] In separate reactions, α,β-unsaturated aldehydes were hydrogenated at the C=C bond by the in-situ-formed rhodium complex and the reduction of the carbonyl group was performed by the in-situ-formed ruthenium complex (Scheme 3.11.13). The substrates were reduced using a 20–50 bar pressure of molecular hydrogen. Excellent yields and selectivities (95%–99%) were observed even after recycling the catalyst.

The discovery of the chiral 2,2-bis(diphenylphosphanyl)-1,1-binaphthyl (BINAP) ligand contributed significantly to the synthesis of enantiomerically pure chiral compounds useful in various fields such as pharmaceutical or fragrance industry.[17] The BINAP-rhodium(II) complexes achieved highly enantioselective hydrogenation of various β-keto esters to chiral β-hydroxy esters or α,β- and α,γ-unsaturated carboxylic acids to saturated chiral carboxylic acids. Functionalized olefins and ketones were also converted with high level of chemo-and/or enantioselectivity. Even though the synthesis of the chiral ligand is complex, it exhibits an unprecedented chiral ability due to four axially and equatorially oriented *P*-phenyl rings and it allows the use of mild conditions for the catalytic reactions. For instance, the reaction of the olefinic substrate was achieved in ethanol using a substrate/catalyst molar ratio of 100 under only 4 atm of hydrogen at room temperature giving the amino acid in 97% yield (Scheme 3.11.14).[18] The S-enantioselectivity was almost perfect.

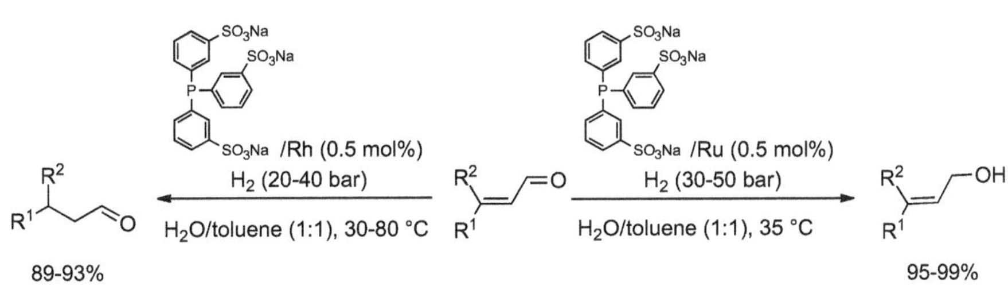

SCHEME 3.11.13

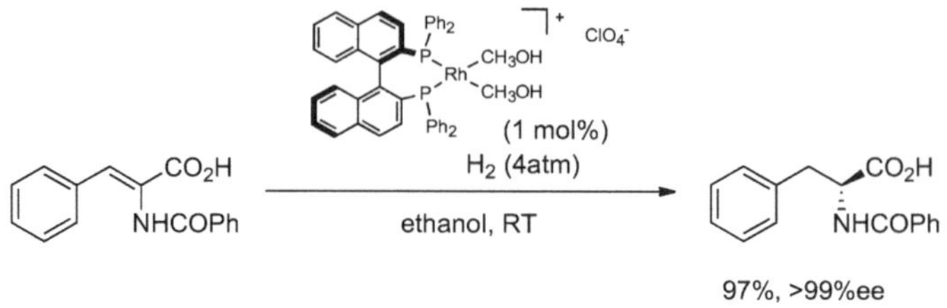

SCHEME 3.11.14

Noyori and his group published other works about the asymmetric hydrogenation catalyzed by chiral ruthenium complexes. They demonstrated that the hydrogenation of ketones and imines with 2-propanol as the hydrogen source led to high yield and up to 99% enantiomeric excess.[19] The reactions were performed in a solution of the inexpensive, environment-friendly, and easy-to-handle alcohol at modest temperature (Scheme 3.11.15). This method avoids the risk caused by presence of the highly combustible hydrogen gas; however, a nonnegligible amount of oxidized reducing agent is generated. A cocatalyst, the solvent's conjugated base [$(CH_3)_2CHOK$], was eventually used to assist in proton dissociation from the hydroxyl group of the alcohol. The authors noticed that the efficiency of the system was affected by the reversibility of the reaction because 2-propanol and the product are both secondary alcohols. This inherent issue highly depends on the redox potential of the hydrogen donor and acceptor present. Even though the reverse process is much slower, it may prevent complete conversion of the substrate or decrease the enantiomeric purity of the desired product.

The asymmetric transfer hydrogenation in aqueous media is a powerful alternative for asymmetric reduction. The water-soluble Ru-TsDPEN (TsDPEN=*N*-(*p*-toluenesulfonyl)-1,2-diphenylethylenediamine) developed by Noyori, Ikaryia, and coworkers[20] demonstrated its ability to catalyze the reduction of ketones by sodium formate in water (Scheme 3.11.16).[21] The optimal conditions were 40 °C and a neutral pH. The study of the effect of pH on the catalytic reaction revealed that the reaction rate was slower at basic and acidic pH. A large range of aromatic ketones afforded chiral secondary alcohols with high yield and high selectivity.

The possibility of using glycerol as hydrogen donor and solvent has also been explored. Glycerol presents many advantages, namely, it is nontoxic and biodegradable, it can be produced from renewable source, and it is able to dissolve many organic compounds. Wolfson et al. reported the transfer hydrogenation of various unsaturated compounds using glycerol and a dimeric ruthenium catalyst (Scheme 3.11.17).[22] Herein, the addition of a base was

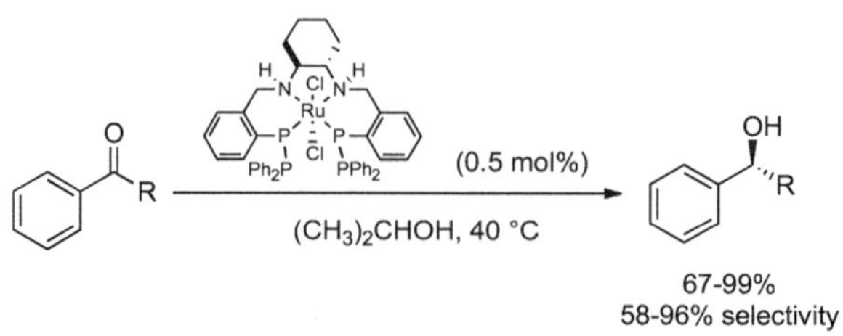

SCHEME 3.11.15

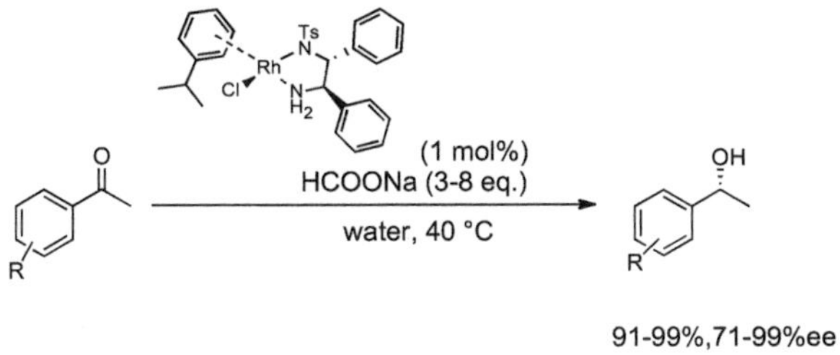

SCHEME 3.11.16

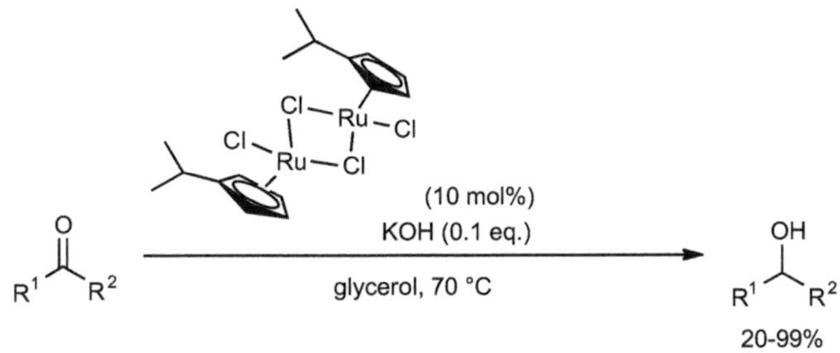

SCHEME 3.11.17

essential to yield the product. Even though the catalyst loading was high (10 mol%), the use of glycerol allowed easy product separation and catalyst recovery. A catalyst recycling test was performed: after extraction of the product and the remaining substrate, the solution was charged with new substrate. A slight decrease of conversion was observed due to the partial dehydrogenation of glycerol during the previous cycle.

A current trend is the use of abundant, inexpensive and nontoxic metals as catalysts to comply with the green chemistry principles. For instance, iron is a good candidate to achieve these goals and it shows competitive activities and selectivities with the appropriate ligands. Milstein et al. developed an efficient iron catalyst with phosphine-based ligands for the hydrogenation of ketones under mild conditions (Scheme 3.11.18).[23] The iron(II) pincer complex with 2,6-bis(diisopropylphosphinomethyl)pyridine reduced aliphatic and aromatic ketones to the corresponding alcohols in no more than 24 h with moderate to excellent yields. The reaction was performed at room temperature and 4.1 atm of hydrogen pressure.

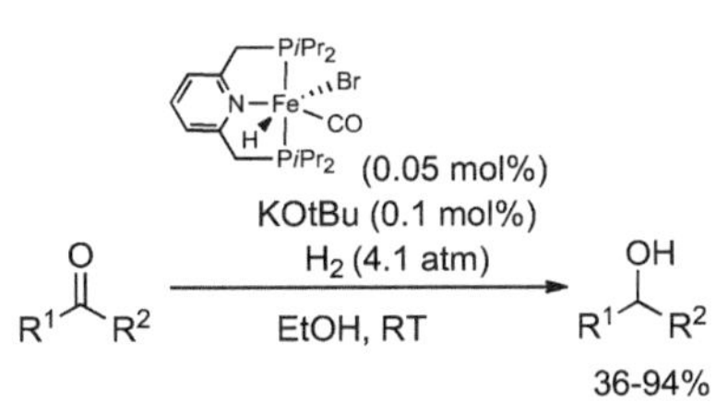

SCHEME 3.11.18

The authors believe that the reaction proceeds through a dearomatized state of the ligand. The dearomatization/rearomatization of the ligand plays a key role in the mechanism and reveals cooperation between the ligand and the metal center.

The same authors extended their work to other substrates such as aldehydes and esters with a similar catalytic process based on the reversible dearomatization/aromatization of different heteroaromatic ligand cores caused by deprotonation/protonation of the ligand.[24]

In the field of iron-catalyzed reduction with phosphine ligands, the hydrogenation of nitroarenes to anilines with formic acid as the reducing agent has been successfully achieved (Scheme 3.11.19).[25] Anilines are of great importance in the synthesis of agrochemicals, pharmaceuticals, dyes, and many more. Primary anilines were obtained in the absence of an additional base using the in-situ-generated catalyst from a 1:1 mixture of $Fe(BF_4)_2 \cdot 6H_2O$ and tris [2-(diphenylphosphino)ethyl]-phosphine [$P(CH_2CH_2PPh_2)_3$]. The catalytic process proceeded through a hydrogen transfer in ethanol using formic acid as an easy-to-handle and inexpensive hydrogen donor. The reactions were energy efficient; they were completed in less than 2 h at 40 °C.

Iron has also been shown to be effective with different types of ligands for the catalytic hydrogenation of olefins.[26] Efficient catalytic activity with low catalyst loading was observed for the dinitrogen- and silane iron-based complexes at ambient temperature in nonpolar media (Scheme 3.11.20). The hydrogenation of simple olefins such as 1-hexene and styrene or diolefins such as 1,5-hexadiene proceeded to completion (>98% yield) within minutes under 4 atm of hydrogen gas.

Among the alternatives to precious metals, zinc proved to reduce imines in the presence of hydrogen gas (Scheme 3.11.21).[27] A catalytic amount of zinc triflate was able to reduce different ortho-, meta-, and para-substituted N-aromatic imines (with both

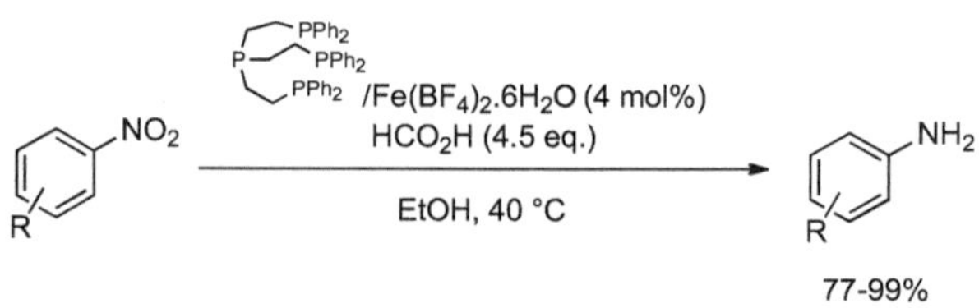

SCHEME 3.11.19

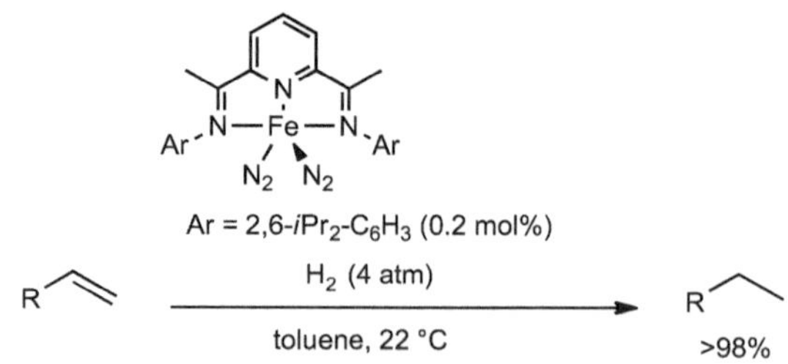

SCHEME 3.11.20

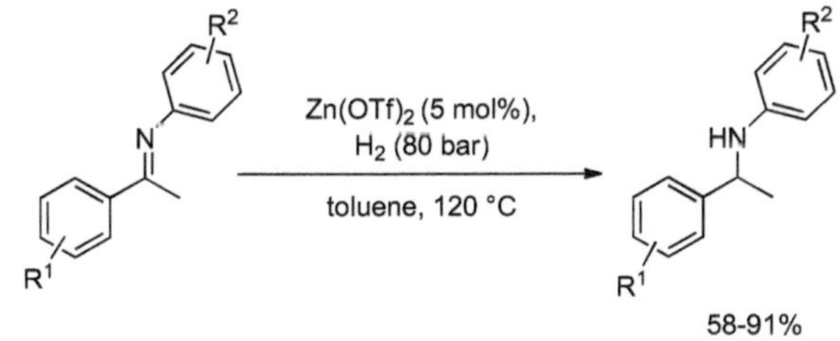

SCHEME 3.11.21

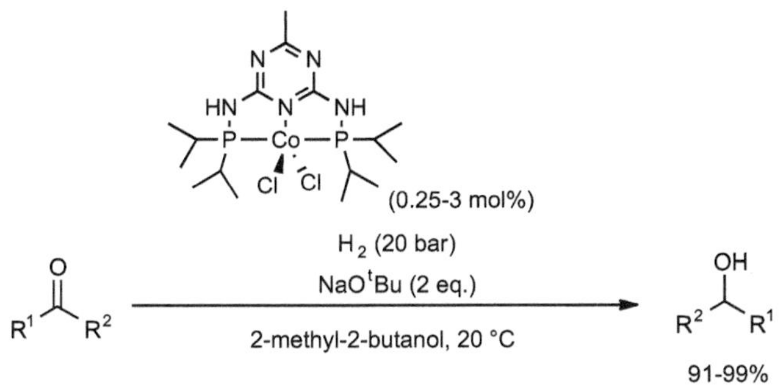

SCHEME 3.11.22

electron-withdrawing and electron-donating groups) to the desired amines in good yield. The advantage of this easy-to-perform protocol is the use of an inexpensive catalyst without additional expensive ligands.

Recently, a system with an easy-to-synthesize cobalt catalyst was developed for the reduction of carbonyl compounds with the same goal of substituting noble metal catalysts (Scheme 3.11.22).[28] The system proceeded under mild conditions with a low catalyst loading and reduced a range of substrates from dialkyl, diaryl, to aryl alkyl ketones. It was selective toward C=O bonds in the presence of C=C bonds. The most effective precatalyst tested was a cobalt-triazine complex. The activation of the precatalyst occurred in solution via salt elimination by addition of 2 equivalents of base.

3.11.2.3 Cross-Coupling Reactions

The formation of C–C and C–X bonds has been the topic of interest of many chemical syntheses. The discovery of cross-coupling reactions provided a powerful tool for the direct formation of these types of bonds. They are typically carried out with a transition metal catalyst and an organometallic precursor. For several years, efforts have been put into the development of sustainable cross-coupling reactions to search for methodologies alternative to traditionally wasteful and energy-intensive protocols.

The Sonogashira coupling is an essential tool to couple terminal alkynes with aryl or vinyl halides. The original Sonogashira reaction is performed in the presence of palladium catalyst

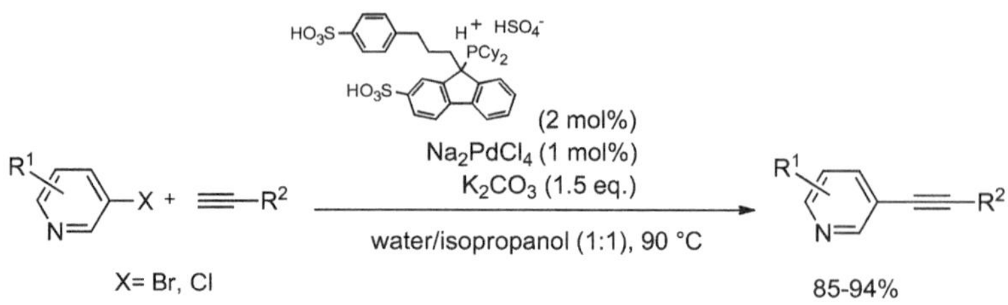

SCHEME 3.11.23

and copper(I) iodide as a cocatalyst in organic solvents, which are economically and environmentally costly. Reactions in aqueous media have recently gained much attention. They offer clear advantages over reactions performed in organic solvents with respect to sustainability. Indeed, organic solvents and additives account for an important part of the environmental impact. Plenio et al. developed a water/isopropanol system for the Sonogashira cross-coupling reaction (Scheme 3.11.23).[29] Although water appears to be a good alternative, the poor solubility of some reactants in water creates a problem. Isopropanol is a suitable cosolvent; it is inexpensive, safe, and biodegradable, and it significantly increases the solubility of organic substrates. It was observed that 1 mol% of the in-situ-formed water-soluble Pd complex with a sulfonic-acid-bearing phosphine ligand effectively catalyzed the cross-coupling of aryl halides with acetylenes. The authors found that water was essential for the catalytic activity as the formation of hydrogen bonding with the nitrogen moieties of the N-heterocyclic substrate prevented the inhibition of the catalyst. The reactions ran at 90 °C from 3 to 20 h with addition of a base. Performed at larger scale, the organic product could be separated from the water solution without additional organic solvent. Thus it eliminates the major waste usually generated during the separation step.

The Suzuki-Miyaura coupling is widely used to synthesize polyolefins, styrenes, and substituted biphenyls. A palladium catalyst is generally the unique choice for this type of cross-coupling reactions due to its incomparable activity. With the aim of reducing the chemical waste and increasing the safety and efficiency of the reaction, a few aspects can be improved. Among them are the use of greener solvent such as water or the design of ligands with specific properties such as stability, efficiency, or nontoxicity. Environment-friendly methods for the Suzuki-Miyaura cross-coupling using water as solvent have been reported. One of these methods is a palladium-phosphine-catalyzed reaction conducted at room temperature and an optimized pH of 11 (Scheme 3.11.24).[30] It was applied to a broad range of substrates, combining aryl halides and boronic acids to form substituted biphenyls. The

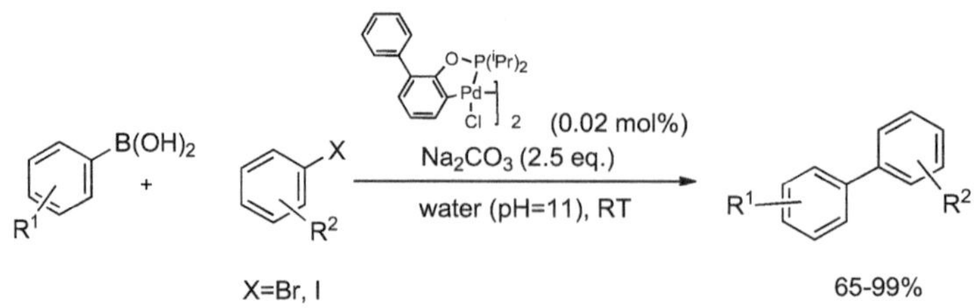

SCHEME 3.11.24

use of water as the only solvent enabled the filtration of the product and the recovery of the catalyst. Recycling experiments showed that the loss of catalytic activity was significant only after the fourth run. It was demonstrated that the addition of an organic solvent deactivated the catalyst, showing that water might be a superior solvent for cross-coupling reactions. Water is not only interesting for its green properties but also for the interesting catalytic activity happening in this medium.[31] In this example, water seems to have a crucial role in the formation of the catalytically active species.

Despite the advantages of the method described previously, the sensitivity of the phosphine ligands remains an important drawback. It is therefore desirable to develop ligand-free methods or using efficient and more stable ligands. Kostas et al. proved that palladium with porphyrin ligands successfully catalyzed the Suzuki-Miyaura cross-coupling with aryl halides and arylboronic acids in aqueous media (Scheme 3.11.25A).[32] Potassium carbonate was required to improve the reactivity of the reactants producing the coupling products in 80%–99% yield depending on the substrate. The best yields were obtained for the halide with an electron-withdrawing group on the aryl ring. The catalyst could be recycled after extraction of the product with an organic solvent; however, a loss of activity was observed after the third run. A similar study, reported by Bora et al., is a Suzuki-Miyaura cross-coupling in water at room temperature with one more advantage, namely, the palladium catalyst is ligand free (Scheme 3.11.25B).[33] The in-situ-generated catalytic system based on

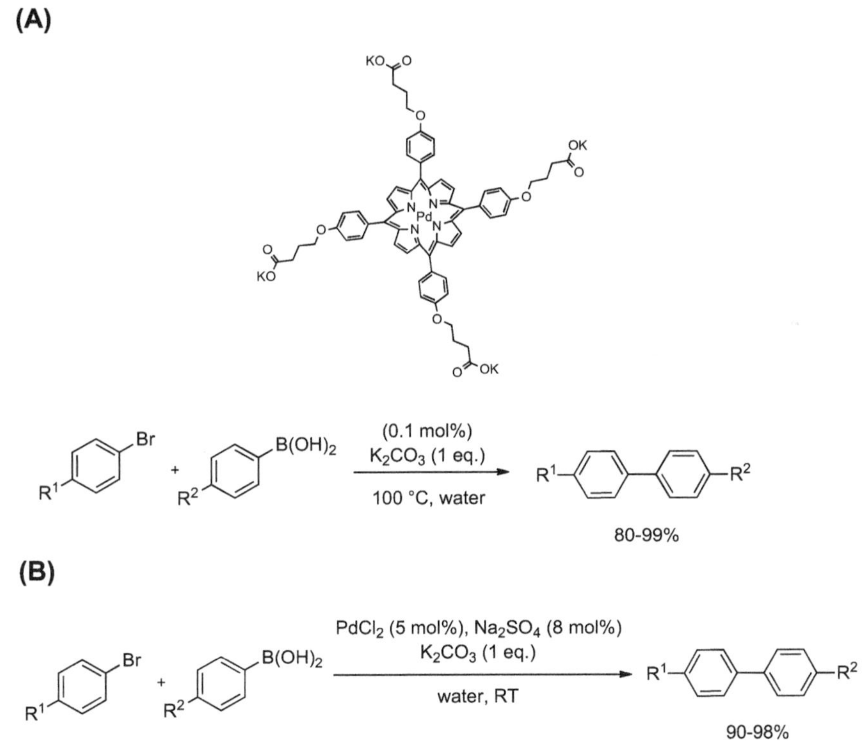

SCHEME 3.11.25

$PdCl_2$ and sodium sulfate exhibited excellent catalytic activity in the coupling of aryl halides with arylboronic acids. The complexation of $PdCl_2$ and sodium sulfate affords the formation of a water-soluble species $Na_2PdCl_2(SO_4)_2$ which is the actual active species that catalyzes the Suzuki-Miyaura reaction. The addition of sodium chloride or sodium acetate offered equal results. Also, in this system, potassium carbonate was necessary for the catalytic reaction to be efficient.

A highly efficient and uncommon protocol was proposed for the Suzuki-Miyaura reaction in water extract of banana (WEB).[34] It offered a mild, economical, and green alternative to the existing methods since it ran at room temperature with a ligand-free palladium acetate catalyst in short time (Scheme 3.11.26). The WEB was prepared by drying the banana peels, and after burning them to ashes, water was added. The filtrate was used as reaction media without any additive. A series of aryl bromides were reacted with arylboronic acids with 0.5 mol% $Pd(OAc)_2$ at room temperature under aerobic condition. Both electron-rich and electron-deficient aryl bromides furnished the products in high yields in 5–20 min only. The role of WEB in the reaction mechanism is not well understood, but it is believed that the sodium and potassium carbonates of banana peels act as a base and sodium and potassium chlorides act as promoters.

The Heck reaction is a key transformation allowing the coupling of an aromatic ring and an alkene. Like other cross-coupling reactions, it suffers from a major drawback: the release of salt waste. Indeed, the use of halogenated compounds requires the use of a stoichiometric amount of base to complete the catalytic cycle, a considerable amount of halide salt is therefore generated. Jacobs et al. proposed a waste-free method for the synthesis of Heck coupling products (Scheme 3.11.27).[35] They employed an appropriate pressure of oxygen (0.8 MPa) to reoxidize the palladium(0) to the active palladium(II) and therefore close the catalytic cycle. This atom-efficient system was solvent free and released water as the only by-product. The reaction was performed with an excess of arene at 90 °C to yield the coupling products in high yields after 22–24 h. For some substrates, the addition of benzoic acid was needed to reduce the induction period. This method, however, is not selective toward one of the *ortho*, *meta*, or *para* positions.

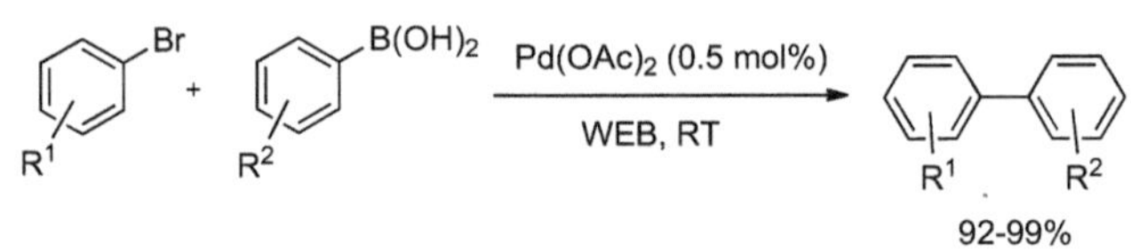

SCHEME 3.11.26

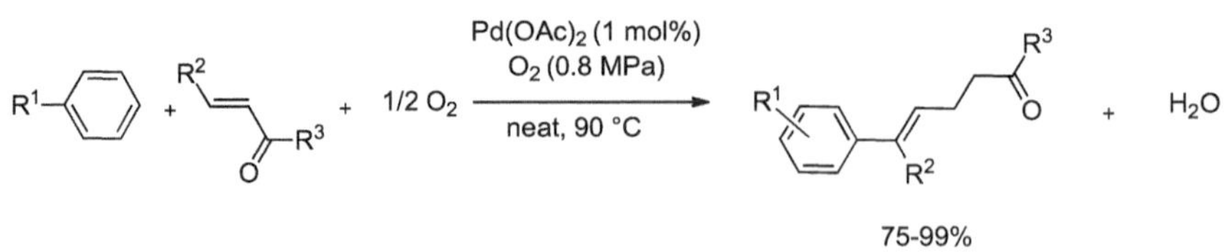

SCHEME 3.11.27

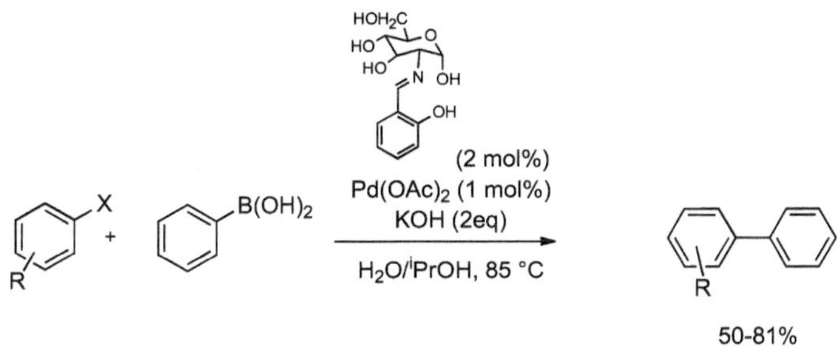

SCHEME 3.11.28

Recently, the capability of sugar molecules as ligands for the Heck and Suzuki coupling reactions in water has been explored.[36] The inexpensive and environment-friendly nature of sugars makes them an interesting option. A system consisting of $Pd(OAc)_2$ and a glucose-based ligand was applied for the preparation of polysubstituted alkenes by the Suzuki-Miyaura cross-couplings and the Heck cross-couplings (Scheme 3.11.28). The in-situ-generated catalyst led to good to excellent yields in the presence of a base at 80–85 °C using pure water or a mixture of water/isopropanol as the solvent. Aryl halides and phenylboronic acids were converted to the corresponding biphenyl compound within 4 h with $Pd(OAc)_2$ and the *N*-salicylidene-D-glucosamine ligand as catalyst.

The Heck reaction was performed with the same complex that successfully coupled iodobenzene and bromobenzene with various olefins in less than 2 h (Scheme 3.11.29).

Beside the carbon-carbon bond formation, cross-coupling reactions can also be employed in carbon-heteroatom bond formation. Likewise, green protocols have been investigated in this field. Tsai et al. proposed the synthesis of carbon-sulfur bonds catalyzed by a recyclable iron-based complex in water under reflux (Scheme 3.11.30).[37] A variety of aryl iodides and thiols were efficiently coupled using environment-friendly and inexpensive $FeCl_3 \cdot 6H_2O$. Even though the catalyst loading was quite high (10 mol%), the authors proved that the catalyst could be reused six times with only a slight decrease in activity. The water-soluble cationic 2,2-bipyridyl ligand was chosen to stabilize the iron salt in the aqueous phase and made the recycling of the catalyst possible. The ionic part of the ligand also played a surfactant role, which facilitated the reaction.

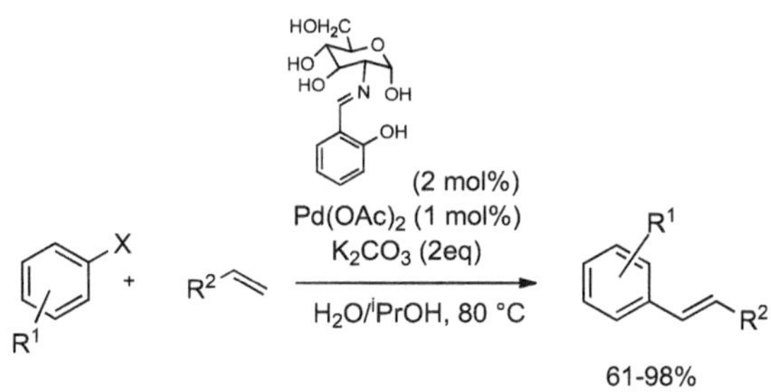

SCHEME 3.11.29

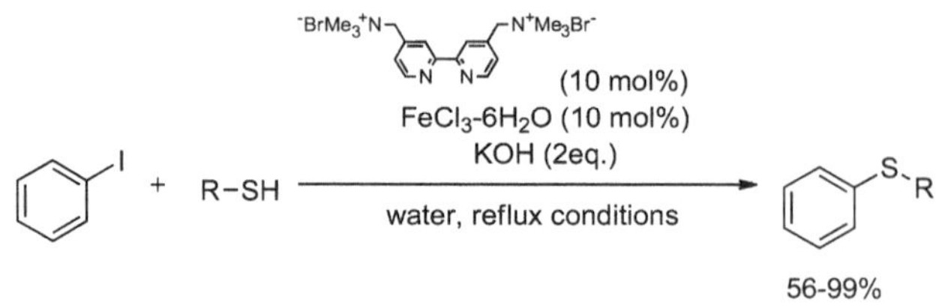

SCHEME 3.11.30

The carbon-nitrogen coupling reaction is an important path for the synthesis of complex nitrogen-containing molecules. It is known that copper is able to catalyze arylation reactions of many nucleophilic species, even though palladium-catalyzed reactions are often more efficient. Efforts have been made to increase the efficiency of copper-mediated processes for economic and environmental reasons. An efficient Ullman-type reaction was developed in aqueous medium using a copper catalyst and the ligand (1*E*, 2*E*)-oxalaldehyde dioxime (Scheme 3.11.31), chosen for its availability and its activity.[38] The nature of the catalyst and the ligand is advantageous when compared with the usual palladium catalyst often combined with complex ligands. However, addition of a base and a transfer agent were necessary. Nitrogen-containing cyclic and aliphatic amines were coupled with good yield at 100 °C under nitrogen atmosphere. The protocol showed good tolerance toward various functional groups.

An Ullmann-type coupling reaction at room temperature was described by Buchwald and his coworkers.[39] The catalyst was a combination of CuI with cyclic β-diketone ligand such as the commercially available 2-acetylcyclohexanone. It allowed the transformation of aryl halides and amines in good yield in no more than 4 h in the presence of Cs_2CO_3 as the base (Scheme 3.11.32). Both activated and nonactivated aryl halides underwent smooth coupling at room temperature.

The same type of reaction, also known as the Buchwald-Hartwig reaction was performed in the nonconventional medium, liquid polyethylene glycol (PEG).[40] PEGs have clear advantages as solvent in the organic synthesis because they are inexpensive, nontoxic, and available. In

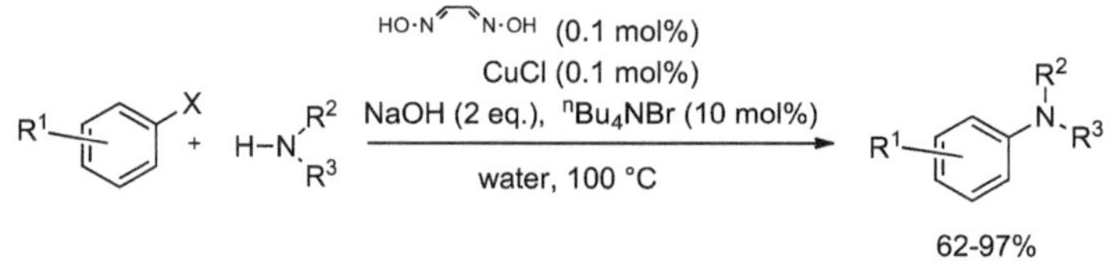

SCHEME 3.11.31

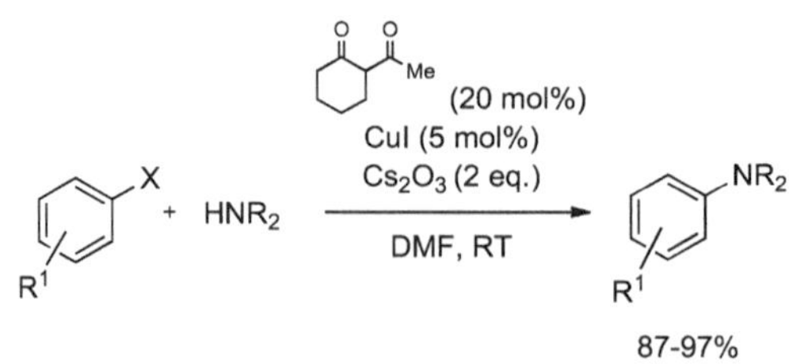

SCHEME 3.11.32

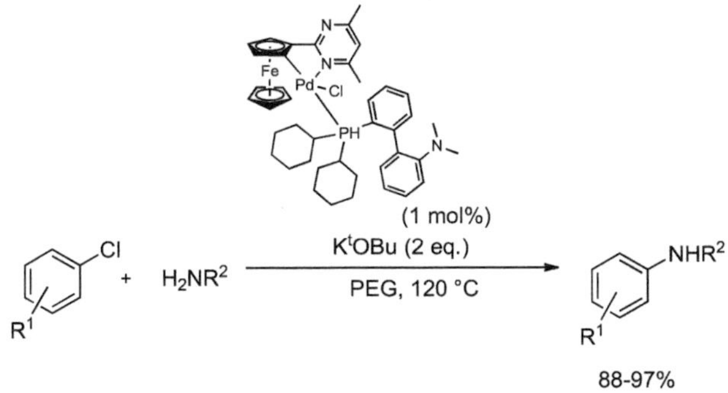

SCHEME 3.11.33

addition, they make the recycling of the catalyst simple.[41] The monophosphine-cyclopalladated ferrocenylpyrimidine complex proved its efficiency in the Buchwald-Hartwig amination of a range of sterically hindered aryl chlorides in PEG (Scheme 3.11.33). The complex combines the stability induced by the palladium framework and the activity associated with the phosphine ligands. This reusable catalytic system led to excellent yields. Thanks to the PEG's properties, the recycling of the catalyst was a simple process: after extraction, the mixture of PEG and catalyst was solidified by cooling followed by evaporation, to finally charge the mixture with new substrate. Only a slight decrease in yields was observed after three runs.

3.11.2.4 Cycloisomerization Reactions

Cycloisomerization is, by nature, atom economical; nothing is wasted as every atom in the starting material is present in the product. However, depending on the reaction protocol, the sources of waste are multiple. Solvents and additives account for a major part of the waste in organic synthesis. It is therefore important to minimize their amount and their toxicity. Cycloisomerizations allow construction of the carbon framework in mostly one step, which would require multiple steps using traditional transformations. The exploitation of new, greener cycloisomerization reactions is in the interest of ongoing research.

Several methods using nonconventional reaction media to address the toxic waste issue were proposed. A palladium-catalyzed cycloisomerization to afford furans has been carried out in the sustainable solvent glycerol, which is a by-product of biodiesel production (Scheme 3.11.34).[42] It is therefore abundant and cheap and also biodegradable and nontoxic. Several enynols were converted to furans with excellent yields in short reaction time (20 min to 2 h). The authors showed that it was also possible to use water as solvent, but the reactions performed in glycerol were faster and catalyst recycling was easier.

Ozdemir et al. successfully synthesized a water-soluble imidazolidin-2-ylidene ruthenium(II) complex that proved to efficiently catalyze the intramolecular cyclization of (Z)-3-methylpent-2-en-4-yn-1-ol into 2,3-dimethylfuran in water (Scheme 3.11.35).[43] The *p*-dimethylaminobenzyl as an N-substituent was chosen because the presence of the peripheral NMe_2 group improved water solubility through quaternization of the NMe_2 group. In addition, the N substituent is believed to have an important effect on the catalytic activity; in this study the chosen substituent

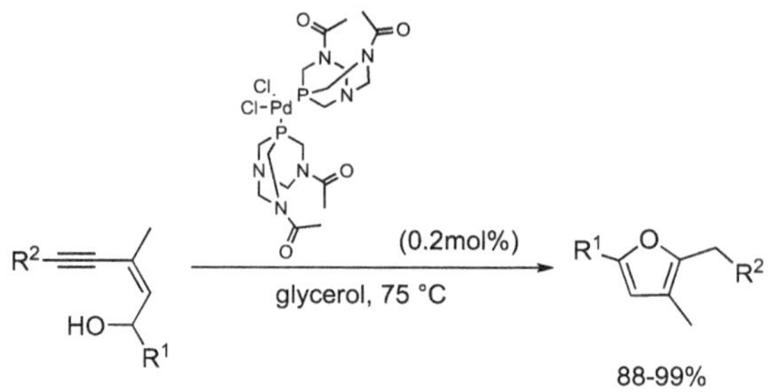

SCHEME 3.11.34

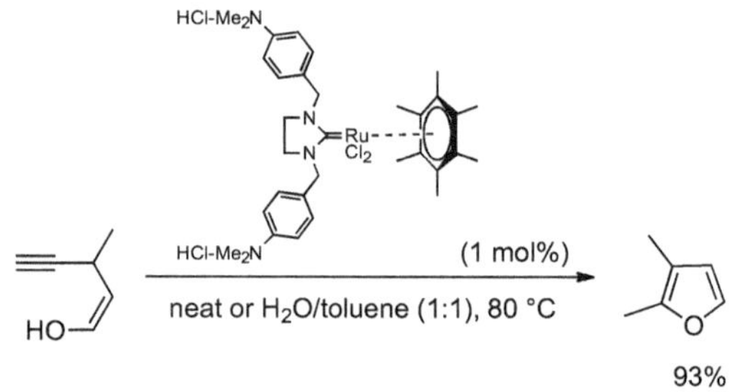

SCHEME 3.11.35

pronounces the activity of the complex. The catalytic reaction was performed either without solvent or in a biphasic system using a mixture of toluene and water. The second system allowed catalyst recovery with a catalytic activity preserved up to the fifth run.

A novel method in the field of gold-catalyzed cycloisomerizations was reported[44] offering, here again, an alternative to the conventional methods using organic solvents such as toluene, dichloromethane, or tetrahydrofuran (THF). The authors carried out the cycloisomerization reactions in water using a catalytic amount of chloroauric acid, which is more water soluble than most of the gold salts (Scheme 3.11.36). A stoichiometric amount of lithium chloride was required to prevent catalyst decomposition. Several functionalized allenes underwent stereoselective cycloisomerization with moderate to good yields at room temperature. The addition of a small amount of diethyl ether was necessary to accelerate the conversion of the non-water-soluble substrates. Despite the obvious benefits that this aqueous system has over organic solvent-based systems, it has some limitations. First, the use of lithium chloride contributes to the waste generated in the process and second, the leaching upon extraction of the products prevents the recycling of the catalyst.

Oshima et al. reported a [4+2] annulation of dienynes at room temperature, in water using a dimeric rhodium catalyst and sodium dodecyl sulfate (SDS) as surfactant (Scheme 3.11.37).[45] The catalyst, di-μ-chloro bis(norbornadiene)dirhodium(I) is believed

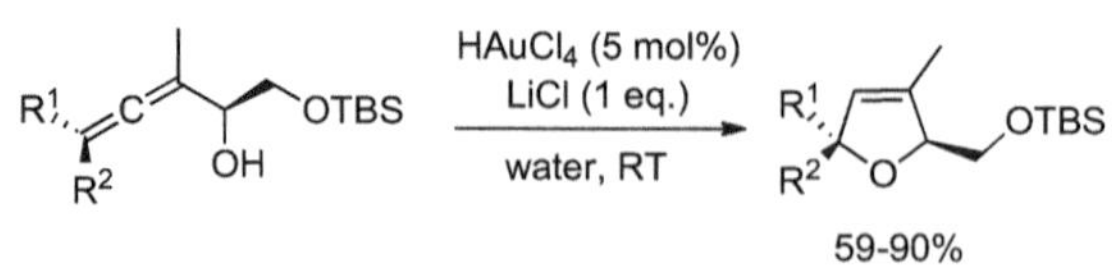

SCHEME 3.11.36

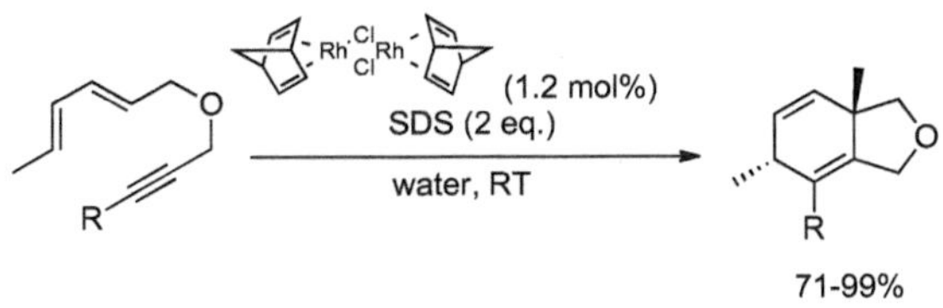

SCHEME 3.11.37

to form a cationic rhodium species after dissociation of the Rh—Cl bond in the presence of SDS. The surfactant fulfilled its role by solubilizing the hydrophobic substrates, and the anionic charge of the micelles seemed to concentrate the cationic rhodium species, which induced rapid conversion of the substrates. This efficient system converted the more hydrophilic dienynes within minutes and the hydrophobic dienynes within hours with good to excellent yields. Although the surfactant clearly improves the catalytic activity and allows the use of a safe solvent, the resulting waste is nonbiodegradable. The treatment of waste is an important factor to take into account from an environmental point of view. Unfortunately, the treatment of surfactants is costly and creates wastewater.[46]

The Pauson-Khand reaction, the carbonylative cycloaddition of an alkene and an alkyne, is a useful method for the preparation of various carbocycles. Highly toxic gaseous carbon monoxide is usually employed to carry out this type of reaction. Chan and his coworkers developed an asymmetric aqueous rhodium-catalyzed method using an aldehyde as CO substitute.[47] The reaction ran in water, which acts as a nonhazardous solvent and eliminates the risks that carbon monoxide presents (Scheme 3.11.38). Various enynes were transformed to the corresponding bicyclic cyclopentenones using cinnamylaldehyde as the source of carbon monoxide. A chiral atropisomeric dipyridyldiphosphane ligand was found to be highly effective in this system. The selected phosphane ligand together with the selected aromatic aldehyde led to good yields and enantiomeric excess. The study of the effect of solvent showed that water provided the best results when compared with organic solvents such as toluene.

The olefin metathesis reaction, usually catalyzed by a ruthenium complex, is a widely used reaction in the production of pharmaceutical compounds and therefore requires high purity of the products obtained. Existing methods lead to ruthenium by-products that are difficult to

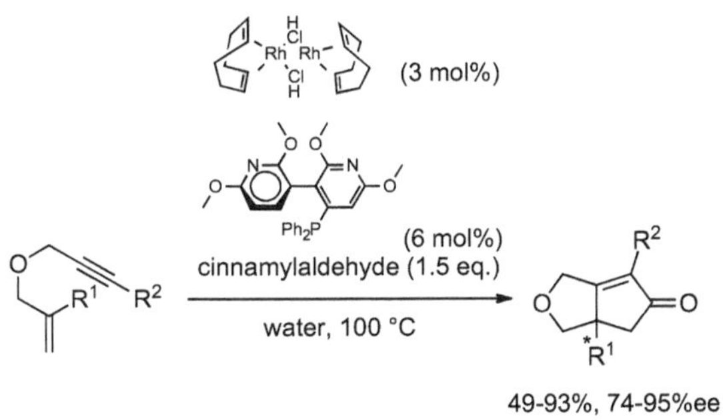

SCHEME 3.11.38

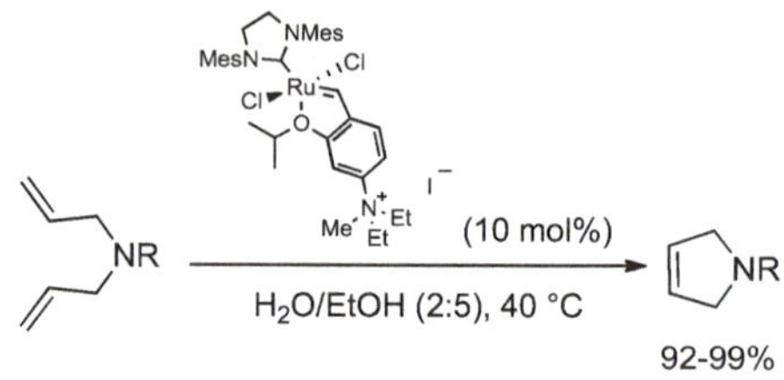

SCHEME 3.11.39

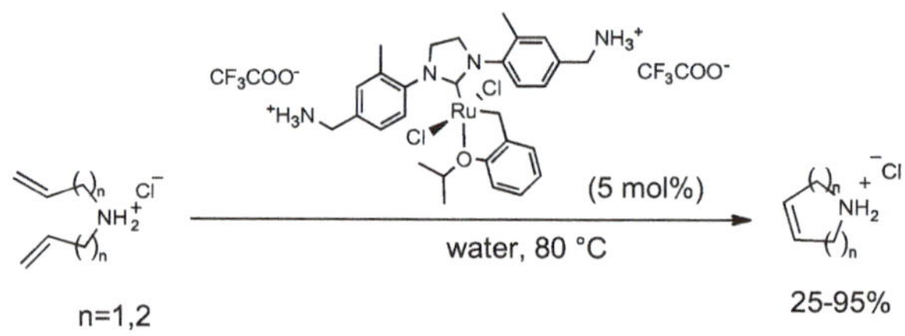

SCHEME 3.11.40

remove from the reaction products.[48] An electron-withdrawing quaternary ammonium ligand has been designed by Grela et al. to improve the metathesis reaction (Scheme 3.11.39).[49] It not only increases the catalytic activity of the ruthenium catalyst but also allows an efficient separation of the catalyst from the desired product. The level of ruthenium contamination exhibited was very low after filtration on silica gel due to the high affinity of the catalyst complex on the silica gel. The reaction was performed under mild conditions in an economically and environmentally beneficial mixture of water and ethanol. A rather high catalyst loading of 10 mol% of the ruthenium complex gave very good yields of the metathesis product.

Recently, a similar study was described, using pure water as solvent and a diammonium-functionalized Ru-alkylidene complex with a lower catalyst loading (5 mol%) than the previously described example (Scheme 3.11.40).[50] However the temperature used was higher and the yields obtained were moderate. In this study, a simple aqueous extraction afforded the products with high purity.

3.11.3 ORGANOCATALYSIS

Even though organocatalysis has its origins in 1971,[51] it is only since the 2000s that the field has experienced a remarkable progress yielding a large number of transformations that can now be performed using this metal-free approach. Recent advances in the field of organocatalysis also include the development of tandem or cascade reactions. Although these are desirable developments from the point of view of green chemistry, many of these reactions still use a high catalyst loading to achieve high conversions and yields. This results in a considerable amount of organic material lost during the workup. This issue becomes even more important considering the fact that the synthesis of many new-generation organocatalysts require a multistep synthesis. As a consequence, to be considered green, organocatalytic reactions should preferentially use an easily accessible catalyst not requiring multiple-step synthesis at low catalyst loadings and comply with other green chemistry principles as well.

3.11.3.1 C—C Bond Formation Reactions

3.11.3.1.1 Amine-Based Catalysts

Amine-based organocatalysts are used in the majority of applications in the field of organocatalysis. Most of these catalysts are based on natural compounds. Some compounds can be used as they appear in nature, whereas others are derivatized before being used as catalyst. Proline and the cinchona alkaloids are prominent examples of compounds that are used directly; the MacMillan type catalysts are a typical example for catalysts obtained after several synthetic steps from a natural compound (Fig. 3.11.1).

Using natural compounds such as proline as catalyst has the advantage that they are abundant and usually can be easily obtained in enantiomerically pure form. The drawback of these molecules is that they often do not perform well for a specific reaction and need high catalyst loading for the reactions to be effective and reach high conversions and enantioselectivities. Efforts have been made to circumvent these issues by structurally modifying the original compounds. One successful approach has been the modification of proline to form diarylprolinol silyl ethers.[52] Although these have been proved to be superior to proline in many regards, from a sustainable point of view their synthesis has to be taken into account. Factoring in these synthetic steps into the environmental balance of a reaction, the design of green reactions with this type of catalysts becomes more challenging. As a result, the examples given here focus mainly on the use of natural compounds such as proline or simple derivatives as catalysts.

One major use of organocatalysis is in the field of aldol reactions. Organocatalysts have been shown to be very effective in directing the outcome of these reactions toward high yield, high diastereoselectivity, and high enantioselectivity. Unfortunately, many of these reactions suffer from some of the common problems such as high catalyst loadings (20–30 mol%) and environmentally harmful organic solvents. To reduce the environmental impact of these reactions, several attempts were made to lower the catalyst loading and use more environment-friendly solvent systems.

FIGURE 3.11.1

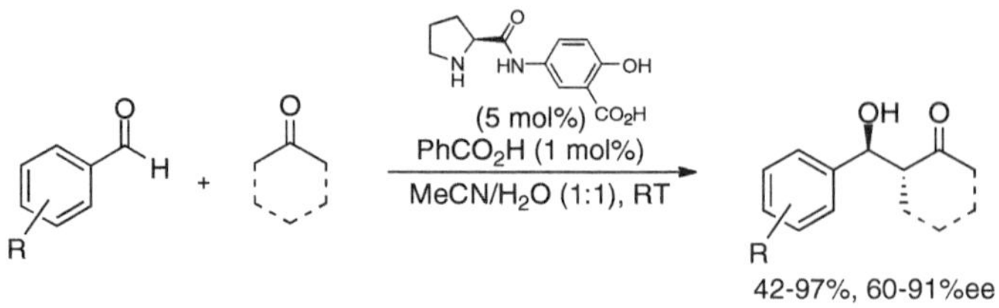

SCHEME 3.11.41

The aldol reaction of aromatic aldehydes and ketones was performed with a proline-derived catalyst in an aqueous system as shown by Bayat et al.[53] The reactions were preferentially performed with an electron-withdrawing group on the aldehyde. The aldol products were obtained in medium to excellent yields, with high enantioselectivities, and an almost exclusive preference for the *anti*-isomer (Scheme 3.11.41).

A very similar approach was taken by Caputo and coworkers. They showed that a sugar-proline catalyst was effective in performing the aldol reaction of aromatic aldehydes with ketones in saturated aqueous NaCl solution (brine). The products were obtained in high yield and enantioselectivities with the *anti*-isomer being the predominant species (Scheme 3.11.42).[54] Careful investigations showed that the catalyst loading could be lowered to 0.1%, albeit with the need of using reduced temperatures (−16 °C) and prolonged reaction times (4 days). Furthermore, the catalyst could be recovered and reused without significant loss of catalytic activity.

A very similar setup for asymmetric aldol reactions in brine was described by Zhang et al.[55] They showed that proline-based dipeptides can serve as effective catalysts for the aldol reaction between aromatic aldehydes and cyclic ketones to give the products in high yields and enantioselectivities. The drawback of their method is that with ketones other than cyclohexanone only poor *syn/anti* selectivity could be obtained.

The same group used a cinchona-derived catalyst in brine for the asymmetric aldol reaction of aromatic aldehydes and cyclohexanone (Scheme 3.11.43).[56] Good yields and enantioselectivities were obtained, although the reaction was only performed with cyclohexanone as the ketone.

Liu et al. showed that a cinchona derivate in combination with a substoichiometric amount of trifluoromethanesulfonic acid was able to perform the reaction of electron-deficient

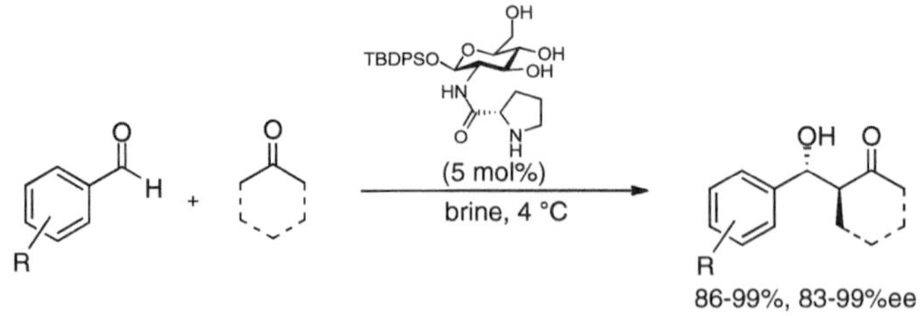

SCHEME 3.11.42

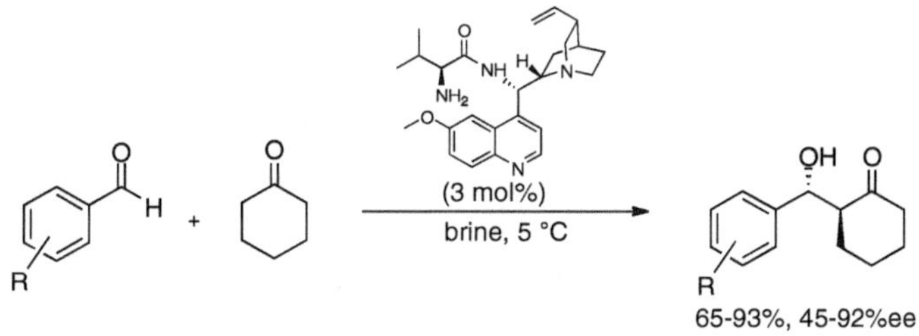

SCHEME 3.11.43

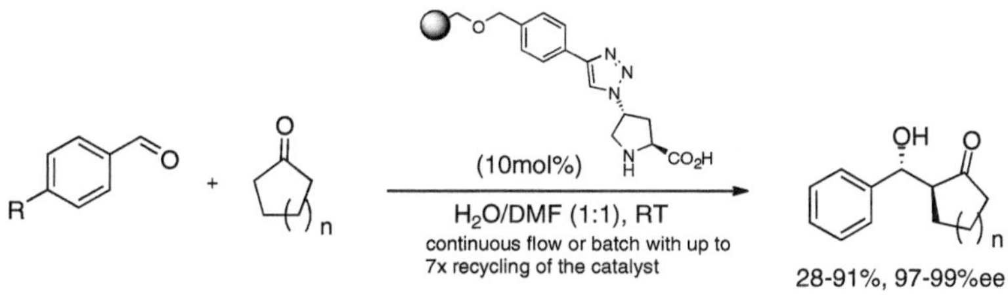

SCHEME 3.11.44

aromatic aldehydes with ketones under neat conditions.[57] These electron-deficient aldehydes are usually less reactive with proline-like catalysts.

The group of Pericàs developed a solid-supported proline derivative for the aldol reaction between aromatic aldehydes and cyclic ketones (Scheme 3.11.44).[58] Aldehydes bearing an electron-withdrawing group gave significantly better results than aldehydes with an electron-donating group. Interestingly, the enantiomeric excess obtained was not affected by the nature of the substituent. The reaction could be run in batch or continuous flow mode. When ran in batch mode, the catalyst could be recycled up to seven times without loss of activity with regard to both yield and enantiomeric excess. The continuous flow reaction allowed the synthesis of up to 5 mmol of product with 0.32 mmol of the catalyst.

Landge and Török described a method for the highly selective aldol reaction between cyclic ketones and ethyl trifluoropyruvate (Scheme 3.11.45).[59] The authors found that the proline-catalyzed reaction yields the corresponding aldol adducts in high yields as well as diastereo- and enantioselectivities. The use of microwave activation at low temperature was crucial to obtain satisfying results.

A solvent-free variation of the proline-catalyzed aldol reaction was described by Bolm et al.[60] A variety of ketones reacted with aromatic aldehydes yielding the *anti*-aldol adduct in medium to high yields and with good to excellent enantioselectivities (Scheme 3.11.46).

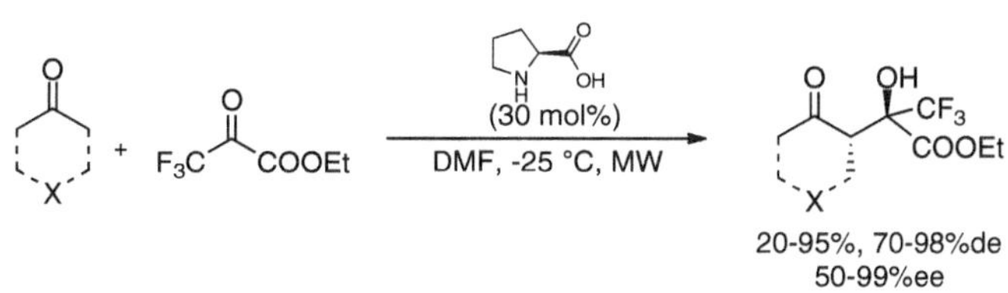

SCHEME 3.11.45

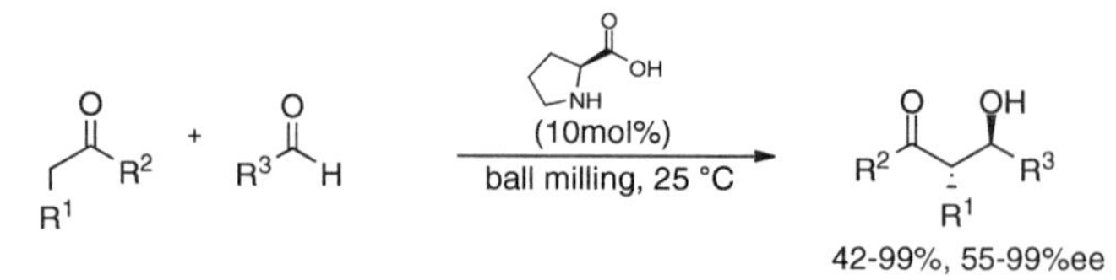

SCHEME 3.11.46

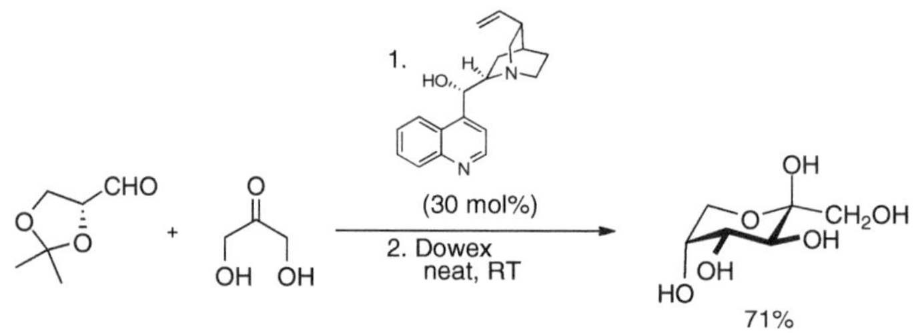

SCHEME 3.11.47

Mahrwald and coworkers showed that cinchonine could be used as a catalyst for the transformation of aliphatic aldehydes with hydroxylated ketones.[61] The reaction between an enantiomerically pure, protected glyceraldehyde and dihydroxyacetone in the presence of cinchonine followed by acidic treatment yielded D-fructopyranose in 71% yield (Scheme 3.11.47).

Conjugate addition to α,β-unsaturated carbonyl compounds is a powerful tool for C–C bond formation. Organocatalysis has been extensively used in asymmetric conjugate addition, as can be seen in a large number of publications covering that field.[62] Still, these methods mostly suffer from a high catalyst loading and/or the use of halogenated solvents. With regard to lowering the environmental footprint of these reactions, Procopio et al. presented a microwave-assisted solvent-free synthesis of optically enriched Michael adducts with proline as catalyst (Scheme 3.11.48).[63] They showed that a variety of α,β-unsaturated ketones reacted with diethylmalonate under solvent-free conditions to provide the corresponding addition product in medium to high yields. During their investigations they found that the steric hindrance of the unsaturated system plays a crucial role in the reaction outcome with a more sterically hindered system giving lower yields and enantioselectivities.

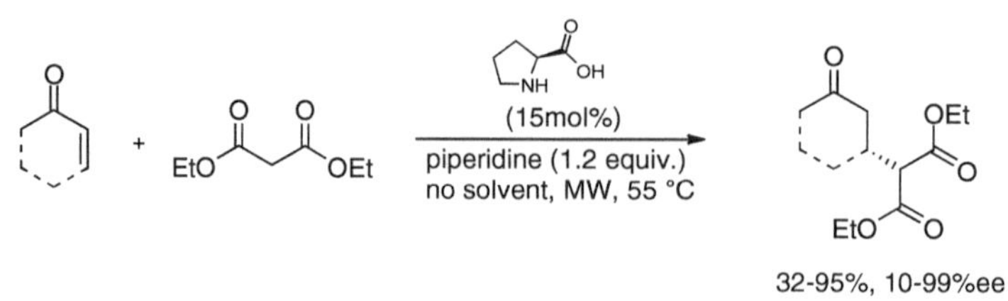

SCHEME 3.11.48

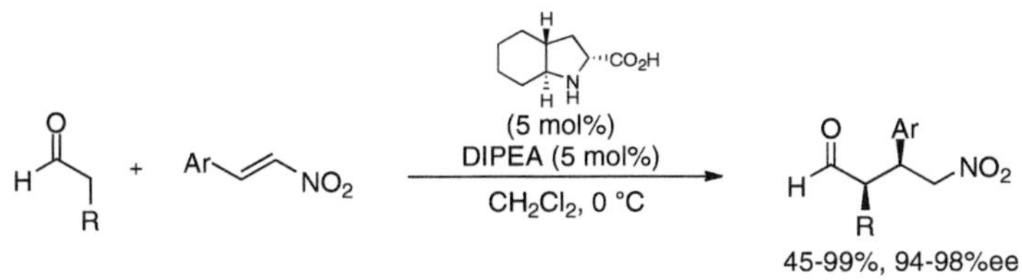

SCHEME 3.11.49

Zhang and his group investigated the Michael addition of aldehydes to aromatic nitroolefins (Scheme 3.11.49).[64] Using a proline-related catalyst, high yields and enantioselectivities of the *syn*-aldol adducts were obtained. The addition of diisopropylethylamine as additive was found to be necessary in achieving good results.

Pericàs et al. performed a nearly identical reaction using polystyrene-supported diarylprolinol as catalyst. Their system could be applied to a large variety of carbonyl and nitro compounds, usually yielding the products in high yields and excellent enantioselectivities.[65]

Multicomponent reactions are an effective way to quickly build up molecular complexity as they allow combining multiple reagents to form a single product in only one step and as such reducing the amount of side products and waste produced. Although the application of proline in multicomponent reactions has been recently reviewed, not all of these reactions can be considered green.[66]

Lee et al. could show that proline is an effective catalyst for the synthesis of functionalized 2*H*-chromenes (Scheme 3.11.50).[67] Combining a salicylaldehyde with a propiolate in refluxing alcohol led to the formation of the 2H-chromene derivatives in good to high yields. They proposed the proline to activate both the aldehyde and the triple bond. They could also show that the propiolate can be replaced by an acetylenedicarboxylate or the salicylaldehyde might be replaced with 2-aminobenzaldehyde. Both variations furnished the desired products in high yields.

A different substitution pattern on the chromene could be obtained when dimedone and different nucleophiles are reacted with salicylaldehyde in refluxing ethanol.[68]

Structurally similar 3-nitrocoumarins have been synthesized from substituted salicylaldehydes and ethyl nitroacetate in high yields using L-proline, as catalyst (Scheme 3.11.51).[69] Depending on the substrate the reaction needed to be performed at either room temperature or 80 °C.

The enantioselective Friedel-Crafts hydroxyalkylation of indols catalyzed by cinchona alkaloids has been shown by Török et al.[70] A variety of indoles were successfully reacted with ethyl trifluoropyruvate to give the corresponding hydroxyalkylated products in high

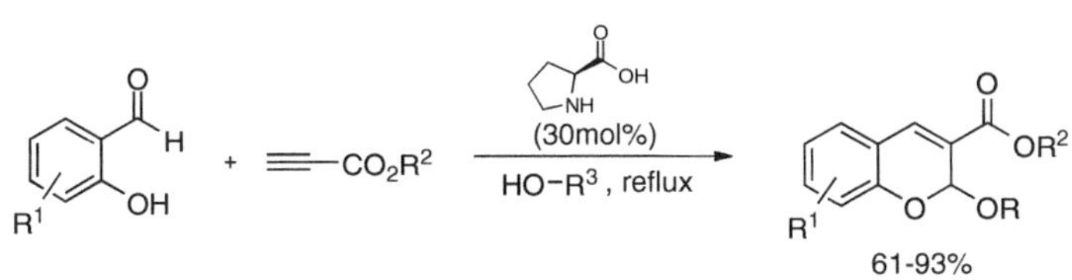

SCHEME 3.11.50

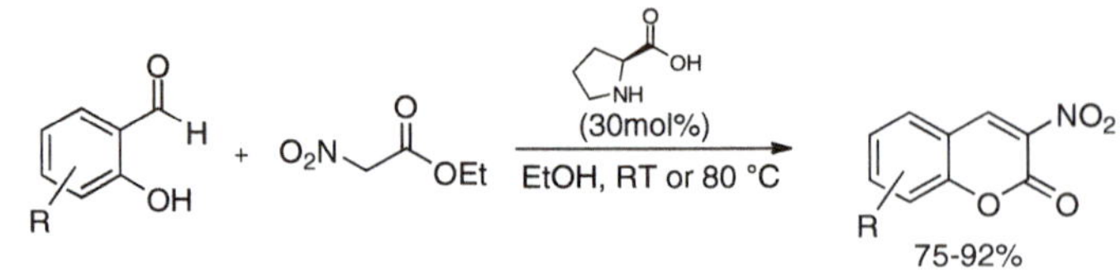

SCHEME 3.11.51

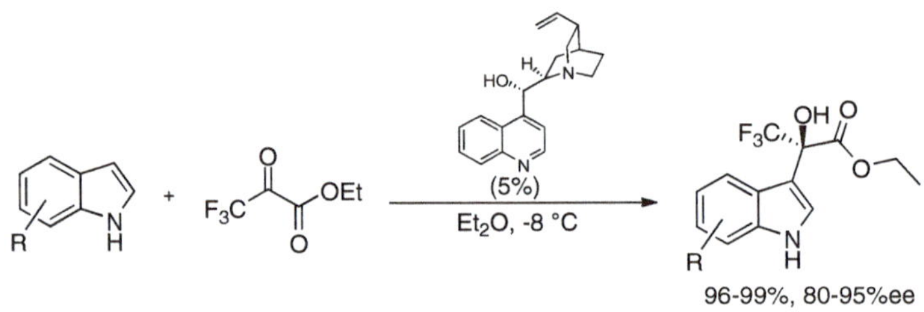

SCHEME 3.11.52

yields and enantioselectivities (Scheme 3.11.52). Using cinchonine or cinchonidine as the catalyst the configuration of the newly formed chiral center can be influenced.

Proline has also been shown to be able to catalyze Biginelli-type reactions. The group of Rimaz described a 3-component-reaction to form pyrimido[4,5-*d*]pyrimidones in high yields using water as solvent with only 2 mol% of L-proline as catalyst (Scheme 3.11.53).[71]

Rajitha et al. showed that proline could be used for the synthesis of highly substituted pyridines from tetralones, benzaldehydes, and ammonium acetate (Scheme 3.11.54).[72] The multicomponent reaction is performed in ethanol and yielded the desired products in high yields

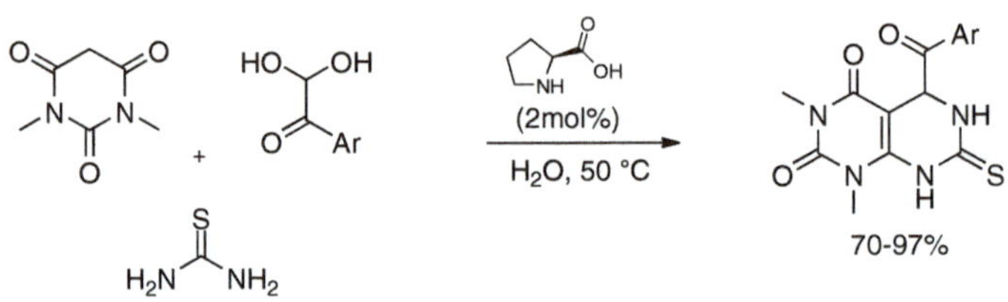

SCHEME 3.11.53

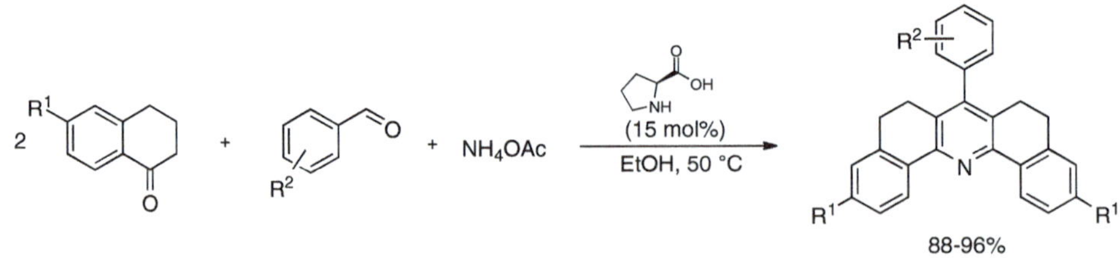

SCHEME 3.11.54

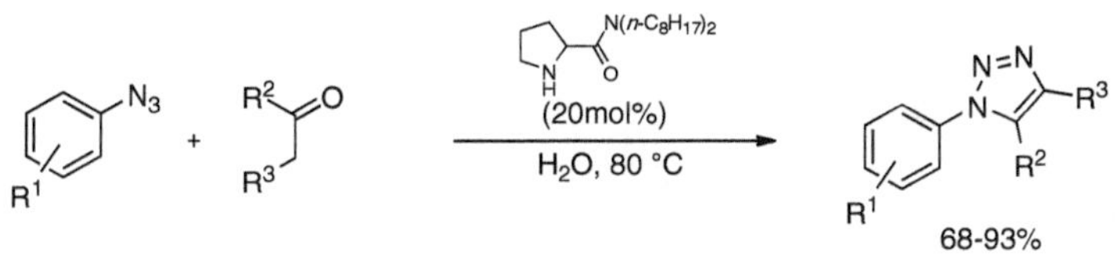

SCHEME 3.11.55

with 15 mol% proline as catalyst. The proline catalyst was easily recovered from the mixture and reused over five cycles without losing its catalytic activity.

Organocatalysis can also be used to perform 1,3-dipolar cycloaddition reactions. Even though the synthesis of triazoles is often known to be copper catalyzed, there are a few examples of performing this reaction using proline as a catalyst; unfortunately, they still require the use of organic solvents.[73] Wang's group could show that with modified proline as a catalyst the reaction could be performed in water as solvent (Scheme 3.11.55).[74] The reaction performed well with a variety of substituted aromatic azides and tolerated different functional groups on the ketone, and the triazoles were obtained in good to high yields.

The concept of organocatalysis has also been employed in the field of C–H activation chemistry. C–H activation is of great interest as it allows the formation of, for example, C–C bonds without prior functionalization of the carbon atom, thus reducing the number of steps needed and waste produced. Most contributions to the field including an organocatalyst use a combination of metal and organocatalysis.[75] An exemption from this is the work of Cui and Jiao (Scheme 3.11.56).[76] Using McMillan's catalyst, they could achieve the α-alkylation of aldehydes with xanthene derivatives. The reaction yielded the desired products in medium to good yields and high enantioselectivities using molecular oxygen as the oxidant allowing the reaction. The addition of a small amount of water was also found to be necessary for the reaction to proceed.

3.11.3.1.2 N-Heterocyclic-Carbene-Based Catalysts

Since the first reports of stable carbenes by Bertrand et al.[77] and Arduengo et al.[78] these compounds have had a considerable impact on homogeneous catalysis. This is to a great part due to their distinct properties allowing them to replace phosphines as ligands in metal catalysis. However, there is also increasing use of stable carbenes, especially N-heterocyclic carbenes (NHCs) as organocatalysts. A seminal report in this field was published by the groups of Enders and Teles.[79] Even though most reactions use NHCs with a fairly complex structure that need to be synthesized, their positive environmental impact lies in the fact that,

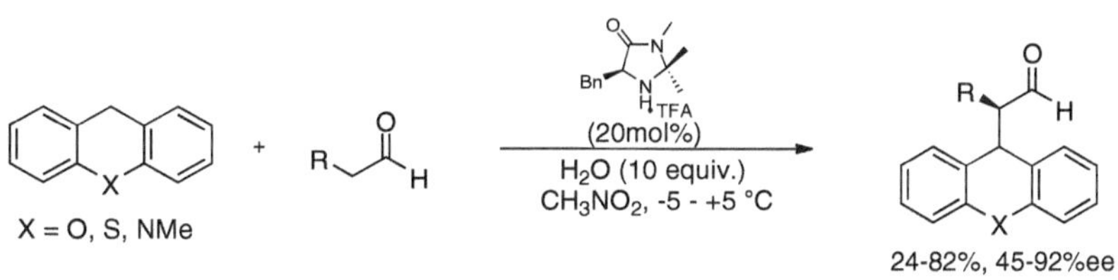

SCHEME 3.11.56

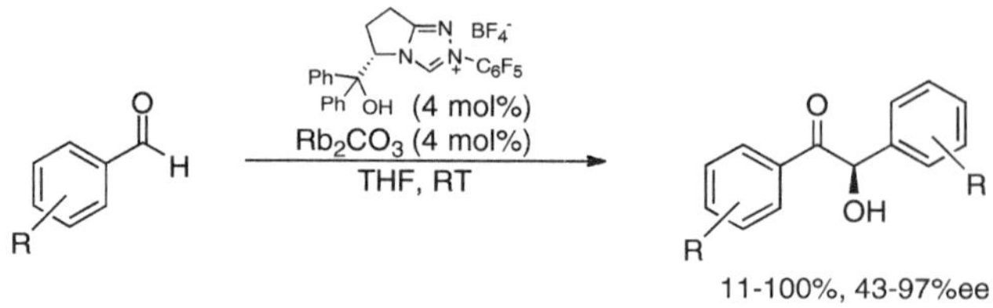

SCHEME 3.11.57

due to their particular properties, they allow reactions that otherwise require multiple steps or allow the substitution of highly toxic species (e.g., CN^- in Umpolung reactions).

NHCs have been proved to be very effective in enantioselective benzoin reactions. Connon and his colleagues have shown that aromatic aldehydes undergo benzoin reaction with a low loading of an NHC catalyst in mostly high yields and high enantioselectivities (Scheme 3.11.57).[80]

The reaction is not limited to homocoupling of aldehydes. It was shown that the cross-benzoin reaction between two aldehydes can be catalyzed by NHCs, albeit the reaction needed higher catalyst loading and the enantiomeric excess obtained was low.

Using a different catalyst, aldehyde-ketone cross-benzoin reactions have been successfully performed. The reaction needed a relatively high catalyst loading but resulted in the formation of a bicyclic system with two new quaternary stereocenters (Scheme 3.11.58).[81] The yield and enantioselectivity obtained depends highly on the ring sizes constructed with the 6-6 systems giving the best yields.

NHCs can be effective catalysts for the hydroacylation of cyclopropenes. Glorius et al. showed that with 5 mol% of an NHC catalyst the synthesis of α-keto-cyclopropanes could be achieved from aromatic aldehydes and aryl-cyclopropenes (Scheme 3.11.59).[82] The products were usually obtained in high diastereoselectivities and high yields with the carbonyl and the aromatic group in *trans*-position. Although the first attempts in performing the

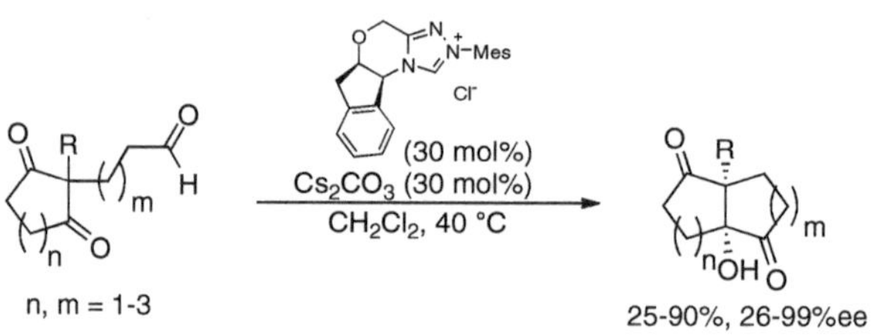

SCHEME 3.11.58

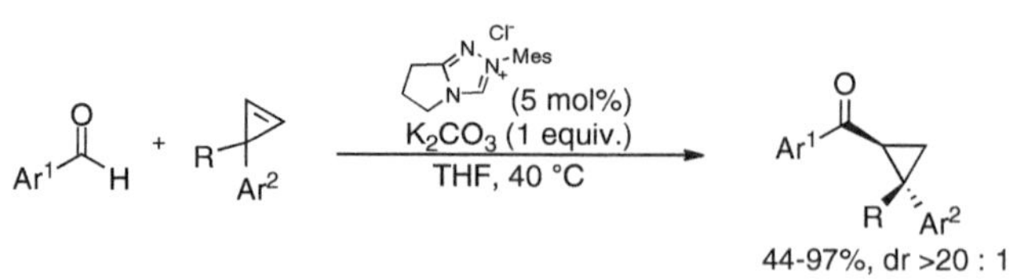

SCHEME 3.11.59

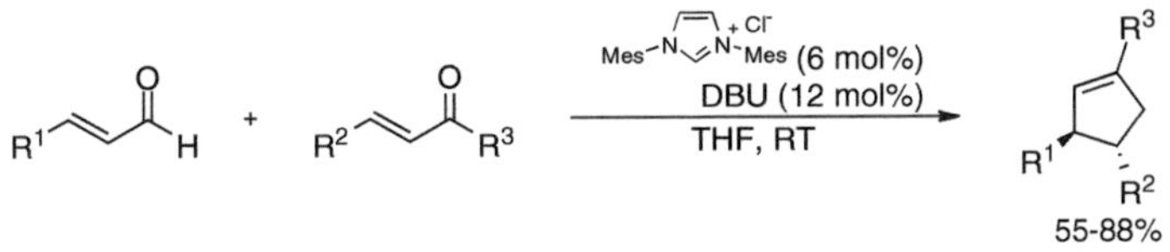

SCHEME 3.11.60

reaction enantioselectively resulted in high enantioselectivities but low yields, the modification of the catalyst made it possible to obtain the products in high yields and high enantioselectivities.[83]

Nair's group published the synthesis of trisubstituted cyclopentenes using an imidazole-based NHC catalyst (Scheme 3.11.60).[84] Only 6 mol% of the catalyst were needed for the reaction during which an aryl enal and a chalcone formed the cyclopentenes with absolute selectivity for the *trans*-isomer. Yields were usually high for the reaction.

The use of NHCs has been extended to transesterification and transamidation as well. Although these reactions are usually performed under relatively harsh acidic or basic conditions, NHCs allow performing these reactions under much milder conditions. The groups of Nolan and Waymouth/Hedrick reported almost simultaneously the use of *N*-alkyl and *N*-arylimidazolium salts for the transesterification of various compounds (Scheme 3.11.61).[85] Under these conditions, a broad range of sensitive and challenging alcohols and esters smoothly underwent transesterification.

In a similar approach, Movassaghi and Schmidt describe the transamidation of esters (Scheme 3.11.62).[86] Using 5 mol% of the catalyst a series of differently substituted esters underwent transamidation smoothly giving the products in high yields. The drawback of the reaction is that it is limited to 1,2- and 1,3-aminoalcohols, probably due to the mechanism involving an O to N acyl shift.

3.11.3.1.3 Chiral Phosphoric Acids Catalysts

A different type of organocatalysts is the chiral phosphoric acids. Their mode of activation usually relies on their Brønsted acidity. When chiral moieties are incorporated as substituents in their structures they can be valuable catalysts for performing enantionselective reactions.[87]

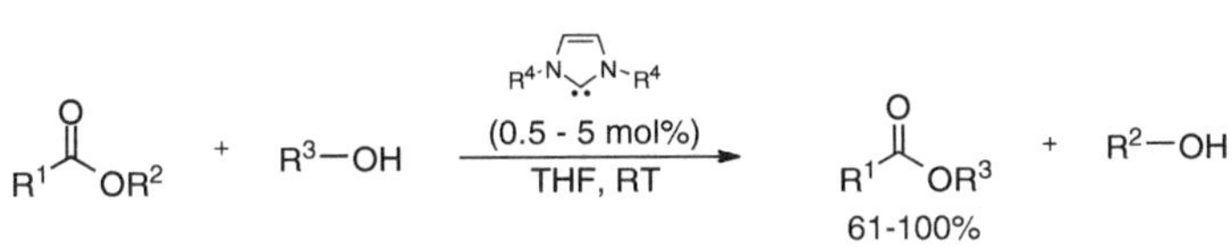

SCHEME 3.11.61

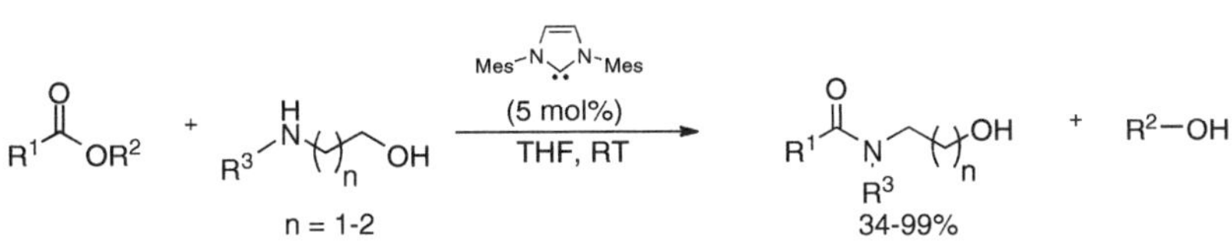

SCHEME 3.11.62

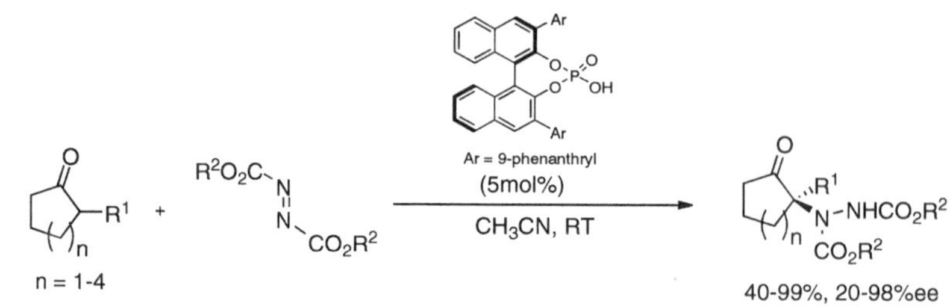

SCHEME 3.11.63

In 2015, List's group showed that chiral phosphoric acids can effectively catalyze the α-amination of α-branched ketones, generating a chiral quaternary center that cannot easily be achieved by amine catalysis (Scheme 3.11.63).[88] A variety of cyclic α-branched ketones were aminated with diazocarboxylates in the α-position, resulting in mostly high yields and good enantioselectivities.

Toste et al. could show that with a differently substituted phosphoric acid the reaction can be performed without any solvent.[89] After dissolving the starting materials in CH_2Cl_2 to ensure proper mixing the solvent was evaporated and the residue was heated at 45 °C for 40–60 h. The α-aminated products are obtained in high yields and high enantioselectivities.

The Friedel-Crafts addition of benzoxazinones with substituted pyrroles could be catalyzed by chiral phosphoric acids, as was shown by Lin's group (Scheme 3.11.64).[90] The reaction proceeds smoothly to form the products in high yields and enantioselectivities, generating a quaternary stereocenter. The authors claim the presence of the CF_3 group to be essential as it contributes to hydrogen bonding in the activation mode of the reaction.

Chiral phosphoric acids were shown to be able to catalyze the asymmetric transfer hydrogenation for a range of imines. Rueping and his coworkers described that different substituted quinolines could be enantioselectively hydrogenated using only 0.5 mol% of chiral phosphoric acid as catalyst (Scheme 3.11.65).[91] Although the reaction could be performed in a traditional batch reactor, they also reported that performing it under continuous flow condition significantly improved the yield of the reaction (67% vs. 97%).

Using a very similar catalyst, More and Bhanage published that asymmetric transfer hydrogenations are also possible in the sustainable solvent diethyl carbonate (Scheme 3.11.66).[92] Quinolines could be reduced into the corresponding tetrahydroquinolines in high yields and enantiomeric excess. As in the work of Rueping, a Hantzsch ester was used as the hydrogen donor.

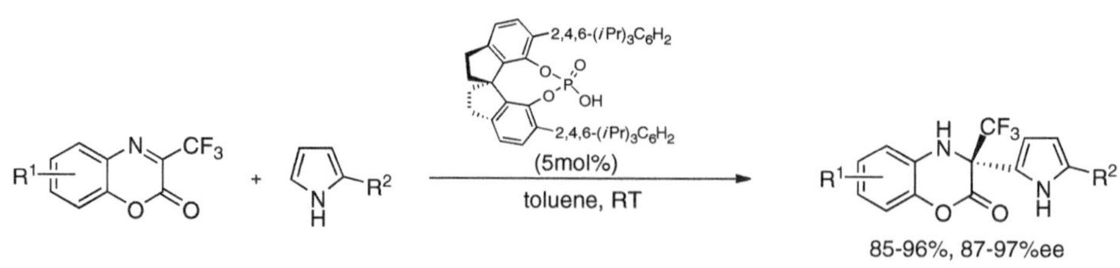

SCHEME 3.11.64

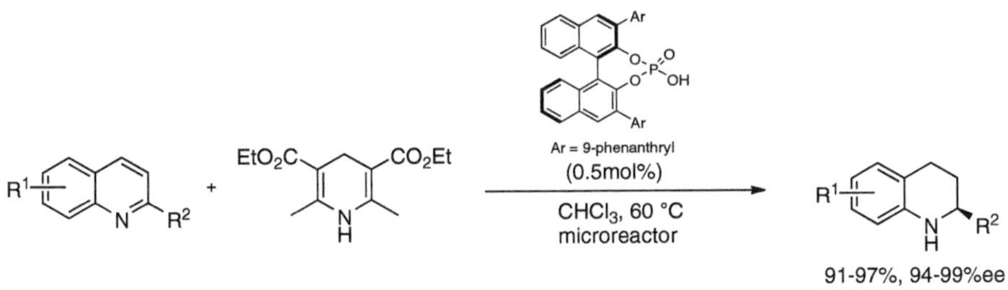

SCHEME 3.11.65

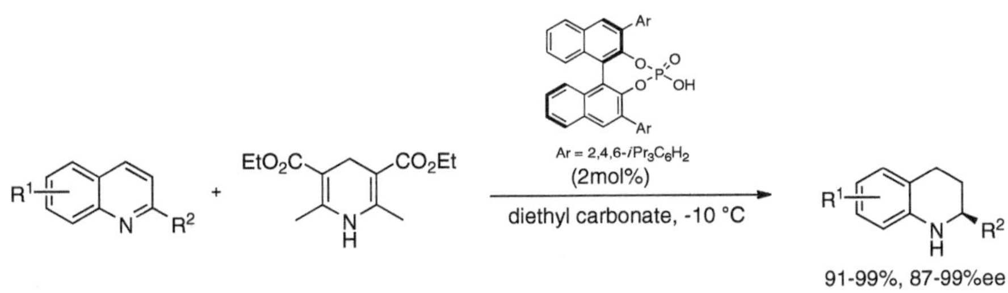

SCHEME 3.11.66

3.11.3.1.4 Carbohydrate-Based Catalysts

Carbohydrates and closely related compounds can easily be derived from natural sources and are the major component of plant-derived biomass. They possess several contiguous stereogenic centers and usually only appear as one of the possible diastereomers. Due to these properties they have been used as building blocks for ligands in metal-catalyzed reactions. With the rise of organocatalysis, there is an increased interest toward their application as direct catalyst. The usually high water solubility makes them attractive for the use in reactions following the principles of green chemistry.

Despite these interesting properties the use of unaltered carbohydrates as catalysts is only sporadically explored. One example is the synthesis of 2-arylbenzimidazoles from *o*-phenylenediamines and substituted benzaldehydes (Scheme 3.11.67).[93] The products are obtained in excellent yields by simple heating of the starting materials in 1M D-glucose solution.

It has been shown by Zhang and coworkers that aldol reactions between isatins and ketones can be successfully catalyzed by protected glucosamines (Scheme 3.11.68).[94] They found that the presence of the unprotected amino group is crucial. Although the reactions were mostly performed in dichloromethane, the authors showed that with cyclohexanone and isatin the reaction could be performed in water.

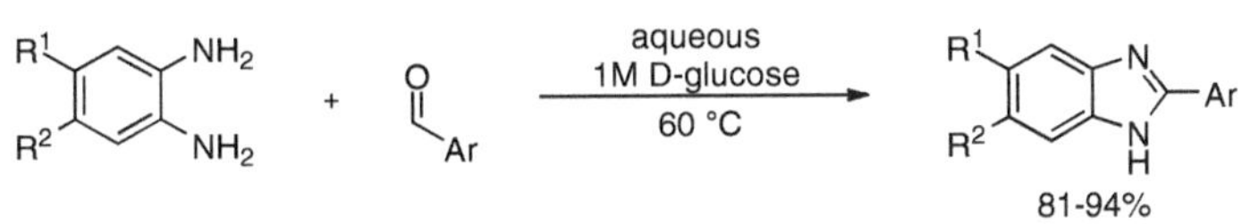

SCHEME 3.11.67

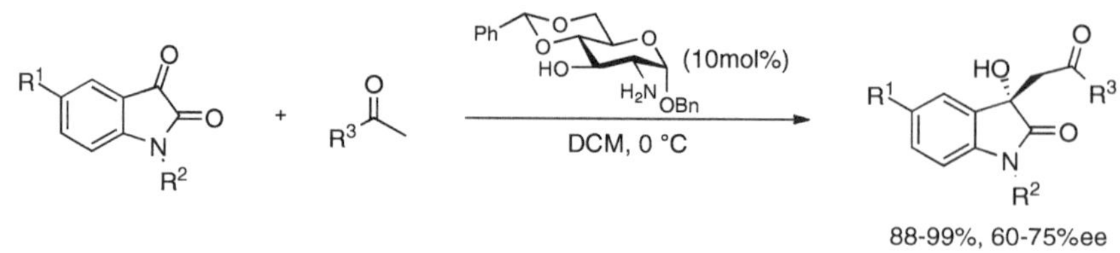

SCHEME 3.11.68

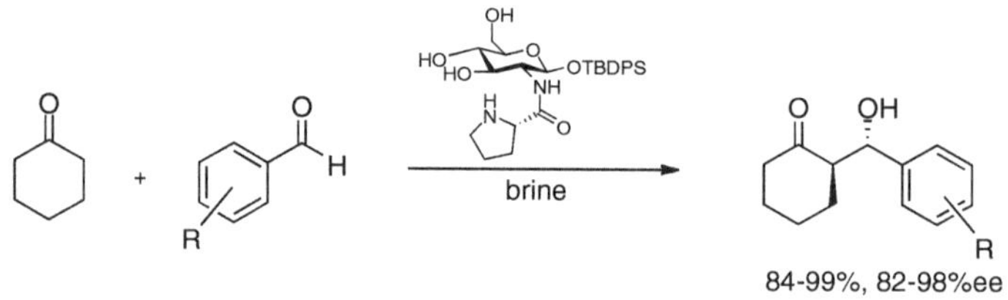

SCHEME 3.11.69

The use of water as a solvent for aldol reactions has been shown to be problematic as the hydrogen bonds needed in the transition state to maintain enantioselectivity are broken up.[95] Using modified carbohydrate-based catalysts, these limitations could be bypassed. When, for example, glucosamides are modified with proline, the reaction between cyclohexanone and benzaldehydes could be performed in water or brine with excellent yields and high enantioselectivities (Scheme 3.11.69).[54,96] The products were obtained as the *anti*-isomer mainly.

3.11.3.2 Oxidation Reactions

Oxidation reactions are traditionally in the realm of metal catalysis, and there are only a few examples of metal-free transformations. One exception to this is the synthesis of epoxides from alkenes, which is usually achieved using stoichiometric amounts of an organic peracid. *m*-Chloroperbenzoic acid is one commonly used peracid to perform epoxidation. To bypass the need for a stoichiometric peracid, recent developments in this area focus on the application of catalytic variations for the epoxidation of alkenes.

Miller et al. showed that in-situ-generated peracids can be successfully used for asymmetric epoxidation reactions. It was found that protected aspartic acid derivatives catalyze the epoxidation of alkenes with H_2O_2 as oxidizing agent; however, the products were obtained in racemic form. When the aspartic acid is incorporated into a small peptide backbone the reaction proceeded with asymmetric induction (Scheme 3.11.70).[97] The drawback of the reaction is that 2 equivalents of diisopropylcarbodimide need to be used.

Another type of reagent commonly used for epoxidation is dioxiranes. While the initial work suffered from the need for a stoichiometric use of the oxirane and low enantioselectivities, a breakthrough was made by Shi and coworkers.[98] They used modified fructose as catalyst and oxone ($KHSO_5$) as reoxidant to generate the oxirane. Although their initial findings were limited to *trans*-alkenes and the reaction required stoichiometric amounts of the

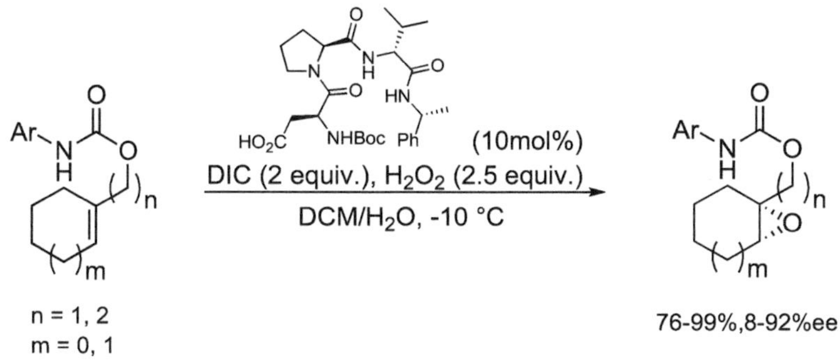

SCHEME 3.11.70

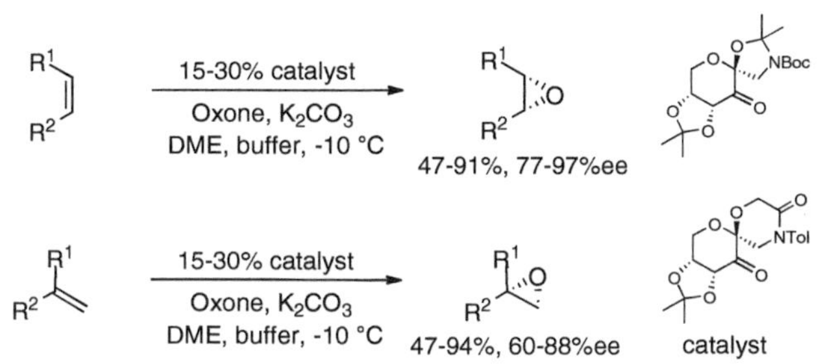

SCHEME 3.11.71

fructose, later modifications of the catalyst allowed the use of *cis*- and terminal olefins (Scheme 3.11.71).[99]

In a similar approach, Yang and coworkers showed that methyl(trifluoromethyl)dioxirane prepared in situ from trifluoroacetone and oxone can effectively perform epoxidation.[100] Although these first attempts needed nearly stoichiometric amounts of the ketone, the same group developed a catalytic enantioselective version of the reaction.[101] Using 10 mol% of a chiral ketone derived from binaphthalene as catalyst, they obtained the epoxides of a variety of different substituted alkenes in high yields and high enantioselectivities (Scheme 3.11.72).

In a 2014 publication by Kokotos, it was shown that with 2,2,2-trifluoroacetophenone as a catalyst H_2O_2 could be used as terminal oxidant in *tert*-butanol as solvent (Scheme 3.11.73).[102] A variety of mono-, di-, and trisubtituted alkenes were transformed into the corresponding epoxides in excellent yields in short reaction times.

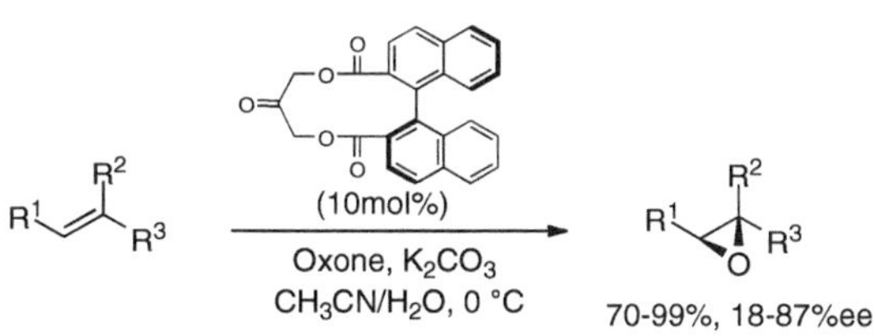

SCHEME 3.11.72

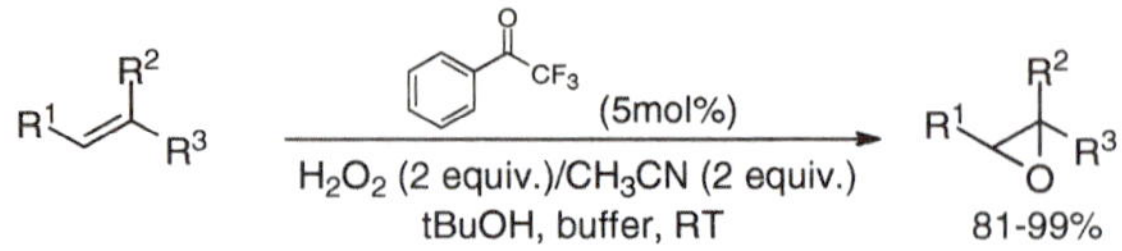

SCHEME 3.11.73

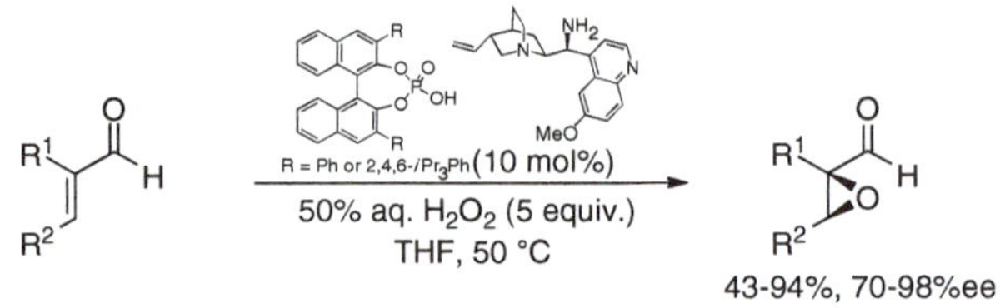

SCHEME 3.11.74

Hydrogen peroxide was also used as an oxidant in the epoxidation of α,β-unsaturated aldehydes. The catalytic system consisted of an amine catalyst activating the aldehyde and a phosphoric acid performing the epoxidation.[103] List et al. showed that the combination of a binaphthol-derived phosphoric acid and a cinchona-alkaloid-derived amine was an effective system yielding the products in moderate to high yield and high enantioselectivities (Scheme 3.11.74).

The same group could show later that the use of phosphoric acid is not necessary when α,β-unsaturated ketones are used as substrate (Scheme 3.11.75).[104] The amine was used in the form of its trifluoroacetic acid salt and the solvent had to be changed to 1,4-dioxane.

Depending on the structure of the substrate, minor modifications to the reaction conditions needed to be made.

A very similar approach has been taken by Hayashi et al.: α-substituted acroleins were made to undergo asymmetric epoxidation using diphenylprolinol diphenylmethylsilylether as catalyst and H_2O_2 as oxidant (Scheme 3.11.76).[105] The reaction tolerated a variety of substituents and could be performed at room temperature in hexane as solvent.

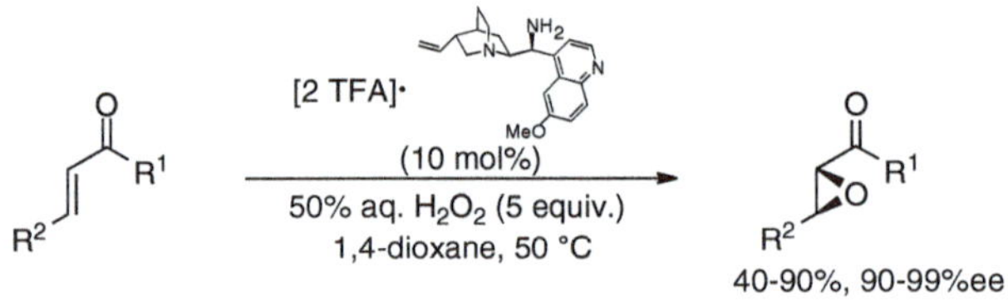

SCHEME 3.11.75

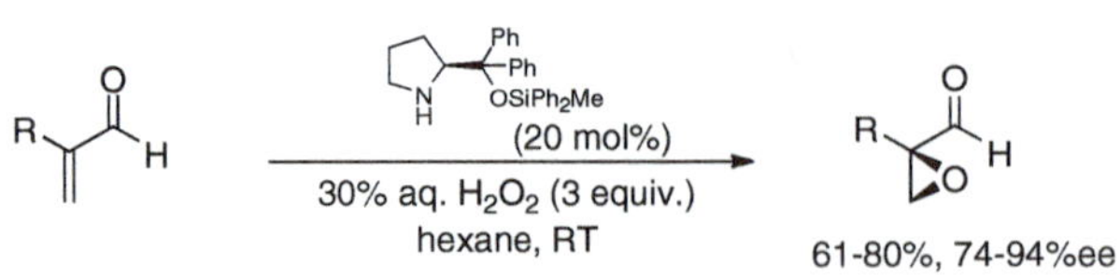

SCHEME 3.11.76

3.11.4 CONCLUSION

As the above examples show, extensive investigations have been done to improve reactions and reaction conditions with the goal to comply with the principles of green chemistry. These include the development of new catalysts that are abundant, environment friendly, easy to obtain, and require only low catalyst loadings; the use of environment-friendly solvents and reagents, and also the development of new reactions that allow improvement of atom economy and reduction in the number of steps needed to obtain a certain product.

Although there have been significant advances in the green direction, many of the examples also show that there are still several drawbacks. The improvements made often impact only one or two aspects of a given reaction or transformation, leaving room for further improvement.

Concluding, it can be stated that although the first steps into the direction of a greener homogeneous catalysis are very promising, there is still more work needed for the full implication of the principles of green chemistry in this field.

References

1. Anastas PT, Warner JC. *Green chemistry: theory and practice*. New York: Oxford University Press; 1998. p. 30.
2. Chadwick JC, van Leeuwen PWNM. *Homogeneous catalysts: activity – stability – deactivation*. Weinheim, Germany: Wiley-VCH; 2001.
3. Buffin B, Clarkson J, Belitz N, Kundu A. Pd(II)-biquinoline catalyzed aerobic oxidation of alcohols in water. *J Mol Catal A-Chem* 2005;**225**:111–6.
4. Ebner D, Bagdanoff T, Ferreira M, McFadden M, Caspi D, Trend M, Stoltz B. The palladium-catalyzed aerobic kinetic resolution of secondary alcohols: reaction development, scope, and applications. *Chem Eur J* 2009;**15**:12978–92.
5. Weerasiri C, Gorden V. Cu(II) 2-quinoxalinol salen catalyzed oxidation of propargylic, benzylic, and allylic alcohols using tert-butyl hydroperoxide in aqueous solutions. *Tetrahedron* 2014;**70**:7962–8.
6. Biradar V, Dongare K, Umbarkar B. Selective oxidation of aromatic primary alcohols to aldehydes using molybdenum acetylide oxo-peroxo complex as catalyst. *Tetrahedron Lett* 2009;**50**:2885–8.
7. Barooah N, Sharma S, Sarma B, Baruah J. Catalytic oxidative reactions of organic compounds by nitrogen-containing copper complexes. *Appl Organomet Chem* 2004;**18**:440–5.
8. Gamez P, Arends I, Reedijk J, Sheldon R. Copper(II)-catalysed aerobic oxidation of primary alcohols to aldehydes. *Chem Commun* 2003:2414–5.
9. Ahmad U, Raisanen M, Leskela M, Repo T. Copper catalyzed oxidation of benzylic alcohols in water with H_2O_2. *Appl Catal A-Gen* 2012;**411**:180–7.
10. Han M, Kim S, Kim T, Lee C. Bismuth tribromide catalyzed oxidation of alcohols with aqueous hydrogen peroxide. *Synlett* 2015;**26**:2434–6.
11. Marui K, Higashiura Y, Kodama S, Hashidate S, Nomoto A, Yano S, Ueshima M, Ogawa A. Vanadium-catalyzed green oxidation of benzylic alcohols in water under air atmosphere. *Tetrahedron* 2014;**70**:2431–8.
12. Du Z, Ma J, Ma H, Gao J, Xu J. Synergistic effect of vanadium-phosphorus promoted oxidation of benzylic alcohols with molecular oxygen in water. *Green Chem* 2010;**12**:590–2.
13. Patil D, Adimurthy S. Copper-catalyzed aerobic oxidation of amines to imines under neat conditions with low catalyst loading. *Adv Synth Catal* 2011;**353**:1695–700.
14. Kang Q, Zhang Y. Copper-catalyzed highly efficient aerobic oxidative synthesis of imines from alcohols and amines. *Green Chem* 2012;**14**:1016–9.
15. Marui K, Nomoto A, Ueshima M, Ogawa A. Eco-friendly copper sulfate-catalyzed oxidation of amines to imines by hydrogen peroxide in water. *Tetrahedron Lett* 2015;**56**:1200–2.

16. Grosselin J, Mercier C, Allmang G, Grass F. Selective hydrogenation of alpha, beta-unsaturated aldehydes in aqueous organic 2-phase solvent systems using ruthenium or rhodium complexes of sulfonated phosphines. *Organometallics* 1991;**10**:2126–33.
17. Noyori R. Asymmetric catalysis: science and opportunities (Nobel lecture). *Angew Chem Int Ed* 2002;**41**:2008–22.
18. Noyori R. Facts are the enemy of truth-reflections on serendipitous discovery and unforeseen developments in asymmetric catalysis. *Angew Chem Int Ed* 2013;**52**:79–92.
19. Noyori R, Hashiguchi S. Asymmetric transfer hydrogenation catalyzed by chiral ruthenium complexes. *Acc Chem Res* 1997;**30**:97–102.
20. Hashiguchi S, Fujia A, Takehara J, Ikariya T, Noyori R. Asymmetric transfer hydrogenation of aromatic ketones catalyzed by chiral ruthenium(ii) complexes. *J Am Chem Soc* 1995;**117**:7562–3.
21. Wu X, Li X, Zanotti-Gerosa A, Pettman A, Liu J, Mills J, Xiao J. Rh(III)- and Ir(III)-catalyzed asymmetric transfer hydrogenation of ketones in water. *Chem Eur J* 2008;**14**:2209–22.
22. Wolfson A, Dlugy C, Shotland Y, Tavor D. Glycerol as solvent and hydrogen donor in transfer hydrogenation-dehydrogenation reactions. *Tetrahedron Lett* 2009;**50**:5951–3.
23. Langer R, Leitus G, Ben-David Y, Milstein D. Efficient hydrogenation of ketones catalyzed by an iron pincer complex. *Angew Chem Int Ed* 2011;**50**:2120–4.
24. Zell T, Milstein D. Hydrogenation and dehydrogenation iron pincer catalysts capable of metal ligand cooperation by aromatization/dearomatization. *Acc Chem Res* 2015;**48**:1979–94.
25. Wienhoefer G, Sorribes I, Boddien A, Westerhaus F, Junge K, Junge H, Llusar R, Beller M. General and selective iron-catalyzed transfer hydrogenation of nitroarenes without base. *J Am Chem Soc* 2011;**133**:12875–9.
26. Bart S, Lobkovsky E, Chirik P. Preparation and molecular and electronic structures of iron(0) dinitrogen and silane complexes and their application to catalytic hydrogenation and hydrosilation. *J Am Chem Soc* 2004;**126**:13794–807.
27. Werkmeister S, Fleischer S, Zhou S, Junge K, Beller M. Development of new hydrogenations of imines and benign reductive hydroaminations: zinc triflate as a catalyst. *Chemsuschem* 2012;**5**:777–82.
28. Roesler S, Obenauf J, Kempe R. A highly active and easily accessible cobalt catalyst for selective hydrogenation of C–O bonds. *J Am Chem Soc* 2015;**137**:7998–8001.
29. Fleckenstein CA, Plenio H. Aqueous/organic cross coupling: sustainable protocol for Sonogashira reactions of heterocycles. *Green Chem* 2008;**10**:563–70.
30. Marziale AN, Jantke D, Faul SH, Reiner T, Herdtweck E, Eppinger J. An efficient protocol for the palladium-catalysed Suzuki-Miyaura cross-coupling. *Green Chem* 2011;**13**:169–77.
31. Simon M, Li C. Green chemistry oriented organic synthesis in water. *Chem Soc Rev* 2012;**41**:1415–27.
32. Kostas D, Coutsolelos G, Charalambidis G, Skondra A. The first use of porphyrins as catalysts in cross-coupling reactions: a water-soluble palladium complex with a porphyrin ligand as an efficient catalyst precursor for the Suzuki-Miyaura reaction in aqueous media under aerobic conditions. *Tetrahedron Lett* 2007;**48**:6688–91.
33. Mondal M, Bora U. An efficient protocol for palladium-catalyzed ligand-free Suzuki-Miyaura coupling in water. *Green Chem* 2012;**14**:1873–6.
34. Oruah R, Ali A, Saikia B, Sarma D. A novel green protocol for ligand free Suzuki-Miyaura cross-coupling reactions in WEB at room temperature. *Green Chem* 2015;**17**:1442–5.
35. Dams M, De Vos D, Celen S, Jacobs P. Toward waste-free production of Heck products with a catalytic palladium system under oxygen. *Angew Chem Int Ed* 2003;**42**:3512–5.
36. Bagherzadeh M, Amini M, Derakhshandeh G, Haghdoost M. An efficient glucose-based ligand for Heck and Suzuki coupling reactions in aqueous media. *J Iran Chem Soc* 2014;**11**:441–6.
37. Wu W, Wang J, Tsai F. A reusable $FeCl_3 \cdot 6H_2O$/cationic 2,2′-bipyridyl catalytic system for the coupling of aryl iodides with thiols in water under aerobic conditions. *Green Chem* 2009;**11**:326–9.
38. Li X, Yang D, Jiang Y, Fu H. Efficient copper-catalyzed N-arylations of nitrogen-containing heterocycles and aliphatic amines in water. *Green Chem* 2010;**12**:1097–105.
39. Shafir A, Buchwald SL. Highly selective room-temperature copper-catalyzed C-N coupling reactions. *J Am Chem Soc* 2006;**128**:8742–3.
40. Xu C, Wang Z, Fu W, Lou X, Li Y, Cen F, Ma H, Ji B. Synthesis and structural characterization of monophosphine-cyclopalladated ferrocenylpyrimidine complexes and reusable catalytic system for amination of hindered aryl chlorides in PEG-400. *Organometallics* 2009;**28**:1909–16.
41. Chen J, Spear S, Huddleston J, Rogers R. Polyethylene glycol and solutions of polyethylene glycol as green reaction media. *Green Chem* 2005;**7**:64–82.

42. Francos J, Cadierno V. Palladium-catalyzed cycloisomerization of (Z)-enynols into furans using green solvents: glycerol vs. water. *Green Chem* 2010;**12**:1552–5.
43. Ozdemir I, Yigit B, Cetinkaya B, Ulku D, Tahir M, Arici C. Synthesis of a water-soluble carbene complex and its use as catalyst for the synthesis of 2,3-dimethylfuran. *J Organomet Chem* 2001;**633**:27–32.
44. Winter C, Krause N. Towards sustainable homogeneous gold catalysis: cycloisomerization of functionalized allenes in water. *Green Chem* 2009;**11**:1309–12.
45. Motoda D, Kinoshita H, Shinokubo H, Oshima K. Phosphane-free rhodium catalyst in an anionic micellar system for [4+2] annulation of dienynes. *Angew Chem Int Ed* 2004;**43**:1860–2.
46. Chitra S, Paramasivan K, Shanmugamani K, Paul R, Paul B. Advanced oxidation processes for the treatment of surfactant wastes. *J Chem Eng Chem Res* 2014;**1**:163–73.
47. Kwong Y, Li M, Lam W, Qiu Q, Lee W, Yeung H, Chan S, Chan C. Rhodium-catalyzed asymmetric aqueous Pauson-Khand-type reaction. *Chem Eur J* 2005;**11**:3872–80.
48. Nicola T, Brenner M, Donsbach K, Kreye P. First scale-up to production scale of a ring closing metathesis reaction forming a 15-membered macrocycle as a precursor of an active pharmaceutical ingredient. *Org Process Res Dev* 2005;**9**:513–5.
49. Michrowska A, Gulajski L, Kaczmarska Z, Mennecke K, Kirschning A, Grela K. A green catalyst for green chemistry: synthesis and application of an olefin metathesis catalyst bearing a quaternary ammonium group. *Green Chem* 2006;**8**:685–8.
50. Wang J, Jackson R, Robinson J. A simple and practical preparation of an efficient water soluble olefin metathesis catalyst. *Green Chem* 2015;**17**:3407–14.
51. a. Eder U, Sauer G, Wiechert R. New type of asymmetric cyclization to optically active steroid CD partial structures. *Angew Chem Int Ed* 1971;**10**:496–7;
    b. Hajos ZG, Parrish DR. *German patent DE 2102623*. 1971.
52. Donslund BS, Johansen TK, Poulsen PH, Halskow KS, Jorgensen KA. The diarylprolinol silyl ethers: ten years after. *Angew Chem Int Ed* 2015;**54**:13860–74.
53. Bayat S, Salleh AB, Abd Malek E, Yousefi S, Abdul Rahman MB. Design of a simple organocatlyst for asymmetric direct aldol reactions in aqueous medium. *Catal Lett* 2015;**145**:1750–5.
54. Pedatella S, De Nisco M, Mastroianni D, Naviglio D, Nucci A, Caputo R. Diastereo- and enantioselective direct aldol reactions in aqueous medium: a new highly efficient proline-sugar chimeric catalyst. *Adv Synth Catal* 2011;**353**:1443–6.
55. Tian H, Gao JL, Xu H, Zheng LY, Huang WB, Liu QW, Zhang SQ. Proline-based dipeptides as efficient organocatalysts for asymmetric aldol reactions in brine. *Tetrahedron Asymmetry* 2011;**22**:1074–80.
56. Huang WB, Liu QW, Zheng LY, Zhang SQ. Novel primary amine organocatalysts derived from cinchona alkaloids for asymmetric direct aldol reactions in brine. *Catal Lett* 2011;**141**:191–7.
57. Zheng BL, Liu QZ, Guo CS, Wang XL, He L. Highly enantioselective direct aldol reaction catalyzed by cinchona derived primary amines. *Org Biomol Chem* 2007;**5**:2913–5.
58. Ayats C, Henseler AH, Pericàs MA. A solid-supported organocatalyst for continuous-flow enantioselective aldol reactions. *ChemSusChem* 2012;**5**:320–5.
59. Landge SM, Török B. Highly enantioselective organocatalytic addition of ethyl trifluoropyruvate to ketones with subzero temperature microwave activation. *Catal Lett* 2009;**131**:432–9.
60. a. Rodriguez B, Ratanen T, Bolm C. Solvent-free asymmetric organocatalysis in a ball-mill. *Angew Chem Int Ed* 2006;**45**:6924–6;
    b. Rodriguez B, Bruckmann A, Bolm C. A highly efficient asymmetric organocatalytic aldol reaction in a ball mill. *Chem Eur J* 2007;**13**:4710–22.
61. Markert M, Mulzer M, Schetter B, Mahrwald R. Amine-catalyzed direct aldol addition. *J Am Chem Soc* 2007;**129**:7258–9.
62. a. Perlmutter A. *Conjugative additions in organic synthesis*. Oxford: Pergamon Press; 1992;
    b. Yamaguchi M. Conjugate addition of stabilized carbanions. In: Jacobsen EN, Pfaltz A, Yamamoto H, editors. *Comprehensive asymmetric catalysis*, Vol. 3. Berlin: Springer; 1999;
    c. Sibi MP, Manyem S. Enantioselective conjugate additions. *Tetrahedron* 2000;**56**:8033–61;
    d. Krause N, Hoffmann-Röder A. Recent advances in catalytic enantioselective Michael additions. *Synthesis* 2001;**2**:171–96.
63. Procopio A, De Nino A, Nardi M, Oliverio M, Paonesa R, Pasceri R. A new microwave-assisted organocatalytic solvent-free synthesis of optically enriched Michael adducts. *Synlett* 2010;**12**:1849–53.

64. Zhao L, Shen J, Liu D, Liu Y, Zhang W. Highly efficient asymmetric Michael addition of aldehyde to nitroolefin using perhydroindolic acid as a chiral organocatalyst. *Org Biomol Chem* 2012;**10**:2840–6.
65. Alza E, Sayalero S, Kasaplar P, Almasi D, Pericàs MA. Polystyrene-supported diarylprolinol ethers as highly efficient organocatalysts for Michael-type reactions. *Chem Eur J* 2011;**17**:11585–95.
66. Khandelwal S, Kumar Tailor Y, Kumar M. L-proline catalyzed multicomponent reactions. *Curr Organocatlysis* 2016;**3**:176–204.
67. Cai H, Xia L, Lee RK, Shim JJ, Kim SH. Construction of diverse and functionalized 2H-chromenes by organocatalytic multicomponent reactions. *Eur J Org Chem* 2015;**23**:5212–20.
68. Li M, Zhang B, Gu Y. Facile construction of densely functionalized 4H-chromenes via three-component reactions catalyzed by L-proline. *Green Chem* 2012;**14**:2421–8.
69. Sharma RK, Priyanka KD. L-Proline catalyzed condensation of salicylaldehydes with ethyl nitroacetate: an efficient access to 3-nitrocoumarins. *Monatsh Chem* 2016:1–5.
70. Török B, Abid M, London G, Esquibel J, Török M, Mhadgut SC, Yan P, Prakash GKS. Highly enantioselective organocatalytic hydroxyalkylation of indoles with ethyl trifluoropyruvate. *Angew Chem Int Ed* 2005;**44**:3086–9.
71. Rimaz M, Khalafy J, Mousavi H. A green organocatalyzed one-pot protocol for efficient synthesis of new substituted pyrimido[4,5- d]pyrimidinones using a Biginelli-like reaction. *Res Chem Intermed* 2016:1–16.
72. Janardhan B, Ravibabu V, Crooks PA, Rajitha B. L-proline catalyzed an efficient multicomponent one-pot synthesis of poly substituted pyridines. *Org Commun* 2012;**5**:186–95.
73. a. Ramachary DB, Ramakumar K, Narayana VV. Amino acid-catalyzed cascade [3+2]-cycloaddition/hydrolysis reactions based on the push–pull dienamine platform: synthesis of highly functionalized NH-1,2,3-triazoles. *Chem Eur J* 2008;**14**:9143–7;
b. Belkheira M, El Abed D, Pons JM, Bressy C. Organocatalytic synthesis of 1,2,3-triazoles from unactivated ketones and arylazides. *Chem Eur J* 2011;**17**:12917–21.
74. Yeung DKJ, Gao T, Huang J, Sun S, Guo H, Wang J. Organocatalytic 1,3-dipolar cycloaddition reactions of ketones and azides with water as a solvent. *Green Chem* 2013;**15**:2384–8.
75. Zhao YL, Wang Y, Luo YC, Fu XZ, Xu PF. Asymmetric C–H functionalization involving organocatalysis. *Tetrahedron Lett* 2015;**56**:3703–14.
76. Zhang B, Xiang SK, Zhang LH, Cui Y, Jiao N. Organocatalytic asymmetric intermolecular dehydrogenative α-alkylation of aldehydes using molecular oxygen as oxidant. *Org Lett* 2011;**13**:5112–5.
77. Igau A, Grutzmacher H, Baceiredo A, Bertrand G. Analogous α,α′-bis-Carbenoid, triply bonded species: synthesis of a stable λ3- phosphinocarbene–λ^5-phosphaacetylene. *J Am Chem Soc* 1988;**110**:6463–6.
78. Arduengo III AJ, Harlow RL, Kline M. A stable crystalline carbene. *J Am Chem Soc* 1991;**113**:361–3.
79. Enders D, Breuer K, Raabe G, Runsink J, Teles JH, Melder JP, Ebel K, Brode S. Preparation, structure, and reactivity of 1,3,4-Triphenyl-4,5-dihydro-1H-1,2,4-triazol-5-ylidene, a new stable carbene. *Angew Chem, Int Ed Engl* 1995;**34**:1021–3.
80. Baragwanath L, Rose CA, Zeitler K, Connon SJ. Highly enantioselective benzoin condensation reactions involving a bifunctional protic pentafluorophenyl-substituted triazolium precatalyst. *J Org Chem* 2009;**74**:9214–7.
81. a. Ema T, Oue Y, Akihara K, Miyazaki Y, Sakai T. Stereoselective synthesis of bicyclic tertiary alcohols with quaternary stereocenters via intramolecular crossed benzoin reactions catalyzed by N-heterocyclic carbenes. *Org Lett* 2009;**11**:4866–9;
b. Ema T, Akihara K, Obayashi R, Sakai T. Construction of contiguous tetrasubstituted carbon stereocenters by intramolecular crossed benzoin reactions catalyzed by N-heterocyclic carbene (NHC) organocatalyst. *Adv Synth Catal* 2012;**354**:3283–90.
82. Bugaut X, Liu F, Glorius F. N-heterocyclic carbene (NHC)-catalyzed intermolecular hydroacylation of cyclopropenes. *J Am Chem Soc* 2011;**133**:8130–3.
83. a. Schedler M, Fröhlich R, Daniliuc CG, Glorius F. 2,6-Dimethoxyphenyl-substituted N-heterocyclic carbenes (NHCs): a family of highly electron-rich organocatalysts. *Eur J Org Chem* 2012;**22**:4164–71;
b. Liu F, Bugaut X, Schedler M, Fröhlich R, Glorius F. Designing N-heterocyclic carbenes: simultaneous enhancement of reactivity and enantioselectivity in the asymmetric hydroacylation of cyclopropenes. *Angew Chem, Int Ed* 2011;**50**:12626–30.
84. Nair V, Vellalath S, Poonoth M, Suresh E. N-heterocyclic carbene-catalyzed reaction of chalcones and enals via homoenolate: an efficient synthesis of 1,3,4-trisubstituted cyclopentenes. *J Am Chem Soc* 2006;**128**:8736–7.

85. a. Grasa GA, Kissling RM, Nolan SP. N-heterocyclic carbenes as versatile nucleophilic catalysts for transesterification/acylation reactions. *Org Lett* 2002;**4**:3583–6;
b. Nyce GW, Lamboy JA, Connor EF, Waymouth RM, Hedrick JL. Expanding the catalytic activity of nucleophilic N-heterocyclic carbenes for transesterification reactions. *Org Lett* 2002;**4**:3587–90;
c. Grasa GA, Güveli T, Singh R, Nolan SP. Efficient transesterification/acylation reactions mediated by N-heterocyclic carbene catalysts. *J Org Chem* 2003;**68**:2812–9.
86. Movassaghi M, Schmidt MA. N-heterocyclic carbene-catalyzed amidation of unactivated esters with amino alcohols. *Org Lett* 2005;**7**:2453–6.
87. a. Connon SJ. Chiral phosphoric acids: powerful organocatalysts for asymetric addition reactions to imines. *Angew Chem Int Ed* 2006;**45**:3909–12;
b. Terada M. Binaphthol-derived phosphoric acid as a versatile catalyst for enantioselective carbon–carbon bond forming reactions. *Chem Commun* 2008;**35**:4097–112;
c. Kampen D, Reisinger CM, List B. Chiral bronsted acids for asymmetric organocatalysis. *Top Curr Chem* 2010;**291**:395–456.
88. Shevchenko GA, Pupo G, List B. Catalytic asymmetric α-amination of α-branched ketones via enol catalysis. *Synlett* 2015;**26**:1413–6.
89. Yang X, Toste FD. Direct asymmetric amination of α-branched cyclic ketones catalyzed by a chiral phosphoric acid. *J Am Chem Soc* 2015;**137**:3205–8.
90. Lou H, Wang Y, Jin E, Lin X. Organocatalytic asymmetric synthesis of dihydrobenzoxazinones bearing trifluoromethylated quaternary stereocenters. *J Org Chem* 2016;**81**:2019–26.
91. Rueping M, Bootwicha T, Sugiono E. Continuous-flow catalytic asymmetric hydrogenations: reaction optimization using FTIR inline analysis. *Beilstein J Org Chem* 2012;**8**:300–7.
92. More GV, Bhanage BM. Chiral phosphoric acid catalyzed asymmetric transfer hydrogenation of quinolines in a sustainable solvent. *Tetrahedron Asymmetry* 2015;**26**:1174–9.
93. Rostamizadeh S, Aryan R, Ghaieni HR. Aqueous 1 M glucose solution as a novel and fully green reaction medium and catalyst for the oxidant-free synthesis of 2-arylbenzimidazoles. *Synth Commun* 2011;**41**:1794–804.
94. Shen C, Shen F, Xia H, Zhang P, Chen X. Carbohydrate-derived alcohols as organocatalysts in enantioselective aldol reactions of isatins with ketones. *Tetrahedron Asymmetry* 2011;**22**:708–12.
95. a. Butler RN, Coyne AG. Water: nature's reaction enforcers comparative effects for organic synthesis "in-water" and "on-water". *Chem Rev* 2010;**110**:6302–37;
b. Lindström UM. Stereoselective oragnic reactions in water. *Chem Rev* 2002;**102**:2751–72.
96. Shen C, Shen F, Zhou G, Xia X, Chen X, Liu X, Zhang P. Novel carbohydrate-derived prolinamide as a highly efficient, recoverable catalyst for direct aldol reactions in water. *Catal Commun* 2012;**26**:6–10.
97. Peris G, Jakobsche CE, Miller SJ. Aspartate-catalyzed asymmetric epoxidation reactions. *J Am Chem Soc* 2007;**129**:8710–1.
98. Tu Y, Wang ZX, Shi Y. An efficient asymmetric epoxidation method for trans-olefins mediated by a fructose-derived ketone. *J Am Chem Soc* 1996;**118**:9806–7.
99. a. Tian H, She X, Shu L, Yu H, Shi Y. Highly enantioselective epoxidation of *cis*-olefins by chiral dioxirane. *J Am Chem Soc* 2000;**122**:11551–2;
b. Wang B, Wong OA, Zhao MX, Shi Y. Asymmetric epoxidation of 1,1-disubstituted terminal olefins by chiral dioxirane *via* a planar-like transition state. *J Org Chem* 2008;**73**:9539–43.
100. Yang D, Wong MK, Yip YC. Epoxidation of olefins using methyl(trifluoromethyl)dioxirane generated *in situ*. *J Org Chem* 1995;**60**:3887–9.
101. a. Yang D, Yip YC, Tang MW, Wong MK, Zheng JH, Cheng KK. A C_2 symmetric chiral ketone for catalytic asymmetric epoxidation of unfunctionalized olefins. *J Am Chem Soc* 1996;**118**:491–2;
b. Yang D, Wong MK, Yip YC, Wang XC, Tang MW, Zheng JH, Cheng KK. Design and synthesis of chiral ketones for catalytic asymmetric epoxidation of unfunctionalized olefins. *J Am Chem Soc* 1998;**120**:5943–52.
102. Limnios D, Kokotos CG. 2,2,2-Trifluoroacetophenone: an organocatalyst for an environmentally friendly epoxidation of alkenes. *J Org Chem* 2014;**79**:4270–6.
103. Lifchits O, Reisinger CM, List B. Catalytic asymmetric epoxidation of α-branched enals. *J Am Chem Soc* 2010;**132**:10227–9.

104. Lifchits O, Mahlau M, Reisinger CM, Lee A, Farès C, Polyak I, Gopakumar G, Thiel W, List B. The cinchona primary amine-catalyzed asymmetric epoxidation and hydroperoxidation of α,β-unsaturated carbonyl compounds with hydrogen peroxide. *J Am Chem Soc* 2013;**135**:6677–93.
105. Bondzic BP, Urushima T, Ishikawa H, Hayashi Y. Asymmetric epoxidation of α-substituted acroleins catalyzed by diphenyl- prolinol silyl ether. *Org Lett* 2010;**12**:5434–7.

CHAPTER

3.12

Heterogeneous Catalysis: A Fundamental Pillar of Sustainable Synthesis

István Pálinkó

University of Szeged, Szeged, Hungary

3.12.1 INTRODUCTORY REMARKS

According to the principles of green chemistry,[1] catalysis belongs to one of the fundamental pillars:

> Principle no. 9:
> Catalytic reagents (as selective as possible) are superior to stoichiometric reagents.

The current emphasis on green chemistry is preventing pollution at its source. Rather than accepting waste generation and disposal as unavoidable, green chemistry seeks new technologies, which are cleaner and economically competitive. Catalysis offers numerous green chemistry benefits including lower energy requirements, catalytic versus stoichiometric amounts of materials, increased selectivity, decreased use of processing and separation agents, and the use of less toxic materials. Heterogeneous catalysis, in particular, addresses the goals of green chemistry by providing easy separation of product and catalyst, thereby eliminating the need for separation through distillation or extraction.[1]

Nevertheless, one must pay attention to the environmentally benign synthesis and degradation of heterogeneous catalysts at the end of their lifetime (Principle no. 10). The heterogeneous catalysts should be stable; no leaching of potentially toxic material should occur during the reactions; the best is no leaching of any material at all.

There are hundreds of articles in the literature describing catalysts meeting these criteria; however, there are numerous other publications, which describe green catalysts or processes without even mentioning these characteristics. The large number of such articles further increases the relevant body of literature. Furthermore, additional catalytic applications

Green Chemistry
http://dx.doi.org/10.1016/B978-0-12-809270-5.00017-0

with some auxiliary conditions like the solvent or the need for heating should also be taken into account when evaluating this research field.

In this chapter, the author have tried to keep out those works in which reasonable doubt can be raised whether the process, even if it is catalytic, is green indeed. Only those heterogeneous catalysts are selected for discussion in which the process itself is green, that is, the synthesis of the catalysts does not require toxic materials; they do not form such waste during degradation, the catalyzed reactions do not need harmful solvents, the best scenario being if they work without using any solvent (Principle no. 5); and the use of catalysts results in energy saving (Principle no. 6). Based on this, for being environmentally benign, it is not enough to claim that the catalysts can be used three, four or more times without significant loss in activity and/or selectivity; a holistic evaluation of the process is necessary.

Even adhering to these criteria, many important catalyst preparations and genuinely green processes could be identified. They are displayed and discussed in the following sections using a somewhat arbitrary classification.

3.12.2 PREPARATION OF CATALYSTS AND THEIR USE IN VARIOUS CHEMICAL REACTIONS

3.12.2.1 Metals, Supported Metals, Metal Nanoparticles, Supported Metal Nanoparticles

Transition metals have been applied for promoting various chemical reactions both in academic research and large-scale chemical processes for over a century. Since it is a fundamental knowledge that only the surface metal atoms are active, the bulk metals are often dispersed over supports, generally, of large specific surface areas. Sometimes, this is the only attribute of the support, but there are instances when the support has catalytic activity as well, thus the support and the metal particles cooperate. Even before the rise of nanoscience, catalytic chemists tried to make highly dispersed supported catalysts, thus they approached the nanometer size range. Since then, novel auxiliary materials (supports, additives, etc.) appeared and the preparation methods were also developed concomitantly. Notable examples are described.

3.12.2.1.1 New Emerging Supports

One of the emerging new group of support materials is the metal organic frameworks (MOFs). MOF is a new class of materials consisting of metal atoms or ions or their clusters connected by coordinatively bonded organic ligands. They can form one-dimensional, two-dimensional (2D), or three-dimensional (3D) polymers of which the 3D structures are the most useful as supports. The latter versions are usually porous materials reaching specific surface areas of $\sim 1000\ m^2/g$. A highly efficient Ru-Pt bimetallic catalyst was prepared applying a simple colloidal deposition method on a zeolite-like MOF (MIL-101) and was used for neat benzene hydrogenation to cyclohexane. The well-characterized catalyst system was shown to be able to provide excellent activities with high selectivity for cyclohexane (up to >99% yield). The enhanced reactivity was thought to be due to the synergy of Ru and Pt.[2]

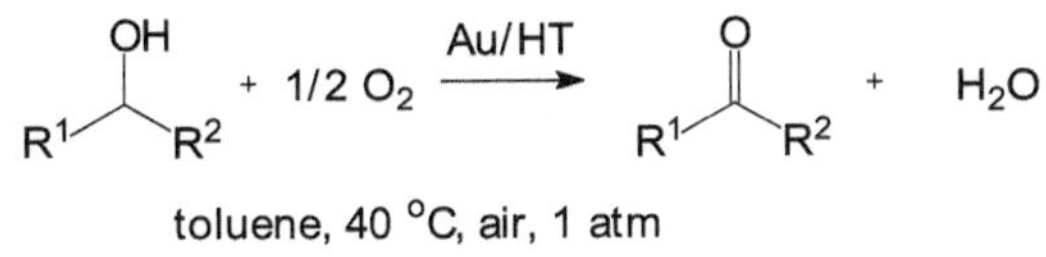

SCHEME 3.12.1 Oxidation reaction of alcohols over Au/hydrotalcite (HT) (yield: over 80% in most cases; 20 examples).

Layered double hydroxides (LDHs) are inexpensive and easily accessible supports, they even can be found in nature. Platinum/gold alloy nanoparticles (NPs) supported on hydrotalcite (HT), a type of LDH, were prepared by using starch as a green reducing and stabilizing agent. The catalyst was found to be an effective heterogeneous catalyst for the selective aerobic oxidation of glycerol and 1,2-propanediol in a base-free aqueous solution using molecular oxygen at atmospheric pressure at room temperature.[3] Similar HT-supported gold NPs (Au/HT) were found to be a highly efficient heterogeneous catalyst for the aerobic oxidation of alcohols under mild reaction conditions (40 °C in air; Scheme 3.12.1).[4] This catalyst system does not require any additives and is applicable to a wide range of alcohols, like the less-reactive cyclohexanol derivatives and 1-phenylethanol under neat conditions.

Au/HT was also active in the synthesis of azo arenes (Scheme 3.12.2).[5] Nitroarenes were deoxygenated and linked selectively through the formation of N=N bonds using molecular H_2 without external additives and tolerating a large variety of functional groups. A facile surface-mediated condensation of nitroso and hydroxylamine intermediates was enabled, and the desired transformation proceeded in a highly selective manner under mild reaction conditions.

Carbon nanotubes [both multiwalled carbon nanotubes (MWCNTs) and single-walled carbon nanotubes (SWCNTs)] also found application as catalyst supports. The Cu-, Ag-, Co- and Pt-MWCNT catalysts[6] and the Fe-SWCNT[7] catalyst work efficiently in different multicomponent one-pot reactions under solvent-free conditions (Scheme 3.12.3). Additional noteworthy

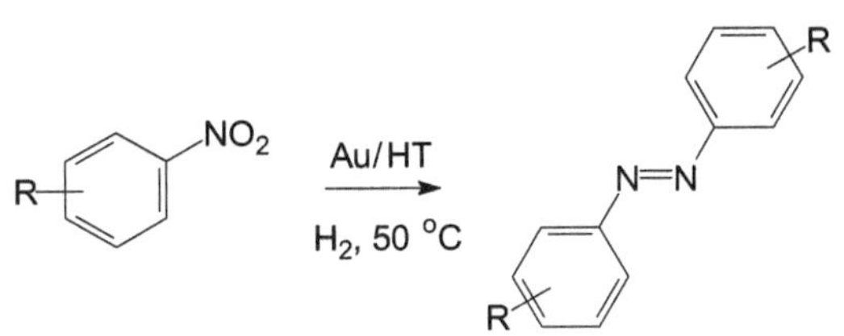

SCHEME 3.12.2 Synthesis of azo arenes over Au/hydrotalcite (HT) (yield: over 80% in most cases; 20 examples).

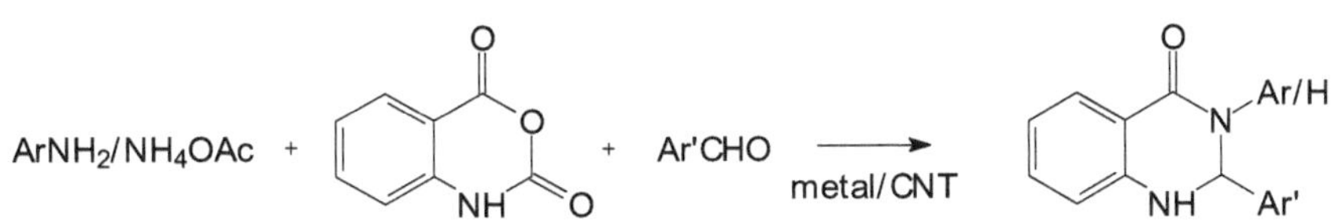

SCHEME 3.12.3 Synthesis of mono- and disubstituted dihydroquinazolinones over multiwalled carbon nanotubes (MWCNTs) decorated with metal (Cu, Ag, Co, Pt) nanoparticles.[6]

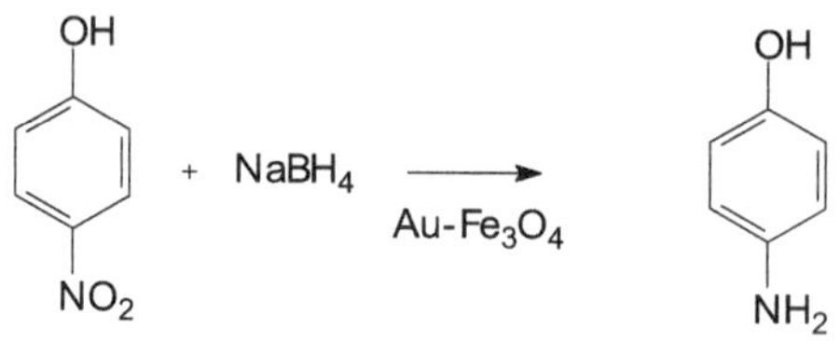

SCHEME 3.12.4 Reduction of *p*-nitrophenol over magnetically separable catalyst.[9]

features are the clean reaction profile, the easy use of accessible starting materials, the high atom economy, the reusability of the catalyst, and the mild reaction conditions.

When the support or part of it is magnetic or superparamagnetic then the catalysts can be easily separated from the reaction mixture by using a magnet. The magnetic or superparamagnetic component is commonly Fe_3O_4 or Fe_2O_3, and the catalytically active part can be a wide range of transition metals in a mono- or bimetallic fashion (e.g., Pt,[8] Ag,[3] Au,[9] Pd,[10,11] Au-Pd [12]). The additional or co-support can also be varied starting from functionalized carbon nanotubes[8] through reduced graphite oxide[12] to hydroxyapatite.[11] The reductions of nitrophenol derivatives[8,9,12,13] were the reactions in which these catalysts were probed most frequently (Scheme 3.12.4). Occasionally, they could be readily separated and recycled by a magnet, and reused in the next reactions with high efficiency for at least 15 successive cycles.[8]

Composite supports can be helpful in complicated reactions such as coupling reactions. Polycarbosilane (an organic-inorganic hybrid polymer containing both Si and C in its backbone structure)[14] and mesoporous cobalt oxide-carbon nanocomposites[15] were applied for anchoring Pd ions. The activity of these catalysts were studied in the Heck and the water-mediated Suzuki coupling reactions, respectively. The catalysts were air-stable and also thermally stable to allow easy use and storage without any precautions, and showed high recyclability. It is highly probable that the coordinatively unsaturated Pd was responsible for the activation of the halogen-containing reactants even in the absence of any additives or ligands. They performed much better than when the Pd ions were supported on more traditional supports, such as SBA-15 (Santa Barbara Amorphous material with hexagonal array of pores with pore diameters in the 4.6–30 nm range), activated charcoal, or amorphous silica.

Natural and many synthetic organic polymers as supports may have the advantage of having several functional groups useful for anchoring the metal particles, easy degradation, and not placing much burden on the environment. Aryl sulfones were synthesized using a highly active and easily recoverable heterogeneous Cu catalyst, which was prepared by simply stirring an aqueous suspension of chitosan (CS) in water with copper salts.[16] The Cu/CS catalyst catalyzed the coupling reactions of aryl halides with sodium sulfinates to readily give the corresponding sulfones in good to excellent yields (Scheme 3.12.5). The highly active catalyst can be reused many times without losing its catalytic activity.

Most of the synthesis methods are solution-based techniques. They use solvent in one or more steps. If it is water without acidic, basic or organic additives, we can talk about an environmentally benign procedure. It is certainly environment-friendly if a solvent is not used at all. For instance, the hydroconversion of cinnamaldehyde as an α,β-unsaturated compound

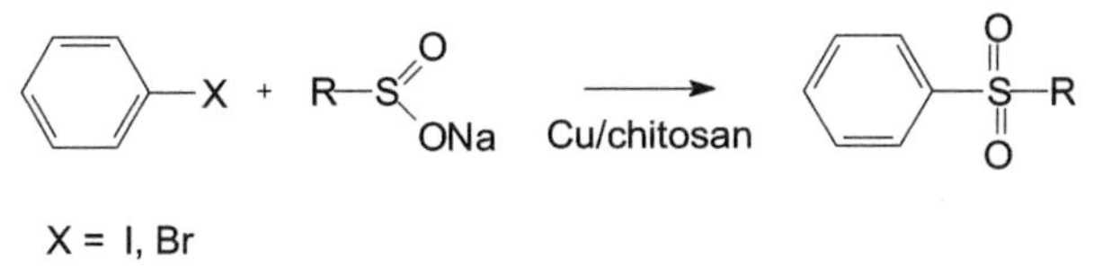

SCHEME 3.12.5 The synthesis of aryl sulfone over Cu/chitosan catalyst (yield: 60%–95% in 48 h; 25 examples).

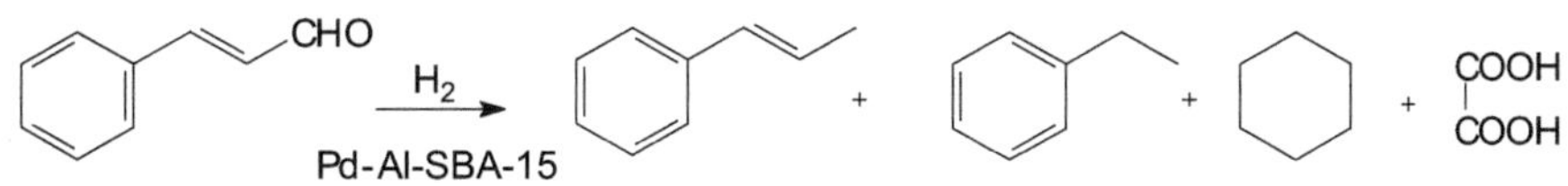

SCHEME 3.12.6 Unusual products in the hydrogenative transformation of cinnamaldehyde over Pd-Al/SBA-15 bifunctional catalyst.

was studied using a simple and efficient hydrogen-donating protocol catalyzed by mechanochemically synthesized bifunctional Pd/Al-SBA-15 catalysts.[17] In contrast to potentially expected products, an interesting selectivity to products including ethylbenzene (via Pd-catalyzed hydrodeformylation reactions) and oxalic acid (via Pd-catalyzed hydrocarboxylation) were observed (Scheme 3.12.6).

A procedure for the stereoretentive direct hydrogenation of amino acids to the respective chiral vicinal amino alcohols was developed (Scheme 3.12.7).[18] The protocol is characterized by using highly active heterogeneous bimetallic Ru/Re sponge catalysts, which allow the process to occur at relatively low temperatures. These conditions are crucial to prevent any racemization. The catalyst system was improved from a practical point of view by supporting the metals on charcoal. In the context of green chemistry, the atom efficiency of this process is worth mentioning along with the fact that water is used as the solvent.

The successful fabrication of silver NPs in situ grown on magnetically separable alginate-based biohydrogels (Ag/AMH) by an environment-friendly light-driven approach was described.[19] In the presence of alginate biopolymers, silver ions can be readily adsorbed and subsequently photoreduced to metallic Ag NPs on the biohydrogels. The resulting Ag/AMH exhibited excellent and durable activity for the catalytic reduction of 4-nitrophenol to 4-aminophenol with $NaBH_4$ in aqueous solution.

Cu-Co catalyst prepared by the oxalate sol-gel method can selectively convert furfural to cyclopentanol in aqueous solution[20] (Scheme 3.12.8). The product distribution was influenced

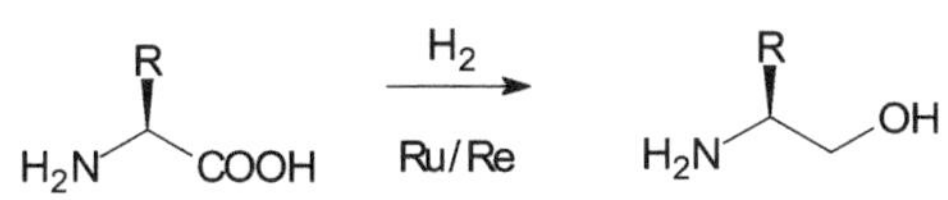

SCHEME 3.12.7 Direct hydrogenation of amino acid to amino alcohol while keeping the configuration of the stereocenter.

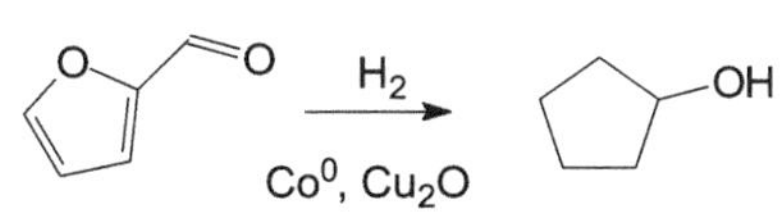

SCHEME 3.12.8 Hydrogenation of furfural to cylopentanol in water over a Cu-Co catalyst.

by the catalyst support. It was found that there was a strong interaction between Cu and Co. Cu^0, Cu_2O, and Co^0 were the main active catalytic species on the surface of the catalyst. Cu^0, Co^0, and Cu_2O were the active hydrogenation species, and Cu_2O also played the role of an electrophile or Lewis acid to polarize the C-O bond via connection to the lone electron pairs of the oxygen atom. The hydrogenation activity of the catalyst was high, and the major product was the fully hydrogenated product cyclopentanol, due to the dominating presence of Co^0 and Cu_2O on the surface of the catalyst.

Pt NPs with size between 2 and 5 nm supported on CeO_2 NPs with size between 30 and 60 nm were prepared by the hydrothermal method in the presence of the surfactant cetyltrimethyl ammonium bromide and polyvinylpyrrolidone.[21] It was found that the catalyst was highly active for the chemoselective hydrogenation of nitro compounds in aqueous medium in the presence of molecular hydrogen at room temperature.

3.12.2.2 Oxides, Mixed/Supported Oxides

While metal-based catalysts are used for hydrogenation reaction most frequently, metal oxide-based ones, either supported or unsupported or in the form of mixed oxides display great versatility; they can be efficient catalysts from simple (but difficult) decarboxylation reactions via oxidations to multicomponent reactions aiming at the syntheses of fine chemicals.

A commercially available Cu_2O catalyst was used for the production of benzene from benzoic acid in water with very high yield and selectivity.[22] This was the first time the reaction from benzoic acid to benzene was completed using an environment-friendly, efficient, and economical method. Cu_2O appeared to be an effective and stable catalyst, and the process was driven by the unique properties of subcritical water [or pressurized or superheated water—the temperature is between 100 °C (atmospheric boiling point) and 374 °C (critical temperature) and the pressure is at least the vapor pressure of water at the given temperature]: the high ion product, the enhanced solubilization of products, and the high diffusivity.

An efficient and versatile practical protocol for the chemoselective *N-tert*-butoxycarbonylation (BOC) of amines using nano-γ-Fe_2O_3 and $(BOC)_2O$ was elaborated (Scheme 3.12.9).[23] Nano-γ-Fe_2O_3 proved to be an efficient, green, heterogeneous, and reusable catalyst at ambient temperature in water. The method was generally applicable for the preparation of *N*-Boc derivatives of aliphatic, heterocyclic, aromatic, as well as amino acid derivatives.

Macroporous transition metal oxides (CuO, NiO, CoO, Mn_2O_3, Cr_2O_3, and ZnO) were synthesized and used as efficient, heterogeneous, reusable, and ecofriendly catalysts in four different organic transformations: *N*-formylation, *N*-acylation, *O*-acylation, and Friedel-Crafts acylation under solvent-free conditions in good (from 50% to 99%) yields.[24] Due to the presence of large internal surface area and pore volume, the porous metal oxides had good control over the diffusion of reagents and products into and out of the porous medium.

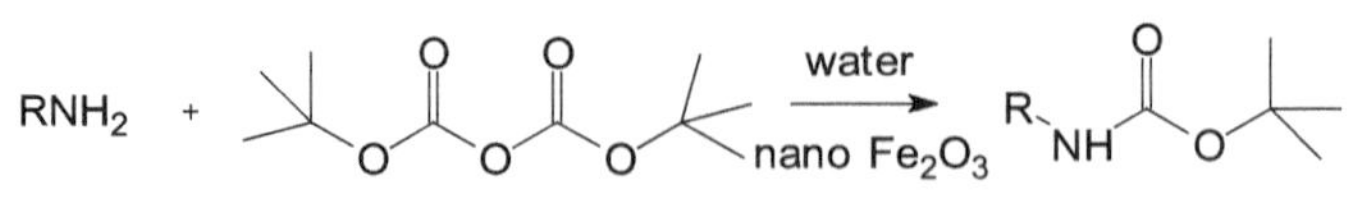

SCHEME 3.12.9 *N-tert*-butoxycarbonylation of various amino compounds (yield: 78%–96% in 1–2 h; 19 examples).

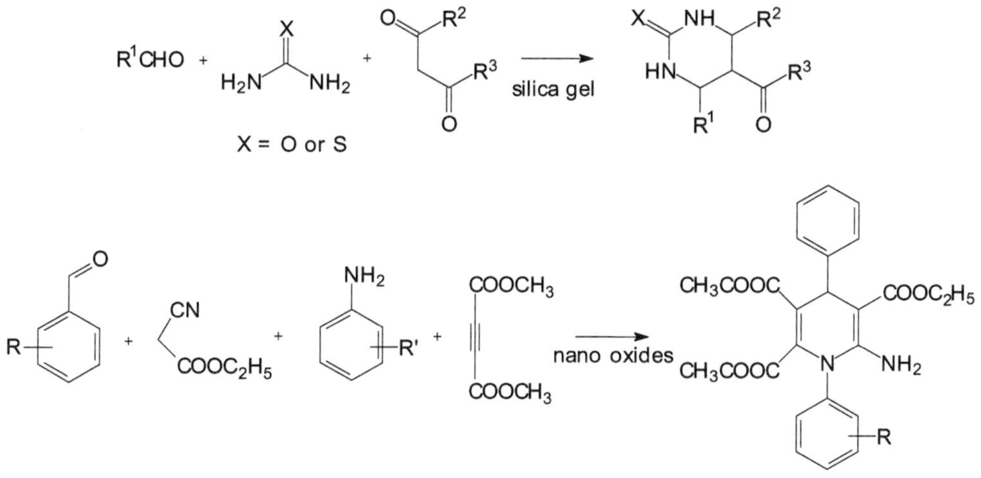

SCHEME 3.12.10 Three- and four-component one-pot reactions providing valuable N-heterocyclic systems.[26]

Three-component (aldehydes, carbamide or thiocarbamide, and 1,4-dioxo compounds) and four-component (aromatic aldehydes, ethyl cyanoacetate, arylamines, and dimethyl acetylenedicarboxylate) reactions could be performed in the presence silica gel[25] and various oxide (ZnO, CuO, CeO_2, SnO, MgO or CaO) NPs,[26] respectively (Scheme 3.12.10), producing polysubstituted, partially saturated, N-containing heterocyclic compounds,[26] while γ-Fe_2O_3 NPs were active in the chemoselective formation of *N-tert*-BOC amines.[23]

CeO_2/Al_2O_3- and CeO_2/Nb_2O_5-supported oxides were active in the carboxylation of glycerol to glycerol carbonate[27] (Scheme 3.12.11). The carbonate, obtained without the use of phosgene, was converted into, e.g., epichlorohydrin having large-scale application. This is a simple and safe synthetic route. The process uses two common abundant starting materials, glycerol and carbon dioxide.

Highly selective propylene formation could be achieved via glycerol hydrodeoxygenation reactions over Fe-Mo oxide catalysts supported on carbon black or activated carbon.[28] Characterization results indicated that molybdenum suboxide (i.e., MoO_2) was the active catalytic species in these reactions.

Silica-supported $LaFeO_3$ could be efficiently applied in the acetylation of amines, alcohols, and phenols to the corresponding acetates using acetic anhydride under solvent-free conditions.[29] The catalytic activity of the $LaFeO_3/SiO_2$ nanocomposite was higher than that of the pure $LaFeO_3$ NPs. The method was high yielding (from 80% to 92% as was demonstrated

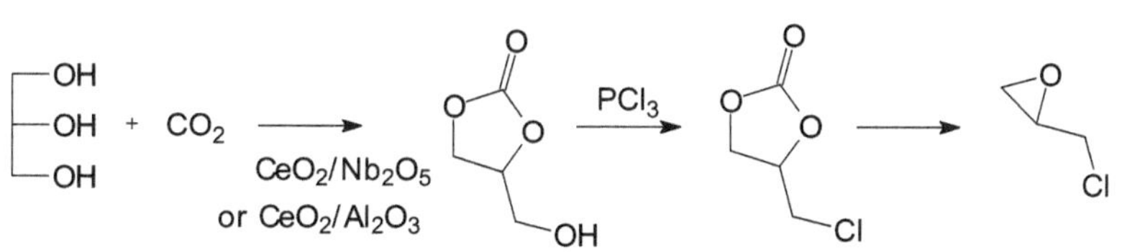

SCHEME 3.12.11 Catalytic carboxylation of glycerol and possible follow-up transformations.

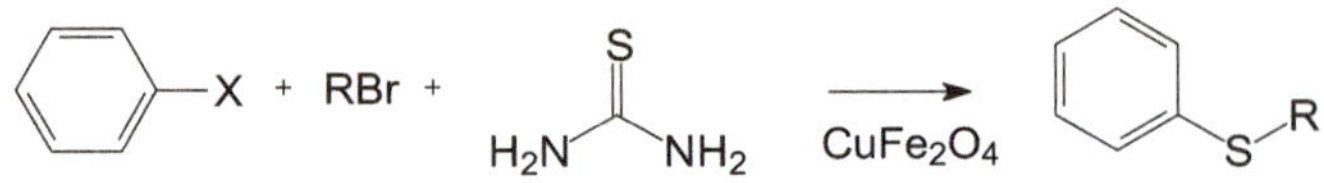

SCHEME 3.12.12 The thioetherification reaction (yield: 73%–93% in 10–24 h; 23 examples).

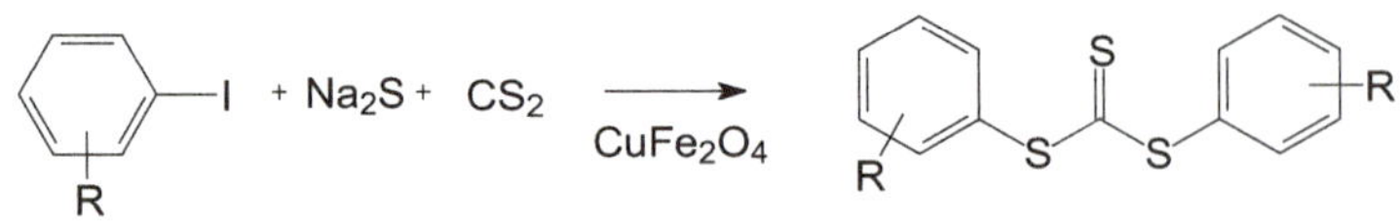

SCHEME 3.12.13 Synthesis of symmetrical trithiocarbonates (yield: 81%–89% in 12 h; 6 examples).

through 18 examples), clean, cost-effective, compatible with the substrates having other functional groups, and very suitable for practical organic synthesis.

A mixed oxide, copper ferrite NPs ($CuFe_2O_4$), was introduced for the one-pot production of aryl alkyl thioethers using thiourea and alkyl bromides in wet polyethylene glycol as a green solvent (Scheme 3.12.12).[30] The catalyst was also successfully applied for one-pot synthesis of symmetric diaryl trithiocarbonates via the reaction of sodium sulfide, carbon disulfide, and aryl iodides under heterogeneous reaction condition (Scheme 3.12.13). The catalyst was recycled using simple magnetic separation.

3.12.2.3 Pristine (Nonfunctionalized) Micro- and Mesoporous Materials

Many of the numerous zeolites and zeotypes are used in large-scale petrochemical industrial technologies producing vast quantities of fuels as well as commodity chemicals. However, simple metal ion-doped or exchanged varieties and related porous materials find applications in the fine chemical industry. They perform superbly even in complicated reactions. Some of the environmentally benign applications are highlighted here.

Sc(III)-exchanged USY (ultrastabilized—very high Si/Al ratio—faujasite type) zeolites were used in aza-Diels-Alder reaction, and this heterogeneous new version provided a green route for quantitative synthesis of tetrahydropyridines (Scheme 3.12.14).[31]

Sc(III)-containing zeolites [Sc(III)-Y, Sc(III)-USY, Sc(III)-β, Sc(III)-ZSM-5, and Sc(III)-mordenite] were active in the Mukaiyama-type aldolization reaction too.[32]

Preparation of pharmaceutically important 2-amino-4*H*-chromene derivatives (Scheme 3.12.15) has been achieved using nano powder of natural clinoptilolite zeolite in aqueous medium with yield reaching 98%.[33] The catalyst was prepared mechanically by planetary ball mill.

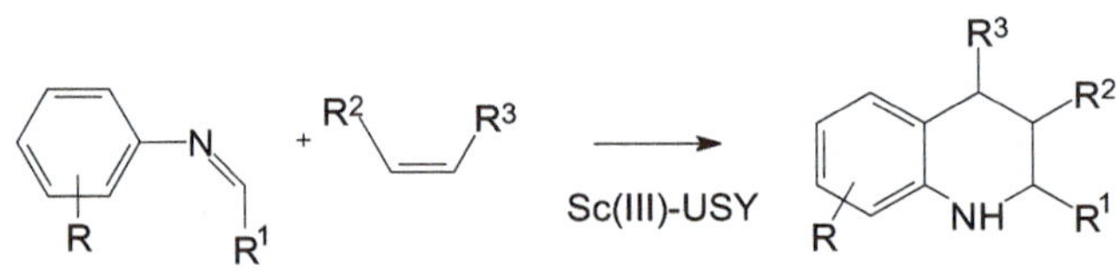

SCHEME 3.12.14 Sc(III)-USY-catalyzed [4 + 2] cyclization producing tetrahydropyridine derivatives.

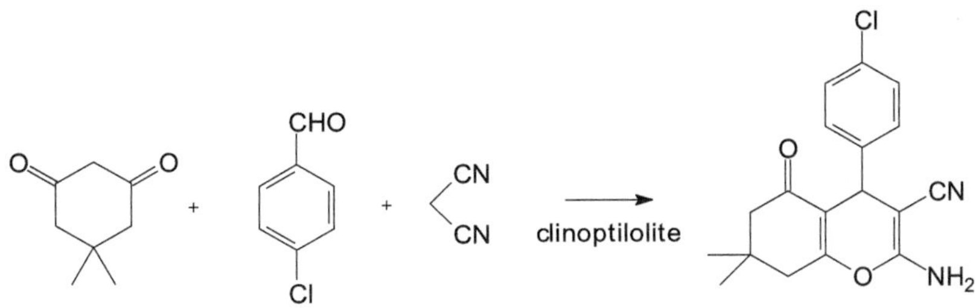

SCHEME 3.12.15 A three-component reaction yielding biologically active 2-amino-4*H*-chromene derivatives.

Cu(I)-exchanged zeolites [Cu(I)-Y, Cu(I)-USY, Cu(I)-β, Cu(I)-ZSM-5, and Cu(I)-mordenite] could catalyze a variety of transformations like 1,3-dipolar reactions or Mannich condensation.[32] These catalysts could be easily prepared, were stable for months and were conveniently recovered by filtration, and they could be recycled. They could be used in environment-friendly solvents and even under solvent-free conditions.

A series of non-noble transition metal (Mn, Zn, Fe, Co, Ni, Ru, V, Cu, and Cr) ion-exchanged zeolite (13X, β, USY, ZSM-5) or zeotype (SAPO-5, SAPO-34—silico-alumino-phosphate molecular sieves) catalysts were prepared and applied in the catalytic oxidation of cyclohexanol and other alcohols with *tert*-butyl hydroperoxide (TBHP) to yield aldehydes or ketones in the absence of any solvent and additive.[34] Among the catalysts tested, Cr^{3+}-exchanged 13X zeolite has exhibited the best activity (close to 100% conversion in 10 h reaction time) with the highest efficiency (always over 77%) of TBHP utilization.

W-AlPO-5, having the tungsten atoms incorporated into the AlPO-5 (aluminophosphate molecular sieve) skeleton, could be used as an efficient heterogeneous catalyst in the synthesis of 5-substituted 1H-tetrazoles by [3 + 2] cycloaddition from nitriles and sodium azide (Scheme 3.12.16).[35] The catalyst exhibited high activity, superior recycling ability, and high substrate tolerance. The significant advantages of W-AlPO-5, such as the simple procedure, the mild reaction conditions, and as an alternative to those corrosive, hazardous, and polluting homogeneous catalysts, warrant its potential application in industrial processes.

A novel hollow-structured Mn-titanosilicate-1 (Mn-HTS) catalyst was reported as a non-nitric acid route for adipic acid production from oxidative cleavage of cyclohexanone (Scheme 3.12.17).[36] The method generates adipic acid in high yields with molecular oxygen under organic solvent- and promoter-free conditions. The hollow nature of the catalyst with large intraparticle voids facilitates the diffusion of bulky molecules to the internal catalytic site, and increases the catalyst activity.

Mesoporous zirconosilicate, stannosilicate, and titanosilicate with the BEA (β zeolite) structure framework were prepared from the commercially available β zeolite via acid-alkaline

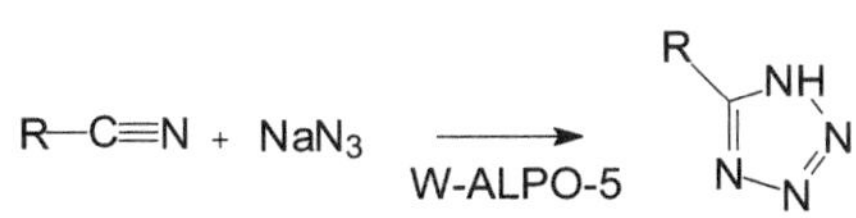

SCHEME 3.12.16 A [3 + 2] cycloaddition leading to 5-substituted 1H-tetrazoles (yield: 74%–95% in 24 h; 9 examples).

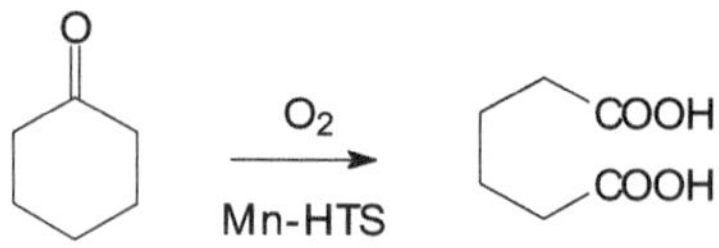

SCHEME 3.12.17 Synthesis of adipic acid over hollow-structure Mn-titanosilicate (yield: ~60% in 9 h, selectivity to adipic acid: ~85%). *Mn-HTS*, hollow-structured Mn-titanosilicate-1.

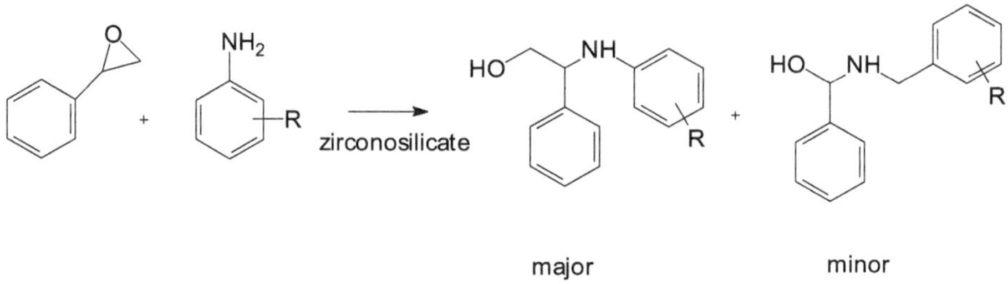

SCHEME 3.12.18 The ring-opening ammonolysis of styrene oxide over mesoporous Zr-β (yield 80.7% in 0.5 h, selectivity: 94.7%).

treatments and subsequent dry impregnation with appropriate organometallic precursors.[37] The as-prepared mesoporous Zr-β exhibited a remarkable catalytic activity and regioselectivity toward the formation of β-amino alcohols in the ring-opening aminolysis of epoxides, and the presence of mesopores can promote the reaction to a great extent through enhanced mass transfer (Scheme 3.12.18).

The high-surface-area 3D porous Ga-TUD-1 (a mesoporous gallosilicate) catalyst was applied in the simple, efficient, and solventless synthesis of α-aminonitriles (Strecker synthesis) (Scheme 3.12.19).[38] The catalyst performed excellently in the synthesis of a range of compounds with variable functionality.

Mesoporous zirconium phosphonates were demonstrated as highly effective catalysts for the heterogeneous catalytic cycloaddition reaction between aziridines and CO_2 to yield oxazolidinones (Scheme 3.12.20) in a solvent-free system without introducing any co-catalysts or halogen species, exhibiting outstanding activity and selectivity as well as excellent recyclability.[39]

Beside zeolites and zeotypes, less frequently though, other 3D [molecular organic framework (MOF)] and 2D (clays capable of ion exchange) materials can also be used as efficient and environmentally benign catalysts even in complicated reactions.

ZIF-67, a cobalt-based zeolitic imidazolate framework having acid-base functionalities was reported for its application in the artificial fixation of CO_2 via cyclic carbonate synthesis

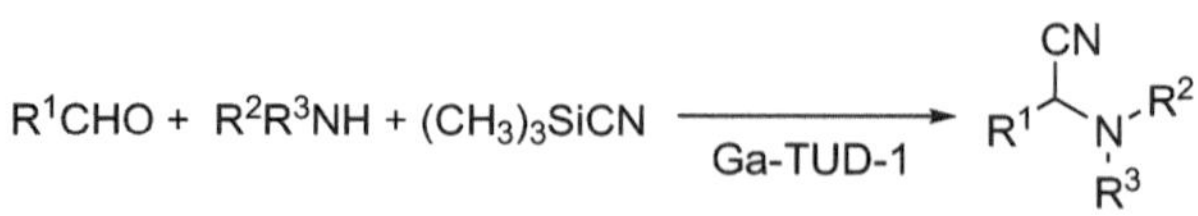

SCHEME 3.12.19 Solventless synthesis of α-aminonitriles over Ga-TUD-1 (yield: 83%–95% in 0.5–1 h; 10 examples).

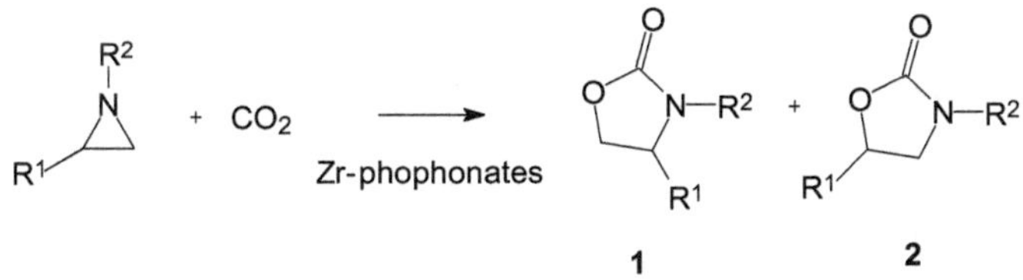

SCHEME 3.12.20 The solvent- and additive-free reaction between substituted aziridine and CO_2 over a Zr-phosphonate (yield: 86%–100% in 8 h, with 98% regioselectivitiy toward product **1**; 5 examples).

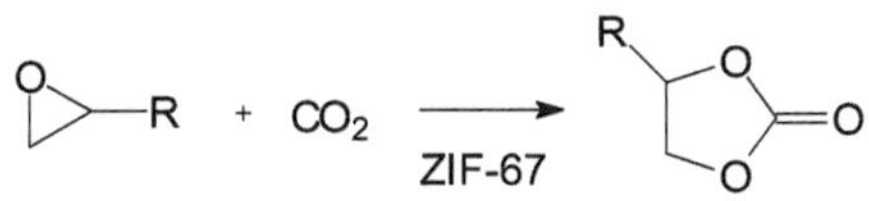

SCHEME 3.12.21 CO_2 fixation in a cyclic carbonate catalyzed by ZIF-67 (yield: 7%–97% in 6 h; 5 examples).

(Scheme 3.12.21).[40] ZIF-67 can be synthesized in water at room temperature with inexpensive precursors in less than 3 h. The heterogeneous ZIF-67 was found to be catalytically efficient toward CO_2-oxirane coupling under moderate reaction conditions with nearly complete selectivity toward five-membered cyclic carbonates under solvent-free and co-catalyst-free conditions.

The catalytic potential of the Cu(I)-exchanged montmorillonite [Cu(I)-MMT] was evaluated in the Huisgen [3 + 2] cycloaddition (Scheme 3.12.22).[41] This catalytic system proved to be efficient for this click chemistry-type transformation. The catalyst exhibited excellent recycling abilities.

Thermally treated (>500 °C) iron-containing MMT was used as catalyst in the conversion of tetrahydrofuran to butyrolactone with hydrogen peroxide as the oxidizing agent (Scheme 3.12.23).[42] Formation of active oxidizing species on the surface occurred on contact of the dislodged Fe(III) oxide with H_2O_2. These active species could promote the oxidation of tetrahydrofuran with high conversion (~60%) and selectivity (~60%) to butyrolactone, whereas the iron-containing clay treated at lower temperatures (<500 °C) performed Fenton-like oxidation with lower tetrahydrofuran conversions and nonselective products.

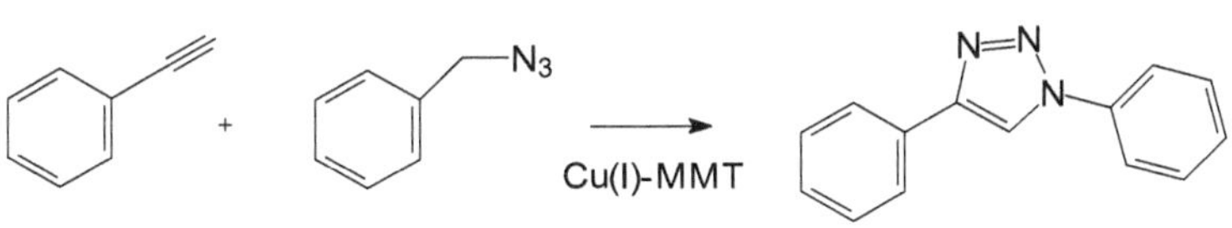

SCHEME 3.12.22 A click reaction among the layers of Cu(I)-exchanged montmorillonite [Cu(I)-MMT] (yield: 88%–98%; 9 examples).

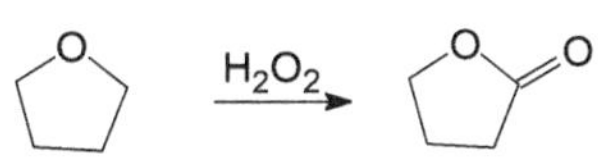

SCHEME 3.12.23 Oxidation of tetrahydrofuran to butyrolactone by Fe(III) species dislodged from Fe(III)-montmorillonite.

3.12.2.4 Immobilized Hybrid Materials

Homogeneous complexes/organometallic materials, organocatalysts, are most often more selective catalysts than the heterogeneous counterparts. However, they are mostly sacrificed, since their recovery is tedious if at all possible. Therefore, it can hardly be stated that their use is benign to the environment. Nevertheless, it is possible to unite the advantageous features of the homogeneous and heterogeneous worlds through immobilizing the homogeneous catalyst on a solid support. This way, probably one loses part of the activity and the selectivity of the homogeneous catalyst, but its recovery is going to be feasible. In fortunate instances, the support material also possesses independent catalytic activity, then, either bifunctional catalysis or, if the catalytic effect of both counterparts are of similar type, synergism may appear. There are several ways of immobilization. It can be done by building primary (covalent or electrostatic) or secondary (anchoring via hydrogen bond or van der Waals forces) bonding interactions between the moiety to be immobilized and the support. The former method is more frequently applied than the latter as it is highlighted by the examples taken from recent literature.

3.12.2.4.1 Immobilization via Covalent Bonds

$Sm[N(SiMe_3)_2]_3$ was immobilized inside the pores of mesoporous silicate SBA-15 leading to the formation of an organolanthanide/inorganic hybrid material $[SBA\text{-}15]Sm[N(SiMe_3)_2]_x$ by releasing $N(SiMe_3)_2$ due to surface reaction with surface silanol groups of SBA-15.[43] It was used as a promising heterogeneous catalyst for the Tishchenko reaction, where it is superior to the homogeneous counterpart ($Sm[N(SiMe_3)_2]_3$) in terms of deactivation behavior, reusability, relative tolerance to oxygen and selectivity control in mixed Tishchenko reaction (Scheme 3.12.24) due to the steric hindrance and the diffusion control derived from the surface confinement.

A series of heterogenized amines on silica, MCM-41 (amorphous silica with regular arrangement of cylindrical mesopores) and delaminated zeolite ITQ-2 (Scheme 3.12.25) were highly active in Michael addition of ethyl 2-oxocycloalkanecarboxylates to acrolein (Scheme 3.12.26).[44] New heterogenized base catalysts are practical alternatives to soluble analogs in view of the following advantages: high stability of the catalysts, high activity under mild conditions, free election of solvent, easy separation of the catalysts by simple filtration, and reuse without the loss of activity.

The reaction sequences shown in Scheme 3.12.25 are very often applied in the covalent anchoring of organocatalysts and complexes.

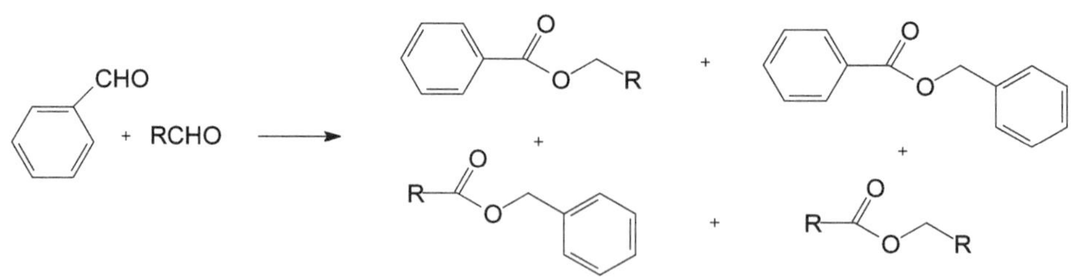

SCHEME 3.12.24 The mixed Tishchenko reaction catalyzed by $[SBA\text{-}15]Sm[N(SiMe_3)_2]_x$.

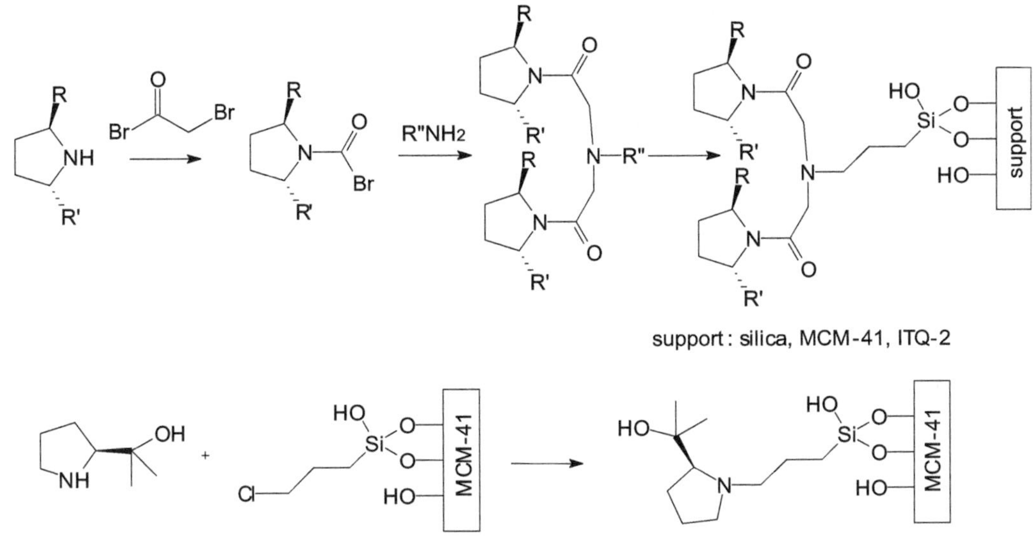

SCHEME 3.12.25 The syntheses of the homogeneous organocatalysts and their heterogenized versions.

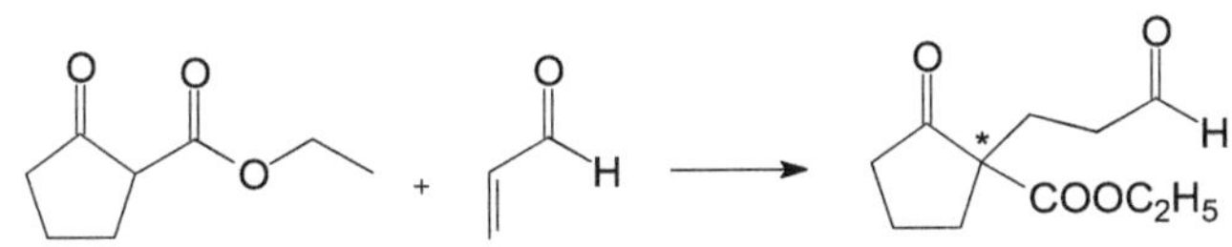

SCHEME 3.12.26 The Michael reactions used as probe for the immobilized organocatalysts.

Sulfonic acid[45] and sulfonic acid monopropylamide (nPr-$NHSO_2OH$)[46] were covalently immobilized in SBA-15[45] and MCM-41,[46] respectively. These catalysts were highly efficient in the one-pot syntheses of triazoloquinazolinones and benzimidazoquinazolinones from the reaction of aromatic aldehydes with 3-amino-1,2,4-triazole (or 2-aminobenzimidazole) and dimedone and of imidazole derivatives. None of these transformations require the presence of solvent (Scheme 3.12.27).

Polypropylene fiber containing covalently grafted sulfonic acid was capable of acting as a supported catalyst in the Biginelli reaction.[47] Excellent results, in terms of chemical yields, catalyst dosage, workup procedure, and catalytic recyclability, were reported for the metal-free catalyzed one-pot synthesis of a number of substituted 3,4-dihydropyrimidin-2-(1H)-ones/-thiones (Scheme 3.12.28). The catalyst can be prepared by a simple process from inexpensive and readily available polypropylene fiber. After completion of the catalytic reaction, the catalyst was simply recovered with common tweezers and recycled with excellent stability and activity even after 11 runs.

L-Proline[49] and propylamine modified with imidazolium ionic moieties,[49] sulfonic acid,[50] sulfamic acid,[51] biguanide[52] or propylsulfamic acid[53] were grafted onto functionalized magnetic Fe_3O_4,[48,49,51] γ-Fe_2O_3,[50] core-shell Fe_3O_4/SiO_2,[52] or hydroxyapatite NPs,[53] and evaluated as catalysts in the condensation reaction of indoles and aldehydes for the synthesis of bis(indolyl)methanes,[48] the Knoevenagel condensation,[49] the Hantzsch reaction,[50] the

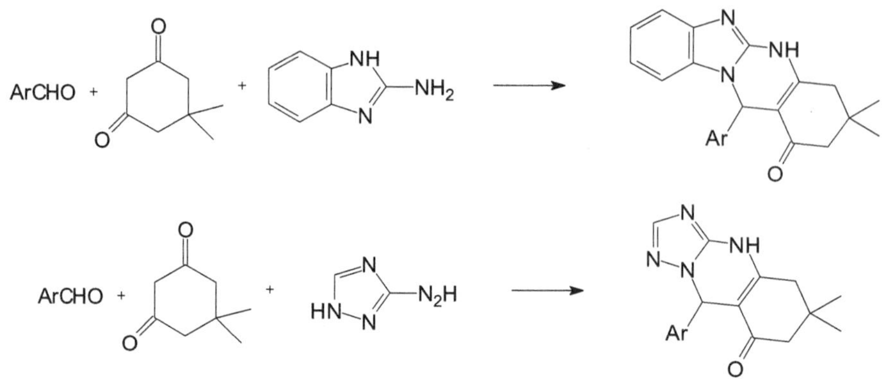

SCHEME 3.12.27 Syntheses of benzimidazoquinazolinones and triazoloquinazolinones over sulfonic acid derivatives immobilized over SBA-15 or MCM-41 (yield: ~80% in ~50 min).

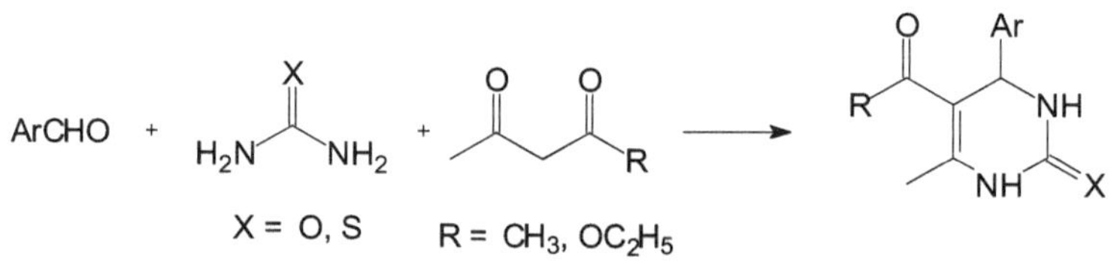

SCHEME 3.12.28 The one-pot Biginelli reaction leading to substituted 3,4-dihydropyrimidin-2-(1H)-ones/-thiones catalyzed by sulfonic acid grafted onto polypropylene fiber (yield: 68%–89% in 8 h; 25 examples).

one-pot synthesis of N-substituted pyrroles from γ-diketones and primary amines,[51] the nitroaldol and domino Knoevenagel condensation/Michael addition/cyclization reactions,[52] and the Friedlander reaction of 2-aminoaryl ketones with ketones[53] using water as solvent[48,49,51] or without solvent[50,52,53] under mild reaction conditions. The heterogeneous catalysts could be recovered easily by applying an external magnetic device and reused many times without significant loss of the catalytic activity.

Surface modification of Fe_3O_4 NPs with triethoxyethylcyanide groups was used for the immobilization of palladium NPs to produce Fe_3O_4/ethyl-CN/Pd.[54] The cyano groups played an important role as capping agent. The catalytic behavior of Fe_3O_4/ethyl-CN/Pd NPs was evaluated in the Suzuki-Miyaura cross-coupling reaction (Scheme 3.12.29). The reaction was carried out in 1:1 water-ethanol, and the yields varied between 88% and 92% when iodo- or bromobenzene derivatives were the reactants, respectively, but remained around 20% when the halogen substituent was chlorine.

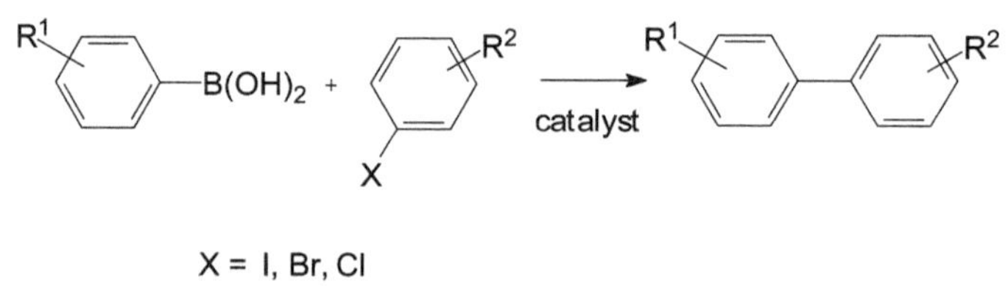

SCHEME 3.12.29 The general scheme of the Suzuki-Miyaura coupling reaction.

The Suzuki-Miyaura reaction was performed using a non-symmetrical salen-type palladium(II) catalyst immobilized onto Merrifield resin and functionalized silica.[55] The catalysts combined high activity with the recoverability and reusability offered by the heterogeneous system without the need for phosphine ligands or additives. Leaching of the metal into solution from the supported catalyst proved to be negligible. The yields were between 91% and 100% for most of the bromobenzene derivatives (they were only used—11 examples) in 30:1 aqueous toluene.

A magnetically separable Fe_3O_4@poly(ethylene glycol) (PEG)-iminophosphine palladium complex was successfully prepared by reacting the palladium complex containing the ligand 2-(diphenylphosphino)benzaldehyde with amino-functionalized PEG-coated iron oxide NPs.[56] It displayed high activity for the Suzuki-Miyaura coupling reaction in water. The yields were between 80% and 99% (19 examples in water as solvent) in all except one chlorobenzene derivative where it was 35%.

By grafting an *N*-heterocyclic carbene palladium complex and ionic liquid on the mesoporous cagelike material SBA-16, an efficient and highly recyclable heterogeneous catalyst for the Suzuki (yield: 87%–99% in 1:1 water-ethanol mixture as solvent; 7 examples) and Heck reactions was prepared (Scheme 3.12.30—yield: 76%–97%).[57] Its recyclability was much better than that of the catalyst prepared with amorphous silica. The significantly enhanced recyclability could be attributed to the isolated nanocages of SBA-16, which could efficiently prevent the aggregation and agglomeration of Pd particles formed during the catalytic reaction into the less active large particles.

An *N*-heterocyclic carbene palladium complex was covalently anchored to silica and was found to be a very efficient and reusable catalyst in the Suzuki and Sonogashira (Scheme 3.12.31) coupling reactions.[58] The catalysts showed excellent performance in these two reactions including various aryl halide derivatives (except the aryl chloride derivative, where the yield was 54%) with phenylboronic acid (yield: 86%–94% via 8 examples) and phenylacetylene (79%–95% via 7 examples) in water as the solvent.

A Cu-*N*-heterocyclic carbene complex was grafted to SBA-15 silica, and was used for the catalytic hydrothiolation of alkynes and electron-deficient alkenes under mild reaction conditions (Scheme 3.12.32).[59] The catalyst displayed higher activity (conversion:98%) and an unusual *Z* stereoselectivity (*Z*:*E* = 89:11) compared with the homogeneous analog (conversion: 41%, *Z*:*E* = 51:49) with phenylacetylene and thiophenol reactants, under otherwise identical reaction conditions.

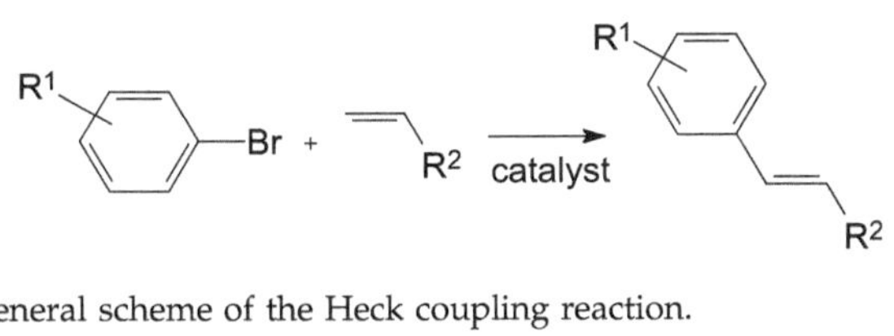

SCHEME 3.12.30 The general scheme of the Heck coupling reaction.

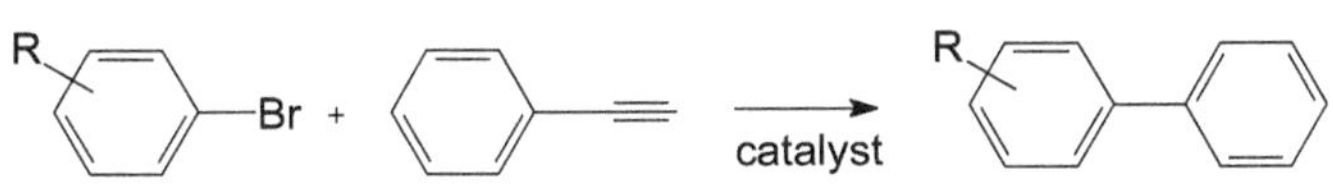

SCHEME 3.12.31 The general scheme of the Sonogashira coupling reaction.

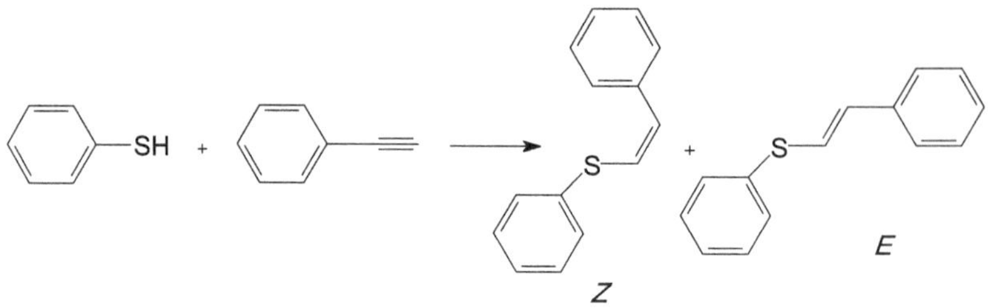

SCHEME 3.12.32 The hydrothiolation reaction of phenylacetylene.

The homogeneous Mn(salen)Cl complex was covalently anchored to cellulose, and this heterogeneous catalyst was able to catalyze the oxidation of aromatic sulfides in good yield under mild conditions.[60] It could also alter the ratio of sulfoxide/sulfone products compared to the unsupported complex. The key advantages of this process are cost-effectiveness of the catalyst, reusability of the catalyst, and easy workup.

Cobalt(II) phthalocyanine covalently immobilized on cellulose was also an excellent oxidation catalyst applied for the aerobic oxidation of alkylarenes and alcohols under relatively mild reaction conditions.[61]

3.12.2.4.2 *Immobilization via Electrostatic Interactions*

A CS-based/MMT/palladium (CS/MMT/Pd) hybrid microsphere catalyst was developed with improved properties for use in Sonogashira coupling reaction (Scheme 3.12.31).[62] An intercalation structure was formed between CS macromolecules and MMT layers with Pd species of different valence states dispersed in nanolevel both in the CS matrix and the interlayers of MMT. CS/MMT/Pd hybrid microspheres were highly active for the Sonogashira reactions of aryl iodides and alkynes. It was concluded that introducing MMT into the CS matrix improved the thermal stability and Pd leaching resistance of the hybrid microsphere catalysts.

Titanium-pillared bentonite was produced by modification of natural bentonite clay using titanium pillaring solution through ion exchange.[63] Esterification reactions were carried out in batch mode. The pillared clay material was shown to be more active in the conversion of acetic acid than the unpillared counterpart.

3.12.2.4.3 *Immobilization via Secondary Bonding Interactions*

Impregnation, where the interaction between the support and the immobilized moieties is ensured by weak forces (hydrogen bonding and/or van der Waals interactions), was also used for immobilization of homogeneous catalysts several times. Triflic acid,[64] sulfuric acid,[65] sulfonic acid,[66] 1-(4-sulfonic acid)butylpyridinium hydrogen sulfate acidic ionic liquid,[67] and ammonium dihydrogenphosphate[68] were supported on silica and the latter on alumina. They were used in various reactions such as the addition of various β-dicarbonyl compounds to a series of alcohols and alkenes[64]; the condensation of benzyl or benzoin, aldehyde, and ammonium acetate for the preparation of 2,4,5-triaryl-1H-imidazoles,[65]; the Biginelli-type reaction between various aldehydes, acetophenones, and urea[66]; in the synthesis of 2-aryl-2,3-dihydroquinazolin-4(1H)-ones by cyclocondensation reaction of

2-aminobenzamide with aromatic aldehydes[67]; and in the synthesis of 1,2,4,5-tetrasubstituted imidazoles by a one-pot, four-component condensation of benzyl, aromatic aldehydes, primary amines, and ammonium acetate.[68] The reactions proceeded under solvent-free conditions at room or moderately elevated temperatures. Under these conditions, leaching of the mostly harmful catalytic components was not a serious problem.

3.12.2.5 Functionalized Hybrid Materials

It is not always easy to distinguish between immobilization and functionalization when a hybrid (composites consisting of two or more constituents at the nanometer or molecular level) catalytic material is prepared. When the moiety is a catalytically active molecule/ion/atom or coordination compound itself, and anchoring to a solid support mainly aims at easing the workup procedure, increasing the stability of the catalyst, and enabling recycling, then the term immobilization is used, and the hybrid material obtained is classified as immobilized hybrid material. When the catalytically active material is constructed on the solid support, then it is termed and classified as functionalized hybrid material. It is true that the borderline between the two classes is somewhat blurred; however, the two material types can most often be distinguished.

Amine-functionalized MCM-41 was found to be an efficient, heterogeneous and recyclable catalyst for the coupling of carbon dioxide (CO_2) and aziridines providing high conversion with excellent regioselectivity toward 5-aryl-2-oxazolidinones[69] under mild and solvent-free conditions (Scheme 3.12.33). The material was equally efficient in the synthesis of a wide variety of quinazoline-2,4(1H,3H)-dione derivatives from 2-aminobenzonitriles and carbon dioxide in water as the solvent once again.[70]

Other mesoporous substances like SBA-15 after amino functionalization through a propyl group linker worked in aqueous medium as well, and were active in the one-pot synthesis of complicated spiro-oxindole derivatives in a three-component condensation reaction.[71]

Poly(4-vinylpyridinium butane sulfonic acid) hydrogen sulfate, a novel polymeric acidic catalyst, could be easily prepared through reaction of cross-linked poly(4-vinylpyridine)

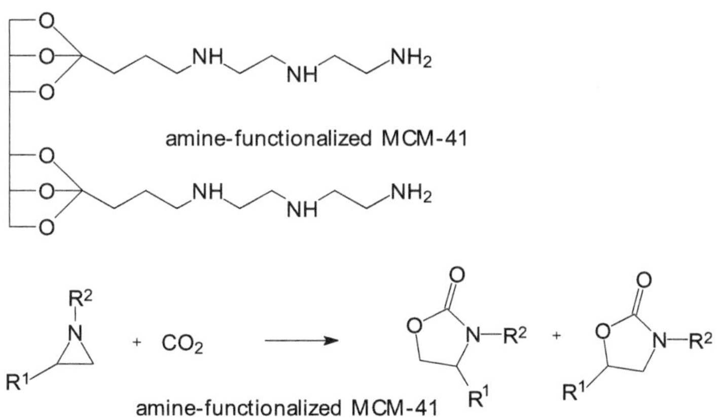

SCHEME 3.12.33 Amine-functionalized MCM-41 in the solvent-free coupling reaction of aziridines and CO_2.

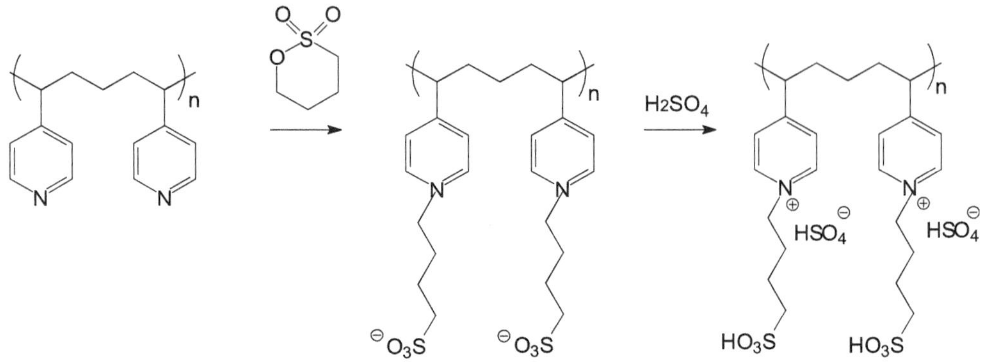

SCHEME 3.12.34 The synthesis of the poly(ionic liquid) catalyst.

with 1,4-butane sultone and further acidification of the catalyst with sulfuric acid[72] (Scheme 3.12.34). The poly(ionic liquid) catalyst was shown to be an efficient dual acidic catalyst for one-pot preparation of 1,8-dioxo-octahydroxanthenes under solvent-free conditions.

After the modification of the surface of magnetic polymer carriers with 1-(2,6-diisopropylphenyl)-1H-imidazole (1-arylimidazole), palladium diacetate was employed to synthesize the Pd-N heterocyclic carbene complex.[73] This magnetic catalyst showed high catalytic activity in the Suzuki-Miyaura reaction of phenylboronic acids with aryl bromides (Scheme 3.12.29) in the ethanol-water solution. After 21 recycling runs, its catalytic activity only decreased slightly, and no leaching of palladium was found either in products or in the reaction residue. When other sources of palladium and ligand ($PdCl_2$ and 3-Cl-pyridine) were employed to construct the palladium complex, the stability of the magnetic catalyst was further improved.

A heterogeneous and recyclable Pd(II)-amino complex constructed on diamine-functionalized ZnAl-LDH proved to be an efficient catalyst for the oxidation of primary alcohol using water as a solvent and molecular O_2 as a terminal oxidant.[74] The *N*-[3-(trimethoxysilyl)-propyl] ethylenediamine ligand was covalently attached to the surface of the LDH. Pd(II) was stabilized by the π-acidic nitrogen-containing ligand.

Similar procedures were applied in the construction of other complexes, such as Cu(I)-amino[75] or copper ferrite-amino and Pd-amino complexes[76] over the surface of 3-aminopropyltriethoxysilane-functionalized KIT-5 (mesoporous silica with a cagelike structure)[75] or silica.[76] The former catalyst was active in the click reaction of terminal alkynes, alkyl halides, and sodium azide using water as the solvent, whereas the latter, magnetically separable one, catalyzed the Sonogashira coupling reactions of aryl iodides and bromides under low Pd loading and mild, phosphine-free reaction conditions.

A heterogeneous bifunctional catalyst composed of OsO_4^{2-}-WO_4^{2-} and a trifunctional catalyst composed of $PdCl_4^{2-}$-OsO_4^{2-}-WO_4^{2-} were designed and prepared by an ion-exchange technique using LDH as an ion-exchanger and their homogeneous bifunctional analog, K_2OsO_4-Na_2WO_4, and trifunctional analog, Na_2PdCl_4-K_2OsO_4-Na_2WO_4.[77] These bifunctional and trifunctional catalysts performed superbly in the synthesis of vicinal diols, the one-pot reaction of asymmetric dihydroxylation-*N*-oxidation and Heck coupling-asymmetric dihydroxylation-*N*-oxidation, respectively, using H_2O_2 as the terminal oxidant.

A robust heterotrimetallic $Ru^{IV}Mn^{IV}Mn^{IV}$ species coordinated to a HT surface as a macroligand facilitated the highly efficient aerobic oxidation of alcohols.[78] This is a fine example of cooperative action among high-valence metal cations in a heterometallic species on a solid surface.

The design, characterization and performance of single-site nanoporous catalysts were described in Ref. 79, and their application in (1) aerobic, shape-selective oxidation of cyclohexane to adipic acid; (2) regiospecific oxidation of linear alkanes; (3) bromine-free synthesis of terephthalic acid from *p*-xylene using air as oxidant; (4) the in situ generation of hazardous reagents for Baeyer-Villiger oxidation of ketones to lactones and for the epoxidation of olefins; and (5) the single-step production of ε-caprolactam from cyclohexanone were elaborated. Multinuclear NP (bimetallic) catalysts engineered at the atomic level were also described, and their application in (1) one-step, solvent-free hydrogenation of polyenes, (2) asymmetric synthesis of pharmaceutical intermediates, (3) regioselective and stereoselective allylic aminations, (4) single-step hydrogenation of dimethyl terepthalate to 1,4-cyclohexane-dimethanol, and (5) renewable nylons from glucose and its biocatalytic derivatives was summarized. The important properties of these solid catalysts beside their easy separation from the products are that (1) they dispense with the use of corrosive reagents (and often of solvents), (2) they permit the replacement of aggressive oxidants (e.g., HNO_3, CrO_3, etc.) or even H_2O_2 and alkyl hydroperoxides with molecular oxygen as the oxidant, (c) they are generally stable and recyclable and require no chemical initiators, and (d) they offer significant processing advantages in large-scale and industrial operations when compared with homogeneous catalysts.

3.12.2.6 Miscellaneous Methods and Catalysts

3.12.2.6.1 Auxiliary Methods Making the Heterogeneous Catalytic Reactions Greener

Microwave or ultrasonic irradiations applied together with solid catalysts provide a green alternative mainly by increasing the efficiency of catalytic processes via either shortening the reaction time or sometimes, in a parallel manner increasing the conversion and selectivity of the reaction.

The highly diastereoselective microwave-assisted three-component synthesis of azabicyclo [2.2.2]octan-5-ones by a silicotungstic acid-catalyzed aza-Diels-Alder cyclization was reported (Scheme 3.12.35).[80] The short reaction times, good yields, and excellent diastereoselectivity made this annulation a practical and environmentally attractive method for this

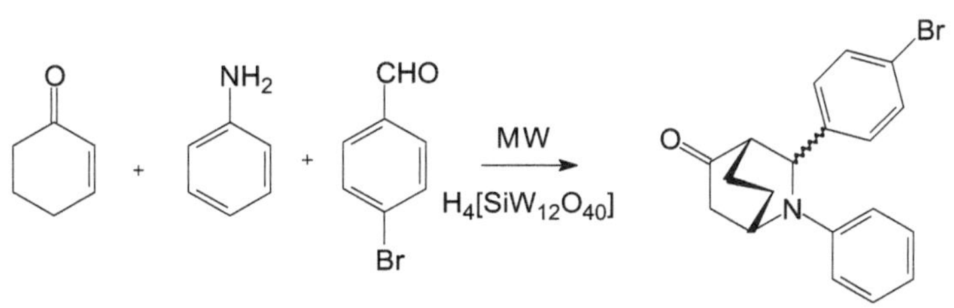

SCHEME 3.12.35 A microwave-assisted three-component synthesis of azabicyclo[2.2.2]octan-5-ones over silicotungstic acid (yield: 59% in 10 min, *endo-exo*: 93:7).

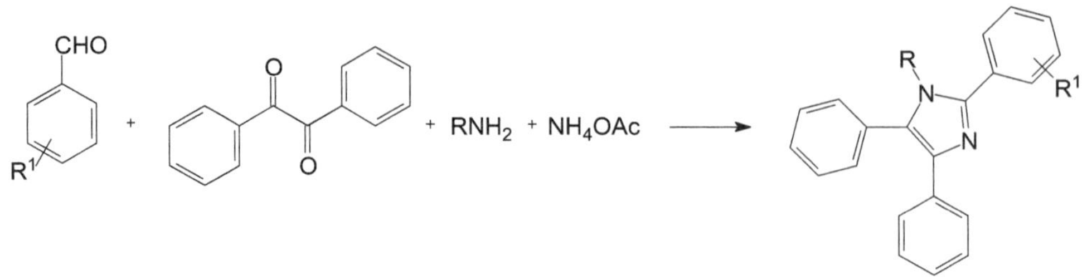

SCHEME 3.12.36 Ultrasonic irradiation accelerated the one-pot synthesis of 1,2,4,5-tetrasubstituted imidazoles over 1-methyl-3-(3-trimethoxysilylpropyl) imidazolium chloride immobilized on Fe_3O_4 NPs.

synthesis. Shortening the reaction time is a crucial asset, since heteropoly acids are excellent catalysts in general, but their stabilities are low. A similar effect of microwave irradiation benefited the phase-transfer catalytic (the catalyst was Aliquat 336) formation of long-chain fatty acid esters used in the cosmetic industry.[81]

Applying the continuous-flow hydration-condensation protocol and microwave irradiation together tremendously increased the efficiency of the Amberlyst 15 catalysts in the synthesis of α,β-unsaturated ketones from alkynes and aldehydes.[82] The procedure presents a straightforward, convenient, and environmental-friendly access to valuable differently substituted chalcones, and can be applied on multigram scale.

Application of ultrasonic irradiation at room temperature accelerated the one-pot synthesis of 1,2,4,5-tetrasubstituted imidazoles catalyzed by the ionic liquid 1-methyl-3-(3-trimethoxysilylpropyl) imidazolium chloride immobilized on Fe_3O_4 NPs (Scheme 3.12.36).[83] The increase in reaction rate is an additional advantage beside the fact that the ionic liquid catalyst could be easily separated from the reaction medium by applying an external magnetic field.

A substantial enhancing effect in the yield of the Claisen-Schmidt condensation between benzaldehyde and acetophenone was observed when the Na- or Cs-Norit (activated carbon) catalysts were activated under ultrasonic conditions.[84] High conversions to chalcones were experienced under solvent-free conditions.

Biodiesel was successfully prepared applying ultrasonic irradiation in a short reaction and at low temperature over $KF/\gamma\text{-}Al_2O_3$ catalyst.[85] The transesterification of soybean oil with methanol was performed by using both low-frequency ultrasonic reactor (20 kHz) and mechanical stirring in the presence of the catalyst. Irradiation with ultrasound significantly accelerated the transesterification reaction when compared with mechanical stirring.

Ultrasonic irradiation shortened the reaction time and increased the selectivity in the cyclization reaction of various primary amines, sodium azide, and triethyl orthoformate leading to 1-substituted tetrazoles (Scheme 3.12.37).[86] The reaction was catalyzed by ZnS NPs instead

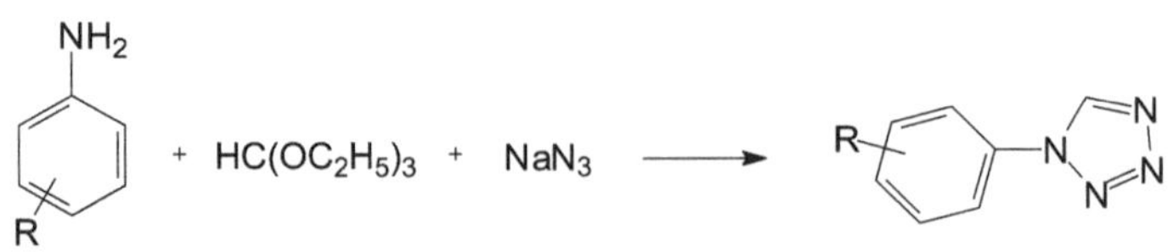

SCHEME 3.12.37 Ultrasonic irradiation-assisted cyclization reaction of primary amines, sodium azide, and triethyl orthoformate leading to 1-substituted tetrazoles over ZnS NPs.

of mineral acids, as efficient, recoverable and reusable catalyst. Compared to conventional methods, this method has considerable advantages such as shorter reaction times, easier workup, purer products with high yields and mild conditions.

3.12.2.6.2 *Ionic Liquids as Catalysts*

Ionic liquids, materials containing cations and anions only and being liquid around room temperature, found applications mainly as solvents in various reactions. Since they have no vapor pressure, they do not evaporate, and in this sense they are considered to be green solvents. (It is to be noted, however, that when it comes to degradation many of them are not so environment-friendly.) Beside their role as solvent, they can act as highly efficient catalysts in various reactions.

Copper(II)-substituted polyoxometalate-based ionic liquids, for example, $[(nC_7H_{15})_4N]6$ $[\alpha\text{-SiW}_{11}O_{39}Cu]$, were successfully developed as halogen-free bifunctional catalysts for the carboxylative cyclization of propargylic amines with CO_2.[87] Such a CO_2 fixation protocol proceeded smoothly at atmospheric pressure under solvent-free conditions, in an environmentally benign and low-energy-input manner.

A sulfonyl-containing ammonium-based Brønsted acid ionic liquid was prepared and used as a liquid heterogeneous catalyst for organic reactions (Scheme 3.12.38).[88] The unique macroscopic phase heterogeneity of the ionic liquid in the reaction system not only ensured excellent catalytic activity of the ionic liquid catalyst but also avoided the use of organic solvents. The catalyst system is applicable for a wide range of reactions like the arylmethylation of phenylacetylene with benzhydrol, the reaction between benzaldehyde and 1,1-diphenylethanol and tandem dehydration/Michael addition.

SO_3H-functionalized ionic liquid was successfully applied[89] as a water-tolerant acidic catalyst for the one-pot domino reaction starting from 2-aminoaryl ketones and α-ketoesters/ketones leading to various quinoline derivatives in aqueous medium. The quinoline products could be conveniently separated from the reaction mixture by filtration, and the catalyst could be easily recycled many times without significant loss of activity.

A 3D network polymer-supported Brønsted acid ionic liquid was synthesized via two postfunctionalization steps.[90] Initially, the active homogeneous catalyst was chemically immobilized onto a polymeric support based on calix^4 resorcinarene by silylation of the hydroxyl groups to form a cationic polymer containing the imidazolium moieties. Subsequently, the HSO_4^- anion was incorporated into the polymer along the imidazolium pendant groups via ion-exchange reaction. The immobilized acidic ionic liquid effectively catalyzed the one-pot synthesis of symmetric bisamides by the multicomponent condensation of 2 mol of amides with aldehydes.

3.12.2.6.3 *Molecular Organic Framework-Based Catalysts*

MOF materials gained importance not only as substances for hydrogen storage but also as catalysts, alone or in combination with other material types.[91]

A cation-exchanged anionic MOF was prepared and was used in the solvent-free Michael addition of pyrrole to electron-deficient β-nitrostyrenes (Scheme 3.12.39).[92]

A zirconium metal-organic framework catalyst was synthesized and investigated in the ring opening of epoxides with nucleophiles such as amines, alcohols, and thiophenol.[93] The catalyst exhibited high activity and excellent recyclability.

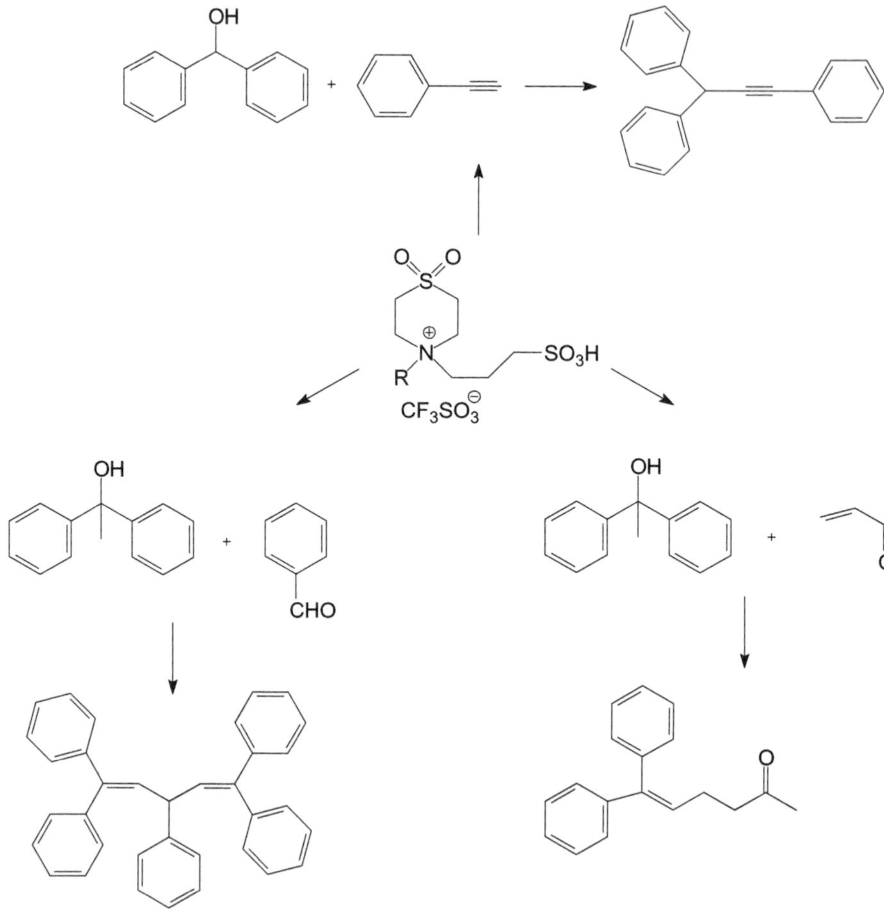

SCHEME 3.12.38 A sulfone-containing ammonium-based Brønsted acid ionic liquid, a liquid heterogeneous catalyst for organic reactions.

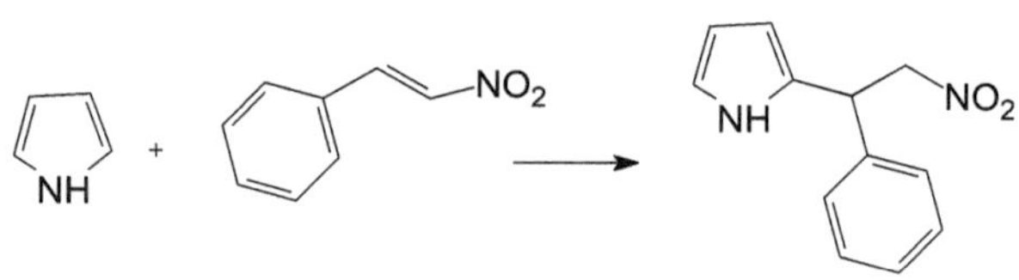

SCHEME 3.12.39 Solvent-free Michael addition of pyrrole to β-nitrostyrene catalyzed by anionic MOFs (yields: 21%–100% in 4 h depending on the specific MOF).

A dual-porous MOF, UMCM-1-NH_2, was used in CO_2 fixation at room temperature in the synthesis of various five-membered cyclic carbonates under solvent-free conditions.[94] Mesopores allowed easy guest diffusion and molecular accessibility, whereas micropores were predominantly helpful in regulating the catalytic interactions in active centers.

Cellulose and cellobiose were selectively converted into sorbitol over water-tolerant phosphotungstic acid (PTA)/MOF hybrid-supported ruthenium catalysts, Ru-PTA/MIL-100(Cr), under aqueous hydrogenation conditions.[95] The goal was to investigate the relationship between the acid/metal balance in the bifunctional catalyst through the effective control of encapsulated PTA loading in MIL-100(Cr). This design and preparation method led to an appropriately balanced Ru-PTA/MIL-100(Cr) in terms of Ru dispersion and hydrogenation capacity on the one hand and acid site density of PTA/MIL-100(Cr) (responsible for acid-catalyzed hydrolysis) on the other hand.

3.12.3 CATALYTIC CONVERSION OF THE BIOMASS

Lignocellulose can be taken as the prototype of biomass. It can be transformed via various pathways (Scheme 3.12.40). Either the raw lignocellulose or the fossilized products are processed; the key steps are catalytic.

A continuous reactive distillation process was developed using solid acid catalysts from prehydrolyzed hemicellulose containing xylose[96] to produce furfural (Scheme 3.12.41). The reactive distillation process provided the unique advantage of immediate furfural separation from the reaction zone as it is formed, thus minimizing the formation of undesired by-products (e.g., oligomers of furfural/humins) from subsequent furfural condensation reactions. Among various solid acid catalysts screened for their ability to produce furfural from dehydration of xylose, the zeolite H-mordenite (Si/Al = 10) gave the highest furfural yield (>75%). During this process, the aqueous feed was introduced at the top of the column, and xylose was dehydrated to furfural, which was co-distilled with water out of the top of the column.

n-Pentane, pentanols, and xylitol could be separately produced from hemicellulose over an Ir-ReOx/SiO_2 catalyst combined with acids by simply adjusting the reaction conditions.[97]

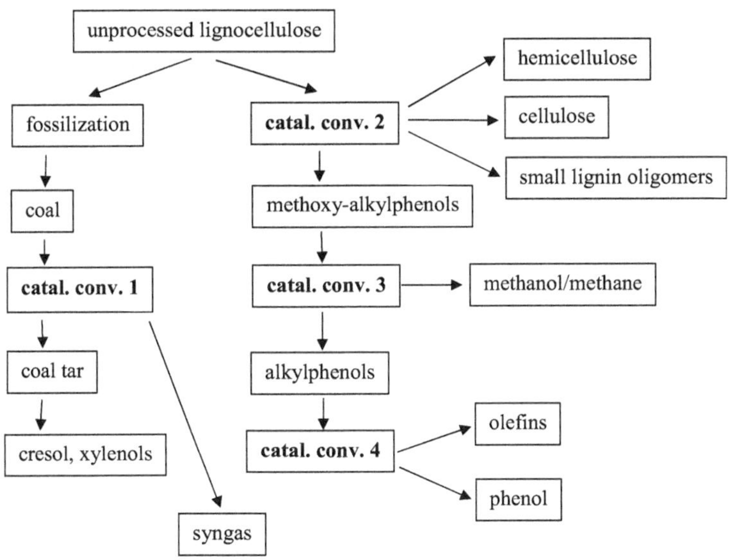

SCHEME 3.12.40 Transformation possibilities of lignocellulose.

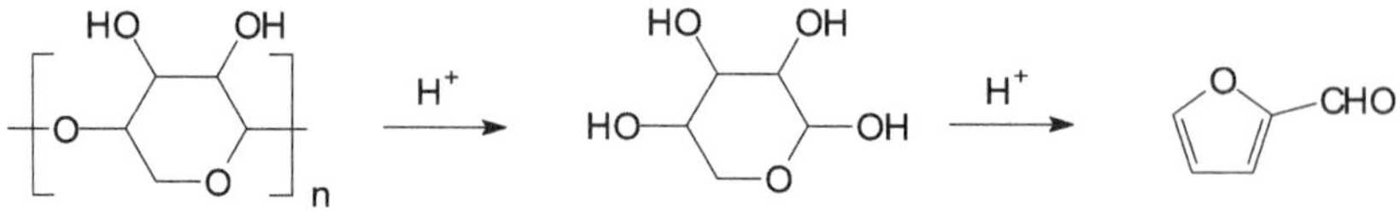

SCHEME 3.12.41 Hydrolysis of a hemicellulose to pentose and its dehydration to furfural.

n-Pentane could be obtained by using Ir-ReOx/SiO_2 combined with HZSM-5 + H_2SO_4 in a biphasic solvent system (4 mL n-dodecane + 9.5 mL H_2O) with a reaction temperature of 463 K for 24 h. Pentanols could be produced by using Ir-ReOx/SiO_2 combined with H_2SO_4 in a biphasic solvent system (20 mL n-dodecane + 9.5 mL H_2O) with a reaction temperature of 413 K for 144 h. Xylitol was gained by using Ir-ReOx/SiO_2 combined with H_2SO_4 in the aqueous phase with a reaction temperature of 413 K for 12 h. The highest yields of n-pentane, pentanols, and xylitol could reach 70%, 32%, and 79%, respectively. The reuse of the catalyst was feasible when the catalyst was regenerated by calcination at 773 K for 3 h. The calcination step was for removing the humins, which were formed in the hydrolysis + hydrogenation step during the conversion of hemicellulose.

The oxidative transformation of lignin into aromatic compounds is an attractive route for upgrading lignocellulosic biomass. It is advantageous that in these reactions, no consumption of expensive hydrogen is required. However, only limited success was achieved for the oxidative conversion of lignin. As a promising result, it was shown that cerium oxide-supported palladium NPs (Pd/CeO_2) could efficiently catalyze the one-pot oxidative conversion of 2-phenoxy-1-phenylethanol, a lignin model compound (Scheme 3.12.42) containing a β-O-4 bond and a Cα-hydroxyl group, in methanol in the presence of O_2, producing phenol, acetophenone, and methyl benzoate as the major products.[98]

Catalytic hydrogenolysis is another transformation possibility of lignin. Sulfided NiMo and CoMo[99] on different acidic and basic supports were studied as hydrogenolysis catalysts in the absence of a solvent. Experiments were carried out in a batch reactor at a reaction temperature of 350 °C for 4 h at 100 bar initial H_2 pressure. Aromatics, alkylphenolics, and alkanes (Scheme 3.12.43) were formed. High lignin oil yields were obtained using the sulfided NiMo supported on activated carbon and MgO-La_2O_3. The highest total monomer yield including alkylphenolics was obtained using the sulfided NiMo/MgO-La_2O_3 catalyst. Others, such as supported NiW[100] or the supported monometallic Ni, proved to be efficient catalysts as well.[101]

Lignin-derived phenolic compounds are important feedstocks for the sustainable production of alkane fuels with C6-C9 carbons. Hydrodeoxygenation (HDO) is the main chemical process to remove oxygen-containing functionalities. HDO reaction of phenols was performed in a biphasic H_2O/n-dodecane system.[102] A series of supported Ru catalysts were prepared and explored for the HDO reaction of phenols, among which Ru/MWCNT showed the highest catalytic activity toward the production of alkanes. The model reaction with eugenol achieved high conversion and high alkane selectivity, much higher than the results from the monophasic aqueous system. It was revealed that eugenol was first hydrogenated to 4-propylguaiacol and then deoxygenated into 4-propylcyclohexanol, which was the main detected intermediate of the reaction. After that, 4-propylcyclohexanol was dehydrated

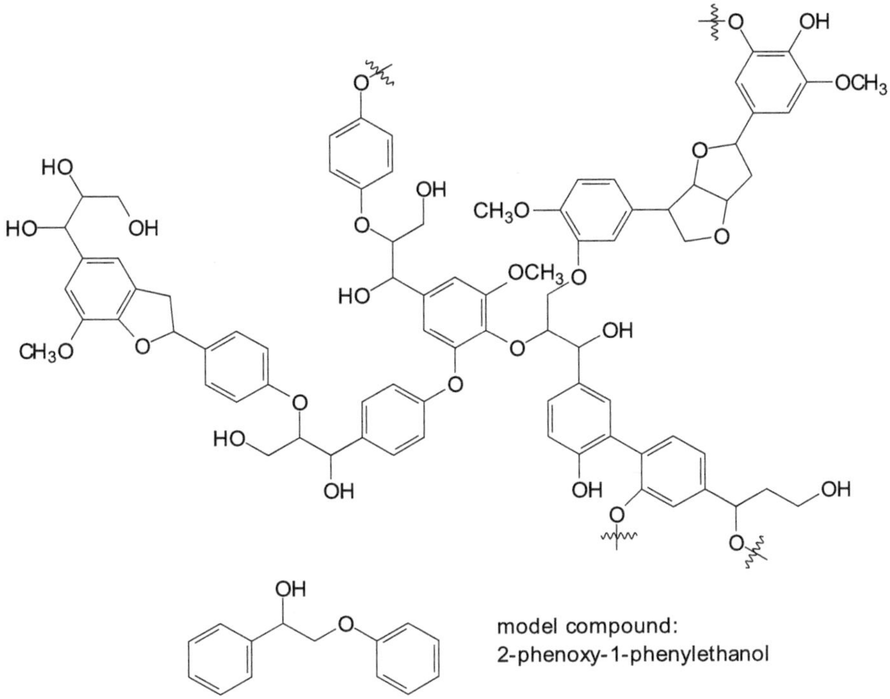

SCHEME 3.12.42 A representative lignin fragment and a model compound.

and hydrogenated into propylcyclohexane (Scheme 3.12.44). Moreover, various phenols and dimeric lignin model compounds were also successfully converted into alkanes in the biphasic systems.

Following the processing of either raw or pyrolyzed lignocellulose, one can arrive at a mixture of alkylphenols. They are often transformed to phenol and olefins. This is an important reaction in upgrading raw and fossilized lignocellulose. An exceptional and stable dealkylation performance was achieved by the application of an acidic ZSM-5 zeolite, in which co-feeding of water was crucial to maintain catalytic activity.[103] The role of water is attributed to competitive adsorption of water and phenol.

Palm oil is another important representative of biomass produced in large quantities. A highly active catalyst, hierarchical nano-sized Ni/H-β, was developed for stearic acid (a model of palm oil) HDO in dodecane.[104] H-β after base treatment developed homogeneously dispersed open intermesopores. The modified H-β largely contained the Ni nanoclusters in the newly formed intercrystalline mesopores, which provide higher accessibility toward heavy molecules, as well as restrict the Ni particle growth. This novel catalyst showed a high initial rate for producing a mixture of n-C17/C18 and iso-C17/C18 alkanes. The HDO route follows the major pathway of sequential hydrogenation and dehydration steps affording a highly atom-economical process and a suitable diesel oil ingredient (with certain branched alkanes).

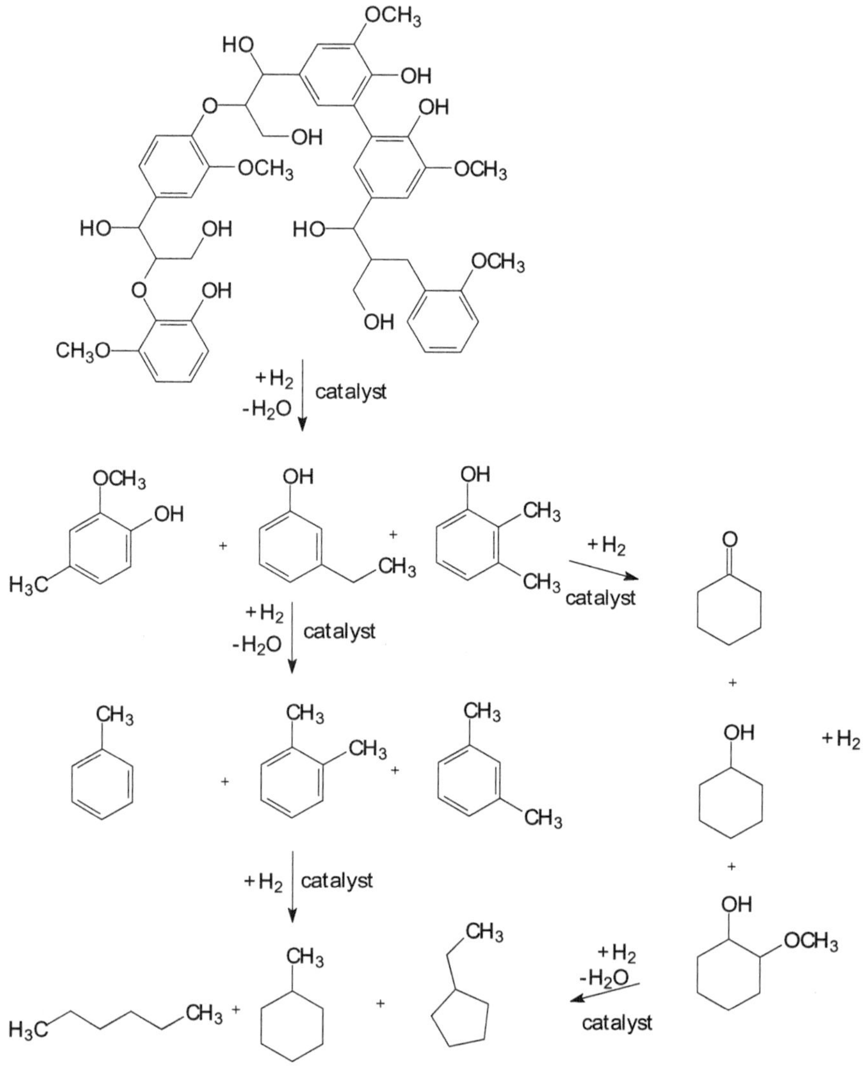

SCHEME 3.12.43 Hydrogenolysis transformation pathways of lignin oligomers.

3.12.4 SELECTED RECENT REVIEWS CONCERNING THE ADVANCES IN PERFORMING REACTION TYPES IN A GREEN WAY AND TRANSFORMING THE BIOMASS

Performing reactions in a green way became a catch phrase in recent years. It was easier to publish the results of research, when it was stated, with or without reason that the requirements of, or at least certain requirements of green chemistry have been met. In filtering out green processes, critical reviews are needed from time to time. Some of the most important

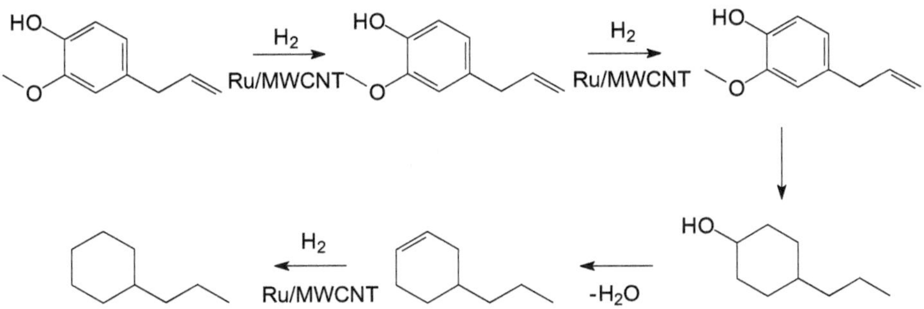

SCHEME 3.12.44 The proposed reaction pathway of transforming phenolic compounds to saturated hydrocarbons over a Ru/multiwalled carbon nanotube catalyst.

ones are collected as follows. One of the most important summarizing the fundamentals of green chemistry in a tutorial way was published by Sheldon in the themed issue of Chemical Society Reviews in 2012.[105] Advances in the development of catalysts for environment-friendly fine chemical syntheses in general[106–116] and more specifically for Baeyer-Villiger oxidation,[117] click reaction,[118] selective oxidation of alcohols,[119] syntheses of heterocyclic compounds,[120] esterification and transesterification,[121] and hydrogenation reaction in the flow mode[122] have also been published.

Large amount of biomass is continuously accumulating during everyday activities of mankind. After processing, it is an important source of valuable compounds, among others, straight-chain as well as branched hydrocarbons, that is, fuel can be produced. While the source is renewable and as often said, a green source of energy, it is not necessarily correct, since, e.g., it has to be transported to the processing plant, which obviously requires non-renewable fossil fuels. Nevertheless, protocols are needed for processing the biomass. They include catalytic reactions, which require heterogeneous catalysts. Research in this area is summarized in the following reviews.[123–126]

3.12.5 CONCLUSIONS AND OUTLOOK

It is beyond doubt that tremendous effort will go into research to find processes, which meet all the principles of green chemistry, since it is our only choice to keep a sustainable world. It is also beyond doubt that the majority of these processes will apply heterogeneous catalysts. Finding new materials and new ways of producing and tailoring heterogeneous catalysts will increase not only their activities but also more importantly, their selectivities. Our present knowledge provides suitable foundations for this development, which will give work for future generations of scholars too.

References

1. Anastas PT, Kirchhoff MM, Williamson TC. Catalysis as a foundational pillar of green chemistry. *Appl Catal A* 2001;**21**:3–13.
2. Liu H, Fang R, Li Z, Li Y. Solventless hydrogenation of benzene to cyclohexane over a heterogeneous Ru–Pt bimetallic catalyst. *Chem Eng Sci* 2015;**122**:350–9.

3. Tongsakul D, Nishimura S, Ebitani K. Platinum/gold alloy nanoparticles-supported hydrotalcite catalyst for selective aerobic oxidation of polyols in base-free aqueous solution at room temperature. *ACS Catal* 2013;**3**:2199–207.
4. Mitsudome T, Noujima A, Mizugaki T, Jitsukawa K, Kaneda K. Efficient aerobic oxidation of alcohols using a hydrotalcite-supported gold nanoparticle catalyst. *Adv Synth Catal* 2009;**351**:1890–6.
5. Liu X, Li H-Q, Ye S, Liu Y-M, He H-Y, Cao Y. Gold-catalyzed direct hydrogenative coupling of nitroarenes to synthesize aromatic azo compounds. *Angew Chem Int Ed* 2014;**53**:7624–8.
6. Safari J, Gandomi-Ravandi S. Application of the ultrasound in the mild synthesis of substituted 2,3-dihydroquinazolin-4(1H)-ones catalyzed by heterogeneous metal–MWCNTs nanocomposites. *J Mol Struct* 2014;**1072**:173–8.
7. Sharghi H, Ebrahimpourmoghaddam S, Doroodmand MM. Iron-doped single walled carbon nanotubes as an efficient and reusable heterogeneous catalyst for the synthesis of organophosphorus compounds under solvent-free conditions. *Tetrahedron* 2013;**69**:4708–24.
8. He H, Gao C. Synthesis of Fe_3O_4/Pt nanoparticles decorated carbon nanotubes and their use as magnetically recyclable catalysts. *J Nanomater* 2011:1–10, 93510.
9. Lin F, Doong R. Bifunctional Au–Fe_3O_4 heterostructures for magnetically recyclable catalysis of nitrophenol reduction. *J Phys Chem C* 2011;**115**:6591–8.
10. Ungaro VA, Liria CW, Romagna CD, Costa NJS, Philippot K, Rossi LM, et al. A green route for the synthesis of a bitter-taste dipeptide combining biocatalysis, heterogeneous metal catalysis and magnetic nanoparticles. *RSC Adv* 2015;**5**:36449–55.
11. Zhang Z, Zhen J, Liu B, Lv K, Deng K. Selective aerobic oxidation of the biomass-derived precursor 5-hydroxymethylfurfural to 2,5-furandicarboxylic acid under mild conditions over a magnetic palladium nanocatalyst. *Green Chem* 2015;**17**:1308–17.
12. Changotra A, Gupta R, Paul S. Reduction of carbon–carbon double bond in α,β-unsaturated ketones over bimetallic Pd and Au nanoparticles immobilized on amine-functionalized magnetite nanoparticles: a novel and highly active bimetallic catalyst. *Der Pharma Chem* 2014;**6**:385–93.
13. Wang Y, Li H, Zhang J, Yan X, Chen Z. Fe_3O_4 and Au nanoparticles dispersed on graphene support as a highly active catalyst toward the reduction of 4-nitrophenol. *PCCP* 2016;**18**:615–23.
14. Mangala K, Sinija PS, Sreekumar K. Palladium(II) supported on polycarbosilane: application as reusable catalyst for Heck reaction. *J Mol Catal A* 2015;**407**:87–92.
15. Duan L, Fu R, Xiao Z, Zhao Q, Wang J-Q, Chen S, et al. Activation of aryl chlorides in water under phase-transfer agent-free and ligand-free Suzuki coupling by heterogeneous palladium supported on hybrid mesoporous carbon. *ACS Catal* 2015;**5**:575–86.
16. Shen C, Xu J, Yu W, Zhang P. A highly active and easily recoverable catalyst for the C–S coupling and its application in the synthesis of zolimidine. *Green Chem* 2014;**16**:3007–12.
17. Yepez A, Hidalgo JM, Pineda A, Černý R, Jíša P, Garcia A, et al. Mechanistic insights into the hydroconversion of cinnamaldehyde using mechanochemically synthesized Pd/Al-SBA-15 catalysts. *Green Chem* 2015;**17**:565–72.
18. Mägerlein W, Dreisbach C, Hugl H, Tse MK, Klawonn M, Bhor S, et al. Homogeneous and heterogeneous ruthenium catalysts in the synthesis of fine chemicals. *Catal Today* 2007;**121**:140–50.
19. Ai L, Yue H, Jiang J. Environmentally friendly light-driven synthesis of Ag nanoparticles in situ grown on magnetically separable biohydrogels as highly active and recyclable catalysts for 4-nitrophenol reduction. *J Mater Chem* 2012;**22**:23447–53.
20. Li X-L, Deng J, Shi J, Pan T, Yu C-G, Xu H-J, Fu Y. Selective conversion of furfural to cyclopentanone or cyclopentanol using different preparation methods of Cu–Co catalysts. *Green Chem* 2015;**17**:1038–46.
21. Shukla A, Singha RK, Sasaki T, Bal R. Nanocrystalline Pt-CeO_2 as an efficient catalyst for a room temperature selective reduction of nitroarenes. *Green Chem* 2015;**17**:785–90.
22. Zheng Q, Morimoto M, Takanohashi T. An efficient production of benzene from benzoic acid in subcritical water using a copper(I) oxide catalyst. *Green Chem* 2015;**17**:791–4.
23. Medisetti V, Parimi U, Anagani RB, Satyanarayana KVVV. Nano-γ-Fe_2O_3: efficient, reusable and green catalyst for *N-tert*-butoxycarbonylation of amines in water. *Green Sustain Chem* 2014;**4**:95–101.
24. Thakuria H, Borah BM, Das G. Macroporous metal oxides as an efficient heterogeneous catalyst for various organic transformations – a comparative study. *J Mol Catal A* 2007;**274**:1–10.

25. Agarwal S, Aware U, Patil A, Rohera V, Ghate M, Jain M, et al. Silica-gel catalyzed facile synthesis of 3,4-dihydropyrimidinones. *Bull Korean Chem Soc* 2012;**33**:377–8.
26. Safaei-Ghomi J, Heidari-Baghbahadorani E, Shahbazi-Alavi H, Asgari-Kheirabadi M. A comparative study of the catalytic activity of nanosized oxides in the one-pot synthesis of highly substituted dihydropyridines. *RSC Adv* 2015;**5**:18145–52.
27. Dibenedetto A, Angelini A, Aresta M, Ethiraj J, Fragale C, Nocito F. Converting wastes into added value products: from glycerol to glycerol carbonate, glycidol and epichlorohydrin using environmentally friendly synthetic routes. *Tetrahedron* 2011;**67**:1308–13.
28. Zacharopoulou V, Vasiliadou ES, Lemonidou AA. One-step propylene formation from bio-glycerol over molybdena-based catalysts. *Green Chem* 2015;**17**:903–12.
29. Farhadi S, Jahanara K, Sepahdar A. Sol–gel derived $LaFeO_3/SiO_2$ nanocomposite: synthesis, characterization and its application as a new, green and recoverable heterogeneous catalyst for the efficient acetylation of amines, alcohols and phenols. *J Iran Chem Soc* 2014;**11**:1103–12.
30. Gholinejad M, Karimi B, Mansouri F. Synthesis and characterization of magnetic copper ferrite nanoparticles and their catalytic performance in one-pot odorless carbon-sulfur bond formation reactions. *J Mol Catal A* 2014;**386**:20–4.
31. Olmos A, Rigolet S, Louis B, Pale P. Design of scandium-doped USY zeolite: an efficient and green catalyst for aza-Diels–Alder reaction. *J Phys Chem C* 2012;**116**:13661–70.
32. Chassaing S, Alix A, Olmos A, Keller M, Sommer J, Pale P. Metal-doped zeolites as green catalysts for organic synthesis. *Z Naturforsch B* 2010;**65**:783–90.
33. Baghbanian SM, Rezaei N, Tashakkorian H. Nanozeolite clinoptilolite as a highly efficient heterogeneous catalyst for the synthesis of various 2-amino-4*H*-chromene derivatives in aqueous media. *Green Chem* 2013;**15**:3446–58.
34. Li K, Zhou D, Deng J, Lu X, Xia Q. Highly efficient catalytic oxidation of cyclohexanol with TBHP over Cr-13X catalysts in a solvent-free system. *J Mol Catal A* 2014;**387**:31–7.
35. Kong D, Liu Y, Zhang J, Li H, Wang X, Liu G, et al. Hierarchically porous AlPO-5-based microspheres as heterogeneous catalysts for the synthesis of 5-substituted 1H-tetrazoles via [3+2] cycloaddition. *New J Chem* 2014;**38**:3078–83.
36. Zou G, Zhong W, Mao L, Xu Q, Xiao J, Yin D, et al. A non-nitric acid method of adipic acid synthesis: organic solvent- and promoter-free oxidation of cyclohexanone with oxygen over hollow-structured Mn/TS-1 catalysts. *Green Chem* 2015;**17**:1884–92.
37. Tang B, Dai W, Sun X, Wu G, Guan N, Hunger M, et al. Mesoporous Zr-Beta zeolites prepared by a post-synthetic strategy as a robust Lewis acid catalyst for the ring-opening aminolysis of epoxides. *Green Chem* 2015;**17**:1744–55.
38. Karmakar B, Sinhamahapatra A, Panda AB, Banerji J, Chowdhury B. Ga-TUD-1: a new heterogeneous mesoporous catalyst for the solventless expeditious synthesis of α-aminonitriles. *Appl Catal A* 2011;**392**:111–7.
39. Lin X-Z, Yang Z-Z, He L-N, Yuan Z-Y. Mesoporous zirconium phosphonates as efficient catalysts for chemical CO_2 fixation. *Green Chem* 2015;**17**:795–8.
40. Kuruppathparambil RR, Jose T, Babu R, Hwang G-Y, Kathalikkattil AC, Kim D-W, et al. A room temperature synthesizable and environmental friendly heterogeneous ZIF-67 catalyst for the solvent less and co-catalyst free synthesis of cyclic carbonates. *Appl Catal B* 2016;**182**:562–9.
41. Mekhzoum MEM, Benzeid H, Qaiss AEK, Essassi EM, Bouhfid R. Copper(I) confined in interlayer space of montmorillonite: a highly efficient and recyclable catalyst for click reaction. *Catal Lett* 2016;**146**:136–43.
42. Ausavasukhia A, Sooknoi T. Oxidation of tetrahydrofuran to butyrolactone catalyzed by iron-containing clay. *Green Chem* 2015;**17**:435–41.
43. Chen Y, Zhu Z, Zhang J, Shen J, Zhou X. Modification of mesoporous silicate SBA-15 with tris[bis(trimethylsilyl)amido]samarium and its utility in Tishchenko reaction. *J Organomet Chem* 2005;**690**:3783–9.
44. Fuerte A, Corma A, Sánchez F. Heterogenised chiral amines as environmentally friendly base catalysts for enantioselective Michael addition. *Catal Today* 2005;**107–108**:404–9.
45. Ziarani GM, Badiei A, Aslani Z, Lashgari N. Application of sulfonic acid functionalized nanoporous silica ($SBA\text{-}Pr\text{-}SO_3H$) in the green one-pot synthesis of triazoloquinazolinones and benzimidazoquinazolinones. *Arab J Chem* 2015;**8**:54–61.

46. Ghorbani-Choghamarani A, Ghorbani F, Yousofvand Z, Azadi G. Mesoporous MCM-41-nPr-NHSO3H as novel and effective nanoreactor catalyst for the synthesis of multi-substituted imidazoles under solvent-free conditions. *J Porous Mater* 2015;**22**:665–73.
47. Shi X-L, Yang H, Tao M, Zhang W. Sulfonic acid-functionalized polypropylene fiber: highly efficient and recyclable heterogeneous Brønsted acid catalyst. *RSC Adv* 2013;**3**:3939–45.
48. Khalafi-Nezhad A, Nourisefat M, Panahi F. L-proline-modified magnetic nanoparticles (LPMNP): a novel magnetically separable organocatalyst. *RSC Adv* 2014;**4**:22497–500.
49. Ying A, Qiu F, Wu C, Hu H, Yang J. Ionic tagged amine supported on magnetic nanoparticles: synthesis and application for versatile catalytic Knoevenagel condensation in water. *RSC Adv* 2014;**4**:33175–83.
50. Koukabi N, Kolvari E, Zolfigol MA, Khazaei A, Shaghasemi BS, Fasahati B. A magnetic particle-supported sulfonic acid catalyst: tuning catalytic activity between homogeneous and heterogeneous catalysis. *Adv Synth Catal* 2012;**354**:2001–8.
51. Veisia H, Mohammadia P, Gholami J. Sulfamic acid heterogenized on functionalized magnetic Fe_3O_4 nanoparticles with diaminoglyoxime as a green, efficient and reusable catalyst for one-pot synthesis of substituted pyrroles in aqueous phase. *Appl Organomet Chem* 2014;**28**:868–73.
52. Alizadeh A, Khodaei MM, Beygzadeh M, Kordestani D, Feyzi M. Biguanide-functionalized Fe_3O_4/SiO_2 magnetic nanoparticles: an efficient heterogeneous organosuperbase catalyst for various organic transformations in aqueous media. *Bull Korean Chem Soc* 2012;**33**:2546–52.
53. Sheykhan M, Mámani L, Ebrahimi A, Heydari A. Sulfamic acid heterogenized on hydroxyapatite-encapsulated γ-Fe_2O_3 nanoparticles as a magnetic green interphase catalyst. *J Mol Catal A* 2011;**335**:253–61.
54. Khakiani BA, Pourshamsian K, Veisi H. A highly stable and efficient magnetically recoverable and reusable Pd nanocatalyst in aqueous media heterogeneously catalyzed Suzuki C–C cross-coupling reactions. *Appl Organomet Chem* 2015;**29**:259–65.
55. Phan NTS, Styring P. Supported phosphine-free palladium catalysts for the Suzuki–Miyaura reaction in aqueous media. *Green Chem* 2008;**10**:1055–60.
56. Liu X, Zhao X, Lub M. A highly water-dispersible and magnetically separable palladium catalyst based on functionalized poly(ethylene glycol)-supported iminophosphine for Suzuki–Miyaura coupling in water. *Appl Organomet Chem* 2015;**29**:419–24.
57. Yang H, Han X, Li G, Wang Y. *N*-Heterocyclic carbene palladium complex supported on ionic liquid-modified SBA-16: an efficient and highly recyclable catalyst for the Suzuki and Heck reactions. *Green Chem* 2009;**11**:1184–93.
58. Ghiaci M, Zarghani M, Khojastehnezhad A, Moeinpour F. Preparation, characterization and first application of silica supported palladium-N-heterocyclic carbene as a heterogeneous catalyst for C–C coupling reactions. *RSC Adv* 2014;**4**:15496–501.
59. Yang Y, Rioux RM. Highly stereoselective anti-Markovnikov hydrothiolation of alkynes and electron-deficient alkenes by a supported Cu-NHC complex. *Green Chem* 2014;**16**:3916–25.
60. Mohammadinezhad A, Nasseri MA, Salimi M. Cellulose as an efficient support for Mn(salen)Cl: application for catalytic oxidation of sulfides to sulfoxides. *RSC Adv* 2014;**4**:39870–4.
61. Shaabani A, Keshipour S, Hamidzad M, Shaabani S. Cobalt(II) phthalocyanine covalently anchored to cellulose as are coverable and efficient catalyst for the aerobic oxidation of alkylarenes and alcohols. *J Mol Catal A* 2014;**395**:494–9.
62. Zeng M, Yuan X, Zuo S, Qi C. Novel chitosan-based/montmorillonite/palladium hybrid microspheres as heterogeneous catalyst for Sonogashira reactions. *RSC Adv* 2015;**5**:37995–8000.
63. Peter OI, Chidi O, Iheanacho MA. The preparation and application of environmentally benign titanium pillared clay catalyst for esterification of ethanol and acetic acid. *Am Chem Sci J* 2012;**2**:45–59.
64. Liu PN, Xia F, Wang QW, Ren YJ, Chen JQ. Triflic acid adsorbed on silica gel as an efficient and recyclable catalyst for the addition of β-dicarbonyl compounds to alcohols and alkenes. *Green Chem* 2010;**12**:1049–55.
65. Maleki B, Shirvan HK, Taimazi F, Akbarzadeh E. Sulfuric acid immobilized on silica gel as highly efficient and heterogeneous catalyst for the one-pot synthesis of 2,4,5-triaryl-1h-imidazoles. *Int J Org Chem* 2012;**2**:93–9.
66. Ziarani GM, Azizi M, Hajiabbasi P, Badiei A. Application of sulfonic acid functionalized nanoporous silica (SBA-Pr-SO_3H) for the preparation of 4,6-diarylpyrimidin-2(1*h*)-ones. *Quim Nova* 2015;**38**:466–70.

67. Yassaghi G, Davoodnia A, Allameh S, Zare-Bidaki A, Tavakoli-Hoseini N. Preparation, characterization and first application of aerosil silica supported acidic ionic liquid as a reusable heterogeneous catalyst for the synthesis of 2,3-dihydroquinazolin-4(1h)-ones. *Bull Korean Chem Soc* 2012;**33**:2724–30.
68. Emrani A, Davoodnia A, Tavakoli-Hoseini N. Alumina supported ammonium dihydrogenphosphate ($NH_4H_2PO_4/Al_2O_3$): preparation, characterization and its application as catalyst in the synthesis of 1,2,4,5-tetrasubstituted imidazoles. *Bull Korean Chem Soc* 2011;**32**:2385–90.
69. Nale DB, Rana S, Parida K, Bhanage BM. Amine functionalized MCM-41 as a green, efficient, and heterogeneous catalyst for the regioselective synthesis of 5-aryl-2-oxazolidinones, from CO_2 and aziridines. *Appl Catal A* 2014;**469**:340–9.
70. Nale DB, Rana S, Parida K, Bhanage BM. Amine functionalized MCM-41: an efficient heterogeneous recyclable catalyst for the synthesis of quinazoline-2,4(1H,3H)-diones from carbon dioxide and 2-aminobenzonitriles in water. *Catal Sci Technol* 2014;**4**:1608–14.
71. Ziarani GM, Badiei A, Mousavi S, Lashgari N, Shahbazi A. Application of amino-functionalized SBA-15 type mesoporous silica in one-pot synthesis of spirooxindoles. *Chin J Catal* 2012;**33**:1832–9.
72. Kiasat AR, Mouradzadegun A, Saghanezhad SJ. Synthesis, characterization, and application of poly(4-vinylpyridinium butane sulfonic acid) hydrogen sulfate as a novel poly(ionic liquid) and heterogeneous solid acid catalyst for the preparation of 1,8-dioxo-octahydroxanthenes. *Res Chem Intermed* 2015;**41**:319–26.
73. Wang Z, Yu Y, Zhang YX, Li SZ, Qian H, Lin ZY. A magnetically separable palladium catalyst containing a bulky N-heterocyclic carbene ligand for the Suzuki–Miyaura reaction. *Green Chem* 2015;**17**:413–20.
74. Sahoo M, Parida KM. Pd(II) loaded on diamine functionalized LDH for oxidation of primary alcohol using water as solvent. *Appl Catal A* 2013;**460–461**:36–45.
75. Mirsafaei R, Heravi MM, Ahmadi S, Moslemin MH, Hosseinnejad T. In situ prepared copper nanoparticles on modified KIT-5 as an efficient recyclable catalyst and its applications in click reactions in water. *J Mol Catal A* 2015;**402**:100–8.
76. Gholinejad M, Ahmadi J. Assemblies of copper ferrite and palladium nanoparticles on silica microparticles as a magnetically recoverable catalyst for Sonogashira reaction under mild conditions. *ChemPlusChem* 2015;**80**:973–9.
77. Choudary BM, Chowdari NS, Madhi S, Kantam ML. A trifunctional catalyst for one-pot synthesis of chiral diols via Heck coupling-*N*-oxidation-asymmetric dihydroxylation: application for the synthesis of diltiazem and Taxol side chain. *J Org Chem* 2003;**68**:1736–46.
78. Ebitani K, Motokura K, Mizugaki T, Kaneda K. Heterotrimetallic RuMnMn species on a hydrotalcite surface as highly efficient heterogeneous catalysts for liquid-phase oxidation of alcohols with molecular oxygen. *Angew Chem Int Ed* 2005;**44**:3423–6.
79. Thomas JM, Raja R. The advantages and future potential of single-site heterogeneous catalysts. *Top Catal* 2006;**40**:3–17.
80. Borkin D, Morzhina E, Datta S, Rudnitskaya A, Sood A, Török M, et al. Heteropoly acid-catalyzed microwave-assisted three-component aza-Diels–Alder cyclizations: diastereoselective synthesis of potential drug candidates for Alzheimer's disease. *Org Biomol Chem* 2011;**9**:1394–401.
81. Villa C, Mariani E, Loupy A, Grippo C, Grossi GC, Bargagna A. Solvent-free reactions as green chemistry procedures for the synthesis of cosmetic fatty esters. *Green Chem* 2003;**5**:623–6.
82. Rueping M, Bootwicha T, Baars H, Sugiono E. Continuous-flow hydration–condensation reaction: synthesis of α,β-unsaturated ketones from alkynes and aldehydes by using a heterogeneous solid acid catalyst. *Beilstein J Org Chem* 2011;**7**:1680–7.
83. Safari J, Zarnegar Z. Magnetic nanoparticle supported ionic liquid as novel and effective heterogeneous catalyst for synthesis of substituted imidazoles under ultrasonic irradiation. *Monatsh Chem* 2013;**144**:1389–96.
84. Durán-Valle CJ, Fonseca IM, Calvino-Casilda V, Picallo M, López-Peinado AJ, Martín-Aranda RM. Sonocatalysis and alkaline-doped carbons: an efficient method for the synthesis of chalcones in heterogeneous media. *Catal Today* 2005;**107–108**:500–6.
85. Shahraki H, Entezari MH, Goharshadi EK. Sono-synthesis of biodiesel from soybean oil by $KF/\gamma\text{-}Al_2O_3$ as a nano-solid-base catalyst. *Ultrason Sonochem* 2015;**23**:266–74.
86. Naeimi H, Kiani F. Ultrasound-promoted one-pot three component synthesis of tetrazoles catalyzed by zinc sulfide nanoparticles as a recyclable heterogeneous catalyst. *Ultrason Sonochem* 2015;**27**:408–15.

87. Wang M-Y, Song Q-W, Ma R, Xie J-N, He L-N. Efficient conversion of carbon dioxide at atmospheric pressure to 2-oxazolidinones promoted by bifunctional Cu(II)-substituted polyoxometalate-based ionic liquids. *Green Chem* 2016;**18**:282–7.
88. Taheri A, Pan X, Liu C, Gu Y. Brønsted acid ionic liquid as a solvent-conserving catalyst for organic reactions. *ChemSusChem* 2014;**7**:2094–8.
89. Akbari J, Heydari A, Kalhor HR, Kohan SA. Sulfonic acid functionalized ionic liquid in combinatorial approach, a recyclable and water tolerant-acidic catalyst for one-pot Friedlander quinoline synthesis. *J Comb Chem* 2010;**12**:137–40.
90. Mouradzadegun A, Elahi S, Abadast F. Synthesis of a 3D-network polymer supported Brønsted acid ionic liquid based on calix^4-resorcinarene via two post-functionalization steps: a highly efficient and recyclable acid catalyst for the preparation of symmetrical bisamides. *RSC Adv* 2014;**4**:31239–48.
91. Zhao M, Yuan K, Wang Y, Li G, Guo J, Hu W, Zhao H, Tang Z. Metal-organic frameworks as selectivity regulators for hydrogenation reactions. *Nature* 2016;**539**:76–80.
92. Beheshti S, Morsali A. Post-modified anionic nano-porous metal–organic framework as a novel catalyst for solvent-free Michael addition reactions. *RSC Adv* 2014;**4**:37036–40.
93. Rani P, Srivastava R. Nucleophilic addition of amines, alcohols, and thiophenol with epoxide/olefin using highly efficient zirconium metal organic framework heterogeneous catalyst. *RSC Adv* 2015;**5**:28270–80.
94. Babu R, Kathalikkattil AC, Roshan R, Tharun J, Kim D-W, Park D-W. Dual-porous metal organic framework for room temperature CO_2 fixation via cyclic carbonate synthesis. *Green Chem* 2016;**18**:232–42.
95. Chen J, Wang S, Huang J, Chen L, Ma L, Huang X. Conversion of cellulose and cellobiose into sorbitol catalyzed by ruthenium supported on a polyoxometalate/Metal–Organic Framework hybrid. *ChemSusChem* 2013;**6**:1545–55.
96. Metkar PS, Till EJ, Corbin DR, Pereira CJ, Hutchenson KW, Sengupta SK. Reactive distillation process for the production of furfural using solid acid catalysts. *Green Chem* 2015;**17**:1453–66.
97. Liu S, Okuyama Y, Tamura M, Nakagawa Y, Imai A, Tomishige K. Selective transformation of hemicellulose (xylan) into n-pentane, pentanols or xylitol over a rhenium-modified iridium catalyst combined with acids. *Green Chem* 2016;**18**:165–75.
98. Deng W, Zhang H, Wu X, Li R, Zhang Q, Wang Y. Oxidative conversion of lignin and lignin model compounds catalyzed by CeO_2-supported Pd nanoparticles. *Green Chem* 2015;**17**:5009–18.
99. Kumar CR, Anand N, Kloekhorst A, Cannilla C, Bonura G, Frusteri K, et al. Solvent free depolymerization of Kraft lignin to alkyl-phenolics using supported NiMo and CoMo catalysts. *Green Chem* 2015;**17**:4921–30.
100. Narani A, Chowdari RK, Cannilla C, Bonura G, Frusteri HHJ, et al. Efficient catalytic hydrotreatment of Kraft lignin to alkylphenolics using supported NiW and NiMo catalysts in supercritical methanol. *Green Chem* 2015;**17**:5046–57.
101. Kasakov S, Shi H, Camaioni DM, Zhao C, Baráth E, Jentys A, et al. Reductive deconstruction of organosolv lignin catalyzed by zeolite supported nickel nanoparticles. *Green Chem* 2015;**17**:5079–90.
102. Chen M-Y, Huang Y-B, Pang H, Liu X-X, Fu Y. Hydrodeoxygenation of lignin-derived phenols into alkanes over carbon nanotube supported Ru catalysts in biphasic systems. *Green Chem* 2015;**17**:1710–7.
103. Verboekend D, Liao Y, Schutyser W, Sels BF. Alkylphenols to phenol and olefins by zeolite catalysis: a pathway to valorize raw and fossilized lignocellulose. *Green Chem* 2016;**18**:297–306.
104. Ma B, Zhao C. High-grade diesel production by hydrodeoxygenation of palm oil over a hierarchically structured Ni/HBEA catalyst. *Green Chem* 2015;**17**:1692–701.
105. Sheldon RA. Fundamentals of green chemistry: efficiency in reaction design. *Chem Soc Rev* 2012;**41**:1437–51.
106. Kwak BS. Applications of heterogeneous catalytic processes to the environmentally friendly synthesis of fine chemicals. *Catal Surv Asia* 2005;**9**:103–16.
107. Thomas JM. Heterogeneous catalysis: Enigmas, illusions, challenges, realities, and emergent strategies of design. *J Chem Phys* 2008;**128**:1–19, 182502.
108. Fukuoka A, Dhepe PL. Sustainable green catalysis by supported metal nanoparticles. *Chem Rec* 2009;**9**:224–35.
109. Polshettiwar V, Varma RS. Green chemistry by nano-catalysis. *Green Chem* 2010;**12**:743–54.
110. Zhou CH. An overview on strategies towards clay-based designer catalysts for green and sustainable catalysis. *Appl Clay Sci* 2011;**53**:87–96.
111. Li H, Bhadury PS, Song B, Yang S. Immobilized functional ionic liquids: efficient, green, and reusable catalysts. *RSC Adv* 2012;**2**:12525–51.

112. Barbaro P, Liguori F, Linares N, Marrodan CM. Heterogeneous bifunctional metal/acid catalysts for selective chemical processes. *Eur J Inorg Chem* 2012:3807–23.
113. Calvete MJF, Silva M, Pereira MM, Burrows HD. Inorganic helping organic: recent advances in catalytic heterogeneous oxidations by immobilised tetrapyrrolic macrocycles in micro and mesoporous supports. *RSC Adv* 2013;**3**:22774–89.
114. Garg B, Bisht T, Ling Y-C. Graphene-based nanomaterials as heterogeneous acid catalysts: a comprehensive perspective. *Molecules* 2014;**19**:14582–614.
115. Wen C, Yin A, Dai W-L. Recent advances in silver-based heterogeneous catalysts for green chemistry processes. *Appl Catal B* 2014;**160–161**:730–41.
116. Li C, Wei M, Evans DG, Duan X. Recent advances for layered double hydroxides (LDHs) materials as catalysts applied in green aqueous media. *Catal Today* 2015;**247**:163–9.
117. Ten Brink G-J, Arends IWCE, Sheldon RA. The Baeyer-Villiger reaction: new developments toward greener procedures. *Chem Rev* 2004;**104**:4105–23.
118. Alonso F, Moglie Y, Radivoy G. Copper nanoparticles in click chemistry. *Acc Chem Res* 2015;**48**:2516–28.
119. Davis SE, Ide MS, Davis RJ. Selective oxidation of alcohols and aldehydes over supported metal nanoparticles. *Green Chem* 2013;**15**:17–45.
120. Dastan A, Kulkarni A, Török B. Environmentally benign synthesis of heterocyclic compounds by combined microwave-assisted heterogeneous catalytic approaches. *Green Chem* 2012;**14**:17–37.
121. Narkhede N, Singh S, Patel A. Recent progress on supported polyoxometalates for biodiesel synthesis via esterification and transesterification. *Green Chem* 2015;**17**:89–107.
122. Irfan M, Glasnov TN, Kappe CO. Heterogeneous catalytic hydrogenation reactions in continuous-flow reactors. *ChemSusChem* 2011;**4**:300–16.
123. Corberán VC, González-Pérez ME, Martínez-González S, Gómez-Avilés A. Green oxidation of fatty alcohols: challenges and opportunities. *Appl Catal A* 2014;**474**:211–23.
124. Wang L, Xiao F-S. Nanoporous catalysts for biomass conversion. *Green Chem* 2015;**17**:24–39.
125. Wang J, Xia J, Wang Y. Recent advances in the catalytic production of glucose from lignocellulosic biomass. *Green Chem* 2015;**17**:737–51.
126. Beerthuis R, Rothenberg G, Shiju NR. Catalytic routes towards acrylic acid, adipic acid and ε-caprolactam starting from biorenewables. *Green Chem* 2015;**17**:1341–61.

CHAPTER

3.13

Phase Transfer Catalysis: A Tool for Environmentally Benign Synthesis

Manisha Mishra, Béla Török

University of Massachusetts Boston, Boston, MA, United States

Despite being developed in the 1940s, phase transfer catalysis (PTC) has been continuing to grow significantly and remains a very important and useful tool in organic synthesis featuring operational efficiency, decreasing product costs, and reducing environmental impacts.[1–3] PTC has been widely employed in the synthesis of an extensive variety of organic compounds. PTC can be divided into two major groups: soluble PTC when the catalyst is soluble in one or both solvents and solid-phase PTC when the effective catalyst is bound to a solid matrix forming a third immiscible solid phase between the organic and aqueous phases. The principles of PTC have already been discussed in Chapter 2.2. This chapter focuses on practical applications of the technique in various bond formation reactions using different types of contemporary phase transfer catalysts.

3.13.1 ASYMMETRIC PHASE TRANSFER CATALYSIS

3.13.1.1 Catalysts

Over the past decades a broad range of reactions have been carried out using chiral phase transfer catalysts with great success.[4] In particular, catalysts based on chiral quaternary ammonium salts derived from cinchona alkaloids have found the most widespread applications.[5] Pioneering works by the groups of O'Donnell,[6] Lygo,[7] and Corey[8] are important examples of cinchona alkaloid-based phase transfer catalysts for the construction of chiral α-substituted and α,α-disubstituted amino acids.

Representative examples of novel cinchona-based phase transfer catalysts are summarized in Fig. 3.13.1.

The structurally rigid, chiral spiro ammonium salts derived from commercially available (*S*)- or (*R*)-1,1′-bi-2-naphthol (BINOL) have been designed by Maruoka et al. as new C_2-symmetric chiral phase transfer catalysts (Fig. 3.13.2) and successfully applied to the highly efficient,

Green Chemistry
http://dx.doi.org/10.1016/B978-0-12-809270-5.00018-2

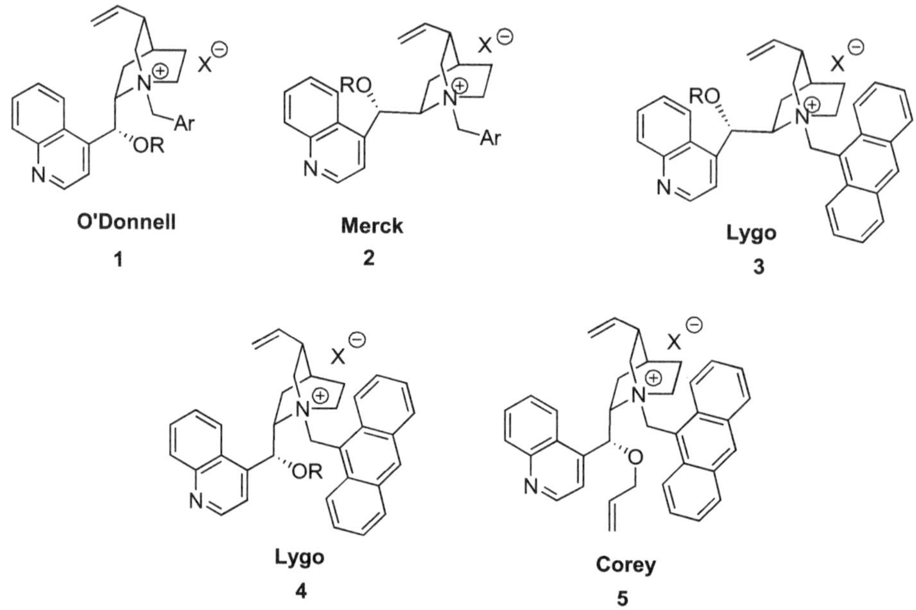

FIGURE 3.13.1 Representative examples of cinchona alkaloid-based phase transfer catalysts.

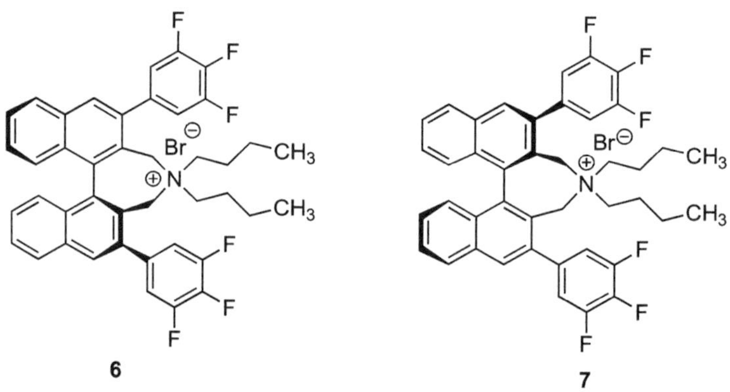

FIGURE 3.13.2 Commercially available examples of chiral phase transfer catalysts developed by Maruoka.

catalytic enantioselective synthesis of various α-amino acids under mild conditions.[9–12] Some of these catalysts (**6**, **7**), shown in Fig. 3.13.2, are called the "Maruoka catalysts" and are commercially available.

Other notable examples in this context are Lygo's biphenyl-based spirocyclic catalysts,[13] Shibasaki's tartaric acid-derived bidentate phase transfer catalysts,[14] Waser's TADDOL (α,α,α,α-tetraaryl-1,3-dioxolane-4,5-dimethanol)-derived N-spiro catalysts,[15] and Denmark's tricyclic ammonium salts[16] (Fig. 3.13.3).

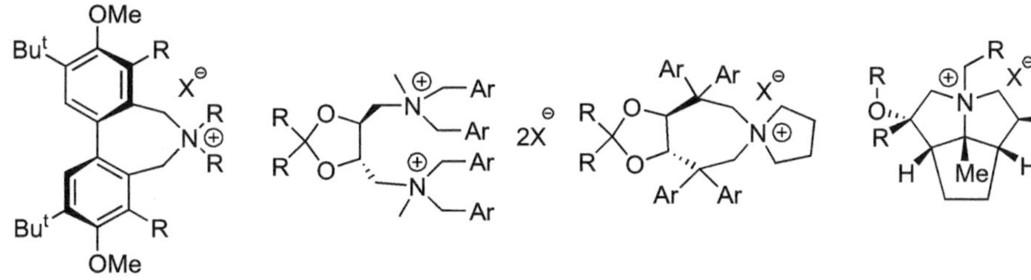

Lygo 8 | Shibasaki 9 | Waser 10 | Denmark 11

FIGURE 3.13.3 Examples of phase transfer catalysts based on alternative chiral backbones.

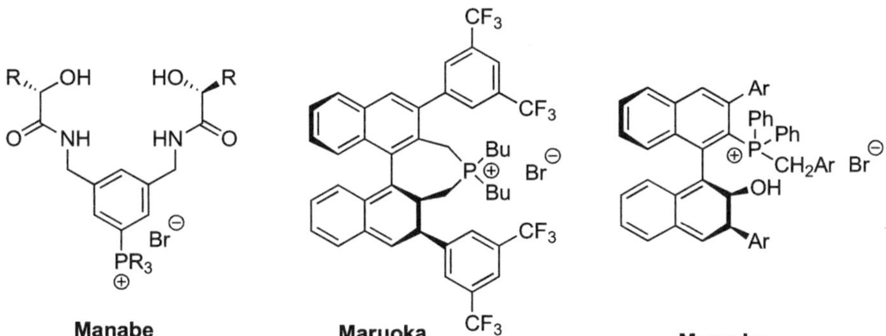

Manabe 12 | Maruoka 13 | Maruoka 14

FIGURE 3.13.4 Phase transfer catalysts based on chiral phosphonium salts.

Apart from the chiral ammonium salt catalysts discussed, the use of phosphonium ion-based catalysts (Fig. 3.13.4) developed by Manabe[17] and Maruoka[18,19] were also investigated in asymmetric transformations.

It was highlighted in numerous case studies that chiral phase transfer catalysts are one of the most important classes of catalysts in different asymmetric reactions where other activation modes provide less than desirable selectivities. Thus these catalysts represent a highly powerful asymmetric induction method that complements other strategies. Some of the selective reactions using the aforementioned catalysts in the synthesis of various biologically active compounds are described in the following sections.

3.13.1.2 Asymmetric Alkylation

Asymmetric phase transfer catalytic alkylation of glycine-based Schiff base **15** with benzyl bromide can be achieved by using cinchona alkaloid-derived phase transfer catalysts **4** and **5** (Scheme 3.13.1). While the reactions provided the products in moderate to good yield, the enantioselectivity was found to be excellent.[20]

Chiral spiro ammonium salts of types **6** and **7** derived by Maruoka from commercially available (*S*)- or (*R*)-1,1′-bi-2-naphthol were successfully applied for the enantioselective

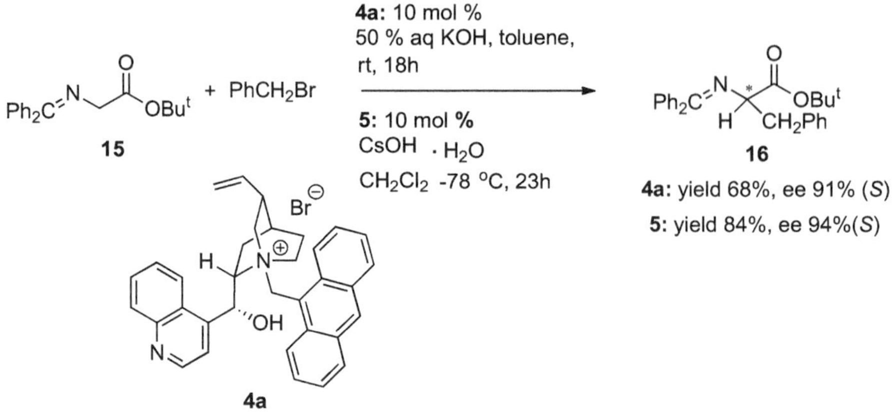

SCHEME 3.13.1

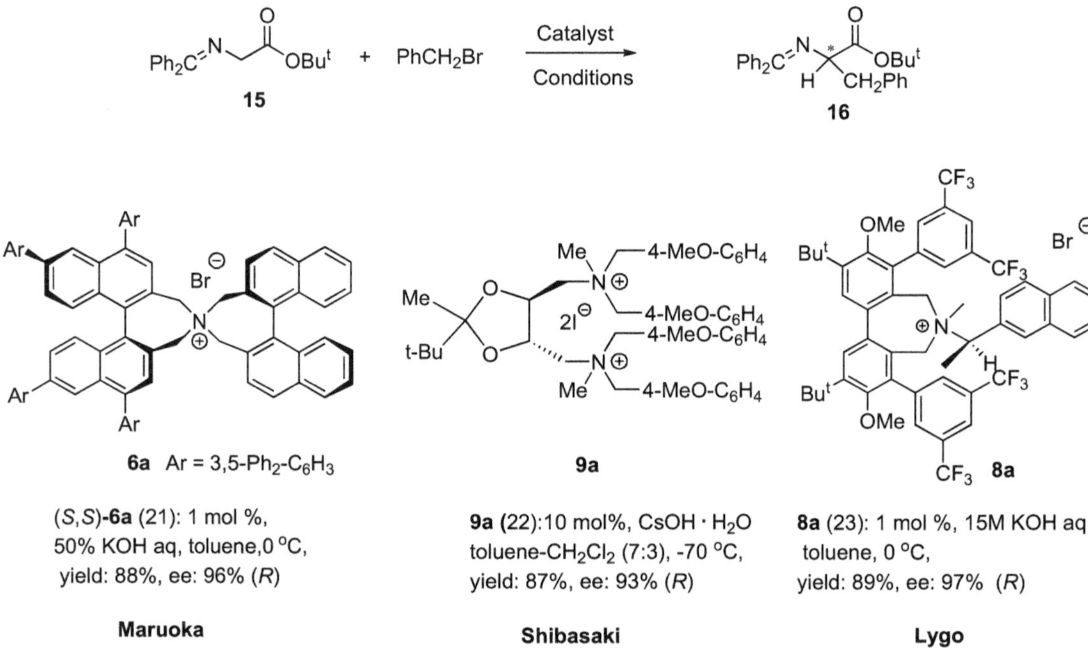

SCHEME 3.13.2

alkylation of **15**, which opened the way for the preparation of a variety of essentially enantiopure α-amino acids.[12] The performances of Maruoka, Lygo, and Shibasaki catalysts[21–23] are summarized in Scheme 3.13.2.

Maruoka developed a powerful chiral quaternary ammonium bromide (**6b**)-type catalyst, which partially possesses flexible unbranched alkyl groups instead of using two of the commonly applied rigid binaphthyl moiety (e.g., in **6a**) and functions as a highly active chiral

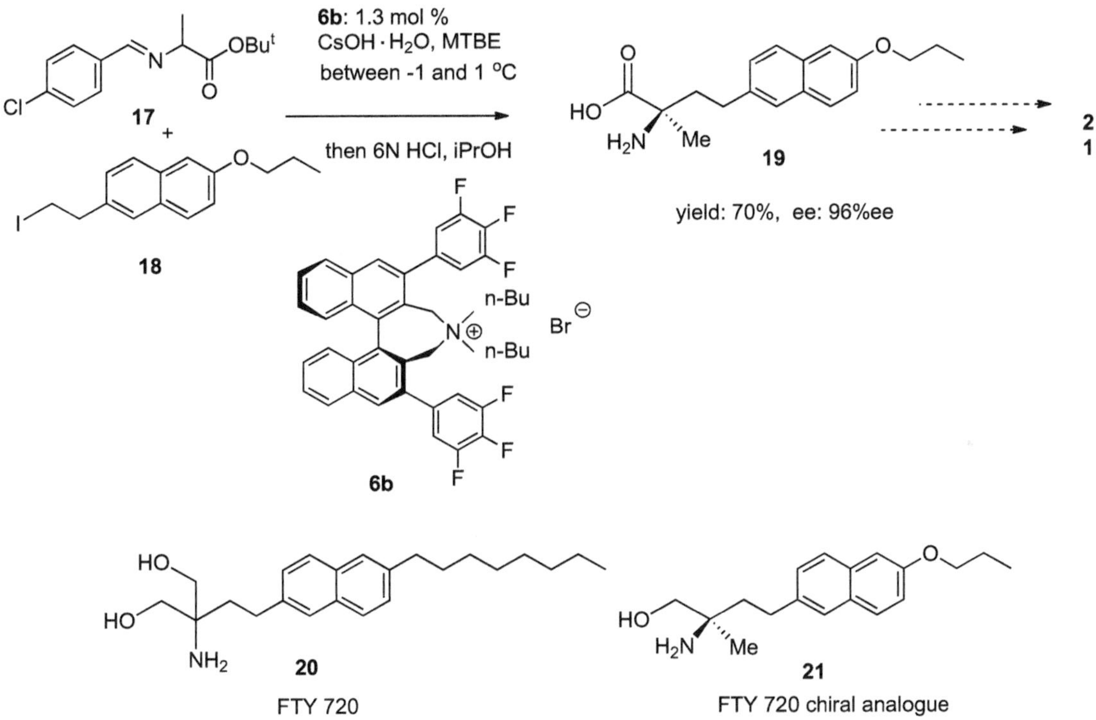

SCHEME 3.13.3

phase transfer catalyst.[11] The aforementioned alkylation reaction of **15** was successfully carried out under mild conditions in the presence of only 0.01–0.05 mol% of **6b** producing the corresponding alkylation product **16** with excellent enantioselectivity (98%ee, yield 92%). FTY 720 (**20**), a spinghosine-1-phosphate receptor agonist and a potential immunomodulatory compound, was originally synthesized by Yoshitomi Pharmaceuticals.[24] An alternative, improved route to synthesize this receptor agonist from its chiral analog **19** has been developed at a large scale. The enantioselective alkylation reaction is the key step to obtain **19** starting from glycine ester **17** catalyzed by **6b**[25] (Scheme 3.13.3).

3.13.1.3 Conjugate Addition

The application of asymmetric conjugate addition reactions is a widely employed strategy to construct functionalized carbon frameworks. The enantioselective Michael addition of glycine derivatives in the presence of chiral phase transfer catalysts is a key tool for obtaining α-alkyl α-amino acids.[22,26,27] A typical example is depicted in Scheme 3.13.4.

The asymmetric conjugate addition of glycine ester **24** to dione **25** was catalyzed by **26** and provided the conjugate adduct **27**. The adduct, upon subsequent reductive amination, produced octahydroindolizine core structure **28**, which served as a key intermediate to furnish (+)-monomorine, in 61% yield without loss of enantioselectivity (Scheme 3.13.5).[28]

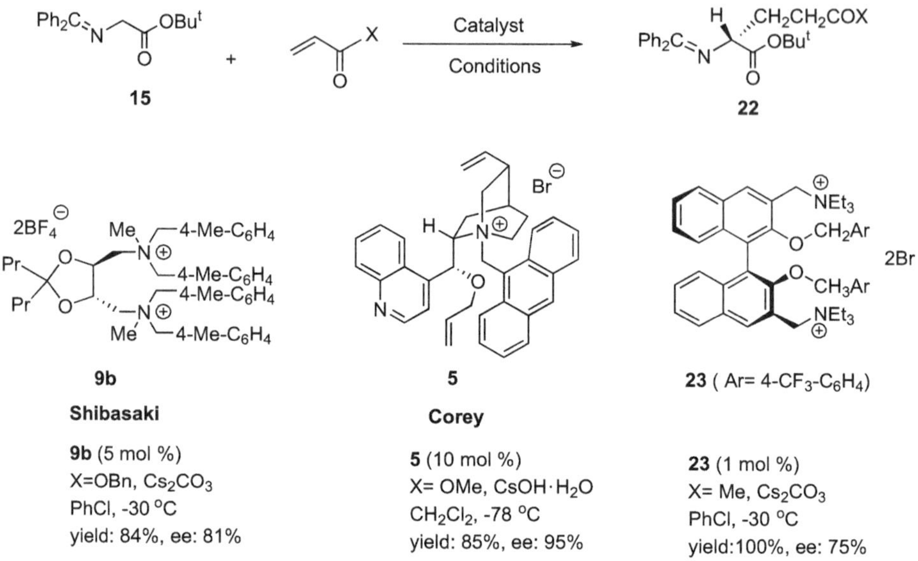

SCHEME 3.13.4

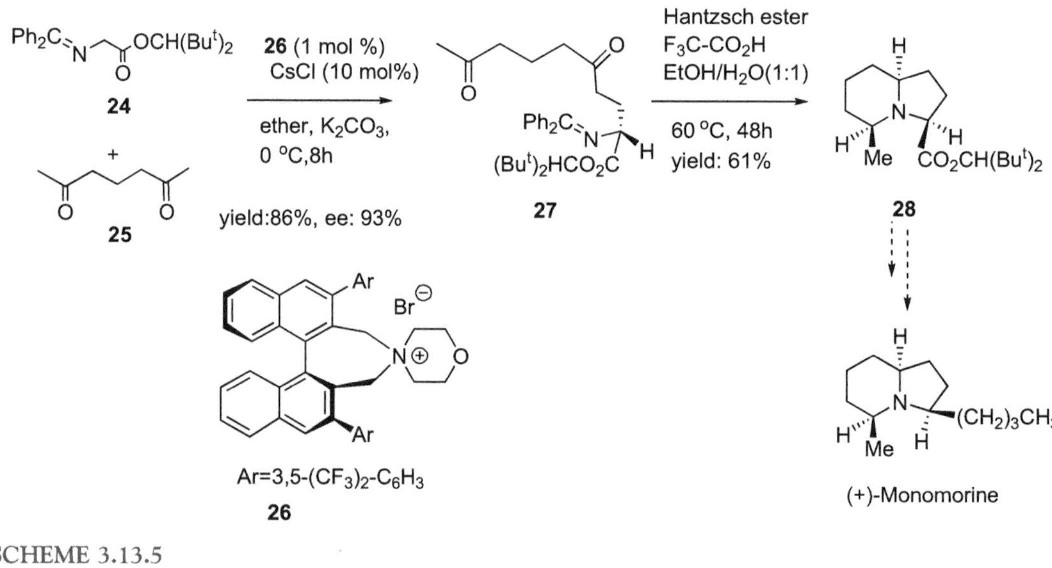

SCHEME 3.13.5

A binaphthyl-modified chiral phase transfer catalyst **26a** has been used in an asymmetric conjugate addition (Scheme 3.13.6) in water-organic biphasic medium under base-free conditions.[29] The reaction provided the product in 93% yield with 90% enantioselectivity.

The bicyclic amino acid **33** is the core structure of telaprevir, a hepatitis C virus protease inhibitor. The highly enantioselective synthesis of this compound has been accomplished by

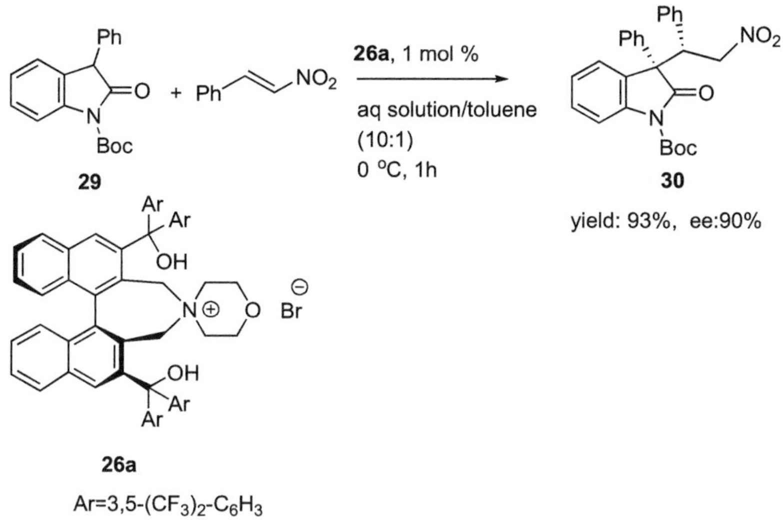

SCHEME 3.13.6

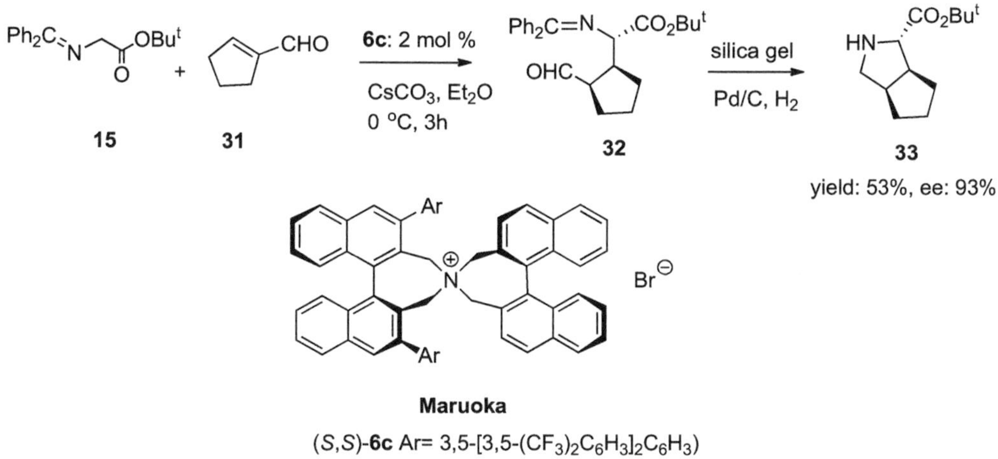

SCHEME 3.13.7

Maruoka and coworkers[30] by using a phase transfer catalytic stereoselective conjugate addition of **15** to cyclopent-1-enecarbaldehyde **31** (Scheme 3.13.7).

3.13.1.4 Cyclization Reactions

Chiral heterocyclic and carbocyclic compounds are of fundamental importance as they are key structural elements and building blocks in the synthesis of different compounds of biological and pharmaceutical importance. Representative methodologies to obtain such compounds with the use of asymmetric phase transfer catalysts are illustrated in the following discussion.

3.13.1.4.1 Epoxidation

The asymmetric epoxidation of electron-deficient olefins with high enantioselectivity was first reported by Wynberg in 1976[31] using hydrogen peroxide (30%) in the presence of different chiral cinchona alkaloid-derived quaternary ammonium salts. Over the years, this reaction has been the subject of great interest and a number of methodologies have been developed including the application of chiral phase transfer catalysts to facilitate the epoxidation reaction[32–34] (Scheme 3.13.8).

In 2013, Shibata and coworkers developed a highly enantioselective aerobic epoxidation of enones **36** catalyzed by cinchona alkaloid-based ammonium salt **38** in the presence of base, oxygen, and methyl hydrazine.[35] The reaction is believed to proceed via in situ generation of hydrogen peroxide from base, oxygen, and methyl hydrazine (Scheme 3.13.9).

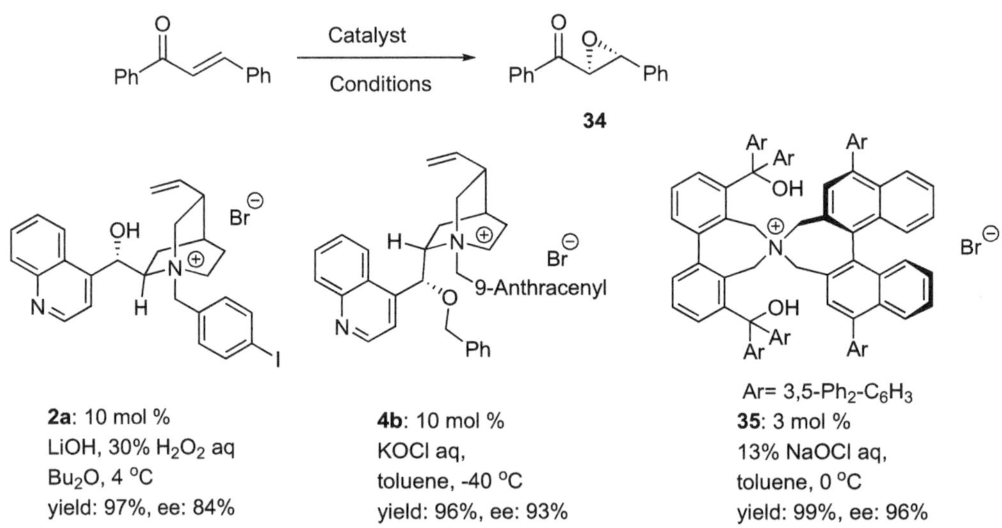

SCHEME 3.13.8

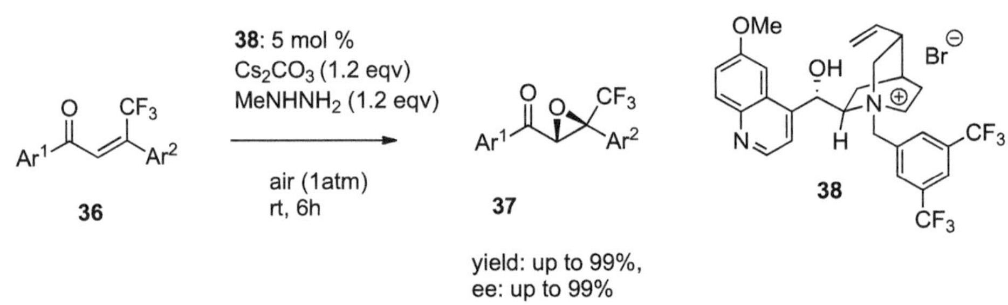

SCHEME 3.13.9

3.13.1.4.2 *Aziridination and Michael Addition*

Chiral aziridines are versatile building blocks for the synthesis of biologically active species such as amino acids, β-lactams, and alkaloids.[36] Murugan and Siva reported new chiral phase transfer catalysts, **1a** and **2b**, derived from cinchonine and cinchonidine, respectively, for the asymmetric aziridination to produce **39** (Scheme 3.13.10).[37]

Minakata and coworkers[38] developed an alternative method for the asymmetric aziridination where *N*-choro-*N*-sodium carbamate has been used as the aziridination agent. The yield is moderate with a good enantiomeric excess (Scheme 3.13.11).

The asymmetric Michael addition reaction by using a chiral phosphonium salt was reported by Maruoka et al. The Michael addition of oxindole **41** with acrylaldehyde in the presence of catalyst **14a** produced the adduct **42** (Scheme 3.13.12), which could be used to obtain **43**, an important core structure of natural alkaloids, after a three-step sequence of transformations.[39]

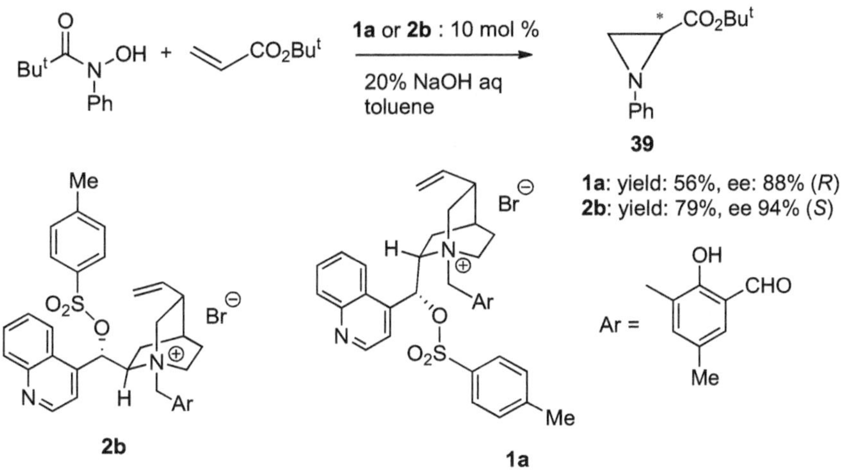

SCHEME 3.13.10

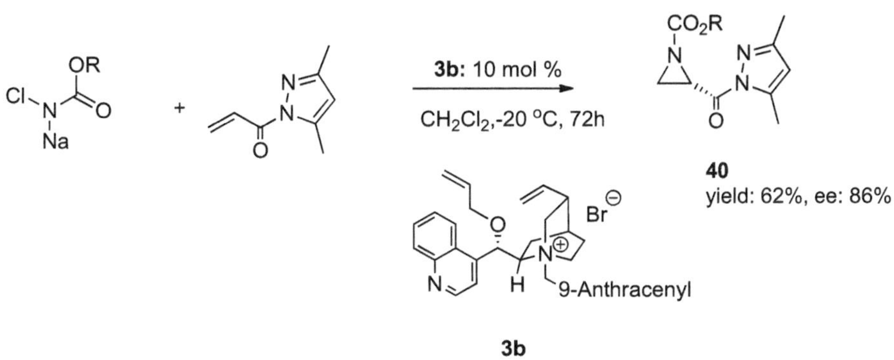

SCHEME 3.13.11

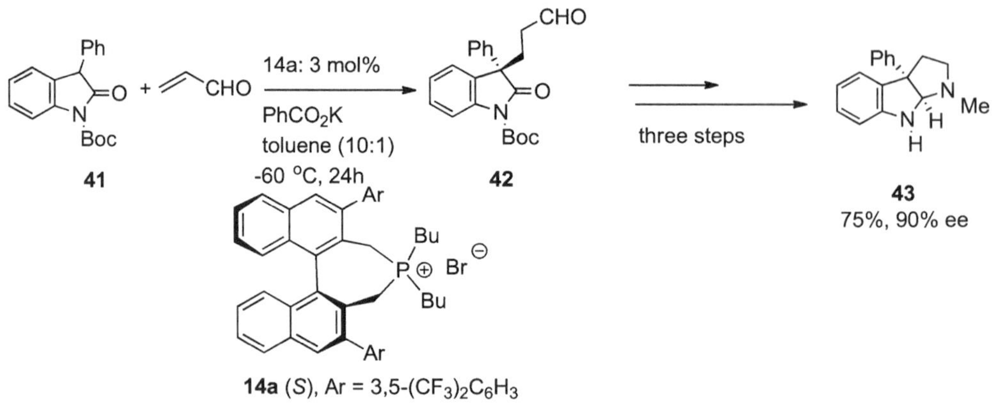

SCHEME 3.13.12

3.13.1.4.3 Synthesis of Pyrazolidine Derivatives

The phase transfer catalytic conjugate addition of N-Boc-hydrazine **44** to chalcone **45** produced 3,5-diaryl pyrazolines such as **46** with high enantioselectivity[40] (Scheme 3.13.13).

3.13.1.4.4 Synthesis of Triazolines

Jørgensen and coworkers[41] developed an asymmetric synthesis of 1,2,4-triazolines **49** starting from isocyano esters **47** and azodicarboxylate **48** with good yield and moderate enantioselectivity (Scheme 3.13.14).

3.13.1.4.5 Synthesis of Carbocycles

Asymmetric PTC is commonly applied for the synthesis of carbocyclic derivatives as these compounds are versatile building blocks in the preparation of many natural products and compounds of biological importance. Chiral cyclopropanes are particularly useful synthons to provide access to complex molecular scaffolds often used in medicinal chemistry

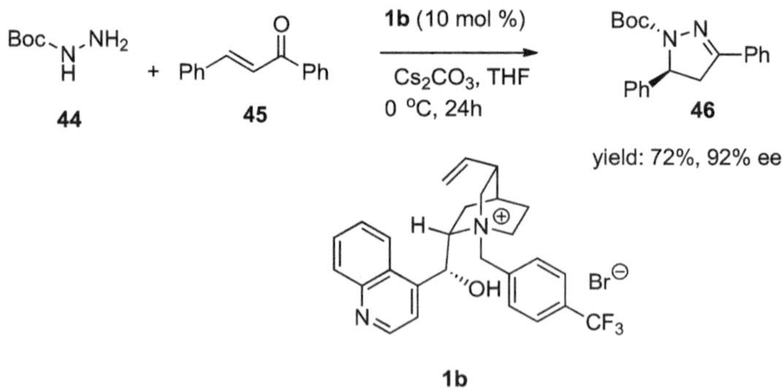

SCHEME 3.13.13

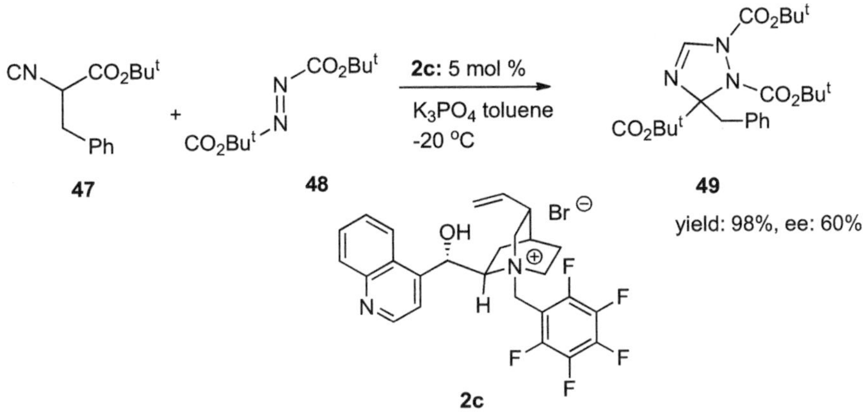

SCHEME 3.13.14

applications.[42] Some recent developments in the synthesis of chiral three-, five-, and six-membered rings are illustrated in Schemes 3.13.15 and 3.13.16.

Cyclopropanation reaction of chalcones **50** with bromomalonate **51** has been developed by Waser and coworkers.[43] It was pointed out that electron-deficient activated chalcones resulted in moderate to high yield (up to 98%) with high enantioselectivity (Scheme 3.13.15). Interestingly, the commercially available Maruoka catalyst **6** does not catalyze this reaction successfully.

A five-membered ring formation reaction was reported by Cobb and coworkers[44] involving intramolecular enantioselective cyclization of **54** catalyzed by **1c** to obtain the *trans*-configured five-membered ϒ-nitro-ester **55** with quantitative yield, but only moderate enantioselectivity (Scheme 3.13.16).

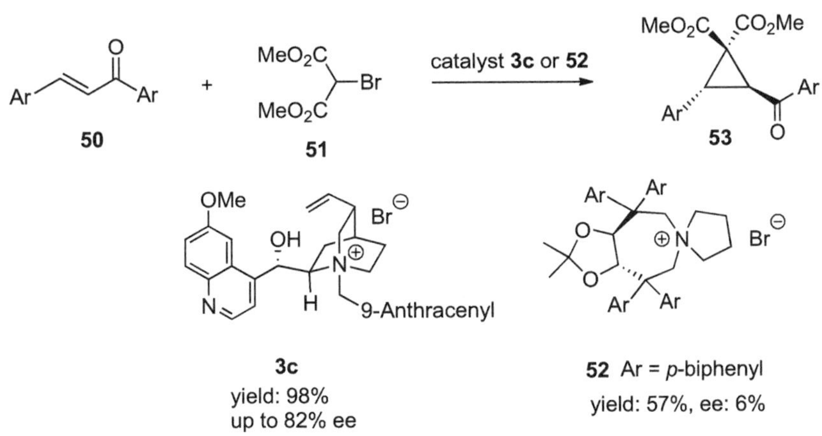

SCHEME 3.13.15

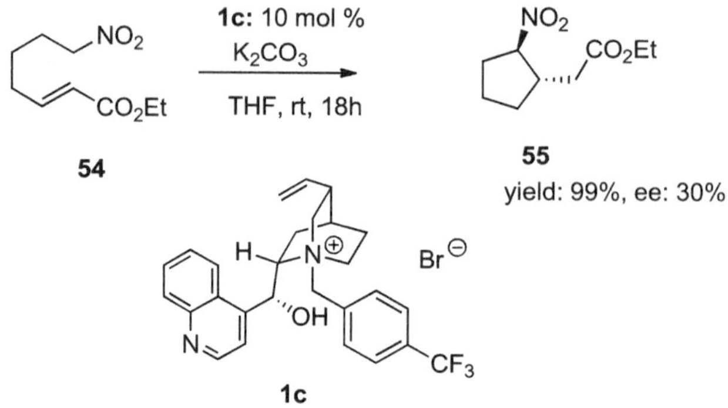

SCHEME 3.13.16

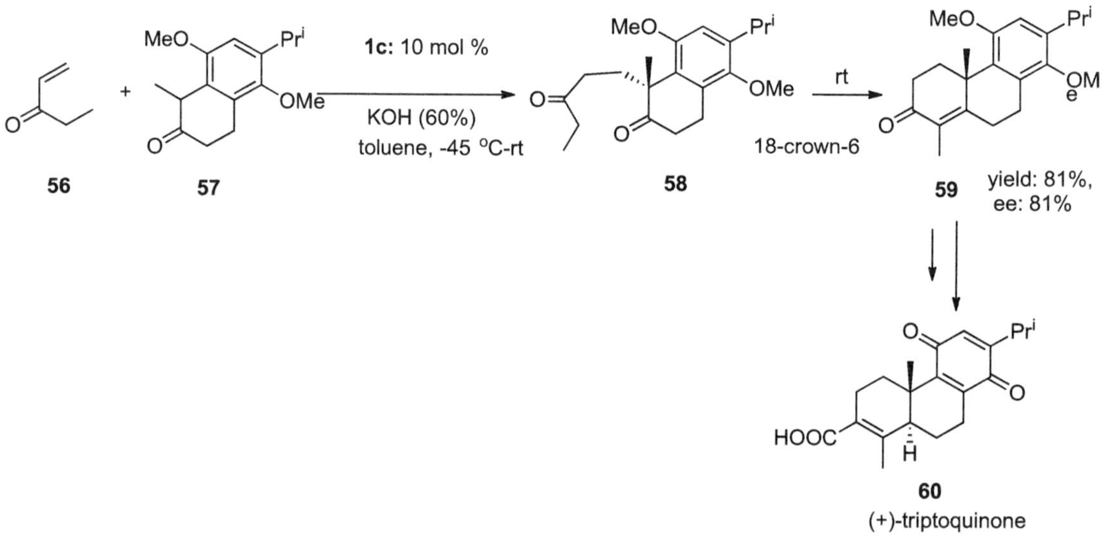

SCHEME 3.13.17

Shishido and coworkers[45] developed a method to synthesize a precursor to a diterpenoid quinone (+)-triptoquinone-A by using the cinchona-based catalyst via the Robinson annulation obtaining the product **59** in good yield and enantioselectivity (Scheme 3.13.17).

3.13.2 POLYMER-ANCHORED AND MULTISITE PHASE TRANSFER CATALYSTS

Similar to asymmetric PTC the use of polymer-supported phase transfer catalysts has attracted significant attention due to their effective utility in organic synthesis. From the industrial point of view, polymer-anchored catalysts are desirable for allowing the use of flow systems, for simplifying catalyst separation from the reaction mixture, for reducing the waste production and energy consumption, and for their potential recyclability. Regen and coworkers[46] were the first to report a phase transfer catalyst linked to a polymer backbone, and suggested the name triphase catalysis. Using nonrecyclable small molecule phase transfer catalysts is unacceptable under some circumstances in terms of cost and environmental impact as these compounds cause the eutrophication of water if they are discharged into the environment due to the presence of nitrogen and phosphorus in the catalysts. The triphase catalysis involves the use of a supported catalyst with two immiscible liquid phases. Polymer-supported ammonium salts, phosphonium salts, macrocyclic polyethers, and poly(-ethylene glycols) (PEGs) are effective and reusable solid-phase phase transfer catalysts. A majority of the work reported is related to quaternary salts supported on polymeric supports; polystyrene-based matrix cross-linked divinylbenzene was found to be the most applicable.

Polymer-bound quaternary ammonium salts are the most widely used heterogeneous phase transfer catalysts in the industry.[47,48] The polymer-supported phase transfer catalysts, prepared by Wang and coworkers[49] using free radical polymerization of styrene and 4-vinylbenzyl chloride were then reacted with various tertiary amines to form quaternary ammonium salts, which have been successfully applied in an etherification reaction (Scheme 3.13.18).

Poly-[*N*-(2-aminoethyl)-acrylamide]-trimethyl ammonium chloride **66** was prepared by the reaction of poly[*N*-(2-aminoethyl)-acrylamide] **65** with an excess of methyl iodide in dioxane.[50] This heterogeneous catalyst can be used in the conversion of alkyl halides to their corresponding thiocyanates, cyanides, azides, and alkyl aryl ethers (Scheme 3.13.19). The catalyst can be recovered and is reusable several times. The reaction has been carried out in aqueous medium, and pure products can be isolated without the need for any purification.

It has been reported that the activity of these catalysts depended on the structure of the active site, which limits their industrial application. In addition, a large amount of single-site phase transfer catalyst is required for a reaction to proceed in a reasonable time. As a result, polymer anchored multisite phase transfer catalysts have been developed, and the salient features of these catalysts include the low energy requirements and high activity under mild reaction conditions.

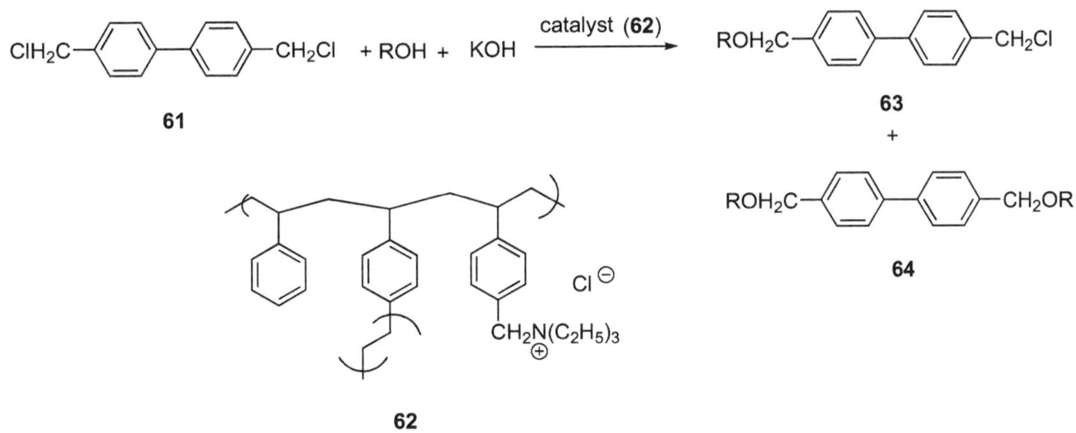

SCHEME 3.13.18

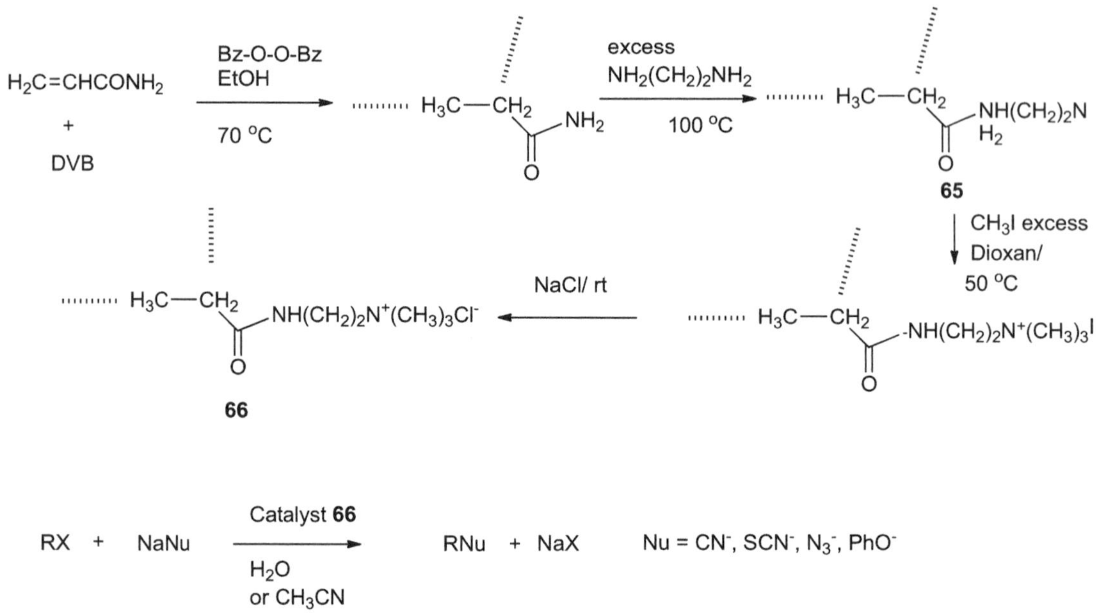

SCHEME 3.13.19

Polymer-based 2-benzyl-2-phenyl-1,3-bis(triethylmethyleneammonium chloride) (PABPBTAC) was designed by Vivekanand et al. as a catalyst, having multiple active sites using polystyrene-based cross-linked beads, for triphase reactions (Scheme 3.13.20). This catalyst (PABPBTAC) was successfully used in dichlorocyclopropanation reaction of indene[51] (Scheme 3.13.21).

1,3,5-Tris(benzyltriethylammonium bromide)-benzene is another important multisite catalyst for triphase reactions designed and synthesized by Vivekanand and coworkers (Scheme 3.13.22) and applied in cyclopentanation reaction of indene[52] (Scheme 3.13.23).

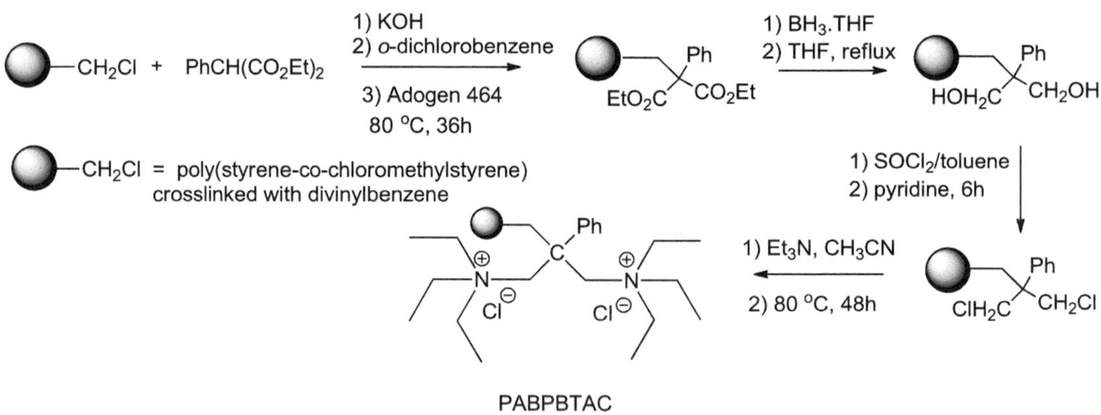

SCHEME 3.13.20

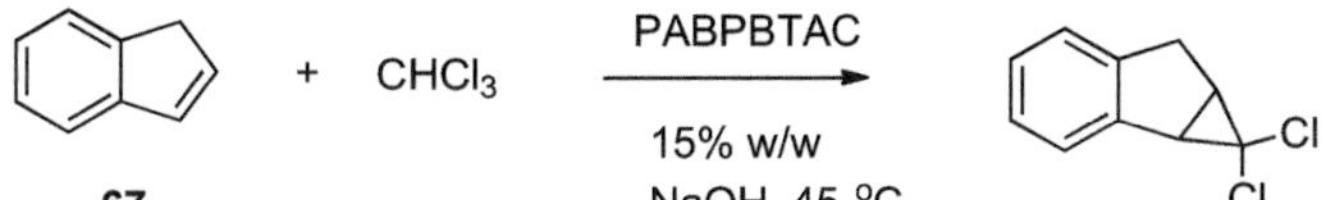

67 68

SCHEME 3.13.21

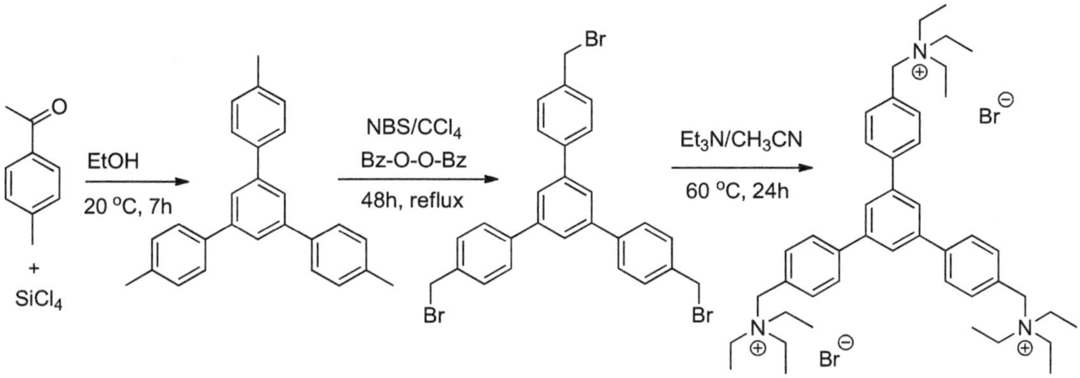

TBTABB

SCHEME 3.13.22

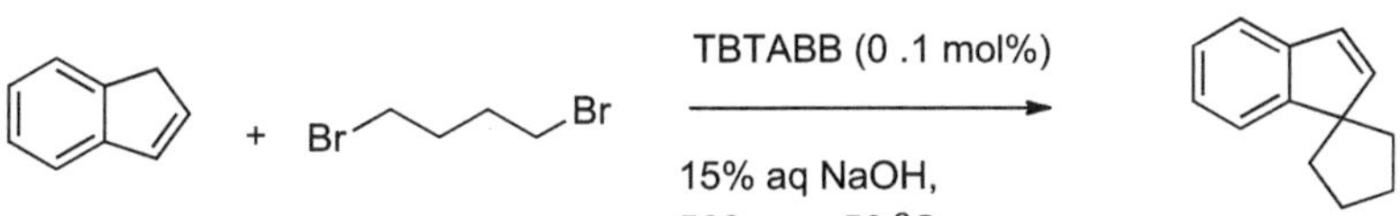

67 69

SCHEME 3.13.23

The advantages of the multisite phase transfer catalysts in the alkylation of indene to its spiro derivative are the increased reaction rate, hydrophilic conditions, and low energy requirement.

Alkylation of phenylacetonitrile (PAN) is one of the most widely studied reactions in the presence of phase transfer catalysts as it has great importance in the pharmaceutical industry. Bead-shaped multisite phase transfer catalysts, prepared by Murugan and coworkers,[53] possessing two, four, and six active sites (Scheme 3.13.24) were studied in the C-alkylation reaction of PAN (Scheme 3.13.25).

Various experiments show that the presence of a large number of active sites in polymer-supported multisite phase transfer catalysts increases the degree of hydrophilicity, which results in enhanced catalytic efficiency. The regioselective formation of β-azido alcohols from azidolysis of epoxides is an important reaction as these are useful synthetic intermediates for the preparation of β-amino alcohols, or in the chemistry of natural products,

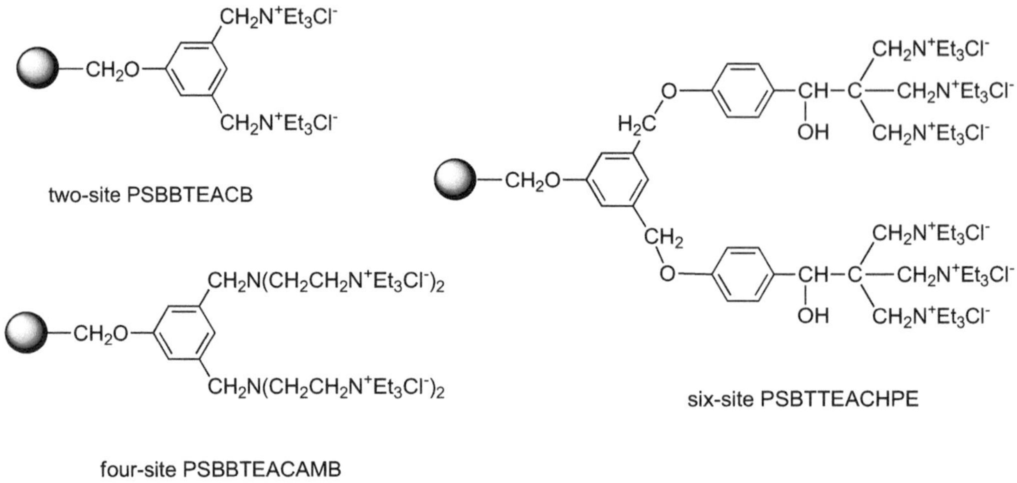

SCHEME 3.13.24

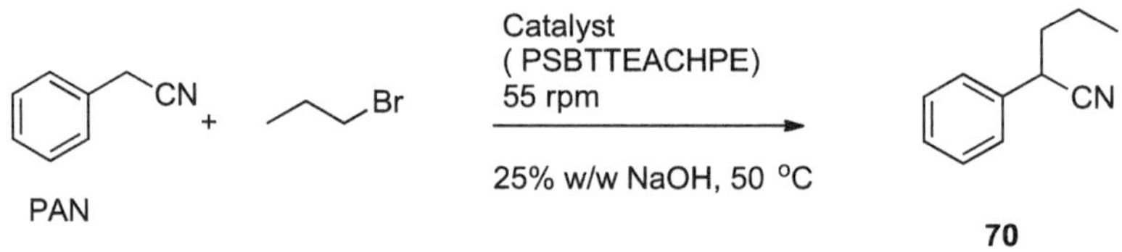

SCHEME 3.13.25

nucleosides, lactams, and oxazolines.[54] Insoluble polymer-supported PTC by simple grafting of PEG onto Dowex resin to form a PTC complex of Dowex-PEG has been designed by Kiasat et al.[55] and applied in various transformations such as azidolysis of epoxides (Scheme 3.13.26).

Porous ionic organic polymers have also been developed to serve as a new type of platform for highly efficient heterogeneous PTC. Such a hierarchical porous organic polymer has been prepared by Sun et al. from vinyl-functionalized quaternary phosphonium salt monomer followed by anion exchange with peroxotungstates. This catalyst showed excellent performance in the epoxidation of olefins in an environment-friendly approach when compared with a conventional method[56] (Scheme 3.13.27).

3.13.3 NANOPARTICLE-SUPPORTED PHASE TRANSFER CATALYSTS

New and novel nanoparticle-supported phase transfer catalysts are being designed to achieve enhanced catalyst recovery and separation from the reaction mixture. Magnetic

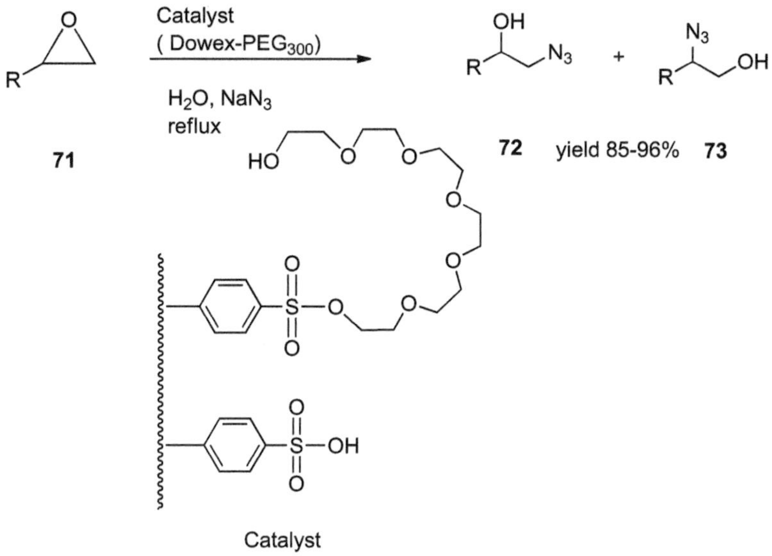

SCHEME 3.13.26

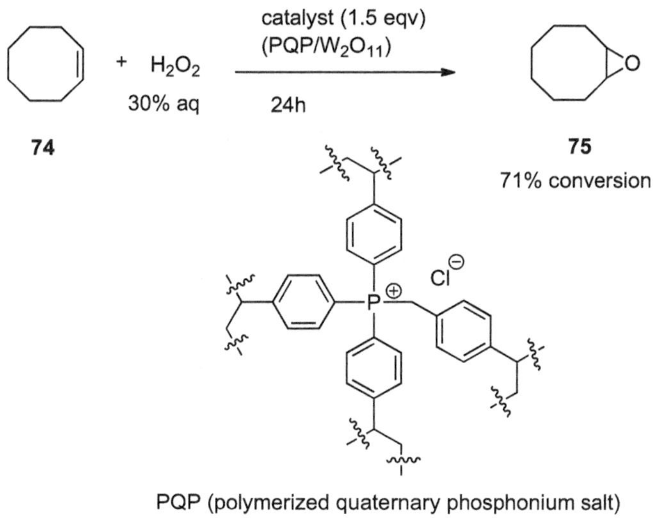

SCHEME 3.13.27

nanoparticles have now been extensively employed in water-mediated reactions as alternative catalyst supports given their high surface area, high catalyst loading capacity, high dispersion, outstanding stability, and convenient catalyst recycling, as these nanoparticles are readily separated from the product mixture by using an external magnet. Kawamura prepared the first example of magnetic nanoparticle-supported phase transfer catalysts through

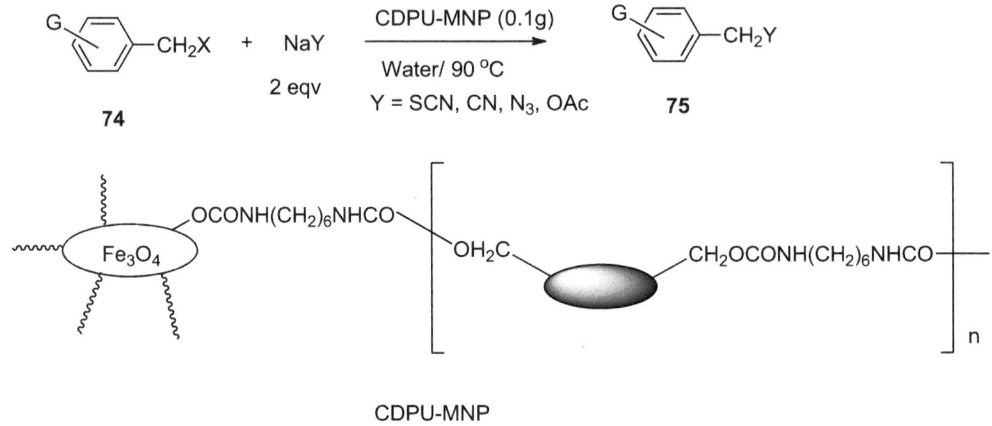

SCHEME 3.13.28

the immobilization of small molecule quaternary ammonium and phosphonium salts on magnetic nanoparticles in 2006.[57]

Kiasat and coworkers[58] designed magnetic nanoparticle-supported polymeric PTC through the immobilization of β-cyclodextrin-polyurethane polymer on magnetite, and used the catalyst in nucleophilic substitution reactions (Scheme 3.13.28). In all cases a very clean reaction was observed without the formation of any by-products, and after the reaction the catalyst could be recovered quantitatively by using an external magnet.

Zhang et al. reported magnetic nanoparticle-supported phase transfer catalysts, which were prepared by depositing and quaternizing poly(glycidyl-methacrylate-ethyleneglycol dimethacrylate) on Fe_3O_4 nanoparticle surfaces.[59] The activity of this catalyst has been studied in a nucleophilic substitution reaction between benzyl alcohol and benzyl bromide (Scheme 3.13.29).

A facile method for one-pot synthesis of β-azido alcohols **78** and β-nitro alcohols **79**[60] has been developed by Kiasat and coworkers applying nano-Fe_3O_4-copoly[(styrene/acrylic acid)/grafted ethylene oxide as a phase transfer catalyst in water (Schemes 3.13.30 and 3.13.31). The simple workup, short reaction time, aqueous medium, and ecofriendly and green nature of the reaction are the noteworthy advantages of this procedure.

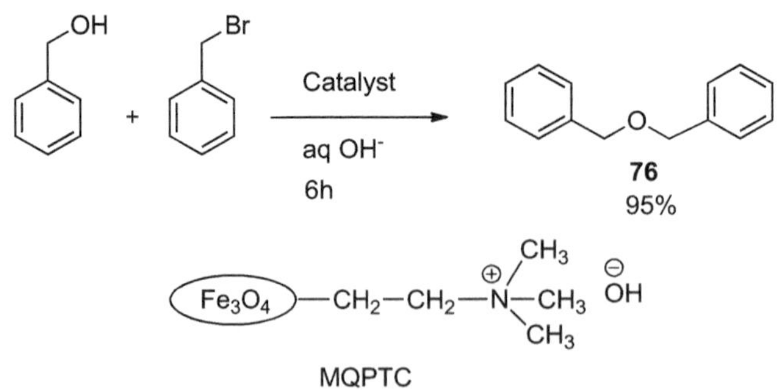

SCHEME 3.13.29

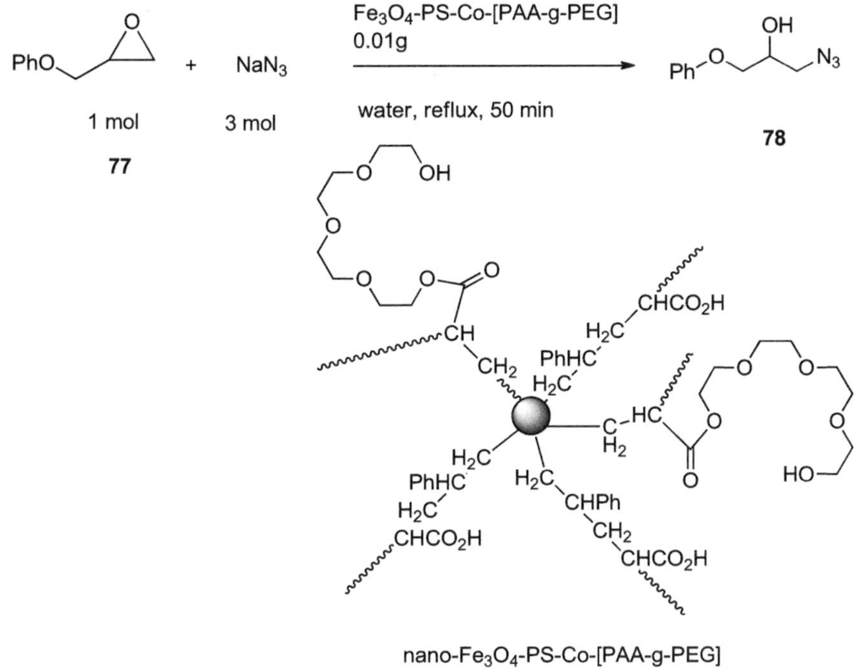

SCHEME 3.13.30

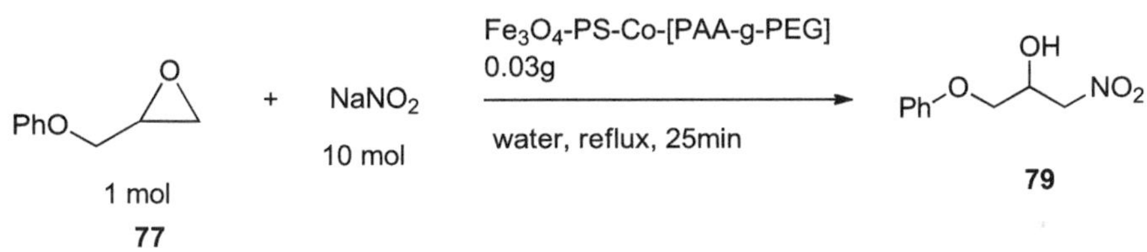

SCHEME 3.13.31

3.13.4 CONCLUSIONS AND OUTLOOK

PTC was a useful tool for sustainable synthesis from the very beginning of the movement, and by continuous changes and reinventing the field, it successfully adjusted to current requirements. As a result, PTC remained a strong contributor to green synthesis. In addition to carrying out asymmetric synthesis in a more and more efficient manner with ever-increasing yields and enantioselectivities, the development of the heterogeneous triphase PTC systems, especially the multisite solid phase transfer catalysts, ensures that PTC will continue to serve as an effective tool for future sustainable synthesis design.

References

1. a. Dehmlow EV, Dehmlow SS. *Phase transfer catalysis*. 3rd ed. Weinheim: VCH; 1993;
   b. Starks CM, Liotta CL, Halpern M. *Phase transfer catalysis*. New York: Chapman & Hall; 1994.
2. Hashimoto T, Maruoka K. Recent development and application of chiral phase-transfer catalysts. *Chem Rev* 2007;**107**:5656–82.
3. Ilia G, Iliescu S, Popa A. Polymer supported phase transfer catalysis in green organic synthesis. *Curr Green Chem* 2015;**2**:264–73.
4. Maruoka K. Practical aspect of recent asymmetric phase-transfer catalysis. *Org Proc Res Dev* 2008;**12**:679–97.
5. a. O'Donnell MJ. The enantioselective synthesis of α-amino acids by phase-transfer catalysis with achiral Schiff-base esters. *Acc Chem Res* 2004;**37**:506–17;
   b. Shirakawa S, Maruoka K. Recent developments in asymmetric phase transfer reactions. *Angew Chem Int Ed* 2013;**52**:4312–48;
   c. Novacek J, Waser M. Bifunctional chiral quaternary ammonium salt catalysts: a rapidly emerging class of powerful asymmetric catalysts. *Eur J Org Chem* 2013:637–48.
6. a. O'Donnell MJ, Bennett WD, Wu S. The stereoselective synthesis of α-amino acids by phase-transfer catalysis. *J Am Chem Soc* 1989;**111**:2353–5;
   b. O'Donell MJ, Wu S, Huffman JC. A new active catalyst species for enantioselective alkylation by phase transfer catalysis. *Tetrahedron* 1994;**50**:4507–18;
   c. Lipkowitz KB, Cavanaugh MW, Becker B, O'Donnell MJ. Theoretical studies in molecular recognition: asymmetric induction of benzophenoneimine ester enolate by the bezylcinchoninium ion. *J Org Chem* 1991;**56**:5181–92.
7. a. Lygo B, Wainwright PG. A new class of asymmetric phase transfer catalysts derived from cinchona alkaloids-application in the enantioselective synthesis of α-amino acids. *Tetrahedron Lett* 1997;**38**:8595–8;
   b. Lygo B, Crosby J, Peterson JA. Enantioselective synthesis of bis-α-amino acid esters via asymmetric phase-transfer catalysis. *Tetrahedron Lett* 1999;**40**:1385–8.
8. a. Corey EJ, Xu F, Noe MC. A Rational approach to catalytic enantioselective enolate alkylation using a structurally rigidified and defined chiral quaternary ammonium salt under phase transfer conditions. *J Am Chem Soc* 1997;**119**:12414–5;
   b. Corey EJ, Noe MC, Xu F. Highly enantioselective synthesis of cyclic and functionalized α-amino acids by means of a chiral phase transfer catalyst. *Tetrahedron Lett* 1998;**39**:5347–50;
   c. Corey EJ, Bo Y, Busch-Peterson J. Highly enantioselective phase transfer catalyzed alkylation of a 3-oxygenated propionic ester equivalent; applications and mechanism. *J Am Chem Soc* 1998;**120**:13000–1.
9. Ooi T, Kameda M, Maruoka KJ. Molecular design of a C_2-symmetric chiral phase-transfer catalyst for practical asymmetric synthesis of α-amino acids. *J Am Chem Soc* 1999;**121**:6519–20.
10. Hashimoto T, Maruoka K. Recent development and application of chiral phase-transfer catalysts. *Chem Rev* 2007;**107**:1556–82.
11. Kitamura M, Shirakawa S, Maruoka K. Powerful chiral phase transfer catalysts for the asymmetric synthesis of α-alkyl and α, α-dialkyl-α-amino acids. *Angew Chem Int Ed* 2005;**44**:1549–51.
12. Ooi T, Arimura Y, Hiraiwa Y, Yuan LM, Kano T, Inoue T, Matsumoto J, Maruoka K. Highly enantioselective monoalkylation of *p*-cholorobenzaldehyde imine of glycine tert-butyl ester under mild phase transfer conditions. *Tetrahedron Asymmetry* 2006;**17**:603–6.
13. Lygo B, Allbutta B, Kirton EHM. Asymmetric michael addition of glycine imines via quaternary ammonium ion catalysis. *Tetrahedron Lett* 2005;**46**:4461–4.
14. Shibuguchi T, Mihara H, Kuramochi A, Ohshima T, Shibasaki M. Catalytic asymmetric phase-transfer Michael reaction and Mannich-type reaction of glycine Schiff bases with tartrate-derived diammonium salts. *Chem Asian J* 2007;**2**:794–801.
15. a. Waser M, Gratzer K, Herchl R, Mueller N. Design, synthesis, and application of tartaric acid derived N-spiro quaternary ammonium salts as chiral phase-transfer catalysts. *Org Biomol Chem* 2012;**10**:251–4;
    b. Gururaja GN, Herchl R, Pichler A, Gratzer K, Waser M. Application scope and limitations of TADDOL-derived chiral ammonium salt phase-transfer catalysts. *Molecules* 2013;**18**:4357–72.
16. Denmark SE, Gould ND, Wolf LM. A systematic investigation of quaternary ammonium ions as asymmetric phase-transfer catalysts. Synthesis catalyst libraries evaluation catalyst activity. *J Org Chem* 2011;**76**:4260–336.

17. Manabe K. Synthesis of novel chiral quaternary phosphonium salts with multiple hydrogen-bonding sites and their application to asymmetric phase-transfer alkylation. *Tetrahedron* 1998;**54**:14465–76.
18. Wang L, Shirakawa S, Maruoka K. Asymmetric neutral amination of nitroolefins catalyzed by chiral bifunctional ammonium salts in water-rich biphasic solvent. *Angew Chem Int Ed* 2011;**50**:5327–30.
19. Shirakawa S, Kasai A, Tokuda T, Maruoka K. Efficient Approach for the design of effective chiral quaternary phosphonium salts in asymmetric conjugate additions. *Chem Sci* 2013;**4**:2248–52.
20. Lygo B, Crosby J, Lowdon TR, Peterson JA, Wainwright PG. Studies on the enantioselective synthesis of α-amino acids via asymmetric phase-transfer catalysis. *Tetrahedron* 2001;**57**:2403–9.
21. Hashimoto T, Maruoka K. Substituent effect of binaphthyl-modified spiro-type chiral phase-transfer catalysts. *Tetrahedron Lett* 2003;**44**:3313–6.
22. Oshima T, Shibuguchi T, Fukuta Y, Shibasaki M. Catalytic asymmetric phase-transfer reactions using tartrate-derived asymmetric two-center organocatalysts. *Tetrahedron* 2004;**60**:7743–54.
23. Lygo B, Allbut B, James SR. Identification of a highly effective asymmetric phase-transfer catalyst derived from α-methylnaphthylamine. *Tetrahedron Lett* 2003;**44**:5629–32.
24. Albert R, Hinterding K, Brinkmann V, Guerini D, Muller-Hartwieg C, Knecht H, Simeon C, Streiff M, Wagner T, Welzenbach K, Zecri F, Zollinger M, Cooke N, Francotte E. Novel immunomodulator FTY720 Is phosphorylated in rats and humans to form a single stereoisomer. Identification, chemical proof, and biological characterization of the biologically active species and its enantiomer. *J Med Chem* 2005;**48**:5373–7.
25. Jiang X, Gong B, Prasad K, Repic O. A practical synthesis of a chiral analogue of FTY720. *Org Process Res Dev* 2008;**12**:1164–9.
26. Arai S, Tokumaru K, Aoyama T. Asymmetric Michael reaction promoted by new chiral phase-transfer catalysts. *Chem Pharm Bull* 2004;**52**:646–8.
27. Zhang F-Y, Corey EJ. Highly enantioselective Michael reactions catalyzed by a chiral quaternary ammonium salt. Illustration by asymmetric syntheses of (S)-ornithine and chiral 2-cyclohexenones. *Org Lett* 2000;**2**:1097–100.
28. Maruoka K. Designer chiral phase-transfer catalysts for green sustainable chemistry. *Pure Appl Chem* 2012;**84**:1575–85.
29. Shirakawa S, Maruoka K. Asymmetric phase-transfer reactions under base-free neutral conditions. *Tetrahedron Lett* 2014;**55**:3833–9.
30. Shirakawa S, Liu Y, Usui A, Maruoka K. Efficient asymmetric synthesis of a bicyclic amino acid as a core structure of telaprevir. *ChemCatChem* 2012;**4**:980–2.
31. Helder R, Hummelen JC, Laane RWPM, Wiering JS, Wynberg H. Catalytic asymmetric induction in oxidation reactions. Synthesis Optically Active Epoxides. *Tetrahedron Lett* 1976;**21**:1831–4.
32. Arai S, Tsuge H, Oku M, Miura M, Shioiri T. Catalytic asymmetric epoxidation of enones under phase-transfer catalyzed conditions. *Tetrahedron* 2002;**58**:1623–30.
33. Corey EJ, Zhang FY. Mechanism and conditions for highly enantioselective epoxidation of α,β-enones using charge-accelerated catalysis by a rigid quaternary ammonium salt. *Org Lett* 1999;**1**:1287–90.
34. Ooi T, Ohara D, Tamura M, Maruoka K. Design of new chiral phase-transfer catalysts with dual functions for highly enantioselective epoxidation of α,β-unsaturated ketones. *J Am Chem Soc* 2004;**126**:6844–5.
35. Kawai H, Okusu S, Yuan Z, Tokunaga E, Yamano A, Shiro M, Shibata N. Enantioselective Synthesis of epoxides having a tetrasubstituted trifluoromethylated carbon center: methylhydrazine-induced aerobic epoxidation of β,β-disubstituted enones. *Angew Chem Int Ed* 2013;**52**:2221–5.
36. McCoull W, Davis FA. Recent synthetic applications of chiral aziridines. *Synthesis* 2000;**10**:1347–65.
37. Murugan E, Siva A. Synthesis of asymmetric N-arylaziridine derivatives using a new chiral phase-transfer catalyst. *Synthesis* 2005;**12**:2022–8.
38. a. Minakata S, Murakami Y, Tsuruoka R, Kitanaka S, Komatsu M. Catalytic aziridination of electron-deficient olefins with an N-chloro-N-sodiocarbamate and application of this novel method to asymmetric synthesis. *Chem Commun* 2008;**47**:6363–5;
    b. Murakami Y, Takeda Y, Minakata S. Diastereoselective aziridination of chiral electron-deficient olefins with N-chloro-N-sodiocarbamates catalyzed by chiral quaternary ammonium salts. *J Org Chem* 2011;**76**:6277–85.
39. He R, Ding C, Maruoka K. Phosphonium salts as chiral phase-transfer catalysts: asymmetric Michael and Mannich reactions of 3-aryloxindoles. *Angew Chem Int Ed* 2009;**48**:4559–61.
40. Mahé O, Dez I, Levacher V, Briére J-F. Enantioselective synthesis of bio-relevant 3,5-diarylpyrazolines. *Org Biomol Chem* 2012;**10**:3946–54.

41. Monge D, Jensen KL, Marín I. Jørgensen KA Synthesis of 1,2,4-Triazolines: base-catalyzed hydrazination/cyclization cascade of α-isocyano esters and amides. *Org Lett* 2011;**13**:328–31.
42. a. Zhang D, Song H, Qin Y. Total synthesis of indoline alkaloids: a cyclopropanation strategy. *Acc Chem Res* 2011;**44**:447–57;
    b. Brandi A, Cicchi S, Cordero FM, Gotti A. Heterocycles from alkylidenecyclopropanes. *Chem Rev* 2003;**103**: 1213–70.
43. Herchl R, Waser M. Asymmetric cyclopropanation of chalcones using chiral phase-transfer catalysts. *Tetrahedron Lett* 2013;**54**:2472–5.
44. Nodes WJ, Shankland K, Rajkumar S, Cobb AJA. Asymmetric phase-transfer-catalyzed synthesis of five-membered cyclic γ-amino acid precursors. *Synlett* 2010:3011–4.
45. Shishido K, Goto K, Miyoshi S, Takaishi Y, Shibuya M. Synthetic studies on diterpenoid quinones with interleukin-1 inhibitory activity. Total synthesis of (±)- and (+)-triptoquinone A. *J Org Chem* 1994;**59**:406–14.
46. Yang HM, Wu HS. Interfacial mechanism and kinetics of phase-transfer catalysis. *Catal Rev* 2003;**45**:463–540.
47. Senthamizh Selvi R, Nanthini R, Sukanya G. The basic principle of phase-transfer catalysis, some mechanistic aspects and important applications. *Int J Sci Tech Res* 2012;**1**:61–3.
48. Jones RA. *Quaternary ammonium salts: their use in phase-transfer catalysis (best synthetic methods)*. Amsterdam: Elsevier; 2001.
49. Wang ML, Lee ZF, Wang F. Phase-transfer catalyzed etherification of 4-4′-bis(chloromethyl)-1,1′-biphenyl with 1- butanol by polymer supported catalysis. *Ind Eng Chem Res* 2005;**44**:5417–26.
50. Tamami B, Ghasemi S. Nucleophilic substitution reactions using polyacrylamide-based phase transfer catalyst in organic and aqueous media. *J Iran Chem Soc* 2008;**5**:S26–32.
51. Vivekanand PA, Balakrishnan T. Catalytic potential of a new polymer-anchored multisite phase transfer catalyst in the dichlorocarbene addition to indene. *Catal Lett* 2009;**131**:587–96.
52. Vivekanand PA, Balakrishnan T. Synthesis and characterization of a novel multi-site phase-transfer catalyst and a kinetic study of the intramolecular cyclopentanation of indene. *Appl Catal A Gen* 2009;**364**:27–34.
53. Murugan E, Gopinath P. Heterogeneous catalysis by new bead-shaped polymer-supported multisite phase transfer catalysts. *J Polym Sci A Polym Chem* 2009;**47**:771–85.
54. Chen S, Thakur SS, Li W, Shin CK, Kawthekar GJ, Kim JW. Efficient catalytic synthesis of optically pure 1,2-azido alcohols through enantioselective epoxide ring opening with HN_3. *J Mol Catal* 2006;**259**:116–20.
55. Kiasat AR, Badri R, Zargar B, Sayyahi S. Poly(ethylene glycol) grafted onto Dowex resin: an efficient, recyclable, and mild polymer-supported phase transfer catalyst for the regioselective azidolysis of epoxides in water. *J Org Chem* 2008;**73**:8382–5.
56. Sun Q, Ma S, Dai Z, Meng X, Xiao F-S. A hierarchical porous ionic organic polymer as a new platform for heterogeneous phase-transfer catalysis. *J Mater Chem A* 2015;**3**:23871–5.
57. Kawamura M, Sato K. Magnetically separable phase-transfer catalysts. *Chem Commun* 2006:4718–9.
58. Kiasat AR, Nazari S. β-Cyclodextrin conjugated magnetic nanoparticles as novel magnetic microvessel and phase-transfer catalyst: synthesis and applications in nucleophilic substitution reaction of benzyl halides. *J Incl Phenom Macrocycl Chem* 2013;**76**:363–8.
59. Jia X, Fan X, Liu Y, Li W, Tian L, Fan L, Zhang B, Zhang H, Zhang Q. Quaternary ammonium functionalized Fe_3O_4@P(GMA-EGDMA) composite particles as highly efficient and dispersible catalysts for phase transfer reactions. *RSC Adv* 2015;**5**:60691–7.
60. Kiasat AR, Daei M, Saghanezhad SJ. Synthesis and characterization of a novel nano-Fe_3O_4-copoly[(styrene/acrylic acid)/grafted ethylene oxide and its application as a magnetic and recyclable phase-transfer catalyst in the preparation of β-azido alcohols and β-nitro alcohols. *Res Chem Intermed* 2016;**42**:581–94.

CHAPTER

3.14

Biocatalysis: Nature's Chemical Toolbox

Julie A. Himmelberger[1], *Kathryn E. Cole*[2], *Daniel P. Dowling*[3]

[1]**DeSales University, Center Valley, PA, United States;** [2]**Christopher Newport University, Newport News, VA, United States;** [3]**University of Massachusetts Boston, Boston, MA, United States**

3.14.1 INTRODUCTION

Nature is filled with examples of challenging chemical transformations performed by biocatalysts. Simply put, biocatalysis is chemical catalysis using biological components, such as proteins or cells. These systems involve highly specific, catalyzed reactions that usually occur in aqueous solvent at near-neutral pH values, which make them ideal systems to produce chemicals and pharmaceuticals with minimal impact on the environment. Indeed, the isolation of secondary metabolites, referred to as natural products, from environmental samples has served as a source of inspiration for pharmaceutical developments[1–5] and given us many important antibiotic and anticancer agents, including penicillin, teixobactin, daptomycin, and epothilones (Fig. 3.14.1).[6–8] These natural product examples highlight how natural systems are capable of generating compounds with complex chemical structures, the total syntheses of which are quite advanced and less environment friendly.

The most common catalysts in biology are enzymes and nucleic acids, although their component amino acids, cofactors, and other small molecules can also serve as organic catalysts in vitro.[9–12] Both enzymes and nucleic acids provide an adaptable framework that can be evolved through directed evolution or structurally and biochemically guided bioengineering to favor different chemical reactions.[13] Enzymes are polymers of amino acids, arranged into functional, three-dimensional (3D) structures that facilitate specific chemical transformations. These amino acids consist of chemical functionalities, termed side chains, that are commonly grouped into basic, acidic, polar, aromatic, hydrophobic, and thiol moieties. The diversity of these protein building blocks enables a broad range of chemical reactions, which include carbonyl, oxidation-reduction, isomerization, polymerization, and

Green Chemistry
http://dx.doi.org/10.1016/B978-0-12-809270-5.00019-4

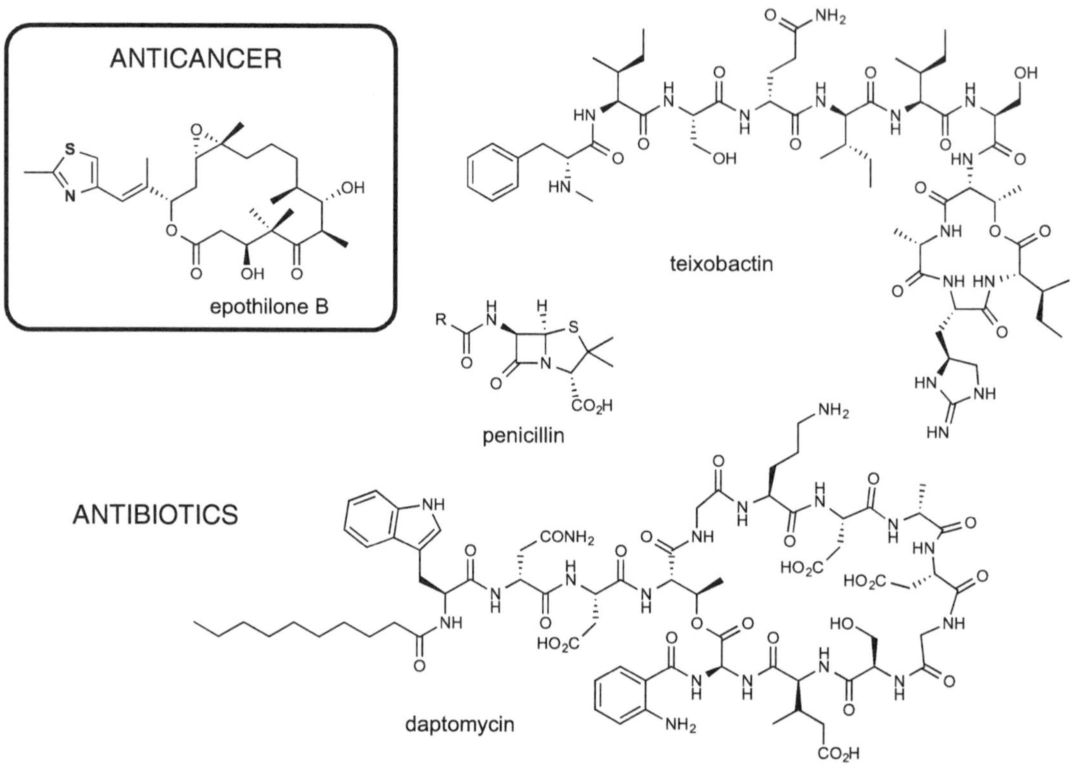

FIGURE 3.14.1 **Representative chemical structures of pharmaceutically important natural products.** Penicillin, daptomycin, and teixobactin are antibiotics, and epothilone B is an anticancer agent.

general acid-general base chemistries; yet, the limited types of amino acid functional groups would seemingly pose a barrier for many types of reactions. To circumvent this, Nature has adapted certain enzymes to utilize chemical cofactors that expand their chemical reactivities, including metallocenters (e.g., Zn, Fe, Mn, Mg, and Ca ions), coenzymes [e.g., flavin mononucleotide and nicotinamide adenine dinucleotide phosphate ($NADP^+$/NADPH)], and cosubstrates [e.g., *S*-adenosyl-L-methionine (SAM) and acetyl coenzyme A (acetyl-CoA)]. Another flavor of protein catalysis involves the development of catalytic antibodies.[14] Antibodies are proteins constituted by two protein chains, a heavy and light chain, forming a dimer-of-dimers structure that contains two epitope-binding regions. As part of the immune system, antibodies typically recognize foreign proteins or small molecules such as sugars. Examples of catalytic antibody generation include utilizing transition state mimics or larger, covalent product adducts of organic reactions as antigens to develop monoclonal antibodies that facilitate organic reactions.[15,16] In addition to these enzyme catalysts, nucleic acids are a form of biocatalysts that will be discussed further in the third case study. When compared with the proteinogenic amino acid possibilities, natural ribonucleic acid and deoxyribonucleic acid polymers (RNA and DNA, respectively) contain only four building blocks: adenosine, guanosine, cytosine, and either uridine for RNA or thymidine for DNA.

The study of biological reactions is of particular interest for the development and use of biocatalysts as greener alternatives to standard chemical syntheses. Biocatalytic reaction types are diverse (Table 3.14.1), making them attractive targets to replace or couple with organic syntheses. The breadth of chemical reactivity is both amazing and exciting, as we consider their use for the production of chemicals and pharmaceuticals.

3.14.1.1 Bioengineering of Biocatalysts

Considerable advancements in the development of biocatalysts for industry have been extensively reviewed.[13,18,27–30] This growth is critical because many industrial applications introduce variables that normally would have a negative impact on biocatalyst function. For example, enzymes are found to be most stable and active at specific temperature and pH values (usually near-neutral pH), and they normally function under physiological

TABLE 3.14.1 Examples of Biocatalytic Reactions

Type	Enzyme	Function	References
Hydrolases	Epoxide hydrolases	Catalyze conversion of epoxides to diols	17
Lipases/Esterases	Lipase	Purification of enantiopure compounds from racemic mixtures	18
Transferases	Lovastatin acyltransferase	Formation of acyl esters	19
Condensation	6-Deoxyerythronolide B synthase	Synthesis of 6-deoxyerythronolide B, the macrocyclic precursor to antibiotics such as erythromycin	20
	Farnesyl diphosphate synthase	Condensation of geranyl diphosphate and isoprenyl diphosphate	21
Amidase/ Aminopeptidase	Rec amidase	Purification of L- or D-amino acids	18
Aldolases	2-Keto-3-deoxygluconate aldolase	Condensation of pyruvate and D-glyceraldehyde	22,23
Oxidoreductases	Ketoreductase	Used to reduce keto functional groups, such as in atorvastatin biosynthesis	24
Oxygenases	CYB5 cytochrome P450	Hydroxylation reaction in artemisinic acid synthesis	25
Skeletal Rearrangement/ Mutases	Taxadiene synthase	Carbocation-mediated complex cyclization of geranylgeranyl diphosphate to taxadiene	21
	Lysine-2,3-aminomutase	Interconversion of L- and β-amino acids	26

concentrations of substrates and products (nanomolar to millimolar ranges). Industrial production of chemicals, however, often requires substrate concentrations upward of 1 M to drive reactions forward and generate sufficient amounts of product to make biocatalysis a reasonable alternative to other, less green synthetic processes. Furthermore, these processes often require substrates that are only soluble in organic solvents. One method to bypass solubility issues is to use a two-phase system (aqueous + organic) that relies on the partition coefficient of the organic substrate in aqueous buffer and the enzyme's high affinity for the substrate. However, because the organic solvent layer can inactivate enzymes, this does not work for all systems. A second issue is substrate and/or product inhibition at the high concentrations necessary for industrial applications.

How then can enzymes be used for efficient syntheses? The answer is in bioengineering and development of enzymes that exhibit the necessary catalytic functions to make them useful industrial biocatalysts. Through bioengineering, the thermal stability of an enzyme, its solubility in organic solvents, and perhaps most importantly, its reaction rate can be optimized for industrial applications. Generally, a turnover rate of >500 per minute with stringent enantio-, regio-, and chemoselectivity under the conditions required for synthesis is desirable for biocatalysts.[31]

The goals of enzyme engineering typically involve increasing enzyme stability, reaction rate, and solubility in organic solvents if desired, all while maintaining the specificity of the reaction (Fig. 3.14.2). Alternatively, less specific systems such as lipases[32,33] or acylases[19] have been explored to generate enzymes that have broad substrate spectrums to serve as biocatalysts that can be used under different synthetic circumstances. Rational bioengineering has been made possible in part due to the rise of macromolecular X-ray crystallography, which provided atomic resolution detail to characterize enzyme structure and function, as well as due to attempts to engineer novel enzymes (reviewed in 2014, Ref. 34). A commonly used method to engineer a biocatalyst is to start from an example enzyme that has a reaction similar to the one that is desired, or a promiscuous enzyme that turns over multiple substrates at different levels, and rationally mutate this enzyme until it demonstrates viability for industrial purposes. A more high-throughput methodology for the development of biocatalysts involves directed evolution,[28] which requires engineered cell lines and error-prone DNA synthesis that facilitate random incorporation of mutations within desired genes that may affect function. This in vitro form of Darwinian natural selection has been employed to increase enzyme stability and substrate tolerance, or completely alter substrate specificity.[28] Importantly, a methodology to screen samples in a high-throughput format is critical for the success of directed evolution techniques. Thus, a *handle* for monitoring the enzyme reaction (e.g., spectroscopy to monitor product levels, or antibiotic resistance marker to directly screen for survival of the organism) is necessary to identify variants that demonstrate the desired increases in enzyme function. Fluorescence-activated cell sorting is one example of a screening technique often used to identify cellular transformants from directed evolution experiments.

During biocatalyst development, where do enzyme mutations most commonly occur from structure-guided and directed evolution experiments? To generate single mutants to screen all 20 amino acids at each position of a 250-amino-acid protein, a laboratory would need to generate 20^{250} different enzyme variants, an unreasonable goal. The obvious answer is to consider only those residues that constitute the enzyme active site that directly bind the substrate and/or product and catalytic residues (first-sphere residues). Mutations of these

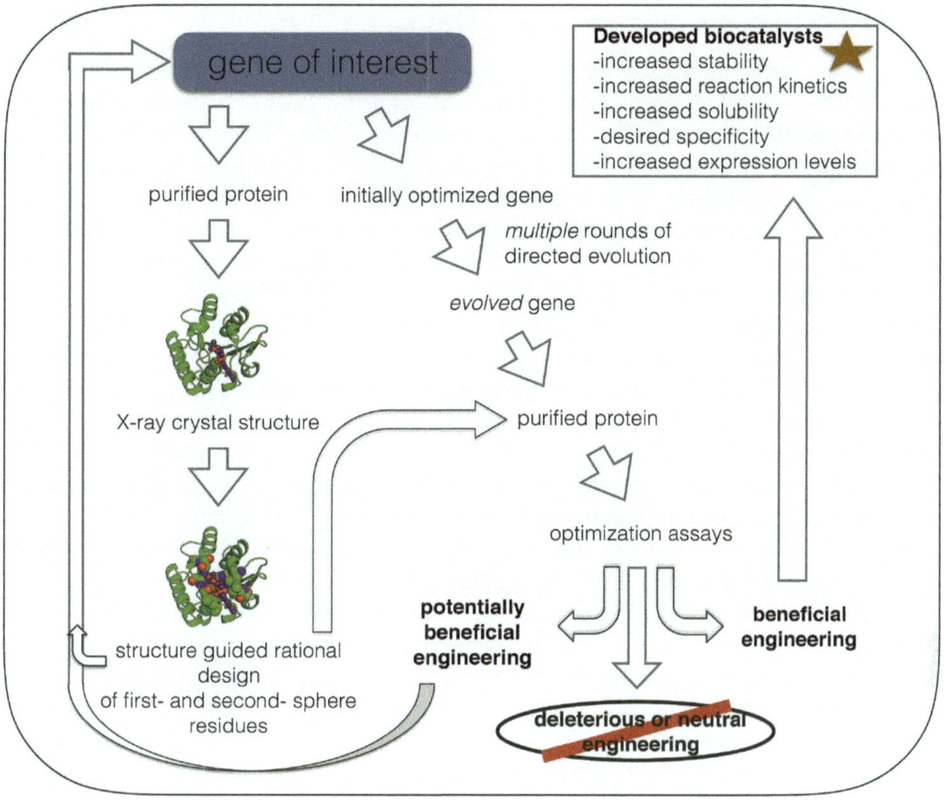

FIGURE 3.14.2 **Scheme of biocatalyst engineering.** Here we consider two methods: structure-guided engineering and directed evolution protocols for the development of optimized biocatalysts. The structure of lipase A33 is used as an example for identifying first- and second-sphere residues (PDB ID 1R4Z).

residues are nearly certain to have a direct effect on enzyme activity, although the exact outcome may sometimes be hard to predict a priori. Still, experiments designed to test all 20 amino acids at just 10 identified first-sphere residues (20^{10} possibilities) may be unrealistic. Therefore scientists often restrict which amino acid mutations to study (e.g., 3^{10}; however, this is still a large experiment!). Importantly, residues that surround the active site but are not directly interacting with the substrate or product (second sphere), in addition to residues near protein channels or dynamic portions of the enzyme structure, may also affect enzyme activity.[35,36] Therefore, although it may be most intuitive to begin making mutations within the active site, one should also be cognizant that second-sphere residues may also play important roles when attempting to engineer enzyme function. An additional complication in bioengineering is the observation that enzyme mutations do not always have an additive effect on enzyme activity.[37] Therefore, it is important to perform exhaustive experiments to identify optimally active enzymes using rational design, highlighting how important of a tool directed evolution has become for the field of bioengineering.

3.14.1.2 Hybrid Enzymatic/Synthetic Methods

Currently, there are many examples of mixed enzymatic/synthetic protocols for chemical production,[13,18,29,30] in addition to examples of developing host expression platforms for the production of specific chemicals.[38–40] One famous example includes protocols developed for the generation of statin drugs, which inhibit cholesterol biosynthesis (reviewed in Ref. 41). Statins function by inhibiting hydroxymethylglutaryl coenzyme A (HMG-CoA) reductase, which catalyzes the committed step of cholesterol biosynthesis. Statins are an extremely large and important segment of the drug market in the United States, where an estimated 71 million adults (from data collected between 2005 and 2008) have high amounts of low-density lipoprotein cholesterol linked to hypercholesterolemia.[42] In the 1970s, the discovery of the statin compactin from *Penicillium citrinum* fermentation broth by Japanese microbiologist Akira Endo provided the first example of an HMG-CoA reductase inhibitor.[43] However, due to complications of extremely upregulated HMG-CoA reductase levels in experiments with rats, studies of compactin and other statins were delayed until these puzzling results were clarified and better model systems were identified.[41] The product lovastatin (first designated mevinolin), became the first US Food and Drug Administration (FDA)-approved statin treatment in 1987. Lovastatin was discovered by researchers at Merck Research Laboratories as a compound produced by *Aspergillus terreus* and has served as a chemical framework for the synthetic development of statins for treating hypercholesterolemia.[44] Lovastatin is produced by an iterative fungal polyketide synthase (PKS) (Fig. 3.14.3A) [see the second case study for an introduction to polyketide (PK) biosynthesis]. Considering the sizable amounts of statin drugs that would need to be synthesized to meet the estimated patient population, several companies including Pfizer and Codexis Inc. have explored the development of biocatalysts to reduce the environmental footprint of statin synthesis.[13] Two examples that we will consider involve atorvastatin and simvastatin.

Biocatalysis has been linked to chemical synthesis in the generation of an intermediate in the synthesis of atorvastatin, the active ingredient in the statin Lipitor produced by Pfizer. In this commonly used example,[13,45] engineering to obtain optimized reaction conditions has allowed for the generation of enzymes with high activity appropriate for chemical synthesis (Fig. 3.14.3B). More specifically in this example, three enzymes were engineered: a halohydrin dehalogenase, which catalyzes replacement of a halide substituent from a halohydrin with a cyano group (reaction was increased ~4000-fold); a ketoreductase (KR), which reduces a ketone using two electrons from the cofactor NADPH (reaction increased sevenfold and maintained >99.5% enantioselectivity); and a glucose dehydrogenase, which couples oxidation of glucose and $NADP^+$ reduction to yield NADPH (reaction increased 13-fold).[24,45] When using a KR, a biocatalyst will require an influx of electrons in the form of NADPH, a technical challenge for the in vitro use of many biocatalysts involving redox chemistry. To overcome this synthetic barrier, the reaction was beautifully coupled to a secondary enzyme reaction (glucose dehydrogenase), to maintain adequate levels of the reduced cofactor (Fig. 3.14.3B).[24] This green alternative in atorvastatin synthesis resulted in a three-enzyme system that generated the chiral hydroxynitrile intermediate of atorvastatin synthesis with an environmental factor of 5.8 (or 18 if including water solvent).[24] A second notable example of enzyme engineering in statin synthesis is simvastatin, for which Prof. Yi Tang and Codesix Inc. received the Presidential Green Chemistry Challenge Award in

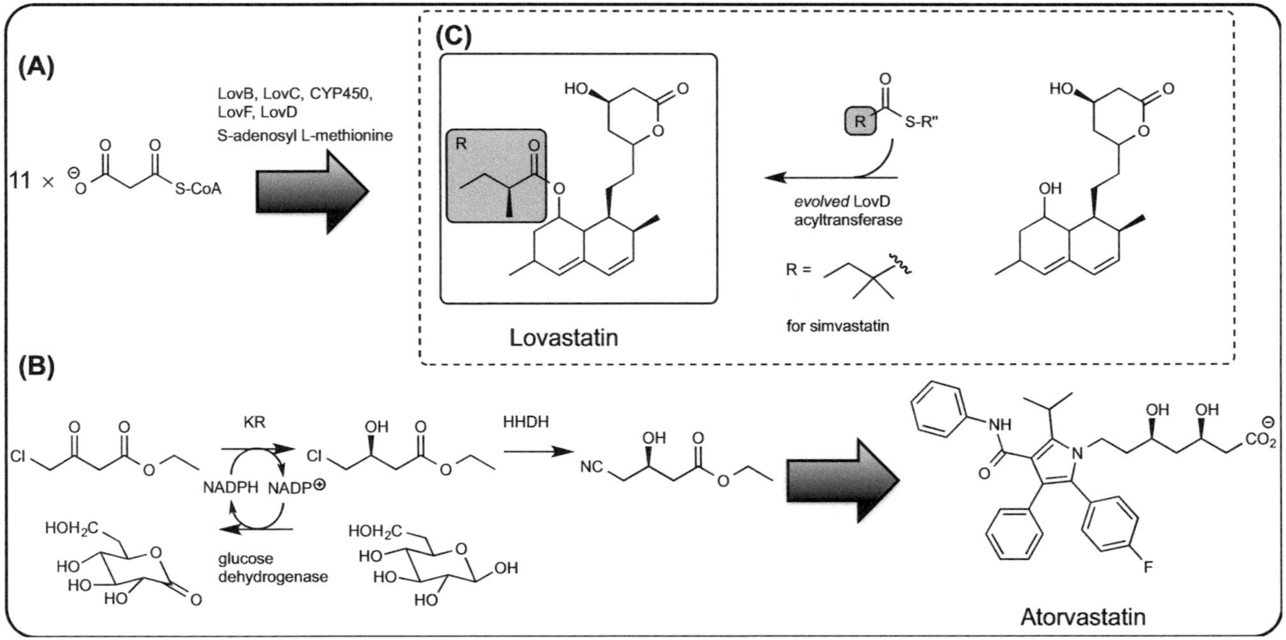

FIGURE 3.14.3 **Industrial production of statins has utilized the development of biocatalysts for mixed enzyme/synthesis methods.** (A) The biosynthesis of lovastatin within *Aspergillus terreus* converts 11 molecules of malonyl-CoA into the final statin product, using five enzymes and *S*-adenosyl L-methionine. (B) Scheme for biocatalytic production of an atorvastatin intermediate entails bioengineered ketoreductase (KR) and halohydrin dehalogenase (HHDH). Maintenance of NADPH levels is achieved by coupling the reaction to a glucose dehydrogenase. (C) A bioengineered LovD acyltransferase has enabled acylation of a lovastatin intermediate with variable acyl groups (labeled R). R″ can be CoA, *N*-acetylcysteamine, or methyl thioglycolate.

2012. Simvastatin is a derivative of lovastatin with a varied R acyl group (Fig. 3.14.3C). The biosynthetic enzyme LovD, an acyltransferase involved in lovastatin biosynthesis, was engineered using rational design and directed evolution to have altered substrate affinity and improved stability for industrial, large-scale applications (Fig. 3.14.3C).[19,46] Whole-cell experiments with LovD, and a knockout of a competing hydrolase (BioH) that would deplete substrate concentrations, have achieved >99% conversion to simvastatin at concentrations of 4–6 g/L (10–15 mM).[47–49]

3.14.2 BENEFITS AND DRAWBACKS OF BIOCATALYST DEVELOPMENT

The positive aspects of using biocatalysts are immediately noticeable: reactions can be performed with extreme specificity and decreased waste. Highly active enzymes can be developed, expressed recombinantly in a genetically tractable system such as *Escherichia coli* or *Streptomyces coelicolor*, or the eukaryotic expression system *Saccharomyces cerevisiae*, purified to homogeneity and then used at catalytic levels during necessary synthetic steps. These methods make it easy to remove the enzyme catalyst and purify samples for the next step of the reaction. Enzymes also can be immobilized on (or "anchored to") solid supports both to ease purification and enable recycling of the biocatalyst (reviewed in Ref. 50). Additionally, multistep reactions can be performed and coupled within a cellular environment. By utilizing host platforms, organisms can be cultured to produce chemicals and pharmaceuticals from feed and components of general metabolism, much like a one-pot synthesis. Genetically modified host organisms have been developed to facilitate their use as expression systems, such as by enriching amounts of necessary building blocks including acetyl- or malonyl-CoA, mevalonate (MVA), and different amino acids.[51–55]

The use of biocatalysts either in vitro with purified samples or in vivo within host systems for production of chemicals is a promising avenue for decreasing the waste footprint of total chemical synthesis for many complex reactions; however, there are several complications and/or drawbacks that must be considered when employing these systems. Foremost, the stability of biocatalysts often complicates their utility. To develop enzyme and nucleic acid catalysts, sample integrity is vitally important. As the 3D structures of these polymeric catalysts is imperative for function, the misfolding and/or truncation of proteins and nucleic acids can be detrimental to chemical activity. Therefore identifying conditions for producing properly folded and functional catalysts in an expression system is an initial hurdle in the development of biocatalysts. The choice of an expression system, identifying proper codon usage, the optimization of expression levels,[56] and correct cellular compartmentalization are fundamentally important factors in recombinant expression of protein targets.[57] For in vitro use, these samples ideally will be highly pure to avoid complications from contaminants that may interfere with their function (e.g., enzymes that catalyze competing reactions) or cellular components that may inhibit or alter the desired reaction, although industrial applications do not always require high purity as long as the specific activity is great enough and there are no competing reactions.[58] After these conditions are met, the temperature and buffer composition must be considered to determine the optimal conditions for each biocatalyst. Analogously, different cellular organelles, such as mitochondria or lysosomes, can be targeted within eukaryotic host systems; each compartment features differences in compound availability and pH, and therefore may affect overall product yields.[57]

Kinetic parameters for biocatalysts must be characterized to determine the conditions that will yield optimal protein expression levels and activity. For both in vitro and in vivo activities, concentrations of substrates and possible cosubstrates and/or cofactors must be optimal, and the possibility of inhibition, by a product or other cellular metabolites, must be considered. Modified host systems that increase concentrations of building block substrates and cosubstrates offer methods to sustain chemical production,[51–55] whereas the level of these molecules will need to be monitored for optimizing in vitro reactions. Lastly, the financial cost of developing a biocatalyst and implementing it for industry must be analyzed relative to the demand of the chemical to make it economically viable.

3.14.3 CASE STUDIES OF BIOCATALYSTS

In this chapter we will consider three biosynthetic systems that explore regio- and stereospecific C–C bond formation, in addition to complex rearrangements, that hold promise for the development of biocatalysts: terpenoid synthases, PKS and nonribosomal peptide synthetase (NRPS) systems, and nucleic acid catalysts. The selection of these examples is by no means meant to be exhaustive, but merely serves as an introduction into the exciting realm of biocatalysis.

3.14.4 CASE STUDY 1: TERPENES

3.14.4.1 Introduction

Terpenoids (isoprenoids) are a family of natural products produced by plants, animals, insects, bacteria, and fungi. This diverse group of molecules includes many chemicals that provide protection against pests and bestow unique flavors and fragrances. A remarkable diversity of enzymes have evolved that are responsible for the biosynthesis of terpenoids; *terpenoid synthases* transform linear polyisoprenoid pyrophosphates into thousands of structurally and stereochemically diverse products. What one of these enzymes can achieve in less than a second is often challenging, or nearly impossible, to replicate in the chemistry laboratory. Current advances in molecular biology, engineering of *S. cerevisiae*, and plant cell culture processes for the synthesis of biomolecules provide opportunities for scientists to utilize the specificity and efficiency of these powerful enzymes to produce large quantities of terpenoid natural products.[59–61] In this section, we will provide an introduction to terpenoids, and explore four examples where scientists exploit enzymes to assist in the synthesis of bioisoprene, menthol, and the pharmaceuticals artemisinin and paclitaxel.[59,62–64]

3.14.4.2 Terpenoids

Terpenoids are the most structurally, stereochemically, and biologically diverse family of natural products. Greater than 55,000 unique terpenoid natural products have been

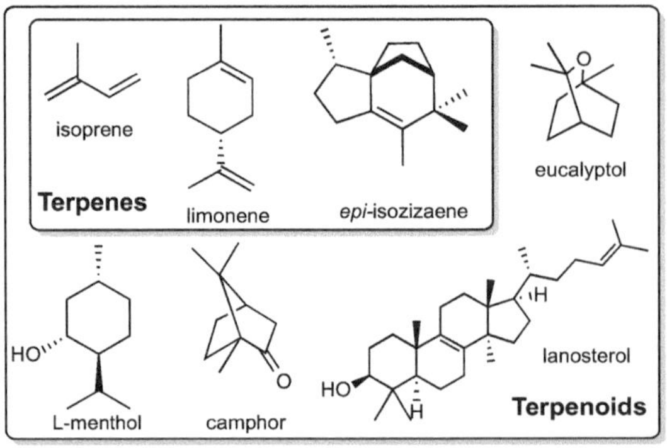

FIGURE 3.14.4 **Structural diversity of terpenoids.** Terpenoids that contain only carbon and hydrogen are known as terpenes; examples include isoprene, the monoterpene limonene, and the sesquiterpene *epi*-isozizaene. Examples of terpenoids that contain oxygen atoms include the monoterpenoids eucalyptol, L-menthol, and camphor, and the triterpenoid lanosterol.

discovered, all of which ultimately derive from the universal 5-carbon precursors isopentenyl pyrophosphate (IPP) and dimethylallyl pyrophosphate (DMAPP).[21] Terpenoids serve a variety of specific functions; some terpenoids are critical for plant survival, including plant primary metabolites such as sterols, whereas many others are secondary metabolites that assist in growth and development but are not absolutely required for survival. For example, secondary metabolites bestow unique flavors and fragrances, provide chemical defense against pests, facilitate interactions between plants and other organisms, and even contribute to photosynthesis.[65]

Traditionally, terpene natural products contain only carbon and hydrogen atoms (e.g., isoprene, limonene, and *epi*-isozizane), whereas the term terpenoid is all-encompassing and includes terpenes that have been further modified by oxidation reactions to form alcohols, aldehydes, ketones, or carboxylic acids, as well as additional reactions that incorporate heteroatoms (e.g., eucalyptol, camphor, and lanosterol) (Fig. 3.14.4). The diverse family of terpene or terpenoid products is generated by an enzyme family known as the terpenoid synthases, which catalyze the modification of linear polyisoprenoid pyrophosphate (diphosphate) substrates. These linear substrates are synthesized via chain elongation reactions of IPP and DMAPP, catalyzed by isoprenoid-coupling enzymes known as prenyl diphosphate synthases (or prenyltransferases).[66] When DMAPP is linked to one or more units of IPP in a head-to-tail fashion, the following molecules are formed: geranyl pyrophosphate (C10), farnesyl pyrophosphate (C15), geranylgeranyl pyrophosphate (C20), and geranylfarnesyl pyrophosphate (C25); these molecules are the linear precursors to the mono-, sesqui-, di-, and sesterterpenes, respectively (Fig. 3.14.5).[21,67] Additionally, two molecules of farnesyl diphosphate can be coupled together in a head-to-head manner (i.e., at C1 of each molecule) to form the 30-carbon (triterpene) molecule squalene, which is the linear precursor of steroids and their hormone derivatives.[68] The variable chain length of linear precursors adds to the diversity of this product family.

Most plants produce complex mixtures of terpenoid natural products depending on their environmental variables, and during the past century a large number of terpenoid products

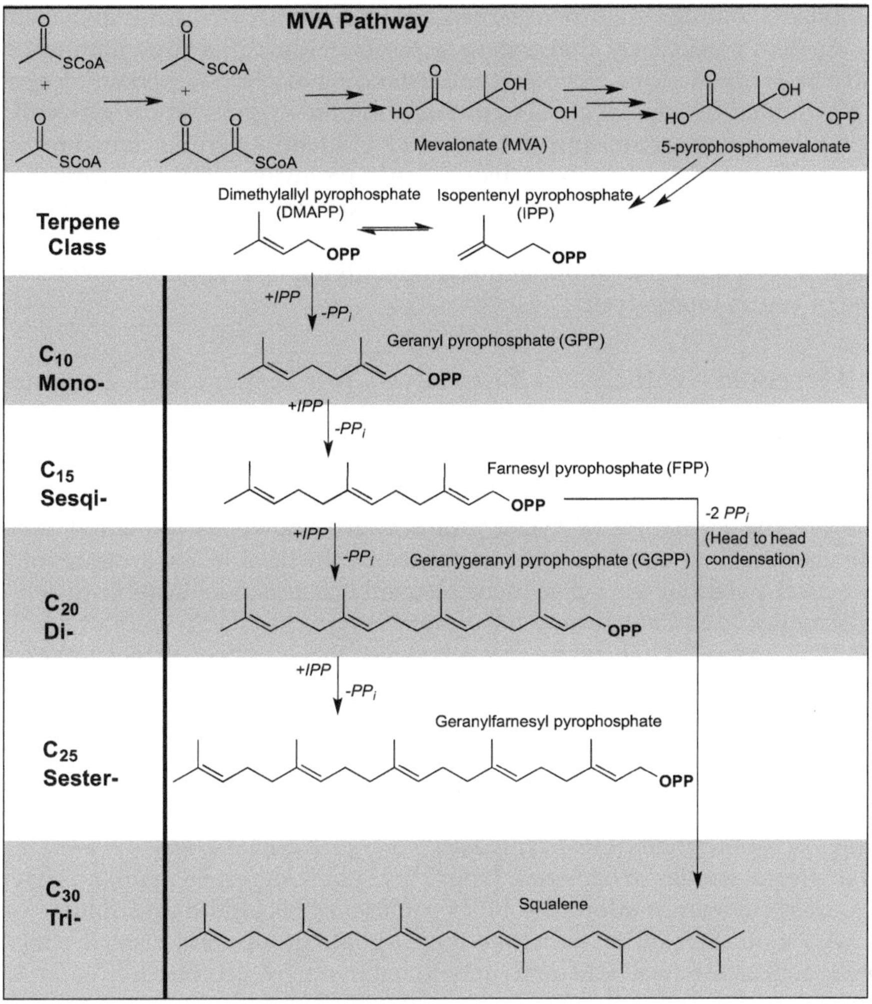

FIGURE 3.14.5 **General scheme of terpenoid biosynthesis and nomenclature.** The mevalonate (MVA) pathway of isoprenoid biosynthesis occurs in the cytoplasm of plant cells. The activated isoprene units IPP and DMAPP produced by this pathway ultimately derive from acetyl-CoA. The activated 5-carbon units are joined by prenyl diphosphate synthases to form increasingly longer linear isoprenoid pyrophosphates. *OPP*, pyrophosphate; *PPi*, inorganic pyrophosphate.

have been extracted and characterized. Prominent examples of well-known plant terpenoids include: the monoterpenes menthol, isolated from wild mint plants; limonene, found in the rind of many citrus fruits; and camphor, produced by several types of trees (Fig. 3.14.4). Historically, the most common method for obtaining plant secondary metabolites, including terpenoids, has been through natural harvest. However, direct extraction of terpenoids from plant sources is limited by the compound's bioavailability. It is rare for a specific secondary metabolite to represent greater than 1% of the dry weight of a plant.[61] Furthermore, many plants synthesize a complex mixture of terpenoids with similar structures and chemical properties, making them particularly challenging to purify. Although a large number of plant terpenoids have been identified, a thorough understanding of plant terpenoid biosynthesis and metabolism is limited by several challenges including the expression of eukaryotic plant enzymes in traditional host organisms, the difficulty in identifying and obtaining appropriate terpenoid substrates for the terpenoid biosynthetic pathways under investigation, and on the genomic level, the absence of functional gene clustering in plants.[69] However, with the increase in high-quality plant genomic sequence information made available since the mid-2000s, there has been a great deal of interest in identifying novel plant biosynthetic enzymes that may be of use as biocatalysts.

3.14.4.3 Terpenoid Synthases as Biocatalysts for Terpene and Terpenoid Production

Terpenoid synthases convert linear polyisoprenoid pyrophosphates into linear or cyclic terpenoid products, often with one or more stereocenters (terpenoid synthases that predominantly catalyze the formation of cyclic products are known as terpenoid cyclases). The potential diversity of C—C bond formation afforded by the flexible linear isoprenoid substrate, and the chemical potential for subsequent biosynthetic functionalization of terpenoids by modifying enzymes (e.g., cytochrome P450 monooxygenases, etc.), account for the tremendous diversity in terpenoid natural products and also make terpenoid biosynthesis an attractive system for engineering novel compounds.[66] Additionally, although some terpene synthases accept only one substrate, others accept multiple substrates. In 1997, the first crystal structure of a terpenoid synthase was determined to 2.6-Å resolution. This structure for the sesquiterpene synthase pentalene synthase from *Streptomyces exfoliatus* revealed that terpene synthases share the same terpenoid synthase α-helical fold as the prenyl diphosphate synthase farnesyl pyrophosphate (FPP) synthase.[70]

Further structural studies of bacterial, fungal, and plant terpenoid synthases have revealed two classes; class I enzymes adopt the FPP synthase α-helical fold and initiate catalysis by metal-triggered ionization of the substrate diphosphate group, and class II enzymes adopt an unrelated double α-barrel fold and initiate catalysis by protonation of an epoxide or olefin.[71] In this section we will only explore examples of type I terpenoid synthases. The 3D structures of known class I terpene synthases share several common features, including a conserved α-helical terpenoid synthase fold, and the presence of two metal-binding motifs, which coordinate the catalytically obligatory metal ions (typically Mg^{2+} or Mn^{2+}) near the opening of the active site (Fig. 3.14.6A).[66] The metal-binding motifs typically consist of an aspartate-rich motif (**DD**xx**D**, where bold residues indicate side chains involved in metal coordination, and x represents any amino acid), which coordinates to two metal ions, known as ${Mg^{2+}}_A$ and ${Mg^{2+}}_C$, and an NSE/DTE motif (**N**xxx**S**xxx**E** or **D**xxx**T**xxx**E**), which coordinates to a third metal ion, known as ${Mg^{2+}}_B$. When the pyrophosphate moiety of the substrate coordinates to the Mg ions, the active site closes to sequester the hydrophobic tail of the substrate away from solvent, forming an enclosed pocket that defines a structural template for the resulting product. Crystal structures of terpenoid synthases have been determined in both "open"/unliganded and "closed"/liganded conformations (Fig. 3.14.6B),[72] defining differences upon substrate binding within the hydrophobic active site cavity. The chemical

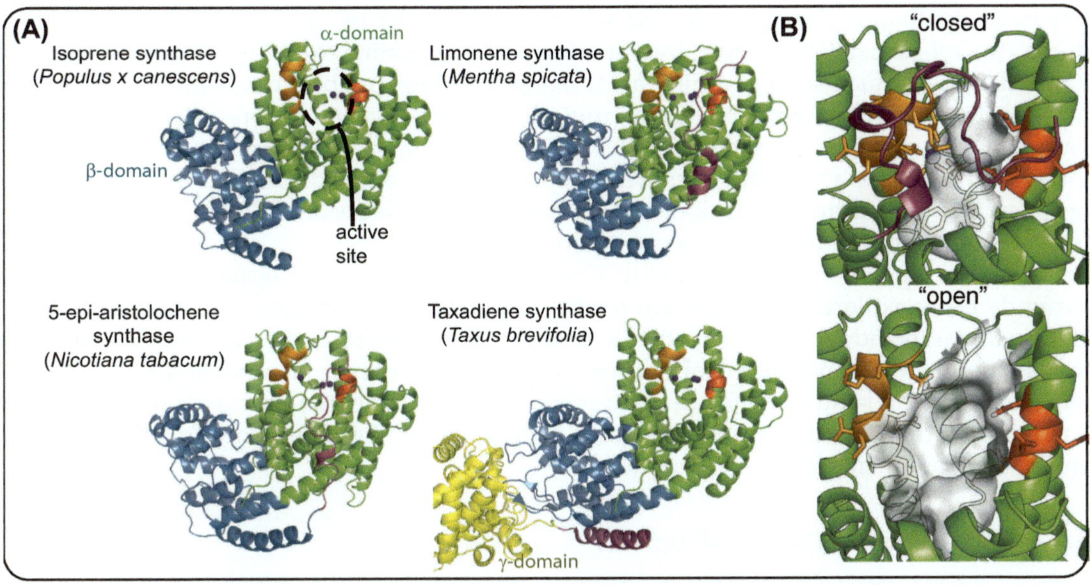

FIGURE 3.14.6 **The conserved class I terpenoid synthase fold.** (A) The X-ray crystal structures of plant terpenoid synthases isoprene synthase (PDB ID 3N0G), limonene synthase (PDB ID 2ONG), 5-*epi*-aristolochene synthase (PDB ID 5EAT), and taxadiene synthase (PDB ID 3P5R) illustrate conservation among plant species of the C-terminal catalytically active class I terpenoid cyclase fold (α-domain, green), containing metal-binding motifs DDxxD and NSD/DTE (red and orange, respectively), which coordinate to the trinuclear metal cluster (dark blue). Additionally, the synthases share a smaller vestigial domain (β-domain, blue). Taxadiene synthase also contains a catalytically inactive class II terpenoid cyclase fold (γ-domain, yellow). An N-terminal α-helix that serves as a "cap" to the active site (purple) is visible in the limonene and 5-*epi*-aristolochene synthase structures. (B) Structures of a bacterial sesquiterpene synthase, *epi*-isozizane synthase, in both the "closed"/liganded and "open"/unliganded forms, depicting how closing of the enzyme active site around the pyrophosphate and metal ions generates an active site template to drive the carbocation rearrangement reaction (PDB IDs 3KB9 and 3LGK). The active site is displayed as a gray surface representation, and the pyrophosphate and bound hydrophobic molecule are displayed in sticks, with carbon in yellow, phosphates in orange, oxygen in red, nitrogen in blue, and Mg^{2+} displayed as in panel (A).

transformation of substrate to product is triggered by the promoted release of the pyrophosphate leaving group; the resulting reactive carbocation quickly undergoes a series of intramolecular C–C bond formation reactions and hydride shifts to form the final product.[73]

Some terpenoid synthases are described as high-fidelity enzymes, meaning they produce almost exclusively one product, whereas others produce a mixture of products. This fidelity is determined by the active site of a terpenoid synthase, which is predominately lined with nonpolar "inert" amino acids, enabling it to direct the precise regio- and stereospecificity of the cyclization reactions without quenching of carbocation intermediates. The terpenoid synthase active site cavity also typically contains an unusually high number of aromatic residues. The aromatic residues are thought to offer cation-π electrostatic stabilization to the carbocation intermediates that form throughout the reaction pathway. The specific sequence of C–C bond-forming reactions and hydride shifts is guided by the shape and polarity of the surface of the active site cavity and culminates in deprotonation or addition of water to the final carbocation intermediate.[74] Numerous studies have shown the remarkable plasticity of terpenoid synthases. For example, the sesquiterpene synthase *epi*-isozizane synthase produces 79% *epi*-isozizane; however, by mutating a phenylalanine residue in the active site cavity to alanine, a much smaller amino acid, the mutant enzyme produces 70% β-farnesene, a new product for this enzyme. Similarly, mutation of a tryptophan to phenylalanine yields a variant of the enzyme that produces 47% (Z)-γ-bisabolene.[72] Therefore just one amino acid mutation within the active site can affect the distribution of products, and the combination of a few amino acid mutations, including residues from the second sphere, can completely transform the identity of the terpenoid synthase.[75]

The following four examples provide a brief introduction to how terpenoid synthases are currently being explored and exploited for the production of chemicals and pharmaceuticals. The examples include microbial synthesis of isoprene, the basic 5-carbon building block of the terpenoids; chemical and microbial synthesis of the monoterpenes menthol and limonene; engineering yeast to synthesize the sesquiterpene amorphadiene, the precursor to the antimalarial drug artemisinin; and heterologous expression of the diterpene Taxol, a popular anticancer drug, using plant cell culture.

3.14.4.4 Production of Bio-Isoprene

Isoprene (2-methyl-1,3-butadiene) is the smallest terpene unit. It is a colorless, volatile liquid with a boiling point of 34 °C. Although isoprene is produced by many plants, it is also a by-product of the thermal cracking of oil and a side product of ethylene production. In industry, isoprene is used to produce synthetic rubber (*cis*-1,4-polyisoprene), with more than 800,000 metric tons produced annually. Limited long-term petroleum reserves have brought an increased interest in engineering microbes to manufacture bio-isoprene as a sustainable alternative to petroleum-based isoprene.[76] There are two pathways for the biosynthesis of DMAPP, the substrate of isoprene synthase to form isoprene. The methylerythritol 4-phosphate (MEP) pathway is used in many Eubacteria, green algae, and chloroplasts of higher plants, and the MVA pathway is common to most eukaryotes, archaea, and cytosols of higher plants.

The production of bio-isoprene has been explored in an engineered *E. coli* strain that is optimized to produce isoprenoid metabolites, serving as an excellent example of how engineered host expression systems can be utilized for the cellular production of fine chemicals. This strain has been manipulated to use a hybrid MVA pathway with proteins from two different bacteria to increase the cellular level of DMAPP. The hybrid MVA pathway consisted of genes from the "upper MVA pathway" from *Enterococcus faecalis* (mvaE, mcaS, and mcaE), which converts acetyl-CoA into MVA followed by genes from the "lower MVA pathway" from *S. cerevisiae* (ERG12, ERG8, ERG19, IDI1), which are responsible for the conversion of MVA to DMAPP. To favor the production of increased levels of isoprene, these cells also contained recombinant isoprene synthase (ispSPa) from *Populus alba* (white poplar).[59] The final *E. coli* strain produced up to 6.3 g/L of isoprene after 40 h of growth using glucose as the carbon feedstock, which corresponds to 28% of the theoretical limit for glucose to isoprene conversion efficiency (gram to gram).[59]

Isoprene synthase is an interesting enzyme. It is the first noncyclase terpenoid synthase found to share the same catalytic metal-binding motif as the terpenoid cyclases, namely, an aspartate-rich motif (DDxxD) and NSE/DTE motif. Although the crystal structure of

P. alba isoprene synthase has yet to be determined, the crystal structure of isoprene synthase from *Populus* × *canescens* (PcISPS) indicated that isoprene synthase is composed of two domains, an inactive N-terminal α-barrel class II terpenoid synthase domain, and the catalytically active C-terminal α-helical class I terpenoid synthase fold (Fig. 3.14.6A).[77] Isoprene synthase is not the only terpenoid synthase that contains inactive domains; in fact, it is hypothesized that many plant terpenoid cyclases have evolved to contain additional inactive domains that may have come from bacterial terpenoid cyclases, including the diterpene cyclase taxadiene synthase (TS), a key enzyme in paclitaxel biosynthesis.[77,78]

3.14.4.5 Chemical and Biocatalytic Synthesis of Menthol and Limonene

Menthol and limonene are both monoterpenes, which share a common carbon skeleton. In fact, the same enzyme, limonene synthase, catalyzes the first step in the biosynthesis of both compounds. L-Menthol [(−)-menthol] is the major component of the essential oil distilled from corn mint (*Mentha arvensis* L.), at greater than 60% (w/v).[79] Its minty taste and cooling effects (when applied to the skin or taken orally) make it a very popular flavoring in personal care products, foods, and nonprescription health care products.[79] As with most terpenoids, L-menthol was first isolated from the natural plant source, and corn mint and peppermint plants continue to supply the majority of the 27,000 metric tons of L-menthol consumed annually. In mint plants, limonene is converted to L-menthol in six enzymatic steps.[80] However, the increased worldwide demand and weather-related fluctuations in production have pushed synthetic production to an all-time high. More than 30 years ago a "green" industrial process for L-menthol synthesis was developed and introduced that uses the renewable monoterpene product myrcene obtained from gum rosin (pine sap) as the starting material (Fig. 3.14.7). The key step in the synthesis involves asymmetric isomerization of geranyldiethylamine catalyzed by the chiral catalyst complex (*S*)-BINAP-Rh. This breakthrough in industrial asymmetric catalytic manufacturing was the result of a collaboration between the Japanese chemical company *Takasago* and several prominent chemists including Dr. Ryoji Noyori, who in 2001 shared the Nobel Prize in Chemistry for his

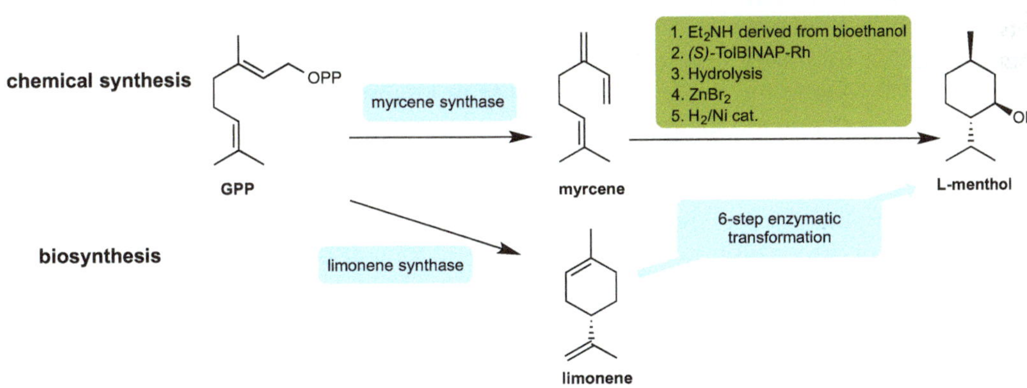

FIGURE 3.14.7 **Synthesis of L-menthol.** The hybrid biocatalyst/chemical synthesis and strict biosynthesis of L-menthol begin with enzymatic conversion of geranyl pyrophosphate (GPP) into myrcene and limonene, respectively.

contributions to chirally catalyzed hydrogenation reactions.[81] It is possible to produce either enantiomer of menthol by changing the configuration of the BINAP-Rh catalyst. Although pine sap is considered renewable, microbial synthesis of myrcene using engineered *E. coli* expressing the monoterpene synthase myrcene synthase is currently under investigation as an alternative green biosynthetic approach.[82]

Despite the successful chemical synthesis of L-menthol, alternative biocatalytic synthesis methods are still being explored. For example, limonene and perillyl alcohol have been successfully produced in *E. coli* using a heterologous MVA pathway together with the limonene synthase and a cytochrome P450 gene, which specifically hydroxylates limonene to produce perillyl alcohol.[40] Transgenic plants offer an alternative approach to the biosynthesis of monoterpenoids. The (−)-limonene synthase gene from *Perilla frutescens* was transformed into tobacco plants, but with poor overall product yields.[79] Also, a biosynthetic method has been developed that uses a cell-free one-pot biotransformation to synthesize menthol isomers from the monoterpenoid pulegone using *E. coli* cell extracts. This method does not require expression of a terpenoid synthase, since the starting material already contains the correct carbon skeleton; however, it does make use of an "ene-reductase" and menthone dehydrogenases.[62]

3.14.4.6 Bioengineering Yeast to Produce Artemisinic Acid for the Treatment of Malaria

Malaria is an infectious disease caused by the parasitic protozoan *Plasmodium*. The number of malarial cases worldwide has decreased since 2000, but in 2015, there were 214 million new cases of malaria according to the World Health Organization, 90% of which occurred in Africa.[83] The most effective treatment for malaria is a combination treatment (known as ACT) that includes the sesquiterpene lactone endoperoxide *artemisinin*, a natural product isolated from *Artemisia annua*, the sweet wormwood plant, native to Asia (Fig. 3.14.8A).[60] In the late 1960's, the Chinese government funded a national project to discover new treatments for malaria. In 1972, after screening thousands of herbal remedies, pharmacologist Youyou Tu and her colleagues at the China Academy of Traditional Chinese Medicine isolated and identified pure artemisinin from *A. annua*. Their discovery was recognized in 2015 when one-half of the Nobel Prize in Physiology or Medicine was awarded to Youyou Tu in recognition of her discovery that revolutionized antimalarial treatment and helped save millions of lives.[84,85] Due to the structural complexity of artemisinin, chemical synthesis is not a financially practical method of production, making extraction of artemisinin from sweet wormwood plants the exclusive method of production. Although efficient and modestly cost-effective, plant-derived artemisinin production depends on crop yields, which can be affected by drought, climate changes, and pests.[86] Natural harvest is estimated to yield from 6 to 14 kg of artemisinin per hectare (i.e., 1.5–2 tons of dried *A. annua*).[87] In an effort to stabilize the international supply of this powerful drug and minimize annual cost-fluctuations, a semisynthetic approach to artemisinin synthesis was championed by Amyris, Inc. since the mid-2000s. To help advance the use of semisynthetic approaches to the synthesis of medically relevant natural products, the intellectual property rights to the semisynthetic product of artemisinin have been provided by Amyris, Inc. free of charge, and are detailed in a series of articles.[60,63]

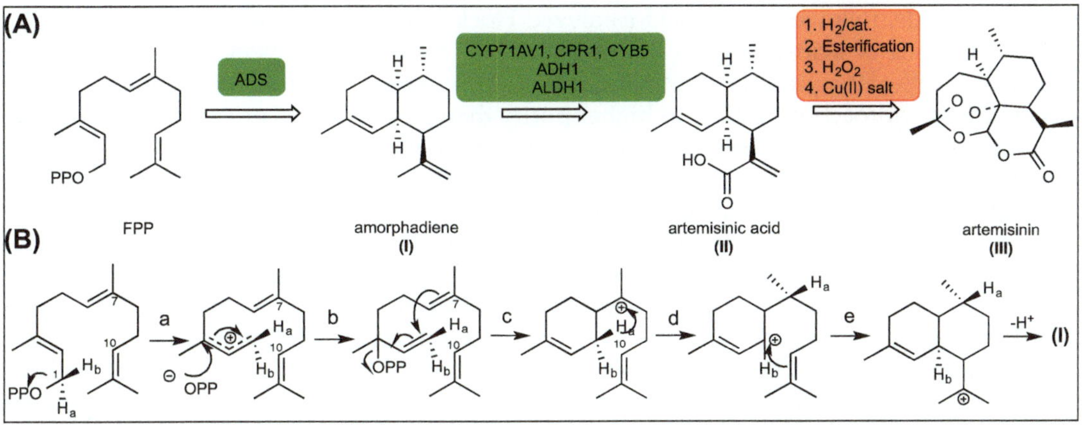

FIGURE 3.14.8 **Semisynthetic production of artemisinin.** (A) The biosynthetic pathway for the production of artemisinic acid in *Artemesia annua* was engineered in baker's yeast. Farnesyl pyrophosphate (FPP) is converted to amorphadiene by the terpenoid synthase amorphadiene synthase (ADS). Amorphadiene is oxidized to artemisinic alcohol by CYP71AV1, CPR1, and CYB5. ADH1 and ALDH1 further oxidize artemisinic alcohol to the aldehyde and carboxylic acid, respectively. A four-step chemical reaction completes the semisynthetic production of artemisinin. (B) Proposed mechanism for the cyclization of FPP by ADS89.

The semisynthetic route to artemisinin production involves engineering yeast to produce artemisinic acid, the precursor of artemisinin. The first committed step in the biosynthetic pathway of artemisinin in *A. annua* is the conversion of FPP almost exclusively into the bicyclic product amorpha-4,11-diene (>95%) by amorphadiene synthase (ADS).[88] The cyclization mechanism of ADS has been investigated using a series of deuterium-labeled FPP substrates. ^{1}H and ^{2}H nuclear magnetic resonance (NMR) studies suggest that the most likely mechanism involves a bisabolyl intermediate and a 1,3-hydride shift (Fig. 3.14.8B).[88]

The engineered artemisinic acid production pathway in *S. cerevisiae* involves overexpressing several genes involved in the MVA pathway to increase IPP and DMAPP production, a similar methodology as was utilized in *E. coli* for the production of bio-isoprene discussed earlier, with incorporation of the ADS gene from *A. annua* to produce the natural ADS synthase for cellular production of amorpha-4,11-diene, the precursor of artemisinic acid. To increase the levels of produced amorpha-4,11-diene, endogenous levels of squalene synthase were suppressed by the downregulation of the *ERG9* gene, which encodes squalene synthase, using a methionine-repressible promoter to reduce the flow of FPP toward the biosynthesis of sterols. The enzymatic oxidation of amorpha-4,11-diene to form artemisinic acid was achieved via coexpression of four additional *A. annua* genes: the cytochrome P450 amorphadiene oxidase (*CYP71AV1*) and its cognate reductase (*CPR1*), a cytochrome *b5* (*CYB5*), and artemisinic aldehyde dehydrogenase (*ALDH1*) (Fig. 3.14.8A).[63] Optimization of the *A. annua* artemisinic acid biosynthetic pathway in yeast resulted in fermentation titers of 25 g/L artemisinic acid.[63] Several synthetic approaches to convert artemisinic acid into artemisinin have been explored. Currently, the most efficient process involves catalytic hydrogenation of the Δ11 double bond to give dihydroartemisinic acid, esterification of the carboxylic acid, an "ene-type" reaction of the C4–C5 double bond with singlet oxygen to form an allylic

3-hydroperoxide, followed by an acid-catalyzed Hock fragmentation and rearrangement that leads to a series of acid-catalyzed cyclizations. This four-step synthetic process ultimately results in a 40%–50% yield of the final product artemisinin. Artemisinin obtained from the semisynthetic approach typically has higher purity than plant-sourced artemisinin[63]; however, despite the initial promise that the semisynthetic approach would lower the cost of malaria treatment, several years of good *A. annua* crops resulted in a decrease in antimalarial drug prices and demand. In 2015, no semisynthetic artemisinin was produced.[89]

To date, there is no experimental structural data available for ADS, although homology modeling has been used to predict its 3D structure.[90] Since ADS has 39% sequence identity with *5-epi*-aristolochene synthase (TEAS) from *Nicotiana tabacum* (cultivated tobacco), a sesquiterpene cyclase whose structure was first determined in 1997, it was used as a basis for homology modeling. TEAS catalyzes the cyclization of FPP to form *5-epi*-aristolochene, the committed step in the biosynthesis of the terpenoid *capsidiol*, a phytoalexin that protects the tobacco plant from fungal infections. The first crystal structure of a plant terpenoid cyclase, the TEAS structure, revealed a unique two-domain architecture later also identified in isoprene synthase (Fig. 3.14.6A): a catalytically active C-terminal domain that adopts the α-helical class I terpenoid synthase fold and an N-terminal domain of unknown function that exhibits an α-helical fold that resembles that of a class II terpenoid synthase.[91]

3.14.4.7 Plant Cell Fermentation of the Potent Antitumor Agent Paclitaxel

One of the most famous terpenoids is paclitaxel (commonly known by its trademark name Taxol), a diterpene isolated from the bark of the Pacific yew tree. The complex structure of paclitaxel consists of four fused rings and contains 11 stereocenters (Fig. 3.14.9). The FDA approved the use of Taxol in 1992 for the treatment of ovarian cancer, more than 20 years after paclitaxel was first isolated and its structure determined. Its mechanism of action involves the stabilization of microtubules, resulting in cell arrest that leads to programmed cell death, or apoptosis. Paclitaxel has been prescribed to over 1 million patients, and two paclitaxel derivatives, taxotere and cabazitaxel, have also been approved by the FDA for the treatment of breast cancer and hormone-refractory prostate cancer, respectively.[64]

Very low levels (0.0004%) of paclitaxel in the bark of Pacific yew trees provide protection against insects and fungi. Unfortunately, the trees take up to 200 years to mature, and the process of stripping the bark to extract paclitaxel is fatal to the tree.[92] More than 38,000 trees (340,000 kg of bark) are needed to supply the annual 25 kg demand for paclitaxel.[61] In the late 1980s as the demand for paclitaxel grew, so did the large-scale destruction of Pacific yew trees, forcing scientists to explore other methods of paclitaxel production.[64,92] The total synthesis of paclitaxel has been achieved; however, it is not an efficient or economical means of large-scale production.[93] A semisynthetic route to the production of paclitaxel was developed and patented in 1992 (Fig. 3.14.9C). The semisynthetic route begins with the compound 10-deacetylbaccatin III (10-DAB), which can be extracted from leaves and twigs of the European yew tree (*Taxus baccata*). A unique collaboration between the National Cancer Institute and the pharmaceutical company Bristol-Myers Squibb (BMS) developed a commercially viable, albeit energy-intensive, synthetic route[92] from 10-DAB to paclitaxel in 1993.

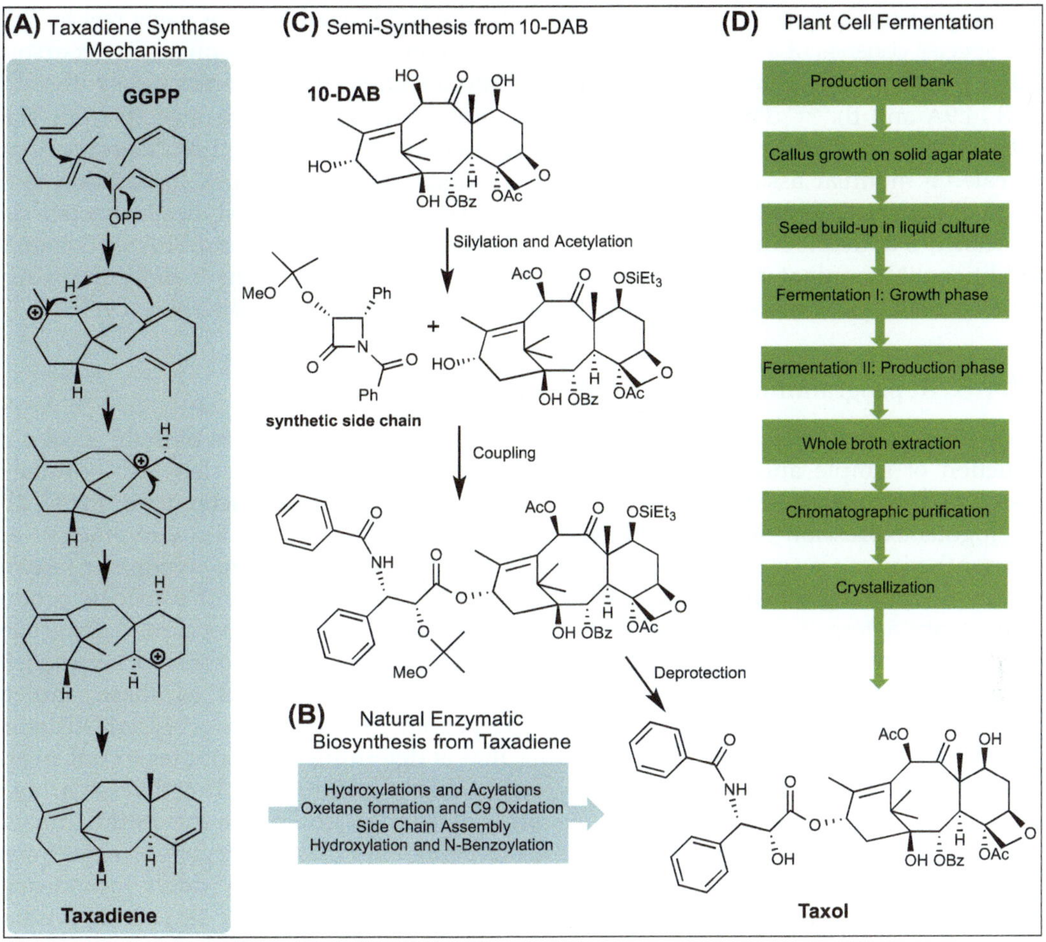

FIGURE 3.14.9 **Synthesis of Taxol.** (A) The cyclization cascade for the conversion of geranylgeranyl pyrophosphate (GGPP) into taxadiene by taxadiene synthase. (B) A summary of the biosynthetic conversion of taxadiene into Taxol. The complete biosynthetic pathway is not yet fully understood. (C) The semisynthetic synthesis of Taxol begins with 10-deacetylbaccatin III (10-DAB). (D) A summary of the current green method of Taxol synthesis using plant cell fermentation.

In 2002, BMS discontinued commercial semisynthetic production of paclitaxel and replaced it with an environment-friendly aqueous-based plant cell fermentation (PCF) process, regarded as the largest commercial application of PCF to date.[92] The process uses cells cultured from the needles of the Chinese yew tree (*Taxus chinensis*) and begins by growing a "callus" of undifferentiated cells on a solid agar medium. These cells are then transferred to a unique liquid growth medium for fermentation whereby the cells are induced to produce the secondary metabolite paclitaxel. The fermentation phase is followed by extraction of the whole cell broth and purification with chromatography and crystallization.[64,92] Overall, the PCF process eliminates the use of many harmful organic solvents including tetrahydrofuran, results in no solid biomass waste, and is significantly less expensive.[92]

In comparison to these semisynthetic methods, the full biosynthesis of paclitaxel in *Taxus* (yew) species consists of 19 steps beginning with the conversion of the universal precursor of the diterpenes, geranylgeranyl pyrophosphate, into taxadiene by the terpenoid synthase TS (Fig. 3.14.9A and B).[77,94] The X-ray crystal structure of a truncated variant of TS from *Taxus brevifolia* was published in 2011. It was the first diterpene cyclase structure determined, and it revealed a modular assembly of three α-helical domains (Fig. 3.14.6A). In addition to the catalytically active C-terminal class I α terpenoid cyclase domain and catalytically inactive class II-associated β domain found in isoprene synthase, limonene synthase, and TEAS, TS contains an additional N-terminal domain identified as a class II γ domain.[77] The β and γ domains in plant terpenoid synthases are thought to have evolved from bacterial diterpene cyclases.[95]

3.14.4.8 Reprogramming Terpenoid Synthases

As we have seen, several terpenoid synthases have been exploited as biocatalysts for the production of simple and complex terpenoid natural products. As scientists continue to investigate the biosynthetic pathways of terpenoid natural products, it is likely that heterologous expression of terpenoid synthases will continue to contribute to efficient and green synthetic routes to the production of novel pharmaceuticals. One avenue of research involves determining whether it is possible to increase the production of a specific product or reprogram a terpenoid synthase entirely to produce alternative products.[96] To achieve this goal, it is necessary to have a greater understanding of the relationship between the structure of the hydrophobic active site pocket of the terpenoid synthase and the corresponding products. Only a handful of terpenoid synthase X-ray crystal structures have been determined, but each new structure that is determined contributes to our overall understanding of how this family of enzymes is able to produce such a diverse array of products. However, structural data alone are not enough to enable successful directed engineering of terpenoid synthases. Numerous terpenoid synthase mutagenesis experiments have demonstrated the enzyme plasticity, yet the altered product array rarely provides any generalizable data that can be applied to assist in engineering other terpenoid synthase systems. Quantum chemical calculations are being used to better understand the level of control exerted by a terpenoid synthase enzyme over a reaction pathway, for example, by studying the electrostatic influence of the pyrophosphate coproduct on the catalyzed reaction.[95] Until the ultimate goal of rational engineering of catalytic activity is realized, terpenoid synthases can continue to be exploited, as seen in the examples mentioned earlier, for their natural products.

3.14.5 CASE STUDY 2: POLYKETIDE AND NONRIBOSOMAL PEPTIDE NATURAL PRODUCTS

3.14.5.1 Introduction

Two families of natural products, the PKs and nonribosomal peptides (NRPs), are found in the three kingdoms of life[97] where they produce a wealth of beneficial chemicals. Many PK and NRP products exhibit a range of therapeutic benefits including antibacterial, anticancer, antiviral, antifungal, and immunosuppressant activities,[3,4] in addition to compounds that target other human diseases such as hypercholesterolemia.[6,7,98] The PKS and NRPS biosynthetic machinery follow similar assembly-line methodologies, so much so that Nature has mixed these two systems together to produce a class of hybrid PK-NRP natural products. By linking NRPS peptides to PK units from PKSs, Nature has expanded the types of compounds that can be generated by these systems[99] and made them attractive targets for the development of biocatalysts.

3.14.5.2 Biosynthesis of Polyketide Synthase and Nonribosomal Peptide Synthetase Products

PKS and NRPS systems follow assembly-line logic: they consist of proteins that are divided into distinct, multidomain modules, each of which incorporates a specific substrate into the growing molecular chain. Let us consider a system divided into three modules (loading module, modules 1, and 2), in which each module, respectively, incorporates the substrates X, Y, and Z to generate a final product X-Y-Z (Fig. 3.14.10A). Transfer of substrates between the modules is facilitated by a protein domain known as the carrier protein, which contains a conserved serine residue. This serine must be posttranslationally modified with a phosphopantetheine moiety (Ppant) derived from CoA, a reaction carried out by the Ppant transferase enzyme family.[100] The Ppant cofactor serves as an ~20-Å flexible "arm" that contains a terminal thiol, which mediates transferring molecules as activated thioesters between the enzyme active sites of different modules in the assembly line. This modularity of PKS and NRPS systems has made them attractive targets for bioengineering,[101] with the idea that by mixing and matching different modules, libraries of different compounds could be produced. These modules may exist on the same protein chain, in which the linearity of the assembly line is easily interpretable, or they may be expressed as separate proteins that must interact in a specific order in trans. This order of interaction is responsible for guiding the assembly line, and thus determining what type of molecule is produced. Such interactions are mediated by communication and docking domains,[102–105] which facilitate protein-protein interactions within NRPS-NRPS, PKS-PKS, NRPS-PKS, or PKS-NRPS junctions.

PKS synthesis is reminiscent of fatty acid synthesis by the enzyme machine fatty acid synthase (FAS).[106] There are three types of bacterial PKS systems that function in either an iterative or noniterative manner to produce a PK.[107] Here, we will focus on the type 1 noniterative multimodular systems. The conventional PKS module contains three tethered domains: β-acyl ketosynthase (KS), acyl transferase (AT), and acyl carrier protein (ACP) (Fig. 3.14.10B). The AT domain recognizes and transfers malonyl or methylmalonyl from soluble (methyl)malonyl-CoA to its respective ACP domain, using a conserved serine within the AT domain active site as a nucleophile to first attack (methyl)malonyl-CoA and generate an acyl enzyme intermediate, and then transfer the acyl group onto the carrier protein Ppant prosthetic group as a thioester. The KS domain serves three main functions: (1) the growing linear chain is transferred from the preceding ACP domain to an internal, activated cysteine residue; (2) it catalyzes decarboxylation of the ACP-bound malonyl substrate; and (3) it facilitates condensation with the KS-bound PK chain donor in a Claisen condensation. Interestingly, condensation by the KS domain inverts the stereochemistry at the C2 carbon of the

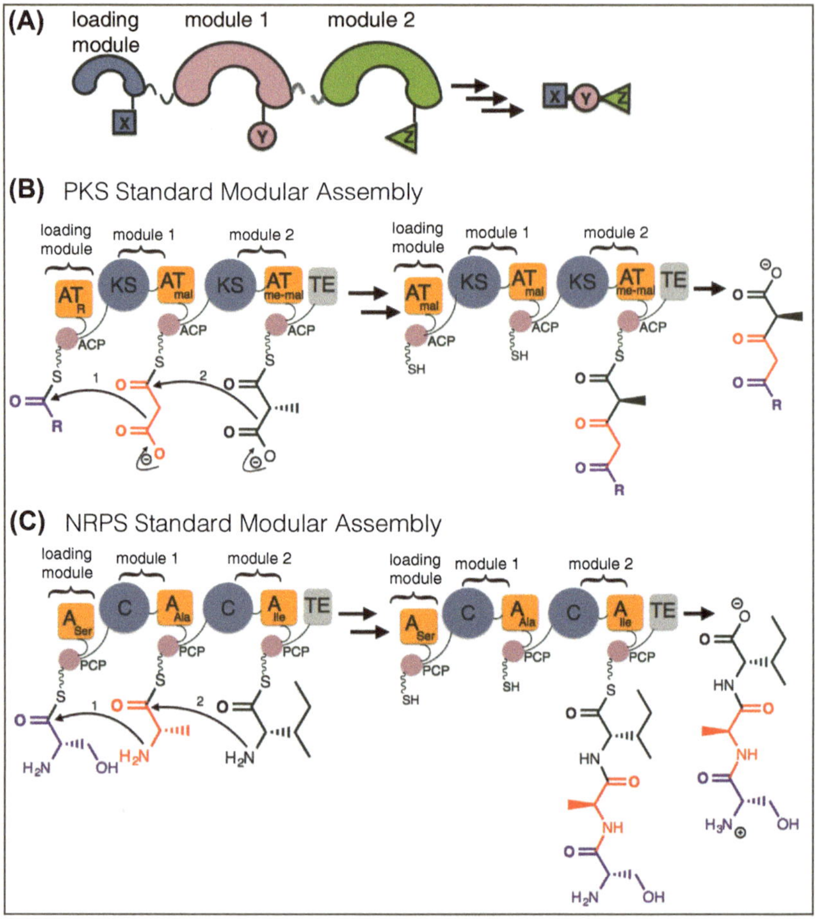

FIGURE 3.14.10 **The modular nature of polyketide synthase (PKS) and nonribosomal peptide synthetase (NRPS) systems determines the assembly line produced product.** (A) Diagram of a modular assembly line containing three separate modules, each a separate protein. Modules communicate *in trans* with each other, using complementary docking domains to define the order of the assembly line. The end result is an ordered condensation of three substrates, yielding a linear product. (B) Diagrams of a PKS protein and (C) an NRPS protein containing two standard modules and a loading module, described in the main text. *A*, adenylation domain; *ACP*, acyl carrier protein; *AT*, acyltransferase; *C*, condensation domain; *KS*, ketosynthase; *mal*: malonyl moiety; *me-mal*: methylmalonyl moiety; *PCP*, peptidyl carrier protein; *TE*, thioesterase.

L-methylmalonyl substrate.[105] Additional tailoring domains define the oxidation state of the growing chain [e.g., KR, enoylreductase (ER), and dehydratase (DH) domains], as well as possible epimerization of the C2 methyl group by certain KR domains. For example, a module containing the KR + DH pair generates an olefin, whereas the KR domain alone yields a β-hydroxyl moiety. Combining the KR, ER, and DH domains generates a fully reduced methylene unit. The final domain of a PKS system usually is the thioesterase (TE) domain, which is responsible for hydrolyzing the final PK product from its Ppant thioester; this domain also can catalyze macrocyclization of molecules at this stage. Additional reductase domains are also found that use NADPH to reduce the final product to an alcohol or aldehyde.

In NRPSs, the biosynthetic enzymes are divided into modules that are responsible for catalyzing peptide synthesis.[105,108,109] Standard NRPS modules typically consist of three domains (Fig. 3.14.10C): the adenylation (A) domain, which uses ATP to form a specific aminoacyl adenylate reminiscent of an aminoacyl synthetase enzyme; a peptidyl carrier protein (PCP) domain, which is similar to the ACP discussed earlier and contains the Ppant cofactor that reacts with the activated aminoacyl adenylate to form a thioester; and a condensation (C) domain, which recognizes two PCP-bound aminoacyl thioesters, termed the upstream and downstream donors, and catalyzes nucleophilic attack of the upstream thioester by the downstream amino acid's amino group. The product dipeptide remains tethered to the downstream PCP, and therefore could serve as a substrate for the next NRPS module. As in PKSs, NRPSs usually terminate with a TE domain that catalyzes product release and possible macrocylization. The structure of a terminal NRPS module with Ppant arm bound has been reported in 2016 (Fig. 3.14.11).[110] In this structure, a Ppant cofactor loaded onto a PCP domain is bound within its module's C domain, providing an excellent visualization of how the carrier protein must move to access all three active sites within this module: it must access the C, A, and TE domain active sites. In addition to the core domains that are responsible for normal chain elongation steps, NRPS natural products display many ornate, chemical features that are not possible without the aid of additional enzyme activities, commonly referred to as tailoring enzymes. Examples of tailoring reactions include epimerization, methylation, cyclization, reduction, and oxidation activities.

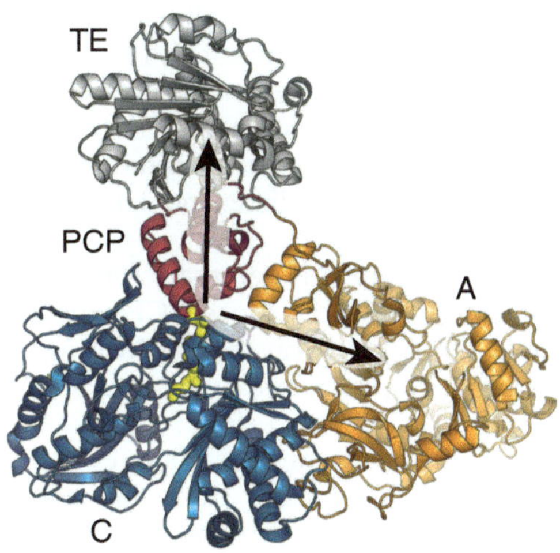

FIGURE 3.14.11 **Structure of the terminal nonribosomal peptide synthetase (NRPS) module AB3403 including its Ppant arm.** The AB3403 module includes four domains: Condensation, C; adenylation, A; peptidyl carrier protein, PCP; and thioesterase, TE (PDB ID 4ZXH). The Ppant arm, displayed as yellow sticks, must be able to reach all three active sites as indicated by the two additional cyan arrows. To achieve this, the PCP domain must be highly mobile. See Fig. 3.14.10C for a description of how NRPS systems function.

Here we will focus on two model systems, the PKS 6-deoxyerythronolide B synthase (DEBS) and the hybrid PKS-NRPS yersiniabactin (Ybt) synthetase. DEBS is known to be an archetypical modular/noniterative type I PKS, and therefore it will provide an excellent example of how modular PKSs function to yield specific PK products of pharmaceutical interest. The added discussion of Ybt synthetase will then expand on PKS systems, adding features of a hybrid PKS-NRPS system that contains deviations from standard PKS and NRPS modules, providing an introduction into how these enzyme systems can be modified in Nature to yield libraries of natural products.

3.14.5.3 6-Deoxyerythronolide B Synthase

The natural product erythromycin, produced by *Saccharopolyspora erythraea* and first isolated in 1948,[111] is an important antibiotic drug that is still used today to battle bacterial infections. It is synthesized from the aglycone precursor 6-deoxyerythronolide B (6dEB) by three, large, multimodular proteins known as DEBS1-3 (Fig. 3.14.12) that were discovered in the early 1990s.[112,113] These three separate proteins each contain two modules (in addition to a loading module within DEBS1), resulting in six overall modules that are responsible for incorporating six ketide units and an initial propionyl group to yield the final, 15-carbon macrolide 6dEB. The presence of three separate proteins, each with multiple modules, has provided an excellent PKS model system for studying assembly-line function, in addition to the presence of β-carbon processing domains within DEBS module 4.

DEBS1-3 proteins are homodimers in solution, similar to most PKS proteins.[106] Recent cryoelectron microscopy structures of an entire protein module within the pikromycin synthase provide an initial understanding of the structure of standard homodimeric PKS modules,[115] which is reminiscent of FAS structures.[116–118] Dimerization and the inherent flexibility of these modular systems enable cross-talk between the ACP and multiple active sites from each protein chain within the dimer.[20,106] Protein dimers of DEBS contain sequences at their N- and C-termini that are responsible for directed communication between respective proteins, termed communication domains. The specificity of these domains for defining specific protein partners is critical for maintaining fidelity within the assembly line: DEBS1 will preferably interact with DEBS2 and DEBS2 will preferably interact with DEBS3, and not other PKS proteins.[119]

For the assembly line of DEBS to begin, it is necessary for each ACP domain to be activated with its cofactor, the PPant moiety from CoA. Once ACP contains its PPant cofactor, the AT domain of each module within DEBS is responsible for loading a methylmalonyl group as a thioester onto the PPant arm of its respective module's ACP, except for the loading module, which catalyzes transfer of a propionyl group (Fig. 3.14.12). Assembly begins between the loading module and module 1, in which the active site cysteine of the KS of module one will accept the propionyl group as a thioester from the upstream loading module ACP domain. The KS domain also has a substrate channel for the PPant arm of the downstream ACP from module 1, enabling the propionyl group to be brought proximal to the downstream methylmalonyl moiety. At this point, the KS domain facilitates the decarboxylative Claisen condensation between the methylmalonyl and propionyl groups, yielding the 5-carbon, 2-methyl β-ketothioester. The KR domain of module 1 facilitates

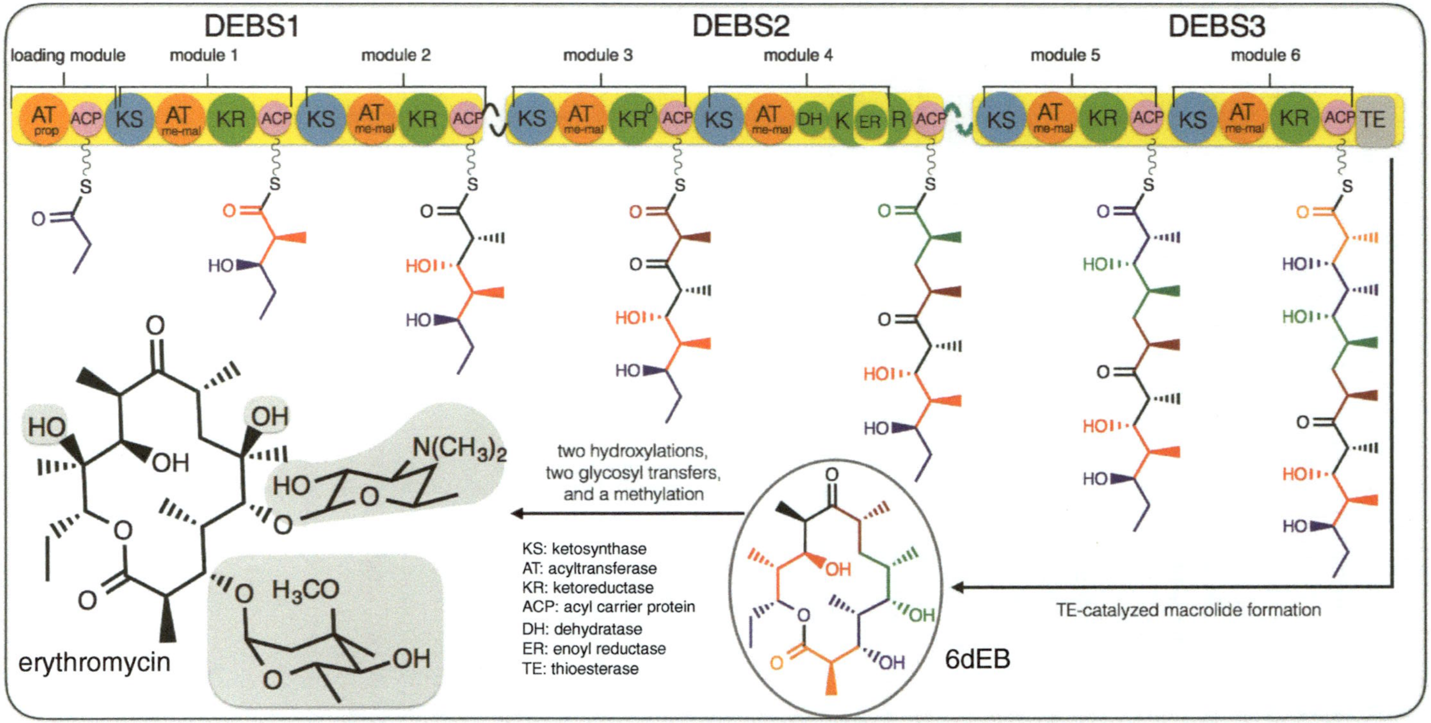

FIGURE 3.14.12 **Scheme of 6-deoxyerythronolide B synthase (DEBS) polyketide synthase (PKS) assembly line and subsequent reactions to generate erythromycin in *Saccharopolyspora erythraea*.** Domains are color coded, with β-carbon-modifying domains colored green. The ketide units incorporated into the 6-deoxyerythronolide B (6dEB) are color coded based on the module responsible for their incorporation. Modifications of 6dEB to generate erythromycin are outlined in gray. *Figure is based on Weissman KJ, Leadlay PF. Combinatorial biosynthesis of reduced polyketides.* Nat Rev Microbiol *2005;3:925–36.*

NADPH-dependent reduction of the β-ketothioester to a hydroxyl moiety, after which the product is passed to the next member of the assembly line, module 2. As the product is passed along from module 1 through module 6, it is extended through the decarboxylative Claisen condensation by a ketide unit from each downstream module. Modules 1, 2, 4, 5, and 6 each have functional KR domains, yielding reduced hydroxyl groups in the growing PK molecule. Module 3 has an inactivated KR domain (KR^0), in which the characteristic NADPH-binding sequence is altered. Module 4 has all three additional β-processing domains (the KR, DH, and ER domains), resulting in a fully reduced ketide unit that generates the deoxy portion of 6dEB. Lastly, the TE domain catalyzes lactonization and product release of the macrolide.[20]

The complexity of multiprotein PKS systems suggests a possible physical barrier for achieving rate constants on the order of simpler biocatalysts, such as a lipase from *Geobacillus stearothermophilus* (k_{cat} on the order of 10^3 per second).[37] To form one 6dEB molecule, access to 28 active sites is necessary, and therefore rates not only involve binding of substrates and product release but also extensive protein-protein interactions between different domains, as well as different proteins.[20] An analysis of reaction rates for individual, isolated DEBS domains suggests that the rate-limiting reaction step is likely to occur within the KS domain.[20] DEBS is one of the first PKS systems that has demonstrated success in heterologous expression systems to produce the final macrolide, with a yield greater than 1 g/L of 6dEB from *E. coli* cells after optimization.[120]

3.14.5.4 Yersiniabactin Synthetase

Ybt synthetase is an example of a modified hybrid PKS-NRPS system. This system is present in *Yersinia pestis* and other *Yersinia* species and enterobacteria.[121] *Y. pestis* is a pathogenic bacteria, most known for causing the bubonic plague.[122] It produces the siderophore Ybt, which chelates Fe^{3+} with a very tight formation constant of $\sim 4 \times 10^{34}$ (Fig. 3.14.13).[123] Production of Ybt has been reported as a virulence factor in *Y. pestis* because the organism needs high amounts of iron for growth, particularly during an infection.[124] Siderophore biosynthesis is prevalent in bacteria, including the prototype *E. coli* and marine organisms, in which environmental concentrations of metal ions may be extremely low. These biosynthetic systems also contain genes responsible for the selective uptake of siderophore-metal ion complexes.[123] The product Ybt is formed by four proteins: two large, multimodular proteins named "high-molecular-weight proteins 1 and 2" (HMWP1/irp1 and HMWP2/irp2), YbtE, and YbtU (Fig. 3.14.13).[125] The structure of the molecule is chemically interesting, as it contains three cyclized cysteine residues in different oxidation states. This pathway contains three heterocyclization (Cy) domains (YbtCy1-3) in place of the standard C domains of NRPS modules; each Cy domain catalyzes peptide bond formation between a downstream L-cysteine residue and a different upstream substrate, followed by cyclization to afford the thiazoline moiety.

The beginning of the assembly line[125] involves the interaction between YbtE, an adenylation domain specific for salicylic acid, and the aryl carrier protein (ArCP) from the 229-kDa HMWP2 protein (Fig. 3.14.13). Here, salicylic acid is loaded as a thioester onto the HMWP2 ArCP, generating Sal-*S*-ArCP. Now the ArCP is primed and ready to enter the assembly line. The second A domain of the system within HMWP2 is specific for L-cysteine. This is a feature

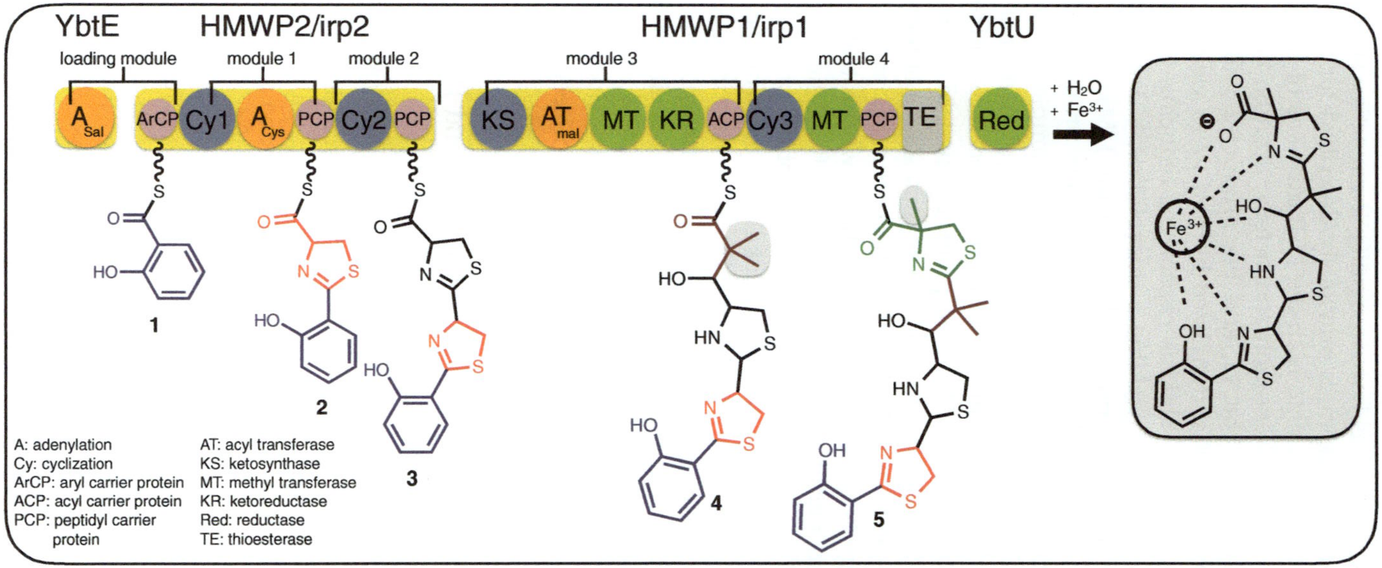

FIGURE 3.14.13 **Scheme of yersiniabactin biosynthesis in *Yersinia pestis*.** Four polyketide synthease (PKS)-nonribosomal peptide synthetase (NRPS) proteins are responsible for the biosynthesis of yersiniabactin: YbtE, high-molecular-weight protein (HMWP)2/irp2, HMWP1/irp1, and YbtU. The components of the yersiniabactin molecule are color coded based on the module responsible for their incorporation. Methyl groups from *S*-adenosyl-L-methionine are *circled*.

of A domains, and mutational studies in related systems have identified the key residues that define amino acid specificity, as well as how to modify these active sites to alter substrate specificity.[126] Using the same mechanism as previously described for salicylic acid loading of the ArCP, L-cysteine will be loaded onto both the PCP domains of HMWP2 by the same A domain. Now, with **1** present, condensation and heterocylization between salicylic acid and L-cysteine will be catalyzed by YbtCy1, and the product **2** will then be condensed with L-cysteine from the terminal PCP and cyclized by YbtCy2 to yield molecule **3**. At this stage, current research indicates that YbtU is capable of reducing the second thiazoline ring in an NADPH-dependent fashion, yielding a thiazolidine.[125] For the assembly line to continue, there must be handoff of the product between the HMWP2 PCP domain and a conserved cysteine residue within the KS domain of the next protein in the pathway, the 348-kDa HMWP1 protein. HMWP1 is an interesting protein example that contains a mixture of PKS and NRPS domains. The acyltransferase domain catalyzes loading of a malonyl group onto the HMWP1 ACP, and the KS domain then facilitates a decarboxylative Claisen condensation of this malonyl substrate with **3**. At some point, the first MT domain donates two methyl groups from two SAM molecules to yield **4**. The final stages of the synthesis involve loading of L-cysteine onto the PCP of HMWP1 (note, this occurs in trans by the A domain of HMWP2 and therefore suggests interaction between these two proteins), and subsequent condensation and cyclization performed by YbtCy3. After final methylation by the second MT domain to yield **5**, the TE domain catalyzes release of the produced Ybt molecule.

Many questions remain regarding molecular details of the Ybt pathway, including protein-protein interactions and the interesting modification reactions. None-the-less, the genes necessary to produce Ybt have been successfully incorporated into an *E. coli* expression strain[127] to yield 67 ± 21 mg of product per litre of cells, with a productivity of 1.1 ± 0.3 mg/L h. Even though this yield is less than that observed for DEBS, optimization would hopefully increase output. Such production has many benefits: complex cyclizations and modifications can be introduced into these molecules with relative ease because they are constantly tethered to carrier proteins.

3.14.5.5 Manipulating Polyketide Synthase/Nonribosomal Peptide Synthetase Systems

Since the discovery of their modular nature, there has been interest in developing PKS and NRPS systems for the production of fine chemicals and pharmaceuticals.[20,114,128] Researchers are approaching this system from many different angles.[55] Originally, after identification of domain boundaries through sequence alignments, simplistic domain swapping was performed. Surprisingly, some of these experiments were successful, supporting the concept that replacement of domains within a large assembly line would not completely shut down the production.[129–131] However, there are limitations, and as we obtain more structural data for PKS and NRPS domains and modules, it is becoming clear that we must consider not only domain replacement but also the size and features of enzyme active sites. A modified system must be able to accommodate a certain substrate, and then every subsequent active site in the remaining assembly line must also be able to function with it, ideally without a loss in catalytic rate. Additionally, it is becoming clear that domain linkers and surface residues define interactions between domains. An analysis of surface residues of ACP domains has provided a nice representation of how different areas of a protein surface may mediate interactions with specific PKS or NRPS domains. Alekseyev et al. have analyzed the different ACP domains of DEBS1-3 proteins[20] by generating homology models based on the NMR structure of an ACP from DEBS module 2. Clear differences in surface electrostatics are observed that allow interpretation of kinetics data from domain-swapping studies within DEBS.

Although there is great interest to develop systems for efficient in vitro production of chemicals, the large sizes of PKS and NRPS systems has made this challenging. Recent structures, however, of entire PKS and NRPS modules are providing critical information toward the development and bioengineering of these systems.[106,110,115,132,133] The fact that many PKS and NRPS modules are split into separate proteins that contain docking domains has provided an initial area of study regarding how to manipulate and engineer these systems to produce novel chemicals. By swapping and exchanging protein communication/docking domains between different proteins, there has been success in generating new compounds.[131,134–136] Swapping of docking domains between the DEBS1-3 proteins led to altered product formation without a large penalty in reaction rates.[119] There has also been success in using nonnatural starter and extender units by replacing the AT domain with homologous domains from other PKS systems, resulting in diversification of produced products.[20] However, due to the extensive protein-protein interactions between different domains, and the likely differences in binding sites for different types and sizes of substrates, a simple swapping of the order of these systems will not likely succeed in generating a highly efficient system for the production of chemicals. Therefore, additional studies at the molecular level to characterize the necessary interactions of these systems will be required to enable their redesign into effective biocatalysts in the future.

Lastly, we would like to consider how directed evolution-based techniques can be applied to PKS and NRPS systems.[137,138] The complexities and size of these systems make them a challenge for standard directed evolution of a single protein. For example, inactivation of just one domain may interrupt the entire assembly line if it is early enough in the pathway or it will greatly reduce its efficiency. There has been success in optimizing A domains by directed evolution, which has been reviewed in 2016,[138] and Fischbach et al. have demonstrated success in using directed evolution methods to restore and improve activity of a chimeric NRPS system.[139] In 2015, work by Chemler et al. took a different approach to generating genetic libraries of PKS systems. In this work, the authors used homologous recombination to generate hybrid libraries of PKS genes from the pikromycin and erythromycin pathways in *S. cerevisiae*.[140]

There is a large interest in understanding how PKS and NRPS systems function at the atomic level and in developing methods to bioengineer these systems for production of novel chemicals. Although the complexities of these systems may make them challenging, it is also their complexity and modular nature that make them attractive as an area of future biocatalysts, able to create complex molecules with regio- and stereoselectivity.

3.14.6 CASE STUDY 3: RIBOZYMES AS BIOCATALYSTS

Ribonucleic enzymes, commonly referred to as ribozymes, are catalytic RNA molecules. The discovery of ribozymes in 1982 first introduced the possibility that catalytic

TABLE 3.14.2 Selected Ribozyme-Catalyzed Reactions

Reaction Catalyzed	Bond Formed or Broken	Rate Enhancement
Aldol reaction	C-C	4×10^3
Alcohol oxidation	C-H	1×10^7
Aldehyde reduction	C-H	3×10^6
Acyl transfer	C-O	1×10^{10}
Aminoacylation	C-O	2×10^5
Porphyrin metalation	Cu-N	500
Palladium nanoparticle formation	Pd-Pd	Nd

Adapted from Silverman SK. Nucleic acid enzymes (ribozymes and deoxyribozymes): in vitro selection and application. In: Wiley Encyclopedia of chemical biology. *John Wiley & Sons, Inc.: Hoboken, NJ, USA; 2008. pp. 1–17.*

biomolecules extend beyond just proteins[141,142] and gave further evidence to the "RNA World" hypothesis, which theorizes that RNA is the common precursor to all life on the earth.[143,144] Naturally occurring ribozymes have been found that are capable of catalyzing both phosphodiester cleavage and ligation, and possibly the formation of peptide bonds; however, their range of catalytic application is somewhat limited. Artificial ribozymes identified from synthetic combinatorial libraries, on the other hand, display a much wider range of catalytic activity, including acyl transfer, aminoacylation, porphyrin metalation, and palladium nanoparticle formation, among others. An extensive list of the reactions catalyzed by artificial ribozymes, and their respective references, has been compiled,[145] with a few representatives listed in Table 3.14.2. From this list, the Diels-Alder C—C bond-forming reactions stand out as being particularly important for both industrial and pharmaceutical applications. The Diels-Alder reaction is a [4 + 2] cycloaddition between a diene and alkene, resulting in a 6-membered cyclohexene system. These reactions are particularly valuable in synthetic organic chemistry because they allow for regio- and stereocontrol over the products. As such, this last case study will discuss ribozymes as related to Diels-Alder reactions (Diels-Alderases), and specifically focus on the detailed evolution of ribozyme discovery and mechanistic investigation.

3.14.6.1 Diels-Alderases: Ribozymes That Catalyze Diels-Alder Reactions

Some of the first Diels-Alderase ribozymes were discovered using an in vitro selection method designed to identify RNA molecules with Diels-Alder activity,[146,147] somewhat analogous to the directed evolution methods described earlier. In a general selection process, a synthetically generated library is applied to an affinity matrix (Fig. 3.14.14A).[148] Molecules expressing affinity to the matrix are retained, whereas those lacking affinity are washed away. The retained molecules are then eluted from the matrix and amplified by polymerase chain reaction. This process is repeated several times until high-affinity molecules are isolated, at which point the individual sequences can be compared to identify conserved oligonucleotide sequences and structural motifs.

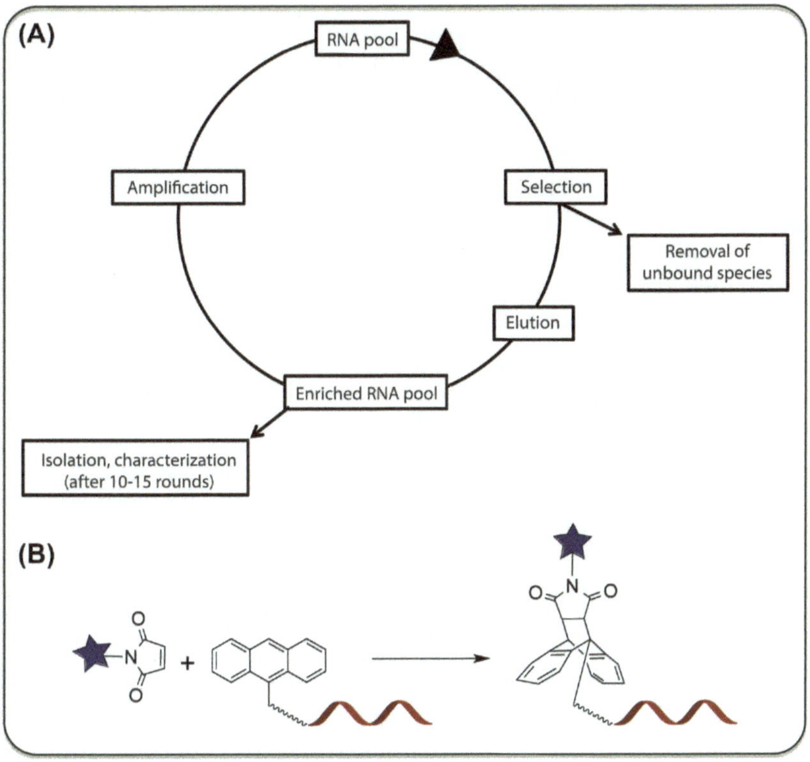

FIGURE 3.14.14 **Overview of RNA selection process.** (A) General scheme of in vitro selection of engineered ribozymes. (B) Scheme of catalytic RNA selection process. Biotinylated maleimide reacts with the RNA-tagged anthracene to generate a biotinylated-RNA-linked Diels-Alder product. The *blue star* represents biotin, and the *red ribbon* represents fused RNA.

In the Seelig study,[147] anthracene (the diene) was tagged to the 5′ end of each RNA molecule in a randomized pool by a polyethylene linker; the starting library included $\sim 2 \times 10^{14}$ sequence variants with 120 randomized nucleotides. The RNA-tagged anthracene molecules were incubated with maleimide (the dienophile) containing biotin at the side chain position, which resulted in the formation of a Diels-Alder product (Fig. 3.14.14B). The resulting product was both biotinylated and linked to the catalytic RNA such that, after immobilization on streptavidin agarose to remove any unreacted anthracene, the selected RNA molecules could be identified and amplified. Approximately 90% of the 13 isolated active ribozymes included two consensus sequences: UGCCA and AAUACU. By identifying other similarities from the 13 isolated ribozymes, three optimized sequences were developed; all showed catalytic activity comparable to the parent sequences, with a synthetic 49-mer being the most active ($K_{app} \sim 144\ M^{-1}\ s^{-1}$).[147] The proposed general sequence and overall structure matched nicely with a subsequently solved X-ray crystal structure of a Diels-Alderase ribozyme (Fig. 3.14.15).[149] There are three helical segments, an internal loop containing the two consensus sequences, and a 5′ terminal GGAG sequence in which the first G is the attachment site for anthracene. The ribozyme adopts an overall λ-shaped fold and, unlike other naturally

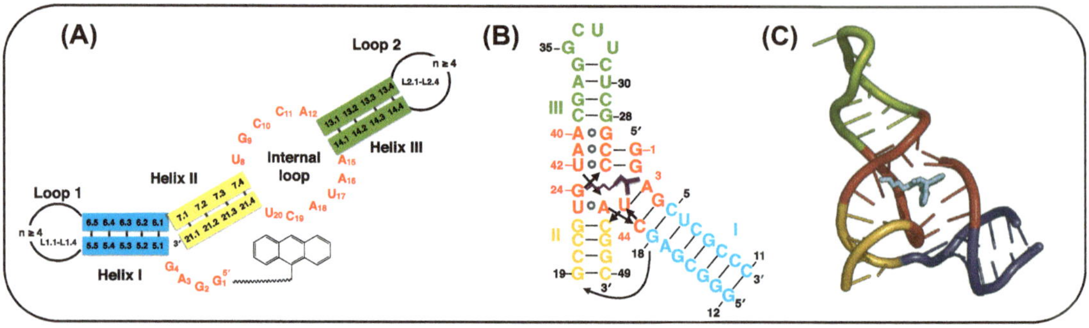

FIGURE 3.14.15 **Structure elucidation of Diels-Alderase ribozyme.** (A) Proposed general structure and sequence of the Diels-Alderase ribozyme. The two consensus sequences are located around the internal loop of the structure. (B) Tertiary fold and (C) X-ray crystal structure of ribozyme-product complex. *(A) Adapted from Elsevier: Keiper S, Bebenroth D, Seelig B, Westhof E, Jäschke A. Architecture of a Diels-Alderase ribozyme with a preformed catalytic pocket. Chem Biol 2004;11:1217–27, copyright (2004). (C) Adapted from Macmillan Publishers Ltd: Serganov A, Keiper S, Malinina L, Tereshko V, Skripkin E, Höbartner C, et al. Structural basis for Diels-Alder ribozyme-catalyzed carbon-carbon bond formation. Nat Struct Mol Biol 2005;12:218–24 copyright (2005).*

occurring ribozymes, the three helices pack against the catalytic pocket to form a pseudoknot (Fig. 3.14.15B and C). Interestingly, the hepatitis delta helper virus (HDV) ribozyme also adopts a pseudoknot fold[150]; the HDV and Diels-Alder ribozymes are two of the fastest known ribozymes. This pseudoknot is formed, in part, by base pairing of the 5′ terminal GGAG sequence. The X-ray crystal structure also revealed eight Mg^{2+}-binding sites; six of these ions aid in stabilizing the tertiary structure, while two interact with other ribozymes in the crystal lattice.[151]

To further understand the overall structure and active site of the synthetic 49-mer, mutagenesis and enzymatic studies were performed.[152] Mutagenic studies found that the highest activity was observed with intact and properly paired helices 1 and 2 of the structure (see Fig. 3.14.15), even though some mutations within these two helices were tolerated. Mutations to helix 3, on the other hand, which had a nonuniform length and sequence in the 13 originally selected ribozymes, did not strongly affect catalysis, and mismatches were better tolerated.[152] Mutagenic studies of the internal loop determined that the upper pentanucleotide region strongly favored conservation compared to the lower hexanucleotide region, and that U_8, C_{10}, C_{11}, and U_{20} were required for proper catalytic function. Functional studies also determined that the 5′ terminal sequence GGAG was required for ribozyme activity; no activity was observed with mutations in positions 1 or 2, and a 50% decrease in activity was observed for mutations at positions 3 and 4. Taken together, these results highlight the importance of the pseudoknot structure of the ribozyme: nucleotides G1 and G2 pair with two cytosines on one strand of the internal loop (from the upper pentanucleotide regions), whereas nucleotides A3 and G4 pair with uracil and cytosine, respectively, from the lower hexanucleotide region on the opposite strand.

An initial question regarding the structure and function of small ribozymes, as well as their DNAzyme counterparts, is whether the active site is preformed for catalysis. Structures of RNA aptamers, which are evolved to bind small molecules within an RNA structure, demonstrate structural changes upon binding and release of ligands; therefore, it was somewhat surprising to consider the active site structure of the developed Diels-Alderase ribozymes. The X-ray crystal structure of a Diels-Alderase ribozyme was effectively identical both with and without substrate,[151] in support of previous chemical studies that found no major structural changes upon substrate binding.[145] Together, these observations suggest that the ribozyme assumes a preformed catalytic pocket.[152] Notably, the Mg^{2+} ions within the crystal structure are not positioned near the active site, indicating that they do not participate directly in catalysis. Furthermore, the active site lacks backbone phosphates, and the sugar rings do not appear to hydrogen bond with the Diels-Alder product, indicating a hydrophobic binding site. Such a binding pocket is unusual for RNA, yet optimal for orienting the largely hydrophobic substrates of a Diels-Alder reaction.

Although these ribozymes were undeniably effective catalysts and accelerated the anthracene-maleimide reaction nearly 20,000-fold, they were initially assessed based on the anthracene being linked to RNA. However, such a system would display altered reaction kinetics compared with a "true" RNA catalyst, in which both organic reactants would ideally be free in solution. As such, the previously constructed synthetic 49-mer was used to catalyze the reaction between untethered anthracene and maleimide derivatives; kinetic rates were monitored with ultraviolet spectroscopy.[149] The 49-mer was indeed catalytically active with untethered reactants, which led to a more in-depth study of a series of anthracene and maleimide derivatives (see later). From these preliminary studies, it appeared that the maleimide moiety and alkyl side chain were the structural features required for proper ribozyme recognition and activity. Remarkably, whereas the uncatalyzed Diels-Alder reaction resulted in a racemic mixture of products, the reaction catalyzed by the 49-mer produced a 20:1 ratio of enantiomers, which corresponds to an overall >90% enantiomeric excess. Furthermore, when the 49-mer was synthesized with L-, rather than D-nucleotides, the opposite product ratio (1:20) was obtained.[149] Importantly, this was the first example of a catalytically active L-nucleic acid.

To better characterize substrate recognition and the 49-mer active site, a series of 44 anthracene and maleimide derivatives were used.[153] A brief discussion of a few key compounds will be presented here, but a thorough reading of this detailed mechanistic study is highly recommended. Briefly, methyl or hydroxymethyl groups were added to the five unique positions around the anthracene rings (Fig. 3.14.16). A hexa(ethylene glycol) was added to some derivatives to improve substrate solubility. Overall, these derivatives revealed that the addition of groups in the 1, 4, and 5 positions (compounds **8**, **9**, and **10**) did not hinder overall ribozyme activity, despite displaying slower kinetics than parent compounds **6** and **7** (only ~1.8%–8% decreased rate enhancements). In contrast, substituents in the 2 and 3 positions (compound **11**) dramatically decreased ribozyme activity, suggesting that the active site is only large enough to accommodate three unsubstituted rings (~10 Å). The two most reactive anthracene derivatives had a hydroxymethyl and carboxylic acid at position 1 (compounds **12** and **13**); although the catalyzed rates were similar to parent compounds **6** and **7**, the uncatalyzed rates were significantly less, leading to an overall rate enhancement of more than 600- and 300-fold, respectively.

With respect to the dienophile, a five-membered maleimidyl ring with at least an ethyl side chain was required for reactivity; maleimide ring opening resulted in activity comparable to that of the background, while maximal activity was observed with a pentyl side chain (compounds **15** and **14**, respectively).[153] The addition of a methyl group on the reactive double

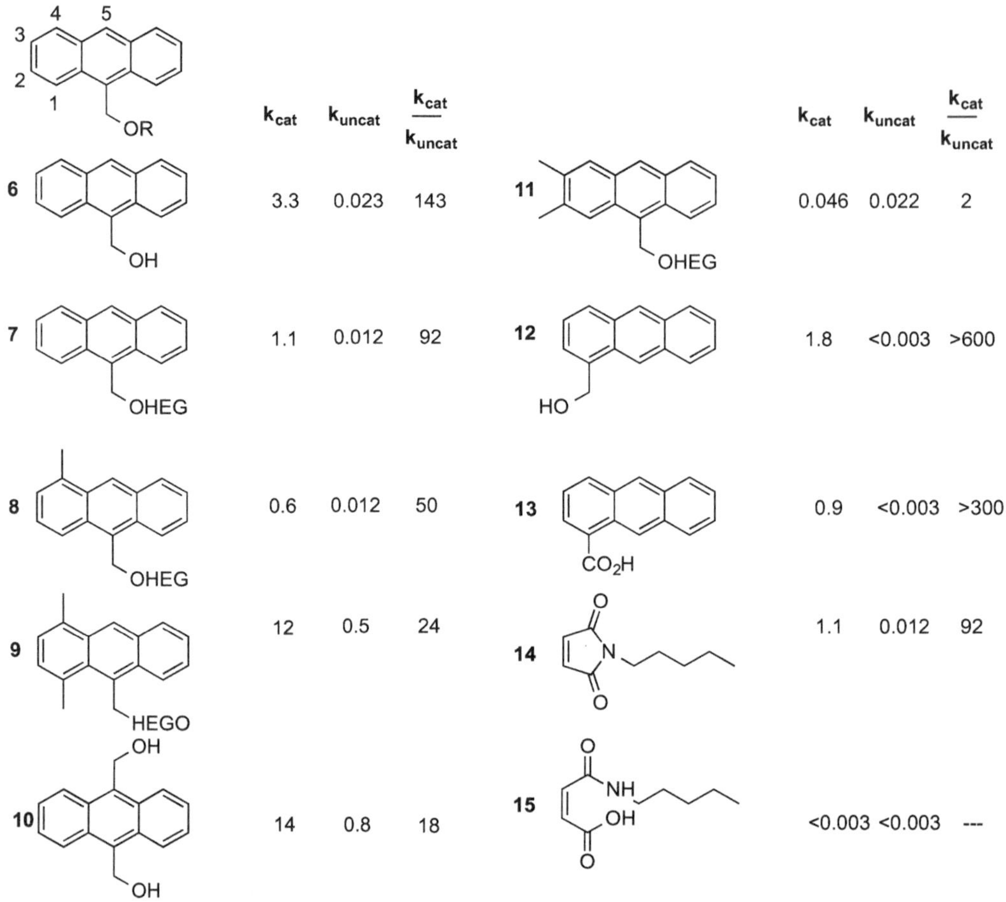

FIGURE 3.14.16 **Representative anthracene and maleimide derivatives and their respective rate constants.** Compound numbering is adjusted from the original reference to exclude omitted structures. *Adapted from Stuhlmann F, Jäschke A. Characterization of an RNA active site: interactions between a Diels–Alderase ribozyme and its substrates and products.* J Am Chem Soc *2002;**124**:3238–44.*

bond, or a single bond in place of the reactive double bond, showed no catalysis over the background reaction (compounds **14** and **7**). As predicted, these observations confirmed that an unsubstituted double bond was required for both catalysis and ribozyme recognition. Finally, the addition of polar groups on the alkyl side chains resulted in an overall decrease in reactivity, confirming that the binding pocket favors hydrophobic substrates and/or van der Waals interactions.[153] Taken together, these data support a general Diels-Alderase binding pocket, where the active site orients the two reactants in a stacked, coplanar fashion, as expected for traditional Diels-Alder chemistry.[153]

These functional studies provide important activity data to relate with the X-ray crystal structure of the Diels-Alderase ribozyme.[153] The physical dimensions of the active site correlate with the effects of anthracene and maleimide derivatives.[153] Substituents along the shorter edges of anthracene (i.e., compound **11**) do not properly fit into the active site, and for example, result in a loss of activity. A longer alkyl side chain on maleimide (i.e., pentyl vs. ethyl) results in higher activity, however, likely due to increased interactions with the hydrophobic binding pocket. Furthermore, the proper positioning of the maleimide side chain also allows formation of only one product enantiomer, as discussed earlier.

The initial discovery, mechanistic studies, and structural elucidation of these Diels-Alderases have led to the following proposed mechanism: the ribozyme likely binds the substrates such that anthracene makes stacking interactions with nearby nucleotides, and maleimide, which is oriented by both its alkyl side chain and hydrogen bonding interactions, is positioned on top of the anthracene. Together, these interactions hold the substrates in proper alignment to facilitate the Diels-Alder reaction. Specific stacking interactions between anthracene and nearby nucleotides increase electron density on the diene, while hydrogen bonding causes maleimide to become more electron deficient, thus enhancing catalysis. Computational studies also predict that the active site is complementary to the transition state, further lowering the energy barrier of the reaction.[151] A detailed comparison of the proposed Diels-Alderase mechanism with other known protein-catalyzed reactions has been reported.[151]

This final case study was intended to present the mechanistic assessment of a series of known Diels-Alderase ribozymes. However, this short review represents only some of the initial studies in this field. A number of additional studies have been published, including advanced techniques for ribozyme selection,[154] control of product stereoselectivity,[155] and further detailed analyses of structure-function relationships,[156] among others.

3.14.7 BENEFITS AND DRAWBACKS OF RNA CATALYSTS

There are many advantages, as well as some challenges, to working with RNA in both industrial and therapeutic capacities.[148,157] The main challenges of working with RNA include solvent systems and handling. Although most organic reactions are carried out in nonpolar solvents, RNA is hydrophilic, which can lead to problems with solubility and/or stability. The ability to add hydrophobic moieties to nucleotides, as well as incorporate stabilizing metal cations, however, allows for increased solubility and overall reactivity. Additionally, using organic-nonorganic solvent mixtures in the selection process may identify sequences that form stable structures under such conditions. Furthermore, RNA is susceptible to degradation by highly active enzymes called RNases, which are present in the air, most aqueous solutions, and hair and skin. For these reasons, strict laboratory protocols are required, including working in RNA-only designated areas, using RNAse-free solutions and/or RNase inhibitors, using RNase-free plasticware and/or maintaining stringent procedures for cleaning and sterilizing glassware, and changing gloves often.

One of the key benefits of working with ribozymes, particularly when compared with traditional protein enzymes, however, is that RNA can be easily synthesized and manipulated by automated solid-phase techniques. Such automated synthesizers allow for the rapid production of randomized sequences and can produce up to 4^N sequences (where N represents the number of nucleotides), creating huge combinatorial libraries for selection processes (as described earlier). RNA is also considered to be thermodynamically stable, particularly compared with DNA, and can adopt various folds based on nucleotide pairing. Because RNA can form aromatic stacking, hydrogen bonds and electrostatic interactions with small molecules, and bind metal cations, it is promising for many catalytic applications.

3.14.8 CONCLUSIONS

Nature has provided an extensive chemical toolbox teeming with diverse biocatalysts for industrial applications. Tapping into this kit has enabled the development of green synthetic methods that have dramatically lowered waste and enabled the formation of complex natural products that are extremely difficult to obtain by standard organic syntheses, and future development of biocatalysts will continue to provide green alternatives to chemical synthesis. With the boom of available genomic information, there is much excitement in researching complex natural product biosynthetic pathways and the bioengineering of known enzymes and ribozymes to achieve competent and efficient biocatalysts for industrial applications. The continued development of bioengineering methods, heterologous expression platforms, and work toward de novo design of biocatalysts will be exciting areas to watch in the future of green chemistry.

References

1. Dias DA, Urban S, Roessner U. A historical overview of natural products in drug discovery. *Metabolites* 2012;**2**:303–36.
2. Katz L, Baltz RH. Natural product discovery: past, present, and future. *J Ind Microbiol Biotechnol* 2016;**43**:155–76.
3. Newman DJ, Cragg GM. Natural products as sources of new drugs over the last 25 years. *J Nat Prod* 2007;**70**:461–77.
4. Koehn FE, Carter GT. The evolving role of natural products in drug discovery. *Nat Rev Drug Discov* 2005;**4**:206–20.
5. Beutler JA. Natural products as a foundation for drug discovery. In: *Current protocols in pharmacology*. Hoboken, NJ, USA: John Wiley & Sons, Inc; 2009. pp. 9.11.1–9.11.21.
6. von Nussbaum F, Brands M, Hinzen B, Weigand S, Häbich D. Antibacterial natural products in medicinal chemistry—exodus or revival? *Angew Chem Int Ed Engl* 2006;**45**:5072–129.
7. Gordaliza M. Natural products as leads to anticancer drugs. *Clin Transl Oncol* 2007;**9**:767–76.
8. Ling LL, Schneider T, Peoples AJ, Spoering AL, Engels I, Conlon BP, et al. A new antibiotic kills pathogens without detectable resistance. *Nature* 2015;**517**:455–9.
9. Motwani HV, Fred C, Haglund J, Golding BT, Törnqvist M. Cob(I)alamin for trapping butadiene epoxides in metabolism with rat S9 and for determining associated kinetic parameters. *Chem Res Toxicol* 2009;**22**:1509–16.
10. Argueta EA, Amoh AN, Kafle P, Schneider TL. Unusual non-enzymatic flavin catalysis enhances understanding of flavoenzymes. *FEBS Lett* 2015;**589**:880–4.
11. Krampitz L. Catalytic functions of thiamin diphosphate. *Annu Rev Biochem* 1969;**38**:213–40.
12. Xu L-W, Lu Y. Primary amino acids: privileged catalysts in enantioselective organocatalysis. *Org Biomol Chem* 2008;**6**:2047–53.
13. Bornscheuer UT, Huisman GW, Kazlauskas RJ, Lutz S, Moore JC, Robins K. Engineering the third wave of biocatalysis. *Nature* 2012;**485**:185–94.
14. Barbas CF. Organocatalysis lost: modern chemistry, ancient chemistry, and an unseen biosynthetic apparatus. *Angew Chem Int Ed* 2008;**47**:42–7.

15. Ishikawa F, Uno K, Nishikawa M, Tsumuraya T, Fujii I. Antibody-catalyzed decarboxylation and aldol reactions using a primary amine molecule as a functionalized small nonprotein component. *Bioorg Med Chem* 2013;**21**:7011–7.
16. Zhong G, Lerner RA, Barbas CF. Broadening the aldolase catalytic antibody repertoire by combining reactive immunization and transition state theory: new enantio- and diastereoselectivities. *Angew Chem Int Ed* 1999;**38**:3738–41.
17. Van Loo B, Kingma J, Arand M, Wubbolts MG, Janssen DB. Diversity and biocatalytic potential of epoxide hydrolases identified by genome analysis. *Appl Environ Microbiol* 2006;**72**:2905–17.
18. Schulze B, Wubbolts MG. Biocatalysis for industrial production of fine chemicals. *Curr Opin Biotechnol* 1999;**10**:609–15.
19. Xie X, Watanabe K, Wojcicki WA, Wang CCC, Tang Y. Biosynthesis of lovastatin analogs with a broadly specific acyltransferase. *Chem Biol* 2006;**13**:1161–9.
20. Khosla C, Tang Y, Chen AY, Schnarr NA, Cane DE. Structure and mechanism of the 6-deoxyerythronolide B synthase. *Annu Rev Biochem* 2007;**76**:195–221.
21. Christianson DW. Unearthing the roots of the terpenome. *Curr Opin Chem Biol* 2008;**12**:141–50.
22. Windle CL, Müller M, Nelson A, Berry A. Engineering aldolases as biocatalysts. *Curr Opin Chem Biol* 2014;**19**:25–33.
23. Royer SF, Haslett L, Crennell SJ, Hough DW, Danson MJ, Bull SD. Structurally informed site-directed mutagenesis of a stereochemically promiscuous aldolase to afford stereochemically complementary biocatalysts. *J Am Chem Soc* 2010;**132**:11753–8.
24. Ma SK, Gruber J, Davis C, Newman L, Gray D, Wang A, et al. A green-by-design biocatalytic process for atorvastatin intermediate. *Green Chem* 2010;**12**:81–6.
25. Renault H, Bassard J, Hamberger B, Werck-Reichhart D. Cytochrome P450-mediated metabolic engineering: current progress and future challenges. *Curr Opin Plant Biol* 2014;**19**:27–34.
26. Rudat J, Brucher BR, Syldatk C. Transaminases for the synthesis of enantiopure beta-amino acids. *AMB Express* 2012;**2**:11.
27. Dunn PJ, Wells AS, Williams (Eds.) MT. *Green chemistry in the pharmaceutical industry*. Weinheim, Germany: Wiley-VCH Verlag GmbH & Co. KGaA; 2010.
28. Turner NJ. Directed evolution drives the next generation of biocatalysts. *Nat Chem Biol* 2009;**5**:567–73.
29. Schmid A, Dordick JS, Hauer B, Kiener A, Wubbolts M, Witholt B. Industrial biocatalysis today and tomorrow. *Nature* 2001;**409**:258–68.
30. Schoemaker HE. Dispelling the myths—biocatalysis in industrial synthesis. *Science* 2003;**299**:1694–7.
31. Fox RJ, Clay MD. Catalytic effectiveness, a measure of enzyme proficiency for industrial applications. *Trends Biotechnol* 2009;**27**:137–40.
32. Qian Z, Fields CJ, Lutz S. Investigating the structural and functional consequences of circular permutation on lipase B from *Candida Antarctica*. *ChemBioChem* 2007;**8**:1989–96.
33. Dröge MJ, Boersma YL, van Pouderoyen G, Vrenken TE, Rüggeberg CJ, Reetz MT, et al. Directed evolution of *Bacillus subtilis* lipase A by use of enantiomeric phosphonate inhibitors: crystal structures and phage display selection. *ChemBioChem* 2006;**7**:149–57.
34. Zanghellini A. *De novo* computational enzyme design. *Curr Opin Biotechnol* 2014;**29**:132–8.
35. Tomchick DR, Phan P, Cymborowski M, Minor W, Holman TR. Structural and functional characterization of second-coordination sphere mutants of soybean lipoxygenase-1. *Bioechemistry* 2001;**40**:7509–17.
36. Leferink NGH, Antonyuk SV, Houwman JA, Scrutton NS, Eady RR, Hasnain SS. Impact of residues remote from the catalytic centre on enzyme catalysis of copper nitrite reductase. *Nat Commun* 2014;**5**:4395.
37. Dror A, Shemesh E, Dayan N, Fishman A. Protein engineering by random mutagenesis and structure-guided consensus of *Geobacillus stearothermophilus* lipase T6 for enhanced stability in methanol. *Appl Environ Microbiol* 2014;**80**:1515–27.
38. Nielsen J. Yeast cell factories on the horizon. *Science* 2015;**349**:1050–1.
39. Sørensen HP, Mortensen KK. Soluble expression of recombinant proteins in the cytoplasm of *Escherichia coli*. *Microb Cell Fact* 2005;**4**:1.
40. Alonso-Gutierrez J, Chan R, Batth TS, Adams PD, Keasling JD, Petzold CJ, et al. Metabolic engineering of *Escherichia coli* for limonene and perillyl alcohol production. *Metab Eng* 2013;**19**:33–41.

41. Tobert JA. Lovastatin and beyond: the history of the HMG-CoA reductase inhibitors. *Nat Rev Drug Discov* 2003;**2**:517–26.
42. Kuklina EV, Shaw KM, Hong Y. Vital signs: prevalence, treatment, and control of high levels of low-density lipoprotein cholesterol—United States, 1999–2002 and 2005–2008. *MMWR Morb Mortal Wkly Rep* 2011;**60**:109–14.
43. Endo A, Kuroda M, Tanzawa K. Competitive inhibition of 3-hydroxy-3-methylglutaryl coenzyme A reductase by ML-236A and ML-236B fungal metabolites, having hypocholesterolemic activity. *FEBS Lett* 1976;**72**:323–6.
44. Alberts AW, Chen J, Kuron G, Hunt V, Huff J, Hoffman C, et al. Mevinolin: a highly potent competitive inhibitor of hydroxymethylglutaryl-coenzyme A reductase and a cholesterol-lowering agent. *Proc Natl Acad Sci USA* 1980;**77**:3957–61.
45. Fox RJ, Davis SC, Mundorff EC, Newman LM, Gavrilovic V, Ma SK, et al. Improving catalytic function by Pro-SAR-driven enzyme evolution. *Nat Biotechnol* 2007;**25**:338–44.
46. Gilson L, Collier SJ, Sukumaran J, Yeo WL, Alviso O, Teo EL, et al. *Variant LovD polypeptides and their uses U.S. Patent No. WO/2011/041231*. Washington DC: U.S. Patent and Trademark Office; 2011.
47. Xie X, Pashkov I, Gao X, Guerrero JL, Yeates TO, Tang Y. Rational improvement of simvastatin synthase solubility in *Escherichia coli* leads to higher whole-cell biocatalytic activity. *Biotechnol Bioeng* 2009;**102**:20–8.
48. Xie X, Wong WW, Tang Y. Improving simvastatin bioconversion in *Escherichia coli* by deletion of bioH. *Metab Eng* 2007;**9**:379–86.
49. Xie X, Tang Y. Efficient synthesis of simvastatin by use of whole-cell biocatalysis. *Appl Environ Microbiol* 2007;**73**:2054–60.
50. Sheldon RA. Enzyme immobilization: the quest for optimum performance. *Adv Synth Catal* 2007;**349**:1289–307.
51. Krivoruchko A, Zhang Y, Siewers V, Chen Y, Nielsen J. Microbial acetyl-CoA metabolism and metabolic engineering. *Metab Eng* 2015;**28**:28–42.
52. Qin J, Zhou YJ, Krivoruchko A, Huang M, Liu L, Khoomrung S, et al. Modular pathway rewiring of *Saccharomyces cerevisiae* enables high-level production of L-ornithine. *Nat Commun* 2015;**6**:8224.
53. Zhou YJ, Gao W, Rong Q, Jin G, Chu H, Liu W, et al. Modular pathway engineering of diterpenoid synthases and the mevalonic acid pathway for miltiradiene production. *J Am Chem Soc* 2012;**134**:3234–41.
54. Billingsley JM, DeNicola AB, Tang Y. Technology development for natural product biosynthesis in *Saccharomyces cerevisiae*. *Curr Opin Biotechnol* 2016;**42**:74–83.
55. Wu MC, Law B, Wilkinson B, Micklefield J. Bioengineering natural product biosynthetic pathways for therapeutic applications. *Curr Opin Biotechnol* 2012;**23**:931–40.
56. Stevens DC, Hari PA, Boddy CN, Hari TPA. The role of transcription in heterologous expression of polyketides in bacterial hosts. *Nat Prod Rep* 2013;**30**:1391–411.
57. DeLoache WC, Dueber JE. Compartmentalizing metabolic pathways in organelles. *Nat Biotechnol* 2013;**31**:320–1.
58. Halling PJ, Gupta MN. Measurement and reporting of data in applied biocatalysis. *Perspect Sci* 2014;**1**:98–109.
59. Yang J, Xian M, Su S, Zhao G, Nie Q, Jiang X, et al. Enhancing production of bio-isoprene using hybrid MVA pathway and isoprene synthase in *E. coli*. *PLoS One* 2012;**7**:e33509.
60. Ro D-K, Paradise EM, Ouellet M, Fisher KJ, Newman KL, Ndungu JM, et al. Production of the antimalarial drug precursor artemisinic acid in engineered yeast. *Nature* 2006;**440**:940–3.
61. Wilson SA, Roberts SC. Recent advances towards development and commercialization of plant cell culture processes for the synthesis of biomolecules. *Plant Biotechnol J* 2012;**10**:249–68.
62. Toogood HS, Ní Cheallaigh A, Tait S, Mansell DJ, Jervis A, Lygidakis A, et al. Enzymatic menthol production: one-pot approach using engineered *Escherichia coli*. *ACS Synth Biol* 2015;**4**:1112–23.
63. Paddon CJ, Westfall PJ, Pitera DJ, Benjamin K, Fisher K, McPhee D, et al. High-level semi-synthetic production of the potent antimalarial artemisinin. *Nature* 2013;**496**:528–32.
64. Wani MC, Horwitz SB. Nature as a remarkable chemist: a personal story of the discovery and development of Taxol. *Anticancer Drugs* 2014;**25**:482–7.
65. Aharoni A, Jongsma MA, Bouwmeester HJH. Volatile science? Metabolic engineering of terpenoids in plants. *Trends in Plant Sci* 2005;**10**:594–602.
66. Aaron JA, Christianson DW. Trinuclear metal clusters in catalysis by terpenoid synthases. *Pure Appl Chem* 2010;**82**:1585–97.
67. Ogura K, Koyama T. Enzymatic aspects of isoprenoid chain elongation. *Chem Rev* 1998;**98**:1263–76.

68. Abe I. Enzymatic synthesis of cyclic triterpenes. *Nat Prod Rep* 2007;**24**:1311–31.
69. Kitaoka N, Lu X, Yang B, Peters R. The application of synthetic biology to elucidation of plant mono-, sesqui-, and diterpenoid metabolism. *Mol Plant* 2015;**8**:6–16.
70. Lesburg CA, Zhai G, Cane DE, Christianson DW. Crystal structure of pentalenene synthase: mechanistic insights on terpenoid cyclization reactions in biology. *Science* 1997;**277**:1820–4.
71. Thulasiram HV, Erickson HK, Poulter CD. Chimeras of two isoprenoid synthases catalyze all four coupling reactions in isoprenoid biosynthesis. *Science* 2007;**316**:73–6.
72. Aaron JA, Lin X, Cane DE, Christianson DW. Structure of *epi*-isozizaene synthase from *Streptomyces coelicolor* A3(2), a platform for new terpenoid cyclization templates. *Biochemistry* 2010;**49**:1787–97.
73. Lesburg CA, Caruthers JM, Paschall CM, Christianson DW. Managing and manipulating carbocations in biology: terpenoid cyclase structure and mechanism. *Curr Opin Struct Biol* 1998;**8**:695–703.
74. Chen M, Al-lami N, Janvier M, D'Antonio EL, Faraldos JA, Cane DE, et al. Mechanistic insights from the binding of substrate and carbocation intermediate analogues to aristolochene synthase. *Biochemistry* 2013;**52**:5441–53.
75. Li J-XX, Fang X, Zhao Q, Ruan J-X, Yang C-QQ, Wang L-JJ, et al. Rational engineering of plasticity residues of sesquiterpene synthases from *Artemisia annua*: product specificity and catalytic efficiency. *Biochem J* 2013;**451**:417–26.
76. Ye L, Lv X, Yu H. Engineering microbes for isoprene production. *Metab Eng* 2016;**38**:125–38.
77. Köksal M, Jin Y, Coates RM, Croteau R, Christianson DW. Taxadiene synthase structure and evolution of modular architecture in terpene biosynthesis. *Nature* 2011;**469**:116–20.
78. Cao R, Zhang Y, Mann FM, Huang C, Mukkamala D, Hudock MP, et al. Diterpene cyclases and the nature of the isoprene fold. *Proteins* 2010;**78**:2417–32.
79. Lange BM. Biosynthesis and biotechnology of high-value p-menthane monoterpenes, including menthol, carvone, and limonene. In: Schrader J, Bohlmann J, editors. *Biotechnology of Isprenoids*. Springer International Publishing; 2015. p. 319–53.
80. Croteau RB, Davis EM, Ringer KL, Wildung MR. (−)-Menthol biosynthesis and molecular genetics. *Naturwissenschaften* 2005;**92**:562–77.
81. Emura M, Matsuda H. A green and sustainable approach: celebrating the 30th anniversary of the asymmetric l-menthol process. *Chem Biodivers* 2014;**11**:1688–99.
82. Kim E-M, Eom J-H, Um Y, Kim Y, Woo HM. Microbial synthesis of myrcene by metabolically engineered *Escherichia coli*. *J Agric Food Chem* 2015;**63**:4606–12.
83. WHO. *Number of malaria cases*. WHO; 2016.
84. Tu Y. The discovery of artemisinin (qinghaosu) and gifts from Chinese medicine. *Nat Med* 2011;**17**:1217–20.
85. Callaway E, Cyranoski D. Anti-parasite drugs sweep Nobel prize in medicine 2015. *Nature* 2015;**526**:174–5.
86. Enserink M. Infectious diseases. Source of new hope against malaria is in short supply. *Science* 2005;**307**:33.
87. Kindermans J-M, Pilloy J, Olliaro P, Gomes M. Ensuring sustained ACT production and reliable artemisinin supply. *Malar J* 2007;**6**:125.
88. Kim SH, Heo K, Chang YJ, Park SH, Rhee SK, Kim SU. Cyclization mechanism of amorpha-4,11-diene synthase, a key enzyme in artemisinin biosynthesis. *J Nat Prod* 2006;**69**:758–62.
89. Peplow M. Synthetic biology's first malaria drug meets market resistance. *Nature* 2016;**530**:389–90.
90. Alam P, Kiran U, Ahmad MM, Kamaluddin, Khan MA, Jhanwar S, et al. Isolation, characterization and structural studies of amorpha - 4, 11-diene synthase (ADS(3963)) from *Artemisia annua* L. *Bioinformation* 2010;**4**:421–9.
91. Starks CM, Back K, Chappell J, Noel JP, Guedes MEM, Kuc J, et al. Structural basis for cyclic terpene biosynthesis by tobacco 5-epi-aristolochene synthase. *Science* 1997;**277**:1815–20.
92. Mountford PG. The Taxol® story- development of a green synthesis via plant cell fermentation. In: Dunn PJ, Wells AS, Williams MT, editors. *Green chemistry in the pharmaceutical industry*. Weinheim, Germany: Wiley-VCH Verlag GmbH & Co. KGaA; 2010. p. 145–60.
93. Nicolaou KC, Yang Z, Liu JJ, Ueno H, Nantermet PG, Guy RK, et al. Total synthesis of taxol. *Nature* 1994;**367**:630–4.
94. Croteau R, Ketchum REB, Long RM, Kaspera R, Wildung MR. Taxol biosynthesis and molecular genetics. *Phytochem Rev* 2006;**5**:75–97.
95. Gao Y, Honzatko RB, Peters RJ. Terpenoid synthase structures: a so far incomplete view of complex catalysis. *Nat Prod Rep* 2012;**29**:1153–75.

96. Li R, Chou WKW, Himmelberger JA, Litwin KM, Harris GG, Cane DE, et al. Reprogramming the chemodiversity of terpenoid cyclization by remolding the active site contour of *epi*-isozizaene synthase. *Biochemistry* 2014;**53**:1155–68.
97. Wang H, Fewer DP, Holm L, Rouhiainen L, Sivonen K. Atlas of nonribosomal peptide and polyketide biosynthetic pathways reveals common occurrence of nonmodular enzymes. *Proc Natl Acad Sci USA* 2014;**111**:9259–64.
98. Endo A. The discover and development of HMG-CoA reductase inhibitors. *J Lipid Res* 1992;**33**:1569–82.
99. Fischbach MA, Walsh CT. Assembly-line enzymology for polyketide and nonribosomal peptide antibiotics: logic, machinery, and mechanisms. *Chem Rev* 2006;**106**:3468–96.
100. Lambalot RH, Gehring AM, Flugel RS, Zuber P, LaCelle M, Marahiel MA, et al. A new enzyme superfamily—the phosphopantetheinyl transferases. *Chem Biol* 1996;**3**:923–36.
101. Cane DE, Walsh CT, Khosla C. Harnessing the biosynthetic code: combinations, permutations, and mutations. *Science* 1998;**282**:63–8.
102. Richter CD, Nietlispach D, Broadhurst RW, Weissman KJ. Multienzyme docking in hybrid megasynthetases. *Nat Chem Biol* 2008;**4**:75–81.
103. Weissman KJ, Müller R. Protein–protein interactions in multienzyme megasynthetases. *ChemBioChem* 2008;**9**:826–48.
104. Broadhurst RW, Nietlispach D, Wheatcroft MP, Leadlay PF, Weissman KJ. The structure of docking domains in modular polyketide synthases. *Chem Biol* 2003;**10**:723–31.
105. Weissman KJ. The structural biology of biosynthetic megaenzymes. *Nat Chem Biol* 2015;**11**:660–70.
106. Weissman KJ. Uncovering the structures of modular polyketide synthases. *Nat Prod Rep* 2015;**32**:436–53.
107. Hertweck C. The biosynthetic logic of polyketide diversity. *Angew Chem Int Ed* 2009;**48**:4688–716.
108. Strieker M, Tanović A, Marahiel MA. Nonribosomal peptide synthetases: structures and dynamics. *Curr Opin Struct Biol* 2010;**20**:234–40.
109. Finking R, Marahiel MA. Biosynthesis of nonribosomal peptides. *Annu Rev Microbiol* 2004;**58**:453–88.
110. Drake EJ, Miller BR, Shi C, Tarrasch JT, Sundlov JA, Allen CL, et al. Structures of two distinct conformations of holo-non-ribosomal peptide synthetases. *Nature* 2016;**529**:235–8.
111. Lewis K. Platforms for antibiotic discovery. *Nat Rev Drug Discov* 2013;**12**:371–87.
112. Donadio S, Staver M, McAlpine J, Swanson S, Katz L. Modular organization of genes required for complex polyketide biosynthesis. *Science* 1991;**252**:675–9.
113. Cortes J, Haydock SF, Roberts GA, Bevitt DJ, Leadlay PF. An unusually large multifunctional polypeptide in the erythromycin-producing polyketide synthase of *Saccharopolyspora erythraea*. *Nature* 1990;**348**:176–8.
114. Weissman KJ, Leadlay PF. Combinatorial biosynthesis of reduced polyketides. *Nat Rev Microbiol* 2005;**3**:925–36.
115. Dutta S, Whicher JR, Hansen DA, Hale WA, Chemler JA, Congdon GR, et al. Structure of a modular polyketide synthase. *Nature* 2014;**510**:512–7.
116. Asturias FJ, Chadick JZ, Cheung IK, Stark H, Witkowski A, Joshi AK, et al. Structure and molecular organization of mammalian fatty acid synthase. *Nat Struct Mol Biol* 2005;**12**:225–32.
117. Maier T, Jenni S, Ban N. Architecture of mammalian fatty acid synthase at 4.5 Å resolution. *Science* 2006;**311**:1258–62.
118. Brignole EJ, Smith S, Asturias FJ. Conformational flexibility of metazoan fatty acid synthase enables catalysis. *Nat Struct Mol Biol* 2009;**16**:190–7.
119. Tsuji SY, Wu N, Khosla C. Intermodular communication in polyketide synthases: comparing the role of protein-protein interactions to those in other multidomain proteins. *Biochemistry* 2001;**40**:2317–25.
120. Lau J, Tran C, Licari P, Galazzo J. Development of a high cell-density fed-batch bioprocess for the heterologous production of 6-deoxyerythronolide B in *Escherichia coli*. *J Biotechnol* 2004;**110**:95–103.
121. Bearden SW, Fetherston JD, Perry RD. Genetic organization of the yersiniabactin biosynthetic region and construction of avirulent mutants in *Yersinia pestis*. *Infect Immun* 1997;**65**:1659–68.
122. Perry RD, Fetherston JD. *Yersinia pestis*—etiologic agent of plague. *Clin Microbiol Rev* 1997;**10**:35–66.
123. Perry RD, Balbo PB, Jones HA, Fetherston JD, DeMoll E. Yersiniabactin from *Yersinia pestis*: biochemical characterization of the siderophore and its role in iron transport and regulation. *Microbiology* 1999;**145**:1181–90.
124. Gehring AM, DeMoll E, Fetherston JD, Mori I, Mayhew GF, Blattner FR, et al. Iron acquisition in plague: modular logic in enzymatic biogenesis of yersiniabactin by *Yersinia pestis*. *Chem Biol* 1998;**5**:573–86.

125. Miller DA, Luo L, Hillson N, Keating TA, Walsh CT. Yersiniabactin synthetase: a four-protein assembly line producing the nonribosomal peptide/polyketide hybrid siderophore of *Yersinia pestis*. *Chem Biol* 2002;**9**:333–44.
126. Challis GL, Ravel J, Townsend CA. Predictive, structure-based model of amino acid recognition by nonribosomal peptide synthetase adenylation domains. *Chem Biol* 2000;**7**:211–24.
127. Pfeifer BA, Wang CCC, Walsh CT, Khosla C. Biosynthesis of yersiniabactin, a complex polyketide-nonribosomal peptide, using *Escherichia coli* as a heterologous host. *Appl Environ Microbiol* 2003;**69**:6698–702.
128. Walsh CT. Polyketide and nonribosomal peptide antibiotics: modularity and versatility. *Science* 2004;**303**:1805–10.
129. Gregory MA, Kaja AL, Kendrew SG, Coates NJ, Warneck T, Nur-e-Alam M, et al. Structure guided design of improved anti-proliferative rapalogs through biosynthetic medicinal chemistry. *Chem Sci* 2013;**4**:1046.
130. Oliynyk M, Brown MJ, Cortés J, Staunton J, Leadlay PF. A hybrid modular polyketide synthase obtained by domain swapping. *Chem Biol* 1996;**3**:833–9.
131. Menzella HG, Reid R, Carney JR, Chandran SS, Reisinger SJ, Patel KG, et al. Combinatorial polyketide biosynthesis by de novo design and rearrangement of modular polyketide synthase genes. *Nat Biotechnol* 2005;**23**:1171–6.
132. Tanovic A, Samel SA, Essen LO, Marahiel MA. Crystal structure of the termination module of a nonribosomal peptide synthetase. *Science* 2008;**321**:659–63.
133. Whicher JR, Dutta S, Hansen DA, Hale WA, Chemler JA, Dosey AM, et al. Structural rearrangements of a polyketide synthase module during its catalytic cycle. *Nature* 2014;**510**:560–4.
134. McDaniel R, Thamchaipenet A, Gustafsson C, Fu H, Betlach M, Ashley G. Multiple genetic modifications of the erythromycin polyketide synthase to produce a library of novel 'unnatural' natural products. *Proc Natl Acad Sci USA* 1999;**96**:1846–51.
135. Tang L, Fu H, McDaniel R. Formation of functional heterologous complexes using subunits from the picromycin, erythromycin and oleandomycin polyketide synthases. *Chem Biol* 2000;**7**:77–84.
136. Whicher JR, Smaga SS, Hansen DA, Brown WC, Gerwick WH, Sherman DH, et al. Cyanobacterial polyketide synthase docking domains: a tool for engineering natural product biosynthesis. *Chem Biol* 2013;**20**:1340–51.
137. Kries H. Biosynthetic engineering of nonribosomal peptide synthetases. *J Pept Sci* 2016;**22**:564–70.
138. Winn M, Fyans JK, Zhuo Y, Micklefield J. Recent advances in engineering nonribosomal peptide assembly lines. *Nat Prod Rep* 2016;**33**:317–47.
139. Fischbach MA, Lai JR, Roche ED, Walsh CT, Liu DR. Directed evolution can rapidly improve the activity of chimeric assembly-line enzymes. *Proc Natl Acad Sci USA* 2007;**104**:11951–6.
140. Chemler JA, Tripathi A, Hansen DA, O'Neil-Johnson M, Williams RB, Starks C, et al. Evolution of efficient modular polyketide synthases by homologous recombination. *J Am Chem Soc* 2015;**137**:10603–9.
141. Kruger K, Grabowski PJ, Zaug AJ, Sands J, Gottschling DE, Cech TR. Self-splicing RNA: autoexcision and autocyclization of the ribosomal RNA intervening sequence of Tetrahymena. *Cell* 1982;**31**:147–57.
142. Guerrier-Takada C, Gardiner K, Marsh T, Pace N, Altman S. The RNA moiety of ribonuclease-P is the catalytic subunit of the enzyme. *Cell* 1983;**35**:849–57.
143. Gilbert W. The RNA world. *Nature* 1986;**319**:618.
144. Orgel LE. Prebiotic chemistry and the origin of the RNA world. *Crit Rev Biochem Mol Biol* 2004;**39**:99–123.
145. Silverman SK. Nucleic acid enzymes (ribozymes and deoxyribozymes): in vitro selection and application. In: *Wiley encyclopedia of chemical biology*. Hoboken, NJ, USA: John Wiley & Sons, Inc; 2008. p. 1–17.
146. Tarasow TM, Tarasow SL, Eaton BE. RNA-catalysed carbon-carbon bond formation. *Nature* 1997;**389**:54–7.
147. Seelig B, Jäschke A. A small catalytic RNA motif with Diels-Alderase activity. *Chem Biol* 1999;**6**:167–76.
148. Jäschke A, Frauendorf C, Hausch F. In vitro selected oligonucleotides as tools in organic chemistry. *Synlett* 1999;**6**:825–33.
149. Seelig B, Keiper S, Stuhlmann F, Jäschke A. Enantioselective ribozyme catalysis of a bimolecular cycloaddition reaction. *Angew Chem Int Ed* 2000;**39**:4576–9.
150. Ferré-D'Amaré AR, Zhou K, Doudna JA. Crystal structure of a hepatitis delta virus ribozyme. *Nature* 1998;**395**:567–74.
151. Serganov A, Keiper S, Malinina L, Tereshko V, Skripkin E, Höbartner C, et al. Structural basis for Diels-Alder ribozyme-catalyzed carbon-carbon bond formation. *Nat Struct Mol Biol* 2005;**12**:218–24.
152. Keiper S, Bebenroth D, Seelig B, Westhof E, Jäschke A. Architecture of a Diels-Alderase ribozyme with a preformed catalytic pocket. *Chem Biol* 2004;**11**:1217–27.

153. Stuhlmann F, Jäschke A. Characterization of an RNA active site: interactions between a Diels–Alderase ribozyme and its substrates and products. *J Am Chem Soc* 2002;**124**:3238–44.
154. Agresti JJ, Kelly BT, Jäschke A, Griffiths AD. Selection of ribozymes that catalyse multiple-turnover Diels–Alder cycloadditions by using in vitro compartmentalization. *Proc Natl Acad Sci USA* 2005;**102**:16170–5.
155. Wombacher R, Keiper S, Suhm S, Serganov A, Patel DJ, Jäschke A. Control of stereoselectivity in an enzymatic reaction by backdoor access. *Angew Chem Int Ed* 2006;**45**:2469–72.
156. Kraut S, Bebenroth D, Nierth A, Kobitski AY, Nienhaus GU, Jäschke A. Three critical hydrogen bonds determine the catalytic activity of the Diels-Alderase ribozyme. *Nucleic Acids Res* 2012;**40**:1318–30.
157. Guo P, Haque F, Hallahan B, Reif R, Li H. Uniqueness, advantages, challenges, solutions, and perspectives in therapeutics applying RNA nanotechnology. *Nucleic Acid Ther* 2012;**22**:226–45.

CHAPTER

3.15

Organic Solvents in Sustainable Synthesis and Engineering

Levente Cseri[1], *Mayamin Razali*[1], *Peter Pogany*[2], *Gyorgy Szekely*[1]

[1]**The University of Manchester, Manchester, United Kingdom;** [2]**Gedeon Richter Plc., Budapest, Hungary**

3.15.1 THE ROLE OF ORGANIC SOLVENTS IN CHEMISTRY AND CHEMICAL ENGINEERING

Solvents, and dissolution in general, have been an integral part of numerous industries for hundreds of years. As early as the 15th century, alchemists were already in search of solvents to aid in dissolution processes. In particular, they were seeking to find the so-called *menstruum universal*, the universal solvent that possessed the powers to remove all seeds of disease from the human body.[1] Throughout the centuries solvents have gained increasing attention, and nowadays they encompass our lives. From the production of life-saving drugs to the manufacture of coffee machines, solvents play a vital role in modern society. Particularly in the chemical industries, solvents are important auxiliaries in processes, and most of the manufacturing methods require various kinds of solvents.

The word "solvent" originates from Latin, in which the verb "solvo" means to loosen or untie. This corresponds to the key property of solvents, which is loosening the intermolecular bonds between the molecules of the solute separating them from each other. The definition of solvents is a subject of debate. The International Union of Pure and Applied Chemistry nomenclature provides the following definition for solutions, solutes, and solvents: "[A solution is] a liquid or solid phase containing more than one substance, when for convenience one (or more) substance, which is called the solvent, is treated differently from the other substances, which are called solutes".[2] Solid solutions (e.g., alloys or plasticizer-containing polymers) are materials in which one compound (solute) is randomly incorporated in the crystal structure of another compound or in an amorphous solid (solvent). The Oxford Dictionary and the Encyclopedia Britannica give a different, more straightforward definition, which narrows down the concept of solvents by excluding solids:

Green Chemistry
http://dx.doi.org/10.1016/B978-0-12-809270-5.00020-0

"[A solvent is a] liquid in which other materials dissolve to form a solution."[3,4] Ignoring the solids in the definition can be justified by the fact that the dissolution phenomenon always takes place in liquid phase, except for one particular example: at room temperature and atmospheric pressure, solid palladium can absorb and dissolve hydrogen up to 900 times its own volume. An ultimate definition can be worded in respect of this chapter as the following: Solvents are substances that are present in a liquid form under the conditions of their applications and in which other substances can dissolve, and from which they can be recovered unchanged on removal of the solvent. Many substances conform to this definition—practically all those that can be liquefied under given conditions—and subsequently the definition is not very helpful, unless the word "application" is stressed, meaning that the solvents and the solutions in them ought to be applicable for some purpose.[5] The essential solvent of Nature is water. Life is unimaginable without water, and it can be found almost everywhere on the earth. It also has been used by the human race as a solvent for many applications since the dawn of mankind. The present chapter will focus on organic solvents, which are excessively used in industrial production. The synthesis and production of organic solvents was one of the biggest successes of the early chemical science. Since the Second Industrial Revolution at the end of the 1800s, organic solvents gained importance in various industrial processes. However, during the 20th century, their adverse effects on health and the environment saw the light of day. In parallel with the increasing emphasis on sustainability and safety, organic solvents have been becoming a concern.[6]

Typical applications of organic solvents are summarized in Table 3.15.1. Nowadays, organic solvents are widely used in various industries including paint, pharmaceutical, and oil industries and also in agriculture and food industry (Fig. 3.15.1). In particular, the largest

TABLE 3.15.1 Common Applications of Organic Solvents

Application	Description	Solvent Example
Absorbent	Absorbs gases from industrial processes or power plants	PEGs
Antifreeze	Added to radiator fluids or motor vehicles to prevent freezing at low outside temperature	Ethylene glycol
Binder	Achieves cohesiveness in composite materials	PEGs
Blowing agent	Used in polymer industry to create/increase porosity	Pentane
Cleaning agent	Removes dirt and impurities from surfaces and equipment, prevents contamination	Acetone, tetrachloroethene, perchloroethene, trichloroethene
Cosmetic	Solvents used in nail polish and nail polish removers	Acetone, methyl ethyl ketone, ethyl acetate, methyl acetate, ethyl acetate
Degreaser	A cleaner that specifically removes grease and oily materials	Benzyl alcohol
Dehydrator	Removes water from materials	Triethylene glycol
Deodorizer	Removes unpleasant smells	1,4-Dichlorobenzene
Diluent	Used in mixtures with other solvents to improve their chemical or physical properties	Acetone, naphtha, toluene

TABLE 3.15.1 Common Applications of Organic Solvents—cont'd

Application	Description	Solvent Example
Dispersant	Forms suspensions of materials without dissolving them	Polysorbates, PEGs
Disinfectant	A chemical liquid that destroys bacteria	Ethanol, isopropanol
Eluent	In chromatographic columns either for analytical or preparative use	Acetonitrile, methanol, tetrahydrofuran
Excipient	An inactive substance that serves as the vehicle or medium for a drug or other active substance.	Butanol, DMSO
Extractant	Used to separate solutes by partitioning between two phases	Hexane, ethyl acetate
Fire suppression agent	Used in fire extinguishers for fire classes A, B, C, E	Perfluoro(2-methyl-3-pentanone)
Food additive	Substance added to food to enhance its flavor or appearance or to preserve it	Ethanol, acetic acid, butyl acetate
Fuel	The chemical energy stored in the intramolecular bonds is liberated by combustion	Methyl *tert*-butyl ether, butanol, toluene
General solvent	Used to bring into solution many kinds of materials (adhesives or pigments), and this usually implies that it is produced as an industrial solvent in large bulk quantities	See Table 3.15.4
Humectant	A hygroscopic material that avidly retains water absorbed from a humid atmosphere, and may be used in cosmetics and pharmaceuticals	Glycerol, DMSO
Lubricant	Used by virtue of its viscosity to reduce friction between moving parts in machinery	Mineral oil
Plasticizer	Used to confer on polymers suitable mechanical elastic properties	Dibutyl phthalate, triethyl citrate
Propellant	Used as a rocket or jet fuel	Kerosene
Reactant	Used to produce or modify other chemicals	See Table 3.15.4
Reaction media	Brings the reactants together in one phase, provides enhanced heat and mass transport properties	See Table 3.15.4
Thermal fluid	Heat transfer media in heating or cooling systems	Methanol, propylene glycol, diphenyl ether
Transport media	Brings compounds together in liquid phase to facilitate the transport and dosing through pipelines and valves	Toluene, xylene
Working fluid	Used to transmit mechanical forces, a required property being low compressibility	Pentane, toluene

A great variety of solvents are used for many different purposes signifying the importance of organic solvents. *DMSO*, dimethyl sulfoxide; *PEG*, polyethylene glycol.

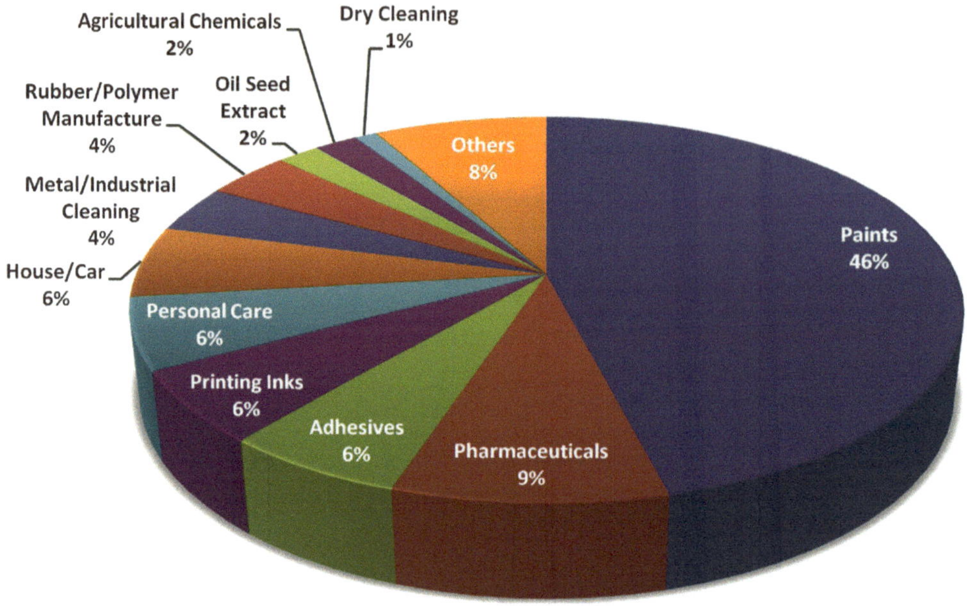

FIGURE 3.15.1 Organic solvent usage of different industrial segments. The paint industry uses by far the most solvents, almost half of the total consumption. The pharmaceutical sector accounts for nearly one-tenth of the overall usage.[10,15]

demand for organic solvents comes from the paint and coating industry, accounting for about 2 million tons per year. Many of the enamel-based paints employ mild petroleum-based solvents with an alkyd vehicle and have long drying and curing times. Conversely, lacquer-based paints require harsher solvents for rapid drying times. The most widely used aromatic hydrocarbon solvents in paint are benzene, toluene, ethylbenzene, mixed xylenes (BTEX), and high-flash aromatic naphthas; aliphatic hydrocarbons include hexanes, heptane, and naphtha.[7] Interestingly, artists are also looking at ways to minimize solvent use and exposure through various means. Some websites describe possible solvent substitutions or alterations in techniques for making art.[8,9] The pharmaceutical industry is a growing market with increasing solvent demand for obtaining high-purity medicines via downstream processing.[10] In general, fine chemical industries use large amounts of solvent per mass of final products as both reaction and purification media. The breakdown of solvent usage by class reveals that the major contributors are alcohol at 30%, followed by aromatic at 20% and ether at 13%.[11] In the face of these facts, it is imperative for solvents to have minimal impact on our health and environment.[12,13] Solvents define a major part of the environmental performance of a process and also affect process efficiency, yield, kinetics, energy consumption, cost, toxicity, safety, and health issues (Fig. 3.15.2).[14]

The pharmaceutical and fine chemical industries manufacture the majority of their products in batch processes, which often requires multiple reaction and purification steps. Most active pharmaceutical ingredients (APIs) are produced using liquid-phase organic reactions, which often utilize excessive quantities of organic solvents to enable and facilitate reactions and downstream processes, as well as to ensure the integrity of intermediates or final products.

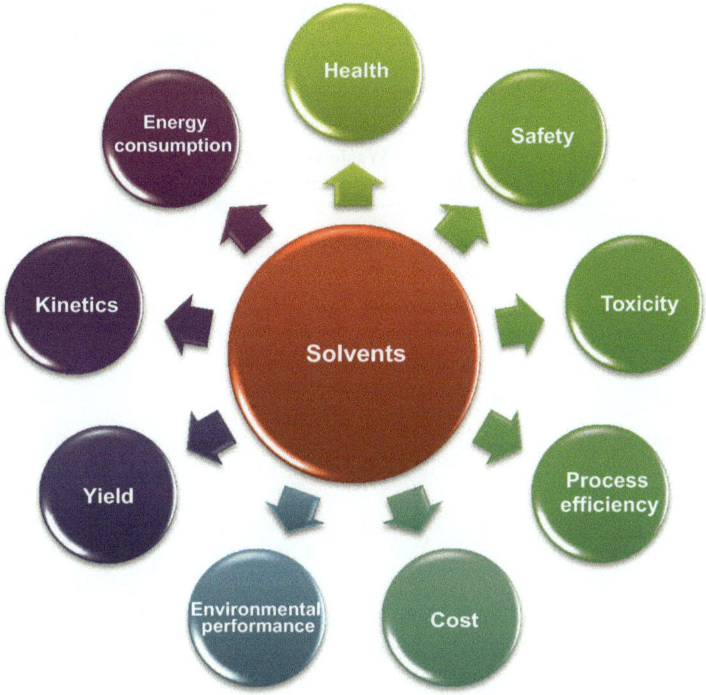

FIGURE 3.15.2 The major factors of an industrial process affected by solvent usage. Although in most cases solvents are not present in the product, they play important roles in most processes, which needs to be considered during process design.

Depending on the type of chemical reactions performed and the physicochemical properties of the catalysts, reagents, and products, both the types and the amounts of solvents can vary widely. Solvents are also used in the pharmaceutical industry for cleaning process equipment and in a plethora of analytical equipment employed for process control and quality monitoring, for instance, high-pressure liquid chromatography. Solvent use can account for as much as 80%–90% (30% water/60% organic solvents) of the total mass in a typical pharmaceutical batch process.[16] Consequently, organic solvents are also a major contributor to the overall toxicity potential associated with these industrial processes,[17] and to the waste generation of chemical industries. Trends about the amount of waste generated by pharmaceutical and medicinal/botanical sectors are shown in Table 3.15.2, demonstrating the major contribution of organic solvents. Therefore process intensification through solvent mitigation (e.g., neat reactions), minimization, recovery, and recycling is of utmost importance in the 21st century. The number of articles about solvent recycling is booming (Fig. 3.15.3). In spite of this, more than half of the chemical production-related waste still does not get recycled, and the trend has not changed in the past decade (Fig. 3.15.4).

Typically, the amount of waste generated from solvents in a pharmaceutical chemical synthetic processing step ranges from 25 to considerably more than 100 kg of solvent per kilogram of API produced. For a commercial process, the American Chemical Society (ACS) Green Chemistry Institute Pharmaceutical Roundtable (GCI-PR) 2007 benchmarking

TABLE 3.15.2 Amount of Waste Generated by Pharmaceutical and Medicinal/Botanical Sectors According to United States Environmental Protection Agency's TRI in 1995 and 2006[16]

	1995		2006	
Solvent	**Rank**	**Waste Generated ($\times 10^6$ kg/year)**	**Rank**	**Waste Generated ($\times 10^6$ kg/year)**
Methanol	1	62.5	1	44.8
Dichloromethane	2	21	2	22.3
Toluene	3	20	3	12.1
Acetonitrile	6	7.39	4	7.9
Hydrochloric acid	4	17.4	5	7.03
Nitrate compounds	12	1.99	6	5.21
Chloroform	21	0.61	7	3.71
n-Hexane	11	2.3	8	2.99
n-Butyl alcohol	13	1.53	9	2.86
N,N-dimethylformamide	8	4.6	10	2.79
Formic acid	9	3.24	11	2.42
N-methyl-2-pyrrolidone	36	0.17	12	2.02
Xylene	16	1.27	13	1.47
Arsenic compounds	83	0.004	14	1.26
1,1,2-Trichloroethane	N/A	0	15	1.23
Methyl *tert*-butyl ether	19	0.77	16	1.2
Ammonia	7	6.27	17	1.01
Ethylene glycol	18	0.9	18	0.82
Sulfuric acid	5	8.79	19	0.71
Certain glycol ethers	31	0.28	20	0.63
Total (top 20 in 2006)		161		124
Total (for all TRI chemicals)		175		128

The top two compounds (methanol and dichloromethane) account for roughly the half of the overall waste in both years. It is also noteworthy that the total amount of solvent waste decreased by nearly 30% during this 10-year period. *TRI*, toxic release inventory.

study showed that the median amount of materials used to make 1 kg of API was 46 kg. Fig. 3.15.5 shows that 56% of the mass used is solvent, i.e., 22 kg of solvent are needed to make 1 kg of API, which results in low atom economy.[17]

Historically, the generation of solvent waste has usually been due to poor solvent selection and processing inefficiencies. The waste generated by pharmaceutical companies has

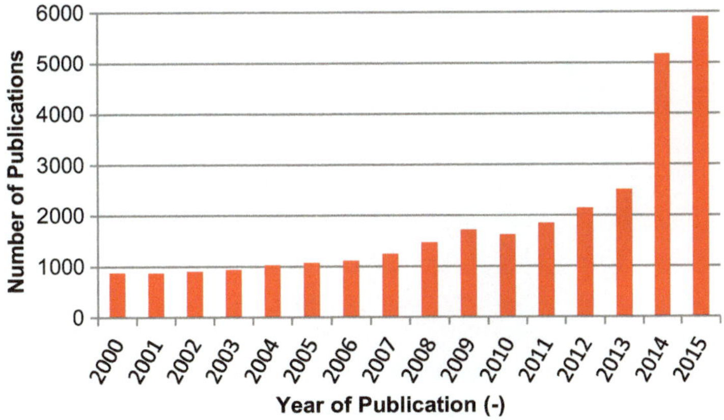

FIGURE 3.15.3 Number of articles per year including the concept of "solvent recovery" throughout the first 16 years of the 21st century. The monotonic increase in the number of articles discussing solvent recovery demonstrates the growing significance of minimizing solvent waste and maximizing product lifetime of raw materials, which contributes to sustainable production. *The data were obtained through Web of Science^{TM}. https://www.webofknowledge.com.*

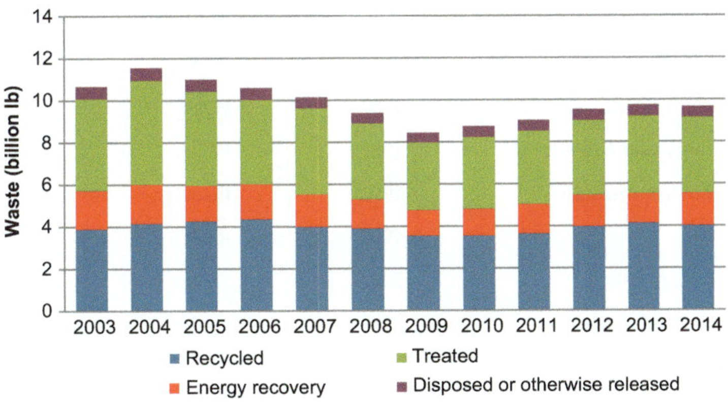

FIGURE 3.15.4 Management of chemical production-related waste in the United States. More than half of the waste is not yet recycled, and the trend has not changed significantly over the past decade.[19]

increased concerns about environmental and human safety. Direct release of treated solvent wastes, hazardous work conditions, and accidental releases of toxic chemicals into the environment have led to the implementation of many laws and regulations. These governmental regulations, in addition to many others, have created a widespread interest in green chemistry and technology.[16,21] As a result of this growing interest the number of published articles has grown sharply and several new journals have been established in this field (Table 3.15.3).

It is also noteworthy that not all the solvents emitted into the environment are of anthropogenic origin. A significant proportion of common organic solvents occur in nature. Ethanol and methanol are produced in the metabolic processes of yeast and bacteria. The esters of

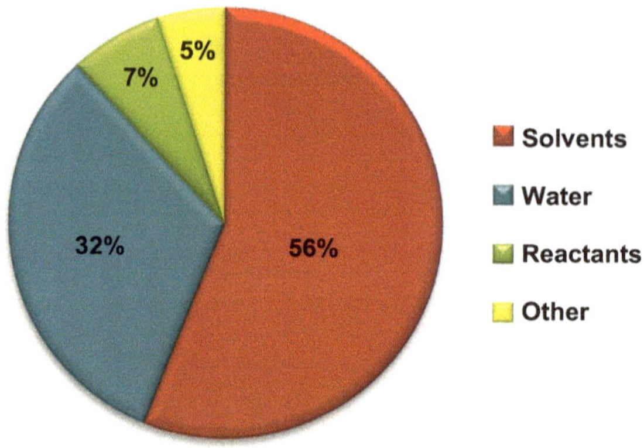

FIGURE 3.15.5 Composition by mass of the type of materials used to manufacture pharmaceuticals. Solvents account for more than half of the total mass, which demonstrates that solvents have a crucial effect on the overall sustainability of pharmaceutical production.[20]

TABLE 3.15.3 Journals Dedicated to Green and Sustainable Chemistry and Engineering

Journal	Year Established	Publisher
Environmental Science & Technology	1967	ACS
Environmental Progress & Sustainable Energy	1982	Wiley-VCH
Environmental Toxicology and Chemistry	1982	Wiley-VCH
Applied Catalysis B: Environmental	1992	Elsevier
Green Chemistry	1999	RSC
Advanced Synthesis & Catalysis	2001	Wiley-VCH
Green Chemistry Letters and Reviews	2007	Taylor & Francis
CLEAN—Soil, Air, Water	2007	Wiley-VCH
International Journal of Sustainable Engineering	2008	Taylor & Francis
Journal of Sustainable Development	2008	Canadian Center of Science and Education
Energy and Environmental Sciences	2008	RSC
ChemSusChem	2008	Wiley-VCH
International Journal of Green Nanotechnology	2009	Taylor & Francis
Sustainability	2009	Multidisciplinary Digital Publishing Institute
Green and Sustainable Chemistry	2011	Scientific Research
Green Processing and Synthesis	2012	De Gruyter
ACS Sustainable Chemistry & Engineering	2013	ACS

TABLE 3.15.3 Journals Dedicated to Green and Sustainable Chemistry and Engineering—cont'd

Journal	Year Established	Publisher
Sustainable Chemical Processes	2013	Springer
Green Materials	2013	Institution of Civil Engineers
Environmental Science: Processes & Impacts	2013	RSC
Current Green Chemistry	2014	Bentham science
Sustainable Materials and Technologies	2014	Elsevier
Sustainable Chemistry and Pharmacy	2015	Elsevier
Current Opinion in Green and Sustainable Chemistry	2016	Elsevier

The booming number and the great variety of these journals correspond to widespread interest in green chemistry and technology. Due to the increasing interest in green and sustainable tools for chemists and engineers, nowadays the main publishers have at least one but usually more Journals dedicated to this field. *ACS*, American Chemical Society; *RSC*, The Royal Society of Chemistry.

acetic acid are released by various plants, providing the scent of the ripe fruits. Pyridine occurs in the leaves and roots of some plants. Organic solvents, such as ethanol, methanol, acetone, or dimethyl sulfoxide, are present in a low concentration even in the metabolic system of healthy humans.[22] A study supported by Euro Chlor, an organization of the European chlorine producers, declared that 90% of the chloroform flux through the environment is of natural origin. Marine flora, especially seaweeds, were identified as the main source of the natural chloroform.[23,24]

3.15.2 RATIONALE FOR SOLVENT SELECTION

The appropriate selection of solvents depends to a large extent on the application, more specifically on what needs to be dissolved and under what conditions. Other considerations such as solvent recovery and solvent release are gaining increasing importance. In the pharmaceutical and specialty chemical industries, the use of solvents to promote reactions is very common. Here, solvents may be used as a reaction medium to bring reactants together, as a reactant to react with a solute, or as a carrier to deliver chemical compounds in solutions to their point of use in the required amounts. In addition, solvents are used to isolate a product, as well as in the cleaning or washing steps of the reaction stage. Besides these general considerations, each application has unique requirements for solvent selection. An overview of these selection criteria is given in the following sections.

3.15.2.1 Solvents as Reaction Media

There are a plethora of criteria to be taken into account for the proper choice of solvents when they are used as reaction media. Examples include maximum working range between melting and boiling point, chemical and thermal stability in this temperature range, good solubility of reactants and products (or sometimes insolubility of the products), compatibility with the analytical methods employed, appropriate degree of purity, price, flammability, explosiveness, low viscosity, miscibility, toxicity, corrosive action, and feasibility of recycling.[25] Table 3.15.4 summarizes some of the main parameters that need to be considered when attempting to select a solvent for a particular application. The first six columns contain relative rankings in different categories for each solvent between 1 (worst) and 10 (best) introduced in the GlaxoSmithKline (GSK) solvent selection guide (SSG).[20] These categories are meant to cover the different aspects of greenness. The "Waste" category covers the recycling, incineration, volatility, and biotreatment issues. The "Environmental Impact" column ranks the fate and effects of solvents on the environment. The acute and chronic effects on human health and the potential for exposure are covered by the "Health" category. "Flammability & Explosion" covers issues affecting handling and storage of solvents. The "Stability" score concerns factors affecting the stability of a solvent. Life cycle impacts from the production of the solvents are covered by the "Life cycle assessment (LCA) ranking." The next column contains the boiling point of each solvent, which can be a major obstacle for sustainable applicability if it is either too low or too high. The most common reaction types for each solvent class are also indicated in the table. The flash point and acute toxicity values—shown in the next two columns—provide numerical data for the comparison of risk when working with different solvents. The short-term exposure limit values and the International Council for Harmonisation of Technical Requirements for Registration of Pharmaceuticals for Human Use (ICH) guideline classification published by the European Medicines Agency give some insight into the regulatory environment of the 2010s.[26]

Solvents in ICH Class 1 are not to be used in the production of APIs and excipients due to their unacceptable toxicity, or their adverse environmental effects. However, if their use is unavoidable to produce a drug product with a significant therapeutic advantage, then their residual levels have to be in the sub-parts per million level to meet the strict pharmaceutical regulations.[27] Solvents in Class 2 are limited in APIs because of their inherent toxicity. Solvents in Class 3 may be regarded as less toxic, and of lower risk to human health. Class 3 contains solvents that do not pose any human health hazard at levels normally accepted in APIs. Nonetheless, there are no long-term toxicity and carcinogenicity studies for many of the solvents in Class 3. Available data indicate that they are less toxic in acute or short-term studies, and negative in genotoxicity studies. It is suggested that the amounts of these residual solvents of 50 mg per day or less would pose negligible toxicity risks.[28]

3.15.2.2 Solvents for Crystallization

The general requirements for organic solvents to be used for recrystallization of organic compounds can be summarized as follows: there should be high solubility at high temperatures and low solubility at low temperatures of the crude product; impurities ought to be either highly soluble or completely insoluble; the solvent must have high boiling and flash point; the solvent should be chemically inert; the solvent must favor crystal growth, and it should be easily separated from the purified crystals by washing and/or drying steps. Subsequently, it should be either volatile or very soluble in a more volatile solvent and not form clathrates or solvates.[25] In the first instance, solvents for crystallization can be selected

TABLE 3.15.4 Classification and Properties of the Most Important Industrial Solvents[20,25,29,30]

SSG Class	Solvent	Waste	Environmental Impact	Health	Flammability and Explosion	Stability	LCA Ranking	bp (°C)	Reaction Types	fp (°C)	LD_{50}	STEL (mg/m³)	EMA ICH[28]
	Water	4	10	10	10	10	10	100	A,C,F,H, J,P,Q	–	–	–	
Supercritical fluid	sc $CO_2^{\dagger}$	10	5	6	6	10	N/A	N/A	C,E,F,J,L	–	N/A	27,400	
Ionic liquid	[bmim][$AlCl_4$]†	4	4	6	10	5	N/A	N/A	F,L	199	N/A	–	
	[bmim][BF_4]†	4	4	6	10	10	N/A	N/A	A, E,F,J,L	288	50-300	–	
Alcohols	Ethylene glycol	5	8	7	10	9	9	197	A,D,F,H, I,J,P,Q	111	4,700	104	Class 2
	1-Butanol	5	7	5	8	9	5	117		35	790	154	Class 3
	Isoamyl alcohol	6	6	7	9	10	6	130		43	>5,000	458	Class 3
	2-Ethylhexanol	9	5	6	9	10	6	185		75	3,700	–	
	2-Butanol	4	6	8	4	9	6	98		27	2,193	462	Class 3
	1-Propanol	4	7	5	7	10	7	97		22	8,038	625	Class 3
	Ethanol	3	8	8	6	9	9	78		14	10,470	5,760	Class 3
	2-Propanol	3	9	8	6	8	4	82		12	5,045	1,250	Class 3
	t-Butanol	3	9	6	6	10	8	83		11	2,743	462	
	Methanol	4	9	5	5	10	9	65		9,7	1,187	333	Class 2
	2-Methoxyethanol	3	8	2	7	6	7	124		40	2,257	9	
Esters	*t*-Butyl acetate	6	9	8	6	10	8	97	C,D,F	4	4,100	1210	
	Butyl acetate	7	7	8	8	10	5	125		23	10,760	966	Class 3
	n-Propyl acetate	5	7	8	6	10	4	102		14	9,370	1,060	Class 3
	Isopropyl acetate	5	7	7	6	9	7	88		4	6,750	849	Class 3
	Ethyl acetate	4	8	8	4	8	6	77		-3	5,620	400	Class 3
	Methyl acetate	3	9	7	4	9	7	58		-13	6,482	770	
	Ethyl lactate	7	5	4	8	10	N/A	154		46	8,200	–	
	Dimethyl carbonate	4	8	8	4	8	6	90		16	>5,000	–	

(Continued)

TABLE 3.15.4 Classification and Properties of the Most Important Industrial Solvents[20,25,29,30]—cont'd

SSG Class	Solvent	Waste	Environmental Impact	Health	Flammability & Explosion	Stability	LCA Ranking	bp (°C)	Reaction Types	fp (°C)	LD_{50}	STEL (mg/m^3)	EMA ICH[28]
Aromatics	Benzene	5	6	1	3	10	7	80	C,E,G,H, I,J,L,P	–11	5,960	9	Class 1
	p-Xylene	7	2	6	5	10	7	138		25	3,523	442	Class 2
	Toluene	6	3	4	4	10	7	110		4	5,580	384	Class 2
	Fluorobenzene	5	3	6	5	9	1	85		15	4,399	–	
Ketones	Methyl isobutyl ketone	6	6	6	7	8	2	117	B,E,J	14	2,080	280	Class 3
	Acetone	3	9	8	4	9	7	56		–17	5,800	3,620	Class 3
	Cyclohexanone	6	8	6	8	9	6	155		44	1,534	81,6	
	Methyl ethyl ketone	3	7	8	4	8	3	80		–3	2,737	900	Class 3
Polar aprotics	1-Methyl-2-pyrrolidone	5	6	3	9	8	4	202	B,H,F,I,J,P	91	3,914	80	Class 2
	Dimethylacetamide	5	6	2	10	8	2	165		70	5,680	72	Class 2
	Dimethylformamide	4	6	2	9	9	7	153		58	2,800	30	Class 2
	Dimethylpropylene urea	7	7	4	9	7	3	146 (5.9 kPa)		120	N/A	–	
	Dimethyl sulfoxide	5	5	7	9	2	6	189		87	14,500	–	Class 3
	Formamide	4	7	2	10	8	8	210		175	5,325	56	Class 2
	Nitromethane	3	8	4	7	2	N/A	101		36	1,478	381	Class 2
	Acetonitrile	2	6	6	6	10	3	81		2	1,320	102	Class 2
Acids	Propionic acid	4	8	6	8	8	7	141	A,C,D,E,F,H, M,N,O,P,Q	54	3,455	46	
	Acetic acid	4	8	6	8	7	8	117		40	3,310	75	Class 3
Alkanes	Cyclohexane	5	5	7	2	10	7	81	F,L,O	–18	12,705	1,050	Class 2
	Methylcyclohexane	6	5	8	3	10	7	101		–4	>3,200	–	Class 2
	Heptane	6	3	8	3	10	7	98		–4	>5,000	6,255	Class 3
	2-Methylpentane	5	4	7	2	10	7	62		–7	N/A	–	
	Hexane	5	3	4	2	10	7	69		–26	25,000	219	Class 2
	Petroleum ether	6	2	2	3	10	7	40-60		–40	N/A	–	

SSG Class	Solvent	Waste	Environmental Impact	Health	Flammability and Explosion	Stability	LCA Ranking	bp (°C)	Reaction Types	fp (°C)	LD_{50}	STEL (mg/m^3)	EMA ICH[28]
Chlorinated	Chloroform	3	6	3	6	9	6	61	C,D,E,I,J, L,M,N,O	–	908	30	Class 2
	Carbon tetrachloride	4	5	3	4	10	7	76		–	2,350	39	Class 1
	1,2-Dichloroethane	4	4	2	6	10	7	83		13	670	63	Class 1
	Chlorobenzene	6	6	4	8	10	8	132		27	1,110	70	Class 2
	1,2-Dichlorobenzene	7	4	6	10	9	8	179		66	500	306	
	Dichloromethane	3	6	4	6	9	7	40		–	2,000	1,060	Class 2
Ethers	Methyl t-butyl ether	4	5	5	3	9	8	55	F,G,H,I, J,K,O	–33	4,000	367	Class 3
	2-Methyltetrahydrofuran	4	5	4	3	6	4	78		–10	300	–	
	1,2-Dimethoxyethane	4	5	2	4	4	7	85		–2	5,370	–	Class 2
	Tetrahydrofuran	3	5	6	3	4	4	66		–17	1,650	300	Class 2
	1,4-Dioxane	3	4	4	4	5	6	101		12	4,200	219	Class 2
	Bis(2-methoxyethyl) ether	4	5	2	8	4	6	162		57	5,400	–	
	Diethyl ether	4	4	5	2	4	6	35		–40	1,215	616	Class 3
	Diisopropyl ether	4	3	8	1	1	9	68		–29	8,470	1,310	
Basics	Triethylamine	4	5	3	4	8	7	89	H,Q	–15	730	12,6	
	Pyridine	3	4	4	7	9	2	115		17	891	33	Class 2

Reactions: *A*, SN_1; *B*, SN_2; *C*, oxidation; *D*, ozonization; *E*, epoxidation; *F*, catalytic hydrogenation; *G*, hydride reduction; *H*, aldol reaction; *I*, Wittig reaction; *J*, Diels-Alder cycloaddition; *K*, Grignard reaction; *L*, Friedel-Crafts reaction; *M*, halogenation; *N*, nitration; *O*, sulfonation; *P*, diazotization; *Q*, diazo coupling. Acute toxicity LD_{50} values are expressed in mg/kg rat.

†These solvents were not part of the original GlaxoSmithKline solvent selection guide. The values were established for this work by the authors of the original guideline.[29] Note that that their methodology was not designed to be used with these types of solvents, and consequently the results must be treated with an even greater degree of caution than the rankings for conventional solvents. These scores were provided on the condition that it is not relied on in environmental practice and GlaxoSmithKline and its employees/agents will accept no liability for any claims arising from reliance on or use of it.

based on the rule of thumb, *like dissolves like*. When the crude product is too soluble in one solvent, and not soluble enough in another, a binary mixture of the two can be applied as recrystallization medium. Examples for binary mixtures useful in crystallization include acetic acid/water, ethanol/water, ethanol/benzene, acetone/petroleum ether, and chloroform/petroleum ether. For a detailed guide on solvent selection for crystallization processes, see the literature from Baumann, Davey, and Craig.[31–33] Besides changing the temperature (decreasing solubility with decreasing temperature) crystallization can also be controlled by changing the solvent polarity at constant temperature. A cold crystallization technique using pairs of solvents of different polarity (e.g., methanol/water, acetone/water) has been described by Bierne et al.[34] The mechanism and rate of the crystallization are determined by a number of kinetic, thermodynamic, and molecular recognition factors (Fig. 3.15.6).[35] Many of these factors depend on the solvent, and consequently, solvent selection is a crucial task that has to be carried out at the beginning of the crystallization processes.

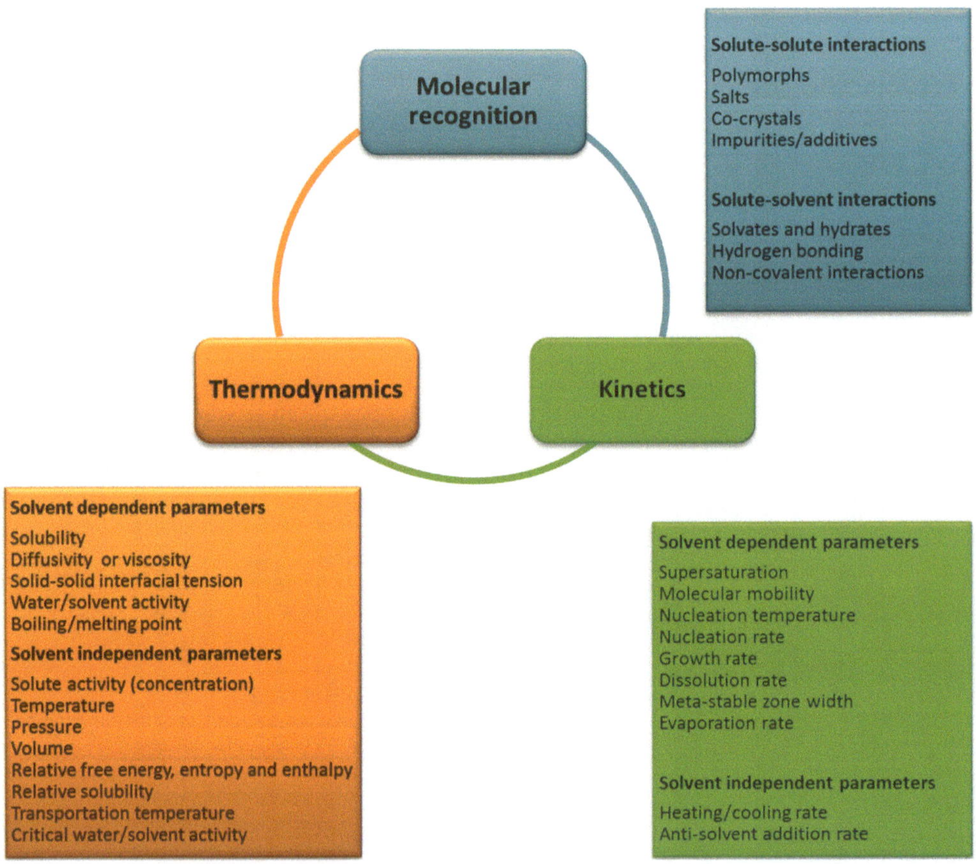

FIGURE 3.15.6 Schematic diagram illustrating the roles of solvent on the kinetic, thermodynamic, and molecular recognition aspects of crystallization. The complex balance between these solvent-dependent factors has to be optimized when selecting solvents for crystallization processes.[35]

3.15.2.3 Solvents for Adsorption

The accumulation of organic species on the surface of adsorbents is determined by the polarity, polarizability, and molecular size of the compound. This is a valid statement for solvents as well: the closer a solvent's polarity is to the polarity of an adsorbent, the stronger the solvent is adsorbed to the surface. The molecules of the adsorbed species and the solvent compete for the active sites on the surface of the adsorbent. Consequently, organic compounds will be adsorbed on a polar adsorbent more strongly from a nonpolar solvent than from a polar one, and vice versa. Conversely, a previously adsorbed material can only be replaced by a solvent when it has a higher affinity for the surface. General requirements for the selection of eluent can be summarized as high purity, solubility of the crude mixture, low viscosity, greenness and cost of regeneration, and suitability for any analytical methods employed (e.g., ultraviolet detection during elution requires a solvent that does not absorb at the wavelength employed in the method). Moreover, the eluent needs to be chemically inert to the adsorbate, otherwise side reactions can occur. For instance, acetone and ethyl acetate on the surface of alkaline adsorbents (e.g., Al_2O_3) are readily transformed into diacetone alcohol and acetic acid, respectively.[25] Eluents can be ranked based on their increasing eluting power by empirical determination of the retention times for a constant adsorbent and various test mixtures. The shorter the retention time of the compound on a polar adsorbent, the higher the eluting power and consequently higher the polarity of the solvent. Oxide-type adsorbents such as aluminum oxide and silica gel give similar eluotropic series, whereas for hydrophobic adsorbents such as charcoal and polyamides, the eluotropic series is almost completely reversed. Although the key factor for determining the elution power of a solvent is polarity, other factors, such as hydrogen bonding or molecular size, can play some role. In some cases, mixtures of two or three solvents and the use of acidic or basic additives (e.g., formic acid or triethylamine) of different polarity results in better separation than a single solvent system. These multicomponent eluents can be ordered in eluotropic series.[25,36,37]

3.15.2.4 Solvents for Extraction and Partitioning

The partitioning of a substance between two liquid phases and the extraction of solids require similar properties of a solvent. A solvent system featuring limited miscibility of the components is required for partitioning because the compound has to dissolve to a different extent in the two phases. The greater the chemical differences between two solvents, the more likely their immiscibility. Further requirements for the solvent system include partition coefficients between 0.2 and 5, large capacity, high selectivity, a high separation factor of at least 1.5, linearity (concentration independence of the partition isotherm), no tendency for emulsion formation, and rapid phase separation, which requires low viscosity, significant density difference, and sufficiently high surface tension. The selection of solvent for partitioning needs to also consider any possible irreversible reactions between the solvent and solute and ease of recovery of the substance and the solvent. The simultaneous optimization of these various requirements is practically impossible, and usually a compromise between these competing factors must be made.[25]

3.15.2.5 Solvents for Membrane Processes

In recent years, application of nanofiltration in organic media has started to rapidly spread.[38] In the pharmaceutical industry, for instance, nanofiltration has been proposed as a sustainable technology to perform concentration, purification, salt/solvent exchange, catalyst recovery, and solvent recycling.[39] Membrane separation performance is much less predictable in the presence of organic solvents than in aqueous solutions.[40] The factors affecting the molecular interactions at the molecular level can be classified in terms of solvent-membrane, solvent-solute, and solute-membrane affinity interactions. These interactions determine the separation factor, permeability, and solute rejection. Hence the selection of solvent has a significant effect on the overall performance of membrane processes. Both viscosity and surface tension are major parameters influencing the flux during membrane processes. Detailed investigation of the effect of solvent on membrane separation can be found in the literature.[41–44] The most important parameters for selecting a solvent for a membrane process include stability and swelling of the membrane in the solvent, solubility of the solutes, and solvent viscosity. The recently marketed solvent-resistant nanofiltration membranes can withstand most solvents with the exceptions of polar aprotic solvents. High solubility of the species to be separated is required for membrane processes, and operation at the solubility limit needs to be avoided. Precipitation of the solutes can lead to clogging of the equipment and fouling on the surface of the membrane. Fouling decreases the membrane performance over time. High-viscosity solvents such as isopropyl alcohol should be avoided as they exhibit low flux through the membrane, which results in increased operation time and thus higher energy consumption. The solvent-resistant membrane manufacturers—Evonik MET, GMT Membrantechnik, SolSep, and Novamem—provide guidance on the applicability and limitations of their membranes including solvent selection.

3.15.2.6 Recent Trends

Global issues are deeply changing the face of the chemical industry. As a result of the endeavor to fulfill the requirements of sustainable and green chemistry, intensive research has begun on solvent use. One segment of this focuses on the design and production of novel solvents. In the first wave, supercritical fluids (SCFs) were the subject of extensive research and found various industrial applications.[45] Owing to its advantageous properties compared with conventional solvents, supercritical CO_2 is now more widely used for extractions,[46,47] polymerization,[48] catalysis,[49] processing polymer melts,[50] and membrane fabrication.[51] Some typical applications include caffeine extraction from coffee beans[52] and oil extraction from plants.[46] In the 2000s, two-phase solvent systems having temperature-dependent mutual miscibility of the two compounds have been applied as reaction media. Having different solubilities for educts, products, reagents, and catalysts, biphasic solvent combinations can enable and facilitate the separation of crude reaction mixtures. Since perfluorohydrocarbons[53] and room temperature ionic liquids (ILs)[54] are immiscible with most common organic solvents, they are particularly suitable for the formation of these biphasic solvent systems. Most recently, deep eutectic solvents (DESs) have become the center of attention.[55–57] These innovations have an important common feature. Conventional solvent selection is about choosing the most appropriate one from solvents with distinct properties. However, using SCFs, ILs, or DESs creates the opportunity to tune the properties according to the purpose.

One of the key factors in sustainable production of solvents is the source of raw materials. Renewable agricultural biomass is a sustainable source for various commodity chemicals replacing fossil resources.[58,59] The global bio-based chemical market is rapidly expanding in both size and importance. Bio-based, and consequently environmentally benign, solvents such as glycerol, ethyl lactate, and 2-methyltetrahydrofuran are often discussed as important additions to the conventional repertoire of solvents.[60,61] The European Commission in the 2012 publication on "Innovating for Sustainable Growth: A Bioeconomy for Europe" as part of the Europe 2020 Strategy encourages smart and green growth, in particular by increasing the use of bio-based products.[62] Nigiz et al. proposed a process for the synthesis of a green solvent, ethyl lactate, using biomass-based sources employing a pervaporation biocatalytic membrane reactor system (PVBCMR).[63] This method benefits from all the advantages of enzymes. It is an energy-intensive, cost-effective, and environment-friendly technique that serves as a reactive separation system. In this study, lipase-loaded biocatalytic sodium alginate membranes were prepared and employed in PVBCMR to synthesize ethyl lactate in mild operation conditions. Ethyl lactate from biorefineries has started to be used in various processes in the paint industry, replacing solvents such as *N*-methylpyrrolidone, xylene, toluene, acetone, and halogenated solvents.[15,64] The utilization of sugar cane bagasse has been explored for methanol to gasoline, biochemical butanol, as well as ethanol production.[65,66] Lactic acid has successfully been used as a reaction medium for ultrasound-assisted scalable synthesis of pyrrole derivatives[67]; for the three-component reactions of styrenes, formaldehyde, and active phenolic compounds or *N*,*N*-dialkylacetoacetamides; for the three-component reactions of diethyl acetylenedicarboxylate, anilines, and aromatic aldehydes; and for Friedländer reactions.[68]

Another segment of the research on solvents focuses on creating reliable computational methods. Using in silico experiments, an adequate decision can be made even before a single drop of solvent has been consumed. In some approaches, novel green solvents were designed. Weis and Visco focused on the green properties of organic molecules and applied a computer-aided molecular design technique to design potentially new environment-friendly solvents.[69] In their work, quantitative structure-property relationships were created to rank designed solvents based on the GSK SSG (see Table 3.15.4). Moity et al. focused rather on the green production of the organic compounds and developed a Computer-Aided Organic Synthesis program. They used it to explore synthetic pathways from a bio-building block (itaconic acid) toward potential solvents.[70] Predictive methods for particular case studies are also available.[71] Simpler group interaction methods such as UNIFAC (UNIQUAC Functional-group Activity Coefficients) and more accurate surface charge interaction methods such as COSMO-RS (Conductor-like Screening Model for Realistic Solvents) have been used in many case studies to minimize the experimental effort and to enlarge the dataspace for optimization of solvent selection.[72,73]

One critical assessment states that academic research in the area of green solvents is focused on neither the largest solvent-consuming industries nor the types of solvents that the research community believes have the best hope of reducing solvent-related environmental damage.[74] Aside from the regulatory and economic pressure on manufacturing companies, academic researchers are also encouraged by editors and publishers to consider using green solvents and solvent management methodologies before submitting their manuscript for publication, and in certain cases, authors are warned that they risk having papers rejected unless environmental impact and green chemistry principles are considered.[75]

3.15.3 CARBON FOOTPRINT OF ORGANIC SOLVENTS

Because many solvents are inherently highly volatile, toxic, considerably persistent, and stable, the handling of organic solvents in the chemical industry represents a high-priority environmental issue. Government regulations play a significant role in the protection of the environment, and solvent emission controls have been used for over 150 years and will continue as with the European Union's Registration, Evaluation, Authorization and Restriction of Chemicals initiative.[76–78] Certain classes of chemical compounds and solvents have been banned; for instance, the Montreal Protocol urges the elimination of substances that deplete the ozone layer.[79] Although early-stage technoeconomic evaluation assessment tools are relatively well established and widely spread, measures of environmental sustainability are less well developed. The most common green metrics used for describing the environmental impact of processes are summarized in Table 3.15.5. Most of the metrics include or are specifically aimed at solvent usage. Mass of solvent, number of solvents used in the process, and ease of solvent recovery are all considered in green metrics analysis.[11]

Global warming and climate change are commonly used terms to describe the increase of average temperature on the earth. The main contributors to global warming are the emissions of greenhouse gases such as carbon dioxide, methane, nitrous oxide, hydrofluorocarbons, and perfluorocarbons. These are increasingly released by human activities. Some anticipated results of global warming are an increase in sea level and sea surface temperature, changes in precipitation, and increases in storm intensity. One of the techniques used to assess the environmental impact of an activity is carbon footprint analysis. Carbon footprint is a measure of the total amount of carbon dioxide emissions directly and indirectly caused by an activity, or the life stages of a product.[80] LCA is a common method to estimate carbon footprint. There are four major steps in LCA: goal and scope definition, inventory analysis, impact assessment, and result interpretation. Fig. 3.15.7 compares the results of carbon footprint analysis of six different types of recycled and virgin solvent.[81] In every case solvent recycling results in significant carbon footprint reduction, which can be as high as 90%. Welton comprehensively reviewed the sustainability aspects of solvents with several industrial examples demonstrating that environmental sustainability must go hand in hand with commercial sustainability.[78]

Excessive solvent use is a major contributor to a chemical company's carbon footprint. The disposal of excessive solvent waste significantly contributes to the release of greenhouse gases and other emissions. It has been estimated that incineration alone creates 6.7 kg of carbon dioxide per kilogram of organic carbon treated.[16] The most widely employed waste disposal tool nowadays in the chemical industry is incineration. The environmental legislation is continuously getting stricter, demonstrating pressure from regulatory agencies. At the same time, price increases of pure solvents are making solvent recovery and recycling a more competitive alternative to incineration. Solvent recovery and recycling can offer significant benefits to both the industry and the environment with regard to reduced purchase, storage, transportation and waste costs, human exposure, increased compliance with environmental legislation, and reduced emission of greenhouse gases.[39] The *Ecosolvent Tool*, developed by the Safety & Environmental Technology Group at Eidgenössische Technische Hochschule Zurich in collaboration with seven chemical companies, is an LCA software that allows for the quantification of the environmental impact of waste solvent treatment.[82] Options for solvent recovery and recycling are discussed in section 3.15.5 book.

TABLE 3.15.5 Summary of the Green Metrics Used for Describing the Environmental Impact of Processes

	Category		Units
MASS			
1	$\frac{\text{Total mass (kg)}}{\text{Mass of product (kg)}}$	Mass intensity	$\frac{kg}{kg}$
2	$\frac{\text{Total mass solvent (gross) (kg)}}{\text{Mass of product (kg)}}$	Solvent intensity	$\frac{kg}{kg}$
3	$\frac{\text{Mass of isolated product (kg)} \times 100}{\text{Total mass of reactants used inreaction (kg)}}$	Reaction mass intensity	%
4	$\frac{\text{FW (g/mol) product} \times 100}{\text{FW of all reactants used in reaction}}$	Atom economy	%
5	$\frac{\text{Mass of waste generated (kg)}}{\text{Mass of product (kg)}}$	E-factor	$\frac{kg}{kg}$
ENERGY			
6	$\frac{\text{Total process energy (MJ)}}{\text{Mass of product (kg)}}$		
7	$\frac{\text{Total solvent recovery energy (MJ)}}{\text{Mass of product (kg)}}$		
PHOTOCHEMICAL OZONE CREATION POTENTIAL			
8	$\frac{\text{Total [mass of solvent (kg)} \times \text{POCP value} \times \text{vapour pressure (mm)]}}{\text{Mass of product (kg)} \times \text{vapour pressure} \times \text{POCP}}$		$\frac{kg}{kg}$ (as toluene)
GREENHOUSE GAS EMISSIONS			
9	$\frac{\text{Total mass of green house gas from energy [askgCO}_2\text{equiv.]}}{\text{Mass of product (kg)}}$		$\frac{kg}{kg}$ (as CO_2)
10	$\frac{\text{Greenhouse gas, kg CO}_2\text{ equivalent, ex energy for solvent recovery}}{\text{Mass of product (kg)}}$		$\frac{kg}{kg}$
SOLVENT			
11	Number of different solvents		Number
12	Overall estimated recovery efficiency		%
13	Energy for solvent recovery		$\frac{MJ}{kg}$
14	Mass intensity net of solvent recovery		$\frac{kg}{kg}$

Most of the metrics include or are specifically aimed at solvent usage. Mass of solvent, number of solvents used in the process, and ease of solvent recovery are all considered in green metrics analysis.[11]

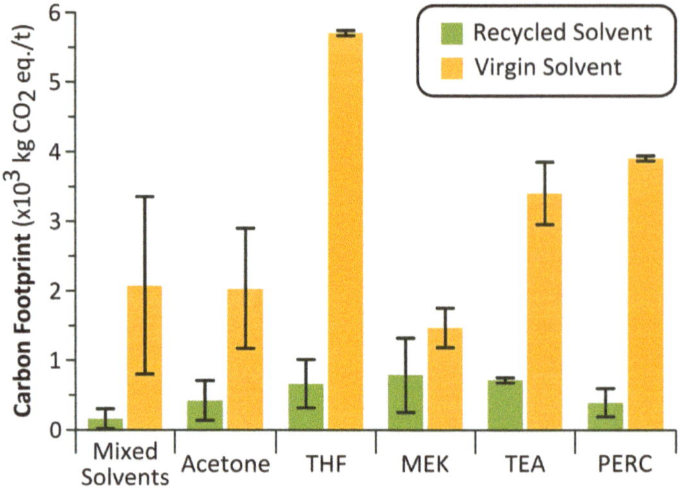

FIGURE 3.15.7 Comparison of the carbon footprints of recycled and virgin solvents: mixed solvents, acetone, tetrahydrofuran (THF), methyl ethyl ketone (MEK), triethylamine (TEA), and perchloroethene (PERC). In every case, solvent recycling results in significant carbon footprint reduction, which can be as high as 90%.[81]

The free-of-charge CCaLC carbon footprinting methodology is a life cycle analysis and decision support tool for calculating and reducing the carbon footprints of different industrial sectors along complete supply chains.[83] It contains over 50 case studies spanning the biofeedstock, biofuel, chemical, food, drinks, and packaging sectors. It has been demonstrated that recovery and reuse of organic solvents have a significant effect on both the economy and environmental impact of the overall process.

3.15.4 SOLVENTS FOR SUSTAINABLE CHEMISTRY

Despite abundant precautions, the use of solvents inevitably contaminates our environment, including air, soil, and water, because they are inherently difficult to contain and recycle. Both academic and industrial researchers have therefore focused on minimizing solvent consumption through the development of solvent-free processes and more efficient solvent recovery and recycling methodologies. However, these approaches have their drawbacks and limitations, necessitating a contamination preventive approach and the search for environmentally benign so-called green solvents. Opportunities for the practical implementation of such solvents in sustainable chemistry are reviewed by DeSimone.[1]

Sustainable chemistry is a concept of increasing interest in the scientific and manufacturing community. Major pharmaceutical companies such as GSK or Pfizer are making increasing efforts to minimize their environmental impact and protect people in the workplace. To share the same vision of more sustainable and green syntheses of APIs, several companies joined working groups, such as the ACS GCI-PR, formed in 2005.[84] Their mission is to facilitate the implementation of the principles of green chemistry and sustainable engineering in the global pharmaceutical industry. Their strategic priorities include informing and influencing the research agenda, developing tools for sustainable innovation, and providing an education resource through global collaboration. This has been achieved through the production of guidelines, literature surveys, disclosure of internal practices and decision making, distribution of grants to universities, and providing open access to sustainable technologies through publications.[85]

The first edition of the GSK SSG was published in 1999,[86] with a further version incorporating LCA assessments published in 2005.[87] A subsequent expansion revised the assessments of factors impacting process safety, significantly increased the number of organic solvents considered, and added a customized version designed with medicinal chemistry and analytical laboratories in mind.[20] The most recent update and further expansion with 44 new solvents was published in 2016.[29] Several other pharmaceutical companies (including Pfizer[88] and Sanofi Aventis[89]) also have their own solvent guides. These publications from the industrial sector are essential to understanding how green chemistry and engineering are approached by chemical manufacturers. Moreover, the challenges and concerns raised by the industrial sector greatly assist in setting the right academic research directions. Collaborative groups such as the ACS GCI-PR and Innovative Medicines Initiative's Chemical Manufacturing Methods for the 21st Century Pharmaceutical Industries[90] are producing guides representing the input of multiple stakeholders. Although differing business priorities and factors such as geography and regulatory environments led to some differences in the final conclusions, all these efforts tend to agree on the majority of solvent assessments.[85]

The SSG established the GSK approach to solvent selection that is based on a relative ranking (1–10) of the inherent environmental, health, and safety issues associated with each solvent (see Table 3.15.4). The objective of the SSG is to highlight potential risks that might arise when a particular solvent is employed in a process. Its main advantage is to permit the user to select the best green alternative based on the general requirements of the chemical process and the properties of the solvent. For instance, using methanol in a synthetic step could afford excellent yield but may have a variety of difficult environment, health, and safety (EHS) issues associated with storing, handling, using, recovering, or disposing that might be eliminated, minimized, or safely managed if replaced by a higher boiling point alcohol. The guide does not state that methanol cannot or should not be used; rather, it attempts to make the various EHS trade-offs readily apparent. This approach provides scientists with the desired degree of freedom and necessary understanding to make the best decisions for their own business and manufacturing environment.[87]

Fig. 3.15.8 shows some green solvent replacement options and their prices—green being the desired solvent and red being the solvents that need to be avoided—based on the Pfizer SSG and recent literature. Prices of 1L CHROMASOLV grade solvents were taken from the Sigma-Aldrich catalog (June 16, 2016). The figure demonstrates that there is no direct correlation between the greenness and the price of the solvents. Replacing a conventional solvent with a greener alternative that is also cheaper seems an obvious decision, but in many cases the routine or the difficult licensing process remains in the way of sustainable development. Furthermore, consistent quality and supply of solvents are important factors for the chemical industry and can hinder the spread of green solvents, whose availability is limited. For instance, piperylene sulfone has been proposed as a green alternative to dimethyl sulfoxide as a solvent for nucleophilic displacement reactions.[91] Although it is not yet commercially

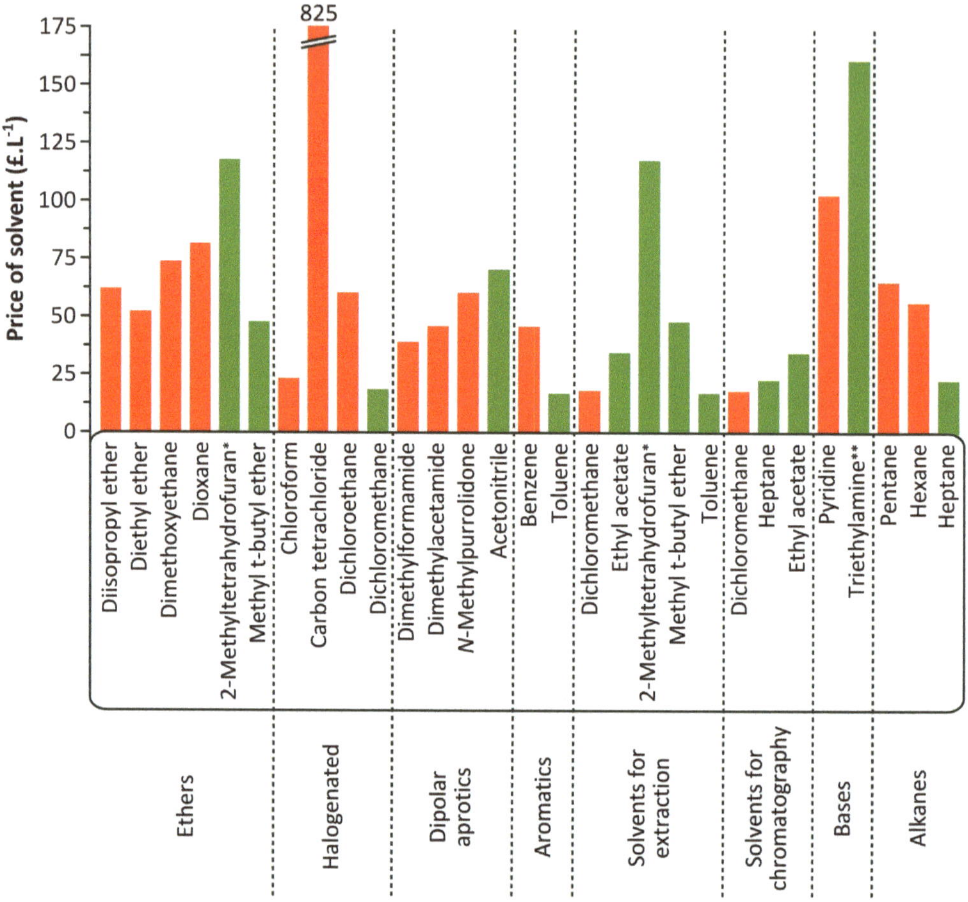

FIGURE 3.15.8 Prices of undesirable solvents (*red bars*) and their alternatives (*green bars*) in 2016. *, anhydrous grade; **, BioUltra grade. The figure demonstrates that there is no direct correlation between the greenness and the price of the solvents. Replacing a conventional solvent with a greener alternative, which is also cheaper seems an obvious decision, but in many cases the routine or the difficult licensing process remains in the way of sustainable development.

available, its sustainable and scalable synthesis has been developed by Marus et al.[92] In some cases, solvent production does not follow the needs of the customers because the solvent is only a by-product of a production process. For instance, as a result of the Olympic games the production of acetonitrile stopped in China in October, 2008, which led to worldwide shortage and subsequently increased price of this particular solvent. Furthermore, a US factory was damaged in Texas during Hurricane Ike. Due to the global economic slowdown, the manufacture of acrylonitrile used for acrylic fibers and acrylonitrile-butadiene-styrene resins decreased. Acetonitrile is generated as a by-product during the production of acrylonitrile, so its availability has also decreased, which prolonged the global shortage of acetonitrile throughout 2009.[93]

The developed production process of pregabalin, a medication used for the treatment of central nervous system disorders such as epilepsy and anxiety, is a good example of considering

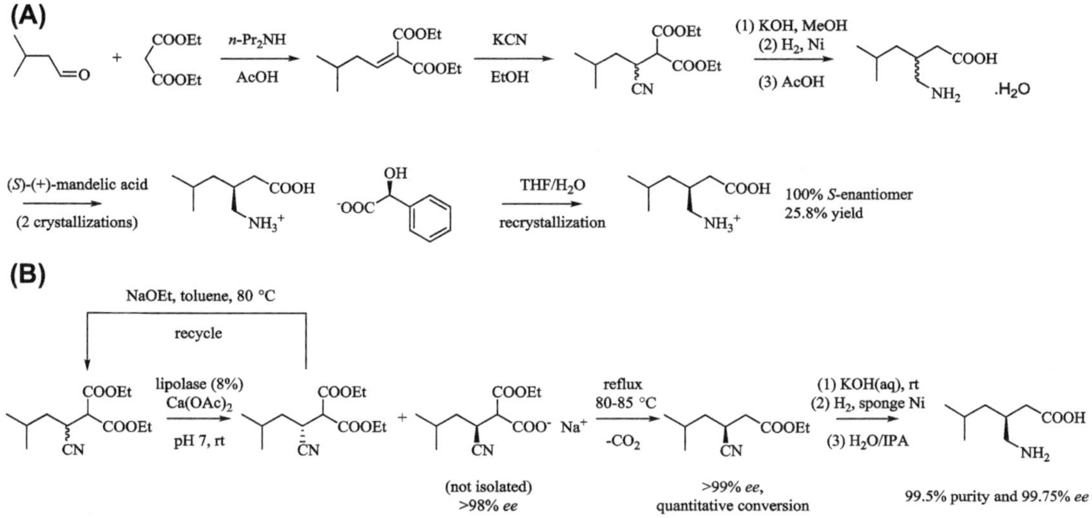

FIGURE 3.15.9 Original (A) and chemoenzymatic (B) synthesis of pregabalin. The improved process reduced the E-factor from 86 to 17, while the solvent use was reduced from 50 to 6.2 kg per kilogram of product.

sustainable process development in the pharmaceutical industry. The original commercial synthesis of pregabalin (Fig. 3.15.9A) started with a Knoevenagel condensation, followed by the addition of a racemic chiral moiety via cyanation, then hydrolysis, decarboxylation, and hydrogenation in methanol solvent to yield a γ-amino acid.[94] Enantiopure mandelic acid was used to form a diastereomeric salt in aqueous isopropyl alcohol for chiral resolution via recrystallization from aqueous tetrahydrofuran, followed by recrystallization from isopropyl alcohol, giving the final pregabalin product. This malonate route was compared with a γ-isobutylglutaric acid route at pilot plant scale, and it was found that the price, the production rate, and the amount of waste generated were similar. However, the latter route using the chlorinated solvent chloroform was rejected due to the higher capital outlay caused by the necessary control measures. This example demonstrates how the avoidance of undesirable solvents may result in lower production costs. Nonetheless, the diastereomeric resolution inherently involved the use of large quantities of solvent and the loss of the APIs itself. Pfizer developed a chemoenzymatic process (Fig. 3.15.9B) where a β-cyano diester is hydrolyzed.[95] The unreacted diester is racemized in toluene. The process yields oil that phase separates from water containing the majority of the impurities. The improved process reduced the E-factor from 86 to 17, while the solvent use was reduced from 50 to 6.2 kg per kilogram of product.

3.15.5 SOLVENT RECOVERY AND RECYCLING

The cost of ownership associated with the continuously increasing use of water and organic solvents is no longer a minor issue. Most of the world's supply of solvents eventually ends up being destroyed or dispersed into the biosphere.[96] There is a negligible accumulation

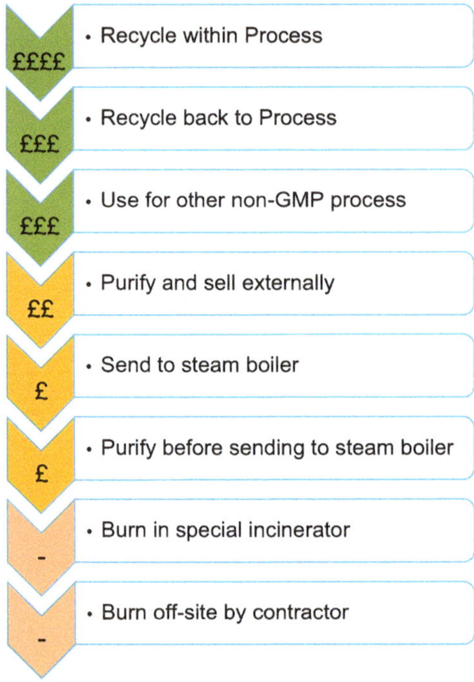

FIGURE 3.15.10 Handling waste process solvent streams and their money-saving potential. From a sustainability point of view, the most desirable product of solvent recovery is one that can be used in situ in the process in which it was used in the first place.

of solvents in long-term artifacts, and consequently the annual production of the solvent industry equates closely to the discharge. In today's economic environment, solvent recovery and recycling play an increasingly important role. Chemical industries need to constantly improve their processes to (1) meet increasingly stringent regulations and (2) increase their profit margins.[97] The design of manufacturing processes determines whether the consumption of organic solvents will be a significant production cost factor. In particular, in pharmaceutical manufacturing processes solvent costs can account for 80%–90% of the total manufacturing cost.[16]

Based on Lonza group's suggestion, Fig. 3.15.10 shows the saving potential of various routes after their intended usage.[98] From a sustainability point of view, the most desirable product of solvent recovery is one that can be used in situ in the process where it was used in the first place. This does not necessarily mean that the recovered solvent meets and has to meet the same specification as the virgin material. Waste solvents produce a significant carbon footprint and associated expenditure independently of being incinerated on-site or disposed of through outsourced services. Preferably solvents need to be recycled on-site using distillation, adsorption, or membrane processing. Due to the importance of rapid time-to-market speed in chemical businesses, solvent recovery and recycling is not considered in the early stages of process development. On the other hand, once a process is well established, solvent recovery is often one of the key upgrades considered to improve profit margins. The University of Minnesota has created a website that discusses the industrial factors to consider with regard to reusing recovered solvents for the same process, which could be useful for companies willing to embark on solvent recycling.[99] The most common technologies for solvent recovery and reuse include: distillation processes such as pressure swing, azeotropic, or extractive distillation; adsorption; and membrane processes such as organophilic pervaporation and solvent-resistant nanofiltration.[96] The following subsections will discuss these three main processes in detail. Less common techniques for solvent recovery include cocurrent or countercurrent extraction and steam or gas stripping. A unique example of microwave-assisted solvent recovery has also been reported. Keller Crescent, a printing company, replaced a distillation unit with a microwave solvent recovery system, which has eliminated all hazardous waste costs associated with both print towels and recovered blanket wash.[100] The solvent recovery rate improved from 50% to 98%, and the company's solvent purchase was reduced by 25%.

3.15.5.1 Distillation Processes

Distillation is a separation process of liquid mixtures based on their boiling points or relative volatility. Distillation processes may have begun as early as 2000 BC in China, Egypt, and Mesopotamia, where different drinks such as *tarasun* were produced by distillation and fermentation of rice.[101] In ancient times, oil essences were produced through distillation and fermentation of cedar, cypress, ginger, and myrrh.[102] Nowadays, there are many distillation techniques for separating solvents from residues, water, or other mixtures. The most commonly used techniques are simple distillation, fractional distillation, steam distillation, and vacuum distillation.

In simple distillation process, a volatile compound is evaporated and channeled through a distillation column into a condenser, where it is eventually captured. This technique can be used to separate mixtures containing nonvolatile compounds such as particles and mixtures with differences of at least 70 °C in boiling points. On the other hand, fractional distillation is used to separate mixtures with nearly equal relative volatility, and as small a difference in boiling points as 25 °C. Fractional distillation columns consist of an array of trays, in which the lowest and highest boiling liquids are collected at the top and bottom of the column, respectively. This process is commonly used in petroleum and food industries due to its better efficiency compared with simple distillation. Vacuum distillation separates mixtures at a temperature much lower than their atmospheric boiling point, and thus it is mainly employed for high-boiling-point solvents such as dimethyl sulfoxide, benzyl alcohol, ethylene glycols, and glycerol.

For separating heat-sensitive compounds, it is recommended to use steam distillation techniques, in which steam is introduced to the mixture causing vaporization at lower temperature than the decomposition temperature of the heat-sensitive compound. This process is commonly used in the production of perfumes, essential oils, and waxes. Azeotropic distillation is used in multipurpose solvent recovery systems for mixtures, whose separation is thermodynamically limited by the presence of azeotrope mixtures. Often, the addition of another component called an "entrainer" facilitates separation. An entrainer is a mass-separating agent that alters the relative volatility of mixtures and thus facilitates the separation process. For instance, benzene is often used as an entrainer to ease the separation of ethanol and water. A residue curve map is a geometric representation of the vapor-liquid equilibrium phase behavior of multicomponent mixtures, and it is used to choose the entrainer.[103]

Distillation is the most well-established technology for solvent recycling, but it consumes a large amount of energy. One of the most effective ways to reduce the energy consumption of distillation is through heat integration.[104] There are two approaches to develop heat-integrated distillation systems: consider new designs of distillation or retrofit heat integration of individual columns.

Industrial case studies demonstrate that distillation is a viable option for solvent recovery. Crumrine, a small business in the United States, recycles 95% of their *N*-methyl-2-pyrrolidone (NMP) solvent using the Re-solv Model RS-7 vacuum solvent recovery system manufactured by SRS Engineering, which led to a savings of $24,000 per year by avoiding waste management and material purchase costs.[105] Bayer Corporation first began to recycle organic solvents used in manufacturing processes in 1974. Total cost savings in their Clayton facility were calculated to be approximately $7 million per year from solvent recycling and the associated purchase costs saved.[106] Pfizer recovers acetonitrile and isopropanol solvent from small-volume waste and plans to recover chlorinated solvents in the future. Through the solvent recovery process Pfizer was able to save $65 million and reduced their carbon emissions by 677,000 kg per year.[107]

3.15.5.2 Adsorption Processes

Adsorption processes are useful and versatile tools when it comes to solvent recovery as they can be applied with high efficiency at relatively low cost in cases in which the desired component presents either a fairly small or a fairly high proportion of the stream. The applicable adsorbents vary according to different purposes.[108,109] Adsorbents with low polarity (activated carbon, etc.) tend to adsorb nonpolar compounds, whereas ones with high polarity (e.g., silica, alumina) have higher affinity to adsorb polar substances. However, some adsorbents operate via specific binding sites (e.g., molecular sieves, molecularly imprinted polymers) rather than simple hydrophilic-hydrophobic interactions. It is worth mentioning that adsorption cannot easily be installed in a continuous configuration and is usually either a one-bed batch process or a twin-bed process with one bed in the adsorption, whereas the other one in the regeneration phase.

In organic solvent recycling, the most frequent issue is the removal of water content. Even traces of water can cause unexpected solubility problems, side reactions, or the decomposition of a reactant. There are various processes to recover wet solvents such as distillation methods or fractional freezing, whereas adsorptive methods are advantageous due to their low energy consumption. Molecular sieves (with pore size 3 or 4 Å), silica, and alumina are widely used for solvent drying.[110,111] The polarity of the solvent affects the efficiency of water removal, which decreases with increasing polarity of the solvent. With the proper choice of adsorption technique, residual water content between 1 and 100 ppm is usually a realistic target.

In the regeneration stage of adsorption, high volumes of gas containing organic solvent are produced. Other processes in the chemical industry, such as paint drying or the drying of solid pharmaceutical intermediates or products, also generate a significant amount of solvent vapor.[112] This raises another issue, as the recovery of this solvent is highly desired to minimize solvent loss and the environmental burden, as urged by the increasingly strict regulatory environment. For example, the recycling of chlorofluorocarbons has gained a lot of attention since the Montreal protocol.[113–115] Incineration of solvent vapors is a widely used solution since it makes use of the solvent's latent heat. However, incineration likely needs supporting fuel to reach the required efficiency and needs continuous solvent vapor feed, not to mention that nonflammable halogenated solvent cannot be eliminated in this manner. Adsorptive systems have proved to be good alternatives. This field of adsorption is dominated by activated carbon adsorbents,[116] but molecular sieve zeolites are also employed.[117] Polymeric adsorbents are seldom employed in such processes, mainly because of their high price compared with activated carbon and zeolites.[118] The choice of adsorbent regeneration technique has a significant effect on the quality of the recovered solvent. Examining the efficiency and applicability of various regeneration processes has been the aim of several studies.[112,119] A typical system utilizing activated carbon adsorption to recover solvents from air emissions is shown in Fig. 3.15.11. Steam regeneration is employed to strip solvents from the activated carbon followed by condensation of the steam/solvent mixture through cooling. Eventually the solvent layer is separated by simple decantation.

Besides waste gases, industrial wastewaters can also be contaminated by significant amount of organic solvents. Activated carbon or charcoal beds are widely employed in water treatment as supplementary units after air-stripping processes.[96] However, for the recovery of less volatile solvents, adsorption is the most reasonable choice as the main unit. Some processes in the chemical industry (e.g., membrane fabrication) produce a considerable amount of wastewater contaminated by high-boiling, polar, aprotic solvents. Razali et al. examined different adsorbents regarding their ability to recover and in situ reuse such waste water.[120] Fig. 3.15.12 shows the adsorption capacity of different adsorbents for the polar, aprotic solvents *N,N*-dimethylformamide (DMF) and NMP. The continuous wastewater treatment process based on MIP7 reduced the waste generation by 99%, and it was demonstrated to be an effective solution to mitigate organic solvent contamination down to 10 ppm level, allowing either safe disposal or in situ reuse.

The color of a pharmaceutical product is an important indicator of the quality and therefore is the subject of regulations. Old or recycled solvents often contain colorful trace impurities. Their removal is essential as they can influence the color of the product. Activated carbon adsorbents are frequently used in this field as they effectively adsorb high-molecular-weight pigments.[121]

3.15.5.3 Membrane Processes

Membrane processes are one of the key technologies to drive process intensification due to their inherent simplicity and energy efficiency.[39,122] The separation proceeds in most cases with a simple pressure gradient, thereby avoiding undesirable phase changes, unlike distillation. The assessment of P.G Jessop particularly mentions that one of the grand challenges in the field of green solvents is the elimination of distillation.[74] Fig. 3.15.13 reveals the comparison of the carbon footprint of distillation-, adsorption-, and nanofiltration-based solvent recovery processes, whereas Table 3.15.6 compares both the carbon footprint and the energy required for solvent recovery by distillation and nanofiltration as a function of boiling point. Distillation processes have a higher carbon footprint than adsorption processes up to 70% solvent recovery due to the high energy consumption, and due to the fact that recovery

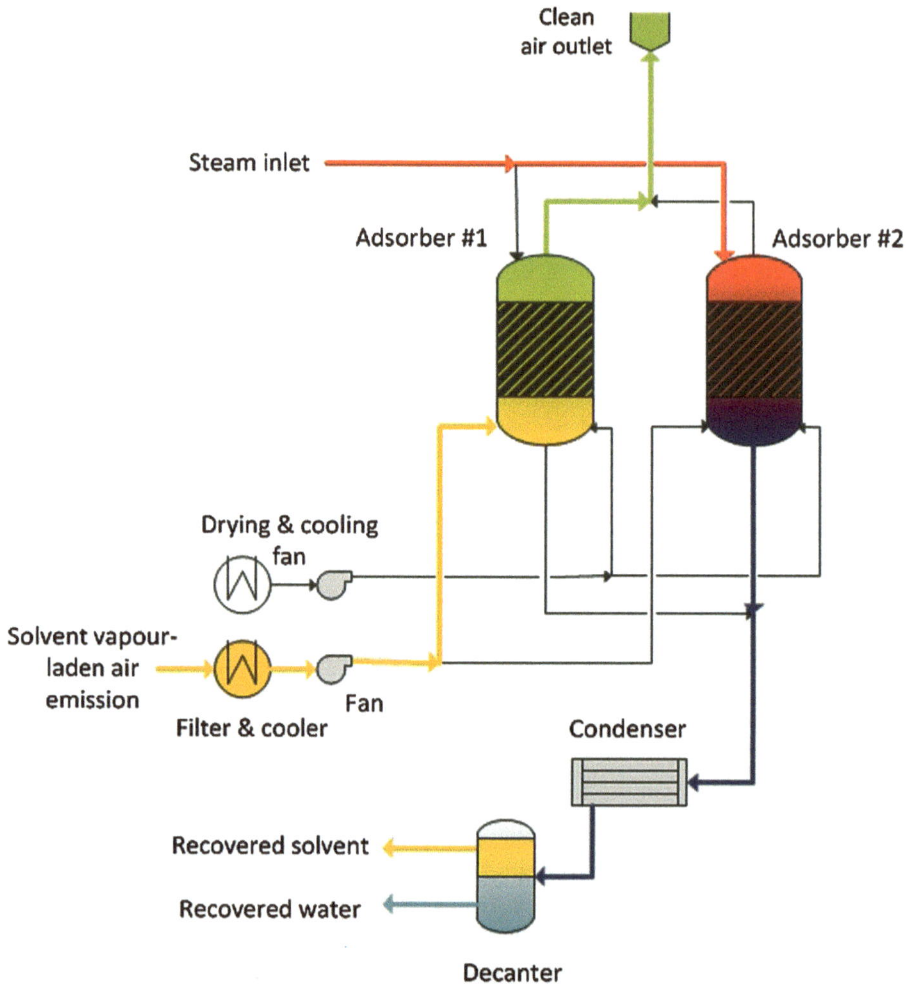

FIGURE 3.15.11 A typical system layout using activated carbon adsorption to recover solvents from air emissions. Steam regeneration is employed to strip solvents from the activated carbon followed by condensation of the steam/solvent mixture through cooling. Eventually the solvent layer is separated by simple decantation. The twin-bed arrangement allows for continuous operation.

of the last 30% of the waste solvent by adsorption requires excessive amounts of adsorbent because of the highly concentrated solution. Over 70% solvent recovery the solid waste adsorbent begins to outweigh the achieved solvent savings.[123]

The modular nature of membrane operations is often regarded as an advantage over other conventional processes since it allows easy integration with existing processes such as distillation.[124] Furthermore, several membrane units can be advantageously arranged into different configurations, such as membrane cascades, to reduce solvent consumption via solvent recovery and in situ solvent recycling.[122,125,126] Recent advancements in the field include membranes that are now stable in a variety of harsh conditions, such as extreme pH and

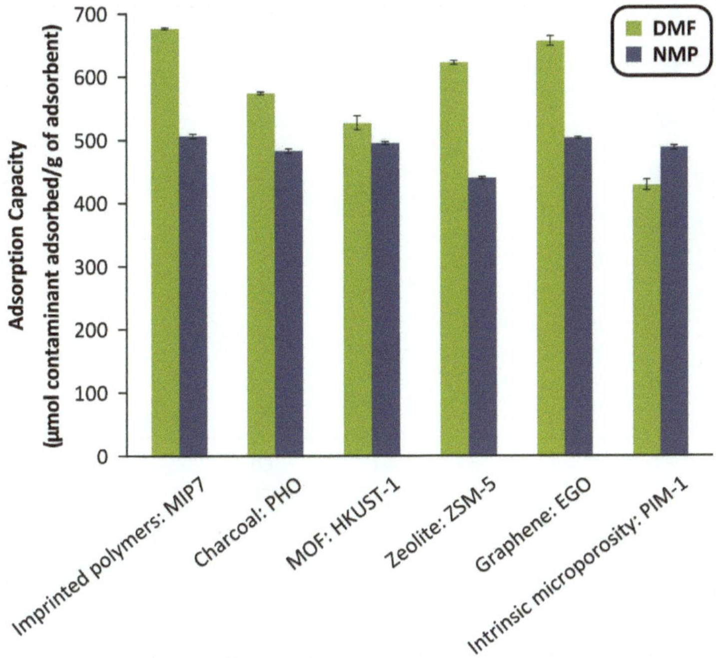

FIGURE 3.15.12 Adsorption capacity of water purification materials from different adsorbent classes: imprinted polymers, charcoals, metal-organic frameworks (MOFs), zeolites, graphene-based materials, and polymers of intrinsic microporosity.[120] The continuous wastewater treatment process based on MIP7 reduced the waste generation by 99%, and it was demonstrated to be an effective solution to mitigate solvent contamination caused by *N,N*-dimethylformamide (DMF) or *N*-methyl-2-pyrrolidone (NMP) down to 10 ppm level, allowing either safe disposal or in situ reuse.

polar aprotic solvents such as DMF and NMP, and are able to remove small impurities from solvents with molecular weights as low as 100 g/mol.[28] Experimentally validated examples of nanofiltration for solvent recovery are presented in Table 3.15.7. The field of solvent-resistant nanofiltration has gone through a significant advancement during the past decade, and improved membranes and process are now available enabling solvent recovery with small impurities at the lower end of the nanofiltration range (100 g/mol). Pervaporation is a membrane process employed for the separation of liquid mixtures. In dehydration applications, water is removed from solvent-water mixtures by selective permeation through a dense hydrophilic pervaporation membrane. The most relevant application of pervaporation is the separation of liquid azeotropes and close-boiling-point solvent-water mixtures. Successfully demonstrated examples include recovery of diisopropyl alcohol from a pharmaceutical effluent; valorization of an industrial ketonic effluent; recovery of 2,3-butanediol, butanol, acetone, and ethanol from water; and dehydration of ethyl acetate.[127–130] Pervaporation has been proposed as a green drying process for tetrahydrofuran recovery during pharmaceutical synthesis.[131] The integration of the pervaporation unit with a constant-volume distillation process resulted in a hybrid system that allows for the recovery and reuse of tetrahydrofuran with 93%, 80%, and 99% reduction in process waste, energy, and operating cost, respectively.

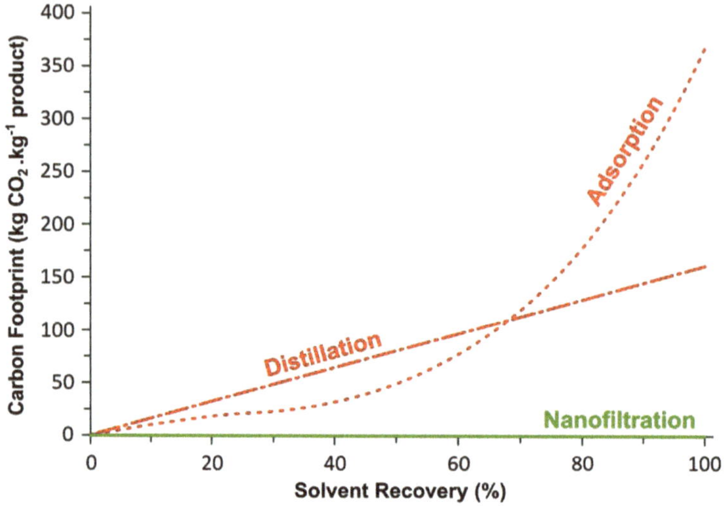

FIGURE 3.15.13 Comparison of CO_2 footprint of common solvent recovery technologies calculated for methanol.[122] Distillation processes have a higher carbon footprint than adsorption processes up to 70% solvent recovery due to the high energy consumption, and due to the fact that recovery of the last 30% of the waste solvent by adsorption requires excessive amounts of adsorbent because of the highly concentrated solution. Over 70% solvent recovery the solid waste adsorbent begins to outweigh the achieved solvent savings.

TABLE 3.15.6 Comparison of Energy, and Carbon Footprint Required for Solvent Recovery by Distillation and Nanofiltration[39]

Solvent	Waste Solvent Generated in 2006 (10^6 kg/year)	Q_{dist} (kWh)	Q_{OSN} (kWh)	Q_{dist}/Q_{OSN}	CO_2 Footprint (10^6 kg/year)
Methanol	44.8	150	0.023	6,453	18
Dichloromethane	22.3	111	0.014	8,010	3
Toluene	12.1	197	0.021	9,278	12
Acetonitrile	7.9	141	0.023	6,029	3
Chloroform	3.71	131	0.012	10,543	0.4
n-Hexane	2.99	149	0.028	5,300	3
n-Butyl alcohol	2.86	223	0.023	9,788	2
N,N-Dimethylformamide	2.79	244	0.019	12,569	2
N-Methyl-2-pyrrolidone	2.02	303	0.018	16,930	1
Xylene	1.47	208	0.021	9,748	1
1,1,2-Trichloroethane	1.23	194	0.013	15,090	0.2
Methyl *tert*-butyl ether	1.2	126	0.025	5,062	1
Ethylene glycol	0.82	337	0.017	20,285	0.3

Q_{dist}, and Q_{OSN} represent energy required for solvent recovery by distillation and OSN processes. The carbon footprint of the waste solvent is estimated based on their disposal using incineration. *OSN*, organic solvent nanofiltration.

TABLE 3.15.7 Experimentally Validated Examples of Solvent Recovery Using Nanofiltration

Year	Application	Solute Molecular Weight (g/mol)	Rejection (%)	Membrane	Solvent
2000	Lube oil dewaxing[132]	500	95	Polyimide based	Methyl ethyl ketone, toluene
2010	Pharmaceutical[123]	>1000, 675	>99	DuraMem 300, 1000	Tetrahydrofuran
2013	Catalyst recovery[133]	1044, 300	>99	DuraMem 300, 500	Tetrahydrofuran
2012	Crystallization[134]	600	>99	PuraMem 280	Isopropyl alcohol
2014	Pharmaceutical[122]	260	>99	DuraMem 150	Methanol
2016	Fine chemical[125]	110	>99	Polyimide-poly (ether-ether-ketone)	Acetone

The field of solvent-resistant nanofiltration has gone through a significant advancement during the past decade, and improved membranes and process are now available enabling solvent recovery with small impurities at the lower end of the nanofiltration range (100 g/mol).

3.15.6 ADVERSE IMPACT OF ORGANIC SOLVENTS

Organic solvents are the source of about 35% of the volatile organic compounds (VOCs) entering the atmosphere in the United Kingdom[96] Their contribution to the total is similar in magnitude to all the VOCs arising from the fueling and use of motor vehicles. The latter source is continuously being substantially reduced by developments in catalysis and car manufacturing and in the fuel distribution system, and thus increased pressure will be brought to bear on solvent users to cut the harm done to the environment by their discharges. This section briefly discusses the exposure to organic solvents and their impact on our health and the environment.

Besides human and environmental exposure, due to the high flammability and low flash point accidents involving organic solvents occasionally occur. In October 2007, at Barton Solvents distribution facility in the United States, a static electrical spark caused by a pressure relief device ignited ethyl acetate solvent, which was being loaded into a 330-gallon square tank.[135] Thanks to the successful evacuation nobody was seriously injured but a large portion of the facility was consumed by the flames. The accident was a result of inadequate electrical bonding and grounding during the filling of the portable steel tank. In November 2006, there was an explosion also in the United States at a solvent plant and ink manufacturer (CAI) due to inadvertent overheating of solvents left stirring overnight in an unsealed mixing tank, releasing flammable vapor that accumulated and ignited. The explosion damaged over 90 homes, and 10 people reported minor injuries.[136] In April 2002, Distillex in the United Kingdom was operating a chemical recovery service, annually recycling thousands of tons of hydrocarbon and halogenated solvents.[137] An angle grinder caused a fire through sparks that ignited solvent-contaminated rags in a waste skip spreading rapidly to a storage area holding almost half a million liters of chemicals. The fire was compounded by chemicals mixing with melting plastic from intermediate bulk containers. Accidents can occur not only at industrial sites but also in university and research environments. In July 2001, a benzene vapor explosion occurred at the University of California, resulting in an injury to a doctoral student and a fire that caused damage worth approximately $3.5 million.[138] During the purification of benzene with a reflux distillation apparatus the system became overpressurized, disassembled, and the solvent vapor was ignited, resulting in an explosion and subsequent fire.

Working with organic solvents can have adverse effects on the environment and on human health and can lead to serious accidents. Therefore the production and consumption of organic solvent needs to be minimized, the recovery and recycle of organic solvents need to be promoted, and all safety regulations must be adhered to at all times to avoid accidents.

3.15.6.1 Exposure and Health Effects of Organic Solvents

Most of the solvents manufactured in bulk quantities were of low quality until the mid-20th century, when chromatography emerged for the purification and analysis of solvents. After their intended use these solvents were dumped in pits and ponds or burnt. Only at the end of the century was it realized that they must be used with caution, and legislation started to emerge to understand and control human and environmental exposure to organic solvents and their vapors. In the United States, new and existing chemicals including solvents are regulated under Toxic Substances Control Act (TSCA), and release of VOCs into the atmosphere are monitored by the Environmental Protection Agency (EPA).[139] In Europe, the European Union Solvents Emission Directive 1999/13/EC16 effective from October 2007 restricts the "loss of solvent to atmosphere," especially carcinogenic, mutagenic, and reprotoxic substances.[140]

Exposure to organic solvents can occur at workplaces either manufacturing or using them, or through consumer products.[141] Consumer products such as paint and coatings contain aliphatic organic solvents (e.g., hexane), adhesives and printing inks contain aromatic organic solvents (e.g., toluene, xylene, benzene), paint strippers contain dichloromethane, dry cleaning agents have perchloroethylene, nail polishes and polish removers have ethyl acetate and acetone, and cleaning products and cosmetics can contain glycol ethers, alcohols, and cyclopentasiloxane. Manufacturing exposure takes place either through evaporation of solvents or due to leaks of normally closed systems.

Two types of effects of organic solvents can be distinguished: acute effects and chronic effects. Acute effects are short-term effects that last for a few hours or days and the effects are reversible. On the other hand, chronic effects are long-term effects due to repeated exposure and the effects are irreversible. Some organic solvents are potential carcinogens; some may cause harmful effects to eyes, skin, and the respiratory, nervous, heart, and cardiovascular systems.[27] The risk assessment of emitted solvents is difficult to ascertain, and solvent exposure is regulated by setting threshold limit values. Exposure and exposure values can be monitored by defined methods such as ambient and biological monitoring.[142]

Interestingly, several solvents have depressant or narcotic effects, and hence, some solvents are used as anesthetics.[143] The anesthetic quality of diethyl ether was first described by Paracelsus, who observed chickens to fall asleep and awaken harmlessly after inhaling ether. Its use in medication began in the mid-1800s and lasted until the end of the 1900s.[144] Diethyl ether also used to be drunk as a recreational drug during the height of its popularity. The use of chloroform in anesthesia and analgesia also emerged in the 19th century. Although doubt had been cast upon its safety since the beginning, chloroform was used in 80%–95% of all anesthesia performed in the United Kingdom and German-speaking countries between around 1865 and 1920.[145] Chloroform has become widely recognized by the public as an incapacitating agent due to the popularity of crime fiction authors having criminals use chloroform-soaked rags to render victims unconscious. However, it is nearly impossible to incapacitate someone using chloroform in this manner.[146]

The most common mild effect of exposure to solvents is nonspecific irritations of skin and mucous membrane structures, causing irritant contact dermatitis.[147] Intact skin structures can be destroyed by solvents that dissolve grease and fat. Typically, the dermatitis on the hands is characterized by dryness, scaling, and fissuring. It is often caused by handling solvent-contaminated products or by cleaning procedures.[143]

Unspecific irritation of mucous membranes is often caused by solvent vapors and results in irritation of the eyes and various sections of the airways. A causal association between exposure to benzene, toluene, or styrene and an increased risk of leukemia, lung cancer, and nasopharyngeal cancer has been established by a large number of studies.[148–150] Increased risks of various gastrointestinal cancers have been suggested following exposure to toluene. Occupation as a painter has consistently been associated with a 40% increased risk of lung cancer.[151] The liver, kidney, and lung are the most common target sites for chlorinated compounds such as trichloroethylene, chloroform, and methylene chloride. It has also been demonstrated that organic solvents have significant toxic effects on petrol-filling workers exposed for longer durations, such as decreases in lung volumes and capacities.[152]

3.15.6.2 Impact on the Environment

3.15.6.2.1 Organic Solvents in Water and Mitigating Technologies

Water miscibility of organic solvents is one of the most important parameters for solvent recovery, environmental impact, and human exposure. The higher the miscibility of a solvent in water, the more likely it is that it will be carried into various segments of the environment and eventually become part of the hydrologic cycle. Miscibility and solubility of solvents and solvent wastes in water can affect the extent of leaching into surface water such as rivers and lakes. Aqueous solubility also determines the efficacy of removal from the atmosphere through dissolution into precipitation and into surface waters. Miscible organic solvent in aqueous media is very difficult to separate and thus causes further problems in solvent recycle, waste disposal, and discharge.

Chlorinated solvents are denser than water. Therefore, these toxic compounds sink and cause harm to aquatic organisms. The most frequently found chlorinated solvents in the environment are dichloromethane, chloroform, carbon tetrachloride, perchloroethylene, trichloroethylene, and trichloroethane. In 2012, the US EPA Toxic Release Inventory reported a total of approximately 10,000,000 chlorinated solvent releases in the country.[153] A 2010 study suggested that groundwater polluted with chlorinated solvents resulted in an increase in liver cancer.[154] Wastewater or municipal water supply treatment systems that utilize coagulation, sedimentation, precipitative softening, filtration, and chlorination have been found ineffective for reducing concentrations of some chlorinated solvents to nonhazardous levels.[155]

Crude oil and petroleum products are lighter than water. Therefore they partition in the top layer of water. However, there are some fractions of these compounds that dissolve in water and may cause harm to aquatic organisms. Mesitylene, *p*-xylene, toluene, heptane, hexane, diisopropyl ether, and fluorobenzene are toxic to aquatic life with long-lasting effects.[30] Therefore these compounds should not be discharged into the environment, and any form of spillage or leakage must be contained immediately. In April 2010, an explosion at the Deepwater Horizon oil rig in the Gulf of Mexico caused the largest oil spill in the earth's history, killing 11 people, spilling 200 million gallons of oil, and directly affecting thousands of species of fish, sea turtles, and birds. Groundwater contamination caused by industrial discharge of chlorinated solvents has been successfully mitigated by activated charcoal treatment.[156] Molecularly imprinted polymers, metal organic frameworks, zeolites, and graphene-based materials, among other advanced adsorbents were used to remove NMP and DMF from industrial wastewater.[120] It has been demonstrated that over 99% of these polar aprotic solvents in the wastewater can be successfully removed and the recovered water can be recycled into the original process. The proposed wastewater treatment technique can reduce the process mass intensity of the manufacturing process by about 99%. Apart from that, a pilot-scale activated sludge system has been established, having acetone, methyl ethyl ketone, cyclohexanone, tetrahydrofuran, carbon tetrachloride, 1,1,1-trichloroethane, 1,1,2-trichloroethane, trichloroethylene, tetrachloroethylene, chlorobenzene, and ethylbenzene solvents to produce an acclimated biomass. Within 7 weeks, 9 of the 11 contaminants were reduced by more than 94% with the estimated biodegradation up to 93.4%.[157] In another study, postprimary municipal wastewater contaminated with acetone, tetrahydrofuran, and 1-butanol was exposed to an experimental microcosm subsurface constructed wetland system consisting of replicates of *Juncus effusus*, *Carex lurida*, *Iris pseudacorus*, and *Pondeteria cordata*. The microbial bioremediation process resulted in 90% removal of 1-butanol, acetone, and tetrahydrofuran in 3, 5–10, and >10 days, respectively. Initial experiments confirmed that the majority of solvent removal was via microbial bioremediation. Biodegradation is identified as the dominant removal mechanism in the model systems, but as much as 20% of the total removal may be attributed to sorption.[158]

3.15.6.2.2 Organic Solvents in Air and Mitigating Technologies

Organic solvents enter the air via evaporation, after which they can slowly photodegrade or react with gas-phase radicals to initiate catalytic oxidation. Once organic solvents enter the air they can be removed via adsorption or thermal oxidizers. Thermal oxidizers are designed to accomplish 95%–99% destruction of virtually all VOCs at about 800 °C with a residence time of less than 1 s.[159] Catalytic oxidation systems directly combust VOCs in a manner similar to thermal oxidizers but at 300–500 °C. This significant reduction in temperature and energy costs is made possible by the use of catalysts that reduce the activation energy for combustion reaction, and thus the overall energy requirement. A reverse flow reactor (RFR) is an adiabatic packed bed reactor in which the direction of the feed flow is reversed periodically. Consequently, the reactor is forced to operate under transient conditions. RFR is an emerging alternative for the removal of VOCs from polluted air because unsteady-state reactor operation can be profitable for the chemical process. Biofiltration is a process in which contaminated air is passed through a porous packed medium that supports a thriving population of microorganisms.[159] The contaminants are first adsorbed from the air onto the water/biofilm phase of the medium and eventually converted into carbon dioxide, water, inorganic products, and biomass. The success of biofiltration depends on the degradability of the contaminants. Ease of degradability of organic solvents follows the order hydrogen sulfide > aromatics > aldehydes and ketones > chlorinated hydrocarbons. Absorber towers are used to remove VOCs from gas streams by contacting the contaminated air with a liquid solvent. Any soluble VOCs will transfer to the liquid phase, and the air stream is scrubbed with VOC removal efficiencies of 95%–98%. Flexible and inexpensive carbon adsorption systems are common tools for VOC emission control. VOCs are removed from the inlet air by physical adsorption onto the surface of the carbon. Drawbacks of such systems include humidity control, high flammability, difficult regeneration for high boiling solvents, and polymerization or oxidation of some solvents to toxic or insoluble derivatives. Therefore, there have been several attempts to replace activated carbon with more advanced materials such as zeolites, metal organic frameworks, and graphene to overcome such barriers.[160–162] Gas separation membranes have been proposed for the recovery of VOCs from air.[163] However, this technique is in the early stages of experimental research and has yet to be commercialized, although a few pilot plants have been developed and are continuously monitored for performance.[159]

3.15.6.2.3 *Organic Solvents in Soil and Mitigating Technologies*

Volatilization from soil may be an important mechanism for the movement of solvents from spills or from the disposal of solvent-containing wastes. The efficacy and rate of volatilization from soil depends on the solvent's vapor pressure, water solubility, and the properties of the soil such as soil water content, airflow rate, humidity, temperature, and the adsorption and diffusion characteristics of the soil. Furthermore, the rate of movement of dissolved solvents through porous geological material may be retarded by adsorption-desorption reactions between the solvents and the solid phases. Solvents may be degraded in soil by the same mechanisms as those in water. In biodegradation, microorganisms utilize the carbon of the solvents for cell growth and maintenance. Phytoremediation is emerging as a cost-effective and green environmental restoration technology that uses plants and associated soil microbes to remove, degrade, or contain toxic chemicals in soils, sediments, groundwater, surface water, and air.[164] It can be used as a stand-alone remediation alternative or as part of a broader site management alternative composed of a number of remediation technologies in which the concentrations of contaminants are not toxic to the plants employed. Although limited to the root zone of plants, phytoremediation is a sustainable alternative to conventional engineering processes that are more destructive to the soil.

Acknowledgment

The authors wish to thank Dr. Helen Sneddon and Dr. Richard Henderson from GlaxoSmithKline plc for their calculations to include supercritical carbon dioxide and the ionic liquids in Table 3.15.4.

References

1. DeSimone JM. Practical approaches to green solvents. *Science* 2002;**297**:799–803.
2. IUPAC. *Compendium of chemical terminology* (The "Gold Book"). Compiled by McNaught AD, Wilkinson A. 2nd ed. Oxford: Blackwell Scientific Publications; 1997, ISBN: 0-9678550-9-8.
3. https://en.oxforddictionaries.com.
4. https://www.britannica.com.

5. Marcus Y, editor. *The properties of solvents*, vol. 1. Weinheim, Germany: Wiley; 1998, ISBN: 978-0-471-98369-9.
6. Kirschner EM. Environment, health concerns force shift in use of organic solvents. *Chem Eng News* 1994;**72**(25):13.
7. Sigma-Aldrich Corporation. http://www.sigmaaldrich.com/analytical-chromatography/air-monitoring/applications/paints-and-coatings/solvents.html.
8. Bissett R. The secret to oil painting without solvents. www.buildart.com/secret_paint_without_solvent.htm.
9. Nontoxic print: nontoxic printmaking, safe painting & printed art – safe solvents. http://www.nontoxicprint.com/safesolvents.htm.
10. AAE Chemie. http://www.aaechemie.com/html/aae_solvent_information.html.
11. Curzons AD, Constable DJC, Mortimer DN, Cunningham VL. So you think your process is green, how do you know?—Using principles of sustainability to determine what is green–a corporate perspective. *Green Chem* 2001;**3**:1–6.
12. Nelson WM, editor. *Green solvents for chemistry: perspectives and practice*. Oxford, UK: Oxford University Press; 2003, ISBN: 0-19-515736-2.
13. Anastas PT, Williamson TC, editors. *Green chemistry: frontiers in benign chemical syntheses and processes*. Oxford, UK: Oxford University Press; 1998, p. 1–26. 0-19-850170-6.
14. Capello C, Fischer U, Hungerbuhler K. What is a green solvent? A comprehensive framework for the environmental assessment of solvents. *Green Chem* 2007;**9**:927–34.
15. Pereira CSM, Silva VMTM, Rodrigues AE. Ethyl lactate as a solvent: properties, applications and production processes – a review. *Green Chem* 2011;**13**:2658–71.
16. Slater CS, Savelski MJ, Carole WA, Constable DJC. Solvent use and waste issues [Chapter 3]. In: Dunn PJ, Wells AS, Williams MT, editors. *Green chemistry in the pharmaceutical industry*. Weinheim, Germany: Wiley; 2010, ISBN: 978-3-527-32418-7.
17. Constable DJC, Gonzalez CJ, Henderson RK. Perspective on solvent use in the pharmaceutical industry. *Org Process Res Dev* 2007;**11**:133–7.
18. Web of Science^{TM}. https://www.webofknowledge.com.
19. Toxics Release Inventory (TRI) Program. Retrieved from: https://www.epa.gov/toxics-release-inventory-tri-program.
20. Henderson RK, Gonzalez CJ, Constable DJC, Alston SR, Inglis GGA, Fisher G, et al. Expanding GSK's solvent selection guide – embedding sustainability into solvent selection starting at medicinal chemistry. *Green Chem* 2011;**13**:854–62.
21. Gonzalez CJ, Curzons AD, Constable DJC, Overcash MR, Cunningham VL. How do you select the "greenest" technology? Development of guidance for the pharmaceutical industry. *Clean Prod Process* 2001;**3**:35–41.
22. Conkle JP, Camp BJ, Welch BE. Trace composition of human respiratory gas. *Arch Environ Health* 1975;**30**(6):290–5.
23. McCulloch A. Chloroform in the environment: occurrence, sources, sinks and effects. *Chemosphere* 2003;**50**:1291–308.
24. Nightingale PD, Malin G, Liss PS. Production of chloroform and other low-molecular-weight halocarbons by some species of macroalgae. *Limnol Oceanogr* 1995;**40**(4):680–9.
25. Reichardt C, editor. *Solvents and solvent effects in organic chemistry*. Weinheim, Germany: Wiley; 2003, ISBN: 3-527-30618-8.
26. *EH40/2005 Workplace exposure limits*. Health and Safety Executive; 2011, ISBN: 978-0-7176-6446-7. http://www.hse.gov.uk/pUbns/priced/eh40.pdf.
27. Szekely G, Amores de Sousa MC, Gil M, Ferreira FC, Heggie W. Genotoxic impurities in pharmaceutical manufacturing: sources, regulations, and mitigation. *Chem Rev* 2015;**115**(16):8182–229.
28. European Medicines Agency. *ICH guideline Q3C (R5) on impurities: guideline for residual solvents, EMA/CHMP/ICH/82260/2006*. 2015.
29. Alder CM, Hayler JD, Henderson RK, Redman AM, Shukla L, Shuster LE, et al. Updating and further expanding GSK's solvent sustainability guide. *Green Chem* 2016;**18**:3879–90.
30. Sigma-Aldrich and MSDS forms. Retrieved from Sigma Aldrich website http://www.sigmaaldrich.com.
31. Baumann JB. Solvent selection for recrystallization: an undergraduate organic experiment. *J Chem Educ* 1979;**56**:64.
32. Davey RJ. Solvent effects in crystallization processes. *Curr Top Mater Sci* 1982;**8**:429–79.

33. Craig RER. Rapid, efficient determination of recrystallization solvents at the microscale level. *J Chem Educ* 1989;**66**:88–9.
34. Bierne D, Smith S, Hoogenboom BE. Recrystallization without tears. *J Chem Educ* 1974;**51**:602.
35. Miller JM, Rodriguez-Hornedo N, Blackburn AC, Macikenas D, Collman BM. Solvent systems for crystallization and polymorph selection [Chapter 3]. In: Augustijns P, Brewster ME, editors. *Solvent systems and their selection in pharmaceutics and biopharmaceutics*. New York, USA: Springer-Verlag; 2007, ISBN: 978-0-387-69149-7.
36. Snyder LR, editor. *Principles of adsorption chromatography. The separation of non-ionic organic compounds, chromatographic science series*. New York, USA: Marcel Dekker; 1968.
37. Snyder LR, Kirkland JJ. *Introduction to modern liquid chromatography*. 2nd ed. New York, USA: Wiley-Interscience; 1979, ISBN: 0-471-03822-9.
38. Marchetti P, Jimenez-Solomon MF, Szekely G, Livingston AG. Molecular separation with organic solvent nanofiltration: a critical review. *Chem Rev* 2014;**114**(21):10735–806.
39. Szekely G, Jimenez-Solomon MF, Marchetti P, Kim JF, Livingston AG. Sustainability assessment of organic solvent nanofiltration: from fabrication to application. *Green Chem* 2014;**16**(10):4440–73.
40. Yang XJ, Livingston AG, Freitas dos Santos L. Experimental observations of nanofiltration with organic solvents. *J Membr Sci* 2001;**190**:45–55.
41. Machado DR, Hasson D, Semiat R. Effect of solvent properties on permeate flow through nanofiltration membranes. Part I: investigation of parameters affecting solvent flux. *J Membr Sci* 1999;**163**:93–102.
42. Machado DR, Hasson D, Semiat R. Effect of solvent properties on permeate flow through nanofiltration membranes: Part II. Transport model. *J Membr Sci* 2000;**166**:63–9.
43. Zhao Y, Yuan Q. Effect of membrane pretreatment on performance of solvent resistant nanofiltration membranes in methanol solutions. *J Membr Sci* 2006;**280**:195–201.
44. Darvishmanesh S, Degrève J, Van der Bruggen B. Mechanisms of solute rejection in solvent resistant nanofiltration: the effect of solvent on solute rejection. *Phys Chem Chem Phys* 2010;**12**(40):13333–42.
45. Knez Z, Markocic E, Leitgeb M, Primozic M, Hrncic MK, Skerget M. Industrial applications of supercritical fluids: a review. *Energy* 2014;**77**:235–43.
46. Rizvi H, editor. *Supercritical fluid processing of food and biomaterials*. London, UK: Blackie Academic and Professional; 1994.
47. Levy JM, editor. *Practical supercritical fluid extraction*. Heidelberg: Hüthig; 1996.
48. Duarte ARC, Casimiro T, Ricardo AA, Simplicio AL, Duarte CMM. Supercritical fluid polymerisation and impregnation of molecularly imprinted polymers for drug delivery. *J Supercrit Fluids* 2006;**39**:102–6.
49. Leitner W. Supercritical carbon dioxide as a green reaction medium for catalysis. *Acc Chem Res* 2002;**35**:746–56.
50. Nalawade SP, Picchioni F, Janssen LPBM. Supercritical carbon dioxide as a green solvent for processing polymer melts: processing aspects and applications. *Prog Polym Sci* 2006;**31**:19–43.
51. Temtem M, Casimiro T, Ricardo AA. Solvent power and depressurization rate effects in the formation of polysulfone membranes with CO_2-assisted phase inversion method. *J Membr Sci* 2006;**283**:244–52.
52. Peker H, Srinivasan MP, Smith JM, McCoy BJ. Caffeine extraction rates from coffee beans with supercritical carbon dioxide. *AIChE J* 1992;**38**(5):761–70.
53. Cavazzini M, Montanari F, Pozzi G, Quici S. Perfluorocarbon-soluble catalysts and reagents and the application of FBS (fluorous biphase system) to organic synthesis. *J Fluor Chem* 1999;**94**:183–93.
54. Earle MJ, Seddon KR. Ionic liquids. Green solvents for the future. *Pure Appl Chem* 2000;**72**:1391–8.
55. Smith EL, Abbott AP, Ryder KS. Deep eutectic solvents (DESs) and their applications. *Chem Rev* 2014;**114**(21):11060–82.
56. Francisco M, Bruinhorst AVD, Kroon MC. Low-transition-temperature mixtures (LTTMs): a new generation of designer solvents. *Angew Chem Int Ed* 2013;**52**:3074–85.
57. Alonso DA, Baeza A, Chinchilla R, Guillena G, Pastor IM, Ramon DJ. Deep eutectic solvents: the organic reaction medium of the century. *Eur J Org Chem* 2016;**4**:612–32.
58. Lomba L, Giner B, Bandres I, Lafuente C, Pino MR. Physicochemical properties of green solvents derived from biomass. *Green Chem* 2011;**13**:2062–70.
59. Pleissner D, Qi Q, Gao C, Rivero CP, Webb C, Lin CSK, Venus J. Valorization of organic residues for the production of added value chemicals: a contribution to the bio-based economy. *Biochem Eng J* 2015. http://dx.doi.org/10.1016/j.bej.2015.12.016.
60. Clark JH, Farmer TJ, Hunt AJ, Sherwood J. Opportunities for bio-based solvents created as petrochemical and fuel products transition towards renewable resources. *Int J Mol Sci* 2015;**16**:17101–59.

61. Gu Y, Jerome F. Bio-based solvents: an emerging generation of fluids for the design of eco-efficient processes in catalysis and organic chemistry. *Chem Soc Rev* 2013;**42**:9550–70.
62. *Innovating for sustainable growth: a bioeconomy for Europe*. Brussels: European Commission; 2012. http://ec.europa.eu/research/bioeconomy/pdf/official-strategy_en.pdf.
63. Nigiz FU, Hilmioglu ND. Green solvent synthesis from biomass based source by biocatalytic membrane reactor. *Int J Energy Res* 2016;**40**:71–80.
64. Bardole JA, White DL. Paint and finish removers. In: *Kirk-othmer Encyclopedia of chemical technology*. Wiley; 2015. p. 1–15.
65. Michailos S, Parker D, Webb C. A multicriteria comparison of utilizing sugar cane bagasse for methanol to gasoline and butanol production. *Biomass Bioenerg* 2016. http://dx.doi.org/10.1016/j.biombioe.2016.06.019.
66. Michailos S, Parker D, Webb C. Simulation studies on ethanol production from sugar cane residues. *Ind Eng Chem Res* 2016;**55**(18):5173–9.
67. Wang SF, Guo CL, Cui KK, Zhu YT, Ding JX, Zou XY, et al. Lactic acid as an invaluable green solvent for ultrasound-assisted scalable synthesis of pyrrole derivatives. *Ultrason Sonochem* 2015;**26**:81–6.
68. Yang J, Tana JN, Gu Y. Lactic acid as an invaluable bio-based solvent for organic reactions. *Green Chem* 2012;**14**:3304–17.
69. Weis DC, Visco DP. Computer-aided molecular design using the Signature molecular descriptor: application to solvent selection. *Comput Chem Eng* 2010;**34**:1018–29.
70. Moity L, Molinier V, Benazzouz A, Barone R, Marionc P, Aubry JM. In silico design of bio-based commodity chemicals: application to itaconic acid based solvents. *Green Chem* 2014;**16**:146–60.
71. Struebing HZ, Ganase Z, Karamertzanis PG, Siougkrou E, Haycock P, Piccione PM, et al. Computer-aided molecular design of solvents for accelerated reaction kinetics. *Nat Chem* 2013;**5**:952–7.
72. Fermeglia M, Braiuca P, Gardossi L, Pricl S, Halling PJ. In silico prediction of medium effects on esterification equilibrium using the COSMO-RS method. *Biotechnol Prog* 2006;**22**:1146–52.
73. Peters MA. *Computational approach to solvent selection for biphasic reaction systems* [Thesis]. Germany: RWTH Aachen University.
74. Jessop PG. Searching for green solvents. *Green Chem* 2011;**13**:1391–8.
75. Laird T. Green chemistry is good process chemistry. *Org Process Res Dev* 2012;**16**:1–2.
76. Towns Improvement Clauses Act 1847. http://www.legislation.gov.uk/ukpga/Vict/10-11/34.
77. REACH. http://echa.europa.eu/web/guest/regulations/reach.
78. Welton T. Solvents and sustainable chemistry. *Proc R Soc A* 2015;**471**:20150502. http://www.dx.doi.org/10.1098/rspa.2015.0502.
79. Ozone Secretariat. http://ozone.unep.org/en/treaties-and-decisions/montreal-protocol-substances-deplete-ozone-layer.
80. Weidema BP, Thrane M, Christensen P, Schmidt J, Lokke S. Carbon footprint – a catalyst for life cycle assessment? *J Ind Ecol* 2008;**12**:3–6.
81. Azapagic A, Amienyo D, Franca RMC, Jeswani HK. Carbon footprints of recycled solvents. *ETHOS Res* 2013. http://esrg.de/media/PDF/Study_print_090514.pdf.
82. Capello C, Hellweg S, Hungerbühler K. Ecosolvent: a tool for the environmental assessment of waste-solvent treatment in chemical industry. http://www.sust-chem.ethz.ch/tools/ecosolvent.
83. http://www.ccalc.org.uk/.
84. ACS GCI Pharmaceutical Roundtable. https://www.acs.org/content/acs/en/greenchemistry/industry-business/pharmaceutical.html.
85. Prat D, Hayler J, Wells A. A survey of solvent selection guides. *Green Chem* 2014;**16**:4546–51.
86. Curzons AD, Constable DC, Cunningham VL. Solvent selection guide: a guide to the integration of environmental, health and safety criteria into the selection of solvents. *Clean Prod Proc* 1999;**1**(2):82–90.
87. Gonzalez CJ, Curzons AD, Constable DJC, Cunningham VL. Expanding GSK's Solvent Selection Guide—application of life cycle assessment to enhance solvent selections. *Clean Tech Environ Policy* 2005;**7**:42–50.
88. Alfonsi K, Colberg J, Dunn PJ, Fevig T, Jennings S, Johnson TA, et al. Green chemistry tools to influence a medicinal chemistry and research chemistry based organisation. *Green Chem* 2008;**10**:31–6.
89. Prat D, Pardigon O, Flemming HW, Letestu S, Ducandas V, Isnard P, et al. Sanofi's solvent selection guide: a step toward more sustainable processes. *Org Process Res Dev* 2013;**17**(12):1517–25.
90. Prat D, Wells A, Hayler J, Sneddon H, McElroy CR, Shehada SA, et al. CHEM21 selection guide of classical and less classical-solvents. *Green Chem* 2016;**18**:288–96.

91. Vinci D, Donaldson M, Hallett JP, John EA, Pollet P, Thomas CA, et al. Piperylene sulfone: a labile and recyclable DMSO substitute. *Chem Commun* 2007:1427–9.
92. Marus GA, Vyhmeister E, Pollet P, Donaldson ME, Mestre VL, Faltermeier S, et al. Sustainable and scalable synthesis of piperylene sulfone: a "volatile" and recyclable DMSO substitute. *Ind Eng Chem Res* 2011;**50**(1):23–7.
93. Tullo A. A solvent dries up. *Chem Eng News* 2008;**86**(47):27. http://pubs.acs.org/doi/abs/10.1021/cen-v086n047.p027.
94. Hoekstra MS, Sobieray DM, Schwindt MA, Mulhern TA, Grote TM, Huckabee BK, et al. Chemical development of CI-1008, an enantiomerically pure anticonvulsant. *Org Process Res Dev* 1997;**1**:26–38.
95. Martinez CA, Hu S, Drumond Y, Tao J, Kelleher P, Tully L. Development of a chemoenzymatic manufacturing process for pregabalin. *Org Process Res Dev* 2008;**12**:392–8.
96. Smallwood IM, editor. *Solvent recovery handbook*. 2nd ed. Florida, USA: Blackwell Science; 2002, ISBN: 0-632-05647-9.
97. Gonzalez CJ, Constable DJC, Ponder CS. Evaluating the "Greenness" of chemical processes and products in the pharmaceutical industry—a green metrics primer. *Chem Soc Rev* 2012;**41**:1485–98.
98. *Solvent recovery*. Lonza Engineering Ltd; May 2011. Retrieved from: http://bio.lonza.com/uploads/tx_mwaxmarketingmaterial/Lonza_Newsletter_May_2011__English.pdf.
99. Selecting a still for on-site solvent recycling, minnesota technical assistance programme. http://www.mntap.umn.edu/mach/resources/62-Still.pdf.
100. Jones G. Keller Crescent Company's solvent recovery solution, Printers' National Environmental Assistance Center. http://www.pneac.org/sheets/litho/Kellercrescent.pdf.
101. Fairley T. *The early history of distillation, annual meeting – yorkshire and north eastern section*. 1906. p. 559–82.
102. History of alcohol distillation. Retrieved from: http://www.copper-alembic.com/en/page/history-of-alcohol-distillation.
103. Julka V, Chiplunkar M, O'Young L. Selecting entrainers for azeotropic distillation. *React Sep* 2009:47–53.
104. Smith R, Jobson M, Farrokhpanah S. Heat integrated distillation system design. *Chem Eng Trans* 2010;**21**:19–24.
105. Solvent recycling at Crumrine. http://unrbep.org/wp-content/uploads/2009/11/Crumrine.pdf.
106. Case study, Bayer Corporation – Clayton facility. http://infohouse.p2ric.org/ref/07/06123.pdf.
107. Thomas P, Solvent recovery at Pfizer: a continuous solution for small waste streams. *Pharm Manuf* Retrieved from: http://www.pharmamanufacturing.com/articles/2011/150/?start=1.
108. Henning KD. Solvent recycling, removal and degradation [Chapter 22]. In: Wypych G, editor. *Handbook of solvents*. New York, USA: ChemTec Publishing; 2001, ISBN: 1895198240.
109. Ghoshal AK, Manjare SD. Selection of appropriate adsorption technique for recovery of VOCs: an analysis. *J Loss Prev Proc* 2002;**15**:413–21.
110. Williams DBG, Lawton M. Drying of organic solvents: quantitative evaluation of the efficiency of several desiccants. *J Org Chem* 2010;**75**:8351–4.
111. Pahl C, Pasel C, Luckas M, Bathen D. Adsorptive water removal from organic solvents in the ppm-region. *Chem Ing Tech* 2011;**83**:177–82.
112. Fuderer A. *Adsorption process for recovery of solvents*. 1998. Patent # US5958109A.
113. Yokota H. *Solvent adsorber and solvent recovery system*. 1990. Patent # US5110328A.
114. Tsai WT, Chen HP, Hsien WY. A review of uses, environmental hazards and recovery/recycle technologies of perfluorocarbons (PFCs) emissions from the semiconductor manufacturing processes. *J Loss Prev Proc* 2002;**15**:65–75.
115. Tsai WT. A review of environmental hazards and adsorption recovery of cleaning solvent hydrochlorofluorocarbons (HCFCs). *J Loss Prev Proc* 2002;**15**:147–57.
116. Bansal RC, Goyal M, editors. *Activated carbon adsorption*. Florida, USA: CRC Press; 2005, ISBN: 978-1-4200-2881-2.
117. Meininghaus CKW, Prins R. Sorption of volatile organic compounds on hydrophobic zeolites. *Micropor Mesopor Mat* 2000;**35–36**:349–65.
118. Harper M. Sorbent trapping of volatile organic compounds from air. *J Chromatogr A* 2000;**885**:129–51.
119. Liu Y, Ritter JA. Pressure swing Adsorption–Solvent vapor Recovery: process dynamics and parametric study. *Ind Eng Chem Res* 1996;**35**:2299–312.
120. Razali M, Kim JF, Attfield M, Budd PM, Drioli E, Lee YM, Szekely G. Sustainable wastewater treatment and recycling in membrane manufacturing. *Green Chem* 2015;**17**:5196–205.

121. Urban P. *Removal of color impurities from organic compounds*. 1956. Patent # US2744938A.
122. Kim JF, Szekely G, Schaepertoens M, Valtcheva IB, Jimenez-Solomon MF, Livingston AG. In situ solvent recovery by organic solvent nanofiltration. *ACS Sustain Chem Eng* 2014;**2**(10):2371–9.
123. Sereewatthanawut I, Lim FW, Bhole YS, Ormerod D, Horvath A, Boam AT, Livingston AG. Demonstration of molecular purification in polar aprotic solvents by organic solvent nanofiltration. *Org Process Res Dev* 2010;**14**:600–11.
124. Micovic J, Werth K, Lutze P. Hybrid separations combining distillation and organic solvent nanofiltration for separation of wide boiling mixtures. *Chem Eng Res Des* 2014;**9**(2):2131–47.
125. Schaepertoens M, Didaskalou C, Kim JF, Livingston A, Szekely G. Solvent recycle with imperfect membranes: a semi-continuous workaround for diafiltration. *J Membr Sci* 2016;**514**:646–58.
126. Abejon R, Garea A, Irabien A. Organic solvent recovery and reuse in pharmaceutical purification processes by nanofiltration membrane cascades. *Chem Eng Trans* 2015;**43**:1057–62.
127. Urtiaga AM, Gorri ED, Gomez P, Casado C, Ibanez R, Ortiz I. Pervaporation technology for the dehydration of solvents and raw materials in the process industry. *Dry Technol* 2007;**25**(11):1819–28.
128. Pervatech case study: an energy efficient way to retrieve ethyl acetate from a waste stream. http://pervaporation-membranes.com/wp-content/uploads/2014/05/Case-Study-Ethyl-Acetate-recovery-Final-17-04-2014.pdf.
129. Shao P, Kumar A. Recovery of 2,3-butanediol from water by a solvent extraction and pervaporation separation scheme. *J Membr Sci* 2009;**329**:160–8.
130. Qureshi N, Meagher MM, Huang J, Hutkins RW. Acetone butanol ethanol (ABE) recovery by pervaporation using silicalite–silicone composite membrane from fed-batch reactor of *Clostridium acetobutylicum*. *J Membr Sci* 2001;**187**:93–102.
131. Slater CS, Savelski MJ, Moroz TM, Raymond MJ. Pervaporation as a green drying process for tetrahydrofuran recovery in pharmaceutical synthesis. *Green Chem Lett Rev* 2012;**5**(1):55–64.
132. White LS, Nitsch AR. Solvent recovery from lube oil filtrates with a polyimide membrane. *J Membr Sci* 2000;**179**:267–74.
133. Siew WE, Ates C, Merschaert A, Livingston AG. Efficient and productive asymmetric Michael addition: development of a highly enantioselective quinidine-based organocatalyst for homogeneous recycling via nanofiltration. *Green Chem* 2013;**15**:663–74.
134. Rundquist EM, Pink CJ, Livingston AG. Organic solvent nanofiltration: a potential alternative to distillation for solvent recovery from crystallisation mother liquors. *Green Chem* 2012;**14**:2197–205.
135. *CSB investigation at Barton solvents Des Moines, Iowa, facility progressing; immediate cause was ignition of spraying ethyl acetate during loading operation - agency also continues examination of July, 2007 Barton accident in Wichita, Kansas*. U.S. Chemical Safety Board, CSB News; November 1, 2007. http://www.csb.gov/news/?SID=33.
136. In Preliminary Findings, CSB Investigators Say. *Danvers, Massachusetts, explosion caused by solvent vapour accumulation, lack of ventilation inside building; flammable liquid safety standards were not implemented, CSB Investigators find*. U.S. Chemical Safety Board, CSB News; 2006. http://www.csb.gov/news/.
137. *Tyneside chemical alert over*. BBC News; January 2002. Retrieved from: http://news.bbc.co.uk/1/hi/england/1790942.stm.
138. University of California. *Irvine independent accident investigation*. Final report. January 24, 2002. http://www.ehs.ucsb.edu/files/docs/ls/UCI_fire.pdf.
139. US Environmental Protection Agency, Summary of the Toxic Substances Control Act [Last updated: July 11, 2016]. Retrieved from: https://www.epa.gov/laws-regulations/summary-toxic-substances-control-act.
140. European Commisions, The VOC solvents emission directive [Last updated: April 20, 2016]. Retrieved from: http://ec.europa.eu/environment/archives/air/stationary/solvents/legislation.htm.
141. Harriman L. *Massachusetts toxics use reduction Institute, six classes – a Webinar series on chemicals of concern*. October 2013. http://www.sixclasses.org/organic-solvents.
142. Baker EL. A review of recent research on health effects of human occupational exposure to organic solvents – a critical review. *J Occup Med* 1994;**36**:1079–92.
143. Riihimaki V, Ulfvarson U, editors. *Safety and health aspects of organic solvents: proceedings of the International Course on safety and health aspects of organic solvents*. New York, USA: Liss AR; 1986, ISBN: 0-845-15070-7.
144. Jacob AK, Kopp SL, Douglas RB, Hugh MS. The history of Anaesthesia [Chapter 1]. In: Barash PG, Cullen BF, Stoelting RK, Cahalan MK, Stock MC, editors. *Clinical Anaesthesia*. Lippincott Williams & Wilkins; 2009. ISBN: 0-7817-8763-5.

145. Wawersik J. History of chloroform anesthesia. *Anaesthesiol Reanim* 1997;**22**(6):144–52.
146. Payne JP. The criminal use of chloroform. *Anaesthesia* 1998;**53**:685–90.
147. Skov H, Allen V, Norval B, Simon B. Contrasting effects of ultraviolet A1 and ultraviolet B exposure on the induction of tumour necrosis factor-α in human skin. *Br J Dermatol* 1998;**138**(2):216–20.
148. Svensson BG, Nise G, Englander V, Attewell R, Skerfving S, Moller T. Deaths and tumours among rotogravure printers exposed to toluene. *Br J Ind Med* 1990;**47**(6):372–9.
149. Walker JT, Bloom TF, Stern FB, Okun AH, Fingerhut MA, Halperin WE. Mortality of workers employed in shoe manufacturing. *Scand J Work Environ Health* 1993;**19**:89–95.
150. Wong O, Trent LS, Whorton MD. An updated cohort mortality study of workers exposed to styrene in the reinforced plastics and composites industry. *Occup Environ Med* 1994;**51**:386–96.
151. Lynge E, Anttila A, Hemminki K. Organic solvents and cancer. *Cancer Causes Control* 1997;**8**(3):406–19.
152. Uzma N, Salar BM, Kumar BS, Aziz N, David MA, Reddy VD. Impact of organic solvents and environmental pollutants on the physiological function in petrol filling workers. *Int J Environ Res Public Health* 2008;**5**:139–46.
153. US EPA TRI. https://iaspub.epa.gov/triexplorer/tri_release.chemical.
154. Lee LJH, Chen CH, Chang YY, Liou SH, Wang JD. An estimation of the health impact of groundwater pollution caused by dumping of chlorinated solvents. *Sci Total Environ* 2010;**408**:1271–5.
155. Robeck G, Love O. *Removal of volatile organic contaminants from groundwater*. Retrieved from U.S.EPA website, https://cfpub.epa.gov/si/si_public_record_Report.cfm?dirEntryID=45829; 1983.
156. Yu JJ, Choub SY. Contaminated site remedial investigation and feasibility removal of chlorinated volatile organic compounds from groundwater by activated carbon fiber adsorption. *Chemosphere* 2000;**41**(3):371–8.
157. Bhattacharya SK, Madura RL, Dobbs RA, Angara RVR, Tabak H. Fate of selected RCRA compounds in a pilot-scale activated sludge system. *Water Environ Res* 1996;**68**(3):260–9.
158. Grove JK, Stein OR. Polar organic solvent removal in microcosm constructed wetlands. *Water Res* 2005;**39**(16):4040–50.
159. Khan FI, Ghoshal AK. Removal of volatile organic compounds from polluted air. *J Loss Prev Proc* 2000;**13**:527–45.
160. Zhao XS, Ma Q, Lu GQ. VOC removal: comparison of MCM-41 with hydrophobic zeolites and activated carbon. *Energy Fuels* 1998;**12**(6):1051–4.
161. Barea E, Montoro C, Navarro JAR. Toxic gas removal – metal–organic frameworks for the capture and degradation of toxic gases and vapours. *Chem Soc Rev* 2014;**43**:5419–30.
162. Wang S, Sun H, Ang HM, Tade MO. Adsorptive remediation of environmental pollutants using novel graphene-based nanomaterials. *Chem Eng J* 2013;**226**:336–47.
163. Drioli E, Barbieri G. *Membrane engineering for the treatment of gases*. RSC Publisher; 2011, ISBN: 1849732442.
164. Greipsson S. Phytoremediation. *Nat Educ Knowl* 2011;**3**(10):7.

CHAPTER

3.16

Ionic Liquids as Novel Media and Catalysts for Electrophilic/Onium Ion Chemistry and Metal-Mediated Reactions

Gopalakrishnan Aridoss[1], *Kenneth K. Laali*[2]

[1]LG Life Sciences Ltd, Daejeon, South Korea; [2]University of North Florida, Jacksonville, FL, United States

3.16.1 INTRODUCTION

Room temperature ionic liquids (RTILs) are continuing to play a major role in advancing synthetic chemistry and catalysis. Steady research progress in the application of ionic liquids (ILs) as solvents and catalysts in synthetic/catalytic transformations is manifested in the publication of a large number of reviews, highlights, commentaries, and accounts.[1–26]

The goal of the present chapter is to summarize in a selective way some of the most recent progress made in the application of RTILs as solvents and catalysts in organic and metal-mediated transformations, with an emphasis on electrophilic/onium ion chemistry.

ILs have had a transformative effect on fundamentally important/textbook reactions such as alkylation, acylation, nitration, and halogenation, and they have also led to notable advancements in metal-mediated transformations as well as those involving onium salts as reagents. Although the focus of the chapter is on electrophilic and metal-mediated reactions, other transformations in which IL coordination polarizes the reactant [as in Diels-Alder (DA)] or an IL solvent helps to solubilize an onium reagent (as in Wittig reaction) are also included.

3.16.2 ELECTROPHILIC ALKYLATION AND ACYLATION REACTIONS

Alkylation and acylation were among the first type of electrophilic transformations to be studied in ILs. Application of ILs as solvent and catalyst for this chemistry, and the progress

Green Chemistry
http://dx.doi.org/10.1016/B978-0-12-809270-5.00021-2

made over the years on improving these transformations, have been summarized in various review articles dealing with application of ILs in organic synthesis.[1,3,4,12,18,19]

3.16.2.1 Alkylation, Adamantylation, Alkenylation, and Benzylation

The importance of IL counterion in catalytic activity of $In(OTf)_3$ was shown in alkylation of aromatic hydrocarbons by using secondary (linear/cyclic), tertiary, and benzylic alcohols as alkylating agents under microwave irradiation.[27] Hydrophilic ILs, namely, 1-ethyl-3-methyl-imidazolium tetrafluoroborate [EMIM][BF_4]- and 1-butyl-3-methylimidazolium tetrafluoroborate [BMIM][BF_4]-bearing coordinating counterion were ineffective, whereas hydrophobic systems, namely, 1-ethyl-3-methylimidazolium bis(trifluoromethylsulfonyl)imide [EMIM][NTf_2]- and 1-ethyl-3-methylimidazolium hexafluorophosphate [EMIM][PF_6]-bearing noncoordinating counterion worked well. In particular, [EMIM][NTf_2] proved superior with respect to reaction rate, *para*-selectivity, and the absence of dibenzylether (DBE) (Scheme 3.16.1). The NTf_2 group is stable toward heating (unlike BF_4 and PF_6, it does not hydrolyze at elevated temperature by the in-situ-formed water to form HF), and the negative charge is highly delocalized. It was reported that this positive "ionic liquid effect" helped increase the efficacy of $In(OTf)_3$ for this transformation.

The occurrence of metathesis between $Sc(OTf)_3$ and the SbF_6 anion in Merrifield's resin-bound imidazolium-SbF_6 was observed by solid-state nuclear magnetic resonance (NMR) and fast atom bombardment mass spectrometry.[28] The in situ metathesis with noncoordinating anions in the IL could enhance its catalytic activity.

Diarylalkanes and triarylmethane were synthesized via double alkylation of aromatics with aldehyde by using a Brønsted acidic IL[29] (Scheme 3.16.2).

A 2015 study described the reductive Friedel-Crafts (FC) coupling of cyclic ketones with indole by employing a task-specific IL as catalyst without the need for any external reagent.[30] The reactions were completed within 1 h with 42%–94% isolated yields (Scheme 3.16.3). IL1 was superior to IL2. Metallic triflates [$Fe(OTf)_3$ or $Zn(OTf)_3$], trifluoromethanesulfonic acid

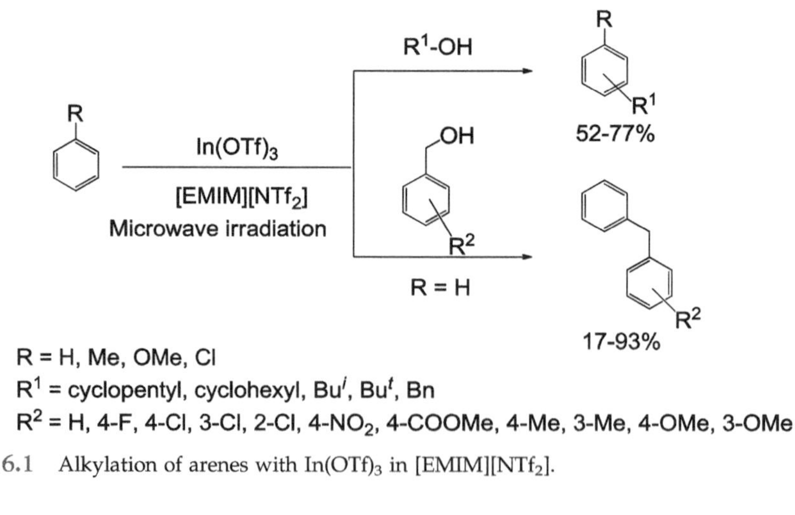

SCHEME 3.16.1 Alkylation of arenes with $In(OTf)_3$ in [EMIM][NTf_2].

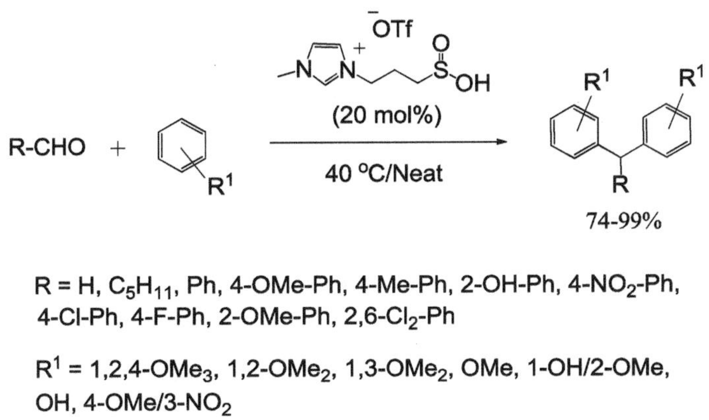

SCHEME 3.16.2 Brønsted acidic ionic liquid-catalyzed double alkylation of arenes with aldehyde.

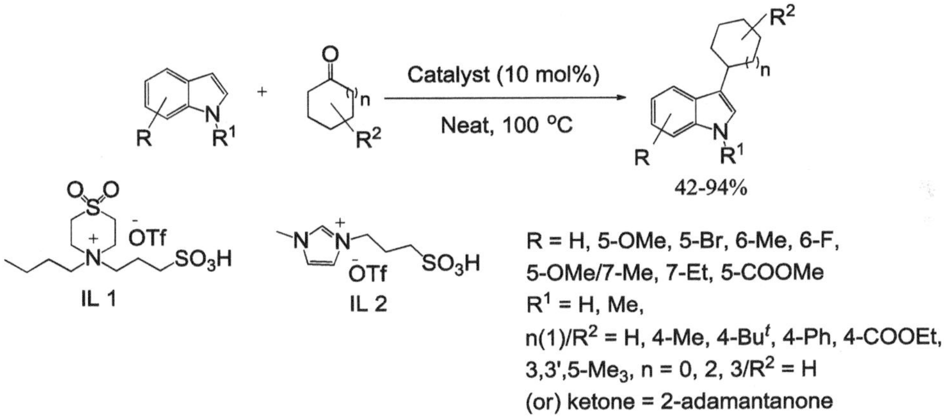

SCHEME 3.16.3 Reductive alkylation of indoles with ketone.

(TfOH), and trifluoroacetic acid (TFAH) at 100 °C were either mildly effective or ineffective for this transformation.

Asymmetric FC reaction of aromatics in ILs is an emerging field. This transformation can be achieved by addition of a chiral reagent and/or catalyst to the IL solvent, or by using the chiral IL itself as a chiral induction agent. This is exemplified in the reaction of ethyl glyoxylate with *N,N*-dimethylanilines in pyridinium-based IL,[31] using $Ti(OPr^i)_4$ as Lewis acid and BINOL/6,6′Br-BINOL as chiral ligand (Scheme 3.16.4).

Adamantylation of arenes with 1-AdaX (X = OH, Cl, Br) was reported by using 1-butyl-3-methylimidazolium trifluoromethanesulfonate [BMIM][OTf]-TfOH.[32] The reactions exhibited high *para*-selectivity, produced little or no adamantane by-product, and the conversions were quantitative or near quantitative (Scheme 3.16.5).

Intramolecular FC reaction in ILs has been applied to medicinal chemistry for the construction of biologically potent heterocyclic frameworks. For example, $Hf(OTf)_4$-catalyzed intramolecular alkenylation[33] allowed the synthesis of 4-phenylcoumarins and

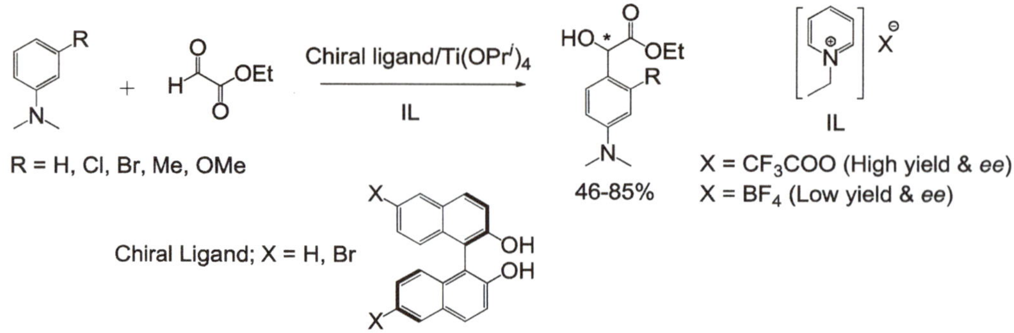

SCHEME 3.16.4 Asymmetric Friedel-Crafts reaction with ethyl glyoxylate.

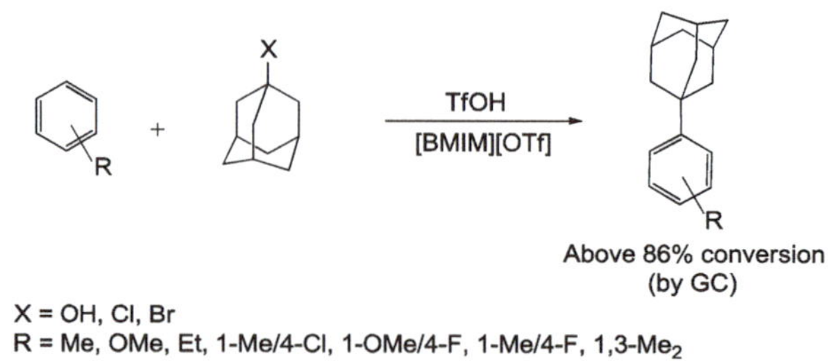

SCHEME 3.16.5 Friedel-Crafts adamantylation of arenes in [BMIM][OTf].

2(1*H*)-quinolinones in 1-butyl-3-methyl imdazolium hexafluoroantimonate [BMIM][SbF_6]/ methylcyclohexane at 85 °C (Scheme 3.16.6). The intramolecular *C*-alkylation of pyrrole with 1(4-bromobutyl)-1*H*-pyrrole has been reported by employing [BMIM][SbF_6] as Lewis acid and $NaHCO_3$ as base, with either CH_3CN or 1-butyl-3-methylimidazolium hexafluorophosphate [BMIM][PF_6] as cosolvent at 100 °C to afford 5,6,7,8-tetrahydropyrrolo[1,2-a] pyridine.[34] However, a low conversion (*c*.23%) was noted when CH_3CN was used as solvent in the absence of IL, at the same temperature.

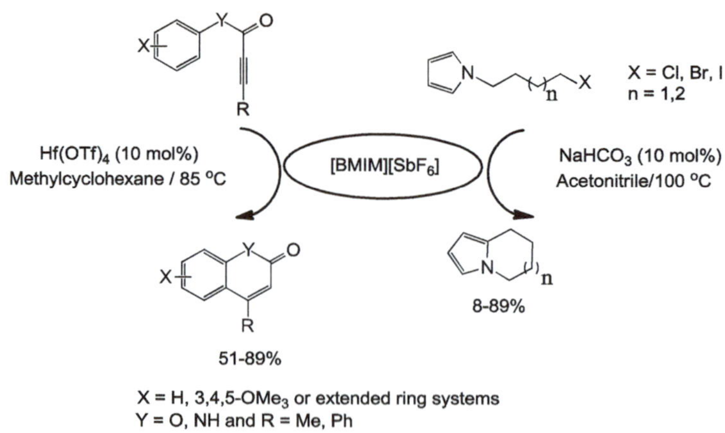

SCHEME 3.16.6 Intramolecular Friedel-Crafts alkyl/alkenylation of arenes.

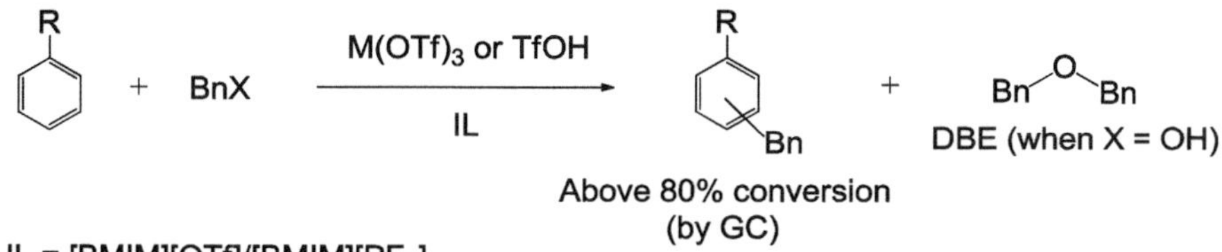

IL = [BMIM][OTf]/[BMIM][PF_6]

BnX = $PhCH_2OH$, $PhCH_2Cl$

R = H, Me, OMe, Et, 1,2,4-Me_3, 1,2-Me_2, 1,3-Me_2, 1,4-Me_2, Ph

$M(OTf)_3$ = $Sc(OTf)_3$; $Yb(OTf)_3 \cdot XH_2O$

SCHEME 3.16.7 Benzylation of aromatics in ionic liquid (IL) catalyzed by $M(OTf)_3$ or TfOH.

Benzylation of arenes with $PhCH_2Cl$ and $PhCH_2OH$ was studied in [BMIM][OTf] and [BMIM][PF_6] as solvent by using TfOH, $Sc(OTf)_3$, or $Yb(OTf)_3 \cdot xH_2O$ as catalysts (Scheme 3.16.7).[35] TfOH was found to be superior for benzylation with BnOH, producing little or no DBE.

A simple method for the synthesis of diphenylmethane with selectivity toward monoalkylated derivatives was developed by using 1-butyl-3-methylimidazolium chloride [BMIM][Cl]−X (X = $ZnCl_2$, $FeCl_3$, $FeCl_2$) as solvent in combination with a Lewis acidic catalyst.[36] Among the IL catalytic systems, [BMIM][Cl]−$ZnCl_2$ showed better recyclability, for at least eight runs, without loss of catalytic activity and selectivity (Scheme 3.16.8).

The IL [BMIM][OTf] served as both solvent and catalyst to promote FC alkylation of indoles and pyrroles with styrene oxide at ambient temperature with short reaction times.[37] The epoxides cleaved selectively at the benzylic position, and alkylation occurred chemoselectively at carbon (Scheme 3.16.9). Interestingly, this reaction did not proceed in [BMIM][PF_6] or in dichloromethane even after prolonged reaction times.

A mild and efficient approach was reported for benzylation of various classes of arenes and heteroarenes into diarylmethanes utilizing Fe(III)-derived Lewis acidic IL 1-butyl-3-methylimidazolium tetrachloroferrate (III) [BMIM][$FeCl_4$] as catalyst and solvent with improved regioselectivity, yield, and recyclability[38] (Scheme 3.16.10).

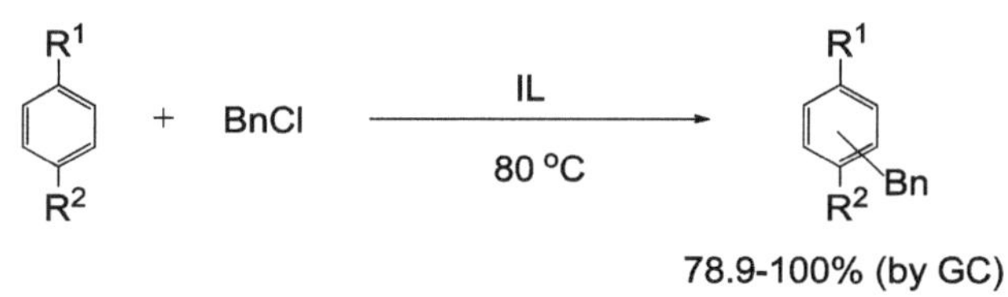

IL = [BMIM][Cl]-X (X = $ZnCl_2$, $FeCl_2$, $FeCl_3$)

R^1 = H, Me, Et, Cl, Br, R^2 = H, Me

SCHEME 3.16.8 Benzylation of aromatics in Lewis acidic ionic liquids (ILs).

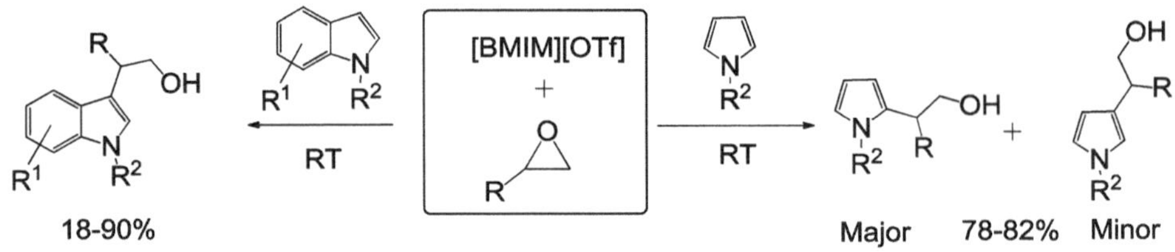

R = Ph, 4-F-Ph, 4-Cl-Ph or Ar = napthalene

R^1 = H, 5-Me, 5-OMe, 2-Me, 1-Me, 5-Br, 4-Br, 5-CN, 5-NO_2; R^2 = H, Me

SCHEME 3.16.9 Friedel-Crafts alkylation of nitrogen heterocycles with styrene oxide.

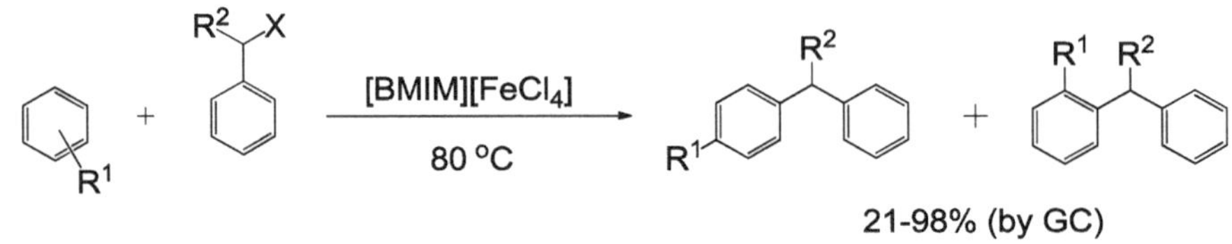

R^1 = H, Me, 2,3-Me_2, 1,4-Me_2, OMe, OH; Ar = 2-naphthol, thiophene, thiazole

R^2 = H, Me, Ph; X = OAc, OH, Cl

SCHEME 3.16.10 Benzylation of arenes in [BMIM][$FeCl_4$].

A series of *N*-methylimidazolium ILs bearing HSO_4^- anions were synthesized and employed as Lewis acids for benzylation of different nucleophiles in acetonitrile. This metal-free approach proved superior in benzylation of thiols, anilines, and indoles, and gave moderate to good yields[39] (Scheme 3.16.11).

Benzylation of diketones was reported in 1-ethyl-3-methylimidazolium trifluoromethanesulfonate [EMIM][OTf] as solvent in combination with a Brønsted acidic IL as catalyst. Suitably substituted benzyl alcohols upon condensation with diketone underwent concomitant benzylation-cyclization-dehydration to furnish functionalized 4*H*-chromenes in respectable yields[40] (Scheme 3.16.12). In another study, a direct C-3 benzylation of 4-hydroxycoumarins in [BMIM][BF_4] or [BMIM][PF_6]/$Bi(NO_3)_3 \cdot 5H_2O$ (BN) system was achieved in excellent isolated yield[41] (Scheme 3.16.13).

3.16.2.2 Acylation in Ionic Liquids

Acetylation and benzoylation are fundamental ways to introduce a keto group into an aryl system, and as such these transformations are of great importance in synthetic chemistry. The main drawbacks in the use of traditional Lewis acids such as $AlCl_3$ are the necessity to use stoichiometric quantities as well as the problems associated with recycling and reuse of the catalyst and/or conventional FC solvents. Modern methodology in combination with the use of ILs has provided solutions to these problems.

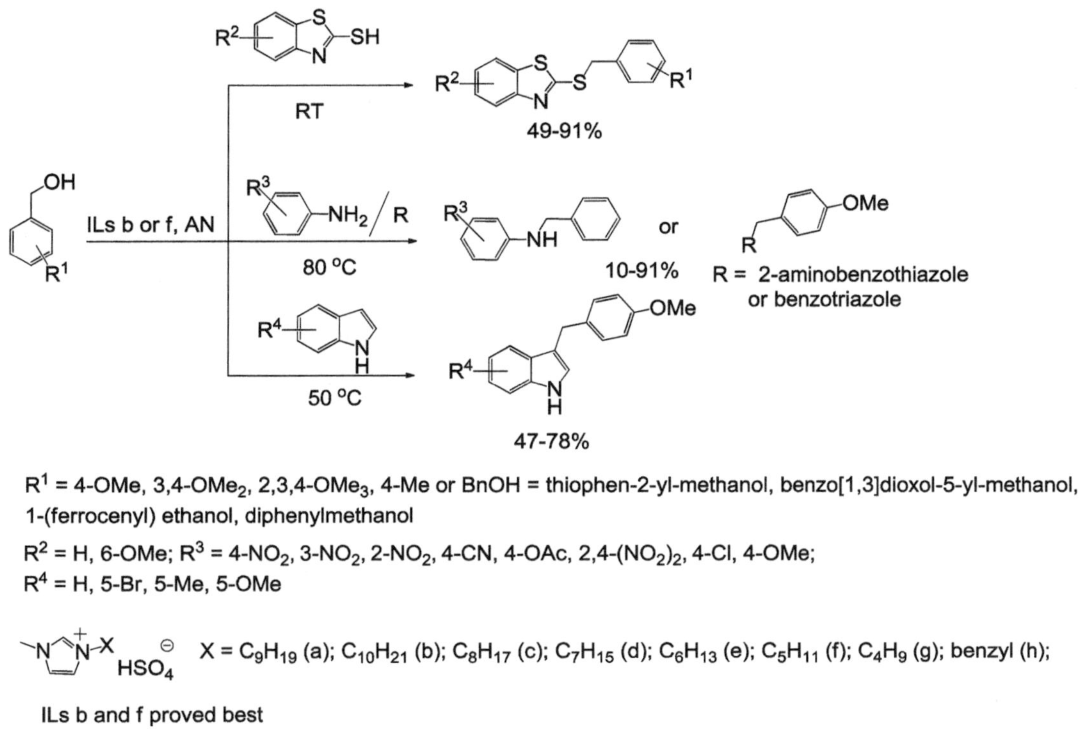

SCHEME 3.16.11 Ionic liquid (IL) catalyzed benzylation of various nucleophiles.

Transacylation constitutes the reverse of FC acylation where the acyl group from a hindered aromatic ketone is transferred to an activated aromatic substrate with the help of an acid catalyst. This transformation was studied in ILs by employing [BMIM][X] as solvent and TfOH as promoter (Scheme 3.16.14).[42] By controlling the temperature, it was possible to bring about transfer acetylation or to simply deacetylate the substrate.

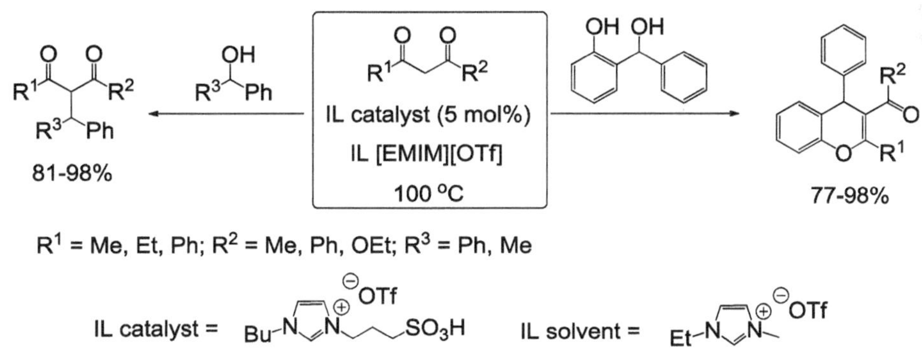

SCHEME 3.16.12 Benzylation of diketones and one-pot synthesis of chromenes catalyzed by Lewis acidic ionic liquid (IL).

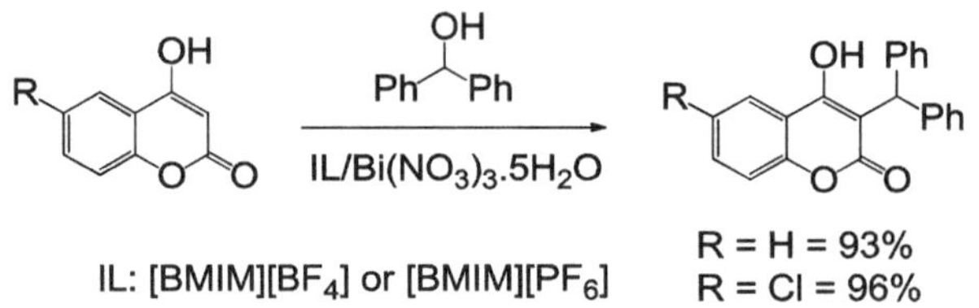

SCHEME 3.16.13 Benzylation of coumarins by $Bi(OTf)_3$/ionic liquid (IL).

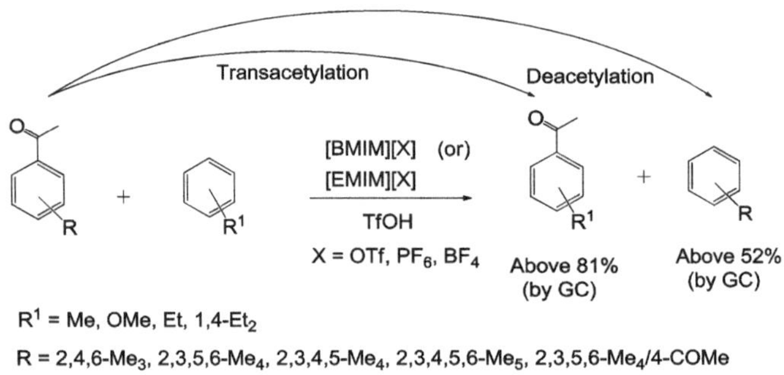

SCHEME 3.16.14 Transacetylation and deacetylation of aromatics.

Although acetyl/benzoyl chlorides or their corresponding anhydrides have traditionally served as electrophilic partners for FC reaction, the use of arylacetic acid/propionic acid/acetonitrile as electrophilic agents for acylation of polyhydroxyaromatics have been shown by using [BMIM][BF_4]-$BF_3 \cdot Et_2O$ or [BMIM][NTf_2]-$HNTf_2$ under microwave irradiation.[43] Reactions were completed in 4 minutes with good chemo- and regioselectivity. The method avoids the need for protection/deprotection sequence for the hydroxyl group (Scheme 3.16.15).

The chloroferrate IL, benzyltributylammonium chloride [BTBA]Cl-0.5$FeCl_3$[44] was used as an alternative to the existing homogeneous or IL catalysts for acetylation and benzoylation of aromatics (Scheme 3.16.16). Reactions proceeded quickly at 50 °C (1–6 min for activated substrates and 12–40 min for deactivated systems such as F/Cl/Br/NO_2-benzene) and exhibited high yields and *para*-selectivity.

Metallic triflates have received considerable attention as catalysts for FC acylation. For example, $Bi(OTf)_3$[45,46] and $In(OTf)_3$[47] in [BMIM][PF_6], [BMIM][OTf], or 1-isobutyl-3-methylimidazolium dihydrogen phosphate [*i*-BMIM][H_2PO_4] were found to exhibit

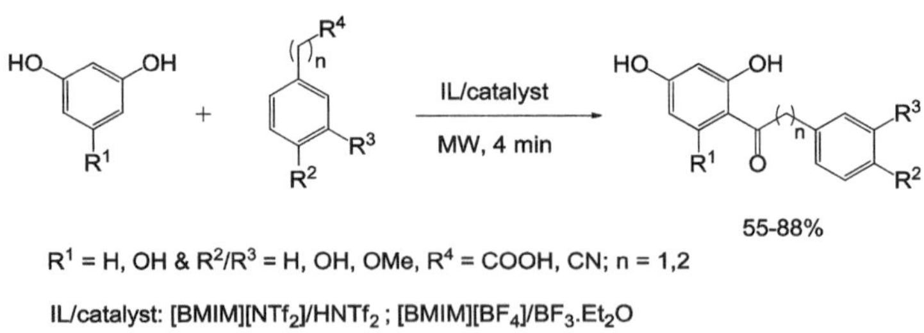

SCHEME 3.16.15 Friedel-Crafts acylation of polyhydroxybenzenes.

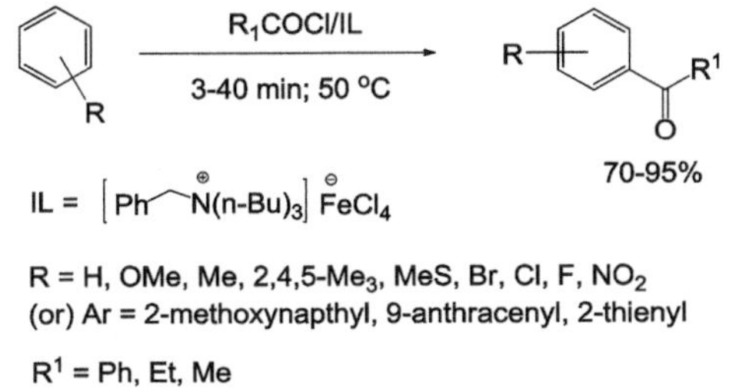

SCHEME 3.16.16 Chloroferrate ionic liquid (IL)-catalyzed Friedel-Crafts acylation of arenes.

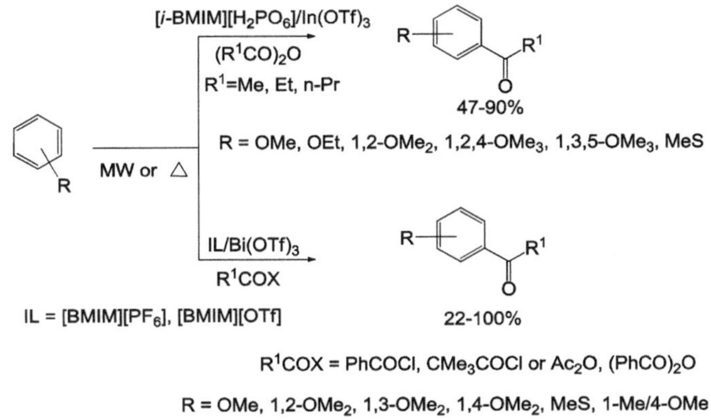

SCHEME 3.16.17 Friedel-Crafts acylation of arenes catalyzed by $Bi(OTf)_3$ or $In(OTf)_3$.

improved catalytic activity in the FC acylation for a range of aromatics by using anhydrides or acid chlorides (Scheme 3.16.17). These reactions have been reported to proceed in a very short time and in good yields under conventional heating or microwave irradiation.

3.16.3 GENERATION OF TAMED PROPARGYLIC AND ALLYLIC CATIONS IN IONIC LIQUIDS FOR FACILE PROPARGYLATION AND ALLYLATION

Propargylic alcohols are efficiently ionized in imidazolium ILs by addition of catalytic amounts of metallic triflates, TfOH, or Brønsted acidic IL to form "tamed" propargylic cations. This approach has enabled the development of a number of IL-based methods for the synthesis of a wide variety of propargylated small-molecule building blocks. Similar activation of allylic and bis-allylic alcohols and reaction with allyl TMS leads to dienes and trienes.

1,3-Diketones were propargylated or allylated by using imidazolium IL as solvent and Brønsted acidic IL as catalyst.[40] Reaction of tertiary allyl-TMS with 1,3-diphenylpropane-1,3-dione, led to the formation of the diene adduct via a tandem Meyer-Schuster rearrangement, aldol condensation, and dehydration sequence (Scheme 3.16.18).

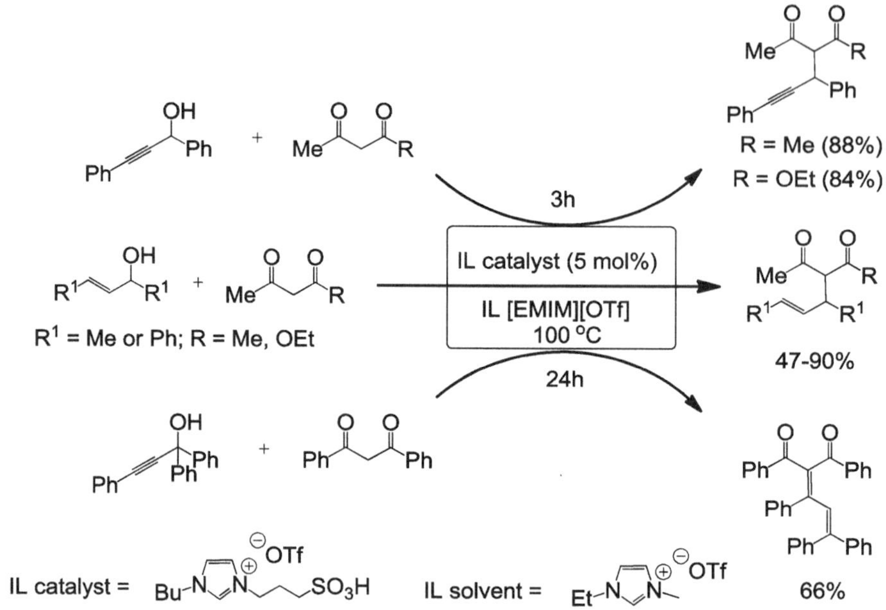

SCHEME 3.16.18 Ionic liquid-catalyzed propargylation and allylation of diketones.

An efficient method for propargylation of arenes and heteroarenes and for the synthesis of a range of symmetrical and unsymmetrical propargylated ethers was developed in [BMIM][X] as solvent along with a Lewis or Brønsted acidic catalyst.[48] Nitroaniline underwent *N*-propargylation instead of ring substitution, while 3-phenyl-2-propyne-1-ol reacted with activated aromatics to form diaryl propanones via concomitant Meyer-Schuster rearrangement and nucleophilic substitution (Scheme 3.16.19).

Condensation of propargylic alcohols or allylic alcohol with 1,3-diketones and β-ketoesters was achieved by using metallic triflate or bismuth nitrate immobilized in imidazolium ILs.[41] Depending on the nature of the substrates, the initially formed propargylated products underwent further cycloisomerization to form highly substituted furans. Bismuth nitrate proved efficient for propargylation of 4-hydroxycoumarins (Scheme 3.16.20).

Indoles and carbazoles were smoothly monopropargylated/allylated in excellent yields by using [BMIM][PF_6] as solvent and bismuth nitrate as catalyst.[49] Triflic acid promoted direct homo bis-propargylation or bis-allylation of carbazoles, while sequential hetero bis-propargylation was achieved by employing $Bi(OTf)_3$[50] (Scheme 3.16.21).

Propargylic alcohols reacted with allyl-tetramethylsilane in [BMIM][BF_4]/$Bi(OTf)_3$ to form 1,5-enynes.[51] Similarly, allylic alcohols were coupled to allyl- and alkynyl silanes. Propargylic, propargylic/allylic, bis-allylic, and allylic alcohols were reduced with Et_3SiH by using $Bi(OTf)_3$/[BMIM][BF_4], affording a variety of functional small molecules (Scheme 3.16.22).

Brønsted acidic IL promoted the Rupe rearrangement of cyclic and acyclic propargylic alcohols in [BMIM][PF_6] to furnish α,β-unsaturated ketones in moderate to good yield.[52] In a typical case, the in-situ-formed enone reacted with benzaldehyde to form indenone through a domino process involving Rupe-Aldol-Nazarov cyclization (Scheme 3.16.23).

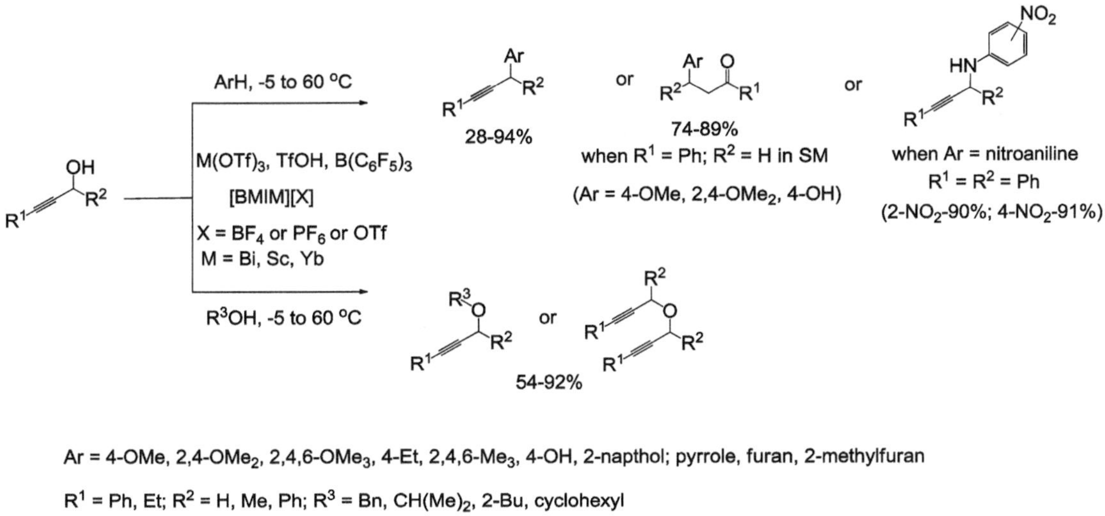

SCHEME 3.16.19 Propargylation of arenes and heteroarenes, and synthesis of dipropargylic ethers.

By using 1-hexyl-3-methylimidazolium hexafluorophosphate [HMIM][PF_6] as solvent/promotor and zinc triflate as catalyst, the 2,4-disubstituted quinolones were synthesized from diversely substituted 2-aminobenzophenones/2-aminoacetophenones and phenyl acetylenes via Meyer-Schuster rearrangement.[53] When compared with [HMIM][PF_6], molecular solvents and other ILs such as [BMIM][BF_4] and [BMIM][PF_6] gave low to moderate yields (Scheme 3.16.24).

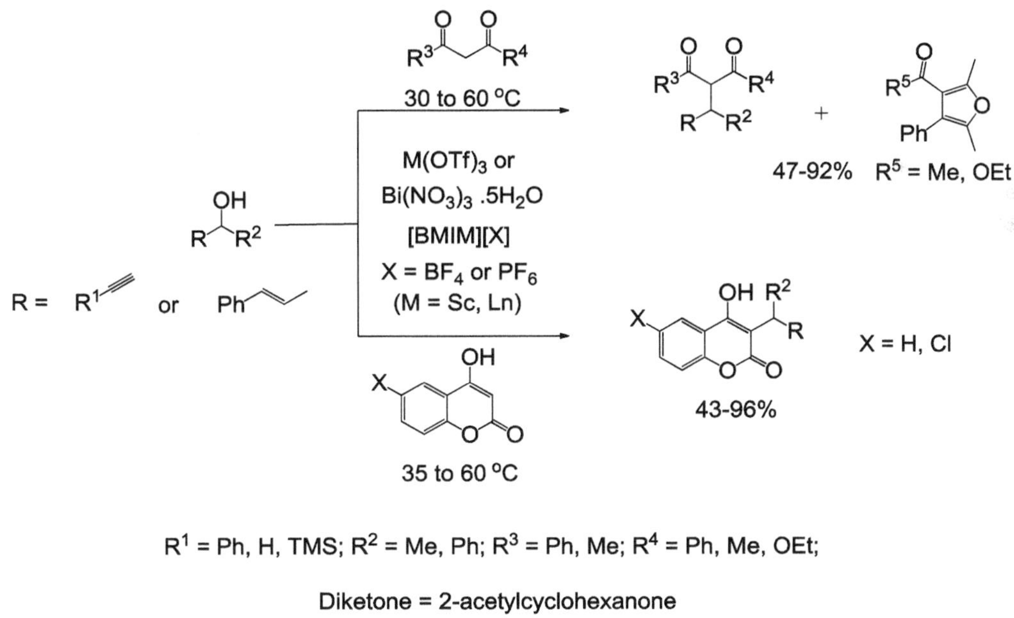

SCHEME 3.16.20 Condensation of propargylic alcohol with nucleophiles in ionic liquid.

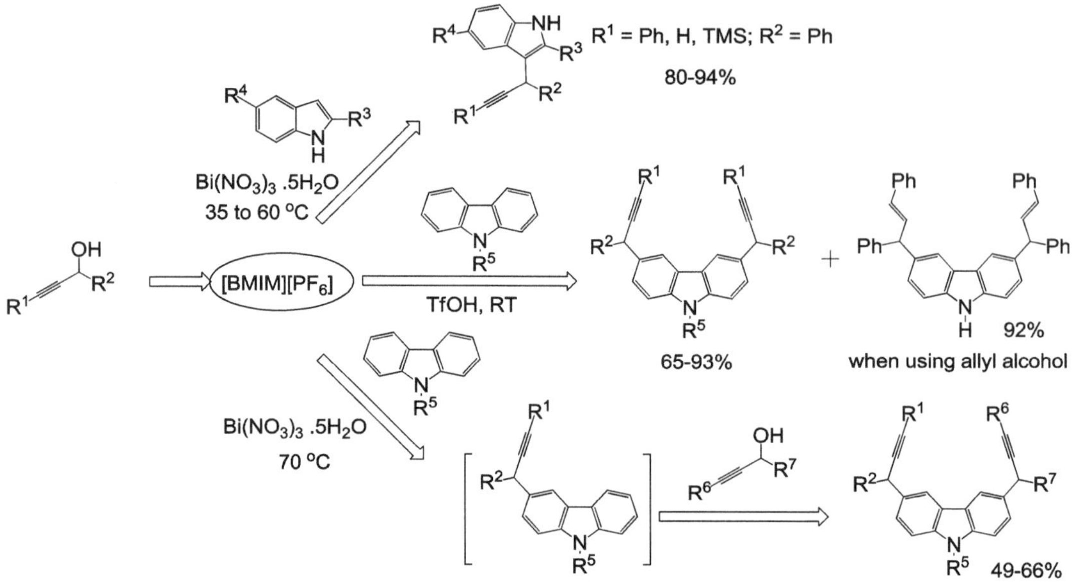

R^1 = Ph, H, Bu^n, TMS; R^2 = Ph, 4-Me-Ph; 4-F-Ph; 4-Cl-Ph; 4-Br-Ph; 4-OMe-Ph; thiophen-2-yl;
R^3 = H, Me; R^4 = H, F, Br, OMe, NO_2, CN; R^5 = H, Me; R^6 = Ph, Bu^n; R^7 = Ph, 4-Me-Ph; 4-Br-Ph

SCHEME 3.16.21 Propargylation of indoles and carbazoles.

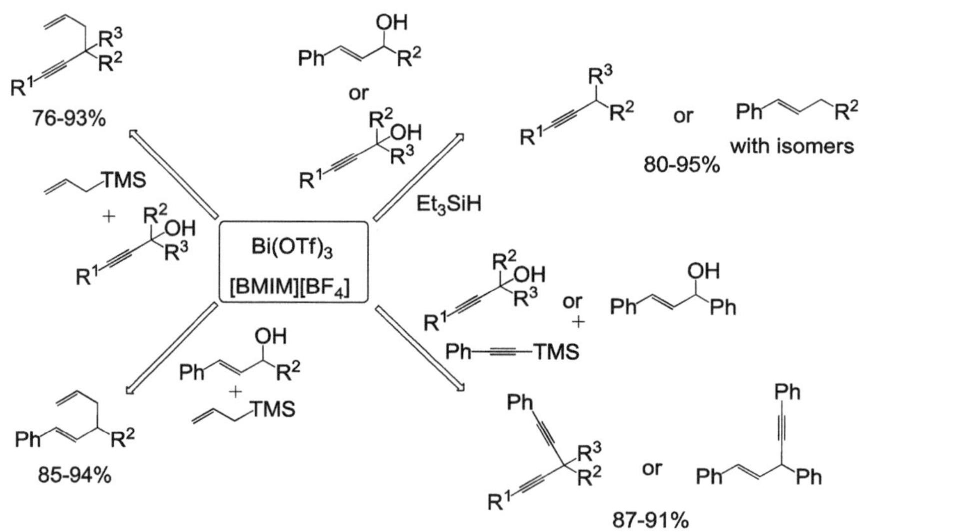

R^1 = aryl, allyl, TMS; R^2 = alkynyl, allyl, Me; R^3 = H, cyclopropyl

SCHEME 3.16.22 Reactions of propargylic and allylic alcohols with allyl-and alkynyl silanes in ionic liquid solvent.

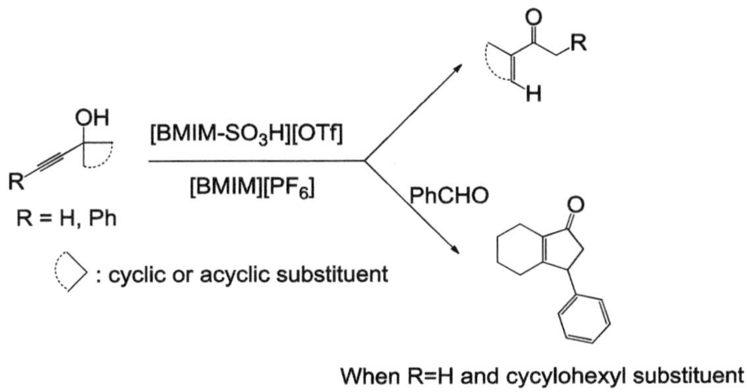

SCHEME 3.16.23 Ionic liquid-catalyzed Rupe rearrangement of propargylic alcohols.

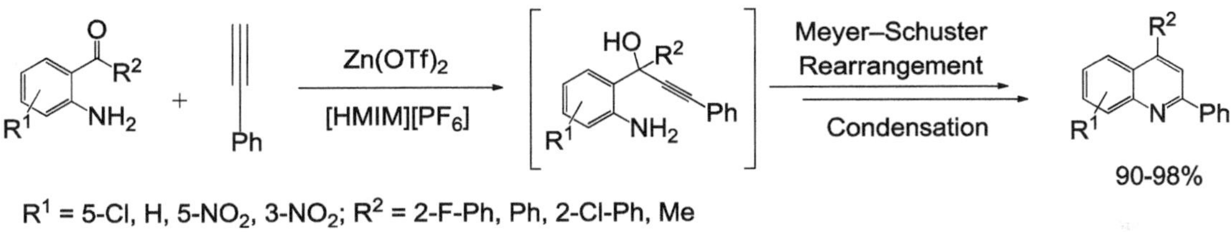

SCHEME 3.16.24 Synthesis of quinolines via Meyer-Schuster rearrangement.

3.16.4 ELECTROPHILIC NITRATION IN IONIC LIQUIDS

In one of the earliest studies aimed at determining the scope of aromatic nitration in RTILs,[54] [EMIM][X] (X = OTf, OAc, NO_3) and diisopropyl ethylammonium acetate [HNEt*i*Pr_2] [CF_3COO] were employed as solvents along with a variety of nitrating systems. Through this survey study, isoamyl nitrate/TfOH, isoamyl nitrate/$BF_3 \cdot Et_2O$, NH_4NO_3/TFAA, $CuNO_3$/TFAA, and $AgNO_3$/Tf_2O were identified as promising systems for aromatic nitration in IL solvents (Scheme 3.16.25). In most cases, the yields and isomer distributions (*ortho/para* ratios) for nitration in ILs were comparable to those employing conventional methods, suggesting similarity in the mechanism.

$Yb(OTf)_3$/HNO_3 and $Cu(OTf)_2$/HNO_3 were employed in another study by using *N*-butyl-*N*-methylpyrrolidinium-NTf_2 as solvent.[55]

Nitration of simple arenes were reported in Brønsted acidic IL *N*-methylimidazolium-*N*-(CH_2)n-SO_3H (where n = 3 and 4) in combination with HNO_3.[56]

The importance of the IL counterion was shown[57] in nitration of arenes in HNO_3-Ac_2O, which led to ring nitration in [BMIM][OTf], halogenation with [BMIM][halide], and side-chain oxidation with [BMIM][mesylate] (Scheme 3.16.26).

Nitration of arenes with HNO_3-Ac_2O system has been studied at room temperature in [BMIM][NTf_2], 1-butyl-2,3-dimethylimidazolium bis(trifluoromethylsulfonyl)imide [BM_2IM] [NTf_2], and 1-butyl-1-methylpyrrolidinium bis(trifluoromethylsulfonyl)imide [BMPy][NTf_2] and [BMPy][OTf] as solvent.[58,59] The [BMPy][NTf_2] proved promising and superior to CH_2Cl_2 in view of the observed reactivity toward deactivated substrates (Scheme 3.16.27).

SCHEME 3.16.25 Nitration of aromatics in imidazolium ionic liquids. *GC*, gas chromatography; *NMR*, nuclear magnetic resonance.

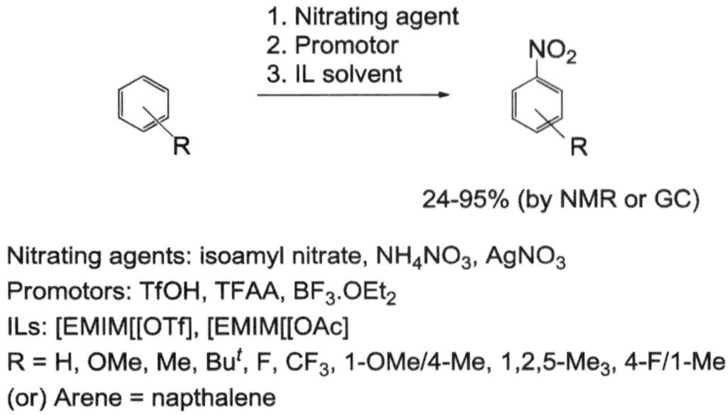

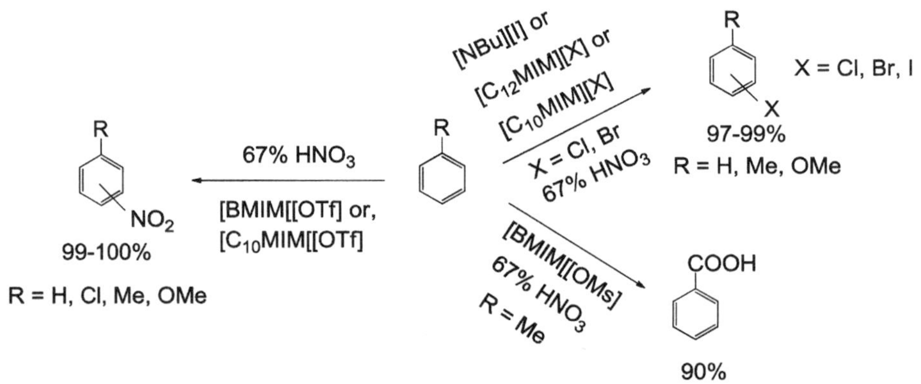

SCHEME 3.16.26 Role of counterion in nitration, halogenation, and oxidation of arenes.

SCHEME 3.16.27 Effect of cations on the nitration of aromatics. *IL*, ionic liquid.

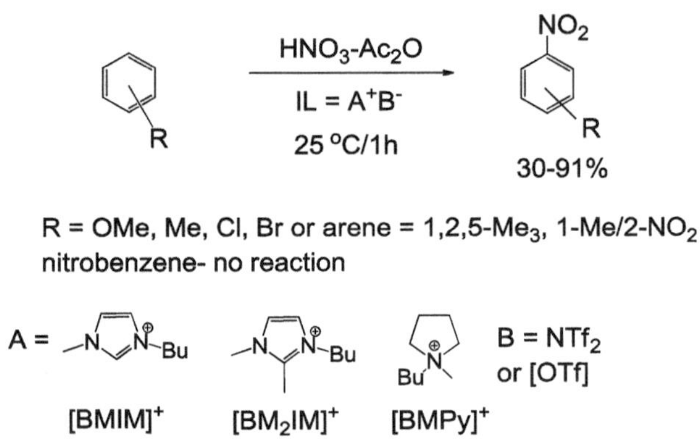

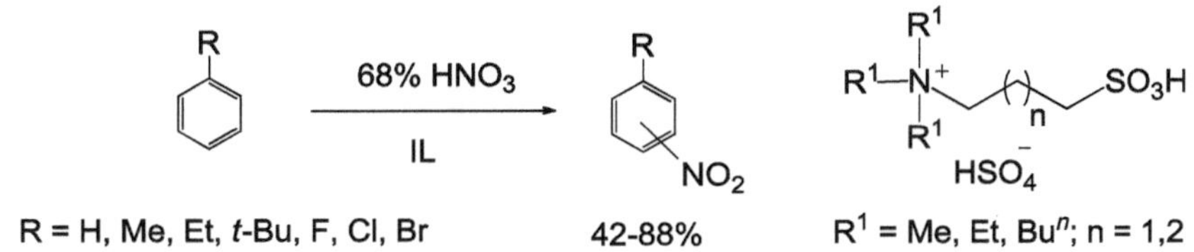

SCHEME 3.16.28 Nitration of arenes in Brønsted acidic ionic liquid (IL).

The acyclic Brønsted acidic IL (Scheme 3.16.28) along with HNO_3 provided a system for mononitration of simple arenes in good to moderate yields.[60]

Triflyl nitrate ($TfONO_2$) and trifluoroacetyl nitrate (CF_3COONO_2) are generated in situ by reacting ethylammonium nitrate (EAN), a readily available and inexpensive IL, with Tf_2O and TFAA, respectively (Scheme 3.16.29). These systems act as in situ sources of nitronium ion and proved effective for mild nitration of a library of arenes. This method provided high regioselectivity and yields, along with the option to recycle and reuse the IL.[61] The system EAN/Tf_2O proved to be a more powerful nitrating agent when compared with EAN/TFAA.

Despite wide synthetic application as versatile oxidant and mild Lewis acid, BN has not been used extensively as a nitrating agent. Suspensions of BN in imidazolium ILs were shown to be effective for ring nitration of activated aromatics under mild conditions without the need for external promoters (Scheme 3.16.30).[62] For activated substrates, reactions were completed faster and in better yields in BN/IL when compared with BN/1,2-dichloroethane (DCE) under comparable conditions. By contrast, 1,2-DCE was a better choice as solvent for deactivated substrates. Since BN is readily available and cheap, and with the possibility of recycling and reuse of IL and absence of promoters, the method provides an attractive alternative to classical nitration methods for activated arenes.

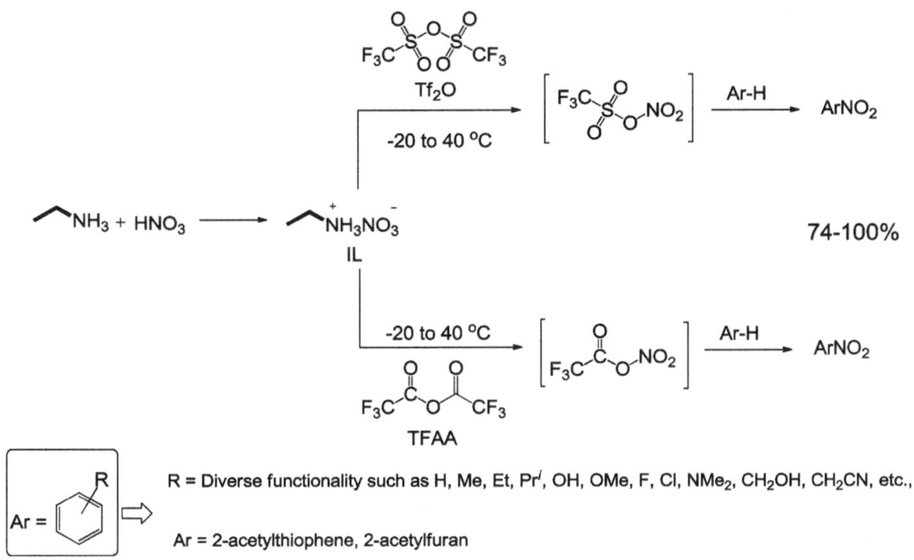

SCHEME 3.16.29 Nitration of aromatics with TFAA/ethylammonium nitrate (EAN) and Tf_2O/EAN.

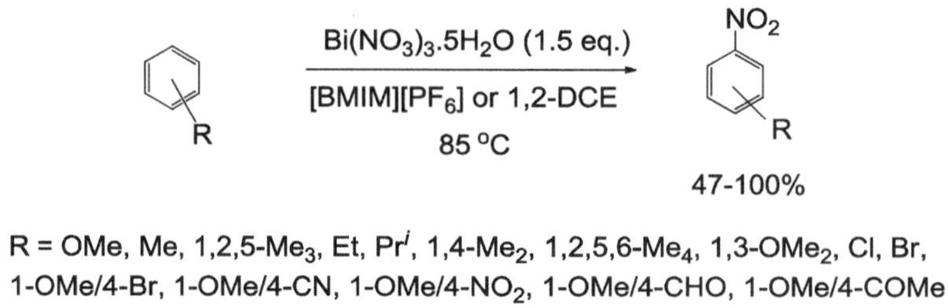

SCHEME 3.16.30 $Bi(NO_3)_3{\cdot}5H_2O$-promoted nitration of arenes.

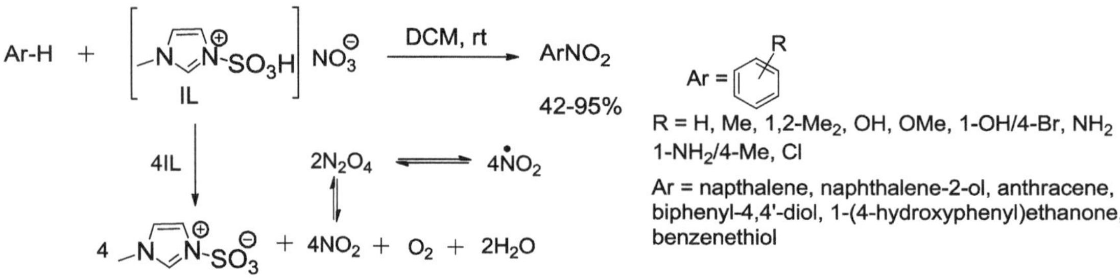

SCHEME 3.16.31 Nitration of aromatics in $[MSIM][NO_3]$-DCM.

The Brønsted acidic IL bearing nitrate counterion 3-methyl-1-sulfonic acid imidazolium nitrate $[MSIM][NO_3]$ was successfully employed for nitration of arenes at room temperature in dichloromethane (DCM) as solvent in good to moderate yields (Scheme 3.16.31).[63] Mechanistic study with radical scavengers suggested the formation of NO_2 radical.

3.16.5 HALOFUNCTIONALIZATION OF ARENES IN IONIC LIQUIDS

ILs have played a significant role in site-selective halofunctionalization of organic compounds. The progress in this area was summarized in a 2009 Tetrahedron Report.[17]

3.16.5.1 Fluorofunctionalization

Due to the remarkable physical and chemical properties that fluorinated compounds offer, the demand for fluorinated organics and building blocks is at an all-time high, and consequently the development of selective fluorination methods is experiencing a huge comeback. Performing mild fluorination methods with onium salts as reagents in ILs offers advantages, not only for recycling and reuse of the IL solvent but also because the reactions can be carried out under homogeneous conditions.

The classical Balz-Schiemann reaction for the synthesis of fluoroaromatics suffers from drawbacks with regard to reproducibility and inconsistent yields. This transformation can be conveniently carried out in imidazolium ILs [EMIM][X] ($X = BF_4$, PF_6) and $[BMIM][PF_6]$ as solvent. The reaction could also be carried out in one pot starting from the anilines by in situ diazotization with the nitrosonium salts (Scheme 3.16.32).[64]

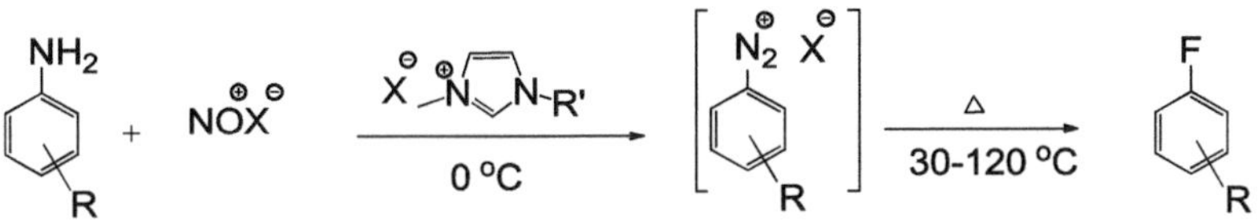

X = BF_4, PF_6; R' = Et, Bu^n
R = H, 4-Cl, 4-Br, 4-OMe, 2-OMe, 2,4,6-Me_3, 4-Bu^t
2-NO_2, 3-NO_2, 4-NO_2, 4-CN, 2,4,6-Br_3

SCHEME 3.16.32 One-pot diazotization-fluorodediazoniation in ionic liquid solvents.

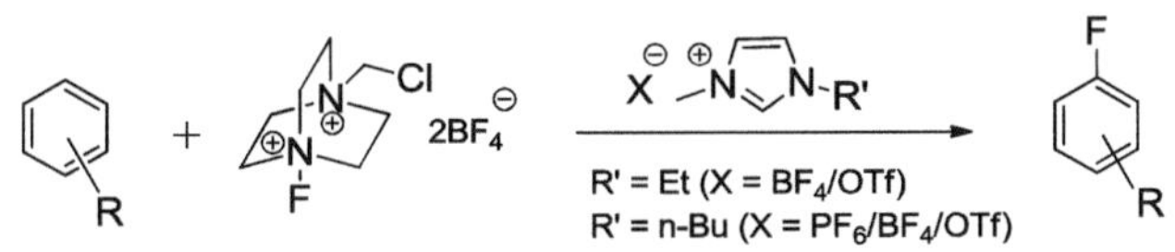

R = OMe, 1-OMe/4-Me, 1-OMe/4-Cl, 1-OMe/4-F, 1-Me/4-Me, 2,4,6-Me_3, 1-Me/4-NO_2
(or) Ar = napthalene, 1-Me-napthalene, dibenzofuran

SCHEME 3.16.33 Electrophilic fluorination of arenes using Selectfluor.

Imidazolium ILs are suitable media for fluorination with Selectfluor (F-TEDA-BF_4) because this N-F reagent (an onium dication salt) can be dissolved in the IL solvent, and this process is aided by sonication, providing a convenient medium for arene fluorination with the possibility to recycle and reuse the solvent. By using this approach the scope of arene fluorination was investigated,[65] and the corresponding fluoroaromatics were obtained under mild conditions in reasonable yields as determined by NMR and gas chromatography. This survey study was subsequently extended to fluorination of bicyclic and polycyclic arenes (Scheme 3.16.33).

The one-pot electrophilic fluorocyclization of alkenols (*trans* isomer) using Selectfluor or *N*-flurobenzenesulfonimide in hydrophilic ILs [EMIM][OTf] and [EMIM][BF_4] as solvent led to preferential formation of the thermodynamically more stable *trans*-diastereomer (Scheme 3.16.34). The reaction was not effective in the hydrophobic IL [BMIM][PF_6],

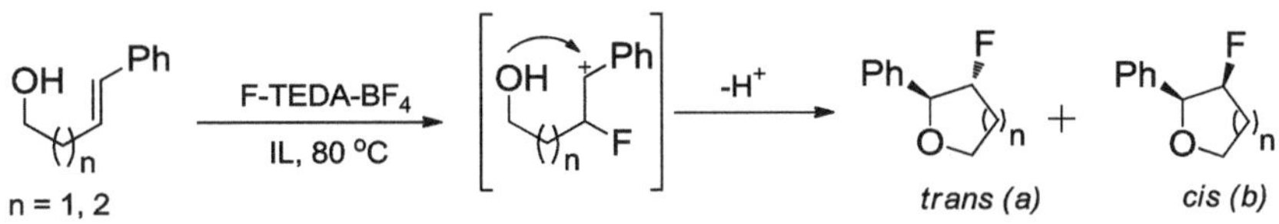

IL = [EMIM][OTf]; [EMIM][BF_4]
[BMIM][OTf]; [BMIM][PF_6]

	Time	Yield (a+b)	a/b	n
[EMIM][OTf] =	40h	80%	5.6	1
$MeNO_2$ =	3h	80%	1.1	1
[EMIM][OTf] =	40h	77%	7.1	2
$MeNO_2$ =	3h	83%	1.2	2
[EMIM][OTf] =	32h	77%	2.1	2 (NFSI)

SCHEME 3.16.34 Stereospecific fluorocyclization of alkenols by N-F reagents in ionic liquid (IL).

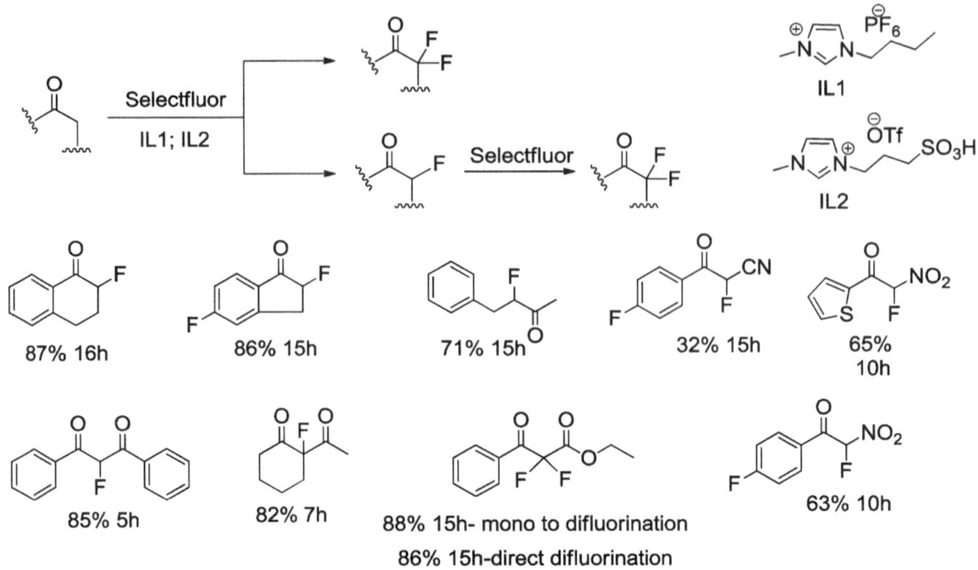

SCHEME 3.16.35 Fluorination of active methylene carbon by Selectfluor in ionic liquid (IL).

seemingly due to lack of solvation of the OH group. Reactions were faster in $MeNO_2$ as solvent, but stereoselectivity was low.[66]

Facile α-monofluorination of ketones, 1,3-diketones, and beta-ketoesters was achieved by using Selectfluor along with the Brønsted acidic IL (IL2) as promoter and [BMIM][PF_6] (IL1) as solvent (Scheme 3.16.35). The method was also applicable to fluorination of beta-nitroketones and with limited success to beta-ketonitriles.[67]

The Umemoto reagent[68] has been used extensively in the recent literature for trifluoromethylation. A study on the influence of structural variations of ILs on trifluoromethylation of aniline using various trifluoromethyl sulfonium salts indicated that the hydrophobic ILs with short alkyl chain were superior for this reaction. The observed regioselectivities (*o*/*p* ratios) measured in different [BMIM][X] solvents were comparable with those observed in DMF (Scheme 3.16.36).

3.16.5.2 Chloro-, Bromo-, and Iodofunctionalization

Trichloroisocyanuric acid (TCICA) in combination with the Brønsted acidic IL [BMIM](SO_3H)] [OTf] was used as an atom-economic chlorinating agent for a variety of aromatics.[69] The reaction was suggested to proceed via a protosolvated trication (Scheme 3.16.37). The mono- to dichlorination ratios were relatively well controlled by adjusting the aromatic to TCICA molar ratios.

Iodination of arenes with *N*-iodosaccharin in refluxing acetonitrile afforded poor yield and low regioselectivity, whereas greatly improved yields and regioselectivity were observed in [BMIM][BF_4] or [BMIM][NTf_2] as solvent[70] (Scheme 3.16.38). This method was not suitable for polycyclic substrates owing to their poor solubility in IL.

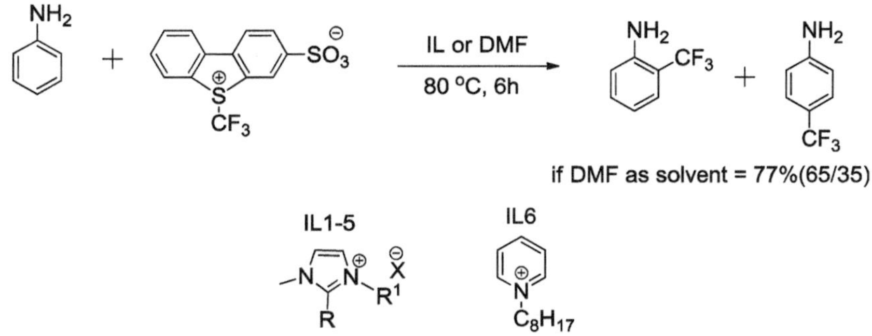

Yield % (o/p ratio)

X ⇨	BF_4	NTf_2	OTf	Br
IL1 ($R=H/R^1=C_4H_9$)	71(61/39)	67(62/38)	59(57/43)	61(57/43)
IL2 ($R=Me/R^1=C_4H_9$)	76(61/39)	69(60/40)	65(58/42)	65(57/43)
IL3 ($R=Me/R^1=C_8H_{17}$)	68(59/41)	68(64/36)	58(58/42)	58(55/45)
IL4 ($R=H/R^1=C_8H_{17}$)	60(61/39)	-	-	-
IL5 ($R=H/R^1=C_{14}H_{29}$)	57(61/39)	-	-	-
IL6	58(55/45)	-	-	-

SCHEME 3.16.36 Trifluoromethylation of aniline with trifluoromethylsulfonium salts in ionic liquids (ILs).

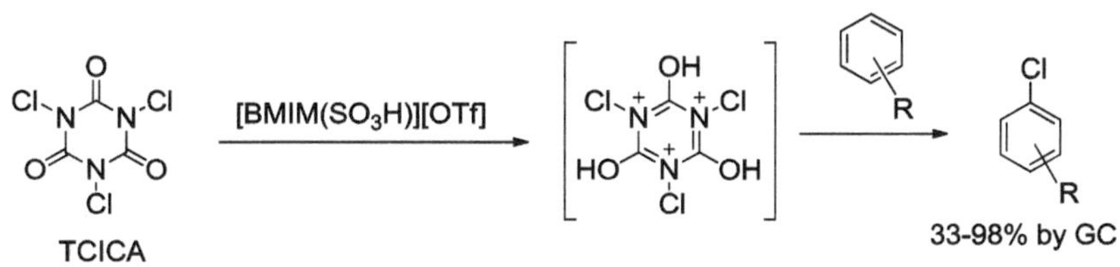

R = H, Me, OMe, Et, 1,3,5-Me_3, 1,4-Me_2, 1,2,4,5-Me_4
1,4-Et_2, 1-F/4-OMe, Cl, 1-OMe/4-NO_2

SCHEME 3.16.37 Acidic ionic liquid-catalyzed aryl chlorination by trichloroisocyanuric acid.

Activated arenes and active methylene compounds were smoothly iodinated by using 30% H_2O_2 or urea-H_2O_2 as mediators in [BMIM][X] (X = BF_4 or PF_6) as solvent. Despite the presence of activated arene in aryl alkyl ketones, iodination occurred selectively at the α-carbon to carbonyl group (Scheme 3.16.39a).[71] Mild halofunctionalization of the same class of compounds was subsequently achieved by using N-halosuccinimide (NXS) (X = Cl, Br, I) along with [BMIM(SO_3H)][OTf] (Scheme 3.16.39b).[72] The IL was recycled and reused in eight successive runs. The degree of halogen introduction was tuned based on the amount of NXS.

Direct chlorination and bromination of free anilines[73] were achieved in 1-hexyl-3-methylimidazolium halide ILs by using copper halides (X = Cl and Br). Enhanced yields were noticed with highly selective *para*-halogenation (Scheme 3.16.40).

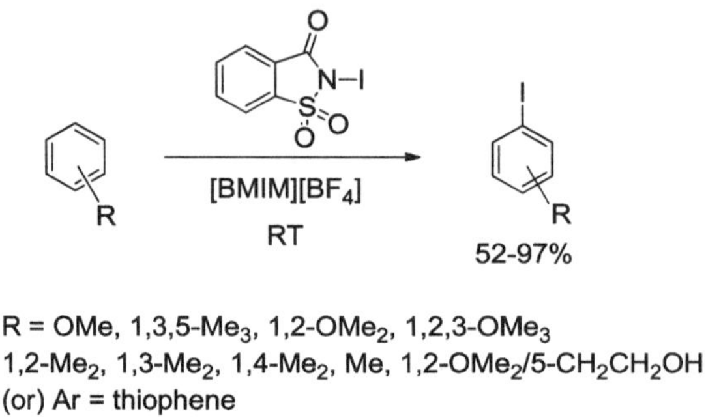

SCHEME 3.16.38 Iodination of arenes by *N*-iodosaccharin in ionic liquid.

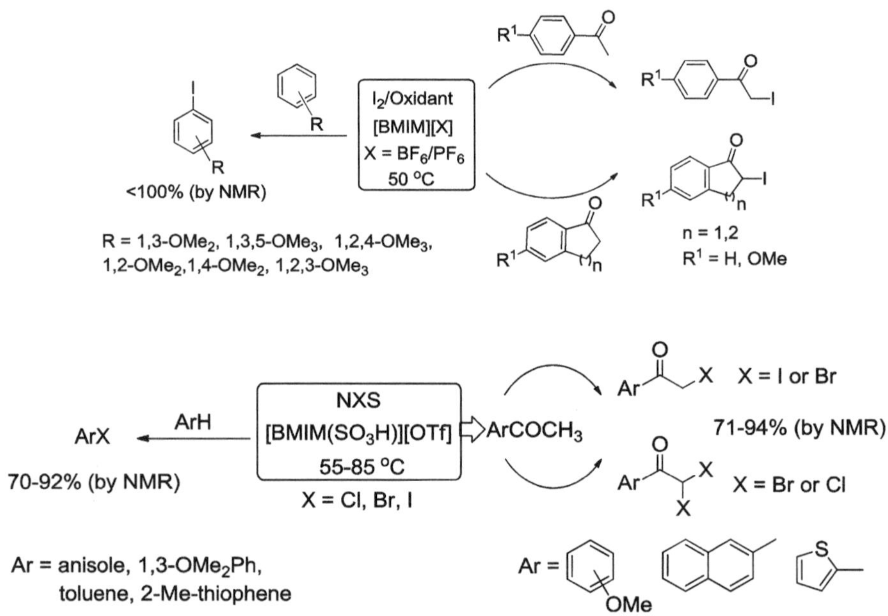

SCHEME 3.16.39 Halogenation of activated organic compounds in ionic liquid.

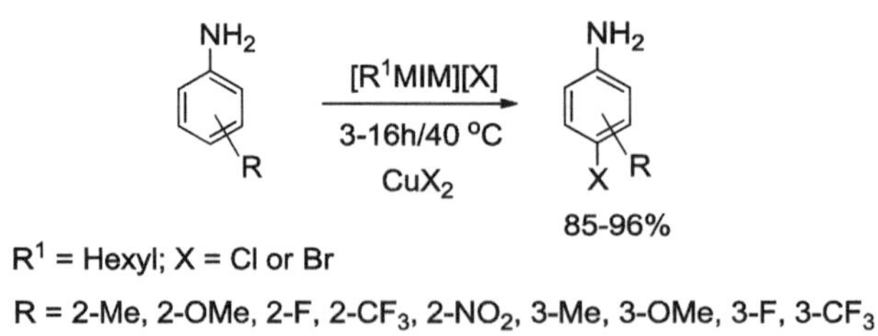

SCHEME 3.16.40 4-Chlorination/bromination of anilines by copper salts in ionic liquid.

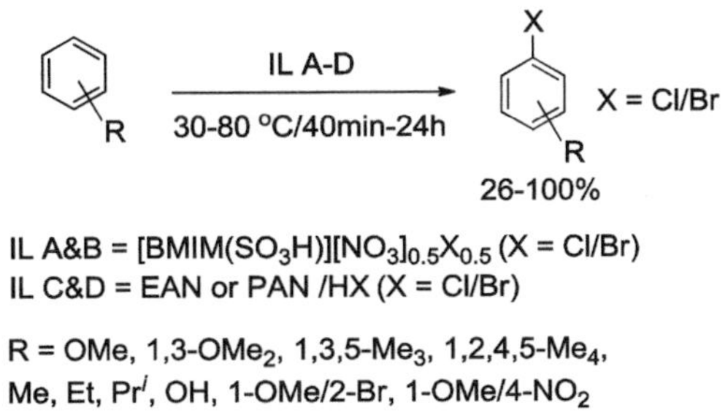

SCHEME 3.16.41 Aerobic oxidative chlorination and bromination of arenes. *IL*, ionic liquid.

Multifunctional ILs $[RNH_3^+][NO_3^-]/HX$ (R = Et or *n*-Pr) and [BMIM(SO_3H)][(NO_3)x(X)y] (X = Br, Cl) synthesized in situ from the corresponding nitrate salts by addition of HBr or HCl proved highly efficient for halofunctionalization of arenes by aerobic oxidation[74] (Scheme 3.16.41).

Chemo- and diastereoselective vicinal dibromination of diverse alkenes were demonstrated in water by employing an amine-tethered imidazolium IL (Scheme 3.16.42).[75]

An IL-mediated palladium-catalyzed one-pot cascade process was investigated for efficient construction of biologically relevant lactones from alkynes and enoic acid.[76] Diversely functionalized alkynes reacted smoothly to furnish the lactones in good yields with high regio- and diastereoselectivities (Scheme 3.16.43).

The IL-ONO/[BMIM][I] system was used for iodofunctionalization of arenes from the anilines by in situ diazotization (Scheme 3.16.44). Electron-deficient anilines proceeded to furnish the desired iodoanilines, whereas the electron-rich counterparts did not move further beyond the diazotization step.[77]

3.16.6 SYNTHESIS OF HIGH-VALUE SMALL MOLECULES VIA DEDIAZONIATIVE FUNCTIONALIZATION IN IONIC LIQUIDS

Highly useful synthons such as halo- and azidoarenes were synthesized from $ArN_2^+BF_4^-$ salts in [BMIM][PF_6]/TMSX (X = Br, I, N_3), with minimal or negligible formation of ArF (Schiemann product) and ArH (via homolytic dediazoniation).[78] The reaction could also be performed via in situ diazotization of arylamines with the nitrosonium salt. NMR monitoring

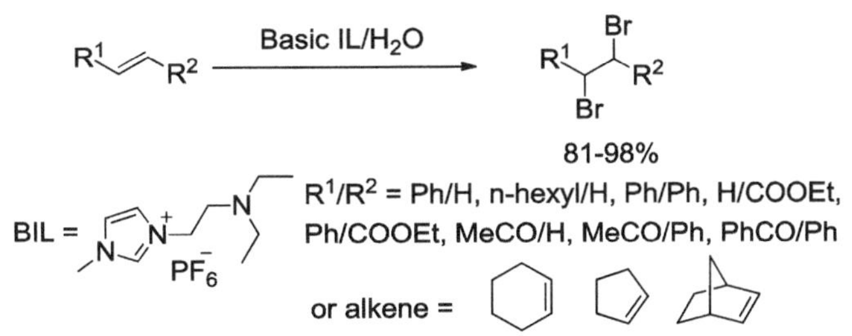

SCHEME 3.16.42 Bromination of alkenes promoted by basic ionic liquid (IL)/water system.

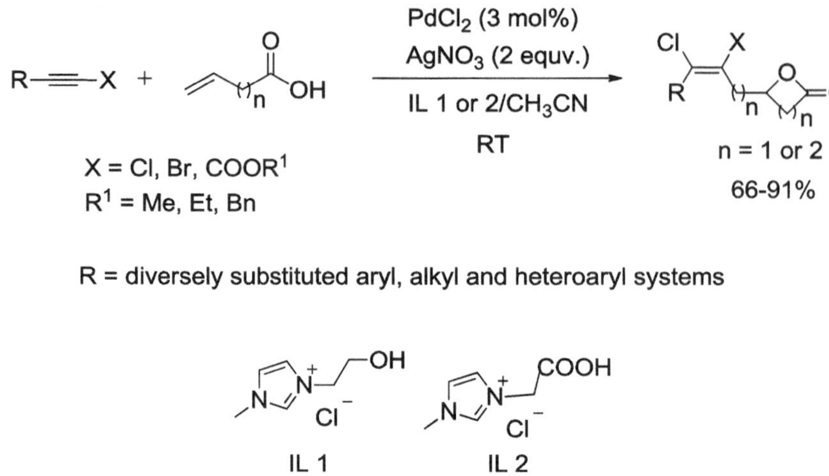

SCHEME 3.16.43 Cascade annulation-halogenation of alkynes and enoic acid to lactones. *IL*, ionic liquid.

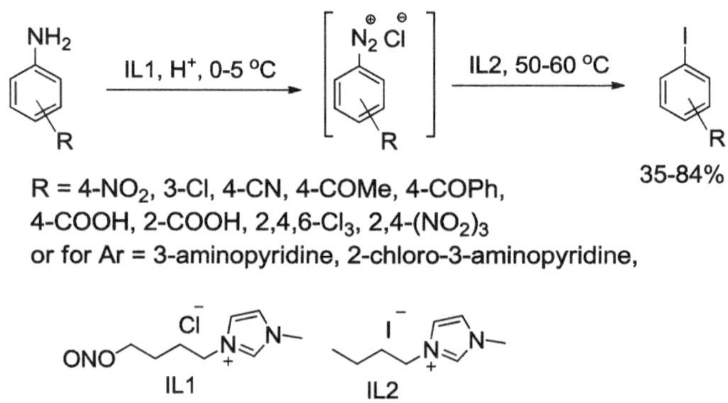

SCHEME 3.16.44 One-pot iodination of anilines via diazotization in ionic liquid (IL).

of the reaction progress provided evidence for the formation of trimethylsilyl fluoride and $[ArN_2][X]$ from TMSX/[BMIM]$[PF_6]$ via metathesis, and subsequent dediazoniation gave the corresponding ArX (Scheme 3.16.45).

Efficient dediazoniative metathesis of $[ArN_2][BF_4]$ in [BMIM]$[NTf_2]$ under thermal or photolytic conditions led to the synthesis of the otherwise highly challenging NTf_2 derivatives along with ArF as a minor product (Scheme 3.16.46).[79] Formation of these products highlights the nucleophilic characteristic of NTf_2 counterion (an ambident nucleophile) toward phenyl cation, despite its well-known noncoordinating and nonnucleophilic properties.

Conversion of arylamines into aryliodides was reported by using the Brønsted acidic IL 3-carboxypyridinium hydrogen sulfate [Hcpy]HSO_4 along with $NaNO_2$/NaI, by a simple grinding method at room temperature[80] (Scheme 3.16.47).

A Brønsted acidic IL immobilized on silica-coated magnetite nanoparticles (Fe_3O_4 at SILnP) was utilized to convert $ArNH_2$ into ArI. The method was extended to pyrazolidine-3,5-dione and pyrazoline-5-one amines bearing a sugar tag, where formation of the

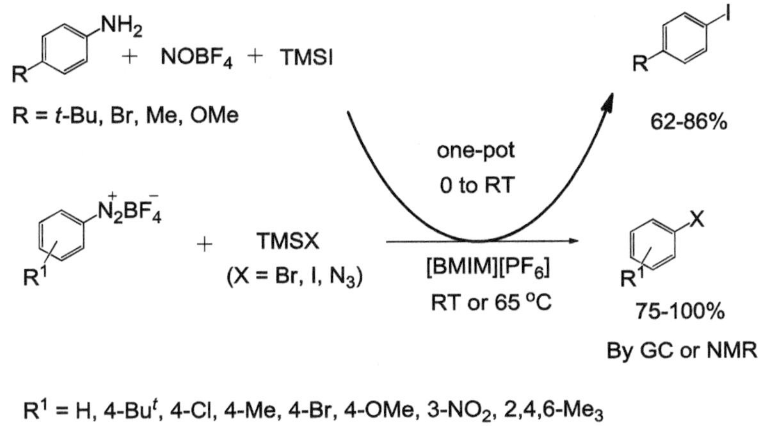

SCHEME 3.16.45 Halo- and azidofunctionalization of arenes via dediazoniation.

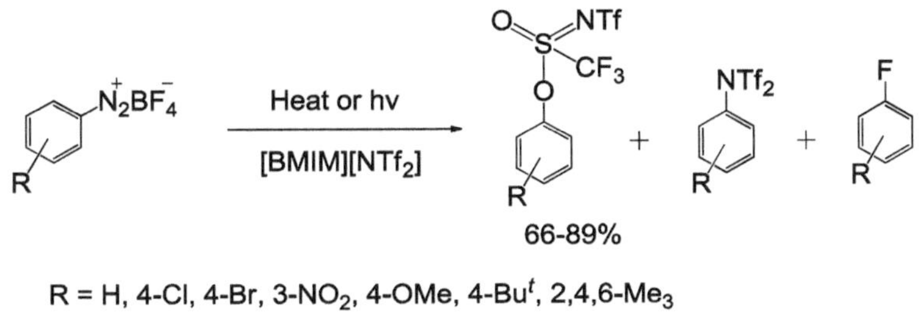

SCHEME 3.16.46 Dediazoniative metathesis of arenediazonium salts in [BMIM][NTf_2].

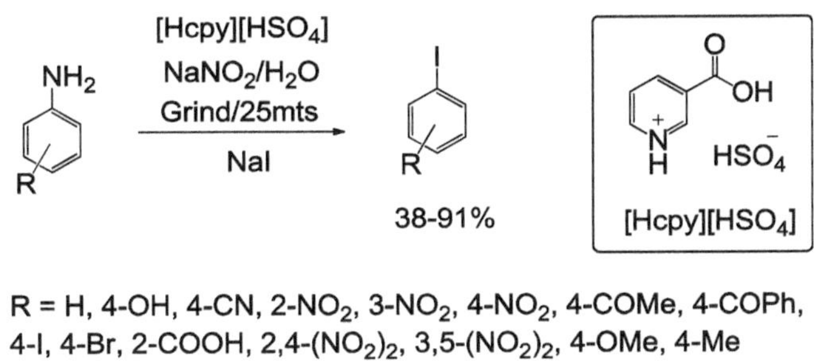

SCHEME 3.16.47 Synthesis of aryl iodides via dediazoniative iodination.

corresponding phenol by-products was minimal. In comparison, when water was used as a solvent, phenol formation increased, whereas the amount of the desired iodo derivatives decreased[81] (Scheme 3.16.48).

A nitrite-anchored IL was employed as a nitrosonium ion source for in situ diazotization of aromatic amines followed by azo coupling with tertiary anilines, phenols (at fourth position), and naphthols (at first position) to prepare azo dyes[82] (Scheme 3.16.49).

SCHEME 3.16.48 One-pot iodination of anilines/amines via dediazoniation using an immobilized acidic ionic liquid (IL).

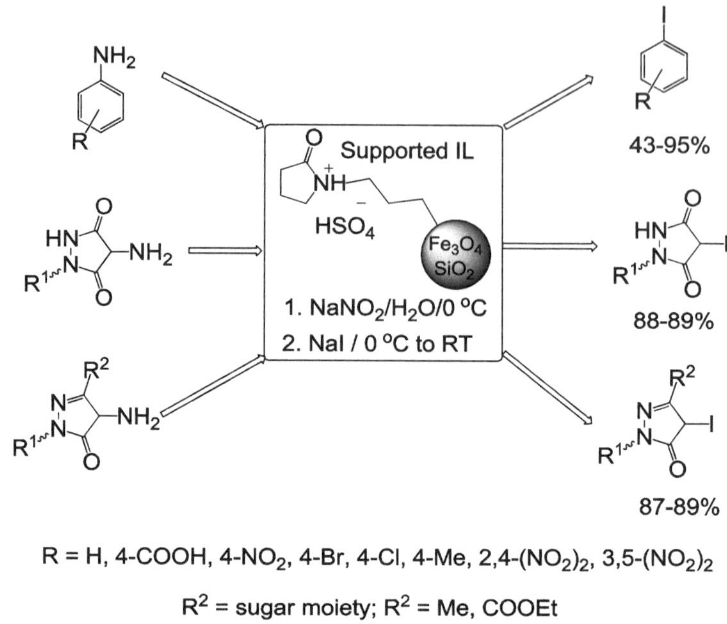

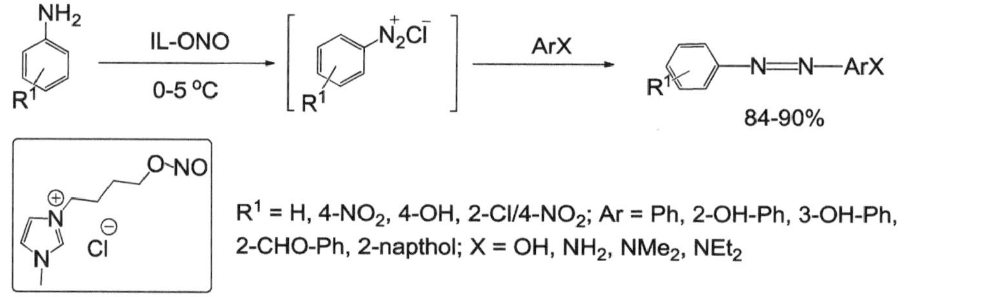

SCHEME 3.16.49 One-pot synthesis of azo dyes from anilines. *IL*, ionic liquid.

3.16.7 IONIC LIQUIDS AS SOLVENT AND CATALYST FOR THE SYNTHESIS OF HETEROCYCLES

Heterocyclic aromatic compounds constitute the core structures of a vast number of pharmaceuticals, agrochemicals, and materials. Application of ILs in the synthesis of various classes of heterocyclic compounds was summarized in a 2009 review.[14]

The $RC(OR)_2^+$ carbocations generated in situ by ionization of the corresponding *ortho*-carboxylic esters in EAN $[EtNH_3][NO_3]$ (IL1) or in Brønsted acidic 1-methyl-3-(3-sulfopropyl)-1H-imidazol-3-ium trifluoromethanesulfonate $[PMIM(SO_3H)][OTf]$ (IL2) reacted with anilines bearing diverse *ortho* substituents to afford a host of benzimidazoles, benzoxazoles, benzothiazoles, and quinazolinones.[83] A convenient method for the synthesis of 1*H*-tetrazoles was

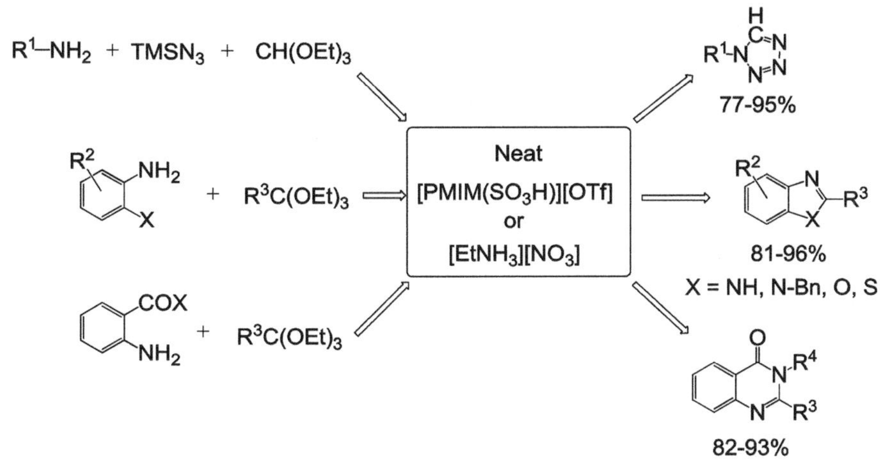

R^1 = substituted phenyl group (H, 4-Cl, 3,5-Cl_2, 3-Br, 4-Br, 4-F, 3-CF_3, 4-CF_3, 4-Me, 2-OMe, 3-OMe, 4-OMe, 4-CH_2COOH, 2-NO_2, 3-NO_2, 4-NO_2, 4-COMe, 2-COOH, 3-COOH), pyridin-2-yl, Bn, thiophen-2-methyl

R^2 = H, 4-Cl, 4-F, 4-Cl/5-NO_2

R^3 = H, Et; R^4 = H, Et, 2-COOH-Ph; X = OH, NH_2

SCHEME 3.16.50 Ionic liquid-catalyzed synthesis of diverse heterocycles.

developed by reacting the IL-generated $CH(OEt)_2^+$ carbocation with $ArNH_2$ and $TMSN_3$.[83] Interestingly, [$EtNH_3$][NO_3] not only acted as catalyst but also acted as substrate to furnish 2-ethylquinazolin-4(3*H*)-ones[83] (Scheme 3.16.50).

Highly substituted imidazoles were synthesized via a one-pot four-component strategy employing [BMIM][Br] as solvent and catalyst. Under conventional or microwave irradiation, a wide range of aryl aldehydes and amines bearing diverse substituents reacted smoothly to furnish imidazoles in excellent yield[84] (Scheme 3.16.51).

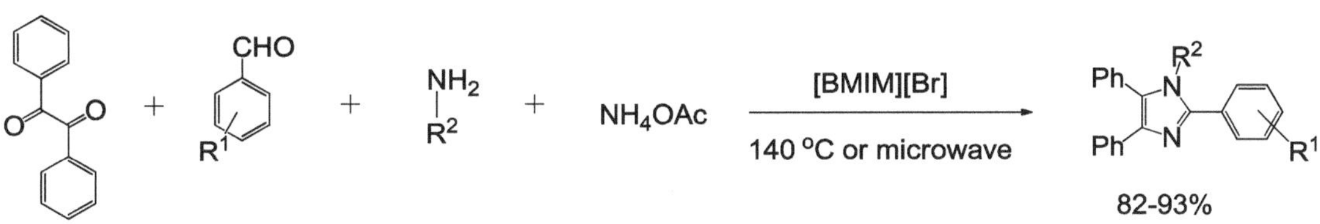

R^1 = H, 4-Me, 4-Br, 4-Cl, 3-NO_3, 4-NO_3, 3-OMe, 4-OH, 4-CN, 3-OH, 4-$CH(Me)_2$;

Aldehyde: 2-furyl, 2-thienyl, 3-indolyl

R^2 = Ph, Me, Bn, Et, 4-Cl-Ph, 4-Me-Ph, 4-F-Ph, cyclohexyl, 4-OH-Ph

SCHEME 3.16.51 One-pot synthesis of imidazoles.

A bifunctional Brønsted acidic IL was employed as catalyst for the three-component synthesis of pyrazolines from aldehydes, hydrazines, and dimethyl acetylenedicarboxylate under solvent-free conditions, with rather remarkable diastereoselectivity and yields. In comparison, lower yields and diastereoselectivity were observed with Lewis acidic metal catalysts, underscoring the superior performance of the 1,4-diazabicyclo[2.2.2]octane (DABCO)-based IL[85] (Scheme 3.16.52).

A convenient metal-free methodology enabled the assembly of 2,3-disubstituted and 3-substituted quinolines from the respective anilines and phenyl acetaldehyde by using [BMIM][BF_4] as solvent and catalyst. A host of anilines bearing diverse functionalities reacted smoothly to furnish quinolines in moderate to good yield with short reaction times, under either conventional or microwave heating[86] (Scheme 3.16.53).

Employing CO_2 as a C1 carbon feedstock, a new green protocol was introduced for the synthesis of benzothiazole from 2-aminothiophenol and triethoxysilane in [BMIM][OAc]. NMR studies showed that the IL played a critical role in the generation of formoxysilane (silyl formate) from CO_2 and hydrosilane, besides activating the 2-aminothiophenol via hydrogen bonding. This method also proved effective in the synthesis of benzimidazoles from *o*-phenylenediamine[87] (Scheme 3.16.54).

Synthesis of the fungicide Fenpropimorph was carried out in 1-butyl-1-methylpyrolidinium bis{(trifluoro-methyl)sulfonyl}amide ([bmpyrr][NTf_2]) IL, utilizing either Heck coupling or

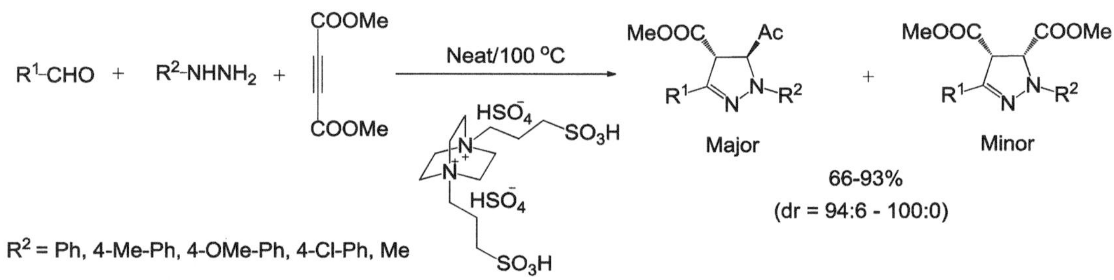

R^2 = Ph, 4-Me-Ph, 4-OMe-Ph, 4-Cl-Ph, Me

R^1 = Ph, 4-Br-Ph, 4-CN-Ph, 4-OMe-Ph, 2,4-Cl_2-Ph, 2-Cl-6-F-Ph, 4-Cl-Ph, thiophen-2-yl, C_3H_7, n-C_6H_{13}

SCHEME 3.16.52 Synthesis of pyrazolines catalyzed by Brønsted acidic IL.

R^1 = H, 2,3-Me_2, 3-Me, 4-Me, 3,4-Me_2, 2,5-Me_2, 3,4,5-OMe_3, 3,4-OMe_2, 3-OMe, 4-OCF_3, 4-Br, 4-Cl, 2-Br
R^2 = Ph, 4-Cl-Ph, 4-Br-Ph, 4-F-Ph, n-C_3H_7, $PhCH_2CH_2$

SCHEME 3.16.53 Synthesis of quinolines in [BMIM][BF_4].

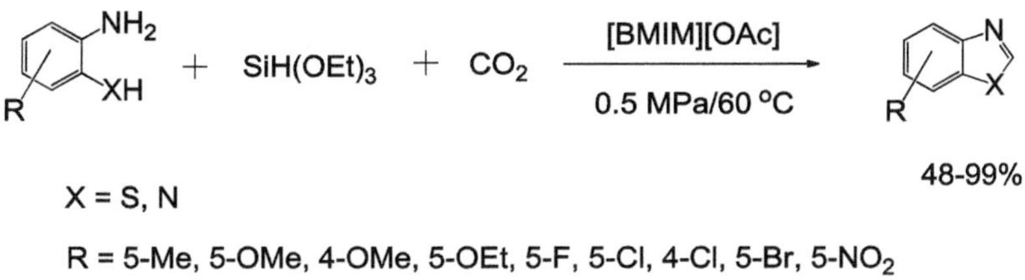

SCHEME 3.16.54 Synthesis of benzothiazoles and benzimidazoles by [BMIM][OAc].

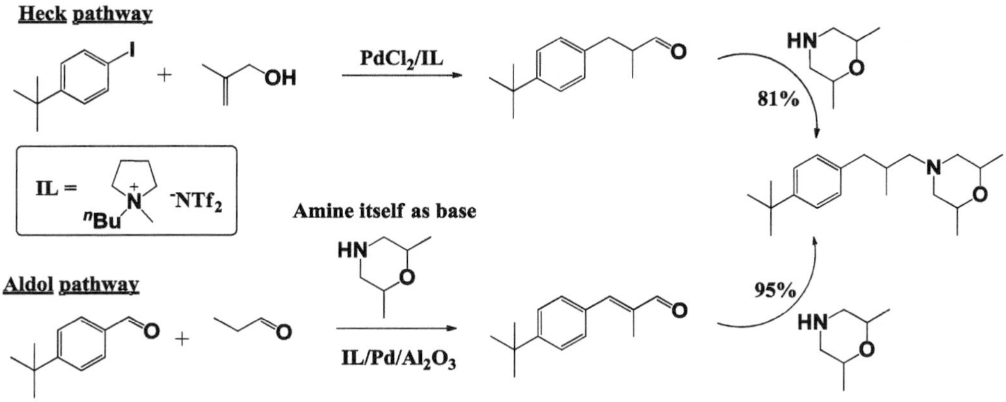

SCHEME 3.16.55 Synthesis of Fenpropimorph in ionic liquid (IL).

aldol condensation/reductive amination sequence in one-pot. Although both routes proved successful, high atom efficiency, better recycling, and improved yield were realized through the aldol pathway[88] (Scheme 3.16.55).

3.16.8 RITTER REACTION

The Ritter reaction of nitriles with tertiary and secondary alcohols was carried out in a number of imidazolium and ammonium-based Brønsted acidic ILs. The survey study showed that the ammonium-tethered Brønsted acidic ILs (Scheme 3.16.56) was superior in its catalytic activity and for recycling/reuse.[89]

Employing the Brønsted acidic IL [BMIM(SO_3H)][OTf] as catalyst, a variety of amides were synthesized from alcohols via the Ritter reaction. Alternatively, NO^+-induced Ritter reaction with Bu^{t}Br and Ada-Br or with Ada-H in [BMIM][PF_6] also furnished the corresponding amides in moderate yields[90] (Scheme 3.16.57).

The *N*-butylammonium carboxylate/Tf_2O-based system proved efficient for the synthesis of unsymmetrical amides from nitriles through a Ritter-type reaction in moderate yields[91] (Scheme 3.16.58).

$R^1\text{-CN} + R^2\text{-OH} \xrightarrow{\text{IL/70 °C}} R^1\text{-CONH-}R^2$ 70-95%

R^1 = Ph, Me, CH_2=CH, 4-Me-Ph, 4-OMe-Ph, 4-F-Ph, 4-Cl-Ph, 4-NO_2-Ph, 4-CF_3-Ph, 2-Me-4-Br-Ph, Ph-$(CH_2)_3$, 1-napthyl

R^2 = Bu^t, cyclohexyl, adamantyl

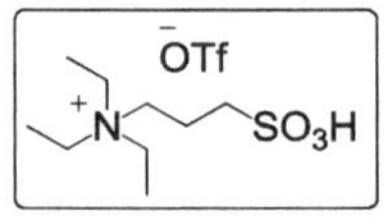

SCHEME 3.16.56 Synthesis of amides via Ritter reaction.

$R^1\text{-CN} + R^2\text{-OH} \xrightarrow[\text{50-80 °C}]{[BMIM(SO_3H)][OTf]} R^1\text{-CONH-}R^2$ 61-98%

R^1 = Me, Et, cyclohexyl, adamantyl, 2-furyl, 4-pyridyl

R^2 = Bu^t, methylcyclohexyl, adamantyl

$R^3\text{-CN} + R^4\text{-X} \xrightarrow[\text{0-90 °C}]{[BMIM[PF_6]/[NO][PF_6]} R^3\text{-CONH-}R^4$ 59-77%

R^3 = Me, Et, cyclohexyl, 2-furyl, Ph

R^4/X = Bu^t/Br, adamantyl/Br, adamantyl/H

SCHEME 3.16.57 Ritter reaction catalyzed by Lewis acidic ionic liquid.

$\text{R-CN} \xrightarrow[[n\text{-Bu-}NH_3][X]]{Tf_2O} \text{R-CONH-X}$ 42-93%

R = 4-Cl-$PhCH_2$, Bn, 4-NO_2-$PhCH_2$, Ph, 4-Cl-Ph, 4-Br-Ph, Me

X = MeCOO, Pr^nCOO, $CH(CH_3)_2CH_2COO$

SCHEME 3.16.58 Ritter reaction of nitriles to amides.

3.16.9 SCHMIDT REACTION

Acetanilides were synthesized in moderate yields from aryl methyl ketones by reaction with sodium azide through the Schmidt reaction in 1-ethyl-3-methylimidazolium triflate/TfOH system (Scheme 3.16.59).[92]

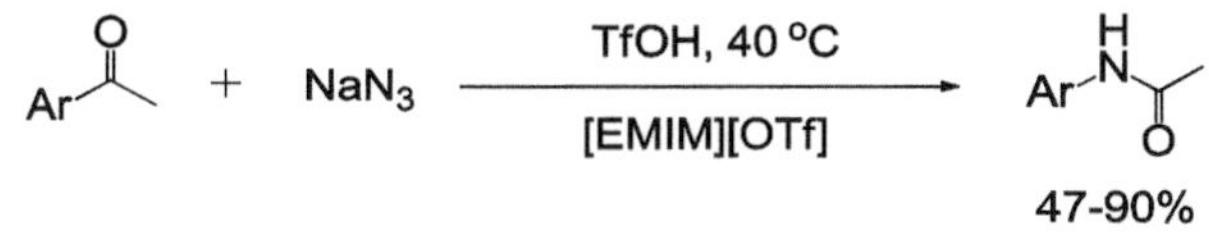

Ar = Ph, 4-Me-Ph, 4-Et-Ph, 2,4-Me$_2$-Ph, 3-OMe-Ph, 4-Cl-Ph, 3-Br-Ph, 4-Br-Ph, 3-NO$_2$-Ph, 2-naphthyl, 2-fluorenyl

SCHEME 3.16.59 Schmidt reaction of acetophenones to acetanilides.

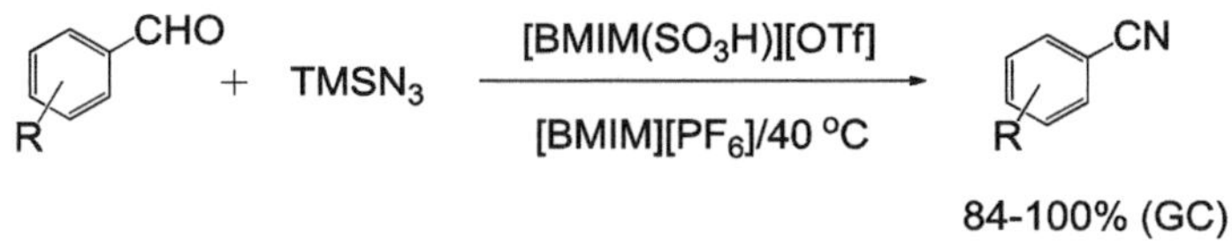

R = H, 2-OMe, 2-F, 3-NO$_2$, 3-Br, 4-Me, 4-OMe, 4-Br, 4-NO$_2$, 4-F, 3-F-4-Cl-Ph, 2,5-OMe$_2$-Ph, 2,3-OMe$_2$-Ph, styryl,
Heteroaldehyde: pyrrol-2-yl, N-Me-pyrrol-2-yl, 2-furyl, N-Me-indol-3yl etc.

SCHEME 3.16.60 Schmidt reaction of aldehydes to nitriles.

Structurally diverse aromatic and heteroaromatic aldehydes were converted to nitriles in a high-yielding Schmidt reaction by using $TMSN_3$, with [BMIM(SO_3H)][OTf] as catalyst and [BMIM][PF_6] as solvent, and with recycling and reuse of [BMIM][PF_6][93] (Scheme 3.16.60).

Employing the task-specific IL [BMIM][N_3] as an azide source, Schmidt reaction of aryl aldehydes and ketones were carried out in the presence of AcOH/H_2SO_4 with good isolated yields. Depending on the nature of substituents on the aldehyde, either nitrile or amide or a mixture of nitrile and amide products were obtained, whereas with ketones only amides were isolated[94] (Scheme 3.16.61).

3.16.10 METAL-MEDIATED CROSS-COUPLING AND CYCLIZATION REACTIONS IN IONIC LIQUIDS

ILs are ideal media for metal-mediated C–C coupling reactions such as Heck, Sonogashira, Suzuki, and others. Progress in this rapidly growing area has been examined in several authoritative reviews.[19,95–98] Representative examples from the recent literature are highlighted here.

3.16.10.1 Heck Cross-Coupling

The pyrazolyl-functionalized *N*-heterocyclic carbene complex of palladium(II) efficiently promoted the Mizoroki-Heck coupling of alkenes with aryl halides in [BMIM][PF_6] as solvent[99] (Scheme 3.16.62). The recycling and reuse of the catalyst and the IL was demonstrated

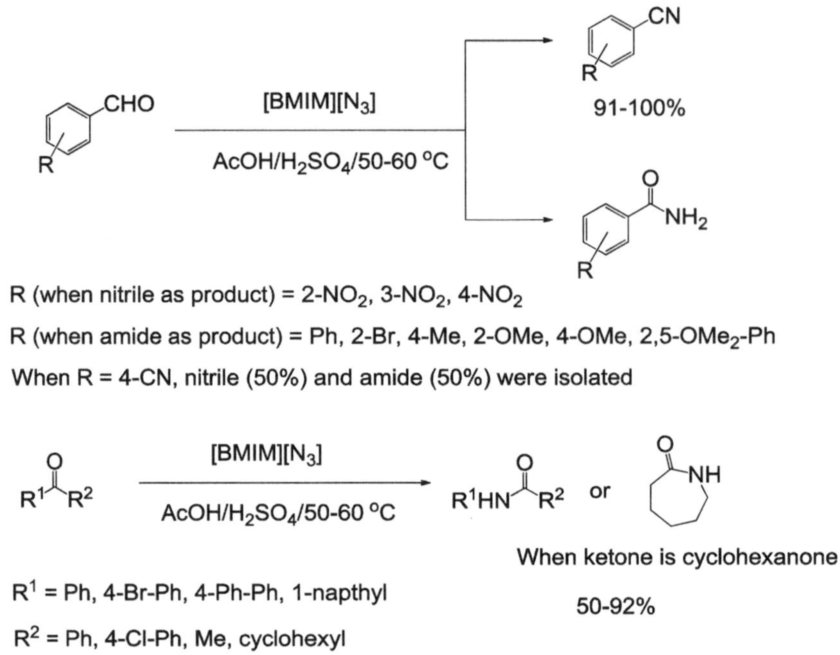

SCHEME 3.16.61 Schmidt reaction of aldehydes and ketones using task-specific ionic liquid.

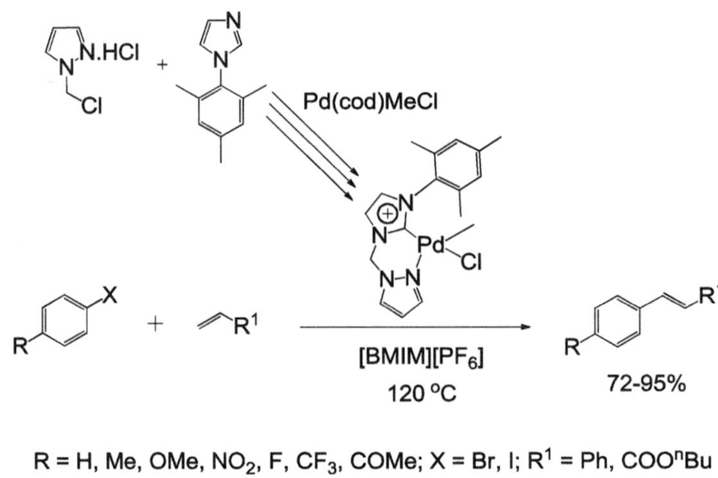

SCHEME 3.16.62 Pd-NHC-catalyzed Heck coupling in ionic liquid.

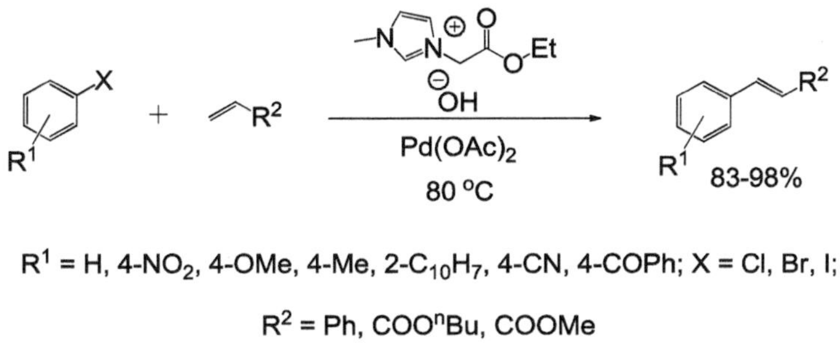

SCHEME 3.16.63 Heck coupling in basic ionic liquid with $Pd(OAc)_2$.

in five reaction cycles, by using the same or different reactants, with no significant loss in activity and yield.

The 3-methyl-1-(ethoxycarbonylmethyl)imidazolium hydroxide IL acted as base, ligand, and reaction medium in the $Pd(OAc)_2$-catalyzed Heck coupling of alkenes with haloarenes[100] (Scheme 3.16.63). The catalytic activity of this IL/Pd system was attributed to the in-situ-formed Pd nanoparticles.

Arenediazonium salts are suitable as coupling partners for the Matsuda-Heck reaction because both the diazonium salt and the Pd catalyst can be dissolved in the IL, and the IL can be recycled and reused. The efficacy of $[ArN_2][X]$ salts as coupling partners was demonstrated in reaction with styrenes and with aliphatic alkenes with $[BMIM]X/Pd(OAc)_2$ ($X = BF_4$ or PF_6).[101] This chemistry can also be performed via in situ diazotization-arylation protocol starting from the aniline (Scheme 3.16.64).

By utilizing the hydrophobic fluorous IL 1-octyl-3-methylimidazolium nonaflurobutanesulfonate as solvent and $Pd(OAc)_2$ as catalyst, a mild and ligand-free method was developed for olefin synthesis, where a host of aryl halides and alkenes reacted to afford moderate to good yields of coupled products[102] (Scheme 3.16.65). In the case of ethyl vinyl ether, the corresponding aryl methyl ketones were synthesized through a concomitant Heck and hydrolysis pathways.

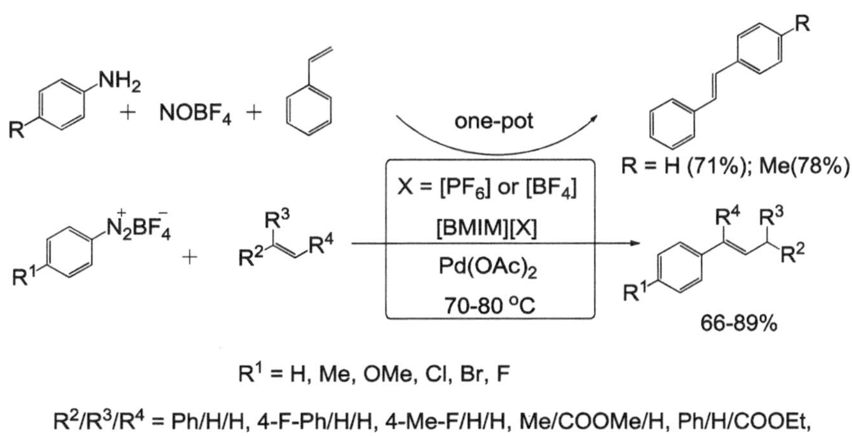

SCHEME 3.16.64 Matsuda-Heck coupling of arenediazonium salts with alkenes.

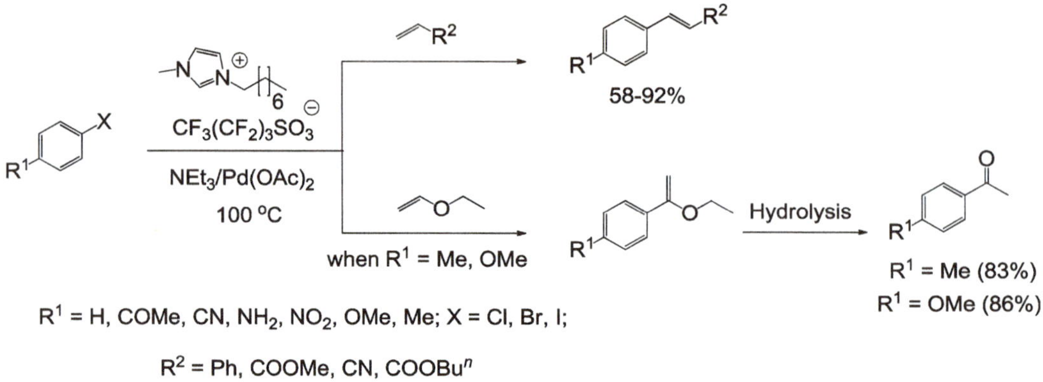

SCHEME 3.16.65 Heck cross-coupling of aryl halides and terminal alkenes in fluorous ionic liquid (IL).

Phosphorylated triethyl or tributylammonium chloride (IL1 and IL2) served as solvent and ligand for $PdCl_2$-catalyzed Heck cross coupling reaction with terminal alkenes[103] (Scheme 3.16.66).

A zwitterionic Pd(II) complex synthesized from pyridine and imidazolium IL bearing different *N*-alkyl chains was employed for arylation of olefins in [BMIM][BF_4] with PPh_3 as ligand and NEt_3 as base.[104] The yield and recyclability were diminished with increasing the alkyl chain length of the IL ligand (Scheme 3.16.67).

3.16.10.2 Sonogashira Cross-Coupling

An efficient copper-free Sonogashira coupling of diverse activated and deactivated aryl bromides with phenyl or n-octyl acetylene in [BMIM][BF_4] at a low loading (1 mol%) of $[Pd(allyl)Cl]_2$ catalyst with PPh_3 as ligand gave the desired products in good to excellent

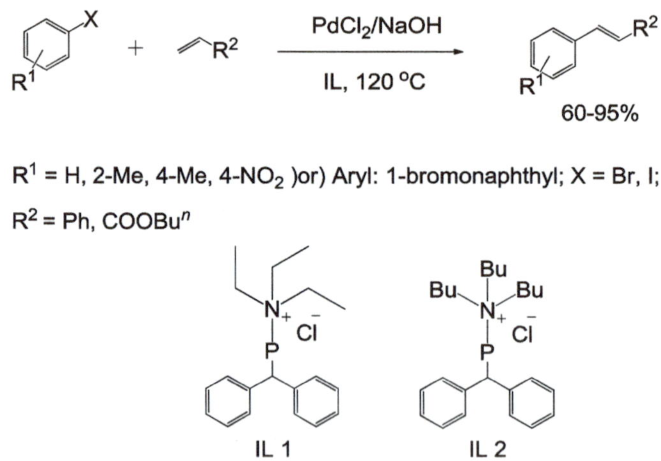

SCHEME 3.16.66 Heck coupling in phosphorylated quaternary ammonium chloride.

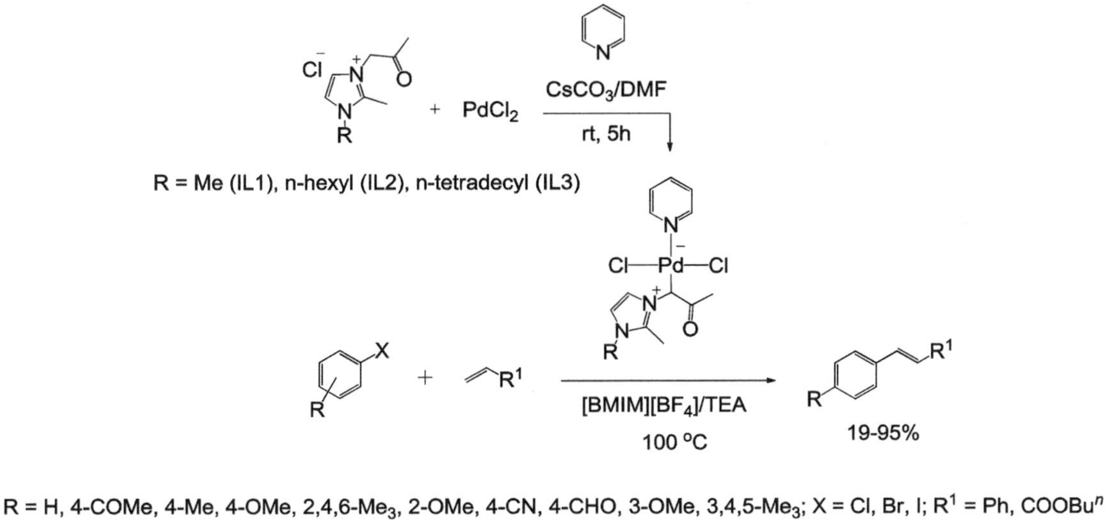

SCHEME 3.16.67 Mizoroki-Heck coupling in [BMIM][BF_4] catalyzed by zwitterionic palladium complex.

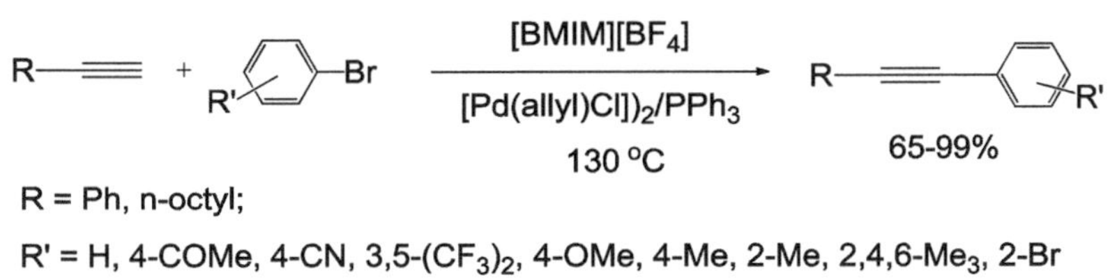

SCHEME 3.16.68 Sonogashira coupling in ionic liquid catalyzed by [Pd(η^3-C_3H_5)Cl]$_2$/PPh_3.

yields[105] (Scheme 3.16.68). The same group reported an extension of this protocol for the mono- and double arylation of terminal alkynes and diynes, respectively, by making extensive use of heterocyclic halides as coupling partners.[106]

Biodegradable ILs derived from esters of nicotinic acid, namely, 3-alkoxycarbonyl-1-methylpyridinium bis(trifluoromethanesulfonyl)imides (ILs 1−4) were screened for Sonogashira coupling reactions using $PdCl_2/NEt_3$ under ultrasonic irradiation at room temperature[107] (Scheme 3.16.69).

Carboxylate-based amine salts (ILs 1−5) acted as ligand, reducing agent, base, and reaction solvent for the copper- and phosphine-free synthesis of diaryl acetylenes in the presence of $PdCl_2$.[108] Additionally, the phosphorylated analog of salt 1 (i.e., IL6) was also equally effective for the coupling reaction with improved yield and excellent recyclability (10 runs) (Scheme 3.16.70).

Functioning as dual solvent and base, the piperidine-incorporated imidazolium IL efficiently promoted the high yield synthesis of diaryl- and aryl alkyl acetylenes including SF_5-bearing analogs and alkyne homodimers.[109] The method employed $PdCl_2(PPh_3)_2$ as catalysts without the need for copper, external base, or additive (Scheme 3.16.71).

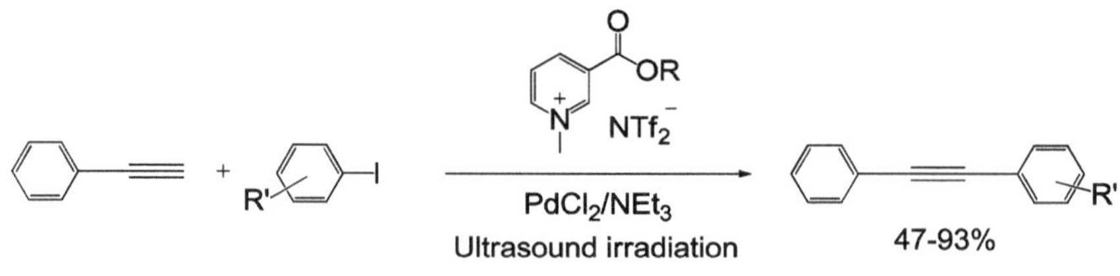

R = Bu^n (IL1); Bu^i (IL2); Bu^s (IL3); Bu^t (IL4)

R' = H, 4-OMe, 4-COMe, 4-Me, 4-COOEt, 4-CH_2OH, 4-OCOMe, 4-OCOPh, 4-NO_2, 4-CH_2OCOMe, 3-COOMe/4-OCOPh, 2-OMe/5-OAc; (or) Ar = 2 or 3-iodothiophene

SCHEME 3.16.69 Aryl-alkynylation in biodegradable ionic liquid catalyzed by $PdCl_2$.

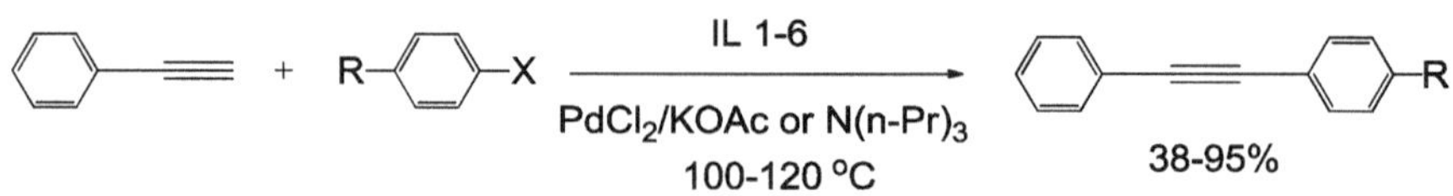

R = H, Me, NO_2, OMe, Me, Cl, (or) Ar = 3-bromopyridine, 3-bromothiophene

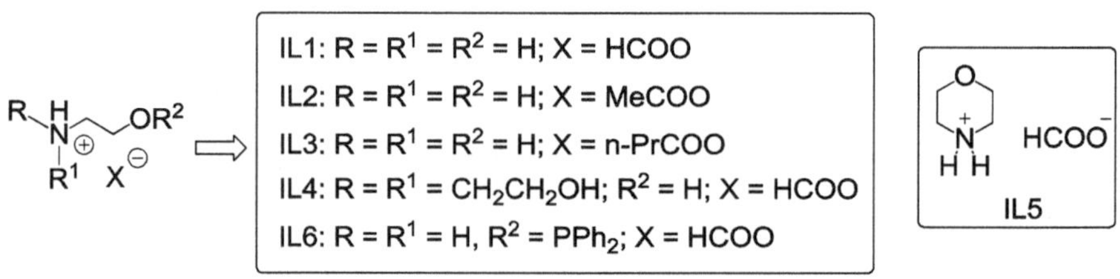

SCHEME 3.16.70 Synthesis of bis aryl-alkyne in ionic liquid by Sonogashira coupling.

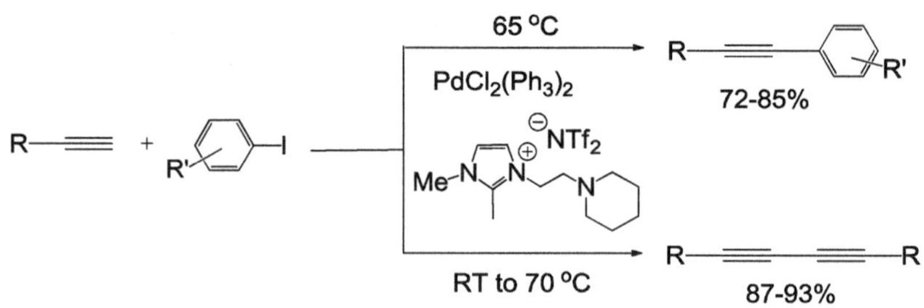

R = Ph, 4-F-Ph, cyclopentyl, TMS, 4-Bu^t-Ph, cyclohexyl-1-ol

R' = 3-OMe, 4-Me, 4-CF_3, 4-COMe, 4-NO_2, 4-OMe, 4-SF_5 (or) Ar = 2-thiophene

SCHEME 3.16.71 Sonogashira coupling in basic ionic liquid.

3.16.10.3 Suzuki Cross-Coupling

A one-pot chemo-/biocatalyzed cascade process involving Pd-catalyzed Suzuki coupling and biosynthesis of enantiopure biaryl alcohols in biphasic IL-water system has been reported.[110] The water-immiscible [BMIM][NTf_2] was recognized as a promising and compatible solvent system for this domino reaction in a survey study. Diarylacetophenone obtained by Suzuki coupling upon subsequent reduction gave alcohols with good yield and exceptional enantioselectivity (Scheme 3.16.72).

New triazole-appended imidazolium ILs were synthesized and explored as ligand and solvent in Suzuki-Miyaura cross-coupling reactions employing $PdCl_2$ as catalyst.[111] Due to the synergetic effect of triazolyl and imidazolium moieties, the mononuclear Pd complex formed with the IL was reported to exert better catalytic stability and excellent recyclability. This was also supported by the high isolated yields of biaryl compound even after five runs (Scheme 3.16.73).

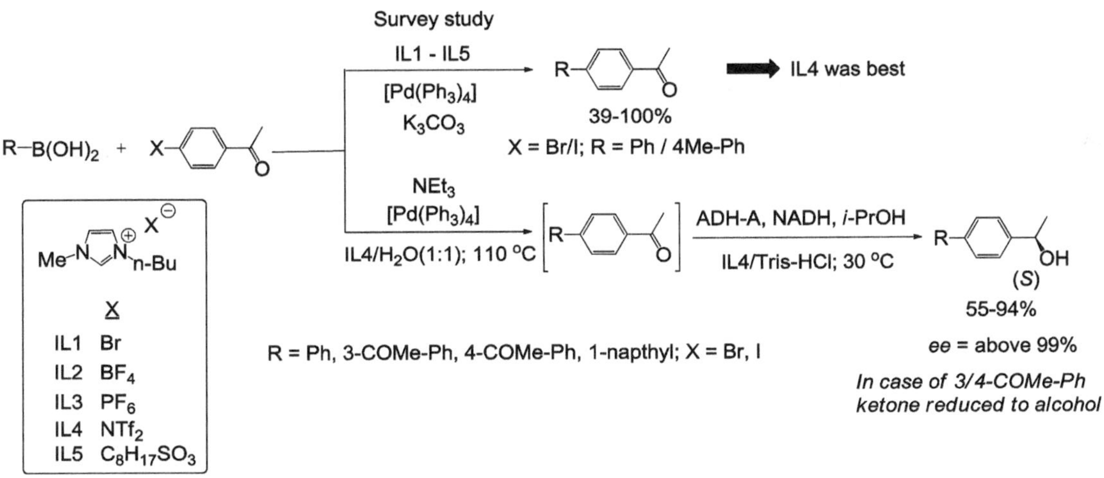

SCHEME 3.16.72 One-pot chemo-/biocatalyzed domino Suzuki and reduction reactions in ionic liquid (IL).

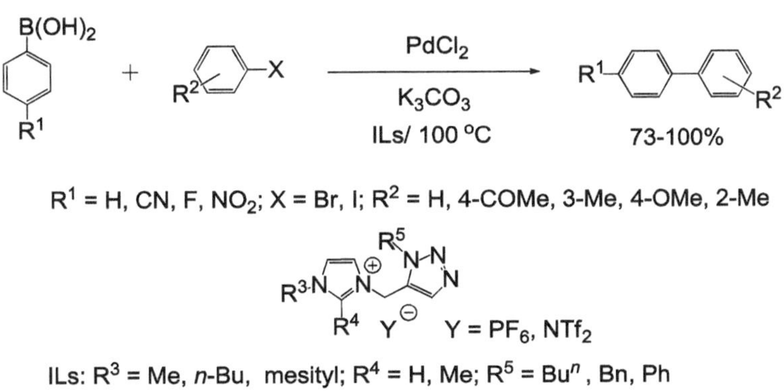

SCHEME 3.16.73 Suzuki cross-coupling in triazole-incorporated imidazolium ionic liquid (IL).

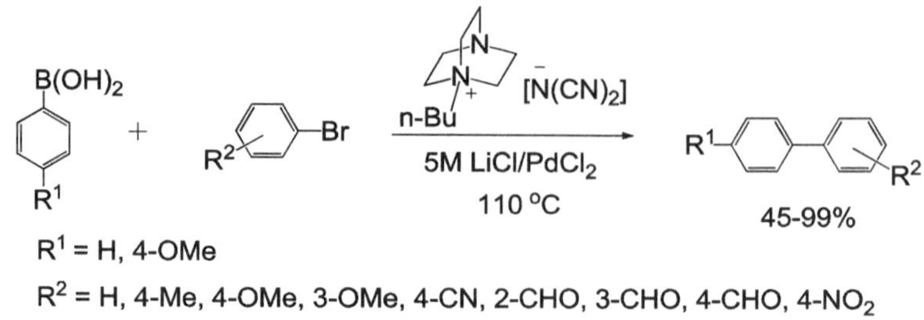

SCHEME 3.16.74 Synthesis of bis-aryls via Suzuki coupling in 1,4-diazabicyclo[2.2.2]octane-based ionic liquid.

A base- and ligand-free approach was developed for the Suzuki coupling using 1-butyl-1,4-diazabicyclo[2.2.2]octandicyanamide, [C-4 DABCO][dca] as IL, with LiCl as promotor and $PdCl_2$ as catalyst, with moderate to good yields[112] (Scheme 3.16.74).

A regio- and stereoselective allyl-aryl coupling reaction was described employing [BMIM][PF_6]/H_2O or [BMIM][SbF_6] with $Pd(OAc)_2$ as catalyst at ambient temperature.[113] Diverse aryl boronic acids and allyl carbonates including the substrates that are prone to β-H elimination in Pd-catalyzed reactions such as 1-benzyl-3-phenylallylic carbonates were tolerated in this system, leading to the desired coupling products in good yields. It is also interesting to note that the same reactions in molecular solvents gave poor conversion, whereas in IL, >99% was achieved (Scheme 3.16.75).

An efficient system consisting of [HMIM][NTf_2]/supercritical CO_2 biphasic medium with $Pd_2(dba)_3$ and K_2CO_3 promoted ligand-free Suzuki cross-coupling reactions of a broad range of aryl halides (X = Cl, Br or I) and arylboronic acids, in respectable yields and with good recyclability[114] (Scheme 3.16.76).

3.16.10.4 Some Featured Coupling and Cyclization Reactions

A simple method was devised for the Stille cross-coupling reaction of aryl/alkenyl bromides by using a wide range of organotin-incorporated task-specific ILs in the presence of palladium catalyst in moderate to good yield.[115] The reaction was straightforward and did

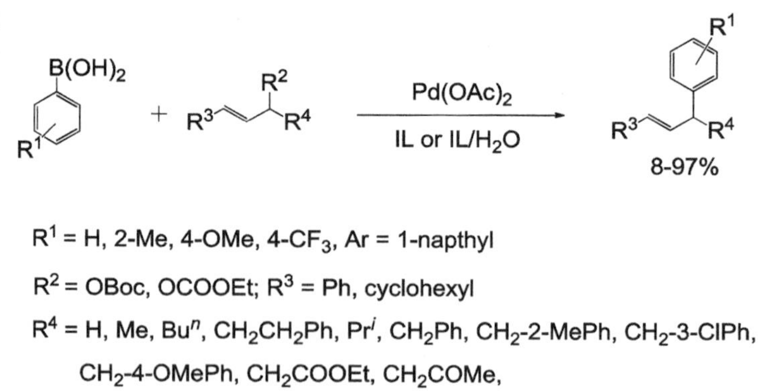

SCHEME 3.16.75 Aryl allyl coupling in ionic liquid (IL) or IL/water system.

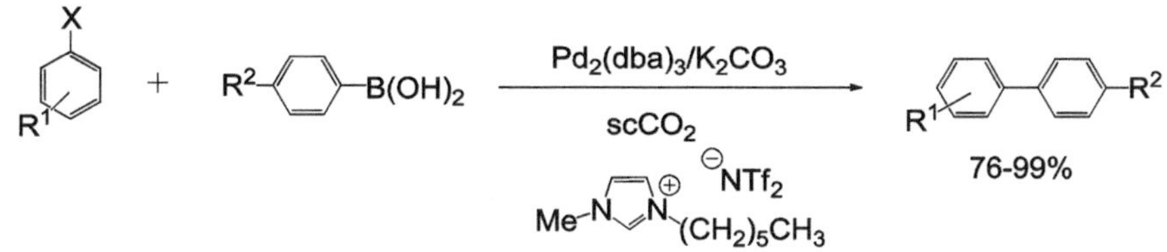

R^1 = H, 4-Me, 2-OMe, 4-OMe, 4-OH, 2-NH_2, 4-NO_2, 4-CHO, 4-F, 4-CN

X = Cl, Br, I; R^2 = H, Me, NO_2, F

SCHEME 3.16.76 Suzuki cross-coupling reaction.

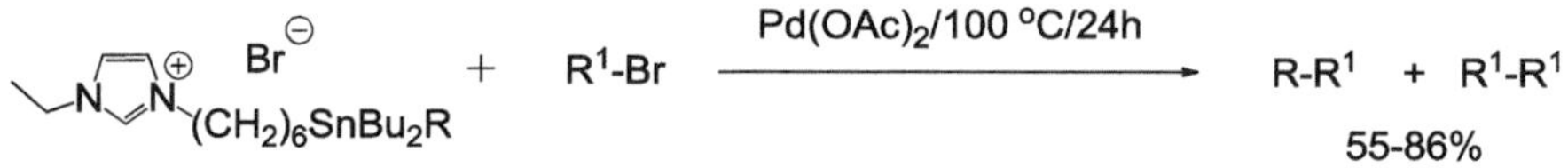

R = Ph, vinyl, allyl, 2-thiophene, 4-OMe-Ph, 4-F-Ph

R^1 = 3-pyridyl, 2-CF_3-pyridyl, 4-COMe-Ph, 4-CHO-Ph, 4-CN-Ph, 3-benzothiozolyl, 2 and 1-napthyl, 2-vinylbenzene, 4-COOEt-Ph

SCHEME 3.16.77 Stille cross-coupling of aryl bromides with ionic liquid-tagged stannanes.

not require additional solvent, base, or additives. The IL was recycled and showed low tin leaching (less than 3 ppm) (Scheme 3.16.77).

Gold nanoparticle-catalyzed reductive Ullman coupling of aryl halides in the presence of tetrabutylammonium acetate (IL1) or tetrabutylammonium hydroxide (IL2)-water micellar condition was reported[116] employing glucose as renewable reductant. Serving as solvent and base, both type of ILs promoted the homocoupling reaction well, with IL2-water being an excellent system based on the isolated yields (Scheme 3.16.78).

An efficient $Pd(OAc)_2$-catalyzed oxidative cross-coupling of polyfluoroarenes/heteroarenes with simple aromatic substrates was demonstrated in imidazolium ILs.[117] The method did not require external oxidant or additives, but a catalytic quantity of AcOH was critical for the success of this C–C coupling transformation (Scheme 3.16.79).

The [BMIM][X] (X = BF_4/PF_6)/$Pd(OAc)_2$ system was employed for efficient synthesis of symmetrical biaryls from readily accessible arenediazonium salts in the absence of ligand and oxidant[118] (Scheme 3.16.80).

A green approach enabling rapid access to a diverse group of benzoxazoles and benzothiazoles from their Schiff's bases was reported in readily accessible imidazolium ILs, employing $Pd(OAc)_2$ as catalyst without the use of additives[119] (Scheme 3.16.81).

A host of 2,3-difunctionalized benzofuran derivatives were synthesized by an atom-economic cascade annulation approach starting from 2-alkynylphenol and β,γ-unsaturated alkenoic acid in the presence of $Pd(TFA)_2$ and $Cu(TFA)_2 \cdot xH_2O$.[120] A survey study revealed that whereas DMF, dioxane, and toluene were not suitable for this transformation resulting in low yields (trace to 9%), a remarkable yield increase was observed (94%) with 1-(2-hydroxyethyl)-3-methyl imidazolium chloride [C_2OHmim]Cl. The study was then

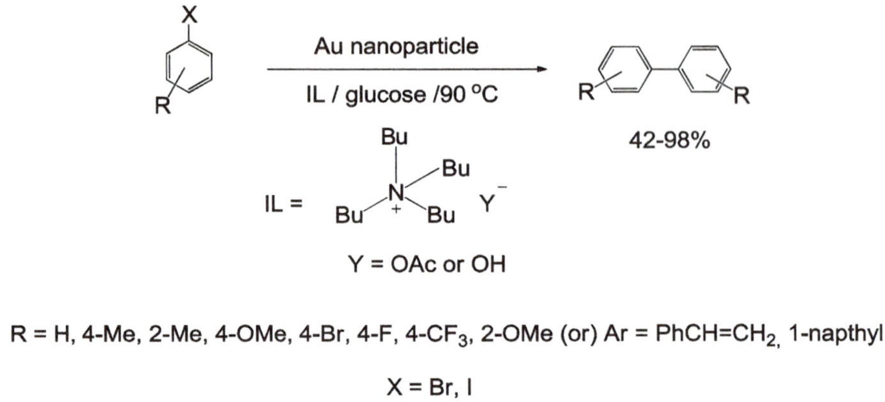

SCHEME 3.16.78 Ullman cross-coupling of aryl halides catalyzed by Au nanoparticles. Il, ionic liquid.

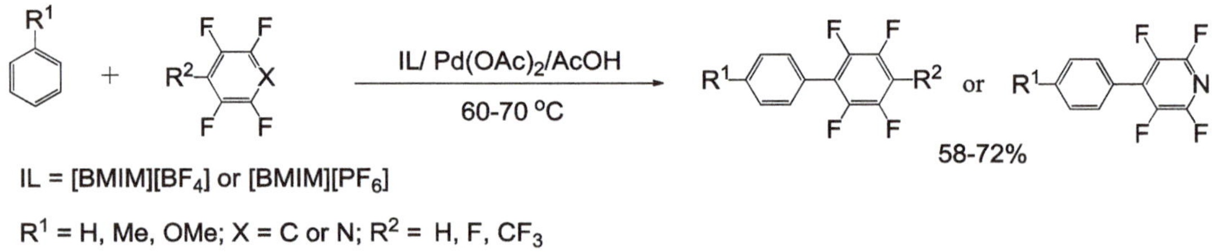

SCHEME 3.16.79 Direct oxidative coupling of polyfluorinated arenes with simple arenes in ionic liquid (IL).

SCHEME 3.16.80 Homocoupling of diazonium salts catalyzed by $Pd(OAc)_2$. *IL*, ionic liquid.

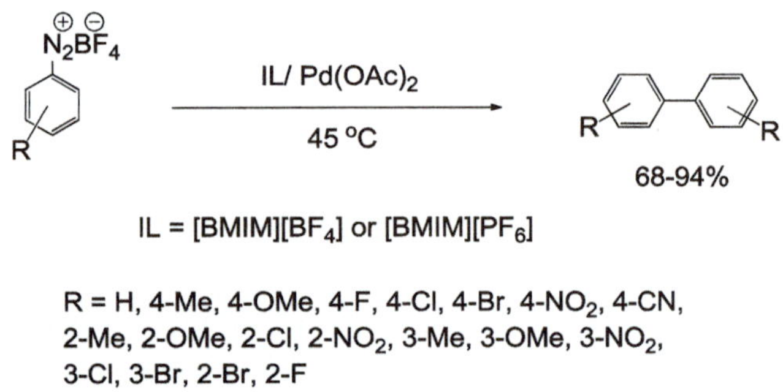

extended to other substrates to synthesize bis-heterocycles in moderate to good yields (56%–89%) (Scheme 3.16.82).

3.16.10.5 Hydroformylation of Alkenes

Hydroformylation is a fundamentally important metal-mediated reaction that is practiced in industry on large scale for the synthesis of aldehydes from alkene by using syngas

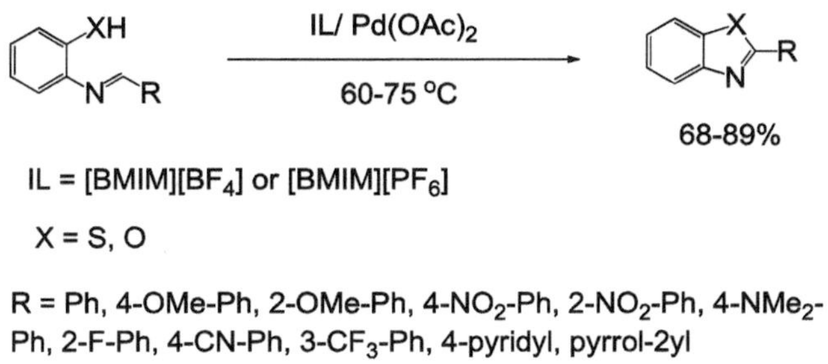

SCHEME 3.16.81 Synthesis of benzothiazoles and benzoxazoles. *IL*, ionic liquid.

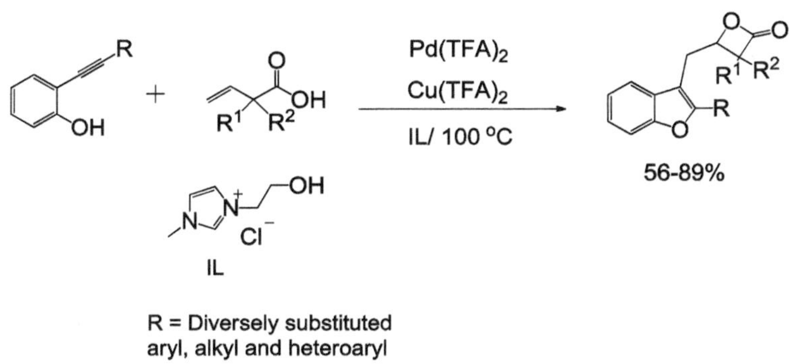

SCHEME 3.16.82 Cyclization of 2-alkynylphenol to 2,3-difunctionalized benzofurans. *IL*, ionic liquid.

($CO + H_2$) under homogeneous catalysis using rhodium catalyst. Problems associated with isolation and recycling of the catalyst provided the impetus for the development of immobilization concepts employing ILs as solvent or hybrid materials.[121,122]

Rh-tri(*m*-sulfonylphenyl)phosphine (TPPTS) immobilized in guanidinium-based IL served as a good catalytic system for the biphasic hydroformylation of higher alkenes (Scheme 3.16.83). Excellent activity and chemoselectivity were reported by using this system for 35 cycles without noticeable metal/ligand leaching.[123]

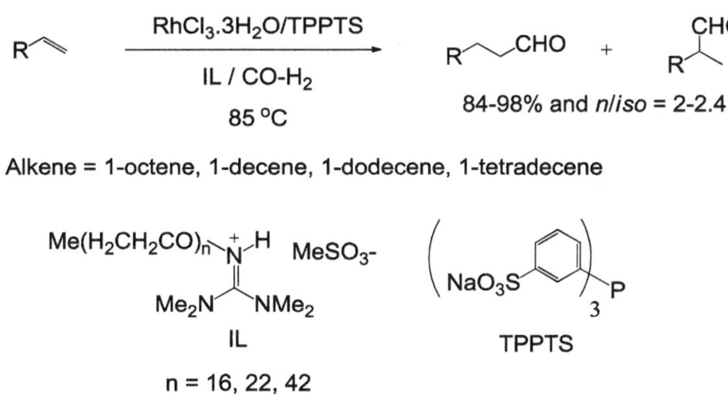

SCHEME 3.16.83 Hydroformylation of alkenes in ionic liquid (IL) catalyzed by Rh/tri(*m*-sulfonylphenyl) phosphine (TPPTS).

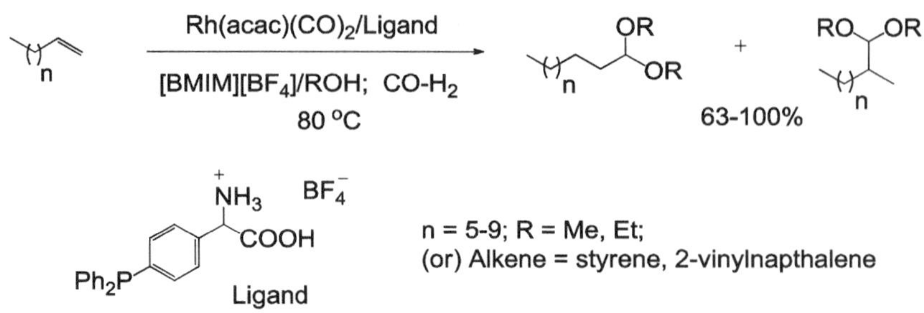

SCHEME 3.16.84 Hydroformylation-acetylation of alkenes in ionic liquid catalyzed by Rh/ligand.

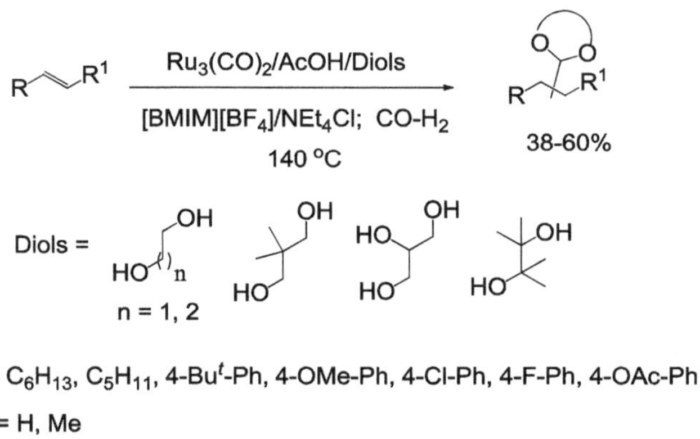

SCHEME 3.16.85 Hydroformylation-acetylation of alkenes in ionic liquid catalyzed by Ru/AcOH.

A tandem hydroformylation-acetylation of olefins was demonstrated in imidazolium IL-MeOH system using zwitterionic phosphine ligand/Rh bifunctional catalyst.[124] Alkyl-substituted alkenes gave better conversion and good selectivity for linear acetal formation, whereas for aryl alkenes, selectivity was reversed, leading mainly to branched acetals (Scheme 3.16.84).

In another study,[125] ruthenium was used in place of rhodium as catalyst for the one-pot hydroformylation and acetylation of alkenes in [BMIM][BF_4] by using acetic acid and NEt_4Cl as additives. Diverse diols and olefins (including aryl alkenes) reacted smoothly to afford modest isolated yields of cyclic acetals with selectivity toward formation of linear aldehydes in the hydroformylation step (Scheme 3.16.85).

A supported IL-phase catalyst was developed for the gas-phase hydroformylation of C4 mixtures (1-butene, *cis/trans* 2-butene, etc.) by using Rh-diphospite ligand.[126,127] This system exhibited exceptional selectivity for isomerization/hydroformylation, as evidenced by the formation of *n*-pentanal in significant yield with a high linear to branched ratio (Scheme 3.16.86).

3.16.10.6 Formylation of Amines and Alcohols

An efficient and chemoselective formylation method for amines has been reported by using formic acid with tetramethylguanidine trifluoroacetate as catalyst and solvent.[128]

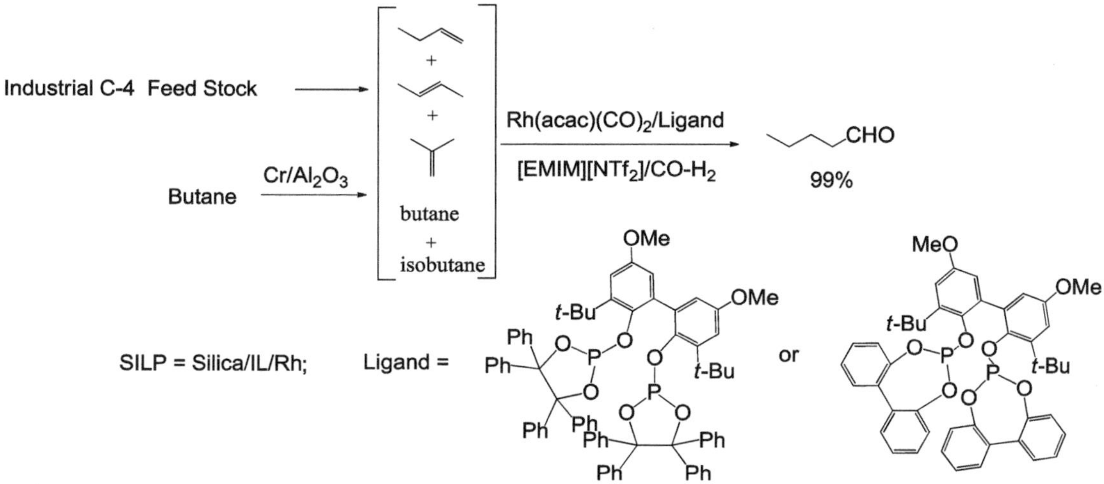

SCHEME 3.16.86 Hydroformylation of alkenes by Rh-supported ionic liquid phase.

Various amines including OH-bearing analogs such as aminoacids and hydroxylamines were converted to the *N*-formylated derivatives with excellent reported yields (Scheme 3.16.87).

Alcohols underwent mild *O*-formylation with ethyl formate at room temperature by using [BMIM][HSO_4] as catalyst and afforded moderate to good yields of the corresponding formates.[129] The ability to selectively formylate a primary alcohol in the presence of a tertiary alcohol or a phenol is a positive feature of this protocol (Scheme 3.16.88).

Imidazolium-based ILs served as catalyst for the formylation of aliphatic/aromatic amines and/or alcohols[130,131] with either CO or formic acid as formylating agent, in moderate to excellent yields (Scheme 3.16.89).

A metal- and acid-free room temperature formylation method for amines that employs CO_2 and phenylsilane and [BMIM]Cl as catalyst has been reported.[132] Aromatic amines underwent smooth monoformylation, whereas the aliphatic counterpart gave both mono- and diformylated products (Scheme 3.16.90). NMR studies of the IL and phenylsilane mixture suggested that IL activates the silane's Si-H bond for CO_2 insertion, and this leads to the formation of formamide.

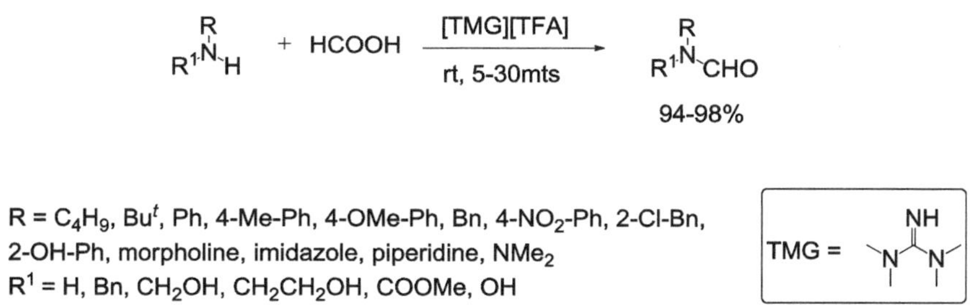

SCHEME 3.16.87 *N*-Formylation of amines by tetramethylguanidine trifluoroacetate ([TMG][TFA]).

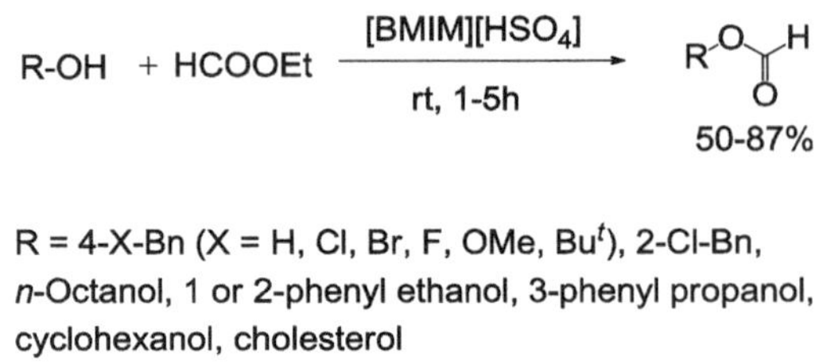

SCHEME 3.16.88 *O*-Formylation of alcohols by imidazolium ionic liquid.

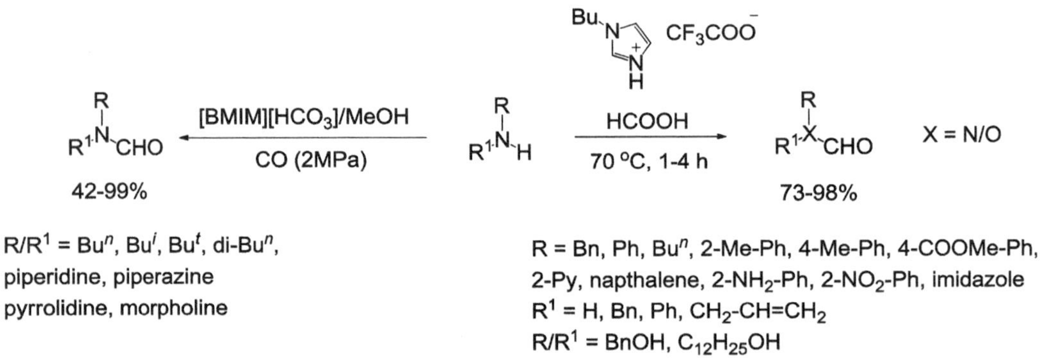

SCHEME 3.16.89 *N*/*O*-Formylation of amines by imidazolium ionic liquid.

SCHEME 3.16.90 *N*-Formylation of amines by [BMIM][Cl] using $PhSiH_3$ and CO_2.

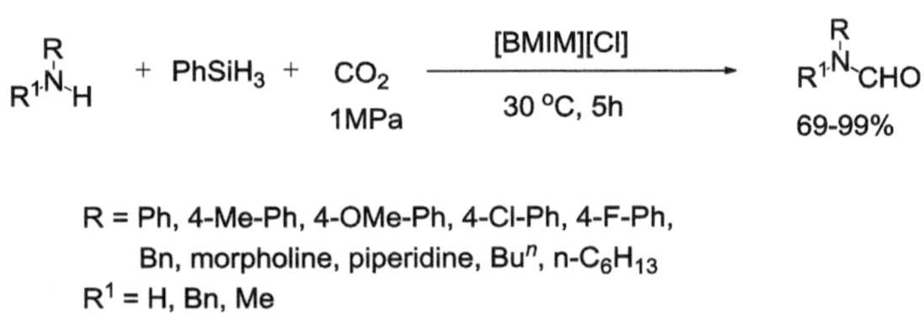

3.16.11 DIELS-ALDER REACTION IN IONIC LIQUIDS

Given the highly fundamental nature of the DA reaction and its paramount importance in synthetic chemistry, application of ILs to this transformation as a way to improve yields and stereoselectivity is a very worthwhile endeavor.[3]

A 2010 authoritative review[133] examined the solvent effect on DA reactions in ILs. Based on multiparameter linear solvation energy relationships and theoretical analysis, it was suggested that increase in selectivity and reaction rate in DA reactions stems from the ability of the IL solvents to hydrogen bond with the dienophile.

Scandium triflate proved to be a highly effective catalyst for DA reactions performed in imidazolium ILs, resulting in high *endo*-selectivity, and also offering the potential for

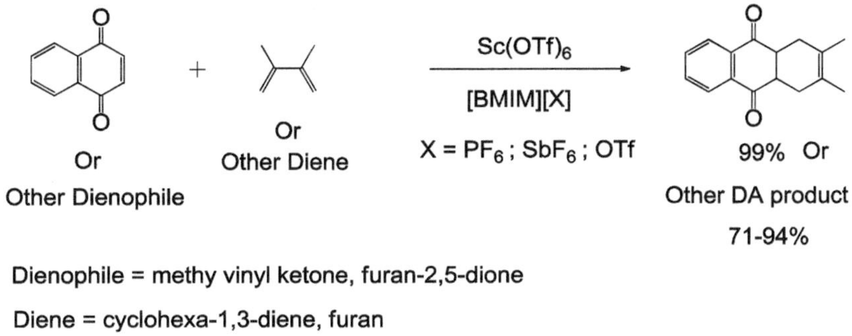

SCHEME 3.16.91 $Sc(OTf)_3$-catalyzed Diels-Alder reaction in ionic liquid.

recycling and reuse.[134] In comparative studies, DCM proved to be a poor solvent (22% yield) when compared with the IL (99% yield) (Scheme 3.16.91).

The influence of ILs on stereoselectivity in DA reactions was demonstrated in cycloaddition reactions between cyclopentadiene and substituted methyl methacrylate.[135] Reactions carried out in organic solvents were *exo*-selective, whereas the ratio was reversed when the reactions were performed in the chloroaluminate *N*-1-butylpyridinium chloride-IL1 (47%–57% yield) or in 1-ethyl-3-methyl-1*H*-imidazolium chloride-IL2 (58%–68% yield) as solvent, where predominant *endo*-selectivity was observed (*endo*:*exo* = 3:1) (Scheme 3.16.92).

A host of Lewis acids have been tested for their catalytic and *endo*-selectivity in the DA reaction between cyclopentadiene and different dienophiles in 1-hexyl-3-methylimidazolium tetrafluoroborate [HMI][BF_4]. Metal salts such as Ce^{IV}, Sc^{III}, and Y^{III} triflates showed promise, and in particular, cerium triflate proved superior to other catalysts. The IL-catalyst system were recycled and reused without significant loss in yield and selectivity[136] (Scheme 3.16.93).

Among various cationic Pd-phosphinooxazolidine catalysts and ILs that were tested, the combination of chiral catalyst with SF_6 counterion and [BMIM][BF_4] was found to be effective for asymmetric DA reaction. Reactions were carried out at −40 °C to room temperature by using a mixture of IL and DCM as solvent. These reactions resulted in better chemical and optical yields (46%–99%) along with excellent recyclability[137] (Scheme 3.16.94).

The DA reaction of cyclopentadiene with alkyl acrylates were carried out in pyrrolidinium IL $Pyrr_{1.4}[NTf_2]$ ($Pyrr_{1.4}$ = 1-methyl-1-butyl-pyrrolidinium cation) as solvent and metallic triflates and metallic chlorides as catalyst.[138] Excellent stereoselectivity and higher conversions

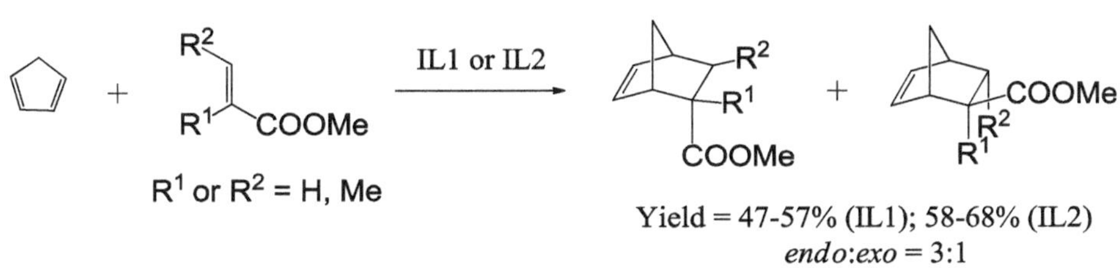

SCHEME 3.16.92 Stereoselective Diels-Alder reaction in the chloroaluminate ionic liquid (IL).

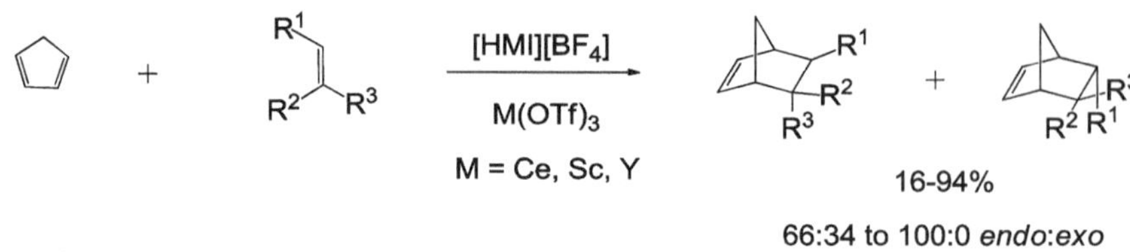

R^1 = COMe, CHO, COEt, H, CN, COOMe,

R^2 = H, Me; R^3: H, CHO

Dienophile = furan-2,5-dione, 1-phenyl-1H-pyrrole-2,5-dione, benzoquinone, 2-methylcyclohexa-2,5-diene-1,4-dione, napthalene-1,4-dione

SCHEME 3.16.93 Diels-Alder reaction in $[HMI][BF_4]$ catalyzed by $M(OTf)_3$.

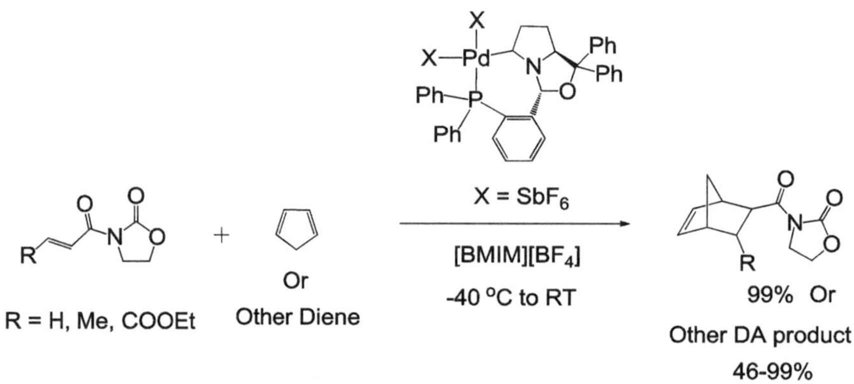

Diene = cyclohexa-1,3-diene, furan, 1-phenoxycarbonyl-1,2-dihydropyridine

SCHEME 3.16.94 Asymmetric Diels-Alder reaction in ionic liquid.

were achieved with metal chlorides ($ScCl_3$, YCl_3, and $YbCl_3$) when compared with the corresponding metal triflates (Scheme 3.16.95).

Hydrogen bond-rich ILs, derived from renewable starting material such as D-glucose, have been employed as solvent and organocatalyst in the DA reaction. Dienes such as cyclopentadiene or isoprene reacted with various dienophiles in high yields and with high *endo*-selectivity[139] (Scheme 3.16.96).

The 1,2-diaryl-2,3-dihydro-4-pyridones were synthesized via the aza-DA reaction by employing 8-ethyl-1,8-diazobicyclo[5,4,0]-7-undecenium trifluoromethanesulfonate or 1-ethyl-3-methyl-1*H*-imidazolium trifluoromethanesulfonate as solvent and microencapsulated $Sc(OTf)_3$ as Lewis acid catalyst.[140] The IL-catalyst system was recovered to about 98% and recycled without loss in yield (Scheme 3.16.97).

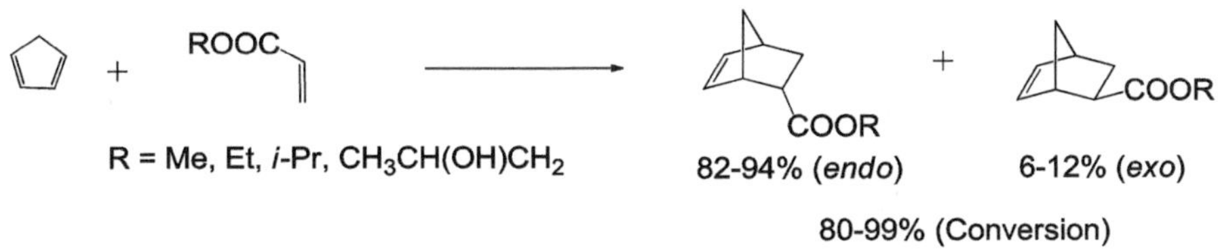

SCHEME 3.16.95 Stereoselective Diels-Alder reaction in pyrrolidinium ionic liquid.

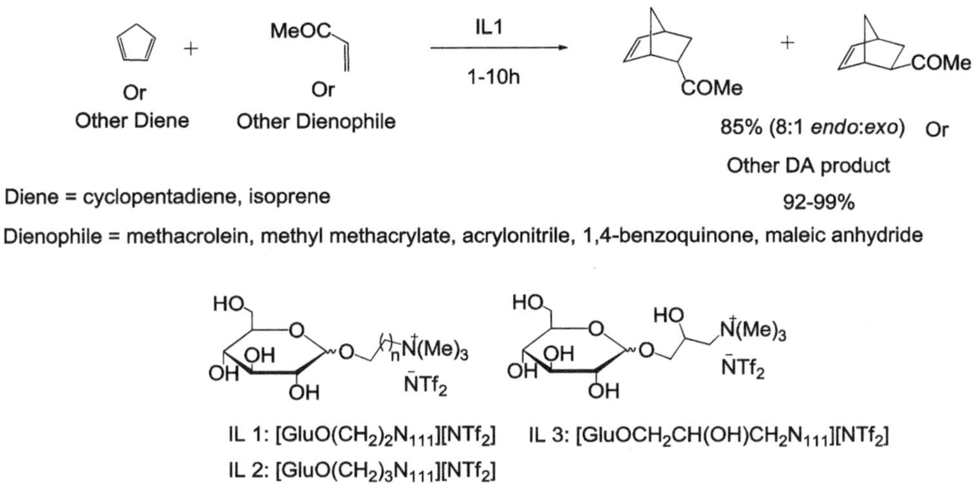

SCHEME 3.16.96 Stereoselective Diels-Alder reaction in glucose-based ionic liquids.

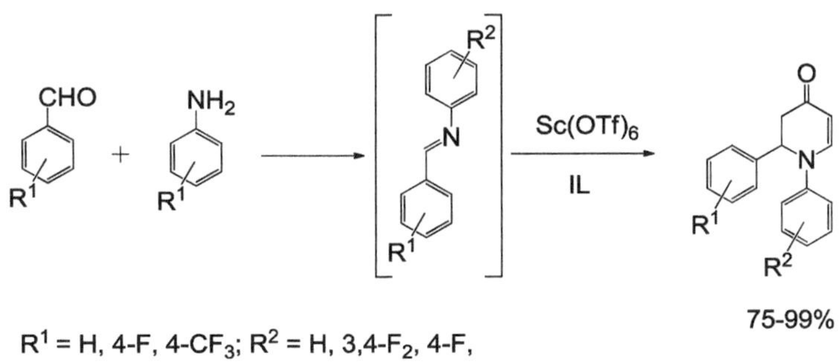

SCHEME 3.16.97 $Sc(OTf)_3$-catalyzed synthesis of 1,2-diaryl-2,3-dihydro-4-pyridones.

The [BMIM][BF_4] was used as solvent and catalyst to promote the three-component aza DA reaction between anilines, aldehydes, and enol ethers at ambient conditions, to furnish pyrano- and furanoquinolines in high yields and with high *endo*-selectivity[141] (Scheme 3.16.98).

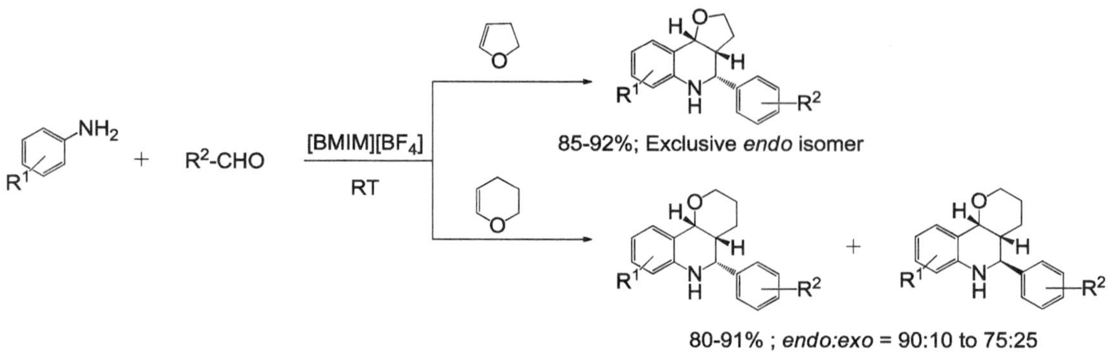

R^1 = H, 4-OMe, 4-Me, 3,5-$(OMe)_2$, 2-Me, 4-F or aniline = 1-napthyl amine

R^2 = Ph, 4-Cl-Ph, 4-F-Ph, 4-OMe-Ph, H, C_8H_5, C_6H_6

SCHEME 3.16.98 Aza Diels-Alder reaction in ionic liquid.

3.16.12 WITTIG REACTION IN IONIC LIQUIDS

The utility of ILs as solvent and/or promoter in the Wittig reaction has been demonstrated in several reported studies.

The Wittig reaction of aldehydes including enals with stabilized ylides was carried out successfully in [BMIM][BF_4] at 60 °C with excellent *E*-stereoselectivity.[142] Successive extraction of the reaction mixture with Bu^tOMe and toluene selectively removed the products and the PPh_3O by-product, respectively, leaving behind the IL for reuse (Scheme 3.16.99).

Metal-free methylenation of ketones with $CH_2(ZnI)_2$ was achieved by using [BMIM][PF_6] as promoter in tetrahydrofuran (THF)-hexane.[143] Various ketones were converted into alkenes at 60 °C in moderate to good yields (Scheme 3.16.100).

The iron complex Fe(tetraphenylporphyrin)Cl-catalyzed olefination of diverse aldehydes was achieved by employing PPh_3 and ethyl diazoacetate in [BMIM][PF_6] as solvent.[144] Depending on the nature and reactivity of aldehydes, the reaction temperature was varied between 50 and 60 °C to improve the yield and selectivity (Scheme 3.16.101).

The Wittig reagent (Scheme 3.16.102) was generated in the phosphonium IL (PhosIL-$C_9H_{19}COO$) by using PhMgBr,[145] and subsequent reaction of this phosphorene with aldehydes afforded the corresponding alkenes in good yield. The IL's quaternary phosphonium

SCHEME 3.16.99 Reaction of aldehydes with stabilized ylides in [BMIM][BF_4].

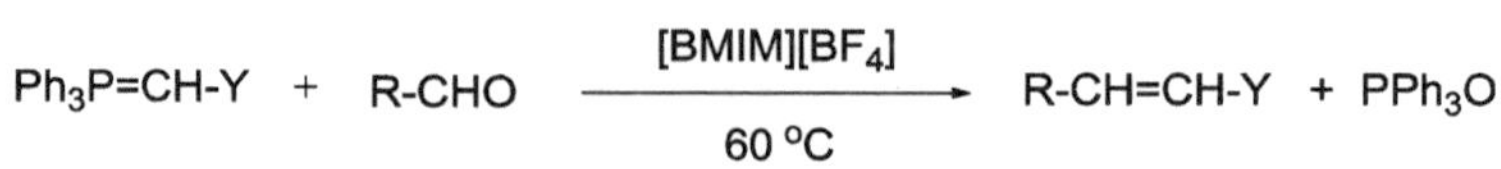

Y = COMe, CO_2Me, CN

R = 4-Cl-Ph, 4-NO_2-Ph, 4-OMe-Ph, 2-Me-Ph, cyclohexyl, C_5H_{11}, Bu^i, (*E*) Ph-CH=CH- , (*E*) Me-CH=CH-

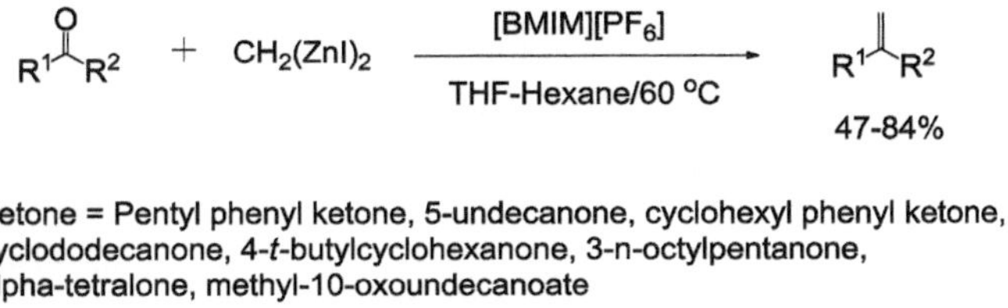

SCHEME 3.16.100 Synthesis of alkene from ketone using ionic liquid.

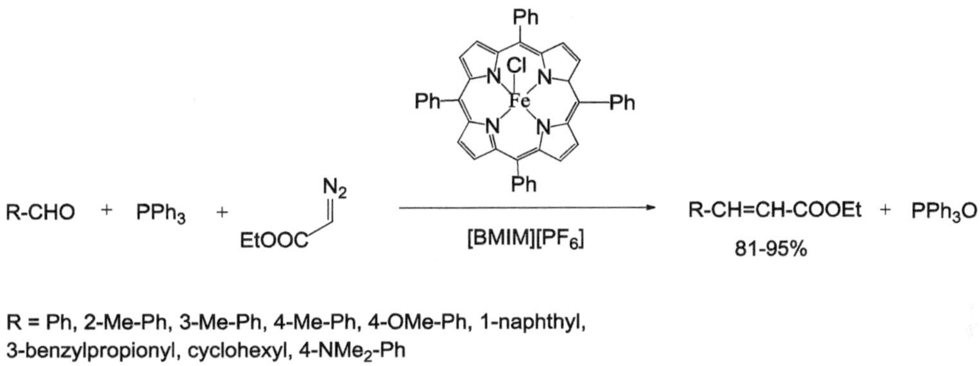

SCHEME 3.16.101 Synthesis of alkene from ethyl diazoacetate in [BMIM][PF_6].

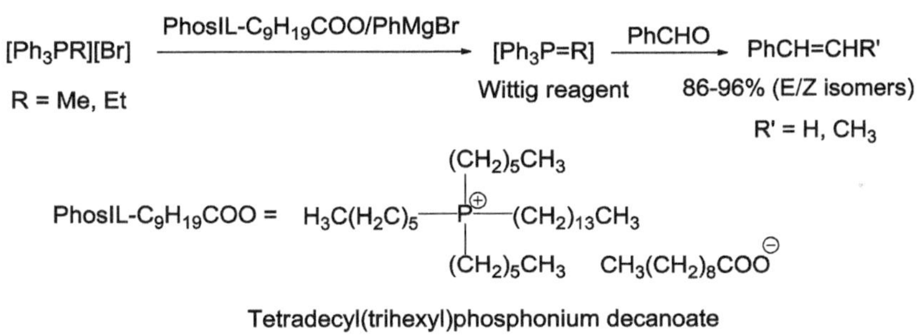

SCHEME 3.16.102 Wittig reaction in phosphonium ionic liquid.

cation is believed to activate carbonyl compounds through effective hydrogen bonding with the proton alpha to the phosphonium ion as evidenced by NMR and X-ray crystallography.

An efficient one-pot protocol has been developed for the synthesis of 1-substituted-9*H*-pyrido[3,4-*b*]indoles from iminophosphorane via an aza-Wittig electrocyclic ring-closure reaction in [BMIM][BF_4] under microwave irradiation.[146] In situ deprotection of *N*-methoxymethyl group proceeded smoothly under these reaction conditions (Scheme 3.16.103).

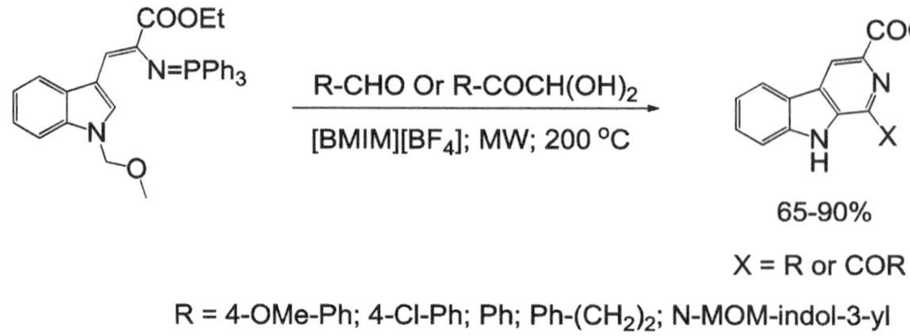

SCHEME 3.16.103 Synthesis of 1-substituted β-carbolines in ionic liquid.

3.16.13 CONCLUDING REMARKS

The foregoing progress review account reflects the truly remarkable role that RTILs have played in advancing synthetic methods and catalysis. Application of ILs as solvent/catalyst in fundamentally important/textbook transformations such as nitration, halogenation, alkylation, and acylation, as well as those involving oniums salts as reagent (as in fluorination with Selectfluor or transformations involving arenediazonium salts) have resulted in major advances. It is also clear that ILs are making significant impact on metal-mediated bond-forming reactions. A major advance has been the design and application of ILs that perform multiple tasks, for example, act as solvent/Brønsted acid, solvent/Lewis acid, solvent/catalyst/base, and other multitasks. The "ionic liquid effect" on improving yields and in controlling/reversing stereoselectivity in DA reactions are also remarkable. Solubility of metal triflates, palladium catalysts, and onium salts such as Selectfluor and arenediazonium salts in imidazolium ILs coupled to the prospect for recycling and reuse have provided added incentives. Clearly, this is a fast moving and highly dynamic research area on a positive trajectory that will likely continue in the next decade.

Acknowledgment

K.L. acknowledges the award of Presidential Professorship at UNF.

References

1. Welton T. Room-temperature ionic liquids. Solvents for synthesis and catalysis. *Chem Rev* 1999;**99**:2071–84.
2. Hagiwara R, Ito Y. Room temperature ionic liquids of alkylimidazolium cations and fluoroanions. *J Fluor Chem* 2000;**105**:221–7.
3. Sheldon R. Catalytic reactions in ionic liquids. *Chem Commun* 2001:2399–407.
4. Zhao H, Malhotra SV. Application of ionic liquids in organic synthesis. *Aldrichim Acta* 2002;**35**:75–83.
5. Dupont J, de Souza RF, Suarez PA. Ionic liquid (molten salt) phase organometallic catalysis. *Chem Rev* 2002;**102**:3667–92.
6. Olivier-Bourbigou H, Magna L. Ionic liquids: perspectives for organic and catalytic reactions. *J Mol Catal A Chem* 2002;**182–183**:419–37.
7. Dzyuba SV, Bartsch RA. Recent advances in applications of room-temperature ionic liquid-supercritical carbon dioxide systems. *Angew Chem Int Ed* 2003;**42**:148–50.
8. Forsyth AA, Pringle JM, MacFarlane DR. Ionic liquids-An overview. *Aust J Chem* 2004;**57**:113–9.
9. Chiappe C, Pieraccini D. Ionic liquids: solvent properties and organic reactivity. *J Phys Org Chem* 2005;**18**:275–97.

10. Ding J, Armstrong DW. Chiral ionic liquids: synthesis and applications. *Chirality* 2005;**17**:281–92.
11. Xue H, Verma R, Shreeve JM. Review of ionic liquids with fluorine containing anions. *J Fluor Chem* 2006;**127**:159–76.
12. Borodkin GI, Shubin VG. Electrophilic reactions of aromatic and heteroaromatic compounds in ionic liquids. *Russ J Org Chem* 2006;**42**:1745–70.
13. Enders F, El Adedin SZ. Air and water stable ionic liquids in physical chemistry. *Phys Chem Chem Phys* 2006;**8**:2101–16.
14. Martins MAP, Frizzo CP, Moreira DN, Zanatta N, Bonacorso HG. Ionic liquids in heterocyclic synthesis. *Chem Rev* 2008;**108**:2015–50.
15. Plechkova NV, Seddon KR. Application of ionic liquids in the chemical industry. *Chem Soc Rev* 2008;**37**:123–50.
16. Greaves TL, Drummond CJ. Protic ionic liquids: Properties and application. *Chem Rev* 2008;**108**:206–37.
17. Pavlinac J, Zupan M, Laali KK, Stavber S. Halogenation of organic compounds in ionic liquids. *Tetrahedron* 2009;**65**:5625–62.
18. Olivier-Bourbigou H, Magna L, Morvan D. Ionic liquids and catalysis: recent progress from knowledge to applications. *Appl Cat A Gen* 2010;**373**:1–56.
19. Hallett JP, Welton T. Room-temperature ionic liquids: solvents for synthesis and catalysis. *Chem Rev* 2011;**111**:3508–76.
20. Chiappe C, Rajamani S. Structural effects on the physico-chemical and catalytic properties of acidic ionic liquids: an overview. *Eur J Org Chem* 2011;**28**:5517–39.
21. Pereiro AB, Araujo JMM, Martinho S, Alves F, Nunes S, Matias A, Duarte CMM, Rebelo LPN, Marrucho IM. Fluorinated ionic liquids: properties and applications. *ACS Sustain Chem Eng* 2013;**1**:427–39.
22. Smiglak M, Pringle JM, Lu X, Han L, Zhang S, Gao H, MacFarlane DR, Rogers RD. Ionic liquids for energy, materials and medicine. *Chem Commun* 2014;**50**:9228–50.
23. Tran AN, Van Do T-N, My Le L-P, Le TG. Synthesis of new fluorinated imidazolium ionic liquids and their prospective function as the electrolytes for lithium-ion batteries. *J Fluor Chem* 2014;**164**:38–43.
24. Chiappe C, Pomelli CS. Point-functionalization of ionic liquids: an overview of synthesis and applications. *Eur J Org Chem* 2014;**28**:6120–39.
25. Garcia-Alvarez J, Hevia E, Capriati V. Reactivity of polar organometallic compounds in unconventional reaction media: challenges and opportunities. *Eur J Org Chem* 2015;**31**:6779–99.
26. Laali KK. Ionic liquids as novel media for electrophilic/onium ion chemistry and metal-mediated reactions: a progress summary. *ARKIVOC* 2016;**i**:150–71.
27. Mack R, Askins G, Lowry J, Hurley N, Reeves PC. Indium(III) triflate—a catalyst for greener aromatic alkylation reactions. *Can J Chem* 2013;**91**:1262–5.
28. Kim JH, Lee JW, Shin US, Lee JY, Lee S-G, Song CE. Activation of Lewis acid catalysts in the presence of an organic salt containing a non-coordinating anion: its origin and application potential. *Chem Commun* 2007:4683–5.
29. Wang A, Zheng X, Zhao Z, Li C, Cui Y, Zheng X, Yin J, Yang G. Brønsted acid ionic liquids catalyzed Friedel—Crafts Alkylations of electron-rich arenes with aldehydes. *Appl Cata A Gen* 2014;**482**:198–204.
30. Taheri A, Lai B, Cheng C, Gu Y. Brønsted acid ionic liquid-catalyzed reductive Friedel—Crafts alkylation of indoles and cyclic ketones without using an external reductant. *Green Chem* 2015;**17**:812–6.
31. Malhotra SV, Xiao Y. Enantioselective Friedel—Crafts reactions of aromatic amines with ethyl glyoxylate in pyridinium-based ionic liquids. *Aust J Chem* 2006;**59**:468–72.
32. Laali KK, Sarca VD, Okazaki T, Brock A, Der P. Triflic acid-catalyzed adamantylation of aromatics in [BMIM][OTf] ionic liquid; synthetic scope and mechanistic insight. *Org Biomol Chem* 2005;**3**:1034–42.
33. Song CE, Jung D-U, Choung SY, Roh EJ, Lee S-G. Dramatic enhancement of catalytic activity in an ionic liquid: highly practical Friedel—Crafts alkenylation of arenes with alkynes catalyzed by metal triflates. *Angew Chem Int Ed* 2004;**43**:6183–5.
34. Jorapur YR, Chi DY. Intramolecular cycloalkylation of pyrrole in ionic liquids and immobilized ionic liquids. *Bull Korean Chem Soc* 2011;**32**:3130–2.
35. Sarca VD, Laali KK. Facile benzylation of aromatics in ionic liquid solvents promoted by TfOH, $Sc(OTf)_3$, and $Yb(OTf)_3 \cdot xH_2O$; New life for a classic transformation. *Green Chem* 2006;**8**:615–20.
36. Yin D, Li C, Tao L, Yu N, Hu S, Yin D. Synthesis of diphenylmethane derivatives in Lewis acidic ionic liquids. *J Mol Catal A Chem* 2006;**245**:260–5.

37. Kantam ML, Chakravarti R, Sreedhar B, Bhargava S. Friedel-Crafts alkylation of nitrogen heterocycles using [bmim][OTf] as a catalyst and reaction medium. *Synlett* 2008;**10**:1449–54.
38. Gao J, Wang J-Q, Song Q-W, He L-N. Iron(III)-based ionic liquid-catalyzed regioselective benzylation of arenes and heteroarenes. *Green Chem* 2011;**13**:1182–6.
39. Chu X-Q, Jiang R, Fang Y, Gu Z-Y, Meng H, Wang S-Y, Ji S-J. Acidic-functionalized ionic liquid as an efficient, green, and metal-free catalyst for benzylation of sulfur, nitrogen, and carbon nucleophiles to benzylic alcohols. *Tetrahedron* 2013;**69**:1166–74.
40. Funabiki K, Komeda T, Kubota Y, Matsui M. Acidic-functionalized ionic liquid as an efficient, green, and metal-free catalyst for benzylation of sulfur, nitrogen, and carbon nucleophiles to benzylic alcohols. *Tetrahedron* 2009;**65**:7457–63.
41. Aridoss G, Laali KK. Condensation of propargylic alcohols with 1,3-dicarbonyl compounds and 4-hydroxycoumarins in ionic liquids (ILs). *Tetrahedron Lett* 2011;**52**:6859–64.
42. Sarca VD, Laali KK. Triflic acid-promoted transacylation and deacylation reactions in ionic liquid solvents. *Green Chem* 2004;**6**:245–8.
43. Hakala U, Wähälä K. Microwave-promoted synthesis of polyhydroxydeoxybenzoins in ionic liquids. *Tetrahedron Lett* 2006;**47**:8375–8.
44. Khodaei MM, Bahrami K, Shahbazi F. An efficient method for aromatic Friedel–Crafts acylation reactions. *Chem Lett* 2008;**37**:844–5.
45. Tran PH, Duus F, Le TN. Friedel–Crafts acylation using bismuth triflate in [BMI][PF_6]. *Tetrahedron Lett* 2012;**53**:222–4.
46. Tran PH, Do NBL, Le TN. Improvement of the Friedel–Crafts benzoylation by using bismuth trifluoromethanesulfonate in 1-butyl-3-methylimidazolium trifluoromethanesulfonate ionic liquid under microwave irradiation. *Tetrahedron Lett* 2014;**55**:205–8.
47. Tran PH, Hansen PE, Hoang HM, Chau D-KN, Le TN. Indium triflate in 1-isobutyl-3-methylimidazolium dihydrogen phosphate: an efficient and green catalytic system for Friedel–Crafts acylation. *Tetrahedron Lett* 2015;**56**:2187–92.
48. Aridoss G, Sarca VD, Ponder Jr JF, Crowe J, Laali KK. Electrophilic chemistry of propargylic alcoholsin imidazolium ionic liquids: propargylation ofarenes and synthesis of propargylic etherscatalyzed by metallic triflates [$Bi(OTf)_3$, $Sc(OTf)_3$, $Yb(OTf)_3$], TfOH, or $B(C_6F_5)_3$. *Org Biomol Chem* 2011;**9**:2518–29.
49. Kumar GGKSN, Aridoss G, Laali KK. Condensation of propargylic alcohols with indoles and carbazole in [bmim][PF_6]/$Bi(NO_3)_3 \cdot 5H_2O$: a simple high yielding propargylation method with recycling and reuse of the ionic liquid. *Tetrahedron Lett* 2012;**53**:3066–9.
50. Kumar GGKSN, Laali KK. Condensation of propargylic alcohols with *N*-methylcarbazole and carbazole in [bmim]PF_6 ionic liquid; synthesis of novel dipropargylic carbazoles using TfOH or $Bi(NO_3)_3 \cdot 5H_2O$ as catalyst. *Tetrahedron Lett* 2013;**54**:965–9.
51. Kumar GGKSN, Laali KK. Facile coupling of propargylic, allylic and benzylic alcohols with allylsilane and alkynylsilane, and their deoxygenation with Et_3SiH, catalyzed by $Bi(OTf)_3$ in [BMIM][BF_4] ionic liquid (IL), with recycling and reuse of the IL. *Org Biomol Chem* 2012;**10**:7347–55.
52. Nandi GC, Rathman BM, Laali KK. Mild conversion of propargylic alcohols to α,β-unsaturated enones in ionic liquids (ILs); a new 'metal free' life for the Rupe rearrangement. *Tetrahedron Lett* 2013;**54**:6258–63.
53. Sarma R, Prajapati D. Ionic liquid - an efficient recyclable system for the synthesis of 2,4-disubstituted quinolines via Meyer-Schuster rearrangement. *Synlett* 2008:3001–5.
54. Laali KK, Gettwert VJ. Electrophilic nitration of aromatics in ionic liquid solvents. *J Org Chem* 2001;**66**:35–40.
55. Handy ST, Egrie CR. In: Rogers RD, Seddon KR, editors. *Green Synthesis: aromatic nitrations in room-temperature ionic liquids ACS Symposium Series e-Books 818; Chapter 11*; 2002. p. 134–46.
56. Qiao K, Yokoyama C. Nitration of aromatic compounds with nitric acid catalyzed by ionic liquids. *Chem Lett* 2004;**33**(7):808–9.
57. Earle MJ, Katdare SP, Seddon KR. Paradigm confirmed: the first use of ionic liquids to dramatically influence the outcome of chemical reactions. *Org Lett* 2004;**6**:707–10.
58. Lancaster NL, Llopis-Mestre V. Aromatic nitrations in ionic liquids: the importance of cation choice. *Chem Commun* 2003:2812–3.
59. Dal E, Lancaster NL. Acetyl nitrate nitrations in [bmpy][$N(Tf)_2$] and [bmpy][OTf], and the recycling of ionic liquids. *Org Biomol Chem* 2005;**3**:682–6.

60. Fang D, Shi R-R, Cheng J, Gong K, Liu Z-L. Regioselective mononitration of aromatic compounds using Brønsted acidic ionic liquids as recoverable catalysts. *Appl Catal A Gen* 2008;**345**:158–63.
61. Aridoss G, Laali KK. Ethylammonium nitrate (EAN)/Tf_2O and EAN/TFAA: ionic liquid based systems for aromatic nitration. *J Org Chem* 2011;**76**:8088–94.
62. Jacoway J, Kumar GGKSN, Laali KK. Aromatic nitration with bismuth nitrate in ionic liquids and in molecular solvents: a comparative study of $Bi(NO_3)_3 \cdot 5H_2O$/[bmim][PF_6] and $Bi(NO_3)_3 \cdot 5H_2O$/1,2-DCE systems. *Tetrahedron Lett* 2012;**53**:6782–5.
63. Zolfigol MA, Khazaei A, Moosavi-Zare AR, Zare A, Kruger HG, Asgari Z, Khakyzadeh V, Kazem-Rostami M. Design of ionic liquid 3-methyl-1-sulfonic acid imidazolium nitrate as reagent for the nitration of aromatic compounds by in Situ generation of NO_2 in acidic media. *J Org Chem* 2012;**77**:3640–5.
64. Laali KK, Gettwert VJ. Fluorodediazoniation in ionic liquid solvents: new life for the Balz–Schiemann reaction. *J Fluor Chem* 2001;**107**:31–4.
65. Laali KK, Borodkin GI. First application of ionic liquids in electrophilic fluorination of arenes; Selectfluor™ (F-TEDA-BF_4) for "green" fluorination. *J Chem Soc Perkin Trans* 2002;**2**:953–7.
66. Lourie LF, Serguchev YA, Ponomarenko MV, Rusanov EB, Vovk MV, Ignat'ev NV. Electrophilic fluorocyclization of unsaturated alcohols in ionic liquids. *Tetrahedron* 2013;**69**:833–8.
67. Reddy AS, Laali KK. Mild and selective α-fluorination of carbonyl compounds (ketones, 1,3-diketones, β-ketoesters, α-nitroketones, and β-ketonitriles) with Selectfluor (F-TEDA-BF_4) in imidazolium ILs [BMIM/PF_6 or BMIM/NTf_2] with Brønsted-acidic IL [PMIM(SO_3H)/OTf] as promoter. *Tetrahedron Lett* 2015;**56**:5495–9.
68. Pégot B, Macé Y, Urban C, Diter P, Blazejewski J-C, Magnier E. Ionic liquids as new media for electrophilic trifluoromethylation reactions. *J Fluor Chem* 2012;**134**:156–9.
69. Hubbard A, Okazaki T, Laali KK. Chlorination of aromatics with trichloroisocyanuric acid (TCICA) in Brønsted-acidic imidazolium ionic liquid [BMIM(SO3H)][OTf]: an economical, green protocol for the synthesis of chloroarenes. *Aust J Chem* 2007;**60**:923–7.
70. Bailey L, Handy ST. Aromatic iodination using *N*-iodosaccharin in room temperature ionic liquids. *Tetrahedron Lett* 2011;**52**:2413–4.
71. Pavlinac J, Laali KK, Zupan M, Stavber S. Iodination of organic compounds with elemental iodine in the presence of hydrogen peroxide in ionic liquid media. *Aust J Chem* 2008;**61**:946–55.
72. Vražič D, Jereb M, Laali KK, Stavber S. Brønsted acidic ionic liquid accelerated halogenation of organic compounds with *N*-halosuccinimides (NXS). *Molecules* 2013;**18**:74–96.
73. Wang H, Wen K, Nurahmat N, Shao Y, Zhang H, Wei C, Li Y, Shen Y, Sun Z. Regioselective chlorination and bromination of unprotected anilines under mild conditions using copper halides in ionic liquids. *Beilstein J Org Chem* 2012;**8**:744–8.
74. Prebil R, Laali KK, Stavber S. Metal and H_2O_2 free aerobic oxidative aromatic halogenation with $[RNH_3^+][NO_3^-]$/HX and [BMIM(SO3H)][NO3)x(X)y] (X = Br, Cl) as multifunctional ionic liquids. *Org Lett* 2013;**15**:2108–11.
75. Primerano P, Cordaro M, Scala A. Aromatic iodination using *N*-iodosaccharin in room temperature ionic liquids. *Tetrahedron Lett* 2013;**54**:4061–3.
76. Li J, Yang W, Yang S, Huang L, Wu W, Sun Y, Jiang H. Palladium-catalyzed cascade annulation to construct functionalized β- and γ-lactones in ionic liquids. *Angew Chem Int Ed* 2014;**53**:7219–22.
77. Valizadeh H, Noorshargh S, Shomali A. Selective iododeamination of electron-deficient anilines using task-specific nitrite- and iodide-based ionic liquids. *Synth Commun* 2015;**45**:1094–100.
78. Hubbard A, Okazaki T, Laali KK. Halo- and azidodediazoniation of arenediazonium tetrafluoroborates with trimethylsilyl halides and trimethylsilyl azide and Sandmeyer-type bromodediazoniation with Cu(I)Br in [BMIM][PF_6] ionic liquid. *J Org Chem* 2008;**73**:316–9.
79. Laali KK, Okazaki T, Bunge S. *N*-(trifluoromethylsulfonyl)aryloxytrifluoromethylsulfoximines [ArO–SO(CF_3)=NTf] and *N*-aryltriflimides Ar–N(Tf)$_2$ by thermal and photolytic dediazoniation of [ArN_2][BF_4] in [BMIM][Tf_2N] ionic liquid: Exploiting the ambident nucleophilic character of a "nonnucleophilic" anion. *J Org Chem* 2007;**72**:6758–62.
80. Hajipour AR, Seddighi M. Application of [Hcpy]HSO4 Brønsted acidic ionic liquid for the synthesis of aryl iodides from aromatic amines. *Org Prep Proc Int* 2011;**43**:292–6.
81. Isaad J. Acidic ionic liquid supported on silica-coated magnetite nanoparticles as a green catalyst for one-pot diazotization–halogenation of the aromatic amines. *RSC Adv* 2014;**4**:49333–41.

82. Valizadeh H, Shomali A. A new nitrite ionic liquid (IL-ONO) as a nitrosonium source for the efficient diazotization of aniline derivatives and *in-situ* synthesis of azo dyes. *Dyes Pigment* 2012;**92**:1138–43.
83. Aridoss G, Laali KK. Building heterocyclic systems with $RC(OR)_2^+$ carbocations in recyclable Brønsted acidic ionic liquids: facile synthesis of 1-substituted 1*H*-1,2,3,4-tetrazoles, benzazoles and other ring systems with $CH(OEt)_3$ and $EtC(OEt)_3$ in $[EtNH_3][NO_3]$ and $[PMIM(SO_3H)][OTf]$. *Eur J Org Chem* 2011:2827–35.
84. Hasaninejad A, Zare A, Shekouhy M, Rad JA. Catalyst-free one-pot four component synthesis of polysubstituted imidazoles in neutral ionic liquid 1-butyl-3-methylimidazolium bromide. *J Comb Chem* 2010;**12**:844–9.
85. Safaei S, Mohammadpoor-Baltork I, Khosropour AR, Moghadam M, Tangestaninejad S, Mirkhania V. Diastereoselective synthesis of pyrazolines using a bifunctional Brønsted acidic ionic liquid under solvent-free conditions. *Adv Synth Catal* 2012;**354**:3095–104.
86. Bharate JB, Bharate SB, Vishwakarma RA. Metal-free, ionic liquid-mediated synthesis of functionalized quinolines. *ACS Comb Sci* 2014;**16**:624–30.
87. Gao X, Yu B, Yang Z, Zhao Y, Zhang H, Hao L, Han B, Liu Z. Ionic liquid-catalyzed C–S bond construction using CO_2 as a C1 building block under mild conditions: a metal-free route to synthesis of benzothiazoles. *ACS Catal* 2015;**5**:6648–52.
88. Forsyth SA, Gunaratne HQN, Hardacre C, McKeown A, Rooney DW. One-pot multistep synthetic strategies for the production of Fenpropimorph using an ionic liquid solvent. *Org Process Res Dev* 2006;**10**:94–102.
89. Feng J, Ying-jie L, Hai-feng D, Zong-han L, Jun-gang C, Da-peng L. Brønsted acidic ionic liquids: novel efficient, cost-effective and reusable solvent-catalysts for Ritter reaction. *Chem Res Chin U* 2010;**26**(3):384–8.
90. Kalkhambkar RG, Waters SN, Laali KK. Highly efficient synthesis of amides via Ritter chemistry with ionic liquids. *Tetrahedron Lett* 2011;**52**:867–71.
91. Khodaei MM, Nazari E. *n*-Butylammonium carboxylates/Tf_2O: ionic liquid based systems for the synthesis of unsymmetrical imides via a Ritter-type reaction. *Tetrahedron Lett* 2012;**53**:2881–4.
92. Epishina MA, Kulikov AS, Ignat̆ev NV, Schulte M, Makhova N. The first example of the Schmidt reaction in ionic liquids. *Mendeleev Commun* 2010;**20**:335–6.
93. Nandi GC, Laali KK. *n*-Butylammonium carboxylates/Tf_2O: ionic liquid based systems for the synthesis of unsymmetrical imides via a Ritter-type reaction. *Tetrahedron Lett* 2013;**54**:2177–9.
94. Valizadeh H, Gholipour H, Ahmadi M, Vaghefi S. $[Bmim]N_3$ as an efficient reagent for the Schmidt reactions of ketones, arylaldehydes and aromatic carboxylic acids. *J Iran Chem Soc* 2014;**11**:1287–94.
95. Bellina F, Chiappe C. The Heck reaction in ionic liquids: progress and challenges. *Molecules* 2010;**15**:2211–45.
96. Prechtl MHG, Scholten JD, Dupont J. Carbon-carbon cross coupling reactions in ionic liquids catalyzed by palladium metal nanoparticles. *Molecules* 2010;**15**:3441–61.
97. Lamblin M, Hardy LN, Hierso J-C, Fouquet E, Felpin F-X. Recyclable heterogeneous palladium catalysts in pure water: sustainable developments in Suzuki, Heck, Sonogashira and Tsuji–Trost reactions. *Adv Synth Catal* 2010;**352**:33–79.
98. Santos CIM, Barata JFB, Faustino MAF, Lodeiro C, Neves MGPMS. Revisiting Heck–Mizoroki reactions in ionic liquids. *RSC Adv* 2013;**3**:19219–38.
99. Wang R, Twamley B, Shreeve JM. A highly efficient, recyclable catalyst for C–C coupling reactions in ionic liquids: Pyrazolyl-functionalized N-heterocyclic carbene complex of palladium(II). *J Org Chem* 2006;**71**:426–9.
100. Sawant AD, Raut DG, Darvatkar NB, Desai UV, Salunkhe MM. An ester appending multifunctional ionic liquid for Pd(II) catalyzed Heck reaction. *Catal Commun* 2010;**12**:273–6.
101. Kalkhambkar RG, Laali KK. Arenediazonium salts immobilized in imidazolium ionic liquids as electrophilic partners in the $Pd(OAc)_2$-catalyzed Matsuda–Heck arylation. *Tetrahedron Lett* 2011;**52**:1733–7.
102. Gaikwad DS, Park YK, Pore DM. A novel hydrophobic fluorous ionic liquid for ligand-free Mizoroki–Heck reaction. *Tetrahedron Lett* 2012;**53**:3077–81.
103. Nowrouzi N, Tarokh D, Motevalli S. An ester appending multifunctional ionic liquid for Pd(II) catalyzed Heck reaction. *J Mol Catal A* 2014;**385**:13–7.
104. Wang Y-P, Lee HM. An ester appending multifunctional ionic liquid for Pd(II) catalyzed Heck reaction. *J Organomet Chem* 2015;**791**:90–8.
105. Hierso J-C, Boudon J, Picquet M, Meunier P. The first catalytic method for Heck alkynylation of unactivated aryl bromides (copper-free Sonogashira) in an ionic liquid: 1 mol-% palladium/triphenylphosphane/pyrrolidine in $[BMIM][BF_4]$ as a simple, inexpensive and recyclable system. *Eur J Org Chem* 2007:583–7.

106. Hierso J-C, Beauperin M, Saleh S, Job A, Andrieu J, Picquet M. Uncommon perspectives in palladium- and copper-catalyzed arylation and heteroarylation of terminal alkynes following Heck or Sonogashira protocols: interactions copper/ligand, formation of diynes, reaction and processes in ionic liquids. *C R Chim* 2013;**16**:580–96.
107. Harjani JR, Abraham TJ, Gomez AT, Garcia MT, Singer RD, Scammells PJ. Sonogashira coupling reactions in biodegradable ionic liquids derived from nicotinic acid. *Green Chem* 2010;**12**:650–5.
108. Iranpoor N, Firouzabadi H, Ahmadi Y. Carboxylate-based, room-temperature ionic liquids as efficient media for palladium-catalyzed homocoupling and Sonogashira–Hagihara reactions of aryl halides. *Eur J Org Chem* 2012:305–11.
109. Reddy AS, Laali KK. Sonogashira cross-coupling in a designer ionic liquid (IL) without copper, external base, or additive, and with recycling and reuse of the IL. *Tetrahedron Lett* 2015;**56**:4807–10.
110. Gauchot V, Kroutil W, Schmitzer AR. Highly recyclable chemo-/biocatalyzed cascade reactions with ionic liquids: one-pot synthesis of chiral biaryl alcohols. *Chem Eur J* 2010;**16**:6748–51.
111. Li L, Wang J, Wu T, Wang R. Click Ionic Liquids: a family of promising tunable solvents and application in Suzuki–Miyaura cross-coupling. *Chem Eur J* 2012;**18**:7842–51.
112. Boruah PR, Koiri MJ, Bora U, Sarma D. A new recyclable/reusable ionic liquid/LiCl system for Suzuki–Miyaura cross coupling reactions. *Tetrahedron Lett* 2014;**55**:2423–5.
113. Xu J, Zhai X, Wub X, Zhang YJ. Allyl–aryl coupling of allylic carbonates with arylboronic acids catalyzed by palladium nanoparticles in ionic liquid. *Tetrahedron* 2015;**71**:1712–7.
114. Wang HB, Hu Y-L, Li D-J. Facile and efficient Suzuki–Miyaura coupling reaction of aryl halides catalyzed by $Pd_2(dba)_3$ in ionic liquid/supercritical carbon dioxide biphasic system. *J Mol Liq* 2016;**218**:429–33.
115. Louaisil N, Pham PD, Boeda F, Faye D, Castanet A-S, Legoupy S. Ionic liquid supported organotin reagents: green tools for Stille cross-coupling reactions with brominated substrates. *Eur J Org Chem* 2011:143–9.
116. Monopoli A, Cotugno P, Palazzo G, Ditaranto N, Mariano B, Cioffi N, Ciminale F, Naccia A. Ullmann homocoupling catalyzed by gold nanoparticles in water and ionic liquid. *Adv Synth Catal* 2012;**354**:2777–88.
117. Kalkhambkar RG, Laali KK. $Pd(OAc)_2$-catalyzed cross-coupling of polyfluoroarenes with simple aromatics in imidazolium ionic liquids (ILs) without oxidant and additive and with recycling/reuse of the IL. *Tetrahedron Lett* 2011;**52**:5525–9.
118. Savanur HM, Kalkhambkar RG, Laali KK. $Pd(OAc)_2$ catalyzed homocoupling of arenediazonium salts in ionic liquids: synthesis of symmetrical biaryls. *Tetrahedron Lett* 2016;**57**:663–7.
119. Kalkhambkar RG, Laali KK. $Pd(OAc)_2$ catalyzed homocoupling of arenediazonium salts in ionic liquids: synthesis of symmetrical biaryls. *Tetrahedron Lett* 2012;**53**:4212–5.
120. Li J, Zhu Z, Yang S, Zhang Z, Wu W, Jiang H. Palladium-catalyzed tandem annulation: a strategy to construct 2,3-difunctionalized benzofuran derivatives in ionic liquids. *J Org Chem* 2015;**80**:3870–9.
121. Haumann M, Riisager A. Hydroformylation in room temperature ionic liquids (RTILs): catalyst and process developments. *Chem Rev* 2008;**108**:1474–97.
122. Jain N, Kumar A, Chauhan S, Chauhan SMS. Chemical and biochemical transformations in ionic liquids. *Tetrahedron* 2005;**61**:1015–60.
123. Jin X, Yang D, Xu X, Yang Z. Super long-term highly active and selectivehydroformylation in a room temperature-solidifiable guanidinium ionic liquid with a polyether tag. *Chem Commun* 2012;**48**:9017–9.
124. Jin X, Zhao K, Cui F, Kong F, Liu Q. Highly effective tandem hydroformylation–acetalization of olefins using a long-life Brønsted acid–Rh bifunctional catalyst in ionic liquid–alcohol systems. *Green Chem* 2013;**15**:3236–42.
125. Norinder J, Rodrigues C, Borner A. Tandem hydroformylation–acetalization with a ruthenium catalyst immobilized in ionic liquids. *J Mol Catal a Chem* 2014;**391**:139–43.
126. Walter S, Haumann M, Wasserscheid P, Hahn H, Franke R. *n*-Butane carbonylation to *n*-pentanal using a cascade reaction of dehydrogenation and SILP-catalyzed hydroformylation. *AIChE* 2015;**61**:893–7.
127. Jakuttis M, Schönweiz A, Werner S, Franke R, Wiese K-D, Haumann M, Wasserscheid P. Rhodium–phosphite SILP catalysis for the highly selective hydroformylation of mixed C_4 feedstocks. *Angew Chem Int Ed* 2011;**50**:4492–5.
128. Akbari J, Hekmati M, Sheykhan M, Heydari A. Guanidine derived ionic liquids: catalyst free medium for N-Formylation of amines. *ARKIVOC* 2009;**xi**:123–9.
129. Niknam K, Zolfigol MA, Saberi D, Khonbazi M. 1-Butyl-3-methylimidazolium hydrogen sulfate [bmim]HSO_4: an efficient reusable acidico Ionic liquid for the formylation of alcohols. *Chin J Chem* 2009;**27**:1548–52.

130. Choi Y-S, Shim YN, Lee J, Yoon JH, Hong CS, Cheong M, Kim HS, Jang HG, Lee JS. Ionic liquids as benign catalysts for the carbonylation of amines to formamides. *Appl Catal a Gen* 2011;**404**:87–92.
131. Majumdar S, De J, Hossain J, Basak A. $Pd(OAc)_2$ catalyzed homocoupling of arenediazonium salts in ionic liquids: synthesis of symmetrical biaryls. *Tetrahedron Lett* 2013;**54**:262–6.
132. Hao L, Zhao Y, Yu B, Yang Z, Zhang H, Han B, Gao X, Liu Z. Imidazolium-based ionic liquids catalyzed formylation of amines using carbon dioxide and phenylsilane at room temperature. *ACS Catal* 2015;**5**(9):4989–93.
133. Chiappe C, Malvaldi M, Pomelli CS. The solvent effect on the Diels–Alder reaction in ionic liquids: multiparameter linear solvation energy relationships and theoretical analysis. *Green Chem* 2010;**12**:1330–9.
134. Song CE, Shim WH, Roh EJ, Lee S-G, Choi JH. Ionic liquids as powerful media in scandium triflate catalysed Diels–Alder reactions: significant rate acceleration, selectivity improvement and easy recycling of catalyst. *Chem Commun* 2001:1122–3.
135. Kumar A, Pawar SS. Converting *exo*-selective Diels–Alder reaction to *endo*-selective in chloroloaluminate ionic liquids. *J Org Chem* 2004;**69**:1419–20.
136. Silvero G, Arévalo MJ, Bravo JL, Ávalos M, Jiménez JL, López I. An in-depth look at the effect of Lewis acid catalysts on Diels–Alder cycloadditions in ionic liquids. *Tetrahedron* 2005;**61**:7105–11.
137. Takahashi K, Nakano H, Fujita R. Reuse of chiral cationic Pd–phosphinooxazolidine catalysts in ionic liquids: highly efficient catalytic asymmetric Diels–Alder reactions. *Chem Commun* 2007:263–5.
138. Stefaniak W, Janus E, Milchert E. Diels–Alder Reaction of cyclopentadiene and alkyl acrylates in the presence of pyrrolidinium ionic liquids with various anions. *Catal Lett* 2011;**141**:742–7.
139. Erfurt K, Wandzik I, Walczak K, Matuszek K, Chrobok A. Hydrogen-bond-rich ionic liquids as effective organocatalysts for Diels–Alder reactions. *Green Chem* 2014;**16**:3508–14.
140. Zulfiqar F, Kitazume T. One-pot aza-Diels–Alder reactions in ionic liquids. *Green Chem* 2000;**2**:137–9.
141. Yadav JS, Reddy BVS, Reddy JSS, Rao RS. Aza-Diels–Alder reactions in ionic liquids: a facile synthesis of pyrano- and furanoquinolines. *Tetrahedron* 2003;**59**:1599–604.
142. Boulaire VL, Grée R. Wittig reactions in the ionic solvent [bmim][BF_4]. *Chem Commun* 2000:2195–6.
143. Yoshino H, Kobata M, Yamamoto Y, Oshima K, Matsubara S. Wittig type methylenation of ketones with bis(iodozincio)methane and ionic liquid. *Chem Lett* 2004;**33**:1224.
144. Sun W, Kühn FE. Efficient aldehyde olefination reactions catalyzed by an iron porphyrin complex in an ionic liquid. *Tetrahedron Lett* 2004;**45**:7415–8.
145. Ramnial T, Taylor SA, Bender ML, Gorodetsky B, Lee PTK, Dickie DA, McCollum BM, Pye CC, Walsby CJ, Clyburne JAC. Carbon-centered strong bases in phosphonium ionic liquids. *J Org Chem* 2008;**73**:801–12.
146. Fresneda PM, Blázquez JA. Synthesis of β-carbolines using microwave-assisted aza-Wittig methodology in ionic liquids. *Tetrahedron Lett* 2012;**53**:2618–21.

CHAPTER

3.17

Solvent-Free Synthesis of Nanoparticles

Shainaz Landge, Debanjana Ghosh, Karelle Aiken

Georgia Southern University, Statesboro, GA, United States

3.17.1 INTRODUCTION

In the past few decades, nanosize (10^{-9} m) has created its own unique niche in multidisciplinary fields. The "nanoworld" encompasses nanoparticles, nanotechnology, nanoscience, and much more. They have gained importance due to their potential to bridge the gap between the atomic/molecular states and the bulk phases through novel physical, chemical, and optoelectronic properties.[1] Particles ranging in size from 1 to 100 nm are utilized in many applications, for example as catalysts,[2] semiconductors,[3] magnetic materials,[4] and in medicinal applications.[5] Since nanoparticles are of great interest, their syntheses and properties are well studied. The size of nanoparticles will affect their physical and chemical properties and are thus manipulated for functional needs. Reagents (reducing agents, stabilizing agents), solvents, additives, and other reaction conditions are strong determinants of particle size.[6]

Typically nanoparticles are synthesized using a variety of methods. In the chemical method, the respective metal salts or composites are mixed with solvent or surfactants and reduced to generate the desired particles.[6] The physical approach tends to use evaporation, condensation, direct heating, or laser ablation. Most of these techniques require large volumes of solvent to carry out the processes. As such, solvent-free methods provide a greener and more economical route to these materials. The present analysis targets solvent-free synthesis of nanoparticles and nanocomposites dispersed in the matrices such as ceramics and polymers.

Solvent-free synthesis of nanoparticles can be broadly divided into (1) mechanochemistry and (2) thermal treatment. The mechanical approach uses pressure and grinding, most commonly ball milling and mortar and pestle. Thermal treatment involves decomposition/thermolysis by conventional and microwave (MW) heating.

Decomposition process is well explored and is the most commonly used method for solvent-free synthesis of nanomaterials. High temperature is required to produce stable monodispersed nanoparticles. It is further categorized into (1) direct heating of the precursor

Green Chemistry
http://dx.doi.org/10.1016/B978-0-12-809270-5.00022-4

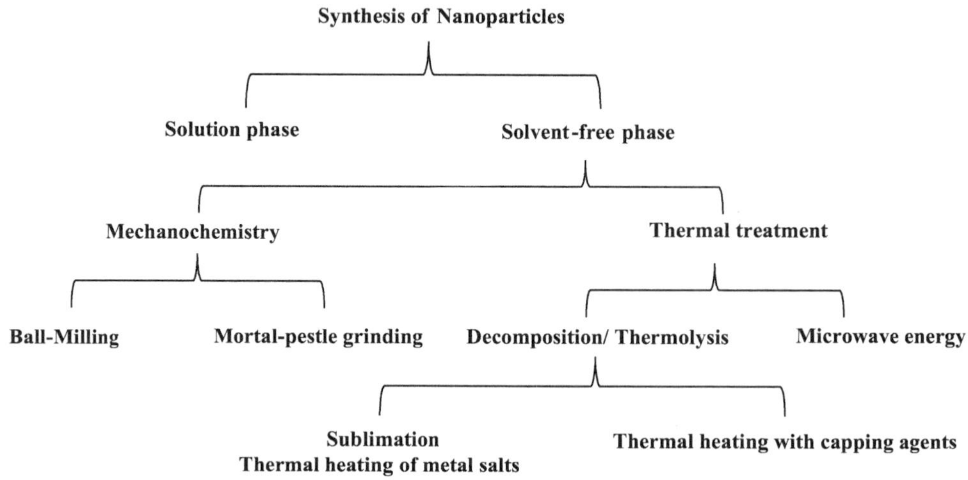

SCHEME 3.17.1 Methods used for the solvent-free synthesis of nanoparticles.

metal salts, e.g., metal acetate and (B) heating of the precursor salts using capping or stabilizing agents such as carboxylate, oleate, and amines.

MW energy as a heating source is still a less-explored area in solvent-free nanoparticle synthesis. It is generally referred to as a combustion method and uses high temperature to heat metal salts. Scheme 3.17.1 depicts the methods used for the solvent-free synthesis of nanoparticles. The common techniques used to characterize the nanoparticles are scanning electron microscopy (SEM), transmission electron microscopy (TEM), X-ray powder diffraction (XRD), X-ray photoelectron spectroscopy (XPS), thermogravimetric analysis (TGA), field emission scanning electron microscopy, etc., to name a few.

The solvent-free synthesis of nanomaterials has paved the way for an ecofriendly technique pertaining to the interests in green chemistry. Based on this viewpoint, several researchers are developing innovative synthetic routes. A few of the advantages of solvent-free synthesis over using solvent-based reactions are as follows:

1. Atom economy: In majority of cases, most components of reactants and reagents are fully converted to products and no solvent is utilized, hence, high atom economy is maintained.
2. Lack of solvent: Most solvents are harmful for the environment, and the solvent-based reactions tend to use excessively large volumes, which has a negative impact on the ecosystem.
3. Catalysis: Use of capping agents helps to inhibit nanoparticle aggregation or overgrowth and also maintains the structural characteristics of the synthesized materials.
4. Reduce by-products: The thermolysis and mechanochemistry methods control the size of the nanoparticles with less impurity in the synthesized product; allowing direct access to extremely clean products without the need for additional purification.

The following discussion considers the synthesis of nanoparticles particularly in solvent-free conditions. Section 3.17.2 addresses mechanochemistry approach and Section 3.17.3 is devoted to thermolysis methods.

3.17.2 MECHANOCHEMISTRY

Ball milling or mortar-and-pestle grinding are the most well-known techniques in mechanochemistry. However, a mechanochemical procedure is any process that involves grinding of solids to produce a chemical reaction or physical change in a substance.[7] Simply put, mechanochemistry is a solvent-free, mechanical activation process. It has been utilized in many areas, and of course, its emergence in nanoparticle/material synthesis has provided a more economical, environment-friendly alternative to traditional, solvent-based methods. The examples that follow illustrate the variety of ways in which mechanochemistry has been used to enhance the properties of nanoparticles and provide more practical, greener, and scalable synthetic avenues for the production of these materials.

3.17.2.1 Ball Milling and Rheomixing

Gold nanoparticles (AuNP) have become a hot topic, so much so, that AuNP is the most recognized nanosystem in the science arena. The expanse of areas with interest in AuNP synthesis is incredibly wide—a few of these are catalysis,[8,9] sensors,[10] and medicine.[11] Reaction scalability is an ongoing challenge for the AuNP field, which is limited by the use of large volumes of solvents and the need for additives.[12] The first mechanochemical, solvent-free approach to AuNP was reported by Debnath and colleagues.[13] The researchers obtained polydispersed samples in the range of 6–30 nm by milling potassium gold(III) chloride, $KAuCl_4$, with sodium borohydride, $NaBH_4$, and poly(vinylpyrrolidone). Rak et al. later reported a "tunable" modification to Debnath's procedure.[14] In the latter work, particle size was better controlled by the type and ratio of the ligand and milling time. The authors explored the use of amines, pyridines, imidazoles, and a carboxylic acid (citric acid) and found the most promising results with the amines. Very small diameters and narrow monodispersity were obtained from chloroauric acid ($HAuCl_4$). The optimal ratio for $HAuCl_4$-to-ligand was 1:5 at a milling speed of 29.5 Hz for 90 min. Various amines yielded an average size of 4.2 nm (pentadecylamine), 1.8 nm (hexadecylamine), 1.5 nm (heptadecylamine), and 1.3 nm (octadecylamine) (Scheme 3.17.2).

Apart from pentadecylamine, which melted during the milling process and produced substantially larger particles, the amines exhibited a clear trend: longer carbon chains gave rise to smaller particles. Importantly, the researchers noted that ultrasmall diameters like

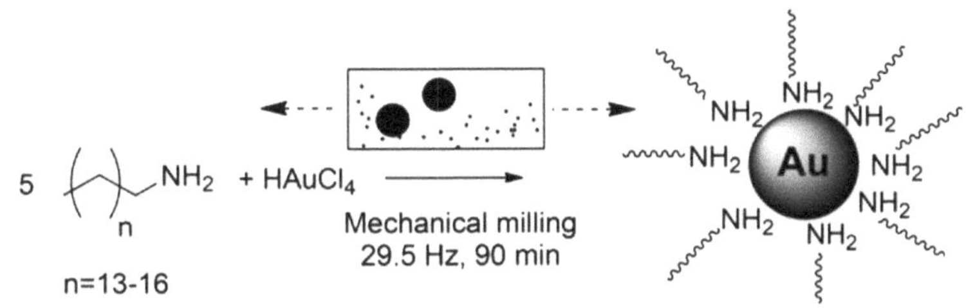

SCHEME 3.17.2 Mechanosynthesis of amine gold nanoparticles. *Reproduced from Rak MJ, Saade NK, Friscic T, Moores A. Mechanosynthesis of ultra-small monodisperse amine-stabilized gold nanoparticles with controllable size.* Green Chem *2014;**16**:86–89, with permission of The Royal Society of Chemistry.*

those in the 1.5 nm range have not been achieved with the traditional solvent-based procedures. Up to 863 mg of the material was made with Rak's "tunable" process.[14] In addition to using common tools such as ultraviolet (UV)-vis, XRD, XPS and TEM were also used for analyses. This was the first report in which matrix-assisted laser desorption/ionization-time of flight mass spectrometry (MS) was employed to assess particle size (Fig. 3.17.1).[14] Fragments that differed in mass by one Au atom, 197.09 units, were observed in the spectra.

The researchers also noted that the type of milling ball was important.[14] Interestingly, Rak and colleagues proved that the steel balls acted as reductants in the reactions. XPS analyses of the nanoparticles confirmed that they were not contaminated with iron. Two 7-mm milling balls were used per 200 mg of Au salt/amine mixture. The reaction was run for 90 min at 30 Hz in a 10-mL stainless steel milling jar. A control study with aluminum balls failed to produce any trace of AuNP.

The most effective catalyst for the conversion of CO_2 to methanol is the copper-zinc system (CuO/ZnO) in a hydrogenation reaction.[15] Carbon dioxide is a major greenhouse gas, and both methanol and CO_2 are inexpensive, renewable resources for a variety of applications.[16] The addition of other metals such as aluminum (Al) can improve the performance of the Cu-Zn catalyst.[15,17–20] Therefore, Lei and coworkers targeted the synthesis of CuO/ZnO/Al_2O_3 (CZA) in a solvent-free, milling-combustion procedure.[21] The researchers employed oxalic acid, citric acid, or urea as fuel to control and enhance the catalyst's physicochemical properties. Previously, less direct methods for the synthesis of the Cu-based systems involved sol-gel,[22] impregnation[18] and coprecipitation.[18,19,23–27] Lei et al. found

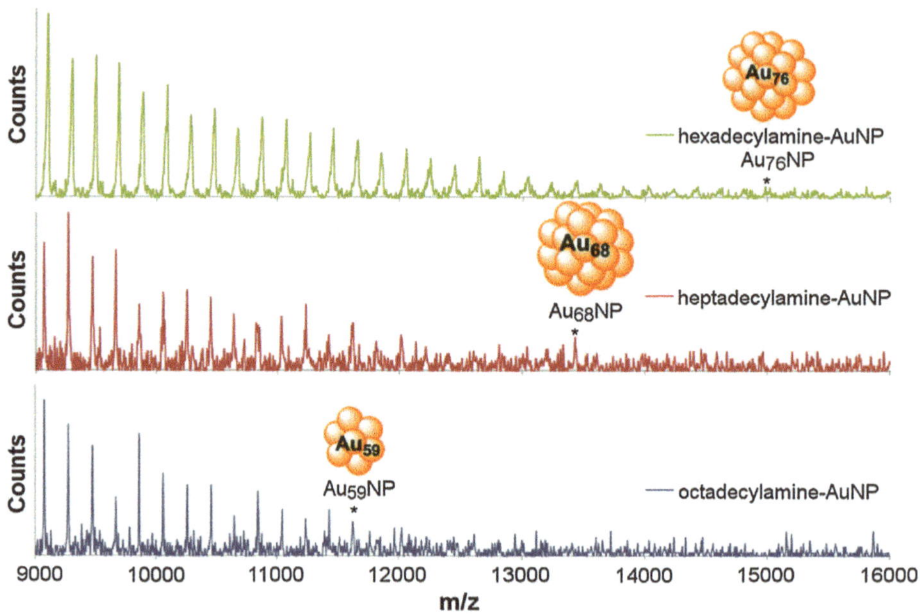

FIGURE 3.17.1 Matrix-assisted laser desorption/ionization-time of flight measurements obtained with hexadecylamine-, heptadecylamine-, and octadecylamine-stabilized gold nanoparticles (AuNPs). *Reproduced from Rak MJ, Saade NK, Friscic T, Moores A. Mechanosynthesis of ultra-small monodisperse amine-stabilized gold nanoparticles with controllable size.* Green Chem *2014;**16**:86–89, with permission of The Royal Society of Chemistry.*

that the choice and equivalents of acid affected the conditions for combustion and the properties of the CZA. They determined that their CZA-C-1.25 was the most effective catalyst for the reduction of CO_2 to methanol with "-C-" symbolizing "citric acid" and "1.25" indicating the molar ratio of citric acid to the salts. The highest yield was 10.3%, a 16.2% conversion of CO_2 to methanol. In general, the solvent-free method with 1.25 equivalents of citric acid resulted in greater surface area and increased CuO-dispersion and Cu-exposure on the surface of the material compared with other results.

The synthesis of the catalysts began with the milling of a premixed, dried powder of $Cu(NO_3)\cdot 3H_2O$, $Zn(OAc)_2\cdot 2H_2O$, $Al(OAc)_3$, and the acid, which the authors called fuel, in an agate vial under air.[21] Reaction conditions involved a 10:1 weight ratio of agate milling ball to powder, a speed of 300 rpm, and a reaction time of 2 h. Heating the milled powders at 450 °C under a flow of air for 4 h yielded the final CZA products. Reduction of these materials with hydrogen gas, the conversion of CuO to Cu(0), activated the catalysts for the CO_2-to-methanol reaction. Six samples were prepared with acid-to-salts molar ratios of 0.50, 0.75, 1.00, and 1.25 for citric acid, and a 1.00 M ratio each for oxalic acid and urea. Samples prepared with urea and oxalic acid suffered from significant agglomeration, lower surface area, and larger CuO particle size compared with their citric acid counterpart. The CZA prepared with one equivalent of citric acid showed uniform, disclike nanoparticles of 170 nm. Fig. 3.17.2 shows the SEM images of the catalysts prepared with 1 equivalent of

FIGURE 3.17.2 Scanning electron microscopic images of CuO/ZnO/Al2O3 (CZA) catalysts prepared with different acids (A) CZA-C-1.00; (B) CZA-O-1.00; (C) CZA-U-1.00. *Reprinted from Lei H, Hou Z, Xie J. Hydrogenation of CO_2 to CH_3OH over $CuO/ZnO/Al_2O_3$ catalysts prepared via a solvent-free routine.* Fuel 2016;**164**:191–8, *with permission from Elsevier.*

the different acids. Changing the equivalents of citric acid from 0.5 to 1.25 resulted in decreased particle sizes for CuO in the initial product, 16.5−8.1 nm, and for Cu(0) in the activated catalyst, 25.1−13.6 nm. Additionally, increasing the equivalents of citric acid also improved the overall CZA surface area, 26.7−77.6 m^2/g; the Cu(0)-dispersion, 3.6%−7.3%; and the exposed Cu(0) surface area, 24.4−49.4 m^2/g.[21]

For comparison, the researchers investigated a CZA sample made by more conventional, solvent-based means through carbonate coprecipitation. In all aspects, the citric acid, solvent-free CZA was physically and chemically superior to the coprecipitation product.

Another application for a Cu-Zn system addresses the need for earth-abundant and thus, more economical alternatives for photovoltaic materials. As such, approaches toward the synthesis of Cu_2ZnSnS_4 (CZTS) and $Cu_2ZnSn(S_{1-x}Se_x)_4$ have garnered extensive attention.[28] Due to the photoelectronic properties of quaternary CZTS, it is seen as a potential replacement for $Cu(In_{1-x}Ga_x)(S_{1-y}Se_y)_2$ (CIGSSe). CZTS is quite comparable to CIGSSe in performance. Unlike Lei's work in which they used salts,[21] procedures for obtaining the solar cell powders from a solvent-free, mechanochemical process involved elemental metals.[29−32] In contrast to solvent-based processes, the "dry" method for the photovoltaic applications affords relatively short reaction times, high yields, and no need of other additives. The first report demonstrated that the CZTS grain size could be controlled by milling time. Longer time led to smaller size.[29] For this case, it was found that the formation of nanocrystalline CZTS was completed within 20 h. Further milling for 25, 30, and 35 h produced grain sizes of 10.6, 9.2, and 8.9 nm, respectively. The investigators used a planetary ball mill (ball mills which are smaller in size than common ball mills and are used in research laboratories for grinding samples), with 50 Hz frequency (frequency refers to speed regulation), 300 rpm revolution (revolution refers to the sun wheel motion) and 600 rpm rotation (rotation refer to the motion of the milling jars), and a 5:1 weight ratio for the ball to powder. The elemental powders were added in accordance with a Cu:Zn:Sn:S ratio of 2:1:1:4. Work by Takahiro targeted the synthesis of CZTSe, $Cu_{2(1-x)}ZnSnSe_4$, particles with planetary ball milling of Cu:Zn:Sn:Se, 2($1-x$):1:1:4, at 800 rpm for a period of 20 min.[31] In this case, the milling was followed by annealing at 500 °C for 5 h under nitrogen gas.

Many solvent-free approaches, although successful in obtaining the Cu-Zn materials, were not able to report similar properties to CIGSSe.[29−31] Park and colleagues, however, developed a relatively large-scale synthesis of precursor CZTS nanocrystals, as much as 20 g, from its elemental precursors.[32] Subsequent conversion to CZTSe afforded the fabrication of highly efficient solar cells. The researchers determined that the reaction for CZTS occurred via a self-propagation process that required 2.5−3 h at 500 rpm for completion. The progress of the reaction was monitored at different time intervals using XRD and Raman spectroscopy (Fig. 3.17.3). The nanoparticles resulting from a 5 h milling were crystalline with an average particle size of 5 nm in 100-nm agglomerates.

The researchers emphasized that the material obtained in their study was far purer than those from previous works, which reported the presence of oxidized species.[29,32,33] The investigators also noted that the products were "bench-top" stable for up to 1 year.[32] Their ball milling process involved the mixing of powders of elemental Cu, Zn, Sn and sulfur in the ratio of 1.7/1.2/1.0/4.0, in that order. The reaction was carried out under argon with zirconia balls at a speed of 500 rpm for 30 min to 5 h. The weight ratio for ball:powder was kept at 5:1 or 5:2 (Scheme 3.17.3).

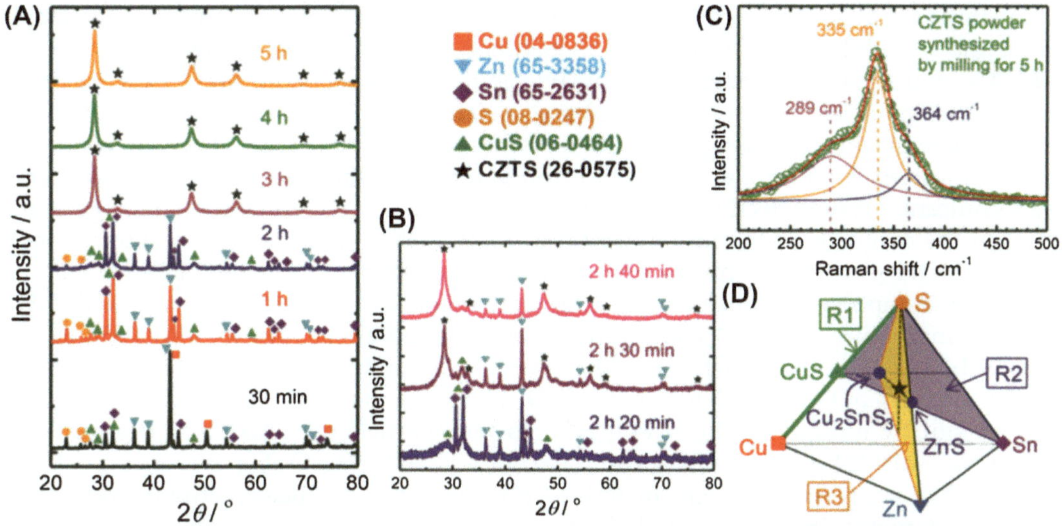

FIGURE 3.17.3 Phase evolution of Cu, Zn, Sn, and S powder mixture to Cu_2ZnSnS_4 (CZTS): X-ray diffraction patterns for the precursors milled at 500 rpm (A) for 30 min, 1, 2, 3, 4, and 5 h and (B) for 2 h 20 min, 2 h 30 min, and 2 h 40 min. (C) Raman spectrum (using 514.5 nm excitation) of CZTS powder synthesized by ball milling for 5 h; and (D) a schematic of the quaternary phase diagram and reaction pathway. Note that the numerals in the parentheses of the legend stand for the JCPDS card numbers. *Reproduced from Park B-I, Hwang Y, Lee SY, Lee J-S, Park J-K, Jeong J, Kim JY, Kim B, Cho SH, Lee DK. Solvent-free synthesis of Cu_2ZnSnS_4 nanocrystals: a facile, green, up-scalable route for low cost photovoltaic cells.* Nanoscale *2014;**6**:11703–11, with permission of The Royal Society of Chemistry.*

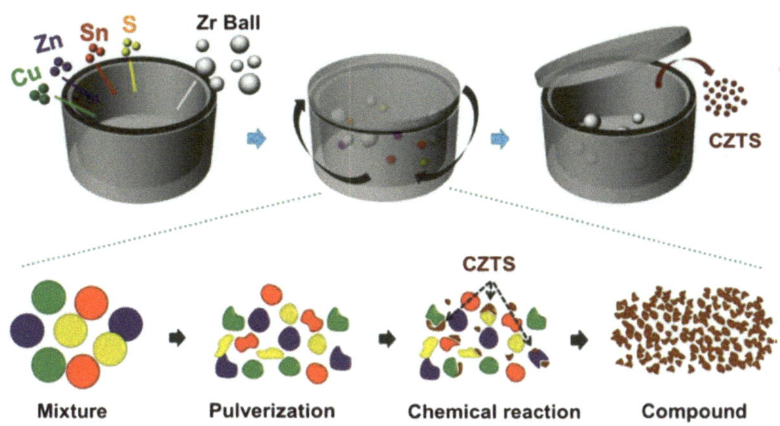

SCHEME 3.17.3 A schematic illustration of the synthetic procedure of Cu_2ZnSnS_4 (CZTS) nanocrystals by a mechanochemical process. *Reproduced from Park B-I, Hwang Y, Lee SY, Lee J-S, Park J-K, Jeong J, Kim JY, Kim B, Cho SH, Lee DK. Solvent-free synthesis of Cu_2ZnSnS_4 nanocrystals: a facile, green, up-scalable route for low cost photovoltaic cells.* Nanoscale *2014;**6**:11703–11, with permission of The Royal Society of Chemistry.*

Mallampati et al. synthesized nano-Fe/CaO, Fe/Ca/CaO, and Fe/Ca/CaO/[PO_4] composites for the immobilization and magnetic separation of Cs from soil.[34] The radioactive ^{137}Cs and ^{134}Cs, which have alarmingly high bioavailability and long half-lives,[35,36] are major contaminants from nuclear power plant disasters.[37–39] The researchers probed the effectiveness of the nanomaterials in scavenging ^{133}Cs and its radioactive counterparts.[34] The nanocomposites were synthesized from Fe, Ca, and CaO through a ball milling process. Fe/Ca/CaO/[PO_4] was significantly more effective than Fe/CaO or Fe/Ca/CaO in decontamination of soil. Fe provided the magnetic properties, while Ca or the Ca/PO_4 system served as precursors for $CaCO_3$/$Ca(OH)_2$, which performed the actual Cs encapsulation (Scheme 3.17.4).

Granular components of the nanoparticles were mixed in a planetary ball mill under Ar(g) at room temperature for 1 h.[34] The speed was maintained at 600 rpm for a rotation to revolution of 1–2 and the researchers used 1 ball per 32 g of mixture. The ratio of the components was 2:5 for Ca/CaO and 2:2:5 for Fe/Ca/CO. The third material, Fe/Ca/CaO/[PO_4], was made in situ by addition of NaH_2PO_4[PO_4] to nanoparticle soil mixtures. The maximum particle size was 221 nm for Fe/CaO and 206 nm for Fe/Ca/CaO (Fig. 3.17.4).

In another study of magnetic materials, this time for magnetorecording applications, researchers targeted iron-platinum (FePt) nanoparticles.[40] These ferromagnetic particles are quite stable under ambient conditions and can be obtained in size ranges as small as ~3 nm.

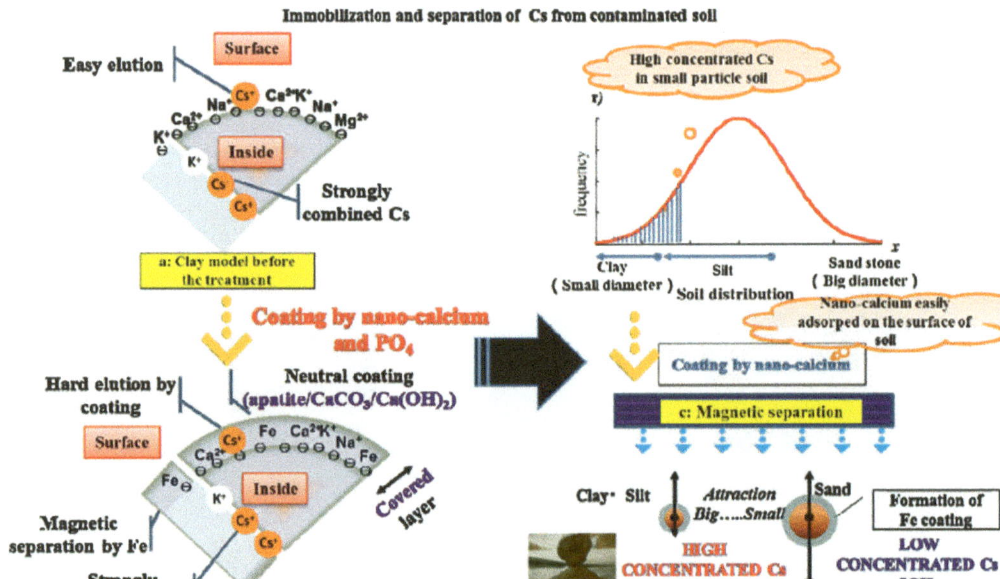

SCHEME 3.17.4 A possible pathway for immobilization with the nano-Fe/Ca/CaO/[PO_4] treatment. (A) ^{133}Cs is adsorbed onto the soil particles, (B) Cs encapsulation through the formation of immobile salts, and (C) solid (small/finer or larger/aggregate) soil fraction separation. *Reprinted from Mallampati SR, Mitoma Y, Okuda T, Simion C, Lee BK. Solvent-free synthesis and application of nano-Fe/Ca/CaO/[PO$_4$] composite for dual separation and immobilization of stable and radioactive cesium in contaminated soils.* J Hazard Mater *2015;***297***:74–82, with permission from Elsevier.*

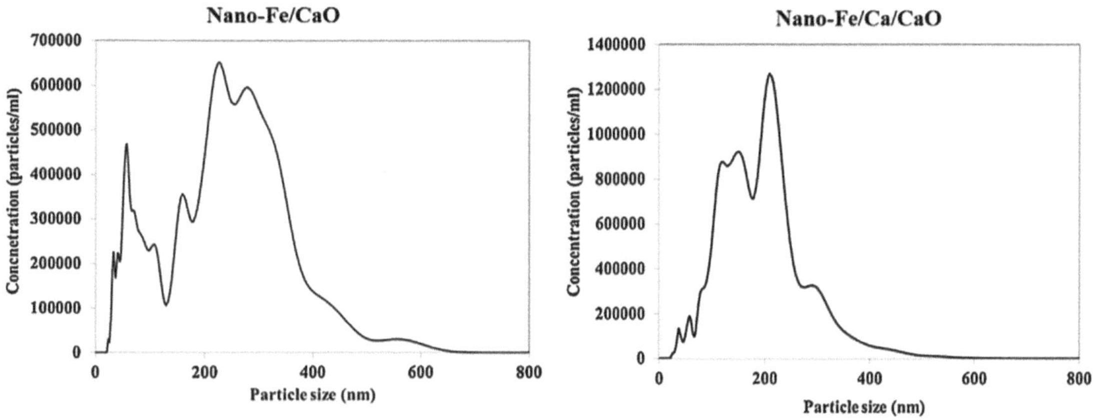

FIGURE 3.17.4 Particle size distribution of nano-Fe/CaO and nano-Fe/Ca/CaO. *Reprinted from Mallampati SR, Mitoma Y, Okuda T, Simion C, Lee BK. Solvent-free synthesis and application of nano-Fe/Ca/CaO/[PO$_4$] composite for dual separation and immobilization of stable and radioactive cesium in contaminated soils.* J Hazard Mater *2015;***297***:74–82, with permission from Elsevier.*

The two most commonly used methods for the preparation of the FePt nanoparticles are (1) physical vapor deposition (PVD)[40–42] and (2) reduction of iron pentacarbonyl, $Fe(CO)_5$, and platinum(II) acetylacetone [Pt(acac)$_2$] salts.[43] Both methods suffer from drawbacks such as high temperatures/energy consumption during the annealing process, such as 650–800 °C, the need for additives and agglomeration.[40–43] Hu et al. introduced a mechanochemical procedure that is an improvement to a previously reported aqueous synthesis by the same group.[44] In a more environment-friendly, i.e., solvent-free manner, they were able to obtain FePt nanoparticles from the planetary milling. Conditions included a speed of 250 rpm for 5 h during which the iron-platinum salt, $Fe(H_2O)_6PtCl_6$, was milled with sodium chloride (NaCl).[45] This process was followed by thermal annealing to reduce the resulting powder under a flow of 5% H_2 in Ar at a relatively low temperature of 400 °C, 250–400 °C less than the PVD and other methods. The nanoparticles were purified by magnetic separation, and washing was carried out with just water. Through varying the equivalent of NaCl (20 g NaCl to 10, 25, and 50 mg of the iron-platinum salt) particle sizes of 6.2, 12.1, and 13.2 nm, respectively, were obtained.

With commercial applications in personal care, coatings/polishing, and antimicrobial products, previous methods for the synthesis of ceramic nanoparticles are already high yielding and provide good control over particle size.[46] The typical approach is a plasma-based method that employs extremely high temperatures. As such, the drawback of this method is its high energy consumption. Shahid and coworkers have developed a mechanical shear approach for the synthesis of carboxylate-alumoxane nanoparticles.[47] In their procedure, particle size is controlled with temperature and the roller rate of a Rheomixer. It is a device that would mix paste/mortar in a closed vessel. It can measure the torque at a certain rotational speed. The authors were successful in synthesizing three composites: L-lysine, stearate, and *p*-hydroxybenzoate-alumoxane. The yields with the shear mixer were higher than those of solution-based methods, and the average particle size for reaction temperatures above 80 °C were significantly smaller, $1–2 \pm 1$ nm, compared with 25 ± 3 nm when the reaction was carried out in solution.[48] The yields obtained with shear mixing were higher than the solution yields as well: 23% versus 35% for *p*-hydroxybenzoate-alumoxane, 18% versus 23% for L-lysine-alumoxane, and 8% versus 28% for stearate-alumoxane.

In the shearing procedure, the carboxylic acid and boehmite were subjected to 60–110 rpm roller rates at temperatures of 60–180 °C for 120 min.[47] The reaction times were monitored for the different variables via measuring the torque reduction. TGA was used to assess the particle size.

3.17.2.2 Mortar and Pestle Milling

Sorbitol, obtained from the reduction of glucose, is a renewable source of highly useful compounds such as ethylene glycol, 1,3-propanediol, and glycerol.[49] The hydrogenolysis of sorbitol can be accomplished with metal catalysts promoted by acidic or basic conditions.[50] Du et al. developed a base-promoted $Ni/Mg_{1.29}Al_{0.06}O_{1.38}$ system initially prepared via thermal decomposition and reduction of nickel-magnesium-aluminum (Ni-Mg-Al) hydrotalcites.[51] Their solvent-free approach to this trimetallic catalyst involved the synthesis of Ni-Mg-Al layered double hydroxide (LDH) crystalline plates, $Ni_{3.6}/Mg_{2.4}Al_2O_{5.4}$ (solvent-free).[52] For the latter procedure an agate mortar was employed in which the nitrate salts of the metals along with sodium carbonate and sodium hydroxide were ground and thermally treated at 393 K for 24 h to produce the precursor $Ni_{3.6}Mg_{2.4}Al_2(OH)_{16}CO_3$ (solvent-free). Upon purification, calcination at 973 K resulted in $Ni_{3.6}Mg_{2.4}Al_2O_9$ (solvent-free) after 4 h. Further reduction under hydrogen gas at 1123 K gave the active catalyst denoted as $Ni_{3.6}/Mg_{2.4}Al_2O_{5.4}$ (solvent-free) (Scheme 3.17.5).

The stability and hydrogenolytic activity of $Ni_{3.6}/Mg_{2.4}Al_2O_{5.4}$ (solvent-free) was somewhat better than that of the same system made by coprecipitation method,[53,54] $Ni_{3.6}/Mg_{2.4}Al_2O_{5.4}$ (coprecipitation). The solvent-free precursor $Ni_{3.6}Mg_{2.4}Al_2(OH)_{16}CO_3$ exhibited high crystallinity. The uniform, hexagonal plates had diameters of 25–30 nm as determined by TEM (22.4 nm by XRD) with a 7.0–10.5 nm thickness. The coprecipitation precursor catalyst, on the other hand, exhibited extensive agglomeration. The researchers attributed the improved performance of the $Ni_{3.6}/Mg_{2.4}Al_2O_{5.4}$-SF to more accessible basic sites and Ni(0)-cluster hydrogenation sites, and smaller particles with uniform distribution of the active Ni(0) (Table 3.17.1).

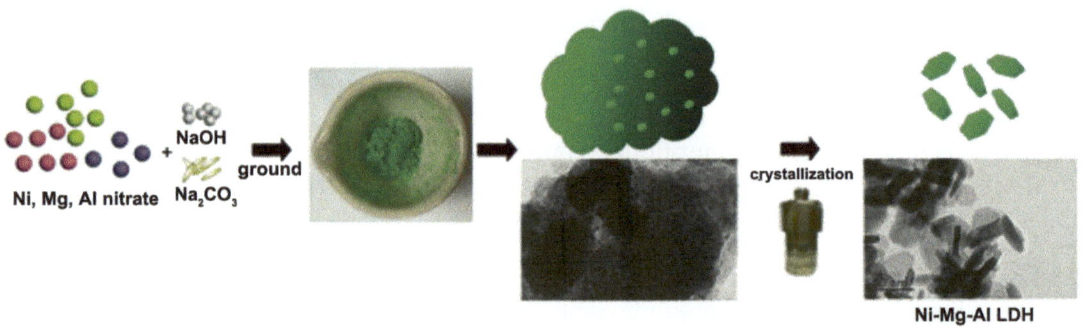

SCHEME 3.17.5 Solvent-free synthesis Ni-Mg-Al layered double hydroxide (LDH). *Reprinted from Du W, Zheng L, Li X, Fu J, Lu X, Hou Z. Plate-like Ni-Mg-Al layered double hydroxide synthesized via a solvent-free approach and its application in hydrogenolysis of D-sorbitol.* App Clay Sci *2016;***123***:166–72, with permission from Elsevier.*

TABLE 3.17.1 The Surface Physicochemical Properties of Catalyst

Catalyst	Specific Surface Area (m^2/G)	Pore Volume (cm^3/G)	Pore Diameter (nm)	Surface Basicity (μmol-CO_2, g^{-1})[a]	Surface Basicity (μmol-CO_2, m^2)[a]	Ni Dispersion (%)[b]	Ni Metallic Surface Area ($m^2/g_{\text{-Ni}}$)[b]
$Ni_{3.6}Mg_{2.4}Al_2O_9$-SF	222	0.57	13.6	371.4	1.7	38.5	215.9
$Ni_{3.6}Mg_{2.4}Al_2O_9$-CP	201	0.50	10.6	322.8	1.6	22.5	130.8

CP, coprecipitation; *SF*, solvent-free.
[a]*Calculated from desorbed amount of CO_2.*
[b]*Calculated from H_2 pulse study after the catalyst was reduced in H_2 at 1123K.*
Reprinted from Du W, Zheng L, Li X, Fu J, Lu X, Hou Z. Plate-like Ni–Mg–Al layered double hydroxide synthesized via a solvent-free approach and its application in hydrogenolysis of D-sorbitol. App Clay Sci 2016;**123**:166–72, *with permission from Elsevier.*

Similar to sorbitol, nanomaterial applications for reactions with ethanol, another renewable compound, are also extensive.[55] In particular, a substantial amount of effort is focused on the development of efficient metal catalysts for the oxidation of ethanol to acetic acid. The stability of the catalysts is a limitation as nanoparticles are prone to aggregation. To address this issue, the metals can be encapsulated in zeolites (M@zeolite).[56–58] Zhang and coworkers have developed a solvent-free process in which the gold-palladium@silica (AuPd@SiO_2), SiO_2-encapsulated bimetallic Au-Pd nanoparticles, is milled with tetrapropylammonium hydroxide.[59] The resulting mixture was thermally treated for 2–4 days at 180 °C, purified, and then calcined for 4 h at 550 °C to yield the S-1 zeolite-encapsulated particles, AuPd@S-1. The Au/Pd content of the particles were varied according to Au_xPd_y@S-1 where x and y reflected the ratio of Au to Pd in the sample, respectively. The materials had high, 349–413 m^2/g, surface areas and 0.16–0.18 cm^3/g pore volumes. The particles were robust, i.e., highly stable under oxidative conditions in the presence and absence of water. These nanoparticles also exhibited exceptional catalytic activity with 80%–100% conversion and remarkable 95%–99% selectivity for the oxidation of ethanol to acetic acid as opposed to stopping at acetaldehyde. Unlike traditional hydrothermal methods,[56–58] the solvent-free process allowed for more than 96% incorporation of the Au-Pd nanoparticle into the zeolite, i.e., the weight-percent of Au and Pd in the precursor Au_xPd_y@SiO_2 was closely maintained in the Au_xPd_y@S-1 product.[59] With the hydrothermal process, metal incorporation is relatively low, about 34%, and the by-products include heavy-metal-contaminated water. There are quite a few reports for the solvent-free encapsulation of metals with zeolites, and Zhang's work exemplifies the general process.[59–63]

In other applications with Ni, nickel hydroxide ($Ni(OH)_2$) has emerged as a promising component of supercapacitors.[64] The nanostructure is quite attractive as it possesses relatively high surface area that allows for optimal contact for electrochemical reactions with the electrolyte.[65–67] The challenge in this area lies in the scalability of the synthesis. Traditional approaches tend to employ high temperatures, elevated pressures, and difficult procedures.[65–67] In the solvent-free approach developed by Cui et al., α-$Ni(OH)_2$ nanosheets are made on a large scale using a four-step process.[68] The authors noted that the scale-up of their method is *only* limited by the size of the reaction vessel. The reaction itself is relatively short as it involves grinding nickel nitrate hexahydrate with morpholine in a mortar for 5 min followed by 30 min of "aging" (Fig. 3.17.5). At least 50 g of the nanosheets can be obtained

FIGURE 3.17.5 Optic image of the α-$Ni(OH)_2$ nanosheets produced from the solvent-free reaction. *Reprinted from Cui H, Xue J, Wang M. Synthesis of high electrochemical performance* $Ni(OH)_2$ *nanosheets through a solvent-free reaction for application in supercapacitor.* Adv Powd Tech *2015;***26***:434–8, with permission from Elsevier.*

from a one batch reaction and the resulting material displayed excellent electrochemical properties. TEM analysis showed that the material consisted of sheets and flakes, the flake nanosheets were 100 nm, laterally, and the flakes were 4–10 nm in size.

Bimetallic, core-shell or alloy, copper-based nanoparticles with noble metals exhibit superior properties in various applications when compared with the monometallic counterparts.[69–72] They have been employed in antibacterial studies,[73] electronic devices,[74] and catalyst-mediated procedures.[75] Solution-based syntheses of these nanoparticles, although efficient in producing uniformity and control over size and morphology, are limited by multistep synthesis, postreaction purifications, and challenges in scale-up. Choi and colleagues have developed a solvent-free, one-pot, scalable "mix-bake-wash" method.[76,77] Their procedure uses thermal annealing to control the structure morphology.[76] Alloys were obtained at 400 °C, whereas core-shell materials were obtained at 300 °C. Prior to heating under nitrogen atmosphere, the samples were prepared by the milling of copper(II) formate tetrahydrate ($Cu(HCO_2)_2 \cdot 4H_2O$), silver nitrate ($AgNO_3$), a 1:5-volume ratio of hexylamine/xylene, and sodium sulfate in a mortar. The sodium sulfate, mimicking the role of a capping agent, prevented aggregation (Scheme 3.17.6).

The Ag@Cu core-shell nanoparticle had a silver core and copper shell with 8.2 ± 2.1 nm shell thickness, 21.4 ± 5.9 nm core size, and an average diameter of 37.8 ± 8.3 nm. The alloys were polycrystalline, somewhat spherical particles with a diameter of 27.7 ± 5.4 nm. The core shells consisted of elemental copper and silver and traces of CuO, whereas the alloy consisted solely of the zero-valent metals.

As discussed earlier, magnetic nanoparticles are useful in many areas such as data storage, cancer treatment, drug delivery, imaging, and more.[78–80] The morphology of the nanoparticles, such as particle size, pore size, and crystallinity, will affect their magnetic properties. Various approaches for the preparation of these materials involve coprecipitation,[81] aging,[82] sol-gel,[83] ultrasound irradiation,[84] laser pyrolysis,[85] a tandem thermal decomposition-oxidation,[86] and

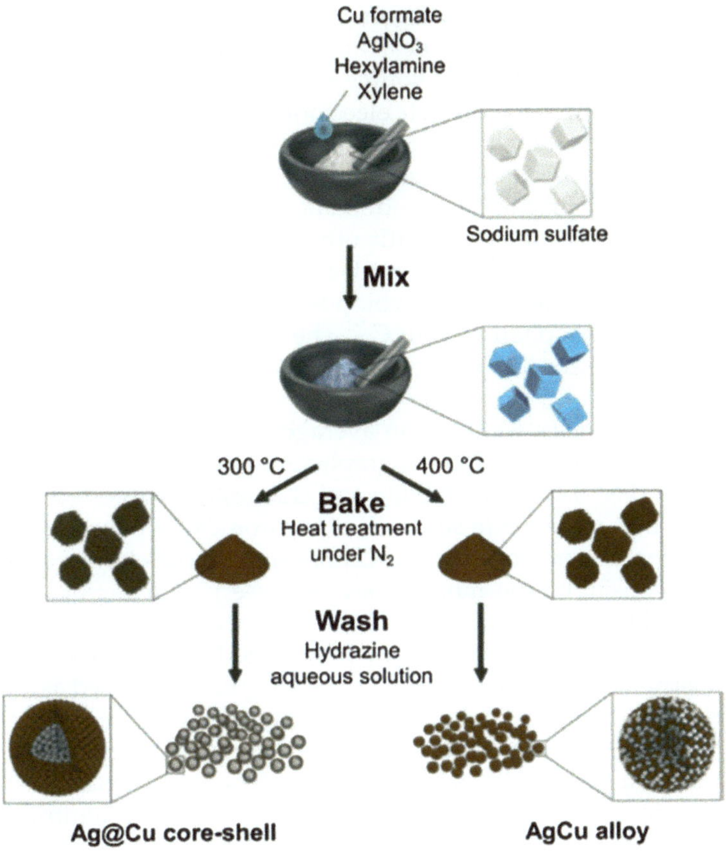

SCHEME 3.17.6 Schematic of the synthetic procedure to prepare Ag-Cu nanoparticles with different bimetallic nanostructures. *Reproduced from Choi E, Lee S, Piao Y. A solventless mix-bake-wash approach to the facile controlled synthesis of core-shell and alloy Ag-Cu bimetallic nanoparticles.* Cryst Eng Comm *2015;***17***:5940–46, with permission of The Royal Society of Chemistry.*

water-in-oil microemulsion also called reverse micelles.[87] Although different techniques are effective in producing a narrow range of small particle sizes, drawbacks involve the need to provide extremely high temperatures, large amounts of solvent and surfactants, toxicity of reaction components, and issues of solubility. Daraio et al. were the first to develop a solvent-free approach to prepare Fe_3O_4 nanoparticles using surfactants and Fe(III) and Fe(II) salts as precursors.[88] Their method involved mortar-and-pestle milling of iron salts ($FeCl_3\cdot6H_2O$ and $FeCl_2\cdot4H_2O$), oleic acid-oleylamine as surfactants, and sodium hydroxide (NaOH) at room temperature. Inert atmosphere was maintained with nitrogen gas. Purification of the resulting particles was performed by extractions with hexanes and ethanol, washings with water, followed by centrifugation for separation of nanoparticles at different points. They demonstrated that the reaction is indeed scalable, synthesizing up to 30 g. The size distribution was quite narrow with an average diameter of 5 nm, and the material was highly crystalline with uniformly spherical particle structures. The authors also performed the synthesis in the absence of surfactant. In this case, the particles were larger (8 nm) with wider size dispersion. Agglomeration was observed in the mixture of spherical- and cubic-shaped structures.

Lithium ion batteries are the leading source of energy for virtually all portable devices. Nanomaterials have been probed for their electronic properties in the race to develop batteries with longer life spans, better charge/discharge rates, and greater cycling stability.[89] As such, researchers are exploring the use of metal oxide/carbon composites. These composites have demonstrated promise as anode material due to their high capacity and relatively low charge/discharge rates. As with other nanoparticles, there are complications with procedures and scalability.[90] Jang and colleagues have developed a solvent-free process in which the iron oxide-graphene composites were prepared by thermal decomposition under a nitrogen atmosphere at 600 °C.[91] The thermal process is preceded by mortar-and-pestle mixing of the iron(III) oleate [from $Fe(acac)_3$ and oleic acid] with graphene (Scheme 3.17.7).

The iron oxide nanoparticles were, on an average, 30 nm in size. They were found to be highly crystalline and well dispersed in the graphene layer; both materials prevented the aggregation of the other (Fig. 3.17.6). In stability and charge/discharge rate, the composite outperformed materials that were made solely of either iron oxide nanoparticles or graphene.

The electronic properties of zinc stannate ($ZnSnO_3$) materials have been exploited in gas sensors.[92–97] Not surprisingly, the morphology of these nanomaterials has a huge impact on the sensing capabilities.[98–100] To control the structures of these particles, most synthetic techniques incorporate organic surfactants or polymers. In their solvent-free approach, Lu and coworkers determined that potassium chloride (KCl) could be used as an additive to control the morphological outcome, cubic or spherical-shaped aggregates.[101] Increasing the equivalents of the salt resulted in a more cubic-shaped morphology. The complete absence of KCl yielded the spherical-like structure. The C-sensors (cubic) outperformed the S-sensors (spherical) in the detection of various vapors, such as acetonitrile, ethanol, chlorobenzene, benzene, formaldehyde, methanol, methylbenzene, acetone, and ammonia. For both sensors, the strongest response was observed with ethanol. The C-sensor aggregates ranged from 50 to 200 nm in size with nanoparticle size of c. 1 nm. The aggregate and nanoparticle sizes were the same for the spherical structures as well. Completely spherical structures were obtained with 1:1:0 = $ZnCl_2$:$SnCl_4 \cdot 5H_2O$:KCl, whereas the optimum conditions for producing cubic aggregates occurred at a ratio of 1:1:6 = $ZnCl_2$:$SnCl_4 \cdot 5H_2O$:KCl (Fig. 3.17.7).

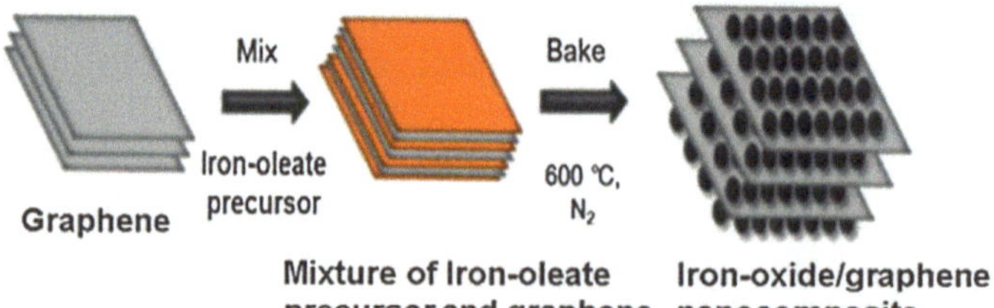

SCHEME 3.17.7 Schematic representation of the direct preparation of iron oxide/graphene nanocomposites by the solvent-free thermal decomposition method. *Reproduced from Jang B, Chae OB, Park S-K, Ha J, Oh SM, Na HB, Piao Y. Solventless synthesis of an iron-oxide/graphene nanocomposite and its application as an anode in high-rate Li-ion batteries.* J Mater Chem A *2013;***1***:15442–46, with permission of The Royal Society of Chemistry.*

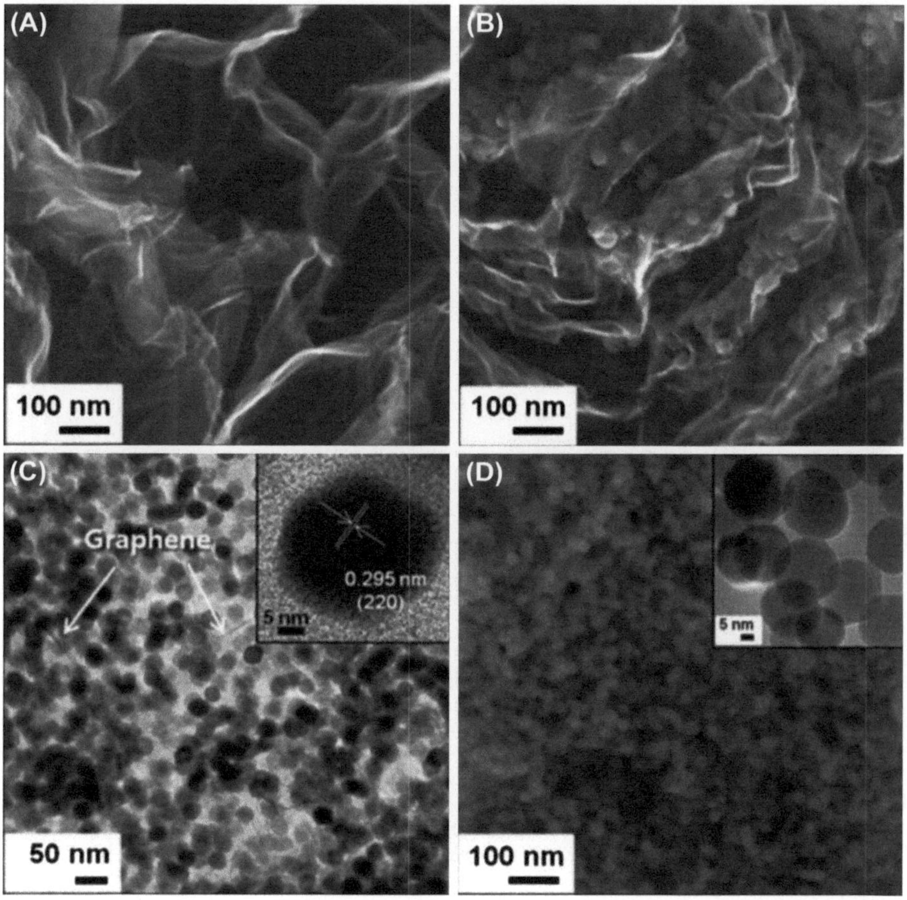

FIGURE 3.17.6 (A) Field emission scanning electron microscopic (FESEM) image of the as-prepared graphene; (B) FESEM image and (C) transmission electron microscopic (TEM) image of the iron-oxide/graphene nanocomposite; and (D) scanning electron microscopic image of the iron oxide nanoparticle. The inset represents a high-resolution TEM image of the iron oxide nanoparticle. *Reproduced from Jang B, Chae OB, Park S-K, Ha J, Oh SM, Na HB, Piao Y. Solventless synthesis of an iron-oxide/graphene nanocomposite and its application as an anode in high-rate Li-ion batteries.* J Mater Chem A *2013;***1***:15442–46, with permission of The Royal Society of Chemistry.*

With a focus on noble metal catalysts, many groups have developed mechanochemical, solvent-free methods for the synthesis of Ir(0) nanoparticles. Both ball milling and mortar-and-pestle mixing have been employed.[102–107] For example, Redón et al. used a standard mortar-and-pestle approach in which the reduction of $IrCl_3$ proceeds through milling with sodium borohydride ($NaBH_4$).[102] The authors have explored the importance of the $NaBH_4$-equivalents, the annealing temperature, and the need for washing the crude product with water prior to annealing. Optimal results were obtained with 5.3 equivalents of the reductant, washing multiple times with water, and annealing at 200 °C. In all their trials, the average particle size was 10 nm. Temperatures above 200 °C and higher $IrCl_3$ to $NaBH_4$ ratios resulted in larger particle sizes.

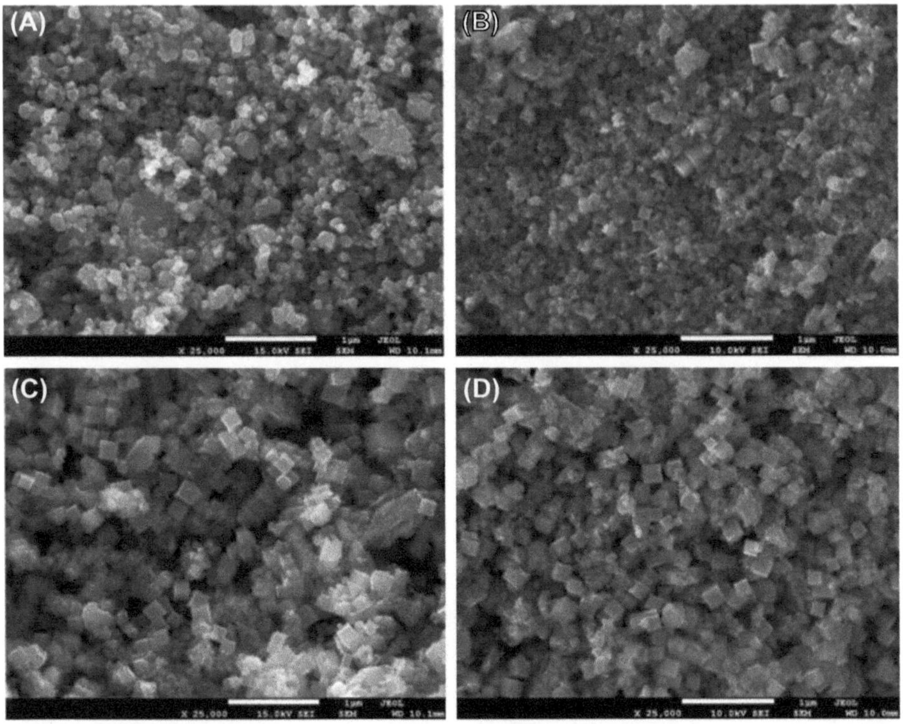

FIGURE 3.17.7 Field emission scanning electron microscopic images of $ZnSnO_3$ particles prepared by adding KCl crystals at different molar ratio of $ZnCl_2$ to $SnCl_4 \cdot 5H_2O$ and to KCl at (A) 1:1:1, (B) 1:1:2, (C) 1:1:4, and (D) 1:1:6. The scale bars in the images are 1 μM. *Reprinted from Lu L, Zhang A, Xiao Y, Gong F, Jia D, Li F. Effect of solid inorganic salts on the formation of cubic-like aggregates of $ZnSnO_3$ nanoparticles in solvent-free, organic-free reactions and their gas sensing behaviors.* Mater Sci Eng B *2012;***177***:942–8, with permission from Elsevier.*

García-Peñia and colleagues were the first to report the solvent-free synthesis of zero-valent ruthenium (Ru) nanoparticles.[108] Ru(0) is used as a catalyst in many reactions, for example, oxidation, reduction, and coupling processes. The investigators noted that their mechanochemical approach requires less time, 5 min, and is devoid of by-products that can sometimes impede the performance of the nanoparticle. The authors synthesized their Ru(0) nanoparticles from the reduction of the chloride salts with sodium borohydride ($NaBH_4$). Particle size was controlled by varying the molar ratio of the reductant. Ten milligrams of samples were prepared with various ratios of $RuCl_3 \cdot nH_2O$ to $NaBH_4$ from 1:0.75 to 1:9.75. Optimum results were obtained with 1:7.75, i.e., particle size decreased with increasing the equivalents of $NaBH_4$ up this point. Beyond this, the particle size increased again. While they were able to detect particles as small as 5 nm with 1:7.75, agglomeration greater than 100 nm was still an issue. As a result, an average particle size could not be determined. Both XRD and TEM analyses showed that hexagonal, close-packed crystallite structures of Ru(0) were present. When compared with nanoparticles made by the colloidal method,[109–112] the solvent-free products were not uniform in shape. The researchers noted that this irregularity was an advantage because it would provide a larger surface area. Overall, they were able to conclude that the equivalents of $NaBH_4$ could be used to control particle size. The Ru(0) nanoparticles were synthesized in an agate mortar and purified through a series of washes with water and ethanol or acetone. Rigorous analyses by XRD, TEM, high-resolution TEM, and XPS were used to measure particle size and to verify the presence of Ru(0).

Cryptomelane (OMS-2), composed of a mixture of Mn(IV) and, to a lesser extent, Mn(III) is an open-tunnel, porous structure that has been probed for applications in heterogeneous redox catalysis, sensors, and batteries.[113] The material's morphology (particle size, surface area, and pore size) and purity affect its physiochemical properties. As such, there has been a focus of developing high-yielding synthetic techniques that provide control and uniformity in the structural characteristics of the cryptomelane. The aforementioned techniques have issues with high temperatures; purity of the resulting material; long reactions times, up to 1–2 days; toxicity of reagents; low surface areas ($\sim$10–90 m^2/g); and more. Ding and coworkers have successfully developed a solvent-free, mortar-and-pestle procedure for the synthesis of pure K-OMS-2.[114] The material synthesized by their method was obtained at a relatively low temperature of 80 °C, with 1 h reaction time that was preceded by the milling of $KMnO_4$ and $Mn(OAc)_2 \cdot 4H_2O$ in a 2:3 ratio, in a mortar. The resulting nanorods had a high surface area of $\sim$160 m^2/g, relatively uniform pore sizes of $\sim$12 nm, and average diameter and rod length of 10 and 50 nm, respectively. The investigators were able to prove that their synthetic K-OMS-2 was stable at temperatures as high as 550 °C and that they had better catalytic activity for the oxidation of alcohols than the material made by a conventional, solvent-based, reflux approach.

Sulfated zirconia is limited in its applications in catalysis by its low surface area.[115,116] Although a number of approaches have yielded uniform particle structures, monoclinic and tetragonal with high surface areas, the procedures are often complicated and require large volumes of solvent.[116] Sun and colleagues have developed a simple solvent-free process in which $ZrOCl_2 \cdot 8H_2O$ is milled with ammonium sulfate [$(NH_4)_2SO_4$] in a carnelian mortar at room temperature followed by calcination at 600 °C for 5 h.[117] The nanoparticles had an average size of 7 nm with a high 165–193 m^2/g surface area. For comparison, the researcher also synthesized a sulfated sample by conventional methods using an aqueous solution of $(NH_4)_2SO_4$. The two samples were probed for their activity in cracking, esterification, and alkane isomerization. Interestingly, the solvent-free product outperformed the "conventional" product in all three processes. It was concluded that the superior activity of the catalyst formed by the mortar approach was due to an increased number of Brønsted acid sites.

The stability of cerium oxide nanoparticles at high temperatures is important due to their use as gas sensors and cracking catalysts.[118] Yu et al. prepared CeO_2 nanoparticles from $(NH_4)Ce(NO_3)_6$ or $Ce(NO_3)_3 \cdot 6H_2O$ and sodium hydroxide at room temperature.[119] The particles were $\sim$3 nm in size with a surface area of 96.2 m^2/g and fluorite structure, with a lattice parameter of $a = 5.42$ Å. The particles from either salt were found to be stable up to 550 °C. Above this temperature, there was a dramatic increase in particles size, from 5 nm to 20–50 nm, and a huge loss in surface area from 96.2 to 113.8 m^2/g to 9.96 m^2/g to "below detectable levels."

3.17.3 SOLVENT-FREE SYNTHESIS OF NANOPARTICLES THROUGH THERMAL TREATMENT

Due to the extensive demand, novel methods of synthesis for metal nanoparticles have found prominence in contemporary research. This part of the chapter focuses on the solvent-free synthesis of the nanostructured materials through thermal treatment of metal salts. Thermolysis has been a well-accepted method that is able to produce high-quality monodispersed nanostructured materials. A typical thermolysis process entails treating precursors in a high-temperature tube furnace. Subsequent cooling to room temperature followed by precipitation produces the desired nanoparticles. Examples below discuss some of the recent advances brought upon by the thermolysis method. The decomposition methods are divided into three categories, those with metal salt precursors (Section 3.17.3.1.1), reactions of the salt precursors with capping agents (Section 3.17.3.1.2), and decomposition using MW energy (Section 3.17.3.2). The former two processes employ conventional heating. In many cases, the salt precursors are acetates.

3.17.3.1 Thermal Decomposition/Thermolysis of Metal Salt Precursor

3.17.3.1.1 Thermal Decomposition of Metal Salt Precursor Nanostructure

Mohadesi et al. reported a novel processing route for mercury oxide (HgO) nanoparticles by solid-state thermal decomposition.[120] The precursor, mercury(II) acetate nanostructures, was obtained by sublimation of powdered sample of mercury(II) acetate in a vertical quartz pipe set. Under vacuum, the system was gradually heated to 150 °C and subjected to a constant water flow to solidify the product vapors. The obtained mercury(II) acetate nanostructures were loaded into a silicon boat that was later transferred into a high-temperature tube furnace and kept there at 350 °C for 2 h. After the thermal treatment, the system was cooled to room temperature, and the HgO nanoparticles were collected for characterization. SEM and TEM images (Fig. 3.17.8) revealed spherical HgO nanostructures with an average particle

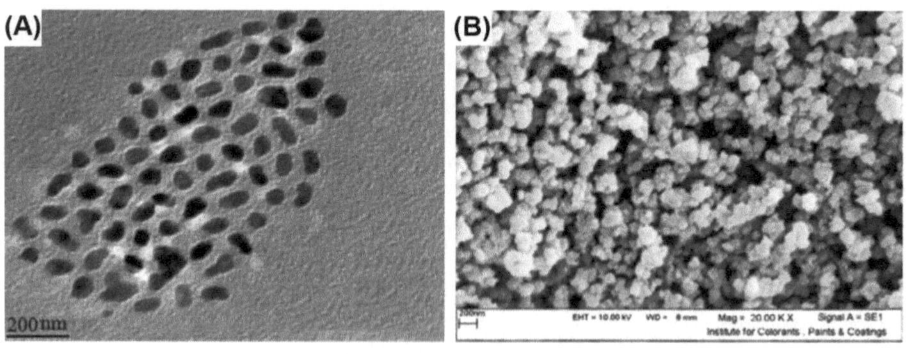

FIGURE 3.17.8 (A) Transmission electron microscopic image and (B) scanning electron microscopic image of the HgO nanoparticles synthesized from the $Hg(OAc)_2$ sublimated at 150 °C as precursor. *Reprinted from Mohadesi A, Ranjbar M, Hosseinpour-Mashkani SM. Solvent-free synthesis of Mercury oxide nanoparticles by a simple thermal decomposition method.* Superlattices Microstruct *2014;**66**:48–53, with permission from Elsevier.*

size of 100 nm. The fine nature of the HgO nanoparticles was confirmed by the broad reflection peaks in XRD analysis. Advantages of this method included the ability to control the size of the nanoparticles, less impurity in the synthesized product, and the inexpensive nature, when compared with other solvent-based procedures.

In a similar method, ZnO nanoparticles were synthesized by the solid-state thermal decomposition from zinc(II) acetate nanostructures of various sizes as starting materials.[121] As in the previous example,[120] the $Zn(OAc)_2$ powder was first sublimed at 150 °C for 2 h to give rise to precursor nanostructures before thermal decomposition to ZnO.[121] With the fixed reaction time of 2 h, the effect of the sublimation temperature on the morphology of the precursor $Zn(OAc)_2$ nanoparticles was investigated by SEM.[122] It was observed that increasing the temperature from 120 °C to 180 °C resulted in larger $Zn(OAc)_2$ nanoparticles.[121] Thermal decomposition of the precursor acetate was carried out at 350 °C for 120 min.

The thermal decomposition technique has been further explored by Sam's group to synthesize NiO nanostructures.[122] Synthesis involved the sublimation of the nickel(II) acetate powder to the corresponding nanomaterials followed by the thermal decomposition at 350 °C for 2 h. Temperatures in the range of 110, 120, 130, and 140 °C were maintained to obtain Ni acetate nanostructures A_1, A_2, A_3, and A_4 of desired particle sizes, respectively. The NiO nanoparticles B_1, B_2, B_3, and B_4 synthesized from the respective precursors A_1, A_2, A_3, and A_4 were pure cubic structures with an average particle diameter of 20–50 nm. The nanoparticles revealed photocatalytic activity with methylene blue (MB) dye. UV-vis absorption was employed to monitor the degradation rate for the decomposition of MB by observing the changes in the absorbance of the dye over 120 min. Fig. 3.17.9 represents the plot of absorbance ratio (A/A_0) against time. A negligible amount of MB degradation was observed in the absence of NiO nanoparticles. However, treatment of MB with the catalyst dramatically increased the rate of photodecolorization indicating efficient photocatalytic activity of the NiO nanoparticles.

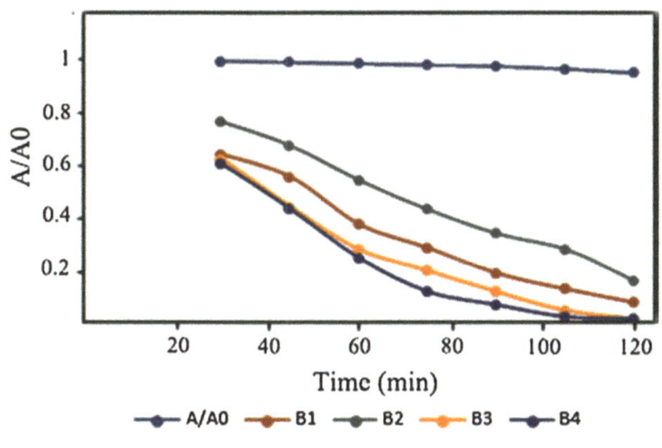

FIGURE 3.17.9 Photodecolorization of methylene blue under ultraviolet illumination by blank, sample B_1, sample B_2, sample B_3, and sample B_4. *Reproduced from Ranjbar M, Taher MA, Sam A. NiO nanostructures: novel solvent-less solid-state synthesis, characterization and MB photocatalytic degradation.* J Mater Sci Mater Electron *2015;**26**:8029–34, with permission of The Springer Science.*

A slight modification in the precursor salt from acetate to the bis(acetylacetonate) counterpart was made by Soofivand et al. while synthesizing one-dimensional CuO nanoparticles.[123] As in other studies, sublimation followed by thermal decomposition of the as-prepared crystalline $Cu(acac)_2$ at varying temperatures produced CuO nanoparticles of different sizes.

Deviating from the acetate salt precursors, ZnO nanoparticles were synthesized from the [*N*,*N*-bis(salicylaldehydo-ethylene diamine]zinc(II) or Zn(salen) in a solvent-free synthetic procedure.[124] ZnO nanoparticles of 10–20 nm were produced by subjecting the Zn(salen) complex to thermal treatment at a temperature of 500 °C in air for 5 h. Room temperature photoluminescence spectra of ZnO nanostructures were dominated by green emission, which was attributed to oxygen vacancy-related donor-acceptor transition.

An environment-friendly synthetic route for ZnO nanoparticles was also executed by Jayanthi's using glycerol as an organic dispersant.[125] The zinc nitrate hexahydrate precursor was made into a paste with two drops of glycerol in a silica crucible. Initially, the temperature was kept at 50 °C before slowly increasing it to 300 °C in a muffle furnace. The resultant product was then calcined at 300 °C for 4 h yielding a colorless crystalline zinc oxide powder. The synthesized nanoparticles revealed a hexagonal shape through XRD analysis, and the average particle size was in the range of 70 nm.

In another work, solvent-free synthesis of metallic Ni and Fe nanoparticles was reported at low temperatures by the reduction of metal organic salt using calcium hydride as the reducing agent.[126] The synthetic method involved the reduction of silica-coated iron oxide nanoparticles by CaH_2 at low temperature (>300 °C). The strong reducing ability of CaH_2 enabled the development of a one-pot synthetic route for the preparation of transition metal nanoparticles.

Solvent-free synthesis of Ag_2S nanoparticles was carried out by Zhang et al.[127] This synthetic route did not involve surfactants, and it was easy to control nanoparticle size via the thermolysis of silver xanthates as the only precursors. In a typical synthesis, the silver octyl xanthate was heated to 200 °C at a rate of 20 °C/min in an oil bath. Upon cooling to room temperature, excess amount of absolute ethanol was added allowing gradual formation of Ag_2S nanoparticles. The as-prepared Ag_2S nanoparticles were redispersed in various nonpolar solvents. The particle size of the synthesized nanoparticles was varied with the variation in the alkyl chain length of the precursors.

Iron oxide nanoparticles are always in demand due to their applicability as magnetic nanomaterials. So synthesis of ferrite nanoparticles using different salts has been a prevalent research question. Among many other materials, ferrocene $(C_5H_5)_2Fe$ is an important precursor for preparing iron oxide nanostructures by thermal decomposition due to its easy sublimating properties.[128,129] Bhattacharjee et al. reported the synthesis of single-phase nanostructured iron oxide nanoparticles by thermal decomposition of ferrocene.[130] Ferrite nanoparticles were prepared by the thermal decomposition of {bis-(cyclopentadienyl)iron}-ferrocene in the presence of oxalic acid through one-step synthesis. Ferrocene and oxalic acid were kept in a porcelain boat, which was loaded into a furnace. The furnace temperature was maintained at 752 °C for 10 h. The thermally decomposed materials were collected at room temperature. The magnetic characterization of the synthesized particles obtained through Mössbauer spectroscopy revealed the formation of hematite nanostructures.

In a follow-up work by the same group,[131] hematite synthesis was reported by changing the precursor components from the previous study.[130] The thermal decomposition of ferrocene in the presence of a guest molecule, iron(II) acetylacetonate, was investigated to determine the correlation between the thermal decomposition reaction kinetic parameters, morphology and physical characteristics of the reaction products, and nature of the guest molecules. TGA analysis revealed that ferrocene alone had suffered about 100% weight loss on heating, whereas when mixed with Fe(II) acetylacetonate, the mixture gave rise to decomposed material below 327 °C. The decomposition of the mixture of ferrocene and Fe(II) acetylacetonate proceeded through two steps and resulted in the formation of ~20% end product. The obtained product was hematite. The formation of hematite is summarized in the chemical Eqs. (3.17.1–3.17.3).

$$(C_5H_5)_2\,Fe \rightarrow Fe + H_2(g) + CH_4(g) + C_5H_6(g) \qquad 3.17.1$$

$$[C_{10}H_{14}FeO_4] \rightarrow Fe_2O_3 + CO_2(g) + 3CO(g) + C_2H_2(g) + H_2(g) \qquad 3.17.2$$

$$\left[(C_5H_5)_2Fe\right] + C_{10}H_{14}FeO_4 \rightarrow Fe_2O_3(s) + CO_2(g) + CO(g) + H_2(g) \qquad 3.17.3$$

It is interesting to note that neat Fe(II) acetylacetonate produces magnetite upon thermal decomposition, whereas hematite was formed in the presence of ferrocene. Several characterization studies corroborated the formation of hematite nanostructures unequivocally through the thermal decomposition at lower temperatures.

The thermal decomposition for synthesis of metal oxide nanoparticles is gradually gaining popularity and has been extended to the synthesis of superparamagnetic nanomaterials. These nanomaterials showed advancement in biomedical applications such as magnetic separation, drug delivery, cancer hyperthermia, and magnetic resonance imaging (MRI) enhancement.[132,133] Iron oxides such as magnetite (Fe_3O_4) and maghemite (γ-Fe_2O_3) are the popular forms of the superparamagnetic materials in this field due to their nontoxicity, biocompatibility, and high stability. However, the difficulty lies in the synthesis of ultrafine particles of controllable size at the nanoscale. Nearly monodispersed superparamagnetic maghemite nanoparticles were synthesized by Kluchova et al. through one-step thermal decomposition of iron(II) acetate.[134] Their work was focused on bioapplications of the iron oxide nanoparticles. In general, these particles are used as a magnetic solid support to immobilize biomolecules such as enzymes, antibodies, other proteins, and oligonucleotides. Immobilization of trypsin is an important example that plays a key role in MS-driven proteomics.[135,136] However, low thermal stability and rapid autolysis of enzymes at basic pH limits the use of trypsin in medical and biotechnological applications.[137] The thermal stability has been enhanced through the conjugation of water-soluble polymers or oligosaccharides, e.g., raffinose. The autolysis of trypsin has been controlled by utilizing nanoparticles as a magnetic carrier for trypsin.[135,138]

With their particles, Kluchova et al. targeted the preparation of an oral negative contrast agent for MRI and a magnetic carrier for trypsin.[134,139,140] The authors performed solid-state isothermal decomposition of iron(II) acetate dihydrate in air at 400 °C. The iron(II) acetate was homogenized by grinding prior to the decomposition. The resulting size distribution of the precursor particles was 1–5 μm. The superparamagnetic maghemite nanoparticles were incorporated into a layered aluminosilicate mineral (bentonite) to form a highly efficient negative contrast agent, maghemite-bentonite (MB), for MRI. The MB nanocomposite was prepared by mixing the maghemite nanoparticles with bentonite powder in a petridish. A homogenous red-brown gel layer was obtained after continuous stirring.

Realizing the necessity for solventless synthesis, Beyki et al. prepared magnetic $MnFe_2O_4$ nanoparticles and $MnFe_2O_4$-reinforced polyamide resin through solid-state synthesis at room temperature.[141] The $MnFe_2O_4$ nanoparticles were obtained with addition of sodium hydroxide to $Mn(NO_3)_2 \cdot 4H_2O$ and $FeCl_3 \cdot 6H_2O$ in desired weighted amounts. The mixture was agitated for 5 min and then dried at 100 °C for 10 min. The polyamide resin and magnetic nanohybrid were fabricated by a solvent-free route with solid reagents. In a typical synthesis, citric acid and 1,4-phenylenediamine were mixed together with ferrite followed by heating at 155 °C for 3 h. After cooling to room temperature and grinding, the obtained particles were stored for characterization. The magnetic-resin nanohybrid was synthesized for the removal of a dye, Congo red, from aqueous solution through adsorption on the resin surface.

3.17.3.1.1.1 OTHER SYNTHETIC ROUTES FOR PREPARATION OF FUNCTIONALIZED NANOPARTICLES USING THERMAL TREATMENT

Sensing and removal of metal ions are attaining importance due to their toxic nature and harmful repercussions when they reach the food chain and are absorbed by living systems. Mercury in this regard is one of the most toxic metals. Research is in progress to synthesize functionalized carbon nanotubes (CNTs) using thermal treatment as the synthetic route. Wang et al. synthesized sulfur and nitrogen co-doped carbon nanoparticles (SNCNs) from glutathione for selective sensing of Mercury(II) ions.[142] The authors used a solvent-free one-pot solid-phase thermal treatment method for synthesis (Scheme 3.17.8). In a typical synthesis of SNCNs, glutathione was heated in a stainless-steel autoclave for 1 h at 260 °C. The formation of SNCNs included four stages: dehydration, polymerization, carbonization, and surface passivation. The glutathione molecules were first aromatized by intermolecular dehydration along with the polymerization process. Then the intermediate formed was carbonized to form carbon seeds followed by functionalization with sulfur and nitrogen

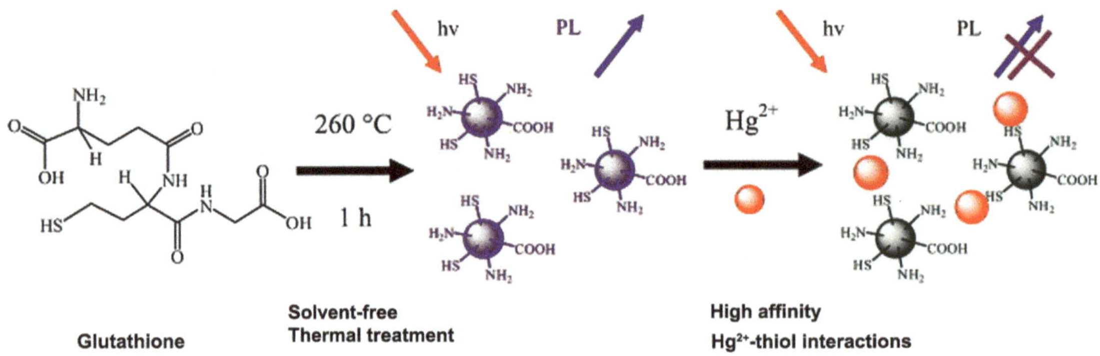

SCHEME 3.17.8 Schematic illustration of the preparation process for sulfur and nitrogen codoped carbon nanoparticles and their application for Hg^{2+} detection. *Reprinted from Wang W, Lu Y-C, Huang H, Wang A-J, Chen J-R, Feng J-J. Solvent-free synthesis of sulfur- and nitrogen-co-doped fluorescent carbon nanoparticles from glutathione for highly selective and sensitive detection of Mercury(II) ions.* Sens Actuators *2014;***202***:741–7, with permission from Elsevier.*

moieties to form SNCNs. The product formed was dissolved in water, centrifuged, and the supernatant was collected for characterization. The synthesized SNCNs were used as the template for detection of Hg^{2+} ions.

Under visible light the SNCN solutions were yellow, and under UV lamp (excitation wavelength = 365 nm) a bright blue fluorescence was observed. The fluorescence spectrum yielded a broad and intense emission band indicating a high quantum yield at ~450 nm. The bright fluorescence of SNCNs is significantly quenched by the addition of trace amount of Hg^{2+} compared with the addition of other metal salts (Fig. 3.17.10).

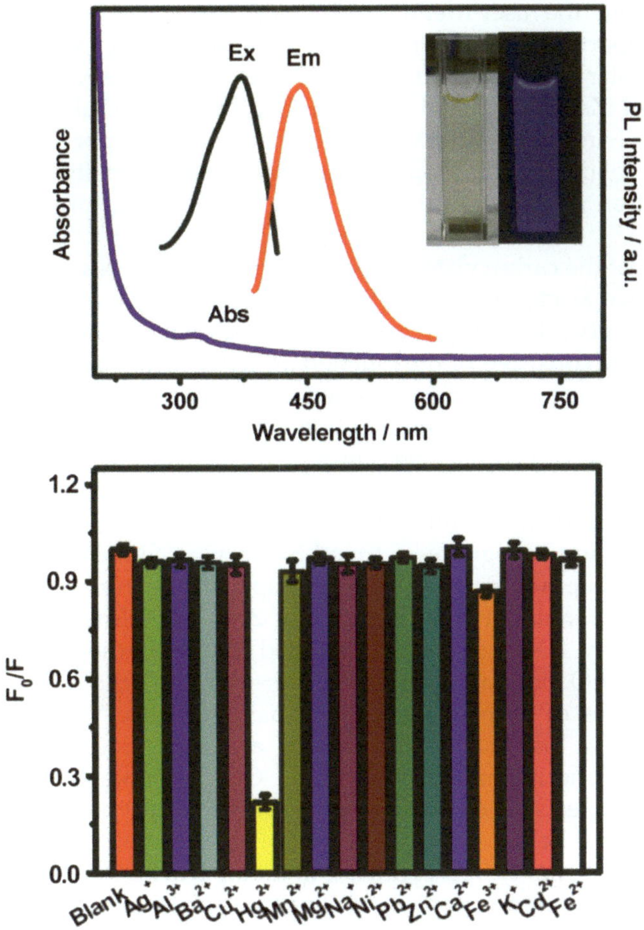

FIGURE 3.17.10 (A) Ultraviolet (UV)-visible absorption, photoluminescence excitation, and emission spectra of sulfur and nitrogen codoped carbon nanoparticles (SNCNs). Inset shows the photographs taken under visible light (left) and UV light of 365 nm (right). (B) Fluorescence responses of SNCNs in the presence of different metal ions (excitation = 371 nm; $[M^{n+}] = 50.0\ \mu M$). *Reprinted from Wang W, Lu Y–C, Huang H, Wang A-J, Chen J-R, Feng J–J. Solvent-free synthesis of sulfur- and nitrogen-co-doped fluorescent carbon nanoparticles from glutathione for highly selective and sensitive detection of Mercury(II) ions.* Sens Actuators *2014;***202***:741–7, with permission from Elsevier.*

3.17.3.1.2 Thermal Decomposition of the Metal Acetate Precursor Nanostructures With Capping Agents

Synthesis of metal oxide nanoparticles through the sublimation process of the precursor nanostructured salts is the commonly employed method for the formation of a myriad of nanoparticles. However, processability and stability of the nanoparticles demand improvement. The use of capping agents has played an important role in enhancing the nanoparticle quality and stability through surface modifications. For instance, carboxylate capping of metal oxide nanostructures has an enriched impact on the processability of the material after the nanoparticle has been synthesized.[143]

Estruga et al. reported a one-step synthesis of carboxylate-capped ZnO nanoparticles at low temperature from the user-friendly solid precursors such as zinc acetate dehydrate (ZAD).[144] The nanoparticles were synthesized by mixing ZAD with a desired carboxylic acid at a definite molar ratio. The authors used two alternative routes to form the mixture (carboxylic acid and ZAD): (1) hand grinding both the chemicals in an agate mortar for 30 min and (2) dissolving both the reagents in ethanol and evaporating the solution to dryness. The obtained solid mixtures from both routes were thermally treated at 90–120 °C in a forced air ventilation furnace for 30–144 h to induce the transformation of the precursors to capped ZnO.

The morphology of the carboxylate-capped ZnO nanostructures was influenced by the nature of the ligand used as capping agent. Spherical aggregates were observed when benzoic, piperonylic, phenylacetic, 4-fluorophenylacetic, and cinnamic acids were used as capping agents. Each aggregate consisted of the junction of many tiny hardly distinguishable crystallites (Bz-ZnO). Strong π-π interactions influenced this structural pattern. The tendency to form aggregates was reduced by using an aliphatic acid. Acids with an aromatic ring sufficiently distant from the carboxylic acid function, as in the case of phenylvaleric acid, did not lead to the formation of strong aggregates. In these cases, the individual nanoparticles were loosely grouped and highly porous. Essentially, the nanomaterial properties, namely, shape and aggregate morphology, were tuned by varying the length of the carboxylic acid chains. The capped nanopowders revealed high dispersibility in several solvents, which facilitated processing of the material and formation of ZnO thin films under mild conditions.

In a different study, Estruga et al. extended their synthetic procedure to prepare carboxylate-capped CuO nanoparticles from user-friendly solid precursors and explored the magnetic properties of each of the capped nanoparticles.[145] The laurate-capped CuO nanoparticles showed a paramagnetic behavior at room temperature. This indicated a possible correlation of the magnetic properties of CuO nanoparticles at 298K with surface chemistry, nature of defects, and concentration. On the other hand, a weak ferromagnetic component was detected below 40K.

The use of capping agents for the synthesis of nanoparticles has been explored in making core-shell particles to provide a new dimension to the solvent-free synthesis of nanomaterials. The fabricated core-shell particles bear unique and tailored properties for various applications in materials science such as adsorbents for high-pressure liquid chromatography, calibration standards, spacers for liquid crystals, inks, catalysis, etc. The colloidal microspheres with magnetic properties have become increasingly applicable in technology due to the tunable anisotropic interaction they exhibit. For instance, the composite magnetic core-shell particles could combine the properties of the individual magnetic particles and nonmagnetic sphere dispersions. In the absence of an applied magnetic field the particles have an isotropic sphere dispersion, whereas in an external magnetic field the particles form anisotropic structures.[146] Looking into the brighter application side of the colloidal nanoparticles, Amara et al. synthesized superparamagnetic poly(divinylbenzene)/iron oxide (PDVB/iron oxide) composite microspheres of narrow size distribution by entrapping sodium acetate and ferrocene within the pores of uniform porous PDVB microspheres.[147] It was then followed by solventless thermal decomposition at 300 °C under ambient atmosphere in a sealed cell. Uniform ferromagnetic C/iron oxide and $C/Fe_3O_4/Fe$ composite microspheres were then formed by annealing the superparamagnetic PDVB/iron oxide particles at 500 and 700 °C, respectively, under argon atmosphere, see Scheme 3.17.9 and Fig. 3.17.11 below.

In 2010, another group synthesized colloidal palladium nanoparticles, Pd(0), PdS, and Pd@PdO, by the solvated metal atom dispersion (SMAD) method.[148] Using SMAD, palladium-butanone-4-tert-butyltoluene-thiol colloid was prepared, where the crucible was loaded with Pd foil. Dodecanethiol and 4-tert-butyltoluene were collected at the bottom of the reactor after purging with argon for 30 min. The crucible was heated in vacuum, and the metal vapor was condensed with 2-butanone on the walls of the reactor. After complete vaporization, dodecanethiol and 4-tert-butyltoluene was stirred with Pd-butanone resulting in the formation of Pd-butanone-4-tert-butyltoluene-thiol colloid. The Pd-thiolate complex was formed by the complete removal of dodecanethiol (Scheme 3.17.10).

Pd(0) nanoparticles were obtained when the Pd-thiolate complex was heated in air at 298 °C for 3 h. Pd@PdO core-shell nanoparticles were collected when the same thiolate complex was heated at 325 °C and 425 °C for 3 h. Finally, PdS nanoparticles were synthesized by the solventless thermolysis of Pd-thiolate complex at 430 °C (Scheme 3.17.11).

In magnetic nanomaterials research, the development of monodispersed size-controlled nanoparticles became a very important issue in their application to high-performance permanent magnets, targeted delivery of drugs, and the ultrahigh-density magnetic storage

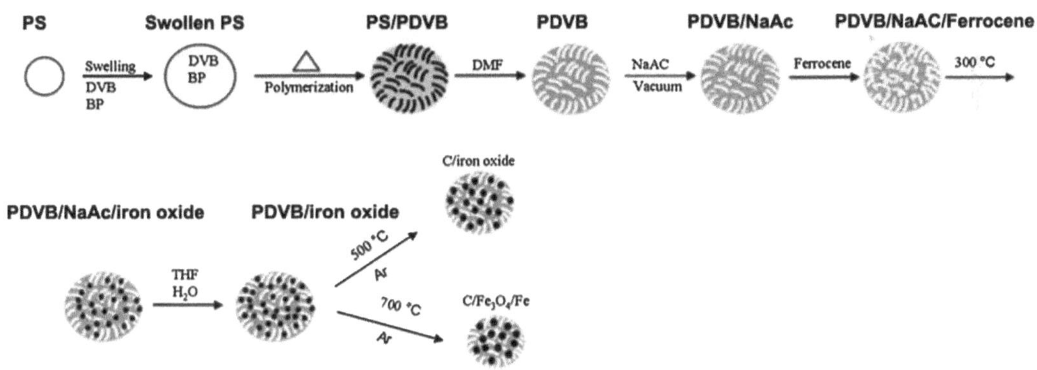

SCHEME 3.17.9 Synthesis of poly(divinylbenzene)/iron oxide (PDVB)/iron oxide, C/iron oxide, and $C/Fe_3O_4/Fe$ composite microspheres from polystyrene (PS). THF (Tetrahydrofuran) and DMF (Dimethylformamide) are the solvents used. *Reproduced from Amara D, Margel S. Solventless thermal decomposition of ferrocene as a new approach for the synthesis of porous superparamagnetic and ferromagnetic composite microspheres of narrow size distribution.* J Mater Chem *2011;**21**:15764–72, with permission of The Royal Society of Chemistry.*

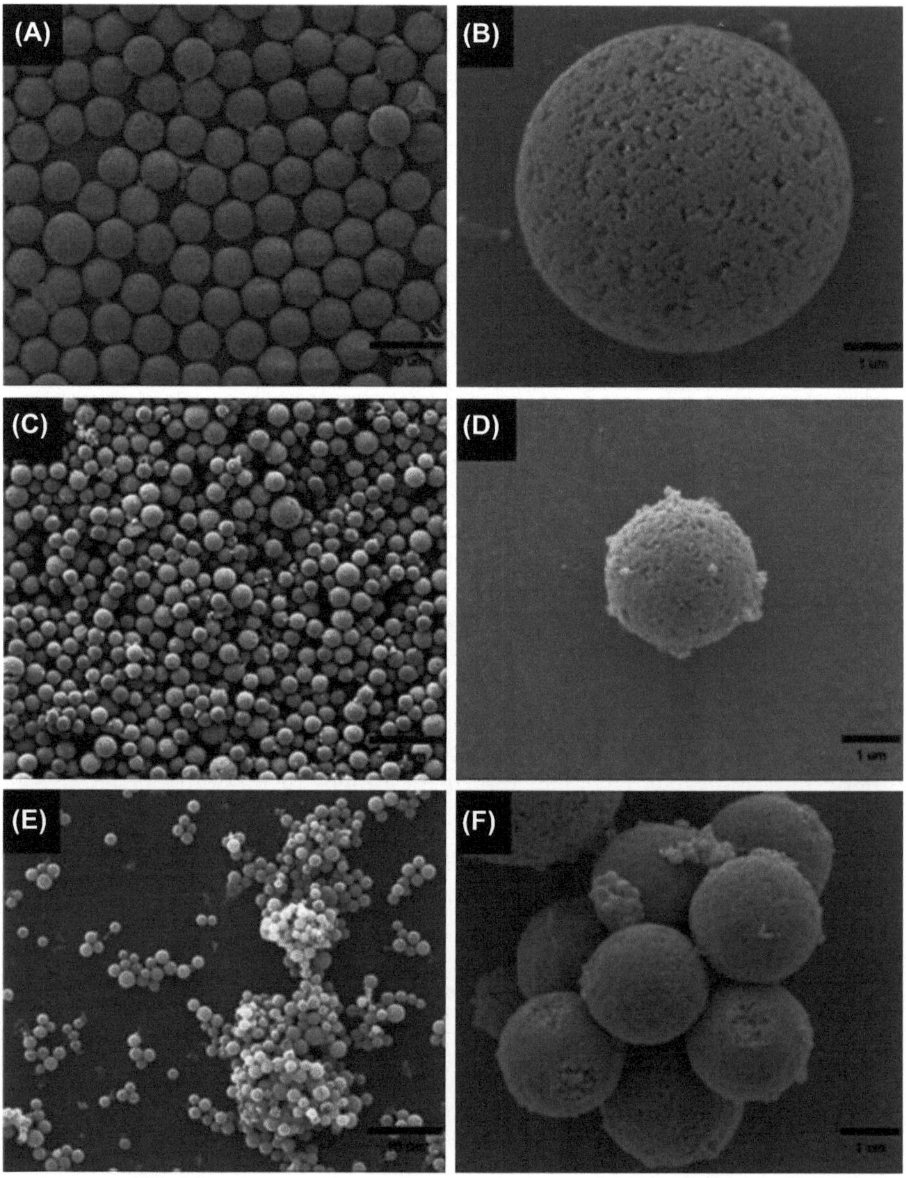

FIGURE 3.17.11 Low- and high-magnification scanning electron microscopic images of the composite microspheres annealed at 300 °C (A and B), 500 °C (C and D), and 700 °C (E and F). *Reproduced from Amara D, Margel S. Solventless thermal decomposition of ferrocene as a new approach for the synthesis of porous superparamagnetic and ferromagnetic composite microspheres of narrow size distribution.* J Mater Chem *2011;***21***:15764–72, with permission of The Royal Society of Chemistry.*

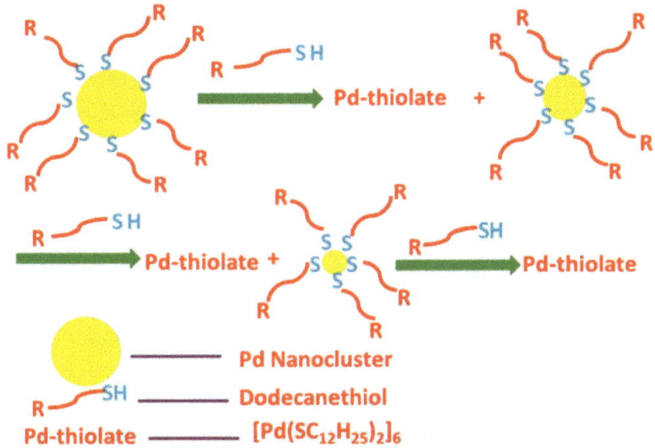

SCHEME 3.17.10 Transformation of Pd nanoparticles into the Pd-thiolate complex $[Pd(SC_{12}H_{25})_2]_6$. *Reprinted from Jose D, Jagirdar BR. Synthesis and characterization of Pd(0), PdS, and Pd@PdO core–shell nanoparticles by solventless thermolysis of a Pd–thiolate cluster.* J Solid State Chem *2010;***183***:2059–67, with permission from Elsevier.*

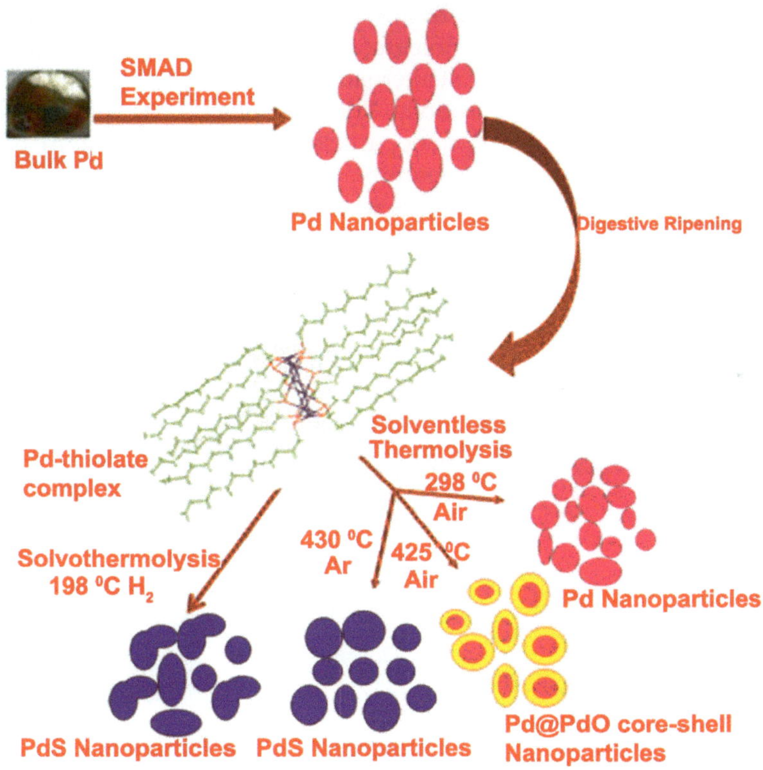

SCHEME 3.17.11 Formation of various Pd nanophases from Pd nanoparticles *SMAD*, solvated metal atom dispersion. *Reprinted from Jose D, Jagirdar BR. Synthesis and characterization of Pd(0), PdS, and Pd@PdO core–shell nanoparticles by solventless thermolysis of a Pd–thiolate cluster.* J Solid State Chem *2010;***183***:2059–67, with permission from Elsevier.*

device. Owing to higher magnetic properties, α-Fe has gained particular interest due to its ability to improve magnetic performance in various fields.[43,149]

Taking into consideration the need for magnetic nanomaterials, Kang and coworkers reported the synthesis of iron and cobalt magnetic materials through solventless thermal decomposition of metal salts with capping agents such as oleate.[150,151] They have synthesized α-Fe nanoparticles from the Fe(II)-oleate complex with sodium chloride (NaCl) as separating media. The Fe(II)-oleate complex was synthesized from iron(II) chloride tetrahydrate and sodium oleate. Sodium oleate was used as a capping agent because oleate has a C18 tail with a *cis* double bond in the middle, forming a kink that is necessary for effective stabilization (Scheme 3.17.12). Each Fe_3O_4 nanoparticle was separated (using NaCl) during the course of reduction by H_2 gas and annealed under a high vacuum system of 1.8×10^{-5} Torr to obtain a high saturation magnetization value of α-Fe nanoparticles. If the ratio between NaCl and Fe_3O_4 nanoparticles was kept lower than 1:40, the effect of managing the size and shape during the reduction and annealing of the NaCl powder for the separating media was decreased. However, when the ratio is higher than 1:40, it is difficult to manage the powder mixture during washing and annealing.

Following the synthetic route of Kang et al.,[150–152] a similar technique was utilized by Wang et al. to synthesize Fe_3O_4 nanoparticles.[153] Highly monodispersed particles with superparamagnetic properties were obtained by the thermal decomposition of the iron(II)-oleate complex. In this route, the iron-oleate precursor was prepared by using oleic acid instead of sodium oleate. This provided a low-cost alternative and a possibility for large-scale synthesis of Fe_3O_4 nanoparticles for future applications. Trioctylamine (TOA) was used as stabilizer to enhance monodispersity of the Fe_3O_4 nanoparticles. The sizes (<20 nm) of nanoparticles were tuned by optimizing reaction temperature, time, the concentration of precursors, and the amount of TOA. The synthesized particles revealed a superparamagnetic behavior, which is useful for many potential applications in the biotechnology field.

In a novel synthetic procedure, Nakaya et al. had synthesized wüstite nanoparticles with cubic shape in the presence of a large amount of oleylamine and oleic acid against an iron

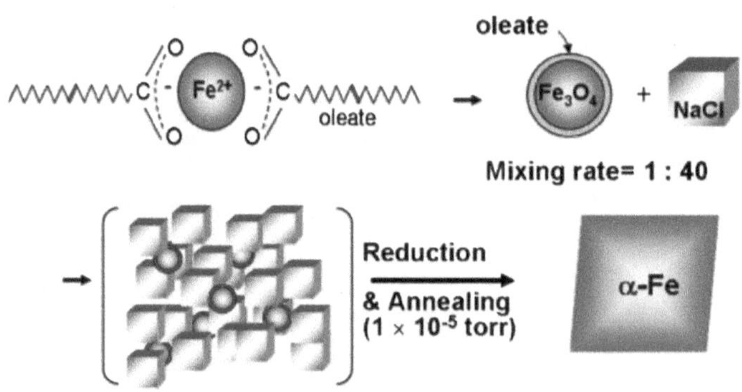

SCHEME 3.17.12 Overall scheme for a modified thermal decomposition method of α-Fe nanoparticle synthesis. *Reprinted with permission from Han YC, Cha HG, Kim CW, Kim YH, Kang YS. Synthesis of Highly Magnetized Iron Nanoparticles by a Solventless Thermal Decomposition Method.* J Phys Chem C;***111****(17):6275–80. Copyright (2007) American Chemical Society.*

precursor.[154] The unique property of wüstite (FeO) in showing paramagnetism at room temperature and antiferromagnetism at temperatures below 203–211K, led them to develop a promising initial precursor of α-Fe (by reduction) and of γ-Fe_2O_3 and Fe_3O_4 nanoparticles (by oxidation). The FeO nanoparticles were synthesized using tris(acetylacetonato) iron(III) as a starting material in the mixture of equal molar concentrations of oleylamine (distilled) and oleic acid. The solution was heated up under a reduced pressure using an oil-sealed rotary vacuum pump with a liquid nitrogen trap. An excess amount of ligand enhanced the particle growth rate by suppressing the nucleation. After the solution was cooled down to room temperature, the resulting precipitate was collected.

The authors further extended their work and synthesized the magnetite (Fe_3O_4) nanoparticles using oleylamine and oleic acid acting as both stabilizing agent and ligand.[155] They followed a similar procedure as previously taking iron(III) acetylacetonate precursor and using 1,2-hexadecanediol as Fe^{3+}-reducing agent. Then the mixture was added to oleylamine and oleic acid and the solution was heated at 130 °C for 30 min with vigorous stirring under reduced pressure. The particle size of the synthesized nanostructures was controlled by varying the reaction conditions and temperatures. It was found that the particle size gradually increased from 5.3 to 20.4 nm with the increase in the reaction temperature from 200 to 300 °C which is in accordance with the previous studies by several other researchers.[121]

The oleylamine and oleic acid as stabilizing agents provided a convenient method for synthesis of nanoparticles. This synthetic route was explored widely by several researchers. In 2013, Abbas et al. had used this solvent-free environmentally benign technique for the synthesis of oleylamine capped silver nanoparticles.[156] The synthesis required heating of 10 equivalents of oleylamine and silver nitrate at 165 °C for 30 min followed by a precipitation (Scheme 3.17.13). The synthesized Ag nanoparticles were dispersed in dicyclopentadiene and cured for preparing a silver nanoparticle-based antimicrobially equipped duroplastic material via the ring-opening metathesis polymerization technique.

Maity et al. also reported a novel solvent-free thermal decomposition method for the synthesis of Fe_3O_4 nanoparticles from $Fe(OAc)_3$ in the presence of oleic acid and oleylamine surfactant mixture.[157] The mixture was magnetically stirred under a flow of argon and then dehydrated at 120 °C for 1 h. The sample was heated for a certain interval of time at a fixed temperature, and the resultant solution was cooled to room temperature. After the completion of reaction, the particles were precipitated in ethanol and then dispersed in hexane.

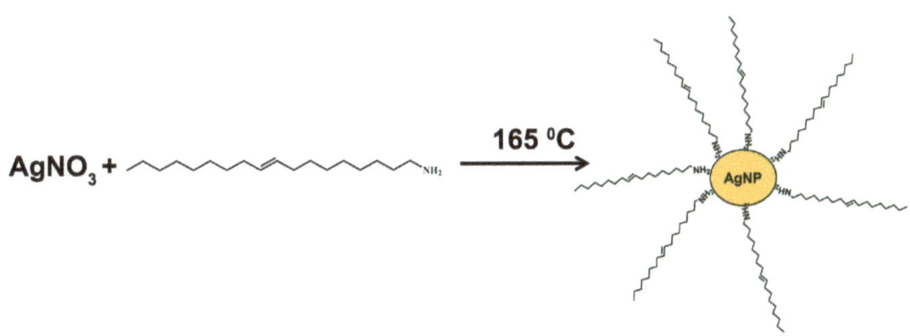

SCHEME 3.17.13 Synthesis of Ag nanoparticles stabilized by oleylamine.

Without the use of solvent, the thermal decomposition temperature and reaction time can be varied easily to vary the particle size. The crystallinity of the magnetite nanoparticles can also be tailored more effectively. The magnetic properties of the magnetite nanoparticles were characterized by using vibrating sample magnetometer and superconducting quantum interference device (SQUID). All samples exhibited superparamagnetic characteristics at 300K, including zero coercivity and remanence. When the reaction time was increased from 2 to 24 h, the saturation magnetization gradually increased from 58 until 71 emu/g. The saturation magnetization of the magnetite nanoparticles also increased to 76 emu/g when the temperature of the reaction was adjusted to 573K for 4 h (Fig. 3.17.12). The obvious increase in the saturation magnetization with reaction temperature and time is attributed to the increase in the crystallinity of the magnetite nanoparticles.

Size-controlled synthesis of AuNPs has been explored using solvent-free synthetic technique. For the purpose, thermolysis of AuCl(SMe_2) in the presence of alkylamines was carried out utilizing one-pot synthesis.[158] In an illustrative synthesis, a mixture of gold(I) sulfide complex [AuCl(SMe_2)] and *n*-hexadecylamine (HDA) was heated to 120 °C for 1 h. The white suspension of AuCl(SMe_2) in HDA progressively turned into a transparent and colorless solution at around 65 °C. Gradual formation of purple color at 120 °C indicated formation of AuNPs (Scheme 3.17.14). The nanoparticles were purified by centrifugation in mixed solvent system (acetone/methanol). The purified AuNPs (HDA-AuNP) were isolated as a brown powder.

The size of the synthesized AuNPs were regulated by varying the alkyl chain lengths of the primary alkylamines. The average diameter of the nanoparticles was measured as 4.3 nm for the olelylamine-capped nanoparticles; however, by increasing the alkyl chain length from HDA to *n*-dodecylamine (DDA) the particle size increased to an average of 6.1 nm for the

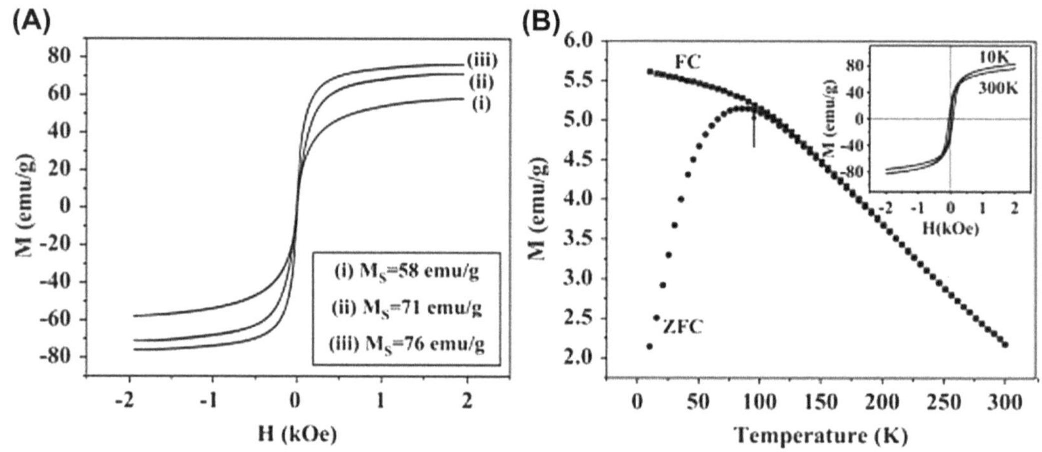

FIGURE 3.17.12 (A) Vibrating sample magnetometer (VSM) results of Fe_3O_4 nanoparticles synthesized at (1) 300 °C for 2 h, (2) 300 °C for 24 h, and (3) 330 °C for 4 h. (B) The temperature dependence of the zero field-cooled/field-cooled (ZFC/FC) magnetization of the Fe_3O_4 nanoparticles synthesized at 330 °C for 4 h. *Reprinted from Maity D, Choo S-G, Yi J, Ding J, Xue JM. Synthesis of magnetite nanoparticles via a solvent-free thermal decomposition route.* J Magn Magn Mater *2009;***321***:1256–9, with permission from Elsevier.*

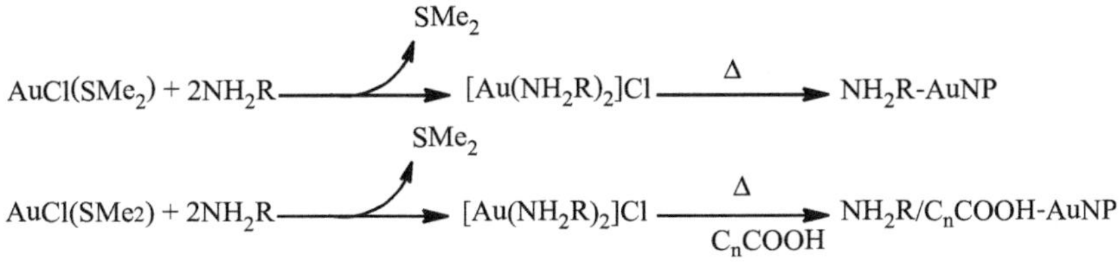

SCHEME 3.17.14 Synthesis of alkylamine capped gold nanoparticles.

HDA- and DDA-capped nanoparticles with a corresponding increase in the yield of nanoparticle formation. All these functionalized AuNPs capped with alkylamines are well dispersed in nonpolar solvents thereby indicating a potential candidate for nanoparticle inks applicable to printed electronics.

3.17.3.2 Solvent-Free Synthesis by Heating via Microwave Energy

MW irradiation as an energy source is different from conventional heating. MW irradiation allows the heat to be generated within the material (local) rather than obtaining energy via conventional external sources.[159] MW heating was found to lead to the effective synthesis of nanoparticles and CNTs by solvent-free approach.

Ecofriendly MW-assisted combustion method (see the following equation) was utilized by Gupta et al. to synthesize homogenous and porous ZnO nanoparticles. Zinc nitrate salt was mixed with two to three drops of ethylene glycol (organic dispersant) and MW energy was provided to heat the paste. Dehydration and decomposition of the mixture released gases, and the combustion of the dispersant released heat to yield a white powder.[160]

$$Zn(NO_3)_2 \rightarrow ZnO_2 + 2NO_2 \text{ (at lower temperature)}$$
$$2ZnO_2 \rightarrow 2ZnO + O_2$$

The use of MW energy is advantageous because of the negligible amount of by-products and short time needed to complete the reaction (seconds vs. hours). The characterizations of the ZnO particles were carried out by XRD, Fourier transform infrared (FTIR), UV-Vis, SEM, etc., in which the XRD and the FTIR confirmed the formation of hexagonal wurtzite structure. The ZnO nanoparticles have shown to have band gap energy of 3.4 eV.

Use of MW irradiation in the synthesis of inorganic nanomaterials has gained significant attention[161] but has mainly been conducted in liquid medium (water, polar organic solvents, ionic liquids, etc.). These solvents are known for absorbing MW energy to induce dielectric heating needed to carry out the reaction.

In 2011, Connell et al. utilized MW energy to synthesize various metal oxide nanoparticles with the direct solid-state Joule heating of CNTs or graphene with organic metal salts.[162] A typical reaction involved the direct mixing of metal salts (e.g., silver acetate) with multiwalled CNTs using mortar and pestle and then irradiating the mixture with MWs for an hour at appropriate temperature and power to give the desired particles. After the reaction was completed the product was dried in air to remove nontoxic volatile components. This method was shown to be widely applicable in the formation of a variety of substrate-supported metals such as Ag, Au, Co, Ni, Pd, or Pt, or metal oxides, e.g., Fe_3O_4, MnO, or TiO_2, all of which formed nanoparticles in high yields. The process was further applied to large substrates such as nanotube yarns.

The major advantage of this method is that CNTs and graphene provided MW good absorption capacity and allowed local heating of the substrates. The Joule heating from CNTs and graphene gave the necessary temperature to decompose the metal salts for instantaneous formation of nanoparticles within seconds.

3.17.4 CONCLUSIONS

Currently, nanoparticles are a hot topic in various disciplines of science. A tremendous number of potential applications have been reported for nanoparticles, e.g., drug delivery; medicine; bio-, chemo-, and electrochemical sensors; antibacterials; antimicrobials; and magnetic materials. The present chapter summarized various ways to synthesize nanoparticles utilizing solvent-free approaches that conform to the principles of green chemistry. Thermolysis including conventional heating and MW energy, and mechanochemistry (for example, ball milling and mortar-pestle) are the primary synthetic methods described in this chapter. Using no solvent in the synthesis makes it greener and safer, and at the same time it can provide easy access to ultrafine nanoparticles.

References

1. Kelly KL, Coronado E, Zhao LL, Schatz GC. The optical properties of metal nanoparticles: the influence of size, shape, and dielectric environment. *J Phys Chem B* 2003;**107**:668–77.
2. Sharma N, Ojha H, Bharadwaj A, Pathak DP, Sharma RK. Preparation and catalytic applications of nanomaterials: a review. *RSC Adv* 2015;**5**:53381–403.
3. Bruchez M, Moronne M, Gin P, Weiss S, Alivisatos AP. Semiconductor nanocrystals as fluorescent biological labels. *Science* 1998;**281**:2013–6.
4. Lu A-H, Salabas EL, Schüth F. Magnetic nanoparticles: synthesis, protection, functionalization, and application. *Angew Chem Int Ed* 2007;**46**:1222–44.
5. De M, Ghosh PS, Rotello VM. Applications of nanoparticles in biology. *Adv Mater* 2008;**20**:4225–41.
6. Abe H, Miyamoto Y, Umetsu M, Uchikoshi T, Okubo T, Naito M. Chapter 4-Control of nanostructure of materials. In: *Nanoparticle technology handbook*. Amsterdam: Elsevier; 2008. p. 177–265.
7. Balaz P, Achimovicova M, Balaz M, Billik P, Cherkezova-Zheleva Z, Criado JM, Delogu F, Dutkovo E, Gaffet E, Gotor FJ, Kumar R, Mitov I, Rojac T, Senna M, Streletski A, Wieczorek-Ciurowa K. Hallmarks of mechanochemistry: from nanoparticles to technology. *Chem Soc Rev* 2013;**42**:7571–637.
8. Stratakis M, Garcia H. Catalysis by supported gold nanoparticles: beyond aerobic oxidative processes. *Chem Rev* 2012;**112**:4469–506.
9. Mikami Y, Dhakshinamoorthy A, Alvaro M, Garcia H. Catalytic activity of unsupported gold nanoparticles. *Catal Sci Tech* 2013;**3**:58–69.
10. Saha K, Agasti SS, Kim C, Li X, Rotello VM. Gold nanoparticles in chemical and biological sensing. *Chem Rev* 2012;**112**:2739–79.
11. Yeh Y-C, Creran B, Rotello VM. Gold nanoparticles: preparation, properties, and applications in bionanotechnology. *Nanoscale* 2012;**4**:1871–80.
12. Dahl JA, Maddux BLS, Hutchison JE. Toward greener nanosynthesis. *Chem Rev* 2007;**107**:2228–69.
13. Debnath D, Kim SH, Geckeler KE. The first solid-phase route to fabricate and size-tune gold nanoparticles at room temperature. *J Mat Chem* 2009;**19**:8810–6.
14. Rak MJ, Saade NK, Friscic T, Moores A. Mechanosynthesis of ultra-small monodisperse amine-stabilized gold nanoparticles with controllable size. *Green Chem* 2014;**16**:86–9.

15. Liao F, Huang Y, Ge J, Zheng W, Tedsree K, Collier P, Hong X, Tsang SC. Morphology-dependent interactions of ZnO with Cu nanoparticles at the materials' interface in selective hydrogenation of CO_2 to CH_3OH. *Angew Chem Int Ed* 2011;**50**:2162–5.
16. Goeppert A, Czaun M, Jones J-P, Surya Prakash GK, Olah GA. Recycling of carbon dioxide to methanol and derived products – closing the loop. *Chem Soc Rev* 2014;**43**:7995–8048.
17. Dibenedetto A, Angelini A, Stufano P. Use of carbon dioxide as feedstock for chemicals and fuels: homogeneous and heterogeneous catalysis. *J Chem Tech Biotech* 2014;**89**:334–53.
18. Angelo L, Kobl K, Tejada LMM, Zimmermann Y, Parkhomenko K, Roger A-C. Study of CuZnMOx oxides (M = Al, Zr, Ce, CeZr) for the catalytic hydrogenation of CO_2 into methanol. *C R Chim* 2015;**18**:250–60.
19. Jun K-W, Shen W-J, Rama Rao KS, Lee K-W. Residual sodium effect on the catalytic activity of $Cu/ZnO/Al_2O_3$ in methanol synthesis from CO_2 hydrogenation. *Appl Catal A* 1998;**174**:231–8.
20. Zangeneh FT, Sahebdelfar S, Ravanchi MT. Conversion of carbon dioxide to valuable petrochemicals: an approach to clean development mechanism. *J Nat Gas Chem* 2011;**20**:219–31.
21. Lei H, Hou Z, Xie J. Hydrogenation of CO_2 to CH_3OH over $CuO/ZnO/Al_2O_3$ catalysts prepared via a solvent-free routine. *Fuel* 2016;**164**:191–8.
22. Yang R, Yu X, Zhang Y, Li W, Tsubaki N. A new method of low-temperature methanol synthesis on $Cu/ZnO/Al_2O_3$ catalysts from $CO/CO_2/H_2$. *Fuel* 2008;**87**:443–50.
23. Witoon T, Permsirivanich T, Donphai W, Jaree A, Chareonpanich M. CO_2 hydrogenation to methanol over Cu/ZnO nanocatalysts prepared via a chitosan-assisted co-precipitation method. *Fuel Process Technol* 2013;**116**:72–8.
24. Yang G, Tsubaki N, Shamoto J, Yoneyama Y, Zhang Y. Confinement effect and synergistic function of H-ZSM-5/Cu-ZnO-Al_2O_3 capsule catalyst for one-step controlled synthesis. *J Am Chem Soc* 2010;**132**:8129–36.
25. Behrens M. Meso- and nano-structuring of industrial $Cu/ZnO/(Al_2O_3)$ catalysts. *J Catal* 2009;**267**:24–9.
26. Bae JW, Kang S-H, Lee Y-J, Jun K-W. Synthesis of DME from syngas on the bifunctional Cu–ZnO–Al_2O_3/Zr-modified ferrierite: effect of Zr content. *Appl Catal B* 2009;**90**:426–35.
27. Shishido T, Yamamoto Y, Morioka H, Takaki K, Takehira K. Active Cu/ZnO and $Cu/ZnO/Al_2O_3$ catalysts prepared by homogeneous precipitation method in steam reforming of methanol. *Appl Catal A* 2004;**263**:249–53.
28. Katagiri H, Sasaguchi N, Hando S, Hoshino S, Ohashi J, Yokota T. Preparation and evaluation of Cu_2ZnSnS_4 thin films by sulfurization of E-B evaporated precursors. *Sol En Mat Sol Cells* 1997;**49**:407–14.
29. Wang Y, Gong H. Cu_2ZnSnS_4 synthesized through a green and economic process. *J Alloys Compd* 2011;**509**:9627–30.
30. Takeshi N, Tsuyoshi M, Takahiro W. Fabrication of Cu_2SnS_3 solar cells by screen-printing and high-pressure sintering process. *Jpn J App Phys* 2014;**53**. 05FW1.
31. Feng G, Seiji Y, Tsuyoshi M, Takahiro W. Structural study of Cu-deficient $Cu_{2(1-x)}ZnSnSe_4$ solar cell materials by X-ray diffraction and X-ray absorption fine structure. *Jpn J App Phys* 2012;**51**:10NC28.
32. Park B-I, Hwang Y, Lee SY, Lee J-S, Park J-K, Jeong J, Kim JY, Kim B, Cho SH, Lee DK. Solvent-free synthesis of Cu_2ZnSnS_4 nanocrystals: a facile, green, up-scalable route for low cost photovoltaic cells. *Nanoscale* 2014;**6**:11703–11.
33. Azanza Ricardo CL, Su'ait MS, Müller M, Scardi P. Production of Cu_2(Zn,Fe)SnS4 powders for thin film solar cell by high energy ball milling. *J Power Sources* 2013;**230**:70–5.
34. Mallampati SR, Mitoma Y, Okuda T, Simion C, Lee BK. Solvent-free synthesis and application of nano-Fe/Ca/CaO/[PO_4] composite for dual separation and immobilization of stable and radioactive cesium in contaminated soils. *J Hazard Mater* 2015;**297**:74–82.
35. Tsukamoto M, Sugiyama D, Yamamoto T, Kawanish M, Saito N. Review of characteristics of post-accident waste generated in Fukushima Daiichi nuclear power plant site and issues to be addressed in in processing and disposal stages. *J Nuc Eng Sci Power Gen Techn* 2013:S001–2.
36. Zakaib GD. Radiation risks unknown. *Nature* 2011;**471**:419.
37. Yoshida N, Takahashi Y. Land-surface contamination by radionuclides from the Fukushima Daiichi nuclear power plant accident. *Elements* 2012;**8**:201–6.
38. Endo S, Kimura S, Takatsuji T, Nanasawa K, Imanaka T, Shizuma K. Measurement of soil contamination by radionuclides due to the Fukushima Dai-ichi nuclear power plant accident and associated estimated cumulative external dose estimation. *J Environ Radio* 2012;**111**:18–27.
39. Morino Y, Ohara T, Watanabe M, Hayashi S, Nishizawa M. Episode analysis of deposition of radiocesium from the fukushima Daiichi nuclear power plant accident. *Environ Sci Technol* 2013;**47**:2314–22.

40. Miyazaki T, Kitakami O, Okamoto S, Shimada Y, Akase Z, Murakami Y, Shindo D, Takahashi YK, Hono K. Size effect on the ordering of $L1_0$FePt nanoparticles. *Phys Rev B* 2005;**72**:1–5. 144419.
41. Bader SD. Colloquium: opportunities in nanomagnetism. *Rev Mod Phys* 2006;**78**:1–15.
42. Yuji I, Takehiro A, Arata T, Katsuji N, Akiyoshi I, Toshikazu K. Structural and magnetization properties of island FePt produced by rapid thermal annealing. *Jpn J App Phys* 2004;**43**:8040–3.
43. Sun S, Murray CB, Weller D, Folks L, Moser A. Monodisperse FePt nanoparticles and ferromagnetic FePt nanocrystal superlattices. *Science* 2000;**287**:1989–92.
44. Capobianchi A, Colapietro M, Fiorani D, Foglia S, Imperatori P, Laureti S, Palange E. General strategy for direct synthesis of $L1_0$ nanoparticle alloys from layered precursor: the case of FePt. *Chem Mater* 2009;**21**: 2007–9.
45. Hu XC, Agostinelli E, Ni C, Hadjipanayis GC, Capobianchi A. A low temperature and solvent-free direct chemical synthesis of $L1_0$ FePt nanoparticles with size tailoring. *Green Chem* 2014;**16**:2292–7.
46. Surmenev RA. A review of plasma-assisted methods for calcium phosphate-based coatings fabrication. *Surf Coat Tech* 2012;**206**:2035–56.
47. Shahid N, Barron AR. Solvent free synthesis of carboxylate-alumoxane nanoparticles using mechanical shear. *J Mater Chem* 2004;**14**:1235–7.
48. Vogelson CT, Koide Y, Alemany LB, Barron AR. Inorganic–organic hybrid and composite resin materials using carboxylate-alumoxanes as functionalized cross-linking agents. *Chem Mater* 2000;**12**:795–804.
49. Anand A, Kulkarni RD, Patil CK, Gite VV. Utilization of renewable bio-based resources, viz. sorbitol, diol, and diacid, in the preparation of two pack PU anticorrosive coatings. *RSC Adv* 2016;**6**:9843–50.
50. Liu X, Wang X, Yao S, Jiang Y, Guan J, Mu X. Recent advances in the production of polyols from lignocellulosic biomass and biomass-derived compounds. *RSC Adv* 2014;**4**:49501–20.
51. Du W-C, Zheng L-P, Shi J-J, Xia S-X, Hou Z-Y. Production of C2 and C3 polyols from d-sorbitol over hydrotalcite-like compounds mediated bi-functional Ni–Mg–Al–Ox catalysts. *Fuel Process Technol* 2015;**139**:86–90.
52. Du W, Zheng L, Li X, Fu J, Lu X, Hou Z. Plate-like Ni–Mg–Al layered double hydroxide synthesized via a solvent-free approach and its application in hydrogenolysis of D-sorbitol. *App Clay Sci* 2016;**123**:166–72.
53. An Z, Zhang W, Shi H, He J. An effective heterogeneous l-proline catalyst for the asymmetric aldol reaction using anionic clays as intercalated support. *J Catal* 2006;**241**:319–27.
54. Xia S, Zheng L, Ning W, Wang L, Chen P, Hou Z. Multiwall carbon nanotube-pillared layered $Cu_{0.4}/Mg_{5.6}Al_2O_{8.6}$: an efficient catalyst for hydrogenolysis of glycerol. *J Mater Chem A* 2013;**1**:11548–52.
55. Chen W-H, Peng J, Bi XT. A state-of-the-art review of biomass torrefaction, densification and applications. *Renew Sus Energ Rev* 2015;**44**:847–66.
56. Laursen AB, Højholt KT, Lundegaard LF, Simonsen SB, Helveg S, Schüth F, Paul M, Grunwaldt J-D, Kegnaes S, Christensen CH, Egeblad K. Substrate size-selective catalysis with zeolite-encapsulated gold nanoparticles. *Angew Chem* 2010;**122**:3582–5.
57. Esken D, Turner S, Lebedev OI, Van Tendeloo G, Fischer RA. Au@ZIFs: stabilization and encapsulation of cavity-size matching gold clusters inside functionalized zeolite imidazolate frameworks, ZIFs. *Chem Mater* 2010;**22**:6393–401.
58. Farrusseng D, Tuel A. Perspectives on zeolite-encapsulated metal nanoparticles and their applications in catalysis. *New J Chem* 2016;**40**:3933–49.
59. Zhang J, Wang L, Zhu L, Wu Q, Chen C, Wang X, Ji Y, Meng X, Xiao F-S. Solvent-free synthesis of zeolite crystals encapsulating gold–palladium nanoparticles for the selective oxidation of bioethanol. *ChemSusChem* 2015;**8**:2867–71.
60. Ren L, Wu Q, Yang C, Zhu L, Li C, Zhang P, Zhang H, Meng X, Xiao F-S. Solvent-free synthesis of zeolites from solid raw materials. *J Am Chem Soc* 2012;**134**:15173–6.
61. Jin Y, Sun Q, Qi G, Yang C, Xu J, Chen F, Meng X, Deng F, Xiao F-S. Solvent-free synthesis of silicoaluminophosphate zeolites. *Angew Chem Int Ed* 2013;**52**:9172–5.
62. Wu Q, Wang X, Qi G, Guo Q, Pan S, Meng X, Xu J, Deng F, Fan F, Feng Z, Li C, Maurer S, Muller U, Xiao F-S. Sustainable synthesis of zeolites without addition of both organotemplates and solvents. *J Am Chem Soc* 2014;**136**:4019–25.
63. Jin Y, Chen X, Sun Q, Sheng N, Liu Y, Bian C, Chen F, Meng X, Xiao FS. Solvent-free syntheses of hierarchically porous aluminophosphate-based zeolites with AEL and AFI Structures. *Chem Eur J* 2014;**20**:17616–23.

64. Wang G, Zhang L, Zhang J. A review of electrode materials for electrochemical supercapacitors. *Chem Soc Rev* 2012;**41**:797–828.
65. Bastakoti BP, Huang H-S, Chen L-C, Wu KCW, Yamauchi Y. Block copolymer assisted synthesis of porous α-$Ni(OH)_2$ microflowers with high surface areas as electrochemical pseudocapacitor materials. *Chem Commun* 2012;**48**:9150–2.
66. Tang Z, Tang C-H, Gong H. A high energy density asymmetric supercapacitor from nano-architectured $Ni(OH)_2$/carbon nanotube electrodes. *Adv Funct Mater* 2012;**22**:1272–8.
67. Cai F-S, Zhang G-Y, Chen J, Gou X-L, Liu H-K, Dou S-X. $Ni(OH)_2$ Tubes with mesoscale dimensions as positive-electrode materials of alkaline rechargeable batteries. *Angew Chem Int Ed* 2004;**43**:4212–6.
68. Cui H, Xue J, Wang M. Synthesis of high electrochemical performance $Ni(OH)_2$ nanosheets through a solvent-free reaction for application in supercapacitor. *Adv Powd Tech* 2015;**26**:434–8.
69. Greeley J, Mavrikakis M. Alloy catalysts designed from first principles. *Nat Mater* 2004;**3**:810–5.
70. Ferrando R, Jellinek J, Johnston RL. Nanoalloys: from theory to applications of alloy clusters and nanoparticles. *Chem Rev* 2008;**108**:845–910.
71. He L, Liu Y, Liu J, Xiong Y, Zheng J, Liu Y, Tang Z. Core–Shell noble-metal@metal-organic-framework nanoparticles with highly selective sensing property. *Angew Chem Int Ed* 2013;**52**:3741–5.
72. Ghosh Chaudhuri R, Paria S. Core/Shell nanoparticles: classes, properties, synthesis mechanisms, characterization, and applications. *Chem Rev* 2012;**112**:2373–433.
73. Paszkiewicz M, Golabiewska A, Rajaski L, Kowal E, Sajdak A, Zaleska-Medynska A. The antibacterial and antifungal textile properties functionalized by bimetallic nanoparticles of Ag/Cu with different structures. *J Nanomater* 2016;**2016**:1–13.
74. Yang A, Li S, Wang Y, Wang L, Bao X, Yang R. Synthesis of Ag@Cu_2O core-shell metal-semiconductor nanoparticles and conversion to Ag@Cu core-shell bimetallic nanoparticles. *Sci China Technol S. C* 2015;**58**:881–8.
75. Gawande MB, Goswami A, Felpin F-X, Asefa T, Huang X, Silva R, Zou X, Zboril R, Varma RS. Cu and Cu-Based nanoparticles: synthesis and applications in catalysis. *Chem Rev* 2016;**116**:3722–811.
76. Choi E, Lee S, Piao Y. A solventless mix-bake-wash approach to the facile controlled synthesis of core-shell and alloy Ag-Cu bimetallic nanoparticles. *Cryst Eng Comm* 2015;**17**:5940–6.
77. Park Y, Piao Y, Lee N, Yoo B, Kim BH, Choi SH, Hyeon T. Transformation of hydrophobic iron oxide nanoparticles to hydrophilic and biocompatible maghemite nanocrystals for use as highly efficient MRI contrast agent. *J Mater Chem* 2011;**21**:11472–7.
78. Pankhurst QA, Connolly J, Jones SK, Dobson J. Applications of magnetic nanoparticles in biomedicine. *J Phys D Appl Phys* 2003;**36**. R167.
79. Berry CC, Curtis SGA. Functionalisation of magnetic nanoparticles for applications in biomedicine. *J Phys D Appl Phys* 2003;**36**. R198.
80. Feng S-S, Chien S. Chemotherapeutic engineering: application and further development of chemical engineering principles for chemotherapy of cancer and other diseases. *Chem Eng Sci* 2003;**58**:4087–114.
81. Fried T, Shemer G, Markovich G. Ordered two-dimensional arrays of ferrite nanoparticles. *Adv Mater* 2001;**13**:1158–61.
82. Rümenapp C, Wagner FE, Gleich B. Monitoring of the aging of magnetic nanoparticles using Mössbauer spectroscopy. *J Magn Magn Mater* 2015;**380**:241–5.
83. Lemine OM, Omri K, Zhang B, El Mir L, Sajieddine M, Alyamani A, Bououdina M. Sol–gel synthesis of 8 nm magnetite (Fe_3O_4) nanoparticles and their magnetic properties. *Superlattices Micros* 2012;**52**:793–9.
84. Azizi K, Karimi M, Shaterian HR, Heydari A. Ultrasound irradiation for the green synthesis of chromenes using L-arginine-functionalized magnetic nanoparticles as a recyclable organocatalyst. *RSC Adv* 2014;**4**:42220–5.
85. Bomatí-Miguel O, Leconte Y, Morales MP, Herlin-Boime N, Veintemillas-Verdaguer S. Laser pyrolysis preparation of SiO_2-coated magnetic nanoparticles for biomedical applications. *J Magn Magn Mater* 2005; **290–291**(Part 1):272–5.
86. Wu W, Wu Z, Yu T, Jiang C, Kim W-S. Recent progress on magnetic iron oxide nanoparticles: synthesis, surface functional strategies and biomedical applications. *Sci Tech Adv Mater* 2015;**16**. 023501.
87. Liu C, Zou B, Rondinone AJ, Zhang ZJ. Reverse micelle synthesis and characterization of superparamagnetic $MnFe_2O_4$ spinel ferrite nanocrystallites. *J Phys Chem B* 2000;**104**:1141–5.
88. Ye XR, Daraio C, Wang C, Talbot JB, Jin S. Room temperature solvent-free synthesis of monodisperse magnetite nanocrystals. *J Nanosci Nanotechnol* 2006;**6**:852–6.

89. Poizot P, Laruelle S, Grugeon S, Dupont L, Tarascon JM. Nano-sized transition-metal oxides as negative-electrode materials for lithium-ion batteries. *Nature* 2000;**407**:496–9.
90. Etacheri V, Marom R, Elazari R, Salitra G, Aurbach D. Challenges in the development of advanced Li-ion batteries: a review. *Energ Environ Sci* 2011;**4**:3243–62.
91. Jang B, Chae OB, Park S-K, Ha J, Oh SM, Na HB, Piao Y. Solventless synthesis of an iron-oxide/graphene nanocomposite and its application as an anode in high-rate Li-ion batteries. *J Mater Chem A* 2013;**1**: 15442–6.
92. Hiratsuka N, Kobayashi H, Uchida H, Katsube T. Gas sensing characteristics of zinc-tin complex oxide thin films with spinel-type structure. *J Ceram Soc Jpn* 1996;**104**:1048–51.
93. Yu-Sheng S, Tian-Shu Z. Preparation, structure and gas-sensing properties of ultramicro $ZnSnO_3$ powder. *Sens Actuators B* 1993;**12**:5–9.
94. Wang Z, Liu J, Wang F, Chen S, Luo H, Yu X. Size-controlled synthesis of $ZnSnO_3$ cubic crystallites at low temperatures and their HCHO-sensing properties. *J Phys Chem C* 2010;**114**:13577–82.
95. Singh RC, Singh O, Singh MP, Chandi PS, Thangaraj R. Sensing behaviour of nanosized zinc–tin composite oxide towards liquefied petroleum gas and ethanol. *Mater Res Bull* 2010;**45**:1162–4.
96. Men H, Gao P, Zhou B, Chen Y, Zhu C, Xiao G, et al. Fast synthesis of ultra-thin $ZnSnO_3$ nanorods with high ethanol sensing properties. *Chem Commun* 2010;**46**:7581–3.
97. Jackson WB, Hoffman RL, Herman GS. High-performance flexible zinc tin oxide field-effect transistors. *Appl Phys Lett* 2005;**87**:3. 193503.
98. Bârsan N, Weimar U. Understanding the fundamental principles of metal oxide based gas sensors; the example of CO sensing with SnO_2 sensors in the presence of humidity. *J Phys Condens Matter* 2003;**15**. R813.
99. Xu J, Jia X, Lou X, Shen J. One-step hydrothermal synthesis and gas sensing property of $ZnSnO_3$ microparticles. *Solid-state Electron* 2006;**50**:504–7.
100. Wang L, Zhang X, Liao X, Yang W. A simple method to synthesize single-crystalline Zn_2SnO_4 (ZTO) nanowires and their photoluminescence properties. *Nanotechnology* 2005;**16**:2928–31.
101. Lu L, Zhang A, Xiao Y, Gong F, Jia D, Li F. Effect of solid inorganic salts on the formation of cubic-like aggregates of $ZnSnO_3$ nanoparticles in solventless, organic-free reactions and their gas sensing behaviors. *Mater Sci Eng B* 2012;**177**:942–8.
102. Redón R, Ramírez-Crescencio F, Fernández-Osorio AL. Solventless synthesis of iridium(0) nanoparticles. *J Nanopart Res* 2011;**13**:5959–65.
103. Yang H, Zhang X, Tang A. Mechanosynthesis and gas-sensing properties of In_2O_3/SnO_2 nanocomposites. *Nanotechnology* 2006;**17**:2860.
104. Delogu F, Orrù R, Cao G. A novel macrokinetic approach for mechanochemical reactions. *Chem Eng Sci* 2003;**58**:815–21.
105. Šepelák V, Feldhoff A, Heitjans P, Krumeich F, Menzel D, Litterst FJ, et al. Nonequilibrium cation distribution, canted spin arrangement, and enhanced magnetization in nanosized $MgFe_2O_4$ prepared by a one-step mechanochemical route. *Chem Mater* 2006;**18**:3057–67.
106. Ennas G, Marongiu G, Marras S, Piccaluga G. Mechanochemical route for the synthesis of cobalt ferrite–silica and iron–cobalt alloy–silica nanocomposites. *J Nanopart Res* 2004;**6**:99–105.
107. Ivanov E, Suryanarayana C. Materials and process design through mechanochemical routes. *J Mater Synth Process* 2000;**8**:235–44.
108. García-Peña NG, Redón R, Herrera-Gomez A, Fernández-Osorio AL, Bravo-Sanchez M, Gomez-Sosa G. Solventless synthesis of ruthenium nanoparticles. *Appl Surf Sci* 2015;**340**:25–34.
109. Yang J, Deivaraj TC, Too H-P, Lee JY. Acetate stabilization of metal nanoparticles and its role in the preparation of metal nanoparticles in ethylene glycol. *Langmuir* 2004;**20**:4241–5.
110. Harpeness R, Peng Z, Liu X, Pol VG, Koltypin Y, Gedanken A. Controlling the agglomeration of anisotropic Ru nanoparticles by the microwave–polyol process. *J Colloid Interface Sci* 2005;**287**:678–84.
111. Park K-W, Choi J-H, Kwon B-K, Lee S-A, Sung Y-E, Ha H-Y, et al. Chemical and electronic effects of Ni in Pt/Ni and Pt/Ru/Ni alloy nanoparticles in methanol electrooxidation. *J Phys Chem B* 2002;**106**:1869–77.
112. Yang J, Lee JY, Deivaraj TC, Too H-P. Preparation and characterization of positively charged ruthenium nanoparticles. *J Colloid Interface Sci* 2004;**271**:308–12.
113. Suib SL. A review of open framework structures. *Annu Rev Mater Sci* 1996;**26**:135–51.

114. Ding Y-S, Shen X-F, Sithambaram S, Gomez S, Kumar R, Crisostomo VMB, Suib SL, Aindow M. Synthesis and catalytic activity of cryptomelane-type manganese dioxide nanomaterials produced by a novel solvent-free method. *Chem Mater* 2005;**17**:5382–9.
115. Corma A. Inorganic solid acids and their use in acid-catalyzed hydrocarbon reactions. *Chem Revs* 1995;**95**:559–614.
116. Yadav GD, Nair JJ. Sulfated zirconia and its modified versions as promising catalysts for industrial processes. *Microporous Mesoporous Mater* 1999;**33**:1–48.
117. Sun Y, Ma S, Du Y, Yuan L, Wang S, Yang J, et al. Solvent-free preparation of nanosized sulfated zirconia with Brønsted acidic sites from a simple calcination. *J Phys Chem B* 2005;**109**:2567–72.
118. Ge C, Xie C, Cai S. Preparation and gas-sensing properties of Ce-doped ZnO thin-film sensors by dip-coating. *Mater Sci Eng B* 2007;**137**:53–8.
119. Yu X, Li F, Ye X, Xin X, Xue Z. Synthesis of cerium(IV) oxide ultrafine particles by solid-state reactions. *J Am Ceram Soc* 2000;**83**:964–6.
120. Mohadesi A, Ranjbar M, Hosseinpour-Mashkani SM. Solvent-free synthesis of mercury oxide nanoparticles by a simple thermal decomposition method. *Superlattices Microstruct* 2014;**66**:48–53.
121. Ranjbar M, Taher MA, Sam A. Solvent-free synthesis of ZnO nanoparticles by a simple thermal decomposition method. *J Clust Sci* 2014;**25**:1657–64.
122. Ranjbar M, Taher MA, Sam A. NiO nanostructures: novel solvent-less solid-state synthesis, characterization and MB photocatalytic degradation. *J Mater Sci Mater Electron* 2015;**26**:8029–34.
123. Soofivand F, Salavati-Niasari M. Novel solvent-less synthesis of CuO nanoparticles by using sublimated precursors. *Mater Lett* 2013;**106**:83–6.
124. Babaheydari AK, Salavati-Niasari M, Khansari A. Solvent-less synthesis of zinc oxide nanostructures from Zn(salen) as precursor and their optical properties. *Particuology* 2012;**10**:759–64.
125. Saravanan M, Dhivakar S, Jayanthi SS. An eco friendly and solvent free method for the synthesis of Zinc oxide nano particles using glycerol as organic dispersant. *Mater Lett* 2012;**67**:128–30.
126. Seinberg L, Yamamoto S, Gallage R, Tsujimoto M, Kobayashi Y, Isoda S, et al. Low temperature solventless synthesis and characterization of Ni and Fe magnetic nanoparticles. *Chem Commun* 2012;**48**: 8237–9.
127. Zhang C, Zhang S, Yu L, Zhang Z, Zhang P, Wu Z. Size-controlled synthesis of monodisperse Ag_2S nanoparticles by a solventless thermolytic method. *Mater Lett* 2012;**85**:77–80.
128. Barreiro A, Hampel S, Rümmeli MH, Kramberger C, Grüneis A, Biedermann K, et al. Thermal decomposition of ferrocene as a method for production of single-walled carbon nanotubes without additional carbon sources. *J Phy Chem B* 2006;**110**:20973–7.
129. Elihn K, Landstrom L, Alm O, Boman M, Heszler P. Size and structure of nanoparticles formed via ultraviolet photolysis of ferrocene. *J Appl Phys* 2007;**101**:034311–5.
130. Bhattacharjee A, Rooj A, Roy M, Kusz J, Gütlich P. Solventless synthesis of hematite nanoparticles using ferrocene. *J Mater Sci* 2013;**48**:2961–8.
131. Bhattacharjee A, Rooj A, Madhusudan R, Kusz J. A solventless method to prepare hematite using thermal decomposition of ferrocene. *Int J Exp Spectrosc Tech* 2016;**1**:1–6.
132. Tartaj P, del Puerto Morales M, Veintemillas-Verdaguer S, González-Carreño T, Carlos JS. The preparation of magnetic nanoparticles for applications in biomedicine. *J Phys D Appl Phys* 2003;**36**. R182.
133. Gupta AK, Gupta M. Synthesis and surface engineering of iron oxide nanoparticles for biomedical applications. *Biomaterials* 2005;**26**:3995–4021.
134. Kluchova K, Zboril R, Tucek J, Pecova M, Zajoncova L, Safarik I, et al. Superparamagnetic maghemite nanoparticles from solid-state synthesis – their functionalization towards peroral MRI contrast agent and magnetic carrier for trypsin immobilization. *Biomaterials* 2009;**30**:2855–63.
135. Liu J, Lin S, Qi D, Deng C, Yang P, Zhang X. On-chip enzymatic microreactor using trypsin-immobilized superparamagnetic nanoparticles for highly efficient proteolysis. *J J Chromatogr A* 2007;**1176**:169–77.
136. Jeng J, Lin M-F, Cheng F-Y, Yeh C-S, Shiea J. Using high-concentration trypsin-immobilized magnetic nanoparticles for rapid in situ protein digestion at elevated temperature. *Rapid Commun Mass Spectrom* 2007;**21**:3060–8.
137. Šebela M, Štosová Tá HJ, Wielsch N, Thomas H, Zdráhal Z, et al. Thermostable trypsin conjugates for high-throughput proteomics: synthesis and performance evaluation. *Proteomic* 2006;**6**:2959–63.

138. Wang C, Oleschuk R, Ouchen F, Li J, Thibault P, Harrison DJ. Integration of immobilized trypsin bead beds for protein digestion within a microfluidic chip incorporating capillary electrophoresis separations and an electrospray mass spectrometry interface. *Rapid Commun Mass Spectrom* 2000;**14**:1377–83.
139. Belessi V, Zboril R, Tucek J, Mashlan M, Tzitzios V, Petridis D. Ferrofluids from magnetic–chitosan hybrids. *Chem Mater* 2008;**20**:3298–305.
140. Li G-Y, Jiang Y-R, Huang K-L, Ding P, Yao L-L. Kinetics of adsorption of saccharomyces cerevisiae mandelated dehydrogenase on magnetic Fe_3O_4–chitosan nanoparticles. *Colloid Surf A* 2008;**320**:11–8.
141. Beyki MH, Alijani H, Fazli Y. Solvent free synthesized $MnFe_2O_4$@polyamid resin as a novel green nanohybrid for fast removing Congo red. *J Mol Liq* 2016;**216**:6–11.
142. Wang W, Lu Y-C, Huang H, Wang A-J, Chen J-R, Feng J-J. Solvent-free synthesis of sulfur- and nitrogen-co-doped fluorescent carbon nanoparticles from glutathione for highly selective and sensitive detection of mercury(II) ions. *Sens Actuators B* 2014;**202**:741–7.
143. Kauffman DR, Ohodnicki PR, Kail BW, Matranga C. Selective electrocatalytic activity of ligand stabilized copper oxide nanoparticles. *J Phys Chem Lett* 2011;**2**:2038–43.
144. Estruga M, Domingo C, Ayllon JA. Solution-processable ZnO nanoparticles obtained by low-temperature solventless synthesis. *J Mater Chem* 2011;**21**:4408–15.
145. Estruga M, Roig A, Domingo C, Ayllón JA. Solution-processable carboxylate-capped CuO nanoparticles obtained by a simple solventless method. *J Nanopart Res* 2012;**14**:1–9.
146. Philipse AP, van Bruggen MPB, Pathmamanoharan C. Magnetic silica dispersions: preparation and stability of surface-modified silica particles with a magnetic core. *Langmuir* 1994;**10**:92–9.
147. Amara D, Margel S. Solventless thermal decomposition of ferrocene as a new approach for the synthesis of porous superparamagnetic and ferromagnetic composite microspheres of narrow size distribution. *J Mater Chem* 2011;**21**:15764–72.
148. Jose D, Jagirdar BR. Synthesis and characterization of Pd(0), PdS, and Pd@PdO core–shell nanoparticles by solventless thermolysis of a Pd–thiolate cluster. *J Solid State Chem* 2010;**183**:2059–67.
149. Park J, Joo J, Kwon SG, Jang Y, Hyeon T. Synthesis of monodisperse spherical nanocrystals. *Angew Chem Int Ed* 2007;**46**:4630–60.
150. Cha HG, Kim YH, Kim CW, Lee DK, Moon SD, Kwon HW, et al. Synthesis of α-Fe nanoparticles by solventless thermal decomposition. *J Nanosci Nanotechnol* 2006;**6**:3412–6.
151. Han YC, Cha HG, Kim CW, Kim YH, Kang YS. Synthesis of highly magnetized iron nanoparticles by a solventless thermal decomposition method. *J Phys Chem C* 2007;**111**:6275–80.
152. Li Y, Zhang XL, Kim YH, Kang YS. Synthesis and characterization of Co nanoparticles by solventless thermal decomposition. *SSP* 2007:71–4.
153. Wang D, Ma Q, Yang P. Synthesis of Fe_3O_4 nanoparticles with tunable and uniform size through simple thermal decomposition. *J Nanosci Nanotechnol* 2012;**12**:6432–8.
154. Nakaya M, Nishida R, Muramatsu A. *Chem Lett* 2013;**42**:863–5.
155. Nakaya M, Nishida R, Muramatsu A. Size control of magnetite nanoparticles in excess ligands as a function of reaction temperature and time. *Molecules* 2014;**19**:11395–403.
156. Abbas M, Kienberger J. Solvent-free synthesis of silver-nanoparticles and their use as additive in poly (dicyclopentadiene). *J Chem Soc Pak* 2013;**35**:360–3.
157. Maity D, Choo S-G, Yi J, Ding J, Xue JM. Synthesis of magnetite nanoparticles via a solvent-free thermal decomposition route. *J Magn Magn Mater* 2009;**321**:1256–9.
158. Yamamoto M, Kashiwagi Y, Nakamoto M. Size-controlled synthesis of gold nanoparticles by thermolysis of a gold(I)-sulfide complex in the presence of alkylamines. *Z Naturforsch B* 2009;**64**:1305–11.
159. Kappe CO, Stadler A. Introduction: microwave synthesis in perspective. In: *Microwaves in organic and medicinal chemistry*. Wiley-VCH Verlag GmbH & Co. KGaA; 2005. p. 1–7.
160. Chandore V, Carpenter G, Sen R, Gupta N. Synthesis of nano crystalline ZnO by microwave assisted combustion method: an eco-friendly and solvent free route. *Int J Environ Sci Dev Monit* 2013;**4**:45–7.
161. Polshettiwar V, Nadagouda MN, Varma RS. Microwave-assisted chemistry: a rapid and sustainable route to synthesis of organics and nanomaterials. *Aust J Chem* 2009;**62**:16–26.
162. Lin Y, Baggett DW, Kim J-K, Siochi EJ, Connell JW. Instantaneous formation of metal and metal oxide nanoparticles on carbon nanotubes and graphene via solvent-free microwave heating. *ACS Appl Mater Interfaces* 2011;**3**:1652–64.

CHAPTER

3.18

Application of Microwaves in Sustainable Organic Synthesis

Abid Shaikh

Georgia Southern University, Statesboro, GA, United States

3.18.1 INTRODUCTION

The application of microwaves in organic synthesis when compared with conventional heating methods is a valuable tool for acceleration of a wide variety of organic reactions.[1] Modern discovery chemistry also places new demands on synthetic routes, requiring rapid access to related structures through simple modifications. The various synthetic methods that had already been reported in the literature, however, are often revisited due to strengthening environmental regulations and safety concerns. Thus the development of novel environmentally benign and energy-efficient synthetic methods to modernize or replace these well-performing but in many ways outdated methods is in great demand. To address this important issue, various new ecofriendly synthetic procedures have been designed.[2] These methods rely on the use of benign reagents or recyclable catalysts and aqueous or minimal solvent use systems, and address the energy consumption/reaction time issue by using nonconventional activation methods (ultrasonic or microwave activations).[3]

The pioneering steps in microwave-assisted organic synthesis (MAOS) were carried out by Gedye and Giguere in 1986 using domestic microwave ovens.[4] They independently reported several successful microwave-assisted reactions and proved that microwave heating could substantially accelerate the rate of reactions. The initial development of the microwave technology was relatively slow due to various reasons, mainly safety concerns, temperature control, and reproducibility. However, in the 1990s, the gradual availability of commercial microwave instruments, specifically designed for synthetic applications, resulted in explosionlike developments in this area.[5] These state-of-the-art instruments are easy to use and provide accurate temperature and pressure control, which critically enhances the reproducibility of the work. Currently, the broad variety of microwave instruments allows performing reactions under open, closed, solvent-free, pressurized conditions and in continuous flow systems. More importantly, the increased interest in MAOS is due to the sometimes

Green Chemistry
http://dx.doi.org/10.1016/B978-0-12-809270-5.00023-6

astonishingly short reaction times, which has a significant contribution in revolutionizing this field in industry. After a relatively slow start, thousands of articles have been published in this research field clearly showing the widespread interest in its application possibilities.[6]

Microwaves interact directly with the organic molecules creating rotational motion, which instantaneously generates heat inside the reaction vessel. The absorption of microwaves depends on the polarity; polar molecules such as water, ethanol, acetonitrile, etc., interact efficiently with microwaves ensuring rapid heat generation.[7] High-energy molecules under microwave irradiation have enhanced collisions and generate heat, which then gets transferred to the entire reaction vessel by convection. On the other hand, conventional heating provides energy to reactants with an external heating source. Heat transfer occurs through the wall of the vessel to the reaction medium, which slows down the energy transfer and thus the reaction rate.

MAOS has been utilized in various organic transformations including one-pot multicomponent reactions, cycloaddition reactions, metathesis, and for the synthesis of heterocycles.[8] In this account, we will provide a brief overview of the advancements reported in MAOS.

3.18.2 MULTICOMPONENT REACTIONS

Multicomponent reactions have attracted considerable attention in the fine chemicals and pharmaceutical industry due to their high atom economy.[9] These are usually not concerted reactions, but rather tandem or domino processes in which multiple reactions are combined into one synthetic operation.[10] This can significantly reduce the time and the amount of waste when compared with alternative multistep synthetic processes. The first step creates an intermediate that need not be isolated, which undergoes a sequence of reactions until the stable product is formed. Careful design of substrates with predetermined functionalities often results in products that are difficult to obtain otherwise in traditional multistep syntheses.

One of the most well-known multicomponent reactions is the Ugi condensation reaction.[11] In 2008, Andreana and coworkers[12] reported the synthesis of 1,4-benzodiazepin-3-ones utilizing the Ugi four-component coupling reaction initiated by microwave irradiation (Scheme 3.18.1). This reaction involves the condensation of an aldehyde or ketone, an amine, an isocyanide, and a carboxylic acid. The authors have attempted the reaction with

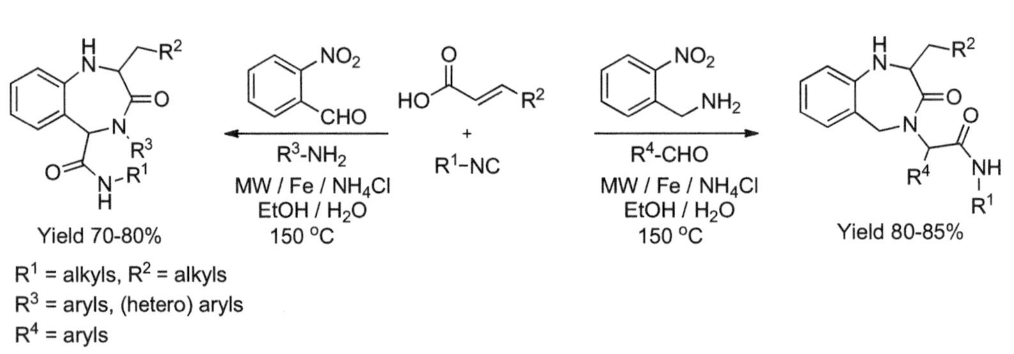

SCHEME 3.18.1 A four-component Ugi condensation for the synthesis of 1,4-benzodiazepin-3-ones.

o-nitrobenzaldehyde to form C2, N4, C5 product (80%–85%), while *o*-nitrobenzylamine provided C2, N4 product (70%–80%). Fe/NH_4Cl was found to be the most effective reagent for the reduction of $-NO_2$ group, which later underwent the aza-Michael reaction forming a doubly conjugated olefin and finally providing the corresponding benzodiazepine-3-ones.

Another variation of the Ugi reaction was described by Dai and coworkers[13] for the rapid synthesis of highly functionalized benzoxazines. The authors uncovered an intramolecular *O*-alkylation sequence starting from 2-aminophenols in combination with α-bromoalkanoic acids, aldehydes, and isocyanides under microwave condensation (Scheme 3.18.2). The reaction was carried out in MeOH, and then aqueous K_2CO_3 was added to facilitate the intermolecular *O*-alkylation in the same reaction vessel without isolating any intermediates. Room temperature stirring for 8–12 h provided a 61%–95% yield. In contrast, microwave heating substantially reduced the reaction times to 20–45 min with excellent yields (up to 90%) under the same reaction conditions.

The quinazolinone scaffold has been extensively utilized in medicinal chemistry.[14] Liu et al.[15] described a microwave-assisted three-component one-pot synthesis of these important heterocycles from readily available anthranilic acids, styryl carboxylic acids, and amines (Scheme 3.18.3). The tandem reaction involves the formation of an amide bond between anthranilic acid and styryl carboxylic acid, and the addition of amine forms the second amide bond with–COOH group of anthranilic acid, which further cyclizes to form the corresponding quinazolinone derivative in excellent yields under short reaction times. The authors successfully demonstrated the use of this methodology in the concise synthesis of the quinazolinone natural product-templated library for high-throughput screening as potent cytotoxic agents.

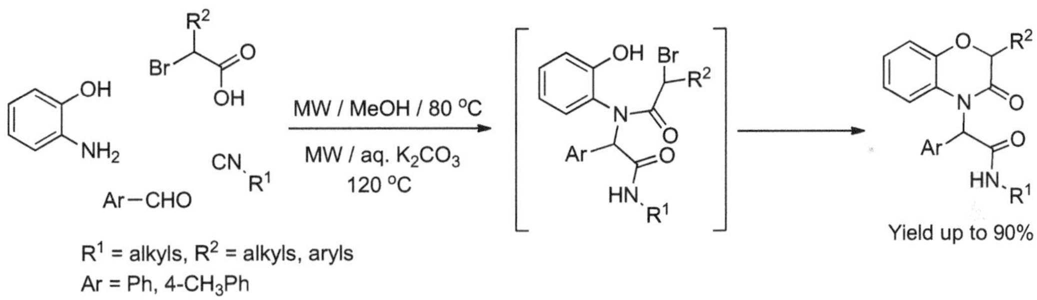

SCHEME 3.18.2 A four-component microwave-assisted synthesis of quinazolinone scaffolds.

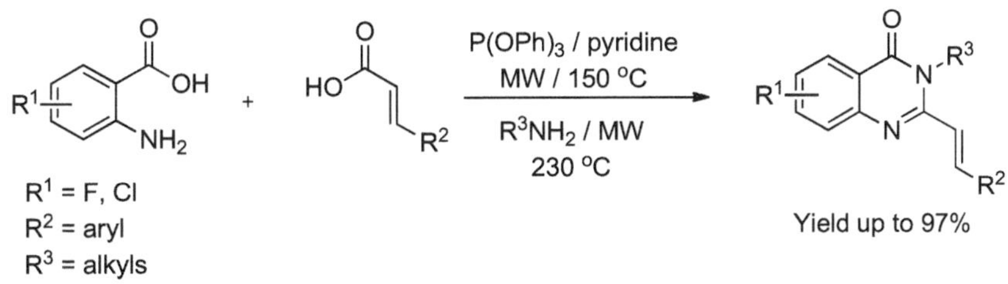

SCHEME 3.18.3 A tandem one-pot synthesis of quinazolinone derivatives using microwaves.

A one-pot microwave-assisted protocol for the synthesis of canthine alkaloid core was reported by Lindsley and coworkers.[16] This protocol involved an indole-tethered acyl hydrazide treated with a 1,2-dicarbonyl compound, such as benzyl, in acetic acid with excess ammonium acetate at 220 °C for 40 min under microwave conditions to form the canthine core in up to 80% yield (Scheme 3.18.4). This methodology was further extended for diversity-oriented synthesis of various substituted canthine derivatives that were difficult to synthesize using traditional reaction pathways. These synthetic derivatives exhibited better activity as kinase inhibitors when compared with the natural canthines.

An efficient one-pot total synthesis of quinazolinobenzodiazepine alkaloid sclerotigenin was reported by Liu et al.[17] via a novel microwave-assisted domino sequence. This synthesis was achieved via a three-component reaction by coupling anthranilic acid with *N*-Boc-glycine in the presence of triphenyl phosphite, $P(OPh)_3$, in pyridine under microwave irradiation at 150 °C for 10 min, followed by the addition of methyl anthranilate at 250 °C for 15 min (Scheme 3.18.5). This method provided 60% yield of the expected product in short reaction time without forming any by-products. The authors extended the scope of this methodology to the synthesis of sclerotigenin analogs by using substituted anthranilic acids.

Microwave-assisted multicomponent reactions have been widely recognized for their utility in generating libraries for biological screening. *N*-substituted 4-aryl-4,6,7,8-tetrahydroquinoline-2,5(1H,3H)-diones were synthesized by Tu and coworkers[18] through a rapid one-pot four-component condensation of aldehydes, Meldrum's acid, dimedone, and amines under microwave conditions (Scheme 3.18.6). This reaction was carried out in 95% EtOH without any catalyst to obtain excellent yields (82%–96%) and with short reaction times (4–9 min).

Multicomponent reactions based on cyclocondensation of 5-aminopyrazoles with pyruvic acid derivatives and aldehydes have been studied for the synthesis of various *N*-heterocycles.

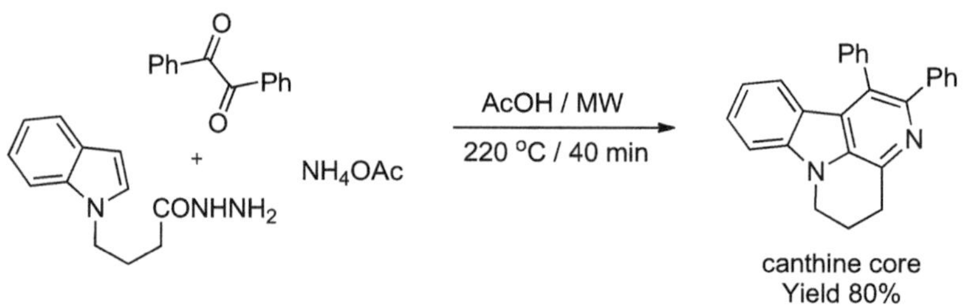

SCHEME 3.18.4 A one-pot microwave-assisted synthesis of canthine alkaloid core.

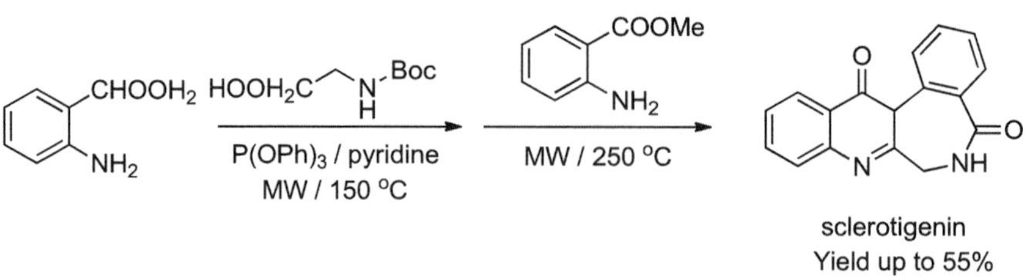

SCHEME 3.18.5 An efficient one-pot synthesis of sclerotigenin alkaloid.

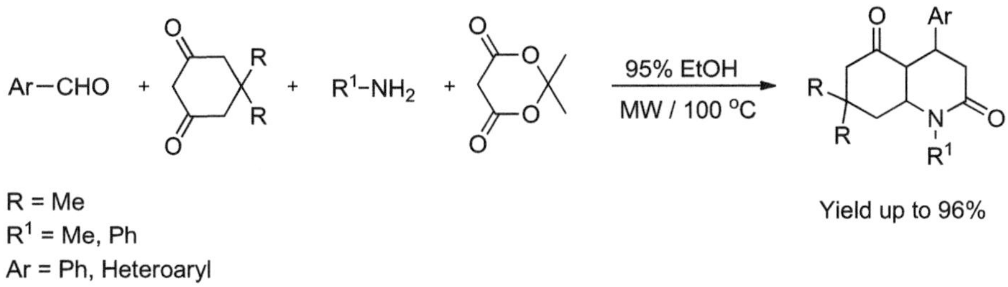

SCHEME 3.18.6 Microwave-assisted multicomponent synthesis of tetrahydro quinolinone derivatives.

Chebanov et al.[19] have investigated the synthesis of fused pyrimidinecarboxylic acids by treatment of 3-amino-1,2,4-triazole, 5-aminotetrazole and 2-aminobenzimidazole with either arylidenepyruvic acids or with pyruvic acid and aldehydes under various reaction conditions (Scheme 3.18.7). The authors observed that this condensation reaction could be carried out efficiently under microwave heating in acetic acid or EtOH in the presence of HCl as catalyst at 150 °C. The microwave-assisted protocol significantly decreased the reaction time from 120 to 10 min and considerably increased the product yields (from 18%–25% to 35%–48%). As the major advantage of this method, it made the product isolation convenient since the product formed a crystalline precipitate in the aforementioned solvent system.

Multicomponent reactions combined with fluorous tagging had simplified the product isolation. The product bearing a fluorous tag can be easily separated from nonfluorous components by fluorous liquid-liquid extraction or solid-phase extraction. Zhang and Chen reported[20] a three-component 1,3-dipolar cycloaddition followed by the Suzuki coupling to synthesize biaryl-substituted proline analogs using benzaldehyde labeled with a fluorous tag as the starting material (Scheme 3.18.8). Amino ester, benzaldehyde, and maleimide were condensed under microwave conditions at 150 °C for 15 min to produce bicyclic proline derivative as a single diastereomer. Products with fluorous components were isolated from the mixture by solid-phase extraction. The isolated products were later subjected to coupling with boronic acids under microwave irradiation at 120 °C for 12 min using bis(diphenylphosphino)ferrocene palladium chloride Pd(dppf)Cl_2 as catalyst, K_3PO_4 as base, and toluene-acetone-H_2O as solvent.

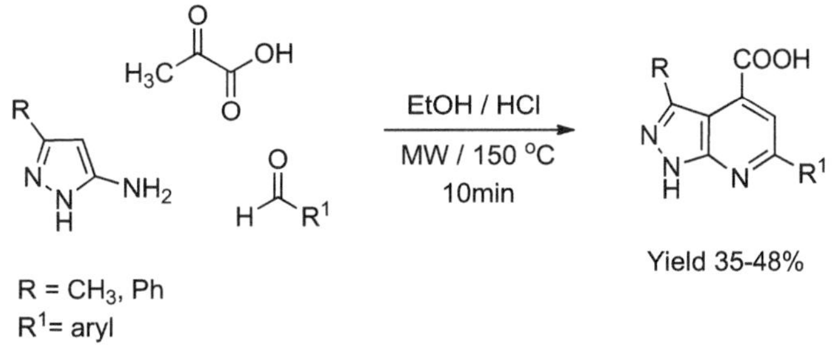

SCHEME 3.18.7 Synthesis of pyrimidinecarboxylic acids using hydrochloric acid as catalyst under microwave irradiation.

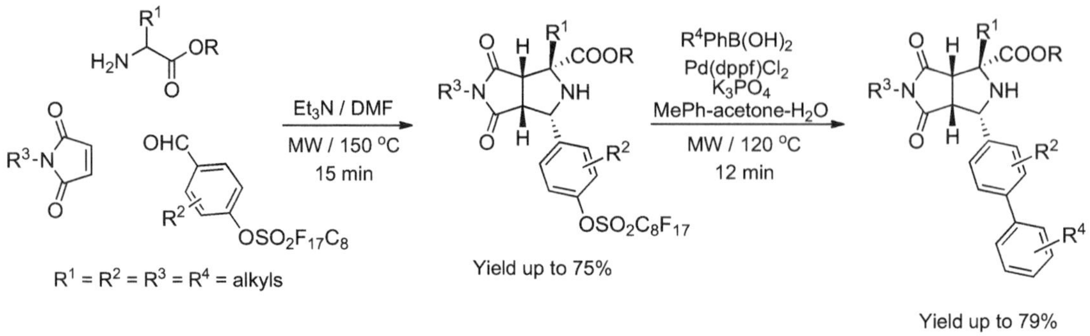

SCHEME 3.18.8 A three-component 1,3-diploar cycloaddition reaction for the synthesis of complex proline derivatives.

3.18.3 CYCLIZATION/CYCLOADDITION REACTIONS

Cycloaddition reactions are one of the pioneering transformations carried out in the history of MAOS. These reactions require activation energy, either by photochemically or heat/high temperatures. This initiated the applications of microwave irradiation in these transformations. These investigations revealed tremendous rate enhancement. Cycloaddition reactions in general represent an important tool for the formation of C–C bonds. Diels-Alder cycloaddition is a well-known strategy for the synthesis of cyclic materials containing multiple stereocenters. Aube's group reported[21] a method that combines the Diels-Alder reaction with an imide acylation reaction, to afford octahydroisoquinol-1-one-8-carboxylic acids. Thermal reactions of aminodienes and maleic anhydride were initially studied, and the high internal pressures and temperatures associated with microwave irradiation facilitated the overall process in dichloroethane at 165 °C (Scheme 3.18.9). This tandem reaction has two possible pathways: intermolecular or intramolecular cyclization followed by acylation. However, the exact mechanism is not known since there were no intermediates detected. The methodology afforded scaffolds containing isoquinolone-based library of products.

The pyrrolidine ring is frequently present in the structures of numerous bioactive molecules, hence, there is a wide interest in the development of synthetic routes to pyrrolidine-containing heterocycles. One of the novel routes is the intramolecular azomethine ylide [3 + 2] cycloaddition. This reaction revealed considerable versatility in the synthesis of pyrrolidines. Kurth and coworkers[22] employed this tandem condensation-cycloaddition strategy for the synthesis of polycyclic pyrrolidines (Scheme 3.18.10). Azomethine ylide cycloadditions carried out under classical reaction conditions required longer reaction times and often afforded products in low yields and/or with lower diastereoselectivity compared with microwave-assisted variants.

Aqueous media represent environmentally benign alternatives for organic synthesis. Fang et al.[23] demonstrated the use of aqueous ammonia for direct conversion of aldehydes and alcohols to nitriles in the presence of iodine, which were heated with dicyandiamide and sodium azide with microwave irradiation to form the corresponding triazines and tetrazoles by a [2 + 3] cycloaddition, respectively (Scheme 3.18.11A). This tandem formation of triazines and tetrazoles requires a prolonged heating period (12–48 h, 100 °C) under

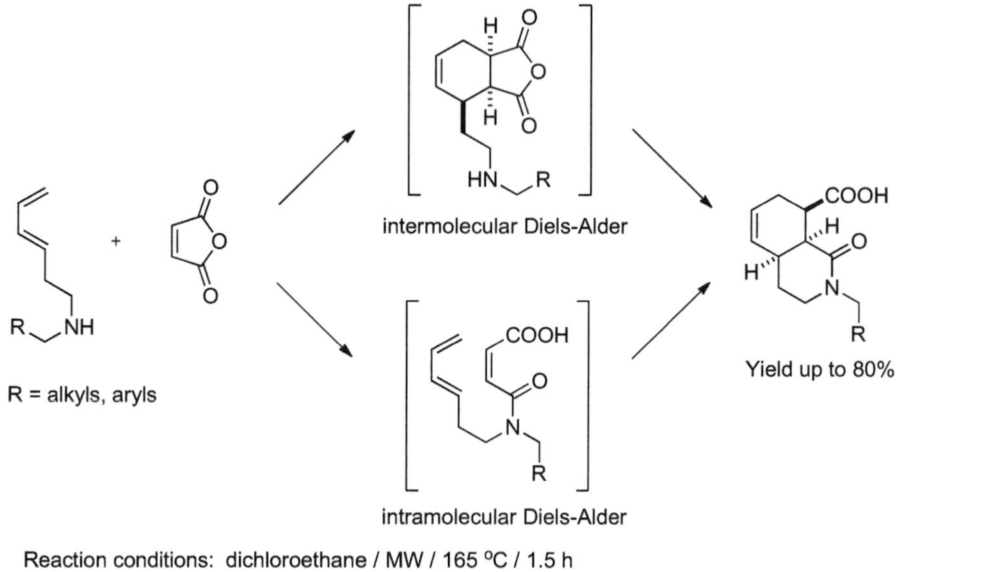

SCHEME 3.18.9 A microwave-assisted Diels-Alder reaction for the synthesis of carboxylic acid derivatives.

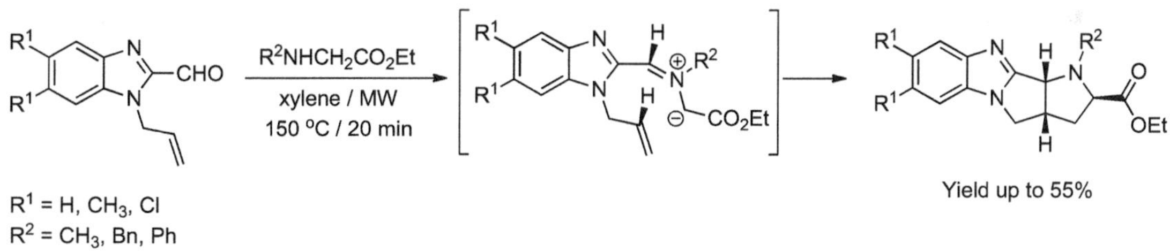

SCHEME 3.18.10 A tandem cyclization/condensation strategy for the synthesis of pyrolidines.

traditional reaction conditions. The microwave approach circumvents the problem of preparation of toxic cyanides, since the nitriles are generated in situ and do not require isolation. Also, the one-pot microwave method is free of KOH, which is required under traditional conditions to accelerate the formation of triazines.

In 2007, Fokin and coworkers[24] described a simple and efficient ruthenium-catalyzed cycloaddition of azides and alkynes for the synthesis of 1,5-disubstituted 1,2,3-triazoles (Scheme 3.18.11B). The authors reported that pentamethylcyclopentadienyl ruthenium(II) chloride tetramer $[Cp^*RuCl]_4$ in dimethylformamide (DMF) under microwave conditions performed significantly better than other ruthenium-based catalysts. Various substituted azides and acetylenes underwent cycloaddition in short reaction times (10–40 min) providing good to excellent yields (52%–92%).

Five-membered *N*-heterocycles such as triazoles, pyrazoles, and imidazoles are found in several pharmacologically active compounds. Yeung and coworkers[25] described a convenient, base-catalyzed one-pot synthesis of 3,5-disubstituted 1,2,4-triazoles by condensation

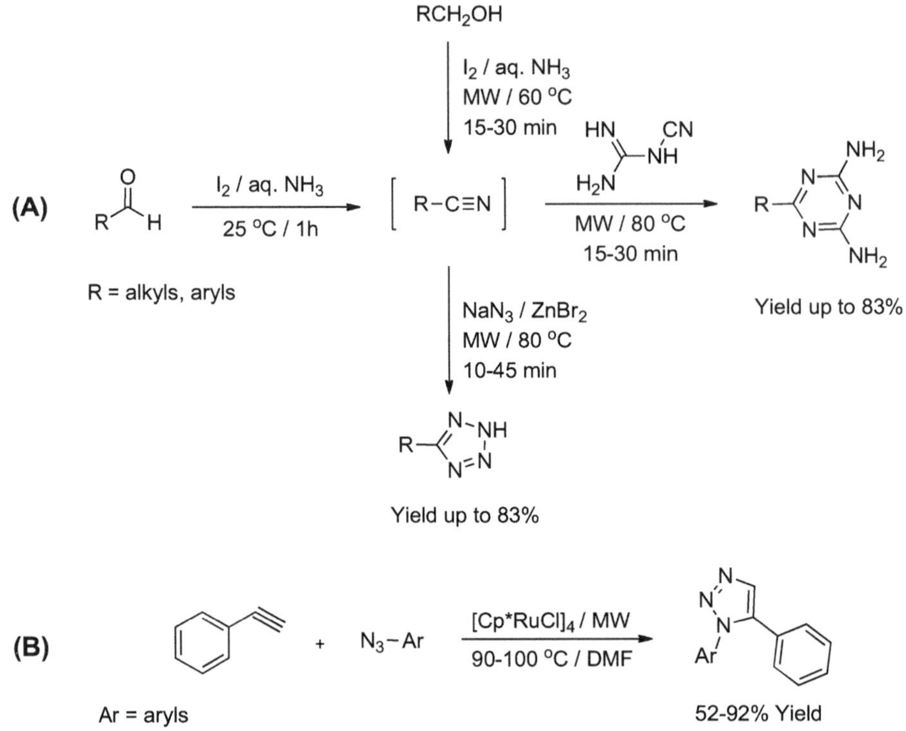

SCHEME 3.18.11 (A) A [2 + 3] cycloaddition reaction for the synthesis of triazines and triazoles using aqueous ammonia. (B) Ruthenium-catalyzed cycloaddition of azides and alkynes to form triazoles.

of a nitrile and a hydrazine (Scheme 3.18.12A). A catalytic amount of K_2CO_3 (0.1 mol%) in *n*-BuOH provided the best results.

In 2008, Lam et al.[26] reported a microwave-assisted protocol for the construction of di- and monosubstituted 2-aminoimidazoles (Scheme 3.18.12B). The two-step reaction involved the synthesis of *N*-(1H-imidazol-2-yl)acetamides from readily available α-haloketones and *N*-acetylguanidine, followed by deacetylation. Significant rate enhancement was observed in both steps of the protocol, and the overall reaction time was shortened to 20 min compared with 48 h of the conventional procedures. Varma and Ju[27] described a similar approach for the synthesis of 4,5-dihydropyrazole, pyrazolidine, and 1,2-dihydrophthalazine derivatives via double alkylation of hydrazine derivatives with alkyl dihalides or ditosylates in aqueous medium under microwave conditions (Scheme 3.18.12C). This methodology provided short reaction time using unprotected hydrazine derivatives and alkyl dihalides or ditosylates to construct heterocycles in a single step, a technique that has never been fully realized under conventional heating conditions.

Benzoxazoles and azaindoles are important heterocycles that are found in many natural product-based drugs, as well as in synthetic pharmaceuticals. Existing methods for the synthesis of benzoxazoles utilize 2-aminophenols as precursors. Batey and coworkers[28] designed a novel one-pot domino annulation strategy involving the reaction of 2-bromoanilines with

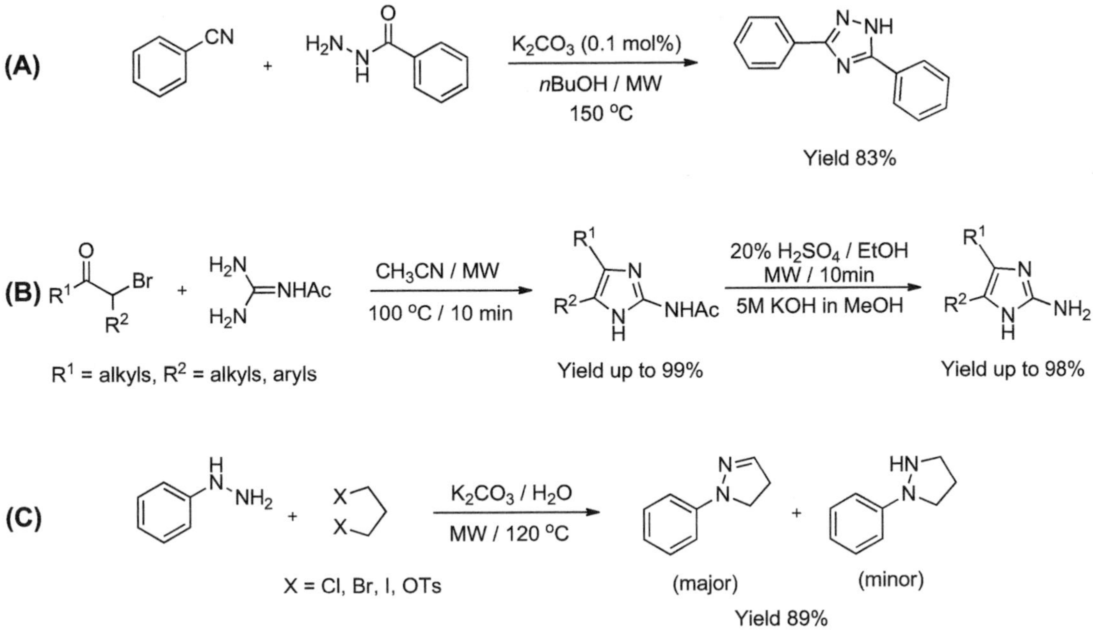

SCHEME 3.18.12 (A) Base-catalyzed one-pot synthesis of triazoles. (B) Microwave-assisted construction of substituted imidazoles. (C) Synthesis of *N*-heterocycles via double alkylation of hydrazines and dihalides.

acyl chlorides in the presence of Cs_2CO_3, catalytic amount of CuI, and the nonacylatable ligand 1,10-phenanthroline (Scheme 3.18.13A). Under these conditions the initial acylation of the aniline is followed by copper-catalyzed intramolecular cyclization of the 2-haloanilide formed to provide the Ar–O bond of the benzoxazole ring. The application of microwave irradiation provided reasonably shorter reaction times compared with conventional heating (i.e., 210 °C for 15 min vs. 95 °C for 24 h) and was applied for the synthesis of a small library of benzoxazoles.

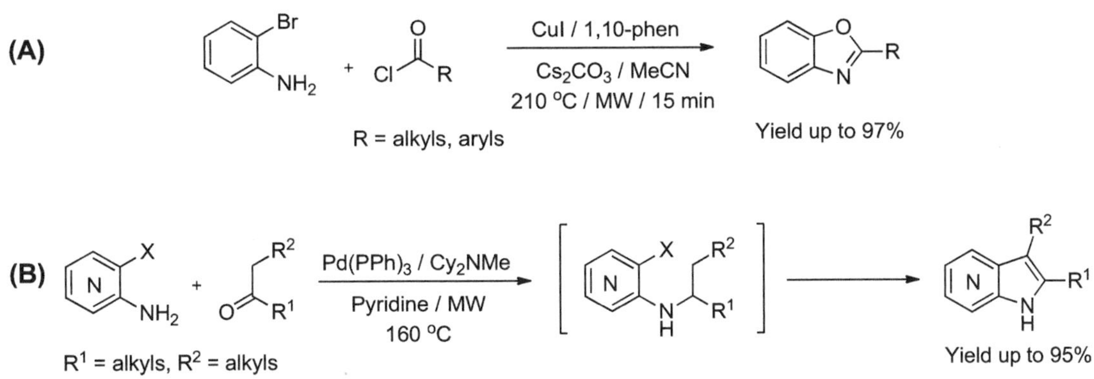

SCHEME 3.18.13 (A) A one-pot domino annulation of 2-bromoanilines with acyl chlorides. (B) A one-pot two-step domino process for the synthesis of 7-azaindoles.

Lachance et al.[29] described a one-pot two-step domino process for the synthesis of 4-, 5-, 6-, or 7-azaindoles via the Hegedus-Mori-Heck reaction (intramolecular Heck reaction) (Scheme 3.18.13B). In this two-step process, the haloaminopyridine first reacts with the ketone to afford enamine, which subsequently undergoes an intramolecular palladium-catalyzed Heck reaction to produce the corresponding azaindole. The reaction of enamines with Cy_2NMe, in the presence of 5 mol% of $Pd(PPh_3)_4$ in pyridine under microwave conditions (160 °C, 20 min) resulted in azaindoles in up to 95% isolated yields.

Tandem microwave-assisted reactions are targets of increasing interest. This tool has been widely utilized for the synthesis of a variety of *N*-heterocycles. Synthesis of pyridines and their fused derivatives have become a major area of interest due to the importance of these heterocycles and their abundance in alkaloids. Wang and coworkers[30] described a rapid and efficient synthesis of polysubstituted quinolines through the Friedländer condensation of 2-aminoarylketones or 2-aminoarylaldehydes with carbonyl compounds (Scheme 3.18.14). Commercially available and inexpensive *p*-toluenesulfonic acid was used as catalyst under microwave conditions. The use of solvent such as toluene provided low yields. In contrast, solvent-free conditions resulted in good to excellent yields. Comparative studies between conventional and microwave heating under solvent-free conditions provided comparable results with slight variation in time. Microwave irradiation required 15–60 sec, whereas conventional heating required 3–10 min.

A tandem microwave-assisted gold(I)-catalyzed hydroamination-hydroarylation strategy for the synthesis of 1,2-dihydroquinolines and quinolines was developed by Che and coworkers in 2007.[31] Au(I) and Au(II) complexes have been widely used as catalysts for a variety of organic transformations. However, most of these complexes are not commercially available. As a test reaction the authors examined the cyclization of *m*-anisidine with phenylacetylene at 80–100 °C in the presence of 5 mol% catalyst. The reaction gave 1,2-dihydroquinoline derivatives in up to 80% yield under normal heating for 12–24 h. After subjecting this reaction to microwave irradiation the reaction times were considerably shortened to 25 min with excellent yields up to 94% (Scheme 3.18.15).

Synthesis of substituted pyridines with molecular complexity was pursued in 2007 by Trost and Gutierrez.[32] This one-pot synthesis involved the reaction between α,β,γ,δ-unsaturated aldehyde and hydroxylamine or methoxyamine (Scheme 3.18.16). Dramatic rate enhancement was observed under microwave conditions, which afforded the desired pyridines in 98% yield.

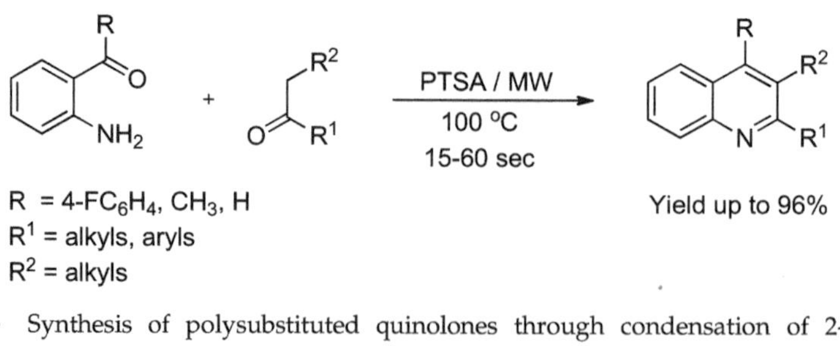

SCHEME 3.18.14 Synthesis of polysubstituted quinolones through condensation of 2-aminoketones with carbonyl compounds.

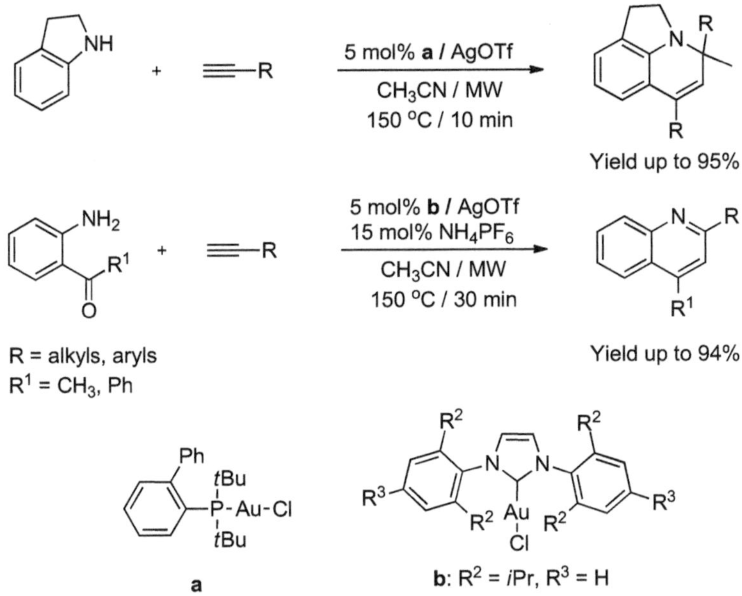

SCHEME 3.18.15 A tandem microwave-assisted gold(I)-catalyzed hydroamination-hydroarylation for the synthesis of quinolines.

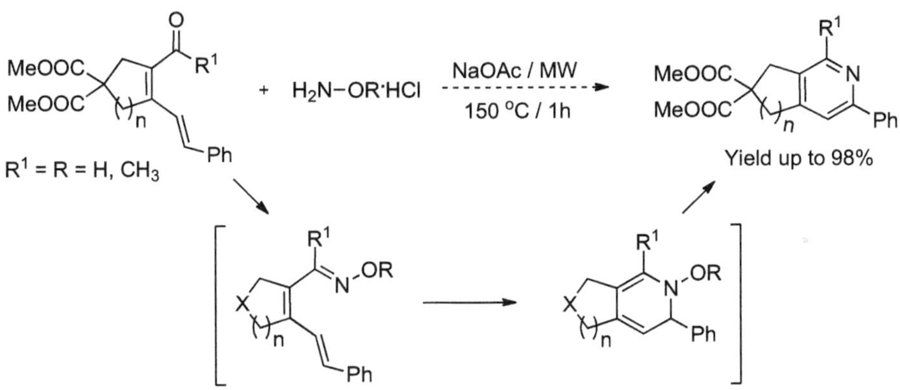

SCHEME 3.18.16 One-pot synthesis of pyridines using aldehydes and hydroxylamine.

Movassaghi and Hill developed a mild and convergent methodology for the synthesis of pyrimidine and pyridine derivatives in one step from amides and nucleophiles (Scheme 3.18.17).[33] The authors discussed the comparison of conventional heating with microwave irradiation for this transformation. Various amides readily underwent the azaheterocycle formation at ambient temperature or with mild heating at 45 °C, whereas recalcitrant amides, including electron-deficient *N*-aryl amides, led to low yields of the desired product under traditional heating conditions. In contrast, microwave heating provided satisfactory yields in this case as well, at relatively high temperature (140 °C) and short reaction time (20 min).

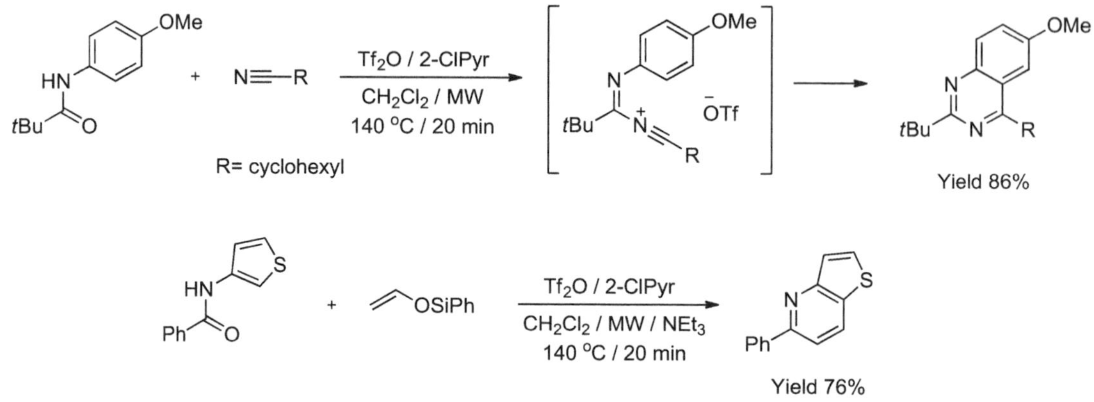

SCHEME 3.18.17 Microwave-assisted convergent synthesis of pyrimidine and pyridine derivatives.

3.18.4 RADICAL CYCLIZATIONS

Radical reactions under microwave conditions are relatively rare in literature and mostly deal with the enhancement of reaction rate in transformations that require relatively long times under conventional heating conditions.[34] Reiser and coworkers reported[35] a new approach for the microwave-assisted synthesis of indolines. The strategy was based on radical cyclization mediated by tributyltin hydride involving a 1,6-H transfer followed by ring closure under microwave heating in an open vessel (Scheme 3.18.18). While traditional heating in toluene provided good yields (62%–76%, 5 h), performing the cyclization under microwave conditions in an open vessel resulted in improved yields (76%–89%) and short reaction times (3.5 h).

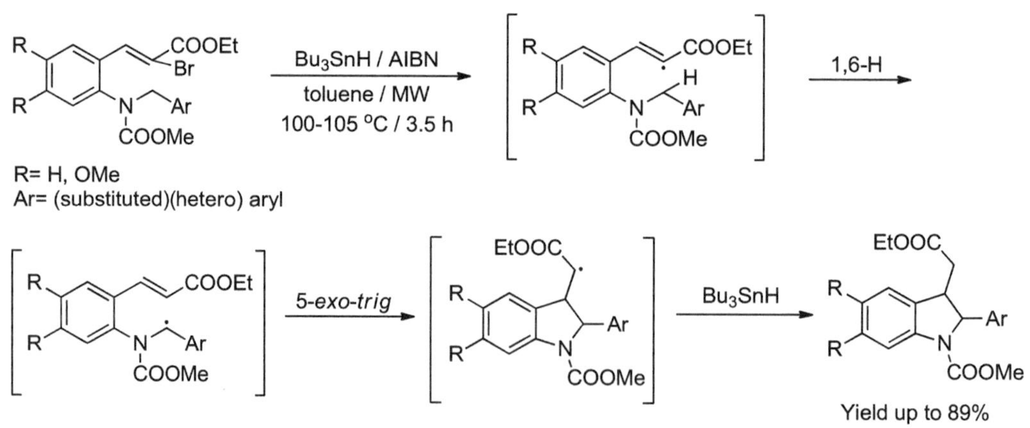

SCHEME 3.18.18 Microwave-assisted radical cyclization strategy for the synthesis of indolines.

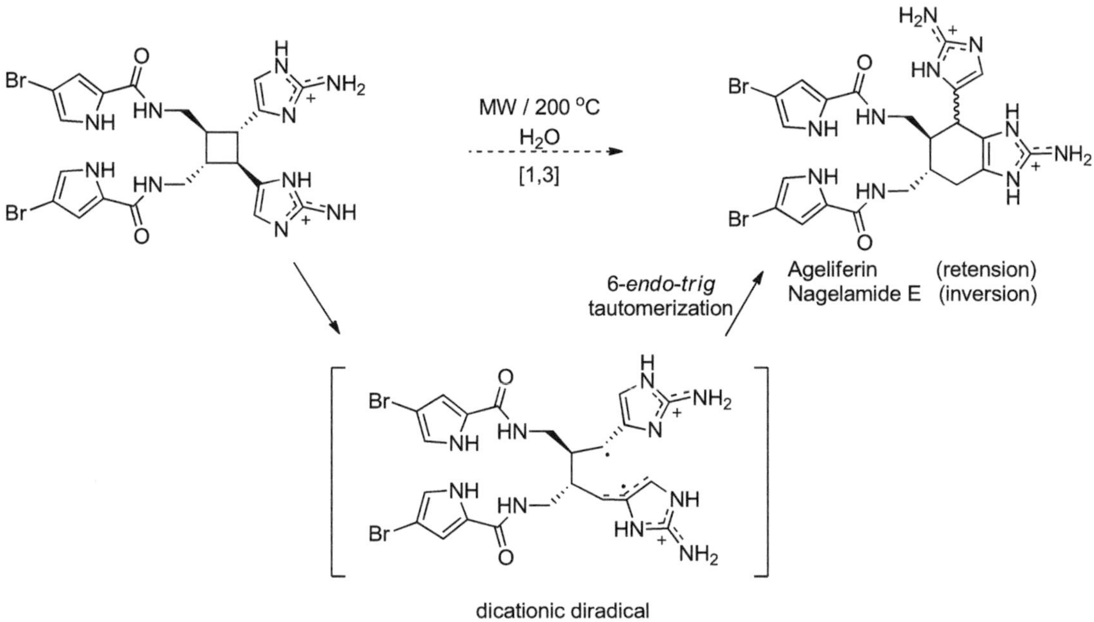

SCHEME 3.18.19 Conversion of sceptrin to ageliferin using microwave heating.

A significant example of the microwave effect was described by Baran and coworkers[36] in the process of converting sceptrin to ageliferin. The authors made several attempts to utilize photochemical [1,3]-sigmatropic rearrangement. However, after several days, only a gradual decomposition occurred. In another attempt, heating sceptrin in methanol at 150 °C with microwave irradiation led to decomposition into a variety of products that were difficult to isolate and characterize. However, when sceptrin was dissolved in water and heated at 200 °C for 1 min using microwave irradiation, the reaction provided ageliferin in 40% yield, along with 52% recovered starting material. Later the authors proposed the mechanism of this rearrangement, suggesting the formation of a diradical intermediate starting from dicationic sceptrin (Scheme 3.18.19).[37]

3.18.5 REACTIONS BY SOLID CATALYSTS

Solvent-free reactions under microwave irradiation have gained significant attention.[38] In many cases, microwave energy is directly absorbed by the solid support or by reagents, allowing homogeneous heat distribution in the sample. This would not be possible under convective heating conditions since the dry media do not have good thermal conductivity.[39] Various solid supports have been utilized as catalysts, such as zeolites and acidic clays.[40] We will focus on some of the advantages of these catalysts in MAOS.

Solid acid catalysis and microwave irradiation have emerged as important tools in organic synthesis as indicated by the extensive number of publications.[41] Use of clay catalysts in organic syntheses is preferable as they are inexpensive, readily available, safe, easy to handle, stable, recyclable, and produce no waste after the reaction. In contrast, the traditional acid catalysts are corrosive, difficult to work with, and produce a substantial amount of waste. Clay catalysts are easy to separate from products and usually can be reused. Montmorillonite K-10 is one of the most widely used solid acids in organic synthesis. K-10 has several advantages such as strong acidity with a Hammett constant of $H_o = \sim -8$, which is comparable to that of concentrated HNO_3; a high surface area of $\sim 250\ m^2/g$ [Brunauer-Emmett-Teller (BET) surface]; and high thermal stability up to 700 °C.[42] It is also safer than traditional mineral acids and noncorrosive. K-10 combined with microwave irradiation has many applications in various acid-catalyzed transformations including heterocyclic chemistry.[1] Our own investigations also revealed the usefulness of K-10 in the synthesis of a broad variety of heterocycles.

The synthesis of substituted pyrroles, indoles, and carbazoles was achieved by involving a solid acid-catalyzed one-step technique.[43] This approach was based on a new pathway, namely, a solid acid-catalyzed intramolecular Friedel-Crafts cyclialkylation/annelation. The intramolecular pathway involved a substituted aromatic substrate that undergoes cyclization due to the close spatial arrangement of two reacting electrophiles. Primary amines readily underwent cyclization on the nitrogen with 2,5-hexanedione to form substituted pyrroles. This process is an efficient modification of the Paal-Knorr synthesis of pyrroles providing nearly quantitative yields (up to 98%) in short reaction times. We have extended the scope of this methodology to synthesize higher analogs such as indoles and carbazoles (Scheme 3.18.20). Pyrroles undergo annelation on the C-2 and C-3 positions to form substituted indoles in excellent yields, however, in longer reaction times. To further widen the applications of the

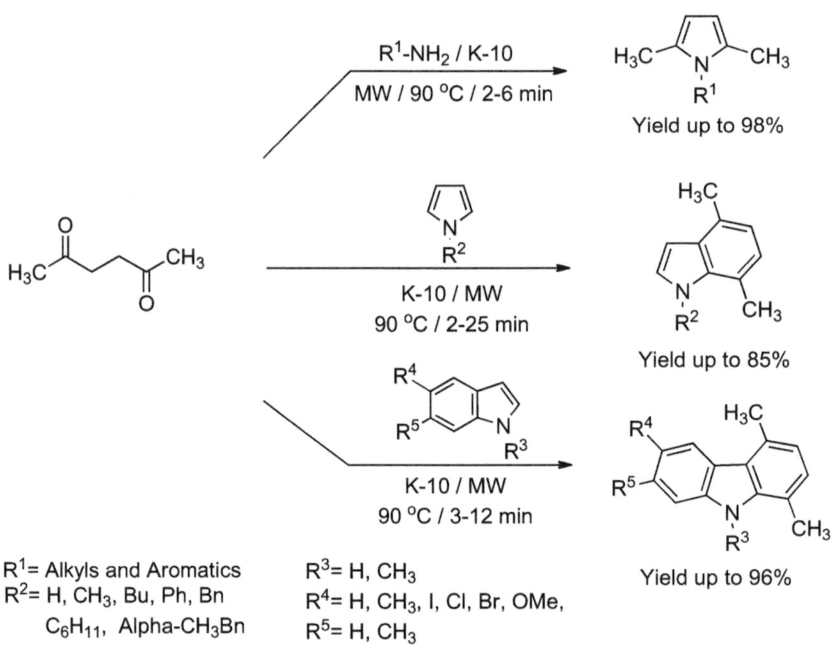

SCHEME 3.18.20 Solid acid-catalyzed microwave-assisted synthesis of *N*-heterocycles.

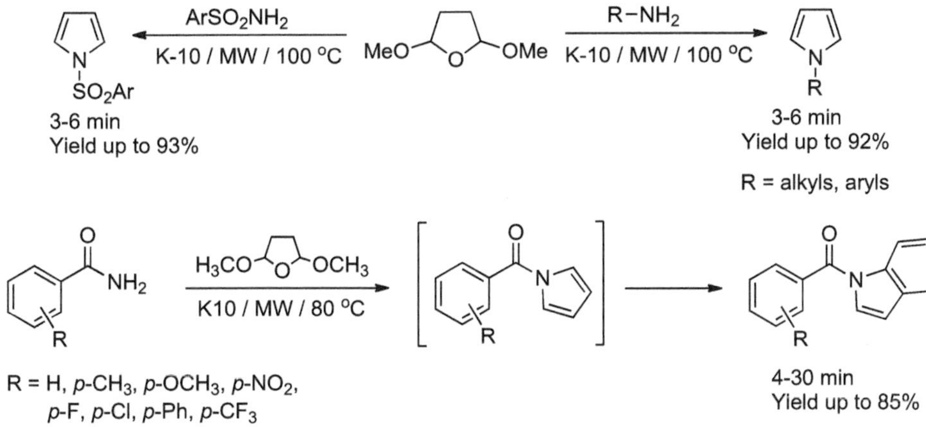

SCHEME 3.18.21 K-10-catalyzed microwave-assisted synthesis of *N*-sulfonyl pyrroles and *N*-acyl indoles.

process, several substituted indoles were used as starting materials to provide the corresponding carbazole derivatives.

Primary amines and sulfonamides also readily underwent cyclization with 2,5-dimethoxytetrahydrofuran with K-10 catalysis under microwave conditions to give the corresponding *N*-alkyl and *N*-sulfonyl pyrroles, respectively.[44] The reactions were completed within 3–6 min. Despite short reaction times, the isolated yields were excellent with 100% selectivity; no by-product formation was observed. Aromatic amides, however, did not provide the expected *N*-acylpyrroles; instead they underwent further cyclization to form *N*-acylindoles under K-10-catalyzed microwave-assisted conditions.[45] This is an efficient one-pot synthesis of *N*-acylindoles. Their synthesis usually requires a multistep process. All the reactions were completed within 4–30 min, providing up to 85% yield (Scheme 3.18.21).

As an important extension of solid acid catalytic technique, K-10 was combined with various metals, such as Pd, to form a bifunctional catalyst. This catalyst enables the synthesis of various heterocycles, which involves a multistep cyclization and dehydrogenation (aromatization) steps. Using the bifunctional catalyst, multistep processes were achieved in one-pot domino reactions.

Many efficient methods had been developed for the synthesis of pyrazoles; however, they usually involve separate processes, cyclization and the oxidative aromatization of the cyclized product, which often requires a stoichiometric amount of oxidant. A novel strategy described the combination of cyclization and oxidation reactions into a two-step/one-pot domino approach utilizing a Pd/C/K-10 bifunctional catalyst.[46] Various substituted hydrazines and chalcones were subjected to reactions under microwave conditions using bifunctional catalysis to provide the corresponding pyrazoles in excellent yields and in short reaction times (Scheme 3.18.22). The cyclization readily occurred at a lower temperature (~100 °C), whereas the oxidative aromatization required the use of elevated temperatures (160 °C). Despite the high temperatures, the short reaction times ensured virtually no by-product formation.

The same methodology of successive cyclization and oxidative aromatization has been extended to other multicomponent reactions such as the Hantzsch pyridine synthesis.[47] Classical one-pot condensation of aldehydes with ethyl acetoacetate and ammonia is either carried out in acetic acid or by refluxing in alcohol. It involves long reaction times, harsh

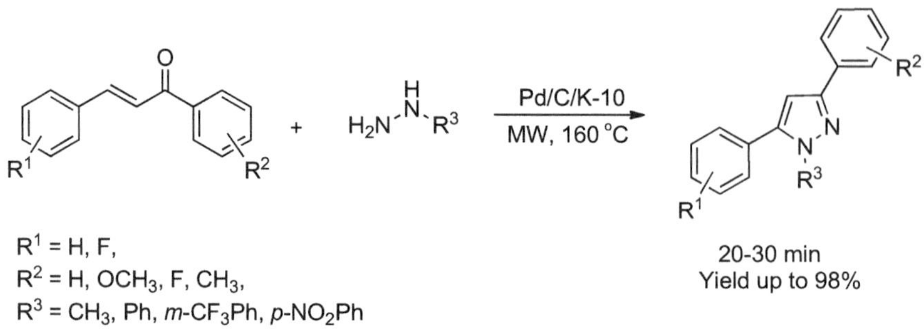

SCHEME 3.18.22 A two-step one-pot cyclization and oxidative aromatization strategy for the synthesis of substituted pyrazoles.

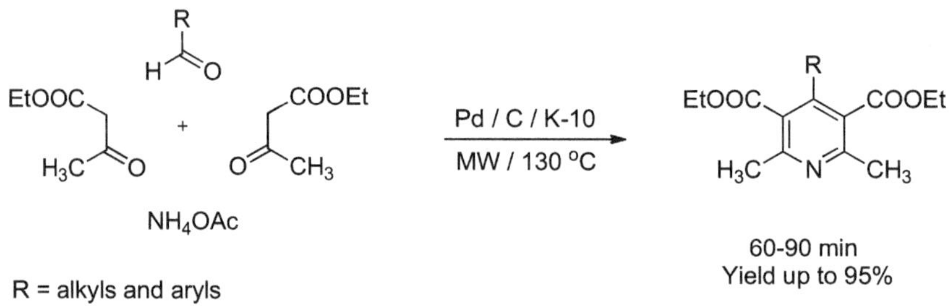

SCHEME 3.18.23 A multicomponent cyclization and dehydrogenation strategy for the synthesis of substituted pyridines.

reaction conditions, and the use of a large quantity of volatile organic solvents and only provides the 1,4-dihydropyridines. This intermediate requires further treatment with an oxidizing agent to form the aromatic product. The aforementioned bifunctional catalyst is capable of carrying out the cyclization and successive dehydrogenation in one-pot condition without isolating the 1,4-dihydropyridine intermediate (Scheme 3.18.23). This method requires reasonable reaction times (60–90 min), which is preferable to traditional methods.

The Pictet-Spengler condensation is one of the most powerful methods for the construction of tetrahydro-β-carboline ring systems. To show the versatility of our bifunctional catalyst, we further extended this approach for the synthesis of β-carbolines under microwave-assisted one-pot conditions.[48] Condensation of tryptamine with aromatic aldehydes provided excellent results depicted in Scheme 3.18.24. The reactions were completed within 45 min at 130 °C providing excellent yields with high selectivities.

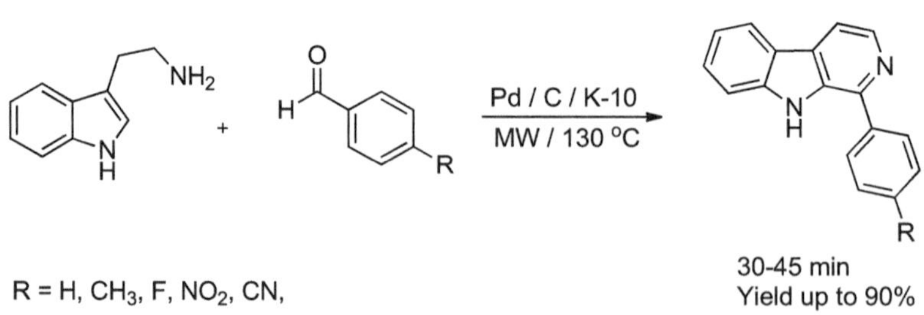

SCHEME 3.18.24 A microwave-assisted Pictet-Spengler synthesis of *N*-heterocycles.

3.18.6 METATHESIS

Ring-closing metathesis (RCM) has been established as a powerful and effective tandem method for the construction of many functionalized heterocycles from acyclic diene precursors.[49] Many useful transformations have been reported for industrial applications as well as for the synthesis of complex molecules due to major advances in catalyst design.[50] Some examples of tandem processes involve an RCM isomerization sequence, a one-pot three-component tandem RCM-Diels-Alder reaction, a tandem RCM-Kharasch addition, an RCM-dihydroxylation sequence, a ring-opening metathesis combined with RCM reaction, a closing metathesis (CM) allylic alcohol isomerization process, a tandem enyne metathesis/ Claisen rearrangement, or a tandem enyne metathesis followed by a cyclopropanation of the newly formed exo-cyclic double bond.[51] Fustero et al.[52] designed a CM-aza-Michael tandem process, leading to the formation of *β*-amino carbonyl units, which are very important structural motifs present in a large number of natural products and pharmaceutical ingredients. The Hoveyda-Grubbs catalyst (5 mol%) with BF_3OEt_2 (1 mol%) as an additive provided Cbz-protected *β*-homoproline and *β*-pipecolic acid derivatives in 99% isolated yield at 45 °C in a sealed tube after 4 days (Scheme 3.18.25). The reaction time was dramatically reduced to 20 min when microwave irradiation was employed.

Enyne cross-metathesis of propargylamines with ethyl vinyl ether was reported for the synthesis of substituted pyrroles.[53] A series of substituted pyrroles having aryl, alkyl, and heteroaryl substituents had been synthesized in good yields under microwave irradiation (Scheme 3.18.26).

Van der Eycken and coworkers[54] developed efficient strategies for the synthesis of *N*-shifted and ring-expanded buflavine analogs having *α*-adrenolytic and antiserotonin activities. This methodology is based on the Suzuki-Miyaura biaryl coupling followed by a RCM reaction for the construction of target molecules (Scheme 3.18.27). The combination of a second-generation Grubbs catalyst and microwave irradiation proved to be highly beneficial in generating the otherwise difficult-to-synthesize eight- and nine-membered ring systems.

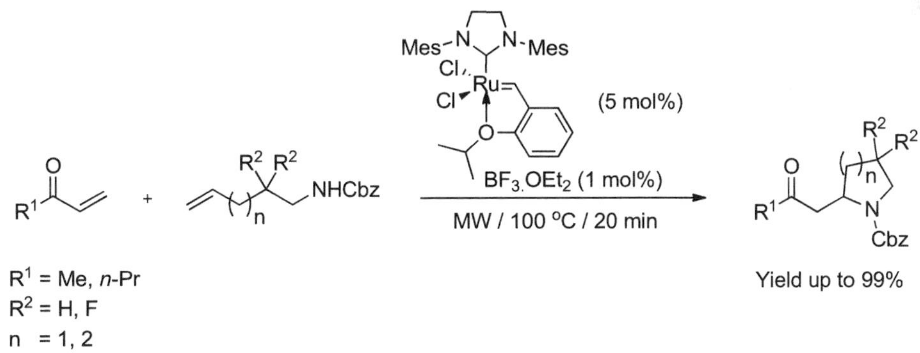

SCHEME 3.18.25 Alkene metathesis combined with aza-Michael cyclization for the formation of animo carbonyl compounds.

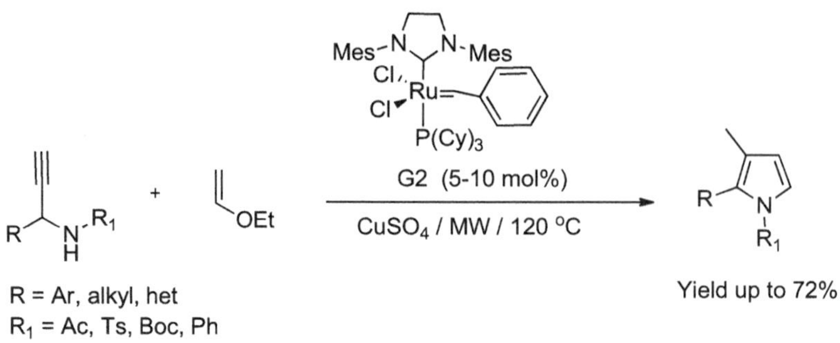

SCHEME 3.18.26 Enyne cross-metathesis of propargylamines with ethyl vinyl ether for the synthesis of substituted pyrroles.

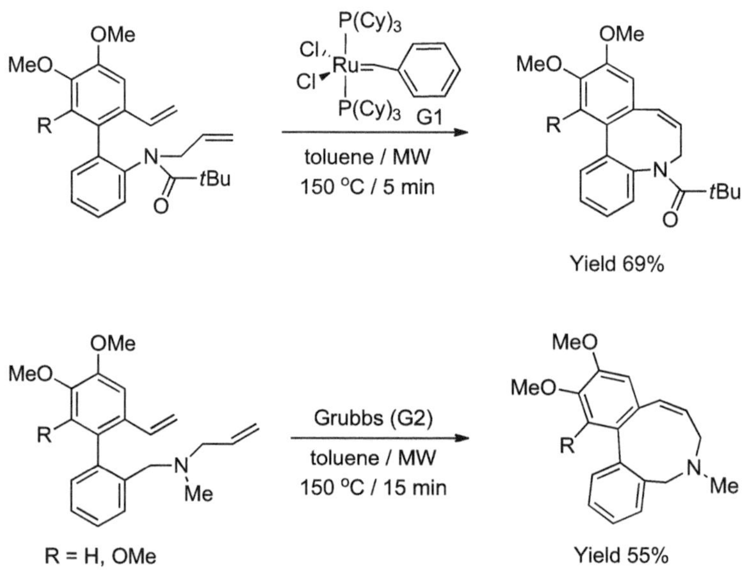

SCHEME 3.18.27 Synthesis of buflavine analogs via successive Suzuki-Miyaura coupling and ring-closing metathesis.

3.18.7 SOLID-PHASE SYNTHESIS ON POLYMER SUPPORTS

Solid-phase synthesis strategy is based on the immobilization of the starting material on a solid support, and then a sequence of reactions is carried out with the various intermediates that remain attached to the solid support. When the synthesis is complete, the products are detached from the support. This technique has many potential advantages, e.g., the excess reagents and by-products can be simply removed from the system.[55] The most common applications of solid-phase methods are in the field of polypeptides and oligonucleotides synthesis.[56] These syntheses can be carried out by sequentially adding amino acids and coupling reagents. Although these reactions can be carried out using conventional methods, they are now synthesized using automated synthesizers. Solid-phase synthesis has also found applications in combinatorial chemistry and library development.[57]

Automated solid-phase synthesis is often coupled with the use of microwave irradiation to address problems associated with the traditional methods, such as slow rates and degradation of polymer support. Microwave-assisted solid-phase synthesis has attracted interest in research laboratories for high-throughput synthesis of small molecule libraries. The discovery of solid-phase synthetic routes toward heterocyclic derivatives is an area of continued creativity for organic chemists. In particular, low-molecular-weight heterocyclic compounds are important as pharmaceutical lead compounds because they are highly functionalizable scaffolds. In solid-phase organic synthesis, resin-bound multifunctional compounds are key intermediates for the production of heterocyclic compound libraries. Giacomelli and coworkers[58] described an interesting example of combinatorial synthesis of *N*-heterocycles using aminofunctionalized support to obtain 1-substituted 4-imidazole carboxylates through a "catch-and-release" methodology (Scheme 3.18.28). Commercially available *N*-methyl aminomethylated polystyrene was charged with alkyl isocyanoacetate and *N*-formyl imidazole diethylacetal. This reaction mixture was heated at 80 °C in a sealed tube under microwave irradiation for 10 min. The resin was then washed thoroughly with portions of DMF, hexane, tetrahydrofuran (THF), hexane, and *n*-butanol in sequence. Treatment of the resin with an amine afforded the isomerically pure alkyl 1-substituted-4-imidazolecarboxylate.

Another attempt utilizing a one-pot tandem microwave-assisted solid-phase protocol was reported by Botta et al.[59] Various substituted benzoxazoles were synthesized from different polymer-bound esters as solid-supported reagents. The authors previously designed a solid support acylating agent, which was reacted with 2-aminophenols in the presence of CH_2Cl_2 for 10 min under microwave irradiation at 160 °C (Scheme 3.18.29). After removal of the solid support by filtration, polymer-bound *p*-toluenesulfonic acid (PTSA) was added to the intermediates mentioned earlier and irradiated for another 10 min at 180 °C to form the corresponding

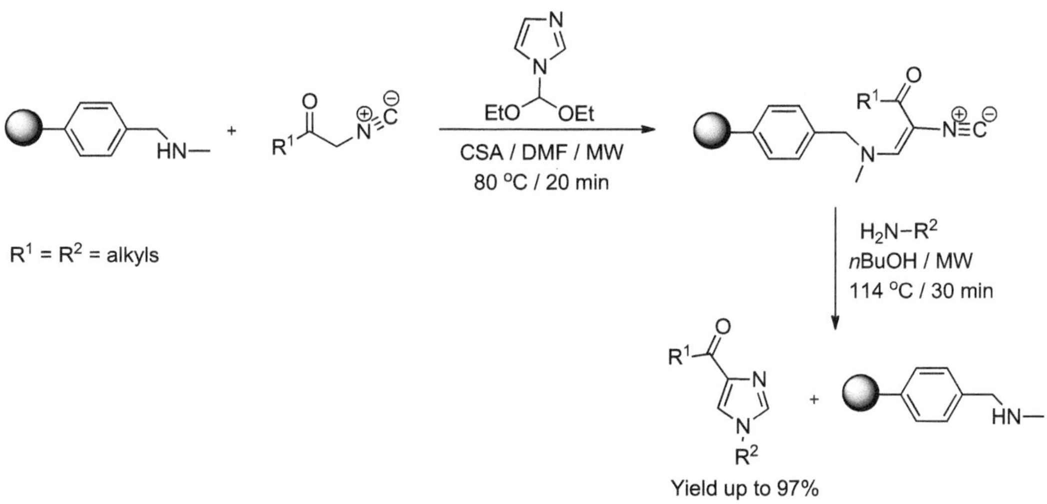

SCHEME 3.18.28 Combinatorial synthesis of *N*-heterocyles using amino-functionalized polymer support.

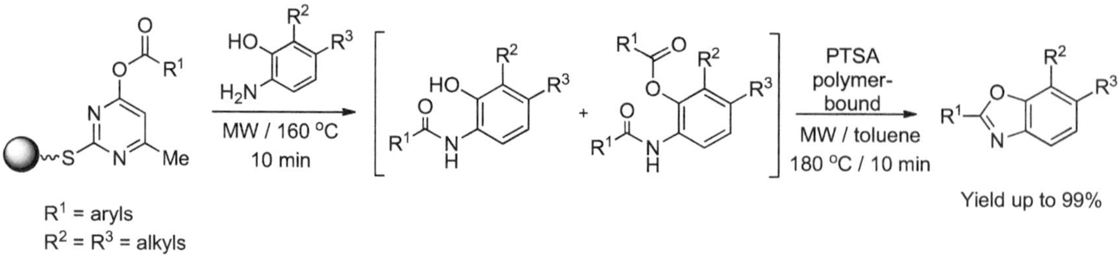

SCHEME 3.18.29 A one-pot tandem microwave-assisted solid-phase synthesis of benzoxazoles.

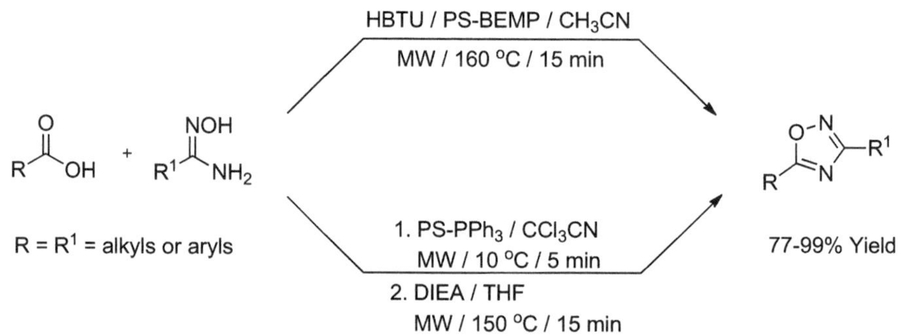

SCHEME 3.18.30 Microwave-assisted condensation of amidoximes with carboxylic acids to form 1,2,4-oxadiazoles.

benzoxazoles in moderate to excellent yields (up to 99%). This protocol is amenable for automation and could be used for the preparation of combinatorial libraries in drug discovery programs.

Rapid and efficient synthesis of 1,2,4-oxadiazoles was reported by Wang and coworkers[60] from a variety of readily available carboxylic acids and amidoximes (Scheme 3.18.30). This methodology was based on the condensation of amidoximes with carboxylic acids in the presence of *O*-(benzotriazol-1-yl)-*N*,*N*,*N*,*N*-tetramethyluronium hexafluorophosphate, 2-*t*-butylimino-2-diethylamino-1,3-dimethyl-perhydro-1,3,2-diazaphosphorine on polystyrene coupling reagent, and microwave heating. The tandem coupling occurred under one-pot conditions providing excellent yields in short reaction times. Another method, reported by the same group, described the conversion of an acid to acid chloride using PS-PPh_3/CCl_3CN reagent under microwave conditions at 100 °C for 5 min. This acid chloride then further reacted with amidoximes to form the corresponding oxadiazoles in excellent yields.

3.18.8 COMBINATION OF IONIC LIQUIDS AND MICROWAVE IRRADIATION

The uses of ionic liquids as solvent, additive, and/or catalyst in microwave-assisted organic syntheses have gained popularity as indicated by a large number of publications. Due to the ionic nature, the catalytic properties of ionic liquids accelerate many organic reactions using conventional organic synthesis methodologies. These compounds can also absorb microwave irradiation very efficiently and can increase the rate of organic reactions, even those involving low-polarity reaction media. In this section, the results describing the innovative applications of ionic liquids will be reviewed.

Portela-Cubillo et al.[61] have described the synthesis of a wide range of biologically active compounds containing the quinazoline ring system. This method utilizes the free-radical-based method that relies on microwave-promoted reactions of *o*-phenyl oximes with aldehydes. 2-Aminoaryl alkanone *o*-phenyl oximes were prepared, which produce dihydroquinazolines when mixed with an aldehyde in toluene and subjected to microwave heating in 1-ethyl-3-methylimidazolium hexafluorophosphate [emim]PF_6. When $ZnCl_2$ was included in the reaction mixture, fully aromatic quinazolines were produced in high yields of up to 94% over a short reaction time of 30 min (Scheme 3.18.31). The method worked well with alkyl, aryl, and heterocyclic aldehydes and for a variety of substituents in the aromatic part of the molecule.

An efficient and fast procedure for the synthesis of 2-(2-pyridyl)azoles, from *o*-substituted aromatic amines and picolinic acid, was described by Likhanova and coworkers using ionic liquid 1-butyl-3-methylimidazolium chloride [bmim]Cl as a catalyst under microwave irradiation (Scheme 3.18.32).[62] These compounds were evaluated as corrosion inhibitors for oil refinery equipment.

Applications of ionic liquids and microwave technology have resulted in the development of a new, environment-friendly technology, such as the aqueous [bmim]BF_4 as a green solvent in the microwave-assisted clean synthesis of 1*H*-benzo[f]chromene derivatives. A high yielding and rapid method for the conversion of substituted 1-(2-hydroxyphenyl)-3-phenyl-1,3-propanediones under microwave irradiation was achieved. This methodology provides high yields in short reaction times (i.e., seconds), using ethyl ammonium nitrate [$EtNH_3$] NO_3 as solvent and catalyst. The catalyst can be recycled and reused several times (Scheme 3.18.33).[63]

The aforementioned examples clearly demonstrate the potential of microwaves using ionic liquids as solvent and/or catalyst. Many well-known and new reactions have been discovered by this technology with evident advantages in most of the cases. The synergism arising

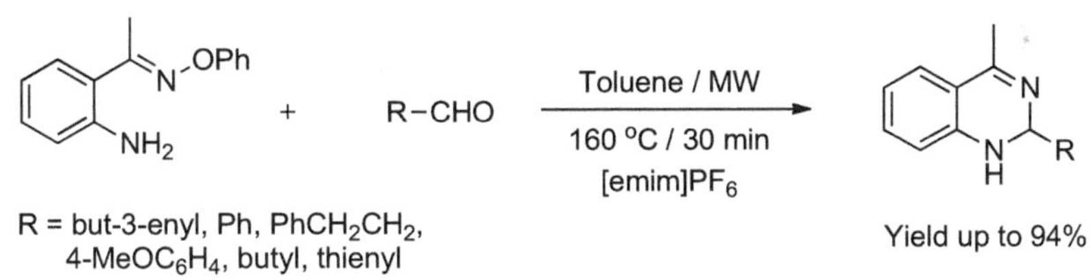

SCHEME 3.18.31 Ionic liquid [emim]PF_6 combined with microwaves for the synthesis of quinazoline ring systems.

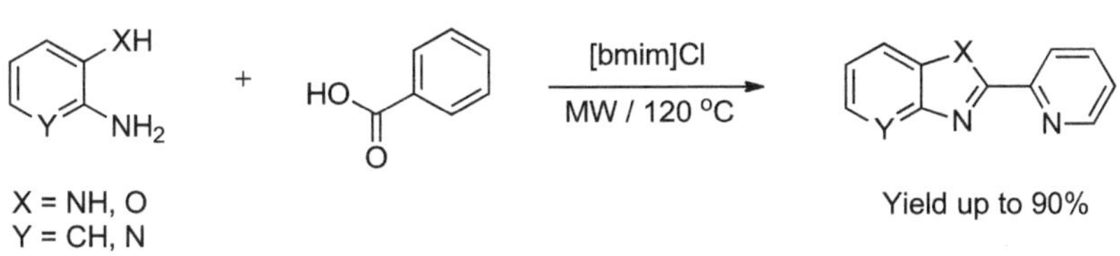

SCHEME 3.18.32 [bmim]Cl-catalyzed synthesis of azoles under microwave irradiation.

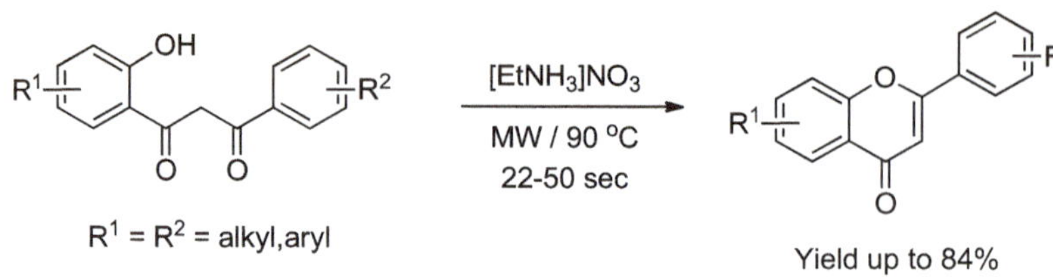

SCHEME 3.18.33 Use of [bmim]BF_4 as a green solvent for microwave-based synthesis.

from the combined use has enormous potential to meet the increasing demand for environmentally benign chemical processes.

3.18.9 MICROWAVE HEATING EFFECT

Reactions described in this chapter clearly outline the dramatic rate enhancement using microwave irradiation when compared with classical heating methods. Microwave heating is often known as dielectric heating.[64] Specific microwave dielectric heating effects result in rapid temperature increase that cannot be easily accomplished using thermal heating.[65] Dielectric heating refers to heating by the electric field component of the high-frequency electromagnetic radiation, owing to the presence of electric dipoles in polar molecules.[66] The rapid heating can be enhanced using solvents such as ionic liquid that have a large dielectric constant.[67] Microwaves are also associated with magnetic field component, which also couples with some materials to induce heating. Experiments performed using microwaves therefore lead to short reaction times and high yields.[68]

3.18.10 CONCLUSIONS AND FUTURE OUTLOOK

Microwave irradiation has become a mainstream activation method of choice for organic synthetic chemists. The early progress made in this area prompted the relevant industry to develop specifically designed, reliable, and controllable instruments that made the development of the MAOS possible. Today, a broad range of such instruments are readily available providing a boost for synthetic applications. In fact, there are very few synthetic laboratories that do not house at least one microwave reactor. Microwave irradiation can be applied to activate a broad variety of chemical transformations, thus providing exceptional importance for MAOS in the development of new synthetic processes. Strengthening environmental regulations and safety concerns generated even further interest in the application of microwave chemistry. Significant developments in several areas of MAOS include the scale-up that would provide a possibility for its extension to industrial processes. In synthesis development most likely, research in the green chemistry areas will dominate in the foreseeable future. These include, but are not limited to the multistep one-pot processes commonly known as tandem reactions; heterogeneous methods such as the application of solid catalysts and solid-phase synthesis are other key areas, as addressed in this work.

References

1. Kappe CO, Stadler A. *Microwaves in organic and medicinal chemistry*. Weinheim: Wiley-VCH Verlag; 2005.
2. Mikami K. *Green reaction media in organic synthesis*. Blackwell: Oxford; 2005.
3. Polshettiwar V, Varma RS. Microwave-assisted organic synthesis and transformations using benign reaction media. *Acc Chem Res* 2008;**41**:629–39.
4. a. Gedye RN, Smith F, Westay K, Ali H, Baldistera H, Baberge L, Rousell J. The use of microwave ovens for rapid organic synthesis. *Tetrahedron Lett* 1986;**27**:279–82;
b. Giguere RJ, Bray TL, Duncan SM, Majetich G. Application of commercial microwave ovens to organic synthesis. *Tetrahedron Lett* 1986;**27**:4945–8.
5. a. Kappe CO, Dallinger D. The impact of microwave synthesis on drug discovery. *Nat Rev Drug Discov* 2006;**5**:51–63;
b. Lidström P, Tierney J, Wathey B, Westman J. Microwave-assisted organic synthesis – a review. *Tetrahedron* 2001;**57**:9225–83.
6. Kappe CO. Controlled microwave heating in modern organic synthesis. *Angew Chem Int Ed* 2004;**43**:6250–84.
7. Loupy A. *Microwaves in organic synthesis*. Weinheim: Wiley-VCH; 2006.
8. a. Tierney JP, Lidström P. *Microwave-assisted organic synthesis*. Blackwell: Oxford; 2005;
b. Kappe OC, Dallinger D, Murphree SS. *Practical microwave synthesis for organic Chemists-stategies, instruments, and protocols*. Weinheim: Wiley-VCH; 2008.
9. Zhu J, Bienaymé H. *Multicomponent reactions*. Weinheim: Wiley-VCH; 2004.
10. Vilotijevic I, Jamison TF. Epoxide-opening cascades promoted by water. *Science* 2007;**317**:1189–92.
11. Gokel G, Lüdke G, Ugi I. *Isonitrile chemistry*. New York: Academic Press; 1971.
12. De Silva RA, Santra S, Andreana PR. A tandem one-pot, microwave-assisted synthesis of regiochemically differentiated 1,2,4,5-tetrahydro-1,4-benzodiazepin-3-ones. *Org Lett* 2008;**10**:4541–4.
13. Xing X, Wu J, Feng G, Dai WM. Microwave-assisted one-pot U-4CR and intermolecular *O*-alkylation towards heterocyclic scaffolds. *Tetrahedron* 2006;**62**:6774–81.
14. Michael JP. Quinoline, quinazoline and acridone alkaloids. *Nat Prod Rep* 2004;**21**:650–68.
15. Liu JF, Kaselj M, Isome Y, Ye P, Sargent K, Sprague K, Cherrak D, Wilson CJ, Si Y, Yohannes D, Ng NC. Design and synthesis of a quinazolinone natural product-templated library with cytotoxic activity. *J Comb Chem* 2006;**8**:7–10.
16. Lindsley CW, Bogusky MJ, Leister WH, McClain RT, Robinson RG, Barnett SF, Defeo-Jones D, Ross III CW, Hartman GD. Synthesis and biological evaluation of unnatural canthine alkaloids. *Tetrahedron Lett* 2005;**46**:2779–82.
17. Liu JF, Kaselj M, Isome Y, Chapnick J, Zhang B, Bi G, Yohannes D, Yu L, Baldino CM. Mirowave-assisted Conside total synthesis of quinazolinobenzodiazepine alkaloids. *J Org Chem* 2005;**70**:10488–93.
18. Tu S, Zhu X, Zhang J, Xu J, Zhang Y, Wang Q, Jia R, Jiang B, Zhang J, Yao C. New potential biologically active compounds: design and an efficient synthesis of *N*-Substituted 4-aryl-4,6,7,8-tetrahydroquinoline-2,5(1*H*,3*H*)-diones under microwave irradiation. *Bioorg Med Chem Lett* 2006;**16**:2925–8.
19. Chebanov VA, Sakhno YI, Desenko SM, Chernenko VN, Musatov VI, Shishkina SV, Shishkin OV, Kappe CO. Cyclocondensation reactions of 5-aminopyrazoles, pyruvic acids and aldehydes. Multicomponent approaches to pyrazolopyridines and related products. *Tetrahedron* 2007;**63**:1229–42.
20. Zhang W, Chen CHT. Fluorous synthesis of biaryl-substituted proline analogs by 1,3-dipolar cycloaddition and Suzuki coupling reactions. *Tetrahedron Lett* 2005;**46**:1807–10.
21. Frankowski KJ, Hirt EE, Zeng Y, Neuenswander B, Fowler D, Schoenen F, Aubé J. Synthesis of *N*-Alkyl-octahydroisoquinolin-1-one-8-carboxamide libraries using a tandem Diels–Alder/acylation sequence. *J Comb Chem* 2007;**9**:1188–92.
22. Meng L, Fettinger JC, Kurth MJ. Intramolecular cycloaddition of azomethine ylides in the preparation of Pyrrolidino[2′,3′:3,4]pyrrolidino[1,2-*a*]benzimidazoles. *Org Lett* 2007;**9**:5055–8.
23. Chie JJ, Fang JM. Microwave-assisted one-pot tandem reactions for direct conversion of primary alcohols and aldehydes to triazines and tetrazoles in aqueous media. *J Org Chem* 2007;**72**:3141–4.
24. Rasmussen LK, Boren BC, Fokin VV. Ruthenium-catalyzed cycloaddition of aryl azides and alkynes. *Org Lett* 2007;**9**:5337–9.
25. Yeung KS, Farkas ME, Kadow JF, Meanwell NA. A base-catalyzed, direct synthesis of 3,5-disubstituted 1,2,4-triazoles from nitriles and hydrazides. *Tetrahedron Lett* 2005;**46**:3429–32.

26. Soh CH, Chui WK, Lam Y. An efficient and expeditious synthesis of di- and monosubstituted 2-aminoimidazoles. *J Comb Chem* 2008;**10**:118–22.
27. Ju Y, Varma RS. Microwave-assisted cyclocondensation of hydrazine derivatives with alkyl dihalides or ditosylates in aqueous media: synthesis of pyrazole, pyrazolidine and phthalizine derivatives. *Tetrahedron Lett* 2005;**46**:6011–4.
28. Viirre RD, Evindar G, Batey RA. Copper-catalyzed domino annulation approaches to the synthesis of benzoxazoles under microwave-accelerated and conventional thermal conditions. *J Org Chem* 2008;**73**:3452–9.
29. Lachance N, April M, Joly MA. Rapid and efficient microwave-assisted synthesis of 4-, 5-, 6- and 7-azaindoles. *Synthesis* 2005:2571–7.
30. Jia CS, Zhang Z, Tu SJ, Wang GW. Rapid and efficient synthesis of poly-substituted quinolines assisted by *p*-toluene sulphonic acid under solvent-free conditions: comparative study of microwave irradiation *versus* conventional heating. *Org Biomol Chem* 2006;**4**:104–10.
31. Liu XY, Ding P, Huang JS, Che CM. Synthesis of substituted 1,2-dihydroquinolines and quinolines from aromatic amines and alkynes by gold(I)-catalyzed tandem hydroamination–hydroarylation under microwave-assisted conditions. *Org Lett* 2007;**9**:2645–8.
32. Trost BM, Gutierrez AC. Ruthenium-catalyzed cycloisomerization–6π-cyclization: a novel route to pyridines. *Org Lett* 2007;**9**:1473–6.
33. Movassaghi M, Hill MD. Synthesis of densely substituted pyrimidine derivatives. *J Am Chem Soc* 2006;**128**: 14254–5.
34. Stella L. *Radicals in organic synthesis*. Weinheim: Wiley-VCH; 2001.
35. Prediger I, Weiss T, Reiser O. Facile access to 2-arylindolines and 2-arylindoles by microwave-assisted tandem radical cyclization. *Synthesis* 2008:2191–8.
36. Baran PS, O'Malley DP, Zografos AL. Sceptrin as a potential biosynthetic precursor to complex pyrrole–imidazole alkaloids: the total synthesis of ageliferin. *Angew Chem Int Ed* 2004;**43**:2674–7.
37. Northrop BH, O'Malley DP, Zografos AL, Baran PS, Houk KN. Mechanism of vinylcyclobutane rearrangement of sceptrin to ageliferin and nagelamide. *Angew Chem Int Ed* 2006;**45**:4126–30.
38. Balogh M, László P. *Organic chemistry using clays*. Berlin, Heidelberg: Springer-Verlag; 1993.
39. Dasgupta S, Török B. Application of clay catalysts in organic synthesis. A review. *Org Prep Proced Int* 2008;**40**:1–65.
40. Gates BC. *Catalysis by solid acids in encyclopedia of catalysis*. New York: Wiley; 2003.
41. Ebitani K, Ide M, Mitsudome T, Mizugaki T, Kaneda K. Creation of a chain-like cationic iron species in montmorillonite as a highly active heterogeneous catalyst for alkane oxygenations using hydrogen peroxide. *Chem Commun* 2002:690–1.
42. a. Szöllösi G, Török B, Baranyi L, Bartók M. Chemoselective hydrogenation of cinnamaldehyde to cinnamyl alcohol over Pt/K-10 catalyst. *J Catal* 1998;**179**:619–23;
b. Török B, London G, Bartók M. Reduction of carbonyl compounds to hydrocarbons by catalytic hydrogenation: a novel one-pot method using Pt/K-10 montmorillonite catalyst. *Synlett* 2000:631–2.
43. Abid M, Spaeth A, Török B. Solvent-free solid acid-catalyzed electrophilic annelations: a new green approach for the synthesis of substituted five-membered *N*-Heterocycles. *Adv Synth Catal* 2006;**348**:2191–6.
44. Abid M, Landge SM, Török B. An efficient and rapid synthesis of *N*-Substituted pyrroles by microwave-assisted solid acid catalysis. *Org Prep Proced Int* 2006;**35**:495–500.
45. Abid M, De Paolis O, Török B. A novel one-pot synthesis of *N*-Acylindoles from primary aromatic amides. *Synlett* 2008:410–2.
46. Landge SM, Schmidt A, Outerbridge V, Török B. Synthesis of pyrazoles by a one-pot tandem cyclization-dehydrogenation approach on Pd/C/K-10 catalyst. *Synlett* 2007:1600–4.
47. De Paolis O, Baffoe J, Landge SM, Török B. Multicomponent domino cyclization-oxidative aromatization on a bifunctional Pd/C/K-10 catalyst: an environmentally benign approach toward the synthesis of pyridines. *Synthesis* 2008:3423–8.
48. Kulkarni A, Abid M, Török B, Huang X. A direct synthesis of β-carbolines via a three-step one-pot domino approach with bifunctional Pd/C/K-10 catalyst. *Tetrahedron Lett* 2009;**50**:1791–4.
49. a. Aitken SG, Abell AD. Olefin metathesis: catalyst development, microwave catalysis, and domino application. *Aust J Chem* 2005;**58**:3–13;
b. Deiters A, Martin SF. Synthesis of oxygen- and nitrogen-containing heterocycles by ring-closing metathesis. *Chem Rev* 2004;**104**:2199–238.

50. La DS, Alexander JB, Cefalo DR, Graf DD, Hoveyda AH, Schrock RR. Mo-catalyzed asymmetric synthesis of dihydrofurans. Catalytic kinetic resolution and enantioselective desymmetrization through ring-closing metathesis. *J Am Chem Soc* 1998;**120**:9720–1.
51. Seigal BA, Fajardo C, Snapper ML. Tandem catalysis: generating multiple contiguous carbon–carbon bonds through a ruthenium-catalyzed ring-closing metathesis/Kharasch addition. *J Am Chem Soc* 2005;**127**:16329–32.
52. Fustero S, Jiménez D, Sánchez-Roselló M, Del Pozo C. Microwave-assisted tandem cross metathesis intramolecular aza-Michael Reaction: an easy entry to cyclic β-amino carbonyl derivatives. *J Am Chem Soc* 2007;**129**:6700–1.
53. Chachignon H, Scalacci N, Petricci E, Castagnolo D. Synthesis of 1,2,3-substituted pyrroles from propargylamines via a one-pot tandem enyne cross metathesis-cyclization reaction. *J Org Chem* 2015;**80**:5287–95.
54. Appukuttan P, Dehaen W, Van der Eycken E. Microwave-assisted transition-metal-catalyzed synthesis of *N*-Shifted and ring-expanded buflavine analogues. *Chem Eur J* 2007;**13**:6452–60.
55. Blaney P, Grigg R, Sridharan V. Traceless solid-phase organic synthesis. *Chem Rev* 2002;**102**:2607–24.
56. Dörwald FC. *Organic synthesis on solid phase*. Weinheim: Wiley-VCH; 2002.
57. Bräse S. *Combinatorial chemistry on solid supports*. Berlin, Heidelberg: Springer-Verlag; 2007.
58. De Luca L, Giacomelli G, Porcheddu A. Synthesis of 1-Alkyl-4-imidazolecarboxylates: a catch and release strategy. *J Comb Chem* 2005;**7**:905–8.
59. Radi M, Saletti S, Botta M. A one-pot, two-step microwave-assisted synthesis of highly functionalized benzoxazoles using solid-supported reagents. *Tetrahedron Lett* 2008;**49**:4464–6.
60. Wang Y, Miller RL, Sauer DR, Djuric SW. Rapid and efficient synthesis of 1,2,4-oxadiazoles utilizing polymer-supported reagents under microwave heating. *Org Lett* 2005;**7**:925–8.
61. Portela-Cubillo F, Scott JS, Walton JC. Microwave promoted synthesis of quinazolines and dihydroquinazolines from 2-aminoarylalkanone *O*-phenyl oximes. *J Org Chem* 2009;**74**:4934–42.
62. Likhanova NV, Veloz MA, Höpfl H, Matias DJ, Reyes-Cruz VE, Olivares O, Martínez-Palou R. Microwave-assisted synthesis of 2-(2-pyridyl)azoles. Study of their corrosion inhibiting properties. *J Heterocycl Chem* 2007;**44**:145–53.
63. Sarda SR, Pathan MY, Paike VV, Pachmase PR, Jadhav WN, Pawar RP. A facile synthesis of flavones using recyclable ionic liquid under microwave irradiation. *Arkivoc* 2006;**XVI**:43–8.
64. Razzaq T, Kremsner JM, Kappe CO. Investigating the existence of nonthermal/specific microwave effects using silicon carbide heating elements as power modulators. *J Org Chem* 2008;**73**:6321–9.
65. Rosanna MR, Tao Y, Steigman AE, Dudley GB. On the rational design of microwave-actuated organic reactions. *Chem Sci* 2012;**3**:1240–4.
66. Kappe CO, Pieber B, Dallinger D. Microwave effects in organic synthesis: myth or reality? *Angew Chem Int Ed* 2013;**52**:1088–94.
67. de la Hoz A, Diaz-Ortiz A, Moreno A. Microwaves in organic synthesis. Thermal and non-thermal microwave Effetcs. *Chem Soc Rev* 2005;**34**:164–8.
68. Cho H, Török F, Török B. Energy efficiency of heterogeneous catalytic microwave-assisted organic reactions. *Green Chem* 2014;**16**:3623–34.

CHAPTER

3.19

Application of Sonochemical Activation in Green Synthesis

Clifford J. Ellstrom, Béla Török

University of Massachusetts Boston, Boston, MA, United States

The application of ultrasounds for the activation of chemical reactions dates back to 1927 when the first reports about the effect of high-intensity sound waves on physical, chemical, and biological systems were published.[1] The field enjoyed much attention, particularly after the 1950s, including both mechanistic interpretations of the ultrasonic effect and synthetic applications. Several books[2–6] and comprehensive reviews[7–10] provided periodic updates on the development of the field.

Similar to microwave-assisted synthesis, the use of ultrasounds is often associated with sustainable or green synthesis.[11,12] Although several of these applications are indeed environmentally benign, one cannot use the automatic green label on either method. Several publications emphasize that a case-by-case holistic evaluation is required before a process can be claimed to be green.[13–17]

The applications of sonochemistry have become more diversified in recent years as it branches out to emerging fields. With the advent of increasing attention toward the development of approaches for green technologies and extended general awareness of the need for preservation of the environment, sonication has continued to grow as a versatile activation method for chemical reactions. With the fundamentals of ultrasonic activation already discussed in Chapter 2.2, here we solely focus on applications. Although the length of this work does not allow a fully comprehensive treatment of the relevant literature, we intended to provide an as-broad-as possible overview in two major fields: the traditional applications in organic synthesis and the emerging area of preparation of nanoparticles.

3.19.1 ORGANIC SYNTHESIS

The applications of sonochemistry in organic synthesis include many areas, such as homogeneous and heterogeneous sonochemistry in aqueous or nonaqueous medium. Fields that

Green Chemistry
http://dx.doi.org/10.1016/B978-0-12-809270-5.00024-8

enjoyed the most success are organometallic chemistry, phase transfer catalysis, or heterogeneous metal-catalyzed reactions.

Homogeneous/liquid-phase applications include the aldol reaction.[18] This reaction has a long history in organic chemistry due to its broad applicability. It can be optimized to provide either the aldol product or the enone that may form as the result of a secondary dehydration. Cravotto et al. used aqueous NaOH to react two carbonyl compounds to create the aldol product (Scheme 3.19.1). After optimization of both the frequency and intensity of the ultrasounds the aldol product was obtained selectively in 73% yield using 18 kHz and 280 W sonication for 15 min. These conditions resulted in the exclusive formation of the aldol product.

In the cases of vanillin and ferrocenecarboxaldehyde a greater amount of catalyst and extended sonication were needed to perform the reaction and subsequently obtain the enone product. This observation initiated the authors to carry out further optimization on this system and developed conditions that favored the exclusive formation of the enone (Scheme 3.19.2). The application of a phase transfer catalyst, Aliquat336, yielded the enone in 82% yield.

Although traditional ozonolysis is a useful reaction, its conditions violate almost every principle of green synthesis. A sonochemical application focused on an oxidative cleavage reaction, which mimics the results of ozonolysis (under oxidative conditions). The process utilizes ultrasounds to accelerate the reaction while maintaining high yields (>85%) (Scheme 3.19.3).[19] This reaction uses $RuCl_3$ that is converted in situ with $NaIO_4$ to form RuO_4, the actual active catalytic species. This catalyst performs the reaction, and the excess $NaIO_4$ regenerates the RuO_4. This is a catalytic process carried out in an aqueous medium, thus it is a greener, easier, and safer way to perform oxidative cleavage when compared with traditional ozonolysis. However, although these conditions are certainly greener than the traditional approach, it could be improved by replacing the $NaIO_4$ with a greener oxidant, such as oxygen or air.

Due to extensive interest in bis(indolyl)methanes, many different syntheses have been developed for the preparation of these compounds, but typically they are not green processes. The use of dodecylbenzenesulfonic acid (DBSA) as a catalyst appears to be a satisfactory solution; the reaction takes place in water, and sonicating the reaction mixture allows for short reaction times at room temperature.[20] Using a 25-kHz-frequency ultrasonic bath, with a 2:1 indole to *p*-chlorobenzaldehyde ratio with a 5% catalyst loading, 98% yield was obtained in 10 min (Scheme 3.19.4). These variables were determined to be the optimal conditions and were tested on other reactants. Of those conditions tested none required longer than 30 min sonication to achieve 85% yield, and the reactions were significantly faster than those with conventional stirring. With these modifications, the reaction is far more efficient and greener than the available alternatives.

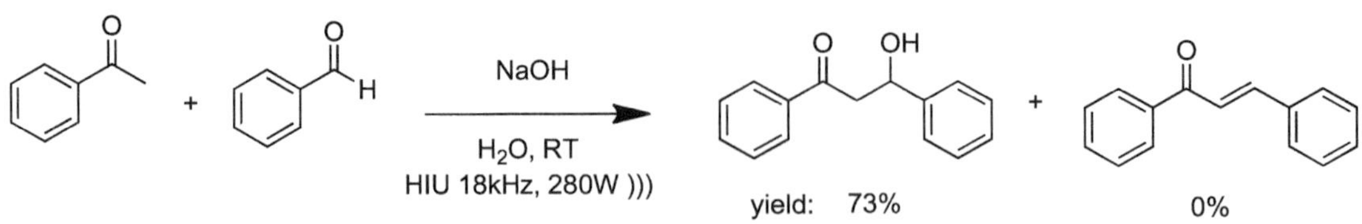

SCHEME 3.19.1

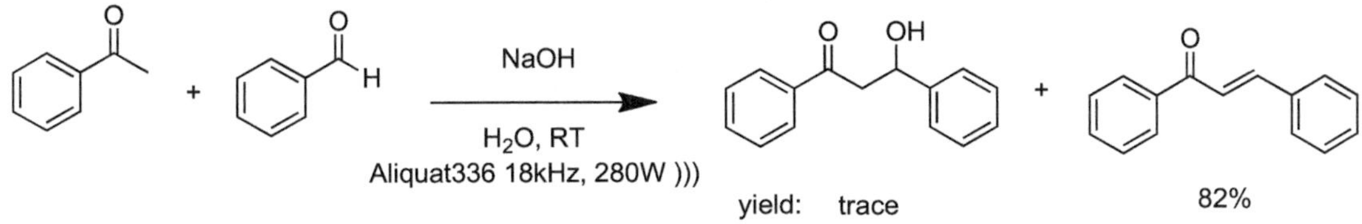

SCHEME 3.19.2

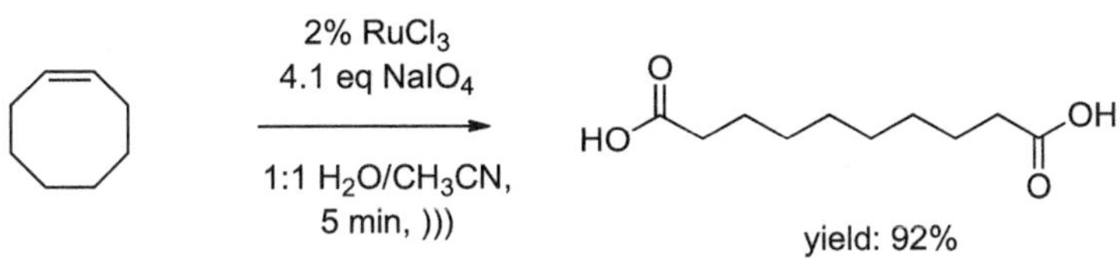

SCHEME 3.19.3

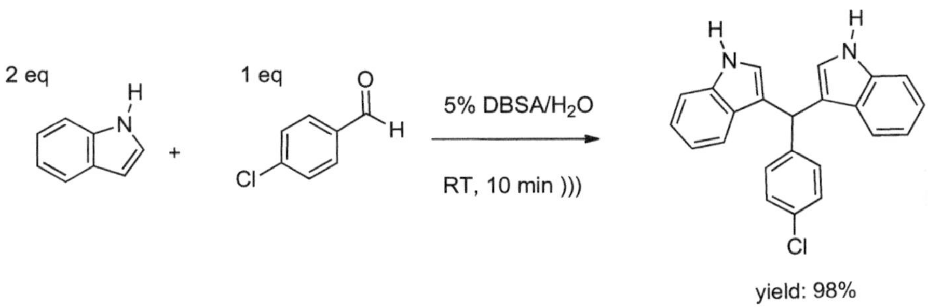

SCHEME 3.19.4

Another reaction utilizing the same DBSA catalyst is the synthesis of amidinohydrazone derivatives.[21] These compounds are versatile and can be used as antimalarial agents. Typically, the synthesis of these compounds required toxic and corrosive reagents and produced low yields in long reactions. The synthesis involving refluxing ethanol with a strong acid for 4–16 hours only resulted in 30%–70% yield. In contrast, utilizing DBSA allowed for the synthesis of these materials in water at room temperature, while sonication drastically decreased reaction times. The optimized system was a 1:1.1 ratio of 1,5-diphenyl-1,4-pentadien-3-one to aminoguanidine hydrochloride with 50% DBSA catalyst load, and it resulted in 94% isolated yield after 2 hours of sonication (Scheme 3.19.5). This system was then tested using

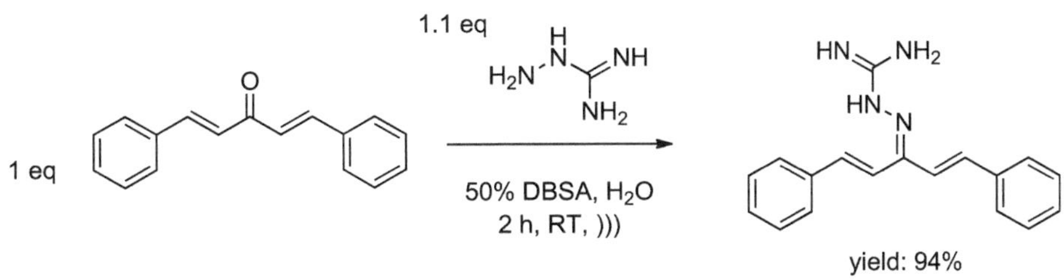

SCHEME 3.19.5

several different analogs, and all the substrates gave 84% or higher isolated yields within 3 h of reaction time. Overall this method is a high-yielding, rapid, environmentally benign process compared to its predecessors, and although it does suffer from a high catalyst loading, this does not take away from the improvements it makes on the current state of this reaction.

A common inorganic base, K_2CO_3, readily catalyzed the ultrasound-promoted synthesis of 4-pyrimidinols in aqueous medium.[22] The cyclocondensation of β-ketoesters and amidines yielded the pyrimidinols (Scheme 3.19.6). While the product yields are mostly moderate (in a few cases excellent) in the 50%–80% range, the reaction conditions are environmentally benign. The authors used sonochemical activation in the next step of the synthesis; however, this step does not qualify as green due to the use of dichloromethane as a solvent and tosyl chloride, which generates HCl as a by-product.

Multicomponent reactions (MCRs) are another common type of reactions that utilize sonochemistry. These reactions typically result in the synthesis of heterocyclic products often without the need for catalysts or complex starting materials. The first example is the synthesis of 1,4-dihydropyridine derivatives.[23] In this four-component, catalyst-free synthesis, multiple 1,4-dihydropyridine derivatives were synthesized utilizing malononitrile, dimethylacetylenedicarboxylate, various benzaldehyde derivatives, and aniline (Scheme 3.19.7). Both catalyzed and catalyst-free reactions were tested; the catalyst-free system provided superior results. When the method of activation was examined, conventional room temperature conditions were tested against ultrasonic irradiation. The conventional reaction took 12 h and resulted in 88% yield. Ultrasonic irradiation, however, improved the yield to 96% while shortening the reaction time to 24 min. Yields of at least 89% were obtained for the 11 derivatives shown. The reagent and catalyst-free nature, the high yields, and the short reaction times make this reaction a green alternative for the synthesis of the products.

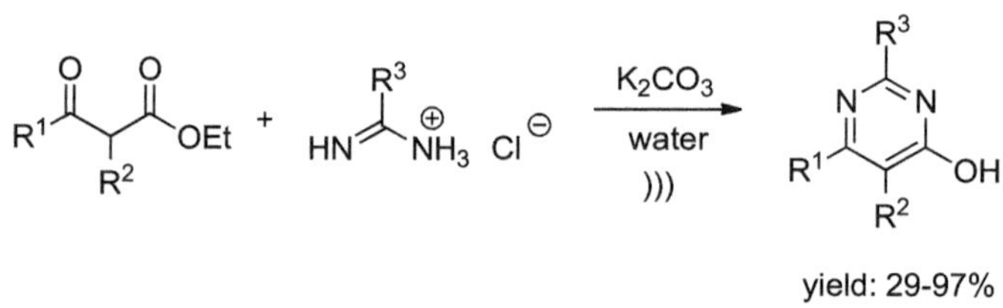

SCHEME 3.19.6

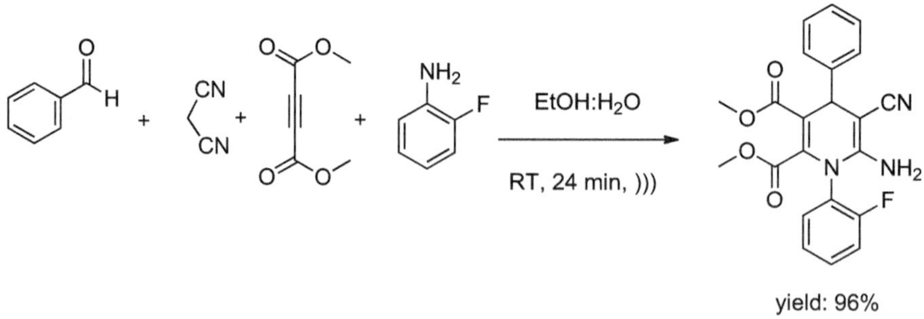

SCHEME 3.19.7

A related method was described by Tabassum et al. using the same starting materials and obtaining the same products with similar yield, applying CuI catalysis and sonication.[24] The catalyst was recyclable for four times with minimal loss of activity.

A similar MCR for the synthesis of indenopyridopyrimidine and pyrimidoquinoline derivatives was described by Mamaghani et al.[25] This process is a three-component reaction requiring slightly elevated temperatures at 65 °C. The solvent dependence of the reaction was examined, and ethylene glycol was found to be the best solvent producing 90% yield in a 3 h reaction. Sonication reduced the reaction time to 25 min, and concurrently, increased the yield to 95%. Water was also tested and good results were obtained (83% yield, 41 min), but clearly the reaction was best run in ethylene glycol (Scheme 3.19.8).

Shukla et al. described the multicomponent one-pot solvent-free sonochemical synthesis of 2H-indazolo[2,1-b]phthalazine-1,6,11-triones and 1H-pyrazolo[1,2-b]phthalazine-5,10-diones.[26] The reaction of phthalhydrazide, 1,3-diketones, and aromatic/aliphatic aldehydes occurred in a domino fashion (Scheme 3.19.9). The syntheses were carried out under both solvent-free conditions at 80 °C and solvent-free system with ultrasonic irradiation at room temperature. Several acid catalysts have been tested, and (*S*)-camphorsulfonic acid was found to be the best promoter. Although the product yields and reaction times were similar, the lower temperature in the sonochemical synthesis is perhaps more energy efficient.

The solvent and catalyst-free sonochemical synthesis of benzoxazoles from azo-linked salicylic acid derivatives and 2-amino-4-chlorophenol has been developed by Nikpassand et al.[27] The reaction occurred at room temperature and provided high yields (85%–96%) under mild conditions. The authors carried out an extensive optimization study to find the best reaction parameters. Several variables, such as temperature, time, ultrasound power, catalyst, etc., were studied. Interestingly, the catalyst-free sonochemical reaction

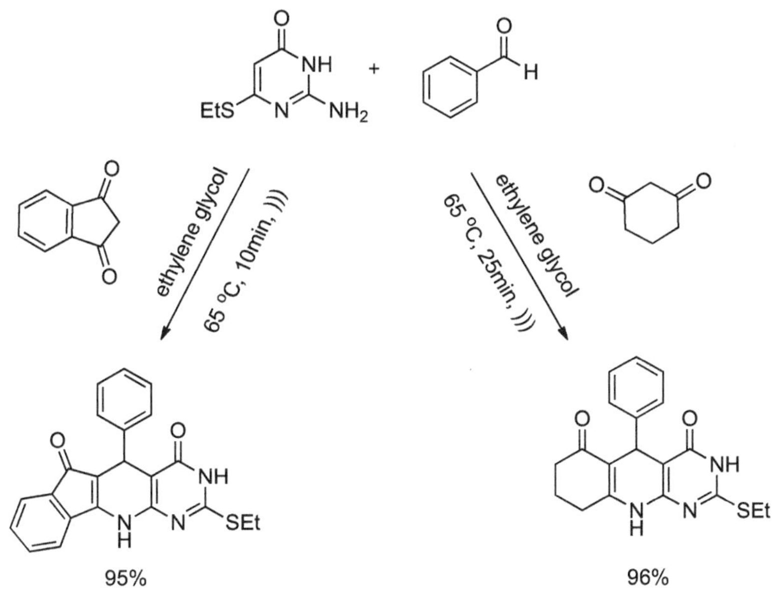

SCHEME 3.19.8

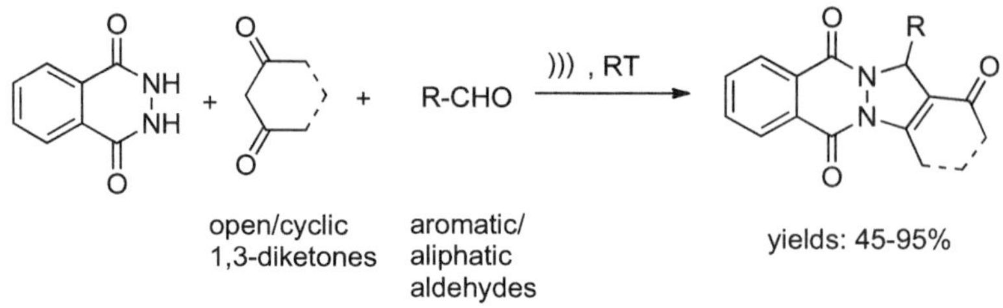

SCHEME 3.19.9

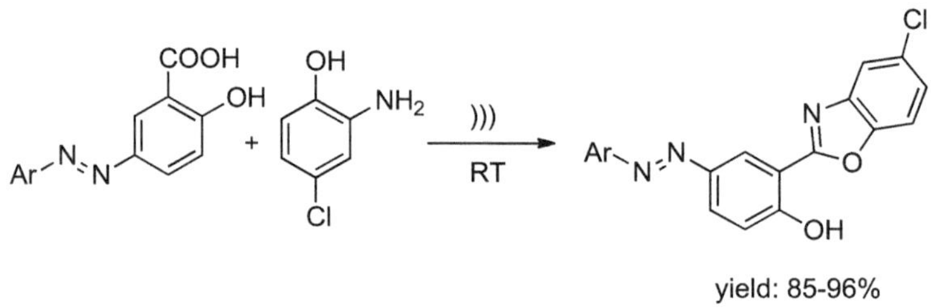

SCHEME 3.19.10

yielded the most products in the shortest times providing a real green alternative for the synthesis of the target compounds (Scheme 3.19.10). When the mixture of two different azo-salicylic acids was used the substituents on the aryl group determined the outcome of the reaction, e.g., the product ratio for the Ar = 2,4-dinitro versus 2,4-dimethyl is 96/0, thus the presence of strongly electron-withdrawing groups remarkably increases the reaction rates.

The environmentally benign sonochemical synthesis of pyrrolin-2-ones has been achieved by Ahankar and coworkers.[28] The MCR of aniline, aldehydes, and diethyl acetylenedicarboxylate was carried out in an ethanolic solution using a green acid catalyst, citric acid. The reaction was complete in less than an hour (mostly 10–15 min) of sonication providing 80%–92% yields. The nonsonicated reaction exhibited similar performance regarding yields; however, in a much longer 10 h reaction time (Scheme 3.19.11).

Reddy et al. applied sonochemical activation in the synthesis of aminopyridine derivatives.[29] Commercially available reactants such as acetophenone, benzaldehyde, *o*-toluidine, and malononitrile and catalyst (FeF_3) were applied in the four-component MCR

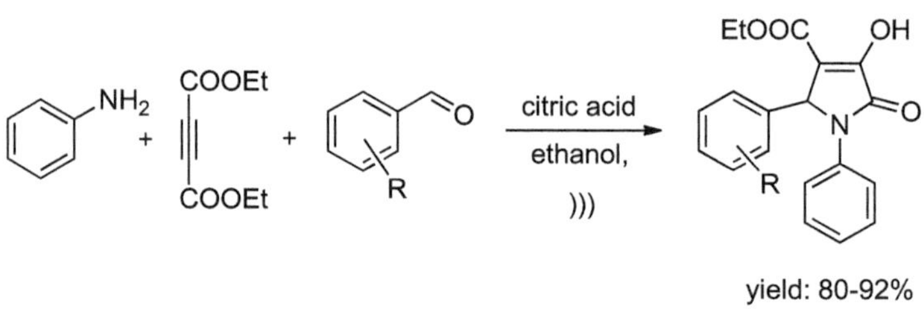

SCHEME 3.19.11

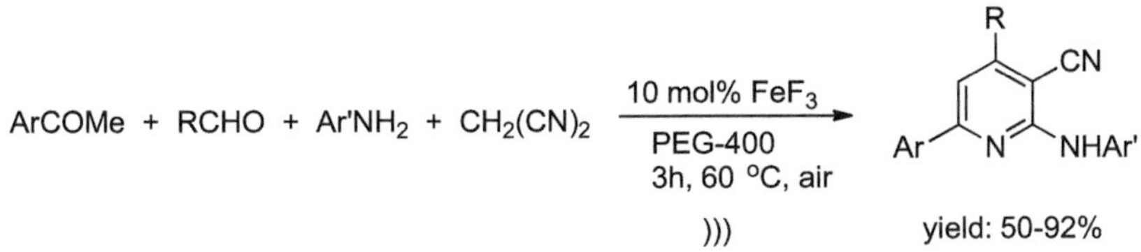

SCHEME 3.19.12

(Scheme 3.19.12). In most cases the reaction occurred with excellent yields, only with a few exceptions.

Heterogeneous catalytic hydrogenation reactions were among the pioneering transformations in *heterogeneous sonochemistry*. Ultrasounds have been used in the hydrogenation of simple C=C bonds[30] or the dehydrogenation of cyclic compounds[31] and were found to highly improve the chemoselectivity of the hydrogenation of α,β-unsaturated compounds[32,33] (Scheme 3.19.13).

However, the most successful examples are found among enantioselective hydrogenations on modified noble metal catalysts. The presonication of the catalysts in the presence of the chiral modifiers resulted in unprecedented, nearly 100% ee values (Scheme 3.19.14). The most successful examples are related to the application of the tartaric acid-modified Raney Ni catalyst for the enantioselective hydrogenation of β-ketoesters (ee up to 98%)[34,35] and the cinchona alkaloid-modified Pt and Pd catalysts for the enantioselective hydrogenation of α-ketoesters[36–39] and trifluoroacetophenone (ee up to 98%).[40] Although the presonication did not result in close-to-perfect enantioselection, it significantly enhanced (over 25%ee increase) the enantioselection of proline-modified Pd catalysts for the hydrogenation of a cyclic α,β-unsaturated ketone, isophorone.[41]

There is an agreement that the role of ultrasounds is in the modification of the nonchiral metal surface. Based on the well-known surface-cleaning effect of ultrasounds, it was proposed that the oxide-contaminated Pt or Raney Ni surface is cleaned and only zero-valence metal atoms are on the surface. These metal atoms will have a high affinity to bind the O-containing tartaric acid and the basic N-containing cinchona alkaloids, hence bringing about the chiral modification of the metal surface. This chirally modified surface will result in the

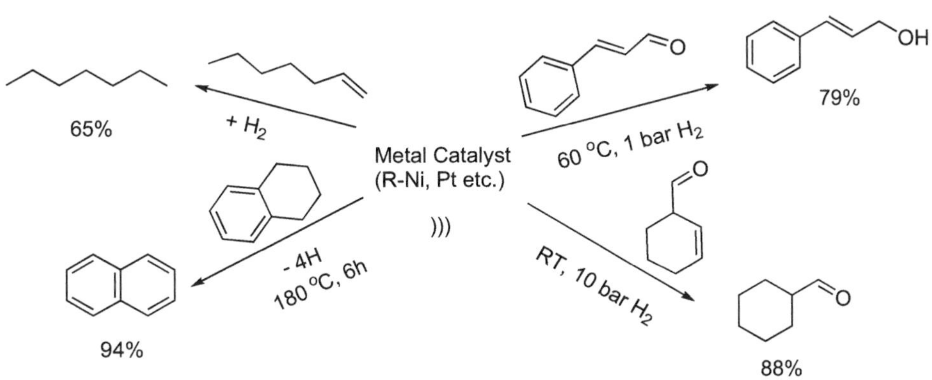

SCHEME 3.19.13

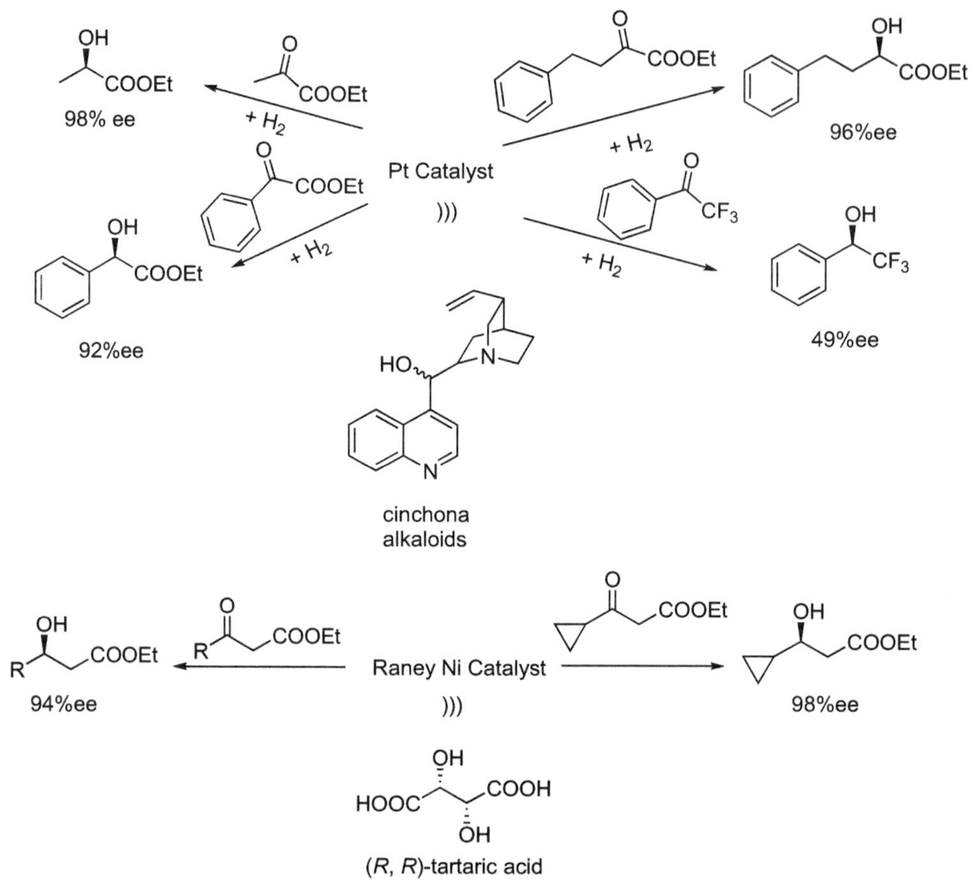

SCHEME 3.19.14

formation of the product with high enantiomeric purity. In the case of the sonochemically modified Pt catalysts, it was observed that the metal particle size became significantly smaller and the product enantioselectivity showed an optimum as a function of the particle size and the presonication time.

Ultrasounds have also been recently applied in the activation of Raney-type Ni-Al alloy that have been used in a variety of reductive transformations (Scheme 3.19.15).[42–44] Typically, the role of ultrasounds is to "clean" the surface of Al, e.g., to remove the gradually accumulating Al_2O_3 impurities. This cleaning effect allowed for a more reactive Al surface, thus improving reaction rate of Al with water, hence enhancing the in situ production of H_2. Once the aluminum component of the Raney alloy began to dissolve the Ni component of the original alloy gradually transformed to a Raney Ni-type hydrogenation catalyst that used the formed hydrogen efficiently. Overall, the direct hydrogen generation resulted in safer, easier, and greener reductions that need neither organic solvents nor dangerous high-pressure hydrogen. It appears that the catalyst is interchangeable as long as it can utilize the hydrogen. This allows for both a Ni-based and a Pd-based variety of these reductions.

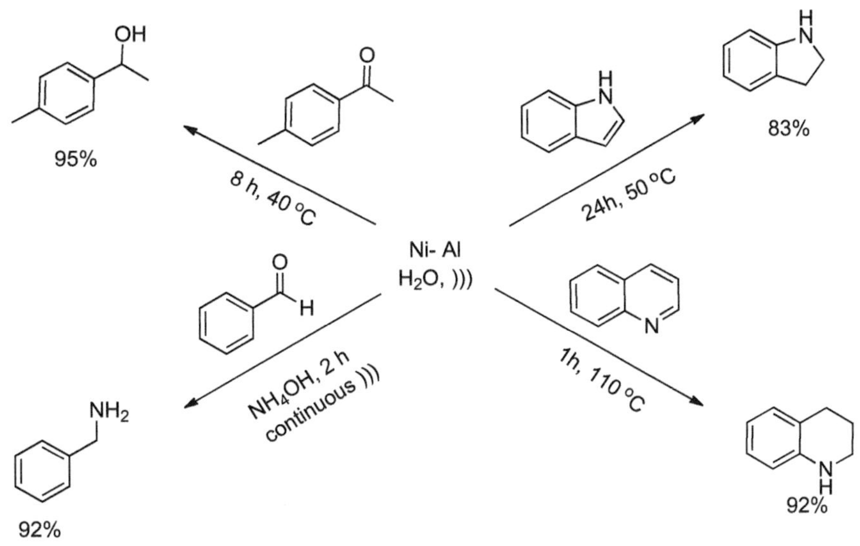

SCHEME 3.19.15

In a subsequent study a commercial Pd catalyst and aluminum powder were used for the easy and fast reduction of multiple functional groups while selectively keeping the aromatic rings intact at room temperature, as shown in Scheme 3.19.16.[45] It was also observed that the appropriate selection of the reaction temperature allowed the successive reduction of the aromatic rings as well. Although Pd is a more expensive catalyst than Ni, it can be used in lower amounts and it does give better yields in more selective reactions under mild conditions in many reactions.

In heterogeneous sonochemistry, *organometallic reactions* can be considered as well-known and popular applications.[46] The aforementioned surface cleaning as well as particle-size-decreasing effect appears to be useful in these reactions as well, as the surface impurities (e.g., oxide or carbonate layers) inhibit the formation of organometallic compounds. When discussing green reactions, it is rare to refer to the Grignard, or the Barbier reaction. These reactions are fundamentally not green; the reaction requires potentially hazardous organic solvents (diethyl ether, tetrahydrofuran) that could undergo peroxidation and both the synthesis and further applications of organometallics produce significant amount of waste. However, beyond doubt these are very useful reactions and synthetic chemists will continue using them. Thus any improvement toward a more efficient protocol for these reactions can yield tremendous environmental benefits. Switching the activation method from microwave or conventional heating to sonication allowed for a more consistent synthesis of the Mg-halide complex.[47] The reactions of several chloro- or bromo-substituted aromatic compounds were tested using various ultrasonic devices (Scheme 3.19.17) to improve the formation of the Grignard reagent. The best performance was observed with the 200-W Cup Horn reactor, operating at 300 kHz, which was able to yield 100% conversion in all cases except for chlorobenzene. The chloro-substituted reactants all exhibited diminished and delayed activities, resulting in induction periods of up to 2 h. The obtained Grignard reagents

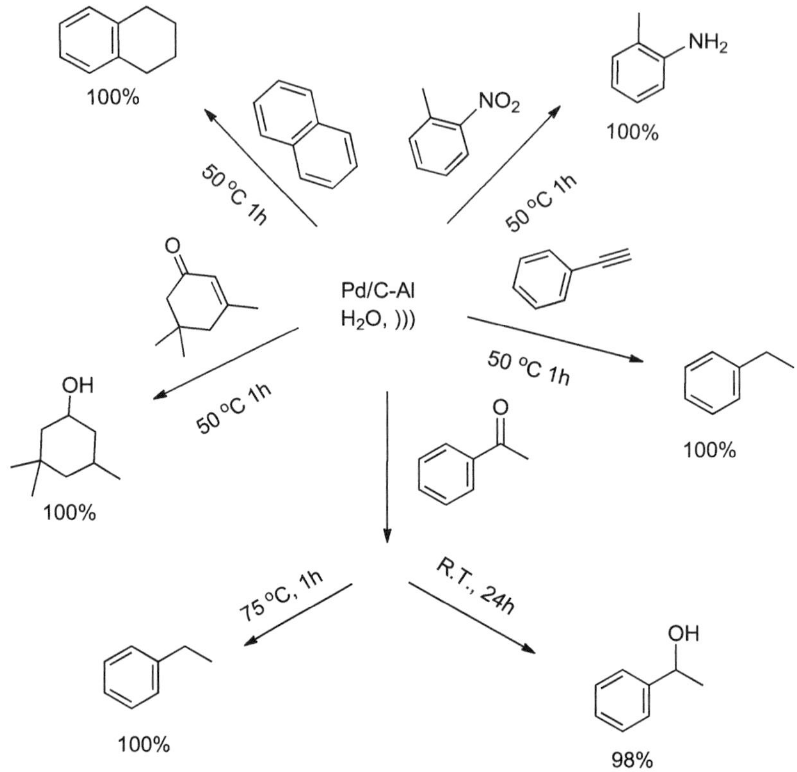

SCHEME 3.19.16

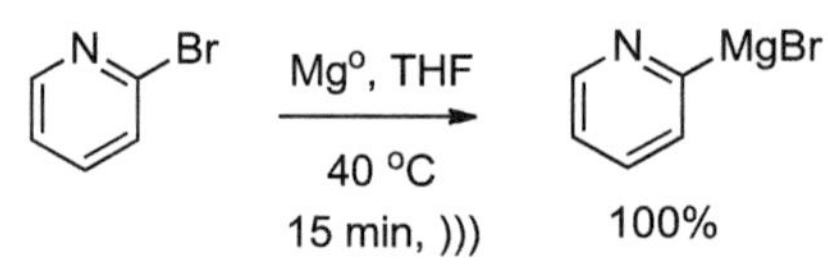

SCHEME 3.19.17

were applied to perform the Grignard reaction with benzophenone to verify the reactivity of the species created. All of the tested intermediates had high conversion rates, and the lowest isolated yield observed was 84%.

A new solvent mixture was studied using ultrasounds for a typical Barbier reaction. The subcritical mixture of carbon dioxide and water was applied as a solvent for a Barbier-type allylation of aldehydes using Zn and Sn as the metals (Scheme 3.19.18).[48] The reaction

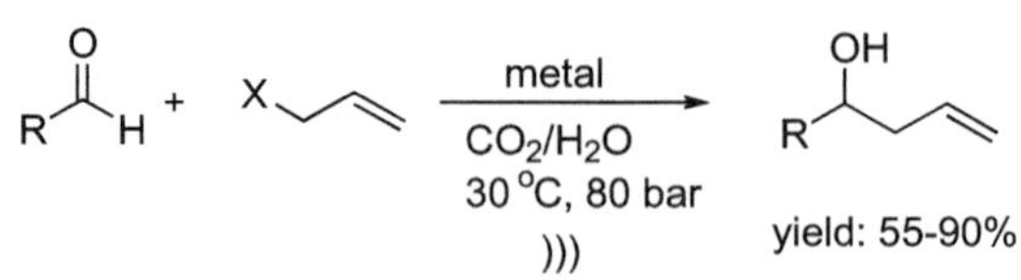

SCHEME 3.19.18

provided the homoallyl alcohols in moderate to excellent yields, however, with almost exclusive chemoselectivity, which was achieved by the addition of a biocompatible nonionic surfactant. Other ultrasound-enhanced examples of the Barbier reaction and Stille coupling have also been published.[49,50]

The ultrasonic activation appears to be applicable for enhancing *Pd-catalyzed cross-coupling* reactions as well. The Suzuki cross-coupling reaction is one of the most important C—C-bond-forming reactions for the synthesis of biaryl compounds. A new variant of this reaction was developed utilizing sonication to replace the microwave-assisted synthesis that was initiated by the ambient conditions that the sonication allows for (Scheme 3.19.19).[51] It was performed utilizing a flow reactor to increase the yield and create a scalable setup of the reaction. The sonicator was a bath operating at 20 kHz built into the flow reactor. Pd^{II} was reduced to Pd^0, which catalyzed the reaction in the sonicator section of the flow reactor. Thirteen aryl halides were tested and showed a clear trend of reduced yield as the size of the halogen decreased and electron-withdrawing groups were added to the ring. Most aryl halides tested resulted in yields of 50% or higher after 3 h, with all iodo-substituted compounds exceeding 85% yield. Ultimately the Pd could be recycled up to 5 times without loss of activity or decrease in yield. If combined with the yields obtained with iodoaryl compounds the reaction provided a real alternative to many other synthetic methods, which is capable of being scaled up to industrial scales.

The surface cleaning and surface modification phenomena described earlier have been the objective of several investigations that targeted the directed surface modification of materials for various purposes from catalysis to coatings.[52]

The sonochemical activation was found to be useful in several heterogeneous catalytic solid acid or metallic nanoparticle-catalyzed synthesis and reactions of heterocycles. Graphene oxide has been applied as a bifunctional catalyst for the synthesis of a broad variety of benzimidazoles and benzothiazoles using ultrasonic activation.[53] The catalyst acted as a solid acid to facilitate the formation of the Schiff base and addition of the NH or SH groups to the C=N double bond (Scheme 3.19.20). Once the nonaromatic product formed the catalyst also assisted in the dehydrogenation/aromatization process. The catalyst remained active after five successive reactions, and it could be reactivated. In a similar application, graphite oxide was used to catalyze the ultrasound-assisted direct oxidative amidation of benzyl alcohols with amines.[54] The reaction occurred under mild conditions providing the amides in moderate to excellent yields (69%—95%).

MCRs carried out in one pot are favored designs in green chemistry applications as there is no need for the isolation and purification of the intermediate products. Several heterogeneous catalytic alternatives have been published, for example, the synthesis of polysubstituted imidazoles.[55] This is a four component reaction, but unlike the homogeneous, catalyst-free

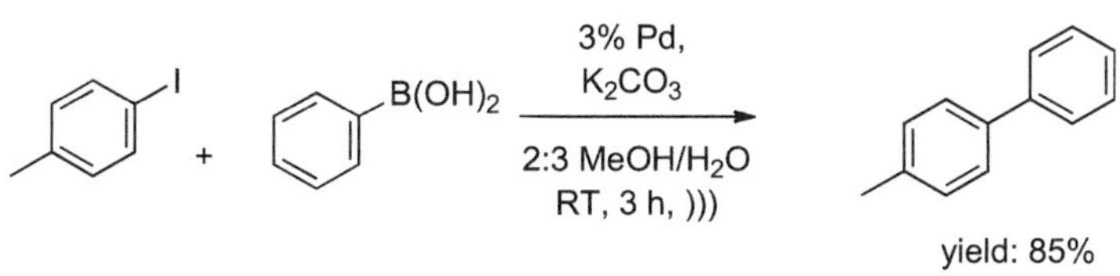

SCHEME 3.19.19

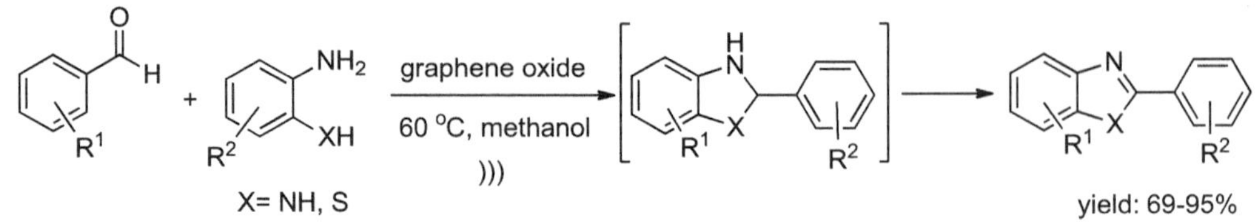

SCHEME 3.19.20

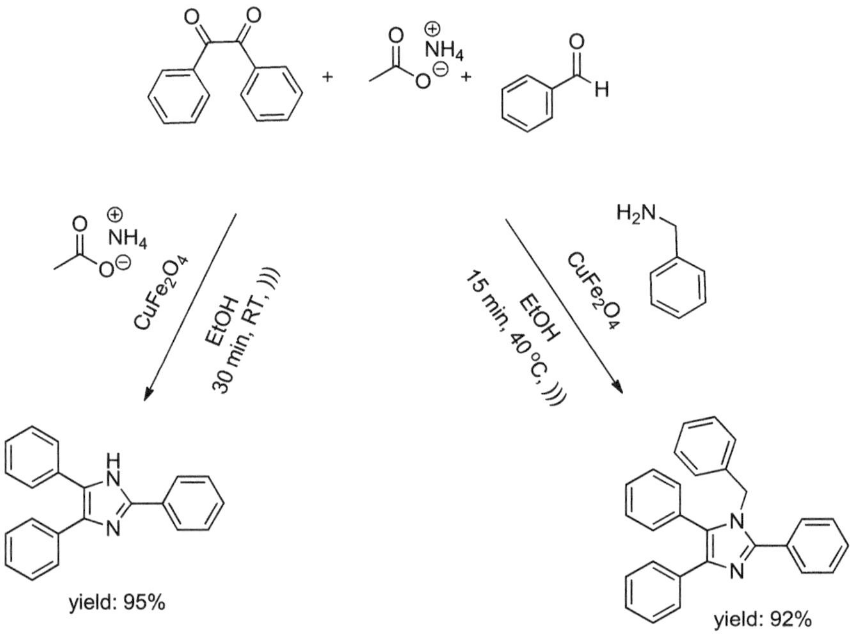

SCHEME 3.19.21

example in Schemes 3.19.7–3.19.10, this reaction requires a catalyst (Scheme 3.19.21). The catalyst used was $CuFe_2O_4$ due to its ability to allow for ambient reaction conditions and decreased reaction time while still obtaining equal or greater yields than either of the other catalysts studied. Further improvements were made with the additional application of sonication, which increased the yields from 45% to 96% in a 30 min timespan. This reaction is a green process with simple and commercially available reactants, which makes the synthesis of imidazoles far more simple and cost-effective than that achieved with most of the available methods. Eidi et al. described the same process using a $CoFe_2O_4$ catalyst obtaining similar yield.[56]

Reddy et al. has developed an ultrasound-assisted synthesis of substituted bezoxanthenes by silica-supported tungstic acid catalysis under solvent-free conditions (Scheme 3.19.22).[57] The lack of solvent, the inexpensive and recyclable catalyst, and the elimination of chromatographic purification are the main advantages of the process. The authors compared the sonochemical method to the conventional synthesis and found that the sonicated reaction took place, on average, in 20 min less time (on a 20–240 min timescale) providing 25% higher yields.

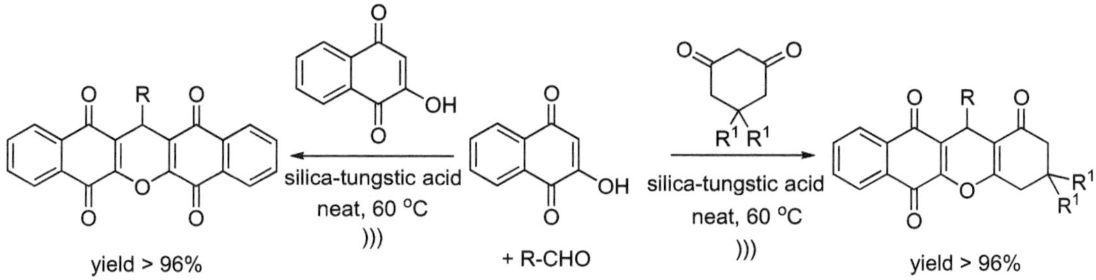

SCHEME 3.19.22

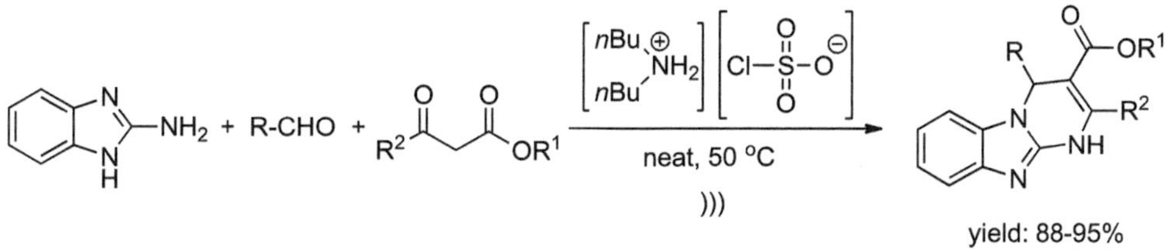

SCHEME 3.19.23

The same group described the solvent-free MCR of aldehydes, 1H-benzo[d]imidazol-2-amine, and β-ketoesters, which provided 1,4-dihydrobenzo[4,5]imidazo[1,2-a]pyrimidine-3-carboxylates in excellent yields (88%–95%).[58] The reaction was catalyzed by a recyclable ionic liquid catalyst, di-*n*-butyl ammonium chlorosulfonate (Scheme 3.19.23).

Clays are versatile solid acid catalysts for a variety of reactions.[59] Safari and coworkers applied an amine-functionalized Cloisite 30B nanoclay [NH_2-montmorillonite (MMT)] catalyst for the ultrasound-assisted synthesis of 3-methyl-4-arylmethylene isoxazole-5(4H)-ones through a three component MCR of hydroxylamine hydrochloride, ethyl acetoacetate, and benzaldehyde derivatives (Scheme 3.19.24).[60] The modified clay appeared to be an effective catalyst for the reaction, and its combination with ultrasonic irradiation yielded the products in excellent yields (80%–97%) in short reaction times (13–55 min).

A broad variety of spiroheterocyclic compounds was synthesized by Singh et al. applying a new organic-inorganic hybrid nanocatalyst in aqueous medium by ultrasonic irradiation.[61] These organo-nanocatalysts were prepared by encapsulating magnetic Fe_2O_3/SiO_2 nanoparticles with thiamine hydrochloride. The synthetic protocol was efficient, economical, and green. The catalyst was mostly recyclable, losing about 10% of its activity in five-consecutive runs.

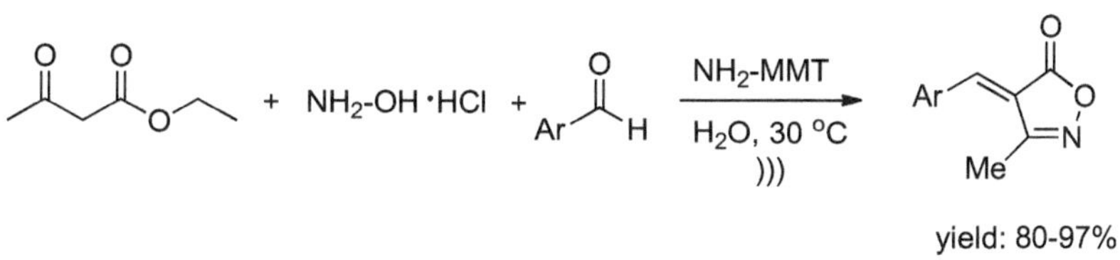

SCHEME 3.19.24

3.19.2 SYNTHESIS OF NANOPARTICLES AND NANOSTRUCTURES

Due to the transient cavitation mechanism, through which ultrasonic irradiation provides the energy, it can perform tasks that other methods of activation are not capable of accomplishing. One of the common emerging applications is based on the particle-size-decreasing effect of sonication. Namely, sonochemistry is uniquely capable of producing nanoparticles from bulk materials. The process and specifics differ from one application to another and even from one solvent to the next, but the final product remains the same. It is believed that the main advantage that the application of ultrasounds represents is in the special role of cavitation that results in significant particle size decrease. When the cavitation bubbles collapse in the crevices of the particles, the extreme pressure and energy result in the breaking of the particle into two smaller units. This effect is well documented and most commonly applied in the activation of heterogeneous metal catalysts.[31–41] In many examples the metal particle size decreased to a few nanometers. The sonication also results in a more homogenous particle size distribution. Based on these earlier observations, it is only natural that the application of ultrasonics enhances the preparation of nanoparticles. A simplified illustration of this process can be seen in Scheme 3.19.25.

An example of this methodology is the creation of nanoparticles of traditional, first- or second-generation antibiotics bringing about their revival against the so-called "super bugs." There are several studies that suggest that using the nanoparticle form of antibiotics may increase their efficacy against nonresistant and even resistant bacteria alike.[62] In a representative study the possible usage of penicillin in the form of nanoparticles was investigated.[63] To prepare the nanoparticles, powdered penicillin was suspended in deionized water and was then sonicated at 20 kHz for 10 min, which resulted in increased efficiency of the antibiotic. The sample was tested against *Staphylococcus aureus*, and its minimum inhibitory concentration (MIC) was determined. The MIC significantly decreased from 33 units/mL for the bulk solution to 0.2 units/mL for the sonicated sample—an increase of more than two orders of magnitude in efficiency. The data suggest that this method could provide a new way to prolong the effectiveness of antibiotics that are no longer efficient in traditional formulation. The increase in efficiency can also allow for smaller treatment doses, reducing the environmental waste as well as the side effects of the antibiotics. Although the procedure carries great promise, the mechanistic explanation is yet to be established. A common issue is that in most publications there was only one antibiotic tested at a time and the procedures are not standardized, making the comparison difficult. A further problem is that several studies examining the effects of nanoparticle formulation on the potency of antibiotics mostly use

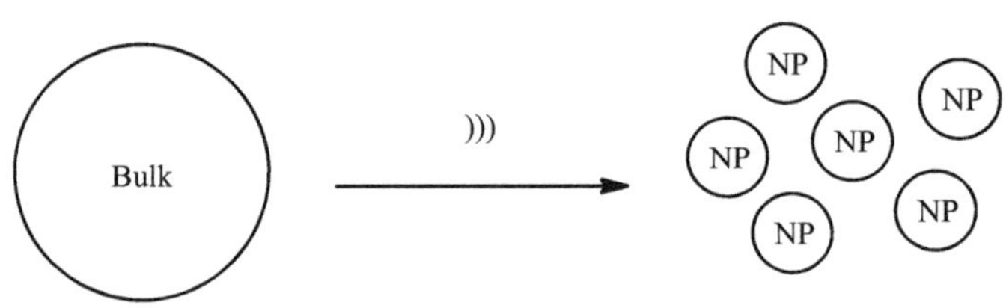

SCHEME 3.19.25

different antibiotic assays as well. Therefore more standardization needs to be performed to allow for easier and better comparisons to confirm the effect. The sonochemical part of the studies are not detailed enough either to make clear conclusions.

Sonication can also allow for the deposition of the in-situ-formed nanoparticles onto different surfaces. After the ultrasonic irradiation created the nanoparticles, the conditions originating from the ongoing irradiation initiated the newly formed nanoparticles to adhere strongly to the surface of a much larger local object. There are many examples of this general procedure ranging from inorganics to proteins being deposited onto different substances.[64–66] Organic nanoparticles were applied to replace the inorganic nanoparticles that are more commonly used in antibacterial cloth. Tannic acid (TA) was also used as the antibacterial agent.[67] As a way to stabilize and properly utilize the nanoparticles created, these TA nanoparticles were deposited on normal cotton bandages. A cotton-based cloth was immersed into the TA/water mixture and sonicated using a 20-kHz sonicator at 30 °C for 1 hour. The antibiotic properties were then examined using *S. aureus* and *Pseudomonas aeruginosa* strains by determining the extent of the reduction of bacterial growth on the TA-impregnated cloth. A significant decrease in growth was observed, ranging from 88% to 100% depending on the preparation method of the cloth, namely, the concentrations of the TA/water mixture used in the sonication. Similar to the aforementioned antibiotic study more data are needed to understand the mechanism that is responsible for the activity enhancement. The well-known antibacterial effect of silver has also been utilized in an effective way when chitosan-capped Ag nanoparticles were prepared and coupled with Cu nanoparticles to obtain antimicrobial nanobunches against *Escherichia coli* by an ultrasound assisted-method.[68]

Another application of nanoparticle sonochemistry is in the field of solar cells. The preparation of solar cells is notoriously wasteful to be considered a green application. This is mainly due to the large-scale synthesis of many of the components used in the solar panels having low yields and being part of toxic synthetic processes. Given the recent popularity and long-term importance of solar cells, improvements to the current methodology would help in reducing the produced waste (and with this, the cost) and environmental impact of solar cells. Solution-based thin-film solar cells are one form of solar cells that aim to eliminate some of the more environmentally damaging aspects of these devices. Both $CuIn_xGa_{1-x}Se_2$ (CIGSe) and $CuIn_xGa_{1-x}S_2$ (CIGS) are common types of nanocrystals that were being proposed to be used in these solution-based thin-layer solar cells.[69] However, the synthesis of these materials requires a large excess of starting materials due to the low overall yield of the long multistep synthesis. This is especially troublesome in industrial-scale synthesis, which creates more waste than useable materials it produces. The new synthesis only requires Se powder, CuCl, and In(OAc) be mixed and sonicated in an ethylene glycol/hydrazine hydrate mixture. This process results in the basic CISe nanocrystal, which is a form of the CIGSe nanocrystal and can, with little modification, also produce the other four variants of the nanocrystals (Scheme 3.19.26).

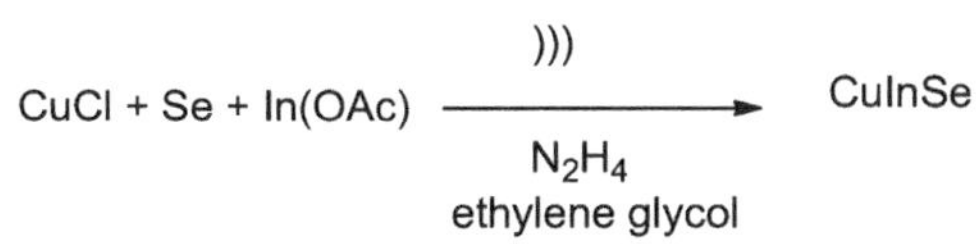

SCHEME 3.19.26

Ultimately, the sonication allows for the elimination of surfactants in the synthesis of these nanocrystals, which is an advantage since the surfactants act as insulators in the final product. Thus both the amount of material and the number of steps required can be reduced. This allows for higher purity and also quantity of the crystals, enabling the solar cell to be more efficient.

Another attempt to reduce the waste created by the synthesis of solar cells is the use of polymer solar cells. These fall under the branch of thin-layer solar cells such as the CIGS and CIGSe cells that were mentioned earlier. These difference, however, due to the use of polymers to act as dyes instead of the nanocrystal structures like the CIGS or CIGSe. Instead of covering the poly(3,4-ethylenedioxythiophene):poly(styrene sulfate)(PEDOT:PSS), which is typically used as the dye for these cells, research focuses on solving the stability issues that PSS cells are often characterized by. This can be achieved by replacing the indium tin oxide (ITO) with a more stable alternative. CuO was chosen due to its ability to potentially withstand the degradation of the PEDOT:PSS as well as for making the entire cell far more environmentally benign.[70] For this synthesis, $CuCl_2$ was dissolved in ethanol to create a 0.01 M solution, then tetramethyl ammonium hydroxide was added until the pH turned to 10, and the solution was sonicated for 45 min at ambient temperature. This mixture was then centrifuged to yield the CuO crystals from the solution. These nanocrystals were found to be in the 3–12 nm diameter range and increased the maximum efficiency of the cell from 6.00% to 6.44% (Scheme 3.19.27). This improved efficiency implies that there is a realistic possibility in CuO replacing ITO in many if not all PSS cells, which would be desirable due to the reduced price, environmentally benign design, as well as greater efficiency.

As outlined earlier sonochemical synthesis has been found to be helpful in the preparation of nanoparticles, with other applications as well. Dutta and coworkers used sonochemical synthesis for the preparation of $CuWO_4$ and $Cu_3Mo_2O_9$ nanoparticles and applied these materials as catalysts for the photocatalytic degradation of harmful pollutants such as certain dyes[71] (Scheme 3.19.28).

$$CuCl_2 \xrightarrow[\text{EtOH} \\ )))]{N(CH_3)_4\,OH} CuO$$

SCHEME 3.19.27

$$Na_2WO_4 \cdot 2\,H_2O + Cu(NO_3)_2 \cdot 3\,H_2O \xrightarrow[\text{water; 40 kHz-100 W/cm}^2\text{; 1.5 h}]{\text{pulse sonication}} CuWO_4 \text{ nanoparticles}$$

$$Na_2MoO_4 \cdot 2\,H_2O + Cu(NO_3)_2 \cdot 3\,H_2O \xrightarrow[\text{water; 40 kHz-100 W/cm}^2\text{; 1.5 h}]{\text{sonication}} Cu_3Mo_2O_9 \text{ nanoparticles}$$

SCHEME 3.19.28

The obtained nanoparticles have shown excellent sorption capacity for cationic dyes, such as rhodamine B, methylene blue, or malachite green. The subsequent decomposition of the dyes was carried out by photocatalysis. The nanoparticle-based sorbents were found to be recyclable in several consecutive experiments. Although the direct aqueous sonolytic oxidative degradation of these dyes can be achieved without the application of the nanoparticles,[72] the use of the nanoparticles is useful for achieving this goal selectively. Other nanoparticles, such as chromium-doped zinc oxide were also found to be effective in the removal of toxic organic dyes (malachite green, eosin yellow, and Auramine O). The ZnO-based nanoparticles were synthesized via an ultrasound-assisted hydrothermal method.[73] The sonochemical synthesis was also useful for the preparation of ZnO quantum dots.[74]

Basavedowda et al. used a green, partially biosynthetic, method applying ultrasounds to synthesize ferromagnetic Fe_3O_4 nanoparticles. These nanoparticles were then used as catalysts in the synthesis of pyrrolo[3,4-c]quinoline-1,3-dione derivatives. The magnetic nanoparticles were found to be easily recyclable and showed negligible activity drop even after five repeated reaction cycles.[75]

Wu and coworkers applied an ultrasound- and microwave-assisted method for the preparation of lead-free palladium nanoparticle catalysts. Similar catalysts, such as the Pb-containing Lindlar catalyst, are frequently used in the hydrogenation of alkynes to alkenes. The sonication of the aqueous dispersion of $Pd(OAc)_2$, the surfactant/capping agent, and the boehmite support provided the Pd nanoparticles and reduced the pores larger than 4 nm in the support. The obtained catalysts exhibited high activity and selectivity in the hydrogenation of diphenylacetylene toward (Z)-stilbene.[76]

Besides actual nanoparticles, the ultrasound-assisted synthesis of nanostructured systems also attracted significant attention. The preparation of metal-organic-framework (MOF) materials and nanocomposites (NCs) are in the forefront of materials science research. These materials are important as catalytic materials, catalyst supports, or in hydrogen storage.[77] The sonochemical synthesis of microporous, partially fluorinated Zn(II) paddle wheel MOF was described by Dhankhar et al.[78] The organic Zn-ligand complex was synthesized by ultrasonic activation. The product adopted a three-dimensional microporous framework and was later used for the preparation of ZnO-C NCs. Similar synthesis conditions were also found applicable for the preparation of zinc(II) amidic pillar-structured[79] or $\{Zn_4O(C_{24}H_{15}N_6O_6)_2(H_2O)_2\ 6H_2O\ DMF\}$ MOFs.[80]

3.19.3 CONCLUSION

Sonochemistry remains a major contributor to the development of green synthetic methods with great variety of applications in several areas, including organic synthesis and the emerging field of nanoparticles. Its ability to not only increase yields and reduce reaction times but also initiate the formation of advanced materials has been thoroughly demonstrated by many applications. Although ultrasounds continue to positively impact the field of organic chemistry, it is currently creating a revolution for the synthesis of nanoparticles. The use of sonochemistry in nanoparticle research is still in its infancy; however, a steady stream of promising applications are published each year. It will continue to grow, and so will the design of equipment used for research and industrial production.

References

1. a. Wood RW, Loomis AL. Physical and biologic effects of high frequency sound waves of great intensity. *Philos Mag* 1927;**4**:414–36;
b. Richards WT, Loomis AL. The chemical effects of high frequency sound waves I. Preliminary survey. *J Am Chem Soc* 1927;**49**:3086–100.
2. Mason TJ, Lorimer JP. *Applied sonochemistry – the uses of power ultrasound in chemistry and processing*. Weinheim: Wiley-VCH; 2002.
3. Grieser F, Choi PK, Enomoto N, Harada H, Okitsu K. *Sonochemistry and the acoustic bubble*. Amsterdam-Oxford-Waltham: Elsevier; 2015.
4. Pankaj AM, editor. *Theoretical and experimental sonochemistry involving inorganic systems*. Dordrecht: Springer; 2011.
5. Luche J-L, editor. *Synthetic organic sonochemistry*. New York: Springer; 1998.
6. Suslick KS, editor. *Ultrasound, its physical, chemical and biological effects*. Weinheim: VCH; 1988.
7. Fernandez RD, Cintas P, Gardeniers HJGE. Merging microfluidics and sonochemistry: towards greener and more efficient micro-sono-reactors. *Chem Commun* 2012;**48**:10935–47.
8. Baig RBN, Varma RS. Alternative energy input: mechanochemical, microwave and ultrasound-assisted organic synthesis. *Chem Soc Rev* 2012;**41**:1559–84.
9. Cravotto G, Borretto E, Oliverio M, Procopio A, Penoni A. Organic reactions in water or biphasic aqueous systems under sonochemical conditions. A review on catalytic effects. *Catal Commun* 2015;**63**:2–9.
10. Török B, Balázsik K, Felföldi K, Bartók M. Asymmmetric reactions in sonochemistry. *Ultrason Sonochem* 2001;**8**:191–200.
11. Polshettiwar V, Varma RS. Non-conventional energy sources for green synthesis in water (microwave, ultrasound, and photo). In: Anastas PT, editor. *Handbook of green chem.***5**; 2010. p. 273–90.
12. Estager J. Integrating ultrasound with other green technologies: toward sustainable chemistry. In: Chen D, Sharma SK, Mudhoo A, editors. *Handbook on applications of ultrasound*. Boca Raton, FL: CRC Press; 2012. p. 675–96.
13. Cintas P. Ultrasound and green chemistry – further comments. *Ultrason Sonochem* 2016;**28**:257–8.
14. Mason TJ. Sonochemistry and the environment – providing a "green" link between chemistry, physics and engineering. *Ultrason Sonochem* 2007;**14**:476–83.
15. Varma RS. Greener and sustainable trends in synthesis of organics and nanomaterials. *ACS Sustain Chem Eng* 2016;**4**:5866–78.
16. Cho H, Török F, Török B. Energy efficiency of heterogeneous catalytic microwave-assisted organic reactions. *Green Chem* 2014;**16**:3623–34.
17. Moseley JD, Kappe CO. A critical assessment of the greenness and energy efficiency of microwave-assisted organic synthesis. *Green Chem* 2011;**13**:794–806.
18. Cravotto G, Demetri A, Nano G, Palmisano G, Penoni A, Tagliapietra S. The aldol reaction under high-intensity ultrasound: a novel approach to an old reaction. *Eur J Org Chem* 2003:4438–44.
19. Rup S, Sindt M, Oget N. Catalytic oxidative cleavage of olefins by RuO_4 organic solvent-free under sonic irradiation. *Tetrahedron Lett* 2010;**51**:3123–6.
20. Li J, Sun M, He G, Xu X. Efficient and green synthesis of bis(indolyl)methanes catalyzed by ABS in aqueous media under ultrasound irradiation. *Ultrason Sonochem* 2011;**18**:412–4.
21. Li J, Du C, Xu X, Chen G. Synthesis of 2-(1,5-diaryl-1,4-pentadien-3-ylidene)-hydrazinecarboximidamide hydrochloride catalyzed by *p*-dodecylbenzenesulfonic acid in aqueous media under ultrasound irradiation. *Ultrason Sonochem* 2012;**19**:1033–8.
22. Vidal M, García-Arriagada M, Rezende MC, Dominguez M. Ultrasound-promoted synthesis of 4-pyrimidinols and their tosyl derivatives. *Synthesis* 2016;**48**:4246–52.
23. Shabalala N, Maddila S, Jonnalagadda S. Catalyst-free, one-pot, four-component green synthesis of functionalized 1-(2-fluorophenyl) 1,4-dihydropyridines under ultrasound irradiation. *New J Chem* 2016;**40**:5107–12.
24. Tabassum S, Govindaraju S, Khan RR, Pasha MA. Ultrasound mediated, green innovation for the synthesis of polysubstituted 1,4-dihydropyridines. *RSC Adv* 2016;**6**:29802–10.
25. Mamaghani M, Tabatabaeian K, Araghi R, Fallah A, HosseinNia R. An efficient, clean, and catalyst-free synthesis of fused pyrimidines using sonochemistry. *Org Chem Int* 2014:1–9.

26. Shukla G, Verma RK, Verma GK, Singh MS. Solvent-free sonochemical one-pot three-component synthesis of 2H-indazolo[2,1-b]phthalazine-1,6,11-triones and 1H-pyrazolo[1,2-b]phthalazine-5,10-diones. *Tetrahedron Lett* 2011;**52**:7195–8.
27. Nikpassand M, Fekri LZ, Farokhian P. An efficient and green synthesis of novel benzoxazoles under ultrasound irradiation. *Ultrason Sonochem* 2016;**28**:341–5.
28. Ahankar H, Ramazani A, Ślepokura K, Lis T, Joo SW. Synthesis of pyrrolidinone derivatives from aniline, an aldehyde and diethyl acetylenedicarboxylate in an ethanolic citric acid solution under ultrasound irradiation. *Green Chem* 2016;**18**:3582–93.
29. Reddy DNK, Chandrasekhar KB, Ganesh YSS, Reddy GR, Kumar JP, Kapavarapu RK, Pal M. FeF_3-catalyzed MCR in PEG-400: ultrasound assisted synthesis of N-substituted 2-aminopyridines. *RSC Adv* 2016;**6**:67212–7.
30. Cains PW, McCausland LJ, Bates DM, Mason TJ. Sonochemical hydrogenation over metal catalysts. *Ultrason Sonochem* 1994;**1**:S45–6.
31. Mason TJ, Lorimer JP, Paniwnyk L, Wright PW, Harris AR. The influence of sonication on the palladium-catalysed dehydrogenation of tetrahydronaphthalene. *J Catal* 1994;**14**:1–4.
32. Szöllösi G, Török B, Szakonyi G, Kun I, Bartók M. Ultrasonic irradiation as activity and selectivity improving factor in the hydrogenation of cinnamaldehyde over Pt/SiO_2 catalysts. *Appl Catal A* 1998;**172**:225–32.
33. Szöllösi G, Kun I, Török B, Bartók M. Ultrasonics in chemoselective heterogeneous metal catalytic processes. Sonochemical hydrogenation of α,β-unsaturated carbonyl compounds over platinum catalysts. *Ultrason Sonochem* 2000;**7**:173–6.
34. Nakagawa S, Sugimura T, Tai A. Almost perfect enantio-differentiating hydrogenation of methyl 3-cyclopropyl-3-oxopropanoate over tartaric acid modified Raney nickel catalyst. *Chem Lett* 1997:859–60.
35. Tai A, Kikukawa T, Sugimura T, Inoue Y, Osawa T, Fujii S. Highly efficient enantio-differentiating hydrogenation over an ultrasonicated Raney nickel catalyst modified with tartaric acid. *J Chem Soc, Chem Commun* 1991:795–6.
36. Török B, Felföldi K, Szakonyi G, Balázsik K, Bartók M. Enantiodifferentiation in asymmetric sonochemical hydrogenations. *Catal Lett* 1998;**52**:81–4.
37. Török B, Balázsik K, Török B, Szöllösi G, Bartók M. Asymmetric sonochemical hydrogenation of α-ketoesters over cinchona-modified platinum catalysts. *Ultrason Sonochem* 2000;**7**:151–5.
38. Török B, Balázsik K, Török M, Felföldi K, Bartók M. Enantiodifferentiation in cinchona-modified platinum-catalyzed sonochemical hydrogenations. *Catal Lett* 2002;**81**:55–62.
39. Kun I, Török B, Felföldi K, Bartók M. Asymmetric hydrogenation of 2-methyl-2-pentenoic acid over cinchona modified Pd/Al_2O_3 catalysts. *Appl Catal A General* 2000;**203**:71–9.
40. Balázsik K, Török B, Felföldi K, Bartók M. Sonochemical enantioselective hydrogenation of trifluoromethyl ketones over platinum catalysts. *Ultrason Sonochem* 1999;**5**:149–55.
41. Mhadgut SC, Bucsi I, Török M, Török B. Sonochemical asymmetric hydrogenation of isophorone on proline modified Pd/Al_2O_3 catalysts. *Chem Commun* 2004:984–5.
42. Cho H, Török F, Török B. Selective reduction of condensed N-heterocycles using water as a solvent and a hydrogen source. *Org Biomol Chem* 2013;**11**:1209–15.
43. Tomin A, Lazarev A, Bere MP, Redjeb H, Török B. Selective reduction of ketones using water as a hydrogen source under high hydrostatic pressure. *Org Biomol Chem* 2012;**10**:7321–6.
44. Schäfer C, Nisanci B, Bere MP, Dastan A, Török B. Heterogeneous catalytic reductive amination of carbonyl compounds with Ni-Al alloy in water as solvent and hydrogen Source. *Synthesis* 2016;**48**:3127–33.
45. Schäfer C, Ellstrom CJ, Cho H, Török B. Pd/C-Al-water facilitated selective reduction of a broad variety of functional groups. *Green Chem* 19, 2017, 1230–1234.
46. Cintas P, Palmisano G, Cravotto G. Power ultrasound in metal-assisted synthesis: from classical Barbier-like reactions to click chemistry. *Ultrason Sonochem* 2011;**18**:836–41.
47. Cravotto G, Procopio A, Oliverio M, Orioa L, Carnaroglioa D. Simple sonochemical protocols for fast and reproducible Grignard reactions. *Green Chem* 2011;**13**:2806–9.
48. Cenci SM, Cox LR, Leeke GA. Ultrasound-induced CO_2/H_2O emulsions as a medium for clean product formation and separation: the Barbier reaction as a synthetic example. *ACS Sustain Chem Eng* 2014;**2**:1280–8.
49. Ocampo RA, Koll LC, Mandolesi SD. Ultrasound promoted Barbier reactions and Csp3–Csp2 Stille coupling for the synthesis of diarylmethanes and substituted benzophenones. *Ultrason Sonochem* 2013;**20**:40–6.

50. Luong M, Domini CE, Silbestri GF, Chopa AB. Ultrasound-assisted synthesis of benzophenones by Stille cross-coupling reactions. Optimization via experimental design. *J Organomet Chem* 2013;**723**:43–8.
51. Shil A, Guha N, Sharma D, Das P. A solid supported palladium(0) nano/microparticle catalyzed ultrasound induced continuous flow technique for large scale Suzuki reactions. *RSC Adv* 2013;**3**:13671–6.
52. a. Cobley A. Ultrasound sonochemistry – a more sustainable approach to surface modification? *Surf Eng* 2009;**25**:559–64;
b. Cobley AJ, Mason TJ, Robinson J. Sonochemical surface modification: a route to lean, green and clean manufacturing? *Plat Surf Finish* 2008;**95**:36–42.
53. Dhopte KB, Zambare RS, Patwardhana AV, Nemade PR. Role of graphene oxide as a heterogeneous acid catalyst and benign oxidant for synthesis of benzimidazoles and benzothiazoles. *RSC Adv* 2016;**6**:8164–72.
54. Mirza-Aghayan M, Ganjbakhsh N, Tavana MM, Boukherroub R. Ultrasound-assisted direct oxidative amidation of benzyl alcohols catalyzed by graphite oxide. *Ultrason Sonochem* 2016;**32**:37–43.
55. Sanasi P, Majji R, Bandaru S, Bassa S, Pinninti S, Vasamsetty S, et al. Nano copper ferrite catalyzed sonochemical, one-pot three and four component synthesis of poly substituted imidazoles. *Mod Res Catal* 2016;**5**:31–44.
56. Eidi E, Kassaee MZ, Nasresfahani Z. Synthesis of 2,4,5-trisubstituted imidazoles over reusable $CoFe_2O_4$ nanoparticles: an efficient and green sonochemical process. *Appl Organomet Chem* 2016;**30**:561–5.
57. Reddy AVS, Reddy MV, Jeong YT. Silica tungstic acid (STA) as a highly efficient and reusable catalyst for the synthesis of benzoxanthenes under solvent-free conditions in ultrasonication. *Res Chem Intermed* 2016;**42**:5209–18.
58. Reddy MV, Reddy AVS, Jeong YT. Di-n-butyl ammonium chlorosulfonate ionic liquids as an efficient and recyclable catalyst for the synthesis of 1,4-dihydrobenzo[4,5]imidazo[1,2-a]pyrimidine-3-carboxylates under solvent-free ultrasound irradiation. *Res Chem Intermed* 2016;**42**:4893–906.
59. a. Dastan A, Kulkarni A, Török B. Environmentally benign synthesis of heterocyclic compounds by combined microwave-assisted heterogeneous catalytic approaches. *Green Chem* 2012;**14**:17–37;
b. Dasgupta S, Török B. Environmentally benign contemporary Friedel-Crafts chemistry by solid acids. *Curr Org Synth* 2008;**5**:321–42;
c. Dasgupta S, Török B. Application of clay catalysts in organic synthesis. A review. *Org Prep Proced Int* 2008;**40**:1–65.
60. Safari J, Ahmadzadeh M, Zarnegar Z. Sonochemical synthesis of 3-methyl-4-arylmethylene isoxazole-5(4H)-ones by amine-modified montmorillonite nanoclay. *Catal Commun* 2016;**86**:91–5.
61. Singh NG, Lily M, Devi SP, Rahman N, Ahmed A, Chandra AK, Nongkhlaw R. Synthetic, mechanistic and kinetic studies on the organo-nanocatalyzed synthesis of oxygen and nitrogen containing spiro compounds under ultrasonic conditions. *Green Chem* 2016;**18**:4216–27.
62. Mishra R, Segal E, Lipovsky A, Natan M, Banin E, Gedanken A. New life for an old antibiotic. *ACS Appl Mater Interfaces* 2015;**7**:7324–33.
63. Yariv I, Lipovsky A, Gedanken A, Lubart R, Fixler D. Enhanced pharmacological activity of vitamin B12 and penicillin as nanoparticles. *Int J Nanomed* 2015;**10**:3593–601.
64. Perkas N, Gedanken A, Wehrschuetz-Sigl E, Guebitz G, Applerot G. Innovative inorganic nanoparticles with antibacterial properties attached to textiles by sonochemistry. In: Altavilla C, Ciliberto E, editors. *Inorganic nanoparticles : synthesis, applications, and perspectives*. Boca Raton, FL: CRC Press; 2011. p. 367–91.
65. Meridor D, Gedanken A. Preparation of enzyme nanoparticles and studying the catalytic activity of the immobilized nanoparticles on polyethylene films. *Ultrason Sonochem* 2013;**20**:425–31.
66. Meridor D, Gedanken A. Enhanced activity of immobilized pepsin nanoparticles coated on solid substrates compared to free pepsin. *Enzyme Microb Tech* 2014;**67**:67–76.
67. Perelshtein I, Ruderman E, Francesko A, Fernandes M, Tzanov T, Gedanken A. Tannic acid NPs – synthesis and immobilization onto a solid surface in a one-step process and their antibacterial and anti-inflammatory properties. *Ultrason Sonochem* 2014;**21**:1916–20.
68. Byeon JH. Rapid green assembly of antimicrobial nanobunches. *Sci Rep* 2016;**6**:27006. http://dx.doi.org/10.1038/srep27006.
69. Lee J, Chang J, Cha J, Lee Y, Han J, Jung D, et al. Large-scale, surfactant-free solution syntheses of $Cu(In,Ga)(S,Se)_2$ nanocrystals for thin film solar cells. *Eur J Inorg Chem* 2011;**1**:647–51.
70. Zhang J, Wang J, Fu Y, Zhanga B, Xie Z. Sonochemistry-synthesized CuO nanoparticles as an anode interfacial material for efficient and stable polymer solar cells. *RSC Adv* 2015;**5**:28786–93.

71. Dutta DP, Rathore A, Ballal A, Tyagi AK. Selective sorption and subsequent photocatalytic degradation of cationic dyes by sonochemically synthesized nano $CuWO_4$ and $Cu_3Mo_2O_9$. *RSC Adv* 2015;**5**:94866–78.
72. Merouani S, Hamdaoui O, Boutamine Z, Rezgui Y, Guemini M. Experimental and numerical investigation of the effect of liquid temperature on the sonolytic degradation of some organic dyes in water. *Ultrason Sonochem* 2016;**28**:382–92.
73. Jamshidi M, Ghaedi M, Dashtian K, Hajati S, Bazrafshan AA. Sonochemical assisted hydrothermal synthesis of ZnO: Cr nanoparticles loaded activated carbon for simultaneous ultrasound-assisted adsorption of ternary toxic organic dye: derivative spectrophotometric, optimization, kinetic and isotherm study. *Ultrason Sonochem* 2016;**32**:119–31.
74. Yang W, Zhang B, Ding N, Ding W, Wang L, Yu M, et al. Fast synthesis of ZnO quantum dots via ultrasonic method. *Ultrason Sonochem* 2016;**30**:103–12.
75. Basavegowda N, Mishra K, Lee YR. Sonochemically synthesized ferromagnetic Fe_3O_4 nanoparticles as a recyclable catalyst for the preparation of pyrrolo[3,4-c]quinoline-1,3-dione derivatives. *RSC Adv* 2014;**4**:61660–6.
76. Wu Z, Cherkasov N, Cravotto G, Borretto E, Ibhadon AO, Medlock J, et al. Ultrasound- and microwave-assisted preparation of lead-free palladium catalysts: effects on the kinetics of diphenylacetylene semi-hydrogenation. *ChemCatChem* 2015;**7**:952–9.
77. Zhao M, Yuan K, Wang Y, Li G, Guo J, Hu W, et al. Metal-organic frameworks as selectivity regulators for hydrogenation reactions. *Nature* 2016;**539**:76–80.
78. Dhankhar SS, Kaur M, Nagaraja CM. Green synthesis of a microporous, partially fluorinated ZnII paddlewheel metal–organic framework: H_2/CO_2 adsorption behavior and solid-state conversion to a ZnO–C nanocomposite. *Eur J Inorg Chem* 2015:5669–76.
79. Bigdeli M, Morsali A. Sonochemical synthesis of a nano-structured zinc(II) amidic pillar metal-organic framework. *Ultrason Sonochem* 2015;**27**:416–22.
80. Moradpour T, Abbasi A, Daftari-Besheli M. Ultrasonic synthesis of $\{Zn_4O(C_{24}H_{15}N_6O_6)_2(H_2O)_2 6\ H_2O \cdot DMF\}$ 3D MOF nanostructures: new precursor to prepare zinc oxide nanoparticles and their application for α-cypermethrin adsorption. *Res Chem Intermed* 2016;**42**:7923–33.

CHAPTER

3.20

Principles of Electrocatalysis

Meaghan McKinnon, Jonathan Rochford

University of Massachusetts Boston, Boston, MA, United States

3.20.1 FUNDAMENTALS OF CYCLIC VOLTAMMETRY

Electrochemistry is an extremely versatile science to study a variety of chemical systems. In particular, the development of molecular redox catalysts benefits tremendously from the suite of potentiometric techniques available to probe kinetic and mechanistic information for a particular chemical transformation or catalytic cycle. Cyclic voltammetry (CV) is typically the method of choice to study a homogeneous catalyst. As such, a brief introduction to the CV method is provided here to aid interpretation of the electrocatalytic data reviewed later in this chapter. The reader is referred to the prominent texts for a more advanced discussion of voltammetry techniques.[1,2]

The CV experiment consists of a three-electrode system [working electrode (WE), counter electrode (CE), and reference electrode (RE)] controlled by a potentiostat, interfaced with a user-friendly software on a computer workstation. The polished disc WE is typically 2–3 mm in diameter and composed of a highly conductive material (e.g., glassy carbon, platinum, or gold). During a CV scan the current amplitude (I, A) is monitored at the WE as the applied potential bias (E, V) is ramped at a designated scan rate (υ, V/s). During a CV scan the I-E applied to the reaction solution at the WE is balanced with an opposite bias (e.g., +1 V vs. −1 V) at the CE typically composed of platinum wire or carbon rod. The applied potential bias in CV is actually a potential difference determined with respect to the reference redox couple of the RE. The RE is isolated via a fine Vycor glass or Teflon frit, which is porous to the supporting electrolyte but nonporous to the analyte, thus maintaining the integrity of the reference redox reaction in situ. A summary of standard aqueous-based REs is provided in Eqs. (3.20.1)–(3.20.3) with potentials referenced to the standard hydrogen electrode [SHE (1.0 M H^+, 1 atm H_2), Eq. 3.20.1], including the mercuric chloride-based saturated calomel electrode (SCE, Eq. 3.20.2) and the silver/silver chloride-based electrode (Eq. 3.20.3). It is recommended to calibrate the RE using an internal standard. The ferricyanide/ferrocyanide Fe(III/II) redox couple is suitable under aqueous conditions (Eq. 3.20.4).

Green Chemistry
http://dx.doi.org/10.1016/B978-0-12-809270-5.00025-X

$$2H^{+}{}_{(aq)} + 2e^{-} \rightleftharpoons H_{2(g)} \quad E^{\circ} = 0.00\ \text{V (vs. SHE, pH} = 0) \tag{3.20.1}$$

$$2Hg_2{}^{2+}{}_{(aq)} + 2e^{-} \rightleftharpoons Hg_{(l)} \quad E^{\circ} = +0.241\ \text{V (vs. SHE)} \tag{3.20.2}$$

$$Ag^{+}{}_{(aq)} + e^{-} \rightleftharpoons Ag_{(s)} \quad E^{\circ} = +0.197\ \text{V (vs. SHE)} \tag{3.20.3}$$

$$[Fe(CN)_6]^{3-}{}_{(aq)} + e^{-} \rightleftharpoons [Fe(CN)_6]^{4-}{}_{(aq)} \quad E^{\circ} = +0.436\ \text{V (vs. SHE, pH} = 7) \tag{3.20.4}$$

Because of the poor solubility (and sometimes instability) of molecular catalysts in aqueous electrolytes, a nonaqueous Ag^{+}/Ag RE is recommended for analysis conducted in organic supporting electrolytes, with the IUPAC recommended[3] ferricenium/ferrocene ($Fc^{+/0}$) redox couple used as an internal pseudoreference; ferrocene being bis(η^5-cyclopentadienyl)iron(II) (Eq. 3.20.5). The $Fc^{+/0}$ redox couple is reported to occur at +0.40 V versus SCE in a 0.1 M Bu_4NPF_6 acetonitrile-based supporting electrolyte[4]; however, the redox properties of nonaqueous electrocatalysts should be reported versus the $Fc^{+/0}$ pseudoreference as recommended by Appel and Helm.[5]

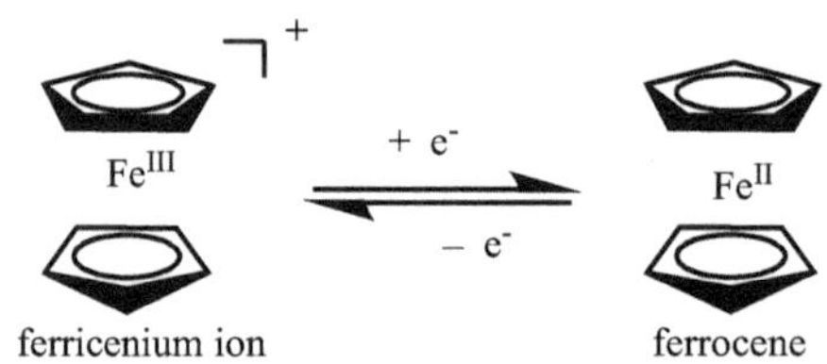

(3.20.5)

A CV scan of ferrocene is illustrated in Fig. 3.20.1 alongside the corresponding potential excitation waveform applied by the potentiostat at the WE (note: this data was recorded using a Ag wire RE in 0.1 M Bu_4NPF_6 acetonitrile supporting electrolyte and subsequently the *x*-axis for the potential data was corrected such that the $Fc^{+/0}$ pseudoreference occurs at 0 V). At the initial potential E_a in Fig. 3.20.1 (point *a*), only background (i.e., ohmic) current is observable. As the WE potential is ramped to a more positive potential (forward scan from point *a* to *c*) the work function of the electrode material is biased sufficiently positive to facilitate electron transfer (observable as faradaic current) from the Fe(II) center of ferrocene, thus forming the oxidized Fe(III)-based ferricenium derivative. The observed anodic current response increases in amplitude with increasing potential (and driving force) until point *d* where the diffusion-controlled concentration of Fc at the surface of the WE decreases, causing the anodic current to peak (i_{pa}) at E_{pa} (the diffusion-controlled peak shapes are discussed later in more detail). Consequently, the anodic current decreases in amplitude from point *d* to *e* as a result of the rapidly decreasing concentration of the Fc analyte, as it is surpassed by the increasing concentration of Fc^{+}. Upon leveling of the anodic faradaic current to pure ohmic current, the scan is reversed at a designated switching potential (point *f*). At this stage the experiment may be terminated if only a half-reaction component of a redox couple need be analyzed. This constitutes the technique of *linear sweep voltammetry* (LSV).

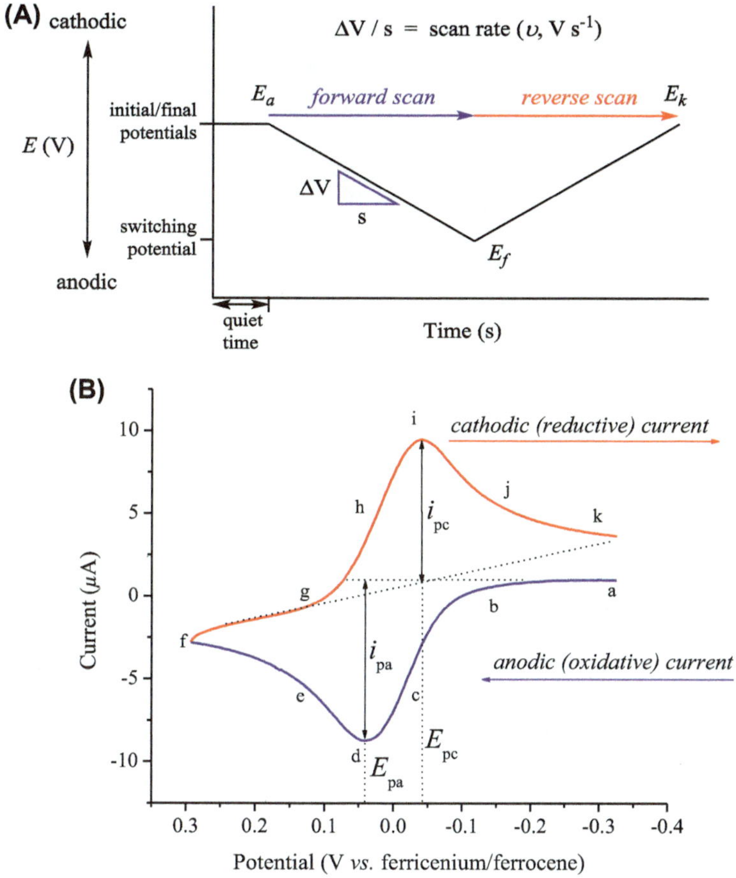

FIGURE 3.20.1 (A) An example of a potential excitation waveform as dictated by user input variables for a cyclic voltammogram experiment and (B) a corresponding cyclic voltammogram of 1 mM ferrocene in 0.1 M Bu_4NPF_6 acetonitrile supporting electrolyte. Experimental conditions: 3 mm glassy carbon disc working electrode, Pt-wire counter electrode, nonaqueous Ag^+/Ag reference electrode, and scan rate $\upsilon = 0.10$ V/s.

Repeating the latter process (*a* through *f*) in situ but in the opposite direction (reverse scan, *f* through *k*) allows analysis of the time-dependent reversibility of the probed redox process, hence the term *cyclic voltammetry*. Scanning back toward negative potential in Fig. 3.20.1, from point *f* to *k*, the work function of the WE is raised to facilitate electron transfer from the electrode surface, reducing the homogeneous Fe(III) center of Fc^+ back to Fe(II) and regenerating Fc. The peak cathodic current (i_{pc}) is observed at point *i* in Fig. 3.20.1 corresponding to the cathodic peak potential E_{pc} beyond which the cathodic current decreases back to baseline as Fc^+ is quantitatively reduced back to Fc (assuming complete reversibility), thus completing a complete CV scan of the $Fc^{+/0}$ redox couple. It is important to appreciate that during this redox transformation the concentration gradient of the homogeneous analyte is determined by the outer Helmholtz plane (OHP), where heterogeneous electron transfer is thought to occur, and the diffuse layer extending from the electrode surface (typically less than 100 Å) to the bulk solution (Fig. 3.20.2).

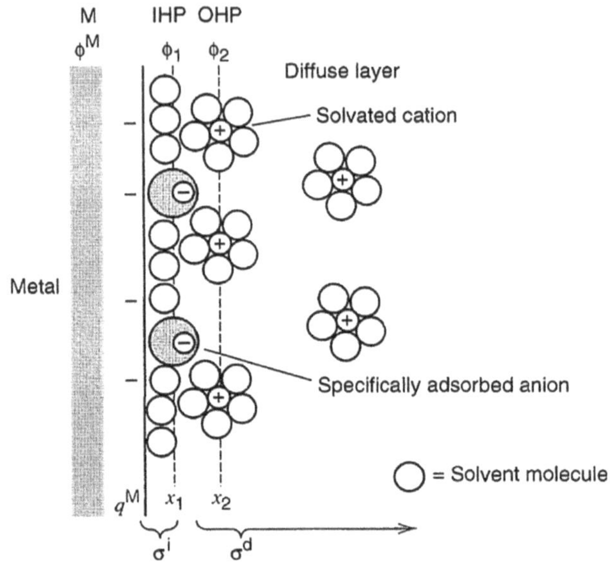

FIGURE 3.20.2 Illustration of the electrochemical double layer at an electrode surface composed of the inner Helmholtz plane (IHP) and outer Helmholtz plane (OHP), as well as the diffuse layer. *Reprinted with permission from Bard AJ, Faulkner LR. Electrochemical methods: fundamentals and applications. In: Harris D, Swain E, Robey C, Aiello E, editor.* Electrochemistry. I. *Faulkner, Larry R. 2nd ed. New York: Wiley; 2000.*

For this reason a quiescent undisturbed solution must be maintained during a CV (or LSV) scan such that mass transfer of analyte to the electrode surface from the diffuse layer is governed only by diffusion in accordance with Fick's law.[1] Accordingly, the supporting electrolyte is typically in 100-fold excess compared to the analyte concentration (typically 0.1 M vs. 1 mM) such that the excess of electrolyte compensates for any electrostatic diffusion of ions in solution to balance charge at the WE and CE surfaces (observable as ohmic current). The characteristic peak profile observed for a reversible redox couple in CV (as illustrated in Fig. 3.20.1) is thus a consequence of mass transfer diffusion of analyte at the electrode surface. For example, maintaining a quiescent solution, the continuous depletion of Fc concentration at the WE surface upon scanning in a positive direction beyond the $Fc^{+/0}$ redox couple has the effect of expanding the diffuse layer thickness (δ) as mass transfer from the bulk solution compensates for the concentration gradient. Mass transfer diffusion therefore determines the peak profile, which is governed by the diffusion coefficient of the analyte (D, cm^2/s). The peak faradaic current (i_p) observed by LSV for a single electron ($n_p = 1$) oxidation (anodic) or reduction (cathodic) event, that is, either the forward *or* reverse scan as illustrated in Fig. 3.20.1, is described by the Randles-Sevcik equation (Eq. 3.20.6):

$$i_p = 0.4463\, n_p FAC\sqrt{\frac{n_p F v D}{RT}} \tag{3.20.6}$$

where F is Faraday's constant (96,485 A s/mol), A is the active electrode surface area (cm^2), C is the bulk analyte concentration (M), R is the gas constant (8.314 V A s/K mol), and T is the temperature (K). As the LSV scan proceeds, the diffuse layer thickness grows in proportion to the square root of the analyte diffusion coefficient and the timescale (t) of the scan, that is, $\delta \sim (Dt)^{1/2}$, where t is inversely proportional to the scan rate ($t \sim RT/Fv$). Accordingly, δ narrows with increasing scan rate.

A useful application of Eq. (3.20.6) is to determine the diffusion coefficient (D) of an analyte. By plotting the peak faradaic current i_p (A) versus the square root of the corresponding scan rate ($v^{1/2}$), D can be extracted from the slope (m) using Eq. (3.20.7):

$$D = \frac{m^2 RT}{(0.4463AC)^2 (n_p F)^3} \tag{3.20.7}$$

The diffusion coefficient of the $Fc^{+/0}$ redox couple, for example, has been reported[6] as 2×10^{-5} cm^2/s, which is the typical order of magnitude for a homogeneous molecular analyte. The time- and distance-dependent concentration gradients of the oxidized and reduced forms of a redox couple during a cyclic voltammogram, from the electrochemical double layer (IHP + OHP) to the diffuse layer, are illustrated in Fig. 3.20.3.

Whether the analyte exists in its oxidized or reduced form, or a mixture of both, is determined by the applied potential bias (E) at the WE with respect to the standard reduction potential of the redox couple E^o following the Nernst relationship (Eq. 3.20.8). The relative concentrations of oxidized to reduced species, at the electrode surface and in the diffuse layer, are also described by the Nernst equation:

$$E = E^\circ + \frac{RT}{nF} \ln\left(\frac{\alpha_{ox}}{\alpha_{red}}\right) \tag{3.20.8}$$

where α_{ox} and α_{red} are the activity coefficients of the oxidized and reduced components of the redox couple, respectively. At the midpoint between E_{pa} and E_{pc} the activity coefficients, and hence concentrations, of both the oxidized and reduced species are equal, the second term of Eq. (3.20.8) reduces to zero, and E° can be derived directly from a voltammogram according to Eq. (3.20.9) (under standard conditions). Another useful consequence of the Nernst equation is how the anodic and cathodic peak separation (ΔE_p) is predicted to be 59 mV for a

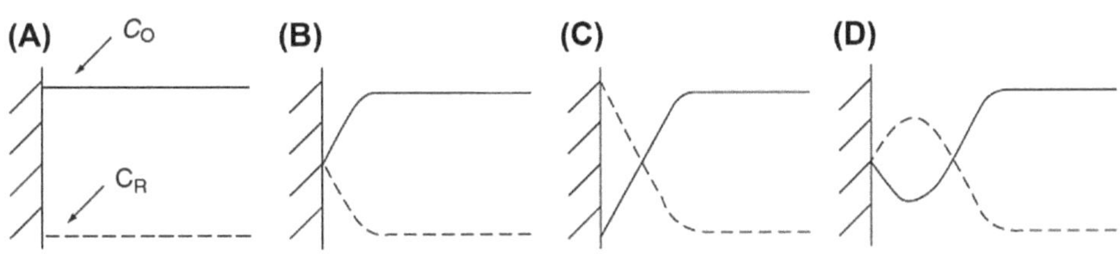

FIGURE 3.20.3 Illustration of the time- and distance-dependent concentration profiles of the oxidized (C_O) and reduced (C_R) forms of a redox couple during a cyclic voltammogram scan where (A) corresponds to the initial potential, (B) to the formal potential during the forward scan, (C) to the achievement of a zero-reactant surface, and (D) to the formal potential during the reverse scan. *Reproduced with permission from Wang J.* Analytical electrochemistry. *3rd ed. Wiley-VCH; 2006.*

one-electron redox process and just 30 mV for a two-electron redox reaction, according to Eq. (3.20.10).

$$E^\circ = \frac{E_{pa} + E_{pc}}{2} \tag{3.20.9}$$

$$\Delta E_p = \frac{59}{n_p} \text{ mV} \tag{3.20.10}$$

3.20.2 ELECTROCATALYSIS

A homogeneous electrocatalyst is typically a solution-phase molecular species that accelerates electron transfer from the WE surface to or from a solution-phase substrate and is also capable of facilitating bond cleavage and/or bond formation to generate a chemical product. As such, time-dependent voltammetry techniques such as LSV and CV are particularly useful to study the thermodynamics and kinetics involved in such mechanistic pathways. To extract rates of reaction from voltammetry data, it is necessary to initially characterize the electrochemical properties of a catalyst under inert noncatalytic conditions in the absence of any substrate, as discussed in Section 3.20.1. Depending on the nature of the molecular catalyst and its intended catalytic application, the solvent, electrolyte, and atmospheric conditions are chosen to create an environment where only the faradaic response of the catalyst complex is observed. After understanding the basic electrochemical properties of the catalyst, the next step is to screen its catalytic activity by exposing it to the necessary reaction conditions. This, again, depends on the desired application. For example, catalytic activity could be triggered by changing the purge gas from argon to CO_2 or by simply changing the supporting electrolyte pH such as in catalytic H_2 evolution.

Before discussing any examples of electrocatalysis, it is important to understand the typical electrocatalytic parameters often utilized to define the efficiency of a catalyst, that is, turnover number (TON), turnover frequency (TOF), and overpotential (η). While the concept of TON and TOF is synonymous in all fields of catalysis, the evaluation of a catalyst TOF using voltammetry techniques is not a straightforward process and deserves significant attention in this chapter. Another topic very relevant for the application of electrocatalysis in fuel-forming reactions, such as H_2O oxidation and the reduction of H^+, O_2, or CO_2, is the proton-coupled electron transfer (PCET), which will be briefly discussed. The term overpotential (η) is unique to electrocatalysis and will be discussed first.

3.20.2.1 Overpotential

The overpotential (η) of an electrocatalytic transformation is a property of the molecular electrocatalyst, and the reaction conditions, employed. First let us consider a simple noncatalyzed reversible electrochemical transformation (E_r) of substrate A to product C characterized by the standard reduction potential, $E^\circ_{A/C}$ (Eq. 3.20.11).

$$A + ne^- \rightleftharpoons C \quad E^\circ_{A/C} \tag{3.20.11}$$

Now let us consider an alternative E_rC_i' mechanism for the same transformation where reversible electrochemical (E_r) activation of a catalyst P, characterized by a standard reduction potential ($E^\circ_{P/Q}$, Eq. 3.20.12), is followed by an irreversible chemical step (C_i'), characterized by a homogeneous electron transfer rate constant (k_e, Eq. 3.20.13), catalytically converting substrate A to product C while regenerating the precatalyst P (note C_i' distinguishes a following regenerative chemical step from a nonregenerative, i.e., noncatalytic, chemical reaction C).[1]

$$P + n_{cat}e^- \rightleftharpoons Q \quad E^\circ_{P/Q} \tag{3.20.12}$$

$$Q + A \rightarrow P + C \quad k_e \tag{3.20.13}$$

Here n_{cat} represents the number of electrons transferred in the rapid heterogeneous reduction (or conversely oxidation for oxidative catalysis) of P to Q. Often n_{cat} is equivalent to n_p [the number of electrons responsible for the peak current (i_p) during reduction of P to Q under noncatalytic conditions], but in rare cases they differ and should be distinguished from each other.

The overpotential (η) is quantitatively defined by the potential difference between the standard reduction potential for the noncatalyzed redox transformation of substrate A to product C, that is, $E^\circ_{A/C}$, and the potential of the catalytic current observed in the presence of catalyst, all other experimental conditions being identical. For many years, there has been inconsistency in the literature as to the most appropriate potential to correlate to the catalytic current. The onset of catalytic current has often been used but biases the system to have a lesser overpotential, whereas using the catalytic current peak potential (E_{cat}) likely overestimates the necessary overpotential. While neither of these methods are incorrect per se, a perspective article[5] on this topic justifiably advocates the use of the catalytic wave half-potential ($E_{cat/2}$) to determine the electrocatalyst overpotential (η, Eq. 3.20.14) to establish consistency moving forward in this growing research field.

$$\eta = E^\circ_{A/C} - E_{cat/2} \tag{3.20.14}$$

Taking what, at face value, appears to be one of the simplest examples, let us consider electrocatalytic proton reduction to evolve hydrogen gas, that is, the H^+/H_2 redox couple (Eq. 3.20.1). Under standard conditions at pH = 0, that is, 1.0 M H^+ and 1 atm H_2, the catalyst overpotential is merely the excess negative potential beyond $E^\circ_{A/C}$ (in this case 0 V vs. SHE, Eq. 3.20.1), which corresponds to $E_{1/2}$ of the cathodic catalytic peak. Although this may seem straightforward, unfortunately the majority of H_2-evolving electrocatalysts are not water soluble and must be investigated in nonaqueous electrolytes. This restriction adds significant challenges to overpotential determination because of the inconsistencies/uncertainties in the reference potential and the pH of a nonaqueous supporting electrolyte. The application of Eq. (3.20.14) for the determination of η for the $Ni(P^{Ph}{}_2N^{Ph}{}_2)_2(BF_4)_2$ catalyst[7] in 0.2 M Bu_4NPF_6 acetonitrile electrolyte is illustrated in Fig. 3.20.4.

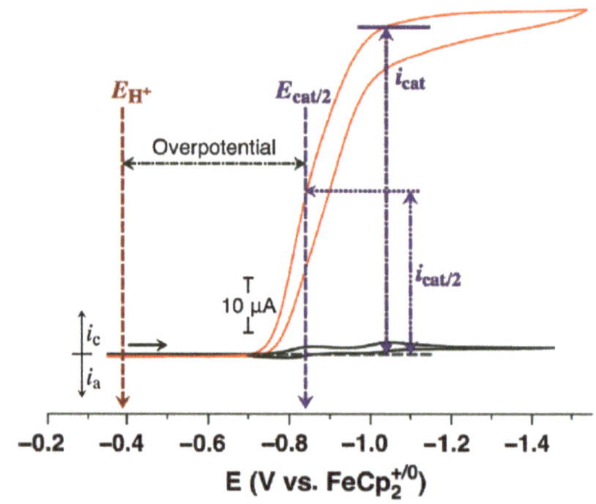

FIGURE 3.20.4 Overlay of the faradaic and catalytic response of the $Ni(P^{Ph}{}_2N^{Ph}{}_2)_2(BF_4)_2$ catalyst in the absence and presence of H^+, respectively, recorded by cyclic voltammetry. Determination of the catalytic half-wave potential, $E_{cat/2}$, is illustrated as a method to consistently report the catalyst overpotential. Note: the H^+/H_2 couple occurs at −0.39 V versus $Fc^{+/0}$ in acetonitrile[8] at pH 6.1 {0.26 M 1:1 DMF:[(DMF)H]$^+$}. *Reprinted with permission from Appel AM, Helm ML. Determining the overpotential for a molecular electrocatalyst.* ACS Catal *2014;***4***(2):630–3.*

Following the recommended method in Eq. (3.20.14), one key to reporting an accurate value for an electrocatalyst overpotential therefore lies in the accurate determination of $E_{A/C}$, that is, under nonstandard conditions. In some cases the value of $E_{A/C}$ can be accurately determined by an open circuit potential experiment.[8] Alternatively, $E_{A/C}$ can be estimated from the standard reduction potential $E^\circ_{A/C}$ (should $E^\circ_{A/C}$ be known in the same nonaqueous solvent) and the nonstandard pH of the supporting electrolyte solution according to the pH-dependent Nernst equation (Eq. 3.20.15):

$$E = E^\circ + \frac{RT}{F}\left(\frac{a_{A*}}{a_{HA}}\right) - 2.303\frac{RT}{F}\text{pH} \tag{3.20.15}$$

In such cases, it is advisable to use at least a 0.1 M concentration of 1:1 acid/base buffer as the supporting electrolyte to stabilize the solution pH during catalysis, and also to better define the pH, as under such conditions the solution pH is equal to the acid pK_a in the nonaqueous solvent. Fortunately, the pK_a scale for many organic acids and their conjugate bases is now well documented in acetonitrile.[9] Furthermore, the standard reduction potential for CO_2 to CO conversion is now established in acetonitrile (Eq. 3.20.16).[10]

$$CO_{2(CH_3CN)} + 2H^+{}_{(CH_3CN)} + 2e^- \rightleftharpoons CO_{(CH_3CN)} + H_2O_{(CH_3CN)} \quad E^\circ = -0.130\text{ V}(\text{vs. }Fc^{+/0}, \text{pH} = 0) \tag{3.20.16}$$

3.20.2.2 Proton-Coupled Electron Transfer

With the goal of this chapter to present the principles of electrocatalysis relevant to CO_2 reduction (and small-molecule activation for sustainable fuels production in general), it is prudent to introduce the concept of proton-coupled redox events.[11–13] More commonly known as PCET, this phenomenon is abundant in nature and is critical in lowering the activation energies of most metalloenzyme-based catalytic transformations. Nature has optimized the orientation and pK_a of pendant acidic and/or basic cofactors at enzymatic catalytic centers to provide for facile accessibility to critical transition states, often inaccessible through pure electron transfer alone. Key to the success of PCET is that by coupling reduction (or oxidation) events in a concerted manner with protonation (or deprotonation) events there is no net change in the charge of the product relative to the substrate. PCET enables catalysts to avoid the generation of highly charged intermediates along their reaction coordinate, allowing access to an alternative low-transition-state pathway. Thus, PCET enables catalytic turnover at reduced overpotential, although often with a cost of kinetic efficiency due to a concerted multisite mechanism. It is also noteworthy that by accessing a low-overpotential pathway, PCET also improves catalyst stability, thereby enhancing the TON by avoiding side reactions often associated with high-energy-reactive intermediates. PCET transformations are most conveniently represented by a square reaction scheme where electron transfer is followed by proton transfer [electron-proton transfer (EPT) pathway] or proton transfer is followed by electron transfer [proton-electron transfer (PET) pathway].

A generic example for a one-electron one-proton PCET reaction of an acid HA is illustrated in Scheme 3.20.1. Electron transfer steps (*horizontal arrows* in Scheme 3.20.1) are characterized by their standard reduction potential, whereas the deprotonation steps (*vertical arrows* in Scheme 3.20.1) are characterized by the pK_a of the specific redox state of the molecule. Of particular interest in the field of electrocatalytic small-molecule activation, because of its thermodynamic advantage, is the concerted proton-electron transfer (CPET) pathway (*diagonal arrow* in Scheme 3.20.1), which avoids the high energy intermediates associated with the stepwise EPT and PET pathways and is characterized by the H–A bond dissociation energy (BDE) according to Eq. (3.20.17):

$$\text{BDE} = RT \ln K_a + FE^\circ + C \tag{3.20.17}$$

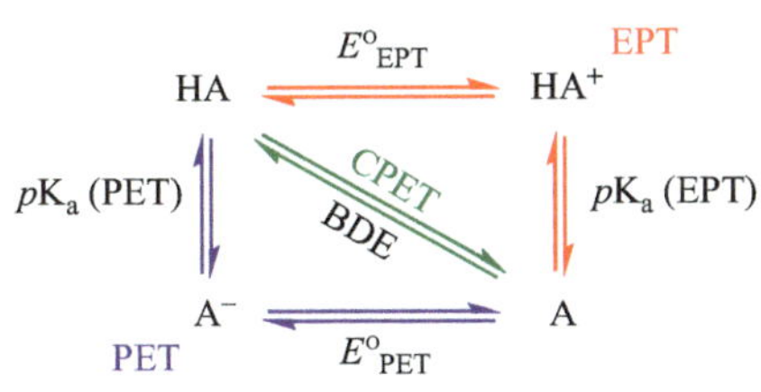

SCHEME 3.20.1 Stepwise and concerted proton-coupled electron transfer (PCET) pathways for a generic acid molecule HA, where EPT is stepwise electron-proton transfer, PET is stepwise proton-electron transfer, CPET is concerted proton-electron transfer, and BDE is bond dissociation energy.

where C is an empirical constant dependent on the experimental conditions (e.g., $C = 59.5$ kcal/mol in acetonitrile).[14] Such pH-dependent redox transformations are often presented in the form of a Pourbaix diagram. With reduction potential plotted as the y-axis versus pH on the x-axis, the following holds true in a Pourbaix diagram: a pH-independent electron transfer process maintains a slope of 0 mV/pH, an n-electron n-proton CPET event (where n typically equals 1 or 2) maintains a slope of 59 mV/pH, a two-electron one-proton CPET event maintains a slope of 30 mV/pH, and a one-electron two-proton CPET maintains a slope of 118 mV/pH. This CPET behavior is summarized by Eq. (3.20.18).

$$\Delta E = -59\left(\frac{\#\ \text{protons}}{\#\ \text{electrons}}\right)\text{mV/pH} \tag{3.20.18}$$

Presented in Fig. 3.20.5 is the Pourbaix diagram for a mononuclear $[Ru(OH_2)(Q)(tpy)]^{2+}$ analogue[15] of the Tanaka dimer water oxidation electrocatalyst.[16] Although not catalytically active itself, this Pourbaix diagram of the mononuclear species played an important role in understanding the PCET mechanisms at play in the active dimer catalyst. For example, critical to water oxidation activity displayed by Tanaka's dimer is formation of the oxyl radical species, abbreviated as $[Ru^{II}(O^{\bullet-})(Q)]^{+}$ in Fig. 3.20.5, which is generated via two-proton, one-electron-coupled oxidation of $[Ru(OH_2)(Q)]^{2+}$ consistent with a pH-dependent slope of −118 mV/pH in the pH range 3–6.

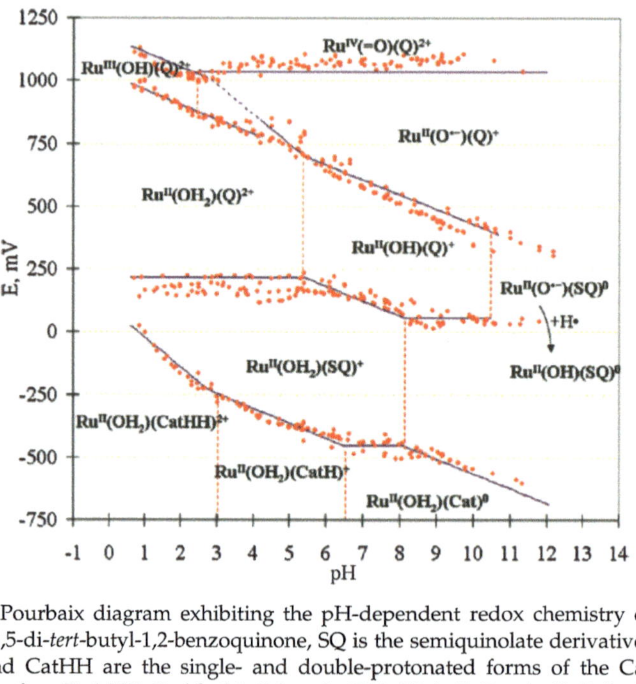

FIGURE 3.20.5 The Pourbaix diagram exhibiting the pH-dependent redox chemistry of $[Ru(OH_2)(Q)(tpy)]^{2+}$ (V vs. SCE), where Q is 3,5-di-*tert*-butyl-1,2-benzoquinone, SQ is the semiquinolate derivative, Cat is the catecholate derivative, and CatH and CatHH are the single- and double-protonated forms of the Cat ligand, respectively. *Reproduced with permission from Tsai MK, Rochford J, Polyansky DM, Wada T, Tanaka K, Fujita E, Micherman J. Characterization of redox states of $Ru(OH_2)(Q)(tpy)^{2+}$ (Q = 3,5,di-tert-butyl-1,2-benzoquinone, tpy =2,2′:6′,2″-terpyridine) and related species through experimental and theoretical studies.* Inorg Chem 2009;**48**(10):4372–83.

3.20.2.3 Rate Analysis of a Homogeneous Electrocatalyst

The maximum TOF (TOF_{max}) of a homogeneous electrocatalytic E_rC_i' reaction is determined by the rate-determining step of the relevant catalytic cycle, which is responsible for the observed rate constant (k_{obs}). Thus, direct measurement of k_{obs} (equivalent to TOF_{max}) by voltammetry can give excellent mechanistic insight to a catalytic process. The direct measurement of k_{obs} by voltammetry, however, is far from a straightforward process. Consider the definition of TOF—the number of moles of product produced per mole of catalyst per unit time—and recall that only the homogeneous catalyst molecules within the diffuse layer at the electrode surface can be activated. Even in ideal situations, as the diffuse layer grows over the course of a voltammetry scan, the concentration of active catalyst varies during the timescale of the scan. More often than not, when studying a homogeneous electrocatalyst, nonideal behavior is observed because of substrate consumption within the diffuse layer and undesired side reactions of the active catalyst. Only with absolute consideration of ideal versus nonideal behavior can k_{obs} be correctly determined and correlated to the true catalyst TOF_{max}.

Recall again the E_rC_i' one-electron catalytic transformation of substrate A to product C by the electrochemically activated catalyst Q, from precatalyst P (Eqs. 3.20.12 and 3.20.13). For a noncatalytic E_rC_i reaction, the peak current i_p of the initial electrochemical step E is not perturbed by a subsequent chemical step C. In contrast, for a catalytic E_rC_i' reaction where the precatalyst P is continuously replenished by the C' reaction step, the peak catalytic current i_{cat} should increase in amplitude relative to the peak noncatalytic current i_p. The amplitude of i_{cat}/i_p will increase as the TONs accessed within the timeframe of the CV scan increase (assuming excess substrate is available, vide infra). Apart from its obvious concentration dependence on the reaction order, with respect to catalyst P and substrate A, the nature of the i_{cat} response depends strongly on the duration of the scan and hence on each of the time-dependent variables involved in the experiment, such as scan rate, diffuse layer thickness (δ), diffusion coefficients of the precatalyst P (D_P) and active catalyst Q (D_Q) (often assumed to be identical), diffusion coefficient of the substrate (D_A, A is often consumed within the diffuse layer at high values of k_{obs} and slow scan rates under nonideal conditions), and rate of homogeneous electron transfer (k_e). Heterogeneous electron transfer at the electrode surface is typically rapid and not rate determining.

Based on these considerations for an E_rC_i' electrocatalytic CV response, a qualitative understanding of the various *I-E* profiles accessible has been modeled by Saveant and Su,[17] where k_e is the rate-determining step. A total of eight catalytic zones are predicted, three of which (KS, KD, and KT2) can be theoretically modeled to allow extraction of k_{obs}.[18] A collection of the simulated CV profiles is illustrated in the zone diagram developed by Saveant and Su (Fig. 3.20.6) where the log of a dimensionless kinetic term (λ, Eq. 3.20.19) is plotted against the log of the excess factor, that is, ratio of substrate versus precatalyst bulk concentrations $\left(C_A^0/C_P^0\right)$, abbreviated as γ.

$$\lambda = \frac{RT}{n_pF}\left(\frac{k_eC_P^0}{v}\right) \tag{3.20.19}$$

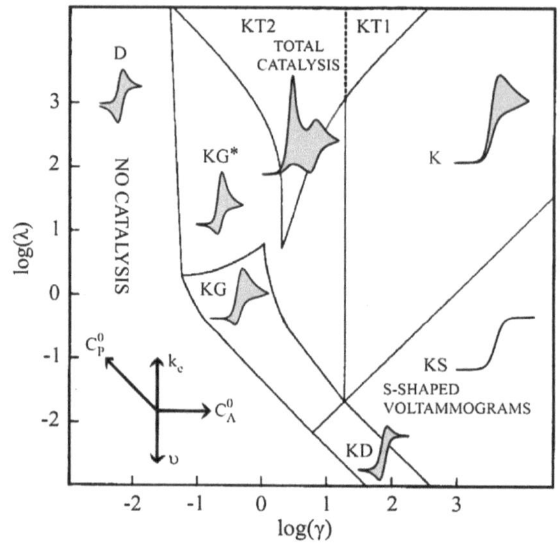

FIGURE 3.20.6 A kinetic zone diagram for an E_rC_i' electrocatalytic cyclic voltammetry response illustrating the predicted current-voltage profiles as a function of the kinetic term λ (here k_e is the homogeneous electron transfer rate constant and υ is the scan rate, see Eq. 3.20.9) and the excess factor γ, that is, the ratio of substrate versus precatalyst bulk concentrations (C_A^0/C_P^0). The "parameter compass" inset on the bottom left of the plot is used to identify how the parameters C_A^0, C_P^0, k_e, and/or υ can be modified to navigate toward a different zone of the diagram with the aim of observing a desired current-voltage response. *Reprinted with permission from Rountree ES, McCarthy BD, Eisenhart TT, Dempsey JL. Evaluation of homogeneous electrocatalysts by cyclic voltammetry.* Inorg Chem *2014;*53*(19):9983–10002.*

While the zone diagram in Fig. 3.20.6 was developed for a one-electron E_rC_i' mechanism, under limiting conditions, this figure can be extended to describe a multielectron multisubstrate electrocatalytic reaction. Qualitative extension of the E_rC_i' zone diagram to describe the *ECEC'* electrocatalytic proton reduction by the cobaloxime catalyst $Co(dmgBF_2)_2(CH_3CN)_2$ has been elegantly demonstrated by Dempsey and coworkers.[19] This precedent will likely provide a sound basis for the future evaluation of electrocatalytic reactions relevant to sustainable fuels production. However, caution should be taken when applying Fig. 3.20.6 to more complex mechanisms where the *I-E* response may be compromised by electrochemically active reaction intermediates, and/or the rate-determining step is not consistent across all values of log (λ) versus log (γ) due to the necessity of a global rate constant ($k_{obs} \propto k_1k_2...k_n$).

An ideal scenario exists with respect to ease of k_{obs} extraction where pure kinetic conditions (zone KS) corresponding to a significant kinetic factor (λ) and fast, but not too fast, k_{obs}, thus maintaining pseudo-zero-order conditions with respect to the substrate concentration, C_A (high excess factor, γ). An abundance of substrate A within the diffuse layer over the course of the scan, that is, negligible decrease in C_A at the electrode surface ($C_A \sim C_A^0$) as catalysis propogates, and negligible side reactions of the active catalyst Q, give rise to an

S-shaped wave where forward and reverse waves trace each other's *I-E* profile as described by Eq. (3.20.20).

$$i_{cat} = n_{cat}FAC_P^0\left(\frac{D_P}{D_Q}\right)\sqrt{D_Q k_{obs}} \quad (3.20.20)$$

where n_{cat} represents the number of electrons transferred in the rapid (non-rate determining) heterogeneous redox conversion of P to Q (Eq. 3.20.12). Assuming that the diffusion coefficients of P and Q are identical, Eq. (3.20.20) is simplified to its more common form, given in Eq. (3.20.21):

$$i_{cat} = n_{cat}FAC_P^0\sqrt{D_P k_{obs}} \quad (3.20.21)$$

With prior knowledge of the bulk concentration of C_P^0, and the precatalyst diffusion coefficient D_P, k_{obs} can be directly determined from the maximum current i_{cat} in zone KS. The characteristic plateau current in zone KS arises because of a steady state between the rates of consumption and regeneration of P by Eqs. (3.20.12) and (3.20.13), respectively. For the same reason, i_{cat} is independent of scan rate under steady-state conditions. In fact, scan rate independence of i_{cat} is an expedient method to experimentally verify that steady-state electrocatalytic conditions have been identified.

Conveniently, to avoid predetermination of D_P and C_P^0 to extract k_{obs} under steady-state conditions, dividing Eq. (3.20.21) by the Randles-Sevcik equation (Eq. 3.20.6), which describes the peak current i_p corresponding to the P/Q standard reduction potential $E^\circ_{A/C}$ under noncatalytic conditions, k_{obs} can be determined from knowledge of the i_{cat}/i_p ratio as described by Eq. (3.20.22). Typically, when n_p is equivalent to n_{cat}, Eq. (3.20.22) can be simplified to Eq. (3.20.23). The latter equation is more commonly represented in terms of k_{obs} (Eq. 3.20.24). It should be emphasized that while i_{cat} is independent of scan rate under steady-state conditions, the noncatalytic peak current, i_p, is not, and thus the i_{cat}/i_p ratio will vary with scan rate while k_{obs} should remain relatively constant if a plateau current is observed.

$$\frac{i_{cat}}{i_p} = \frac{1}{0.4463}\left(\frac{n_{cat}}{n_p^{3/2}}\right)\sqrt{\left(\frac{RT}{Fv}\right)k_{obs}} \quad (3.20.22)$$

$$\frac{i_{cat}}{i_p} = \frac{1}{0.4463}\sqrt{\left(\frac{RT}{n_pFv}\right)k_{obs}} \quad (3.20.23)$$

$$k_{obs} = 0.1991\left(\frac{n_pFv}{RT}\right)\left(\frac{i_{cat}}{i_p}\right)^2 \quad (3.20.24)$$

Notably, pure kinetic conditions are best identified by designing experimental parameters with a high excess factor (e.g., low catalyst concentration and high substrate concentration), and a suitably fast scan rate relative to the rate-determining homogeneous electron transfer step k_e. Without a prior knowledge of k_e, trial and error is best guided by the log (λ) versus log (γ) navigational compass provided in Fig. 3.20.6. When such conditions are identified, and a plateau steady-state current is obtained, such that the substrate is not consumed within the diffuse layer, k_{obs} depends solely on the rate-determining homogeneous electron transfer step (k_e, Eq. 3.20.13), which can thus be interpreted as the maximum catalyst TOF (TOF_{max}, Eq. 3.20.25).

$$TOF_{max} = k_{obs} = k_e C_A^0 \tag{3.20.25}$$

Conversely, should the homogeneous electron transfer rate k_e be very slow, a peak catalytic current is observed (zone KD, Fig. 3.20.6), as opposed to a plateau current, as the rate of precatalyst regeneration (P) is faster than the rate of substrate consumption. This may be accompanied by evidence of anodic current (CV reverse scan) belonging to the P/Q redox couple due to the slow reactivity of Q with the analyte. In this case a slower scan rate should be employed to allow the catalytic current to plateau, satisfying the requirements to allow k_{obs} interpretation by any equation of Eqs. (3.20.21)–(3.20.24). Unfortunately, this is a common problem when screening new structures for electrocatalytic activity and it raises the added complication that slower scan rate may enhance substrate consumption within the diffuse layer (i.e., shifting from zone KD directly to zone K, bypassing zone KS) precluding the application of Eqs. (3.20.21)–(3.20.24) to extract a trustworthy value for k_{obs}. Ideal pure kinetic observation of zone KS should always be confirmed before determination of TOF_{max} for homogeneous molecular electrocatalysts. Unfortunately, this practice is not always followed in the literature because of poor execution or knowledge of the methods discussed here. Where the pure kinetic zone KS cannot be identified, alternative methods, such as foot-of-the-wave analysis (FOWA), should be conducted.

3.20.2.4 Foot-of-the-Wave Analysis

For more complicated reaction mechanisms, that is, those not amenable to the E_rC_i' model described earlier, modeling of the catalyst's *I-E* response is necessary. This requires an in-depth knowledge of the reaction mechanism to determine k_{obs}. As previously highlighted, in the case of a multistep mechanism (e.g., *ECEC'*) described by a global rate constant, prevailing reaction conditions can be identified such that k_{obs} is dependent on a single rate-determining step. Thus, on observation of a steady-state plateau current, an *EC'* mechanism can be assumed allowing application of Eqs. (3.20.21)–(3.20.24). However, for scenarios falling outside the pure kinetic profile of zone KS, Eqs. (3.20.21)–(3.20.24) are not applicable. In the pure kinetic region with a high kinetic factor (λ) (Fig. 3.20.6, zone K), or with low substrate concentration where consumption becomes a problem within the diffuse layer ($C_A < C_A^0$), a wave is again observed for the *I-E* profile and Eq. (3.20.26) should be applied. Eq. (3.20.26) can also be normalized using 3.20.6 with respect to electrode surface area (A), the bulk concentration of precatalyst (C_P^0), diffusion coefficient of precatalyst (D_P), and scan rate as expressed in Eq. (3.20.27).

$$i_{cat} = \frac{n_{cat} F A C_P^0 \sqrt{D_P k_{obs}}}{1 + \exp\left[\frac{F}{RT}\left(E - E^{\circ}_{P/Q}\right)\right]} \tag{3.20.26}$$

$$\frac{i_{cat}}{i_p} = \frac{2.24\sqrt{\left(\frac{RT}{Fv}\right)k_{obs}}}{1+\exp\left[\left(\frac{F}{RT}\right)(E - E^\circ_{P/Q})\right]}\left(\frac{n_{cat}}{n_p^{3/2}}\right) \tag{3.20.27}$$

Fortunately, Savéant and Costentin have developed a practical analysis to provide a means to estimate pure kinetic information of a nonideal catalytic response using Eq. (3.20.27). This method is called FOWA, as it limits kinetic analysis to the foot of the catalytic wave where no other side phenomena exist. Specifically, substrate consumption, catalyst deactivation (e.g., by cocatalyst), and inhibition by-products (e.g., surface adsorption) are common occurrences that occur beyond the foot of the catalytic wave as the overpotential is increased. These processes ultimately disrupt the pure electrocatalytic response precluding an accurate determination of TOF_{max}. When the catalytic response is perturbed and falls outside the pure kinetic KS zone, and when navigating the zone diagram (Fig. 3.20.6) to obtain steady-state conditions is unsuccessful, FOWA provides an alternative method to estimate TOF_{max}. An example of FOWA, which requires a plot of i_{cat}/i_p versus $1\Big/\left(1+\exp\left[\frac{F}{RT}(E - E^\circ_{P/Q})\right]\right)$, is illustrated in Fig. 3.20.7 where a plateau catalytic current is not observed because of substrate consumption within the reaction-diffuse layer.[20]

While Eq. (3.20.24) provides for the most straightforward access to k_{obs} within the pure kinetic KS zone, Eq. (3.20.27) is used in Fig. 3.20.7a′ to illustrate the power of FOWA with respect to a nonideal kinetic response in Fig. 3.20.7b′. Plotting the ratio of i_{cat}/i_p versus $1\Big/\left(1+\exp\left[\frac{F}{RT}(E - E^\circ_{P/Q})\right]\right)$ gives a straight line in Fig. 3.20.7a′ for zone KS profile *I-E* response with a slope of $2.24\sqrt{\left(\frac{RT}{Fv}\right)2k_{obs}}\left(\frac{n_{cat}}{n_p^{3/2}}\right)$. Thus, the value of k_{obs} is directly accessible from the slope of this plot. By fitting only the linear portion of the plot in Fig. 3.20.7b′ with Eq. (3.20.27), at low values of $1\Big/\left(1+\exp\left[\frac{F}{RT}(E - E^\circ_{P/Q})\right]\right)$ a slope of $2.24\sqrt{\left(\frac{RT}{nFv}\right)2k_{obs}}\left(\frac{n_{cat}}{n_p^{3/2}}\right)$ applies equally well, allowing k_{obs} to be extracted with confidence. With knowledge of the catalyst TOF in hand, a Tafel analysis can be plotted as log (TOF) versus the applied electrochemical overpotential (η, Eq. 3.20.14) as described by Eq. (3.20.28). Tafel analysis routinely allows identification of the optimum applied overpotential (η) to reach TOF_{max}, with respect to the applied experimental conditions for a given catalyst. In addition, an arguably more relevant metric obtainable from a Tafel analysis is the intrinsic TOF, TOF_0, of a catalyst at zero overpotential, that is, $E = E^\circ_{A/C}$, observable as the *y*-intercept in Fig. 3.20.8.[21]

$$TOF = \frac{TOF_{max}}{1+\exp\left[\frac{F}{RT}(E^\circ_{A/C} - E_{cat/2} - \eta)\right]} \tag{3.20.28}$$

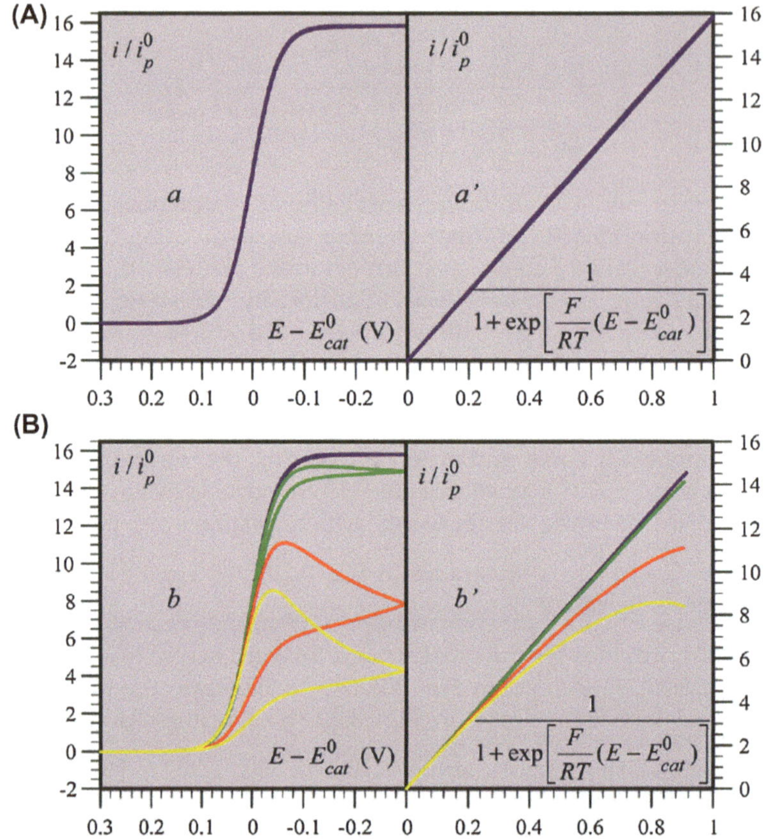

FIGURE 3.20.7 Catalytic cyclic voltammetry responses (left) and foot-of-the-wave analyses (right, FOWAs) for an E_rC_i' reaction. A linear FOWA plot is only observed for the S-shaped steady-state current-voltage cyclic voltammetry response (*blue line*). FOWA allows TOF_{max} to be estimated under non-steady-state conditions (green, red, and yellow cyclic voltammetry plots) by fitting only the linear portion of their FOWA plots. Experimental parameters: $v = 0.1$ V/s, $D_P = 10^{-5}$ cm^2/s, $C_P^0 = 1$ mM, $T = 298$K; (a, a′) no substrate consumption, $C_A^0 = 1$ M, $2k = 100$ s^{-1}; (b, b′) substrate consumption, $C_A^0 = 1$ M (blue), 0.1 M (green), 0.01 M (red), 0.001 M (yellow), $2k = 100$ s^{-1}. *Reprinted with permission from Costentin C, Robert M, Savéant J-M. Catalysis of the electrochemical reduction of carbon dioxide.* Chem Soc Rev *2013;**42**(6):2423–36.*

3.20.3 A CASE STUDY IN HOMOGENEOUS ELECTROCATALYTIC CO_2 REDUCTION

As underlined at the beginning of this chapter, electrocatalysis is capable of addressing a variety of critical sustainable energy concerns. Having now discussed the theoretical basis for electrocatalyst characterization, an overview of one of the primary challenges in electrocatalysis, namely, electrocatalytic CO_2 reduction, is presented. A detailed account of

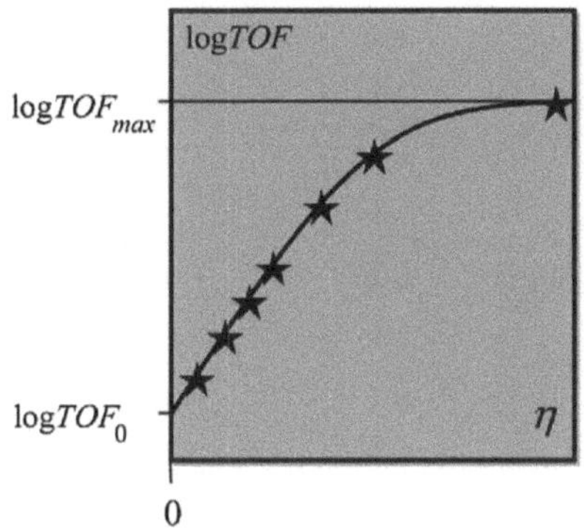

FIGURE 3.20.8 A generic Tafel plot for an E_rC_i' reaction derived from cyclic voltammetry (CV) analysis as described by Eq. (3.20.28). The data points highlighted by the original authors are suggested as points of reference to compare TOF values obtained by CV and bulk electrolysis data for verification of Eq. (3.20.28). *Reprinted with permission from Artero V, Saveant J. Toward the rational benchmarking of homogenous H_2-evolving catalysts.* Energy Environ Sci *2014;7:3808–14.*

homogeneous electrocatalytic CO_2 reduction is beyond the scope of this chapter; therefore, focus will be placed on recent developments in the field that have utilized the "greener," more abundant, and economically viable first-row transition metal-based catalysts.

The conversion of CO_2 into clean fuels (e.g., methanol) and fuel precursors (e.g., carbon monoxide or formic acid) is a critical strategy to achieve a sustainable, carbon-neutral, global energy technology. A highly desired approach to the catalytic conversion of CO_2 to reduced forms of carbon is to use electrical potential energy, viz. electrocatalysis, ideally derived from renewable sources such as solar energy. The one-electron reduction of free CO_2 to generate its one-electron reduced radical anion $CO_2^{\bullet -}$ (Eq. 3.20.29) is a thermodynamically demanding reaction, which occurs at an equilibrium potential of −1.99 V versus the SHE because of the large reorganization energy involved.[22] Through the application of bioinspired PCET catalysis, the thermodynamic requirements for CO_2 reduction can be improved significantly, producing, for example, HCO_2H at a more modest potential of −0.61 V versus SHE in water at pH 7 (Eq. 3.20.31), or CO at −0.52 V versus SHE in water at pH 7 (Eq. 3.20.32).[23] To illustrate the diversity and proton dependency of CO_2 redox chemistry, a summary of aqueous equilibrium potentials (vs. SHE) for the pure redox (Eqs. 3.20.29 and 3.20.30) and proton-coupled (Eqs. 3.20.31–3.20.36) conversions of CO_2 to its reduced derivatives is provided below for the following reaction conditions: 1 atm CO_2, 25 °C, and pH = 7.

$$CO_{2(g)} + e^- \rightleftharpoons CO_2{}^{\bullet -}{}_{(aq)} \quad -1.99\ \text{V} \tag{3.20.29}$$

$$CO_{2(aq)} + e^- \rightleftharpoons CO_2{}^{\bullet -}{}_{(aq)} \quad -1.90\ \text{V} \tag{3.20.30}$$

$$CO_{2(g)} + 2H^+{}_{(aq)} + 2e^- \rightleftharpoons HCO_2H_{(aq)} \quad -0.61\ \text{V} \tag{3.20.31}$$

$$CO_{2(g)} + 2H^+{}_{(aq)} + 2e^- \rightleftharpoons CO_{(g)} + H_2O \quad -0.52\ V \tag{3.20.32}$$

$$3CO_{2(g)} + H_2O_{(l)} + 2e^- \rightleftharpoons CO_{(g)} + 2HCO_3^- \quad -0.56\ V \tag{3.20.33}$$

$$CO_{2(g)} + 4H^+{}_{(aq)} + 4e^- \rightleftharpoons HC(OH)_{2(aq)} + H_2O_{(l)} \quad -0.49\ V \tag{3.20.34}$$

$$CO_{2(g)} + 6H^+{}_{(aq)} + 6e^- \rightleftharpoons CH_3OH_{(aq)} + H_2O_{(l)} \quad -0.38\ V \tag{3.20.35}$$

$$CO_{2(g)} + 8H^+{}_{(aq)} + 8e^- \rightleftharpoons CH_{4(g)} + 2H_2O_{(l)} \quad -0.24\ V \tag{3.20.36}$$

The majority of the literature reporting on homogeneous catalytic CO_2 reduction has focused predominantly on the two-electron two-proton coupled conversion of CO_2 to CO and H_2O (Eq. 3.20.32). While this is often by default, as CO is frequently the sole product of homogeneous transition metal CO_2 reduction catalysis, the catalytic transformation of CO_2 to CO is nonetheless highly desired because CO is a key raw material alongside hydrogen gas for liquid fuel production by the Fischer-Tropsch reaction.[24] Reports of CO_2 to HCO_2H conversion (Eq. 3.20.31), also a two-electron two-proton coupled reaction, by homogeneous first-row transition metal complexes are less common. Formic acid is also a highly valued product, as it is less toxic than CO and nonflammable. Moreover, formic acid is a liquid-phase product that can be directly used more efficiently in a fuel cell than the methanol fuel cell.[25,26] The global production of formic acid is approximately 720,000 tons/annum,[27] used mostly in feed preservation, leather processing, textile processing, and flue gas desulfurization. The principal method of formic acid production is hydrolysis of methyl formate (ironically a product of MeOH and CO).[28] Thus, there is great potential for formic acid to be invested in the future as a renewable economy.

One of the major challenges in identifying a successful homogeneous catalyst for the proton-coupled conversion of CO_2 to CO or HCO_2H is competitive proton reduction and H_2 evolution (Eq. 3.20.31). This is especially challenging for electrocatalytic HCO_2H production, which shares a metal hydride intermediate with the H_2 evolution mechanistic pathway. A general comparison of electrocatalytic CO, HCO_2H, and H_2 production is illustrated in Scheme 3.20.2. It is noteworthy that both CO and HCO_2H reaction pathways share tautomeric intermediates in the metallocarboxylic acid and metalloformate species. Such tautomerization has not been experimentally observed, but has been suggested based on indirect experimental and computational analysis.[29] Although CO is more commonly observed as a product of nonaqueous electrocatalytic CO_2 reduction, the formic acid product is typically favored for aqueous electrolytes.

The engineering of a homogeneous catalyst to orchestrate higher-order PCET conversions of CO_2 (Eqs. 3.20.34–3.20.36), while highly desired, is an extremely challenging and hitherto unsuccessful task. The remainder of this chapter will focus on the two-electron two-proton coupled reduction of CO_2 to CO and/or HCO_2H by homogeneous, molecular first-row transition metal electrocatalysts.

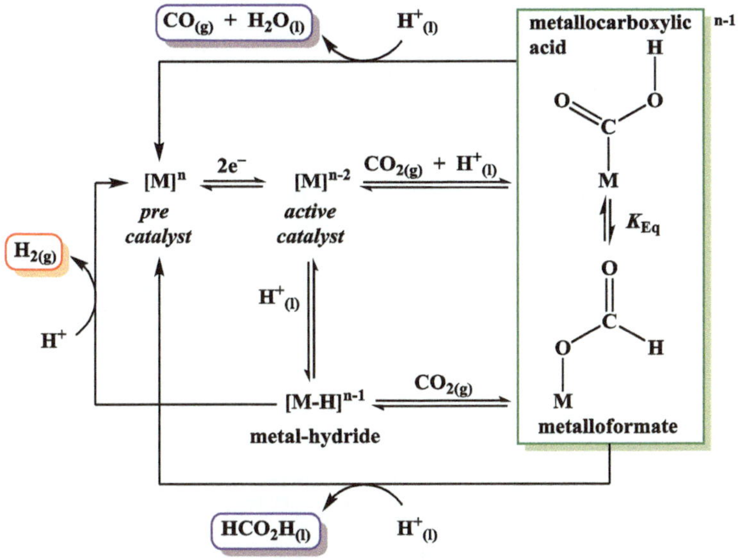

SCHEME 3.20.2 Reaction pathways for CO and HCO_2H formation from CO_2 using the same active catalyst, as well as competitive H_2 evolution. Both pathways have a strong proton (pK_a) dependence and share isostructural metallocarboxylic acid and metalloformate intermediates. (Note: this scheme is not intended to describe the complete mechanistic pathways; deviations may occur depending on the catalyst properties.)

3.20.4 HIGHLIGHTS IN HOMOGENEOUS ELECTROCATALYTIC CO_2 REDUCTION BY FIRST-ROW TRANSITION METALS

Transition metal complexes are prime candidates for the development of selective CO_2 reduction electrocatalysts, as they offer easy access to a diverse range of metal oxidation states and ligand structures to allow electronic fine-tuning of the metal center and optimization of hydrogen bonding in the second coordination sphere. This is no more evident than in nature with the [NiFe] CO dehydrogenase enzyme that promotes the proton-coupled reduction of CO_2 to CO. Scheme 3.20.3 illustrates how the Ni center behaves as the active site for CO_2 coordination, with the adjacent Fe and histidine cofactors acting in tandem to control the local pH environment to promote protonation of the metallocarboxylate intermediate and ultimate cleavage of the C–OH bond. The resulting Ni–CO bond readily dissociates, opening up the unsaturated Ni center for another cycle.[30]

Molecular homogeneous transition metal-based catalysts have long been utilized for electrocatalytic CO_2 reduction,[31,32] as the metallocarboxylate intermediate reduces the reorganization energy for CO_2 activation relative to the formation of the one-electron reduced radical anion (Eq. 3.20.32) by stabilizing a bent configuration of the carboxylate anion.[33] Catalysts have varied in design using electron-rich late transition metal complexes such as cobalt and nickel cyclam or tetraamine complexes[34–37]; cobalt and iron pincer complexes[29,38] and tetrapyrroles[39–41]; polypyridyl complexes of cobalt,[42–44] rhenium,[45–48] ruthenium,[49–53] osmium,[54] rhodium, and iridium[55,56]; and phosphine complexes of cobalt,[57] nickel,[58] rhodium, and palladium.[59,60] Select highlights from the early and more recent literature

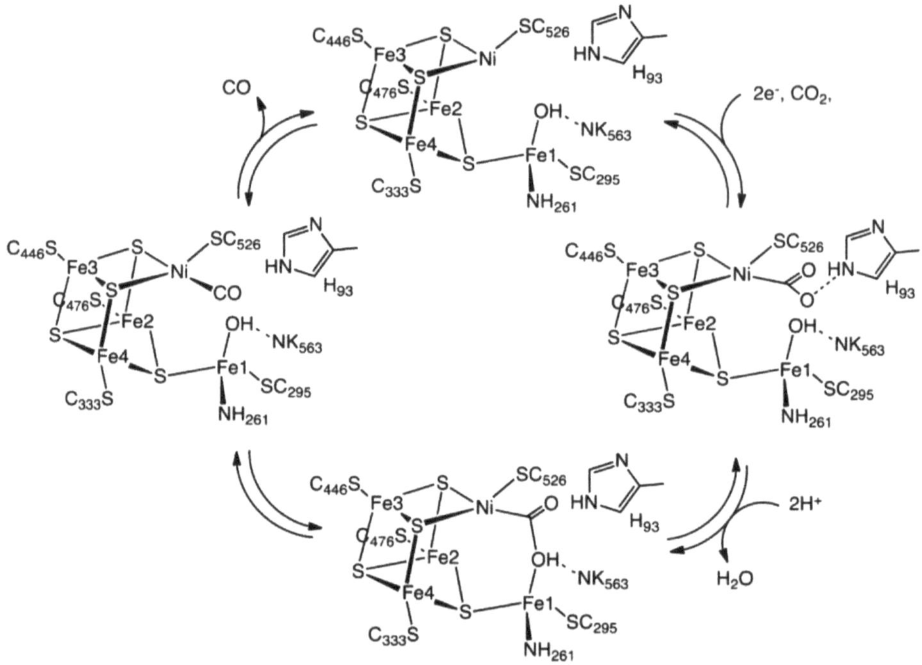

SCHEME 3.20.3 Catalytic cycle for the 2-electron 2-proton coupled reduction of CO_2 to CO and H_2O by the [NiFe] CO dehydrogenase enzyme. *Reproduced with permission from Appel AM, Bercaw JE, Bocarsly AB, Dobbek H, Dubois DL, Dupuis M, et al. Frontiers, opportunities, and challenges in biochemical and chemical catalysis of CO_2 fixation.* Chem Rev *2013;**113**(8):6621–58.*

involving homogeneous molecular first-row transition metal catalysts are discussed here. For a more comprehensive account of homogeneous electrocatalytic CO_2 reduction the reader is directed to a number of excellent reviews on this topic.[20,23,30,32,61–70]

3.20.4.1 Cyclam and Pincer Complexes

The Co(II) and Ni(II) cyclam complexes (where cyclam is 1,4,8,11-tetraazacyclotetradecane) introduced by Eisenberg and coworkers in 1980 were the first of their kind to exhibit high faradaic yields (FYs) for CO production.[36,63] In particular, $[Ni(cyclam)]^{2+}$ exhibits a high selectivity (>90%) for CO production with a high TOF at −1.4 V versus SCE at a pH of 4 in aqueous electrolyte.[71] The Ni(I) species, $[Ni(cyclam)]^{+}$, generated upon one-electron reduction of $[Ni(cyclam)]^{2+}$, is thought to be the active catalyst because it initiates CO_2 activation via formation of the metallocarboxylate intermediate $[Ni(cyclam)(CO_2)]^{+}$ (Scheme 3.20.4). Subsequent protonation generates the corresponding metallocarboxylic acid intermediate $[Ni(cyclam)(CO_2H)]^{2+}$. This is followed by the rate-determining C−OH bond cleavage via further proton-coupled reduction generating H_2O and $[Ni(cyclam)(CO)]^{2+}$, which dissociates CO regenerating the precatalyst $[Ni(cyclam)]^{2+}$.

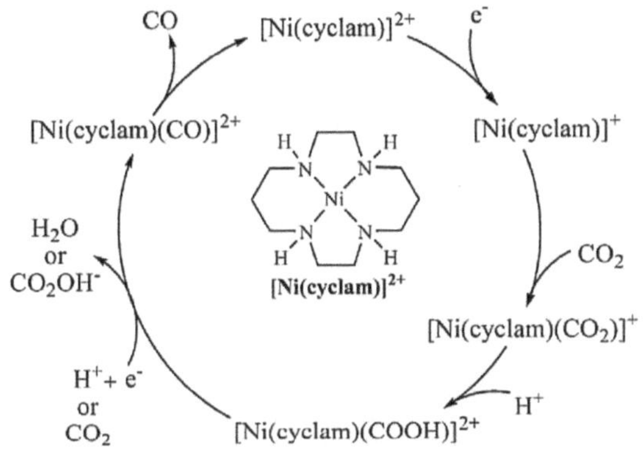

SCHEME 3.20.4 Electrocatalytic CO_2 to CO cycle for $[Ni(cyclam)]^{2+}$ as proposed by Sauvage. *Reproduced with permission from Froehlich JD, Kubiak CP. The homogeneous reduction of CO_2 by $[Ni(cyclam)]^{+}$: increased catalytic rates with the addition of a CO scavenger.* J Am Chem Soc *2015;***137***(10):3565–73.*

Interestingly, when the water content in the electrolyte was decreased, the production of formate (HCO_2^-) took over at ~75% faradaic efficiency.[71] Sauvage and coworkers later demonstrated that $[Ni(cyclam)]^{2+}$ requires the use of a mercury WE (a static mercury dropping electrode is typically used), also that the anion in the supporting electrolyte affected the catalysis selectivity.[34,71,72] It is now well established that this efficient catalytic activity was in fact derived from a heterogeneous $Hg|[Ni(cyclam)]^{+}$ interfacial assembly. While divergent from the homogeneous focus of this chapter, a valuable lesson was learned by the catalysis community, that is, homogeneous catalytic activity must be proven and cannot be assumed.

In 2012, Schneider et al.[37] published a comprehensive study of $[Ni(cyclam)]^{2+}$ and related macrocycles (Fig. 3.20.9) for the selective electrocatalytic reduction of CO_2 to CO at a mercury WE in aqueous electrolyte under various pH conditions. Catalytic activity was found to vary tremendously with respect to the structural isomer of the catalyst, attributed to favorable

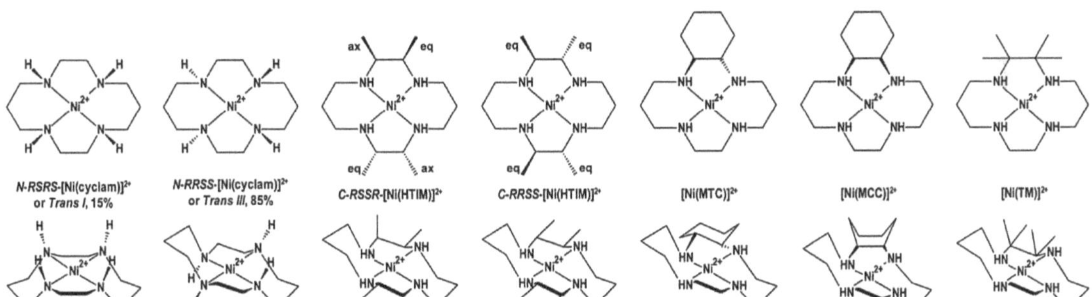

FIGURE 3.20.9 Structures of the Ni(II) cyclam complexes investigated by Schneider et al., including illustrations of the two primary isomers of $[Ni(cyclam)]^{2+}$ (trans I and trans III) present in solution at equilibrium, two structural isomers of $[Ni(HTIM)]^{2+}$ (C-RSSR and C-RRSS), two isomers having either a *trans*-cyclohexane (MTC) or a *cis*-cyclohexane (MCC) group, and a tetramethyl substituted (TM) cyclam analogue. *Reproduced with permission from Schneider J, Jia H, Kobiro K, Cabelli DE, Muckerman J. Nickel(II) macrocycles: highly efficient electrocatalysts for the selective reduction of CO_2 to CO.* Energy Environ Sci *2012;5(11):9502–10.*

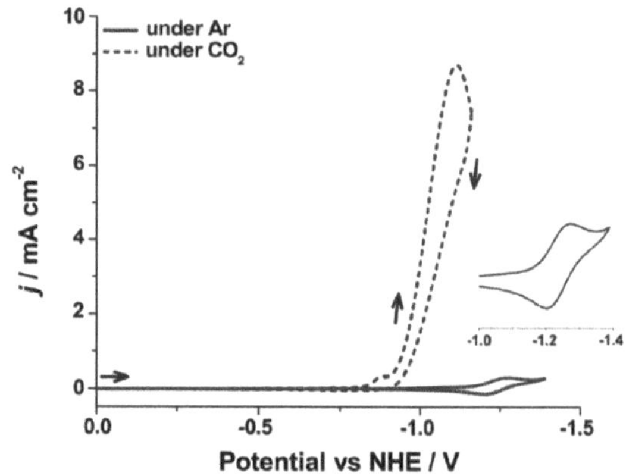

FIGURE 3.20.10 Cyclic voltammetry of a 1.6 mM solution of $[Ni(MTC)]^{2+}$ in 0.1 M $NaClO_4$, pH 5, purged with Ar (*solid line*) and CO_2 (*dashed line*) exhibiting strong catalytic current, scanning at 100 mV/s, static mercury dropping electrode working electrode (area = 0.0104 cm^2). The inset is an expansion of the Ni(II/I) couple, observed under Ar. *Arrows* indicate the direction of the applied potential. *Reproduced with permission from Schneider J, Jia H, Kobiro K, Cabelli DE, Muckerman J. Nickel(II) macrocycles: highly efficient electrocatalysts for the selective reduction of CO_2 to CO.* Energy Environ Sci *2012;5(11):9502–10.*

versus unfavorable geometries for effective adsorption onto the mercury electrode surface. With the suitable isomer present, all catalysts performed excellently in pH 5 aqueous electrolyte at an overpotential of $\eta = -0.55$ V, exhibiting high rate constants ($k_e = 3.3 - 7.1 \times 10^9\ M^{-1}\ s^{-1}$) and faradaic efficiencies (FY = 84%–92%) for the selective reduction of CO_2 to CO, with $[Ni(MTC)]^{2+}$ exhibiting one of the best activities (Fig. 3.20.10). Notably, under more acidic conditions, below the p*K*a of the Ni(H) metal-hydride species (pH < 2), competitive H_2 evolution became dominant.

Kubiak has since demonstrated homogeneous electrocatalytic activity for $[Ni(cyclam)]^+$ in 0.1 M KCl aqueous electrolyte at the inert glassy carbon WE, albeit with an order of magnitude drop in current density (1 mA/cm^2; 1 mM $[Ni(cyclam)]^{2+}$, $\upsilon = 100$ mV/s) relative to the heterogeneous Hg|$[Ni(cyclam)]^+$ system (11 mA/cm^2).[73] A major problem with the $[Ni(cyclam)]^{2+}$ system is catalyst poisoning by the formation of the $[Ni^0(cyclam)(CO)]^0$ species.[63,71] Kubiak has addressed this problem by taking advantage of a CO scavenger in the $[Ni(TMC)]^+$ species (where TMC is 1,4,8,11-tetramethyl-1,4,8,11-tetraazacyclotetradecane) by generating a 10-fold increase in catalytic current at a glassy carbon WE.[73,74] Lacking any secondary amine protons, the tetra-*N*-methylated TMC complex is a poor CO_2 reduction catalyst but exhibits a greater binding affinity for CO [$K_{CO} = (1.2 \pm 0.4) \times 10^5$].[75] In a rare report of a modified nickel cyclam-based catalyst, Cowan and coworkers[76] demonstrated improved catalytic activity for the pendant carboxylated cyclam system $[Ni(cyclam\text{-}CO_2H)]^{2+}$ (cyclam-CO_2H is 1,4,8,11-tetraazacyclotetradecane-6-carboxylic acid), which maintains selectivity for CO production even at low pH.

Beyond the cyclam ligand, reports of nickel- or cobalt-mediated electrocatalytic CO_2 reduction are rare. Abruna and coworkers[77] reported on the nonaqueous (0.1 M Bu_4NClO_4 in DMF) electrocatalytic conversion of CO_2 to HCO_2H by a series of pseudo-octahedral Ni(II),

Fe(II), and Co(II) complexes. As these precatalysts were coordinatively saturated with a series of bis(ketimino)pyridine NNN pincer ligands, the electrochemical reduction-induced ligand dissociation is less likely, thus generating an unsaturated low-valence active catalyst for CO_2 activation. A study by Chang and coworkers reported on the nonaqueous electrocatalytic and photocatalytic activity for a series of tetrachelated bis(2-isoquinoline)carbene complexes at Ni(II)[78,79] in acetonitrile electrolyte. Although electrocatalytic currents remained low, a high selectivity was observed for CO production, with faradaic efficiencies up to 90%. Subsequently, a tris-coordinated Ni(II) bis(carbene)pyridyl CNC pincer complex was also reported to show electrocatalytic activity for CO_2 reduction in acetonitrile.[80] Again poor catalytic currents translated into a low TOF_{max} of just 90 s^{-1}. CO was identified by gas chromatography (GC); however, the FY or the presence of other products (e.g., HCO_2H or H_2) was never reported. Narayanan et al. demonstrated that evaluation of FY is critical when investigating electrocatalytic CO_2 reduction. In the latter study, two $Ni(NNN)X_2$ pincer complexes [where $X = Cl^-$ or Br^- and NNN = *N,N′*-(2,6-diisopropylphenyl)bis-aldiminopyridine] exhibited catalytic current upon purging an acetonitrile electrolyte with 1 atm CO_2. However, the primary reaction product determined by GC was H_2, with negligible CO detected. Similar ambiguous behavior was reported previously for a pentacoordinate cyclic bis(ketimino)pyridine Mn carbonyl complex, for which Mukhopadhyay et al. [81] demonstrated that the presence of CO_2 was critical to generate carbonic acid in situ as a viable proton source for proton reduction and H_2 evolution.

Somewhat related to the cyclam ligand, Marinescu and coworkers[44] demonstrated electrocatalytic CO_2 to CO conversion for a [NiFe] CODH inspired Co(II) aminopyridine macrocycle, which also has the capability of hydrogen bonding to the metallocarboxylate intermediate via peripheral secondary amine ligand groups. An FY of 98% was reported, demonstrating high selectivity for CO production; however, weak catalytic current again translated to a low TOF_{max} of 360 s^{-1}. Another interesting example of cobalt-mediated electrocatalytic CO_2 reduction is by Peters and coworkers[38] who reported nonaqueous (10 M H_2O in acetonitrile) electrocatalytic CO_2 to CO conversion by the formally Co(I) complex $[CoN_4H(CH_3CN)]^+$ (where N_4H is 2,12-dimethyl-3,7,11,17-tetraazabicyclo-[11.3.1]-heptadeca-1(7),2,11,13,15-pentaene). Although the reported FY of 45% was relatively low because of competitive H_2 evolution (FY = 30%), this study is notable in that it demonstrated a redox noninnocent contribution of the tetradentate N_4H ligand to the CO_2 reduction cycle. Based on a combination of crystallographic and computational analysis, the $[CoN_4H(CH_3CN)]^+$ precatalyst was ascribed as having a low-spin Co(II) ion antiferromagnetically coupled to a ligand radical-anion ($N_4H^{\bullet -}$), which was deemed critical for favorable CO_2 over H^+ reduction. Perhaps the most intriguing report of electrocatalytic CO_2 conversion by a pincer type complex is that by Chen et al.,[29] where a pentacoordinate cyclic bis(ketimino)pyridine ligand showed high selectivity for either CO (FY = 82%) or HCO_2H (FY = 80%), which could be selectively controlled based on whether the catalyst was Co or Fe based, Fig. 3.20.11a and b, respectively. It is noteworthy that selective HCO_2H production by the Fe complex was attributed to tautomerization of an Fe(III)-CO_2H metallocarboxylate intermediate to and Fe(III)-O(CHO) metalloformate intermediate.

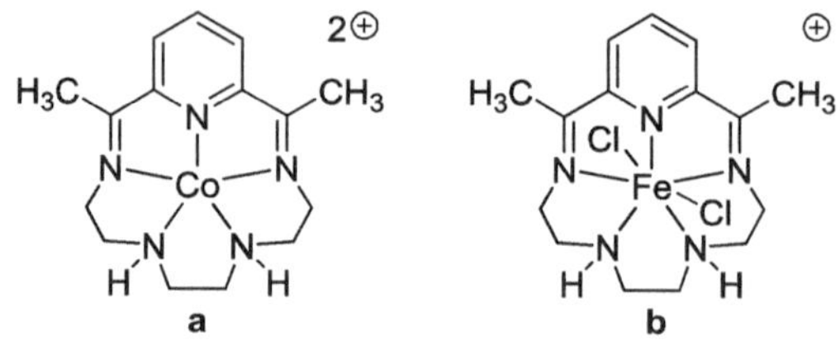

FIGURE 3.20.11 Co(II) and Fe(III) chloride complexes of the pentacoordinate cyclic bis(ketimino)pyridine ligand reported by Chen et al.[29] exhibit high electrocatalytic CO_2 reduction product selectivity for CO and HCO_2H, respectively.

3.20.4.2 Polypyridyl Complexes

The group VII *fac*-M^I(NN)$(CO)_3$X [M = Re(I) or Mn(I)] class of catalysts for electrocatalytic CO_2 to CO formation, where NN is a polypyridyl ligand and X is a monodentate ligand, have been intensely studied owing to their high product selectivity for CO formation. Since the first report of photocatalytic CO_2 reduction with *fac*-Re(bpy)$(CO)_3$Cl (bpy = 2,2′-bipyridine) by Hawecker, Lehn, and Ziessel, there have been many literature reports of related photo-/electrocatalytic systems, with numerous reviews written on the topic.[23,32,45,61–68] A development pioneered by Deronzier[82] has been the successful application of *fac*-Mn^I(NN)$(CO)_3$Br complexes as electrocatalysts (where NN is 2,2′-bipyridine or 4,4′-dimethyl-2,2′-bipyridine), taking advantage of the more abundant and economical first-row transition metal, manganese (Fig. 3.20.12a and b).

In contrast to their Re(I) analogues, Mn(I) polypyridyl catalysts typically require an excess source of Brønsted acid for binding of CO_2 and formation of the manganese carboxylic acid intermediate. This prerequisite has been investigated computationally by Carter who compared the catalytic cycles for both Re and Mn systems.[83] Before entering the catalytic cycle (Scheme 3.20.5), two-electron reduction is required to generate the five-coordinate anionic active catalyst in situ (Scheme 3.20.5d). The neutral five-coordinate one-electron reduced species (Scheme 3.20.5b) is prone to dimerization.[84] While a side reaction of the desired two-electron activation, this dimer can itself produce two equivalents of the active catalyst upon two-electron reduction. The active catalyst then reacts with CO_2 in the presence of a proton source to generate the neutral six-coordinate metallocarboxylic acid intermediate (Scheme 3.20.5e). One-electron reduction to a metallocarboxylic acid anion (Scheme 3.20.5f) is then required, at the expense of a high overpotential, to promote proton-coupled hydroxide abstraction, a rate-determining step that involves C–OH bond cleavage, eliminating H_2O and generating a Mn(I) tetracarbonyl intermediate (Scheme 3.20.5g). Subsequent reduction eliminates CO regenerating the five-coordinate active catalyst.

Inspired by the work of Deronzier et al. many reports of Mn(I) CO_2 reduction electrocatalysts have appeared in the recent years.[83–103] One notable example reported by Kubiak et al. [87] demonstrated how the steric bulk of the 6,6′-dimesityl-2,2′-bipyridine ligand (Fig. 3.20.12d and e) hinders competitive formation of a Mn^0-Mn^0 dimer complex, enhancing catalytic TOF for CO formation. Gobetto et al.[86] and Agarwal et al.[97] each reported on the introduction of a pendant hydroxyphenyl group to *fac*-Mn^I(NN)$(CO)_3$Br electrocatalysts (Fig. 3.20.12g and h respectively). Interestingly, Gobetto et al.[86] reported a 22% faradaic

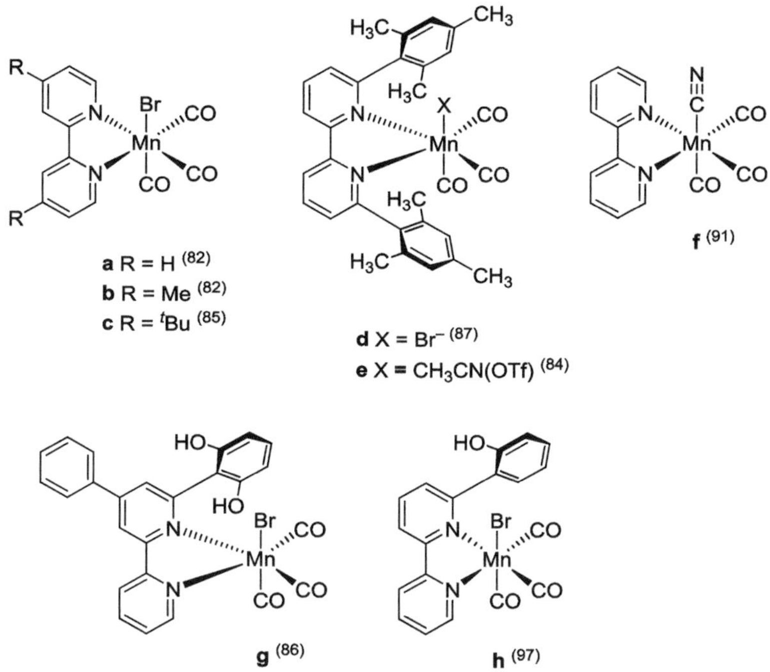

FIGURE 3.20.12 Molecular structures of [*fac*-Mn^{I}(NN)$(CO)_3X]^n$ CO_2 reduction catalysts discussed in this chapter, where N^N is a bipyridine ligand. Citations are provided in parentheses.

efficiency for HCO_2H production in acetonitrile electrolyte in the absence of a sacrificial Brønsted acid, with 70% of the faradaic current being attributed to CO.

Recent progress in catalyst design has seen the introduction of asymmetric bidentate pyrid-2-yl ligand scaffolds in place of the prevalent 2,2′-bipyridine backbone. First reported by Agarwal et al.,[91] the NHC-pyridyl Mn(I) tricarbonyl catalysts containing the *N*-methyl-*N*′-2-pyridyl imidazol-2-ylidine and benzimidazol-2-ylidine ligands (Fig. 3.20.13a and b, respectively) exhibited electrocatalytic activity for CO_2 reduction. While each complex demonstrated low TOFs ($<0.1\ s^{-1}$) and poor faradaic efficiencies (<35%), Agarwal improved this catalytic activity by engineering a more π-acidic ligand.[101] Replacing the pyridyl ring with pyrimidine (Fig. 3.20.13c vs. d) caused a 160-mV anodic shift in the catalyst reduction potential with a concomitant increase in TOF ($4.5–42.0\ s^{-1}$) and faradaic efficiency for CO production (48%–72%).

Spall et al.[102] reported on the electrocatalytic CO_2 reduction properties for a series of five asymmetric 2-(phenylimino)pyridine-containing Mn(I) tricarbonyl complexes (Fig. 3.20.13e–i). In this study the imine substituent (Fig. 3.20.13, R_1) was used to tune catalyst electronic properties, whereas the 2,6-phenyl substituents at the α-imine (Fig. 3.20.13, R_2 and R_3) were used to impart a steric influence to hinder Mn^0-Mn^0 dimer formation. Both CO and HCO_3^- were observed as products of electrocatalysis in wet (4.7% H_2O) acetonitrile. Although the reported i_{cat}/i_p ratios were relatively low [<0.5 relative to $Mn(bpy)(CO)_3Br$], the authors highlighted the true versatility of this ligand platform, which may provide for significant enhancement

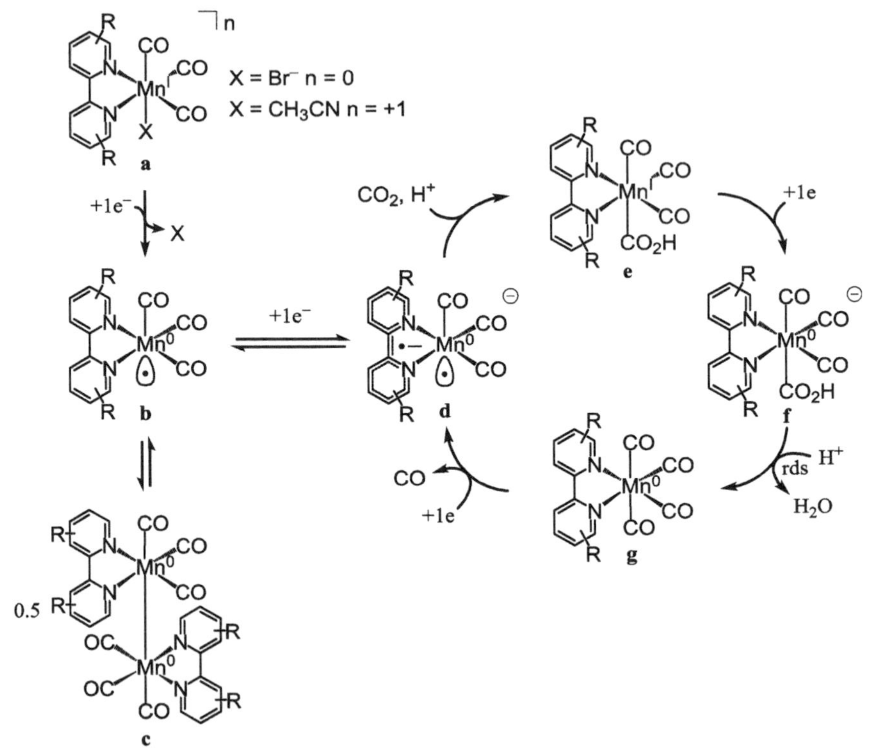

SCHEME 3.20.5 Electrocatalytic CO_2 to CO cycle for Mn(I) polypyridyl complexes including precatalyst activation and competitive Mn-Mn dimer formation.

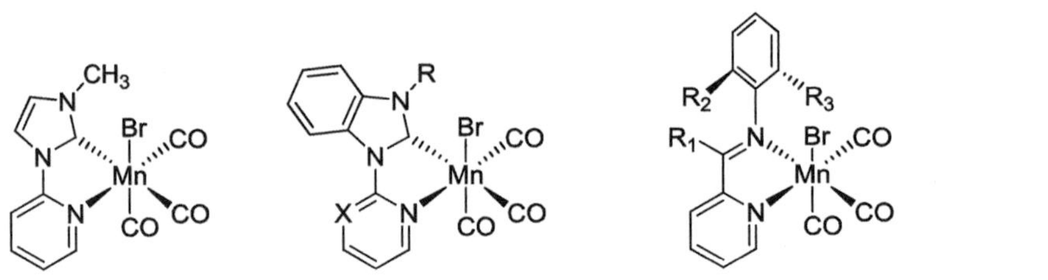

a R = Me (91)

b R = Me, X = C-H (101)
c R = Et, X = C-H (101)
d R = Et, X = N (101)

e $R_1 = R_2 = R_3 = H$ (102)
f $R_1 = H$, $R_2 = {}^iPr$, $R_3 = H$ (102)
g $R_1 = H$, $R_2 = R_3 = {}^iPr$, (102)
h $R_1 = H$, $R_2 = {}^tBu$, $R_3 = H$ (102)
i $R_1 = Me$, $R_2 = {}^tBu$, $R_3 = H$ (102)

FIGURE 3.20.13 Molecular structures of Mn(I) CO_2 reduction catalysts discussed in this chapter utilizing an asymmetric bidentate pyridyl ligand.

in catalytic activity should ideal electronic and steric factors be identified. However, stability still remains a problem, consistent with two related independent studies[92,103] of Mn(I) tricarbonyl complexes containing a nonaromatic α-diimine backbone. Both these studies demonstrated that electrocatalytic CO_2 reduction activity still persists in the absence of a nonaromatic ligand platform generating both CO and HCO_3^-, albeit with significantly reduced catalyst stability. Perhaps most departed from the 2,2′-bipyridine standard is the recent communication by Richeson who demonstrated selective CO production (FY = 96%) for a tridentate *N,N′*-bis(diphenylphosphino)-2,6-di(methylamino)pyridine (PNP) pincer ligand at an Mn(I) tricarbonyl center. Interestingly, this catalyst exhibited catalytic current and CO evolution in the absence of excess Brønsted acid, in contrast to the asymmetric *N*-(diphenylphosphino)-2-(methylamino)pyridine bidentate analogue that required wet (5% H_2O) acetonitrile.[99]

3.20.4.3 Iron Porphyrin Catalysts

Some of the more notable CO_2 reduction electrocatalysts are the series of iron(III) chloride tetraaryl porphyrins reported by the Savéant group.[63,104,105] First reported in 1988,[106] early studies identified Fe(0) tetraphenylporphyrin as an active catalyst for CO_2 to CO conversion, generated in situ upon sequential two-electron reduction of the Fe(III) chloride precursor FeTPP (Fig. 3.20.14).

FeTPP shows modest catalytic current with selective CO production in 0.1 M Et_4NClO_4 DMF electrolyte under 1 atm of CO_2; however, it suffers from rapid degradation due to outer sphere reactivity at the porphyrin macrocycle. Introduction of weak Lewis acids significantly improves the efficiency and stability of FeTPP without compromising selectivity for CO evolution ($FY_{CO} = 70\%$),[110–112] with Mg^{2+} in particular exhibiting an i_{cat}/i_p ratio close to 20 (recorded at 0.1 V/s). It was later proven that Lewis acid cations assist in C—O bond cleavage to form an Fe^{II}(CO) intermediate and the corresponding carbonate salt of the Lewis acid.[63,104,112] Deposition of carbonate salts at the electrode surface ultimately inhibits electrocatalysis, therefore replacing the Lewis acid with a Brønsted acid made for a logical progression. Introduction of the 1-propanol, 2-pyrrolidone, or trifluoroethanol Brønsted acids saw a significant improvement in catalytic current, with the latter exhibiting an impressive i_{cat}/i_p ratio of 81[105,111] and a much improved bulk electrolysis

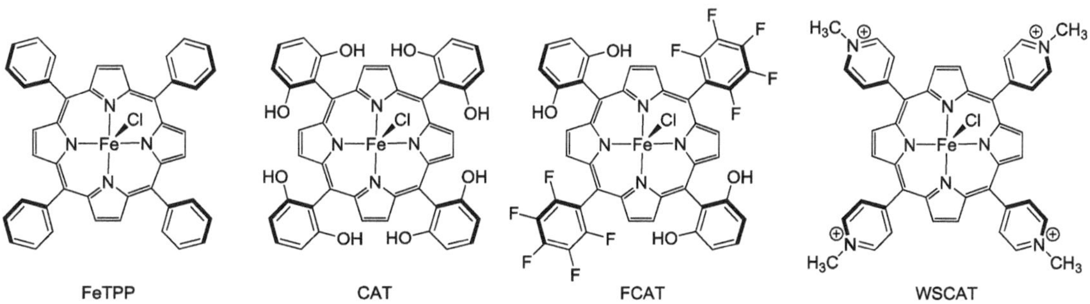

FIGURE 3.20.14 Molecular structures of the *meso*-tetraarylporphyrin iron(III) chloride CO_2 reduction electrocatalysts reported by Savéant and coworkers.[106–109]

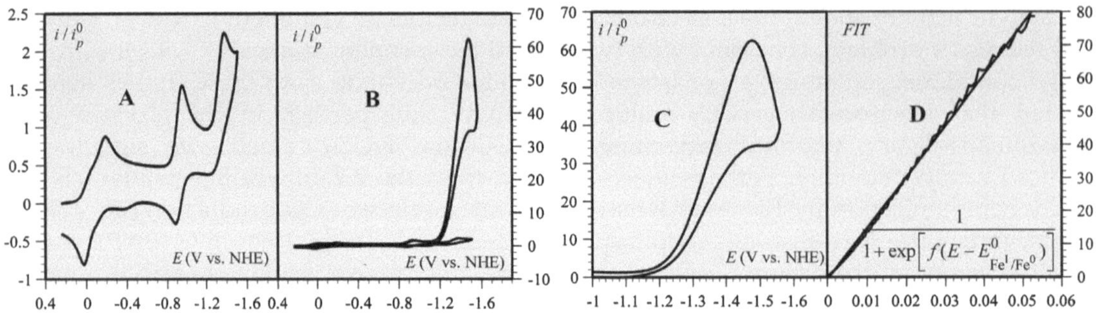

FIGURE 3.20.15 Cyclic voltammetry of 1 mM CAT recorded in 0.1 M $^{n}Bu_4NPF_6$ DMF electrolyte at 0.1 V/s with all current normalized with respect to i_p at the Fe(II/I) couple. Strong catalytic current is exhibited under 1 atm of CO_2 + 2 M H_2O at the onset of the Fe(I/0) redox couple. Reaction conditions are (A) + 2 M H_2O under 1 atm Ar, (B and C) + 2 M H_2O in the presence (upper trace, 0.23 M) and absence (lower trace) of CO_2, and (D) FOWA of the voltammogram in (C). *Reproduced with permission from Costentin C, Drouet S, Robert M, Savéant J. A local proton source enhances CO_2 electroreduction to CO by a molecular Fe catalyst.* Science *2012;***338***(6103):90–4.*

stability with TONs of 350 h^{-1}.[63,104,105,113] A faradaic efficiency of 100% has since been reported for CO evolution in the presence of 0.1 M PhOH (1 mM FeTPP).[40] Founded upon the positive influence of Brønsted acids on the electrocatalytic response of FeTPP, Savéant and coworkers[107] subsequently introduced eight pendant hydroxyl substituents, via four *meso*-2,6-dihydroxyphenyl substituents, to provide a large local concentration of protons at the Fe center (Fig. 3.20.14, CAT). It was in the same report that Savéant also introduced the novel concept of FOWA to accurately determine TOF_{max} in the absence of an ideal plateau current response. An impressive i_{cat}/i_p ratio of 60 (0.1 V/s scan rate) was reported for CAT in the presence of 2 M H_2O (Fig. 3.20.15). Furthermore, extracted from Tafel analysis, a log TOF range of 2.3–4.2 (200–15,849 s^{-1}) was reported across the overpotential range of $\eta = 0.41–0.56$ V.

A study by Costentin et al.[108] has provided an in-depth mechanistic study of CAT versus the slightly improved FCAT system, with log TOF_{max} values reported[114] as 3.8 (6.3×10^3 s^{-1}) and 4.0 (1.0×10^4 s^{-1}), respectively, in the presence of 3.0 M PhOH. Perhaps more impressive is the performance of the water-soluble FeTPP analogue WSCAT (Fig. 3.20.14) whose product selectivity can be tuned from 100% CO (pH = 6.7, no buffer) to 50%:50% CO:H_2 (pH = 6.7, 0.1 M phosphate buffer) to 100% H_2 (pH = 3.7, 0.1 M formic acid buffer).[109]

3.20.5 CONCLUSIONS

Hopefully, this chapter has provided a coherent and inclusive introduction to the theory, application, and practice of electrocatalysis in the area of sustainable fuels production, specifically toward CO_2 reduction. Not only is molecular engineering of efficient, selective, and stable homogeneous catalysts a challenging task but also the task of comprehensive and qualitative analysis of a newly discovered catalyst. Only with a true appreciation of the intricacies detailed in this chapter (and the literature cited herein) will it be possible to truly make progress toward the further reduction of CO_2 toward higher reduced derivatives. It is imperative to realize the urgency with which chemists need to address the challenges discussed here. It is also worth considering that electrocatalysis alone may not mitigate the imminent environmental issues associated with global CO_2 emissions. Photovoltaics and photocatalysis are also capable of making a meaningful contribution, as discussed in Chapters 3.21 and 3.25.

References

1. Bard AJ, Faulkner LR. Electrochemical methods: fundamentals and applications. In: Harris D, Swain E, Robey C, Aiello E, editors. *Electrochemistry. I.* 2nd ed. New York: Wiley; 2000. Faulkner, Larry R.
2. Wang J. *Analytical electrochemistry*. 3rd ed. Wiley-VCH; 2006.
3. Gritzner G, Kuta J. Recommendations on reporting electrode potentials in non-aqueous solvents. *Pure Appl Chem* 1984;**56**(4):461–6.
4. Connelly NG, Geiger WE. Chemical redox agents for organometallic chemistry. *Chem Rev* 1996;**96**(2):877–910.
5. Appel AM, Helm ML. Determining the overpotential for a molecular electrocatalyst. *ACS Catal* 2014;**4**(2):630–3.
6. Janisch JJ, Ruff A, Speiser B, Wolff C, Zigelli J, Benthin S, Feldmann V, Mayer H. Consistent diffusion coefficients of ferrocene in some non-aqueous solvents: electrochemical simultaneous determination together with electrode sizes and comparison to pulse-gradient spin-echo NMR results. *J Solid State Electrochem* 2011;**15**:2083.
7. Wilson AD, Newell RH, McNevin MJ, Mucjerman JT, Rakowski DuBois M, DuBois D. Hydrogen oxidation and production using nickel-based molecular catalysts with positioned proton relays. *J Am Chem Soc* 2006;**128**(1):358–66.
8. Roberts JAS, Bullock R. Direct determination of equilibrium potentials for hydrogen oxidation/production by open circuit potential measurements in acetonitrile. *Inorg Chem* 2013;**52**(7):3823–35.
9. McCarthy BD, Martin DJ, Rountree ES, Ullman AC, Dempsey JL. Electrochemical reduction of Brønsted acids by glassy carbon in acetonitrile: implications for electrocatalytic hydrogen evolution. *Inorg Chem* 2014;**53**(16):8350–61.
10. Matsubara Y, Grills DC, Kuwahara Y. Thermodynamic aspects of electrocatalytic CO_2 reduction in acetonitrile and with an ionic liquid as solvent or electrolyte. *ACS Catal* 2015;**5**(11):6440–52.
11. Huynh MHV, Meyer T. Proton-coupled electron transfer. *Chem Rev* 2007;**107**(11):5004–64.
12. Costentin C, Drouet S, Passard G, Robert M, Saveant J. Proton-coupled electron transfer cleavage of heavy-atom bonds in electrocatalytic processes. Cleavage of a C–O bond in the catalyzed electrochemical reduction of CO_2. *J Am Chem Soc* 2013;**135**(24):9023–31.
13. Siewart I. Proton-coupled electron transfer reactions catalyzed by 3*d* metal complexes. *Chem Eur J* 2015;**21**(43):15078–91.
14. Liu WZ, Bordwell F. Gas-phase and solution-phase homolytic bond dissociation energies of $H{-}N^+$ bonds in the conjugate acids of nitrogen bases. *J Org Chem* 1996;**61**(14):4778–83.
15. Tsai MK, Rochford J, Polyansky DM, Wada T, Tanaka K, Fujita E, Micherman J. Characterization of redox states of $Ru(OH_2)(Q)(tpy)^{2+}$ (Q = 3,5,di-tert-butyl-1,2-benzoquinone, tpy =2,2′:6′,2″-terpyridine) and related species through experimental and theoretical studies. *Inorg Chem* 2009;**48**(10):4372–83.
16. Muckerman JT, Polyansky DE, Wada T, Tanaka K, Fujita E. Water oxidation by a ruthenium complex with non-innocent quinone ligands: possible formation of an O–O bond at a low oxidation state of the metal. *Inorg Chem* 2008;**47**(6):1787–802.
17. Saveant JM, Su K. Homogeneous redox catalysis of electrochemical reaction. Part VI. Zone diagram representation of the kinetic regimes. *J Electroanal Chem Interfacial Electrochem* 1984;**171**(1–2):341–9.
18. Rountree ES, McCarthy BD, Eisenhart TT, Dempsey JL. Evaluation of homogeneous electrocatalysts by cyclic voltammetry. *Inorg Chem* 2014;**53**(19):9983–10002.
19. Martin DJ, McCarthy BD, Rountree ES, Dempsey J. Qualitative extension of the *EC′* zone diagram to a molecular catalyst for a multi-electron, multi-substrate electrochemical reaction. *Dalt Trans* 2016;**45**:9970–6.
20. Costentin C, Robert M, Savéant J-M. Catalysis of the electrochemical reduction of carbon dioxide. *Chem Soc Rev* 2013;**42**(6):2423–36.
21. Artero V, Saveant J. Toward the rational benchmarking of homogenous H_2-evolving catalysts. *Energy Environ Sci* 2014;**7**:3808–14.
22. Schmidt M. The thermodynamics of CO_2 conversion. In: Paul J, Pradier C, editors. *Carbon dioxide chemistry environmental issues*. The Royal Society of Chemistry; 1994.

23. Arakawa H, Aresta M, Armor JN, Barteau MA, Beckman EJ, Bell AT, et al. Catalysis research of relevance to carbon management: progress, challenges, and opportunities. *Chem Rev* 2001;**101**(4):953–96.
24. Zhang QH, Kang JC, Wang Y. Development of novel catalysts for Fischer-Tropsch synthesis: tuning the product selectivity. *ChemCatChem* 2010;**2**(9):1030–58.
25. Jiang H-L, Singh SK, Yan J-M, Zhang X-B, Xu Q. Liquid-phase chemical hydrogen storage: catalytic hydrogen generation under ambient conditions. *ChemSusChem* 2010;**3**(5):541–9.
26. Loges B, Boddien A, Gärtner F, Junge H, Beller M. Catalytic generation of hydrogen from formic acid and its derivatives: useful hydrogen storage materials. *Top Catal* 2010;**53**(13–14):902–14.
27. IHS. *Formic acid. Chemical economics handbook.* 2016. Available from: https://www.ihs.com/products/formic-acid-chemical-economics-handbook.html.
28. *Ullmann's encyclopedia of industrial chemistry*. Wiley-VCH Verlag GmbH & Co. KGaA; 2000.
29. Chen L, Guo Z, Wei X-G, Gallenkamp C, Bonin J, Anxolabehere-Mallart E, Lau K-C, Lau T-C, Robert M. Molecular catalysis of the electrochemical and photochemical reduction of CO_2 with earth-abundant metal complexes. Selective production of CO vs HCOOH by switching of the metal center. *J Am Chem Soc* 2015;**137**(34):10918–21.
30. Appel AM, Bercaw JE, Bocarsly AB, Dobbek H, Dubois DL, Dupuis M, et al. Frontiers, opportunities, and challenges in biochemical and chemical catalysis of CO_2 fixation. *Chem Rev* 2013;**113**(8):6621–58.
31. Qiao JL, Liu YY, Hong F, Zhang J, Qiao J, Liu Y, Hong F, Zhang J. A review of catalysts for the electroreduction of carbon dioxide to produce low-carbon fuels. *Chem Soc Rec* 2014;**43**:631–75.
32. Keene FR, Sullivan B. Thermodynamic, kinetic, and product considerations in carbon dioxide reactivity. In: Sullivan BP, Krist K, Guard H, editors. *Electrochemical and electrocatalytic reactions of carbon dioxide*. Elsevier; 1993. p. 1–18.
33. Leitner W. The coordination chemistry of carbon dioxide and its relevance for catalysis: a critical survey. *Coord Chem Rev* 1996;**153**:257–84.
34. Beley M, Collin JP, Ruppert R, Sauvage J. Electrocatalytic reduction of carbon dioxide by nickel cyclam^{2+} in water: study of the factors affecting the efficiency and the selectivity of the process. *J Am Chem Soc* 1986;**108**:7461–7.
35. Schneider J, Jia H, Muckerman JT, Fujita E. Thermodynamics and kinetics of CO_2, CO, and H^+ binding to the metal center of CO_2 reduction catalysts. *Chem Soc Rev* 2012;**41**(6):2036–51.
36. Fisher B, Eisenberg R. Electrocatalytic reduction of carbon dioxide by using macrocycles of nickel and cobalt. *J Am Chem Soc* 1980;**102**:7361–3.
37. Schneider J, Jia H, Kobiro K, Cabelli DE, Muckerman J. Nickel(II) macrocycles: highly efficient electrocatalysts for the selective reduction of CO_2 to CO. *Energy Environ Sci* 2012;**5**(11):9502–10.
38. Lacy DC, Mccrory CCL, Peters JC. Studies of cobalt-mediated electrocatalytic CO_2 reduction using a redox-active ligand. *Inorg Chem* 2014;**53**(10):2–10.
39. Manbeck GF, Fujita E. A review of iron and cobalt prophyrins, phthalocyanines and related complexes for electrochemical and photochemical reduction of carbon dioxide. *J Porphyr Phthalocyanines* 2015;**19**(1–3):45–64.
40. Costentin C, Robert M, Saveant J. Current, issues in molecular catalysis illustrated by iron porphyrins as catalysts of the CO_2-to-CO electrochemical conversion. *Acc Chem Res* 2015;**48**(12):2996–3006.
41. Lin S, Diercks CS, Zhang YB, Kornienko N, Nichols EM, Zhao Y, et al. Covalent organic frameworks comprising cobalt porphyrins for catalytic CO_2 reduction in water. *Science* 2015;**349**(6253):1208–13.
42. Chan SLF, Lam TL, Yang C, Yan SC, Cheng N. A robust and efficient cobalt molecular catalyst for CO_2 reduction. *Chem Commun* 2015;**51**(37):7799–801.
43. Elgrishi N, Chambers MB, Fontecave M. Turning it off! Disfavouring hydrogen evolution to enhance selectivity for CO production during homogeneous CO_2 reduction by cobalt–terpyridine complexes. *Chem Sci* 2015;**6**(4):2522–31.
44. Chapovetsky A, Do TH, Haiges R, Takase MK, Marinescu S. Proton-assisted reduction of CO_2 by cobalt aminopyridine macrocycles. *J Am Chem Soc* 2016;**138**(18):5765–8.
45. Hawecker J, Lehn J-M, Ziessel R. Electrocatalytic reduction of carbon dioxide mediated by $Re(bipy)(CO)_3Cl$ (bipy = 2,2′-bipyridine). *J Chem Soc Chem Commun* 1984;**984**(6):328–30.
46. Keith JA, Grice KA, Kubiak CP, Carter EA. Elucidation of the selectivity of proton-dependent electrocatalytic CO_2 reduction by *fac*-$Re(bpy)(CO)_3Cl$. *J Am Chem Soc* 2013;**135**(42):15823–9.
47. Manbeck GF, Muckerman JT, Szalda DJ, Himeda Y, Fujita E. Push or pull? Proton responsive ligand effects in rhenium tricarbonyl CO_2 reduction catalysts. *J Phys Chem B* 2015;**119**(24):7457–66.

48. Franco F, Cometto C, Garino C, Minero C, Sordello F, Nervi C, et al. Photo- and electrocatalytic reduction of CO_2 by $Re(CO)_3${a,a′-diimine-(4-piperidinyl-1,8-napthalimide}Cl complexes. *Eur J Inorg Chem* 2015;(2):296–304.
49. Ishida H, Tanaka K, Tanaka T. Electrochemical CO_2 reduction catalyzed by $[Ru(bpy)_2(CO)_2]^{2+}$ and $[Ru(bpy)(CO)Cl]^+$ – the effect of pH on the formation of CO and $HCOO^-$. *Organometallics* 1987;**6**(1):181–6.
50. Nagao H, Mizukawa T, Tanaka K. Carbon-carbon bond formation in the electrochemical reduction of carbon-dioxide catalyzed by a ruthenium complex. *Inorg Chem* 1994;**33**(15):3415–20.
51. Tanaka K, Ooyama D. Multi-electron reduction of CO_2 via Ru–CO_2, –C(O)OH, –CO, –CHO, and –CH2OH species. *Coord Chem Rev* 2002;**226**(1–2):211–8.
52. Chen Z, Chen C, Weinberg DR, Kang P, Concepcion JJ, Harrison D, et al. Electrocatalytic reduction of CO_2 to CO by polypyridyl ruthenium complexes. *Chem Commun* 2011;**47**(47):12607–9.
53. Johnson BA, Maji S, Agarwala H, White TA, Mijangos E, Ott S. Activating a low overpotential CO_2 reduction mechanism by a strategic ligand modification on a ruthenium polypyridyl catalyst. *Angew Chem Int Ed* 2016;(5):55.
54. Bruce MRM, Megehee E, Sullivan BP, Thorp HH, Otoole TR, Downard A, et al. Electrocatalytic reduction of carbon dioxide based on 2,2′-bipyridyl complexes of osmium. *Inorg Chem* 1992;**31**(23):4864–73.
55. Bolinger CM, Story N, Sullivan BP, Meyer T. Electrocatalytic reduction of carbon dioxide by 2,2′-bipyridine complexes of rhodium and iridium. *Inorg Chem* 1988;**27**(25):4582–7.
56. Sypaseuth FD, Matlachowski C, Weber M, Schwalbe M, Tzschucke C. Electrocatalytic carbon dioxide reduction by using cationic pentamethylcyclopentadienyl-Iridium complexes with unsymmetrically substituted bipyridine ligands. *Chem Eur J* 2015;**21**(17):6564–71.
57. Haines RJ, Wittrig RE, Kubiak C. Electrocatalytic reduction of carbon dioxide by the binuclear copper complex $[Cu_2(6\text{-diphenylphosphine})\text{-}2,2'\text{-bipyridyl})_2(MeCN)_2](PF_6)_2$. *Inorg Chem* 1994;**33**(21):4723–8.
58. Dalaet DL, Delrosario R, Fanwick PE, Kubiak C. Carbon dioxide chemistry and electrochemistry of a binuclear cradle of Ni(0), $Ni_2(\mu\text{-}CNMe)(CNMe)_2(PPh_2CH_2PPh_2)_2$. *J Am Chem Soc* 1987;**109**(3):754–8.
59. Slater S, Wagenknecht J. Electrochemical reduction of CO_2 catalyzed by $Rh(diphos)_2Cl$. *J Am Chem Soc* 1984;**106**(18):5367–8.
60. Raebiger JW, Turner JW, Noll BC, Curtis CJ, Miedaner A, Cox B, et al. Electrochemical reduction of CO_2 to CO catalyzed by a bimetallic palladium complex. *Organometallics* 2006;**25**(14):3345–51.
61. Fujita E, Brunschwig B. Homogeneous redox catalysis in CO2 fixation. In: Balzani V, editor. *Catalysis, heterogeneous systems, gas phase systems, electron transfer in chemistry*, vol. 4. Wiley-VCH; 2001. p. 88–126.
62. Morris AJ, Meyer GJ, Fujita E. Molecular approaches to the photocatalytic reduction of carbon dioxide for solar fuels. *Acc Chem Res* 2009;**42**(12):1983–94.
63. Benson EE, Kubiak CP, Sathrum AJ, Smieja JM. Electrocatalytic and homogeneous approaches to conversion of CO_2 to liquid fuels. *Chem Soc Rev* 2009;**38**(1):89–99.
64. Doherty MD, Grills DC, Muckerman JT, Polyansky DE, Fujita E. Toward more efficient photochemical CO_2 reduction: use of $scCO_2$ or photogenerated hydrides. *Coord Chem Rev* 2010;**254**(21–22):2472–82.
65. Takeda H, Ishitani O. Development of efficient photocatalytic systems for CO_2 reduction using mononuclear and multinuclear metal complexes based on mechanistic studies. *Coord Chem Rev* 2010;**254**(3–4):346–54.
66. Mikkelsen M, Jorgensen M, Krebs F. The teraton challenge. A review of fixation and transformation of carbon dioxide. *Energy Environ Sci* 2010;**3**(1):43–81.
67. Hawecker J, Lehn JM, Ziessel R. Photochemical and electrochemical reduction of carbon dioxide to carbon monoxide mediated by (2,2′-bipyridine)tricarbonylchlororhenium(I) and related complexes as homogeneous catalysts. *Helv Chim Acta* 1986;**69**(8):1990–2012.
68. Grills DC, Fujita E. New directions for the photocatalytic reduction of CO_2: supramolecular, $scCO_2$ or biphasic ionic liquid-$scCO_2$ systems. *J Phys Chem Lett* 2010;**1**(18):2709–18.
69. Grice KA. Carbon dioxide reduction with homogenous early transition metal complexes: opportunities and challenges for developing CO_2 catalysis. *Coord Chem Rev* 2017;**336**:78–95.
70. Elgrishi N, Chambers MB, Wang X, Fontecave M. Molecular polypyridine-based metal complexes as catalysts for the reduction of CO_2. *Chem Soc Rev* 2017;**46**:761–96.
71. Deronzier A, Moutet J. 9.10 Electrochemical reactions catalyzed by transition metal complexes. *Compr Coord Chem II* 2004;**9**:471–507.
72. Collin JP, Jouaiti A, Sauvage J. Electrocatalytic properties of $Ni(cyclam)^{2+}$ and $Ni_2(biscyclam)^{4+}$ with respect to CO_2 and H_2O reduction. *Inorg Chem* 1988;**27**:1986–90.

73. Froehlich JD, Kubiak CP. Homogeneous CO_2 reduction by Ni(cyclam) at a glassy carbon electrode. *Inorg Chem* 2012;**51**(7):3932–4.
74. Froehlich JD, Kubiak CP. The homogeneous reduction of CO_2 by $[Ni(cyclam)]^+$: increased catalytic rates with the addition of a CO scavenger. *J Am Chem Soc* 2015;**137**(10):3565–73.
75. Szalda DJ, Fujita E, Sanzenbacher R, Paulus H, Elias H. Syntheses and properties of nickel(I) and nickel(II) complexes of a series of macrocyclic N4 ligands: crystal structures of *C-RSSR*-$[Ni^{I}HTIM](ClO_4)$, *C-RSSR*-$[Ni^{II}HTIM](ClO_4)_2$, *C-RSSR*-$[Ni^{II}HTIM](ClO_4)_2$, and $[Ni^{II}TIM](ClO_4)_2$ (HTIM = 2,3,9,10-Tetramethyl-1,4,8,11-tetraazacyclotetradecane, TIM = 2,3,9,10-Tetramethyl-1,4,8,11-tetraazacyclotetradeca-1,3,8,10-tetraene). *Inorg Chem* 1994;**33**(25):5855–63.
76. Neri G, Aldous IM, Walsh JJ, Hardwick LJ, Cowan A. A highly active nickel electrocatalyst shows excellent selectivity for CO_2 reduction in acidic media. *Chem Sci* 2016;**7**:1521–6.
77. Chiericato Jr G, Arana CR, Casado C, Cuadrado I, Abruna H. Electrocatalytic reduction of carbon dioxide mediated by transition metal complexes with terdentate ligands derived from diacetylpyridine. *Inorg Chim Acta* 2000;**300–302**:32–42.
78. Thoi VS, Chang C. Nickel N-heterocyclic carbene-pyridine complexes that exhibit selectivity for electrocatalytic reduction of carbon dioxide over water. *Chem Commun* 2011;**47**:6578–80.
79. Thoi VS, Kornienko N, Margarit CG, Yang P, Chang C. Visible-light photoredox catalysis: selective reduction of carbon dioxide to carbon monoxide by a nickel N-heterocyclic carbene-isoquinoline complex. *J Am Chem Soc* 2013;**135**(38):14413–24.
80. Sheng M, Jiang N, Gustafson S, You B, Ess DH, Sun Y. A nickel complex with a biscarbene pincer-type ligand shows high electrocatalytic reduction of CO_2 over H_2O. *Dalton Trans* 2015;**44**:16247–50.
81. Mukhopadhyay TK, MacLean NL, Gan L, Ashley DC, Groy TL, Baik M-H, et al. Carbon dioxide promoted H^+ reduction using a bis(imino)pyridine manganese electrocatalyst. *Inorg Chem* 2015;**54**(9):4475–82.
82. Bourrez M, Molton F, Chardon-Noblat S, Deronzier A. $[Mn(bipyridyl)(CO)_3Br]$: an abundant metal carbonyl complex as efficient electrocatalyst for CO_2 reduction. *Angew Chem Int Ed* 2011;**50**(42):9903–6.
83. Riplinger C, Carter EA. Influence of weak Brønsted acids on electrocatalytic CO_2 reduction by manganese and rhenium bipyridine catalysts. *ACS Catal* 2015;**5**(2):900–8.
84. Grills DC, Farrington JA, Layne BH, Lymar SV, Mello BA, Preses JM, Wishart J. Mechanism of the formation of a Mn-based CO_2 reduction catalyst revealed by pulse radiolysis with time-resolved infrared detection. *J Am Chem Soc* 2014;**136**(15):5563–6.
85. Smieja JM, Sampson MD, Grice KA, Benson EE, Froehlich JD, Kubiak P. Manganese as a substitute for rhenium in CO_2 reduction catalysts: the importance of acids. *Inorg Chem* 2013;**52**(5):2484–91.
86. Franco F, Cometto C, Ferrero Vallana F, Sordello F, Priola E, Minero C, et al. A local proton source in a $[Mn(bpy-R)(CO)_3Br]$-type redox catalyst enables CO_2 reduction even in the absence of Brønsted acids. *Chem Commun* 2014;**50**(93):14670–3.
87. Sampson MD, Nguyen AD, Grice KA, Moore CE, Rheingold AL, Kubiak CP. Manganese catalysts with bulky bipyridine ligands for the electrocatalytic reduction of carbon dioxide: eliminating dimerization and altering catalysis. *J Am Chem Soc* 2014;**136**(14):5460–71.
88. Riplinger C, Sampson MD, Ritzmann AM, Kubiak CP, Carter EA. Mechanistic contrasts between manganese and rhenium bipyridine electrocatalysts for the reduction of carbon dioxide. *J Am Chem Soc* 2014;**136**(46):16285–98.
89. Takeda H, Koizumi H, Okamoto K, Ishitani O. Photocatalytic CO_2 reduction using a Mn complex as a catalyst. *Chem Commun* 2014;**50**(12):1491–3.
90. Bourrez M, Orio M, Molton F, Vezin H, Duboc C, Deronzier A, et al. Pulsed-EPR evidence of a manganese(II) hydroxycarbonyl intermediate in the electrocatalytic reduction of carbon dioxide by a manganese bipyridyl derivative. *Angew Chem Int Ed* 2014;**53**(1):240–3.
91. Agarwal J, Shaw TW, Stanton CJ, Majetich GF, Bocarsly AB, Schaefer HF. NHC-containing manganese(I) electrocatalysts for the two-electron reduction of CO_2. *Angew Chem Int Ed* 2014;**53**(20):5152–5.
92. Zeng Q, Tory J, Hartl F. Electrocatalytic reduction of carbon dioxide with a manganese(I) tricarbonyl complex containing a nonaromatic α-diimine ligand. *Organometallics* 2014;**33**(18):5002–8.
93. Walsh JJ, Smith CL, Neri G, Whitehead GFS, Robertson CM, Cowan A. Improving the efficiency of electrochemical CO_2 reduction using immobilized manganese complexes. *Faraday Discuss* 2015;**183**:147–60.
94. Grice K, Kubiak C. Recent studies of rhenium and manganese bipyridine carbonyl catalysts for the electrochemical reduction of CO_2. *Adv Inorg Chem* 2014;**66**:163–88.

95. Machan CW, Stanton CJ, Vandezande JE, Majetich GF, Schaefer HF, Kubiak C, et al. Electrocatalytic reduction of carbon dioxide by Mn(CN)(2,2′-bipyridine)(CO)3: CN coordination alters mechanism. *Inorg Chem* 2015;**54**(17):8849–56.
96. Lam YC, Nielsen RJ, Gray HB, Goddard WAA. Mn bipyrimidine catalyst predicted to reduce CO_2 at lower overpotential. *ACS Catal* 2015;**5**(4):2521–8.
97. Agarwal J, Shaw TW, Schaefer HF, Bocarsly AB. Design of a catalytic active site for electrochemical CO_2 reduction with Mn(I)-tricarbonyl species. *Inorg Chem* 2015;**54**(11):5285–94.
98. Sampson MD, Kubiak CP. Manganese electrocatalysts with bulky bipyridine ligands: utilizing Lewis acids to promote carbon dioxide reduction at low overpotentials. *J Am Chem Soc* 2016:3–6.
99. Rao GK, Pell W, Korobkov I, Richeson D. Electrocatalytic reduction of CO_2 using Mn complexes with unconventional coordination environments. *Chem Commun* 2016;**52**(51):8010–3.
100. Rawat KS, Mahata A, Choudhuri I, Pathak B. N-Heterocyclic carbene-based Mn electrocatalyst for two-electron CO_2 reduction over proton reduction. *J Phys Chem C* 2016;**120**(16):8821–31.
101. Stanton III CJ, Vandezande JE, Majetich GF, Schaefer III HF, Agarwal J. Mn-NHC electrocatalysts: increasing pi acidity lowers the reduction potential and increases the turnover frequency for CO_2 reduction. *Inorg Chem* 2016;**55**(19):9509–12.
102. Spall SJP, Keane T, Tory J, Cocker DC, Adams H, Fowler H, Meijer AJHM, Hartl F, Weinstein J. Manganese tricarbonyl complexes and asymmetric 2-iminopyridine ligands: toward decoupling steric and electronic factors in electrocatalytic CO_2 reduction. *Inorg Chem* 2016;**55**(24):12568–82.
103. Vollmer MV, Machan CW, Clark ML, Antholine WE, Agarwal J, Schaefer III HF, Kubiak CP, Walensky J. Synthesis, spectroscopy, and electrochemistry of (a-diimine)$M(CO)_3Br$, M = Mn, Re, complexes: ligands isoelectronic to bipyridyl show differences in CO_2 reduction. *Organometallics* 2015;**34**(1):3–12.
104. Hammouche M, Lexa D, Momenteau M, Saveant JM. Chemical catalysis of electrochemical reactions. Homogeneous catalysis of the electrochemical reduction of carbon dioxide by iron(0) porphyrins. Role of the addition of magnesium cations. *J Am Chem Soc* 1991;**113**(22):8455–66.
105. Bhugun I, Lexa D, Saveant J. Catalysis of the electrochemical reduction of carbon dioxide by iron(0) porphyrins: synergystic effect of weak Brønsted acids. *J Am Chem Soc* 1996;**118**:1769–76.
106. Hammouche M, Lexa D, Saveant J. Catalysis of the electrochemical reduction of carbon dioxide by iron(0) porphyrins. *J Electroanal Chem Interfacial Electrochem* 1988;**249**(1–2):347–51.
107. Costentin C, Drouet S, Robert M, Savéant J. A local proton source enhances CO_2 electroreduction to CO by a molecular Fe catalyst. *Science* 2012;**338**(6103):90–4.
108. Costentin C, Passard G, Robert M, Savéant J-M. Pendant acid-base groups in molecular catalysis: H-bond promoters or proton relays? Mechanisms of the conversion of CO_2 to CO by electrogenerated iron(0) porphyrins bearing prepositioned phenol functionalities. *J Am Chem Soc* 2014;**136**(0):11821–9.
109. Costentin C, Robert M, Savéant J-M, Tatin A. Efficient and selective molecular catalyst for the CO_2-to-CO electrochemical conversion in water. *Proc Natl Acad Sci* 2015;**112**(22):6882–6.
110. Bordwell FG, Ji G. Effects of structural changes on acidities and homolytic bond dissociation energies of the hydrogen-nitrogen bonds in amidines, carboxamides, and thiocarboxamides. *J Am Chem Soc* 1991;**113**(22):8398–401.
111. Bhugun I, Lexa D, Saveant J. Ultraefficient selective homogeneous catalysis of the electrochemical reduction of carbon dioxide by an iron(0) porphyrin associated with a weak Brønsted acid cocatalyst. *J Am Chem Soc* 1994;**116**(11):5015–6.
112. Bhugun I, Lexa D, Saveant J. Catalysis of the electrochemical reduction of carbon dioxide by iron(0) porphyrins. Synergistic effect of Lewis acid cations. *J Phys Chem* 1996;**100**(51):19981–5.
113. Costentin C, Drouet S, Robert M, Savéant JM. Turnover numbers, turnover frequencies, and overpotential in molecular catalysis of electrochemical reactions. Cyclic voltammetry and preparative-scale electrolysis. *J Am Chem Soc* 2012;**134**(27):11235–42.
114. Costentin C, Passard G, Robert M, Savéant J-M. Ultraefficient homogeneous catalyst for the CO_2–CO electrochemical conversion. *Proc Natl Acad Sci* 2014;**111**(42):14990–4.

CHAPTER

3.21

Principles of Photochemical Activation Toward Artificial Photosynthesis and Organic Transformations

Ken T. Ngo, Jonathan Rochford

University of Massachusetts Boston, Boston, MA, United States

3.21.1 INTRODUCTION

Minimizing the environmental impacts caused by human activities is one of the greatest challenges of the 21st century. Goals have been set to address the problems, and these goals have been driving the development of sustainable chemistry to reduce pollutants from reaching the environment, to diminish the amount of greenhouse gases that is already in the atmosphere, and to utilize the principle of conservation of energy for cost-effective products.[1,2] A major part of addressing this challenge is the development of catalysts for a variety of applications. Catalysts are typically divided into two main classes, that is, heterogeneous and homogeneous. Heterogeneous catalysts are known for their robustness and ease of recovery, but they often suffer from poor efficiency and selectivity. Homogeneous catalysts, in particular molecular transition metal catalysts, can be tailored to be highly efficient and selective but often lack the robustness of heterogeneous systems. Essential in the development of sustainable catalytic systems, whether they are heterogeneous or homogeneous in nature, is the utilization of solar energy to provide the requisite driving force to initiate catalysis and promote a desired chemical transformation. Photocatalysis utilizes light as a source of energy to drive chemical reactions ranging from complex organic transformations to small-molecule activation, such as water splitting and CO_2 reduction; topics that will be highlighted in this chapter as prototypical examples. From

Green Chemistry
http://dx.doi.org/10.1016/B978-0-12-809270-5.00026-1

the perspective of sustainable fuels production, it is highly desired to drive photocatalytic small-molecule activation chemistry using solar energy. This is motivated by the fact that more solar energy is incident on the earth's surface in 2 h (178,600 TW h) than is consumed globally in an entire year (157,768 TW h).[3] By taking advantage of the large quantity of "free" solar energy available, photoexcited state species can be readily generated to initiate the desired catalytic transformation. A major advantage of homogeneous molecular catalysts is that they offer a high level of tunability through tailored ligand design (inner and outer coordination sphere) as well as tunability of the redox properties of the central metal. Beyond providing for highly selective and efficient catalysis, homogeneous catalysts also play a critical role in developing a detailed mechanistic understanding behind many chemical transformations, which are difficult to probe experimentally via a heterogeneous pathway. Indeed nature has adapted a molecular approach to harvest solar energy and transform it into chemical energy by splitting water and reducing carbon dioxide. This work will outline some of the basic principles of photocatalysis. As a comprehensive overview of the photocatalysis literature is beyond the scope of this chapter, we will just highlight some prototypical examples. First and foremost will be examples of artificial photosynthesis, namely, photocatalytic CO_2 reduction and water oxidation. Finally, an introduction to the growing field of photocatalytic organic transformations will be presented.

3.21.2 SOLAR ENERGY DISTRIBUTION

An understanding of solar irradiance and its properties is fundamental for the field of photocatalysis. Solar irradiance is defined as the power per unit area received from the sun. Due to a blackbody radiation effect from the sun, the energy that reaches the earth is not distributed uniformly across the spectrum. Approximately, 5% of the power is in the ultraviolet (UV) region, 43% is in the visible region, and 52% is in the near-infrared region. Before reaching the earth's surface, solar radiation also interacts with a variety of atmospheric light-absorbing molecules such as O_3, H_2O, CO_2, and a wide array of volatile organic carbons. This gives rise to an unevenly distributed solar radiation curve as shown in Fig. 3.21.1. Despite the highly energized photons that the UV region provides, the solar spectrum has motivated the development of many new materials capable of light harvesting and solar energy conversion using primarily visible and near-infrared radiation. The unit of solar irradiance is watt per square meter (W/m^2). Solar fluence can also be expressed in suns where one sun is equivalent to 1000 W/m^2. Time can be integrated into this unit to give watt-hour per square meter (Wh/m^2) so 1 W/m^2 equals 24 $Wh/m^2/day$, and this is a common unit for solar irradiance used in the solar cell industry. The total power coming from the sun that hits the earth in 1 year is 3×10^{24} J per year, or enough energy to power the earth for 10,000 years at the current level of global energy consumption.[4]

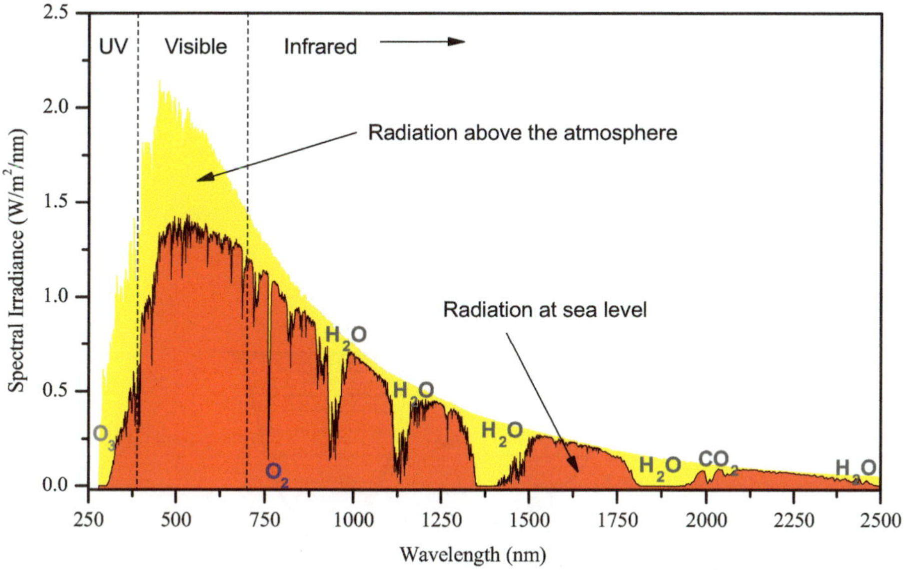

FIGURE 3.21.1 Solar radiation curve with most of the radiation energy distributed in the visible region. *Data was downloaded from the ASTM International homepage;* Standard tables for reference solar spectral irradiances: direct normal and hemispherical on 37 degree tilted surface. *ASTM International; 2012;* Standard solar constant and zero air mass solar spectral irradiance tables. *ASTM International; 2014.*

3.21.3 THE JABLONSKI DIAGRAM

Aleksander Jablonski (1898–1980), a Polish physicist, was born in Imperial Russia and later became a member of the Polish Academy of Sciences. He was the first to propose the concept of describing fluorescence schematically.[7] Any photochemical process, such as absorption and emission, can be described by the Jablonski diagram (Fig. 3.21.2).[8]

The Jablonski diagram describes the possible events that occur after absorbance of a photon by a molecule. Singlet (S) and triplet (T) electronic energy states are depicted as parabolas, while vibrational energy levels are shown as horizontal lines. The first event in the Jablonski diagram is the absorption of a photon with a resonant energy of $h\nu$ (where h is Planck's constant and ν is the frequency of light) to excite one electron from the electronic ground state S_0 to an excited state S_n ($n \geq 1$). This event is restricted to a vertical transition in accordance with the Born-Oppenheimer approximation. Based on the Boltzmann distribution of electrons in the ground state, the electron almost always starts from the lowest vibrational energy level from S_0, but it does not necessarily end up at the lowest vibrational energy level ν_0 of S_n. In fact, the final state of the excited electron depends on the degree of overlap of the two vibrational wave functions, that is, the Frank-Condon factor, describing the initial state and the final state. Typically, the initial excited state is not ν_0 of S_n, and this short-lived state is called the Frank-Condon state. This quickly relaxes down to ν_0 of S_n through internal conversion (ic) and/or vibrational relaxation (vr), releasing vibrational

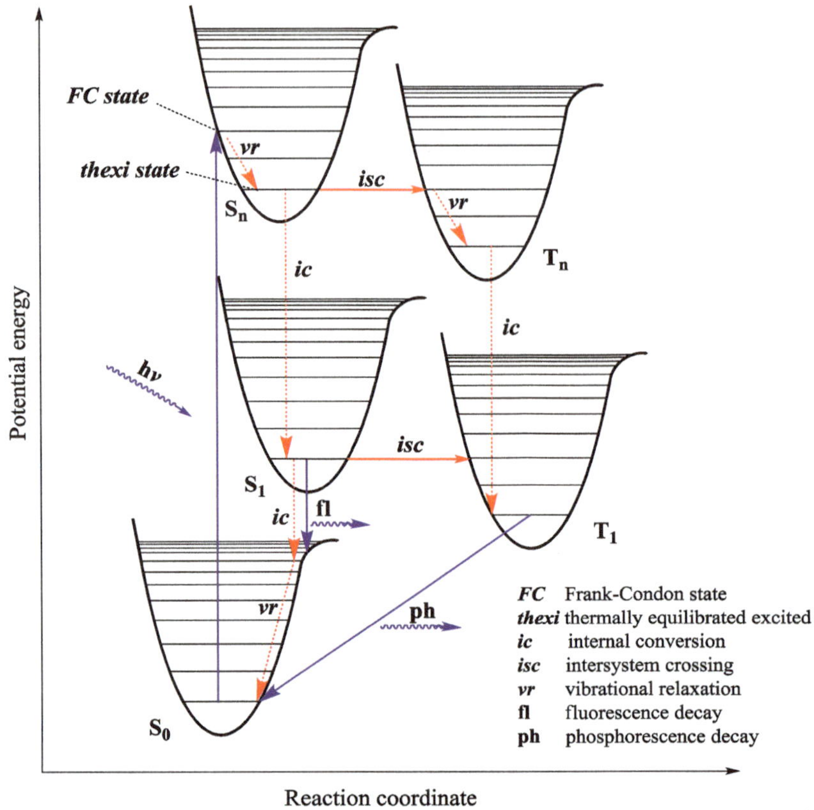

FIGURE 3.21.2 The Jablonski diagram shows the possible processes after the initial excitation event by a photon.

energy as heat to the surrounding solvent. This is now at the thermally equilibrated excited (thexi) state. Internal conversion is dependent on resonant mixing of vibrational wave functions between two closely lying electronic states. This occurs rapidly (attoseconds to femtoseconds) from S_n ($n \geq 2$) excited states until the S_1v_0 level is reached (a.k.a. Kasha's rule). A number of processes can occur from the S_1v_0 level including additional ic and vr, intersystem crossing (isc), fluorescence decay (fl), and/or phosphorescence decay (ph). Both ic and vr release the excited state energy as heat, whereas fluorescence and phosphorescence release the excited state energy in the emission of a photon. Similar to internal conversion, intersystem crossing also depends on resonance of two vibrational wave functions between two electronic states; however, for intersystem crossing this is a spin-forbidden lateral transition (enhanced by spin-orbit coupling via heavy atom relativistic effects) resulting in a change of multiplicity. Fluorescence, typically $S_1 \rightarrow S_0$, is spin-allowed and typically occurs on the nanosecond timescale. In contrast, phosphorescence, for example, $T_1 \rightarrow T_0$, is spin-forbidden and as such triplet states are longer lived, ranging from nanoseconds to microseconds.

3.21.4 PRINCIPLES OF PHOTOCHEMICAL ACTIVATION

Homogeneous photocatalysis is typically conducted at ambient atmospheric pressure and room temperature. The most basic of photocatalytic systems consists of a light-absorbing photosensitizer (PS), a sacrificial electron donor (SED) or sacrificial electron acceptor (SEA), a catalyst (Cat), and the substrate (Sub). Although photocatalysis is initiated on photon absorption, the subsequent redox chemistry that pursues in the "dark" reactions requires an in-depth knowledge of the constituent redox properties of all components of the system. A general schematic is provided in Scheme 3.21.1 to illustrate how a PS may catalyze oxidative or reductive chemical transformations. Choice of the PS, Cat, Sub, and SEA or SED, as well as their concentrations, will dictate the relevant rates of electron transfer and the particular mechanism that is followed. For example, in photooxidative catalysis the excited state PS (PS*) may undergo oxidation by the SEA if $k_{SEA} > k_{ox}$ (oxidative quenching), thus generating the oxidized PS (PS^+), which is subsequently regenerated via the catalyst activation step (Scheme 3.21.1(i)). However, should $k_{ox} > k_{SEA}$ hold true, electron transfer from Cat to PS* (reductive quenching) is kinetically favored generating PS^- as a reactive intermediate. In either case the same catalytic cycle is followed and it is merely the order of electron transfer events and the nature of the PS^+ or PS^- intermediate that varies albeit ideally inconsequential to product formation. A similar nature of events and kinetic competition can be described for photoreductive catalysis as illustrated in Scheme 3.21.1.

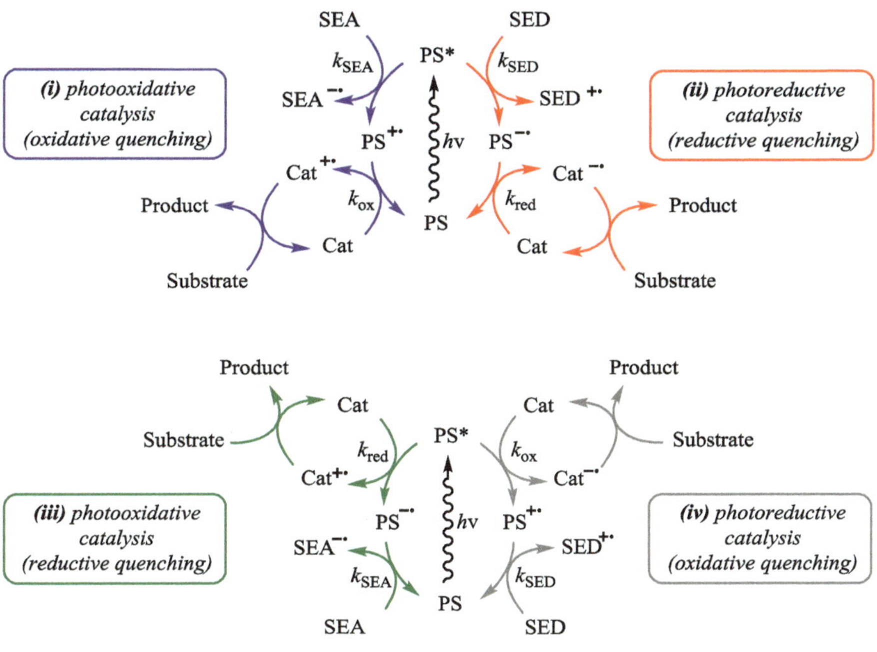

SCHEME 3.21.1 Generic schemes illustrating how a photosensitizer may catalyze oxidative or reductive chemical transformations through judicious choice of the catalyst (Cat), substrate, and sacrificial electron acceptor (SEA) *or* sacrificial electron donor (SED). Whether oxidative or reductive quenching pathways are followed, (i) versus (iii) *or* (ii) versus (iv), is dictated by competition between electron transfer rates (k).

Of course, Scheme 3.21.1 is simplified significantly if a single species can carry out the combined tasks of the PS and catalyst—such species are indeed true photocatalysts. This modification is often achieved through the synthesis of supramolecular systems where a typically independent PS is covalently bound to the required catalyst. This option removes the mass transfer diffusion dependence of k_{ox} and k_{red}, minimizing the concentration of PS required. For a more detailed discussion of photoactivated catalysis the reader is directed to a number of comprehensive reviews outlining molecular approaches to organic photoredox catalysis[9–11] and artificial photosynthesis.[12,13]

The efficiency of the oxidative quenching process to generate PS^+ (Scheme 3.21.1(i)) depends on the relative energy separation between the reduction potential of the SEA, $E(SEA/SEA^-)$, and the excited state oxidation potential of the PS, $E(PS^+/PS^*)$. For the SEA to accept one electron from PS* the SEA/SEA^- redox couple must be more positive than the PS^+/PS^* redox couple to provide an exergonic driving force. The PS excited state oxidation potential PS^+/PS^* can be calculated using the ground state oxidation potential (PS^+/PS) and the zero-zero energy of the electronic excited state $E_{0\text{-}0}$ as shown in Eq. (3.21.1).

$$E(PS^+/PS^*) = E(PS^+/PS) - E_{0\text{-}0} \tag{3.21.1}$$

The energy $E_{0\text{-}0}$ may correspond to the optical band-gap $S_1\nu_0 \rightarrow S_0\nu_0$ for a singlet excited state, or $T_1\nu_0 \rightarrow S_0\nu_0$ for a triplet excited state (typical of transition metal-based PSs). The PS excited state reduction potential, $E(PS^*/PS^-)$, can be calculated in a similar manner using the PS ground state reduction potential $E(PS/PS^-)$ according to Eq. (3.21.2).

$$E(PS^*/PS^-) = E(PS/PS^-) + E_{0\text{-}0} \tag{3.21.2}$$

Similarly, the reduction potential of the PS excited state (PS^*/PS^-) needs to be lower in energy than the oxidation potential of the SED (SED^+/SED) in the reductive quenching pathway (Scheme 3.21.1(ii)). A similar argument can be made for Scheme 3.21.1(iii) and (iv) where Eqs. (3.21.1 and 3.21.2) equally apply with respect to the catalyst oxidation (Cat^+/Cat) and reduction (Cat/Cat^-) redox couples. The quenching rate constant (k_q) for electron transfer is typically determined by the Stern-Volmer analysis where a ratio of the steady state emission intensity (I) or lifetime (τ) of the PS with incrementing quencher (SED, SEA, or Cat) concentration is determined with respect to the emission intensity or lifetime in the absence of quencher (I_0 or τ_0) according to Eq. (3.21.3).

$$\frac{I_0}{I} = \frac{\tau_0}{\tau} = 1 + K_{SV}[Q] \tag{3.21.3}$$

where Q is the quenching agent (SEA, SED, or Cat), K_{SV} is the Stern-Volmer constant that is equal to k_q, and τ_0 and k_q can be correlated to k_{red}, k_{ox}, or k_{Cat} in Scheme 3.21.1 depending on the pathway followed. Typically, an efficient PS is chosen based on the following photophysical and electrochemical properties:

1. Excellent light-harvesting properties; the PS should ideally exhibit strong absorption across the visible and near-infrared regions of the solar spectrum.
2. Competitive absorption by other components of the photocatalytic system (SEA, SED, Cat, substrate, product) should be avoided.

3. The PS* excited state lifetime should be long-lived enough to ensure an efficient oxidative or reductive quenching process to follow.
4. The PS should display excellent redox stability to allow for large catalytic turnover numbers (TONs).

Transition metal complexes are capable of meeting each of these requirements because of their redox stability, high molar extinction coefficients, and the long excited state lifetimes typically exhibited by triplet metal-to-ligand charge-transfer (^{3}MLCT) excited states, at least for second- and third-row complexes. The two most popular transition metal-based PSs for application in homogeneous photocatalysis are $[Ru(bpy)_3]^{2+}$ and $Ir(ppy)_3$, where bpy is 2,2′-bipyridine and ppy is 2-phenylpyridine. A summary of their photophysical and electrochemical properties is provided in Scheme 3.21.2 where λ_{max} is the wavelength of the lowest

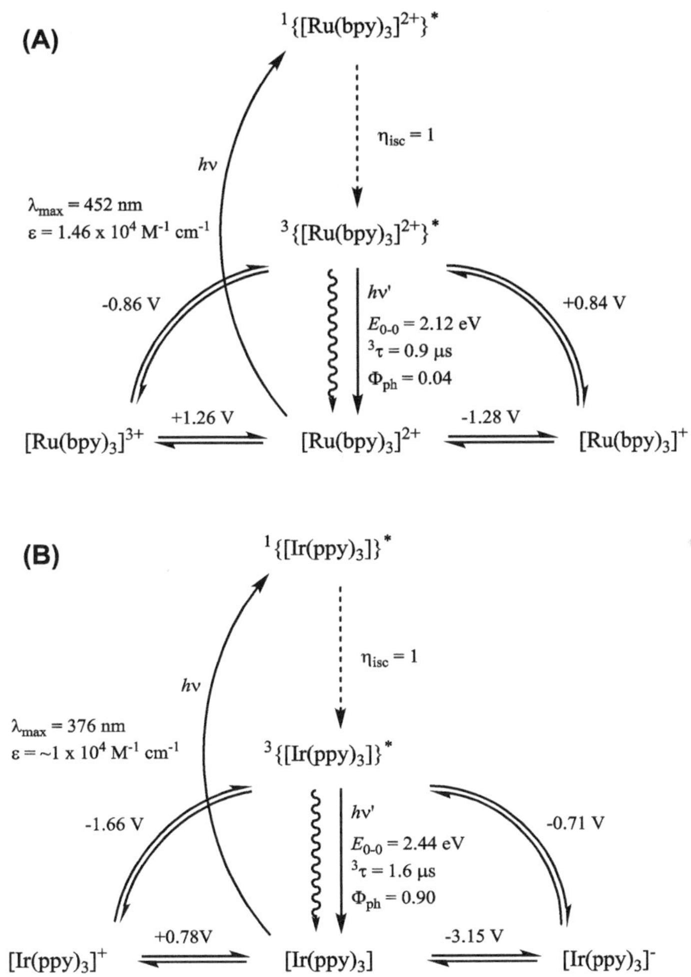

SCHEME 3.21.2 A summary of photophysical and electrochemical (vs. SCE) properties for both ground and excited states of (A) $[Ru(bpy)_3]^{2+}$ and (B) $Ir(ppy)_3$.

energy maximum absorption band, ε is the molar extinction coefficient at λ_{max}, η_{isc} is the intersystem conversion efficiency between the singlet and triplet excited states (maximum is 1), ${}^3\tau$ is the triplet lifetime, and Φ_{ph} is the phosphorescence emission quantum yield.

Of course, selection of the SEA or SED also has a big impact on the outcome of the electron transfer events and overall quantum efficiency in a photocatalytic reaction as quantified by the Stern-Vollmer constant for a particular set of experimental conditions. Back electron transfer events are of course undesirable; however, the sacrificial nature of the SEA or SED species ultimately results in the accumulation of SEA^- or SED^+ that are often radical in nature and prone to H atom side reactions, which may result in PS or Cat decomposition and reduced photocatalytic TONs. For this reason, there is a growing trend to replace the homogeneous SEA or SED species with a heterogeneous electrode surface in a charge-balanced photovoltaic cell.[4]

3.21.5 EVALUATING THE EFFICIENCY OF A PHOTOCATALYTIC SYSTEM

While there are typically many components and independent steps in any photocatalytic system, quantitative analysis is often evaluated by three common terms each of which depend on the product selectivity, that is, quantum yield of product formation (Φ), TON, and turnover frequency (TOF). Product selectivity itself is a useful metric for reporting photocatalytic efficiency and is determined by calculating the ratio of desired product with respect to the total molar quantity of all reaction products (Eq. 3.21.4).[14]

$$\text{Product selectivity} = \frac{\text{Moles of desired product formed}}{\text{Total moles of all products formed}} \tag{3.21.4}$$

The quantum yield of a photocatalytic reaction (Φ) is evaluated using Eq. (3.21.5) where a measure of the molar fraction between desired products and incident photons is required.

$$\Phi = \frac{\text{Moles of desired product formed}}{\text{Moles of incident photons}} \tag{3.21.5}$$

It is important to appreciate here that the maximum Φ for a one-electron catalytic process is 1, as photoexcitation leads to one-electron transfer event and eventually one molecule of product at a time. However, if the catalytic cycle requires n electrons where $n > 1$ then the maximum Φ for that process is $1/n$. The TON carries information on the stability of a photocatalytic system and is equivalent to the total number of moles of product produced, irrespective of reaction time, per mole of catalyst (Eq. 3.21.6). A stable photocatalytic system should have a very high TON. In practice, catalysts used for industrial applications usually have TON in the range of 1 million. It is however quite common that catalysts in their developing stages may exhibit lower TONs in the range 1–500.[15,16]

$$\text{TON} = \frac{[\text{Desired product}]}{[\text{Catalyst}]} \tag{3.21.6}$$

TON should be recorded after the activity of Cat is completely ceased; however, it is quite common practice to compare TONs for various catalysts after a fixed period before reaction completion. Finally, TOF refers to the rate of a photocatalytic cycle. TOF is determined by the ratio of moles of desired product produced per mole of catalyst per unit time (Eq. 3.21.7, equivalent to TON per unit time). TOF is typically reported in units of inverse hours (h^{-1}); however, catalysts with low TONs are commonly reported in units of inverse minutes or even inverse seconds because of short reaction times.

$$\mathrm{TOF} = \frac{\mathrm{TON}}{\text{Reaction time}} \tag{3.21.7}$$

3.21.6 EXAMPLES OF PHOTOCATALYTIC SYSTEMS

Inspired by natural photosynthesis, researchers have developed artificial photosynthetic systems with an ultimate goal to harvest and store solar energy in the form of chemical bonds, that is, solar-to-fuel conversion. One famous early example reported by Fujishima and Honda[17] in 1972 was the photocatalytic water-splitting activity of TiO_2, which demonstrated the huge potential of solar energy conversion. Since then many research programs have developed a better understanding of different semiconductor systems as well as a variety of molecular catalysts and nanomaterials to increase reaction efficiencies and catalyst stabilities, although significant progress is still needed.[18] Inspired by natural photosynthesis, the fundamental aspects of artificial photosynthesis related to photocatalytic CO_2 reduction and photocatalytic H_2O oxidation will be emphasized here. Finally, the growing field of photocatalytic organic synthesis will also be discussed.

3.21.6.1 Principles of Artificial Photosynthesis: Photocatalytic CO_2 Reduction and H_2O Oxidation

The reduction of CO_2 by one electron to form $CO_2^{\bullet-}$ has a formal reduction potential of −1.99 V versus standard hydrogen electrode.[19] This reduction potential can be minimized through proton-assisted multiple electron transfer pathways as shown in Table 3.21.1, where the reduction of CO_2 to useful chemical feedstock is still an energetically unfavorable process. It is challenging to advance catalysts to efficiently drive these proton-coupled multielectron transformation reactions. Another challenge is the development of a photocatalytic system capable of multielectron multiproton coupled redox events driven by single-photon single-electron transfer events.

The simplest, and most successful, cases where only two electrons are needed to reduce CO_2 to isolable products are those of photocatalytic carbon monoxide (CO) and formic acid (HCO_2H) production. The first example was reported by Lehn and coworkers in 1982.[20] In this study, $[Ru(bpy)_3]^{2+}$ was used as a PS, $CoCl_2$ as the CO_2-coordinating catalyst, and triethanolamine (TEOA) as an SED in aqueous solution. A simplified photocatalytic scheme is illustrated in Scheme 3.21.3.[20]

TABLE 3.21.1 Standard Reduction Potentials for CO_2 and H^+ Reduction[19]

Reaction	E^o (V) Versus SHE[a]
$CO_{2(g)} + e^- \rightleftharpoons CO_2^{\bullet -}{}_{(aq)}$	−1.99
$CO_{2(g)} + 2H^+{}_{(aq)} + 2e^- \rightleftharpoons HCO_2H_{(aq)}$	−0.61
$CO_{2(g)} + 2H^+{}_{(aq)} + 2e^- \rightleftharpoons CO_{(g)} + H_2O_{(l)}$	−0.52
$3CO_{2(g)} + H_2O_{(l)} + 2e^- \rightleftharpoons CO_{(g)} + 2HCO_3^-{}_{(aq)}$	−0.56
$CO_{2(g)} + 4H^+{}_{(aq)} + 4e^- \rightleftharpoons HC(OH)_{2(aq)} + H_2O_{(l)}$	−0.49
$CO_{2(g)} + 6H^+{}_{(aq)} + 6e^- \rightleftharpoons CH_3OH_{(aq)} + H_2O_{(l)}$	−0.38
$CO_{2(g)} + 8H^+{}_{(aq)} + 8e^- \rightleftharpoons CH_{4(g)} + 2H_2O_{(l)}$	−0.24

SHE, standard hydrogen electrode.
[a]*E^o potentials are reported at 1 atm CO_2; 25 °C; pH = 7.*

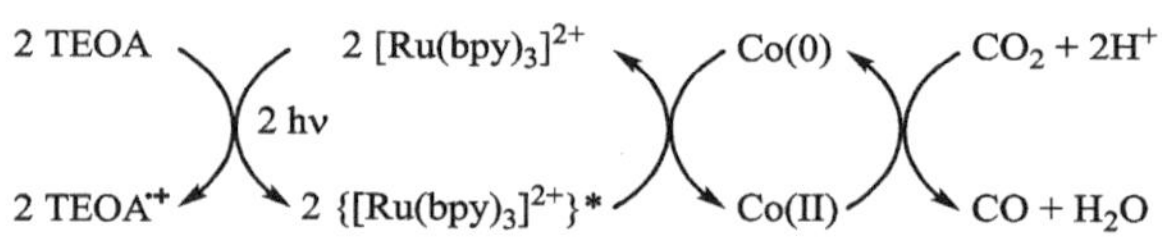

SCHEME 3.21.3 Simplified photocatalytic scheme of CO_2 to CO transformation using a $[Ru(bpy)_3]^{2+}$ molecular photosensitizer and triethanolamine (TEOA) sacrificial electron donor and $CoCl_2$ catalyst.[20]

During the same period, pioneering work in heterogeneous photocatalytic CO_2 reduction was also reported for semiconductor catalysts. Hemminger et al.[21] reported the photocatalytic reduction of CO_2 and gaseous H_2O to methane on the surface of $SrTiO_3$. Halmann[22] has shown that a *p*-type GaP photocathode could reduce CO_2 to HCO_2H in the presence of water with a fairly high conversion efficiency in a photoelectrochemical cell. Honda and coworkers[23] had demonstrated both photocatalytic and photoelectrocatalytic reduction of CO_2 in aqueous suspensions of a number of semiconductors including WO_3, TiO_2, ZnO, CdS, GaP, and SiC. These heterogeneous photocatalytic reactions are thought to occur according to Eq. (3.21.8).

$$Cat + h\nu \rightarrow Cat^*(e^-{}_{cond} + p^+{}_{val}) \quad (3.21.8)$$

where $e^-{}_{cond}$ is a conduction band electron and $p^+{}_{val}$ is a positive hole in the valance band of the semiconducting catalyst, with Cat* denoting the electronically excited state of the semiconductor. Upon exciton formation, both conduction band and valence band levels carry the reduction and oxidation potentials, respectively, where both half reactions are described by Eqs. (3.21.9 and 3.21.10):

$$H_2O + 2p^+{}_{val} \rightarrow \tfrac{1}{2}O_2 + 2H^+ \quad (3.21.9)$$

$$CO_{2(aq)} + 2H^+ + 2e^-_{cond} \rightarrow HCOOH \quad (3.21.10)$$

For a homogeneous molecular approach to photocatalytic CO_2 reduction, a large body of literature has focused on the *fac*-Re(CO)$_3$(NN)X class of photocatalysts (exhibiting both PS + Cat activity) where NN is a polypyridyl ligand, for example, 2,2′-bipyridine (bpy). More elaborate supramolecular complexes, metal tetraaza macrocyclic compounds with nickel and cobalt,[24–31] and metalloporphyrins and related metallomacrocycles[32–38] have also garnered much attention. These systems can favorably bind to CO_2 and in the presence of available protons are capable of maintaining a high selectivity for CO and/or HCO_2H formation, despite the potential for competitive hydrogen evolution via H^+ reduction. In particular, the pioneering work of Lehn and coworkers on photocatalytic CO_2 reduction using *fac*-Re(bpy)(CO)$_3$X (X = Cl^- or Br^-) has attracted a tremendous amount of attention.[39] Ishitani and coworkers[40] have developed impressive photocatalytic systems utilizing an independent PS in a two-component (PS + Cat) or dyad supramolecular (PS – Cat) photocatalytic system inspired by Lehn's original work, as shown in Fig. 3.21.3.

Both the two-component system and supramolecular photocatalyst reported by Ishitani and coworkers include a PS that can reductively transfer one electron at a time to the catalyst where two sequential one-electron transfer events result in the multielectron reduction required for CO formation. The two-component system presented in Fig. 3.21.3 consists of a trinuclear rhenium(I)-based PS *fac*-{Re[4,4′-(MeO)$_2$bpy](CO)$_3$[P(OEt)$_3$]}$^+$ initiating the photochemical electron transfer by its stability and strong reducing power, which ultimately generates the two-electron reduced five-coordinate active catalyst *fac*-[Re(bpy)(CO)$_3$]$^-$. However, the two-component system still relies on mass transfer diffusion and the probability of heteronuclear collision to initiate electron transfer, which ultimately restricts the catalytic TOF.[40] To avoid this rate-determining diffusion, the supramolecular photocatalyst model was developed with the PS and the catalyst being linked together by a hydrocarbon chain allowing the electron transfer step to take place intramolecularly. In this example, [Ru(dmbpy)$_3$]$^{2+}$ was used as the PS for its ability to absorb a wide range of wavelengths in the visible spectrum, and *cis,trans*-{Re(NN)(CO)$_2$[P(*p*-F-Ph)$_3$]$_2$}$^+$ was employed as the

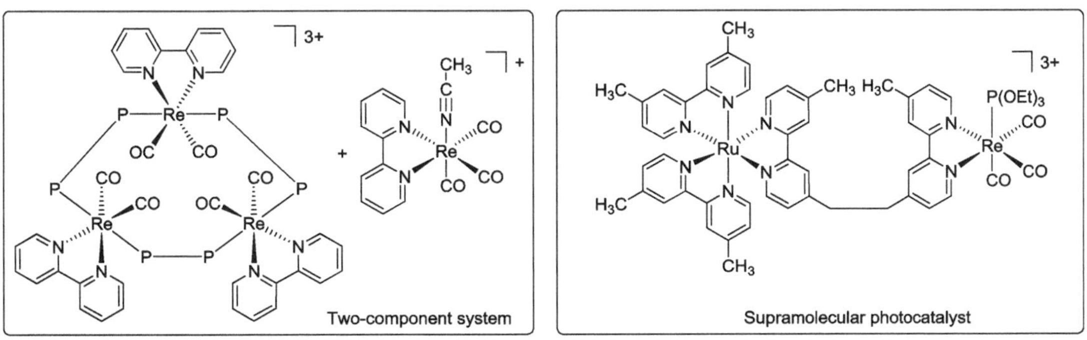

FIGURE 3.21.3 Examples of (A) a two-component (PS + Cat) system and (B) a supramolecular (PS – Cat) system for photocatalytic CO_2 reduction.[40]

catalyst with two peripheral P(*p*-F-Ph)$_3$ ligands that help increase the photocatalytic activity.[41] Different lengths of the bridging ligand were studied leading to the conclusion that a two carbon bridge is the most effective in this case. In these examples, 1-benzyl-1,4-dihydronicotinamide (BNAH) and 1,3-dimethyl-2-phenyl-2,3-dihydro-1*H*-benzo[*d*]imidazole (BIH) were chosen as SEDs to quench the ^{3}MLCT excited state of the PS. The use of TEOA was also required as a base to abstract a proton from BNAH or BIH subsequent to electron transfer to suppress the back electron transfer pathway.[42] The oxidation process of these SEDs is shown in Scheme 3.21.4.[43]

Supramolecular photocatalysts are more advantageous than the traditional two-component system; however, they often still rely on sacrificial reductants to regenerate the PS. In an effort to address this issue a heterogeneous system was developed where the supramolecular catalyst design and semiconductor were combined to make a hybrid Z-scheme-type photocatalyst (Scheme 3.21.5).[44] In this example the electron donor selected was methanol, which was no longer sacrificial in nature, as upon oxidation, by the photoexcited nitrogen-doped tantalum oxide (TaON) semiconductor valence band, formaldehyde was produced as a desirable product.

The use of SEDs has helped simplify such photocatalytic systems to focus on elucidation of the mechanism of the catalytic cycle and the properties of the PS. However, this also limits such reactions, as the TON relies on a bulk concentration of SED for practical applications (although it is often SED$^+$ side reactions that prevail). To overcome this problem, an abundant source of electron donor such as water needs to be utilized coupling CO_2 reduction with H_2O oxidation, truly mimicking natural photosynthesis. Similar to CO_2 reduction, the water-splitting reaction is also an uphill process that has a standard Gibbs free energy of 238 kJ/mol (Eq. 3.21.11).

$$2H_2O \rightarrow 2H_2 + O_2 \quad \Delta G^o = 238 \text{ kJ/mol} \tag{3.21.11}$$

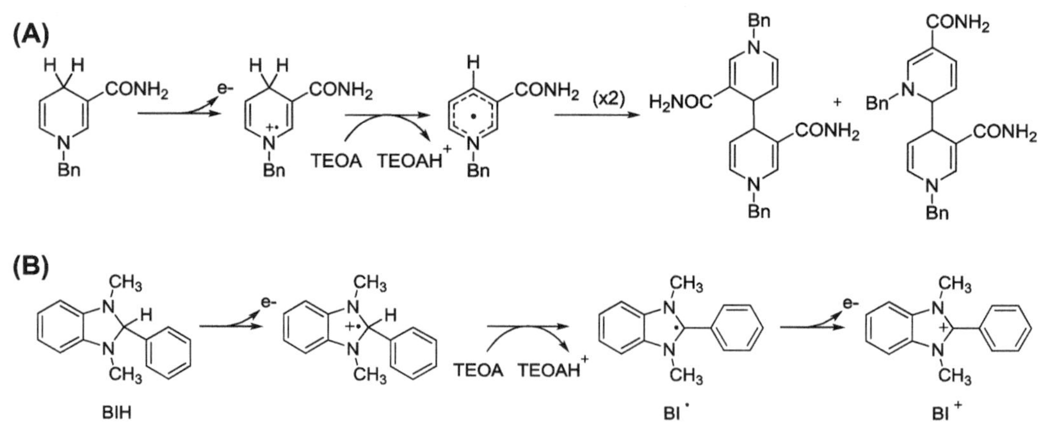

SCHEME 3.21.4 Oxidation and possible dimerization process of the common sacrificial electron donors (A) 1-benzyl-1,4-dihydronicotineamide (BNAH) and (B) 1,3-dimethyl-2-phenyl-2,3-dihydro-1*H*-benzo[*d*]imidazole (BIH).[43]

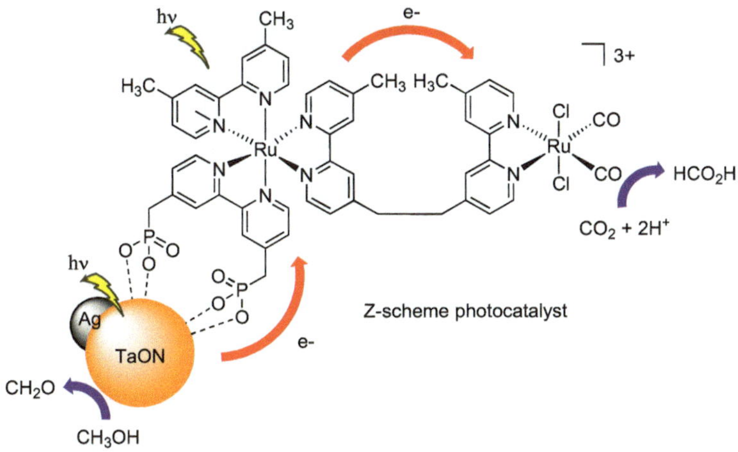

SCHEME 3.21.5 Artificial photosynthetic Z-scheme for photocatalytic CO_2 reduction that utilizes both TaON/Ag and $[Ru(dmbpy)_3]^{2+}$ (where dmbpy is 4,4′-dimethyl-2,2′-bipyridine) as photosensitizers and $[Ru(dmbpy)(CO)_2Cl_2]^{2+}$ as the catalyst.[44]

The full water-splitting reaction can be described by two half reactions from an electrochemical perspective (Eqs. 3.21.12 and 3.21.13).

$$2H^+ + 2e^- \rightarrow H_2 \quad E^\circ = 0.0 \text{ V versus SHE (pH} = 0) \qquad (3.21.12)$$

$$2H_2O + 4\text{ hole}^+ \rightarrow O_2 + 4H^+ \quad E^\circ = +1.23 \text{ V versus SHE (pH} = 0) \qquad (3.21.13)$$

The first homogeneous catalyst reported for water oxidation by Meyer and coworkers[45] is famously referred to as the "Blue Dimer." Homogeneous ruthenium-based transition metal complexes have been the primary focus for the development of many electrocatalytic water oxidation catalysts since catalytic activity of the blue dimer was first established[46–53]; however, other transition metals such Mn, Ir, Fe, and Co[54] metal centers have also contributed greatly to the overall mechanistic understanding of water oxidation. A notable paradigm shift in homogeneous water oxidation catalyst was reported by Concepcion et al. using $[Ru^{II}(tpy)(bpm)(H_2O)]^{2+}$ (tpy is 2,2′:6′,2″-terpyridine and bpm is 2,2′-bipyrimidine),[55] indicating that a dimeric or higher-ordered structure is not strictly required and one metal center is sufficient for catalytic water oxidation (Fig. 3.21.4). This catalyst has since inspired other monomer designs for successful water oxidation catalyst via electrocatalysis,[56] surface-bound heterogeneous electrocatalysis,[57] and photoelectrocatalysis.[58]

As highlighted earlier, there is increasing interest and motivation to replace the sacrificial redox species so often employed in photocatalysis systems with a heterogeneous electrode surface in a charge-balanced photovoltaic cell. One prominent example was reported by Meyer and coworkers[59] in which a dye-sensitized photoelectrosynthesis cell was engineered based on the well-established dye-sensitized solar cell device structure. The surface of the TiO_2 semiconductor oxide was modified with a molecular PS, thus establishing a photoanode capable of mediating electron and hole transfer that promotes the photocatalytic oxidation of H_2O, evolving oxygen with concomitant proton reduction and H_2 evolution occurring at the cathode (Fig. 3.21.5).

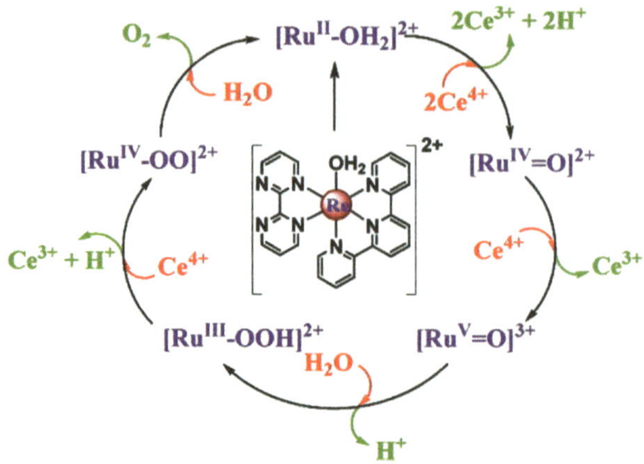

FIGURE 3.21.4 Proposed mechanism for water oxidation by $[Ru^{II}(tpy)(bpm)(H_2O)]^{2+}$ under aqueous acidic condition, with Ce(IV/II) as a sacrificial chemical oxidant. *Reprinted with permission from Concepcion JJ, Jurss JW, Templeton JL, Meyer TJ. One site is enough. Catalytic water oxidation by $[Ru(tpy)(bpm)(OH_2)]^{2+}$ and $[Ru(tpy)(bpz)(OH_2)]^{2+}$.* J Am Chem Soc *2008;**130**:16462–3.*

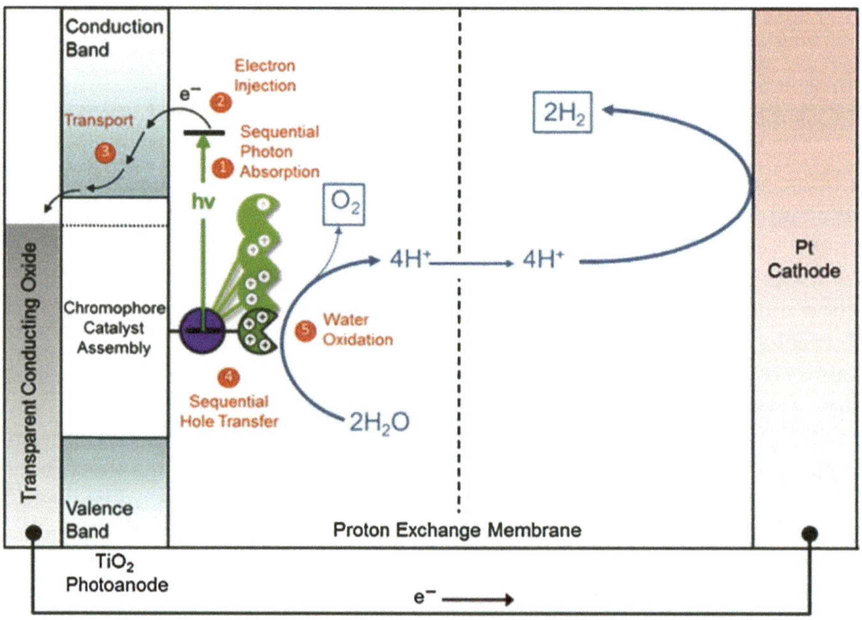

FIGURE 3.21.5 A dye-sensitized photoelectrosynthesis cell engineered for photocatalytic water, evolving hydrogen and oxygen without the need of any sacrificial redox species. *Reprinted with permission from Ashford DL, Gish MK, Vannucci AK, Brennaman MK, Templeton JL, Papanikolas JM, et al. Molecular chromophore–catalyst assemblies for solar fuel applications.* Chem Rev *2015;**115**:13006–49.*

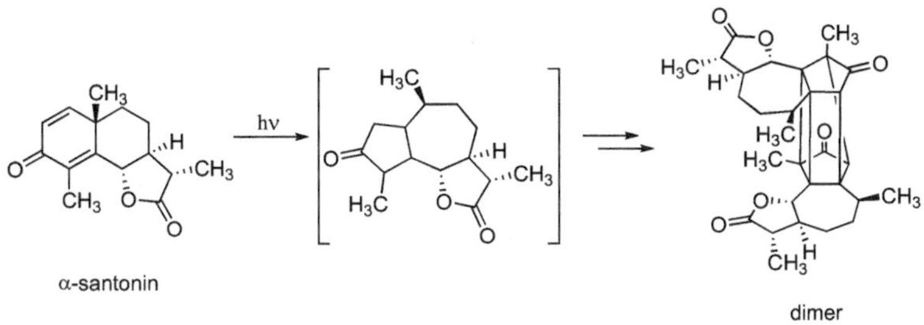

FIGURE 3.21.6 Colorless crystal of α-santonin turns into a yellow dimer upon exposure to sunlight.[60]

3.21.6.2 Photocatalysis in Organic Synthesis

Harnessing solar energy to perform chemical transformations in a controlled fashion has long been a goal that involves an interdisciplinary effort. Just as nature has been using solar energy to transform CO_2 and H_2O into organic compounds, it is possible for humans to utilize a similar approach for a large scope of chemical transformations in organic synthesis. One of the earliest examples of this was shown by Trommsdorff[60] in 1834 when upon exposure to sunlight the color of an anthelminthic α-santonin drug was changed drastically from a colorless crystal to yellow (Fig. 3.21.6).

Curiosity and effort have driven the field to gain a better understanding about this transformation and other related photochemical reactions. Throughout the years, a substantial wealth of knowledge has been developed garnering a broad understanding of the mechanism for many photochemical processes that sequentially lead to notable applications in photoassisted organic synthesis.[8,61–64] One of the principal challenges in this field is the weak absorption of visible light by small or poorly conjugated organic molecules, which tend to be colorless and only absorb in the UV region of the solar spectrum. This has limited the early application of photoassisted organic synthesis to UV light, which carries only 5% of the energy of the solar spectrum, and UV excitation also promotes bond cleavage, radical formation, and unwanted side products. One way to avoid the use of UV light is to increase the degree of conjugation in an organic molecule, which can increase its absorbance in the visible region. This, however, severely limits the scope of a photocatalytic approach to organic synthesis. An alternative method is to use a photoredox catalyst or a PS that can absorb solar energy from the desirable region of visible light and transfer the absorbed energy to a less photoreactive reactant toward a desired chemical transformation. The latter approach (as outlined in Scheme 3.21.1) is a rapidly growing area of green chemistry and deserves ample discussion in this chapter.

Photoredox catalysts for organic synthesis have also been receiving extended attention in the past four decades.[65] The first example of the use of photoredox catalyst for modern organic synthesis was reported in 1978 by Kellogg,[66,67] in which $[Ru(bpy)_3]Cl_2$ was used to accelerate the transformation of sulfonium salt and dihydropyridine into the corresponding ketone (Fig. 3.21.7). In 1984, Cano-Yelo and Deronzier[68] showed that $[Ru(bpy)_3]Cl_2$ could effectively convert the tetrafluoroborate salt of the stilbenediazonium ion into the corresponding phenanthrene (the Pschorr reaction) by visible light. This photoredox catalyst was also explored by two independent groups, Tanaka and Sakurai groups, in the late

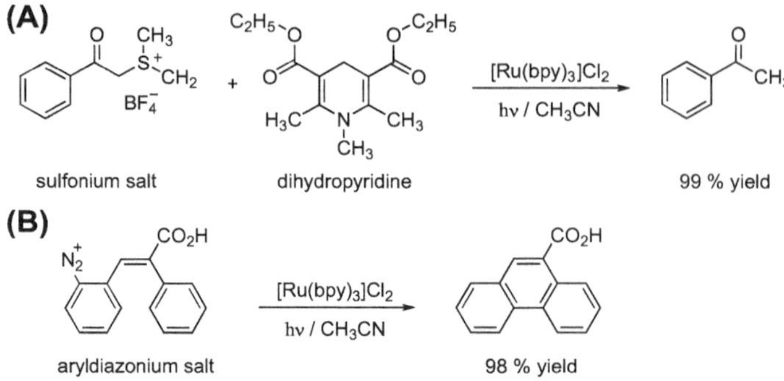

FIGURE 3.21.7 The first few examples of photoassisted organic synthesis in modern chemistry. (A) Reductive desulfuration by Kellogg and coworkers[66] and (B) photocatalytic Pschorr's reaction by Deronzier and coworkers.[68]

1980s and early 1990s for a number of reactions including hydride transfer from NADH to *p*-benzoquinone, 1-benzyl-1,4-dihydronicotinamide with benzyl bromide, electron-deficient olefins, aromatic ketones, and reduction of phenacyl halides.[69–74]

Upon visible light absorption, the ^{3}MLCT excited state of $[Ru(bpy)_3]Cl_2$ generates a unique environment that provides for both a strong oxidant, $E(PS^+/PS^*) = +0.84$ V versus SCE, and a strong reductant, $E(PS^*/PS^-) = -0.86$ V versus SCE, in the same medium. Single-electron transfer (SET) can then occur between the excited state PS and a suitable reactant to initiate the catalytic cycle. Synthetic methodologies have been utilized to modify the photophysical properties of ruthenium(II) polypyridyl complexes and their related Ir(III) cyclometalled counterparts, as well as to develop new classes of organic PSs. Some common photoredox catalysts include $[Ru(bpy)_3]^{2+}$, $[Ru(bpz)_3]^{2+}$, $Ir(ppy)_3$, $[Ir(ppy)_2(dtbpy)]^+$, $[Ir(dF(CF_3)ppy)_2(dtbpy)]^+$, rose bengal, eosin Y, 2,4,6-triphenylpyrylium (TPP^+), 9-mesityl-10-methylacridinium ($Mes\text{-}Acr^+$), and riboflavin (Fig. 3.21.8, where bpz is 2,2′-bipyrazine and dtbpy is 4,4′-di-*tert*-butyl-2,2′-bipyridyl).[75]

Development in the field of photoredox catalysis has had a significant impact on promoting green methodologies for organic synthesis, including but not limited to C-H functionalization of amines, hydrogen transfer, C-C, and C-X coupling.[75] Hydrogen-atom transfer (HAT) is a critical step in many organic syntheses and industrial and biological processes. Several important reactions such as combustion, aerobic oxidation, and reduction to enzymatic catalysis involve a HAT step.[76] One example of oxidative HAT is the initiation of a combustion process when oxygen or other oxidants remove a hydrogen atom from a hydrocarbon to form a carbon radical. Such chemistry is conducted on an enormous scale in a range of industrial partial oxidation processes.[77] In the case of reductive HAT, transfer hydrogenation occurs from a non-H_2 hydrogen source such as isopropanol. These latter reactions are very attractive because high pressure H_2 gas is not required and the source of the hydrogen atom is also inexpensive. However, most of these reactions do require the presence of a metal catalyst.[78] Although these reactions could be driven by traditional thermally driven homogeneous or heterogeneous catalysis, visible light-driven photoredox catalysts may also provide enough driving force for these transformations.[79,80] The photoassisted oxidative and reductive cases of HAT reactions are discussed separately in the following sections.

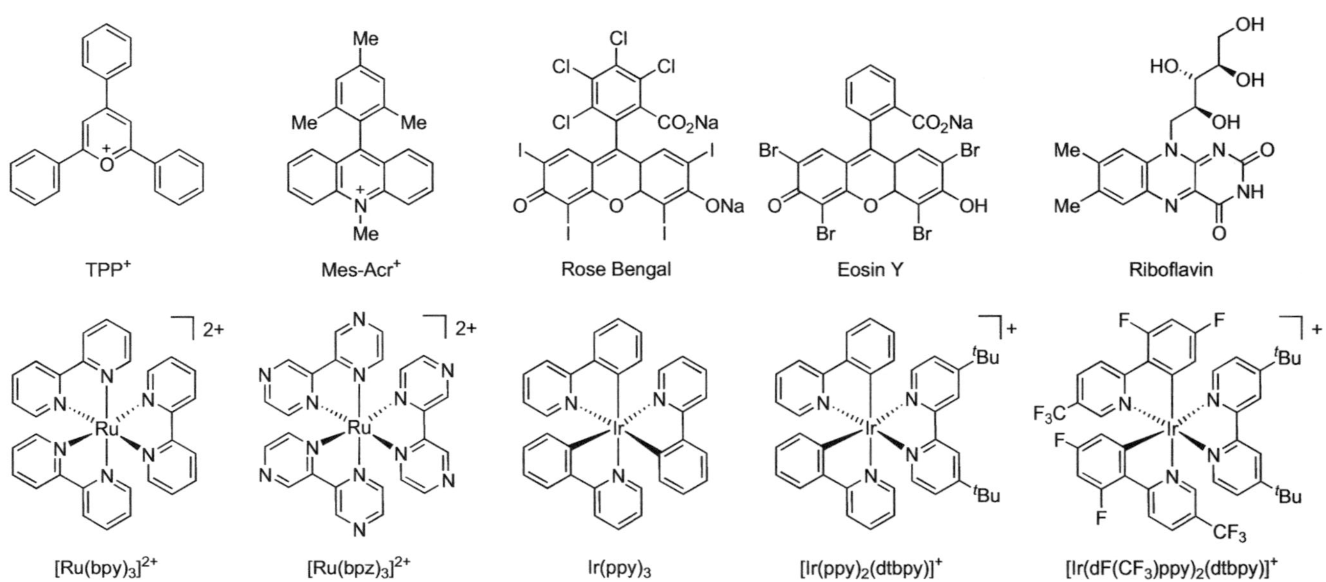

FIGURE 3.21.8 Chemical structures of some common photoredox catalysts.[65]

The critical step in oxidative HAT is cleavage of the C—H bond to form a carbon radical that ultimately leads to product formation, such as a new C—C bond. In traditional catalysis, C—H bond cleavage is usually carried out by a metal complex through oxidative addition, electrophilic activation, or σ-bond metathesis to form the carbon radical. In contrast, photocatalysis takes advantage of the strong reducing power that the photogenerated electron carries to reduce the substrate and to generate the reactive carbon radical center. In 2011, Sanford and coworkers[81] combined the palladium-catalyzed C—H functionalization reactivity and visible light ruthenium photoredox catalysis to perform a C—H arylation at room temperature with aryldiazonium salts ($[ArN_2]BF_4$). Subsequent studies involving theoretical calculations agree on the role of the catalyst in the catalytic cycle, as shown in Fig. 3.21.9.[65,82,83]

The reaction utilizes $[Ru(bpy)_3]Cl_2$ as the photocatalyst. The ^{3}MLCT excited state of $[Ru(bpy)_3]Cl_2$ reduces $[ArN_2]BF_4$ to generate the Ar• radical. 2-arylpyridine chelates $Pd^{II}(OAc)_2$ to form the palladium complex which in turn interacts with the Ar• radical to generate the Pd(III)-aryl species. After an SET event to regenerate the $[Ru(bpy)_3]Cl_2$ photocatalyst, the Pd(IV)-aryl species is formed. This species undergoes reductive elimination to yield the arylated C—C coupling product and to regenerate the Pd(II) catalyst.

Another interesting example of C—C coupling assisted by photoredox catalysis is the functionalization of the α-amino C—H bond. Many biologically active compounds and pharmaceuticals on the market contain amino group and this makes the functionalization around the amine center an attractive task for organic chemists. In 2010 Stephenson and coworkers

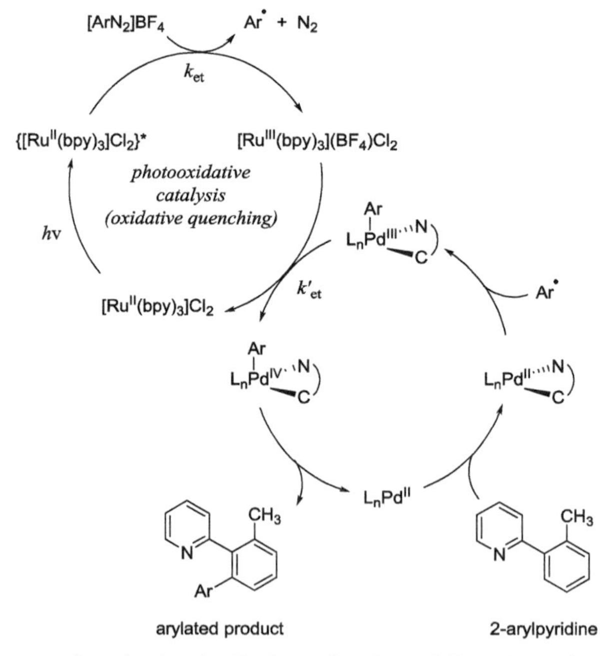

FIGURE 3.21.9 The proposed mechanism for C—C coupling by aryldiazonium salts and ruthenium photoredox catalysis.[65]

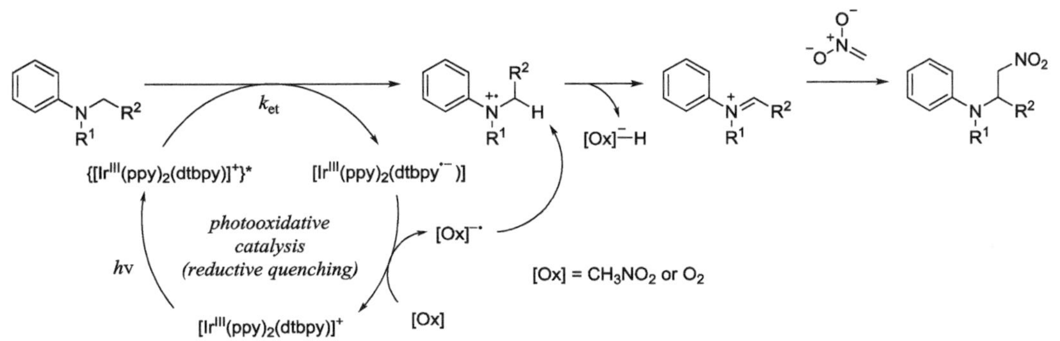

FIGURE 3.21.10 Proposed mechanism for the formation of C—C bonds between tertiary *N*-arylamines and nitroalkanes via an oxidative aza-Henry reaction, with $[Ir(ppy)_2(dtbpy)]PF_6$ photoredox catalyst.[82]

had shown that the α-amino C—H bond of *N*-arylamines could be activated by using the $[Ir(ppy)_2(dtbpy)]PF_6$ photocatalyst in an aza-Henry reaction (Fig. 3.21.10).[82] Upon photoexcitation by visible light, the excited state of the iridium chromophore is reductively quenched by the tertiary *N*-arylamine generating an amine radical cation. This increases the acidity of the α-amino C—H bond, and the H atom could be abstracted by an oxidant [O]• in the medium to generate an α-amino radical, which in turn is oxidized to an iminium ion. Nucleophilic attack at the iminium ion center by a nitroalkane generates the desired product (Fig. 3.21.10).

A wide range of synthetic strategies was built based on this work, and thus this approach was applied to many coupling reactions including but not limited to aerobic cross-dehydrogenative coupling, oxidative C—H functionalization of tertiary amines, the Friedel-Crafts amidoalkylation, carbon-phosphorus bond formation, α-arylation of α-amino carbonyl compounds with indoles, and cyclization of amino alcohols.[84–90] Photoassisted organic synthesis also led to other useful transformations such as sp^2 and sp C—C and C—X bond formation, where X is a halogen, nitrogen, or oxygen. These areas have been reviewed comprehensively in the literature.[85–90]

Similar to the mechanism of oxidative HAT, the excited state of the photocatalyst reduces a substrate to generate a reactive radical in reductive HAT reactions. However, instead of forming a coupled product such as a new C—C bond, this radical species is converted back to a neutral species by accepting an H atom after participating in the desired chemistry.[75] Nicewicz and coworkers[91] had shown that 2-phenylmalonitrile could act as an organocatalyst to mediate HAT in the presence of an organic photocatalyst 9-mesityl-10-methylacridinium perchlorate ($[Mes\text{-}Acr]ClO_4$). This reaction is driven by visible light, and the intramolecular cyclization product of this reaction also shows anti-Markovnikov selectivity (Fig. 3.21.11).[92]

Photoexcitation of $Mes\text{-}Acr^+$ generates an excited electron that could reduce the organic substrate. This reduction step gives a tertiary carbon radical center that drives an internal nucleophilic addition between the hydroxyl lone pair of electrons and the secondary carbocation next to the radical center. Subsequently 2-phenylmalonitrile mediates HAT from the hydroxyl group to the tertiary carbon radical center to give the anti-Markovnikov product and to regenerate the organic photoredox catalyst by SET. The same strategy was applied to a variety of transformations including hydroacetoxylation, hydroamination,

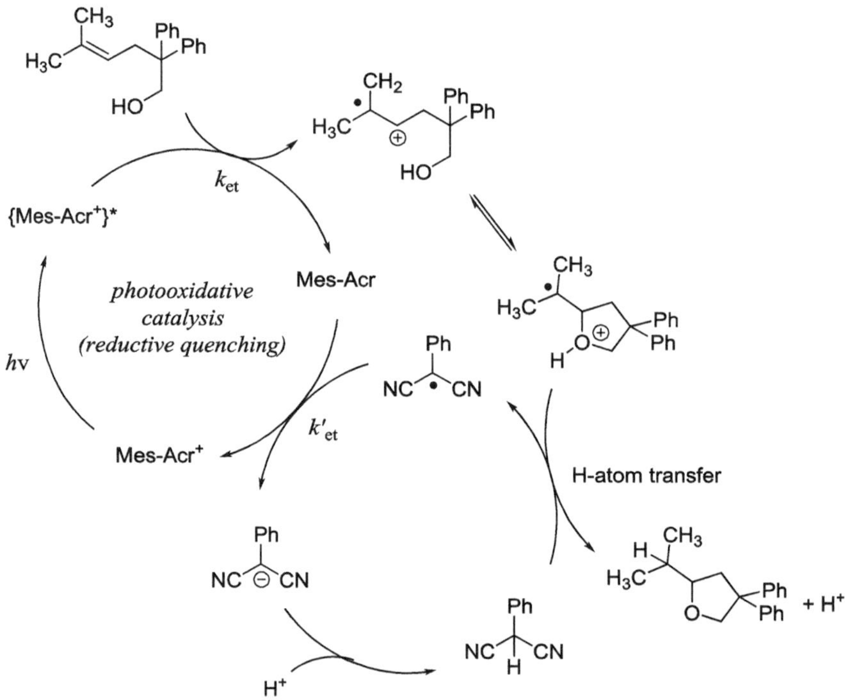

FIGURE 3.21.11 Anti-Markovnikov product formation driven by visible light organic photocatalyst and 2-phenylmalonitrile H-atom transfer agent.[92]

hydrochlorination, hydrofluorination, and cycloaddition of alkenes.[75] The subsequent work of Nicewicz and coworkers[93–95] also unveiled that thiophenol could carry out the HAT step at a faster rate than 2-phenylmalonitrile.

The activation of C—H bond by metal complex PS and HAT agent for the aforementioned transformations is just one example of the active areas in photoassisted organic synthesis literature. Other active areas include but not limited to semiconductor,[80,96] supramolecular systems,[97–99] and polymer photocatalysts.[100,101]

References

1. Anastas PT, Kirchhoff MM. Origins, current status, and future challenges of green chemistry. *Acc Chem Res* 2002;**35**:686–94.
2. Galian RE, Pérez-Prieto J. Catalytic processes activated by light. *Energy Environ Sci* 2010;**3**:1488.
3. *Statistical review of world energy*. 2015.
4. Brennaman MK, Dillon RJ, Alibabaei L, Gish MK, Dares CJ, Ashford DL, et al. Finding the way to solar fuels with dye-sensitized photoelectrosynthesis cells. *J Am Chem Soc* 2016;**138**:13085–102.
5. *Standard tables for reference solar spectral irradiances: direct normal and hemispherical on 37 degree tilted surface*. ASTM International; 2012.
6. *Standard solar constant and zero air mass solar spectral irradiance tables*. ASTM International; 2014.
7. Jablonski A. Efficiency of anti-stokes fluorescence in dyes. *Nature* 1933;**131**:839–40.
8. Turro NJ, Scaiano JC, Ramamurthy V. *Principles of molecular photochemistry: an introduction*. University Science Books; 2008. p. 530.

9. Schultz DM, Yoon TP. Solar synthesis: prospects in visible light photocatalysis. *Science* 2014;**343**:985–94.
10. Prier CK, Rankic DA, MacMillan DWC. Visible light photoredox catalysis with transition metal complexes: applications in organic synthesis. *Chem Rev* 2013;**113**:5322–63.
11. Douglas JJ, Sevrin MJ, Stephenson CRJ. Visible light photocatalysis: applications and new disconnections in the synthesis of pharmaceutical agents. *Org Process Res Dev* 2016;**20**:1134–47.
12. Alstrum-Acevedo JH, Brennaman MK, Meyer TJ. Chemical approaches to artificial photosynthesis. 2. *Inorg Chem* 2005;**44**:6802–27.
13. Berardi S, Drouet S, Francas L, Gimbert-Surinach C, Guttentag M, Richmond C, et al. Molecular artificial photosynthesis. *Chem Soc Rev* 2014;**43**:7501–19.
14. Yui T, Tamaki Y, Sekizawa K, Ishitani O. Photocatalytic reduction of CO_2: from molecules to semiconductors. *Top Curr Chem* 2011;**303**:151–84.
15. Takeda H, Ishitani O. Development of efficient photocatalytic systems for CO_2 reduction using mononuclear and multinuclear metal complexes based on mechanistic studies. *Coord Chem Rev* 2010;**254**:346–54.
16. Fujita E. Photochemical carbon dioxide reduction with metal complexes. *Coord Chem Rev* 1999;**185**:373–84.
17. Fujishima A, Honda K. Electrochemical photolysis of water at a semiconductor electrode. *Nature* 1972;**238**:37–8.
18. Navarro Yerga RM, Álvarez Galván MC, del Valle F, Villoria de la Mano JA, Fierro JLG. Water splitting on semiconductor catalysts under visible-light irradiation. *ChemSusChem* 2009;**2**:471–85.
19. Matsubara Y, Grills DC, Kuwahara Y. Thermodynamic aspects of electrocatalytic CO_2 reduction in acetonitrile and with an ionic liquid as solvent or electrolyte. *ACS Catal* 2015;**5**:6440–52.
20. Lehn J-M, Ziessel R. Photochemical generation of carbon monoxide and hydrogen by reduction of carbon dioxide and water under visible light irradiation. *Proc Natl Acad Sci USA* 1982;**79**:701–4.
21. Hemminger JC, Carr R, Somorjai GA. The photoassisted reaction of gaseous water and carbon dioxide adsorbed on the $SrTiO_3$ (111) crystal face to form methane. *Chem Phys Lett* 1978;**57**:100–4.
22. Halmann M. Photoelectrochemical reduction of aqueous carbon dioxide on p-type gallium phosphide in liquid junction solar cells. *Nature* 1978;**275**:115–6.
23. Inoue T, Fujishima A, Konishi S, Honda K. Photoelectrocatalytic reduction of carbon dioxide in aqueous suspensions of semiconductor powders. *Nature* 1979;**277**:637–8.
24. Fujita E, Furenlid LR, Renner MW. Direct XANES evidence for charge transfer in Co–CO_2 complexes. *J Am Chem Soc* 1997;**119**:4549–50.
25. Tinnemans AHA, Koster TPM, Thewissen DHMW, Mackor A. Tetraaza-macrocyclic cobalt(II) and nickel(II) complexes as electron-transfer agents in the photo(electro)chemical and electrochemical reduction of carbon dioxide. *Recl Des Trav Chim Des Pays Bas* 1984;**103**:288–95.
26. Ogata T, Yanagida S, Brunschwig BS, Fujita E. Mechanistic and kinetic studies of cobalt macrocycles in a photochemical CO_2 reduction system: evidence of Co–CO_2 adducts as intermediates. *J Am Chem Soc* 1995;**117**:6708–16.
27. Fujita E, Creutz C, Sutin N, Brunschwig BS. Carbon dioxide activation by cobalt macrocycles: evidence of hydrogen bonding between bound CO_2 and the macrocycle in solution. *Inorg Chem* 1993;**32**:2657–62.
28. Fujita E, Haff J, Sanzenbacher R, Elias H. High electrocatalytic activity of *RRSS*-[Ni^{II}HTIM](ClO_4)$_2$ and [Ni^{II}DMC](-ClO_4)$_2$ for carbon dioxide reduction (HTIM = 2,3,9,10-Tetramethyl-1,4,8,11-tetraazacyclotetradecane, DMC = C-meso-5,12-Dimethyl-1,4,8,11-tetraazacyclotetradecane). *Inorg Chem* 1994;**33**:4627–8.
29. Kelly CA, Mulazzani QG, Venturi M, Blinn EL, Rodgers MAJ. The thermodynamics and kinetics of CO_2 and H^+ binding to Ni(cyclam)$^+$ in aqueous solution. *J Am Chem Soc* 1995;**117**:4911–9.
30. Fujita E, Creutz C, Sutin N, Szalda DJ. Carbon dioxide activation by cobalt(I) macrocycles: factors affecting carbon dioxide and carbon monoxide binding. *J Am Chem Soc* 1991;**113**:343–53.
31. Fujita E, Szalda DJ, Creutz C, Sutin N. Carbon dioxide activation: thermodynamics of carbon dioxide binding and the involvement of two cobalt centers in the reduction of carbon dioxide by a cobalt(I) macrocycle. *J Am Chem Soc* 1988;**110**:4870–1.
32. Hammouche M, Lexa D, Momenteau M, Saveant JM. Chemical catalysis of electrochemical reactions. Homogeneous catalysis of the electrochemical reduction of carbon dioxide by iron("0") porphyrins. Role of the addition of magnesium cations. *J Am Chem Soc* 1991;**113**:8455–66.
33. Dhanasekaran T, Grodkowski J, Neta P, Hambright P, Fujita E. p-Terphenyl-sensitized photoreduction of CO_2 with cobalt and iron porphyrins. Interaction between CO and reduced metalloporphyrins. *J Phys Chem A* 1999;**103**:7742–8.

34. Grodkowski J, Neta P, Fujita E, Mahammed A, Simkhovich L, Gross Z. Reduction of cobalt and iron corroles and catalyzed reduction of CO_2. *J Phys Chem A* 2002;**106**:4772–8.
35. Grodkowski J, Neta P. Cobalt corrin catalyzed photoreduction of CO_2. *J Phys Chem* 2000;**104**:1848–53.
36. Grodkowski J, Dhanasekaran T, Neta P, Hambright P, Brunschwig BS, Shinozaki K, et al. Reduction of cobalt and iron phthalocyanines and the role of the reduced species in catalyzed photoreduction of CO_2. *J Phys Chem A* 2000;**104**:11332–9.
37. Behar D, Dhanasekaran T, Neta P, Hosten CM, Ejeh D, Hambright P. Cobalt porphyrin catalyzed reduction of CO_2. Radiation chemical, photochemical, and electrochemical studies. *J Phys Chem A* 1998;**5639**:2870–7.
38. Grodkowski J, Behar D, Neta P, Hambright P. Iron porphyrin-catalyzed reduction of CO_2. Photochemical and radiation chemical studies. *J Phys Chem A* 1997;**101**:248–54.
39. Grice KA. Carbon dioxide reduction with homogenous early transition metal complexes: opportunities and challenges for developing CO_2 catalysis. *Coord Chem Rev* 2017;**336**:78–95.
40. Sahara G, Ishitani O. Efficient photocatalysts for CO_2 reduction. *Inorg Chem* 2015;**54**:5096–104.
41. Tamaki Y, Watanabe K, Koike K, Inoue H, Morimoto T, Ishitani O. Development of highly efficient supramolecular CO_2 reduction photocatalysts with high turnover frequency and durability. *Faraday Discuss* 2012;**155**:115–27.
42. Tamaki Y, Morimoto T, Koike K, Ishitani O. Photocatalytic CO_2 reduction with high turnover frequency and selectivity of formic acid formation using Ru(II) multinuclear complexes. *Proc Natl Acad Sci USA* 2012;**109**:15673–8.
43. Tamaki Y, Koike K, Morimoto T, Ishitani O. Substantial improvement in the efficiency and durability of a photocatalyst for carbon dioxide reduction using a benzoimidazole derivative as an electron donor. *J Catal* 2013;**304**:22–8.
44. Sekizawa K, Maeda K, Domen K, Koike K, Ishitani O. Artificial Z-scheme constructed with a supramolecular metal complex and semiconductor for the photocatalytic reduction of CO_2. *J Am Chem Soc* 2013;**135**:4596–9.
45. Gersten SW, Samuels GJ, Meyer TJ. Catalytic oxidation of water by an oxo-bridged ruthenium dimer. *J Am Chem Soc* 1982;**104**:4029–30.
46. Wang Y, Ahlquist MSG. Where does the water go? A computational study on the reactivity of a ruthenium(v) oxo complex (bpc)(bpy)$Ru^{V}O$. *Phys Chem Phys* 2014;**16**:11182–5.
47. Tamaki Y, Vannucci AK, Dares CJ, Binstead RA, Meyer TJ. One-electron activation of water oxidation catalysis. *J Am Chem Soc* 2014;**136**:6854–7.
48. Romero I, Rodríguez M, Sens C, Mola J, Rao Kollipara M, Francàs L, et al. Ru complexes that can catalytically oxidize water to molecular dioxygen. *Inorg Chem* 2008;**47**:1824–34.
49. Neudeck S, Maji S, López I, Meyer S, Meyer F, Llobet A. New powerful and oxidatively rugged dinuclear Ru water oxidation catalyst: control of mechanistic pathways by tailored ligand design. *J Am Chem Soc* 2014;**136**:24–7.
50. Muckerman JT, Kowalczyk M, Badiei YM, Polyansky DE, Concepcion JJ, Zong R, et al. New water oxidation chemistry of a seven-coordinate ruthenium complex with a tetradentate polypyridyl ligand. *Inorg Chem* 2014;**53**:6904–13.
51. Li T-T, Chen Y, Li F-M, Zhao W-L, Wang C-J, Lv X-J, et al. Efficient water oxidation catalyzed by mononuclear ruthenium(ii) complexes incorporating Schiff base ligands. *Chem A Eur J* 2014;**20**:8054–61.
52. Kärkäs MD, Åkermark T, Chen H, Sun J, Åkermark B. Back cover: a tailor-made molecular ruthenium catalyst for the oxidation of water and its deactivation through poisoning by carbon monoxide. *Angew Chem Int Ed* 2013;**52**:4274.
53. Ghosh S, Baik M-H. Redox properties of Tanaka's water oxidation catalyst: redox noninnocent ligands dominate the electronic structure and reactivity. *Inorg Chem* 2011;**50**:5946–57.
54. Blakemore JD, Crabtree RH, Brudvig GW. Molecular catalysts for water oxidation. *Chem Rev* 2015;**115**:12974–3005.
55. Concepcion JJ, Jurss JW, Templeton JL, Meyer TJ. One site is enough. Catalytic water oxidation by $[Ru(tpy)(bpm)(OH_2)]^{2+}$ and $[Ru(tpy)(bpz)(OH_2)]^{2+}$. *J Am Chem Soc* 2008;**130**:16462–3.
56. Chen Z, Concepcion JJ, Hu X, Yang W, Hoertz PG, Meyer TJ. Concerted O atom–proton transfer in the O–O bond forming step in water oxidation. *Proc Natl Acad Sci USA* 2010;**107**:7225–9.

57. Chen Z, Concepcion JJ, Jurss JW, Meyer TJ. Single-site, catalytic water oxidation on oxide surfaces. *J Am Chem Soc* 2009;**131**:15580–1.
58. Kohl SW, Weiner L, Schwartsburd L, Konstantinovski L, Shimon LJW, Ben-David Y, et al. Consecutive thermal H_2 and light-induced O_2 evolution from water promoted by a metal complex. *Science* 2009;**324**:74–7.
59. Ashford DL, Gish MK, Vannucci AK, Brennaman MK, Templeton JL, Papanikolas JM, et al. Molecular chromophore–catalyst assemblies for solar fuel applications. *Chem Rev* 2015;**115**:13006–49.
60. Trommsdorff H. Ueber santonin. *Ann Pharm* 1834;**11**:190–207.
61. Koike T, Akita M. Visible-light radical reaction designed by Ru- and Ir-based photoredox catalysis. *Inorg Chem Front* 2014;**1**:562–76.
62. Corrigan N, Shanmugam S, Xu J, Boyer C. Photocatalysis in organic and polymer synthesis. *Chem Soc Rev* 2016;**45**:6165–212.
63. Arora A, Weaver JD. Visible light photocatalysis for the generation and use of reactive Azolyl and Polyfluoroaryl intermediates. *Acc Chem Res* 2016;**49**:2273–83.
64. Kozlowski M, Yoon T. Editorial for the special issue on photocatalysis. *J Org Chem* 2016;**81**:6895–7.
65. Shaw MH, Twilton J, MacMillan DWC. Photoredox catalysis in organic chemistry. *J Org Chem* 2016;**81**:6898–926.
66. Hedstrand DM, Kruizinga WH, Kellogg RM. Light induced and dye accelerated reductions of phenacyl onium salts by 1,4-dihydropyridines. *Tetrahedron Lett* 1978;**19**:1255–8.
67. Van Bergen TJ, Hedstrand DM, Kruizinga WH, Kellogg RM. Chemistry of dihydropyridines. 9. Hydride transfer from 1,4-dihydropyridines to sp3-hybridized carbon in sulfonium salts and activated halides. Studies with NAD(P)H models. *J Org Chem* 1979;**44**:4953–62.
68. Cano-Yelo H, Deronzier A. Photocatalysis of the Pschorr reaction by tris-(2,2′-bipyridyl)ruthenium(II) in the phenanthrene series. *J Chem Soc Perkin Trans* 1984;**2**:1093–8.
69. Pac C, Miyauchi Y, Ishitani O, Ihama M, Yasuda M, Sakurai H. Redox-photosensitized reactions. 11. $Ru(bpy)_3^{2+}$-photosensitized reactions of 1-benzyl-1,4-dihydronicotinamide with aryl-substituted enones, derivatives of methyl cinnamate, and substituted cinnamonitriles: electron-transfer mechanism and structure-reactivity. *J Org Chem* 1984;**49**:26–34.
70. Pac C, Ihama M, Yasuda M, Miyauchi Y, Sakurai H. Tris(2,2′-bipyridine)ruthenium(2+)-mediated photoreduction of olefins with 1-benzyl-1,4-dihydronicotinamide: a mechanistic probe for electron-transfer reactions of NAD(P)H-model compounds. *J Am Chem Soc* 1981;**103**:6495–7.
71. Fukuzumi S, Mochizuki S, Tanaka T. Photocatalytic reduction of phenacyl halides by 9,10-dihydro-10-methylacridine: control between the reductive and oxidative quenching pathways of tris(bipyridine)ruthenium complex utilizing an acid catalysis. *J Phys Chem* 1990;**94**:722–6.
72. Fukuzumi S, Koumitsu S, Hironaka K, Tanaka T. Energetic comparison between photoinduced electron-transfer reactions from NADH model compounds to organic and inorganic oxidants and hydride-transfer reactions from NADH model compounds to *p*-benzoquinone derivatives. *J Am Chem Soc* 1987;**109**:305–16.
73. Ishitani O, Pac C, Sakurai H. Redox-photosensitized reactions. 10. Formation of a novel type of adduct between an NADH model and carbonyl compounds by photosensitization using $Ru(bpy)_3^{2+}$. *J Org Chem* 1983;**48**:2941–2.
74. Hironaka K, Fukuzumi S, Tanaka T. Tris(bipyridyl)ruthenium(II)-photosensitized reaction of 1-benzyl-1,4-dihydronicotinamide with benzyl bromide. *J Chem Soc Perkin Trans* 1984;**2**:1705–9.
75. Yang H, Cui X, Dai X, Deng Y, Shi F. Carbon-catalysed reductive hydrogen atom transfer reactions. *Nat Commun* 2015;**6**:6478.
76. Wang D, Astruc D. The golden age of transfer hydrogenation. *Chem Rev* 2015;**115**:6621–86.
77. Roth J, Lovell S, Mayer JM. Intrinsic barriers for electron and hydrogen atom transfer reactions of biomimetic iron complexes. *J Am Chem Soc* 2000;**122**:5486–98.
78. Ravelli D, Dondi D, Fagnoni M, Albini A. Photocatalysis. A multi-faceted concept for green chemistry. *Chem Soc Rev* 2009;**38**:1999–2011.
79. Fagnoni M, Dondi D, Ravelli D, Albini A. Photocatalysis for the formation of the C–C bond. *Chem Rev* 2007;**107**:2725–56.
80. Kalyani D, McMurtrey KB, Neufeldt SR, Sanford MS. Room-temperature C–H arylation: merger of Pd-catalyzed C–H functionalization and visible-light photocatalysis. *J Am Chem Soc* 2011;**133**:18566–9.

81. Neufeldt SR, Sanford MS. Combining transition metal catalysis with radical chemistry: dramatic acceleration of palladium-catalyzed C−H arylation with diaryliodonium salts. *Adv Synth Catal* 2012;**354**:3517−22.
82. Condie AG, González-Gómez JC, Stephenson CRJ. Visible-light photoredox catalysis: aza-Henry reactions via C−H functionalization. *J Am Chem Soc* 2010;**132**:1464−5.
83. Maestri G, Malacria M, Derat E. Radical Pd(III)/Pd(I) reductive elimination in palladium sequences. *Chem Commun* 2013;**49**:10424−6.
84. Wang Z-Q, Hu M, Huang X-C, Gong L-B, Xie Y-X, Li J-H. Direct α-arylation of α-amino carbonyl compounds with indoles using visible light photoredox catalysis. *J Org Chem* 2012;**77**:8705−11.
85. Pan Y, Wang S, Kee CW, Dubuisson E, Yang Y, Loh KP, et al. ChemInform abstract: graphene oxide and Rose Bengal: oxidative C−H functionalization of tertiary amines using visible light. *ChemInform* 2012;**43**. http://dx.doi.org/10.1002/chin.201217163.
86. Liu Q, Li Y-N, Zhang H-H, Chen B, Tung C-H, Wu L-Z. Reactivity and mechanistic insight into visible-light-induced aerobic cross-dehydrogenative coupling reaction by organophotocatalysts. *Chem Eur J* 2012;**18**:620−7.
87. Reckenthäler M, Griesbeck AG. Photoredox catalysis for organic syntheses. *Adv Synth Catal* 2013;**355**:2727−44.
88. Rueping M, Zhu S, Koenigs RM. Photoredox catalyzed C−P bond forming reactions-visible light mediated oxidative phosphonylations of amines. *Chem Commun* 2011;**47**:8679−81.
89. Mathis CL, Gist BM, Frederickson CK, Midkiff KM, Marvin CC. Visible light photooxidative cyclization of amino alcohols to 1,3-oxazines. *Tetrahedron Lett* 2013;**54**:2101−4.
90. Dai C, Meschini F, Narayanam JMR, Stephenson CRJ. Friedel−crafts amidoalkylation via thermolysis and oxidative photocatalysis. *J Org Chem* 2012;**77**:4425−31.
91. Zeller MA, Riener M, Nicewicz DA. Butyrolactone synthesis via polar radical crossover cycloaddition reactions: diastereoselective syntheses of methylenolactocin and protolichesterinic acid. *Org Lett* 2014;**16**:4810−3.
92. Hamilton DS, Nicewicz DA. Direct catalytic anti-Markovnikov hydroetherification of alkenols. *J Am Chem Soc* 2012;**134**:18577−80.
93. Wilger DJ, Grandjean J-MM, Lammert TR, Nicewicz DA. The direct anti-markovnikov addition of mineral acids to styrenes. *Nat Chem* 2014;**6**:720−6.
94. Nguyen TM, Nicewicz DA. Anti-markovnikov hydroamination of alkenes catalyzed by an organic photoredox system. *J Am Chem Soc* 2013;**135**:9588−91.
95. Perkowski AJ, Nicewicz DA. Direct catalytic anti-markovnikov addition of carboxylic acids to alkenes. *J Am Chem Soc* 2013;**135**:10334−7.
96. Hoffmann MR, Martin ST, Choi W, Bahnemann DW. Environmental applications of semiconductor photocatalysis. *Chem Rev* 1995;**95**:69−96.
97. Zoltan T, Vargas F, Rivas C, López V, Perez J, Biasutto A. Synthesis, photochemical and photoinduced antibacterial activity studies of *meso*-tetra(pyren-1-yl)porphyrin and its Ni, Cu and Zn complexes. *Sci Pharm* 2010;**78**:767−89.
98. Sun H, Blatter F, Frei H. Selective oxidation of toluene to benzaldehyde by O_2 with visible light in barium(2+)- and calcium(2+)-exchanged zeolite Y. *J Am Chem Soc* 1994;**116**:7951−2.
99. Corma A, Garcia H. Zeolite-based photocatalysts. *Chem Commun* 2004:1443−59.
100. Zhang M, Rouch WD, McCulla RD. Conjugated polymers as photoredox catalysts: visible-light-driven reduction of aryl aldehydes by poly(*p*-phenylene). *Eur J Org Chem* 2012;**2012**:6187−96.
101. Wada Y, Ogata T, Hiranaga K, Yasuda H, Kitamura T, Murakoshi K, et al. Visible light-induced photofixation of CO_2 into benzophenone: roles of poly(*p*-phenylene) as photocatalyst and two-electron mediator in the presence of quaternary onium salts. *J Chem Soc Perkin Trans* 1998;**2**:1999−2004.

CHAPTER

3.22

Biopolymers: Biodegradable Alternatives to Traditional Plastics

Christopher Brigham

University of Massachusetts Dartmouth, North Dartmouth, MA, United States

3.22.1 INTRODUCTION

The Merriam-Webster dictionary offers several definitions for the word "plastic," including but not limited to "made or consisting of a plastic," "capable of being deformed," and "having a quality consisting of mass-produced plastic goods."[1] Although it is debatable whether these definitions are of use to the biotechnologist or the biochemist, one can see that the term "plastic" can potentially refer to many different types of materials. We traditionally think of plastic as material from which household items are made, such as toothpaste tubes, shampoo bottles, water bottles, and children's toys, among many other things. Traditionally, these types of plastics are produced synthetically, using refined petroleum products. Plastics are polymers, long chains of molecules joined together by covalent bonds. The molecules that make up the plastic polymer chain are called monomers. Traditionally, petroleum-based plastics consist of high-molecular-weight (HMW) chains, suggesting that many monomers are required for making a single polymer chain. Most petroleum-based (i.e., synthetic) plastics are tough, flexible, and resistant to the majority of environmental phenomena. However, the toughness and general resistance to breakdown of synthetic plastics means that, when disposed, they persist in the environment for a long, indefinite period of time, contaminating the soil and aqueous ecosystems. The continued acceptance and prevalence of synthetic plastics as a raw material has resulted in the presence of plastic waste in rivers and oceans, with microscopic plastic fibers finding their way into the tissues and organs of marine life. In the past couple of decades, recycling of synthetic plastic has become more universally accepted. Although recycling plastic offers some mitigation to the problem of plastic introduced into the environment, the process consumes energy and less than 50% of the plastic used in the United States each year is actually recycled, including 31.8% of all plastic bottles used in 2014.[2] Furthermore, each community seems to have its own rules about recycling (i.e., what forms of plastic can be recycled, allowable dirt and grease content, etc.), so there is

Green Chemistry
http://dx.doi.org/10.1016/B978-0-12-809270-5.00027-3

not enough consistency in the process. In short, while we are moving toward more responsible use and disposal of petroleum-based plastics, there is still a pressing need for bio-based, biodegradable alternatives for traditional, petroleum-based plastics for many applications.

Plastics are ubiquitous because of their mechanical properties. They are flexible, moldable, and tough. They can be injection molded and melt-processed, allowing them to take many shapes. It is these properties that have come to define the very word "plastic." Since much of a plastic's utility lies in its structure, if we are to look for alternatives to synthetic plastics, we must examine the structural polymers in nature. There are several organisms that produce substances that have properties similar to petroleum-based plastics. Like plastics, these substances are polymers that have structural integrity, toughness, and flexibility. Many, if not all of these bio-based polymers, are biodegradable and thus represent sources of raw material that could potentially replace traditional synthetic plastics.

In terms of polymer production as an industrial process, there is also a major difference between bio-based polymers and petroleum-based polymers, the polymerization process itself. As mentioned above, synthetic chemical processes produce traditional plastics like polyethylene, polypropylene, and nylon. Biochemical processes, often carried out in an organism in vivo, produce bio-based polymers. For bio-based plastics, we can consider the properties and utility of polymers made from different biological monomers, such as amino acids, sugars, and organic acids. In this chapter, you will learn about some representative bio-based polymers, their relevant material properties, and some of their commercial applications.

3.22.2 PROTEIN: A UBIQUITOUS BIOPOLYMER

Proteins are natural polymers that consist of one or more polypeptide chains. Proteins are made up of amino acids as monomers, and polymerization proceeds by the formation of peptide bonds that occur via an enzyme-mediated condensation reaction (Fig. 3.22.1). Proteins

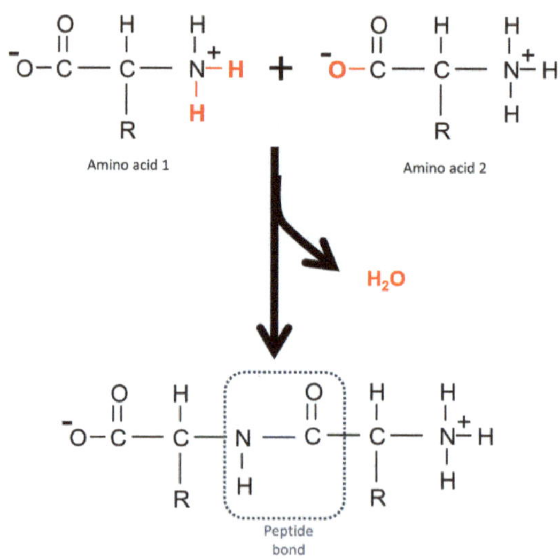

FIGURE 3.22.1 To form a peptide bond, a condensation reaction must occur where water (red) is liberated.

are ubiquitous throughout the domains of life. There are two main types of proteins: enzymes and structural proteins. For each protein, the primary amino acid sequence dictates, in part, the secondary, tertiary, and quaternary structures of the polypeptides. Enzymes tend to be globular in nature, whereas structural proteins are more ordered.[3] For the purposes of proteins as replacement for petrochemical polymers, we will consider the ordered structural proteins. Structural proteins like collagen, gelatin, and silk possess mechanical properties that make these proteins useful in a variety of applications.

3.22.2.1 Collagen and Gelatin

The structural proteins collagen and gelatin constitute >30% of the total proteins of most animals. These two polymers are used in a variety of applications, many of which are medical due mainly to their biodegradability and biocompatibility.[4]

A collagen molecule consists of a trimer of α-chains, each of which has a molecular mass of approximately 100,000. The average amino acid molecular weight in collagen is low, compared with most proteins, because of the high content of the amino acid glycine.[4,5] Depending on the origin of collagen, the individual subunits of the collagen trimer can make up a homotrimer of identical chains or a heterotrimer of different chains.[4] Collagen fibers tend to have high tensile strength and stability, aspects that make the molecule useful in many applications. Individual α-chains of collagen can form left-handed helical structures by themselves, and three of these chains will then intertwine to form a right-handed "superhelix."[4] The helical regions of the collagen α-chains possess a "Gly-X-Y" repeating triplet motif, where a glycine residue is flanked by two amino acids "X" and "Y," which are predominantly proline and hydroxyproline, respectively (Fig. 3.22.2). The presence of the glycine at every third residue in the collagen α-chain is necessary for the helical structure. As is evident, the architecture of collagen is due to its primary amino acid sequence.

In the body, collagen forms elastic molecular networks that can strengthen tendons, as well as sheets that support skin and internal organs.[6] Studies in rats and clinical trials in humans have demonstrated that type II collagen, commonly isolated from chicken combs, reduces cartilage destruction in patients suffering from osteoarthritis.[7–9] As collagen is a very strong molecule, it is often used in bone grafts.[10] Collagen has also been widely used in the cosmetic surgery and as tissue engineering scaffold material.[11] For nonmedical uses, collagen is also used as sausage casings. When collagen is denatured by heating, the three strands separate (either partially or completely) into globular domains and random coils, thus becoming gelatin.

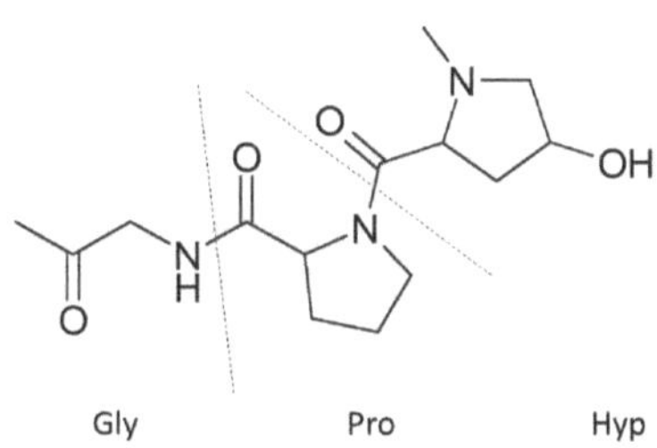

FIGURE 3.22.2 A schematic representation of the Gly-Pro-Hyp motif found in collagen α-chains.

Gelatin is a mixture of peptides derived from collagen by breakage of cross-linkages and some peptide bonds. Typically, this breakage of bonds to convert collagen to gelatin is performed by enzymatic degradation.[4] Any protease could be used to convert collagen to gelatin peptides, such as pepsin, trypsin, papain, and other enzymes. Hydrolysis to peptides is performed to improve the functional as well as nutritional properties of the protein molecules. Several fish-based gelatin peptides have been shown to possess antioxidant properties.[4] The molecular weight of gelatin is significantly lower than that of collagen, typically in the range of 1400–26,000.[4] Just as glycine, proline and hydroxyproline are predominant amino acid residues in collagen, these three amino acids are often found in gelatin peptides.

There are many uses for gelatin peptides. When one thinks of gelatin, the typical applications associated are food related. Indeed, gelatin is a constituent of "gummy" candy products, and is also used as a thickening agent in foods like ice cream, yogurt, and marshmallow (synthetic).[12] However, in the pharmaceutical industry, gelatin has been used for decades as coatings for pills and capsules to aid in swallowing. Also, ballistic gelatin is used as a medium for testing guns and ammunition, due to its similar consistency to muscle tissue. Gelatin can also be used as a binder in a variety of applications, including match and sandpaper making.[13]

3.22.2.2 Silk

Silks are protein polymers that are spun into fibers by some insects. Commonly, we think of silkworms and spiders as silk producers, but scorpions as well as some fly and mite larvae are also capable of making silk fibers.[14,15] Silk fibers, like those of collagen and gelatin, are biodegradable and biocompatible. Silk possesses superior mechanical properties (e.g., tensile strength, toughness, elasticity). Several types of silk can be produced by spiders. Each different silk has a different composition, structure, and properties, depending on the insect producing it and the function for which it is used by the producing organism. The most-well-characterized silk is known as *major ampullate*, or dragline silk. Dragline silk is used as lifeline support for some spiders.

Silks, like collagen, are characterized by a repetitive primary amino acid sequence. Like collagen, the amino acid sequence of silks tends to be glycine rich. The glycine-rich regions of spider silk proteins are characterized by the tripeptide motif Gly-Gly-X, where X represents one of a subset of nonglycine amino acids. These glycine-rich regions are followed by stretches of the amino acid alanine (poly-A regions). Several Gly-Gly-X motifs and poly-A stretches are repeated on a silk polypeptide, sandwiched between N- and C-terminal nonrepeated regions (Fig. 3.22.3). These poly-A regions are thought to form crystalline, hydrophobic domains that are responsible for the high tensile strength of silk fiber.[16,17] Secondary structures of silk include β-sheet structures separated by random coils.[15] Spider silk proteins (often called spidroins) have very HMW, from 200,000–350,000 mass units.

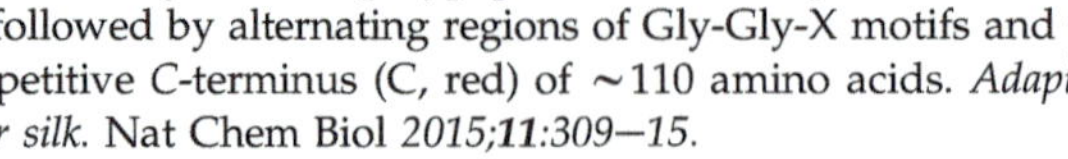

FIGURE 3.22.3 Schematic drawing of a silk polypeptide. The nonrepetitive *N*-terminus (N, blue) consisting of ~130 amino acid residues is followed by alternating regions of Gly-Gly-X motifs and poly-Ala regions (yellow and orange), followed by a nonrepetitive *C*-terminus (C, red) of ~110 amino acids. *Adapted from Rising A, Johansson J. Toward spinning artificial spider silk.* Nat Chem Biol *2015;***11***:309–15.*

Given the heterogeneity of silk proteins synthesized by insects and arachnids, a recombinant source of silk proteins has been sought. Recombinant silk proteins also exhibit more processability and can be formed into morphologies like films, hydrogels, particles, or nonwoven meshes.[15] Silk is commonly used as a material for clothing manufacture, but other, more specialized uses exist, due to the high strength of silk. One such use that relies on the strength of silk is its use in parachutes.

3.22.3 POLYSACCHARIDES

Polysaccharides are produced in nature as carbon and energy storage molecules and also as structural molecules. Bacterial, fungal, and plant cell walls are composed of polysaccharides. Polysaccharides such as cellulose are partly responsible for the rigidity of plant structures like stems and grass blades. Chitin is a strong structural polysaccharide that comprises the shells of several crustaceans, including lobsters, crabs, and shrimp. The use of these polysaccharides in nature suggests that they would have utility for several household, industrial, and other applications, as will be discussed in the following pages.

3.22.3.1 Starch

Starch, a homopolymer of α-D-glucose (Fig. 3.22.4A), is the principal form of stored carbon and energy in plants. Starch is composed of two structurally distinct polymers, the linear amylose and the branched amylopectin (Fig. 3.22.5). Starches from most food crops contain 20%–30% amylose.[18] Amylopectin is the main crystalline component in granular starch. Starch is synthesized in the plastids of plant cells, either in the chloroplast or in the dedicated amyloplast (e.g., in tubers like potatoes). Gluconeogenesis produces hexoses like glucose and fructose, which are phosphorylated. The hexose phosphates are then isomerized (if necessary) and combined with ATP to produce ADP-glucose (ADP-GLC), a key intermediate in starch synthesis. Starch synthase enzymes utilize the glucose moiety of ADP-GLC to produce α-1,4-linked chains of varying length (amylose). Starch branching enzymes create α-1,6 linkages to produce branched polysaccharides (amylopectin). Activities of starch branching and debranching enzymes dictate the branching pattern and thus the crystallinity of starch.[19] A schematic pathway of starch biosynthesis is shown in Fig. 3.22.6.

The presence of different types and abundances of crystal structures in starch granules are influenced by the ratio of amylose to amylopectin. The branching of amylopectin, as well as the length of the outer branches, influence the crystallinity of starch. Starch can be used as a thermoplastic, and when it is processed in a temperature window of 100–200 °C, several types of products can be made. To enhance the properties of starch as a bioplastic, different plasticizers can be added, including the natural products glycerol, sucrose, urea, maltose, and even water. The greater the amount of plasticizer in the starch blend, the more rubbery the mechanical properties of the plastic. However, the interaction between the chains of amylose and amylopectin is important for starch as a plastic, and the presence of too much plasticizer results in a very soft, gel-like material.[19]

Starch is often used as a thickening and/or binding agent in foods, and it is also used as a binder in pills and tablets in the pharmaceutical industry. In textile manufacturing, addition

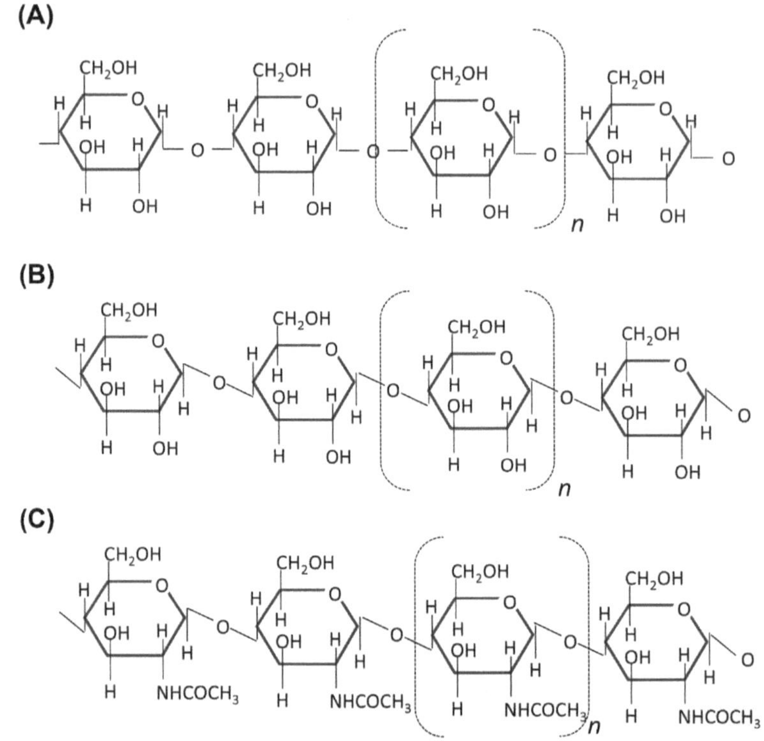

FIGURE 3.22.4 Schematic drawing of the structure of starch (A), cellulose (B), and chitin (C). Note that the difference between (A) and (B) is an α-linkage of glucose monomers in starch and a β-linkage of glucose monomers in cellulose.

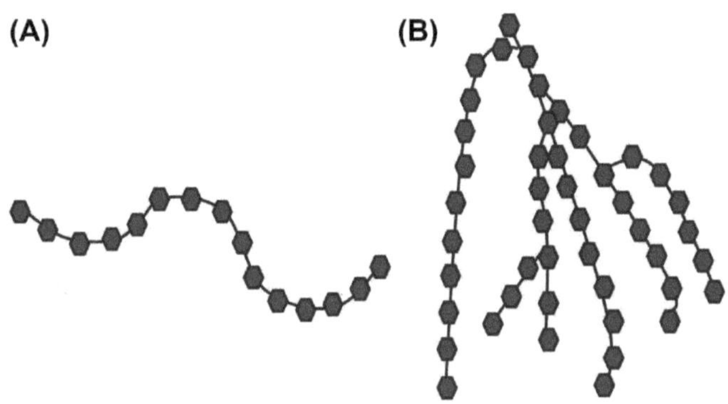

FIGURE 3.22.5 Structure of starch polysaccharides. Amylose (A) is a linear chain of glucose monomers. Amylopectin (B) is a highly branched chain of glucose monomers.

of starch reduces breakage of yarns during weaving. Starch has been used as a bio-based plastic in disposable flatware and also for packing material (i.e., "packing peanuts"). Starch can also be the carbon source used by bioplastic-producing microorganisms to produce monomers and/or biopolymers.

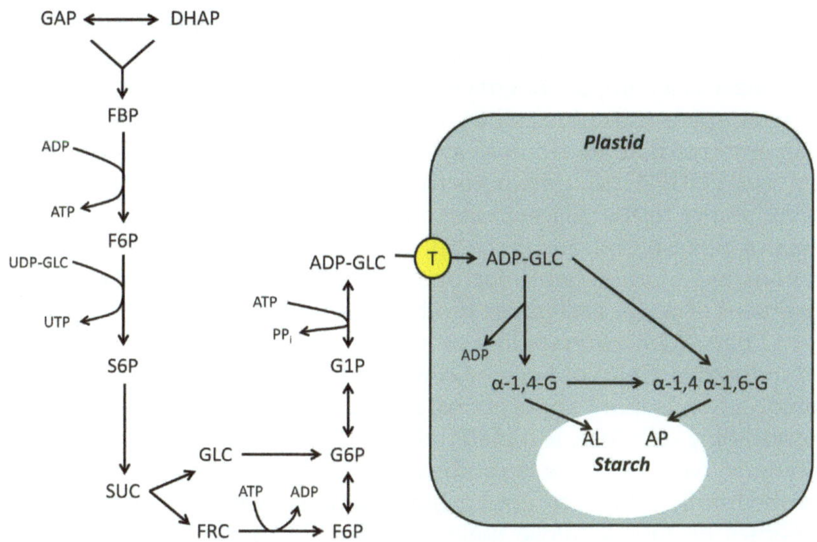

FIGURE 3.22.6 The starch biosynthesis pathway in plants. *AL*, amylose; *AP*, amylopectin; *DHAP*, dihydroxyacetone phosphate; *F6P*, fructose 6-phosphate; *FBP*, fructose bisphosphate; *FRC*, fructose; *G1P*, glucose 1-phosphate; *G6P*, glucose 6-phosphate; *GAP*, glyceraldehyde 3-phosphate; *GLC*, glucose; *S6P*, sucrose 6-phosphate; *SUC*, sucrose. T is a transport protein that shuttles ADP-GLC across the plastid membrane.

3.22.3.2 Cellulose

Cellulose is one of the most abundant biomaterials on the earth. It is generally synthesized by plants, but it is also produced by some bacteria. Like starch, cellulose is a homopolymer of glucose, and yet unlike starch, glucose monomers are joined by β-1,4 linkages (Fig. 3.22.4B). Cellulose, a tough, fibrous, and water-insoluble polysaccharide, plays an integral role in keeping the structure of plant cell walls stable.[20] Cellulose chains are arranged in microfibrils or bundles of polysaccharide that are arranged in fibrils (bundles of microfibrils), which in turn make up the plant cell wall. This arrangement not only aids in the stability of plant structures but also suggests that cellulose is a biomaterial with high strength and other superior mechanical properties.

It is not only plants that synthesize cellulose. Bacteria are capable of producing the polysaccharide as well. Bacterial cellulose synthesis has been most extensively studied in *Acetobacter xylinum*. It is thought that the biological role of cellulose produced by bacteria is to aid in flocculation or to maintain certain environment, such as aerobic conditions or allowing attachment to plants.[21] Bacterial cellulose is arranged similarly to plant cellulose, as polysaccharide chains form microfibrils and bundles of microfibrils form ribbons.[22,23] Unlike plant-based cellulose, bacterial cellulose is highly pure and does not need to be separated from lignin in processing. Also unlike plant cellulose, bacterial cellulose has superior water retention properties; plant cellulose exhibits water retention values of 60%, whereas bacterial cellulose can exhibit retention of water up to 1000% of the cellulose sample weight.[24] Superior water retention of bacterial cellulose allows the polymer to possess high crystallinity, yet it is smooth and moldable, thus adding to its suitability in medical applications, such as a structural component of artificial organs and blood vessels.[22]

Cellulose synthase is a complex of enzymes that spans the cell membrane in plants and, in the case of bacteria, spans the entire cell wall. In both plants and bacteria, UDP-glucose (UDP-GLC) is the key intermediate in cellulose synthesis. The cellulose synthase complex uses the glucose moiety from UDP-GLC, transports the glucose across the cell membrane or cell wall, and adds the monomer to the nascent extracellular cellulose chain.[23,25]

Applications of plant-based and bacterial cellulose are many, and, as can be expected, some applications are more suitable for one type of cellulose over the other. Cellulose is the main constituent of paper and paper products as well as textiles made from cotton, linen, and other plant fibers.[24] Microcrystalline cellulose is used as a filler, both in the pharmaceutical and food industries. Cellulose can be converted into cellophane or nitrocellulose for use in photographic and household applications, respectively, as well as gunpowder. Nanocellulose, or nanocrystalline cellulose, is a material that is becoming increasingly more valued for several applications. Nanocellulose is produced by acid treatment of plant or bacterial cellulose. As the name implies, nanocellulose exhibits nanoscale structures, such as fibers or crystals. The properties of nanocellulose make it suited for applications such as photonics or optoelectronics. Crystallized nanocellulose has been proposed as a raw material for food packaging and electronics. In addition, nanocellulose can be derivatized with different chemical groups, which confers different properties to the polymer. Esterification of nanocellulose increases the hydrophobicity and strength of the material, while cationization confers pH sensitivity to nanocellulose, as well as allows for CO_2-controlled flocculation. Currently, like with many biopolymers, nanocellulose applications are limited because of availability and cost.[26] With such promising applications, it seems as though nanocellulose is a potential value target for metabolic engineering of microorganisms.

3.22.3.3 Chitin/Chitosan

While cellulose is the most abundant polysaccharide in nature, chitin is the second most abundant. The primary structure of chitin is very similar to that of cellulose, in that monosaccharides are joined by β-1,4 linkages (Fig. 3.22.4C). The main difference between chitin and cellulose is the monosaccharide of which the polymer is composed of. Chitin is a polymer of the amino sugar *N*-acetylglucosamine (NAG), an amino sugar. Chitosan is a polymer similar to chitin, except that chitosan is a heteropolymer composed of NAG and D-glucosamine monosaccharide groups. Chitin is found in the shells of crustaceans, insects, and mollusks, as well as in fungal cell walls. Chitin is synthesized via UDP-NAG as the principal precursor.[27] As with cellulose, the chitin biosynthetic machinery is set up to secrete the polysaccharide to the exterior of the cell. The structure of naturally occurring chitin is found in three main configurations, based on the alignment of the polysaccharide chains with respect to each other (Fig. 3.22.7). The most common configuration of chitin is the α configuration, which is found in crustacean and insect shells. β-Chitin is found in squid pens and beaks, while γ-chitin is found in fungi and yeasts.[28,29] The different configurations of chitin result in differences in mechanical properties. Experiments have shown that β-chitin is more soluble in solvents than α-chitin, suggesting that the packing between the chains is more "open" in β-chitin to allow solvent access.[28]

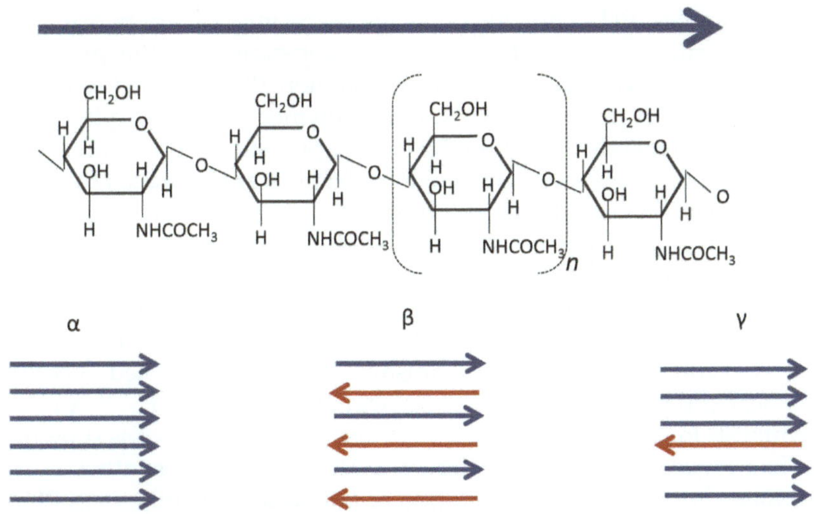

FIGURE 3.22.7 Alignment of chitin strands in different configurations. (Top) Image shows the directionality of a single chitin strand. (Bottom) α-Chitin is aligned monodirectionally, β-chitin is aligned bidirectionally in alternating fashion, and γ-chitin includes chains oriented in both directions, but the majority of the chitin strands face one direction. *Adapted from Kumirska J, Weinhold MX, Thoeming J, Stepnowski P. Biomedical activity of chitin/chitosan based materials—influence of physicochemical properties apart from molecular weight and degree of N-acetylation.* Polymers 2011;3:1875–901.

Chitin can be purified from crustacean or insect shells by deproteination and demineralization. The removal of protein and minerals from crustacean shells has traditionally been performed in two steps: (1) protein separation and removal by alkali treatment and (2) demineralization using high temperature and acidity.[29,30] The harsh conditions of the treatments often result in a degree of deacetylation in chitin, thus producing chitosan. Furthermore, if we are interested in chitin or chitosan as commercial biomaterials, then these traditional pretreatments leave something to be desired. Treatment of large quantities of crustacean shells in this way would result in large volumes of highly toxic wastewater requiring neutralization and treatment. Ideally, a more "natural" method of deproteination and demineralization of shells should be sought. Deproteination can be performed by microorganisms possessing a high amount of proteolytic activity. This would allow for wastewater that not only does not require significant treatment/neutralization but also contains plentiful nutrients (i.e., proteins and peptides). In the case of lobster shell treatment, enzymatic proteolysis could result in liberation of the molecule astaxanthin, which has potential pharmaceutical use.[30]

3.22.4 POLYHYDROXYALKANOATE—A NATURAL AND DIVERSE POLYESTER

Polyhydroxyalkanoate (PHA) is a family of polymers that are produced intracellularly as carbon and energy storage molecules for many types of microorganisms. Many bacteria produce PHA under conditions in which carbon is plentiful, but concentrations of other key nutrients (e.g., nitrogen phosphorus) are low. These polymers are accumulated in intracellular bodies known as granules. PHA monomers, as the name implies, are hydroxyalkanoic acids, and the monomer composition of PHA can vary, depending on the producing organism and the PHA synthase enzyme used. The most common PHA studied, indeed perhaps the most common type of PHA produced, is the homopolymer polyhydroxybutyrate (PHB). The type of PHA produced depends in part on the type of polymer synthase present in the biocatalyst organism. PHB and other short chain length (scl) PHAs are produced by *Ralstonia eutropha* (also called *Cupriavidus necator*), the model organism of PHA biosynthesis. *R. eutropha* and other similar microorganisms possess a Class I PHA synthase. Generally, organisms with this class of synthase produce scl-PHA like PHB and the copolymer poly (hydroxybutyrate-*co*-hydroxyvalerate) [P(HB-*co*-HV)]. PHA monomer content is also determined by the carbon source consumed by the biocatalyst. P(HB-*co*-HV) is typically only synthesized in *R. eutropha* and other organisms when they are fed 3-carbon (C3) and 5-carbon (C5) linear compounds, like propionic acid and valeric acid.[31,32] A schematic depiction of PHA synthesis in organisms that possess Class I PHA synthases is shown in Fig. 3.22.8.

Organisms like *Pseudomonas* use a different route for PHA biosynthesis. These bacteria utilize intermediates from fatty acid β-oxidation or fatty acid biosynthesis as PHA precursors. *Pseudomonas* has two dedicated enzymes that convert fatty acid β-oxidation or synthesis monomers [typically in the (*S*) configuration] to an (*R*)-hydroxyacyl-CoA precursor, which is then polymerized by a Class II PHA synthase enzyme. As a result, the PHA produced by these organisms contains multiple different monomers, usually four or more. Commonly, these monomers are much longer than the 4-carbon (C4) 3-hydroxybutyrate (3HB). The monomers present in PHA synthesized by *Pseudomonas* sp are of 6-carbon (C6) length and above. These

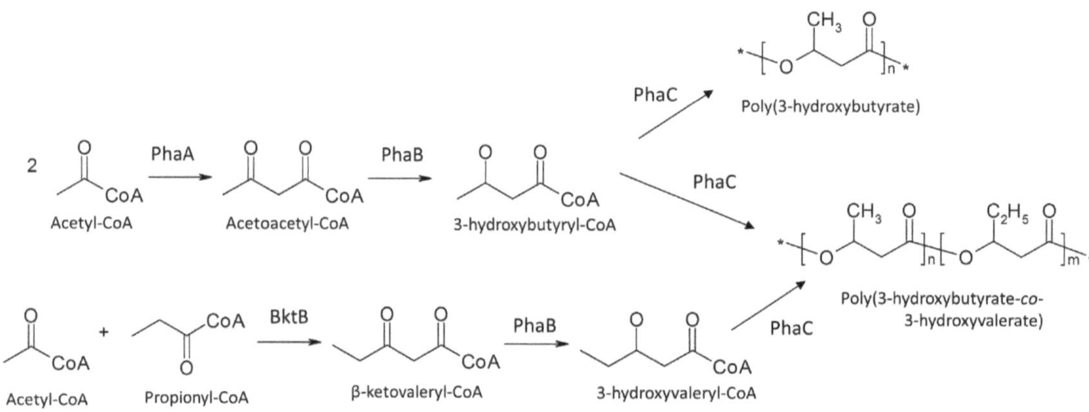

FIGURE 3.22.8 Polyhydroxyalkanoate (PHA) biosynthesis in organisms that contain a Class I PHA synthase (e.g., *Ralstonia eutropha*). Depending on the carbon sources fed to the organism, the polyhydroxybutyrate (PHB) polymer or the poly(hydroxybutryate-*co*-hydroxyvalerate) [P(HB-*co*-HV)] polymer can be made. For PHB: two molecules of acetyl-CoA are condensed by the β-ketothiolase enzyme PhaA to produce acetoacetyl-CoA; the acetoacetyl-CoA is reduced to 3-hydroxybutyryl-CoA by the reductase PhaB; and 3-hydroxybutyryl-CoA is the precursor for the PHA synthase PhaC. For P(HB-*co*-HV): a molecule of acetyl-CoA and a molecule of propionyl-CoA are condensed by the β-ketothiolase enzyme BktB to produce β-ketovaleryl-CoA; the ketovaleryl-CoA is reduced to 3-hydroxyvaleryl-CoA by the reductase PhaB; and 3-hydroxyvalerryl-CoA is the precursor for the PHA synthase PhaC.

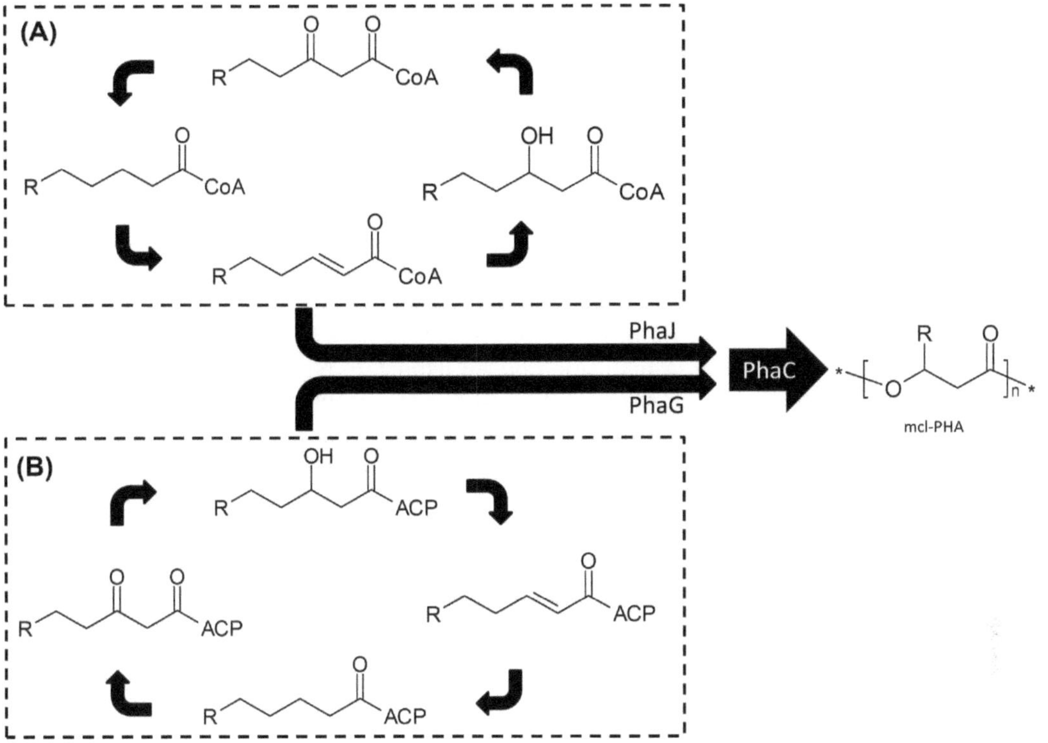

FIGURE 3.22.9 Synthesis of mcl-PHA. The fatty acid β-oxidation pathway (A) produces precursors for PHA synthesis as enoyl-CoA is syphoned off of the pathway by the enzyme PhaJ to produce monomers for PHA synthesis using the Class I PHA synthase, PhaC. The fatty acid biosynthesis pathway (B) produces hydroxyacyl-ACP, which is converted to PHA precursor by the 3-hydroxyacyl-ACP:CoA transacylase enzyme PhaG. The PHA synthase PhaC uses the 3-hydroxyacyl-CoA precursors as substrates for PHA synthesis. *mcl-PHA*, medium chain length polyhydroxyalkanoate; *ACP*, acyl carrier protein.

are called medium chain length (mcl) PHAs. Fig. 3.22.9 shows a schematic pathway of the mcl-PHA synthesis. There are also Class III and Class IV PHA synthases, which are less common than Class I and II synthases.[33,34] Thus far, those synthase enzymes have been shown to make only PHB. It should be noted that regardless of the class of PHA synthases, all enzymes require an active site cysteine moiety for function.

From a molecular viewpoint, PHA polymerization is a well-studied process. PHA synthase enzymes have been purified and their activities tested in vitro. For Class I synthases, like those produced by *R. eutropha*, time for initiation of polymerization is significantly greater than time for extension. Loading of the synthase enzyme might be the cause of this disparity in time. Purified PHA synthase from *R. eutropha* cells was shown to consist of three separate species: (1) a large portion of synthase monomer not bound with substrate, (2) a small synthase dimer portion not bound with substrate, and (3) an HMW fraction that consists of PHA synthase containing oligomeric PHA bound to the active site. In addition, the HMW fraction consisted of some PHA granule-associated proteins.[35] The enzymology of PHA synthesis is surprisingly simple, once the precursor molecules, 3-hydroxyacyl-CoA, are provided to the system. In each microbial system for synthesizing PHA, a dimer of active PHA synthase molecules is required for polymerization. In the case of Class III and Class IV synthases, which are heterodimers,[33,36] a heterotetramer consisting of two aforementioned heterodimers is required for chain elongation. The biochemical mechanism of chain initiation and elongation is of great interest to biopolymer researchers. The mechanism for initiation and elongation for the Class III PHA synthase from *Allochromatium vinosum* has been elucidated. As alluded to above, one synthase from *A. vinosum* is a heterodimer, consisting of PhaC, which contains the active site of the enzyme, and an accessory protein, PhaE. There are two models for activity for the *A. vinosum* PhaEC enzyme: one consists of a hydroxybutyrate monomer binding to the active site cysteine, while the nascent PHB chain is stabilized in the active site of the molecule by noncovalent interactions with other amino acid residues found in the active site, namely, a histidine and an aspartate residue; the other model takes into account the dimeric (or in the case of Class III synthases, heterotetrameric) nature of the PHA synthase. The nascent PHB chain is "swapped" back and forth between active site cysteine moieties, covalently binding in each active site, resulting in one cysteine covalently bound with polymer and the other unbound. Studies using active site mutant synthase enzymes and radiolabeled substrate showed that the most likely mechanism for chain elongation is the mechanism in which the polymer is unbound from the active site, stabilized by other amino acid residues, and rapidly rebound upon introduction of a new monomer.[37] PHA chain termination is also of interest, especially when the goal is to create custom-made PHA polymers for different applications. Several unique substrates were synthesized for testing the chain elongation mechanism of PHA synthases from different organisms.[38,39] One synthetic "monomer" possessed a chemical change in which the sulfur in 3HB-CoA (i.e., 3HB-SCoA) was converted to a methyl group ($3HBCH_2CoA$) and was found to be a good chain terminator.[39] Using a radiolabeled priming molecule for PHA synthase, the $3HBCH_2CoA$ was shown to promote chain termination, leaving a hydroxybutyryl moiety attached to the enzyme, suggesting that the chemical mechanism of chain termination is distinct from that of chain removal from the active site during elongation.

From a material property standpoint, PHAs that are composed of both scl and mcl monomers have been shown to have superior flexibility, elasticity, and toughness.[40–42] A small subset of microorganisms that are capable of synthesizing PHAs can produce copolymers consisting of an scl monomer, typically the C4 monomer hydroxybutyrate, and an mcl monomer, such as the C6 monomer hydroxyhexanoate, the C8 monomer hydroxyoctanoate, or others. The synthase enzymes required to polymerize these copolymers are very similar to Class I enzymes, but with key amino acid residue changes that broaden the substrate specificity. These enzymes, and the bacteria that produce them, are very desirable for production of PHA with favorable mechanical properties for diverse applications.[43,44]

A commonly studied PHA copolymer containing both an scl- and an mcl-HA monomer is poly(3-hydroxybutyrate-*co*-3-hydroxyhexanoate) [P(HB-*co*-HHx)]. The P(HB-*co*-HHx) copolymer can be produced by wild-type strains of bacteria (e.g., *Aeromonas caviae*) and recombinant strains of bacteria (e.g., *R. eutropha* and *Escherichia coli*). It has been shown that recombinant *R. eutropha* strains produce up to 80% of their cell dry weight in P(HB-*co*-HHx) when cultivated on palm oil,[43,45] crude palm kernel oil, coconut oil,[46] and waste animal fat,[47] among others, as the sole carbon source. The recombinant strains express a broad substrate range PHA synthase from *Rhodococcus aetherivorans*, and the *PhaJ* gene, encoding an (*R*)-specific enoyl-CoA hydratase.[34,48] The PhaJ enzyme utilizes fatty acid β-oxidation intermediates as substrates and converts them to PHA precursors.[34,42] The specific PhaJ enzyme expressed in the recombinant *R. eutropha* strains is able to produce HHx-CoA as the main product when suitable carbon sources (e.g., lipids and fatty acids) are present. This is not to suggest that other, longer-chain-length HA-CoA molecules cannot be produced. It is possible that the heterologously expressed PhaC enzyme is unable to incorporate monomers that are longer than C6. As of the writing of this chapter, the biochemistry of this enzyme has not been studied in detail. P(HB-*co*-HHx) exhibits thermal and mechanical properties that make it a suitable biodegradable alternative to many petroleum-based polymers. The melting temperature has been shown to range from 52 °C to 150 °C, depending on the hydroxyhexanoate (HHx) monomer content present in the PHA.[40,46,49,50] The elongation to break of many P(HB-*co*-HHx) preparations varies from 6.5% to 1075%, again depending in a large part on the monomer concentrations.[46,49–52] A PHA homopolymer that has been shown to have favorable mechanical properties is poly(4-hydroxybutyrate) [P(4HB)]. This polymer can be produced by wild-type organisms containing the Class I PHA synthase; however, the cultures must be fed a suitable precursor molecule, like butyrolactone.[53,54]

PHA polymers have been examined for many different applications. PHB and some of its copolymers have been examined in packaging, agricultural, and industrial use. A chief impediment to using these biopolymers in place of petroleum-based plastics is the high cost of fermentation and recovery of product. Thus, with PHA, typically higher value applications are sought. One such high-value application for PHA that has been studied by several research groups and biotechnology companies is the use of PHA as raw material for medical implants, sutures, drug delivery matrices, and other health care applications. PHAs of many types have been shown to be biocompatible in many different animal model systems.[49,55] Homopolymers like P(4HB) and copolymers like P(3HB-*co*-4HB) or P(HB-*co*-HHx) have suitable material properties to allow these polymers to be used as sutures. To be used as suture material, a polymer must be flexible and tough. In addition, a material used as a biodegradable (resorbable) suture must maintain the desirable properties long enough to allow for wound closure and healing, prior to breakdown in the body. Currently, Tepha, Inc., a biotechnology company in Lexington, MA, USA, produces sutures and surgical meshes that are flexible, strong, and sufficiently biodegradable, fabricated from P(4HB). The biodegradability of PHA allows for use as a drug delivery carrier. PHA matrixes of different types have been examined for their degradation and concomitant release of drugs like antibiotics or eukaryotic cell growth inhibitors.

One aspect of PHAs that makes the family of polymers desirable as replacement raw materials for petroleum-based plastics is their biodegradability. As PHA is a form of carbon storage for several microorganisms, many species that synthesize the polyester also express dedicated intracellular PHA depolymerase enzymes.[34,56] However, when PHA has been removed from intracellular granules, it loses its completely amorphous character and becomes semicrystalline.[57,58] Intracellular PHA depolymerases from microbial species discovered thus far are unable to act on semicrystalline polymer. However, many microbes possess extracellular enzymes capable of breaking down semicrystalline PHA polymer. These extracellular enzymes include dedicated PHA depolymerases,[59] as well as lipase and esterase enzymes that act in a nonspecific manner to break down extracellular PHA.[60] Thus higher the concentration of microbes in an environmental niche, the more likely rapid degradation of PHA matrixes (and products) will occur. This has been demonstrated by incubating PHA films in many different environments, from river water to soil to sewage. The environments that contain the largest concentration of microbes (i.e., sewage and compost) were able to rapidly degrade the films to completion (from 10 to 50 weeks).[61] Different soil communities, containing different microbial makeup, were examined for PHA degradation. It was shown that, as the composition and quantity of the microbiota of root zones of different trees changed over time, so too did the ability to decompose PHA matrixes.[59] The same trend was also shown in eutrophic zones of river reservoirs.[62]

Degradation of PHA by lipases and other enzymes nonspecific to the polymer is the key phenomenon allowing for use of these materials as surgical implants, tissue scaffolds, sutures, and other medical products. PHA implants have been shown to degrade in different animal systems, including rats,[63] mice,[64] and human cell culture.[65] For materials like PHA to be used for medical purposes, they must be shown to be biocompatible. PHB, for example, has been shown to be a natural constituent of mammalian cells (including human cells) as part of ion channels.[55,66–68] Regardless of this, are PHA copolymers and their potential degradation products nontoxic and nonimmunogenic in animals? Research groups have characterized biocompatibility of different PHA polymers in different animal systems. Polymers like P(4HB), P(3HB-*co*-4HB), and P(HB-*co*-HHx) have been demonstrated to be nontoxic and generally biocompatible in rats, rabbits, and dogs.[55] It should be noted, however, that great care must be taken in preparation of PHA matrixes for implantation or other contact with animal tissue. Since PHA is a bacterial product, there is a high likelihood that isolation of the polymer will also bring along the presence of contaminants like the potentially immunogenic lipopolysaccharide (LPS). Thus, extra purification steps are usually warranted if the PHA is to be used as a medical polymer. Typically, isolated PHA is redissolved in solvent and then reprecipitated to produce a purer polymer. This can be done numerous times to remove potentially toxic LPS and other contaminants.[64,69] PHA has also been examined as a carrier for time-release drugs. Ideally, the porosity of PHA and the biodegradation in tissues will result in the release over time of the embedded drug, thus delivering drug over an extended period of time. This application has been examined using antibiotics as the type of time-release drug of choice. Many different types of PHA have been tested, and all have shown the ability of drug release in solution over time.[55,70,71]

To use PHA for household and industrial purposes, data must be gathered on how aging and degradation affects the material itself. This is especially true if the purpose of the PHA material is to degrade, as in its use in surgical implants and sutures. An important aspect of PHA polymers is the monomer composition. As mentioned before, the presence of scl and mcl monomers in PHA results in a polymers that is more flexible and tough.[49] In such copolymers, regions of mcl-HA monomers disrupt the crystalline packing of the scl-HA monomer chains, in turn decreasing the overall crystallinity of the polymer sample.[72,73] The effect of decreasing the overall crystallinity of PHA is the enhancement of biodegradability. The increase of HHx content in P(HB-*co*-HHx) has been shown to make the polymer easier to degrade, presumably because disruption of crystalline packing allows for easier access of enzymes involved in degradation.[72,73]

3.22.5 CONCLUSION AND OUTLOOK

We are gaining an increased understanding about how biomaterials can be used for a variety of applications, especially those in which petroleum-based polymers have been the typical raw material. In many cases, these polymers play structural roles in their native organisms and thus have the material properties necessary for fabrication of a durable product. The two key aspects of the polymers mentioned in this chapter, however, are their bio-based origin and potential for biodegradability. Due to the high price of production and isolation of raw material, it is likely that biopolymers will never fully replace petrochemical polymers in all applications. However, given the advances that have been made in biomaterials over the past 20 years, bio-based polymers are an excellent alternative for many applications. Also, given the amount of waste we produce due to consumption of food and other goods, we have a pressing need for biodegradable alternatives in packaging and other applications. The hope is that many natural products will be available to replace plastics that end up as waste.

References

1. *Plastic*. 2011. Available at: http://www.merriam-webster.com/dictionary/plastic.
2. APPR, ACC. *2014 United States national postconsumer plastics bottle recycling report*. 2015. p. 1–15.
3. Walsh G. *Proteins: biochemistry and biotechnology*. Singapore: John Wiley & Sons, Inc.; 2014.
4. Liu D, Nikoo M, Boran G, Zhou P, Regenstein JM. Collagen and gelatin. *Annu Rev Food Sci Technol* 2015;**6**:527–57.
5. Subhan F, Ikram M, Shehzad A, Ghafoor A. Marine collagen: an emerging player in biomedical applications. *J Food Sci Technol* 2015;**52**:4703–7.
6. Di Lullo GA, Sweeney SM, Korkko J, Ala-Kokko L, San Antonio JD. Mapping the ligand-binding sites and disease-associated mutations on the most abundant protein in the human, type I collagen. *J Biol Chem* 2002;**277**:4223–31.
7. Klatt AR, Paul-Klausch B, Klinger G, Kuhn G, Renno JH, Banerjee M, et al. A critical role for collagen II in cartilage matrix degradation: collagen II induces pro-inflammatory cytokines and MMPs in primary human chondrocytes. *J Orthop Res* 2009;**27**:65–70.
8. Schauss AG, Stenehjem J, Park J, Endres JR, Clewell A. Effect of the novel low molecular weight hydrolyzed chicken sternal cartilage extract, BioCell Collagen, on improving osteoarthritis-related symptoms: a randomized, double-blind, placebo-controlled trial. *J.Agric Food Chem* 2012;**60**:4096–101.
9. Xu D, Shen W. Chicken collagen type II reduces articular cartilage destruction in a model of osteoarthritis in rats. *West Indian Med J* 2007;**56**:202–7.
10. Cunniffe G, O'Brien F. Collagen scaffolds for orthopedic regenerative medicine. *J Miner Metal Mater Soc* 2011;**63**:66–73.
11. Oliveira SM, Ringshia RA, Legeros RZ, Clark E, Yost MJ, Terracio L, et al. An improved collagen scaffold for skeletal regeneration. *J Biomed Mater Res A* 2010;**94**:371–9.
12. Djagny VB, Wang Z, Xu S. Gelatin: a valuable protein for food and pharmaceutical industries: review. *Crit Rev Food Sci Nutr* 2001;**41**:481–92.
13. Ward AG, Courts A, editors. *The science and technology of gelatin*. New York: Academic Press; 1977.
14. Altman GH, Diaz F, Jakuba C, Calabro T, Horan RL, Chen J, et al. Silk-based biomaterials. *Biomaterials* 2003;**24**:401–16.
15. Tokareva O, Michalczechen-Lacerda VA, Rech EL, Kaplan DL. Recombinant DNA production of spider silk proteins. *Microb Biotechnol* 2013;**6**:651–63.
16. Rising A, Johansson J. Toward spinning artificial spider silk. *Nat Chem Biol* 2015;**11**:309–15.
17. Ayoub NA, Garb JE, Tinghitella RM, Collin MA, Hayashi CY. Blueprint for a high-performance biomaterial: full-length spider dragline silk genes. *PLoS One* 2007;**2**:e514.
18. Beckles DM, Thitisaksakul M. Use of biotechnology to engineer starch in cereals. In: Heldman DR, Hoover DG, Wheeler MB, editors. *Encyclopedia of biotechnology in agriculture and food*. CRC Press; 2010. p. 1–8.
19. van Soest JJ, Vliegenthart JF. Crystallinity in starch plastics: consequences for material properties. *Trends Biotechnol* 1997;**15**:208–13.
20. Bringmann M, Landrein B, Schudoma C, Hamant O, Hauser MT, Persson S. Cracking the elusive alignment hypothesis: the microtubule-cellulose synthase nexus unraveled. *Trends Plant Sci* 2012;**17**:666–74.

21. Jones R, Farah LF. Production and application of microbial cellulose. *Polym Degrad Stab* 1998;**59**:101–6.
22. Klemm D, Schumann D, Udhardt U, Marsch S. Bacterial synthesized cellulose – artificial blood vessels for microsurgery. *Prog Polym Sci* 2001;**26**:1561–603.
23. Endler A, Sanchez-Rodriguez C, Persson S. Glycobiology: cellulose squeezes through. *Nat Chem Biol* 2010;**6**:883–4.
24. Klemm D, Heublein B, Fink HP, Bohn A. Cellulose: fascinating biopolymer and sustainable raw material. *Angew Chem Int Ed Engl* 2005;**44**:3358–93.
25. Brown Jr RM, Willison JH, Richardson CL. Cellulose biosynthesis in *Acetobacter xylinum*: visualization of the site of synthesis and direct measurement of the in vivo process. *Proc Natl Acad Sci USA* 1976;**73**:4565–9.
26. Abitbol T, Rivkin A, Cao Y, Nevo Y, Abraham E, Ben-Shalom T, et al. Nanocellulose, a tiny fiber with huge applications. *Curr Opin Biotechnol* 2016;**39**:76–88.
27. Merzendorfer H, Zimoch L. Chitin metabolism in insects: structure, function and regulation of chitin synthases and chitinases. *J Exp Biol* 2003;**206**:4393–412.
28. Kumirska J, Weinhold MX, Thoeming J, Stepnowski P. Biomedical activity of chitin/chitosan based materials—influence of physicochemical properties apart from molecular weight and degree of *N*-acetylation. *Polymers* 2011;**3**:1875–901.
29. Younes I, Rinaudo M. Chitin and chitosan preparation from marine sources. Structure, properties and applications. *Mar Drugs* 2015;**13**:1133–74.
30. Arbia W, Arbia L, Adour L, Amrane A. Chitin extraction from Crustacean shells by biological methods – a review. *Food Technol Biotechnol* 2012;**51**:12–25.
31. Yang YH, Brigham CJ, Budde CF, Boccazzi P, Willis LB, Hassan MA, et al. Optimization of growth media components for polyhydroxyalkanoate (PHA) production from organic acids by *Ralstonia eutropha*. *Appl Microbiol Biotechnol* 2010;**87**:2037–45.
32. Yang YH, Brigham CJ, Song E, Jeon JM, Rha CK, Sinskey AJ. Biosynthesis of poly(3-hydroxybutyrate-*co*-3-hydroxyvalerate) containing a predominant amount of 3-hydroxyvalerate by engineered *Escherichia coli* expressing propionate-CoA transferase. *J Appl Microbiol* 2012;**113**:815–23.
33. Rehm BH. Polyester synthases: natural catalysts for plastics. *Biochem J* 2003;**376**:15–33.
34. Brigham CJ, Sinskey AJ. Polyhydroxyalkanoate production enzymes: a survey and biological perspective. *J Sib Fed Univ* 2012;**3**:220–42.
35. Cho M, Brigham CJ, Sinskey AJ, Stubbe J. Purification of polyhydroxybutyrate synthase from its native organism, *Ralstonia eutropha*: implications for the initiation and elongation of polymer formation in vivo. *Biochemistry* 2012;**51**:2276–88.
36. Yuan W, Jia Y, Tian J, Snell KD, Muh U, Sinskey AJ, et al. Class I and III polyhydroxyalkanoate synthases from *Ralstonia eutropha* and *Allochromatium vinosum*: characterization and substrate specificity studies. *Arch Biochem Biophys* 2001;**394**:87–98.
37. Li P, Chakraborty S, Stubbe J. Detection of covalent and noncovalent intermediates in the polymerization reaction catalyzed by a C149S class III polyhydroxybutyrate synthase. *Biochemistry* 2009;**48**:9202–11.
38. Buckley RM, Stubbe J. Chemistry with an artificial primer of polyhydroxybutyrate synthase suggests a mechanism for chain termination. *Biochemistry* 2015;**54**:2117–25.
39. Zhang W, Shrestha R, Buckley RM, Jewell J, Bossmann SH, Stubbe J, et al. Mechanistic insight with HBCH2CoA as a probe to polyhydroxybutyrate (PHB) synthases. *ACS Chem Biol* 2014;**9**:1773–9.
40. Fukui T, Abe H, Doi Y. Engineering of *Ralstonia eutropha* for production of poly(3-hydroxybutyrate-*co*-3-hydroxyhexanoate) from fructose and solid-state properties of the copolymer. *Biomacromolecules* 2002;**3**:618–24.
41. Insomphun C, Mifune J, Orita I, Numata K, Nakamura S, Fukui T. Modification of beta-oxidation pathway in *Ralstonia eutropha* for production of poly(3-hydroxybutyrate-*co*-3-hydroxyhexanoate) from soybean oil. *J Biosci Bioeng* 2013;**117**:184–90.
42. Riedel SL, Lu J, Stahl U, Brigham CJ. Lipid and fatty acid metabolism in *Ralstonia eutropha*: relevance for the biotechnological production of value-added products. *Appl Microbiol Biotechnol* 2014;**98**:1469–83.
43. Budde CF, Riedel SL, Willis LB, Rha C, Sinskey AJ. Production of poly(3-hydroxybutyrate-*co*-3-hydroxyhexanoate) from plant oil by engineered *Ralstonia eutropha* strains. *Appl Environ Microbiol* 2011;**77**:2847–54.
44. Fukui T, Doi Y. Cloning and analysis of the poly(3-hydroxybutyrate-*co*-3-hydroxyhexanoate) biosynthesis genes of *Aeromonas caviae*. *J Bacteriol* 1997;**179**:4821–30.

45. Riedel SL, Bader J, Brigham CJ, Budde CF, Yusof ZA, Rha C, et al. Production of poly(3-hydroxybutyrate-*co*-3-hydroxyhexanoate) by *Ralstonia eutropha* in high cell density palm oil fermentations. *Biotechnol Bioeng* 2012;**109**:74–83.
46. Wong YM, Brigham CJ, Rha C, Sinskey AJ, Sudesh K. Biosynthesis and characterization of polyhydroxyalkanoate containing high 3-hydroxyhexanoate monomer fraction from crude palm kernel oil by recombinant *Cupriavidus necator*. *Bioresour Technol* 2012;**121**:320–7.
47. Riedel SL, Jahns S, Koenig S, Bock MC, Brigham CJ, Bader J, et al. Polyhydroxyalkanoates production with *Ralstonia eutropha* from low quality waste animal fats. *J Biotechnol* 2015;**214**:119–27.
48. Hisano T, Tsuge T, Fukui T, Iwata T, Miki K, Doi Y. Crystal structure of the (R)-specific enoyl-CoA hydratase from *Aeromonas caviae* involved in polyhydroxyalkanoate biosynthesis. *J Biol Chem* 2003;**278**:617–24.
49. Kehail A, Foshey M, Chalivendra V, Brigham C. Thermal and mechanical characterization of solvent-cast poly(3-hydroxybutyrate-*co*-3-hydroxyhexanoate). *J Polym Res* 2015;**22**:216.
50. Doi Y, Kitamura S, Abe H. Microbial synthesis and characterization of poly(3-hydroxybutyrate-*co*-3-hydroxyhexanoate). *Macromolecules* 1995;**28**:4822.
51. Jeon JM, Brigham CJ, Kim YH, Kim HJ, Yi DH, Kim H, et al. Biosynthesis of poly(3-hydroxybutyrate-*co*-3-hydroxyhexanoate) (P(HB-*co*-HHx)) from butyrate using engineered *Ralstonia eutropha*. *Appl Microbiol Biotechnol* 2014;**98**:5461–9.
52. Asrar J, Valentin HE, Berger PA, Tran M, Padgette SR, Garbow JR. Biosynthesis and properties of poly(3-hydroxybutyrate-*co*-3-hydroxyhexanoate) polymers. *Biomacromolecules* 2002;**3**:1006–12.
53. Hein S, Sohling B, Gottschalk G, Steinbuchel A. Biosynthesis of poly(4-hydroxybutyric acid) by recombinant strains of *Escherichia coli*. *FEMS Microbiol Lett* 1997;**153**:411–8.
54. Moore T, Adhikari R, Gunatillake P. Chemosynthesis of bioresorbable poly(gamma-butyrolactone) by ring-opening polymerisation: a review. *Biomaterials* 2005;**26**:3771–82.
55. Brigham CJ, Sinskey AJ. Applications of polyhydroxyalkanoates in the medical industry. *Int J Biotech Wellness Ind* 2012;**1**:53–60.
56. Jendrossek D, Handrick R. Microbial degradation of polyhydroxyalkanoates. *Annu Rev Microbiol* 2002;**56**:403–32.
57. Jendrossek D, Pfeiffer D. New insights in the formation of polyhydroxyalkanoate granules (carbonosomes) and novel functions of poly(3-hydroxybutyrate). *Environ Microbiol* 2014;**16**:2357–73.
58. Papageorgiou AC, Hermawan S, Singh CB, Jendrossek D. Structural basis of poly(3-hydroxybutyrate) hydrolysis by PhaZ7 depolymerase from *Paucimonas lemoignei*. *J Mol Biol* 2008;**382**:1184–94.
59. Boyandin AN, Prudnikova SV, Filipenko ML, Khrapov EA, Vasilev AD, Volova TG. Biodegradation of polyhydroxyalkanoates by soil microbial communities of different structures and Detection of PHA degrading microorganisms. *Appl Biochem Microbiol* 2012;**48**:28–36.
60. Shang L, Fei Q, Zhang YH, Wang XZ, Fan DD, Chang HN. Thermal properties and biodegradability studies of poly(3-hydroxybutyrate-*co*-3-hydroxyvalerate). *J Polym Environ* 2012;**20**:23–8.
61. Matavulj M, Molitoris HP. Biodegradation of polyhydroxyalkanoate-based plastic (BIOPOL) under different environmental conditions: I. weight loss of substrate. *Denkschr Regensb Bot Ges* 2000;**61**:735–49.
62. Volova TG, Gladyshev MI, Trusova MY, Zhila NO. Degradation of polyhydroxyalkanoates in eutrophic reservoir. *Polym Degrad Stab* 2007;**92**:580–6.
63. Ying TH, Ishii D, Mahara A, Murakami S, Yamaoka T, Sudesh K, et al. Scaffolds from electrospun polyhydroxyalkanoate copolymers: fabrication, characterization, bioabsorption and tissue response. *Biomaterials* 2008;**29**:1307–17.
64. Shishatskaya EI. Biomedical investigations of biodegradable PHAs. *Macromol Symp* 2008;**269**:65–81.
65. Xu XY, Li XT, Peng SW, Xiao JF, Liu C, Fang G, et al. The behaviour of neural stem cells on polyhydroxyalkanoate nanofiber scaffolds. *Biomaterials* 2010;**31**:3967–75.
66. Reusch RN. Biological complexes of poly-beta-hydroxybutyrate. *FEMS Microbiol Rev* 1992;**9**:119–29.
67. Reusch RN. Physiological importance of poly-(R)-3-hydroxybutyrates. *Chem Biodivers* 2012;**9**:2343–66.
68. Reusch RN. Poly-(R)-3-hydroxybutyrates (PHB) are atherogenic components of lipoprotein Lp(a). *Med Hypotheses* 2015;**85**:1041–3.
69. Riedel SL, Brigham CJ, Budde CF, Bader J, Rha C, Stahl U, et al. Recovery of poly(3-hydroxybutyrate-*co*-3-hydroxyhexanoate) from *Ralstonia eutropha* cultures with non-halogenated solvents. *Biotechnol Bioeng* 2012;**110**:461–70.

70. Murueva AV, Shishatskaya EI, Kuzmina AM, Volova TG, Sinskey AJ. Microparticles prepared from biodegradable polyhydroxyalkanoates as matrix for encapsulation of cytostatic drug. *J Mater Sci Mater Med* 2013;**24**:1905–15.
71. Shishatskaya EI, Goreva AV, Voinova ON, Inzhevatkin EV, Khlebopros RG, Volova TG. Evaluation of antitumor activity of rubomycin deposited in absorbable polymeric microparticles. *Bull Exp Biol Med* 2008;**145**:358–61.
72. Morse MC, Liao Q, Criddle CS, Frank CW. Anaerobic biodegradation of the microbial copolymer poly(3-hydroxybutyrate-*co*-3-hydroxyhexanoate): effects of comonomer content, processing history, and semi-crystalline morphology. *Polymer* 2011;**52**:547–56.
73. Numata K, Abe H, Iwata T. Biodegradability of poly(hydroxyalkanoate) materials. *Materials* 2009;**2**:1104–26.

CHAPTER

3.23

Modern Applications of Green Chemistry: Renewable Energy

John Collins[1], *Gerald Gourdin*[2], *Deyang Qu*[3]

[1]IBM Thomas J Watson Research Center, Yorktown Heights, NY, United States; [2]Georgia Institute of Technology, Atlanta, GA, United States; [3]University of Wisconsin Milwaukee, Milwaukee, WI, United States

3.23.1 The Static Concentration of Energy in Chemical Bonds and the Physical Double Layer: Modern Methods of Energy Storage

3.23.1.1 INTRODUCTION

New energy generation systems such as solar and wind systems have great potential for eliminating our dependency on real-time, on-demand energy *generation* systems. However, these new-generation systems are intermittent and periodic, so there is a need for new forms of efficient, on-demand, energy *storage* systems to fully exploit green generation technologies. It is these technological demands that are driving the need for specialized electrochemical energy storage devices. However, before we can delve into discussing those specialized electrochemical energy storage devices in this and the following sections of this chapter, it is necessary to provide a short review of the ways that energy is stored in chemical bonds and how it can be harnessed.

There are two fundamental ways that the energy stored in chemical bonds can be captured or harnessed. The first is to break the bonds, through a chemical or thermal process, and then capture that energy that is released, in the form of heat, and put it to work. Releasing the energy stored in covalent bonds by breaking those bonds is the domain of combustion. The other way to harness a chemical bond's stored energy is through the direct transfer of

Green Chemistry
http://dx.doi.org/10.1016/B978-0-12-809270-5.00028-5

electrons from a chemical species or a bond to another chemical species for bond formation. Although the net effect is the same—one or more chemical species is converted to another chemical species—it can be shown that the electron transfer process is a much more efficient method of harnessing a chemical bond's energy. This process is the province of electrochemistry and electrochemical energy storage devices.

Since the breaking of chemical bonds is an endothermic process, new bonds have to be formed and any excess energy released and captured for use. The classic example of combustion is the consumption of a hydrocarbon in the presence of oxygen (the reactants), which forms water and CO_2 as the products.

$$C_nH_{2n+2} + \frac{1}{2}(3n+1)O_2 \rightarrow (n+1)H_2O + nCO_2 \tag{3.23.1.1}$$

The energy that is produced during combustion of a hydrocarbon or other chemical species is released in the form of heat. Alone, however, the energy that is released *only* provides heat and is not in a form that is very useful for doing work. Therefore a system has to be designed so that the released energy can be captured in such a form to harness it to do work, and the system (or device) that accomplishes this is the heat engine.

3.23.1.2 HEAT ENGINE

Under the basic definition, a heat engine is a device or system that converts heat or thermal energy—and chemical energy—to mechanical energy, which can then be used to do mechanical work. Work is defined as a force that enables the displacement of mass (e.g., fuel combustion that results in an increase of pressure/heat—where the increased pressure displaces the pistons of an engine, which consequently turns wheels via an axle). A heat engine can harness the energy that is stored in chemical bonds, through the breaking of those bonds and the forming of new bonds, and then utilize the excess energy that is generated from that process to do work. Unfortunately, the theoretical amount of energy that is actually available is limited by a certain constraint summarized in Carnot's theorem, which states that *no engine operating between two heat reservoirs can be more efficient than a Carnot engine operating between those same reservoirs*. Hence, since it is not possible to use heat energy taken from the source (high-temperature reservoir) to *lower* the temperature of the sink (low-temperature reservoir) more than it already is, the amount of heat available to do work will be limited by the energy difference between the source and the sink. This concept is graphically illustrated in Fig. 3.23.1.1.

To understand the limitations of Carnot's theorem, we can evaluate the example of a device that harnesses this method of energy utilization: the internal combustion engine (ICE). The ICE is a Carnot heat engine that is designed to convert the stored chemical energy in a fuel to mechanical work through the combustion of that fuel. Unfortunately, this process is an inefficient one, and, when accounting for thermodynamic considerations, there is a maximum limit to the theoretical efficiency of a Carnot heat engine. To put it simply, the excess energy that is released from combustion is used to expand a gas (i.e., the atmosphere within the cylinder), which is then used to move a piston. The mechanical motions of the moving pistons are then used to move and power a vehicle. The chemicals combusted in

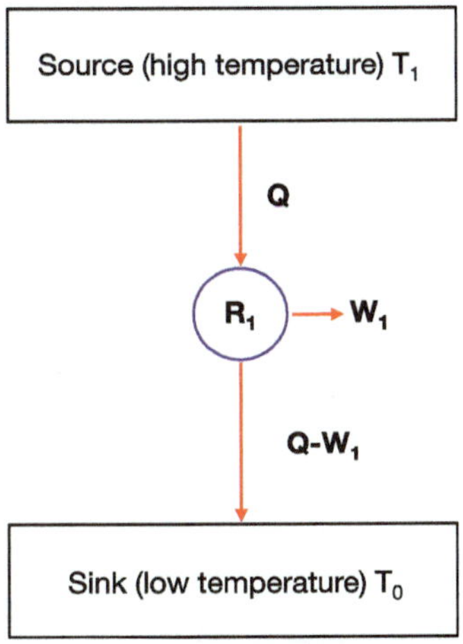

FIGURE 3.23.1.1 The Carnot cycle. Q is the total available heat taken from the source, Q-W_1 is the net heat that is lost to the sink that is unavailable to do work, and W_1 is the actual work that can be done by the system.

an ICE result in an increase in the kinetic energy ($^3\!/_2 k_b T_1$) of the gaseous products of those reactions and in the surrounding atmosphere. A part of that kinetic energy is transferred to do the work of moving that piston, and the remainder ($^3\!/_2 k_b T_2$) is released to the environment (the sink). From Carnot's theorem, the maximum theoretical efficiency can be calculated from the following expression:

$$\varepsilon = \frac{\frac{3}{2}kT_1 - \frac{3}{2}kT_2}{\frac{3}{2}kT_1} = \frac{T_1 - T_2}{T_1} \qquad (3.23.1.2)$$

The maximum limit to the efficiency of the engine is determined by the difference in temperature between the heat source and the heat sink. For gasoline, the ignition temperature in air is approximately 542K. Assuming the heat source is near that temperature, the ideal maximum efficiency of a heat engine would be about 49%. Although this may seem relatively high, there is, in addition, a loss of energy in the form of heat being transferred to the surroundings (i.e., the walls) through convection, solid-state conduction, and the friction associated with moving the pistons. This loss of energy decreases the kinetic energy of the gas particles that are formed from the combustion of the fuel and the heat energy also raises the temperature (T_2) of the sink/surroundings (the engine), therefore the actual amount of work that can be performed through gas expansion is decreased even further. In the case of the ICE operating at the thermodynamic limit, the efficiency can be decreased by up to half, or to about 25%–39%. A diagram illustrating these additional losses is presented in Fig. 3.23.1.2.

In the previous paragraphs, utilization of the energy stored in chemical bonds via a conventional heat engine was illustrated in the context of the ICE and its application to supplying energy for transportation. Conventional heat engines can also be used to produce electricity from chemical energy through an intermediate mechanical process that converts the heat

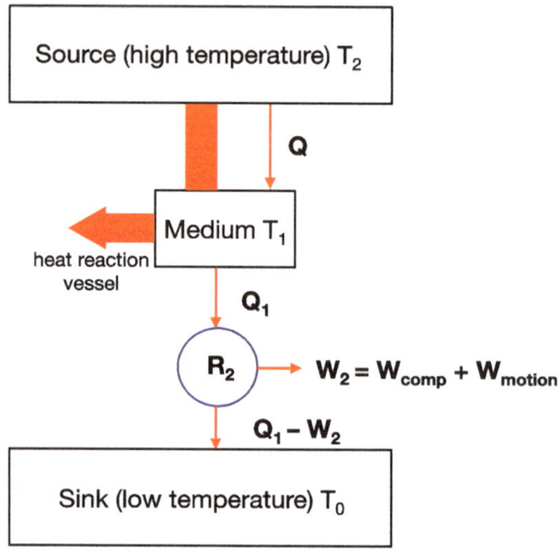

FIGURE 3.23.1.2 The internal combustion engine as a modified Carnot heat engine. T_1 is the temperature of the medium (surroundings, chamber walls, etc.) between the source (hot) and sink (cold) of the engine. R_2 is the amount of work done by the system after compensating for heat lost to the reaction vessel.

energy into mechanical energy that drives a device that produces electrical energy, which is the basis of operation of fossil fuel-based power plants. The same discussion regarding the theoretical and actual limitations on the efficiency of the process can then be applied to the combustion of natural gas in the conversion of water to steam to drive an electricity-producing turbine. A certain amount of energy is applied to raise the temperature of the natural gas sufficiently for ignition to take place, but the theoretical maximum amount of energy available to do work (i.e., produce steam) is also likewise limited by the lower temperature of the heat sink. Furthermore, the actual efficiency is limited by losses to the environment, friction, etc. This, of course, results in an even greater reduction in efficiency.[1] In the end, the same limits on the maximum efficiency apply.

3.23.1.3 ENERGY AND ELECTRON TRANSFER OR "COLD COMBUSTION"

The preceding discussion described one way of accessing the energy stored in chemical bonds, but as was shown, there are significant limitations to the use of that energy in terms of the maximum theoretical efficiency of the process. However, it should be recalled that the breakage/formation of chemical bonds during combustion was just one way to access the energy that is stored in chemical bonds. The direct transfer of electrons from a chemical species or a bond to another chemical species or for bond formation was the other way to access the energy stored in chemical bonds. As the reader will soon see, the direct transfer of electrons from one medium to another is a more efficient way to access that stored energy. One way this "alternate" process can be illustrated and understood is through a rather interesting example: the introduction of sodium metal to an atmosphere of chlorine gas. When the sodium metal and chlorine gas come into contact at the right ratios, with a little bit of water, the sodium burns, the chlorine-chlorine bond breaks, and sodium chloride is formed as the product, as spectacularly illustrated in this online video (https://www.youtube.com/watch?v=tbPxwDiX1NU).

$$2Na_{(s)} + Cl_{2\,(g)} \rightarrow NaCl_{(s)} \quad (3.23.1.3)$$

What is not directly shown in that reaction is that there was a transfer of electrons, but if we can break the complete reaction into its component steps, it can be shown that, as a molecule of sodium loses an electron to become a cation, the chlorine molecule picks up those electrons to become anions.

$$\begin{aligned} Na &\rightarrow Na^+ + e^- \\ Cl_2 + 2e^- &\rightarrow Cl^- \end{aligned} \quad (3.23.1.4)$$

In this reaction, there is a direct transfer of an electron from a sodium atom to a chlorine atom to form ions, which is the second way that energy can be harnessed from chemical bonds. The reduction of one chemical species (e.g., chlorine) paired with the oxidation of another species (e.g., sodium) is a more efficient way of electrons being transferred between species. In the case of ionic bonds, the energy associated with that bond comes in the form of electron transfer: an electron is transferred from one species to another and the resultant electrostatic attraction between the now charged species forms the bond. The practical efficiency of an electron transfer process used to produce or store energy is significantly higher than that for a Carnot heat engine, since it is not subjected to those thermodynamic constraints. We can illustrate this by using the combustion of methanol as an example.

$$CH_3OH_{(l)} + {}^3\!/_2O_{2\,(g)} \rightarrow 2H_2O_{(l)} + CO_{2\,(g)} \quad (3.23.1.5)$$

The enthalpy change for this reaction is −726 kJ/mol, but the actual available energy, the Gibbs energy, will be −702 kJ/mol. It is necessary for a heat engine to convert all of the free energy from the combustion reaction to work. However, as was described in Section 3.23.1.2, a portion of the free energy is irretrievably consumed in heating of the vessel and fuel and what remains is further limited by Carnot efficiency constraints. While it may not be intuitive, the conversion of methanol and oxygen to water and carbon dioxide through an electron transfer process still results in releasing the same amount of energy to do work, but it would not be subject to Carnot efficiency constraints.

However, just as with the translation of combustion to its application in the ICE, redox reactions performed in isolation are not very useful for doing work. The question then arises, what would happen if we could separate the redox active species while still allowing the charge transfer reaction to take place? A system has to be therefore designed such that the transfer of electrons between two species can be made to do useful work. Rather than having the species that comprise the redox couple come into direct contact with each other, as in the example of sodium metal and chlorine gas, the species are physically separated.

That temporarily "free" electron could be then controlled and directed so that devices could be engineered to take advantage of the now isolated electron transfer processes between redox species to store and release energy. By redirecting the "freed" electron through an external circuit with a load (some device that consumes energy from the system), such as an electric motor, the process of reducing and oxidizing chemical species can therefore be made to do work. The reduction and oxidation reactions could then occur through intermediaries in the form of their corresponding ions dissolved in a supporting medium. This is the functioning principal of electrochemical energy devices.

While a portion of the Gibbs energy that is potentially released from these spatially separated redox reactions may be released in the form of heat, the majority of the released energy is in the form of electron transfer/movement and the resultant current flow. Since the majority of the released energy is not utilized in the form of heat, a device that is designed to capture that released energy, such as a fuel cell, does not suffer the same constraints on its operational efficiency as does a Carnot heat engine. The remainder of this chapter will focus on introducing the fundamentals of the various electrochemical cells and how they are employed as energy generation and storage devices.

3.23.1.4 ELECTROCHEMICAL ENERGY STORAGE

3.23.1.4.1 Fundamentals

To effectively discuss the fundamentals of electrochemical cells, it is necessary to introduce some of the basic principles of electrochemistry. Electrochemistry is, as the name implies, a combination of electricity and chemistry where chemical reactions are induced or spontaneously occur through the application and control of a flow of electrons. This process can take place whenever a solution, aqueous or otherwise, comes into contact with a metal surface or any polarizable surface that allows for the transfer of electrons across the interface. In a controlled laboratory setting or a device, the aqueous solution is an electrolyte of known ionic strength, and an electrode provides the polarizable solid surface. In the bulk of the electrolyte, the arrangements of the ions and solvent molecules is more or less random and homogeneous because the forces acting on those particles are, on average, equal in all directions. However, since an interface with an electrode is different from the interactions with the bulk electrolyte, the forces that act upon the electrolyte particles at the interface are anisotropic, which gives rise to new arrangements of solvent molecules and ionic species.[2] The moment the electrolyte contacts the electrode surface, there is an ordering of the electrolyte ions and solvent molecules with dipoles at the interface. This phenomenon is illustrated in Fig. 3.23.1.3.

Across the electrode/electrolyte interface, a potential difference will develop, which results in the development of an electric field. In addition, if the electrode itself is further charged via an external power source, both sides of the interface become polarized and form a structure that is referred to as the double layer: ions of a specific charge are arranged in a layer parallel to the interface, whereas particles of opposite charge within the electrode material itself become arranged near the interface. The formation of the induced double layer at a charged interface is illustrated in Fig. 3.23.1.4.

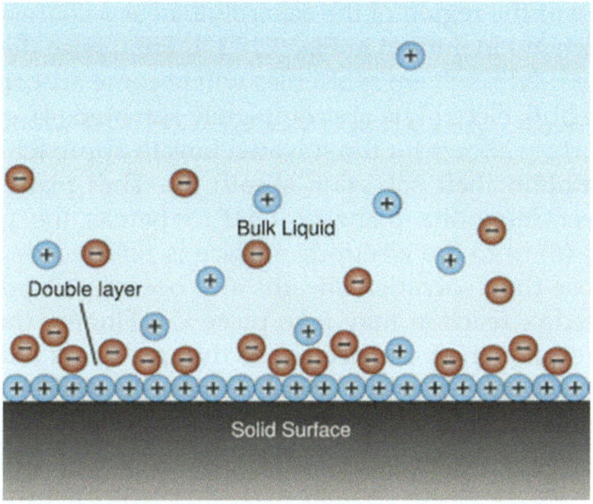

FIGURE 3.23.1.3 The electrode/electrolyte interface.[3] A diagram of the electric double layer that forms at the surface of a charged surface. The double layer is the structure that forms from the instantaneous ordering of electrolyte ions and solvent molecules with dipoles at the electrode/electrolyte interface. A potential difference will develop across the interface as a result of the formation of the double layer. *From © Snubcube, used under Creative Commons (CC-BY-SA-3.0).*

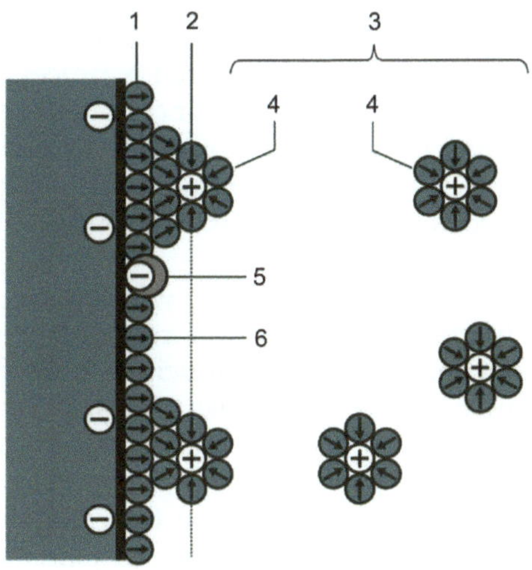

FIGURE 3.23.1.4 The double layer.[4] Structure of the double layer and the components of diffusion layer. The charging of the electrode results in both sides of the interface becoming polarized and the formation of the structured double layer and diffusion zone. The components that constitute the double layer and zone of diffusion are identified in the figure as: (1) the plane at which solvent molecules are adsorbed onto the electrode surface or the inner Helmholtz plane, (2) the closet point at which solvated ions still maintain their solvation sheath or the outer Helmholtz plane, (3) the zone of diffusion between the double layer and the bulk electrolyte, which consists of (4) completely solvated ions, (5) specifically adsorbed ions that have lost their solvation sheaths, and (6) solvent molecules that have become adsorbed onto the electrode surface. *From ©Tosaka, used under Creative Commons (CC-BY-3.0).*

The complex structure of the region of the electrolyte near a charged interface is composed of several elements, which are identified in Fig. 3.23.1.4. First, once the electrode, or any other surface, contacts the electrolyte, solvent molecules will become adsorbed onto the surface (6). Although the ions in the bulk electrolyte are completely solvated (4), electrostatic attraction to the charged electrode surface allows for the solvated ions to approach to the closest point that allows them to still maintain their solvation sheath (2). That plane of closest approach is referred to as the outer Helmholtz plane or OHP, whereas the plane at which solvent molecules are adsorbed (6) onto the electrode surface is referred to as the inner Helmholtz plane (1).[5,6] Ions may lose their solvation sheaths and become adsorbed onto the surface of the electrode where a redox reaction may take place (5). Finally, there is a transition zone between the double layer and the bulk electrolyte that produces a distribution of ions and charge as a function of distance from the metal surface. In this zone, which is referred to as the diffuse layer, the electric potential decreases exponentially away from the interface.

From the preceding discussion, it can be shown that there are two processes that readily occur at the interface of the electrode surface and the electrolyte that provide methods for charge (energy) storage. Through the use of an external power source, the electrode can be charged, which not only results in the accumulation of charged particles (i.e., electrons) within the electrode but also induces an accumulation of charge (ions) at the interface within the electrolyte. The ability of a material to accumulate and hold charge in this manner is a measure of a material's capacitance, which is one form of storing electrical energy. Devices that employ this form of charge storage are called capacitors, and their more modern, high-capacity variants are referred to as supercapacitors. Capacitance is a measure of an electrode's ability to store charged particles on the surface of electrodes, in a static fashion, without the transfer of electrons between the charged ion and the surface of the electrode. The second form of charge storage occurs through redox processes. A species in the electrolyte may gain or lose electrons through a reduction or oxidation process, and if that process is reversible under the right conditions, then that stored energy can be recaptured to do work. The redox charge storage process is a chemical form of charge storage due to the transfer of electrons between a surface species and an adsorbed ion species. This is the functional basis of operation for batteries and fuel cells.

Typically, a chemistry student's first exposure to a working electrochemical cell is in a general or physical chemistry teaching laboratory. A common electrochemical cell that has routinely been used for illustrating some of the principles of electrochemistry is the Zn/Cu cell. Zinc and copper metal are employed as the electrodes, and both are submerged in their corresponding sulfate solutions to form one half of the cell, or what is commonly referred to as a half-cell. The two half-cells are joined through a salt bridge, or junction, of NaCl or $NaSO_4$ that allows for counter ions to cross between the cell halves to maintain electroneutrality in each half-cell. At each electrode, a reduction or oxidation reaction can take place, with the expressions for the reduction reactions being as follows:

$$\text{Copper electrode:}\ Cu^{2+}{}_{(aq)} + 2e^- \rightarrow Cu_{(s)}$$

$$\text{Zinc electrode:}\ Zn^{2+}{}_{(aq)} + 2e^- \rightarrow Zn_{(s)}$$

Since only one species can be reduced, whereas the other must be oxidized, we can combine these two reactions to obtain the potential overall cell reaction shown in Eq. (3.23.1.6). At this point, the arrow in the reaction is somewhat arbitrary since we have not yet determined if this reaction is spontaneous as it is currently written.

$$Cu^{2+}{}_{(aq)} + Zn_{(s)} \rightarrow Cu_{(s)} + Zn^{2+}{}_{(aq)} \tag{3.23.1.6}$$

Spontaneity of a chemical reaction is determined by calculating the Gibbs energy for the reaction, and if that calculated value is negative, then the reaction is spontaneous. The change in the Gibbs energy for the Cu-Zn reaction is calculated in Eq. (3.23.1.7):

$$\begin{aligned} \Delta_r G^\phi &= \sum_{\text{products}} n\Delta_f G^\phi - \sum_{\text{reactants}} n\Delta_f G^\phi \\ \Delta_r G^\phi &= \left(\Delta_f G^\phi_{Cu^{2+}} + \Delta_f G^\phi_{Zn}\right) - \left(\Delta_f G^\phi_{Zn^{2+}} + \Delta_f G^\phi_{Cu}\right) \\ \Delta_r G^\phi &= (-147.06 \text{ kJ/mol} + 0) - (-65.49 \text{ kJ/mol} + 0) \\ \Delta_r G^\phi &= -212.55 \text{ kJ/mol} \end{aligned} \tag{3.23.1.7}$$

In this example, $\Delta_r G^\phi = -212.55$ kJ/mol, and so the reaction is spontaneous as written: the copper ion is reduced, whereas the zinc metal is oxidized. A similar analysis can be made using standard reduction potentials. As the reader may recall, the reduction potential can be thought of as a measure of how easy or difficult it would be for a particular species to be reduced. For two species to spontaneously undergo a combined redox reaction, one species must be capable of donating one or more electrons, and therefore become oxidized, whereas the other, paired, species must be capable of accepting those electrons and consequently become reduced. To put it simply, if two species come into contact with each other and one of those species is more readily reduced with respect to the other, then a charge transfer reaction will take place. Whether that charge transfer reaction can proceed spontaneously as written will depend on the cell potential, which is calculated from the difference between the standard reduction potentials of the cell half-reactions. In the case of the Cu/Zn cell, the standard reduction potential of $Cu/Cu^{2+} = +0.34$ V and the standard reduction potential for $Zn/Zn^{2+} = -0.76$ V.

$$E^\phi = E^\phi_{Cu} - E^\phi_{Zn} = +0.34 \text{ V} - (-0.76 \text{ V}) = 1.1 \text{ V} \tag{3.23.1.8}$$

The calculated potential for the Cu/Zn cell is 1.1 V, and this value can be related to the Gibbs energy for the reactions using Eq. (3.23.1.9), where ν is number of electrons transferred and F is Faraday's constant.

$$\Delta_r G^\phi = -\nu F E^\phi = -(2)(96{,}485 \text{ C/mol})(1.1 \text{ V}) = -214.2 \text{ kJ/mol} \tag{3.23.1.9}$$

The calculated value of −214.2 kJ/mol compares quite well to the value that was calculated using the standard Gibbs free energy of formation values. Since we had already established that the reaction is spontaneous as written, it can therefore be concluded that, if the calculated cell potential is positive, then the cell reaction is spontaneous, whereas if the calculated cell potential is negative, then the reverse reaction would be spontaneous. A cell that generates electricity as a result of a spontaneous electrochemical reaction is referred to as a galvanic cell. If a cell reaction is determined not to be spontaneous, it can still be driven in that direction through the application of electricity. This scenario arises when chemical species are formed through an electrochemical process that is referred to as *electrolysis*. An example of this is the electrolysis of water to form hydrogen and oxygen gasses. A cell that drives a nonspontaneous electrochemical reaction through the application of electricity is referred to as an electrolytic cell.

3.23.1.4.2 Ideal vs. Real Behavior: Energy Losses

In real electrochemical devices, there are four major factors that lower the open cell potential and the operational potential of the device. These major factors are (1) activation losses, (2) internal current losses, (3) mass transport/concentration losses, and (4) ohmic losses. Activation losses arise from the need to overcome the activation energy barriers associated with the charge transfer reactions and to ensure that the electrochemical reaction will be driven toward the formation of products. These losses will have the most significant impact at low current densities. The second source of loss is internal or parasitic current losses. Depending on the nature of charge storage or generation processes associated with the device, these losses can be attributed to either the reverse electrochemical reaction, side redox reactions with either components of the electrolyte, electrolyte contaminants, functional groups on the electrode surface, or unreacted fuel that can be attributed to physical losses. The third type of loss is mass transport/concentration losses, which are most prevalent at high current densities. These losses result from the effect of having a less-than-sufficient concentration of the electroactive species at an electrode's surface. It essentially occurs because the redox reactions involved in the discharge of stored energy or the electricity generation process occur faster than the involved electroactive species can be brought to the electrode surface through mass transport. These losses result in a maximum limit of the current output of the device. The final type and most common source of loss in any electrical device is ohmic losses. This type of loss occurs because of the impact on the flow of electrons due to the combined resistances within the cell (separator, electrodes, current collectors) and the interfaces between those components. All ohmic losses are directly proportional to the current. All these losses contribute in their own way to decrease the operational voltages in electrochemical devices.

3.23.1.5 ELECTROCHEMICAL ENERGY DEVICES

The remainder of this section will briefly introduce the fundamentals of the three general electrochemical devices that are used for the storage and generation of electrical energy. This overview will focus on the generalized principals of the operation of these devices, whereas Section 3.23.2 will provide a more thorough discussion of the associated current state and challenges concerning electrochemical devices.

3.23.1.5.1 Batteries

Since we have already introduced the Cu/Zn example cell, which is representative of the operation of a battery, it is perhaps easiest to start with those devices. In the normal operation of a battery, it functions just like a galvanic cell in that electricity is generated as the result of a spontaneous chemical reaction. If a battery is recharged, it is operating as an electrolytic cell, which uses electricity to reverse a chemical reaction. Although a single electrochemical cell may constitute a battery, typical commercial batteries are usually a combination of multiple cells arranged in such a way as to obtain the desired battery voltage and current output.

Although the simple example of the redox reaction between copper and zinc might be considered reversible, the circumstances that allow for reversibility are not always present in batteries. The energy that is obtained from a battery during discharge is typically less than what the theoretical calculations would predict. Physical changes that the electrode may undergo during discharge and the losses described previously (activation, parasitic, mass transport, and ohmic losses) will diminish the energy that is available for work. Those energy losses during discharge have to be compensated for during recharging of the battery if a battery is to be considered potentially rechargeable. A battery would not be considered rechargeable if the energy required to reverse the chemical reaction and to compensate for those energy losses is too high to be practical. Batteries can therefore be separated into two general classes: primary and secondary batteries. A primary battery is assembled in the fully charged state and once discharged, is not rechargeable in a practical sense, so it is considered a single-use device. A secondary battery is typically assembled in the discharged state and must therefore be charged to be useful, but it is also rechargeable and is thus considered to be a multiuse device.

In the simplest cell model, the electrodes are the materials that are reduced or oxidized (the electroactive species), which means the electrodes will generate the ions that conduct the charge through the electrolyte (ionic conduction). The electrodes also act as the current collectors passing the temporarily "freed" electrons into and out of the external circuit (electronic conduction). The electrolyte used is frequently the same for both halves of a battery cell, and a separator is placed between the electrodes. The separator may be composed of a polymer or other inert material (e.g., cellulose), and its function is to allow for ion conduction between electrodes while inhibiting electronic conduction through direct contact of the electrodes. A diagram illustrating a representative battery is provided in Fig. 3.23.1.5.

As stated previously, the oxidation or reduction of a chemical species is the operating principal of a battery. In the Zn/Cu cell example in Section 3.23.1.4.1, the open cell potential was calculated to be 1.1 V. If that cell were connected to an external circuit and a load through which the current flowed, copper ions would be reduced and the zinc metal oxidized to its corresponding cation. As long as there were still copper ions in the electrolyte and zinc metal remaining, the cell potential would theoretically still be 1.1 V. The operating potential of a cell, or a battery, is determined by the redox potentials of the species that are being reduced or oxidized. From a strictly theoretical perspective, the cell potential will not change during the charge and discharge cycles. Since all the reducible species have the same reduction

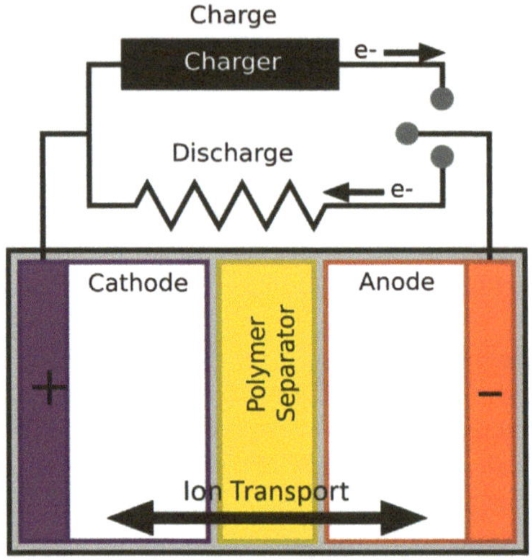

FIGURE 3.23.1.5 Battery diagram illustrating the directional flow of electrons during charge (toward anode) and discharge (toward cathode).[7] A battery is composed of three main elements: the electrodes (anode and cathode), the electrolyte, and the separator. During the charge and discharge processes, ions in the electrolyte migrate to either the anode (charging) or the cathode (discharging) through the separator, which is permeable to ions but does not allow the conduction of electrons. *From © Tkarcher, used under Creative Commons (CC-BY-SA-3.0).*

potential and all the oxidizable species have the same oxidation potential, the cell potential will not change until all of those species are either reduced or oxidized. Once the last copper ion is reduced and zinc atom oxidized, then the cell potential would drop to zero because the potential energy that drove those coupled reduction/oxidation reactions would be gone (Figs. 3.23.1.5).

However, due to the various losses described in the preceding section (activation, parasitic, mass transport, and ohmic losses), the energy required to fully charge the battery will be greater than the energy that can be obtained from the battery during discharge, and this results in a hysteresis in the charge-discharge profile. Hysteresis occurs when the energetics of the forward process does not precisely match the energetics of the reverse process, thereby producing a gap between the cell potential measured during charge and discharge as opposed to an overlap. An example charge-discharge potential curve that illustrates this imbalance is shown in Fig. 3.23.1.6. It should be pointed out that, under real-world operating conditions and in real manufactured batteries, there are other factors besides those discussed energy losses that will have a significant effect on the charge/discharge curves.

3.23.1.5.2 Fuel Cells

The operational principals of fuel cells are similar to those of batteries in that redox processes are involved in the generation of electrical energy. Since the output of electricity results from a chemical reaction, fuel cells can be considered as functioning like galvanic cells. From that perspective, fuel cells are similar to primary batteries with one significant

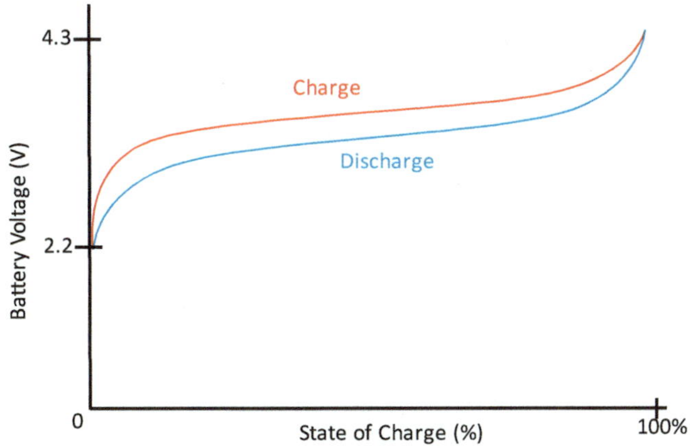

FIGURE 3.23.1.6 Charge-discharge profile. Shown are representative charge (*red*) and discharge (*blue*) curves for a typical battery. The hysteresis, or gap between the curves, arises from differences in the energetics of the forward process versus the energetics of the reverse process. The energy required to fully charge a battery is typically greater than the energy that is released during discharge.

difference: the electroactive species are not contained within a fuel cell but are supplied from external reservoirs. Therefore from an operational or user point of view, the fuel cell could be considered to be most similar to the ICE in that a fuel is fed into the device and power is generated in the form of electricity. That generated electricity can be used to power electric motors to move a vehicle, and once the fuel supply has been exhausted, power generation (and motion) stops until the fuel source can be replenished.

A prototypical example of a fuel cell is the hydrogen fuel cell, which, through the reverse hydrolysis of water, converts hydrogen and oxygen directly into electricity, water, and some heat: $2H_{2\,(g)} + O_{2\,(g)} \rightarrow 2H_2O_{(g)}$. Hydrogen fuel is supplied at the anode where it is oxidized, and the resultant proton migrates to the cathode, where it is combined with oxygen in a reduction reaction to form liquid water. The electrodes are composed of porous materials that are permeable to the gasses and contain catalysts, such as platinum, that facilitate the corresponding reduction (cathode) or oxidation (anode) reaction. A diagram that illustrates this process is shown in Fig. 3.23.1.7. Hydrogen gas permeates through the anode where it is oxidized by a platinum catalyst, releasing the proton into the electrolyte and the electron into the external circuit: $H_{2\,(g)} \rightarrow 2H^{+}{}_{(aq)} + 2e^{-}$. At the cathode, the electron from the external circuit and the proton from the electron combine with oxygen in a reduction reaction to form liquid water: $O_{2\,(g)} + 2H^{+}{}_{(aq)} + 2e^{-} \rightarrow H_2O_{(l)}$.

Unlike a battery, the electroactive species are within the electrolyte; the electrodes themselves do not directly participate in the redox processes but act as facilitators. The electrodes function strictly as the interface at which the redox reactions occur and as the current collectors that transfer the charge from the electroactive species into the external circuit. Since no combustion is involved in this process, the efficiencies of fuel cells are not subjected to Carnot limitations. Therefore the resulting efficiency can be between 50% and 80%; approximately double that of an ICE.[9] Fuel cells can be thought of as combining the

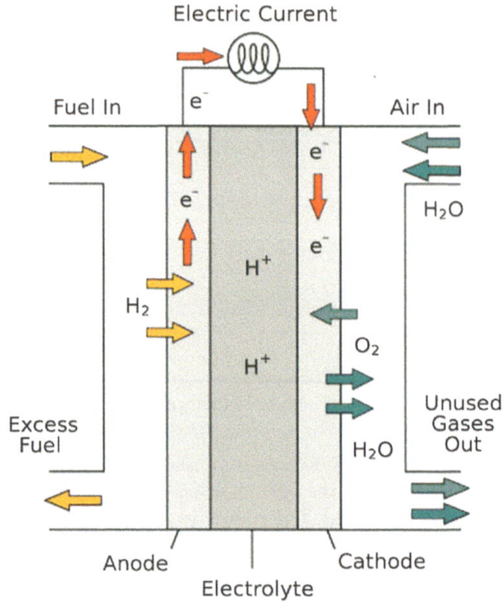

FIGURE 3.23.1.7 Fuel cell diagram illustrating the flow of current due to the oxidation of hydrogen and reduction of oxygen.[8] The hydrogen fuel cell converts hydrogen and oxygen directly into electricity, water, and some heat. Hydrogen is oxidized at the anode, $H_{2\,(g)} \rightarrow 2H^{+}{}_{(aq)} + 2e^{-}$, releasing a proton that migrates to the cathode, where it is reduced in the presence of oxygen to form liquid water, $O_{2\,(g)} + 2H^{+}{}_{(aq)} + 2e^{-} \rightarrow H_2O_{(l)}$. The electron(s) travel through the external circuit where it is available to do work. *From ©R. Dervisoglu, used under Creative Commons (CC0 1.0).*

best features of engines and batteries; like an engine they can operate for as long as fuel is available without any intermediate mechanical energy conversion, but the characteristics are similar to a battery when under load conditions.

Just as with batteries, the maximum energy output of fuel cells is subjected to the types of losses described in Section 3.23.1.4.2. A loss that is specific to fuel cells is a contribution to the internal current losses due to fuel or electron crossover. Separating the anode from the cathode is an ion-conducting membrane that allows for ions to cross from the anode side of the cell to the cathode side, while preventing other potentially reactive species from crossing. However, this selective barrier is not ideal and either unreacted fuel (e.g., hydrogen gas) or free electrons can cross over to the cathode. In either situation, electrons that normally would have flowed through the external circuit now directly combine with the oxygen at the cathode, which results in a decrease of the energy output of the device. Fig. 3.23.1.8 illustrates the effect that these losses have on the cell potential with increasing current densities. The ideal performance of a fuel cell (i.e., no energy losses) would exhibit an operating potential that would be constant at all levels of output current densities, a measure of energy output, and is represented by the blue line in Fig. 3.23.1.8. However, a fuel cell operating under realistic conditions that experiences those previously discussed energy losses would show

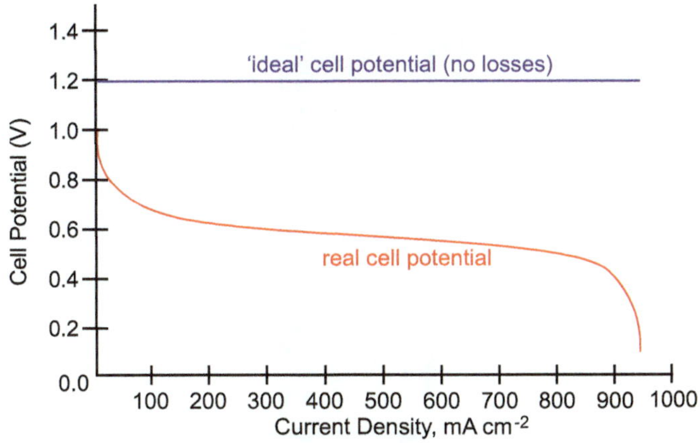

FIGURE 3.23.1.8 Fuel cell losses of cell potential with increasing current densities. The ideal performance (*blue line*) of a fuel cell exhibits a constant operating potential at all levels of output current densities. A fuel cell operating under realistic conditions (*red line*) experiences energy losses that result in a decrease in its operating potential as the output current density increases.

a decrease in its operating potential as the energy output increases. As the energy demand increases (increasing current density), the energy losses that depend on current density (mass transport limitations, ohmic losses) will have increasing impact on the fuel cell's operating potential, eventually resulting in a precipitous drop in the operating potential, as is represented by the red line in Fig. 3.23.1.8.

3.23.1.5.3 Electrochemical Capacitors

Another method of storing energy in an electrochemical cell was introduced in the discussion of the electrode/electrolyte interface in Section 3.23.1.4. It is the storage of energy through double-layer capacitance. Double-layer capacitance has been known and exploited for several decades in the form of electrostatic and electrolytic capacitors. These devices are composed of two metal plates separated by a dielectric. Electrical energy is stored by introducing a charge to the electrodes (metal plates), which in turn induces a polarization of the dielectric between the plates. This buildup of electrical charges results in the development of an electric field between the plates and has historically been referred to as *dielectric storage*. A diagram illustrating a parallel plate capacitor is shown in Fig. 3.23.1.9.

The amount of energy that can be stored, or the device's *capacitance*, is determined by Eq. (3.23.1.10), where C is the capacitance, ε is the dielectric constant of the dielectric, A is the area of the plate, and δ is the distance between the plates.

$$C = \varepsilon \frac{A}{\delta} \tag{3.23.1.10}$$

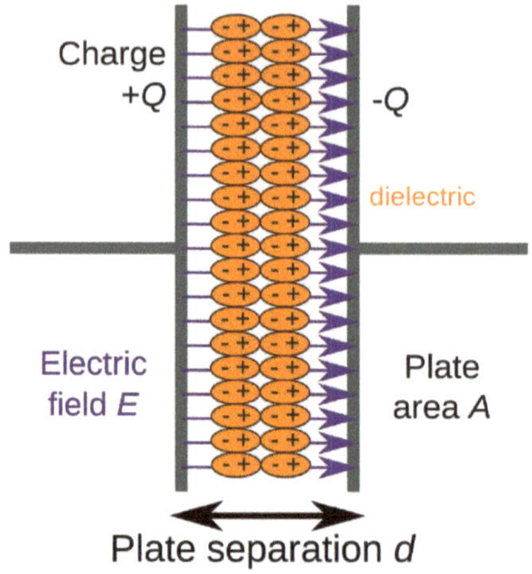

FIGURE 3.23.1.9 Parallel plate capacitor illustrating the buildup of opposite, polarized charges at the dielectric interfaces as the electrodes are charged.

Electrostatic storage of energy offers an advantage over storing energy in chemical bonds in that the stored energy can be released very rapidly since no redox reactions are involved. If it is assumed that a fraction, λ, of the maximum stored charge, q, can be established on the electrode surface, then the electrode potential at that point is $\lambda q/C$. The addition of an infinitely small amount of charge, δq, results in the element of energy of charging being defined as:

$$dG = \frac{\lambda q}{C} \cdot dq = \frac{\lambda q}{C} \cdot q d\lambda \tag{3.23.1.11}$$

The energy of charging can be derived through integration and is shown in Eq. (3.23.1.12), where G is the energy of charging, λ is the fraction of stored charge, q is the electric charge, C is the capacitance, and V_0 is the operating potential.[10]

$$G = \int_0^q dG = \int_0^1 \frac{\lambda q}{C} \cdot q d\lambda = \frac{q^2}{C} \int_0^1 \lambda d\lambda$$

$$G = \frac{1}{2}\frac{q^2}{C} \equiv E = \frac{1}{2} C V_0^2 \quad \text{given } q = C V_0 \tag{3.23.1.12}$$

The work, or free energy, that is required to establish the double layer increases as the population of ions in the OHP and the population of charge particles just beneath the electrode surface increases. Each additional charged particle that is to be added to either of

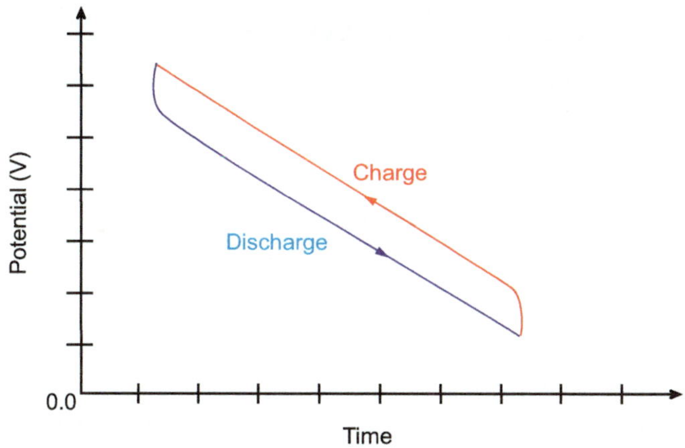

FIGURE 3.23.1.10 Charge/discharge hysteresis of supercapacitor. The work that is required to establish the double layer increases as the population of charged particles in solution (ions) and within the electrode at the interface increases. This increasing energy results in linear increases in the cell potential during charge (red) and a linear decrease during discharge (blue).

those two regions must overcome the electrostatic repulsion from the previously accumulated charges of the same polarity that are already present. Therefore, as the charging of the double layer proceeds, more energy is required to accumulate additional charge at the interface. This increasing energy results in a linear increase in the cell potential during charge and a corresponding linear decrease in the cell potential during discharge. The cell potential as a function of time over the charge and discharge cycles is illustrated in Fig. 3.23.1.10. The hysteresis (gap) between the curves is a result of those energy losses discussed in Section 3.23.1.4.2, which results in more energy required to charge the cell than can be obtained during discharge.

In parallel plate capacitors, the thickness of the double layer that forms between the polarized molecules of the dielectric and the charged plate will be limited by the distance between the charged parallel plates. In addition, the metal plate electrodes will typically have a very small surface area and so plate capacitors, or their modern-day equivalents, will typically have storage capacities in the picofarad range (1–25 pF). Given that some capacitors can support up to 2000 V and that $1\ F = 1\ A\ s/V$, the storage capacity for those devices would be in the 5–10 μAh range. Increasing this capacity by changing the electrode materials to porous materials and by replacing the dielectric with a liquid electrolyte is what has given rise to a new class of energy storage device: the supercapacitor.

3.23.2 The Movement of Electrons in Batteries, Fuel Cells, and Supercapacitors: Methods of Energy Delivery

3.23.2.1 INTRODUCTION

Section 3.23.1 provided a brief introduction to the technologies that generate and store electrical energy and that can enable society to transition from heat engine-type devices to devices that access that same stored energy in chemical bonds through electrochemistry. The electrochemical devices employed for the storage and generation of electricity that arise from that transition utilize the energy that is stored in chemical bonds more efficiently and would thus have a significant impact on realizing a clean energy future. Lastly, that section provided a short overview of the fundamentals that govern the operation of electrochemical generation and storage devices: batteries, fuel cells, and supercapacitors. This section will continue that discussion by providing an overview of current technologies and an introduction to some of the basic challenges to fully exploiting their potential.

To have a constructive discussion on these storage and generation technologies, it is important to establish how they relate to each other in a meaningful way, which can be done based on two metrics that are the most important and the most useful for comparison: (1) energy density (Wh/L), or specific energy (Wh/kg), and (2) power density (W/L), or specific power (W/kg). A device's specific energy indicates its capacity, which is defined as the amount of energy that the device can release before it has to be replaced or recharged, and a device's specific power indicates its loading, or how rapidly that stored energy can be released. A higher power density means that the energy can be released in a shorter amount of time. The capabilities of different electrochemical energy storage devices are typically compared and contrasted quickly by plotting the values for those two metrics in a Ragone plot.

Although the simplified Ragone plot in Fig. 3.23.2.1 just compares the generalized classes of energy devices (both storage and generation), it does allow for a broader understanding of the relationship among the various classes of devices. From that perspective, it is possible to talk about a spectrum of electrochemical energy storage devices classified by those two important capabilities. At the high end of the energy density scale are fuel cells and batteries, such as lithium ion batteries, whereas the much lower energy densities of plate capacitors place them at the lowest end of that scale. However, in terms of power densities, capacitors are superior devices since they can release, and recapture, their stored energy very quickly, whereas batteries take much longer to charge and discharge. It is also clear from the Ragone plot that devices that are optimized for maximizing one energy storage attribute are likely to be poorly optimized for the other attribute. Much of the current research into electrochemical energy storage is directed toward the development of devices that improve on one or more of those attributes, while attempting to bridge the gap between high-energy-density and high-power-density devices. The main focus of the bridging research is centered on the basic components of all electrochemical energy storage devices: electrolytes, electrode materials,

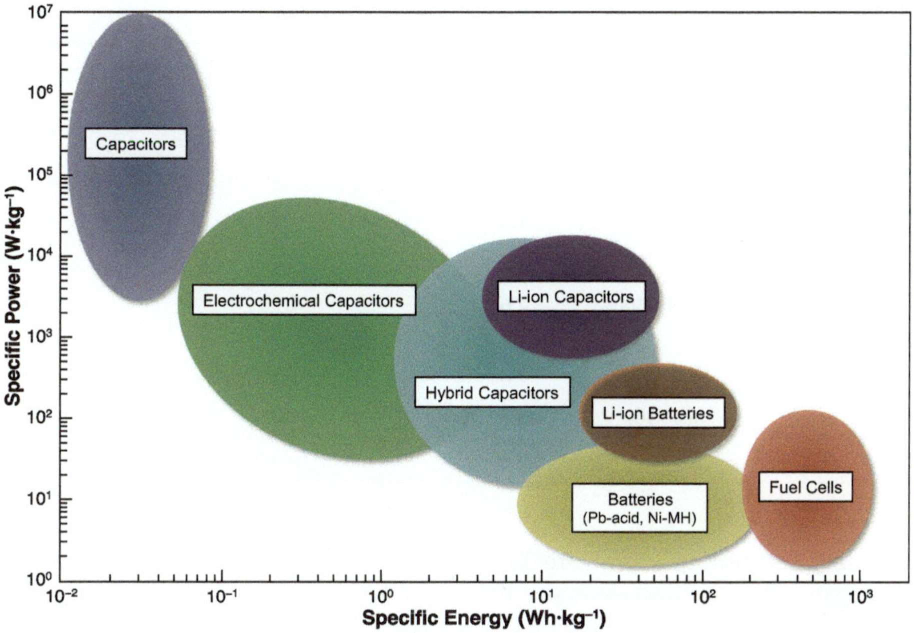

FIGURE 3.23.2.1 Ragone plot. The Ragone plot illustrates the relationship between energy density and power density among different energy storage devices. In general, devices that are optimized for maximizing one attribute are poorly optimized for the other attribute. Fuel cells and batteries are placed at the higher end of the energy density scale, but take much longer to charge and discharge: lower power density. Parallel plate capacitors and super-capacitors have much lower energy densities but are superior devices in terms of power density, since they can release, and recapture, their stored energy very quickly.

and separators. In the remainder of this section, each class of energy storage device will be discussed in terms of the limitations of the current generations of those components and what is being done to address those limitations.

3.23.2.2 BATTERIES

3.23.2.2.1 Battery Types

As discussed in the previous section, batteries store energy in the form of chemical changes that the two electrodes of a battery cell will undergo during charge and discharge operations. As a battery releases its stored electrical energy, the spontaneous redox reactions that take place result in each electrode material undergoing a bulk chemical and physical change that allows it to reach a lower and more stable energy state. For rechargeable, or secondary, batteries, energy can then be applied to drive the reverse chemical processes to change the electrode material's composition back to the form and state that it was in prior to discharge. In general, these processes are multielectron transfer processes, and as would be expected, the bulk changes in the electrode material are slow. However, due to that more substantial degree of change, the energy storage density can be very high. An abbreviated listing of

TABLE 3.23.2.1 Characteristics of Common Battery Types

Type	Cycles	Cell Potential (V)	Specific Energy (Wh/kg)	Specific Power (W/kg)
Alkaline	N/A	1.5	85–190	50
Lead-acid	500	2.1	30–40	180
Nickel-cadmium (NiCd)	~1500	1.2	50	150
Nickel metal hydride	700–1200	1.2	100	250–1000
Sodium-nickel chloride (Ni-NaCl)	1500–3000	2.6	90	150
Lithium ion	400–2000	3.2–3.9	100–265	250–400
Lithium ion polymer	300–400	3.3–3.7	250	600
Zinc-air	>300	1.65	470–1360	100

common battery technologies that are in current use today is provided in Table 3.23.2.1. Given the length of history and diversity of battery technologies in use, a comprehensive review of each of these technologies is well outside the scope of this chapter. However, an overview of the electrochemical reactions involved in the operation of some of these battery types is provided. It is perhaps easiest to begin with primary batteries, since they were the earliest batteries put to some type of practical use.

3.23.2.2.1.1 Alkaline Battery

The alkaline battery is one such primary battery that is still widely used today. One of the more common chemistries for this type of battery is zinc/manganese dioxide with a basic potassium hydroxide electrolyte. Zinc metal functions as the negative electrode following the half-reaction during discharge shown in Eq. (3.23.2.1a), whereas MnO_2 acts as the positive electrode following the half-reaction shown in Eq. (3.23.2.1b).

$$Zn_{(s)} + 2OH^-_{(aq)} \rightleftharpoons ZnO_{(s)} + H_2O_{(l)} + 2e^- \tag{3.23.2.1a}$$

$$2MnO_{2\,(s)} + H_2O_{(l)} + 2e^- \rightleftharpoons Mn_2O_{3\,(s)} + 2OH^-_{(aq)} \tag{3.23.2.1b}$$

When combined, the overall net reaction is as shown in Eq. (3.23.2.1c).

$$Zn_{(s)} + 2MnO_{2\,(s)} \rightleftharpoons ZnO_{(s)} + Mn_2O_{3\,(s)} \tag{3.23.2.1c}$$

The typical alkaline batteries in use are not considered rechargeable because the energy required to reverse the chemical processes shown in Eqs. (3.23.2.1a,b,c) would be so high as to not only make it impractical but also too dangerous to recharge since it would result in a substantial increase in the device's temperature.

3.23.2.2.1.2 *Lead Acid Battery*

The lead acid battery is another highly used battery but one that is also rechargeable. The battery uses an acidic electrolyte (H_2SO_4), and in the fully charged state, the negative electrode is composed of lead, whereas the positive electrode is lead(IV) oxide. During discharge, the negative electrode donates two electrons to the external circuit forming Pb^{2+}, following the reaction shown in Eq. (3.23.2.2a), whereas at the positive electrode Pb^{4+} is reduced to Pb^{2+}, following the reaction shown in Eq. (3.23.2.2b).

$$Pb_{(s)} + HSO_4^-(aq) \rightleftharpoons PbSO_{4\,(s)} + H^+{}_{(aq)} + 2e^- \tag{3.23.2.2a}$$

$$PbO_{2\,(s)} + HSO_4^-(aq) + 3H^+{}_{(aq)} + 2e^- \rightleftharpoons PbSO_{4\,(s)} + 2H_2O_{(l)} \tag{3.23.2.2b}$$

Therefore, in the discharged state, both the positive and negative electrodes generate lead(II) sulfate ($PbSO_4$), and the electrolyte loses much of its dissolved sulfuric acid, becoming primarily water. The overall net reaction is shown in Eq. (3.23.2.2c).

$$Pb_{(s)} + PbO_{2\,(s)} + 2H_2SO_{4\,(aq)} \rightleftharpoons 2PbSO_{4\,(s)} + 2H_2O_{(l)} \tag{3.23.2.2c}$$

3.23.2.2.1.3 *Ni Metal Hydride Battery*

The nickel metal hydride (NiMH) battery is a widely used rechargeable battery that uses nickel oxyhydroxide (NiOOH) as the positive electrode and a hydrogen-absorbing metal alloy as the negative electrode. This battery technology was introduced in the 1990s, and due to its superior capacity and overall lower cost, it significantly reduced the use of the NiCd battery. In addition, cadmium is a toxic metal and its disposal presented a significant environmental hazard. The most common composition for the metal alloy (M) is AB_5, where A is a mixture of rare earths, such as lanthanum, cerium, neodymium, or praseodymium, and B is nickel, cobalt, manganese, or aluminum. In addition, higher capacity materials for the negative electrode have been based on AB_2 compounds, where A is titanium or vanadium and B is zirconium or nickel, modified with chromium, cobalt, iron, or manganese. During discharge, the positive electrode (NiOOH) will produce nickel(II) hydroxide according to the reaction shown in Eq. (3.23.2.3a), whereas the metal hydride (negative electrode) will lose hydrogen to form the alloy following the reaction shown in Eq. (3.23.2.3b).

$$NiO(OH)_{(s)} + H_2O_{(l)} + e^- \rightleftharpoons Ni(OH)_{2\,(s)} + OH^-{}_{(aq)} \tag{3.23.2.3a}$$

$$MH_{(s)} + OH^-{}_{(aq)} \rightleftharpoons M_{(s)} + H_2O_{(l)} + e^- \tag{3.23.2.3b}$$

The overall net cell reaction is shown in Eq. (3.23.2.3c).

$$NiO(OH)_{(s)} + MH_{(s)} \rightleftharpoons Ni(OH)_{2\,(s)} + M_{(s)} \tag{3.23.2.3c}$$

3.23.2.2.1.4 Lithium Ion Battery

Our discussion now arrives at what is now the most commonly used secondary battery: the lithium ion battery. There are two properties of lithium that make it very attractive as an anode material: it is both the least electronegative and the lightest metal. These properties make lithium the element with the most negative redox potential and an anode of high specific capacity (3.86 Ah/g) when it is matched with an appropriate cathode.[11,12] However, the use of pure lithium metal as the anode in a secondary battery proved to be too dangerous for practical use. Inhomogeneities in the surface of the Li metal produce regions that pass charge at different rates, which results in an uneven deposition of lithium during the charge cycle. Over repeated charge-discharge cycles of a Li metal battery, thicker deposits of lithium, with larger surface areas, allow for more charge transfer to occur and therefore continue the growth of those lithium deposits or "dendrites." The formation of dendritic lithium results in a decrease in the energy density, due to the loss of free lithium, and continued growth could become sufficient to pierce the separator, resulting in an internal short circuit and thermal runaway.[11,13] Instead, lithium ions are employed as the alternative with an electrode material that can internally accommodate them through a process called *intercalation*. The structure of an intercalation material is such that the material allows for the insertion and deinsertion of "guest" ions without the ions participating in any chemical bonds, unlike typical battery electrode materials.

Lithium ion batteries are the primary example of an energy storage device that employs intercalation materials as the means for storing energy. The redox reactions that occur during injection or extraction of electrons take place on the host lattice. For the positive electrode, the majority of the intercalation materials are transition metal oxides or chalcogenides (i.e., sulfides, selenides, tellurides), whereas carbonaceous materials are typically employed as the intercalation host for the negative electrode.[11,14,15] The intercalated lithium is stored in the carbon matrix in its ionic form, which eliminates the possibility for the formation of dendritic lithium during normal operations, as would be the case for lithium metal.[11,13] Furthermore, the use of ion intercalation materials incurs an additional advantage with regard to cycle life. Unlike nonintercalation electrode materials that undergo chemical and physical changes during charge/discharge, ion intercalation materials possess a two-dimensionally stable (i.e., "low strain") structure, with volume changes during intercalation/deintercalation typically $\leq 10\%$. This added stability greatly adds to the longer cycle life attributed to intercalation devices.[16]

A diagram that illustrates the operation of a lithium ion battery is shown in Fig. 3.23.2.2. $LiCoO_2$ is the positive electrode and graphite is the negative electrode, which during discharge, act as the cathode and anode, respectively. The lithiated transition metal oxide material of the positive electrode undergoes the reaction shown in Eq. (3.23.2.4a), whereas the carbon (graphite) negative electrode undergoes the reaction shown in Eq. (3.23.2.4b).

$$Li_{1-x}M_yO_z + xLi^+ + xe^- \rightleftharpoons LiM_yO_z \tag{3.23.2.4a}$$

$$Li_xC_6 \rightleftharpoons xLi^+ + xe^- + 6C \tag{3.23.2.4b}$$

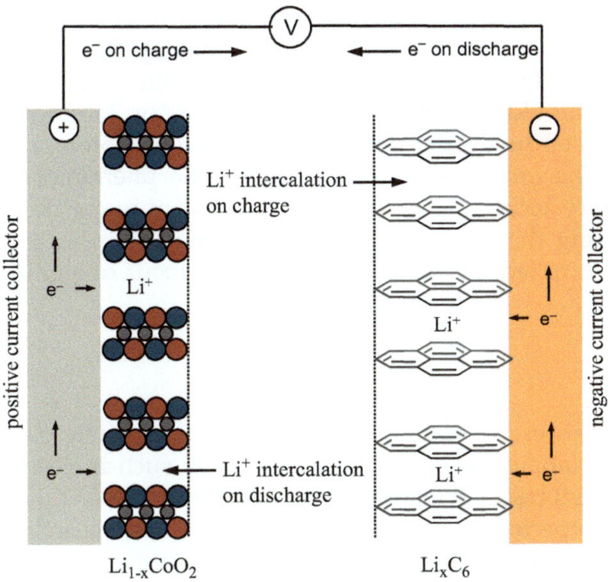

FIGURE 3.23.2.2 Lithium ion battery ($LiCoO_2$) illustrating the flow of electrons and intercalation of Li^+ on charge and discharge. During charging of the battery, Li ions are released from the positive electrode ($LiCoO_2$) and migrate to the negative electrode (graphene) where they are intercalated. During discharge, the reverse process holds. Li ions deintercalate from the matrix of the graphene and reinsert into the structure of the positive electrode.

In a completely charged battery, all the stored charge is in the negative electrode. As the battery is used, charge is taken from out of the negative electrode, and, in the example shown in Fig. 3.23.2.2, Li ions deintercalate from the graphite layers and electrons are released into the external circuit. This is the discharge process, and at this point in the use of the battery, the negative electrode (graphite) is the anode—because the anode has all of the charge (negatively charged electrons) and is delivering it toward the cathode. As the positively charged Li ions migrate through the electrolyte toward the $LiCoO_2$ cathode, negative charge, in the form of electrons, is delivered from the graphite anode, through the external circuit, to meet the positively charged Li ions at the surface of the positive electrode (cathode). It is during this electron transfer process that it is possible to utilize the current flowing as a charge source to power other devices. Once discharge is complete, the positive electrode is "full" of lithium and the negative electrode is "depleted" of lithium. At this point, the battery can no longer supply useful energy and needs to be charged. Now, the $LiCoO_2$ (positive electrode) acts as the anode (since it now contains all of the stored charge) and Li ions will diffuse and intercalate into the matrix of the graphite (acting as the cathode). The overall reaction is shown in Eq. (3.23.2.4c), where M is an oxide of a transition metal, such as Co ($y = 1$, $z = 2$) or Mn ($y = 2$, $z = 4$).

$$Li_{1-x}M_yO_z + Li_xC_6 \rightleftharpoons LiM_yO_z + 6C \qquad (3.23.2.4c)$$

The carbon-based intercalation materials typically employed as the negative electrode possess a low insertion/deinsertion potential, so the operating potential range of these devices will exceed the decomposition potential of the electrolytes. At certain cell potentials during the lithium intercalation process, components of the electrolyte will react at active sites on the carbon surface and will reductively decompose. Over the initial cycling of the device, these decomposition reactions will form what is referred to as the "solid electrolyte interphase" (SEI) layer. Ideally, this layer should be both an electronic insulator, inhibiting any charge transfer reactions, and an ion conductor that allows for the conduction of cations to/from the carbon host material. The formation of an SEI layer is a phenomenon that is not limited to current Li ion battery technologies, but also a factor in many next-generation battery chemistries. Determining the mechanisms that form the SEI layer and how that formation can be optimized, controlled, or even eliminated, is a focus of significant research.[17–21]

3.23.2.2.2 Challenges

Lithium batteries have become the workhorse of rechargeable batteries in portable electronic applications and have demonstrated viability in larger scale applications, such as electric vehicles (EVs). The majority of the more recent research (starting from 2000) is on the so-called beyond lithium technologies. As might be expected, the most common goal of this research is to increase the energy density of batteries, but the real challenge is to maintain the desirable high energy density while under high rates of discharge (high power delivery capability). The direction of recent research can be viewed as falling under three general approaches: (1) higher performing positive and negative electrode materials, (2) "beyond Li ion" batteries, or alternative, higher capacity cathode materials, and (3) "beyond lithium," or alternative intercalation ions. A brief overview of these new battery chemistries is provided next.

3.23.2.2.3 Current Research

The phrase "higher performing" used with regard to positive and negative electrode materials generally refers to materials with increased capacity and/or power capabilities that still function within the existing spectrum of lithium ion battery technologies. The most common materials used as positive electrodes have been layered oxides of the form $LiMO_2$ (where M = Co, Ni, or Mn). Even today, the positive electrode material that is most typically found in lithium batteries is lithium cobalt oxide ($LiCoO_2$), which offers relatively high energy density. However, with the focus on reducing cost and improving safety and energy density, chemical substitution at the metal (M) site has been extensively studied. Several viable combinations containing either two or three 3d metals were produced, such as $Li[Ni_{1-y-z}Mn_yCo_z]O_2$. An example of such a material with this composition is $LiNi_{0.33}Mn_{0.33}Co_{0.33}O_2$, which has a reported reversible specific capacity of 153 mAh/g. Other cathode materials that have been developed and follow the same composition patterns have even higher capacities (up to 200 mAh/g).[22] Other oxides that have also been developed, such as lithium iron phosphate ($LiFePO_4$), lithium manganese oxide ($LiMn_2O_4$, $LiMnO_3$), and lithium nickel manganese cobalt oxide ($LiNiMnCoO_2$), offer lower energy density, but they also provide longer cycle life and increased safety and so are also widely used. More recent research into new positive electrode materials have focused on what has been referred to as "layered-layered" metal oxides like $x LiMnO_2 \cdot (1-x) LiMO_2$, where

M is a mixture of Ni, Co, and Mn. This material has often been referred to as high-energy $LiNi_xMn_yCo_zO_2$ (NMC) but has so far only demonstrated ~100 cycles with 20% capacity loss.[22]

Research into higher capacity negative electrode materials is aimed at addressing the inherent capacity limitation of graphite. Referring back to Eq. (3.23.2.4c) which presents the charge/discharge of the negative electrode, fully lithiated graphite has a capacity of one Li atom for six carbon atoms or LiC_6. The capacity can be calculated as follows:

$$1\ \text{Li} = 1\text{e}^- = 96485.33289\ \text{C/mol} = 96485.33289\ \text{Ah/mol}$$

$$6\ \text{carbons} = 72.065\ \text{g}$$

$$\frac{96485.33289\ \text{Ah}}{72.065\ \text{gC}} = 371.91\ \text{mAh/g}$$

Given that the maximum theoretical capacity of graphite to store lithium is only 372 mAh/g, any research that produces higher capacity positive electrode materials will still be limited by the capacity of graphite. However, silicon is considered a very attractive choice as an anode material in lithium ion batteries and has become a strong focus of research.[23] Negative electrodes of silicon have a significantly greater theoretical capacity than graphite because silicon has been shown to be capable of forming $Li_{22}Si_5$ and $Li_{15}Si_4$.[24,25] Based on a similar calculation as mentioned earlier, silicon has a theoretical capacity of either 4200 mAh/g ($Li_{22}Si_5$) or 3580 mAh/g ($Li_{15}Si_4$), which is ~10 times higher than that of commercial graphitic materials. However, silicon undergoes significant volume expansion during the charging process because, unlike graphite, silicon is not an intercalation material but instead forms alloys. As a consequence of that alloying, there is a significant physical change in the composition of the electrode material, which results in substantial volumetric expansion. In addition, just as with the lithium ion batteries, an SEI layer will form during the initial charge/discharge cycles, and, with repeated expansions and contractions during the charge and discharge cycles, the coherency of the SEI layer is destroyed and gives rise to newly exposed and activated surfaces that continually react with and decompose the electrolyte. A significant loss of capacity over time results—the major issue that has so far hindered the commercialization of silicon anodes. Strategies for addressing this issue include exploiting suitable nanostructures, such as nanowires, nanotubes, or hollow particles, and using silicon-carbon nanocomposites, which have been shown to be better at buffering the volume changes and, in the case of the carbon composites, offer increased electronic conductivity.[26] Similar studies have also been performed with Sn-Li alloys as anode materials, which also suffer from similar expansion issues.[26]

In the category of "beyond Li ion" batteries are two cathode materials that, while still involving lithium ion as the charge carrier, make use of chemistries significantly different enough to warrant a separate discussion: lithium-air and lithium-sulfur batteries. Theoretically these batteries can offer higher capacity, but they both suffer from issues that have so far inhibited their commercialization potential. In addition, current research has shown that lithium metal will need to be used as the negative electrode for these two chemistries to achieve their highest practical capacities. Therefore the issues with dendrite formation that accompanies the use of lithium metal also have to be addressed.[27]

The Li-air battery is a promising candidate for future high-density storage applications, such as EVs. Li-air systems have significantly higher energy density relative to lithium ion batteries, as oxygen from the atmosphere is incorporated into the battery and is used as a reactant with the intercalating lithium on the surface of the cathode. Since oxygen is the most electronegative element that could be potentially employed in a battery, the Li-air battery has a predicted theoretical capacity of 3840 mAh/g[28] and, given an operating potential of 2.9 V, a theoretical specific energy density of 11,140 Wh/kg. Primary Li-air batteries have been available for years; however, the development of a Li-air secondary battery is very exciting as it is projected to maintain a three to four times increase in the gravimetric energy density (charge per unit mass) compared with Li ion batteries.

The positive electrode must be composed of a material that is permeable to oxygen, is conductive, and possesses a relatively high active surface area. Mesoporous carbon has typically been employed as the positive electrode, which is also catalyzed with either Mn, Co, or another transition metal to enhance the reduction kinetics and to increase the electrode's specific capacity.[29,30] The generalized charge/discharge processes for a Li-air battery are shown in Fig. 3.23.2.3.

The overall discharge (anodic) reaction at the negative electrode is $Li \rightarrow Li^+ + e^-$, whereas the overall discharge (cathodic) reaction at the positive electrode is $2Li^+ + 2e^- + O_2 \rightarrow Li_2O_2$. However, there is evidence that points to LiO_2 as an intermediate in the reaction that occurs at the positive electrode.[22,32–34] The oxygen molecule can be reduced to peroxide, which reacts with a lithium ion to form lithium peroxide, which can be reduced further by reacting with another lithium ion to form Li_2O_2. The reactions at the positive electrode that involve the LiO_2 intermediate are as follows, where "*" denotes a surface site on which Li_2O_2 growth could proceed.

$$\begin{aligned} Li^+ + e^- + O_2 &\rightleftharpoons LiO_2^* \\ Li^+ + e^- + LiO_2^* &\rightarrow Li_2O_2^* \end{aligned} \tag{3.23.2.5}$$

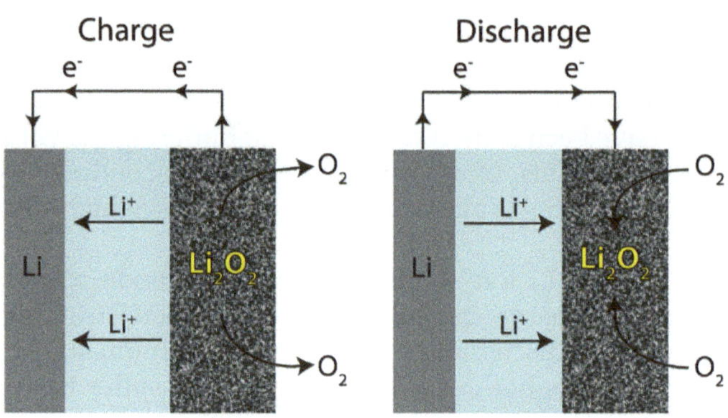

FIGURE 3.23.2.3 Charge/discharge of Li-air battery.[31] During discharge, Li ions migrate to the positive electrode where they are reduced in the presence of O_2 to form Li_2O_2. During charging of the battery, the Li_2O_2 is oxidized releasing O_2 to the atmosphere and the Li ions in the electrolyte are reduced at the negative electrode to Li metal. *From ©Na9234, used under Creative Commons (CC-BY-3.0).*

Some of the issues involving lithium-air batteries are similar to those in lithium ion batteries. There are side reactions during both charge and discharge that have a negative effect on the storage capacity. Specifically, during discharge, solvents can react to produce formates and acetates; the carbons of the positive electrode and gas-phase CO_2 have been found to form lithium carbonate, Li_2CO_3, whereas trace water leads to the formation of large Li_2O_2 crystals. During charging of the battery, alkyl carbonate-based electrolytes were found to react and impede charge storage. However, glycol ethers, or "glymes," are relatively stable against the highly reactive superoxide molecule that forms under standard charging conditions. Unfortunately, glymes are prone to autoxidation, which leads to other degradation products.[22,33,35] The search for a stable electrolyte is probably the most critical challenge for Li-O_2 battery research and commercialization to progress.

Sulfur is another potential cathode material that has the capacity to host two lithium atoms. This trait gives sulfur cathode batteries a very high theoretical specific capacity of 1660 mAh/g, although in practice, this capacity is significantly reduced. With an operating potential of between 1.7 and 2.5 V, a theoretical specific energy density of 500 Wh/kg is obtained. Therefore, the lithium-sulfur battery is a very promising energy storage device since it could potentially provide a substantial gravimetric energy density and cost advantage over lithium ion batteries. As with the lithium-air battery, lithium metal is typically the material of choice for the negative electrode since it provides the highest capacity, but silicon nanowires have also been employed.[22,23,36,37] The positive electrode is typically a mixture of sulfur dispersed onto a porous carbon matrix such as mesoporous carbons, disordered carbon nanotubes, or a graphene oxide nanocomposite.[38–40] However, the overall discharge reaction at the positive electrode is more complicated than other chemistries since elemental sulfur can exist in several polyatomic forms, with S_8 being the most common form that exists in the Li-S battery.[41] During discharge, solid-phase S_8 in the positive electrode first dissolves into the electrolyte and is then reduced in sequence to sulfide ions with progressively lower oxidation states according to the following electrochemical series.

$$\begin{aligned}
&{}^{1}\!/_{2}S_8 + e^- \rightarrow {}^{1}\!/_{2}S_8^{2-}\\
&{}^{3}\!/_{2}S_8^{2-} + e^- \rightarrow 2S_6^{2-}\\
&S_6^{2-} + e^- \rightarrow {}^{3}\!/_{2}S_4^{2-}\\
&{}^{1}\!/_{2}S_4^{2-} + e^- \rightarrow S_2^{2-}\\
&{}^{1}\!/_{2}S_2^{2-} + e^- \rightarrow S^{2-}
\end{aligned} \tag{3.23.2.6}$$

Chemical precipitation/dissolution reactions with lithium ions will also occur during the discharge process according to the following generalized reaction scheme: $2Li^+ + S_n^{2-} \rightarrow Li_2S_n$, where $n = 1, 2, 4$, or 8.[40,42–44] Just as with lithium-oxygen chemistry, lithium-sulfur batteries suffer from cycle stability issues related to the decomposition of the electrolyte, which so far have limited the LiS battery's potential for commercialization. Specifically, these issues are largely due to the formation of highly soluble, long-chain polysulfides that can diffuse/migrate to the negative electrode. There, these polysulfides (Li_2S_x, with $x \geq 4$) are further reduced to shorter chain polysulfides (Li_2S_x, with $x = 3$–6), thus enabling a process in which these reducible species are shuttled between the two electrodes at which they can be reduced in what is referred to as a redox shuttle. This results in a decrease in the coulombic efficiency.

Last we turn to alternative intercalation ions to the lithium ion, and the most promising one is sodium. Lithium is not a ubiquitous resource, so there is always the potential for regional scarcity, which will result in unpredictable costs. However, sodium does not suffer from that localization problem and is also a very cheap resource. A sodium ion battery functions the same as a lithium ion battery in that electrical energy is stored through the intercalation of the ions into the materials of the electrodes.

As with lithium, single metal and mixed metal oxides have commonly been used as materials for the positive electrode and the corresponding reaction that takes place is $Na_{1-x}M_yO_z + xNa^+ + xe^- \rightleftharpoons NaM_yO_z$, where M = Co, Mn, Cr, or Fe or a combination of Ni, Mn, Co, Fe, and Mg. In addition, sulfides/sulfates, phosphates, fluorides, and cyanoferrates have also been evaluated as cathode materials.[45–48] Unlike in lithium ion batteries, the use of graphite as a negative electrode is not feasible since sodium ions are both heavier and larger than lithium ions, which leads to larger volume expansion of the electrode material during intercalation. In addition, the larger size of sodium ions leads to relatively slow diffusion within a solid electrode during cycling of sodium ion batteries. Consequently, hard carbons or nanostructured carbons—designed in such a way as to accommodate the larger ions—have been employed as negative electrode materials.[45,49,50] The electrochemical reaction that takes place at the negative electrode is $Na_xC_n \rightleftharpoons xNa^+ + xe^- + nC$. Metal alloys and metal oxides that have also been evaluated as potential negative electrode materials include silicon, antimony, and copper and titanium oxides.[45,51,52] Sodium ion batteries typically have poor electrochemical activity when compared with their lithium ion counterparts. Lower ion activity is due, in part, to the large volumetric expansion that is intrinsic to the intercalation of Na ions (steric issues) and because Na has a lower ionization potential than Li. This leads to lower operating potentials and thus lower energy densities when compared with lithium ion batteries.

3.23.2.3 FUEL CELLS

Although the energy generated from a fuel cell is stored in chemical bonds, the actual electroactive species are not contained within the device itself. However, as long as sufficient fuel is supplied to the cell, it can be consumed during operation to convert the energy of the chemical bonds comprising the fuel into useful work. Just as with batteries, there are a variety of different fuel cell technologies that differ in their chemistries and operating conditions, and as a result of those differences, their applications are diverse as well. A list of the some of the more common types of fuel cells, and their associated characteristics, is provided in Table 3.23.2.2.[1]

Unlike rechargeable batteries, fuel cells always operate in energy-producing mode, similar to a galvanic cell, and so the terms "cathode" and "anode" are synonymous with the terms "positive electrode" and "negative electrode," respectively. Diagrams that illustrate the operation of the different fuel cells are provided in Fig. 3.23.2.4 followed by a brief description of each fuel cell and its operation.

TABLE 3.23.2.2 Characteristics of Different Fuel Cell Types

Type	T (°C)	Anode Reaction	Cathode Reaction
Alkaline	<100	$H_2 + 2OH^- \rightarrow 2H_2O + 2e^-$	$\frac{1}{2}O_2 + H_2O + 2e^- \rightarrow 2OH^-$
Proton exchange membrane	<120	$H_2 \rightarrow 2H^+ + 2e^-$	$\frac{1}{2}O_2 + H^+ + 2e^- \rightarrow H_2O$
Direct methanol	<120	$CH_3OH + H_2O \rightarrow CO_2 + H^+ + 6e^-$	$\frac{3}{2}O_2 + 6H^+ + 6e^- \rightarrow 3H_2O$
Phosphoric acid	150–200	$H_2 \rightarrow 2H^+ + 2e^-$	$\frac{1}{2}O_2 + H^+ + 2e^- \rightarrow H_2O$
Molten carbonate	600–700	$H_2 + CO_3^{2-} \rightarrow H_2O + CO_2 + 2e^-$	$\frac{1}{2}O_2 + CO_2 + 2e^- \rightarrow CO_3^{2-}$
Solid oxide	500–1000	$H_2 + 2O^{2-} \rightarrow H_2O + 2e^-$	$\frac{1}{2}O_2 + 2e^- \rightarrow O^{2-}$

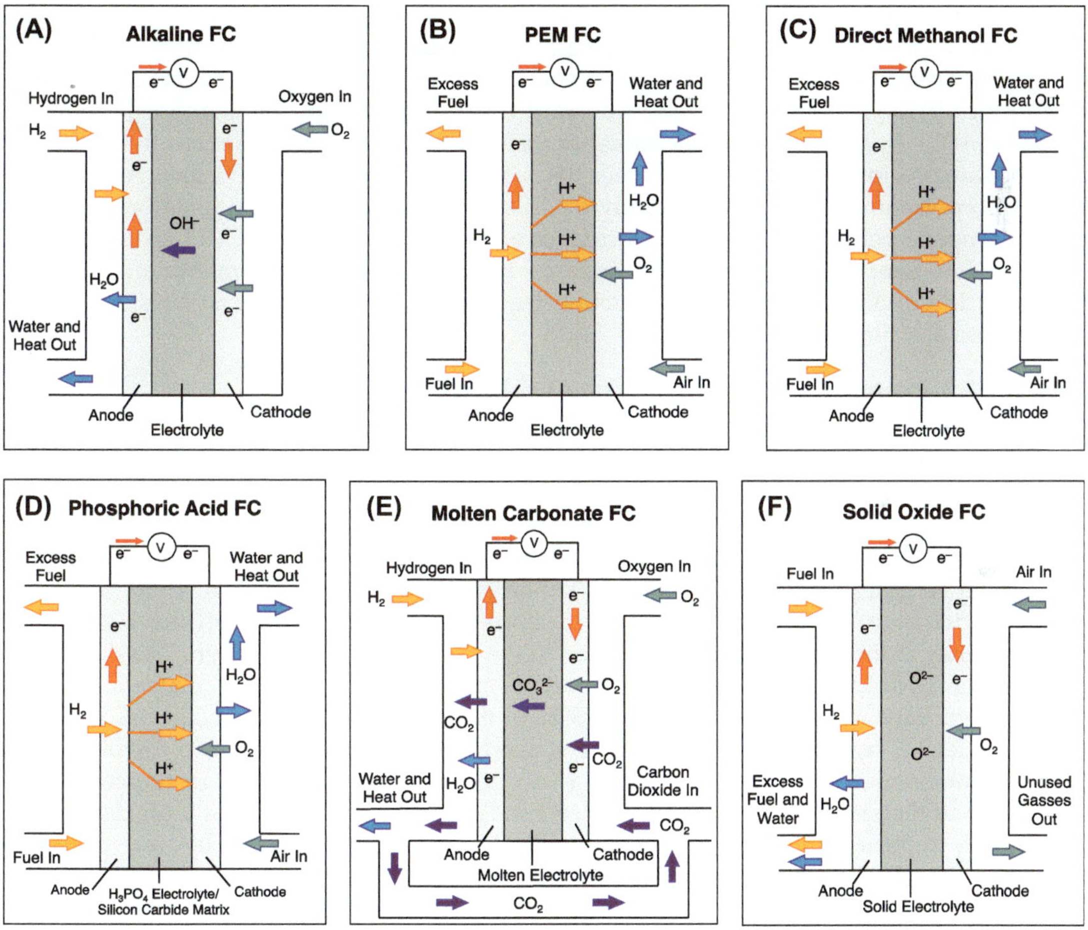

FIGURE 3.23.2.4 Types of fuel cells (FCs).[53] (A) Alkaline, (B) proton exchange membrane (PEM), (C) direct methanol, (D) phosphoric acid, (E) molten carbonate, (F) solid oxide. *Adapted from Types of fuel cells. U.S. Department of Energy; n.d. http://energy.gov/eere/fuelcells/types-fuel-cells.*

3.23.2.3.1 Fuel Cell Types

3.23.2.3.1.1 Alkaline Fuel Cell

In the alkaline fuel cell (AFC), hydrogen (the fuel) is oxidized at the anode, whereas oxygen (the oxidant) is reduced at the cathode. The AFC uses an aqueous electrolyte consisting of 35%–50% KOH, the electrodes are composed of porous carbons that have been catalyzed with platinum, and the hydroxide ion acts as the charge carrier. Basic electrolytes offer two primary advantages over acidic ones: (1) the kinetics of the reduction reactions at the cathode and anode are more facile at higher pH, which (2) allows for nonprecious metal catalysts to be used to efficiently enhance cell reactions.[54,55] The chemical reactions that occur at the respective electrodes are shown in the first row of Table 3.23.2.2, and the diagram that illustrates the operation of an AFC is shown in Fig. 3.23.2.4A.

A key challenge for AFCs has been their susceptibility to poisoning by carbon dioxide. CO_2, present in the supplied air, reacts with the potassium hydroxide electrolyte, following the reaction: $2KOH + CO_2 \rightarrow K_2CO_3 + H_2O$. While this conversion process is relatively slow, even a small amount of CO_2 in the air can dramatically affect cell performance and durability. To address this problem, researchers have moved to polymeric anion exchange membranes (AEMs) as the hydroxide transport medium in anion exchange membrane fuel cells (AEMFC). In an AEM, cationic moieties fixed to polymer chains prevent the formation of carbonate precipitates[56] and so fuel cells that use those membranes have lower susceptibility to CO_2 poisoning than liquid electrolyte AFCs do. However, even small amounts of CO_2 can still impede performance, and performance and durability of the AEMFCs still lag those of other low-temperature fuel cells, such as the proton exchange membrane fuel cell (PEMFC). Improving the oxidation-reduction reaction by increasing the activity of the catalysts is also a major focus of research.[55]

3.23.2.3.1.2 Proton Exchange Membrane Fuel Cell

The PEMFC has also been referred to as the polymer electrolyte fuel cell because it utilizes a solid-state electrolyte membrane that facilitates the transfer of cations. This exchange membrane separates the electrodes and is a hydrated solid that promotes the conduction of protons, which act as the charge carriers. A diagram that illustrates the operation of a PEMFC is shown in Fig. 3.23.2.4B. Although many different types of membranes are used, the most common is Nafion produced by DuPont. Nafion is composed of a carbon chain layer that has a fluorine atom layer attached to it, with branches that are formed off this basic chain. The frequency of these side chains is reflected in the different types of Nafion. The side chains increase the hydrophilic nature of the material, and it is this property that allows for a Nafion membrane to take up water up to 50% of the membrane's dry weight. As indicated in Table 3.23.2.2, hydrogen is oxidized at the anode releasing an electron to travel through the external circuit, doing work, whereas a solvated proton travels though the conductive membrane to the cathode. At the cathode, oxygen is reduced and combines with the electron and proton to form water. Since high temperatures are not necessary to hydrate the membrane, the PEMFC can run at very low temperatures relative to some other classes of fuel cells. The electrodes are typically composed of porous composites of polymer electrolyte binder and catalyst nanoparticles supported on carbon particles.

For the electrochemical reactions to occur at the electrodes, protons, electrons, and the reactant gases (hydrogen or oxygen) must be able to gain access to the very small surfaces that support the catalyst in the electrodes. In addition, the water that is produced must be able to permeate from the catalyst to the gas outlet. Because water is a product of the electrochemical reaction, the management of water is an issue for the PEMFC. It is important not to flood the electrodes since that will cause a decrease in surface area at which the separation of hydrogen or the formation of water takes place. The excess water at the cathode is also managed to some extent, since the flow of air by the cathode can remove some of that water through evaporation. Unfortunately, water also tends to move toward the cathode, and this "electroosmotic drag" can result in the removal of water at the anode at high current densities. Such a process results in the loss of fuel pressure since no water would be present to transfer new protons. However, the polymer membrane also needs to stay hydrated, so a balance among these competing factors must be achieved. Another issue with PEMFC is that the Nafion separator is unstable above 70 °C and is also expensive. Accordingly, there has been a significant amount of research into alternative membranes that are more stable at higher temperatures, which would improve the efficiency and inhibit the crossover of water while still allowing for proton transfer. Some of the other types of membranes that are being researched include polymer-zeolite nanocomposite proton exchange membranes and sulfonated polyphosphazene-based membranes.[57–60]

While hydrogen is an ideal fuel for a fuel cell, it has to initially be produced from either electrolyzing water or from steam reforming of hydrocarbons following this two-stage reaction: $CH_4 + H_2O \rightarrow CO + 3H_2 \rightarrow CO + H_2O \rightarrow CO_2 + H_2$. The steam reformation process is energy intensive, so there are advantages to obtaining the hydrogen directly from hydrocarbons within the fuel cell itself. This is the basis of the direct methanol fuel cell (DMFC), which utilizes methanol as the fuel.

3.23.2.3.1.3 *Direct Methanol Fuel Cell*

The use of methanol provides an advantage in that it is easier to store the fuel in liquid form rather than as compressed gas or as a hydride as with hydrogen. Like the PEMFA, the DMFC uses a polymer electrolyte membrane and also operates at relatively low temperatures (<100 °C). The methanol reacts with the catalysts on the anode to produce electrons, protons, and CO_2. The reaction is a multistep mechanism with the appearance of catalyst-adsorbed species as well as HCOH and HCOOH:

$$\begin{aligned} &CH_3OH + x\text{Pt} \rightarrow \text{Pt}{-}CH_2OH + H^+ + e^- \\ &\text{Pt}{-}CH_2OH + x\text{Pt} \rightarrow \text{Pt}{-}CHOH + H^+ + e^- \\ &\text{Pt}{-}CHOH + x\text{Pt} \rightarrow \text{Pt}{-}CO + 2H^+ + 2e^- + x\text{Pt} \end{aligned} \tag{3.23.2.7}$$

The reaction can also result in the formation of CO, which is a poison for platinum catalysts, and so catalysts are typically Pt-Ru or Pt-WO_3 alloys. Use of these alloys allows for the curing of Pt and prevents poisoning of the catalysts:[61,62]

$$\begin{aligned} &M + H_2O \rightarrow M - OH + H^+ + e^- \quad \text{where } M = \text{Pt, Ru} \\ &\text{Pt}{-}CHOH + \text{Pt}{-}OH \rightarrow HCOOH + H^+ + e^- + \text{Pt} \\ &\text{Pt}{-}CO + \text{RuOH} \rightarrow CO_2 + H^+ + e^- + x\text{Pt} + \text{Ru} \end{aligned} \tag{3.23.2.8}$$

At the cathode, the protons are recombined with the electrons and oxygen to form water. A diagram that illustrates the operation of a DMFC is shown in Fig. 3.23.2.4C.

There are, however, several issues with the DMFC that make it a less attractive option then a hydrogen fuel cell, and they can generally be attributed to the inability to get the full potential out of the anode and cathode.[54] First, acid electrolytes must be used because carbonate formation is a serious problem in alkaline solutions, but the corrosive tendency of the acid limits the type of catalysts that can be used. This has to be addressed through appropriate catalysts that are stable in acidic environments.[63] Second, since methanol oxidation is a multistep and more complex reaction, as is shown in Eq. (3.23.2.7), it proceeds much more slowly than hydrogen oxidation, as would be with the PEMFC. This results in a fuel cell that has a far lower power output for a given size.[63] In addition, although the cell's characteristics are similar to those of the PEMFC, methanol has a tendency to crossover from the anode to the cathode, and this results in a situation where it is possible to have "chemical short circuits." The short circuits result in decreased efficiencies since the same catalyst materials are typically used at both the cathode and the anode.[63] Although platinum or its alloys are considered to be ideal, the scarcity and expense of platinum ultimately limits its widespread application and so new catalysts are currently being researched. Nitrogen-doped carbon materials exhibit high catalytic activities in the oxygen reduction reaction,[64,65] while templated gold nanowires demonstrate promising activity in the oxidation of carbon monoxide, the aforementioned catalyst poison.[66]

3.23.2.3.1.4 Phosphoric Acid Fuel Cell

The phosphoric acid fuel cell (PAFC) is similar to the PEMFC in that the charge carrier is H^+ and so it uses a proton-conducting electrolyte, phosphoric acid. Phosphoric acid provides numerous advantages: the required thermal, chemical, and electrochemical stability; low volatility (which other inorganic acids do not possess); and the absence of the CO_2 reaction to form carbonate ions. Unfortunately, phosphoric acid has a freezing point of 42 °C and so the electrolyte of the fuel cell must be kept well above that temperature, which leads to a typical operating range of 150−200 °C. The electrolyte is held in a silicon carbide matrix that is composed of particles approximately 1 μm in size, which allows for the matrix to be about 0.1−0.2 mm thick and provides for considerably low ohmic losses, while still being thick enough to prevent crossover of the reactant gasses. The PAFC uses gas diffusion electrodes that are composed of a porous carbon matrix catalyzed by platinum. The porous carbon matrix must provide the following useful functions: (1) disperse the Pt or Pt-alloy catalyst to ensure good utilization of the catalytic metal, (2) provide micropores in the electrode for maximum gas diffusion to the catalyst and electrode/electrolyte interface, and (3) increase the electrical conductivity of the catalyst. The PAFC operation is shown in Fig. 3.23.2.4D.

The PAFC is considered to be the "first generation" of modern fuel cells, and so it is one of the more mature cell types and the first to be used commercially. Given the high operating temperature required for this type of fuel cell, it is typically used for stationary power generation. However, some PAFCs have been used to power large vehicles such as city buses.[53] Unfortunately, the cost of this technology is still too high to be economically competitive with alternative power generation systems, so research is being directed toward increasing the power and energy densities of the cells and thereby reducing the costs.

3.23.2.3.1.5 Molten Carbonate Fuel Cell

The molten carbonate fuel cell (MCFC) employs a molten mixture of alkali metal carbonates as the electrolyte and the carbonate ion (CO_3^{2-}) acts as the charge carrier. The anodes are composed of a porous, sintered Ni-Cr/Ni-Al alloy, whereas the cathodes are made from nickel oxide. The electrolyte is held in a ceramic matrix of $LiAlO_2$ and is usually a mixture of lithium and potassium or lithium and sodium carbonates that form a highly conductive molten salt at very high temperatures (approximately 600–700 °C). Because these devices operate at high temperatures, nonprecious metals can be used as catalysts at the anode and cathode, thereby reducing costs. The carbonate ion is oxidized in the presence of hydrogen to form water and carbon dioxide, whereas at the cathode the carbon dioxide is then reduced and reacts with oxygen to reform the carbonate. The operation of an MCFC is shown in Fig. 3.23.2.4E.

Unlike alkaline, phosphoric acid, and PEMFCs, MCFCs do not require an external reformer to convert fuels such as natural gas to hydrogen. At the high temperatures at which MCFCs operate, methane and other light hydrocarbons are converted to hydrogen within the fuel cell itself by a process referred to as internal reforming, which reduces cost. However, since MCFCs operate at temperatures high enough to keep the electrolyte molten, they are more suited for stationary applications as opposed to mobile applications. In addition, the high temperatures and the corrosive electrolyte accelerate component breakdown and corrosion, significantly decreasing cell life.

3.23.2.3.1.6 Solid Oxide Fuel Cell

The solid oxide fuel cell (SOFC) is part of a broader category of ceramic fuel cells. These cells have solid-state electrolytes, as well as a solid-state anodes, cathodes, and cell interconnects.[67,68] The SOFC in particular uses a hard, nonporous ceramic compound, such as yttria-stabilized zirconia, as the electrolyte. Ni ceramics and lanthanum manganite ($LaMnO_3$) have typically been used as the anode and cathode, respectively, but other solid-state materials such as layered perovskites have also been employed.[68] As with the MCFC, these cells operate at high temperatures (>500 °C) and therefore do not require precious metals on the anode or cathode. An added benefit to operating at very high temperatures is that SOFCs are around 60% efficient at converting fuel to electricity and are able to reform fuels internally. Therefore, SOFCs are able to use pure H_2 or even CO, as well as methane, diesel, gasoline, or coal gas.[53] The oxide anion (O^{2-}) acts as the charge carrier that will react with H_2 that has been oxidized at the anode to form water and free electrons. At the cathode, the oxygen from air will be reduced to reform the oxide anion. The operation of the SOFC is shown in Fig. 3.23.2.4F.

Operating at high temperatures also has its disadvantages, such as a slow startup time and the need for significant thermal shielding to retain heat. Just as with the MCFC, the high operating temperature also has a detrimental impact on the durability of the materials. Hence the development of low-cost materials with high durability and high activity at cell operating temperatures is a key technical challenge.[69]

3.23.2.3.2 Current Research

While the discussions covering the different fuel cell types also included a brief introduction to the technical challenges that are facing each specific device, there is one challenge that is common to almost all fuel cells: the catalysts. Platinum is a rare and expensive metal, which limits its uses. Key to the efficient use of platinum as a catalyst or as part of the catalyst is understanding the stability of platinum catalysts[70,71] and developing different strategies to lower the costs of its use. Those strategies include decreasing the actual amount of platinum needed by enhancing its catalytic activity, for example, through the use of carbon nanotubes as the support substrate,[72] or the use of binary and ternary alloys that not only exhibit increased activity but also demonstrate an increase in stability.[73] As discussed in the sections on the PEMFC and DMFC, platinum catalysts are also extremely sensitive to carbon monoxide poisoning, making it necessary to employ an additional reactor to reduce carbon monoxide in the fuel gas if the hydrogen is derived from a hydrocarbon fuel. In addition, platinum is susceptible to poisoning by other atmospheric contaminants, such as NO_x, which would be a concern if the fuel cell will take in air to obtain oxygen. Therefore the development of durable, ideally metal-free, and low-cost catalysts that are applicable in the oxygen reduction and fuel oxidation reactions is a significant challenge facing most modern fuel cells.

Lastly, in the introduction to the discussion of fuel cells, it was stated that those devices are energy producers, whereas batteries and supercapacitors are energy storage devices. However, there can be reversible fuel cells. Those fuel cell systems would produce electricity from hydrogen and oxygen and generate heat and water as by-products, just like other fuel cells, but the reversible fuel cells would also be able to use electricity from solar power, wind power, or other sources to split water into oxygen and hydrogen fuel through electrolysis. Reversible fuel cells could then provide power when needed. During times of high power production from other technologies, reversible fuel cells can store the excess energy in the form of hydrogen. This energy storage capability could be a key enabler for intermittent renewable energy technologies, such as wind and solar power.

3.23.2.4 SUPERCAPACITORS

As stated in Section 3.23.2.2 of this chapter, the energy storage mechanism of a capacitor derives from the formation of an electric double layer at the electrode/dielectric interface.[10] The charge carriers (ions or electrons) accumulate near the electrode/dielectric interface during charging and release during discharge. Since no chemical reactions take place, charging and discharging of a capacitor is very rapid, providing the high power capability of these devices. Since electrostatic charge storage does not involve chemical or physical changes to the electrode materials, it also provides a capacitor with an exceptionally high rate of recyclability—hundreds of thousands of cycles—without any noticeable degradation in performance. The charge storage capacity, or capacitance, is directly proportional to the surface area of the electrode, as was shown in Eq. (3.23.1.10), but this dependence on the surface area also limits the maximum storage capacity of the device.

Recall from the previous section the design of the simple dielectric capacitor consisting of two parallel plates separated by a dielectric (Fig. 3.23.1.9) and the resulting limitations that were imposed by that design. If the solid dielectric is replaced with a liquid electrolyte, it would allow ions to move to within molecular-sized distances (0.3–0.8 nm) of the charged surface of the electrode. The double layer that spontaneously forms at the interface between an electrolyte and a metal (or any conductive solid) surface has been well studied, and the capacitance of this polarizable interface can be as much as 200 times greater than that in a dielectric capacitor.[62] In addition, replacement of the metal plate electrodes with electrodes composed of high-surface-area powders will result in a substantial increase in the amount of energy that can be stored. Porous carbon materials can have total specific surface areas ranging from 1000 to 3000 m^2/g. Electrochemical capacitors (ECs) that use a liquid electrolyte in combination with high-surface-area electrodes to achieve high specific capacitance values are commonly referred to as *supercapacitors*.

As with batteries and fuel cells, supercapacitors come in different classes. The different types of the main components (electrolytes and electrode materials) give rise to different classes of supercapacitors. Since the specific properties of the electrode materials, for example, the pore structure and pore distribution, have to be matched to the physical properties of the ions within the electrolyte, the different electrolytes and electrode materials will be briefly discussed.

3.23.2.4.1 Electrolytes and Electrodes

The most common electrolytes for supercapacitors can be classified as either aqueous or non-aqueous, which indicate that the electrolyte is composed of ionic salts dissolved in either water or a nonaqueous solvent. Aqueous electrolytes have certain inherent advantages since water itself is slightly conductive (5.5 μS/m), easily polarizable, and has a relatively low viscosity (1.002 mPa s). Water is also a good solvent for ionic compounds, so electrolyte solutions can be created from acids such as sulfuric acid, from bases like potassium hydroxide, or from solutions of salts such as quaternary phosphonium salts, sodium perchlorate, lithium perchlorate, or lithium hexafluoride arsenate ($LiAsF_6$). Aqueous electrolytes generally have relatively high conductivity values ranging from 100 to 1000 mS/cm. The drawbacks to aqueous electrolytes are that they have a low dissociation potential (~1.15 V) per electrode and a relatively limited operating temperature range. The dissociation potential is the limit at which the capacitor can no longer be further charged without resulting in irreversible decomposition of the electrolyte. With these limitations, aqueous electrolytes are more suitable in supercapacitors that do not require high specific energy but where high specific power is more desirable.

Nonaqueous electrolytes are composed of organic solvents, such as acetonitrile, propylene carbonate, tetrahydrofuran, diethyl carbonate, or γ-butyrolactone, and quaternary ammonium salts or alkyl ammonium salts such as tetraethylammonium tetrafluoroborate [$N(Et)_4BF_4$] or triethyl(methyl)tetrafluoroborate [$NMe(Et)_3BF_4$]. Organic solvents will generally have a lower innate conductivity (10–60 mS/cm) and can have higher viscosities (0.334–2.507 mPa s), but they also have higher dissociation potentials, ranging from 1.35 V to ~4.0 V per electrode. In addition, nonaqueous electrolytes will have a higher flexibility in their operating temperature range. While the lower electrical conductivity of organic solvents leads to devices with a lower specific power relative to aqueous electrolytes, non-aqueous supercapacitors will generally have a higher specific energy since the specific energy increases with the square of the voltage (Eq. 3.23.1.12) and that, in general, will only be limited by the decomposition potential of the electrolyte.

Finally, there are ionic liquids (ILs). ILs are pure ionic materials that are a liquid at or very near to room temperature. Higher temperature ILs have been employed for other applications historically and are often referred to as molten salts. The temperature of those salts must remain extremely high (>500 °C) to remain liquid, whereas room-temperature ILs do not need to be so. The use of ILs in electrochemical energy storage offers numerous advantages. First, since ILs are not being employed as a solvent and have extremely low vapor pressures, there is no risk of loss of viscosity or ion mobility due to solvent evaporation. Second, the low vapor pressure also aids in improving the operation of the device at the upper temperature range since ILs will still be liquid. Third, although their conductivity may be quite moderate, in the context of a supercapacitor that may be less of an issue since the thickness of the double layer that is formed during charging is of the order of angstroms. Last, ILs have high dissociation potentials that provide a very wide voltage stability window (up to 6.0 V). However, there are drawbacks and these include extremely low viscosities and consequently a limit on the low end of the operating temperature range. Also, just as with nonaqueous electrolytes, it is very important to keep the ILs free of water as that can cause degradation and will result in decomposition if an IL-based supercapacitor is charged to high potentials.

The choice of the electrode material will depend on what mode of charge storage will be utilized for the device. The historical basis of charge storage for electrostatic and electrolytic capacitors is the double layer within the dielectric that forms during polarization of the electrodes. The type of supercapacitor that has directly descended from the electrolytic or electrostatic capacitors, and the more common type, is the *electrochemical double-layer capacitor* (EDLC). As the name implies, electrical energy is stored in the double layer that forms at the electrode/electrolyte interface. This is accomplished by changing the potential of each electrode so that the electrode obtains either a negative or a positive charge. Ions in the electrolyte are then attracted to the oppositely polarized electrode and accumulate in their respective double layers at the OHP. The charging and discharging processes are illustrated in Fig. 3.23.2.5, where Fig. 3.23.2.5A is the EDLC in the discharged state and Fig. 3.23.2.5B is the EDLC in the charged state.

In the discharged state (Fig. 3.23.2.5A), the electrodes are not charged and the ions of the electrolyte are randomly distributed. Although not explicitly illustrated in this diagram, the electrodes are composed of a high-surface-area material, and so there are channels and pores whose surfaces are accessible by the ions of the electrolyte. During charging of the device (Fig. 3.23.2.5B), the electrodes take on a positive or negative character and the ions of the electrolyte are attracted to the oppositely charged surfaces. An electrode with a surface area of about 1000 m^2/g results in a typical double-layer capacitance of about 10 $\mu F/cm^2$ and a specific capacitance of 100 F/g. In general, carbon materials for double-layer capacitive electrodes must have three characteristics: (1) high specific areas (>1000 m^2/g), (2) good intra- and interparticle conductivity in porous matrices, and (3) good electrolyte accessibility to the intrapore space of carbon materials. The proportion of the total surface area of the electrode material that is electrochemically accessible by the ions of the electrolyte limits

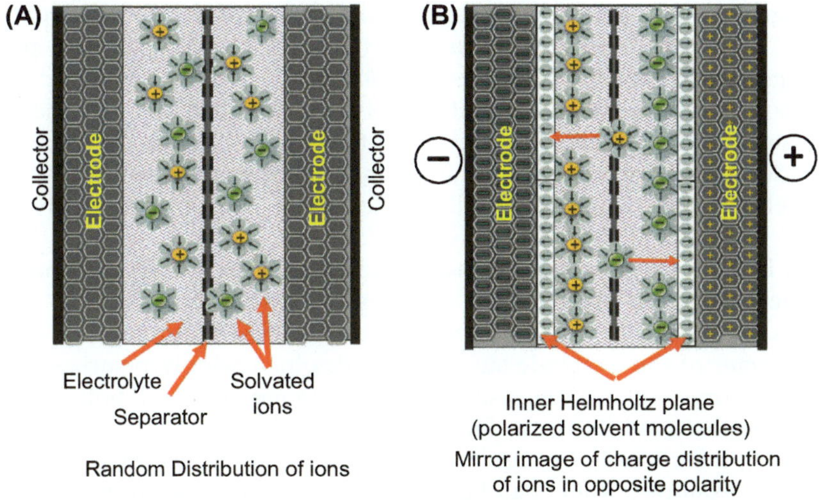

FIGURE 3.23.2.5 Double-layer capacitor diagram in (A) discharged and (B) charged states.[74] The electrodes are not charged in the discharged state (A), so the ions of the electrolyte are randomly distributed in the electrolyte. Charging of the device (B) introduces a positive or negative character to the electrodes, which attracts the ions of the electrolyte to the oppositely charged surfaces. *From ©Elcap, used under Creative Commons (CC0 1.0).*

the total amount of stored charge. Optimizing the pore structure of the electrode material so that more of the electrolyte ions have ready access to those surfaces is a major focus of research.[75]

It should be noted that heteroatoms and surface functionalities that may exist on the external surfaces of carbon materials used for electrodes could also undergo redox reactions. These surface functionalities may be the result of incomplete carbonization of the precursor or may intentionally have been added during preparation. The typical heteroatoms present in appreciable quantities include oxygen, hydrogen, and nitrogen. Depending on the nature of the functionality, those heteroatoms can either add to or detract from the total capacitance through their participation in Faradaic reactions, changing the electronic conductivity of the bulk material or affecting the electroactive surface area by influencing wettability.[76] In addition, there is an increased risk of electrolyte decomposition induced by active surface functional groups (especially oxygen-containing acidic groups), particularly for nonaqueous electrolytes. This will depend on the concentration and type of functional groups, the electrode surface area, and the operating potential.[77] For a more detailed discussion on the topic of surface functionality and their influence on electrochemical performance, the reader is directed to the excellent review articles by Hall et al.[76] and Wang et al.[77]

Lastly, in Fig. 3.23.1.4 of Section 3.23.2.2 it was indicated that there is a possibility that some of the ions in the electrolyte might lose their solvation sheaths and become adsorbed onto the surface of the electrode, where those specifically adsorbed ions could undergo redox reactions. If that redox reaction is highly reversible, then that process gives rise to an additional method of storing energy in what has been referred to as *psuedocapacitance*.[78] Psuedocapacitance is the electrochemical storage of electricity through very fast and reversible charge transfer (Faradaic) processes on the surface of electrodes prepared with suitable materials. Electrochemical supercapacitors that rely on fast Faradaic processes as a means of charge storage are referred to as *pseudocapacitors*. With pseudocapacitors, charge storage does not involve a chemical reaction between the adsorbed ion and the atoms of the electrode since only a single charge transfer takes place. It should be pointed out that although pseudocapacitance is the predominant method of storing electrical energy for pseudocapacitors, electrostatic capacitance still makes some contribution to the total energy stored.

Just as with EDLCs, the reversible Faradaic processes are a surface-dependent phenomenon (no chemical change is taking place within the electrode bulk) and so as each charge transfer process occurs, the energy required to effect the next redox reaction increases, since the overall character of the electrode surface becomes increasingly polarized. Therefore the charge/discharge curve still exhibits a similar linear change in the cell potential as a function of time to the one that was shown in Fig. 3.23.1.10 in Section 3.23.2.2. Certain noble metal oxides (e.g., RuO_2, IrO_3, MnO_2, Co_3O_4)[10,77,79] are well suited as electrode materials for pseudocapacitors. For example, RuO_2 has been employed as the electrode material using an aqueous electrolyte following the redox reaction shown in Eq. (3.23.2.9).[10]

$$RuO_x(OH)_y + \delta e^- + 8H^+ \rightarrow RuO_{x-8}(OH)_{y+\delta} \tag{3.23.2.9}$$

In addition to metal oxides, another type of pseudocapacitive electrode material that has been applied to ECs is intrinsically conducting polymers. These polymers not only possess π-orbital conjugation that provides electronic conductivity but also give those materials the ability to be reduced or oxidized, allowing them to store and release charge.[76,77,80] During oxidation, ions are transferred to the polymer backbone, and during reduction, the ions are released back into solution. Polyaniline was one of the earliest materials to be studied, and, due to its environmental stability, controllable electrical conductivity, and high specific capacitance, it continues to be intensively researched as an electrode material.[80] Another advantage to conducting polymers is that they are synthesized, which offers a potential method for tailoring the properties of the electrode material to fit specific applications at relatively low cost. As such, low-cost conducting polymers with good electrical conductivity and large pseudocapacitance continue to attract significant interest as electrode materials.

3.23.2.4.2 Classifying Supercapacitors

Based on the different modes of charge storage (i.e., double-layer capacitance versus pseudocapacitance) it is possible to develop a simple classification of supercapacitors with the first being the symmetric supercapacitor. A symmetric supercapacitor has electrodes that are of exactly the same material and utilize the same mode of charge storage. Although this is the simplest assemblage of components, it may not necessarily be the most efficient, as can be shown in the example of a double-layer EC manufactured using the same electrode material. Since solvated cations are typically smaller than anions (due to the larger electron cloud), a particular material that relies on double-layer capacitance for charge storage will be able to better accommodate cations than anions. Therefore the positive electrode would have to be prepared with more material (more mass) to provide sufficient surface area to balance the amount of charge storage on both electrodes, which would lower the specific capacitance for the cell. This issue is alleviated with an asymmetric supercapacitor.

An asymmetric supercapacitor has electrodes that are composed of different materials that nevertheless still utilize the same mode of charge storage. Returning to the example of the double-layer EC, the device is now fabricated using two different carbon materials. Such a cell allows for the material used at a particular electrode to be tailored to better accommodate cations or anions, depending on if it is the negative or positive electrode, respectively. In addition, the electrochemical accessibility of surfaces and how quickly those surfaces can be accessed are both equally important. Balancing the properties of each electrode so that they will have the same time constant related to the accessibility of ions interacting with electrode surfaces will determine the rate performance, or power capabilities, of the device. Specific to the asymmetric pseudocapacitor is that the electrodes will be composed of materials that should have corresponding reduction/oxidation potentials. Hence, while reduction is taking place at one electrode, oxidation is also taking place at the opposite electrode.

3.23.2.4.3 Challenges

Recall that in the Ragone plot shown in Fig. 3.23.2.1, dielectric and ECs dominated the higher end of the power density scale. Among the energy storage devices that utilize capacitive charge storage, supercapacitors are placed higher on the energy density scale, but even with the use of high-surface-area materials as the electrode materials, the energy densities of supercapacitors are at least two or more orders of magnitude lower than those obtained by batteries. Therefore it is not a coincidence that the biggest challenge for supercapacitors is the same as batteries: increased energy density. Additionally, size, weight, flexibility, cost, and resources are significant factors that present a constant challenge for supercapacitors. The following sections will further discuss some of those challenges.

3.23.2.4.4 Current and Future Research

3.23.2.4.4.1 Electrode Materials

One approach for increasing the electrostatic capacity in electrode materials is to increase the electrochemical accessibility to the surfaces. This can be accomplished by controlling the porosity and pore size distribution of the materials. With that in mind, forms of carbon that provide more control over those characteristics other than activated carbons have also been researched as electrode materials. These include carbon fiber cloths, carbide-derived carbons, carbon aerogels, graphite (in the form of graphene), graphane, and carbon nanotubes.[76,77,81–84]

With carbon fiber cloths, it is the diameters of the carbon fibers and how they are woven that are used to control the porosity of the material. Carbon fiber cloths have been used as the electrodes in IL-based and solid-state EDLCs,[85,86] as well as the conducting substrate for metal oxide, pseudocapacitive electrodes.[87] Carbide-derived carbons are materials that are derived from the chemical or physical etching of metal or metalloid atoms from binary (e.g., SiC, TiC) or ternary (e.g., Ti_2AlC, Ti_3SiC_2) carbides, and it is the structure of the original carbide that determines the porosity characteristics of the resultant carbon material. Carbon aerogels are synthetic, porous, and ultralight materials that are produced from the carbonization of organic gels, which results in a network structure of interconnected nanosized particles. They are composed of covalently bonded particles that possess high electrical conductivity, relatively high surface area (400–1000 m^2/g), and controllable pore structure (<100 nm). The carbonization process produces micropores within the particles, and mesopores and macropores are produced in the interparticle structure formerly occupied by the solvent. The distribution of micropores and mesopores will depend on the original ingredients, the curing and drying methods of the organic gel, and the conditions employed for the carbonization process.[75,88] However, they are predominantly mesoporous, and so microporosity is introduced using a postsynthesis activation step.[75] For example, Zhu et al. applied a CO_2 activation step to carbon aerogels to increase the surface area from 245 to 1418 m^2/g, which resulted in an increase of capacitance from 78 to 146 F/g.[89] Carbon nanotubes as capacitor electrode materials can only supply very moderate capacitance, but the addition of a thin amorphous carbon layer provides larger specific surface area and better conditions for increased charge accumulation. In addition, dramatic improvements in capacitance can be achieved by introducing pseudocapacitance effects through electrically conducting polymers[77] or through the formation of composites with transition metal oxides.[72]

3.23.2.4.4.2 *Hybrid Electrochemical Capacitors*

Another approach for addressing energy density limitations that has arisen in more recent research is the hybrid EC. Although it may seem similar to an asymmetric EC, each electrode material in a hybrid EC differs not only in its composition but also in its mode of storing charge. These different electrode materials are not just limited to electrostatic or pseudocapacitive materials but can also include intercalation or batterylike (i.e., bulk change) materials. Distinct from pseudocapacitive materials that are surface-dependent charge transfer phenomenon, batterylike materials will undergo both electrochemical (Faradaic) and physical changes during the charge and discharge processes, whereas intercalation materials will undergo a single charge transfer involving just the charge carrier (ion), and there is little effect on the composition of the electrode material. These different materials can be paired to produce an electrochemical energy storage device with certain energy densities and rate capabilities that are designed to fit a specific application. Typically for hybrid capacitors, the material of one electrode will use electrostatic energy storage. Two examples of hybrid capacitors will now be briefly introduced: (1) an electrostatic/pseudocapacitive EC and (2) the lithium ion capacitor (LIC).

The first example from Xu et al. is a hybrid capacitor that uses a high-surface-area carbon, with a Brunauer-Emmett-Teller (BET) surface area of >2000 m^2/g as the negative electrode and a positive electrode composed of MnO_2 combined with carbon black or carbon nanotubes to enhance conductivity.[90,91] An aqueous solution (pH ~6) of $Mg(NO_3)_2$, $Ca(NO_3)_2$, or $Ba(NO_3)_2$ was evaluated as the electrolyte.[90] In neutral medium, the reaction at the MnO_2 electrode follows the equation: $MnO_x(H_2O)_y \rightleftharpoons MnO_{(x+\delta)}(H_2O)_{(y-\delta)} + 2\delta H^+ + 2\delta e^-$, where $MnO_x(H_2O)_y$ and $MnO_{(x+\delta)}(H_2O)_{(y-\delta)}$ indicate interfacial manganese oxide at high and low oxidation states, respectively. It was expected that the stability window for a pH 6.4 aqueous electrolyte would be approximately be −0.4 to 0.85 V. Therefore the positive polarization of the manganese oxide would be limited by the evolution of O_2, but instead a high oxygen overpotential was measured. It was determined that the manganese oxide composite could be polarized up to higher potential values without exhibiting signs of electrolyte decomposition. This result was attributed to H^+ species being generated on the manganese oxide electrode, producing a shift in the pH down to lower values. Decreasing the potential cutoff to more negative values led to an increase in the electrode capacitance from 170 to 230 F/g that was attributed to a pseudocapacitive contribution produced as a result of reversible hydrogen adsorption on the activated carbon.

A hybrid LIC combines a lithium intercalation negative electrode with a high-surface-area capacitive charge storage positive electrode to increase the energy density relative to EDLCs while maintaining higher power densities than lithium ion batteries.[92–95] During charging of an LIC, Li ions are stored in the matrix of the intercalation material electrode, while at the same time the counterions establish a double layer at the interface of the capacitive electrode. During discharge, the intercalation material loses electrons and the Li ions deintercalate, while the double layer is broken at the positive electrode. The operation of an LIC is illustrated in Fig. 3.23.2.6.

Two issues face LICs, which are specifically manufactured having all of the Li ions present only in the Li salt of the electrolyte. The first issue is that since the potential required for intercalation generally exceeds the decomposition potential of the electrolyte, an SEI layer will be formed, and this results in a decrease in the energy storage capacity due to the irretrievable loss of Li ions into the SEI layer. The second is that during the charging process, the majority of the anions and Li cations will become localized at their respective electrode and this creates a decrease in the ionic conductivity of the electrolyte.[27,97] A way to address these potential problems is to prelithiate the negative electrode, which provides three advantages: (1) the total capacity of the LIC does not solely depend on the Li salt from the electrolyte; (2) the concentrations of the anion and cation do not change substantially during the charge and

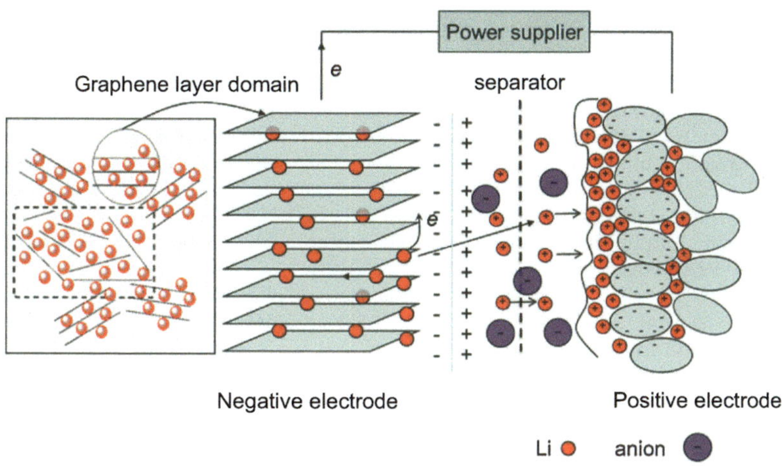

FIGURE 3.23.2.6 Lithium ion capacitor charge/discharge mechanisms. Li ions from the electrolyte are consumed to form the solid electrolyte interphase layer during the initial charging of the device. During charging of an lithium ion capacitor, Li ions insert into the matrix of the intercalation material electrode, while a double -layer is established with the counter ions at the interface of the capacitive (positive) electrode. During discharge, the intercalation material loses electrons and Li ions, while the double layer is broken at the positive electrode. *Adapted from Gourdin G, Smith PH, Jiang T, Tran TN, Qu D. Lithiation of amorphous carbon negative electrode for Li ion capacitor.* J Electroanal Chem *2013;688:103–12. http://dx.doi.org/10.1016/j.jelechem.2012.08.029.*

discharge processes, which would result in a decrease in conductivity; and (3) the portion of the irreversible capacity that arises out of formation of the SEI layer during the lithiation process when the capacitor is initially assembled can be minimized.[96,98]

As with Li ion batteries and supercapacitors in general, a major focus of the research involving LICs is the electrode material. Intercalation materials that have been exploited as the negative electrode include hydroxides,[99] transition metal oxides,[97,100,101] polyanions,[102] and graphitic carbons,[103] whereas materials for the positive electrode include layered oxides,[104,105] spinels,[106,107] phosphates,[108,109] and silicates.[110] In a comprehensive article, Aravindan et al. provided an excellent review covering the variety of intercalation materials that have been or could be used in LICs.[102]

References

1. Kirubakaran A, Jain S, Nema RK. A review on fuel cell technologies and power electronic interface. *Renew Sust Energy Rev* 2009;**13**:2430–40. http://dx.doi.org/10.1016/j.rser.2009.04.004.
2. Bockris J, Reddy A, Gamboa-Aldeco M. Chapter 1: The electrified interface. In: Bockris J, editor. *Modern electrochemistry 2A*. Springer; 2000.
3. Snubcube. *Double Layer 2*. Wikimedia Foundation; 2013.
4. Tosaka. *Electric double-layer (BMD model) NT*. vol. 540 × 640 pixels. Wikimedia Foundation; 2008.
5. Bockris JO, Devanathan MAV, Muller K. On the structure of charged interfaces. *Proc R Soc Lond A Math Phys Eng Sci* 1963;**274**:55–79. http://dx.doi.org/10.1098/rspa.1963.0114.
6. Grahame DC. The electrical double layer and the theory of electrocapillarity. *Chem Rev* 1947;**41**:441–501. http://dx.doi.org/10.1021/cr60130a002.
7. Tkarcher. *Diagram of a battery with a polymer separator*. vol. 570 × 609 pixels. Wikimedia Foundation; 2012.
8. Dervisoglu R. *Solid oxide fuel cell protonic*. vol. 508 × 570 pixels. Wikimedia Foundation; 2012.
9. Cook B. Introduction to fuel cells and hydrogen technology. *Eng Sci Educ J* 2002;**11**:205–16. http://dx.doi.org/10.1049/esej:20020601.
10. Conway BE. *Electrochemical Supercapacitors: Scientific Fundamentals and Technological Applications*. 1st ed. New York: Springer; 1999.
11. Xu K. Nonaqueous liquid electrolytes for lithium-based rechargeable batteries. *Chem Rev* 2004;**104**:4303–417. http://dx.doi.org/10.1021/cr030203g.
12. Haynes WM, editor. *CRC Handbook of Chemistry and Physics, 93rd edition*. 93rd ed. CRC Press; 2012.
13. Nagaura T, Tozawa K. Lithium ion rechargeable battery. *Prog Batteries Sol Cells* 1990;**9**:209–17.
14. Auborn JJ, Barberio YL. Lithium intercalation cells without metallic lithium: $MoO_2/LiCoO_2$ and $WO_2/LiCoO_2$. *J Electrochem Soc* 1987;**134**:638–41. http://dx.doi.org/10.1149/1.2100521.
15. Scrosati B. Lithium rocking chair batteries: an old concept? *J Electrochem Soc* 1992;**139**:2776–81. http://dx.doi.org/10.1149/1.2068978.
16. Winter M, Besenhard JO, Spahr ME, Novák P. Insertion electrode materials for rechargeable lithium batteries. *Adv Mater* 1998;**10**:725–63. http://dx.doi.org/10.1002/(SICI)1521-4095(199807)10:10<725::AID-ADMA725>3.0.CO;2-Z.
17. Xu K. Electrolytes and interphases in Li-ion batteries and beyond. *Chem Rev* 2014;**114**:11503–618. http://dx.doi.org/10.1021/cr500003w.
18. Bhatt MD, Geaney H, Nolan M, O'Dwyer C. Key scientific challenges in current rechargeable non-aqueous $Li-O_2$ batteries: experiment and theory. *Phys Chem Chem Phys* 2014;**16**:12093–130. http://dx.doi.org/10.1039/C4CP01309C.
19. Lu P, Li C, Schneider EW, Harris SJ. Chemistry, impedance, and morphology evolution in solid electrolyte interphase films during formation in lithium ion batteries. *J Phys Chem C* 2014;**118**:896–903. http://dx.doi.org/10.1021/jp4111019.
20. Collins J, Gourdin G, Foster M, Qu D. Carbon surface functionalities and SEI formation during Li intercalation. *Carbon* 2015;**92**:193–244. http://dx.doi.org/10.1016/j.carbon.2015.04.007.
21. Gourdin G, Collins J, Zheng D, Foster M, Qu D. Spectroscopic compositional analysis of electrolyte during initial SEI layer formation. *J Phys Chem C* 2014;**118**:17383–94. http://dx.doi.org/10.1021/jp504181b.

22. Gröger O, Gasteiger HA, Suchsland J-P. Review—Electromobility: batteries or fuel cells? *J Electrochem Soc* 2015;**162**:A2605–22. http://dx.doi.org/10.1149/2.0211514jes.
23. Erickson EM, Markevich E, Salitra G, Sharon D, Hirshberg D, la Llave de E, et al. Review—development of advanced rechargeable batteries: a continuous challenge in the choice of suitable electrolyte solutions. *J Electrochem Soc* 2015;**162**:A2424–38. http://dx.doi.org/10.1149/2.0051514jes.
24. Liu XH, Zheng H, Zhong L, Huang S, Karki K, Zhang LQ, et al. Anisotropic swelling and fracture of silicon nanowires during lithiation. *Nano Lett* 2011;**11**:3312–8. http://dx.doi.org/10.1021/nl201684d.
25. Hatchard TD, Dahn JR. In situ XRD and electrochemical study of the reaction of lithium with amorphous silicon. *J Electrochem Soc* 2004;**151**:A838–42. http://dx.doi.org/10.1149/1.1739217.
26. Hassoun J, Scrosati B. Review—advances in anode and electrolyte materials for the progress of lithium-ion and beyond lithium-ion batteries. *J Electrochem Soc* 2015;**162**:A2582–8. http://dx.doi.org/10.1149/2.0191514jes.
27. Goodenough JB, Kim Y. Challenges for rechargeable batteries. *J Power Sources* 2011;**196**:6688–94. http://dx.doi.org/10.1016/j.jpowsour.2010.11.074.
28. Luntz AC, McCloskey BD. Nonaqueous Li–Air batteries: a status report. *Chem Rev* 2014. http://dx.doi.org/10.1021/cr500054y. 141107065747008.
29. Harding JR, Lu Y-C, Tsukada Y, Shao-Horn Y. Evidence of catalyzed oxidation of Li_2O_2 for rechargeable Li–Air battery applications. *Phys Chem Chem Phys* 2012;**14**:10540. http://dx.doi.org/10.1039/c2cp41761h.
30. Cao Y, Wei Z, He J, Zang J, Zhang Q, Zheng M, et al. α-MnO_2 nanorods grown in situ on graphene as catalysts for Li–O_2 batteries with excellent electrochemical performance. *Energy Environ Sci* 2012;**5**:9765. http://dx.doi.org/10.1039/c2ee23475k.
31. Na9234. *Li-air-charge-discharge*. vol. 3789 × 2184 pixels. Wikimedia Foundation; 2011.
32. Lu Y-C, Crumlin EJ, Veith GM, Harding JR, Mutoro E, Baggetto L, et al. In situ ambient pressure X-ray photoelectron spectroscopy studies of lithium-oxygen redox reactions. *Sci Rep* 2012:2. http://dx.doi.org/10.1038/srep00715.
33. Li F, Zhang T, Zhou H. Challenges of non-aqueous Li–O_2 batteries: electrolytes, catalysts, and anodes. *Energy Environ Sci* 2013;**6**:1125–41. http://dx.doi.org/10.1039/C3EE00053B.
34. Adams BD, Radtke C, Black R, Trudeau ML, Zaghib K, Nazar LF. Current density dependence of peroxide formation in the Li–O_2 battery and its effect on charge. *Energy Environ Sci* 2013;**6**:1772–8. http://dx.doi.org/10.1039/C3EE40697K.
35. Geaney H, O'Dwyer C. Examining the role of electrolyte and binders in determining discharge product morphology and cycling performance of carbon cathodes in Li-O_2 batteries. *J Electrochem Soc* 2016;**163**:A43–9. http://dx.doi.org/10.1149/2.1011514jes.
36. Bresser D, Passerini S, Scrosati B. Recent progress and remaining challenges in sulfur-based lithium secondary batteries – a review. *Chem Commun* 2013;**49**:10545–62. http://dx.doi.org/10.1039/C3CC46131A.
37. Xu G, Ding B, Pan J, Nie P, Shen L, Zhang X. High performance lithium–sulfur batteries: advances and challenges. *J Mater Chem A* 2014;**2**:12662–76. http://dx.doi.org/10.1039/C4TA02097A.
38. Zhu L, Zhu W, Cheng X-B, Huang J-Q, Peng H-J, Yang S-H, et al. Cathode materials based on carbon nanotubes for high-energy-density lithium–sulfur batteries. *Carbon* 2014;**75**:161–8. http://dx.doi.org/10.1016/j.carbon.2014.03.049.
39. Ji X, Lee KT, Nazar LF. A highly ordered nanostructured carbon–sulphur cathode for lithium–sulphur batteries. *Nat Mater* 2009;**8**:500–6. http://dx.doi.org/10.1038/nmat2460.
40. Manthiram A, Fu Y, Chung S-H, Zu C, Su Y-S. Rechargeable lithium–sulfur batteries. *Chem Rev* 2014. http://dx.doi.org/10.1021/cr500062v. 140715153614001.
41. Yamin H, Gorenshtein A, Penciner J, Sternberg Y, Peled E. Lithium sulfur battery oxidation/reduction mechanisms of polysulfides in THF solutions. *J Electrochem Soc* 1988;**135**:1045–8. http://dx.doi.org/10.1149/1.2095868.
42. Kumaresan K, Mikhaylik Y, White RE. A mathematical model for a lithium–sulfur cell. *J Electrochem Soc* 2008;**155**:A576–82. http://dx.doi.org/10.1149/1.2937304.
43. Scrosati B, Hassoun J, Sun Y-K. Lithium-ion batteries. A look into the future. *Energy Environ Sci* 2011;**4**:3287–95. http://dx.doi.org/10.1039/C1EE01388B.
44. Cañas NA, Fronczek DN, Wagner N, Latz A, Friedrich KA. Experimental and theoretical analysis of products and reaction intermediates of lithium–sulfur batteries. *J Phys Chem C* 2014;**118**:12106–14. http://dx.doi.org/10.1021/jp5013208.

45. Sawicki M, Shaw LL. Advances and challenges of sodium ion batteries as post lithium ion batteries. *RSC Adv* 2015;**5**:53129–54. http://dx.doi.org/10.1039/C5RA08321D.
46. Li X, Wu D, Zhou Y-N, Liu L, Yang X-Q, Ceder G. O3-type $Na(Mn_{0.25}Fe_{0.25}Co_{0.25}Ni_{0.25})O_2$: a quaternary layered cathode compound for rechargeable Na ion batteries. *Electrochem Commun* 2014;**49**:51–4. http://dx.doi.org/10.1016/j.elecom.2014.10.003.
47. Doubaji S, Valvo M, Saadoune I, Dahbi M, Edstrom K. Synthesis and characterization of a new layered cathode material for sodium ion batteries. *J Power Sources* 2014;**266**:275–81. http://dx.doi.org/10.1016/j.jpowsour.2014.05.042.
48. Yoshida J, Guerin E, Arnault M, Constantin C, de Boisse BM, Carlier D, et al. New P2-$Na_{0.70}Mn_{0.60}Ni_{0.30}Co_{0.10}O_2$ layered oxide as electrode material for Na-ion batteries. *J Electrochem Soc* 2014;**161**:A1987–91. http://dx.doi.org/10.1149/2.0121414jes.
49. Irisarri E, Ponrouch A, Palacin MR. Review—hard carbon negative electrode materials for sodium-ion batteries. *J Electrochem Soc* 2015;**162**:A2476–82. http://dx.doi.org/10.1149/2.0091514jes.
50. Suryawanshi A, Mhamane D, Nagane S, Patil S, Aravindan V, Ogale S, et al. Indanthrone derived disordered graphitic carbon as promising insertion anode for sodium ion battery with long cycle life. *Electrochim Acta* 2014;**146**:218–23. http://dx.doi.org/10.1016/j.electacta.2014.09.052.
51. Kim H, Hong J, Park K-Y, Kim H, Kim S-W, Kang K. Aqueous rechargeable Li and Na ion batteries. *Chem Rev* 2014. http://dx.doi.org/10.1021/cr500232y.
52. Yu P, Li C, Guo X. Sodium storage and pseudocapacitive charge in textured $Li_4Ti_5O_{12}$ thin films. *J Phys Chem C* 2014;**118**:10616–24. http://dx.doi.org/10.1021/jp5010693.
53. *Types of fuel cells*. U.S. Department of Energy; n.d. http://energy.gov/eere/fuelcells/types-fuel-cells.
54. Winter M, Brodd RJ. What are batteries, fuel cells, and supercapacitors? *Chem Rev* 2004;**104**:4245–70. http://dx.doi.org/10.1021/cr020730k.
55. Sun W, Hsu A, Chen R. Carbon-supported tetragonal MnOOH catalysts for oxygen reduction reaction in alkaline media. *J Power Sources* 2011;**196**:627–35. http://dx.doi.org/10.1016/j.jpowsour.2010.07.082.
56. Miyazaki K, Sugimura N, Kawakita K-I, Abe T, Nishio K, Nakanishi H, et al. Aminated perfluorosulfonic acid ionomers to improve the triple phase boundary region in anion-exchange membrane fuel cells. *J Electrochem Soc* 2010;**157**:A1153–7. http://dx.doi.org/10.1149/1.3483105.
57. Holmberg BA, Wang X, Yan Y. Nanocomposite fuel cell membranes based on Nafion and acid functionalized zeolite beta nanocrystals. *J Membr Sci* 2008;**320**:86–92.
58. Peighambardoust SJ, Rowshanzamir S, Amjadi M. Review of the proton exchange membranes for fuel cell applications. *Int J Hydrogen Energy* 2010;**35**:9349–84. http://dx.doi.org/10.1016/j.ijhydene.2010.05.017.
59. Luo T, Zhang Y, Xu H, Zhang Z, Fu F, Gao S, et al. Highly conductive proton exchange membranes from sulfonated polyphosphazene-graft-copolystyrenes doped with sulfonated single-walled carbon nanotubes. *J Membr Sci* 2016;**514**:527–36. http://dx.doi.org/10.1016/j.memsci.2016.04.071.
60. Dong Y, Zhang YX, Xu HL, Luo TW, Fu FY, Zhu CJ. Proton exchange membrane based on crosslinked sulfonated polyphosphazene containing pendent perfluorosulfonic acid groups with sulfonated poly(ether ether ketone). *J Appl Polym Sci* 2016:133. http://dx.doi.org/10.1002/app.43492.
61. Jayaraman S, Jaramillo TF, Baeck SH, McFarland EW. Synthesis and characterization of Pt–WO_3 as methanol oxidation catalysts for fuel cells. *J Phys Chem B* 2005;**109**:22958–66. http://dx.doi.org/10.1021/jp053053h.
62. Bockris J, Reddy A. Chapter 4: Conversion and storage of electrochemical energy. In: Bockris JO'M, editor. *Modern Electrochemistry 2B. Electrodics in Chemistry*. Springer; 2000. p. 1789–901.
63. Direct Methanol Fuel Cells. *Fuel cell systems explained*. West Sussex, England: John Wiley & Sons Ltd; 2003.
64. Guo D-J. Electrooxidation of ethanol on novel multi-walled carbon nanotube supported platinum-antimony tin oxide nanoparticle catalysts. *J Power Sources* 2011;**196**:679–82. http://dx.doi.org/10.1016/j.jpowsour.2010.07.075.
65. Sun Y, Wu Q, Shi G. Graphene based new energy materials. *Energy Environ Sci* 2011;**4**:1113–32. http://dx.doi.org/10.1039/C0EE00683A.
66. Lee Y, Kim J, Yun DS, Nam YS, Shao-Horn Y, Belcher AM. Virus-templated Au and Au–Pt core–shell nanowires and their electrocatalytic activities for fuel cell applications. *Energy Environ Sci* 2012;**5**:8328–34. http://dx.doi.org/10.1039/c2ee21156d.
67. Goodenough JB, Singh P. Review—solid electrolytes in rechargeable electrochemical cells. *J Electrochem Soc* 2015;**162**:A2387–92. http://dx.doi.org/10.1149/2.0021514jes.
68. Kim YN, Manthiram A. Layered LnBaCoCuO ($0 \leq x \leq 1.0$) perovskite cathodes for intermediate-temperature solid oxide fuel cells. *J Electrochem Soc* 2011;**158**:B276–82. http://dx.doi.org/10.1149/1.3527006.

69. Mutoro E, Crumlin EJ, Pöpke H, Luerssen B, Amati M, Abyaneh MK, et al. Reversible compositional control of oxide surfaces by electrochemical potentials. *J Phys Chem Lett* 2012;**3**:40–4. http://dx.doi.org/10.1021/jz201523y.
70. Carlton CE, Chen S, Ferreira PJ, Allard LF, Shao-Horn Y. Sub-nanometer-resolution elemental mapping of "Pt 3Co" nanoparticle catalyst degradation in proton-exchange membrane fuel cells. *J Phys Chem Lett* 2012;**3**:161–6. http://dx.doi.org/10.1021/jz2016022.
71. Sheng W, Chen S, Vescovo E, Shao-Horn Y. Size influence on the oxygen reduction reaction activity and instability of supported Pt nanoparticles. *J Electrochem Soc* 2012;**159**:B96. http://dx.doi.org/10.1149/2.009202jes.
72. Lota G, Fic K, Frackowiak E. Carbon nanotubes and their composites in electrochemical applications. *Energy Environ Sci* 2011;**4**:1592–605. http://dx.doi.org/10.1039/C0EE00470G.
73. Mani P, Srivastava R, Strasser P. Dealloyed binary PtM_3 (M = Cu, Co, Ni) and ternary $PtNi_3M$ (M = Cu, Co, Fe, Cr) electrocatalysts for the oxygen reduction reaction: performance in polymer electrolyte membrane fuel cells. *J Power Sources* 2011;**196**:666–73. http://dx.doi.org/10.1016/j.jpowsour.2010.07.047.
74. Elcap. *EDLC-charge-distribution*. Wikimedia Foundation; 2013.
75. Sevilla M, Mokaya R. Energy storage applications of activated carbons: supercapacitors and hydrogen storage. *Energy Environ Sci* 2014;**7**:1250. http://dx.doi.org/10.1039/c3ee43525c.
76. Hall PJ, Mirzaeian M, Fletcher SI, Sillars FB, Rennie AJR, Shitta-Bey GO, et al. Energy storage in electrochemical capacitors: designing functional materials to improve performance. *Energy Environ Sci* 2010;**3**:1238. http://dx.doi.org/10.1039/c0ee00004c.
77. Wang G, Zhang L, Zhang J. A review of electrode materials for electrochemical supercapacitors. *Chem Soc Rev* 2012;**41**:797. http://dx.doi.org/10.1039/c1cs15060j.
78. Conway BE, Birss V, Wojtowicz J. The role and utilization of pseudocapacitance for energy storage by supercapacitors. *J Power Sources* 1997;**66**:1–14. http://dx.doi.org/10.1016/S0378-7753(96)02474-3.
79. Sharma P, Bhatti TS. A review on electrochemical double-layer capacitors. *Energy Convers Manage* 2010;**51**:2901–12. http://dx.doi.org/10.1016/j.enconman.2010.06.031.
80. Xu G, Wang W, Qu X, Yin Y, Chu L, He B, et al. Electrochemical properties of polyaniline in p-toluene sulfonic acid solution. *Eur Polym J* 2009;**45**:2701–7. http://dx.doi.org/10.1016/j.eurpolymj.2009.05.016.
81. Egashira M, Sawada N, Ueda K, Yoshimoto N, Morita M. Capacitance of porous carbon electrode in mixed salt non-aqueous electrolytes. *J Power Sources* 2010;**195**:1761–4. http://dx.doi.org/10.1016/j.jpowsour.2009.09.041.
82. Tien C-P, Teng H-S. Efficient ion transport in activated carbon capacitors assembled with gelled polymer electrolytes based on poly(ethylene oxide) cured with poly(propylene oxide) diamines. *J Taiwan Inst Chem Eng* 2009;**40**:452–6. http://dx.doi.org/10.1016/j.jtice.2008.11.005.
83. Simon P, Gogotsi Y. Materials for electrochemical capacitors. *Nat Mater* 2008;**7**:845–54. http://dx.doi.org/10.1038/nmat2297.
84. Jänes A, Tõnurist K, Thomberg T, Lust E. Comparison of electrospun and commercially available separator materials for supercapacitors. In: *ECS 215th meeting*, vol. 19. ECS; 2009. p. 23–32. http://dx.doi.org/10.1149/1.3253360.
85. Palm R, Kurig H, Tonurist K, Jänes A, Lust E. Electrical double layer capacitors based on 1-ethyl-3-methylimidazolium tetrafluoroborate with small addition of acetonitrile. *Electrochim Acta* 2012;**85**:139–44. http://dx.doi.org/10.1016/j.electacta.2012.08.030.
86. Aoki M, Tadanaga K, Tatsumisago M. All-solid-state electric double-layer capacitor using ion conductive inorganic-organic hybrid membrane based on 3-glycidoxypropyltrimethoxysilane. *Electrochem Solid St* 2010;**13**:A52–4. http://dx.doi.org/10.1149/1.3305323.
87. Xu J, Wang Q, Wang X, Xiang Q, Liang B, Chen D, et al. Flexible asymmetric supercapacitors based upon Co_9S_8 nanorod//Co_3O_4@RuO_2 nanosheet arrays on carbon cloth. *ACS Nano* 2013. http://dx.doi.org/10.1021/nn401450s. 130508150416007.
88. Moreno-Castilla C. Carbon gels in catalysis. In: Serp P, Figueiredo JL, editors. *Carbon materials for catalysis*. Hoboken, New Jersey: John Wiley & Sons, Inc.; 2009. p. 373–400.
89. Zhu Y, Hu H, Li W-C, Zhang X. Cresol–formaldehyde based carbon aerogel as electrode material for electrochemical capacitor. *J Power Sources* 2006;**162**:738–42.
90. Xu C, Du H, Li B, Kang F, Zeng Y. Asymmetric activated carbon-manganese dioxide capacitors in mild aqueous electrolytes containing alkaline-earth cations. *J Electrochem Soc* 2009;**156**:A435–41. http://dx.doi.org/10.1149/1.3106112.
91. Khomenko VG, Raymundo-Pinero E, Béguin F. Optimisation of an asymmetric manganese oxide/activated carbon capacitor working at 2 V in aqueous medium. *J Power Sources* 2006;**153**:183–90. http://dx.doi.org/10.1016/j.jpowsour.2005.03.210.

92. Amatucci GG, Badway F, Pasquier Du A, Zheng T. An asymmetric hybrid nonaqueous energy storage cell. *J Electrochem Soc* 2001;**148**:A930–9. http://dx.doi.org/10.1149/1.1383553.
93. Pasquier AD, Plitz I, Gural J, Badway F, Amatucci GG. Power-ion battery: bridging the gap between Li-ion and supercapacitor chemistries. *J Power Sources* 2004;**136**:160–70. http://dx.doi.org/10.1016/j.jpowsour.2004.05.023.
94. Choi N-S, Chen Z, Freunberger SA, Ji X, Sun Y-K, Amine K, et al. Challenges facing lithium batteries and electrical double-layer capacitors. *Angew Chem Int Ed* 2012;**51**:9994–10024. http://dx.doi.org/10.1002/anie.201201429.
95. Yoo HD, Markevich E, Salitra G, Sharon D, Aurbach D. On the challenge of developing advanced technologies for electrochemical energy storage and conversion. *Mater Today* 2014;**17**:110–21. http://dx.doi.org/10.1016/j.mattod.2014.02.014.
96. Gourdin G, Smith PH, Jiang T, Tran TN, Qu D. Lithiation of amorphous carbon negative electrode for Li ion capacitor. *J Electroanal Chem* 2013;**688**:103–12. http://dx.doi.org/10.1016/j.jelechem.2012.08.029.
97. Naoi K, Ishimoto S, Isobe Y, Aoyagi S. High-rate nano-crystalline $Li_4Ti_5O_{12}$ attached on carbon nano-fibers for hybrid supercapacitors. *J Power Sources* 2010;**195**:6250–4. http://dx.doi.org/10.1016/j.jpowsour.2009.12.104.
98. Gourdin G, Smith P, Jiang T, Tran T. Lithiation of anode for lithium-ion capacitor. In: *ECS 220th meeting*; 2011.
99. Cheng L, Li H-Q, Xia Y-Y. A hybrid nonaqueous electrochemical supercapacitor using nano-sized iron oxyhydroxide and activated carbon. *J Solid State Electrochem* 2005;**10**:405–10. http://dx.doi.org/10.1007/s10008-005-0016-8.
100. Cai Y, Zhao B, Wang J, Shao Z. Non-aqueous hybrid supercapacitors fabricated with mesoporous TiO_2 microspheres and activated carbon electrodes with superior performance. *J Power Sources* 2014;**253**:80–9. http://dx.doi.org/10.1016/j.jpowsour.2013.11.097.
101. Aravindan V, Shubha N, Ling WC, Madhavi S. Constructing high energy density non-aqueous Li-ion capacitors using monoclinic TiO_2-B nanorods as insertion host. *J Mater Chem A* 2013;**1**:6145–51. http://dx.doi.org/10.1039/C3TA11103B.
102. Aravindan V, Gnanaraj J, Lee Y-S, Madhavi S. Insertion-type electrodes for nonaqueous Li-ion capacitors. *Chem Rev* 2014. http://dx.doi.org/10.1021/cr5000915. 141118084758002.
103. Kim J-H, Kim J-S, Lim Y-G, Lee J-G, Kim Y-J. Effect of carbon types on the electrochemical properties of negative electrodes for Li-ion capacitors. *J Power Sources* 2011;**196**:10490–5. http://dx.doi.org/10.1016/j.jpowsour.2011.08.081.
104. Park M-S, Lim Y-G, Park J-W, Kim J-S, Lee J-W, Kim JH, et al. Li_2RuO_3 as an additive for high-energy lithium-ion capacitors. *J Phys Chem C* 2013;**117**:11471–8. http://dx.doi.org/10.1021/jp4005828.
105. Yoon JH, Bang HJ, Prakash J, Sun YK. Comparative study of $Li[Ni_{1/3}Co_{1/3}Mn_{1/3}]O_2$ cathode material synthesized via different synthetic routes for asymmetric electrochemical capacitor applications. *Mater Chem Phys* 2008;**110**:222–7. http://dx.doi.org/10.1016/j.matchemphys.2008.01.032.
106. Li H, Cheng L, Xia Y. A hybrid electrochemical supercapacitor based on a 5 V Li-ion battery cathode and active carbon. *Electrochem Solid St* 2005;**8**:A433–6. http://dx.doi.org/10.1149/1.1960007.
107. Ma S-B, Nam K-W, Yoon W-S, Yang X-Q, Ahn K-Y, Oh K-H, et al. A novel concept of hybrid capacitor based on manganese oxide materials. *Electrochem Commun* 2007;**9**:2807–11. http://dx.doi.org/10.1016/j.elecom.2007.09.015.
108. Wu XL, Jiang LY, Cao FF, Guo Y-G, Wan L-J. $LiFePO_4$ nanoparticles embedded in a nanoporous carbon matrix: superior cathode material for electrochemical energy-storage devices. *Adv Mater* 2009;**21**:2710–4. http://dx.doi.org/10.1002/adma.200802998.
109. Karthikeyan K, Amaresh S, Kim KJ, Kim SH, Chung KY, Cho BW, et al. A high performance hybrid capacitor with Li_2CoPO_4F cathode and activated carbon anode. *Nanoscale* 2013;**5**:5958–64. http://dx.doi.org/10.1039/C3NR00760J.
110. Karthikeyan K, Aravindan V, Lee SB, Jang IC, Lim HH, Park GJ, et al. Electrochemical performance of carbon-coated lithium manganese silicate for asymmetric hybrid supercapacitors. *J Power Sources* 2010;**195**:3761–4. http://dx.doi.org/10.1016/j.jpowsour.2009.11.138.

3.23.3 Applications of Renewable Energy From Green Chemistry: Electric Vehicles, Smart Grids, Smart Roadways, and Smart Buildings/Homes

3.23.3.1 INTRODUCTION

Due to the potentially catastrophic consequences of modern society's past and present reliance on fossil fuel consumption, greenhouse gas (GHG) emissions can now be considered the paramount human influence on future lifestyles and economies. GHG emissions are progressively becoming a fundamental metric by which we should be monitoring common practices in all economic sectors. Transitioning to renewable energy sources is a major method of suppressing GHG emissions while maintaining and growing the current world economy and way of life. As such, it is extremely important to understand the impact of specific economic sectors and their associated fossil fuel-based influence on the environment. More importantly, it is imperative for humans to accurately and accountably replace their GHG-emitting fuel sources with the sustainable, noncarbon emitting use of renewable energy.

To compare anthropogenic GHG emissions from general economic sectors, all GHG emission types (e.g., CH_4, N_2O) are converted to greenhouse warming equivalents of CO_2 (CO_2-eq).[1] Direct and indirect CO_2 equivalent emissions produced from various major economic sectors are shown in Fig. 3.23.3.1A.

In the first decade of the 21st century, cumulative GHGs have increased by 10 Gt CO_2-eq. To completely understand the sources of the net emissions, indirect CO_2 emissions in the form of electricity and heat production (Fig. 3.23.3.1A) have also been deeply investigated. Through accounting for both direct and indirect GHG emissions, it can be seen that the industry (net 32%) and buildings (net ~19%) sectors increased significantly. Combined with cumulative transport (~14%) and "other energy" (~10%) emission contributions, the interrelated economic sectors of transport, buildings, and industry (with indirect emissions contributing the majority) have the highest potential for reduction of emission contributions when replaced with renewable energy sources.

The relative trends of various specific direct sources of global CO_2 emissions between the years 1970 and 2004, as reported from the Intergovernmental Panel on Climate Change's Fourth Assessment report, are shown in Fig. 3.23.3.1B.[5] The specific economic sources of CO_2 emissions displaying increasing rate trends for over 30 years are electricity plants and road transportation, with refineries and residences also showing positive upward trends to a smaller degree. A more accurate understanding of the most significant emission sector CO_2 contributors can be obtained when combining this assessment with the assertions reached from Fig. 3.23.3.1A. The production of electricity from nonrenewable sources (e.g., from coal, natural gas), road transportation, and residential/service energy use are the most serious direct emission sector threats, whereas electricity and heat production from buildings and industry are the heaviest indirect contributors to global warming.

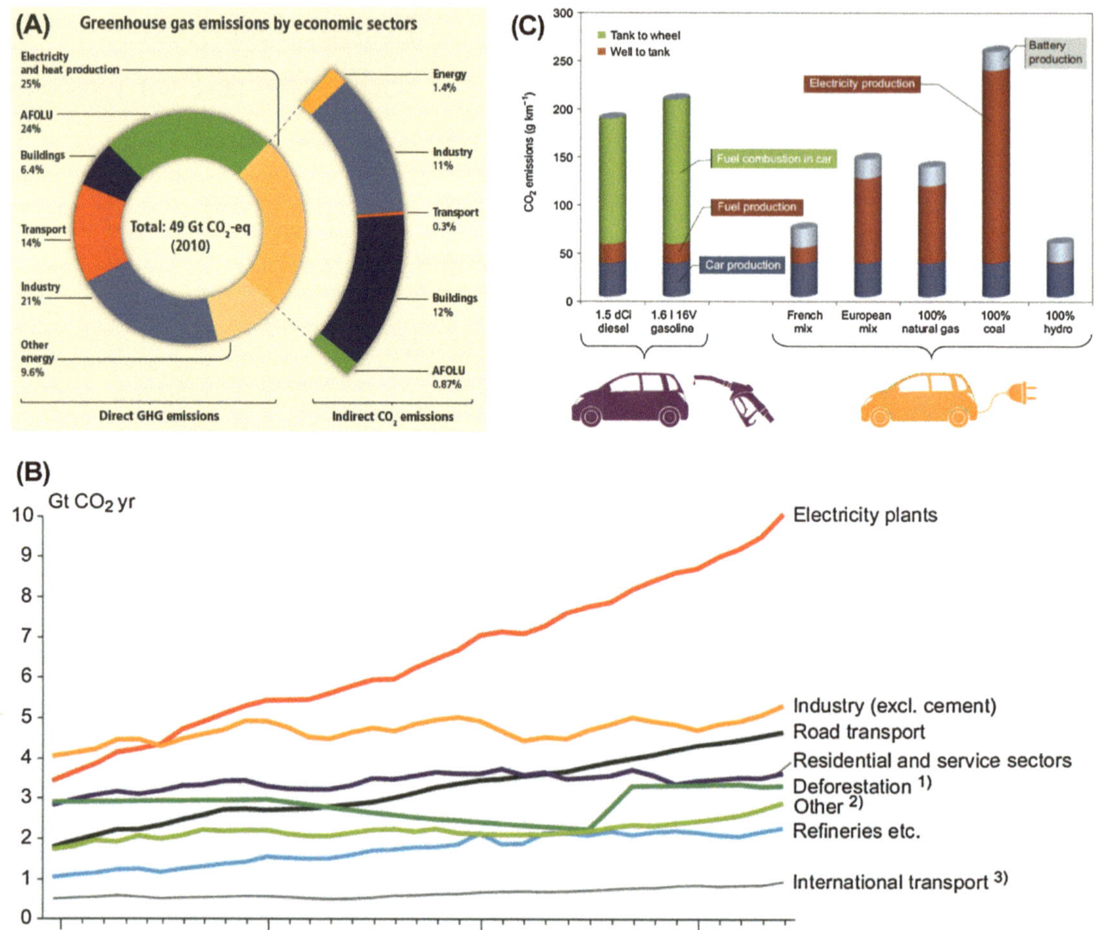

FIGURE 3.23.3.1 (A) Total anthropogenic greenhouse gas (GHG) emissions [gigatons of CO_2-equivalent per year (Gt CO_2-eq/year)] from economic sectors in 2010. The *circle* represents both direct and indirect GHG emissions. The direct GHG emissions relate to direct fuel consumption and consequential emissions (e.g., combustion of fossil fuels for automobiles). The pullout further delineates sources of indirect CO_2 emissions from electricity and heat production, which are attributed to sectors of later-stage energy use. AFOLU refers to agriculture, forestry, and other land use-based CO_2 emissions. Other energy here means energy sector GHGs (e.g., fuel production and transport, petroleum refining) other than electricity and heat production. (B) Sources of global CO_2 emissions, 1970–2004 (only direct emissions by sector). (C) The CO_2 benefit of electric vehicles (EVs) relies heavily on the origin of the electricity. Total CO_2 emissions (g/km) for internal combustion engine cars (left) and for full EVs (right) with various electricity origins. The "well-to-tank" step refers to the fuel production and delivery, whereas the "tank-to-wheel"' step deals with the car operation (i.e., fuel combustion). *(A) Adapted from IPCC publications, specifically IPCC. Climate change 2014: synthesis report. In: Pachauri RK, Meyer LA, editor.* Contribution of working Groups I, II and III to the Fifth Assessment Report of the Intergovernmental Panel on Climate Change 2014; *2014. (B) Adapted from Fig. 1.2 in Section 1.3 of Intergovernmental Panel on Climate Change. Climate change 2007: mitigation of climate change. In: Metz B, Davidson OR, Bosch PR, Dave R, Meyer LA, editors. Contribution of working group III to the Fourth Assessment Report of the Intergovernmental Panel on Climate Change. Cambridge, New York: Cambridge University Press; 2007 and originally from Olivier JGJ, Aardenne JAV, Dentener FJ, Pagliari V, Ganzeveld LN, Peters JAHW. Recent trends in global greenhouse gas emissions: regional trends 1970–2000 and spatial distribution of key sources in 2000.* Environ Sci *2005;2:81–99. http://dx.doi.org/10.1080/15693430500400345. (C) Adapted from Larcher D, Tarascon J-M. Towards greener and more sustainable batteries for electrical energy storage.* Nat Chem *2015;7:19–29. http://dx.doi.org/10.1038/nchem.2085.*

The single human innovation that can enable the change of all the aforementioned emission sources to renewable energy dependence is the battery (or, more generally, electrochemical energy storage devices). One possibility for streamlined renewable energy source integration is through the harvesting, storing, and utilizing of solar energy by individual consumers (e.g., harvested energy stored in their home wall sized batteries).[6] Such devices could charge and run appliances/electronics, heat needs, EVs, etc. Modern home batteries and EVs are based on a smart, bidirectional grid system (which will be covered later in this section) that allows for the reversible charge/discharge between car, home, office building, public utility/battery storage site, etc. The flexibility of integrating renewable energy-based sources with electrochemical energy storage has the real potential to curb the dominant portions of the major direct and indirect CO_2 emission sources. By delivering energy from renewable energy source locations (office, home, hydroelectric plant, public spaces, wind farm, etc.) through a closed loop energy distribution system via primarily the smart grid and EVs (home → EV/smart grid → office → EV/smart grid → public space → EV/smart grid → home...), the efficient storage and distribution of renewable energy will enable a practical and competitive market edge for the replacement of the fossil fuels.

CO_2 emissions from various sources of electricity are compared with traditional fossil fuels in Fig. 3.23.3.1C. It should be noted that the average European electricity mix is composed of fossil fuels and waste combustion (53%), nuclear (25%), and renewable hydro/wind (21%), whereas the French electricity mix is composed of nuclear (80%), renewable hydro/wind (11%), and fossil fuels and waste combustion (9%). Obviously, all electricity mixes, except for 100% coal, show significant CO_2 emission decreases over conventional diesel/gasoline sources. However, the percentage of renewable energies in each of these electricity mixes is quite low. The case of 100% renewable hydro-produced electricity has the least CO_2 emissions of any electricity source. Although the world's limited number of hydroenergy plants restricts their usage, hydroelectric power serves as an important example of the potential CO_2 reduction obtainable with 100% renewable energy electricity sources. The French mix of electricity sources displays the lowest CO_2 emissions from mixed sources due to the high contribution of nuclear energy. However, this low CO_2-producing energy source is also not a plausible option for effective, long-term sustainable worldwide distribution. Nuclear energy is currently impeded by a multitude of sustainable and environmental considerations such as those expressed by the International Atomic Energy Agency's classification of nuclear energy scenario sustainability.[7] Such considerations highlight the possibility of immediate dangers to humans (e.g., Chernobyl, Fukushima) and nuclear waste's potential chronic hazard to organic life. However, advancing technologies such as nuclear waste recycling and upcycling have been at the modern forefront of nuclear energy interest.[8,9] Consequently, the low contribution of renewable energy to both electricity mixes directly implies that a significant CO_2 emission reduction via renewable energy has yet to be tapped.

This section investigates some particulars of EVs, smart grids, and smart homes/buildings as the major adaptable technologies and essential economic sectors required for the sustainable integration of renewable energy sources in the modern world. Special attention is paid to EVs and various battery types with specific uses, as they are viewed as essential to the replacement of fossil fuels with renewable energy sources.

3.23.3.2 THE HISTORY OF ELECTRIC VEHICLES

EVs have had a tumultuous history from their conception. Although it is not commonly known, EVs were the major automobile of choice in America during the 1830s,[10,11] after Robert Anderson developed the first primary battery (nonrechargeable) prototype. However, EV popularity dropped due to the lack of practical rechargeable batteries and effectively high, overall vehicle energy efficiencies—leading to high cost with low payoffs. EVs briefly resurfaced between 1850 and 1880 through great advancements of direct current (DC) electric engines and rechargeable batteries due to the contributions of individuals such as Siemens and Gramme.[10,11] By the year 1900, EV taxis accounted for 28% of road vehicles and were a popular vehicle of choice.[12,13] Yet once again, cost, inefficiencies, limited range, and lack of consumer access suppressed EV growth and sustainability. Highly available and advertised, the invention of the gas-powered and affordable Model T by Henry Ford starting in 1908, along with the invention of the electric starter by Charles Kettering in 1912, resulted in lower cost and more accessible fossil fuel-based automobiles dominating the past century of consumer locomotion. The short traveling distance of the EV in combination with the emergence of cheap, highly available gasoline fuel sources allowed gas-powered vehicles to begin their dominance of the global automobile market.

EVs were virtually eliminated from the global consumer market until the early 2000s, when maturing battery technologies rapidly evolving toward higher efficiencies and more direct integration with renewable energy sources led to a major increase in the type and quantity of EV production and sales. Similar to the booms of the early and mid-20th century, next-generation technological advancements are again shepherding market changes—in both the automotive and energy sectors.

Fig. 3.23.3.2 displays the change in EV sales since 2010.[14,15] EV types are classified foremost as either hybrid electric vehicles (HEVs), plug-in hybrid electric vehicles (PHEVs), nonhybrid, all electric battery electric vehicles (BEVs), or fuel cell electric vehicles (FEVs). HEVs are still the dominant consumer-advertised and available EVs on the market. HEVs (red-Fig. 3.23.3.2) have displayed a near steady state since the year 2012. However, more advanced BEVs and PHEVs (green and blue-Fig. 3.23.3.2) have shown a significant increase in sales relative to total EVs (from 0.12% to 22.9%) between 2010 and 2015.[14–16] Additionally, the relative to total US car sales as well as total PHEV and BEV US car sales have demonstrated an increase from 0.0062% to 1.5% between 2010 and 2015[14–16]—a very noteworthy increase for the new-generation EV types. The emergences of more advanced and cheaper EV batteries, powering stations, and power train technologies have had a significant impact outside of the United States—as seen in Fig. 3.23.3.2 with the noteworthy uptrend of total non-US EV car sales in 2014/2015.

The Tesla model S was the world's best-selling plug-in EV during 2015 and was the second best in 2014.[17] The Model S was ranked by the Environmental Protection Agency as having the highest driving range in 2012, with 265 miles from one full charge.[18] This broke down to an 89 miles per gallon gasoline equivalent when the EV was equipped with an 85 kWh pack, yielding an annual fuel/energy cost of $750.[19] In April of 2016, Tesla unveiled its new Model 3 fully EV, which achieves 215 miles per full charge and is priced competitively at the consumer level, with their base model starting at $35,000. One week after the car's unveiling,

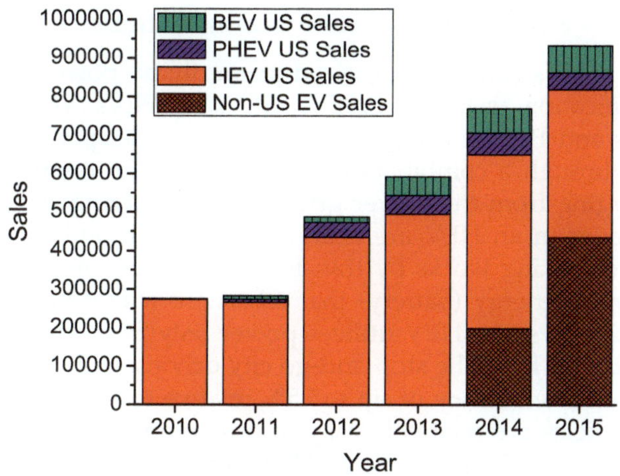

FIGURE 3.23.3.2 Average battery electric vehicle (BEV), plug-in hybrid electric vehicle (PHEV), and hybrid electric vehicle (HEV) US sales (number of vehicles per year) between 2010 and 2015. Total non-US electric vehicle (EV) sales are displayed for 2014 and 2015 only. The trends show the relative upward mobility of US BEV and PHEVs along with non-US EV in recent years—a result from affordability of advanced EV battery, charging stations, and power train technology. Values calculated from the Electric Drive Transportation Association[14] and the monthly plug-in sales scorecard.[15]

Tesla had received over 325,000 reservations for the future purchase of Model 3 vehicles planned for delivery at the end of 2017.[20] This number is more than triple the number of Model S cars Tesla had sold by the end of 2015. The unveiling of the Model 3 is part of Tesla's three-step strategy of starting at the higher priced consumer bracket and moving progressively toward lower cost, whereby past and present revenues from higher priced vehicles such as the Tesla Roadster and Model S primarily pay for the research and development of cheaper battery and drivetrain technologies that are priced to be available to a wider market.[21] Such innovation mandates the direct cooperation and feedback loop between consumer sales and technology development and is at the core of successful modern EV development. These types of strategies work to directly compete with the dominance of lower priced, gas-powered vehicles that are manufactured through a nearly 75-year-old mature infrastructure of vehicle manufacturing lines that have long produced vehicles in a highly efficient and relatively cheap fashion. Increasing consumer confidence and invested production of EVs are the result of modern "Henry Fords," as exemplified by EV innovators such as Elon Musk and Franz von Holzhausen of Tesla.

The primary factors impeding the integration of renewable energy with EVs are storage capacity and cost. Storage capacity refers to the concentration of stored charge within a given EV. The higher the capacity of energy storage, the longer EVs can run before requiring a charging station. An additional benefit to longer-running EVs is the reduction in the frequency of charging stations in nation-wide charging station networks. The cost of EV production/sales, and its competition with the gas-powered automobile industry, is all that would remain in streamlining the adoption of zero-emission transportation vehicles in the consumer world. If these two primary hurdles are overcome, the establishment of a modern EV consumer market is inevitable.

3.23.3.2.1 Electric Vehicle Power Train Configurations

The diverse array of EVs can be categorized primarily by power train and battery type, which leads to classification as HEVs, PHEVs, BEVs, or FEVs. The most common power trains for both HEVs and PHEVs are series, parallel, or series-parallel power trains. HEVs are supplied power through a combination of an ICE and an electric motor—they do not receive electrical charging from the power grid through charging stations or other external energy sources. Batteries in an HEV are charged by the ICE and/or through regenerative braking. Regenerative braking works by transforming kinetic energy from braking mechanisms to stored chemical energy (battery) that otherwise would be lost as heat from wear and tear on brake pads. A series HEV utilizes power only from an electric motor, supplied by regenerative braking or ICE. The stop-and-go city driving with this vehicle configuration enables an increase in the efficiency of up to 25% relative to conventional ICE vehicles.[13,22] Series power train vehicles are relatively easy to design and are popular for large vehicles such as city buses. In a parallel power train HEV configuration (e.g., Ford Escape, Honda Civic Hybrid), the electric motor and ICE are both coupled to the transmission independently. The efficiency of this vehicle configuration is therefore increased up to 40% through improvements in both city and highway driving. A series-parallel HEV also has both the electric motor and ICE coupled to the transmission/wheels—however, a generator that couples the ICE and electric motor allows for the system to run in either series or parallel mode, which makes these power train setups more complicated and costly. Modern consumer model examples are the Toyota Prius, the Lexus LS 600h, and the Nissan Tino.

PHEVs can be externally charged at stations or home plug-in sites (typically overnight). They can run solely on electric power longer than regular hybrids, due to a high-energy-density on-board storage system, resulting in better fuel economy. The power train configuration for PHEVs, which are very similar to HEVs with the addition of an on-board, plug-in battery charger, is shown in Fig. 3.23.3.3. In a series PHEV (Fig. 3.23.3.3A), power is supplied only to the transmission (which turns the wheels) from the electric motor. This electric motor is supplied with energy from either a rechargeable, plug-in battery or an ICE. Versatility (e.g., stop-and-go city driving and highway driving) and efficiency are increased with the parallel PHEV configuration (Fig. 3.23.3.3B), where power can be supplied independently from either the ICE or battery pack/electric motor. The series-parallel PHEV is shown in

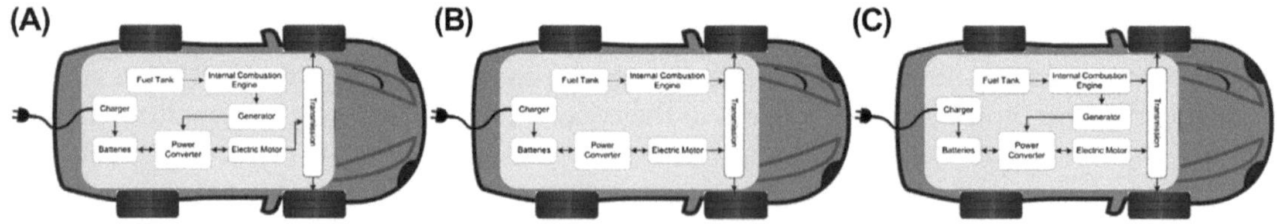

FIGURE 3.23.3.3 Modern power train configurations of plug-in hybrid electric vehicles (PHEVs): (A) series PHEV, (B) parallel PHEV, (C) series-parallel PHEV. All power train configurations displayed require a combination of internal combustion and battery-stored energy sources. *Adapted from Yong JY, Ramachandaramurthy VK, Tan KM, Mithulananthan N. A review on the state-of-the-art technologies of electric vehicle, its impacts and prospects.* Renew Sustain Energy Rev *2015;**49**:365–85. http://dx.doi.org/10.1016/j.rser.2015.04.130.*

Fig. 3.23.3.3C where both series and parallel modes are available to power the transmission and the electric motor interchangeably.

PHEVs offer bidirectional alternating current-DC power electronic converters, which can be used as the utility interface converter, allowing for grid-to-vehicle or vehicle-to-grid power transfer—thereby enabling compatibility with smart grid, 2-way energy storage.[23] There are two operating modes for PHEVs: charge-depleting (CD) and charge-sustaining (CS) modes. In the CD mode, the engine is off and all power is supplied by the battery pack. As the vehicle is driven, the state of charge (SOC) of the battery system decreases. The all-electric power supply from the CD mode continues until the battery SOC reaches a predefined minimum, at which point the engine turns on and the vehicle then operates in a hybrid, CS mode. PHEVs allow for both grid electricity and liquid fuels—increasing the potential diversity of energy sources, including renewable energy from grid sources. Hence, the PHEV can be viewed as a pivot vehicle toward renewable energy consumption EVs—surpassing the on-board, electrically regenerative, yet solely fossil-fuel-consuming energy consumption of HEVs.

BEVs have excellent environmental and efficiency benefits as they have zero tailpipe emissions and are the *only* EVs that have the potential to be powered completely from renewable energies. The only energy source for a BEV comes from a battery-powered electric motor that is powered by a plug-in battery and charger system. Both regenerative braking and external sources of energy charge the battery system. Since BEVs are all powered by electric batteries, they always run in the CD mode. BEVs have a considerably high torque produced from the electric motor, creating smoother acceleration/deceleration and an almost silent operating noise compared with ICE vehicles.[24] Battery capacity is the sole factor that determines the travel range of BEVs. Examples of BEVs are the Nissan Leaf, the Tesla Model S, and the Tesla Model 3. Fig. 3.23.3.4 displays a typical power train configuration for a BEV, which illustrates the direct power train energy sourced solely from the plug-in charging process.[13] Since the BEV uses only electric propulsion systems, it has traditionally been considered only suitable

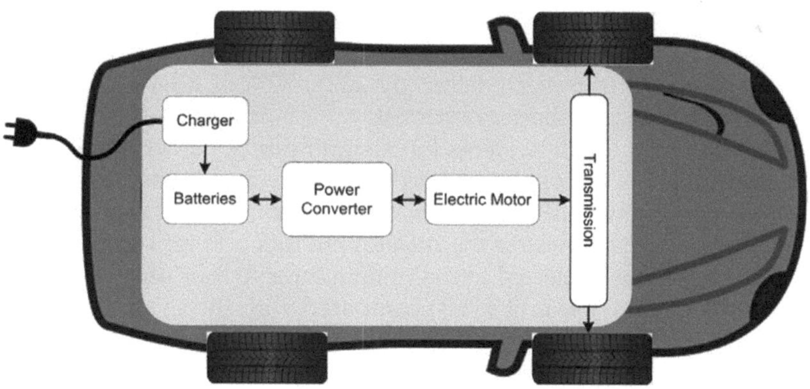

FIGURE 3.23.3.4 Typical power train of a battery electric vehicle. This vehicle's sole power source is from on-board electrochemical energy storage. *Adapted from Yong JY, Ramachandaramurthy VK, Tan KM, Mithulananthan N. A review on the state-of-the-art technologies of electric vehicle, its impacts and prospects.* Renew Sustain Energy Rev *2015;**49**:365–85. http://dx.doi.org/10.1016/j.rser.2015.04.130.*

for city driving due to its limited range. An example is the Nissan Leaf, which is powered by 24 kWh lithium ion battery packs and has a range of just 160 km (~99 miles). However, with the implementation of battery charging stations throughout the globe, BEVs are quickly achieving driving ranges similar to ICEs, HEVs, and PHEVs. As discussed, the Tesla Model 3 is an up-and-coming BEV that allows for up to 215 miles before requiring a recharge, and has fast-charging capability.[25] Past, contemporary, and future designs of electrochemical energy storage are a critical component of EV advancement and societal integration.

3.23.3.2.2 Advancements in Electric Vehicle Batteries

Batteries are an essential component of all EV propulsion systems. EV battery storage capacity is, of course, the primary factor in determining a vehicle's driving range and therefore also determines the progression of the EV consumer market. The selection of a suitable drivetrain and the implementation of mechanisms like regenerative braking, as discussed in the previous section, are the secondary factors that help to extend the runtime of a given battery. Batteries are made up of one or more electrochemical cells that store energy in a specific chemical form, and then transform that chemical energy into electrical energy when the user requires it. Although battery capacity (amount of charge stored) is typically reported in ampere-hours (Ah), batteries are often measured in watt-hours (Wh), which signifies the capacity stored in a battery multiplied by the average voltage during battery use. The SOC of a battery denotes the percentage of charge currently available in the battery—like the battery percentage indicator on many smartphones. Provided in Table 3.23.3.1 is a listing of various battery technologies integrated with EVs to varying degrees throughout the years. Please refer to Section 3.23.3.2.2 for a detailed discussion of the chemistries and operation of the more common battery types.

Compared with other battery storage technologies, Li ion batteries are the most efficient EV batteries due to their high storage capacity and power capability per unit of battery mass—allowing for relatively lighter and smaller rechargeable energy storage devices. Currently, the majority of EVs commercially available utilize NiMH batteries, but the relatively higher efficiency and lower toxicity of Li ion batteries project it to become the dominant EV battery source. One study projected that by 2025, 70% of HEVs, 100% of PHEVs, and 100% of BEVs will run on Li ion battery chemistry.[32] Li-air batteries are a promising candidate for future EV batteries. Li-air systems have significantly higher energy density—as oxygen from the atmosphere is incorporated into the battery and is used as a reactant with the intercalating lithium on the surface of the cathode. Primary (nonrechargeable) Li-air batteries have been available for years; however the development of a Li-air secondary (rechargeable) battery is very exciting as it is projected to maintain a three to four times increase in the gravimetric energy density (charge per unit mass) compared with the Li ion battery. The development of better Li-based energy storage is inevitable, and next-generation production is projected to create severe strain on the world's limited lithium reserves. For this reason, materials with high global reserves and lower cost (e.g., Zn and Na) are potential EV battery storage sources.

TABLE 3.23.3.1 EV Battery Development for the Past 100 Years

Battery	Description	Advantages	Disadvantages
Lead-acid (Pb-acid)	• Lead sulfate in sulfuric acid electrolyte • Oldest rechargeable battery used in both ICEs and EVs (est. 1859) • Low-use (starters) batteries in ICEs • Commonly used for golf carts and utility vehicles; being replaced by Li ion-based batteries	• Inexpensive to produce • High power/weight ratio	• Toxic • Very heavy (low energy/weight ratio) • Reaches equilibrium (becomes ineffective) when not in use for extended periods or when used for too long in one sitting (listening to music/using headlights when car is turned off)
Nickel-cadmium (NiCd)	• Est. 1899 • Rechargeable Ni oxide hydroxide and metallic cadmium	• Relatively high lifespan • Relatively high discharge/power: 1 h discharge	• Toxic—limited by European Union • Progressive reductions in storage capacity ("memory effect") after successive partial discharges
NiMH	• Est. 1986 • Similar to NiCd batteries in manufacturing and operation, except anode is H-adsorbing lanthanum/rare earth metal • Vehicles: Toyota Rav4EV-I (1997–2003), Toyota Prius I-III	• 50% higher energy density and less of a memory effect when compared with NiCd • 2000 cycles • Robust battery materials • Can be recharged quickly	• High self-discharge rate • Rate of discharge not as good as that of NiCd • Memory effect, yet not as bad as NiCd • Limited supply of lanthanum/rare earth metals (China)
Sodium-nickel chloride (Ni-NaCl)	• Est. 1985, known as "ZEBRA battery," Zeolite battery research Africa • Utilizes a molten salt at high temperatures up to 350 °C during operation • Nickel and molten sodium electrodes • Vehicles: smart EV, Think City, Smith electric vehicles	• High energy density (5× lead acid) • 1500–3000 cycles depending on the number of cells in the system	• Liquid electrolyte freezes at 157 °C, normal operating temperature is 270–350 °C • Limited range of available sizes and capacities • Preheat required to 270 °C • High self-discharge (15% capacity/day when not in use to maintain necessary functioning high temps)
Lithium ion (Li ion)	• Carbon-based negative electrode • Metal oxide positive electrode • Electrolyte is typically Li salt in organic solvent • Vehicles: Li-Ni-Co-Al: Think City, Mitsubishi-MiEV, GM Volt • Li-Ni-Mn-Co: Nissan Leaf, Ampera, Tesla	• Relatively large power, energy density and specific energy storage capacity • Highly versatile for its physical shape and uses—portable devices, power tools, and EVs • Possibility for battery recycling • Much less toxic than lead or cadmium	• High cost, limited recharging cycles, and overheating danger • Pure lithium is highly reactive • Strict temperature range to preserve battery lifetime (5–45 °C) and avoid degradation • Limited supply of lithium in world (can be expensive)
Lithium-ion polymer (LIP)	• Similar to a Li-ion battery but it uses a solid polymer electrolyte (SPE) instead of Li-salt/organic solvent electrolyte	• Lighter and smaller battery materials with higher charge storage capacity than conventional Li ion batteries	• Instability upon being overloaded and discharge below minimum values: cells can expand greatly and rupture cell lining—fire hazard

(Continued)

TABLE 3.23.3.1 EV Battery Development for the Past 100 Years—cont'd

Battery	Description	Advantages	Disadvantages
	• Vehicles: Hyundai HEVs, Kia Soul	• Even more flexible than Li ion batteries and allows for almost any desired shape of battery—great applications in radio-controlled devices, thin digital devices, phones, and EVs	
Zinc-air (Zn-air)	• Powered by the oxidation of zinc from oxygen (ideally from the air) • Highly energy dense, relatively inexpensive new-generation battery	• Promising development; Zn-air batteries have replaced toxic mercury batteries in small applications (camera, hearing aids, etc.) • Scalable to EV battery and grid-scale energy storage • Does not require rare earth metals	• Must be sealed during storage to preserve shelf life • Self-discharge (from water and CO_2) when exposed to air • Cell can be destroyed (quenching the cathode) in environments with high temperature/humidity • Zinc corrosion can concentrate explosive hydrogen
Lithium-air (Li-air)	• Next-generation Li ion batteries, that rely on oxidation of lithium at the anode and reduction of oxygen at the cathode • Optimal, controlled chemistry of Li and O_2 has highest theoretical specific energy of all modern batteries	• Energy density superior to any other EV battery (5–15 times specific energy of Li ion batteries) • Only battery where the theoretical energy density could compete with energy delivered through the power train system by gasoline (1.7 kWh/kg)	• Cost and very limited cycle life constrain commercial usage • Li oxides form precipitates on electrode surface—limiting the capacity of extended cycling • Cycle life of Li-air batteries requires significant improvement before commercialization (still in R&D phase)
Lithium-sulfur (Li-S)	• Li-dissolution/plating at the anode. Polysulfides are reduced at the cathode during discharge. S-polymer formation at cathode during charging	• Another next-generation Li ion-based-battery also with much higher energy density: (sulfur can host two Li ions where ~0.6 Li ions are accommodated in Li ion batteries) • Light weight and very low cost due to the use of earth-abundant sulfur—mass production possibility • 50–1500 cycle life	• High mechanical stress on the cell as the sulfur in cathode adsorbs Li and expands nearly double in size • Many polysulfides that form are insoluble in electrolyte—causing irreversible passivation of active sulfur and decreasing capacity • Not yet commercially available—still in R&D phase • Rapid discharge hazard possibility due to high energy density

est., established; *EV*, electric vehicle; *HEV*, hybrid electric vehicles; *ICE*, internal combustion engine; *LIP*, lithium ion polymer; *NiMH*, nickel metal hydride; *R&D*, research and development.

Compiled from Manzetti S, Mariasiu F. Electric vehicle battery technologies: from present state to future systems. Renew Sustain Energy Rev *2015;***51***:1004–12. http://dx.doi.org/10.1016/j.rser.2015.07.010; Helmers E, Marx P. Electric cars: technical characteristics and environmental impacts.* Environ Sci Eur *2012;***24***:14. http://dx.doi.org/10.1186/2190-4715-24-14; Dunn B, Kamath H, Tarascon J-M. Electrical energy storage for the grid: a battery of choices.* Science *2011;***334***:928–35. http://dx.doi.org/10.1126/science.1212741; Daniel C, Besenhard JO, editors.* Handbook of battery materials: DANIEL: HBK BATTERY MAT E2 O-BK. *Weinheim, Germany: Wiley-VCH Verlag GmbH & Co. KGaA; 2011; Girishkumar G, McCloskey B, Luntz AC, Swanson S, Wilcke W. Lithium–air battery: promise and challenges.* J Phys Chem Lett *2010;***1***:2193–203. http://dx.doi.org/10.1021/jz1005384; Balaish M, Kraytsberg A, Ein-Eli Y. A critical review on lithium–air battery electrolytes.* Phys Chem Chem Phys *2014;***16***:2801. http://dx.doi.org/10.1039/c3cp54165g; Yekini Suberu M, Wazir Mustafa M, Bashir N. Energy storage systems for renewable energy power sector integration and mitigation of intermittency.* Renew Sustain Energy Rev *2014;***35***:499–514. http://dx.doi.org/10.1016/j.rser.2014.04.009.*

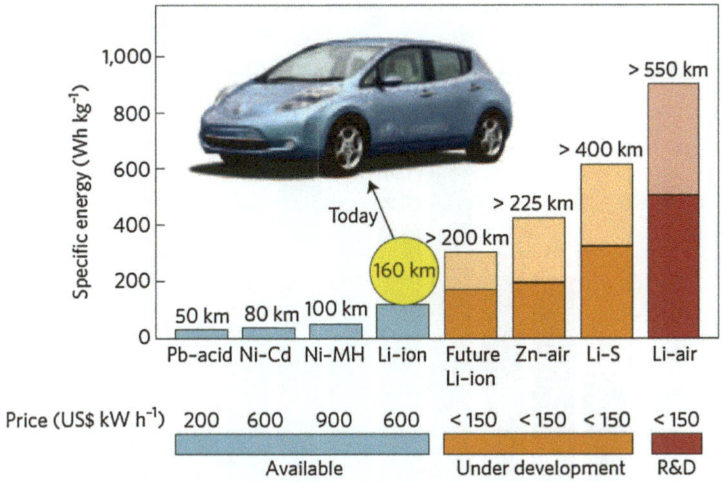

FIGURE 3.23.3.5 Practical specific energies for some rechargeable batteries, along with estimated driving distances and pack prices. For future technologies, a range of anticipated specific energies are given as shown by the *lighter shaded region* on the bars in the chart for rechargeable batteries under development and research and development. *Darker shaded regions* for these batteries are practical estimations of existing technology. The values for driving ranges are based on the minimum specific energy for each technology and scaled on the specific energy of the Li ion cells (140 Wh/kg) and driving range (160 km) of the Nissan Leaf.[34] The prices for technologies under development represent targets set by the US Advanced Battery Consortium.[35] *Adapted from Bruce PG, Freunberger SA, Hardwick LJ, Tarascon J-M. Li-O_2 and Li-S batteries with high energy storage.* Nat Mater *2012;**11**:19–29. http://dx.doi.org/10.1038/nmat3191.*

The pattern of EV battery development has been repeated throughout its commercial development: theoretical limits of a new battery's charge storage capability are evaluated through a multitude of experiments, and eventually, the practical, specific battery formulation is optimized and commercialized. As history has demonstrated again and again, theoretical predictions rarely meet practical reality. Fig. 3.23.3.5 displays a leading 2012 tabulation and estimations of fully developed and developing EV batteries as well as their driving ranges and pricing.[33] The commercially available batteries displayed in blue are fully developed and illustrate minimal energy density and therefore driving distance at mid- to high-level production cost. Relative to the range and cost efficiency of Li ion batteries powering the 2012 Nissan Leaf, Li ion, Zn-air, Li-S, and Li-air batteries are all predicted to make significant improvements over 2012 commercial capabilities. Li-S and Li-air batteries have been seen as particularly promising prospects for the future of energy storage for nearly a decade. Both battery technologies have particularly high theoretical energy densities given that the common forms of stored charge in Li-S or Li-air cathodes—Li_2S, or Li_2O_2 and LiOH, respectively—all store more charge per unit mass, which results in correspondingly high specific energy. Additionally, the Li metal anode stores more charge per unit mass compared with traditional graphitic anodes utilized in $LiCoO_2$ batteries.

To illustrate the theoretical versus practical development of EV batteries, current battery performance (Fig. 3.23.3.6) can be compared with theoretical predictions from 2012 (Fig. 3.23.3.5). A critical component of battery performance and usability is cycle life (the

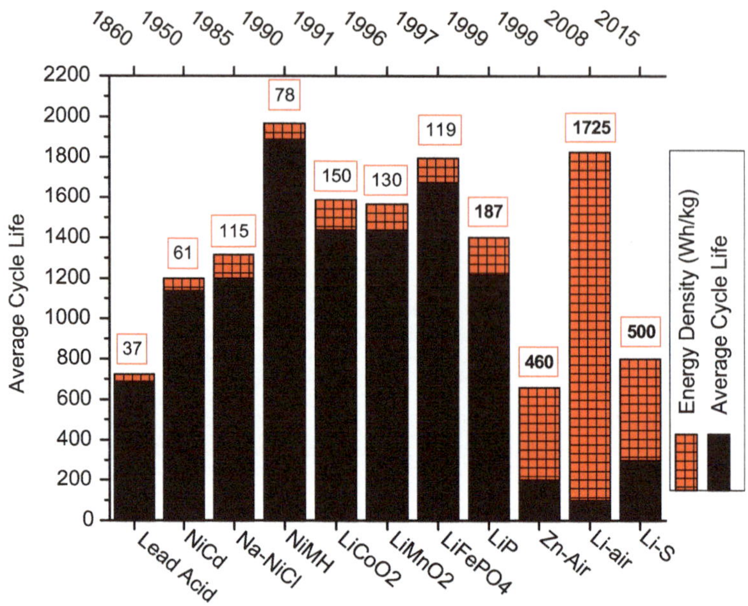

FIGURE 3.23.3.6 Average number of rechargeable cycles (cycle life) and average energy density (Wh/kg) of commercially utilized batteries since 1860. Numerical values of average energy density are displayed above each battery type column. Dates of initial commercial availability are displayed for each battery type. Average values calculated from Diouf and Pode,[36] Yong et al.,[13] Manzetti and Mariasiu,[24] Yekini Seberu et al.,[31] Tie and Tan.[22]

number of rechargeable cycles within a battery). Fig. 3.23.3.6 displays the timeline for every EV battery's initial year of commercial development, along with the energy density and cycle life, as averaged from several sources.[13,22,24,31,36] The average life cycle is displayed as the black column in Fig. 3.23.3.6 and is quite high for all mature battery technologies prior to 1999. After 1999, new-generation batteries: Zn-air, Li-air, and Li-S, all have poor life cycle limits in commercial EV applications. However, this was also the case for several of the now mature EV battery types prior to their optimization. Over time, the battery cell, electrolyte formulations, and material surface properties of electrodes developed,[37–39] allowing for significant advances in cycle life for every conventional battery type. However, with the advent of the next-generation, extremely energy-dense EV batteries, the challenge of increasing cycle life is once again the primary hurdle to commercial development. This is apparent with the stark color contrast displayed in Fig. 3.23.3.6 between mature and new-generation EV batteries. The future advancement of widespread EV use is once again a matter of battery cycle life research and development. However, even if the cycle life for new-generation Li-based batteries is significantly increased, the matter of cost and earth abundance remain limiting factors—harboring the increased development of non-rare earth metals (e.g., Zn, Na) in metal-air and alternative electrode composition (e.g., S, Si, O_2) well into the future of commercially available batteries.

3.23.3.3 SMART GRIDS

Traditional power grids operate in a very outdated and risky manner. Fox-Penner likens the traditional power grid to a network of large ponds, connected via canals, across a vast, flat landscape as depicted in Fig. 3.23.3.7.[40] Each pond's level is lowered by houses surrounding the pond, which pump water out, and raised by water towers that add water to individual ponds. In this manner, every pond maintains the same water level. Maintaining the same, balanced water level between the grid of networked ponds is "the single most important aspect of power systems' architecture."[40] In terms of real power grid systems, as represented by the simple pond system (where the ponds represent power generators and the channels are the transmission lines), if a single pond or channel were to overflow, the system imbalance would trip the power grid and lead to a blackout. Modern power systems in the United States and much of the world are still maintained by individuals who sit in control centers

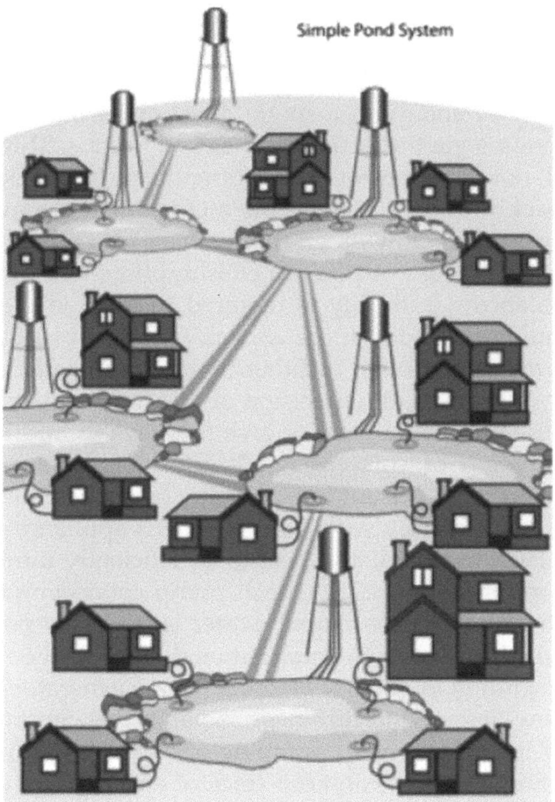

FIGURE 3.23.3.7 The pond system analog to the power grid. *Ponds* represent power generators and the *channels* are the transmission lines. Each residence pumps water in a unidirectional or bidirectional manner. The ease by which multiple, connected pond systems can maintain the same water level is analogous to the evolution of power grid efficiency. *Reproduced from Fox-Penner PS.* Smart power: climate change, the smart grid, and the future of electric utilities. *Washington: Island Press; 2010.*

throughout the country and monitor the power consumption of the entire nation. The name given for the system operators and the control centers is the balancing authority. The balancing authority turns off and on the power supply as needed, via generators (mostly coal, oil, gas, and nuclear power plants) all over the country. The power generators throughout the United States are both old and new, and therefore vary in efficiency, GHG production, and operation costs. Individuals must monitor the balancing authority control centers 24-h a day and make changes instantly, in real time, to avoid system blackouts. Even with the help of advanced computers, the fate of the US national power grid still comes down to people being present to program and push buttons every minute of the day.

Since the early days of the electrical grid system, technology has advanced to connect cities and states on the national level. The first days of the electrical grid system were made up of individual, isolated power generators for single towns or cities, called microgrids. These were largely unreliable since the imbalance of a single, isolated generator (or pond system) could not be aided by other surrounding generators (or ponds). Microgrids were also expensive to maintain without the aid of a network. With time, the power grid was developed through advancements in energy transmission, which revolutionized the advent of "endless and cheap energy" for everyone, yet without high regard for the production of GHGs.

What may perhaps be the largest inefficiency of the conventional power grid system is that it only allows for the flow of energy in a single direction—from the "water tower," into a "pond," and then into houses via transmission lines and pumps. This means that the energy system is unidirectional, not allowing for any informational feedback or power contribution from the energy user back into the "ponds." As a consequence, power companies did not know which consumer was using how much energy in a given power grid sector until power meters were installed at each house or place of consumption. During peak power consumption times, when the balancing authority is required to turn on more or all of the power generating plants, including the older, more expensive-to-run plants, the power companies could now know who was using power at either night or day and therefore charge the consumer more accordingly. However, the precision of "night" or "daytime" consumption was still not enough to bill the individual user, or give feedback to inform that user.

Today, the implementation of smart meters[41] allows for power companies to record customer power use in a far more precise way: hour by hour or minute by minute. Smart meters can now communicate wirelessly with the user, smart appliances, and the power grid—allowing energy use, GHG emissions, and cost to be efficiently minimized in real time. To return to our simple pond system analogy, each pump connecting every house and pond is now monitored, with respect to the overall water level of the pond network as well as the preference, real-time adjustments, or informational input of the user. In addition, smart meters allow for the integration of small-scale local power generators, such as personal consumer, public, or company-based wind turbines or solar panels. Smart grids and smart microgrids additionally allow for the two-way pumping of water in/out of a residence, pond, or nearby pond via canal. Smart meters can keep track of energy use from the utility company vs. energy use from the on-site generator and maintain the efficient use and storage of energy based on the time of day, the rate of solar or wind power generation, and the cost of energy from the utility grid based on demand, availability, etc. The ability for the smart meter to turn off an externally provided energy source and turn on an on-site energy source allows for both on-site and off-site energy usage optimizations.

Fig. 3.23.3.8A illustrates the differences in power and information flow between traditional power grids and new-generation smart grids.[42] Unidirectional power/information flow is characteristic of traditional grids where the power generation sources are largely made up of coal, nuclear, and hydroelectric energy generator sources. Bidirectional power/information flow is a key component of smart grids where modern microgrids at consumers' residences and places of business can not only communicate to the main power grid but also contribute power generation to the entire network. Smart grids are essential to the

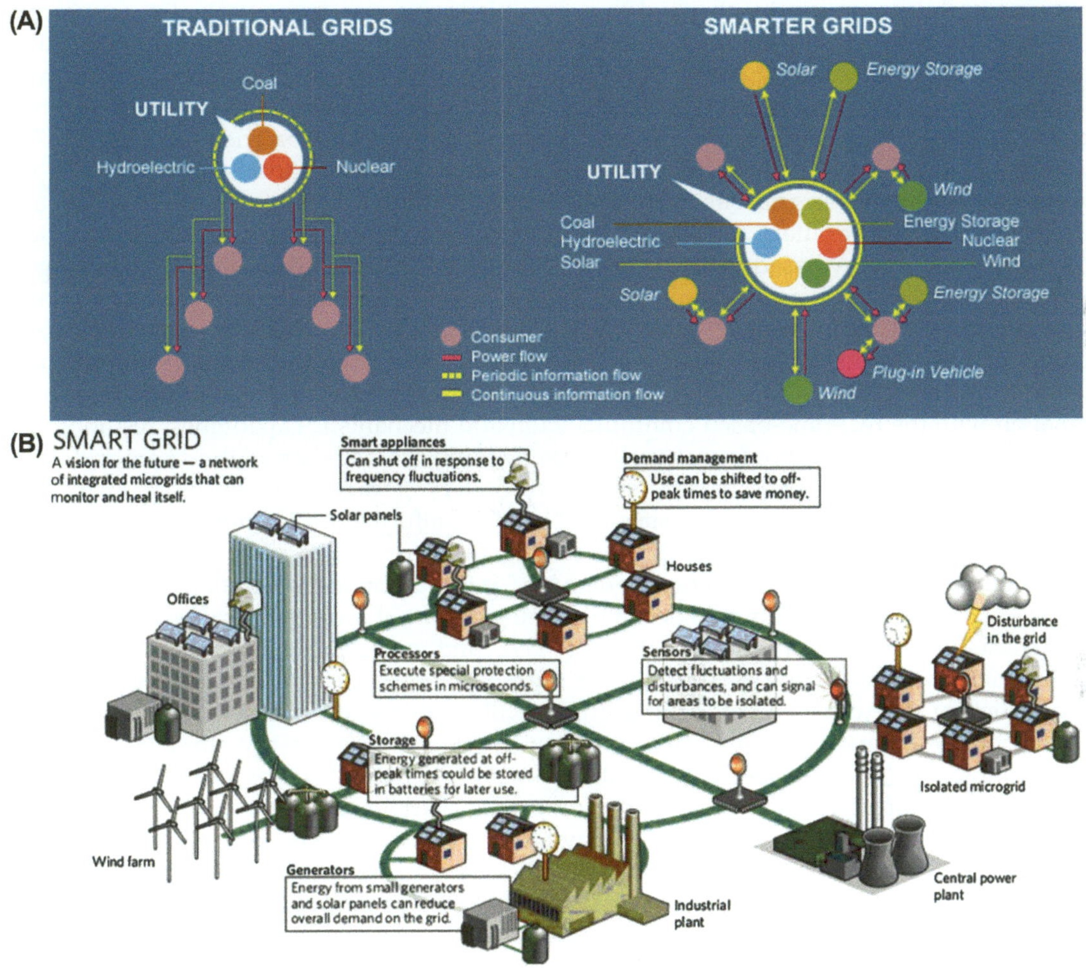

FIGURE 3.23.3.8 (A) Power flow and information flow between power generation sources and consumers for both traditional power grids and smarter power grids. (B) Smart grid: a "network of integrated microgrids that can monitor and heal itself". *(A) Courtesy of International Business Machines Corporation, © International Business Machines Corporation, IBM – Electric cars – United States 2010. http://www.ibm.com/smarterplanet/us/en/smart_grid/article/electric_cars.html. (B) Reproduced from Wolsink M. The research agenda on social acceptance of distributed generation in smart grids: renewable as common pool resources.* Renew Sustain Energy Rev *2012;**16**:822–35. http://dx.doi.org/10.1016/j.rser.2011.09.006.*

establishment of a network of consumer-based microgrids as they can allow integration of renewable sources of energy such as solar and wind and for the possibility of a more efficient expansion of the balancing authority. Customers will have far more control over their power use, generation, and costs. Advancing information technology will allow "control stations" to take the form of cell phone apps. Individuals will be able to turn off power when it is at its highest price/highest demand and turn it on when the cost drops. Via the touch of a button, consumers will be able to program the energy use for their residence, place of business, plug-in EV, appliances, energy storage, and contribution of energy to the grid network.

The illustration in Fig. 3.23.3.8B shows how the infrastructure of the current centralized power supply system can be changed to a possible smart grid framework. The framework consists of a network of integrated microgrids that are regulated internally, automatically, and manually from different sources and key nodes—such as individual users, smart appliances, smartphones, sensors, processors, etc.—thereby greatly lowering the risk of power outages and the demanding control of the traditional power grid balancing authority. The advent of such a two-way feedback loop allows individual users to also contribute to the regulation of the power grid balance (pond-levels), thereby increasing the effectiveness and decreasing the power-outage risks of the power grid balance authority as a whole. Additionally, renewable energy based microgrids allow for individual energy users to become largely independent of the utility grid, while allowing for "extra" energy to be sold back to the main power grid if it is not immediately needed. The other primary drawback of the unidirectional power grid is that power consumers cannot contribute to the total power generation (water level of the pond). The ability of bidirectional power flow (consumer ↔ power grid), in combination with the real-time-based communication and mechanistic capabilities, resulting in a more efficient, integrated power network, is the defining characteristic of the smart grid. With the implementation of smart meters, stationary battery storage, the significant increase of local, renewable power generation sites, and modern information/communication technology, designs based on microgrids are beginning to be reimplemented. Such advances in renewable energy harvesting, energy storage, and information technology are quickly allowing every energy grid consumer to become an integral part of the energy balancing authority.

Vehicle-to-grid (V2G), vehicle-to-home (V2H), and vehicle-to-vehicle (V2V) advanced technologies will also have a large impact on the potential to distribute and increase the production of renewable energy.[43] Plug-in EVs will be able to contribute energy to the local energy micro- or macrogrid, greatly impacting the sale of EVs, energy storage capacity of households, and grid storage of renewable energy. Fig. 3.23.3.9 maps the energy sources, carriers, and corresponding vehicles that utilize the various energy sources. Electricity is the key for all sources of renewable energy-based vehicle propulsion. In addition to this, electricity is also the primary energy carrier for V2G as well as renewable energy grid storage from both micro- and macrogrids. The actual storage of the renewable energy, as depicted in Fig. 3.23.3.8, relies on stationary battery technology.

As discussed in the previous section, battery EVs are expanding rapidly. The implementation of smart grids is contributing to the capability of widespread EV production and charging-station-enabled drivability throughout the globe. Related to the recent technology boom of EV batteries, next-generation stationary batteries are also currently under development. Such batteries would allow for potentially enormous energy storage capacity for next-generation smart grids and alternative energy harvesting. Combined solar and wind energy

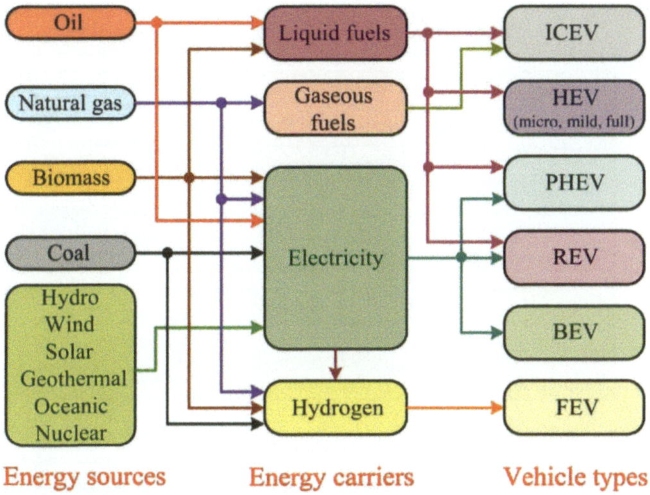

FIGURE 3.23.3.9 Energy diversification of electric vehicles (EVs). Internal combustion engine vehicle (ICEV), hybrid electric vehicle (HEV), plug-in hybrid electric vehicle (HEV battery electric vehicle), range-extended electric vehicle (REV), battery electric vehicle (BEV), and fuel cell electric vehicle (FEV). *Reproduced from Liu C, Chau KT, Wu D, Gao S. Opportunities and challenges of vehicle-to-home, vehicle-to-vehicle, and vehicleto-grid technologies. Proc IEEE 2013;101:2409e27. http://dx.doi.org/10.1109/JPROC.2013.2271951.*

harvested throughout the daytime could be stored and then utilized at night for entire cities. Vehicles charged from renewable sources could deliver energy back to older, nearby nonrenewable energy-sourced sectors via bidirectional chargers. To make such technological feats a reality, electrical energy storage (EES) will be required on a vast scale. The same smart grid technologies that enable micro-/macrogrid generation and energy transfer will accommodate EES units. If such energy storage becomes a low-cost reality, the entire conventional balancing authority will be fundamentally altered for the better. The balancing authority will now become smart grid enabled through the use of stationary battery/grid-integrated technology. Instead of a human control center or individual authorities balancing the power grid, the batteries and smart grid technology, combined with individual energy user real-time input, will become the balancing authority. There will cease to be critical need to turn power plants on or off to control energy flow and prevent blackouts. Instead, battery stations will progressively become efficient at discharging when more energy is demanded and renewable energy generation is low and at charging when less energy is needed and renewable energy generation is high. Everyone on the grid will be able to create a reserve of stored energy, primarily sourced from renewables. Individual users can become the primary authorities in determining how, when, and where energy is delivered.

EES technologies can be classified into six categories based on what form the energy is stored in (Fig. 3.23.3.10A). Since the implementation of grid storage is to improve grid reliability and inefficiency, EES technology must allow for a very high rate of delivery of electrical energy as soon as it is required to prevent a blackout. Hence, EES materials are often rated based on their power (W/s). The higher the power density, the more energy can be delivered when it is immediately needed. Fig. 3.23.3.10B illustrates the majority of the grid storage

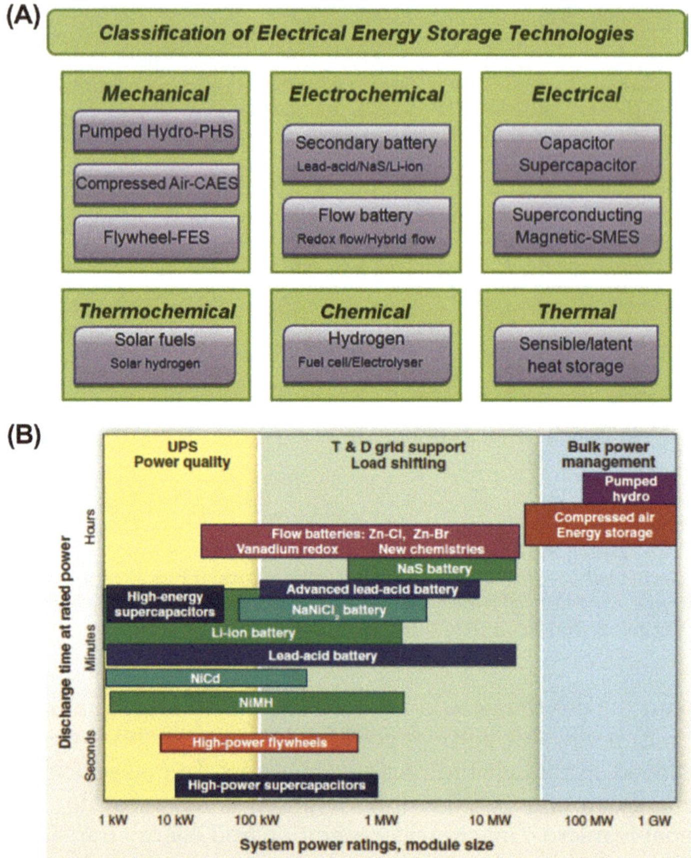

FIGURE 3.23.3.10 (A) Classification of electrical energy storage (EES) technologies based on form of stored energy. (B) Comparison of discharge time and power rating for various EES technologies. The comparisons are of a general nature because several of the technologies have broader power ratings and longer discharge times than illustrated. *(A) Adapted from Luo X, Wang J, Dooner M, Clarke J. Overview of current development in electrical energy storage technologies and the application potential in power system operation.* Appl Energy *2015;***137***:511–36. http://dx.doi.org/10.1016/j.apenergy.2014.09.081. (B) Adapted from Dunn B, Kamath H, Tarascon J-M. Electrical energy storage for the grid: a battery of choices.* Science *2011;***334***:928–35. http://dx.doi.org/10.1126/science.1212741.*

devices currently available or under research and development with each storage device's relative size and discharge duration at a given power output. An uninterruptible power supply is a device that can switch on automatically when there is a power outage, and it is typically used for low-load-drawing uses (e.g., computers servers, backup emergency lighting). Transmission and distribution grid support energy storage would supplement part of the grid energy being consumed in real time, thereby taking some of the energy demand off of traditional power generators. Bulk power management devices, with very high system power ratings, are currently the only EES devices that have the potential to completely, temporarily replace constant sources of power generation. Pumped hydro, for example, currently stores over 99% (127,000 MW) of the total worldwide storage capacity for electrical energy.[44]

3.23.3.4 SMART ROADWAYS

Intelligent transportation systems, or smart roads, are just beginning to come onto market. Both solar-to-electricity and solar-to-thermal energy conversion concepts and designs are currently being implemented in both experimental and initial commercial capacities for the generation of renewable energy systems integrated with transportation roadways. An example of this type of technology is the solar-powered hexagonal road panels developed by the company Solar Roadways.[46] These panels are produced from recycled waste materials with transparent driving surfaces, underlying photovoltaic (PV) cells, and sensors. Fig. 3.23.3.11A shows three of the 4 ft^2 hexagonal PV-capable road panels with three different colored embedded light-emitting diodes (LEDs). The various LED lights can be utilized not only to illuminate the road for drivers and pedestrians, but also to integrate with sidewalk sensors to warn drivers and pedestrians of impending accidents by displaying messages (Fig. 3.23.3.11B). Components include the base conduit and cable corridor, which also can filter and retain storm water for irrigation, gray water, landscaping, etc. (Fig. 3.23.3.11C). The Solar Roadways founders, Scott and Julie Brusaw, are pictured with their prototype parking lot in Fig. 3.23.3.11D. Initial trials and research have yielded an estimate that the SR2 panels shown here can generate ~302 MWh per year, per lane-mile of solar roadway panels. That is

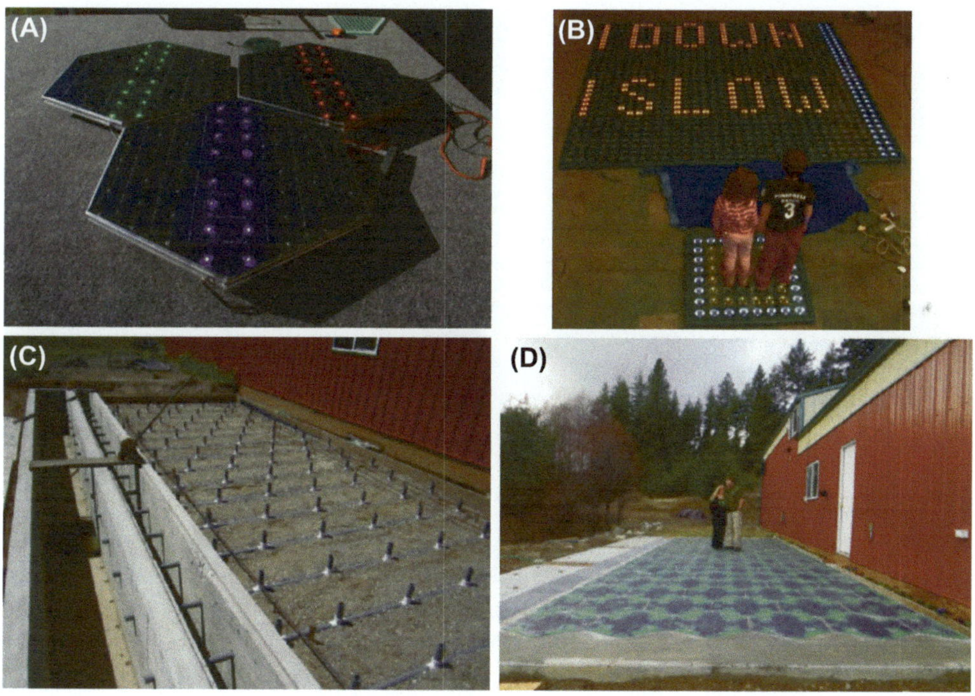

FIGURE 3.23.3.11 (A) Actual solar roadways, third-generation prototype; (B) solar roadways first-generation panel illustrating the pressure-sensitive response panels that illuminate light-emitting diode warning signals embedded in the road; (C) solar roadway conduit, cable corridor and drainage prototype; (D) completed prototype parking lot. *Adapted from Solar Roadways; n.d. http://www.solarroadways.com/.*

enough energy to power 86 homes from the power grid on a two-lane solar roadway.[47] In 2015, Solar Roadways was granted Phase II USDOT funding (2-year; $750,000) to further develop their research into commercialization by advancing properties such as freeze/thaw cycling, moisture conditioning, shear testing, and higher weight loads. Solar Roadways is planning to install solar road panels on the famous Route 66 near Conway, Missouri, in the near future.[48]

France announced in 2016 that it would be building over 600 miles of solar-paneled roads over the next 5 years, where the renewable energy yielded will serve 5 million people. Deemed the "Wattway," the roads are being constructed by the National Institute of Solar Energy and the French road construction company Colas. These specific road panels are much thinner than Solar Roadway's—only a quarter of an inch thick (Fig. 3.23.3.12A). They also have the added benefit of being installed directly on top of existing roadways—meaning that these systems can be integrated into the existing infrastructure without having to tear out old roadways. Wattway panels are textured similarly to typical pavement to provide traction. The manufacturer claims that 1000 m of Wattway panels can generate renewable energy for lighting 5000 residences, with 20 m^2 providing enough electricity for an average household.[49] Fig. 3.23.3.12B shows part of a test solar roadway developed by the Colas company and an EV charging from the energy harvested via the solar roadway. In

FIGURE 3.23.3.12 (A) The height and asphaltlike texture of "Wattway" solar panels installed at the top of existing road pavement, next to a coin for scale. (B) Test site of textured PV road surface and charging station at Colas site. (C) The first kilometer of Wattway panels during its inauguration in Tourouvre, France. (D) Glowing roadway lines charged by solar roadways in the Netherlands. *(B) Adapted from Wattway by Colas | Paving the way to tomorrow's energy; n.d. http://www.wattwaybycolas.com/en/. (C) Material is copyrighted by Ars Technica, Anthony S. World's first solar road opens in France: it's ridiculously expensive.* Ars Tech *2016. http://arstechnica.com/cars/2016/12/worlds-first-solar-road-opens-in-france/. (D) Reproduced with permission from Studio Roosegaarde – Smart highway info; n.d. https://www.studioroosegaarde.net/project/smart-highway/info/.*

December 2016, France inaugurated its first continuous kilometer of the Wattway, as shown in Fig. 3.23.3.12C. Colas claims that theoretically, France could become energy independent by utilizing Wattway panels on roughly a quarter of the country's roads.[50] Additional applications of energy harvested from solar roadways include providing illumination or directional signage (Fig. 3.23.3.12D). Here the road is illuminated through glowing LEDs powered by solar roadways—a function currently being utilized in the Netherlands. The glowing lines are powered by energy stored during the daytime and glow at night for up to 10 h—creating day-to-day energy-neutral roadway illumination.

Electricity generation is not the only method of energy utilization from solar energy capture. Solar thermal energy is also a significant source of energy produced by the bombardment of solar photons on roads. When exposed to solar radiation for extended periods, asphalt dissipates the sun-generated thermal energy as heat throughout its bulk structure. Asphalt can reach temperatures higher than 70 °C during a hot summer day due to its high absorption coefficient. Generally, hot days in the summer are viewed as a negative impact on the asphalt—degrading the structural integrity and durability of the pavement by creating deformations (cracks, thermal oxidation degradation, etc.) with the added negative force effects of traffic. Initial research on how to "extract" the thermal energy was done simply to improve asphalt lifetime and performance. However, like much of scientific discovery, separate applications and opportunities spawned out of the initial aims of scientific interest. The harnessing of hot asphalt pavement into an energy source through various means has become increasingly popular.

Asphalt solar collector systems are the most modern popular forms of thermal energy solar harvesting. They come in two forms: porous asphalt layers between two impermeable layers and asphalt concrete layers with a network of piping through them, referred to as hydronic asphalt pavement (HAP). HAP is a developing renewable energy application that yields a unique method of solar energy utilization via asphalt thermal radiation.[53] HAP systems have networks of piping where high thermal energy-retaining fluid is circulated below the surface of the pavement. Energy from high surface temperatures from asphalt can be transferred to a circulating fluid in the pipes beneath the asphalt. This benefits the structural integrity of the road in the summer time by cooling the pavement. The now high-thermal-energy fluid can be stored in insulated, underground tanks. In the summer time, this process protects the structural integrity and reduces the degradation of roadways. In the wintertime, the thermal energy from sunlight during the day can be stored in underground tanks, allowing for the heating of the roads at night and/or when snow and ice begins to form on the roadway. Fig. 3.23.3.13 illustrates the principles of both work functions of the HAP system. Cold fluid is pumped up from an insulated underground storage medium and cools the pavement as the fluid flows at a slow rate parallel to the surface top layer (Fig. 3.23.3.13A). The warm fluid is then circulated back to the storage medium where it can be stored for later use. During the deicing phase of HAP systems, warm fluid from the insulated storage medium is laterally pumped underneath the frozen pavement, allowing for heated pavement and snow/ice melt (Fig. 3.23.3.13B). Aside from these applications, HAP systems are also utilized for localized heating and recharging of battery systems.

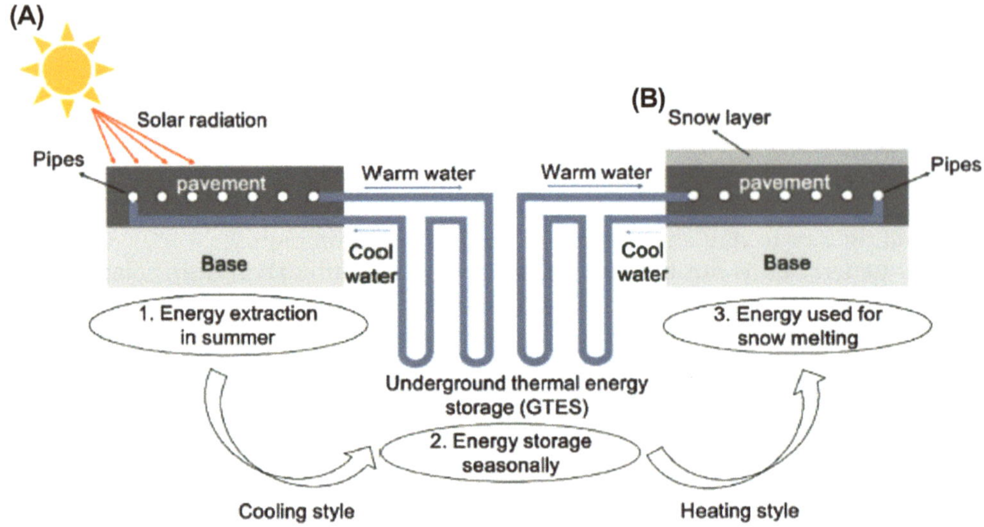

FIGURE 3.23.3.13 Working principle of hydronic asphalt pavement (HAP) system. The heating mechanisms in HAPs are based on a closed loop system—typically made up of flexible polymer tubing, water, and propylene glycol. Fluid is warmed and thermally stored underground for pavement heating at the user's discretion. *Reproduced from Pan P, Wu S, Xiao Y, Liu G. A review on hydronic asphalt pavement for energy harvesting and snow melting.* Renew Sustain Energy Rev *2015;***48***:624–34. http://dx.doi.org/10.1016/j.rser.2015.04.029.*

3.23.3.5 SMART BUILDINGS (GREEN BUILDINGS AND HOMES)

Recent times have seen the establishment of a multitude of organizations and assessment methods to assist and rate the construction and energy use of buildings of all types. Among the leading organizations are Leadership in Energy and Environmental Design (LEED) in the United States, BRE Environmental Assessment Method (BREEAM) in the United Kingdom, Green Building Council of Australia Green Star (GBCA) in Australia, Green Mark Scheme in Singapore, Pearl Rating System for Estidama in Abu Dhabi, Green Building Index in Malaysia, Hong Kong Building Environmental Assessment Method (HK BEAM), Germany Sustainable Building Council (DGNB) in Germany, and Comprehensive Assessment System for Built Environmental Efficiency (CASBEE) in Japan. Every organization has similar assessment tools and categories,[54] such as energy, transport, water, material, land use, ecology, and emissions. It is important to note that all these green building assessment organizations are voluntary organizations rather than mandates or government-run. However, the World Green Building Council (another voluntary organization) monitors and coordinates the efforts of the green building organizations of the world. Even with all of these various organizations that assess the efficiencies of development and energy use, there is still no clear definition or common validation of a "green" building or residence. One paramount consideration is that these organizations, such as LEED, often do not include source energy in their life cycle assessments (LCAs) and instead only focus on energy consumed at the building site.[55] However, any comprehensive and accurate assessment of a building's energy use and emissions should include the assessment of all energy components at every point in the full lifetime of the building's components. Therefore, any designation of "Green," "Gold," or "Silver" rankings by these organizations should be closely examined to validate the assessment findings and honorary labels from *private* assessment entities.

The total life cycle of a building includes both embodied energy (EE) and operating energy (OE), where EE is defined as the amount of energy that goes into the building materials, and during all phases of construction and demolition, whereas OE is the amount of energy consumed during operation and inhabitation of the building. Only relatively recently have LCAs required both, as it was long thought that the vast majority of energy consumption and GHG emissions were carried out during the building's operating stage.[56] Energy-efficient materials, insulations, advanced appliances and equipment, however, have been thought to tip the scales both ways for the total EE in a building's LCA.[57] For example, low-energy buildings (building designs that demand less OE and total life cycle energy compared with conventional building materials) are made of reduced-OE materials via both active and passive technologies. However, as Fig. 3.23.3.14 illustrates with the OE and EE of four case studies,[56] a decrease in OE due to advanced building materials is often accompanied by a semiproportional increase in EE. Hence, the cumulative OE and EE of every green building must be analyzed in great detail to accurately assess the relative LCA of a building. The following case study illustrates the fallibility in LEED-rated buildings, via a thorough examination of only EE, or "source energy," when not enough analysis is included in rating schemes.

Fig. 3.23.3.15 displays an objective assessment of almost 975 office buildings in New York City (NYC)[55] that includes 21 LEED-rated office buildings with respect to their energy

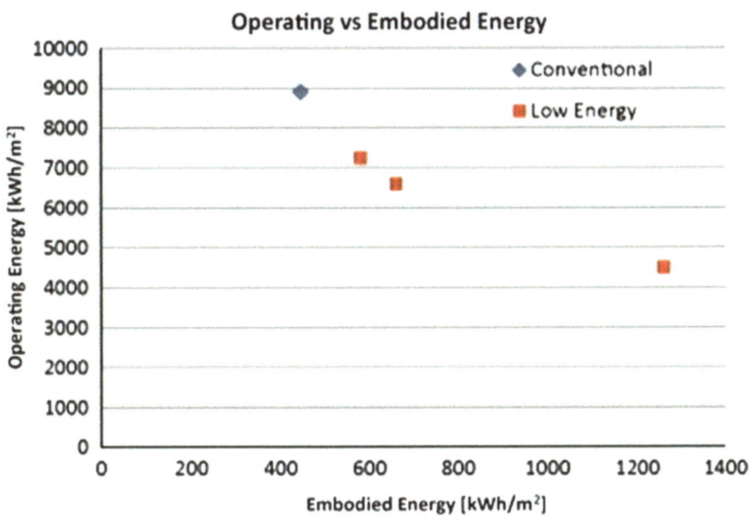

FIGURE 3.23.3.14 Interplay between operating and embodied energy for case studies. Newer, more advanced building materials are often lower in operating energy yet require larger magnitudes of embodied energy to produce. *Reproduced from Cabeza LF, Rincón L, Vilariño V, Pérez G, Castell A. Life cycle assessment (LCA) and life cycle energy analysis (LCEA) of buildings and the building sector: a review.* Renew Sustain Energy Rev *2014;***29***:394–416. http://dx.doi.org/10.1016/j.rser.2013.08.037.*

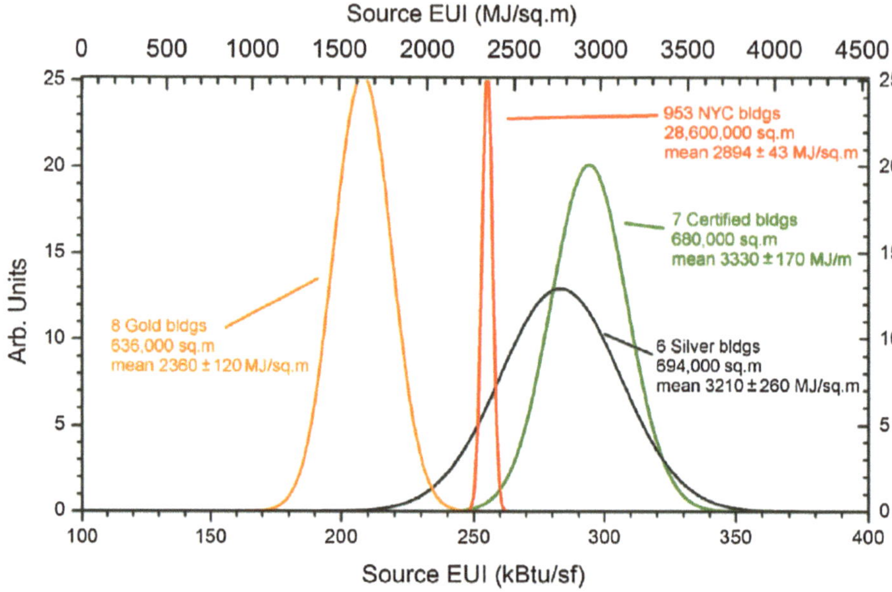

FIGURE 3.23.3.15 Normal distributions with mean and standard deviations for Certified (*green*), Silver (*black*), and Gold (*orange*) LEED office buildings and for 953 conventional, non-LEED-rated, office buildings (*red*). Source energy utilization intensities are presented in two metrics (top axis: MJ/m^2 and bottom axis: kBtu/sf). *Adapted from Scofield JH. Efficacy of LEED-certification in reducing energy consumption and greenhouse gas emission for large New York City office buildings.* Energy Build *2013;***67***:517–24. http://dx.doi.org/10.1016/j.enbuild.2013.08.032.*

utilization intensities (EUIs). EUIs are categorized by energy utilization, which includes both primary and source contributions, very similar to the OE and EE classifications discussed previously. The comparison presented in Fig. 3.23.3.15 also appropriately accounts for the weight and size of buildings in calculating accurate source, or embodied, EUIs. When the full accountability of source EUI metrics are taken into account, only the LEED Gold-rated buildings use less energy (18%) than conventional office buildings—as represented by the very precise distribution of averaged data assessed from 953 NYC office buildings. Such tight distributions are often consistently observed when a large number of buildings are surveyed—representing a strong conventional building assessment for a given city. In fact, LEED Silver and LEED Certified buildings contain, respectively, 15% and 11% more source energy than conventional buildings. When compared with conventional office buildings (93 kg/m^2), less GHGs were emitted from LEED Gold buildings (76 kg/m^2), whereas more GHGs were emitted by both LEED Certified (116 kg/m^2) and Silver (100 kg/m^2) buildings. Obviously, this case study illustrates that the apparent "green" certifications of buildings do not necessarily reflect their rating labels—an observation that has been used for scrutiny of "green" certification-based entities for quite a while. Hence, any LCA assessment should be conducted in an extremely thorough manner to accurately assess how much energy is being utilized/saved and the resulting GHGs suppressed throughout the building's manufacture and operation processes.

3.23.3.6 HOME AND COMMUNITY SOLAR HARVESTING AND ENERGY STORAGE SYSTEMS

Even though the aforementioned assessment of "green" energy-efficient buildings does not paint an authentic picture for the energy/environmental "certification" of the building economic sector, there are certainly many who take a far more effective approach working and integrating their efforts at the local level. Individual homes and communities have some of the biggest opportunities to replace fossil fuel energy with renewables. The implementation of a locally integrated smart grid, powered by home-based renewables and transferred with V2G and grid-to-vehicle (G2V) technology, is one very feasible option currently available to energy consumers.

The renewable energy source that has the potential to make the most significant difference for fossil fuel replacement at the local level is the utilization of home-based solar energy harvesting. The use of solar PV energy was the fastest growing energy technology in the world between 2000 and 2010, increasing from 1.5 to 40 GW of cumulative power capacity.[58] Germany is the single most solar PV-progressing country—contributing 7 GW of the 17 GW added globally in 2010.[59] As discussed in Section 3.23.3.3, electricity from solar PV local sources is very close to being cost competitive with retail prices of traditional grid electricity, especially during high peak power demand and their associated high prices, but in many markets, a considerable price difference still exists, making widespread use limited. However, the base cost of solar PV cells and systems directly determines the potential for widespread consumer use. Fortunately, the cost of PV systems has been consistently dropping for over three decades (Fig. 3.23.3.16), exhibiting a reduction in cost by ~20% (learning rate = 19.3%) for every doubling in PV cell capacity. Such a continuing, advancing trend has shown significant changes at national levels, year after year. China, for example, is now the world's largest manufacturer of solar modules, with its global production share increasing from 39% in 2009 to 55% in 2010. In that same year, PV module production more than doubled from 7.5 to 18.1 GW.[59] However, such a significantly increased share of the global market did not lead PV manufacturers in other countries to keep pace with global production. Due to a near-stagnant level of production during the same 2009 to 2010 transition (~1.7 vs. ~2.4 GW), the US share of global manufacturing decreased from 23% to 13% in response to the shift in the global production market. Additionally, massive and rapid growth of the PV market has been having profound effects on beneficial subsidies and significant policy changes, such as the accelerated growth spurts of "PV bubbles"[59] that appeared in Spain in 2008 and the Czech Republic in 2010. Similar trends in higher production and lower cost are expected to continue, along with increases in capacity efficiency and production scale-up processes. Hence it is only going to become easier and more cost effective with the passing of time to replace fossil fuel use for individual consumers at home or in the office through PV integration.

At this point in global manufacture and development, solar PV systems are extremely modular, allowing for fast and easy installation by the general consumer public. Such ease of access will only support the steadily growing trend of consumer solar energy harvesting and grid storage contribution. The market availability is rapidly expanding throughout the world as the cost of solar PV systems is becoming progressively affordable to all consumer

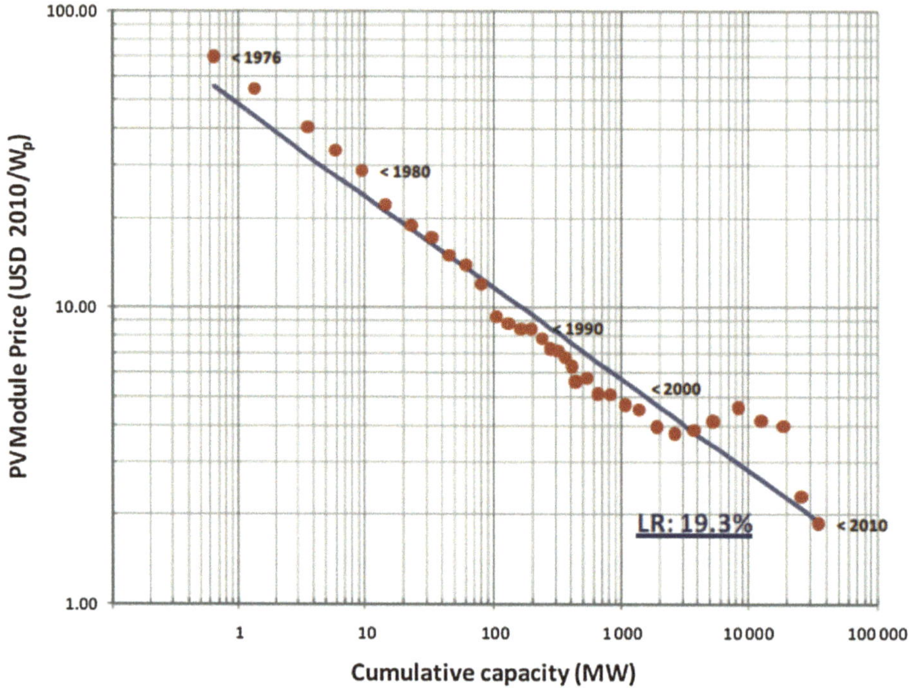

FIGURE 3.23.3.16 Cost degression of solar photovoltaic modules, 1976–2010. *Adapted from Brown A, Muller SS, Dobrotkova Z. Renewable energy: markets and prospects by technology; n.d; © OECD/IEA 2011 Renewable Energy: Markets and Prospects by Technology, IEA Publishing. License: www.iea.org/t&c with permission from Breyer and Gerlach.*

groups. Perspectives on the feasibility of worldwide solar PV integration are widespread.[60–64] Many current studies promote and praise the rapidly growing trends of the affordable PV market, claiming its growth has already outpaced traditional fossil fuels or will soon become more affordable than nonrenewable sources of electricity. Other studies claim that the solar PV and renewable energy market is still dramatically far from global replacement of traditional fossil fuels for sources of electricity. A notable reason for the disparity of solar PV integration is the price of building and utilizing such systems in different parts of the world—with geographic, climate, and local production factors weighing in heavily. Levelized cost of electricity (LCOE) is one method to compare sources of electricity through the use of consistent variables. The LCOE assesses the average total cost to build and operate a new electricity power plant over its lifetime. The method reflects the expected annual electricity generation while also considering all the costs including initial capital, investment return, maintenance, and expected lifetime. The LCOE also represents the cost at which the resulting electricity must be sold in order for the electricity plant to break even over its lifetime. Hence, for all renewable and nonrenewable sources, LCOE is one metric of valid cost comparison.

Fig. 3.23.3.17 displays LCOEs for both nonrenewable (coal and natural gas) and renewable (on-shore wind and solar PV) at the global scale. LCOEs were averaged from Europe, China, and the United States to generate current and predicted operating costs (US dollars per MWh)

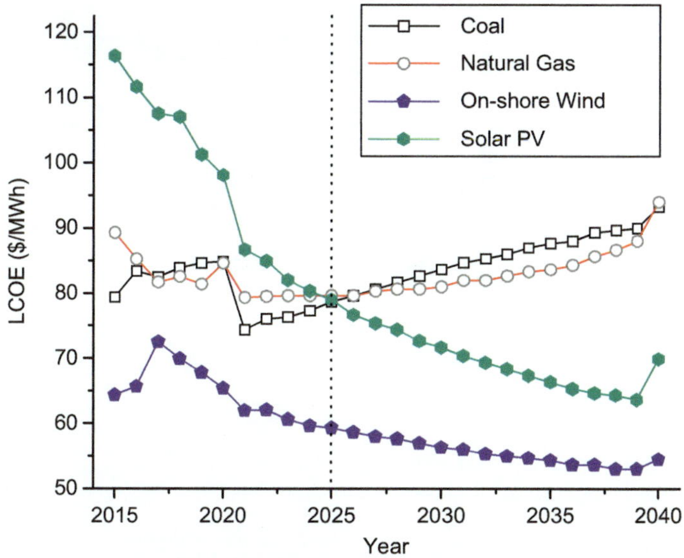

FIGURE 3.23.3.17 Global average and predicted averaged levelized cost of electricity (LCOE) of major nonrenewable (coal and natural gas) and renewable [onshore wind and solar photovoltaic (PV)] sources of electricity. Major renewable sources are predicted to outcompete nonrenewable energy options when the solar PV becomes less expensive than major competitors at the worldwide level—as predicted here with the dotted line shown at year 2025. Values were averaged from: Bloomberg New Energy Outlook 2015 (data used for 2015–40),[60] US Energy Information Agency Annual Energy Outlooks from years 2012, 2013, 2014 and 2015 using averages from: (1) conventional coal, advanced coal, and carbon capture processed coal; (2) conventional combined cycle and conventional combustion natural gas; (3) on-shore wind; and (4) solar PV (data used for 2017–20, 2022 and 2040),[64] Lazard levelized cost of energy analysis 2014 and 2015 (data used for 2015–17),[62,65–67] and US DOE reports (data used for 2016).[68]

up to the year 2040 at the global level. As displayed, onshore wind power is one source of renewable energy that already beats out coal and natural gas sources and continues to decease in its LCOE with time. However, due to the geographic and climate limitations of onshore wind power, its worldwide integration is limited. Solar PVs, with progressively decreasing production costs as shown in Fig. 3.23.3.16, are expected to outcompete coal and natural gas electricity generation sources by the year 2025, as displayed with the vertical dashed line in Fig. 3.23.3.17. Every year beyond 2025, the global cost of solar PV (and onshore wind)-based electricity generation is projected to be more affordable than coal and natural gas. Hence, even though many studies presently show some economies in a fortuitous position to outcompete nonrenewable energy sources, Fig. 3.23.3.17 illustrates that it is just a matter of time before both major renewable energy sources dominate the global electricity market.

Fig. 3.23.3.18A illustrates that the peak solar harvesting time is typically in the middle of the day, when most people are not at home, whereas there is some electricity demand in the morning hours and a large demand in the evening. Interestingly, the peak solar timing is beneficial to energy storage as many working consumers will not use energy during a time that would normally charge large-scale stationary storage devices. However, the capacity

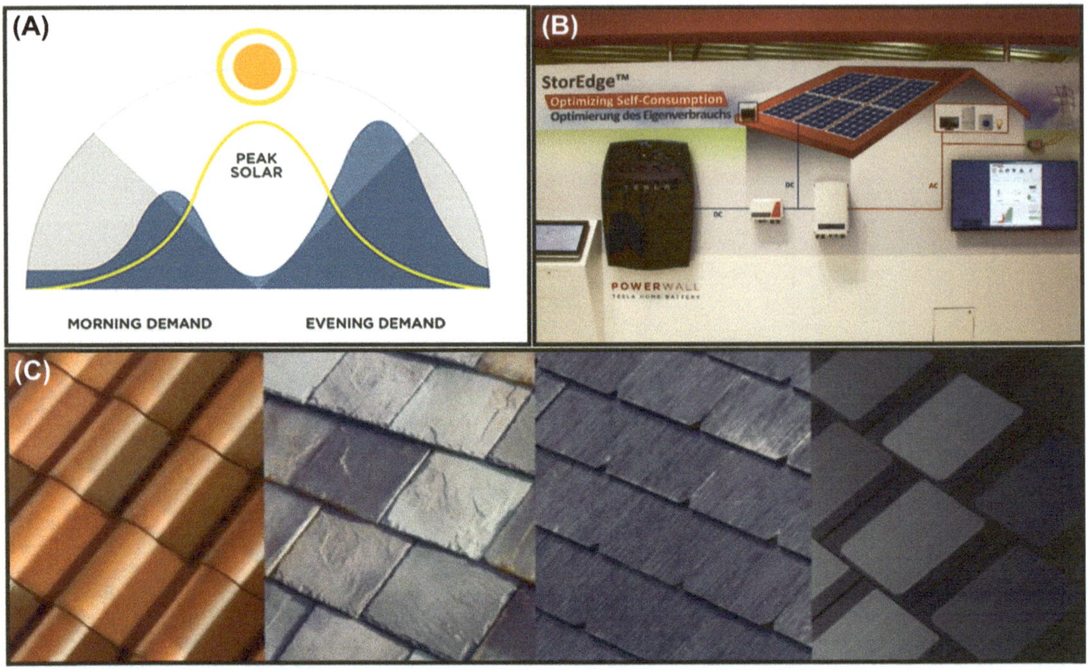

FIGURE 3.23.3.18 (A) Illustration of peak solar radiation during the daytime and the morning and evening energy demands of a typical home. (B) Tesla energy Powerwall battery system diagram illustrating solar panels, hybrid inverter (manages on-grid and off-grid energy and battery grid energy), meter, and auto transformer (regulates direct current and alternating current voltage/current). (C) The four solar roofing types unveiled in 2016 by Tesla/solar city. The solar panels are made to function and look like a fashionable roof while integrating with a smart home or smart community's energy storage smart grid. *(A) Reproduced from* Tesla is planning to release the Powerwall version 2.0 this summer. *TreeHugger; n.d. http://www.treehugger.com/clean-technology/tesla-planning-release-powerwall-version-20-summer.html. (B) Reproduced from* Tesla to launch Powerwall 2.0 home battery this summer; *n.d. http://inhabitat.com/powerwall-2-0-an-improved-tesla-battery-pack-to-be-unveiled-this-summer/. (C) Adapted from* Tesla's solar roof: solar shingles reimagined. *EnergySage; 2016. http://news.energysage.com/tesla-solar-panel-roof-the-next-solar-shingles/ and Tesla Powerwall: https://www.tesla.com/en_AU/powerwall?redirect=no.*

of single home stationary storage devices is somewhat limited, and therefore much of the peak solar energy is essentially untapped due to lack of on-site stationary storage capacity. One increasingly popular home solar PV storage system relies heavily on products such as Tesla's Powerwall battery packs (Fig. 3.23.3.18B). Such home battery packs come in a variety of sizes and capacities—allowing consumers to directly regulate their home-based fossil fuel consumption and GHG emissions. A significant impact on the residential economic sector was the recent acquisition of Solar City by Tesla. This high-quality battery and production-based home solar energy manufacturer is now one of the significant entities allowing for the streamlining of more advanced solar PV/home battery kits for grid-meter-storage integration and user-directed implementation. As Fig. 3.23.3.18B illustrates, in one such kit, solar energy is converted to unidirectional DC and transferred to an inverter. The inverter allows for the transfer of unidirectional DC to multidirectional DC to be stored in a home battery system. When stored energy is either required for the home's needs or if the home owner would like to sell or contribute his or her unneeded stored energy to the community grid, the inverter can then convert the unidirectional DC stored in the home battery system into alternating current—allowing for the home owner's preferred usage. Cloud-based monitoring platforms have also become quite popular for the control of smart home systems. The ability to monitor the real-time status of your current battery state/capacity, turn on and off in-home appliances/heat, and direct energy back to the grid from your smartphone or digital device will help in filling the gap between energy conservation and minimizing GHG emissions. Fig. 3.23.3.18C illustrates Tesla's new solar roof shingles. Made of material as durable as quartz and being esthetically attractive, these efficient solar roof shingles allow for extremely robust and easy addition to any residence seeking alternative energy home integration and renewable energy offsetting of GHGs. Similar to the same business model discussed earlier with Tesla's car sales, these high-quality solar roof shingles are expected to initially be sold at a relatively high price, where the revenue will be reinvested for research and production of more affordable models for a larger consumer base. Companies like Tesla and SolarCity continue to produce renewable energy-utilizing products that appeal to every market sector and progressively grow in sales distribution through combined integration of superior function, fashion, and affordability.

Fig. 3.23.3.19A shows a community energy storage (CES) system composed of several different solar energy harvesting dwellings utilizing solar PV/battery storage setups such as the ones described previously. The figure also displays a necessary CES inverter to properly manage the bidirectional capability of the localized grid system. In general, inverter costs are also directly proportional to solar PV capacity—another important factor in CES cost analyses. CES systems have been shown to make very significant differences based on the size of the community in the system and the PV consumer market penetration (% of home-solar energy harvesters). Many models and predictions have been assessed based on current trends, technology advancement, and cost decrease. For example, one investigation found the predicted PV penetration to be 7.6% for the year 2020.[58] Fig. 3.23.3.19B illustrates the effect of a 7.6% predicted PV penetration for various community sizes. As the figure illustrates, smaller communities are more likely to form CES systems all by themselves. As the community gets larger, more and more homes tap into the CES systems—where many of the new CES home additions do not have PV arrays because the community is larger than the PV penetration percentage. This is illustrated with the figure to show that for a 7.6% solar penetration, the PV percentage and therefore the CES system changes from 100% for a single home to 7.6% for a community of 100. Hence, over time PV penetration percentages will increase due to decreasing solar PV cost—necessitating the optimal CES size to be progressively growing as a time-dependent process.

One way to maximize the storage of solar PV-based electricity is to seek out ways to charge all solar energy-storing batteries at peak solar times. EVs have a direct hand in this possibility, as EVs would be the one battery source that is not physically located at the solar PV-equipped smart home. Since EVs are used for transport only 4% of the time, this allows for the battery to be utilized for renewable energy storage 96% of their lifetimes.[72] While the EV sits all day at work or the office, it is a solar PV energy storage site that is inefficiently being utilized. Work and office buildings that have solar powered charging stations for vehicles could also utilize the harvested electricity to power the business with non-GHG-emitting

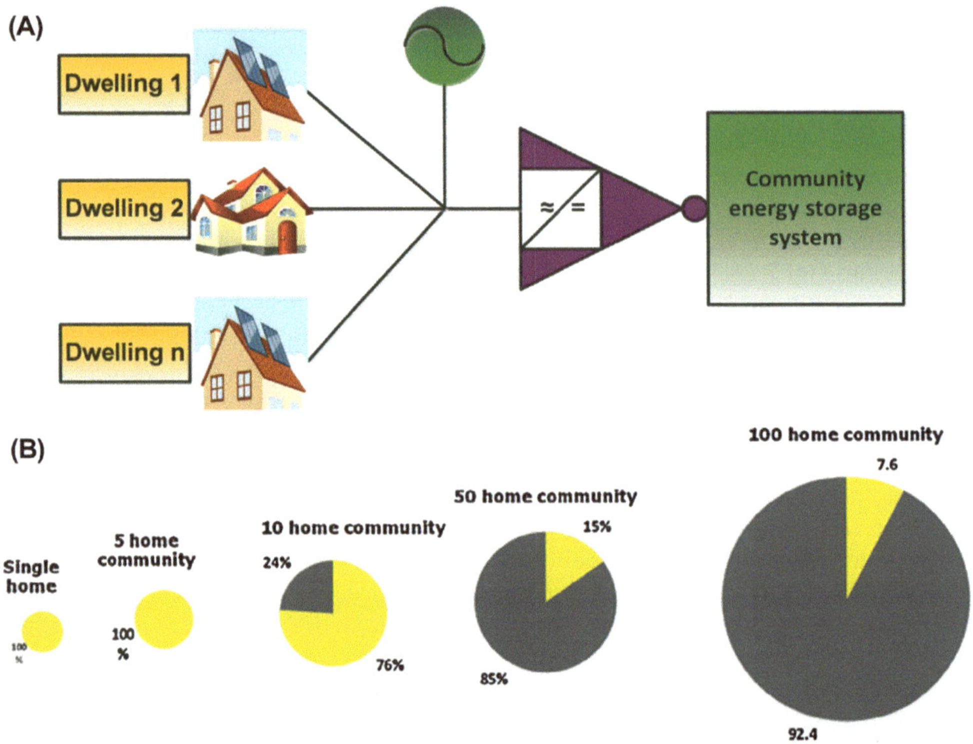

FIGURE 3.23.3.19 (A) Schematic representation of the bidirectional inverter connected to a community energy storage system. (B) Evolution of the community photovoltaic (PV) percentage (*yellow* area in the pie charts) as a function of the size of the community in 2020 when the PV penetration in the United Kingdom was assumed to be 7.6%. *Adapted from Parra D, Gillott M, Norman SA, Walker GS. Optimum community energy storage system for PV energy time-shift.* Appl Energy *2015;**137**:576–87. http://dx.doi.org/10.1016/j.apenergy.2014.08.060.*

renewable energy sources. As we discussed in the previous smart grid section (Section 3.23.3.3), V2G, V2H or vehicle-to- office, and V2V energy transfer technology make it possible to utilize the smart grid at the workplace. Such an implementation would allow for a more advanced and efficient distribution of renewable solar power from and for the energy uses of every element connected to the smart grid: EVs, homes, offices, etc. The more onsite and renewable energy-integrated EV charging stations a business, office building, or residence has, the higher the energy, emissions, and transportation rating of the building economic sector. Fig. 3.23.3.20A displays one such business-located charging station.[73] Such green business implementations allow for the carbon footprint of their daily energy needs to be reduced through direct business site-facilitated renewable energy sources and a corresponding decrease in fossil fuel emissions. Fig. 3.23.3.20B displays typical simulated PV production for a single home throughout a summer day in Croatia. As discussed previously, the middle of the day is when the most solar energy is available for harvesting and storage. Fig. 3.23.3.20C illustrates eight different EV simulation tests that illustrate the number of

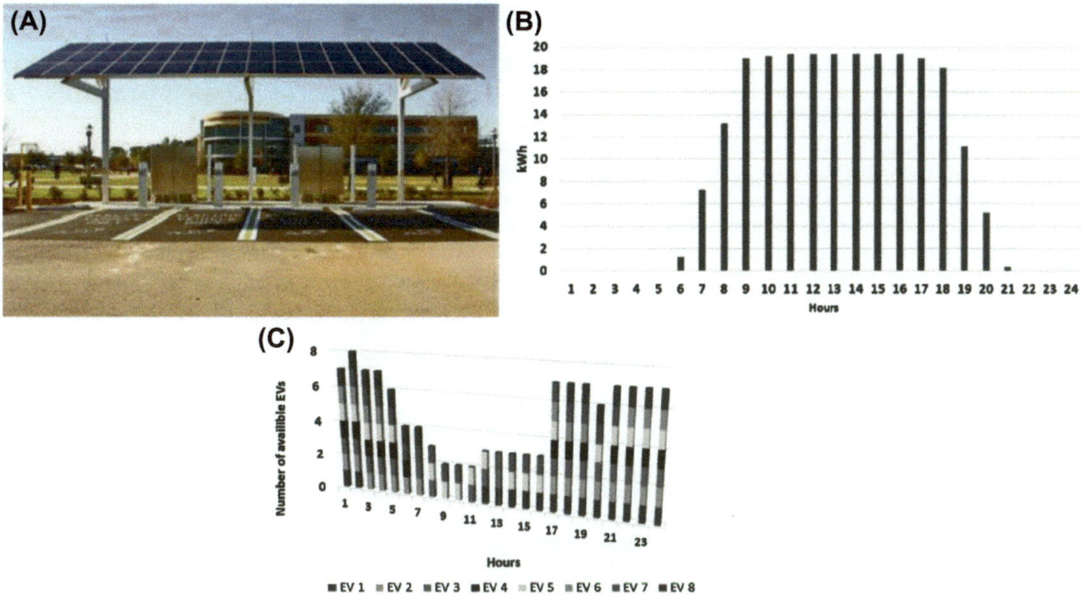

FIGURE 3.23.3.20 (A) An example of photovoltaic (PV)-based parking lots for electric vehicle (EV) charging. (B) PV production for typical summer day in Croatia displaying the middle portion of the day with the highest solar activity. (C) EV availability for a typical summer day. EV use is maximized in the middle of the day, when solar activity is also at its highest magnitude, illustrating the necessity of public and workspace PV charging stations. *(A) Reproduced from Bhatti AR, Salam Z, Aziz MJBA, Yee KP, Ashique RH. Electric vehicles charging using photovoltaic: status and technological review.* Renew Sustain Energy Rev *2016;**54**:34–47. http://dx.doi.org/10.1016/j.rser.2015.09.091. (C) Adapted from Mesarić P, Krajcar S. Home demand side management integrated with electric vehicles and renewable energy sources.* Energy Build *2015;**108**:1–9. http://dx.doi.org/10.1016/j.enbuild.2015.09.001.*

EVs available for home charging, which is shown to be quite low during the peak PV production hours. Hence, the installation of solar-powered EV charging stations at work, at shopping malls, at parks, and at public spaces and parking lots allows for the EV charging gap to be filled, and for solar EV charging to be efficiently maximized and distributed as a renewable energy source throughout the grid.

The conventional electricity market is just beginning to fundamentally transform due to the generation, storage, and distribution of renewable energy sources, similar to what Tesla has initiated with the EV market in the past decade. To take one last note from the inventor of the Model T, Henry Ford: "If everyone is moving forward together, then success takes care of itself." Just as the unstoppable momentum for the ICE automobile defined modern transportation and mechanics more than a century ago, the motivation from global climate change and the perpetual benefits of renewable energy use are moving society forward together on a global scale—progressing humanity into the development of novel energy and transportation sectors, creating a successful, renewable energy-based sustainable infrastructure and world.

References

1. IPCC. Climate change 2014: synthesis report. In: Pachauri RK, Meyer LA, editors. *Contribution of working Groups I, II and III to the Fifth Assessment Report of the Intergovernmental Panel on Climate Change 2014*; 2014. http://www.globalchange.gov/browse/reports/ipcc-climate-change-2014-synthesis-report.
2. Intergovernmental Panel on Climate Change. Climate change 2007: mitigation of climate change. In: Metz B, Davidson OR, Bosch PR, Dave R, Meyer LA, editors. *Contribution of working group III to the Fourth Assessment Report of the Intergovernmental Panel on Climate Change*. Cambridge, New York: Cambridge University Press; 2007.
3. Olivier JGJ, Aardenne JAV, Dentener FJ, Pagliari V, Ganzeveld LN, Peters JAHW. Recent trends in global greenhouse gas emissions: regional trends 1970–2000 and spatial distribution of key sources in 2000. *Environ Sci* 2005;**2**:81–99. http://dx.doi.org/10.1080/15693430500400345.
4. Larcher D, Tarascon J-M. Towards greener and more sustainable batteries for electrical energy storage. *Nat Chem* 2015;**7**:19–29. http://dx.doi.org/10.1038/nchem.2085.
5. *1.3 Energy, emissions, trends in research and development — are we on track? — AR4 WGIII Chapter 1: Introduction*; n.d. https://www.ipcc.ch/publications_and_data/ar4/wg3/en/ch1s1-3.html.
6. Powerwall | The Tesla Home Battery; n.d. https://www.tesla.com/powerwall.
7. Sustainable energy: renewable energy. World Nuclear Association; n.d. http://www.world-nuclear.org/information-library/energy-and-the-environment/sustainable-energy.aspx.
8. Bristol U of November: diamond-power | News | University of Bristol; n.d. http://www.bristol.ac.uk/news/2016/november/diamond-power.html.
9. Payne L, Heard PJ, Scott TB. Examination of surface deposits on Oldbury reactor core graphite to determine the concentration and distribution of 14C. *PLoS One* 2016;**11**:e0164159. http://dx.doi.org/10.1371/journal.pone.0164159.
10. Guarnieri M. When cars went electric, Part 1 [historical]. *IEEE Ind Electron Mag* 2011;**5**:61–2. http://dx.doi.org/10.1109/MIE.2011.940248.
11. Guarnieri M. When cars went electric, Part 2 [historical]. *IEEE Ind Electron Mag* 2011;**5**:46–53. http://dx.doi.org/10.1109/MIE.2011.941122.
12. Publication: Global EV Outlook; n.d. https://www.iea.org/publications/freepublications/publication/global-ev-outlook.html.
13. Yong JY, Ramachandaramurthy VK, Tan KM, Mithulananthan N. A review on the state-of-the-art technologies of electric vehicle, its impacts and prospects. *Renew Sustain Energy Rev* 2015;**49**:365–85. http://dx.doi.org/10.1016/j.rser.2015.04.130.
14. Electric Drive Sales; n.d. http://electricdrive.org/index.php?ht=d%2Fsp%2Fi%2F20952%2Fpid%2F20952.
15. Monthly Plug-In Sales Scorecard; n.d. http://insideevs.com/monthly-plug-in-sales-scorecard/.
16. U.S. car sales 1951–2015 | Statistic. Statista; n.d. http://www.statista.com/statistics/199974/us-car-sales-since-1951/.
17. Tesla model S was World's best-selling plug-in car in 2015; n.d. http://www.hybridcars.com/tesla-model-s-was-worlds-best-selling-plug-in-car-in-2015/.
18. Green Car Congress: EPA rating for 85 kWh Tesla Model S: 89 MPGe, 265-mile range; n.d. http://www.greencarcongress.com/2012/06/models-20120621.html.
19. 2012 Tesla Model S; n.d. http://www.fueleconomy.gov/feg/Find.do?action=sbsid=32557.
20. Danahull DH. Tesla says it received more than 325,000 model 3 reservations. Bloombm; n.d. http://www.bloomberg.com/news/articles/2016-04-07/tesla-says-model-3-pre-orders-surge-to-325-000-in-first-week.
21. The secret tesla motors master plan (just between you and me); 2006. https://www.teslamotors.com/blog/secret-tesla-motors-master-plan-just-between-you-and-me.
22. Tie SF, Tan CW. A review of energy sources and energy management system in electric vehicles. *Renew Sustain Energy Rev* 2013;**20**:82–102. http://dx.doi.org/10.1016/j.rser.2012.11.077.
23. Emadi A. Transportation 2.0. *IEEE Power Energy Mag* 2011;**9**:18–29. http://dx.doi.org/10.1109/MPE.2011.941320.
24. Manzetti S, Mariasiu F. Electric vehicle battery technologies: from present state to future systems. *Renew Sustain Energy Rev* 2015;**51**:1004–12. http://dx.doi.org/10.1016/j.rser.2015.07.010.
25. Model 3 | Tesla Motors; n.d. https://www.teslamotors.com/model3.
26. Helmers E, Marx P. Electric cars: technical characteristics and environmental impacts. *Environ Sci Eur* 2012;**24**:14. http://dx.doi.org/10.1186/2190-4715-24-14.

27. Dunn B, Kamath H, Tarascon J-M. Electrical energy storage for the grid: a battery of choices. *Science* 2011;**334**:928–35. http://dx.doi.org/10.1126/science.1212741.
28. Daniel C, Besenhard JO, editors. *Handbook of battery materials: DANIEL: HBK BATTERY MAT E2 O-BK*. Weinheim, Germany: Wiley-VCH Verlag GmbH & Co. KGaA; 2011.
29. Girishkumar G, McCloskey B, Luntz AC, Swanson S, Wilcke W. Lithium–air battery: promise and challenges. *J Phys Chem Lett* 2010;**1**:2193–203. http://dx.doi.org/10.1021/jz1005384.
30. Balaish M, Kraytsberg A, Ein-Eli Y. A critical review on lithium–air battery electrolytes. *Phys Chem Chem Phys* 2014;**16**:2801. http://dx.doi.org/10.1039/c3cp54165g.
31. Yekini Suberu M, Wazir Mustafa M, Bashir N. Energy storage systems for renewable energy power sector integration and mitigation of intermittency. *Renew Sustain Energy Rev* 2014;**35**:499–514. http://dx.doi.org/10.1016/j.rser.2014.04.009.
32. Deutsche Bank. Autos & auto parts electric cars: plugged in 2, 2009; n.d. From: https://www.db.com/company/index.htm.
33. Bruce PG, Freunberger SA, Hardwick LJ, Tarascon J-M. Li-O_2 and Li-S batteries with high energy storage. *Nat Mater* 2012;**11**:19–29. http://dx.doi.org/10.1038/nmat3191.
34. Nissan LEAF® Electic car features. Nissan USA; n.d. http://www.nissanusa.com/electric-cars/leaf/features.
35. *US Advanced Battery Consortium USABC Goals for Advanced Batteries for EVs*. 2006. Available at: http://uscar.org/commands/files_download.php?files_id=27.
36. Diouf B, Pode R. Potential of lithium-ion batteries in renewable energy. *Renew Energy* 2015;**76**:375–80. http://dx.doi.org/10.1016/j.renene.2014.11.058.
37. Collins J, Ngo T, Qu D, Foster M. Spectroscopic investigations of sequential nitric acid treatments on granulated activated carbon: effects of surface oxygen groups on π density. *Carbon* 2013;**57**:174–83. http://dx.doi.org/10.1016/j.carbon.2013.01.061.
38. Collins J, Zheng D, Ngo T, Qu D, Foster M. Partial graphitization of activated carbon by surface acidification. *Carbon* 2014;**79**:500–17. http://dx.doi.org/10.1016/j.carbon.2014.08.009.
39. Collins J, Gourdin G, Foster M, Qu D. Carbon surface functionalities and SEI formation during Li intercalation. *Carbon* 2015;**92**:193–244. http://dx.doi.org/10.1016/j.carbon.2015.04.007.
40. Fox-Penner PS. *Smart power: climate change, the smart grid, and the future of electric utilities*. Washington: Island Press; 2010.
41. Wolsink M. The research agenda on social acceptance of distributed generation in smart grids: renewable as common pool resources. *Renew Sustain Energy Rev* 2012;**16**:822–35. http://dx.doi.org/10.1016/j.rser.2011.09.006.
42. IBM – Electric cars – United States 2010. http://www.ibm.com/smarterplanet/us/en/smart_grid/article/electric_cars.html.
43. Liu C, Chau KT, Wu D, Gao S. Opportunities and challenges of vehicle-to-home, vehicle-to-vehicle, and vehicle-to-grid technologies. *Proc IEEE* 2013;**101**:2409–27. http://dx.doi.org/10.1109/JPROC.2013.2271951.
44. EPRI | Product Abstract; n.d. http://www.epri.com/abstracts/pages/productabstract.aspx?ProductID=000000000001020676.
45. Luo X, Wang J, Dooner M, Clarke J. Overview of current development in electrical energy storage technologies and the application potential in power system operation. *Appl Energy* 2015;**137**:511–36. http://dx.doi.org/10.1016/j.apenergy.2014.09.081.
46. Solar Roadways; n.d. http://www.solarroadways.com/.
47. R&D - Solar Roadways; n.d. http://www.solarroadways.com/Research/Research.
48. GCR – News – Solar paving to make first public US appearance on Route 66; n.d. http://www.globalconstructionreview.com/news/solar-paving-ma8ke-first-pub4lic-us-appear5ance/.
49. Wattway by Colas | Paving the way to tomorrow's energy; n.d. http://www.wattwaybycolas.com/en/.
50. Small French town's "Wattway" solar road converts sunshine to power street lights. *Jpn Times Online* December 23, 2016. http://www.japantimes.co.jp/news/2016/12/23/business/small-french-towns-wattway-solar-road-converts-sunshine-power-street-lights/#.WUB2GOvytaQ.
51. Anthony S. World's first solar road opens in France: it's ridiculously expensive. *Ars Tech* 2016. http://arstechnica.com/cars/2016/12/worlds-first-solar-road-opens-in-france/.
52. Studio Roosegaarde – Smart highway info; n.d. https://www.studioroosegaarde.net/project/smart-highway/info/.

53. Pan P, Wu S, Xiao Y, Liu G. A review on hydronic asphalt pavement for energy harvesting and snow melting. *Renew Sustain Energy Rev* 2015;**48**:624–34. http://dx.doi.org/10.1016/j.rser.2015.04.029.
54. Zuo J, Zhao Z-Y. Green building research—current status and future agenda: a review. *Renew Sustain Energy Rev* 2014;**30**:271–81. http://dx.doi.org/10.1016/j.rser.2013.10.021.
55. Scofield JH. Efficacy of LEED-certification in reducing energy consumption and greenhouse gas emission for large New York City office buildings. *Energy Build* 2013;**67**:517–24. http://dx.doi.org/10.1016/j.enbuild.2013.08.032.
56. Cabeza LF, Rincón L, Vilariño V, Pérez G, Castell A. Life cycle assessment (LCA) and life cycle energy analysis (LCEA) of buildings and the building sector: a review. *Renew Sustain Energy Rev* 2014;**29**:394–416. http://dx.doi.org/10.1016/j.rser.2013.08.037.
57. Spence R, Mulligan H. Sustainable development and the construction industry. *Habitat Int* 1995;**19**:279–92. http://dx.doi.org/10.1016/0197-3975(94)00071-9.
58. Parra D, Gillott M, Norman SA, Walker GS. Optimum community energy storage system for PV energy time-shift. *Appl Energy* 2015;**137**:576–87. http://dx.doi.org/10.1016/j.apenergy.2014.08.060.
59. Brown A, Muller SS, Dobrotkova Z. Renewable energy: markets and prospects by technology; n.d.
60. 2015 New Energy Outlook. Bloomberg; n.d. http://bloom.bg/1IT8ezN.
60a. *International Technology Roadmap for Photovoltaic (ITRPV)*, 7th Ed. 2016. http://www.itrpv.net/Reports/Downloads/2016/.
61. *Levelized cost of new electricity generating technologies*. IER; 2009. http://instituteforenergyresearch.org/studies/levelized-cost-of-new-generating-technologies/.
62. Comparing the levelized cost of energy technologies. *Energy Innov Policy Technol* 2015. http://energyinnovation.org/2015/02/07/levelized-cost-of-energy/.
63. Photovoltaic system pricing trends: Historical, recent, and near-term projections 2015 edition - photovoltaic system pricing trends 2015 edition.xlsx — OpenEI Datasets; n.d. http://en.openei.org/datasets/dataset/photovoltaic-system-pricing-trends-historical-recent-and-near-term-projections-2015-edition/resource/f0000000-58cc-4372-a567-000000000046.
64. U.S. Energy Information Administration (EIA) — Source; n.d. http://www.eia.gov/outlooks/aeo/electricity_generation.cfm.
65. *Levelized cost of energy analysis 9.0*. Lazard.com; n.d. https://www.lazard.com/perspective/levelized-cost-of-energy-analysis-90/.
66. *Levelized cost of energy analysis 10.0*. Lazard.com; n.d. https://www.lazard.com/perspective/levelized-cost-of-energy-analysis-100/.
67. *Levelized cost of energy v8 abstract*. Lazard.com; n.d. https://www.lazard.com/perspective/levelized-cost-of-energy-v8-abstract/.
68. *Renewable energy sources: cost comparison*; n.d. http://www.renewable-energysources.com/.
69. *Tesla is planning to release the Powerwall version 2.0 this summer*. TreeHugger; n.d. http://www.treehugger.com/clean-technology/tesla-planning-release-powerwall-version-20-summer.html.
70. *Tesla to launch Powerwall 2.0 home battery this summer*; n.d. http://inhabitat.com/powerwall-2-0-an-improved-tesla-battery-pack-to-be-unveiled-this-summer/.
71. *Tesla's solar roof: solar shingles reimagined*. EnergySage; 2016. http://news.energysage.com/tesla-solar-panel-roof-the-next-solar-shingles/.
72. Mesarić P, Krajcar S. Home demand side management integrated with electric vehicles and renewable energy sources. *Energy Build* 2015;**108**:1–9. http://dx.doi.org/10.1016/j.enbuild.2015.09.001.
73. Bhatti AR, Salam Z, Aziz MJBA, Yee KP, Ashique RH. Electric vehicles charging using photovoltaic: status and technological review. *Renew Sustain Energy Rev* 2016;**54**:34–47. http://dx.doi.org/10.1016/j.rser.2015.09.091.

3.23.4 Innovations and Future Opportunities for Renewable Energy From Green Chemistry

3.23.4.1 INTRODUCTION

The future of renewable energy integration relies directly on the combined scientific development and economic integration of energy storage, smart grid, and smart infrastructure technologies. As illustrated by the market advancements of companies such as Tesla, renewable energy-utilizing products and infrastructure have been integrated into a very fossil-fuel-dependent economy and lifestyle through the continual reinvestment in smarter, more energy efficient, and environment-friendly technology. Historically, there is a slim chance that any product will be able to directly replace its corresponding fossil-fueled commodity, no matter what new products are brought to the market. However, modernized investment and marketing strategies, such as those explained in the previous section, have been proved to result in affordable and sustainable replacement products for traditional energy sources and fossil-fuel-powered consumer products. Tesla's example of initially developing very high-end automobiles with very high performance (e.g., Tesla Roadster) enabled enough revenue to reinvest in product diversification (e.g., Tesla Model X and S). Eventually, the *invest, develop, sell, reinvest method* has allowed for the development of high-performing and cost-effective EV product lines suitable for the status quo (e.g., Tesla Model 3). Currently, this modernized business model is being applied to significantly diversify Tesla's current up-and-coming product line, strengthening the renewable energy/environment-friendly market by developing fully electric buses, trucks, and semi-trucks to reach the mass consumer populace. Tesla is not stopping there, as the company is applying the same strategy toward home solar systems. As discussed in the previous section, Tesla's home energy storage technology, such as the Powerwall, is now being developed for streamlined integration with local and community-based solar PV energy harvesting. Tesla's recent acquisition of SolarCity will eventually allow for the affordable use of home-based solar systems to power not only the local community and grid but also the majority of individual consumer's home and transportation electricity needs. Furthermore, Elon Musk's achievements with SpaceX and Hyperloop technologies[1,2] are showing us that this "reach for the stars" type of technological investment and innovation is progressively successful in modern times—where nearly every new successful mass-consumer-utilized technology contains a core component of renewable or sustainable energy. There are several other technology companies that are in the process of implementing what, at one time, may have been considered incredible but unlikely innovations. This section summarizes many of the highlights of the chapter while also presenting some newly designed or developed technologies, which are still far from any real market/consumer integration.

3.23.4.2 FUTURE ELECTRIC VEHICLE BATTERIES: THE HYBRIDIZATION OF BATTERY AND SUPERCAPACITOR TECHNOLOGY

Significant improvements in stored energy density as well as power density have been made, especially during the past two decades. The modern advances that have allowed for these two types of energy storage to be commercialized for EV use have progressed in the form of secondary batteries (Section 3.23.2.2) and supercapacitors (Section 3.23.2.4) to allow for advances in both constant speed/idling (high-energy-density batteries) and acceleration (high-power-density supercapacitors), respectively. The previous section discussed the most recent advancements in battery technology, where storage advancements, primarily in the form of energy density, have been shown to increase as a function of materials. However, energy density is only part of the necessary energy storage required for effective EV implementation. In reality, high and long-lasting specific power is far more in demand than high specific energy. The power factor has always been a necessary selling point and capability required in consumer vehicles. The problem in developing long-lasting, high-power-storage devices is centered on the electrode materials utilized for energy storage. One major area of research is focused on diffusion of ions inserted/intercalated and deinserted/deintercalated in the materials comprising the electrodes. Another significant area of research involves the suppression or solubility of precipitates (e.g., polysulfides, Li oxides) formed during the charge and discharge processes. Smooth, long surface structure domains allow for effective electron transfer and ion transport, yet lack the very highly accessible surface areas required for extended high-power charge storage. Ion diffusion through large, ordered structures creates very slow rates of charge and discharge—thereby diminishing maximum power output, yet providing extensive discharge time. Highly amorphous electrode structure domains, however, can store a large amount of charge that can be released very quickly, yet are limited by the total discharge time due to diffusion of ions out of amorphous layers deep within the bulk of the electrode. From these traits, it is easy to see that extended high-power charge storage capability quickly becomes a balance of the overall structural order versus electrochemically accessible active surface area. The low kinetic barriers of high-power supercapacitor materials are currently being developed toward more batterylike devices through specific molecular changes at the nanoscale, thereby allowing for the discharge capacity to increase. The same alterations are taking place with energy-dense battery materials in efforts to better enable the transition from first-generation EVs toward the next-generation of integrated renewable energy source EVs.[3–5]

Traditional EV production has enabled the effective delivery of energy in EVs via separate energy storage sources—where a rechargeable battery system is separate from a supercapacitor, or power box, system. This design decreases the overall efficiency of EVs, as the two energy storage sources must be coordinated together to most effectively deliver stored energy based on the needs of the driver. More importantly, two energy storage systems significantly increase the weight, cost, and carbon footprint of every EV produced. The Ragone plot in Section 3.23.2.1 (Fig. 3.23.2.1) illustrates the modern advancements of both battery- and supercapacitor-type materials, where the integration of both chemical and physical forms of stored energy in a single material can be enabled (e.g., Li-ion capacitors)—bringing EV storage technology closer to sustainable, weight, and cost-efficient EV storage devices.

The evolution of the major generations of battery- and capacitor-type electrode materials are pictured in Fig. 3.23.4.1. Material properties and composition examples are described for both positive and negative electrodes. The first generation of both batteries and capacitors display the least advanced charge storage capabilities. Battery materials are highly ordered—allowing for high durability and long lifetimes. However, kinetically, charge/discharge rates are very poor for insertion-type batteries as the intercalation/deintercalation processes are extremely slow. First-generation capacitors were also highly ordered parallel plates, placed very close to each other and having extremely high kinetics and lifetimes. However, low surface area, and therefore low charge storage, limited conventional uses. Second-generation batteries and capacitors are more disordered in their three-dimensional structure and smaller in single domain particle size. This alteration allows for more accessible surface areas for both battery and capacitor functions and therefore more charge storage by either physical (electrostatic) or a chemical means of charge storage adsorption. The relative increase in disorder incorporates some capacitor-type domains into second-generation battery structures—creating relatively flat, open surfaces between the ordered layered-type storage domains. At this point, electrostatic charge storage (observed in Fig. 3.23.4.1 as stored charge particles, in single layer or pyramidal shape, between ordered domains) begins to become prevalent in second-generation

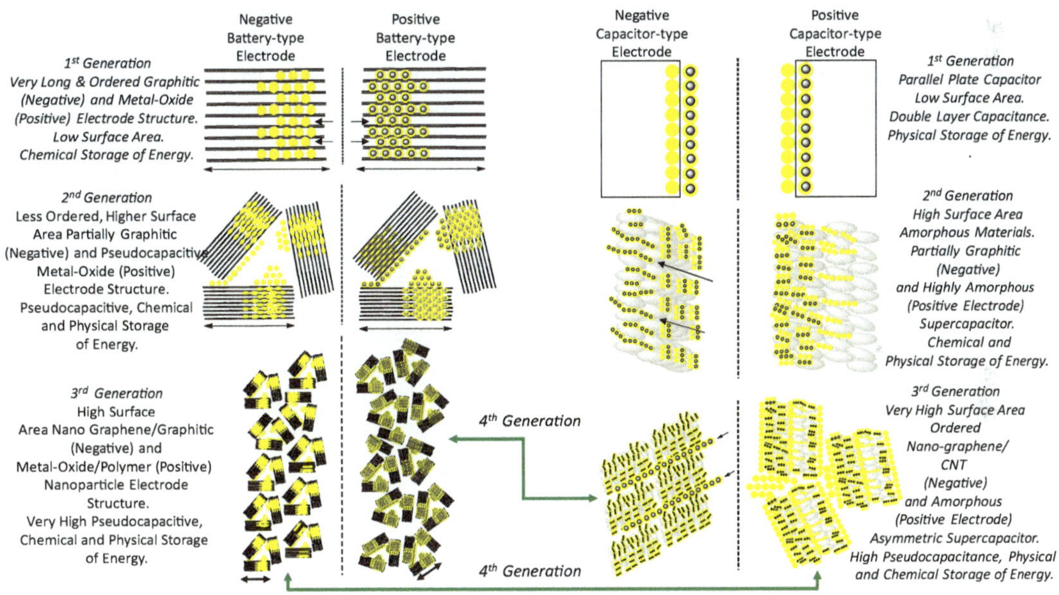

FIGURE 3.23.4.1 The evolution of hybrid battery-type (left) and capacitor-type (right) electrodes. *Yellow* and *black/yellow dots* represent relative negatively and positively charged particles. Single layered and stacked charged particles physically (electrostatically) adsorbed to surfaces located in between graphitic or metal oxide positive/negative highly ordered electrodes (this type of charge storage is typical of capacitor-type forms of charge storage) illustrate the evolving charge storage capabilities of modern battery-type electrodes. Two-way *black arrows* shown with battery-type electrodes display approximate, relative sizes of electrode domains. Single-way *arrows* shown with evolving capacitor-type electrodes indicate the incorporation of intercalation processes (typical of a battery-type form of charge storage) of charging an electrode with ion species. Two-way *green arrows* display the fourth generation of both battery- and capacitor-type electrodes, where hybrid battery/capacitor devices are created through various combinations of hybrid chemical (battery) and physical (capacitor) charge storing electrodes.

battery-type electrodes. The increased disorder in second-generation capacitor-type electrodes breaks down the single, flat surface structure of the parallel plate capacitor and incorporates higher surface area where charge can be stored through intercalation of charged particles (a seemingly battery-type form of charge storage). Such textural changes in battery electrode materials allow for increases in specific power where increased intercalation battery-type charge storing observed with the structural changes of second-generation supercapacitors enables significant increases in specific energy. This is the basis for hybrid capacitor and battery-type energy storage in a single electrode material. Pseudocapacitance, which is a phenomenon known to occur in the presence of nanosurfaces as well as nano-metal oxide particles, enables extremely high magnitudes of combined Faradaic and electrostatic charge storage (Section 3.23.2.4.1). Pseudocapacitance has been shown to increase capacitance, relative to its initial double-layer capacitance (the static charge adsorption we observe in first-generation capacitors), well over 100 times.[6] Both second-generation hybrid battery/capacitor electrode material types can display pseudocapacitance properties. Third-generation batteries and capacitors have extremely high surface area nanostructured surfaces where the electrochemical accessibility of charge storage particles in the high-surface-area nanostructures can diffuse through porous networks with very little resistance. Both battery and supercapacitor third-generation hybrids display their largest magnitudes of combined chemical and physical electrochemical storage. Hybrid supercapacitors have increased relative specific energy by incorporating specific asymmetric molecular and ordered stacked particle structure into each electrode. Many current studies are investigating methods to convert battery-type electrode material into capacitor-type material, and vice versa, allowing for tunable properties from a single source material.[7–9] This enables the development of relatively high specific energy through enhanced battery-type storage combined with a more traditional high-power supercapacitor-type electrode.[10–12] The asymmetric hybrid supercapacitor is a natural transition into the fourth generation of evolving electrode materials. Fourth-generation hybrid electrodes are, in fact, optimized combinations of hybrid-designed battery and capacitor electrodes, enabling a diverse array of possible battery-type capacitor or capacitor-type battery system combinations. The ability to put hybrid "types" of both capacitor or battery electrodes into a single charge-storing electrochemical cell, in a single lightweight device, with low electrochemical resistance from diffusion of charged particles in the electrode nanostructure is the keystone of fourth-generation hybrid energy storage devices and the leading design concept of advanced energy storage materials.

3.23.4.3 FUTURE SMART GRID AND SMART ROAD OPPORTUNITIES

The advent of the smart grid in the past decade has produced a wide array of sustainable energy advancements through the integration of modern renewable energy sources and digital communications. Section 3.23.3.3 described the key components of the smart grid—describing why stationary EES is essential to the development of a workable, renewable energy-based smart grid. Zinc-air grid storage is one example of a large-scale grid storage option currently being utilized in the world and growing throughout the worldwide market.

FIGURE 3.23.4.2 A picture the Eos Aurora 1000|4000, a low-cost direct current battery system designed to meet requirements of grid scale energy storage. *Reproduced with permission from Eos Energy Storage | Technology and Products; n.d. http://www.eosenergystorage.com/technology-and-products/.*

Eos is a company that is currently making headlines for its development of commercial-grade grid-storage battery systems based on the novel design and manufacture of Zn-air electrode materials.[13] One product making headlines (the Eos Aurora 1000|4000) is a high market competitor for future grid storage (Fig. 3.23.4.2). Utilizing four 250 kW submodules, the Aurora DC battery system yields 4 MWh of energy capacity, enabling 4 full hours of discharge capacity with a high response time and a relatively long battery lifetime of 5000 cycles and 15 years. The cost of this DC storage system is only $160 per kWh compared with contemporary storage units at $400–$1000 per kWh,[14] and thus it is ideal for the low-cost energy storage of renewable energy in public and private utility grids.

As we have seen in Section 3.23.3.4, smart roads and solar roadways are progressively being developed in a very diverse manner, adding multiple pathways and applications for renewable energy sources, especially solar energy. However, the most recent innovations for smart roads rely on integrated renewable energy—where energy harvesting for an energy-neutral transportation system originates from a variety of combined sources. These systems are still in their initial development stages and have limited deployment in real-world test sites at proprietary protected locations. However, when this is the case, one method by which to learn more about such next-generation scientific developments is through patents. Patents often yield valuable information about new discoveries and designs because they must be very specific and functional to be considered patentable. A scientific diagram of an integrated renewable energy smart roadway entitled "System and method for creating a networked infrastructure roadway distribution platform of solar energy gathering devices"[15] was patented and published in 2009 (Fig. 3.23.4.3). The patent describes the key components that are integral for the production and function of this particular integrated renewable energy smart road system, and enough key information is given so any knowledgeable reader can understand the particulars relevant to their own scientific developments in PV solar or wind turbine energy harvesting systems. This particular patent describes designs for wind turbines that are combined with solar panels in a number of manners. The patent not only integrates wind and solar energy from turbines on the side of roadways but also utilizes miniaturized wind/solar systems that are attached to solar-paint-coated EVs that travel on smart, bidirectional information and energetic roadways. Hence the system utilizes all the possible surface-available sites on the roadway, as well as on the vehicle and wind turbine surfaces, to implement both renewable energy sources present on a given roadway—

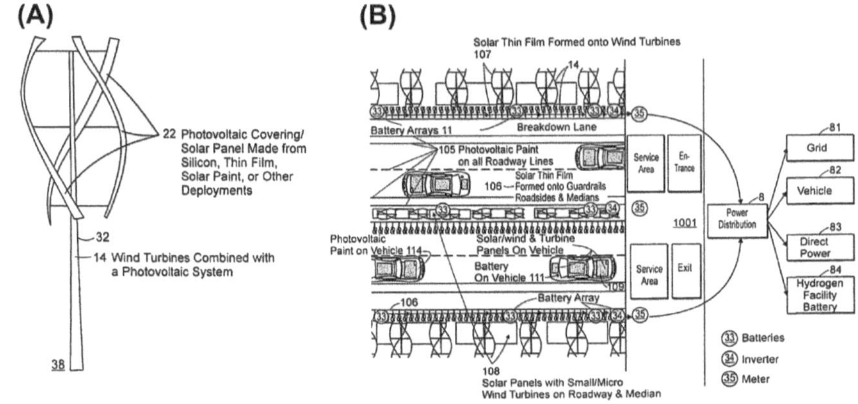

FIGURE 3.23.4.3 Two figures from the patent titled "System and method for creating a networked infrastructure roadway distribution platform of solar energy gathering devices". Fig. 3A from the patent illustrates integrated wind/ solar turbine systems that would be present both on the side of smart roads and on the electric vehicles directly, as shown with Fig. 3B. The patent design's integration of solar and turbine energy generation devices on both vehicles and roadway space, with renewable energy being controlled between both on-board vehicle and road grid storage for the powering of battery electric vehicles is an effective example of renewable smart grid energy harvesting, storage, and delivery. *Adapted from US Pat No. US 7501713 B2, Fein GS, Merritt E, Llc G.* System and method for creating a networked infrastructure roadway distribution platform of solar energy gathering devices; *2007.*

solar power from the sun and wind power generated by the locomotion of EVs. Such a design is a prime example of integrated renewable energy systems for up-and-coming next-generation smart roadways—and a good example of the importance of utilizing the maximum available surface area available (a resounding theme for efficiency) to achieve the best energy harvesting, transfer, or storage performance for smart materials.

3.23.4.4 FUTURE SMART BUILDING AND SMART HOME OPPORTUNITIES

Energy harvesting and storage in the conventional battery/electricity sense is not the only manner by which the building economic sector can improve and implement renewable energy growth. Biomimicry is a concept that enables sustainable innovation and real-world applications based on the patterns and strategies learned from nature.[16] Biomimicry has recently entered into the forefront of green design and manufacturing practices.[17] All scales of chemical and physical engineering, energy use, harvesting, etc., from the nano- to macroscales, have been shown to respond in similar biologically predicted manners, when elements found in biology or nature have been the basis of the design. Aprilli is one such company designing and proposing biomimetic, renewable energy-based buildings and sites.[18] Fig. 3.23.4.4 displays one of Aprilli's conceptual designs (which was awarded first prize for the 2013 Green Dot Design competition) for an urban skyfarm in Seoul, Korea. The conceptual prototype of a vertical farm proposal is based on biomimicry ideas that yield energetic benefits from the environment due to the farm infrastructure's similarities to

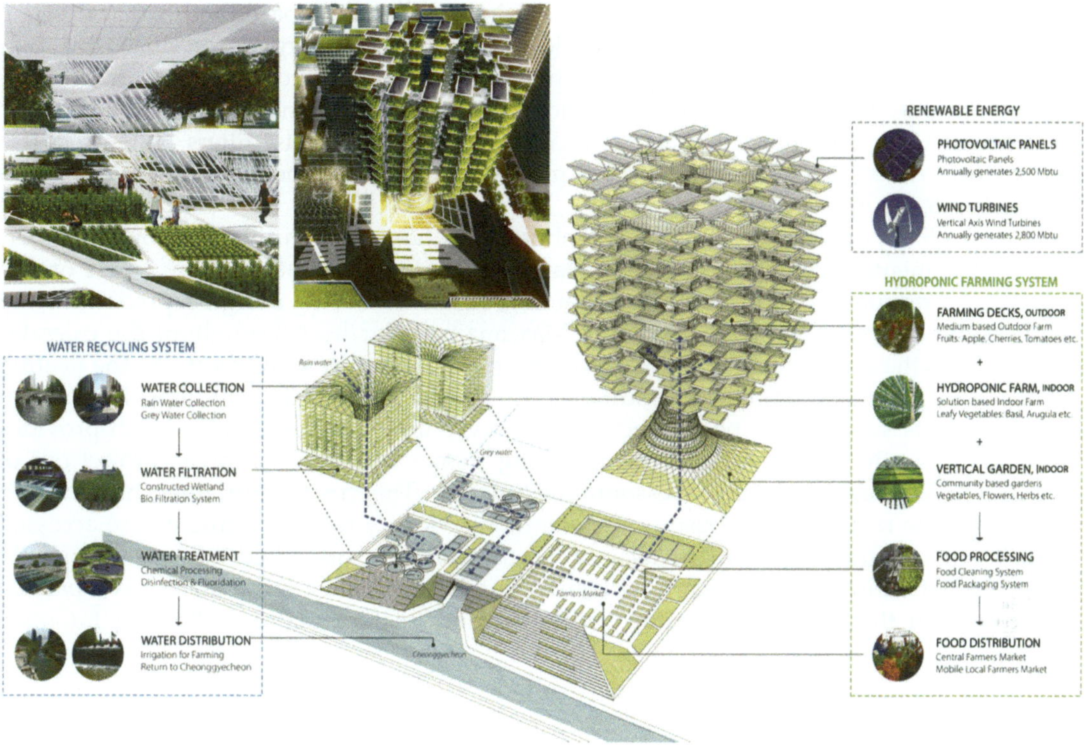

FIGURE 3.23.4.4 A schematic showing the design of an urban skyfarm inspired by the ecological system of giant trees. Water collection, filtration, distribution, efficiency, etc., work in the same gravity-assisted manner as a large, multibranched tree. Photovoltaic panels and wind turbines are optimally positioned throughout the structure to maximize solar and wind energy harvesting. Agriculture for the system is maximized in the tree-mimicked infrastructure by "smart-gridding" the local sourced water and renewable energy into the infrastructure's hydroponic farming system, including indoor/outdoor farming, on-site food processing/distribution, and vertical community gardens. *Adapted from http://www.aprilli.com/urban-skyfarm/.*

organic vegetation in the natural world. This urban skyfarm example was inspired by the ecological system of giant trees. The proposal site is located in a very densely populated area and will effectively support local food agriculture, distribution, water quality, air filtration, and renewable energy production needs. The elevation of the main food production fields allows for vegetation to access more beneficial sunlight and wind—similar to the effect of producing strong top branches or blossoming flowers from a tree. As the elevation increases, the floors extend further laterally, similar to the branches of a tall bush or tree that naturally strengthen toward the direction of maximum sunlight exposure. The lower levels are controlled indoor hydroponic environments, allowing for the minimization of pesticides, herbicides, and fungicides with no environmental runoff. The ground level is shaded and utilized by the public. The major components of this design are the root, trunk, branch, and leaf—where the spatial characteristics of the infrastructure design are optimized for diversity of farming products and techniques—and water, which is utilized for self-irrigation. In all, this particular urban skyfarm supports nearly 70 farming decks, 44,000 m^2

of surface area in the leaf portion, 28,000 m^2 of indoor farming space, and 3200 m^2 of PV panels. The urban skyfarm is a net-zero building in which all energy is produced by solar and wind energy. Additional benefits of having a massive green space in a dense urban environment include reduction of heat, regulation of storm water runoff, and photosynthetic reduction of CO_2. The integrated water processing system collects, filters, and processes rain or storm runoff. Such smart integrated processes in housing construction are adaptable to a multitude of urban housing models.

Fig. 3.23.4.5 is a proposed schematic of a phototropic house to be utilized in lower temperature environments such as San Francisco. Based on the idea of phototropism, where organisms grow in response to light, this self-sustaining, net-zero housing design is predominantly based on the control of sunlight exposure. Maximizing solar heat throughout the calendar year increases the comfortable living atmosphere for inhabitants, while reducing the heating requirements from external energy sources. Each unit in the housing building for the San Francisco site is strategically positioned. Nearly every unit is facing south, which provides the most sun in the northern hemisphere. Living rooms and common areas project out of the building, whereas bedrooms recess into the building. This creates a brighter, heated environment for the major daytime use of the common spaces, whereas bedrooms are accessed primarily at nighttime. The main facades of the building are tilted toward the south, improving light and solar heat gain. As shown in the figure, the central atrium is a passive design element that provides natural ventilation and heat regulation, allowing for open

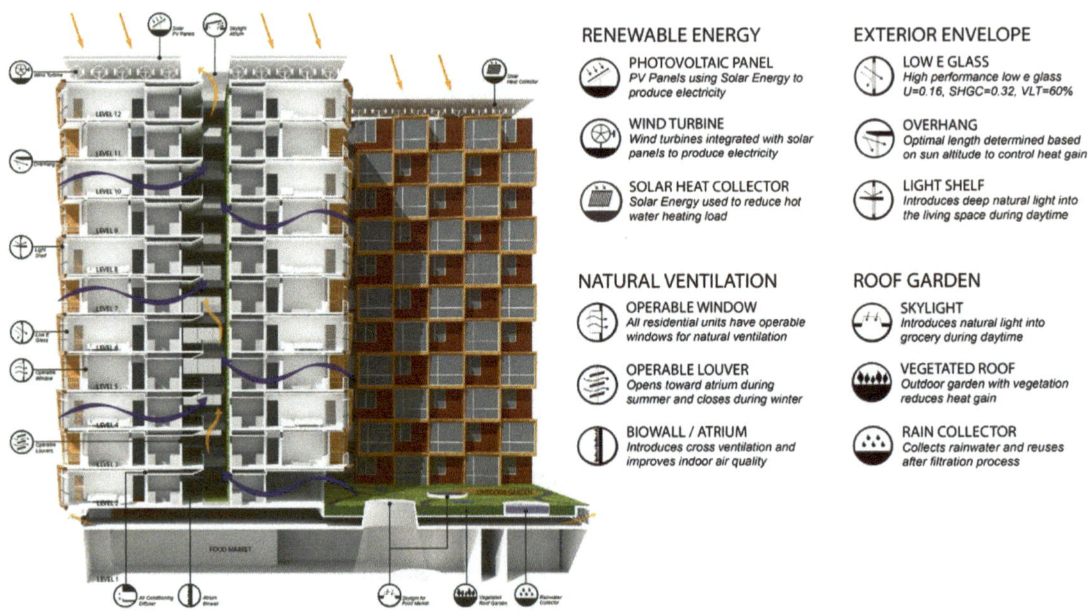

FIGURE 3.23.4.5 Phototropic house proposal schematic for mild temperate climates such as San Francisco. Strategically placed photovoltaic panels, wind turbines, and solar heat collectors enable local renewably sourced electricity and hot water supply. Architectural designs allow for natural ventilation, air conditioning, efficient natural light use, and heat shielding. Roof gardens enable locally produced water collection and food production. *Adapted from Phototropic Housing.* Aprilli; *n.d. http://www.aprilli.com/phototropic-housing/.*

and closed skylights and windows in the summer and winter, respectively. Heat transfer piping is integrated into the ceiling, walls, and ground floors to enable a solar source of heating and cooling water—similar to the smart road system described in Section 3.23.3.4. The net-zero energy use of the phototropic house is achieved through not only passive ventilation and heat transfer loops but also the utilization of renewable resources located on-site, as illustrated in Fig. 3.23.4.5. The full design details for the phototropic house include the following renewable energy source yields: roof PV (8400 ft^2, 654 Mbtu/year), west facade PV (3000 ft^2, 129 Mbtu/year), solar hot water (663 Mbtu/year), and wind turbines (130 units, 887 Mbtu/year). The EUI (see Section 3.23.3.5) of the phototropic building site is 12.8 kBtu/ft^2, with the total renewable energy yield at 13.1 kBtu/ft^2, thereby exceeding the entire building's OE use and preventing all GHG emissions while under normal OE use conditions.

3.23.4.5 FINAL WORDS

The companies briefly discussed in the previous section serve as examples that illustrate the ways in which renewable energy technologies have been and still are making inroads in the efforts to shift the economy toward a society based on renewable energy, and therefore a society lower in GHG emissions. Tesla's innovation and production history is a great example of one manner in which renewable energy can be effectively integrated into the product infrastructure of a fossil-fuel-based economy. However, for renewable energy to become the dominant energy source, far more innovation and economical acceptance is required. The integration of bioinspired designs and technologies, as demonstrated by Aprilli's renewable energy-based buildings, is another example that shows that researchers have many more systems to choose from for new inspirations than humankind's now demonstrably limited technological history. The designs and materials costs associated with renewable energy generation systems have reached a point that allows them to be cost competitive with fossil fuels. The efficient and low-cost storage of renewable-generated energy is what is needed to enable the transition to a green energy economy. Hence, storage technologies, such as the grid storage Zn-air battery systems being developed by Eos, are important for the evolution of human sustainability. The authors hope that the reader will have understood the discussed technological projections, ideas, and innovations for their potential impacts on future renewable energy lifestyles and fossil fuel replacement, while also appreciating that many future technological ideas originate from a common perspective of implausibility, only to be later molded into reality through time and persistent research.

References

1. Musk E. *Hyperloop Alpha*. 2013.
2. spacexcmsadmin. *Hyperloop SpaceX*. 2015. http://www.spacex.com/hyperloop.
3. Schuller A, Flath CM, Gottwalt S. Quantifying load flexibility of electric vehicles for renewable energy integration. *Appl Energy* 2015;**151**:335–44. http://dx.doi.org/10.1016/j.apenergy.2015.04.004.
4. Nunes P, Farias T, Brito MC. Day charging electric vehicles with excess solar electricity for a sustainable energy system. *Energy* 2015;**80**:263–74. http://dx.doi.org/10.1016/j.energy.2014.11.069.
5. Dubal DP, Ayyad O, Ruiz V, Gómez-Romero P. Hybrid energy storage: the merging of battery and supercapacitor chemistries. *Chem Soc Rev* 2015;**44**:1777–90. http://dx.doi.org/10.1039/C4CS00266K.

6. Conway BE. *Electrochemical supercapacitors scientific fundamentals and technological applications*. Boston (MA), US: Springer; 1999.
7. Collins J, Ngo T, Qu D, Foster M. Spectroscopic investigations of sequential nitric acid treatments on granulated activated carbon: effects of surface oxygen groups on π density. *Carbon* 2013;**57**:174–83. http://dx.doi.org/10.1016/j.carbon.2013.01.061.
8. Collins J, Zheng D, Ngo T, Qu D, Foster M. Partial graphitization of activated carbon by surface acidification. *Carbon* 2014;**79**:500–17. http://dx.doi.org/10.1016/j.carbon.2014.08.009.
9. Collins J, Gourdin G, Foster M, Qu D. Carbon surface functionalities and SEI formation during Li intercalation. *Carbon* 2015;**92**:193–244. http://dx.doi.org/10.1016/j.carbon.2015.04.007.
10. Liu J, Zhang L, Wu HB, Lin J, Shen Z, Lou XWD. High-performance flexible asymmetric supercapacitors based on a new graphene foam/carbon nanotube hybrid film. *Energy Environ Sci* 2014;**7**:3709–19. http://dx.doi.org/10.1039/C4EE01475H.
11. Amatucci GG, Badway F, Du Pasquier A, Zheng T. An asymmetric hybrid nonaqueous energy storage cell. *J Electrochem Soc* 2001;**148**:A930. http://dx.doi.org/10.1149/1.1383553.
12. Fan Z, Yan J, Wei T, Zhi L, Ning G, Li T, et al. Asymmetric supercapacitors based on graphene/MnO_2 and activated carbon nanofiber electrodes with high power and energy density. *Adv Funct Mater* 2011;**21**:2366–75. http://dx.doi.org/10.1002/adfm.201100058.
13. Eos Energy Storage | Technology and Products; n.d. http://www.eosenergystorage.com/technology-and-products/.
14. Cardwell D. Battery seen as way to cut heat-related power losses. *N Y Times* 2013.
15. Fein GS, Merritt E, Llc G. *System and method for creating a networked infrastructure roadway distribution platform of solar energy gathering devices*. 2007.
16. Benyus JM. *Biomimicry: innovation inspired by nature*. repr. New York (NY): Harper Perennial; 2008.
17. Benyus J. *Biomimicry in action*; n.d.
18. Products. *Aprilli*; n.d. http://www.aprilli.com/.
19. Urban Skyfarm. *Aprilli*; n.d. http://www.aprilli.com/urban-skyfarm/.
20. Phototropic Housing. *Aprilli*; n.d. http://www.aprilli.com/phototropic-housing/.

CHAPTER

3.24

From Ethanol to Biodiesel: A Survey of Green Fuels

Christopher Brigham[1], *Johannes Bader*[2]

[1]University of Massachusetts Dartmouth, North Dartmouth, MA, United States;
[2]Beuth University of Applied Sciences, Berlin, Germany

3.24.1 INTRODUCTION

Early on in history, wood, charcoal, dung, and other bio-based substances were burned to produce energy. These were the first biofuels ever used by mankind, with wood being the most common and still used today. Also, olive oil and whale oil have been used as bio-based energy sources. Most of these renewable energy sources have been fully or partially replaced by brown and black coal, petroleum, and natural gas due to their high energy contents, widespread availability, and ease of transportation of these sources of energy. Today, most energy is produced from fossil-based sources (i.e., petroleum and coal). One of the principal disadvantages of fossil-based fuels is the liberation of huge amounts of carbon dioxide as the oxidation product, which acts as a greenhouse gas.

There is scientific consensus that the global climate is changing, and the effects can be seen all over the world. Sea levels are increasing, glaciers are melting, and global temperatures are increasing. It seems likely that 2016 could be the hottest year since climate data have been recorded. Furthermore, 9 of the 10 hottest years have occurred during the past 15-year time span.[1] The carbon dioxide concentration has reached a level of more than 400 ppm, which is higher than that in the past 400,000 years.[1] All these data indicate a required shift toward carbon emission neutral industry, in part reflected in the increased utilization of biofuels. Utilization of biofuels offers the opportunity to reduce greenhouse gas emissions. Furthermore, the dependency on fossil-based energy sources can be reduced.

Research on liquid biofuels has already been conducted for more than 100 years. Nicolaus Otto and Rudolf Diesel invented engines that can be run with ethanol or plant oil, respectively.[2] Due to lower prices, mineral oil-based fuels have traditionally substituted these biofuels. However, during the oil crisis in the 1970s, the interest in renewable fuels increased and development of new production processes was started.[2] The first-generation biofuels were

Green Chemistry
http://dx.doi.org/10.1016/B978-0-12-809270-5.00029-7

produced from farm crops, directly. Due to the competition between fuel and food production, new processes were developed and are still being optimized. These second-generation biofuels are based on nonfood substrates such as grass, straw, or wood. Processing of these substrates presents significant challenges. Therefore, the third-generation biofuels, based on algae, are now being developed.

The following sections of this chapter will summarize the actual developments in biofuel production, including the chemistry behind some potentially transformative bioenergy sources.

3.24.2 BIOETHANOL PRODUCTION

Ethanol has been an important consumer product for many centuries. The first proof describes that ethanol was produced by the Chinese in approximately 7000 BC.[3] Wine, beer, mead, and other alcoholic beverages have been developed and are continually being improved around the world. For the majority of the time that societies have been producing alcoholic beverages, humans had no knowledge about the microorganisms that perform the fermentation. It can be assumed that these fermentations started spontaneously by the native microorganisms prevalent on berries or grain used in the generation of the product. The first observation of yeast was achieved by Antonie van Leeuwenhoek in 1680.[4] In 1815, Joseph Louis Gay-Lussac developed the chemical equation for the ethanolic fermentation[5]:

$$C_6H_{12}O_6 \rightarrow 2C_2H_5OH + 2CO_2$$

He also developed a process for the cleavage of cellulose resulting in the production of glucose.

Afterward, it took another 42 years until Louis Pasteur demonstrated in 1857 that alcoholic fermentation is conducted by living yeasts. Again, in 1899, Eduard Buchner proved that yeast cell extracts can cause the alcoholic fermentation.[6] For this observation he was awarded the Nobel Prize in chemistry in 1907. Further insight into the biochemical reactions associated with ethanol metabolism had been achieved by Arthur Harden and W. Young, who found a phosphorylated intermediate, the "Harden-Young Ester," which is otherwise known as fructose-1,6-bisphosphate. In 1929, Arthur Harden and Hans von Euler-Chelpin received the Nobel Prize for their investigation in enzymatic fermentation reactions.[6] In 1936, Otto Heinrich Warburg proved that reduced nicotinamide adenine dinucleotide (NADH) is required as a cofactor for the enzymatic reduction of acetaldehyde to ethanol. One year later, Erwin Negelein and Hans Joachim Wulff purified and crystalized the enzyme alcohol dehydrogenase, which converts acetaldehyde to ethanol.[6]

Already in 1876, ethanol was used to power the first engines by Nicolaus Otto.[2] Also, Henry Ford used ethanol to fuel his first cars. At the beginning of the 20th century, petroleum-based fuels were developed and replaced ethanol as fuel. In World War II, blends of petroleum-based gasoline and ethanol were applied due to fuel shortages. After World War II, the utilization of ethanol for combustion machines was reduced, due to widespread availability, and thus low cost, of petroleum-based fuels. In the 1970s, ethanol production was promoted, again due to a world oil crisis. Increased ecological awareness led to elevated ethanol production for blends of ethanol and petroleum-based fuels.[2] In 2015, approximately 100 million cubic meters of bioethanol were produced worldwide. The largest producers were the United States, Brazil, and Europe with 56 million cubic meters, 26 million cubic meters, and 5 million cubic meters, respectively.[7] The advantage of bioethanol is the possible reduction of carbon footprint compared with crude petroleum-based fuels. The bioethanol production processes that are developed must be cost efficient to compete with petroleum-based fuels. Societal and economic pressure on biofuel production has increased due to the strong reduction of petroleum prices from approximately US$110 per barrel in 2014 to US$40 per barrel in April 2016. Especially, the production of environment-friendly, second-generation bioethanol is affected by the strong decrease in petroleum price, due to the high production costs of the biofuel. Compared with first-generation bioethanol, larger monetary investments and higher processing costs are required for the production of second-generation bioethanol. Reboredo et al.[8] compared prices of first- and second-generation ethanol. First-generation bioethanol costs were US$0.30/L, while costs for second-generation bioethanol were at least US$0.60/L. Due to the reduction of oil prices, second-generation bioethanol is more expensive than fossil fuel at the moment.

However, even at the current low petroleum prices, the production of bioethanol is advantageous due to the reduction of the greenhouse gas carbon dioxide. Burning bioethanol will liberate approximately the same amount of carbon dioxide as had been captured from the atmosphere by the plants that were used as feedstocks (Fig. 3.24.1). Furthermore, utilization of bioethanol as fuel can reduce the societal dependency on fossil fuels, which are a limited source. Reducing dependency of these petroleum-based fuels will thus reduce the amount of petroleum needed per capita in a population that also relies on ethanol as a fuel.

FIGURE 3.24.1 The complete combustion of bioethanol releases carbon dioxide into the atmosphere, which is then taken up by plants via photosynthesis. Ethanologenic microbes take up components of plants via fermentation, beginning the cycle anew.

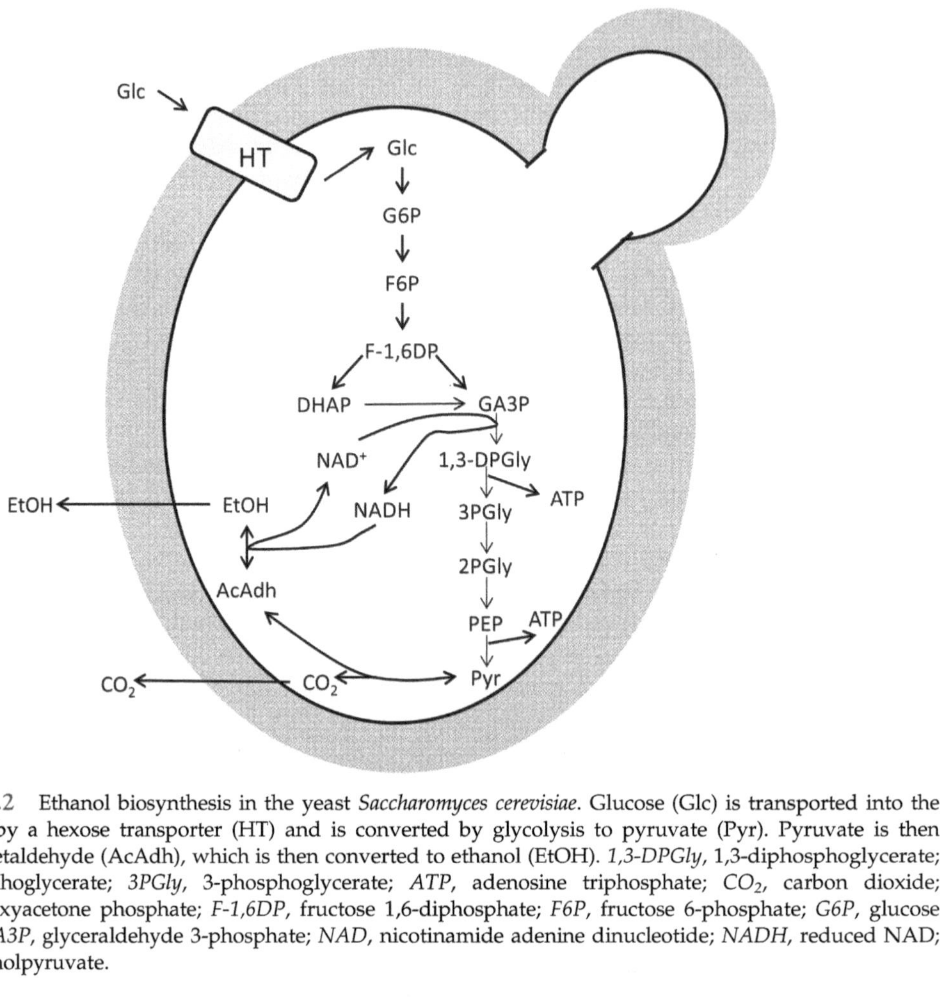

FIGURE 3.24.2 Ethanol biosynthesis in the yeast *Saccharomyces cerevisiae*. Glucose (Glc) is transported into the cell cytoplasm by a hexose transporter (HT) and is converted by glycolysis to pyruvate (Pyr). Pyruvate is then converted to acetaldehyde (AcAdh), which is then converted to ethanol (EtOH). *1,3-DPGly*, 1,3-diphosphoglycerate; *2PGly*, 2-phosphoglycerate; *3PGly*, 3-phosphoglycerate; *ATP*, adenosine triphosphate; *CO_2*, carbon dioxide; *DHAP*, dihydroxyacetone phosphate; *F-1,6DP*, fructose 1,6-diphosphate; *F6P*, fructose 6-phosphate; *G6P*, glucose 6-phosphate; *GA3P*, glyceraldehyde 3-phosphate; *NAD*, nicotinamide adenine dinucleotide; *NADH*, reduced NAD; *PEP*, phosphoenolpyruvate.

Fig. 3.24.2 displays a schematic illustration of the ethanol biosynthesis pathway found in the yeast *Saccharomyces cerevisiae.* First-generation bioethanol is produced by fermenting corn, sugar cane, sugar beet, or grain. All these plants deliver high amounts of fermentable sugars or easily degradable starch. Starch is saccharified by microbial enzymes, yielding the fermentable monosaccharide glucose. In most production plants, ethanol is produced by the yeast *S. cerevisiae* up to an average of 15% (v/v). Compared with possible other ethanol producers like *Zymomonas mobilis* or genetically modified *Escherichia coli* strains, *S. cerevisiae* offers some advantages. Yeast exhibits high acid tolerance, and cell viability in yeast cultures are not negatively affected by phages during fermentation process, avoiding problems such as those that may occur during bacterial fermentation processes. In addition to the desired ethanol product, the same amount of carbon dioxide is formed on a molar ratio when C6 sugars are fermented. The ethanol that is produced is then purified by distillation to up to 96% (v/v) purity, where the remaining water and the ethanol form an azeotrope mixture. Further purification by distillation requires the utilization of cyclohexane. In modern production plants, the remaining 4% (v/v) of water is removed using molecular sieves. These compounds (e.g., zeolite) incorporate water within their crystal structure, while ethanol cannot permeate into the mineral. There are attempts to use membranes for evaporation processes during recovery and purification of ethanol, but up to now, membranes are too expensive for an economically feasible industrial application.

The disadvantage of first-generation bioethanol is the land requirement for farming of sugar-rich crops. Large amounts of agricultural area are required worldwide for energy crops. In the United States, approximately 25% of total corn production is used for ethanol manufacturing. In this calculation, the utilization of corn side products for feeding animals is already considered. Without this consideration, approximately 40% of the total corn production is used for ethanol production.[9] These numbers indicate that there is a strong competition between food production, protection of natural forest areas, and the growth of energy plants for biofuel production. For this reason, the production of first-generation ethanol is limited, due to limited amounts of required substrates. Furthermore, reduction of total carbon dioxide emission is quite low when ethanol produced from cellulose is compared to traditional energy crops like grasses, corn and trees. Jones and Mayfield[10] compare the total greenhouse gas emissions of different first-generation bioethanol processes. Depending on the energy production (based on coal, gas, or plant material) required for harvesting and transport, the first-generation bioethanol pretreatment and production process may result in even higher greenhouse gas emissions than petroleum. Many efforts have been made to overcome the disadvantages of first-generation bioethanol and to achieve a sustainable, environment-friendly, and economically viable biofuel production scenario.

Since crops that produce first-generation bioethanol are specifically grown and harvested for the purpose of making biofuel, the energy required for seeding, fertilization, growth, harvesting, and transport of the resulting biomass must be taken into account in a life cycle analysis of the process. Thus, second-generation bioethanol production processes that use nonfood (i.e., waste) substrates result in a marked decrease in greenhouse gas emissions, compared with their first-generation counterparts. Strong research has been conducted for the development of second-generation ethanol. This ethanol is based on many different lignocellulosic and other sugar-rich substrates to reduce the "food versus fuel" competition. Furthermore, the availability of substrates allows for increased ethanol production. Possible feedstock substrates are: rice straw, corn stover, cotton stalk, wood, forest residues, poplar, switchgrass, bagasse, and paper pulp. According to Saini et al.[11] more than 1 billion ton of agricultural waste products are available per year for ethanol production. Utilization of agricultural waste products offers the possibility to produce food and a substrate for ethanol production simultaneously. Carbon emissions are very low with this strategy.

Utilization of many different types of feedstock requires sophisticated methods for the pretreatment of plant-derived substrates. The compositions of different plant materials are presented in Table 3.24.1.[11] During pretreatment of different feedstocks, the amount of crystalline cellulose also must be taken into account, due to the poor bioavailability of these molecules.

Compared with first-generation biofuel production processes, second-generation biofuel productions are much more complicated and pretreatment strategies are more expensive. Cellulose and hemicellulose can be used for ethanol production, while hydrolysis of lignin does not result in fermentable sugars. Cleavage of cellulose results in glucose molecules,

TABLE 3.24.1 Composition of Possible Feedstocks That Can Act as Substrates for Bioethanol Production

Material	Cellulose	Hemicellulose	Lignin
Algae (green)	20–40	20–50	–
Cotton	80–95	5–20	–
Grasses	25–40	25–50	10–30
Hardwoods	43–47	25–35	16–24
Hardwood barks	22–40	20–38	30–55
Softwoods	40–44	25–29	25–31
Softwood barks	18–38	15–33	30–60
Cornstalk	39–47	26–31	3–5
Wheat straw	37–41	27–32	13–15
Newspaper	40–55	25–40	18–30
Chemical pulp	60–80	20–30	2–10
Sorghum stalks	27	25	11
Corn stover	38–40	28	7–21
Coir	36–43	0.15–0.25	41–45
Bagasse	32–48	19–24	23–32
Rice straw	28–36	23–28	12–14
Wheat straw	33–38	26–32	17–19
Barley straw	31–45	27–38	14–19
Sorghum straw	32	24	13
Sweet sorghum bagasse	34–45	18–28	14–22

Numbers represent the amount of cellulose, hemicellulose, and lignin in %weight/weight. Quantities of other components, such as protein and minerals, are not presented.
The table is modified according to Saini JK, Saini R, and Tewari L. Lignocellulosic agriculture wastes as biomass feedstocks for second-generation bioethanol production: concepts and recent developments. 3 Biotech *2015;5:337–353.*

while hydrolysis of hemicellulose results in a variety of products such as xylose, arabinose, mannose, galactose, and glucose. Depending on the feedstock, different amounts of toxic lignin-derived cleavage products must be removed before the fermentation process can begin. Lignin is not degradable by the yeast *S. cerevisiae* and contains growth-inhibiting or even toxic components such as acetate, 5-hydroxymethylfurfural, and furfural. These substances can inhibit growth and ethanol formation during the fermentation process. Plant-derived cellulose and hemicellulose can be hydrolyzed by chemical pretreatment or with cellulases. Chemical pretreatment includes methods using sulfuric acid, ammonia, calcium hydroxide, or sulfur dioxide. All pretreatment methods may result in liberation of around 90% of fermentable C5 and C6 sugars.[12] Using enzymes to cleave cellulose offers the advantage of mild and nonoxidizing reaction conditions compared with the application of sulfuric acid. Hence, the requirements on the equipment are lower, resulting in reduced costs. Furthermore, application of enzymes can reduce costs of wastewater treatment or recycling of the applied acid. One challenge with enzyme pretreatment of feedstocks is product inhibition. Cellulase activity is inhibited by produced cellobiose, resulting in low yields of fermentable sugars from the pretreatment process. The solution for the product inhibition during saccharification is a process called simultaneous saccharification and fermentation (SSF). Enzymatic hydrolysis and microbial fermentation are conducted in the same vessel, simultaneously. The technology of SSF was invented and patented 40 years ago.[13] It offers the advantage that liberated sugar molecules are immediately fermented to produce ethanol. Hence the product inhibition of the cellulase or amylase enzyme does not occur. Compared with separate hydrolysis and fermentation processes, higher production rates, increased yield, and elevated concentrations of ethanol can be achieved by SSF.[14] For an optimal SSF process, enzymes and microorganisms must have comparable optima in pH values and temperatures. For hemicellulose hydrolysis, there have also been attempts to use fungi for microbial hemicellulose breakdown without the addition of pure enzymes.[15]

The C6 sugars that are liberated from lignocellulosic biomass can be readily fermented by *Saccharomyces* sp during ethanol production. To increase product yield and improve the economics of ethanol production, different strategies have been developed to utilize the C5 sugars that are liberated from hemicellulose as well as the C6 sugars.

Genetically modified microorganisms, especially *Saccharomyces* strains, have been developed for efficient utilization of C5 sugars. Bacterial and/or fungal enzymes (xylose reductase and xylitol dehydrogenase from fungi, or xylose isomerase from bacteria) have been incorporated into yeast to extend sugar utilization capabilities. Moysés et al.[16] offer a summary of genetically modified yeast strains that have been developed for xylose fermentation. Besides the genetic incorporation of the aforementioned enzymes, sugar transporter proteins must be expressed or modified in their substrate selectivities or transport rates. Inhibition of xylose uptake in the presence of glucose results in a prolonged fermentation processes. This kind of inhibition must be overcome in cultures growing on lignocellulosic feedstock. Furthermore, undesired side reactions must be reduced, and cofactor recycling has to be adapted to the new pathways in the genetically modified yeast. It was observed that fermenting yeasts are much more sensitive to growth-inhibiting factors from lignin when they ferment xylose, compared with when they ferment glucose. Under industrial fermentation conditions, the yeast strain has to cope with many additional stress factors, such as temperature, high concentrations of acetic acid and ethanol, as well as increases in ionic strength of the fermentation media. Hence, the genetic construction of industrially applicable strains is very challenging, and further improvement is required.[16]

Research is conducted with non-*Saccharomyces* strains that are able to utilize C5 and C6 sugars. Biswas et al.[17] report a fermentation process using the yeast *Pichia stipitis* CBS6054 as a biocatalyst that uses undetoxified cleavage products of steam-exploded bagasse. The authors were able to demonstrate production of ethanol concentrations of up to 18.7 g/L. Further development is required to create an economic process at an industrial scale. Also, genetically modified microorganisms and adapted fermentation strategies are required for the efficient production of second-generation ethanol, due to the high concentration of salts in many substrates. Casey et al.[18] investigated the influence of different ions on the production of ethanol by a genetically modified strain of *Saccharomyces* that is able to utilize both C5 and C6 sugars. This group found that, due to the concentrations of chloride, sulfate, sodium, potassium, and ammonium, growth rate and ethanol production rates decreased due to a reduced consumption rate of xylose. The presence of 0.5 M NaCl resulted in a 65% decrease of xylose utilization.

The bacterium *Z. mobilis* is a very promising candidate for ethanol production. It has been used traditionally for the production of the beverages tequila and African palm wine. Naturally, *Z. mobilis* exhibits a high ethanol tolerance. Ma et al.[19] reported about the utilization of an acid-tolerant strain used for the production of ethanol from food waste. The authors demonstrated production of up to 100 g/L of ethanol under nonsterile conditions.

Commonly applied wild-type strains of *S. cerevisiae* and *Z. mobilis* lack the native ability to utilize pentose sugars, which represent a significant fraction of sugars liberated from hemicellulose. *E. coli* is able to utilize hexoses, pentoses, and also glycerol (a common waste product from biodiesel production), but naturally only very small amounts of ethanol are produced. Genetically engineered strains are able to produce ethanol from different substrates with high yield. Some strains are even able to express the required enzymes for hemicellulose hydrolysis.[20,21] Hence addition of hydrolytic enzymes is not required during pretreatment. Despite extensive research involving recombinant *E. coli* as an ethanol-producing microorganism, the problem of comparatively low resistance to high concentrations of the ethanol product limits the application of these recombinant strains for commercial ethanol production processes.

Besides the utilization of genetically modified microorganisms, the application of controlled cocultures can be beneficial for ethanol production, especially in the production of second-generation bioethanol. Controlled cocultures consist of at least two different microorganisms. Hence the metabolic functions of both microorganisms can be combined during the fermentation process. Utilization of nonrecombinant strains simplifies regulatory specifications and the disposal of biomass produced in countries with strict regulations regarding genetically modified organisms. Much research has been conducted in this field. Specifically, SSF processes can be performed using different microorganisms instead of with pure enzymes. Paschos et al.[15] used a controlled coculture consisting of *Fusarium oxysporum* and *S. cerevisiae* for the simultaneous hydrolysis and fermentation of liquefied wheat straw where *F. oxysporum* provides the lignocellulose hydrolytic capabilities and *S. cerevisiae* converted fermentable sugars to EtOH. These researchers achieved 58 g/L of final ethanol concentration without the use of additional enzymes. In a much older study, Abate et al.[22] reported about the increase of ethanol yield when using a coculture of *S. cerevisiae* and *Z. mobilis*, when compared with yields achieved by monoculture. Coculture fermentation processes offer a cost-efficient possibility of ethanol production, especially for second-generation processes when saccharification and fermentation can be conducted in one simultaneous process.

As mentioned earlier, it is possible for microorganisms in a modified ethanol production process to utilize C5 sugars from lignocellulosic material. After a suitable pretreatment, both sugars can be fermented to ethanol in a single fermentation step. In 2013, the mixed fermentation of C5 and C6 sugars could be demonstrated at an industrial scale of 270 m^3. Compared with single C6 fermentation, a 40% increase in ethanol yield per ton straw could be achieved. Companies, such as DSM and DONG Energy are currently active in this field and have developed the requisite microorganisms and processes.[23]

In the United States, one of the largest plants for cellulosic bioethanol production was built and opened in November 2015, with an annual production of more than 110 million liters. Corn stover is used as the feedstock for production. The pretreatment is conducted using ammonia followed by enzymatic hydrolysis. Ammonia treatment offers several advantages such as decrystallization of cellulose, partial hydrolysis of hemicellulose, and disruption of fiber structure,[24] all of which would allow for better access of hydrolytic enzymes to the polysaccharide structure. Furthermore, most of the ammonia used for this pretreatment process can be recycled. The remaining ammonia serves as a nitrogen source during the fermentation process.[24] In this production process, the following fermentation is performed using the bacterium *Z. mobilis*. Leftover lignocellulose is dried and can be burned for energy production.

Another example of second-generation bioethanol is the utilization of saw dust as substrate. In 2015, it was announced that North European Bio Tech Oy (NEB) will build a plant for the annual production of 10 million liters of bioethanol in Finland. The substrate will be processed via steam explosion and enzymatic hydrolysis to gain the required sugars. Following feedstock treatment, an ethanol fermentation process will be conducted.[25]

Substrates for second-generation bioethanol can also be optimized for the desired degree of hydrolysis of sugars and a reduced liberation of growth-inhibiting substances from lignin. Historically, development of farm crops has always been done to achieve higher yields of a desired product. In this regard, no selection or breeding has been done to achieve easily degradable straw. Rendleman et al.[26] reported that large seed-producing companies are planning to characterize their crop regarding not only the productivity in grain but also in fermentability of the straw. Just by selecting highly fermentable varieties of corn, the yield in fermentation could be increased by 6% when compared with varieties that did not undergo a selection. Torres et al.[27] published results about diversity in cell wall composition and cell wall digestibility of various cultivars of maize. Some cultivars exhibited much better hydrolysis properties of maize straw due to increased hemicellulose concentration and reduced lignin content, resulting in reduced amount of required chemicals and lower temperatures. Using these cultivars enables a reduction of processing costs of up to 15% during pretreatment and saccharification. Also, Barros-Rios[28] compared variations of maize in hydrothermal pretreatment of straw for the purpose of ethanol production. They suggest that breeding of food crops for increased biomass yields higher amounts of substrate for bioethanol production from straw.

Allwright and Taylor[29] discuss possible optimization of energy plants, such as poplar, with the goal of reduced lignin content, low crystallinity of cellulose, and increased growth density of the plant. Additionally, tolerance against drought, salinity, and diseases can be increased. Also, efficient use of water and nitrogen can be achieved by breeding techniques.

For the production of first-generation bioethanol, optimization of the plants is beneficial to reduce production costs. Zabed et al.[30] present results of high-sugar-containing varieties of corn that allow for a reduction in the amount of amylolytic enzymes used during the saccharification process, which leads to a reduction of processing costs. Until now, sweet corn had exhibited lower grain yields when compared with many common varieties of edible corn. Hence, sweet corn is not typically utilized for bioethanol production. Therefore, the authors suggest greater efforts in breeding to combine the high yields of newly developed sweet corn varieties with the reduced amount of required enzymes during ethanol production process.

3.24.3 HIGHER ALCOHOLS

Higher alcohols (i.e., alcohols that possess a longer carbon chain than methanol and ethanol) are more desirable as biofuel compounds because they possess higher energy content and are less hygroscopic. Some of these higher alcohols can be produced by wild-type organisms, whereas the production of others relies on production strain engineering and unique fermentation processes. Although many longer chain higher alcohols are made naturally as by-products and contribute to the complexity of flavors and fragrances,[31] we will focus on the two higher alcohols that have attracted much attention as biofuel molecules over the past decade, *n*-butanol and isobutanol.

3.24.3.1 *n*-Butanol

There are a variety of applications for the industrially significant compound *n*-butanol (referred to as butanol for the remainder of the chapter). Butanol is used to produce methacrylate esters and butyl acrylate for coatings and enamels, and derivatives of butanol are used to make butyl ether plasticizers and other compounds.[32] An important application that is relevant for this chapter is that butanol can be used as a direct replacement of gasoline in existing automobile engines.[32,33] Butanol contains 29% more energy than ethanol (27 MJ/L, compared to 21 MJ/L)[32,34] and is less hygroscopic than ethanol, which results in less engine corrosion.[35] Butanol can be produced synthetically from petroleum or fermentatively by wild-type or engineered microorganisms. Due to instability in petroleum costs, the fermentative route for butanol production has been examined with an eye upon improving yields and productivities in processes involving wild-type or engineered organisms. Butanol production by microbial fermentation was first reported by Louis Pasteur in 1861.[36] Butanol is produced by microorganisms using the acetone-butanol-ethanol (ABE) fermentation scheme. As the name of the pathway implies, acetone and ethanol are coproducts in butanol production. The typical ratio of acetone, ethanol, and butanol in this process is 3:6:1.[32] Fig.3.24.3 shows a schematic pathway of ABE production from acetyl-CoA as a key intermediate. It is believed that the only wild-type organisms known to produce butanol are a subset of bacteria known as solventogenic clostridia. These microbes make up a subset of the genus *Clostridium*, with *Clostridium acetobutylicum* and *Clostridium biejerinckii* being the best-studied examples.[37–39] ABE fermentation typically occurs as a result of anaerobic breakdown of sugar substrates by these solventogenic clostridia under the right conditions.[33] These clostridia are able to utilize a variety of sugar substrates, including glucose, mannose, galactose, and xylose, although the last of these is not utilized efficiently.[39] However, given the range of substrate utilization by these organisms, much research has been performed on the utilization of waste carbon substrates like lignocellulosic biomass for butanol fermentation. Starch has also been used as a substrate for butanol fermentation. However, starch is not truly a waste carbon source, so processes involving starch utilization for biofuels are likely to be nonviable. Some clostridia, namely, *Clostridium ljungdahlii* and *Clostridium carboxidivorans*, have even been shown to utilize the carbon in syngas (a mixture of CO, CO_2, and H_2 produced by coal-fired power plants). However, both species must be further engineered to produce significant quantities of butanol.[40]

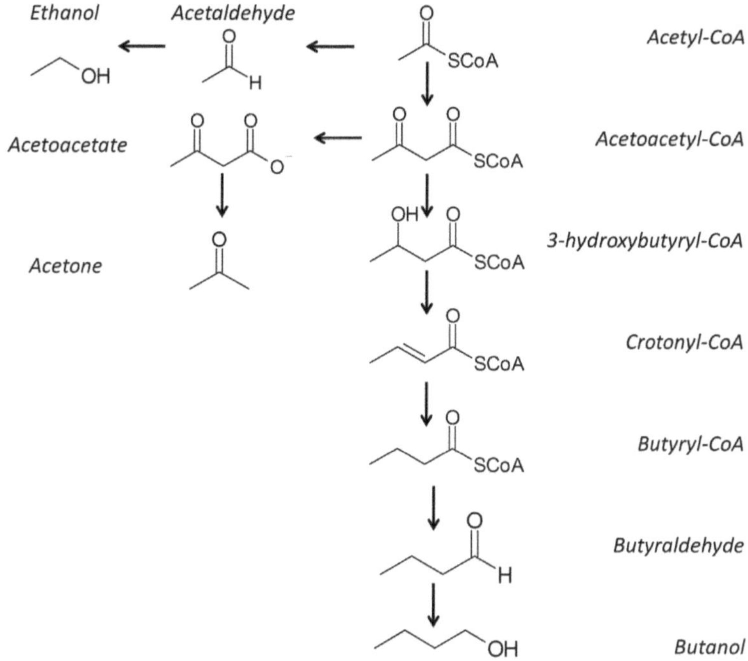

FIGURE 3.24.3 Schematic of the acetone-butanol-ethanol fermentation pathway.

Since the butanol yield from solventogenic clostridia is not high enough to justify the organisms' use as biofuel production biocatalysts, metabolic engineering must be performed. For the syngas-utilizing *C. ljungdahlii*, which does not produce butanol, a heterologous butanol synthesis pathway from *C. acetobutylicum* was inserted, resulting in production of ~0.5 mM butanol from syngas by the engineered *C. ljungdahlii* strain.[40,41] Metabolic engineering of a biocatalyst with the goal of increasing production typically begins with the elimination or downregulation of competing pathways. *C. acetobutylicum* was engineered to produce higher titers of butanol by removal of the genes that encode the first four steps of the short-chain fatty acid production pathway (both pathways use acetyl-CoA as a key intermediate). A titer of 16 g/L was produced from a multiple gene knockout strain.[42,43] Another aspect to consider when performing strain engineering for alcohol-producing microorganisms is product tolerance. Butanol tolerance has been sought by researchers looking to increase the alcohol titer in solventogenic clostridial strains. Butanol is a lipophilic solvent, and in large quantities, it can literally dissolve cell membranes of the strain that produces it. In the laboratory, ABE fermentation with clostridia has been shown to produce a maximum of 20 g/L butanol. Thus more tolerant production strains are sought to be able to withstand higher butanol production levels. Using evolutionary engineering with the help of mutagenesis, a strain of *C. acetobutylicum* was isolated that could tolerate up to 4% butanol.[44]

In general, clostridia lack sufficient genetic tools and tractability, making them less-than-ideal targets for metabolic engineering. Many other species have been examined for butanol production, including *Saccharomyces cerevisiae*, *Pseudomonas putida*, *Ralstonia eutropha*, among others.[38,45,46] The genetically tractable *E. coli* has been engineered for butanol production. Several genes were introduced to allow for butanol production from acetyl-CoA. Furthermore, three different competing pathways for acetyl-CoA or NADH were hindered or eliminated to increase butanol titer.[47] Some species, like *R. eutropha*, already possess part of the pathway that can lead to butanol synthesis, although more pathways competing for acetyl-CoA or NADH could hinder production titers.

3.24.3.2 Isobutanol

Another higher alcohol of interest as a biofuel compound is the branched-chain isobutanol. Unlike butanol, wild-type microorganisms generally do not synthesize isobutanol in significant quantities. However, engineered strains that express a heterologously expressed production pathway have been developed. A representative isobutanol production pathway is illustrated in Fig.3.24.4. As a fuel, isobutanol has an energy density that is 95% that of gasoline (32 MJ/L, compared with 36 MJ/L for gasoline) and, like butanol, can be delivered to automobiles by already-existing fuel delivery infrastructure.[48]

The design of a fermentative route for isobutanol production relies on intermediates fed from the branched-chain amino acid biosynthesis pathways. Many host organisms have been engineered to produce isobutanol, including *E. coli*, *R. eutropha*, *Corynebacterium glutamicum*, *S. cerevisiae*, and others.[40,49] For the majority of reported metabolic engineering efforts, the key intermediate is a 2-ketoacid that is then converted to a primary aldehyde, which is

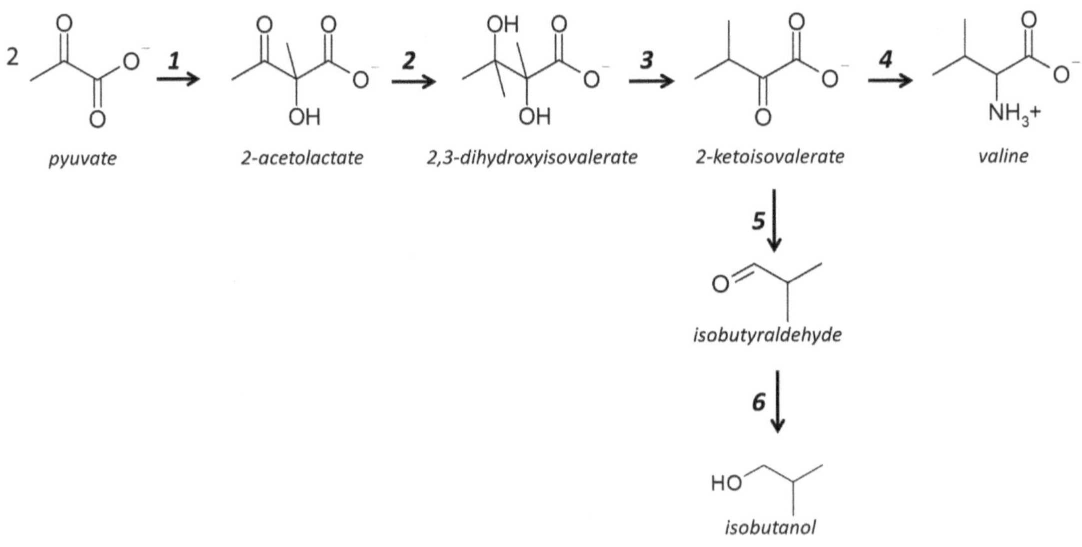

FIGURE 3.24.4 Design of an isobutanol synthesis pathway in a biocatalyst microorganism utilizes valine biosynthesis intermediates, specifically 2-ketoisovalerate. The key intermediate 2-ketoisovalerate is decarboxylated by a heterologously expressed ketoisovalerate decarboxylase (KivD, 5) to produce isobutyraldehyde. The isobutyraldehyde is then converted to isobutanol by a native or heterologously expressed alcohol dehydrogenase (Adh, 6). Other enzymes in the pathway are acetohydroxyacid synthase (AHAS, 1), acetohydroxyacid isomereductase (AHAIR, 2), dihydroxyacid dehydratase (DHAD, 3), and transaminase (TA, 4).

then converted to the alcohol molecule of interest.[35,43] For enzymatic conversion of the 2-ketoacid to an aldehyde, the gene that is typically expressed, *kivD*, encodes a ketoacid decarboxylase. This gene was isolated and characterized from *Lactococcus lactis*, and plays a central role in the native organism in the cheese-making process.[35,50] Following the aldehyde synthesis by KivD, an alcohol is produced by the action of an alcohol dehydrogenase enzyme.[35] *R. eutropha*, traditionally a biopolymer-producing organism, was engineered to synthesize isobutanol. The polyhydroxyalkanoate biosynthesis pathway genes were removed, and *kivD* was expressed heterologously. Since *R. eutropha* appears to possess several short-chain alcohol dehydrogenase homologs, a native Adh was used to complete the isobutanol biosynthesis pathway.[49] The *adh* gene used for the strain construction in this case was found to be expressed in *R. eutropha* only under anaerobic conditions. However, mutant strains that constitutively express Adh were isolated previously,[51] so these strains were used as the basis for construction of an isobutanol production variant of *R. eutropha*. The resulting strain was able to produce isobutanol using fructose as the main carbon source. Ketoisovalerate from the valine biosynthesis pathway was used to produce the final product. A maximum titer of 14 g/L isobutanol was obtained. Concomitantly, smaller quantities of 3-methyl-1-butanol were also produced, via the 2-ketoisocaproate precursor from the isoleucine synthesis pathway.[49] The resulting production strain then evolved to tolerate larger quantities of extracellular isobutanol.[52] Some of the mutations that occurred in the tolerant strains were analogous to those seen in isobutanol-tolerant *E. coli* strains, as published earlier,[53] suggesting that mechanisms of tolerance are similar in both organisms. *R. eutropha* isobutanol tolerance was improved from 1.0% to 2.0%.[52]

As with butanol or any biofuel, to make a fermentative production process viable, costs of operation must be minimal and productivities must be high. Thus researchers look to utilize growth and production substrates from waste streams as a plentiful supply of often-underutilized carbon. *R. eutropha* is capable of growth on carbon substrates like CO_2 and formic acid, suggesting that it could act as a biocatalyst for conversion of these inexpensive carbon sources to isobutanol. Indeed, laboratory-scale fermentations have been performed using CO_2 and electrically split water (supply of O_2 and H_2) to produce formate, which is then fed to engineered *R. eutropha* cells to produce isobutanol.[54] *R. eutropha* as a biofuel producer has been a key player in the "electrofuels" trend that has been supported by the US Department of Energy. The question that remains now for electrofuels researchers, with proof of principle established, is how to significantly improve isobutanol productivity.

3.24.4 BIODIESEL: LIPID-DERIVED BIOFUEL

As mentioned at the beginning of the chapter, many early biofuels are plant based (e.g., wood). Also, plant oils have been used as a biofuels, due to their high energy content. These lipid-based fuels are generally referred to as biodiesel, as the combustion of these oils could be used to power Rudolf Diesel's eponymous engine. These days, commercially produced biodiesel is typically made by transesterification of lipids using chemical and enzymatic catalysts and ethanol or methanol. Chemical catalysts used are typically strong bases like sodium or potassium hydroxide.[55] The chemical treatment of lipids with strong base and alcohol to produce biodiesel esters is highly energy intensive,[56] and thus not ideal as a biofuel synthesis method. Enzymatic transesterification is generally preferable as a production method, as it minimizes the synthesis of contaminating by-products.[57] The use of plant oils as the carbon feedstock from which to make biodiesel is controversial in that plant oils themselves are valuable as cooking oil. Also, virgin plant oils generally are too expensive to be used as a biodiesel substrate and are generally prized as cooking oils. Thus it is much more cost-effective to utilize waste cooking oil as the feedstock for biodiesel production.[57]

A schematic of conversion of lipids to fatty acid methyl esters (FAMEs), the biodiesel molecules, is shown in Fig. 3.24.5. The lipase enzyme will catalyze the hydrolysis of the bond between the fatty acid and the glycerol backbone of a triacylglycerol (TAG), creating a new covalent bond with an alcohol substrate, producing a fatty acid alcohol ester and a diacylglycerol (DAG). For biodiesel production, large quantities of lipase can be immobilized to allow for large-scale synthesis of FAMEs.[55]

One group of microorganisms that has attracted much attention as a biocatalyst in biodiesel production is microalgae. These organisms are favorable for several reasons: (1) they can produce TAGs that can be used as feedstock for biodiesel production, (2) several species can grow in saline (brackish) water or waste water, and (3) they can grow and produce TAGs photosynthetically using CO_2 as the main carbon source. Some species of microalgae can produce up to 60% of their dry weight as lipid, whereas oilseed crops are often only 5%–10% oil

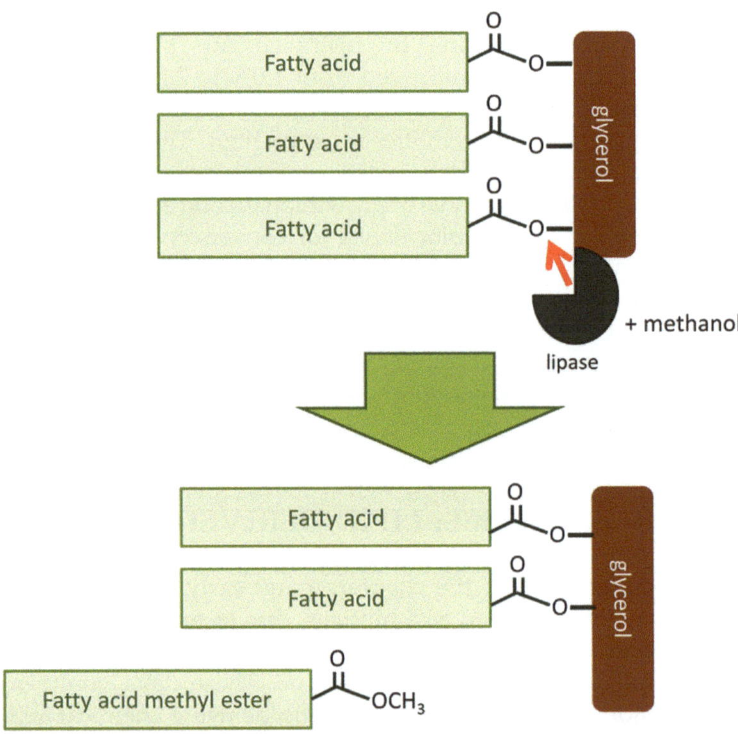

FIGURE 3.24.5 Lipase will cleave a fatty acid off of a triacylglycerol to produce a diacylglycerol (DAG) and a free fatty acid. In the presence of an alcohol like methanol, the products would be a DAG and a fatty acid methyl ester, which is a biofuel molecule.

by weight.[58] However, even with all these positive aspects of microalgal TAG production, there are still many challenges to overcome before a cost-effective process is introduced. Many conditions that can be applied to increase the yield of lipid to >50% involve some kind of nutrient limitation. This is a production condition that is used sometimes in bacterial polyhydroxyalkanoate bioplastic production.[59,60] However, nutrient limitation of microalgae cultures can severely limit the amount of biomass produced, in spite of increasing the cellular yield of TAG. One way to overcome this deficiency is to perform a two-stage culture in which biomass is accumulated and then nutrient limitation is applied to allow for production of TAG. Generally, however, the two-stage culture is less cost effective, so additional means must be taken to increase biomass and TAG yield, thus producing costcompetitive biodiesel.[58] Another issue with photosynthetic cultivation of microalgae is slow growth. Production of a significant amount of TAG for biodiesel requires a lot of time, decreasing the cost-effectiveness. This concern can be alleviated by using some species of algae that can grow on heterotrophic carbon sources. Specifically, if microalgae can grow on waste carbon sources, like organic acids, batch cultivation time can be shortened and TAG productivity can be increased. Such is the case of *Chlorella protothecoides*, which can be cultivated heterotrophically using mixed organic acids as the main carbon source.[61] Another challenge is the harvesting of algal biomass from large ponds. Centrifugation could be employed for this purpose, but this unit operation would clearly drive up process costs. To increase efficiency of biomass harvest, flocculation can be performed. However, chemical flocculants like $FeCl_3$ or $AlCl_3$ are harmful to the environment. Researchers have had some success with natural flocculants like chitosan.[58] Chitosan is a charged polysaccharide obtained from the shells of marine crustaceans.[62] Use of chitosan has been shown to be effective for flocculation of freshwater *Chlorella* sp, among others.[63] For algae cultivated in photosynthetic cultures, large surface area is required for the production of significant amounts of biomass and thus significant amounts of TAG. Traditionally, photosynthetic cultures are grown in raceway ponds that allow exposure of maximum surface area to sunlight. However, this presents another challenge in microalgae cultivation, in that raceway ponds require a lot of land usage. This has been addressed by design of algae cultivation systems that rely on synthetic lighting of cultures in photobioreactors that can use less water without sacrificing surface area.[64]

Some heterotrophic microorganisms are capable of producing a large amount of intracellular TAG that can be used for biodiesel production. One such organism is the Gram-positive, oleaginous bacterium *Rhodococcus opacus*. *R. opacus* has been shown to accumulate over 80% of its cell dry weight as TAGs, which are stored intracellularly in lipid bodies.[65] The bacterium can grow on a wide variety of carbon sources, including monosaccharides, disaccharides, organic acids, alkanes, and phenolic compounds.[65,66] *R. opacus* can grow in the presence of high concentrations of carbon source, resulting in a high-cell-density culture.[67] Also, *R. opacus* has exhibited some amenability to metabolic engineering as the carbon substrate range has been broadened to include xylose,[68] cellobiose,[69] and glycerol.[70,71] Research groups have been developing a TAG production scheme using *R. opacus* as the biocatalyst to convert waste carbon like lignocellulose or chitin to cost-effective biodiesel.[68,70,72]

While the production of bio-based diesel fuel is attractive, how do the combustion properties of biodiesel compare with traditional, petroleum-based diesel fuel? Cetane number (CN) is a dimensionless indicator of the ignition quality of a diesel fuel. The scale of CN is arbitrary, but most American engine manufacturers designate an acceptable range for fuels of a CN of around 50.[73] Petroleum-based diesel fuel has an average CN of 50.[73] Based on the fatty acid contents of *R. opacus* TAGs, the predicted CN for biodiesel produced from these TAGs is ~62.[72] This number is within the acceptable range of operation for biodiesel.

Biodiesel is prized because it could be an acceptable replacement for petroleum should reserves run low. The general consensus on biodiesel is that it burns "cleaner" with less toxic by-products being produced. In general, higher the biodiesel content of a fuel mixture, the less is the particulate matter, carbon monoxide, and hydrocarbons produced.[73] This suggests that use of biodiesel as an additive in fuel can result in a substantial reduction in emissions. Not all emissions are reduced when biodiesel is added to the fuel mixture. Oxides of nitrogen are found in increased concentrations when biodiesel is burned.[73] A study of exposure of *Salmonella typhimurium* cells to FAME combustion particulates demonstrated lower incidents of mutagenicity.[73] In more recent studies, particulate matter produced by combustion of 100% biodiesel was shown to cause elevated immunogenic effects in mice.[74] In general, more study is needed to evaluate the health effects of exposure to biodiesel combustion by-products.

3.24.5 CONCLUSION AND FUTURE PROSPECTS

The main question with biofuels does not center on a novel process, on a necessarily energy-dense molecule, or on a specific biocatalyst. The question that is most important in these biofuels research endeavors is "will it matter?" We can design an efficient ethanol production process. We can design a butanol or isobutanol production process. We can produce TAGs using waste carbon or photosynthesis. Will a microbial-centered biofuels synthesis process ever produce enough fuel at a low-enough cost to compete with petroleum-based fuels? Will energy sources produced by fermentation even be able to compete with other renewable fuel sources like solar or wind power? To begin to answer these questions, we must examine these processes and perform life cycle analyses that help shed light on the economic competitiveness and the environmental impacts of biofuel productions.

One specific problem with biofuels research that seems to pop up is the inability to assess the impact of biofuel use. A pessimist may be quick to point out the lack of clear impact of a biofuel in a national and/or global fuel market. In response to this, and perhaps a bit contradictively, there is an argument against thinking too big in regard to fledgling biofuels research. Especially with a process of waste carbon conversion to biofuel, the argument is in favor of acting locally. Local business and agriculture ecosystems have their own specific waste streams. Tailoring a bioproduct synthesis process to effectively use these locally produced waste streams as feedstock should be the main focus. National incorporation of biofuel usage should only be considered once a suitable infrastructure is in place (e.g., Brazilian ethanol production).

Overall, there is an uncertainty in how much of the energy "pie" will consist of bio-based energy-like biofuels. The considerations we as individuals must make is, are we really in a position to deny the potential impact of a new prospective energy source? True, there is an argument about cost-effectiveness of specific energy sources, but if one should appear promising, the answer to "how much of the pie?" should be, "who cares, it's pie!!!"[75]

References

1. Shaftel H. *Global climate change*. 2016. http://climate.nasa.gov/.
2. Songstad DD, Lakshmanan P, Chen J. Historical perspective of biofuels : learning from the past to rediscover the future. *In Vitro Cell Dev Biol – Plant* 2009;**45**:189–92.
3. McGovern PE, Zhang J, Tang J, Zhang Z, Hall GR, Moreau RA, et al. Fermented beverages of pre- and proto-historic China. *Proc Natl Acad Sci USA* 2004;**101**:17593–8.
4. Suh S-O, Blackwell M, Kurtzman CP, Lachance M-A. Phylogenetics of Saccharomycetales, the ascomycete yeasts. *Mycologia* 2006;**98**:1006–17.
5. Barnett JA. A history of research on yeasts 1: work by chemists and biologists 1789–1850. *Yeast* 1998;**14**:1439–51.
6. Barnett JA. A history of research on yeasts 5: the fermentation pathway. *Yeast* 2003;**20**:509–43.
7. *World fuel ethanol production*. 2016. http://www.ethanolrfa.org/resources/industry/statistics/#1454098996479-8715d404-e546.
8. Reboredo FH, Lidon F, Pessoa F, Ramalho JC. The fall of oil prices and the effects on biofuels. *Trends Biotechnol* 2016;**34**:3–6.
9. Mumm RH, Goldsmith PD, Rausch KD, Stein HH. Land usage attributed to corn ethanol production in the United States: sensitivity to technological advances in corn grain yield, ethanol conversion, and co-product utilization. *Biotechnol Biofuels* 2014;**7**:61.
10. Jones CS, Mayfield SP. *Our energy future: introduction to renewable energy and biofuels*. Oakland, California: University of California Press; 2016.
11. Saini JK, Saini R, Tewari L. Lignocellulosic agriculture wastes as biomass feedstocks for second-generation bioethanol production: concepts and recent developments. *3 Biotech* 2015;**5**:337–53.
12. Galbe M, Zacchi G. *Biofuels*. Berlin, Heidelberg: Springer Berlin Heidelberg; 2007. http://dx.doi.org/10.1007/978-3-540-73651-6.
13. Gauss WF, Suzuki S, Takagi M. United States patent 3,990,944. 1976:3–5.
14. Neves Ma, Kimura T, Shimizu N, Nakajima M. State of the art and future trends of bioethanol production. *Dyn Biochem Process Biotechnol Mol Biol* 2007;**1**:1–14.
15. Paschos T, Xiros C, Christakopoulos P. Simultaneous saccharification and fermentation by co-cultures of *Fusarium oxysporum* and *Saccharomyces cerevisiae* enhances ethanol production from liquefied wheat straw at high solid content. *Ind Crops Prod* 2015;**76**:793–802.
16. Moysés DN, Reis VCB, de Almeida JRM, de Moraes LMP, Torres FAG. Xylose fermentation by *Saccharomyces cerevisiae*: challenges and prospects. *Int J Mol Sci* 2016;**17**:1–18.
17. Biswas R, Uellendahl H, Ahring B. Conversion of C6 and C5 sugars in undetoxified wet exploded bagasse hydrolysates using *Scheffersomyces* (*Pichia*) *stipitis* CBS6054. *Amb Express* 2013;**3**:42.
18. Casey E, Sedlak M, Ho NWY, Mosier NS. Effect of acetic acid and pH on the cofermentation of glucose and xylose to ethanol by a genetically engineered strain of *Saccharomyces cerevisiae*. *FEMS Yeast Res* 2010;**10**:385–93.
19. Ma K, Ruan Z, Shui Z, Wang Y, Hu G, He M. Open fermentative production of fuel ethanol from food waste by an acid-tolerant mutant strain of *Zymomonas mobilis*. *Bioresour Technol* 2016;**203**:295–302.
20. Ryu S, Karim MN. A whole cell biocatalyst for cellulosic ethanol production from dilute acid-pretreated corn stover hydrolyzates. *Appl Microbiol Biotechnol* 2011;**91**:529–42.
21. Muñoz-Gutiérrez I, Oropeza R, Gosset G, Martinez A. Cell surface display of a β-glucosidase employing the type v secretion system on ethanologenic *Escherichia coli* for the fermentation of cellobiose to ethanol. *J Ind Microbiol Biotechnol* 2012;**39**:1141–52.
22. Abate C, Callieri D, Rodriguez E, Garro O. Ethanol production by a mixed culture of flocculent strains of *Zymomonas mobilis* and *Saccharomyces* sp. *Appl Microbiol Biotechnol* 1996;**45**:580–3.
23. *DSM proves cellulosic bio-ethanol fermentation on industrial scale with 40% higher yield*. 2013. http://www.dsm.com/corporate/media/informationcenter-news/2013/12/2013-12-04-dsm-proves-cellulosic-bio-ethanol-fermentation-on-industrial-scale-with-40-percent-higher-yield.html.
24. Yang B, Wyman CE. Pretreatment: the key to unlocking low-cost cellulosic ethanol. *Biofuels Bioprod Biorefining* 2008;**2**:26–40.
25. Snook A. *Cellunolix ethanol plant to be built in Finland*. 2015. http://www.canadianbiomassmagazine.ca/biofuel/cellunolix-ethanol-plant-to-be-built-in-finland-5188.
26. Rendleman CM, Shapouri H. *New technologies*. 2007. http://dx.doi.org/10.1016/S0959-8049(15)30013-7.

27. Torres AF, Slegers PM, Noordam-Boot CMM, Dolstra O, Vlaswinkel L, Boxtel AJB, et al. Maize feedstocks with improved digestibility reduce the costs and environmental impacts of biomass pretreatment and saccharification. *Biotechnol Biofuels* 2016;**9**:1.
28. Barros-rios J. Second-generation bioethanol of hydrothermally pretreated stover biomass from maize genotypes. *Biomass Bioenergy* 2016;**90**:42–9.
29. Allwright MR, Taylor G. Molecular breeding for improved second generation bioenergy crops. *Trends Plant Sci* 2016;**21**:43–54.
30. Zabed H, Faruq G, Sahu JN, Boyce AN, Ganesan P. A comparative study on normal and high sugary corn genotypes for evaluating enzyme consumption during dry-grind ethanol production. *Chem Eng J* 2016;**287**:691–703.
31. Nozzi NE, Desai SH, Case AE, Atsumi S. Metabolic engineering for higher alcohol production. *Metab Eng* 2014;**25**:174–82.
32. Zheng J, Tashiro Y, Wang Q, Sonomoto K. Recent advances to improve fermentative butanol production: genetic engineering and fermentation technology. *J Biosci Bioeng* 2015;**119**:1–9.
33. Jang YS, Malaviya A, Cho C, Lee J, Lee SY. Butanol production from renewable biomass by clostridia. *Bioresour Technol* 2012;**123**:653–63.
34. Fei Q, Fu R, Shang L, Brigham CJ, Chang HN. Lipid production by microalgae *Chlorella protothecoides* with volatile fatty acids (VFAs) as carbon sources in heterotrophic cultivation and its economic assessment. *Bioprocess Biosyst Eng* 2014;**38**:691–700.
35. Brigham CJ, Gai CS, Lu J, Speth DR, Worden RM, Sinskey AJ. Engineering *Ralstonia eutropha* for production of isobutanol from CO_2, H_2, and O_2. In: *Advanced biofuels and bioproducts*. New York, NY: Springer New York; 2013. p. 1065–90.
36. Jones DT, Woods DR. Acetone-butanol fermentation revisited. *Microbiol Rev* 1986;**50**:484–524.
37. Schiel-Bengelsdorf B, Montoya J, Linder S, Dürre P. Butanol fermentation. *Environ Technol* 2013;**34**:1691–710.
38. Xue C, Zhao XQ, Liu CG, Chen LJ, Bai FW. Prospective and development of butanol as an advanced biofuel. *Biotechnol Adv* 2013;**31**:1575–84.
39. Gu Y, Jiang Y, Yang S, Jiang W. Utilization of economical substrate-derived carbohydrates by solventogenic clostridia: pathway dissection, regulation and engineering. *Curr Opin Biotechnol* 2014;**29**:124–31.
40. Lan EI, Liao JC. Microbial synthesis of n-butanol, isobutanol, and other higher alcohols from diverse resources. *Bioresour Technol* 2013;**135**:339–49.
41. Köpke M, Held C, Hujer S, Liesegang H, Wiezer A, Wollherr A, et al. *Clostridium ljungdahlii* represents a microbial production platform based on syngas. *Proc Natl Acad Sci USA* 2010;**107**:13087–92.
42. Jang Y, Lee Y, Lee J, Park H, Im A, Eom M, et al. Enhanced butanol production obtained by reinforcing the direct butanol-forming route in *Clostridium acetobutylicum*. *MBio* 2012;**3**:1–9.
43. Choi YJ, Lee J, Jang Y, Lee SY. Higher alcohols. *MBio* 2014;**5**:1–10.
44. Liu XB, Gu QY, Yu XB. Repetitive domestication to enhance butanol tolerance and production in *Clostridium acetobutylicum* through artificial simulation of bio-evolution. *Bioresour Technol* 2013;**130**:638–43.
45. Steen EJ, Chan R, Prasad N, Myers S, Petzold C, Redding A, et al. Metabolic engineering of *Saccharomyces cerevisiae* for the production of isobutanol and 3-methyl-1-butanol. *Microb Cell Fact* 2008:7. http://dx.doi.org/10.1186/1475-2859-7-36.
46. Machado HB, Dekishima Y, Luo H, Lan EI, Liao JC. A selection platform for carbon chain elongation using the CoA-dependent pathway to produce linear higher alcohols. *Metab Eng* 2012;**14**:504–11.
47. Atsumi S, Cann AF, Connor MR, Shen CR, Smith KM, Brynildsen MP, et al. Metabolic engineering of *Escherichia coli* for 1-butanol production. *Metab Eng* 2008;**10**:305–11.
48. Fei Q, Brigham CJ, Lu J, Fu R, Sinskey AJ. Production of branched-chain alcohols by recombinant *Ralstonia eutropha* in fed-batch cultivation. *Biomass Bioenergy* 2013;**56**:334–41.
49. Lu J, Brigham CJ, Gai CS, Sinskey AJ. Studies on the production of branched-chain alcohols in engineered *Ralstonia eutropha*. *Appl Microbiol Biotechnol* 2012;**96**:283–97.
50. De La Plaza M, Fernández De Palencia P, Peláez C, Requena T. Biochemical and molecular characterization of α-ketoisovalerate decarboxylase, an enzyme involved in the formation of aldehydes from amino acids by *Lactococcus lactis*. *FEMS Microbiol Lett* 2004;**238**:367–74.
51. Jendrossek D, Kruger N, Steinbuchel A. Characterization of alcohol dehydrogenase genes of derepressible wild-type *Alcaligenes eutrophus* H16 and constitutive mutants. *J Bacteriol* 1990;**172**:4844–51.

52. Bernardi AC, Gai CS, Lu J, Sinskey AJ, Brigham CJ. Experimental evolution and gene knockout studies reveal AcrA-mediated isobutanol tolerance in *Ralstonia eutropha*. *J Biosci Bioeng* 2015;**122**:64–9.
53. Minty JJ, Lesnefsky Aa, Lin F, Chen Y, Zaroff Ta, Veloso AB, et al. Evolution combined with genomic study elucidates genetic bases of isobutanol tolerance in *Escherichia coli*. *Microb Cell Fact* 2011;**10**:18.
54. Li H, Opgenorth PH, Wernick DG, Rogers S, Wu T-YT-Y, Higashide W, et al. Integrated electromicrobial conversion of CO_2 to higher alcohols. *Science* 2012;**335**:1596.
55. Hood EE. Plant-based biofuels. *F1000Research* 2016;**5**:1–9.
56. Harding KG, Dennis JS, von Blottnitz H, Harrison STL. A life-cycle comparison between inorganic and biological catalysis for the production of biodiesel. *J Clean Prod* 2008;**16**:1368–78.
57. Du W, Li W, Sun T, Chen X, Liu D. Perspectives for biotechnological production of biodiesel and impacts. *Appl Microbiol Biotechnol* 2008;**79**:331–7.
58. Mallick N, Bagchi SK, Koley S, Singh AK. Progress and challenges in microalgal biodiesel production. *Front Microbiol* 2016;**7**:1019.
59. Brigham CJ, Speth DR, Rha C, Sinskey AJ. Whole-genome microarray and gene deletion studies reveal regulation of the polyhydroxyalkanoate production cycle by the stringent response in *Ralstonia eutropha* H16. *Appl Environ Microbiol* 2012;**78**:8033–44.
60. Riedel SL, Bader J, Brigham CJ, Budde CF, Yusof ZAM, Rha C, et al. Production of poly(3-hydroxybutyrate-co-3-hydroxyhexanoate) by *Ralstonia eutropha* in high cell density palm oil fermentations. *Biotechnol Bioeng* 2012;**109**:74–83.
61. Xiong T, Peng F, Liu Y, Deng Y, Wang X, Xie M. Fermentation of Chinese sauerkraut in pure culture and binary co-culture with *Leuconostoc mesenteroides* and *Lactobacillus plantarum*. *LWT – Food Sci Technol* 2014;**59**:713–7.
62. Kurita K. Chitin and Chitosan: Functional Biopolymers from marine Crustaceans. *Mar Biotechnol* 2006;**8**:203–26.
63. Knuckey RM, Brown MR, Robert R, Frampton DMF. Production of microalgal concentrates by flocculation and their assessment as aquaculture feeds. *Aquac Eng* 2006;**35**:300–13.
64. Zijffers JWF, Janssen M, Tramper J, Wijffels RH. Design process of an area-efficient photobioreactor. *Mar Biotechnol* 2008;**10**:404–15.
65. Alvarez HM, Mayer F, Fabritius D, Steinbüchel A. Formation of intracytoplasmic lipid inclusions by *Rhodococcus opacus* strain PD630. *Arch Microbiol* 1996;**165**:377–86.
66. Holder JW, Ulrich JC, DeBono AC, Godfrey PA, Desjardins CA, Zucker J, et al. Comparative and functional genomics of *Rhodococcus opacus* PD630 for Biofuels development. *PLoS Genet* 2011:7. http://dx.doi.org/10.1371/journal.pgen.1002219.
67. Kurosawa K, Boccazzi P, de Almeida NM, Sinskey AJ. High-cell-density batch fermentation of *Rhodococcus opacus* PD630 using a high glucose concentration for triacylglycerol production. *J Biotechnol* 2010;**147**:212–8.
68. Kurosawa K, Plassmeier J, Kalinowski J, R??ckert C, Sinskey AJ. Engineering l-arabinose metabolism in triacylglycerol-producing *Rhodococcus opacus* for lignocellulosic fuel production. *Metab Eng* 2013;**30**:89–95.
69. Hetzler S, Bröker D, Steinbüchel A. Saccharification of cellulose by recombinant *Rhodococcus opacus* PD630 strains. *Appl Environ Microbiol* 2013;**79**:5159–66.
70. Kurosawa K, Radek A, Plassmeier JK, Sinskey AJ. Improved glycerol utilization by a triacylglycerol-producing *Rhodococcus opacus* strain for renewable fuels. *Biotechnol Biofuels* 2015;**8**:31.
71. Herrero OM, Alvarez HM, Moncalián G. Physiological and genetic differences amongst *Rhodococcus* species for using glycerol as a source for growth and triacylglycerol production. *Microbiology* 2016;**162**:384–97.
72. Palmer JD, Brigham CJ. Feasibility of triacylglycerol production for biodiesel, utilizing *Rhodococcus opacus* as a biocatalyst and fishery waste as feedstock. *Renew Sustain Energy Rev* 2016;**56**:922–8.
73. Knothe G, Van Gerpen JH, Krahl J, et al. *The biodiesel handbook*. 2005. http://dx.doi.org/10.1016/B978-1-893997-62-2.50015-2.
74. Madden MC. A paler shade of green? The toxicology of biodiesel emissions: recent findings from studies with this alternative fuel. *Biochim Biophys Acta – Gen Subj* 2016:1–7.
75. Somerville C, Youngs H. How big is the bioenergy piece of the energy pie? Who cares-it's pie! *Biotechnol Bioeng* 2014;**111**:1717–8.

CHAPTER

3.25

Solar Energy Conversion

Nicholas A. Lee, Gerald E. Gilligan, Jonathan Rochford

University of Massachusetts Boston, Boston, MA, United States

3.25.1 SOLAR ENERGY AND DYE-SENSITIZED SOLAR CELLS

As the world's energy needs continue to increase, the development and implementation of renewable energy technologies is of paramount importance. In particular, solar energy has the highest potential to replace current energy-generation methods. Global energy consumption is presently estimated to be around 17 TW (2014) and is expected to increase with population growth to 24 TW by 2035.[1] The majority of this energy is sourced from the combustion of fossil fuels (coal, natural gas, and oil) that release large amounts of carbon dioxide into the atmosphere. Currently, greener alternatives such as hydroelectric and renewables (hydroelectric, wind, geothermal, tidal, biomass, and solar energy) only account for 10% of global energy consumption.[2] Fig. 3.25.1 shows the stark correlation between increasing atmospheric CO_2 concentrations and increasing global temperature over the past 125 years. The concerns surrounding climate change and carbon emissions are so significant that in 2015 the Obama administration announced the Clean Power Plan, which aims to regulate emissions from existing power plants with the goal of reducing carbon emissions by 32% by the year 2030.[3]

Renewable energy sources have the potential to be carbon neutral and, in particular, many scientists believe solar energy is best suited to satisfy the world's energy needs. An estimated 120,000 TW of sunlight strikes the earth every year.[4] Capturing just a small portion of this available energy could replace that currently produced by fossil fuels, with just 28 TW required annually to meet global demand by 2050. The dominant technology in the market today for solar energy capture remains the crystalline silicon solar cell. Crystalline silicon solar cells were first developed in the 1950s and 1960s to support the space program's satellite energy requirements. At the time, solar panels were very expensive because of the challenges of creating large crystals of silicon. In the past 40 years, however, the cost has dropped as manufacturing techniques have improved and efficiencies increased.

The maximum theoretical efficiency of a single-junction solar cell was calculated by William Shockley and Hans J. Queisser in 1961.[5] This "detailed balance limit of efficiency" is deeply rooted in thermodynamics and considers a number of factors such as blackbody

Green Chemistry
http://dx.doi.org/10.1016/B978-0-12-809270-5.00030-3

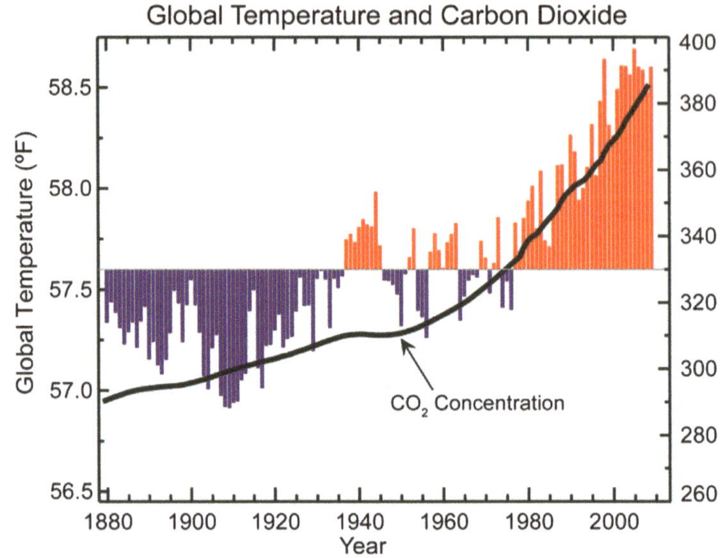

FIGURE 3.25.1 Plot of average global temperature and CO_2 concentration per year. Bar graph baseline is set at the average temperature between 1901 and 2000.[107]

radiation, radiative recombination, and spectral losses. As a result, the calculated maximum efficiency for a single p-n junction silicon solar cell is 30%, considering its band gap of 1.1 eV and assuming the sun is a blackbody at 6000K. Commonly referred to as the Shockley-Queisser limit, this has encouraged investigation into different types of solar cells that can replicate, and ultimately outperform, the efficiency of silicon solar cells but at a lower cost of production. A plot depicting the evolution of solar cell efficiencies since 1975, itemized with respect to the variety of available photovoltaic technologies, is provided in Fig. 3.25.2 (courtesy of data available from the US National Renewable Energy Laboratory, NREL).

Recently, the record solar cell efficiencies have soared beyond the Shockley-Queisser limit as high as 46% using multijunction photovoltaic devices.[6] In particular, the GaAs multijunction photovoltaic cell has gained widespread fame. Unfortunately, this class of solar cell is not particularly well suited for widespread commercial application because of its high materials cost and complexity of fabrication. As such, the majority of current photovoltaic research focuses on reducing materials and fabrication costs while maintaining a positive trajectory in power conversion efficiencies (PCEs). In fact, record single-junction photovoltaic device efficiencies have approached the Shockley-Queisser limit of 30% in the recent years in the case of silicon photovoltaics.[7]

One alternative to silicon-based devices is the dye-sensitized solar cell (DSSC), first introduced by Brain O'Regan and Michael Grätzel.[8] Their preeminent 1991 *Nature* publication reported a solar-to-electric PCE of almost 8%; a significant improvement over the previously studied dye-sensitized photoelectrochemical cells.[9,10] This work was so significant that it jump-started a global research effort in the area of dye-sensitized photovoltaics and has since been cited almost 15,000 times. Utilizing a mesoporous semiconducting TiO_2 electrode with a high surface area that was sensitized with a molecular ruthenium(II) polypyridyl complex,

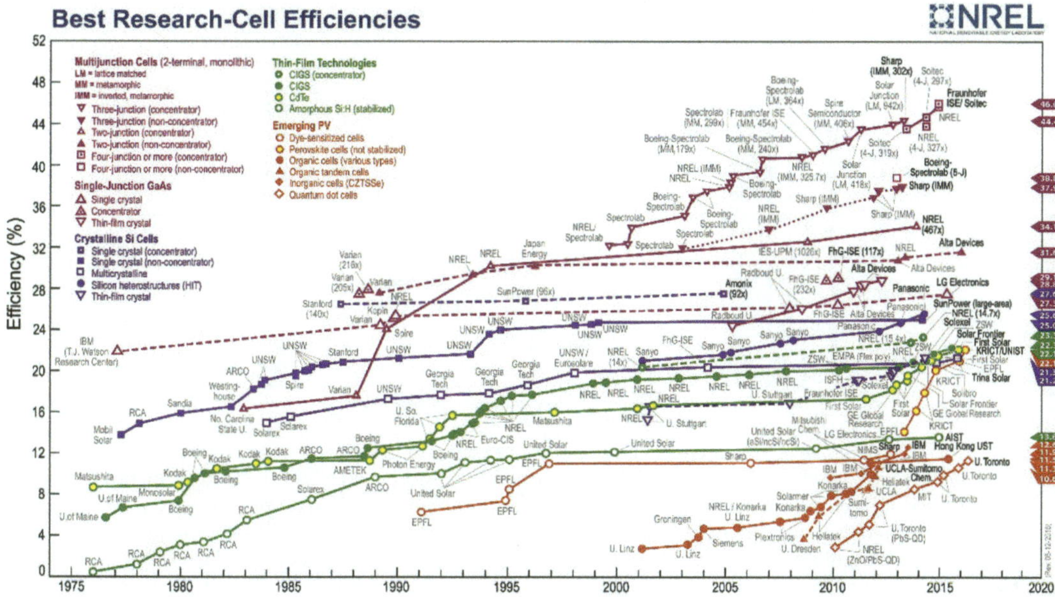

FIGURE 3.25.2 Chart of "best research-cell efficiencies" reported in laboratory settings since 1975, updated to 2016. *This plot is courtesy of the National Renewable Energy Laboratory, Golden, CO; United States Department of Energy, NREL/NCPV.* Best research-cell efficiencies. *http://www.nrel.gov/ncpv/.*

O'Regan and Grätzel improved the light-harvesting ability of the cell along with its long-term stability. Based on this design, recent advances in DSSC design have led to efficiencies of up to 13%.[11,12]

As illustrated in Fig. 3.25.2, DSSCs have yet to reach the record efficiencies of their silicon counterparts. Despite this fact, the many benefits associated with DSSCs have kept this versatile technology especially promising. In comparison to first-generation silicon solar cells and second-generation amorphous silicon, copper indium gallium selenide, and cadmium telluride (CdTe) solar cells, DSSCs have low production costs, flexible fabrication platforms, and enhanced performance under extreme working conditions, for example, low light and high temperature.[13] In addition, the fabrication of DSSCs requires very little special equipment and can be done in an open-air environment on a benchtop. As a result, DSSCs are accessible to many research groups worldwide. This accessibility and potential for commercialization has led to an expansion of DSSC research in the past few decades.

3.25.2 DYE-SENSITIZED SOLAR CELL DESIGN, MECHANISM, AND THERMODYNAMIC CONSIDERATIONS

Unlike their silicon counterparts, conventional DSSCs involve many components that can be tuned, both photophysically and electrochemically, to optimize device performance using conventional synthetic methods. Consisting of a dye-sensitized n-type semiconductor

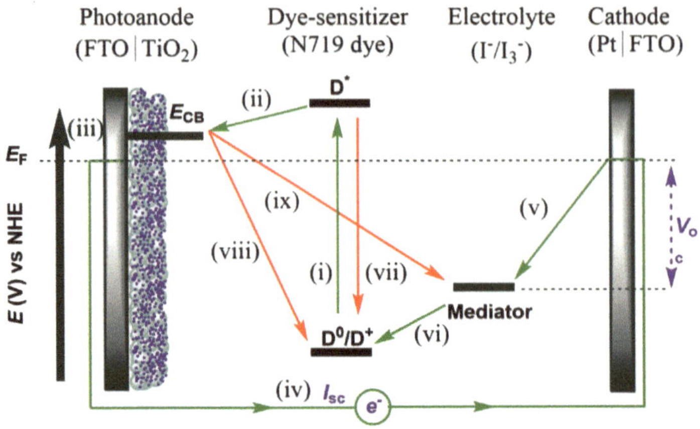

SCHEME 3.25.1 Energy level alignment of components in a conventional n-type DSSC device. *Green arrows* (gray in print versions) represent desired forward electron-transfer processes, *red arrows* (light gray in print versions) represent undesired charge recombination processes, and the *blue arrow* (dark gray in print versions) represents the open circuit potential energy difference (V_{OC}) at the p-n junction. The mechanistic legend is as follows, with typical kinetics provided in parenthesis: (i) dye electronic excitation (fs), (ii) charge injection to semiconductor conduction band (150 ps), (iii) electron equilibration to semiconductor Fermi level and diffusion to the FTO contact (100 μs), (iv) external current load between photoanode and cathode, (v) redox mediator reduction at the cathode, (vi) dye regeneration/mediator oxidation (1 μs), (vii) mass transfer diffusion of redox mediator, (viii) $TiO_2^{\bullet-} \rightarrow Dye^+$ charge recombination (3 μs), and (ix) $TiO_2^{\bullet-}$ (→ (mediator + charge recombination (1 ms).

(the photocathode), a solution-phase redox mediator (RM), and a platinum cathode, the basic schematic of a DSSC based on Grätzel and O'Regan's initial design is illustrated in Scheme 3.25.1.

A mesoporous, n-type-semiconducting metal oxide film (thickness varies from 2 to 20 μm, typically composed of 20-nm-diameter TiO_2 nanoparticles) with a high surface area is grafted to a substrate of transparent conductive oxide glass, commonly fluorine-doped tin oxide (FTO). Covalently anchored to the TiO_2 surface is a light-harvesting dye molecule, for example, N719 [a.k.a. bis(tetrabutylammonium) *cis*-dithiocyanatobis(2,2′-bipyridine-4,4′-di-carboxylate)ruthenium(II)], which on photoexcitation (Scheme 3.25.1(i)) injects an electron into the conduction band of the semiconductor (Scheme 3.25.1(ii)) resulting in an oxidized dye species (D^+). The conduction band electron subsequently diffuses through the mesoporous TiO_2 film to be collected at the FTO|TiO_2 contact (Scheme 3.25.1(iii)) and then is passed through an external load to the Pt|FTO cathode (Scheme 3.25.1(iv)). At the cathode, the electron reduces the oxidized species in the redox-mediator solution, for example, $I_3^- + 2e^- \rightarrow 3I^-$ (Scheme 3.25.1(v)). Meanwhile the oxidized dye molecule is regenerated following mass transfer diffusion of the reduced redox-mediator, for example, I^-, from the cathode to the photoanode completing the DSSC cycle (Scheme 3.25.1(vi)). The overall process results in no change in the chemical nature of the cell, integral to the design of stable, long-lifetime solar cells.

The efficiency of a DSSC is hindered by a number of charge recombination processes. Exciton recombination (Scheme 3.25.1(vii)) involves relaxation of the excited state dye (D*) back to its ground state because of a short excited state lifetime. For example, the triplet metal-to-ligand charge-transfer excited state (^{3}MLCT) of the N3 [a.k.a. *cis*-di(thiocyanato) bis(2,2′-bipyridyl-4,4′-di-carboxylate)ruthenium(II)] dye is 50 ns.[14] Recombination of TiO_2 conduction band electrons with the oxidized dye (Scheme 3.25.1(viii)) and oxidized RM (Scheme 3.25.1(ix)) is also a possible pathway for charge recombination. Not illustrated in Scheme 3.25.1 is the additional possibility, due to any flaws in device fabrication, of charge recombination from any exposed FTO surface sites to the oxidized mediator.[15] Of course, minimizing the possibility of any charge recombination processes is critical to efficient DSSC design and fabrication.

Fundamental to the operation of an *n*-type semiconductor-based DSSC and optimization of electron-transfer kinetics is energy level alignment of the metal oxide conduction band, dye sensitizer highest occupied molecular orbital (HOMO) and lowest unoccupied molecular orbital (LUMO) energies, and RM redox potentials. In addition to "hot-electron" injection, which is an established phenomenon in DSSCs, the high-energy singly occupied molecular orbital (SOMO) of the thermally equilibrated photoexcited state of the dye sensitizer should be sufficiently negative in comparison to the semiconductor conduction band to promote rapid charge transfer. Similarly, the SOMO of the oxidized dye sensitizer should have a redox potential more positive than that of the RM to promote dye regeneration. In these cases, the excited state oxidation potential and excited state reduction potentials are critical, as they provide an estimate of the excited state thermodynamic potentials involved in the forward electron-transfer processes responsible for photocurrent generation. The photosensitizer excited state oxidation, $E(S^+/S^*)$, and excited state reduction, $E(S^*/S^-)$, potentials can be estimated by consideration of the excited state energy (ΔE_{0-0}) according to Eqs. (3.25.1 and 3.25.2).[16]

$$E(S^+/S^*) = E(S^+/S) - \Delta E_{0-0}(S/S^*) \tag{3.25.1}$$

$$E(S^*/S^-) = E(S^-/S) + \Delta E_{0-0}(S/S^*) \tag{3.25.2}$$

Here, "S" symbolizes the ground state photosensitizer; "S^+" and "S^-" are the oxidized and reduced forms of the photosensitizer, respectively; S* is the excited state of the photosensitizer; and ΔE_{0-0} is the 0-0 vibrational energy gap between the ground lowest electronic excited state. For example, $E(T_1,\nu_0) - E(S_0,\nu_0)$ for the N719 dye as intersystem from the S_1 to T_1 excited state is rapid and quantitative.

The driving force for excited state charge-transfer oxidation to the TiO_2 conduction band can be estimated by Eq. (3.25.3) and the driving force for dye regeneration by the RM is given by Eq. (3.25.4), where $E_{redox}(RM)$ corresponds to the equilibrium potential of the oxidized and reduced forms of the RM species. These processes are not necessarily competitive; however, if charge injection to the semiconductor is very fast, the RM will only ever react with the ground state oxidized dye.

$$\Delta G = |E(S^+/S*) - E_{CB}(TiO_2)| \tag{3.25.3}$$

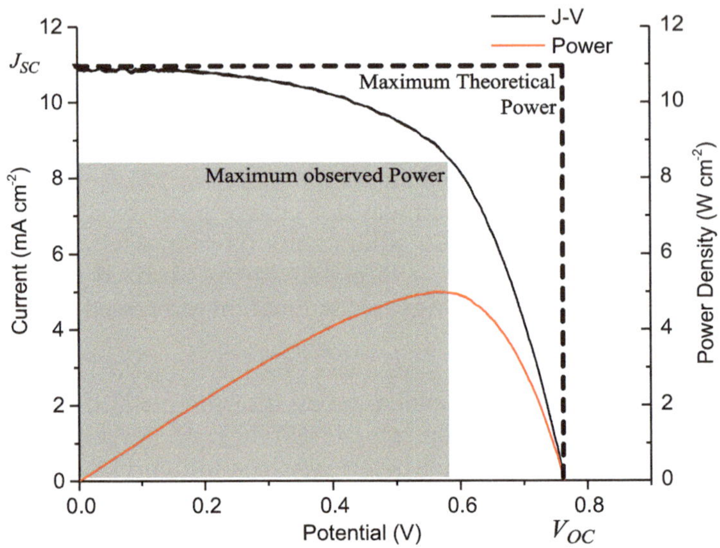

FIGURE 3.25.3 Representative current density-voltage curve of a dye-sensitized solar cell irradiated under 1 sun illumination. The fill factor is the ratio of the shaded region to the region bounded by the *dashed lines* at J_{SC} and V_{OC} (i.e., "maximum observed power/maximum theoretical power"). *Adapted from Ardo S, Meyer GJ. Photodriven heterogeneous charge transfer with transition-metal compounds anchored to TiO_2 semiconductor surfaces.* Chem Soc Rev *2009;38:115–64.*

$$\Delta G = |E(S^*/S^-) - E_{redox}(RM)| \quad (3.25.4)$$

Therefore, it is imperative that when designing dye molecules the HOMO and LUMO potential energies be considered, not only for tuning of the optical absorption band gap but also with respect to energy level alignment adjacent to the metal oxide conduction band and the RM equilibrium potential.

The percentage PCE or η of a DSSC is calculated by measuring the current-voltage relationship for the device under illumination according to Eq. (3.25.5).

$$\eta = \frac{J_{SC} \cdot V_{OC} \cdot FF}{P_{in}} \times 100 \quad (3.25.5)$$

Taken from *J-V* curve measurements as seen in Fig. 3.25.3, the short circuit current density, J_{SC}, is observed at the intercept of the curve and the y-axis at $V_{OC} = 0$ (Fig. 3.25.3). This value relates the maximum rate at which charge can flow through an external circuit. The open circuit potential, V_{OC}, is observed when $J_{SC} = 0$ A/cm^2 and corresponds to the energy gap between the Fermi level (E_{Fermi}) of the semiconductor and the redox potential of the RM (Scheme 3.25.1, Fig. 3.25.3). The fill factor, *FF*, is a value between 0 and 1 that relates the maximum observed power to the theoretical maximum observed power, calculated as the product of J_{SC} and V_{OC}. In essence, the higher the fill factor, the more ideal the device is, reflected by a more rectangular *J-V* curve. The standard power input (P_{in}) is 1.5 Air Mass (AM) solar irradiation (a.k.a. 1 sun = 1000 W/m^2 or 100 mW/cm^2).

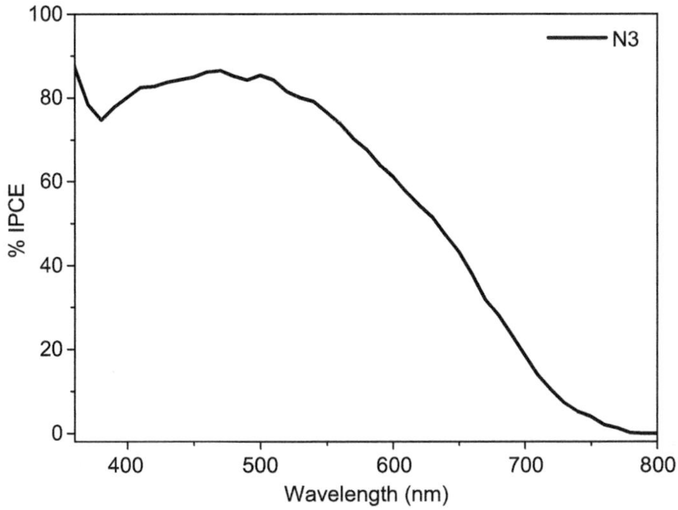

FIGURE 3.25.4 Percentage incident photon-to-current conversion efficiency (IPCE) spectrum for a TiO_2-based dye-sensitized solar cell device sensitized with the N3 dye.

In addition to a *J-V* plot, the spectral response of a DSSC can be characterized by measuring the external quantum efficiency, also known as the *incident photon-to-current conversion efficiency* or %IPCE, which is calculated by Eq. (3.25.6).

$$\mathrm{IPCE}(\lambda) = \frac{I(\lambda)/P_{\mathrm{in}}(\lambda)}{hc/(e\lambda)} \times 100 \tag{3.25.6}$$

where λ is the incident wavelength, $I(\lambda)$ is the photocurrent measured under monochromatic illumination, $P_{\mathrm{in}}(\lambda)$ is the incident monochromatic light fluence, h is Planck's constant, c is the speed of light, and e is the elementary charge. The %IPCE is then plotted as a function of wavelength. A %IPCE plot for a DSSC with the reference N3 dye is shown in Fig. 3.25.4.

3.25.3 OPTIMIZATION OF DYE-SENSITIZED SOLAR CELL DESIGN

The unique structure of the DSSC allows for, and requires, optical and electrochemical tuning of individual components while considering their compatibility for optimization of overall device performance. Although each individual aspect of the DSSC has been investigated from a molecular perspective, research has focused primarily on increasing the light-harvesting properties of the dye sensitizer and more recently on increasing the open circuit voltage by modifying the RM structure. The following sections will serve as a brief overview of studies into each component of a DSSC.

3.25.3.1 Mesoporous Metal Oxide Working Electrode

As used in the original Grätzel cell, titanium dioxide (TiO_2) has been the metal oxide of choice for most DSSCs because of its favorable band gap (3.2 eV) and conduction band energy [$E_{CB} \approx -0.5$ V vs. normal hydrogen electrode (NHE)].[17,18] The anatase crystal phase of TiO_2 with an average particle size ranging from 10 to 20 nm diameter is typically most efficient in terms of conductivity and surface area. Optimum TiO_2 photoanodes in high-performance DSSCs typically consist of up to four different TiO_2 layers[12]:

1. An ultrathin layer of TiO_2 (~50 nm thick) on the FTO substrate to block charge recombination between the FTO back contact and the RM electrolyte (Scheme 3.25.1(vii)).
2. A thick, mesoporous, light-absorbing layer (6–10 μm, 10–25 nm particles) that serves as a high-surface-area substrate for dye molecule attachment as well as a conductive medium for electron diffusion to the FTO back contact.
3. A second ultrathin posttreatment layer of TiO_2 on the mesoporous film to further increase the surface area throughout the mesoporous structure and increase electrical contact between TiO_2 nanoparticles.
4. A light-scattering layer consisting of ~400-nm TiO_2 particles (2–4 μm) to promote back reflection of transmitted photons for increased light absorption.

Ito et al.[19] showed that the addition of a thin blocking layer, deposited on the surface of the FTO glass by spray pyrolysis, and a thin posttreatment layer, deposited via an aqueous $TiCl_4$ solution, resulted in a PCE increase from 9.4% to 10.6% with the N719 dye and I_3^-/I^- RM. This increase was due to not only a reduction of charge recombination at the FTO surface but also an increase in the roughness of the TiO_2 surface, allowing for additional dye loading.

One major drawback of mesoporous TiO_2 films is the "random walk" that electrons must take to reach the FTO back contact. This can result in electrons being "trapped" in TiO_2. More ordered structures such as single crystalline TiO_2 nanotubes or nanowires have also been explored to minimize this random walk process but these materials tend to suffer from smaller surface areas due to the difficulty in engineering stable films of such hierarchical assemblies with appropriate thickness. In a publication by So and colleagues,[20] wide single-walled TiO_2 nanotubes were developed and subsequently treated with $TiCl_4$. This approach implemented the ordered approach of using nanotubes while keeping a large surface area. As a result, they achieved a PCE close to 8% via back-illumination, that is, irradiation through the counter electrode (CE) side of the device. Interestingly, after four depositions of $TiCl_4$, the efficiency of the cell decreased, most likely because of poor diffusion of electrolyte into the nanotube as the porosity decreased (Fig. 3.25.5).

Wurtzite ZnO has also been used as an electrode scaffold in DSSCs. ZnO has a similar band gap (3.2 eV) and conduction band energy ($E_{CB} = -0.5$ V vs. NHE) to TiO_2.[17,18] A number of different ZnO nanostructured materials have been developed for DSSC applications.[21] However, efficiencies of ZnO-based DSSCs are at best half of those based on TiO_2 electrodes. This has been attributed to charge-transfer impedance at the ZnO|dye interface following photoexcitation, which is not observed for TiO_2.[22]

Saito and Fujihara[23] achieved high photocurrent generation using a ZnO photoelectrode sensitized with the N719 dye coupled with an I_3^-/I^- RM. Reducing the working electrode (WE) and CE spacing from 120 to 25 μm, the J_{SC} was increased from 11.94 to 18.11 mA/cm^2

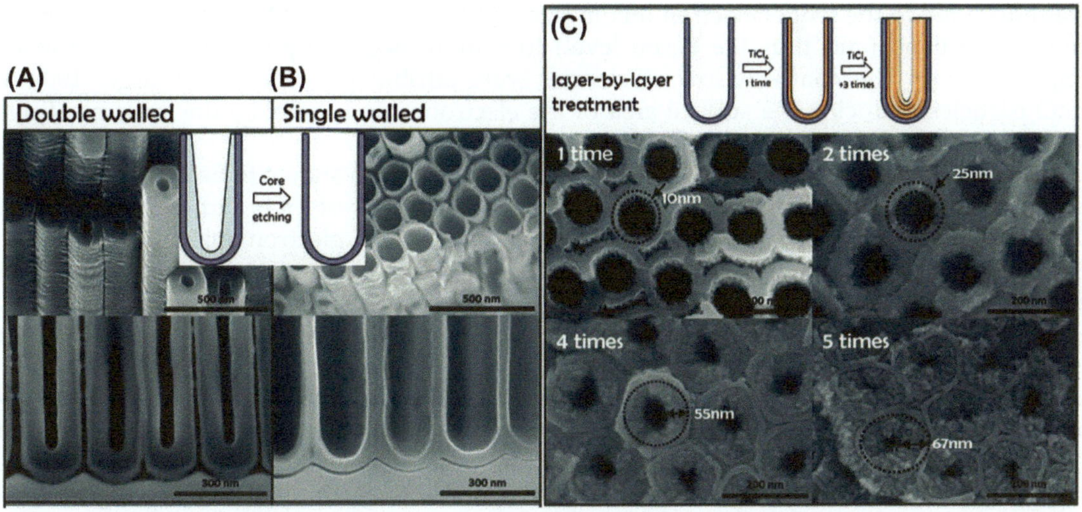

FIGURE 3.25.5 Process to build optimized hierarchical TiO_2 nanotube structures for dye-sensitized solar cells. (A) scanning electron microscopic images of "as-formed" tubes (showing typical double-walled morphology). (B) Tubes after the core removal process, leaving only the outer tube shell present. (C) Layer-by-layer decoration with TiO_2 nanoparticles. *Reprinted with permission from So S, Hwang I, Schmuki P. Hierarchical DSSC structures based on "single walled" TiO_2 nanotube arrays reach a back-side illumination solar light conversion efficiency of 8%.* Energy Environ Sci *2015;8:849–54.*

while maintaining a consistent V_{OC} and FF, resulting in an increase in efficiency from 4.1% to 6.6%. In addition, a short 90 min sensitization time was used to limit the precipitation of Zn^{2+}/dye within the pores of the electrode as described by Keis et al.[24]

Sakai et al.[25] combined ZnO with the commonly used $TiCl_4$ after treatment, resulting in TiO_2-coated ZnO electrodes. These electrodes showed higher J_{SC}, V_{OC}, and PCE compared to bare ZnO, bare TiO_2, and TiO_2-coated TiO_2 electrodes. Efficiencies of 3.6% and 4.9% were reported for the N719 and D149 photosensitizers (Fig. 3.25.6), respectively, using these TiO_2-coated ZnO electrodes in conjunction with the I_3^-/I^- RM. This improved performance

N719 D149

FIGURE 3.25.6 Structures of (A) the ruthenium-based N719 dye and (B) the indoline-based D149 organic dye. N719, in particular, is an industry standard because of its well-studied properties and NCS ligand.[25]

was attributed to the longer electron lifetimes in the TiO_2-coated ZnO electrodes, shifting the conduction band (and thus the Fermi level) to a more negative potential, resulting in an improved V_{OC}. The relatively low efficiencies were attributed to the use of larger 100-nm ZnO particles, reducing the surface area of the electrodes and therefore the amount of dye loading.

Other metal oxides such as SnO_2 and Nb_2O_5 have also been shown to be viable in DSSCs. SnO_2 has a conduction band energy (0.5 eV) lower than that of TiO_2 and therefore can be used with dyes having lower LUMOs to ensure efficient electron injection. Although SnO_2 presents an opportunity for the application of low band-gap, near-infrared (NIR)-absorbing photosensitizers in a DSSC, this approach often leads to a lower V_{OC}, especially when used with an I_3^-/I^- RM, because of the correspondingly lower E_{Fermi} energy. Kumar et al.[26] reported an efficiency of 3% using SnO_2 nanoflowers with a V_{OC} of 700 mV.[26] Nb_2O_5, in contrast, has a conduction band energy (0.2 eV) higher than that of TiO_2. If the photosensitizer excited state oxidation potential is sufficiently negative, this results in an increased Nb_2O_5 E_{Fermi} level and higher observed V_{OC}. Rani et al.[27] developed Nb_2O_5 nanochannels with a film thickness of 10 μm. Using the N3 and the I_3^-/I^- electrolyte, an efficiency of 4.5% was achieved.

3.25.3.2 Counter Electrode

The CE, or cathode, is required to reduce the oxidized form of the RM, regenerating the reduced form of the RM. Ideally, the CE should have a small sheet resistance, high catalytic activity, and low cost.[28] A commonly used CE is platinized FTO conductive glass. Utilized in the original Grätzel cell, the highly conductive and catalytic nature of Pt makes it suitable for CE applications. However, Pt is not cheap and the scaling of Pt CEs could limit its application. Carbon CEs have also been explored. Carbon is a viable alternative to Pt because of its low cost and multitude of structures. Kay and Grätzel[29] implemented a mixture of porous graphite and carbon black for monolithic DSSCs and obtained an efficiency of 6.6%. Wang et al.[30] incorporated a honeycomb-structured graphene as a CE synthesized from LiO_2 and CO. This graphene CE resulted in efficiencies similar to those of Pt (7.8% vs. 8%) for an N719 and I_3^-/I^- system. Other carbon structures such as single-walled nanotubes and multiwalled nanotubes have shown PCEs of 4.5%[31] and 7.7%,[32] respectively.

In addition to platinum and carbon, conductive polymers such as poly(3,4-ethylenedioxythiophene) (PEDOT) and inorganic semiconductors such as cobalt sulfide (CoS) have also been investigated. High-surface-area PEDOT CEs can be easily made by electrochemical polymerization of 3,4-ethylenedioxythiophene (EDOT) on a conductive FTO glass substrate. The thickness and morphology of PEDOT can be modified by the number of cyclic voltammetry cycles during electrodeposition. Tian et al. utilized PEDOT in studying a series of organic redox couples and the TH305 dye. An efficiency of 6% was achieved, which is higher than the efficiency obtained with Pt under the same conditions (Fig. 3.25.7).[33] CoS CEs were also introduced by Wang et al.[34] as a cheaper alternative to Pt and showed increased catalytic activity compared to a Pt CE for I_3^- reduction. The overall efficiency was identical to a Pt CE but showed additional stability over prolonged light soaking and thermal stress. Tian et al.[35] showed that CoS CEs are more efficient than Pt CEs when

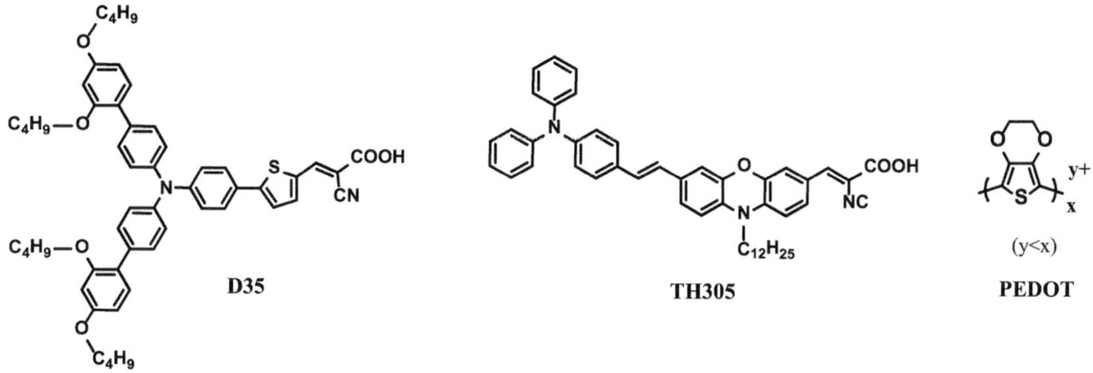

FIGURE 3.25.7 Molecular structures of the D35 and TH305 organic photosensitizers, which perform best in combination with a thiolate/disulfide redox mediator and a CoS or poly(3,4-ethylenedioxythiophene) (PEDOT) counter electrode, respectively.[33,35]

used in devices employing an organic dye, D35, and a thiolate/disulfide RM in ionic liquid solvent (Fig. 3.25.7).

3.25.4 DYE SENSITIZERS AND ANCHORING GROUPS

The dye sensitizer is critical to the operation of an efficient DSSC. Its role is to absorb photons of light across the broad UV-vis-NIR spectrum and inject an excited state electron from the LUMO (or higher-lying energy levels) to the semiconductor conduction band while concurrently driving RM oxidation and limiting unwanted charge recombination processes. Several criteria for an efficient dye sensitizer are described in the following[36]:

1. The dye must be strongly anchored to the WE metal oxide surface to promote efficient electron injection to the conduction band of the semiconductor.
2. The excited state oxidation potential of the dye must be higher in energy than the conduction band of the semiconductor to create enough driving force for charge injection.
3. The HOMO of the oxidized sensitizer must lie at a lower potential energy than the redox potential of the RM to facilitate dye regeneration.
4. The dye must have panchromatic absorption across the UV-vis-NIR solar spectrum.
5. The geometry of the dye must be optimized such that the LUMO is on the anchoring ligand (e.g., close to TiO_2 surface) and the HOMO away from the surface and near the RM for effective dye regeneration.
6. The dye must be stable under a number of different photochemical, thermal, and electrochemical states to be practical in commercial applications.
7. The dye should reduce charge recombination by passivating the photoanode surface through steric bulk, for example, incorporation of steric alkyl chains in the dye structure limits $TiO_2^{\bullet-} \rightarrow$ RM charge recombination of cobalt(III/II) mediators.[37]

A number of different classes of dyes, most notably consisting of metal coordination complexes and organic dyes, have been studied in DSSC settings. Champion dyes from each class will be discussed, with a focus on ruthenium polypyridyl sensitizers.

3.25.4.1 Metal Coordination Complexes

Metal coordination complexes have been investigated as sensitizers for their redox stability, broad absorption from to metal-to-ligand charge-transfer (MLCT) transitions, and tunable redox potentials. In general, the metal complex photosensitizer consists of one central metal atom grafted to the TiO_2 surface by a ligand with at least one anchoring group. By varying the nature of the metal and/or its ligand set, the photophysical and electrochemical properties can be fine-tuned to optimize DSSC performance.

The seminal 1991 O'Regan and Grätzel *Nature* paper utilized a trimeric $Ru(dcbpy)_2(\mu\text{-}(CN)Ru(CN)(bpy)_2)_2$ sensitizer, where dcbpy is 2,2′-bipyridine-4,4′-dicarboxylic acid and bpy is 2,2′-bipyridine, as a strong light absorber and achieved a PCE of just over 7%.[8] However, this dye would later be superseded by N3. The N3 dye first studied by Nazeeruddin et al.[14] in 1993 has since become one of the standards in DSSC testing and analysis. N3 is a favorable sensitizer because of its ability to absorb a large amount of visible light, relatively long excited state lifetime, and inclusion of four carboxylic acid anchoring groups resulting in an efficiency of over 10% with an I_3^-/I^- RM. In addition, it is an excellent example of a dye that meets requirements 1–6 listed earlier.[36] Its HOMO and LUMO energy levels are situated at +1.09 and −0.77 V versus NHE, respectively, and are favorable for dye regeneration and charge injection when used in combination with the I_3^-/I^- RM. The numerous MLCT transitions present in the N3 dye lead to a broad absorption spectrum, and the geometry of the dye creates a dipole that effectively shuttles excited state electrons toward the TiO_2 surface. Lastly, the dye was shown to keep a fairly constant efficiency over a period of 10 months under illumination.[14] The deprotonated form of N3, N719 (Fig. 3.25.8), has also been studied in depth and has been shown to have a slightly improved performance over N3 by minimizing conduction band leveling due to protonation sites at the TiO_2 interface.[38]

Since the report of the N3 dye, a large number of thiocyanate (NCS^-)-based ruthenium sensitizers have been developed. The "black dye," also reported by Nazeeruddin et al., used one terpyridine ligand substituted with three carboxylic acid groups (tctpy) and three isothiocyanato ligands attached to the ruthenium(II) center (Fig. 3.25.8). A decrease in the

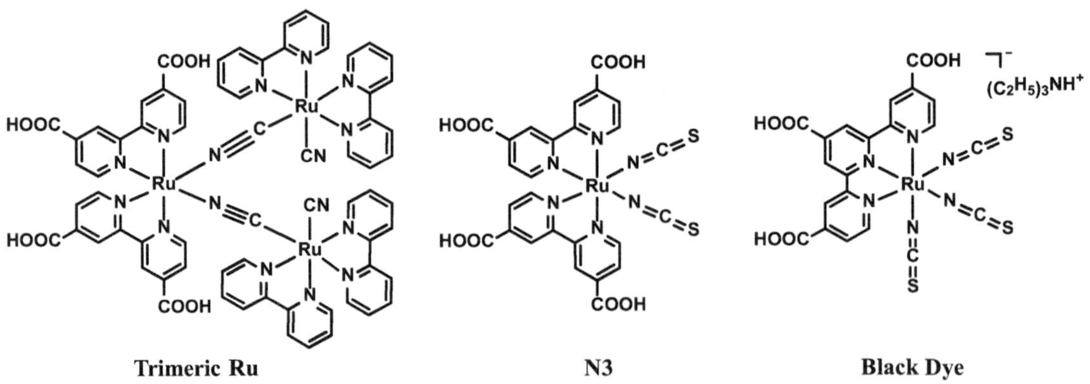

FIGURE 3.25.8 Structures of the trimeric Ru photosensitizer used by O'Regan and Grätzel in their 1991 publication as well as structures of the standard N3 and "black" dye.[39]

energy of tctpy(π^*)-based LUMO shifts the visible absorption to lower energy relative to N3, resulting in photocurrent generation into the NIR region. Overall, the black dye achieved a PCE of 10.4%.[39] The black dye is commonly used as standard for dyes incorporating terpyridine and analogous tridentate ligands, while the N3 or N719 dye is used when studying dyes with bipyridine or analogous bidentate ligands.

A large number of complexes inspired by the N3, N719, and black dyes have been designed in the past 20 years to increase the long-term stability of the sensitizer, increase the molar extinction coefficient, and optimize the redox potential of the sensitizer. One approach is to increase the bulk and hydrophobicity of the sensitizer by introducing long alkyl chains on the bipyridine ligand. The tris-heteroleptic Z907 dye, Ru(H_2dcbpy)(dnbpy)$(NCS)_2$, displayed excellent thermal stability over 1000 h at 80 °C when coadsorbed with 1-decylphosphonic acid on TiO_2.[40] Over this time, the open circuit voltage dropped about 3% while holding similar short-circuit currents. Later studies further investigated this impressive thermal stability imparted by hydrophobic TiO_2 passivation. For example, Nazeeruddin et al.[41] presented a comparative study of the alkyl chain length on DSSC efficiency. This study found that when extending the chain length from methyl to hexyl to nonyl to tridecyl, the efficiency of the DSSC increased to a maximum of 8.6% while sustaining 96% of its initial efficiency under 1000 h irradiation at 80 °C. The passivation of TiO_2 by the long alkyl chains could also be used to limit FTO-to-I_3^- and TiO_2-to-I_3^- charge recombination.

Increasing the light-harvesting ability should lead to higher DSSC efficiencies because of the increased number of excited state sensitizers and potentially reduced thickness of metal oxide WEs. For example, a significant increase in absorption is observed when the bipyridine rings are modified with triarylamine moieties (Fig. 3.25.9, KW1 & KW2). Fig. 3.25.9 shows the increased MLCT absorbance of KW1 and KW2 dyes compared to Z907.[42] A PCE of 10.1% was achieved for both KW1 and KW2 with an I_3^-/I^- mediator compared to a PCE of 8.9% for the Z907 dye under similar conditions. The difference in PCE was attributed to an increase in J_{SC} from 16.7 mA/cm^2 for Z907 to 18.5 mA/cm^2 and 18.9 mA/cm^2 for KW1 and KW2, respectively, because of increased absorption.

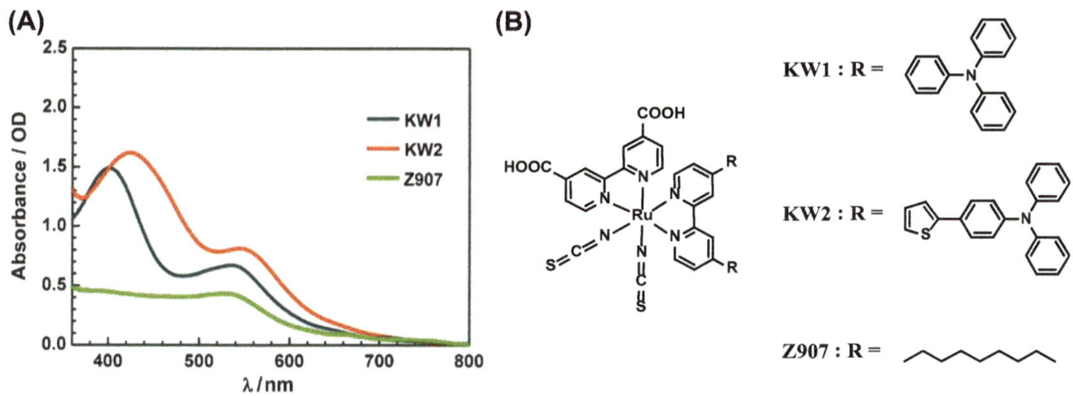

FIGURE 3.25.9 (A) Ruthenium photosensitizers KW1 and KW2 show an increased molar extinction coefficient relative to the Z907 dye because of an added charge-transfer/conjugation effect. (B) Molecular structures of the KW1, KW2, and Z907 dyes.[42]

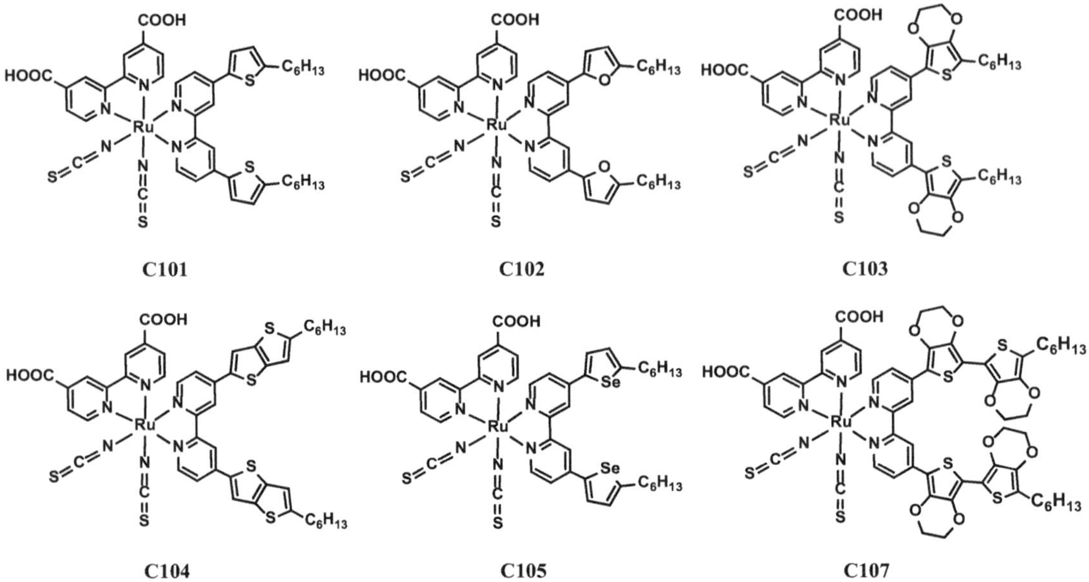

FIGURE 3.25.10 Molecular structures of the C100 class of dyes first developed by Gao and colleagues, which are capable of exceeding 11% dye-sensitized solar cell power conversion efficiency.[43–46]

Fig. 3.25.10 shows similarly structured dyes incorporating alkyl thiophene, alkyl furan, alkyl ethylenedioxythiophene, alkyl thieno[3,2-*b*]thiophene, and alkyl selenophene that have also been explored.[43–46] These dyes, sometimes referred to as the C100 series, have achieved efficiencies of over 11% and occasionally set DSSC benchmarks.[36] The first of these dyes, C101, was introduced by Gao et al. in 2008 and achieved an efficiency of 11.0% because of its strong absorption. It is also very robust, showing reasonable efficiency with low volatility electrolytes and ionic liquid electrolytes.[43] The absorption of these dyes has also been improved by increasing the ligand π-delocalization, such as in C107.[44]

Although many efficient Ru sensitizers have been developed with isothiocyanate ligands, there are some limitations to this approach. First, as the HOMO energy level is delocalized between the metal and isothiocyanate, fine-tuning of this energy level is difficult. If the potential of the HOMO could be changed, a variety of RMs could be used and the band gap could shift the absorbance to optimize the light-harvesting properties of the dye. The lack of potential for variation in isothiocyanate-based photosensitizers limits this possibility. Second, the monodentate nature of isothiocyanate is the dye's weakest point limiting dye stability under harsh working conditions. In particular, the NCS ligand is believed to dissociate and be replaced by a solvent molecule, especially in the dye's oxidized state where Ru-NCS bonding is weakest due to poorer Ru($d\pi$) → NCS(π^*) back donation.

Therefore, efforts have been made to replace all monodentate isothiocyanate ligands with bidentate or tridentate chelating ligands. Wadman et al.[47] first showed that cyclometallated ruthenium complexes could be used in DSSCs. These dyes consisted of two tridentate ligands in a bis-heteroleptic [Ru(CNN)(NNN)] configuration whose absorption was slightly redshifted compared to N719 yet gave similar overall efficiencies, but each being rather

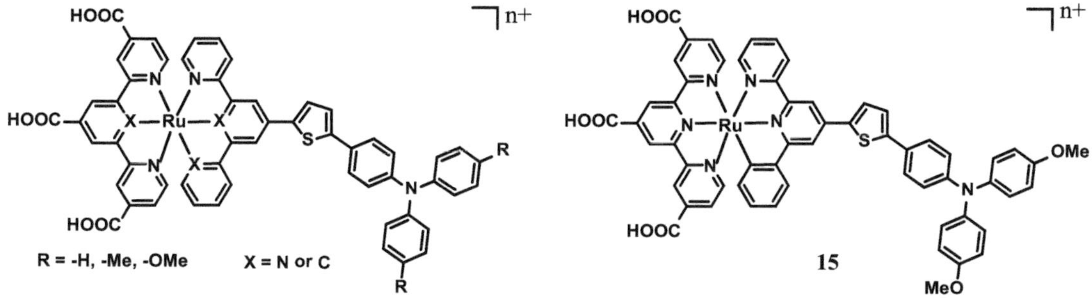

FIGURE 3.25.11 (A) Molecular structures of the bis-heteroleptic tridentate orthometallated dyes studied by Robson et al. (B) Complex **15** exhibited the highest dye-sensitized solar cell power conversion efficiency of 8.0% from this series of photosensitizers.[48]

low at PCEs of 2.1% and 2.3%, respectively. A series of tridentate Ru(II) dyes incorporating triphenylamine (TPA) substituents was studied by Robson et al.[48] The general structure for these dyes is shown in Fig. 3.25.11. Inclusion of the TPA electron donor and a bridging thiophene linkage enhances light absorption in all complexes. This comprehensive study of bis-tridentate complexes resulted in a PCE of 8.0% for complex **15** (Fig. 3.25.11), close to the PCE of 8.6% observed for the N3 dye under the same conditions.

Chen et al.[49] replaced two isothiocyanate ligands in the black dye with an anionic bidentate pyridine pyrazolate (pypz) ligand and observed that similar efficiencies could be achieved for tris-heteroleptic complexes with a combination of mono, bi, and tridentate ligands (Fig. 3.25.12, PRT1). The use of cyclometallated complexes was further validated when Bessho et al.[50] studied the bis-heteroleptic YE-05 dye, Ru(dcbpy)$_2$(2-(2,4-difluorophenyl)pyridine) (Fig. 3.25.12). The absorption spectra of N719 and YE-05 showed stark differences. Not only does YE-05 have higher molar extinction coefficients at similar wavelengths as N719 but it also has a small peak at 490 nm where there is a valley in the N719 spectrum (Fig. 3.25.13). The electron-donating nature of the carbanionic ligand was stabilized by the addition of two electron-withdrawing fluorine atoms adjusting the redox potential for sufficient dye regeneration by the I^- species of the RM. As a result, the overall

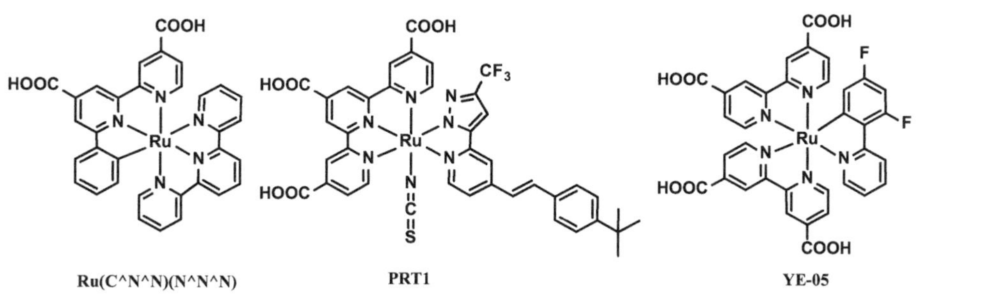

FIGURE 3.25.12 Cyclometallated ruthenium photosensitizers studied by (A) Wadman et al.,[47] (B) Chen et al.,[49] and (C) Bessho et al.[50]

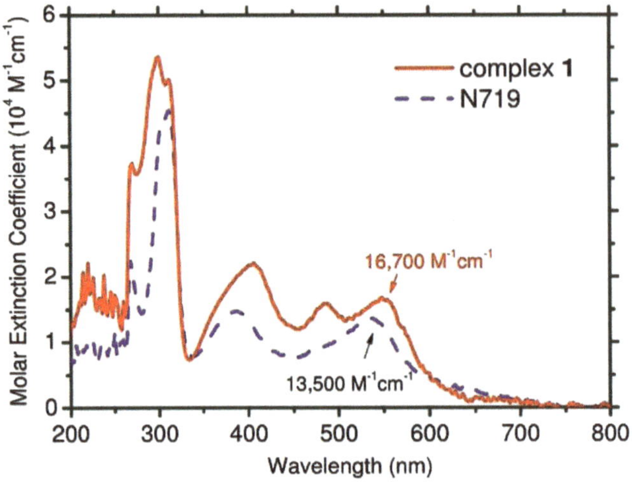

FIGURE 3.25.13 UV-vis spectra of YE-05 (complex 1) and N719. Note the additional peak in the YE-05 spectrum at 490 nm and the increased molar extinction coefficient. *Reprinted with permission from Bessho T, Yoneda E, Yum JH, Guglielmi M, Tavernelli I, Imai H, et al. New paradigm in molecular engineering of sensitizers for solar cell applications.* J Am Chem Soc *2009;**131**:5930–4.*

efficiency was shown to be 10.1% with the I_3^-/I^- RM, the first time a cyclometallated system achieved over 10% efficiency and was comparable to N719.

Another breakthrough in orthometallated complexes was reported by Bomben et al.,[51] where the YE-05 dye was modified by replacing the fluorine atoms on the phenyl pyridine ring with CF_3 groups and one of the dcbpy ligands with a bipyridine ligand substituted with long aliphatic chains to suppress charge recombination at the electrode surface, akin to the Z907 dye (Fig. 3.25.14). The resulting complex had a very positive first oxidation potential at +0.99 V versus NHE, providing sufficient driving force for dye regeneration by an I_3^-/I^- or $[Co^{III/II}(bpy)_3]^{3+/2+}$ RM. In addition, the long-chain aliphatic groups not only

FIGURE 3.25.14 (A) Molecular design elements for proposed by Bomben et al. for designer Ru-based dyes where R^1 is utilized for increased light harvesting and passivation of the photoanode surface, and R^2 for fine-tuning of highest occupied molecular orbital energy level through nature of substituent groups.[51] (B) A bis-hexylthiophene tris-heteroleptic Ru-based dye designed by this approach exhibited a 6.2% dye-sensitized solar cell power conversion efficiency.

improved light absorption but also sufficiently passivated the TiO_2 photoanode. As a result, the new dye showed an increased performance with an I_3^-/I^- mediator and achieved an incredible 6.2% efficiency when paired with the $[Co^{III/II}(bpy)_3]^{3+/2+}$ RM, where N3 was less than 1% efficient under the same conditions. These results helped formulate a general methodology when designing high-efficiency ruthenium dyes where three bidentate ligand are used. Each ligand serves a different purpose: (1) anchoring the dye to the WE (usually dcbpy), (2) enhancing light absorption and passivating the WE through bulky ligand substituents, and (3) tuning the position of the HOMO through its electron-donating/-withdrawing character.

A number of orthometallated Ru(II) complexes are shown in Fig. 3.25.15. The neutral TFRS-2 dye was developed by Wu et al.,[52] who incorporated 2-pyridyl pyrazolate as chelating ligands. This dye showed strong absorption in the visible region ($\varepsilon = 23{,}400\ M^{-1}\ cm^{-1}$ at 424 nm, $\varepsilon = 21{,}900\ M^{-1}\ cm^{-1}$ at 460 nm, and $\varepsilon = 16{,}400\ M^{-1}\ cm^{-1}$ at 533 nm) and a PCE of 9.5% was achieved, surpassing the PCE of 8.6% reported for the N719 dye due to increased J_{SC} (17.15 mA/cm^2 for TFRS-2 and 15.64 mA/cm^2 for N719). Li et al.[53] studied a series of Ru(II) dipyrinates that showed panchromatic light harvesting due to strong $\pi \rightarrow \pi^*$ dipyrrin transitions at ~460 nm and an MLCT transition at ~540 nm. However, the PCE for these dyes was low ($\eta < 1.5\%$) because of poor electron injection as the lowest excited state, $E(S^+/S^*)$, was approximately −0.59 V versus NHE, a mere 90 mV above the TiO_2 conduction band (E_{cb} ~ 0.50 V vs. NHE). Zhao et al.[54] synthesized and studied $[Ru(dcbpy)_2(OQN)]^+$, where

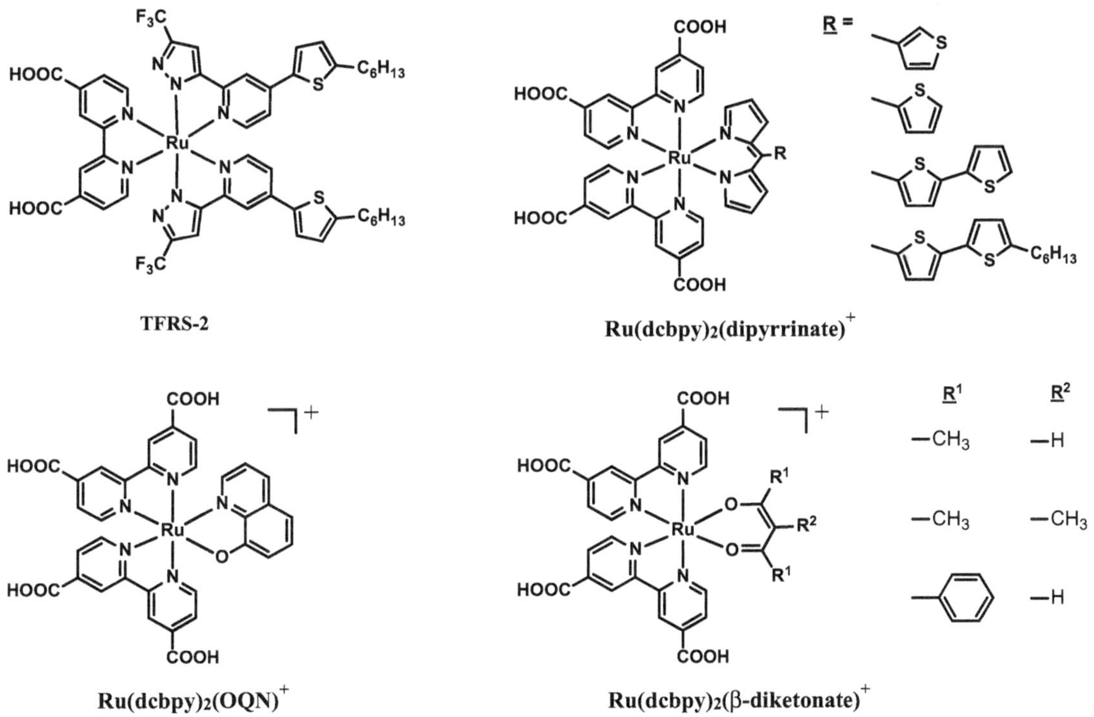

FIGURE 3.25.15 Orthometallated Ru-based photosensitizers studied by (A) Wu et al.,[52] (B) Li et al.,[53] (C) Zhao et al.,[54] and (D) Takahashi et al.[55]

OQN is 8-oxyquinolate. Although the UV-vis spectrum was remarkably similar to that of N3, a PCE of just 1.8% was observed primarily because of a low charge injection yield and poor driving force for $[Ru(dcbpy)_2(OQN)]^{2+}$ reduction. Takahashi et al.[55] studied $[Ru(dcbpy)_2(\beta\text{-diketonate})]^+$ complexes where the electron-donating nature of the β-diketonate ligands redshifted the absorption spectra, showing a shoulder at about 600 nm. A PCE of 6.0% was observed for $[Ru(dcbpy)_2(acac)]^+$ (where acac is acetylacetonato), surpassing the efficiency of N3 (PCE = 5.7%).

In addition to Ru-based dyes, other metal ions such as Os^{II}, Re^{I}, Fe^{II}, Pt^{II}, and Cu^{I} have been used as coordination centers in complex molecular sensitizers (Fig. 3.25.16).[12] Wu et al.[56] developed an Os-centered dye, TF-52, that achieved an efficiency of 8.9%, comparable to the black dye, by extending the absorption spectrum well into the NIR region. Hasselmann and Meyer[57] used $Re(deeb)(CO)_3(X)$, where deeb is 4,4′-$(COOEt)_2$-2,2′-bipyridine and X is I^-, Br^-, Cl^-, or CN^-, to study electron transfer and driving force dependence of charge injection to TiO_2. Ferrere and Gregg[58] replaced the ruthenium atom in the N3 dye with an iron atom and observed a significantly decreased PCE, possibly because of band-selective injection from excited state electrons. Islam et al.[59] studied a series of square planar platinum(II) diamine dithiolate complexes where an overall efficiency of 3% was achieved using Pt(dcbpy)(qdt), where qdt is quinoxaline-2,3-dithiolate. A record efficiency for a copper-centered dye of 4.7% was achieved by Sandroni et al.[60] utilizing a heteroleptic copper(I) polypyridine dye (C3) with chenodeoxycholic acid as a coabsorbent.

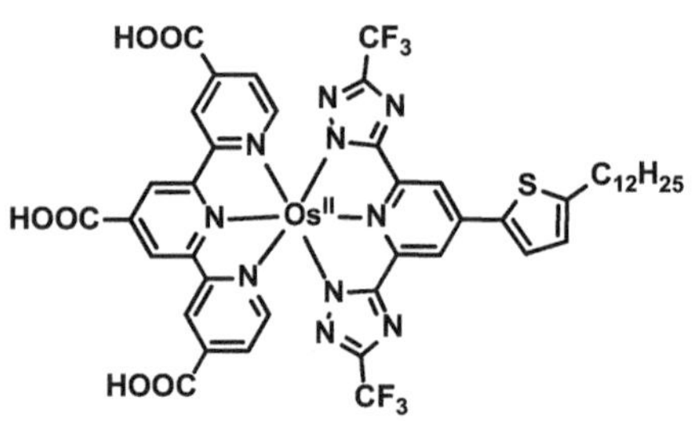

TF-52

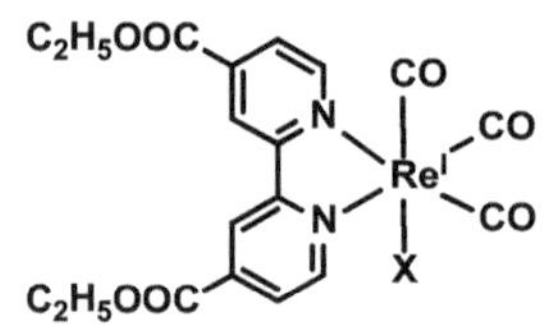

$Re(CO)_3(deeb)X$
X = I^-, Br^-, Cl^-, or CN^-

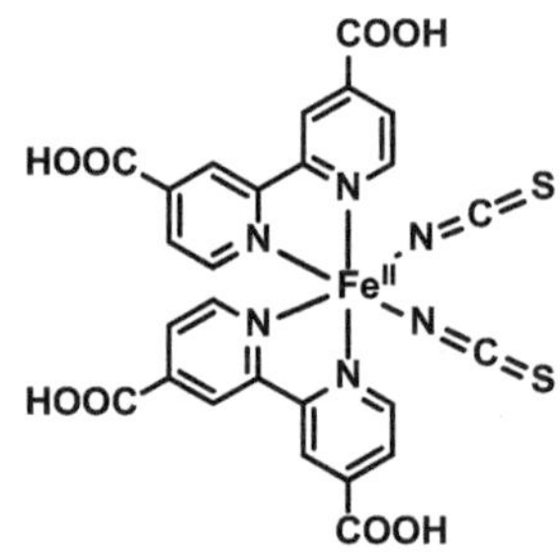

Fe-N3

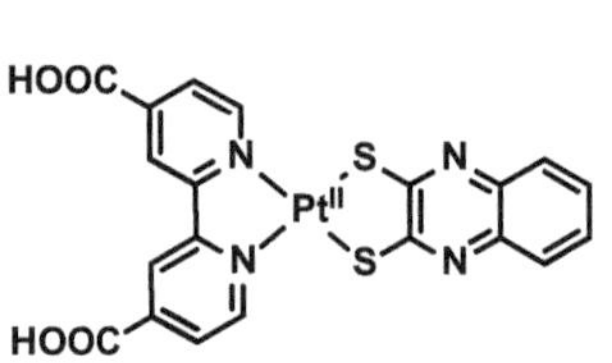

Pt(dcbpy)(qdt)

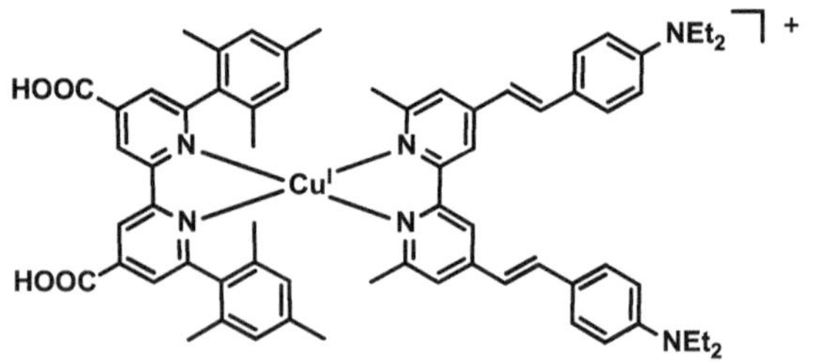

C3

FIGURE 3.25.16 Examples of dye-sensitized solar cell photosensitizers incorporating a variety of transition metal centers alternative to ruthenium including osmium(II),[56] rhenium(I),[57] iron(II),[58] platinum(II),[59] and copper(I).[60]

3.25.4.2 Organic Dyes

Organic dyes have been studied as an alternative to transition metal-based photosensitizers. These dyes have a few advantages compared to their metal counterparts. Firstly, taking advantage of established organic syntheses, a large collection of diverse dyes can be designed. Secondly, organic dyes are generally cheaper and greener than using rare earth metals such as Ru and Os; however, some multistep organic synthesis can be comparatively expensive to rare earth metal sensitizer production. Lastly, organic dyes are capable of exhibiting $\pi \rightarrow \pi^*$ electronic transitions with molar absorption coefficients up to an order of magnitude greater than the ^{1}MLCT transitions of their transition metal complex competitors.[61]

In general, organic sensitizers are typically engineered to have a donor-π-acceptor structure where π-conjugated terminal carboxylic acid groups are often used as both the electron acceptor and the TiO_2 surface-anchoring group. This design strategy imparts a charge-transfer character to the $\pi \rightarrow \pi^*$ electronic transition, which typically has a large transition state dipole moment (μ_{ge}) and difference in ground state-excited state dipole moments ($\mu_{gg}-\mu_{ee}$). So in essence, the dye absorbs a visible photon very efficiently, having a high molar extinction coefficient, and electron density is transferred from the donor moiety to the acceptor through a π-conjugated bridge. Changing the structure and length of the π-conjugated bridge affects the HOMO and LUMO energy levels, giving rise to tunable photophysical and electrochemical properties. This type of design strategy has led to the development of a large number of various metal-free sensitizers. A number of these dyes have been summarized in a 2010 review by Hagfeldt et al. and in a 2015 review by Lee et al.[13,61]

One of the most cited examples of an organic dye is described in a 2006 paper by Koumura et al.[62] who developed a series of carbazole dyes with an oligothiophene spacer consistent with a donor-π-acceptor structure. A major breakthrough occurred when long alkyl groups were substituted on the oligothiophene linker. It was subsequently found that these long alkyl chains, first incorporated to prevent dye aggregation at the surface, also prevent FTO-to-I_3^- and TiO_2-to-I_3^- charge recombination, increasing electron lifetimes in the TiO_2 electrode. The most efficient of these dyes, MK-2 (Fig. 3.25.17), achieved an efficiency of 7.7% with an I_3^-/I^- RM.

Porphyrin systems have been extensively studied due to their ubiquitous presence in nature as light absorbers in photosynthesis. These systems have very high molar extinction coefficients distinguished by the sharp Soret band around 400–450 nm ($\varepsilon > 10^5\ M^{-1}\ cm^{-1}$) and Q-bands in the 500–700 nm region ($\varepsilon > 10^4\ M^{-1}\ cm^{-1}$). While benzoic acid-functionalized porphyrin systems such as ZnTCPP can be directly bound to TiO_2, using them as a light-absorbing π-bridge in a donor-π-acceptor structure has shown to be very effective. Mathew et al. designed the Zn porphyrin system, YD2-*o*-C8, where the donor moiety a diarylamine group, the porphyrin is a 5,10-subsituted diethynyl π-bridge, and the acceptor moiety is a benzoic acid unit (Fig. 3.25.18).[11] When paired with a $[Co^{(III/II)}(bpy)_3]^{3+/2+}$ RM, an efficiency of 11.9% was observed. Furthermore, when cosensitized with another organic dye, Y123, to improve panchromatic absorption for the device, the J_{SC} increased from 17.3 to 17.7 mA/cm^2, the V_{OC} decreased from 0.965 to 0.935 V, and the FF increased from 0.71 to 0.74. Ultimately, the work progressed to the point where a record PCE of 13% was achieved using the same method and the porphyrin SM371, the highest ever recorded for an organic photosensitizer-based DSSC device.

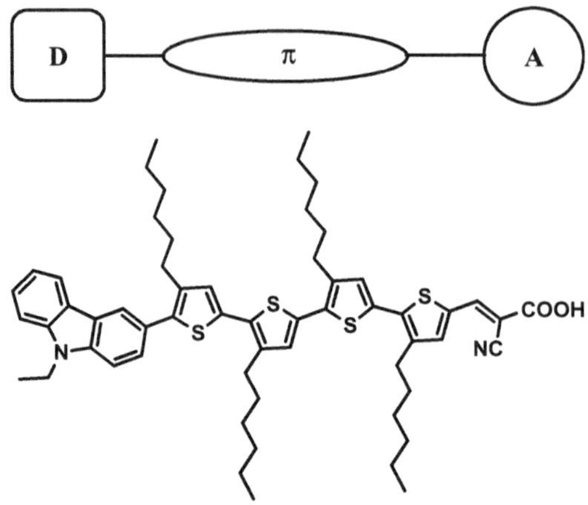

MK-2

FIGURE 3.25.17 Schematic donor-π-acceptor design of organic dyes and the prototypical molecular structure of the MK-2 dye.

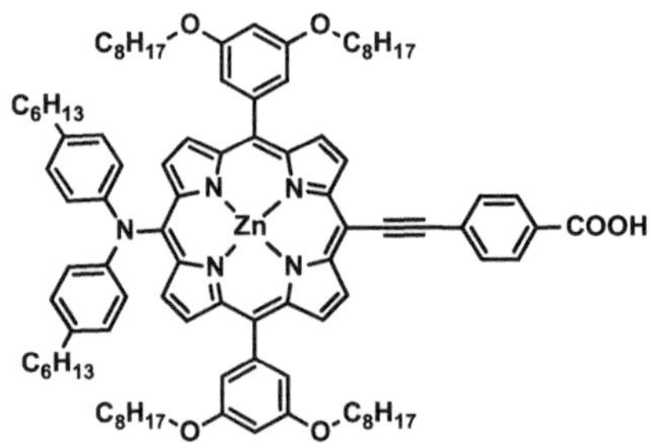

YD2-*o*-C8

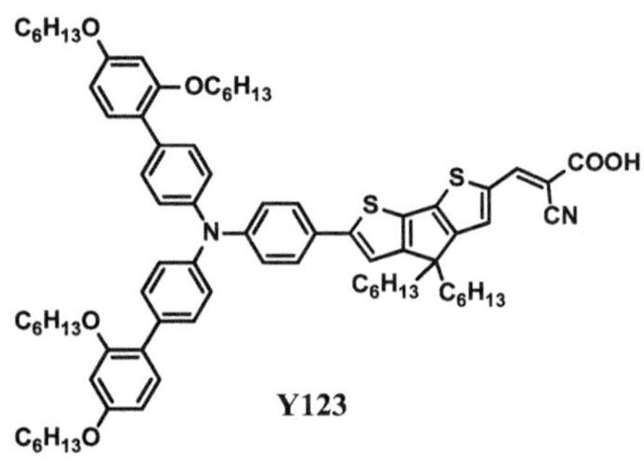

Y123

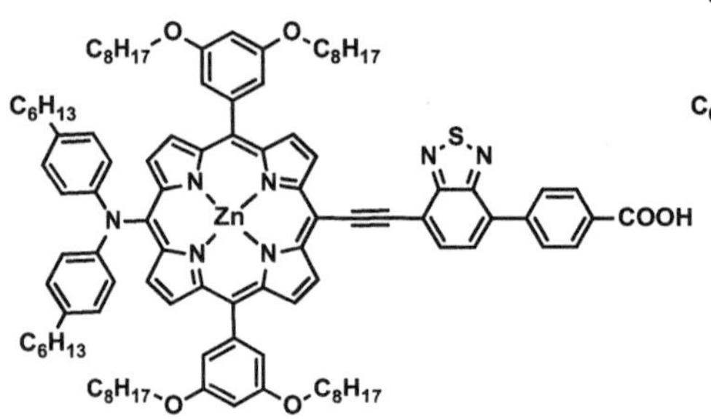

SM371

FIGURE 3.25.18 Molecular structures of the YD2-*o*-C8, Y123, and SM371 dyes used by Mathew et al. where a record 13% dye-sensitized solar cell power conversion efficiency was achieved.[11]

3.25.4.3 Surface-Anchoring Groups

The photosensitizer must be chemically attached to the metal oxide surface to ensure efficient charge injection. Attachment to the electrode surface occurs through anchoring groups, which should allow for rapid charge transfer and strong binding to the electrode surface for increased stability in the DSSC. These anchors chemically bond to the semiconductor surface by reacting with exposed hydroxyl groups and can bind in a number of different modes, which are summarized in Fig. 3.25.19. A commonly used type of anchoring group is carboxylic acid, though other acids such as phosphorous acid, hydroxamic acid, malonic acid, and salicylic acid have been used.[63–65] Carboxylic acids have been shown to be the most efficient at transferring charge to the TiO_2 conduction band, while phosphonic acids tend to be more stable because of their inherently lower pK_a values. Brown et al.[66] developed a unique dye that incorporated both carboxylic acid and phosphorous acid linkers. They showed that electron injection was possible through the carboxylate group and the phosphonate groups simply helped the sensitizer bind to the TiO_2 surface. As a result, dyes with both groups were more strongly bound to TiO_2, increasing the stability of the cell. While the latter dye exhibited a poor efficiency, the study revealed that multiple anchors could be used and that proper anchoring of the dye is key for the longer-term stability of the cell.

The MK-2 dye has been modified to study the effect of various anchoring groups. Koenigsmann et al.[67] replaced the carboxylate-anchoring group with a hydroxamate-anchoring group and saw increased durability with aqueous electrolytes (Fig. 3.25.20). Over the span of 8 days in aqueous environment, no loss in efficiency was observed for the hydroxamate-modified MK-2 device, whereas the device with commercial MK-2 saw a decrease of 50% because of desorption of the dye from the WE. Kakiage et al. designed an alkoxysilyl derivative of MK-2 (ADEKA-1) using trimethoxysilane as the anchoring group (Fig. 3.25.20).[12] In studying the stability of the anchor in an aqueous environment, they observed that 70%–80% of the dye molecules anchored with carboxylate groups desorbed after 120 min, whereas 95% of the dyes anchored by trimethoxysilane remained on the electrode. In addition, a PCE of 12.5% was achieved using ADEKA-1, $[Co(Cl\text{-}phen)_3]^{3+/2+}$ RM, and a novel "multicapping" posttreatment. These examples show how a simple change to the anchoring group can improve stability of the photosensitizer at the photoanode and potentially improve the efficiency and long-term stability of a DSSC device.

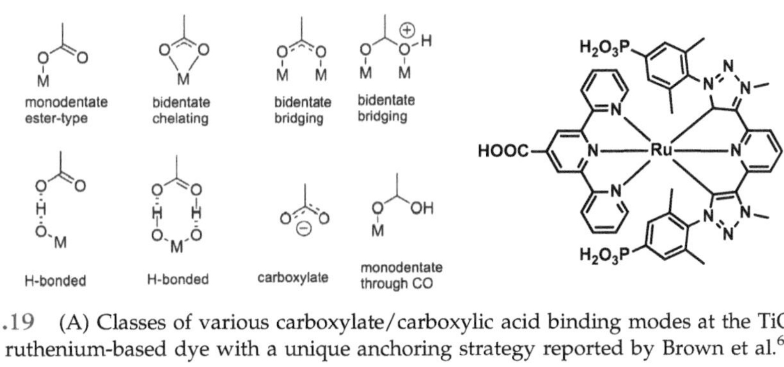

FIGURE 3.25.19 (A) Classes of various carboxylate/carboxylic acid binding modes at the TiO_2 interface. (B) A bis-heteroleptic ruthenium-based dye with a unique anchoring strategy reported by Brown et al.[66]

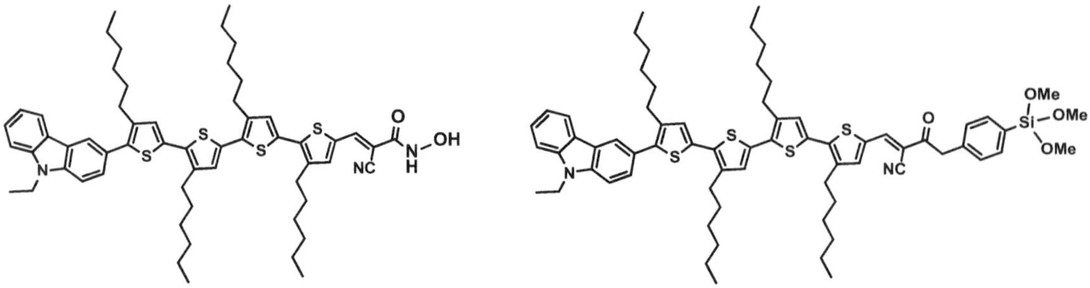

FIGURE 3.25.20 Molecular structures of the MK-2HA[67] and ADEKA-1[11] dyes that incorporate hydroxamic acid and trimethoxysilyl-based anchoring groups, respectively.

3.25.4.4 Redox Mediators and Supporting Electrolyte Formulations

The RM is necessary in a DSSC to transport charge between the cathode and photoanode to complete the circuit of the device. The open circuit potential (V_{OC}) is derived from the difference between the E_{Fermi} level of the semiconductor WE and the electrochemical potential of the RM and is critical to determine the overall efficiency of a DSSC. More specifically, the reduced RM regenerates the oxidized dye molecules by the following mechanism:

$$RM + D^{+} \rightarrow RM^{+} + D \quad (3.25.7)$$

after which the dye can absorb another photon and the oxidized RM^{+} diffuses to the CE to collect an electron and return to its reduced form. The thermodynamics and kinetics of this catalytic cycle are integral to the performance, stability, and longevity of a DSSC device.

A number of different classes of RMs have been used in DSSCs. The triiodide/iodide (I_3^-/I^-) RM is most commonly used primarily because of its ease of preparation and compatibility with a majority of both ruthenium-based and organic photosensitizers. Typically, it consists of roughly a 1:10 mixture of I_3^-:I^- in acetonitrile (e.g., 0.5 M tetrapropylammonium iodide +0.04 M iodide in 80:20 v/v ethylene carbonate:acetonitrile) as incorporated into the first of Grätzel's efficient DSSCs.[6] However, its redox potential and visible light-absorbing qualities can be limiting in a DSSC. Additional RMs include tribromide/bromide (Br_3^-/Br^-), organic systems (e.g., disulfide/thiolate), and inorganic transition metal complexes such as Co(III/II) polypyridyl complexes are discussed in detail in the following.

It is important to note, however, that these redox couples can be paired with a number of different solvents such as water, alcohols, nitriles, esters (e.g., ethylene carbonate), and ionic liquids. Acetonitrile is commonly used for its large electrochemical window, high dielectric constant, low viscosity, and ability to dissolve a number of different compounds. In addition, additives are often used to increase the V_{OC} of a DSSC. For example, *tert*-butylpyridine (TBP) was first used in 1993 by Nazeeruddin et al.[14] and resulted in a notable improvement in V_{OC} after submerging the photoanode in TBP solution for 15 min before cell fabrication. Further studies into the effect of TBP have shown that it reduces the electron recombination rate,[68] attributed to "band bending" of the TiO_2 E_{CB} to a more negative potential,[69] and increased electron lifetimes in TiO_2.[70] Cationic additives, such as Li^+, have also been shown to lower E_{CB} of TiO_2, increasing electron injection rates from the excited state dye to TiO_2 and leading to higher J_{SC}. Liu et al. showed that as the size of the cation increases, the E_{CB} of the electrode is shifted to a more negative potential. While this increases V_{OC}, there is a decrease in J_{SC}, resulting in no significant improvements to PCE.[71] The selection of solvent and additives is key to optimization of any DSSC, but the effects are minimal compared to selection of an appropriate RM. Thus, the remainder of this discussion will focus on RM properties. To summarize, an established and commercially available I_3^-/I^- electrolyte composition now contains a mixture of 1,3-dimethylimidazolium iodide, 1-ethyl-3-methylimidazolium iodide, and iodine in an ionic liquid (Solaronix, Mosalyte TDE-250) or 1,3-dimethylimidazolium iodide and iodine in acetonitrile (Solaronix, Iodolyte HI-30).

3.25.4.4.1 *Triiodide/Iodide*

The most versatile and studied RM is the triiodide/iodide (I_3^-/I^-) couple, which has been the go-to electrolyte for fabrication of many DSSC devices. The I_3^-/I^- RM penetrates into the porous semiconductor because of its small size and quickly regenerates dye molecules if there is sufficient driving force for reduction.[72] In addition, its ability to avoid recombination with electrons in the semiconductor conduction band leads to increased photocurrent, E_{Fermi}, V_{OC}, and ultimately a greater PCE over other redox couples in majority of cases. However, there are a few negative features of the I_3^-/I^- RM that have warranted extensive investigation into alternatives options: (1) iodine is a good oxidizing agent and is thus corrosive, causing deterioration of metal contacts; (2) iodine has a high vapor pressure that can lead to leakage over time; (3) the triiodide ion absorbs visible light, which hinders photoexcitation of dye molecules; and (4) its fixed redox potentials (Eqs. 3.25.10 and 3.25.12) limit the V_{OC} and result in significant thermal loss due to disproportionation of the $I_2^{\bullet-}$ radical anion to generate the I_3^- electron acceptor (Eq. 3.25.11).[73] This difference of approximately 0.4 V theoretically results in an equivalent loss of 0.4 V in the V_{OC} of a DSSC.

$$D^* \rightarrow D^+ + e^-(TiO_2) \tag{3.25.8}$$

$$D^+ + I^- \rightarrow (D\cdots I) \tag{3.25.9}$$

$$(D\cdots I) + I^- \rightarrow D + I_2^{\bullet-} \quad E^0 = 0.79 \text{ V versus NHE}^a \tag{3.25.10}$$

$$2I_2^{\bullet-} \rightarrow I_3^- + I^- \tag{3.25.11}$$

$$I_3^- + 2e^-(CE) \rightarrow 3I^- \quad E^0 = 0.35 \text{ V versus NHE}^a \tag{3.25.12}$$

[a] Standard redox potentials measured in difference acetonitrile by Boschloo et al.[74]

The RM cycle of dye regeneration is summarized as follows: after photoexcitation and electron injection, the oxidized dye is reduced by I^- to form the encounter complex (**D**···**I**). Interaction with a second I^- anion forms the (**D**···$\mathbf{I_2^-}$) complex, which then dissociates to the ground state **D** and $\mathbf{I_2^{\bullet -}}$. Two diiodide radicals subsequently undergo disproportionation to form equal equivalents of the I^- and I_3^- anions. The limited possibility of V_{OC} enhancement when employing the I_3^-/I^- RM in DSSCs is not only a result of the potential loss between the redox potentials of $\mathbf{I^-/I_2^{\bullet -}}$ and I_3^-/I^- but also due to the fact this there is no scope for tuning these redox potentials.

3.25.4.4.2 *Tribromide/bromide*

Similar to the I_3^-/I^- redox couple, a Br_3^-/Br^- RM has been used in conjunction with organic dye molecules. Both being halogen elements, iodine and bromine have similar chemical properties. Owing to the greater electronegativity of the smaller bromine atom, the redox potential of Br_3^-/Br^- is more positive than I_3^-/I^-, which should give higher V_{OC} when paired with appropriate dyes. While the Br_3^-/Br^- RM allows for a significant enhancement in V_{OC} relative to the I_3^-/I^- system, the oxidation potential of Br^- is estimated to be 1.1 V versus NHE, which requires dyes to have lower-lying HOMO energies. This has the negative impact of widening the band gap of dyes compatible with the Br_3^-/Br^- RM, rendering them less efficient at harvesting low-energy visible light.[75] For example, Wang et al.[76] showed that the positive potential of the Br_3^-/Br^- RM limits its effectiveness with particular dyes because of mismatched energy levels and poor dye regeneration. This study did demonstrate, however, that use of the Br_3^-/Br^- RM electrolyte increased PCE to 2.6% from 1.7% when compared to an I_3^-/I^- electrolyte with the eosin Y dye. This improvement was mostly due to the increase in V_{OC} from 0.451 to 0.813 V, consistent with difference in redox potential between I_3^-/I^- and Br_3^-/Br^-.

Overall, the Br_3^-/Br^- RM can lead to DSSCs with a higher V_{OC} but it has seen very limited use because of its positive potential, which limits the short circuit current due to decreased light harvesting of compatible photosensitizers. Br_2 is also a stronger oxidant than I_2, which can lead to some degradation of dyes, especially under light irradiation, limiting the long-term stability of a bromine-based device.

3.25.4.4.3 *Organic Redox Mediators*

Organic RMs are another class of RM that have been studied extensively as an I_3^-/I^- alternative. Fig. 3.25.21 shows several common mediators of this class. One major benefit of organic systems is the possibility of modifying the structure to tune the redox potential and solubility and to introduce steric factors. 2,2,6,6-Tetramethyl-1-piperidinyloxy (TEMPO) was incorporated to a DSSC by Zhang et al. in 2008.[77] It was calculated that the redox potential for a $TEMPO^+/TEMPO^{\bullet}$ was 310 mV more positive than the I_3^-/I^- couple. This difference led to an increased V_{OC} and improved PCE when paired with the D149 dye. In 2010, the first

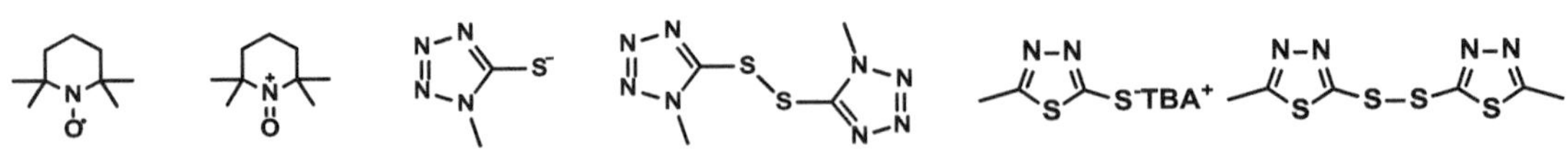

TEMPO TEMPO$^+$ T$^-$ T$_2$ McMT$^-$ BMT

FIGURE 3.25.21 Molecular structures of alternative iodine-free organic redox mediators $TEMPO^+/TEMPO^{\bullet}$,[77] T_2/T^-,[78] and BMT/McMT.[33]

efficient sulfur-based RM, 5-mercapto-1-methyltetrazole and its dimer (T_2/T^-), was introduced by Wang et al.[78] A PCE of 6.4% was obtained using T_2/T^-, matching the efficiency of I_3^-/I^- under the same conditions. Since then, a number of sulfur-based disulfide/thiolate derivatives have been studied. Tian et al. studied a series of RMs derived from mercapto-5-methyl-1,3,4-thiadiazole (McMT) and their dimers (BMT).[33] Incorporating a PEDOT-based cathode, a PCE of 6.0% was achieved.

3.25.4.4.4 Transition Metal Complex Mediators

Inorganic complexes have also been explored as RMs in DSSCs. Inorganic complexes exhibit a greater redox stability compared to organic molecules and their redox potentials can be easily tuned by varying the nature of the chelating ligands or the metal center itself. In addition, inorganic RMs typically exhibit reversible one-electron redox couples that avoid the potential loss as observed in both the I_3^-/I^- and Br_3^-/Br^- RMs. However, each metal-based RM suffers from significant charge recombination at the TiO_2 interface, so dyes or additives must be designed to sufficiently insulate the semiconductor surface.[79] Furthermore, as transition metal-based RMs are typically cationic complexes, there typically exists coulombic repulsion to the formation of the precursor complex to electron transfer between the RM and the oxidized form of the photosensitizer.

Ferrocenium/ferrocene (Fc^+/Fc) and similar iron-based redox couples are known for their stable redox properties and are often used as a reference standard in electrochemical analysis. Daeneke et al.[80] achieved a PCE of 7.5% using the Fc^+/Fc RM and a PCE of 6.1% with I_3^-/I^- when combined with the novel organic dye Carbz-PAHTDTT and chenodeoxycholic acid (cheno) as a coabsorbent (Fig. 3.25.22). This study noted a V_{OC} increase of 100 mV when switching from the I_3^-/I^- RM to Fc^+/Fc RM and suggested that modification of the substituents on the cyclopentadienyl rings could further fine-tune the redox potential of metallocene-based redox couples.

Bai et al.[81] reported a PCE of 7.0% utilizing the tetrahedral Cu(2,9-dimethyl-1,10-phenanthroline)$_2$ redox couple ($[Cu(dmp)_2]^{2+/+}$) and the C218 dye, which incorporates long alkyl chains to protect the TiO_2 surface (Fig. 3.25.23). An open circuit potential of 932 mV was achieved with their cobalt electrolyte, nearly 200 mV more than an analogous device fabricated with an I_3^-/I^- electrolyte. In 2014, Colombo et al. studied Cu(2,9-dimethyl-4,7-diphenyl-

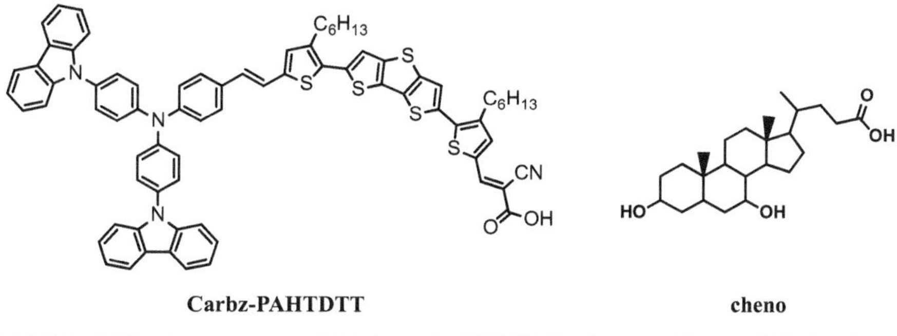

FIGURE 3.25.22 Molecular structures of (A) the carbz-PAHTDTT photosensitizer and (B) the chenodeoxycholic acid (cheno) coabsorbent.[80]

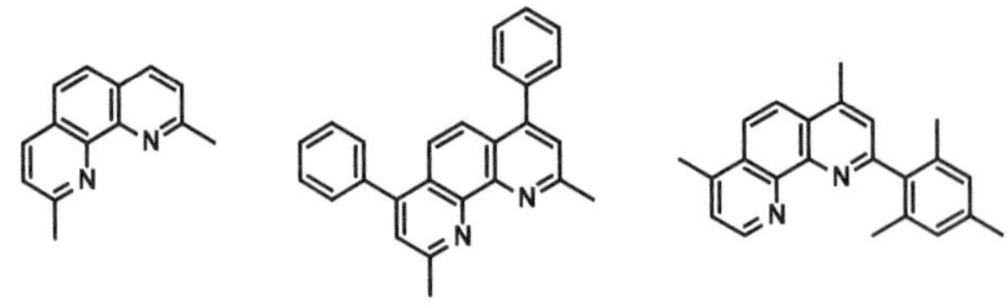

FIGURE 3.25.23 Polypyridyl ligands used in studying Cu(II/I)-based redox mediators: (A) dmp = 2,9-dimethyl-1,10-phenanthroline, (B) dmdpp = 2,9-dimethyl-4,7-diphenyl-1,10-phen-anthroline, and mdmp = 2-mesityl-4,7-dimethyl-1,10-phenanthroline.[81,82]

1,10-phenanthroline)$_2$ ($[Cu(dmdpp)_2]^{3+/2+}$) and Cu(2-mesityl-4,7-dimethyl-1,10-phenanthroline)$_2$ ($[Cu(mdmp)_2]^{3+/2+}$), adding bulky substituents to the efficient $[Cu(dmp)_2]^{2+/+}$ RM (Fig. 3.25.23). They found that compared to $[Cu(dmp)_2]^{2+/+}$, $[Cu(mdmp)_2]^{3+/2+}$ leads to superior dye regeneration, longer conduction band electron lifetimes, and increased efficiency.[82] Copper RMs are still relatively new but could be very promising because of the possibility for high V_{OC} when the redox potential for the copper RM is matched with the HOMO energy level of the photosensitizer for sufficient dye regeneration.

Of all inorganic RMs, cobalt(III/II) polypyridyl-based mediators have garnered the most interest because of their low visible light absorption and redox tenability. These properties are critical when optimizing V_{OC} of a DSSC and tailoring energy level alignment with a specific photosensitizer combination. However, the performance of cobalt(III/II) RMs is restricted by rapid recombination of electrons from the metal oxide conduction band and the Co(III) species, slow dye regeneration, and poor mass transport (depending on the steric bulk of the cobalt coordination environment) through the mesoporous WE. All of this considered, implementation of cobalt(III/II) polypyridyl mediators has shown great promise with a variety of sensitizers.

In a 2002 paper, Sapp et al.[83] studied a number of cobalt complexes with substituted polypyridine ligands as RM with the ruthenium N3 dye. This comprehensive study showed that while a Pt CE is sufficient for regeneration of I^-, it may not be the best for cobalt(III/II) polypyridyl RMs, and carbon-based CEs were identified as a more suitable alternative cathode. Sapp et al. also demonstrated that the addition of Li^+ ions to the electrolyte increased both the J_{SC} and the V_{OC} of the DSSC, possibly due to limited charge recombination. A PCE of 1.3% with $[Co(dtb)_3]^{3+/2+}$ was achieved representing about 80% of the efficiency observed by the I_3^-/I^- RM under identical conditions.

The mass transport properties of cobalt(III/II) polypyridyl RMs were studied by Nelson et al.[84] Working with the bulky $[Co(t\text{-}Bu_2bpy)_3]^{3+/2+}$ RM, a sharp decline in photocurrent was observed within 1 s of illumination, whereas a minimal decline in current was noted for the I_3^-/I^- RM. The rate of diffusion of cobalt(III/II) polypyridyl RMs through a TiO_2 anode was measured using rotating disk electrode voltammetry and was shown to be at least an order of magnitude slower than the I_3^-/I^- RM. This difference can be attributed to the different sizes of the molecules, the opposing charges, and possibly the composition of the TiO_2 electrode. Increasing the porosity of the electrode or the viscosity of the electrolyte could have major effects on the efficiency of a cobalt-based DSSC. Klahr and Hamann[85] reported that cobalt(III/II) polypyridyl RMs are not limited by their ability to regenerate dyes. This study attributed the variance in PCE of cobalt complexes not to the different redox potentials but due to recombination rates that are heavily limited by mass transport. In addition, this study found that a super thin layer of Al_2O_3 at the surface of TiO_2 significantly reduces charge recombination with cobalt(III/II) polypyridyl RMs and improves the overall PCE.

A number of studies have since reported DSSC designs in which efficiencies utilizing cobalt(III/II) polypyridyl RMs rival or surpass that of the I_3^-/I^- RM. Polander et al.[37] developed a very bulky ruthenium orthometallated photosensitizer (Fig. 3.25.24) with long alkoxy chains to passivate the TiO_2 electrode. This dye, when paired with $[Co(phen)_3]^{3+/2+}$, showed a V_{OC} of 0.837 V and a PCE of 8.6%, equivalent to a similar cell with an I_3^-/I^- RM. Wu et al.[86] also have designed Ru sensitizers with bulky ligands. Their dye (TFRS-80c), when paired with the same $[Co(phen)_3]^{3+/2+}$ RM, gave a PCE of 9%, surpassing the 7% PCE observed with an I_3^-/I^- RM.

In 2016, the highest PCE to date for a ruthenium photosensitizer with a cobalt mediator was reported by Wu and Nazeeruddin et al.[87] The dye molecules employed in this study use isoquinazolylpyrazolate ligands further functionalized with thiophene groups to improve light absorption (Fig. 3.25.25). At the terminal end of the thiophene groups, long hanging alkyl chains were incorporated to act as steric buffers to the cobalt electrolyte, thus insulating the semiconductor surface from Co(III) species and preventing charge recombination. Furthermore, the isoquinazolylpyrazolate ligand is further functionalized at the pyrazole ring with fluorinated alkyl chains to also tune the Ru(III/II) redox potential, while again providing steric bulk to insulate the semiconductor surface from undesired charge recombination with Co(III). The best performing photosensitizer from this study is "51-57dht.1," as illustrated in Fig. 3.25.25, exhibiting a PCE of 9.5%.

Some of the most efficient DSSCs have been fabricated using an organic dye and cobalt RM. In 2010, Feldt et al.[88] studied the photophysical and electrochemical properties of four different cobalt(III/II) polypyridyl RMs: $[Co(bpy)_3]^{3+/2+}$, $[Co(dmb)_3]^{3+/2+}$, $[Co(dtb)_3]^{3+/2+}$, and $[Co(phen)_3]^{3+/2+}$ (Fig. 3.25.26). When $[Co(bpy)_3]^{3+/2+}$ was paired with the D35 photosensitizer (Fig. 3.25.7), a PCE of 6.7% was achieved, the highest of any iodide-free DSSC at

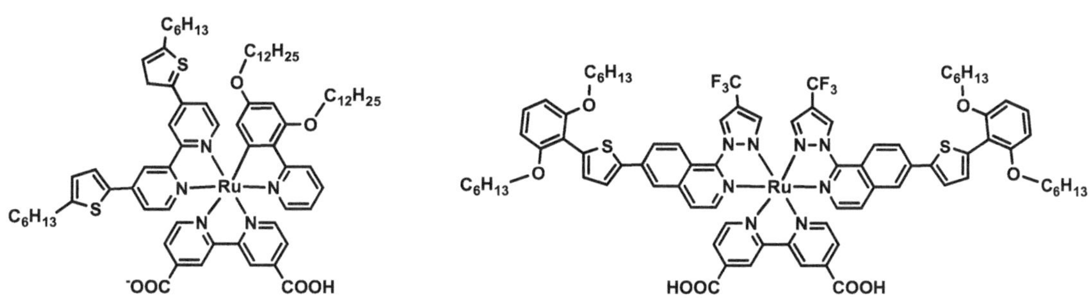

TFRS-80c

FIGURE 3.25.24 Molecular structures of ruthenium-based photosensitizers designed for compatibility with a cobalt-based redox mediator designed by (A) Polander et al.[37] and (B) Wu et al.[86] Note that steric groups play an important role in protecting the metal center from direct reactions with the mediator.

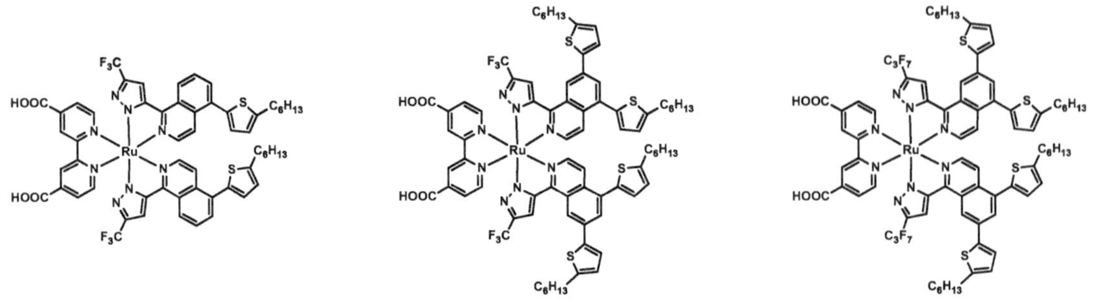

51-5ht **51-57dht** **51-57dht.1**

FIGURE 3.25.25 Record-holding ruthenium photosensitizers for Co(III/II)-based redox mediator dye-sensitized solar cells as reported by Wu and Nazeeruddin et al. where the "51-57dht.1" dye reached a power conversion efficiency of 9.5%.[87]

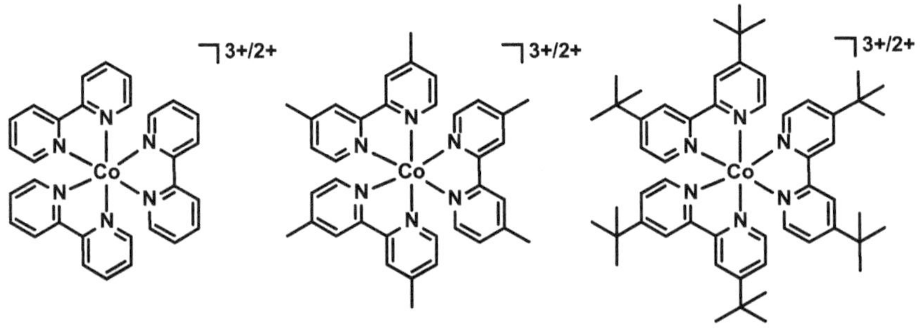

$[Co(bpy)_3]^{3+/2+}$ $[Co(dmb)_3]^{3+/2+}$ $[Co(dtb)_3]^{3+/2+}$

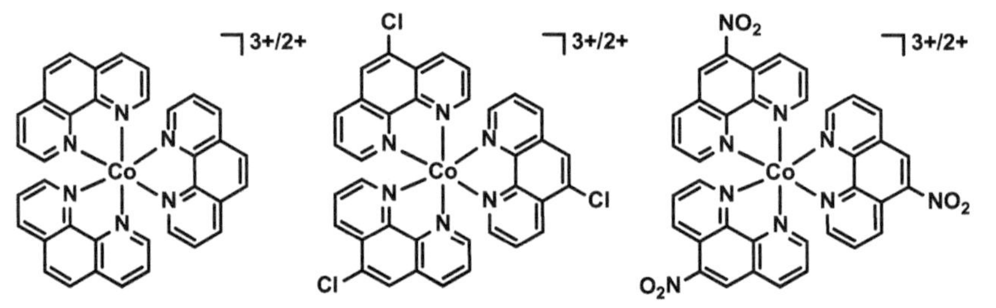

$[Co(phen)_3]^{3+/2+}$ $[Co(Cl\text{-}phen)_3]^{3+/2+}$ $[Co(NO_2\text{-}phen)_3]^{3+/2+}$

FIGURE 3.25.26 Structure of Co(III/II)-based redox mediators studied by Feldt et al. (bpy = 2,2′-bipyridine, dmb = 4,4′-dimethyl-2,2′-bipyridine, dtb = 4,4′-di-*tert*-butyl-2,2′-bipyridine, phen = 1,10-phenanthroline, Cl-phen = 5-chloro-1,10-phenanthroline, NO_2-phen = 5-nitro-1,10-phenanthroline).[88]

that time. Feldt et al. also concluded that the insulating effect of the alkoxyl groups on the D35 dye prevented facile charge recombination and eliminated the need for additional surface-insulting layers. The passivation of TiO_2 by pendant dye alkoxy groups also allowed for the use of less bulky polypyridyl ligands on the Co(III/II) RMs, increasing the mass transfer capability of the RM. A year later, the same authors reported a remarkable V_{OC} of

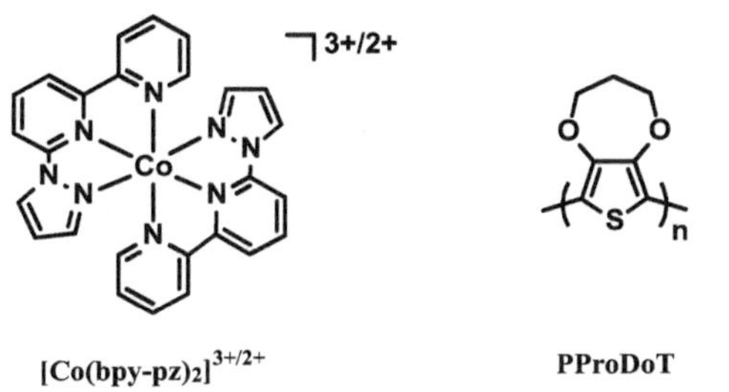

FIGURE 3.25.27 Molecular structures of the $[Co(bpy\text{-}pz)_2]^{3+/2+}$ redox mediator and PProDoT used for fabrication of a Pt-free counter electrode [bpy-pz = 6-(1H-pyrazol-1-yl)-2,2′-bipyridine].[90]

over 1 V for a DSSC with D35 and a cobalt RM.[89] By introducing electron-withdrawing chlorine or nitro substituents to the phenanthroline ligands of $[Co(phen)_3]^{3+/2+}$, the redox potential was shifted to a more positive potential, thus increasing V_{OC}. Unfortunately, this change in redox potential limited the amount of driving force for dye regeneration, thus reducing the photocurrent of the DSSC. This study represented a turning point for DSSC development in that a V_{OC} of over 1 V was achieved and could lead to very-high-performing DSSCs if paired with a suitable photosensitizer having an appropriately aligned oxidation potential (e.g., Mathew et al.[11]).

Yum et al.[90] reported a PCE of over 10% for an organic dye Y123 (Fig. 3.25.27) and a new cobalt RM with a tridentate ligand system bpy-pz. The complex, $[Co(bpy\text{-}pz)_2]^{3+/2+}$, has a redox potential of 0.86 V versus NHE, which leads to a V_{OC} of nearly 1 V, whereas the I_3^-/I^- RM managed a V_{OC} of just 0.75 V. To further optimize the DSSC, a Pt-free CE composed of poly(3,4-propylenedioxythiophene) (PProDoT) layers was also used. This significantly improved the fill factor and led to a PCE of over 10%. This study is an excellent example of how selecting the proper dye and RM pair along with modification to the CE could lead to high-performing DSSCs while avoiding costly materials such as Pt.

3.25.5 PEROVSKITES: EMERGING SOLAR CELL PHOTOSENSITIZERS

Beyond the silicon and dye-sensitized TiO_2 solar cells discussed earlier, a number of interesting alternative solutions have been emerging in the recent years. Fig. 3.25.28 illustrates how quickly various solar cell technologies are experiencing efficiency gains. A prime example of such a photovoltaic technology is the class of perovskite solar cells. Perovskite solar cells have seen notable efficiency increases in the recent years, rising to almost 22% efficiency.[91] Perovskite solar cells have taken particular inspiration from dye-sensitized systems and share common optimization strategies.

3.25.5.1 Perovskite Solar Cells

The general term "perovskite" can refer to a wide range of compounds that share the same crystal structure and the formula ABX_3. The crystal structure of a typical perovskite material

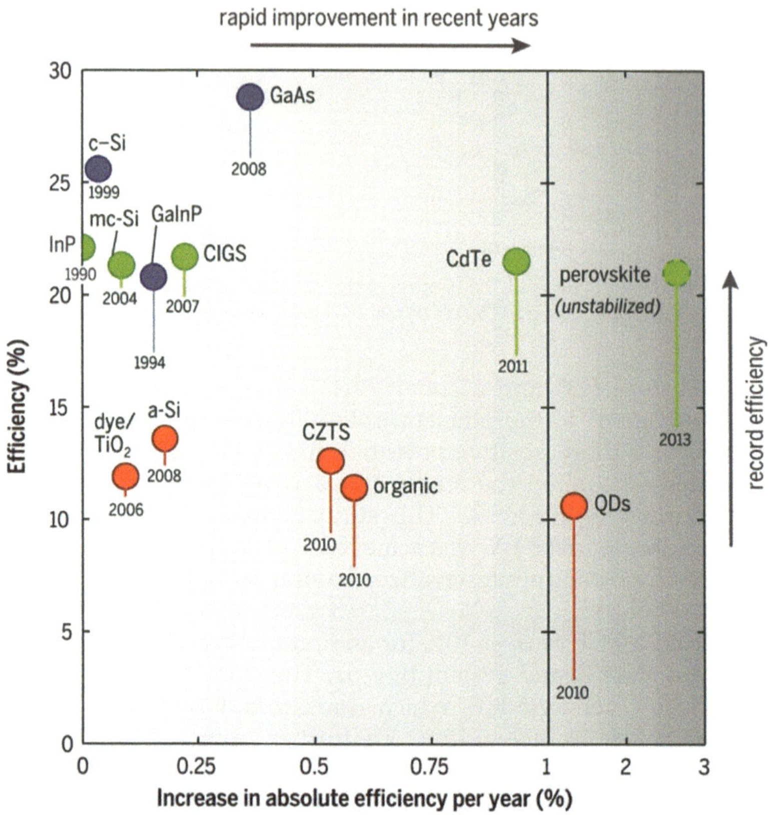

FIGURE 3.25.28 Increase in power conversion efficiencies plotted for various classes of solar cells per year. Perovskites sit in a unique position where reported absolute efficiencies have increased greatly every year for about the past decade. These solar cells are rapidly encroaching on record efficiencies within the field. *Reproduced with permission from Polman A, Knight M, Garnett EC, Ehrler B, Sinke WC. Photovoltaic materials: present efficiencies and future challenges.* Sci Mag *2016;***352***:6283. aad4244-1-10.*

is shown in Fig. 3.25.29. In terms of perovskite solar cells, most arise from a specific series of mixed organic-inorganic halide perovskite materials: those having an organic cation such as methylammonium (A), an inorganic cation such as Pb(II) (B), and a halide or mixed group of halides (X).[91,92]

Mixed organic-inorganic halide perovskite materials were developed and recognized to have potential for solar cell devices as early as in 1995,[92,93] but the materials significantly lagged in efficiency gains behind other solar cell technologies such as those employing ruthenium dye sensitizers. The highest efficiency of a perovskite system was reported using $CH_3NH_3PbBr_3$ at 2.2% by Miyasaka et al.[92,94] This was followed by an important breakthrough by the same group when the bromide ions in the previous perovskite were replaced with iodide ions. This effectively tuned the material's band gap from 3.3 to 1.6 eV in the iodide-containing perovskite and corresponded with an efficiency increase to 3.8%.[95]

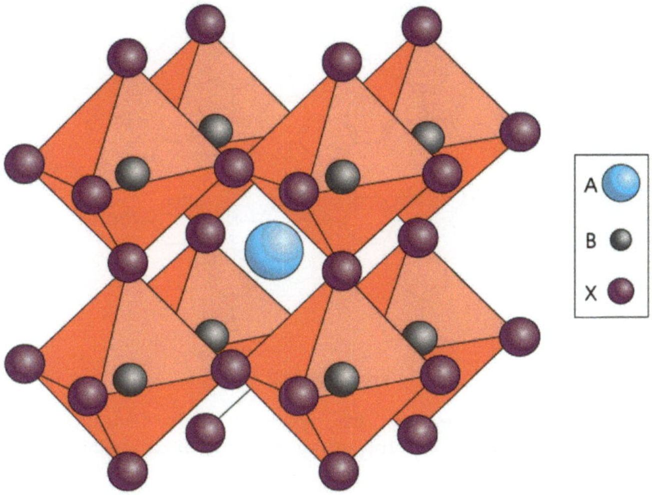

FIGURE 3.25.29 Crystal structure of perovskite materials with the formula ABX_3. In most perovskite solar cell materials, A is the organic cation methylammonium, B is Pb(II), and X is a halide. *Reproduced with permission from Green MA, Ho-Baillie A, Snaith HJ. The emergence of perovskite solar cells.* Nat Photonics *2014;8:506–14.*

Thus, methylammonium lead iodide, $CH_3NH_3PbI_3$, became the dominant perovskite component in such solar cell devices.

Recognizing the importance of $CH_3NH_3PbI_3$ as a potentially high-efficiency perovskite material with a more amenable band gap for solar cell work, researchers set out to optimize the fabrication of perovskite-based cells. Cells had been previously constructed around a simple solution-based organic electrolyte of lithium halide sandwiched between FTO glass and an appropriate metal electrode.[92,96] Merging work from the field of DSSCs with their own, Park et al. optimized TiO_2 nanoparticles deposited on the FTO glass surface for use with the $CH_3NH_3PbI_3$ perovskite. This led to a 6.5% efficiency in 2011[97]; this result is historically important, as it beats the efficiency of the standard DSSC format N719 dye under similar working conditions.[92,96]

One of the more pervasive problems of perovskite materials used in solar cell work is stability. In liquid electrolyte, this is especially pronounced as the perovskite salt readily dissolves, eventually leading to catastrophic performance decrease as the perovskite parts contact with the TiO_2 layer on the FTO.[96] This necessitated a move to a solid-state electrolyte, or a hole transport medium (HTM). Park and Grätzel recognized spiro-OMeTAD (2,2′,7,7′-tetrakis(*N*,*N*-di-*p*-methoxyphenylamine)-9,9′-spirobifluorene, Fig. 3.25.30) as an ideal HTM for perovskites (again inspired by earlier DSSC studies), reporting a better cell stability. The solid-state HTM also resulted in an improved efficiency of 9.7% because of the optimized hole transport across the solar cell.[98] Snaith et al.[99] further improved the efficiency of the $CH_3NH_3PbI_3$ perovskite with a spiro-OMeTAD HTM by replacing conductive TiO_2 with a nonconductive Al_2O_3 scaffold, increasing the efficiency of the solar cell to 10.9% by improving V_{OC}.

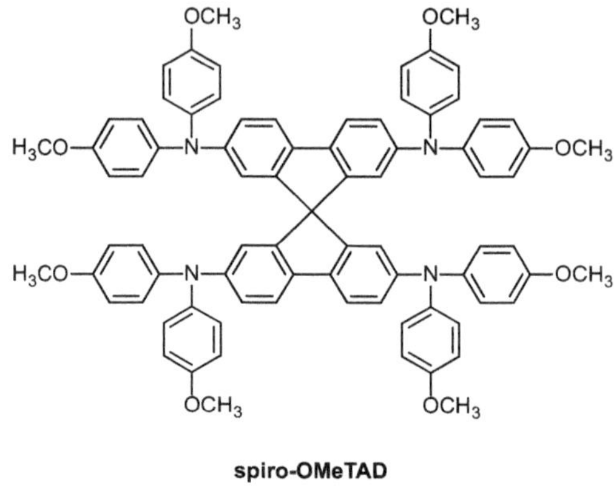

FIGURE 3.25.30 Molecular structure of the popular spiro-OMeTAD [2,2′,7,7′-tetrakis(*N*,*N*-di-*p*-methoxyphenylamine)-9,9′-spirobifluorene] hole-transporting medium for solid-state perovskite and dye-sensitized solar cells.

To this point, the solar cell architectures discussed have utilized TiO_2 nanoparticles acting as a scaffold for thin layers of perovskites, supported between an FTO cathode and HTM-separated reflective metal anode. Grätzel and Seok iterated slightly on this design to improve efficiency to 12%.[100] Two major changes were made to reach this point. First, the spiro-OMeTAD HTM was replaced with poly-triarylamine (Fig. 3.25.31). This improved the efficiency by improving V_{OC} and thus the fill factor.[100] Second, an additional layer of unbound perovskite material was introduced between the HTM layer and the perovskite-coated nanoporous TiO_2. The layer of unbound perovskite material was determined to play an important role in electron-hole transport, and by careful manipulation of the ratio between the TiO_2-supported perovskite layer and the unbound layer (also called the capping layer, or overlayer), Seok achieved an efficiency in excess of 16%.[101]

In the recent years, efficiencies approaching and in excess of 20%[91] have been achieved in perovskite systems. These gains have been facilitated beyond the point of Seok[101] and Grätzel's[100] optimizations of solar cell construction by improvements in cell construction

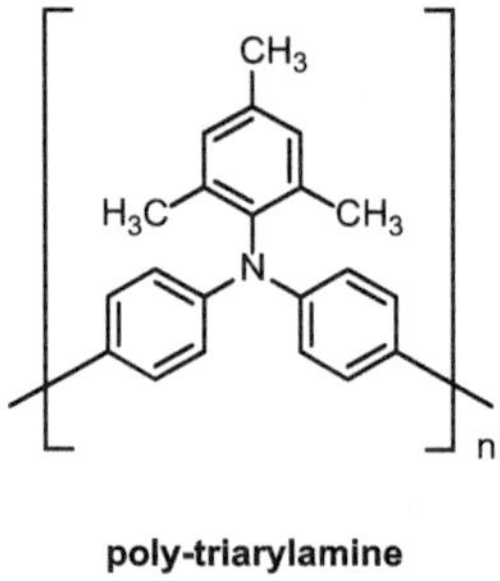

FIGURE 3.25.31 Poly-triarylamine is an important hole-transporting medium for perovskite solar cells and exhibits an improvement in V_{OC} compared to the standard spiro-OMeTAD.[92,100]

and introduction of fractional additives to the perovskite structure in attempts to improve crystal morphology. One of the first and most successful fractional additives employed was formamidinium.[102] Formamidinium replaces some small fraction of the methyl ammonium in the perovskite structure and results in a more symmetrical cubic phase and corresponding efficiency increase.[92] For this reason, formamidinium factions are prevalent throughout high-efficiency perovskites. Along the same lines, Saliba et al. added a third fractional additive to the mix of methylammonium and formamidinium: cesium. By using a ratio of this set of three ions in the A position of the perovskite structure (ABX_3), Saliba et al.[103] were able to increase the solar cell efficiency to 21.1%. Even higher efficiencies have now been reported for cesium-free systems, with the current record for perovskite solar cell efficiency held by Seok et al.[104] at 22% through improvements in cell fabrication.

The main issues that prevent wide implementation of perovskite solar cells are the stability of the perovskite material and of course environmental factors of a Pb-based device in widespread practice. Degradation due to heat, moisture exposure, or slow alteration of crystal structure during operation cycles can all hamper cell efficiency significantly.[96,105] Thus, it is critical to use the most stable perovskite possible rather than just trying to protect it from these destabilizing influences. Snaith et al. made the first advance in this area, discovering that a perovskite formula of $CH_3NH_3PbI_{3-x}Cl_x$, where the perovskite material incorporates some fractional amount of chloride into its structure instead of iodide alone, is significantly more stable and extends operational lifespan.[99] $CH_3NH_3PbI_{3-x}Br_x$ is even more stable, but the widening band gap due to the inclusion of bromide reduces cell efficiency.[106] Additives of formamidinium and cesium were shown by Saliba et al.[103] to improve stability and operating lifetime as well. But even in the case of the more stable cesium/methylammonium/formamidinium perovskite system reported, the operating lifetime of most perovskites remains on the order of hundreds of hours rather than the 10,000+ h of desired lifetime for commercial application.[91,92] So as research into improving efficiency tapers off as perovskites grow toward their Shockley-Quiesser limit, one can expect research on stability improvement to rise steeply. Presently, perovskite solar cell devices are potentially cheap, relatively easy to manufacture, and highly efficient—the lagging characteristics are stability and potentially environmental factors.

Research addressing the fact that lead-based perovskites are not environmentally friendly (they are soluble in water and toxic to human health and the environment) has been steadily increasing. Attempts to replace Pb in perovskite solar cells will be worth watching but have met with mixed success thus far in terms of reproducing their efficiency and stability. Tin has been a potential replacement of interest and has a theoretically higher efficiency limit because of a smaller band gap. But stability for tin compounds is significantly reduced when compared with lead because of easier oxidation of the tin compounds. However, any advances in general perovskite stability in the future may revive tin as a potential alternative to lead if the stability increase is significant.[91,92]

References

1. International Energy Agency. *2016 Key world energy statistics*. Paris: IEA; 2016.
2. British Petroleum. *BP statistical review of world energy*. London: British Petroleum; 2015.
3. The White House. *Climate change and President Obama's action plan*. https://www.whitehouse.gov/climate-change.

4. Lewis NS, Nocera DG. Powering the planet: chemical challenges in solar energy utilization. *Proc Natl Acad Sci USA* 2006;**103**:15729–35.
5. Shockley W, Queisser HJ. Detailed balance limit of efficiency of *p-n* junction solar cells. *J Appl Phys* 1961;**32**:510–9.
6. Dimroth F, Grave M, Beutel P, Fiedeler U, Karcher C, Tibbits TND, et al. Wafer bonded four-junction GaInP/GaAs//GaInAsP/GaInAs concentrator solar cells with 44.7% efficiency. *Prog Photovoltaics Res Appl* 2014;**22**:277–82.
7. United States Department of Energy, NREL/NCPV. *Best research-cell efficiencies*. http://www.nrel.gov/ncpv/.
8. O'Regan B, Grätzel M. A low-cost, high-efficiency solar cell based on dye-sensitized colloidal TiO_2 films. *Nature* 1991;**353**:737–40.
9. Matsumura M, Nomura Y, Tsubomura H. Dye-sensitization on the photocurrent at zinc oxide electrode in aqueous electrolyte solution. *Bull Chem Soc Jpn* 1977;**50**:2533–7.
10. Alonso VN, Beley M, Chartier P, Ern V. Dye sensitization of ceramic semiconducting electrodes for photoelectrochemical conversion. *Rev Phys Appl* 1981;**16**:5–10.
11. Mathew S, Yella A, Gao P, Humphry-Baker R, Curchod BFE, Ashari-Astani N, et al. Dye-sensitized solar cells with 13% efficiency achieved through the molecular engineering of porphyrin sensitizers. *Nat Chem* 2014;**6**:242–7.
12. Kakiage K, Aoyama Y, Yano T, Otsuka T, Kyomen T, Unno M, et al. An achievement of over 12 percent efficiency in an organic dye-sensitized solar cell. *Chem Commun* 2014;**50**:6379–81.
13. Hagfeldt A, Boschloo G, Sun L, Kloo L, Pettersson H. Dye-sensitized solar cells. *Chem Rev* 2010;**110**:6595–663.
14. Nazeeruddin MK, Kay A, Rodicio I, Humphry-Baker R, Mueller E, Liska P, et al. Conversion of light to electricity by cis-X_2bis(2,2′-bipyridyl-4,4′-dicarboxylate)ruthenium(II) charge-transfer sensitizers (X = Cl-, Br-, I-, CN-, and SCN-) on nanocrystalline titanium dioxide electrodes. *J Am Chem Soc* 1993;**115**:6382–90.
15. Cameron PJ, Peter LM. How important is the back reaction of electrons via the substrate in dye-sensitized nanocrystalline solar cells? *J Phys Chem B* 2005;**109**:7392–8.
16. Zhou Y, Lee NA, Ngo KT, Peng X, Feng Y, Rochford J. Rigid triarylamine donor–π–acceptor porphyrin dyes and their application in dye-sensitized solar cells. *RSC Adv* 2015;**5**:41193–202.
17. Tiwana P, Docampo P, Johnston MB, Snaith HJ, Herz LM. Electron mobility and injection dynamics in mesoporous ZnO, SnO_2, and TiO_2 films used in dye-sensitized solar cells. *ACS Nano* 2011;**5**:5158–66.
18. Katoh R, Furube A, Yoshihara T, Hara K, Fujihashi G, Takano S, et al. Efficiencies of electron injection from excited N3 dye into nanocrystalline semiconductor (ZrO_2, TiO_2, ZnO, Nb_2O_5, SnO_2, In_2O_3) films. *J Phys Chem B* 2004;**108**:4818–22.
19. Ito S, Liska P, Comte P, Charvet R, Péchy P, Bach U, et al. Control of dark current in photoelectrochemical (TiO_2/I^--I_3^-) and dye-sensitized solar cells. *Chem Commun* 2005;**34**:4351–3.
20. So S, Hwang I, Schmuki P. Hierarchical DSSC structures based on "single walled" TiO_2 nanotube arrays reach a back-side illumination solar light conversion efficiency of 8%. *Energy Environ Sci* 2015;**8**:849–54.
21. Zhang QF, Dandeneau CS, Zhou XY, Cao GZ. ZnO nanostructures for dye-sensitized solar cells. *Adv Mater* 2009;**21**:4087–108.
22. Němec H, Rochford J, Taratula O, Galoppini E, Kužel P, Polívka T, et al. Influence of the electron-cation interaction on electron mobility in dye-sensitized ZnO and TiO_2 nanocrystals: a study using ultrafast terahertz spectroscopy. *Phys Rev Lett* 2010;**104**:1–4.
23. Saito M, Fujihara S. Large photocurrent generation in dye-sensitized ZnO solar cells. *Energy Environ Sci* 2008;**1**:280–3.
24. Keis K, Lindgren J, Lindquist SE, Hagfeldt A. Studies of the adsorption process of Ru complexes in nanoporous ZnO electrodes. *Langmuir* 2000;**16**:4688–94.
25. Sakai N, Miyasaka T, Murakami TN. Efficiency enhancement of ZnO-based dye-sensitized solar cells by low-temperature $TiCl_4$ treatment and dye optimization. *J Phys Chem C* 2013;**117**:10949–56.
26. Kumar EN, Jose R, Archana PS, Vijila C, Yusoff MM, Ramakrishna S. High performance dye-sensitized solar cells with record open circuit voltage using tin oxide nanoflowers developed by electrospinning. *Energy Environ Sci* 2012;**5**:5401–7.
27. Rani RA, Zoolfakar AS, Subbiah J, Ou JZ, Kalantar-zadeh K. Highly ordered anodized Nb_2O_5 nanochannels for dye-sensitized solar cells. *Electrochem Commun* 2014;**40**:20–3.

28. Wang H, Hu YH. Graphene as a counter electrode material for dye-sensitized solar cells. *Energy Environ Sci* 2012;**5**:8182–8.
29. Kay A, Grätzel M. Low cost photovoltaic modules based on dye sensitized nanocrystalline titanium dioxide and carbon powder. *Sol Energy Mater Sol Cells* 1996;**44**:99–117.
30. Wang H, Sun K, Tao F, Stacchiola DJ, Hu YH. 3D honeycomb-like structured graphene and its high efficiency as a counter-electrode catalyst for dye-sensitized solar cells. *Angew Chem Int Ed* 2013;**52**:9210–4.
31. Park JG, Akhtar MS, Li ZY, Cho DS, Lee W, Yang OB. Application of single walled carbon nanotubes as counter electrode for dye sensitized solar cells. *Electrochim Acta* 2012;**85**:600–4.
32. Lee WJ, Ramasamy E, Lee DY, Song JS. Efficient dye-sensitized solar cells with catalytic multiwall carbon nanotube counter electrodes. *ACS Appl Mater Interfaces* 2009;**1**:1145–9.
33. Tian H, Yu Z, Hagfeldt A, Kloo L, Sun L. Organic redox couples and organic counter electrode for efficient organic dye-sensitized solar cells. *J Am Chem Soc* 2011;**133**:9413–22.
34. Wang M, Anghel AM, Marsan B, Ha NLC, Pootrakulchote N, Zakeeruddin SM, et al. CoS supersedes Pt as efficient electrocatalyst for triiodide reduction in dye-sensitized solar cells. *J Am Chem Soc* 2009;**131**:15976–7.
35. Tian H, Gabrielsson E, Yu Z, Hagfeldt A, Kloo L, Sun L. A thiolate/disulfide ionic liquid electrolyte for organic dye-sensitized solar cells based on Pt-free counter electrodes. *Chem Commun* 2011;**47**:10124–6.
36. Bomben PG, Robson KCD, Koivisto BD, Berlinguette CP. Cyclometalated ruthenium chromophores for the dye-sensitized solar cell. *Coord Chem Rev* 2012;**256**:1438–50.
37. Polander LE, Yella A, Curchod BFE, Astani NA, Teuscher J, Scopelliti R, et al. Towards compatibility between ruthenium sensitizers and cobalt electrolytes in dye-sensitized solar cells. *Angew Chem Int Ed* 2013;**52**:8731–5.
38. Nazeeruddin MK, Zakeeruddin SM, Humphry-Baker R, Jirousek M, Liska P, Vlachopoulos N, et al. Acid–base equilibria of (2,2′-bipyridyl-4,4′-dicarboxylic acid)ruthenium(II) complexes and the effect of protonation on charge-transfer sensitization of nanocrystalline titania. *Inorg Chem* 1999;**38**:6298–305.
39. Nazeeruddin MK, Péchy P, Renouard T, Zakeeruddin SM, Humphry-Baker R, Cointe P, et al. Engineering of efficient panchromatic sensitizers for nanocrystalline TiO_2-based solar cells. *J Am Chem Soc* 2001;**123**:1613–24.
40. Wang P, Zakeeruddin SM, Moser JE, Nazeeruddin MK, Sekiguchi T, Grätzel M. A stable quasi-solid-state dye-sensitized solar cell with an amphiphilic ruthenium sensitizer and polymer gel electrolyte. *Nat Mater* 2003;**2**:402–7.
41. Nazeeruddin MK, Zakeeruddin SM, Lagref JJ, Liska P, Comte P, Barolo C, et al. Stepwise assembly of amphiphilic ruthenium sensitizers and their applications in dye-sensitized solar cell. *Coord Chem Rev* 2004;**248**:1317–28.
42. Cao K, Lu J, Cui J, Shen Y, Chen W, Alemu G, et al. Highly efficient light harvesting ruthenium sensitizers for dye-sensitized solar cells featuring triphenylamine donor antennas. *J Mater Chem A* 2014;**2**:4945–53.
43. Gao F, Wang Y, Shi D, Zhang J, Wang M, Jing X, et al. Enhance the optical absorptivity of nanocrystalline TiO_2 film with high molar extinction coefficient ruthenium sensitizers for high performance dye-sensitized solar cells. *J Am Chem Soc* 2008;**130**:10720–8.
44. Yu QJ, Liu S, Zhang M, Cai N, Wang Y, Wang P. Dye-sensitized solar cells with a high absorptivity ruthenium sensitizer featuring a 2-(hexylthio)thiophene conjugated bipyridine. *J Phys Chem C* 2009;**113**:14559–66.
45. Gao F, Wang Y, Zhang J, Shi D, Wang M, Humphry-Baker R, et al. A new heteroleptic ruthenium sensitizer enhances the absorptivity of mesoporous titania film for a high efficiency dye-sensitized solar cell. *Chem Commun* 2008;**7345**:2635–7.
46. Gao F, Cheng Y, Yu Q, Liu S, Shi D, Li Y, et al. Conjugation of selenophene with bipyridine for a high molar extinction coefficient sensitizer in dye-sensitized solar cells. *Inorg Chem* 2009;**48**:2664–9.
47. Wadman SH, Kroon JM, Bakker K, Lutz M, Spek AL, van Klink GPM, et al. Cyclometalated ruthenium complexes for sensitizing nanocrystalline TiO_2 solar cells. *Chem Commun* 2007;**4**:1907–9.
48. Robson KCD, Koivisto BD, Yella A, Sporinova B, Nazeeruddin MK, Baumgartner T, et al. Design and development of functionalized cyclometalated ruthenium chromophores for light-harvesting applications. *Inorg Chem* 2011;**50**:5494–508.
49. Chen BS, Chen K, Hong YH, Liu WH, Li TH, Lai CH, et al. Neutral, panchromatic Ru(II) terpyridine sensitizers bearing pyridine pyrazolate chelates with superior DSSC performance. *Chem Commun* 2009:5844–6.
50. Bessho T, Yoneda E, Yum JH, Guglielmi M, Tavernelli I, Imai H, et al. New paradigm in molecular engineering of sensitizers for solar cell applications. *J Am Chem Soc* 2009;**131**:5930–4.

51. Bomben PG, Gordon TJ, Schott E, Berlinguette CP. A trisheteroleptic cyclometalated Ru^{II} sensitizer that enables high power output in a dye-sensitized solar cell. *Angew Chem Int Ed* 2011;**50**:10682–5.
52. Wu KL, Hsu HC, Chen K, Chi Y, Chung MW, Liu WH, et al. Development of thiocyanate-free, charge-neutral Ru(II) sensitizers for dye-sensitized solar cells. *Chem Commun* 2010;**46**:5124–6.
53. Li G, Bomben PG, Robson KCD, Gorelsky SI, Berlinguette CP, Shatruk M. Ru complexes of thienyl-functionalized dipyrrins as NCS-free sensitizers for the dye-sensitized solar cell. *Chem Commun* 2012;**48**:8790–2.
54. Zhao HC, Harney JP, Huang Y, Yum J, Nazeeruddin MK, Grätzel M, et al. Evaluation of a ruthenium oxyquinolate architecture for dye-sensitized solar cells. *Inorg Chem* 2012;**51**:1–3.
55. Takahashi Y, Arakawa H, Sugihara H, Hara K, Islam A, Katoh R, et al. Highly efficient polypyridyl-ruthenium(II) photosensitizers with chelating oxygen donor ligands: β-diketonato-bis(dicarboxybipyridine) ruthenium. *Inorg Chim Acta* 2000;**310**:169–74.
56. Wu KL, Ho ST, Chou CC, Chang YC, Pan HA, Chi Y, et al. Engineering of osmium(II)-based light absorbers for dye-sensitized solar cells. *Angew Chem Int Ed* 2012;**51**:5642–6.
57. Hasselmann GM, Meyer GJ. Diffusion-limited interfacial electron transfer with large apparent driving forces. *J Phys Chem B* 1999;**103**:7671–5.
58. Ferrere S, Gregg BA. Photosensitization of TiO_2 by $[Fe^{II}(2,2'\text{-bipyridine-4,4'-dicarboxylic acid})_2(CN)_2]$: band selective electron injection from ultra-short-lived excited states. *J Am Chem Soc* 1998;**120**:843–4.
59. Islam A, Sugihara H, Hara K, Singh LP, Katoh R, Takahashi Y, et al. Dye sensitization of nanocrystalline titanium dioxide with square planar platinum(II) diimine dithiolate complexes. *Inorg Chem* 2001;**40**:5371–80.
60. Sandroni M, Favereau L, Planchat A, Akdas-Kilig H, Szuwarski N, Pellegrin Y, et al. Heteroleptic copper(I)–polypyridine complexes as efficient sensitizers for dye sensitized solar cells. *J Mater Chem A* 2014;**2**:9944–7.
61. Lee CP, Lin RYY, Lin LY, Li CT, Chu TC, Sun S, et al. Recent progress in organic sensitizers for dye-sensitized solar cells. *RSC Adv* 2015;**5**:23810–25.
62. Koumura N, Wang ZS, Mori S, Miyashita M, Suzuki E, Hara K. Alkyl-functionalized organic dyes for efficient molecular photovoltaics. *J Am Chem Soc* 2006;**128**:14256–7.
63. Brewster TP, Konezny SJ, Sheehan SW, Martini LA, Schmuttenmaer CA, Batista VS, et al. Hydroxamate anchors for improved photoconversion in dye-sensitized solar cells. *Inorg Chem* 2013;**52**:6752–64.
64. Heuer WB, Xia HL, Ward W, Zhou Z, Pearson WH, Siegler MA, et al. New dicarboxylic acid bipyridine ligand for ruthenium polypyridyl sensitization of TiO_2. *Inorg Chem* 2012;**51**:3981–8.
65. Gou F, Jiang X, Li B, Jing H, Zhu Z. Salicylic acid as a tridentate anchoring group for azo-bridged zinc porphyrin in dye-sensitized solar cells. *ACS Appl Mater Interfaces* 2013;**5**:12631–7.
66. Brown DG, Schauer PA, Borau-Garcia J, Fancy BR, Berlinguette CP. Stabilization of ruthenium sensitizers to TiO_2 surfaces through cooperative anchoring groups. *J Am Chem Soc* 2013;**135**:1692–5.
67. Koenigsmann C, Ripolles TS, Brennan BJ, Negre CFA, Koepf M, Durrell AC, et al. Substitution of a hydroxamic acid anchor into the MK-2 dye for enhanced photovoltaic performance and water stability in a DSSC. *Phys Chem Chem Phys* 2014;**16**:16629–41.
68. Huang SY, Schlichthörl G, Nozik AJ, Grätzel M, Frank AJ. Charge recombination in dye-sensitized nanocrystalline TiO_2 solar cells. *J Phys Chem B* 1997;**101**:2576–82.
69. Schlichthörl G, Huang SY, Sprague J, Frank AJ. Band edge movement and recombination kinetics in dye-sensitized nanocrystalline TiO_2 solar cells: a study by intensity modulated photovoltage spectroscopy. *J Phys Chem B* 1997;**101**:8141–55.
70. Boschloo G, Häggman L, Hagfeldt A. Quantification of the effect of 4-tert-butylpyridine addition to I^-/I_3^- redox electrolytes in dye-sensitized nanostructured TiO_2 solar cells. *J Phys Chem B* 2006;**110**:13144–50.
71. Liu Y, Hagfeldt A, Xiao XR, Lindquist SE. Investigation of influence of redox species on the interfacial energetics of a dye-sensitized nanoporous TiO_2 solar cell. *Sol Energy Mater Sol Cells* 1998;**55**:267–81.
72. Yu Z, Vlachopoulos N, Gorlov M, Kloo L. Liquid electrolytes for dye-sensitized solar cells. *Dalton Trans* 2011;**40**:10289–303.
73. Boschloo G, Hagfeldt A. Characteristics of the iodide/triiodide redox mediator in dye-sensitized solar cells. *Acc Chem Res* 2009;**42**:1819–26.
74. Boschloo G, Gibson EA, Hagfeldt A. Photomodulated voltammetry of iodide/triiodide redox electrolytes and its relevance to dye-sensitized solar cells. *J Phys Chem Lett* 2011;**2**:3016–20.
75. Cong J, Yang X, Kloo L, Sun L. Iodine/iodide-free redox shuttles for liquid electrolyte-based dye-sensitized solar cells. *Energy Environ Sci* 2012;**5**:9180–94.

76. Wang ZS, Sayama K, Sugihara H. Efficient Eosin Y dye-sensitized solar cell containing Br^-/Br_3^- electrolyte. *J Phys Chem B* 2005;**109**:22449–55.
77. Zhang Z, Chen P, Murakami TN, Zakeeruddin SM, Grätzel M. The 2,2,6,6-Tetramethyl-1-piperidinyloxy radical: an efficient, iodine- free redox mediator for dye-sensitized solar cells. *Adv Funct Mater* 2008;**18**:341–6.
78. Wang M, Chamberland N, Breau L, Moser JE, Humphry-Baker R, Marsan B, et al. An organic redox electrolyte to rival triiodide/iodide in dye-sensitized solar cells. *Nat Chem* 2010;**2**:385–9.
79. Gregg BA, Pichot F, Ferrere S, Fields CL. Interfacial recombination processes in dye-sensitized solar cells and methods to passivate the interfaces. *J Phys Chem B* 2001;**105**:1422–9.
80. Daeneke T, Kwon TH, Holmes AB, Duffy NW, Bach U, Spiccia L. High-efficiency dye-sensitized solar cells with ferrocene-based electrolytes. *Nat Chem* 2011;**3**:211–5.
81. Bai Y, Yu Q, Cai N, Wang Y, Zhang M, Wang P. High-efficiency organic dye-sensitized mesoscopic solar cells with a copper redox shuttle. *Chem Commun* 2011;**47**:4376–8.
82. Colombo A, Dragonetti C, Magni M, Roberto D, Demartin F, Caramori S, et al. Efficient copper mediators based on bulky asymmetric phenanthrolines for DSSCs. *ACS Appl Mater Interfaces* 2014;**6**:13945–55.
83. Sapp SA, Elliott CM, Contado C, Caramori S, Bignozzi CA. Substituted polypyridine complexes of cobalt(II/III) as efficient electron-transfer mediators in dye-sensitized solar cells. *J Am Chem Soc* 2002;**124**:11215–22.
84. Nelson JJ, Amick TJ, Elliott CM. Mass transport of polypyridyl cobalt complexes in dye-sensitized solar cells with mesoporous TiO_2 photoanodes. *J Phys Chem C* 2008;**112**:18255–63.
85. Klahr BM, Hamann TW. Performance enhancement and limitations of cobalt bipyridyl redox shuttles in dye-sensitized solar cells. *J Phys Chem C* 2009;**113**:14040–5.
86. Wu KL, Hu Y, Chao CT, Yang YW, Hsiao TY, Robertson N, et al. Dye sensitized solar cells with cobalt and iodine-based electrolyte: the role of thiocyanate-free ruthenium sensitizers. *J Mater Chem A* 2014;**2**:19556–65.
87. Wu KL, Huckaba AJ, Clifford JN, Yang YW, Yella A, Palomares E, et al. Molecularly engineered Ru(II) sensitizers compatible with cobalt(II/III) redox mediators for dye-sensitized solar cells. *Inorg Chem* 2016;**55**:7388–95.
88. Feldt SM, Gibson EA, Gabrielsson E, Sun L, Boschloo G, Hagfeldt A. Design of organic dyes and cobalt polypyridine redox mediators for high-efficiency dye-sensitized solar cells. *J Am Chem Soc* 2010;**132**:16714–24.
89. Feldt SM, Wang G, Boschloo G, Hagfeldt A. Effects of driving forces for recombination and regeneration on the photovoltaic performance of dye-sensitized solar cells using cobalt polypyridine redox couples. *J Phys Chem C* 2011;**115**:21500–7.
90. Yum JH, Baranoff E, Kessler F, Moehl T, Ahmad S, Bessho T, et al. A cobalt complex redox shuttle for dye-sensitized solar cells with high open-circuit potentials. *Nat Commun* 2012;**3**:631.
91. Polman A, Knight M, Garnett EC, Ehrler B, Sinke WC. Photovoltaic materials: present efficiencies and future challenges. *Sci Mag* 2016;**352**:6283. aad4244-1-10.
92. Green MA, Ho-Baillie A, Snaith HJ. The emergence of perovskite solar cells. *Nat Photonics* 2014;**8**:506–14.
93. Mitzi DB, Wang S, Field CA, Chess CA, Guloy AM. Conducting layered organic-inorganic halides containing <110>-Oriented perovskite sheets. *Science* 1995;**267**:1473–6.
94. Kojima A, Teshima K, Miyasaka T, Shirai Y. Novel photoelectrochemical cell with mesoscopic electrodes sensitized by lead-halide compounds (2). In: *Proc. 210th ECS meeting*; 2006. p. 397.
95. Kojima A, Teshima K, Shirai Y, Miyasaka T. Organometal halide perovskites as visible-light sensitizers for photovoltaic cells. *J Am Chem Soc* 2009;**131**:6050–1.
96. Wang B, Xiao X, Chen T. Perovskite photovoltaics: a high-efficiency newcomer to the solar cell family. *Nanoscale* 2014;**6**:12287–97.
97. Im JH, Lee CR, Lee JW, Park SW, Park NG. 6.5% efficient perovskite quantum-dot-sensitized solar cell. *Nanoscale* 2011;**3**:4088–93.
98. Kim HS, Lee CR, Im KH, Lee KB, Moehl T, Marchioro A, et al. Lead iodide perovskite sensitized all-solid-state submicron thin film mesoscopic solar cell with efficiency exceeding 9%. *Sci Rep* 2012;**2**:591.
99. Lee MM, Teuscher J, Miyasaka T, Murakami TN, Snaith HJ. Efficient hybrid solar cells based on meso-superstructured organometal halide perovskites. *Science* 2012;**338**:643–7.
100. Heo JH, Im SH, Noh JH, Mandal TN, Lim CS, Chang JA, et al. Efficient inorganic–organic hybrid heterojunction solar cells containing perovskite compound and polymeric hole conductors. *Nat Photonics* 2013;**7**:486–91.
101. Jeon NJ, Noh JH, Kim YC, Yang WS, Ryu S, Seok SI. Solvent engineering for high-performance inorganic–organic hybrid perovskite solar cells. *Nat Mat* 2014;**13**:897–903.

102. Pellet N, Gao P, Gregori G, Yang TY, Nazeeruddin MK, Maier J, et al. Mixed-organic-cation perovskite photovoltaics for enhanced solar-light harvesting. *Angew Chem Int Ed* 2014;**53**:3151–7.
103. Saliba M, Matsui T, Seo JY, Domanski K, Correa-Baena JP, Nazeeruddin MK, et al. Cesium-containing triple cation perovskite solar cells: improved stability, reproducibility and high efficiency. *Energy Environ Sci* 2016;**9**:1989–97.
104. Seo J, Noh JH, Seok SI. Rational strategies for efficient perovskite solar cells. *Acc Chem Res* 2016;**49**:562–72.
105. Docampo P, Bein T. A long-term view on perovskite optoelectronics. *Acc Chem Res* 2016;**49**:339–46.
106. Noh JH, Im SH, Heo JH, Mandal TN, Seok SI. Chemical management for colorful, efficient, and stable inorganic–organic hybrid nanostructured solar cells. *Nano Lett* 2013;**13**:1764–9.
107. Karl KR, Melillo JM, Peterson TC. *Global climate change impacts in the United States.* New York: Cambridge University Press; 2009. NOAA/NCDC.
108. Ardo S, Meyer GJ. Photodriven heterogeneous charge transfer with transition-metal compounds anchored to TiO_2 semiconductor surfaces. *Chem Soc Rev* 2009;**38**:115–64.

Recommended Reading

1. Green MA, Emery K, Hishikawa Y, Warta W, Dunlop ED. Solar cell efficiency tables (version 47). *Prog Photovoltaics Res Appl* 2015;**23**:805–12.

CHAPTER

3.26

Toward a Sustainable Carbon Cycle: The Methanol Economy

Alain Goeppert, George A. Olah, G.K. Surya Prakash

University of Southern California, Los Angeles, CA, United States

3.26.1 INTRODUCTION

Since the beginning of the industrial revolution in the 18th century, modern societies have relied increasingly on fossil fuels to fulfill the energy needs of the growing world population. Presently, about 22 million tons of coal, 85 million barrels of oil (around 12 million metric tons), and 10 billion m^3 of natural gas are consumed each and every day to fulfill some 81% of our energy demand.[1] Fossil fuels are also the raw materials for a wide variety of derived hydrocarbon materials and products. These range from gasoline and diesel fuel to varied petrochemical and chemical products including synthetic fibers, plastics, fertilizers, and pharmaceuticals. This "gift" that Nature has given us in the form of fossilized sunshine, that is, fossil fuels, has enabled an unprecedented development of human societies to a level of prosperity and advancement that could only be dreamed of only a few centuries ago. To put it in perspective, it has been determined that a single barrel of oil has an energy content equivalent to 12 humans working all year round or about 25,000 man-hours. Considering that each person in the United States consumes on an average 25 barrels of oil annually, about 300 people working on a yearly basis would be required to power the industries, fuel the transportation sector, and provide for many of their household needs to cover and maintain the current standard of living of just one person. However, beside many benefits, there are also a number of problems associated with the utilization of fossil fuels on such a massive scale. Their combustion contributes to the annual release of more than 35 billion metric tons of carbon dioxide into the environment, far outpacing Nature's carbon cycle. This has resulted in a rapid increase in the atmospheric carbon dioxide concentration from a preindustrial level of about 270 ppm to more than 400 ppm at present. It is now generally accepted that this excessive release of CO_2 along with other greenhouse gases due to human activity are major contributors to the global warming of our planet.[2] They have also caused other significant environmental disruptions including ocean acidification, melting of the ice caps,

http://dx.doi.org/10.1016/B978-0-12-809270-5.00031-5

rising of sea levels, alteration to the hydrological cycles, more frequent events of climate extremes, and unpredictable changes in biodiversity.[3] In fact, the impact of humanity on the earth, not only from the emissions of greenhouse gases but also from many other elements, is now so profound that expert geologist have proposed a new geological epoch to describe it: the *Anthropocene*. This new epoch would begin in about 1950 and follows the *Holocene*, 12,000 years of stable climate since the end of the last ice age. Numerous solutions have been proposed to alleviate the problems associated with the emission of CO_2, the inevitable by-product of fossil fuel combustion. Energy conservation can help but will not solve the problem by itself. The use of renewables as well as nuclear energy to fulfill an increasing part of our energy mix is also being pursued, but they will be hard pressed to replace fossil fuels, at least in the short term. One of the most commonly proposed solution is carbon capture and sequestration (CCS), wherein CO_2 would be captured from various point sources including power plants and industrial facilities, concentrated, pressurized, and then pumped underground into various geological formations. However, CCS has not been demonstrated yet on the immense scale needed, tens of billions of metric tons of CO_2 per year, and we have to make sure that all the CO_2 stored underground also remains there and does not leak overtime, rendering the overall process pointless. Monitoring of stored CO_2 over decades and probably centuries is therefore crucially important but might be difficult to enforce. The storage of large amounts of CO_2 underground has also seen a pushback from local populations, which do not wish to have such facilities located close to or under them. Although in general people agree that something has to be done about CO_2 emissions, on a local level, we are facing a classical "not in my backyard" attitude. Alternatively, the captured CO_2 could be mineralized either on the surface or underground. However, be it for sequestration, mineralization, or recycling, the first step is the same, namely, CO_2 capture. To capture CO_2, different technologies including absorption into liquids, adsorption on solids, and membrane separation could be utilized depending on the CO_2 concentration, temperature, presence of impurities, etc., in the effluent to be treated. Eventually, even the CO_2 present in the air could be captured, and technologies are being developed for this purpose.[4,5] Direct air capture (DAC) is important to address about half the CO_2 emissions that originate from small and dispersed fossil fuel-burning units such as cars, airplanes, trains, stoves, domestic heating, etc., for which CO_2 capture at the source would be impractical and/or economically not viable.

Another issue with fossil fuels is that they were formed naturally millions of years ago and their amount is finite and limited. Once used up, they cannot be renewed on the human timescale. Despite their considerable size, our fossil fuel reserves will therefore be increasingly depleted. Exploitation of unconventional fossil fuel resources including shale oil and shale gas, tight oil, tar sands, and possibly methane hydrate in the future, could significantly increase the availability of affordable fossil fuels. However, the impact of their exploitation on the environment is raising numerous concerns. A number of countries, for example, including France and Germany banned or put altogether a moratorium on the use of hydraulic fracturing, also called "fracking," technologies used to exploit shale resources as long as they are not proven safe for the environment. The utilization of these unconventional resources also exacerbates the problems associated with greenhouse gas emissions as more energy is expended for their extraction compared with that required for conventional resources.

It was thought that the demand for petroleum oil and natural gas would outpace the global production capacity in the relatively near term (the so-called global Hubbert's peak), after which the availability of these resources would decrease and their prices would increase sharply. However, with the mentioned recently found unconventional gas and oil resources and the future possibility of economical exploitation of methane hydrate and still very large reserves of coal, it seems that we will not run out of fossil fuels for quite a while. Instead, if no acceptable and economical solution is found for the CO_2 problem, it could well be that our usage of fossil fuels will be strongly constrained not by their actual availability but by the levels of CO_2 emissions and the accompanying environmental effects that we are willing to endure. Sheik Zaki Yamani, a former oil minister from Saudi Arabia, might have therefore well been correct, when he said that "the stone age didn't end for lack of stone, and the oil age will end long before the world runs out of oil."[6]

In fact we are not facing an energy shortage per se. Almost all of our energy, including fossil fuels (coal, oil, natural gas), wind, hydro, and of course solar comes in one way or another from the energy of the sun.[7] At any given time, sunshine delivers to the earth as light and heat about 10,000 times more energy than the entire world is consuming. As the sun is estimated to last for at least another 4.5 billion years, the challenge is to find more efficient and feasible ways to capture, store, and utilize its energy in suitable energy storage devices and energy carriers.

With fossil fuels still at the heart of its energy system, the world is presently between a rock and a hard place. CO_2 emissions need to be tamed, while at the same time we have to maintain or improve the standard of living of billions of people. Alternatives are thus clearly needed to solve our carbon conundrum. To tackle the CO_2 problem, the mentioned CCS can help in the short term, but burying CO_2 underground does not generate any economic value. At the same time, whether humankind uses up most of the fossil fuel resources (combined with CCS) or relies increasingly on alternative energies, the need for transportation fuels and materials currently obtained from petroleum oil and natural gas will remain. CCS does not provide a solution for these essential carbon-based products ranging from plastics and aviation fuels to medicines. As an alternative to sequestration, CO_2 capture and recycling (CCR) or CO_2 capture and utilization, promises an ultimate solution. With CCR, CO_2 captured from any source can be recycled back to fuels and materials. Initially CO_2 could be obtained from fossil fuel-burning power plants, but to be truly carbon neutral and sustainable in the long term it will have to be eventually captured mainly from the atmosphere. The energy needed for the recycling of CO_2 would come from any alternative energy source, including solar, wind, and hydroelectricity, and also nuclear energy, be it fission of fusion, albeit made safer. As such, CCR also offers a way to store energy generated from renewable energy. The recycling and use of atmospheric CO_2-based materials would constitute an anthropogenic version of Nature's own carbon cycle.

One of the most versatile, simple, and easy-to-obtain liquid product from CO_2 is methanol through hydrogenation, electrochemical reduction, or by other means. Biomass, which is already a form of recycled atmospheric CO_2, can also be transformed efficiently to methanol through gasification. The produced renewable and sustainable methanol can then be further processed into dimethyl ether (DME), ethylene, propylene, gasoline, and all other products currently obtained from petroleum and natural gas. Once methanol and its derivatives are combusted they will release CO_2, which can then be recycled back, effectively closing the carbon loop (Fig. 3.26.1). The goal of this chapter is to give an overview of the conversion of CO_2 to methanol, potentially solving our carbon conundrum.

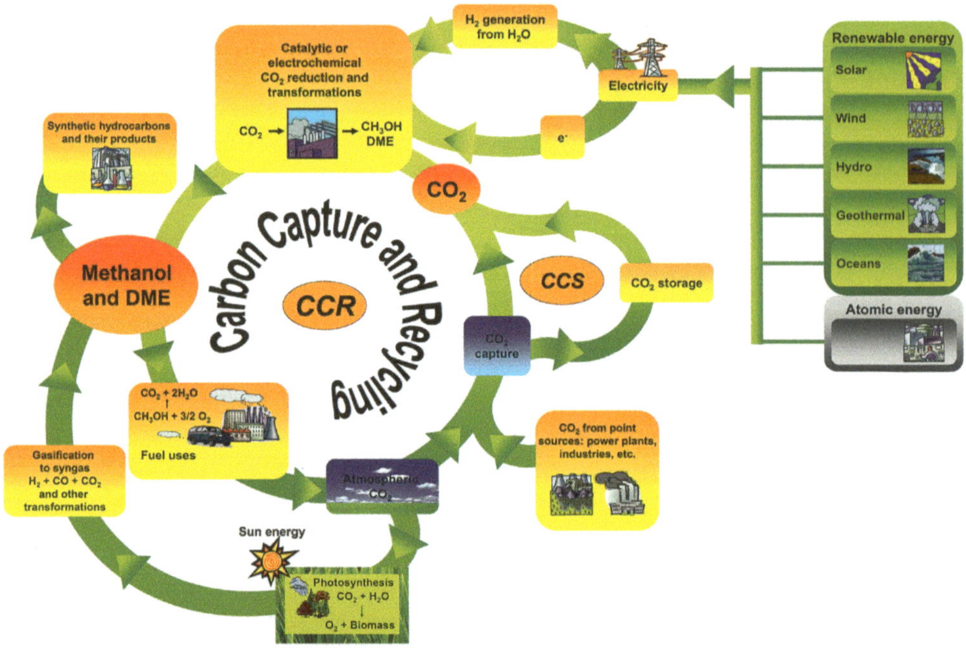

FIGURE 3.26.1 Anthropogenic carbon cycle within the methanol economy. *CCR*, CO_2 capture and recycling; *CCS*, carbon capture and sequestration; *DME*, dimethyl ether. *Reproduced from Goeppert A, Czaun M, Jones J-P, Prakash GKS, Olah GA. Recycling of carbon dioxide to methanol and derived products – closing the loop.* Chem Soc Rev *2014;**43**:7995–8048 by permission of the Royal Society of Chemistry.*

3.26.2 WHY METHANOL?

The primary product of most renewable energy sources including wind, solar, hydro and geothermal, as well nuclear energy that are bound to play an increasing role in the world's energy mix is electricity. Although electrical energy is an excellent way to transfer energy over relatively short distances, its efficient transportation over longer distances and storage on a large scale remains a challenge. Pumped hydro, compressed air storage, flywheels, and batteries are all possible but have limited capacity and/or high costs. This is especially problematic for the two most promising and scalable renewable energy sources, solar and wind, which by nature are intermittent and highly fluctuating. Solar energy produces much less power under cloudy conditions and none at night. Wind, on the other hand, does not blow constantly or consistently at the same speed. Production also varies by season. For these renewable sources to fulfill more than a marginal to small portion of our energy needs, leveling out these fluctuations in production will be essential to safeguard the stability of the electrical grid. In the current system, the production of electrical power follows closely the demand by varying the output of power plants. Although this is possible when using fossil fuels, especially natural gas, it does not seem to be an option for intermittent renewable energy sources. The implementation of "smart grids" can help to alleviate part of the problems, but storing excess capacity for use when needed is a necessity. One way of achieving this is by storing electrical energy in the form of chemical energy (bonds) in compounds such as hydrogen, methanol, methane, and higher hydrocarbons. These chemicals can then be stored and/or transported and used at a later time to generate electricity or in other applications such as transportation, heating, and cooking.[9,10]

The simplest compound that can be produced from electrical energy is hydrogen, by splitting of water. Presently, the sole practical method to split water into hydrogen is electrolysis, but others such as photochemical and thermal splitting as well as high-temperature thermochemical cycles are also under investigation. Electrolysis of water has been practiced for more than 100 years and is efficient with overall system conversions higher than 75%–80%, with further improvements possible.[11,12]

Hydrogen gas has been proposed as an energy storage media and is an excellent and clean fuel that produces, besides energy, only water when combusted. In theory, a "hydrogen economy" based on simple recycling of water to hydrogen would be very attractive.[13] In practice, however, due to its physical and chemical properties, hydrogen has a number of serious drawbacks.[14] Because of its low volumetric density, it requires either compression to high pressures (350–700 bars) or liquefaction at very low temperature (−253 °C), making its storage problematic and energy intensive. It is also highly flammable and explosive and can diffuse through many commonly used metals and materials. The construction of an entirely new infrastructure needed to transport, store and dispense hydrogen safely would therefore be very expensive.

Other, less cumbersome ways of storing electrons in chemical bonds are thus needed. Among the possible chemical storage media, a liquid medium would be logistically preferable to a gaseous one in most applications. In the transportation sector in particular, a transition from liquid fossil fuel-derived products (gasoline, diesel fuel, kerosene, etc.) to a renewable and sustainable liquid fuel would be highly desirable and avoid the construction of an entirely new transportation/storage/distribution network. It would allow for the use of the existing infrastructure with only minor modifications. Among the possible candidates fulfilling these requirements is methanol,[15] a simple compound containing only one carbon, that is liquid a room temperature. Methanol has many advantages, and its use as an energy carrier has been promoted in the past.[10,16–26] Due to its high octane rating, it is an excellent additive or substitute for gasoline in internal combustion engines. Emissions from methanol engines are also clean with low NO_x and SO_x emission and practically no soot formation. Methanol can also be used efficiently in modified diesel engines[27,28] as well as in direct methanol fuel cells converting the chemical energy in methanol directly to electrical power at ambient temperature.[29] DME, produced from methanol by simple one-step bimolecular dehydration, is a gas that can be easily liquefied at moderate pressure, much like liquefied petroleum gas (LPG). DME is an excellent diesel fuel substitute with a high cetane rating producing almost no soot emissions and has also attracted much interest.[30,31] It can also replace LPG in most applications such as heating and cooking. Methanol and DME are also superior fuels for electric power generation in gas turbines.[32–34] If needed, methanol can even be transformed to gasoline through the methanol-to-gasoline process developed by Mobil in the 1970 to the 1980s.[35] Besides its fuel application, methanol is a feedstock for numerous chemicals such as formaldehyde, acetic acid, and methyl-*tert*-butyl ether. Through the methanol-to-olefins process, it can also produce light olefins including ethylene and propylene used in polymers (chiefly polyethylene and polypropylene), as well as any hydrocarbon and product currently obtained from petroleum oil.[10]

Methanol is already one of the most important organic feedstock in the chemical industry with a current worldwide annual production of over 70 million metric tons.[36] Although, for economic reasons, methanol is presently almost exclusively produced from fossil fuels (natural gas and coal), in the future it could be made from any carbon-containing feedstock including biomass and any CO_2 source. Such a "Methanol Economy" concept includes an anthropogenic carbon cycle to produce methanol, which can then be used as a renewable fuel or for the synthesis of nearly all products presently derived from fossil fuels. All aspects of a Methanol Economy have been described in detail in our recent papers and a monograph.[10,16,17,37]

3.26.3 METHANOL PRODUCTION FROM FOSSIL FUELS WITH REDUCED OR NO CO_2 EMISSION

Presently, methanol is still almost exclusively produced from fossil fuels, mainly natural gas and coal. Coal gasification and methane reforming both afford syngas, a mixture of hydrogen and carbon monoxide (H_2/CO), and some carbon dioxide, that can be transformed to methanol (Fig. 3.26.2). While the conversion of H_2/CO mixture to methanol (Eq. 1) and the hydrogenation of CO_2 to methanol (Eq. 2) are exothermic, the reverse water gas shift reaction (RWGS, Eq 3) is endothermic.

All three reactions are reversible, and the optimal reaction conditions were developed to shift these equilibria to the formation of methanol, which require higher pressures and lower temperature according to the Le Chatelier's principle. Although the original methanol synthesis technologies introduced by BASF in the 1920's were operated at high pressures of 250–300 atm and temperatures of 300–400 °C over zinc oxide/chromium oxide catalysts, present technologies require milder reaction conditions (200–300 °C, 50–100 atm) using copper-based catalysts. Methane from any source such as natural gas and shale gas is the preferred feedstock for syngas production due to the relatively low level of impurities (H_2S, COS, and mercaptans) and ease of handling compared with coal-based gasification technologies. Producing syngas with a satisfactory purity level from coal requires generally much more workup and is more capital intensive. It also releases more CO_2 per unit of methanol produced. Using cleaner syngas as a feed allows the application of more active and selective catalysts at lower pressures and temperatures resulting in fewer side products (higher alcohols, DME, and hydrocarbons). At present, almost all methanol production processes are carried out via the gas-phase reaction of H_2 and CO. However, the catalyst arrangement (e.g., tube, fixed bed, or suspension) and the reactor design may differ.[38,39]

$$CO + 2\,H_2 \rightleftharpoons CH_3OH \qquad \Delta H_{298K} = -21.7\ \text{kcal mol}^{-1} \qquad \text{Eq.1}$$

$$CO_2 + 3\,H_2 \rightleftharpoons CH_3OH + H_2O \qquad \Delta H_{298K} = -11.9\ \text{kcal mol}^{-1} \qquad \text{Eq.2}$$

$$CO_2 + H_2 \rightleftharpoons CO + H_2O \qquad \Delta H_{298K} = 9.8\ \text{kcal mol}^{-1} \qquad \text{Eq.3}$$

FIGURE 3.26.2 Methanol synthesis from syngas.

The process for the production of methanol is quite mature, but in a carbon-constrained world the use of fossil fuels such as methane or coal will be increasingly seen as unsustainable and an environmental liability. If the humankind wants to continue using fossil fuels, reducing their carbon footprint will therefore become a necessity. A number of ways exist to achieve partial to almost complete carbon neutrality while still using fossil fuels. Natural gas, because of its high hydrogen/carbon ratio is the fossil fuel that is the most amenable for achieving such a goal.

3.26.3.1 Steam Reforming of Methane

The current syngas production from natural gas, which is mainly composed of methane with traces of higher hydrocarbons, is conducted by reforming reactions. The most commonly applied is steam reforming of methane (SMR, Fig. 3.26.3).

This highly energy demanding process ($\Delta H_{298K} = 49.1$ kcal/mol) is generally carried out on nickel-containing catalysts at temperatures of 800–1000 °C and pressures of 20–40 atm. The resulting H_2/CO ratio of 3:1 ($S = 3$) obtained is not practical for methanol synthesis (Eq. 1), which requires a lower ratio ($S \approx 2–2.2$), and therefore appropriate adjustments need to be made.[40,41]

3.26.3.2 Partial Oxidation of Methane

Partial oxidation of methane (Fig. 3.26.4) is the reaction between methane and oxygen giving ideally a mixture of hydrogen and CO with a H_2/CO ratio of 2:1 ($S = 2$) either in the presence or absence of a catalyst.[42] This mixture would be close to ideal for methanol synthesis. However, practically, the gas mixture contains CO_2 and H_2O due to side reactions such as H_2 or CO oxidation and complete oxidation of methane to CO_2, which cause losses in product gases. The large output of energy due to complete combustion can also lead to safety concerns due to runaway reactions.

3.26.3.3 Dry Reforming of Methane

Carbon dioxide reforming of methane (Fig. 3.26.5), also called "dry" reforming indicating the absence of steam, is the reaction between an equimolar amount of CO_2 and CH_4 giving a

$$CH_4 + H_2O \rightleftharpoons CO + 3\,H_2 \quad \Delta H_{298K} = 49.1\ \text{kcal mol}^{-1}$$

FIGURE 3.26.3 Methane steam reforming.

$$CH_4 + 1/2O_2 \longrightarrow CO + 2\,H_2 \quad \Delta H_{298K} = -8.5\ \text{kcal mol}^{-1}$$

FIGURE 3.26.4 Partial oxidation of methane.

$$CO_2 + CH_4 \rightleftharpoons 2\,CO + 2\,H_2 \quad \Delta H_{298K} = 59.1\ \text{kcal mol}^{-1}$$

FIGURE 3.26.5 Dry reforming of methane.

1:1 ratio of CO and H_2 (S = 1).[43] In commercial processes, this endothermic reaction is conducted on Ni/MgO or Ni/$MgAl_2O_4$ catalyst at temperatures from 800 to 1000 °C. This reaction has the advantage of using very cost-effective raw materials. The fact that one molecule of CO_2 is used for each CH_4 molecule makes it an attractive alternative to mitigate global warming. This process can be environmentally beneficial if the reaction heat comes from renewable energy and the utilized CO_2 is recycled from a CO_2-rich stream.

A disadvantage of dry reforming in the context of methanol production is the low H_2/CO (S = 1) ratio in the product gas, which is not suitable as such for methanol synthesis. Enrichment of the obtained H_2-CO mixture by adding H_2 from other sources could be a solution to adjust the needed H_2/CO ratio (S ≈ 2). Another way to obtain the right H_2/CO ratio is to combine several of these reforming reactions.

3.26.3.4 Bi-reforming of Methane (Natural Gas) for Methanol Production

In the process that has been termed "bi-reforming,"[16,44] SMR and dry reforming of methane are combined[43] to give exclusively metgas with a H_2/CO ratio of 2/1 (S = 2); the needed feed for subsequent methanol synthesis (Fig. 3.26.6).[45] The bireforming of methane is also a practical procedure for natural (shale) gas that contains CO_2 because it does not require the removal and release of CO_2 into the atmosphere and allows the adjustment of the feed ratio according to local conditions. Higher aliphatic hydrocarbons can also be utilized as feedstocks for bi-reforming (Fig. 3.26.6, Eq. 5).[16,44,46]

steam reforming

$$2\,CH_4 + 2\,H_2O \rightleftharpoons 2\,CO + 6\,H_2$$

dry reforming

$$CO_2 + CH_4 \rightleftharpoons 2\,CO + 2\,H_2$$

bi-reforming

$$3\,CH_4 + 2\,H_2O + CO_2 \rightleftharpoons \underset{\text{metgas}}{4\,CO + 8\,H_2} \rightarrow 4\,CH_3OH \quad \text{Eq. 4}$$

bi-reforming of higher hydrocarbons

$$3\,C_nH_{2n+2} + (3n-1)\,H_2O + CO_2 \rightleftharpoons \underset{\text{metgas}}{(3n+1)\,CO + (6n+2)\,H_2} \rightarrow (3n+1)\,CH_3OH \quad \text{Eq. 5}$$

FIGURE 3.26.6 Forming of methane and higher hydrocarbons.

Steam reforming $$3\,CH_4 + 3\,H_2O \rightleftharpoons 3\,CO + 9\,H_2$$

$$3\,CO + 6\,H_2 \rightleftharpoons 3\,CH_3OH$$

CO_2 addition $$CO_2 + 3\,H_2 \rightleftharpoons CH_3OH + H_2O$$

Overall reaction $$3\,CH_4 + 2\,H_2O + CO_2 \rightleftharpoons 4\,CH_3OH$$

FIGURE 3.26.7 Addition of CO_2 to syngas from steam reforming.

3.26.3.5 Addition of CO_2 to Syngas From Methane Steam Reforming

The proper H_2/CO ratio of 2 can be achieved not only by the combination of steam reforming and dry reforming but also by the addition of CO_2 to the syngas flow from steam reforming during the methanol synthesis step (Fig. 3.26.7). The advantage of this method is that the excess hydrogen produced from steam reforming can be also utilized in the same process.[41]

3.26.3.6 Production of H_2 From CH_4 Without CO_2 Formation and the Carnol Process

Although the discussed reforming processes involved the utilization of some CO_2 to reduce the carbon footprint of their products, it would be even more beneficial to eliminate entirely the emission of CO_2. One way of achieving this goal and still use fossil fuels is by combining the reforming reaction with the water gas shift (WGS) reaction. The CO produced during any reforming reaction could be reacted with water to generate CO_2 and more H_2. In this case, however, the CO_2 would have to be captured and sequestered (CCS), reducing substantially or eliminating the emission of CO_2. The pure hydrogen produced would then be combined with non-fossil CO_2 (i.e., from biomass or the atmosphere) to obtain methanol. Applying this process to steam reforming, all the carbon contained in methane would be sequestered in the form of CO_2 (3.26.8). Interestingly, the overall reaction is the same as the one for bi-reforming (Fig. 3.26.6, Eq. 4). A similar reaction can be obtained when dry reforming is combined with the WGS reaction followed by sequestration of CO_2 and methanol synthesis (Fig. 3.26.9).

Another possibility to produce methanol from methane without CO_2 by-product is to thermally decompose methane at high temperatures above 800 °C in the absence of air to hydrogen and carbon (fossil fuel decarbonization).[47,48] In the following step, catalytic hydrogenation of CO_2 (captured from any non-fossil source including the atmosphere) can be carried out to produce methanol. This process is the so-called "Carnol process" (Fig. 3.26.10). It has been developed by the Brookhaven National Laboratory and was also used to produce carbon black as a filler material or pigment.[49]

It is apparent that in the overall reaction all three carbon atoms from methane end up in elemental carbon, while carbon atoms from the added CO_2 are incorporated into methanol. When methanol is combusted, this CO_2 is released back into the atmosphere. Consequently,

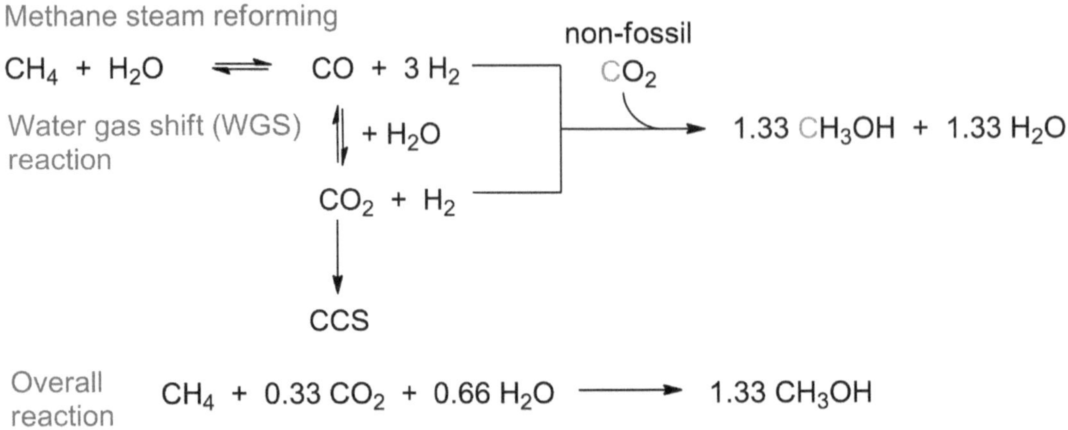

FIGURE 3.26.8 Combination of steam reforming and water gas shift with carbon capture and sequestration (CCS) followed by methanol synthesis.

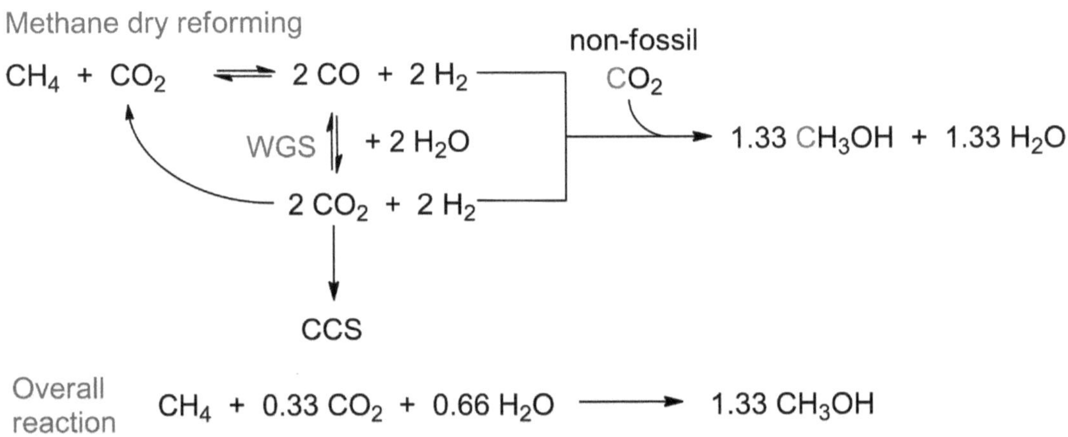

FIGURE 3.26.9 Combination of dry reforming and water gas shift with carbon capture and sequestration (CCS) followed by methanol synthesis.

$$3\,CH_4 \xrightarrow{>800\,^{\circ}C} 3\,C + 6\,H_2$$ CH_4 thermal decomposition

$$2\,CO_2 + 6\,H_2 \longrightarrow 2\,CH_3OH + 2\,H_2O$$ CO_2 hydrogenation to methanol

$$3\,CH_4 + 2\,CO_2 \longrightarrow 2\,CH_3OH + 2\,H_2O + 3\,C$$ Overall Carnol process

FIGURE 3.26.10 Carnol process.

overall, the net emission of CO_2 from this process is close to zero, because CO_2 released when methanol is combusted as a fuel was recycled from existing emission sources. The solid carbon formed as a by-product can be handled and stored much more easily than gaseous CO_2, and be disposed of or used as a commodity material, for example, in tire production, soil conditioning, or as a filler for road construction.

The methane thermal decomposition process is, however, still in its development stage and will require considerable further research to become a mature and efficient technology for commercial application.

Combinations of methane decomposition with reforming reactions or the Boudouard reaction are also possible.[8] Methane decomposition can, for example, be combined with dry reforming. The result is the production of methanol and carbon. For 2 mol of CH_4 used, 1 mol of carbon is formed (Fig. 3.26.11). Because not all the carbon in methane ends up in solid carbon, the environmental benefit is not as high as with the Carnol process, but the economic cost might be lower. Fig.3.26.12 gives a summary of some of the reactions and their combinations to convert CO_2 to methanol while still using abundant methane resources. Depending on the reactions, they offer partial to complete CO_2 emission reduction.

3.26.3.7 Coal to Methanol Without CO_2 Emissions

Because of the enormous global reserves of coal still available, processes to produce hydrogen from coal with concomitant capture and sequestration of the generated CO_2 have also been developed. Such an approach has been proposed for advanced coal-fired power plants[50–52] to produce not methanol but electricity and/or hydrogen. Through coal gasification with steam and WGS reaction, all the carbon in coal can be converted to H_2 and CO_2. To avoid CO_2 emissions, the generated CO_2 would have to be captured and sequestered (CCS). CO_2 from renewable sources such as biomass or the air could then be converted to methanol with the hydrogen generated from coal gasification (Fig. 3.26.13). Overall, this process would be CO_2 neutral as all the carbon in coal would be captured and stored as CO_2, possibly allowing the use of still extensive coal reserves. While this might be possible, the amounts of CO_2 that need to be captured and sequestered are enormous. As with other processes, the carbon sequestration part will probably be the most difficult to solve as

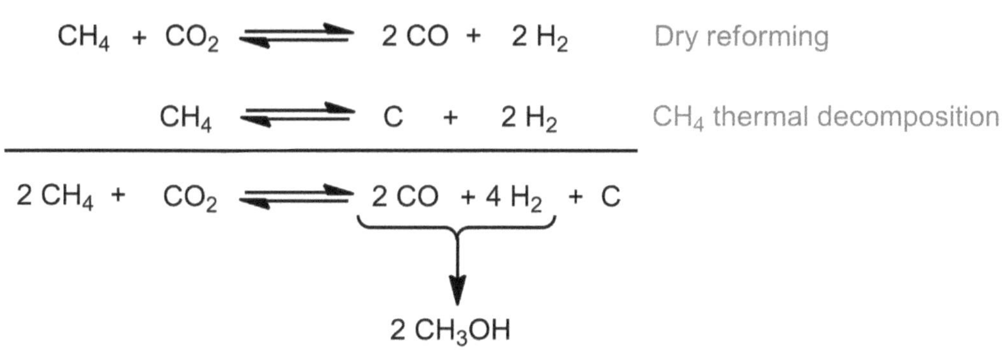

FIGURE 3.26.11 Combination of methane decomposition with dry reforming.

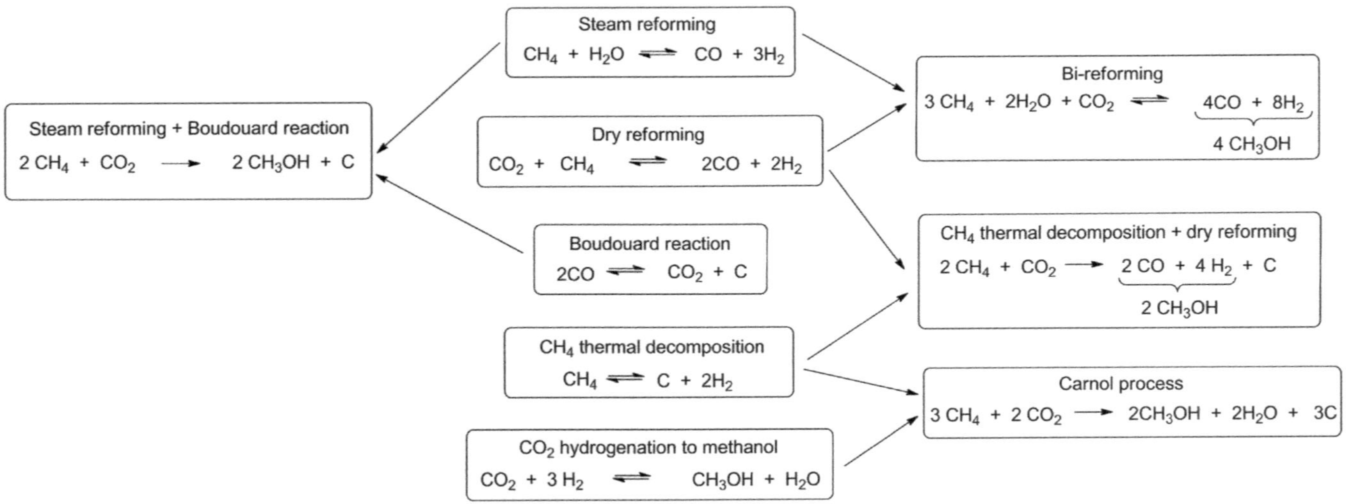

FIGURE 3.26.12 Examples of processes to convert CO_2 to methanol using methane.

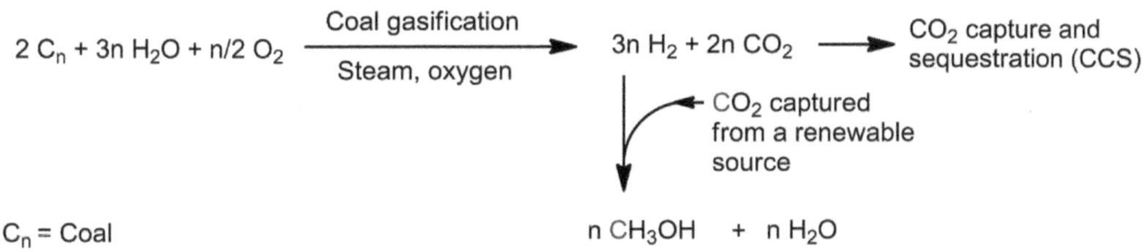

FIGURE 3.26.13 Methanol production from coal without CO_2 emissions.

populations are increasingly pushing back on plans to store large amounts of CO_2 underground due to possible leaks and induced seismic activity.

All the processes based on fossil fuels ultimately depend not only on the availability of these resources but also on their impact on the environment. Even the utilization of methane, the "cleanest" of our fossil fuels, compared with coal or petroleum, still releases large amounts of CO_2. If environmental costs are factored into the equation, it might well be that methanol from fossil fuel sources without the emission of CO_2 will become more expensive than its synthesis from alternative non-fossil sources. Therefore to allow for continued and sustainable development, humanity needs to wean itself from fossil fuels by increasingly using renewable and nuclear energy sources and recycling CO_2.

3.26.4 SUSTAINABLE PRODUCTION OF METHANOL

3.26.4.1 Biomass- and Waste-Based Methanol and Dimethyl Ether: Bio-methanol and Bio-Dimethyl Ether

Nature's photosynthesis uses the sun's energy with chlorophyll in plants as a catalyst to recycle carbon dioxide and water into new plant life, that is, biomass. Converting biomass into methanol offers therefore a way to recycle CO_2 from the air. Biomass is generally referred to as any type of plant or animal material, that is, materials produced by life forms. This includes wood and agricultural crops and their waste by-products, municipal solid waste, animal waste, and aquatic plants and algae. Methanol was originally made exclusively, among many other products, through the thermal destructive distillation of wood (hence the name *wood alcohol* is sometimes given to methanol). However, due to its inefficiency and the advent of synthetic processes via syngas, this route was soon abandoned in the first half of the 20th century. Because biomass itself is a very general term describing a variety of heterogeneous materials (in most cases bulky solids such as wood and agricultural crops and by-products), its conversion to a single, easy-to-trade, and convenient-to-transport liquid product such as methanol could be advantageous. The modern methods for producing bio-methanol from biomass are now different and much more efficient than a century ago.[53,54] Basically, not only wood but also any organic (i.e., carbon-containing) material obtained from living systems can be used in the process.

Depending on the nature of the feedstock, various technologies, namely, pyrolysis, liquefaction, gasification, and combinations thereof can be used (Fig. 3.26.14).[55–59] For solid feedstocks such as wood and other cellulosic materials the technologies are similar to the ones

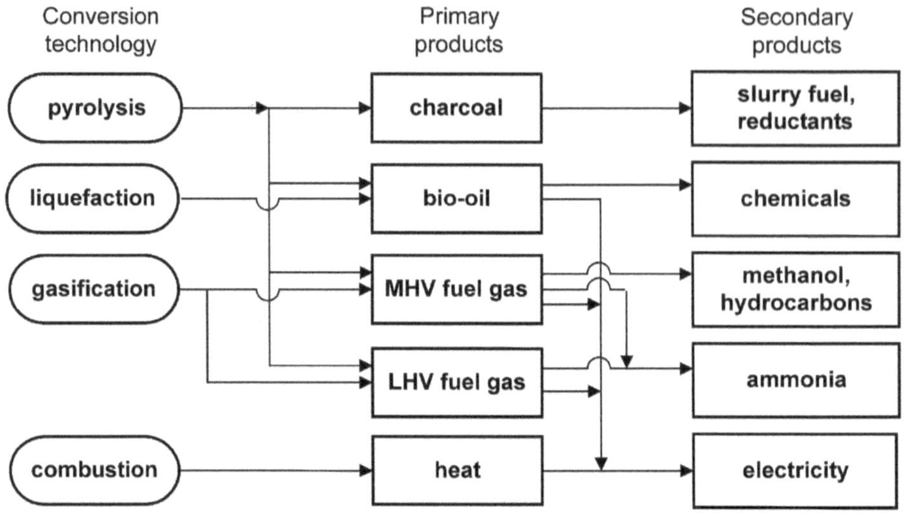

FIGURE 3.26.14 Thermochemical conversion-products and uses. *LHV*, lower heating value; *MHV*, medium heating value. *Reproduced with permission from Elsevier from Chmielniak T, Sciazko M. Co-gasification of biomass and coal for methanol synthesis.* Appl Energ *2003;***74***:393–403.*

used to convert coal to methanol, that is, gasification to syngas followed by methanol synthesis. For some other biomass feedstocks such as animal manure, the conversion to biogas, a mixture rich in methane and CO_2, followed by reforming to syngas and finally methanol synthesis is a possibility. An overview of various options is shown in Fig. 3.26.15. In theory, any organic carbon containing material can be utilized in the process, making it relatively flexible. The direct conversion of biomass to methanol following enzymatic routes is also an attractive alternative, but is still in early research phase.

The production of methanol from biomass suffers from some of the same problems as its production from coal. Due to their chemical composition, both these feedstocks produce a syngas mixture with a low hydrogen/carbon ratio and a high CO_2/CO ratio. For methanol synthesis the optimal ratio $S = (H_2\text{-}CO_2)/(CO + CO_2)$ has to be close to 2. To adjust the ratio, part of the CO_2 has to be separated, generally after gasification.[62] Preferably, the captured CO_2 should be sequestered or used in some other process, but it could also simply be vented. If the production of methanol from biomass generates a lot of CO_2, the apparent conversion of biomass into methanol is significantly reduced.[61] Another option is to react the excess CO_2 with hydrogen produced from some other source.[63] Syngas with a H_2/CO ratio close to 3 obtained from steam reforming of natural gas could be used to compensate for the lack of hydrogen in the syngas produced from biomass.[64,65] Biomass/natural gas hybrid plants would avoid the need for a H_2/CO ratio adjustment and CO_2 addition usually practiced in natural gas-based plants. At the same, it would avoid the need for CO_2 removal encountered in the biomass-based plant. The Hynol process, originally developed at the Brookhaven National Laboratory, was successful in converting materials such as woodchips into methanol on a pilot scale using a combination of biomass and natural gas.[66–68] The combination of biomass gasification products with steel-work off-gases such as coke oven gas

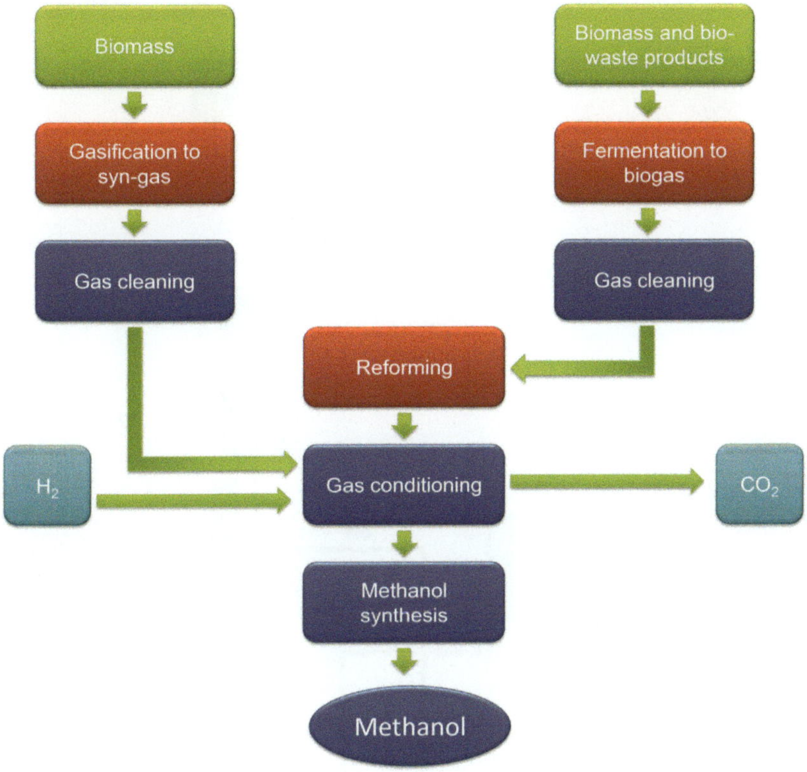

FIGURE 3.26.15 Simplified schematic illustrating methanol production from biomass-based feedstocks. *Reproduced from Goeppert A, Czaun M, Jones J-P, Prakash GKS, Olah GA. Recycling of carbon dioxide to methanol and derived products e closing the loop.* Chem Soc Rev *2014;***43***:7995–8048 by permission of the Royal Society of Chemistry. Taken from Chmielniak T, Sciazko M. Co-gasification of biomass and coal for methanol synthesis.* Appl Energ *2003;***74***:393–403.*

containing typically about 66% hydrogen was also studied.[69] However, although natural gas and other industrial gases can be used, renewable sources of hydrogen would be preferable to lower the carbon footprint of the methanol production.

Hybrid systems based on biomass and alternative energy sources have been proposed for the production of methanol[26,63,70–74] as well as liquid hydrocarbons through the Fischer-Tropsch process.[75] Ouellette et al., for example, discussed the production of bio-methanol from biomass and hydrogen produced via electrolysis of water using hydroelectricity.[70] In Denmark, the combination of hydrogen produced from water in a solid oxide electrolyzer cell (SOEC) and biomass was investigated for the production of methanol and DME (Fig. 3.26.16).[76] This study reported that when compared with a traditional methanol synthesis plant operating on biomass gasification without electrolysis, the output of methanol from the plant is doubled and the methanol production efficiency is boosted from 59% to 71%. The total plant efficiency was 81.6%. With this process, it should be possible to synthesize 1053 t of methanol per day with 1000 t of wood per day and 141 MW of installed SOEC. Oxygen produced as a by-product of water electrolysis could also be used to gasify the biomass, eliminating the need for a stand-alone air separation unit.

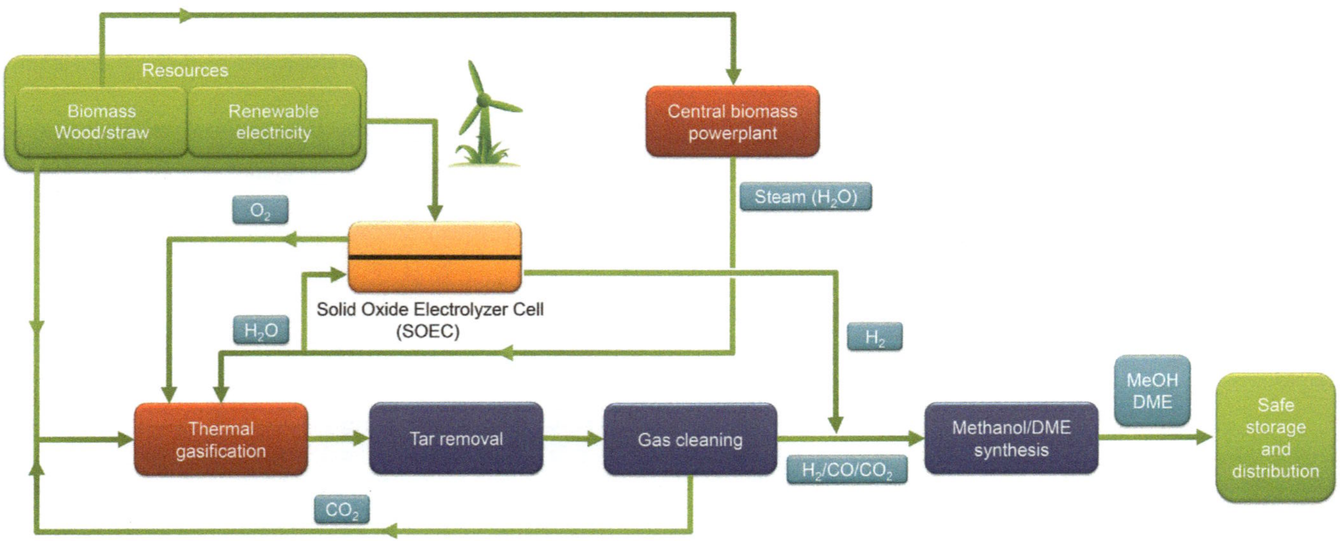

FIGURE 3.26.16 Methanol/dimethyl ether synthesis based on electrolysis-assisted gasification of wood.[76]

The anaerobic digestion of various feedstocks such as animal dung, municipal organic waste, sewage sludge, and wastewater from food, paper and pulp, fiber, meat, milk, brewing, and pharmaceutical plants has been adapted commercially to produce biogas.[77] All types of biomass can be used as substrates for biogas production as long as they contain carbohydrates, proteins, fats, cellulose, and hemicelluloses as main components. However, strong lignified organic substances such as wood are generally not suitable due to their slow anaerobic decomposition. The gas produced in anaerobic digesters, depending on the feedstock and effectiveness of the process, consists of 50%–70% methane with the remainder being mostly CO_2. At present, the gas is mainly used to produce electricity or heat. After purification and the removal of impurities (especially hydrogen sulfide), biogas could also be used for the production of methanol in much the same way as natural gas. In the United States, Oberon Fuels in California is commercializing small-scale skid-mounted plants for the production of DME and methanol from biogas. However, as in the case of syngas from biomass gasification, biogas contains an excess of CO_2, which would have to be removed, and possibly stored, or reacted with hydrogen from another source.[73] Therefore utilization of wind energy to produce H_2 by electrolysis of water in an SOEC and combination with biogas has, for example, been studied in Denmark for the production of methanol and DME (Fig. 3.26.17).[76,78] Bireforming of biogas to produce syngas is also a possibility.[79,80]

3.26.4.1.1 *Limitations of Biomass*

Although waste products from wood processing, agricultural residues and by-products, as well as solid municipal waste represent suitable feedstocks for methanol production, the quantities generated from these resources are limited. In the long term, growing demand for bio-methanol would necessitate larger and reliable sources of raw biomass. There is growing concern that the use of food crops for the production of fuels, such as ethanol

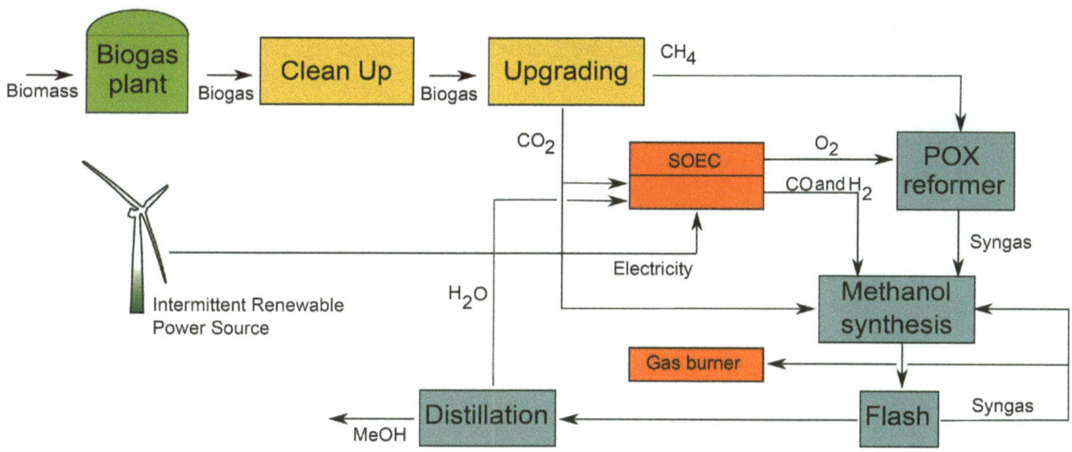

FIGURE 3.26.17 Methanol production concept from biogas and H_2 via solid oxide electrolyzer cell (SOEC) using wind energy. *POX*, partial oxidation of methane. *Reproduced with permission from Pedersen TH, Schultz RH. Technical and economic assessment of methanol production from biogas [Master thesis]. Denmark: Department of Energy Technology, University of Aalborg; 2012.*

from corn (first-generation biofuels), contributes to increasing food prices by competing for the same agricultural resources.[81] For the production of bio-methanol, non-food crops selected specifically for energy purposes would therefore have to be cultivated on a large scale if a significant amount of methanol were to be produced from biomass resources. Suitable "energy crops" are being identified, and the most promising—essentially fast-growing grasses and trees—are being field tested. However, in most scenarios, these energy crops would still compete for the same arable lands as the ones used for food production, limiting their growth potential. Furthermore, crop production requires sufficient water, a suitable temperature range, fertilizers (mostly derived from fossil fuel resources), as well as agricultural land and time. Energy crops should also not compete with food crops for these essential resources. Increased pollution due to fertilizer run-off leading to such problems as "dead zones" in the ocean is also a concern[82] as is the carbon footprint of biomass production.[83,84] Taking these and other factors into account, it has been estimated that biomass produced in a renewable and sustainable manner can cover at most about 20% of our energy needs of the future.[85–88]

3.26.4.2 Methanol Through CO_2 Recycling

The limitations of the natural photosynthesis-based carbon cycle can be overcome by supplementing with a feasible *anthropogenic chemical carbon cycle* based on CCR fuels and materials through methanol.

3.26.4.2.1 Methanol From CO_2 and H_2

It has long been known that CO_2 can be converted to methanol by catalytic hydrogenation (Eq. 2).[89] This is the most straightforward way to produce methanol from CO_2 and in fact, some of the earliest methanol plants operating in the 1920s and the 1930s in the United States used CO_2 and H_2 obtained as by-products of fermentation processes.[38,90] Both homogeneous and heterogeneous catalysts have been studied for the hydrogenation of CO_2. The latter is, however, preferable in terms of cost, stability, separation, handling and reuse of the catalyst, as well as reactor design and is also the one currently practiced for the industrial production of methanol from syngas.

3.26.4.2.1.1 HETEROGENEOUS CATALYSTS FOR THE PRODUCTION OF METHANOL FROM CO_2 AND H_2

Efficient catalysts, based on metals and their oxides, notably copper and zinc, have been developed for the conversion of CO_2 to methanol.[91,92] These catalysts are very similar to the ones currently used for the production of methanol from syngas based on Cu/ZnO/Al_2O_3. In view of our present understanding of the mechanism of methanol synthesis from syngas, this is not unexpected. It is now generally accepted that methanol is most probably almost exclusively formed by hydrogenation of CO_2 contained in syngas on the catalyst's surface. To be converted to methanol, the CO in the syngas first undergoes a WGS reaction (Fig. 3.26.18) to form CO_2 and H_2. The formed CO_2 then reacts with hydrogen to yield

$$CO + H_2O \rightleftharpoons H_2 + CO_2 \quad \Delta H_{298K} = -9.8\ \text{kcal mol}^{-1}$$

FIGURE 3.26.18 Water gas shift reaction.

methanol.[93–95] CO would therefore serve essentially as a CO_2 source as well as a scavenger for the oxygen atoms in water, which acts as an inhibitor for the active metal sites.[96,97] It has been shown that reacting a CO/H_2 mixture carefully purified from CO_2 and water on a commercial methanol catalyst produces no or very little methanol.

In an industrial setting, an adequate catalyst should remain active for several years, to sustain high plant output. Deactivation through sintering (clustering of copper sites), carbon deposition, and other phenomena should thus be minimal to avoid premature catalyst replacement and associated costs. Improvement to the catalyst system's activity and stability over time is therefore of prime importance for the economics of any methanol plant.[98]

On a commercial standard catalyst (for example, from Süd Chemie), CO_2 hydrogenation was found to be slower than CO hydrogenation.[99] However, with CO_2 as the carbon source, the selectivity was higher, with less by-product formation. More problematic was that conventional Cu/ZnO-based methanol synthesis catalysts exhibit a tendency to deactivate prematurely at higher CO_2 partial pressures.[98] It appears, however, that the effect of high CO_2 on the methanol catalysts is substantially due to the presence of water formed during the synthesis reaction and not the CO_2 itself. In the presence of CO, this water can react via the WGS reaction to form more CO_2 and H_2.[100] The presence of water can inhibit the metal active sites of the catalyst and lead to poorer performance and deactivation in conventional catalysts.[101–104] Due to the lack of long-term stability of conventional syngas to methanol catalysts in the presence of high levels of CO_2, more stable catalysts have been studied and developed.

Lurgi AG, a leader in methanol synthesis process technology, in collaboration with Süd-Chemie developed and thoroughly tested a high activity catalyst for methanol production from CO_2 and H_2.[91] Operating at a temperature around 260 °C, slightly higher than that used for conventional methanol synthesis catalysts, the methanol selectivity was excellent. Other companies commercializing methanol synthesis catalysts such as Sinetix, Haldor Topsøe, and Mitsubishi Gas Chemical have also developed over the years more stable catalysts for the CO_2 to methanol process.

A large number of studies have been devoted to the development of efficient and stable heterogeneous catalysts for the production of methanol from CO_2 hydrogenation. Most of the catalysts described still rely on Cu on a support such as ZrO_2 or/and ZnO and diverse additives to improve the activity, stability, and other parameters (Fig. 3.26.19). They were summarized recently in one of our review article.[8] Beside the composition of the catalyst, the preparation method also played a significant role. Some of these catalysts are already used in pilot and demonstration plants.

Catalysts based on various metals including palladium, platinum, and molybdenum have shown limited applicability for CO_2 hydrogenation to methanol. Recently, however, the discovery of a Ni-Ga-based catalyst for this reaction was reported.[105] In comparison to conventional $Cu/ZnO/Al_2O_3$ catalysts, a similar methanol synthesis activity was observed with, however, a much lower CO generation. This is of interest for the production of methanol at lower pressures and possibly even ambient pressure.

3.26.4.2.1.2 REDUCTION OF CO_2 TO METHANOL WITH HOMOGENEOUS CATALYSTS

Although numerous heterogeneous systems were described for the direct hydrogenation of carbon dioxide to methanol, only a limited number of homogeneous catalysts have been reported and this field has been reviewed recently.[8,106]

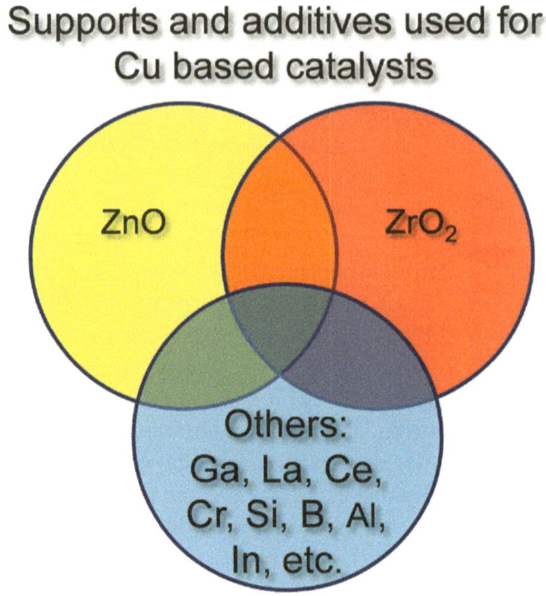

FIGURE 3.26.19 Most common combinations of supports and additives used for Cu-based heterogeneous catalysts for the hydrogenation of CO_2 to methanol. *Reproduced from Goeppert A, Czaun M, Jones J-P, Prakash GKS, Olah GA. Recycling of carbon dioxide to methanol and derived products — closing the loop.* Chem Soc Rev *2014;**43**:7995–8048 by permission of the Royal Society of Chemistry.*

The first example of direct hydrogenation of CO_2 to methanol under homogeneous conditions was reported in the 1990s using a $Ru_3(CO)_{12}$ catalyst precursor in the presence of KI. In addition to methanol, the formation of CO, CH_4, and C_2H_6 was observed.[107] The catalytic performance of the $Ru_3(CO)_{12}$-KI system surpassed that of other transition metal carbonyls such as $Ir_4(CO)_{12}$, $Rh_4(CO)_{12}$, $W(CO)_6$, $Mo(CO)_6$, $Fe_2(CO)_9$, and $Co_2(CO)_8$.[108] The catalysts based on ruthenium seem to be especially active in this reaction.[109] Recently, the capture of CO_2 by bases and its transformation to methanol in a single pot was also demonstrated on a ruthenium catalyst.[110] We have also shown that CO_2 captured from air with an amine can be efficiently converted to methanol using a homogeneous ruthenium catalyst (Fig.3.26.20).[111]

Another approach is to divide the six electron reduction of CO_2 to methanol into several consecutive steps, in a so-called cascade process.[112] The three steps for the homogeneous reduction of CO_2 to methanol encompass (1) reduction of CO_2 to formic acid (FA), (2) esterification of FA to formate esters, and (3) hydrogenation of formate ester to methanol. For each

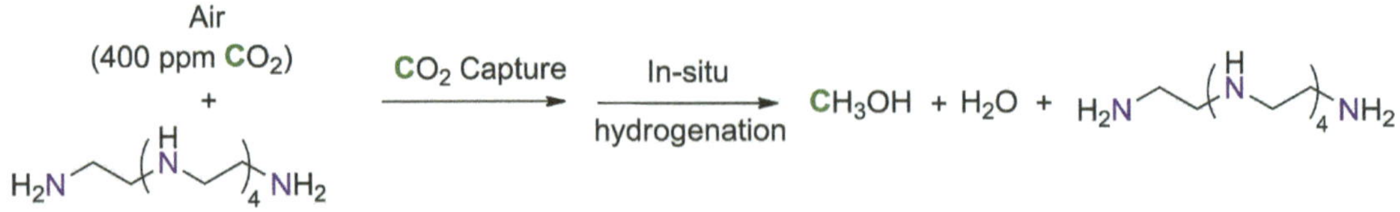

FIGURE 3.26.20 Carbon dioxide capture from the air and conversion to methanol in one pot.

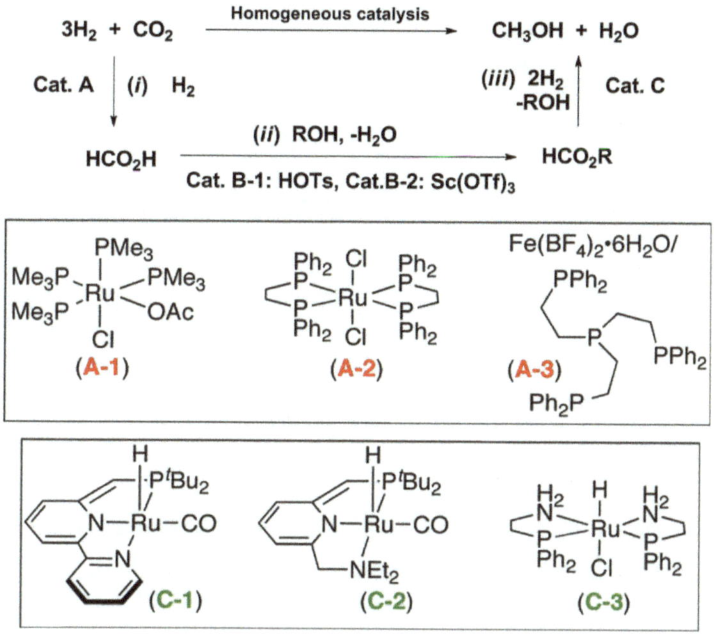

FIGURE 3.26.21 Hydrogenation of CO_2 to methanol in a cascade system.[112]

step a different catalyst is required, which has to coexist and must be tolerant to the reaction conditions and relatively high temperature (135 °C or more) and pressure (40 bar or more). Such systems have been described and tested for the homogeneous hydrogenation of CO_2 to methanol, but with limited success (Fig. 3.26.21).[112] The fact that three different catalysts have to be used is also not very practical/economical. Usually, simple and robust systems are preferred especially for industrial applications.

In addition to the direct hydrogenation of CO_2 to methanol, the hydrogenation of CO_2 derivatives such as carbonates, polycarbonates, carbamates, urea derivatives, and formates[113] has gained some attention[114] because the activation barriers of these reactions are smaller and these molecules can be readily synthesized from CO_2 (Fig. 3.26.22).[114] In a second step, these molecules are hydrogenated to methanol. However, adding steps to the preparation of methanol also adds costs to the overall process.

Reduction of CO_2 by a variety of reducing agents other than hydrogen is also possible. Silanes, hydrides, and boranes can, for example, be used for the reduction of CO_2 to methanol in the presence of a catalyst such as *N*-heterocyclic carbenes or with sterically encumbered Lewis acid-Lewis base ion pairs (so-called "frustrated" pairs).[8] The utilization of these reducing agents is, however, in most cases not economical especially compared with direct hydrogenation with H_2.

Despite considerable efforts, a well-defined homogeneous catalyst for the direct conversion of CO_2 to methanol that is also efficient, robust, and selective remains elusive. Heterogeneous catalysts are therefore still the standard for the commercial preparation of methanol from CO_2.

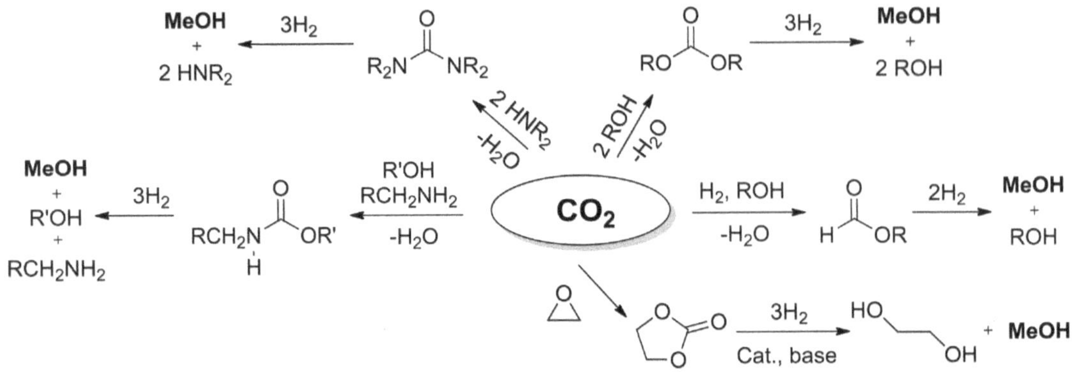

FIGURE 3.26.22 Methanol production by indirect hydrogenation of CO_2. *Reproduced from Goeppert A, Czaun M, Jones J-P, Prakash GKS, Olah GA. Recycling of carbon dioxide to methanol and derived products – closing the loop.* Chem Soc Rev *2014;***43***:7995–8048 by permission of the Royal Society of Chemistry.*

3.26.4.2.1.3 TWO-STEP ROUTE FOR CO_2 HYDROGENATION TO METHANOL

CO_2 to methanol can also be achieved in two steps. In the first step, CO_2 and H_2 are fed to a reactor where part of the CO_2 is converted to CO through the RWGS reaction. After separation of the produced water, the resulting gas composed of $CO/CO_2/H_2$ is directed to a methanol synthesis reactor (Fig. 3.26.23). This process called CAMERE (carbon dioxide hydrogenation to methanol via RWGS reaction) was developed by the Korean Institute of Science and Technology.[115] It was claimed that this two-step process allows a sizable reduction in size of the methanol synthesis reactor compared with the direct CO_2 hydrogenation route. The process is also described as having higher efficiency, with twice the methanol production yield and lower operating costs. The RWGS reaction has to be operated at temperatures higher than 600 °C to obtain a reasonable CO_2 conversion to CO (>60%). The catalyst employed is based on zinc aluminate.[116] The catalyst for the second step, the methanol synthesis, is composed of $Cu/ZnO/ZrO_2/Ga_2O_3$.[115] A pilot plant based on the CAMERE process with a production capacity of 100 kg/day methanol has been built in Korea to test this concept.[116]

3.26.4.2.2 *CO_2 Reduction to CO Followed by Hydrogenation*

In the reaction of CO_2 with H_2, a mole of water is generated for every mole of methanol. One-third of the H_2, and therefore energy needed to produce this H_2, ends up in water

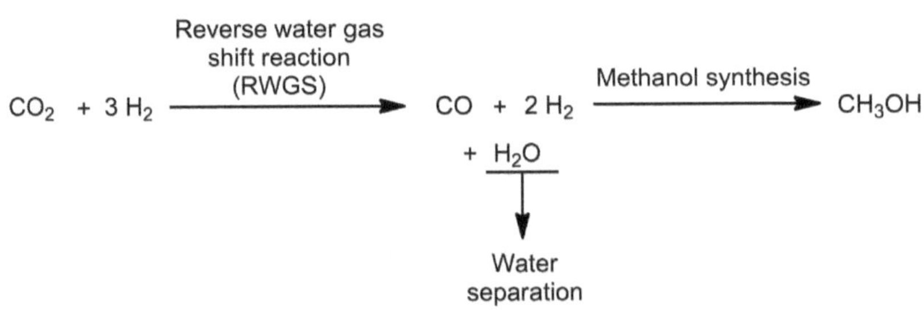

FIGURE 3.26.23 Simplified schematic of the CAMERE (carbon dioxide hydrogenation to methanol via RWGS reaction) process.

$$CO_2 + C \xrightarrow{\Delta} 2\,CO \quad \Delta H_{298K} = 40.8 \text{ kcal mol}^{-1}$$

FIGURE 3.26.24 Reverse Boudouard reaction.

by-product (Eq. 2). Likewise, DME produced via the direct catalytic hydrogenation of CO_2 also generates water as a by-product.[117] Water formed can be reused for hydrogen production or, when needed, be recycled, particularly in arid areas or where the need for pure water would warrant it.

Nevertheless, to utilize hydrogen more efficiently for CO_2 conversion to methanol and DME the initial chemical or electrochemical reduction of CO_2 to CO to minimize hydrogen use in water formation is also feasible. The CO formed can then react with two molecules of H_2 to form methanol without water by-product. Carbon dioxide reduction to CO can be achieved by the reverse Boudouard reaction, that is, by thermal reaction with carbon at temperatures above 800 °C (Fig. 3.26.24). This, however, necessitates solid carbon, and although coal could be used, the process would not be sustainable. Although biomass can also be used as a carbon source, other technologies are probably more efficient at transforming it into fuels.

The direct conversion of CO_2 to CO using a thermochemical cycle and solar energy is also being intensively studied. Due to the high stability of CO_2, its direct splitting (thermolysis) is not favorable, and even at 2000 °C only about 2% of CO_2 is decomposed to CO and O_2.[118] In addition to choosing the proper materials that can resist such high temperatures, other issues such as separation of explosive product gases or fast quenching to avoid recombination upon cooling makes the direct thermal splitting challenging. Two-step thermochemical cycles operating at lower temperatures can overcome the aforementioned issues associated with the direct thermal splitting of carbon dioxide.

Researchers at the Sandia National Laboratories working on the sunshine to petrol project (S2P) have, for example, developed a solar furnace, which heats a device containing cobalt-doped ferrite (Fe_3O_4) to temperatures around 1400–1500 °C, driving off oxygen gas (Fig. 3.26.25). In a second step at a lower temperature, the reduced material FeO is then exposed to CO_2, from which it absorbs oxygen, producing CO and ferrite, which can be recycled.[119–121] Numerous other redox systems have been developed for the thermochemical splitting of CO_2 such as ZnO/Zn,[118,121,122] SnO_2/SnO,[118] nonstoichiometric ceria,[123] or Ni-, Fe-, Mg- and Mn-doped ceria/zirconia solid solutions,[124] which differ in the decomposition temperature of the oxide material. Although this technology shows promise, its viability on an industrial scale is still unproven.

3.26.4.2.3 *Electrochemical Routes From CO_2 to Methanol*

Another way to perform the reduction of CO_2 to CO, which does not require high temperatures, is electrochemical reduction in aqueous or organic media (Fig. 3.26.26).

$$Fe_3O_4 \xrightarrow{1400-1500\ °C} 3\,FeO + 1/2\,O_2$$
$$3\,FeO + CO_2 \xrightarrow{<1000\ °C} Fe_3O_4 + CO$$
$$CO_2 \longrightarrow CO + 1/2\,O_2 \quad \text{Overall reaction}$$

FIGURE 3.26.25 Thermochemical cycle for the production of CO from CO_2 at high temperature using solar heat.

$$CO_2 \xrightarrow{e^-} CO + 1/2\ O_2$$

FIGURE 3.26.26 Electrochemical reduction of CO_2 to CO.

This approach has been investigated using various metal electrodes in aqueous media as well as in organic solvent media, in particular methanol.[125,126]

Besides CO, other products can be generated by electrochemical reduction of CO_2. For many decades, the electrochemical reduction of CO_2 has been practiced at the laboratory scale but has not yet successfully made the transition to an industrial process. Part of this reluctance by industry to adopt electrochemical CO_2 processing technologies has been the lack of a driving force to utilize CO_2. Part of it has been due to the absence of a practical technology to convert CO_2 electrochemically into useful products. Both of these areas are, however, rapidly changing, and progress is being made in solving the problems encountered in CO_2 electrolysis.

Depending on the conditions and electrocatalyst used, a variety of reduced products can be generated electrochemically from CO_2, including FA, carbon monoxide, formaldehyde, methanol, and methane (Table 3.26.1).[127]

Part of the problem is that none of these reactions are well separated by potential, meaning that the control between various products must emanate from a choice in catalyst and conditions rather than potential. Although all these standard potentials are close to the one for the hydrogen evolution, generation of the first intermediate (carbon dioxide radical anion, $CO_2^{\bullet -}$) has been estimated to occur only at −1.89 V versus standard hydrogen electrode.[128] Even though it can be stabilized, this high-energy intermediate needs to be generated first for CO_2 reduction to occur on a traditional metal electrode. Therefore, to achieve CO_2 reduction at a reasonable rate, significant overpotentials are generally required. During the electrochemical CO_2 reduction, water is typically present as a proton source (and electrolyte) and thus the competing hydrogen evolution reaction (HER) must also be taken into account. For this reason, many metals reported to be active for CO_2 reduction have also relatively high HER overpotentials. Because of these competing reactions, a balancing act must be conducted to identify the optimal CO_2 reduction electrode able to reduce CO_2 selectively at high rates and low overpotentials without reducing water simultaneously. In an effort to limit hydrogen evolution, similar CO_2 reduction reactions in some organic solvent media were also studied. Methanol in particular, used industrially as a physical absorber for CO_2 in the Rectisol process, has been extensively studied as a medium for the electrochemical reduction of CO_2.[129–131]

TABLE 3.26.1 Standard Potentials for CO_2 Reduction[128]

Half-Cell Reaction	E° Versus SHE
$CO_2 + 2H^+ + 2e^- \rightarrow HCOOH$	−0.11
$CO_2 + 2H^+ + 2e^- \rightarrow CO + H_2O$	−0.10
$CO_2 + 4H^+ + 4e^- \rightarrow CH_2O + H_2O$	−0.028
$CO_2 + 6H^+ + 6e^- \rightarrow CH_3OH + H_2O$	+0.031
$CO_2 + 8H^+ + 8e^- \rightarrow CH_4 + 2H_2O$	+0.17

SHE, standard hydrogen electrode.

Overcoming the large overpotential to CO_2 reduction also results in low to moderate energy efficiency.[8] Energy efficiency is potentially the biggest issue facing the commercialization of electrochemical CO_2 reduction because the cost of electricity will likely be the main factor influencing the cost of the product. In alkaline water electrolysis, probably the closest commercial process, the cost of electricity is estimated to represent 80% of the total cost of the hydrogen produced. Improvements in this area are critical if electrochemical CO_2 reduction is ever to become a viable choice for energy storage since a loss of energy efficiency must be made up with additional renewable energy.

3.26.4.2.3.1 DIRECT ELECTROCHEMICAL CO_2 REDUCTION TO METHANOL

Converting CO_2 to methanol directly has the distinct advantage of producing a useful product that can be directly utilized by many energy-consuming devices. There is considerable interest in this field, given its allure from a practical standpoint. Table 3.26.2 illustrates some of the conditions reported in the literature to obtain methanol directly from CO_2.

The most striking observation that can be made when looking at Table 3.26.2 is that very few metals have been used for the direct generation of methanol. The only example is in entry 4, with the remainder being semiconductors (entries 1–3 and 8), oxides (entries 5 and 9), alloys (entry 6), or electrolytes containing a homogeneous catalyst (entries 7 and 8). Canfield and Frese first determined that semiconductors have some ability to form methanol from CO_2 back in 1983,[132] albeit at extremely low faradaic efficiencies (FEs) and current densities. Subsequent efforts by a number of research groups increased both the FE of the process as well as the current density. Bocarsly et al.[136] introduced the concept of adding a homogeneous cocatalyst

TABLE 3.26.2 Direct Electrochemical Reduction of CO_2 to Methanol

Entry	Electrode	E Versus NHE (V)	Major Product	FE (%)	Current Density (mA/cm^2)	References	Electrolyte
1	n-GaAs	−1.06	MeOH	1.0	0.16	132	Sat. Na_2SO_4
2	*p*-GaAs	−1.06	MeOH	0.52	0.08	132	Sat. Na_2SO_4
3	*p*-InP	−1.06	MeOH	0.8	0.06	132	Sat. Na_2SO_4
4	Mo	−0.56	MeOH	84	0.12	133	0.2 M Na_2SO_4
5	CuO	−1.3	MeOH	28	6.9	134	0.5 M $KHCO_3$
6	Pt-Ru/C	−0.06	MeOH	7.5	0.4	135	Flow cell
7	Pd	−0.51	MeOH	30	0.04	136	0.5 M $NaClO_4$ + pyridine
8[a]	n-GaP	−0.06	MeOH	90	0.27	137	10 mM pyridine pH 5.2
9	RuO_2/TiO_2 nanotubes	−0.6	MeOH	60	1	138	0.5 M $NaHCO_3$

NHE, normal hydrogen electrode; *FE*, faradaic efficiency.
[a]*Solar irradiation of the working electrode.*

(pyridine) and then improved their results by switching to a semiconductor irradiated with light.[137] The latter example has the highest FE, coupled with the lowest applied potential (entry 8) in Table 3.26.2. Recently, using pyridine (which is protonated in situ to form the active pyridinium species) as a cocatalyst has been explored extensively, particularly with the goal of determining the mechanism of activation precisely.[139–144] Entry 8 also involves an irradiated semiconductor, meaning that some of the energy for CO_2 reduction is coming from light.

As illustrated by these examples of direct CO_2 to methanol reduction, one of the main problems remains the relatively low current density, especially when compared with two electron reduction products such as CO and formate, likely due to the increased complexity of reducing CO_2 with six electrons. The direct electrochemical reduction of CO_2 to methanol certainly holds much promise. However, with such low current densities, it would not be practical on a large scale since it has been shown that the investment cost is almost proportional to the surface area of the electrode.[12] Therefore, to minimize these costs and be economical current density must be in the range of at least hundreds of milliamperes per square centimeter. This is several orders of magnitude higher than anything reported in Table 3.26.2. When the current density is increased, the selectivity generally suffers. Besides methanol, depending on the electrodes and conditions, varying amounts of FA, carbon monoxide, formaldehyde, and methane can be produced in two-, four-, or eight-electron processes, respectively.[145] Although most of these mixtures of products could be further reacted to increase the yield of methanol, it would add to the complexity and cost of the process. Avoiding additional steps in methanol synthesis remains the most desirable goal.

3.26.4.2.3.2 METHODS FOR HIGH RATE ELECTROCHEMICAL CO_2 REDUCTION

Although the electrochemical reduction of CO_2 directly to methanol would be ideal, it is currently outside the scope of industrially viable processes. For this reason, the electrochemical reduction of CO_2 by two electrons to CO or formate/FA, which has been demonstrated at relatively high current densities and selectivity followed by subsequent conversion to methanol using more traditional chemical processes, as described vide supra, might be a more feasible route.

During the electrochemical reduction of CO_2 in water, hydrogen formation generally competes with CO_2 reduction, thereby reducing the FE of the CO_2 reduction. Progress is being made to suppress hydrogen formation.

However, in our studies of electrochemical CO_2 recycling, instead of considering H_2 formation as a problem, it was found advantageous to generate CO and H_2 concomitantly at the cathode preferably in a H_2:CO ratio close to 2. This syngas mixture ("metgas") can then be further transformed into methanol, allowing for a more energy-efficient CO_2 reduction (Fig. 3.26.27).[146,147] An additional advantage is the valuable pure oxygen produced at the anode.

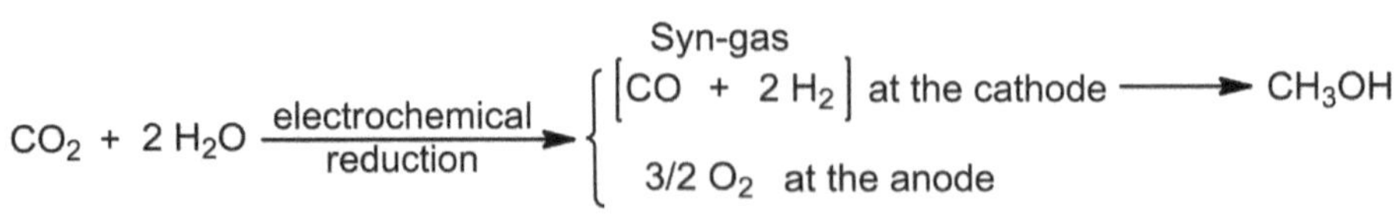

FIGURE 3.26.27 Two-electron reduction of CO_2 to CO with concomitant H_2 evolution for subsequent production of methanol.

Thus, methanol and DME can be produced selectively from CO_2 via electrochemically generated syngas (metgas) in much the same way as it is done from natural gas or coal. The advantages are that no purification step is required and no impurities such as sulfur, which could deactivate the methanol production catalyst, are present. The reaction is preferably run under pressure to feed the metgas directly into the methanol synthesis reactor and avoid compression costs.

For the coproduction of CO and H_2, proton exchange membrane electrolyzer cells can be used. The pressurization of CO_2 was found to increase the FE and rate of the reduction reaction. More recently, SOECs have been used for CO_2 reduction[148,149] coupled with steam electrolysis. Instead of relying on electricity for all of the energy required to reduce CO_2, SOECs operate at very high temperatures (typically ~800–900 °C) where some of the energy for CO_2 reduction can be supplied by heat. SOECs have the advantage of being capable of producing extremely high current densities (up to 1500 mA/cm^2) compared with other types of cells while simultaneously operating at reasonable cell potentials (2.65 V or less).[150] Research is still very active in understanding the mechanisms at play in SOECs used for CO_2 reduction.[151] A drawback or selectivity advantage, depending on the point of view, is that SOECs are essentially limited to CO as the product, since other products decompose at such high temperatures. Another related technology inspired by SOEC for CO_2 reduction is the use of solid proton conductors, which have been reported to yield both CO and CH_4 electrochemically.[150] Although alkaline electrolyzers are the most mature electrolyzer technology, with the first demonstration in 1800,[152] they can only be used to produce H_2 from water. CO_2 would unfortunately react with the basic electrolyte of the alkaline electrolyzer cell and render it useless in a very short time.

As an alternative to the preferred electrochemical direct syngas production for methanol synthesis, CO_2 can also be first reduced to FA, another two-electron reduction product. The obtained FA can then be converted into syngas for methanol synthesis. This is possible because of a unique feature of FA that can be decomposed via two distinct pathways: decarbonylation and decarboxylation. Decarbonylation of FA leads to CO and H_2O and was first reported by Sabatier in 1912 over heterogeneous catalysts.[153,154] On the other hand, the decarboxylation of FA gives CO_2 and H_2[155] and is mainly carried out in the presence of homogeneous catalysts based on Ru and Ir.[156–161] After the separation of water (in the decarbonylation pathway) and carbon dioxide (in the decarboxylation pathway), the combined metgas streams with a H_2/CO ratio of 2:1 can be subjected to methanol synthesis (Fig. 3.26.28).

Although the decomposition of FA generally proceeds through decarboxylation and decarbonylation, it can also undergo decomposition by disproportionation to methanol, CO_2, and water (Fig. 3.26.29). This was first shown by Miller et al. in 2013 using a homogeneous catalyst based on iridium, albeit with low selectivity.[162] The selectivity and yield were greatly improved by Cantat et al. using ruthenium phosphine complexes.[163] Recently, Laurenczy et al. achieved both the hydrogenation of CO_2 to FA and the subsequent disproportionation of FA to methanol in an aqueous acidic solution with an iridium catalyst.[164] The acidification of the medium resulted in complete (98%) and selective (96%) disproportionation of FA into methanol.

In addition, FA can be esterified to methyl formate and then hydrogenated (with hydrogen from water electrolysis) to yield methanol, as shown in Fig. 3.26.30. Although hydrogenation of methyl formate is not commonly practiced industrially due to the higher value of FA and methyl formate compared to methanol, it can be accomplished using heterogeneous catalysts.[165,166]

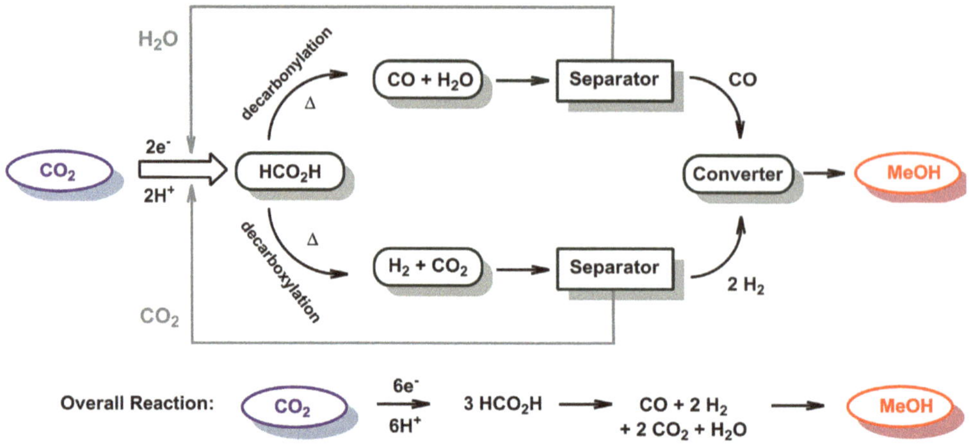

FIGURE 3.26.28 Methanol production from CO_2 by sequential electrochemical reduction to formic acid (FA), FA decomposition, and metgas conversion to methanol.

$$3\,(CO_2 + 2\,H^+ + 2e^-) \xrightarrow{\text{Catalyst}} 3\,HCOOH \xrightarrow{\text{Catalyst}} CH_3OH + 2\,CO_2 + H_2O$$

FIGURE 3.26.29 Disproportionation of formic acid to methanol.

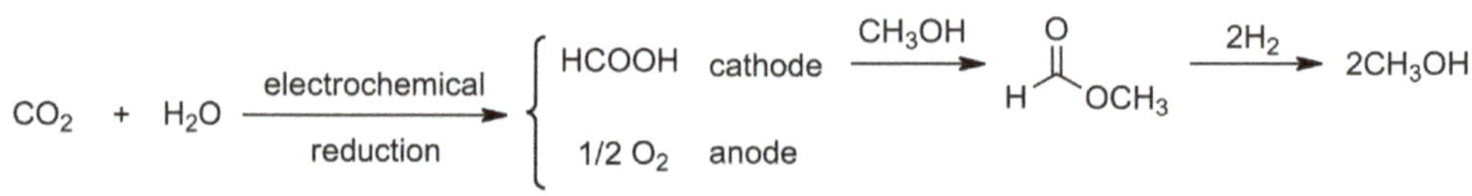

FIGURE 3.26.30 Two-electron reduction of CO_2 to formic acid with subsequent esterification to methyl formate and further hydrogenation to methanol.

3.26.4.2.4 Photochemical Reduction of CO_2 to Methanol

The photochemical reduction of CO_2 to methanol shares some similarities with the electrochemical CO_2 reduction, especially when it comes to the molecular catalysts used in both cases and product distribution. Although the selectivity for methanol is quite low, the direct conversion of CO_2 to methanol using photochemistry has been demonstrated. Typically, photochemical (or photocatalytic) CO_2 reduction gives either formate or CO as a major product. This area has been reviewed in 2010.[167] One of the major limitations to photochemical CO_2 reduction as it is generally practiced is that a sacrificial hydride source (in most cases an amine, ascorbic acid, or 1-benzyl-1,4-dihydronicotinamide) must be added to the solution to substitute for the anode, which would generally be used in electrochemical CO_2 reduction. Photochemical regeneration of hydride donor remains a major challenge.[168] Furthermore, the activity and lifespan of the photochemical catalysts used is still very limited and will require significant improvement if these processes are to be considered for industrial application. In addition to the molecular catalyst approach, semiconductors and metal oxides such as silicon carbide,[169] WO_3,[170] $InVO_4$,[171] TiO_2,[172–175] $InTaO_4$,[176] NiO,[175] and ZnO[175] either by themselves or in combination with various other heterogeneous catalysts have also been tested. A challenge to producing methanol on semiconductors irradiated with light is that the reaction can be reversible,[169,175] so strategies to mitigate methanol oxidation are essential to achieve a practical process.

Instead of directly reducing CO_2 photochemically, one could produce hydrogen using a photochemical cell[177,178] and take this renewable hydrogen to hydrogenate CO_2. Although this approach would not be as straightforward as direct photochemical CO_2 reduction, it could still reduce the number of steps required for CO_2 reduction by combining light harvesting with hydrogen evolution.[179] Compared to photochemical CO_2 reduction, water splitting can occur at higher efficiencies (>12%) and current densities (120 mA/cm^2).[180] A significant amount of research is ongoing in this area, with centers like the Joint Center for Artificial Photosynthesis focused entirely on photochemical reactions and their potential commercialization. The road for practical application, however, remains challenging.[179]

3.26.4.2.5 Practical Applications of CO_2 to Methanol

Besides the experience of Lurgi AG described vide supra, methanol synthesis from CO_2 and H_2 has also been demonstrated on a laboratory pilot scale in Japan, where a 50 kg CH_3OH/day production with 99.8% selectivity for methanol was achieved.[92,181] A liquid-phase methanol synthesis process was also developed, which allowed for a CO_2 and H_2 conversion to methanol of about 95% with very high selectivity in a single pass.[182]

The first contemporary commercial CO_2 to methanol recycling, tapping into locally produced, cheap geothermal energy, is presently being operated in Iceland by the company, Carbon Recycling International. The demonstration plant, named after one of the authors of this book chapter (G. A. Olah), has an annual capacity of 4500 m^3 of methanol (10 metric ton/day), and is based on the conversion of CO_2 accompanying the readily available local geothermal energy (hot water and steam) sources (Fig. 3.26.31). The required H_2 is being generated by water electrolysis using cheap geothermal electricity.[183] Iceland embarked on this development as a way to exploit and possibly export its cheap and clean renewable

FIGURE 3.26.31 The "George Olah Renewable CO_2-to-Methanol Plant" of Carbon Recycling International (CRI) in Iceland. Based on local renewable energy and CO_2. *Courtesy: CRI.*

electrical energy. The produced methanol, called Vulcanol, is currently mixed with gasoline locally and is being sent to continental Europe to be used as a fuel either directly or for the preparation of biodiesel.

In Japan, Mitsui chemicals has also constructed of a 100 metric ton/year demonstration plant producing methanol from CO_2, obtained as an industrial by-product, and hydrogen generated by photochemical splitting of water using solar energy.[184] Blue Fuel Energy in Canada is planning to use hydroelectricity and concentrated CO_2 emissions from natural gas processing to produce preferably methanol.[185] In Germany, Sunfire is also working on converting CO_2 and H_2 from water and renewable electricity sources to liquid fuels including methanol, gasoline, and diesel.

Various improvements for producing methanol from CO_2 and H_2 have been reported in the literature.[91,92,186,187] An electricity-to-fuel efficiency of about 70% in a state-of-the-art methanol plant operating on CO_2 and H_2 feedstocks has been claimed.[188] The capital investment for a methanol synthesis unit using CO_2 and H_2 is estimated to be about the same as that for a conventional syngas-based plant.[91] The limiting factor for large scale-up of such carbon neutral processes is the availability and price of CO_2 and H_2 sources, the price of H_2 depending mainly on the necessary electrical energy cost.

3.26.4.3 Production of Dimethyl Ether From CO_2

Methanol obtained using any process described in this chapter can be transformed easily to DME by conventional bimolecular dehydration (Fig. 3.26.32). This reaction is readily carried out catalytically over varied solid acids such as alumina or phosphoric acid-modified γ-Al_2O_3.[99,189]

Like methanol, DME can also be produced by direct catalytic hydrogenation of CO_2. Similar to the route from syngas to DME, CO_2 hydrogenation to DME can be achieved on a hybrid catalyst system consisting of a combination of methanol synthesis and dehydration catalysts.[117,190] These "bifunctional" catalytic systems thus contain two complementary catalysts, one for the methanol synthesis and another for the dehydration. Combinations of Cu/ZnO/Al_2O_3/Ga_2O_3/MgO for the methanol synthesis and γ-Al_2O_3 or ZrO_2/Al_2O_3 for the dehydration step have been studied, and it was shown that the catalyst bed was the most effective when the methanol synthesis catalyst was placed in a layer upstream of the dehydration catalyst layer.[117,191] A number of other systems were also tested. They include (Cu/ZnO/Al_2O_3/ZrO_2)/HZSM-5,[192,193] (Cu/ZnO/Al_2O_3/La_2O_3)/HZSM-5,[194] (Cu/ZnO/Al_2O_3)/γ-Al_2O_3,[195,196] (Cu/ZnO/Al_2O_3)/HZSM-5,[195] (Cu/ZnO/ZrO_2)/HZSM-5,[197] and (Cu/TiO_2/ZrO_2)/HZSM-5.[198]

A core-shell type catalyst with a CuO/ZnO/Al_2O_3 core and an outer shell composed of a membrane of metal-doped silica-alumina was also recently proposed for DME synthesis by CO_2 reduction.[199]

$$2\,CH_3OH \longrightarrow CH_3OCH_3 + H_2O$$

FIGURE 3.26.32 Dehydration of methanol to dimethyl ether.

3.26.5 WHERE SHOULD THE CO_2 COME FROM?

3.26.5.1 Capture of CO_2 From Any Source

To produce fuels from CO_2, the first step is to capture it from natural and anthropogenic sources as efficiently and economically as possible. Carbon dioxide for chemical recycling to methanol or DME can come from various natural and industrial emissions, and eventually from the CO_2 content of the atmosphere. As pointed out, presently, worldwide, more than 35 billion metric tons of CO_2 related to human activities are released into the atmosphere every year. To significantly reduce emissions, capture of CO_2 from industrial and natural sources is becoming essential. In the short term, the capture of CO_2 from fossil fuel-burning power plants will probably be the most economical because of their relatively high CO_2 concentration (typically between 5% and 15% by volume). Technologies to capture CO_2 emissions from such sources and other industrial sources have been described extensively in the literature.[200] The removal and capture of CO_2 from gas streams can be achieved by a range of separation techniques depending on factors such as CO_2 concentration, pressure, and temperature. These separation technologies are based on various physical and chemical processes including absorption into a liquid solution system, adsorption onto a solid, cryogenic separation, and permeation through membranes (Fig. 3.26.33). The description of these techniques for the separation of CO_2 from relatively concentrated sources have been reviewed in the recent past.[200–208]

3.26.5.2 CO_2 From Biomass and the Atmosphere

To be not only renewable but also sustainable, sources of CO_2 decoupled from fossil fuel resources need to be increasingly and eventually exclusively utilized for the synthesis of

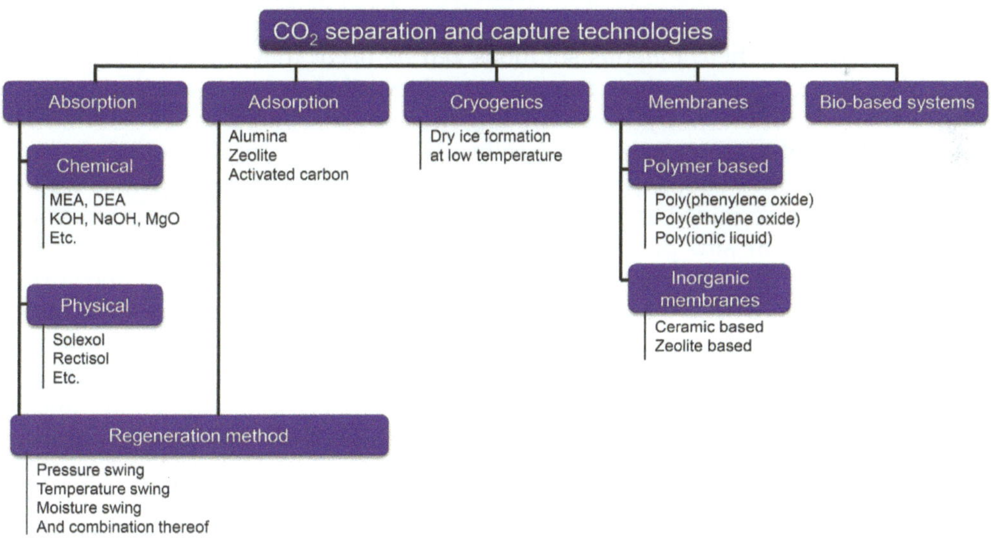

FIGURE 3.26.33 CO_2 separation and capture technologies. *DEA*, diethanolamine; *MEA*, monoethanolamine. *Reproduced from Goeppert A, Czaun M, Jones J-P, Prakash GKS, Olah GA. Recycling of carbon dioxide to methanol and derived products – closing the loop.* Chem Soc Rev *2014;***43***:7995–8048 by permission of the Royal Society of Chemistry.*

carbon-containing material. Biomass could be used and CO_2 captured from their combustion with technology similar to the ones applied for fossil fuels. However, even though the resources of biomass are enormous, the amounts of CO_2 that they can generate in a sustainable manner are limited. This points to another source of CO_2, namely, the atmosphere, which contains for all practical purpose, an essentially inexhaustible amount of CO_2. The capture of CO_2 from the air, generally termed as direct air capture, is feasible and has a number of advantages. Because the concentration of CO_2 in the atmosphere is similar all around the world (~400 ppm presently), CO_2 extraction plants could be placed anywhere, ideally, close to the sequestration site or hydrogen production sites for recycling. This makes the CO_2 collection independent from the CO_2 source, meaning that CO_2 originating from any source, small or big, static or moving (car, airplanes, trains, etc.) can be captured. The atmosphere would thus essentially serve as a "CO_2 conveyor belt" to transport CO_2 at no cost from the source point to the site of capture. No CO_2 transportation costs from a capture to a sequestration or recycling site would be needed, eliminating the need for a vast and costly infrastructure. DAC has been proposed for a while now,[19,26,209–223] and technologies to make it a reality are being developed.

Due to the low CO_2 concentration in air (~400 ppm), the presence of moisture, and the necessity to operate close to room temperature and ambient pressure, many of the technologies for gases containing higher CO_2 concentration have to be ruled out. At atmospheric pressure, physical adsorbents such as zeolites, activated carbon, and alumina can be excluded due to their low selectivity for CO_2 in the presence of moisture and very low heat of adsorption resulting in low adsorption capacities. Physical adsorption in liquids for the capture of CO_2 at rather high pressure is also not applicable. In practice, only sorbents having a chemical interaction can be used effectively for the capture of CO_2 from the air. Widely used monoethanolamine-based sorbents suffer from stability problems, when contacted with air, as well as evaporation issues due to their relatively high volatility and the large volume of gas to be handled. Although strong bases such as calcium hydroxide, potassium hydroxide, sodium hydroxide, and their aqueous solutions can be used, they bind CO_2 strongly, thus requiring high temperature for their regeneration. Nevertheless, their application for DAC is being explored, notably by Carbon Engineering in Canada.[224–226] Amine- and polyamine-based solid sorbents either physically adsorbed or chemically bound on a support such as silica, mesoporous solids (MCM-41, MCM-48, SBA-15, etc.), polymers, as well as carbon fibers have been recognized as potential candidates for DAC.[227–240] Hyperbranched aminosilicas prepared by in situ polymerizing of aziridine on porous solids have also been reported for DAC.[241,242] Adsorbents based on modified metal-organic frameworks[243,244] and porous polymer networks[245] as well as quaternary ammonium-functionalized anionic exchange resin that can be regenerated by moisture swing protocols are another possibility.[214,246–248] Technologies for DAC have been reviewed recently.[4,249] Practical applications have been developed for the essential removal of CO_2 from submarines and spacecraft for life support,[237,250] whereas the separation and recovery of CO_2 from ambient air on a larger scale is still in its infancy.[4] And while there is no question that the capture of CO_2 from air is possible, more research and development is clearly needed to optimize this technology and determine its economic viability. Using current technologies, the cost of removing a metric ton of CO_2 from point sources such as a coal-burning power plant that contains 10%–15% CO_2 has been estimated at between \$30 and \$100. The estimated cost of DAC, on the other hand, still varies greatly from about $20 to more than $1000 per metric ton of CO_2 depending on the assumptions.[4] The cost of a commercial plant will depend on many factors including the process used, labor cost, materials, and energy consumption. Only with the construction of demonstration and pilot plants will we have a better understanding of the total cost associated with DAC. Examples of proposed devices and prototypes for DAC are represented in Fig. 3.26.34.

It should be pointed out that air also contains other essentials for humankind's sustainable future in considerably higher concentration than the low (0.040%) CO_2 content. Pure water vapor, essential to life, in concentrations ranging from 1% to 6% depending on the moisture content of the air in varied location, could be separated from the air at the same time as CO_2 and provide clean water as an added value. This inexhaustible and renewable source of water could also be the source for the hydrogen needed in the reduction of CO_2.

Regardless of the method used to separate CO_2 from various sources, once CO_2 has been purified and compressed to the right pressure, it can be converted to methanol using any of the processes and catalysts described earlier.

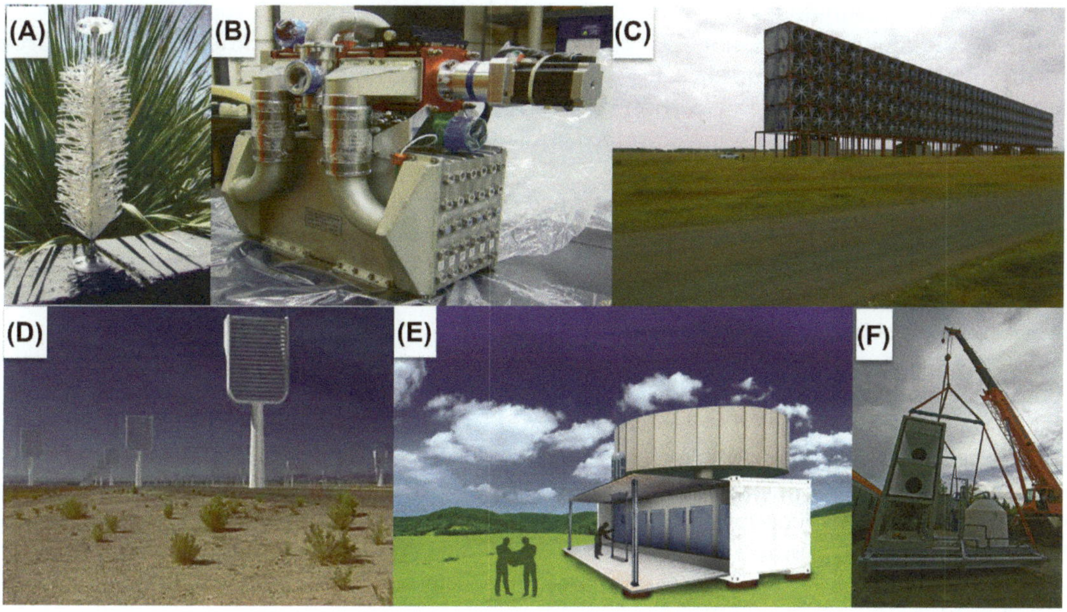

FIGURE 3.26.34 Examples of prototypes and proposed designs for the separation of CO_2 from the air. (A) Prototype for CO_2 capture from the air. (B) Solid amine-based swing bed for CO_2 removal in human spaceflights. (C) Artist rendering of an atmospheric CO_2 capture contactor. (D) Artist rendering of an array of atmospheric CO_2 capture units also known as Synthetic Trees. (E) Artist rendering of a prototype for CO_2 capture from the air using an anionic exchange resin and regeneration by moisture swing. (F) Construction of a prototype for CO_2 capture from the air in Alberta, Canada. *(A) Reproduced by permission of Kilimanjaro Energy. (B) Reproduced by permission of NASA. Courtesy: JC Graf. (C) Reproduced by permission of Carbon Engineering. (D) Reproduced by permission of Stonehaven Production. (E) Reproduced by permission of Kilimanjaro Energy. (F) Reproduced by permission of Carbon Engineering. Reproduced from Goeppert A, Czaun M, Prakash GKS, Olah GA. Air as the renewable carbon source of the future: an overview of CO_2 capture from the atmosphere.* Energ Environ Sci *2012;5:7833–53 with permission from The Royal Society of Chemistry.*

3.26.6 THE PATH TOWARD AN ANTHROPOGENIC CARBON CYCLE

The energy harvested through the burning of fossil fuels has allowed a rapid and unprecedented development of human society. However, the utilization of what amounts to fossilized sunshine has come at the expense of adverse global environmental changes, the full extent of which we are only starting to grasp. As proposed presently, CO_2 captured from fossil fuel emissions could be stored/sequestered in depleted gas and oil fields, deep aquifers, underground cavities, or at the bottom of the seas (CCS). This approach, however, neither provides a permanent solution nor does it help any way in fulfilling humankind's future needs for fuels, hydrocarbons, and their products. On the other hand, even if reserves of fossil fuels, either conventional or unconventional, remain considerable, they will eventually be depleted, be too expensive, or their environmental impact will be unacceptable. As these are complex issues, there will not be a single solution. The discussed chemical recycling of carbon dioxide (CCR) to produce carbon-neutral renewable methanol-based fuels and derived materials offers, however, a feasible and powerful new alternative to tackle both problems: global climate change and the inevitable depletion of fossil fuels.

The production of methanol, DME, hydrocarbons, and other synthetic materials from anthropogenic sources including CO_2 in flue gases of various industries could be the first steps toward an anthropogenic carbon cycle. As fossil fuels become less abundant and their use regulated by stricter emission standards, related CO_2 emissions will eventually diminish. Biomass sources could play a role, but even if the amounts of biomass that can be generated in a sustainable way are large, they are, nevertheless, limited and will not be able to cover all our needs by themselves. These limitations imply that methanol and derived products would be increasingly produced from CO_2 captured from the air, which offers a nearly inexhaustible carbon source for humankind. The energy required for the recycling of CO_2 will be provided by renewable energy sources or atomic energy, be it nuclear fission or fusion.

Processes and catalysts for the efficient conversion of CO_2 are in various stages of development. The most mature technology for CO_2 to methanol relies on hydrogenation with H_2 on heterogeneous catalysts. The hydrogen required would be obtained from water by electrolysis. Upon combustion and use, methanol and its derived product will be transformed back to water and CO_2, closing the anthropogenic carbon cycle. This would constitute humankind's artificial version of Nature's own photosynthesis-based carbon cycle (Fig. 3.26.1). Recycling water and CO_2 contained in the air with alternative energy sources represents pretty much the end game in terms of sustainability.

It should, however, be mentioned that significant bottlenecks remain, such as the cost of electricity from renewable sources needed to produce hydrogen. The amount of energy required to replace petroleum oil and other fossil fuels, in the order of 150 petawatt-hour per year, is also staggering. Nevertheless, these problems are not unique to methanol and all alternative fuels face similar economic and scale hurdles. The costs for renewables such as solar and wind have already come down dramatically, and continued effort will eventually make them competitive with fossil fuels in most markets.

Methanol from non-renewable sources such as natural gas and coal is already competitive with gasoline and diesel fuel, and could therefore constitute a bridge toward renewable

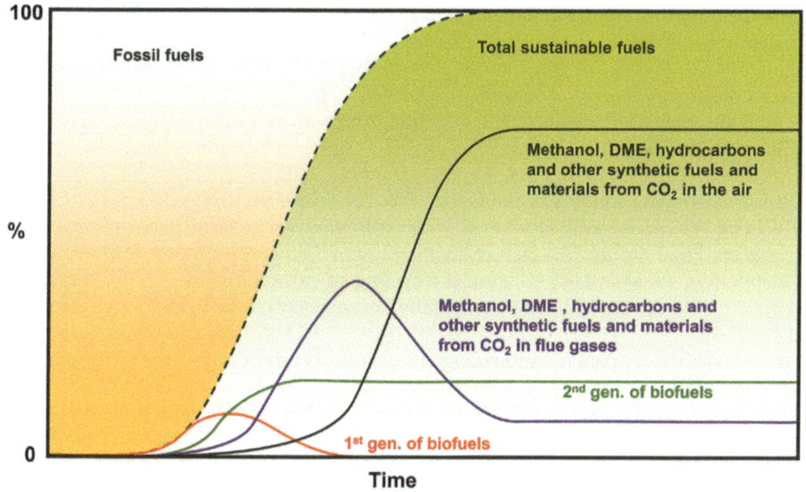

FIGURE 3.26.35 Possible transition to a sustainable fuel future including methanol and dimethyl ether (DME) as key components. *Reproduced from Goeppert A, Czaun M, Jones J-P, Prakash GKS, Olah GA. Recycling of carbon dioxide to methanol and derived products – closing the loop.* Chem Soc Rev *2014;**43**:7995–8048 by permission of the Royal Society of Chemistry.*

methanol. Hybrid systems using both renewable and fossil fuels with less or no CO_2 emissions to produce methanol could also be used during a transition period. Once the infrastructure for the distribution of methanol is in place, it could be seamlessly shifted to sustainable renewable methanol in the future. From a chemical point of view, there is no difference between methanol from fossil sources and renewable methanol.

Depending on a number of factors, including policies, state of development, locally available resources, and competing technologies, the timeline and layout for the transition to sustainable fuels and products could vary from country to country, and location to location. A possible outline for the transition to a sustainable future for carbon fuels and hydrocarbon products is shown in Fig. 3.26.35.

References

1. *Key world energy statistics 2014*. Paris: International Energy Agency (IEA); 2014.
2. *IPCC fourth assessment report: climate change 2007*. Intergovernmental Panel on Climate Change; 2007.
3. Synthesis Report. Summary for Policymaker *IPCC fifth assessment report: climate change 2014*. Cambridge, U.K: Cambridge University Press; 2014. Available at: http://www.ipcc.ch/.
4. Goeppert A, Czaun M, Prakash GKS, Olah GA. Air as the renewable carbon source of the future: an overview of CO_2 capture from the atmosphere. *Energ Environ Sci* 2012;**5**:7833–53.
5. Meth S, Goeppert A, Prakash GKS, Olah GA. Silica nanoparticles as supports for regenerable CO_2 sorbents. *Energ Fuel* 2012;**26**:3082–90.
6. Maass P. *The breaking point*. N.Y.T.; August 21, 2005.
7. Smil V. *Energy at the crossroads, global perspectives and uncertainties.* , Cambridge: MIT Press; 2003.
8. Goeppert A, Czaun M, Jones J-P, Prakash GKS, Olah GA. Recycling of carbon dioxide to methanol and derived products – closing the loop. *Chem Soc Rev* 2014;**43**:7995–8048.
9. Schlögl R. *Chemical energy storage*. Berlin/Boston: Walter de Gruyter; 2013.

10. Olah GA, Goeppert A, Prakash GKS. *Beyond oil and gas: the methanol economy*. 2nd ed. Weinheim, Germany: Wiley-VCH; 2009.
11. Chisholm G, Cronin L. Hydrogen from water electrolysis. In: Letcher TM, editor. *Storing energy: with special reference to renewable energy sources*. Amsterdam: Elsevier; 2016.
12. Ursua A, Gandia LM, Sanchis P. Hydrogen production from water electrolysis: current status and future trends. *Proc IEEE* 2012;**100**:410–26.
13. Rifkin J. *The hydrogen economy*. New York: Tarcher/Putnam; 2002.
14. Bossel U. Does a hydrogen economy make sense? *Proc IEEE* 2006;**94**:1826.
15. Bromberg L, Cheng WK. *Methanol as an alternative transportation fuel in the us: options for sustainable and/or energy-secure transportation*. Final report: Massachussets Institute of Technology report PSFC/RR-10-12. 2010.
16. Olah GA, Goeppert A, Prakash GKS. Chemical recycling of carbon dioxide to methanol and dimethyl ether: from greenhouse gas to renewable, environmentally carbon neutral fuels and synthetic hydrocarbons. *J Org Chem* 2009;**74**:487–98.
17. Olah GA, Prakash GKS, Goeppert A. Anthropogenic chemical carbon cycle for a sustainable future. *J Am Chem Soc* 2011;**133**:12881–98.
18. *Methanol as an energy carrier*. Forschungszentrum Jülich GmbH, Institut für Werkstoffe und Verfahren der Energietechnik; 2006.
19. Asinger F. *Methanol, Chemie- und Energierohstoff. Die Mobilisation der Kohle*. Heidelberg: Springer-Verlag; 1987.
20. Prepared for the Methanol Institute, Available at *On the road with methanol: the present and future benefits of methanol fuel*. 1994. http://www.methanol.org.
21. Reed TB, Lerner RM. Methanol: a versatile fuel for immediate use. *Science* 1973;**182**:1299–304.
22. Perry JH, Perry CP. *Methanol, bridge to a renewable energy future*. Lanham, Maryland: University Press of America; 1990.
23. Moffat AS. Methanol-powered cars get ready to hit the road. *Science* 1991;**251**:514–5.
24. Gray Jr CL, Alson JA. *Moving America to methanol*. Ann Arbor: The University of Michigan Press; 1985.
25. *Methanol as an alternative fuel choice: an assessment*. Washington, DC: The Johns Hopkins University; 1990.
26. Specht M, Bandi A. *"The methanol cycle" - sustainable supply of liquid fuels*. Stuttgart, Germany: Center for Solar Energy and Hydrogen Research (ZSW); 1999.
27. Bromberg L, Cohn DR. *Alcohol fueled heavy duty vehicles using clean, high efficiency engines, PSFC/JA-09-31*. Cambridge: MIT; 2009.
28. Bromberg L, Cohn DR. *Heavy duty vehicles using clean, high efficiency alcohol engines, PSFC/JA-10-43*. Cambridge: MIT; 2010.
29. McGrath KM, Prakash GKS, Olah GA. Direct methanol fuel cells. *J Ind Eng Chem* 2004;**10**:1063–80.
30. Semelsberger TA, Borup RL, Greene HL. Dimethyl ether (DME) as an alternative fuel. *J Power Sources* 2006;**156**:497–511.
31. Arcoumanis C, Bae C, Crookes R, Kinoshita E. The potential of di-methyl ether (DME) as an alternative fuel for compression-ignition engines: a review. *Fuel* 2008;**87**:1014–30.
32. Temchin J. *Analysis of market characteristics for conversion of liquid fueled turbines to methanol: prepared for the methanol foundation and methanex by electrotek concepts*. 2003.
33. Basu A, Wainwright JM. DME as a power generation fuel: performance in gas turbines. In: *New Dehli, India, January: Presented at the PETROTECH-2001 Conf*; 2001.
34. *GE position paper: feasibility of methanol as gas turbine fuel: general electric*. 2001.
35. Chang CD. Methanol to gasoline and olefins. In: Cheng W-H, Kung HH, editors. *Methanol production and use*. New York: Marcel Dekker; 1994. p. 133.
36. Methanol Institute. www.methanol.org.
37. Olah GA. After oil and gas: methanol economy. *Catal Lett* 2004;**93**:1–2.
38. Cheng W-H, Kung HH, editors. *Methanol production and use*. New York: Marcel Dekker; 1994.
39. Bertau M, Offermanns H, Plass L, Schmidt F, Wernicke H-J, editors. *Methanol: the basic chemical and energy feedstock of the future, Berlin/Heidelberg*. Springer-Verlag; 2014.
40. Subramani V, Sharma P, Zhang L, Liu K. *Catalytic steam reforming technology for the production of hydrogen and syngas. Hydrogen and syngas production and purification technologies*. John Wiley & Sons, Inc; 2009. p. 14–126.
41. Rostrup-Nielsen J, Christiansen LJ. *Concepts in syngas manufacture*. London: Imperial College Press; 2011.
42. Choudhary TV, Choudhary VR. Energy-efficient syngas production through catalytic oxy-methane reforming reactions. *Angew Chem Int Ed* 2008;**47**:1828–47.

43. Bradford MCJ, Vannice MA. CO_2 reforming of CH_4. *Catal Rev Sci Eng* 1999;**41**:1–42.
44. Olah GA, Prakash GKS. *Conversion of carbon dioxide to methanol and/or dimethyl ether using bi-reforming of methane or natural gas*. 2008. International Patent Application WO 2008157673 A1.
45. Olah GA, Prakash GKS. *Conversion of carbon dioxide to methanol and/or dimethyl ether using bi-reforming of methane or natural gas*. US Patent 7,906,559. 2011.
46. Olah GA, Prakash GKS. *Conversion of carbon dioxide to dimethyl ether using bi-reforming of methane or natural gas*. US Patent 8,133,926. 2012.
47. Steinberg M. Fossil fuel decarbonization technology for mitigating global warming. *Int J Hydrog Energ* 1999;**24**:771–7.
48. Halmann MM, S M. *Greenhouse gas carbon dioxide mitigation. Greenhouse gas carbon dioxide mitigation; Science and technology*. Boca Raton: CRC Press; 1999.
49. Steinberg M. The Carnol process system for CO_2 mitigation and methanol production. *Energy* 1997;**22**:143–9.
50. Report to the Congress *FutureGen. Integrated hydrogen, electric power production and carbon sequestration research initiative*. US: Department of Energy; 2004. Available at: http://www.energy.gov.
51. *Office of clean coal, strategic plan. Moving America towards an affordable "zero" emissions coal energy option*. U.S. Department of Energy; 2006.
52. External Draft *Hydrogen from coal program. Research development, and demonstration plan. For the period 2010 through 2016*. U.S. Department of Energy; 2010.
53. Spath PL, Dayton DC. *Preliminary Screening—technical and economic assessment of synthesis gas to fuels and chemicals with emphasis on the potential for biomass-derived syngas, NETL/TP-510-34929*. Golden, Colorado: National Renewable Energy Laboratory; 2003.
54. Hamelinck CN, Faaij APC. Future prospects for production of methanol and hydrogen from biomass. *J Power Sources* 2002;**111**:1–22.
55. Bridgwater AV. Catalysis in thermal biomass conversion. *Appl Catal* 1994;**116**:5–47.
56. Stevens C, editor. *Thermochemical processing of biomass: conversion into fuels, chemicals and power*. Wiley; 2011.
57. Bulushev DA, Ross JRH. Catalysis for conversion of biomass to fuels via pyrolysis and gasification: a review. *Catal Today* 2011;**171**:1–13.
58. Larson ED. Technology for electricity and fuels from biogas. *Annu Rev Energy Environ* 1993:567–630.
59. Bain RL. *Material and energy balances for methanol from biomass using biomass gasifiers. NREL/TP-510-17098*. Golden, Colorado: National Renewable Energy Laboratory; 1992.
60. Chmielniak T, Sciazko M. Co-gasification of biomass and coal for methanol synthesis. *Appl Energ* 2003;**74**:393–403.
61. Reschetilowski W. Alternative resources for the methanol economy. *Russ Chem Rev* 2013;**82**:624–34.
62. Ribeiro AM, Santos JC, Rodrigues AE. PSA design for stoichiometric adjustment of bio-syngas for methanol production and co-capture of carbon dioxide. *Chem Eng J* 2010;**163**:355–63.
63. Specht M, Bandi A, Baumgart F, Murray CN, Gretz J. Synthesis of methanol from biomass/CO_2 resources. In: Eliasson B, Riemer PWF, Wokaum A, editors. *Greenhouse gas control technologies*. Amsterdam: Pergamon; 1999. p. 723.
64. Borgwardt RH. Methanol production from biomass and natural gas as transportation fuel. *Ind Eng Chem Res* 1998;**37**:3760–7.
65. Li H, Hong H, Jin H, Cai R. Analysis of a feasible polygeneration system for power and methanol production taking natural gas and biomass as materials. *Appl Energ* 2010;**87**:2846–53.
66. Dong Y, Steinberg M. Hynol-an economical process for methanol production form biomass and natural gas with reduced CO_2 emission. *Int J Hydrog Energ* 1997;**22**:971–7.
67. Dong Y, Borgwardt RH. Biomass reactivity in gasification by the hynol process. *Energ Fuel* 1998;**12**:479–84.
68. Norbeck JM, Johnson K. *The hynol process: a promising pathway for renewable production of methanol*. Riverside: College of Engineering, Center for Environmental Research and Technology, University of California; 2000.
69. Lundgren J, Ekbom T, Hulteberg C, Larsson M, Grip CE, Nilsson L, et al. Methanol production from steel-work off-gases and biomass based synthesis gas. *Appl Energ* 2013;**112**:431–9.
70. Ouellette N, Rogner HH, Scott DS. Hydrogen from remote excess hydroelectricity. Part II: hydrogen peroxide or biomethanol. *Int J Hydrog Energ* 1995;**20**:873–80.
71. Ng KS, Sadhukhan J. Process integration and economic analysis of bio-oil platform for the production of methanol and combined heat and power. *Biomass Bioenerg* 2011;**35**:1153–69.
72. Galindo Cifre P, Badr O. Renewable hydrogen utilization for the production of methanol. *Energ Convers Manage* 2007;**48**:519–27.

73. Clausen LR, Houbak N, Elmegaard B. Technoeconomic analysis of a methanol plant based on gasification of biomass and electrolysis of water. *Energy* 2010;**35**:2338–47.
74. Mignard D, Pritchard C. On the use of electrolytic hydrogen from variable renewable energies for the enhanced conversion of biomass to fuels. *Chem Eng Res Des* 2008;**86**:473–87.
75. Agrawal R, Singh NR, Ribeiro FH, Delgass WN. Sustainable fuel for the transportation sector. *P Natl Acad Sci* 2007;**104**:4828–33.
76. Final Project Report. EUDP Project Journal Number: 64010-0011 *GreenSynFuels. Economical and technological statement regarding integration and storage of renewable energy in the energy sector by production of green synthetic fuels for utilization in fuel cells.* Danish Technological Institute; 2011.
77. Weiland P. Biogas production: current state and perspectives. *Appl Microbiol Biotechnol* 2010;**85**:849–60.
78. Pedersen TH, Schultz RH. *Technical and economic assessment of methanol production from biogas* [Master thesis]. Denmark: Department of Energy Technology, University of Aalborg; 2012.
79. Olah GA, Goeppert A, Czaun M, Prakash GKS. Bi-reforming of methane from any source with steam and carbon dioxide exclusively to metgas ($CO-2H_2$) for methanol and hydrocarbon synthesis. *J Am Chem Soc* 2013;**135**:648–50.
80. Olah GA, Prakash GKS, Goeppert A, Czaun M, Mathew T. Self-sufficient and exclusive oxygenation of methane and its source materials with oxygen to methanol via metgas using oxidative Bi-reforming. *J Am Chem Soc* 2013;**135**:10030–1.
81. Proceedings of the high level conference on world food security: the challenges of climate change and bioenergy. Soaring food prices: facts, perspectives, impacts and actions required (HLC/08/INF/1). In: *Conference organized by the food and agricultural organization of the united Nations, Rome, 3–5 June*; 2008.
82. Donner SD, Kucharik CJ. Corn-based ethanol production compromises goal of reducing nitrogen export by the Mississippi River. *Proc Nat Acad Sci USA* 2008;**105**:4513–8.
83. Fargione J, Hill J, Tilman D, Polasky S, Hawthorne P. Land clearing and biofuel carbon debt. *Science* 2008;**319**:1235–8.
84. Searchinger T, Heimlich R, Houghton RA, Fengxia D, Elobeid A, Fabiosa J, et al. Use of U.S. Croplands for biofuels increases greenhouse gases through emissions from land use change. *Science* 2008;**319**:1238–40.
85. Doornbosch R, Steenblik R. *Biofuels: is the cure worse than the Disease?: report by the organisation for economic co-operation and development.* 2007.
86. Hedegaard K, Thyø KA, Wenzel H. Life cycle assessment of an advanced bioethanol technology in the perspective of constrained biomass availability. *Environ Sci Technol* 2008;**42**:7992–9.
87. *World energy outlook 2004.* Paris: International Energy Agency; 2004.
88. Pearson RJ, Eisaman MD, Turner JWG, Edwards PP, Zheng J, Kuznetsov VL, et al. Energy storage via carbon-neutral fuels made from CO_2, water, and renewable energy. *Proc IEEE* 2012;**100**:440–60.
89. Chorkendorff I, Niemantsverdriet JW. *Concepts of modern catalysis and kinetics.* Weinheim: Wiley; 2007.
90. Paul JK. *Methanol technology and application in Motor fuels.* Park Ridge: Noyes Data Corporation; 1978.
91. Goehna H, Koenig P. Producing methanol from CO_2. *Chemtech* June 1994;**39**.
92. Saito M. R&D activities in Japan on methanol synthesis from CO_2 and H_2. *Catal Surv Jpn* 1998;**2**:175–84.
93. Hansen JB. Methanol synthesis. In: Ertl G, Knözinger H, Weitkamp J, editors. *Handbook of heterogeneous catalysis*, Vol. 4. Weinheim: Wiley-VCH; 1997. p. 1856.
94. Ostrovskii VE. Mechanism of methanol synthesis from hydrogen and carbon oxides at Cu-Zn containing catalysts in the context of some fundamental problems of heterogeneous catalysis. *Catal Today* 2002;**77**:141–60.
95. Rozovskii AY, Lin GI. Fundamentals of methanol synthesis and decomposition. *Top Catal* 2003;**22**:137–50.
96. Chinchen GC, Denny PJ, Parker DG, Spencer MS, Waugh KC, Whan DA. Mechanism of methanol synthesis from $CO_2/CO/H_2$ mixtures over copper/zinc oxide/alumina catalysts: use of ^{14}C-labelled reactants. *Appl Catal* 1987;**30**:333–8.
97. Bart JCJ, Sneeden RPA. Copper-zinc oxide-alumina methanol catalysts revisited. *Catal Today* 1987;**2**:1–124.
98. Tijm PJA, Waller FJ, Brown DM. Methanol technology developments for the new Millenium. *Appl Catal A-Gen* 2001;**221**:275–82.
99. Pontzen F, Liebner W, Gronemann V, Rothaemel M, Ahlers B. CO_2-based methanol and DME – efficient technologies for industrial scale production. *Catal Today* 2011;**171**:242–50.
100. Clarke DB, Bell AT. An infrared study of methanol synthesis from CO_2 on clean and potassium-promoted Cu/SiO_2. *J Catal* 1995;**154**:314–28.

101. Saito M, Fujitani T, Takeuchi M, Watanabe T. Development of copper/zinc oxide-based multicomponent catalysts for methanol synthesis from carbon dioxide and hydrogen. *Appl Catal A-Gen* 1996;**138**:311–8.
102. Inui T, Hara H, Takeguchi T, Kim J-B. Structure and function of Cu-based composite catalysts for highly effective synthesis of methanol by hydrogenation of CO_2 and CO. *Catal Today* 1997;**36**:25–32.
103. Fujita S, Usui M, Ito H, Takezawa N. Mechanisms of methanol synthesis from carbon dioxide and from carbon monoxide at atmospheric pressure over Cu/ZnO. *J Catal* 1995;**157**:403–13.
104. Arena F, Barbera K, Italiano G, Bonura G, Spadaro L, Frusteri F. Synthesis, characterization and activity pattern of Cu–ZnO/ZrO_2 catalysts in the hydrogenation of carbon dioxide to methanol. *J Catal* 2007;**249**:185–94.
105. Studt F, Sharafutdinov I, Abild-Pedersen F, Elkjaer CF, Hummelshøj JS, Dahl S, et al. Discovery of a Ni-Ga catalyst for carbon dioxide reduction to methanol. *Nat Chem* 2014;**6**:320–4.
106. Li Y-N, Ma R, He L-N, Diao Z-F. Homogeneous hydrogenation of carbon dioxide to methanol. *Catal Sci Technol* 2014;**4**:1498–512.
107. Tominaga K, Sasaki Y, Kawai M, Watanabe T, Saito M. Ruthenium complex catalyzed hydrogenation of carbon-dioxide to carbon-monoxide, methanol and methane. *J Chem Soc Chem Comm* 1993:629–31.
108. Tominaga K, Sasaki Y, Watanabe T, Saito M. Homogeneous hydrogenation of carbon-dioxide to methanol catalyzed by ruthenium cluster anions in the presence of halide anions. *B Chem Soc Jpn* 1995;**68**:2837–42.
109. Wesselbaum S, vom Stein T, Klankermayer J, Leitner W. Hydrogenation of carbon dioxide to methanol by using a homogeneous ruthenium-phosphine catalyst. *Angew Chem Int Ed* 2012;**51**:7499–502.
110. Rezayee NM, Huff CA, Sanford MS. Tandem amine and ruthenium-catalyzed hydrogenation of CO_2 to methanol. *J Am Chem Soc* 2015;**137**:1028–31.
111. Kothandaraman J, Goeppert A, Czaun M, Olah GA, Prakash GKS. Conversion of CO_2 from air into methanol using a polyamine and a homogeneous ruthenium catalyst. *J Am Chem Soc* 2016;**138**:778–81.
112. Huff CA, Sanford MS. Cascade catalysis for the homogeneous hydrogenation of CO_2 to methanol. *J Am Chem Soc* 2011;**133**:18122–5.
113. Balaraman E, Gunanathan C, Zhang J, Shimon LJW, Milstein D. Efficient hydrogenation of organic carbonates, carbamates and formates indicates alternative routes to methanol based on CO_2 and CO. *Nat Chem* 2011;**3**:609–14.
114. Choudhury J. New strategies for CO_2-to-methanol conversion. *ChemCatChem* 2012;**4**:609–11.
115. Joo O-S, Jung K-D, Moon I, Rozovskii AY, Lin GI, Han S-H, et al. Carbon dioxide hydrogenation to form methanol via a reverse-water-gas-shift reaction (the CAMERE process). *Ind Eng Chem Res* 1999;**38**:1808–12.
116. Joo O-S, Jung K-D, Jung Y. Camere Process for methanol synthesis from CO_2 hydrogenation. In: Sang-Eon Park J-SC, Kyu-Wan L, editors. *Stud. Surf. Sci. Catal.*, Vol. 153. Elsevier; 2004. p. 67–72.
117. Hirano M, Imai T, Yasutake T, Kuroda K. Dimethyl ether synthesis from carbon dioxide by catalytic hydrogenation (Part 2) hybrid catalyst consisting of methanol synthesis and methanol dehydration catalysts. *J Jpn Petrol Inst* 2004;**47**:11–8.
118. Abanades S, Chambon M. CO_2 dissociation and upgrading from two-step solar thermochemical processes based on ZnO/Zn and SnO_2/SnO redox pairs. *Energ Fuel* 2010;**24**:6667–74.
119. Miller JE, Allendorf MD, Diver RB, Evans LR, Siegel NP, Stuecker JN. Metal oxide composites and structures for ultra-high temperature solar thermochemical cycles. *J Mater Sci* 2008;**43**:4714.
120. Steinfeld A, Sanders S, Palumbo R. Design aspects of solar thermochemical engineering – a case study: two-step water-splitting cycle using the Fe_3O_4/FeO redox system. *Sol Energ* 1999;**65**:43–53.
121. Galvez ME, Loutzenhiser PG, Hischier I, Steinfeld A. CO_2 splitting via two-step solar thermochemical cycles with Zn/ZnO and FeO/Fe_3O_4 redox reactions: thermodynamic analysis. *Energ fuel* 2008;**22**:3544–50.
122. Loutzenhiser PG, Galvez ME, Hischier I, Graf A, Steinfeld A. CO_2 splitting in an aerosol flow reactor via the two-step Zn/ZnO solar thermochemical cycle. *Chem Eng Sci* 2010;**65**:1855–64.
123. Chueh WC, Falter C, Abbott M, Scipio D, Furler P, Haile SM, et al. High-flux solar-driven thermochemical dissociation of CO_2 and H_2O using nonstoichiometric ceria. *Science* 2010;**330**:1797–801.
124. Kang M, Zhang J, Wang C, Wang F, Zhao N, Xiao F, et al. CO_2 splitting via two step thermochemical reactions over doped ceria/zirconia solid solutions. *RSC Adv* 2013;**3**:18878–85.
125. Jitaru M, Lowy DA, Toma M, Oniciu L. Electrochemical reduction of carbon dioxide on flat metallic cathodes. *J Appl Electrochem* 1997;**27**:875.
126. Gattrell M, Gupta N, Co A. A review of the aqueous electrochemical reduction of CO_2 to hydrocarbons at copper. *J Electroanal Chem* 2006;**594**:1–19.

127. Hori Y. Electrochemical CO_2 reduction on metal electrodes. In: Vayenas CG, White RE, Gamboa-Aldeco ME, editors. *Modern aspects of electrochemistry*, Vol. 42. New York: Springer; 2008. p. 89–189.
128. *Carbon dioxide chemistry: environmental issues*. Woodhead Publishing; 1994.
129. Kaneco S, Iiba K, Suzuki SK, Ohta K, Mizuno T. Electrochemical reduction of carbon dioxide to hydrocarbons with high faradaic efficiency in LiOH/methanol. *J Phys Chem B* 1999;**103**:7456.
130. Kaneco S, Iwao R, Iiba K, Itoh SI, Ohata K, Mizuno T. Electrochemical reduction of carbon dioxide on an indium wire in a KOH/methanol-based electrolyte at ambient temperature and pressure. *Environ Eng Sci* 1999;**16**:131.
131. Kaneco S, Katsumata H, Suzuki T, Ohta K. Photoelectrochemical reduction of carbon dioxide at p-type gallium arsenide and p-type indium phosphide electrodes in methanol. *Chem Eng J* 2006;**116**:227.
132. Canfield D, Frese KW. Reduction of carbon dioxide to methanol on n- and p-GaAs and p-InP. Effect of crystal face, electrolyte and current density. *J Electrochem Soc* 1983;**130**:1772–3.
133. Frese Jr KW, Leach SC, Summers DP. *Electrochemical reduction of aqueous carbon dioxide to methanol*. US Patent 4,609,441 A. USA: Gas Research Institute; 1986.
134. Le M, Ren M, Zhang Z, Sprunger PT, Kurtz RL, Flake JC. Electrochemical reduction of CO_2 to CH_3OH at copper oxide surfaces. *J Electrochem Soc* 2011;**158**:E45–9.
135. Shironita S, Karasuda K, Sato K, Umeda M. Methanol generation by CO_2 reduction at a Pt–Ru/C electrocatalyst using a membrane electrode assembly. *J Power Sources* 2013;**240**:404–10.
136. Seshadri G, Lin C, Bocarsly AB. A new homogeneous electrocatalyst for the reduction of carbon dioxide to methanol at low overpotential. *J Electroanal Chem* 1994;**372**:145–50.
137. Barton EE, Rampulla DM, Bocarsly AB. Selective solar-driven reduction of CO_2 to methanol using a catalyzed p-GaP based photoelectrochemical cell. *J Am Chem Soc* 2008;**130**:6342–4.
138. Qu J, Zhang X, Wang Y, Xie C. Electrochemical reduction of CO_2 on RuO_2/TiO_2 nanotubes composite modified Pt electrode. *Electrochim Acta* 2005;**50**:3576–80.
139. Morris AJ, McGibbon RT, Bocarsly AB. Electrocatalytic carbon dioxide activation: the rate-determining step of pyridinium-catalyzed CO_2 reduction. *ChemSusChem* 2011;**4**:191–6.
140. Barton Cole E, Lakkaraju PS, Rampulla DM, Morris AJ, Abelev E, Bocarsly AB. Using a one-electron shuttle for the multielectron reduction of CO_2 to methanol: kinetic, mechanistic, and structural insights. *J Am Chem Soc* 2010;**132**:11539–51.
141. Lim C-H, Holder AM, Musgrave CB. Mechanism of homogeneous reduction of CO_2 by pyridine: proton relay in aqueous solvent and aromatic stabilization. *J Am Chem Soc* 2012;**135**:142–54.
142. Keith JA, Carter EA. Theoretical insights into pyridinium-based photoelectrocatalytic reduction of CO_2. *J Am Chem Soc* 2012;**134**:7580–3.
143. Kamrath MZ, Relph RA, Johnson MA. Vibrational predissociation spectrum of the carbamate radical anion, $C_5H_5N\text{-}CO_2-$, generated by reaction of pyridine with $(CO_2)m-$. *J Am Chem Soc* 2010;**132**:15508–11.
144. Ertem MZ, Konezny SJ, Araujo CM, Batista VS. Functional role of pyridinium during aqueous electrochemical reduction of CO_2 on Pt(111). *J Phys Chem Lett* 2013;**4**:745–8.
145. Halmann MM, Steinberg M. *Greenhouse gas carbon dioxide mitigation: science and technology*. CRC Press; 1999.
146. Olah GA, Prakash GKS. *Electrolysis of carbon dioxide in aqueous media to carbon monoxide and hydrogen for production of methanol*. US Patent 8,138,380. 2012.
147. Jensen SH, Larsen PH, Mogensen M. Hydrogen and synthetic fuel production from renewable energy sources. *Int J Hydrogen Energy* 2007;**32**:3253–7.
148. Stoots CM, Hartvigsen JJ, O'Brien JE, Herring JS. Syngas production via high-temperature coelectrolysis of steam and carbon dioxide. *J Fuel Cell Sci Technol* 2008;**6**:011014.
149. Zhan Z, Zhao L. Electrochemical reduction of CO_2 in solid oxide electrolysis cells. *J Power Sources* 2010;**195**:7250–4.
150. Xie K, Zhang Y, Meng G, Irvine JTS. Electrochemical reduction of CO_2 in a proton conducting solid oxide electrolyser. *J Mater Chem* 2011;**21**:195–8.
151. Shi Y, Luo Y, Cai N, Qian J, Wang S, Li W, et al. Experimental characterization and modeling of the electrochemical reduction of CO_2 in solid oxide electrolysis cells. *Electrochim Acta* 2013;**88**:644–53.
152. Kreuter W, Hofmann H. Electrolysis: the important energy transformer in a world of sustainable energy. *Int J Hydr Energ* 1998;**23**:661–6.
153. Sabatier P, Mailhe A. Sur la decomposition catalytique de l'acide formique. *Compt Rend* 1911;**152**:1212–5.
154. Loges B, Boddien A, Gartner F, Junge H, Beller M. Catalytic generation of hydrogen from formic acid and its derivatives: useful hydrogen storage materials. *Top Catal* 2010;**53**:902–14.

155. Joo F. Breakthroughs in hydrogen storage-formic acid as a sustainable storage material for hydrogen. *ChemSusChem* 2008;**1**:805–8.
156. Boddien A, Loges B, Junge H, Gartner F, Noyes JR, Beller M. Continuous hydrogen generation from formic acid: highly active and stable ruthenium catalysts. *Adv Synth Catal* 2009;**351**:2517–20.
157. Loges B, Boddien A, Junge H, Beller M. Controlled generation of hydrogen from formic acid amine adducts at room temperature and application in H_2/O_2 fuel cells. *Angew Chem Int Ed* 2008;**47**:3962–5.
158. Sponholz P, Mellmann D, Junge H, Beller M. Towards a practical setup for hydrogen production from formic acid. *ChemSusChem* 2013;**6**:1172–6.
159. Laurenczy G, Fellay C, Dyson PJ. A viable hydrogen-storage system based on selective formic acid decomposition with a ruthenium catalyst. *Angew Chem Int Ed* 2008;**47**:3966–8.
160. Czaun M, Goeppert A, May R, Haiges R, Prakash GKS, Olah GA. Hydrogen generation from formic acid decomposition by ruthenium carbonyl complexes. Tetraruthenium dodecacarbonyl tetrahydride as an active intermediate. *ChemSusChem* 2011;**4**:1241–8.
161. Himeda Y. Highly efficient hydrogen evolution by decomposition of formic acid using an iridium catalyst with 4,4′-dihydroxy-2,2′-bipyridine. *Green Chem* 2009;**11**:2018–22.
162. Miller AJM, Heinekey DM, Mayer JM, Goldberg KI. Catalytic disproportionation of formic acid to generate methanol. *Angew Chem Int Ed* 2013;**52**:3981–4.
163. Savourey S, Lefèvre G, Berthet J-C, Thuéry P, Genre C, Cantat T. Efficient disproportionation of formic acid to methanol using molecular ruthenium catalysts. *Angew Chem Int Ed* 2014;**53**:10466–70.
164. Sordakis K, Tsurusaki A, Iguchi M, Kawanami H, Himeda Y, Laurenczy G. Carbon dioxide to methanol: the aqueous catalytic way at room temperature. *Chem – A Eur J* 2016;**22**:15605–8.
165. Gormley RJ, Rao VUS, Soong Y, Micheli E. Methyl formate hydrogenolysis for low-temperature methanol synthesis. *Appl Catal A-Gen* 1992;**87**:81–101.
166. Iwasa N, Terashita M, Arai M, Takezawa N. New catalytic functions of Pd and Pt catalysts for hydrogenolysis of methyl formate. *React Kinet Catal Lett* 2001;**74**:93–8.
167. Doherty MD, Grills DC, Muckerman JT, Polyansky DE, Fujita E. Toward more efficient photochemical CO_2 reduction: use of $scCO_2$ or photogenerated hydrides. *Coord Chem Rev* 2010;**254**:2472–82.
168. Dibenedetto A, Stufano P, Macyk W, Baran T, Fragale C, Costa M, et al. Hybrid technologies for an enhanced carbon recycling based on the enzymatic reduction of CO_2 to methanol in water: chemical and photochemical NADH regeneration. *ChemSusChem* 2012;**5**:373–8.
169. Gondal M, Ali M, Dastageer M, Chang X. CO_2 conversion into methanol using granular silicon carbide (α6H-SiC): a comparative evaluation of 355 nm laser and xenon mercury broad band radiation sources. *Catal Lett* 2013;**143**:108–17.
170. Chen X, Zhou Y, Liu Q, Li Z, Liu J, Zou Z. Ultrathin, single-crystal WO^3 nanosheets by two-dimensional oriented attachment toward enhanced photocatalytic reduction of CO_2 into hydrocarbon fuels under visible light. *ACS Appl Mater Inter* 2012;**4**:3372–7.
171. Jia ZF, Wang FM, Bai Y, Liu N. Photocatalytic reduction of CO_2 to methanol using the $InVO_4$-based photocatalysts. *Adv Mater Res* 2011:2033–7.
172. Li X, Liu H, Luo D, Li J, Huang Y, Li H, et al. Adsorption of CO_2 on heterostructure $CdS(Bi_2S_3)/TiO_2$ nanotube photocatalysts and their photocatalytic activities in the reduction of CO_2 to methanol under visible light irradiation. *Chem Eng J* 2012;**180**:151–8.
173. Luo D, Chen C, Zhang N, Hong S, Wu H, Liu Z. Characterization, DFT Research of Nd/TiO2: photocatalyst for synthesis of methanol from CO_2 and H_2O. *Z Phys Chem* 2009;**223**:1465–76.
174. Wu JCS, Lin H-M, Lai C-L. Photo reduction of CO_2 to methanol using optical-fiber photoreactor. *Appl Catal A-Gen* 2005;**296**:194–200.
175. Yahaya AH, Gondal MA, Hameed A. Selective laser enhanced photocatalytic conversion of CO_2 into methanol. *Chem Phys Lett* 2004;**400**:206–12.
176. Tsai C-W, Chen HM, Liu R-S, Asakura K, Chan T-S. Ni@NiO core–shell structure-modified nitrogen-doped $InTaO_4$ for solar-driven highly efficient CO_2 reduction to methanol. *J Phys Chem C* 2011;**115**:10180–6.
177. Rajeshwar K. Hydrogen generation from irradiated semiconductor-liquid interfaces. In: Rajeshwar K, McConnell R, Licht S, editors. *Solar hydrogen generation*. New York: Springer; 2008. p. 167–228.
178. Grimes CA, Varghese OK, Ranjan S. *Light, water, hydrogen. The solar generation of hydrogen by water photoelectrolysis*. New York: Springer; 2008.
179. McKone JR, Lewis NS, Gray HB. *Chem Mater* 2013;**26**:407–14.

180. Khaselev O, Turner JA. A monolithic photovoltaic-photoelectrochemical device for hydrogen production via water splitting. *Science* 1998;**280**:425–7.
181. Ushikoshi K, Moria K, Watanabe T, Takeuchi M, Saito MA. 50 kg/day class test plant for methanol synthesis from CO_2 and H_2. *Stud Surf Sci Catal* 1998;**114**:357–62.
182. Final report: prepared by air products liquid phase conversion company for the us DOE national energy technology laboratory *Commercial-scale demonstration of the liquid phase methanol (LPMEOHTM) process*. 2003.
183. Shulenberger AM, Jonsson FR, Ingolfsson O, Tran K-C. *Process for producing liquid fuel from carbon dioxide and water*. US Patent Application 2007/0244208 A1. 2007.
184. Tremblay J-F. CO_2 as feedstock. Mitsui will make methanol from the greenhouse gas. *Chem Eng News* 2008;**86**:13.
185. Blue Fuel Energy. http://bluefuelenergy.com/.
186. Xiaoding X, Moulijn JA. Mitigation of CO_2 by chemical conversion: plausible chemical reactions and promising products. *Energy & Fuels* 1996;**10**:305.
187. Saito M, Murata K. Development of high performance Cu/ZnO-Based catalysts for methanol synthesis and water-gas shift reaction. *Catal Surv Asia* 2004;**8**:285–94.
188. Graves C, Ebbesen SD, Mogensen M, Lackner KS. Sustainable hydrocarbon fuels by recycling CO_2 and H_2O with renewable or nuclear energy. *Renew Sust Energ Rev* 2011;**15**:1–23.
189. Xu M, Lunsford JH, Goodman DW, Bhattacharyya A. Synthesis of dimethyl ether (DME) from methanol over solid-acid catalysts. *Appl Catal A-Gen* 1997;**149**:289–301.
190. Arena F, Spadaro L, Di Blasi O, Bonura G, Frusteri F. Integrated synthesis of dimethylether via CO_2 hydrogenation. In: Xinhe B, Yide X, editors. *Stud. Surf. Sci. Catal*, Vol. 147. Elsevier; 2004. p. 385–90.
191. Hirano M, Imai T, Yasutake T, Kuroda K. Dimethyl ether synthesis from carbon dioxide by catalytic hydrogenation (Part 1). Activities of methanol dehydration catalysts. *J Jpn Petrol Inst* 2002;**45**:169–74.
192. An X, Zuo Y-Z, Zhang Q, Wang D-Z, Wang J-F. Dimethyl ether synthesis from CO_2 hydrogenation on a $CuO-ZnO-Al_2O_3-ZrO_2$/HZSM-5 bifunctional catalyst. *Ind Eng Chem Res* 2008;**47**:6547–54.
193. Sun K, Lu W, Wang M, Xu X. Low-temperature synthesis of DME from CO_2/H_2 over Pd-modified $CuO-ZnO-Al_2O_3-ZrO_2$/HZSM-5 catalysts. *Catal Comm* 2004;**5**:367–70.
194. Gao W, Wang H, Wang Y, Guo W, Jia M. Dimethyl ether synthesis from CO_2 hydrogenation on La-modified $CuO\text{-}ZnO\text{-}Al_2O_3$/HZSM-5 bifunctional catalysts. *J Rare Earths* 2013;**31**:470–6.
195. Naik SP, Ryu T, Bui V, Miller JD, Drinnan NB, Zmierczak W. Synthesis of DME from CO_2/H_2 gas mixture. *Chem Eng J* 2011;**167**:362–8.
196. Ereña J, Sierra I, Aguayo AT, Ateka A, Olazar M, Bilbao J. Kinetic modelling of dimethyl ether synthesis from ($H_2 + CO_2$) by considering catalyst deactivation. *Chem Eng J* 2011;**174**:660–7.
197. Bonura G, Cordaro M, Spadaro L, Cannilla C, Arena F, Frusteri F. Hybrid $Cu-ZnO-ZrO_2$/H-ZSM5 system for the direct synthesis of DME by CO_2 hydrogenation. *Appl Catal B-Environ* 2013;**140–141**:16–24.
198. Wang S, Mao D, Guo X, Wu G, Lu G. Dimethyl ether synthesis via CO_2 hydrogenation over $CuO-TiO_2-ZrO_2$/HZSM-5 bifunctional catalysts. *Catal Comm* 2009;**10**:1367–70.
199. Zha F, Ding J, Chang Y, Ding J, Wang J, Ma J. Cu–Zn–Al oxide cores packed by metal-doped amorphous silica–alumina membrane for catalyzing the hydrogenation of carbon dioxide to dimethyl ether. *Ind Eng Chem Res* 2011;**51**:345–52.
200. Markewitz P, Kuckshinrichs W, Leitner W, Linssen J, Zapp P, Bongartz R, et al. Worldwide innovations in the development of carbon capture technologies and the utilization of CO_2. *Energ Environ Sci* 2012;**5**:7281–305.
201. Pires JCM, Martins FG, Alvim-Ferraz MCM, Simões M. Recent developments on carbon capture and storage: an overview. *Chem Eng Res Des* 2011;**89**:1446–60.
202. Choi S, Drese JH, Jones CW. Adsorbent materials for carbon dioxide capture from large anthropogenic point sources. *ChemSusChem* 2009;**2**:796–854.
203. D'alessandro DM, Smit B, Long JR. Carbon dioxide capture: prospects for new materials. *Angew Chem Int Ed* 2010;**49**:6058–82.
204. MacDowell N, Florin N, Buchard A, Hallett J, Galindo A, Jackson G, et al. An overview of CO_2 capture technologies. *Energy Environ Sci* 2010;**3**:1645–69.
205. Wang Q, Luo J, Zhong Z, Borgna A. CO_2 capture by solid adsorbents and their applications: current status and new trends. *Energy Environ Sci* 2011;**4**:42–55.
206. Allam RJ, Bredesen R, Drioli E. Carbon dioxide separation technologies. In: Aresta M, editor. *Carbon dioxide recovery and utilization*. Dordrecht, The Netherlands: Kluwer Academic Publishers; 2003. p. 53.

207. Boot-Handford ME, Abanades JC, Anthony EJ, Blunt MJ, Brandani S, Mac Dowell N, et al. Carbon capture and storage update. *Energ Environ Sci* 2014;**7**:130–89.
208. Kohl A, Nielsen R. *Gas purification*. 5th ed. Houston: Gulf Publishing Company; 1997.
209. Specht M, Bandi A. *Herstellung von Fluessigen Kraftstoffen aus Atmosphaerischem Kohlendioxid. Forshungsverbund Sonnenenergie, Themen 1994–1995*. Energiespeicherung; 1995. p. 41.
210. Pasel J, Peters R, Specht M. *Methanol Herstellung und Einsatz als Energietraeger fuer Brennstoffzellen. Forshungsverbund Sonnenenergie, Themen 1999–2000*. Berlin, Germany: Zukunftstechnologie Breenstoffzelle; 2000. p. 46.
211. Specht M, Bandi A. "The methanol cycle" – sustainable supply of liquid fuels. Stuttgart, Germany: Center for Solar Energy and Hydrogen Research (ZSW); 1999.
212. Lackner K.S., Grimes P., and Ziock H-J.. Carbon dioxide extraction from the air: is it an option? Proceedings of the 24th annual technical conference on coal utilization and fuel systems, March 8–11, 1999. Clearwater, Florida 1999:885–896.
213. Lackner KS, Ziock H-J, Grimes P. The case for carbon dioxide extraction from air. *SourceBook* 1999;**57**:6.
214. Lackner KS. Capture of carbon dioxide from ambient air. *Eur Phys J-Spec Top* 2009;**176**:93–106.
215. Lackner KS. Washing carbon out of the air. *Sci Am* June 2010;66–71.
216. Keith DW, Ha-Duong M. In: Gale J, Kaya Y, Pergamon, editors. *Sixth international conference on greenhouse gas control technologies*; 2003. p. 187–97. Oxford, UK, Kyoto, Japan.
217. Keith DW, Ha-Duong M, Stolaroff JK. Climate strategy with CO_2 capture from the air. *Clim Change* 2006;**74**:17–45.
218. Stolaroff JK, Keith DW, Lowry GV. Carbon dioxide capture from atmospheric air using sodium hydroxide spray. *Environ Sci Technol* 2008;**42**:2728–35.
219. Mahmoudkhani M, Heidel KR, Ferreira JC, Keith DW, Cherry RS. Low energy packed tower and caustic recovery for direct capture of CO_2 from air. *Energ Procedia* 2009;**1**:1535–42.
220. Mahmoudkhani M, Keith DW. Low-energy sodium hydroxide recovery for CO_2 capture from atmospheric air-thermodynamic analysis. *Int J Greenh Gas Con* 2009;**3**:376–84.
221. Steinberg M, Dang VD. *Use of controlled thermonuclear reactor fusion power for the production of synthetic methanol from the air and water: report from the Brookhaven national laboratory*. 1975.
222. Dang VD, Steinberg M. Production of synthetic methanol from air and water using controlled thermonuclear reactor power—II. Capital investment and production costs. *Energ Convers* 1977;**17**:133–40.
223. Steinberg M, Dang V-D. Production of synthetic methanol from air and water using controlled thermonuclear reactor power—I. Technology and energy requirement. *Energ Convers* 1977;**17**:97–112.
224. Keith DW, Mahmoudkhani M. *Carbon dioxide capture*. US Patent Application 2010/0034724. 2010.
225. Keith DW, Mahmoudkhani M, Biglioli A, Hart B, Heidel K, Foniok M. *Carbon dioxide capture method and facility*. US Patent Application 2010/0064890 A1. 2010.
226. Carbon Engineering Inc. www.carbonengineering.com.
227. Belmabkhout Y, Serna-Guerrero R, Sayari A. Amine-bearing mesoporous silica for CO_2 removal from dry and humid air. *Chem Eng Sci* 2010;**65**:3695–8.
228. Gebald C, Wurzbacher JA, Steinfeld A. Eur. Pat. Appl. EP2266680A1; 2010.
229. Belmabkhout Y, Serna-Guerrero R, Sayari A. Adsorption of CO_2-containing gas mixtures over amine-bearing pore-expanded MCM-41 silica: application for gas purification. *Ind Eng Chem Res* 2010;**49**:359–65.
230. Olah GA, Goeppert A, Meth S, Prakash GKS. *Nano-structure supported solid regenerative polyamine and polyamine polyol absorbents for the separation of carbon dioxide from gas mixtures including the air*. US Patent 7 795 175. 2010.
231. Wurzbacher JA, Gebald C, Steinfeld A. Separation of CO_2 from air by temperature-vacuum swing adsorption using diamine-functionalized silica gel. *Energ Environ Sci* 2011;**4**:3584–92.
232. Stuckert NR, Yang RT. CO_2 capture from the atmosphere and simultaneous concentration using zeolites and amine-grafted SBA-15. *Environ Sci Technol* 2011;**45**:10257–64.
233. Chaikittisilp W, Khunsupat R, Chen TT, Jones CW. Poly(allylamine) mesoporous silica composite materials for CO_2 capture from simulated flue gas or ambient air. *Ind Eng Chem Res* 2011;**50**:14203–10.
234. Gebald C, Wurzbacher JA, Tingaut P, Zimmermann T, Steinfeld A. Amine-based nanofibrillated cellulose as adsorbent for CO_2 capture from air. *Environ Sci Technol* 2011;**45**:9101–8.
235. Didas SA, Kulkarni A, Sholl DS, Jones CW. Role of amine structure on carbon dioxide adsorption from ultra-dilute gas streams such as ambient air. *ChemSusChem* 2012;**5**:2058–64.
236. Sculley JP, Zhou H-C. Enhancing amine-supported materials for ambient air capture. *Angew Chem Int Ed* 2013;**51**:12660–1.

237. Satyapal S, Filburn T, Trela J, Strange J. Performance and properties of a solid amine sorbent for carbon dioxide removal in space life support applications. *Energ Fuel* 2001;**15**:250–5.
238. Chen Z, Deng S, Wei H, Wang B, Huang J, Yu G. Polyethylenimine-impregnated resin for high CO_2 adsorption: an efficient adsorbent for CO_2 capture from simulated flue gas and ambient air. *ACS Appl Mater Interfaces* 2013;**5**:6937–45.
239. Goeppert A, Czaun M, May RB, Prakash GKS, Olah GA, Narayanan SR. Carbon dioxide capture from the air using a polyamine based regenerable solid adsorbent. *J Am Chem Soc* 2011;**133**:20164–7.
240. Goeppert A, Zhang H, Czaun M, May RB, Prakash GKS, Olah GA, et al. Easily regenerable solid adsorbents based on polyamines for carbon dioxide capture from the air. *ChemSusChem* 2014;**7**:1386–97.
241. Choi S, Drese JH, Eisenberger PM, Jones CW. Application of amine-tethered solid sorbents for direct CO_2 capture from the ambiant air. *Environ Sci Technol* 2011;**45**:2420–7.
242. Choi S, Drese JH, Chance RR, Eisenberger PM, Jones CW. *Application of amine-tethered solid sorbents to CO_2 fixation from air*. US Patent Application 2011/0179948A1. 2011.
243. Choi S, Watanabe T, Bae T-H, Sholl DS, Jones CW. Modification of the Mg/DOBDC MOF with amines to enhance CO_2 adsorption from ultradilute gases. *J Phys Chem Lett* 2012;**3**:1136–41.
244. McDonald TM, Lee WR, Mason JA, Wiers BM, Hong CS, Long JR. Capture of carbon dioxide from air and flue gas in the alkylamine-appended metal–organic framework mmen-Mg_2(dobpdc). *J Am Chem Soc* 2012;**134**:7056–65.
245. Lu W, Sculley JP, Yuan D, Krishna R, Zhou H-C. Carbon dioxide capture from air using amine-grafted porous polymer networks. *J Phys Chem* 2013;**117**:4057–61.
246. Wang T, Lackner KS, Wright AB. Moisture-swing sorption for carbon dioxide capture. *Phys Chem Chem Phys* 2013;**15**:504–14.
247. Wang T, Lackner KS, Wright A. Moisture swing sorbent for carbon dioxide capture from ambient air. *Environ Sci Technol* 2011;**45**:6670–5.
248. He H, Li W, Zhong M, Konkolewicz D, Wu D, Yaccato K, et al. Reversible CO_2 capture with porous polymers using the humidity swing. *Energ Environ Sci* 2013;**6**:488–93.
249. Sanz-Pérez ES, Murdock CR, Didas SA, Jones CW. Direct capture of CO_2 from ambient air. *Chem Rev* 2016;**116**:11840–76.
250. Huang Z, Chen ZB, Ren NQ, Hu DX, Zheng DH, Zhang ZP. A novel application of the SAWD-Sabatier-SPE integrated system for CO_2 removal and O_2 regeneration in submarine cabins during prolonged voyages. *J Zhejiang Univ-Sci A* 2009;**10**:1642–50.

CHAPTER

3.27

Natural and Nature-Inspired Synthetic Small Molecule Antioxidants in the Context of Green Chemistry

William Horton, Marianna Török

University of Massachusetts Boston, Boston, MA, United States

3.27.1 INTRODUCTION

Reactive oxygen species (ROS) are produced by enzymatic/nonenzymatic metabolic redox reactions starting with the partial reduction of oxygen to superoxide $(O_2^{\bullet-})$ or hydrogen peroxide (H_2O_2) followed by further secondary reactions of the products.[1] Often, transition metal ions, such as Cu^{2+}, Co^{2+}, Ni^{2+}, or Fe^{2+}, are also involved in these reactions.[1] Similarly, reactive nitrogen species (RNS) are derived from various reactions of the free radical nitrogen oxide (NO•) that is synthetized from arginine by nitrogen oxide synthases.[1] The ROS/RNS family includes both free radicals and nonradical species, with superoxide $(O_2^{\bullet-})$, hydroxyl (OH•) radicals, hydroperoxyl (HOO•) radicals, the peroxynitrite ($OONO^-$) ion, the paramagnetic singlet oxygen ($^1O_2$), nitrogen oxide (NO•) radical, hydrogen peroxide (H_2O_2), ozone (O_3), and hypochlorous acid (HOCl) molecules being the most frequently mentioned members.[1–4] The production of ROS/RNS can be both harmful and beneficial in living systems. At physiological concentration, they play significant roles in cell survival by regulating signaling pathways or fighting infections.[3,4] At high concentrations, however, they react with proteins, lipids, and nucleic acids and may modify their biological function.[4] It is a consequence of an imbalance in the redox homeostasis due to the overproduction of ROS/RNS or inadequate activity of the cellular antioxidant defenses and is referred to as oxidative (or sometimes nitrosative) stress. The cellular damage caused by superoxide and other ROS/RNS has been implicated in the aging process and numerous diseases

Green Chemistry
http://dx.doi.org/10.1016/B978-0-12-809270-5.00032-7

including cancer, cardiovascular diseases, neurodegenerative diseases (e.g., Alzheimer's and Parkinson's diseases, amyotrophic lateral sclerosis, multiple sclerosis), macular degeneration, rheumatoid arthritis, and diabetes.[1,4,5] The collective term "antioxidants" refers to a diverse group of bioactive molecules that protect against the cellular damage caused by oxidative stress through various mechanisms (e.g., scavenging free radicals, regenerating other antioxidants, chelating metal ions, regulating enzyme activities, repairing oxidative damage).[3–5] Endogenous antioxidants (Table 3.27.1) include protective enzymes (e.g., superoxide dismutase, catalase, glutathione peroxidase), nonenzymatic peptides/proteins (e.g., glutathione, ferritin, transferrin, ceruloplasmin, albumin), enzyme cofactors (e.g., coenzyme Q, lipoic acid), and metabolites (e.g., bilirubin, uric acid, melatonin).[4,5] A diet rich in fruits and vegetables usually contains low-molecular-weight exogenous antioxidants (Table 3.27.1), (e.g., vitamin C, vitamin E, β-carotene, resveratrol, curcumin), contributing to the cellular antioxidant defense.[4,5]

Beside the age and state of health of a person, lifestyle issues and environmental factors may contribute to oxidative stress and result in different pathological conditions.[6–8] A 2016 study in China warns of the increasing cases of lung, colorectal and breast cancers, blaming rapid industrialization and urbanization combined with unhealthy lifestyle changes (heavy smoking, poor diet and obesity) in an aging population.[9]

As mentioned earlier, excessive ROS/RNS can form not only as natural by-products of metabolic redox reactions but also in response to environmental stress (e.g., pollution or radiation). We have less control over the amount and activity of endogenous antioxidants in our body than those of the exogenous antioxidants in our daily diet. Therefore, the dietary natural antioxidants and their synthetic analogs garner extensive attention, as these compounds are potential candidates for preventing and/or treating many diseases.[10]

From the viewpoint of green chemistry, small molecule antioxidants have been the focus of interest for several reasons. The investigation of their protective/repairing role against oxidative damage triggered by environmental exposures,[6–10] design and preferably green synthesis of novel antioxidants,[11] the extraction of natural antioxidants from inexpensive sustainable resources,[12–17] and the development of environment-friendly technologies to do so[18–21] are all active current topics in this research area.

TABLE 3.27.1 Endogenous and Exogenous Antioxidants

Source	Antioxidants	Examples
Endogenous	Protective enzymes	Superoxide dismutase, catalase, glutathione peroxidase
	Nonenzymatic peptides/proteins	Glutathione, ferritin, transferrin, ceruloplasmin, albumin
	Enzyme cofactors	Coenzyme Q, lipoic acid
	Metabolites	Bilirubin, uric acid, melatonin
Exogenous	Dietary small molecule antioxidants	Vitamin C, vitamin E, β-carotene, resveratrol, curcumin

3.27.2 IDENTIFICATION, ISOLATION, AND STRUCTURAL CHARACTERIZATION

Instead of the costly isolation of individual bioactive compounds, researchers rather focus on extracts from natural sources in search for new antioxidants. Antioxidants can be found in various inexpensive, renewable, and widely abundant sources including several plants and plant-derived biomass. A majority of antioxidants are soluble in organic solvents, and their simple extraction can be carried out using ether, ethanol, or methanol, for example; however, this process may result in the loss of many water-soluble antioxidants.[18] Polyphenols, known for their antioxidant properties, are widely available in plants and are generally present in their organic extracts. These extracts frequently contain what is known as free polyphenols; the polyphenol content of the source that is not bound to complex sugars, proteins, and other large biomolecules. However, bound phenols are more complicated to extract, as the molecules they are bound to are often part of the cell membrane or water-soluble entities not easily obtained by extraction with organic solvents.[22–24] Extracts are widely used by the food industry as flavorings, preservatives, and supplements to our daily diet. Numerous types and varieties of vegetables and fruits[25] (vide infra), especially red grapes, cranberries, and blueberries have been of specific interest due to their high concentrations of antioxidants and associated potential health benefits.[13,16,26,27] Polyphenol-rich natural extracts are isolated from unprocessed foods, food waste, and other sources of biomass. It is very important to acknowledge that the age of the sample, cooking (pressure, heat and shearing forces), and digestion can all change the chemical makeup and hence the antioxidant properties of the original material.[14,28,29]

Several waste products from food processing and other biomass sources, including rice husk, coffee silverskin, grape skin, and barley green biomass, are rich in polyphenols and provide a simple and plentiful source of antioxidants.[12,14,15,19,21] By using different, sometimes relatively harsh extraction procedures the bound polyphenols can also be isolated.[15,17,19] Extraction of antioxidants, for example, under extreme pH and temperature conditions yields different antioxidant extracts from plant and biomass sources.[15] It has been shown that bound polyphenols are typically higher in antioxidant activity than the free polyphenols extracted through organic phase extraction.[17] Utilizing abundant biomass that is considered as waste to obtain antioxidant-rich extracts is an excellent application of green chemistry principles.

There are also significant efforts to develop green techniques to obtain antioxidant-rich samples from natural sources. Some of these environment-friendly processes include microwave-assisted extraction (MAE), supercritical fluid extraction, and ultrasound-assisted extraction.[14,15,17–21] MAE allows for maximum solvent penetration into the sample by breaking down the cellular matrix to extract any residual polyphenols.[18] MAE, however, may degrade the polyphenols during their extraction due to the significant temperature increase during the process.[18] Supercritical fluid extraction, particularly with supercritical CO_2, is an excellent method for obtaining polyphenols from biomass. Although supercritical water can be used in place of a solvent, its generation requires high temperature (374 °C) and pressure (22.1 MPa), making it less desirable than supercritical CO_2, which can be used under milder conditions (31.1 °C and 7.39 MPa).[18,30] Extraction with supercritical CO_2 also eliminates the exposure to oxygen, which reduces the possibility that the polyphenols extracted will be oxidized before analysis.[18] Supercritical CO_2 has added benefits, such as being easily attainable, nontoxic, and nonflammable, which makes it a very useful solvent for green extraction of antioxidants from natural sources.[18] An additional environmentally benign method is the ultrasound-assisted extraction, which uses sonication to disrupt cell walls and enhance membrane permeability.[18] The aforementioned are only a few examples of extraction techniques used to obtain natural antioxidants from inexpensive, abundant, and sustainable resources.

To identify common structural motifs responsible for biological activity, the antioxidant property of a sample must be quantified with standardized assays that are applicable to not only natural extracts but also individual compounds. However, there are extensive variations in the structures and mechanisms of antioxidants, and as such the physiological properties are widely different. Furthermore, extracts must be further tested and refined to determine the individual compounds and other factors responsible for their activity. As a result, multiple extraction methods and antioxidant assays should be used when trying to assess any source. There is a broad array of different antioxidant assays currently used in research, and these assays are summarized in excellent recent reviews.[31–34] Here we only mention a few examples without providing a comprehensive overview. There are simple, widely used ultraviolet (UV)-Vis assays, including the DPPH assay utilizing the relatively stable, purple-colored radical 2,2-diphenyl-1-picrylhydrazyl (DPPH) or the ABTS assay using 2,2′-azino-bis(3-ethylbenzthiazoline-6-sulfonic acid) (ABTS). Both of these monitor the loss of signal intensity over time as the antioxidant scavenges the radical.[31,32] The radicals used in the DPPH and ABTS assays are bulky, are biologically nonrelevant, and their poor reaction kinetics may yield lower antioxidant activity.[31,32] The DPPH assay uses an organic solvent, which makes its biochemical applicability questionable, as opposed to the ABTS assay, which utilizes a phosphate buffer.[31,32] Other common methods, such as the oxygen radical absorbance capacity (ORAC) assay, usually operate with smaller, biologically significant ROS/RNS generated in situ.[31,32] In the ORAC assay, a radical damages a fluorescent probe molecule, such as fluorescein, that is measured spectroscopically. The antioxidant protection of the fluorescent molecule prevents its damage and thus the decrease in the fluorescence intensity.[31] Lipid peroxidation is another common biomarker of radical damage as lipids are widely accessible and a constant target for radical species in the body.[3,31,32] Although the lipid peroxidation assays are biologically relevant, the chemical nature of the lipid radicals formed varies (e.g., in length) and depends on the conditions of the assay, making the interpretation of the results difficult sometimes. The total oxyradical scavenging capacity assay monitors the oxidation of alpha-keto-gamma-(methylthio)butyric acid to ethylene by OH•, ROO•, and $ONOO^-$ radicals using gas chromatography.[33]

Other assay methods such as cyclic voltammetry, ferric-reducing antioxidant power, total reactive antioxidant potential, cupric-reducing antioxidant capacity, and the Folin-Ciocalteu method assess the reducing rather than the radical scavenging ability of an antioxidant.[31–34]

Electron paramagnetic resonance (EPR) or electron spin resonance spectroscopy is a unique method for both qualitative and quantitative analysis of paramagnetic species such as free radicals and is excellent for assessing antioxidant properties. EPR antioxidant assays usually directly monitor stable free radicals (e.g., DPPH) or use the spin-trapping technique, in which short-lived free radicals (e.g., superoxide or hydroxyl) are identified and quantified based on the spectra of their stable paramagnetic nitroxide adducts with diamagnetic spin traps.[34,35]

In cell-based assays, the capacity of an antioxidant can be determined by adding the tested compound directly to the growth medium and exposing the culture to an oxidative stressor, such as H_2O_2.[36–38] Cell-based assays provide biologically relevant data on how the antioxidant protects the living cell under oxidative stress conditions.[36–38] Although cell-based assays provide the most data for high-throughput screening, these assays require sterile systems and time to culture the cells for use in the assays and the antioxidants tested may be toxic to the cells, which adds more variables for uncertainty.[36–38]

Many of these assays utilize standardization, so the data between different assays can be easily compared. It is common practice in laboratories to apply well-known antioxidants as standard molecules, such as Trolox, glutathione, gallic acid, and catechin (Fig. 3.27.1).

Some of these assays, such as ORAC, ABTS, and cell-based antioxidant assays, can be used in aqueous buffer, whereas others require organic solvent.[31,32] The characteristics of the assays (type of radical and solvents used, mechanism of action, reproducibility, etc.) must be taken into careful consideration in the selection process, and the antioxidant assay should be standardized if possible. High-performance liquid chromatography-tandem mass spectrometry is a useful technique for further separation and identification of individual compounds responsible for the major antioxidant capacity of an extract.[39] Although this method can identify new or already known individual bioactive molecules, often there are other cofactors and bulk proteins that may help increase the antioxidant capacity of a sample.

Antioxidants are structurally diverse molecules (Figs. 3.27.2–3.27.5) with different modes of action including (1) chain-breaking antioxidants, which act via direct ROS/RNS scavenging, and (2) preventative antioxidants, which function through indirect actions for example, by transition metal ion chelation, enzyme modulation, and manipulation of gene expression.[22,40] The largest class of natural antioxidants in our diet is the earlier mentioned phenols/polyphenols that can be divided into flavonoids and nonflavonoids; these can also be categorized into further structural subgroups (Figs. 3.27.2 and 3.27.3). However, there are several other exogenous (e.g., some vitamins/provitamins) and endogenous small molecule antioxidants present in Nature. A few representative compounds are illustrated in Figs. 3.27.2–3.27.5, with some of their common dietary sources for the exogenous antioxidants, to demonstrate their structural diversity and abundance.[4,5,22–24,41–46]

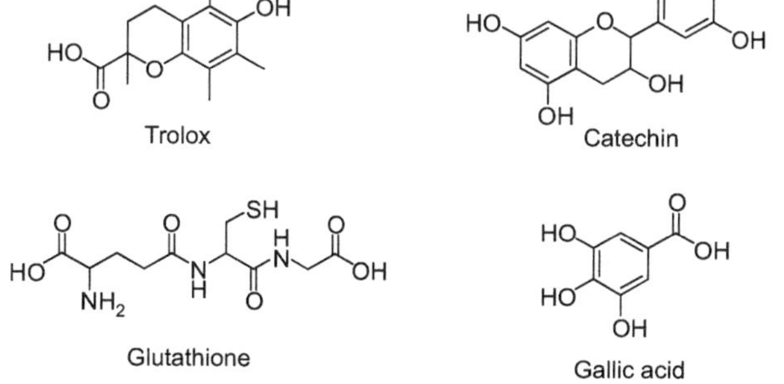

FIGURE 3.27.1 Traditional antioxidant assay standards.

Subgroup	Representative compound	Example of dietary source
Flavones	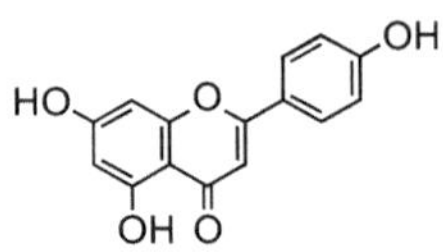 Apigenin	parsley celery
Flavonols	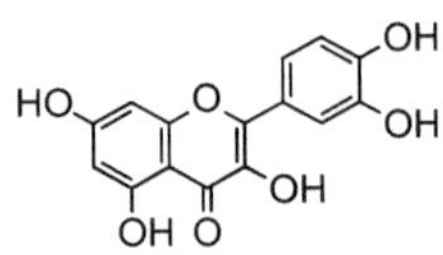 Quercetin	onion berries broccoli red wine tea
Flavanols	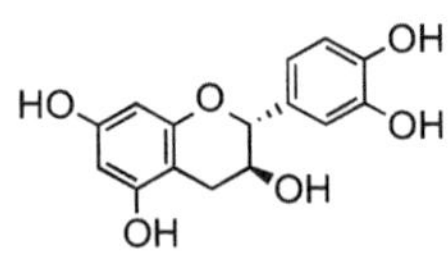 Catechin	apricot chocolate green tea red wine
Isoflavones	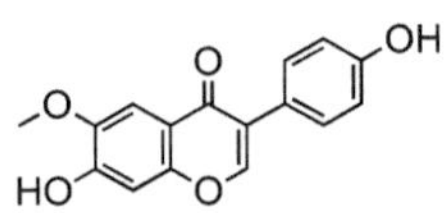 Glycitein	soybeans black beans green peas
Flavanones	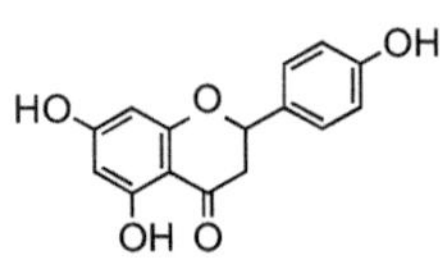 Naringenin	citrus fruits
Anthocyanidins	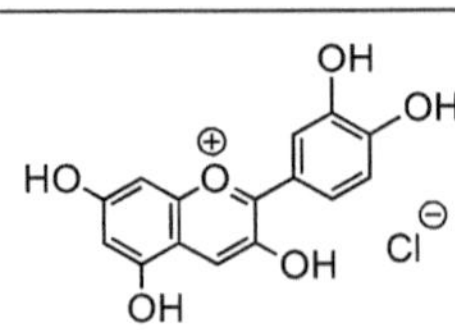 Cyanidin (chloride)	cherries rasberries strawberries grapes

FIGURE 3.27.2 Selected structural subgroups of flavonoids with examples.

Subgroup	Representative compound	Example of dietary source
Lignans	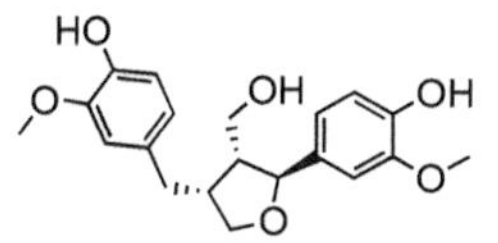 Lariciresinol	flaxseeds sesame seeds sunflower seeds pistachios
Coumarins	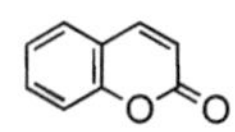 Coumarin	green tea cinnamon bark oil chicory cloudberry
Stilbenes	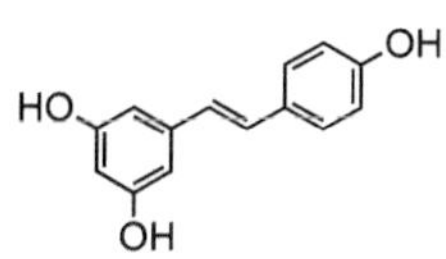 Resveratrol	grapes peanuts soy red wine berries red cabbage
Curcumins	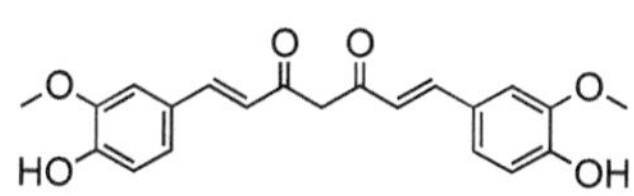 Curcumin	turmeric nuts
Phenolic acids	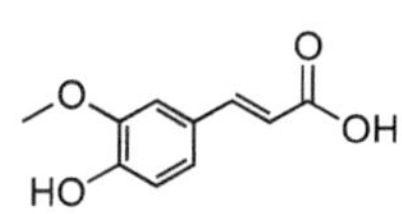 Ferulic acid	cereal grains wheat oats rice
Tannins	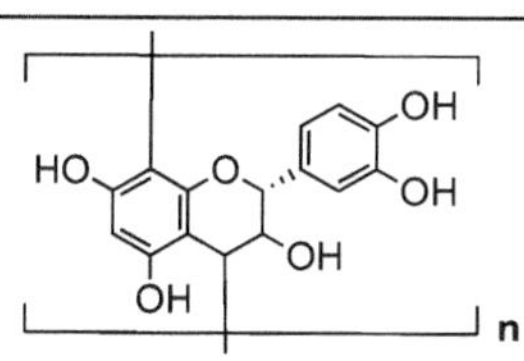 Proanthocyanidins	tea berries cocoa wine beer nuts

FIGURE 3.27.3 Selected structural subgroups of nonflavonoid phenolic antioxidants with examples.

Name	Representative structure	Example of dietary source
Preformed vitamin A retinoids	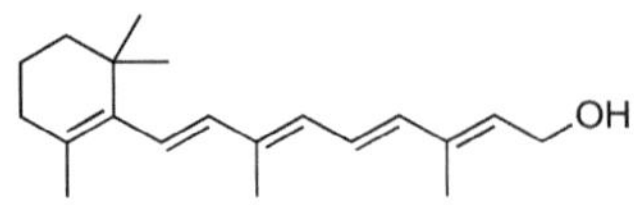 Retinol	liver fish fish oil milk eggs
Provitamin A carotenoids	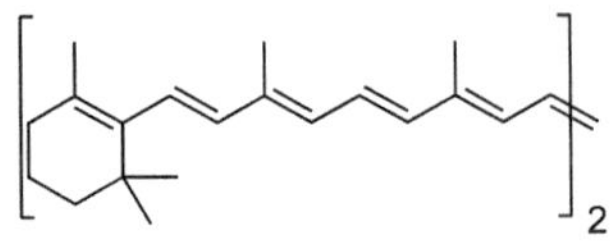 Beta-carotene	carrots spinach apricot sweet potato mango
Vitamin C	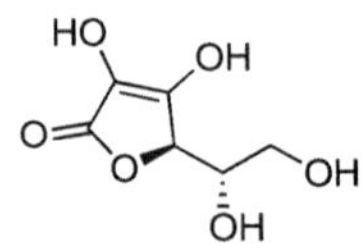 Ascorbic acid	citrus fruits broccoli potato greens strawberries
Vitamin E	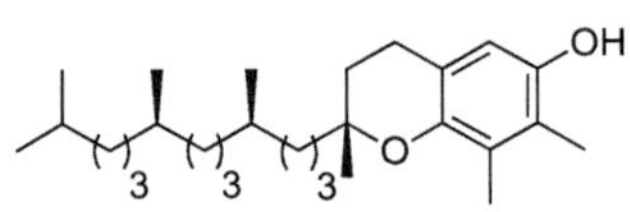 alpha-Tocopherol	plant oils seeds nuts asparagus wheat germ

FIGURE 3.27.4 Examples of vitamins/provitamins with antioxidant activity.

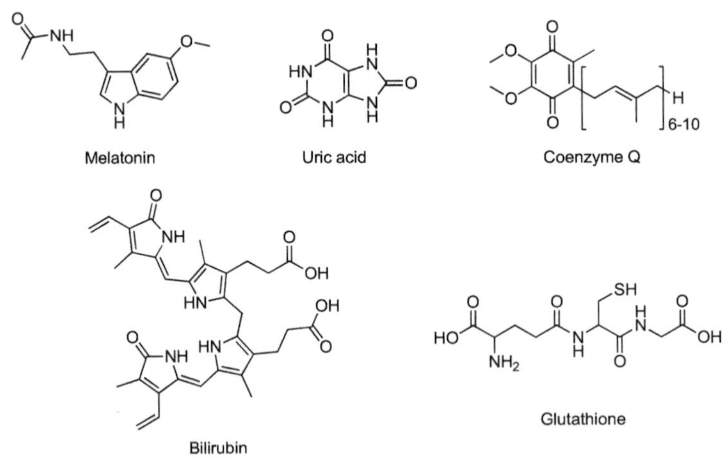

FIGURE 3.27.5 Chemical structures of a few endogenous small molecule antioxidants.

One common structural feature in most of these molecules is the continuous electron conjugation across the entire molecule, which contributes to radical scavenging.[41] Many of these antioxidant species have also been shown to have other biological effects contributing to their anticancer, anti-inflammatory, neuro- and cardioprotective, or other physiological activity.[4,5,13,16,39,42,44,47]

3.27.3 LIMITATIONS OF THERAPEUTIC APPLICATIONS

Researchers often face difficulties with translating the results of high-throughput antioxidant assays directly into biological activity. The absorption, distribution, metabolism and excretion properties (or pharmacokinetic properties) of antioxidants are significant factors in their application as therapeutic agents. If an antioxidant is too hydrophobic, it becomes embedded into membranes, whereas molecules that are too hydrophilic are not absorbed well.[48–50] The major location for free radical species in the cell is the mitochondria, where superoxide is produced as a by-product of the electron transport chain.[1,4] If the antioxidant cannot pass through the membranes of the mitochondria, e.g., because it is too hydrophilic, it cannot scavenge the free radicals inside.[48] Bioavailability of antioxidants, such as polyphenols, is affected in vivo as they are constantly interacting with proteins found throughout the body. After ingestion of antioxidant-rich food, polyphenols have been found to bind to salivary proteins.[51] In the blood stream, antioxidants are bound to serum proteins, such as albumin.[51] Polyphenols have shown good stability in gastric juice, allowing for absorption by diffusion across the stomach lining and also by the bilitranslocase transporter, which is responsible for the movement of bilirubin as well.[51] Metabolic enzymes and membrane transporters participate in directing the antioxidants to their site of action.[51] By exploring these mechanisms, a more efficient delivery of antioxidants to the target tissues and cellular compartments can be developed. Also, there is a large gap in our current understanding on the metabolism of antioxidants in the body. Dietary antioxidants and orally administered supplements undergo a first-pass metabolism in the liver carried out by a family of enzymes known as cytochrome P450s, as well as by other enzymes present, mostly with the ultimate purpose of reducing toxicity and improve water solubility.[51–54] Easy excretion is important for removal of the damaged or unnecessary antioxidants from the circulation.[50] Using resveratrol as an example, studies have shown that resveratrol is readily metabolized by glucuronidation and sulfation after absorption.[51,55,56] This leads to lower concentration of resveratrol available in the blood, whereas the concentrations of resveratrol's metabolites are still high.[52,54,56] Further research is necessary to determine exactly how the activity of an antioxidant is affected after consumption by comparing the data gathered from in vitro high-throughput screening assays with that obtained from animal or human models.[57–59]

There are a number of cases when a powerful antioxidant becomes a prooxidant. A prooxidant is a compound that inhibits natural antioxidant systems in the body, or generates ROS, causing oxidative stress.[4] Several studies have investigated the role of α-tocopherol and β-carotene (Fig. 3.27.4) in preventing or treating cancer. High concentrations of β-carotene, however, have been shown to actually stimulate cancer development in patients.[60,61] Antioxidants have been found to become prooxidants at high concentrations or elevated O_2 pressures.[4] The exchange of electrons, with or without proton transfer, can easily occur not only between the antioxidant and a radical but also between two antioxidant molecules, which can lead to molecular damage or polymerization of the antioxidant. When an antioxidant becomes damaged, it needs to be removed from the body or recycled. Meanwhile, its metabolites still can generate further radicals working against the overall antioxidant effect of the parent compound. β-carotene (Fig. 3.27.4) specifically is a long-chain antioxidant that breaks down quickly after it quenches a radical. Its degradation products have been shown to be prooxidative and cause more oxidative damage in tissues rather than preventing it.[62,63] Antioxidants taken in high concentrations can interact with metal ions and oxygen in the body producing radical species causing oxidative damage.[64] When high concentrations of ascorbic acid (Fig. 3.27.4), a water-soluble compound, were given to patients with cancer, the cancer cells proliferated.[65–67] The question remains open: what can chemists do to mitigate the negative effects of antioxidants while trying to use them to reduce the oxidative stress, which is an underlying factor in many different diseases?

3.27.4 CHEMICAL MODIFICATIONS AND FORMULATIONS TO IMPROVE DRUGLIKE PROPERTIES AND THERAPEUTIC POTENTIAL

With so much potential for antioxidants to help combat radical damage that contributes to aging, various cancers, neurodegenerative and other diseases, extensive research is being directed towards the modification and formulation of antioxidants to increase their potency, reduce prooxidative effects, and improve the delivery of the antioxidants to specific target sites.

Coadministration of antioxidants with each other or with other synergists, readily available through extraction from renewable resources, have allowed for the development of new drug therapies, showing promise in reducing cancer risk and treating diseases that are known to progress due to radical species.[68–71] In combined antioxidant therapy, one antioxidant can regenerate the other. Furthermore, as different antioxidants target unique tissues and cellular compartments, their combination allows for a broader coverage than a single antioxidant alone.[72] Studies on the coadministration of vitamins C and E, for example, indicate promising results.[73,74]

There have also been extensive efforts to encapsulate antioxidants in a biodegradable lipid shell or encase the antioxidants in nanoparticles.[51,75–78] These techniques are designed to protect the antioxidant from unwanted metabolic degradation and also help target-specific delivery.[51,79] Curcumin has been encapsulated in poly(lactic-co-glycolic acid) nanoparticles to increase its solubility and oral bioavailability, allowing for smaller doses of curcumin to be just as effective as larger doses.[80] Resveratrol has been also loaded into solid nanoparticles and nanostructured lipids to increase its bioavailability.[81] The shelf-life of the nanoparticles used in the formulation of the antioxidants needs to be considerably long. If the nanoparticles leaked any antioxidant or degraded before they could be administered, then the encapsulation would not be successful. In both of these cases, the nanoparticles were found to be stable for several hours.[80,81] Similarly, the antioxidant activity of compounds isolated from plant extracts, such as curcumin from turmeric, is being boosted by encapsulation of the antioxidant within a nanoparticle.[82,83] These nanoparticles are allowing for specific delivery of antioxidants to target sites requiring lower concentrations of the antioxidant and providing less exposure to O_2.[51] This decreases the risk of the antioxidant becoming degraded and promoting oxidative damage. Encapsulation of antioxidants allows for improved potency and bioavailability, as well as, longer shelf life.[75,78,80]

The chemical modifications of natural antioxidants to achieve better potency and decreased toxicity must retain conjugation, make the antioxidant more membrane permeable, or prevent degradation during metabolism or after radical scavenging.[41,61,62,84] A quantitative structure-activity relationship (QSAR) study coupled with a high-throughput screening method is one of the best ways to develop potent synthetic derivatives of a natural antioxidant. QSAR uses a large variety of data inputs such as those from structural information, electrochemical properties, and antioxidant assay data to determine what modifications to the core structure improve the pharmacological properties of the compound.[85–89] QSAR studies require as many physicochemical parameters as can be determined so that the data and conclusions derived from the study can be applied to future modifications of antioxidant core structures. The goal is to identify important features that increase the potency of an antioxidant, to further aid in the design and synthesis of new antioxidant derivatives.

Finally, we demonstrate a few current examples for the development of Nature-inspired novel antioxidants based on the structure of three well-known antioxidants resveratrol, β-carotene, and curcumin.

Resveratrol (Fig. 3.27.6), a stilbene, is a powerful antioxidant; however it suffers from relatively low aqueous solubility and is not photostable because of its ethylene bridge that isomerizes between *cis-* to *trans-*isomers when UV light is applied.[90] Resveratrol has a semiquinone structure with unpaired electrons localized on the oxygen atoms in the *ortho-* and *para-*positions, and the 4-OH is more acidic, which most likely reacts with a radical.[91] This bridge is an essential part of the stilbene structure, and *trans-*stilbenes have shown better activity in antioxidant studies.[90] Increasing the bridge size in resveratrol increases its ε_{HOMO} value and also appears to increase its antioxidant activity.[91] Cyclodextrins and bile acids can be used to improve the aqueous solubility and photostability of stilbenes.[90] Other

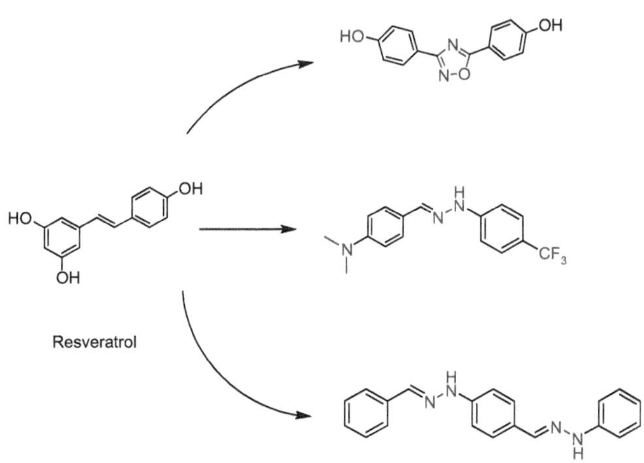

FIGURE 3.27.6 Synthetic analog of the natural antioxidant resveratrol and resveratrol-inspired derivatives.

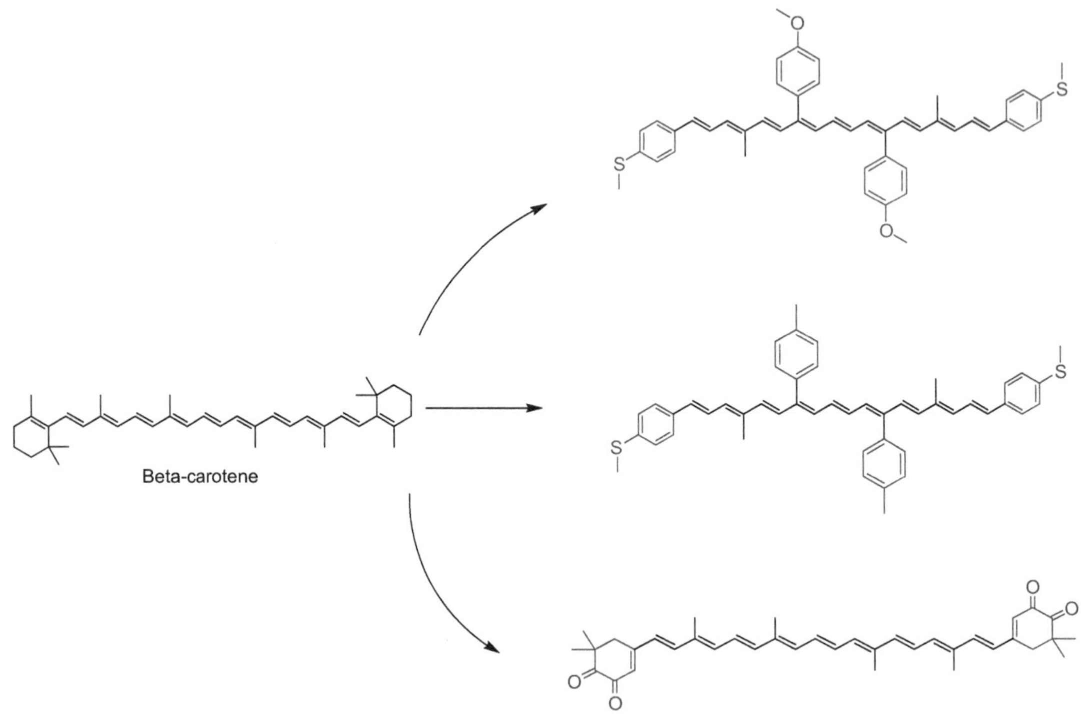

FIGURE 3.27.7 Synthetic derivatives of the natural antioxidant β-carotene.

synthetic derivatives of resveratrol have been shown to have multifunctional activity including inhibition of amyloid-β self-assembly and radical scavenging.[11,92]

β-Carotene (Fig. 3.27.7) has been thoroughly discussed previously, and even though it has significant side effects, it is still a commonly used antioxidant. Carotenoids are known to interact in solution with other carotenoids and proteins (e.g., via tyrosine or cysteine), which affects their bioavailabilities.[93] Carotenoids have been stabilized, their bioavailabilities have been increased, and the formation of some prooxidative by-products have been prevented by the development of novel synthetic derivatives of the original β-carotene structure.[62,94,95]

Curcumin (Fig. 3.27.8) is an abundant antioxidant in turmeric, which, in powdered form, is a major constituent of the popular spice curry. Curcumin suffers from poor solubility and rapid degradation in the body, and at high concentrations, curcumin has been shown to exhibit a prooxidant effect.[96,97] The stability of curcumin is affected by a host of factors such as pH, temperature, light, and structural modifications.[97] The primary cause of curcumin degradation is alkaline hydrolysis in buffer solutions.[96] To improve its bioavailability, the structure of curcumin has been modified, conjugated with other biomolecules, or even encapsulated in nanoparticles.[96,98,99] Many variations on the curcumin core structure have been synthesized, and most have typically lower antioxidant activities[100] than that of the parent compound. Other research groups have encapsulated the curcumin molecules into liposomes, nanoparticles, microspheres, and hydrogels.[1,97,99]

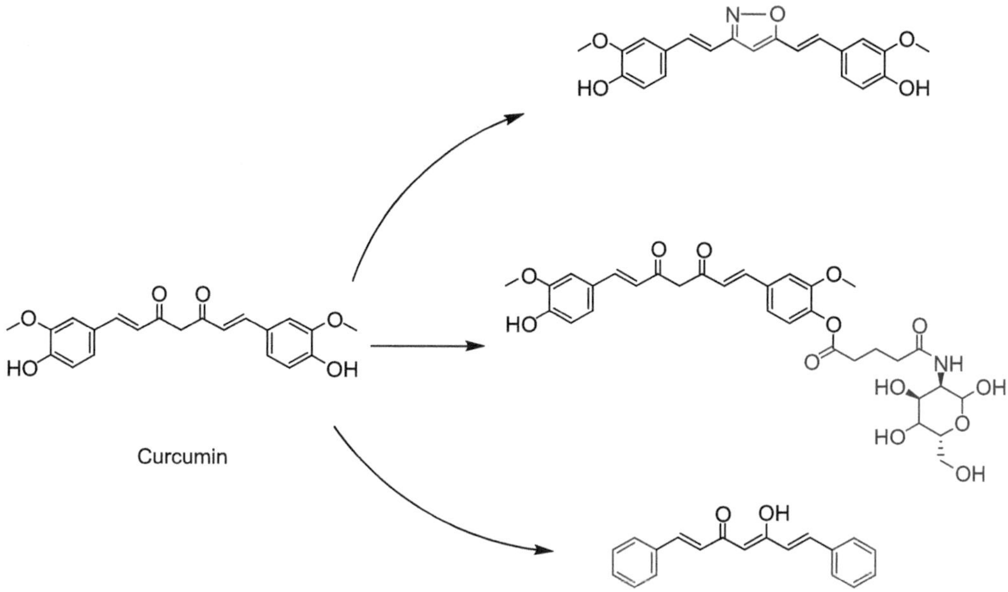

FIGURE 3.27.8 Synthetic derivatives of the natural antioxidant curcumin.

Modifications to the delivery method, chemical structure, or supplementation with another antioxidant are all designed to enhance the antioxidant capacity of the original molecule. The goal is to lower the concentration of the antioxidant used to reduce the negative side effects in the body while retaining, or even increasing, its desired activity. Several of these modifications can be considered as environmentally benign procedures.[11,96,99,100]

3.27.5 CONCLUSIONS

Antioxidants play important roles in several areas of the current green chemistry research ranging from understanding their bioactivity to seeking inexpensive, renewable sources and developing green techniques for isolating or synthetizing potent antioxidants. These findings may lead to promising novel antioxidant therapies that alleviate the effects of environmental pollution and contribute to the prevention or treatment of many human diseases.

References

1. Valko M, Leibfritz D, Moncol J, Cronin MTD, Mazur M, Telser J. Free radicals and antioxidants in normal physiological functions and human disease. *Int. J. Biochem. Cell Biol.* 2007;**39**:44–84.
2. Forman HJ, Augusto O, Brigelius-Flohe R, Dennery PA, Kalyanaraman B, Ischiropoulos H, et al. Even free radicals should follow some rules: a guide to free radical research terminology and methodology. *Free Radic. Biol. Med.* 2015;**78**:233–5.
3. Brender DA. Free radicals and antioxidant nutrients. In: Murray RK, Bender DA, Botham KM, Kennely PJ, Rodwell VW, Weil PA, editors. *Harper's illustrated biochemistry*. McGraw-Hill Companies, Inc.; 2012. p. 543–8.

4. Pisoschi AM, Pop A. The role of antioxidants in the chemistry of oxidative stress: a review. *Eur. J. Med. Chem.* 2015;**97**:55–74.
5. Kelsey NA, Wilkins HM, Linseman DA. Nutraceutical antioxidants as novel neuroprotective agents. *Molecules* 2010;**15**:7792–814.
6. Pryor WA. Biological effects of cigarettes smoke, wood smoke, and the smoke from plastics – the use of electron spin resonance. *Free Radic. Biol. Med.* 1992;**13**:659–76.
7. Riley PA. Free radicals in biology – oxidative stress and the effects of ionizing radiation. *Int. J. Radiat. Biol.* 1994;**65**:27–33.
8. Benedetti M, Giuliani ME, Regoli F. Oxidative metabolism of chemical pollutants in marine organisms: molecular and biochemical biomarkers in environmental toxicology. *Ann. N. Y Acad. Sci.* 2015;**1340**:8–19.
9. Bode1 AM, Dong Z, Wang H. Cancer prevention and control: alarming challenges in China. *Natl. Sci. Rev.* 2016;**3**:117–27.
10. Poljšak B, Fink R. The protective role of antioxidants in the defence against ROS/RNS-mediated environmental pollution. *Oxid. Med. Cell. Longev.* 2014;**2014**:1–22.
11. Török B, Sood A, Bag S, Tulsan R, Ghosh S, Borkin D, et al. Diaryl hydrazones as multifunctional inhibitors of amyloid self-assembly. *Biochemistry* 2013;**52**:1137–48.
12. Gupta GK, De S, Franco A, Balu AM, Luque R. Sustainable biomaterials: current trends, challenges and applications. *Molecules* 2016;**21**:48–59.
13. Neto CC. Cranberries: ripe for more cancer research? *J. Sci. Food Agric.* 2011;**91**:2303–7.
14. Belcredi NR, Ehrenbergerova J, Fiedlerova V, Belakova S, Vaculova K. Antioxidant vitamins in barley green biomass. *J. Agric. Food Chem.* 2010;**58**:11755–61.
15. Narita Y, Inouye K. High antioxidant activity of coffee silverskin extracts obtained by the treatment of coffee silverskin with subcritical water. *Food Chem.* 2012;**135**:943–9.
16. Neto CC. Cranberry and blueberry: evidence for protective effects against cancer and vascular diseases. *Mol. Nutr. Food Res.* 2007;**51**:652–64.
17. Lingua MS, Fabani MP, Wunderlin DA, Baroni MV. From grape to wine: changes in phenolic composition and its influence on antioxidant activity. *Food Chem.* 2016;**208**:228–38.
18. Talmaciu AI, Volf I, Popa VIA. Comparative analysis of the 'green' techniques applied for polyphenols extraction from bioresources. *Chem. Biodivers.* 2015;**12**:1635–50.
19. da Costa Lopes AM, Brenner M, Fale P, Roseiro LB, Bogel-Łukasik R. Extraction and purification of phenolic compounds from lignocellulosic biomass assisted by ionic liquid, polymeric resins, and supercritical CO_2. *ACS Sustainable Chem. Eng.* 2016;**4**:3357–67.
20. Aizpurua-Olaizola O, Ormazabal M, Vallejo A, Olivares M, Navarro P, Etxebarria N, et al. Optimization of supercritical fluid consecutive extractions of fatty acids and polyphenols from *Vitis vinifera* grape wastes. *J. Food Sci.* 2015;**80**:101–7.
21. Jeong KM, Zhao J, Jin Y, Heo SR, Han SY, Yoo DE, et al. Highly efficient extraction of anthocyanins from grape skin using deep eutectic solvents as green and tunable media. *Arch. Pharm. Res.* 2015;**38**:2143–52.
22. Sandoval-Acuña C, Ferreira J, Speisky H. Polyphenols and mitochondria: an update on their increasingly emerging ROS-scavenging independent actions. *Arch. Biochem. Biophys.* 2014;**559**:75–90.
23. Shahidi F, Ambigaipalan P. Phenolics and polyphenolics in foods, beverages and spices: antioxidant activity and health effects – a review. *J. Funct. Foods* 2015;**18**:820–97.
24. Manach C, Scalbert A, Morand C, Rémésy C, Jiménez L. Polyphenols: food sources and bioavailability. *Am. J. Clin. Nutr.* 2004;**79**:727–47.
25. Brewer MS. Natural antioxidants: sources, compounds, mechanisms of action, and potential applications. *Compr. Rev. Food Sci. Food Saf.* 2011;**10**:221–47.
26. Xia E, Deng G, Guo Y, Li H. Biological activities of polyphenols from grapes. *Int. J. Mol. Sci.* 2010;**11**:622–46.
27. Cranberries and health. Cranberry Health Research Center University of Massachusetts Dartmouth. http://www.umassd.edu/chrc/cranberriesandhealth/.
28. Sarawong C, Schoenlechner R, Sekiguchi K, Berghofer E, Ng PK. Effect of extrusion cooking on the physiochemical properties, resistant starch, phenolic content and antioxidant capacities of green banana flour. *Food Chem.* 2014;**143**:33–9.
29. Slavova-Kazakova A, Karamac M, Kancheva V, Amaroxicz R. Antioxidant activity of flaxseed extracts in lipid systems. *Molecules* 2016;**21**:17–28.

30. Plaza M, Turner C. Pressurized hot water extraction of bioactivities. *Trends Anal. Chem.* 2015;**71**:39–54.
31. Apak R, Özyürek M, Güclü K, Capanoglu E. Antioxidant activity/capacity measurement. 1. Classification, physicochemical principles, mechanisms, and electron transfer (ET)-based assays. *J. Agric. Food Chem.* 2016;**64**:997–1027.
32. Apak R, Özyürek M, Güclü K, Capanoglu E. Antioxidant activity/capacity measurement. 2. Hydrogen atom transfer (HAT)-based, mixed-mode (electron transfer (ET)/HAT), and lipid peroxidation assays. *J. Agric. Food Chem.* 2016;**64**:1028–45.
33. Amorati R, Valgimigli L. Advantages and limitations of common testing methods for antioxidants. *Free Radic. Res.* 2015;**49**:633–49.
34. Antolovich M, Prenzler PD, Patsalides E, McDonald S, Robards K. Methods for testing antioxidant activity. *Analyst* 2002;**127**:183–98.
35. Rohn S, Kroh LW. Electron spin resonance – a spectroscopic method for determining the antioxidative activity. *Mol. Nutr. Food Res.* 2005;**49**:898–907.
36. Halliwell B, Whiteman M. Measuring reactive species and oxidative damage in vivo and in cell culture: how should you do it and what do the results mean? *Brit. J. Pharmacol.* 2004;**142**:231–55.
37. Pietta PG. Flavonoids as antioxidants. *J. Nat. Prod.* 2000;**63**:1035–42.
38. Jensen GS, Wu X, Patterson KM, Barnes J, Carter SG, Scherwitz L, et al. In vitro and in vivo antioxidant and anti-inflammatory capacities of an antioxidant-rich fruit and berry juice blend. Results of a pilot and randomized, double-blinded, placebo-controlled, crossover study. *J. Agric. Food Chem.* 2008;**56**:8326–33.
39. Kang J, Thakali KM, Jensen GS, Wu X. Phenolic acids of the two major blueberry species in the US market and their antioxidant and anti-inflammatory activities. *Plant Foods Hum. Nutr.* 2015;**70**:56–62.
40. Valko M, Morris H, Cronin MT. Metals, toxicity and oxidative stress. *Curr. Med. Chem.* 2005;**12**:1161–208.
41. Halake K, Birajdar M, Lee J. Structural implications of polyphenolic antioxidants. *J. Ind. Eng. Chem.* 2016;**35**:1–7.
42. Burns J, Yokota T, Ashihara H, Lean M, Crozier A. Plant foods and herbal sources of resveratrol. *J. Agric. Food Chem.* 2002;**50**:3337–40.
43. Thompson L, Boucher B, Liu Z, Cotterchio M, Kreiger N. Phytoestrogen content of foods consumed in Canada, including isoflavones, lignans, and coumestan. *Nutr. Cancer* 2006;**54**:184–201.
44. Landete JM. Dietary intake of natural antioxidants: vitamins and polyphenols. *Crit. Rev. Food Sci. Nutr.* 2012;**53**:706–21.
45. Byrd-Bredbenner C, Beshgetoor D, Moe G, Berning J. *Wardlaw's perspectives in nutrition.* 8th ed. New York: McGraw-Hill Companies, Inc.; 2009.
46. Lake BG. Coumarin metabolism, toxicity and carcinogenicity: relevance for human risk assessment. *Food Chem. Toxicol.* 1999;**37**:423–53.
47. Yu HY, Chen ZY, Sun B, Liu J, Meng FY, Liu Y, Tian T,Jin A, Ruan HL. Lignans from the fruit of *Schisandra glaucescens* with antioxidant and neuroprotective properties. *J. Nat. Prod.* 2014;**77**:1311–20.
48. Lipinski CA. Drug-like properties and the causes of poor solubility and poor permeability. *J. Pharmacol. Toxicol.* 2000;**44**:235–49.
49. Ekins S, Waller CL, Swaan PW, Cruviani G, Wrighton SA, Wikel JH. Progress in predicting human ADME parameters in silico. *J. Pharmacol. Toxicol.* 2000;**44**:251–72.
50. Hansch C, Leo A, Mekapati SB, Kurup A. QSAR and ADME. *Bioorg. Med. Chem.* 2004;**12**:3391–400.
51. Lewandowska U, Szewczyk K, Hrabec E, Janecka A, Gorlach S. Overview of metabolism and bioavailability enhancement of polyphenol. *J. Agric. Food Chem.* 2013;**61**:12183–99.
52. Klaassen C, Casarett LJ, Doull J. *Casarett & Doull's toxicology: the basic science of poisons.* 8th ed. New York: McGraw-Hill Companies, Inc.; 2013.
53. Azzolini M, Mattarei A, Spina M, Marotta E, Zoratti M, Paradisi C, et al. Synthesis and evaluation as prodrugs of hydrophilic carbamate ester analogues of resveratrol. *Mol. Pharm.* 2015;**12**:3441–54.
54. Guengerich PF. Common and uncommon cytochrome P450 reactions related to metabolism and chemical toxicity. *Chem. Res. Toxicol.* 2001;**14**:611–50.
55. Kapetanovic IM, Muzzio M, Huang A, Thompson TN, McCormick DL. Pharmacokinetics, oral bioavailability, and metabolic profile of resveratrol and its dimethylether analog, pterostilbene, in rats. *Cancer Chemother. Pharmacol.* 2011;**68**:593–601.
56. Deng L, Li Y, Zhang X, Chen B, Deng Y, Li Y. UPLC–MS method for quantification of pterostilbene and its application to comparative study of bioavailability and tissue distribution in normal and Lewis lung carcinoma bearing mice. *J. Pharm. Biomed.* 2015;**114**:200–7.

57. Nebert DW, Russell DW. Clinical importance of the cytochromes P450. *Lancet* 2002;**360**:1155–62.
58. Moon YJ, Wang X, Morris ME. Dietary flavonoids: effects on xenobiotic and carcinogen metabolism. *Toxicol. In Vitro* 2006;**20**:187–210.
59. Anzenbacher P, Anzenbacherova E. Cytochrome P450 and metabolism of xenobiotics. *Cell. Mol. Life Sci.* 2001;**58**:737–47.
60. Albanes D, Heinonen OP, Taylor PR, Virtamo J, Edwards BK, Rautalahti M, et al. Alpha-tocopherol and beta-carotene supplements and lung cancer incidence in the alpha-tocopherol, beta-carotene cancer prevention study: effects of base-line characteristics and study compliance. *J. Natl. Cancer Inst.* 1996;**88**:1560–70.
61. Heinonen OP, Albanes D, Virtamo J, Taylor PR, Huttunen JK, et al. Prostate cancer and supplementation with alpha-tocopherol and beta-carotene: incidence and mortality in a controlled trial. *J. Natl. Cancer Inst.* 1998;**90**:440–6.
62. You JS, Jeon S, Byun YJ, Koo SS, Choi S. Enhanced biological activity of carotenoids stabilized by phenyl groups. *Food Chem.* 2015;**177**:339–45.
63. de la Lastra CA, Villegas I. Resveratrol as an antioxidant and prooxidant agent: mechanisms and clinical implications. *Biochem. Soc. Trans.* 2007;**35**:1156–60.
64. Sen C, Packer L, Hänninen O. *Handbook of oxidants and antioxidants in exercise*. 1st ed. Elsevier Science; 2000.
65. Du J, Martin SM, Levine M, Wagner BA, Buettner GR, Wang SH, et al. Mechanisms of ascorbate-induced cytotoxicity in pancreatic cancer. *Clin Cancer Res* 2010;**16**:509–20.
66. Stephenson CM, Levin RD, Spector T, Lis CG. Phase I clinical trial to evaluate the safety, tolerability, and pharmacokinetics of high-dose intravenous ascorbic acid in patients with advanced cancer. *Cancer Chemother. Pharmacol.* 2013;**72**:139–46.
67. Padayatty SJ, Sun H, Wang Y, Riordan HD, Dewitt SM, Katz A, et al. Vitamin C pharmokinetics: implications for oral and intravenous use. *Ann. Intern. Med.* 2004;**140**:533–7.
68. Hecht F, Pessoa CF, Gentile LB, Rosenthal D, Carvalho DP, Fortunato RS. The role of oxidative stress on breast cancer development and therapy. *Tumour Biol.* 2016;**37**:4281–91.
69. Vicetti R, Ishitani T, Salas A, Ayala M. Use of alpha-tocopherol combined with synergists and compared to other antioxidants on the oxidative stability of sardine skin lipids. *J. Food Compos. Anal.* 2005;**18**:131–7.
70. Yun JI, Gong SP, Song YH, Lee ST. Effects of combined antioxidant supplementation on human sperm motility and morphology during sperm manipulation in vitro. *Fertil. Steril.* 2013;**100**:373–8.
71. Koyuturk M, Yanardag R, Bolkent S, Tunali S. Influence of combined antioxidants against cadmium induced testicular damage. *Environ. Toxicol. Pharmacol.* 2006;**21**:235–40.
72. Hagymasi K, Egresi A, Lengyel G. Antioxidants – antioxidative stress: facts and questions, 2015. *Orv. Hetil.* 2015;**156**:1884–7 (in Hungarian).
73. Kim Y, Chongviriyapan N, Liu C, Russell RM, Wang X. Combined α-tocopherol and ascorbic acid protect again smoke-induced metaplasia in ferrets. *Lung Cancer* 2012;**75**:15–23.
74. Salonen RM, Nyyssönen K, Kaikkonen J, Porkkala-Sarataho E, Voutilainen S, Rissanenet TH, et al. Six-year effect of combined vitamin C and E supplementation on atherosclerotic progression. The antioxidant supplementation in atherosclerosis prevention (ASAP) study. *Circulation* 2003;**107**:947–53.
75. Weissig V, Guzman-Villanueva V. Nanocarrier-based antioxidant therapy: promise or delusion? *Expert Opin. Drug Deliv.* 2015;**12**:1783–90.
76. Ibañez IL, Notcovich C, Catalano PN, Bellino MG, Durán H. The redox-active nanomaterial toolbox for cancer therapy. *Cancer Lett.* 2015;**359**:9–19.
77. Richard PU, Duskey JT, Stolarov S, Spulber M, Palivan CG. New concepts to fight oxidative stress: nanosized three-dimensional supramolecular antioxidant assemblies. *Expert Opin. Drug Deliv.* 2015;**12**:1527–45.
78. Rzigalinski BA, Meehan K, Davis RM, Xu Y, Miles WC, Cohen CA. Radical nanomedicine. *Nanomedicine (Lond.)* 2006;**1**:399–412.
79. Chorny M, Hood E, Levy RJ, Muzykantov VR. Endothelial delivery of antioxidant enzymes loaded into non-polymeric magnetic nanoparticles. *J. Control. Release* 2010;**146**:144–51.
80. Shaikh J, Ankola DD, Beniwal V, Singh D, Kumar MN. Nano particle encapsulation improves oral bioavailability of curcumin by at least 9-fold when compared to curcumin administered with piperine as absorption enhancer. *Eur. J. Pharm. Sci.* 2009;**37**:223–30.
81. Neves AR, Lucio M, Martins S, Lima J, Reis S. Novel resveratrol nanodelivery systems based on lipid nanoparticles to enhance its oral bioavailability. *Int. J. Nanomed.* 2013;**8**:177–87.

82. Malik P, Ameta RK, Singh M. Preparation and characterization of bionanoemulsions for improving and modulating the antioxidant efficacy of natural phenolic antioxidant curcumin. *Chem. Biol. Interact.* 2014;**222**:77–86.
83. Yen FL, Wu TH, Tzeng CW, Lin LT, Lin CC. Curcumin nanoparticles improve the physicochemical properties of curcumin and effectively enhance its antioxidant and antihepatoma activities. *J. Agric. Food Chem.* 2010;**58**:7376–82.
84. Rosini M, Simoni E, Milelli A, Minarini A, Melchiorre C. Oxidative stress in Alzheimer's disease: are we connecting the dots? *J. Med. Chem.* 2014;**57**:2821–31.
85. Bors W, Michel C. Chemistry of the antioxidant effect of polyphenols. *Ann. N. Y Acad. Sci.* 2002;**957**:57–69.
86. Mitra I, Saha A, Roy K. Quantification of contributions of different molecular fragments for antioxidant activity of coumarin derivatives based on QSAR analyses. *Can. J. Chem.* 2013;**91**:428–41.
87. Khlebnikov AI, Schepetkin IA, Domina NG, Kirpotina LN, Quinn MT. Improved quantitative structure-activity relationship models to predict antioxidant activity of flavonoids in chemical, enzymatic, and cellular systems. *Bioorg. Med. Chem.* 2007;**15**:1749–70.
88. Rastija V, Medić-Šarić M. QSAR study of antioxidant activity of wine polyphenols. *Eur. J. Med. Chem.* 2009;**44**:400–8.
89. Amić D, Davidović-Amić D, Bešlo D, Rastija V, Lučić B, Trinajstić N. SAR and QSAR of the antioxidant activity of flavanoids. *Curr. Med. Chem.* 2007;**14**:827–45.
90. Silva F, Figuiras A, Gallardo E, Nerín C, Dominges FC. Strategies to improve the solubility and stability of stilbene antioxidants: a comparative study between cyclodextrins and bile acids. *Food Chem.* 2014;**145**:115–25.
91. Benayahoum A, Amira-Guebailia H, Houache O. On the role of ethylene bridge elongation in the antioxidant activity of polyhydroxylated stilbenes: a theoretical approach. *C. R. Chim.* 2015;**18**:149–59.
92. Peerannawar S, Horton W, Kokel A, Török F, Török M, Török B. Theoretical and experimental analysis of the antioxidant features of diarylhydrazones. *Struct. Chem.* 2017;**28**:391–402.
93. Fiedor J, Burda K. Potential role of carotenoids as antioxidants in human health and disease. *Nutrients* 2014;**6**:466–88.
94. Nowicka B, Kruk J. Occurrence, biosynthesis and function of isoprenoid quinones. *Biochim. Biophys. Acta* 2010;**1797**:1587–605.
95. Beutner S, Boledorn B, Frixel S, Blanco I, Hoffmann T, Martin H, et al. Quantitative assessment of antioxidant properties of natural colorants and phytochemicals: carotenoids, flavonoids, phenols and indigoids. The role of β-carotene in antioxidant functions. *J. Sci. Food Agric.* 2001;**81**:559–68.
96. O'Toole MG, Soucy PA, Chauhan R, Raju MV, Patel DN, Nunn BM, et al. Release-modulated antioxidant activity of a composite curcumin-chitosan polymer. *Biomacromolecules* 2016;**17**:1253–60.
97. Betbeder D, Lipka E, Howsam M, Carpentie R. Evolution of availability of curcumin inside poly-lactic-co-glycolic acid nanoparticles: impact on antioxidant and antinitrosant properties. *Int. J. Nanomed.* 2015;**10**:5355–66.
98. Nimiya Y, Wang W, Du Z, Sukamtoh E, Zhu J, Decker E, Zhang G. Redox modulation of curcumin stability: redox active antioxidants increase chemical stability of curcumin. *Mol. Nutr. Food Res.* 2016;**60**:487–94.
99. Suwannateep N, Wanichwecharungruang S, Haag S, Devahastin S, Groth N, Fluuhr J, Lademann J, Meinke M. Encapsulated curcumin results in prolonged curcumin activity in vitro and radical scavenging activity ex vivo on skin after UVB-irradiation. *Eur. J. Pharm. Biopharm.* 2012;**82**:485–90.
100. Sherin DR, Rajasekharan KN. Mechanochemical synthesis and antioxidant activity of curcumin-templated azoles. *Arch. Pharm. Chem. Life Sci.* 2015;**348**:908–14.

CHAPTER

3.28

The Value-Adding Connections Between the Management of Ecoinnovation and the Principles of Green Chemistry and Green Engineering

Philip Coish[1], Enda McGovern[2], Julie B. Zimmerman[1], Paul T. Anastas[1]

[1]Yale University, New Haven, CT, United States; [2]Sacred Heart University, Fairfield, CT, United States

3.28.1 INTRODUCTION

3.28.1.1 Importance of Sustainability to Business and the Connections to the Principles of Green Chemistry and Green Engineering

The challenges facing society today threaten the sustainability of its current practices. According to the Department of Economic and Social Affairs of the United Nations Secretariat, the world population is projected to increase by more than 1 billion people within the next 15 years, reaching 8.5 billion by 2030, and to increase further to 9.7 billion by 2050.[1] Such increases in world population will require the use of constrained resources in ways that are sustainable. Another significant challenge facing today's societies is climate change, which the United Nations Secretary-General has called the "greatest collective challenge facing humankind today."[2] According to Moon, (society) must turn this challenge into the greatest opportunity for common progress toward a sustainable future. Accepting this challenge was the focus of the 195 nations who attended the December, 2015 United Nations Conference on Climate Change in Paris. Global challenges, such as climate change and

Green Chemistry
http://dx.doi.org/10.1016/B978-0-12-809270-5.00033-9

population growth, require the adoption of sustainable practices across all sectors of industry, and societies will look to the economic marketplace for solutions, in addition to the disciplines of science and technology.[3,4]

In 1987, the Brundtland Commission published a report in which sustainable development was defined as "development that seeks to meet the needs and aspirations of the present without compromising the ability to meet those of the future."[5] The Brundtland definition confirms that the continued economic and social well-being of the global population cannot be maintained by measures that are outdated and continue to destroy the natural environment. The concept of sustainability has played a critical role in shaping the international community's response toward economic, social, and environmental development.[6] Importantly, the concept has influenced the way companies conduct business affairs. As a result, companies are now adopting the three pillars of sustainability—profit, people, and planet—as an accounting framework that more accurately measures their performance.[7,8] These dimensions relate to the corporate "triple-bottom-line" framework proposed by Elkington.[9]

Since the Brundtland report, academics have built on the concept of business sustainability and its impact on competitiveness. Dyllick and Hockerts have transposed the Brundtland definition to the business level by defining corporate sustainability "as meeting the needs of a firm's direct and indirect stakeholders without compromising its ability to meet the needs of future stakeholders as well."[10] In 1999, Lovins and coauthors published the theory of "natural capitalism" as a new approach to protecting the biosphere while improving company profits and competitiveness.[11] The approach advocated for the implementation of whole-system designs that introduced alternative, environment friendly technologies. According to these authors, companies need to productively use and reinvest in the capital of people and nature, in addition to financial and physical capital. Porter and Kramer later described the business concept of creating shared value (CSV)[12] that was defined as "policies and operating practices that enhance the competitiveness of a company while simultaneously advancing the economic and social conditions in the communities in which it operates." According to Porter and Kramer, CSV is integral to a company's profitability and competitive position and differs from corporate social responsibility programs that focus mostly on reputation and have only a limited connection to the business. Other authors have also noted that sustainability has the potential to deliver new sources of competitive advantage.[13] Simply put, the concept of business sustainability refers to "business models and managerial decisions grounded in financial, environmental and social concerns."[14]

There are many reasons for companies to focus on sustainability. Companies engaged in environmental management and green innovation can lower costs by minimizing waste and increasing productivity.[15] According to Porter and van der Linde, "pollution represents economic waste and involves unnecessary, inefficient or incomplete utilization of resources, or resources not used to generate highest value."[16] When a company avoids waste, they avoid non-value-creating activities such as waste handling, storage, and disposal that cost the company both time and money. Companies can also possibly contribute to growth in revenues by creating a positive corporate reputation with environmentalists and green-minded consumers and thereby enhance corporate competitiveness while avoiding protests.[15] Sustainable development that incorporates green innovation also enables companies to recognize new business opportunities with first mover advantage, thus creating the opportunity to develop unique competencies that would be difficult to match by competitors or imitators.[16,17] Importantly, companies that comply with local, national, and international regulations of environmental protection can avoid expensive litigations and the potential burden of heavy financial penalties.[15]

3.28.1.1.1 Connections Between Sustainability, Value Creation Levers, and the Principles of Green Chemistry and Green Engineering

The value creation of green initiatives is further supported by the results of a survey in 2009 of more than 1500 worldwide executives and managers about their perspectives on business sustainability.[13] The study provided evidence that sustainability has the potential to affect multiple value creation levers over both the short and long terms. For example, sustainability efforts can result in a stronger brand and greater pricing power (a value creation lever). This can lead to increased margins, producing greater profits and maximizing total shareholder return. Sustainability can also help improve customer loyalty and reduce the rate of churn, thereby boosting market share and revenue growth. If enabling these value creation levers can help increase profits, free up cash flow, and ultimately improve total shareholder return, then assisting leaders and managers of companies to implement sustainable policies should be highly desirable. Such assistance can be provided by adopting the Principles of Green Chemistry and Green Engineering (GC&GE) as these principles offer companies a set of methodologies, or framework, to implement sustainability that can have a positive impact on business.

Despite these advantages, many executives are of the opinion they have to choose between the social benefits of developing sustainable products or processes and the financial costs attributed to such developments.[17] The thinking behind this trade-off, however, has been challenged. In the article entitled "The Transformative Innovations Needed by Green Chemistry for Sustainability," it was stated that the greening of the environment and the green color of the US dollar can be coupled and that the trade-offs are not inherent and necessary.[18] An economy exists within a society, and the society exists within the environment.[19] Balancing the three elements of environment, economy, and society will result in less-than-optimal trade-offs. Instead, the long-term approach must be to ensure that the goals of environment, society, and economy are working in concert via a synergistic approach.[19] Similarly, Porter and van der Linde state that the policy focus should be on "relaxing the tradeoff between competitiveness and the environment rather than accepting it as a given."[16]

3.28.1.1.2 Connections Between Sustainability Literacy and the Principles of Green Chemistry and Green Engineering

According to a 2010 McKinsey & Company study of more than 1,500 worldwide executives and managers, greater than 50% of the sample consider sustainability very or extremely important in areas such as new product development and corporate strategy.[20] Yet only approximately 30% of executives state that their companies actively seek opportunities to invest in sustainable related projects or embed it in their business practices. According to the study authors, a potential reason for the lack of engagement in sustainability may be that companies have no clear definition of sustainability, which suggests the need for improved sustainability literacy.

A lack of understanding among business leaders of what sustainability actually means to a company was also observed by Berns and coauthors from the results of an extensive survey.[13]

Almost 70% of survey respondents said that their company has not developed a clear business case for sustainability. The authors concluded that "shortcomings were due to several underlying information gaps, and emphasized a critical need for gathering and sharing of basic facts about sustainability" as a first step toward helping managers make decisions that lead to sustainable development. The authors note that there is a "high correlation between the depth of a business leader's experience with sustainability and the drivers and benefits that he or she perceived, suggesting that the more people know about sustainability, the more thoughtfully they evaluate it and the more opportunity."

Today's companies are facing sustainability-related issues such as resource volatility, customer demands, and impending regulations, and there are also mounting pressures from multiple stakeholders that include employees, customers, investors, insurers, nongovernmental organizations, government, and society.[13] As a further complication, companies that are becoming globalized and extensively networked are finding it difficult to ensure that inputs into their value chain are green.[21] The leaders and managers of organizations that are facing sustainability-related issues and challenges will require new core skills and competencies to successfully manage business sustainability and the related activity of ecoinnovation. One such competency is sustainability literacy that can address the informational gaps.[22,23] It has been stated that the "green core competency" of a company is defined as the collective learning and capabilities about green innovation and environmental management within an organization.[15] Consistent with this definition is the notion that knowledge resources and human skills, in addition to provisions and access to finance, are essential drivers to green innovation.[24] Green core competencies can be enhanced with knowledge and/or an awareness of two sets of guiding principles that offer frameworks to implement sustainability. The two frameworks are the Principles of Green Chemistry, published by Anastas and Warner,[25,25a,25b] and the Principles of Green Engineering, published by Anastas and Zimmerman.[26] The relevance of these principles to the management of business sustainability, and specifically to the management of ecoinnovation, is discussed in this chapter.

3.28.1.2 What Are Principles of Green Chemistry and Green Engineering in the Context of Business Sustainability?

The Principles of GC&GE (Boxes 3.28.1 and 3.28.2) provide guidelines for the design of new materials, products, processes, and systems that are benign to human health and the environment. A design that is based on these principles moves beyond baseline engineering, chemical, and safety specifications to consider the environmental, economic, and social factors of sustainability.[26] As such, the principles can be organized into categories that are based on sustainability goals: "maximize resource efficiency," "eliminate and minimize hazards and pollution," and "design systems holistically and use life cycle thinking."[27] The relevance of these goals to sustainability is why the principles are useful to managers and leaders as they implement sustainability initiatives or programs in their organizations, or as they pursue green innovation strategies that can create competitive advantages in new markets or extend advantages in old ones. The principles are intended to have a positive impact on society (people), earnings per share (profit), and the environment (planet).

BOX 3.28.1

PRINCIPLES OF GREEN ENGINEERING

1. Inherent rather than circumstantial: Designers need to strive to ensure that all materials and energy inputs and outputs are as inherently nonhazardous as possible.
2. Prevention instead of treatment: It is better to prevent waste than to treat or clean up waste after it is formed.
3. Design for separation: Separation and purification operations should be designed to minimize energy consumption and materials use.
4. Maximize efficiency: Products, processes, and systems should be designed to maximize mass, energy, space, and time efficiency.
5. Output pulled versus input pushed: Products, processes, and systems should be "output pulled" rather than "input pushed" through the use of energy and materials.
6. Conserve complexity: Embedded entropy and complexity must be viewed as an investment when making design choices on recycle, reuse, or beneficial disposition.
7. Durability rather than immortality: Targeted durability, not immortality, should be a design goal.
8. Meet need, minimize excess: Design for unnecessary capacity or capability (e.g., "one size fits all") solutions should be considered a design flaw.
9. Minimize material diversity: Material diversity in multicomponent products should be minimized to promote disassembly and value retention.
10. Integrate material and energy flows: Design of products, processes, and systems must include integration and interconnectivity with available energy and materials flows.
11. Design for commercial "afterlife": Products, processes, and systems should be designed for performance in a commercial "afterlife."
12. Renewable rather than depleting: Material and energy inputs should be renewable rather than depleting.

ACS principles of green engineering. Retrieved from http://www.acs.org/content/acs/en/greenchemistry/what-is-green-chemistry/principles/12-principles-of-green-engineering.html.

3.28.1.3 What Do the Principles Offer to Managers, Leaders, and Their Companies?

As Hopkins has pointed out, "You don't need to find sustainability. Sustainability will find you."[30] A close inspection of a firm's value chain will reveal inputs that require chemistry and/or engineering. For example, the dyes found in clothing are chemicals sourced at some point in the manufacturing process of the garment. Given such an input, the Principles of GC&GE can be used to implement sustainability via tangible actions and can be used to achieve sustainability in a company and its supply chain on multiple levels from local to

BOX 3.28.2

PRINCIPLES OF GREEN CHEMISTRY

1. Prevention: It is better to prevent waste than to treat or clean up waste after it has been created.
2. Atom economy: Synthetic methods should be designed to maximize the incorporation of all materials used in the process into the final product.
3. Less hazardous chemical syntheses: Wherever practicable, synthetic methods should be designed to use and generate substances that possess little or no toxicity to human health and the environment.
4. Designing safer chemicals: Chemical products should be designed to affect their desired function while minimizing their toxicity.
5. Safer solvents and auxiliaries: The use of auxiliary substances (e.g., solvents, separation agents, etc.) should be made unnecessary wherever possible and innocuous when used.
6. Design for energy efficiency: Energy requirements of chemical processes should be recognized for their environmental and economic impacts and should be minimized. If possible, synthetic methods should be conducted at ambient temperature and pressure.
7. Use of renewable feedstock: A raw material or feedstock should be renewable rather than depleting whenever technically and economically practicable.
8. Reduce derivatives: Unnecessary derivatization (use of blocking groups, protection/deprotection, temporary modification of physical/chemical processes) should be minimized or avoided if possible, because such steps require additional reagents and can generate waste.
9. Catalysis: Catalytic reagents (as selective as possible) are superior to stoichiometric reagents
10. Design for degradation: Chemical products should be designed so that at the end of their function they break down into innocuous degradation products and do not persist in the environment.
11. Real-time analysis for pollution prevention: Analytical methodologies need to be further developed to allow for real-time, in-process monitoring and control prior to the formation of hazardous substances.
12. Inherently safer chemistry for accident prevention: Substances and the form of a substance used in a chemical process should be chosen to minimize the potential for chemical accidents, including releases, explosions, and fires.

ACS principles of green chemistry. Retrieved from http://www.acs.org/content/acs/en/greenchemistry/what-is-green-chemistry/principles/12-principles-of-green-chemistry.html.

global. For example, the manufacturer of the garment can use a dye that is made by a less hazardous chemical synthesis as per Green Chemistry Principle 3. Importantly, the principles are innovative, nonregulatory, and economically driven toward achieving a sustainable future and offer companies an opportunity to move away from the "command-and-control" regulatory system of hazardous substances or practices.

The value of GC&GE to companies can be best illustrated through looking at examples from industry. The types of products and processes that are being offered by companies adopting the Principles of Chemistry include medicine, food production, energy production, packaging, household and commercial cleaning products, electronics, automotive products, and a wide range of consumer goods.[19] For example, as highlighted by Nike Inc. in their 2016 Sustainable Business Report, the company is reducing the use of petroleum-derived solvents (PDSs) across the manufacturing process of their footwear product lines. Since 1995, there has been an actual 96% reduction in the use of PDSs per pair of shoes through the adoption of water-based adhesives.[31] Another example in the dry cleaning sector of industry is the development of supercritical carbon dioxide as a viable alternative to tetrachloroethylene in dry cleaning applications. Supercritical carbon dioxide is environment friendly, nontoxic, and biodegradable and requires no hazardous waste removal.[19]

The fourth Principle of Green Engineering states that products, processes, and systems should be designed to maximize mass, energy, space, and time efficiency. This is related to the "Porter hypothesis,"[16] which says, in part, that environmentally benign innovations can lead to an increase in a firm's performance through a reduction of energy or materials use. A good example from industry is displayed by Unilever. As a result of their sustainability initiatives, Unilever factories are releasing 37% less emissions than in 2008, and waste is down 85%.[32] The example highlights how innovators are attempting to utilize matter and energy in a way that enhances performance and value while protecting the environment.[26]

Managers do not have to be scientists to appreciate the value of the GC&GE Principles. Just as managers do not need to be experts in all aspects of business, such as finance and marketing, managers can acquire a basic understanding of principles while effectively implementing the principles as part of a company's sustainability strategy. However, employees whose job function is directly related to sustainability, such as a sustainability officer, might benefit from a more in-depth knowledge of the principles. Similarly, designers may be more successful if they have a thorough understanding of the Principles of GC&GE. In addition to a knowledge of the principles, managers will need the ability to apply *systems thinking* to implement the principles across value chains. Systems thinking and external collaboration, like sustainability literacy, are core competencies needed by those who manage business sustainability.[22]

3.28.2 DISCUSSION

3.28.2.1 What Is Ecoinnovation? What Is the Relevance of Green Chemistry and Green Engineering Principles to Ecoinnovation?

An important component of sustainability is innovation.[33] According to Porter and van der Linde,[16] the competitive advantage of a company neither depends on static efficiency nor on optimizing within fixed constraints, but rather on the capacity of the company for innovation and improvement, which shifts the constraints. Also, within the context of ecoinnovation, Nidumolu and coauthors state that "smart companies are now treating sustainability as innovation's new frontier."[17]

There are three stages to technological advancement, namely ideation, innovation, and diffusion.[34,35] While an invention is an idea or a model for a new improved model or process, an invention becomes an innovation *in an economic sense* when the improved product or process is first introduced to the market.[36] A further distinction of invention and innovation was highlighted by Kemp and Pearson who note that "While the concept of invention refers to discovery, discovery may not be needed for innovation."[37]

The definition of innovation in an economic sense is different from the Webster definition that states that innovation is a new idea, device, or method, or the act or process of introducing new ideas, devices, or methods.[38] In this definition, innovation is not separate from invention. For the purposes of this chapter, and for the discussion of the connections between the Principles of GC&GE and the management of ecoinnovation, the invention stage is included in reference to ecoinnovation. The reason for this inclusion is that the principles can impact the inherent design of new products and processes, and design can require, or lead to, invention and discovery. This is aligned with the statement that a future challenge in sustainability is to shift from improving to inventing.[39]

Based on the Organization for Economic Co-operation and Development definition of innovation, Kemp and Pearson propose the following definition for ecoinnovation:[37]

"Eco-innovation is the production, assimilation or exploitation of a product, production process, service, or management or business method that is novel to the organization (developing it or adopting it) and which results, throughout its life cycle, in a reduction of environmental risk, pollution and other negative impacts of resources use (including energy use) compared to relevant alternatives."[37]

As Remmings pointed out in his seminal article in 1998, ecoinnovations differ from normal innovations because they produce two types of positive externalities, the usual knowledge externalities in the research and innovation phases and externalities in the adoption and diffusion phases due to the positive impact upon the environment. Ecoinnovation develops products and services which themselves cause external benefits.[36]

In the context of sustainable development, four notions of innovation have evolved and are used synonymously according to Schiederig and coauthors.[40] These notions are: eco-, green, sustainable, and environmental innovation. The main difference between sustainable innovation and the other three notions is that the latter three notions include economic and ecological aspects, whereas sustainable innovation, in its original meaning, includes a social dimension as well. However, as the authors note, this difference has been removed by some scholars and today, the notions are used interchangeably. Yet, Klewitz and Hansen distinguished ecoinnovation (related to environmental issues) from sustainability-orientated innovations that have a broader focus on environmental, social, and economic dimensions.[41]

Based on their review, Schiederig and coauthors concluded that the four notions of ecoinnovation have six common aspects (Box 3.28.3).[40]

Given these six aspects that are common to the four notions of green innovation, the relevance of the Principles of GC&GE to the notion of ecoinnovation are easily observed. The Principles of GC&GE provide a framework that can guide scientists and engineers when designing new materials, products, processes, and systems that are benign to human health and the environment (Aspects 1 and 5). The Principles are relevant to market orientation or pull (Aspect 2) as many customers in today's market are demanding and judging sustainable based products more positively, which provide benefits to human health and the environment as an integral part of a products performance. Ecoinnovations based on the principles can set a new standard not only for the firm but also for each market sector (Aspect 6). To be

BOX 3.28.3

SIX ASPECTS COMMON TO THE FOUR NOTIONS OF ECO-, GREEN, SUSTAINABLE, AND ENVIRONMENTAL INNOVATION

1. Innovation object: Product, process, service, method;
2. Market orientation: Satisfy needs/be competitive on the market;
3. Environmental aspect: Reduce negative impact (optimum: zero impact);
4. Phase: Full life cycle must be considered (for material flow reduction);
5. Impulse: Intention for reduction may be economical or ecological; and
6. Level: Setting a new innovation/green standard to the firm.

Adopted from Schiederig T, Tietze F, and Herstatt C. Green innovation in technology and innovation management – an exploratory literature review. R&D Manage *2012;**42**:180–192.*

truly sustainable, the principles should be implemented across multiple scales (local to global). Lastly, the principles address two fundamental concepts that designers should strive to integrate into their design of products or processes, namely, life cycle considerations (Aspect 4) and inherently benign by design[42] (Aspect 5). Clearly, the Principles of GC&GE can guide sustainable research and development and are applicable to ecoinnovation.

A focus of this chapter is "technological eco-innovation" as opposed to organizational, institutional, or marketing-based innovation. Technological ecoinnovation can be separated into curative and preventative technologies.[36] "Curative technologies" repair damage (e.g., the remediation of contaminated soil), whereas "preventative technologies" attempt to avoid creating any damage. Preventative technologies can be further subdivided into "integrated" and "additive technologies." Additive (or end-of-pipe) technologies include measures like disposal and recycling occurring after the actual production and consumption process. In contrast, integrated technologies directly address the cause and comprise "all measures leading to a reduction in input materials, energy inputs and emissions during production and consumption."[36] Integrated technologies remove the need for curative technologies that repair damage. Integrated technologies can also avoid command and control regulation, which can present regulatory barriers to implementation.[43] Thus integrated technologies are preferred to additive (or end-of-pipe) technologies. The goals of the GC&GE Principles are to minimize waste and to totally eliminate the toxicity of waste through *incorporated inherency that is benign by design via integrated technologies*.

Tidd noted that innovations can be further defined by the type of innovation, namely, continuous incremental, complex, radical, and disruptive innovation.[44] It has been noted that in the context of environmental innovation, there is a need for both radical and incremental change.[45] Others have emphasized the need for radical or disruptive innovation with the view that a shift is needed to fast-track business strategies that address pressing global issues like climate change and resource depletion. An analytical framework of scoring of ecoinnovation has been developed by Carrillo-Hermosilla, del Río, and Könnölä.[46] The type of innovation that is pursued by a firm varies with the organizational capacity that is required to effect the type of innovation. This capacity is defined by three factors: resources (money, people, and technology), process, and values (as it relates to prioritization decisions).[47] Bernauer and coauthors posited that the growing maturity of a firm's green capabilities increases the ability to ecoinnovate and general innovativeness increases the probability that firms will also ecoinnovative but concludes that more research is needed on this topic.[48]

An example of a shift from incremental innovation is by Nike Inc. who report that the company has disrupted the traditional method of manufacturing shoes with the Flyknit Lunar 1+ running shoe and is reducing waste by about 60% on an average compared with cut-and-sew footwear.[31] Nike's breakthrough innovation significantly reduced waste and condensed non-value-creating activities such as waste handling, storage, and disposal that incurred more costs in time and money and, at the same time, reduced the number of negative environmental impacts.

3.28.2.2 Who Are the Actors Involved in the Management of Ecoinnovation, and Who Can Implement or Use the Principles of Green Chemistry and Green Engineering?

According to Kemp and Pearson, there are four types of ecoinnovators at the firm level that show different behaviors toward ecoinnovation.[35,37] The most active innovators are the *strategic ecoinnovators* and *strategic ecoadapters* that intentionally develop or implement ecoinnovations, in contrast to the *passive ecoadapters* and *non-ecoinnovators*. Bossink discusses ecoinnovators at the individual or actor level in his book entitled *Eco-innovation and Sustainability Management*.[49] Bossink presents a model in which ecoinnovation is conceptualized as a concentric system that consists of three interdependent and interacting levels of cooperative activity. According to the model, there are three managerial levels of coideation, coinnovation, and coinstitutionalization. The inner coideation level consists of leaders, champions, and entrepreneurs. The middle coinnovation level consists of teams, projects, businesses, and public-private partnerships. The outer coinstitutionalization level consists of market and society, knowledge and technology, and policy and regulation. Bossink contends that it is the individuals of the inner coideation level that take the lead and have a central position in this ecoinnovation and sustainability model.

An awareness of the Principles of GC&GE should not be limited to the members of the design teams of sustainable and innovative products and processes. Ecoinnovation requires top-down implementation of sustainable innovation from managers and leaders in addition to the bottom-up (emergence) implementation from designers (followers). The managers and leaders of ecoinnovators must also be aware of the principles if an organization is truly going to implement sustainable approaches across multiple scales. At the leadership level, the Principles of GC&GE can better assist leaders to develop business models and strategies that are sustainable (green) in the long term. Related to strategy is the corporate vision and leaders who incorporate sustainability into their vision play a key role in ensuring that sustainability is part of the organization's creative process.[21] According to Berns and coauthors, some thought leaders have suggested that leadership in sustainability might be viewed as a proxy for management quality.[13] At the managerial level, the Principles of GC&GE can assist in managing ecoinnovators involved in research and development. Other discipline managers, for example, in marketing and operations, could also benefit from an awareness of sustainability[50] and, in particular, the Principles of GC&GE. According to Cooper and Edgett, the most effective methods of ideation for product innovation are customer visit teams, ethnographic research, leader user analysis, and focus groups.[34] As ideas for new inventions and innovations are fed back to design teams, the translation might be more effective if the communicators (the marketing staff) have high levels of sustainability literacy. The overarching concept is that the actors in all phases of technology advancement (ideation, innovation, and diffusion) have to make decisions related to sustainability and these actors would benefit from an understanding of the Principles of GC&GE.

3.28.2.3 Management of Ecoinnovation and the Value-Adding Connections to the Principles of Green Chemistry and Green Engineering

3.28.2.3.1 Overview

As early as 1998, Remmings wrote that "managing eco-innovation is an increasingly important issue for many firms."[36] This point was made again in 2008 by Chen who wrote "environmental management is gathering importance within organizations and it is increasingly becoming an important part of management agendas."[15] There is unlikely to be "one best way" to manage ecoinnovation since there are multiple factors that affect the management of innovation, including type, stage, scope, and type of organization.[44] Despite this variance, the Network for Business Sustainability has published a report that presents a sustainability roadmap for business leaders, including a three-stage framework for assessing which stage(s) of the sustainability continuum an organization currently occupies.[14] The report provides 38 practices for fostering innovation that are binned into eight categories such as "Leadership and Governance," "Knowledge Management," and "Tools and Platforms." It has also been stated that integrating sustainability and innovation is really about understanding the future direction of the business and managing it inclusively.[21] To that end, one can ask how to apply or include the Principles of GC&GE in the management of ecoinnovation. One approach is to look at a framework for managing innovation and identify the opportunities for implementing the principles.

3.28.2.3.2 Framework of the Innovation Management Process: Opportunities to Apply the Principles

Adams and coauthors reviewed the literature pertaining to the measurement of innovation management at the company level and developed a framework of the innovation management process consisting of seven categories.[51] The framework incorporates a diverse set of literature and provides "a holistic framework covering the range of activities required to turn ideas into useful and marketable products." In this chapter, this framework of the innovation management process will be used to identify where leaders and managers can implement the Principles of GC&GE or, in some cases, highlight the potential impact that the principles can have on these categories. The seven categories as identified are listed in Box 3.28.4, along with added commentary.

BOX 3.28.4

FRAMEWORK OF THE INNOVATION MANAGEMENT PROCESS: OPPORTUNITIES TO APPLY THE PRINCIPLES

1. Inputs

"Inputs management is concerned with the resourcing of innovation activities and includes factors ranging from finance, to human and physical resources, to generating new ideas."[51]

Human: Ecoinnovation and GC&GE Principles require unique staffing requirements and networks. The need for systems thinking and a multidisciplinary approach requires innovative teams with diverse skill sets and knowledge. Actors require the ability to network and collaborate within the departments of an organization and beyond.

Physical: Specific Principles of Green Engineering consider material and energy inputs in the life cycle of an innovation.

Natural: The GC&GE Principles can guide scientists and engineers when designing new materials, products, processes, and systems that are benign to human health and the environment.

Financial: The GC&GE Principles are innovative, nonregulatory, economically driven approaches toward sustainability. The principles can have a positive impact on value creation levers that can lead to increased total shareholder return.

2. Innovation Strategy

An emerging field of management is dealing with the natural environment as it affects corporate strategy.[15] The innovation strategy can be part of the larger corporate strategy that evolves from the business model. The Principles of GC&GE can help leaders develop strategies that are sustainable. The principles offer tangible and actionable guidance on pursuing sustainability (specifically decision making) within the company's value chain. Boons and Ludeke-Freund noted that both Lovins and Hart envisioned changing the company's business model as a way to reduce negative social and ecological impacts, or even as a way to purposefully achieve sustainable development.[52] Notably, Boons and Ludeke-Freund reviewed the current literature on business models as it relates to innovation and proposed a conceptual definition of a sustainable business model in the context of technological innovation.[52] Innovation attempts are related to the innovation strategy, and the implementation of the GC&GE Principles could be an integral part of a firm's innovation strategy. Certainly, if a business model is a way to reduce negative social and ecological impact, then the principles can be used as a tangible means to execute the model and derive strategy. While doing so, companies that follow a sustainable business should be able to derive competitive advantages that are based on a sustainable value proposition. For example, by reducing the costs associated with waste, companies can improve operational efficiencies and gain a competitive advantage via a cost advantage. Alternatively, a firm may gain a differentiation advantage by offering to the market a "greener" product and perhaps demand a higher price, especially if the product is certified as sustainable. A continued effort by companies to grow their sustainable product range can result in developing a stronger brand image and greater pricing power (a value creation lever). This can lead to an improvement in margin and improved profits, and subsequently a greater total shareholder return.

BOX 3.28.4 *(cont'd)*

3. Knowledge Management

According to Adams and coauthors, knowledge management is concerned with obtaining and communicating ideas and information that is the foundation of innovation competencies.[51] The leaders and managers of sustainability will require new core competencies to be able to manage ecoinnovation. Such core competencies include sustainability literacy, external collaboration, and systems thinking.[22,23,53] While managers and leaders may possess an expert knowledge of sustainability, they can build organizational capabilities by hiring experts who are educated on the Principles of GC&GE. "Green core competence" has been defined as the collective learning and capabilities about green innovation and environmental management within an organization.[15] The collective learning, or knowledge repository, is considered by Adams and colleagues to be an important area for innovation management.[51] Other important areas are idea generation (the ideation or invention stage) and information flows. As mentioned earlier, the GC&GE Principles can guide designers during idea creation and innovation phases as they create new products, services, or processes. As for information flow, the achievements gained from adopting green chemistry in the design of new products, services, or processes will take place at the interface of numerous disciplines and will require information exchange between multiple collaborators engaged in systems thinking. Ecoinnovations and green chemistry technologies are being developed at the nexus of resources issues, such as the water-energy nexus, and will require cross-disciplinary approaches. Project teams that pursue sustainability initiatives, or are involved in the creation of a new product or process, will require diverse skill sets.

4. Organization and Culture

As noted earlier in this chapter, leaders with a vision that incorporates sustainability play a key role in ensuring that sustainability is part of the organization's creative process.[21] Leaders and managers can create the value system of a company that can impact the type of innovation (e.g., incremental vs. radical) and the extent of innovation that is pursued by the company (e.g., strategic ecoinnovators vs nonecoinnovators). Christensen and Overdorf defined an organization's values as "the standards by which employees set priorities that enable them to judge whether an idea for a new product is attractive or marginal, and so on."[47] At the executive level, prioritization decisions based on values take the form of decisions to invest, or not, in new products, services, and processes. These values can be influenced by the guiding Principles of GC&GE. Similarly, when an employee, perhaps a designer, is faced with a prioritization decision, that employee can utilize the Principles of GC&GE to make decisions that lead to sustainable innovation and ultimately sustainable development.

Related to culture, the sustainable practices of a company has become an important means of motivating top talent, and a small but growing population of "sustainability enthusiasts" pay close attention to a firm's behavior according to a 2013 survey by Bain and Company.[23] No doubt the Principles of GC&GE would be attractive to such enthusiasts and could reduce employee turnover or churn. Moreover, sustainability practices

Continued

BOX 3.28.4 *(cont'd)*

are an important statement in attracting the right talent[23] and speaks to the potential impact of sustainability efforts on the value creation lever of employee recruitment and engagement.[13,54]

Finally, a review by Tidd suggests that "the complexity and uncertainty of the environment affects the degree, type, organization and management of innovation" and that a coherent organization with the proper fit to specific technologies and markets will lead to greater performance.[44] Irrespective of the complexity of a business case (high or low complexity), it is thought that the principles can facilitate managers and leaders in finding the proper fit with respect to sustainability issues.

5. Project Management

The Principles of GC&GE can assist managers with the management of ecoinnovators who design sustainable and innovative products and services. The manager may be a research director interacting with a project team, or he or she could be the project manager. The project managers are required to interact with personnel across multiple levels of an organization, including middle, upper manager, and executive levels (e.g., Vice President of Research). This is the type of interaction Bossink recognized in his ecomanagement model consisting of interdependent and interacting levels of cooperative activity. An awareness of the Principles of GC&GE can guide project managers as they operate within the multiple levels of reporting and interaction during the ideation, innovation, and diffusion stages. This relates to an important aspect of project management, which is life cycle analysis (from concept to end-of-life). Research staff involved with innovation, including project managers, have to make decisions related to sustainability, for example, decisions related to the environmental impact of a new product. If the product is designed to be environmentally benign but is manufactured using hazardous or nonrenewable substances, the impacts have simply been shifted to another part of the overall life cycle.[26] Another example relates to energy. If a product or process is energy efficient, or even energy generating, but the manufacturing process consumes energy to a degree that offsets any energy gains, there is no net sustainability advantage. Accordingly, designers and managers should consider the entire life cycle of a product, including those of the materials and energy inputs.[26]

6. Portfolio Management (Both Leader and Manager Levels)

Leaders and managers can evaluate the sustainability of the projects that make up a portfolio in the context of the principles. If a company's portfolio of projects consists of projects that result from ecoinnovation, then the company will be following a path of sustainable development. The evaluation of the portfolio should consider the project details as it evolves through each phase of the innovation process, from ideation to diffusion. An assessment framework for a sustainable portfolio that can be used to sustain business value has been developed by Graham and Bertels.[55] A challenge for leadership is the development of performance metrics for sustainability efforts, and some have developed green metrics such as an "e-factor," that are different from the usual metrics of portfolio management, such as

BOX 3.28.4 *(cont'd)*

patent portfolio size and R&D expenditure.[44,56] Performance metrics for sustainability, and for GC&GE, is an evolving area of interest within corporate sustainability.

7. Commercialization (With a Focus on Marketing and Sales)

As highlighted earlier, it was determined by Cooper and Edgett that the most effective methods of ideation for product innovation are customer visit teams, ethnographic research, leader user analysis, and focus groups.[34] Feedback from marketing and sales teams involved in the diffusion stage will help the designers of new products in the ideation phase since these teams will best understand market orientation or pull. Customers are currently searching for products that provide benefits to improving human health and the environment as an integral part of a product's performance. It would be wise that customer preferences for green products be transferred to the designers (and managers) in the ideation phase. It is therefore very important that marketing personnel and development managers be aware of sustainability principles, such as those of GC&GE, as they interact with customers and the designers of new products.[50]

Adopted from Adams R, Bessant J, and Phelps R. Innovation management measurement: a review. Int J Manag Rev *2006;8:21–47.*

3.28.3 CONCLUSIONS AND SUMMARY

In this chapter, the value-adding connections between the management of ecoinnovation and the Principles of GC&GE have been examined. The chapter began with a discussion of sustainability and sustainable development, then explored the definition of ecoinnovation, and finally discussed a framework for managing ecoinnovation and the relevance of the Principles of GC&GE. This chapter should be of interest to business leaders, managers across various departments, innovation experts, sustainability officers, directors of research and development, product designers, engineers, and other people from across all disciplines who have an interest in business sustainability.

It is the premise of this perspective that green core competencies can be enhanced with knowledge and/or an awareness of the Principles of GC&GE. The principles provide guidelines for the design of new materials, products, processes, and systems that are benign to human health and the environment. Such innovative designs move beyond baseline engineering and chemical and safety specifications to fully embrace the environmental, economic, and social factors of sustainability.[26]

Although leaders need not be experts, there is a need for expert knowledge within the organization (knowledge repositories), and the organizational capabilities of a company can be enhanced by hiring experts who are educated in the Principles of GC&GE. Managers who are intimately involved with the implementation and operation of sustainability programs would benefit from a knowledge of the principles. Certainly the designers, such as chemists or chemical engineers, can apply the principles in their daily functions, as can employees within the supply chain organizational chart who develop sustainability criteria for environmentally preferable purchasing, and the marketing personnel who increasingly engage green-minded customers. Enabling the development of sustainability literacy throughout an organization, in particular knowledge of the Principles of GC&GE, would facilitate the advancement of green technologies through the stages of ideation, innovation, and diffusion.

In summary, there are many reasons for companies to focus on sustainability, including the value creation of green initiatives and the subsequent potential to deliver new sources of competitive advantage. Nurturing company policies in support of sustainable development can be guided more efficiently by embracing the Principles of GC&GE in the product research and development phases. The principles not only provide innovative, nonregulatory, economically driven approaches toward sustainability but also are expected to make a significant and positive impact on the three pillars of sustainability—profit, people, and planet.

Funding Sources

This work was supported in part by the NSF Division of Chemistry and the EPA through a program of Networks for Sustainable Molecular Design and Synthesis. Grant No. 1339637.

References

1. World population prospects: the 2015 revision. Key findings and advance tables. Retrieved from http://esa.un.org/unpd/wpp/publications/files/key_findings_wpp_2015.pdf.
2. Ki-Moon B. Big idea 2014: the year for climate action. Retrieved from http://www.un.org/climatechange/summit/2013/12/big-idea-2014-the-year-for-climate-action-by-ban-ki-moon/.
3. The imperative of disruptive innovation. Retrieved from http://www.innosight.com/innovation-resources/upload/Disruptive-Innovation-Primer.pdf.
4. El-Kafafi S, Liddle S. Innovative sustainable practices: are they commercially viable? *World J Entrep Manag Sustain Dev* 2010;**6**:19–28.
5. Brundtland G, Khalid M, Agnelli S, et al. Report of the world commission on environment and development: our common future. Retrieved from http://www.un-documents.net/our-common-future.pdf.
6. UNECE sustainable development – concept and action. Retrieved from http://www.unece.org/oes/nutshell/2004-2005/focus_sustainable_development.html.
7. Oehlberg L, Shelby R, Agogino A. Sustainable product design: designing for diversity in engineering education. *Int J Eng Educ* 2010;**26**:489–98.
8. Slaper T, Hall T. The triple bottom line: what is it and how does it work? *Indiana Bus Rev* 2011;**86**:4–8.
9. Elkington J. Partnerships from cannibals with forks: the triple bottom line of 21st century business. *Environ Qual Manag* 1998:37–51.
10. Dyllick T, Hockerts K. Beyond the business case for corporate sustainability. *Bus Strateg Environ* 2002;**11**:130–41.
11. Lovins A, Hunter L, Hawken P. Road map for natural capitalism. *Harv Bus Rev* 1999;**77**:145–58.
12. Porter M, Kramer M. Creating shared value. *Harv Bus Rev* 2011;**89**:62–77.
13. Berns M, Townsend A, Khayat Z, Balagopal B, Reeves M, Hopkins M, Kruschwitz N. Sustainability and competitive advantage. *Sloan Manage Rev* 2009;**51**:19–26.
14. Innovating for sustainability: a guide for executives. Retrieved from www.nbs.net/knowledge.
15. Chen Y-S. The driver of green innovation and green image – green core competence. *J Bus Ethics* 2008;**81**:531–43.
16. Porter M, van der Linde C. Toward a new conception of the environment-competitiveness relationship. *J Econ Perspect* 1995;**9**:97–118.

17. Nidumolu R, Prahalad C, Rangaswami M. Why sustainability is now a key driver of innovation. Innovating for Sustainability: a Guide for Executives. *Harv Bus Rev* 2009;**87**:57–64.
18. Anastas P. The transformative innovations needed by green chemistry for sustainability. *ChemSusChem* 2009;**2**:391–2.
19. Manley J, Anastas P, Cue Jr B. Frontiers in green chemistry: meeting the grand challenges for sustainability in R&D and manufacturing. *J Clean Prod* 2008;**16**:743–50.
20. How companies manage sustainability: McKinsey global survey results. Retrieved from http://www.mckinsey.com/insights/sustainability/how_companies_manage_sustainability_mckinsey_global_survey_results.
21. Dormann J, Holliday C. Innovation, technology, sustainability and society. Retrieved from http://www.wbcsd.org/pages/edocument/edocumentdetails.aspx?id=88&nosearchcontextkey=true.
22. Strandberg C. *Sustainability talent management: the new business imperative. Five sustainability leadership competencies for your competency model*. 2015.
23. Davis-Peccoud J, Artabane M. *The big green talent machine. How sustainability can help hone your talent agenda*. 2013.
24. Cuerva M, Triguero-Cano Á, Córcoles D. Drivers of green and non-green innovation: empirical evidence in low-tech SMEs. *J Clean Prod* 2014;**68**:104–13.
25. Anastas P, Warner J. *Green chemistry: theory and practice*. New York: Oxford University Press; 1998.
25a. Anastas PT, Heine L, Williamson C. *Green chemical syntheses and processes: Introduction*. 2000. p. 1–6.
25b. Anastas PT, Williamson C, Hjeresen D. Breen JJ. Promoting Green Chemistry Initiatives. *Environ Sci Technol* 1999:116–9.
26. Anastas PT, Zimmerman JB. Design through the 12 principles of green engineering. *Environ Sci Technol* 2003:95A–101A.
27. ACS. ACS annual green chemistry & engineering conference brochure. In: *Proccedings of conference ACS annual green chemistry & engineering conference brochure*; 2015.
28. ACS principles of green chemistry. Retrieved from http://www.acs.org/content/acs/en/greenchemistry/what-is-green-chemistry/principles/12-principles-of-green-chemistry.html.
29. ACS principles of green engineering. Retrieved from http://www.acs.org/content/acs/en/greenchemistry/what-is-green-chemistry/principles/12-principles-of-green-engineering.html.
30. Hopkins M. 8 reasons sustainability will change management (that you never thought of). *Sloan Manag Rev* 2009;**51**:27–30.
31. *Nike sustainable business report*. 2016.
32. Gellesnov D. Unilever finds that shrinking its footprint is a giant task. Retrieved from http://www.nytimes.com/2015/11/22/business/unilever-finds-that-shrinking-its-footprint-is-a-giant-task.html?_r=0.
33. Baregheh A, Rowley J, Sambrook S. Towards a multidisciplinary definition of innovation. *Manag Decis* 2009;**47**:1323–39.
34. Cooper R, Edgett S. Ideation for product Innovation: what are the best methods? 2008.
35. Ozusaglam S. Environmental innovation: a concise review of the literature. *Vie Sci Entre* 2012:15–38.
36. Rennings K. Towards a theory and policy of eco-innovation – neoclassical and (Co)evolutionary perspectives. 1998.
37. Kemp R, Pearson P. *Deliverable 15. Final report MEI project about measuring eco-innovation*. 2007.
38. Merriam-Webster definition of innovation. Retrieved from http://beta.merriam-webster.com/dictionary/innovation.
39. Anastas P. Green chemistry next: moving from evolutionary to revolutionary. *Aldrichimica Acta* 2015;**48**:3–4.
40. Schiederig T, Tietze F, Herstatt C. Green innovation in technology and innovation management – an exploratory literature review. *R&D Manag* 2012;**42**:180–92.
41. Klewitz J, Hansen E. Sustainability-oriented innovation of SMEs: a systematic review. *J Clean Prod* 2014;**65**:57–75.
42. Farris CA, Anastas PT. *Benign by design*. Washington, DC: American Chemical Society Books; 1994.
43. Matus K, Clark W, Anastas P, Zimmerman J. Barriers to the implementation of green chemistry in the United States. *Environ Sci Technol* 2012;**46**:10892–9.
44. Tidd J. Innovation management in context: environment, organization, and performance. *Int J Manag Rev* 2001;**3**:169–83.
45. Hall J. Introduction. Special issue: environmental innovation. *J Clean Prod* 2003;**11**:343–6.
46. Carrillo-Hermosilla J, Del Río P, Könnölä T. Diversity of eco-innovations: reflections from selected case. *J Clean Prod* 2010;**18**:1073–83.
47. Christensen C, Overdorf M. Meeting the challenge of disruptive change. *Harv Bus Rev* 2000:66–76.

48. Bernauer T, Engels S, Kammerer D, Seijas J. Explaining green innovation. Ten years after Porter's win-win proposition: how to study the effects of regulation on corporate environmental innovation?. In: *Comparative and international studies working paper, no. 17*. ETH Zurich; 2006.
49. Bossink B. *Eco-innovation and sustainability management*. Routledge; 2013.
50. Visser R, Jongen M, Zwetsloot G. Business-driven innovations towards more sustainable chemical products. *J Clean Prod* 2008;**16**:S85–94.
51. Adams R, Bessant J, Phelps R. Innovation management measurement: a review. *Int J Manag Rev* 2006;**8**:21–47.
52. Boons F, Lüdeke-Freund F. Business models for sustainable innovation: state-of-the-art and steps towards a research agenda. *J Clean Prod* 2013;**45**:9–19.
53. Fiksel J. Sustainability and resilience: toward a systems approach. *Sustain Sci Pract Policy* 2006;**2**:14–22.
54. Lovins A. What executives don't get about sustainability (and further notes on the profit motive). In: Hopkins M, editor. *Sloan management review*, vol. 51; 2009. p. 35–40.
55. Graham R, Bertels S. Achieving sustainable value: sustainability portfolio assessment. *Green Manag Int* 2008;**54**:57–67.
56. Constable D, Curzonsb A, Cunningham V. Metrics to green chemistry—which are the best? *Green Chem* 2002;**4**:521–7.

CHAPTER

3.29

The International Chemicals Regime: Protecting Health and the Environment

Natalia Escobar-Pemberthy, Maria Ivanova, Gabriela Bueno
University of Massachusetts Boston, Boston, MA, United States

Traditionally, the study of natural sciences has been separated from the worlds of policy and government. Scientists have advanced the pursuit of knowledge in numerous ways through both curiosity-driven research and applied scientific inquiry. Basic science, even though it may not provide immediate discoveries or solutions to social problems, has often generated new knowledge and new approaches leading to innovation and breakthroughs and can provide important input to policy-making processes. It often takes years, however, to integrate scientific knowledge into policy development.

Take climate change for example—as early as in the 1800s, scientists Joseph Fourier, John Tyndall, and Svante Arrhenius all independently discovered and explained the dynamics of the earth's atmosphere, elucidating the link between carbon dioxide and temperature. Policy commitments to address greenhouse gas emissions came over a 100 years later, in the 1990s, with the adoption of the UN Framework Convention on Climate Change in 1992. Implementation has been fraught with difficulties and only in 2015, after many international negotiations, did *all* the world's governments commit to undertaking national actions to reduce greenhouse gas emissions and to reach net zero emissions by year 2050. Indeed, only when science is fully utilized in the policy process, can we effectively explore the nature of complex problems and identify appropriate responses. It is therefore important to understand the systems of governance, their fundamental principles, and the various ways in which they do or do not incorporate science.

Often themselves the product of scientific discovery, chemicals are critical to all aspects of modern life. They play an important role in agriculture, industry, energy, and medicine. Through the different stages of their life cycle, from extraction to disposal, chemicals might pose various threats to human health and the environment.[1] According to the Global Chemicals Outlook, "Exposure to toxic chemicals can cause or contribute to a broad range of health

Green Chemistry
http://dx.doi.org/10.1016/B978-0-12-809270-5.00034-0

outcomes. These include eye, skin, and respiratory irritation; damage to organs such as the brain, lungs, liver or kidneys; damage to the immune, respiratory, cardiovascular, nervous, reproductive or endocrine systems; and birth defects and chronic diseases, such as cancer, asthma, or diabetes."[2]

The number of chemicals available on the market is increasing every year, consumption is rising across countries, and the international chemicals industry is growing dramatically. For example, there are currently more than 140,000 chemicals on the European Union (EU) market alone and another 30,000 are expected to be registered by 2018.[2] The global chemical industry output was valued at US$171 billion in 1970 and, in 2010, had increased more than 20-fold to US$4.12 trillion.[2] By 2020, chemicals are expected to represent one-third of the overall global consumption.[3]

Environmental and human exposure to chemicals occurs through the manufacturing process, the use of products containing chemicals, and the disposal of chemicals. Chemical pollution has become a transboundary issue as many hazardous substances are transported through air and water across the globe. Moreover, the disposal of hazardous wastes has also become an international concern. National regulatory systems are critical to the safe management of chemical substances but are often insufficient as trade volumes increase, opportunities for illegal dumping appear in places around the globe with weak regulatory systems, and the cost for chemical management in industrialized countries increases. As governments have noted, "the challenges posed by chemicals and wastes are global, enduring and constantly evolving and ... are interrelated with crucial environmental issues such as environment-dependent human health, the health of ecosystems and better ecosystem management, the preservation of biodiversity, and the link between poverty and environment, environmental disasters, climate change and sustainable consumption."[2,4]

For decades, the system of global environmental governance has been concerned with hazardous chemicals and pollutants and their management and effects on health and the environment. Through various international commitments, countries have consistently articulated a clear commitment to reduce the generation of toxic substances, improve their management, and reduce the environmental and health risks associated with them.[5–7] The Sustainable Development Goals (SDGs) connected the issue of chemicals and waste to various targets in terms of human health, water management, and sustainable consumption and production.[8]

In this chapter, we provide an overview of the international regulatory system for chemicals and waste and the mechanisms through which it regulates the production, use, and trade of chemical substances worldwide. We explain the origins of this system and its functions and examine its applications at the national level. We seek to provide the reader with a broad understanding of the international regulation on chemicals. Although chemicals regulation at the international level spans a number of specific treaties, from nuclear power to long-range transboundary air pollution, we cover in detail four issues that directly impact, and are informed by, green chemistry: persistent organic pollutants (POPs), hazardous chemicals, hazardous waste, and mercury. Green chemistry seeks to address many of the threats that chemical substances could pose to humans and ecosystems through the "design of chemical products and processes that reduce or eliminate the generation of hazardous substances."[9] Green chemistry offers alternatives to avoid deleterious impacts on the health of humans, animals, plants, and entire ecosystems, which corresponds to the objective of the treaties we analyze later. We conclude this chapter with a discussion of the ways in which the new SDGs deal with chemicals and their impacts on health and the environment.

3.29.1 THE INTERNATIONAL REGIME FOR REGULATION OF CHEMICALS AND HAZARDOUS WASTE

In 1962, with the publication of *Silent Spring*, Rachel Carson sounded an alarm about the then conventional wisdom on the benefits of industrial chemicals. The book presented a compelling case about the deleterious effects, especially on birds, of the use of pesticides and DDT [i.e., 1,1′-(2,2,2-trichloroethane-1,1-diyl)bis(4-chlorobenzene), or dichlorodiphenyltrichloroethane] in particular. It spurred widespread environmental activism; led to a ban on the use of DDT in agriculture in the United States, enabling the recovery of endangered species, such as the bald eagle; and catalyzed an environmental movement. It also triggered the creation of the US Environmental Protection Agency and the establishment of an environmental regulatory system and heightened international awareness about environmental concerns.

In the 1970s, the United States took the lead in the creation of the first international body for the protection of the environment—the United Nations Environment Programme (UNEP). In 1972, at the first United Nations Conference on the Human Environment in Stockholm, Sweden, governments recognized that human activities caused the degradation of important environmental resources and harmed the "physical, mental, and social health of man" (see Figs. 3.29.1 and 3.29.4).[6] They committed to preventing pollution (Principle 7 of the Stockholm Declaration) and collaborating with other states to address common challenges.

FIGURE 3.29.1 **Evolution of the system of global environmental governance.** The system of global environmental governance has evolved through global summits on environmental issues that served as the framework for designing global strategies toward sustainable development. One of these strategies is the articulation of global environmental conventions to address specific environmental problems such as chemical pollution and waste management or climate change.

FIGURE 3.29.2 **The International Environment House in Geneva (Switzerland).** Host of the executive secretariats of the chemicals conventions, the International Environment House, houses the secretariats of the chemicals conventions and the secretariat of the Convention on International Trade in Endangered Species (CITES), as well as other institutions working on environmental issues, including the UN Environment Programme and its chemicals branch (Figs. 3.29.2 and 3.29.3).

FIGURE 3.29.3 **Delegates meet in Geneva in May 2015 for the joint sessions of the Conferences of the Parties of the Basel, Rotterdam, and Stockholm Conventions.** Since 2010, and as part of the synergies process, the three conventions hold a joint session of their Conferences of the Parties and then back-to-back sessions on each convention to discuss issues related to their specific regulations. *Photo by IISD/ENB*

Twenty years later, at the second United Nations Conference on Environment and Development in 1992 (also known as the Rio Earth Summit), states agreed to "effectively cooperate to discourage or prevent the relocation and transfer to other States of any activities and substances that cause severe environmental degradation or are found to be harmful to human health" (Principle 14).[10] Over time, as new environmental issues emerged, the international

FIGURE 3.29.4 **Negotiations for the Minamata Convention.** On October 10, 2013, Nobuteru Ishihara, Minister of Environment of Japan; Achim Steiner, UNEP Executive Director; and local authorities from the city of Minamata marked the completion of the negotiations of the mercury convention.

community negotiated treaties to address each of them. These treaties gave rise to what is called the international chemicals and waste regime. In International Relations, a regime is a set of "principles, norms, rules, and decision-making procedures around which actor expectations converge in a given issue-area."[11] In broader terms, we understand regimes to be collections of norms and rules contained in documents such as international treaties, resolutions, and declarations, as well as the organizational framework that oversees and creates international regulation and convenes international participants. However, before delving into the specific norms and institutions, it is important to understand how the international chemicals regime emerged.

Recognizing that economic development has traditionally relied on the heavy use of chemicals and that these chemicals pose threats to critical ecosystem services, countries proposed specific strategies for the environmentally sound management of toxic chemicals.[5] Agenda 21, the plan of action established at the Earth Summit, acknowledged two major problems regarding chemicals and waste, particularly in developing countries: (1) lack of sufficient scientific information to assess the risks entailed by the use of chemicals and (2) lack of resources to evaluate the chemicals for which data is available.[5] The 2002 Plan for Implementation of the World Summit on Sustainable Development followed up on previous decisions and established as one of its objectives the minimization, by 2020, of the adverse effects of chemicals.[12] Throughout this period (1972–2002), these decisions resulted in two international treaties: the Basel Convention on the Transboundary Movement of Hazardous Waste[13] and the Stockholm Convention on Persistent Organic Pollutants.[14] The SDGs that governments adopted in 2015 maintain this intent and define different patterns for sustainable consumption and production, calling for the reduction and elimination of wastes and the control of hazardous substances across different policy areas.

Thus in a span of about four decades, the international community has been actively working toward bridging the science-policy gap, promoting international cooperation, and increasing awareness about environmental issues related to hazardous chemicals and waste. In addition to the broader political efforts to include chemicals in the international agenda, countries have also created specific policy instruments to address the challenge of hazardous substance management in an environmentally sound manner. According to an international lawyer Philippe Sands, there are four different approaches by which international agreements define hazardous substances and activities:

- by reference to their inherent characteristics, including their toxicity, flammability, explosiveness, and oxidization;
- by reference to a listing system that identified certain activities or projects on the basis that they are, per se, likely to have significant effects on the environment;
- by reference to national laws;
- by regulating specific substances, instead of establishing definitions.[15]

In the following, we focus on four international instruments for regulating chemicals and waste (see Table 3.29.1). These treaties span several decades, from the 1980s, with the Basel Convention, to the 2010s, with the Mercury Convention, and are particularly relevant for this chapter because they deal with issues important to green chemistry. Three global environmental conventions, Basel, Rotterdam, and Stockholm, regulate multiple aspects of the use, management, and disposal of hazardous substances through their life cycle. The latest one, the Minamata Convention, regulates mercury. Following a brief overview of each convention, we turn to their institutional arrangements, which comprise a secretariat and a decision-making body called the Conference of the Parties (COP).

3.29.1.1 Chemicals and Hazardous Waste Regulation

In the 1970s and 1980s, the number of sites for disposal of hazardous substances in industrialized countries was inadequate for the storage and safe treatment of the chemicals to be disposed. Sites in developing countries therefore became more appealing, as there were no domestic regulations in those countries to hinder acceptance of hazardous chemicals and no international regulation to prevent their transboundary movement. In addition, developing countries lacked proper disposal procedures and therefore became extremely vulnerable to hazardous substances.[16–18] Chemical pollution and other misuses of hazardous substances generated environmental harm, health consequences, and economic costs that called for the design of international mechanisms to regulate these procedures.[19,20] The movement of hazardous waste and the control of pollutant substances became some of the most contentious issues in global environmental governance.[21]

Increasing "international concern about environmental accidents and the discovery of toxic waste deposits in African countries imported from abroad that led to pollution of global resources"[22] brought momentum for the negotiation of different multilateral environmental agreements to address the challenge of managing, reducing, and eliminating chemicals and waste. Initially, UNEP decided to tackle the issue with regulation to stabilize the transportation and disposal of toxic wastes, leading to the establishment in 1985 of the Cairo Guidelines and Principles for the Environmentally Sound Management of Hazardous Wastes.[23] From there, a joint proposal by the governments of Switzerland and Hungary mandated the Executive Director to convene a working group for the elaboration of a global convention to control the transboundary movements of hazardous wastes. The *Basel Convention on the Transboundary Movement of Hazardous Waste* was signed, in the city that gives it its name, on March 22, 1989, with 53 original state parties and was entered into force in 1992.[13] The Basel Convention was at that point perceived as the most comprehensive solution to the problem of hazardous waste at different levels. The Convention "discourages exports of

TABLE 3.29.1 Environmental Agreements in the Chemicals and Waste Regime Related to Green Chemistry

	Year	Parties	Main Objectives
BASEL CONVENTION *the world environmental agreement on wastes*	1989	185	• Protect human health and the environment against the adverse effects of hazardous waste • Control of the transboundary movement of hazardous wastes as well as responsible trade in hazardous chemicals • Support for countries to strengthen their capacity for the sound management of chemicals and waste • Protect and preserve the marine environment from all sources of pollution • Promote safe radioactive and nuclear waste management. • Reduce the transboundary movement of hazardous wastes • Restrict those movements of hazardous wastes that are perceived in discordance with the principles of environmental sound management • Regulate the transboundary movements when they are permissible • Promote the environmentally sound management of hazardous wastes and adequate disposal activities
ROTTERDAM CONVENTION Share Responsibility	1998	156	• Promote shared responsibility and cooperative efforts among parties in the international trade of certain hazardous chemicals to protect human health and the environment from potential harm • Contribute to the environmentally sound use of those hazardous chemicals
STOCKHOLM CONVENTION	2001	180	• Regulate the sound management of chemicals throughout their life cycle, including persistent organic pollutants and heavy metals, and of waste
UNEP MINAMATA CONVENTION ON MERCURY	2013	35	• Protect human health and the environment from anthropogenic emissions and releases of mercury and mercury compounds • Ban new mercury mines and phaseout existing ones • Phaseout and phase down of mercury use in a number of products and processes • Control measures on emissions to air and on releases to land and water • Regulate the informal sector or artisanal and small-scale mining • Ensure the environmentally sound interim storage of mercury and its disposal once it becomes waste

Key global environmental conventions relevant to the field of green chemistry. Starting in 1989 with the Basel Convention on the Transboundary Movement of Hazardous Wastes, these conventions address the transport of hazardous substances and the management of specific pollutants. The Stockholm Convention addresses the phaseout of persistent organic pollutants (POPs) and the Minamata Convention tackles mercury. All of them have the common goal of reducing the harmful effects of chemicals on human health and the environment.

Basel Convention. UNEP/WG.180/3 report of the meeting – Ad Hoc working group of legal and technical experts with a mandate to Prepare a global convention on the control of the transboundary movements of hazardous wastes. *Budapest (Hungary): UNEP; 1987; Kummer K. The Basel Convention: ten years on.* Rev Eur Community Int Environ Law *1998;7:227–36; Porta M, Zumeta E. Implementing the Stockholm treaty on persistent organic pollutants.* Occup Environ Med *2002;59:651–2; Rotterdam Convention.* History of the negotiations of the Rotterdam convention. *Geneva (Switzerland)/Rome (Italy): UNEP/FAO; 2010; Stockholm Convention.* Overview of the Stockholm convention. *Geneva (Switzerland): UNEP; 2008; From United Nations Environment Programme.* Minamata convention on mercury; *2013.*

hazardous and other wastes, which should only be allowed if the exporting state does not have the capacity, facilities or suitable sites to dispose of them in an environmentally sound or efficient manner, or if the wastes are required as a raw material for recycling or recovery in the importing state, or in accordance with other criteria decided by the parties. Moreover, parties may not transfer to importing or transit states their obligation under the Convention to carry out environmentally sound management, and can impose additional requirements consistent with the Convention to better protect human health and the environment."[24] It also provides specific rules for the international movement and transport of waste, including packaging and labeling guidelines.

A key component of the regulations established by the Basel Convention is the Ban Amendment adopted by the second meeting of the COP in 1994. It was conceived as an instrument to address the challenges of developing countries and economies in transition regarding import of wastes that they were not prepared to receive and manage. The amendment (Annex VII) prohibits Organisation for Economic Co-operation and Development and the EU countries from all transboundary movements of hazardous wastes covered by the convention to countries not listed in that Annex if the intention is final disposal and disposal of specific wastes destined for reuse, recycling, and recovery operations.

An additional key step in the regulation of the movement of hazardous wastes had to do with information and notification processes among exporters and recipients of these substances. The 1998 Rotterdam Convention on Prior Informed Consent (PIC) procedure for Certain Hazardous Chemicals and Pesticides in International Trade[25] regulates this matter. Specifically, it creates legally binding obligations to ensure that governments respect certain rules in the distribution of chemicals, particularly having all the information required to assess and take informed decisions on export and import transactions. The overall objective of the convention is to facilitate information exchange and to promote shared responsibility and cooperation among parties in the international trade of hazardous chemicals.[26]

3.29.1.2 Persistent Organic Pollutants

Responding to increasing international concern about pollution and hazardous substances, in 1995, UNEP developed an international assessment of the effects of POPs. Based on the alarming results, in 1997, UNEP received a mandate to negotiate a binding international agreement to identify, regulate, and control the effects of POPs.[27–29] In 2001, after five negotiation rounds, 92 countries signed the *Stockholm Convention on Persistent Organic Pollutants (POPs)*, which entered into force in 2004.[14] The Stockholm Convention includes detailed provisions to eliminate POP releases and the associated risks. It also obliges countries to submit a National Implementation Plan that is to be included in the national sustainable development strategy, designs instruments to establish a clear route for countries to advance in the implementation of their commitments under the convention,[30] and recognizes the need to work with developing countries to strengthen their capacities to achieve this objective.

3.29.1.3 Mercury

The most recent development in the chemicals and waste regime has to do with the regulation of the anthropogenic emissions and releases of mercury and mercury compounds. These substances have long been identified as toxic to humans and other organisms. Back in the 1950s, a specific disease associated with the bioaccumulation of mercury in chemical waste that was transferred to seafood was identified as having caused mercury poisoning to the population of the Minamata Bay and the Shiranui Sea in Japan. Since then, more than 2000 victims have been identified. In 2003, different countries started calling for voluntary commitments to decrease emissions. But it was only in 2009 that UNEP adopted the decision "to initiate action to manage mercury in an efficient, effective, and coherent manner."[31] Through five intergovernmental negotiation committees, the chemicals branch of UNEP, Division of Technology, Industry and Economics, led countries to complete their negotiations and finalize the convention in 2013. The resulting *Minamata Convention on Mercury* also aims at protecting human health and the environment, in this case from the adverse effects of this toxic metal. It draws global attention to a substance that is broadly used and releases hazardous components in the atmosphere, soil, and water. The main objective of the convention is to control the anthropogenic releases of mercury throughout its life cycle. As of December 2016, the convention had 128 signatories and 35 countries had ratified it. It will enter into force after ratification from 50 countries.[32]

3.29.1.4 Institutional Framework

Each agreement within the chemicals and waste regime addresses a specific element of the larger environmental issue with the common goal of protecting human health and the environment. Each agreement has its own objectives and carries separate legal, political, and practical implications.[1,33] However, the four agreements also complement each other. The core of those interlinkages lies in the coordination of their functions, obligations, and objectives to guarantee that the overall goal of protecting human health and the environment is achieved. In the following, we explain the different parts of the institutional arrangements of the chemicals and waste agreements and how they work.

A key component of each convention, and consequently of the chemicals and waste regime, is the institutional arrangements established to support state parties in the implementation of each agreement.[34–36] Each convention establishes a secretariat, a COP, and other subsidiary bodies, which may include scientific or implementation bodies, for example.

3.29.1.4.1 Secretariats

When the Basel Convention was first drafted, Article 16 assigned to the parties the task of designating "the secretariat from among those existing competent intergovernmental organizations, which have signified their willingness to carry out the secretariat functions."[13] Furthermore, the Rotterdam Convention established a joint secretariat between UNEP and the United Nations Food and Agriculture Organization (Art. 19), whereas the Stockholm Convention (Art. 20) and the Minamata Convention (Art. 16) assigned this role to the executive director of UNEP.[14,25,32]

Each convention also established specific functions of its governing bodies and the extent of its interaction with other institutional arrangements established for the operation of each agreement. In case of the Basel Convention, for example, the role of the secretariat mainly focuses on information and coordination, serving as a bridge to inform countries about the management of hazardous substances and assisting countries in the process of implementing the convention, technology, assessment, and monitoring (Art. 16).[37] For the Rotterdam Convention, secretariat functions include assisting parties, ensuring coordination, developing reports, and establishing additional institutional arrangements to guarantee the operation of the agreement (Art. 19).[25] Similar functions were also established by the Stockholm Convention (Art. 20) and the Minamata Convention (Art. 16).[14,32]

3.29.1.4.2 *Conference of the Parties*

Conventions also establish a COP, a body that brings together all the countries that have signed and ratified the agreement and that constitutes the main decision-making body for all the regulations regarding each treaty. Each COP establishes its own schedule for meetings and the specific rules for operations. It can also establish other subsidiary bodies and define rules of procedure for operation and for financial participation of the parties under each convention. COPs are also the institutional body in charge of reviewing and evaluating the effectiveness of the conventions and of promoting the definition of national policies, strategies, and measures that enable parties to achieve the objectives of each agreement at the national level.

In some cases, the evolution of the agreements and of the issue areas they address requires the conventions to consider and adopt amendments and protocols to undertake additional actions to achieve the purposes of the conventions.[14,25] The COPs also analyze and approve these instruments, define strategies and mechanisms for cooperation, connect with international governmental and nongovernmental organizations, regulate the provision of information, and coordinate operations to guarantee the implementation of the conventions.[14] Serving as decision-making, information, dispute-settlement, and monitoring hubs, the COPs are critical governance mechanisms in environmental conventions. The conventions also have mechanisms for scientific data management and interaction with the scientific community. Those are discussed in the next section.

Considering these similarities, and recognizing the overarching common goal among the Basel, Rotterdam, and Stockholm Conventions of protecting human health and the environment and promoting sustainable development, in 2008/2009, governments mandated the three COPs to enhance the coordination and cooperation among the three conventions, to strengthen their implementation, and to increase coherence, efficiency, and effectiveness. This approach identified synergies and led to the adoption of different decisions to coordinate organizational, administrative, technical, informational, and decision-making practices and improve efficiency and implementation through joint activities.[38–40] A fundamental consequence of this process, also known as the synergies process, was the establishment of a joint executive secretariat to oversee the three agreements. The executive secretariat, based in Geneva, Switzerland, is currently led by Rolph Payet, former Minister for Environment and Energy of Seychelles.

3.29.1.4.3 *Strategic Approach to International Chemicals Management*

The Strategic Approach to International Chemicals Management (SAICM) is a key global policy framework that followed the discussions on this environmental issue in the Johannesburg Plan of Action[12] and the First International Conference on Chemicals Management,[41] and a consultative process with the actors involved in the programs associated with chemicals and their effects on human health and the environment. Representatives from relevant sectors called on governing bodies and organizations to incorporate a series of strategies into their objectives and mandates and requested UNEP to assume the administrative responsibility for this Strategic Approach. SAICM's main objective is to achieve the goal that, by

2020, "chemicals will be produced and used in ways that minimize significant adverse impacts on the environment and human health."[42] The spirit behind this strategy, outlined in the Dubai Declaration on International Chemicals Management and the Overarching Policy Strategy, is the need for coordinated action at the national level that supports chemicals management and safety concern in all countries and all types of economic sectors. Furthermore, it promotes effective governance for the management of chemicals, public participation, and accountability and acknowledges the relevance of this approach for sustainable development. "The sound management of chemicals," the Dubai Declaration establishes, "is essential if we are to achieve sustainable development, including the eradication of poverty and disease, the improvement of human health and the environment and the elevation and maintenance of the standard of living in countries at all levels of development."[41]

3.29.1.5 Scientific Input to the Chemicals and Waste Conventions

A key feature of the chemicals and waste regime is the relevance of scientific data for all the conventions. Science-based evaluations and management procedures are essential to identify hazardous substances and to manage, reduce, and eliminate them. Scientific information clearly contributes to the policy-making processes within the chemicals and waste regime, and conventions need assessment and monitoring mechanisms to support their efforts—with scientific evidence—for the sound management of hazardous substances, particularly in developing countries.[1,12] Beyond the assessment of the global state of chemicals and waste, science is also central to assess the effectiveness of the conventions and to evaluate the information submitted by state parties, which is often incomplete and unverified, specifically with regard to the assessment of inventories, stockpiles, generation, and movement of hazardous wastes and POPs.[43,44]

Part of the importance that each convention assigns to science is reflected in their institutional arrangements. Conventions establish subsidiary bodies that support their interaction with the scientific community; contribute to the monitoring, assessment, and evaluation functions of the different agreements; and bring science-based input to the decision-making processes. In the chemicals and waste regime, each convention has specific bodies to bridge the existing gap on technical and scientific data necessary to assess the presence of hazardous substances in the environment and the problem of pollution.

In the case of the *Basel Convention*, scientific functions rest with the Open Ended Working Group, one of the convention subsidiary bodies with the mandate to consider and advise the COPs on different issues in connection to the technical guidelines for the environmentally sound management of wastes, the distinctions between waste and nonwaste substances, the management of hazardous waste on land, and physiochemical treatment. The COP requires permanent access to available scientific information to achieve the purposes of the convention.

The *Stockholm Convention* is one of the agreements with strong relationship to scientific analysis. Scientific assessments form the basis for determining the presence of POPs and their effects on human health and the environment. Different bodies in the convention have the responsibility to address scientific uncertainty about the existence and consequences of chemicals in the environment and classify each substance. Furthermore, implementing the convention requires countries to strengthen their national scientific and technical research so that they can generate and exchange data and analysis on pollutant substances.

The Persistent Organic Pollutants Review Committee (POPRC) evaluates the listing of the substances under one of three annexes that indicate the treatment for each POP: elimination (Annex A), restriction (Annex B), and reduction (Annex C). The committee considers proposals submitted by the parties for listing new chemicals, examines them based on the criteria established by the agreement, develops a risk profile, and determines whether a specific chemical is likely to generate significant harmful effects for human health and the environment. After a pollutant has been determined as unhealthy, the committee takes into consideration the socioeconomic implications for state parties, conducts a risk management evaluation, and recommends to the COPs in which Annex it should be included. The COPs should take into account not only the recommendations of the POPRC but also any scientific uncertainty. When a chemical is listed, related control measures are to be established.[45] The 31 experts that form the committee are government-designated and confirmed by the COPs on the basis of equitable geographical distribution and the need to balance different types of expertise on chemical assessment and management.

The scientific subsidiary body of the *Rotterdam Convention* is the Chemicals Review Committee (CRC). It was established to review chemicals and pesticide formulations and make recommendations to the COPs on their management. When state parties notify the secretariat about chemicals that meet the requirements for PIC but such consent is not completed, the CRC has to review if the chemical should be subject to the PIC procedure. The CRC also reviews information provided on severely hazardous pesticide formulations proposed by developing countries or economies in transition as subject to the PIC procedure.

The latest chemical convention, the *Minamata Convention on Mercury*, has a strong scientific foundation. The convention establishes scientific information exchange as a critical factor to reduce or eliminate the production, use, trade, emissions, and release of mercury and other mercury compounds. Nonetheless, the actual structure of scientific institutional arrangements is still to be determined, as the convention has not yet entered into force.

3.29.2 IMPLEMENTING THE BASEL AND STOCKHOLM CONVENTIONS

As international environmental agreements operate in a system without hierarchical authority, there is no direct effort to enforce them. It is therefore critical to understand how they perform as instruments of global governance and how they contribute to the solution of the environmental problems they are designed to address. The existing literature offers no definitive arguments or evidence about the extent of implementation of international environmental agreements. Although some experts argue that conventions are effective instruments,[46–48] others contend that they are not very effective in resolving the problems they were designed to address and are highly dependent on countries' capacity, political will, and resources.[11] The impact of international conventions can only be measured through the actions of the member states and it is necessary to understand the extent to which state parties are "domesticating" the conventions, complying with the expectations as signatories and adopting regulations to facilitate implementation.[47,49–54] In the chemicals and waste regime, national laws need to be enacted to comply with international requirements.

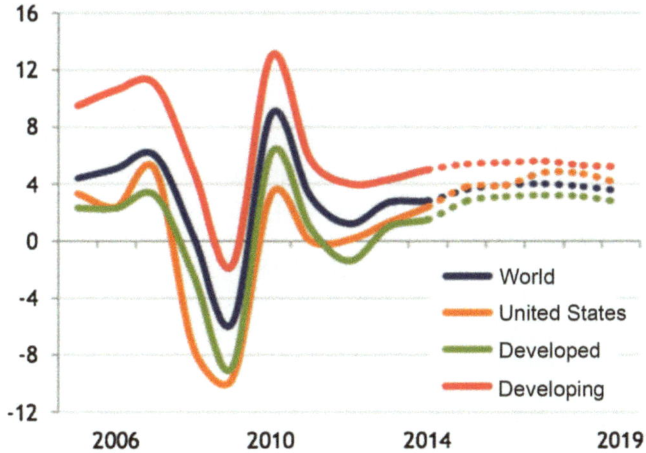

FIGURE 3.29.5 **Global chemistry business output percentage of annual change.** The global chemistry business is growing significantly in developing countries and in the United States. Developed countries are projected to maintain a stable change rate. Importantly, chemicals production has shifted from developed countries to countries such as Brazil, Russia, India, China, and developing states more broadly as sales of new types of chemicals have also doubled. *UNEP. Global environmental outlook 5. Nairobi: United Nations; 2012 and American Chemistry Council. Year-end 2014 chemical industry situation and outlook, p. 5. http://files.clickdimensions.com/americanchemistrycom-avo5d/files/year-end2014situationandoutlookf6c2.pdf.*

As Fig. 3.29.5 illustrates, the output of the chemistry business has increased most in developing countries and regulation and legislation need to keep pace to develop national and local standards, reflect the reality of national chemical industries, and establish monitoring, inspection, reporting, and enforcement mechanisms. Conventions also require national action on the appointment of focal points and other institutional and strategic arrangements.

There is no empirical evidence about the extent to which signatory parties to any of the international conventions are acting on their obligations at the national level. Indeed, measuring implementation is a particularly difficult task because of the vagueness of legal obligations and the lack of regular and reliable reporting. Common understanding of terminology simply does not exist, and key concepts for assessing the state of the environment and of the effectiveness of measures taken are missing.[55,56] Only recently, the Center for Governance and Sustainability at the University of Massachusetts Boston launched a research project, the Environmental Conventions Initiative, to address this gap and offer analytical input on the implementation of six global environmental conventions.[a] Drawing on data from the research effort, this section offers some insights into implementation of the Basel and Stockholm conventions.[b]

[a]This project assesses the extent to which countries implement their commitments under international environmental conventions and is currently under development. The results presented here are therefore preliminary. For more information, see http://environmentalgovernance.org/research/environmental-conventions-initiative/.

[b]Data for the other conventions in the chemicals and waste cluster is not included, as they are not part of the group of conventions studied by the Environmental Conventions Initiative.

To understand implementation, it is important to first define the signatory parties for each convention. Legislation, strategies, institutional arrangements, and monitoring efforts differ among developed and developing countries. Furthermore, the nature of the problem of chemicals production and pollution also varies across different countries and regions. According to UNEP's flagship environmental assessment GEO-5, the production of chemicals has shifted in the past decade, moving from developed countries to the BRIC countries (Brazil, Russia, India, and China) and other developing countries that are also increasing sales and the generation of new chemicals.[3,57]

Membership in the Basel Convention reached 185 countries in 2015. About 26% of them are developed countries and 74% are developing countries. The Rotterdam Convention reached 156 state parties in early 2016, in which 27% are developed countries and 73% are developing countries. In a similar pattern, 24% of the 180 parties to the Stockholm Convention are developed countries and 76% are developing countries. Figs. 3.29.6 and 3.29.7 shows the evolution of membership in each of the conventions. The Stockholm Convention had approximately three times more original signatories than the Basel Convention, and twice as many as the Rotterdam Convention, illustrating the evolution of the regime from the 1990s when the Basel and Rotterdam conventions entered into force and increased the awareness of the international community about the importance of international cooperation to address chemical pollution and wastes. Expanding membership also confirms the success

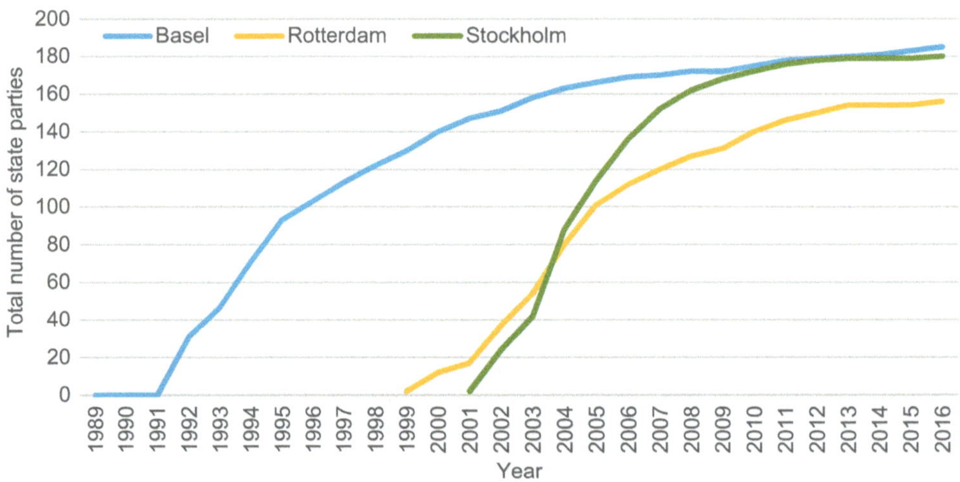

FIGURE 3.29.6 **Historical evolution of the membership in the Basel and Stockholm Conventions.** The number of countries parties to the Basel, Rotterdam, and Stockholm Conventions has increased since the conventions were signed in 1989, 1998, and 2001, respectively. In the case of the Basel Convention, 42 countries signed the convention originally and ratifications started 3 years later (1992) going from 31 to the current 185 state parties.[90] Sierra Leone was the last country to join the convention in November 2016. For the Rotterdam Convention, 53 countries were the original signatories of the convention. Slovenia was the first country to ratify the convention in 1999, and the number of state parties has since increased to 156. Sierra Leone was the last country to ratify the convention as of the writing of this chapter in 2016. For the Stockholm Convention, ratifications started sooner going from 2 the same year that the convention was concluded (2001) to the current 180 state parties. Iraq was the last country to join the convention in March 2016.[91]

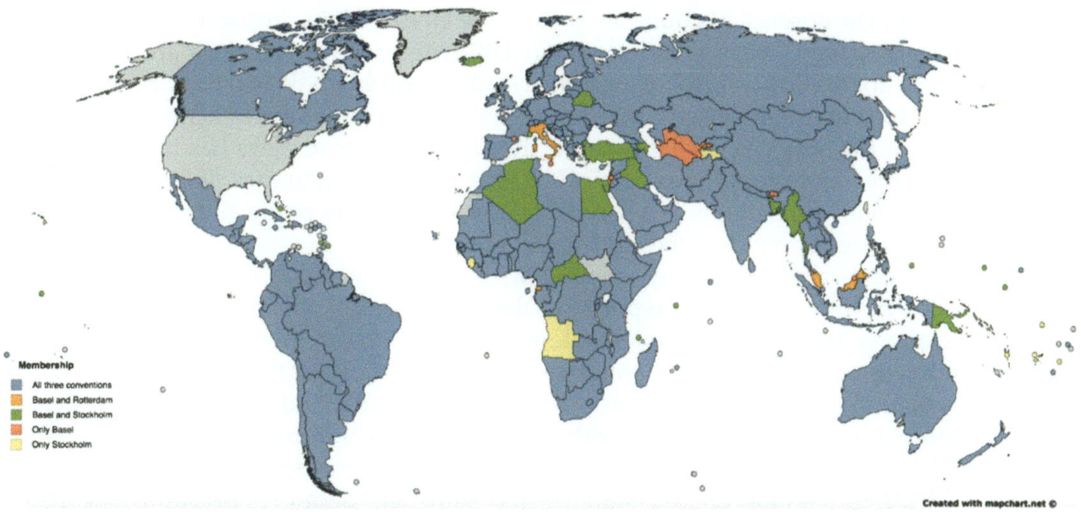

FIGURE 3.29.7 **Status of ratifications of the Basel, Rotterdam, and Stockholm Conventions.** Map indicates the status of each country regarding the ratification of the Basel, Stockholm, and Rotterdam Conventions. Although the majority of countries are parties to the three conventions, there are important gaps in membership. Some countries are not members of any of the three agreements. Others are not involved in one or two of the agreements. Asymmetry in membership presents challenges at the international level, as some countries can be seen as free riders and at the national level where harmonized pollution policies will be harder to achieve.

of the 2001 Johannesburg Plan of Action[12] in its objective to increase the ratification and implementation of the conventions.[c] Given the transboundary nature of the threat of wastes and pollution, it is critical that all countries engage in the regulation of chemicals and waste.

All the conventions, however, have notable gaps in membership. Seven countries are not members in either convention: Grenada, Haiti, San Marino, South Sudan, Timor-Leste, the United States, and the Vatican City. Other nonmembers include Angola for the Basel Convention; Algeria, Egypt, Iceland, Turkmenistan, and Uzbekistan for the Rotterdam Convention; and Israel, Italy, Malaysia, and Uzbekistan for the Stockholm Convention. The absence of some countries then leads to lower levels of implementation. Italy, for example, shows lower implementation of the Stockholm Convention when compared to other European countries,[58] and ratifying the convention will be critical to improve national chemicals regulations. Other countries, such as Malaysia, have designed and executed policies consistent with the issues addressed by the conventions but have not ratified the agreements yet, which leads to illegal use of some substances listed as POPs and their detection in the environment.[59] Yet others, such as the United States, possess advanced cradle-to-grave hazardous waste management systems and have signed both conventions but have not ratified, largely because of domestic

[c]Since 2001, 38 countries have joined the Basel Convention, which represents 21% of its current membership. About 95% of those new members are developing countries. For the Rotterdam Convention, 139 countries joined after 2001, which represents 89% of its current membership, and 75% of those countries are developing. Membership in the Stockholm Convention has also increased. A total of 156 countries ratified the convention after 2002, in which 79% (123) are developing countries and 21% (33) are developed.

political gridlock. The absence of countries such as the United States impacts the legitimacy of the conventions and limits their global coverage. Ultimately, the conventions are instruments of international cooperation to resolve a particular global problem. When the United States is not a party, it does not engage with the larger community of countries acting to achieve a common goal. At the national level, however, the United States already has in place regulations that attain the objectives of the chemicals conventions. The reticence of the United States to ratify international agreements is a direct consequence of the political deadlock between the executive and legislative branches in the US government.

The fundamental measure of implementation is reporting. Each convention secretariat requires parties to submit national reports on the fulfillment of their obligations under the agreement. National reporting on implementation is critical to monitor compliance and effectiveness and to assess the status of chemicals in each country. Both the Basel and Stockholm Conventions have specific reporting systems, requesting annual and periodic (every 4 years) reports, respectively. In both cases, reports include specific information on the measures taken to implement the convention, the effectiveness of those measures, designation of focal points to address convention-related matters, and statistical data on hazardous substances production, import, export, movement, and impact on human health and the environment.[13,14] Reporting rates for the conventions are relatively low. Not all countries submit the national reports they are required to submit, and, of the ones that do, some delay submission, inhibiting the prompt availability of data to assess performance. In the Basel Convention, countries have reported on average 50% of the time they were required to report since 2001, whereas for the Stockholm Convention they have only fulfilled this obligation 38% of the time since 2004.[60,61] Only 22 countries have a 100% reporting rate for the Basel Convention. Most of them (16) are developed countries, but Bahrain, Barbados, Madagascar, Malaysia, Singapore, and Thailand are also part of this group. However, 21 countries, all of them developing, have never submitted a report. For the Stockholm Convention, only 32 countries (18% of the members) have submitted all the reports they were required to submit, including the Central African Republic, Costa Rica, Mali, and Nepal, whereas 72 countries (40% of the parties) have never submitted a report.

Historical analysis of reporting behavior illustrates a key challenge in the process of implementation. For the Basel Convention, the number of countries submitting a report each year has decreased from 74% in 2001 to 39% in 2013. The Stockholm Convention exhibits a more positive trend, with countries' reporting increasing from 39% in 2002–06 to 56% in 2006–10. However, there is still a significant group of countries for which data is not available, and it includes both the developed and developing countries.

Regarding the implementation of the rest of the obligations under the agreements, a group of countries including Switzerland, Austria, Belgium, Canada, Japan, Finland, Norway, Australia, and the Czech Republic have implemented most of the obligations for control of the transboundary movement of hazardous waste and control and eradication of POPs. Several developing countries have also achieved positive results. In the case of the Basel Convention, countries such as Colombia, Malaysia, Morocco, and Tunisia have consistently reported their success on implementing the obligations regarding the legal definitions of the different types of hazardous wastes and the control of their harmful effects on human health and the environment. In the case of the Stockholm Convention, developed countries seem to have stronger performance. However, countries such as The Gambia, El Salvador,

Zambia, Chile, and Colombia evidence positive results despite some challenges in the reporting process.

Analysis of implementation in both conventions points to causality between countries' level of development and their compliance with the internationally agreed goals on the environmentally sound management of chemicals and waste. Developing countries seem to be struggling the most to translate their international commitments into national laws. In the case of the Basel Convention, countries such as Barbados, Equatorial Guinea, Turkmenistan, Saint Vincent and the Grenadines, Nauru, and Burkina Faso consistently report failure in putting in place the necessary domestic measures required to comply with the convention, due in part to low state capacities and resource availability. Similarly, Barbados, a country that has reported to the convention 100% of the time, shows significant difficulties in implementation, which raises questions about the connection between the complex, transboundary nature of the environmental threat and the difficulties of implementation for many developing countries.[62] Detailed analysis of the country's national reports shows that Barbados informs about low progress on obligations regarding legislation, information, and regulation of transactions in hazardous wastes, as it fails to submit information of the actual generation, export, and import of these substances. There is a clear need for systematic and scientific monitoring mechanisms to support developing countries in fulfilling all obligations under the convention.

In the case of the Stockholm Convention, most countries with poor results are developing countries, including Myanmar, Malawi, Ethiopia, Cote d'Ivoire, Cameroon, Cambodia, Laos, Central African Republic, and Mauritania. Collection of and access to detailed scientific information is required to identify, regulate, and eliminate POPs,[1] and there is an important gap that undermines developing countries' potential to fully implement their obligations under the Stockholm Convention. However, countries such as Russia, the United Arab Emirates, and Monaco also register low performance; such cases demand closer examination and analysis, and perhaps greater openness about the challenges these countries face.

The evolution of the process of implementation for both the Basel and the Stockholm conventions reflects multiple realities not only for specific countries but also for the conventions' institutional structures and leadership. First, it is evident that additional efforts are required to improve compliance with national reporting. Second, implementation results call for analysis to determine which factors, besides the level of development, act as the main obstacles to progress, so that both conventions and countries can address them. Data are also required to connect the definition of national policies with the effectiveness of the conventions. According to the Global Chemicals Outlook, for example, more than 70% of the countries that have submitted information on hazardous waste generation to the Basel Convention reported an increase of 12% in these substances.[2] Furthermore, even though developing countries and economies in transition are decreasing the amount of hazardous waste imported, the amount exported has increased considerably. However, as data are incomplete, additional information is necessary to evaluate the extent to which the conventions are effectively addressing the threat of chemical pollution and its effects on human health and the environment.

Environmental conventions have defined "high-priority substances" for countries to establish national policies and baselines.[2] However, existing disparities in the implementation across types of countries and across regions call for "chemicals policy instruments and approaches that are appropriate to the economic conditions and strategies" of specific countries.[2] Additional research will help determine best practices that can be replicated and the type of information collection mechanisms and integrated approaches at the national level to implement the required chemicals policies.

As the chemicals and waste conventions advance in the process to develop joint, synergistic operations at the global and national levels, it is important to determine the extent to which countries are following the guidlines and objectives established by the conventions. Greater cooperation and coordination between the chemicals and waste conventions provide an opportunity for capacity building, knowledge transfer, enhanced awareness, and efficiency, as well as improved implementation. Furthermore, the chemicals and waste regime is being integrated with other global agendas. The SDGs—specifically Goal 12 "Responsible Production and Consumption"—aim to achieve the environmentally sound management of chemicals and all wastes throughout their life cycle, a goal that goes hand in hand with the framework of the conventions. Attaining this target will require the full implementation of the conventions, which now have both the opportunity and the responsibility to coordinate their efforts and integrate chemical management strategies in ways that contribute to the ultimate objective of sustainable development at the global, national, and local levels.

3.29.3 INTERNATIONAL CHEMICALS AND WASTE REGULATION AT THE NATIONAL LEVEL: COUNTRY CASE STUDIES

Countries translate their global environmental commitments into national policy and strategies through regulations, initiatives, and institutional arrangements with the ultimate goal of protecting human health and the environment from the pervasive effects of hazardous substances. Domestic legislation and policy on chemicals management specify the norms and procedures to be followed by persons and companies, as well as government agencies in the national context. In this section, we address examples of national policies that draw attention to some of the characteristics of the global regime. The case studies were selected to exemplify some key characteristics of the regime, such as relevance of scientific information and interagency cooperation, and the range of different factors that influence the translation of the global commitments into national law and policy. The case of the United States demonstrates how a country acts on chemicals and waste regulations without having ratified the agreements.

3.29.3.1 Canada: Using the Input of the Scientific Community and Other Stakeholders

Canada bases its national legislation on chemicals on the "Chemicals Management Plan" (CMP), an initiative designed to reduce the risk posed by chemicals. Originally, the plan identified 4300 chemicals as priorities. The first stage (2011–16), divided into three phases, aimed to address approximately 1700 substances. Currently, in the third phase of the first stage, the plan includes 1550 substances. The goal is to address all the chemicals by the year 2020.[63,64]

An important element of the Canadian approach to chemicals management is the significant input from the scientific community, so as to ensure a strong scientific foundation for the management of chemicals and hazardous waste. A CMP Science Committee, created by the federal offices in charge of Health and Environment and Climate Change, brings together experts on the scientific considerations for the implementation of the CMP. Members of the committee are selected based on their expertise and do not represent their individual affiliations or interests.[65] The plan also establishes a Stakeholder Advisory Council. Stakeholders can offer advice and input to the government on the implementation of chemicals regulation and foster dialogue on issues regarding risk assessment, risk management, risk monitoring, communications, and integrated activities that contribute to the policy goals that the government has established.[66,67] Members represent national aboriginal organizations, consumer groups, nongovernmental organizations, and industries. The Stakeholder Committee interacts with government officials periodically to review the progress reports on the implementation of the CMP. It discusses risk management, designs outreach strategies to increase public participation in chemicals management, and brings into the analysis and policy process issues that should be connected to chemicals strategies such as climate change and human rights.[67]

3.29.3.2 European Union: Regional Approaches to Chemicals Management and Safety

In the EU, the European Chemicals Agency (ECHA) is in charge of the management of chemicals and hazardous wastes.[68] The agency is at the center of the regulatory authorities implementing chemicals legislation in each of the 28 countries of the Union. As an institution, ECHA supports governments and companies to comply with the legislation, advances the safe use of chemicals, provides information on chemicals, and addresses chemicals of concern. Two initiatives are particularly important.

The REACH (Regulation for Registration, Evaluation, Authorisation and Restriction of Chemicals) regulatory program aims at protecting human health and the environment while enhancing the competitiveness of the EU chemicals industry. It establishes the registration, evaluation, authorization, and restriction processes to manage unambiguous substances and regulate their use within the European market. The initiative also defines alternative methods for the hazard assessment of substances.[69] In principle, the REACH program is applicable to all chemical substances and places the burden of proof on companies, which must demonstrate how substances can be safely used, and the risk management measures to the users. If risks cannot be managed, authorities can restrict the use of substances in different ways.[70]

Another key characteristic of the EU chemicals regulation is the integration with consumers and workers' safety standards. The Classification, Labeling and Packaging Regulation guarantees that before chemicals are placed on the market, industries should identify the potential risks to human health and the environment, classify them, label them, and inform workers and consumers about their effects before they handle them.[71,72] The EU also has its own PIC Regulation administering the transboundary movement of hazardous chemicals to non-EU countries,[72] based on the regulations established by the Rotterdam Convention.

3.29.3.3 Colombia: Calling for Additional Knowledge and Technical Capacity

After becoming party to the Basel Convention in 1989, and in the process of ratifying the convention, Colombia approved Law 253 (1996) to acquire specific obligations regarding trade, management, reduction, and eradication of hazardous waste.[73] This complemented previous guidelines and regulations developed before the convention and the measures established by the 1991 Constitution stating that no individual or organization had the right to dispose of hazardous substances or chemicals into the environment.[74] A core component of the legislation in the chemicals and waste regime relates to water management policies, specifically regulating water pollution and its negative effects on biodiversity and other ecosystems such as wetlands.[75] Colombia was also one of the latest state parties to ratify the Ban Amendment to this convention in 2015.

The country, however, exemplifies some of the different challenges in the implementation process faced by developing countries and emerging economies. One is the balance between the need for economic expansion and the definition of policies, strategies, and technical infrastructure for the adequate management of chemicals and waste.[74] For example, the current debate about the ratification of the Minamata Convention on Mercury brings up the pressure for expansion of the mining industry, one of the pillars of economic development and growth.[75] In addition, even when the country is aware of the impact and risks of inadequately managing hazardous waste, it also acknowledges that scientific input and knowledge of their production, composition, and impact in the medium and long term are limited and need to be improved.[75]

3.29.3.4 United States: Nonparties and Their Domestic Regulation on Chemicals and Waste

The United States is not a party of the Basel, Stockholm, or Rotterdam conventions. However, the national legislation on chemicals and waste is well developed and often seen as more stringent than the regulations of these international treaties.[76,77] Specifically, regarding the transboundary movement of hazardous waste, previous agreements between countries are required to trade these substances between the United States as a nonparty and parties to the Basel Convention. Separate agreements have been established with Canada, Mexico, Costa Rica, Malaysia, and the Philippines for importing and exporting waste. In the case of Canada, the agreement aims, among other goals, to develop low-cost options for waste management in the cases in which either country lacks the domestic capacity or technology required to appropriately handle these substances. Agreements with developing countries also specify that the United States may receive their waste for disposal or recycling but may not export waste to these countries.

In the case of POPs, the Stockholm Convention has influenced US policies to control chemicals, helping the country's efforts to reduce the generation of these substances. The US government has also developed bilateral efforts with Canada and the UN Commission for Europe to eliminate persistent toxic substances included in the convention and others that are not regulated under the convention. The United States also provides financial and technical support to other countries for POP reduction.

Perhaps somewhat surprisingly, the United States was the first country to sign and ratify the Minamata Convention on Mercury.[78] Global sources are estimated to cause 70% of the mercury deposited in the United States. Thus, even when the United States is significantly decreasing its use and emissions of mercury, its domestic efforts are not sufficient to significantly reduce mercury pollution. The country also engages the scientific community in the development of an understanding of the effects of mercury pollution on human health, the measuring and monitoring of mercury presence, the cleaning up of contaminated sites, and the development of technologies to prevent mercury emissions into the air.[78] Importantly, however, the ratification of the Minamata Convention did not require the approval of Congress and the Senate. Because it contained provisions that were already part of domestic laws in the United States, the president could ratify the convention by executive order.

3.29.3.5 Brazil: Legislation and Interagency Partnerships for Implementation

In the area of chemicals and waste, Brazil's Ministry of Environment has developed a National Policy of Chemical Safety, which is to be fully implemented by 2021. The main objective of this initiative is to guarantee that chemicals are produced and used responsibly, minimizing the harmful effects on the environment and human health. Specific instruments have also been established to address the commitments defined by each convention.

Brazil ratified the Basel Convention in 1992 and has developed a range of legal instruments, including specific standards to regulate waste movement. Brazil also plans to prevent and respond to accidents involving hazardous chemical products.[57] Recently, the country approved its National Policy for Solid Waste for the prevention and reduction of waste generation and the promotion of sustainable consumption.[57,79]

The Stockholm Convention is also important for Brazil, given the significance of the agricultural sector for the country. Brazil joined the agreement in 2001 and ratified it in 2004. Since then, the Ministry of Foreign Affairs has been working in partnership with other participants on implementation. The country has established technical groups with representatives from the federal and the state governments in agriculture, health, and environment; nongovernmental organizations; industrial associations; and academia. These technical groups work on the definition of the country's National Implementation Plan, develop the national inventory of sources and emissions of POPs, and identify areas that require plans of action to fulfill the regulations established by the convention.

Despite some challenges in implementation and enforcement, Brazil is an example of the importance of national laws to consolidate the country's international commitments in the chemicals and waste regime.[80]

3.29.4 SUSTAINABLE DEVELOPMENT GOALS: FUTURE FOR CHEMICALS AND WASTE REGULATION?

Since its inception, the chemicals and waste regime has been at the core of the international agenda to address the threat of hazardous wastes and pollution. The conventions discussed in this chapter have achieved important progress in addressing not only the different threats but also awareness about the harmful effects of pollutant substances for health and the environment. However, there is still work to be done in four core areas.

First, conventions need to assess and show some progress in the achievement of the existing global policy objectives. Otherwise, state parties and stakeholders will question the role of the regime not only as a governance instrument but also in the protection of human health and the environment.[3] Second, provision of assistance is critical to the ability of developing countries to strengthen their national capacities and implement the agreements to achieve the objectives of the conventions.[3,81] Third, to reduce the level of uncertainty that negatively affects implementation and evaluation effectiveness, convention secretariats and parties need to work with the scientific community and academia in the collection and generation of quality data and information regarding the uses, emissions, and effects of chemicals.[3,17,82] Finally, the connection between the conventions and the global sustainable development agenda should be deepened and strengthened through the implementation of the SDGs.

In 2015 the United Nations adopted the 2030 Sustainable Development Agenda and 17 ambitious and universal global goals—the SDGs. The SDGs aim to address the challenges of the economic, social, and environmental dimensions of development in an integrated and comprehensive manner. Pollution control and the environmentally sound management of chemicals and wastes are connected to several goals in areas such as health, water sanitation, the protection of life on land and oceans, and the sustainability of cities. However, the chemicals conventions are particularly relevant to SDG12, Responsible Production and Consumption, and are explicitly included in the specific targets for the goal. Target 12.4 posits that countries should "By 2020, achieve the environmentally sound management of chemicals and all wastes throughout their life cycle, in accordance with agreed international frameworks, and significantly reduce their release to air, water and soil in order to minimize their adverse impacts on human health and the environment."[8]

Ultimately, implementing the chemicals and waste conventions will be the key not only to achieve their overall objectives but also to contribute to the attainment of the SDGs. Scholars argue that the challenges of the conventions include both ratification and implementation[83] and that even when levels of ratification are positive, the chemicals and waste agreements have not "always transposed into national legislation in a comprehensive manner" and countries still "face considerable difficulties establishing effective policies and administrative structures for managing hazardous chemicals."[1,3,84] Factors affecting the level of implementation include the level of development, the level of technical capacity, the debate between sovereignty and cooperation, and the fact that the policies do not seem adaptable to the different national characteristics of state parties.[19,21,83,85,86] Only by addressing the root causes of low implementation and noncompliance will it be possible to achieve comprehensive regulation—using scientific knowledge as a foundation—to fully translate global environmental commitments into national policies.

References

1. Selin H. *Global governance of hazardous chemicals*. MIT Press; 2010.
2. UNEP. *Global chemicals outlook: towards sound management of chemicals*. Nairobi: United Nations; 2013.
3. UNEP. *Global environmental outlook 5*. Nairobi: United Nations; 2012.
4. UNEP. *UNEP/GC.27/17 Proceedings of the governing council/global ministerial environment forum at its first universal session*. Nairobi, Kenya: United Nations; 2013.

5. UNCED. *Agenda 21*. New York: United Nations; 1992.
6. United Nations. A/CONF.48/14 Declaration of the United Nations conference on the human environment (Stockholm declaration). In: *United Nations conference on the human environment*; 1972. Stockholm.
7. United Nations. A/CONF.199/20 Johannesburg declaration on sustainable development. In: *World Summit on Sustainable Development*; 2002.
8. UN General Assembly. *A/RES/70/1 Transforming our world: the 2030 agenda for sustainable development*. New York, NY: United Nations; 2015.
9. US Environmental Protection Agency. *Basics of green chemistry*. 2016.
10. United Nations. A/CONF.151/26 Rio declaration on environment and development. In: *United Nations Conference on environment and development*; 1992. Rio de Janeiro.
11. Krasner SD. *International regimes*. Ithaca: Cornell University Press; 1983.
12. WSSD. *A/CONF.199/L.1 Draft plan of implementation of the world summit on sustainable development*. Johannesburg, South Africa: United Nations; 2002.
13. UNEP. *Basel Convention on the control of transboundary movements of hazardous wastes and their disposal*. 1989.
14. United Nations. *Stockholm convention on persistent organic pollutants (POPs)*. 2001.
15. Sands P, Peel J. *Principles of international environmental law*. 3rd ed. Cambridge University Press; 2012.
16. Critharis M. Third World nations are down in the dumps: the exportation of hazardous waste. *Brook J Int'l L* 1990;**16**:311.
17. Krueger J. The Basel Convention and the international trade in hazardous wastes. In: Stokke OS, Thommessen ØB, editors. *Yearbook of international co-operation on environment and development*. London (United Kingdom): Earthscan; 2001. p. 43–51.
18. Lucier CA, Gareau BJ. Obstacles to preserving precaution and equity in global hazardous waste regulation: an analysis of contested knowledge in the Basel Convention. *Int Environ Agreements Polit Law Econ* 2014:1–16.
19. Krueger J. What's to become of trade in hazardous wastes?: the Basel Convention one decade later. *Environ Sci Policy Sustain Dev* 1999;**41**:10–21.
20. Waugh T. Where do we go from here: legal controls and future strategies for addressing the transportation of hazardous wastes across international borders. *Fordham Environ Law Rev* 1999;**11**:477.
21. Krueger J. Prior informed consent and the Basel Convention: the hazards of what isn't known. *J Environ Dev* 1998;**7**:115–37.
22. Basel Convention. *Overview of the Basel Convention*. Geneva (Switzerland): UNEP; 2011.
23. UNEP. *UNEP/GC decision 14/30 Cairo guidelines for the environmentally sound management of hazardous waste*. Nairobi (Kenya): United Nations; 1987.
24. Sands P. *Principles of international environmental law*. 2nd ed. Cambridge University Press; 2003.
25. FAO/UNEP. *Rotterdam convention on the prior informed consent procedure for certain hazardous chemicals and pesticides in international trade*. 1998.
26. Rotterdam Convention. *History of the negotiations of the Rotterdam convention*. Geneva (Switzerland)/Rome (Italy): UNEP/FAO; 2010.
27. Stockholm Convention. *Overview of the Stockholm convention*. Geneva (Switzerland): UNEP; 2008.
28. Hagen PE, Walls MP. The Stockholm Convention on persistent organic pollutants. *Nat Resour Environ* 2005:49–52.
29. UNEP. *UNEP/GC Decision 19/13C international action to protect human health and the environment through measures which will reduce and/or eliminate emissions and discharges of persistent organic pollutants, including the development of an international legally binding instrument*. Nairobi (Kenya): United Nations; 1997.
30. Lu Y, Giesy J, Holliday L. *Implementing the Stockholm convention on persistent organic pollutants*. Washington, DC: The National Academies Press; 2007.
31. UNEP. *UNEP/GC.25/5 Decision 25/5: chemicals management, including mercury*. Nairobi (Kenya): United Nations; 2009.
32. UNEP. *Minamata convention on mercury*. 2013.
33. Krueger J, Selin H. Governance for sound chemicals management: the need for a more comprehensive global strategy. *Glob Gov* 2002:323–42.
34. Abbott KW, Snidal D. Why states act through formal international organizations. *J Confl Resolut* 1998;**42**:3–32.
35. Barnett MN, Finnemore M. The politics, power, and pathologies of international organizations. *Int Organ* 1999;**53**:699–732.
36. List M, Rittberger V. The role of intergovernmental organizations in the formation and evolution of international environmental regimes. In: Underdal A, editor. *The politics of international environmental management*. Strasbourg (France): Springer; 1998. p. 67–82.

37. Basel Convention. *UNEP/WG.180/3 report of the meeting — Ad Hoc working group of legal and technical experts with a mandate to Prepare a global convention on the control of the transboundary movements of hazardous wastes*. Budapest (Hungary): UNEP; 1987.
38. Basel Convention. *COP decision BC VII/8 cooperation and coordination between the Basel, Rotterdam and Stockholm conventions*. Nairobi (Kenya): UNEP; 2006.
39. Stockholm Convention. *COP decision SC 4/34 enhancing cooperation and coordination among the Basel, Rotterdam and Stockholm conventions*. Geneva (Switzerland): UNEP; 2006.
40. Rotterdam Convention. *COP decision RC 4/11 enhancing cooperation and coordination among the Basel, Rotterdam and Stockholm conventions*. Rome (Italy): FAO; 2008.
41. ICCM. *Dubai declaration on international chemicals management*. Dubai (UAE): United Nations; 2006.
42. UNEP. *Strategic Approach to International Chemicals Management: SAICM texts and resolutions of the International Conference on Chemicals Management*. Geneva (Switzerland): UNEP; 2006.
43. Basel Convention. *UNEP/CHW.10/28 report of the conference of the parties to the Basel Convention on the control of transboundary movements of hazardous wastes and their disposal on its tenth meeting*. Cartagena (Colombia): UNEP; 2011.
44. Stockholm Convention. *UNEP/POPS/COP.5/36 report of the conference of the parties to the Stockholm convention on persistent organic pollutants on the work of its fifth meeting*. Geneva (Switzerland): UNEP; 2011.
45. Stockholm Convention. *Persistent organic pollutants review committee (POPRC)*. Geneva (Switzerland): UNEP; 2008.
46. Crossen TE. Multilateral environmental agreements and the compliance continuum. *Bepress Legal Series* 2003.
47. Chayes A, Chayes AH. On compliance. *Int Organ* 1993;**47**:175—205.
48. Downs GW, Rocke DM, Barsoom PN. Is the good news about compliance good news about cooperation? *Int Organ* 1996;**50**:379—406.
49. Hasenclever A, Mayer P, Rittberger V. *Theories of international regimes*. Cambridge; New York: Cambridge University Press; 1997.
50. Jacobson HK, Brown-Weiss E. Compliance with international environmental accords: achievements and strategies. In: Rolen M, Sjöberg H, Svedin U, editors. *International governance on environmental issues*. Dordrecht; Boston, MA: Kluwer Academic Publishers; 1997. p. 78—110.
51. Simmons BA. International law and state behavior: commitment and compliance in international monetary affairs. *Am Political Sci Rev* 2000;**94**:819—35.
52. Young OR. *Compliance and public authority : a theory with international applications*. Baltimore, MD: Johns Hopkins University Press; 1979.
53. Jacobson HK, Brown-Weiss E. Strengthening compliance with international environmental accords: preliminary observations from collaborative project. *Glob Gov* 1995;**1**:119.
54. Mitchell RB. Institutional aspects of implementation, compliance, and effectiveness. In: Luterbacher U, Sprinz DF, editors. *International relations and global climate change*. Cambridge, MA: MIT Press; 2001. p. 221—44.
55. Helm C, Sprinz DF. Measuring the effectiveness of international environmental regimes. *J Confl Resolut* 2000;**44**:630—52.
56. Levy MA. Is the environment a national security issue? *Int Secur* 1995;**20**:35—62.
57. Ghosh SK, Debnath B, Baidya R, et al. Waste electrical and electronic equipment management and Basel Convention compliance in Brazil, Russia, India, China and South Africa (BRICS) nations. *Waste Manag Res* 2016;**34**:693—707.
58. Miniero R, De Felip E, Magliuolo M, Ferri F, Di Domenico A. Selected persistent organic pollutants (POPs) in the Italian environment. *Ann Ist Super Sanità* 2005;**41**:487—92.
59. IPEN. In: Project IPE, editor. *Malaysia country situation report*; 2005.
60. Basel Convention. *National reporting*. Geneva (Switzerland): UNEP; 2016.
61. Stockholm Convention. *National reporting*. Geneva (Switzerland): UNEP; 2016.
62. Kummer K. The international regulation of transboundary traffic in hazardous wastes: the 1989 Basel Convention. *Int Comp Law Q* 1992;**41**:530—62.
63. Government of Canada. *Chemicals management plan*. 2011.
64. Gauthier J, Gosselin A, Eggleton M. Ecological risk assessments of metal-containing substances under Canada's chemical management plan. In: *E3S Web of Conferences*. EDP Sciences; 2013.
65. Government of Canada. *Chemicals management plan science committee*. 2014.

66. Government of Canada. *Chemicals management plan stakeholder advisory council.* 2011.
67. Edge S, Eyles J. Message in a bottle: claims disputes and the reconciliation of precaution and weight-of-evidence in the regulation of risks from Bisphenol A in Canada. *Health Risk Soc* 2013;**15**:432–48.
68. European Chemicals Agency. *The European chemicals agency: working for the safe use of chemicals.* 2013.
69. European Parliament, European Council. *Regulation (EC) No 1907/2006 concerning the Registration, Evaluation, Authorisation and Restriction of Chemicals (REACH).* 2006.
70. Christensen FM, Eisenreich SJ, Rasmussen K, Sintes JR, Sokull-Kluettgen B, EJvd P. European experience in chemicals management: integrating science into policy. *Environ Sci Technol* 2010;**45**:80–9.
71. European Parliament, European Council. *Regulation (EC) No 1272/2008 on classification, labelling and packaging of substances and mixtures.* 2008.
72. Molander L, Rudén C. Narrow-and-sharp or broad-and-blunt—regulations of hazardous chemicals in consumer products in the European Union. *Regul Toxicol Pharmacol* 2012;**62**:523–31.
73. Congreso de Colombia. *LEY 253 DE 1996 Aprobación del Convenio de Basilea sobre el control de los movimientos transfronterizos de los desechos peligrosos y su eliminación.* 1996. Bogotá (Colombia).
74. O'Brien PJ. Participation and sustainable development in Colombia. *Revista Eur de Estudios Latinoamericanos y del Caribe/Eur Rev Lat Am Caribb Stud* 1995:7–35.
75. Murillo Chavarro J. Legal protection of areas of ecological importance such as Paramo in Colombia. *IUCN Acad Environ Law E J* 2011:81–9.
76. U.S. Government. *International agreements on transboundary shipments of waste.* U.S. Environmental Protection Agency; 2016.
77. Crook JR. Contemporary practice of the United States relating to international law. *Am J Int Law* 2013;**107**:431–77.
78. Rotondi J, Smaczniak K. The Minamata convention on mercury: what it does and does not mean for the United States. *Nat Resour Environ* 2014;**29**:19.
79. Government of Brazil. Política Nacional de Resíduos Sólidos. *Ministério do Meio Ambiente* 2010.
80. Ziglio L. Industrial solid waste management in Brazil and the Basel Convention. *Novos Estudos Jurídicos* 2014;**19**:585–606.
81. Perrez FX. *Building an effective future-proof international chemicals and waste regime. Chemicals and wastes policy & practice guest articles.* Winnipeg (Canada): IISD; 2015.
82. Lallas PL. The Stockholm Convention on persistent organic pollutants. *Am J Int Law* 2001;**95**:692–708.
83. Selin H. *Managing hazardous chemicals: longer-range challenges.* Boston, MA: Boston University – The Frederick S. Pardee Center for the Study of the Longer-Range Future; 2009.
84. UNCSD. *E/CN.17/2011/6 report of the secretary general – policy options and actions for expediting progress in implementation: waste management.* New York, NY: UN Economic and Social Council; 2011.
85. Schneider W. Basel Convention ban on hazardous waste exports: paradigm of efficacy or exercise in futility. *Suffolk Transnat'l L Rev* 1996;**20**:247.
86. Walsh MT. Global trade in hazardous wastes: domestic and international attempts to cope with a growing crisis in waste management. *Cathol Univ Law Rev* 1992;**42**:103.
87. Kummer K. The Basel Convention: ten years on. *Rev Eur Community Int Environ Law* 1998;**7**:227–36.
88. Porta M, Zumeta E. Implementing the Stockholm treaty on persistent organic pollutants. *Occup Environ Med* 2002;**59**:651–2.
89. Council AC. *Year-end 2014 chemical industry situation and outlook: American chemistry builds momentum.* 2014. Washington, DC.
90. Basel Convention. *Status of ratifications.* Geneva (Switzerland): UNEP; 2015.
91. Stockholm Convention. *Status of ratifications.* Geneva (Switzerland): UNEP; 2016.

Index

'*Note*: Page numbers followed by "f" indicate figures, "t" indicate tables, and "b" indicate boxes.'

B

N

Q

R

S